HOW TO USE
the Pearson Nurse's Drug Guide 2013

Classifications and Prototype Drugs

The classifications used in this book are based on the system used by the American Hospital Formulary Service (AHFS). This book further classifies drugs by therapeutic uses, enabling the nurse to identify drugs in the same class that have similar indications for use. Thus, the book provides a framework for understanding how drugs in a given class are used in clinical practice. The pharmacologic classification appears immediately after the **Classi-**

> **AMIODARONE HYDROCHLORIDE**
> (a-mee'oh-da-rone)
> **Cordarone, Amio-Aqueous, Nexterone, Pacerone**
> **Classification:** ANTIARRHYTHMIC, CLASS III
> **Therapeutic:** CLASS III ANTIARRHYTHMIC; ANTIANGINAL
> **Pregnancy Category:** D

fication heading, followed by the **Therapeutic** classification. In general, all drugs in a class will have similar actions, uses, adverse effects, and nursing implications. Therefore, we have selected certain drugs that are representative of a classification or its subclassification—**prototype drugs**—to aid the nurse in understanding the classification of drugs. Prototype drug monographs are identified with a small icon. The user can refer to the prototype drug to develop a better understanding of drugs that belong within the same classification or subclassification. When a drug belongs to a classification that has a designated prototype drug, that prototype is identified directly below the therapeutic classification. The Classification Scheme on pages xi–xviii identifies the drug prototype considered to be representative of each class. All prototype drugs are highlighted in **bold** type in the index for quick identification. Some drugs have a unique mechanism of action or therapeutic effect. In these cases, there is no prototype drug to be identified.

Pregnancy Category

Drugs may be described as category A, B, C, D, or X according to the risk to the fetus, with A being the lowest and X the highest risk. If the FDA pregnancy category is known, it is indicated after the Therapeutic classifications. Refer to Appendix C, *FDA Pregnancy Categories*, for a more complete description of each category.

Controlled Substances

In the United States, controlled substances are classified as belonging to one of five Schedules (I to V) according to abuse potential. If a drug is a controlled substance, then its Schedule number is listed below the

Pregnancy Category. Schedule I has the highest and Schedule V has the lowest potential for abuse. Refer to Appendix B, *U.S. Schedules of Controlled Substances*, for a more complete description of each schedule.

Availability

Because drugs come in a variety of dosages and forms, the authors include a section devoted to Availability in each monograph. This section identifies the available forms (e.g., tablets, capsules) and the available dosage strengths for every drug.

> **AVAILABILITY** 100 mg, 200 mg, 400 mg tablets; 50 mg/mL injection

Action and *Therapeutic Effect*

Each monograph describes the mechanism by which the specific drug produces physiologic and biochemic changes at the cellular, tissue, and organ levels. This information helps the user understand how the drug works in the body and makes it easier to learn its adverse reactions, and cautious uses. The *therapeutic effects*, which are set in italics for clarity and ease of use, are the reasons why a drug is prescribed. Therapeutic effectiveness of the drug can be determined by monitoring improvement in the condition for which the drug is prescribed.

> **ACTION & *THERAPEUTIC EFFECT***
> Class III antiarrhythmic that has antianginal and antiadrenergic properties. Acts directly on all cardiac tissues by prolonging duration of action potential and refractory period. Slows conduction time through the AV node and can interrupt the reentry pathways through the AV node. *Effective in prevention or suppression of cardiac arrhythmias.*

Uses and Unlabeled Uses

The therapeutic applications of each drug are described in terms of approved (i.e., FDA-labeled) uses and unlabeled uses. An unlabeled use is one that does not appear on the drug label or in the manufacturer's literature. Although currently supported by medical literature, unlabeled uses are not currently approved by the FDA.

> **USES** Prophylaxis and treatment of life-threatening ventricular arrhythmias and supraventricular arrhythmias, particularly with atrial fibrillation.
> **UNLABELED USES** Treatment of nonexertional angina, conversion of atrial fibrillation to normal sinus rhythm, paroxysmal supraventricular tachycardia, ventricular rate control due to accessory pathway conduction in pre-excited atrial arrhythmia, after defibrillation and epinephrine in cardiac arrest, AV nodal reentry tachycardia.

Contraindications and Cautious Use

Many drugs have contraindications and therefore should not be used in specific conditions, such as during pregnancy or pathologic disorders. In other cases, the drug should be used with great caution because of a greater than average risk of untoward effects.

CONTRAINDICATIONS Hypersensitivity to amiodarone, or benzyl alcohol; cardiogenic shock, severe sinus bradycardia, advanced AV block unless a pacemaker is available, severe sinusnode dysfunction or sick sinus syndrome, bradycardia, congenital or acquired QR prolongation syndromes, or history of torsades de pointes; severe liver disease, pregnancy (category D), lactation.
CAUTIOUS USE Hepatic disease, cirrhosis; Hashimoto's thyroiditis, goiter, thyrotoxicosis, or history of other thyroid dysfunction; mild to moderate hepatic toxicity; CHF, left ventricular dysfunction; hypersensitivity to iodine; older adults; Fabry disease, especially with visual disturbances; electrolyte imbalance, hypokalemia, hypomagnesemia, hypovolemia; preexisting lung disease, COPD; open heart surgery.

Arrhythmias

Adult: **PO Loading Dose**
800–1600 mg/day in 1–2 doses for 1–3 wk **PO Maintenance Dose** 400–600 mg/day in 1–2 doses **IV Loading Dose** 150 mg over 10 min followed by 360 mg over next 6 h **IV Maintenance Dose** 540 mg over 18 h (0.5 mg/min), may continue at 0.5 mg/min
Convert IV to PO Duration of infusion less than 1 wk use 800–1600 mg PO, 1–3 wk use 600–800 mg PO, greater than 3 wk use 400 mg PO
Child: **IV** 5 mg/kg then repeat to max of 300 mg total

Hepatic Impairment Dosage Adjustment

Adjustment only suggested in severe hepatic impairment

Route and Dosage

The routes and dosages are highlighted in a gray box for easy access. Route of administration is specified as subcutaneous, IM, IV, PO, PR, nasal, ophthalmic, vaginal, topical, aural, intradermal, or intrathecal. Dosages are listed according to indication or FDA-approved labeled use(s). One of the hallmarks of this drug guide is the comprehensive dosage information it provides. The guide includes adult, geriatric, and pediatric dosages, as well as dosages for neonates and infants whenever applicable. This section also indicates dosage adjustments for renal impairment (based on creatinine clearance), hepatic impairment, patients undergoing hemodialysis, and obese patients (based on ideal body weight). Information about the need for dosage adjustments based on pharmacogenetic variables [e.g., cytochrome (CYP) system of enzymes] is also provided as available.

ADMINISTRATION
- Note: Correct hypokalemia and hypomagnesemia prior to initiation of therapy.

Oral
- Give consistently with respect to meals. Avoid grapefruit juice.
- Note: Only a prescriber experienced with the drug and treatment of life-threatening arrhythmias should give loading doses.
- Note: GI symptoms commonly occur during high-dose therapy, especially with loading doses. Symptoms usually respond to dose reduction or divided dose given with food, including milk.

Administration

Drug administration is an important primary role for the nurse. Organized by different routes, this section lists comprehensive instructions for administering, handling, and storing medications.

Intravenous Drug Administration

Within the **Administration** section of appropriate monographs, the authors highlight intravenous drugs, indicated by a vertical red bar. This section provides users with comprehensive instructions on how to **Prepare** and **Administer** direct, intermittent, and continuous intravenous medications. When different from adults, intravenous administration and preparation for pediatric patients is provided. It also includes **Solution/Additive** and **Y-Site** incompatibility for every monograph, where appropriate, to indicate which drugs and solutions should not be mixed with the intravenous drug. This is crucial information for drug administration. A chart for Y-Site compatibility for common intravenous drugs is located inside the back cover of this drug guide. These enhancements eliminate the need for additional resources for intravenous administration.

Intravenous

PREPARE: **IV Infusion: First rapid loading dose infusion:** Add 150 mg (3 mL) amiodarone to 100 mL D5W to yield 1.5 mg/mL. **Second infusion during first 24 h (slow loading dose and maintenance infusion):** Add 900 mg (18 mL) amiodarone to 500 mL D5W to yield 1.8 mg/mL. **Maintenance infusions after the first 24 h:** Prepare concentrations of 1–6 mg/mL amiodarone. Note: Use central line to give concentrations greater than 2 mg/mL.

ADMINISTER: **IV Infusion:** Rapidly infuse initial 150 mg dose over the first 10 min at a rate of 15 mg/min. ▪ Over next 6 h, infuse 360 mg at a rate of 1 mg/min. ▪ Over the remaining 18 h, infuse ▪ 540 mg at a rate of 0.5 mg/min..

INCOMPATIBILITIES Solution/additive: Aminophylline, amoxicillin/clavulanic acid, cefazolin, floxacillin, furosemide, quinidine. Y-site: Aminocaproic acid, aminophylline, Amoxicillin, penem, fludarabine, fluorouracil, gemtuzumab, heparin, imipenem/cilastatin, levofloxacin, mechlorethamine, magnesium sulfate, methotrexate, micafungin, paclitaxel.

Adverse Effects

Virtually all drugs have adverse side effects that may be bothersome to some individuals but not to others. Adverse effects with an incidence of ≥1% are listed by body system or organs. The most common adverse effects appear in *italic* type, whereas those that are life-threatening are underlined. Users of the drug guide will find a key at the bottom of every page as a quick reminder.

Diagnostic Test Interference

This section describes the effect of the drug on various diagnostic tests and alerts the nurse to possible misinterpretations of test results when applicable. The name of the specific test altered is highlighted in ***bold italic*** type.

ADVERSE EFFECTS (≥1%) **CNS:** Peripheral neuropathy (*muscle weakness,* wasting numbness, tingling), *fatigue,* abnormal gait, dyskinesias, *dizziness,* paresthesia, headache. **CV:** Bradycardia, *hypotension* (IV), sinus arrest, cardiogenic shock, CHF, arrhythmias; AV block. **Special Senses:** *Corneal microdeposits,* blurred vision, optic neuritis, optic neuropathy, permanent blindness, corneal degeneration, macular degeneration, photosensitivity. **GI:** *Anorexia, nausea, vomiting, constipation,* hepatotoxicity. **Metabolic:** Hyperthyroidism or hypothyroidism; may cause neonatal hypo- or hyperthyroidism if taken during pregnancy. **Respiratory:** (Pulmonary toxicity) Alveolitis, pneumonitis (fever, dry cough, dyspnea), interstitial pulmonary fibrosis, *fatal gasping syndrome* with IV in children. **Skin:** Slate-blue pigmentation, *photosensitivity,* rash. **Other:** With chronic use, angioedema.

INTERACTIONS Drug: Significantly increases **digoxin** levels; enhances pharmacologic effects and toxicities of **disopyramide, procainamide, quinidine, flecainide, lidocaine, lovastatin, simvastatin;** anticoagulant effects of ORAL ANTICOAGULANTS enhanced; **verapamil, diltiazem,** BETA-ADRENERGIC BLOCKING AGENTS may potentiate sinus bradycardia, sinus arrest, or AV block; may increase **phenytoin** levels 2- to 3-fold; **cholestyramine** may decrease amiodarone levels; **fentanyl** may cause bradycardia, hypotension, or decreased output; may increase **cyclosporine** levels and toxicity; **cimetidine** may increase amiodarone levels; **ritonavir** may increase risk of amiodarone toxicity, including cardiotoxicity; **simvastatin** doses over 20 mg increase risk of rhabdomyolysis; **loratadine** use may increase risk of QT prolongation.

Interactions

When applicable, this section lists individual drugs, drug classes, foods, and herbs that interact with the drug discussed in the monograph. Drugs may interact to inhibit or enhance one another. Thus, drug interactions may improve the therapeutic response, lead to therapeutic failure, or produce specific adverse reactions. Only drugs that have been shown to cause clinically significant and documented interactions with the drug discussed in the monograph are identified. Note that generic drugs appear in **bold** type, and drug classes appear in SMALL CAPS.

Pharmacokinetics

This section identifies how the drug moves throughout the body. It lists the mechanisms of absorption, distribution, metabolism, elimination, and half-life when known. It also provides information about onset, peak, and duration of the drug action. Where appropriate, information appears for protein-binding and CYP450 metabolism.

> **PHARMACOKINETICS Absorption:** 22–86% absorbed. **Onset (PO):** 2–3 days to 1–3 wk. **Peak:** 3–7 h. **Distribution:** Concentrates in adipose tissue, lungs, kidneys, spleen; crosses placenta; 96% protein bound. **Metabolism:** Extensively in liver; undergoes some enterohepatic cycling; via CYP2C8 and 3A4. **Elimination:** Excreted chiefly in bile and feces; also in breast milk. **Half-Life:** Biphasic, initial 2.5–10 days, terminal 40–55 days.

Nursing Implications

Under the headings **Assessment & Drug Effects** and **Patient & Family Education**, the nurse can quickly and easily identify needed information and incorporate it into the appropriate steps of the nursing process. Before administering a drug, the nurse should read both sections to determine the assessments that should be made before and after administration of the drug, the indicators of drug effectiveness, laboratory tests recommended for individual drugs, and the essential patient and/or family education related to the drug.

Therapeutic Effectiveness

Therapeutic effectiveness of a drug can be determined by monitoring improvement in the condition for which the drug is prescribed, and by using the **Assessment & Drug Effects** section. Drugs have multiple uses or indications. Therefore, it is important to know why a drug is being prescribed for a specific patient (**Uses** and **Unlabeled Uses**). In the italicized sentences at the end of the **Action & *Therapeutic Effect*** section in all monographs, specific indicators of the effectiveness of the drug are provided. Additionally, in the **Route & Dosage** table for each drug, the dosages are listed according to the indications for FDA-labeled use(s) of the drug. Furthermore, the **Therapeutic** classifications listed within the red box at the beginning of the monograph provides the nurse with further assistance in determining and evaluating the therapeutic effectiveness of the drug.

> **NURSING IMPLICATIONS**
>
> **Assessment & Drug Effects**
> - Monitor BP carefully during infusion and slow the infusion if significant hypotension occurs; bradycardia should be treated by slowing the infusion or discontinuing if necessary. Monitor heart rate and rhythm and BP until drug response has stabilized; report promptly symptomatic bradycardia. Sustained monitoring is essential because drug has an unusually long half-life.
>
> **Patient & Family Education**
> - Check pulse daily once stabilized, or as prescribed. Report a pulse less than 60.
> - Take oral drug consistently with respect to meals. Do not drink grapefruit juice while taking this drug.

PEARSON
NURSE'S
DRUG GUIDE
2013

Billie Ann Wilson, BSN, MSN, PhD

Professor Emerita
School of Nursing
Loyola University New Orleans
New Orleans, Louisiana

Margaret T. Shannon, BSN, MSN, PhD

Professor Emeritus of Nursing
Our Lady of Holy Cross College
New Orleans, Louisiana

Kelly M. Shields, PharmD

Associate Professor of Pharmacy Practice
Director of Drug Information Center
Raabe College of Pharmacy
Ohio Northern University
Ada, Ohio

Pearson

Boston Columbus Indianapolis New York San Francisco
Upper Saddle River Amsterdam Cape Town Dubai London
Madrid Milan Munich Paris Montreal Toronto Delhi
Mexico City São Paulo Sydney Hong Kong Seoul
Singapore Taipei Tokyo

Notice: The authors and the publisher of this volume have taken care to make certain that the doses of drugs and schedules of treatment are correct and compatible with the standards generally accepted at the time of publication. Nevertheless, as new information becomes available, changes in treatment and in the use of drugs become necessary. The reader is advised to carefully consult the instruction and information material included in the package insert of each drug or therapeutic agent before administration. This advice is especially important when using, administering, or recommending new and infrequently used drugs. The authors and publisher disclaim all responsibility for any liability, loss, injury, or damage incurred as a consequence, directly or indirectly, of the use and application of any of the contents of this volume.

www.pearsonhighered.com/drugguides

12   13   /   10   9   8   7   6   5   4   3   2   1

ISBN 0-13-296489-9 / 978-0-13-296489-0
[retail] 0-13-297485-1 / 978-0-13-297485-1

PRINTED IN THE UNITED STATES OF AMERICA

CONTENTS

To
Alvin, Theresa, Ellen, and
Michael, Rick, Kris, and Leah

without whom this work would not have been possible

♦

ABOUT THE AUTHORS

Billie Ann Wilson is Professor Emerita in the School of Nursing at Loyola University in New Orleans, Louisiana. Prior to entering nursing, she taught natural and physical sciences at the secondary and collegiate levels. She holds a BS in Biology from Boston College, an MS in Biology from Purdue University, a BS in Nursing from Northwestern State University of Louisiana, an MSN from Louisiana State University Health Sciences Center, and a PhD in Curriculum and Instruction from the University of New Orleans.

Margaret T. Shannon is Professor Emeritus of Nursing at Our Lady of Holy Cross College, New Orleans, Louisiana. She holds a BS and an MS in Chemistry, both from Saint Louis University; an MA in Teaching Biology from Saint Mary's University, a BS in Nursing from Northwestern State University of Louisiana, an MSN from Louisiana State University Health Sciences Center, and a PhD in Curriculum and Instruction from the University of New Orleans. Prior to entering nursing, she taught physical science, natural science, and mathematics at the secondary and collegiate levels.

Kelly M. Shields is currently Associate Professor of Pharmacy Practice at Ohio Northern University's Raabe College of Pharmacy. She holds a Doctor of Pharmacy from Butler University and completed a fellowship in Natural Product Information and Research at University of Missouri-Kansas City. She has practiced pharmacy in retail, community, and academic settings and has worked as a freelance medical writer.

EDITORIAL REVIEW PANEL

We wish to thank the following individuals for conducting thorough reviews of the drug information in this book for its accuracy, currency, relevance, presentation, accessibility, and use. Their feedback guided us in developing a better book for nurses.

Nurse's Drug Guide Reviewers

Carol Agana, MSNc, RNP, APN
University of Arkansas
Fayetteville, AR

Charlene M. Chapman, RN, BSN
Pennsylvania Institute
of Technology
Media, PA

Patti Christy, DNP, FNP, APRN-BC
Memorial Medical Group
Lake Charles, LA

Christy L. Henry, RN, BSN
Texas County Technical College
Houston, MO

Mary Catherine Rawls, MS, BSN, RN-BC, ONC
Dartmouth-Hitchcock
Medical Center
Lebanon, NH

Marianne Swihart, RN, MEd, MSN
Pasco-Hernandez
Community College
New Port Richey, FL

Nancy Lynn Whitehead, MS, FNP, C, CSN, CLNC
Milwaukee Area Technical
College
Milwaukee, WI

PHARMACY CONSULTANT

A special acknowledgment to **Marc Harrold, PhD, RPh,** Professor of Medicinal Chemistry, School of Pharmacy at Duquesne University in Pittsburgh, Pennsylvania, who is a tremendous addition to the author team as a contributor for the monographs of the new drugs in this edition. We are grateful for his expertise and for his valued input.

PREFACE

Pearson Nurse's Drug Guide 2013 is a current and reliable reference designed to provide comprehensive information needed to make appropriate decisions regarding drug administration. This new edition includes 21 monographs for new drugs recently approved by the Food and Drug Administration (FDA) and over 350 updates to drug indications, available forms, adverse effects, dosages, and more. Revised IV administration information has been added and/or updated regarding adults as well as children. These revisions include subheadings for IV preparation and administration of IV medications for children, which add to the ease of locating appropriate information by age.

On pages xi–xviii, the user will find a current listing of drug classifications and their associated drug prototypes. Prototype drugs are representative of all drugs in a particular classification or subclassification. The classification scheme serves as a valuable tool, especially for students learning pharmacology and familiarizing themselves with drug families and prototype drugs for classes of drugs.

Each drug monograph provides the necessary information for safe and effective drug administration. The user should read all the information provided. Occasionally, the user will be referred to Appendix F, *Glossary of Key Terms, Clinical Conditions, and Associated Signs and Symptoms*. This unique glossary provides valuable information regarding common assessment findings related to therapeutic effectiveness or ineffectiveness of specific drugs.

The authors recognize that the decision-making process related to drug administration is a cyclical one. For example, assessments are made both prior to and after drug administration. Thus, nursing diagnoses and interventions may change as a result of an *achieved therapeutic effect, therapeutic failure, manifestation of an adverse effect,* or *demonstration of a learning need*. The authors believe that the users of this drug reference will find that the clear and logical design of the drug monographs facilitates decision making and supports the nursing process.

Since physicians, advanced practice nurses, and other health professionals now have prescriptive privileges, the term *prescriber* is used throughout this book.

ORGANIZATION

The **Pearson Nurse's Drug Guide 2013** is user friendly. Nurses in clinical practice, nursing professors, pharmacists, and nursing students from across the country reviewed the content of this handbook and provided helpful suggestions on how it could be made more useful. Based on their comments, the authors updated and added new information to the monographs. To help readers better understand how to use the drug guide, the authors illustrate and describe all the components of a drug monograph in the **Guide to Using the Pearson Nurse's Drug Guide 2013**, at the beginning of the book.

In this drug guide, all drugs are listed alphabetically according to their generic names. Pharmacologic classifications are paired with therapeutic classifications for every drug monograph for ease of use by nurse clinicians and students alike. Each drug is indexed by both its generic and trade names in the back of the guide to make it easier for the user to locate individual drug monographs. Trade names followed by a maple leaf indicate that brand of the drug is available in Canada.

If a drug is not listed in the alphabetical section, it may be a combination drug, which is a drug made up of more than one generic component. These combination drugs are listed under their trade names in the index and in Appendix E, *Prescription Combination Drugs*. The appendix identifies the generic components and the amount of each generic drug contained in the combination. Users of this drug guide will find the page numbers for monographs of the component drugs in this appendix to make access to this information easier and faster.

Appendixes

This edition of the drug guide includes several helpful tables and charts in the appendixes, including Appendix A, *Ocular Medications, Low Molecular Weight Heparins, Inhaled Corticosteroids, and Topical Corticosteroids*; Appendix B, *U.S. Schedules of Controlled Substances*; Appendix C, *FDA Pregnancy Categories*; Appendix D, *Oral Dosage Forms That Should Not Be Crushed*; Appendix E, *Prescription Combination Drugs*; Appendix F, *Glossary of Key Terms, Clinical Conditions, and Associated Signs and Symptoms*; Appendix G, *Abbreviations*; and Appendix H, *Herbal and Dietary Supplement Table*. Appendix I, *Look-Alike, Sound-Alike Medications*, which highlights medications that are commonly involved in medication errors; and Appendix J, *Vaccines*, which highlights vaccines commonly seen/administered in practice.

Index

The index in the ***Pearson Nurse's Drug Guide 2013*** is perhaps the most often-used section in the entire book. All generic, trade, and combination drugs are listed in this index. Whenever a trade name is listed, the generic drug monograph is listed in parentheses. Additionally, classifications are listed and identified in SMALL CAPS, whereas all prototype drugs are highlighted in **bold** type. Drugs belonging to various classifications and subclassifications, including therapeutic classes, are also cross-referenced in this index. As a special feature, the index includes entries for combination drugs (e.g. Tylenol with Codeine) with index references to component drugs as well as the combination drug reference to Appendix E. Medications listed in Appendix A (*Ocular Medications, Low Molecular Weight Heparins, Inhaled Corticosteroids, and Topical Corticosteroid*) or Appendix J (*Vaccines*) are also cross-referenced in the index.

ONLINE COMPANION

The ***Pearson Nurse's Drug Guide 2013*** also comes with an online companion designed to assist nurses in providing drug information and nursing implications for patients in hospitals, clinics, and all community settings. The online companion provides access to many more resources:

- monographs for newly approved drugs,
- calculators to help nurses do conversions or calculate dosages and IV drip rates,
- a link to access and purchase mobile versions of this drug guide,
- access to drug updates,
- links to drug-related sites,
- drug-related tools,
- medication administration techniques,
- drug classifications,
- common herbal remedies,
- list of look-alike, sound-alike drugs, and
- most commonly administered vaccinations.

You can also send the authors your feedback about the drug guide through this website. To access this online resource, go to www.pearsonhighered.com/drugguides.

ACKNOWLEDGMENTS

We wish to express our appreciation to our past and present students who have provided the inspiration for this work. It is for these individuals and all who strive for excellence in patient care that this work was undertaken.

Billie Ann Wilson, BSN, MSN, PhD
Margaret T. Shannon, BSN, MSN, PhD
Kelly M. Shields, PharmD

Classifications	Prototype
ANTIGOUT AGENT	Probenecid

ANTIHISTAMINES

ANTIHISTAMINES (H1-RECEPTOR ANTAGONIST)	Diphenhydramine HCl
NON-SEDATING	Loratadine
ANTIPRURITIC	Hydroxyzine HCl
ANTIVERTIGO AGENT	Meclizine HCl

ANTIINFECTIVES

ANTIBIOTICS

AMEBICIDE	Paromomycin Sulfate
ANTHELMINTIC	Mebendazole
AMINOGLYCOSIDES	Gentamicin Sulfate
ANTIFUNGALS	Amphotericin B
AZOLE ANTIFUNGAL	Fluconazole
ALLYLAMINE ANTIFUNGAL	Terbinafine
ECHINOCARDIN ANTIFUNGAL	Caspofungin
BETA-LACTAM	Imipenem-Cilastatin
CEPHALOSPORIN	
FIRST GENERATION	Cefazolin Sodium
SECOND GENERATION	Cefaclor
THIRD GENERATION	Cefotaxime Sodium
CLINDAMYCIN	Clindamycin HCl
MACROLIDES	Erythromycin
PENICILLIN	
AMINOPENICILLIN	Ampicillin
EXTENDED SPECTRUM PENICILLIN	Piperacillin/Taxobactum
PENICILLASE-RESISTANT PENICILLIN	Oxacillin Sodium
NATURAL PENICILLIN	Penicillin G Potassium
QUINOLONES	Ciprofloxacin HCl
SULFONAMIDES	Sulfisoxazole
TETRACYCLINE	Tetracycline HCl
URINARY TRACT ANTI-INFECTIVE	Trimethoprim
ANTILEPROSY (SULFONE) AGENT	Dapsone
ANTIMALARIAL	Chloroquine Phosphate
ANTIPROTOZOAL	Metronidazole
ANTITUBERCULOSIS AGENTS	Isoniazid
ANTITUBERCULOSIS AGENT, ANTIMYCOBACTERIAL	Rifampin

---

*Based on the American Hospital Formulary Service Pharmacologic–Therapeutic Classification.
†Prototype drugs are highlighted in tinted boxes in this book.
Complete list of drugs for each classification found in classification index starting on p. 1662.

Classifications	Prototype

ANTIVIRAL AGENTS.................................Acyclovir
   ADAMANTANES ....Amantadine
ANTIRETROVIRAL AGENTS
   NUCLEOSIDE REVERSE TRANSCRIPTASE
      INHIBITOR....................................Lamivudine
   NONNUCLEOSIDE REVERSE
      TRANSCRIPTASE INHIBITOR............Efavirenz
   PROTEASE INHIBITOR..........................Saquinavir
URINARY TRACT ANTI-INFECTIVE ....Trimethoprim

ANTINEOPLASTICS
ALKYLATING AGENT.................................Cyclophosphamide
ANTHRACYCLINE (ANTIBIOTIC)...............Doxorubicin HCl
ANTIANDROGEN....................................Flutamide
ANTIMETABOLITES
   ANTIMETABOLITE (ANTIFOLATE) ....Methotrexate
   ANTIMETABOLITE
   (PURINE ANTAGONIST) 6-Mercaptopurine
   ANTIMETABOLITE (PYRIMIDINE) 5-Fluorouracil
AROMATASE INHIBITOR..........................Anastrozole
DNA TOPOISOMERASE INHIBITOR
   (CAMPTOTHECIN) ....Topotecan HCl
EPIDERMAL GROWTH FACTOR
   RECEPTOR-TYROSINE
   KINASE INHIBITOR (EGFR-TKI)..............Gefitinib
HORMONE, SELECTIVE ESTROGEN
   RECEPTOR MODIFIERS (SERMS) ....Tamoxifen Citrate
MITOTIC INHIBITOR.................................Vincristine Sulfate
TAXANE (TAXOID)...................................Paclitaxel

ANTITUSSIVES, EXPECTORANTS, & MUCOLYTICS
ANTITUSSIVE...Benzonatate
EXPECTORANT ....Guaifenesin
MUCOLYTIC ....Acetylcysteine

AUTONOMIC NERVOUS SYSTEM AGENTS
ADRENERGIC AGONISTS (SYMPATHOMIMETICS)
   ALPHA-ADRENERGIC AGONIST ....Dexmedetomide HCl
   ALPHA- & BETA-ADRENERGIC
      AGONIST.....................................Epinephrine
   BETA-ADRENERGIC AGONIST ....Isoproterenol HCl

Classifications   Prototype

ADRENERGIC ANTAGONISTS (SYMPATHOLYTICS)
    ALPHA-1 ADRENERGIC ANTAGONIST
      (GENITOURINARY SMOOTH
      MUSCLE RELAXANT) .... Tamsulosin
    ALPHA ANTAGONISTS
      (BLOCKING AGENT) .... Prazosin HCl
    BETA ANTAGONISTS .... Propranolol HCl
    ERGOT ALKALOID .... Ergotamine Tartrate
    5-HT1 SEROTONIN AGONISTS .... Sumatriptan
ANTICHOLINERGICS (PARASYMPATHOLYTICS)
    ANTIMUSCARINIC .... Atropine Sulfate
    ANTISPASMODIC
    [GENITOURINARY (GU)] .... Oxybutynin
CHOLINERGICS (PARASYMPATHOMIMETICS)
    CHOLINESTERASE INHIBITOR .... Neostigmine
    CENTRAL-ACTING .... Donepezil
    DIRECT-ACTING CHOLINERGIC .... Bethanechol Chloride
AUTONOMIC DRUGS, MISC .... Nicotine

BENZODIAZEPINE ANTAGONIST .... Flumazenil

BIOLOGICAL RESPONSE MODIFIERS
FUSION PROTEIN .... Alefacept
IMMUNOSUPPRESSANT .... Cyclosporine
IMMUNOGLOBULIN .... Immune Globulin
IMMUNOMODULATORS .... Immune Globulin
    INTERFERON .... Peginterferon Alfa-2a
    TUMOR NECROSIS FACTOR MODIFIER ... Etanercept
MONOCLONAL ANTIBODY .... Basiliximab

**BISPHOSPHONATE (REGULATOR, BONE
    METABOLISM)** .... Etidronate Disodium

**BLOOD DERIVATIVE, PLASMA VOLUME
    EXPANDER** .... Normal Serum Albumin

**BLOOD FORMERS, COAGULATORS, &
    ANTICOAGULANTS**
ANTICOAGULANT .... Heparin Sodium
    DIRECT THROMBIN INHIBITOR .... Lepirudin
    LOW MOLECULAR WEIGHT HEPARIN ... Enoxaparin
ANTIPLATELET AGENTS .... Clopidogrel
    GLYCOPROTEIN IIb/IIIa INHIBITOR .... Abciximab

Classifications	Prototype
COLONY STIMULATING FACTOR	Filgrastim
HEMATOPOIETIC GROWTH FACTOR	Epoetin Alpha
HEMOSTATIC (COAGULATOR)	Aminocaproic Acid
IRON PREPARATION	Ferrous Sulfate
THROMBOLYTIC ENZYME	Alteplase

BRONCHODILATORS (RESPIRATORY SMOOTH MUSCLE RELAXANT)

BETA-ADRENERGIC AGONIST	Albuterol
LEUKOTRIENE INHIBITOR	Zafirlukast
XANTHINE	Theophylline

CARDIOVASCULAR AGENTS

ANGIOTENSIN II RECEPTOR ANTAGONISTS	Losartan Potassium
ANGIOTENSIN-CONVERTING ENZYME INHIBITORS	Enalapril
ANTIARRHYTHMIC AGENTS	
CLASS IA	Procainamide HCl
CLASS IB	Lidocaine HCl
CLASS IC	Flecainide
CLASS II	Propranolol HCl
CLASS III	Amiodarone HCl
ANTILIPEMICS	
BILE ACID SEQUESTRANT	Cholestyramine
FIBRATES	Fenofibrate
HMG-CoA REDUCTASE INHIBITOR (STATIN)	Lovastatin
CALCIUM CHANNEL BLOCKERS	
1,4 DIHYDROPYRIDINE	Nifedipine
MISCELLANEOUS	Verapamil
CARDIAC GLYCOSIDE	Digoxin
CENTRAL-ACTING ANTIHYPERTENSIVE	Methyldopa
INOTROPIC AGENT	Milrinone Lactate
NITRATE VASODILATOR	Nitroglycerin
NONNITRATE VASODILATOR	Hydralazine HCl
PROSTAGLANDIN (PULMONARY ANTIHYPERTENSIVE)	Epoprostenol sodium
RAUWOLFIA ALKALOID	Reserpine
RENIN-ANGIOTENSION RECEPTOR ANTAGONIST	
ANGIOTENSIN II RECEPTOR ANTAGONIST	Losartan Potassium

Classifications   Prototype

CENTRAL NERVOUS SYSTEM AGENTS
ANALGESICS, ANTIPYRETICS
    NARCOTIC (OPIATE) AGONISTS .... Morphine
    NARCOTIC (OPIATE)
        AGONIST-ANTAGONIST .... Pentazocine HCl
    NARCOTIC (OPIATE) ANTAGONIST ..... Naloxone HCl
    NONNARCOTIC ANALGESICS............ Acetaminophen
    NONSTEROI DAL ANTI-INFLAMMATORY
        DRUGS (NSAIDS)
            COX-1 .... Ibuprofen
            COX-2 .... Celecoxib

        SALICYLATE .... Aspirin
ANESTHETIC
    LOCAL (ESTER TYPE) .... Procaine HCl
    LOCAL (AMIDE TYPE) .... Lidocaine HCl
ANTIPARKINSON AGENTS
    CENTRALLY ACTING CHOLINERGIC
        RECEPTOR ANTAGONISTS .... Benztropine
    CATECHOLAMINE O-METHYL TRANSFERASE
        (COMT) INHIBITORS...................... Tolcapone
    DOPAMINE RECEPTOR AGONISTS....... Levodopa
ANTICONVULSANTS
    BARBITURATE...................................... Phenobarbital
    BENZODIAZEPINE .... Diazepam
    GABA INHIBITOR .... Valproic Acid Sodium
    GABA ANALOG .... Gabapentin
    HYDANTOIN....................................... Phenytoin
    SUCCINIMIDE..................................... Ethosuximide
    SULFONAMIDE.................................... Zonisamide
    TRICYCLIC .... Carbamazepine
ANXIOLYTICS, SEDATIVE-HYPNOTICS
    BARBITURATE...................................... Secobarbital
    BENZODIAZEPINE .... Lorazepam
    CARBAMATE .... Meprobamate
    NONBENZODIAZEPINE .... Zolpidem
PSYCHOTHERAPEUTIC
    ANTIDEPRESSANTS
        MONOAMINE OXIDASE (MAO)
            INHIBITORS .... Phenelzine Sulfate
        SELECTIVE SEROTONIN REUPTAKE
            INHIBITORS (SSRIS) .... Fluoxetine HCl

Classifications	Prototype

      SEROTONIN NOREPINEPHRINE
         REUPTAKE INHIBITORS .... Venlafaxine
      TETRACYCLIC ANTIDEPRESSANTS .... Mirtazapine
      TRICYCLIC ANTIDEPRESSANTS .... Imipramine HCl
    ANTIPSYCHOTIC AGENT
      ATYPICAL .... Clozapine
      BUTYROPHENONE .... Haloperidol
      MOOD STABILIZER .... Lithium Carbonate
      PHENOTHIAZINE............................ Chlorpromazine
CEREBRAL STIMULANT
    AMPHETAMINE .... Amphetamine Sulfate
    XANTHINE .... Caffeine

ELECTROLYTIC & WATER BALANCE AGENTS
DIURETIC
    LOOP.. Furosemide
    OSMOTIC .... Mannitol
    POTASSIUM-SPARING......................... Spironolactone
    THIAZIDE... Hydrochlorothiazide
    VASOPRESSIN ANTAGONIST .... Conivaptan
PHOSPHATE BINDER .... Sevelamer HCl
REPLACEMENT SOLUTION .... Calcium Gluconate

ENZYMES
ENZYME REPLACEMENT.......................... Pancrelipase
ENZYME INHIBITOR .... Alpha$_1$-Proteinase Inhibitor

EYE, EAR, NOSE, & THROAT (EENT) PREPARATIONS
ANTIHISTAMINE, OCULAR .... Emedastine
CARBONIC ANHYDRASE INHIBITOR .... Acetazolamide
CYCLOPLEGIC (MYDRIATIC)..................... Atropine Sulfate
MIOTIC (ANTIGLAUCOMA AGENT) .... Pilocarpine HCl
PROSTAGLANDIN .... Latanoprost
VASOCONSTRICTOR, DECONGESTANT .. Naphazoline HCl

GASTROINTESTINAL AGENTS
ANORECTANT... Diethylpropion HCl
ANTACID, ADSORBENT .... Aluminum Hydroxide
ANTIDIARRHEAL..................................... Loperamide
ANTIDIARRHEAL, ADSORBENT .... Bismuth Subsalicylate
ANTIEMETIC.. Prochlorperazine
ANTIEMETIC (5-HT3 ANTAGONIST) .... Ondansetron HCl

Classifications   Prototype

ANTISECRETORY (H2-RECEPTOR
    ANTAGONIST) ....Cimetidine
BULK LAXATIVE ....Psyllium Hydrophilic Mucilloid
MUCOUS MEMBRANE
    ANTIINFLAMMATORY ....Mesalamine
PROKINETIC AGENT (GI STIMULANT) ....Metoclopramide HCl
PROTON PUMP INHIBITORS....................Omeprazole
SALINE CATHARTIC...................................Magnesium Hydroxide
STIMULANT LAXATIVE ....Bisacodyl
STOOL SOFTENER ....Docusate Calcium

GOLD COMPOUND..................................Auranofin

HORMONES & SYNTHETIC SUBSTITUTES

ADRENAL CORTICOSTEROID
    GLUCOCORTICOSTEROID...................Prednisone
    MINERALOCORTICOID ....Fludrocortisone Acetate
ANDROGEN/ANABOLIC STEROIDS ....Testosterone
ANTIANDROGENS
    5-ALPHA REDUCTASE INHIBITORS........Finasteride
ANTIDIABETIC AGENTS
    ALPHA-GLUCOSIDASE INHIBITOR........Acarbose
    BIGUANIDES ....Metformin
    DD$_4$ INHIBITOR INCRETIN MIMETIC ....Exenatide
    INCRETIN MODIFIER
        (DPP-4 INHIBITOR) ....Sitagliptan
    INSULIN ....Insulin Injection
    MEGLITINIDES ....Repaglinide
    SULFONYLUREAS ....Glyburide
    THIAZOLIDINEDIONES........................Rosiglitazone
ESTROGENS ....Estradiol
GONADOTROPIN-RELEASING
    HORMONE ANALOGS ....Leuprolide Acetate
GONADOTROPIN-RELEASING
    HORMONE ANTAGONIST ....Ganirelix Acetate
GROWTH HORMONE............................Somatropin
OXYTOCIC...Oxytocin Injection
PITUITARY (ANTIDIURETIC) ....Vasopressin Injection
PROGESTINS (INJECTABLE PRODUCTS) ....Progesterone
PROGESTINS (ORAL PRODUCTS) ....Norethindrone
PROSTAGLANDIN (OXYTOCIC) ....Carboprost
RAUWOLFIA ALKALOID ....Reserpine
SOMASTATIN ANALOG ....Octreotide

Classifications   Prototype

THYROID AGENTS
    ANTITHYROID AGENT..........................Propylthiouracil
    THYROID ....Levothyroxine Sodium
VITAMIN D ANALOG ....Calcitriol

IMPOTENCE AGENT
PHOSPHDIESTERASE (PDE-5)
    INHIBITOR ....Sildenafil

RESPIRATORY AGENTS
MAST CELL STABILIZER ....Cromolyn Sodium

SKIN & MUCOUS MEMBRANE AGENTS
ANTIACNE (RETINOID) ....Isotretinoin
ANTI-INFLAMMATORY STEROID ....Hydrocortisone
PEDICULICIDE ....Permethrin
PSORALEN ....Methoxsalen
SCABICIDE ....Lindane

SOMATIC NERVOUS SYSTEM AGENTS
SKELETAL MUSCLE RELAXANTS
    CENTRAL-ACTING ....Cyclobenzaprine HCl
    DEPOLARIZING ....Succinylcholine Chloride
    NONDEPOLARIZING ....Atracurium

ABACAVIR SULFATE

(a-ba′ca-vir)

Ziagen

Classifications: ANTIRETROVIRAL;
NUCLEOSIDE REVERSE TRANSCRIPTASE
INHIBITOR (NRTI)
Therapeutic: ANTIRETROVIRAL
(NRTI)
Prototype: Lamivudine
Pregnancy Category: C

AVAILABILITY 300 mg tablets; 20
mg/mL oral solution

ACTION & *THERAPEUTIC EFFECT*
Abacavir is a synthetic nucleoside
analog with inhibitory activity against
HIV. It inhibits the activity of viral
reverse transcriptase (RT) both by
competing with natural DNA nucleoside and by incorporation into viral
DNA. Abacavir prevents the formation of viral DNA replication. *Viral
load decreases as measured by an
increased CD_4 lymphocyte cell count
and suppression of HIV RNA, indicated by decreased HIV RNA copies,
in HIV-positive individuals with little
or no exposure to zidovudine (AZT).*

USES Treatment of HIV infection in
combination with other antiretroviral agents.

CONTRAINDICATIONS Hypersensitivity to abacavir (fatal rechallenge reactions reported); lactic acidosis; creatinine clearance of less than 50 mL/
min; severe hepatomegaly; moderate
to severe hepatic impairment; patients
with HLA-B*5701 allele (high risk for
hypersensitivity reaction); lactation.
CAUTIOUS USE Prior resistance to
another nucleoside reverse transcriptase inhibitor (NRTI); history
of cardiac disease; older adults;
pregnancy (category C). Safe use in
children younger than 3 mo has not
been established.

ROUTE & DOSAGE

HIV Infection

Adult: **PO** 300 mg b.i.d.
Child (3 mo–16 y): **PO** 8 mg/kg
b.i.d. (max: 300 mg b.i.d.)
Patients weighing 14–21 kg, give
150 mg PO twice daily; *21–30
kg,* give 150 mg PO in the morning
and 300 mg in the evening;
over 30 kg, give 300 mg PO
twice daily

Hepatic Impairment Dosage Adjustment

Mild (Child-Pugh score 5–6): 200
mg b.i.d.

ADMINISTRATION

Oral

- Tablets and oral solution are interchangeable on a mg-for-mg basis.
- Store tablets and liquid at 20°–
  25° C (68°–77° F). Liquid may be
  refrigerated.

**ADVERSE EFFECTS (≥1%) Body as
a Whole:** Hypersensitivity reactions
(including fever, skin rash, fatigue,
nausea, vomiting, diarrhea, abdominal pain); malaise; lethargy; myalgia; arthralgia; paresthesia; edema;
shortness of breath. **CNS:** Insomnia,
headache, fever. **CV:** Hypotension
(associated with hypersensitivity
reaction), <u>heart attack</u>. **GI:** <u>Hepato-
megaly</u> with steatosis, *nausea, vomiting, diarrhea, anorexia,* pancreatitis, increased GGT, increased liver
function tests. **Skin:** *Rash.* **Other:**
Lactic acidosis, renal insufficiency.

INTERACTIONS Drug: Alcohol may
increase abacavir blood levels.

PHARMACOKINETICS Absorption:
Rapidly absorbed, 83% bioavailable. **Distribution:** Distributes into

Common adverse effects in *italic*, life-threatening effects <u>underlined</u>; generic names
in **bold;** classifications in SMALL CAPS; ✛ Canadian drug name; ☉ Prototype drug

1

extravascular space and erythrocytes; 50% protein bound. **Metabolism:** Metabolized by alcohol dehydrogenase and glucuronyl transferase to inactive metabolites. **Elimination:** 84% in urine, primarily as inactive metabolites;16% in feces. **Half-Life:** 1.5 h.

NURSING IMPLICATIONS

Assessment & Drug Effects

- Monitor for S&S of hypersensitivity: fever, skin rash, fatigue, GI distress (nausea, vomiting, diarrhea, abdominal pain). Withhold drug and immediately notify prescriber if hypersensitivity develops.
- Lab tests: Periodic liver function, tests, BUN and creatinine, CBC with differential, triglyceride levels, and blood glucose, especially in diabetics.
- Withhold drug and immediately notify prescriber for S&S of acidosis, hepatotoxicity, or renal insufficiency.

Patient & Family Education

- Take drug exactly as prescribed at indicated times. Missed dose: Take immediately, then resume dosing schedule. Do not double a dose.
- Withhold drug immediately and notify prescriber at first sign of hypersensitivity reaction (see Assessment & Drug Effects).
- Carry Warning Card provided with drug at all times.

ABATACEPT

(a-ba-ta'sept)

Orencia

Classifications: BIOLOGIC AND IMMUNOLOGICAL; IMMUNOMODULATOR; DISEASE-MODIFYING ANTIRHEUMATIC DRUG (DMARD)

Therapeutic: ANTIRHEUMATIC (DMARD); ANTI-INFLAMMATORY

Pregnancy Category: C

AVAILABILITY 250 mg lyophilized powder for injection

ACTION & *THERAPEUTIC EFFECT*

Abatacept inhibits T-cell (T lymphocyte) proliferation and inhibits production of tumor necrosis factor (TNF)-alpha, interferon-gamma, and interleukin-1 (IL-2, IL-6, and IL-15). It suppresses inflammation, decreases anticollagen antibody production, and reduces antigen-specific production of interferon-gamma. *Reduces the number of activated T lymphocytes found in synovial fluid of rheumatoid arthritis patients. It relieves RA symptoms and slows progression of structural damage. It improves RA physical function in adults with active RA who have had an inadequate response to other drugs.*

USES Treatment of moderate to severe rheumatoid arthritis or juvenile rheumatoid arthritis.

CONTRAINDICATIONS Known hypersensitivity to abatacept, live vaccines; active infections; co-administration with anakinra; TNF antagonists; other biologic RA therapy; lactation. **CAUTIOUS USE** COPD; malignancies; pregnancy (category C); children younger than 6 y.

ROUTE & DOSAGE

Rheumatoid Arthritis

Adult: **IV** Initial dose: *Weight less than 60 kg,* 500 mg; *60–100 kg,* 750 mg; *weight greater than 100 kg,* 1000 mg. Give dose at weeks 2 and 4, then monthly.

Juvenile Idiopathic Arthritis

Child (at least 6 y): **IV** *Weight less than 75 kg,* 10 mg/kg, repeat at weeks 2 and 4, then monthly (max: 1000 mg); *weight 75–100 kg,* 750 mg at weeks 2 and 4, then

Common adverse effects in *italic*, life-threatening effects underlined; generic names in **bold;** classifications in SMALL CAPS; ♣ Canadian drug name; ⊘ Prototype drug

monthly; *weight at least 100 kg,* 1000 mg at weeks 2 and 4, then monthly

ADMINISTRATION

Subcutaneous

- Abatacept for subcutaneous injection is supplied in a 125mg/mL prefilled syringe.
- Do not remove the small bubble of air in the syringe and do not pull back on the plunger head before injection.
- Ensure that the full contents of the syringe are injected. Rotate injection sites.
- Prefilled syringes must be stored between 2°–8° C (36°–46° F). Do not freeze.

Intravenous

- Note: The prefilled syringe is for subcutaneous injection only. Do not use for IV infusion.

PREPARE: **IV Infusion:** Use the supplied silicone-free disposable syringe with an 18–21 gauge needle to reconstitute the vial. ▪ Add 10 mL sterile water to each 250 mL to yield 25 mg/mL. ▪ To avoid foaming, gently swirl until completely dissolved. Do not shake or vigorously agitate. ▪ After dissolving, vent the vial with a needle to dissipate any foam. ▪ The reconstituted solution **must be** further diluted to a total of 100 mL as follows: From a 100 mL NS IV bag remove a volume equal to the total volume of abatacept in the reconstituted vials (e.g., for 2 vials, remove 20 mL). Using the supplied silicone-free syringe, slowly add the reconstituted abatacept to the IV bag and gently mix. ▪ The final concentration of the IV solution will be approximately 5, 7.5, or 10 mg/mL, depending on whether 2, 3, or 4 vials are used. ▪ Discard any unused abatacept.

ADMINISTER: **IV Infusion:** Use a 0.2–1.2 micron low-protein-binding filter. Infuse over 30 min.

INCOMPATIBILITIES **Solution/additive:** Should not be infused in the same intravenous line with other agents. **Y-site:** Should not be infused in the same intravenous line with other agents.

- Store at 2°–8° C (36°–46° F).

ADVERSE EFFECTS (≥1%) **Body as a Whole:** Infusion-related reactions, malignancies, cough, hypersensitivity reactions. **CNS:** *Headache,* dizziness. **CV:** Hypertension. **GI:** *Nausea,* dyspepsia. **Musculoskeletal:** Back pain, pain in extremity. **Respiratory:** *Upper respiratory tract infection, nasopharyngitis,* sinusitis, influenza, bronchitis. **Skin:** Rash. **Urogenital:** Urinary tract infection.

INTERACTIONS Drug: TNF ANTAGONISTS increase the risk of serious infections. Avoid use of anakinra.

PHARMACOKINETICS Half-Life: 13.1 days.

NURSING IMPLICATIONS

Assessment & Drug Effects

- Prior to initiating treatment with abatacept, screen for latent TB infection with a TB skin test.
- Monitor for S&S of hypersensitivity (e.g., hypotension, urticaria, and dyspnea); discontinue infusion and notify prescriber if any of these occur.
- Monitor for S&S of infection. Withhold drug and notify prescriber if patient develops a serious infection.
- Monitor for deterioration of respiratory status in patients with COPD.

Patient & Family Education

- Report any of the following to a health care provider: Any type of

Common adverse effects in *italic,* life-threatening effects underlined; generic names in **bold;** classifications in SMALL CAPS; ♣ Canadian drug name; ⊘ Prototype drug

3

infection, a positive TB skin test, a recent vaccination, a persistent cough, unexplained weight loss, fever, sore throat, or night sweats.

- Report S&S of an allergic reaction that may develop within 24 h of receiving abatacept (e.g., hives swollen face, eyelids, lips, tongue, throat, or trouble breathing).

- Do not accept immunizations with live vaccines while taking or within 3 mo of discontinuing abatacept.

ABCIXIMAB ⊕

(ab-cix′i-mab)
ReoPro
Classifications: ANTIPLATELET; GLYCOPROTEIN IIB/IIIA INHIBITOR
Therapeutic: PLATELET AGGREGA-TION INHIBITOR
Pregnancy Category: C

AVAILABILITY 2 mg/mL solution

ACTION & *THERAPEUTIC EFFECT*
Abciximab is a human-murine monoclonal antibody Fab (fragment antigen binding) fragment that binds to the glycoprotein IIb/IIIa (GPIIb/IIIa) receptor sites of platelets. *Abciximab inhibits platelet aggregation by preventing fibrinogen, von Willebrand's factor, and other molecules from adhering to GPIIb/IIIa receptor sites of the platelets.*

USES Adjunct to aspirin and heparin for the prevention of acute cardiac ischemic complications in patients undergoing percutaneous transluminal coronary angioplasty (PTCA).
UNLABELED USES Acute MI, Kawasaki disease.

CONTRAINDICATIONS Hypersensitivity to abciximab or to murine proteins; active internal bleeding; GI or

GU bleeding within 6 wk; history of CVA within 2 y or a CVA with severe neurologic deficit; administration of oral anticoagulants unless PT less than 1.2 times control; thrombocytopenia (less than 100,000 cells/mL); recent major surgery or trauma; intracranial neoplasm, aneurysm, severe hypertension; history of vasculitis; use of dextran before or during PTCA.

CAUTIOUS USE Patients weighing less than 75 kg; history of previous GI disease; recent thrombolytic therapy; PTCA within 12 h of MI; unsuccessful PTCA; PTCA procedure lasting longer than 70 min; older adults; pregnancy (category C); lactation. Safe use in children has not been established.

ROUTE & DOSAGE

PTCA

Adult: **IV** 0.25 mg/kg bolus 10–60 min prior to angioplasty, followed by continuous infusion of 0.125 mcg/kg/min (up to 10 mcg/min) for next 12–24 h

ADMINISTRATION

Intravenous
Do not shake vial. Discard if visible opaque particles are noted.

- Use a nonpyrogenic low protein-binding 0.2- or 0.22-micron filter when withdrawing drug into a syringe from the 2 mg/mL vial and when infusing as continuous IV.

PREPARE: Direct: No dilution required. **Continuous:** Inject 5 mL of abciximab (10 mg) into 250 mL of NS or D5W.
ADMINISTER: Direct: Give undiluted bolus dose over 5 min. **Continuous:** Infuse at no more than 15 mL/h (10 mcg/min) via an infusion pump over 12 h to 24 h.

Common adverse effects in *italic*, life-threatening effects underlined; generic names in **bold**; classifications in SMALL CAPS; ◆ Canadian drug name; ⊕ Prototype drug

INCOMPATIBILITIES **Solution/additive:** Infuse through separate IV line. **Y-site:** Infuse through separate IV line.

▪ Discard any unused drug at the end of the 12 h infusion as well as any unused portion left in vial. ▪ Store vials at 2°–8° C (36°–46° F).

ADVERSE EFFECTS (≥1%) Hematologic: <u>*Bleeding*</u>, including intracranial, retroperitoneal, and hematemesis; <u>thrombocytopenia</u>.

INTERACTIONS Drug: ORAL ANTICOAGULANTS, NSAIDS, **dipyridamole, ticlopidine, dextran** may increase risk of bleeding. **Herbal:** Gingko can increase bleeding risk.

PHARMACOKINETICS **Onset:** Greater than 90% inhibition of platelet aggregation within 2 h. **Duration:** Approximately 48 h. **Half-Life:** 30 min.

NURSING IMPLICATIONS

Assessment & Drug Effects
▪ Monitor for S&S of: Bleeding at all potential sites (e.g., catheter insertion, needle puncture, or cutdown sites; GI, GU, or retroperitoneal sites); hypersensitivity that may occur any time during administration.
▪ Lab tests: Baseline platelet count, PT, aPTT, and ACT, then repeat every 2–4 h during first 24 h; aPTT or ACT prior to arterial sheath removal (do not remove unless aPTT is 50 sec or less or ACT is 75 sec or less).
▪ Avoid or minimize unnecessary invasive procedures and devices to reduce risk of bleeding.
▪ Elevate head of bed 30° or less and keep limb straight when femoral artery access is used; following sheath removal, apply pressure for 30 min.

▪ Stop infusion immediately and notify prescriber if bleeding or S&S of hypersensitivity occurs.

Patient & Family Education
▪ Report any S&S of bleeding immediately.

ABIRATERONE ACETATE
(a'-bir-a'-ter-one as'-e-tate)
Zytiga
Classifications: ANTIANDROGEN; ANDROGEN BIOSYNTHESIS INHIBITOR
Therapeutic: ANTIANDROGEN
Pregnancy Category: X

AVAILABILITY 250 mg tablets

ACTION & *THERAPEUTIC EFFECT* Inhibits the enzyme required for androgen biosynthesis in testicular, adrenal, and prostatic tumor tissues. Enzyme inhibition may also result in increased mineralocorticoid production in the adrenal glands. *Decreased levels of serum testosterone and other androgens slow the growth of androgen-sensitive carcinomas.*

USES Metastatic castration-resistant prostate cancer, in combination with prednisone, in those previously treated with docetaxel.

CONTRAINDICATIONS Severe hepatic impairment (Child-Pugh class C; pregnancy (category X).
CAUTIOUS USE History of CV disease (e.g., heart failure, hypertension, recent MI, ventricular arrhythmias); hypokalemia; fluid retention; concurrent steroid therapy, especially during dosage adjustment or with concurrent infection or stress; moderate hepatic impairment (Child-Pugh class B).

Common adverse effects in *italic*, life-threatening effects <u>underlined</u>; generic names in **bold**; classifications in SMALL CAPS; ♣ Canadian drug name; ⊙ Prototype drug

5

ROUTE & DOSAGE

Metastatic Prostate Cancer

Adult: **PO** 1000 mg once daily in combination with PO prednisone 5 mg b.i.d.

Hepatic Impairment Dosage Adjustment

Moderate impairment (Child-Pugh class B): **PO** 250 mg once daily

Severe impairment (Child-Pugh class C): Do not use

ADMINISTRATION

Oral

- Give on an empty stomach 2 h before or 1 h after food.
- Tablets should be swallowed whole with water.
- Women who are or may be pregnant must use gloves to handle abiraterone.
- Store at 15°–30° C (59°–86° F).

ADVERSE EFFECTS (≥1%) **CV:** *Arrhythmia, cardiac failure,* chest pain or discomfort, *hot flush, hypertension.* **GI:** *Diarrhea,* dyspepsia. **Metabolic:** *Edema,* elevated ALT and AST, elevated total bilirubin, elevated triglycerides, hypokalemia, hypophosphatemia. **Musculoskeletal:** *Joint discomfort and swelling, muscle discomfort.* **Respiratory:** *Cough,* upper respiratory tract infection. **Urogenital:** Nocturia, *urinary frequency, urinary tract infection.*

INTERACTIONS Drug: Abiraterone can increase the levels of drugs requiring CYP2D6 (e.g., **dextromethorphan, thioridazine**). Strong inhibitors of CYP3A4 (e.g., **ketoconazole, itraconazole, clarithromycin, atazanavir,** nefazodone, saquinavir, telithromycin, ritonavir, indinavir, nelfinavir, voriconazole)** or inducers of CYP3A4 (e.g., **phenytoin, carbamazepine, rifampin, rifabutin, rifapentine, phenobarbital**) may increase or decrease, respectively, the levels of abiraterone.

PHARMACOKINETICS: 2 h. **Distribution:** Greater than 99% Plasma protein bound. **Metabolism:** In the liver to an active metabolite. **Elimination:** Fecal (88%) and renal (5%). **Half-Life:** 7–17 h.

NURSING IMPLICATIONS

Assessment & Drug Effects

- Monitor BP and cardiac function especially with a history of CV disease.
- Monitor for and report signs of fluid retention (e.g., sudden weight gain, peripheral edema).
- Monitor for and report S&S of hypokalemia or hepatotoxicity (see Appendix F). Withhold drug and notify prescriber if AST/ALT is above 5 × ULN or bilirubin above 3 × ULN.
- Lab tests: Monitor ALT, AST, and bilirubin at baseline, then q2wk for first 3 mo, then monthly thereafter; monitor serum electrolytes (especially potassium).

Patient & Family Education

- Do not take this drug within 2 h before or 1 h after consuming food.
- A condom should be used during sexual intercourse with a woman who is or could become pregnant.
- Report any of the following to a health care provider: Sudden weight gain, swelling of feet or legs, palpitations, unusual weakness,

muscle pain, S&S of a urinary tract infection.

ACAMPROSATE CALCIUM
(a-cam-pro′sate)

Campral
Classifications: SUBSTANCE ABUSE DETERRENT
Therapeutic: SUBSTANCE ABUSE INHIBITOR
Pregnancy Category: C

AVAILABILITY 333 mg delayed release tablets

ACTION & *THERAPEUTIC EFFECT*
A neurotransmitter analog that may interact with CNS glutamate and GABA neurotransmitter systems and help restore normal balance between neuronal excitation and inhibition. *Reduces craving for alcohol intake due to chronic use, but does not cause alcohol aversion or a disulfiram-like reaction as a result of ethanol ingestion.*

USES Maintenance of abstinence from alcohol in patients with alcoholism.

CONTRAINDICATIONS Hypersensitivity to acamprosate calcium or any of its components; suicidal ideation; severe renal impairment (CrCl less than 30 mL/min).
CAUTIOUS USE Moderate renal impairment; depression; pregnancy (category C); lactation. Safety and efficacy of acamprosate have not been established in adolescents or children younger than 18 y.

ROUTE & DOSAGE

Maintenance of Alcohol Abstinence
Adult: **PO** 666 mg t.i.d.

Renal Impairment Dosage Adjustment
CrCl 30–50 mL/min: 333 mg t.i.d.;
less than 30 mL/min: Do not use

ADMINISTRATION

Oral
- Ensure that the drug is not chewed or crushed. It **must be** swallowed whole.
- Store at 15°–30° C (59°–86° F).

ADVERSE EFFECTS (≥1%) **Body as a Whole:** Flu syndrome, chills. **CNS:** Depression, anxiety, insomnia, asthenia, dizziness, paresthesia, headache, somnolence, decreased libido, amnesia, abnormal thinking, tremor. **CV:** Palpitation, syncope. **GI:** *Diarrhea,* nausea, vomiting, anorexia, flatulence, dry mouth, abdominal pain, dyspepsia, constipation, increased appetite. **Metabolic:** Peripheral edema, weight gain. **Musculoskeletal:** Musculoskeletal pain. **Respiratory:** Rhinitis, cough, dyspnea, pharyngitis, bronchitis. **Skin:** Pruritus, diaphoresis, rash. **Special Senses:** Abnormal vision, taste perversion. **Urogenital:** Impotence.

INTERACTIONS Drug: None reported.

PHARMACOKINETICS Absorption: 11% bioavailability. **Metabolism:** Not metabolized. **Elimination:** Renal. **Half-Life:** 20–33 h.

NURSING IMPLICATIONS

Assessment & Drug Effects
- Monitor for S&S of depression or suicidal thinking.
- Monitor for: Impaired judgment or thinking; dizziness or impaired motor skills. Take appropriate protective measures.

Common adverse effects in *italic*, life-threatening effects underlined; generic names in **bold**; classifications in SMALL CAPS; ♣ Canadian drug name; ☺ Prototype drug

7

Patient & Family Education

- Report any alcohol consumption while taking acamprosate.
- Report promptly any of the following: Unusual anxiousness or nervousness; depression or suicidal thoughts; burning or tingling sensations in arms, legs, hands, or feet; chest pains or palpitations; difficulty urinating.
- Do not drive or engage in other hazardous activities until reaction to the drug is known.

ACARBOSE ⦿

(a-car′bose)

Precose

Classifications: ANTIDIABETIC; ALPHA-GLUCOSIDASE INHIBITOR

Therapeutic: ANTIDIABETIC

Pregnancy Category: B

AVAILABILITY 25 mg, 50 mg, 100 mg tablets

ACTION & *THERAPEUTIC EFFECT*
Acarbose is an oral alpha-glucosidase inhibitor that delays absorption of sugars from the intestinal tract. The inhibitory effect varies according to the enzymes involved; from most to least inhibited are glucoamylase, sucrase, maltase, and isomaltase. *Acarbose reduces blood sugar by interfering with carbohydrate absorption from the GI tract.*

USES In conjunction with diet and exercise for type 2 diabetes.
UNLABELED USES Adjunctive treatment of type 1 diabetes.

CONTRAINDICATIONS Inflammatory bowel disease, colon ulcers, partial bowel obstruction, predisposition for obstruction; lactation.

CAUTIOUS USE GI distress or liver disorders, pregnancy (category B); children younger than 18 y.

ROUTE & DOSAGE

Type 2 Diabetes Mellitus
Adult: **PO** Start with 25 mg daily to t.i.d. with meals, titrate to individual response (max: 150 mg/day for 60 kg or less, 300 mg/day for greater than 60 kg)

ADMINISTRATION

Oral

- Remove drug from foil wrapper immediately before administration.
- Give drug with first bite at each of the three main meals.
- Do not store above 25° C (77° F). Keep tightly closed and protect from moisture.

ADVERSE EFFECTS (≥1%) **CNS:** Sleepiness, weakness, dizziness, headache, vertigo (may be due to poor diabetic control). **Endocrine:** Hypoglycemia (especially in combination with sulfonylureas and insulin). **GI:** *Diarrhea, flatulence, abdominal distention,* borborygmi, increased liver function tests. **Hematologic:** Anemia (especially iron deficiency). **Skin:** Erythema, exanthema, urticaria.

INTERACTIONS Drug: SULFONYLUREAS may increase hypoglycemic effects. Drugs that induce hyperglycemia (e.g., THIAZIDES, CORTICOSTEROIDS, PHENOTHIAZINES, ESTROGENS, **phenytoin, isoniazid**) may decrease effectiveness of acarbose. May decrease the effect of **digoxin**. **Herbal:** Ginseng may increase hypoglycemic effects.

PHARMACOKINETICS Absorption: 0.5–2% is absorbed intact from GI tract. **Peak:** Peak blood glucose reduction approximately 70 min

---

after dose. **Metabolism:** In GI tract by intestinal bacteria and digestive enzymes. **Elimination:** 35% in urine, 51% in feces, 5% in air as CO_2. **Half-Life:** 2 h.

NURSING IMPLICATIONS

Assessment & Drug Effects

- Lab tests: Frequent blood glucose and periodic HbA1C; periodic liver enzymes; Hct and Hgb if anemia suspected.
- Treat hypoglycemia with dextrose; not with sucrose (table sugar).

Patient & Family Education

- Note: Acarbose prevents the breakdown of table sugar. Have a source of dextrose, such as dextrose paste, available to treat low blood sugar.
- Monitor closely blood glucose, especially following dosage changes.
- Report abdominal distress; dietary adjustment or dosage reduction may be warranted.
- Monitor weight and report significant changes.

ACEBUTOLOL HYDROCHLORIDE

(a-se-byoo-toe'lole)

Sectral

Classifications: BETA-ADRENERGIC ANTAGONIST; ANTIHYPERTENSIVE; CLASS II ANTIARRHYTHMIC

Therapeutic: ANTIHYPERTENSIVE; CLASS II ANTIARRYTHMIC

Prototype: Propranolol

Pregnancy Category: B

AVAILABILITY 200 mg, 400 mg capsules

ACTION & *THERAPEUTIC EFFECT*

Beta$_1$-selective adrenergic blocking agent with mild sympathomimetic activity (partial beta-agonist activity). Produces negative chronotropic and inotropic activity (i.e., decreases exercise-induced heart rate, inhibits reflex orthostatic tachycardia, and decreases cardiac output at rest and during exercise). *Decreases both systolic and diastolic BP at rest and during exercise. Exhibits antiarrhythmic activity (class II antiarrhythmic agent).*

USES Treatment of mild to moderate hypertension. Management of recurrent stable ventricular arrhythmias.

UNLABELED USES Supraventricular arrhythmias, chronic stable angina pectoris.

CONTRAINDICATIONS Overt CHF, second- or third-degree AV block, severe bradycardia, cardiogenic shock; acute bronchospasm, pulmonary edema; lactation.

CAUTIOUS USE Impaired cardiac function, well-compensated CHF, mesenteric or peripheral vascular disease; cerebrovascular disease; patients undergoing major surgery involving general anesthesia; renal or hepatic impairment; labile diabetes mellitus; hyperthyroidism; bronchospastic disease (asthma, emphysema); avoid abrupt withdrawal; pregnancy (category B); children younger than 12 y.

ROUTE & DOSAGE

Hypertension

Adult: **PO** 400–800 mg/day in 1–2 divided doses (max: 1200 mg/day)
Geriatric: **PO** 200–400 mg/day (max: 800 mg/day)

Ventricular Arrhythmias

Adult: **PO** 200 mg b.i.d. increased to 600–1200 mg/day

Common adverse effects in *italic*, life-threatening effects underlined; generic names in **bold**; classifications in SMALL CAPS; ♦ Canadian drug name; ⊘ Prototype drug

9

Renal Impairment Dosage Adjustment

CrCl less than 50 mL/min: Reduce dose by 50%; less than 25 mL/min: Reduce dose by 75%; Intermittent hemodialysis: Reduce dose by 75%

ADMINISTRATION

Oral

- Check BP and apical pulse before administration. If heart rate is less than 60 bpm or other ordered parameter, consult prescriber.
- Drug is usually discontinued gradually over a period of 2 wk.
- Store at 15°–30° C (59°–86° F).

ADVERSE EFFECTS (≥1%) **Body as a Whole:** *Fatigue.* **CNS:** Dizziness, insomnia, drowsiness, confusion, fainting. **CV:** *Bradycardia,* hypotension, CHF. **GI:** Nausea, *diarrhea, constipation,* flatulence. **Hematologic:** Agranulocytosis, *antinuclear antibodies (ANA).* **Metabolic:** Hypoglycemia (may mask symptoms of a hypoglycemic reaction). **Respiratory:** Bronchospasm, pulmonary edema, dyspnea. **Urogenital:** Decreased libido; impotence.

DIAGNOSTIC TEST INTERFERENCE False-negative test results possible (see **propranolol**).

INTERACTIONS Drug: Other HYPO-TENSIVE AGENTS, DIURETICS increase hypotensive effect; with **albuterol, metaproterenol, terbutaline,** or **pirbuterol,** there is mutual antagonism with acebutolol; NSAIDS blunt hypotensive effect; decreases hypoglycemic effect of **glyburide;** increases bradycardia and sinus arrest with **amiodarone.**

PHARMACOKINETICS Absorption: Average bioavailability of 40%. (In geriatric patients, bioavailability increases twofold.) **Peak:** 3 h.

Distribution: Minimally into CSF; crosses placenta; is excreted in breast milk. **Metabolism:** In liver. **Elimination:** In urine, feces. **Half-Life:** 3–4 h.

NURSING IMPLICATIONS

Assessment & Drug Effects

- Monitor BP and cardiac status throughout therapy. Observe for and report marked bradycardia or hypotension, especially when patient is also receiving a catecholamine-depleting drug (e.g., reserpine).
- Monitor I&O ratio and pattern and report changes to prescriber (e.g., dysuria, nocturia, oliguria, weight change).
- Monitor for S&S of CHF, especially peripheral edema, dyspnea, activity intolerance.
- Lab tests: Monitor for drug induced positive ANA titer during long-term therapy, especially in women and older adults; periodic CBC with long-term therapy.

Patient & Family Education

- Know parameters for withholding drug (e.g., pulse less than 60).
- Note: Common adverse effects include insomnia, drowsiness, and confusion.
- Do not drive or engage in potentially hazardous activities until response to drug is known.
- Do not increase, decrease, omit, or discontinue drug regimen without advice from the prescriber. Abrupt withdrawal may worsen angina or precipitate MI in patient with heart disease.
- Contact prescriber promptly at the first signs or symptoms of CHF (see Appendix F).
- Report muscle and joint pain to prescriber. Discontinuation of drug therapy usually reverses these adverse effects.

Common adverse effects in *italic*, life-threatening effects <u>underlined</u>; generic names in **bold**; classifications in SMALL CAPS; ✦ Canadian drug name; ⊙ Prototype drug

- Monitor for loss of glycemic control if diabetic.
- Note: Drug may mask symptoms of hypoglycemia (see Appendix F) and potentiate insulin-induced hypoglycemia in diabetics.
- Avoid use of OTC oral cold preparations and topical nasal decongestants unless approved by the prescriber.

ACETAMINOPHEN, PARACETAMOL ⊕

(a-seat-a-mee'noe-fen)

Abenol ♦, **A'Cenol, Acephen, Anacin-3, Anuphen, APAP, Atasol** ♦, **Campain** ♦, **Dolanex, Exdol** ♦, **Halenol, Liquiprin, Ofirmev, Panadol, Pedric, Robigesic** ♦, **Rounox** ♦, **Tapar, Tempra, Tylenol, Tylenol Arthritis, Valadol**

Classifications: NONNARCOTIC ANALGESIC, ANTIPYRETIC
Therapeutic: NONNARCOTIC ANALGESIC; ANTIPYRETIC
Pregnancy Category: B

AVAILABILITY 80 mg, 120 mg, 125 mg, 300 mg, 325 mg, 650 mg suppositories; 80 mg, 160 mg, 325 mg, 500 mg tablets/caplets; 650 mg extended release tablets/capsules; 80 mg/0.8 mL, 80 mg/2.5 mL, 80 mg/5 mL, 120 mg/5 mL, 160 mg/5 mL, 500 mg/5 mL liquid; 1000 mg/100 mL injection

ACTION & *THERAPEUTIC EFFECT* Produces analgesia by unknown mechanism, but it is centrally acting in the CNS by increasing the pain threshold by inhibiting cyclooxygenase. Reduces fever by direct action on hypothalamus heat-regulating center with consequent peripheral vasodilation, sweating, and dissipation of heat. *It provides temporary analgesia for mild to moderate pain.*

In addition, acetaminophen lowers body temperature in individuals with a fever.

USES Fever reduction. Temporary relief of mild to moderate pain. Generally as substitute for aspirin when the latter is not tolerated or is contraindicated.

CONTRAINDICATIONS Hypersensitivity to acetaminophen or phenacetin; use with alcohol.
CAUTIOUS USE Repeated administration to patients with anemia, G6PD deficiency, renal or hepatic disease; arthritic or rheumatoid conditions affecting children younger than 12 y; alcoholism; malnutrition; thrombocytopenia; bone marrow depression, immunosuppression; pregnancy (category B).

ROUTE & DOSAGE

Mild to Moderate Pain, Fever

Adult: **PO** 325–650 mg q4–6h (max: 4 g/day) **PR** 650 mg q4–6h (max: 4 g/day) **IV** 1000 mg q6h or 650 q4h prn
Child: **PO** 10–15 mg/kg q4–6h **PR** 2–5 y, 120 mg q4–6h (max: 720 mg/day); 6–12 y, 325 mg q4–6h (max: 2.6 g/day)
Child (2 yr or more): **IV** 15 mg/kg/dose q6h or 12.5 mg/kg/dose q4h prn
Neonate: **PO** 10–15 mg/kg q6–8h

ADMINISTRATION

Oral
- Ensure that extended release tablets are not crushed or chewed. These must be swallowed whole.
- Chewable tablets should be thoroughly chewed and wetted before they are swallowed.

- Do not coadminister with a high carbohydrate meal; absorption rate may be significantly retarded.
- Store in light-resistant containers at room temperature, preferably between 15°–30° C (59°–86° F).

Rectal
- Insert suppositories beyond the rectal sphincter.

Intravenous

PREPARE: **Intermittent:** For adults and adolescents weighing 50 kg (110 lb) or more, give without dilution by attaching a vented IV set directly to the 100 mL (1000 mg) vial. For patients weighing less than 50 kg (110 lb), withdraw the needed dose from a sealed 1000 mg vial and place in an empty sterile container (e.g., plastic IV bag, syringe) for infusion.
ADMINISTER: **Intermittent:** Infuse over 15 min. For small volume pediatric doses up to 60 mL, use a syringe pump to administer over 15 min. Store at controlled temperature and use within 6 h after opening.

ADVERSE EFFECTS (≥1%) **Body as a Whole:** Negligible with recommended dosage; rash. **Acute poisoning:** Anorexia, nausea, vomiting, dizziness, lethargy, diaphoresis, chills, epigastric or abdominal pain, diarrhea; onset of <u>hepatotoxicity</u>: elevation of serum transaminases (ALT, AST) and bilirubin; hypoglycemia, <u>hepatic coma, acute renal failure</u> (rare). **Chronic ingestion:** Neutropenia, pancytopenia, <u>leukopenia</u>, <u>thrombocytopenic purpura</u>, *hepatotoxicity in alcoholics,* renal damage.

DIAGNOSTIC TEST INTERFERENCE False increases in ***urinary 5-HIAA*** (5-hydroxyindoleacetic acid) by-product of serotonin; false decreases in ***blood glucose*** (by ***glucose oxidase–peroxidase procedure***); false increases in urinary glucose (with certain instruments in glucose analyses); and false increases in ***serum uric acid*** (with ***phosphotungstate method***). High doses or long-term therapy: hepatic, renal, and hematopoietic function (periodically).

INTERACTIONS Drug: Cholestyramine may decrease acetaminophen absorption. With chronic coadministration, BARBITURATES, **carbamazepine, phenytoin,** and **rifampin** may increase potential for chronic hepatotoxicity. Chronic, excessive ingestion of **alcohol** will increase risk of hepatotoxicity.

PHARMACOKINETICS Absorption: Rapid and almost complete absorption (PO) less complete absorption from rectal suppository. **Peak:** 0.5–2 h. **Duration:** 3–4 h. **Distribution:** In all body fluids; crosses placenta. **Metabolism:** Extensively in liver. **Elimination:** 90–100% of drug excreted as metabolites in urine; excreted in breast milk. **Half-Life:** 1–3 h.

NURSING IMPLICATIONS
Assessment & Drug Effects
- Monitor for S&S of: Hepatotoxicity, even with moderate acetaminophen doses, especially in individuals with poor nutrition or who have ingested alcohol (3 or more alcoholic drinks daily) over prolonged periods; poisoning, usually from accidental ingestion or suicide attempts; potential abuse from psychological dependence (withdrawal has been associated with restless and excited responses).

Patient & Family Education

- Do not take other medications (e.g., cold preparations) containing acetaminophen without medical advice; overdosing and chronic use can cause liver damage and other toxic effects.
- Do not self-medicate adults for pain more than 10 days (5 days in children) without consulting a prescriber.
- Do not use this medication without medical direction for: Fever persisting longer than 3 days, fever over 39.5° C (103° F), or recurrent fever.
- Do not give children more than 5 doses in 24 h unless prescribed by prescriber.

ACETAZOLAMIDE ℗⊕

(a-set-a-zole'a-mide)

Acetazolam ♣, **Apo-Acetazola-mide** ♣, **Diamox Sequels**
Classifications: CARBONIC ANHYDRASE INHIBITOR
Therapeutic: DIURETIC; ANTICONVULSANT; ANTIGLAUCOMA
Pregnancy Category: C

AVAILABILITY 125 mg, 250 mg tablets; 500 mg sustained release capsules; 500 mg powder for injection

ACTION & THERAPEUTIC EFFECT
The mechanism of anticonvulsant action is thought to involve inhibition of CNS carbonic anhydrase, which retards abnormal transmission from CNS neurons. Diuretic effect is due to inhibition of carbonic anhydrase activity in proximal renal tubule, preventing formation of carbonic acid, and therefore the formation of H^+ and HCO_3^-. Inhibition of carbonic anhydrase in eye reduces rate of aqueous humor formation with consequent lowering of intraocular pressure. *Reduces seizure activity and intraocular pressure. Additionally, it has a diuretic effect.*

USES Focal Absence seizures; reduction of intraocular pressure in open-angle glaucoma and secondary glaucoma; preoperative treatment of acute closed-angle glaucoma; edema.
UNLABELED USES Hydrocephalus; familial periodic paralysis, metabolic alkalosis, nystagmus, urinary alkalinization.

CONTRAINDICATIONS Hypersensitivity to carbonic anhydrase inhibitors, marked renal and hepatic disease; Addison's disease or other types of adrenocortical insufficiency; hyponatremia, hypokalemia, hyperchloremic acidosis; prolonged administration to patients with hyphema; chronic noncongestive angle-closure glaucoma.
CAUTIOUS USE Hypersensitivity to sulfonamides and derivatives (e.g., thiazides), history of hypercalciuria; diabetes mellitus, elderly, gout, patients receiving digitalis, obstructive pulmonary disease, respiratory acidosis; pregnancy (category C).

ROUTE & DOSAGE

Glaucoma

Adult: **PO** 250 mg 1–4 times/day; 500 mg sustained release b.i.d., up to 1 g/day **IM/IV** 500 mg, may repeat in 2–4 h

Absence Seizures

Adult: **PO/IV** 8–30 mg/kg/day in 1–4 doses

Edema

Adult: **PO/IV** 250–375 mg every a.m. (5 mg/kg); may be given every other day if condition improves

---

Common adverse effects in *italic*, life-threatening effects underlined; generic names in **bold**; classifications in SMALL CAPS; ♣ Canadian drug name; ⊕ Prototype drug

13

Altitude Sickness

Adult: **PO** 250 mg q6–12h or 500 mg sustained release q12–24h, starting 24–48 h before climb and continuing for 48 h at high altitude

Renal Impairment Dosage Adjustment

CrCl 10–50 mL/min: Extend interval to q12h; *less than 10 mL/min:* Use not recommended

Hemodialysis Dosage Adjustment
Administer post-dialysis

ADMINISTRATION

Oral
- Administer diuretic dose in morning to avoid interrupted sleep.
- Give with food or meals to minimize GI upset.
- Note: If tablet(s) cannot be swallowed, soften tablet(s) (not sustained release form) in 2 tsp of hot water and add to 2 tsp of honey/syrup to disguise bitter taste; avoid syrups containing alcohol or glycerin, or crush tablet(s) and suspend in syrup (250–500 mg/5 mL syrup). Prepare just before administration. Drug does not dissolve in fruit juices.
- Store oral preparations at 15°–30° C (59°–86° F) unless otherwise directed.

Intramuscular
- Reconstitute as for IV administration. See PREPARE Direct.
- Give IM for rapid lowering of intraocular pressure or in patients unable to take oral dosage.
- Note: The intramuscular dosage is not the route of choice because the alkalinity of the solution makes the injection painful.

Intravenous

PREPARE: **Direct:** Reconstitute each 500 mg vial with at least 5 mL of sterile water for injection to yield approximately 100 mg/mL. ▪ May be used as prepared or further diluted. **IV Infusion:** Dilute reconstituted solution with D5W or NS. Use within 24 h of reconstitution.

ADMINISTER: **Direct:** Give at a rate of 500 mg or fraction thereof over 1 min. **IV Infusion:** Give as a continuous infusion over 4–8 h.

INCOMPATIBILITIES **Solution/additive:** Amino acid, multivitamin **Y-site:** Diltiazem, TPN.

ADVERSE EFFECTS (≥1%) **CNS:** Paresthesias, sedation, malaise, disorientation, depression, fatigue, muscle weakness, flaccid paralysis. **GI:** Anorexia, nausea, vomiting, weight loss, dry mouth, thirst, diarrhea. **Hematologic:** Bone marrow depression with agranulocytosis, hemolytic anemia, aplastic anemia, leukopenia, pancytopenia. **Metabolic:** Increased excretion of calcium, potassium, magnesium, and sodium, metabolic acidosis, hyperglycemia, hyperuricemia. **Ocular:** Transient myopia. **Urogenital:** Glycosuria, urinary frequency, polyuria, dysuria, hematuria, crystalluria. **Other:** Exacerbation of gout, hepatic dysfunction, Stevens-Johnson syndrome.

DIAGNOSTIC TEST INTERFERENCE Monitor for false-positive *urinary protein* determinations; falsely high values for *urine urobilinogen;* depressed *iodine uptake* values (exception: hypothyroidism).

INTERACTIONS Drug: Renal excretion of AMPHETAMINES, **ephedrine, flecainide, quinidine, procainamide,** TRICYCLIC ANTIDEPRESSANTS may be decreased, thereby enhancing or prolonging their effects. Renal excretion of **lithium, phenobarbital** may be increased. **Amphotericin B** and CORTICOSTEROIDS may accelerate **potassium** loss. **Digoxin** may predispose persons with hypokalemia to **digitalis** toxicity; puts patients on high doses of SALICYLATES at high risk for SALICYLATE toxicity.

PHARMACOKINETICS Absorption: Well absorbed from GI tract. **Onset:** 1 h regular release; 2 h sustained release; 2 min IV. **Peak:** 2–4 h reg; 8–18 h sustained; 15 min IV. **Duration:** 8–12 h reg; 18–24 h sustained; 4–5 h IV. **Distribution:** Distributed throughout body; crosses placenta. **Elimination:** In urine. **Half-Life:** 2.4–5.8 h.

NURSING IMPLICATIONS

Assessment & Drug Effects

- Establish baseline weight before initial therapy and weigh daily thereafter when used to treat edema.
- Monitor for S&S of: Mild to severe metabolic acidosis; potassium loss which is greatest early in therapy (see hypokalemia in Appendix F).
- Monitor I&O especially when used with other diuretics.
- Lab tests: Blood pH, blood gases, urinalysis, CBC, and serum electrolytes (initially and periodically during prolonged therapy or concomitant therapy with other diuretics or digitalis).

Patient & Family Education

- Maintain adequate fluid intake (1.5–2.5 L/24 h; 1 liter is approximately equal to 1 quart) to reduce risk of kidney stones.

- Report any of the following: Numbness, tingling, burning, drowsiness, and visual problems, sore throat or mouth, unusual bleeding, fever, skin or renal problems.
- Eat potassium-rich diet and take potassium supplement when taking this drug in high doses or for prolonged periods.
- Use caution when engaging in hazardous activities until reaction to drug is known.

ACETYLCYSTEINE ⊕

(a-se-til-sis'tay-een)
Acetadote, N-Acetylcysteine, Mucomyst, Parvolex ♦
Classifications: MUCOLYTIC; ANTIDOTE
Therapeutic: MUCOLYTIC; ANTIDOTE
Pregnancy Category: B

AVAILABILITY 10%, 20% solution for inhalation; 20% solution for injection

ACTION & *THERAPEUTIC EFFECT* Acetylcysteine probably acts by disrupting disulfide linkages of mucoproteins in purulent and nonpurulent bronchial secretions. In acetaminophen overdose, it helps to prevent hepatotoxicity by serving as a substrate for the toxic metabolites of acetaminophen. *Lowers viscosity and facilitates the removal of secretions. Removes the toxic metabolites of acetaminophen.*

USES Adjuvant therapy in patients with abnormal mucous secretions in acute and chronic bronchopulmonary diseases, and in pulmonary complications of cystic fibrosis and surgery, tracheostomy, and atelectasis. Also used in diagnostic bronchial studies and as an antidote for acute acetaminophen poisoning.

Common adverse effects in *italic*, life-threatening effects underlined; generic names in **bold**; classifications in SMALL CAPS; ♦ Canadian drug name; ⊕ Prototype drug

15

UNLABELED USES Meconium ileus; prevention of radiocontrast-induced renal dysfunction.

CONTRAINDICATIONS Hypersensitivity to acetylcysteine; patients at risk of gastric hemorrhage.

CAUTIOUS USE Patients with asthma, older adults, severe hepatic disease, esophageal varices, peptic ulcer disease; debilitated patients with severe respiratory insufficiency; pregnancy (category B), lactation.

ROUTE & DOSAGE

Mucolytic

Adult: **Inhalation** 1–10 mL of 20% solution q4–6h or 2–20 mL of 10% solution q4–6h **Direct Instillation** 1–2 mL of 10–20% solution q1–4h
Child: **Inhalation** 3–5 mL of 20% solution or 6–10 mL of 10% solution 3–4 times/day
Infant: **Inhalation** 1–2 mL 20% solution or 2–4 mL of 10% solution 3–4 times/day

Acetaminophen Toxicity

Adult/Child: **PO** 140 mg/kg followed by 70 mg/kg q4h for 17 doses (use a 5% solution)
Adult/Adolescent/Child: **IV** 150 mg/kg infused over 60 min, followed by 50 mg/kg over 4 h, then 100 mg/kg over 16 h; total dose 300 mg/kg over 21 h

ADMINISTRATION

Inhalation and Instillation

- Prepare dilution within 1 h of use; drug does not contain an antimicrobial agent. A light purple discoloration does not significantly impair drug's effectiveness.
- Dilute the 20% solution with NS or water for injection. The 10% solution may be used undiluted.
- Give by direct instillation into tracheostomy (1–2 mL of 10–20% solution).
- Instruct patient to clear airway, if possible, coughing productively prior to aerosol administration to ensure maximum effect.
- Store opened vial in refrigerator to retard oxidation; use within 96 h.
- Store unopened vial at 15°–30° C (59°–86° F), unless otherwise directed.

Oral

- Dilute the 20% solution 1:3 with cola, orange juice, or other soft drink to make a 5% solution. If administered via a gastric tube, water may be used as the diluent.
- Freshly prepare all diluted solutions and use within 1 h of preparation.

Intravenous

PREPARE: IV Infusion: Acetylcysteine reacts with certain metals and rubber; use IV equipment made of plastic or glass.
- Dilute all required doses in D5W as follows: For loading dose, add a dose equal to 150 mg/kg to 200 mL; for first maintenance dose, add a dose equal to 50 mg/kg to 500 mL; for second maintenance dose, add a dose equal to 100 mg/kg to 1000 mL. ▪ Note: The total IV volume should be reduced for patients less than 40 kg and for those with fluid restriction. In small children, individualize the total IV volume to avoid water intoxication and hyponatremia.

ADMINISTER: IV Infusion: Give loading dose over 60 min, maintenance dose 1 over 4 h, maintenance dose 2 over 16 h. Complete all infusions over 24 h.
INCOMPATIBILITIES Y-site: Ceftazidime.

- Store reconstituted solution for up to 24 h at 15°–30° C (59°–86° F).

ADVERSE EFFECTS (≥1%) **CNS:** Dizziness, drowsiness. **GI:** Nausea, *vomiting*, stomatitis, hepatotoxicity (urticaria). **Respiratory:** <u>Broncho-spasm</u>, rhinorrhea, burning sensation in upper respiratory passages, epistaxis.

PHARMACOKINETICS Onset: 1 min. **Peak:** 5–10 min. **Metabolism:** Deacetylated in liver to cysteine.

NURSING IMPLICATIONS

Assessment & Drug Effects

- During IV infusion, carefully monitor for fluid overload and signs of hyponatremia (i.e., changes in mental status).
- Monitor for S&S of aspiration of excess secretions, and for bronchospasm (unpredictable); withhold drug and notify prescriber immediately if either occurs.
- Lab tests: ABGs, pulmonary functions and pulse oximetry as indicated; baseline serum acetaminophen level (for toxicity), LFTs, bilirubin, serum electrolytes, BUN, and plasma glucose.
- Have suction apparatus immediately available. Increased volume of respiratory tract fluid may be liberated; suction or endotracheal aspiration may be necessary to establish and maintain an open airway. Older adults and debilitated patients are particularly at risk.
- Nausea and vomiting may occur, particularly when face mask is used, due to unpleasant odor of drug and excess volume of liquefied bronchial secretions.

Patient & Family Education

- Report difficulty with clearing the airway or any other respiratory distress.
- Report nausea, as an antiemetic may be indicated.

- Note: Unpleasant odor of inhaled drug becomes less noticeable with continued use.

ACITRETIN
(a-ci-tree'tin)
Soriatane
Classifications: RETINOID
Therapeutic: ANTIPSORIATIC
Prototype: Isotretinoin
Pregnancy Category: X

AVAILABILITY 10 mg, 17.5 mg, 22.5 mg, 25 mg capsules

ACTION & *THERAPEUTIC EFFECT* Acitretin binds to the retinoic acid receptors in the skin, thus modifying gene expression, epithelial cell growth, and cell differentiation. *Acitretin is a highly toxic metabolite of retinol (vitamin A).*

USES Treatment of severe, recalcitrant psoriasis in adults.
UNLABELED USES Eczema, lichen planus.

CONTRAINDICATIONS Sensitivity to parabens, papilledema, severe renal impairment or renal failure, development of psychiatric symptoms (depression, etc); pregnancy (category X) for at least 3 y after use, lactation.
CAUTIOUS USE Patients with impaired hepatic function, hepatitis, diabetes mellitus, obesity, alcoholism, history of pancreatitis, hypertriglyceridemia, hypercholesterolemia, coronary artery disease, retinal disease, degenerative joint disease.

ROUTE & DOSAGE

Psoriasis
Adult: **PO** 25–50 mg daily with main meal

Common adverse effects in *italic*, life-threatening effects <u>underlined</u>; generic names in **bold**; classifications in SMALL CAPS; ♣ Canadian drug name; ☺ Prototype drug

17

ADMINISTRATION

Oral

- Administer as single dose with main meal because food enhances absorption.
- Store at 15°–25° C (59°–77° F) and protect from light. After opening, avoid exposure to high temperatures and humidity.

ADVERSE EFFECTS (≥1%) Body as a Whole: *Hyperesthesia, paresthesias, arthralgia, progression of existing spinal hyperostosis, rigors,* back pain, hypertonia, myalgia, fatigue, hot flashes, increased appetite. **CNS:** Headache, depression, aggressive feelings and thoughts of self-harm, insomnia, somnolence. **CV:** Flushing, edema. **GI:** *Dry mouth, increased liver function tests, increased triglycerides and cholesterol,* hepatitis, gingival bleeding, gingivitis, increased saliva, stomatitis, thirst, ulcerative stomatitis, abdominal pain, diarrhea, nausea, tongue disorder. **Special Senses:** Blurred vision, blepharitis, conjunctivitis, decreased night vision/night blindness, eye pain, photophobia; earache, tinnitus; taste perversion. **Respiratory:** Sinusitis. **Skin:** *Alopecia, skin peeling, dry skin, nail disorders, pruritus, rash, cheilitis, skin atrophy, paronychia,* abnormal skin odor and hair texture, cold/clammy skin, increased sweating, purpura, seborrhea, skin ulceration, sunburn. **Other:** *Rhinitis, epistaxis, xerophthalmia.*

INTERACTIONS Drug: Use with **ethanol** causes longer half-life than acitretin; interferes with the contraceptive efficacy of **progestin**-only ORAL CONTRACEPTIVES. Use with **methotrexate** increases the risk of hepatitis. **Food:** Excess vitamin A.

PHARMACOKINETICS Absorption: Rapidly from GI tract, optimal absorption when taken with food.

Peak: 2–5 h. **Distribution:** Crosses placenta, distributed into breast milk. **Metabolism:** Active metabolite, *cis*-acitretin. **Elimination:** In both urine and feces. **Half-Life:** 49 h acitretin, 63 h *cis*-acitretin.

NURSING IMPLICATIONS

Assessment & Drug Effects

- Monitor for S&S of pancreatitis or loss of glycemic control in diabetics. Report either condition immediately to prescriber.
- Lab tests: Baseline and at 1- to 2-wk intervals until response to drug is known, do lipid profile and liver function tests. Monitor blood glucose and HbA1C periodically.

Patient & Family Education

- Note: Transient worsening of psoriasis may occur during early therapy.
- Review common adverse effects of drug; lag time of 2–3 mo may be necessary before drug effect is evident.
- Discontinue drug and report immediately to prescriber if visual problems develop.
- Note: Dry eyes with decreased tolerance for contact lenses may occur.
- Do not drink alcohol while taking this drug; it increases risk of hepatotoxicity and hypertriglyceridemia; females should avoid alcohol during and for 2 mo following therapy.
- Avoid excessive amounts of vitamin A (consult prescriber).
- Do not donate blood for 3 y following therapy.
- Avoid excessive exposure to sunlight or UV light.
- Use two forms of effective contraception for 1 mo before and at least 3 y following therapy because of the serious risk of fetal deformities that could result from exposure to this medication.

Common adverse effects in *italic*, life-threatening effects underlined; generic names in **bold**; classifications in SMALL CAPS; ♣ Canadian drug name; ⊙ Prototype drug

ACRIVASTINE/ PSEUDOEPHEDRINE

(a-cri-vas'teen)

Semprex-D (combination with pseudoephedrine)
Classifications: H₁-RECEPTOR ANTAGONIST; DECONGESTANT
Therapeutic: ANTIHISTAMINE; DECONGESTANT
Prototype: Diphenhydramine
Pregnancy Category: B

AVAILABILITY Acrivastine 8 mg/ pseudoephedrine 60 mg capsules

ACTION & *THERAPEUTIC EFFECT*
Acrivastine is an H₁-receptor histamine antagonist that controls histamine-mediated symptoms and acts on sympathetic nerve endings. Pseudoephedrine shrinks swollen nasal mucous membranes and reduces nasal congestion of the mucosa. *It is effective in allergic rhinitis by reducing nasal congestion and decreasing respiratory mucosa swelling.*

USES Seasonal and perennial allergic rhinitis with nasal congestion.

CONTRAINDICATIONS Hypersensitivity to acrivastine, triprolidine, pseudoephedrine, or ephedrine; severe hypertension or severe coronary artery disease; patients on MAO inhibitor drugs; uncontrolled hypertension; tachycardia, acute cardiac arrhythmias; closed-angle glaucoma.

CAUTIOUS USE Renal insufficiency, hypertension, DM, ischemic heart disease, increased intraocular pressure, hyperthyroidism, BPH, GI disorders, older adults, pregnancy (category B), lactation. Safety and effectiveness in children younger than 12 y have not been established.

ROUTE & DOSAGE

Allergic Rhinitis
Adult: **PO** 1 cap (8 mg acrivastine/60 mg pseudoephedrine) q4–6h

Renal Impairment Adjustment
CrCl less than 48 mL/min: Do not use

ADMINISTRATION

Oral
- Do not give to patients with a creatinine clearance of 48 mL/min or less.
- Store at 15°–25° C (59°–77° F); protect from light and moisture.

ADVERSE EFFECTS (≥1%) **CNS:** Headache, vertigo, dizziness, insomnia, jitteriness, *drowsiness*. **GI:** Nausea, diarrhea, dry mouth, dyspepsia.

INTERACTIONS Drug: Alcohol may increase psychomotor impairment.

PHARMACOKINETICS Absorption: Rapidly from GI tract. **Onset:** 1 h. **Duration:** Approximately 12 h. **Metabolism:** In liver. **Elimination:** Approximately 65% excreted unchanged in urine. **Half-Life:** 1.5 h.

NURSING IMPLICATIONS

Assessment & Drug Effects
- Monitor for dizziness, sedation, urinary obstruction, and hypotension, especially in older adults.
- Assess for significant drowsiness, which may necessitate drug discontinuation.

Patient & Family Education
- Do not use this drug in combination with other OTC antihistamines or decongestants.
- Do not drive or engage in potentially hazardous activities until response to drug is known.
- Do not take alcohol or other CNS depressants while taking this drug.

Common adverse effects in *italic*, life-threatening effects underlined; generic names in **bold**; classifications in SMALL CAPS; ♣ Canadian drug name; ⊕ Prototype drug

19

√ACYCLOVIR, ACYCLOVIR SODIUM ℗℞

(ay-sye'kloe-ver)

Zovirax

Classifications: ANTIVIRAL
Therapeutic: ANTIVIRAL; ANTI-HERPES
Pregnancy Category: B

AVAILABILITY 200 mg capsules; 400 mg, 800 mg tablets; 200 mg/5 mL oral suspension; 50 mg/mL injection; 5% ointment, cream

ACTION & *THERAPEUTIC EFFECT*
Acyclovir is a synthetic nucleoside analog of guanine. It preferentially interferes with DNA synthesis of herpes simplex virus types 1 and 2 (HSV-1 and HSV-2) and varicella-zoster virus, thereby inhibiting viral replication. *Acyclovir reduces viral shedding and formation of new lesions and speeds healing time. It demonstrates antiviral activity against herpes virus simiae (B virus), Epstein-Barr (infectious mononucleosis), varicella-zoster and cyto-megalovirus, but does not eradicate the latent herpes virus.*

USES Parenterally for treatment of viral encephalitis, treatment of herpes simplex, and treatment of varicella-zoster virus (shingles/chickenpox). Used orally for treatment of herpes simplex treatment, and prophylaxis and treatment of varicella-zoster virus (shingles/chickenpox). Used topically for herpes labialis (cold sores), initial episodes of herpes genitalis and in non-life-threatening mucocutaneous herpes simplex virus infections in immunocompromised patients.

UNLABELED USES Treatment of eczema herpeticum caused by HSV localized and disseminated herpes zoster, CMV prophylaxis, pharyngitis, stomatitis, varicella prophylaxis.

CONTRAINDICATIONS Hypersensitivity to acyclovir and valacyclovir.
CAUTIOUS USE Renal insufficiency, dehydration, seizure disorders, or neurologic disease; immunocompromised individuals; older adults; pregnancy (category B).

ROUTE & DOSAGE

Cold Sores

Adult/Adolescents (12 y or older): **Topical** Apply 5 times/day for 4 days

Genital Herpes Simplex

Adult: **PO** 400 mg t.i.d. for 5–14 day cycle **IV** 5 mg/kg q8h × 7 days **Topical** Apply q3h 6 times/day × 7 days

Herpes Simplex Immunocompromised Patient

Adult: **IV** 5 mg/kg q8h × 7 days
Child: **IV** 10 mg/kg q8h × 7 days

Prophylaxis for Genital Herpes Simplex

Adult: **PO** 400 mg b.i.d.

Herpes Zoster

Adult: **PO** 800 mg q4h 5 times/day × 7–10 days
Child: **PO** 80 mg/kg/day in 5 divided doses

Herpes Zoster (in immunocompromised patients)

Adult/Adolescent: **IV** 10 mg/kg q8h × 7 days
Child (under 12 years): **IV** 20 mg/kg q8h × 7–10 days

Viral Encephalitis

Adult: **IV** 10 mg/kg q8h × 10 days
Child (3 mo–12 y): **IV** 20 mg/kg q8h × 10 days

Neonate (younger than 3 mo):
IV 10 mg/kg q8h × 10 days

Varicella Zoster

Child/Adolescent: **PO** 20 mg/kg
(max: 800 mg) q.i.d. for 5 day
cycle initiated within 24 h of onset
of rash
Adult: **IV** 10 mg/kg q8h × 7 days
Child: **IV** 20 mg/kg q8h × 7 days

Obesity Dosage Adjustment

Patient dose should be calculated
using IBW.

Renal Impairment Dosage Adjustment

CrCl 25–50 mL/min: Standard dose
q12h; *10–25 mL/min:* Standard
dose q24h; *less than 10 mL/min:*
Give half normal dose q24h (see
package insert for neonatal renal
impairment adjustment)

Hemodialysis Dosage Adjustment

Administer dose after dialysis

ADMINISTRATION

Oral

- Shake suspension well prior to use.
- Store capsules in tight, light-resistant containers at 15°–30° C (59°–86° F) unless otherwise directed.

Topical

- Wash hands thoroughly before and after treatment of lesions and after handling and disposition of secretions.
- Apply approximately ½ inch of cream or ointment ribbon for each 4 square inches of surface area. Use sufficient ointment or cream to completely cover lesions.
- Apply topical preparation with finger cot or surgical glove.
- Store at 15°–25° C (59°–78° F) unless otherwise directed.

Intravenous

PREPARE: **Intermittent:** Reconstitute by adding 10 mL sterile water for injection to 500-mg vial to yield 50 mg/mL. Note: Do not use bacteriostatic water for injection containing benzyl alcohol. Shake well. ▪ Further dilute to 7 mg/mL or less to reduce risk of renal injury and phlebitis. Example: Add 1 mL of reconstituted solution to 9 mL of diluent to yield 5 mg/mL. ▪ Use standard electrolyte and glucose solutions (e.g., NS, LR, D5W) for dilution.

ADMINISTER: **Intermittent:** Administer over at least 1 h to prevent renal tubular damage. Rapid or bolus IV administration **must be** avoided. ▪ Monitor IV flow rate carefully; infusion pump or microdrip infusion set preferred.

INCOMPATIBILITIES **Solution/additive:** Bacteriostatic water for injection, **dobutamine, dopamine. Y-site:** Amifostine, aminocaproic acid, amsacrine, ampho-tericin B, ampicillin/sulbac-tam, aztreonam, cefepime, chlorpromazine, ciprofloxa-cin, codeine, daptomycin, diaz-epam, dobutamine, dopamine, doxorubicin, epinephrine, epi-rubicin, eptfibatide, esmolol, fenoldopam, fludarabine, fos-carnet, gemcitabine, gemtu-zumab, haloperidol, hydrala-zine, hydroxyzine, idarubicin, irinotecan, ketamine, ketorol-ac, labetolol, levofloxacin, lido-caine, methyldopa, midazo-lam, mycophenolate, nitroprus-side, ondansetron, palonosetron, pentamidine, phenylephrine, phenytoin, piperacillin/tazobac-tam, potassium phosphate, pro-cainamide, prochlorperazine, promethazine, quinupristin/ dalfopristin, sargramostim, strep-tozocin, tacrolimus, ticarcillin/

Common adverse effects in *italic*, life-threatening effects underlined; generic names in **bold;** classifications in SMALL CAPS; ✤ Canadian drug name; ✪ Prototype drug

21

clavulanate, TPN, vecuronium, verapamil, vinorelbine.

- Refrigerated reconstituted solution may precipitate; however, crystals will redissolve at room temperature. • Store acyclovir powder and reconstituted solutions at controlled room temperature, preferably at 15°–30° C (59°–86° F) unless otherwise directed by manufacturer. • Use reconstituted solution within 12 h. Use diluted solution within 24 h.

ADVERSE EFFECTS (≥1%) **Body as a Whole:** Generally minimal and infrequent. **CNS:** *Headache,* lightheadedness, lethargy, fatigue, tremors, confusion, seizures, dizziness. **GI:** *Nausea, vomiting, diarrhea.* **Urogenital:** Glomerulonephritis, renal pain, renal tubular damage, acute renal failure. **Skin:** Rash, urticaria, pruritus, burning, stinging sensation, irritation, sensitization. **Other:** Inflammation or phlebitis at IV injection site, sloughing (with extravasation), thrombocytopenic purpura/hemolytic uremic syndrome.

INTERACTIONS Drug: Probenecid decreases acyclovir elimination; **zidovudine** may cause increased drowsiness and lethargy.

PHARMACOKINETICS Absorption: Oral dose is 15–30% absorbed. **Peak:** 1.5–2 h after oral dose. **Distribution:** Into most tissues with lower levels in the CNS; crosses placenta. **Metabolism:** Drug is primarily excreted unchanged. **Elimination:** Renally eliminated; also excreted in breast milk. **Half-Life:** 2.5–5 h.

NURSING IMPLICATIONS

Assessment & Drug Effects
- Observe infusion site during infusion and for a few days following infusion for signs of tissue damage.

- Monitor I&O and hydration status. Keep patient adequately hydrated during first 2 h after infusion to maintain sufficient urinary flow and prevent formation of renal stones. Consult physician about amount and length of time oral fluids need to be pushed after IV drug treatment.
- Monitor for S&S of: Reinfection in pregnant patients; acyclovir-induced neurologic symptoms in patients with history of neurologic problems; drug resistance in immunocompromised patients receiving prolonged or repeated therapy; acute renal failure with concomitant use of other nephrotoxic drugs or preexisting renal disease.
- Lab tests: Baseline and periodic renal function tests, particularly with IV administration. Elevations of BUN and serum creatinine and decreases in creatinine clearance indicate need for dosage adjustment, discontinuation of drug, or correction of fluid and electrolyte balance.
- Monitor for adverse effects and viral resistance with long-term prophylactic use of the oral drug.

Patient & Family Education
- Start therapy as soon as possible after onset of S&S for best results.
- Do not exceed recommended dosage, frequency of drug administration, or specified duration of therapy. Contact prescriber if relief is not obtained or adverse effects appear.
- Cleanse affected areas with soap and water 3–4 times daily prior to topical application; dry well before application. With application to genitals, wear loose-fitting clothes over affected areas.
- Refrain from sexual intercourse while herpes lesions are present;

neither topical nor systemic drug prevents transmission to other individuals.

- Avoid topical drug contact in or around eyes. Report unexplained eye symptoms to prescriber immediately (e.g., redness, pain); untreated infection can lead to corneal keratitis and blindness.

ADALIMUMAB

(a-da-lim′u-mab)

Humira

Classifications: BIOLOGICAL RESPONSE MODIFIER; IMMUNOMODULATOR; TUMOR NECROSIS FACTOR (TNF) MODIFIER; DISEASE-MODIFYING ANTIRHEUMATIC DRUG (DMARD)
Therapeutic: ANTIRHEUMATIC; DMARD; ANTI-INFLAMMATORY
Prototype: Etanercept
Pregnancy Category: B

AVAILABILITY 40 mg/0.8 mL injection

ACTION & *THERAPEUTIC EFFECT*
Adalimumab is a human recombinant IgG1 monoclonal antibody. It neutralizes the effects of tumor necrosis factor (TNF)-alpha, a cytokine, by blocking its interaction with cell surface TNF receptors. This mechanism blocks the normal inflammatory and immune responses controlled by TNF-alpha. In the presence of complement, adalimumab may also lyse TNF-expressing cells. *Reduces the levels of acute phase inflammatory reactants (C-reactive protein, ESR, interleukin-6) thus decreasing overall joint inflammation; also reduces levels of enzymes that produce tissue remodeling responsible for cartilage destruction. In RA, adalimumab reduces the numerous inflammatory events of polyarthritis. It reduces the overproduction of TNF-alpha principally by macrophages in rheumatoid joints.*

USES Treatment of moderate to severe rheumatoid arthritis or psoriatic arthritis and to reduce progression of the disease in patients with or without a disease-modifying antirheumatic drug (DMARD), treatment of Crohn's disease.

CONTRAINDICATIONS Hypersensitivity to adalimumab or mannitol; active infection, either chronic or acute; neoplastic disease; sepsis; lactation.
CAUTIOUS USE History of recurrent infection or conditions predisposing to infection; recurrent history of sensitivity to monoclonal antibodies; cardiovascular disease; neurologic disease; patients residing in areas with endemic TB or histoplasmosis; active or latent TB infection prior to therapy; demyelinating disorders; Crohn's disease; ulcerative colitis; surgery; pregnancy (category B). Safe use in children except for juvenile idiopathic arthritis has not been established.

ROUTE & DOSAGE

Rheumatoid Arthritis

Adult: **Subcutaneous** 40 mg every other wk (may use 40 mg every wk if not on concomitant methotrexate)

Crohn's Disease

Adult: **Subcutaneous** Initial dose of 160 mg (dose can be administered as 4 injections in 1 day or as 2 injections/day for 2 consecutive days), then 80 mg at wk 2, followed by 40 mg every other wk beginning at wk 4

Common adverse effects in *italic*, life-threatening effects underlined; generic names in **bold;** classifications in SMALL CAPS; ◆ Canadian drug name; ⊕ Prototype drug

23

ADMINISTRATION
Subcutaneous
- Do not administer to persons with active infections. Evaluate for latent TB with TB skin test prior to initiation of therapy.
- Inspect prefilled syringe for particulate matter and discoloration prior to subcutaneous injection.
- Rotate injection sites and do not inject into skin that is red, bruised, tender, or hard. After injecting the drug, do not rub the site.
- Discard any remaining solution in prefilled syringe, as it contains no preservatives.
- Store in original carton at 2°–4° C (38°–48° F). Protect from light. Do not use beyond the expiration date.

ADVERSE EFFECTS (≥1%) **Body as a Whole:** Infections (especially reactivation of latent tuberculosis), sepsis, may see increase in malignancies (lymphoma), back pain, fever, allergic reactions (including <u>anaphylactic shock</u>), *flu-like symptoms, fatigue.* **CNS:** *Headache.* **CV:** Hypertension. **GI:** *Nausea, vomiting,* abdominal pain. **Hematologic:** Development of ANA antibodies. **Respiratory:** *Upper respiratory infection, sinusitis.* **Skin:** *Injection site reactions (erythema, itching, hemorrhage, pain, swelling), rash,* urticaria, fixed drug reaction. **Urogenital:** Urinary tract infection.

INTERACTIONS Drug: Do not give live virus vaccines to patient on adalimumab; not recommended for use with other TNF BLOCKERS (**etanercept, infliximab, anakinra**).

PHARMACOKINETICS Absorption: 64% absorbed from subcutaneous injection site. **Peak:** 131 h. **Distribution:** Minimal beyond vascular/synovial space. **Elimination:** Higher

clearance in presence of anti-adalimumab antibodies, lower clearance with increasing age. **Half-Life:** 11.8 days (10–20 days).

NURSING IMPLICATIONS
Assessment & Drug Effects
- Monitor for and report lupus-like syndrome (e.g., joint pain, rash on cheeks or arms that is sensitive to sun).
- Monitor for and report promptly S&S of infection, including TB. Monitor known HBV carriers for signs of active HBV infection during therapy and for several months after therapy is discontinued.
- Monitor neurologic status closely. Report any change in status such as blurred vision or paresthesia.

Patient & Family Education
- Live vaccines should not be accepted by persons taking this drug.
- Report promptly any of the following to the prescriber: Unexplained joint pain, rash on cheeks or arms, fever, sore throat or other signs of infection, changes in vision, numbness or tingling in extremities.

ADAPALENE
(a-da′pa-leen)
Differin
Classifications: ANTIACNE; RETINOID
Therapeutic: ANTIACNE
Prototype: Isotretinoin
Pregnancy Category: C

AVAILABILITY 0.1%, 0.3% gel; 0.1% cream, 0.1% lotion, 0.1% topical solution

ACTION & *THERAPEUTIC EFFECT*
Adapalene is a topical retinoid-like compound that modulates cellular differentiation, keratinization, and inflammatory processes related to the

pathology of acne vulgaris. Topical adapalene may normalize the differentiation of epithelial follicular cells. *Adapalene decreases the inflammatory process and acne formation.*

USES Treatment of acne vulgaris.

CONTRAINDICATIONS Hypersensitivity to adapalene or any of the components of the gel, irritating topical products, and sunburn; skin abrasion, eczema, seborrheic dermatitis. **CAUTIOUS USE** Pregnancy (category C), lactation. Safety and effectiveness in children younger than 12 y are not established.

ROUTE & DOSAGE

Acne

Adult: Apply once daily to affected areas in evening

ADMINISTRATION

Topical
- Apply a thin film to clean skin, avoiding eyes, lips, mucous membranes, cuts, abrasions, eczematous or sunburned skin.
- Do not apply to skin recently treated with preparations containing sulfur, resorcinol, or salicylic acid.
- Store at 20°–25° C (68°–77° F).

ADVERSE EFFECTS (≥1%) **Skin:** *Erythema, scaling, dryness, pruritus, burning,* skin irritation, stinging, acne flares.

PHARMACOKINETICS Absorption: Minimal through intact skin. **Elimination:** Primarily in bile.

NURSING IMPLICATIONS

Assessment & Drug Effects
- Monitor therapeutic effectiveness, which is indicated by improvement after 8–12 wk of treatment;

early therapy may be marked by apparent worsening of acne.
- Note: Cutaneous reactions (e.g., erythema, scaling, pruritus) are common and normally diminish after first month of therapy.

Patient & Family Education
- Apply only as directed; excessive application will not result in faster healing but will cause marked redness, peeling, and discomfort.
- Minimize exposure to sunlight and sunlamps, and use sunscreen and protective clothing as needed.

ADEFOVIR DIPIVOXIL

(a-de′fo-vir)

Hepsera

Classifications: ANTIVIRAL; NUCLEOTIDE ANALOG

Therapeutic: ANTIVIRAL

Pregnancy Category: C

AVAILABILITY 10 mg tablets

ACTION & *THERAPEUTIC EFFECT* Inhibits human hepatitis virus (HBV) DNA polymerase (reverse transcriptase) by competing with its DNA and by causing DNA chain termination after its incorporation into viral DNA. This results in inhibition of HBV DNA replication. *A nucleotide analog with activity against human hepatitis B virus (HBV).*

USES Treatment of chronic hepatitis B.

CONTRAINDICATIONS Hypersensitivity to adefovir; untreated or unknown human immunodeficiency virus (HIV); exacerbations of hepatitis B, especially in patients who have discontinued anti-hepatitis B therapy; children younger than 2 y; lactation.

Common adverse effects in *italic*, life-threatening effects underlined; generic names in **bold;** classifications in SMALL CAPS; ♣ Canadian drug name; ◎ Prototype drug

25

CAUTIOUS USE Decreased cardiac function due to concomitant disease or other drug therapy; elderly; concomitant use of highly nephrotoxic drugs; renal dysfunction; co-administration with drugs that reduce renal function or compete for active tubular secretion; pregnancy (category C). Appropriate infant immunizations should be used to prevent neonatal acquisition of the hepatitis B virus.

ROUTE & DOSAGE

Hepatitis B

Adult: **PO** 10 mg daily

Renal Impairment Dosage Adjustment

CrCl 20–49 mL/min: 10 mg q48h; *10–19 mL/min:* 10 mg q72h

Hemodialysis Dosage Adjustment

10 mg q7days following dialysis

ADMINISTRATION

Oral

▪ May be given without regard to food.
▪ Store in original container at 15°–30° C (59°–86° F).

ADVERSE EFFECTS (≥1%) **CNS:** *Asthenia,* headache. **GI:** Abdominal pain, nausea, flatulence, diarrhea, dyspepsia, exacerbation of hepatitis after discontinuation of therapy, hepatomegaly. **Metabolic:** *Increased ALT, AST,* increased creatine kinase, amylase, lactic acidosis. **Urogenital:** *Hematuria,* glycosuria, increased serum creatinine, nephrotoxicity. **Other:** HIV resistance in patient with unrecognized HIV, hematuria.

INTERACTIONS Drug: Risk of lactic acidosis when used with NUCLEOSIDE ANALOGS. **Ibuprofen** increases bioavailability of adefovir.

PHARMACOKINETICS Absorption: Adefovir dipivoxil is a prodrug. 59% of dose is absorbed as active drug. **Peak:** 1–4 h. **Distribution:** Minimal protein binding. **Metabolism:** Adefovir dipivoxil is rapidly converted to active adefovir. **Elimination:** Primarily in urine. **Half-Life:** 7.5 h.

NURSING IMPLICATIONS

Assessment & Drug Effects

▪ Lab tests: Baseline and periodic renal function tests (monitor more often with preexisting impairment or other risk factors for renal impairment); periodic liver function tests, creatinine kinase, serum amylase, and routine blood chemistries including serum electrolytes.
▪ Withhold drug and notify prescriber if lactic acidosis is suspected [e.g., hyperventilation, lethargy, plasma pH less than 7.35 and lactate greater than 5–6 mol/L (mEq/L)].

Patient & Family Education

▪ Report any of the following to prescriber: Blood in urine, unexplained weakness, or exacerbation of S&S of hepatitis.
▪ Patients who discontinue adefovir should be monitored at repeated intervals over a period of time for hepatic function.

ADENOSINE

(a-den'o-sin)

Adenocard, Adenoscan
Classifications: ANTIARRHYTHMIC
Therapeutic: ANTIARRHYTHMIC
Pregnancy Category: C

AVAILABILITY 3 mg/mL

ACTION & *THERAPEUTIC EFFECT*
Slows conduction through the atrioventricular (AV) and sinoatrial

Common adverse effects in *italic*, life-threatening effects underlined; generic names in **bold**; classifications in SMALL CAPS; ♦ Canadian drug name; ⊕ Prototype drug

(SA) nodes. Can interrupt the reentry pathways through the AV node. *Restores normal sinus rhythm in patients with paroxysmal supraventricular tachycardia.*

USES Conversion to sinus rhythm of paroxysmal supraventricular tachycardia (PSVT) including PSVT associated with accessory bypass tracts (Wolff-Parkinson-White syndrome). "Chemical" thallium stress test.

UNLABELED USES Afterload-reducing agent in low-output states; to prevent graft occlusion following aortocoronary bypass surgery; to produce controlled hypotension during cerebral aneurysm surgery.

CONTRAINDICATIONS AV block, preexisting second- and third-degree heart block or sick sinus rhythm without pacemaker, since a heart block may result.

CAUTIOUS USE Asthmatics, unstable angina, stenotic valvular disease, hypovolemia; hepatic and renal failure; pregnancy (category C).

ROUTE & DOSAGE

Supraventricular Tachycardia

Adult/Adolescent (Weight 50 kg or more): **IV** 6 mg bolus initially; after 1–2 min may give two additional 12 mg bolus doses for a total of 3 doses. Do not exceed 12 mg in any one dose.
Neonate/Infant/Child: **IV** 0.05–1 mg/kg bolus; additional doses may be increased by 0.05–1 mg/kg q2min (max: 12 mg/dose)

Stress Thallium Test

Adult: **IV** 140 mcg/kg/min × 6 min (max: 0.84 mg/kg total dose)

ADMINISTRATION

Intravenous
Make sure solution is clear at time of use.
▪ Discard unused portion (contains no preservatives).

PREPARE: **Direct:** No dilution is required.
ADMINISTER: **Direct:** *Supraventricular Tachycardia:* Give rapid bolus over 1–2 sec. *Thallium Stress Test:* Give bolus over 6 min.
▪ If given by IV line, administer as proximally as possible, and follow with a rapid saline flush.

▪ Store at room temperature 15°–30° C (59°–86° F). Do not refrigerate, as crystallization may occur. If crystals do form, dissolve by warming to room temperature.

ADVERSE EFFECTS (≥1%) **CNS:** Headache, light-headedness, dizziness, tingling in arms (from IV infusion), apprehension, blurred vision, burning sensation (from IV infusion). **CV:** *Transient facial flushing,* sweating, palpitations, chest pain, atrial fibrillation or flutter. **Respiratory:** Shortness of breath, transient *dyspnea,* chest pressure. **GI:** Nausea, metallic taste, tightness in throat. **Other:** Irritability in children.

INTERACTIONS Drug: Dipyridamole can potentiate the effects of adenosine; **theophylline** will block the electrophysiologic effects of adenosine; **carbamazepine** may increase risk of heart block.

PHARMACOKINETICS Absorption: Rapid uptake by erythrocytes and vascular endothelial cells after IV administration. **Onset:** 20–30 sec. **Metabolism:** Rapid uptake into cells; degraded by deamination to inosine, hypoxanthine, and

Common adverse effects in *italic,* life-threatening effects underlined; generic names in **bold;** classifications in SMALL CAPS; ♣ Canadian drug name; ⊘ Prototype drug

27

adenosine monophosphate. **Elimination:** Route unknown. **Half-Life:** 10 sec.

NURSING IMPLICATIONS

Assessment & Drug Effects

- Monitor for S&S of bronchospasm in asthma patients. Notify prescriber immediately.
- Use a hemodynamic monitoring system during administration; monitor BP and heart rate and rhythm continuously for several minutes after administration.
- Note: Adverse effects are generally self-limiting due to short half-life (10 sec).
- Note: At the time of conversion to normal sinus rhythm, PVCs, PACs, sinus bradycardia, and sinus tachycardia, as well as various degrees of AV block, are seen on the ECG. These usually last only a few seconds and resolve without intervention.

Patient & Family Education

- Note: Flushing may occur along with a feeling of warmth as drug is injected.

AGALSIDASE BETA

(a-gal´si-dase)

Fabrazyme

Classification: ENZYME REPLACEMENT
Therapeutic: ENZYME REPLACEMENT
Prototype: Pancrelipase
Pregnancy Category: B

AVAILABILITY 35 mg/vial injection

ACTION & *THERAPEUTIC EFFECT*
Fabry disease is caused by a deficiency of alpha-galactosidase A resulting in accumulation of glycosphingolipids in body tissues causing cardiomyopathy, renal failure, and CVA. Agalsidase beta provides an exogenous source of K-galactosidase A that catalyzes the breakdown of glycosphingolipids including GL-3. *Reduces globotriaosylceramide (GL-3) deposition in capillary endothelium of the kidney and certain other cell types.*

USES Treatment of Fabry disease.

CAUTIOUS USE Hypersensitivity reaction to agalsidase beta or mannitol; compromised cardiac function, mild to severe hypertension; renal impairment; pregnancy (category B). Safety and efficacy in children younger than 16 y have not been established; lactation.

ROUTE & DOSAGE

Fabry Disease
Adult: **IV** 1 mg/kg q2wk

ADMINISTRATION

Intravenous
Give antipyretics prior to infusion.

PREPARE: **Infusion:** Bring agalsidase vials and supplied sterile water for injection to room temperature prior to reconstitution.
- Reconstitute each 35 mg vial slowly injecting 7.2 mL of sterile water for injection down inside wall of vial. Roll and tilt vial gently to mix but do not shake. ▪ Reconstituted vial contains 5.0 mg/mL of clear, colorless solution. Do not use if there is particulate matter or if discolored. ▪ **Must be** further diluted in NS to a final total volume of 500 mL; prior to adding the volume of reconstituted agalsidase required for the dose, remove an equal volume of NS from the 500 mL infusion bag.
ADMINISTER: **Infusion:** Initial rate should not exceed 0.25 mg/min (15 mg/h; give more slowly if infusion-associated reactions occur). ▪ After tolerance

to infusion is established, may increase rate in increments of 0.05–0.08 mg/min (increments of 3 to 5 mg/h) for each subsequent infusion.
INCOMPATIBILITIES **Solution/additive:** Do not infuse with other products.

- Store refrigerated until needed. Vials are for single use. Discard any unused portion. Do NOT use after expiration date.

ADVERSE EFFECTS (≥1%) **Body as a Whole:** *Fever, skeletal pain, pallor, rigors, temperature change sensation,* ataxia, stroke. **CNS:** *Dizziness, headache, paresthesia, anxiety, depression,* vertigo. **CV:** *Chest pain, cardiomegaly, hypertension, hypotension, dependent edema,* bradycardia, <u>heart failure</u>, exacerbation of preexisting arrhythmias. **GI:** *Dyspepsia, nausea, abdominal pain.* **Metabolic:** *Antibody development.* **Musculoskeletal:** *Arthrosis, skeletal pain.* **Respiratory:** *Bronchitis,* bronchospasm, laryngitis, *pharyngitis, rhinitis,* sinusitis, dyspnea. **Skin:** Pruritus, urticaria. **Special Senses:** Hearing loss. **Urogenital:** Testicular pain, nephrotic syndrome.

INTERACTIONS Drug: Coadministration with **amiodarone, chloroquine, hydroxychloroquine, gentamicin** is not recommended due to theoretical decreased response to agalsidase beta therapy.

PHARMACOKINETICS Metabolism: Degraded through peptide hydrolysis. **Elimination:** Renal elimination expected to be a minor pathway. **Half-Life:** 45–102 min.

NURSING IMPLICATIONS
Assessment & Drug Effects
- During infusion, monitor for infusion-related reactions such as hypertension or hypotension, chest

pain or chest tightness, dyspnea, fever and chills, headache, abdominal pain, pruritus and urticaria.
- Slow infusion and notify prescriber immediately if infusion reaction occurs. Note that additional antipyretic and/or an antihistamine and oral steroid may reduce the symptoms.
- Monitor cardiac status closely, especially with preexisting heart disease.

Patient & Family Education
- Notify prescriber if you have experienced an unusual reaction to agalsidase beta, agalsidase alfa, mannitol, other drugs, foods, or preservatives.
- Report any of the following to prescriber immediately: Chest pain or chest tightness, rapid heartbeat, shortness of breath or difficulty breathing; depression; dizziness; skin rash, hives or itching; throat tightness; swelling of the face, lips, neck, ears, or extremities.
- Do not drive or engage in other hazardous activities until reaction to drug is known.

ALBENDAZOLE
(al-ben'da-zole)
Albenza
Classifications: ANTHELMINTIC
Therapeutic: ANTHELMINTIC
Prototype: Mebendazole
Pregnancy Category: C

AVAILABILITY 200 mg tablets

ACTION & *THERAPEUTIC EFFECT*
A broad-spectrum oral anthelmintic agent. It is the only anthelmintic drug active against all stages of the helminth life cycle (ova, larvae, and adult worms). Its mechanism of action appears to cause selective degeneration of cytoplasmic microtubules in the intestinal cells of the helminths and larvae. *Albendazole ultimately*

Common adverse effects in *italic*, life-threatening effects <u>underlined</u>; generic names in **bold**; classifications in SMALL CAPS; ♥ Canadian drug name; ☯ Prototype drug

29

causes decreased ATP production in the helminths, resulting in energy depletion, which kills the worms.

USES Treatment of neurocysticercosis caused by pork tapeworm (*Taenia solium*), hydatid disease caused by the larval form of dog tapeworm (*Echinococcus granulosus*).

UNLABELED USES Giardiasis, pinworm infection, hookworm infection, microsporidosis.

CONTRAINDICATIONS Hypersensitivity to the benzimidazole class of compounds or any components of albendazole; hepatic enzymes twice the ULN; clinically significant decreases in blood cell counts.

CAUTIOUS USE Hepatic dysfunction; retinal lesions, pregnancy (category C), lactation; children younger than 6y.

ROUTE & DOSAGE

Neurocysticercosis

Adult/Adolescent/Child: **PO** *Older than 6 y, weight less than 60 kg,* 15 mg/kg/day divided b.i.d. for 8–30 day cycle (max: 800 mg/day); *weight 60 kg or more,* 400 mg b.i.d. for 8–30 days

Hydatid Disease

Adult/Child: **PO** *Older than 6 y, weight less than 60 kg,* 15 mg/kg/day divided b.i.d. x 28 days, then 14 days drug free and repeat for 2 more cycles (max: 800 mg/day); *weight 60 kg or more,* 400 mg b.i.d. for 28-day cycle (then 14 days without drug and repeat regimen for 3 cycles)

ADMINISTRATION

Oral

- Give with meals. Absorption is significantly increased with a fatty meal.

- Do not exceed maximum total daily dose of 800 mg.
- Store at 20°–25° C (68°–77° F).

ADVERSE EFFECTS (≥1%) **Body as a Whole:** Hypersensitivity reactions. **CNS:** *Headache,* dizziness, vertigo, increased intracranial pressure, meningeal signs, fever. **GI:** *Abnormal liver function tests,* abdominal pain, nausea, *vomiting.* **Hematologic:** (Rare) Reversible <u>leukopenia</u>, granulocytopenia, pancytopenia, <u>agranulocytosis</u>. **Skin:** Rash, urticaria, reversible alopecia.

INTERACTIONS Drug: Cimetidine, praziquantel increase levels. **Food:** Avoid **grapefruit juice**.

PHARMACOKINETICS Absorption: Poorly absorbed, absorption enhanced with a fatty meal. **Peak:** 2–5 h. **Distribution:** 70% protein bound; widely distributed, including cyst fluid and CSF; secreted into animal breast milk. **Metabolism:** In liver to active metabolite. **Elimination:** In bile. **Half-Life:** 9 h.

NURSING IMPLICATIONS

Assessment & Drug Effects

- Lab tests: WBC count, absolute neutrophil count, and liver function tests at start of each 28-day cycle and q2wk during cycle.
- Withhold drug and notify prescriber if WBC count falls below normal or liver enzymes are elevated.
- Note: Patients should be concurrently treated with appropriate steroid and anticonvulsant therapy.

Patient & Family Education

- Take with meals (see ADMINISTRATION), but avoid grapefruit juice while taking this drug.
- Do not become pregnant during or for at least 1 mo after therapy.

ALBUTEROL 💬

(al-byoo′ter-ole)

Accuneb, Novosalmol ♣, Pro-Air HFA, Proventil, Proventil HFA, ReliOn Ventolin HFA, Ventolin, Ventolin HFA, VoSpire ER
Classifications: BRONCHODILATOR (RESPIRATORY SMOOTH MUSCLE RELAXANT); BETA-ADRENERGIC AGONIST
Theurapeutic: BRONCHODILATOR
Pregnancy Category: C

AVAILABILITY 2 mg, 4 mg tablets; 4 mg, 8 mg extended release tablets; 2 mg/5 mL syrup; 200 mcg capsules for inhalation; 0.083%, 0.5% solution for inhalation; 90 mcg/actuation

ACTION & *THERAPEUTIC EFFECT*

Moderately selective beta$_2$-adrenergic agonist that acts prominently on smooth muscles of trachea, bronchi, uterus, and vascular supply to skeletal muscles. Inhibits histamine release by mast cells. Produces bronchodilation by relaxing smooth muscles of bronchial tree. *Bronchodilation decreases airway resistance, facilitates mucous drainage, and increases vital capacity.*

USES To relieve bronchospasm associated with acute or chronic asthma, bronchitis, or other reversible obstructive airway diseases. Also used to prevent exercise-induced bronchospasm.

CONTRAINDICATIONS Albuterol or levalbuterol hypersensitivity; congenital long QT syndrome. Use of oral syrup in children younger than 2 y.
CAUTIOUS USE Cardiovascular disease, renal impairment, hypertension, hyperthyroidism, diabetes mellitus, older adults; history of seizures; hypersensitivity to sympathomimetic amines or to fluorocarbon propellant used in inhalation aerosols; pregnancy (category C).

ROUTE & DOSAGE

Bronchospasm

Adult: **PO** 2–4 mg 3–4 times/day, 4–8 mg sustained release 2 times/day **Inhaled** 1–2 inhalations q4–6h
Child: **PO** 2–6 y, 0.1–0.2 mg/kg t.i.d. (max: 4 mg/dose); 6–12 y, 2 mg 3–4 times/day **Inhaled** 4–12 y, 1–2 inhalations q4–6h

Prevention of Exercise-Induced Brochospasm

Adult: **Inhaled** 2 inhalations 30 min prior to exercise
Child (older than 4 y): **Inhaled** 2 inhalations 30 min prior to exercise

ADMINISTRATION

Oral

- Do not crush extended release tablets. Scored tablets may be broken in half.
- Note: An initial dose of 2 mg t.i.d. or q.i.d. is recommended for older adult patients.
- Store tablets and syrup between 2°–25° C (36°–77° F) in tight, light-resistant container.

Inhalation

- If ordered with beclomethasone, give 20–30 min before beclomethasone unless otherwise directed by prescriber.
- Administer albuterol inhalation aerosol canister only with the actuator provided.
- Store canisters between 15°–30° C (59°–86° F) away from heat and direct sunlight.

ADVERSE EFFECTS (≥1%) **Body as a Whole:** Hypersensitivity reaction. **CNS:** *Tremor,* anxiety, nervousness, restlessness, convulsions, weakness, headache, hallucinations. **CV:** Palpitation, hypertension,

Common adverse effects in *italic*, life-threatening effects <u>underlined</u>; generic names in **bold;** classifications in SMALL CAPS; ♣ Canadian drug name; 💬 Prototype drug

31

hypotension, bradycardia, reflex tachycardia. **Special Senses:** Blurred vision, dilated pupils. **GI:** Nausea, vomiting. **Other:** Muscle cramps, hoarseness.

DIAGNOSTIC TEST INTERFERENCE Transient small increases in *plasma glucose* may occur.

INTERACTIONS Drug: With **epinephrine,** other SYMPATHOMIMETIC BRONCHODILATORS, possible additive effects; MAO INHIBITORS, TRICYCLIC ANTIDEPRESSANTS potentiate action on vascular system; BETA-ADRENERGIC BLOCKERS antagonize the effects of both drugs.

PHARMACOKINETICS Onset: Inhaled: 5–15 min; PO: 30 min. **Peak:** Inhaled: 0.5–2 h; PO: 2.5 h. **Duration:** Inhaled: 3–6 h; PO: 4–6 h (8–12 h with sustained release). **Metabolism:** In liver by CYP3A4; may cross the placenta. **Elimination:** 76% of dose eliminated in urine in 3 days. **Half-Life:** 2.75 h.

NURSING IMPLICATIONS

Assessment & Drug Effects

- Monitor therapeutic effectiveness which is indicated by significant subjective improvement in pulmonary function within 60–90 min after drug administration.
- Monitor for: S&S of fine tremor in fingers, which may interfere with precision handwork; CNS stimulation, particularly in children 2–6 y (hyperactivity, excitement, nervousness, insomnia), tachycardia, GI symptoms. Report promptly to prescriber.
- Lab tests: Periodic ABGs, pulmonary functions, and pulse oximetry.
- Consult prescriber about giving last albuterol dose several hours before bedtime, if drug-induced insomnia is a problem.

Patient & Family Education

- Review directions for correct use of medication and inhaler (see ADMINISTRATION).
- Avoid contact of inhalation drug with eyes.
- Do not increase number or frequency of inhalations without advice of prescriber.
- Notify prescriber if albuterol fails to provide relief because this can signify worsening of pulmonary function and a reevaluation of condition/therapy may be indicated.
- Note: Albuterol can cause dizziness or vertigo; take necessary precautions.
- Do not use OTC drugs without prescriber approval. Many medications (e.g., cold remedies) contain drugs that may intensify albuterol action.

ALCLOMETASONE DIPROPIONATE

(al-clo-met′a-sone)

Aclovate
Pregnancy Category: C
See Appendix A-4.

ALEFACEPT ℗

(a-le′fa-cept)

Amevive
Classifications: BIOLOGICAL RESPONSE MODIFIER; IMMUNOLOGIC
Therapeutic: IMMUNOSUPPRESSANT
Pregnancy Category: B

AVAILABILITY 15 mg vials

ACTION & *THERAPEUTIC EFFECT* Alefacept is thought to bind to CD2 receptors found on all peripheral T cells and to immunoglobulin receptors on cytotoxic cells, such as natural killer cells. Alefacept blocks further activation of T cells and reduces cellular-mediated apoptosis of T cells. Activation of T cells plays a role in chronic plaque psoriasis. *Alefacept modulates the immune response by decreasing activation of T cells that are believed to be the key mediators of psoriasis.*

USES Treatment of moderate to severe chronic plaque psoriasis.
UNLABELED USES Treatment of psoriatic arthritis.

CONTRAINDICATIONS Hypersensitivity to alefacept; CD4+ T lymphocyte count below normal; patients with HIV, history of systemic malignancies; patients with a clinically important infection; serious infections; live or attenuated vaccines; lactation.
CAUTIOUS USE Patients at high risk for malignancies; older adults; pregnancy (category B).

ROUTE & DOSAGE

Chronic Plaque Psoriasis

Adult: **IM** 15 mg once/wk × 12 wk, may repeat course after 12 wk of therapy if CD4 count is above 250 cells/mcL **IV** 7.5 mg once/wk × 12 wk, may repeat course after 12 wk of therapy if CD4 count is above 250 cells/mcL

ADMINISTRATION

- Administer only if CD4+ T lymphocyte count is 250 cells/mcL or more.
- Reconstituted alefacept should be clear and colorless to slightly yellow.

Intramuscular

- Reconstitute the 15 mg vial for IM administration with 0.6 mL of the supplied diluent to yield 15 mg/0.5 mL. Gently swirl vial for about 2 min to mix, but do not shake.

- Rotate the injection sites and space at least 1 inch from an old site.
- Never inject into areas where the skin is tender, bruised, red, or hard.

Intravenous

PREPARE: **Direct:** Use supplied needles for preparation and administration. ▪ Reconstitute the 7.5 mg vial for IV administration with 0.6 mL of the supplied diluent to yield 7.5 mg/0.5 mL. ▪ Keeping the needle pointed at the sidewall of the vial, slowly inject the diluent then gently swirl vial for about 2 min to mix, but do not shake.
ADMINISTER: **Direct:** Prime supplied infusion set with 3.0 mL of supplied diluent and insert into vein. ▪ Give reconstituted solution over 5 sec or less and do not use a filter. ▪ Follow with flush using 3 mL of supplied diluent.
INCOMPATIBILITIES **Solution/additive:** Do not add other medications to IV solution.

- Store vials of powder away from light at 15°–30° C (59°–86° F).
- Store reconstituted solution for up to 4 h between 2°–8° C (36°–46° F); discard solution not used within 4 h of reconstitution.

ADVERSE EFFECTS (≥1%) **Body as a Whole:** Secondary malignancies, serious infections, chills, *injection site pain,* injection site inflammation. **CNS:** Dizziness, headache. **GI:** Nausea, vomiting. **Hematologic:** *Lymphopenia,* alefacept antibody formation. **Musculoskeletal:** Myalgia. **Respiratory:** Pharyngitis, increased cough. **Skin:** Pruritus.

INTERACTIONS Drug: Additive immunosuppression with other immunosuppressant drugs (e.g., CORTICOSTEROIDS); LIVE VACCINES increase risk of secondary transmission of infection.

Common adverse effects in *italic*, life-threatening effects underlined; generic names in **bold**; classifications in SMALL CAPS; ✦ Canadian drug name; ⊙ Prototype drug

33

PHARMACOKINETICS Absorption: 63% from IM injection. **Metabolism:** Presumed to be broken down in plasma. **Half-Life:** 270 h after IV.

NURSING IMPLICATIONS

Assessment & Drug Effects

- Discontinue drug immediately and institute supportive measures if a serious hypersensitivity reaction occurs.
- Note: Drug should be discontinued if CD4+ T lymphocyte counts remain below 250 cells/mcL for 1 mo.
- Lab tests: Weekly WBC with differential during 12-wk dosing period; periodic liver enzymes.
- Monitor for and promptly report S&S of infection.

Patient & Family Education

- Report any of the following promptly: Chest pain or tightness, rapid or irregular heart beat; difficulty breathing or swallowing; swelling of face, tongue, hands, feet or ankles; rapid weight gain; signs of infection (e.g., fever, chills, cough, sore throat, pain or difficulty passing urine); skin rash or itchy skin; severe stomach pain.
- Do not accept live or live-attenuated vaccines while taking this drug.
- Notify prescriber if you become pregnant while taking this drug or within 8 wk of discontinuing drug.

ALEMTUZUMAB

(a-lem'tu-zu-mab)

Campath

Classifications: BIOLOGICAL RESPONSE MODIFIER; MONOCOLONAL ANTIBODY; ANTINEOPLASTIC **Therapeutic:** ANTINEOPLASTIC

Pregnancy Category: C

AVAILABILITY 30 mg/mL injection

ACTION & *THERAPEUTIC EFFECT*

Monoclonal antibody that attaches to CD52 cell surface antigens expressed on a variety of leukocytes, including normal and malignant B and T lymphocytes, monocytes, and some granulocytes. Proposed mechanism of action is antibody-dependent lysis of leukemic cells following binding to cell surface antigens. *Initiates antibody-dependent cell lysis, thus inhibiting cell proliferation in chronic lymphocytic leukemia.*

USES Treatment of B-cell chronic lymphocytic leukemia in patients who have failed fludarabine therapy. **UNLABELED USES** Treatment of mycosis fungoides, non-Hodgkin's lymphoma.

CONTRAINDICATIONS Type I hypersensitivity to alemtuzumab or its components, hamster protein hypersensitivity; serious infection or exposure to viral infections (i.e., herpes or chickenpox), HIV infection, dental work; infection; lactation. **CAUTIOUS USE** History of hypersensitivity to other monoclonal antibodies; ischemic cardiac disease, angina, coronary artery disease; dental disease; history of varicella disease; females of childbearing age; pregnancy (category C). Safety and efficacy in children are not established.

ROUTE & DOSAGE

B-Cell Chronic Lymphocytic Leukemia

Adult: **IV** Start with 3 mg/day, when that is tolerated, increase dose over next 3–7 days to 10 mg/day; when 10 mg/day is tolerated, increase to maintenance dose of 30 mg/day (give 30 mg/day 3 times/

Common adverse effects in *italic*, life-threatening effects underlined; generic names in **bold;** classifications in SMALL CAPS; ♣ Canadian drug name; ☯ Prototype drug

wk). Single dose should not exceed 30 mg; cumulative dose should not exceed 90 mg/wk

Toxicity Dosage Adjustment

First time ANC falls below 250/mcL or platelet count falls below 25,000/mcL

• Stop therapy until ANC is at least 500 mcL and platelet count is at least 50,000 mcL, resume at previous dose
• If therapy is stopped for 7 or more days, restart at 3 mg and taper up

Second time ANC below 250/mcL or platelet count falls below 25,000 mcL

• Stop therapy until ANC is at least 500 mcL and platelet count is at least 50,000 mcL, resume at 10 mg dose
• If therapy is stopped for 7 or more days, restart at 3 mg and taper up, do not exceed 10 mg

Third time ANC below 250/mcL or platelet count falls below 25,000/mcL

• Stop therapy permanently
*Patients starting therapy with baseline ANC less than 500 mcL or baseline platelet count less than 25,000 mcL who experience a 50% decrease from baseline should stop therapy until values return to baseline then resume at previous dose. If therapy is stopped for 7 or more days, restart at 3 mg and taper up.

ADMINISTRATION

▪ Note: Premedication with antihistamines, acetaminophen, antiemetics, and corticosteroids prior to infusion may reduce the severity of adverse side effects.

Intravenous

PREPARE: **IV Infusion:** Do NOT shake ampule prior to use. ▪ Withdraw required dose into a syringe with a sterile, low-protein binding, non-fiber releasing 5 micron filter. Inject into 100 mL NS or D5W. ▪ Gently invert bag to mix. Infuse within 8 h of mixing. ▪ Protect from light. Discard any unused solution.

ADMINISTER: **IV Infusion:** Infuse each dose over 2 h. ▪ Do NOT give single doses greater than 30 mg or cumulative weekly doses greater than 90 mg.

INCOMPATIBILITIES **Solution/additive:** Do not infuse or mix with other drugs.

▪ Store at 2°–8° C (36°–46° F). Discard if ampule has been frozen. Protect from direct light.

ADVERSE EFFECTS (≥1%) **Body as a Whole:** *Infusion reactions (rigors, fever, nausea, vomiting, hypotension, rash, shortness of breath, bronchospasm, chills), fatigue, pain, sepsis, asthenia,* edema, herpes simplex, myalgias, malaise, moniliasis, temperature change sensation, <u>coma, seizures</u>. **CNS:** *Headache, dysesthesias, dizziness, insomnia,* depression, tremor, somnolence, <u>cerebrovascular accident, subarachnoid hemorrhage</u>. **CV:** *Hypotension, tachycardia, hypertension,* <u>cardiac failure, arrhythmias, MI</u>. **GI:** *Diarrhea, nausea, vomiting, stomatitis, abdominal pain, dyspepsia, anorexia,* constipation. **Hematologic:** *Neutropenia, anemia,* <u>thrombocytopenia</u>, purpura, epistaxis, <u>pancytopenia</u>. **Respiratory:** *Dyspnea, cough, bronchitis, pneumonia, pharyngitis,* bronchospasm, rhinitis. **Skin:** *Rash,*

Common adverse effects in *italic*, life-threatening effects <u>underlined</u>; generic names in **bold;** classifications in SMALL CAPS; ♣ Canadian drug name; ☯ Prototype drug

35

urticaria, pruritus, increased sweating. **Other:** Risk of opportunistic infections.

INTERACTIONS Drug: Additive risk of bleeding with ANTICOAGULANTS, NSAIDS, PLATELET INHIBITORS, SALICYLATES and increased risk of opportunistic infections with **fludarabine. Herbal: Feverfew, garlic, ginger, ginkgo** may increase risk of bleeding.

PHARMACOKINETICS Peak: Steady-state levels in approximately 6 wk. **Half-Life:** 12 days.

NURSING IMPLICATIONS

Assessment & Drug Effects

- Discontinue infusion and notify prescriber immediately if any of the following occurs: Hypotension, fever, chills, shortness of breath, bronchospasm, or rash.
- Withhold drug and notify prescriber if absolute neutrophil count less than 250/mcL or platelet count of 25,000/mcL or less.
- Monitor BP closely during infusion period. Careful monitoring of BP and hypotensive symptoms is especially important in patients with ischemic heart disease and those on antihypertensives.
- Withhold drug during any serious infection. Therapy may be reinstituted following resolution of the infection.
- Lab tests: CBC with differential and platelet counts weekly or more frequently in the presence of anemia, thrombocytopenia, or neutropenia; periodic blood glucose, serum electrolytes, and alkaline phosphatase.
- Monitor diabetics closely for loss of glycemic control.
- Monitor for S&S of dehydration especially with severe vomiting.

Patient & Family Education

- Do not accept immunizations with live viral vaccines during therapy or if therapy has been recently terminated.
- Use effective methods of contraception to prevent pregnancy during therapy and for at least 6 mo following therapy.
- Report any of the following to prescriber immediately: Unexplained bleeding, fever, sore throat, flu-like symptoms, S&S of an infection, difficulty breathing, significant GI distress, abdominal pain, fluid retention, or changes in mental status.
- Diabetics should monitor blood glucose levels carefully since loss of glycemic control is a possible adverse reaction.

ALENDRONATE SODIUM
(a-len'dro-nate)

Fosamax, Fosamax D (with 2800 International Units Vitamin D3)
Classifications: BISPHOSPHONATE; BONE METABOLISM REGULATOR
Therapeutic: BONE METABOLISM REGULATOR
Prototype: Etidronate
Pregnancy Category: C

AVAILABILITY 5 mg, 10 mg, 35 mg, 40 mg, 70 mg tablets; 70 mg/75 mL oral solution

ACTION & *THERAPEUTIC EFFECT*
A bisphosphonate that inhibits osteoclast-mediated bone resorption. Antiresorption mechanism is thought to be localized to resorption sites of active bone turnover and to have minimal to no interference with bone mineralization. *Alendronate decreases bone resorption, thus minimizing loss of bone density.*

USES Prevention and treatment of osteoporosis in postmenopausal women or in men, Paget's disease. Treatment of glucocorticoid-induced osteoporosis.

CONTRAINDICATIONS Hypersensitivity to alendronate or other bisphosphonates; achalasia, esophageal stricture, severe renal impairment (CrCl less than 35 mL/min); hypocalcemia; abnormalities; lactation.
CAUTIOUS USE Renal impairment, CHF, hyperphosphatemia, liver disease, fever or infection, active upper GI problems; pregnancy (category C).

ROUTE & DOSAGE

Treatment of Osteoporosis
Adult: **PO** 10 mg once/day (max: 40 mg/day) or 70 mg qwk

Prevention of Osteoporosis
Adult: **PO** 5 mg daily or 35 mg qwk

Treatment of Steroid-Induced Osteoporosis
Adult: **PO** 5 mg daily or 35 mg qwk

Treatment of Paget's Disease
Adult: **PO** 40 mg once/day for 6 mo

Renal Impairment Dosage Adjustment
CrCl less than 35 mL/min: Use not recommended

ADMINISTRATION

Oral
- Correct hypocalcemia before administering alendronate.
- Administer in the morning at least 30 min before the first food, beverage, or medication. Do not administer within 2 h of calcium-containing foods, beverages, or medications. At least 30 min should

elapse after alendronate dose before taking any other drugs.
- *Oral Solution:* Use oral syringe for accurate dosage. Give with at least 60 mL (2 oz) of plain water.
- *Tablet:* Give with 8 oz of plain water.
- Keep patient sitting up or ambulating for 30 min after taking drug.
- Store according to manufacturer's directions.

ADVERSE EFFECTS (≥1%) **Endocrine:** Hypocalcemia, hypophosphatemia. **GI:** Esophageal irritation and *ulceration, nausea, vomiting, abdominal pain, dyspepsia,* diarrhea, constipation, flatulence. **Other:** Arthralgias, myalgias, headache, rash, alopecia.

INTERACTIONS Drug: Ranitidine increases alendronate availability. **Food: Calcium** and food (especially dairy products) reduce alendronate absorption.

PHARMACOKINETICS Absorption: 0.5–1% from GI tract (absorption significantly decreased by calcium and food). **Onset:** 3–6 wk. **Duration:** 12 wk after discontinuation. **Distribution:** Rapid skeletal uptake. **Metabolism:** Not metabolized. **Elimination:** Up to 50% excreted unchanged in urine. **Half-Life:** Up to 10 h.

NURSING IMPLICATIONS
Assessment & Drug Effects
- Lab tests: Baseline and periodic albumin-adjusted serum calcium, serum phosphate, serum alkaline phosphatase, fasting and 24 h urinary calcium, and serum electrolytes. Periodically monitor renal and liver functions.
- Diagnostic test: Bone density scan every 12–18 mo.
- Discontinue drug if the CrCl less than 35 mL/min.

Common adverse effects in *italic*, life-threatening effects <u>underlined</u>; generic names in **bold**; classifications in SMALL CAPS; ♣ Canadian drug name; ⊘ Prototype drug

37

Patient & Family Education
- Review directions for taking drug correctly (see ADMINISTRATION).
- Report fever, especially when accompanied by arthralgia and myalgia.

ALFENTANIL HYDROCHLORIDE

(al-fen'ta-nill)

Alfenta

Classifications: NARCOTIC OPIATE AGONIST ANALGESIC; GENERAL ANESTHETIC

Therapeutic: NARCOTIC ANALGESIC; GENERAL ANESTHETIC

Prototype: Morphine

Pregnancy Category: C

Controlled Substance: Schedule II

AVAILABILITY 500 mcg/mL injection

ACTION & *THERAPEUTIC EFFECT*
Alfentanil is a narcotic agonist analgesic with CNS effects that appear to be related to interaction of drug with opiate receptors. *Analgesia is mediated through changes in the perception of pain at the spinal cord and at higher levels in the CNS.*

USES General anesthesia induction and maintenance, and sedation maintenance.

UNLABELED USES Severe pain.

CONTRAINDICATIONS Coagulation disorders, bacteremia, infection at injection site; lactation.

CAUTIOUS USE Older adults, history of pulmonary disease; pregnancy (category C). Safety in children younger than 12 y is not established.

ROUTE & DOSAGE

Anesthesia

Adult: **IV** induction 130–245 mcg/kg; maintenance 0.5–1.5 mcg/kg/min

Conscious Sedation

Adult: **IV** 3–8 mcg/kg then 3–5 mcg/kg q5–20 min or continuous infusion of 0.25–1 mcg/kg/min (total 3–40 mcg/kg)

Obesity Dosage Adjustment
Dose based on IBW

Hepatic Impairment Dosage Adjustment
Maintenance dosage adjustment recommended

ADMINISTRATION

Intravenous

***PREPARE:* Direct or Continuous:** Alfentanil is available in a concentration of 500 mcg/mL. Small volumes may be given direct IV undiluted or diluted in 5 mL of NS.
- For IV infusion, add 20 mL of alfentanil to 230 mL of compatible IV solution to yield 40 mcg/mL. Compatible IV solutions include NS, D5/NS, D5W, and LR. ■ Note: Alfentanil may be diluted to concentrations of 25–80 mcg/mL.

***ADMINISTER:* Direct** Administer over at least 3 min. Do not administer more rapidly. **Continuous:** Administer at a rate of 0.25–1 mcg/kg/min. Note: Dose may be individualized.

***INCOMPATIBILITIES* Y-site: Amphotericin B, amphotericin B (lipid), dantrolene, diazepam, diazoxide, lansoprazole, pantoprazole, phenytoin, sulfamethoxazole/trimethoprim, thiopental.**

- Store at 15°–30° C (59°–86° F). Avoid freezing.

ADVERSE EFFECTS (≥1%) **Body as a Whole:** Thoracic muscle rigidity, flushing, diaphoresis; extremities feel heavy and warm. **CNS:** Dizziness, euphoria, drowsiness. **CV:** Hypotension, hypertension, tachycardia, bradycardia. **GI:** *Nausea,* vomiting, anorexia, constipation, cramps. **Respiratory:** Apnea, respiratory depression, dyspnea.

INTERACTIONS Drug: BETA-ADRENERGIC BLOCKERS increase incidence of bradycardia; CNS DEPRESSANTS such as BARBITURATES, TRANQUILIZERS, NEUROMUSCULAR BLOCKING AGENTS, OPIATES, and INHALATION GENERAL ANESTHETICS may enhance the cardiovascular and CNS effects of alfentanil in both magnitude and duration; enhancement or prolongation of postoperative respiratory depression also may result from concomitant administration of any of these agents with alfentanil.

PHARMACOKINETICS Onset: 2 min. **Duration:** Injection 30 min; continuous infusion 45 min. **Distribution:** Crosses placenta. **Metabolism:** In liver by CYP3A4. **Elimination:** Excreted in breast milk. **Half-Life:** 46–111 min.

NURSING IMPLICATIONS

Assessment & Drug Effects

- Monitor for S&S of increased sympathetic stimulation (arrhythmias) and evidence of depressed postoperative analgesia (tachycardia, pain, pupillary dilation, spontaneous muscle movement) if a narcotic antagonist has been administered to overcome residual effects of alfentanil.
- Evaluate adequacy of spontaneous ventilation carefully during postoperative period.
- Monitor vital signs carefully during postoperative period; check for bradycardia, especially if patient is also taking a beta-blocker.
- Note: Dizziness, sedation, nausea, and vomiting are common when drug is used as a postoperative analgesic.

Patient & Family Education

- Report unpleasant adverse effects when drug is used for patient-controlled analgesia.

ALFUZOSIN
(al-fuz′o-sin)

UroXatral, Xatral ♦

Classifications: ALPHA-ADRENERGIC ANTAGONIST

Therapeutic: GENITOURINARY SMOOTH MUSCLE RELAXER

Prototype: Tamsulosin

Pregnancy Category: B

AVAILABILITY 10 mg extended release tablet

ACTION & *THERAPEUTIC EFFECT*

Alfuzosin is a short-acting, selective antagonist at alpha-1 receptors with a low incidence of hypotension and sexual dysfunction. Alpha-1 receptors cause contraction of smooth muscle in the prostate, prostatic capsule, prostatic urethra, bladder base, and bladder neck. Both alpha-1a (70%) and alpha-1b receptors exist in the prostate. *Blockade of alpha-1 receptors by alfuzosin causes smooth muscles in the bladder neck and prostate to relax, thereby reducing pressure on the urethra and improving urine flow rate. This results in a reduction in BPH symptoms.*

USES Treatment of symptomatic benign prostatic hypertrophy (BPH). **UNLABELED USE** Erectile dysfunction (with sildenafil).

CONTRAINDICATIONS Hypersensitivity to alfuzosin; severe hepatic insufficiency; concurrent treatment with potent CYP3A4 inhibitors (e.g., ketoconazole, itraconazole, and ritonavir); angina; QT prolongation.

Common adverse effects in *italic*, life-threatening effects <u>underlined</u>; generic names in **bold**; classifications in SMALL CAPS; ♦ Canadian drug name; ☉ Prototype drug

39

CAUTIOUS USE Coronary artery disease, cardiac arrhythmias; hepatic disease; dizziness, light-headedness, orthostatic hypotension; pregnancy (category B).

ROUTE & DOSAGE

Benign Prostatic Hypertrophy
Adult: **PO** 10 mg daily

ADMINISTRATION

Oral
- Give immediately after same meal each day.
- Ensure that extended release tablet is not crushed or chewed. It **must be** swallowed whole.
- Store at 15°–30° C (59°–86° F). Protect from light and moisture.

ADVERSE EFFECTS (≥1%) **Body as a Whole:** Fatigue, pain. **CNS:** Dizziness, headache. **GI:** Abdominal pain, dyspepsia, constipation, nausea. **CV:** Orthostatic hypotension. **Respiratory:** Upper respiratory infection, bronchitis, sinusitis, pharyngitis. **Urogenital:** Impotence, priapism.

INTERACTIONS Drug: Increased risk of hypotension with other ANTIHYPERTENSIVE AGENTS or PDE5 INHIBITORS **ketoconazole, itraconazole,** PROTEASE INHIBITORS may increase alfuzosin levels and toxicity. Contraindicated with ANTIRETROVIRAL PROTEASE INHIBITORS or potent CYP3A4 inhibitors.

PHARMACOKINETICS **Absorption:** 80% protein bound **Peak:** 8 h. **Metabolism:** In liver by CYP3A4. **Elimination:** 69% in feces, 24% in urine. **Half-Life:** 10 h.

NURSING IMPLICATIONS

Assessment & Drug Effects
- Monitor CV status and BP, especially with concurrent antihypertensive drugs or inhibitors of CYP3A4. See INTERACTIONS.
- Check postural vital signs for orthostatic hypotension within a few hours following administration.
- Withhold drug and report new or worsening angina to prescriber.
- Lab tests: Baseline and periodic LFTs.

Patient & Family Education
- Inform prescriber about all other prescription, nonprescription, or herbal drugs being taken.
- Make position changes slowly to minimize dizziness.
- Do not drive or engage in other hazardous activities until reaction to drug is known.

ALISKIREN
(a-lis'ki-ren)

Tekturna
Classifications: RENIN ANGIOTENSIN SYSTEM ANTAGONIST; ANTIHYPERTENSIVE
Therapeutic: DIRECT RENIN INHIBITOR; ANTIHYPERTENSIVE
Pregnancy Category: C first trimester; D second and third trimester

AVAILABILITY 150 mg, 300 mg tablets

ACTION & *THERAPEUTIC EFFECT*
A direct renin inhibitor that reduces plasma renin activity and inhibits the conversion of angiotensinogen to angiotensin I (Ang I) and subsequent production of angiotensin II (Ang II). *Lowers blood pressure by decreasing vasoconstriction and aldosterone production, thus reducing sodium reabsorption and fluid retention.*

USES Treatment of hypertension, either as monotherapy or in combination with other antihypertensive agents.

Common adverse effects in *italic*, life-threatening effects <u>underlined</u>; generic names in **bold;** classifications in SMALL CAPS; ♣ Canadian drug name; ⊕ Prototype drug

CONTRAINDICATIONS Hypersensitivity to aliskiren; hyperkalemia; hypercalcemia; dehydration; pregnancy (category D second and third trimester); lactation.

CAUTIOUS USE Cautious use in patients with CrCl less than 30 mL/min; older adults; history of angioedema; respiratory disorders; history of airway surgery; diabetes mellitus; pregnancy (category C first trimester); children younger than 18 y.

ROUTE & DOSAGE

Hypertension
Adult: PO 150–300 mg once daily

ADMINISTRATION
Oral
- Give consistently with regard to meals.
- Store at 15°–30° C (59°–86° F) and protect from light.

ADVERSE EFFECTS (≥1%) **CNS:** *Headache, dizziness.* **GI:** *Diarrhea.* **Metabolic:** Hyperkalemia. **Skin:** Angioedema, rash.

INTERACTIONS Drug: Enhances effects of other ANTIHYPERTENSIVE AGENTS. **Atorvastatin** and **ketoconazole** increase the plasma level of aliskiren, while **irbesartan** decreases its plasma level. Aliskiren decreases the plasma level of **furosemide.**

PHARMACOKINETICS Absorption: 2.5%. **Peak:** 1–3 h. **Metabolism:** Less than 10% via liver. **Elimination:** Primarily in stool. **Half-Life:** 24 h.

NURSING IMPLICATIONS
Assessment & Drug Effects
- Monitor for hypotension after the initiation of therapy and following dosage increase.

- Monitor for angioedema, which may occur any time during treatment. Withhold drug and immediately report to prescriber.
- Lab tests: Periodic serum electrolytes, especially with concurrent ACE inhibitor.

Patient & Family Education
- Full therapeutic effect is usually obtained by 2 wk of therapy.
- Report immediately any of the following: Swelling about the face, lips, tongue; difficulty breathing or swallowing; swelling of hands or feet.
- High fat meals interfere with the absorption of this drug. Do not take drug following a high fat meal.
- Do not use salt substitutes or potassium supplements without consulting prescriber.
- Lab tests: Periodic serum potassium.

ALITRETINOIN (9-*cis*-RETINOIC ACID)
(a-li-tre'ti-noyne)

Panretin
Classifications: ANTIACNE (RETINOID)
Therapeutic: ANTIACNE
Prototype: Isotretinoin
Pregnancy Category: D

AVAILABILITY 0.1% gel

ACTION & *THERAPEUTIC EFFECT*
Naturally occurring retinoid that binds to and activates all known retinoid receptors in cells, which regulate cellular differentiation and proliferation in both healthy and neoplastic cells. *Inhibits the growth of Kaposi's sarcoma (KS) in HIV patients. It does not prevent the development of new KS lesions.*

USES Treatment of cutaneous lesions of AIDS-related Kaposi's sarcoma.
UNLABELED USES Cutaneous T-cell lymphomas.

Common adverse effects in *italic*, life-threatening effects <u>underlined</u>; generic names in **bold;** classifications in SMALL CAPS; ♣ Canadian drug name; ☻ Prototype drug

41

CONTRAINDICATIONS Hypersensitivity to alitretinoin or other retinoids including vitamin A; when systemic anti-KS therapy is required; pregnancy (category D), lactation.
CAUTIOUS USE Cutaneous T-cell lymphoma. Safety and efficacy in children younger than 18 y, or older adults 65 y or older, are unknown.

ROUTE & DOSAGE

Cutaneous Kaposi's Sarcoma

Adult: **Topical** Apply sufficient gel to cover lesions b.i.d., may increase application to 3–4 times daily if tolerated

ADMINISTRATION

Topical
- Apply gel liberally over lesions; avoid unaffected skin and mucus membranes.
- Dry 3–5 min before covering with clothes. Do not cover with occlusive dressing.
- Store at 15°–30° C (59°–86° F).

ADVERSE EFFECTS (≥1%) **Skin:** Erythema, edema, vesiculation, *rash, burning pain,* pruritus, <u>exfoliative dermatitis</u>, excoriation, paresthesia.

INTERACTIONS Drug: Increased toxicity with insect repellents containing DEET.

PHARMACOKINETICS Absorption: Minimal.

NURSING IMPLICATIONS

Assessment & Drug Effects
- Monitor for S&S of dermal toxicity (e.g., erythema, edema, vesiculation).

Patient & Family Education
- Allow up to 14 wk for therapeutic response.

- Avoid exposure of medicated skin to sunlight or sun lamps.
- Contact prescriber if inflammation, swelling, or blisters appear on medicated areas.

ALLOPURINOL

(al-oh-pure′i-nole)
Alloprin A ♦, Apo-allopurinol-A ♦, Zyloprim
Classifications: ANTIGOUT
Therapeutic: ANTIGOUT
Pregnancy Category: C

AVAILABILITY 100 mg, 300 mg tablets; 500 mg vial

ACTION & *THERAPEUTIC EFFECT*
Allopurinol reduces endogenous uric acid by selectively inhibiting action of xanthine oxidase, the enzyme responsible for converting hypoxanthine to xanthine and xanthine to uric acid (end product of purine catabolism). *Urate pool is decreased by the lowering of both serum and urinary uric acid levels, and hyperuricemia is prevented.*

USES To control hyperuricemia, gout, nephrolithiasis, renal calculus, uric acid nephropathy.

CONTRAINDICATIONS Hypersensitivity to allopurinol; as initial treatment for acute gouty attacks; idiopathic hemochromatosis (or those with family history); HLA-B*5801 genotype (strongly associated with allopurinol induced severe cutaneous reactions).
CAUTIOUS USE Impaired hepatic or renal function, history of peptic ulcer, lower GI tract disease, bone marrow depression; pregnancy (category C).

Common adverse effects in *italic*, life-threatening effects <u>underlined</u>; generic names in **bold**; classifications in SMALL CAPS; ♦ Canadian drug name; ☯ Prototype drug

ROUTE & DOSAGE

Treatment of Hyperuricemia

Adult/Adolescent/Child (over 10):
PO 600–800 mg per day (in divided doses) **IV** 200–400 mg/m²/day (max: 600 mg/day) in 1–4 divided doses
Child: **PO** 6–10 years, 300 mg/day *(less than 6 years)* **PO** 150 mg/day **IV** 200 mg/m²/day in 1–4 divided doses

Treatment of Recurrent Renal Calculi

Adult: **PO** 200–300 daily (may divide dose)

Renal Impairment Dosage Adjustment

CrCl 10–20 ml/min: **PO** 200 mg/day **IV** 100 mg; *3–9 mL/min:* 100 mg daily; *less than 3 ml/min:* 100 mg with extended interval between doses (24 h or more)

Hemodialysis Dosage Adjustment

Administer dose after dialysis or use 50% supplemental dose.

ADMINISTRATION

Oral

- Give after meals for best toleration; tablet may be crushed and taken with fluid or mixed with food.
- Store at 15°–30° C (59°–86° F) in a tightly closed container.

Intravenous

PREPARE: **Intermittent:** Reconstitute a single dose vial (500 mg) with 25 mL of sterile water for injection to yield 20 mg/mL.
- **Must be** further diluted with NS or D5W to a concentration of 6 mg/mL or less. - Note: Adding 2.3 mL of diluent yields 6 mg/mL.
ADMINISTER: **Intermittent** Usually administered over 30–60 min.

INCOMPATIBILITIES **Solution/additive:** Amikacin, amphotericin B, carmustine, cefotaxime, chlorpromazine, cimetidine, clindamycin, cytarabine, dacarbazine, daptomycin, daunorubicin, diltiazem, diphenhydramine, doxorubicin, doxycycline, droperidol, epirubicin, ertapenem, etoposide, floxuridine, gentamicin, haloperidol, hydroxyzine, idarubicin, imipenem-cilastatin, irinotecan, mechlorethamine, meperidine, methylprednisolone, metoclopramide, metoprolol, minocycline, mycophenolate, nalbuphine, netilmicin, ondansetron, palonosetron, pancuronium, potassium acetate, prochlorperazine, promethazine, sodium bicarbonate, streptozocin, tacrolimus, tobramycin, vecuronium, vinorelbine.

ADVERSE EFFECTS (≥1%) **CNS:** Drowsiness, headache, vertigo. **GI:** Nausea, vomiting, diarrhea, abdominal discomfort, indigestion, malaise. **Hematologic:** (Rare) <u>Agranulocytosis, aplastic anemia, bone marrow depression,</u> thrombocytopenia. **Skin:** Urticaria or pruritus, pruritic maculopapular rash, toxic epidermal necrolysis. **Other:** <u>Hepatotoxicity,</u> renal insufficiency.

DIAGNOSTIC TEST INTERFERENCE Possibility of elevated blood levels of *alkaline phosphatase* and *serum transaminases (AST, ALT),* and decreased blood *Hct, Hgb, leukocytes*.

INTERACTIONS Drug: Alcohol may inhibit renal excretion of uric acid; **ampicillin, amoxicillin** increase risk of skin rash; enhances anticoagulant effect of

---

warfarin; toxicity from **azathio-prine, mercapto-purine, cyclophosphamide, cyclosporin** increased; increases hypoglycemic effects of **chlorpropamide;** THI-AZIDES increase risk of allopurinol toxicity and hypersensitivity (especially with impaired renal function); ACE INHIBITORS increase risk of hypersensitivity; high dose **vitamin C** increases risk of kidney stone formation.

PHARMACOKINETICS Absorption: 80–90% from GI tract. **Onset:** 24–48 h. **Peak:** 2–6 h. **Metabolism:** 75–80% to the active metabolite oxypurinol. **Elimination:** Slowly excreted in urine; excreted in breast milk. **Half-Life:** 1–3 h; oxypurinol, 18–30 h.

NURSING IMPLICATIONS

Assessment & Drug Effects

▪ Monitor for therapeutic effectiveness which is indicated by normal serum and urinary uric acid levels usually by 1–3 wk, gradual decrease in size of tophi, absence of new tophaceous deposits (after approximately 6 mo), with consequent relief of joint pain and increased joint mobility.

▪ Monitor for S&S of an acute gouty attack which is most likely to occur during first 6 wk of therapy.

▪ Lab tests: Serum uric acid levels q1–2wk to check adequacy of dosage; baseline CBC, liver and kidney function tests and then monthly, particularly during first few months. Check urinary pH at regular intervals.

▪ Monitor patients with renal disorders more often; they tend to have a higher incidence of renal stones and drug toxicity problems.

▪ Report onset of rash or fever immediately to prescriber; withhold drug. Life-threatening toxicity syndrome can occur 2–4 wk after initiation of therapy (more common with impaired renal function) and is generally accompanied by malaise, fever, and aching, a diffuse erythematous, desquamating rash, hepatic dysfunction, eosinophilia, and worsening of renal function.

Patient & Family Education

▪ Drink enough fluid to produce urinary output of at least 2000 mL/day (fluid intake of at least 3000 mL/day). (Note that 1000 mL is approximately equal to 1 quart.) Report diminishing urinary output, cloudy urine, unusual color or odor to urine, pain or discomfort on urination.

▪ Report promptly the onset of itching or rash. Stop drug if a skin rash appears, and report to prescriber.

▪ Minimize exposure of eyes to ultraviolet or sunlight, which may stimulate the development of cataracts.

▪ Do not drive or engage in potentially hazardous activities until response to drug is known.

ALMOTRIPTAN

(al-mo-trip′tan)

Axert

Classifications: SEROTONIN 5-HT₁ RECEPTOR AGONIST

Therapeutic: ANTIMIGRAINE

Prototype: Sumatriptan

Pregnancy Category: C

AVAILABILITY 6.25 mg, 12.5 mg tablets

ACTION & THERAPEUTIC EFFECT Selective agonist that binds with high affinity to serotonin receptors found on extracerebral and intracranial blood vessels. Due primarily to its effects on 5-HT$_{1D}$ and 5-HT$_{1B}$ serotonin receptors on cranial blood vessels, it causes vasoconstriction and decreases inflammation and neurotransmission. *This results in constriction of cranial vessels that become dilated during a migraine attack and reduces signal transmission in the pain pathways.*

USES Treatment of migraine headache with or without aura.

CONTRAINDICATIONS Hypersensitivity to almotriptan malate; significant cardiovascular disease such as ischemic heart disease, coronary artery vasospasms, MI, angina, arteriosclerosis, cardiac arrhythmias, history of cerebrovascular events, or uncontrolled hypertension; stroke, Wolff-Parkinson-White syndrome, within 24 h of receiving another 5-HT$_1$ agonist or an ergot-amine-containing or ergot-type drug; basilar or hemiplegic migraine.

CAUTIOUS USE Significant risk factors for coronary artery disease unless a cardiac evaluation has been done; hypertension; risk factors for cerebrovascular accident; diabetes; colitis; smoking; obesity; peripheral vascular disease, impaired liver or kidney function, Raynaud's disease, older adults; pregnancy (category C), lactation; children.

ROUTE & DOSAGE

Migraine Headache

Adult: **PO** 6.25–12.5 mg; if headache returns, may repeat after at least 2 h (max: 2 tabs/24 h)

Renal Impairment Dosage Adjustment

CrCl less than 10 mL/min: 6.25 mg (max: 12.5 mg/day)

Hepatic Impairment Dosage Adjustment

6.25 mg (max: 12.5 mg/day)

ADMINISTRATION

Oral

- Do not give within 24 h of an ergot-containing drug.
- Administer any time after symptoms of migraine appear.
- Do not administer a second dose without consulting the prescriber for any attack during which the FIRST dose did NOT work.
- Give a second dose if headache was relieved by first dose but symptoms return; however, wait at least 2 h after the first dose before giving a second dose.
- Do not give more than two doses in 24 h.
- Store at 15°–30° C (59°–86° F).

ADVERSE EFFECTS (≥1%) **Body as a Whole:** Flushing. **CNS:** Drowsiness, headache, paresthesia. **CV:** Palpitations, tachycardia, serious cardiac events (including <u>MI</u>) have been reported within a few hours after administration. **GI:** Nausea, vomiting, dry mouth.

INTERACTIONS Drug: Dihydroergotamine, methysergide, other 5-HT$_1$ AGONISTS may cause prolonged vasospastic reactions; SSRIS, **sibutramine** have rarely caused weakness, hyperreflexia, and incoordination; MAOIS should not be used with 5-HT$_1$ AGONISTS.

PHARMACOKINETICS Absorption: Well absorbed, 70% reaches systemic

Common adverse effects in *italic*, life-threatening effects <u>underlined</u>; generic names in **bold**; classifications in SMALL CAPS; ♣ Canadian drug name; ⊙ Prototype drug

45

circulation. **Peak:** 1–3 h. **Distribution:** 35% protein bound. **Metabolism:** 27% metabolized by monoamine oxidase. **Elimination:** 75% renally, 13% in feces. **Half-Life:** 3–4 h.

NURSING IMPLICATIONS

Assessment & Drug Effects

- Monitor cardiovascular status carefully following first dose in patients at relatively high risk for coronary artery disease (e.g., postmenopausal women, men over 40 years old, persons with known CAD risk factors) or who have coronary artery vasospasms.
- Report to prescriber immediately chest pain or tightness in chest or throat that is severe or does not quickly resolve following a dose of almotriptan.
- Pain relief usually begins within 10 min of ingestion, with complete relief in approximately 65% of all patients within 2 h.
- Monitor BP, especially in those being treated for hypertension.

Patient & Family Education

- Notify prescriber immediately if symptoms of severe angina (e.g., severe or persistent pain or tightness in chest, back, neck, or throat) or hypersensitivity (e.g., wheezing, facial swelling, skin rash, or hives) occur.
- Do not take any other serotonin receptor agonist (e.g., Imitrex, Maxalt, Zomig, Amerge) within 24 h of taking almotriptan.
- Advise prescriber of any drugs taken within 1 wk of beginning almotriptan.
- Check with prescriber regarding drug interactions before taking any new OTC or prescription drugs.
- Report any other adverse effects (e.g., tingling, flushing, dizziness) at next prescriber visit.

ALOSETRON

(a-lo′se-tron)

Lotronex

Classifications: SEROTONIN 5-HT$_3$ RECEPTOR ANTAGONIST
Therapeutic: GI ANTIMOTILITY
Pregnancy Category: B

AVAILABILITY 0.5 mg, 1 mg tablets

ACTION & *THERAPEUTIC EFFECT*

Potent and selective serotonin (5-HT$_3$) receptor antagonist. Serotonin 5-HT$_3$ receptors are extensively located on enteric neurons of the GI tract. Activation of these receptors affects amount of visceral pain experienced, transit time in the colon, and GI secretions. *Alosetron significantly controls GI pain, and severe diarrhea related to irritable bowel syndrome.*

USES Treatment of severe chronic irritable bowel syndrome (IBS) in women whose predominant symptom is diarrhea and whose symptoms have lasted longer than 6 mo and have failed to respond to conventional therapy.

CONTRAINDICATIONS Constipation, ischemic colitis, development of ischemic bowel symptoms such as sudden onset of rectal bleeding, bloody diarrhea, new or sudden worsening of abdominal pain; history of chronic or severe constipation, intestinal obstruction, toxic megacolon, GI adhesions, GI perforation, active diverticulitis, history of, or current Crohn's disease or ulcerative colitis; hypersensitivity to alosetron; thrombophlebitis, hypercoagulable state, inability to comply with Patient–Prescriber Agreement, severe hepatic impairment.

Common adverse effects in *italic*, life-threatening effects <u>underlined</u>; generic names in **bold**; classifications in SMALL CAPS; ✦ Canadian drug name; ⊘ Prototype drug

CAUTIOUS USE Hepatic insufficiency, renal impairment; older adults; pregnancy (category B), lactation. Safety and efficacy in children are not established.

ROUTE & DOSAGE

Irritable Bowel Syndrome
Adult: **PO** Start with 0.5 mg b.i.d. for 4 wk, may increase to 1 mg b.i.d. if tolerated

ADMINISTRATION

Oral
- Ensure that the patient has signed the Patient–Prescriber Agreement prior to administering alosetron.
- Do not give this drug if the patient has constipation.
- Review the contraindications for this drug and ensure that the patient has none of the conditions for which the drug is contraindicated.
- Store at 25° C (77° F).

ADVERSE EFFECTS (≥1%) **Body as a Whole:** Malaise, fatigue, cramps, pain. **CNS:** Anxiety. **CV:** Tachyarrhythmias. **GI:** *Constipation*, abdominal pain, nausea, distention, reflux, hemorrhoids, hyposalivation, dyspepsia, <u>ischemic colitis</u>. **Skin:** Sweating, urticaria. **Urogenital:** Urinary frequency.

INTERACTIONS Drug: Fluvoxamine increases alosetron serum level.

PHARMACOKINETICS Absorption: Rapidly absorbed, average bioavailability of 50–60%. **Peak:** 1 h. **Distribution:** 82% protein bound. **Metabolism:** Extensively in liver by CYP2C9. **Elimination:** 73% in urine, 24% in feces. **Half-Life:** 1.5 h.

NURSING IMPLICATIONS

Assessment & Drug Effects
- Monitor for and report immediately signs of ischemic colitis such as new or worsening abdominal pain, bloody diarrhea, or blood in the stool.
- Withhold drug and notify prescriber if patient has not had adequate control of IBS symptoms after 4 wk of treatment with 1 mg twice a day.
- Monitor carefully patients who have decreased GI motility (e.g., older adults, persons receiving other drugs which may decrease GI motility) as they may be at greater risk of serious complications of constipation.
- Monitor carefully patients with any degree of hepatic insufficiency as they may be more susceptible to adverse drug effects.
- Monitor periodically for cardiac arrhythmias, especially with preexisting cardiovascular disease.

Patient & Family Education
- Read the Medication Guide before starting alosetron and each time you refill your prescription.
- Do not start taking alosetron if you are constipated.
- Discontinue alosetron immediately and contact your prescriber if you experience any of the following: Constipation, new or worsening abdominal pain, bloody diarrhea, or blood in the stool.
- Contact your prescriber immediately if constipation does not resolve after discontinuation of alosetron. Resume alosetron again only if constipation has resolved and your prescriber directs you to begin taking the medication again.
- Stop taking alosetron and contact your prescriber if IBS symptoms are not adequately controlled after 4 wk of taking 1 tablet twice a day.

Common adverse effects in *italic*, life-threatening effects <u>underlined</u>; generic names in **bold**; classifications in SMALL CAPS; ♣ Canadian drug name; ⊘ Prototype drug

47

ALPHA₁-PROTEINASE INHIBITOR (HUMAN) ℗⊕

(pro′ten-ase)

Aralast, Prolastin, Zemaira

Classification: RESPIRATORY ENZYME INHIBITOR

Therapeutic: RESPIRATORY ENZYME REPLACEMENT

Pregnancy Category: C

AVAILABILITY Prolastin: 25 mg/mL; **Aralast:** 16 mg/mL; **Zemaira:** 50 mg/mL

ACTION & *THERAPEUTIC EFFECT*
Alpha₁-proteinase inhibitor (alpha₁-PI; alpha₁-antitrypsin) is extracted from plasma and used in patients with panacinar emphysema who have alpha₁-antitrypsin deficiency. Alpha₁-antitrypsin deficiency is a chronic, hereditary, and usually fatal autosomal recessive disorder that results in a slowly progressive, panacinar emphysema. *Prevents the progressive breakdown of elastin tissues in the alveoli, thus slowing panacinar emphysema progression.*

USES Indicated for chronic replacement therapy in patients with alpha₁-antitrypsin deficiency and demonstrable panacinar emphysema.

CONTRAINDICATIONS Individuals with selective IgA deficiencies; lactation.

CAUTIOUS USE Patients with significant heart disease or other conditions that may be aggravated with slight increases in plasma volume; pregnancy (category C). Safety and efficacy in children are not established.

ROUTE & DOSAGE

Panacinar Emphysema
Adult: **IV** 60 mg/kg once/wk

ADMINISTRATION

Intravenous

Give hepatitis B vaccine prior to utilizing this drug.

***PREPARE:* IV Infusion:** Warm unopened diluent and concentrate to room temperature. ▪ Use the supplied, double needle transfer device to reconstitute with sterile water for injection (supplied by manufacturer) to yield a concentration of 20 mg/mL.

***ADMINISTER:* IV Infusion:** Give within 3 h after reconstitution. ▪ Give alone, without mixing with other agents. ▪ Administer at rate of 0.08 mL/kg/min or more slowly as determined by response and comfort of the patient. ▪ Note: The recommended dosage takes about 30 min to administer to a 70 kg person.

▪ Store unreconstituted drug at 2°–8° C (35°–46° F). Do not refrigerate after reconstitution. Discard unused solution.

ADVERSE EFFECTS (≥1%) **Hematologic:** Leukocytosis. **CNS:** Dizziness, fever (may be delayed). **Respiratory:** Upper and lower respiratory tract infections. **Other:** Hepatitis B if not immunized.

PHARMACOKINETICS Distribution: Crosses placenta; distributed into breast milk. **Metabolism:** Undergoes catabolism in the intravascular space; approximately 33% is catabolized per day. **Half-Life:** 4.5–5.2 days.

NURSING IMPLICATIONS

Assessment & Drug Effects

▪ Administer with caution in patients at risk for circulatory overload. Monitor cardiac status.
▪ Monitor respiratory status (rate, dyspnea, lung sounds) throughout therapy.

Common adverse effects in *italic*, life-threatening effects underlined; generic names in **bold;** classifications in SMALL CAPS; ♣ Canadian drug name; ⊕ Prototype drug

- Lab tests: Monitor serum alpha₁-PI level; periodic pulmonary functions and ABGs.

Patient & Family Education

- Avoid smoking and notify prescriber of any changes in respiratory pattern.

ALPRAZOLAM

(al-pray'zoe-lam)
Niravam, Xanax, Xanax XR
Classifications: ANXIOLYTIC; SEDATIVE-HYPNOTIC; BENZODIAZEPINE
Therapeutic: ANTIANXIETY; SEDATIVE-HYPNOTIC
Prototype: Lorazepam
Pregnancy Category: D
Controlled Substance: Schedule IV

AVAILABILITY 0.25 mg, 0.5 mg, 1 mg, 2 mg tablets; 0.5 mg 1 mg, 2 mg, 3 mg sustained release tabs; 1 mg/mL oral solution; 0.25 mg, 0.5 mg, 1 mg, 2 mg orally disintegrating tabs

ACTION & *THERAPEUTIC EFFECT*
A CNS depressant that appears to act at the limbic, thalamic, and hypothalamic levels of the CNS. *Has antianxiety and sedative effects with addictive potential.*

USES Management of anxiety disorders or for short-term relief of anxiety symptoms. Also used as adjunct in management of anxiety associated with depression and agitation, and for panic disorders, such as agoraphobia.
UNLABELED USES Alcohol withdrawal.

CONTRAINDICATIONS Sensitivity to benzodiazepines; acute narrow angle glaucoma; pulmonary disease; use alone in primary depression or psychotic disorders, bipolar disorders, organic brain disorders; myasthenia gravis; pregnancy (category D), lactation.
CAUTIOUS USE Impaired hepatic function; history of alcoholism; renal impairment, hepatic disease; geriatric and debilitated patients ; children younger than 18 y. Effectiveness for long-term treatment (greater than 4 mo) not established.

ROUTE & DOSAGE

Anxiety Disorders
Adult: **PO** 0.25–0.5 mg t.i.d. (max: 4 mg/day)
Geriatric: **PO** 0.125–0.25 mg b.i.d.

Panic Attacks
Adult: **PO** 1–2 mg t.i.d. (max: 8 mg/day); sustained release: Initiate with 0.5–1 mg once/day. Depending on the response, the dose may be increased at intervals of 3 to 4 days in increments of no more than 1 mg/day. Target range 3–6 mg/day (max: 10 mg/day).

Hepatic Impairment Dosage Adjustment
Reduce dose by 50% in hepatic impairment.
Do not discontinue abruptly.

ADMINISTRATION
Oral

- Reduce drug gradually when discontinuing drug.
- Store in light-resistant containers at 15°–30° C (59°–86° F), unless otherwise directed.

ADVERSE EFFECTS (≥1%) **CNS:** *Drowsiness, sedation,* lightheadedness, dizziness, syncope,

Common adverse effects in *italic*, life-threatening effects underlined; generic names in **bold;** classifications in SMALL CAPS; ♣ Canadian drug name; ☻ Prototype drug

49

depression, headache, confusion, insomnia, nervousness, fatigue, clumsiness, unsteadiness, rigidity, tremor, restlessness, paradoxical excitement, hallucinations. **CV:** Tachycardia, hypotension, ECG changes. **Special Senses:** Blurred vision. **Respiratory:** Dyspnea.

INTERACTIONS Drug: Alcohol and other CNS DEPRESSANTS, ANTICONVULSANTS, ANTIHISTAMINES, BARBITURATES, NARCOTIC ANALGESICS, BENZODIAZEPINES compound CNS depressant effects; **cimetidine, disulfiram, fluoxetine,** TRICYCLIC ANTIDEPRESSANTS increase alprazolam levels (decreased metabolism); ORAL CONTRACEPTIVES may increase or decrease alprazolam effects. **Herbal: Kava, valerian** may potentiate sedation; **St. John's wort** decreases serum level of alprazolam.Cigarette smoking may decrease serum level of alprazolam by 50%.

PHARMACOKINETICS Absorption: Rapidly absorbed. **Peak:** 1–2 h. **Distribution:** Crosses placenta. **Metabolism:** Oxidized in liver to inactive metabolites by CYP3A4. **Elimination:** Renal elimination. **Half-Life:** 12–15 h.

NURSING IMPLICATIONS
Assessment & Drug Effects

- Monitor for S&S of drowsiness and sedation, especially in older adults or the debilitated; they may require supervised ambulation and/or fall precautions.
- Lab tests: Periodic blood counts, urinalyses, and blood chemistry studies, particularly during continuing therapy.

Patient & Family Education

- Make position changes slowly and in stages to prevent dizziness.
- Do not use alcohol, other CNS depressants, or OTC medications containing antihistamines (e.g.,

sleep aids, cold, hay fever, or allergy remedies) without consulting prescriber.
- Do not drive or engage in potentially hazardous activities until response to drug is known.
- Taper dosage following continuous use; abrupt discontinuation of drug may cause withdrawal symptoms: Nausea, vomiting, abdominal and muscle cramps, sweating, confusion, tremors, convulsions.

ALPROSTADIL (PGE₁)

(al-pross'ta-dil)
Caverject, Edex, Muse, Prostin VR Pediatric
Classification: PROSTAGLANDIN
Therapeutic: PROSTAGLANDIN
Prototype: Epoprostenol
Pregnancy Category: C

AVAILABILITY 500 mcg/mL injection; 10 mcg/mL, 20 mcg/mL, 40 mcg/mL powder for injection; 125 mcg, 250 mcg, 500 mcg, 1000 mcg; urethral suppository

ACTION & *THERAPEUTIC EFFECT*
Preserves ductal patency by relaxing smooth muscle of ductus arteriosus. Alprostadil induces penile erection by relaxing the smooth muscles of the corpus cavernosum and dilating the cavernosal arteries and their penile arterioles. Sufficient rigidity of the penis also requires increased venous outflow resistance. *Preserves ductal patency by relaxing smooth muscle of ductus arteriosus. It induces penile rigidity and erection by penile blood engorgement.*

USES Temporary measure to maintain patency of ductus arteriosus in infants with ductal-dependent congenital heart defects until corrective surgery can be performed. Also used in erectile dysfunction.

CONTRAINDICATIONS Ductus arteriosus respiratory distress syndrome (hyaline membrane disease); neonates with respiratory distress syndrome; hypersensitivity to alprostadil; patients with penile implants. *Muse, Edex:* Women, children, and newborns; lactation. *Muse:* Patients with urethral stricture, inflammation/ infection of glans of penis, severe hypospadias, acute or chronic urethritis; sickle cell anemia or trait, thrombocytopenia, thrombocytosis; polycythemia, multiple myeloma.

CAUTIOUS USE Ductus arteriosus; bleeding tendencies; cardiovascular disease; erectile dysfunction; hypersensitivity to alprostadil; leukemia; penile anatomic deformations; patients on anticoagulants, vasoactive or antihypertensive drugs.

ROUTE & DOSAGE

To Maintain Patency of Ductus Arteriosus

Neonate: **IV** 0.05–0.1 mcg/kg/ min, may increase gradually (max: 0.4 mcg/kg/min)

Erectile Dysfunction of Vasculogenic, Psychogenic, or Mixed Etiology

Adult: **Intracavernosal** Initiate with 2.5 mcg; if inadequate response, increase dose by 2.5 mcg. May then increase dose in 5 mcg increments until a suitable erection occurs, not exceeding 1 h in duration (max: 60 mcg). *Adult:* **Intraurethral (Muse)** 125 mcg or 250 mcg; dose adjusted to patient satisfaction (max: 2 x/24h)

Erectile Dysfunction of Pure Neurogenic Etiology

Adult: **Intracavernosal** Initiate with 1.25 mcg; if inadequate response, increase dose by 1.25 mcg, then increase by 2.5 mcg, may then increase dose in 5 mcg increments until a suitable erection occurs, not exceeding 1 h in duration; wait 24 hrs between doses (max: 60 mcg)

ADMINISTRATION

Intracavernosal Injection

- Administer only after proper training in the penile injection technique. Refer to information on administration provided to the patient by the manufacturer.
- Use reconstituted solutions immediately.
- Store dry powder at or below 25° C (77° F) for up to 3 mo. Do not freeze.

Transurethral Insertion

- Refer to information on insertion of urethral suppository into the urethra provided to the patient by the manufacturer.

Intravenous

PREPARE: Continuous: Dilute 500 mcg alprostadil with NS or D5W to volume appropriate for pump delivery system. ▪ Prepare fresh solution q24h. Discard unused portions. ▪ A 500 mcg ampule diluted in 250 mL yields a concentration of 2 mcg/mL.

ADMINISTER: Continuous: Infuse at rate of 0.05–0.1 mcg/ kg/min up to a maximum of 0.4 mcg/kg/min. ▪ Reduce infusion rate immediately if arterial pressure drops significantly or if fever occurs. ▪ Discontinue promptly, if apnea or bradycardia occurs.

Common adverse effects in *italic*, life-threatening effects underlined; generic names in **bold**; classifications in SMALL CAPS; ✦ Canadian drug name; ☺ Prototype drug

51

- Store at 2°–8° C (36°–46° F) unless otherwise directed by manufacturer. Protect from freezing.

ADVERSE EFFECTS (≥1%) **CNS:** *Fever,* seizures, lethargy. **CV:** *Flushing,* bradycardia, hypotension, syncope, tachycardia; CHF, <u>ventricular fibrillation, shock.</u> **GI:** Diarrhea, gastric regurgitation. **Hematologic:** <u>Disseminated intravascular coagulation</u> (DIC), <u>thrombocytopenia.</u> **Respiratory:** Apnea. **Urogenital:** Oliguria, anuria. *Penile pain,* prolonged erection, priapism, penile fibrosis, injection site hematoma/ecchymosis, penile rash and edema, prostatitis, perineal pain. **Skin:** Rash on face and arms, alopecia. **Other:** Leg pain.

INTERACTIONS Drug: May increase anticoagulant properties of **warfarin;** ANTIHYPERTENSIVE AGENTS increase risk of hypotension.

PHARMACOKINETICS Onset: 15 min to 3 h. **Metabolism:** Rapidly in lungs. **Elimination:** Through kidneys. **Half-Life:** 5–10 min.

NURSING IMPLICATIONS
Assessment & Drug Effects
Ductus Arteriosus
- Monitor therapeutic effectiveness which is indicated by increased blood oxygenation (Po₂), usually evident within 30 min, in infants with cyanotic heart disease; increased pH in those with acidosis, increased systemic BP and urinary output, return of palpable pulses, and decreased ratio of pulmonary artery to aortic pressure in infants with restricted systemic blood flow.
- Monitor arterial pressure, ECG, heart rate, BP, respiratory rate, and temperature, throughout the infusion.

- Lab tests: Monitor arterial blood gases and blood pH throughout the infusion.

Patient & Family Education
Erectile Dysfunction
- Follow carefully directions for penile injection provided by the manufacturer.
- Do not change dose without consulting the prescriber.
- Do not use intracavernosal injection more often than 3 times/wk; allow at least 24 h between uses.
- Do not use more than 2 urethral suppository systems in a 24 h period.
- Report promptly any of the following: Nodules or hard tissue in penis; penile pain, redness, swelling, tenderness; or curvature of the erect penis.
- Seek immediate medical attention if an erection persists longer than 6 h.

ALTEPLASE RECOMBINANT 🅿
(al′te-plase)
Actilyse, Activase, Cathflo Activase
Classification: THROMBOLYTIC, TISSUE PLASMINOGEN ACTIVATOR
Therapeutic: THROMBOLYTIC ENZYME
Pregnancy Category: C

AVAILABILITY 50 mg, 100 mg vials

ACTION & *THERAPEUTIC EFFECT* This recombinant DNA-derived form of human tissue-type plasminogen activator (t-PA) is a thrombolytic agent. t-PA promotes thrombolysis by forming the active proteolytic enzyme, plasmin. *Plasmin is capable of degrading fibrin, fibrinogen, and factors V, VIII, and XII.*

USES Indicated in selective cases of acute MI, preferably within 6 h of attack for recanalization of the coronary artery; lysis of acute pulmonary emboli; acute ischemic stroke or thrombotic stroke (within 3 h of onset); treatment of acute coronary artery thrombosis in the setting of percutaneous coronary intervention (PCI); reestablishing patency of occluded IV catheter.

UNLABELED USES Lysis of arterial occlusions in peripheral and bypass vessels; DVT.

CONTRAINDICATIONS Active internal bleeding, history of cerebrovascular accident, aneurysm, recent (within 2 mo) intracranial or interspinal surgery or trauma, intracranial neoplasm, increased intracranial pressure; arteriovenous malformation, bleeding disorders, severe uncontrolled hypertension, likelihood of left heart thrombus, acute pericarditis, bacterial endocarditis, severe liver or renal dysfunction, septic thrombophlebitis.

CAUTIOUS USE Recent major surgery (within 10 days), cerebral vascular disease, recent GI or GU bleeding, recent trauma, renal impairment, hypertension, hemorrhagic ophthalmic conditions; age greater than 75 y; pregnancy (category C), lactation.

ROUTE & DOSAGE

Acute MI

Adult: IV *Weight 65 kg or more, 60 mg over first hour, 20 mg/h over second hour, and 20 mg over third hour (for a total of 100 mg over 3 h; weight less than 65 kg, 1.25 mg/kg over 3 h (60% of dose over first hour, 20% of* dose over second hour, and 20% of dose over third hour).
Accelerated schedule (with heparin and aspirin): Weight greater than 67 kg, 15 mg bolus, then 50 mg over next 30 min, then 35 mg over next 60 min.
Accelerated schedule (with heparin and aspirin): Weight 67 kg or less, 15 mg bolus, then 0.75 mg/kg (not to exceed 50 mg) over next 30 min, then 0.5 mg/kg (not to exceed 35 mg) over next 60 min

Acute Ischemic Stroke/ Thrombotic Stroke

Adult: IV *0.9 mg/kg over 60 min with 10% of dose as an initial bolus over 1 min (max: 90 mg)*

Pulmonary Embolism

Adult: IV *100 mg infused over 2 h*

Reopen Occluded IV Catheter

Adult/Child (greater than 30 kg): IV *Instill 2 mg/2 mL into dysfunctional catheter for 2 h. May repeat once if needed.*
Child: IV *2 y or older, weight 10–29 kg. Instill 110% of internal lumen volume with 1 mg/mL concentration (max: 2 mg). May repeat if function not restored within 2 h.* IV *Younger than 2 y, weight less than 10 kg, 0.5 mg diluted in a volume to fill the lumen of the catheter*

ADMINISTRATION

Intravenous ——

PREPARE: IV Infusion: Reconstitute the *50 mg vial* as follows: Do not use if vacuum in vial has been broken. Use a large-bore needle (e.g., 18 gauge) and do not prime needle with air. ▪ Dilute contents of vial with sterile water for injection

Common adverse effects in *italic*, life-threatening effects underlined; generic names in **bold**; classifications in SMALL CAPS; ♣ Canadian drug name; ✪ Prototype drug

53

supplied by manufacturer. ▪ Direct stream of sterile water into the lyophilized cake. Slight foaming is usual. Allow to stand until bubbles dissipate. Resulting concentration is 1 mg/mL. ▪ Reconstitute the *100-mg vial* using supplied transfer device for reconstitution. Follow manufacturer's directions.

ADMINISTER: **IV Infusion:** ▪ Start IV infusion as soon as possible after the thrombolytic event, preferably within 6 h. ▪ Administer drug as reconstituted (1 mg/mL) or further diluted with an equal volume of NS or D5W to yield 0.5 mg/mL. **Acute MI:** Administer 60% of total dose in the first hour for acute MI, with 6–10% given as a bolus dose over 1–2 min and remainder of first dose infused over hour 1. Follow with second dose (20% of total) over hour 2, and third dose (20% of total) over hour 3. ▪ For patients weighing less than 65 kg calculate dose using 1.25 mg/kg over 3 h. See accelerated schedule under Route & Dosage. **Pulmonary embolism:** Administer entire dose over a 2 h period. **Acute ischemic stroke:** Give 5 mg as an initial bolus over 1 min, then give the remainder of the 0.75 mg/kg dose over 60 min. ▪ Do not exceed a total dose of 100 mg. Higher doses have been associated with intracranial bleeding. ▪ Follow infusion of drug by flushing IV tubing with 30–50 mL of NS or D5W. ▪ Reconstituted drug is stable for 8 h in above solutions at room temperature (2°–30° C; 36°–86° F). Since there are no preservatives, discard any unused solution after that time.

INCOMPATIBILITIES **Solution/additive: Dobutamine, dopamine, heparin. Y-site: Bivalirudin, dobutamine, dopamine, heparin, nitroglycerin.**

▪ Store above reconstituted solutions at room temperature 2°–30° C (36°–86° F) for no longer than 8 h. Discard any unused solution after that time.

ADVERSE EFFECTS (≥1%) **Hematologic:** Internal and superficial bleeding (cerebral, retroperitoneal, GU, GI).

PHARMACOKINETICS Peak: 5–10 min after infusion completed. **Duration:** Baseline values restored in 3 h. **Metabolism:** In liver. **Elimination:** In urine. **Half-Life:** 26.5 min.

NURSING IMPLICATIONS

Assessment & Drug Effects

▪ Monitor for S&S of excess bleeding q15min for the first hour of therapy, q30min for second to eighth hour, then q8h.

▪ Monitor neurologic checks throughout drug infusion q30min and qh for the first 8 h after infusion.

▪ Protect patient from invasive procedures because spontaneous bleeding occurs twice as often with alteplase as with heparin. IM injections are contraindicated. Minimize physical manipulation of patient during thrombolytic therapy to prevent bruising.

▪ Lab tests: Coagulation tests including APTT, bleeding time, PT, TT, INR, **must be** done before administration of drug. Also check *baseline* Hct, Hgb, and platelet counts. Draw Hct following drug administration to detect possible blood loss.

▪ Check vital signs frequently. Be alert to changes in cardiac rhythm.

▪ Report signs of bleeding: Gum bleeding, epistaxis, hematoma, spontaneous ecchymoses, oozing at catheter site, increased pain from internal bleeding. Stop the infusion, then resume when bleeding stops.

- Use the radial artery to draw ABGs. Pressure to puncture sites, if necessary, should be maintained for up to 30 min.
- Continue monitoring vital signs until laboratory reports confirm anticoagulant control; patient is at risk for postthrombolytic bleeding for 2–4 days after intracoronary alteplase treatment.

Patient & Family Education
- Report promptly any of the following: Sudden severe headache, blood in urine, bloody or tarry stool, any sign of bleeding or oozing from injection/insertion sites.
- Remain quiet and on bedrest while receiving this medicine.

ALTRETAMINE

(al-tre′ta-meen)
Hexalen
Classifications: ANTINEOPLASTIC; ALKYLATING
Therapeutic: ANTINEOPLASTIC
Prototype: Cyclophosphamide
Pregnancy Category: D

AVAILABILITY 50 mg capsules

ACTION & *THERAPEUTIC EFFECT*
Altretamine is a synthetic cytotoxic antineoplastic drug. Its metabolites have cytotoxic properties. *Altretamine has demonstrated neoplastic activity in patients resistant to alkylating agents.*

USES Ovarian cancer.
UNLABELED USES Breast, cervical, colon, endometrial, head, and neck cancer; small-cell lung cancers and lymphomas.

CONTRAINDICATIONS Hypersensitivity to altretamine, severe bone marrow depression, neurologic toxicity, neurologic disease; pregnancy (category D), lactation.

CAUTIOUS USE History of viral infections (i.e., herpes simplex); radiation therapy. Safety and efficacy in children are not established.

ROUTE & DOSAGE

Ovarian Cancer
Adult: **PO** 260 mg/m²/day for 14 or 21 consecutive days in a 28-day cycle

ADMINISTRATION
Oral
- Give only under supervision of a qualified prescriber experienced in the use of antineoplastics.
- Give in 3-4 divided doses after meals and at bedtime.
- Altretamine is usually discontinued for 14 days or longer and restarted at 200 mg/m²/day if any of the following occur: Severe GI intolerance; WBC count less than 2000/mm³, granulocyte count less than 1000/mm³; or progressive neurotoxicity.
- Store at room temperature, 15°–30° C (59°–86° F).

ADVERSE EFFECTS (≥1%) **CNS:** *Paresthesias, hyporeflexia, muscle weakness, peripheral numbness, ataxia, Parkinson-like tremors.* **GI:** *Nausea, vomiting.* **Hematologic:** Leukopenia, thrombocytopenia. **Urogenital:** Slight increase in serum creatinine. **Skin:** Alopecia and eczema.

INTERACTIONS Drug: Concomitant administration of TRICYCLIC ANTIDEPRESSANTS (**imipramine, amitriptyline**), MONOAMINE OXIDASE INHIBITORS, or **selegiline** result in orthostatic hypotension. Avoid use with **sargramostim, filgramstim**.

PHARMACOKINETICS Absorption: Rapidly ·from GI tract. Approximately 25% reaches systemic

Common adverse effects in *italic*, life-threatening effects underlined; generic names in **bold**; classifications in SMALL CAPS; ♣ Canadian drug name; ❂ Prototype drug

55

circulation. **Metabolism:** Rapidly demethylated in the liver. **Elimination:** 62% of the dose is excreted in the urine in 24 h. **Half-Life:** 13 h.

NURSING IMPLICATIONS

Assessment & Drug Effects

- Lab tests: Monitor blood counts at least monthly and prior to each course of therapy.
- Perform a neurologic examination regularly; assess for the presence of paresthesias, hypoesthesias, muscle weakness, peripheral numbness, ataxia, decreased sensations, and alterations in mood or consciousness.
- Withhold medication if neurologic symptoms fail to resolve with dose reduction. Notify prescriber.
- Monitor for nausea and vomiting, which are related to the cumulative dose of altretamine. After several weeks some patients develop tolerance to the GI effects. Antiemetics may be required to control GI distress.

Patient & Family Education

- Taking altretamine after meals or with food or milk may decrease nausea.
- Report symptoms indicative of neurotoxicity to prescriber (paresthesias, hypoesthesias, muscle weakness, peripheral numbness, ataxia, decreased sensations, and alterations in mood or consciousness).

ALUMINUM HYDROXIDE ⊕
(a-lu′mi-num)
ALternaGEL, Alu-Cap, Alugel, Alu-Tab, Amphojel, Dialume

ALUMINUM CARBONATE, BASIC
Basaljel

ALUMINUM PHOSPHATE
Phosphaljel
Classifications: ANTACID; ADSORBENT
Therapeutic: ANTACID
Pregnancy Category: C

AVAILABILITY Aluminum Hydroxide: 300 mg, 400 mg, 500 mg, 600 mg tablets; 300 mg, 400 mg, 500 mg, 600 mg capsules; 320 mg/5 mL, 450 mg/5 mL, 600 mg/5 mL, 675 mg/5 mL suspension; **Aluminum Carbonate, Basic:** 608 mg tablets; 608 mg capsules; 400 mg/5 mL suspension; **Aluminum Phosphate:** 608 mg tablets; 608 mg capsules; 400 mg/5 mL suspension

ACTION & *THERAPEUTIC EFFECT* Nonsystemic antacid with moderate neutralizing action. Reduces acid concentration and pepsin activity by raising pH of gastric and intra-esophageal secretions. *Reduces gastric acidity by neutralizing the stomach acid content. Aluminum carbonate lowers serum phosphate by binding dietary phosphate to form insoluble aluminum phosphate, which is excreted in feces.*

USES Symptomatic relief of gastric hyperacidity associated with gastritis, esophageal reflux, and hiatal hernia; adjunct in treatment of gastric and duodenal ulcer. More commonly used in combination with other antacids. Aluminum carbonate is used primarily in conjunction with a low phosphate diet to reduce hyperphosphatemia in patients with renal insufficiency and for prophylaxis and treatment of phosphatic renal calculi.

CONTRAINDICATIONS Prolonged use of high doses in presence of low serum phosphate.
CAUTIOUS USE Renal impairment; gastric outlet obstruction; older

adults; decreased bowel activity (e.g., patients receiving anticholinergic, antidiarrheal, or antispasmodic agents); patients who are dehydrated or on fluid restriction; pregnancy (category C).

ROUTE & DOSAGE

Antacid (Hydroxide and Phosphate)
Adult: PO 600 mg t.i.d. or q.i.d.

Antacid (Carbonate)
Adult: PO 10–30 mL of regular suspension or 5–15 mL of extra strength suspension or 2 capsules or tablets q2h

Phosphate Lowering (Carbonate)
Adult: PO 10–30 mL of regular suspension or 5–15 mL of extra strength suspension or 2–6 capsules or tablets 1 h p.c. and at bedtime

ADMINISTRATION

Oral

- Tablet **must be** chewed until it is thoroughly wetted before swallowing.
- Note for antacid use: Follow well-chewed tablet with one-half glass of water or milk; follow liquid preparation (suspension) with water to ensure passage into stomach. For phosphate lowering: Follow tablet, capsule, or suspension with full glass of water or fruit juice.
- Store between 15°–30° C (59°–86° F) in tightly closed container.

ADVERSE EFFECTS (≥1%) GI:
Constipation, fecal impaction, intestinal obstruction. **CNS:** Dialysis dementia (thought to be due to aluminum intoxication). **Metabolic:** Hypophosphatemia, hypomagnesemia.

INTERACTIONS Drug: Aluminum will decrease absorption of **chloroquine, cimetidine, ciprofloxacin, digoxin, isoniazid,** IRON SALTS, NSAIDS, **norfloxacin, ofloxacin, phenytoin, phenothiazines, quinidine, tetracycline, thyroxine. Sodium polystyrene sulfonate** may cause systemic alkalosis.

PHARMACOKINETICS Absorption: Minimal absorption. **Peak:** Slow onset. **Duration:** 2 h when taken with food; 3 h when taken 1 h after food. **Elimination:** In feces as insoluble phosphates.

NURSING IMPLICATIONS

Assessment & Drug Effects

- Note number and consistency of stools. Constipation is common and dose related. Intestinal obstruction from fecal concretions has been reported.
- Lab tests: Periodic serum calcium and phosphorus levels with prolonged high-dose therapy or impaired renal function.

Patient & Family Education

- Increase phosphorus in diet when taking large doses of these antacids for prolonged periods; hypophosphatemia can develop within 2 wk of continuous use of these antacids. The older adult in a poor nutritional state is at high risk.
- Antacid may cause stools to appear speckled or whitish.
- Report epigastric or abdominal pain; it is a clinical guide for adjusting dosage. Keep prescriber informed. Pain that persists beyond 72 h may signify serious complications.
- Seek medical help if indigestion is accompanied by shortness of breath, sweating, or chest pain, if stools are dark or tarry, or if

Common adverse effects in *italic*, life-threatening effects underlined; generic names in **bold**; classifications in SMALL CAPS; ♣ Canadian drug name; ⊘ Prototype drug

57

symptoms are recurrent when taking this medication.
- Seek medical advice and supervision if self-prescribed antacid use exceeds 2 wk.

ALVIMOPAN
(al-vi-mo′pan)
Entereg
Classifications: PERIPHERAL OPIOID RECEPTOR ANTAGONIST; GI MOTILITY STIMULANT
Therapeutic: GI MOTILITY STIMULANT
Pregnancy Category: B

AVAILABILITY 12 mg capsules

ACTION & *THERAPEUTIC EFFECT*
Morphine and other post-op analgesics are mu-opioid receptor agonists known to inhibit GI motility and prolong the duration of postoperative ileus. Alvimopan is a selective antagonist of mu-opioid receptors. *It competitively antagonizes the effect of morphine on contractility, shortening the duration of post-op ileus.*

USES To accelerate the time to upper and lower gastrointestinal recovery after partial large or small bowel resection surgery with primary anastomosis.
UNLABELED USES Constipation, opioid-induced constipation.

CONTRAINDICATIONS Therapeutic doses of opioids for greater than 7 consecutive days immediately preoperative; end-stage renal disease; severe hepatic impairment (Child-Pugh class C).
CAUTIOUS USE Recent exposure to opioids; surgery for complete bowel obstruction; history of CAD or MI; pregnancy (category B); lactation. Safety and efficacy in children not established.

ROUTE & DOSAGE

Acceleration of Postoperative GI Recovery
Adult: **PO** 12 mg 0.5–5 h preoperative; then 12 mg b.i.d. up to 7 days

ADMINISTRATION
Oral
- Give pre-op dose 30 min–5 h before surgery.
- Do not exceed 15 doses (maximum allowed).
- Store at 15°–30° C (59°–86° F).
- Note: Hospitals **must be** registered in and have met all of the requirements for the **Entereg** Access Support and Education (E.A.S.E.) program in order to use alvimopan.

ADVERSE EFFECTS (≥3 %) **GI:** Constipation, dyspepsia, flatulence. **Hematologic:** Anemia. **Metabolic:** Hypokalemia. **Musculoskeletal:** Back pain. **Urogenital:** Urinary retention.

INTERACTIONS Food: Decreased extent and rate of absorption if taken with a high fat meal.

PHARMACOKINETICS Absorption: Bioavailability 6%. **Peak:** 2 h. **Distribution:** 90–94% plasma protein bound. **Metabolism:** By intestinal flora. **Elimination:** Fecal (primary) and renal (35%). **Half-Life:** 10–18 h.

NURSING IMPLICATIONS
Assessment & Drug Effects
- Monitor frequently for return of bowel sounds and ability to pass flatus.
- Monitor closely patients with impaired renal function for adverse effects.
- Report to prescriber increasing abdominal pain, diarrhea, nausea and vomiting.

Common adverse effects in *italic*, life-threatening effects <u>underlined</u>; generic names in **bold**; classifications in SMALL CAPS; ♣ Canadian drug name; ⊘ Prototype drug

▪ Lab tests: Serum potassium in patients predisposed to hypokalemia.

Patient & Family Education

▪ Report promptly increasing abdominal pain and nausea.

AMANTADINE HYDROCHLORIDE ⊕

(a-man'ta-deen)

Symmetrel

Classifications: ANTIVIRAL; CENTRAL-ACTING CHOLINERGIC RECEPTOR ANTAGONIST; ANTIPARKINSON

Therapeutic: ANTIVIRAL; ANTIPARKINSON

Pregnancy Category: C

AVAILABILITY 100 mg capsules; 50 mg/5 mL syrup

ACTION & *THERAPEUTIC EFFECT*
Because amantadine does not suppress antibody formation, it can be administered for interim protection in combination with influenza A virus vaccine until antibody titer is adequate or to augment prophylaxis in a previously vaccinated individual. Mechanism of action in parkinsonism may be related to release of dopamine from neuronal storage sites. *Active against several strains of influenza A virus. Effective in management of symptoms of parkinsonism when used in conjunction with other antiparkinson agents.*

USES Treatment of idiopathic parkinsonism or Parkinson's disease. Also used for prophylaxis and symptomatic treatment of influenza A infections.

UNLABELED USES Neuroleptic malignant syndrome (NMS), management of cocaine dependency, fatigue.

CONTRAINDICATIONS Hypersensitivity to amantadine or rimantadine, closed angle glaucoma; suicidal ideation; lactation.

CAUTIOUS USE History of epilepsy or other types of seizures; CHF, peripheral edema, orthostatic hypotension; recurrent eczematoid dermatitis; psychoses, severe psychoneuroses; hepatic disease; renal impairment; older adults, cerebral arteriosclerosis; pregnancy (category C). Safety in children younger than 1 y for Influenza A is not established.

ROUTE & DOSAGE

Influenza A Treatment

Adult (younger than 65 y)/Child (older than 9 y): **PO** 200 mg once/day or 100 mg q12h
Adult (65 y or older): **PO** 100 mg once/day
Child (1–9 y): **PO** 5 mg/kg in 2–3 equal doses (max: 150 mg/day)

Influenza A Prevention

Adult (younger than 65 y): **PO** 200 mg/day or 100 mg q12h; begin as soon as possible after initial exposure and continue for at least 10 days after exposure
Adult (older than 65 y): **PO** 100 mg once daily
HIV-Infected Adult/Adolescent/Child (at least 10 y): **PO** 100 mg twice daily
Child (1–9 y): **PO** 5 mg/kg/day (up to 150 mg/day) given in 2 divided doses (not more than 150 mg/day)

Parkinsonism

Adult: **PO** 100 mg 1–2 times/day, start with 100 mg/day if patient is on other antiparkinsonism medications

Drug-Induced Extrapyramidal Symptoms

Adult: **PO** 100 mg b.i.d. (max: 400 mg/day if needed)

Common adverse effects in *italic*, life-threatening effects <u>underlined</u>; generic names in **bold**; classifications in SMALL CAPS; ♣ Canadian drug name; ⊕ Prototype drug

Renal Impairment Dosage Adjustment
CrCl 30–50 mL/min: 200 mg PO for 1st day, then 100 mg PO daily; 15–29 mL/min: 100 mg PO for 1st day, then 100 mg PO on alternate days; *less than 15 mL/min:* 200 mg q7days

ADMINISTRATION

Oral

- Give with water, milk, or food.
- Use supplied calibrated device for measuring syrup formulation.
- Influenza prophylaxis: Drug should be initiated when exposure is anticipated and continued for at least 10 days.
- Used in conjunction with influenza A vaccine (generally in high-risk patients who have not been vaccinated previously) until protective antibodies develop (10–21 days) after vaccine administration.
- Schedule medication in the morning or, with q12h dosing, schedule 2nd dose several hours before bedtime. If insomnia is a problem, suggest patient limit number of daytime naps.
- Store in tightly closed container preferably at 15°–30° C (59°–86° F) unless otherwise directed by manufacturer. Avoid freezing.

ADVERSE EFFECTS (≥1%) **CNS:** *Dizziness, light-headedness,* headache, ataxia, irritability, anxiety, *nervousness, difficulty in concentrating,* mood or other mental changes, confusion, visual and auditory hallucinations, *insomnia,* nightmares, convulsions. **CV:** Orthostatic hypotension, peripheral edema, dyspnea. **Special Senses:** Blurring or loss of vision. **GI:** Anorexia, *nausea,* vomiting, dry mouth. **Hematologic:** Leukopenia, agranulocytosis.

INTERACTIONS Drug: Alcohol enhances CNS effects; may potentiate effects of ANTICHOLINERGICS.

PHARMACOKINETICS Absorption: Almost completely absorbed from GI tract. **Onset:** Within 48 h. **Peak:** 1–4 h. **Distribution:** Through body fluids. **Metabolism:** Not metabolized. **Elimination:** 90% unchanged in urine. **Half-Life:** 9–37 h (prolonged in renal insufficiency).

NURSING IMPLICATIONS

Assessment & Drug Effects

- Monitor effectiveness. Note that with parkinsonism, maximum response occurs within 2 wk–3 mo. Effectiveness may wane after 6–8 wk of treatment; report change to prescriber.
- Monitor and report: Mental status changes; nervousness, difficulty concentrating, or insomnia; loss of seizure control; S&S of toxicity, especially with doses above 200 mg/day.
- Monitor for and report promptly suicidal ideation, especially in those with a history of psychiatric disorders.
- Establish a baseline profile of the patient's disabilities to accurately differentiate disease symptoms and drug-induced neuropsychiatric adverse reactions.
- Monitor vital signs for at least 3 or 4 days after increases in dosage; also monitor urinary output.
- Lab tests: pH and serum electrolytes.
- Monitor for and report reduced salivation, increased akinesia or rigidity, and psychological disturbances that may develop within 4–48 h after initiation of therapy and after dosage increases with parkinsonism.

Patient & Family Education

- Note: For influenza within 24 h but no later than 48 h after onset of symptoms.
- Make all position changes slowly, particularly from recumbent to upright position, in order to minimize dizziness.
- Report any of the following to prescriber: Shortness of breath, peripheral edema, significant weight gain, dizziness or lightheadedness, inability to concentrate, and other changes in mental status, suicidal ideation, difficulty urinating, and visual impairment.
- Do not drive and exercise caution with potentially hazardous activities until response to the drug is known.
- Note: People with Parkinson's disease should not discontinue therapy abruptly; doing so may precipitate a parkinsonian crisis with severe akinesia, rigidity, tremor, and psychic disturbances. Adhere to established dosage regimen.

AMBENONIUM CHLORIDE

(am-be-noe′nee-um)

Mytelase

Classification: CHOLINESTERASE INHIBITOR

Therapeutic: CHOLINESTERASE INHIBITOR

Prototype: Neostigmine

Pregnancy Category: C

AVAILABILITY 10 mg tablets

ACTION & *THERAPEUTIC EFFECT*
Inhibits destruction of acetylcholine (ACh) by cholinesterase, thereby prolonging effects of ACh (neurotransmitter) at postsynaptic receptor sites. Has direct stimulant effect on striated muscles. *Improves muscular strength in myasthenia gravis.*

USES Symptomatic treatment of myasthenia gravis.

CONTRAINDICATIONS Intestinal or urinary tract obstruction; patients receiving mecamylamine; lactation.

CAUTIOUS USE Epilepsy, bradycardia, cardiac arrhythmias, recent coronary occlusion; bronchial asthma; hyperthyroidism; older adults; vagotonia; peptic ulcer, megacolon; pregnancy (category C).

ROUTE & DOSAGE

Myasthenia Gravis
Adult: **PO** 5 mg t.i.d. or q.i.d., may titrate q48 hours
Child: **PO** 0.3 mg/kg/day in 3–4 divided doses, may need up to 1.5 mg/kg/day in 3–4 divided doses

ADMINISTRATION

Oral

- Give with food or milk to minimize adverse effects.
- Schedule larger doses when patient experiences the most fatigue or muscle weakness; to improve ability to eat, give drug 30–45 min before meals.
- Store at 15°–30° C (59°–86° F) unless otherwise directed.

ADVERSE EFFECTS (≥1%) **CNS:** Exaggerated cholinergic (muscarinic) effects; muscle cramps, headache, confusion, dizziness, incoordination, fasciculations, agitation, restlessness, muscle weakness, paralysis, slurred speech, convulsions, respiratory depression. **CV:** Bradycardia. **GI:** Nausea, vomiting, diarrhea, abdominal cramps, excessive salivation. **Special Senses:** Blurred vision, lacrimation. **Respiratory:** Bronchospasm,

Common adverse effects in *italic*, life-threatening effects underlined; generic names in **bold**; classifications in SMALL CAPS; ♣ Canadian drug name; ✪ Prototype drug

61

increased bronchial secretions, dyspnea. **Other:** Diaphoresis.

INTERACTIONS Drug: Demecarium and other CHOLINESTERASE INHIBITORS possibly compound toxicity; **mecamylamine, succinylcholine, procainamide, quinidine,** AMINOGLYCOSIDES increase neuromuscular blocking effects with possibility of respiratory depression; atropine antagonizes effects of ambenonium.

PHARMACOKINETICS Absorption: Poorly absorbed from GI tract. **Onset:** 20–30 min. **Duration:** 3–8 h.

NURSING IMPLICATIONS

Assessment & Drug Effects

- Therapeutic effect may not be apparent for several days after initiation of therapy.
- Keep atropine sulfate immediately available to treat severe cholinergic reactions.
- Monitor for S&S of overdosage (muscle weakness within 1 h; headache, weakness of muscles of neck, chewing, and swallowing, increased salivation) and inadequate ventilation (unusual apprehension, restlessness, rapid pulse and respirations, rising BP).
- Monitor vital signs during dosage adjustment periods.
- Note: Muscle weakness beginning 3 h or more after drug administration is probably due to underdosage or drug resistance.

Patient & Family Education

- Learn to recognize adverse effects, how to modify the doses accordingly, and when to take atropine.
- Note: During long-term therapy the drug may become ineffective; responsiveness usually returns when dosage is reduced or drug is withdrawn for several days.

- Carry medical identification indicating medical diagnosis and medication(s) being taken.

AMCINONIDE
(am-sin'oh-nide)
Pregnancy Category: C
See Appendix A-4.

AMIFOSTINE
(am-i-fos'teen)
Ethyol
Classification: CYTOPROTECTIVE
Therapeutic: CYTOPROTECTIVE
Pregnancy Category: C

AVAILABILITY 500 mg vial

ACTION & *THERAPEUTIC EFFECT*
Amifostine reduces cytotoxic damage induced by radiation or antineoplastic agents in well-oxygenated cells. Protective effects appear to be mediated by the formation of a metabolite of amifostine that removes free radicals from normal cells exposed to cisplatin. *Amifostine is cytoprotective in the kidney, bone marrow, and GI mucosa, but not in the brain or spinal cord. The cytoprotection results in decreased myelosuppression and peripheral neuropathy.*

USES Reduction of the cumulative renal toxicity associated with cisplatin, xerostomia.
UNLABELED USES Reduction of paclitaxel toxicity, bone marrow suppression prophylaxis.

CONTRAINDICATIONS Sensitivity to aminothiol compounds or mannitol, patients with potentially curable malignancies, hypotensive patients or those who are dehydrated, exfoliated dermatitis; lactation.

CAUTIOUS USE Patients at risk for hypocalcemia, cardiovascular disease (i.e., arrhythmias, CHF, TIA, CVA); radiation therapy; renal disease; pregnancy (category C).

ROUTE & DOSAGE

Renal Protection
Adult: IV 910 mg/m² once daily prior to chemotherapy

Reduction of Xerostomia
Adult: IV 200 mg/m² prior to radiation therapy

ADMINISTRATION

Intravenous
Give antiemetics, adequately hydrate, and defer antihypertensives for 24 h prior to administration. Do not administer if patient is hypotensive or dehydrated. Consult prescriber.

PREPARE: **IV Infusion:** Reconstitute by adding 9.7 mL of NS injection to a single-dose vial to yield 50 mg/mL. ▪ May be further diluted with NS to a concentration as low as 5 mg/mL.

ADMINISTER: **IV Infusion:** Infuse over no more than 15 min, beginning 30 min before chemotherapy; place patient in supine position prior to and during infusion. ▪ For xerostomia, infuse over 3 min; begin 15–30 min before radiation.

INCOMPATIBILITIES **Solution/additive:** Do not mix with any solutions other than NS. **Y-site: Acyclovir, amphotericin B, amphotericin B lipid, cefoperazone, chlorpromazine, cisplatin, ganciclovir, hydroxyzine, minocycline, mycophenolate, prochlorperazine, quinupristin-dalfopristin.**

▪ Store reconstituted solution at 15°–30° C (59°–86° F) for 5 h or refrigerate up to 24 h.

ADVERSE EFFECTS (≥1%) **CV:** *Transient reduction in blood pressure.* **GI:** *Nausea, vomiting.* **Other:** Infusion reactions (flushing, feeling of warmth or coldness, chills, dizziness, somnolence, hiccups, sneezing), hypocalcemia, hypersensitivity reactions.

INTERACTIONS Drug: ANTIHYPERTENSIVES could cause or potentiate hypotension.

PHARMACOKINETICS Onset: 5–8 min. **Metabolism:** In liver to active free thiol metabolite. **Elimination:** Renally excreted. **Half-Life:** 8 min.

NURSING IMPLICATIONS

Assessment & Drug Effects
▪ Monitor for S&S of hypocalcemia and fluid balance if vomiting is significant.
▪ Monitor BP every 5 min during infusion. Stop infusion if systolic BP drops significantly from baseline (e.g., 20% drop in systolic BP) and place patient flat with legs raised. Restart infusion if BP returns to normal in 5 min.

Patient & Family Education
▪ Know and understand adverse effects.

AMIKACIN SULFATE
(am-i-kay′sin)
Amikin
Classification: AMINOGLYCOSIDE ANTIBIOTIC
Therapeutic: ANTIBIOTIC
Prototype: Gentamicin
Pregnancy Category: C

AVAILABILITY 250 mg/mL, 50 mg/mL injection

Common adverse effects in *italic*, life-threatening effects underlined; generic names in **bold**; classifications in SMALL CAPS; ♣ Canadian drug name; ☯ Prototype drug

63

ACTION & *THERAPEUTIC EFFECT*
Appears to inhibit protein synthesis in bacterial cells and is usually bactericidal. *Effective against a wide range of gram-negative bacteria, including many strains resistant to other aminoglycosides. Also effective against penicillinase- and non-penicillinase-producing* Staphylococcus.

USES Primarily for short-term treatment of serious infections of respiratory tract, bones, joints, skin, and soft tissue, CNS (including meningitis), peritonitis burns, recurrent urinary tract infections (UTIs).
UNLABELED USES Intrathecal or intraventricular administration, in conjunction with IM or IV dosage.

CONTRAINDICATIONS History of hypersensitivity or toxic reaction with an aminoglycoside antibiotic; lactation.
CAUTIOUS USE Impaired renal function; eighth cranial (auditory) nerve impairment; preexisting vertigo or dizziness, tinnitus, or dehydration; fever; older adults, premature infants, neonates and infants; myasthenia gravis; parkinsonism; hypocalcemia; pregnancy (category C).

ROUTE & DOSAGE

Moderate to Severe Infections
Adult: **IV/IM** 5–7.5 mg/kg loading dose, then 7.5 mg/kg q12h (max: 15 mg/kg/day) for 7–10 days
Child: **IV/IM** 5–7.5 mg/kg loading dose, then 5 mg/kg q8h or 7.5 mg/kg q12h for 7–10 days (max: 1.5 g/day)
Neonate: **IV/IM** 10 mg/kg loading dose, then 7.5 mg/kg q12h for 7–10 days

Uncomplicated UTI
Adult: **IV/IM** 250 mg q12h

Obesity Dosage Adjustment
Calculate dose based on IBW

Renal Impairment Dosage Adjustment
CrCl greater than 60 mL/min: Normal dose q8h; *40–60 mL/min:* Normal dose q12h; *20–39 mL/min:* Half dose q24h; *less than 20 mL/min:* Administer loading dose then monitor closely

Hemodialysis Dosage Adjustment
Administer dose post-dialysis or give ⅔ dose as supplemental dose

ADMINISTRATION

Intramuscular
- Use the 250 mg/mL vials for IM injection. Calculate the required dose and withdraw the equivalent number of mLs from the vial.
- Give deep IM into a large muscle.

Intravenous
Verify correct IV concentration and rate of infusion with prescriber for neonates, infants, and children.

PREPARE: Intermittent: Add contents of 500 mg vial to 100 or 200 mL D5W, NS injection, or other diluent recommended by manufacturer. ■ For pediatric patients, volume of diluent depends on patient's fluid tolerance. ■ Note: Color of solution may vary from colorless to light straw color or very pale yellow. Discard solutions that appear discolored or that contain particulate matter.
ADMINISTER: Intermittent: Give a single dose (including loading

dose) over at least 30–60 min by IV infusion. ▪ Increase infusion time to 1–2 h for infants. ▪ Monitor infusion rate carefully. A rapid rise in serum amikacin level can cause respiratory depression (neuromuscular blockade) and other signs of toxicity.

INCOMPATIBILITIES **Solution/additive: Aminophylline, amphotericin B, ampicillin, CEPHALOSPORINS, chlorothiazide, heparin, PENICILLINS, phenytoin, thiopental, vitamin B complex with C. Y-site: Allopurinol, amphotericin B, azithromycin, hetastarch, propofol, thiopental.**

▪ Store at 15°–30° C (59°–86° F) unless otherwise directed.

ADVERSE EFFECTS (≥1%) **CNS:** Neurotoxicity: Drowsiness, unsteady gait, weakness, clumsiness, paresthesias, tremors, convulsions, peripheral neuritis. **Special Senses:** *Auditory–ototoxicity,* high-frequency hearing loss, complete hearing loss (occasionally permanent); tinnitus; ringing or buzzing in ears; *Vestibular:* Dizziness, ataxia. **GI:** Nausea, vomiting, hepatotoxicity. **Metabolic:** Hypokalemia, hypomagnesemia. **Skin:** Skin rash, urticaria, pruritus, redness. **Urogenital:** Oliguria, urinary frequency, hematuria, tubular necrosis, azotemia. **Other:** Superinfections.

INTERACTIONS Drug: ANESTHETICS, SKELETAL MUSCLE RELAXANTS have additive neuromuscular blocking effects; **acyclovir, amphotericin B, bacitracin, capreomycin, cephalosporins, colistin, cisplatin, carboplatin, methoxyflurane, polymyxin B, vancomycin, furosemide, ethacrynic acid** increase risk of ototoxicity and nephrotoxicity.

PHARMACOKINETICS Peak: 30 min IV; 45 min to 2 h IM. **Distribution:** Does not cross blood–brain barrier; crosses placenta; accumulates in renal cortex. **Elimination:** 94–98% renally in 24 h, remainder in 10–30 days. **Half-Life:** 2–3 h in adults, 4–8 h in neonates.

NURSING IMPLICATIONS

Assessment & Drug Effects

▪ Baseline tests: Before initial dose, C&S; renal function and vestibulocochlear nerve function (and at regular intervals during therapy; closely monitor in the older adult, patients with documented ear problems, renal impairment, or during high dose or prolonged therapy).

▪ Monitor peak and trough amikacin blood levels: Draw blood 1 h after IM or immediately after completion of IV infusion; draw trough levels immediately before the next IM or IV dose.

▪ Lab tests: Periodic serum creatinine and BUN, complete urinalysis. Monitor serum creatinine or creatinine clearance (generally preferred) more often, in the presence of impaired renal function, in neonates, and in the older adult; note that prolonged high trough (greater than 8 mcg/mL) or peak (greater than 30–35 mcg/mL) levels are associated with toxicity.

▪ Monitor for and promptly report S&S of: Ototoxicity [primarily involves the cochlear (auditory) branch; high-frequency deafness usually appears first and can be detected only by audiometer]; indicators of declining renal function; respiratory tract infections and other symptoms indicative of superinfections.

Common adverse effects in *italic*, life-threatening effects underlined; generic names in **bold;** classifications in SMALL CAPS; ♣ Canadian drug name; ◐ Prototype drug

65

- Monitor for and report auditory symptoms (tinnitus, roaring noises, sensation of fullness in ears, hearing loss) and vestibular disturbances (dizziness or vertigo, nystagmus, ataxia).
- Monitor and report any changes in I&O, oliguria, hematuria, or cloudy urine. Keeping patient well hydrated reduces risk of nephrotoxicity; consult prescriber regarding optimum fluid intake.

Patient & Family Education
- Report immediately any changes in hearing or unexplained ringing/roaring noises or dizziness, and problems with balance or co-ordination.

AMILORIDE HYDROCHLORIDE
(a-mill′oh-ride)

Midamor
Classification: DIURETIC, POTASSI-UM-SPARING
Therapeutic: DIURETIC, POTASSIUM-SPARING; ANTIHYPERTENSIVE
Prototype: Spironolactone
Pregnancy Category: B

AVAILABILITY 5 mg tablets

ACTION & *THERAPEUTIC EFFECT*
Potassium-sparing diuretic with mild diuretic and antihypertensive action. Diuretic action is independent of aldosterone and carbonic anhydrase. Induces urinary excretion of sodium and reduces excretion of potassium and hydrogen ions by direct action on distal renal tubules. *Lowers blood pressure by excretion of sodium ion and water from the kidney while sparing potassium excretion.*

USES Adjunctive treatment of heart failure, hypertension, hypokalemia.
UNLABELED USES Ascites.

CONTRAINDICATIONS Elevated serum potassium (greater than 5.5 mEq/L) anuria, acute or chronic renal insufficiency; evidence of diabetic nephropathy; type 1 diabetes mellitus; metabolic or respiratory acidosis; lactation.
CAUTIOUS USE Debilitated patients; diet-controlled or uncontrolled diabetes mellitus; cardiopulmonary disease; hepatic disease; older adult; pregnancy (category B). Safe use in children not established.

ROUTE & DOSAGE

HTN, HF, Hypokalemia
Adult: **PO** 5 mg/day, may increase up to 20 mg/day in 1–2 divided doses

ADMINISTRATION

Oral
- Give once/day dose in the morning and schedule the second b.i.d. dose early to avoid interrupting sleep.
- Give with food to reduce possibility of gastric distress.
- Store at 15°–30° C (59°–86° F) in a tightly closed container unless otherwise directed.

ADVERSE EFFECTS (≥1 %) **Body as a Whole:** Generally well tolerated. **CNS:** *Headache,* dizziness, nervousness, confusion, paresthesias, drowsiness. **CV:** Cardiac arrhythmias. **Metabolic:** Hyperkalemia, hyponatremia, positive Coombs' test. **Hematologic:** Aplastic anemia. **Special Senses:** Tinnitus; nasal congestion. Visual disturbances, increased intraocular pressure. **GI:** *Diarrhea* or constipation, anorexia, *nausea,* vomiting, abdominal cramps, dry mouth, thirst. **Urogenital:** Polyuria, dysuria, bladder spasms, urinary

Common adverse effects in *italic*, life-threatening effects underlined; generic names in **bold**; classifications in SMALL CAPS; ♣ Canadian drug name; ☻ Prototype drug

frequency, impotence, decreased libido. **Respiratory:** Dyspnea, shortness of breath. **Skin:** Rash, pruritus, photosensitivity reactions. **Other:** Weakness, fatigue, muscle cramps.

DIAGNOSTIC TEST INTERFERENCE
Discontinue at least 3 days before glucose tolerance test.

INTERACTIONS Drug: Blood
from blood banks, ACE INHIBITORS (e.g., **captopril**), **spironolactone, triamterene,** POTASSIUM SUPPLEMENTS may cause hyperkalemia with cardiac arrhythmias; possibility of increased **lithium** toxicity (decreased renal elimination); possibility of altered **digoxin** response; NSAIDS may attenuate antihypertensive effects. **Food:** POTASSIUM-CONTAINING SALT SUBSTITUTES or foods high in **potassium** increase risk of hyperkalemia.

PHARMACOKINETICS Absorption: 50% from GI tract. Onset: 2 h. Peak: 6–10 h. Duration: 24 h. Elimination: 20–50% unchanged in urine, 40% in feces. Half-Life: 6–9 h.

NURSING IMPLICATIONS

Assessment & Drug Effects
- Monitor for S&S of hyperkalemia and hyponatremia (see Appendix F). Hyperkalemia occurs in about 10% of patients receiving amiloride and serum potassium can rise suddenly and without warning. It is more common in older adults and patients with diabetes or renal disease.
- Lab tests: Serum potassium levels, particularly when therapy is initiated, whenever dosage adjustments are made, and during any illness that may affect kidney function; periodic BUN, creatinine, for patients with renal or hepatic dysfunction, diabetes mellitus, older adults, or the debilitated.
- Monitor ECG as warranted.

Patient & Family Education
- Learn S&S of hyperkalemia and hyponatremia (see Appendix F) and report to prescriber immediately.
- Do not take potassium supplements, salt substitutes, high intake of dietary potassium unless prescribed by prescriber.
- Do not drive or engage in potentially hazardous activities until response to drug is known.

AMINOCAPROIC ACID ⓟ
(a-mee-noe-ka-proe′ik)
Amicar
Classifications: COAGULATOR; SYSTEMIC HEMOSTATIC
Therapeutic: ANTIHEMORRHAGIC; ANTIFIBRINOLYTIC
Pregnancy Category: C

AVAILABILITY 250 mg/mL injection; 500 mg, 1000 mg tablets; 250 mg/mL syrup

ACTION & *THERAPEUTIC EFFECT*
Synthetic hemostatic with specific antifibrinolysis action. Inhibits plasminogen activator substance, and to a lesser degree plasmin (fibrinolysin), which is concerned with destruction of clots. *Acts as an inhibitor of fibrinolytic bleeding.*

USES To control excessive bleeding resulting from systemic hyperfibrinolysis; also used in urinary fibrinolysis associated with severe trauma, anoxia, shock, urologic surgery, and neoplastic diseases of GU tract.
UNLABELED USES To prevent hemorrhage in hemophiliacs undergoing dental extraction; as a specific antidote for streptokinase or urokinase toxicity; to prevent recurrence of subarachnoid hemorrhage, especially when surgery is delayed; for management of amegakaryocytic thrombocytopenia; and

Common adverse effects in *italic*, life-threatening effects underlined; generic names in **bold**; classifications in SMALL CAPS; ♣ Canadian drug name; ⓟ Prototype drug

67

to prevent or abort hereditary angiooedema episodes.

CONTRAINDICATIONS Severe renal impairment; active disseminated intravascular clotting (DIC); upper urinary tract bleeding (hematuria); hemophilia; benzyl alcohol hypersensitivity, especially in neonates; paraben hypersensitivity; lactation.
CAUTIOUS USE Cardiac, renal, or hepatic disease; renal impairment; history of pulmonary embolus or other thrombotic diseases; hypovolemia; pregnancy (category C).

ROUTE & DOSAGE

Hemostatic
Adult: **PO/IV** 4–5 g during first hour, then 1–1.25 g qh for 8 h or until bleeding is controlled (max: 30 g/24h)
Child: **PO/IV** 100 mg/kg or 3 g/m² during first hour, then 33.3 mg²/kg qh (max: 18 g/m²/24 h)
Renal Impairment Dosage Adjustment
Reduce dose to 15–25% of normal dose

ADMINISTRATION

Oral
- Note: May need to give patient as many as 10 tablets or 4 tsp for a 5 g dose during the first hour of treatment.

Intravenous
PREPARE: **IV Infusion:** Dilute parenteral aminocaproic acid before use. ■ Each 4 mL (1 g) is diluted with 50 mL of NS, D5W, or LR.
ADMINISTER: **IV Infusion:** Prescriber orders specific IV flow rate. ■ Usual rate is 5 g or a fraction thereof over first hour (5 g/250 mL). ■ Give each additional gram over 1 h.

Avoid rapid infusion to prevent hypotension, faintness, and bradycardia or other arrhythmias.

INCOMPATIBILITIES **Solution/additive: Fructose solution.**

■ Store in tightly closed containers at 15°–30° C (59°–86° F) unless otherwise directed. Avoid freezing.

ADVERSE EFFECTS (≥ 1%) CNS: Dizziness, malaise, headache, seizures. **CV:** Faintness, orthostatic hypotension; dysrhythmias; thrombophlebitis, thromboses. **Special Senses:** Tinnitus, nasal congestion. Conjunctival erythema. **GI:** Nausea, vomiting, cramps, diarrhea, anorexia. **Urogenital:** Diuresis, dysuria, urinary frequency, oliguria, reddish-brown urine (myoglobinuria), <u>acute renal failure</u>. Prolonged menstruation with cramping. **Skin:** Rash.

DIAGNOSTIC TEST INTERFERENCE *Serum potassium* may be elevated (especially in patients with impaired renal function).

INTERACTIONS Drug: ESTROGENS, ORAL CONTRACEPTIVES may cause hypercoagulation.

PHARMACOKINETICS Absorption: Rapidly from GI tract. **Peak:** 2 h. **Distribution:** Readily penetrates RBCs and other body cells. **Elimination:** 80% as unmetabolized drug in 12 h.

NURSING IMPLICATIONS
Assessment & Drug Effects
- Check IV site at frequent intervals for extravasation.
- Observe for signs of thrombophlebitis. Change site immediately if extravasation or thrombophlebitis occurs (see Appendix F).
- Monitor and report S&S of myopathy: Muscle weakness, myalgia, diaphoresis, fever, reddish-brown urine (myoglobinuria), oliguria, as

well as thrombotic complications: Arm or leg pain, tenderness or swelling, Homans' sign, prominence of superficial veins, chest pain, breathlessness, dyspnea. Drug should be discontinued promptly.
- Monitor vital signs and urine output.
- Lab tests: With prolonged therapy, monitor creatine phosphokinase activity and urinalyses for early detection of myopathy.

Patient & Family Education
- Report difficulty urinating or reddish-brown urine.
- Report arm or leg pain, chest pain, or difficulty breathing.

AMINOPHYLLINE (THEOPHYLLINE ETHYLENEDIAMIDE)
(am-in-off'i-lin)

Corophyllin ♥, Paladron ♥
Classifications: BRONCHODILATOR; RESPIRATORY SMOOTH MUSCLE RELAXANT; XANTHINE
Therapeutic: BRONCHODILATOR
Prototype: Theophylline
Pregnancy Category: C

AVAILABILITY 100 mg, 200 mg tablets; 105 mg/5 mL oral liquid; 25 mg/mL injection

ACTION & *THERAPEUTIC EFFECT*
A xanthine derivative that relaxes smooth muscle in the airways of the lungs and suppresses the response of the airways to stimuli that constrict them. *It is a respiratory smooth muscle relaxant that results in bronchodilation.*

USES To prevent and relieve symptoms of acute bronchial asthma and treatment of bronchospasm associated with chronic bronchitis and emphysema.
UNLABELED USES As a respiratory stimulant in Cheyne-Stokes respiration; for treatment of apnea and bradycardia in premature infants; as cardiac stimulant and diuretic in treatment of CHF.

CONTRAINDICATIONS Hypersensitivity to xanthine derivatives or to ethylenediamine component; cardiac arrhythmias; lactation.
CAUTIOUS USE Severe hypertension, cardiac disease, arrhythmias; impaired hepatic function; diabetes mellitus; hyperthyroidism; glaucoma; prostatic hypertrophy; fibrocystic breast disease; history of peptic ulcer; neonates and young children, patients over 55 y; COPD, acute influenza or patients receiving influenza immunization; pregnancy (category C).

ROUTE & DOSAGE

Bronchospasm

Adult: **IV Loading Dose** 6 mg/kg over 30 min **IV Maintenance Dose** *Nonsmoker,* 0.5 mg/kg/h; *smoker,* 0.8 mg/kg/h; *CHF or cirrhosis,* 0.1–0.2 mg/kg/h **PO** *Nonsmoker,* 0.5 mg/kg/h times 24 h in 4 divided doses; *smoker,* 0.75 mg/kg/h times 24 h in 4 divided doses; *CHF or cirrhosis,* 0.25 mg/kg/h times 24 h in 4 divided doses
Child: **IV Loading Dose** 6 mg/kg IV over 30 min **IV Maintenance Dose** *1–9 y,* 1 mg/kg/h; *older than 9 y,* 0.8 mg/kg/h **PO** *1–9 y,* 1 mg/kg/h times 24 h in 4 divided doses; *older than 9 y,* 0.75 mg/kg/h times 24 h in 4 divided doses
Infant: **PO/IV** *6–11 mo,* 0.7 g/kg/h; *2–6 mo,* 0.5 mg/kg/h

Neonatal Apnea

Neonate: **PO/IV Loading Dose** 5 mg/kg **PO/IV Maintenance Dose** 5 mg/kg/day divided q12h

Common adverse effects in *italic,* life-threatening effects underlined; generic names in **bold**; classifications in SMALL CAPS; ♥ Canadian drug name; ● Prototype drug

69

Geriatric Patients: **PO** 6.25 mg/kg loading dose, then 2.5 mg/kg q8h

Obesity Dosage Adjustment
Calculate dose based on IBW

ADMINISTRATION

Oral
- Give with a full glass of water on an empty stomach (30 min–1 h before or 2 h after meals) for faster absorption, which is delayed but is not reduced with food.
- Minimize GI symptoms by taking immediately after a meal or with food.
- Extended (controlled) release preparations should not be crushed or chewed; however, if tablet is scored, it can be broken in half, then swallowed.
- Contents of extended release capsules may be mixed with soft, moist food to promote swallowing.

Intravenous
Verify correct IV concentration and rate of infusion with prescriber for neonates, infants, and children.

PREPARE: **IV Infusion:** Dilute loading dose in 100–200 mL NS, D5W,D5/NS, or LR. For continuous or intermittent infusion dilute in 500–1000 mL. • Do not use aminophylline solutions if discolored or if crystals are present.

ADMINISTER: **IV Infusion:** Infuse at a rate not to exceed 25 mg/min.

INCOMPATIBILITIES **Solution/additive:** Amikacin, bleomycin, CEPHALOSPORINS, **chlorpromazine, ciprofloxacin, clindamycin, dimenhydrinate, dobutamine, doxorubicin, epinephrine, hydralazine, hydroxyzine, insulin, isoproterenol, meperidine, methylprednisolone, morphine, nafcillin, norepine-** phrine, papaverine, penicillin G, pentazocine, procaine, prochlorperazine, promazine, promethazine, verapamil, vitamin B complex with C, zinc. **Y-site: Amiodarone, ciprofloxacin, clarithromycin, dobutamine, fenoldopam, hydralazine, lansoprazole, ondansetron, TPN, vinorelbine, warfarin.**

- Store at 15°–30° C (59°–86° F) in tightly closed containers unless otherwise directed.

ADVERSE EFFECTS (≥1%) CNS: *Nervousness,* restlessness, depression, insomnia, irritability, headache, dizziness, muscle hyperactivity, convulsions. **CV:** Cardiac arrhythmias, tachycardia (with rapid IV), hyperventilation, chest pain, severe hypotension, cardiac arrest. **GI:** *Nausea, vomiting, anorexia,* hematemesis, diarrhea, epigastric pain.

INTERACTIONS Drug: Increases **lithium** excretion, lowering **lithium** levels; **cimetidine,** high-dose **allopurinol** (600 mg/day), **ciprofloxacin, erythromycin, troleandomycin** can significantly increase **theophylline** levels. **Herbal: St. John's wort** may decrease effect.

PHARMACOKINETICS Absorption: Most products are 100% absorbed from GI tract. **Peak:** IV 30 min; uncoated tablet 1 h; sustained release 4–6 h. **Duration:** 4–8 h; varies with age, smoking, and liver function. **Distribution:** Crosses placenta. **Metabolism:** Extensively in liver; by CYP1A2. **Elimination:** Parent drug and metabolites excreted by kidneys; excreted in breast milk. **Half-Life:** 3.7 h (child); 7.7 h (adult).

NURSING IMPLICATIONS

Assessment & Drug Effects
- Monitor for S&S of toxicity (generally related to theophylline serum

levels over 20 mcg/mL). Observe patients receiving parenteral drug closely for signs of hypotension, arrhythmias, and convulsions until serum theophylline stabilizes within the therapeutic range.

- Monitor and record vital signs and I&O. A sudden, sharp, unexplained rise in heart rate may indicate toxicity.
- Lab tests: Monitor serum theophylline levels. Blood samples should be drawn 1–2 h after a dose at steady state.
- Note: Older adults, acutely ill, and patients with severe respiratory problems, liver dysfunction, or pulmonary edema are at greater risk of toxicity due to reduced drug clearance.
- Note: Children appear more susceptible to CNS stimulating effects of xanthines (nervousness, restlessness, insomnia, hyperactive reflexes, twitching, convulsions). Dosage reduction may be indicated.

Patient & Family Education

- Note: Use of tobacco tends to increase elimination of this drug (shortens half-life), necessitating higher dosage or shorter intervals than in nonsmokers.
- Report excessive nervousness or insomnia. Dosage reduction may be indicated.
- Note: Dizziness is a relatively common side effect, particularly in older adults; take necessary safety precautions.
- Do not take OTC remedies for treatment of asthma or cough unless approved by prescriber.

AMINOSALICYLIC ACID (*PARA*-AMINOSALICYLIC ACID)

(a-mee-noe-sal-i-sil'ik)

Paser
Classification: ANTITUBERCULOSIS

Therapeutic: ANTITUBERCULOSIS
Prototype: Isoniazid
Pregnancy Category: C

AVAILABILITY 4 g packets

ACTION & *THERAPEUTIC EFFECT*
Aminosalicylic acid and salts are highly specific bacteriostatic agents that suppress growth and multiplication of *Mycobacterium tuberculosis* by preventing folic acid synthesis. Aminosalicylates reportedly have potent hypolipemic action. *Aminosalicylates are an effective antiinfective alone or in combined therapy and reduce serum cholesterol and triglycerides by lowering LDL and VLDL.*

USES With **streptomycin** or **isoniazid** or both in treatment of pulmonary and extrapulmonary tuberculosis to delay drug resistance.

CONTRAINDICATIONS Hypersensitivity to aminosalicylates, salicylates, or to compounds containing *para*-aminophenyl groups (e.g., sulfonamides, certain hair dyes); G6PD deficiency, use of the sodium salt in patients on sodium restriction or CHF; lactation.
CAUTIOUS USE Impaired renal and hepatic function; blood dyscrasias; goiter; gastric ulcer; pregnancy (category C).

ROUTE & DOSAGE

Tuberculosis
Adult/Child: **PO** 150 mg/kg/day in 3 divided doses (max: 12 g)

ADMINISTRATION

Oral

- Mix granules in apple sauce or yogurt, or suspend in an acidic drink such as fruit juice or tomato juice. Do not administer granules that have lost their tan color.

Common adverse effects in *italic*, life-threatening effects underlined; generic names in **bold**; classifications in SMALL CAPS; ♣ Canadian drug name; ⊘ Prototype drug

71

- Give with or immediately following meals to reduce irritative gastric effects.
- Store in tight, light-resistant containers in a cool, dry place, preferably at 15°–30° C (59°–86° F), unless otherwise directed.

ADVERSE EFFECTS (≥1%) **Body as a Whole:** Fever, chills, generalized malaise, joint pain, rash, fixed-drug eruptions, pruritus; vasculitis; Loeffler's syndrome. **CNS:** Psychotic reactions. **GI:** *Anorexia, nausea, vomiting, abdominal distress, diarrhea*, peptic ulceration, acute hepatitis, malabsorption. **Hematologic:** Leukopenia, agranulocytosis, eosinophilia, lymphocytosis, thrombocytopenia, hemolytic anemia; (G6PD deficiency); prothrombinemia. **Urogenital:** Renal (irritation), crystalluria. **Other:** With long-term administration, goiter.

DIAGNOSTIC TEST INTERFERENCE Aminosalicylates may interfere with urine ***urobilinogen*** determinations (using ***Ehrlich's reagent***), and may cause false-positive ***urinary protein*** and ***VMA*** determinations (with ***diazoreagent***); false-positive ***urine glucose*** may result with ***cupric sulfate tests*** (e.g., ***Benedict's solution***), but reportedly not with ***glucose oxidase reagents*** (e.g., ***TesTape, Clinistix***). Reduces ***serum cholesterol,*** and possibly ***serum potassium, serum PBI,*** and 24-hour ***I-131 thyroidal uptake*** (effect may last almost 14 days).

INTERACTIONS Drug: Increases hypoprothrombinemic effects of ORAL ANTICOAGULANTS; increased risk of crystalluria with **ammonium chloride, ascorbic acid;** decreased intestinal absorption of **cyanocobalamin, folic acid, digoxin;** ANTIHISTAMINES may inhibit PAS absorption; may increase or decrease **phenytoin** levels; **probenecid, sulfinpyrazone** decrease PAS elimination.

PHARMACOKINETICS Absorption: Almost completely from GI tract; sodium form more rapidly absorbed than the acid. **Peak:** 1.5–2 h. **Duration:** 4 h. **Distribution:** Well distributed to tissue and body fluids except CSF unless meninges are inflamed. **Metabolism:** In liver. **Elimination:** greater than 80% in urine in 7–10 h. **Half-Life:** 1 h.

NURSING IMPLICATIONS

Assessment & Drug Effects

- Monitor for abrupt onset of fever, particularly during the early weeks of therapy, and clinical picture resembling that of infectious mononucleosis (malaise, fatigue, generalized lymphadenopathy, splenomegaly, sore throat), as well as minor complaints of pruritus, joint pains, and headache, which strongly suggest hypersensitivity; report these symptoms promptly.
- Monitor I&O and encourage fluids. High concentrations of drug are excreted in urine, and this can cause crystalluria and hematuria.

Patient & Family Education

- Note: Hypersensitivity reactions may occur after a few days, but most commonly in the fourth or fifth week; report promptly.
- Notify prescriber if sore throat or mouth, malaise, unusual fatigue, bleeding or bruising occurs.
- Note: Therapy generally lasts about 2 y. Adhere to the established drug regimen, and remain under close medical supervision to detect possible adverse drug effects during the treatment period. Resistant TB strains develop more rapidly when drug regimen is interrupted or is sporadic.
- Do not take aspirin or other OTC drugs without prescriber's approval.
- Discard drug if it discolors (brownish or purplish); this signifies decomposition.

AMIODARONE HYDROCHLORIDE ⊙

(a-mee'oh-da-rone)

Cordarone, Amio-Aqueous, Nexterone, Pacerone
Classification: ANTIARRHYTHMIC, CLASS III
Therapeutic: CLASS III ANTIARRHYTHMIC; ANTIANGINAL
Pregnancy Category: D

AVAILABILITY 100 mg, 200 mg, 400 mg tablets; 50 mg/mL injection

ACTION & *THERAPEUTIC EFFECT*

Class III antiarrhythmic that has antianginal and antiadrenergic properties. Acts directly on all cardiac tissues by prolonging duration of action potential and refractory period. Slows conduction time through the AV node and can interrupt the reentry pathways through the AV node. *Effective in prevention or suppression of cardiac arrhythmias.*

USES Prophylaxis and treatment of life-threatening ventricular arrhythmias and supraventricular arrhythmias, particularly with atrial fibrillation.

UNLABELED USES Treatment of nonexertional angina, conversion of atrial fibrillation to normal sinus rhythm, paroxysmal supraventricular tachycardia, ventricular rate control due to accessory pathway conduction in pre-excited atrial arrhythmia, after defibrillation and epinephrine in cardiac arrest, AV nodal reentry tachycardia.

CONTRAINDICATIONS Hypersensitivity to amiodarone, or benzyl alcohol; cardiogenic shock, severe sinus bradycardia, advanced AV block unless a pacemaker is available, severe sinus-node dysfunction or sick sinus syndrome, bradycardia, congenital or acquired QR prolongation syndromes, or history of torsades de pointes; severe liver disease, pregnancy (category D), lactation.

CAUTIOUS USE Hepatic disease, cirrhosis; Hashimoto's thyroiditis, goiter, thyrotoxicosis, or history of other thyroid dysfunction; mild to moderate hepatic toxicity; CHF, left ventricular dysfunction; hypersensitivity to iodine; older adults; Fabry disease, especially with visual disturbances; electrolyte imbalance, hypokalemia, hypomagnesemia, hypovolemia; preexisting lung disease, COPD; open heart surgery.

ROUTE & DOSAGE

Arrhythmias

Adult: **PO Loading Dose** 800–1600 mg/day in 1–2 doses for 1–3 wk **PO Maintenance Dose** 400–600 mg/day in 1–2 doses **IV Loading Dose** 150 mg over 10 min followed by 360 mg over next 6 h **IV Maintenance Dose** 540 mg over 18 h (0.5 mg/min), may continue at 0.5 mg/min **Convert IV to PO** Duration of infusion less than 1 wk use 800–1600 mg PO, 1–3 wk use 600–800 mg PO, greater than 3 wk use 400 mg PO
Child: **IV** 5 mg/kg then repeat to max of 300 mg total

Hepatic Impairment Dosage Adjustment

Adjustment only suggested in severe hepatic impairment

ADMINISTRATION

- Note: Correct hypokalemia and hypomagnesemia prior to initiation of therapy.

Oral

- Give consistently with respect to meals. Avoid grapefruit juice.

Common adverse effects in *italic*, life-threatening effects underlined; generic names in **bold**; classifications in SMALL CAPS; ♣ Canadian drug name; ⊙ Prototype drug

73

- Note: Only a prescriber experienced with the drug and treatment of life-threatening arrhythmias should give loading doses.
- Note: GI symptoms commonly occur during high-dose therapy, especially with loading doses. Symptoms usually respond to dose reduction or divided dose given with food, including milk.

Intravenous

PREPARE: **IV Infusion: First rapid loading dose infusion:** Add 150 mg (3 mL) amiodarone to 100 mL D5W to yield 1.5 mg/mL. **Second infusion during first 24 h (slow loading dose and maintenance infusion):** Add 900 mg (18 mL) amiodarone to 500 mL D5W to yield 1.8 mg/mL. **Maintenance infusions after the first 24 h:** Prepare concentrations of 1–6 mg/mL amiodarone. Note: Use central line to give concentrations greater than 2 mg/mL.

ADMINISTER: **IV Infusion:** Rapidly infuse initial 150 mg dose over the first 10 min at a rate of 15 mg/min. ▪ Over next 6 h, infuse 360 mg at a rate of 1 mg/min. ▪ Over the remaining 18 h, infuse 540 mg at a rate of 0.5 mg/min. ▪ After the first 24 h, infuse maintenance doses of 720 mg/24 h at a rate of 0.5 mg/min.

INCOMPATIBILITIES **Solution/additive: Aminophylline, amoxicillin/clavulanic acid, cefazolin, floxacillin, furosemide, quinidine. Y-site: Aminocaproic acid, aminophylline, amoxicillin, ampicillin/sulbactam, argatroban, atenolol, bivalirudin, cefamandole, cefazolin, ceftazidime, ceftopribole, cytarabine, digoxin, doxorubicin, drotrecogin, ertapenem, fludarabine, fluorouracil, gemtuzumab, heparin, imipenem/cilastatin,** **levofloxacin, magnesium sulfate, meclorethamine, methotrexate, micafungin, paclitaxel, piperacillin, piperacillin/tazobactam, potassium acetate, potassium phosphate, sodium bicarbonate, sodium phosphate, thiotepa, tigecycline.**

- Store at 15°–30° C (59°–86° F) protected from light, unless otherwise directed.

ADVERSE EFFECTS (≥1%) **CNS:** Peripheral neuropathy (*muscle weakness,* wasting numbness, tingling), *fatigue,* abnormal gait, dyskinesias, *dizziness,* paresthesia, headache. **CV:** Bradycardia, *hypotension* (IV), sinus arrest, cardiogenic shock, CHF, arrhythmias; AV block. **Special Senses:** *Corneal microdeposits,* blurred vision, optic neuritis, optic neuropathy, permanent blindness, corneal degeneration, macular degeneration, photosensitivity. **GI:** *Anorexia, nausea, vomiting, constipation,* hepatotoxicity. **Metabolic:** Hyperthyroidism or hypothyroidism; may cause neonatal hypo- or hyperthyroidism if taken during pregnancy. **Respiratory:** (Pulmonary toxicity) Alveolitis, pneumonitis (fever, dry cough, dyspnea), interstitial pulmonary fibrosis, *fatal gasping syndrome* with IV in children. **Skin:** Slate-blue pigmentation, *photosensitivity,* rash. **Other:** With chronic use, angioedema.

INTERACTIONS Drug: Significantly increases **digoxin** levels; enhances pharmacologic effects and toxicities of **disopyramide, procainamide, quinidine, flecainide, lidocaine, lovastatin, simvastatin;** anticoagulant effects of ORAL ANTICOAGULANTS enhanced; **verapamil, diltiazem,** BETA-ADRENERGIC BLOCKING AGENTS may potentiate sinus bradycardia, sinus arrest, or AV block; may

Common adverse effects in *italic*, life-threatening effects underlined; generic names in **bold**; classifications in SMALL CAPS; ♣ Canadian drug name; ⦿ Prototype drug

increase **phenytoin** levels 2- to 3-fold; **cholestyramine** may decrease amiodarone levels; **fentanyl** may cause bradycardia, hypotension, or decreased output; may increase **cyclosporine** levels and toxicity; **cimetidine** may increase amiodarone levels; **ritonavir** may increase risk of amiodarone toxicity, including cardiotoxicity; **simvastatin** doses over 20 mg increase risk of rhabdomyolysis; **loratadine** use may increase risk of QT prolongation. **Food: Grapefruit juice** may increase amidarone concentrations. **Herbal: Echinacea** may increase hepatotoxicity, **St. John's wort** may decrease efficacy. **Lab Test:** Affects thyroid function tests, causing an increase in serum T_4 and serum reverse T_3 levels, and a decline in serum T_3 levels.

PHARMACOKINETICS **Absorption:** 22–86% absorbed. **Onset (PO):** 2–3 days to 1–3 wk. **Peak:** 3–7 h. **Distribution:** Concentrates in adipose tissue, lungs, kidneys, spleen; crosses placenta; 96% protein bound. **Metabolism:** Extensively in liver; undergoes some enterohepatic cycling; via CYP2C8 and 3A4. **Elimination:** Excreted chiefly in bile and feces; also in breast milk. **Half-Life:** Biphasic, initial 2.5–10 days, terminal 40–55 days.

NURSING IMPLICATIONS

Assessment & Drug Effects

- Monitor BP carefully during infusion and slow the infusion if significant hypotension occurs; bradycardia should be treated by slowing the infusion or discontinuing if necessary. Monitor heart rate and rhythm and BP until drug response has stabilized; report promptly symptomatic bradycardia. Sustained monitoring is essential because drug has an unusually long half-life.

- Monitor for S&S of: Adverse effects, particularly conduction disturbances and exacerbation of arrhythmias, in patients receiving other antiarrhythmic drugs; drug-induced hypothyroidism or hyperthyroidism (see Appendix F), especially during early treatment period; pulmonary toxicity (progressive dyspnea, fatigue, cough, pleuritic pain, fever) throughout therapy.

- Lab tests: Baseline and periodic assessments should be made of liver, lung, thyroid, neurologic, and GI function. Drug may cause thyroid function test abnormalities in the absence of thyroid function impairment.

- Monitor for elevations of AST and ALT. If elevations persist or if they are 2–3 times above normal baseline readings, reduce dosage or withdraw drug promptly to prevent hepatotoxicity and liver damage.

- Auscultate chest periodically or when patient complains of respiratory symptoms. Check for diminished breath sounds, rales, pleuritic friction rub; observe breathing pattern. Drug-induced pulmonary function problems **must be** distinguished from CHF or pneumonia. Keep prescriber informed.

- Anticipate possible CNS symptoms within a week after amiodarone therapy begins. Proximal muscle weakness, a common side effect, intensified by tremors presents a great hazard to the ambulating patient. Assess severity of symptoms. Supervision of ambulation may be indicated.

Patient & Family Education

- Check pulse daily once stabilized, or as prescribed. Report a pulse less than 60.

- Take oral drug consistently with respect to meals. Do not drink grapefruit juice while taking this drug.

Common adverse effects in *italic*, life-threatening effects underlined; generic names in **bold;** classifications in SMALL CAPS; ♣ Canadian drug name; ⊘ Prototype drug

75

- Become familiar with potential adverse reactions and report those that are bothersome to the prescriber.
- Use dark glasses to ease photophobia; some patients may not be able to go outdoors in the daytime even with such protection.
- Follow recommendation for regular ophthalmic exams, including funduscopy and slit-lamp exam.
- Wear protective clothing and a barrier-type sunscreen that physically blocks penetration of skin by ultraviolet light to prevent a photosensitivity reaction (erythema, pruritus); avoid exposure to sun and sunlamps.

AMITRIPTYLINE HYDROCHLORIDE

(a-mee-trip′ti-leen)

Apo-Amitriptyline ✦, Levate ✦, Novotriptyn ✦
Classification: TRICYCLIC ANTIDEPRESSANT
Therapeutic: ANTIDEPRESSANT
Prototype: Imipramine
Pregnancy Category: C

AVAILABILITY 10 mg, 25 mg, 50 mg, 75 mg, 100 mg, 150 mg tablets

ACTION & *THERAPEUTIC EFFECT*
A tricyclic antidepressant (TCA) that inhibits the reuptake of serotonin (5-HT) and norepinephrine from the synaptic gap; also inhibits norepinephrine reuptake to a moderate degree. Restoration of the levels of these neurotransmitters is a proposed mechanism of its antidepressant action. *Interference with the reuptake of serotonin and norepinephrine results in the antidepressant activity of amitriptyline.*

USES Endogenous depression.

UNLABELED USES Prophylaxis for cluster, migraine, and chronic tension headaches; neuropathic pain, to increase muscle strength in myotonic dystrophy, enuresis, fibromyalgia, insomnia, panic disorder, social anxiety disorder, and as sedative for nondepressed patients.

CONTRAINDICATIONS TCA hypersensitivity; acute recovery period after MI, cardiac arrhythmias, AV block, long-QT prolongation; suicidal ideation; history of seizure disorders; lactation, children younger than 12 y.

CAUTIOUS USE Prostatic hypertrophy, history of urinary retention or obstruction; angle-closure glaucoma; diabetes mellitus; history of hematologic disorders; history of alcoholism; GERD, BPH; hyperthyroidism; patient with cardiovascular, hepatic, or renal dysfunction; patient with suicidal tendency, electroshock therapy; elective surgery; schizophrenia; respiratory disorders; Parkinson's disease; seizure disorders; older adults, adolescents; pregnancy (category C).

ROUTE & DOSAGE

Antidepressant

Adult: **PO** 25–75 mg/day, may gradually increase to 150–300 mg/day
Adolescent/Geriatric: **PO** 10 mg t.i.d. with 20 mg at bedtime (max: 150–200 mg/day)

Pharmacogenetic Dosage Adjustment

CYP2D6 poor metabolizers: Dose at 60–75% of normal dose

ADMINISTRATION

Oral
- Give with or immediately after food to reduce possibility of GI

irritation. Tablet may be crushed if patient is unable to take it whole; administer with food or fluid.

- Give increased doses preferably in late afternoon or at bedtime due to sedative action that precedes antidepressant effect.
- Note that dose is usually tapered over 2 wk at discontinuation to prevent withdrawal symptoms (headache, nausea, malaise, musculoskeletal pain, panic attack, weakness).

ADVERSE EFFECTS (≥1%) **CNS:** *Drowsiness, sedation, dizziness,* nervousness, restlessness, fatigue, headache, insomnia, abnormal movements (extrapyramidal symptoms), seizures. **CV:** *Orthostatic hypotension,* tachycardia, palpitation, ECG changes. **Special Senses:** Blurred vision, mydriasis. **GI:** *Dry mouth,* increased appetite especially for sweets, *constipation,* weight gain, sour or metallic taste, nausea, vomiting. **Urogenital:** *Urinary retention.* **Other:** (Rare) <u>Bone marrow depression</u>.

INTERACTIONS Drug: Avoid drugs affecting QT interval. CNS DEPRESSANTS, **alcohol,** HYPNOTICS, BARBITURATES, SEDATIVES potentiate CNS depression; ANTICOAGULANTS, ORAL, may increase hypoprothrombinemic effect; **levodopa,** SYMPATHOMIMETICS (e.g., **epinephrine, norepinephrine**), possibility of sympathetic hyperactivity with hypertension and hyperpyrexia; MAO INHIBITORS, possibility of severe reactions, toxic psychosis, cardiovascular instability; **methylphenidate** increases plasma TCA levels; THYROID DRUGS may increase possibility of arrhythmias; **cimetidine** may increase plasma TCA levels. **Herbal: St. John's wort** may cause serotonin syndrome.

PHARMACOKINETICS Absorption: Rapidly from GI tract. **Peak:** 2–12 h.

Distribution: Crosses placenta. **Metabolism:** In liver (CYP2D6). **Elimination:** Primarily in urine; enters breast milk. **Half-Life:** 10–50 h.

NURSING IMPLICATIONS

Assessment & Drug Effects
- Monitor therapeutic effectiveness. It may take 1–6 wk to reduce attacks when used for migraine prophylaxis.
- Monitor for S&S of drowsiness and dizziness (initial stages of therapy); institute measures to prevent falling. Also monitor for overdose or suicide ideation especially in children and adolescents and in patients who use excessive amounts of alcohol.
- Lab tests: Baseline and periodic leukocyte and differential counts; renal and hepatic function tests.
- Eye examinations (including glaucoma testing) are recommended particularly for older adults, adolescents, and patients receiving high doses/prolonged therapy.
- Monitor BP and pulse rate in patients with preexisting cardiovascular disease. Assess for orthostatic hypotension especially in older adults. Withhold drug if there is a rise or fall in systolic BP (by 10–20 mm Hg), or a sudden increase or a significant change in pulse rate or rhythm. Notify prescriber.
- Monitor I&O, including bowel elimination pattern.

Patient & Family Education
- Monitor weight; drug may increase appetite or a craving for sweets.
- Understand that tolerance/adaptation to anticholinergic actions (see Appendix F) usually develops with maintenance regimen. Keep prescriber informed.

Common adverse effects in *italic*, life-threatening effects <u>underlined</u>; generic names in **bold**; classifications in SMALL CAPS; ✤ Canadian drug name; ⊙ Prototype drug

77

- Relieve dry mouth by taking frequent sips of water and increasing total fluid intake.
- Make position change slowly and in stages to prevent dizziness.
- Do not drive or engage in potentially hazardous activities until response to drug is known.
- Do not use OTC drugs without consulting prescriber while on TCA therapy; many preparations contain sympathomimetic amines.
- Note: Amitriptyline may turn urine blue-green.

AMLEXANOX
(am-lex'a-nox)

Aphthasol, OraDisc A
Classification: ANTI-INFLAMMATORY
Therapeutic: ANTI-INFLAMMATORY
Pregnancy Category: B

AVAILABILITY 5% paste; 2 mg mucoadhesive disc

ACTION & THERAPEUTIC EFFECT
Amlexanox is a potent inhibitor of inflammatory mediators (e.g., leukotrienes, IgE, IgG). Its mechanism of healing is unknown. *Amlexanox reduces healing time and pain related to aphthous ulcers or canker sores.*

USES Treatment of aphthous ulcers in patients with normal immune systems.

CONTRAINDICATIONS Sensitivity to amlexanox or benzyl alcohol.
CAUTIOUS USE Immunosuppressed patients; pregnancy (category B), lactation. Safety and efficacy in children are not established.

ROUTE & DOSAGE

Aphthous Ulcers
Adult: **Topical** Apply ¼ in. (0.5 cm) to finger and dab onto each mouth ulcer q.i.d. (after oral hygiene p.c.

and at bedtime) for 10-day cycle; apply disc to each mouth ulcer and allow to dissolve

ADMINISTRATION
Topical
- Apply after oral hygiene following each meal and before bedtime.
- Avoid prolonged contact with skin and wash off skin if contact occurs.
- Store at 15°–30° C (59°–86° F) away from heat and moisture. Do not freeze.

ADVERSE EFFECTS (≥1%) **Body as a Whole:** Transient pain, stinging, or burning at application site.

PHARMACOKINETICS Absorption: Minimally absorbed through ulcer. **Onset:** Approximately 3 days. **Elimination:** 17% in urine. **Half-Life:** 3.5 h.

NURSING IMPLICATIONS
Assessment & Drug Effects
- Discontinue use if rash or inflamed membranes develop.

Patient & Family Education
- Use at first sign of canker sore. Wash hands before and immediately after application.
- Flush eyes immediately with large amount of cold water if paste accidentally comes in contact with eyes or eye area.
- Contact prescriber if healing does not result after 10 days of therapy.

AMLODIPINE
(am-lo'di-peen)

Norvasc
Classifications: CALCIUM CHANNEL BLOCKER; ANTIHYPERTENSIVE
Therapeutic: ANTIHYPERTENSIVE; ANTIANGINAL
Prototype: Nifedipine
Pregnancy Category: C

AVAILABILITY 2.5 mg, 5 mg, 10 mg tablets

ACTION & *THERAPEUTIC EFFECT*
Amlodipine is a calcium channel blocking agent that selectively blocks calcium influx across cell membranes of cardiac and vascular smooth muscle without changing serum calcium concentrations. It reduces coronary vascular resistance and increases coronary blood flow. Additionally, amlodipine decreases peripheral vascular resistance, increases oxygen delivery to myocardial tissue, and increases cardiac output. *Amlodipine reduces systolic, diastolic, and mean arterial blood pressure. It also decreases pain due to angina.*

USES Treatment of mild to moderate hypertension and stable angina.

CONTRAINDICATIONS Hypersensitivity to amlodipine; hypotension; severe obstructive coronary artery disease; severe aortic stenosis.

CAUTIOUS USE Liver disease; concomitant use with hypotension; CHF, ventricular dysfunction; lactation; older adults; GERD; hepatic disease; pregnancy (category C); children younger than 6 y.

ROUTE & DOSAGE

Hypertension

Adult: PO 5–10 mg once daily
Geriatric: Start with 2.5 mg, adjust dose at intervals of not less than 2 wk

Adolescent/Child (at least 6 y old): PO 2.5–5 mg daily (max: 10 mg)

Stable/Vasospastic Angina

Adult: PO 5–10 mg daily (usually 10 mg)

Hepatic Impairment Dosage Adjustment

Start with 2.5 mg, adjust dose at intervals of not less than 2 wk

ADMINISTRATION

Oral
- Give drug without regard to meals.
- Note: Doses are usually titrated upward over a period of 14 days or more rapidly if warranted.
- Store at 15°–30° C (59°–86° F).

ADVERSE EFFECTS (≥1%) **CV:** Palpitations, flushing tachycardia, *peripheral or facial edema*, bradycardia, chest pain, syncope, postural hypotension. **CNS:** Light-headedness, fatigue, *headache*. **GI:** Abdominal pain, nausea, anorexia, constipation, dyspepsia, dysphagia, diarrhea, flatulence, vomiting. **Urogenital:** Sexual dysfunction, frequency, nocturia. **Respiratory:** Dyspnea. **Skin:** Flushing, rash. **Other:** Arthralgia, cramps, myalgia.

INTERACTIONS Drug: Adenosine may increase the risk of bradycardia; **bosentan** may decrease efficacy of amlodipine; additive hypotensive effects with other ANTIHYPERTENSIVE AGENTS; AZOLE ANTIFUNGALS (e.g., **fluconazole, itraconazole**) may inhibit metabolism of amlodipine; **itraconazole** may increase edema. **Food: Grapefruit juice** may increase amlodipine levels. **Herbal: Ephedra, ma huang, melatonin** may antagonize antihypertensive effects.

PHARMACOKINETICS Absorption: Greater than 90% absorbed from GI tract. **Peak:** 6–9 h. **Duration:** 24 h. **Distribution:** Greater than 95% protein bound. **Metabolism:** In liver (CYP3A4) to inactive metabolites. **Elimination:** In urine (less than 5–10% excreted unchanged),

Common adverse effects in *italic*, life-threatening effects underlined; generic names in **bold**; classifications in SMALL CAPS; ✦ Canadian drug name; ⊙ Prototype drug

79

20–25% in feces. **Half-Life:** Less than 45 y: 28–69 h; greater than 60 y: 40–120 h.

NURSING IMPLICATIONS

Assessment & Drug Effects

- Monitor BP for therapeutic effectiveness. BP reduction is greatest after peak levels of amlodipine are achieved 6–9 h following oral doses.
- Monitor for S&S of dose-related peripheral or facial edema that may not be accompanied by weight gain; rarely, severe edema may cause discontinuation of drug.
- Monitor BP with postural changes. Report postural hypotension. Monitor more frequently when additional antihypertensives or diuretics are added.
- Monitor heart rate; dose-related palpitations (more common in women) may occur.

Patient & Family Education

- Report significant swelling of face or extremities.
- Exercise caution when standing and walking due to possible dose-related light-headedness/dizziness.
- Report shortness of breath, palpitations, irregular heartbeat, nausea, or constipation to prescriber.

AMMONIUM CHLORIDE
(ah-mo'ni-um)
Classification: ELECTROLYTIC BALANCE AGENT
Therapeutic: ACIDFIER; ELECTROLYTE REPLACEMENT
Pregnancy Category: B

AVAILABILITY 26.75% or 5 mEq/mL solution; 500 mg tablets; 486 mg enteric-coated tablets

ACTION & *THERAPEUTIC EFFECT*
Acidifying property is due to conversion of ammonium ion (NH_4^+) to urea in liver with liberation of H^+ and Cl^-. Potassium excretion also increases acid, but to a lesser extent. *Effective as a systemic acidifier in metabolic alkalosis by releasing H^+ ions which lower pH.*

USES Treatment of hypochloremic states and metabolic alkalosis. Diuretic or urinary acidifying agent.

CONTRAINDICATIONS Severe renal or hepatic insufficiency; primary respiratory acidosis.

CAUTIOUS USE Cardiac edema, cardiac insufficiency, pulmonary insufficiency; pregnancy (category B), lactation.

ROUTE & DOSAGE

Urine Acidifier, Diuretic

Adult: **PO** 4–12 g/day divided q4–6h
Child: **PO** 75 mg/kg/day in 4 divided doses

Metabolic Alkalosis and Hypochloremic States

Adult/Child: **IV** Dose calculated on basis of CO_2 combining power or serum Cl deficit, 50% of calculated deficit is administered slowly

ADMINISTRATION

Oral

- Give after meals for best tolerance or use enteric-coated tablets. Tablets should be swallowed whole.
- Store in airtight container.

Intravenous —
Check with prescriber for slower rate for infants.

PREPARE: **Intermittent:** Dilute each 20 mL vial in 500–1000 mL NS. Do not exceed a concentration of 1–2%.

ADMINISTER: **Intermittent:** Give slowly to avoid serious adverse effects (ammonia toxicity) and

local irritation and pain. ▪ Give at a rate not to exceed 5 mL/min.
INCOMPATIBILITIES **Solution/additive: Codeine phosphate, dimenhydrinate. Y-site: Warfarin.**

▪ Avoid freezing. ▪ Concentrated solutions crystallize at low temperatures. ▪ Crystals can be dissolved by placing intact container in a warm water bath and warming to room temperature.

ADVERSE EFFECTS (≥1%) **Body as a Whole:** Most secondary to ammonia toxicity. **CNS:** Headache, depression, drowsiness, twitching, excitability; EEG abnormalities. **CV:** Bradycardia and other arrhythmias. **GI:** Gastric irritation, nausea, vomiting, anorexia. **Metabolic:** Metabolic acidosis, hyperammonia. **Respiratory:** Hyperventilation. **Skin:** Rash. **Urogenital:** Glycosuria. **Other:** Pain and irritation at IV site.

DIAGNOSTIC TEST INTERFERENCE Ammonium chloride may increase *blood ammonia* and *AST,* decrease *serum magnesium* (by increasing urinary magnesium excretion), and decrease *urine urobilinogen.*

INTERACTIONS Drug: Aminosalicylic acid may cause crystalluria; increases urinary excretion of AM-PHETAMINES, **flecainide, mexiletine, methadone, ephedrine, pseudoephedrine;** decreased urinary excretion of SULFONYLUREAS, SALICYLATES.

PHARMACOKINETICS Absorption: Completely absorbed in 3–6 h. **Metabolism:** In liver to HCl and urea. **Elimination:** Primarily in urine.

NURSING IMPLICATIONS

Assessment & Drug Effects
▪ Assess IV infusion site frequently for signs of irritation. Change site as warranted.
▪ Monitor for S&S of: Metabolic acidosis (mental status changes in-

cluding confusion, disorientation, coma, respiratory changes including increased respiratory rate and depth, exertional dyspnea); ammonium toxicity (cardiac arrhythmias including bradycardia, irregular respirations, twitching, seizures).
▪ Monitor I&O ratio and pattern. The diuretic effect of ammonium chloride is compensatory and lasts only 1–2 days.
▪ Lab tests: Baseline and periodic determinations of CO_2 combining power, serum electrolytes, and urinary and arterial pH during therapy to avoid serious acidosis.

Patient & Family Education
▪ Report pain at IV injection site.

AMOXAPINE
(a-mox′a-peen)
Classification: TRICYCLIC ANTIDEPRESSANT
Therapeutic: ANTIDEPRESSANT
Prototype: Imipramine
Pregnancy Category: C

AVAILABILITY 25 mg, 50 mg, 100 mg, 150 mg tablets

ACTION & THERAPEUTIC EFFECT Tricyclic antidepressant (TCA) with mixed antidepressant and tranquilizing properties. Antidepressant activity is thought to be due to reduced reuptake of norepinephrine and serotonin at the cell membrane of the neuron, thus increasing the level of both neurotransmitters. *Enhancement of neurotransmitters results in its antidepressant activity.*

USES Neurotic and endogenous depression accompanied by anxiety or agitation.

CONTRAINDICATIONS Hypersensitivity to other tricyclic compounds;

Common adverse effects in *italic*, life-threatening effects underlined; generic names in **bold**; classifications in SMALL CAPS; ✤ Canadian drug name; Prototype drug

81

acute recovery period after MI; AV block; MAOI therapy, QT prolongation; suicidal ideation; lactation.

CAUTIOUS USE History of convulsive disorders, schizophrenia, manic depression, electroshock therapy; alcohol abuse; history of urinary retention, benign prostatic hypertrophy; angle-closure glaucoma or increased intraocular pressure; cardiovascular disorders; impaired renal or hepatic function; elective surgery; pregnancy (category C); children younger than 16 y.

ROUTE & DOSAGE

Antidepressant

Adult: **PO** Start at 50 mg b.i.d. or t.i.d., may increase on third day to 100 mg t.i.d. Maintenance doses less than 300 mg/day as single dose at bedtime.
Geriatric: **PO** 25 mg at bedtime, may increase q3–7days to 50–150 mg/day in divided doses (max: 300 mg/day)

ADMINISTRATION

Oral
- Give with or after food to reduce GI irritation; tablet may be crushed and taken with fluid or mixed with food.
- Give maintenance dose as a single dose at bedtime to minimize daytime sedation and other annoying drug adverse effects.
- Do not abruptly discontinue drug. Doses should be tapered over 2 wk.
- Store at 15°–30° C (59°–86° F) in tightly closed container unless otherwise directed.

ADVERSE EFFECTS (≥1%) **CNS:** *Drowsiness,* dizziness, headache, fatigue, *sedation,* lethargy; extra-

pyramidal effects (acute dystonic reactions, panic attacks, parkinsonism, tardive dyskinesia), seizures (overdosage). **CV:** Orthostatic hypotension; arrhythmias. **GI:** Constipation, diarrhea, flatulence, *dry mouth,* peculiar taste, nausea, heartburn. **Special Senses:** Blurred vision, dry eyes. **Urogenital:** Nephrotoxicity (overdosage).

INTERACTIONS Drug: May decrease response to ANTIHYPERTENSIVES; CNS DEPRESSANTS, **alcohol,** HYPNOTICS, BARBITURATES, SEDATIVES potentiate CNS depression; may increase hypoprothrombinemic effect of ORAL ANTICOAGULANTS; **ethchlorvynol,** transient delirium; with **levodopa,** SYMPATHOMIMETICS (e.g., **epinephrine, norepinephrine**), possibility of sympathetic hyperactivity with hypertension and hyperpyrexia; with MAO INHIBITORS, possibility of severe reactions: toxic psychosis, cardiovascular instability; **methylphenidate** increases plasma TCA levels; thyroid drugs may increase possibility of arrhythmias; **cimetidine** may increase plasma TCA levels. **Herbal: Ginkgo** may decrease seizure threshold, **St. John's wort** may cause serotonin syndrome.

PHARMACOKINETICS Absorption: Rapidly absorbed. **Peak:** 1–2 h. **Distribution:** Probably crosses placenta; distributed into breast milk. **Metabolism:** Via CYP2D6; active metabolite. **Elimination:** 60% in urine in 6 days; 7–18% in feces. **Half-Life:** 8 h parent drug, 30 h metabolite.

NURSING IMPLICATIONS

Assessment & Drug Effects
- Monitor therapeutic effectiveness. Initial antidepressant effect (mild euphoria, increased energy) may

occur within 4–7 days; however, in most patients clinical response does not occur until after 2–3 wk of drug therapy.

- Supervise patient closely during therapy for suicidal ideation and potential serious adverse effects.
- Report immediately signs of neuroleptic malignant syndrome: Fever, sweating, rigidity (catatonia), unstable BP, rapid, irregular pulse; changes in level of consciousness, coma. Although rare, it can be life-threatening if drug is not stopped immediately. Death can result from acute respiratory, renal, or cardiovascular failure.
- Report immediately the onset of signs of tardive dyskinesia (see Appendix F); careful observation/reporting may prevent irreversibility.
- Monitor I&O ratio and bowel elimination pattern. Report continuing constipation.

Patient & Family Education
- Report promptly suicidal ideation. Children and adolescents may be especially vulnerable to suicidal thoughts.
- Do not abruptly discontinue drug. Dosage should be tapered over 2 wk. Maintain established dosage regimen. Do not skip, reduce, or double doses or change dose intervals.
- Minimize alcohol intake as it may potentiate drug effects, thus increasing the dangers of overdosage or suicidal ideation.
- Drink at least 2000 mL (approximately 2 qts) fluid daily and eat foods with high fiber content (if allowed) to provide needed roughage.
- Monitor weight at least weekly and report significant weight gain.
- Do not drive or engage in potentially hazardous tasks until response to drug is known.

- Rinse mouth frequently with clear water, especially after eating, to relieve mouth dryness.
- Do not take any prescription or OTC drugs without consulting prescriber.

AMOXICILLIN
(a-mox-i-sill'in)

Amoxil, Apo-Amoxi ♦, Moxa-tag, Trimox
Classifications: ANTIBIOTIC; AMINOPENICILLIN
Therapeutic: ANTIBIOTIC
Prototype: Ampicillin
Pregnancy Category: B

AVAILABILITY 125 mg, 200 mg, 250 mg, 400 mg, 500 mg tablets; 250 mg, 500 mg capsules; 50 mg/mL, 125 mg/5 mL, 200 mg/5 mL, 250 mg/5 mL, 400 mg/5 mL powder for suspension; 775 mg extended release tabs

ACTION & THERAPEUTIC EFFECT Broad-spectrum semisynthetic aminopenicillin and analog of ampicillin. Like other penicillins, amoxicillin inhibits the final stage of bacterial cell wall synthesis by binding to specific penicillin-binding proteins (PBPs) located inside the cell wall of rapidly multiplying bacteria. It results in bacterial cell lysis and death. *Active against both aerobic gram-positive and aerobic gram-negative bacteria.*

USES Infections of ear, nose, throat, GU tract, skin, and soft tissue caused by susceptible bacteria. Also used in uncomplicated gonorrhea.

CONTRAINDICATIONS Hypersensitivity to penicillins; infectious mononucleosis.
CAUTIOUS USE History of or suspected atopy or allergy (hives, eczema, hay fever, asthma); history

Common adverse effects in *italic*, life-threatening effects underlined; generic names in **bold**; classifications in SMALL CAPS; ♦ Canadian drug name; ⊙ Prototype drug

83

of cephalosporin or carbapenem hypersensitivity; colitis, dialysis, diarrhea, GI disease; viral infection, syphilis, severe hepatic impairment; renal impairment or failure, diabetes mellitus, leukemia, pregnancy (category B); lactation.

ROUTE & DOSAGE

Mild to Moderate Infections

Adult: **PO** 250–500 mg q8h
Child/Infant (at least 3 mo): **PO** 25–50 mg/kg/day (max: 60–80 mg/kg/day) divided q8h or 200–400 mg q12h
Neonate/Infant (up to 3 mo): **PO** 20–30 mg/kg/day divided q12h

Gonorrhea

Adult: **PO** 3 g as single dose with 1 g probenecid
Child (2 y or older): **PO** 50 mg/kg as single dose with probenecid 25 mg/kg

Tonsillitis/Pharyngitis (extended release tabs)

Adult/Adolescent: **PO** 775 mg daily with food

Renal Impairment Dosage Adjustment

CrCl 10–30 mL/min: 250–500 mg q12h; *less than 10 mL/min:* 250–500 mg q24h

Hemodialysis Dosage Adjustment

Administer extra dose after dialysis

ADMINISTRATION

Oral

- Ensure that chewable tablets are chewed or crushed before being swallowed with a liquid.
- Do not crush or chew extended release tablets.

- Place reconstituted pediatric drops directly on child's tongue or add to formula, milk, fruit juice, water, ginger ale, or other soft drink. Have child drink all the prepared dose promptly.
- Store in tightly covered containers at 15°–30° C (59°–86° F) unless otherwise directed. Reconstituted oral suspensions are stable for 7 days at room temperature.

ADVERSE EFFECTS (≥1%) **Body as a Whole:** As with other penicillins. Hypersensitivity (rash, anaphylaxis), superinfections. **GI:** Diarrhea, nausea, vomiting, pseudomembranous colitis (rare). **Hematologic:** Hemolytic anemia, eosinophilia, agranulocytosis (rare). **Skin:** Pruritus, urticaria, or other skin eruptions. **Special Senses:** Conjunctival ecchymosis.

INTERACTIONS TETRACYCLINES may inhibit activity of amoxicillin; **probenecid** prolongs the activity of amoxicillin.

PHARMACOKINETICS Absorption: Nearly complete absorption. **Peak:** 1–2 h. **Distribution:** Diffuses into most tissues and body fluids, except synovial fluid and CSF (unless meninges are inflamed); crosses placenta; distributed into breast milk in small amounts. **Metabolism:** In liver. **Elimination:** 60% in urine. **Half-Life:** 1–1.3 h.

NURSING IMPLICATIONS

Assessment & Drug Effects

- Determine previous hypersensitivity reactions to penicillins, cephalosporins, and other allergens prior to therapy.
- Lab tests: Baseline C&S tests prior to initiation of therapy, start drug pending results; periodic assessments of renal, hepatic, and

hematologic functions with prolonged therapy.

- Monitor for S&S of an urticarial rash (usually occurring within a few days after start of drug) suggestive of a hypersensitivity reaction. If it occurs, look for other signs of hypersensitivity (fever, wheezing, generalized itching, dyspnea), and report to prescriber immediately.
- Report onset of generalized, erythematous, maculopapular rash (ampicillin rash) to prescriber. Ampicillin rash is not due to hypersensitivity; however, hypersensitivity should be ruled out.
- Monitor for and report diarrhea which may indicate pseudomembranous colitis.

Patient & Family Education

- Take drug around the clock, do not miss a dose, and continue therapy until all medication is taken, unless otherwise directed by prescriber.
- Report to prescriber onset of diarrhea and other possible symptoms of superinfection (see Appendix F).

AMOXICILLIN AND CLAVULANATE POTASSIUM

(a-mox-i-sill'in)

Amoclan, Augmentin, Augmentin-ES600, Augmentin XR, Clavulin ♦
Classifications: BETA-LACTAM ANTIBIOTIC; AMINOPENICILLIN
Therapeutic: ANTIBIOTIC
Prototype: Ampicillin
Pregnancy Category: B

AVAILABILITY 250 mg, 500 mg, 875 mg tablets; 125 mg, 200 mg, 400 mg chewable tablets; 125 mg/5 mL, 200 mg/5 mL, 250 mg/5 mL, 400 mg/5 mL, 600 mg/5 mL oral suspension; 1000 mg amoxicillin/62.5 mg clavulanate sustained release tablets

ACTION & *THERAPEUTIC EFFECT*
As a beta-lactam antibiotic, amoxicillin is bactericidal. It inhibits the final stage of bacterial cell wall synthesis by binding with specific penicillin-binding proteins (PBPs) that are located inside the bacterial cell wall that leads to bacterial cell lysis and death. *Effectivenes of ampicillin is synergistic in combination with clavulanic acid. Clavulanic acid in combination with ampicillin inhibits enzyme (beta-lactamase) degradation of amoxicillin and by synergism extends both spectrum of activity and bactericidal effect of amoxicillin against many strains of beta-lactamase-producing bacteria resistant to amoxicillin alone.*

USES Infections caused by susceptible beta-lactamase-producing organisms: Lower respiratory tract infections, acute bacterial sinusitis, community acquired pneumonia, otitis media, sinusitis, skin and skin structure infections, and UTI.

CONTRAINDICATIONS Hypersensitivity to penicillins; infectious mononucleosis; patient with previous history of drug-induced cholestasis, jaundice, or other hepatic dysfunction; severe renal impairment.

CAUTIOUS USE Allergic disorders; cephalosporin hypersensitivity; GI disorders; colitis; hepatic or renal disease; elderly; pregnancy (category B), lactation.

ROUTE & DOSAGE

Mild to Moderate Infections

Adult: **PO** 250 or 500 mg tablet (each with 125 mg clavulanic acid) q8–12h; Sustained release

---

Common adverse effects in *italic*, life-threatening effects <u>underlined</u>; generic names in **bold**; classifications in SMALL CAPS; ♦ Canadian drug name; ⊘ Prototype drug

85

tabs: 2 tablets (2000 mg amoxicillin/125 mg clavulanate) q12h × 7–10 days
Child: **PO** *Less than 40 kg,* 20–40 mg/kg/day (based on amoxicillin component) divided q8–12h; *older than 3 mo,* 90 mg/kg/day of 600 ES divided q12h × 10 days
Neonate: Infant (younger than 3 mo): **PO** 30 mg/kg/day (amoxicillin) divided q12h

ADMINISTRATION

Oral

- Give at the start of a meal to minimize GI upset and enhance absorption.
- Reconstitute oral suspension by adding amount of water specified on container to provide a 5 mL suspension. Tap bottle before adding water to loosen powder, then add water in 2 portions, agitating suspension well before each addition.
- Agitate suspension well just before administration of each dose.
- Give dialysis patient an additional 2 doses on the day of dialysis; one dose during and another dose after dialysis.
- Store tablets in tight containers at less than 24° C (71° F). Reconstituted oral suspension should be refrigerated at 2°–8° C (36°–46° F), then discarded after 10 days.

ADVERSE EFFECTS (≥1%) **GI:** *Diarrhea,* nausea, vomiting. **Skin:** Rash, urticaria. **Other:** Candidal vaginitis; moderate increases in serum ALT, AST; hypersensitivity reactions, glomerulonephritis; <u>agranulocytosis</u> (rare).

DIAGNOSTIC TEST INTERFERENCE May interfere with **urinary glucose** determinations using **cupric sulfate, Benedict's solution, Clinitest;** does not affect **glucose oxidase methods** (e.g., Clinistix, TesTape). Positive direct **antiglobulin** (**Coombs'**) test results may be reported, a reaction that could interfere with **hematologic studies** or with **transfusion cross-matching** procedures.

INTERACTIONS Drug: TETRACYCLINES may inhibit activity of amoxicillin; **probenecid** prolongs the activity of amoxicillin.

PHARMACOKINETICS Absorption: Nearly complete absorption. **Peak:** 1–2 h. **Distribution:** Diffuses into most tissues and body fluids, except synovial fluid and CSF (unless meninges are inflamed); crosses placenta; distributed into breast milk in very small amounts. **Metabolism:** In liver. **Elimination:** 50–73% of the amoxicillin and 25–45% of the clavulanate dose excreted in urine in 2 h. **Half-Life:** Amoxicillin 1–1.3 h, clavulanate 0.78–1.2 h.

NURSING IMPLICATIONS

Assessment & Drug Effects

- Determine previous hypersensitivity reactions to penicillins, cephalosporins, and other allergens prior to therapy.
- Lab tests: Baseline C&S tests prior to initiation of therapy; start drug pending results.
- Monitor for S&S of an urticarial rash (usually occurring within a few days after start of drug) suggestive of a hypersensitivity reaction. If it occurs, look for other signs of hypersensitivity (fever, wheezing, generalized itching, dyspnea), and report to prescriber immediately.
- Monitor for and report diarrhea which may indicate pseudomembranous colitis.

Common adverse effects in *italic,* life-threatening effects <u>underlined</u>; generic names in **bold;** classifications in SMALL CAPS; ✚ Canadian drug name; ⊘ Prototype drug

Patient & Family Education

- Female patients should report onset of symptoms of *Candidal vaginitis* (e.g., moderate amount of white, cheesy, nonodorous vaginal discharge; vaginal inflammation and itching; vulvar excoriation, inflammation, burning, itching). Therapy may have to be discontinued.
- Report onset of diarrhea and other possible symptoms of superinfection to prescriber (see Appendix F).

AMPHETAMINE SULFATE ⊙

(am-fet′a-meen)

Adderall, Adderall XR
Classification: CEREBRAL STIMULANT; ANOREXIANT
Therapeutic: CEREBRAL STIMULANT
Pregnancy Category: C
Controlled Substance: Schedule II

AVAILABILITY 5 mg, 10 mg tablets; **Adderall:** 5 mg, 10 mg, 20 mg, 30 mg tablets; 5 mg, 15 mg, 20 mg, 25 mg, 30 mg sustained release capsules

ACTION & *THERAPEUTIC EFFECT*

Marked stimulant effect on CNS thought to be due to action on cerebral cortex and possibly the reticular activating system. Acts indirectly on adrenergic receptors by increasing synaptic release of norepinephrine in the brain and by blocking reuptake of norepinephrine at presynaptic membranes. *CNS stimulation results in increased motor activity, diminished sense of fatigue, alertness, wakefulness, and mood elevation. Anorexigenic effect thought to result from direct inhibition of hypothalamic appetite center as well as mood elevation.*

USES Narcolepsy, attention deficit disorder in children and adults (hyperkinetic behavioral syndrome, minimal brain dysfunction). Use as short-term adjunct to control exogenous obesity not generally recommended because of its potential for abuse.

CONTRAINDICATIONS Hypersensitivity to sympathomimetic amines; history of drug abuse; severe agitation; hyperthyroidism; diabetes mellitus; moderate to severe hypertension, advanced arteriosclerosis, angina pectoris or other cardiovascular disorders; Gilles de la Tourette disorder; glaucoma; during or within 14 days after treatment with MAOIs; lactation.
CAUTIOUS USE Mild hypertension; pregnancy (category C).

ROUTE & DOSAGE

Narcolepsy
Adult: **PO** 5–60 mg/day divided q4–6h in 2–3 doses
Child: **PO** *Older than 12 y,* 10 mg/day, may increase by 10 mg at weekly intervals; *6–12 y,* 5 mg/day, may increase by 5 mg at weekly intervals
Attention Deficit Disorder
Adult/Adolescent: **PO** 10 mg extended release once daily in a.m.; may increase by 5–10 mg at weekly intervals if needed to max of 30 mg/day.
Child: **PO** *6 y,* 5 mg 1–2 times/day, may increase by 5 mg at weekly intervals (max: 40 mg/day); *3–5 y,* 2.5 mg 1–2 times/day, may increase by 2.5 mg at weekly intervals; 10 mg extended release once daily in a.m.; may increase by 5–10 mg at weekly intervals if needed to max of 30 mg/day.

Common adverse effects in *italic*, life-threatening effects underlined; generic names in **bold**; classifications in SMALL CAPS; ♣ Canadian drug name; ⊙ Prototype drug

87

Obesity Dosage Adjustment
Adult: **PO** 5–10 mg 1 h before meals

ADMINISTRATION

Oral
- Give first dose on awakening or early in a.m. when prescribed for narcolepsy.
- Give last dose no later than 6 h before patient retires to avoid insomnia.
- Ensure that sustained release capsules are not crushed or chewed.
- Store at 15°–30° C (59°–86° F) unless otherwise directed.

ADVERSE EFFECTS (≥1%) Body as a Whole: Allergy, urticaria, <u>sudden death</u> (reported in children with structural cardiac abnormalities). **CNS:** *Irritability,* psychosis, *restlessness,* nervousness, headache, *insomnia,* weakness, *euphoria,* dysphoria, drowsiness, trembling, hyperactive reflexes. **CV:** *Palpitation,* elevated BP; tachycardia, vasculitis. **Urogenital:** Impotence and change in libido with high doses. **GI:** Dry mouth, anorexia, unusual weight loss, nausea, vomiting, diarrhea, or constipation.

DIAGNOSTIC TEST INTERFERENCE Elevations in *serum thyroxine (T₄)* levels with high amphetamine doses.

INTERACTIONS Drug: Acetazolamide, sodium bicarbonate decrease amphetamine elimination; **ammonium chloride, ascorbic acid** increase amphetamine elimination; effects of both amphetamine and BARBITURATES may be antagonized if given together; **furazolidone** may increase BP effects of amphetamines, and interaction may persist for several weeks after furazolidone is discontinued;

guanethidine antagonizes antihypertensive effects; because MAO INHIBITORS, **selegiline** can precipitate hypertensive crisis (fatalities reported), do not administer amphetamines during or within 14 days of these drugs; PHENOTHIAZINES may inhibit mood elevating effects of amphetamines; TRICYCLIC ANTIDEPRESSANTS enhance amphetamine effects through increased **norepinephrine** release; BETA AGONISTS increase cardiovascular adverse effects.

PHARMACOKINETICS Absorption: Rapid. **Peak effect:** 1–5 h. **Duration:** Up to 10 h. **Distribution:** All tissues, especially CNS. **Metabolism:** In liver. **Elimination:** Renal; excreted into breast milk. **Half-Life:** 10–30 h.

NURSING IMPLICATIONS

Assessment & Drug Effects
- Monitor for S&S of toxicity in children. Response to this drug is more variable in children than adults; acute toxicity has occurred over a wide dosage range.
- Monitor for S&S of insomnia or anorexia. Report complaints to prescriber. Dosage reduction may be required.
- Monitor BP and HR, especially in those with hypertension.
- Monitor diabetics closely for loss of glycemic control.
- Monitor growth in children; drug may be discontinued periodically to allow for normal growth.
- Note: Drug's excitatory and euphoric effects are associated with a high abuse potential.

Patient & Family Education
- Keep prescriber informed of clinical response and persistent or bothersome adverse effects. This drug exerts a stimulating effect that

Common adverse effects in *italic*, life-threatening effects <u>underlined</u>; generic names in **bold**; classifications in SMALL CAPS; ✤ Canadian drug name; ⊚ Prototype drug

masks fatigue; after exhilaration disappears, fatigue and depression are usually greater than before, and a longer period of rest is needed.

- Report insomnia or undesired weight loss.
- Do not drive or engage in potentially hazardous tasks until response to drug is known.
- Rinse mouth frequently with clear water, especially after eating, to relieve mouth dryness; increase fluid intake, if allowed; chew sugarless gum or sourballs.
- Note: Meticulous oral hygiene is required because decreased saliva encourages demineralization of tooth surfaces and mucosal erosion.
- Avoid caffeine-containing beverages because caffeine increases amphetamine effects.
- Note that drug is usually tapered gradually following prolonged administration of high doses. Abrupt withdrawal may result in lethargy, profound depression, or other psychotic manifestations that may persist for several weeks.

AMPHOTERICIN B 💊

(am-foe-ter'i-sin)

Amphocin, Fungizone
Classification: ANTIFUNGAL
Therapeutic: ANTIFUNGAL
Pregnancy Category: B; C (oral suspension)

AVAILABILITY 50 mg powder for injection; 100 mg/mL suspension; 3% cream, lotion, ointment

ACTION & *THERAPEUTIC EFFECT*
Fungistatic antibiotic that exerts antifungal action on both resting and growing cells at least in part by selectively binding to sterols in fungus cell membrane resulting in cell death. *Fungicidal at higher*

concentrations, depending on sensitivity of fungus.

USES Used intravenously for a wide spectrum of potentially fatal systemic fungal (mycotic) infections. Has been used to potentiate antifungal effects of flucytosine (*Ancobon*) and to provide anticandidal prophylaxis in certain susceptible patients receiving immunosuppressive therapy. Used topically for cutaneous and mucocutaneous infections caused by *Candida* (monilia).

UNLABELED USES Treatment of candiduria, fungal endocarditis, meningitis, septicemia; fungal infections of urinary bladder and urinary tract; amebic meningoencephalitis, and paracoccidioidomycosis.

CONTRAINDICATIONS Hypersensitivity to amphotericin; lactation.
CAUTIOUS USE Severe bone marrow depression; renal function impairment; anemia; pregnancy (category B), oral supsension (category C).

ROUTE & DOSAGE

**Systemic Infections
[Amphocin, Fungizone]**

Adult: **IV Test Dose** 1 mg dissolved in 20 mL of D5W by slow infusion (over 10–30 min) **IV Maintenance Dose** 0.25–0.3 mg/kg/day infused over 4–6 h, may gradually increase by 0.125–0.25 mg/kg/day up to 1–1.5 mg/kg/day (max dose: 1.5 mg/kg)
Child: **IV Test Dose** 0.1 mg/kg up to 1 mg dissolved in 20 mL of D5W by slow infusion (over 10–30 min) **IV Maintenance Dose** 0.4 mg/kg/day infused over 4–6 h, may increase by 0.25 mg/kg/day to target dose of 0.25–1 mg/kg/day infused over 2–6 h (max: 1.5 mg/kg)

Common adverse effects in *italic*; life-threatening effects <u>underlined</u>; generic names in **bold**; classifications in SMALL CAPS; ♣ Canadian drug name; 💊 Prototype drug

89

Renal Impairment Dosage Adjustment

The dose can be reduced or interval extended

ADMINISTRATION

Intravenous ─────────

PREPARE: Typically prepared by pharmacy service due to complex technique required for IV solution preparation. Each brand of amphotericin is prepared differently according to manufacturer's directions. ▪ Refer to specific manufacturer's guidelines for preparation of IV solutions.

ADMINISTER: Intermittent: Use a 1-micron filter. ▪ Infuse total daily dose over 2–6 h. Use longer infusion time for better tolerance. ▪ Alert: Rapid infusion of any amphotericin can cause cardiovascular collapse. If hypotension or arrhythmias develop interrupt infusion and notify prescriber. ▪ Protect IV solution from light during administration. ▪ Note incompatibilities. When given through an existing IV line, flush before and after with D5W. ▪ Initiate therapy using the most distal vein possible and alternate sites with each dose if possible to reduce the risk of thrombophlebitis. ▪ Check IV site frequently for patency.

INCOMPATIBILITIES Solution/additive: Any **saline**-containing solution (precipitate will form), PARENTERAL NUTRITION SOLUTIONS, **amikacin, calcium chloride, calcium gluconate, chlorpromazine, cimetidine, ciprofloxacin, diphenhydramine, dopamine, edetate calcium disodium, gentamicin, kanamycin, magnesium sulfate, meropenem, metaraminol, methyldopate, penicillin G, polymyxin, potassium chloride, prochlorperazine, ranitidine, streptomycin, verapamil.** Y-site: AMINOGLYCOSIDES, PENICILLINS, PHENOTHIAZINES, **allopurinol, amifostine, amsacrine, aztreonam, bivalirudin, cefepime, cefpirome, cisatracurium, dexmedetomidine, docetaxel, doxorubicin liposome, enalaprilat, etoposide, fenoldopam, filgrastim, fluconazole, fludarabine, foscarnet, gemcitabine, granisetron, heparin** (flush lines with D5W, not NS), **hetastarch, lansoprazole, linezolid, melphalan, meropenem, ondansetron, paclitaxel, pemetrexed, piperacillin/tazobactam, propofol, TPNs, vinorelbine.**

▪ Store according to manufacturer's recommendations for reconstituted and unopened vials.

ADVERSE EFFECTS (≥1%) **Body as a Whole:** Hypersensitivity (pruritus, urticaria, skin rashes, fever, dyspnea, <u>anaphylaxis</u>); *fever, chills.* **CNS:** Headache, sedation, muscle pain, arthralgia, weakness. **CV:** Hypotension, <u>cardiac arrest</u>. **Special Senses:** Ototoxicity with tinnitus, vertigo, loss of hearing. **GI:** Nausea, vomiting, diarrhea, epigastric cramps, anorexia, weight loss. **Hematologic:** Anemia, <u>thrombocytopenia</u>. **Metabolic:** *Hypokalemia, hypomagnesemia.* **Urogenital:** <u>Nephrotoxicity</u>, urine with low specific gravity. **Skin:** Dry, erythema, pruritus, burning sensation; allergic contact dermatitis, exacerbation of lesions. **Other:** Pain; arthralgias, thrombophlebitis (IV site), superinfections.

INTERACTIONS Drug: AMINOGLYCOSIDES, **capreomycin, cisplatin, carboplatin, colistin, cyclosporine, mechlorethamine,**

furosemide, vancomycin increase the possibility of nephrotoxicity; CORTICOSTEROIDS potentiate hypokalemia; with DIGITALIS GLYCOSIDES, hypokalemia increases the risk of **digitalis** toxicity.

PHARMACOKINETICS Peak: 1–2 h after IV infusion. **Duration:** 20 h. **Distribution:** Minimal amounts enter CNS, eye, bile, pleural, pericardial, synovial, or amniotic fluids; similar plasma and urine concentrations. **Elimination:** Excreted renally; can be detected in blood up to 4 wk and in urine for 4–8 wk after discontinuing therapy. **Half-Life:** 24–48 h.

NURSING IMPLICATIONS

Assessment & Drug Effects

- Lab tests: Baseline C&S tests prior to initiation of therapy; start drug pending results. Baseline and periodic BUN, serum creatinine, creatinine clearance; during therapy periodic CBC, serum electrolytes (especially K^+, Mg^{++}, Na^+, Ca^{++}), and LFTs.
- Monitor for S&S of local inflammatory reaction or thrombosis at injection site, particularly if extravasation occurs.
- Monitor cardiovascular and respiratory status and observe patient closely for adverse effects during initial IV therapy. If a test dose (1 mg over 20–30 min) is given, monitor vital signs every 30 min for at least 4 h. Febrile reactions (fever, chills, headache, nausea) occur in 20–90% of patients, usually 1–2 h after beginning infusion, and subside within 4 h after drug is discontinued. The severity of this reaction usually decreases with continued therapy. Keep prescriber informed.
- Monitor I&O and weight. Report immediately: Oliguria, any change in I&O ratio and pattern, or

appearance of urine [e.g., sediment, pink or cloudy urine (hematuria)], abnormal renal function tests, unusual weight gain or loss. Generally, renal damage is reversible if drug is discontinued when first signs of renal dysfunction appear.

- Report to prescriber and withhold drug, if BUN exceeds 40 mg/dL or serum creatinine rises above 3 mg/dL. Dosage should be reduced or drug discontinued until renal function improves.
- Consult prescriber for guidelines on adequate hydration and adjustment of daily dose as a possible means of avoiding or minimizing nephrotoxicity.
- Report promptly any evidence of hearing loss or complaints of tinnitus, vertigo, or unsteady gait. Tinnitus may not be a complaint in older adults or the very young. Other signs of ototoxicity (i.e., vertigo or hearing loss) are more reliable indicators of ototoxicity in these age groups.

Patient & Family Education

- Maintain a high fluid intake of 2–3 liters (approximately 2–3 quarts) of fluid daily if not directed otherwise.
- Report promptly chills, fever, or unusual weakness.

AMPHOTERICIN B LIPID-BASED

Abelcet, Amphotec, AmBisome
Classification: ANTIFUNGAL
Therapeutic: ANTIFUNGAL
Prototype: AMPHOTERICIN B
Pregnancy Category: B; C (oral suspension)

AVAILABILITY Abelcet: 100 mg/20 mL suspension for injection; **Amphotec:** 50 mg, 100 mg powder for

Common adverse effects in *italic*, life-threatening effects underlined; generic names in **bold**; classifications in SMALL CAPS; ♣ Canadian drug name; ✪ Prototype drug

91

injection; **AmBisome:** 50 mg powder for injection

ACTION & *THERAPEUTIC EFFECT*
Fungistatic antibiotic that exerts antifungal action on both resting and growing cells at least in part by selectively binding to sterols in fungus cell membrane. This results in fungal cell death. *Fungicidal at higher concentrations, depending on sensitivity of fungus.*

USES Used intravenously for a wide spectrum of potentially fatal systemic fungal (mycotic) infections.

UNLABELED USES Treatment of candiduria, fungal endocarditis, meningitis, septicemia; fungal infections of urinary bladder and urinary tract; amebic meningoencephalitis, and paracoccidioidomycosis.

CONTRAINDICATIONS Hypersensitivity to amphotericin; lactation.
CAUTIOUS USE Severe bone marrow depression; renal function impairment; anemia; pregnancy (category B).

ROUTE & DOSAGE

Systemic Infections [Abelcet]
Adult/Child: **IV** 5 mg/kg/day
[Amphotec]
Adult/Child: **IV Test Dose** 10 mL (1.6–8.3 mg) of initial dose infused over 10–30 min **IV Maintenance Dose** 3–4 mg/kg/day (max: 7.5 mg/kg/day) infused at 1 mg/kg/h
[AmBisome]
Adult/Child: **IV** 3–5 mg/kg/day infused over 1–2 h

Cryptococcal Meningitis in HIV [AmBisome]
Adult: **IV** 6 mg/kg/day infused over 2 h

Leishmaniasis [AmBisome]
Adult: **IV** *Immunocompetent patient:* 3 mg/kg/day on days 1–5, 14, and 21; may repeat if necessary. *Immunocompromised:* 4 mg/kg/day on days 1–5, 10, 17, 24, 31, and 38

ADMINISTRATION
Intravenous

PREPARE: Each brand of amphotericin is prepared differently according to manufacturer's directions. ▪ Refer to specific manufacturer's guidelines for preparation of IV solutions.
ADMINISTER: **Abelcet Intermittent:** Flush existing IV line with D5W before infusion. ▪ Use 5 micron in-line filter. Infuse total daily dose at 2.5 mg/kg/h. ▪ Shake IV bag at least q2h to evenly mix solution.
Amphotec Intermittent: Do not use an in-line filter. ▪ Infuse total daily dose at 1 mg/kg/h. Infusion time may be shortened but should never be less than 2 h. ▪ Infusion time may also be extended for better tolerance.
AmBisome Intermittent: Do not use an in-line filter. ▪ Infuse total daily dose over 2 h. Infusion time may be shortened but should never be less than 1 h. ▪ Alert: Rapid infusion of any amphotericin can cause cardiovascular collapse. If hypotension or arrhythmias develop interrupt infusion and notify prescriber. ▪ Protect IV solution from light during administration. ▪ Note incompatibilities. When given through an existing IV line, flush before and after with D5W. ▪ Initiate therapy using the most distal vein possible and alternate sites with each

Common adverse effects in *italic*, life-threatening effects <u>underlined</u>; generic names in **bold**; classifications in SMALL CAPS; ✚ Canadian drug name; ⊙ Prototype drug

dose if possible to reduce the risk of thrombophlebitis. ▪ Check IV site frequently for patency.

INCOMPATIBILITIES **Solution/additive:** Any **saline**-containing solution (precipitate will form), PARENTERAL NUTRITION SOLUTIONS. **Y-site:** AMINOGLYCOSIDES, PENICILLINS, PHENOTHIAZINES, **alfentanil, amikacin, ampicillin, ampicillin/sulbactam, atenolol, aztreonam, bretylium, buprenorphine, butorphanol, calcium salts, carboplatin, cefazolin, cefepime, ceftazidime, ceftriaxone, chlorpromazine, cimetidine, cisatracurium, cyclophosphamide, cyclosporine, cytarabine, diazepam, digoxin, diphenhydramine, dobutamine, dopamine, doxorubicin, doxorubicin liposome, droperidol, enalaprilat, esmolol, etoposide, famotidine, fluconazole, fluorouracil, haloperidol, heparin** (flush lines with D5W, not NS), **hetastartch, hydromorphone, hydroxyzine, imipenem/cilastatin, labetalol, leucovorin, lidocaine, magnesium sulfate, meperidine, mesna, metoclopramide, midazolam, mitoxantrone, morphone, nalbuphine, naloxone, netilmicin, ofloxacin, ondansetron, paclitaxel, phenytoin, piperacillin, piperacillin/tazobactam, potassium chloride, prochlorperazine, promethazine, propranolol, ranitidine, remifentanil, sodium bicarbonate, ticarcillin/clavulanate, vecuronium, verapamil, vinorelbine.** ▪ Do not mix Abelcet or Amphotec with any other drugs.

▪ Store according to manufacturer's recommendations for reconstituted and unopened vials.

ADVERSE EFFECTS (≥1%) **Body as a Whole:** Hypersensitivity (pruritus, urticaria, skin rashes, fever, dyspnea, <u>anaphylaxis</u>); *fever, chills.* **CNS:** Headache, sedation, muscle pain, arthralgia, weakness. **CV:** Hypotension, <u>cardiac arrest</u>. **Special Senses:** Ototoxicity with tinnitus, vertigo, loss of hearing. **GI:** Nausea, vomiting, diarrhea, epigastric cramps, anorexia, weight loss. **Hematologic:** Anemia, <u>thrombocytopenia</u>. **Metabolic:** *Hypokalemia, hypomagnesemia.* **Urogenital:** <u>Nephrotoxicity</u>, urine with low specific gravity. **Skin:** Dry, erythema, pruritus, burning sensation; allergic contact dermatitis, exacerbation of lesions. **Other:** Pain; arthralgias, thrombophlebitis (IV site), superinfections.

INTERACTIONS Drug: AMINOGLYCOSIDES, **capreomycin, cisplatin, carboplatin, colistin, cyclosporine, mechlorethamine, furosemide, vancomycin** increase the possibility of nephrotoxicity; CORTICOSTEROIDS potentiate hypokalemia; with DIGITALIS GLYCOSIDES, hypokalemia increases the risk of **digitalis** toxicity.

PHARMACOKINETICS Peak: 1–2 h after IV infusion. **Duration:** 20 h. **Distribution:** Minimal amounts enter CNS, eye, bile, pleural, pericardial, synovial, or amniotic fluids; similar plasma and urine concentrations. **Elimination:** Excreted renally; can be detected in blood up to 4 wk and in urine for 4–8 wk after discontinuing therapy. **Half-Life:** 24–48 h.

NURSING IMPLICATIONS

Assessment & Drug Effects

▪ Lab tests: Baseline C&S tests prior to initiation of therapy; start drug pending results. Baseline and periodic BUN, serum creatinine,

---

Common adverse effects in *italic*, life-threatening effects <u>underlined</u>; generic names in **bold**; classifications in SMALL CAPS; ♣ Canadian drug name; ⊕ Prototype drug

creatinine clearance; during therapy periodic CBC, serum electrolytes (especially K^+, Mg^{++}, Na^+, Ca^{++}), and LFTs.

- Monitor for S&S of local inflammatory reaction or thrombosis at injection site, particularly if extravasation occurs.
- Monitor cardiovascular and respiratory status and observe patient closely for adverse effects during initial IV therapy. If a test dose (1 mg over 20–30 min) is given, monitor vital signs every 30 min for at least 4 h. Febrile reactions (fever, chills, headache, nausea) occur in 20–90% of patients, usually 1–2 h after beginning infusion, and subside within 4 h after drug is discontinued. The severity of this reaction usually decreases with continued therapy. Keep prescriber informed.
- Monitor I&O and weight. Report immediately oliguria, any change in I&O ratio and pattern, or appearance of urine [e.g., sediment, pink or cloudy urine (hematuria)], abnormal renal function tests, unusual weight gain or loss. Generally, renal damage is reversible if drug is discontinued when first signs of renal dysfunction appear.
- Report to prescriber and withhold drug if BUN exceeds 40 mg/dL or serum creatinine rises above 3 mg/dL. Dosage should be reduced or drug discontinued until renal function improves.
- Consult prescriber for guidelines on adequate hydration and adjustment of daily dose as a possible means of avoiding or minimizing nephrotoxicity.
- Report promptly any evidence of hearing loss or complaints of tinnitus, vertigo, or unsteady gait. Tinnitus may not be a complaint in older adults or the very young.

Other signs of ototoxicity (i.e., vertigo or hearing loss) are more reliable indicators of ototoxicity in these age groups.

Patient & Family Education

- Maintain a high fluid intake of 2–3 liters (approximately 2–3 quarts) of fluid daily if not directed otherwise.
- Report promptly chills, fever, or unusual weakness.

AMPICILLIN 🅿

(am-pi-sill′in)

Novo-Ampicillin ◆, Principen

AMPICILLIN SODIUM

Ampicin ◆, Penbritin ◆
Classifications: ANTIBIOTIC; AMINOPENICILLIN
Therapeutic: ANTIBIOTIC
Pregnancy Category: B

AVAILABILITY 250 mg, 500 mg capsules; 125 mg/5 mL, 250 mg/5 mL oral suspension; 125 mg, 250 mg, 500 mg, 1 gm, 2 gm vials

ACTION & THERAPEUTIC EFFECT
A broad-spectrum, semisynthetic aminopenicillin that is bactericidal but is inactivated by penicillinase (beta-lactamase). Like other penicillins, ampicillin inhibits the final stage of bacterial cell wall synthesis by binding to specific penicillin-binding proteins (PBPs) located inside the bacterial cell wall resulting in lysis and death of bacteria. *Effective against gram-positive bacteria as well as some gram-negative bacteria.*

USES Infections of GU, respiratory, and GI tracts and skin and soft tissues; also gonococcal infections, bacterial meningitis, otitis media, sinusitis, and septicemia and for

Common adverse effects in *italic*, life-threatening effects underlined; generic names in **bold**; classifications in SMALL CAPS; ◆ Canadian drug name; 🅿 Prototype drug

prophylaxis of bacterial endocarditis. Used parenterally only
for moderately severe to severe
infections.

CONTRAINDICATIONS Hypersensitivity to penicillin derivatives; infectious mononucleosis.

CAUTIOUS USE History of hypersensitivity to cephalosporins; GI
disorders; renal disease or impairment; pregnancy (category B),
lactation.

ROUTE & DOSAGE

Systemic Infections

Adult: **PO/IV/IM** 250–500 mg
q6h
Child (under 40 kg): **PO/IV** 25–50
mg/kg/day divided q6–8h
Neonate: **IV/IM** *Up to 7 days,
weight up to 2000 g,* 50 mg/kg/
day divided q12h; *up to 7 days,
weight greater than 2000 g,*
75 mg/kg/day divided q8h; *older
than 7 days, weight less than
1200 g,* 50 mg/kg/day divided
q12h; *older than 7 days, weight
1200–2000 g,* 75 mg/kg/day
divided q8h; *older than 7 days,
weight greater than 2000 g,*
100 mg/kg/day divided q6h

Meningitis

Adult/Child: **IV** 150–200 mg/kg/
day divided q3–4h
Neonate: **IV/IM** *Up to 7 days,
weight up to 2000 g,* 100 mg/kg/
day divided q12h; *up to 7 days,
weight greater than 2000 g,* 150
mg/kg/day divided q8h; *older
than 7 days, weight less than
1200 g,* 100 mg/kg/day divided
2h; *older than 7 days, weight
1200–2000 g,* 150 mg/kg/day
divided q8h; *older than 7 days,*

weight greater than 2000 g,
200 mg/kg/day divided q6h

Gonorrhea

Adult: **PO** 3.5 g with 1 g
probenecid × 1 **IV/IM** 500 mg
q8–12h

Bacterial Endocarditis Prophylaxis

Adult: **IV** 2 g 30 min before
procedure
Child: **IV** 50 mg/kg 30 min before
procedure (max: 2 g)

Group B Strep Prophylaxis

Adult: **IV** 2 g, then 1 g q4h until
delivery

Renal Impairment Dosage Adjustment

CrCl 10–30 mL/min: Give
q6–12h; *less than 10 mL/min:*
Give q12h

Hemodialysis Dosage Adjustment

Dose should be given after dialysis

ADMINISTRATION

Oral

- Give with a full glass of water on
an empty stomach (at least 1 h before or 2 h after meals) for maximum absorption. Food hampers
rate and extent of oral absorption.

Intramuscular

- Reconstitute each vial by adding
the indicated amount of sterile water for injection or bacteriostatic
water for injection (1.2 mL to 125
mg; 1 mL to 250 mg; 1.8 mL to 500
mg; 3.5 mL to 1 g; 6.8 mL to 2 g).
All reconstituted vials yield 250
mg/mL except the 125 mg vial
which yields 125 mg/mL. Administer within 1 h of preparation.
- Withdraw the ordered dose and
inject deep IM into a large muscle.

Common adverse effects in *italic*, life-threatening effects underlined; generic names
in **bold**; classifications in SMALL CAPS; ✦ Canadian drug name; ✪ Prototype drug

95

Intravenous

Verify correct IV concentration and rate of infusion with prescriber for administration to neonates, infants, and children.

PREPARE: **Direct/Intermittent:** Reconstitute as follows with sterile water for injection: Add 5 mL to 500 mg or fraction thereof; add 7.4 mL to 1 g; add 14.8 mL to 2 g. Final concentration **must be** 30 mg/mL or less; may be given direct IV as prepared or further diluted in 50 mL or more of NS, D5W, D5/NS, D5W/0.45NaCl, or LR. ▪ Stability of solution varies with diluent and concentration of solution. Solutions in NS are stable for up to 8 h at room temperature; other solutions should be infused within 2–4 h of preparation. Give direct IV within 1 h of preparation. ▪ Wear disposable gloves when handling drug repeatedly; contact dermatitis occurs frequently in sensitized individuals.

ADMINISTER: **Direct/Intermittent:** Infuse 500 mg or less slowly over 3–5 min. Give 1–2 g over at least 15 min. ▪ With solutions of 100 mL or more, set rate according to amount of solution, but no faster than direct IV rate. ▪ Convulsions may be induced by too rapid administration.

INCOMPATIBILITIES **Solution/additive:** Do not add to a **dextrose**-containing solution unless entire dose is given within 1 h of preparation. **Aztreonam, cefepime, hydrocortisone, prochlorperazine.** **Y-site:** **Amphotericin B, epinephrine, fenoldopam, fluconazole, hydralazine, lansoprazole, midazolam, nicardipine, ondansetron, sargramostim, TPN, verapamil, vinorelbine.**

▪ Store capsules and unopened vials at 15°–30° C (59°–86° F) unless otherwise directed. Keep oral preparations tightly covered.

ADVERSE EFFECTS (≥1%) **Body as a Whole:** Similar to those for penicillin G. Hypersensitivity (pruritus, urticaria, eosinophilia, hemolytic anemia, interstitial nephritis, <u>anaphylactoid reaction</u>); superinfections. **CNS:** Convulsive seizures with high doses. **GI:** *Diarrhea,* nausea, vomiting, <u>pseudomembranous colitis</u>. **Other:** Severe pain (following IM); phlebitis (following IV). **Skin:** *Rash.*

DIAGNOSTIC TEST INTERFERENCE Elevated *CPK* levels may result from local skeletal muscle injury following IM injection. *Urine glucose:* High urine drug concentrations can result in false-positive test results with *Clinitest* or *Benedict's* [enzymatic *glucose oxidase methods* (e.g., *Clinistix, Diastix, Tes-Tape*) are not affected]. *AST* may be elevated (significance not known).

INTERACTIONS Drug: Allopurinol increases incidence of rash. Effectiveness of the AMINOGLYCOSIDES may be impaired in patients with severe end-stage renal disease. **Chloramphenicol, erythromycin,** and **tetracycline** may reduce bactericidal effects of ampicillin; this interaction is primarily significant when low doses of ampicillin are used. Ampicillin may interfere with the contraceptive action of oral contraceptives (**estrogens**). Female patients should be advised to consider nonhormonal contraception while on antibiotics. **Food:** Food may decrease absorption of ampicillin, so it should be taken 1 h before or 2 h after meals.

PHARMACOKINETICS Absorption: Oral dose is 50% absorbed. Peak effect: 5 min IV, 1 h IM, 2 h PO. **Duration:** 6–8 h. **Distribution:** Most body tissues; high CNS concentrations only with inflamed meninges; crosses the placenta. **Metabolism:** Minimal hepatic metabolism. **Elimination:** 90% in urine; excreted into breast milk. **Half-Life:** 1–1.8 h.

NURSING IMPLICATIONS

Assessment & Drug Effects

- Determine previous hypersensitivity reactions to penicillins, cephalosporins, and other allergens prior to therapy. Monitor closely for signs of hypersensitivity during first 30 min after administration.
- Lab tests: Baseline C&S tests prior to initiation of therapy; start drug pending results. Baseline and periodic assessments of renal, hepatic, and hematologic functions, particularly during prolonged or high-dose therapy.
- Note: Sodium content of IV drug should be considered in patients on sodium restriction.
- Inspect skin daily and instruct patient to do the same. The appearance of a rash should be carefully evaluated to differentiate a nonallergenic ampicillin rash from a hypersensitivity reaction. Report rash promptly to prescriber.
- Monitor for and report diarrhea which may indicate pseudomembranous colitis.

Patient & Family Education

- Report diarrhea to prescriber; do not self-medicate. Give a detailed report to the prescriber regarding onset, duration, character of stools, associated symptoms, temperature and weight loss to help rule out the possibility of drug-induced, potentially fatal pseudomembranous colitis (see Appendix F).
- Report S&S of superinfection (onset of black, hairy tongue; oral lesions or soreness; rectal or vaginal itching; vaginal discharge; loose, foul-smelling stools; or unusual odor to urine).
- Notify prescriber if no improvement is noted within a few days after therapy is started.
- Take medication around the clock; continue taking medication until it is all gone (usually 10 days) unless otherwise directed by prescriber or pharmacist.

AMPICILLIN SODIUM AND SULBACTAM SODIUM

(am-pi-sill'in/sul-bak'tam)

Unasyn

Classifications: ANTIBIOTIC; AMINOPENICILLIN

Therapeutic: ANTIBIOTIC

Prototype: Ampicillin

Pregnancy Category: B

AVAILABILITY 1.5 g, 3 g vials

ACTION & *THERAPEUTIC EFFECT*
Ampicillin inhibits the final stage of bacterial cell wall synthesis by binding to specific penicillin-binding proteins (PBPs) located inside the bacterial cell wall, thus destroying the cell wall. Sulbactam inhibits beta-lactamases, most frequently responsible for transferred drug resistance. Thus the spectrum of drugs affected by the combination of the two is increased. *Effective against both gram-positive and gram-negative bacteria including those that produce beta-lactamase and non-beta-lactamase producers. Ampicillin without sulbactam is not effective against beta-lactamase producing strains.*

USES Treatment of infections due to susceptible organisms in skin

---

Common adverse effects in *italic*, life-threatening effects <u>underlined</u>; generic names in **bold;** classifications in SMALL CAPS; ♣ Canadian drug name; ❶ Prototype drug

97

and skin structures, intra-abdominal infections, and gynecologic infections.

CONTRAINDICATIONS Hypersensitivity to penicillins; mononucleosis. **CAUTIOUS USE** Hypersensitivity to cephalosporins; GI disorders; renal disease or impairment; pregnancy (category B) or lactation.

ROUTE & DOSAGE

Systemic Infections

Adult/Child (weight greater than 40 kg): **IV/IM** 1.5–3 g q6h (max: 4 g sulbactam/day)
Child (1 y or older): **IV** 300 mg/kg/day (200 mg/kg ampicillin and 100 mg/kg sulbactam) divided q6h

Renal Impairment Dosage Adjustment

CrCl greater than 30 mL/min: Give q6–8h; *15–29 mL/min:* Give q12h; *5–14 mL/min:* Give q24h
Dialysis: Give dose after dialysis

ADMINISTRATION

Intramuscular

▪ Reconstitute solution with sterile water for injection by adding 6.4 mL diluent to a 3 g vial. Each mL contains 250 mg ampicillin and 125 mg sulbactam.
▪ Give deep IM into a large muscle. Rotate injection sites.

Intravenous

PREPARE: Direct/Intermittent: Reconstitute each 1.5 g vial with 3.2 mL of sterile water for injection to yield 375 mg/mL (250 mg ampicillin/125 mg sulbactam); must further dilute with NS, D5W, D5/NS, D5W/0.45NS, or LR to a final concentration within the range of 3–45 mg/mL.

ADMINISTER: Direct: Give slowly over at least 10–15 min. **Intermittent:** Infuse solutions of less than 50 mL over 10–15 min and solutions of 50–100 mL over 15–30 min. With solutions of 100 mL or more, set rate according to amount of solution but no faster than direct IV rate (e.g., 100 mL over 30 min). ▪ Convulsions may be induced by too rapid administration. ▪ Use only freshly prepared solution; administer within 1 h after preparation.

INCOMPATIBILITIES Solution/additive: Do not add to a **dextrose**-containing solution unless entire dose is given within 1 h of preparation. **Ciprofloxacin. Y-site: Amiodarone, amphotericin B, ciprofloxacin, idarubicin, lansoprazole, nicardipine, ondansetron, sargramostim.**

▪ Store powder for injection at 15°–30° C (59°–86° F) before reconstitution. Storage times and temperatures vary for different concentrations of reconstituted solutions; consult manufacturer's directions.

ADVERSE EFFECTS (≥1%) **Body as a Whole:** Hypersensitivity (rash, itching, <u>anaphylactoid reaction</u>), fatigue, malaise, headache, chills, edema. **GI:** *Diarrhea, nausea,* vomiting, abdominal distention, candidiasis. **Hematologic:** Neutropenia, <u>thrombocytopenia</u>. **Urogenital:** Dysuria. **CNS:** Seizures. **Other:** Local pain at injection site; thrombophlebitis.

INTERACTIONS Drug: Allopurinol increases incidence of rash; effectiveness of the AMINOGLYCOSIDES may be impaired in patients with severe end stage renal disease; **chloramphenicol, erythromycin, tetracycline** may reduce bactericidal effects of ampicillin—this interaction is primarily significant when low doses

Common adverse effects in *italic*, life-threatening effects <u>underlined</u>; generic names in **bold**; classifications in SMALL CAPS; ✤ Canadian drug name; ✪ Prototype drug

are used; ampicillin may interfere with the contraceptive action of ORAL CONTRACEPTIVES—female patients should be advised to consider nonhormonal contraception while on antibiotics.

PHARMACOKINETICS Peak: Immediate after IV. **Duration:** 6–8 h. **Distribution:** Most body tissues; high CNS concentrations only with inflamed meninges; crosses placenta; appears in breast milk. **Metabolism:** Minimal hepatic metabolism. **Elimination:** In urine. **Half-Life:** 1 h.

NURSING IMPLICATIONS

Assessment & Drug Effects

- Determine previous hypersensitivity reactions to penicillins, cephalosporins, and other allergens prior to therapy.
- Lab tests: Baseline C&S tests prior to initiation of therapy; start drug pending results. Baseline and periodic assessments of renal, hepatic, and hematologic functions, particularly during prolonged or high-dose therapy.
- Report promptly unexplained bleeding (e.g., epistaxis, purpura, ecchymoses).
- Monitor patient carefully during the first 30 min after initiation of IV therapy for signs of hypersensitivity and anaphylactoid reaction (see Appendix F). Serious anaphylactoid reactions require immediate use of emergency drugs and airway management.
- Monitor for and report diarrhea which may indicate pseudomembranous colitis. Observe for and report other S&S of superinfection.
- Monitor I&O ratio and pattern. Report dysuria, urine retention, and hematuria.

Patient & Family Education

- Report chills, wheezing, pruritus (itching), respiratory distress, or palpitations to prescriber immediately.
- Report diarrhea to prescriber; do not self-medicate.

AMYL NITRITE

(am'il)

Amyl Nitrite

Classifications: NITRATE VASODILATOR; ANTIDOTE

Therapeutic: ANTIDOTE; NITRATE VASODILATOR

Prototype: Nitroglycerin

Pregnancy Category: C

AVAILABILITY 0.3 mL ampules

ACTION & *THERAPEUTIC EFFECT*
Short-acting vasodilator and smooth muscle relaxant. It is converted to nitric oxide, which causes the vasodilation. Action in treatment of cyanide poisoning is based on ability of amyl nitrite to convert hemoglobin to methemoglobin, which forms a nontoxic complex with cyanide ion. *Effective for immediate treatment of cyanide poisoning.*

USES As an adjunct antidote in the immediate treatment of cyanide poisoning. (Because of adverse effects, unpleasant odor, and expense, infrequently used to treat angina pectoris.)

UNLABELED USES Change intensity of heart murmurs.

CONTRAINDICATIONS Hypersensitivity to nitrites or nitrates; severe anemia; uncontrolled hyperthyroidism; acute alcoholism.

CAUTIOUS USE Older adults; recent increase in intracranial pressure; cerebral hemorrhage; head trauma; hypotension; glaucoma; recent MI; pregnancy (category C), lactation, children.

Common adverse effects in *italic*, life-threatening effects <u>underlined</u>; generic names in **bold**; classifications in SMALL CAPS; ♣ Canadian drug name; ❂ Prototype drug

ROUTE & DOSAGE

Acute Angina
Adult: **Inhalation** 0.18–0.3 mL prn

Cyanide Poisoning
Adult/Child: **Inhalation** 0.3 mL perle crushed every minute and inhaled for 15–30 sec until sodium nitrite infusion is ready

ADMINISTRATION

Inhalation

- Crush ampule between fingers to prepare (amyl nitrite is available in 0.18 mL and 0.3 mL perles, which are thin, friable glass ampules enveloped with woven fabric cover).
- Instruct patient to sit during and immediately after drug administration.
- Note: Amyl nitrite is volatile and highly flammable; when mixed with air or oxygen, it forms a mixture that can explode if ignited.
- Store at 8°–15° C (46°–59° F), unless otherwise directed. Protect from light.

ADVERSE EFFECTS (≥1%) **Body as a Whole:** Transient flushing, weakness. **CV:** Orthostatic hypotension, palpitation, <u>cardiovascular collapse</u>, tachycardia. **GI:** Nausea, vomiting. **Hematologic:** <u>Methemoglobinemia (large doses)</u>. **CNS:** *Headache,* dizziness, syncope. **Respiratory:** <u>Respiratory depression</u>.

PHARMACOKINETICS Absorption: Rapidly from mucous membranes. **Onset:** 10–30 sec. **Duration:** 3–5 min.

NURSING IMPLICATIONS

Assessment & Drug Effects

- Monitor for S&S of syncope, due to a sudden drop in systolic BP, which

sometimes follows drug inhalation, particularly in older adults.
- Monitor vital signs until stable. Rapid pulse, which usually lasts for a brief period, is an expected response to the fall in BP produced by the drug.
- Chart length of time required for angina to subside after administration of drug.
- Note: Tolerance may develop with repeated use over prolonged periods.

Patient & Family Education

- Note: Drug has a strongly fruity odor.
- Go to the emergency room immediately or consult prescriber if no relief from angina is experienced after 3 doses 5 min apart.

ANAGRELIDE HYDROCHLORIDE

(a-na′gre-lyde)
Agrylin
Classifications: ANTICOAGULANT; ANTIPLATELET
Therapeutic: ANTIPLATELET; REDUCER OF PLATELET COUNT
Pregnancy Category: C

AVAILABILITY 0.5 mg, 1 mg capsules

ACTION & THERAPEUTIC EFFECT
Anagrelide action appears to be related to a selective inhibition of platelet production. It inhibits platelet aggregation by affecting several aggregating agents (e.g., thrombin and arachidonic acid, ADP, and collagen). *Anagrelide is associated with significant decreases in platelet counts and is thought to prevent early changes in shape of platelets.*

USES Essential thrombocytemia.
UNLABELED USES Polycythemia vera.

CONTRAINDICATIONS Hypotension, severe hepatic impairment, females of childbearing age; lactation.

CAUTIOUS USE Cardiovascular disease, renal and hepatic function impairment, jaundice; pregnancy (category C); children.

ROUTE & DOSAGE

Essential Thrombocythemia
Adult (16 y or older): **PO** Start with 0.5 mg q.i.d. or 1 mg b.i.d. × 1 wk, may increase by 0.5 mg/day qwk until platelet count is less than 600,000/mcL (max: 10 mg/day)

Hepatic Impairment Dosage Adjustment
0.5 mg daily for 1 wk

ADMINISTRATION

Oral

- Make sure dosage increments do not exceed 0.5 mg/day in any 1 wk.
- Store at 15°–25° C (59°–77° F) in a light-resistant container.

ADVERSE EFFECTS (≥1%) **Body as a Whole:** *Asthenia, pain, edema (general),* paresthesia, back pain, malaise, fever, chills, photosensitivity. **CNS:** Headache, *dizziness,* CVA, syncope, seizures. **CV:** *Palpitations,* chest pain, tachycardia, peripheral edema, CHF, MI, cardiomyopathy, heart block, atrial fibrillation, pericarditis, arrhythmia, hemorrhage. **GI:** *Diarrhea, abdominal pain, nausea,* flatulence, vomiting, dyspepsia, anorexia, pancreatitis, constipation, GI hemorrhage, and ulceration. **Hematologic:** Anemia, thrombocytopenia, ecchymoses, lymphedema. **Respiratory:** *Dyspnea,* pulmonary infiltrates, pulmonary fibrosis, pulmonary hypertension. **Skin:** Rash, urticaria. **Other:** Dysuria.

PHARMACOKINETICS Absorption: 70% from GI tract. Food reduces bioavailability. **Onset:** 7–14 days. **Duration:** Increased platelet counts were observed 4 days after discontinuing drug. **Metabolism:** Extensively metabolized. **Elimination:** Primarily in urine as metabolites. **Half-Life:** 1.3–1.8 h.

NURSING IMPLICATIONS

Assessment & Drug Effects

- Monitor for therapeutic effectiveness which is indicated by reduction of platelets for at least 4 wk to 600,000/mcL or less or 50% from baseline.
- Monitor for S&S of CHF or myocardial ischemia.
- Monitor for S&S of renal toxicity in patients with renal insufficiency (creatinine 2 mg/dL or more).
- Monitor for S&S of hepatic toxicity in patients with liver functions greater than 1.5 times upper limit of normal.
- Lab tests: Monitor platelet count q2days for first wk, weekly thereafter until maintenance dose reached; closely monitor Hgb, WBC count, LFTs, and BUN and creatinine while platelet count is being lowered.

Patient & Family Education

- Contact prescriber if palpitations, fluid retention, breathing difficulty, or any other distressful symptoms develop.
- Avoid excessive exposure to sunlight or UV light.

ANAKINRA
(an-a-kin′ra)
Kineret
Classifications: IMMUNOMODULATOR; INTERLEUKIN-1 RECEPTOR ANTAGONIST; DISEASE-MODIFYING ANTIRHEUMATIC DRUG (DMARD)

Common adverse effects in *italic*, life-threatening effects underlined; generic names in **bold**; classifications in SMALL CAPS; ♣ Canadian drug name; ❷ Prototype drug

101

Therapeutic: ANTIRHEUMATIC; DMARD
Pregnancy Category: B

AVAILABILITY 100 mg prefilled syringes

ACTION & *THERAPEUTIC EFFECT*
Anakinra is a recombinant human interleukin-1 (IL-1) receptor antagonist (IL-1R1). It blocks the biological activity of IL-1 by inhibiting it from binding to interleukin receptors that are present in both bone and cartilage as well as other kinds of tissues. Interleukin-1 is produced in response to inflammation. IL-1 mediates various responses of tissues, including inflammatory and immunologic responses. *Anakinra competes with interleukin-1 (IL-1) by inhibiting it from binding to its receptor sites in tissues.*

USES Treatment of rheumatoid arthritis in patients failing other disease-modifying antirheumatic drugs (DMARDs). Usually given in combination with another DMARD.

CONTRAINDICATIONS Hypersensitivity to anakinra, *E. coli*–derived products, latex; active infections; live vaccines.
CAUTIOUS USE Neutropenia, immunosuppressed patients, or patients with frequent, serious infections; asthmatics; older adults; renal impairment; pregnancy (category B), lactation, children.

ROUTE & DOSAGE

Rheumatoid Arthritis
Adult: **Subcutaneous** 100 mg daily

ADMINISTRATION

Subcutaneous
- Do not give anakinra if the patient has an active infection.

- Note that anakinra should not ordinarily be given with tumor necrosis factor (TNF) blocking agents.
- Discard any unused portions as the drug contains no preservative.
- Check expiration date and do not use if expired or if an excessive number of translucent particles appears in the syringe.
- Store in the refrigerator at 2°–8° C (36°–46° F). DO NOT FREEZE OR SHAKE. Protect from light.

ADVERSE EFFECTS (≥1%) **Body as a Whole:** *Bacterial infections (URI,* sinusitis, flu, *other).* **CNS:** Headache. **GI:** Nausea, diarrhea, abdominal pain. **Hematologic:** Decreased neutrophil count, antibody formation. **Other:** *Injection site reactions (erythema, ecchymosis, edema, inflammation, pain).*

INTERACTIONS Drug: Increased risk of infection with live virus vaccine, **etanercept, infliximab.** Increased risk of neutropenia as well as infection with **etanercept** and **infliximab.**

PHARMACOKINETICS Absorption: 95% absorbed subcutaneous site. **Peak:** 3–7 h. **Elimination:** In urine. **Half-Life:** 4–6 h.

NURSING IMPLICATIONS

Assessment & Drug Effects
- Monitor for S&S of infection (e.g., pneumonia or other URI, cellulitis). Withhold drug and notify prescriber if these appear.
- Lab tests: Monitor absolute neutrophil count (ANC) prior to initiating anakinra, monthly for 3 mo, and q3mo thereafter for 1 y; monitor periodically WBC and platelet counts.
- Monitor closely patients with impaired renal function for S&S of adverse drug reactions.

Common adverse effects in *italic*, life-threatening effects underlined; generic names in **bold**; classifications in SMALL CAPS; ♣ Canadian drug name; ● Prototype drug

- Assess for injection site reactions manifested by erythema, ecchymosis, inflammation, and pain.

Patient & Family Education
- Review carefully the "Information for Patients and Caregivers" leaflet for detailed instructions on handling and injecting anakinra.
- Give the injection at approximately the same time every day.
- Administer only 1 dose (the entire contents of 1 prefilled glass syringe) per day. Discard any unused portions as the drug contains no preservative. Do not save unused drug.
- Do not permit vaccination with live vaccines while taking anakinra.
- Withhold drug and notify prescriber for S&S of upper respiratory, skin, or other infection(s).

ANASTROZOLE ☻

(a-nas'tro-zole)

Arimidex
Classifications: ANTINEOPLASTIC; NONSTEROIDAL AROMATASE INHIBITOR
Therapeutic: ANTINEOPLASTIC
Pregnancy Category: D

AVAILABILITY 1 mg tablets

ACTION & *THERAPEUTIC EFFECT*
Anastrozole is a potent and selective nonsteroidal aromatase inhibitor that converts estrone to estradiol. It lowers serum estrogen levels in postmenopausal women without interfering with adrenal steroid synthesis. *Inhibiting the biosynthesis of estrogens is one way to deprive tumors of estrogens, and thus restrict tumor growth.*

USES Early and advanced breast cancer with hormone receptor positive or hormone status unknown in postmenopausal women.

CONTRAINDICATIONS Premenopausal women, postmenopausal hormone replacement therapy; severe hepatic disease, pregnancy (category D), lactation.
CAUTIOUS USE Mild to moderate hepatic disease; older adults; children.

ROUTE & DOSAGE

Breast Cancer
Adult: **PO** 1 mg once daily

ADMINISTRATION

Oral
- Give on an empty stomach, 1 h before or 2 h after meals, because food affects extent of absorption.
- Store at 20°–25° C (68°–77° F).

ADVERSE EFFECTS (≥1%) **CNS:** Asthenia, headache, hot flushes, pain, dizziness, depression, paresthesia, malaise, insomnia, confusion, anxiety, nervousness. **CV:** Chest pain, hypertension, thrombophlebitis, <u>ischemic cardiac events</u>, edema. **GI:** *Diarrhea,* nausea, vomiting, constipation, abdominal pain, anorexia, dry mouth, increased liver function tests (ALT, AST, GGT). **Respiratory:** Dyspnea, cough, pharyngitis, bronchitis, rhinitis, sinusitis. **Other:** Rash, peripheral edema, pelvic pain, flu-like syndrome.

PHARMACOKINETICS Absorption: Rapidly absorbed from GI tract. **Distribution:** 40% protein bound. **Metabolism:** 85% metabolized in liver to inactive metabolites. **Elimination:** 10% excreted unchanged; 60% as metabolites in urine. **Half-Life:** 50 h.

NURSING IMPLICATIONS
Assessment & Drug Effects
- Lab Tests: Monitor periodically liver enzymes, CBC with differential,

Common adverse effects in *italic*, life-threatening effects <u>underlined</u>; generic names in **bold**; classifications in SMALL CAPS; ♣ Canadian drug name; ☻ Prototype drug

103

alkaline phosphatases, total cholesterol, and lipid profile.

- Assess for hypertension, complications of edema, thrombotic events, and signs of liver toxicity.

Patient & Family Education

- Recognize common adverse effects and seek information on measures to control discomfort.
- Seek medical attention if you experience chest pain, calf pain, or shortness of breath; unexplained loss of appetite or nausea; jaundice.

ANIDULAFUNGIN

(a-ni-dul'a-fun-gin)

Eraxis

Classifications: ECHINOCANDIN ANTIFUNGAL

Therapeutic: ANTIFUNGAL

Prototype: Caspofungin

Pregnancy Category: C

AVAILABILITY 50 mg powder for injection

ACTION & *THERAPEUTIC EFFECT*
Anidulafungin is a semisynthetic echinocandin that inhibits glucan synthase, an enzyme present in fungal cells. Glucan is an essential component of the fungal cell wall; therefore, anidulafungin causes fungal cell death. *Interferes with reproduction and growth of susceptible fungi.*

USES Treatment of candidemia and other *Candida* infections caused by *C. albicans*, *C. glabrata*, *C. parapsilosis*, and *C. tropicalis*. Treatment of esophageal candidiasis.
UNLABELED USES Fungal prophylaxis in immunocompromised children who are hospitalized with neutropenia.

CONTRAINDICATIONS Hypersensitivity to anidulafungin or another echinocandin antifungal; lactation.

CAUTIOUS USE Hepatic impairment; pregnancy (category C). Safety and efficacy in children have not been established.

ROUTE & DOSAGE

Candidemia and Other *Candida* Infections

Adult: **IV** 200 mg loading dose on day 1, then 100 mg IV daily for at least 14 days after last positive culture

For the Treatment of Esophageal Candidiasis

Adult: **IV** 100 mg loading dose on day 1, then 50 mg IV daily for at least 14 days (and for at least 7 days after resolution of symptoms)

ADMINISTRATION

Intravenous

***PREPARE:* IV Infusion:** Reconstitute each vial with the supplied single-use 15 mL vial of diluent [20% (w/w) dehydrated alcohol in water] to yield 3.33 mg/mL. ▪ Each 50 mg reconstituted vial **must be** further diluted with NS for D5W to a total infusion volume of 100 mL with a concentration of 0.5 mg/mL. ▪ For a 50 mg dose, dilute 15 mL reconstituted solution in 100 mL of IV fluid to yield 0.43 mg/mL. ▪ For a 100 mg dose, dilute 30 mL reconstituted solution in 250 mL of IV fluid to yield 0.36 mg/mL. ▪ For a 200 mg dose, dilute 60 mL reconstituted solution in 500 mL of IV fluid to yield 0.36 mg/mL.

***ADMINISTER:* IV Infusion:** Give at a rate **no greater** than 1.1 mg/min. DO NOT give a bolus dose.

***INCOMPATIBILITIES* Y-site: Amphotericin B, ertapenem, sodium bicarbonate.**

Common adverse effects in *italic*, life-threatening effects <u>underlined</u>; generic names in **bold**; classifications in SMALL CAPS; ♣ Canadian drug name; ❂ Prototype drug

- Store unreconstituted vials, reconstituted vials, and companion diluent vials at 15°–30° C (59°–86° F). Reconstituted vials **must be** further diluted and administered within 24 h.

ADVERSE EFFECTS (≥1%) **Body as a Whole:** Hypersensitivity. **CNS:** Headache. **GI:** Diarrhea, nausea. **Hematologic:** Neutropenia. **Metabolic:** Increased alkaline phosphatase, increased ALT, increased gamma-glutamyl transferase, hypokalemia. **Skin:** Rash.

INTERACTIONS Drug: Cyclosporin increases overall systemic exposure (i.e., area under the curve, or AUC).

PHARMACOKINETICS Distribution: 84% protein bound. **Metabolism:** Nonhepatic degradation to inactive metabolites. **Elimination:** Fecal. **Half-Life:** 40–50 h.

NURSING IMPLICATIONS

Assessment & Drug Effects

- Prior to initiating therapy with anidulafungin, obtain specimen for fungal culture.
- Monitor for and report S&S of hypersensitivity (e.g., dyspnea, flushing, hypotension, swelling about the face, pruritus, rash, and urticaria), abnormal LFTs, or liver dysfunction (e.g., jaundice, clay-colored stools).
- Discontinue infusion if signs of hypersensitivity appear.
- Monitor cardiac status especially with a preexisting history of dysrhythmias.
- Lab tests: Baseline and periodic liver function tests; periodic CBC with differential and platelet count; periodic serum electrolytes, amylase, and lipase.
- Monitor for S&S hypokalemia and hepatic toxicity (see Appendix F).
- Monitor diabetics for loss of glycemic control.

Patient & Family Education

- Report any of the following immediately if experienced during or shortly after infusion: Difficulty breathing, swelling about the face, itching, rash.
- Report S&S of jaundice to the prescriber: Clay-colored stool, dark urine, yellow skin or sclera, unexplained abdominal pain, or fatigue.

APOMORPHINE HYDROCHLORIDE

(a-po-mor'feen)

Apokyn

Classifications: ANTIPARKINSON; DOPAMINE RECEPTOR AGONIST

Therapeutic: ANTIPARKINSON

Prototype: Levodopa

Pregnancy Category: C

AVAILABILITY 10 mg/mL injection

ACTION & THERAPEUTIC EFFECT
Apomorphine has central dopamine receptor agonist properties. Its use in treatment of Parkinson's is thought to be related to its stimulation of centrally located postsynaptic dopamine D_2-type receptors. *Diminishes hypomobility associated with "off" episodes ("end-of-dose wearing off" and unpredictable "on/off" episodes) in persons with advanced Parkinson's disease.*

USES Rescue of "off" episodes associated with advanced Parkinson's disease.

CONTRAINDICATIONS Hypersensitivity to the drug or its ingredients (i.e., sodium metabisulfite), benzyl alcohol hypersensitivity; renal failure; QT prolongation; heart failure, or shock; depression, suicidal ideation; decreased alertness, seizures, seizure disorder, unconscious state or coma, decreased alertness; lactation.

Common adverse effects in *italic*, life-threatening effects underlined; generic names in **bold**; classifications in SMALL CAPS; ♣ Canadian drug name; ● Prototype drug

105

CAUTIOUS USE Hypersensitivity to sulfites; cardiovascular, cerebrovascular, respiratory, renal, or hepatic disease; CNS depression, history of (chronic) depression or suicidal ideation; hypotension; vomiting; bradycardia; hypokalemia and hypomagnesemia; older adult; pregnancy (category C).

ROUTE & DOSAGE

"Off" Episodes of Parkinson's Disease

Adult: **Subcutaneous** Start with a test dose where BP can be closely monitored. Escalate test dose no sooner than 2 h after last dose until dose is not tolerated or patient has response. If test dose of 0.2 mL (2 mg) is tolerated and has positive response, continue with 0.2 mL (2 mg); if no response, use test dose of 0.4 mL (4 mg); if tolerated and has a positive response, continue with 0.3 mL (3 mg); if 0.4 mL test dose is not tolerated, try 0.3 mL (3 mg) test dose; if 0.3 mL is tolerated, continue with 0.2 mL (2 mg). May increase by 1 mg every few days, generally should not exceed 0.4 mL (4 mg) as an outpatient. Max dose: 0.6 mL as single injection and max 5 injections/ day. If therapy is interrupted for 1 wk, restart with 2 mg dose.

ADMINISTRATION

Subcutaneous

- Aspirate to avoid intravascular injection and ensure the injection is subcutaneous and not intradermal.
- Rotate subcutaneous sites to reduce skin reactions.
- If the patient has not received apomorphine in more than 1 wk, reinstitute it by starting with the initial test dose and titrating to the desired dose.
- Apomorphine causes nausea and vomiting; thus the recommendation is to give 300 mg of trimethobenzamide PO t.i.d., starting 3 days before the first injection and continued for at least the first 2 mo of treatment.
- Store at 15°–30° C (59°–86° F).

ADVERSE EFFECTS (≥1%) **Body as a Whole:** Weakness, yawning, tiredness. **CNS:** CNS depression, dizziness, drowsiness, headache, lightheadedness, euphoria, restlessness, tremor, depression, *dyskinesias, hallucinations.* **CV:** <u>Acute circulatory failure</u>, bradycardia, hypertension, orthostatic hypotension, QT prolongation, vasovagal response, syncope. **GI:** *Nausea, vomiting,* hypersalivation, taste perversions. **Metabolic:** Peripheral edema. **Respiratory:** Respiratory depression, tachypnea, cough, pharyngitis, rhinitis. **Skin:** Contact dermatitis, *bruising,* granuloma, pruritus, sweating. **Urogenital:** *Frequent penile erections,* painful erections.

INTERACTIONS Drug: Alosetron, dolasetron, granisetron, ondansetron, palonosetron may cause severe hypotension and unconsciousness; **alfuzosin, amoxapine, bepridil, chloroquine, clozapine, cyclobenzaprine, droperidol, flecainide, halofantrine, halothane, levomethadyl,** LOCAL ANESTHETICS, MACROLIDES **(clarithromycin, erythromycin, troleandomycin), maprotiline, mefloquine, methadone, pentamidine,** PHENOTHIAZINES, **probucol, gatifloxacin, gemifloxacin, grepafloxacin, levofloxacin, moxifloxacin, sparfloxacin, tacrolimus,** TRICYCLIC ANTIDEPRESSANTS, **amiodarone,**

clozapine, disopyramide, dofetilide, dolasetron, haloperidol, ibutilide, mesoridazine, palonosetron, pimozide, procainamide, quinidine, thioridazine, sotalol, ziprasidone may exacerbate QT_c prolongation; may increase CNS depression with other CNS depressants, including TRICYCLIC ANTIDEPRESSANTS, ANXIOLYTICS, SEDATIVES, HYPNOTICS, **dronabinol,** GENERAL ANESTHETICS, **mirtazapine, nefazodone,** OPIATE AGONISTS, **pramipexole, ropinirole,** SKELETAL MUSCLE RELAXANTS, **tramadol, trazodone. Herbal: Kava** may increase the symptoms of Parkinson's disease.

PHARMACOKINETICS Absorption: Subcutaneous absorption dependent on site utilized—abdominal injection absorbed faster than thigh; lowering the temperature of the injection site slows absorption. **Onset:** 7–14 min. **Peak:** 40–60 min. **Duration:** Up to 2 h. **Distribution:** 85–90% protein bound. **Metabolism:** Metabolized by glucuronidation, sulfation, and N-demethylation. **Elimination:** Excreted by kidneys. **Half-Life:** 30–60 min.

NURSING IMPLICATIONS

Assessment & Drug Effects
- Periodic ECG, especially in those with known CV disease.
- Withhold drug and notify prescriber for S&S of torsades de pointes (i.e., palpitations and syncope), especially in those with bradycardia or suspected hypokalemia or hypomagnesemia.
- Lab tests: Periodic serum electrolytes.
- Monitor closely for orthostatic hypotension, especially when doses are increased, and in patients taking antihypertensive medications and vasodilators (especially nitrates).

- Monitor orthostatic vital signs. Institute fall precautions, especially if orthostatic hypotension occurs.

Patient & Family Education
- Avoid the use of alcohol while taking this drug.
- Report promptly any of the following: Irregular or fast, pounding heartbeat, or palpitations; dizziness, light-headedness, or fainting; unexplained weakness, tiredness, or sleepiness; confusion, hallucinations, or depression; unusual body movements; vomiting; or prolonged painful erections.
- Do not engage in potentially hazardous activities until reaction to drug is known.

APRACLONIDINE

(a-pra-clo′ni-deen)
Iopidine
Pregnancy Category: C
See Appendix A-1.

APREPITANT

(a-pre′pi-tant)

FOSAPREPITANT

(fos-a-pre′pi-tant)
Emend
Classifications: CENTRAL ACTING ANTIEMETIC; SUBSTANCE P/NEUROKININ 1 (NK_1) RECEPTOR ANTAGONIST
Therapeutic: ANTIEMETIC
Pregnancy Category: B

AVAILABILITY 80 mg, 125 mg capsules; 115 mg powder for injection

ACTION & THERAPEUTIC EFFECT
Aprepitant is a selective substance P/neurokinin 1 (NK_1) receptor antagonist. Substance P and the NK-1 receptors are present in areas in the brain that control the emetic

Common adverse effects in *italic*, life-threatening effects underlined; generic names in **bold**; classifications in SMALL CAPS; ♣ Canadian drug name; ❂ Prototype drug

107

reflex. Aprepitant crosses the blood–brain barrier and occupies brain NK_1 receptors. Peripheral blockade by NK_1 receptor antagonists at receptors located in the GI is an additional hypothesized mechanism of action. *Aprepitant augments the antiemetic activity of the 5-HT$_3$-receptor antagonist, ondansetron, and inhibits both the acute and delayed phases of emesis induced by chemotherapy agents.*

USES Prevention of acute and delayed nausea and vomiting associated with emetogenic chemotherapy.

CONTRAINDICATIONS Hypersensitivity to aprepitant; lactation.

CAUTIOUS USE Chemotherapeutic agents metabolized through CYP3A4; severe hepatic impairment; severe renal impairment without dialysis; pregnancy (category B); children younger than 18 y.

ROUTE & DOSAGE

Chemotherapy-Induced Nausea & Vomiting

Adult: **PO** 125 mg 1 h prior to chemotherapy, then 80 mg q a.m. for the next 2 days in conjunction with other antiemetics **IV** 115 mg substituted for oral dose on day 1

ADMINISTRATION

Oral

- Ensure that capsule is swallowed whole with a full glass of water. Do not crush or sprinkle the contents of the capsule.
- Give 1 h before start of chemotherapy.
- Store at 20°–25° C (68°–77° F). Keep the desiccant in the original bottle.

Intravenous (Fosaprepitant)

PREPARE: Infusion: Inject 5 mL NS onto the inside of either a 115 mg or 150 mg vial to prevent foaming. Swirl gently to dissolve. Withdraw contents of the 115 mg vial and add to 110 mL NS or contents of 150 mg vial and add to 145 mL NS. Both dilutions yield 1 mg/mL. Gently invert bag several times to distribute.
ADMINISTER: Infusion: Infuse the 115 mg dose over 15 min and the 150 mg dose over 20–30 min. Longer infusion times may be used if patient complains of burning.

- Reconstituted drug is stable for 24 h at room temperature.

ADVERSE EFFECTS (≥1%) **Body as a Whole:** *Fatigue,* asthenia, malaise, dehydration, fever. **CNS:** Dizziness, insomnia, headache, peripheral neuropathy, sensory neuropathy, anxiety, confusion, depression. **GI:** *Constipation, diarrhea, anorexia, nausea, hiccups,* abdominal pain, gastritis, gastroesophageal reflux, abnormal or impaired taste (dysgeusia), dyspepsia, dysphagia, flatulence, hypersalivation, increased taste disturbance, increased AST and ALT. **Hematologic:** Neutropenia, anemia. **Musculoskeletal:** Pain, myalgia. **Respiratory:** Cough, dyspnea, upper or lower respiratory infection, pneumonitis, respiratory insufficiency. **Special Senses:** Tinnitus.

INTERACTIONS Drug: Increased risk of cardiovascular toxicity with **dofetilide, pimozide;** may decrease **warfarin** concentrations and INR; may decrease levels and effectiveness of ORAL CONTRACEPTIVES; **carbamazepine, griseofulvin, modafinil, rifabutin, rifapentine, phenobarbital, primidone** may decrease antiemetic efficacy; may increase levels of **dexamethasone.** Because aprepitant is a substrate of CYP3A4, many additional drug interactions are theoretically possible. **Food: Grapefruit juice** may decrease effectiveness of aprepitant. **Herbal: St. John's wort** may decrease effectiveness of aprepitant.

PHARMACOKINETICS Absorption: 60–65% of oral dose reaches systemic circulation. **Peak:** 4 h. **Duration:** 95% protein bound; readily crosses the blood–brain barrier. **Metabolism:** In liver by CYP3A4. **Elimination:** Not renally excreted. **Half-Life:** 9–12 h.

NURSING IMPLICATIONS

Assessment & Drug Effects

- Monitor cardiac status especially with preexisting CV disease or concurrent use of any CYP3A4 substrate drug (e.g., ketoconazole, itraconazole, nefazodone, troleandomycin, clarithromycin).
- Lab tests: Monitor PT/INR 7–10 days after 3-day regimen with concurrent warfarin use; monitor phenytoin level with concurrent use; monitor serum electrolytes, UA, and CBC.

Patient & Family Education

- Report immediately to prescriber any of the following: Skin rash; difficulty breathing or shortness of breath; rapid, slow, or irregular heartbeat; changes in BP; dizziness or confusion; unexplained sharp or severe pain in leg or stomach; rectal bleeding. Inform prescriber of all other drugs or herbal products you are using. Do not take new drugs (prescription, OTC, herbal) without first consulting prescriber.

- Use barrier contraception in addition to oral contraceptives while taking drug.

ARGATROBAN

(ar-ga′tro-ban)

Acova, Novastan

Classifications: ANTICOAGULANT; DIRECT THROMBIN INHIBITOR
Therapeutic: ANTITHROMBOTIC; DIRECT THROMBIN INHIBITOR
Prototype: Lepirudin
Pregnancy Category: C

AVAILABILITY 250 mg/2.5 mL vials

ACTION & *THERAPEUTIC EFFECT* Synthetic derivative of arginine that is a direct thrombin inhibitor. Capable of inhibiting the action of both free and clot-bound thrombin. *Reversibly binds to the thrombin active site, thereby blocking clot-forming activity of thrombin.*

USES Prophylaxis or treatment of thrombosis in patients with heparin-induced thrombocytopenia (HIT); prophylaxis or treatment of coronary artery thrombosis during percutaneous coronary interventions (PCI) in patients at risk for HIT.
UNLABELED USES Treatment of disseminated intravascular coagulation (DIC).

CONTRAINDICATIONS Hypersensitivity to argatroban. Any bleeding including intracranial bleeding, GI bleeding, retroperitoneal bleeding; severe hepatic impairment; lactation.
CAUTIOUS USE Diseased states with increased risk of hemorrhaging; severe hypertension; GI ulcerations, hepatic impairment; spinal anesthesia, stroke, surgery, trauma; pregnancy (category C). Safety and effectiveness in

Common adverse effects in *italic*, life-threatening effects underlined; generic names in **bold**; classifications in SMALL CAPS; ♣ Canadian drug name; ❷ Prototype drug

109

children younger than 18 y are not established.

ROUTE & DOSAGE

Prevention & Treatment of Thrombosis

Adult: **IV** 2 mcg/kg/min, may be adjusted to maintain an aPTT of 1.5–3 times baseline (max: 10 mcg/kg/min)

Hepatic Impairment Dosage Adjustment

0.5 mcg/kg/min, may be adjusted to maintain an aPTT of 1.5–3 times baseline (max: 10 mcg/kg/min)

Prophylaxis or Treatment of Coronary Thrombosis during PCI

Adult: **IV** Initiate at 25 mcg/kg/min, then bolus of 350 mcg/kg administered via a large bore IV line over 3–5 min, then 25 mcg/kg/min by continuous infusion; maintain activated clotting time (ACT) 300–450 sec; if ACT below 300 sec, increase infusion to 30 mcg/kg/min; if ACT over 450 sec, decrease infusion to 15 mcg/kg/min

ADMINISTRATION
Intravenous

Note: Argatroban is supplied in 100 mg/mL vials which **must be** diluted 100-fold prior to infusion.

PREPARE: **Continuous:** Dilute each 2.5 mL vial by mixing with 250 mL of D5W, NS, or LR to yield 1 mg/mL. ▪ Mix by repeated inversion of the diluent bag for 1 min.

ADMINISTER: **Continuous for Heparin-Induced Thrombocytopenia (HIT/HITTS):** Before administration, discontinue heparin and obtain a baseline aPTT. ▪ Give at a rate of 2 mcg/kg/min, or as ordered. Lower initial doses are required with hepatic impairment. ▪ Check aPTT 2 h after initiation of therapy. After the initial dose, adjust dose (not to exceed 10 mcg/kg/min) until the steady-state aPTT is 1.5 to 3 times baseline (not to exceed 100 sec). Adjust dose to maintain aPTT at 1.5–3 times baseline, but not greater than 100 sec. ▪ Check aPTT 2 h after initiation of therapy to confirm desired therapeutic range. **Continuous for Percutaneous Coronary Intervention:** Start an infusion at 25 mcg/kg/min and give a bolus of 350 mcg/kg, via a large bore IV line, over 3–5 min. ▪ Check ACT 5–10 min after the bolus dose. If the ACT is greater than 450 sec, decrease infusion rate to 15 mcg/kg/min. If ACT is less than 300 sec, give an additional bolus of 150 mcg/kg and increase infusion to 30 mcg/kg/min. ▪ Check ACT q5–10min to maintain an ACT level 300–450 sec.

▪ Diluted solutions are stable for 24 h at 25° C (77° F) in ambient indoor light. ▪ Protect from direct sunlight. Store solutions refrigerated at 2°–8° C (36°–46° F) in the dark.

ADVERSE EFFECTS (≥1%) **Body as a Whole:** Fever, sepsis, pain, allergic reactions (rare). **CV:** Hypotension, <u>cardiac arrest</u>, ventricular tachycardia. **GI:** Diarrhea, nausea, vomiting, coughing, abdominal pain. **Hematologic:** <u>Major GI bleed</u>, *minor GI bleeding, hematuria, decrease Hgb/Hct,* groin bleed, hemoptysis, brachial bleed. **Respiratory:** Dyspnea. **Urogenital:** UTI.

INTERACTIONS Drug: Heparin results in increased bleeding; may prolong PT with **warfarin;** may increase risk of bleeding with THROMBOLYTICS. **Herbal: Feverfew,**

Common adverse effects in *italic*, life-threatening effects <u>underlined</u>; generic names in **bold;** classifications in SMALL CAPS; ✚ Canadian drug name; ✪ Prototype drug

garlic, ginger, ginkgo may increase potential for bleeding.

PHARMACOKINETICS Peak: 1–3 h. **Distribution:** In extracellular fluid; 54% protein bound. **Metabolism:** In liver by CYP3A4/5. **Elimination:** Primarily in bile (78%). **Half-Life:** 39–51 min.

NURSING IMPLICATIONS

Assessment & Drug Effects

- **Heparin-Induced Thrombocytopenia:** Monitor aPTT. Dose adjustment may be needed to reach the target aPTT. Check aPTT 2 h after initiation of therapy. After the initial dose, adjust dose (not to exceed 10 mcg/kg/min), until the steady-state aPTT is 1.5 to 3 times baseline (not to exceed 100 sec).
- Monitor cardiovascular status carefully during therapy.
- Monitor for and report S&S of bleeding: Ecchymosis, epistaxis, GI bleeding, hematuria, hemoptysis.
- Note: Patients with history of GI ulceration, hypertension, recent trauma, or surgery are at increased risk for bleeding.
- Monitor neurologic status and report immediately focal or generalized deficits.
- Lab tests: Baseline and periodic ACT (activated clotting time), thrombin time (TT), platelet count, Hgb and Hct; daily INR when argatroban and warfarin are co-administered; periodic stool test for occult blood; urinalysis.

Patient & Family Education

- Report immediately any of the following to prescriber: Unexplained back or stomach pain; black, tarry stools; blood in urine; coughing up blood; difficulty breathing; dizziness or fainting spells; heavy menstrual bleeding; nosebleeds; unusual bruising or bleeding at any site.

ARIPIPRAZOLE

(a-rip′-i-pra-zole)

Abilify, Abilify Discmelt
Classifications: ATYPICAL ANTIPSYCHOTIC; DOPAMINE SYSTEM STABILIZER
Therapeutic: ANTIPSYCHOTIC
Prototype: Clozapine
Pregnancy Category: C

AVAILABILITY 2 mg, 5 mg, 10 mg, 15 mg, 20 mg, 30 mg tablets; 10 mg, 15 mg disintegrating tablets; 1 mg/mL oral solution; 9.75 mg/13 mL injection

ACTION & *THERAPEUTIC EFFECT* Efficacy of aripiprazole may be mediated through a combination of partial agonist activity at D_2 and 5-HT_{1A} receptors and antagonist activity at 5-HT_{2A} receptors. *Partial dopaminergic agonist property of aripiprazole accounts for antipsychotic treatment of schizophrenic and bipolar individuals.*

USES Treatment of schizophrenia, bipolar mania, maintenance in bipolar 1 disorder, adjunct treatment in major depressive disorder, irritability associated with autism.

UNLABELED USES Restless leg syndrome, acute psychosis.

CONTRAINDICATIONS Hypersensitivity to aripiprazole; dementia related psychosis in elderly; QT prolongation; lactation.

CAUTIOUS USE History of seizures or conditions that lower seizure threshold (e.g., Alzheimer's dementia); suicidal ideation, depression; brain tumor; dementia; diabetes mellitus; patients with known cardiovascular disease (history of MI or ischemic heart disease, heart failure, or conduction abnormalities), cerebrovascular disease, or conditions that predispose to hypotension (dehydration, hypovolemia, and

Common adverse effects in *italic*, life-threatening effects underlined; generic names in **bold**; classifications in SMALL CAPS; ◆ Canadian drug name; ⓟ Prototype drug

111

treatment with antihypertensive medications); dysphagia; ethanol intoxication; hyperglycemia, hypothermia; obesity, elderly; pregnancy (category C); children younger than 13 y with schizophrenia, children younger than 10 y with bipolar mania, children younger than 6 y with autistic disorder. Safety and efficacy in children with severe depression is not established.

ROUTE & DOSAGE

Schizophrenia
Adult: **PO** 10–15 mg once daily, may increase at 2-wk intervals to max of 30 mg/day if needed
Adolescent/Child (at least 10 y old): **PO** 2 mg daily, increase to 5 mg after 2 days, increase to 10 mg after 2 more days. Can increase up to 30 mg.

Bipolar Mania
Adult: **PO** 15–30 mg once daily
Adolescent/Child (at least 10 y old): **PO** 2 mg daily, increase to 5 mg after 2 days, increase to 10 mg after 2 more days. Can increase up to 30 mg.

Agitation Associated with Schizophrenia/Bipolar
Adult: **IM** 9.75 mg (range: 5.25–15 mg)

Adjunct in Major Depression
Adult: **PO** 2–5 mg daily

Irritability Associated with Autism
Adolescent/Child (over 6 y old): **PO** 2 mg daily, increase as needed

Pharmacogentic Dosage Adjustment
Reduced CYP2D6 expression (i.e., poor metabolizers): Give 70% of normal starting dose

ADMINISTRATION

Oral
- Remove tablet from blister pack immediately before administration.
- Orally disintegrating tablet should be given without water; however, if needed, liquid may be given. Do not split orally disintegrating tablet.
- Note that dose should be reduced by 50% with concurrent treatment with ketoconazole, quinidine, fluoxetine, or paroxetine.

Intramuscular
- Inject slowly and deeply into a large muscle.
- Ensure that drug is not injected intravenously or subcutaneously.
- Store at 15°–30° C (59°–86° F).

ADVERSE EFFECTS (≥1%) **Body as a Whole:** *Headache,* asthenia, fever, flu-like symptoms, peripheral edema, chest pain, neck pain, neck rigidity. **CNS:** *Anxiety, insomnia, light-headedness, somnolence, akathisia,* tremor, extrapyramidal symptoms, depression, nervousness, increased salivation, hostility, suicidal thought, manic reaction, abnormal gait, confusion, cogwheel rigidity. **CV:** Hypertension, tachycardia, hypotension, bradycardia. <u>Risk of stroke in elderly with dementia-related psychosis.</u> **GI:** *Nausea, vomiting, constipation,* anorexia. **Hematologic:** Ecchymosis, anemia. **Metabolic:** Weight gain, weight loss, hyperglycemia, diabetes mellitus, increased creatine kinase. **Musculoskeletal:** Muscle cramp. **Respiratory:** Rhinitis, cough. **Skin:** Rash. **Special Senses:** Blurred vision.

INTERACTIONS Drug: CYP3A4 inducers (**carbamazepine, phenytoin,** etc.) will decrease aripiprazole levels (may need to double aripiprazole dose); use with

Common adverse effects in *italic*, life-threatening effects <u>underlined</u>; generic names in **bold**; classifications in SMALL CAPS; ♣ Canadian drug name; ● Prototype drug

CYP2D6 or CYP3A4 inhibitors (**ketoconazole, quinidine, fluoxetine, paroxetine,** etc.) may increase aripiprazole levels (reduce dose by ½); may cause additive sedation with other SEDATIVES (alcohol, tramadol, BARBITURATES, etc.); may enhance effects of ANTIHYPERTENSIVE AGENTS. **Herbal: St. John's wort** may decrease aripiprazole levels. **Food:** High fat meals may delay time to peak plasma levels.

PHARMACOKINETICS Absorption: 87% bioavailable. **Peak:** 3–5 h. **Metabolism:** In liver by CYP3A4 and 2D6. Major metabolite, has some activity. **Elimination:** 55% in feces, 25% in urine. **Half-Life:** 75 h (94 h for metabolite); 146 h (poor metabolizers).

NURSING IMPLICATIONS

Assessment & Drug Effects

▪ Monitor for and report immediately worsening depression or suicidal ideation, especially in children, aldolescents, and young adults.

▪ Monitor cardiovascular status. Assess for and report orthostatic hypotension. Take BP supine then in sitting position. Report systolic drop of greater than 15–20 mm Hg. Patients at increased risk are those who are dehydrated, hypovolemic, or receiving concurrent antihypertensive therapy.

▪ Monitor body temperature in situations likely to elevate core temperature (e.g., exercising strenuously, exposure to extreme heat, receiving drugs with anticholinergic activity, or being subject to dehydration).

▪ Monitor for and report signs of tardive dyskinesia.

▪ Monitor for and immediately report S&S of neuroleptic malignant syndrome (NMS) (see Appendix F). Withhold drug if NMS is suspected.

▪ Lab tests: Periodic Hct and Hgb, and blood glucose. Monitor for elevated CPK and myoglobinuria

if NMS is suspected. Monitor CBC frequently in patients with history of low WBC or drug-induced low WBC.

▪ Monitor diabetics for loss of glycemic control.

Patient & Family Education

▪ Report promptly deterioration of mental status or behavior (especially suicidal ideation).

▪ Carefully monitor blood glucose levels if diabetic.

▪ Do not drive or engage in other potentially hazardous activities until reaction to drug is known.

▪ Avoid situations where you are likely to become overheated or dehydrated.

▪ Notify prescriber if you become pregnant or intend to become pregnant while taking this drug.

ASCORBIC ACID (VITAMIN C)

Apo-C ♦, Ascorbicap, Cecon, Cetane, Cevalin, CeVi-Sol ♦, Flavorcee, Redoxon ♦, Schiff Effervescent Vitamin C, Vita-C

ASCORBATE, SODIUM

(a-skor′bate)

Cenolate
Classification: VITAMIN
Therapeutic: VITAMIN SUPPLEMENT; URINARY ACIDIFIER
Pregnancy Category: C

AVAILABILITY 25 mg, 50 mg, 100 mg, 250 mg, 500 mg, 1000 mg tablets; 500 mg/mL injection

ACTION & *THERAPEUTIC EFFECT*
Water-soluble vitamin essential for synthesis and maintenance of collagen and intercellular ground substance of body tissue cells, blood vessels, cartilage, bones, teeth, skin, and tendons. Humans are unable to synthesize ascorbic acid in

Common adverse effects in *italic*, life-threatening effects underlined; generic names in **bold**; classifications in SMALL CAPS; ♦ Canadian drug name; ☻ Prototype drug

113

the body therefore, it **must be** consumed daily. *Increases protective mechanism of the immune system, thus supporting wound healing, and resistance to infection.*

USES Prophylaxis and treatment of scurvy and as a dietary supplement.

UNLABELED USES To acidify urine; to prevent and treat cancer; to treat idiopathic methemoglobinemia; as adjuvant during deferoxamine therapy for iron toxicity; in megadoses will possibly reduce severity and duration of common cold. Widely used as an antioxidant in formulations of parenteral tetracycline and other drugs.

CONTRAINDICATIONS Use of sodium ascorbate in patients on sodium restriction; use of calcium ascorbate in patients receiving digitalis.

CAUTIOUS USE Excessive doses in patients with G6PD deficiency; hemochromatosis, thalassemia, sideroblastic anemia, sickle cell anemia; pregnancy (category C); patients prone to gout or renal calculi.

ROUTE & DOSAGE

Therapeutic

Adult: **PO/IV/IM/Subcutaneous** 150–500 mg/day in 1–2 doses
Child: **PO/IV/IM/Subcutaneous** 100–300 mg/day in divided doses

Prophylactic

Adult: **PO/IV/IM/Subcutaneous** 45–60 mg/day
Child: **PO/IV/IM/Subcutaneous** 30–60 mg/day

Urinary Acidifier

Adult: **PO/IV/IM/Subcutaneous** 4–12 g/day in divided doses
Child: **PO/IV/IM/Subcutaneous** 500 mg q6–8h

ADMINISTRATION

Oral

- Give oral solutions mixed with food.
- Dissolve effervescent tablet in a glass of water immediately before ingestion.

Intramuscular, Subcutaneous

- Open ampules with caution. After prolonged storage, decomposition may occur with release of carbon dioxide and resulting increase in pressure within ampule.
- Be aware that ascorbic acid injection may gradually darken on exposure to light; slight coloration reportedly does not affect its therapeutic action.

Intravenous

Verify correct IV concentration and rate of infusion for children with prescriber.

PREPARE: **Direct/Continuous/Intermittent:** Give undiluted or diluted (preferred) in solutions such as NS, D5W, D5/NS, LR. ▪ Be aware that parenteral vitamin C is incompatible with many drugs. ▪ Consult pharmacist for compatibility information.

ADMINISTER: **Direct:** Give undiluted at a rate of 100 mg or a fraction thereof over 1 min. **Continuous/Intermittent (preferred):** Give at ordered rate determined by volume of solution to be infused.

INCOMPATIBILITIES Solution/additive: **Aminophylline, bleomycin, erythromycin, nafcillin, sodium bicarbonate, theophylline.** Y-site: **Etomidate, thiopental.**

- Store in airtight, light-resistant, nonmetallic containers, away from heat and sunlight, preferably at 15°–30° C (59°–86° F), unless otherwise specified by manufacturer.

Common adverse effects in *italic*, life-threatening effects underlined; generic names in **bold**; classifications in SMALL CAPS; ✚ Canadian drug name; ⊙ Prototype drug

ADVERSE EFFECTS (≥1%) **GI:** Nausea, vomiting, heartburn, diarrhea, or abdominal cramps (high doses). **Hematologic:** Acute hemolytic anemia (patients with deficiency of G6PD); sickle cell crisis. **CNS:** Headache or insomnia (high doses). **Urogenital:** Urethritis, dysuria, crystalluria, hyperoxaluria, or hyperuricemia (high doses). **Other:** Mild soreness at injection site; dizziness and temporary faintness with rapid IV administration.

DIAGNOSTIC TEST INTERFERENCE High doses of ascorbic acid can produce false-negative results for *urine glucose* with *glucose oxidase* methods (e.g., *Clinitest, TesTape, Diastix*); false-positive results with *copper reduction methods* (e.g., Benedict's solution, Clinitest); and false increases in *serum uric acid* determinations (by *enzymatic methods*). Interferes with *urinary steroid* (17-OHCS) determinations (by *modified Reddy, Jenkins, Thorn procedure*), decreases in *serum bilirubin,* and may cause increases in *serum cholesterol, creatinine,* and *uric acid* (methodologic inferences). May produce false-negative tests for *occult blood* in stools if taken within 48–72 h of test.

INTERACTIONS Drug: Large doses may attenuate hypoprothrombinemic effects of ORAL ANTICOAGULANTS; SALICYLATES may inhibit ascorbic acid uptake by leukocytes and tissues, and ascorbic acid may decrease elimination of SALICYLATES; chronic high doses of ascorbic acid may diminish the effects of **disulfiram.**

PHARMACOKINETICS Absorption: Readily absorbed PO; however, absorption may be limited with large doses. **Distribution:** Widely distributed to body tissues; crosses placenta; distributed into breast milk. **Metabolism:** In liver. **Elimination:** Rapidly in urine when plasma level exceeds renal threshold of 1.4 mg/dL.

NURSING IMPLICATIONS

Assessment & Drug Effects
- Lab tests: Periodic Hct and Hgb, serum electrolytes.
- Monitor for S&S of acute hemolytic anemia, sickle cell crisis.

Patient & Family Education
- High doses of vitamin C are not recommended during pregnancy.
- Take large doses of vitamin C in divided amounts because the body uses only what is needed at a particular time and excretes the rest in urine.
- Megadoses can interfere with absorption of vitamin B_{12}.
- Note: Vitamin C increases the absorption of iron when taken at the same time as iron-rich foods.

ASENAPINE

(a-sin′a-peen)

Saphris

Classification: ATYPICAL ANTIPSYCHOTIC; SEROTONIN ANTAGONIST; ANTIMANIC; ANTIDEPRESSANT
Therapeutic: ANTIPSYCHOTIC; ANTIMANIC; ANTIDEPRESSANT
Prototype: Clozapine
Pregnancy Category: C

AVAILABILITY 5 mg, 10 mg sublingual tablets

ACTION & THERAPEUTIC EFFECT Mechanism of action thought to be related to antagonism to certain CNS dopamine (D_2) and serotonin ($5-HT_{2A}$) receptors. *Effect on serotonin and dopamine receptors may account for activity of asenapine against the negative symptoms of psychotic disorders.*

Common adverse effects in *italic*, life-threatening effects underlined; generic names in **bold**; classifications in SMALL CAPS; ♣ Canadian drug name; ⊙ Prototype drug

115

USES Acute treatment of schizophrenia; acute treatment of manic or mixed episodes associated with bipolar disorder (bipolar I disorder) with or without psychotic features.

CONTRAINDICATIONS Dementia-related psychosis; ketoacidosis; severe neutropenia (ANC less than 1000/mm³); patients with history of torsades de pointes related to drugs; suicidal ideation; severe hepatic impairment (Child-Pugh Class C).

CAUTIOUS USE Tardive dyskinesia, cardiovascular disease (history of MI, ischemic heart disease, HF, conduction abnormalities, bradycardia), cerebrovascular disease, dehydration, hypovolemia, diabetes mellitus; history of seizures; Alzheimer's dementia; older adults; history of suicidal tendencies; pregnancy (category C); lactation. Safety and effectiveness in pediatric patients have not been established.

ROUTE & DOSAGE

Schizophrenia

Adult: **SL** 5 mg b.i.d.

Bipolar Disorder

Adult: **SL** Start at 10 mg b.i.d., may decrease to 5 mg b.i.d. if higher dose not tolerated.

ADMINISTRATION

Sublingual

- Tablet must be placed under tongue and allowed to dissolve completely.
- Ensure that tablet is not chewed or swallowed.
- Eating and drinking should be avoided for 10 min after administration.
- Store at 15°–30° C (59°–86° F).

ADVERSE EFFECTS (≥1%) **Body as a Whole:** Fatigue, irritability, pain in extremities. **CNS:** Agitation, akathisia, anxiety, *dizziness*, dysgeusia, *extrapyramidal symptoms, headache, insomnia, somnolence.* **CV:** Hypertension. **GI:** Constipation, dyspepsia, nausea, vomiting. **Metabolic:** Increased appetite, increased weight. **Musculoskeletal:** Arthralgia. **Respiratory:** Dry mouth, oral hypoesthesia, salivary hypersecretion.

INTERACTIONS Drug: Fluvoxamine and **imipramine** can increase asenapine levels.

PHARMACOKINETICS Absorption: Bioavailability 35%. **Peak:** 0.5–1.5 h. **Distribution:** 95% plasma protein bound. **Metabolism:** In liver via CYP1A2. **Elimination:** 50% renal; 40% fecal. **Half-Life:** 24 h.

NURSING IMPLICATIONS

Assessment & Drug Effects

- Monitor BP, HR, and weight. Monitor orthostatic vital signs with concurrent antihypertensive therapy or any condition that predisposes to hypotension (e.g., advanced age, dehydration).
- Monitor for orthostatic hypotension and syncope, especially early in therapy.
- Monitor diabetics or those at risk for diabetes for loss of glycemic control.
- Withhold drug and report promptly S&S of neuroleptic malignant syndrome (see Appendix F).
- Lab tests: Baseline and periodic CBC.

Patient & Family Education

- Be alert for and report worsening of condition, including ideas of suicide.
- Make position changes slowly, especially from lying or sitting to a standing position.
- If diabetic, monitor blood sugar closely for loss of control.
- Stop taking the drug and report immediately any of the following:

High fever, muscle rigidity, altered mental status, or palpitations.
- Avoid engaging in hazardous activities until response to drug is known.
- Avoid alcohol while taking this drug.

ASPARAGINASE
(a-spar′a-gi-nase)
Colaspase, Elspar, Kidrolase A, L-asparaginase
Classifications: ANTINEOPLASTIC; ENZYME; ANTIMETABOLITE
Therapeutic: ANTINEOPLASTIC
Pregnancy Category: C

AVAILABILITY 10,000 international unit vial

ACTION & THERAPEUTIC EFFECT
A highly toxic drug with a low therapeutic index. Catalyzes asparagine to aspartic acid and ammonia, thus depleting extracellular supply of an amino acid essential to synthesis of DNA and other nucleoproteins. *Reduced availability of asparagine causes death of tumor cells, since unlike normal cells, tumor cells are unable to synthesize their own supply. Resistance to cytotoxic action develops rapidly, and it is not an effective treatment for solid tumors.*

USES Primarily in combination regimens with other antineoplastic agents to treat acute lymphocytic leukemia (ALL).
UNLABELED USES Other leukemias, lymphosarcoma, and (intraarterially) treatment of hypoglycemia due to pancreatic islet cell tumor.

CONTRAINDICATIONS Hypersensitivity to *Escherichia coli* protein; history of or existing pancreatitis; chickenpox (existing or recent illness or exposure), herpetic infection; lactation.

CAUTIOUS USE Liver impairment; diabetes mellitus; anticoagulation therapy, coagulopathy; infections; history of urate calculi or gout; antineoplastic or radiation therapy; pregnancy (category C).

ROUTE & DOSAGE

Induction Agent
Adult: **IV** (sole agent) 200 international units/kg/day for 28 days, inject over at least 30 min into running IV OR 1000 international units/kg/day for days 22–32 (along with prednisone and vincristine)
Child: **IV** 1000 units/kg/day × 10 days starting day 22 (along with prednisone and vincristine)

Desensitization Protocol
Adult/Child: **IV** Schedule begins with 1 international unit, then double the dose q10min until the accumulated total matches the planned dose. See package insert for detailed dosing.

ADMINISTRATION

Intravenous
An **intradermal skin test** is usually performed prior to initial dose and when drug is readministered after an interval of a week or more; allergic reactions are unpredictable.

- Observe test site for at least 1 h for evidence of positive reaction (wheal, erythema). A negative skin test, however, does not preclude possibility of an allergic reaction.
- Administer test dose and IV infusion under constant supervision by clinician experienced in cancer chemotherapy.
- Use only clear solutions.

Common adverse effects in *italic*, life-threatening effects underlined; generic names in **bold**; classifications in SMALL CAPS; ♦ Canadian drug name; ❂ Prototype drug

117

***PREPARE:* IV Infusion:** Reconstitute with sterile water or with 0.9 % NaCl. ▪ Each 10,000 international unit vial is diluted with 5 mL of diluent to yield 2000 international units/mL. ▪ Shake vial well to promote dissolution of powder. Avoid vigorous shaking. Ordinary shaking does not inactivate the enzyme or cause foaming of content.

***ADMINISTER:* IV Infusion:** Further dilute reconstituted solution with NS or D5W by administration into tubing of an already free flowing infusion of one of these solutions. ▪ Give over a period of not less than 30 min. ▪ Use a 5-micron filter to remove gelatinous fiber-like particles that can develop in solutions on standing.

▪ Store sealed vial of lyophilized powder below 8° C (46° F) unless otherwise directed by manufacturer. Store reconstituted solutions and solutions diluted for IV infusion at 2°–8° C (36°–46° F) for up to 8 h; then discard.

ADVERSE EFFECTS (≥1%) Body as a Whole: Hypersensitivity (*Skin rashes, urticaria,* respiratory distress, anaphylaxis), chills, fever, fatal hyperthermia, perspiration, weight loss. **CNS:** Depression, fatigue, lethargy, drowsiness, confusion, agitation, hallucinations, dizziness, Parkinson-like syndrome with tremor and progressive increase in muscle tone. **GI:** *Severe vomiting, nausea,* anorexia, abdominal cramps, diarrhea, acute pancreatitis, liver function abnormalities. **Urogenital:** Uric acid nephropathy, azotemia, proteinuria, renal failure. **Hematologic:** *Reduced clotting factors* (especially V, VII, VIII, IX), *decreased circulating platelets and fibrinogen,* leukopenia. **Metabolic:** Hyperglycemia, glycosuria, polyuria, hypoalbuminemia, hypocalcemia, hyperuricemia. **Other:** Flank pain, infections.

DIAGNOSTIC TEST INTERFERENCE
Asparaginase may interfere with ***thyroid function*** tests: Decreased total ***serum thyroxine*** and increased ***thyroxine-binding globulin index;*** pretreatment values return within 4 wk after drug is discontinued.

INTERACTIONS Drug: Decreased hypoglycemic effects of SULFONYLUREAS, **insulin;** increased potential for toxicity if asparaginase is given concurrently or immediately before CORTICOSTEROIDS, **vincristine;** blocks antitumor effect of **methotrexate** if given concurrently or immediately before it.

PHARMACOKINETICS Distribution: Into intravascular space (80%) and lymph; low levels in CSF, pleural and peritoneal fluids. **Elimination:** Small amounts found in urine. **Half-Life:** 8–30 h.

NURSING IMPLICATIONS

Assessment & Drug Effects

▪ Have immediately available: Personnel, drugs, and equipment for treating allergic reaction (which may range from urticaria to anaphylactic shock) whenever drug is administered, including skin testing.

▪ Monitor for S&S of hypersensitivity or anaphylactoid reaction (see Appendix F) during drug administration. Anaphylaxis usually occurs within 30–60 min after dose has been given and is more likely with intermittent administrations, particularly at intervals of 7 days or more.

▪ Monitor I&O and maintain adequate fluid intake.

▪ Evaluate CNS function (general behavior, emotional status, level

Common adverse effects in *italic*, life-threatening effects underlined; generic names in **bold**; classifications in SMALL CAPS; ♣ Canadian drug name; ⊘ Prototype drug

of consciousness, thought content, motor function) before and during therapy.

- Note: Toxicity potential is increased when giving drug immediately before a course of prednisone and vincristine; toxicity appears less when given after these drugs.
- Lab tests: Periodic serum amylase, serum calcium, blood glucose, coagulation factors, ammonia and uric acid levels, renal function tests, peripheral blood counts, and bone marrow function; LFTs at least twice weekly during therapy.
- Monitor diabetics for loss of glycemic control.
- Monitor for and report S&S of hyperammonemia: Anorexia, vomiting, lethargy, weak pulse, depressed temperature, irritability, asterixis, seizures, coma.
- Anticipate possible prolonged or exaggerated effects of concurrently given drugs or their toxicity because of potential serious hepatic dysfunction that reduces metabolism of other drugs. Report incidence promptly.
- Watch for neurotoxic reaction (25% of patients) which usually appears within the first few days of therapy. It is manifested by tiredness and changing levels of consciousness (ranging from confusion to coma).
- Note: Protect from infection during first several days of treatment when circulating lymphoblasts decrease markedly and leukocyte counts may fall below normal. Report promptly S&S of infection: Chill, fever, aches, sore throat.
- Report sudden severe abdominal pain with nausea and vomiting, particularly if these symptoms occur after medication is discontinued (may indicate pancreatitis).

Patient & Family Education

- Note: Therapeutic response will most likely be accompanied by some toxicity in all patients; toxicity is reportedly greater in adults than in children.
- Notify prescriber of continued loss of weight or onset of foot and ankle swelling.
- Notify prescriber without delay if nausea or vomiting makes it difficult to take all prescribed medication.
- Report onset of unusual bleeding, bruising, petechiae, melena, skin rash or itching, yellowed skin and sclera, joint pain, puffy face, or dyspnea.
- Do not drive or operate equipment that requires alertness and skill. Exercise caution with potentially hazardous activities. These effects can continue several weeks after last dose of the drug.

ASPIRIN (ACETYLSALICYLIC ACID) ⊕
(as'pe-ren)

Alka-Seltzer, A.S.A., Aspergum, Astrin ♣, Bayer, Bayer Children's, Cosprin, Easprin, Ecotrin, Empirin, Entrophen ♣, Halfprin, Measurin, Novasen ♣, St. Joseph Children's, Supasa ♣, Triaphen-10 ♣, ZORprin
Classifications: NONNARCOTIC ANALGESIC, SALICYLATE; ANTIPYRETIC; ANTIPLATELET
Therapeutic: ANALGESIC; ANTIPYRETIC; ANTIPLATELET
Pregnancy Category: C in first and second trimester and D in third trimester

AVAILABILITY 81 mg chewable tablets; 325 mg, 500 mg tablets; 81 mg, 165 mg, 325 mg, 500 mg, 650 mg, 975 mg enteric-coated tablets; 650 mg, 800 mg sustained release

Common adverse effects in *italic*, life-threatening effects underlined; generic names in **bold**, classifications in SMALL CAPS; ♣ Canadian drug name; ⊕ Prototype drug

119

tablets; 120 mg, 200 mg, 300 mg, 600 mg suppositories

ACTION & *THERAPEUTIC EFFECT*

Major action is primarily due to inhibiting the formation of prostaglandins involved in the production of inflammation, pain, and fever. **Anti-inflammatory action:** Inhibits prostaglandin synthesis. As an anti-inflammatory agent, aspirin appears to be involved in enhancing antigen removal and in reducing the spread of inflammatory substances. **Analgesic action:** Principally peripheral with limited action in the CNS in the hypothalamus; results in relief of mild to moderate pain. **Antipyretic action:** Suppress the synthesis of prostaglandin in or near the hypothalamus. Aspirin also lowers body temperature in fever by indirectly causing centrally mediated peripheral vasodilation and sweating. **Antiplatelet action:** Aspirin (but not other salicylates) powerfully inhibits platelet aggregation. High serum salicylate concentrations can impair hepatic synthesis of blood coagulation factors VII, IX, and X. *Reduces inflammation, pain, and fever. Also inhibits platelet aggregation, reducing ability of blood to clot.*

USES To relieve pain of low to moderate intensity. Also for various inflammatory conditions, such as acute rheumatic fever, systemic lupus, rheumatoid arthritis, osteoarthritis, bursitis, and calcific tendonitis, and to reduce fever in selected febrile conditions. Used to reduce recurrence of TIA due to fibrin platelet emboli and risk of stroke in men; to prevent recurrence of MI; as prophylaxis against MI in men with unstable angina.

UNLABELED USES As prophylactic against thromboembolism; to prevent cataract and progression of diabetic retinopathy; and to control symptoms related to gluten sensitivity.

CONTRAINDICATIONS History of hypersensitivity to salicylates including methyl salicylate (oil of wintergreen); sensitivity to other NSAIDs; patients with "aspirin triad" (aspirin sensitivity, nasal polyps, asthma); chronic rhinitis; acute bronchospasm; agranulocytosis; head trauma; increased intracranial pressure; intracranial bleeding; chronic urticaria; history of GI ulceration, bleeding, or other problems; hypoprothrombinemia, vitamin K deficiency, hemophilia, or other bleeding disorders; CHF; pregnancy (category D, especially in third trimester); lactation.

CAUTIOUS USE Otic diseases; gout; children with fever accompanied by dehydration; hyperthyroidism; immunosuppressed individuals; asthma; GI disease; history of gout; cardiac disease; renal or hepatic impairment; G6PD deficiency; vitamin K deficiency; anemia; preoperatively; Hodgkin's disease pregnancy (category C in first and second trimester). Prematures, neonates, or children younger than 2 y, except under advice and supervision of prescriber; children or teenagers with chickenpox or influenza-like illnesses because of possible association with Reye's syndrome.

ROUTE & DOSAGE

Mild to Moderate Pain, Fever

Adult: **PO/PR** 350–650 mg q4h (max: 4 g/day)
Child: **PO/PR** 10–15 mg/kg in 4–6 h (max: 3.6 g/day)

Arthritic Conditions

Adult: **PO** 3.6–5.4 g/day in 4–6 divided doses

Child: **PO** 80–100 mg/kg/day in 4–6 divided doses (max: 130 mg/kg/day)

Thromboembolic Disorders

Adult: **PO** 81–325 mg daily

TIA Prophylaxis

Adult: **PO** 650 mg b.i.d.

MI Prophylaxis

Adult: **PO** 80–325 mg/day

ADMINISTRATION

Oral

- Give with a full glass of water (240 mL), milk, food, or antacid to minimize gastric irritation.
- Do not give to children or adolescents with chickenpox or flu-like symptoms.
- Enteric-coated tablets dissolve too quickly if administered with milk and should not be crushed or chewed.

Rectal

- Ensure that suppository is inserted beyond the internal sphincter.
- Store at 15°–30° C (59°–86° F) in airtight container and dry environment unless otherwise directed by manufacturer. Store suppositories in a cool place or refrigerate but do not freeze.

ADVERSE EFFECTS (≥1%) **Body as a Whole:** Hypersensitivity (urticaria, <u>bronchospasm, anaphylactic shock (laryngeal edema)</u>). **CNS:** Dizziness, confusion, drowsiness. **Special Senses:** Tinnitus, hearing loss. **GI:** *Nausea*, vomiting, diarrhea, anorexia, *heartburn, stomach pains,* ulceration, occult bleeding, GI bleeding. **Hematologic:** <u>Thrombocytopenia, hemolytic anemia,</u> prolonged bleeding time. **Skin:** Petechiae,

easy bruising, rash. **Urogenital:** Impaired renal function. **Other:** Prolonged pregnancy and labor with increased bleeding.

DIAGNOSTIC TEST INTERFERENCE Bleeding time is prolonged 3–8 days (life of exposed platelets) following a single 325-mg (5 grains) dose of aspirin. Large doses of salicylates equivalent to 5 g or more of aspirin per day may cause prolonged *prothrombin time* by decreasing prothrombin production; interference with *pregnancy tests* (using mouse or rabbit); decreases in *serum cholesterol, potassium, PBI, T_3 and T_4 concentrations,* and an increase in *T_3 resin uptake. Serum uric acid* may increase when plasma salicylate levels are below 10 and decrease when above 15 mg/dL using colorimetric methods. *Urine 5-HIAA:* Aspirin may interfere with tests using fluorescent methods. *Urine ketones:* Salicylates interfere with Gerhardt test (reaction with ferric chloride produces a reddish color that persists after boiling). *Urine glucose:* Moderate to large doses of salicylates equivalent to an aspirin dosage 2.4 g/day or more may produce false-negative results with glucose oxidase methods (e.g., *Clinistix, TesTape*) and false-positive results with copper reduction methods *(Benedict's solution, Clinitest). Urinary PSP excretion* may be reduced by salicylates. Salicylates may cause *urine VMA* to be falsely elevated (by most tests), or reduced (by Pisano method). Salicylates may interfere with or cause false decreases in plasma theophylline levels using Schack and Waxler method. High plasma salicylate levels may cause abnormalities in *liver function tests.*

Common adverse effects in *italic*, life-threatening effects <u>underlined</u>; generic names in **bold**; classifications in SMALL CAPS; ♣ Canadian drug name; ◎ Prototype drug

121

INTERACTIONS Drug: Aminosalicylic acid increases risk of SALICYLATE toxicity. **Ammonium chloride** and other ACIDIFYING AGENTS decrease renal elimination and increase risk of SALICYLATE toxicity. ANTICOAGULANTS increase risk of bleeding. ORAL HYPOGLYCEMIC AGENTS increase hypoglycemic activity with aspirin doses greater than 2 g/day. CARBONIC ANHYDRASE INHIBITORS enhance SALICYLATE toxicity. CORTICOSTEROIDS add to ulcerogenic effects. **Methotrexate** toxicity is increased. Low doses of SALICYLATES may antagonize uricosuric effects of **probenecid** and **sulfinpyrazone. Herbal:** Feverfew, garlic, ginger, ginkgo, evening primrose oil may increase bleeding potential.

PHARMACOKINETICS Absorption: 80–100% absorbed (depending on formulation), primarily in stomach and upper small intestine. **Peak levels:** 15 min to 2 h. **Distribution:** Widely distributed in most body tissues; crosses placenta. **Metabolism:** Aspirin is hydrolyzed to salicylate in GI mucosa, plasma, and erythrocytes; salicylate is metabolized in liver. **Elimination:** 50% of dose is eliminated in the urine in 2–4 h (low doses) or 15–30 h (high doses). Excreted into breast milk. **Half-Life:** Aspirin 15–20 min; salicylate 2–18 h (dose dependent).

NURSING IMPLICATIONS

Assessment & Drug Effects

- Monitor for loss of tolerance to aspirin. Symptoms usually occur 15 min to 3 h after ingestion: Profuse rhinorrhea, erythema, nausea, vomiting, intestinal cramps, diarrhea.
- Monitor closely the diabetic child for need to adjust insulin dose. Children on high doses of aspirin are particularly prone to hypoglycemia (see Appendix F).
- Monitor for salicylate toxicity. In adults, a sensation of fullness in the ears, tinnitus, and decreased or muffled hearing are the most frequent symptoms.
- Monitor children for S&S of salicylate toxicity manifested by: hyperventilation, agitation, mental confusion, or other behavioral changes, drowsiness, lethargy, sweating, and constipation.

Patient & Family Education

- Use enteric-coated tablets, extended release tablets, buffered aspirin, or aspirin administered with an antacid to reduce GI disturbances.
- Discontinue aspirin use with onset of ringing or buzzing in the ears, impaired hearing, dizziness, GI discomfort or bleeding, and report to prescriber.
- Do not use aspirin for self-medication of pain (adults) beyond 5 days without consulting a prescriber. Do not use aspirin longer than 3 days for fever (adults and children), never for fever over 38.9° C (102° F) in older adults or 39.5° C (103° F) in children and adults under 60 y or for recurrent fever without medical direction.
- Consult prescriber before using aspirin for any fever accompanied by rash, severe headache, stiff neck, marked irritability, or confusion (all possible symptoms of meningitis).
- Avoid alcohol when taking large doses of aspirin.
- Observe and report signs of bleeding (e.g., petechiae, ecchymoses, bleeding gums, bloody or black stools, cloudy or bloody urine).
- Maintain adequate fluid intake when taking repeated doses of aspirin.

Common adverse effects in *italic*, life-threatening effects <u>underlined</u>; generic names in **bold**; classifications in SMALL CAPS; ✤ Canadian drug name; ⊙ Prototype drug

ATAZANAVIR

(a-ta-zan'a-vir)

Reyataz
Classifications: ANTIRETROVIRAL;
PROTEASE INHIBITOR
Therapeutic: PROTEASE INHIBITOR
Prototype: Saquinavir
Pregnancy Category: B

AVAILABILITY 100 mg, 150 mg, 200 mg, 300 mg capsules

ACTION & *THERAPEUTIC EFFECT*

Atazanavir is an HIV-1 protease inhibitor that selectively inhibits the replication of HIV. Protease plays a major role in virus-specific processing of gene products used in the replication enzymes of HIV-1 infected cells. Thus, protease is necessary for the production of mature viruses. Atazanavir reduces the viral load and increases CD4+ cell count. *Protease inhibition renders the virus noninfectious. Because HIV protease inhibitors inhibit the HIV replication cycle midway in the process, they are active in acutely and chronically infected cells.*

USES Treatment of HIV infection in combination with other antiretroviral agents.

CONTRAINDICATIONS Hypersensitivity to atazanavir; severe hepatic insufficiency; lactase deficiency; severe rash; lactation.

CAUTIOUS USE Mild to moderate hepatic impairment, hepatitis B or C; females, diabetes mellitus, diabetic ketoacidosis; hemophilia, hepatic disease; hepatitis; jaundice, hypercholesterolemia, hypertriglyceridemia; preexisting conduction system disease (e.g. marked first-degree AV block or second- or third-degree AV block; lactic acidosis, obesity;

older adults; pregnancy (category B); infants and children younger than 6 y.

ROUTE & DOSAGE

HIV Infection

Adult/Adolescent: Child (older than 6 y and weight greater than 39 kg): **PO** 400 mg once/day with a light meal OR 300 mg once/day with 100 mg of ritonavir
Adolescent/Child (at least 6 y and weight 32–39 kg): **PO** 250 mg plus ritonavir (100 mg) once daily
Adolescent/Child (at least 6 y and weight 25–32 kg): **PO** 200 mg plus ritonavir (100 mg) once daily (i.e., 7 mg/kg plus ritonavir 4 mg/kg)
Adolescent/Child (6 y or older and weight 20 to less than 25 kg): **PO** 150 mg plus ritonavir (80 mg)
Adolescent/Child (6 y or older and weight 15–20 kg): **PO** 150 mg plus ritonavir (80 mg) once daily

Hepatic Impairment Dosage Adjustment

Moderate impairment (Child-Pugh class B): Reduce dose to 300 mg once a day. *Severe impairment:* Not recommended for use.

ADMINISTRATION

Oral

- Give with a light meal, not on an empty stomach.
- When co-administered with didanosine buffered formulations, give atazanavir (with food) 2 h before or 1 h after didanosine.
- Give 2 h before/1 h after antacids or buffered drugs.
- Store at 15°–30° C (59°–86° F).

Common adverse effects in *italic*, life-threatening effects underlined; generic names in **bold**; classifications in SMALL CAPS; ♣ Canadian drug name; ⊙ Prototype drug

123

ADVERSE EFFECTS (≥1%) **Body as a Whole:** *Peripheral neuropathy,* fever, pain, fatigue, allergic reaction, angioedema, asthenia, burning sensation, chest pain, edema, facial atrophy, generalized edema, heat sensitivity, infection, malaise, pallor, peripheral edema, photosensitivity, substernal chest pain, sweating. **CNS:** *Headache,* depression, insomnia, dizziness, abnormal dream, abnormal gait, agitation, amnesia, anxiety, confusion, convulsion, decreased libido, emotional lability, hallucination, hostility, hyperkinesia, hypesthesia, increased reflexes, nervousness, psychosis, sleep disorder, somnolence, <u>suicide attempt</u>, twitch. **CV:** <u>Cardiac arrest</u>, heart block (PR prolongation), hypertension, myocarditis, palpitation, syncope, vasodilatation. **GI:** Hyperbilirubinemia, jaundice, *nausea, vomiting, diarrhea,* abdominal pain, anorexia, aphthous stomatitis, colitis, constipation, dental pain, dyspepsia, enlarged abdomen, esophageal ulcer, esophagitis, flatulence, gastritis, gastroenteritis, gastrointestinal disorder, hepatitis, hepatomegaly, hepatosplenomegaly, increased appetite, liver damage, liver fatty deposit, mouth ulcer, pancreatitis, peptic ulcer. **Endocrine:** Decreased male fertility. **Hematologic:** Ecchymosis, purpura. **Metabolic:** Lipodystrophy syndrome, hypercholesterolemia, hypertriglyceridemia. **Musculoskeletal:** Myalgia, arthralgia. **Respiratory:** Cough, dyspnea, hiccup. **Skin:** *Rash,* alopecia, cellulitis, dermatophytosis, dry skin, eczema, nail disorder, pruritus, seborrhea, urticaria, vesiculobullous rash. **Special Senses:** Otitis, taste perversion, tinnitus. **Urogenital:** Abnormal urine, amenorrhea, crystalluria, gynecomastia, hematuria, impotence, kidney calculus, kidney failure, kidney pain, menstrual disorder, oliguria, pelvic pain, polyuria, proteinuria, urinary frequency, urinary tract infection.

INTERACTIONS Drug: There are extensive drug interactions reported; check the package insert for complete listing. May increase levels and toxicity of **cyclosporine,** systemic **lidocaine, sirolimus, tacrolimus, amiodarone, alfuzosin, dronedarone, eplerenone;** increase risk of myopathy and rhabdomyolysis with **atorvastatin, lovastatin, simvastatin;** may increase risk of heart block with **diltiazem;** ANTACIDS, H₂-RECEPTOR ANTAGONISTS, PROTON PUMP INHIBITORS may decrease absorption of atazanavir; may increase toxicity of **irinotecan;** increased risk of prolonged sedations with BENZODIAZEPINES; **indinavir** may increase risk of hyperbilirubinemia; **didanosine, efavirenz, rifampin** may decrease atazanavir levels; **ergotamine, ergonovine dihydroergotamine, bepridil, pimozide** may cause serious adverse reactions; may increase risk of hypotension, visual changes, and priapism with **sildenafil, tadalafil, vardenafil.** Avoid use of **sildenafil, alfuzosin, salmeterol** and **bosentan. Herbal: St. John's wort, red yeast rice** may decrease atazanavir levels.

PHARMACOKINETICS Absorption: 68% into systemic circulation; taking with food enhances bioavailability. **Peak:** 2–2.5 h. **Metabolism:** In liver by CYP3A4. **Elimination:** 70% in feces, 13% in urine. **Half-Life:** 7 h.

NURSING IMPLICATIONS

Assessment & Drug Effects

- Monitor CV status and ECG closely, especially with concurrent treatment with other drugs known to prolong the PR interval.
- Monitor for and report promptly S&S of lactic acidosis (see metabolic acidosis, Appendix F).

Common adverse effects in *italic,* life-threatening effects <u>underlined</u>; generic names in **bold;** classifications in SMALL CAPS; ♣ Canadian drug name; ◑ Prototype drug

- Monitor neonates and infants exposed to atazanavir in utero for severe hyperbilirubinemia during the first few days of life.
- Lab tests: Baseline and periodic LFTs; total bilirubin if jaundiced; periodic PT/INR with concurrent warfarin therapy; monitor blood glucose closely, especially if diabetic. Monitor CD4+ cell count and HIV RNA viral load.

Patient & Family Education

- Do not alter the dose or discontinue therapy without consulting prescriber.
- Inform prescriber of all prescription, nonprescription, or herbal meds being used.
- Report promptly any of the following: Dizziness or light-headedness; muscle pain (especially with concurrent statin therapy); severe nausea, vomiting (especially if red or "coffee-ground" in appearance), stomach pain, black tarry stools; yellowing of skin or whites of eyes; skin rash or itchy skin; sore throat, fever, or other S&S of infection; unexplained tiredness or weakness.
- If taking both sildenafil and atazanavir, promptly report any of the following sildenafil-associated adverse effects: Hypotension, visual changes, or prolonged penile erection.

ATENOLOL

(a-ten'oh-lole)

Apo-Atenolol ✦, Tenormin
Classifications: BETA-ADRENERGIC ANTAGONIST; ANTIHYPERTENSIVE
Therapeutic: ANTIHYPERTENSIVE; ANTIANGINAL
Prototype: Propranolol
Pregnancy Category: D

AVAILABILITY 25 mg, 50 mg, 100 mg tablets

ACTION & THERAPEUTIC EFFECT Atenolol selectively blocks beta$_1$-adrenergic receptors located chiefly in cardiac muscle. Mechanisms for antihypertensive action include central effect leading to decreased sympathetic outflow to periphery, reduction in renin activity with consequent suppression of the renin-angiotensin-aldosterone system, and competitive inhibition of catecholamine binding at beta-adrenergic receptor sites. *Reduces rate and force of cardiac contractions (negative inotropic action); cardiac output is reduced as well as systolic and diastolic BP. Atenolol decreases peripheral vascular resistance both at rest and with exercise.*

USES Management of hypertension as a single agent or concomitantly with other antihypertensive agents, especially a diuretic, and in treatment of stable angina pectoris, MI.

UNLABELED USES Antiarrhythmic, mitral valve prolapse, adjunct in treatment of pheochromocytoma and of thyrotoxicosis, lithium-induced tremor, and for vascular headache prophylaxis.

CONTRAINDICATIONS Sinus bradycardia, greater than first-degree heart block, uncompensated heart failure, cardiogenic shock, abrupt discontinuation, untreated phenochrormocytoma; pregnancy (category D); lactation.

CAUTIOUS USE Hypertensive patients with CHF controlled by digitalis and diuretics, vasospastic angina (Prinzmetal's angina); bronchospastic disease; asthma, bronchitis, emphysema, and COPD; major depression; diabetes mellitus; impaired renal function, dialysis; myasthenia gravis; pheochromocytoma, hyperthyroidism, thyrotoxicosis; older adults.

Common adverse effects in *italic*, life-threatening effects underlined; generic names in **bold**; classifications in SMALL CAPS; ✦ Canadian drug name; ✪ Prototype drug

125

ROUTE & DOSAGE

Hypertension, Angina

Adult: **PO** 25–50 mg/day, may increase to 100 mg/day
Child: **PO** 0.8–1.5 mg/kg/day (max: 2 mg/kg/day)

MI

Adult: **PO** Start 50 mg/day

Renal Impairment Dosage Adjustment

CrCl 15–35 mL/min: Max dose of 50 mg/day; less than 15 mL/min: Max dose of 25 mg/day

ADMINISTRATION

Oral

- Crush tablets, if necessary, before administration and give with fluid of patient's choice.
- When drug is discontinued, it should be tapered and not stopped abruptly.
- Store in tightly closed, light-resistant container at 15°–30° C (59°–86° F) unless otherwise directed.

ADVERSE EFFECTS (≥1%) **CNS:** Dizziness, vertigo, light-headedness, syncope, fatigue or weakness, lethargy, drowsiness, insomnia, mental changes, depression. **CV:** *Bradycardia, hypotension, CHF,* cold extremities, leg pains, dysrhythmias. **GI:** Nausea, vomiting, diarrhea. **Respiratory:** Pulmonary edema, dyspnea, bronchospasm. **Other:** May mask symptoms of hypoglycemia; decreased sexual ability.

INTERACTIONS Drug: Atropine and other ANTICHOLINERGICS may increase atenolol absorption from GI tract; NSAIDS may decrease hypotensive effects; may mask symptoms of a hypoglycemic reaction induced by **insulin,** SULFONYLUREAS; may increase **lidocaine** levels and toxicity; pharmacologic and toxic effects of both atenolol and **verapamil** are increased. **Prazosin, terazosin** may increase severe hypotensive response to first dose of atenolol.

PHARMACOKINETICS Absorption: 50% of dose absorbed. **Peak:** 2–4 h. **Duration:** 24 h. **Distribution:** Does not readily cross blood–brain barrier. **Metabolism:** No hepatic metabolism. **Elimination:** 40–50% in urine; 50–60% in feces. **Half-Life:** 6–7 h.

NURSING IMPLICATIONS

Assessment & Drug Effects

- Measure trough BP (just prior to scheduled dose) to determine efficacy.
- Check apical pulse before administration in patients receiving digitalis (both drugs slow AV conduction). If below 60 bpm (or other ordered parameter), withhold dose and consult prescriber.
- Monitor BP throughout dosage adjustment period. Consult prescriber for acceptable parameters.
- Monitor diabetics for loss of glycemic control.

Patient & Family Education

- Adhere rigidly to dose regimen. Sudden discontinuation of drug can exacerbate angina and precipitate tachycardia or MI in patients with coronary artery disease, and thyroid storm in patients with hyperthyroidism.
- Make position changes slowly and in stages, particularly from recumbent to upright posture.
- If diabetic, closely monitor blood glucose values.

ATOMOXETINE

(a-to-mox'e-teen)

Strattera

Common adverse effects in *italic*, life-threatening effects underlined; generic names in **bold**; classifications in SMALL CAPS; ♣ Canadian drug name; ⊘ Prototype drug

Classification: MISCELLANEOUS PSYCHOTHERAPEUTIC
Therapeutic: ADHD AGENT
Pregnancy Category: C

AVAILABILITY 10 mg, 18 mg, 25 mg, 40 mg, 60 mg, 80 mg, 100 mg capsules

ACTION & *THERAPEUTIC EFFECT* Selective inhibition of the pre-synaptic norepinephrine trans-porter, resulting in norepineph-rine reuptake inhibition. *Improved attentiveness, ability to follow through on tasks with less distrac-tion and forgetfulness, and dimin-ished hyperactivity.*

USES Acute and maintenance treat-ment of attention deficit/hyperac-tivity disorder (ADHD) in adults and children.

CONTRAINDICATIONS Hypersen-sitive to atomoxetine or any of its constituents; concomitant use or use within 2 wk of MAOIs; narrow angle glaucoma; structural cardiac abnor-malities or other serious heart prob-lems; pheochromocytoma or history of pheochromocytoma; jaundice or liver injury; suicidal ideation; major depressive disorder (MDD); lacta-tion. Safety and efficacy in children younger than 6 y and the older adult have not been established.
CAUTIOUS USE Severe liver in-jury may progress to liver failure or death in a small percentage of patients. History of hypertension, tachycardia, cardiovascular or cere-brovascular disease; any condition that predisposes to hypotension; urinary retention or urinary hesi-tancy; history of bipolar disorder; history of suicidal tendencies; preg-nancy (category C).

ROUTE & DOSAGE

ADHD

Adult/Adolescent/Child (older than 6 y and weight greater than 70 kg): **PO** Start with 40 mg in morning. May increase after 3 days to target dose of 80 mg/day given either once in the morning or as divided dose. May increase to max of 100 mg/day if needed.
Child/Adolescent (weight less than 70 kg): **PO** Start with 0.5 mg/kg/day. May increase after 3 days to target dose of 1.2 mg/kg/day. Administer once daily in morning or divide dose. Max dose is 1.4 mg/kg or 100 mg, whichever is less.

Hepatic Impairment Dosage Adjustment

Child-Pugh class B: Initial and target doses should be reduced to 50% of the normal dose
Child-Pugh class C: Initial dose and target doses should be reduced to 25% of normal dose

Pharmacogenetic Dosage Adjustment /Patients Receiving Concurrent CYP2D6 Inhibitors

CYP2D6 poor metabolizers: In children/adolescents (weight up to 70 kg) start at 0.5 mg/kg, adjust upward only after 4 wk if well tolerated; adults/adolescents (weight greater than 70 kg) start at 40 mg/day, adjust upward only after 4 wk if well tolerated; do not exceed 80 mg

ADMINISTRATION

Oral
- Note that total daily dose in chil-dren and adolescents is based on weight. Determine that ordered dose is appropriate for weight prior to administration of drug.

Common adverse effects in *italic*, life-threatening effects underlined; generic names in **bold**; classifications in SMALL CAPS; ♣ Canadian drug name; ⊙ Prototype drug

- Note manufacturer recommends dosage adjustments with concomitant administration of strong CYP2D6 inhibitors (e.g., paroxetine, fluoxetine, quinidine). Consult prescriber.
- Store at 15°–30° C (59°–86° F).

ADVERSE EFFECTS (≥1%) **Body as a Whole:** Flu-like syndrome, flushing, fatigue, fever, rigors. **CNS:** Dizziness, *headache,* somnolence, crying, tearfulness, irritability, mood swings, *insomnia,* depression, tremor, early morning awakenings, paresthesias, abnormal dreams, decreased libido, sleep disorder, <u>suicidal ideation</u>. **CV:** Increased blood pressure, sinus tachycardia, palpitations. **GI:** *Upper abdominal pain,* constipation, dyspepsia, *nausea, vomiting, decreased appetite,* anorexia, dry mouth, diarrhea, flatulence, <u>severe liver injury (rare)</u>. **Endocrine:** Hot flushes, sexual dysfunction. **Metabolic:** Weight loss. **Hepatic:** Hepatotoxicity. **Musculoskeletal:** Arthralgia, myalgia. **Respiratory:** *Cough,* rhinorrhea, nasal congestion, sinusitis. **Skin:** Dermatitis, pruritus, increased sweating. **Special Senses:** Mydriasis. **Urogenital:** Urinary hesitation/retention, dysmenorrhea, ejaculation dysfunction, impotence, delayed onset of menses, irregular menstruation, prostatitis; priapism, male pelvic pain.

INTERACTIONS Drug: Albuterol may potentiate cardiovascular effects of atomoxetine; CYP2D6 inhibitors **(fluoxetine, paroxetine, quinidine)** may increase atomoxetine levels and toxicity; MAOIS may precipitate a hypertensive crisis; may attenuate effects of ANTIHYPERTENSIVE AGENTS.

PHARMACOKINETICS Absorption: Well absorbed from GI tract. **Distribution:** 98% protein bound. **Peak:** 1–2 h. **Metabolism:** In liver by CYP2D6. **Elimination:** Primarily in urine. **Half-Life:** 5.2 h.

NURSING IMPLICATIONS

Assessment & Drug Effects
- Evaluate for continuing therapeutic effectiveness especially with long-term use.
- Report increased aggression and irritability as these may indicate a need to discontinue the drug.
- Monitor children and adolescents for behavior changes that may indicate suicidal ideation, including aggression and anxiety that may be precursors of it.
- Monitor cardiovascular status especially with preexisting hypertension.
- Monitor HR and BP at baseline, following a dose increase, and periodically while on therapy.
- Lab tests: Periodic LFTs.

Patient & Family Education
- Instruct patients on S&S of liver toxicity.
- Report any of the following to the prescriber: Indicators of suicidal ideation in children and adolescents; chest pains or palpitations, urinary retention or difficulty initiating voiding urine, appetite loss and weight loss, or insomnia.
- Make position changes slowly if you experience dizziness with arising from a lying or sitting position.
- Do not drive or engage in potentially hazardous activities until reaction to the drug is known.

ATORVASTATIN CALCIUM
(a-tor-va′sta-tin)

Lipitor
Classifications: ANTILIPEMIC; HMG-COA; REDUCTASE INHIBITOR (STATIN)
Therapeutic: ANTILIPEMIC; STATIN
Prototype: Lovastatin
Pregnancy Category: X

Common adverse effects in *italic*, life-threatening effects <u>underlined</u>; generic names in **bold**; classifications in SMALL CAPS; ✦ Canadian drug name; ⊙ Prototype drug

AVAILABILITY 10 mg, 20 mg, 40 mg tablets

ACTION & *THERAPEUTIC EFFECT*
Atorvastatin is an inhibitor of reductase 3-hydroxy-3-methyl-glutaryl coenzyme A (HMG-CoA), which is essential to hepatic production of cholesterol. Atorvastatin increases the number of hepatic low-density-lipid (LDL) receptors, thus increasing LDL uptake and catabolism of LDL. HDL cholesterol blood level increases with use of atorvastatin. *Atorvastatin reduces LDL and total triglyceride (TG) production as well as increases the plasma level of high-density lipids (HDL).*

USES Adjunct to diet for the reduction of LDL cholesterol and triglycerides in patients with primary hypercholesterolemia and mixed dyslipidemia, prevention of cardiovascular disease in patients with multiple risk factors.

CONTRAINDICATIONS Hypersensitivity to atorvastatin, myopathy, active liver disease, unexplained persistent transaminase elevations, renal failure, renal impairment, hepatic encephalopathy, hepatitis, hepatic disease; jaundice, rhabdomyolysis; uncontrolled seizure disorders; pregnancy (category X), lactation.
CAUTIOUS USE Hypersensitivity to other HMG-CoA reductase inhibitors, history of liver disease, patients who consume substantial quantities of alcohol. Safety and efficacy in children younger than 10 y have not been established.

ROUTE & DOSAGE

Hypercholesterolemia/ Prevention of Cardiovascular Disease

Adult: **PO** Start with 10–40 mg daily, may increase up to 80 mg/day

Child/Adolescent (10–17 y): **PO** Start with 10 mg daily, may increase up to 20 mg/day

ADMINISTRATION
Oral
- May be given at any time of day.
- Store at 20°–25° C (68°–77° F).

ADVERSE EFFECTS (≥1%) **Body as a Whole:** Back pain, asthenia, hypersensitivity reaction, myalgia, rhabdomyolysis. **CNS:** Headache. **GI:** Abdominal pain, constipation, diarrhea, dyspepsia, flatulence, increased liver function tests. **Respiratory:** Sinusitis, pharyngitis. **Skin:** Rash.

INTERACTIONS Drug: May increase **digoxin** levels 20%, increases levels of **norethindrone** and **ethinyl estradiol** oral contraceptives; **erythromycin** may increase atorvastatin levels 40%; MACROLIDE ANTIBIOTICS, **cyclosporine, delavirdine, gemfibrozil, niacin, clofibrate,** AZOLE ANTIFUNGALS (**ketoconazole, itraconazole**) may increase risk of rhabdomyolysis; **nelfinavir** may increase atorvastatin levels. **Food: Grapefruit juice** (greater than 1 qt/day) may increase risk of myopathy and rhabdomyolysis.

PHARMACOKINETICS Absorption: Rapidly from GI tract. 30% reaches the systemic circulation. **Onset:** Cholesterol reduction—2 wk. **Peak:** Plasma concentration, 1–2 h; effect 2–4 wk. **Distribution:** 98% or greater protein bound. Crosses placenta, distributed into breast milk of animals. **Metabolism:** In the liver by CYP3A4 to active metabolites. **Elimination:** Primarily in bile; less than 2% in urine. **Half-Life:** 14 h; 20–30 h for active metabolites.

Common adverse effects in *italic*, life-threatening effects underlined; generic names in **bold**; classifications in SMALL CAPS; ♣ Canadian drug name; ☺ Prototype drug

129

NURSING IMPLICATIONS

Assessment & Drug Effects

- Monitor for therapeutic effectiveness which is indicated by reduction in the level of LDL-C.
- Lab tests: Monitor lipid levels within 2–4 wk after initiation of therapy or upon change in dosage; monitor LFTs at 6 and 12 wk after initiation or elevation of dose, and periodically thereafter.
- Assess for muscle pain, tenderness, or weakness; and, if present, monitor CPK level (discontinue drug with marked elevations of CPK or if myopathy is suspected).
- Monitor carefully for digoxin toxicity with concurrent digoxin use.

Patient & Family Education

- Report promptly any of the following: Unexplained muscle pain, tenderness, or weakness, especially with fever or malaise; yellowing of skin or eyes; stomach pain with nausea, vomiting, or loss of appetite; skin rash or hives.
- Do not take drug during pregnancy because it may cause birth defects. Immediately inform prescriber of a suspected or known pregnancy.
- Inform prescriber regarding concurrent use of any of the following drugs: Erythromycin, niacin, antifungals, or birth control pills.
- Minimize alcohol intake while taking this drug.

ATOVAQUONE

(a-to′va-quone)

Mepron

Classification: ANTIPROTOZOAL
Therapeutic: ANTIPROTOZOAL
Prototype: Metronidazole
Pregnancy Category: C

AVAILABILITY 750 mg/5 mL suspension

ACTION & *THERAPEUTIC EFFECT*
Atovaquone is an antiprotozoal with antipneumocystic activity, including *Pneumocystis carinii* (PCP) and the *Plasmodium* species. The site of action in PCP is linked to inhibition of the electron transport system in the mitochondria. This results in the inhibition of nucleic acid and ATP synthesis. *Effective against* P. carinii *and the* Plasmodium *species, as well as other protozoans.*

USES Second-line oral therapy of mild to moderate *P. carinii* pneumonia (PCP) in immunocompromised patients intolerant of co-trimoxazole.
UNLABELED USES May be effective in the treatment of cerebral toxoplasmosis.

CONTRAINDICATIONS History of potential life-threatening allergies to atovaquone.
CAUTIOUS USE Severe PCP, concurrent pulmonary diseases, older adults, impaired hepatic function; pregnancy (category C), lactation; neonates and infants.

ROUTE & DOSAGE

Mild to Moderate *Pneumocystis carinii* Pneumonia (PCP)

Adult: **PO** 750 mg (5 mL) suspension b.i.d. for 21 days

ADMINISTRATION

Oral

- Give with meals because food significantly enhances absorption.
- Store at room temperature 15°–30° C (59°–86° F) unless otherwise directed by the manufacturer.

Common adverse effects in *italic*, life-threatening effects <u>underlined</u>; generic names in **bold**; classifications in SMALL CAPS; ♣ Canadian drug name; ⊕ Prototype drug

ADVERSE EFFECTS (≥1%) Body as a Whole: *Fever.* **CV:** Hypotension. **CNS:** *Headache, insomnia, dizziness, strange or vivid dreams, anxiety, depression.* **Hematologic:** Anemia, neutropenia. **Metabolic:** Hyponatremia, hypoglycemia. **GI:** *Nausea, diarrhea, vomiting,* abdominal pain, anorexia, dyspepsia, oral candidiasis, oral ulcers. **Skin:** *Rash,* pruritus, erythema multiforme. **Respiratory:** Cough, sinusitis.

DIAGNOSTIC TEST INTERFERENCE May cause increase in **amylase** and other **liver function tests.**

INTERACTIONS Drug: Zidovudine may increase risk of bone marrow toxicity. **Food:** Oral absorption is increased 3- to 4-fold when administered with food, especially with fatty foods.

PHARMACOKINETICS Absorption: Poor, absorption improved when taken with a fatty meal. **Duration:** 6–23 wk after a 3-wk course of therapy. **Distribution:** Penetrates poorly into cerebrospinal fluid; greater than 99.9% protein bound. **Metabolism:** Not metabolized. **Elimination:** Greater than 94% in feces over 21 days (enterohepatically cycled). **Half-Life:** 2–3 days.

NURSING IMPLICATIONS

Assessment & Drug Effects
- Assess for therapeutic failure in patients with GI disorders that may limit absorption of drug.
- Lab tests: Monitor CBC with differential, blood glucose, serum sodium, creatinine, BUN, and serum amylase periodically. Report abnormal elevations in these values; drug may need to be discontinued.

Patient & Family Education
- Note: It is necessary to take this drug exactly as prescribed because it is slowly eliminated from the body.

ATOVAQUONE/PROGUANIL HYDROCHLORIDE

(a-to′va-quone/pro′gua-nil)

Malarone, Malarone Pediatric
Classification: ANTIMALARIAL
Therapeutic: ANTIMALARIAL
Prototype: Chloroquine HCl and Metronidazole
Pregnancy Category: C

AVAILABILITY Atovaquone 250 mg/proguanil HCl 100 mg (adult dose), atovaquone 62.5 mg/proguanil HCl 25 mg (pediatric dose) tablets

ACTION & *THERAPEUTIC EFFECT* Combination of two antimalarial drugs. Atovaquone inhibits electron transport system in mitochondria of the malaria parasite, thus interfering with nucleic acid and ATP synthesis of the parasite. Proguanil interferes with DNA synthesis of the malaria parasite. *This drug combination has synergistic activity toward malarial treatment because each component has a different mode of action.*

USES Prevention and treatment of malaria due to *P. falciparum,* even in chloroquine-resistant areas.

CONTRAINDICATIONS Known hypersensitivity to atovaquone or proguanil; severe malaria.
CAUTIOUS USE Cerebral malaria, complicated malaria, pulmonary edema; renal failure, renal impairment; hepatic disease; lactation; older adults; African Americans, Chinese, Japanese; diarrhea, emesis, GI disease; hepatic disease, infection, sunlight (UV) exposure;

Common adverse effects in *italic*, life-threatening effects <u>underlined</u>; generic names in **bold**; classifications in SMALL CAPS; ♣ Canadian drug name; ☺ Prototype drug

131

pregnancy (category C). Use in children weighing less than 9 kg is not established.

ROUTE & DOSAGE

Prevention of Malaria

Adult: **PO** 1 tablet daily with food starting 1–2 days before travel to malarial area and continuing for 7 days after return

Child: **PO** *Weight 11–20 kg,* 1 pediatric tablet daily; *weight 21–30 kg,* 2 pediatric tablets daily; *weight 31–40 kg,* 3 pediatric tablets daily; *weight greater than 40 kg,* 1 adult tablet daily with food starting 1–2 days before travel to malarial area and continuing for 7 days after return

Treatment of Malaria

Adult: **PO** 4 tablets as a single daily dose for 3 days

Child: **PO** *Weight 5–8 kg,* 2 pediatric tablets; *weight 9–10 kg,* 3 pediatric tablets; *weight 11–20 kg,* 1 adult tablet; *weight 21–30 kg,* 2 adult tablets; *weight 31–40 kg,* 3 adult tablets; *weight greater than 40 kg,* 4 adult tablets as a single daily dose for 3 days

ADMINISTRATION

Oral

- Give at the same time each day with food or a drink containing milk.
- Give a repeat dose if vomiting occurs within 1 h after dosing.

ADVERSE EFFECTS (≥1%) **Body as a Whole:** Fever, *myalgia,* back pain, asthenia, anorexia. **Digestive:** *Nausea, abdominal pain, diarrhea,* dyspepsia. **CNS:** *Headache.* **Respiratory:** Cough. **Skin:** Pruritus. **Other:** <u>Anaphylactic reaction.</u>

INTERACTIONS Drug: Rifampin, rifabutin, tetracycline may decrease serum levels; **metoclopramide** may decrease absorption.

PHARMACOKINETICS Absorption: Atovaquone (A), Poor, absorption improved when taken with a fatty meal; **Proguanil (P),** Extensively absorbed. **Duration: A,** 6–23 wk after a 3-wk course of therapy. **Distribution: A,** Penetrates poorly into cerebrospinal fluid; greater than 99.9% protein bound; **P,** 75% protein bound. **Metabolism: A,** Not metabolized; **P,** Metabolized by CYP2C19 to cycloguanil. **Elimination: A,** Greater than 94% in feces over 21 days (enterohepatically cycled); **P,** Primarily in urine. **Half-Life: A,** 2–3 days; **P,** 12–21 h.

NURSING IMPLICATIONS

Assessment & Drug Effects

- Lab tests: Monitor AST and ALT periodically, especially with long-term therapy.
- Monitor for S&S of parasitemia in patients receiving tetracycline and in those experiencing diarrhea or vomiting.
- Note: Only use metoclopramide to control vomiting if other antiemetics are not available.

Patient & Family Education

- Take this drug at the same time each day for maximum effectiveness.
- Note: Absorption of this drug may be reduced with diarrhea and vomiting. Consult prescriber if either of these occurs.

ATRACURIUM BESYLATE 💿

(a-tra-kyoor'ee-um)

Common adverse effects in *italic,* life-threatening effects <u>underlined</u>; generic names in **bold**; classifications in SMALL CAPS; ♣ Canadian drug name; 💿 Prototype drug

Classifications: NONPOLARIZING SKELETAL MUSCLE RELAXANT; NEUROMUSCULAR BLOCKER
Therapeutic: SKELETAL MUSCLE RELAXANT
Pregnancy Category: C

AVAILABILITY 10 mg/mL injection

ACTION & *THERAPEUTIC EFFECT* Inhibits neuromuscular transmission by binding competitively with acetylcholine to muscle end plate receptors. Given in general anesthesia only after unconsciousness has been induced by other drugs. *Synthetic skeletal muscle relaxant that produces short duration of neuromuscular blockade, exhibits minimal direct effects on cardiovascular system, and has less histamine-releasing action.*

USES Adjunct for general anesthesia to produce skeletal muscle relaxation during surgery; to facilitate endotracheal intubation. Especially useful for patients with severe renal or hepatic disease, limited cardiac reserve, and in patients with low or atypical pseudocholinesterase levels.

CONTRAINDICATIONS Myasthenia gravis.

CAUTIOUS USE When appreciable histamine release would be hazardous (as in asthma or anaphylactoid reactions, significant cardiovascular disease), neuromuscular disease (e.g., Eaton-Lambert syndrome), carcinomatosis, electrolyte or acid–base imbalances, dehydration, impaired pulmonary function; pregnancy (category C), lactation; children younger than 1 mo.

ROUTE & DOSAGE

Skeletal Muscle Relaxation
Adult/Child (2 y or older): **IV** 0.4–0.5 mg/kg initial dose, then 0.08–0.1 mg/kg bolus 20–45 min after the first dose and q15–25 min thereafter; reduce doses if used with general anesthetics
Child (1 mo–2 y): **IV** 0.3–0.4 mg/kg

Mechanical Ventilation
Adult: **IV** 5–9 mcg/kg/min by continuous infusion

ADMINISTRATION
- Verify correct concentration and rate of infusion for infants and children with prescriber.

Intravenous

***PREPARE:* Direct:** Give initial bolus dose undiluted. **Continuous:** Maintenance dose **must be** diluted with NS, D5W or D5/NS. Maximum concentration should be 0.5 mg/mL. Do not mix in same syringe or administer through same needle as used for alkaline solutions [incompatible with alkaline solutions (e.g., barbiturates)].

***ADMINISTER:* Direct:** Give as bolus dose over 30–60 sec. **Continuous:** Give infusion at rate required to maintain desired effect.

***INCOMPATIBILITIES* Solution/additive: Lactated Ringer's, aminophylline, cefazolin, heparin, nitroprusside quinidine, ranitidine, sodium nitroprusside. Y-site: Aminophylline, amphotericin B, cefonicid, cefoperazone, cefoxitin, ceftazidime, dantrolene, diazepam, diazoxide, furosemide, ganciclovir, indomethacin, pantoprazole, pentobarbital, phenobarbital, phenytoin, propofol, sodium bicarbonate, thiopental.**

- Store at 2°–8° C (36°–46° F) to preserve potency unless otherwise directed. Avoid freezing.

Common adverse effects in *italic*, life-threatening effects underlined; generic names in **bold**; classifications in SMALL CAPS; ♣ Canadian drug name; ● Prototype drug

133

ADVERSE EFFECTS (≥1%) **CV:** Bradycardia, tachycardia. **Respiratory:** Respiratory depression. **Other:** Increased salivation, anaphylaxis.

INTERACTIONS Drug: GENERAL ANESTHETICS increase magnitude and duration of neuromuscular blocking action; AMINOGLYCOSIDES, **bacitracin, polymyxin B, clindamycin, lidocaine, parenteral magnesium, quinidine, quinine, trimethaphan, verapamil** increase neuromuscular blockade; DIURETICS may increase or decrease neuromuscular blockade; **lithium** prolongs duration of neuromuscular blockade; NARCOTIC ANALGESICS present possibility of additive respiratory depression; **succinylcholine** increases onset and depth of neuromuscular blockade; **phenytoin** may cause resistance to or reversal of neuromuscular blockade.

PHARMACOKINETICS Onset: 2 min. **Peak:** 3–5 min. **Duration:** 60–70 min. **Distribution:** Well distributed to tissues and extracellular fluids; crosses placenta; distribution into breast milk unknown. **Metabolism:** Rapid nonenzymatic degradation in bloodstream. **Elimination:** 70–90% in urine in 5–7 h. **Half-Life:** 20 min.

NURSING IMPLICATIONS

Assessment & Drug Effects

- Lab tests: Baseline serum electrolytes, acid–base balance, and renal function as part of preanesthetic assessment.
- Note: Personnel and equipment required for endotracheal intubation, administration of oxygen under positive pressure, artificial respiration, and assisted or controlled ventilation **must be** immediately available.
- Evaluate degree of neuromuscular blockade and muscle paralysis

to avoid risk of overdosage by qualified individual using peripheral nerve stimulator.

- Monitor BP, pulse, and respirations and evaluate patient's recovery from neuromuscular blocking (curare-like) effect as evidenced by ability to breathe naturally or to take deep breaths and cough, keep eyes open, lift head keeping mouth closed, adequacy of hand-grip strength. Notify prescriber if recovery is delayed.
- Note: Recovery from neuromuscular blockade usually begins 35–45 min after drug administration and is almost complete in about 1 h. Recovery time may be delayed in patients with cardiovascular disease, edematous states, and in older adults.

ATROPINE SULFATE

(a′troe-peen)

Atropair ♦

Classifications: ANTICHOLINERGIC; ANTIMUSCARINIC; ANTIARRHYTHMIC

Therapeutic: ANTISECRETORY; ANTIARRHYTHMIC; BRONCHODILATOR

Pregnancy Category: C

AVAILABILITY 0.4 mg tablets; 0.05 mg/mL, 0.1 mg/mL, 0.3 mg/mL, 0.4 mg/mL, 0.5 mg/mL, 0.8 mg/mL, 1 mg/mL injection

ACTION & *THERAPEUTIC EFFECT* Acts by selectively blocking all muscarinic responses to acetylcholine (ACh), whether excitatory or inhibitory. Antisecretory action (vagolytic effect) suppresses sweating, lacrimation, salivation, and secretions from nose, mouth, pharynx, and bronchi. Blocks vagal impuse to heart with resulting decrease in AV conduction time, increase in heart rate, and cardiac output, and shortened PR interval. *Potent bronchodilator when bronchoconstriction has been induced by parasympathomimetics, and decreases bronchial secretions. Decreases GI spasm. Produces mydriasis and cycloplegia by blocking responses of iris sphincter muscle and ciliary muscle of lens to cholinergic stimulation. Increases heart rate and cardiac output.*

USES Adjunct in symptomatic treatment of GI disorders (e.g., peptic ulcer, pylorospasm, GI hypermotility, irritable bowel syndrome) and spastic disorders of biliary tract. Relaxes upper GI tract and colon during hypotonic radiography. *Ophthalmic Use:* To produce mydriasis and cycloplegia before refraction and for treatment of anterior uveitis and iritis. *Preoperative Use:* To suppress salivation, perspiration, and respiratory tract secretions; to reduce incidence of laryngospasm, reflex bradycardia arrhythmia, and hypotension during general anesthesia. *Cardiac Uses:* For sinus bradycardia or asystole during CPR or that is induced by drugs or toxic substances (e.g., pilocarpine, beta-adrenergic blockers, organophosphate pesticides, and *Amanita* mushroom poisoning); for management of selected patients with symptomatic sinus bradycardia and associated hypotension and ventricular irritability; for diagnosis of sinus node dysfunction and in evaluation of coronary artery disease during atrial pacing; for management of chronic symptomatic sinus node dysfunction. *Other Uses:* Oral inhalation for short-term treatment and prevention of bronchospasms associated with asthma, bronchitis, and COPD and as drying agent in upper respiratory infection. Adjunctive therapy for hypermotility of GI tract.

CONTRAINDICATIONS Hypersensitivity to belladonna alkaloids; synechiae; angle-closure glaucoma; parotitis; obstructive uropathy (e.g., bladder neck obstruction caused by prostatic hypertrophy); intestinal atony, paralytic ileus, achalasia, pyloric stenosis, obstructive diseases of GI tract, severe ulcerative colitis, toxic megacolon; tachycardia secondary to cardiac insufficiency or thyrotoxicosis; acute hemorrhage; acute MI; myasthenia gravis.

CAUTIOUS USE Myocardial infarction, hypertension, hypotension; coronary artery disease, CHF, tachyarrhythmias; gastric ulcer, GI infections, hiatal hernia with reflux esophagitis; hyperthyroidism; COPD; autonomic neuropathy; hepatic or renal disease; older adults; debilitated patients; Down syndrome; autonomic neuropathy, spastic paralysis, brain damage in children; patients exposed to high environmental temperatures; patients with fever; pregnancy (category C); infants.

ROUTE & DOSAGE

Preanesthesia

Adult: **IV/IM/Subcutaneous** 0.4–0.6 mg 30–60 min before surgery
Child: **IV/IM/Subcutaneous** Weight less than 5 kg, 0.04 mg/kg; weight greater than 5 kg, 0.03 mg/kg 30–60 min before surgery (max: 0.4 mg)

Bradyarrhythmias

Adult: **IV/IM** 1 mg q2–3min (max: 3 mg)
Child: **IV/IM** 0.01–0.03 mg/kg for 1–2 doses

Common adverse effects in *italic*, life-threatening effects underlined; generic names in **bold;** classifications in SMALL CAPS; ✤ Canadian drug name; ❂ Prototype drug

135

Organophosphate Antidote

Adult: **IV/IM** 1–2 mg q5–60min until muscarinic signs and symptoms subside (may need up to 50 mg)
Child: **IV/IM** 0.05 mg/kg q10–30min until muscarinic signs and symptoms subside

COPD

Adult: **Inhalation** 0.025 mg/kg diluted with 3–5 mL saline, via nebulizer 3–4 times daily (max: 2.5 mg/day)
Child: **Inhalation** 0.03–0.05 mg/kg diluted with 3–5 mL saline, via nebulizer 3–4 times daily

Uveitis

Adult/Child: **Ophthalmic** 1–2 drops of solution or small amount of ointment in eye up to t.i.d.

Cycloplegia

Adult: **Ophthalmic** 1 drop of solution or small amount of ointment in eye 1 h before the procedure
Child: **Ophthalmic** 1–2 drops in eye b.i.d. for 1–3 days prior to procedure or a small amount of ointment in conjunctival sac t.i.d. for 1–3 days prior to procedure with last dose applied several hours before the procedure

ADMINISTRATION

Subcutaneous/Intramuscular:
- Inject indiluted. When using AtroPen, inject into the outer thigh at a 90 degree angle.

Intravenous

PREPARE: **Direct:** Give undiluted or diluted in up to 10 mL of sterile water.

ADMINISTER: **Direct:** Give 1 mg or fraction thereof over 1 min directly into a Y-site.
INCOMPATIBILITIES **Solution/additive: Pantoprazole.**

- Store at room temperature 15°–30° C (59°–86° F) in protected airtight, light-resistant containers unless otherwise directed by manufacturer.

ADVERSE EFFECTS (≥1%) **CNS:** Headache, ataxia, dizziness, excitement, irritability, convulsions, drowsiness, fatigue, weakness; mental depression, confusion, disorientation, hallucinations. **CV:** Hypertension or hypotension, ventricular tachycardia, palpitation, paradoxical bradycardia, AV dissociation, atrial or <u>ventricular fibrillation</u>. **GI:** Dry mouth with thirst, dysphagia, loss of taste; nausea, vomiting, constipation, delayed gastric emptying, antral stasis, paralytic ileus. **Urogenital:** Urinary hesitancy and retention, dysuria, impotence. **Skin:** Flushed, dry skin; anhidrosis, rash, urticaria, contact dermatitis, allergic conjunctivitis, fixed-drug eruption. **Special Senses:** Mydriasis, blurred vision, photophobia, increased intraocular pressure, cycloplegia, eye dryness, local redness.

DIAGNOSTIC TEST INTERFERENCE *Upper GI series:* Findings may require qualification because of anticholinergic effects of atropine (reduced gastric motility and delayed gastric emptying). *PSP excretion test:* Atropine may decrease urinary excretion of PSP (phenolsulfonphthalein).

INTERACTIONS Drug: Amantadine, ANTIHISTAMINES, TRICYCLIC ANTIDEPRESSANTS, **quinidine, disopyramide, procainamide** add to anticholinergic effects. **Levodopa** effects decreased. **Methotrimeprazine** may precipitate extrapyramidal effects. Antipsychotic effects of PHENOTHIAZINES are decreased due to decreased absorption.

PHARMACOKINETICS Absorption: Well absorbed from all administration sites. Peak effect: 30 min IM, 2–4 min IV, 1–2 h subcutaneous, 1.5–4 h inhalation, 30–40 min topical. **Duration:** Inhibition of salivation 4 h; mydriasis 7–14 days. **Distribution:** In most body tissues; crosses blood–brain barrier and placenta. **Metabolism:** In liver. **Elimination:** 77–94% in urine in 24 h. **Half-Life:** 2–3 h.

NURSING IMPLICATIONS

Assessment & Drug Effects

- Monitor vital signs. HR is a sensitive indicator of patient's response to atropine. Be alert to changes in quality, rate, and rhythm of HR and respiration and to changes in BP and temperature.
- Initial paradoxical bradycardia following IV atropine usually lasts only 1–2 min; it most likely occurs when IV is administered slowly (more than 1 min) or when small doses (less than 0.5 mg) are used. Postural hypotension occurs when patient ambulates too soon after parenteral administration.
- Note: Frequent and continued use of eye preparations, as well as overdosage, can have systemic effects. Some atropine deaths have resulted from systemic absorption following ocular administration in infants and children.
- Monitor I&O, especially in older adults and patients who have had surgery (drug may contribute to urinary retention).

- Monitor CNS status. Older adults and debilitated patients sometimes manifest drowsiness or CNS stimulation (excitement, agitation, confusion) with usual doses of drug or other belladonna alkaloids. Supervision of ambulation may be indicated.
- Monitor infants, small children, and older adults for "atropine fever" (hyperpyrexia due to suppression of perspiration and heat loss), which increases the risk of heatstroke.
- Patients receiving atropine via inhalation sometimes manifest mild CNS stimulation with doses in excess of 5 mg and mental depression and other mental disturbances with larger doses.

Patient & Family Education

- Follow measures to relieve dry mouth: Adequate hydration; small, frequent mouth rinses with tepid water; meticulous mouth and dental hygiene; gum chewing or sucking sugarless sourballs.
- Note: Drug causes drowsiness, sensitivity to light, blurring of near vision, and temporarily impairs ability to judge distance. Avoid driving and other activities requiring visual acuity and mental alertness.
- Discontinue ophthalmic preparations and notify prescriber if eye pain, conjunctivitis, palpitation, rapid pulse, or dizziness occurs.

AURANOFIN ⊕

(au-rane′eh-fin)

Ridaura

Classifications: GOLD COMPOUND; IMMUNOLOGIC; ANTI-INFLAMMATORY; ANTI-RHEUMATIC

Therapeutic: ANTI-INFLAMMATORY; ANTIRHEUMATIC

Pregnancy Category: C

Common adverse effects in *italic*, life-threatening effects <u>underlined</u>; generic names in **bold**; classifications in SMALL CAPS; ◆ Canadian drug name; ⊕ Prototype drug

137

AVAILABILITY 3 mg capsules

ACTION & *THERAPEUTIC EFFECT*
Strongly lipophilic and almost neutral in solution, properties that may facilitate transport of agent across cell membranes. Action appears to be immunomodulatory: Serum immunoglobulin concentrations and rheumatoid factor titers are decreased; and anti-inflammatory: Gold is taken up by macrophages with resulting inhibition of phagocytosis and lysosomal enzyme release. *Auranofin is immunomodulatory and anti-inflammatory.*

USES Management of rheumatoid arthritis.

CONTRAINDICATIONS History of gold-induced necrotizing enterocolitis, renal disease, exfoliative dermatitis or bone marrow aplasia; patient who has recently received radiation therapy, history of severe toxicity from previous exposure to gold or other heavy metals; uncontrolled CHF; marked hypertension; SLE; lactation.
CAUTIOUS USE Inflammatory bowel disease, rash, liver disease, renal disease; history of bone marrow depression; older adults; diabetes mellitus, CHF; pregnancy (category C).

ROUTE & DOSAGE

Rheumatoid Arthritis
Adult: **PO** 6 mg/day in 1–2 divided doses, may increase to 6–9 mg/day in 3 divided doses after 6 mo (max: 9 mg/day)

ADMINISTRATION
Oral
- Give capsule with food or fluid of patient's choice.

- Store at 15°–30° C (59°–86° F); protect from light and moisture.
- Note: Expiration date is 4 y after date of manufacture.

ADVERSE EFFECTS (≥1%) **GI:** *Diarrhea, abdominal cramping* and pain; *nausea,* vomiting, anorexia, dysphagia; *stomatitis,* glossitis, metallic taste; flatulence, constipation, GI bleeding, melena. **Hematologic:** Thrombocytopenia, leukopenia, eosinophilia, agranulocytosis, aplastic anemia. **Urogenital:** Proteinuria, hematuria, renal failure. **Skin:** *Rash, pruritus,* dermatitis, urticaria.

DIAGNOSTIC TEST INTERFERENCE
Auranofin may enhance response to a ***tuberculin skin test.***

PHARMACOKINETICS Absorption:
20% from small intestine. **Peak:**
2 h. **Distribution:** Highest concentrations in kidneys, spleen, lungs, adrenals, and liver; unknown if crosses placenta; small amounts distributed into breast milk. **Elimination:** 60% of absorbed gold eliminated in urine, remainder in feces. **Half-Life:** 11–23 days.

NURSING IMPLICATIONS

Assessment & Drug Effects
- Monitor for therapeutic effectiveness which develops slowly and is not usually apparent for 3–4 mo.
- Report any of the following S&S promptly: Unexplained bleeding or bruising, metallic taste, sore mouth; pruritus, rash; diarrhea and melena; yellow skin and sclera; unexplained cough or dyspnea.
- Lab tests: Periodic CBC with differential and platelet count; periodic LFTs and renal function tests.

Patient & Family Education

- Report adverse effects of therapy, especially abdominal cramping and pain; discontinuance of therapy may be necessary.
- Report metallic taste and pruritus with or without rash. These are among earliest symptoms of impending gold toxicity.
- Do not change dosage (dose or dose interval) by omission, increase, or decrease without first consulting prescriber.
- Use antidiarrheal OTC drug and high-fiber diet for drug-induced diarrhea.
- Avoid exposure to sunlight or to artificial ultraviolet light to prevent photosensitivity reaction.
- Rinse mouth with water frequently for symptomatic treatment of mild stomatitis. Avoid commercial mouth rinses; clean teeth with soft tooth brush and gentle brushing to avoid gingival trauma. Floss at least once daily.

AZACITIDINE

(a-za-ci'ti-deen)

Vidaza

Classifications: ANTINEOPLASTIC AGENT; ANTIMETABOLITE (PYRIMIDINE)

Therapeutic: ANTINEOPLASTIC

Prototype: Fluorouracil

Pregnancy Category: D

AVAILABILITY 100 mg powder for injection

ACTION & *THERAPEUTIC EFFECT*
Causes changes in DNA in abnormal blood-forming cells in the bone marrow, resulting in restoration of normal function to tumor-suppressor genes that are responsible for regulating cell differentiation and growth. *Cytotoxic effects of azacitidine cause the death of rapidly dividing cancer cells that are no longer responsive to normal growth control mechanisms.*

USES Treatment of myelodysplastic syndrome, specifically refractory anemia.

UNLABELED USES Refractory acute lymphocytic and myelogenous leukemia.

CONTRAINDICATIONS Hypersensitivity to azacitidine or mannitol; advanced malignant hepatic tumors, myelodysplastic syndrome with hepatic impairment; vaccination; active infection; dental work; intramuscular injections, if platelets less than 50,000 mm³; pregnancy (category D), lactation. Safety and efficacy in children have not been established.

CAUTIOUS USE Hypoalbuminemia (less than 3 g/dL), hepatic disease; elderly; bone marrow depression; dental disease; history of varicella zoster or other herpes infections; renal impairment, renal failure; older adults.

ROUTE & DOSAGE

Myelodysplastic Syndrome

Adult: **Subcutaneous** 75 mg/m² once daily for 7 days every 4 wk; may increase to 100 mg/m² if no beneficial response is seen after 2 treatment cycles and no toxicity other than nausea and vomiting has occurred

Renal Impairment Dosage Adjustment

If unexplained elevations of BUN or creatinine occur, the next cycle should be delayed until the values return to normal or

Common adverse effects in *italic*, life-threatening effects underlined; generic names in **bold**; classifications in SMALL CAPS; ♦ Canadian drug name; ⊘ Prototype drug

139

baseline, and the dose should be reduced by 50% in the next course

ADMINISTRATION

Subcutaneous

- Reconstitute by slowly injecting 4 mL of sterile water for injection into 100 mg vial to yield 25 mg/mL. Invert 2–3 times and gently rotate until a uniform suspension is achieved. The suspension will be cloudy. If not used immediately, see directions for storage.
- Doses greater than 4 mL should be divided equally into 2 syringes and injected into 2 separate sites. Rotate sites for each injection (thigh, abdomen, or upper arm). Give subsequent injections at least 1 in from an old site and never into areas where the site is tender, bruised, red, or hard.
- Storage: Reconstituted suspension may be kept in the vial or syringe. May refrigerate for up to 8 h. Before use, suspension may be kept at room temperature for up to 30 min. Resuspend by inverting the syringe 2–3 times and gently roll between the palms for 30 sec immediately before administration.

ADVERSE EFFECTS (≥1%) **Body as a Whole:** *Fever, fatigue, malaise, weakness, asthenia, limb pain, back pain,* lymphadenopathy, hematoma, night sweats, cellulitis, lethargy. **CNS:** *Dizziness, headache, depression,* syncope. **CV:** *Chest pain,* cardiac murmur, tachycardia, hypotension. **GI:** *Nausea, vomiting, diarrhea, constipation, anorexia, weight loss, abdominal pain,* stomatitis, dyspepsia. **Hematologic:** *Anemia,* thrombocytopenia, leukopenia, *neutropenia, ecchymosis, febrile neutropenia.* **Metabolic:** *Peripheral edema.* **Musculoskeletal:** *Myalgia, arthralgia,* muscle cramps. **Respiratory:** *Cough, dyspnea, pharyngitis, nasopharyngitis, pneumonia,* wheezing, pleural effusion, rhonchi. **Skin:** *Injection site erythema, injection site reactions, rash, pruritus, sweating,* urticaria. **Urogenital:** Dysuria, urinary tract infection.

INTERACTIONS Drug: ANTICOAGULANTS, NSAIDS, ANTIPLATELET AGENTS may increase risk of bleeding; **filgrastim, sargramostim** may interfere with the efficacy of azacitidine if given within 24 h of azacitidine dose.

PHARMACOKINETICS Peak: 30 min. **Metabolism:** In liver. **Elimination:** By kidneys. **Half-Life:** 4 h.

NURSING IMPLICATIONS

Assessment & Drug Effects

- Monitor for S&S of drug toxicity in those with renal insufficiency.
- Lab tests: Baseline LFTs and serum creatinine; monitor CBC with differential before each treatment cycle and prn.
- Withhold drug and notify prescriber for any of the following: S&S of hepatic or renal insufficiency; lab values that indicate leukopenia, neutropenia, thrombocytopenia, or hepatic or renal insufficiency; or serum bicarbonate levels less than 20 mEq/L.

Patient & Family Education

- Promptly report S&S of infection or indication of unusual bleeding tendencies (e.g., dark, tarry stools and easy bruising).
- Women should avoid becoming pregnant and men should not father a child while taking this drug.

AZATHIOPRINE

(ay-za-thye'oh-preen)

Azasan, Imuran

Classifications: IMMUNOSUPPRESSANT; DISEASE-MODIFIYING RHEUMATIC DRUG (DMARD)

Therapeutic: IMMUNOSUPPRESSANT; ANTI-INFLAMMATORY; ANTIRHEUMATIC; DMARD

Prototype: Cyclosporine

Pregnancy Category: D

AVAILABILITY 25 mg, 50 mg, 75 mg, 100 mg tablets; 100 mg vial

ACTION & *THERAPEUTIC EFFECT*

Antagonizes purine metabolism and appears to inhibit DNA, RNA, and normal protein synthesis in rapidly growing cells. *Suppresses T cell effects before transplant rejection. Has immunosuppressant and anti-inflammatory properties.*

USES Adjunctive agent to prevent rejection of kidney allografts, usually with other immunosuppressants. Also used in selective adult patients with severe, active rheumatoid arthritis unresponsive to conventional therapy.

UNLABELED USES SLE, lupus nephritis, psoriatic arthritis; ulcerative colitis, pemphigus, nephrotic syndrome, and other inflammatory and immunologic diseases.

CONTRAINDICATIONS Hypersensitivity to azathioprine or mercaptopurine; clinically active infection; immunization of patient or close family members with live virus vaccines; anuria; pancreatitis; patients previously treated with alkylating agents (increased risk of neoplasms), concurrent radiation therapy; development of GI toxicity to drug; pregnancy (category D), lactation.

CAUTIOUS USE Impaired kidney and liver function; patients receiving cadaver kidney; myasthenia gravis; serious infections. Safety and efficacy in children have not been established.

ROUTE & DOSAGE

Renal Transplantation

Adult: **PO** 3–5 mg/kg/day initially, may be able to reduce to 1–3 mg/kg/day **IV** 3–5 mg/kg/day initially, may be able to reduce to 1–3 mg/kg/day; transfer to PO

Rheumatoid Arthritis

Adult: **PO** 1 mg/kg/day initially, may be increased by 0.5 mg/kg/day at 4–6 wk intervals if needed up to 2.5 mg/kg/day

Obesity Dosage Adjustment

Doses calculated on IBW

Renal Impairment Dosage Adjustment

CrCl 10–50 mL/min: 75% of usual dose; *less than* 10 mL/min: 50% of usual dose

Hemodialysis Dosage Adjustment

Administer after dialysis

ADMINISTRATION

Oral

- Give oral drug in divided doses (as prescribed) with food or immediately after meals to minimize gastric disturbances.

Intravenous

PREPARE: **Direct/Intermittent:** Reconstitute by adding 10 mL sterile water for injection into vial; swirl until dissolved. May be given as prepared or further diluted with 50 mL NS, D5W, or D5/NS. ▪ Reconstituted solution may be stored at room temperature but **must be** used within 24 h after reconstitution (contains no preservatives).

ADMINISTER: **Direct/Intermittent:** May infuse over 30 min to 8 h. Typical infusion time is 30–60 min

Common adverse effects in *italic*, life-threatening effects underlined; generic names in **bold**; classifications in SMALL CAPS; ✦ Canadian drug name; ⊙ Prototype drug

141

or longer. ▪ If longer infusion time is ordered, the final volume of the IV solution is increased appropriately. Check with prescriber.

INCOMPATIBILITIES: **Y-site: amikacin, ampicillin/sulbactam, ascorbic acid, aztreonam, bumetanide, buprenorphine, butophanol, calcium chloride, CEPHALOSPORINS, chloramphenicol, chlorpromazine, cimetidine, clindamycin, dantrolene, diazepam, diazoxide, diphenydramine, dobutamine, dopamine, doxycycline, ephedrine, epinephrine, esmolol, famotidine, ganciclovir, gentamicin, haloperidol, hydralazine, hydrocortisone, hydroxyzine, imipenem/cilastin, isoproterenol, ketorolac, labetolol, lidocaine, magnesium sulfate, meperidine, methyldopate, miconazole, midazolam, milrinone acetate, minocycline, morphine, nafcillin, nalbuphine, nitroprusside, norepinephrine, ondansetron, papaverine, pentazocine, phenylephrine, phenytoin, piperacillin, procainamide, prochlorperazine, promethazine, pyridoxine, quinidine, ritodrine, rocuronium, sodium bicarbonate, streptokinase, succinylcholine, sulfamethaxazole/trimethoprim, tacrolimus, thiamine, ticarcillin, tobramycin, tolazoline, vancomycin, verapamil**

▪ Store at 15°–30° C (59°–86° F) in tightly closed, light-resistant containers unless otherwise directed.

ADVERSE EFFECTS (≥1%) Body as a Whole: Hypersensitivity (skin eruptions, rash, arthralgia). **GI:** Nausea, vomiting, anorexia, esophagitis, diarrhea, steatorrhea, hepatitis with elevations in bilirubin, alkaline phosphatase, AST, ALT, biliary stasis, toxic hepatitis. **Hematologic:** <u>Bone marrow depression, thrombocytopenia, leukopenia,</u> anemia, <u>agranulocytosis,</u> pancytopenia. **Other:** *Secondary infection (immunosuppression);* dysarthria, alopecia; carcinogenic and teratogenic potential reported.

DIAGNOSTIC TEST INTERFERENCE Azathioprine may decrease plasma and urinary ***uric acid*** in patients with gout.

INTERACTIONS Drug: Allopurinol increases effects and toxicity of azathioprine by reducing metabolism of the active metabolite; **allopurinol** doses should be decreased by one third or one fourth; **tubocurarine** and other NONDEPOLARIZING SKELETAL MUSCLE RELAXANTS may reverse or inhibit neuromuscular blocking effects. Ribavirin or peginterferon alfa-2A increase risk of pancytopenia.

PHARMACOKINETICS Absorption: Readily from GI tract. **Distribution:** Crosses placenta. **Metabolism:** Extensively in liver to active metabolite mercaptopurine. **Elimination:** In urine. **Half-Life:** 3 h.

NURSING IMPLICATIONS

Assessment & Drug Effects

▪ Monitor therapeutic effectiveness which usually requires 6–8 wk of therapy for patients with rheumatoid arthritis (improvement in morning stiffness and grip strength). If no improvement has occurred after 12-wk trial period, drug is generally discontinued.

▪ Lab tests: Baseline CBC with differential and platelet count, then weekly during first month of therapy, twice monthly during second and third months, and monthly, or more frequently thereafter, if indicated (e.g., by dosage or therapy changes); periodic LFTs and renal function tests throughout therapy.

- Monitor for toxicity. Drug has a high toxic potential. Because it may have delayed action, dosage should be reduced or drug withdrawn at the first indication of an abnormally large or persistent decrease in leukocyte or platelet count.
- Monitor vital signs. Report signs of infection.
- Monitor I&O ratio; note color, character, and specific gravity of urine. Report an abrupt decrease in urinary output or any change in I&O ratio.
- Monitor for signs of abnormal bleeding [easy bruising, bleeding gums, petechiae, purpura, melena, epistaxis, dark urine (hematuria), hemoptysis, hematemesis]. If thrombocytopenia occurs, invasive procedures should be withheld, if possible.

Patient & Family Education

- Avoid contact with anyone who has a cold or other infection and report signs of impending infection. Exercise scrupulous personal hygiene because infection is a constant hazard of immunosuppressive therapy.
- Practice birth control during therapy and for 4 mo after drug is discontinued. This drug is associated with potential hazards in pregnancy.
- Do not receive/take vaccinations or other immunity-conferring agents during therapy because they may precipitate unusually severe reactions due to the immunosuppressive effects of the drug.

AZELAIC ACID

(a'ze-laic)

Azelex, Finacea

Classification: ANTIACNE
Therapeutic: ANTIACNE
Prototype: Isotretinoin
Pregnancy Category: B

AVAILABILITY 20% cream; 15% gel

ACTION & *THERAPEUTIC EFFECT*
Azelaic acid is a naturally occurring dicarboxylic acid. Antimicrobial action is attributable to inhibition of microbial cellular protein synthesis. A normalization of keratinization of the follicle occurs and it reduces the number of acne lesions. *Reduces the number of inflammatory pustules and papules.*

USES Mild to moderate inflammatory acne vulgaris, mild to moderate rosacea.

CONTRAINDICATIONS Hypersensitivity to any component in the drug.
CAUTIOUS USE Dark complexion, pregnancy (category B), lactation. Safety and efficacy in children younger than 12 y are not established.

ROUTE & DOSAGE

Acne Vulgaris, Rosacea

Adult/Child (older than 12 y):
Topical Apply thin film to clean and dry area b.i.d.

ADMINISTRATION

Topical

- Wash and dry skin thoroughly prior to application of drug.
- Apply by thoroughly massaging a thin film of the cream or gel into

Common adverse effects in *italic*, life-threatening effects <u>underlined</u>; generic names in **bold**; classifications in SMALL CAPS; ◆ Canadian drug name; ⊕ Prototype drug

143

the affected area. Avoid occlusive dressing.
- Wash hands before and after application of cream or gel.
- Store at 15°–30° C (59°–86° F).

ADVERSE EFFECTS (≥1%) **Skin:** Pruritus, burning, stinging, tingling, erythema, dryness, rash, peeling, irritation, contact dermatitis, vitiligo depigmentation, hypertrichosis. **Other:** Worsening of asthma.

PHARMACOKINETICS Absorption: Approximately 4% absorbed through the skin. **Onset:** 4–8 wk. **Distribution:** Into all tissues. **Metabolism:** Partially by beta oxidation in liver. **Elimination:** Primarily in urine. **Half-Life:** 12 h.

NURSING IMPLICATIONS

Assessment & Drug Effects
- Assess for signs of hypopigmentation and report immediately.
- Monitor for sensitivity or severe irritation, which may warrant drug dosage reduction or discontinuation.

Patient & Family Education
- Learn proper application of cream or gel and avoid contact with eyes or mucous membranes.
- Wash eyes with copious amounts of water if contact with medication occurs.
- Note: Transient pruritus, burning, and stinging are common; however, severe skin irritation or hypopigmentation should be reported.

AZELASTINE HYDROCHLORIDE

(a-ze-las′teen)

Astelin, Astepro, Optivar
Classifications: ANTIHISTAMINE; H₁-RECEPTOR ANTAGONIST; NASAL AND OCULAR ANTIHISTAMINE
Therapeutic: ANTIHISTAMINE

Prototype: Diphenhydramine
Pregnancy Category: C

AVAILABILITY 137 mcg/spray nasal spray; 0.05% ophth solution; 0.15% nasal spray

ACTION & THERAPEUTIC EFFECT
Potent histamine H₁-receptor antagonist and inhibitor of mast cell release of histamine. *Effective in the symptomatic treatment of seasonal allergic rhinitis and as a nasal decongestant.*

USES Allergic conjunctivitis, allergic rhinitis, ocular pruritus, vasomotor rhinitis.

CONTRAINDICATIONS Hypersensitivity to azelastine; pregnancy (category C). Safety and efficacy in children younger than 5 y for ophthalmic solution and nasal spray use are not established.

CAUTIOUS USE Hepatic or renal disease; asthmatics; older adults; lactation. **Astepro** nasal spray in children younger than 12 y.

ROUTE & DOSAGE

Allergic Rhinitis

Adult/Adolescent: **Intranasal** 1–2 sprays/nostril b.i.d.
Child (5–11 y): **Intranasal** 1 spray/nostril b.i.d.

Perennial Allergic Rhinitis (Astepro only)

Adult/Adolescent: **Intranasal** 2 sprays/nostril b.i.d.

Ocular Puritus
See Appendix A-1.

ADMINISTRATION

Intranasal
- Prime delivery unit before first use (see manufacturer's instructions).
- Instruct patient to clear nasal passages prior to drug installation;

then tilt head forward slightly and sniff gently when drug is sprayed into each nostril.

- Store the bottle upright at room temperature, 15°–30° C (59°–86° F).

ADVERSE EFFECTS (≥1%) **Body as a Whole:** Fatigue, dizziness. **GI:** Dry mouth, nausea. **Metabolic:** Weight gain. **CNS:** *Headache, somnolence.* **Respiratory:** Pharyngitis, *rhinitis,* paroxysmal sneezing, *cough,* asthma. **Special Senses:** *Bitter taste,* nasal burning, epistaxis, conjunctivitis.

INTERACTIONS Drug: Alcohol and CNS DEPRESSANTS, sedating ANTIHISTAMINES may cause reduced alertness.

PHARMACOKINETICS Absorption: 40% from nasal inhalation. **Peak:** 2–3 h. **Metabolism:** Active metabolites. **Elimination:** Primarily in feces. **Half-Life:** 22 h.

NURSING IMPLICATIONS

Assessment & Drug Effects

- Monitor level of alertness especially in older adults and with concurrent use of other CNS depressants.

Patient & Family Education

- Follow manufacturer's directions for priming the metered dose spray unit before first use and after storage of greater than 3 days.
- Tilt head forward while instilling spray. Avoid getting spray in eyes.
- Do not drive or engage in potentially hazardous activities until response to drug is known.
- Avoid concurrent use of CNS depressants, such as alcohol, while taking this drug.
- Discard spray unit and dispensing package bottle after 3 mo.

AZILSARTAN MEDOXOMIL
(ay′-zil-sar′-tan me-dox′-oh-mil)

Edarbi
Classifications: ANGIOTENSIN II RECEPTOR ANTAGONIST; ANTIHYPERTENSIVE
Therapeutic: ANTIHYPERTENSIVE
Prototype: Losartan
Pregnancy Category: C first trimester; D second and third trimesters

AVAILABILITY 40 mg; 80 mg tablets

ACTION & *THERAPEUTIC EFFECT*
An angiotensin II receptor blocker (ARB) that prevents binding of angiotension II to AT_1 receptors in vascular smooth muscle and other tissues. *Lowers blood pressure thus reducing the risk for fatal and nonfatal cardiovascular events (e.g., stroke and myocardial infarction).*

USES Hypertension, either alone or in combination with other agents.

CONTRAINDICATIONS Pregnancy (2nd and 3rd trimesters).
CAUTIOUS USE Severe CHF, renal artery stenosis, volume depletion, and renal impairment; pregnancy (1st trimester); lactation. Safety and effectiveness in children younger than 18 y have not been established.

ROUTE & DOSAGE

Hypertension

Adult: **PO** 80 mg once daily; consider 40 mg once daily if on high-dose diuretic

ADMINISTRATION

Oral

- May be given without regard to meals.
- Store at 15°–30° C (59°–86° F), and protect from light and moisture.

ADVERSE EFFECTS (≥1%) **CNS:** Dizziness, fatigue. **CV:** Hypotension.

Common adverse effects in *italic*, life-threatening effects <u>underlined</u>; generic names in **bold**; classifications in SMALL CAPS; ✤ Canadian drug name; ⊕ Prototype drug

145

GI: Diarrhea. **Metabolic:** Hyperkalemia.

INTERACTIONS Drug: Concurrent use of azilsartan with NSAIDs may cause deterioration of renal function in patients who are elderly, volume-depleted, on diuretic therapy, or have compromised renal function.

PHARMACOKINETICS Absorption: 60% bioavailability. **Distribution:** Greater than 99% plasma protein bound. **Metabolism:** In the liver to pharmacologically inactive metabolites. **Elimination:** Fecal (55%) and renal (42%). **Half-Life:** 11 h.

NURSING IMPLICATIONS

Assessment & Drug Effects
- Monitor HR and BP at regular intervals.
- Monitor for and report S&S of orthostatic hypotension, especially with volume depletion or with concurrent diuretic use.
- Lab test: Periodic CBC especially with those at risk for anemia; periodic serum creatinine, especially with preexisting renal impairment.

Patient & Family Education
- Report to prescriber if you become or plan to become pregnant.
- Make position changes slowly, especially from lying or sitting to standing.
- Report promptly to prescriber if you experience faintness or dizziness.

AZITHROMYCIN

(a-zi-thro-mye′sin)

AzaSite, Zithromax, Zmax
Classification: MACROLIDE ANTIBIOTIC
Therapeutic: ANTIBIOTIC
Prototype: Erythromycin
Pregnancy Category: B

AVAILABILITY 500 mg, 600 mg tablets; 100 mg/5 mL, 200 mg/5 mL, 1 g/packet oral suspension; 500 mg injection; 1% ophthalmic; extended release: 176 mg/5 mL oral suspension

ACTION & *THERAPEUTIC EFFECT* A macrolide antibiotic that reversibly binds to the 50S ribosomal subunit of susceptible organisms and consequently inhibits protein synthesis. *Effective for treatment of mild to moderate infections caused by pyogenic organisms.*

USES Pneumonia, lower respiratory tract infections, pharyngitis/tonsillitis, gonorrhea, nongonococcal urethritis, skin and skin structure infections due to susceptible organisms, otitis media, *Mycobacterium avium–intracellulare* complex infections, acute bacterial sinusitis. **Zmax:** Acute bacterial sinusitis and community acquired pneumonia. **AzaSite:** Bacterial conjunctivitis.
UNLABELED USES Bronchitis, *Helicobacter pylori* gastritis.

CONTRAINDICATIONS Hypersensitivity to azithromycin, erythromycin, or any of the macrolide antibiotics; viral infection.
CAUTIOUS USE Older adults or debilitated persons, hepatic or renal impairment; GI disease; ventricular arrhythmias, QT prolongation; UV exposure; pregnancy (category B), lactation; children younger than 6 mo.

ROUTE & DOSAGE

Bacterial Infections

Adult: **PO** 500 mg on day 1, then 250 mg q24h for 4 more days **IV** 500 mg daily for at least 2 days, administer 1 mg/mL over 3 h or 2 mg/mL over 1 h
Child (6 mo or older): **PO** 10 mg/kg on day 1, then 5 mg/kg for 4 more days (max: 250 mg/day)

Common adverse effects in *italic*, life-threatening effects underlined; generic names in **bold**; classifications in SMALL CAPS; ◆ Canadian drug name; ❂ Prototype drug

Acute Bacterial Sinusitis

Adult: **PO** 500 mg once daily × 3 days. Zmax: Single one-time dose of 2 g
Child (6 mo or older): **PO** 10 mg/kg once daily × 3 days

Otitis Media

Child (older than 6 mo): **PO** 30 mg/kg as a single dose or 10 mg/kg once daily (not to exceed 500 mg/day) for 3 days or 10 mg/kg as a single dose on day 1 followed by 5 mg/kg/day on days 2–5

Gonorrhea

Adult: **PO** 2 g as a single dose

Chancroid

Adult: **PO** 1 g as a single dose
Child: **PO** 20 mg/kg as single dose (max: 1 g)

Bacterial Conjunctivitis

Adult: **Ophthalmic** 1 drop b.i.d. × 2 days then daily × 5 days

Renal Impairment Dosage Adjustment

CrCl less than 10 mL/min: Use with caution

ADMINISTRATION

Oral

- Give extended-release oral suspension (Zmax) at least 1 h before or 2 h after a meal. Tablets may be taken without regard to food.
- Do not give within 2 h of an aluminum- or magnesium-containing antacid.

Intravenous

PREPARE: **Intermittent:** Reconstitute 500-mg vial with 4.8 mL of sterile water for injection and shake until dissolved. ▪ Final concentration is 100 mg/mL.

▪ Solution **must be** further diluted to 1.0 or 2.0 mg/mL by adding 5 mL of the 100-mg/mL solution to 500 mL or 250 mL, respectively, of D5W, D5/NS, 0.45NaCl, or other compatible solution.
ADMINISTER: **Intermittent:** Administer 1 mg/mL over 3 h. Infuse 2 mg/mL over 1 h. ▪ Note: Do not give a bolus dose.
INCOMPATIBILITIES Y-site: **Amikacin, aztreonam, cefotaxime, ceftazidime, ceftriaxone, cefuroxime, ciprofloxacin, clindamycin, famotidine, fentanyl, furosemide, gemtuzumab, gentamicin, imipenem/cilastatin, ketorolac, levofloxacin, mitoxantrone, morphine, mycophenolate, ondansetron, piperacillin/tazobactam, potassium, quinupristin/dalfopristin, ticarcillin/clavulanate, tobramycin.**

▪ Store drug when diluted as directed for 24 h at or below 30° C (86° F) or for 7 days under 5° C (41° F).

ADVERSE EFFECTS (≥1%) **CNS:** Headache, dizziness. **GI:** Nausea, vomiting, diarrhea, abdominal pain; hepatotoxicity, mild elevations in liver function tests.

DIAGNOSTIC TEST INTERFERENCE Liver function tests: Reversible, asymptomatic elevations in *liver enzymes (AST, ALT, gamma glutamyl transferase, alkaline phosphatase)* have been reported in some patients treated with azithromycin.

INTERACTIONS Drug: ANTACIDS may decrease peak level of azithromycin; may increase toxicity of **digoxin, cyclosporine, phenytoin, dihydroergotamine, ergotamine. Nelfinavir** may increase side effects of azithromycin. Effects of **warfarin** may be potentiated. **Food:** Food will decrease the amount of azithromycin absorbed by 50%.

Common adverse effects in *italic*, life-threatening effects underlined; generic names in **bold;** classifications in SMALL CAPS; ✚ Canadian drug name; ✪ Prototype drug

PHARMACOKINETICS Absorption: 37% of dose reaches the systemic circulation. **Onset:** 48 h. **Peak:** 2.5–4 h. **Distribution:** Extensively into tissues including sputum, blister, and vaginal secretions; tissue concentrations are often higher than serum concentrations. **Metabolism:** In liver. **Elimination:** 5–12% of dose in urine. **Half-Life:** 60–70 h.

NURSING IMPLICATIONS

Assessment & Drug Effects

- Monitor for and report loose stools or diarrhea, since pseudomembranous colitis (see Appendix F) **must be** ruled out.
- Lab tests: Monitor PT and INR closely with concurrent warfarin use.
- Report immediately any S&S of hypersensitivity; though rare, these reactions can be serious.

Patient & Family Education

- Direct sunlight (UV) exposure should be minimized during therapy with drug.
- Take aluminum or magnesium antacids 2 h before or after drug.
- Report onset of loose stools or diarrhea.

AZTREONAM

(az-tree'oh-nam)

Azactam, Cayston
Classifications: ANTIBIOTIC; MONOBACTAM ANTIBIOTIC
Therapeutic: ANTIBIOTIC
Prototype: Imipenem-cilastatin
Pregnancy Category: B

AVAILABILITY 1 g, 2 g vials

ACTION & *THERAPEUTIC EFFECT* Differs structurally from beta-lactam antibiotics (penicillins and cephalosporins) in having a monocyclic rather than a bicyclic nucleus. Acts by inhibiting synthesis of bacterial cell wall by preferentially binding to specific penicillin-binding proteins (PBP) in the bacterial cell wall. *Highly resistant to beta-lactamases and does not readily induce their formation. Spectrum of activity limited to aerobic, gram-negative bacteria.*

USES Gram-negative infections of urinary tract, lower respiratory tract, skin and skin structures, and for intra-abdominal and gynecologic infections, septicemia, and as adjunctive therapy for surgical infections, community acquired pneumonia. Inhalation form for cystic fibrosis.

CONTRAINDICATIONS Hypersensitivity to aztreonam; viral infections. **CAUTIOUS USE** History of hypersensitivity reaction to penicillin, cephalosporins, or to other drugs; impaired renal or hepatic function, older adults; pregnancy (category B); lactation. For Cayston inhalation only: Safety and effectiveness have not been established in pediatric patients younger than 7 y, patients with FEV_1 less than 25% or greater than 75% predicted, or colonized with *Burkholderia cepacia.*

ROUTE & DOSAGE

Urinary Tract Infection

Adult/Adolescent: **IV/IM** 0.5–1 g q8–12h
Child/Infant (over 9 mo): **IV** 30 mg/kg q6–8h

Moderate to Severe Infections

Adult: **IV/IM** 1–2 g q8–12h (max: 8 g/24 h)
Child: **IV** 30 mg/kg/day q6–8h

Community Acquired Pneumonia

Adult: **IV** 1–2 g q8–12h or 2 g q6–8h

Cystic Fibrosis Improvement in Respiratory Symptoms

Adult/Adolescent/Child (over 7): **Inhaled** 75 mg TID × 28 days

Common adverse effects in *italic*, life-threatening effects underlined; generic names in **bold;** classifications in SMALL CAPS; ♣ Canadian drug name; ⊘ Prototype drug

Renal Impairment Dosage Adjustment

CrCl 10–30 mL/min: Reduce dose 50%; less than 10 mL/min: Reduce dose by 75%

Hemodialysis Dosage Adjustment

Reduce dose to 12.5% and give after hemodialysis

ADMINISTRATION

Inhalation

- Administer immediately after reconstitution.
- Administer only via an Altera Nebulizer System after patient has used a bronchodilator. Short-acting bronchodilators can be taken 15 min–4 h before Cayston. Long-acting bronchodilators can be taken 30 min–12 h before Cayston.

Intramuscular

- Reconstitute with at least 3 mL of diluent per gram of drug for IM injection. Immediately and vigorously shake vial to dissolve. Suitable diluents include sterile water for injection; bacteriostatic water for injection (with benzyl alcohol and propyl parabens); NS 0.9% for injection.
- Give IM injections deeply into large muscle mass such as the upper outer quadrant of the gluteus maximus or lateral thigh. Rotate injection sites.

Intravenous

Verify correct IV concentration and rate of infusion/injection with prescriber before giving to neonates, infants, and children.

PREPARE: **Direct** Reconstitute a single dose with 6–10 mL of sterile water for injection. ▪ Immediately shake vial until solution is dissolved. May be given direct IV as prepared or further diluted for IV infusion. ▪ Reconstituted solutions

are colorless to light straw yellow and turn slightly pink on standing. **Intermittent:** Each gram of reconstituted aztreonam **must be** further diluted in at least 50 mL of D5W, NS, or other solution approved by manufacturer to yield a concentration not to exceed 20 mg/mL.

ADMINISTER: **Direct:** Give over 3–5 min. **Intermittent:** Give over 20–60 min through Y-site.

INCOMPATIBILITIES **Solution/additive: Ampicillin, metronidazole, nafcillin. Y-site: Acyclovir, alatrofloxacin, amphotericin B, amphotericin B cholesteryl complex, amsacrine, azathioprine, azithromycin, chlorpromazine, dantrolene, daunorubicin, erythromycin, ganciclovir, milrinone acetate, indomethacin, lansoprazole, lorazepam, metronidazole, mitomycin, mitoxantrone, mycophenolate, oritavancin, pantoprazole, papaverine, pentamidine, pentazocine, pentobarbital, phenytoin, prochlorperazine, quinidine, streptozocin, trastuzumab.**

ADVERSE EFFECTS (≥1%) Body as a Whole: Hypersensitivity (urticaria, eosinophilia, <u>anaphylaxis</u>). **CNS:** Headache, dizziness, confusion, paresthesias, insomnia, seizures. **GI:** Nausea, *diarrhea,* vomiting, elevated liver function tests. **Hematologic:** Eosinophilia. **Special Senses:** Tinnitus, nasal congestion, sneezing, diplopia. **Skin:** Rash, purpura, erythema multiforme, exfoliative dermatitis, diaphoresis; petechiae, pruritus. **Other:** Local reactions (phlebitis, thrombophlebitis (following IV), pain at injection sites), superinfections (gram-positive cocci), vaginal candidiasis.

DIAGNOSTIC TEST INTERFERENCE

Aztreonam may cause transient

Common adverse effects in *italic*, life-threatening effects <u>underlined</u>; generic names in **bold**; classifications in SMALL CAPS; ♣ Canadian drug name; 🅖 Prototype drug

149

elevations of *liver function tests,* increases in *PT* and *PTT,* minor changes in *Hgb,* and positive *Coombs' test.*

INTERACTIONS Drug: Imipenem-cilastatin, cefoxitin may be antagonistic.

PHARMACOKINETICS Peak: 1 h IM. **Distribution:** Widely distributed including synovial and blister fluid, bile, bronchial secretions, prostate, bone, and CSF; crosses placenta; distributed into breast milk in small amounts. **Metabolism:** Not extensively metabolized. **Elimination:** 60–70% in urine within 24 h. **Half-Life:** 1.6–2.1 h.

NURSING IMPLICATIONS

Assessment & Drug Effects

- Lab tests: Obtain baseline C&S test prior to initiation of therapy. Start drug pending results. Baseline and periodic renal function tests, particularly in older adults and in those with history of renal impairment.
- Inspect IV injection sites daily for signs of inflammation. Pain and phlebitis occur in a significant number of patients.
- Monitor for and report loose stools or diarrhea, since pseudomembranous colitis (see Appendix F) **must be** ruled out.
- Monitor for S&S of opportunistic infections (rectal or vaginal itching or discharge, fever, cough) and promptly report onset to prescriber. Overgrowth of nonsusceptible organisms, particularly *staphylococci, streptococci,* and fungi, is a threat, especially in patients receiving prolonged or repeated therapy.

Patient & Family Education

- Determine previous hypersensitivity reactions to penicillins, cephalosporins, and other allergens prior to therapy.

- Report promptly any of the following: Unexplained diarrhea or loose stools, any sign of allergic reaction, any worsening symptoms.
- Note: IV therapy may cause a change in taste sensation. Report interference with eating.

BACITRACIN
(bass-i-tray'sin)
Baci-IM
Classification: ANTIBIOTIC
Therapeutic: ANTIBIOTIC
Pregnancy Category: C

AVAILABILITY 50,000 unit vial; 500 units/g ophthalmic ointment

ACTION & *THERAPEUTIC EFFECT*
Polypeptide antibiotic derived from cultures of *Bacillus subtilis.* Interferes with the bacterial cell membrane by inhibiting cell wall synthesis. *Spectrum of antibacterial activity similar to that of penicillin. Active against many gram-positive organisms. Ineffective against most other gram-negative organisms.*

USES Parenteral therapy restricted to infants with staphylococcal pneumonia and empyema where adequate laboratory facilities and constant supervision are available. Used topically in treatment of superficial infections of skin.
UNLABELED USES Orally for treatment of antibiotic-associated colitis.

CONTRAINDICATIONS Toxic reaction or renal dysfunction associated with bacitracin; pulmonary disease; atopic individuals.
CAUTIOUS USE Hypersensitivity to neomycin; myasthenia gravis or other neuromuscular disease; renal impairment; pregnancy (category C), lactation.

Common adverse effects in *italic,* life-threatening effects underlined; generic names in **bold;** classifications in SMALL CAPS; ♣ Canadian drug name; ☺ Prototype drug

ROUTE & DOSAGE

Systemic Infections

Infant: **IM** *Weight less than 2.5 kg,* up to 900 units/kg/24 h divided q8–12h; *weight greater than 2.5 kg,* up to 1000 units/kg/24h divided q8–12h

Skin Infections

Adult: **Topical** Apply thin layer of ointment b.i.d., t.i.d., as solution of 250–1000 units/mL in wet dressing

ADMINISTRATION

Intramuscular

- Reconstitute with NS containing 2% procaine hydrochloride (prescribed). Do not reconstitute with diluents containing parabens because solution may precipitate or become cloudy.
- Alternate injection sites since injections are painful.
- Powder vials should be stored in refrigerator at 2°–8° C (36°–46° F). Store solution for a maximum of 1 wk if refrigerated. Inactivation occurs at room temperature.

Topical

- Clean affected area prior to application. May be covered with a sterile bandage.
- Store ointments in tightly closed containers at 15°–30° C (59°–86° F) unless otherwise directed.

ADVERSE EFFECTS (≥1%) **GI:** Anorexia, nausea, vomiting, diarrhea, rectal itching and burning. **Hematologic:** Systemic use: Bone marrow depression, blood dyscrasias; eosinophilia. **Body as a Whole:** Hypersensitivity (erythema, <u>anaphylaxis</u>). **Urogenital:** <u>Nephrotoxicity</u>; dose related: Increased BUN, uremia, <u>renal tubular</u> and <u>glomerular necrosis</u> (IM route). **Special Senses:** Tinnitus.

Other: Pain and inflammation at injection site, fever, superinfection, neuromuscular blockade <u>with respiratory depression</u>.

INTERACTIONS Drug: With AMINO-GLYCOSIDES, possibility of additive nephrotoxic and neuromuscular blocking effects; with **tubocurarine** and other NONDEPOLARIZING SKELETAL MUSCLE RELAXANTS, possibility of additive neuromuscular blocking effects.

PHARMACOKINETICS Absorption: Poorly absorbed from intact or denuded skin or mucous membranes. **Peak:** 1–2 h IM. **Duration:** 6–8 h. **Distribution:** Widely distributed including peritoneal and ascitic fluids. **Elimination:** Slow renal excretion (10–40% in 24 h).

NURSING IMPLICATIONS

Assessment & Drug Effects

- Lab tests: Baseline C&S tests prior to initiation of therapy, baseline and periodic kidney function tests.
- Watch for signs of local allergic reaction (itching, burning, redness) with topical skin applications.
- Monitor I&O during parenteral therapy. Adequate urinary output is important to reduce possibility of renal toxicity.
- Inspect urine for turbidity and hematuria, and watch for other S&S of urinary tract dysfunction. Report any changes in urination pattern (e.g., oliguria, urinary frequency, nocturia).
- Note: Prolonged use may result in overgrowth of nonsusceptible organisms, especially *Candida albicans.*

Patient & Family Education

- Report local allergic reactions with topical applications (e.g., itching, burning, redness).

Common adverse effects in *italic*, life-threatening effects <u>underlined</u>; generic names in **bold**; classifications in SMALL CAPS; ✤ Canadian drug name; ☯ Prototype drug

B

BACLOFEN

(bak'loe-fen)

Kemstro, Lioresal

Classifications: CENTRAL-ACTING SKELETAL MUSCLE RELAXANT; GABA AGONIST

Therapeutic: SKELETAL MUSCLE RELAXANT

Prototype: Cyclobenzaprine

Pregnancy Category: C

AVAILABILITY 10 mg, 20 mg tablets; 10 mg, 20 mg orally disintegrating tablets; 50 mcg/mL, 250 mcg/mL ampules

ACTION & *THERAPEUTIC EFFECT*
Centrally acting skeletal muscle relaxant that depresses monosynaptic and polysynaptic afferent reflex activity at spinal cord level. Baclofen stimulates the GABA receptors, which results in decreased excitatory input into alpha-motor neurons. *Reduces skeletal muscle spasm caused by upper motor neuron lesions.*

USES Symptomatic relief of painful spasms in multiple sclerosis and in the management of detrusor sphincter dyssynergia in spinal cord injury or disease.

UNLABELED USES Treatment of trigeminal neuralgia and of tardive dystonia associated with antipsychotic medications, chronic pain.

CONTRAINDICATIONS Coagulopathy, bacteremia, intramuscular or intrathecal administration, subcutaneous administration; lactation.

CAUTIOUS USE Impaired renal and hepatic function; bipolar disorder, psychosis, schizophrenia, seizure disorders, seizures, stroke, cerebral palsy, depression, diabetes mellitus, dialysis, head trauma, PKU, epilepsy; thrombocytopenia; psychiatric or brain disorders; older adults, pregnancy (category C); children younger than 2 y.

ROUTE & DOSAGE

Muscle Spasm

Adult: **PO** 5 mg t.i.d., may increase by 5 mg/dose q3days prn (max: 80 mg/day)

Child: **PO** 2–7 y, 10–15 mg/day divided q8h, may increase by 5–15 mg/day q3days (max: 40 mg/day); *8 y or older,* 10–15 mg/day divided q8h, may increase by 5–15 mg/day q3days (max: 60 mg/day)

Adult: **Intrathecal** Prior to infusion pump implantation, initiate trial dose of 50 mcg/mL bolus administered in intrathecal space by barbotage over 1 min or less. Observe patient over next 4–8 h for significant decrease in muscle spasm. If response is less than desired, administer second bolus of 75 mcg/1.5 mL and observe 4–8 h. May repeat in 24 h with a 100 mcg/2-mL bolus if necessary. *Post-implant titration:* Use screening dose if response lasted longer than 12 h or double screening dose if response lasted less than 12 h and administer over 24 h. After first 24 h, decrease dose by 10–30% q24h until desired response achieved. Maintenance doses range from 12–1500 mcg/day, with most patients maintained on 300–800 mcg/day.

ADMINISTRATION

Oral

- Give with food or milk to avoid GI distress.

Intrathecal

- Give by direct intrathecal injection (via lumbar puncture or catheter) over at least 1 min or longer.
- Dilute *only* with sterile, preservative free NS injection. Baclofen **must be** diluted to a concentration

of 50 mcg/mL when preparing test doses.

- Intrathecal infusion pump: Do not abruptly discontinue as serious adverse effects may develop.
- Store at 15°–30° C (59°–86° F) in tightly closed container unless otherwise directed.

ADVERSE EFFECTS (≥1%) **CNS:** *Transient drowsiness,* vertigo, dizziness, weakness, fatigue, headache, confusion, insomnia; ataxia, loss of seizure control in epileptic patients; abrupt discontinuation of intrathecal administration may result in high fever, altered mental status, exaggerated rebound spasticity, and muscle rigidity, that in rare cases has advanced to rhabdomyolysis, multiple organ-system failure, and death. **CV:** Hypotension. **Special Senses:** Tinnitus, nasal congestion; blurred vision, mydriasis, nystagmus, diplopia, strabismus, miosis. **GI:** Nausea, constipation, vomiting; mild increases in AST, and alkaline phosphatase, jaundice. **Urogenital:** Urinary frequency.

DIAGNOSTIC TEST INTERFERENCE Possibility of increases in *blood-glucose,* serum *alkaline phosphatase,* and *AST levels*.

INTERACTIONS Drug: Alcohol, CNS DEPRESSANTS, MAO INHIBITORS, ANTIHISTAMINES compound CNS depression; baclofen may increase blood **glucose** levels, making it necessary to increase dosage of SULFONYLUREAS, **insulin.**

PHARMACOKINETICS Absorption: Readily from GI tract. **Peak:** 2–3 h. **Duration:** 8 h. **Distribution:** Minimal amounts cross blood–brain barrier; crosses placenta; distribution into breast milk unknown. **Metabolism:** 15% in liver. **Elimination:** 70–85% in urine within 72 h; some elimination in feces. **Half-Life:** 3–4 h.

NURSING IMPLICATIONS

Assessment & Drug Effects

- Supervise ambulation. Initially, the loss of spasticity induced by baclofen may affect patient's ability to stand or walk.
- Lab tests: Baseline and periodic blood sugar, hepatic function tests, and urine.
- Monitor for orthostatic hypotension. Fall precautions may be necessary.
- Observe carefully for CNS side effects: Mental confusion, depression, hallucinations. Older adults are especially sensitive to this drug.
- Monitor patients with epilepsy for possible loss of seizure control.

Patient & Family Education

- Note: CNS depressant effects will be additive to other CNS depressants, including alcohol.
- Monitor blood glucose for loss of glycemic control if diabetic.
- Do not drive or engage in other potentially hazardous activities until the response to drug is known.
- Do not self-dose with OTC drugs without prescriber's approval.
- Do not stop this drug unless directed to do so by prescriber. Drug withdrawal needs to be accomplished gradually over a period of 2 wk or more. Abrupt withdrawal following prolonged administration may cause anxiety, agitated behavior, auditory and visual hallucinations, severe tachycardia, acute exacerbation of spasticity, and seizures.

BALSALAZIDE

(bal-sal'a-zide)

Colazal

Classifications: MUCOUS MEMBRANE AGENT; ANTI-INFLAMMATORY

Therapeutic: ANTI-INFLAMMATORY

Prototype: Mesalamine (5-ASA)

Pregnancy Category: B

Common adverse effects in *italic*, life-threatening effects underlined; generic names in **bold;** classifications in SMALL CAPS; ♣ Canadian drug name; ⊚ Prototype drug

153

B

AVAILABILITY 750 mg capsules

ACTION & *THERAPEUTIC EFFECT* A
prodrug of mesalamine that remains intact until it reaches the lumen of the colon. Thought to decrease inflammation of the mucous lining of the colon by blocking cyclooxygenase and inhibiting prostaglandin synthesis in the lining of the colon. *An anti-inflammatory agent and a prodrug of 5-ASA.*

USES Treatment of mild to moderate active ulcerative colitis.

CONTRAINDICATIONS Prior hypersensitivity to salicylates, balsalazide. **CAUTIOUS USE** Hypersensitivity to mesalamine, sulfasalazine, olsalazine, salicylate. Allergic response to any medications; hepatic or renal impairment; pyloric stenosis; pregnancy (category B), lactation; children younger than 5 y.

ROUTE & DOSAGE

Ulcerative Colitis
Adult: **PO** 2250 mg t.i.d. for 8–12 wk
Adolescent/Child (older than 5 y): 2250 mg (1–3 caps) t.i.d. for up to 8 wk

ADMINISTRATION

Oral
- Give in a consistent manner with respect to food intake (i.e., either always with or always without food).
- Store at room temperature, preferably between 15°–30° C (59°–86° F).

ADVERSE EFFECTS (≥1%) **Body as a Whole:** Arthralgia, fatigue, fever, pain, back pain. **CNS:** Headache, insomnia. **GI:** Abdominal pain, nausea, diarrhea, vomiting, rectal bleeding, flatulence, dyspepsia, coughing, anorexia. **Respiratory:** Rhinitis, pharyngitis.

INTERACTIONS Avoid for 6 wk after VARICELLA VACCINE.

PHARMACOKINETICS Absorption: Low and variable absorption from the colon. **Distribution:** 99% protein bound. **Metabolism:** Metabolized in colon to release 5-aminosalicylic acid. **Elimination:** Feces.

NURSING IMPLICATIONS

Assessment & Drug Effects
- Monitor for S&S of myelosuppression in patients also receiving azathioprine. Monitor S&S of colitis, including rectal bleeding and frequent stools. Report worsening symptoms.
- Lab tests: Closely monitor CBC with concomitant azathioprine therapy; monitor renal and liver functions when used with other aminosalicylates.

Patient & Family Education
- Report worsening of S&S of colitis to prescriber (e.g., diarrhea, abdominal pain, fever, rectal bleeding).

BASILIXIMAB 🔾
(bas-i-lix′i-mab)
Simulect
Classifications: IMMUNOSUPPRESSANT; MONOCLONAL ANTIBODY; INTERLEUKIN-2 RECEPTOR ANTAGONIST
Therapeutic: IMMUNOSUPPRESSANT
Pregnancy Category: B

AVAILABILITY 10 mg, 20 mg vials

ACTION & *THERAPEUTIC EFFECT*
Immunosuppressant agent that is an interleukin-2 (IL-2) receptor monoclonal antibody produced by recombinant DNA technology. Binds to and blocks the interleukin-2R-alpha chain (CD-25 antibodies) on surface of activated T lymphocytes. *Binding to CD-25 antibodies inhibits a critical*

pathway in the immune response of the lymphocytes involved in allograft rejection.

USES Prophylaxis of acute renal transplant rejection.

CONTRAINDICATIONS Hypersensitivity to mannitol or murine protein; serious infection or exposure to viral infections (e.g., chickenpox, herpes zoster); lactation.

CAUTIOUS USE History of untoward reactions to dacliximab or other monoclonal antibodies; pregnancy (category B).

ROUTE & DOSAGE

Transplant Rejection Prophylaxis

Adult/Child (weight greater than 35 kg): IV 20 mg times 2 doses (1st dose 2 h before surgery, 2nd dose 4 days after transplant)
Child (weight less than 35 kg, 2–15 y): IV 12 mg/m² (max: 20 mg/dose) times 2 doses (1st dose 2 h before surgery, 2nd dose 4 days after transplant)

ADMINISTRATION

Intravenous

PREPARE: **Direct/IV infusion:** Add 2.5 mL or 5 mL sterile water for injection to the 10 mg or 20 mg vial, respectively. Rock vial gently to dissolve. ▪ May be given as prepared direct IV as a bolus dose or further diluted in an infusion bag to a volume of 50 mL in NS or D5W. The resulting solution has a concentration of 2.5 mg/mL. ▪ Invert IV bag to dissolve but do not shake. ▪ Discard if diluted solution is colored or has particulate matter. ▪ Use IV solution immediately. *ADMINISTER:* **Direct:** Give bolus over 20–30 sec. **IV Infusion:** Infuse the ordered dose of diluted drug over 20–30 min.

▪ If necessary, the diluted solution may be stored at room temperature for 4 h or at 2°–8° C (36°–46° F) for 24 h. Discard after 24 h. ▪ Store undiluted drug at 2°–8° C (36°–46° F).

ADVERSE EFFECTS (≥1%) **Body as a Whole:** Pain, peripheral edema, edema, fever, viral infection, asthenia, arthralgia, acute hypersensitivity reactions with any dose. **CNS:** Headache, tremor, dizziness, insomnia, paresthesias, agitation, depression. **CV:** Hypertension, chest pain, hypotension, arrhythmias. **GI:** Constipation, nausea, diarrhea, abdominal pain, vomiting, dyspepsia, moniliasis, flatulence, GI hemorrhage, melena, esophagitis, erosive stomatitis. **Hematologic:** Anemia, thrombocytopenia, thrombosis, polycythemia. **Respiratory:** Dyspnea, URI, cough, rhinitis, pharyngitis, bronchospasm. **Skin:** Poor wound healing, acne. **Urogenital:** Dysuria, UTI, albuminuria, hematuria, oliguria, frequency, renal tubular necrosis, urinary retention. **Other:** Cataract, conjunctivitis. **Metabolic:** Hyperkalemia, hypokalemia, hyperglycemia, hyperuricemia, hypophosphatemia, hypocalcemia, increased weight, hypercholesterolemia, acidosis.

PHARMACOKINETICS Duration: 36 days. **Distribution:** Binds to interleukin-2R-alpha sites on lymphocytes. **Half-Life:** 7.2 ± 3.2 days in adults, 11.5 ± 6.3 days in children.

NURSING IMPLICATIONS

Assessment & Drug Effects

▪ Monitor carefully for and immediately report S&S of opportunistic infection or anaphylactoid reaction (see Appendix F).
▪ Monitor CV status with periodic BP measurements. Hypertension is a common adverse effect.
▪ Lab tests: Periodic serum electrolytes, lipid profile, and uric acid.

Common adverse effects in *italic*, life-threatening effects underlined; generic names in **bold**; classifications in SMALL CAPS; ✚ Canadian drug name; ☺ Prototype drug

B

Patient & Family Education

- Report any distressing adverse effects.
- Avoid vaccination for 2 wk following last dose of drug.

BCG (BACILLUS CALMETTE-GUÉRIN) VACCINE

(ba-cil'lus cal'met-te guer'in)

Tice, TheraCys

Classifications: VACCINE; IMMUNOMODULATOR; BIOLOGICAL RESPONSE MODIFIER

Therapeutic: ANTINEOPLASTIC; IMMUNOMODULATOR

Pregnancy Category: C

Vaccine information.

AVAILABILITY 50 mg, 81 mg powder for suspension

ACTION & *THERAPEUTIC EFFECT*
BCG vaccine is an immunization agent for tuberculosis (TB). This vaccine is also active immunotherapy. BCG vaccine stimulates the reticuloendothelial system (RES) to produce macrophages that do not allow mycobacteria to multiply. BCG live is thought to cause a local, chronic inflammatory response involving macrophage and leukocyte infiltration of the bladder. This leads to destruction of superficial tumor cells. *BCG vaccine is an immunization agent for tuberculosis (TB). BCG is active immunotherapy that stimulates the immune mechanism to reject the tumor. It enhances the cytotoxicity of macrophages. BCG live is used intravesically as a biological response modifier for bladder cancer in situ.*

USES Carcinoma in situ of the bladder.

UNLABELED USES Malignant melanoma.

CONTRAINDICATIONS Impaired immune responses, immunosuppressive corticosteroid therapy, active TB, concurrent infections; recent TURP, severe hematuria; asymptomatic carriers with positive HIV serology; fever; UTI; lactation.

CAUTIOUS USE Hypersensitivity to BCG; high risk for HIV; aneurysm or prosthesis; pregnancy (category C).

ROUTE & DOSAGE

Carcinoma of the Bladder

Adult: **Intravesical** 3 vials of **TheraCys** at 27 mg each (81 mg total) of BCG reconstituted with accompanying diluent 7–14 days after biopsies/transurethral resections once/wk for 6 wk plus one treatment at 3, 6, 12, 18, and 24 mo; 1 vial of **Tice** per intravesical instillation once/wk for 6 wk plus one treatment/mo for 6–12 mo

Prevention of Tuberculosis (Tice Only)
See Appendix J.

ADMINISTRATION

Intravesical Instillation

- **TheraCys:** Dilute 3 vials of **TheraCys** in 50 mL of sterile preservative free NS and instill into bladder slowly by gravity flow via urethral catheter. Patient retains suspension for 2 h and then voids.
- **Tice:** Instill 1 vial of **Tice** intravesically once/wk for 6 wk plus one per mo for 6–12 mo.
- Important: Exercise care when handling BCG vaccine to avoid contact with the product.
- Store dry BCG powder, reconstituted vaccine, and diluent refrigerated at 2°–8° C (35°–46° F). Use reconstituted solution within 2 h.

ADVERSE EFFECTS (≥1%) **CNS:** Intravesical administration: *Malaise,* dizziness, headache, weakness. **Endocrine:** Hyperpyrexia. **GI:** Abdominal pain, anorexia, constipation, nausea,

Common adverse effects in *italic*, life-threatening effects <u>underlined</u>; generic names in **bold**; classifications in SMALL CAPS; ♥ Canadian drug name; ⊙ Prototype drug

vomiting, diarrhea; hepatic dysfunction following intratumor injection, granulomatous hepatitis. **Urogenital:** Intravesical administration: Bladder spasms, clot retention, decreased bladder capacity, decreased urine flow, *dysuria, hematuria,* incontinence, nocturia, UTI, cystitis, hemorrhagic cystitis, penile pain, prostatism. **Hematologic:** Thrombocytopenia, eosinophilia, *anemia,* leukopenia, disseminated intravascular coagulation. **Respiratory:** Cough (rare), pulmonary granulomas, pulmonary infection. **Skin:** Abscess with recurrent discharge, red papule that scales or ulcerates in about 5–6 wk, dermatomyositis, granulomas at injection site 4–6 wk after inoculation, keloid formation, lupus vulgaris. **Body as a Whole:** Systemic BCG infection, *chills, flu-like syndrome,* anaphylaxis (rare), allergic reactions, lymphadenitis.

DIAGNOSTIC TEST INTERFERENCE Prior BCG vaccination may result in false-positive *tuberculin skin test (PPD).* Following BCG vaccination, tuberculin sensitivity may persist for months to years.

INTERACTIONS Drug: Concurrent antimycobacterial therapy (**aminosalicylic acid, capreomycin, cycloserine, ethambutol, ethionamide, isoniazid, pyrazinamide, rifabutin, rifampin, streptomycin**) that inhibits multiplication of BCG bacilli has the potential to antagonize or altogether negate the BCG vaccine-mediated immune response. **Cyclosporine** may reduce the immunologic response to BCG vaccine. *Cytomegalovirus* **immune globulin** and other live vaccines (measles/mumps/rubella, oral polio) may interfere with immune response to BCG. Previous vaccination with or other exposure to BCG

may induce variable sensitivity to tuberculin. A greater booster effect following repeat tuberculin testing has been reported in individuals with prior BCG vaccination when compared with individuals without prior vaccination.

NURSING IMPLICATIONS

Assessment & Drug Effects

- Monitor for S&S of systemic BCG infection: Fever, chills, severe malaise, or cough.
- Lab tests: Culture blood and urine, if systemic infection is suspected.
- Assess for regional lymph node enlargement and report fistula formation.

Patient & Family Education

- Report promptly S&S of infections including fever, chills, or unexplained fatigue and weakness.
- Retain instillation in bladder for as long as possible (up to 2 h) before voiding.
- Unless instructed otherwise, increase fluid intake to flush bladder after first voiding.
- Disinfect voided urine with bleach for 15 min before flushing.

BECAPLERMIN
(be-cap′ler-min)

Regranex
Classification: PLATELET-DERIVED GROWTH FACTOR (PDGF)
Therapeutic: GROWTH FACTOR
Pregnancy Category: C

AVAILABILITY 0.01% gel

ACTION & *THERAPEUTIC EFFECT*
Recombinant human platelet-derived growth factor B in a topical gel. It induces fibroblast proliferation in new granulation tissue. *It is effective against diabetic neuropathic ulcers that involve subcutaneous or deeper*

Common adverse effects in *italic*, life-threatening effects underlined; generic names in **bold**; classifications in SMALL CAPS; ✦ Canadian drug name; ⊘ Prototype drug

157

tissue, and also have an adequate blood supply. Hence it promotes wound healing of diabetic ulcers.

USES Lower-extremity diabetic neuropathic ulcers, wound management.

CONTRAINDICATIONS Hypersensitivity to drug or any component in formulation; cresol or paraben hypersensitivity; neoplasms at site of application; wounds that close by primary intention; increased risk of death in patients with diabetes mellitus.

CAUTIOUS USE Concurrent use of corticosteroids, cancer chemotherapy, or other immunosuppressive agents; systemic infection; peripheral vascular disease; ulcer wounds related to arterial or venous insufficiency; thermal, electrical, or radiation burns at wound site; malignancy; elderly; pregnancy (category C), lactation; children younger than 16 y.

ROUTE & DOSAGE

Diabetic Neuropathic Ulcers
Adult/Adolescent (older than 16 y): **Topical** Calculate the length of gel based on ulcer size and apply once/day until healed; reassess if ulcer not completely healed in 20 wk

ADMINISTRATION

Topical

▪ Squeeze calculated length of gel onto clean, firm, nonabsorbable surface.

▪ Apply even layer to ulcer area with clean tongue depressor or cotton swab and cover with saline-moistened dressing. After 12 h, remove dressing, clean ulcer by rinsing with water or saline to remove residual gel, and apply new saline-moistened dressing without becaplermin for next 12 h. Repeat cycle.

▪ Apply only to ulcers with good blood supply.

▪ Dosage calculation: Measure greatest length (*L*) and greatest width (*W*) of ulcer in inches or centimeters; using 15-g tube multiply ($L \times W$) × 0.6 for dose in inches or ($L \times W$)/4 for dose in cm; using 2-g tube multiply ($L \times W$) × 1.3 for dose in inches or ($L \times W$)/2 for dose in cm.

▪ Store at 2°–8° C (36°–46° F). Do not freeze and do not use beyond expiration date.

ADVERSE EFFECTS (≥1%) **Skin:** Erythematous rash.

PHARMACOKINETICS Absorption: Less than 3% absorbed into systemic circulation.

NURSING IMPLICATIONS

Assessment & Drug Effects

▪ Therapeutic effectiveness: 30% decrease in ulcer size after 10 wk or complete healing after 20 wk.

▪ Monitor for and report appearance of erythematous rash.

Patient & Family Education

▪ Consult wound care provider who typically recalculates dosage weekly/biweekly.

▪ Follow directions for application carefully. Gel may be measured out on waxed paper.

▪ Wash hands prior to application and do not allow tip of tube to contact ulcer or any surface.

▪ Report worsening ulceration or development of skin rash.

BECLOMETHASONE DIPROPIONATE

(be-kloe-meth′a-sone)

Beconase AQ, QVAR, Vancenase AQ

Pregnancy Category: C

See Appendix A-3.

Common adverse effects in *italic*, life-threatening effects underlined; generic names in **bold;** classifications in SMALL CAPS; ✚ Canadian drug name; ◑ Prototype drug

BELIMUMAB

(be-lim'-ue-mab)

Benlystra

Classifications: BIOLOGICAL RE-SPONSE MODIFIER; IMMUNOLOGI-CAL AGENT; MONOCLONAL ANTI-BODY

Therapeutic: IMMUNOSUPPRES-SANT

Prototype: Basiliximab

Pregnancy Category: C

AVAILABILITY 120 mg, 400 mg powder for injection

ACTION & *THERAPEUTIC EFFECT* A monoclonal antibody that inhibits the survival of B cells, including autore-active B cells, and reduces the differ-entiation of B cells into immunoglob-ulin-producing plasma cells. *Helps control lupus erythematosus by decreasing the level of a specific im-munoglobulin G responsible for the damaging effects of lupus.*

USES Treatment of active, autoanti-body-positive systemic lupus ery-thematosis (SLE) in adults who are receiving standard therapy.

CONTRAINDICATIONS Previous hy-persensitivity to belimumab.

CAUTIOUS USE New onset or chron-ic infection; psychiatric disorders; suicidal behavior; pregnancy; lacta-tion. Safety and effectiveness have not been established in children.

ROUTE & DOSAGE

Systemic Lupus Erythematosis

Adult: **IV** 10 mg/kg q2wk for 3 doses, then q4wk

ADMINISTRATION

Intravenous

■ Use only in a setting capable of managing hypersensitivity re-actions.

PREPARE: **IV Infusion:** Place vial at room temperature for 10–15 min. Reconstitute with sterile water for injection by adding 1.5 mL or 4.8 mL to the 120 mg or 400 mg vial, respectively, to yield 80 mg/mL. Minimize foam-ing by directing stream to side of vial, then swirl gently for 60 sec. Keep at room temperature, swirling for 60 sec q5min until dissolved. Withdraw the re-quired dose of belimumab, then remove from 250 mL of NS a volume equal to the volume of the dose. Add belimumab to the NS and invert to mix.

ADMINISTER: **IV Infusion:** DO NOT give IV push or bolus. In-fuse over 1 h through a dedi-cated IV line. Do not mix with other solutions or drugs. ■ Mon-itor closely for S&S of an infu-sion reaction. Stop infusion, institute supportive measures and notify prescriber if an infu-sion reaction (or hypersensitivity) is suspected.

INCOMPATIBILITIES **Solution/additive:** Dextrose solutions. **Y-site:** None listed.

■ May store reconstituted solu-tion for up to 7 h at 2°–8° C (36°–46° F). Protect from light. Total time from reconstitution to com-pletion of infusion should not ex-ceed 8 h.

ADVERSE EFFECTS (≥1%) **Body as a Whole:** Cystitis, pain in the ex-tremities, *pyrexia.* **CNS:** Depres-sion, insomnia, migraine headache. **GI:** *Diarrhea,* gastroenteritis, *nau-sea.* **Hematological:** Leukopenia. **Respiratory:** Bronchitis, *nasophar-yngitis,* pharyngitis.

PHARMACOKINETICS **Half-Life:** 19.4 days.

Common adverse effects in *italic*, life-threatening effects underlined; generic names in **bold**; classifications in SMALL CAPS; ♣ Canadian drug name; ☺ Prototype drug

159

NURSING IMPLICATIONS

Assessment & Drug Effects

- Monitor vital signs often during and after infusion.
- Monitor closely for S&S of infusion reaction or hypersensitivity. Stop the infusion and notify prescriber if any of these occur.
- Monitor closely if a new infection develops.
- Assess for depression, anxiety or other manifestations of psychiatric problems. Report immediately suicidal ideation.

Patient & Family Education

- Report immediately any of the following: Difficulty breathing, wheezing, rash, itching, swelling of the face or tongue, chest pain, or other discomfort during drug infusion.
- Inform prescriber immediately of new or worsening mental health problems such as anxiety, depression, or thoughts of suicide.
- Live vaccines received within 30 days of beginning or concurrently with belimumab therapy may not be effective. Seek advice of prescriber.
- Do not breast feed while taking this drug without consulting prescriber.
- Women should use effective means of contraception during and for at least 4 mo following termination of treatment.

BENAZEPRIL HYDROCHLORIDE

(ben-a′ze-pril)

Lotensin

Classifications: ANTIHYPERTENSIVE; RENIN ANGIOTENSIN SYSTEM ANTAGONIST

Therapeutic: ANTIHYPERTENSIVE, ACE INHIBITOR

Prototype: Enalapril

Pregnancy Category: C first trimester, D second and third trimester

AVAILABILITY 5 mg, 10 mg, 20 mg, 40 mg tablets

ACTION & *THERAPEUTIC EFFECT*
Lowers blood pressure by specific inhibition of the angiotensin-converting enzyme (ACE) and thus by decreasing angiotensin II (a potent vasoconstrictor) and aldosterone secretion. *Achieves an antihypertensive effect by suppression of the renin-angiotensin-aldosterone system.*

USES Treatment of mild to moderate hypertension.

UNLABELED USES CHF, reno-protective agent.

CONTRAINDICATIONS Hypersensitivity to benazepril or another ACE inhibitor; pregnancy (category D second and third trimester), lactation; children with a GFR less than 30 mL/h.

CAUTIOUS USE Renal impairment, renal-artery stenosis; patients with hypovolemia, receiving diuretics, undergoing dialysis; patients in whom excessive hypotension would present a hazard (e.g., cerebrovascular insufficiency); CHF; hepatic impairment; diabetes mellitus; pregnancy (category C first trimester); children younger than 6 y.

ROUTE & DOSAGE

Hypertension

Adult/Adolescent: **PO** 10–40 mg/day in 1–2 divided doses (max: 80 mg/day)
Child (older than 6 y): **PO** 0.1–0.6 mg/kg daily (max: 40 mg/day)

Renal Impairment Dosage Adjustment

CrCl less than 30 mL/min: Use 5 mg starting dose. Do not use in children with CrCl less than 30 mL/min.

Common adverse effects in *italic*, life-threatening effects underlined; generic names in **bold**; classifications in SMALL CAPS; ♣ Canadian drug name; ♦ Prototype drug

ADMINISTRATION
Oral
- Consult prescriber about initial dose if patient is also receiving diuretics. Typically an initial dose of 5 mg is used to minimize the risk of hypotension.
- Store at room temperature, but not above 30° C (86° F).

ADVERSE EFFECTS (≥1%) **CV:** Hypotension. **CNS:** *Headache,* dizziness, fatigue, weakness. **Endocrine:** Hyperkalemia (at higher doses). **GI:** Nausea, diarrhea or constipation, gastritis. **Urogenital:** Azotemia, oliguria, renal failure in patients with CHF. **Respiratory:** Cough, rhinitis, bronchitis. **Other:** Back pain.

DIAGNOSTIC TEST INTERFERENCE Elevations in *serum bilirubin* have been observed after benazepril administration. Benazepril inhibits *aldosterone* secretion, which causes an increase in *serum potassium.*

INTERACTIONS Drug: POTASSIUM-SPARING DIURETICS may increase the risk of hyperkalemia. Benazepril may increase **lithium** toxicity. Use with **azathioprine** increases risk of myelosuppression.

PHARMACOKINETICS Absorption: Readily from GI tract; 37% reaches the systemic circulation. **Peak:** 2–6 h. **Duration:** 20–24 h. **Distribution:** Small amounts cross the blood–brain barrier; crosses placenta; small amount excreted in breast milk. **Metabolism:** In liver to active metabolite, benazeprilat. **Elimination:** Benazeprilat is primarily excreted in urine. **Half-Life:** Benazepril 0.6 h; benazeprilat 22 h.

NURSING IMPLICATIONS
Assessment & Drug Effects
- Assess for hypotension, especially in patients who may be volume depleted (e.g., prolonged diuretic therapy, recent vomiting or diarrhea, salt restriction) or who have CHF.
- Lab tests: Monitor serum potassium levels for hyperkalemia (see Appendix F).

Patient & Family Education
- Do not use salt substitutes unless recommended by prescriber.
- Report swelling of face, eyes, lips, or tongue or difficulty breathing immediately to prescriber.

BENDAMUSTINE
(ben-da-mus′teen)
Treanda
Classifications: ANTINEOPLASTIC; ALKYLATING AGENT
Therapeutic: ANTINEOPLASTIC
Prototype: Cyclophosphamide
Pregnancy Category: D

AVAILABILITY 25 mg, 100 mg powder for injection

ACTION & *THERAPEUTIC EFFECT* Alkylating agent that causes the formation of intrastrand and interstrand crosslinks between DNA molecules, thus resulting in inhibition of DNA replication, repair and transcription. *Active against both dividing and resting neoplastic lymphocytes.*

USES Treatment of chronic lymphocytic leukemia, indolent B cell non-Hodgkin's lymphoma.
UNLABELED USES Treatment of mantle cell lymphoma.

CONTRAINDICATIONS Known hypersensitivity (e.g., anaphylaxic reaction) to bendmustine or mannitol); grade 2 or greater nonhematologic toxicity; moderate-to-severe hepatic impairment (ALT or AST 2.5–10 × ULN + total bilirubin 1.5–3 × ULN, or bilirubin greater than 3 × ULN); CrCl of less than 40 mL/min;

Common adverse effects in *italic*, life-threatening effects <u>underlined</u>; generic names in **bold**; classifications in SMALL CAPS; ♣ Canadian drug name; ⊙ Prototype drug

161

B

severe or progressive skin reactions; infusion reaction; pregnancy (category D); lactation.

CAUTIOUS USE Myelosuppression; mild hepatic or mild-to-moderate renal impairment; grade 3 or 4 infusion reaction; infection; poorly managed hypertension; children younger than 6 y.

ROUTE & DOSAGE

Chronic Lymphocytic Leukemia

Adult: **IV** 120 mg/m² on days 1 and 2 of a 28-day cycle, up to 6 cycles

Hematologic Toxicity Dosage Adjustment

Grade 3 or higher: Reduce dose to 50 mg/m² on days 1 and 2 of each cycle; if grade 3 or higher toxicity recurs, reduce the dose to 25 mg/m² on days 1 and 2 of each cycle.

Grade 4 hematologic toxicity: Hold dose until neutrophil above 1000 and platelets above 75,000.

Indolent B Cell Non-Hodgkin's Lymphoma

Adult: **IV** 120 mg/m² on days 1 and 2 q21days

Toxicity Dosage Adjustment

- For grade 3 or greater non-hematologic toxicity or grade 4 hematologic toxicity: Reduce dose to 90 mg/m²
- For recurrent grade 3 or greater non-hematologic toxicity or recurrent grade 4 hematologic toxicity: Reduce dose to 60 mg/m²

ADMINISTRATION

Intravenous

Exercise caution in handling and disposal. Avoid contact with skin. Wash immediately with soap and water if contact occurs.

- Withhold dose for grade 4 hematologic toxicity or clinically significant grade 2 or higher nonhematologic toxicity until nonhematologic toxicity has recovered to grade 1 or less and/or the blood cell counts have improved (ANC count 1000 or more and platelets 75,000 or more).

PREPARE: IV Infusion: Reconstitute with sterile water for injection; add 5 mL to the 25 mg vial or 20 mL to the 100 mg vial to yield 5 mg/mL; should dissolve completely in 5 min. • Withdraw required dose within 30 min of reconstitution and immediately add to 500 mL of NS. **ADMINISTER: IV Infusion:** Infuse over 30 min.

ADVERSE EFFECTS (≥5%) Body as a Whole: Hypersensitivity, infection, *pyrexia.* **CNS:** Asthenia, chills, fatigue. **GI:** Diarrhea, *nausea, vomiting.* **Hematologic:** *Anemia,* leukopenia, lymphopenia, *neutropenia,* thrombocytopenia. **Metabolic:** Hyperuricemia, weight loss. **Respiratory:** Cough, nasopharyngitis. **Skin:** Pruritus, rash.

INTERACTIONS Drug: Compounds that inhibit CYP1A2 (**atazanavir, cimetidine, ciprofloxacin, fluvoxamine, omeprazole, tacrine, zileuton**) will increase levels of bendamustine but decrease levels of its active metabolite. Avoid **clozapine** due to additive bone marrow suppression.

PHARMACOKINETICS Distribution: 95% protein bound. **Metabolism:** Hepatic oxidation by CYP1A2. **Elimination:** Primarily fecal (90%). **Half-Life:** 40 min.

B

NURSING IMPLICATIONS

Assessment & Drug Effects

- Monitor closely for infusion reactions (i.e., chills, fever, pruritus, rash) and signs of anaphylaxis. Reactions are more likely with the second and subsequent cycles. Discontinue infusion immediately and notify prescriber for severe reactions.
- Maintain adequate hydration status to minimize risk of tumor lysis syndrome.
- Monitor for and report S&S of infection.
- Lab tests: Baseline and weekly Hgb, WBC with differential, platelet count during each cycle; frequent serum potassium and uric acid; frequent renal function tests with preexisting renal impairment.

Patient & Family Education

- Men and women should use reliable contraception to avoid pregnancy during and for 3 mo after bendamustine therapy is completed.
- Do not drive or engage in other dangerous activities until reaction to drug is known.
- Report promptly any of the following: Signs of infection, nausea, vomiting, diarrhea, worsening rash or itching.

BENZALKONIUM CHLORIDE

(benz-al-koe'nee-um)

Benza, Benzalchlor-50, Germicin, Pharmatex ♣, Sabol, Zephiran
Classification: TOPICAL ANTIBIOTIC
Therapeutic: ANTIBIOTIC
Pregnancy Category: C

AVAILABILITY 17% concentrate, 1:750 solution, 1:750 tincture/tincture spray

ACTION & *THERAPEUTIC EFFECT*

Bactericidal or bacteriostatic action (depending on concentration), probably due to inactivation of bacterial enzyme. *Effective against bacteria, some fungi (including yeasts) and certain protozoa. Generally not effective against spore-forming organisms.*

USES Antisepsis of intact skin, mucous membranes, superficial injuries, and infected wounds; also for irrigations of the eye and body cavities and for vaginal douching. A component of several contact lens wetting and cushioning solutions, and a preservative for ophthalmic solutions.

CONTRAINDICATIONS Casts, occlusive dressings, anal or vaginal packs, lactation.
CAUTIOUS USE Irrigation of body cavities; pregnancy (category C).

ROUTE & DOSAGE

Minor Wounds or Preoperative Disinfection
Adult: **Topical** 1:750 tincture or spray

Preoperative Disinfection of Denuded Skin and Mucous Membranes
Adult: **Topical** 1:10,000–1:2000 solution

Wet Dressings
Adult: **Topical** 1:5000 solution

Urinary Bladder Irrigation
Adult: **Topical** 1:20,000–1:5000 solution

Urinary Bladder Instillation
Adult: **Topical** 1:40,000–1:20,000 solution

Common adverse effects in *italic*, life-threatening effects underlined; generic names in **bold**; classifications in SMALL CAPS; ♣ Canadian drug name; ⊕ Prototype drug

163

Irrigation of Deep Infected Wounds

Adult: Topical 1:20,000–1:3000 solution

Vaginal Irrigation

Adult: Topical 1:5000–1:2000 solution

Sterile Storage of Instruments, Thermometers, Ampules

Adult: Topical 1:750 solution

ADMINISTRATION

Topical

- Use sterile water for injection as diluent for aqueous solutions to be instilled in wounds or body cavities. For other uses, fresh sterile distilled water is used.
- Irrigate eyes immediately and repeatedly with water if medication solution stronger than 1:5000 enters eyes; see a prescriber promptly.
- Rinse first with water, then with 70% alcohol, before applying benzalkonium for preoperative skin preparation.
- Consult prescriber about proper dilution of solutions used on denuded skin or inflamed or irritated tissues.
- Store at room temperature, preferably between 15°–30° C (59°–86° F) in airtight container, protected from light.

ADVERSE EFFECTS (≥1%) **Body as a Whole:** Few or no toxic effects in recommended dilutions. **Skin:** Erythema, local burning, hypersensitivity reactions.

NURSING IMPLICATIONS

Assessment & Drug Effects

- Monitor wounds carefully. Report increasing signs of infection or lack of healing.

✓BENZOCAINE

(ben'zoe-caine)

Americaine, Americaine Anesthetic Lubricant, Americaine-Otic, Anbesol Cold Sore Therapy, Chigger-Tox, Dermoplast, Foille, Hurricane, Orabase with Benzocaine, Orajel, Solarcaine, T-Caine

Classifications: LOCAL ANESTHETIC (ESTER TYPE); ANTIPRURITIC
Therapeutic: LOCAL ANESTHETIC; ANTIPRURITIC
Prototype: Procaine
Pregnancy Category: C

AVAILABILITY 5% spray, cream, ointment; 6% cream; 8% lotion, 20% spray, ointment, gel, liquid; 20% otic solution

ACTION & *THERAPEUTIC EFFECT*
Produces surface anesthesia by inhibiting conduction of nerve impulses from sensory nerve endings. Almost identical to procaine in chemical structure, but has prolonged duration of anesthetic action. *Temporary relief of pain and discomfort.*

USES Temporary relief of pain and discomfort in pruritic skin problems, minor burns and sunburn, minor wounds, and insect bites. Otic preparations are used to relieve pain and itching in acute congestive and serous otitis media, swimmer's ear, and otitis externa. Preparations are also available for toothache, minor sore throat pain, canker sores, hemorrhoids, rectal fissures, pruritus ani or vulvae, as male genital desensitizer to slow onset of ejaculation, and for use as anesthetic-lubricant for passage of catheters and endoscopic tubes.

CONTRAINDICATIONS Hypersensitivity to benzocaine or other PABA derivatives (e.g., sunscreen preparations), or to any of the components in the formulation; use of ear preparation in patients with perforated eardrum or ear discharge; applications to large areas. **Dental use:** Children younger than 2 y. **Anesthetic lubricant:** Infants younger than 1 y.

CAUTIOUS USE History of drug sensitivity; denuded skin or severely traumatized mucosa; pregnancy (category C).

ROUTE & DOSAGE

Anesthetic

Adult: **Topical** Lowest effective dose

Child: **Topical** Lower strengths

ADMINISTRATION

Topical

- Avoid contact of all preparations with eyes and be careful not to inhale mist when spray form is used.
- Do not use spray near open flame or cautery and do not expose to high temperatures. Hold can at least 12 inches (30 cm) away from affected area when spraying.
- Store at 15°–30° C (59°–86° F) in tight, light-resistant containers unless otherwise specified.

ADVERSE EFFECTS (≥1%) **Body as a Whole:** Low toxicity; sensitization in susceptible individuals; allergic reactions, anaphylaxis. **Hematologic:** Methemoglobinemia reported in infants.

INTERACTIONS Drug: Benzocaine may antagonize antibacterial activity of SULFONAMIDES.

PHARMACOKINETICS Absorption: Poorly absorbed through intact skin; readily absorbed from mucous membranes. **Peak:** 1 min. **Duration:** 15–30 min. **Metabolism:** By plasma cholinesterases and to a lesser extent by hepatic cholinesterases. **Elimination:** In urine.

NURSING IMPLICATIONS

Assessment & Drug Effects

- Assess swallowing when used on oral mucosa, as benzocaine may interfere with second (pharyngeal) stage of swallowing; hold food and liquids accordingly.
- Assess for sensitivity. Local anesthetics are potentially sensitizing to susceptible individuals when applied repeatedly or over extensive areas.

Patient & Family Education

- Use specific benzocaine preparation ONLY as prescribed or recommended by manufacturer.
- Discontinue medication if the condition persists, worsens, or if signs of sensitivity, irritation, or infection occur.

√BENZONATATE (Pr)

(ben-zoe′na-tate)

Tessalon

Classification: ANTITUSSIVE

Therapeutic: COUGH SUPPRESSANT

Pregnancy Category: C

AVAILABILITY 100 mg, 200 mg capsules

ACTION & *THERAPEUTIC EFFECT* Nonnarcotic antitussive activity that acts periperally by anesthetizing the receptors in the respiratory bronchi, lungs, and pleura, thus reducing the cough reflex. *Decreases frequency and intensity of nonproductive cough.*

USES Symptomatic treatment of cough.

UNLABELED USES Hiccups.

CONTRAINDICATIONS Hypersensitivity to benzonatate.

CAUTIOUS USE Pregnancy (category C), lactation; children younger than 10 y.

ROUTE & DOSAGE

Antitussive

Adult/Child (older than 10 y): **PO** 100 mg t.i.d. Max: 600 mg/day

ADMINISTRATION

Oral

- Ensure that soft capsules called perles are swallowed whole.
- Store in airtight containers protected from light.

ADVERSE EFFECTS (≥1%) **CNS:** Drowsiness, sedation, headache, mild dizziness. **GI:** Constipation, nausea. **Skin:** Rash, pruritus.

PHARMACOKINETICS **Onset:** 15–20 min. **Duration:** 3–8 h.

NURSING IMPLICATIONS

Assessment & Drug Effects
- Auscultate lungs anteriorly and posteriorly at scheduled intervals.
- Observe character and frequency of coughing and volume and quality of sputum. Keep prescriber informed.

Patient & Family Education
- Do not chew or allow perle to dissolve in mouth; swallow whole. If perle dissolves in mouth, the mouth, tongue, and pharynx will be anesthetized. Also, it is unpleasant to taste.

BENZPHETAMINE HYDROCHLORIDE

(benz-fet'a-meen)

Didrex

Classifications: CEREBRAL STIMULANT; ANOREXIANT
Therapeutic: ANOREXIANT
Prototype: Amphetamine
Pregnancy Category: X
Controlled Substance: Schedule III

AVAILABILITY 50 mg tablets

ACTION & THERAPEUTIC EFFECT Indirect acting sympathomimetic amine with amphetamine-like actions but with fewer side effects than amphetamine. Anorexiant effect thought to be secondary to stimulation of hypothalamus releasing stored catecholamines in the CNS. *Effective as an appetite suppressant.*

USES Short-term management of exogenous obesity.

CONTRAINDICATIONS Known hypersensitivity to sympathomimetic amines; angle-closure glaucoma; advanced arteriosclerosis, angina pectoris, severe cardiovascular disease, moderate to severe hypertension; hyperthyroidism, agitated states; history of drug abuse; lactation; pregnancy (category X).

CAUTIOUS USE Diabetes mellitus; older adults; psychosis; mild hypertension; children younger than 12 y.

ROUTE & DOSAGE

Obesity

Adult/Adolescent: **PO** 25–50 mg 1–3 times/day

ADMINISTRATION

Oral
- Give as a single daily dose, preferably midmorning or midafternoon, according to patient's eating habits.
- Schedule daily dose no later than 6 h before patient retires to avoid insomnia.

Common adverse effects in *italic*, life-threatening effects underlined; generic names in **bold**; classifications in SMALL CAPS; ♣ Canadian drug name; ⊘ Prototype drug

■ Store in tight, light-resistant containers at 15°–30° C (59°–86° F) unless otherwise directed.

ADVERSE EFFECTS (≥1%) **CNS:** Euphoria, irritability, hyperactivity, nervousness, *restlessness, insomnia,* tremor, headache, light-headedness, dizziness, depression following stimulant effects. Marked insomnia, irritability, hyperactivity, personality changes, psychosis, dermatoses. **CV:** *Palpitation,* tachycardia, elevated BP, irregular heartbeat. **GI:** Xerostomia, nausea, vomiting, diarrhea or constipation, abdominal cramps.

INTERACTIONS Drug: **Acetazolamide, sodium bicarbonate** decrease AMPHETAMINE elimination; **ammonium chloride, ascorbic acid** increase AMPHETAMINE elimination; BARBITURATES may antagonize the effects of both drugs; **furazolidone** may increase BP effects of AMPHETAMINES, and interaction may persist for several weeks after discontinuation of **furazolidone; guanethidine** antagonizes antihypertensive effects; because MAO INHIBITORS, **selegiline** can cause hypertensive crisis (fatalities reported); do not administer AMPHETAMINES during or within 14 days of these drugs; PHENOTHIAZINES may inhibit mood-elevating effects of AMPHETAMINES; BETA AGONISTS increase AMPHETAMINE'S adverse cardiovascular effects. **Herbal:** Do not use with melatonin.

PHARMACOKINETICS Absorption: Readily absorbed from GI tract. **Duration:** 4 h. **Metabolism:** Via CYP3A4. **Elimination:** Renal elimination.

NURSING IMPLICATIONS

Assessment & Drug Effects

■ Assess for signs of excessive CNS stimulation: Insomnia, restlessness, tremor, palpitations. These may indicate need for dosage adjustment.

■ Monitor vital signs; report elevated BP, tachycardia, and irregular heart rhythm.

■ Monitor diabetics for loss of glycemic control.

Patient & Family Education

■ Note: Anorexiant effects are temporary and tolerance may occur; long-term use is not indicated.

■ Do not drive or engage in potentially hazardous activities until response to drug is known.

■ Do not terminate high dosage therapy abruptly; GI distress, stomach cramps, trembling, unusual tiredness, weakness, and mental depression may result.

BENZTROPINE MESYLATE ℞

(benz'troe-peen)

Apo-Benztropine ✦, **Cogentin, PMS Benztropine** ✦
Classifications: CENTRALLY ACTING CHOLINERGIC RECEPTOR ANTAGONIST; ANTIPARKINSON
Therapeutic: ANTIPARKINSON
Pregnancy Category: C

AVAILABILITY 0.5 mg, 1 mg, 2 mg tablets; 1 mg/mL ampules

ACTION & *THERAPEUTIC EFFECT*
Synthetic centrally acting anticholinergic agent that acts by diminishing excess cholinergic effect associated with dopamine deficiency. *Suppresses tremor and rigidity; does not alleviate tardive dyskinesia.*

USES Parkinson's disease or Parkinsonism, drug-induced extrapyramidal symptoms, tremor.

CONTRAINDICATIONS Narrow-angle glaucoma; myasthenia gravis; obstructive diseases of GU and GI tracts;

Common adverse effects in *italic*, life-threatening effects underlined; generic names in **bold**; classifications in SMALL CAPS; ✦ Canadian drug name; ℗ Prototype drug

167

tendency to tachycardia; tardive dyskinesia.

CAUTIOUS USE Older children, older adults or debilitated patients, patients with poor mental outlook, mental disorders; tachycardia; autonomic neuropathy; enlarged prostate; hypertension; history of renal or hepatic disease; pregnancy (category C), lactation; children younger than 3 y.

ROUTE & DOSAGE

Parkinsonism

Adult: **PO/IM** 0.5–1 mg/day, may gradually increase as needed up to 6 mg/day

Extrapyramidal Reactions

Adult: **PO/IM/IV** 1–4 mg once or twice daily

Child (older than 3 y): **PO/IM/IV** 0.02–0.05 mg/kg/dose once or twice daily

Acute Dystonia

Adult: **IV** 1–2 mg daily

ADMINISTRATION

Oral

- Give immediately after meals or with food to prevent gastric irritation. Tablet can be crushed and sprinkled on or mixed with food.
- Initiate and withdraw drug therapy gradually; effects are cumulative.
- Store in tightly covered, light-resistant container at 15°–30° C (59°–86° F) unless otherwise directed.

Intramuscular

- Inject undiluted solution deeply into a large muscle.

Intravenous

IV administration to infants and children: Verify correct IV concentration with prescriber.

PREPARE: Direct: Give undiluted.

ADMINISTER: Direct: Give 1 mg or a fraction thereof over 1 min.

ADVERSE EFFECTS (≥1%) CNS: *Sedation,* drowsiness, dizziness, paresthesias; agitation, irritability, restlessness, nervousness, insomnia, hallucinations, delirium, mental confusion, toxic psychosis, muscular weakness, ataxia, inability to move certain muscle groups. **CV:** Palpitation, tachycardia, flushing. **Special Senses:** Blurred vision, mydriasis, photophobia. **GI:** Nausea, vomiting, *constipation, dry mouth,* distention, paralytic ileus. **Urogenital:** Dysuria.

INTERACTIONS Drug: Amantadine, TRICYCLIC ANTIDEPRESSANTS, MAO INHIBITORS, PHENOTHIAZINES, **procainamide, quinidine** have additive anticholinergic effects and cause confusion, hallucinations, paralytic ileus. **Nifedipine** extended release increases the risk of GI obstruction.

PHARMACOKINETICS Onset: 15 min IM/IV; 1 h PO. **Duration:** 6–10 h.

NURSING IMPLICATIONS

Assessment & Drug Effects

- Monitor I&O ratio and pattern. Advise patient to report difficulty in urination or infrequent voiding. Dosage reduction may be indicated.
- Closely monitor for appearance of S&S of onset of paralytic ileus including intermittent constipation, abdominal pain, diminution of bowel sounds on auscultation, and distention.
- Monitor HR especially in patients with a tendency toward tachycardia.
- Monitor for and report muscle weakness or inability to move certain muscle groups. Dosage reduction may be needed.
- Supervise ambulation and use protective measures as necessary.
- Report immediately S&S of CNS depression or stimulation. These usually require interruption of drug therapy.

Patient & Family Education

- Do not drive or engage in potentially hazardous activities until response to drug is known. Seek help walking as necessary.
- Avoid alcohol and other CNS depressants because they may cause additive drowsiness. Do not take OTC cold, cough, or hay fever remedies unless approved by prescriber.
- Sugarless gum, hard candy, and rinsing mouth with tepid water will help dry mouth.
- Avoid strenuous exercise in hot weather; diminished sweating may require dose adjustments because of possibility of heat stroke.

BETAMETHASONE

(bay-ta-meth'a-sone)
Betnelan ♦, Celestone

BETAMETHASONE ACETATE AND BETAMETHASONE SODIUM PHOSPHATE

Celestone Soluspan

BETAMETHASONE BENZOATE

Beben ♦

BETAMETHASONE DIPROPIONATE

Alphatrex, Diprolene

BETAMETHASONE SODIUM PHOSPHATE (PH 8.5)

Betameth, Betnesol ♦, Celestone S

BETAMETHASONE VALERATE

Betaderm ♦, Beta-Val, Betnovate ♦, Celestoderm ♦, Ectosone Lotion ♦, Luxiq, Metaderm ♦, Novobetamet ♦, Valnac

Classifications: ADRENAL CORTICOSTEROID; GLUCOCORTICOID; ANTI-INFLAMMATORY
Therapeutic: ANTI-INFLAMMATORY; ADRENAL CORTICOSTEROID
Prototype: Hydrocortisone
Pregnancy Category: D in first trimester and C in second and third trimester

AVAILABILITY **Betamethasone:** 0.6 mg tablets; 0.6 mg/5 mL syrup; **Betamethasone Acetate and Betamethasone Sodium:** 3 mg acetate, 3 mg sodium phosphate/mL suspension; **Betamethasone Benzoate and Betamethasone Dipropionate:** 4 mg/mL injection; **Betamethasone Valerate:** 0.1% ointment; 0.01%, 0.05%, 0.1% cream; 0.1% lotion; 1.2 mg/g foam; **Betamethasone Sodium Phosphate:** 0.6 mg/5 mL syrup

ACTION & *THERAPEUTIC EFFECT* Synthetic, long-acting glucocorticoid with minor mineralocorticoid properties but strong immunosuppressive, anti-inflammatory, and metabolic actions. *Relieves anti-inflammatory manifestations and is an immunosuppressive agent.*

USES Reduces serum calcium in hypercalcemia, suppresses undesirable inflammatory or immune responses, produces temporary remission in nonadrenal disease, and blocks ACTH production in diagnostic tests. Topical use provides relief of inflammatory manifestations of corticosteroid-responsive dermatoses.

UNLABELED USES Prevention of neonatal respiratory distress syndrome (hyaline membrane disease).

CONTRAINDICATIONS In patients with systemic fungal infections; acne vulgaris; acne rosacea; Cushing's syndrome; periorbital dermatitis;

Common adverse effects in *italic*, life-threatening effects underlined; generic names in **bold**; classifications in SMALL CAPS; ♦ Canadian drug name; ☻ Prototype drug

vaccines; pregnancy (category D in first trimester); lactation.

CAUTIOUS USE Ocular herpes simplex; concomitant use of aspirin; osteoporosis; diverticulitis, nonspecific ulcerative colitis, abscess or other pyrogenic infection, peptic ulcer disease; asthmatics; diabetes mellitus; hypertension; renal insufficiency; myasthenia gravis; pregnancy (category C in second and third trimester).

ROUTE & DOSAGE

Anti-Inflammatory Agent

Adult: **PO** 0.6–7.2 mg/day **IM/IV** Up to 9 mg/day as sodium phosphate

Topical

See Appendix A-4.
Child: **PO** 0.0175–0.25 mg/kg/day or 0.5–0.75 mg/m²/day divided q6–8h **IM** 0.0175–0.125 mg/kg/day or 0.5–0.75 mg/m²/day divided q6–8h

Respiratory Distress Syndrome

Adult: **IM** 2 mL of sodium phosphate to mother once daily 2–3 days before delivery

ADMINISTRATION

Oral

- Give with food or milk to lessen stomach irritation.

Intra-articular/Intramuscular/Intralesional

- Use Celestone Soluspan for intra-articular, IM, and intralesional injection. The preparation is not intended for IV use. Do not mix with diluents containing preservatives (e.g., parabens, phenol).
- Use 1% or 2% lidocaine hydrochloride if prescribed. Withdraw betamethasone mixture first, then lidocaine; shake syringe briefly.

Intravenous

PREPARE: **Direct/IV Infusion:** Give by direct IV undiluted or further diluted for infusion in D5W or NS.
ADMINISTER: **Direct:** Give at a rate of 1 dose/min. **IV Infusion:** Give at a rate determined by the total amount of IV fluid.
INCOMPATIBILITIES **Solution/additive and Y-site:** Do not infuse with other drugs.

ADVERSE EFFECTS (≥1%) **Body as a Whole:** Hypersensitivity or <u>anaphylactoid reactions; aggravation or masking of infections;</u> malaise, weight gain, obesity. Most adverse effects are dose and treatment duration dependent. **CNS:** Vertigo, headache, nystagmus, ataxia (rare), increased intracranial pressure with papilledema (usually after discontinuation of medication), mental disturbances, aggravation of preexisting psychiatric conditions, insomnia. **CV:** Hypertension; syncopal episodes, thrombophlebitis, thromboembolism or fat embolism, palpitation, tachycardia, necrotizing angitis; CHF. **Endocrine:** Suppressed linear growth in children, decreased glucose tolerance; hyperglycemia, manifestations of latent diabetes mellitus; hypocorticism; amenorrhea and other menstrual difficulties. **Special Senses:** Posterior subcapsular cataracts (especially in children), glaucoma, exophthalmos, increased intraocular pressure with optic nerve damage, perforation of the globe, fungal infection of the cornea, decreased or blurred vision. **Metabolic:** Hypocalcemia; *sodium and fluid retention;* hypokalemia and hypokalemic alkalosis; negative nitrogen balance. **GI:** *Nausea,* increased appetite, ulcerative esophagitis, pancreatitis, abdominal distention, peptic ulcer with perforation and hemorrhage, melena; decreased serum concentration of vitamins A and C. **Hematologic:** Thrombocytopenia. **Musculoskeletal:** Osteoporosis, compression fractures, muscle wasting and weakness, tendon rupture, aseptic necrosis of femoral and humeral heads (all resulting from long-term use). **Skin:** Skin thinning and atrophy, *acne, impaired wound healing;* petechiae, ecchymosis, easy bruising; suppression of skin test reaction; hypopigmentation or hyperpigmentation, hirsutism, acneiform eruptions, subcutaneous fat atrophy; allergic dermatitis, urticaria, angioneurotic edema, increased sweating. **Urogenital:** Increased or decreased motility and number of sperm; urinary frequency and urgency, enuresis. **With parenteral therapy, IV site:** Pain, irritation, necrosis, atrophy, sterile abscess; Charcot-like arthropathy following intra-articular use; burning and tingling in perineal area (after IV injection).

DIAGNOSTIC TEST INTERFERENCE May increase *serum cholesterol, blood glucose, serum sodium, uric acid* (in acute leukemia) and *calcium* (in bone metastasis). It may decrease *serum calcium, potassium, PBI, thyroxin (T_4), triiodothyronine (T_3) and reduce thyroid I 131* uptake. It increases *urine glucose* level and *calcium* excretion; decreases *urine 17-OHCS* and *17-KS* levels. May produce false-negative results with *nitroblue tetrazolium test* for systemic bacterial infection and may suppress reactions to skin tests.

INTERACTIONS Drug: BARBITURATES, **phenytoin, rifampin** may reduce pharmacologic effect of betamethasone by increasing its metabolism.

PHARMACOKINETICS Peak: 1–2 h **Half-Life:** 35–54 h

NURSING IMPLICATIONS

Assessment & Drug Effects
▪ Assess therapeutic effectiveness. Response following intra-articular, intralesional, or intrasynovial administration occurs within a few hours and persists for 1–4 wk. Following IM administration response occurs in 2–3 h and persists for 3–7 days.

Patient & Family Education
▪ Monitor weight at least weekly.
▪ Discontinue slowly after systemic use of 1 wk or longer. Abrupt withdrawal, especially following high doses or prolonged use, can cause dizziness, nausea, vomiting, fever, muscle and joint pain, weakness.

BETAXOLOL HYDROCHLORIDE
(be-tax'oh-lol)

Betoptic, Betoptic-S, Kerlone
Classifications: EYE PREPARATION; MIOTIC (ANTIGLAUCOMA); BETA-ADRENERGIC ANTAGONIST; ANTIHYPERTENSIVE
Therapeutic: ANTIGLAUCOMA (MIOTIC); ANTIHYPERTENSIVE
Prototype: Propranolol
Pregnancy Category: C

AVAILABILITY 10 mg, 20 mg tablets; 0.25%, 0.5% ophthalmic solution

ACTION & *THERAPEUTIC EFFECT*
Acts as a beta$_1$-selective adrenergic receptor blocking agent, especially in the cardioselective beta$_1$ receptors. Its antihypertensive effect is thought to be due to: (1) decreasing cardiac output, (2) reducing sympathetic nervous system outflow to the periphery resulting in

Common adverse effects in *italic*, life-threatening effects <u>underlined</u>; generic names in **bold**; classifications in SMALL CAPS; ✦ Canadian drug name; ☻ Prototype drug

171

vasodilatation, and (3) suppression of renin activity in the kidney. It reduces intraocular pressure within the eye by decreasing the production of aqueous humor. *It has antihypertensive and antiglaucoma effects.*

USES Hypertension. Ocular use for intraocular hypertension, chronic open angle glaucoma (see Appendix A-1).

CONTRAINDICATIONS Hypersensitivity to beta-blockers; sinus bradycardia, AV block greater than first degree, cardiogenic shock, overt cardiac failure; CHF unless secondary to tachyarrhythmia treatable with beta-blockers; glaucoma, angle closure (unless with a miotic); lactation.

CAUTIOUS USE History of heart failure; renal impairment, hepatic impairment; elderly; hyperthyroidism or thyrotoxicosis; diabetes mellitus; with evidence of airflow obstruction or reactive airway disease; depression; pregnancy (category C); children younger than 18 y.

ROUTE & DOSAGE

Hypertension

Adult: **PO** 10–20 mg daily (max: 40 mg/day in 1–2 divided doses)
Elderly: **PO** Start with 5 mg/day and taper up

Renal Impairment Dosage Adjustment

CrCl less than 60 mL/min: Administer 50% of dose (max: 20 mg/day)

Chronic Open-Angle Glaucoma/ Ocular Hypertension

See Appendix A-1.

ADMINISTRATION

Oral

▪ Check pulse before administering betaxolol, oral or ophthalmic. If there are extremes (rate or rhythm), withhold medication and notify prescriber.

ADVERSE EFFECTS (≥1%) **CV:** Bradycardia, hypotension. **CNS:** Depression. **Respiratory:** Increased airway resistance. **Special Senses:** With ophthalmic solution, *mild ocular stinging* and discomfort, tearing.

INTERACTIONS Drug: Reserpine and other CATECHOLAMINE-DEPLETING AGENTS may cause additive hypotensive effects or bradycardia. **Verapamil** may cause additive heart block.

PHARMACOKINETICS Absorption: 90% of PO dose reaches systemic circulation. **Onset:** 0.5–1 h. **Peak:** 2 h. **Duration:** Greater than 12 h. **Metabolism:** In liver. **Elimination:** 30–40% in urine, 50–60% in bile and feces. **Half-Life:** 12–22 h.

NURSING IMPLICATIONS

Assessment & Drug Effects

▪ Monitor pulse rate and BP at regular intervals in patients with known heart disease.
▪ Report promptly onset of bradycardia or signs of CHF.
▪ Monitor therapeutic effectiveness. Some patients develop tolerance during long-term therapy.

Patient & Family Education

▪ Report unusual pulse rate or significant changes to prescriber according to parameters provided.
▪ Adhere to regimen EXACTLY as prescribed. Do not stop drug abruptly; angina may be exacerbated; dosage is reduced over a period of 1–2 wk.

- Report difficulty in breathing promptly to prescriber. Drug withdrawal may be indicated.

BETHANECHOL CHLORIDE ⏺

(be-than′e-kole)

Urecholine
Classification: DIRECT-ACTING CHO-LINERGIC
Therapeutic: CHOLINERGIC
Pregnancy Category: C

AVAILABILITY 5 mg, 10 mg, 25 mg, 50 mg tablets

ACTION & *THERAPEUTIC EFFECT*
Synthetic choline ester with effects similar to those of acetylcholine (ACh). Acts directly on postsynaptic receptors, and since it is not hydrolyzed by cholinesterase, its actions are more prolonged than those of ACh. Produces muscarinic effects primarily on GI tract and urinary bladder. Increases tone and peristaltic activity of esophagus, stomach, and intestine; contracts detrusor muscle of urinary bladder, usually enough to initiate micturition. *Bethanechol is a synthetic parasympathomimetic indicated for the treatment of urinary retention associated with neurogenic bladder.*

USES Acute postoperative and postpartum nonobstructive (functional) urinary retention, and for neurogenic atony of urinary bladder with retention.
UNLABELED USES Anticholinergic syndrome, GERD, ileus.

CONTRAINDICATIONS COPD; history of or active bronchial asthma; hyperthyroidism; recent urinary bladder surgery, cystitis, bacteriuria, urinary bladder neck or intestinal obstruction, peptic ulcer, recent GI surgery, peritonitis; marked vagotonia, pronounced vasomotor instability, AV conduction defects, severe bradycardia, hypotension or hypertension, coronary artery disease, recent MI; epilepsy, parkinsonism; lactation, children younger than 8 y.
CAUTIOUS USE Urinary retention; bacteriemia; patients at risk for syncopy; pregnancy (category C).

ROUTE & DOSAGE

Urinary Retention

Adult: **PO** 5–10 mg hourly until goals attained (max: 50 mg dose), usually 10–50 mg 3–4 times daily

ADMINISTRATION

Oral

- Give on an empty stomach (1 h before or 2 h after meals) to lessen possibility of nausea and vomiting, unless otherwise advised by prescriber.

ADVERSE EFFECTS (≥1%) **Body as a Whole:** (Dose-related) Increased sweating, malaise, headache, substernal pain or pressure, hypothermia. **CV:** Hypotension with dizziness, faintness, flushing, orthostatic hypotension (large doses); mild reflex tachycardia, atrial fibrillation (hyperthyroid patients), transient complete heart block. **Special Senses:** Blurred vision, miosis, lacrimation. **GI:** Nausea, vomiting, abdominal cramps, diarrhea, borborygmi, belching, salivation, fecal incontinence (large doses), urge to defecate (or urinate). **Respiratory:** Acute asthmatic attack, dyspnea (large doses).

DIAGNOSTIC TEST INTERFERENCE

Bethanechol may cause increases in ***serum amylase*** and ***serum lipase,*** by stimulating pancreatic secretions, and may increase *AST, serum bilirubin,* and *BSP retention* by causing spasms in sphincter of Oddi.

INTERACTIONS Drug: Amoxapine, ANTIMUSCARINICS, **maprotiline,** TRICYCLIC ANTIDEPRESSANTS **may decrease the effect of bethanechol. Ambenonium, neostigmine,** other CHOLINESTERASE INHIBITORS compound cholinergic effects and toxicity; **procainamide, quinidine, atropine, epinephrine** antagonize effects of bethanechol.

PHARMACOKINETICS Absorption: Poorly absorbed. **Onset:** 30 min. **Duration:** 1 h. **Metabolism:** Unknown. **Elimination:** Unknown.

NURSING IMPLICATIONS

Assessment & Drug Effects

▪ Monitor BP and pulse. Report early signs of overdosage: Salivation, sweating, flushing, abdominal cramps, nausea.

▪ Monitor I&O. Observe and record patient's response to bethanechol.

▪ Monitor respiratory status. Promptly report dyspnea or any other indication of respiratory distress.

▪ Supervise ambulation as indicated by patient response to drug.

Patient & Family Education

▪ Make position changes slowly and in stages, particularly from lying down to standing.

▪ Do not stand still for prolonged periods; sit or lie down at first indication of faintness.

▪ Do not drive or engage in potentially hazardous activities until response to drug is known.

▪ Note: Drug may cause blurred vision; take appropriate precautions.

BEVACIZUMAB
(be-va-ci-zu′mab)

Avastin
Classifications: ANTINEOPLASTIC; BIOLOGICAL RESPONSE MODIFIER; MONOCLONAL ANTIBODY
Therapeutic: ANTINEOPLASTIC
Pregnancy Category: C

AVAILABILITY 25 mg/mL injection

ACTION & *THERAPEUTIC EFFECT*
Binds to vascular endothelial growth factor (VEGF) and prevents the interaction of VEGF to its receptors on the surface of endothelial cells. This blocks endothelial cell proliferation and new blood vessel formation in tumor cells. *Believed to cause reduction of microvascularization in the tumor inhibiting the progression of metastatic disease.*

USES Metastatic colorectal cancer, non–small-cell lung cancer, malignant glioblastoma.
UNLABELED USES Metastatic renal cell cancer, age-related macular degeneration.

CONTRAINDICATIONS Nephrotic syndrome; active bleeding, recent hemoptysis; GI perforation; nephritic syndrome; leucopenia; surgery within 28 days; dental work within 20 days; severe thromboembolic event due to drug; hypertensive crisis; lactation.
CAUTIOUS USE Hypersensitivity to bevacizumab; renal disease; hypertension, history of arterial thromboembolic, cardiovascular, or cerebrovascular disease; CHF; history of GI bleeding; pregnancy (category C). Safety and effectiveness

in children and infants are not established.

ROUTE & DOSAGE

Metastatic Colorectal Cancer

Adult: IV 5–10 mg/kg q14days until disease progression; in conjunction with other chemotherapy

Non–Small-Cell Lung Cancer

Adult: IV 15 mg/kg q3wk

Glioblastoma

Adult: IV 10 mg/kg q14days in 28-day cycle

ADMINISTRATION

Intravenous

PREPARE: **IV Infusion:** Withdraw the desired dose of 5 mg/kg and dilute in 100 mL of NS injection. ▪ Do not shake and do NOT mix or administer with dextrose solutions. Discard any unused portion.

ADMINISTER: **IV Infusion:** DO NOT administer IV push or bolus. ▪ Infuse first dose over 90 min; if well tolerated, infuse second dose over 60 min; if well tolerated, infuse all subsequent doses over 30 min.

INCOMPATIBILITIES **Solution/additive: Dextrose**-containing solutions. **Y-site: Dextrose**-containing solutions.

▪ Store diluted solution at 2°–8° C (36°–46° F) for up to 8 h. Store vials at 2°–8° C (36°–46° F) and protect from light.

ADVERSE EFFECTS (≥1%) Body as a Whole: *Asthenia,* pain, wound

dehiscence, tracheoesophageal (TE) fistula formation. **CNS:** Syncope, headache, dizziness, confusion, abnormal gait, leukoencephalopathy. **CV:** DVT, *hypertension,* heart failure, intraabdominal thrombosis, cerebrovascular events. **GI:** Abdominal pain, *diarrhea,* constipation, nausea, vomiting, anorexia, stomatitis, dyspepsia, weight loss, flatulence, dry mouth, colitis, gastrointestinal perforation. **Hematologic:** Leukopenia, *neutropenia,* thrombocytopenia, hemorrhage, *thromboembolism.* **Metabolic:** Hypokalemia, hyperbilirubinemia. **Musculoskeletal:** Myalgia. **Respiratory:** Upper respiratory infection, epistaxis, dyspnea, hemoptysis. **Skin:** Exfoliative dermatitis, alopecia. **Special Senses:** Taste disorder, increased tearing. **Urogenital:** *Proteinuria,* urinary frequency/urgency.

PHARMACOKINETICS Half-Life: 20 days (11–50 days).

NURSING IMPLICATIONS

Assessment & Drug Effects

▪ Monitor for S&S of an infusion reaction (hypersensitivity); infusion should be interrupted in all patients with severe infusion reactions and appropriate therapy instituted.

▪ Monitor BP at least every 2–3 wk; if hypertension develops, monitor more frequently, even after discontinuation of bevacizumab.

▪ Withhold drug and promptly notify prescriber for S&S of CHF, hemorrhage (e.g., epistaxis, hemoptysis, or GI bleeding), or unexplained abdominal pain.

▪ Lab tests: Urinalysis for proteinuria and 24 h urine if protein 2+ or greater.

B

- Monitor for dizziness, lightheadedness, or loss of balance. Take appropriate safety measures.

Patient & Family Education

- Report any of the following to the prescriber: Bloody or black, tarry stool; changes in patterns of urination; swelling of legs or ankles; increased shortness of breath; severe abdominal pain; change in mental awareness, inability to talk or move one side of the body.

BEXAROTENE
(bex-a-ro′teen)

Targretin
Classifications: ANTINEOPLASTIC
Therapeutic: ANTINEOPLASTIC
Prototype: Isotretinoin
Pregnancy Category: X

AVAILABILITY 75 mg capsules; 1% gel

ACTION & *THERAPEUTIC EFFECT*
Selectively binds to retinoid X receptors (RXR). Activation of the RXR pathway leads to cell death by interfering with cellular differentiation and proliferation of cells. *Inhibits the growth of tumor cells of squamous (skin) cell origin inducing tumor regression.*

USES Treatment of cutaneous manifestations of cutaneous T-cell lymphoma, mycosis fungoides.

CONTRAINDICATIONS Hypersensitivity to bexarotene; pregnancy (category X), lactation.
CAUTIOUS USE Hypersensitivity to retinoid agents; coronary artery disease; diabetes mellitus; alcoholism, history of pancreatitis, hepatitis; elevated triglycerides, hepatic

impairment. Safety and efficacy in children are not established.

ROUTE & DOSAGE

T-Cell Lymphoma
Adult: **PO** 300 mg/m²/day as a single dose if no response after 8 wk, may increase to 400 mg/m²/day. Adjust dose downward in 100 mg/m²/day increments if toxicity occurs. **Topical** Apply once every other day × 1 week, increase frequency at weekly intervals to once/day, b.i.d., t.i.d., and q.i.d.

ADMINISTRATION

Oral

- Give drug with or immediately following a meal. Ensure that capsules are swallowed whole.
- Do not initiate therapy in a woman of childbearing age until the possibility of pregnancy has been completely ruled out.

Topical

- Apply a generous coating only to skin lesions; avoid normal skin.
- Do not cover with clothing until gel dries.
- Store capsules and gel at 20°–25° C (36°–77° F). Protect from light and avoid high temperatures and humidity after bottle or tube is opened.

ADVERSE EFFECTS (≥1%) **Body as a Whole:** *Headache, asthenia, infection,* chills, fever, flu-like syndrome, back pain, bacterial infection. **CNS:** Insomnia. **CV:** *Peripheral edema.* **GI:** *Abdominal pain, nausea,* diarrhea, vomiting, anorexia. **Endocrine:** *Hyperthyroidism.* **Hematologic:** Leukopenia, anemia, hypochromic anemia. **Metabolic:**

Common adverse effects in *italic*, life-threatening effects underlined; generic names in **bold;** classifications in SMALL CAPS; ♣ Canadian drug name; ☯ Prototype drug

Hyperlipidemia, hypercholesterol-emia, increased LDH. **Skin:** *Rash, dry skin,* exfoliative dermatitis, alopecia, photosensitivity.

INTERACTIONS No clinically significant interactions established.

PHARMACOKINETICS Absorption: Best with a fat-containing meal. **Peak:** 2 h. **Distribution:** Greater than 99% protein bound. **Metabolism:** Metabolized by CYP3A4. **Elimination:** Primarily in bile. **Half-Life:** 7 h.

NURSING IMPLICATIONS

Assessment & Drug Effects

- Monitor (with oral dose) for S&S of: Hypothyroidism, hypertriglyceridemia, hypercholesterolemia, and pancreatitis.
- Lab tests (with oral dose): Baseline blood lipids, then weekly for 2–4 wk, and every 8 wk thereafter; baseline LFTs then repeat at 1, 2, 4 wk, and every 8 wk thereafter; baseline WBC and thyroid function tests, then repeat periodically thereafter; periodic serum calcium; for females, pregnancy test monthly throughout therapy.
- Withhold oral drug and notify prescriber if triglycerides greater than 400 mg/dL or AST, ALT, or bilirubin greater than 3 times upper limit of normal.

Patient & Family Education

- Use effective methods of contraception (both men and women) while taking/using this drug and for at least 1 mo after the last dose of the drug.
- Do not take this drug if you are or could be pregnant.
- Report immediately any of the following: Swelling in the face, lips, or wheezing; persistent

bloating, constipation, diarrhea, vomiting, or stomach pain; persistent headache, severe drowsiness or weakness.
- Report changes in vision to the prescriber. An ophthalmologic evaluation may be needed.
- Limit exposure to sunlight or sun lamps and wear sunscreen.
- Report significant skin irritation.

BICALUTAMIDE

(bi-ca-lu′ta-mide)

Casodex

Classifications: ANTINEOPLASTIC; HORMONE

Therapeutic: ANTINEOPLASTIC; NONSTEROIDAL ANTIANDROGEN

Prototype: Flutamide

Pregnancy Category: X

AVAILABILITY 50 mg tablets

ACTION & *THERAPEUTIC EFFECT*
Bicalutamide is a nonsteroidal antiandrogen that inhibits the pharmacologic effects of androgen by binding to androgen receptors in target tissue. *Prostatic carcinoma is androgen sensitive; it responds to removal of the source of androgen or treatment that counteracts the effects of androgen.*

USES In combination with a luteinizing hormone-releasing hormone (LHRH) analog for advanced prostate cancer.

CONTRAINDICATIONS Hypersensitivity to bicalutamide, pregnancy (category X), hepatic failure; lactation.

CAUTIOUS USE Moderate to severe hepatic impairment; glucose intolerance; diabetes mellitus. Safety and efficacy in children are not established.

Common adverse effects in *italic,* life-threatening effects underlined; generic names in **bold;** classifications in SMALL CAPS; ♣ Canadian drug name; ❸ Prototype drug

177

B

ROUTE & DOSAGE

Advanced Prostate Cancer
Adult: **PO** 50 mg once/day

ADMINISTRATION

Oral
- Give drug at the same time each day.
- Start treatment with bicalutamide at the same time as treatment with a luteinizing hormone-releasing hormone (LHRH) analog.
- Store at 15°–30° C (59°–86° F).

ADVERSE EFFECTS (≥1%) **CNS:** Dizziness, paresthesia, insomnia, anxiety, decreased libido, confusion, neuropathy, somnolence, nervousness, headache. **CV:** *Hot flashes,* hypertension, chest pain, CHF. **GI:** *Constipation, nausea, diarrhea,* vomiting, increased liver function tests, abdominal pain, anorexia, dyspepsia, dry mouth, melena. **Urogenital:** Nocturia, hematuria, UTI, impotence, gynecomastia, incontinence, frequency, dysuria, urinary retention, urgency. **Metabolic:** Peripheral edema, hyperglycemia, weight loss, weight gain, gout. **Musculoskeletal:** Myasthenia, arthritis, myalgia, leg cramps, pathologic fractures. **Skin:** Rash, sweating, dry skin, pruritus, alopecia. **Body as a Whole:** Flu syndrome, bone pain, infection, anemia.

INTERACTIONS Drug: May increase effects of ORAL ANTICOAGULANTS. Bicalutamide concentrations may be increased by PROTEASE INHIBITORS, **fluoxetine, fluvoxamine, erythromycin** and other CYP3A4 inhibitors. Efficacy of bicalutamide may be decreased by **bosentan,** **carbamazepine,** BARBITURATES and other CYP3A4 inducers.

PHARMACOKINETICS Absorption: Readily from GI tract. **Metabolism:** In liver. **Elimination:** In urine and feces. **Half-Life:** 5.8 days.

NURSING IMPLICATIONS

Assessment & Drug Effects
- Monitor for S&S of disease progression.
- Lab tests: Periodic PSA levels, CBC, LFTs, renal functions; with concurrent Coumadin therapy, closely monitor PT and INR.

Patient & Family Education
- Report jaundice or any other troubling adverse effects immediately.

BIMATOPROST
(bi-mat′o-prost)

Lumigan
Pregnancy Category: C
See Appendix A-1.

BISACODYL ⊙
(bis-a-koe′dill)

Apo-Bisacodyl ✦, **Bisacolax, Bisco-Lax** ✦, **Dacodyl, Deficol, Doxidan, Dulcolax, Fleet Bisacodyl, Laxit** ✦, **Theralax**
Classification: STIMULANT LAXATIVE
Therapeutic: LAXATIVE
Pregnancy Category: C

AVAILABILITY 5 mg tablets, enteric coated; 5 mg tablets, delayed release; 10 mg suppository; 10 mg/30 mL enema

ACTION & *THERAPEUTIC EFFECT* Expands intestinal fluid volume by increasing epithelial permeability.

Common adverse effects in *italic*, life-threatening effects <u>underlined</u>; generic names in **bold;** classifications in SMALL CAPS; ✦ Canadian drug name; ⊙ Prototype drug

Induces peristaltic contractions by direct stimulation of sensory nerve endings in the colonic wall.

USES Temporary relief of acute constipation and for evacuation of colon before GI procedures.

CONTRAINDICATIONS Acute surgical abdomen, nausea, vomiting, abdominal cramps, intestinal obstruction, fecal impaction; use of rectal suppository in presence of anal or rectal fissures, ulcerated hemorrhoids, proctitis, bowel obstruction or perforation, ileus; children younger than 1 y.
CAUTIOUS USE Pregnancy (category C), lactation.

ROUTE & DOSAGE

Laxative

Adult: PO 5–15 mg prn (max: 30 mg for special procedures) PR 10 mg prn
Child: PO 6 y or older, 5–10 mg prn PR 2 y or older, 10 mg; younger than 2 y, 5 mg

ADMINISTRATION

Oral
- Give in the evening or before breakfast because of action time required.
- Ensure that enteric-coated tablets are swallowed whole; they should not be crushed or chewed. Do not give within 1 h of antacids or milk.
- Store tablets in tightly closed containers at temperatures not exceeding 30° C (86° F).

Rectal
- Suppository may be inserted at time bowel movement is desired.
- Storage is same as tablets.

ADVERSE EFFECTS (≥1%) Systemic effects not reported. Mild cramping, nausea, diarrhea, fluid and electrolyte disturbances (especially potassium and calcium).

INTERACTIONS Drug: ANTACIDS will cause early dissolution of enteric-coated tablets, resulting in abdominal cramping.

PHARMACOKINETICS Absorption: 5–15% from GI tract. **Onset:** 6–8 h PO; 15–60 min PR. **Metabolism:** In liver. **Elimination:** In urine, bile, and breast milk.

NURSING IMPLICATIONS

Assessment & Drug Effects
- Evaluate periodically patient's need for continued use of drug; bisacodyl usually produces 1 or 2 soft formed stools daily.
- Monitor patients receiving concomitant anticoagulants. Indiscriminate use of laxatives results in decreased absorption of vitamin K.

Patient & Family Education
- Add high-fiber foods slowly to regular diet to avoid gas and diarrhea. Adequate fluid intake includes at least 6–8 glasses/day.

BISMUTH SUBSALICYLATE ⊕

(bis'muth)

Pepto-Bismol
Classifications: ANTIDIARRHEAL; SALICYLATE
Therapeutic: ANTIDIARRHEAL
Pregnancy Category: C

AVAILABILITY 262 mg tablets/caplets; 130 mg/15 mL, 262 mg/15 mL, 524 mg/15 mL liquid

ACTION & *THERAPEUTIC EFFECT*
Hydrolyzed in GI tract to salicylate, which is believed to inhibit synthesis of prostaglandins responsible for GI hypermotility and inflammation.

Common adverse effects in *italic*, life-threatening effects underlined; generic names in **bold**; classifications in SMALL CAPS; ✤ Canadian drug name; ⊕ Prototype drug

179

It acts as a direct mucosal protective agent. Effectiveness as an antidiarrheal appears to be due to direct antimicrobial action and to an antisecretory effect on intestinal secretions exposed to toxins.

USES Prophylaxis and treatment of traveler's diarrhea (turista) and for temporary relief of indigestion.
UNLABELED USES *Helicobacter pylori* associated with peptic ulcer disease.

CONTRAINDICATIONS Hypersensitivity to aspirin or other salicylates; coagulopathy, severe hepatic impairment; use for more than 2 days in presence of high fever or in children younger than 3 y unless prescribed by prescriber; chickenpox or flu; dysentery.
CAUTIOUS USE Diabetes and gout; alcoholism; renal impairment; older adults; smoking; pregnancy (category C), lactation.

ROUTE & DOSAGE

Diarrhea

Adult: **PO** 30 mL or 2 tab q30–60min prn (max: 8 doses/day)
Child: **PO** *3–6 y,* 5 mL or 1/2 tab q30–60min prn (max: 8 doses/day); *6–9 y,* 2/3 tab or 10 mL q30–60min prn (max: 8 doses/day); *9–12 y,* 15 mL or 1 tab q30–60min prn (max: 8 doses/day)

Traveler's Diarrhea

Adult: **PO** 2–4 tab or 15–30 mL q.i.d. for 3 wk

Peptic Ulcer Disease

Adult: **PO** 2 tablets q.i.d. with 2 additional antibiotics for 10–14 days
Child (younger than 10 y): **PO** 15 mL q.i.d. × 6 wk

ADMINISTRATION

Oral

- Ensure chewable tablets are chewed or crushed before being swallowed and followed with at least 8 oz water or other liquid.
- Store at 15°–30° C (59°–86° F) unless otherwise directed.

ADVERSE EFFECTS (≥1%) **GI:** Temporary *darkening of stool* and tongue, metallic taste, bluish gum line; bleeding tendencies. With high doses: Fecal impaction. **CNS:** Encephalopathy (disorientation, muscle twitching). **Hematologic:** Bleeding tendency. **Special Senses:** Tinnitus, hearing loss.

DIAGNOSTIC TEST INTERFERENCE Because bismuth subsalicylate is radiopaque, it may interfere with *radiographic studies* of GI tract.

INTERACTIONS Drug: Bismuth may decrease the absorption of TETRACYCLINES, QUINOLONES **(ciprofloxacin, norfloxacin, ofloxacin).** May increase level of **aspirin.**

PHARMACOKINETICS Absorption: Undergoes chemical dissociation in GI tract to bismuth subcarbonate and sodium salicylate; bismuth is minimally absorbed, but the salicylate is readily absorbed.

NURSING IMPLICATIONS

Assessment & Drug Effects

- Monitor bowel function; note that stools may darken and tongue may appear black. These are temporary effects and will disappear without treatment.
- Lab tests: *H. pylori* breath test when used for peptic ulcers.

Patient & Family Education

- Note: Bismuth contains salicylate. Use caution when taking aspirin and other salicylates. Many OTC medications for colds, fever, and pain contain salicylates.

- Consult prescriber if diarrhea is accompanied by fever or continues for more than 2 days.
- Note: Temporary grayish black discoloration of tongue and stool may occur.

BISOPROLOL FUMARATE

(bis-o-pro'lol fum'a-rate)

Zebeta

Classifications: BETA-ADRENERGIC ANTAGONIST; ANTIHYPERTENSIVE
Therapeutic: ANTIHYPERTENSIVE
Prototype: Propranolol
Pregnancy Category: C

AVAILABILITY 5 mg, 10 mg tablets

ACTION & *THERAPEUTIC EFFECT*
Long-acting cardioselective (beta$_1$) adrenoreceptor blocking agent. To maintain beta$_1$ cardioselectivity, the lowest effective dose is necessary. Bisoprolol decreases heart rate, blood pressure, contractile force, and cardiac workload, which reduces myocardial oxygen consumption and increases blood flow to myocardium. *Bisoprolol has antianginal properties, especially improving exercise tolerance. It reduces both systolic and diastolic blood pressure at rest and with exercise.*

USES Hypertension.
UNLABELED USES Angina.

CONTRAINDICATIONS History of hypersensitivity to bisoprolol, severe sinus bradycardia, second- and third-degree AV block, overt cardiac failure, cardiogenic shock; pulmonary edema.
CAUTIOUS USE Asthma or COPD, peripheral vascular disease, diabetes mellitus, Prinzmetal's angina;

hyperthyroidism, renal or hepatic insufficiency, cerebrovascular disease, stroke; pregnancy (category C), lactation.

ROUTE & DOSAGE

Hypertension
Adult: **PO** 2.5–5 mg once daily, may increase to 20 mg/day if necessary

ADMINISTRATION

Oral

- Note: The half-life of the drug is increased in those with significant liver dysfunction; usual initial dose is 2.5 mg and may be carefully titrated upward if necessary.
- Discontinue drug gradually over a period of 1–2 wk to avoid rebound, withdrawal angina, or hypertension.
- Store at room temperature, 15°–30° C (59°–86° F).

ADVERSE EFFECTS (≥1%) CNS: Dizziness, fatigue, tiredness, vertigo, anxiety, headache, sleep disturbances. **CV:** Bradycardia, orthostatic hypotension, rebound/withdrawal angina or hypertension following abrupt discontinuation, may exacerbate intermittent claudication. **Endocrine:** Increases serum levels of VLDL-C and decreases levels of HDL-C lipoproteins, may cause slight rise in serum potassium. **GI:** Abdominal pain, dyspepsia, nausea, vomiting, constipation, diarrhea. **Respiratory:** Asthma, bronchospasm, cough, dyspnea, pharyngitis, sinusitis. **Skin:** Rash, acne, pruritus, eczema. **Other:** Arthralgia.

INTERACTIONS Drug: Amiodarone may cause significant bradycardia; BETA-BLOCKERS may reduce **glucose**

Common adverse effects in *italic*, life-threatening effects underlined; generic names in **bold**; classifications in SMALL CAPS; ♦ Canadian drug name; ⊘ Prototype drug

181

B

tolerance, inhibit **insulin** secretion, alter rate of recovery from hypoglycemia, reduce peripheral circulation, and suppress hypoglycemic symptoms; **rifampin** decreases bisoprolol blood levels.

PHARMACOKINETICS Absorption: Readily from GI tract; 82–94% reaches systemic circulation. **Peak:** Therapeutic effect 2–4 wk. **Duration:** 24 h. **Distribution:** Some CNS penetration. **Metabolism:** 50% in liver by CYP3A4. **Elimination:** 50–60% unchanged in urine. **Half-Life:** 10–12.4 h.

NURSING IMPLICATIONS

Assessment & Drug Effects

- Monitor BP frequently during periods of dose adjustment or drug withdrawal. Monitor postural vital signs for orthostatic hypotension.
- Monitor for activity-induced angina both during therapy and following discontinuation of drug.
- Monitor for and report severe hypotension and bradycardia. Dosage adjustment may be required.
- Monitor for bronchospasms in patients with a history of asthma or COPD.
- Monitor diabetics for loss of glycemic control.
- Lab tests: Periodic CBC, electrolytes, renal function, LFTs lipid profile.

Patient & Family Education

- Report orthostatic hypotension and dizziness to prescriber.
- Do not discontinue drug abruptly unless specifically instructed to do so.
- Note: Drug-induced nightmares and unpleasant dreams are possible when taking this drug.
- Monitor blood glucose for loss of glycemic control if diabetic.

BIVALIRUDIN
(bi-val'i-ru-den)

Angiomax
Classifications: ANTICOAGULANT; DIRECT THROMBIN INHIBITOR
Therapeutic: ANTITHROMBOTIC
Prototype: Lepirudin
Pregnancy Category: B

AVAILABILITY 250 mg vial

ACTION & *THERAPEUTIC EFFECT* Direct inhibitor of thrombin similar to lepirudin. Capable of inhibiting the action of both free and clot-bound thrombin. *Reversibly binds to the thrombin active site, thereby blocking the thrombogenic activity of thrombin.*

USES Used with aspirin as an anticoagulant in patients undergoing PTCA or PCR, patients at risk for HIT undergoing PCI.
UNLABELED USES DVT prevention.

CONTRAINDICATIONS Hypersensitivity to bivalirudin; cerebral aneurysm, intracranial hemorrhage; patients with increased risk of bleeding (e.g., recent surgery, trauma, CVA, hepatic disease); coagulopathy; lactation.
CAUTIOUS USE Asthma or allergies; blood dyscrasia or thrombocytopenia; GI ulceration, serious hepatic disease; hypertension, renal impairment, pregnancy (category B). Safety and efficacy in children are not established.

ROUTE & DOSAGE

Anticoagulation
Adult: **IV** 0.75 mg/kg bolus (5 min after the bolus, ACT should be performed and 0.3 mg/kg given

if needed) followed by 1.75 mg/kg/h for the duration of the procedure, may continue at 0.2 mg/kg/h up to 20 h if needed; intended for use with aspirin 300–325 mg

Renal Impairment Dosage Adjustment

If *CrCl less than 30 mL/min:* Give maintenance dose of 1 mg/kg/h

Hemodialysis Dosage Adjustment

Give 0.25 mg/kg/h maintenance dose

ADMINISTRATION

Intravenous

PREPARE: **Direct/Continuous:** Direct IV bolus dose and initial 4-h continuous infusion: Reconstitute each 250 mg vial with 5 mL of sterile water for injection; gently swirl until dissolved. ▪ Must further dilute each reconstituted vial in 50 mL of D5W or NS to yield 5 mg/mL. **Continuous:** Subsequent low-dose, continuous infusions: Reconstitute each 250 mg vial as above. ▪ Further dilute each reconstituted vial in 500 mL of D5W or NS to yield 0.50 mg/mL.
ADMINISTER: **Direct:** Give bolus dose over 3–5 sec. **Continuous:** Give 1.75 mg/kg/h for the duration of the PTCA procedure. ▪ Subsequent doses, give 0.2 mg/kg/h for up to 20 h as ordered.
INCOMPATIBILITIES **Y-site:** Alteplase, amiodarone, amphotericin B, chlorpromazine, diazepam, prochlorperazine, reteplase, streptokinase, vancomycin.

▪ Store reconstituted vials refrigerated at 2°–8° C (35.6°–46.4° F) for up to 24 h. Store diluted concentrations between 0.5 mg/mL and 5 mg/mL at room temperature, 15°–30° C (59°–86° F), for up to 24 h.

ADVERSE EFFECTS (≥1%) **Body as a Whole:** *Back pain,* pain, fever. **CV:** *Hypotension,* hypertension, bradycardia. **GI:** *Nausea,* vomiting, dyspepsia, abdominal pain. **Hematologic:** Bleeding. **CNS:** *Headache,* anxiety, nervousness. **Urogenital:** Urinary retention, pelvic pain. **Other:** Injection site pain.

INTERACTIONS No clinically significant interactions established.

PHARMACOKINETICS Duration: 1 h. **Distribution:** No protein binding. **Metabolism:** Proteolytic cleavage and renal metabolism. **Elimination:** Renal. **Half-Life:** 25 min.

NURSING IMPLICATIONS

Assessment & Drug Effects

▪ Monitor cardiovascular status carefully during therapy.
▪ Monitor for and report S&S of bleeding: Ecchymosis, epistaxis, GI bleeding, hematuria, hemoptysis.
▪ Patients with history of GI ulceration, hypertension, recent trauma or surgery are at increased risk for bleeding.
▪ Monitor neurologic status and report immediately: Focal or generalized deficits.
▪ Lab tests: Baseline and periodic ACT, aPTT, PT, platelet count, Hgb and Hct; periodic serum creatinine, stool for occult blood, urinalysis.

Patient & Family Education

▪ Report any of the following immediately: Unexplained back or stomach pain; black, tarry stools; blood in urine, coughing up blood; difficulty breathing; dizziness or fainting spells; heavy menstrual bleeding; nosebleeds; unusual bruising or bleeding at any site.

Common adverse effects in *italic*, life-threatening effects <u>underlined</u>; generic names in **bold;** classifications in SMALL CAPS; ✤ Canadian drug name; ⊘ Prototype drug

183

B

BLEOMYCIN SULFATE

(blee-oh-mye'sin)

Classifications: ANTINEOPLASTIC; ANTIBIOTIC
Therapeutic: ANTINEOPLASTIC
Prototype: Doxorubicin
Pregnancy Category: D

AVAILABILITY 15 units, 30 units powder for injection

ACTION & *THERAPEUTIC EFFECT* A toxic drug with low therapeutic index; intensely cytotoxic. By unclear mechanism, blocks DNA, RNA, and protein synthesis. A cell cycle-phase nonspecific agent widely used in combination with other chemotherapeutic agents because it lacks significant myelosuppressive activity. *This mixture of cytotoxic antibiotics from a strain of* Streptomyces verticillus *has strong affinity for skin and lung tumor cells, in contrast to its low affinity for cells in hematopoietic tissue.*

USES As single agent or in combination with other agents, as adjunct to surgery and radiation therapy. Squamous cell carcinomas of head, neck, penis, testicles, cervix, and vulva; lymphomas (including reticular cell sarcoma, lymphosarcoma, Hodgkin's); malignant pleural effusions.
UNLABELED USES *Mycosis fungoides* and *Verruca vulgaris* (common warts), AIDS-related Kaposi's sarcoma.

CONTRAINDICATIONS History of hypersensitivity or idiosyncrasy to bleomycin; pulmonary infection; concurrent radiation therapy; women of childbearing age, pregnancy (category D), lactation.
CAUTIOUS USE Compromised hepatic, renal, or pulmonary function; peripheral vascular disease; history of tobacco use; previous cytotoxic drug or radiation therapy.

ROUTE & DOSAGE

Squamous Cell Carcinoma, Testicular Carcinoma

Adult/Child: **Subcutaneous/IM/ IV** 10–20 units/m² or 0.25–0.5 units/kg 1–2 times/wk (max: 300–400 units)

Lymphomas

Adult/Adolescent: **Subcutaneous/IM/ IV** 10–20 units/m² 1–2 times/wk after a 1–2 units test dose times 2 doses

Malignant Pleural Effusion

Adult: **Intrapleural** 60 units single dose

Cervical, Penile, Vuvlvar, Head-Neck Cancer

Adult: **IV/IM/Subcutaneous** 5–20 units/m² (0.25–0.5 units/kg)

Renal Impairment Dosage Adjustment

CrCl 40–50 mL/min: Reduce dose 30%; *CrCl 30–39 mL/min:* Reduce dose 40%; *CrCl 20–29 mL/min:* Reduce dose 45%; *CrCl 10–19 mL/min:* Reduce dose 55%; *CrCl 5–10 mL/min:* Reduce dose 60%

ADMINISTRATION

Note: Due to risk of anaphylactoid reaction, give lymphoma patients 2 units or less for first two doses. If no reaction, follow regular dosage schedule.

Subcutaneous/Intramuscular

- Reconstitute with sterile water, NS, or bacteriostatic water by adding 1–5 mL to the 15 units vial or 2–10 mL to the 30 units vial. Amount of diluent is determined by the total volume of solution that will be injected.

- Inject IM deeply into upper outer quadrant of buttock; change sites with each injection.

Intravenous —————————

IV administration to infants and children: Verify correct IV concentration and rate of infusion with prescriber.

PREPARE: **Intermittent:** Dilute each 15 units with at least 5 mL of sterile water or NS. ▪ May be further diluted in 50–100 mL of the chosen diluent. ▪ Do not dilute with any solution containing D5W.

ADMINISTER: **Intermittent:** Give each 15 units or fraction thereof over 10 min through Y-tube of free-flowing IV.

INCOMPATIBILITIES **Solution/additive: Aminophylline, ascorbic acid, cefazolin, diazepam, hydrocortisone, methotrexate, mitomycin, nafcillin, penicillin G, terbutaline Y-site: amphotericin B, dantrolene, diazepam, phenytoin, tigecycline.**

- Store unopened ampules at 15°–30° C (59°–86° F) unless otherwise specified by manufacturer.

ADVERSE EFFECTS (≥1%) **Body as a Whole:** Hypersensitivity (anaphylactoid reaction); *mild febrile reaction.* **CNS:** Headache, mental confusion. **GI:** Stomatitis, ulcerations of tongue and lips, anorexia, nausea, vomiting, diarrhea, weight loss. **Hematologic:** Thrombocytopenia, leukopenia, (rare). **Respiratory:** Pulmonary toxicity (dose and age-related); interstitial pneumonitis, pneumonia, or fibrosis. **Skin:** Diffuse alopecia (reversible), *hyperpigmentation, pruritic erythema,* vesiculation, acne, thickening of skin and nail beds, *patchy hyperkeratosis,* striae, peeling, bleeding. **Other:** Pain at tumor site; phlebitis; necrosis at injection site, shivering.

INTERACTIONS Drug: Other ANTI-NEOPLASTIC AGENTS increase bone marrow toxicity; decreases effects of **digoxin, phenytoin,** avoid use with LIVE VACCINES.

PHARMACOKINETICS Distribution: Concentrates mainly in skin, lungs, kidneys, lymphocytes, and peritoneum. **Metabolism:** Unknown. **Elimination:** 60–70% recovered in urine as parent compound. **Half-Life:** 2 h.

NURSING IMPLICATIONS

Assessment & Drug Effects

- Monitor closely for an acute reaction (hypotension, hyperpyrexia, chills, confusion, wheezing, cardiopulmonary collapse). Anaphylactoid reaction can be fatal (see Appendix F). It may occur immediately or several hours after first or second dose, especially in lymphoma patients.
- Monitor temperature. Febrile reaction (mild chills and fever) is relatively common and usually occurs within the first few hours after administration of a large single dose and lasts about 4–12 h. Reaction tends to become less frequent with continued drug administration, but can recur at any time.
- Monitor for and report any of the following: Unexplained bleeding or bruising; evidence of deterioration of renal function (changed I&O ratio and pattern, decreasing creatinine clearance, weight gain or edema); evidence of pulmonary toxicity (nonproductive cough, chest pain, dyspnea).
- Check weight at regular intervals under standard conditions. Weight loss and anorexia may persist a long time after therapy has been discontinued.
- Report promptly symptoms of skin toxicity (hypoesthesia, urticaria, tender swollen hands) that may develop in second or third

Common adverse effects in *italic*, life-threatening effects underlined; generic names in **bold**; classifications in SMALL CAPS; ✦ Canadian drug name; ❂ Prototype drug

185

week of treatment and after 150–200 units of bleomycin have been administered. Therapy may be discontinued.

Patient & Family Education

- Avoid OTC drugs during antineoplastic treatment period unless approved by prescriber.
- Report skin irritation which may not develop for several weeks after therapy begins.
- Hyperpigmentation may occur in areas subject to friction and pressure, skin folds, nail cuticles, scars, and intramuscular sites.

BOCEPREVIR

(boe-se′-pre-vir)

Victelis
Classifications: ANTIVIRAL AGENT; ANTIHEPATITIS AGENT; NS3/4A PROTEASE INHIBITOR
Therapeutic: ANTIVIRAL; PROTEASE INHIBITOR
Prototype: None
Pregnancy Category: B (alone); X (in combination with ribavirin and peginterferon alfa)

AVAILABILITY 200 mg capsules

ACTION & *THERAPEUTIC EFFECT* A direct-acting antiviral that inhibits an HCV protease enzyme needed for replication of HCV in the host cells. *Effective in treatment of hepatitis C genotype 1.*

USES Treatment of chronic hepatitis C virus (HCV) genotype 1 infection in patients 18 y and older with compensated liver disease. Must be used in combination with ribavirin and peginterferon alfa, and can be used in patients with or without cirrhosis, and in patients who are previously untreated or who have failed previous interferon and ribavirin therapy.

CONTRAINDICATIONS Pregnant women; men whose female partners are pregnant; co-administration with drugs that are highly dependent on CYP3A4/5 for clearance or are potent CYP3A4/5 inducers (see drug interactions); older adult; lactation.
CAUTIOUS USE Anemia, neutropenia, and thromboembolic events. Safety and effectiveness in children not established.

ROUTE & DOSAGE

Chronic Hepatitis C Infection

Adult: **PO** 800 mg t.i.d. (every 7–9 h); begin treatment 4 wk after initiating therapy with ribavirin and peginterferon alfa. Duration of combination drug therapy 28–48 wk

ADMINISTRATION

Oral

- Give with a meal or light snack.
- Doses should be given approximately q7–9 h.
- Boceprevir **MUST** be given in combination with peginterferon alfa and ribavirin (started 4 wk prior to adding boceprevir).
- Store 2°–8° C (36°–46° F). Can be stored at room temperature up to 25° C (77° F) for 3 mo. Avoid excessive heat.

ADVERSE EFFECTS (≥1%) **Body as a Whole:** Arthralgia, chills, decreased appetite, dyspnea on exertion. **CNS:** Asthenia, dizziness, *fatigue, headache,* insomnia, irritability. **GI:** Diarrhea, dry mouth, *dysgeusia, nausea,* vomiting. **Hematological:** *Anemia,* neutropenia. **Metabolic:** Decreased hemoglobin, neutropenia. **Skin:** Alopecia, dry skin, rash.

INTERACTIONS Drug: Strong CYP3A4 inhibitors (e.g., **ketoconazole, itraconazole, voriconazole,**

clarithromycin, nefazodone, ritonavir, saquinavir, nelfinavir, indinavir, atazanavir and telithromycin) can increase boceprevir levels, while potent CYP3A4 inducers (e.g. rifampin, dexamethasone, phenytoin, carbamazepine and phenobarbital) can decrease boceprevir levels. Due to its ability to inhibit CYP3A4/5, boceprevir can increase plasma levels of other compounds requiring CYP3A4/5 (e.g., ketoconazole, itraconazole, voriconazole, posiconazole, budesonide, fluticasone, rifampin and other RIFAMYCINS, alfuzosin, ERGOT ALKALOIDS, HMG CoA REDUCTASE INHIBITORS). Boceprevir can increase adverse effects of desipramine and trazodone. Use of BENZODIAZEPINES with boceprevir may increase the risk of respiratory depression. Boceprevir increases the cardiovascular effects of salmeterol. Combination use with colchicine increases the risk of colchicines toxicity. Boceprevir increases the risk of hyperkalemia if used with drospirenone. Boceprevir can increase digoxin levels. Boceprevir may alter the levels of warfarin. **Food:** Co-administration with food enhances bioavailability. **Herbal: St. John's wort** may decrease the levels of boceprevir.

PHARMACOKINETICS Peak: 2 h. **Distribution:** Approximately 75% plasma protein bound. **Metabolism:** Hepatic metabolism to inactive compounds. **Elimination:** Fecal (79%) and renal (9%). **Half-Life:** 3.4 h.

NURSING IMPLICATIONS

Assessment & Drug Effects
- Monitor for and report S&S of anemia or infection.
- Compile a complete list of all drugs, vitamins, herbals, etc. used by the patient.

- Lab tests: Monitor CBC with differential at baseline and wk 4, 8, 12; HCV-RNA levels at wk 4, 8, 12, 24, end of treatment, and periodically thereafter; monthly pregnancy tests during treatment and for 6 mo thereafter.
- Withhold drug and notify prescriber if pregnancy is suspected.

Patient & Family Education
- Women who are pregnant or men whose female partners are pregnant should not take boceprevir. Notify prescriber immediately if a pregnancy develops.
- Use at least two forms of effective birth control during and for 6 mo after completion of therapy. Systemic hormonal contraceptives may not be effective while taking boceprevir.
- If a dose is missed and it is less than 2 h before next dose is due, skip the missed dose; if it is 2 h or more before the next dose is due, take the missed dose and resume dosing schedule.
- Boceprevir interacts with a large number of drugs. Inform prescriber about any prescription and nonprescription medicines, vitamins, and herbal supplements you are taking.
- Do not breast feed while taking this drug without consulting prescriber.

BORTEZOMIB
(bor-te-zo'mib)

Velcade

Classifications: ANTINEOPLASTIC; BIOLOGICAL RESPONSE MODIFIER; PROTEOSOME INHIBITOR

Therapeutic: ANTINEOPLASTIC; SIGNAL TRANSDUCTION INHIBITOR (STI)

Pregnancy Category: D

Common adverse effects in *italic*, life-threatening effects underlined; generic names in **bold**; classifications in SMALL CAPS; ♣ Canadian drug name; ⊙ Prototype drug

187

AVAILABILITY 3.5 mg powder for injection

ACTION & *THERAPEUTIC EFFECT*

Bortezomib is an inhibitor of proteasome, which is responsible for regulation of protein expression and degradation of damaged or obsolete proteins within the cell; its activity is critical to activation or suppression of cellular functions including the cell cycle, oncogene expression, and apoptosis. *Proteasome inhibition may reverse some of the changes that allow proliferation of malignant cells and suppress apoptosis (programmed cell death) in malignant cells.*

USES Treatment of relapsed or refractory multiple myeloma or mantle cell lymphoma.

UNLABELED USES Myelomatous pleural effusion, non-Hodgkin's lymphoma.

CONTRAINDICATIONS Hypersensitivity to bortezomib, boron, or mannitol; acute diffuse infiltrative pulmonary and pericardial disease; pregnancy (category D), lactation.

CAUTIOUS USE Peripheral neuropathy; history of syncope, dehydration, hypotension; history of allergies, asthma; preexisting electrolyte or acid-base disturbances, especially hypokalemia or hyponatremia; diabetic mellitus; liver disease; myelosuppression, renal impairment; risk factors for cardiac disease; history of peripheral neuropathy or other neurologic disorders; risk factors for pulmonary disorders; GI toxicities. Safety and effectiveness in children are not established.

ROUTE & DOSAGE

Multiple Myeloma/Mantle Cell Lymphoma (failed previous therapy)

Adult: **IV** 1.3 mg/m^2 days 1, 4, 8, and 11 followed by a 10-day rest period; at least 72 h should elapse between consecutive doses; 3 wk period is a treatment cycle

Previously Untreated Multiple Myeloma

Adult: **IV** 1.3 mg/m^2/dose on days 1, 4, 8, and 11 followed by a 10-day rest period (days 12–21) then on days 22, 25, 29, and 32 followed by a 10-day rest period; 6 wk is one course

Toxicity Dosage Adjustment

Withhold dose with grade 3 or 4 hematologic toxicity, when symptoms resolve dose may be reduced 25% and restarted.

Hemodialysis Dosage Adjustment

Administer after hemodialysis

Hepatic Impairment Adjustment

Moderate/severe impairment (bilirubin above 1.5 x ULN) reduce dose to 0.7 mg/m^2

ADMINISTRATION

Intravenous
Wear protective gloves and prevent contact with skin.

***PREPARE*: Direct:** Reconstitute 3.5 mg vial with 3.5 mL of NS for injection to yield 1 mg/mL. ▪ Discard if not clear and colorless. Give within 8 h of reconstitution.
***ADMINISTER*: Direct:** Give as a bolus dose over 3–5 sec. Flush before/after with NS.
INCOMPATIBILITIES **Solution/additive:** Do not recommend mixing or injecting with any other drugs.

Common adverse effects in *italic*, life-threatening effects underlined; generic names in **bold**; classifications in SMALL CAPS; ♣ Canadian drug name; ⊙ Prototype drug

▪ Store unopened vials at 15°–30° C (59°–86° F). Protect from light.
▪ Store reconstituted vials at 15°–30° C (59°–86° F). ▪ Give within 8 h of reconstitution. May store up to 3 h in a syringe; however, total storage time must not exceed 8 h when exposed to normal indoor lighting.

ADVERSE EFFECTS (≥1%) **Body as a Whole:** *Asthenia, weakness, fatigue, malaise, fever, dehydration, peripheral neuropathy, rigors, herpes zoster.* **CNS:** *Insomnia, headache, paresthesia, dizziness, anxiety.* **CV:** *Edema, hypotension, orthostatic hypotension.* **GI:** *Nausea, vomiting, diarrhea, anorexia, abdominal pain, constipation, dyspepsia, dysphagia.* **Hematologic:** <u>Thrombocytopenia, neutropenia, anemia.</u> **Musculoskeletal:** *Arthralgia, musculoskeletal pain, bone pain, myalgia, back pain, muscle cramps.* **Respiratory:** *Dyspnea, cough, upper respiratory infection.* **Skin:** *Rash, pruritus.* **Special Senses:** *Blurred vision, diplopia.*

INTERACTIONS Drug: Hypoglycemia and hyperglycemia have been reported with ANTIDIABETIC AGENTS; ANTIHYPERTENSIVE AGENTS may exacerbate hypotension; ANTICOAGULANTS, **antithymocyte globulin,** NSAIDS, PLATELET INHIBITORS, **aspirin,** THROMBOLYTIC AGENTS may increase risk of bleeding. **Food: Grapefruit juice** may increase drug levels.

PHARMACOKINETICS Distribution: 85% protein bound. **Metabolism:** In the liver (CYP3A4, CYP2C19, CYP1A2). **Half-Life:** 9–15 h.

NURSING IMPLICATIONS
Assessment & Drug Effects
▪ Monitor for and report S&S of neuropathy (e.g., hyperesthesia, hypoesthesia, paresthesia, discomfort or neuropathic pain).

▪ Monitor diabetics for loss of glycemic control.
▪ Monitor postural vital signs for orthostatic hypotension.
▪ Monitor for S&S of developing a pulmonary disorder.
▪ Monitor I&O and assess for S&S of dehydration or electrolyte imbalance if vomiting and/or diarrhea develop.
▪ Monitor for exacerbation of CHF, or acute onset of CHF.
▪ Lab tests: Frequent CBC with platelet count prior to each dose; baseline and periodic LFTs; frequent blood glucose in diabetics.

Patient & Family Education
▪ Report promptly any of the following: Dizziness, light-headedness or fainting spells; numbness, tingling, or other unusual sensations; signs of infection (e.g., fever, chills, cough, sore throat); bruising, pinpoint red spots on the skin; black, tarry stools, nosebleeds, or any other sign of bleeding.
▪ Monitor closely blood glucose level if diabetic.
▪ Report increased S&S of CHF, or acute onset of these S&S.
▪ Do not drive or engage in other hazardous activities until reaction to drug is known.
▪ Report any S&S of respiratory difficulty.
▪ Females should use reliable methods of contraception to avoid pregnancy while on this drug.

BOTULINUM TOXIN TYPE A
(bo'tul-i-num)
Botox, BOTOX Cosmetic
Classifications: SKELETAL MUSCLE RELAXANT; ANTISPASMODIC
Therapeutic: MUSCLE RELAXANT; ANTISPASMODIC
Pregnancy Category: C

Common adverse effects in *italic*, life-threatening effects <u>underlined</u>; generic names in **bold**; classifications in SMALL CAPS; ♣ Canadian drug name; ⊘ Prototype drug

189

AVAILABILITY 100 units, 200 units powder for injection

ACTION & *THERAPEUTIC EFFECT*
Botulinum toxin type A blocks neuromuscular transmission by binding to receptor sites on motor nerve terminals, entering the nerve terminals, and inhibiting the release of acetylcholine. This inhibition occurs as the neurotoxin splits a protein molecule integral to the successful docking and releasing of acetylcholine from storage areas located within nerve endings. *When injected intramuscularly at therapeutic doses, botulinum toxin type A produces partial chemical denervation of the muscle resulting in a localized reduction in muscle activity.*

USES Treatment of blepharospasm, cervical dystonia, strabismus, glabellar frown wrinkles, severe axillary hyperhidrosis, spasticity, chronic migraine.
UNLABELED USES Treatment of other types of wrinkles, achalasia.

CONTRAINDICATIONS Presence of infection at the proposed injection site(s); hypersensitivity to Botox. Patients with dysphagia or respiratory compromise.
CAUTIOUS USE Hypersensitivity to albumin; individuals with peripheral motor neuropathic diseases (e.g., amyotrophic lateral sclerosis, or motor neuropathy), or neuromuscular junctional disorders (e.g., myasthenia gravis or Lambert-Eaton syndrome); neuromuscular disorders; ocular disease; cardiovascular disease; elderly; inflammation at the proposed injection site; weakness in the target muscle(s); pregnancy (category C), lactation; children.

ROUTE & DOSAGE

Blepharospasm
Adult/Child (older than 12 y):
Intradermal 1.25–2.5 units injected at each site, may repeat in 3 mo if needed; cumulative dose should not exceed 200 units in a 30-day period

Cervical Dystonia
Adult/Adolescent (older than 16 y): **IM** 198–300 units divided among affected muscles

Frown Wrinkles
Adult (younger than 65 y): **IM** 20 units divided among affected muscles in 5 step doses, may repeat in 3–4 mo if needed

Spasticity
Adult: **IM** , 75–360 units divided among muscles
Child (2–18 y): **IM** 4 units/kg (max: 200 units per treatment) q3mo

Axillary Hyperhidrosis
Adult: **IM** 50 units per site

Migraine Prophylaxis
Adult: **IM** 155U/treatment across 31 sites (see prescribing information)

ADMINISTRATION

Intramuscular, Intradermal, Subcutaneous
- Slowly inject required amount of nonpreserved NS (see dilution calculation) into vial. Discard vial if a vacuum does not pull diluent into vial. Gently rotate to mix. Discard if not clear, colorless, and free of particulate matter. Dilution calculation: Add 1, 2, 4, or 6 mL of NS to yield, respectively, 10 units/0.1 mL,

5 units/0.1 mL, 2.5 units/0.1 mL, 1.25 units/0.1 mL.

- Store at 2°–8° C (36°–46° F) (refrigerated). Administer within 4 h of reconstitution.

INCOMPATIBILITIES Do not mix with other solutions/additives.

ADVERSE EFFECTS (≥1%) **Body as a Whole:** Injection site reactions (localized pain, tenderness, bruising), neck pain, flu-like symptoms, hypertonia, asthenia, fever. **CNS:** *Headache,* drowsiness. **GI:** *Dysphagia,* dry mouth, fever, nausea, vomiting. **Hematologic:** Ecchymosis. **Musculoskeletal:** Local muscle weakness, dysarthria. **Respiratory:** Cough, rhinitis, upper respiratory infection. **Special Senses:** *Ptosis,* superficial punctate keratitis, dry eyes, ocular irritation, lacrimation, photophobia, keratitis, diplopia.

INTERACTIONS Drug: AMINOGLYCOSIDES, NEUROMUSCULAR BLOCKING AGENTS may potentiate neuromuscular blockade; **chloroquine** may antagonize blocking effects.

NURSING IMPLICATIONS

Assessment & Drug Effects
- Evaluate for therapeutic effectiveness, maximal at about 1–2 wk (lasting 3–4 mo).

Patient & Family Education
- Inform prescriber about all prescription, nonprescription, and herbal drugs being taken.
- Report immediately any of the following: Difficulty breathing or swallowing, problem with speech; unusual bleeding, bruising, or swelling around injection site.
- Note: Effects of the injection generally last 3–4 mo and then repeat treatments may be given.

BRENTUXIMAB
(bren-tuk' see-mab)

Adcetris
Classifications: BIOLOGICAL RESPONSE MODIFIER; MONOCLONAL ANTIBODY; CD-30 SPECIFIC ANTIBODY
Therapeutic: MONOCLONAL ANTIBODY
Prototype: Basiliximab
Pregnancy Category: D

AVAILABILITY 50 mg lyophilized powder for dissolution and injection

ACTION & THERAPEUTIC EFFECT
An antibody-drug complex that binds to a protein (CD 30) on the surface of lymphoma cells and is internalized into these cells where it releases a microtubule disrupting agent. *Causes cell cycle arrest and apoptosis of lymphoma cells.*

USES Treatment of Hodgkin lymphoma in patients non-responsive to autologous stem cell transplant (ASCT) or nonresponsive to at least two multiagent chemotherapy (for patients not ASCT candidates) regimens. Treatment of systemic anaplastic large cell lymphoma (sALCL) in patients nonresponsive to at least one prior multiagent chemotherapy regimen.

CAUTIOUS USE Infusion reactions and hypersensitivity; neutropenia; tumor lysis syndrome; Stevens-Johnson Syndrome; renal and hepatic impairment; older adult (65 and over); pregnancy category D; lactation. Safety and effectiveness in children not established.

Common adverse effects in *italic*, life-threatening effects underlined; generic names in **bold**; classifications in SMALL CAPS; ♥ Canadian drug name; ☻ Prototype drug

191

B

ROUTE & DOSAGE

Hodgkin Lymphoma or Systemic Anaplastic Large Cell Lymphoma

Adult: IV 1.8 mg/kg q3wk for maximum of 16 cycles

Peripheral Neuropathy Dosage Adjustment

Grade 2/3 peripheral neuropathy: Stop treatment until toxicity resolves to grade 1 or better. Reduce dosage to 1.2 mg/kg q3wk
For Grade 4 peripheral neuropathy: Discontinue treatment

Neutropenia Dosage Adjustment

Grade 3/4 neutropenia: Stop treatment until toxicity resolves to baseline or grade 2 or better. Consider the use of growth factors (CSFs) for subsequent cycles
Grade 4 neutropenia (despite the use of growth factors): Discontinue treatment or reduce dosage to 1.2 mg/kg IV q3wk

ADMINISTRATION

Intravenous

- Premedication with acetaminophen, an antihistamine and a corticosteroid are recommended for prior infusion reactions.

PREPARE: **IV Infusion:** Reconstitute each 50 mg vial with 10.5 mL sterile water (SW) for injection to yield 5 mg/mL. Direct the stream of SW on to side of vial and not into powder. Swirl gently to dissolve. Do not shake. Immediately withdraw the required volume of reconstituted solution and add to a minimum of 100 mL of NS, D5W, or LR to produce a final concentration of 0.4–1.8 mg/mL. Invert bag gently to mix.

ADMINISTER: **IV Infusion:** Give over 30 min. Do not give IV push or bolus.

INCOMPATIBILITIES Solution/additive: Do not mix or administer as an IV infusion with any other medicinal products.

- Store refrigerated, if necessary, at 2°–8° C (36°–46° F). Should be used immediately after preparation but no later than 24 h following reconstitution.

ADVERSE EFFECTS (≥10%) **Body as a Whole:** Chills, *fatigue*, pain, peripheral edema, *pyrexia*. **CNS:** Anxiety, dizziness, headache, insomnia, *peripheral neuropathy*. **GI:** *Abdominal pain*, constipation, *diarrhea*, *nausea*, *vomiting*. **Hematological:** Anemia, lymphadenopathy, *neutropenia*, *thrombocytopenia*. **Metabolic:** Decreased appetite, weight loss. **Musculoskeletal:** Arthralgia, back pain, muscle spasms, myalgia, pain in extremity. **Respiratory:** *Cough*, dyspnea, oropharyngeal pain, *upper respiratory tract infection*. **Skin:** Alopecia, dry skin, night sweats, pruritis, *rash*.

INTERACTIONS Drug: Monomethyl auristatin E (MMAE), one of the components of brentuximab is primarily metabolized by CYP3A4. Strong inhibitors of CYP3A4 (e.g., **ketoconazole, delavirdine, indinavir, itraconazole**) may increase the levels of brentuximab. Strong inducers of CYP3A4 (i.e., **phenobarbital, rifampin** may decrease the levels of brentuximab. **Herbal:** **St. John's wort** and **Hypericum perforatum** may reduce the levels of brentuximab.

PHARMACOKINETICS Peak: 1–3 d. **Distribution:** 68-82% plasma protein bound. **Metabolism:** Oxidative metabolism. **Elimination:** Primarily fecal. **Half-Life:** 4–6 d.

B

NURSING IMPLICATIONS

Assessment & Drug Effects

- Discontinue infusion immediately and institute supportive measures if hypersensitivity (see Appendix F) is suspected.
- Monitor closely for and report promptly S&S of an infusion reaction (e.g., chills, nausea, dyspnea, pruritus, fever, cough).
- Monitor cardiac status throughout infusion and during the immediate period thereafter.
- Monitor closely for a report S&S of peripheral neuropathy (e.g., hypo/hyperesthesia, paresthesia, burning sensation, nerve pain, weakness). Withhold drug and notify prescriber for new or worsening grade 2 or 3 neuropathy.
- Lab tests: Baseline and prior to each dose CBC with differential.

Patient and Family Education

- Report any of the following to prescriber: Chills, rash or difficulty breathing within 24 h of infusion; numbness or tingling of hands or feet; muscle weakness; fever of 100.5° F or greater; unexplained cough; or pain on urination.
- Women should use effective means of contraception and avoid breast feeding while being treated with brentuximab.

BRIMONIDINE TARTRATE

(bry-mon′i-deen)

Alphagan P
Pregnancy Category: B
See Appendix A-1.

BRINZOLAMIDE

(brin-zol′a-mide)

Azopt
Pregnancy Category: C
See Appendix A-1.

BROMFENAC

(brom′fen-ac)

Xibrom
Pregnancy Category: C
See Appendix A-1.

BROMOCRIPTINE MESYLATE

(broe-moe-krip′teen)

Cycloset, Parlodel
Classifications: ERGOT ALKALOID; DOPAMINE RECEPTOR AGONIST; ANTIPARKINSON
Therapeutic: ERGOT REPLACEMENT; ANTIDYSKINETIC; ANTIPARKINSON; GLYCEMIC CONTROL AGENT
Prototype: Ergotamine
Pregnancy Category: C

AVAILABILITY 0.8 mg, 2.5 mg tablets; 5 mg capsules

ACTION & *THERAPEUTIC EFFECT*
Semisynthetic ergot alkaloid that is a synthetic dopamine (D_2) receptor agonist. Reduces elevated serum prolactin levels in men and women by activating postsynaptic dopaminergic receptors in hypothalamus to stimulate release of prolactin-inhibiting factor. Activates dopaminergic receptors (D_2 receptors) in neostriatum of CNS, which may explain action in parkinsonism. *Restores ovulation and ovarian function in amenorrheic women, thus correcting female infertility secondary to elevated prolactin levels. Activates dopaminergic receptors in CNS resulting in antiparkinsonism effect. Improves glycemic control in Type 2 diabetics.*

USES Acromegaly, hyperprolactinemia, infertility, Parkinson's disease, pituitary adenoma. As adjunct to diet and exercise in Type 2 diabetes.

Common adverse effects in *italic*, life-threatening effects <u>underlined</u>; generic names in **bold;** classifications in SMALL CAPS; ♣ Canadian drug name; ⊘ Prototype drug

193

B

UNLABELED USES To prevent postpartum lactation, to relieve premenstrual symptoms, mastalgia, and cocaine withdrawal.

CONTRAINDICATIONS Hypersensitivity to ergot alkaloids; uncontrolled hypertension; severe ischemic heart disease or peripheral vascular disease; pituitary tumor; **Cycloset:** Type 1 diabetes mellitus or diabetic ketoacidosis; synopal migraine due to hypertensive episodes. **Parlodel:** Uncontrolled hypertension; postpartum period in women with history of CAD, or severe cardiovascular conditions. Normal prolactin levels, preeclampsia, eclampsia; lactation.

CAUTIOUS USE Hepatic and renal dysfunction; history of psychiatric disorder; history of GI bleeding or peptic ulcer; history of MI with residual arrhythmia; pregnancy (category C); children.

ROUTE & DOSAGE

Amenorrhea or Galactorrhea, Female Infertility

Adult: **PO** 1.25–2.5 mg/day (max: 2.5 mg 2–3 times/day)

Parkinson's Disease

Adult: **PO** 1.25 day (increase by 2.5 mg q14–28 days)

Acromegaly

Adult: **PO** 1.25–2.5 mg/day for 3 days, then increase by 1.25–2.5 mg q3–7days until desired effect is achieved, usually 30–60 mg/day in divided doses

Adjunct in Type 2 Diabetes (Cycloset)

Adult: **PO** 0.8 mg daily in the a.m., titrate up (max dose: 1.6–4.8 mg)

ADMINISTRATION

Oral

- Give with meals, milk, or other food to reduce incidence of GI side effects.
- Have patient in supine position before receiving first dose because dizziness and fainting may occur. For this reason, initial dose is usually prescribed for evening administration.
- Store in tightly closed, light-resistant containers, preferably at 15°–30° C (59°–86° F) unless otherwise directed.

ADVERSE EFFECTS (≥1%) **Body as a Whole:** Mostly dose related. **CNS:** Headache, dizziness, vertigo, light-headedness, fainting, sedation, nightmares, insomnia, dyskinesia, ataxia; mania, nervousness, anxiety, depression. **CV:** *Orthostatic hypotension,* <u>shock</u>, postpartum hypertension, palpitation, extrasystoles, Raynaud's phenomenon, red, tender, hot, edematous extremities (erythromelalgia), exacerbation of angina, arrhythmias, <u>acute MI</u>. **Special Senses:** Blurred vision, burning sensation in eyes, blepharospasm, diplopia. **GI:** *Nausea,* vomiting, abdominal cramps, epigastric pain, constipation (long-term use) or diarrhea; metallic taste, dry mouth, dysphagia, anorexia, peptic ulcers. **Skin:** Urticaria, rash, mottling, livedo reticularis. **Other:** Fatigue, nasal congestion, asthenia.

INTERACTIONS Drug: Possibility of decreased tolerance to **alcohol;** ANTIHYPERTENSIVE AGENTS add to hypotensive effects; ORAL CONTRACEPTIVES, **estrogen, progestins** may interfere with effect of bromocriptine by causing amenorrhea and galactorrhea; PHENOTHIAZINES, TRICYCLIC ANTIDEPRESSANTS, **methyldopa,**

reserpine can cause an increase in **prolactin,** which may interfere with bromocriptine activity.

PHARMACOKINETICS Absorption:
28% from GI tract. **Peak:** 1–2 h. **Duration:** 4–8 h. **Metabolism:** In liver by CYP3A4. **Elimination:** 85% in feces in 5 days; 3–6% in urine. **Half-Life:** 50 h.

NURSING IMPLICATIONS
Assessment & Drug Effects

- Monitor vital signs closely during the first few days and periodically throughout therapy.
- Lab tests: Periodic CBC, liver functions and renal functions with prolonged therapy.
- Monitor for and report psychotic symptoms and other adverse reactions in Parkinson's patients.
- Improvement in Parkinson's disease may be noted in 30–90 min following administration of bromocriptine, with maximum effect in 2 h.

Patient & Family Education

- Make position changes slowly and in stages, especially from lying down to standing, and to dangle legs over bed for a few minutes before walking. Lie down immediately if light-headedness or dizziness occurs.
- Do not drive or engage in other potentially hazardous activities until response to drug is known.
- Avoid exposure to cold and report the onset of pallor of fingers or toes.
- Note: Use barrier-type contraceptive measures until normal ovulating cycle is restored. Oral contraceptives are contraindicated.

BROMPHENIRAMINE MALEATE
(brome-fen-ir′a-meen)
Veltane

Classifications: ANTIHISTAMINE; H$_1$-RECEPTOR ANTAGONIST
Therapeutic: ANTIHISTAMINE
Prototype: Diphenhydramine
Pregnancy Category: C

AVAILABILITY 10 mg/mL injection; 2 mg/5 mL elixir; 4 mg tablet; ingredient in many oral combination products containing a decongestant, expectorant, and/or analgesic

ACTION & *THERAPEUTIC EFFECT*
Antihistamine that competes with histamine for H$_1$-receptor sites on effector cells in the bronchi and bronchioles, thus blocking histamine-mediated responses. *Effective against upper respiratory symptoms and allergic manifestations.*

USES Symptomatic treatment of allergic manifestations. Also used in various cough mixtures and antihistamine-decongestant cold formulations.

CONTRAINDICATIONS Hypersensitivity to antihistamines; acute asthma; newborns.
CAUTIOUS USE Older adults; prostatic hypertrophy; GI obstruction; asthma; narrow-angle glaucoma; COPD, cardiovascular or renal disease; seizure disorders; hyperthyroidism; pregnancy (category C), lactation.

ROUTE & DOSAGE

Allergy

Adult: **PO** 4–8 mg t.i.d. or q.i.d. or 8–12 mg of sustained release b.i.d. or t.i.d. **Subcutaneous/ IM/IV** 5–20 mg q6–12h (max: 40 mg/24 h)

Common adverse effects in *italic;* life-threatening effects <u>underlined;</u> generic names in **bold;** classifications in SMALL CAPS; ♦ Canadian drug name; ⊘ Prototype drug

195

Geriatric: **PO** 4 mg 1–2 times/
day
Child: **PO** *Older than 6 y, 2–4
mg t.i.d. or q.i.d. or 8–12 mg
of sustained release b.i.d.
(max: 12 mg/24 h); younger
than 6 y, 0.5 mg/kg in 3–4
divided doses*

ADMINISTRATION

Oral

- Give with meals or a snack to prevent gastric irritation.

Subcutaneous/Intramuscular

- Give without further dilution or diluted to a 1:10 ratio with NS.

Intravenous ———————

PREPARE: **Direct:** Give undiluted or diluted with 10 mL D5W or NS.
ADMINISTER: **Direct:** Give IV push slowly over 1 min to a recumbent patient.
INCOMPATIBILITIES **Solution/additive: Radio-contrast media (diatrizoate, iothalamate), insulin, pentobarbital.**

- Store in tightly covered container at 15°–30° C (59°–86° F) unless otherwise directed. Elixir and parenteral form should be protected from light. Avoid freezing.

ADVERSE EFFECTS (≥1%) **Body as a Whole:** Hypersensitivity reaction (urticaria, increased sweating, <u>agranulocytosis</u>). **CNS:** *Sedation,* drowsiness, dizziness, headache, disturbed coordination. **GI:** Dry mouth, throat, and nose, stomach upset, constipation. **Special Senses:** Ringing or buzzing in ears. **Skin:** Rash, photosensitivity.

DIAGNOSTIC TEST INTERFERENCE
May cause false-negative **allergy skin tests.**

INTERACTIONS Drug: Alcohol and other CNS DEPRESSANTS add to sedation.

PHARMACOKINETICS Peak: 3–9 h. **Duration:** Up to 48 h. **Distribution:** Crosses placenta. **Elimination:** 40% in urine within 72 h; 2% in feces. **Half-Life:** 12–34 h.

NURSING IMPLICATIONS

Assessment & Drug Effects

- Drowsiness, sweating, transient hypotension, and syncope may follow IV administration; reaction to drug should be evaluated. Keep prescriber informed.
- Note: Older adults tend to be particularly susceptible to drug's sedative effect, dizziness, and hypotension. Most symptoms respond to reduction in dosage.
- Lab tests: Periodic CBC in patients receiving long-term therapy.

Patient & Family Education

- Acute hypersensitivity reaction can occur within minutes to hours after drug ingestion. Reaction is manifested by high fever, chills, and possible development of ulcerations of mouth and throat, pneumonia, and prostration. Patient should seek medical attention immediately.
- Sugarless gum, lemon drops, or frequent rinses with warm water may relieve dry mouth.
- Do not drive or perform other potentially hazardous activities until response to drug is known.
- Do not take alcoholic beverages or other CNS depressants (e.g., tranquilizers, sedatives, pain or sleeping medicines) without consulting prescriber.

BUDESONIDE
(bu-des′o-nide)

Entocort EC, Pulmicort Flexhaler, Rhinocort, Rhinocort Aqua, Rhinocort Turbuhaler

Common adverse effects in *italic*, life-threatening effects <u>underlined</u>; generic names in **bold**; classifications in SMALL CAPS; ♣ Canadian drug name; ⊘ Prototype drug

Classifications: ADRENAL CORTI-COSTEROID GLUCOCORTICOID; ANTI-INFAMMATORY
Therapeutic: ANTI-INFLAMMATORY; ADRENAL CORTICOSTEROID
Prototype: Hydrocortisone
Pregnancy Category: B (inhaled); C (oral)

AVAILABILITY 32 mcg/inhalation; 3 mg capsule

ACTION & *THERAPEUTIC EFFECT*
Has potent glucocorticoid activity. Its anti-inflammatory action on nasal mucosa is thought to be a result of decreased IgE synthesis and decreased arachidonic acid metabolism. *Glucocorticoids have a wide range of inhibitory activities against multiple cell types (e.g., neutrophils, macrophages) and mediators (e.g., histamine, cytokines) involved in allergic and nonallergic/irritant-mediated inflammation.*

USES Treatment of allergic and perennial rhinitis, maintain remission in mild to moderate Crohn's disease; prophylaxis for asthma.

CONTRAINDICATIONS Hypersensitivity to budesonide, status asthmaticus, acute bronchospasms; peptic ulcer disease; lactation.
CAUTIOUS USE Active or quiescent tuberculosis; infections of respiratory tract; in sun-treated fungal, bacterial, or systemic viral infections or ocular herpes simplex; recent nasal septal ulcers; recurrent epistaxis; nasal surgery or trauma; psychosis; myasthenia gravis; diabetes mellitus; seizure disorders; **oral:** pregnancy (category C); **nasal:** pregnancy (category B).

ROUTE & DOSAGE

Crohn's Disease
Adult: **PO** 9 mg once/day in a.m. for up to 8 wk, may taper to 6 mg daily for 2 wk prior to discontinuing. May repeat 8-wk course for recurring episodes of active Crohn's disease.

Asthma Prophylaxis, Rhinitis
See Appendix A-3.

ADMINISTRATION
Oral
- Ensure that capsules are swallowed whole and not chewed.
- Give only in the morning.
- Patients with moderate to severe liver disease should be monitored for increased signs and/or symptoms of hypercorticism. Reducing the dose of Entocort EC capsules should be considered in these patients.
- Store at 25° C (77° F); excursions permitted to 15°–30° C (59°–86° F).

ADVERSE EFFECTS (≥1%) **Body as a Whole:** Arthralgia, fatigue, fever, hyperkinesis, myalgia, asthenia, paresthesia, tremor. **CNS:** Dizziness, emotional lability, facial edema, nervousness, *headache,* agitation, confusion, insomnia, drowsiness. **CV:** Chest pain, hypertension, palpitations, sinus tachycardia. **GI:** Abdominal pain, dyspepsia, gastroenteritis, oral candidiasis, xerostomia, diarrhea, nausea, vomiting, cramps. **Hematologic:** Epistaxis. **Metabolic:** Hypokalemia, weight gain. **Respiratory:** Bronchospasms, *infections,* cough, rhinitis, sinusitis, dyspnea, hoarseness, wheezing. **Skin:** Eczema, pruritus, purpura, rash, alopecia. **Special Senses:** Contact dermatitis, reduced sense of smell,

Common adverse effects in *italic*, life-threatening effects <u>underlined</u>; generic names in **bold**; classifications in SMALL CAPS; ✤ Canadian drug name; ⊙ Prototype drug

197

B

nasal pain. **Urogenital:** Intermenstrual bleeding, dysuria.

INTERACTIONS Drug: Ketoconazole may increase oral budesonide concentrations and toxicity; toxicity may also occur with **anastrozole** (high doses only), **clarithromycin, cyclosporine, danazol, delavirdine, diltiazem, erythromycin, fluconazole, fluoxetine, fluvoxamine, indinavir, isoniazid, INH, itraconazole, mibefradil, nefazodone, nelfinavir, nicardipine, norfloxacin, oxiconazole, quinidine, quinine, ritonavir, saquinavir, troleandomycin, verapamil,** and **zafirlukast. Food:** Grapefruit juice will significantly increase bioavailability of oral budesonide.

PHARMACOKINETICS Absorption: 20% (nasal) dose, 6–13% of (orally inhaled) dose, 9% PO dose reaches systemic circulation; PO form is absorbed from duodenum at pH greater than 5.5; oral bioavailability increases 2.5 times in hepatic cirrhosis. **Onset:** 8–12 h inhaled, 2 wk oral. **Peak:** 2 wk inhaled, 8 wk oral delayed by high-fat meal. **Distribution:** 90% protein bound. **Metabolism:** 85% of absorbed dose undergoes first pass metabolism by CYP3A4. **Elimination:** 60% in urine, 40% in feces. **Half-Life:** 2–3.6 h.

NURSING IMPLICATIONS

Assessment & Drug Effects

- Monitor closely for S&S of hypercorticism if concomitant doses of ketoconazole or other CYP3A4 inhibitors (see Drug Interactions) are being given.
- Monitor patients with moderate to severe liver disease for increased S&S of hypercorticism.
- Lab tests: Periodic serum potassium.

Patient & Family Education

- Notify the prescriber immediately for any of the following: Itching, skin rash, fever, swelling of face and neck, difficulty breathing, or if you develop S&S of infection.
- Do not drink grapefruit juice or eat grapefruit regularly.
- Avoid people with infections, especially those with chickenpox or measles if you have never had these conditions.

BUMETANIDE

(byoo-met′a-nide)

Bumex

Classifications: FLUID AND WATER BALANCE AGENT; LOOP DIURETIC

Therapeutic: DIURETIC; ANTIHYPERTENSIVE

Prototype: Furosemide

Pregnancy Category: C

AVAILABILITY 0.5 mg, 1 mg, 2 mg tablets; 0.25 mg/mL injection

ACTION & *THERAPEUTIC EFFECT* Sulfonamide derivative that causes both potassium and magnesium wastage. Inhibits sodium and chloride reabsorption by direct action on proximal ascending limb of the loop of Henle. Also appears to inhibit phosphate and bicarbonate reabsorption. *Produces only mild hypotensive effects at usual diuretic doses. Controls formation of edema.*

USES Edema, heart failure.

UNLABELED USES Ascites, hypercalcemia, hypertension.

CONTRAINDICATIONS Hypersensitivity to bumetanide or to other sulfonamides; anuria, markedly elevated BUN; hepatic coma; ventricular arrhythmias; severe electrolyte deficiency; lactation.

Common adverse effects in *italic*, life-threatening effects underlined; generic names in **bold**; classifications in SMALL CAPS; ♣ Canadian drug name; ⊘ Prototype drug

CAUTIOUS USE Hepatic cirrhosis; renal impairment; severe renal disease; history of gout; history of pancreatitis; history of hypersensitivity to furosemide; diabetes mellitus; acute MI; older adults; pregnancy (category C).

ROUTE & DOSAGE

Edema

Adult: **PO** 0.5–2 mg once/day, may repeat at 4–5 h intervals if needed (max: 10 mg/day) **IV/IM** 0.5–1 mg over 1–2 min, repeated q2–3h prn (max: 10 mg/day)

ADMINISTRATION

Oral
- Give with food or milk to reduce risk of gastrointestinal irritation.
- Administered in the morning as a single dose, either daily or by intermittent schedule.

Intramuscular
- Use undiluted solution for injection.

Intravenous

PREPARE: **Direct/Continuous:** Give direct IV undiluted (typical) or diluted for infusion with D5W, NS, LR. *ADMINISTER:* **Direct:** Give IV push at a rate of a single dose over 1–2 min. **Continuous:** Give diluted solution over 5 min or at prescribed rate. *INCOMPATIBILITIES* **Solution/additive:** Dobutamine. **Y-site:** Amphotericin B, azathioprine, chlorpromazine, dantrolene, diazepam, diazoxide, fenoldopam, ganciclovir, gemtuzumab, haloperidol, midazolam, minocycline, ofloxacin, oritavancin, papaverine, phenytoin, quinupristin/dalfopristin, sulfamethoxazole/trimethoprim.

- Diluted infusion should be used within 24 h after preparation.

- Store in tight, light-resistant container at 15°–30° C (59°–86° F) unless otherwise directed.

ADVERSE EFFECTS (≥1%) **Body as a Whole:** Sweating, hyperventilation, glycosuria. **CNS:** Dizziness, headache, weakness, fatigue. **CV:** Hypotension, ECG changes, chest pain, *hypovolemia*. **GI:** Nausea, vomiting, abdominal or stomach pain, GI distress, diarrhea, dry mouth. **Metabolic:** *Hypokalemia,* hyponatremia, hyperuricemia, hyperglycemia; *hypomagnesemia;* decreased calcium, chloride. **Musculoskeletal:** Muscle cramps, muscle pain, stiffness or tenderness; arthritic pain. **Special Senses:** Ear discomfort, ringing or buzzing in ears, impaired hearing.

INTERACTIONS **Drug:** AMINOGLYCOSIDES, **cisplatin** increase risk of ototoxicity; bumetanide increases risk of hypokalemia-induced **digoxin** toxicity; NONSTEROIDAL ANTI-INFLAMMATORY DRUGS (NSAIDS) may attenuate diuretic and hypotensive response; **probenecid** may antagonize diuretic activity; bumetanide may decrease renal elimination of **lithium; sotalol** may increase risk of cardiotoxicity.

PHARMACOKINETICS **Absorption:** Readily from GI tract. **Onset:** 30–60 min PO; 40 min IV. **Peak:** 0.5–2 h. **Duration:** 4–6 h. **Distribution:** Distributed into breast milk. **Metabolism:** In liver. **Elimination:** 80% in urine in 48 h, 10–20% in feces. **Half-Life:** 60–90 min.

Common adverse effects in *italic,* life-threatening effects <u>underlined</u>; generic names in **bold**; classifications in SMALL CAPS; ✤ Canadian drug name; ⊘ Prototype drug

NURSING IMPLICATIONS

Assessment & Drug Effects

- Monitor I&O and report onset of oliguria or other changes in I&O ratio and pattern promptly.
- Monitor weight, BP, and pulse rate. Assess for hypovolemia by taking BP and pulse rate while patient is lying, sitting, and standing. Older adults are particularly at risk for hypovolemia with resulting thrombi and emboli.
- Lab tests: Serum electrolytes, blood studies, liver and kidney function tests, uric acid (particularly patients with history of gout), and blood glucose. Determine values initially and at regular intervals.
- Monitor for S&S of hypomagnesemia and hypokalemia (see Appendix F) especially in those receiving digitalis or who have CHF, hepatic cirrhosis, ascites, diarrhea, or potassium-depleting nephropathy.
- Question patient about hearing difficulty or ear discomfort. Patients at risk of ototoxic effects include those receiving the drug IV, especially at high doses, those with severely impaired renal function, and those receiving other potentially ototoxic or nephrotoxic drugs (see Appendix F).
- Monitor diabetics for loss of glycemic control.

Patient & Family Education

- Report promptly to prescriber symptoms of electrolyte imbalance (e.g., weakness, dizziness, fatigue, faintness, confusion, muscle cramps, headache, paresthesias).
- Eat potassium-rich foods such as fruit juices, potatoes, cereals, skim milk, and bananas while taking bumetanide.

- Report S&S of ototoxicity promptly to prescriber (see Appendix F).
- Monitor blood glucose for loss of glycemic control if diabetic.

BUPRENORPHINE HYDROCHLORIDE

(byoo-pre-nor′feen)

Buprenex, Suboxone, Subutex
Classifications: ANALGESIC; NARCOTIC (OPIATE AGONIST–ANTAGONIST)
Therapeutic: NARCOTIC ANALGESIC
Prototype: Pentazocine
Pregnancy Category: C
Controlled Substance: Schedule III

AVAILABILITY 0.3 mg (base)/mL injection; 2 mg, 8 mg sublingual tablets

ACTION & *THERAPEUTIC EFFECT*
Opiate agonist–antagonist with agonist activity approximately 30 times that of morphine and antagonist activity equal to or up to 3 times that of naloxone. Respiratory depression occurs infrequently, probably due to drug's opiate antagonist activity. *Dose-related analgesia results from a high affinity of buprenorphine for mu-opioid receptors and as an antagonist at the kappa-opiate receptors in the CNS. Naloxone is an antagonist at the mu-opioid receptor.*

USES *Injectable* used for moderate to severe pain. *Sublingual tablets* used for treatment of opioid dependence.
UNLABELED USES *Injectable* to reverse fentanyl-induced anesthesia. *Sublingual tablets* may be used to ease cocaine withdrawal.

CONTRAINDICATIONS Hypersensitivity to buprenorphine or hypersensitivity to naloxone; lactation.
CAUTIOUS USE Patient with history of opiate use; compromised respiratory function [e.g., chronic obstructive pulmonary disease (COPD), cor pulmonale, decreased respiratory reserve, hypoxia, hypercapnia, or preexisting respiratory depression]; hypothyroidism, myxedema, Addison's disease; severe renal or hepatic impairment; geriatric or debilitated patients; acute alcoholism, delirium tremens; prostatic hypertrophy, urethral stricture; comatose patient; patients with CNS depression, head injury, or intracranial lesion; biliary tract dysfunction; pregnancy (category C); children less than 2 y.

ROUTE & DOSAGE

Postoperative Pain

Adult/Adolescent (older than 12 y): IV/IM 0.3 mg q6h up to 0.6 mg q4h or 25–50 mcg/h by IV infusion
Geriatric: IV/IM 0.15 mg q6h
Child (2–12 y): IV/IM 2–6 mcg/kg q4–6h prn

Opioid Dependence/Cocaine Withdrawal

Adult: SL Initiate with 8 mg daily on day 1 at least 4 h after last opioid dose, 16 mg daily on day 2, then switch to maintenance therapy at the same buprenorphine dose as day 2 (e.g., 16 mg daily). Adjust dose daily until opiate withdrawal effects are suppressed. Maintenance dose range 4–24 mg/day buprenorphine.

ADMINISTRATION

Sublingual

- Place **Suboxone** and **Subutex** tablets under tongue until dissolved. For doses requiring more than two tablets, place all tablets at once under tongue, or if patient cannot accommodate all tablets, place two tablets at a time under tongue.
- Instruct to hold the tablets under tongue until dissolved; advise not to swallow.

Intramuscular

- Give undiluted, deep IM into a large muscle.

Intravenous

PREPARE: Direct/IV Infusion: May be given undiluted direct IV or further dilute each 1 mL (0.3 mg) ampule in 50 mL of D5W, NS, D5NS, or LR to yield 6 mcg/mL for infusion. ▪ Do not use if discolored or contains particulate matter.
ADMINISTER: Direct: Give slowly at a rate of 0.3 mg over 2 min to a patient in a recumbent position. **IV Infusion:** Give by slow infusion over 3 min or longer depending on volume of IV solution.
INCOMPATIBILITIES **Solution/additive: Diltiazem, floxacillin, furosemide, lorazepam. Y-site: Amphotericin B cholesteryl sulfate complex, doxorubicin liposome, lansoprazole.**

- Store at 15°–30° C (59°–86° F); avoid freezing.

ADVERSE EFFECTS (≥1%) **CNS:** *Sedation, drowsiness,* dizziness, vertigo, *headache,* amnesia, euphoria, asthenia, *insomnia, pain* (when used for withdrawal), *withdrawal symptoms.* **CV:** Hypotension,

Common adverse effects in *italic*, life-threatening effects underlined; generic names in **bold**; classifications in SMALL CAPS; ✦ Canadian drug name; ⊙ Prototype drug

201

vasodilation. **Special Senses:** Miosis. **GI:** *Nausea,* vomiting, diarrhea, *constipation.* **Respiratory:** Respiratory depression, hyperventilation. **Skin:** Pruritus, injection site reactions, *sweating.*

INTERACTIONS Drug: Alcohol, OPIATES, other CNS DEPRESSANTS, BENZODIAZEPINES augment CNS depression; **diazepam** may cause respiratory or cardiovascular collapse; AZOLE ANTIFUNGALS (e.g., **fluconazole**), MACROLIDE ANTIBIOTICS (e.g., **erythromycin**), and PROTEASE INHIBITORS (e.g., **saquinavir**) may increase buprenorphine levels.

PHARMACOKINETICS Absorption: Widely variable sublingual absorption. **Onset:** 10–30 min IM/IV. **Peak:** 1 h IM/IV; 2–6 h SL. **Duration:** 6–10 h. **Metabolism:** Extensively in liver by CYP3A4 to active metabolite norbuprenorphine. **Elimination:** 70% in feces, 30% in urine in 7 days. **Half-Life:** 2.2 h IM/IV; 37 h SL.

NURSING IMPLICATIONS

Assessment & Drug Effects

▪ Monitor respiratory status during therapy. Buprenorphine-induced respiratory depression is about equal to that produced by 10 mg morphine, but onset is slower, and if it occurs, it lasts longer.

▪ Monitor I&O ratio and pattern urinary retention is a potential adverse effect.

▪ Lab tests: Baseline LFTs and renal function.

▪ Supervise ambulation; drowsiness occurs in 66% of patients taking this drug.

Patient & Family Education

▪ Do not drive or engage in other potentially hazardous activities until response to drug is known.

▪ Do not use alcohol or other CNS depressing drugs without consulting prescriber. An additive effect exists between buprenorphine hydrochloride and other CNS depressants including alcohol.

BUPROPION HYDROCHLORIDE

(byoo-pro′pi-on)

Wellbutrin, Wellbutrin SR, Wellbutrin XL, Zyban

BUPROPION HYDROBROMIDE

Aplenzin

Classification: ANTIDEPRESSANT
Therapeutic: ANTIDEPRESSANT
Pregnancy Category: C

AVAILABILITY 75 mg, 100 mg tablets; 100 mg, 150 mg, 200 mg sustained release tablets; 150 mg, 300 mg extended release tablets; Hydrobromide form: 174 mg, 348 mg, 522 mg extended release tablet

ACTION & *THERAPEUTIC EFFECT*
The neurochemical mechanism of bupropion is not fully understood. It selectively inhibits the neuronal reuptake of dopamine. *Its antidepressive effect is related to CNS stimulant effects.*

USES Depression, smoking cessation; seasonal affective disorder.
UNLABELED USES Neuropathic pain, ADHD.

CONTRAINDICATIONS Hypersensitivity to bupropion; history of seizure disorder; current or prior diagnosis of bulimia or anorexia nervosa; suicidal ideation; concurrent administration of an MAO inhibitor; head trauma; seizure disorder; CNS tumor; recent MI; abrupt discontinuation, anorexia nervosa, bulimia nervosa; lactation.

B

CAUTIOUS USE Renal or hepatic function impairment; drug abuse or dependence; cardiac disease, MI, renal impairment; hepatic disease, biliary cirrhosis; hypertension, bipolar disorder, mania, psychosis, diabetes mellitus, older adults, ethanol intoxication, tics, Tourette's syndrome; pregnancy (category C); children younger than 18 y.

ROUTE & DOSAGE

Depression/Seasonal Affective Disorder

Adult: **PO** 100 mg t.i.d. (immediate release) or 150 mg SR b.i.d., or 300 mg XL daily, doses greater than 450 mg/day are associated with an increased risk of adverse reactions including seizures; **Aplenzin:** 174 mg daily can increase to 348 mg daily
Geriatric: **PO** 50–100 mg/day, may increase by 50–100 mg q3–4days (max: 150 mg/dose)

Smoking Cessation

Adult: **PO** Start with 150 mg once daily × 3 days, then increase to 150 mg b.i.d. (max: 300 mg/day) for 7–12 wk

Hepatic Impairment Dosage Adjustment

Start at lower dose, decrease dose or dosage frequency

ADMINISTRATION

Oral

- Give with meals to decrease incidence of nausea and vomiting.
- Ensure that extended release and sustained release tablets are not chewed or crushed. They **must be** swallowed whole.
- Store away from heat, direct light, and moisture.

ADVERSE EFFECTS (≥1%) **Body as a Whole:** Weight loss, weight gain. **CNS:** Seizures. The risk of seizure appears to be strongly associated with dose (especially greater than 450 mg/day) *agitation, insomnia, dry mouth, blurred vision, headache, dizziness, tremor.* **GI:** *Nausea, vomiting, constipation.* **CV:** Tachycardia. **Skin:** Rash.

INTERACTIONS Drug: May increase metabolism of **carbamazepine, cimetidine, phenytoin, phenobarbital,** decreasing their effect; may increase incidence of adverse effects of **levodopa,** contraindicated with MAO INHIBITORS.

PHARMACOKINETICS Absorption: Readily from GI tract. **Onset:** 3–4 wk. **Peak:** 1–3 h. **Metabolism:** In liver to active metabolites by CYP2D6. **Elimination:** 80% in urine. **Half-Life:** 8–24 h.

NURSING IMPLICATIONS

Assessment & Drug Effects

- Monitor for therapeutic effectiveness. The full antidepressant effect of drug may not be realized for 4 or more weeks.
- Monitor for and report promptly worsening of depression or suicidal ideation.
- Monitor for and report promptly suicidal ideation or worsening of mental status.
- Use extreme caution when administering drug to patient with history of seizures, cranial trauma, or other factors predisposing to seizures; during sudden and large increments in dose, seizure potential is increased.
- Report significant restlessness, agitation, anxiety, and insomnia. Symptoms may require treatment or discontinuation of drug.

Common adverse effects in *italic*, life-threatening effects <u>underlined</u>; generic names in **bold**; classifications in SMALL CAPS; ♣ Canadian drug name; ✪ Prototype drug

203

- Monitor for and report delusions, hallucinations, psychotic episodes, confusion, and paranoia.
- Lab tests: Periodic renal function tests and LFTs.

Patient & Family Education
- Monitor weight at least weekly. Report significant changes in weight (±5 lb) to prescriber.
- Minimize or avoid alcohol because it increases the risk of seizures.
- Report promptly suicidal thoughts, especially when treated for depression.
- Do not drive or engage in potentially hazardous activities until response to drug is known because judgment or motor and cognitive skills may be impaired.
- Do not abruptly discontinue drug. Gradual dosage reduction may be necessary to prevent adverse effects.
- Do not take any OTC drugs without consulting prescriber.

BUSPIRONE HYDROCHLORIDE

(byoo-spye'rone)

BuSpar
Classification: ANXIOLYTIC
Therapeutic: ANTIANXIETY
Prototype: Lorazepam
Pregnancy Category: B

AVAILABILITY 5 mg, 10 mg, 15 mg tablets

ACTION & *THERAPEUTIC EFFECT*
An anxiolytic that focuses mainly on the brain D_2-dopamine receptors. It has agonist effects on presynaptic dopamine receptors and also a high affinity for serotonin (5-HT$_{1A}$) receptors. *Antianxiety effect is due to serotonin reuptake inhibition and agonist effects on dopamine receptors of the brain.*

USES Management of anxiety disorders and for short-term treatment of generalized anxiety.

UNLABELED USES Adjuvant for nicotine withdrawal, premenstrual syndrome.

CONTRAINDICATIONS Concomitant use of MAOI therapy; lactation.
CAUTIOUS USE Moderate to severe renal or hepatic impairment, pregnancy (category B); children less than 18 y.

ROUTE & DOSAGE

Anxiety
Adult: **PO** 7.5–15 mg/day in divided doses, may increase by 5 mg/day q2–3days as needed (max: 60 mg/day) *Geriatric:* **PO** 5 mg b.i.d., may increase to max of 60 mg/day

ADMINISTRATION

Oral
- Give with food to decrease nausea.
- Store at 15°–30° C (59°–86° F) in tightly closed container unless otherwise directed.

ADVERSE EFFECTS (≥1%) **CNS:** Numbness, paresthesia, tremors, *dizziness, headache,* nervousness, *drowsiness,* lightheadedness, dream disturbances, decreased concentration, excitement, mood changes. **CV:** Tachycardia, palpitation. **Special Senses:** Blurred vision. **GI:** *Nausea,* vomiting, dry mouth, abdominal/gastric distress, diarrhea, constipation. **Urogenital:** Urinary frequency, hesitancy. **Musculoskeletal:** Arthralgias. **Respiratory:** Hyperventilation, shortness of breath. **Skin:** Rash, edema, pruritus, flushing, easy bruising, hair loss, dry skin. **Other:** Fatigue, weakness.

DIAGNOSTIC TEST INTERFERENCE
Buspirone may increase serum concentrations of *hepatic aminotransferases (ALT, AST)*.

INTERACTIONS Drug: May cause hypertension with MAO INHIBITORS, **trazodone,** possible increase in liver transaminases; increased **haloperidol** serum levels. **Food: Grapefruit juice** may increase drug levels. **Herbal: St. John's wort** may increase drug levels.

PHARMACOKINETICS Absorption: Readily from GI tract, undergoes first pass metabolism. **Onset:** 5–7 days. **Peak:** 1 h. **Metabolism:** In liver. **Elimination:** 30–63% in urine as metabolites within 24 h. **Half-Life:** 2–4 h.

NURSING IMPLICATIONS

Assessment & Drug Effects
- Monitor for therapeutic effectiveness. Desired response may begin within 7–10 days; however, optimal results take 3–4 wk. Reinforce the importance of continuing treatment to patient.
- Monitor for and report dystonia, motor restlessness, and involuntary repetitious movement of facial or cervical muscle.
- Observe for and report swollen ankles, decreased urinary output, changes in voiding pattern, jaundice, itching, nausea, or vomiting.

Patient & Family Education
- Report any of the following immediately: Involuntary, repetitive movements of face or neck; weakness, nervousness, nightmares, headache, or blurred vision; depression or thoughts of suicide.
- Do not use OTC drugs without advice of the prescriber while taking buspirone.

- Do not drive or engage in other potentially hazardous activities until response to drug is known.
- Discuss limits of alcohol intake with prescriber; cautious use is generally advised.

BUSULFAN
(byoo-sul'fan)

Busulfex, Myleran
Classifications: ANTINEOPLASTIC; ALKYLATING AGENT
Therapeutic: ANTINEOPLASTIC
Prototype: Cyclophosphamide
Pregnancy Category: D

AVAILABILITY 2 mg tablets; 6 mg/mL injection

ACTION & *THERAPEUTIC EFFECT*
Potent cytotoxic alkylating agent that may be mutagenic or carcinogenic, and cell cycle nonspecific. Reduces total granulocyte mass but has little effect on lymphocytes and platelets except in large doses. May cause widespread epithelial cellular dysplasia severe enough to make it difficult to interpret exfoliative cytologic examinations. *Causes cell death by acting predominantly on slowly proliferating stem cells by inducing cross linkage in DNA, thus blocking replication.*

USES Bone marrow ablation, chronic myelogenous leukemia, stem cell transplant preparation.

CONTRAINDICATIONS Therapy-resistant chronic lymphocytic leukemia; lymphoblastic crisis of chronic myelogenous leukemia; bone marrow depression, immunizations (patient and household members), chickenpox (including recent exposure), herpetic infections; pregnancy (category D), lactation.
CAUTIOUS USE Men and women in childbearing years; hepatic

Common adverse effects in *italic*, life-threatening effects underlined; generic names in **bold**; classifications in SMALL CAPS; ♦ Canadian drug name; ☺ Prototype drug

205

disease; history of gout or urate renal stones; prior irradiation or chemotherapy.

ROUTE & DOSAGE

Chronic Myelogenous Leukemia

Adult: **PO** 4–8 mg/day or 1.8–4 mg/m² daily until maximal clinical and hematologic improvement, may use 1–4 mg/day if remission is shorter than 3 mo

Child: **PO** 0.06–0.12 mg/kg/day or 1.8–4.6 mg/m² as a daily dose

Stem Cell Transplant Preparation and Bone Marrow Ablation

Adult: **IV** (used with cyclophosphamide) 0.8 mg/kg IBW or ABW (whichever is lower) q6h × 4 days

Obesity Dosage Adjustment

In obese patients, use adjusted ideal body weight = IBW + 0.25 × (actual weight – IBW)

ADMINISTRATION

Oral

- Give at same time each day.
- Give on an empty stomach to minimize nausea and vomiting.
- Store in tightly capped, light-resistant container at 15°–30° C (59°–86° F), unless otherwise specified.

Intravenous

PREPARE: **Intermittent:** Prepare a volume of NS or D5W IV solution that is 10 times the volume of busulfan needed. ▪ Using a 5 micron nylon filter (supplied), withdraw the needed dose of busulfan. Remove needle and filter and use a new, nonfiltered needle to add busulfan to the IV fluid. (Always add busulfan to IV fluid rather than IV fluid to busulfan.) ▪ Mix by inverting the IV bag several times.

ADMINISTER: **Intermittent:** Infuse via a central venous catheter over 2 h. ▪ Flush line before/after infusion with at least 5 mL D5W or NS.

ADVERSE EFFECTS (≥1%) **Hematologic:** Major toxic effects are related to bone marrow failure; agranulocytosis (rare), pancytopenia, thrombocytopenia, leukopenia, anemia. **Urogenital:** Flank pain, renal calculi, uric acid nephropathy, acute renal failure, gynecomastia, testicular atrophy, azoospermia, impotence, sterility in males, ovarian suppression, menstrual changes, amenorrhea (potentially irreversible), menopausal symptoms. **Respiratory:** Irreversible pulmonary fibrosis ("busulfan lung"). **Skin:** Alopecia, hyperpigmentation. **Other:** Endocardial fibrosis, dizziness, cholestatic jaundice, infections.

DIAGNOSTIC TEST INTERFERENCE Busulfan may decrease *urinary 17-OHCS* excretion, and may increase *blood and urine uric acid* levels. Drug-induced cellular dysplasia may interfere with interpretation of *cytologic studies.*

INTERACTIONS Drug: Probenecid, sulfinpyrazone may increase uric acid levels.

PHARMACOKINETICS Absorption: Readily from GI tract. **Peak:** 4 h. **Duration:** 4 h. **Metabolism:** In liver by CYP3A4. **Elimination:** 10–50% in urine within 48 h.

NURSING IMPLICATIONS

Assessment & Drug Effects

- Monitor the following: Vital signs, weight, I&O ratio and pattern. Urge patient to increase fluid intake to 10–12 (8 oz) glasses daily (if allowed) to assure adequate urinary output.

- Monitor for and report symptoms suggestive of superinfection (see Appendix F), particularly when patient develops leukopenia.
- Lab test: Baseline Hgb, Hct, WBC with differential, platelet count, LFTs, kidney function, serum uric acid; repeat at least weekly.
- Avoid invasive procedures during periods of platelet count depression.

Patient & Family Education
- Report to prescriber any of the following: Easy bruising or bleeding, cloudy or pink urine, dark or black stools; sore mouth or throat, unusual fatigue, blurred vision, flank or joint pain, swelling of lower legs and feet; yellowing white of eye, dark urine, light-colored stools, abdominal discomfort, or itching (hepatotoxicity).
- Use contraceptive measures during busulfan therapy and for at least 3 mo after drug is withdrawn.

BUTABARBITAL SODIUM

(byoo-ta-bar'bi-tal)

Butisol Sodium
Classifications: BARBITURATE; ANXIOLYTIC; SEDATIVE-HYPNOTIC
Therapeutic: ANTIANXIETY; SEDATIVE-HYPNOTIC
Prototype: Phenobarbital
Pregnancy Category: C
Controlled Substance: Schedule III

AVAILABILITY 30 mg, 50 mg tablets; 30 mg/5 mL elixir

ACTION & *THERAPEUTIC EFFECT*
Intermediate-acting barbiturate that appears to act at thalamus level of the brain, where it interferes with transmission of impulses to the cerebral cortex. *Preoperative sedative agent that also is an effective antianxiety agent.*

USES Anxiety, insomnia, sedation induction and maintenance.

CONTRAINDICATIONS Porphyria; uncontrolled pain; severe respiratory disease; history of addiction; lactation.

CAUTIOUS USE Severe renal or hepatic impairment; acute abdominal conditions; head trauma, history of seizures; history of herpes infection; older adults or debilitated patients; pregnancy (category C).

ROUTE & DOSAGE

Daytime Sedation
Adult: **PO** 15–30 mg t.i.d. or q.i.d.
Preoperative Sedation
Adult: **PO** 50–100 mg 60–90 min before surgery
Child: **PO** 2–6 mg/kg/dose (max: 100 mg)
Insomnia
Adult: **PO** 100 mg at bedtime

ADMINISTRATION

Oral
- Schedule slow withdrawal following long-term use to avoid precipitating withdrawal symptoms.
- Store in tightly covered containers, preferably at 15°–30° C (59°–86° F), unless otherwise directed.

ADVERSE EFFECTS (≥1%) **CNS:** Drowsiness, *residual sedation* ("hangover"), headache. **GI:** Nausea, vomiting, constipation, diarrhea. **Skin:** Urticaria, skin rash. **Musculoskeletal:** Muscle or joint pain.

INTERACTIONS Drug: Alcohol and other CNS DEPRESSANTS add to CNS and respiratory depression; butabarbital increases the metabolism of ORAL

Common adverse effects in *italic*, life-threatening effects underlined; generic names in **bold**; classifications in SMALL CAPS; ♣ Canadian drug name; ♥ Prototype drug

207

B

ANTICOAGULANTS, BETA-BLOCKERS, CORTICOSTEROIDS, **doxycycline, griseofulvin, quinidine,** THEOPHYLLINES, ORAL CONTRACEPTIVES, decreasing their effectiveness. **Herbal: Kava, valerian** may potentiate sedation.

PHARMACOKINETICS Absorption: Readily from GI tract. **Onset:** 40–60 min. **Peak:** 3–4 h. **Duration:** 6–8 h. **Distribution:** Crosses placenta; distributed into breast milk. **Metabolism:** In liver. **Elimination:** In urine primarily as metabolites. **Half-Life:** Average 100 h.

NURSING IMPLICATIONS

Assessment & Drug Effects

- Assess for adverse effects. Older adults and debilitated patients sometimes manifest excitement, confusion, or depression. Children also may react with paradoxical excitement. Side rails may be advisable. Report these reactions to prescriber.

Patient & Family Education

- Do not drive or engage in other potentially hazardous activities until response to drug is known.
- Do not drink alcoholic beverages while taking this drug. Other CNS depressants may produce additive drowsiness; do not take without approval of prescriber.

BUTENAFINE HYDROCHLORIDE
(bu-ten'a-feen)

Lotrimin Ultra, Mentax
Classification: ANTIFUNGAL ANTIBIOTIC
Therapeutic: ANTIFUNGAL
Prototype: Terbinafine
Pregnancy Category: B

AVAILABILITY 1% cream

ACTION & *THERAPEUTIC EFFECT* Exerts antifungal action by inhibiting fungal sterol synthesis that is needed in formation of the fungal cell membrane. *Antifungal effectiveness against interdigital tinea pedis (athlete's foot), tinea corporis (ringworm), and tinea cruris (jock itch).*

USES Treatment of tinea pedis, tinea corporis, and tinea cruris.

CONTRAINDICATIONS Hypersensitivity to butenafine; ophthalmic or vaginal administration.
CAUTIOUS USE Hypersensitivity to naftifine or tolnaftate; pregnancy (category B); lactation; children younger than 12 y.

ROUTE & DOSAGE

Tinea Pedis

Adult/Child (older than 12 y):
Topical Apply to affected area and surrounding skin b.i.d. × 7 days or daily × 4 wk

Tinea Corporis, Tinea Cruris

Adult/Child (younger than 12 y):
Topical Apply to affected area and surrounding skin once daily

ADMINISTRATION

Topical

- Apply sufficient cream to cover affected skin and surrounding areas.
- Do not use occlusive dressing unless specifically directed to do so.
- Store at 5°–30° C (41°–86° F).

ADVERSE EFFECTS (≥1%) **Skin:** Burning/stinging at application site, contact dermatitis, erythema, irritation, itching.

NURSING IMPLICATIONS

Assessment & Drug Effects

- Note: 2–4 wk of therapy are usually required for effective treatment.

Common adverse effects in *italic*, life-threatening effects <u>underlined</u>; generic names in **bold**; classifications in SMALL CAPS; ♣ Canadian drug name; ⊙ Prototype drug

Patient & Family Education
- Discontinue medication and notify prescriber if irritation or sensitivity develops.
- Avoid contact with mucous membranes.
- Wash hands thoroughly before and after application of cream.

BUTOCONAZOLE NITRATE
(byoo-toe-koe'na-zole)

Femstat 3, Gynazole 1
Classification: AZOLE ANTIFUNGAL ANTIBIOTIC
Therapeutic: ANTIFUNGAL
Prototype: Fluconazole
Pregnancy Category: C

AVAILABILITY 2% cream

ACTION & *THERAPEUTIC EFFECT* Imidazole derivative with antifungal activity. Alters fungal cell membrane permeability, permitting loss of phosphorous compounds, potassium, and other essential intracellular constituents with consequent loss of ability to replicate. Action takes place primarily on medicated infected surface tissues. *Has fungicidal effect as well as effectiveness against some gram-positive bacteria.*

USES Local treatment of vulvovaginal candidiasis.

CAUTIOUS USE Hypersensitivity to azole antifungals; HIV patients; diabetes mellitus; pregnancy (category C), lactation; children less than 12 y.

ROUTE & DOSAGE

Vulvovaginal Candidiasis
Adult: **Topical** 1 applicator full intravaginally at bedtime for 3 days, may be extended another 3 days if needed

Pregnant women: **Topical** 1 applicator full intravaginally at bedtime for 6 days

ADMINISTRATION

Topical Intravaginal
- Continue treatment even during menstruation.
- Store medication at 15°–30° C (59°–86° F); avoid extreme temperature and freezing.

ADVERSE EFFECTS (≥1%) **Urogenital:** Vulvar or vaginal burning, vulvar itching, discharge, soreness, swelling; urinary frequency and burning. **Skin:** Itching of fingers. **CNS:** Headache.

PHARMACOKINETICS Absorption: Small amount absorbed systemically from intravaginal administration. **Distribution:** Crosses placenta in animals. **Metabolism:** In liver. **Elimination:** In both urine and feces within 4–7 days. **Half-Life:** 21–24 h.

NURSING IMPLICATIONS

Assessment & Drug Effects
- Monitor for therapeutic effectiveness. Candidiasis in nonpregnant women is usually controlled in 3 days.

Patient & Family Education
- Take medication exactly as prescribed; do not increase or decrease dosage or discontinue or extend treatment period. Contact prescriber if symptoms (vaginal burning, discharge, or itching) persist; drug may be discontinued if acute irritation occurs.
- Patient's sexual partner should wear a condom during intercourse.

Common adverse effects in *italic*, life-threatening effects <u>underlined</u>; generic names in **bold**; classifications in SMALL CAPS; ♦ Canadian drug name; ◉ Prototype drug

209

B

BUTORPHANOL TARTRATE

(byoo-tor'fa-nole)

Stadol, Stadol NS

Classifications: ANALGESIC; NAR-COTIC (OPIATE AGONIST-ANTAGO-NIST)

Therapeutic: NARCOTIC ANALGESIC
Prototype: Pentazocine
Pregnancy Category: C
Controlled Substance: Schedule IV

AVAILABILITY 1 mg/mL, 2 mg/mL injection; 10 mg/mL spray

ACTION & *THERAPEUTIC EFFECT*
Synthetic, centrally acting analgesic with mixed narcotic agonist and antagonist actions. Acts as agonist on one type of opioid receptor and as a competitive antagonist at others. Site of analgesic action believed to be subcortical, possibly in the limbic system of the brain. Respiratory depression does not increase appreciably with higher doses, as it does with morphine, but duration of action increases. *Narcotic analgesic that relieves moderate to severe pain.*

USES Relief of moderate to severe pain, preoperative or preanesthetic sedation and analgesia, obstetric analgesia during labor, cancer pain, renal colic, burns.

UNLABELED USES Musculoskeletal and post-episiotomy pain.

CONTRAINDICATIONS Narcotic-dependent patients; opiate agonist hypersensitivity.

CAUTIOUS USE History of drug abuse or dependence; emotionally unstable individuals; head injury, increased intracranial pressure; acute MI, ventricular dysfunction, coronary insufficiency, hypertension; patients undergoing biliary tract surgery; respiratory depression, bronchial asthma, obstructive respiratory disease; and renal or hepatic dysfunction; prior to labor, pregnancy (category C). Safe use in children under 18 y not established.

ROUTE & DOSAGE

Pain Relief

Adult: **IM** 1–4 mg q3–4h as needed (max: 4 mg/dose) **IV** 0.5–2 mg q3–4h as needed
Geriatric: **IM/IV** 0.25–1 mg q6–8h **Intranasal** 1 mg (1 spray) in one nostril, may repeat in 90 sec, then may repeat these 2 doses q3–4h prn

Adjunct to Balanced Anesthesia

Adult: **IV** 2 mg before induction or 0.5–1 mg in increments during anesthesia

Labor

Adult: **IV/IM** 1–2 mg may repeat in 4 h

Renal Impairment Dosage Adjustment

For GFR less than 10 mL/min, use 50% of dose

Hepatic Impairment Dosage Adjustment

Use half normal dose and at least 6 h interval

ADMINISTRATION

Intranasal
- Give 1 spray into one nostril only. One spray provides a 1 mg dose.

Intramuscular
- Give preoperative IM injection 60–90 min before surgery.

Intravenous

PREPARE: Direct: Give undiluted.
ADMINISTER: Direct: Give at a rate of 2 mg over 3–5 min.

Common adverse effects in *italic*, life-threatening effects <u>underlined</u>; generic names in **bold**; classifications in SMALL CAPS; ♣ Canadian drug name; ☉ Prototype drug

INCOMPATIBILITIES **Y-site: Amphotericin B cholesteryl, lansoprazole, midazolam.**

- Store at 15°–30° C (59°–86° F) unless otherwise directed. Protect from light.

ADVERSE EFFECTS (≥1%) **CNS:** Drowsiness, *sedation,* headache, vertigo, dizziness, floating feeling, weakness, lethargy, confusion, light-headedness, insomnia, nervousness, <u>respiratory depression</u>. **CV:** Palpitation, bradycardia. **GI:** Nausea. **Skin:** Clammy skin, tingling sensation, flushing and warmth, cyanosis of extremities, diaphoresis, sensitivity to cold, urticaria, pruritus. **Genitourinary:** Difficulty in urinating, biliary spasm.

INTERACTIONS Drug: Alcohol and other CNS DEPRESSANTS augment CNS and respiratory depression.

PHARMACOKINETICS Onset: 10–30 min IM; 1 min IV. **Peak:** 0.5–1 h IM; 4–5 min IV. **Duration:** 3–4 h IM; 2–4 h IV. **Distribution:** Crosses placenta; distributed into breast milk. **Metabolism:** In liver in inactive metabolites. **Elimination:** Primarily in urine. **Half-Life:** 3–4 h.

NURSING IMPLICATIONS

Assessment & Drug Effects
- Monitor for respiratory depression. Do not administer drug if respiratory rate is less than 12 breaths/min.
- Monitor vital signs. Report marked changes in BP or bradycardia.
- Note: If used during labor or delivery, observe neonate for signs of respiratory depression.
- Note: Drug can induce acute withdrawal symptoms in opiate-dependent patients.
- Drug is usually withdrawn gradually following chronic administration. Abrupt withdrawal may produce vomiting, loss of appetite, restlessness, abdominal cramps, increase in BP and temperature, mydriasis, faintness. Withdrawal symptoms peak 48 h after discontinuation of drug.

Patient & Family Education
- Lie down to control drug-induced nausea.
- Do not take alcohol or other CNS depressants with this drug without consulting prescriber because of possible additive effects.
- Do not drive or engage in other potentially hazardous activities until response to drug is known.

CABAZITAXEL
(ka-baz'i tax-el)

Jevtana
Classifications: ANTINEOPLASTIC; TAXANE; MICROTUBULE INHIBITOR
Therapeutic: ANTINEOPLASTIC; ANTIMICROTUBULE
Prototype: Paclitaxel
Pregnancy Category: D

AVAILABILITY 60 mg/1.5 mL solution for injection

ACTIONS & *THERAPEUTIC EFFECT*
Cabazitaxel binds to the microtubule network essential for interphase and mitosis of the cell cycle stabilizing the microtubules involved in cell division and preventing their normal functioning. *This antimicrotubular effect results in inhibition of mitosis in cancer cells.*

USES Treatment of prostate cancer.

CONTRAINDICATIONS Neutrophil count of 1500/mm³ or less; hypersensitivity to cabazitaxel or drugs formulated with polysorbate 80; total bilirubin at or above the ULN or AST/ALT at or above 1.5 × ULN; pregnancy category D; lactation.

Common adverse effects in *italic*, life-threatening effects <u>underlined</u>; generic names in **bold**; classifications in SMALL CAPS; ♣ Canadian drug name; ⓟ Prototype drug

211

CAUTIOUS USE Neutropenia; history of hypersensitivity reactions; GI distress (nausea, vomiting, diarrhea); renal or hepatic impairment; patients 65 yrs or greater. Safety and efficacy in children not established.

ROUTE & DOSAGE

Prostate Cancer

Adult: **IV** 25 mg/m^2 over 1 h, repeat q3wk (max: 10 cycles)

Toxicity Dosage Adjustment

Grade 3 or greater neutropenia: Delay treatment until ANC greater than 1500/mm^3. Reduce dose to 20 mg/m^2. Use G-CSF for secondary prophylaxis.

Patients who develop febrile neutropenia: Delay treatment until improvement or resolution, and until ANC is greater than 1500/mm^3. Reduce dose to 20 mg/m^2. Use G-CSF for secondary prophylaxis.
Grade 3 or greater diarrhea or persisting diarrhea: Delay treatment until improvement or resolution. Reduce dose to 20 mg/m^2. Discontinue cabazitaxel if toxicity persists with 20 mg/m^2 dose.

Hepatic Impairment Dosage Adjustment

Should not be given to patients with a total bilirubin at or above the ULN, or AST/ALT at or above 1.5 x ULN

Renal Impairment Dosage Adjustment

No dosage adjustments required

ADMINISTRATION

Intravenous

- Pre-medicate at least 30 min prior to each dose to avoid severe hypersensitivity: Dexamethasone 8 mg (or equivalent), dexchlorpheniramine 5 mg or diphenhydramine 25 mg, and ranitidine 50 mg (or equivalent).
- Follow institutional or standard guidelines for preparation, handling, and disposal of cytotoxic agents.

PREPARE: **Continuous:** Prepare under aseptic conditions. Do NOT use infusion containers or equipment made with PVC or polyurethane. *First dilution:* Add all of the supplied diluent to the 60 mg vial of cabazitaxel. ▪ Direct flow of diluent onto inside wall of the cabazitaxel vial; inject slowly to limit foaming then invert vial gently for 45 sec to mix. ▪ Allow vial to stand until foam dissipates then inspect to ensure there are no visible particles. ▪ The resulting solution (10 mg/mL of cabazitaxel) should be further diluted within 30 min. *Second dilution:* Withdraw required dose and dilute in 250 mL or more of NS or D5W to yield a concentration no greater than 0.26 mg/mL. Solution should be clear without precipitate. ▪ Use immediately.

ADMINISTER: **Continuous:** Infuse over 1 hr through a 0.22 micron in-line filter. Do NOT use tubing containing PVC or polyurethane. ▪ Monitor closely during infusion for S&S of hypersensitivity. Stop infusion immediately and institute supportive care should a hypersensitivity reaction occur.

INCOMPATIBILITIES **Solution/additive:** None listed. **Y-site:** None listed.

- Store undiluted at 15°–30° C (59°–86° F). First dilution should be used immediately. Second dilution may be stored for 8 hr at room temperature (including 1 hr infusion time) or for a total of 24 hr if refrigerated (including 1 hr infusion time).

ADVERSE EFFECTS (≥5%) **Body as a Whole:** *Alopecia,* mucosal inflammation, pain, peripheral edema, *pyrexia.* **CNS:** *Asthenia,* dizziness, *dysgeusia, fatigue,* headache, *periphery neuropathy.* **CV:** Arrhythmia, hypotension. **GI:** *Abdominal pain, constipation, diarrhea,* dyspepsia, *nausea, vomiting.* **Hematological:** *Anemia,* febrile neutropenia, *leukopenia, neutropenia, thrombocytopenia.* **Metabolic:** *Anorexia,* dehydration, weight loss. **Musculoskeletal:** *Arthralgia, back pain,* muscle spasms. **Respiratory:** *Cough, dyspnea.* **Urogenital:** Dysuria, *hematuria,* urinary tract infections.

INTERACTIONS Drug: Coadministration of CYP3A4 inducers (e.g., **carbamazepine, phenobarbital, phenytoin, rifabutin, rifampin, rifapentine**) can decrease the levels of cabazitaxel. Coadministration of strong CYP3A inhibitors (e.g., **atazanavir, clarithromycin, indinavir, itraconazole, ketoconazole, nefazodone, nelfinavir, ritonavir, saquinavir, telithromycin, voriconazole**) can increase the levels of cabazitaxel. **Food:** Grapefruit juice can increase the levels of cabazitaxel. **Herbal:** **St. John's wort** can decrease the levels of cabazitaxel.

PHARMACOKINETICS Distribution: 89–92% Plasma protein bound. **Metabolism:** Extensively metabolized in the liver. **Elimination:** Primarily fecal elimination. **Half-Life:** 95 h.

NURSING IMPLICATIONS

Assessment & Drug Effects
- Monitor for hypersensitivity reactions especially during cycles 1 and 2. S&S requiring treatment and discontinuation of the drug include: Hypotension, bronchospasm, and generalized rash/erythema. Discontinue immediately and manage symptoms aggressively.
- Monitor vital signs frequently, especially during the first hour of infusion. Cardiac monitoring may be indicated for those with conduction abnormalities.
- Lab tests: Monitor CBC with differential weekly during cycle 1 and prior to each cycle thereafter; baseline and periodic LFTs and kidney function tests; periodic serum electrolytes, especially if diarrhea occurs.
- Monitor for and report promptly severe or persistent diarrhea as it may cause dehydration and electrolyte imbalances.

Patient & Family Education
- Report immediately to prescriber S&S of hypersensitivity during drug infusion: Rash or itching, skin redness, difficulty breathing, chest pain or throat tightness, swelling of face, or feeling faint.
- Report severe or persistent diarrhea or vomiting as additional medications may be required.
- Avoid aspirin, NSAIDs, or alcohol to minimize GI distress.
- Do not drink grapefruit juice while taking this drug.
- Report unusual bruising or bleeding (e.g., blood in urine, or dark tarry stools).
- Use caution with exposure to potential sources of infection during periods when your blood count is low.

Common adverse effects in *italic,* life-threatening effects underlined; generic names in **bold;** classifications in SMALL CAPS; ♣ Canadian drug name; ⊘ Prototype drug

213

CABERGOLINE
(ka-ber′go-leen)

Dostinex
Classifications: ERGOT ALKALOID;
DOPAMINE RECEPTOR AGONIST
Therapeutic: ERGOT ALKALOID;
ANTI-PARKINSON
Prototype: Ergotamine
Pregnancy Category: B

AVAILABILITY 0.5 mg tablets

ACTION & *THERAPEUTIC EFFECT*
Cabergoline is a synthetic ergot
derivative, long-acting dopamine
receptor agonist with a high affinity for D_2 receptors in the anterior pituitary. It also suppresses
prolactin secretion. *Cabergoline
inhibits both puerperal lactation
and pathologic hyperprolactinemia. Exhibits antiparkinsonism
effects due to increased levels of
dopamine.*

USES Treatment of hyperprolactinemia.
UNLABELED USES Parkinson's disease, restless leg syndrome.

CONTRAINDICATIONS Uncontrolled hypertension and hypersensitivity to ergot derivatives;
pregnancy-induced hypertension, preeclampsia, eclampsia,
lactation.
CAUTIOUS USE Hepatic function
impairment; elderly, psychosis;
pregnancy (category B). Safety and
efficacy in pediatric patients are
unknown.

ROUTE & DOSAGE

Hyperprolactinemia
Adult: **PO** Start with 0.25 mg 2
times/wk, may increase by 0.25
mg 2 times/wk to a max of 1 mg
2 times/wk

ADMINISTRATION
Oral
▪ Give on same days each week.

ADVERSE EFFECTS (≥1%) **Body as a
Whole:** Asthenia, fatigue, hot flashes.
CV: Postural hypotension. **GI:** *Nausea, constipation,* abdominal pain,
dyspepsia, vomiting, dry mouth, diarrhea, flatulence. **Endocrine:** Breast
pain, dysmenorrhea. **CNS:** *Headache, dizziness,* paresthesia, somnolence, depression, nervousness.

INTERACTIONS Drug: Concurrent
use with PHENOTHIAZINES, BUTYROPH-
ENONES, THIOXANTHINES, and **metoclopramide** decreases therapeutic
effects of both drugs.

**PHARMACOKINETICS   Absorp-
tion:** Rapidly absorbed in GI tract,
undergoes first-pass metabolism.
Peak: 2–3 h. **Distribution:** 40–42%
protein bound. Crosses placenta.
Metabolism: Extensively metabolized. **Elimination:** Approximately
22% in urine, 60% in feces. **Half-
Life:** 63–69 h.

NURSING IMPLICATIONS
Assessment & Drug Effects
▪ Lab tests: Monitor serum prolactin
levels to assess response to each
dosing level.
▪ Monitor for hypotension, especially when given with other
drugs known to lower BP.

Patient & Family Education
▪ Discontinue this drug once prescriber advises that serum prolactin
level has been maintained for 6 mo.

CAFFEINE ⊘
(kaf-een′)

**Caffedrine, Dexitac, NoDoz,
Quick Pep, S-250, Tirend, Viva-
rin**

CAFFEINE AND SODIUM BENZOATE

CITRATED CAFFEINE
Cafcit
Classifications: RESPIRATORY AND CEREBRAL STIMULANT; XANTHINE
Therapeutic: RESPIRATORY AND CEREBRAL STIMULANT
Pregnancy Category: C

AVAILABILITY 100 mg, 150 mg, 200 mg tablets; 200 mg capsules; 10 mg/mL caffeine citrate oral solution; 10 mg/mL caffeine citrate injection

ACTION & *THERAPEUTIC EFFECT*
Stimulant effect is thought to be related to inhibition of the enzyme phosphodiesterase, which results in higher concentrations of cyclic AMP. Releases epinephrine and norepinephrine from adrenal medulla, producing CNS stimulation. Small doses improve psychic and sensory awareness and reduce drowsiness and fatigue by stimulating cerebral cortex. Higher doses stimulate medullary, respiratory, vasomotor, and vagal centers. Produces smooth muscle relaxation by direct action on vascular musculature. *Effective in managing neonatal apnea, and as an adjuvant for pain control in headaches and following dural puncture. Relief of headache is perhaps due to mild cerebral vasoconstriction action and increased vascular tone. It acts as a bronchodilator in asthma.*

USES Orally as a mild CNS stimulant to aid in staying awake and restoring mental alertness, and as an adjunct in narcotic and nonnarcotic analgesia. Used parenterally as an emergency stimulant in acute circulatory failure, as a diuretic, and for neonatal apnea.

UNLABELED USES Topical treatment of atopic dermatitis; to re-leave spinal puncture headache.

CONTRAINDICATIONS Acute MI, symptomatic cardiac arrhythmias, palpitations; peptic ulcer; pulmonary disease; insomnia, panic attacks.
CAUTIOUS USE Diabetes mellitus; hiatal hernia; psychotic disorders; dementia; depressive disorders; hepatic disease; hypertension with heart disease; pregnancy (category C), lactation.

ROUTE & DOSAGE

Mental Stimulant
Adult: **PO** 100–200 mg q3–4h prn
Circulatory Stimulant
Adult: **IM** 200–500 mg prn
Apnea of Prematurity (Caffeine Citrate Only)
Neonate (28–33 wk gestation): **PO/IV** 20 mg/kg (loading dose); then, after 24 h, 5 mg/kg/day

ADMINISTRATION
Oral
- Powdered form may be dissolved in the patient's liquid of choice.

Intramuscular
- Give deep IM into a large muscle.

Intravenous
Note: IV route reserved for emergency situations only.

***PREPARE*: IV Infusion:** May be diluted for infusion in D5W.
***ADMINISTER*: IV Infusion:** A syringe infusion pump is recommended. ▪ Give loading dose over 30 min and maintenance dose over at least 10 min.
***INCOMPATIBILITIES* Y-site: Acyclovir, furosemide, lorazepam,**

Common adverse effects in *italic*, life-threatening effects <u>underlined</u>; generic names in **bold**; classifications in SMALL CAPS; ✦ Canadian drug name; ⊘ Prototype drug

215

nitroglycerin, oxacillin, pantoprazole.

ADVERSE EFFECTS (≥1%) **CV:** Tingling of face, flushing, palpitation, tachycardia, arrhythmia, angina, ventricular ectopic beats. **GI:** Nausea, vomiting; epigastric discomfort, gastric irritation (oral form), diarrhea, hematemesis, kernicterus (neonates). **CNS:** *Nervousness, insomnia,* restlessness, irritability, confusion, agitation, fasciculations, delirium, twitching, tremors, clonic convulsions. **Respiratory:** Tachypnea. **Special Senses:** Scintillating scotomas, tinnitus. **Urogenital:** Increased urination, diuresis.

DIAGNOSTIC TEST INTERFERENCE Caffeine reportedly may interfere with diagnosis of pheochromocytoma or neuroblastoma by increasing urinary excretion of **catecholamines, VMA,** and **5-HIAA** and may cause false positive increases in **serum urate** (by *Bittner method*).

INTERACTIONS Drug: Increases effects of **cimetidine;** increases cardiovascular stimulating effects of BETA-ADRENERGIC AGONISTS; possibly increases **theophylline** toxicity.

PHARMACOKINETICS Absorption: Rapid. **Peak:** 15–45 min. **Distribution:** Widely throughout body; crosses blood–brain barrier and placenta. **Metabolism:** In liver. **Elimination:** In urine as metabolites; excreted in breast milk in small amounts. **Half-Life:** 3–5 h in adults, 36–144 h in neonates.

NURSING IMPLICATIONS

Assessment & Drug Effects

- Monitor vital signs closely as large doses may cause intensification rather than reversal of severe drug-induced depressions.

- Observe children closely following administration as they are more susceptible than adults to the CNS effects of caffeine.
- Lab tests: Monitor blood glucose and HbA1c levels in diabetics.

Patient & Family Education

- Caffeine in large amounts may impair glucose tolerance in diabetics.
- Do not consume large amounts of caffeine as headache, dizziness, anxiety, irritability, nervousness, and muscle tension may result from excessive use, as well as from abrupt withdrawal of coffee (or oral caffeine). Withdrawal symptoms usually occur 12–18 h following last coffee intake.

CALCIPOTRIENE

(cal-ci′po-tri-een)

Dovonex, Sorilux
Classification: VITAMIN D ANALOG
Therapeutic: VITAMIN D ANALOG
Prototype: Calcitriol
Pregnancy Category: C

AVAILABILITY 0.005% ointment and cream, scalp solution

ACTION & THERAPEUTIC EFFECT Calcipotriene is a synthetic vitamin D_3 analog for the treatment of moderate plaque psoriasis. *Calcipotriene controls psoriasis by inhibiting proliferation of psoiatic skin, reducing the number of polymorphonuclear leukocytes (PMNs) in the skin cells, and decreasing the number of epithelial cells.*

USES Treatment of psoriasis.

CONTRAINDICATIONS Hypersensitivity to calcipotriene, hypercalcemia or vitamin D toxicity, lactation.
CAUTIOUS USE History of nephrolithiasis; dermatoses other than psoriasis; patients older than 65 y;

pregnancy (category C). Safety and efficacy in children not established.

ROUTE & DOSAGE

Adult: **Topical** Apply a thin layer to affected area once or twice daily

ADMINISTRATION

Topical
- A thin layer should be applied to the affected skin and rubbed in gently and completely.
- Calcipotriene should not be applied to the face.
- Wash hands before and after application of medication.

ADVERSE EFFECTS (≥1%) **Skin:** Facial dermatitis, burning, stinging, erythema, folliculitis, mild transient itching.

INTERACTIONS No clinically significant interactions established.

PHARMACOKINETICS Absorption: 6% absorbed systemically. **Onset:** 1 wk. **Peak:** 8 wk. **Duration:** 4 wk. **Metabolism:** Recycled via liver. **Elimination:** In bile.

NURSING IMPLICATIONS

Assessment & Drug Effects
- Observe reductions in scaling, erythema, and lesion thickness indicating a positive therapeutic response.
- Significant reduction in psoriatic lesions usually occurs following 1 wk of treatment. Marked improvement is generally noted by the 8th wk of treatment.
- Lab tests: Monitor periodically serum calcium, phosphate, and calcitriol levels during long-term therapy.

Patient & Family Education
- Treatment with calcipotriene may be indefinite, as reappearance of

psoriatic lesions is common following discontinuation of the drug.
- Adverse effects may include burning and stinging with drug application; these are usually transient.
- Do not mix calcipotriene with any other topical medicine.
- Report appearance of facial dermatitis (redness and scaling around mouth and nose).

CALCITONIN (SALMON)

Fortical, Miacalcin
Classification: BONE METABOLISM REGULATOR
Therapeutic: BONE METABOLISM REGULATOR
Pregnancy Category: C

AVAILABILITY 200 international units/mL injection; 200 international units/spray

ACTION & *THERAPEUTIC EFFECT*
Calcitonin opposes the effects of parathyroid hormone on bone and kidneys, reduces serum calcium by binding to a specific receptor site on osteoclast cell membrane, and alters transmembrane passage of calcium and phosphorus. Promotes renal excretion of calcium and phosphorus. *Effective in osteoporosis due to inhibition of bone resorption. Effective in symptomatic hypercalcemia by rapidly lowering serum calcium.*

USES Symptomatic Paget's disease of bone (osteitis deformans), postmenopausal osteoporosis. Orphan drug approval (calcitonin human): Short-term adjunctive treatment of severe hypercalcemic emergencies.
UNLABELED USES Diagnosis and management of medullary carcinoma of thyroid; treatment of osteogenesis imperfecta.

Common adverse effects in *italic*, life-threatening effects <u>underlined</u>; generic names in **bold;** classifications in SMALL CAPS; ◆ Canadian drug name; ⓟ Prototype drug

217

C

CONTRAINDICATIONS Hypersensitivity to fish proteins or to calcitonin; hypocalcemia.

CAUTIOUS USE Renal impairment; osteoporosis; pernicious anemia; Zollinger-Ellison syndrome; pregnancy (category C), lactation. Safe use in children younger than 12 y not established.

ROUTE & DOSAGE

Paget's Disease
Adult: **Subcutaneous/IM** 100 international units/day, may decrease to 50–100 international units/day or every other day

Hypercalcemia
Adult: **Subcutaneous/IM** 4 international units/kg q12h, may increase to 8 international units/kg q6h if needed

Postmenopausal Osteoporosis
Adult: **Subcutaneous/IM** 100 international units/day **Intranasal** 1 spray (200 international units) daily, alternate nostrils

ADMINISTRATION

Allergy Test Dose
- An allergy skin test is usually done prior to initiation of therapy. The appearance of more than mild erythema or wheal 15 min after intracutaneous injection indicates that the drug should not be given.

Intranasal
- Activate the pump prior to first use; hold bottle upright and depress white side arms 6 times.
- The nasal spray is administered in one nostril daily; alternate nostrils.

Subcutaneous
- Calcitonin human is administered only by subcutaneous injection; calcitonin salmon may be administered by subcutaneous or IM injection.

Intramuscular
- Use IM route when the volume to be injected is greater than 2 mL.
- Rotate injection sites.
- Store calcitonin (human) at or below 25° C (77° F), protected from light, unless otherwise specified by manufacturer.
- Store calcitonin (salmon) in refrigerator, preferably at 2°–8° C (36°–46° F) unless otherwise directed.

ADVERSE EFFECTS (≥1%) **Body as a Whole:** Headache, eye pain, feverish sensation, hypersensitivity reactions, <u>anaphylaxis</u>. Reported for calcitonin human only: Urinary frequency, chills, chest pressure, weakness, paresthesias, tender palms and soles, dizziness, nasal congestion, shortness of breath. **GI:** *Transient nausea,* vomiting, anorexia, unusual taste sensation, abdominal pain, diarrhea. **Skin:** Inflammatory reactions at injection site, flushing of face or hands, pruritus of earlobes, edema of feet, skin rashes. **Urogenital:** Nocturia, diuresis, abnormal urine sediment.

INTERACTIONS Drug: May decrease serum **lithium** levels.

PHARMACOKINETICS Onset: 15 min. **Peak:** 4 h. **Duration:** 8–24 h. **Distribution:** Does not cross placenta; distribution into breast milk unknown. **Metabolism:** In kidneys. **Elimination:** In urine. **Half-Life:** 1.25 h.

NURSING IMPLICATIONS

Assessment & Drug Effects
- Have readily available parenteral calcium, particularly during early therapy. Hypocalcemic tetany is a theoretical possibility.
- Examine urine specimens periodically for sediment with long-term therapy.

- Lab tests: Monitor for hypocalcemia (see Signs & Symptoms, Appendix F). Theoretically, calcitonin can lead to hypocalcemic tetany. Latent tetany may be demonstrated by Chvostek's or Trousseau's signs and by serum calcium values: 7–8 mg/dL (latent tetany); below 7 mg/dL (manifest tetany).
- Examine nasal passages prior to treatment with the nasal spray and anytime nasal irritation occurs.
- Nasal ulceration or heavy bleeding are indications for drug discontinuation.

Patient & Family Education
- Watch for redness, warmth, or swelling at injection site and report to prescriber, as these may indicate an inflammatory reaction. The transient flushing that commonly occurs following injection of calcitonin, particularly during early therapy, may be minimized by administering the drug at bedtime. Consult prescriber.
- Maintain your drug regimen to prevent early relapses even though symptoms have improved.
- Ensure that you feel comfortable using the nasal pump properly. Notify prescriber if significant nasal irritation occurs.
- Consult prescriber before using OTC preparations. Some supervitamins, hematinics, and antacids contain calcium and vitamin D (vitamin may antagonize calcitonin effects).

CALCITRIOL 🅿

(kal-si-trye'ole)

Calcijex, Rocaltrol, Vectical
Classification: VITAMIN D ANALOG
Therapeutic: VITAMIN D ANALOG
Pregnancy Category: C

AVAILABILITY 0.25 mcg, 0.5 mcg capsule; 1 mcg/mL oral solution; 1 mcg/mL injection; 3 mcg/g ointment

ACTION & *THERAPEUTIC EFFECT*
Synthetic form of an active metabolite of ergocalciferol (vitamin D_2). In the liver, cholecalciferol (vitamin D_3) and ergocalciferol (vitamin D_2) are enzymatically metabolized to calcifediol, an activated form of vitamin D_3 in the kidney. Patients with nonfunctioning kidneys are unable to synthesize sufficient calcitriol. *By promoting intestinal absorption and renal retention of calcium, calcitriol elevates serum calcium levels, decreases elevated blood levels of phosphate and parathyroid hormone. Thus it decreases subperiosteal bone resorption and mineralization defects.*

USES Management of hypocalcemia in patients undergoing chronic renal dialysis and in patients with hypoparathyroidism or pseudohypoparathyroidism. Patients with hyperparathyroidism in moderate to severe chronic renal failure not on dialysis; psoriasis; renal osteodystrophy.
UNLABELED USES Selected patients with vitamin D–dependent rickets, familial hypophosphatemia; osteopetrosis; osteoporosis.

CONTRAINDICATIONS Hypersensitivity to calcitriol; hypercalcemia or vitamin D toxicity.
CAUTIOUS USE Hyperphosphatemia, renal failure; sarcoidosis; patients receiving digitalis glycosides; older adults; pregnancy (category C).

ROUTE & DOSAGE

Hypocalcemia
Adult: **PO** 0.25 mcg/day, may be increased by 0.25 mcg/day

Common adverse effects in *italic*, life-threatening effects underlined; generic names in **bold**; classifications in SMALL CAPS; ◆ Canadian drug name; 🅿 Prototype drug

219

q4–8wk for dialysis patients or q2–4wk for hypoparathyroid patients if necessary **IV** 1–2 mcg 3 times/wk at the end of dialysis, may need up to 3 mcg 3 times/wk *Child:* **PO** *On hemodialysis:* 0.25–2 mcg/day

Renal Osteodystrophy (stage 3 or 4 chronic kidney disease)

Adult/Adolescent/Child (older than 2 y): **PO** 0.25 mcg/day may increase; monitor closely and adjust as needed
Child (younger than 3 y): **PO** 0.01–0.015 mcg/kg/day. Monitor closely and adjust as needed

ADMINISTRATION

Oral

- Oral dose can be taken either with food or milk or on an empty stomach. Discuss with prescriber.
- When given for hypoparathyroidism, the dose is given in the morning.
- Capsules should be protected from heat, light, and moisture. Store in tightly closed container.

Topical

- Do not apply to face, lips, or area around eyes.
- Store at 15°–30° C (59°–87° F). Do not refrigerate.

Intravenous

PREPARE: Direct: Give undiluted.
ADMINISTER: Direct: Give IV push over 30–60 sec.

ADVERSE EFFECTS (≥1%) **Body as a Whole:** Muscle or bone pain. **CV:** Palpitation. **GI:** Anorexia, nausea, vomiting, dry mouth, thirst, constipation, abdominal cramps, metallic taste. **Metabolic:** Vitamin D intoxication, hypercalcemia, hypercalciuria, hyperphosphatemia. **CNS:** Headache, weakness. **Special Senses:** Blurred vision, photophobia. **Urogenital:** Increased urination.

INTERACTIONS Drug: THIAZIDE DIURETICS may cause hypercalcemia; calcifediol-induced hypercalcemia may precipitate digitalis arrhythmias in patients receiving DIGITALIS GLYCOSIDES.

PHARMACOKINETICS Absorption: Readily absorbed from GI tract. **Onset:** 2–6 h. **Peak:** 10–12 h. **Duration:** 3–5 days. **Metabolism:** In liver. **Elimination:** Mainly in feces. **Half-Life:** 3–6 h.

NURSING IMPLICATIONS

Assessment & Drug Effects

- Lab tests: Determine baseline and periodic levels of serum calcium, phosphorus, magnesium, alkaline phosphatase, creatinine; measure urinary calcium and phosphorus levels q24h.
- Effectiveness of therapy depends on an adequate daily intake of calcium and phosphate. The prescriber may prescribe a calcium supplement on an as-needed basis.
- Monitor for hypercalcemia (see Signs & Symptoms, Appendix F). During dosage adjustment period, monitor serum calcium levels particularly twice weekly to avoid hypercalcemia.
- If hypercalcemia develops, withhold calcitriol and calcium supplements and notify prescriber. Drugs may be reinitiated when serum calcium returns to normal.

Patient & Family Education

- Oral/IV: Discontinue the drug if experiencing any symptoms of hypercalcemia (see Appendix F) and contact prescriber.
- Oral/IV: Do not use any other source of vitamin D during therapy, since calcitriol is the most potent

form of vitamin D$_3$. This will avoid the possibility of hypercalcemia.

- Oral/IV: Consult prescriber before taking an OTC medication. (Many products contain calcium, vitamin D, phosphates, or magnesium, which can increase adverse effects of calcitriol.)
- Oral/IV: Maintain an adequate daily fluid intake unless you have kidney problems, in which case consult your prescriber about fluids.
- Stop using ointment and contact physician if severe irritation occurs.
- Avoid natural or artificial sunlight when using ointment.
- Limit ointment use to no more than 2 tubes/wk.

CALCIUM CARBONATE

Apo-Cal ♦, BioCal, Calcite-500, Calsan ♦, Cal-Sup, Caltrate ♦, Chooz, Dicarbosil, Equilet, Mallamint, Mega-Cal, Nu-Cal, Os-Cal, Oystercal, Titralac, Tums

CALCIUM ACETATE

PhosLo

CALCIUM CITRATE

Citracal

CALCIUM PHOSPHATE TRIBASIC (TRICALCIUM PHOSPHATE)

CALCIUM LACTATE

Cal-Lac

Classifications: FLUID AND ELECTROLYTIC REPLACEMENT SOLUTION; ANTACID
Therapeutic: NUTRITIONAL SUPPLEMENT; ANTACID
Prototype: Calcium gluconate
Pregnancy Category: B for calcium acetate; other salts not rated

AVAILABILITY Calcium carbonate: 125 mg, 250 mg, 650 mg, 750 mg, 1.25 g, 1.5 g tablets; **Calcium acetate:** 667 mg tablets; **Calcium citrate:** 950 mg, 2376 mg tablets; **Calcium phosphate tribasic:** 1565.2 mg tablets

ACTION & *THERAPEUTIC EFFECT*
Calcium carbonate is a rapid-acting antacid with high neutralizing capacity and relatively prolonged duration of action. Decreases gastric acidity, thereby inhibiting proteolytic action of pepsin on gastric mucosa. All forms of calcium salts are used for calcium replacement therapy. *Effectively relieves symptoms of acid indigestion and useful as a calcium supplement.*

USES Relief of transient symptoms of hyperacidity as in acid indigestion, heartburn, peptic esophagitis, and hiatal hernia. Also as calcium supplement in treatment of mild calcium deficiency states. Control of hyperphosphatemia in chronic renal failure (calcium acetate).
UNLABELED USES For treatment of hyperphosphatemia in patients with chronic renal failure and to lower BP in selected patients with hypertension.

CONTRAINDICATIONS Hypercalcemia and hypercalciuria (e.g., hyperparathyroidism, vitamin D overdosage, decalcifying tumors, bone metastases), calcium loss due to immobilization, severe renal failure, renal calculi, GI hemorrhage or obstruction, dehydration, digitalis toxicity; hypochloremic alkalosis, ventricular fibrillation, cardiac disease.
CAUTIOUS USE Decreased bowel motility (e.g., with anticholinergics, antidiarrheals, antispasmodics), the older adult; **Calcium acetate:** pregnancy (category B).

Common adverse effects in *italic*, life-threatening effects underlined; generic names in **bold**; classifications in SMALL CAPS; ♦ Canadian drug name; 🅟 Prototype drug

221

ROUTE & DOSAGE

All doses are in terms of *elemental calcium:* 1 g calcium carbonate = 400 mg (20 mEq, 40%) elemental calcium; 1 g calcium acetate = 250 mg (12.6 mEq, 25%) elemental calcium; 1 g calcium citrate = 210 mg (12 mEq, 21%) elemental calcium; 1 g tricalcium phosphate = 390 mg (19.3 mEq, 39%) elemental calcium; calcium lactate = 130 mg (6.5 mEq, 13%) elemental calcium

Supplement for Osteoporosis

Adult: **PO** 1–2 g b.i.d. or t.i.d.

Antacid

Adult: **PO** 0.5–2 g 4–6 times/day

Hyperphosphatemia

Adult: **PO** Calcium acetate 2–4 tablets with each meal

Supplement for Mild Hypercalcemia

Child: **PO** 500 mg/kg/day in divided doses (lactate)

ADMINISTRATION

Oral

- When used as antacid, give 1 h after meals and at bedtime. When used as calcium supplement, give 1–1½ h after meals, unless otherwise directed by prescriber.
- Chewable tablet should be chewed well before swallowing or allowed to dissolve completely in mouth, followed with water. Powder form may be mixed with water.
- Ensure that sustained release form of drug is not chewed or crushed. It **must be** swallowed whole.

ADVERSE EFFECTS (≥1%) **GI:** *Constipation* or laxative effect, acid rebound, nausea, eructation, *flatulence,* vomiting, fecal concretions.

Metabolic: Hypercalcemia with alkalosis, metastatic calcinosis, hypercalciuria, hypomagnesemia, hypophosphatemia (when phosphate intake is low). **CNS:** Mood and mental changes. **Urogenital:** Polyuria, renal calculi.

INTERACTIONS Drug: May enhance inotropic and toxic effects of **digoxin; magnesium** may compete for GI absorption; decreases absorption of TETRACYCLINES, QUINOLONES **(ciprofloxacin).**

PHARMACOKINETICS Absorption: Approximately ⅓ of dose absorbed from small intestine. **Distribution:** Crosses placenta. **Elimination:** Primarily in feces; small amounts in urine, pancreatic juice, saliva, breast milk.

NURSING IMPLICATIONS

Assessment & Drug Effects

- Note number and consistency of stools. If constipation is a problem, prescriber may prescribe alternate or combination therapy with a magnesium antacid or advise patient to take a laxative or stool softener as necessary.
- Lab tests: Determine serum and urine calcium weekly in patients receiving prolonged therapy and in patients with renal dysfunction.
- Record amelioration of symptoms of hypocalcemia (see Signs & Symptoms, Appendix F).
- Observe for S&S of hypercalcemia in patients receiving frequent or high doses, or who have impaired renal function (see Appendix F).

Patient & Family Education

- Do not continue this medication beyond 1–2 wk, since it may cause acid rebound, which generally occurs after repeated use for 1 or 2 wk and leads to chronic use. It is potentially dangerous to self-medicate. Do not take antacids longer than 2 wk without medical supervision.

- Avoid taking calcium carbonate with cereals or other foods high in oxalates. Oxalates combine with calcium carbonate to form insoluble, nonabsorbable compounds.

- Do not use calcium carbonate repeatedly with foods high in vitamin D (such as milk) or sodium bicarbonate, as it may cause milk-alkali syndrome: hypercalcemia, distaste for food, headache, confusion, nausea, vomiting, abdominal pain, metabolic alkalosis, hypercalciuria, polyuria, soft tissue calcification (calcinosis), hyperphosphatemia and renal insufficiency. Predisposing factors include renal dysfunction, dehydration, electrolyte imbalance, and hypertension.

CALCIUM CHLORIDE

Classification: FLUID AND ELECTROLYTIC REPLACEMENT SOLUTION
Therapeutic: FLUID AND ELECTROLYTE REPLACEMENT
Prototype: Calcium gluconate
Pregnancy Category: A; C in high doses

AVAILABILITY 10% injection

ACTION & *THERAPEUTIC EFFECT* Ionizes readily and provides excess chloride ions that promote acidosis and temporary (1–2 days) diuresis secondary to excretion of sodium. *Rapidly and effectively restores serum calcium levels in acute hypocalcemia of various origins and an effective cardiac stabilizer under conditions of hyperkalemia or resuscitation.*

USES Treatment of cardiac resuscitation when epinephrine fails to improve myocardial contractions; for treatment of acute hypocalcemia (as in tetany due to parathyroid deficiency, vitamin D deficiency, alkalosis, insect bites or stings, and during exchange transfusions), for treatment of hypermagnesemia, and for cardiac disturbances of hyperkalemia.

CONTRAINDICATIONS Ventricular fibrillation, hypercalcemia, digitalis toxicity, injection into myocardium or other tissue.
CAUTIOUS USE Digitalized patients; sarcoidosis, renal insufficiency, history of renal stone formation; cardiac arrhythmias; dehydration; diarrhea; cor pulmonale, respiratory acidosis, respiratory failure; pregnancy (category A; category C in high doses).

ROUTE & DOSAGE

All doses are in terms of *elemental calcium:* 1 g calcium chloride = 272 mg (13.6 mEq) elemental calcium

Hypocalcemia

Adult: **IV** 0.5–1 g (7–14 mEq) at 1–3 day intervals as determined by patient response and serum calcium levels
Child: **IV** 2.7–5 mg/kg administered slowly
Neonate: **IV** Less than 1 mEq/day

Hypocalcemic Tetany

Adult: **IV** 4.5–16 mEq prn
Child: **IV** 0.5–0.7 mEq/kg t.i.d. or q.i.d.
Neonate: **IV** 2.4 mEq/kg/day in divided doses

CPR

Adult: **IV** 2–4 mg/kg, may repeat in 10 min
Child: **IV** 20 mg/kg, may repeat in 10 min

Common adverse effects in *italic*, life-threatening effects <u>underlined</u>; generic names in **bold**; classifications in SMALL CAPS; ◆ Canadian drug name; ✪ Prototype drug

223

C

ADMINISTRATION

Intravenous

IV administration to neonates, infants, and children: Verify correct IV concentration and rate of infusion with prescriber.

PREPARE: Direct: May be given undiluted or diluted (preferred) with an equal volume of NS for injection. ▪ Solution should be warmed to body temperature before administration.

ADMINISTER: Direct: Give at 0.5–1 mL/min or more slowly if irritation develops. Avoid rapid administration. ▪ Use a small-bore needle and inject into a large vein to minimize venous irritation and undesirable reactions.

INCOMPATIBILITIES Solution/additive: Amphotericin B, chlorpheniramine, dobutamine, concentration-dependent incompatibility with other ELECTROLYTES. **Y-site: Amphotericin B cholesteryl complex, propofol, sodium bicarbonate.**

ADVERSE EFFECTS (≥1%) **Body as a Whole:** Tingling sensation. With rapid IV, sensations of heat waves (peripheral vasodilation), fainting, **CV:** (With rapid infusion) hypotension, bradycardia, cardiac arrhythmias, <u>cardiac arrest</u>. **Skin:** Pain and burning at IV site, severe venous thrombosis, necrosis and sloughing (with extravasation).

INTERACTIONS Drug: May enhance inotropic and toxic effects of **digoxin;** antagonizes the effects of **verapamil** and possibly other CALCIUM CHANNEL BLOCKERS.

PHARMACOKINETICS Distribution: Crosses placenta. **Elimination:** Primarily in feces; small amounts in urine, pancreatic juice, saliva, and breast milk.

NURSING IMPLICATIONS

Assessment & Drug Effects

▪ Monitor ECG and BP and observe patient closely during administration. IV injection may be accompanied by cutaneous burning sensation and peripheral vasodilation, with moderate fall in BP.

▪ Advise ambulatory patient to remain in bed for 15–30 min or more depending on response following injection.

▪ Observe digitalized patients closely since an increase in serum calcium increases risk of digitalis toxicity.

▪ Lab tests: Determine serum pH, calcium, and other electrolytes frequently as guides to dosage adjustments.

Patient & Family Education

▪ Remain in bed for 15–30 min or more following injection and depending on response.

▪ Symptoms of mild hypercalcemia, such as loss of appetite, nausea, vomiting, or constipation may occur. If hypercalcemia becomes severe, call health care provider if feeling confused or extremely excited.

▪ Do not use other calcium supplements or eat foods high in calcium, like milk, cheese, yogurt, eggs, meats, and some cereals, during therapy.

CALCIUM GLUCONATE 🅿

(gloo'koe-nate)

Classification: ELECTROLYTE AND WATER BALANCE
Therapeutic: ELECTROLYTE REPLACEMENT SOLUTION
Pregnancy Category: A; C in high doses

AVAILABILITY 500 mg, 650 mg, 975 mg, 1 g tablets; 10% injection

Common adverse effects in *italic*, life-threatening effects <u>underlined</u>; generic names in **bold**; classifications in SMALL CAPS; ♣ Canadian drug name; 🅿 Prototype drug

ACTION & *THERAPEUTIC EFFECT*

Calcium gluconate acts like digitalis on the heart, increasing cardiac muscle tone and force of systolic contractions (positive inotropic effect). *Rapidly and effectively restores serum calcium levels in acute hypocalcemia of various origins; also effective as a cardiac stabilizer under conditions of hyperkalemia or resuscitation.*

USES Negative calcium balance (as in neonatal tetany, hypoparathyroidism, vitamin D deficiency, alkalosis). Also to overcome cardiac toxicity of hyperkalemia, for cardiopulmonary resuscitation, to prevent hypocalcemia during transfusion of citrated blood. Also as antidote for magnesium sulfate, for acute symptoms of lead colic, to decrease capillary permeability in sensitivity reactions, and to relieve muscle cramps from insect bites or stings. Oral calcium may be used to maintain normal calcium balance and to prevent primary osteoporosis. Also in osteoporosis, osteomalacia, chronic hypoparathyroidism, rickets, and as adjunct in treatment of myasthenia gravis and Eaton-Lambert syndrome.

UNLABELED USES To antagonize aminoglycoside-induced neuromuscular blockage, and as "calcium challenge" to diagnose Zollinger-Ellison syndrome and medullary thyroid carcinoma.

CONTRAINDICATIONS Ventricular fibrillation, metastatic bone disease, injection into myocardium; renal calculi, hypercalcemia, predisposition to hypercalcemia (hyperparathyroidism, certain malignancies); digitalis toxicity.

CAUTIOUS USE Digitalized patients, renal or cardiac insufficiency, arrhythmias; dehydration; diarrhea; hyperphosphatemia; sarcoidosis, history of lithiasis, immobilized patients; pregnancy (category A; category C in high doses).

ROUTE & DOSAGE

All doses are in terms of *elemental calcium:* 1 g calcium gluconate = 90 mg (4.5 mEq, 9.3%) elemental calcium

Supplement for Osteoporosis

Adult: **PO** 1–2 g b.i.d. to q.i.d. **IV** 7 mEq q1–3days
Child: **PO** 45–65 mg/kg/day in divided doses **IV** 1–7 mEq q1–3days
Neonate: **PO** 50–130 mg/kg/day (max: 1 g)

Hypocalcemia

Adult: **IV** 2–15 g/day continuous or divided dose
Child: **IV** 200–500 mg/kg/day (max: 2–3 g/dose)
Neonate: **IV** Not more than 0.93 mEq

Hypocalcemic Tetany

Adult: **IV** 1–3 g prn
Child: **IV** 100–200 mg/kg/dose, may repeat q6–8h
Neonate: **IV** 200 mg followed by 500 mg/kg/day infusion

CPR

Adult: **IV** 2.3–3.7 mEq x 1

Hyperkalemia with Cardiac Toxicity

Adult: **IV** 500–800 mg (max dose: 3 g)

Exchange Transfusions with Citrated Blood

Adult: **IV** 500–1000 mg for each 500 mL of blood
Neonate: **IV** 98 mg for each 100 mL of blood

Common adverse effects in *italic*, life-threatening effects underlined; generic names in **bold**; classifications in SMALL CAPS; ♣ Canadian drug name; ☻ Prototype drug

ADMINISTRATION

Oral

- Ensure that chewable tablets are chewed or crushed before being swallowed with a liquid.
- Give with meals to enhance absorption.

Intravenous

PREPARE: Direct: May be given undiluted. **Intermittent/Continuous:** May be diluted in 1000 mL of NS.

ADMINISTER: Direct: Give direct IV at a rate of 0.5 mL or a fraction thereof over 1 min. • Do not exceed 2 mL/min. **Intermittent/Continuous:** Give slowly, not to exceed 200 mg/min for adults or 100 mg/min for children. Use a small-bore needle into a large vein to avoid possibility of extravasation and resultant necrosis. • With children, scalp veins should be avoided. Avoid rapid infusion. • High concentrations of calcium suddenly reaching the heart can cause fatal cardiac arrest.

INCOMPATIBILITIES Solution/additive: Amphotericin B, cefamandole, dobutamine, methylprednisolone, metoclopramide, concentration-dependent incompatibility with other ELECTROLYTES. **Y-site: Amphotericin B cholesteryl complex, fluconazole, indomethacin, lansoprazole, meropenem.**

- Injection should be stopped if patient complains of any discomfort. • Patient should be advised to remain in bed for 15–30 min or more following injection, depending on response.

ADVERSE EFFECTS (≥1%) **Body as a Whole:** Tingling sensation. With rapid IV, sensations of heat waves (peripheral vasodilation), fainting. **GI:** PO preparation: Constipation, increased gastric acid secretion. **CV:** (With rapid infusion) hypotension, bradycardia, cardiac arrhythmias, <u>cardiac arrest</u>. **Skin:** Pain and burning at IV site, severe venous thrombosis, necrosis and sloughing (with extravasation).

DIAGNOSTIC TEST INTERFERENCE

IV calcium may cause false decreases in *serum and urine magnesium* (by *Titan yellow method*) and transient elevations of *plasma 11-OHCS* levels by *Glenn-Nelson technique.* Values usually return to control levels after 60 min; *urinary steroid values (17-OHCS)* may be decreased.

INTERACTIONS Drug:

May enhance inotropic and toxic effects of **digoxin; magnesium** may compete for GI absorption; decreases absorption of TETRACYCLINES, QUINOLONES **(ciprofloxacin);** antagonizes the effects of **verapamil** and possibly other CALCIUM CHANNEL BLOCKERS (IV administration).

PHARMACOKINETICS Absorption:

30% from small intestine. **Onset:** Immediately after IV. **Distribution:** Crosses placenta. **Elimination:** Primarily in feces; small amounts in urine, pancreatic juice, saliva, and breast milk.

NURSING IMPLICATIONS

Assessment & Drug Effects

- Assess for cutaneous burning sensations and peripheral vasodilation, with moderate fall in BP, during direct IV injection.
- Monitor ECG during IV administration to detect evidence of hypercalcemia: Decreased QT interval associated with inverted T wave.
- Observe IV site closely. Extravasation may result in tissue irritation and necrosis.

- Monitor for hypocalcemia and hypercalcemia (see Signs & Symptoms, Appendix F).
- Lab tests: Determine levels of calcium and phosphorus (tend to vary inversely) and magnesium frequently, during sustained therapy. Deficiencies in other ions, particularly magnesium, frequently coexist with calcium ion depletion.

Patient & Family Education
- Report S&S of hypercalcemia (see Appendix F) promptly to your care provider.
- Milk and milk products are the best sources of calcium (and phosphorus). Other good sources include dark green vegetables, soy beans, tofu, and canned fish with bones.
- Calcium absorption can be inhibited by zinc-rich foods: Nuts, seeds, sprouts, legumes, soy products (tofu).
- Check with prescriber before self-medicating with a calcium supplement.

CALCIUM POLYCARBOPHIL

(pol-ee-kar'boe-fil)

FiberCon
Classifications: BULK LAXATIVE; ANTIDIARRHEAL
Therapeutic: BULK LAXATIVE; ANTIDIARRHEAL
Prototype: Psyllium hydrophilic mucilloid
Pregnancy Category: A; C in high doses

AVAILABILITY 500 mg, 625 mg tablets

ACTION & *THERAPEUTIC EFFECT*
Hydrophilic, bulk-producing laxative that restores normal moisture level and bulk content of intestinal tract. In constipation, retains free water in intestinal lumen, thereby indirectly opposing dehydrating forces of the bowel; in diarrhea, when intestinal mucosa is incapable of absorbing fluid, drug absorbs fecal fluid to form a gel. *Relieves constipation or diarrhea associated with bowel disorders and acute nonspecific diarrhea.*

USES Constipation or diarrhea associated with diverticulitis or irritable bowel syndrome; acute nonspecific diarrhea.

CONTRAINDICATIONS GI obstruction; children younger than 6 y.
CAUTIOUS USE Pregnancy (category A; category C in high doses), lactation.

ROUTE & DOSAGE

Constipation or Diarrhea
Adult: **PO** 1 g q.i.d. as needed (max: 6 g/24 h)
Child: **PO** 6–12 y, 500 mg 1–3 times/day (max: 3 g/24 h); *younger than 6 y,* 500 mg 1–2 times/day (max: 1.5 g/24 h)

ADMINISTRATION

Oral
- Administer with at least 180–240 mL (6–8 oz) water or other fluid of patient's choice when used as a laxative and with at least 60–90 mL (2–3 oz) of fluid when used as an antidiarrheal. Chewed tablets should not be swallowed dry.
- If diarrhea is severe, dose can be repeated every half hour up to maximum daily dose.

ADVERSE EFFECTS (≥1%) **GI:** *Flatulence,* abdominal fullness, <u>intestinal obstruction</u>; laxative dependence (long-term use).

PHARMACOKINETICS Absorption: Not absorbed from the intestine. Bowel movement usually occurs within 12–72 h. **Elimination:** In feces.

NURSING IMPLICATIONS

Assessment & Drug Effects

- Evaluate effectiveness of medication. If it is ineffective as an anti-diarrheal, report to prescriber.
- Report rectal bleeding, very dark stools, or abdominal pain promptly.

Patient & Family Education

- You will likely have a bowel movement within 12–72 h.
- This is an OTC product. Take this drug exactly as ordered. Do not increase the dose if response is inadequate. Consult prescriber. Do not use other laxatives while you are taking calcium polycarbophil.

CANDESARTAN CILEXETIL

(can-de-sar′tan ci-lex′e-til)

Atacand

Classification: ANGIOTENSIN II RECEPTOR ANTAGONIST, ANTIHYPERTENSIVE

Therapeutic: ANTIHYPERTENSIVE

Prototype: Losartan

Pregnancy Category: C first trimester, D second and third trimester

AVAILABILITY 4 mg, 8 mg, 16 mg, 32 mg tablets

ACTION & *THERAPEUTIC EFFECT*

Angiotensin II receptor (type AT_1) antagonist. Angiotensin II is a potent vasoconstrictor and primary vasoactive hormone of the renin–angiotensin–aldosterone system. Candesartan selectively blocks binding of angiotensin II to the AT_1 receptors found in many tissues (e.g., vascular smooth muscle, adrenal glands). *Results in blocking the vasoconstricting and aldosterone-secreting effects of angiotensin II, resulting in an antihypertensive effect. Effectively lowers BP from hypertensive to normotensive range.*

USES Hypertension, heart failure.

CONTRAINDICATIONS Known sensitivity to candesartan or any other angiotensin II (AT_1) receptor antagonist (e.g., losartan, valsartan); primary hyperaldosteronism; bilateral renal artery stenosis; pregnancy (category D second and third trimesters); lactation; children younger than 1 y for hypertension, or children with GFR less than 30 mL/min/1.73 m².

CAUTIOUS USE Unilateral renal artery stenosis; aortic or mitral valve stenosis; hypertrophic cardiomyopathy; CHF; diabetes; moderate hepatic or renal impairment, significant renal failure; pregnancy (category C first trimester); lactation.

ROUTE & DOSAGE

Hypertension

Adult: **PO** Start at 16 mg daily (range 8–32 mg divided once or twice daily)

Adolescent/Child: **PO** At least 6 y, weighing more than 50 kg, 8–16 mg given in single or divided doses; adjust based on response; at least 6 y, weighing less than 50 kg, 4–8 mg given in single or divided doses; adjust based on response

Child (1–6 y): **PO** 0.2 mg/kg/day; adjust based on response

Heart Failure

Adult: **PO** Start at 4 mcg once daily, double the dose at 2 wk intervals as tolerated by the patient until a dose of 32 mg is reached

Hepatic Impairment Dosage Adjustment

For patients with moderate hepatic impairment, initiate therapy at lower dose

Common adverse effects in *italic*, life-threatening effects underlined; generic names in **bold**; classifications in SMALL CAPS; ♣ Canadian drug name; ◐ Prototype drug

ADMINISTRATION

Oral

- Volume depletion should be corrected prior to initiation of therapy to prevent hypotension.
- Dose is individualized and may be given once or twice daily. The daily dose may be titrated up to 32 mg; larger doses are not likely to provide additional benefit.

ADVERSE EFFECTS (≥1%) Body as a Whole: Fatigue, peripheral edema, back pain, arthralgia. **CV:** Chest pain. **GI:** Nausea, abdominal pain, diarrhea, vomiting. **CNS:** Headache, dizziness. **Respiratory:** Cough, sinusitis, upper respiratory infection, pharyngitis, rhinitis. **Urogenital:** Albuminuria.

INTERACTIONS Drug: May increase risk of **lithium** toxicity.

PHARMACOKINETICS Absorption: 15% reaches systemic circulation. **Peak:** Serum concentration, 3–4 h; therapeutic effect, 2–4 wk. **Duration:** 24 h. **Distribution:** Greater than 99% protein bound; crosses placenta; distributed into breast milk. **Metabolism:** Minimally in liver. **Elimination:** Primarily in bile (67%) and urine (33%). **Half-Life:** 9–12 h.

NURSING IMPLICATIONS

Assessment & Drug Effects

- Monitor BP as therapeutic effectiveness is indicated by decreases in systolic and diastolic BP within 2 wk with maximal effect at 4–6 wk.
- Monitor for transient hypotension in volume/salt-depleted patients; if hypotension occurs, place in supine position and notify prescriber.
- Monitor BP periodically; trough readings, just prior to the next scheduled dose, should be made when possible.

- Lab tests: Periodically monitor BUN and creatinine, serum potassium, liver enzymes, and CBC with differential.

Patient & Family Education

- Inform your prescriber immediately if you become pregnant.
- You may not notice maximum pressure-lowering effect for 6 wk.
- Report episodes of dizziness especially when making position changes.

CAPECITABINE

(cap-e-si'ta-been)
Xeloda
Classifications: ANTINEOPLASTIC; ANTIMETABOLITE, PYRIMIDINE
Therapeutic: ANTINEOPLASTIC
Prototype: 5-Fluorouracil (5-FU)
Pregnancy Category: D

AVAILABILITY 150 mg, 500 mg tablets

ACTION & THERAPEUTIC EFFECT Pyrimidine antagonist and cell cycle-specific antimetabolite. Prodrug of 5-FU. Blocks actions of enzymes essential to normal DNA and RNA synthesis. May become incorporated into RNA molecules of tumor cells, thereby interfering with RNA and protein synthesis. *Reduces or stabilizes tumor size in metastatic breast cancer.*

USES Metastatic breast cancer and colorectal cancer.
UNLABELED USES Ovarian cancer.

CONTRAINDICATIONS Hypersensitivity to capecitabine, doxifluridine, 5-FU; myelosuppression; dihydropyrimidine dehydrogenase (DPD) deficiency; females of childbearing age; active infection; jaundice; severe renal failure; pregnancy (category D); lactation, children younger than 18 y. **CAUTIOUS USE** Mild to moderate renal or hepatic dysfunction; bacterial

Common adverse effects in *italic*, life-threatening effects <u>underlined</u>; generic names in **bold**; classifications in SMALL CAPS; ♣ Canadian drug name; ☯ Prototype drug

229

or viral infection; coronary artery disease, angina, cardiac arrhythmias; history of varicella zoster or other herpes infections; older adults.

ROUTE & DOSAGE

Breast Cancer, Colorectal Cancer

Adult: **PO** 2500 mg/m²/day in 2 divided doses × 2 wk (timing of second cycle differs based on other medications used)

Renal Impairment Dosage Adjustment

CrCl 30–50 mL/min: Reduce dose by 25%; less than 30 mL/min: Do not use

ADMINISTRATION

Oral

- Morning and evening doses (about 12 h apart) should be given within 30 min of the end of a meal. Water is the preferred liquid for taking this drug.

ADVERSE EFFECTS (≥1%) **Body as a Whole:** *Fatigue,* pyrexia, pain, myalgia. **CV:** Edema. **GI:** *Severe diarrhea, nausea, vomiting, stomatitis,* abdominal pain, constipation, dyspepsia, *anorexia.* **Hematologic:** Neutropenia, thrombocytopenia, anemia, lymphopenia. **Metabolic:** Dehydration, hyperbilirubinemia. **CNS:** Paresthesias, headache, dizziness, insomnia. **Skin:** Hand-and-foot syndrome, *dermatitis,* nail disorder. **Special Senses:** Eye irritation.

INTERACTIONS Drug: Leucovorin increases concentration and toxicity of **5-FU,** altered coagulation and/or bleeding reported with **warfarin** and **NSAIDs. Food:** Food decreases extent of absorption.

PHARMACOKINETICS Absorption: Absorption significantly reduced by food. **Peak:** 1.5–2 h.

Distribution: Approx 35% protein bound. **Metabolism:** Extensively metabolized to 5-FU. **Elimination:** In urine. **Half-Life:** 45 min.

NURSING IMPLICATIONS

Assessment & Drug Effects

- Lab tests: Monitor periodically CBC with differential and LFTs including bilirubin, transaminases, alkaline phosphatase.
- Monitor carefully for S&S of grade 2 or greater toxicity: Diarrhea greater than 4 BMs/day or at night; vomiting greater than 1 time/24 h; significant loss of appetite or anorexia; stomatitis; hand-and-foot syndrome (pain, swelling, erythema, desquamation, blistering); temperature = 100.5° F; and S&S of infection.
- Withhold drug and immediately report S&S of grade 2 or greater toxicity.
- Monitor for dehydration and replace fluids as needed.
- Monitor carefully patients with coronary artery disease for S&S of cardiotoxicity (e.g., increasing angina).

Patient & Family Education

- Report immediately significant nausea, loss of appetite, diarrhea, soreness of tongue, fever of 100.5° F or more, or signs of infection. Review patient drug package insert carefully for more detail.
- Inform prescriber immediately if you become pregnant.

CAPREOMYCIN

(kap-ree-oh-mye'sin)

Capastat

Classifications: ANTIBIOTIC; ANTITUBERCULOSIS

Therapeutic: ANTITUBERCULOSIS

Prototype: Isoniazid

Pregnancy Category: C

Common adverse effects in *italic,* life-threatening effects underlined; generic names in **bold;** classifications in SMALL CAPS; ♣ Canadian drug name; ⊙ Prototype drug

AVAILABILITY 1 g powder for injection

ACTION & *THERAPEUTIC EFFECT*
Polypeptide antibiotic that is bacteriostatic; action is unclear. Should not be used alone. *Bacteriostatic against human strains of* Mycobacterium tuberculosis *and other species of* Mycobacterium. *Effective second-line antimycobacterial in conjunction with other antitubercular drugs.*

USES Treatment of active tuberculosis when primary agents cannot be tolerated or when causative organism has become resistant.

CONTRAINDICATIONS Lactation.
CAUTIOUS USE Hypersensitivity to antibiotics, including capreomycin, or to other drugs; renal insufficiency (extreme caution); acoustic nerve impairment; history of allergies (especially to drugs); preexisting liver disease; myasthenia gravis; parkinsonism; pregnancy (category C). Safe use in infants and children younger than 2 y is not established.

ROUTE & DOSAGE

Tuberculosis

Adult: IM/IV 1 g/day (not to exceed 20 mg/kg/day) for 60–120 days, then 1 g 2–3 times/wk × 18–24 mo. See prescribing information for dose adjustments for renal insufficiency.

Renal Impairment Dosage Adjustment

CrCl 25–50 mL/min: Reduce dose by 50%; 10–24 mL/min: Reduce dose by 50% and give q48h; *less than* 10 mL/min: Reduce dose by 50% and give twice weekly

ADMINISTRATION

Intramuscular

- Reconstitute by adding 2 mL of NS injection or sterile water for injection to each 1 g vial. Allow 2–3 min for drug to dissolve completely.
- Make IM injections deep into large muscle mass. Superficial injections are more painful and are associated with sterile abscess. Rotate injection sites.
- Solution may become pale straw color and darken with time, but this does not indicate loss of potency.
- After reconstitution, solution may be stored 48 h at room temperature and up to 14 days under refrigeration unless otherwise directed.

Intravenous

PREPARE: **IV Infusion:** Reconstitute by adding 2 mL of NS or sterile water to each 1 g to yield 370 mg/mL. ▪ Allow 2–3 min to dissolve, then add required dose to 100 mL of NS.
ADMINISTER: **IV Infusion:** Give over 60 min. Avoid rapid infusion.

ADVERSE EFFECTS (≥1%) **Skin:** Urticaria, maculopapular rash, photosensitivity. **Hematologic:** Leukocytosis, leukopenia, *eosinophilia.* **CNS:** Neuromuscular blockage (large doses: Skeletal muscle weakness, respiratory depression or arrest). **Urogenital:** Nephrotoxicity (long-term therapy), tubular necrosis. **Special Senses:** *Ototoxicity,* eighth nerve (auditory and vestibular) damage. **Metabolic:** Hypokalemia, and other electrolyte imbalances. **Other:** Impaired hepatic function (decreased BSP excretion); IM site reactions: Pain, induration, excessive bleeding, sterile abscesses.

DIAGNOSTIC TEST INTERFERENCE
BSP and ***PSP*** excretion tests may be decreased.

Common adverse effects in *italic*, life-threatening effects underlined; generic names in **bold**; classifications in SMALL CAPS; ♣ Canadian drug name; ⊙ Prototype drug

231

INTERACTIONS Drug: Increased risk of nephrotoxicity and ototoxicity with AMINOGLYCOSIDES, **amphotericin B, colistin, polymyxin B, cisplatin, vancomycin.**

PHARMACOKINETICS Peak: 1–2 h. **Distribution:** Does not cross blood–brain barrier; crosses placenta. **Elimination:** 52% in urine unchanged in 12 h; small amount in bile. **Half-Life:** 4–6 h.

NURSING IMPLICATIONS

Assessment & Drug Effects

- Observe injection sites for signs of excessive bleeding and inflammation.
- Lab tests: Perform the following as guidelines for therapy before drug is started and at regular intervals during therapy: Appropriate bacterial sensitivity tests; CBC; weekly renal function studies (BUN, NPN), creatinine clearance, sediment); LFTs (periodically); serum potassium levels (monthly).
- Dosage of capreomycin is typically reduced in patients with impaired renal function, as it is cumulative. Follow renal function tests closely.
- Monitor I&O rates and pattern: Report immediately any change in output or I&O ratio, any unusual appearance of urine, or elevation of BUN above 30 mg/dL.
- Evaluate hearing and balance by audiometric measurements (twice weekly or weekly) and tests of vestibular function (periodically).

Patient & Family Education

- Report any change in hearing or disturbance of balance. These effects are sometimes reversible if drug is withdrawn promptly when first symptoms appear.
- Ensure that you know about adverse reactions and what to do about them. Report immediately the appearance of any unusual symptom, regardless of how vague it may seem.

CAPSAICIN
(cap-say'i-sin)

Axsain, Capsaicin, Capsin, Capsacin-P, Dolorac, Qutenza, Trixaicin, Zostrix, Zostrix-HP
Classification: TOPICAL ANALGESIC
Therapeutic: TOPICAL ANALGESIC

AVAILABILITY 0.025%, 0.075% lotion; 0.025%, 0.075%, 0.25% cream; 0.025%, 0.05% gel, 8% topical patch

ACTION & *THERAPEUTIC EFFECT*
Capsaicin depletes and prevents reaccumulation of Substance P, the primary chemical mediator of pain impulses from the periphery to the CNS. *Renders skin and joints insensitive to pain; therefore, it serves as an effective peripheral analgesic.*

USES Temporary relief of pain from arthritis, neuralgias, diabetic neuropathy, and herpes zoster.
UNLABELED USES Phantom limb pain, psoriasis, intractable pruritus.

CONTRAINDICATIONS Hypersensitivity to capsaicin or any ingredient in the cream.
CAUTIOUS USE Patients on ACE inhibitors. Safety and efficacy in children younger than 2 y have not been established.
OTC For children under 18 y, consult with prescriber.

ROUTE & DOSAGE

Analgesia

Adult/Child (older than 2 y):
Topical Apply to affected area not more than 3–4 times/day

Common adverse effects in *italic*, life-threatening effects underlined; generic names in **bold**; classifications in SMALL CAPS; ♦ Canadian drug name; ⊘ Prototype drug

ADMINISTRATION

Topical

- Apply to affected areas only and avoid contact with eyes or broken or irritated skin.
- If applied with bare hand, wash immediately following application.
- Use only nitrile gloves when handling a capsaicin patch.
- Avoid tight bandages over areas of application of the cream.
- If necessary for adherence, clip hair (do not shave) on skin where patch will be applied.
- Patch may be cut (before removing protective liner) to match size and shape of treatment area.
- Leave patch on for 60 min. To ensure contact, a dressing may be applied.
- Following patch removal, apply cleansing gel to treatment area and leave on for at least 1 min.

ADVERSE EFFECTS (≥1%) **CNS:** Concentration greater than 1%: Neurotoxicity, hyperalgesia. **Skin:** *Burning, stinging, redness,* itching. **Other:** Cough.

INTERACTIONS Drug: May increase incidence of cough with ACE INHIBITORS.

PHARMACOKINETICS Onset: Postherpetic neuralgia: 2–6 wk.

NURSING IMPLICATIONS

Assessment & Drug Effects

- Monitor for significant pain relief, which may require 4–6 wk of application three or four times daily.
- Monitor for and report signs of skin breakdown as these generally indicate need for drug discontinuation.

Patient & Family Education

- Report local discomfort at site of application if discomfort is distressing or persists beyond the first 3–4 days of use.

- Use caution in handling contact lenses following application of cream. Wash hands thoroughly before touching lenses.
- Notify prescriber if symptoms do not improve or condition worsens within 14–28 days.

CAPTOPRIL
(kap′toe-pril)
Classifications: RENIN ANGIOTENSIN SYSTEM ANTAGONIST; ANTIHYPERTENSIVE
Therapeutic: ANTIHYPERTENSIVE; ACE INHIBITOR
Prototype: Enalapril
Pregnancy Category: D

AVAILABILITY 12.5 mg, 25 mg, 50 mg, 100 mg tablets

ACTION & *THERAPEUTIC EFFECT* Lowers blood pressure by specific inhibition of the angiotensin-converting enzyme (ACE) utilized by renin in the formation of angiotensin II, a potent vasoconstrictor. ACE inhibition alters hemodynamics without compensatory changes in cardiac output (except in patients with CHF). Inhibition of ACE also leads to decreased circulating aldosterone. In heart failure, captopril administration is followed by a fall in CVP and pulmonary wedge pressure; hypotensive action appears to be unrelated to plasma renin levels. *Effective in management of hypertension, and in congestive heart failure with resulting decreases in dyspnea and improved exercise tolerance.*

USES Hypertension; heart failure, diabetic nephropathy, left ventricular dysfunction post MI.
UNLABELED USES Idiopathic edema.

CONTRAINDICATIONS Angioedema, hypersensitivity to captopril or ACE inhibitors; hypotension; jaundice,

Common adverse effects in *italic*, life-threatening effects underlined; generic names in **bold**; classifications in SMALL CAPS; ♣ Canadian drug name; ● Prototype drug

233

or marked elevations of hepatic enzymes; pregnancy (category D), lactation.

CAUTIOUS USE Impaired renal function, patient with solitary kidney; collagen-vascular diseases (scleroderma, SLE); autoimmune disease, bone marrow suppression, coronary or cerebrovascular disease; cardiomyopathy, aortic stenosis; severe salt/volume depletion; heart failure, renal artery stenosis, renal disease, renal failure, renal impairment; hyperkalemia, elderly.

ROUTE & DOSAGE

Hypertension

Adult/Adolescent: **PO** 12.5–25 mg b.i.d. or t.i.d., may increase to 50 mg t.i.d. (max: 450 mg/day)

Heart Failure

Adult: **PO** 25 mg b.i.d.; may increase to 50 mg t.i.d. if needed (max: 450 mg/day)

Proteinuria with Diabetic Nephropathy

Adult: **PO** 25 mg t.i.d.

Renal Insufficiency Dosage Adjustment

CrCl 10–50 mL/min: 75% of dose; *less than 10 mL/min:* 50% of dose

ADMINISTRATION

Oral

- Give captopril 1 h before meals. Food reduces absorption by 30–40%.
- Store in light-resistant containers at no more than 30° C (86° F) unless otherwise directed.

ADVERSE EFFECTS (≥1%) **Body as a Whole:** Hypersensitivity reactions, serum sickness-like reaction, arthralgia, skin eruptions. **CV:** Slight increase in heart rate, first dose hypotension, dizziness, fainting. **GI:** Altered taste sensation (loss of taste perception, persistent salt or metallic taste); weight loss, intestinal angioedema. **Hematologic:** Hyperkalemia, neutropenia, agranulocytosis (rare). **Respiratory:** *Cough.* **Skin:** *Maculopapular rash,* urticaria, pruritus, angioedema, photosensitivity. **Urogenital:** Azotemia, impaired renal function, nephrotic syndrome, membranous glomerulonephritis. **Other:** Positive antinuclear antibody (ANA) titers.

DIAGNOSTIC TEST INTERFERENCE Elevated *urine protein levels* may persist even after captopril has been discontinued. Possibility of transient elevations of *BUN* and *serum creatinine,* slight increase in *serum potassium,* and *serum prolactin,* increases in *liver enzymes,* and false-positive *urine acetone* (using *sodium nitroprusside reagent*). Captopril may decrease *fasting blood sugar.*

INTERACTIONS Drug: NITRATES, DIURETICS, and ANTIHYPERTENSIVES enhance hypotensive effects. POTASSIUM-SPARING DIURETICS **(spironolactone, amiloride)** increase potassium levels. May increase risk of angioedema when used with **pregabalin. Food:** Decreases absorption; take 30–60 min before meals.

PHARMACOKINETICS Absorption: 60–75% absorbed; food may decrease absorption 25–40%. **Onset:** 15 min. **Peak:** 1–2 h. **Duration:** 6–12 h. **Distribution:** To all tissues except CNS; crosses placenta. **Metabolism:** Some liver metabolism. **Elimination:** Primarily in urine; excreted in breast milk.

NURSING IMPLICATIONS

Assessment & Drug Effects

- Monitor BP closely following the first dose. A sudden exaggerated hypotensive response may occur within 1–3 h of first dose, especially in those with high BP or on a diuretic and restricted salt intake.

- Advise bed rest and BP monitoring for the first 3 h after the initial dose.
- Monitor therapeutic effectiveness. At least 2 wk of therapy may be required before full therapeutic effects are achieved.
- Lab tests: Establish baseline urinary protein levels before initiation of therapy and check at monthly intervals for the first 8 mo of treatment and then periodically thereafter. Perform WBC and differential counts before therapy is begun and at approximately 2-wk intervals for the first 3 mo of therapy and then periodically thereafter.

Patient & Family Education

- Report to prescriber without delay the onset of unexplained fever, unusual fatigue, sore mouth or throat, easy bruising or bleeding (pathognomonic of agranulocytosis).
- Mild skin eruptions are most likely to appear during the first 4 wk of therapy and may be accompanied by fever and eosinophilia.
- Consult prescriber promptly if vomiting or diarrhea occur.
- Report darkening or crumbling of nailbeds (reversible with dosage reduction).
- Taste impairment occurs in 5–10% of patients and generally reverses in 2–3 mo even with continued therapy.
- Use OTC medications only with approval of the prescriber. Inform surgeon or dentist that captopril is being taken. Alert diabetic patient that captopril may produce hypoglycemia. Monitor blood glucose and HbA1C closely during first few weeks of therapy.

CARBACHOL INTRAOCULAR
(kar′ba-kole)

Miostat
Pregnancy Category: C
See Appendix A-1.

CARBAMAZEPINE 🅟
(kar-ba-maz′e-peen)

Apo-Carbamazepine ✤, Carbatrol, Epitol, Equetro, Mazepine ✤, PMS-Carbamazepine ✤, Tegretol, Tegretol XR
Classification: ANTICONVULSANT TRICYCLIC
Therapeutic: ANTICONVULSANT; ANTIMANIA
Pregnancy Category: D

AVAILABILITY 100 mg chewable tablets; 200 mg tablets; 100 mg, 200 mg, 400 mg sustained release tablets; 100 mg, 200 mg, 300 mg sustained release capsules; 100 mg/5 mL suspension

ACTION & *THERAPEUTIC EFFECT*
Anticonvulsant action appears to inhibit sustained repetitive impulses and reduces post-tetanic synaptic transmission in the spinal cord. It limits the spread of seizure activity. Provides relief in trigeminal neuralgia by reducing synaptic transmission within trigeminal nucleus. Unknown mechanism in regard to bipolar disorder. *Effective anticonvulsant for a range of seizure disorders and as an adjuvant reduces depressive signs and symptoms and stabilizes mood. It is effective for pain and other symptoms associated with neurologic disorders.*

USES Partial seizures, bipolar disorder, mania, neuropathic pain, tonic-clonic seizures, trigeminal neuralgia.

Common adverse effects in *italic*, life-threatening effects underlined; generic names in **bold**; classifications in SMALL CAPS; ✤ Canadian drug name; 🅟 Prototype drug

UNLABELED USES Diabetic neuropathy, agitation, postherpetic neuralgia, hiccups.

CONTRAINDICATIONS Hypersensitivity to carbamazepine and to TCAs or MAOI therapy; history of myelosuppression or hematologic reaction to other drugs; leukopenia; bone marrow depression; within 14 d use of MAOI drugs; increased IOP; SLE; hepatic, or renal failure; coronary artery disease; hypertension; petit mal (absent) seizures, atonic or myoclonic seizures; acute intermediate porphyria; pregnancy (category D).

CAUTIOUS USE The older adult; history of cardiac disease, alcoholism; hepatic disease; cardiac arrhythmias; presence of HLA-B*1502 gene increases risk of Stevens-Johnson syndrome or toxic epidermal necrolysis; mixed seizure disorder including atypical absence seizures; children younger than 6 y.

ROUTE & DOSAGE

Seizures

Adult: **PO** 200 mg b.i.d., gradually increased to 800–1200 mg/day in 3–4 divided doses. Tegretol XR dosed b.i.d.
Child: **PO** *Younger than 6 y:* 10–20 mg/kg/day, may gradually increase weekly (recommended max: 35 mg/kg/day in 3–4 divided doses); *6–12 y:* 100 mg b.i.d., gradually increased to 400–800 mg/day in 3–4 divided doses (max: 1 g/day); *younger than 6 y:* 20–30 mg/kg/day in 3–4 divided doses

Trigeminal Neuralgia

Adult: **PO** 100 mg b.i.d., gradually increased by 100 mg increments q12h until relief; usual dose 200–800 mg/day in 3–4 divided doses (max: 1.2 g/day). Tegretol XR dosed b.i.d.

Bipolar Disorder (Equetro)

Adult: **PO** 200 mg b.i.d.

ADMINISTRATION

Oral

- Do not administer within 14 days of patient receiving a MAO inhibitor.
- Give with a meal to increase absorption.
- Ensure that chewable tablets are chewed or crushed before being swallowed with a liquid.
- Ensure that sustained release form of drug is not chewed or crushed. It **must be** swallowed whole.
- Do not administer carbamazepine suspension simultaneously with other liquid medications: A precipitate may form in the stomach.

ADVERSE EFFECTS (≥1%) **Body as a Whole:** Myalgia, arthralgia, leg cramps, carbamazepine-induced SLE. **CV:** Edema, syncope, arrhythmias, heart block. **GI:** Nausea, vomiting, anorexia, abdominal pain, diarrhea, constipation, dry mouth and pharynx, abnormal liver function tests, hepatitis, cholestatic and hepatocellular jaundice, pancreatitis. **Endocrine:** Hypothyroidism, SIADH. **Hematologic:** Aplastic anemia, leukopenia (transient), leukocytosis, agranulocytosis, eosinophilia, thrombocytopenia. **CNS:** Dizziness, vertigo, drowsiness, disturbances of coordination, ataxia, confusion, headache, fatigue, listlessness, speech difficulty, development of minor motor seizures, hyperreflexia, akathisia, involuntary movements, tremors, visual hallucinations, activation of latent psychosis, aggression; agitation, respiratory depression. **Skin:** Skin rashes, urticaria, petechiae, erythema multiforme, Stevens-Johnson syndrome, photosensitivity reactions, altered skin pigmentation, exfoliative dermatitis, alopecia. **Special Senses:** Abnormal hearing acuity, scotomas, conjunctivitis, blurred vision, transient diplopia, oculomotor disturbances, oscillopsia, nystagmus. **Urogenital:** Urinary frequency or retention, oliguria, impotence.

DIAGNOSTIC TEST INTERFERENCE May cause false positive TCA screen; may interact with pregnancy tests.

INTERACTIONS Drug: Serum concentrations of other ANTICONVULSANTS may decrease because of increased metabolism; **verapamil, erythromycin, ketoconazole, nefazodone** may increase carbamazepine levels; decreases hypoprothrombinemic effects of ORAL ANTICOAGULANTS; increases metabolism of ESTROGENS, thus decreasing effectiveness of ORAL CONTRACEPTIVES. Reduces concentration of **delavirdine, etravirine. Food: Grapefruit juice** may increase drug levels. **Herbal: Ginkgo** may decrease anticonvulsant effectiveness.

PHARMACOKINETICS Absorption: Slowly from GI tract. **Peak:** 2–8 h. **Distribution:** Widely distributed; high concentrations in CSF; crosses placenta; distributed into breast milk. **Metabolism:** In liver by CYP3A4; can induce liver microsomal enzymes. **Elimination:** In urine and feces. **Half-Life:** Variable due to autoinduction: 25–65 h then 14–16 h (with repeated use).

NURSING IMPLICATIONS

Assessment & Drug Effects

- Lab tests: Baseline and periodic CBCs including platelets, reticulocytes, serum electrolytes and serum iron, LFTs, BUN, and complete urinalysis.

- At least 3 mo into therapy, it is recommended that prescriber attempt dosage reduction or termination of drug therapy, if possible, in patients with trigeminal neuralgia. Some patients develop tolerance to the effects of carbamazepine.

- Monitor for the following reactions, which commonly occur during early therapy: Drowsiness, dizziness, light-headedness, ataxia, gastric upset. If these symptoms do not subside within a few days, dosage adjustments may be indicated.

- Withhold drug and notify prescriber if any of the following signs of myelosuppression occur: RBC less than 4 million/mm^3, Hct less than 32%, Hgb less than 11 g/dL, WBC less than 4000/mm^3, platelet count less than 100,000/mm^3, reticulocyte count less than 20,000/mm^3, serum iron greater than 150 mg/dL.

- Monitor for toxicity, which can develop when serum concentrations are even slightly above the therapeutic range.

- Monitor I&O ratio and vital signs during period of dosage adjustment. Report oliguria, signs of fluid retention, changes in I&O ratio, and changes in BP or pulse patterns.

- Cardiac syncope may resemble epileptic seizures. Therefore, it is recommended that patients who experience an apparent increase in frequency of seizures or a change in their character should be checked by continuous ECG monitoring for 24 h.

- Doses higher than 600 mg/day may precipitate arrhythmias in patients with heart disease.

- Confusion and agitation may be aggravated in the older adult; therefore, side rails and supervision of ambulation may be indicated.

Common adverse effects in *italic*, life-threatening effects <u>underlined</u>; generic names in **bold**; classifications in SMALL CAPS; ♣ Canadian drug name; ☺ Prototype drug

237

Patient & Family Education

- Discontinue drug and notify prescriber immediately if early signs of toxicity or a possible hematologic problem appear, (e.g., anorexia, fever, sore throat or mouth, malaise, unusual fatigue, tendency to bruise or bleed, petechiae, ecchymoses, bleeding gums, nose bleeds).
- Avoid hazardous tasks requiring mental alertness and physical coordination until reaction to drug is known, since dizziness, drowsiness, and ataxia are common adverse effects.
- Remain under close medical supervision throughout therapy.
- Avoid excessive sunlight, as photosensitivity reactions have been reported. Apply a sunscreen (if allowed) with SPF of 12 or above.
- Carbamazepine may cause breakthrough bleeding and may also affect the reliability of oral contraceptives.
- Be aware that abrupt withdrawal of any anticonvulsant drug may precipitate seizures or even status epilepticus.

CARBIDOPA-LEVODOPA

(kar-bi-doe′pa)

Sinemet, Sinemet-CR, Parcopa

CARBIDOPA

Lodosyn
Classifications: DOPAMINE RECEPTOR AGONIST; ANTIPARKINSON
Therapeutic: ANTIPARKINSON
Pregnancy Category: C

AVAILABILITY Carbidopa: 25 mg tablet; **Carbidopa/Levodopa:** 10 mg /100 mg, 25 mg/100 mg, 25 mg/250 mg tablets; 25 mg/100 mg, 50 mg/200 mg sustained release tablets and orally disintegrating tablets

ACTION & *THERAPEUTIC EFFECT*
When levodopa is given alone, large doses **must be** administered. Carbidopa prevents peripheral metabolism (decarboxylation) of levodopa and thereby makes more levodopa available for transport to the brain. Carbidopa does not cross blood–brain barrier and therefore does not affect metabolism of levodopa within the brain. *Effective in management of symptoms of Parkinson's disease and parkinsonism of secondary origin while improving life expectancy and quality of life.*

USES Symptomatic treatment of idiopathic Parkinson's disease (paralysis agitans), postencephalitic parkinsonism, and parkinsonism following carbon dioxide and manganese intoxication. Carbidopa is available alone from manufacturer, on request by prescriber, for use with levodopa when separate titration of each agent is indicated, and for investigational purposes.

CONTRAINDICATIONS Hypersensitivity to carbidopa or levodopa; narrow-angle glaucoma; history of or suspected melanoma; lactation. Safe use in women of childbearing potential is not established.

CAUTIOUS USE Cardiovascular, hepatic, pulmonary, or renal disorders; history of MI; urinary retention; history of peptic ulcer; psychiatric states; endocrine disease; chronic wide-angle glaucoma; seizure disorders; pregnancy (category C). Safe use in children younger than 18 y is not established.

ROUTE & DOSAGE

Parkinson's Disease in Patients Not Currently Receiving Levodopa

Adult: **PO** 1 tablet containing 10 mg carbidopa/100 mg levodopa or 25 mg carbidopa/100 mg levodopa t.i.d., increased by

---

Common adverse effects in *italic*, life-threatening effects <u>underlined</u>; generic names in **bold**; classifications in SMALL CAPS; ♣ Canadian drug name; ⊘ Prototype drug

1 tablet daily or every other day up to 6 tablets/day

Patients Receiving Levodopa

Adult: **PO** 1 tablet of the 25/250 mixture t.i.d. or q.i.d., adjusted by ½–1 tablet as needed up to 8 tablets/day (start at 20–25% of initial dose of levodopa)

ADMINISTRATION

Oral

- Ensure that sustained release form of drug (Sinemet CR) is not chewed or crushed. It may be broken in half but otherwise swallowed whole.
- Give consistently with respect to food. High protein meals may interfere with absorption of levodopa.
- When patient has been taking levodopa alone, carbidopa-levodopa is usually initiated with a morning dose after patient has been without levodopa for at least 8 h.
- Store in tight, light-resistant containers.

ADVERSE EFFECTS (≥1%) **Body as a Whole:** Hoarseness, unusual breathing patterns, neuroleptic malignant syndrome. **CV:** Orthostatic hypotension, irregular heartbeat, palpitation, arrhythmias, phlebitis, edema. **GI:** Nausea, anorexia, dry mouth, bruxism, vomiting, excess salivation. **Hematologic:** Hemolytic and nonhemolytic anemia, thrombocytopenia, agranulocytosis. **Metabolic:** Abnormal liver function tests, abnormal BUN. **CNS:** *Involuntary movements (dyskinetic, dystonic, choreiform),* ataxia, muscle twitching, increase in hand tremor, numbness, headache, dizziness, euphoria, fatigue, confusion, insomnia, nightmares, mental disturbances, anxiety, depression with suicidal tendencies,

delirium, seizures. **Skin:** Body odor, skin rash, dark sweat, loss of hair. **Special Senses:** Blepharospasm, mydriasis, miosis, blurred vision, diplopia, oculogyric crisis. **Urogenital:** Dark urine, priapism, urinary frequency, retention, incontinence.

DIAGNOSTIC TEST INTERFERENCE
Urine glucose: False-negative tests may result with use of *glucose oxidase methods* (e.g., *Clinistix, TesTape*) and false-positive results with *copper reduction methods* (e.g., *Benedict's, Clinitest*), especially in patients receiving large doses. It is reported that *Clinistix* and *TesTape* may be used if reading is taken at margin of wet and dry tape. There is also the possibility of false-positive tests for *urinary ketones* by *dipstick tests* [e.g., *Acetest* (equivocal), *Ketostix, Labstix*] false elevation of *serum* and *urinary uric acid* levels by *colorimetric methods* (*not* with *uricase*); and interference with *urine PKU test* results.

INTERACTIONS Drug: MAO INHIBITORS may precipitate hypertensive crisis; TRICYCLIC ANTIDEPRESSANTS potentiate postural hypotension; ANTICHOLINERGIC AGENTS may enhance levodopa effects but can exacerbate involuntary movements; **methyldopa, guanethidine** increase hypotensive and CNS effects; PHENOTHIAZINES, haloperidol, **phenytoin, papaverine** may interfere with levodopa effects.

PHARMACOKINETICS Absorption: 40–70% of carbidopa absorbed after PO dose; carbidopa may enhance absorption of levodopa. **Distribution:** Widely distributed in most body tissues except CNS; crosses placenta; excreted in breast milk. **Elimination:** In urine. **Half-Life:** 2 h.

C

NURSING IMPLICATIONS

Assessment & Drug Effects

- Make accurate observations and report promptly adverse reactions and therapeutic effects. Rate of dosage increase is determined primarily by patient's tolerance and response to levodopa.
- Monitor vital signs, particularly during period of dosage adjustment. Report alterations in BP, pulse, and respiratory rate and rhythm.
- Monitor all patients closely for behavior changes. Patients in depression should be closely observed for suicidal tendencies.
- Monitor for changes in intraocular pressure in patients with chronic wide-angle glaucoma.
- Monitor patients with diabetes carefully for alterations in diabetes control. Frequent monitoring of blood sugar is advised.
- Lab tests: Periodic blood glucose, hepatic and renal function tests, CBC with differential, Hgb and Hct.
- Report promptly abnormal involuntary movement such as facial grimacing, exaggerated chewing, protrusion of tongue, rhythmic opening and closing of mouth, bobbing of head, jerky arm and leg movements, and exaggerated respiration.
- Assess for "on-off" phenomenon: Sudden, unpredictable loss of drug effectiveness ("off" effect), which lasts 1 min–1 h. This is followed by an equally abrupt return of function ("on" effect). Sometimes symptoms can be controlled by increasing number of doses per day.
- Monitor therapeutic effects. Some patients manifest increase in bradykinesia ("leg freezing" or slow body movement). The patient is unable to start walking and frequently falls. Reduction of dosage may be indicated in these patients.
- Patients who require more frequent drug administration are most likely to manifest gradual return of parkinsonian symptoms toward the end of a dose period.

Patient & Family Education

- Follow prescriber's directions regarding continuation or discontinuation of levodopa. Both adverse reactions and therapeutic effects occur more rapidly with carbidopa-levodopa combination than with levodopa alone.
- Make positional changes slowly and in stages, particularly from recumbent to upright position, dangle your legs a few minutes before standing, and walk in place before ambulating, as some patients experience weakness, dizziness, and faintness. Tolerance to this effect usually develops within a few months of therapy. Support stockings may help. Consult prescriber.
- Report muscle twitching and spasmodic winking promptly, as these may be early signs of overdosage.
- You may notice elevation of mood and sense of well-being before any objective improvement. Resume activities gradually and observe safety precautions to avoid injury.
- Maintain your prescribed drug regimen. Abrupt withdrawal can lead to parkinsonian crisis with return of marked muscle rigidity, akinesia, tremor, hyperpyrexia, mental changes.
- Avoid driving or other hazardous activities until reaction to drug is determined.
- Levodopa may cause urine to darken on standing and may also cause sweat to be dark-colored. This effect is not clinically significant.
- Wear medical identification. Inform all health care providers that you are taking carbidopa-levodopa.

Common adverse effects in *italic*, life-threatening effects <u>underlined</u>; generic names in **bold**; classifications in SMALL CAPS; ✦ Canadian drug name; ● Prototype drug

CARBINOXAMINE

(car-bi-nox-a-meen)
Arbinoxa, Palgic
Classification: ANTIHISTAMINE;
H₁-RECEPTOR ANTAGONIST
Therapeutic: ANTIHISTAMINE;
SEDATING H₁-ANTAGONIST
Prototype: Diphenhydramine
Pregnancy Category: C

AVAILABILITY 4 mg tablet;
4mg/5mL oral solution

ACTION & *THERAPEUTIC EFFECT*

Carbinoxamine competes for H₁-receptor sites on effector cells thus blocking histamine release. *Has antihistaminic, anticholinergic (drying), antitussive, and sedative properties.*

USES Allergic conjunctivitis, allergic rhinitis, rhinorrhea, pruritus, sneezing, urticaria.

CONTRAINDICATIONS Hypersensitivity to antihistamines of similar structure; lower respiratory tract symptoms (including acute asthma); narrow-angle glaucoma; prostatic hypertrophy, bladder neck obstruction; GI obstruction or stenosis; children younger than 1 y; lactation.
CAUTIOUS USE History of asthma; COPD; convulsive disorders; increased IOP; hyperthyroidism; hypertension, cardiovascular disease; hepatic disease; diabetes mellitus; older adults, infants, and young children; pregnancy (category C).

ROUTE & DOSAGE

Allergic Conjunctivitis, Allergic rhinitis, Rhinorrhea, Pruritus, Sneezing, Urticaria

Adult: **PO** 4–8 mg t.i.d. or q.i.d.
Children: **PO** 1–3 y, 2 mg t.i.d. or q.i.d.; 3–6 y, 2–4 mg t.i.d. or q.i.d.; > 6 y, 4–6 mg t.i.d. or q.i.d.

ADMINISTRATION

Oral
- Give with food or milk if needed to lessen GI distress.
- Store a 15°–30°C (59°–86°F).

ADVERSE EFFECTS (≥1%) **Body as a Whole:** <u>Anaphylactic shock</u>, chills, excessive perspiration, polyuria, photosensitivity, weakness. **CNS:** Acute labyrinthitis, blurred vision, confusion, convulsions, diplopia, disturbed coordination, dizziness, euphoria, excitation, fatigue, headache, hysteria, insomnia, irritability, nervousness, neuritis, paresthesia, restlessness, sedation, sleepiness, tinnitus, tremor, vertigo. **CV:** Extrasystoles, headache, hypotension, palpitations, tachycardia. **GI:** Anorexia, constipation, diarrhea, epigastric distress, heartburn, nausea, vomiting. **Hematological:** <u>Agranulocytosis</u>, hemolytic anemia, <u>thrombocytopenia</u>. **Respiratory:** Dryness of mouth, nose and throat, nasal stuffiness, thickening of bronchial secretions, tightness of chest and wheezing. **Skin:** Drug rash, urticaria. **Urogenital:** Difficult urination, early menses, increased urinary frequency, urinary retention.

INTERACTIONS Drug: MAO INHIBITORS may prolong and intensify the anticholinergic effects of carbinoxamine. Carbinoxamine may enhance the effects of TRICYCLIC ANTIDEPRESSANTS, BARBITURATES, **alcohol**, and other CNS DEPRESSANTS.

PHARMACOKINETICS Onset: 15–30 m. **Peak:** 1 h. **Metabolism:** Extensive hepatic metabolism to inactive compounds. **Elimination:** Primarily renal. **Half-Life:** 10–20 h.

Common adverse effects in *italic*, life-threatening effects <u>underlined</u>; generic names in **bold**; classifications in SMALL CAPS; ✤ Canadian drug name; ❂ Prototype drug

241

NURSING IMPLICATIONS

Assessment & Drug Effects

- Monitor CV status especially with preexisting cardiovascular disease.
- Monitor for adverse effects especially in young children and older adults.
- Supervise ambulation and institute fall precautions as necessary.

Patient & Family Education

- Do not use alcohol and other CNS depressants because of the possible additive CNS depressant effects.
- Do not drive or engage in other potentially hazardous activities until the response to drug is known.
- Increase fluid intake, if not contraindicated; drug has a drying effect (thickens bronchial secretions) that may make expectoration difficult.

CARBOPLATIN

(car-bo-pla'tin)

Classifications: ANTINEOPLASTIC; ALKYLATING AGENT
Therapeutic: ANTINEOPLASTIC
Prototype: Cyclophosphamide
Pregnancy Category: D

AVAILABILITY 150 mg/15 mL, 450 mg/45 mL, 50 mg/5mL, 600 mg/60 mL solution for injection

ACTION & *THERAPEUTIC EFFECT*
Carboplatin is a platinum compound that is a chemotherapeutic agent. It produces interstrand DNA cross-linkages, thus interfering with DNA, RNA, and protein synthesis. Carboplatin is cell-cycle nonspecific and induces programmed cell death. *Full or partial activity against a variety of cancers resulting in reduction or stabilization of tumor size. Useful in patients with impaired renal function, patients unable to accommodate high-volume hydration, or patients at high risk for neurotoxicity and/or ototoxicity.*

USES Ovarian cancer.
UNLABELED USES Combination therapy for breast, cervical, colon, endometrial, head and neck, and lung cancer; leukemia, lymphoma, and melanoma.

CONTRAINDICATIONS History of severe reactions to carboplatin or other platinum compounds, severe bone marrow depression; significant bleeding; impaired renal function; pregnancy (category D), and lactation.
CAUTIOUS USE Use with other nephrotoxic drugs; coagulopathy; previous radiation therapy; renal impairment.

ROUTE & DOSAGE

Ovarian Cancer

Adult: **IV** 360 mg/m^2 once q4wk. May be repeated when neutrophil count is at least 2000 mm^3 and platelet count is at least 100,000 mm^3. If neutrophil and platelet counts are lower, dose of carboplatin should be reduced by 50–75% of initial dose. Alternatively, 400 mg/m^2 as a 24-h infusion for 2 consecutive days can be used.

Renal Impairment Dosage Adjustment

CrCl 41–59 mL/min: Dose 250 mg/m^2; 16–40 mL/min: Dose 200 mg/m^2

Hemodialysis Dosage Adjustment
Initial dose not to exceed 150 mg/m^2

ADMINISTRATION

Intravenous

PREPARE: IV Infusion: Do not use needles or IV sets containing aluminum. ▪ Immediately before use, reconstitute with either sterile water for injection or D5W or NS as follows: 50-mg vial plus 5 mL diluent; 150-mg vial plus 15 mL diluent; 450-mg vial plus 45 mL diluent. All dilutions yield 10 mg/mL. ▪ May be further diluted for infusion with D5W or NS to concentrations as low as 0.5 mg/mL.

ADMINISTER: **IV Infusion:** Give IV solution over 15 min or longer, depending on total amount of solution and patient tolerance.
▪ Lengthening duration of administration may decrease nausea and vomiting. ▪ Premedication with a parenteral antiemetic 30 min before and on a scheduled basis thereafter is normally used. ▪ Do not repeat doses until the neutrophil count is at least 2000/mm^3 and platelet count at least 100,000/mm^3.

INCOMPATIBILITIES **Solution/additive: Sodium bicarbonate, fluorouracil, mesna. Y-site: Amphotericin B cholesteryl complex, lansoprazole.**

▪ Protect from light. Reconstituted solutions are stable for 8 h at room temperature; discard solutions 8 h after dilution.

ADVERSE EFFECTS (≥1%) **Body as a Whole:** Hypersensitivity reactions. **GI:** *Mild to moderate nausea and vomiting,* anorexia, hypogeusia, dysgeusia, mucositis, diarrhea, constipation, elevated liver enzymes. **Hematologic:** <u>Thrombocytopenia, leukopenia, neutropenia, anemia.</u> **Metabolic:** *Mild hyponatremia, hypomagnesemia, hypocalcemia, and hypokalemia.* **CNS:** Peripheral neuropathy. **Skin:** Rash, alopecia. **Special Senses:** Tinnitus. **Urogenital:** Nephrotoxicity.

DIAGNOSTIC TEST INTERFERENCE Decreased **calcium levels;** mild increases in **liver function tests;** decreased levels of **magnesium, potassium,** and **sodium.**

INTERACTIONS Drug: AMINOGLYCOSIDES may increase the risk of ototoxicity and nephrotoxicity. May decrease **phenytoin** levels.

PHARMACOKINETICS Onset: 8 wk (2 cycles). **Duration:** 2–16 mo. **Distribution:** Highest concentration in the liver, lung, kidney, skin, and tumors. Not bound to plasma proteins. **Metabolism:** Hydrolyzed in the serum. **Elimination:** Primarily by the kidneys; 60–80% excreted in urine within 24 h. **Half-Life:** 3 h.

NURSING IMPLICATIONS

Assessment & Drug Effects
▪ Monitor closely during first 15 min of infusion, since allergic reactions have occurred within minutes of carboplatin administration.
▪ Lab tests: Baseline and periodic CBC with differential, platelet count, Hgb and Hct. Monitor kidney function at baseline and prior to each infusion; periodic serum electrolytes.
▪ Monitor results of peripheral blood counts. Leukopenia, neutropenia, and thrombocytopenia are dose related and may produce dose-limiting toxicity.
▪ Monitor for peripheral neuropathy (e.g., paresthesias), ototoxicity, and visual disturbances.
▪ Monitor serum electrolyte studies, because carboplatin has been associated with decreases in sodium, potassium, calcium, and magnesium. Special precautions may be warranted for patients on diuretic therapy.

Patient & Family Education
▪ Learn about the range of potential adverse effects. Strategies for nausea prevention should receive special attention.

Common adverse effects in *italic*, life-threatening effects <u>underlined</u>; generic names in **bold**; classifications in SMALL CAPS; ♣ Canadian drug name; ☻ Prototype drug

243

- During therapy you are at risk for infection and hemorrhagic complications related to bone marrow suppression. Avoid unnecessary exposure to crowds or infected persons during the nadir period.
- Report paresthesias (numbness, tingling), visual disturbances, or symptoms of ototoxicity (hearing loss and/or tinnitus).

CARBOPROST TROMETHAMINE ⓟ

(kar'boe-prost)

Hemabate
Classifications: PROSTAGLANDIN; OXYTOCIC
Therapeutic: OXYTOCIC
Pregnancy Category: D

AVAILABILITY 250 mcg/mL injection

ACTION & *THERAPEUTIC EFFECT*
Synthetic analog of naturally occurring prostaglandin F_2 alpha with longer duration of biological activity. Stimulates myometrial contractions of gravid uterus at term labor. Mean time to abortion 16 h; mean dose required 2.6 mL. *Effectively stimulates uterine contraction and is used to induce abortion. Useful in treatment of postpartum hemorrhage due to uterine atony unresponsive to usual measures.*

USES Pregnancy termination. Also for refractory postpartum bleeding.
UNLABELED USES To reduce blood loss secondary to uterine atony; to induce labor in intrauterine fetal death and hydatidiform mole.

CONTRAINDICATIONS Acute pelvic inflammatory disease; active cardiac, pulmonary, renal, or hepatic disease; pregnancy (category D); lactation.
CAUTIOUS USE History of asthma; adrenal disease; anemia; hypotension; hypertension; diabetes mellitus; epilepsy; history of uterine surgery; cervical stenosis; fibroids.

ROUTE & DOSAGE

Abortion, Postpartum Bleeding

Adult: **IM** Initial: 250 mcg (1 mL) repeated at 1.5–3.5-h intervals if indicated by uterine response. Dosage may be increased to 500 mcg (2 mL) if uterine contractility is inadequate after several doses of 250 mcg (1 mL), not to exceed total dose of 12 mg or continuous administration for more than 2 days.

ADMINISTRATION

- Give deep IM into a large muscle. Aspirate carefully before injecting drug to avoid inadvertent entry into blood vessel which can result in bronchospasm, tetanic contractions, and shock. Do not use same site for subsequent doses.
- Store drug in refrigerator at 2°–4° C (36°–39° F) unless otherwise specified.

ADVERSE EFFECTS (≥1%) **Body as a Whole:** Fever, flushing, chills, cough, headache, pain (muscles, joints, lower abdomen, eyes), hiccups, breast tenderness. **GI:** *Nausea,* diarrhea, vomiting.

PHARMACOKINETICS Peak: 30–90 min. **Elimination:** Renal within 24 h.

NURSING IMPLICATIONS

Assessment & Drug Effects

- Monitor uterine contractions and observe and report excessive vaginal bleeding and cramping pain. Save all clots and tissue for prescriber inspection and laboratory analysis.

- Check vital signs at regular intervals. Carboprost-induced febrile reaction occurs in more than 10% of patients and **must be** differentiated from endometritis, which occurs around third day after abortion.

Patient & Family Education

- Report promptly onset of bleeding, foul-smelling discharge, abdominal pain, or fever.
- Since ovulation may reoccur as early as 2 wk post-abortion, consider appropriate contraception.

CARISOPRODOL

(kar-eye-soe-proe'dole)

Soma

Classification: CENTRALLY-ACTING SKELETAL MUSCLE RELAXANT

Therapeutic: SKELETAL MUSCLE RELAXANT

Prototype: Cyclobenzaprine

Pregnancy Category: C

Controlled Substance: Schedule IV

AVAILABILITY 350 mg tablets

ACTION & *THERAPEUTIC EFFECT*
Centrally acting skeletal muscle relaxant that appears to cause slight reduction in muscle tone leading to relief of pain and discomfort of muscle spasm. *Effective spasmolytic while reducing pain associated with acute musculoskeletal disorders.*

USES Skeletal muscle spasm, stiffness, and pain in a variety of musculoskeletal disorders and to relieve spasticity and rigidity in cerebral palsy.

CONTRAINDICATIONS Hypersensitivity to carisoprodol and related compounds (e.g., meprobamate, carbamate); acute intermittent porphyria.

CAUTIOUS USE Impaired liver or kidney function, addiction-prone individuals; seizure disorder; pregnancy (category C), lactation; children younger than 5 y.

ROUTE & DOSAGE

Muscle Spasm

Adult: **PO** 350 mg t.i.d.
Child (older than 5 y): **PO** 25 mg/kg/day in 4 divided doses

---

ADMINISTRATION

Oral

- Give with food, as needed, to reduce GI symptoms. Last dose should be taken at bedtime.
- Store in tightly closed container.

ADVERSE EFFECTS (≥1%) **Body as a Whole:** Eosinophilia, asthma, fever, anaphylactic shock. **CV:** Tachycardia, postural hypotension, facial flushing. **GI:** Nausea, vomiting, hiccups. **CNS:** *Drowsiness, dizziness,* vertigo, ataxia, tremor, headache, irritability, depressive reactions, syncope, insomnia. **Skin:** Skin rash, erythema multiforme, pruritus.

INTERACTIONS Drug: Alcohol, CNS DEPRESSANTS potentiate CNS effects.

PHARMACOKINETICS Onset: 30 min. **Duration:** 4–6 h. **Distribution:** Crosses placenta. **Metabolism:** In liver by CYP2C19. **Elimination:** By kidneys; excreted in breast milk (2–4 times the plasma concentrations). **Half-Life:** 8 h.

NURSING IMPLICATIONS

Assessment & Drug Effects

- Monitor for allergic or idiosyncratic reactions that generally occur from the first to the fourth dose in patients taking the drug for the first time. Symptoms usually subside after several hours; they are treated by supportive and symptomatic measures.

Patient & Family Education

- Avoid driving and other potentially hazardous activities until

---

Common adverse effects in *italic*, life-threatening effects underlined; generic names in **bold**; classifications in SMALL CAPS; ◆ Canadian drug name; ⊙ Prototype drug

245

response to the drug has been evaluated. Drowsiness is a common side effect and may require reduction in dosage.

- Report to prescriber if symptoms of dizziness and faintness persist. Symptoms may be controlled by making position changes slowly and in stages.
- Do not take alcohol or other CNS depressants (effects may be additive) unless otherwise directed by prescriber.
- Discontinue drug and notify prescriber if skin rash, diplopia, dizziness, or other unusual signs or symptoms appear.

CARMUSTINE
(kar-mus'teen)

BiCNU, Gliadel
Classifications: ANTINEOPLASTIC; ALKYLATING
Therapeutic: ANTINEOPLASTIC
Prototype: Cyclophosphamide
Pregnancy Category: D

AVAILABILITY 100 mg injection; 7.7 mg wafer

ACTION & *THERAPEUTIC EFFECT*
Highly lipid-soluble compound with cell-cycle-nonspecific activity against rapidly proliferating cells. Produces cross-linkage of DNA strands, thereby blocking DNA, RNA, and protein synthesis in tumor cells. *Drug metabolites are thought to be responsible for antineoplastic activities. Full or partial activity against a variety of cancers results in reduction or stabilization of tumor size and increased survival rates.*

USES As single agent or with other antineoplastics in treatment of Hodgkin's disease and other lymphomas, melanoma, primary and metastatic tumors of brain, and GI tract malignancies.

UNLABELED USES Treatment of carcinomas of breast and lungs, Ewing's sarcoma, Burkitt's tumor, malignant melanoma, and topically for mycosis fungoides.

CONTRAINDICATIONS History of pulmonary function impairment; recent illness with or exposure to chickenpox or herpes zoster; infection; severe bone marrow depression, decreased circulating platelets, leukocytes, or erythrocytes; pregnancy (category D), lactation.
CAUTIOUS USE Hepatic and renal insufficiency; patient with previous cytotoxic medication, or radiation therapy; history of herpes infections.

ROUTE & DOSAGE

Previously Untreated Patients— Carcinoma
Adult: **IV** 150–200 mg/m² q6wk in one dose *or* given over 2 days; adjust for hematologic toxicity
Mycosis Fungoides
Adult: **Topical** 0.05–0.4% solution in 30% alcohol to paint entire body (60 mL) or ointment 1–2 times/day for 6–8 wk (10 mg/ day) **(must be** specially compounded)

ADMINISTRATION

Note: When administering IV to infants and children, verify correct IV concentration and rate of infusion with prescriber.

Intravenous

PREPARE: IV Infusion: Wear disposable gloves; contact of drug with skin can cause burning, dermatitis, and hyperpigmentation. ▪ Add supplied diluent to the 100 mg vial. Further dilute with 27 mL of sterile water for injection to yield a concentration

of 3.3 mg/mL. ▪ Each dose is then added to 100–500 mL of D5W or NS. ▪ If possible avoid using PVC IV tubing and bags.
ADMINISTER: **IV Infusion:** Infuse a single dose over at least 1 h. Slow infusion over 1–2 h and adequate dilution will reduce pain of administration. ▪ Avoid starting infusion into dorsum of hand, wrist, or the antecubital veins; extravasation in these areas can damage underlying tendons and nerves leading to loss of mobility of entire limb.
INCOMPATIBILITIES **Solution/additive: Dextrose 5%, sodium bicarbonate. Y-site: Allopurinol.**

▪ Reconstituted solutions of carmustine are clear and colorless and may be stored at 2°–8° C (36°–46° F) for 8 h protected from light. ▪ Store unopened vials at 2°–8° C (36°–46° F), protected from light, unless otherwise directed by manufacturer. ▪ Signs of decomposition of carmustine in unopened vial: Liquefaction and appearance of oil film at bottom of vial. Discard drug in this condition.

ADVERSE EFFECTS (≥1%) **Hematologic:** Delayed <u>myelosuppression</u> (dose-related); <u>thrombocytopenia</u>. **CNS:** Dizziness, ataxia. **Respiratory:** <u>Pulmonary infiltration or fibrosis</u>. **Skin:** Skin flushing and burning pain at injection site, hyperpigmentation of skin (from contact). **Special Senses:** (With high doses) <u>Eye infarctions</u>, retinal hemorrhage, suffusion of conjunctiva. **GI:** Stomatitis, *nausea, vomiting*.

INTERACTIONS Drug: **Cimetidine** may potentiate neutropenia and thrombocytopenia.

PHARMACOKINETICS Distribution: Readily crosses blood–brain barrier; CSF concentrations 15–70% of plasma concentrations. **Metabolism:** Rapidly metabolized. **Elimination:** 60–70% in urine in 96 h; 6% via lungs, 1% in feces; excreted in breast milk.

NURSING IMPLICATIONS

Assessment & Drug Effects

▪ Frequently check rate of flow and blood return; monitor injection site for extravasation. If there is any question about patency, line should be restarted.
▪ Monitor for nausea and vomiting (dose related), which may occur within 2 h after drug administration and persist for up to 6 h.
▪ Lab tests: Baseline CBC with differential and platelet count, repeat blood studies following infusion at weekly intervals for at least 6 wk. Baseline and periodic tests of hepatic and renal function.
▪ Platelet nadir usually occurs within 4–5 wk, and leukocyte nadir within 5–6 wk after therapy is terminated.
▪ Check temperature daily. Avoid use of rectal thermometer to prevent injury to mucosa. An elevation of 0.6° F or more above normal temperature warrants reporting.
▪ Report symptoms of lung toxicity (cough, shortness of breath, fever) to the prescriber immediately.
▪ Be alert to signs of hepatic toxicity (jaundice, dark urine, pruritus, light-colored stools) and renal insufficiency (dysuria, oliguria, hematuria, swelling of lower legs and feet).

Patient & Family Education

▪ Report burning sensation immediately, as carmustine can cause burning discomfort even in the absence of extravasation. Infusion will be discontinued and restarted in another site. Ice application over the area may decrease the discomfort.

Common adverse effects in *italic*, life-threatening effects <u>underlined</u>; generic names in **bold**; classifications in SMALL CAPS; ✦ Canadian drug name; ⊘ Prototype drug

247

- Intense flushing of skin may occur during IV infusion. This usually disappears in 2–4 h.
- You will be highly susceptible to infection and to hemorrhagic disorders. Be alert to hazardous periods that occur 4–6 wk after a dose of carmustine. If possible, avoid invasive procedures (e.g., IM injections, enemas, rectal temperatures) during this period.
- Report promptly the onset of sore throat, weakness, fever, chills, infection of any kind, or abnormal bleeding (ecchymosis, petechiae, epistaxis, bleeding gums, hematemesis, melena).

CARTEOLOL HYDROCHLORIDE
(car′tee-oh-lole)

Cartrol, Ocupress
Classifications: BETA-ADRENERGIC ANTAGONIST; ANTIHYPERTENSIVE
Therapeutic: ANTIHYPERTENSIVE; BETA-ADRENERGIC BLOCKER
Prototype: Propranolol
Pregnancy Category: C

AVAILABILITY 2.5 mg, 5 mg tablets; 1% solution

ACTION & *THERAPEUTIC EFFECT*
Carteolol is a beta-adrenergic blocking agent (antagonist) that competes for available beta receptor sites. It inhibits both beta$_1$ receptors (chiefly in cardiac muscle) and beta$_2$ receptors (chiefly in the bronchial and vascular musculature). It decreases standing and supine hypertension. *Effective antihypertensive agent by reducing BP to normotensive range and useful in managing angina by decreasing myocardial oxygen demand and lowering cardiac work load.*

USES For hypertension, either alone or in combination with other drugs,

particularly a thiazide diuretic (not indicated for hypertensive crisis); chronic open-angle glaucoma.
UNLABELED USES To reduce the frequency of anginal attacks.

CONTRAINDICATIONS Sinus bradycardia, severe CHF; greater than first-degree heart block, cardiogenic shock, CHF secondary to tachycardia treatable with beta-blockers, overt cardiac failure, hypersensitivity to beta-blocking agents, persistent severe bradycardia, bronchial asthma or bronchospasm, and severe COPD; pulmonary edema.
CAUTIOUS USE CHF patients treated with digitalis and diuretics, peripheral vascular disease; diabetes, hypoglycemia, thyrotoxicosis; renal disease; CVA; pregnancy (category C), lactation.

ROUTE & DOSAGE

Hypertension
Adult: **PO** 2.5 mg once/day, may increase to 5–10 mg if needed (max: 10 mg/day)

Open-Angle Glaucoma
Adult: **Ophthalmic** 1 drop in affected eye b.i.d.

ADMINISTRATION

Oral
- Administer capsule or tablet whole. Do not crush or break and instruct patient not to chew before swallowing.
- Store away from heat, light, or moisture.

ADVERSE EFFECTS (≥1%) **Body as a Whole:** Rash, muscle cramps, bronchospasm. **CV:** Increased angina, hypotension, CHF, bradycardia. **GI:** Abdominal pain, diarrhea, nausea. **Endocrine:** Hyperglycemia. **CNS:** *Headache, dizziness,* drowsiness, insomnia, anxiety, tremor, paresthesia, weakness.

INTERACTIONS Drug: DIURETICS and other HYPOTENSIVE AGENTS increase hypotensive effect; carteolol and **albuterol, metoproterenol, terbutaline, pirbuterol** are mutually antagonistic; NSAIDS may blunt hypotensive effect; decreases hypoglycemic effect of **glyburide;** may increase bradycardia and sinus arrest with **amiodarone.**

PHARMACOKINETICS Absorption: Readily from GI tract; 85% reaches systemic circulation. **Peak:** 1–3 h. **Duration:** 24–48 h. **Distribution:** Crosses placenta; distributed into breast milk. **Metabolism:** In liver to active metabolite. **Elimination:** Primarily in urine. **Half-Life:** 4–6 h.

NURSING IMPLICATIONS

Assessment & Drug Effects

- Assess heart rate prior to administration. If pulse is less than 50 bpm, withhold drug and notify prescriber.
- Monitor BP and pulse frequently during period of adjustment and periodically throughout therapy.
- If hypotension (systolic BP less than 90 mm Hg) occurs, withhold the drug and notify prescriber, and carefully assess the hemodynamic status of the patient.
- Monitor daily weight and assess for evidence of fluid overload since drug may precipitate CHF (see Signs & Symptoms, Appendix F).
- Monitor diabetic for loss of diabetic control. Drug may prevent the appearance of early S&S of acute hypoglycemia (see Appendix F).
- Drug may reduce tolerance to cold temperatures in older adults or in those who have circulatory problems.

Patient & Family Education

- Report the first sign or symptom of impending CHF (see Signs & Symptoms, Appendix F) or unexplained respiratory symptoms.
- Do not discontinue medication abruptly, since sudden withdrawal may precipitate or exacerbate angina.
- Report slow pulse rate, confusion or depression, dizziness or lightheadedness, skin rash, fever, sore throat, or unusual bleeding or bruising.
- Be cautious while driving or performing other hazardous activities until response to drug is known.
- Take your pulse before and after taking the medication. If it is much slower than normal rate (or less than 50 bpm), check with your prescriber.

CARVEDILOL
(car-ve-di'lol)
Coreg, Coreg CR, Kredex ✚
Classifications: ALPHA- AND BETA-ADRENERGIC ANTAGONIST; ANTIHYPERTENSIVE
Therapeutic: ANTIHYPERTENSIVE; ADRENERGIC BLOCKER
Prototype: Propanolol HCl
Pregnancy Category: C

AVAILABILITY 3.125 mg, 6.25 mg, 12.5 mg, 25 mg tablets; ; 10 mg, 20 mg, 40 mg, 80 mg extended release capsule

ACTION & *THERAPEUTIC EFFECT*
Adrenergic receptor blocking agent that contributes to blood pressure reduction. Peripheral vasodilation and, therefore, decreased peripheral resistance results from alpha$_1$-blocking activity. *An effective antihypertensive agent reducing BP to normotensive range and useful in managing some angina,*

C

dysrhythmias, and CHF by decreasing myocardial oxygen demand and lowering cardiac workload.

USES Management of essential hypertension, CHF, left ventricular dysfunction post MI.

CONTRAINDICATIONS Patients with class IV decompensated cardiac failure, abrupt cessation in CAD patients; bronchial asthma, or related bronchospastic conditions (e.g., chronic bronchitis and emphysema), pulmonary edema; second- and third-degree AV block, sick sinus syndome, cardiogenic shock or severe bradycardia; lactation.

CAUTIOUS USE Patients on MAOI agents, diabetes, hypoglycemia; patients at high risk for anaphylactic reaction, peripheral vascular disease, cerebrovascular insufficiencies, major depression, hepatic or renal impairment; older adults; pregnancy (category C); children younger than 18 y.

ROUTE & DOSAGE

CHF

Adult: **PO** Start with 3.125 mg b.i.d. × 2 wk, may double dose q2wk as tolerated up to 25 mg b.i.d. if weight less than 85 kg *or* 50 mg b.i.d. if weight greater than 85 kg

Left Ventricular Dysfunction Post MI

Adult: **PO** 6.25 mg bid, can double every 3–10 days as tolerated

Hypertension

Adult: **PO** Start with 6.25 mg b.i.d., may increase by 6.25 mg b.i.d. to max of 50 mg/day

ADMINISTRATION

Oral

- Give with food to slow absorption and minimize risk of orthostatic hypotension.
- Dose increments should be made at 7- to 14-day intervals.

ADVERSE EFFECTS (≥1%) **Body as a Whole:** Increased sweating, fatigue, chest pain, pain, arthralgia. **CV:** Bradycardia, hypotension, syncope, hypertension, AV block, angina. **GI:** Diarrhea, nausea, abdominal pain, vomiting. **Respiratory:** Sinusitis, bronchitis. **Hematologic:** Thrombocytopenia. **Metabolic:** Hyperglycemia, weight increase, gout. **CNS:** *Dizziness,* headache, paresthesias.

INTERACTIONS Drug: Rifampin significantly decreases **carvedilol** levels; **cimetidine** may increase **carvedilol** levels; **clonidine, reserpine,** MAO INHIBITORS may cause hypotension or bradycardia; **carvedilol** may increase **digoxin** levels and may enhance hypoglycemic effects of **insulin** and oral HYPOGLYCEMIC AGENTS, may enhance the effects of ANTIHYPERTENSIVES. **Amiodarone** and **fluconazole** increase side effects.

PHARMACOKINETICS Absorption: Rapidly from GI tract, 25–35% reaches the systemic circulation. **Peak:** Antihypertensive effect 7–14 days. **Distribution:** Greater than 98% protein bound. **Metabolism:** In the liver by CYP2D6 and CYP2C9. **Elimination:** Primarily through feces. **Half-Life:** 7–10 h.

NURSING IMPLICATIONS

Assessment & Drug Effects

- Monitor for therapeutic effectiveness which is indicated by lessening of S&S of CHF and improved BP control.

- Lab tests: Monitor LFTs periodically; at first sign of hepatic toxicity (see Appendix F) stop drug and notify prescriber.
- Monitor for worsening of symptoms in patients with PVD.
- Monitor digoxin levels with concurrent use; plasma digoxin concentration may increase.

Patient & Family Education
- Do not abruptly discontinue taking this drug.
- Make position changes slowly due to the risk of orthostatic hypotension.
- Do not engage in hazardous activities while experiencing dizziness.
- If you have diabetes, the drug may increase effects of hypoglycemic drugs and mask S&S of hypoglycemia.

CASCARA SAGRADA

(kas-kar'a)

Classification: GI STIMULANT, LAXATIVE
Therapeutic: LAXATIVE
Prototype: Bisacodyl
Pregnancy Category: C

AVAILABILITY 325 mg tablets; liquid

ACTION & THERAPEUTIC EFFECT
Acts principally in large intestine by stimulating propulsive movements of colon through direct chemical irritation. Casanthrol is a derivative of cascara sagrada. *Effective laxative with results in 6–12 h. Useful in conditions where straining at stool is to be avoided.*

USES Temporary relief of constipation and to prevent straining at stool in various disease conditions. Sometimes used with milk of magnesia.

CONTRAINDICATIONS Abdominal pain, fecal impaction; GI bleeding, ulcerations; appendicitis, gastroenteritis, intestinal obstruction, CHF.
CAUTIOUS USE Renal impairment; diabetic patients; rectal bleeding; concomitant laxative use; pregnancy (category C); lactation.

ROUTE & DOSAGE

Laxative
Adult: **PO** Tablet: 325–1000 mg/day; fluid extract: 0.5–1.5 mL/day; aromatic fluid extract: 2–6 mL/day *Child:* **PO** 2–12 y, ½ of adult dose; *younger than 2 y,* ¼ of adult dose *Infant:* **PO** Aromatic fluid extract: 1.25 mL/day as single dose

ADMINISTRATION

Oral
- Administer with a full glass of water on an empty stomach for best results. Results may be delayed somewhat by food.
- Store in tightly covered, light-resistant containers, unless otherwise directed by manufacturer.

ADVERSE EFFECTS (≥1%) **GI:** Anorexia, nausea, gripping, abnormally loose stools, constipation rebound, melanosis of colon. **Metabolic:** Hypokalemia, impaired glucose tolerance, calcium deficiency. **Urogenital:** Discoloration of urine.

DIAGNOSTIC TEST INTERFERENCE Possibility of interference with *PSP excretion test* because of urine discoloration.

INTERACTIONS Drug: Decreased effect of ORAL ANTICOAGULANTS.

PHARMACOKINETICS Absorption: Minimal from GI tract. **Onset:** 6–12 h. **Metabolism:** In liver. **Elimination:** In feces and urine; excreted in breast milk.

Common adverse effects in *italic*, life-threatening effects <u>underlined</u>; generic names in **bold**; classifications in SMALL CAPS; ♣ Canadian drug name; ⦿ Prototype drug

251

NURSING IMPLICATIONS

Assessment & Drug Effects

- Monitor electrolyte balance if significant diarrhea occurs, especially with frail older adults.
- Monitor restoration of normal bowel function.

Patient & Family Education

- A single dose taken before retiring usually results in evacuation of soft stool 6–12 h later.
- Frequent or prolonged use of irritant cathartics disrupts normal reflex activity of colon and rectum and can lead to drug dependence for evacuation.
- See bisacodyl for additional information.

CASPOFUNGIN ⏎

(cas-po-fun′gin)

Cancidas

Classifications: ANTIBIOTIC; ECHINOCANDIN ANTIFUNGAL

Therapeutic: ANTIFUNGAL

Pregnancy Category: C

AVAILABILITY 50 mg and 70 mg powder for injection

ACTION & *THERAPEUTIC EFFECT* Caspofungin is an antifungal agent that inhibits the synthesis of an integral component of the fungal cell wall of susceptible species. *Interferes with reproduction and growth of susceptible fungi.*

USES Treatment of invasive aspergillosis in those refractory to or intolerant of other antifungal therapies; empirical therapy for presumed fungal infection with febrile neutropenia; treatment of candidemia and intra-abdominal abscesses, peritonitis, and pleural space infections due to *Candida*.

UNLABELED USES Treatment of esophageal candidiasis with or without oropharyngeal candidiasis (thrush).

CONTRAINDICATIONS Hypersensitivity to any component of this product; mannitol; not studied in patients with ESRF, or children younger than 18 y.

CAUTIOUS USE Patients with moderate hepatic insufficiency; cholestasis; pregnancy (category C); lactation.

ROUTE & DOSAGE

Invasive *Aspergillosis*, Empirical Therapy, *Candida*

Adult: **IV** 70 mg on day 1, then 50 mg daily thereafter

ADMINISTRATION

Intravenous

Allow vial to come to room temperature.

***PREPARE:* IV Infusion:** Reconstitute a 50 mg or 70 mg vial with 10.5 mL of NS, sterile water for injection, or bacteriostatic water for injection to yield 5 mg/mL and 7 mg/mL, respectively. Mix gently until clear. ▪ Withdraw the required dose of reconstituted solution and add to 250 mL of NS, 1/2NS, or 1/4NaCl, or LR. **DO NOT** use diluents or IV solutions containing dextrose.

***ADMINISTER:* IV Infusion:** Give slowly over at least 1 h. Do not co-infuse with any other medication.

***INCOMPATIBILITIES* Solution/additive:** Any **dextrose**-containing solution. Do not mix or co-infuse with any other medications.

▪ Reconstituted solution should be stored at or below 25° C (77° F) for 1 h prior to preparing the IV solution for infusion. ▪ Store IV solution for up to 24 h at or below 25° C (77° F) or 48 h at 2°–8° C (36°–46° F).

ADVERSE EFFECTS (≥1%) **Body as a Whole:** Anaphylaxis, chills, *injection site reaction,* sensation of warmth. **CNS:** Headache. **CV:** Sinus tachycardia. **GI:** *Nausea, vomiting,* diarrhea, abdominal pain. **Hematologic/Lymphatic:** *Phlebitis, thrombophlebitis,* vasculitis, anemia. **Hepatic:** Elevated liver enzymes. **Metabolic:** Anorexia, *hypokalemia.* **Musculoskeletal:** Pain, myalgia. **Respiratory:** <u>Acute respiratory distress syndrome,</u> dyspnea. **Skin:** Rash, facial swelling, pruritus.

INTERACTIONS Drug: Cyclosporine increases overall systematic exposure to caspofungin; inducers of drug clearance or mixed inducer/inhibitors (e.g., **carbamazepine, dexamethasone, efavirenz, nelfinavir, nevirapine, phenytoin, rifampin**) can decrease caspofungin levels; caspofungin decreases the overall systematic exposure to **tacrolimus.**

PHARMACOKINETICS Distribution: 97% protein bound. **Metabolism:** Liver and plasma to inactive metabolites. **Elimination:** Both in urine and feces. **Half-Life:** 9–11 h.

NURSING IMPLICATIONS

Assessment & Drug Effects
- Monitor for S&S of hypersensitivity during IV infusion; frequently monitor IV site for thrombophlebitis.
- Monitor for and report S&S of fluid retention (e.g., weight gain, swelling, peripheral edema), especially with known cardiovascular disease.
- Lab tests: Baseline and periodic LFTs; periodic kidney function tests, serum electrolytes, and CBC with differential, platelet count.
- Monitor blood levels of tacrolimus with concurrent therapy.

Patient & Family Education
- Report immediately any of the following: Facial swelling, wheezing, difficulty breathing or swallowing,

tightness in chest, rash, hives, itching, or sensation of warmth.

CEFACLOR ⊕

(sef′a-klor)

Ceclor ◆

Classifications: ANTIBIOTIC; SECOND-GENERATION CEPHALOSPORIN
Therapeutic: ANTIBIOTIC
Pregnancy Category: B

AVAILABILITY 250 mg, 500 mg capsules; 500 mg sustained release tablets; 125 mg/5 mL, 187 mg/5 mL, 250 mg/5 mL, 375 mg/5 mL suspension

ACTION & *THERAPEUTIC EFFECT*
Semisynthetic, second-generation oral cephalosporin antibiotic. Possibly more active than other oral cephalosporins against gram-negative bacilli, especially beta-lactamase-producing *Haemophilus influenzae,* including ampicillin-resistant strains and certain gram-positive strains. Preferentially binds to one or more of the penicillin-binding proteins (PBPs) located on cell walls of susceptible organisms. This inhibits third and final stage of bacterial cell wall synthesis, thus killing bacterium. *Effective in treating acute otitis media and acute sinusitis where causative agent is resistant to other antibiotics. Useful in treating respiratory and urinary tract infections.*

USES Treatment of otitis media and infections of upper and lower respiratory tract, urinary tract, and skin and skin structures caused by ampicillin-resistant *H. influenzae;* acute uncomplicated UTI.

CONTRAINDICATIONS Hypersensitivity to cephalosporins and related antibiotics.
CAUTIOUS USE History of sensitivity to penicillins or other drug allergies; GI disease, colitis;

Common adverse effects in *italic,* life-threatening effects <u>underlined</u>; generic names in **bold**; classifications in SMALL CAPS; ◆ Canadian drug name; ⊕ Prototype drug

253

markedly impaired renal function; older adults; coagulopathy; pregnancy (category B), lactation.

ROUTE & DOSAGE

Mild to Moderate Infections

Adult: **PO** 250–500 mg q8h, or Ceclor CD 250–500 mg/q12h
Child (older than 1 mo): **PO** 20–40 mg/kg/day divided q8h (max: 2 g/day)

ADMINISTRATION

Oral

- Give sustained release tablets with food to enhance absorption. Food does not affect absorption of capsules.
- Ensure that sustained release tablets are not chewed or crushed. They **must be** swallowed whole.
- After stock oral suspension is prepared, it should be kept refrigerated. Expiration date should appear on label. Discard unused portion after 14 days. Shake well before pouring.
- Store pulvules in tightly closed container unless otherwise directed.

ADVERSE EFFECTS (≥1%) **Body as a Whole:** Serum sickness-like reaction, eosinophilia, joint pain or swelling, fever, superinfections. **GI:** *Diarrhea,* nausea, vomiting, anorexia, pseudomembranous colitis (rare). **Skin:** Urticaria, pruritus, morbilliform eruptions.

DIAGNOSTIC TEST INTERFERENCE May produce positive *direct Coombs' test.* False-positive *urine glucose* determinations may result with use of *copper sulfate reduction methods* (e.g., *Clinitest* or *Benedict's reagent*) but not with *glucose oxidase* (enzymatic) *tests* such as *Clinistix, Diastix, TesTape.*

INTERACTIONS Drug: Avoid live vaccines.

PHARMACOKINETICS Absorption: Well absorbed; acid stable. **Peak:** 30–60 min. **Elimination:** 60% of dose eliminated renally in 8 h; crosses placenta; excreted in breast milk. **Half-Life:** 0.5–1 h.

NURSING IMPLICATIONS

Assessment & Drug Effects

- Determine previous hypersensitivity to cephalosporins, penicillins, and other drug allergies before therapy is initiated.
- Lab tests: Perform culture and sensitivity tests prior to and periodically during therapy.
- Report persistent diarrhea, as interruption of therapy may be necessary.
- Monitor for manifestations of drug hypersensitivity (see Appendix F). Discontinue drug and promptly report them if they appear.
- Monitor for manifestations of superinfection (see Appendix F). Promptly report their appearance.

Patient & Family Education

- Report promptly any signs or symptoms of superinfection (see Appendix F).
- Yogurt or buttermilk (if allowed) may serve as a prophylactic against intestinal superinfections by helping to maintain normal intestinal flora.

CEFADROXIL

(sef-a-drox′ill)

Classifications: ANTIBIOTIC; FIRST-GENERATION CEPHALOSPORIN
Therapeutic: ANTIBIOTIC
Prototype: Cefazolin
Pregnancy Category: B

AVAILABILITY 500 mg capsules; 1 g tablets; 125 mg/5 mL, 250 mg/5 mL, 500 mg/5 mL suspension

ACTION & *THERAPEUTIC EFFECT*
Semisynthetic, first-generation cephalosporin antibiotic. Bactericidal action (similar to that of penicillins): Drug penetrates bacterial cell wall, resists beta-lactamases, and inactivates enzymes essential to cell wall synthesis. *Active against organisms that liberate cephalosporinase and penicillinase (beta-lactamases). Effective in reducing signs and symptoms of urinary tract infections, bone and joint infections, skin and soft tissue infections, and pharyngitis.*

USES Primarily in treatment of urinary tract infections caused by *Escherichia coli, Proteus mirabilis,* and *Klebsiella* sp.; infections of skin and skin structures caused by *Staphylococci* and *Streptococci;* and for treatment of group A beta-hemolytic streptococcal pharyngitis and tonsillitis.

CONTRAINDICATIONS Hypersensitivity to cephalosporins.
CAUTIOUS USE Sensitivity to penicillins or other drug allergies; impaired renal function, older adults; GI disease, history of colitis, coagulopathy; pregnancy (category B).

ROUTE & DOSAGE

Uncomplicated Urinary Tract Infection

Adult: **PO** 1–2 g/day in 1–2 divided doses
Child: **PO** 30 mg/kg/day in 2 divided doses

Skin and Skin Structure Infections, Streptococcal Pharyngitis, or Tonsillitis

Adult: **PO** 1 g/day in 1–2 divided doses

Child: **PO** 30 mg/kg/day in 2 divided doses

Renal Impairment Dosage Adjustment

CrCl less than 25 mL/min:
Adult: **PO** 1 g q24h
Child: **PO** 15 mg/kg q24h

ADMINISTRATION
Oral

- Give with food or milk to reduce nausea. If nausea persists, termination of therapy may be necessary.
- Follow directions for mixing oral suspension found on drug label. Reconstituted suspension contains 125 mg or 250 mg cefadroxil per 5 mL.
- Shake suspension well before use; discard after 14 days.
- Store in tight container unless otherwise directed. Oral suspensions are stable for 14 days under refrigeration at 2°–8° C (36°–46° F). Avoid freezing. Note expiration date on label.

ADVERSE EFFECTS (≥1%) **Body as a Whole:** Hypersensitivity [rash, swollen eyelids (<u>angioedema</u>), pruritus, chills], superinfections. **GI:** Nausea, *diarrhea,* vomiting, heartburn, gastritis, bloating, abdominal cramps.

DIAGNOSTIC TEST INTERFERENCE
False-positive ***urine glucose*** determinations using ***copper sulfate reduction reagents,*** such as ***Clinitest*** or ***Benedict's reagent,*** but not with ***glucose oxidase tests*** (e.g., ***Clinistix, Diastix, TesTape***). **Cefadroxil-induced positive direct Coombs' test** may interfere with ***cross-matching procedures*** and ***hematologic studies.***

INTERACTIONS Drug: Probenecid decreases renal excretion of cefadroxil.

Common adverse effects in *italic*, life-threatening effects <u>underlined</u>; generic names in **bold**; classifications in SMALL CAPS; ♣ Canadian drug name; ☢ Prototype drug

255

PHARMACOKINETICS Absorption: Acid stable; rapidly absorbed from GI tract. **Peak:** 1 h. **Elimination:** 90% unchanged in urine within 8 h; bacterial inhibitory levels persist 20–22 h; crosses placenta; excreted in breast milk. **Half-Life:** 1–12 h.

NURSING IMPLICATIONS

Assessment & Drug Effects

- Determine previous hypersensitivity to cephalosporins, penicillins, and other drug allergies, before therapy is initiated.
- Lab tests: Perform culture and sensitivity testing prior to and periodically during therapy.
- Lab tests: Perform baseline and periodic renal function studies in patients with renal function impairment, and monitor I&O ratio and pattern.
- Monitor for manifestations of drug hypersensitivity (see Signs & Symptoms, Appendix F). Discontinue drug and promptly report them if they appear.
- Monitor for manifestations of superinfection (see Signs & Symptoms, Appendix F). Promptly report their appearance.

Patient & Family Education

- Report promptly the onset of rash, urticaria, pruritus, or fever, as the possibility of an allergic reaction is high, if you are allergic to penicillin.
- Take medication for the full course of therapy as directed by your prescriber.
- Report promptly S&S of superinfections (see Appendix F).

CEFAZOLIN SODIUM ⊕

(sef-a'zoe-lin)

Ancef

Classifications: ANTIBIOTIC; FIRST-GENERATION CEPHALOSPORIN

Therapeutic: ANTIBIOTIC
Pregnancy Category: B

AVAILABILITY 250 mg, 500 mg, 1 g injection

ACTION & *THERAPEUTIC EFFECT*

Semisynthetic, first-generation cephalosporin C with limited activity against gram-negative organisms. Bactericidal action: Preferentially binds to one or more of the penicillin-binding proteins (PBP) located on cell walls of susceptible organisms. This inhibits third and final stage of bacterial cell wall synthesis, thus killing the bacterium. *Effective treatment for bone and joint infections, biliary tract infections, endocarditis prophylaxis and treatment, respiratory tract and genital tract infections, septicemia and skin infections, and surgical prophylaxis.*

USES Severe infections of urinary and biliary tracts, skin, soft tissue, and bone, and for bacteremia and endocarditis caused by susceptible organisms; also perioperative prophylaxis in patients undergoing procedures associated with high risk of infection (e.g., open heart surgery).

CONTRAINDICATIONS Hypersensitivity to any cephalosporin and related antibiotics.

CAUTIOUS USE History of penicillin sensitivity, impaired renal function, patients on sodium restriction; coagulopathy; GI disease, colitis; pregnancy (category B).

ROUTE & DOSAGE

Moderate to Severe Infections

Adult: **IV/IM** 250 mg–2 g q8h, up to 2 g q4h (max: 12 g/day)
Child: **IV/IM** 25–100 mg/kg/day in 3–4 divided doses, up to 100 mg/kg/day (not to exceed adult doses)

Common adverse effects in *italic*, life-threatening effects underlined; generic names in **bold**; classifications in SMALL CAPS; ♦ Canadian drug name; ⊕ Prototype drug

Neonate: **IV** *Younger than 7 days, 40 mg/kg/day divided q12h; 7 days or older, 40–60 mg/kg/day divided q8–12h*

Surgical Prophylaxis

Adult: **IV/IM** 1–2 g 30–60 min before surgery, then 0.5–1 g q8h
Child: **IV/IM** 25–50 mg/kg 30–60 min before surgery, then q8h for 24 h

Renal Impairment Dosage Adjustment

CrCl less than 35 mL/min: Dose q12h; *10 mL/min:* Dose q24h

ADMINISTRATION

Intramuscular

- Preparation of IM solution: Reconstitute with sterile water for injection, bacteriostatic water for injection, or 0.9% sodium chloride injection.
- Reconstituted solutions are stable for 24 hr at room temperature and for 96 hr refrigerated.
- IM injections should be made deep into large muscle mass. Pain on injection is usually minimal. Rotate injection sites.

Intravenous

IV administration to neonates, infants, and children: Verify correct IV concentration and rate of infusion with prescriber.

PREPARE: Direct: Add 2 mL sterile water for injection to the 500 mg vial to yield 225 mg/mL, or add 2.5 mL to the 1 g vial to yield 330 mg/mL. Shake well to dissolve.
- Further dilute with 5 mL sterile water for injection. **Intermittent:** After initial vial reconstitution, add required dose to 50–100 mL of NS or D5W.

ADMINISTER: Direct/Intermittent: Infuse 1 g over 5 min or longer as

determined by the amount of solution. ■ The risk of IV site reactions may be reduced by proper dilution of IV solution, use of small bore IV needle in a large vein, and by rotating injection sites.

INCOMPATIBILITIES Solution/additive: AMINOGLYCOSIDES, **atracurium, bleomycin, cimetidine, clindamycin, lidocaine, ranitidine. Y-site:** Amiodarone, AMINOGLYCOSIDES, **amphotericin B cholesteryl complex, cisatracurium, hydromorphone, idarubicin, pentamidine, high dose vancomycin, vinorelbine.**

ADVERSE EFFECTS (≥1%) **Body as a Whole:** <u>Anaphylaxis</u>, fever, eosinophilia, superinfections, seizure (high doses in patients with renal insufficiency). **GI:** *Diarrhea,* anorexia, abdominal cramps. **Skin:** Maculopapular rash, urticaria.

DIAGNOSTIC TEST INTERFERENCE Because of cefazolin effect on the *direct Coombs' test,* transfusion *cross-matching procedures* and *hematologic studies* may be complicated. False-positive *urine glucose* determinations are possible with use of *copper sulfate tests* (e.g., *Clinitest* or *Benedict's reagent*) but not with *glucose oxidase tests* such as *TesTape, Diastix,* or *Clinistix.*

INTERACTIONS Drug: Probenecid decreases renal elimination of cefazolin.

PHARMACOKINETICS Peak: 1–2 h after IM; 5 min after IV. **Distribution:** Poor CNS penetration even with inflamed meninges; high concentrations in bile and in diseased bone; crosses placenta. **Elimination:** 70% unchanged in urine in 6 h; small amount excreted in breast milk. **Half-Life:** 90–130 min.

Common adverse effects in *italic*, life-threatening effects <u>underlined</u>; generic names in **bold**; classifications in SMALL CAPS; ♣ Canadian drug name; ⊕ Prototype drug

257

NURSING IMPLICATIONS

Assessment & Drug Effects

- Determine history of hypersensitivity to cephalosporins, penicillins, and other drugs, before therapy is initiated.
- Lab tests: Perform culture and sensitivity testing prior to and during therapy. Therapy may be initiated pending results.
- Monitor I&O rates and pattern: Be alert to changes in BUN, serum creatinine.
- Prompt attention should be given to onset of signs of hypersensitivity (see Appendix F).
- Promptly report the onset of diarrhea. Pseudomembranous colitis, a potentially life-threatening condition, starts with diarrhea.

Patient & Family Education

- Report promptly any signs or symptoms of superinfection (see Appendix F).
- Report signs of coagulation problems such as easy bruising and nosebleeds.

CEFDINIR

(cef′di-nir)

Omnicef

Classifications: ANTIBIOTIC; THIRD-GENERATION CEPHALOSPORIN

Therapeutic: ANTIBIOTIC

Prototype: Cefotaxime sodium

Pregnancy Category: B

AVAILABILITY 300 mg capsules; 125 mg/5 mL, 250 mg/5 mL suspension

ACTION & *THERAPEUTIC EFFECT*
Broad-spectrum semisynthetic third-generation beta-lactamase cephalosporin antibiotic. *Effective against a wide variety of gram-positive and gram-negative bacteria.*

USES Community-acquired pneumonia, acute exacerbations of chronic bronchitis, acute maxillary sinusitis, pharyngitis, tonsillitis, uncomplicated skin infections, bacterial otitis media.

CONTRAINDICATIONS Hypersensitivity to cefdinir and other cephalosporins.

CAUTIOUS USE Hypersensitivity to penicillins, penicillin derivatives; renal impairment; ulcerative colitis or antibiotic-induced colitis; bleeding disorders; GI disorders; liver or kidney disease; pregnancy (category B), lactation. Safety and efficacy in neonates and infants younger than 6 mo old not established.

ROUTE & DOSAGE

Community-Acquired Pneumonia, Skin Infections

Adult: **PO** 300 mg q12h × 10 days
Child/Infant (6 mo–12 y): **PO** 7 mg/kg q12h × 10 days

Chronic Bronchitis, Maxillary Sinusitis, Pharyngitis, Tonsillitis

Adult/Adolescent: **PO** 600 mg q24h or 300 mg q12h × 10 days
Child/Infant (6 mo–12 y): **PO** 14 mg/kg q24h or 7 mg/kg q12h × 10 days

Renal Impairment Dosage Adjustment

CrCl less than 10 mL/min: 300 mg daily

Hemodialysis Dosage Adjustment
300 mg or 7 mg/kg dose PO every other day; dose given at the end of each session

ADMINISTRATION

Oral

- Do not give within 2 h of aluminum- or magnesium-containing antacids or iron supplements.

- Reconstitute oral suspension to 125 mg/mL by adding water (38 mL to 60 mL bottle or 63 mL to 100 mL bottle). Shake well before each use.
- Store in tightly closed container. Discard after 10 days.

ADVERSE EFFECTS (≥1%) **GI:** *Diarrhea*, nausea, abdominal pain. **Metabolic:** Increased GGT, increased urine protein, hematuria. **CNS:** Headache. **Skin:** Rash, cutaneous moniliasis. **Urogenital:** Vaginal moniliasis, vaginitis.

DIAGNOSTIC TEST INTERFERENCE False positive for *ketones* or *glucose* in urine using *nitroprusside* or *Clinitest.*

INTERACTIONS Drug: ANTACIDS should be taken at least 2 h before or after cefdinir; **probenecid** prolongs cefdinir elimination; **iron** decreases absorption.

PHARMACOKINETICS Absorption: 16–25% bioavailability. **Peak:** 2–4 h. **Distribution:** 60–70% protein bound; penetrates sinus tissue, blister fluid, lung tissue, middle ear fluid. **Metabolism:** Hepatic. **Elimination:** In urine. **Half-Life:** 1.6 h.

NURSING IMPLICATIONS

Assessment & Drug Effects

- Determine previous hypersensitivity to cephalosporins, penicillins, and other drug allergies before therapy is initiated.
- Carefully monitor for and immediately report S&S of: Hypersensitivity, superinfection, or pseudomembranous colitis (see Appendix F).
- Discontinue drug and notify prescriber if seizures associated with drug therapy occur.

Patient & Family Education

- Allow a minimum of 2 h between cefdinir and antacids containing aluminum or magnesium, or drugs containing iron.
- Immediately contact prescriber if a rash, diarrhea, or new infection (e.g., yeast infection) develops.

CEFDITOREN PIVOXIL
(cef-di-tor′en)

Spectracef
Classifications: ANTIBIOTIC; THIRD-GENERATION CEPHALOSPORIN
Therapeutic: ANTIBIOTIC
Prototype: Cefotaxime sodium
Pregnancy Category: B

AVAILABILITY 200 mg, 400 mg tablets

ACTION & *THERAPEUTIC EFFECT*
Semisynthetic cephalosporin with bactericidal activity resulting from the inhibition of cell wall synthesis through an affinity for penicillin-binding proteins (PBPs). Stable in the presence of a variety of bacterial beta-lactamase enzymes, including penicillinases and some cephalosporinases. *Antibacterial activity is effective against both aerobic gram-positive and aerobic gram-negative bacteria.*

USES Acute exacerbation of bacterial chronic bronchitis, pharyngitis, tonsillitis, pneumonia, uncomplicated skin and skin-structure infections.

CONTRAINDICATIONS Known allergy to cephalosporins or any of the components of cefditoren; carnitine deficiency; milk protein hypersensitivity.
CAUTIOUS USE History of hypersensitivity to penicillin or other drugs; renal or hepatic impairment; poor nutritional status; coagulopathy; diabetes mellitus; colitis, GI disease; older adults; concurrent anticoagulant therapy; pregnancy (category B), lactation. Safety and efficacy in children younger than 12 y are not established.

Common adverse effects in *italic*, life-threatening effects <u>underlined</u>; generic names in **bold**; classifications in SMALL CAPS; ♣ Canadian drug name; ⊙ Prototype drug

ROUTE & DOSAGE

Chronic Bronchitis/Pneumonia

Adult/Adolescent: PO 400 mg b.i.d. × 10–14 days

Pharyngitis, Tonsillitis, Skin Infections

Adult: PO 200 mg b.i.d. × 10 days

Uncomplicated Skin/Soft Tissue Infections

Adult/Adolescent: PO 200–400 mg b.i.d. × 10 days

Renal Impairment Dosage Adjustment

CrCl 30–49 mL/min: 200 mg b.i.d.; *less than 30 mL/min:* 200 mg daily

ADMINISTRATION

Oral

- Give with food to enhance absorption.
- Do not give within 2 h of an antacid or H₂-receptor antagonist (such as cimetidine).
- Store at 15°–30° C (58°–86° F). Protect from light and moisture.

ADVERSE EFFECTS (≥1%) **GI:** *Diarrhea,* nausea, abdominal pain, dyspepsia, vomiting. **Hematologic:** Anemia, leukocytosis. **CNS:** Headache. **Urogenital:** Vaginal moniliasis, hematuria.

INTERACTIONS Drug: ANTACIDS, H₂-RECEPTOR ANTAGONISTS may decrease absorption; **probenecid** will decrease elimination.

PHARMACOKINETICS Absorption: 14% reaches systemic circulation. **Distribution:** 88% protein bound, distributes into blister fluid, tonsils. **Metabolism:** Hydrolyzed. **Elimination:** Primarily in urine. **Half-Life:** 1.6 h.

NURSING IMPLICATIONS

Assessment & Drug Effects

- Obtain history of hypersensitivity to cephalosporins, penicillins, and other drug allergies.
- Lab tests: Baseline C&S tests recommended prior to and periodically during therapy. Baseline and periodic studies of kidney function; frequent PT determinations in patients at risk for increased prothrombin time; as indicated, Hct and Hgb, CBC with differential, urinalysis, serum electrolytes, and liver enzymes.
- Monitor for manifestations of drug hypersensitivity (see Appendix F). Withhold drug and report promptly to prescriber if they appear.
- Monitor for and report promptly manifestations of superinfection (see Appendix F), especially diarrhea. Diarrhea may indicate a change in intestinal flora and development of enterocolitis.
- Monitor for and report immediately signs of seizure activity or loss of seizure control.

Patient & Family Education

- Do not take within 2 h of antacids or other drugs used to reduce stomach acids.
- Discontinue drug and report to prescriber signs of an allergic reaction (e.g., rash, urticaria, pruritus, fever).
- Report promptly S&S of superinfection (see Appendix F), especially unexplained diarrhea. Antibiotic-associated colitis is a superinfection that may occur in 4–9 days or as long as 6 wk after drug is discontinued.
- Use daily yogurt or buttermilk (if allowed) as a prophylactic against intestinal superinfections.

CEFEPIME HYDROCHLORIDE

(cef'e-peem)

Common adverse effects in *italic*, life-threatening effects underlined; generic names in **bold**; classifications in SMALL CAPS; ◆ Canadian drug name; 🚱 Prototype drug

Maxipime
Classifications: ANTIBIOTIC; FOURTH-GENERATION CEPHALOSPORIN
Therapeutic: ANTIBIOTIC; CEPHALOSPORIN
Prototype: Cefotaxime sodium
Pregnancy Category: B

AVAILABILITY 1 g, 2 g vials

ACTION & *THERAPEUTIC EFFECT*
Cefepime, considered to be a fourth-generation cephalosporin antibiotic that preferentially binds to one or more of the penicillin-binding proteins (PBPs) located on cell walls of susceptible organisms. This inhibits the third and final stage of bacterial cell wall synthesis, thus killing the bacteria (bactericidal). *Cefepime is similar to third-generation cephalosporins with respect to broad gram-negative coverage; however, cefepime has broader gram-positive coverage than third-generation cephalosporins. It is highly resistant to hydrolysis by most beta-lactamase bacteria.*

USES Uncomplicated and complicated UTI, skin and soft tissue infections, pneumonia. Empiric monotherapy for febrile neutropenic patients.

CONTRAINDICATIONS Hypersensitivity to cefepime, other cephalosporins, severe reaction to penicillins, or other beta-lactam antibiotics.
CAUTIOUS USE Patients with history of GI disease, particularly colitis; renal insufficiency; pregnancy (category B), lactation.

ROUTE & DOSAGE

Mild to Moderate Infections
Adult: **IV/IM** 0.5–1g q12h for 7–10 days

Moderate to Severe Infections
Adult/Adolescent/Child (over 40 kg): **IV** 1–2g q12h for 10 days
Child (up to 40 kg)/Infant (over 2 mo): **IV** 50 mg/kg q12h for 10 days

Febrile Neutropenia
Adult/Adolescent/Child (over 40 kg): **IV** 2 g q8h for 7 days or until resolution of neutropenia
Child (under 40 kg)/Infant (over 2 mo): **IV** 50 mg/kg q8h until resolution of neutropenia

UTI
Adult/Adolescent/Child (over 40 kg): **IV/IM** 0.5–1g q12h × 7–10 days
Child (up to 40 kg)/Infant (over 2 mo): **IV/IM** 50 mg/kg q12h × 7–10 days

Community-Acquired Pneumonia
Adult: **IV** 1–2g q12h

Renal Impairment Dosage Adjustment
CrCl 30–60 mL/min: Dose q24h; *11–29 mL/min:* Give 50% of normal dose q24h; *less than 10 mL/min:* 250–500 mg q24h

Hemodialysis Dosage Adjustment
Administer dose after dialysis

ADMINISTRATION
Intramuscular
- Reconstitute 500-mg vial and 1-g vial, respectively, with 1.3 or 2.4 mL of one of the following: Sterile water for injection, 0.9% NaCl injection, bacteriostatic water for injection with parabens or benzyl alcohol, or other compatible solution.

Common adverse effects in *italic*, life-threatening effects <u>underlined</u>; generic names in **bold**; classifications in SMALL CAPS; ♣ Canadian drug name; ⊙ Prototype drug

261

C

Intravenous

PREPARE: **Intermittent:** Dilute with 50–100 mL of one of the following: NS, D5W, D5/NS or other compatible solution.

ADMINISTER: **Intermittent:** Infuse over 30 min; with Y-type administration set, discontinue other compatible solutions while infusing cefepime.

INCOMPATIBILITIES **Solution/additive:** AMINOGLYCOSIDES, **aminophylline, metronidazole. Y-site:** **Acetylcysteine, acyclovir, amphotericin B, amphotericin B cholesteryl complex, amphotericin B liposomal, caspofungin, chlordiazepoxide, chlorpromazine, cimetidine, ciprofloxacin, cisplatin, dacarbazine, daunorubicin, diazepam, diltiazem, diphenhydramine, dobutamine, dopamine, doxorubicin, droperidol, enalaprilat, epirubicin, erythromycin, etoposide, famotidine, filgrastim, floxuridine, ganciclovir, gemcitabine, haloperidol, hydroxyzine, idarubicin, ifosfamide, irinotecan, lansoprazole, magnesium sulfate, mannitol, mechlorethamine, meperidine, metoclopramide, mitomycin, mitoxantrone, morphine, nalbuphine, nesiritide, nicardipine, ofloxacin, ondansetron, oxaliplatin, pantoprazole, premetrexed, phenytoin, plicamycin, prochlorperazine, promethazine, quinupristin-dalfopristin, streptozocin, tacrolimus, temocillin, theophylline, vancomycin, vecuronium, vinblastine, vincristine, voriconazole.**

■ Store reconstituted solution at 20°–25° C (68°–77° F) for 24 h or in refrigerator at 2°–8° C (36°–46° F) for 7 days. Protect from light.

ADVERSE EFFECTS (≥1%) **Body as a Whole:** Eosinophilia. **GI:** Antibiotic-associated colitis, diarrhea, nausea, oral moniliasis, vomiting, elevated liver function tests (ALT, AST). **CNS:** Headache, fever. **Skin:** Phlebitis, pain, inflammation, rash, pruritus, urticaria. **Urogenital:** Vaginitis.

DIAGNOSTIC TEST INTERFERENCE Positive *Coombs' test* without hemolysis. May cause false-positive *urine glucose test* with *Clinitest.*

INTERACTIONS Drug: AMINOGLYCOSIDES may increase risk of nephrotoxicity and have additive/synergistic effects. May decrease efficacy of ORAL CONTRACEPTIVES. **Probenecid** may increase levels.

PHARMACOKINETICS Absorption: Well absorbed after IM administration; serum levels significantly lower than after equivalent IV dose. **Distribution:** Widely distributed, may cross inflamed meninges; crosses placenta, secreted into breast milk. **Metabolism:** In liver. **Elimination:** In urine. **Half-Life:** 2 h.

NURSING IMPLICATIONS

Assessment & Drug Effects

■ Determine history of hypersensitivity reactions to cephalosporins, penicillins, or other drugs before therapy is initiated.

■ Lab tests: Perform culture and sensitivity tests before initiation of therapy.

■ Monitor for S&S of hypersensitivity (see Appendix F). Report their appearance promptly and discontinue drug.

■ Monitor for S&S of superinfection or pseudomembranous colitis (see Appendix F); immediately report either to prescriber.

Common adverse effects in *italic*, life-threatening effects underlined; generic names in **bold**; classifications in SMALL CAPS; ✤ Canadian drug name; ⊘ Prototype drug

- With concurrent high-dose aminoglycoside therapy, closely monitor for nephrotoxicity and ototoxicity.

Patient & Family Education
- Promptly report S&S of hypersensitivity (e.g., rash) or superinfection, especially unexplained diarrhea (see Appendix F).

CEFIXIME
(ce-fix'ime)

Suprax

Classifications: BETA-LACTAM ANTIBIOTIC; THIRD-GENERATION CEPHALOSPORIN
Therapeutic: ANTIBIOTIC
Prototype: Cefotaxime sodium
Pregnancy Category: B

AVAILABILITY 100 mg/5 mL, 200 mg/5 mL suspension; 400 mg tablet

ACTION & *THERAPEUTIC EFFECT*
Cefixime is a third-generation cephalosporin. As a beta-lactam antibiotic like the penicillins, it is mainly bactericidal. It inhibits the third and final stage of bacterial cell wall synthesis by preferentially binding to specific penicillin-binding proteins (PBPs) located inside the bacterial cell wall. *Cefixime is highly stable in the presence of beta-lactamases (penicillinases and cephalosporinases) and therefore has excellent activity against a wide range of gram-negative bacteria. It is bactericidal against susceptible bacteria.*

USES Uncomplicated UTI, otitis media, pharyngitis, tonsillitis, and bronchitis.

CONTRAINDICATIONS Patients with known allergy to the cephalosporin group of antibiotics, severe reaction to penicillin.

CAUTIOUS USE Allergy to penicillin, history of colitis, renal insufficiency, GI disease, coagulopathy, pregnancy (category B), lactation. Safety and effectiveness in infants younger than 6 mo have not been established.

ROUTE & DOSAGE

Infection
Adult: PO 400 mg/day in 1–2 divided doses
Child: PO 8 mg/kg/day in 1–2 divided doses

Renal Impairment Dosage Adjustment
CrCl 21–60 mL/min: Give 75% of dose; less than 20 mL/min: Give 50% of dose

ADMINISTRATION

Oral
- Do not substitute tablets for liquid in treatment of otitis media because of lack of bioequivalence.
- After reconstitution, suspension may be kept for 14 days at room temperature or refrigerated. Store away from heat and light. Keep tightly closed and shake well before using.

ADVERSE EFFECTS (≥1%) **GI:** *Diarrhea,* loose stools, nausea, vomiting, dyspepsia, flatulence. **CNS:** Drug fever, headache, dizziness. **Skin:** Rash, pruritus. **Urogenital:** Vaginitis, genital pruritus.

INTERACTIONS Drug: AMINOGLYCOSIDES may increase risk of nephrotoxicity and have additive/synergistic effects. May decrease efficacy of ORAL CONTRACEPTIVES. **Probenecid** may increase levels.

PHARMACOKINETICS Absorption: 40–50% from GI tract. **Peak:** 2–6 h.

Common adverse effects in *italic*, life-threatening effects underlined; generic names in **bold;** classifications in SMALL CAPS; ♦ Canadian drug name; ☻ Prototype drug

263

Distribution: Into breast milk. **Elimination:** 50% in urine, 50% in bile. **Half-Life:** 3–4 h.

NURSING IMPLICATIONS

Assessment & Drug Effects

- Determine previous hypersensitivity reactions to cephalosporins, penicillins, and history of other allergies, particularly to drugs prior to initiation of therapy.
- Lab tests: Perform culture and sensitivity tests prior to initiation of therapy.
- Monitor for superinfections (see Appendix F) caused by overgrowth of nonsusceptible organisms, particularly during prolonged use.
- Monitor I&O rates and pattern: Nephrotoxicity occurs more frequently in patients older than 50 y, with impaired renal function, in the debilitated, and in patients receiving high doses or other nephrotoxic drugs.
- Carefully monitor anyone with a history of allergies. Report manifestations of hypersensitivity (see Appendix F).
- Promptly report loose stools or diarrhea, which may indicate pseudomembranous colitis (see Appendix F). Discontinuation of drug may be necessary.

Patient & Family Education

- Report loose stools or diarrhea during drug therapy and for several weeks after. Older adult patients are especially susceptible to pseudomembranous colitis.

CEFOTAXIME SODIUM ⓟ
(sef-oh-taks′eem)

Claforan
Classifications: BETA-LACTAM ANTIBIOTIC; THIRD-GENERATION CEPHALOSPORIN

Therapeutic: ANTIBIOTIC
Pregnancy Category: B

AVAILABILITY 500 mg, 1 g, 2 g injection

ACTION & *THERAPEUTIC EFFECT*
Broad-spectrum semi-synthetic third-generation cephalosporin antibiotic. Preferentially binds to one or more of the penicillin-binding proteins (PBP) located on cell walls of susceptible organisms. This inhibits third and final stage of bacterial cell wall synthesis, thus killing the bacteria. *Generally active against a wide variety of gram-negative bacteria including most of the Enterobacteriaceae. Also active against some organisms resistant to first- and second-generation cephalosporins, and aminoglycoside antibiotics and penicillins.*

USES Serious infections of lower respiratory tract, skin and skin structures, bones and joints, CNS (including meningitis and ventriculitis), gynecologic and GU tract infections, including uncomplicated gonococcal infections caused by penicillinase-producing *Neisseria gonorrhoeae* (PPNG). Also used to treat bacteremia or septicemia, intra-abdominal infections, and for perioperative prophylaxis.
UNLABELED USES Treatment of disseminated gonococcal infections (gonococcal arthritis-dermatitis syndrome) and as drug of choice for gonococcal ophthalmia caused by PPNG in adults, children, and neonates.

CONTRAINDICATIONS Hypersensitivity to cefotaxime, cephalosporins and other beta-lactam antibiotics.
CAUTIOUS USE History of Type I hypersensitivity reactions to penicillin; history of allergy to other beta-lactam; antibiotics; coagulopathy;

renal impairment; history of colitis or other GI disease; pregnancy (category B).

ROUTE & DOSAGE

Moderate to Severe Infections

Adult: **IV/IM** 1–2 g q8–12h, up to 2 g q4h (max: 12 g/day)
Child: **IV/IM** 1 wk or younger, 50 mg/kg q12h; 1–4 wk, 50 g/kg/q8h; 1 mo–12 y, 50–200 mg/kg/day divided q4–8h (max: 12 g/24 h)

Disseminated Gonorrhea

Adult: **IV** 1 g q8h

Surgical Prophylaxis

Adult: **IV/IM** 1 g 30–90 min before surgery

Renal Impairment Dosage Adjustment

CrCl less than 20 mL/min: Give ½ normal dose

Hemodialysis Dosage Adjustment

Supplemental dose may be needed

ADMINISTRATION

Intramuscular

- Add 3 mL sterile water for injection or bacteriostatic water for injection to vial containing 1 g drug, providing a solution of approximately 300 mg cefotaxime/mL.
- Administer IM injection deeply into large muscle mass (e.g., upper outer quadrant of gluteus maximus). Aspirate to avoid inadvertent injection into blood vessel. If IM dose is 2 g, divide dose and administer into 2 different sites.

Intravenous

IV administration to neonates, infants, and children: Verify correct IV concentration and rate of infusion with prescriber.

- Do not admix cefotaxime with sodium bicarbonate or any fluid with a pH greater than 7.5.
- Risk of phlebitis may be reduced by use of a small needle in a large vein.

PREPARE: **Direct:** Add 10 mL diluent to vial with 1 or 2 g drug providing a solution containing 95 or 180 mg/mL, respectively. **Intermittent:** To 1 or 2 g drug add 50 or 100 mL D5W, NS, D5/NS, D5/.45% NaCl, LR, or other compatible diluent. **Continuous:** Dilute in 500–1000 mL compatible IV solution.

ADMINISTER: **Direct:** Give over 3–5 min. **Intermittent:** Give over 20–30 min, preferably via butterfly or scalp vein-type needles. **Continuous:** Infuse over 6–24 h.

INCOMPATIBILITIES **Solution/additive:** AMINOGLYCOSIDES, **aminophylline. Y-site:** Allopurinol, azithromycin, cisatracurium, filgrastim, fluconazole, gemcitabine, hetastarch, pentamidine, vancomycin.

- Protect from excessive light. Reconstituted solutions may be stored in original containers for 24 h at room temperature; for 10 days under refrigeration at or below 5° C (41° F); or for at least 13 wk in frozen state.

ADVERSE EFFECTS (≥1%) **Body as a Whole:** Fever, nocturnal perspiration, inflammatory reaction at IV site, phlebitis, thrombophlebitis; pain, induration, and tenderness at IM site, superinfections. **GI:** Nausea, vomiting, *diarrhea*, abdominal pain, colitis, <u>pseudomembranous colitis</u>, anorexia. **Metabolic:** Transient increases in serum AST, ALT, LDH, bilirubin, alkaline phosphatase concentrations. **Skin:** Rash, pruritus.

Common adverse effects in *italic*, life-threatening effects <u>underlined</u>; generic names in **bold**; classifications in SMALL CAPS; ♣ Canadian drug name; ⊙ Prototype drug

265

DIAGNOSTIC TEST INTERFERENCE
May cause falsely elevated **serum** or **urine creatinine** values **(Jaffe reaction)**. Positive **direct antiglobulin (Coombs') test** results may interfere with **hematologic studies** and **cross-matching** procedures.

INTERACTIONS Drug: Probenecid decreases renal elimination; **alcohol** produces disulfiram reaction.

PHARMACOKINETICS Peak: 30 min after **IM**; 5 min after **IV**. **Distribution:** CNS penetration except with inflamed meninges; also penetrates aqueous humor, ascitic and prostatic fluids; crosses placenta. **Metabolism:** In liver to active metabolites. **Elimination:** 50–60% unchanged in urine in 24 h; small amount excreted in breast milk. **Half-Life:** 1 h.

NURSING IMPLICATIONS

Assessment & Drug Effects

- Determine previous hypersensitivity reactions to cephalosporins and penicillins, and history of other allergies, particularly to drugs, before therapy is initiated.
- Lab tests: Perform culture and sensitivity tests before initiation of therapy. Serum creatinine, creatinine clearance, BUN should be evaluated at regular intervals during therapy and for several months after drug has been discontinued. Perform periodic hematologic studies (including PT and/or PTT) and evaluation of hepatic functions with high doses or prolonged therapy.
- Monitor I&O rates and patterns, especially with higher doses or concurrent aminoglycoside therapy. Report significant changes in I&O.
- Superinfection due to overgrowth of nonsusceptible organisms may occur, particularly with prolonged therapy.

- Report onset of diarrhea promptly. Check for fever. If diarrhea is mild, discontinuation of cefotaxime may be sufficient.
- If diarrhea is severe, suspect antibiotic-associated pseudomembranous colitis, a life-threatening superinfection (may occur in 4–9 days or as long as 6 wk after cephalosporin therapy is discontinued).

Patient & Family Education

- Report any early signs or symptoms of superinfection promptly. Superinfections caused by overgrowth of nonsusceptible organisms may occur, particularly during prolonged use.
- Yogurt or buttermilk, 120 mL (4 oz) of either (if allowed), may serve as a prophylactic against intestinal superinfection by helping to maintain normal intestinal flora.
- Report loose stools or diarrhea.

CEFOTETAN DISODIUM
(sef'oh-tee-tan)

Cefotan
Classifications: ANTIBIOTIC; SECOND-GENERATION CEPHALOSPORIN
Therapeutic: ANTIBIOTIC
Prototype: Cefotaxime sodium
Pregnancy Category: B

AVAILABILITY 1 g, 2 g, 10 g injection

ACTION & THERAPEUTIC EFFECT Semisynthetic beta-lactam antibiotic, classified as a second-generation cephalosporin. Preferentially binds to one or more of the penicillin-binding proteins (PBP) located on cell walls of susceptible organisms. This inhibits third and final stage of bacterial cell wall synthesis, thus killing the bacterium. *Generally less active against susceptible*

Staphylococci than first-generation cephalosporins, but has broad spectrum of activity against gram-negative bacteria when compared to first- and second-generation cephalosporins. It also shows moderate activity against gram-positive organisms. It is active against the Enterobacteriaceae and anaerobes.

USES Infections caused by susceptible organisms in urinary tract, lower respiratory tract, skin and skin structures, bones and joints, gynecologic tract; also intra-abdominal infections, bacteremia, and perioperative prophylaxis.

CONTRAINDICATIONS Hypersensitivity to cephalosporins and related beta-lactam antibiotics.

CAUTIOUS USE Preexisting coagulopathy; colitis, GI disease; renal impairment; pregnancy (category B); lactation.

ROUTE & DOSAGE

Moderate to Severe Infections

Adult: **IV/IM** 1–2 g q12h

UTI

Adult: **IV** 500 mg q12h or 1–4 g/day

Surgical Prophylaxis

Adult/Adolescent: **IV/IM** 1–2 g 30–60 min before surgery

Renal Impairment Dosage Adjustment

CrCl greater than 30 mL/min: Regular dose q12h; *10–30 mL/min:* Regular dose q24h; *less than 10 mL/min:* Regular dose q48h

Hemodialysis Dosage Adjustment

Give ¼ dose q24h on days between sessions, ½ dose on day of dialysis

ADMINISTRATION

Intramuscular

- For IM reconstitution (follow manufacturer's directions for selection of diluent), add 2 mL diluent to 1 g vial; yields approximately 2.4 mL (375 mg/mL).
- For IM administration, inject well into body of large muscle such as upper outer quadrant of buttock (gluteus maximus).

Intravenous

IV administration to infants and children: Verify correct IV concentration and rate of infusion with prescriber.

PREPARE: Direct: Dilute each 1 g with 10 mL of sterile water for injection. **Intermittent:** Dilute each 1 g with 50–100 mL of D5W or NS.

ADMINISTER: Direct: Give over 3–5 min. **Intermittent:** Give a single dose over 30 min. ▪ For IV infusion, solution may be given for longer period of time through tubing system through which other IV solutions are being given.

INCOMPATIBILITIES Solution/additive: AMINOGLYCOSIDES, **heparin, promethazine,** TETRACYCLINES. **Y-site:** AMINOGLYCOSIDES, **cistracurium, lansoprazole, pemetrexed, promethazine, vancomycin, vinorelbine.**

▪ Protect sterile powder from light; store at or below 22° C (71.6° F); remains stable 24 mo after date of manufacture. May darken with age, but potency is unaffected. ▪ Reconstituted solutions: Stable for 24 h at 25° C (77° F); 96 h when refrigerated at 5° C (41° F); or at least 1 wk when frozen at –20° C (–4° F).

ADVERSE EFFECTS (≥1%) **Body as a Whole:** Fever, chills, injection site pain, inflammation, disulfiram-like reaction. **GI:** Nausea, vomiting, *diarrhea*, abdominal pain, antibiotic-associated

Common adverse effects in *italic*, life-threatening effects underlined; generic names in **bold**; classifications in SMALL CAPS; ✚ Canadian drug name; ⊘ Prototype drug

267

colitis. **Hematologic:** <u>Thrombocytopenia</u>, prolongation of bleeding time or prothrombin time. **Skin:** Rash, pruritus.

DIAGNOSTIC TEST INTERFERENCE
May cause falsely elevated *serum* or *urine creatinine* values *(Jaffe reaction)*. False-positive reactions for *urine glucose* have not been reported using *copper sulfate reduction methods* (e.g., *Benedict's, Clinitest*); however, since it has occurred with other cephalosporins, it may be advisable to use *glucose oxidase tests (Clinistix, TesTape, Diastix)*. Positive *direct antiglobulin (Coombs') test* results may interfere with *hematologic studies* and *cross-matching* procedures.

INTERACTIONS Drug: Probenecid decreases renal elimination of cefotetan; **alcohol** produces disulfiram reaction; **chloramphenicol** may effect therapeutic activity.

PHARMACOKINETICS Peak: 1.5–3 h after IM. **Distribution:** Poor CNS penetration; widely distributed to body tissues and fluids, including bile, sputum, prostatic and peritoneal fluids; crosses placenta. **Elimination:** 51–81% unchanged in urine; 20% in bile; small amount in breast milk. **Half-Life:** 180–270 min.

NURSING IMPLICATIONS
Assessment & Drug Effects
- Determine history of hypersensitivity to cephalosporins and penicillins, and other drug allergies, before therapy begins.
- Lab tests: Perform culture and sensitivity studies before initiation of therapy. Perform periodic hematologic studies (including PT/INR and PTT) and evaluation of renal function, especially if cefotetan dose is high or if therapy is prolonged in order to recognize

symptoms of nephrotoxicity and ototoxicity (see Appendix F).
- Report onset of loose stools or diarrhea. If diarrhea is severe, suspect pseudomembranous colitis (see Appendix F) caused by *Clostridium difficile*. Check temperature. Report fever and severe diarrhea to prescriber; drug should be discontinued.

Patient & Family Education
- Report promptly S&S of superinfection (see Appendix F).
- Report loose stools or diarrhea.

CEFOXITIN SODIUM
(se-fox'i-tin)
Mefoxin
Classifications: ANTIBIOTIC; SECOND-GENERATION CEPHALOSPORIN
Therapeutic: ANTIBIOTIC
Prototype: Cefaclor
Pregnancy Category: B

AVAILABILITY 1 g, 2 g injection

ACTION & THERAPEUTIC EFFECT
Semisynthetic, broad-spectrum beta-lactam antibiotic classified as second-generation cephalosporin; structurally and pharmacologically related to cephalosporins and penicillins. Preferentially binds to one or more of the penicillin-binding proteins (PBP) located on cell walls of susceptible organisms, thus making it bactericidal. *It shows enhanced activity against a wide variety of gram-negative organisms and is effective for mixed aerobic-anaerobic infections.*

USES Infections caused by susceptible organisms in the lower respiratory tract, urinary tract, skin and skin structures, bones and joints; also intra-abdominal endocarditis, gynecologic infections, septicemia, uncomplicated gonorrhea, and peri-operative prophylaxis in prosthetic arthroplasty or cardiovascular surgery.

CONTRAINDICATIONS Hypersensitivity to cephalosporins and related antibiotics.

CAUTIOUS USE History of sensitivity to penicillin or other allergies, particularly to drugs; impaired renal function; coagulopathy; GI disease, colitis; pregnancy (category B).

ROUTE & DOSAGE

Moderate to Severe Infections

Adult: **IV/IM** 1–2 g q6–8h, up to 12 g/day
Child (older than 3 mo): **IV/IM** 80–160 mg/kg/day in 4–6 divided doses (max: 12 g/day)

Surgical Prophylaxis

Adult: **IV/IM** 2 g 30–60 min before surgery, then 2 g q6h for 24 h
Child: **IV/IM** 30–40 mg/kg 30–60 min before surgery, then 30–40 mg q6h for 24 h

Cesarean Surgery

Adult: **IV/IM** 2 g after clamping umbilical cord

Renal Impairment Dosage Adjustment

CrCl 30–50 mL/min: 1–2 g q8–12h; *10–29 mL/min:* 1–2 g q12–24h; *5–9 mL/min:* 0.5–1 g q12–24h; *greater than 5 mL/min:* 0.5–1 g q24–48h

Hemodialysis Dosage Adjustment

Dose of 1–2 g post dialysis

ADMINISTRATION

Intramuscular

- Reconstitute each 1 g with 2 mL sterile water for injection or 0.5 or 1% lidocaine hydrochloride (without epinephrine), used to reduce discomfort of IM injection.

After reconstitution for IM use, shake vial and allow solution to stand until it becomes clear.
- Administer IM injections deep into large muscle mass such as upper outer quadrant of gluteus maximus. Aspirate before injecting drug. Rotate injection sites.

Intravenous

IV administration to neonates, infants and children: Verify correct IV concentration and rate of infusion/injection with prescriber.

PREPARE: **Direct:** Reconstitute each 1 g with 10 mL sterile water, D5W, or NS. **Intermittent:** Following reconstitution, dilute 1–2 g in 50–100 mL of D5W or NS. **Continuous:** Dilute large doses in 1000 mL of D5W or NS.

ADMINISTER: **Direct:** Give over 3–5 min. **Intermittent:** Give over 15 min. **Continuous:** Give at a rate determined by the volume of solution. ▪ Reconstituted solution may become discolored (usually light yellow to amber) if exposed to high temperatures; however, potency is not affected. ▪ Solution may be cloudy immediately after reconstitution; let stand and it will clear.

INCOMPATIBILITIES **Solution/additive:** AMINOGLYCOSIDES, **ranitidine. Y-site: Ampicillin/ sulbactam,** AMINOGLYCOSIDES, **cisatracurium, fenoldopam, filgrastim, hetastarch, lansoprazole, pentamidine, vancomycin.**

▪ After reconstitution, solution is stable for 24 h at 25° C (77° F); 7 days when refrigerated at 4° C (39° F), or 30 wk when frozen at −20° C (−4° F).

ADVERSE EFFECTS (≥1%) **Body as a Whole:** Drug fever, eosinophilia,

Common adverse effects in *italic,* life-threatening effects <u>underlined</u>; generic names in **bold;** classifications in SMALL CAPS; ✦ Canadian drug name; ⦿ Prototype drug

269

superinfections, local reactions: pain, tenderness, and induration (IM site), thrombophlebitis (IV site). **GI:** *Diarrhea,* pseudomembranous colitis. **Skin:** Rash, exfoliative dermatitis, pruritus, urticaria. **Urogenital:** Nephrotoxicity, interstitial nephritis.

DIAGNOSTIC TEST INTERFERENCE
Cefoxitin causes false-positive (black-brown or green-brown color) *urine glucose* reaction with *copper reduction reagents* such as *Benedict's* or *Clinitest,* but not with *enzymatic glucose oxidase reagents (Clinistix, TesTape).* With high doses, falsely elevated *serum and urine creatinine* (with *Jaffe reaction*) reported. False-positive *direct Coombs' test* (may interfere with *cross-matching procedures* and *hematologic studies*) has also been reported.

INTERACTIONS Drug: Probenecid decreases renal elimination of cefoxitin.

PHARMACOKINETICS Peak: 20–30 min after IM; 5 min after IV. **Distribution:** Poor CNS penetration even with inflamed meninges; widely distributed in body tissues including pleural, synovial, and ascitic fluid and bile; crosses placenta. **Elimination:** 85% unchanged in urine in 6 h, small amount in breast milk. **Half-Life:** 45–60 min.

NURSING IMPLICATIONS
Assessment & Drug Effects
- Determine previous hypersensitivity to cephalosporins, penicillins, and other drug allergies before therapy is initiated.
- Lab tests: Perform culture and sensitivity testing prior to therapy; periodic renal function tests.
- Inspect injection sites regularly. Report evidence of inflammation and patient's complaint of pain.

- Monitor I&O rates and pattern: Nephrotoxicity occurs most frequently in patients older than 50 y, in patients with impaired renal function, the debilitated, and in patients receiving high doses or other nephrotoxic drugs.
- Be alert to S&S of superinfections (see Appendix F). This condition is most apt to occur in older adult patients, especially when drug has been used for prolonged period.
- Report onset of diarrhea (may be dose related). If severe, pseudomembranous colitis (see Signs & Symptoms, Appendix F) **must be** ruled out. Older adult patients are especially susceptible.

Patient & Family Education
- Report promptly S&S of superinfection (see Appendix F).
- Report watery or bloody loose stools or severe diarrhea.
- Report severe vomiting or stomach pain.
- Report infusion site swelling, pain, or redness.

CEFPODOXIME
(cef-po-dox′eem)

Vantin
Classifications: ANTIBIOTIC; THIRD-GENERATION CEPHALOSPORIN
Therapeutic: ANTIBIOTIC
Prototype: Cefotaxime sodium
Pregnancy Category: B

AVAILABILITY 100 mg, 200 mg tablets; 10 mg/mL, 20 mg/mL suspension

ACTION & THERAPEUTIC EFFECT
Semisynthetic beta-lactam cephalosporin antibiotic that inhibits the final stage of bacterial cell wall synthesis by preferentially binding to specific penicillin-binding proteins (PBPs) within the bacterial cell

wall. *Highly active against gram-negative bacteria.*

USES Gonorrhea, otitis media, lower and upper respiratory tract infections, urinary tract infections.
UNLABELED USES Skin and soft tissue infections.

CONTRAINDICATIONS Hypersensitivity to cephalosporins and other beta-lactam antibiotics.
CAUTIOUS USE Renal impairment, history of Type I hypersensitivity reactions to penicillins; coagulopathy; history of colitis or other GI disease; pregnancy (category B); lactation.

ROUTE & DOSAGE

Respiratory Tract, Skin, and Soft Tissue Infections
Adult: **PO** 200 mg q12h for 10 days
Child: **PO** 10 mg/kg/day divided q12h

Urinary Tract Infections
Adult: **PO** 100 mg q12h

Gonorrhea
Adult: **PO** 200 mg as single dose

Otitis Media
Child (5 mo–12 y): **PO** 10 mg/kg/day divided q12–24h

ADMINISTRATION

Oral
- Give with food to enhance absorption.
- Give 1 h before or 2 h after an antacid.
- Consult prescriber regarding patients with renal impairment (i.e., creatinine clearance less than 30 mL/min); dosage intervals should be every 12 h.
- Preparation of suspension: To either the 50 mg/5 mL strength or the 100 mg/5 mL strength, add 25 mL of distilled water, then shake vigorously for 15 seconds. Next, to the 50 mg/5 mL strength add 33 mL, or to the 100 mg/5 mL strength add 32 mL, of distilled water, and shake for at least 3 minutes.
- Store suspension for up to 14 days in a refrigerator [2°–8° C (36°–46° F)]. Shake well before using.

ADVERSE EFFECTS (≥1%) **Body as a Whole:** Eye itching, cough, epistaxis, fever, decreased appetite, malaise. **GI:** Diarrhea, nausea, vomiting, abdominal pain, soft stools, flatulence, pseudomembranous colitis (rare). **CNS:** Rare: Headache, asthenia, dizziness, fatigue, anxiety, insomnia, flushing, nightmares, weakness. **Urogenital:** Vaginal candidiasis. **Skin:** Urticaria, rash, scaling, peeling.

INTERACTIONS Drug: ANTACIDS, **ranitidine** may decrease absorption. **Food:** Food may increase the absorption.

PHARMACOKINETICS Absorption: 40–50% absorbed from GI tract. **Onset:** Therapeutic effect in 3 days. **Distribution:** Distributes well into inflammatory, pulmonary, and pleural fluid, and tonsils. Some distribution into prostate. 40% bound to plasma proteins. Distributed into breast milk. **Elimination:** 80% in urine. **Half-Life:** 2–3 h.

NURSING IMPLICATIONS

Assessment & Drug Effects
- Determine history of hypersensitivity reactions to cephalosporins and penicillins, and history of allergies, particularly to drugs, before therapy is initiated.
- Lab tests: Perform culture and sensitivity tests before initiation of therapy. Therapy may be instituted pending test results.

Common adverse effects in *italic*, life-threatening effects underlined; generic names in **bold**; classifications in SMALL CAPS; ✤ Canadian drug name; ⊙ Prototype drug

271

- Report onset of loose stools or diarrhea. Although pseudomembranous enterocolitis (see Appendix F) rarely occurs, this potentially life-threatening complication should be ruled out.
- Monitor for manifestations of hypersensitivity (see Appendix F). Discontinue drug and report S&S of hypersensitivity promptly.
- Monitor I&O (especially with high doses). Report significant changes.

Patient & Family Education
- Report any signs or symptoms of hypersensitivity immediately.
- Report loose stools or diarrhea.

CEFPROZIL
(cef′pro-zil)
Cefzil ✦
Classifications: ANTIBIOTIC; SECOND-GENERATION CEPHALOSPORIN
Therapeutic: ANTIBIOTIC
Prototype: Cefaclor
Pregnancy Category: B

AVAILABILITY 250 mg, 500 mg tablets; 125 mg/5 mL, 250 mg/5 mL suspension

ACTION & THERAPEUTIC EFFECT
Semisynthetic, second-generation cephalosporin antibiotic with drug structure characterized by a beta-lactam ring; generally resistant to hydrolysis by beta-lactamases. Preferentially binds to proteins in cell walls of susceptible organisms, thus killing the bacteria. *Third-generation cephalosporins are more active and have a broader spectrum against gram-negative bacteria than first- or second-generation of cephalosporins.*

USES Upper and lower respiratory tract infections, otitis media, skin infections.

CONTRAINDICATIONS Hypersensitivity to cephalosporin and related antibiotics; severely impaired renal or hepatic function; phenylketonuria (PKU).
CAUTIOUS USE Patients with delayed reaction to penicillin or other drugs; coagulopathy; renal impairment, renal disease; GI disease, especially colitis; pregnancy (category B); infants younger than 6 mo.

ROUTE & DOSAGE

Mild to Moderate Infections
Adult: **PO** 250–500 mg q12–24h for 10–14 days
Child (older than 6 mo): **PO** 15 mg/kg q12h

Renal Impairment Dosage Adjustment
CrCl less than 29 mL/min: Reduce dose 50%

ADMINISTRATION
Oral
- Drug may be given without regard to meals.
- Consult prescriber for patients with impaired renal function. Dose is reduced by 50% when creatinine clearance is 0–29 mL/min.
- Administer after hemodialysis since drug is partially removed by dialysis.
- After reconstitution, oral suspension is refrigerated. Discard unused portion after 14 days.

ADVERSE EFFECTS (≥1%) **Body as a Whole:** Hypersensitivity reactions, superinfections. **GI:** Nausea, vomiting, diarrhea, abdominal pain. **Hematologic:** Eosinophilia. **CNS:** Headache. **Skin:** Rash, diaper rash. **Urogenital:** Genital pruritus, vaginal candidiasis.

DIAGNOSTIC TEST INTERFERENCE
May cause a positive *direct*

Coombs' test; false-positive reactions for *urine glucose* with *copper reduction tests* such as *Benedict's* or *Fehling's solution* or *Clinitest tablets.*

INTERACTIONS Drug: Probenecid prolongs the elimination of cefprozil.

PHARMACOKINETICS Absorption: Readily from GI tract. **Peak:** 1–2 h. **Distribution:** Distributes into blister fluid at 50% of the serum level. **Elimination:** Primarily by kidneys. **Half-Life:** 1–2 h.

NURSING IMPLICATIONS

Assessment & Drug Effects
- Determine previous hypersensitivity to cephalosporins or penicillins before treatment.
- Withhold drug and notify prescriber if hypersensitivity occurs (e.g., rash, urticaria).
- Lab tests: Perform culture and sensitivity tests before therapy. Therapy may be initiated while results are pending.
- Monitor for and report diarrhea, as pseudomembranous colitis is a potential adverse effect.
- Monitor for and report signs of superinfection (see Appendix F).
- When given concurrently with other cephalosporins or aminoglycosides, monitor for signs of nephrotoxicity.

Patient & Family Education
- Report rash or other signs of hypersensitivity immediately.
- Report signs of superinfection (see Appendix F).
- Report loose stools and diarrhea even after completion of drug therapy.

CEFTAROLINE
(cef-tar′o-line)
Teflaro

Classifications: BETA-LACTAM ANTIBIOTIC; THIRD-GENERATION CEPHALOSPORIN
Therapeutic: ANTIBIOTIC; CEPHALOSPORIN
Prototype: Cefotaxime
Pregnancy Category: B

AVAILABILITY 400 mg and 600 mg powder for injection

ACTION & *THERAPEUTIC EFFECT*
A bactericidal broad-spectrum, beta-lactam third-generation cephalosporin antibiotic. It preferentially binds to one or more of the penicillin-binding proteins (PBP) located on the cell walls of susceptible organisms. This inhibits the third and final stage of cell wall synthesis, thus destroying the bacterium. *Effective against certain gram-positive and gram-negative bacteria responsible for complicated, acute skin infections and community-acquired pneumonia.*

USES Treatment of acute bacterial skin and skin structure infections (ABSSSI) and community-acquired bacterial pneumonia (CABP) caused by susceptible microorganisms: *Haemophilus influenzae.*

CONTRAINDICATIONS Known hypersensitivity to ceftaroline or other cephalosporins; *C.difficile*-associated diarrhea.
CAUTIOUS USE Previous hypersensitivity to penicillins or carbapenems; renal impairment; direct Coombs' test seroconversion; pregnancy (category B); lactation. Safety and efficacy in children not established.

Common adverse effects in *italic*, life-threatening effects <u>underlined</u>; generic names in **bold**; classifications in SMALL CAPS; ♣ Canadian drug name; ☺ Prototype drug

273

ROUTE & DOSAGE

Acute Bacterial Skin and Skin Structure Infections (ABSSSI) or Community-Acquired Bacterial Pneumonia (CABP)

Adult: **IV** 600 mg infused over 1 h q12h for 5–14 days

Renal Impairment Dosage Adjustment

CrCl greater than 30 to 50 mL/ min: 400 mg over 1 h q12h; *15–30 mL/min:* 300 mg over 1 h q12h; *less than 15 mL/min:* 200 mg over 1 h q12h

Hemodialysis Dosage Adjustment
Administer dose after hemodialysis.

ADMINISTRATION

Intravenous

PREPARE: **Intermittent:** Reconstitute the 400 mg or 600 mg vial with 20 mL of sterile water to yield 20 mg/mL or 30 mg/mL, respectively. ▪ Mix gently then withdraw the required dose and add to at least 250 mL of NS, D5W, 2.5% DW, 0.45% NaCl, or LR. ▪ Do not mix with or add to solutions containing other drugs. *ADMINISTER:* **Intermittent:** Give at a constant rate over 1 h. ▪ Monitor closely for S&S of hypersensitivity (see Appendix F). If suspected, stop infusion and notify prescriber immediately. ▪ Use within 6 h when stored at room temperature or within 24 h if refrigerated.

INCOMPATIBILITIES **Solution/additive:** Do not mix with another drug. **Y-site:** Do not mix with another drug.

ADVERSE EFFECTS (≥2%) **Body as a Whole:** Phlebitis. **GI:** Constipation, diarrhea, nausea, vomiting.

Metabolic: Elevated ALT and AST levels, hypokalemia. **Skin:** Rash.

DIAGNOSTIC TEST INTERFERENCE
Ceftaroline can cause false positive results for a *Direct Coombs' Test.*

INTERACTIONS Drug: PROBENECID may decrease the renal excretion of ceftaroline.

PHARMACOKINETICS Peak: 1 h. **Distribution:** Approximately 20% plasma protein bound. **Metabolism:** Dephosphorylated to active metabolite; hydrolyzed to inactive metabolite. **Elimination:** Primarily renal (88%) with minor fecal (6%). **Half-Life:** 1.6 h.

NURSING IMPLICATIONS

Assessment & Drug Effects

▪ Determine previous hypersensitivity reactions to cephalosporins and penicillins, and history of other allergies, particularly to drugs, before therapy is initiated.
▪ Lab tests: Perform culture and sensitivity tests before initiation of therapy; baseline and periodic kidney function tests, especially in the older adult; periodic serum electrolytes, CBC with differential and platelet count, and LFTs.
▪ Monitor closely for S&S of hypersensitivity (see Appendix F).
▪ Monitor I&O rates and patterns, especially with concurrent aminoglycoside therapy. Report significant changes in I&O.
▪ Report promptly onset of diarrhea. If fever is present and diarrhea is severe, suspect antibiotic-associated pseudomembranous colitis (may occur during therapy or following discontinuation of ceftaroline).

Patient & Family Education

▪ Report promptly frequent watery stools or bloody diarrhea.

- Yogurt or buttermilk may serve as a prophylactic against mild forms of diarrhea.
- Report any signs of hypersensitivity (see Appendix F).

CEFTAZIDIME
(sef'tay-zi-deem)

Fortaz, Tazicef
Classifications: ANTIBIOTIC; THIRD-GENERATION CEPHALOSPORIN
Therapeutic: ANTIBIOTIC
Prototype: Cefotaxime sodium
Pregnancy Category: B

AVAILABILITY 500 mg, 1 g, 2 g injection

ACTION & THERAPEUTIC EFFECT
Semisynthetic, third-generation broad-spectrum cephalosporin antibiotic. Preferentially binds to one or more of the penicillin-binding proteins (PBP) located on cell walls of susceptible microbes; this inhibits the final stage of bacterial cell wall synthesis, leading to cell death of the bacterium. *More active and has a broader spectrum against aerobic gram-negative bacteria than do either first- or second-generation agents.*

USES To treat infections of lower respiratory tract, skin and skin structures, urinary tract, bones, and joints; also used to treat bacteremia, gynecologic, intra-abdominal, and CNS infections (including meningitis).
UNLABELED USES Surgical prophylaxis.

CONTRAINDICATIONS Hypersensitivity to cephalosporins and related beta-lactam antibiotics; viral disease.
CAUTIOUS USE Pregnancy (category B); elderly; coagulopathy, renal disease, renal impairment; GI disease; colitis.

ROUTE & DOSAGE

Moderate to Severe Infections
Adult: **IV/IM** 1–2 g q8–12h, up to 2 g q6h
Geriatric: **IV/IM** 1–2 g q12h
Child: **IV/IM** 30–50 mg/kg/day q8h (max: 6 g/day)
Neonate (4 wk or younger): **IV** 30 mg/kg q12h

Very Severe Infection
Adult: **IV** 2 g q8h

Renal Impairment Dosage Adjustment
CrCl 30–50 mL/min: Dose q12h; *10–30 mL/min:* Dose q24h; *less than 10 mL/min:* Dose q48–72h

Hemodialysis Dosage Adjustment
Removed by dialysis

ADMINISTRATION

Intramuscular
- Reconstitute by adding 3 mL sterile water or bacteriostatic water for injection or 0.5% or 1% lidocaine HCl injection to 1 g vial to yield 280 mg/mL.
- Inject into large muscle mass (e.g., upper outer quadrant of gluteus maximus or lateral part of thigh).

Intravenous
PREPARE: Direct: Add 10 mL of sterile water for injection to 1 g to yield 280 mg/mL. **Intermittent:** Prepare as for direct injection then further dilute with 50–100 mL of D5W, NS, or LR.
ADMINISTER: Direct: Give over 3–5 min. **Intermittent:** Give over 30–60 min. • If given through a Y-type set, discontinue other solutions during infusion of ceftazidime.
INCOMPATIBILITIES Solution/additive: AMINOGLYCOSIDES, **aminophylline,**

Common adverse effects in *italic*, life-threatening effects <u>underlined</u>; generic names in **bold**; classifications in SMALL CAPS; ♣ Canadian drug name; ⊘ Prototype drug

275

ciprofloxacin, ranitidine, sodium bicarbonate. **Y-site:** Alatrofloxacin, amiodarone, AMINOGLYCOSIDES, **amphotericin B cholesteryl complex, amsacrine, azithromycin, clarithromycin, doxorubicin liposome, fluconazole, idarubicin, midazolam, pentamidine, sargramostim, vancomycin, warfarin.**

- Protect sterile powder from light. Reconstituted solution is stable 7 days when refrigerated at 4°–5° C (39°–41° F); for 18–24 h when stored at 15°–30° C (59°–86° F).

ADVERSE EFFECTS (≥1%) **Body as a Whole:** Fever, phlebitis, pain or inflammation at injection site, superinfections. **GI:** Nausea, vomiting, *diarrhea,* abdominal pain, metallic taste, drug-associated <u>pseudomembranous colitis.</u> **Skin:** Pruritus, rash, urticaria. **Urogenital:** Vaginitis, candidiasis.

DIAGNOSTIC TEST INTERFERENCE False-positive reactions for **urine glucose** have been reported using **copper sulfate** (e.g., *Benedict's solution, Clinitest*). *Glucose oxidase tests (Clinistix, TesTape)* are unaffected. May cause positive **direct antiglobulin (Coombs')** **test** results, which can interfere with **hematologic studies** and **transfusion cross-matching procedures.**

INTERACTIONS Drug: Probenecid decreases renal elimination of ceftazidine.

PHARMACOKINETICS Peak: 1 h. **Distribution:** CNS penetration with inflamed meninges; also penetrates bone, gallbladder, bile, endometrium, heart, skin, and ascitic and pleural fluids; crosses placenta. **Metabolism:** Not metabolized. **Elimination:** 80–90% unchanged in urine in 24 h; small amount in breast milk. **Half-Life:** 25–60 min.

NURSING IMPLICATIONS

Assessment & Drug Effects

- Determine history of hypersensitivity to cephalosporins and penicillins, and other drug allergies, before therapy begins.
- Lab tests: Perform culture and sensitivity studies before initiation of therapy. Therapy may begin pending test results.
- If administered concomitantly with another antibiotic, monitor renal function and report if symptoms of dysfunction appear (e.g., changes in I&O ratio and pattern, dysuria).
- Be alert to onset of rash, itching, and dyspnea. Check patient's temperature. If it is elevated, suspect onset of hypersensitivity reaction (see Appendix F).
- Monitor for superinfection. (See Appendix F.)
- If diarrhea occurs and is severe, suspect pseudomembranous colitis (caused by *Clostridium difficile*). Report severe diarrhea to prescriber.

Patient & Family Education

- Report loose stools or diarrhea promptly.
- Report any signs or symptoms of superinfection promptly (see Appendix F).

CEFTIBUTEN
(sef-ti-bu′ten)
Cedax
Classifications: BETA-LACTAM ANTIBIOTIC; THIRD-GENERATION CEPHALOSPORIN
Therapeutic: ANTIBIOTIC
Prototype: Cefotaxime sodium
Pregnancy Category: B

AVAILABILITY 400 mg capsules; 90 mg/5 mL, 180 mg/mL suspension

ACTION & *THERAPEUTIC EFFECT*

Ceftibuten is a broad-spectrum, third-generation beta-lactam antibiotic. Preferentially binds to one or more of the penicillin-binding proteins located in the cell wall of susceptible organisms. This inhibits the final stage of bacterial cell wall synthesis, killing the bacterium. *It has antibacterial activity against both gram-negative and gram-positive bacteria.*

USES Acute bacterial exacerbations of chronic bronchitis; acute bacterial otitis media; pharyngitis or tonsillitis.

CONTRAINDICATIONS Hypersensitivity to ceftibuten or cephalosporins.
CAUTIOUS USE Renal dysfunction, penicillin hypersensitivity, history of colitis or diabetes; renal impairment; GI disease; coagulopathy; elderly; pregnancy (category B), lactation. Safety and efficacy in infants younger than 6 mo not established.

ROUTE & DOSAGE

Mild to Moderate Infections

Adult: **PO** 400 mg once daily for 10 days

Renal Impairment Dosage Adjustment

CrCl 30–49 mL/min: 200 mg q24h; *less than 30 mL/min:* 100 mg q24h
Child (6 mo–12 y): **PO** 9 mg/kg once daily (max: 400 mg) for 10 days

Renal Impairment Dosage Adjustment

CrCl 30–49 mL/min: 4.5 mg/kg q24h; *less than 30 mL/min:* 2.25 mg/kg q24h

ADMINISTRATION

Oral

- Give oral suspension 1 h before or 2 h after a meal.
- Children weighing more than 45 kg may receive maximum daily dose.
- Hemodialysis patients should receive drug at the end of dialysis.
- Store capsules at 2°–25° C (36°–77° F); keep container tightly closed. Reconstituted oral suspension is stable for 14 days under refrigeration at 2°–8° C (36°–46° F).

ADVERSE EFFECTS (≥1%) **Body as a Whole:** Dyspnea, dysuria, fatigue, vaginitis, moniliasis, urticaria, pruritus, rash, paresthesia, taste perversion. **GI:** Nausea, vomiting, diarrhea, dyspepsia, abdominal pain, anorexia, constipation, dry mouth, eructation, flatulence. **CNS:** Headache, dizziness, nasal congestion, somnolence.

INTERACTIONS Drug: AMINOGLYCOSIDES may increase risk of nephrotoxicity and have additive/synergistic effects. May decrease efficacy of ORAL CONTRACEPTIVES. **Probenecid** may increase levels.

PHARMACOKINETICS Absorption: Rapidly from GI tract. **Peak:** Approx 2–3 h. **Distribution:** Bronchial mucosa levels are approx 37% of plasma levels, middle ear levels approx 50% of plasma levels. **Elimination:** Primarily in urine. **Half-Life:** 1.5–2.5 h.

NURSING IMPLICATIONS

Assessment & Drug Effects

- Determine history of hypersensitivity reactions to cephalosporins, penicillins, or other drugs, before therapy is initiated. Monitor for S&S of hypersensitivity (see

Common adverse effects in *italic*, life-threatening effects <u>underlined</u>; generic names in **bold**; classifications in SMALL CAPS; ♣ Canadian drug name; ⊘ Prototype drug

277

Appendix F); report their appearance promptly and discontinue drug.

- Lab tests: Perform culture and sensitivity tests before initiation of therapy. Dosage may be started pending test results.
- Monitor for S&S of superinfection or pseudomembranous colitis (see Appendix F); immediately report either to prescriber.
- Closely monitor patients with renal impairment; if seizures develop, discontinue drug and notify prescriber.

Patient & Family Education

- If on dialysis treatment, take this drug after dialysis.
- Report any S&S of hypersensitivity, superinfection, and pseudomembranous colitis promptly.

CEFTIZOXIME SODIUM

(sef-ti-zox'eem)

Cefizox

Classifications: ANTIBIOTIC; THIRD-GENERATION CEPHALOSPORIN
Therapeutic: ANTIBIOTIC
Prototype: Cefotaxime sodium
Pregnancy Category: B

AVAILABILITY 1 g, 2 g injection

ACTION & *THERAPEUTIC EFFECT*
Semisynthetic third-generation cephalosporin antibiotic. Preferentially binds to one or more of the penicillin-binding proteins (PBP) located on cell walls of susceptible organisms. This inhibits third and final stage of bacterial cell wall synthesis, thus killing the bacterium. *Its spectrum includes some gram-positive organisms but has predominantly gram-negative coverage.*

USES Infections caused by susceptible organisms in lower respiratory tract, skin and skin structures, urinary tract, bones and joints; also used to treat intra-abdominal infections, pelvic inflammatory disease, uncomplicated gonorrhea, meningitis *(Haemophilus influenzae, Streptococcus pneumoniae),* and for surgical prophylaxis.

UNLABELED USES Meningitis caused by *Neisseria meningitidis* and *E. coli.*

CONTRAINDICATIONS Hypersensitivity to cephalosporins and other beta-lactam antibiotics; viral disease. Safe use in infants younger than 6 mo not established.

CAUTIOUS USE Penicillin hypersensitivity; coagulopathy; GI disease, colitis; elderly; renal disease, renal impairment; pregnancy (category B); lactation.

ROUTE & DOSAGE

Moderate to Severe Infections

Adult: **IV/IM** 1–2 g q8–12h, up to 2 g q4h
Child (6 mo or older): **IV/IM** 50 mg/kg q6–8h, up to 200 mg/kg/day

Life Threatening Infections

Adult: **IV** 3–4 g q8h

Renal Impairment Dosage Adjustment

Use lower dose.

Hemodialysis Dosage Adjustment

Administer dose after dialysis

ADMINISTRATION

Intramuscular

- Reconstitute as follows with sterile water for injection: Add 1.5 mL to 500 mg to yield 280 mg/mL; add 3 mL to 1 g or 6 mL to 2 g to yield 270 mg/mL.
- Give deep IM into a large muscle. Give no more than 1 g into a single injection site.

Common adverse effects in *italic*, life-threatening effects underlined; generic names in **bold;** classifications in SMALL CAPS; ♣ Canadian drug name; ⊘ Prototype drug

Intravenous

PREPARE: **Direct:** Reconstitute each 1 g with 10 mL sterile water. Shake well. **Intermittent:** Further dilute reconstituted solution in 50–100 mL D5W, NS, D5/NS, D5/.45% NaCl, LR, or other compatible IV solution.
ADMINISTER: **Direct:** Give over 3–5 min. **Intermittent:** Give over 30 min.
INCOMPATIBILITIES **Solution/additive: Promethazine. Y-site: Filgrastim, lansoprazole, vancomycin.**

■ Protect from light. Consult manufacturer's directions concerning storage of reconstituted solutions.

ADVERSE EFFECTS (≥1%) **Body as a Whole:** Fever, phlebitis, vaginitis, pain and induration at injection site, paresthesia. **GI:** Nausea, vomiting, diarrhea, <u>pseudomembranous colitis</u>. **Skin:** Rash, pruritus.

DIAGNOSTIC TEST INTERFERENCE Ceftizoxime causes false-positive *direct Coombs' test* (may interfere with *cross-matching procedures* and *hematologic studies*).

INTERACTIONS Drug: Probenecid decreases renal elimination of ceftizoxime.

PHARMACOKINETICS Peak: 1 h. **Distribution:** Crosses placenta. **Metabolism:** Not metabolized. **Elimination:** 80–90% unchanged in urine in 24 h; small amount in breast milk. **Half-Life:** 25–60 min.

NURSING IMPLICATIONS
Assessment & Drug Effects
■ Determine history of hypersensitivity reactions to cephalosporins, penicillin, or other drugs before therapy is instituted. Report to prescriber history of allergy, particularly to drugs.
■ Lab tests: Perform culture and sensitivity tests before initiation of therapy. Therapy may be instituted pending test results.
■ Be alert to symptoms of hypersensitivity reaction (see Appendix F). Serious reactions may require emergency measures.

Patient & Family Education
■ Report loose stools or diarrhea promptly.
■ Report any signs or symptoms of hypersensitivity (see Appendix F) promptly.

CEFTRIAXONE SODIUM
(sef-try-ax′one)
Rocephin
Classifications: ANTIBIOTIC; THIRD-GENERATION CEPHALOSPORIN
Therapeutic: ANTIBIOTIC
Prototype: Cefotaxime sodium
Pregnancy Category: B

AVAILABILITY 250 mg, 500 mg, 1 g, 2 g injection

ACTION & *THERAPEUTIC EFFECT*
Semisynthetic third-generation cephalosporin antibiotic. Preferentially binds to one or more of the penicillin-binding proteins (PBP) located on cell walls of susceptible organisms. This inhibits third and final stage of bacterial cell wall synthesis, thus killing the bacterium. *Similar to other third-generation cephalosporins, ceftriaxone is effective against serious gram-negative organisms and also penetrates the CSF in concentrations useful in treatment of meningitis.*

USES Infections caused by susceptible organisms in lower respiratory tract, skin and skin structures,

Common adverse effects in *italic*, life-threatening effects <u>underlined</u>; generic names in **bold**; classifications in SMALL CAPS; ♣ Canadian drug name; ⊘ Prototype drug

279

urinary tract, bones and joints; also intra-abdominal infections, pelvic inflammatory disease, uncomplicated gonorrhea, meningitis, and surgical prophylaxis.

CONTRAINDICATIONS Hypersensitivity to cephalosporins; viral infections; neonates with hyperbilirubinemia, especially premature neonates; neonates with calcium-containing infusions such as parenteral nutrition; signs and symptoms of gallbladder disease.
CAUTIOUS USE Coagulopathy, hypersensitivity to penicillin or other drugs; impaired vitamin K synthesis; chronic hepatic disease; history of GI disease, colitis; renal disease, renal impairment; pregnancy (category B).

ROUTE & DOSAGE

Moderate to Severe Infections

Adult: **IV/IM** 1–2 g q12–24h × 4–14 days (max: 4 g/day)
Child: **IV/IM** 50–75 mg/kg/day in 2 divided doses × 4–14 days (max: 2 g/day)

Bacterial Otitis Media

Child: **IM** 50 mg/kg (max: 1 g)

Meningitis

Adult: **IV/IM** 2 g q12h
Child: **IV/IM** 100 mg/kg/day in 2 divided doses (max: 4 g/day)

Surgical Prophylaxis

Adult: **IV/IM** 1 g 30–120 min before surgery

Uncomplicated Gonorrhea

Adult: **IM** 250 mg as single dose
Child: **IM** 125 mg as single dose

ADMINISTRATION

Intramuscular

- Reconstitute the 1 g or 2 g vial by adding 2.1 mL or 4.2 mL, respec-

tively, of sterile water for injection. Yields 350 mg/mL. See manufacturer's directions for other dilutions.
- Give deep IM into a large muscle.

Intravenous

IV administration to infants and children: Verify correct IV concentration and rate of infusion with prescriber.

PREPARE: Intermittent: Reconstitute each 250 mg with 2.4 mL of sterile water, D5W, NS, or D5/NS to yield 100 mg/mL. ▪ Further dilute with 50–100 mL of the selected IV solution.
ADMINISTER: Intermittent: Give over 30 min.
INCOMPATIBILITIES Solution/ additive: AMINOGLYCOSIDES, **aminophylline, clindamycin, lidocaine, linezolid, metronidazole, theophylline, calcium**-containing products such as parenteral nutrition. **Y-site:** AMINOGLYCOSIDES, **amphotericin B cholesteryl complex, amsacrine, azithromycin, calcium**-containing products, **filgrastim, fluconazole, labetalol, pentamidine, vancomycin, vinorelbine.**

- Protect sterile powder from light. Store at 15°–25° C (59°–77° F).
- Reconstituted solutions: Diluent, concentration of solutions are determinants of stability. See manufacturer's instructions for storage.

ADVERSE EFFECTS (≥1%) Body as a Whole: Pruritus, fever, chills, pain, induration at IM injection site; phlebitis (IV site). **GI:** *Diarrhea,* abdominal cramps, <u>pseudomembranous colitis</u>, biliary sludge. **Urogenital:** Genital pruritus; moniliasis.

DIAGNOSTIC TEST INTERFERENCE Causes prolonged *PT/INR* during therapy.

INTERACTIONS Drug: Probenecid decreases renal elimination of ceftriaxone; **alcohol** produces disulfiram reaction; effect of **warfarin** may be increased.

PHARMACOKINETICS Peak: 1.5–4 h after IM; immediately after IV. **Distribution:** Widely in body tissues and fluids; good CNS penetration; crosses placenta. **Metabolism:** Not metabolized. **Elimination:** 33–65% unchanged in urine; also in bile and breast milk. **Half-Life:** 5–10 h.

NURSING IMPLICATIONS

Assessment & Drug Effects

- Determine history of hypersensitivity reactions to cephalosporins and penicillins and history of other allergies, particularly to drugs, before therapy is initiated.
- Lab tests: Perform culture and sensitivity tests before initiation of therapy. Dosage may be started pending test results. Periodic coagulation studies (PT and INR) should be done when on concurrent warfarin.
- Inspect injection sites for induration and inflammation. Rotate sites. Note IV injection sites for signs of phlebitis (redness, swelling, pain).
- Monitor for manifestations of hypersensitivity (see Appendix F). Report promptly.
- Watch for and report: Petechiae, ecchymotic areas, epistaxis, or any unexplained bleeding. Ceftriaxone appears to alter vitamin K–producing gut bacteria; therefore, hypoprothrombinemic bleeding may occur.
- Report promptly development of diarrhea. The incidence of antibiotic-produced pseudomembranous colitis (see Appendix F) is higher than with most cephalosporins.

Patient & Family Education

- Report any signs of bleeding.
- Report loose stools or diarrhea promptly.

CEFUROXIME SODIUM
(se-fyoor-ox'eem)
Zinacef

CEFUROXIME AXETIL
Ceftin
Classifications: ANTIBIOTIC; SECOND-GENERATION CEPHALOSPORIN
Therapeutic: ANTIBIOTIC
Prototype: Cefaclor
Pregnancy Category: B

AVAILABILITY 125 mg, 250 mg, 500 mg tablets; 125 mg/5 mL, 250 mg/5 mL suspension; 750 mg, 1.5 g injection

ACTION & *THERAPEUTIC EFFECT*
Semisynthetic second-generation cephalosporin beta-lactam antibiotic. Preferentially binds to one or more of the penicillin-binding proteins (PBP) located on cell walls of susceptible organisms. This inhibits third and final stage of bacterial cell wall synthesis, thus killing the bacterium. *Similar to other second-generation cephalosporins, cefuroxime is more active against gram-negative bacteria than are first-generation cephalosporins but not as active as third-generation cephalosporins.*

USES Infections caused by susceptible organisms in the lower respiratory tract, urinary tract, skin, and skin structures; also used for treatment of meningitis, gonorrhea, and otitis media and for perioperative prophylaxis (e.g., open-heart surgery), early Lyme disease.

CONTRAINDICATIONS Hypersensitivity to cephalosporins and related antibiotics; viral infections.

Common adverse effects in *italic*, life-threatening effects <u>underlined</u>; generic names in **bold**; classifications in SMALL CAPS; ♣ Canadian drug name; ⊘ Prototype drug

281

CAUTIOUS USE History of allergy, particularly to drugs; penicillin sensitivity; renal insufficiency; history of colitis or other GI disease; potent diuretics; pregnancy (category B), lactation.

ROUTE & DOSAGE

Moderate to Severe Infections
Adult: **PO** 250–500 mg q12h **IV/IM** 750 mg–1.5 g q6–8h
Child (3 mo–12 y): **PO** 10–15 mg/kg (125–250 mg) q12h **IV/IM** 50–100 mg/kg/day divided q8h (max: 6 g/day)

Bacterial Meningitis
Adult: **IV/IM** 1.5–3 g q8h
Child/Infant (older than 3 mo): **IV/IM** 200–240 mg/kg/day divided q6–8h; reduced to 100 mg/kg/day upon improvement

Surgical Prophylaxis
Adult/Adolescent: **IV/IM** 1.5 g 30–60 min before surgery, then 750 mg q8h for 24 h

Renal Impairment Dosage Adjustment
CrCl 10–20 mL/min: Give q12h; less than 10 mL/min: Give q24h

Hemodialysis Dosage Adjustment
Give supplemental dose

ADMINISTRATION

Oral
- Cefuroxime tablets and oral suspension are not substitutable on a mg-for-mg basis.
- The oral suspension is for infants and children 3 mo to 12 y. Each teaspoon (5 mL) contains the equivalent of 125 mg cefuroxime. Shake oral suspension well before each use.

Intramuscular
- Shake IM suspension gently before administration. IM injections should be made deeply into large muscle mass. Rotate injection sites.

Intravenous
IV administration to neonates, infants and children: Verify correct IV concentration and rate of infusion/injection with prescriber.

PREPARE: Direct: Dilute each 750 mg with 8 mL sterile water, D5W, or NS. **Intermittent:** Further dilute in 50–100 mL of compatible solution. **Continuous:** May be added to 1000 mL of IV compatible solution.
ADMINISTER: Direct: Give slowly over 3–5 min. **Intermittent:** Give over 30 min. **Continuous:** Give over 6–24 h.
INCOMPATIBILITIES Solution/additive: AMINOGLYCOSIDES, **ciprofloxacin, ranitidine. Y-site:** AMINOGLYCOSIDES, **azithromycin, cisatracurium, clarithromycin, filgrastim, fluconazole, midazolam, vancomycin, vinorelbine.**

- Store powder protected from light unless otherwise directed. After reconstitution, store suspension at 2°–30° C (36°–86° F). Discard after 10 days.

ADVERSE EFFECTS (≥1%) **Body as a Whole:** Thrombophlebitis (IV site); pain, burning, cellulitis (IM site); superinfections, positive Coombs' test. **GI:** *Diarrhea,* nausea, antibiotic-associated colitis. **Skin:** Rash, pruritus, urticaria. **Urogenital:** Increased serum creatinine and BUN, decreased creatinine clearance.

DIAGNOSTIC TEST INTERFERENCE Cefuroxime causes false-positive (black-brown or green-brown color) *urine glucose* reaction with copper reduction reagents (e.g., **Benedict's** or **Clinitest**) but not with **enzymatic glucose oxidase reagents** (e.g., **Clinistix, TesTape**). False-positive **direct Coombs' test** (may interfere with **cross-matching procedures** and **hematologic studies**) has been reported.

INTERACTIONS Drug: Probenecid decreases renal elimination of cefuroxime, thus prolonging its action.

PHARMACOKINETICS Absorption: Well absorbed from GI tract; hydrolyzed to active drug in GI mucosa. **Peak:** PO 2 h; IM 30 min. **Distribution:** Widely distributed in body tissues and fluids; adequate CNS penetration with inflamed meninges; crosses placenta. **Elimination:** 66–100% in 24 h; in breast milk. **Half-Life:** 1–2 h.

NURSING IMPLICATIONS

Assessment & Drug Effects
▪ Determine history of hypersensitivity reactions to cephalosporins, penicillins, and history of allergies, particularly to drugs, before therapy is initiated.
▪ Lab tests: Perform culture and sensitivity tests before initiation of therapy. Therapy may be instituted pending test results. Monitor periodically BUN and creatinine clearance.
▪ Report onset of loose stools or diarrhea. Pseudomembranous colitis (see Signs & Symptoms, Appendix F) should be ruled out as the cause of diarrhea during and after antibiotic therapy.
▪ Monitor for manifestations of hypersensitivity (see Appendix F). Discontinue drug and report their appearance promptly.

Patient & Family Education
▪ Report loose stools or diarrhea promptly.

▪ Report any signs or symptoms of hypersensitivity (see Appendix F).

CELECOXIB ⊕
(cel-e-cox'ib)
Celebrex
Classifications: ANALGESIC, NONSTEROIDAL ANTI-INFLAMMMATORY DRUG (NSAID); CYCLOOXYGENASE-2 (COX-2) INHIBITOR; ANTI-INFLAMMATORY
Therapeutic: ANALGESIC, NSAID; COX-2 INHIBITOR; ANTI-INFLAMMATORY; ANTIRHEUMATIC
Pregnancy Category: C first and second trimester; D third trimester

AVAILABILITY 50 mg, 100 mg, 200 mg, 400 mg capsules

ACTION & THERAPEUTIC EFFECT Although an NSAID, unlike ibuprofen celecoxib inhibits prostaglandin synthesis by inhibiting cyclooxygenase-2 (COX-2), but does not inhibit cyclooxygenase-1 (COX-1). *Exhibits anti-inflammatory, analgesic, and antipyretic activities. Reduces or eliminates the pain of rheumatoid and osteoarthritis.*

USES Relief of S&S of osteoarthritis and rheumatoid arthritis. Treatment of acute pain and primary dysmenorrhea; ankylosing spondylitis, juvenile rheumatoid arthritis.

CONTRAINDICATIONS Hypersensitivity to celecoxib, salicylate, or sulfonamide; asthmatic patients with aspirin triad; GI bleeding; advanced renal disease; development of S&S of renal impairment due to drug; severe hepatic impairment; development of S&S of hepatic impairment due to drug; anemia; pain from CABG surgery; pregnancy (category D third trimester); lactation.

Common adverse effects in *italic*, life-threatening effects underlined; generic names in **bold**; classifications in SMALL CAPS; ✦ Canadian drug name; ⊕ Prototype drug

283

CAUTIOUS USE Patients who are CYP2C9 poor metabolizers; patients who weigh less than 50 kg; mild or moderate hepatic impairment; elevated LTFs; renal insufficiency; prior history of GI bleeding or peptic ulcer disease; alcoholics; asthmatics; bone marrow suppression; CVA; PVD; fluid retention and/or HF; known risks for cardiovascular disease; kidney disease; hypertension; fluid retention; older adults; pregnancy (category C first and second trimester); children with systemic-onset juvenile rheumatoid arthritis younger than 2 y.

ROUTE & DOSAGE

Osteoarthritis/Arthritis/ Ankylosing Spondylitis

Adult: **PO** 100 mg b.i.d. or 200 mg daily

Rheumatoid Arthritis

Adult: **PO** 100–200 mg b.i.d.

Acute Pain, Dysmenorrhea

Adult: **PO** 400 mg 1st dose, then 200 mg same day if needed, then 200 mg b.i.d. prn

FAP

Adult: **PO** 400 mg b.i.d.

Juvenile Rheumatoid Arthritis

Adolescent/Child (older than 2 y, weight greater than 25 kg): **PO** 100 mg b.i.d.
Child (older than 2 y, weight 10–25 kg): **PO** 50 mg b.i.d.

Hepatic Dosage Adjustment

Child-Pugh class B: Reduce dose by 50%

Pharmacogenetic Dosage Adjustment

Poor CYP2C9 metabolizers: Start with ½ normal dose

ADMINISTRATION

Oral

- Give 2 h before/after magnesium- or aluminum-containing antacids.
- Store in tightly closed container and protect from light.

ADVERSE EFFECTS (≥1%) **Body as a Whole:** Back pain, peripheral edema. Increased risk of cardiovascular events. **GI:** Abdominal pain, diarrhea, dyspepsia, flatulence, nausea. **CNS:** Dizziness, headache, insomnia. **Respiratory:** Pharyngitis, rhinitis, sinusitis, URI. **Skin:** Rash.

INTERACTIONS Drug: May diminish effectiveness of ACE INHIBITORS; **fluconazole** increases celecoxib concentrations; may increase **lithium** concentrations; may increase INR in older patients on **warfarin.**

PHARMACOKINETICS Peak: 3 h. **Distribution:** 97% protein bound; crosses placenta. **Metabolism:** In liver by CYP2C9. **Elimination:** Primarily in feces (57%), 27% in urine. **Half-Life:** 11.2 h.

NURSING IMPLICATIONS

Assessment & Drug Effects

- Lab tests: Periodically monitor Hct and Hgb, LFTs, BUN and creatinine, and serum electrolytes.
- Monitor closely lithium levels when the two drugs are given concurrently.
- Monitor closely PT/INR when used concurrently with warfarin.
- Monitor for fluid retention and edema especially in those with a history of hypertension or CHF.

Patient & Family Education

- Promptly report any of the following: Unexplained weight gain, edema, skin rash.
- Stop taking celecoxib and promptly report to prescriber if any of

the following occurs: S&S of liver dysfunction including nausea, fatigue, lethargy, itching, jaundice, abdominal pain, and flu-like symptoms; S&S of GI ulceration including black, tarry stools and upper GI distress.

- Avoid using celecoxib during the third trimester of pregnancy.

CEPHALEXIN

(sef-a-lex'in)

Ceporex A, Keflex, Novolexin A
Classifications: BETA-LACTAM ANTIBIOTIC; FIRST-GENERATION CEPHALOSPORIN
Therapeutic: ANTIBIOTIC
Prototype: Cefazolin
Pregnancy Category: B

AVAILABILITY 250 mg, 500 mg capsules; 250 mg, 500 mg, 1 g tablets; 125 mg/5 mL, 250 mg/5 mL suspension

ACTION & THERAPEUTIC EFFECT Semisynthetic beta-lactam cephalosporin. Preferentially binds to one or more of the penicillin-binding proteins (PBP) located on cell walls of susceptible organisms. This inhibits third and final stage of bacterial cell wall synthesis, thus killing the bacterium. *Broad-spectrum, first-generation cephalosporin antibiotic active against many gram-positive aerobic cocci and much less active against gram-negative bacteria or anaerobic organisms.*

USES To treat infections caused by susceptible pathogens in respiratory and urinary tracts, middle ear, skin, soft tissue, and bone.

CONTRAINDICATIONS Hypersensitivity to cephalosporin antibiotics; viral infections.
CAUTIOUS USE History of hypersensitivity to penicillin or other

drug allergy; severely impaired renal function; GI disease, colitis; hepatic disease; coagulopathy; pregnancy (category B), lactation. Safe use in infants younger than 1 mo not established.

ROUTE & DOSAGE

Mild to Moderate Infection
Adult: **PO** 250–500 mg q6h
Child: **PO** 25–100 mg/kg/day in 4 divided doses
Skin and Skin Structure Infections
Adult: **PO** 500 mg q12h
Otitis Media
Child: **PO** 75–100 mg/kg/day in 4 divided doses

ADMINISTRATION

Oral

- Cephalexin oral suspension should be refrigerated; discard unused portions 14 days after preparation. Label should indicate expiration date. Keep tightly covered. Shake suspension well before pouring.

ADVERSE EFFECTS (≥1%) **Body as a Whole:** Angioedema, <u>anaphylaxis</u>, superinfections. **GI:** *Diarrhea* (generally mild), nausea, vomiting, anorexia, abdominal pain. **CNS:** Dizziness, headache, fatigue. **Skin:** Rash, urticaria.

DIAGNOSTIC TEST INTERFERENCE False-positive **urine glucose** determinations using **copper sulfate reagents** (e.g., **Clinitest, Benedict's reagent**), but not with **glucose oxidase (enzymatic) tests** (e.g., **Tes-Tape, Diastix, Clini-stix**). Positive **direct Coombs' test** may complicate transfusion **cross-matching procedures** and **hematologic studies**.

Common adverse effects in *italic*, life-threatening effects <u>underlined</u>; generic names in **bold**; classifications in SMALL CAPS; ♣ Canadian drug name; ☻ Prototype drug

285

INTERACTIONS Drug: Probenecid decreases renal elimination of cephalexin.

PHARMACOKINETICS Absorption: Rapidly from GI tract; stable in stomach acid. **Peak:** 1 h. **Distribution:** Widely distributed in body fluids with highest concentration in kidney; crosses placenta. **Elimination:** 80–100% unchanged in urine in 8 h; excreted in breast milk. **Half-Life:** 38–70 min.

NURSING IMPLICATIONS

Assessment & Drug Effects

- Determine history of hypersensitivity reactions to cephalosporins and penicillin and history of other drug allergies before therapy is initiated.
- Lab tests: Evaluate renal and hepatic function periodically in patients receiving prolonged therapy.
- Monitor for manifestations of hypersensitivity (see Signs & Symptoms, Appendix F). Discontinue drug and report their appearance promptly.

Patient & Family Education

- Keep prescriber informed if adverse reactions appear.
- Be alert to S&S of superinfections (see Appendix F). These symptoms should be reported promptly and appropriate therapy instituted.

CERTOLIZUMAB PEGOL

(cer-to'li-zu-mab)

Cimzia, Pegol

Classifications: BIOLOGICAL RESPONSE MODIFIER; IMMUNOMODULATOR; TUMOR NECROSIS FACTOR (TNF) MODIFIER

Therapeutic: TNF MODIFIER; IMMUNOLOGIC; ANTIRHEUMATIC

Prototype: Etanercept

Pregnancy Category: B

AVAILABILITY 400 mg lyophilized powder for injection; 200 mg/mL injection

ACTION & *THERAPEUTIC EFFECT* A fragment of an antibody Fab fragment with specificity for tumor necrosis factor (TNF)-alpha. This causes a reduction in the production of proinflammatory cytokines including interleukin-1 beta as well as TNF-alpha. Increased levels of TNF-alpha are found in the bowel wall areas that are affected by Crohn's disease and RA. *Reduces inflammatory cytokine production in Crohn's disease. It also decreases the serum level of C-reactive protein, a direct measure of the inflammatory process related to Crohn's disease. Effective for treatment of adults with moderately to severely active rheumatoid arthritis.*

USES Reduction of signs and symptoms, as well as maintenance of clinical response, in patients with moderately to severely active Crohn's disease. For treatment of moderate to severely active rheumatoid arthritis.

UNLABELED USES Psoriasis.

CONTRAINDICATIONS Active chronic or localized infections (e.g., TB, histoplasmosis, other fungal infections); HBV reactivation; lupus-like syndrome.

CAUTIOUS USE History of recurrent infection; concurrent immunosuppressive therapy; past/current residence in region where TB and histoplasmosis are endemic; CNS demyelinating disease; neurologic disorders, including seizure disorder, optic neuritis, peripheral neuropathy; recurrent/previous hematologic disorders; heart failure; hypersensitivity response to other TNF blocker(s); older adults; pregnancy (category B); lactation.

Safety and efficacy in children not established.

ROUTE & DOSAGE

Crohn's Disease
Adult: **Subcutaneous** 400 mg (two 200 mg injections) at weeks 0, 2, and 4, then 400 mg q4wk if clinical response occurs

Rheumatoid Arthritis
Adult: **Subcutaneous** Two 200 mg injections, at weeks 0, 2, and 4, then 200 mg every other week

ADMINISTRATION

Subcutaneous
- Reconstitute two 200 mg vials by adding 1 mL sterile water to each using a 20-gauge needle. Swirl gently then allow to sit to dissolve (may require up to 30 min); yields 200 mg/mL. Use within 2 h of reconstitution.
- Use two separate syringes with 20-gauge needles; withdraw 1 mL from each vial. Change to 23-gauge needles and inject into two separate sites on the abdomen or thigh.
- Store reconstituted solution. May be kept at room temperature for no longer than 2 h and refrigerated for up to 24 h.

ADVERSE EFFECTS (≥1%) **Body as a Whole:** Erythema nodosum, injection-site pain, pain in extremity, peripheral edema, pneumonia, *upper respiratory infection, urinary tract infection,* viral infections. **GI:** Abdominal pain. **Musculoskeletal:** *Arthralgia.* **Urogenital:** Pyelonephritis.

DIAGNOSTIC TEST INTERFERENCE Certolizumab may cause erroneously elevated *activated partial thromboplastin time (aPTT)* assay results.

INTERACTIONS Drug: Coadministration of **anakinra, abatacept** may cause increased risks of serious infections and neutropenia. Do not use with TNF ALPHA BLOCKERS.

PHARMACOKINETICS Absorption: 80% bioavailable. **Peak:** 54–171 h. **Half-Life:** 14 days.

NURSING IMPLICATIONS

Assessment & Drug Effects
- Prior to initiating therapy, patient should be evaluated for TB risk factors and latent TB. Monitor for S&S of TB throughout therapy.
- Report promptly any S&S of infection or hypersensitivity reaction. (See Appendix F for S&S.)
- Monitor closely carriers of HBV for signs of active infection. If suspected, withhold injection and notify prescriber.
- Monitor closely patients with heart failure for worsening cardiac status.
- Monitor for and report promptly any abnormal neurologic finding or unexplained bruising or bleeding.
- Lab tests: Baseline TB test; periodic CBC with platelet count.

Patient & Family Education
- Report promptly any of the following: S&S of infections, such as persistent fever; signs of an allergic reaction (e.g., hives, itching, swelling); unexplained bleeding or bruising.
- Do not accept vaccination with live (or attenuated) vaccines while on certolizumab.

✓CETIRIZINE
(ce-tir′i-zeen)
Reactine ✦, Zyrtec⊘

LEVOCETIRIZINE
(lev-o-ce-tir′i-zeen)

Common adverse effects in *italic*, life-threatening effects underlined; generic names in **bold**; classifications in SMALL CAPS; ✦ Canadian drug name; ⊙ Prototype drug

287

C

✓Xyzal
✓**Classifications:** ANTIHISTAMINE;
H_1-RECEPTOR ANTAGONIST; NON-
SEDATING
Therapeutic: ANTIHISTAMINE,
NON-SEDATING
Prototype: Loratadine
Pregnancy Category: B

AVAILABILITY 5 mg, 10 mg tablets;
5 mg, 10 mg chewable tablets; 5
mg/5 mL, 2.5 mg/5 mL syrup; **Levocetirizine:** 2.5 mg/mL syrup; 5
mg tablet

ACTION & *THERAPEUTIC EFFECT*
Cetirizine is a potent H_1-receptor
antagonist and thus an antihistamine without significant anticholinergic or CNS activity. Low
lipophilicity combined with its
H_1-receptor selectivity probably
accounts for its relative lack of
anticholinergic and sedative properties. *Effectively treats allergic
rhinitis and chronic urticaria by
eliminating or reducing the local
and systemic effects of histamine
release.*

USES Seasonal and perennial allergic rhinitis and chronic idiopathic
urticaria.

CONTRAINDICATIONS Hypersensitivity to H_1-receptor antihistamines
or hydroxyzine; lactation, children
younger than 2 y.
CAUTIOUS USE Moderate to severe
renal impairment, hepatic impairment, pregnancy (category B), children.

ROUTE & DOSAGE

Allergic Rhinitis

Adult: **PO** 5–10 mg once/day
Child: **PO** 2–5 y, 2.5 mg daily
(max: 5 mg/day); 6 y or older,
5–10 mg daily

Allergic Rhinitis (Levocetirizine)

*Adult/Adolescent/Child (older
than 6 y):* **PO** 2.5–5 mg once/day
Child (2–5 y): **PO** 1.25 mg each
evening

Chronic Urticaria

Adult: **PO** 10 mg daily or b.i.d.

Chronic Urticaria (Levocetirizine)

*Adult/Adolescent/Child (older
than 6 y):* **PO** 2.5–5 mg each
evening
Child (6 mo–5 y): **PO** 1.25 mg
each evening

***Renal Impairment Dosage
Adjustment (Levocetirizine)***

CrCl 51–80 mL/min: 2.5 mg
daily; *30–50 mL/min:* 2.5 mg
every other day; *10–29 mL/min:*
2.5 mg twice a week; *less than 10
mL/min:* Do not use

ADMINISTRATION

Oral
▪ Consult prescriber about dosage
if significant adverse effects appear. As elimination half-life is
prolonged in the older adult,
dosage adjustments may be
warranted.

ADVERSE EFFECTS (≥1%) **GI:** Constipation, diarrhea, dry mouth.
CNS: *Drowsiness, sedation, headache,* depression. **CV:** Syncope.

**INTERACTIONS Drug: Theophyl-
line** may decrease cetirizine clearance leading to toxicity. Use with
scopolamine or **atropine** may
cause anticholinergic effects.

PHARMACOKINETICS Absorption:
Readily from GI tract. **Peak:** 1 h.
Distribution: 93% protein bound;
minimal CNS concentrations. **Me-
tabolism:** Minimal (by CYP3A4).
Elimination: 60% unchanged in

Common adverse effects in *italic*, life-threatening effects underlined; generic names
in **bold;** classifications in SMALL CAPS; ♥ Canadian drug name; ☺ Prototype drug

urine within 24 h, 5% in feces. **Half-Life:** 7.4 h (cetirizine), 8–9 h (levocetirizine).

NURSING IMPLICATIONS

Assessment & Drug Effects

- Monitor for drug interactions. As the drug is highly protein bound, the potential for interactions with other protein-bound drugs exists.
- Monitor for sedation, especially the older adult.

Patient & Family Education

- Do not use in combination with OTC antihistamines.
- Do not engage in driving or other hazardous activities, before experiencing your responses to the drug.

CETRORELIX

(ce-tro-re'lix)

Cetrotide
Classification: GONADOTROPIN-RELEASING HORMONE (GnRH) ANTAGONIST
Therapeutic: LUTEINIZING HORMONE-RELEASING HORMONE RECEPTOR ANTAGONIST
Pregnancy Category: X

AVAILABILITY 0.25 mg, 3 mg injection

ACTION & *THERAPEUTIC EFFECT* Cetrotide is a luteinizing hormone-releasing hormone antagonist. *Prevents premature LH surges in patients undergoing controlled ovarian hyperstimulation for assisted reproduction.*

USES Treatment of infertility as part of an assisted reproduction program.
UNLABELED USES BPH, endometriosis.

CONTRAINDICATIONS Hypersensitivity to cetrorelix, extrinsic peptide hormones, mannitol, gonadotropin-releasing hormone analogs; primary ovarian failure; renal failure; pregnancy (category X); known or suspected pregnancy; lactation.
CAUTIOUS USE Hepatic insufficiency; polycystic ovary syndrome.

ROUTE & DOSAGE

Infertility

Adult: **Subcutaneous** 0.25 mg/days during early to mid follicular phase of the cycle (stimulation day 5 or 6) following the initiation of FSH or 3 mg as a single dose is administered when the serum estradiol level is indicative of an appropriate stimulation response, usually on FSH stimulation day 7 (range day 5–9). If HCG has not been administered within 4 days after the injection of 3 mg, then 0.25 mg should be administered once daily until HCG administration.

ADMINISTRATION

Subcutaneous

- Reconstitute the 0.25 or 3 mL vial with 1 or 3 mL, respectively, of sterile water for injection.
- Inject into lower abdominal wall following reconstitution. Rotate injection sites.
- Store the 3 mg dose at room temperature, 15°–30° C (59°–86° F). Store the 0.25 mg dose in the refrigerator.

ADVERSE EFFECTS (≥1%) CNS: Headache. **GI:** Nausea, vomiting, abdominal pain. **Endocrine:** Hot flashes. **Skin:** Pruritus at injection site. **Urogenital:** Ovarian enlargement, ovarian hyperstimulation syndrome, pelvic pain.

INTERACTIONS Drug: Cimetidine, methyldopa, metoclopramide, reserpine, PHENOTHIAZINES may

Common adverse effects in *italic*, life-threatening effects underlined; generic names in **bold**; classifications in SMALL CAPS; ♣ Canadian drug name; ⊘ Prototype drug

289

interfere with fertility efforts. **Herbal: Black cohosh, DHEA** may antagonize fertility efforts.

PHARMACOKINETICS Absorption: 85% absorbed from subcutaneous injection site. **Peak:** 1–2 h. **Metabolism:** Metabolized by peptidases. **Elimination:** 2–4% in urine, 5–10% in bile. **Half-Life:** 62 h after single dose, 20 h after multiple doses.

NURSING IMPLICATIONS

Assessment & Drug Effects

- Lab test: Monitor routine blood chemistries.
- Monitor weight and report development of edema and/or shortness of breath.

Patient & Family Education

- Contact prescriber immediately for any of the following: Abdominal or stomach pain, persistent or severe nausea, vomiting or diarrhea; decreased urination; pelvic pain; moderate to severe bloating, rapid weight gain; shortness of breath; swelling of lower legs.

CETUXIMAB

(ce-tux'i-mab)

Erbitux

Classifications: ANTINEOPLASTIC; MONOCLONAL ANTIBODY; EPIDERMAL GROWTH FACTOR RECEPTOR (EGFR) INHIBITOR

Therapeutic: ANTINEOPLASTIC

Prototype: Gefitinib

Pregnancy Category: C

AVAILABILITY 100 mg/50 mL injection

ACTION & *THERAPEUTIC EFFECT*

Cetuximab is a recombinant, monoclonal antibody that binds specifically to the epidermal growth factor receptor (EGFR, HER1, c-ErbB-1)

on both normal and tumor cells. Binding to the EGFR results in inhibition of cell growth, induction of apoptosis, and decreased vascular endothelial growth factor production. *Overexpression of EGFR is detected in many human cancers, including those of the colon and rectum. Cetuximab inhibits the growth and survival of tumor cells that overexpress the EGFR.*

USES Treatment of EGFR-expressing metastatic colorectal cancer in combination with irinotecan in patients who are refractory to irinotecan-based chemotherapy or as monotherapy in patients who are intolerant to irinotecan-based chemotherapy. Used in combination with radiation for squamous cell cancer of head and neck.

CONTRAINDICATIONS Lactation within 60 days of using cetuximab; worsening of preexisting pulmonary edema or interstitial lung disease; serious infusion reaction to drug. Safety and efficacy in children have not been established.

CAUTIOUS USE Infusion reaction, especially with first-time users; history of hypersensitivity to murine proteins or cetuximab; cardiac disease, coronary artery disease, CHF, arrhythmias; pulmonary disease, pulmonary fibrosis; UV exposure, radiation therapy; older adults; pregnancy (category C).

ROUTE & DOSAGE

Colorectal Cancer/Head and Neck Cancer
Adult: **IV** Start with 400 mg/m² over 2 h; continue with 250 mg/m² over 1 h weekly

Common adverse effects in *italic*, life-threatening effects underlined; generic names in **bold**; classifications in SMALL CAPS; ♣ Canadian drug name; ⊘ Prototype drug

ADMINISTRATION

Intravenous

Administer with full resuscitation equipment available and under the supervision of a prescriber experienced with chemotherapy.

- Premedication with an H_1-receptor antagonist (e.g., diphenhydramine 50 mg IV) is recommended.
- Monitor for an infusion reaction for at least 1 h following completion of infusion.

PREPARE: IV Infusion: • Do not shake or further dilute vial. Do not mix with other medication. • Inject cetuximab solution into a sterile, evacuated container or bag (i.e., glass, polyolefin, ethylene vinyl acetate, DEHP plasticized PVC, or PVC); repeat until needed dose has been added to container, using a new needle for each vial. • Attach to infusion set with a low-protein-binding 0.22-micron filter and prime line with cetuximab. May also administer by syringe and syringe pump; use a new needle and filter for each vial.
ADMINISTER: IV Infusion: Do NOT administer a bolus dose. • Give IV infusion via an infusion pump or syringe pump at 5 mL/min or less; piggyback into the patient's IV line. • Flush line with NS after infusion. • Note: Slow infusion rate by 50% if a prior, mild infusion reaction occurred.
***INCOMPATIBILITIES* Solution/additive:** Do not mix with other additives.

- Store unopened vials at 2°–8° C (36°–46° F). Note: Vials may contain a small amount of easily visible, white particles. • Cetuximab in IV bag is stable for up to 12 h refrigerated and up to 8 h at 20°–25° C (68°–77° F).

ADVERSE EFFECTS (≥1%) **Body as a Whole:** Infusion reactions (allergic reaction, anaphylactoid reaction, fever, chills, dyspnea, bronchospasm stridor, hoarseness, urticaria, hypotension), *fever,* sepsis, *asthenia, malaise,* pain, infection. **CNS:** *Headache,* insomnia, depression. **CV:** Cardiopulmonary arrest. **GI:** *Nausea, vomiting, diarrhea, abdominal pain, constipation,* stomatitis, dyspepsia. **Hematologic:** Leukopenia, anemia. **Metabolic:** Weight loss, peripheral edema, dehydration, hypomagnesemia, hypokalemia. **Respiratory:** Pulmonary embolism, pulmonary fibrosis (rare), *dyspnea,* cough. **Skin:** *Rash,* alopecia, pruritus. **Urogenital:** Kidney failure.

PHARMACOKINETICS Half-Life: 114 h (75–188 h).

NURSING IMPLICATIONS

Assessment & Drug Effects
- Discontinue infusion and notify prescriber for S&S of a severe infusion reaction: Chills, fever, bronchospasm, stridor, hoarseness, urticaria, and/or hypotension. Carefully monitor until complete resolution of all S&S.
- Monitor pulmonary status and report onset of acute or worsening pulmonary symptoms.
- Lab tests: Periodic CBC with differential, electrolytes, Hct and Hgb. Closely monitor serum electrolytes, including serum magnesium, potassium, and calcium, during and for 1 h after administration of this drug. Additionally, electrolytes need to be monitored for 8 wk after completion of therapy.

Patient & Family Education
- Report immediately: Difficulty breathing, wheezing, shortness of breath, hives, faintness and/or dizziness anytime during IV infusion.

Common adverse effects in *italic*, life-threatening effects underlined; generic names in **bold**; classifications in SMALL CAPS; ♣ Canadian drug name; ⊘ Prototype drug

291

- Report promptly any of the following: Eye inflammation, mouth sores, skin rash, redness, or severe dry skin.
- Wear sunscreen and a hat and limit sun exposure while being treated with this drug.

CEVIMELINE HYDROCHLORIDE

(cev-i-may′leen)

Evoxac

Classifications: CHOLINERGIC AGONIST; CHOLINERGIC ENHANCER
Therapeutic: CHOLINERGIC RECEPTOR ENHANCER
Pregnancy Category: C

AVAILABILITY 30 mg capsules

ACTION & THERAPEUTIC EFFECT
Cholinergic agent that binds to muscarinic receptors. *Increases secretion of exocrine glands, such as salivary and sweat glands. It relieves severe dry mouth.*

USES Treatment of dry mouth in patients with Sjögren's syndrome.

CONTRAINDICATIONS Hypersensitivity to cevimeline; uncontrolled asthma; acute iritis; narrow-angle glaucoma; lactation.
CAUTIOUS USE Controlled asthma; chronic bronchitis, COPD; cardiac disease, cardiac arrhythmias, myocardial infarction; history of nephrolithiasis or cholelithiasis; older adults; pregnancy (category C). Safety and effectiveness in children are not established.

ROUTE & DOSAGE

Dry Mouth
Adult: **PO** 30 mg t.i.d.

ADMINISTRATION

Oral
- Give without regard to food.
- Store refrigerated at 2°–8° C (35.6°–46.4° F) with occasional fluctuations between 15°–30° C (59°–86° F).

ADVERSE EFFECTS (≥1%) **Body as a Whole:** *Excessive sweating, headache,* back pain, dizziness, fatigue, pain, hot flushes, rigors, tremor, hypertonia, myalgia, fever, eye pain, earache, flu-like symptoms.

CNS: Insomnia, anxiety, vertigo, depression, hyporeflexia. **CV:** Peripheral edema, chest pain. **GI:** *Nausea, diarrhea,* excessive salivation, dyspepsia, abdominal pain, coughing, vomiting, constipation, anorexia, dry mouth, hiccup. **Respiratory:** *Rhinitis, sinusitis, upper respiratory tract infection,* pharyngitis, bronchitis. **Skin:** Rash, conjunctivitis, pruritus. **Special Senses:** Abnormal vision. **Urogenital:** Urinary tract infection.

INTERACTIONS Drug: BETA-ADRENERGIC AGONISTS may cause conduction disturbances; PARASYMPATHOMIMETIC DRUGS may have additive effects.

PHARMACOKINETICS Absorption: Rapidly absorbed. **Peak:** 1.5–2 h. **Distribution:** Less than 20% protein bound. **Metabolism:** In liver by CYP2D6 and 3A3/4. **Elimination:** Primarily in urine. **Half-Life:** 5 h.

NURSING IMPLICATIONS

Assessment & Drug Effects
- Monitor for S&S of increased airway resistance, especially in patient with asthma, bronchitis, emphysema, or COPD.
- Lab tests: Routine blood chemistry during long-term therapy.
- Report S&S of excess cholinergic activity (e.g., diaphoresis, frequent

Common adverse effects in *italic*, life-threatening effects underlined; generic names in **bold**; classifications in SMALL CAPS; ♣ Canadian drug name; ⊘ Prototype drug

urge to urinate, nausea and/or diarrhea).

Patient & Family Education
- Do not drive or engage in potentially hazardous activities until response to drug is known.
- Consult prescriber if confusion, dizziness, or faintness occur.
- Report diminished night vision or depth perception.
- Drink fluids liberally (2000–3000 mL/day) in the event of excessive sweating.

CHARCOAL, ACTIVATED (LIQUID ANTIDOTE)

Actidose, CharcoAid, Charco-caps, Charcodote, Insta-Char
Classifications: ANTIDOTE; AD-SORBENT
Therapeutic: ANTIDOTE
Pregnancy Category: C

AVAILABILITY 208 mg/mL, 15 g, 30 g, 50 mg liquid/suspension

ACTION & *THERAPEUTIC EFFECT*
Activated charcoal (carbon) is a chemically inert, odorless, tasteless, fine black powder with wide spectrum of adsorptive activity. Acts by binding (adsorbing) toxic substances, thereby inhibiting their GI absorption, enterohepatic circulation, and thus bioavailability. *Action appears to result from drug diffusion from plasma into GI tract where it is adsorbed by activated charcoal. Effectively adsorbs toxins in the gut preventing their systemic absorption and impact.*

USES General purpose emergency antidote in the treatment of poisonings by most drugs and chemicals. Gastric dialysis (repetitive doses) in uremia to adsorb various waste products from GI tract; severe acute poisoning. Has been used to adsorb intestinal gases in treatment of dyspepsia, flatulence, and distention (value in these conditions not established). Sometimes used topically as a deodorant for foul-smelling wounds and ulcers.

CONTRAINDICATIONS Reportedly not effective for poisonings by cyanide, mineral acids, caustic alkalis, organic solvents, iron, ethanol, methanol; gag reflex depression, coma; GI obstruction; quinidine or quinine hypersensitivity.
CAUTIOUS USE Pregnancy (category C); lactation.

ROUTE & DOSAGE

Acute Poisonings
Adult: **PO** 30–100 g in at least 180–240 mL (6–8 oz) of water or 1 g/kg
Child (1–12 y): **PO** 1–2 g/kg or 15–30 g in at least 6–8 oz of water
Infant (younger than 1 y): **PO** 1 g/kg

Gastric Dialysis
Adult: **PO** 20–40 g q6h for 1 or 2 days

GI Disturbances
Adult: **PO** 520–975 mg p.c. up to 5 g/day

ADMINISTRATION

Oral
- In an emergency, dose may be approximated by stirring sufficient activated charcoal into tap water to make a slurry the consistency of soup (about 20–30 g in at least 240 mL of water).
- Activated charcoal can be swallowed or given through a nasogastric tube. If administered too rapidly, patient may vomit.
- Store in tightly covered container.

Common adverse effects in *italic*, life-threatening effects underlined; generic names in **bold**; classifications in SMALL CAPS; ♣ Canadian drug name; ⊙ Prototype drug

293

ADVERSE EFFECTS (≥1%) **GI:** Vomiting (rapid ingestion of high doses), constipation, diarrhea (from sorbitol).

INTERACTIONS Drug: May decrease absorption of all other oral medications—administer at least 2 h apart.

PHARMACOKINETICS Absorption: Not absorbed. **Elimination:** In feces.

NURSING IMPLICATIONS

Assessment & Drug Effects

▪ Record appearance, color, consistency, frequency, and relative amount of stools. Inform patient that activated charcoal will color feces black.

CHLORAL HYDRATE

(klor′al hye′drate)

Aquachloral Supprettes, Noctec, Novochlorhydrate ♣
Classifications: ANXIOLYTIC; SEDATIVE-HYPNOTIC
Therapeutic: ANTIANXIETY; SEDATIVE-HYPNOTIC
Prototype: Secobarbital
Pregnancy Category: C
Controlled Substance: Schedule IV

AVAILABILITY 500 mg capsules; 250 mg/5 mL, 500 mg/5 mL syrup; 324 mg, 500 mg, 648 mg suppositories

ACTION & *THERAPEUTIC EFFECT* Produces "physiologic sleep" by mild cerebral depression with little effect on respirations or BP and little or no hangover. *Chloral hydrate in low doses is a sedative-hypnotic that does not affect sleep physiology (e.g., REM sleep).*

USES Short-term management of insomnia, general sedation (especially in the young and the older adult), sedation before and after surgery, to reduce anxiety associated with drug withdrawal, and alone or with paraldehyde to prevent or suppress alcohol withdrawal symptoms.

CONTRAINDICATIONS Known hypersensitivity to chloral hydrate or chloral derivatives; severe hepatic, renal, or cardiac disease; rectal dosage form in patients with proctitis; oral use in patients with esophagitis, gastritis, gastric or duodenal ulcers.
CAUTIOUS USE History of intermittent porphyria, asthma, history of or proneness to drug dependence, depression, suicidal tendencies; pregnancy (category C).

ROUTE & DOSAGE

Sedative

Adult: **PO/PR** 250 mg t.i.d. p.c.
Child: **PO/PR** 25–50 mg/kg/day divided q6–8h (max: 500 mg/dose)

Hypnotic

Adult: **PO/PR** 500 mg–1 g 15–30 min before bedtime or 30 min before surgery
Geriatric: **PO/PR** 250 mg at bedtime
Child: **PO/PR** 50 mg/kg 15–30 min before bedtime or 30 min before surgery (max: 1 g)

EEG Premedication

Child: **PO/PR** 20–25 mg/kg 30–60 min prior to procedure

ADMINISTRATION

Oral

▪ Dilute liquid preparations in chilled fluids to minimize unpleasant taste.
▪ Watch to see that drug is not cheeked and hoarded.

Rectal

▪ Moisten suppository with a water-based lubricant, such as K-Y jelly, prior to insertion.

Common adverse effects in *italic*, life-threatening effects underlined; generic names in **bold;** classifications in SMALL CAPS; ♣ Canadian drug name; ❷ Prototype drug

ADVERSE EFFECTS (≥1%) **Body as a Whole:** Angioedema, eosinophilia, breath odor, leukopenia, ketonuria, renal and hepatic damage, sudden death. **CV:** Arrhythmias, cardiac arrest. **GI:** *Nausea, vomiting, diarrhea,* severe gastritis. **CNS:** Dizziness, motor incoordination, headache. **Skin:** Purpura, urticaria, erythematous rash, eczema, erythema multiforme, fixed drug eruptions. **Special Senses:** Conjunctivitis.

DIAGNOSTIC TEST INTERFERENCE False-positive results for *urine glucose* with *Benedict's solutions,* and *Clinitest.* Possible interference with fluorometric test for *urine catecholamines* and *urinary 17-OHCS* determinations.

INTERACTIONS Drug: Alcohol, BARBITURATES, **paraldehyde,** other CNS DEPRESSANTS potentiate CNS depression; tachycardia may also occur with **alcohol;** increases anticoagulant effect of ORAL ANTICOAGULANTS; **furosemide** IV can produce flushing, diaphoresis, BP changes.

PHARMACOKINETICS Absorption: Readily from oral or rectal administration. **Onset:** 30–60 min. **Peak:** 1–3 h. **Duration:** 4–8 h. **Distribution:** Well distributed to all tissues; 70–80% protein bound; crosses placenta. **Metabolism:** In liver to the active metabolite trichloroethanol. **Elimination:** Primarily by kidneys; small amount excreted in feces via bile. **Half-Life:** 8–11 h.

NURSING IMPLICATIONS

Assessment & Drug Effects
- Do not discontinue abruptly following prolonged use. Sudden withdrawal from dependent patients may produce delirium, mania, or convulsions.

- Monitor for S&S of allergic skin reactions, which may occur within several hours or as long as 10 days after drug administration.

Patient & Family Education
- Do not ambulate without assistance until response to drug is known.
- Avoid concomitant use of alcoholic beverages.
- Avoid driving and other potentially hazardous activities while under the influence of chloral hydrate.

CHLORAMBUCIL
(klor-am′byoo-sil)
Leukeran
Classifications: ANTINEOPLASTIC; ALKYLATING AGENT
Therapeutic: ANTINEOPLASTIC; NITROGEN MUSTARD
Prototype: Cyclophosphamide
Pregnancy Category: D

AVAILABILITY 2 mg tablets

ACTION & *THERAPEUTIC EFFECT* Potent aromatic derivative of the alkylating agent nitrogen mustard which is slowest acting and least toxic of the nitrogen mustards. A cell-cycle nonspecific drug (kills both resting and dividing cells), it causes cytotoxic cross linkage in DNA, thus preventing synthesis of DNA, RNA, and proteins. *Lymphocytic effect is marked; thus it is effective in treatment of various lymphomas.*

USES As single agent or with other antineoplastics in treatment of chronic lymphocytic leukemia, malignant lymphomas including lymphosarcoma, Hodgkin's disease, and giant follicular lymphoma, and in treatment of carcinoma of the ovary, breast, and testes.
UNLABELED USES Nonneoplastic conditions: Vasculitis complicating rheumatoid arthritis, autoimmune

Common adverse effects in *italic*, life-threatening effects underlined; generic names in **bold;** classifications in SMALL CAPS; ◆ Canadian drug name; ⊘ Prototype drug

295

hemolytic anemias associated with cold agglutinins, lupus glomerulonephritis, idiopathic nephrotic syndrome, polycythemia vera, macroglobulinemia.

CONTRAINDICATIONS Hypersensitivity to chlorambucil or to other alkylating agents; administration within 4 wk of a full course of radiation or chemotherapy; full dosage if bone marrow is infiltrated with lymphomatous tissue or is hypoplastic; smallpox and other vaccines; pregnancy (category D), lactation.

CAUTIOUS USE Excessive or prolonged dosage, pneumococcus vaccination, history of seizures or head trauma.

ROUTE & DOSAGE

Malignant Diseases (Lymphomas, Hodgkin's Disease, etc.)
Adult: **PO** 0.1–0.2 mg/kg/day (usual dose 4–10 mg/day)
Child: **PO** 0.1–0.2 mg/kg/day in single or divided doses

ADMINISTRATION

Oral
- Control nausea and vomiting by giving entire daily dose at one time, 1 h before breakfast or 2 h after evening meal, or at bedtime. Consult prescriber.
- Store in tightly closed, light-resistant container.

ADVERSE EFFECTS (≥1%) **Body as a Whole:** Drug fever, skin rashes, papilledema, alopecia, peripheral neuropathy, sterile cystitis, pulmonary complications, seizures (high doses). **GI:** Low incidence of gastric discomfort, hepatotoxicity. **Hematologic:** Bone marrow depression: *Leukopenia,* thrombocytopenia, anemia. **Metabolic:** Sterility, hyperuricemia.

INTERACTIONS Drug: May have to adjust dose of **allopurinol, colchicine** because of chlorambucil-associated hyperuricemia.

PHARMACOKINETICS Absorption: Rapidly and completely from GI tract. **Peak:** 1 h. **Distribution:** Extensively bound to plasma and tissue proteins; crosses placenta. **Metabolism:** In liver. **Elimination:** 60% in urine as metabolites within 24 h. **Half-Life:** 1.5–2.5 h.

NURSING IMPLICATIONS

Assessment & Drug Effects
- Lab tests: CBC, Hgb, total and differential leukocyte counts, and serum uric acid initially and at least once weekly during treatment.
- Leukopenia usually develops after the third week of treatment; it may continue for up to 10 days after last dose, then rapidly return to normal.
- Avoid or reduce to minimum injections and other invasive procedures (e.g., rectal temperatures, enemas) when platelet count is low.

Patient & Family Education
- Notify prescriber if the following symptoms occur: Unusual bleeding or bruising, sores on lips or in mouth; flank, stomach, or joint pain; fever, chills, or other signs of infection, sore throat, cough, dyspnea.
- Report immediately the onset of a skin reaction.
- Drink at least 10–12 glasses [240 mL (8 oz) each] of fluid per day, if not contraindicated.

CHLORAMPHENICOL

(klor-am-fen′i-kole)

Chloromycetin, Novo-chloro-cap ✦

CHLORAMPHENICOL SODIUM SUCCINATE

Chloromycetin Sodium Succinate

Classification: ANTIBIOTIC
Therapeutic: BROAD-SPECTRUM ANTIBIOTIC
Pregnancy Category: C

AVAILABILITY 250 mg capsules; 100 mg/mL injection; 5 mg/mL ophth solution; 10 mg/g ointment

ACTION & *THERAPEUTIC EFFECT*

Synthetic broad-spectrum antibiotic believed to act by binding to the 50S ribosome of bacteria and thus interfering with protein synthesis. *Effective against a wide variety of gram-negative and gram-positive bacteria and most anaerobic microorganisms.*

USES Severe infections when other antibiotics are ineffective or are contraindicated. Particularly effective against *Salmonella typhi* and other *Salmonella* sp., *Streptococcus pneumoniae, Neisseria,* meningeal infections caused by *H. influenzae,* and infections involving *Bacteroides fragilis* and other anaerobic organisms, *Rickettsia rickettsii* (cause of Rocky Mountain spotted fever) and other rickettsiae, the lymphogranuloma-psittacosis group *(Chlamydia),* and *Mycoplasma.* Also used in cystic fibrosis anti-infective regimens and topically for infections of skin, eyes, and external auditory canal.

CONTRAINDICATIONS History of hypersensitivity or toxic reaction to chloramphenicol; treatment of minor infections, prophylactic use; typhoid carrier state, history or family history of drug-induced bone marrow depression, concomitant therapy with drugs that produce bone marrow depression; lactation.

CAUTIOUS USE Impaired hepatic or renal function, premature and full-term infants, children; intermittent porphyria; patients with G6PD deficiency; patient or family history of drug-induced bone marrow depression; pregnancy (category C).

ROUTE & DOSAGE

Serious Infections

Adult: **PO/IV** 50 mg/kg/day in 4 divided doses
Infant/Child: **PO/IV** 50–75 mg/kg/day divided q6h (max: 4 g/day)
Neonate: **IV** 25–50 mg/kg/day divided q12–24h

Meningitis

Adult/Child: **IV** 75–100 mg/kg/day divided q6h

ADMINISTRATION

Oral

- Give preferably with a full glass of water on an empty stomach, at least 1 h before or 2 h after a meal, to achieve optimum blood levels.

Ophthalmic

- Apply light pressure to lacrimal duct after instillation for 1–2 min to prevent drainage into nasopharynx and systemic absorption. This is an extremely important step to decrease absorption. Several cases of aplastic anemia have been associated with use of ophthalmic preparations.

Intravenous

IV administration to neonates, infants, children: Verify correct IV concentration and rate of infusion with prescriber.

PREPARE: Direct: Dilute each 1 g with 10 mL of sterile water or D5W. **Intermittent:** Further dilute in 50–100 mL of D5W.

Common adverse effects in *italic*, life-threatening effects underlined; generic names in **bold**; classifications in SMALL CAPS; ♣ Canadian drug name; ✪ Prototype drug

297

ADMINISTER: **Direct:** Give slowly over a period of at least 1 min. **Intermittent:** Give over 30–60 min.
INCOMPATIBILITIES **Solution/additive:** Chlorpromazine, glycopyrrolate, metoclopramide, polymyxin B, prochlorperazine, promethazine, TETRACYCLINES, vancomycin. **Y-site:** Fluconazole.

▪ Solution for infusion may form crystals or a second layer when stored at low temperatures. Solution can be clarified by shaking vial. ▪ Do not use cloudy solutions.

▪ Store topical ophthalmic, otic, and skin preparations, PO forms, and unopened ampuls at room temperature and protected from light unless otherwise directed by manufacturer.

ADVERSE EFFECTS (≥1%) **Body as a Whole:** Hypersensitivity, angioedema, dyspnea, fever, anaphylaxis, superinfections, Gray syndrome. **GI:** Nausea, vomiting, diarrhea, perianal irritation, enterocolitis, glossitis, stomatitis, unpleasant taste, xerostomia. **Hematologic:** Bone marrow depression (dose-related and reversible): Reticulocytosis, leukopenia, granulocytopenia, thrombocytopenia, increased plasma iron, reduced Hgb, hypoplastic anemia, hypoprothrombinemia. Non-dose-related and irreversible pancytopenia, agranulocytosis, aplastic anemia, paroxysmal nocturnal hemoglobinuria, leukemia. **CNS:** Neurotoxicity: Headache, mental depression, confusion, delirium, digital paresthesias, peripheral neuritis. **Skin:** Urticaria, contact dermatitis, maculopapular and vesicular rashes, fixed-drug eruptions. **Special Senses:** Visual disturbances, optic neuritis, optic nerve atrophy, contact conjunctivitis.

DIAGNOSTIC TEST INTERFERENCE
Possibility of false-positive results

for **urine glucose** by **copper reduction methods** (e.g., **Benedict's solution, Clinitest**).

INTERACTIONS Drug: The metabolism of **chlorpropamide, dicumarol, phenytoin, tolbutamide** may be decreased, prolonging their activity. **Phenobarbital** decreases chloramphenicol levels. The response to **iron** preparations, **folic acid,** and **vitamin B$_{12}$** may be delayed.

PHARMACOKINETICS **Absorption:** Rapidly from GI tract. **Peak:** PO: 1–3 h; IV: 1 h. **Distribution:** Widely distributed to most body tissues including saliva and ascitic, pleural and synovial fluid; concentrates in liver and kidneys; penetrates CNS; crosses placenta. **Metabolism:** Primarily inactivated in liver. **Elimination:** Much longer in neonates; metabolite and free drug excreted in urine; excreted in breast milk. **Half-Life:** 1.5–4.1 h.

NURSING IMPLICATIONS

Assessment & Drug Effects

▪ Lab tests: Perform bacterial culture and susceptibility tests prior to first dose. Baseline CBC, platelets, serum iron, and reticulocyte cell counts before initiation of therapy, at 48 h intervals during therapy, and periodically. Monitor chloramphenicol blood levels weekly or more frequently with hepatic dysfunction and in patients receiving therapy for longer than 2 wk. Desired concentrations: Peak 10–20 mcg/mL; through 5–10 mcg/mL.

▪ Check temperature at least q4h. Usually chloramphenicol is discontinued if temperature remains normal for 48 h.

▪ Monitor I&O ratio or pattern: Report any appreciable change.

Common adverse effects in *italic*, life-threatening effects underlined; generic names in **bold**; classifications in SMALL CAPS; ✚ Canadian drug name; ⊘ Prototype drug

- Monitor for S&S of gray syndrome, which has occurred 2–9 days after initiation of high dose chloramphenicol therapy in premature infants and neonates and in children 2 y or younger. Report early signs: Abdominal distention, failure to feed, pallor, changes in vital signs.

Patient & Family Education

- A bitter taste may occur 15–20 sec after IV injection; it usually lasts only 2–3 min.
- Report immediately sore throat, fever, fatigue, petechiae, nose bleeds, bleeding gums, or other unusual bleeding or bruising, or any other suspicious sign of symptom.
- Watch for S&S of superinfection (see Appendix F).
- Notify prescriber immediately if signs of hypersensitivity reaction (see Appendix F), irritation, superinfection, or other adverse reactions appear.

CHLORDIAZEPOXIDE HYDROCHLORIDE

(klor-dye-az-e-pox'ide)

Librium, Solium ♦
Classifications: ANXIOLYTIC; SEDATIVE-HYPNOTIC; BENZODIAZEPINE
Therapeutic: ANTIANXIETY; SEDATIVE-HYPNOTIC
Prototype: Lorazepam
Pregnancy Category: D
Controlled Substance: Schedule IV

AVAILABILITY 5 mg, 10 mg, 25 mg capsules

ACTION & *THERAPEUTIC EFFECT*
Benzodiazepine derivative that acts on the limbic, thalamic, and hypothalamic areas of the CNS. Has long-acting hypnotic properties.

Causes mild suppression of REM sleep and of deeper phases, particularly stage 4, while increasing total sleep time. *Produces mild anxiolytic (reduces anxiety), sedative, anticonvulsant, and skeletal muscle relaxant effects.*

USES Relief of various anxiety and tension states, preoperative apprehension and anxiety, and for management of alcohol withdrawal.
UNLABELED USES Essential, familial, and senile action tremors.

CONTRAINDICATIONS Hypersensitivity to chlordiazepoxide and other benzodiazepines; narrow-angle glaucoma, prostatic hypertrophy, shock, comatose states, primary depressive disorder or psychoses, acute alcohol intoxication; pregnancy (category D), lactation.
CAUTIOUS USE Anxiety states associated with impending depression, history of impaired hepatic or renal function; addiction-prone individuals, blood dyscrasias; in the older adult, debilitated patients, children; aggressive or hyperactive children; hyperkinesis; oral use in children younger than 6 y, parenteral use in children younger than 12 y.

ROUTE & DOSAGE

Mild Anxiety, Preoperative Anxiety
Adult: **PO** 5–10 mg t.i.d. or q.i.d.
Geriatric: **PO** 5 mg b.i.d. to q.i.d.
Child: **PO** 5 mg b.i.d. to q.i.d.; may be increased to 10 mg t.i.d.

Severe Anxiety and Tension
Adult: **PO** 20–25 mg t.i.d. or q.i.d.

Alcohol Withdrawal Syndrome
Adult: **PO** 50–100 mg prn up to 300 mg/day

Common adverse effects in *italic*, life-threatening effects <u>underlined</u>; generic names in **bold**; classifications in SMALL CAPS; ♦ Canadian drug name; ⊘ Prototype drug

299

ADMINISTRATION

Oral

- Give with or immediately after meals or with milk to reduce GI distress. If an antacid is prescribed, it should be taken at least 1 h before or after chlordiazepoxide to prevent delay in drug absorption.
- Store in tight, light-resistant containers at room temperature unless otherwise specified by manufacturer.

ADVERSE EFFECTS (≥1%) **Body as a Whole:** Edema, pain in injection site, jaundice, hiccups, <u>respiratory depression</u>. **CV:** Orthostatic hypotension, tachycardia, changes in ECG patterns seen with rapid IV administration. **GI:** Nausea, dry mouth, vomiting, constipation, increased appetite. **CNS:** *Drowsiness,* dizziness, *lethargy,* changes in EEG pattern; vivid dreams, nightmares, headache, vertigo, syncope, tinnitus, confusion, hallucinations, parodoxic rage, depression, delirium, ataxia. **Skin:** Photosensitivity, skin rash. **Urogenital:** Urinary frequency.

DIAGNOSTIC TEST INTERFERENCE Chlordiazepoxide increases *serum bilirubin, AST* and *ALT;* decreases *radioactive iodine uptake;* and may falsely increase readings for *urinary 17-OHCS* (modified *Glenn-Nelson* technique).

INTERACTIONS Drug: Alcohol, CNS DEPRESSANTS, ANTICONVULSANTS potentiate CNS depression; **cimetidine** increases **chlordiazepoxide** plasma levels, thus increasing toxicity; may decrease antiparkinson effects of **levodopa;** may increase **phenytoin** levels; smoking decreases sedative and antianxiety effects. **Herbal: Kava, valerian** may potentiate sedation.

PHARMACOKINETICS Absorption: Well absorbed from GI tract. **Peak:** 1–4 h. **Distribution:** Widely distributed throughout body; crosses placenta. **Metabolism:** In liver via CYP3A4 to long-acting active metabolite. **Elimination:** Slowly excreted in urine (may last several days); excreted in breast milk. **Half-Life:** 5–30 h.

NURSING IMPLICATIONS

Assessment & Drug Effects

- Monitor for S&S of orthostatic hypotension and tachycardia, observe closely and monitor vital signs.
- Check BP and pulse before giving benzodiazepine in early part of therapy. If blood pressure falls 20 mm Hg or more or if pulse rate is above 120 bpm, notify prescriber.
- Lab tests: Periodic blood cell counts and LFTs are recommended during prolonged therapy.
- Monitor I&O until drug dosage is stabilized. Report changes in I&O ratio and dysuria to prescriber.
- Monitor for S&S of paradoxic reactions—excitement, stimulation, disturbed sleep patterns, acute rage—which may occur during first few weeks of therapy in psychiatric patients and in hyperactive and aggressive children receiving chlordiazepoxide. Withhold drug and report to prescriber.
- Assess patient's sleep pattern. If dreams or nightmares interfere with rest, notify prescriber.
- Supervise ambulation, especially with older adults & debilitated patients.

Patient & Family Education

- Abrupt discontinuation of drug in patients receiving high doses for long periods (4 mo or longer) has precipitated withdrawal symptoms, but not for at least

5–7 days because of slow elimination.

- Long-term use of this drug may cause mouth soreness. Good oral hygiene can alleviate the discomfort.
- Avoid activities requiring mental alertness until reaction to the drug has been evaluated.
- Avoid drinking alcoholic beverages. When combined with chlordiazepoxide, effects of both are potentiated.
- Avoid excessive sunlight. Use sunscreen lotion (SPF 12 or above) if allowed.

CHLOROQUINE PHOSPHATE ⊕

(klor'oh-kwin)

Aralen
Classification: ANTIMALARIAL
Therapeutic: ANTIMALARIAL; AMEBICIDE
Pregnancy Category: C

AVAILABILITY 250 mg (150 mg base), 500 mg (300 mg base) tablets

ACTION & THERAPEUTIC EFFECT Antimalarial activity is believed to be based on its ability to form complexes with DNA of parasite, thereby inhibiting replication and transcription to RNA and nucleic acid synthesis. *Acts as a suppressive agent in patient with* P. vivax *or* P. malariae *malaria; terminates acute attacks and increases intervals between treatment and relapse of malaria. Abolishes the acute attack of* P. falciparum *malaria but does not prevent the infection.*

USES Suppression and treatment of malaria caused by *P. malariae, P. ovale, P. vivax,* and susceptible forms of *P. falciparum,* and in the treatment of extraintestinal amebiasis. Concomitant therapy with primaquine is necessary for radical cure of *P. vivax* and *P. malariae* malarias.

UNLABELED USES Discoid and systemic lupus erythematosus, porphyria cutanea tarda, solar urticaria, polymorphous light eruptions, and in rheumatoid arthritis (as second-line therapy).

CONTRAINDICATIONS Hypersensitivity to 4-aminoquinolines, psoriasis; porphyria, renal disease, 4-aminoquinoline-induced retinal or visual field changes; long-term therapy in children. Safe use in women of childbearing potential not established.

CAUTIOUS USE Impaired hepatic function, alcoholism, eczema, patients with G6PD deficiency, infants and children, hematologic, GI, and neurologic disorders; pregnancy (category C).

ROUTE & DOSAGE

Doses are expressed in terms of chloroquine base.

Acute Malaria

Adult: **PO** 600 mg base followed by 300 mg base at 6, 24, and 48 h
Child: **PO** 10 mg base/kg, then 5 mg base/kg at 6, 24, and 48 h

Malaria Suppression

Adult: **PO** 300 mg base the same day each week starting 2 wk before exposure and continuing for 4–6 wk after leaving the area of exposure (max: 300 mg base/wk)
Child: **PO** 5 mg base/kg the same day each week starting 2 wk before exposure and continuing for 4–6 wk after leaving the area of exposure (max: 300 mg base/wk)

Common adverse effects in *italic,* life-threatening effects underlined; generic names in **bold;** classifications in SMALL CAPS; ♣ Canadian drug name; ⊕ Prototype drug

301

Extraintestinal Amebiasis

Adult: **PO** 600 mg base/day for 2 days, then 300 mg base/day for 2–3 wk
Child: **PO** 10 mg base/kg/day for 2–3 wk

Rheumatoid Arthritis, SLE

Adult: **PO** 150 mg base/day with evening meal

ADMINISTRATION

Oral

- Give immediately before or after meals to minimize GI distress.
- Monitor child's dose closely. Children are extremely susceptible to overdosage.

ADVERSE EFFECTS (≥1%) **Body as a Whole:** Slight weight loss, myalgia, lymphedema of upper limbs. **CV:** Hypotension; ECG changes. **GI:** *Diarrhea,* abdominal cramps, *nausea,* vomiting, anorexia. **Hematologic:** Hemolytic anemia in patients with G6PD deficiency. **CNS:** Mild transient headache, fatigue, irritability, confusion, nightmares, skeletal muscle weakness, paresthesias, reduced reflexes, vertigo. **Skin:** Bleaching of scalp, eyebrows, body hair, and freckles, pruritus, patchy alopecia (reversible). **Special Senses:** (Usually reversible): Blurred vision, disturbances of accommodation, night blindness, scotomas, visual field defects, photophobia, corneal edema, opacity or deposits, ototoxicity (rare).

INTERACTIONS Drug: Aluminum- and **magnesium**-containing ANTACIDS and LAXATIVES decrease chloroquine absorption, so separate administration by at least 4 h; chloroquine may interfere with response to **rabies vaccine. Food:** Taking **lemon juice** decreases therapeutic effect.

PHARMACOKINETICS Absorption: Rapidly and almost completely absorbed. **Peak:** 1–2 h. **Distribution:** Widely distributed; concentrates in lungs, liver, erythrocytes, eyes, skin, and kidneys; crosses placenta. **Metabolism:** Partially in liver to active metabolites. **Elimination:** In urine; excreted in breast milk. **Half-Life:** 70–120 h.

NURSING IMPLICATIONS

Assessment & Drug Effects

- Lab tests: CBC and ECG are advised before initiation of therapy and periodically thereafter in patients on long-term therapy.
- Monitor for changes in vision. Retinopathy (generally irreversible) can be progressive even after termination of therapy. Patient may be asymptomatic or complain of night blindness, scotomas, visual field changes, blurred vision, or difficulty in focusing. Withhold drug and report immediately to prescriber.

Patient & Family Education

- Report promptly visual or hearing disturbances, muscle weakness, or loss of balance, symptoms of blood dyscrasia (fever, sore mouth or throat, unexplained fatigue, easy bruising or bleeding).
- Use of dark glasses in sunlight or bright light may provide comfort (because of photophobia) and reduce risk of ocular damage.
- Avoid driving or other potentially hazardous activities until reaction to drug is known.
- May cause rusty yellow or brown discoloration of urine.
- Do not drink lemon juice along with chloroquine. It decreases the drug's effectiveness.

Common adverse effects in *italic,* life-threatening effects <u>underlined</u>; generic names in **bold**; classifications in SMALL CAPS; ✚ Canadian drug name; ✪ Prototype drug

CHLOROTHIAZIDE

(klor-oh-thye'a-zide)

CHLOROTHIAZIDE SODIUM

Diuril
Classifications: ELECTROLYTE AND
WATER BALANCE AGENT; THIAZIDE
DIURETIC; ANTIHYPERTENSIVE
Therapeutic: THIAZIDE DIURET-
IC; ANTIHYPERTENSIVE
Prototype: Hydrochlorothiazide
Pregnancy Category: C

AVAILABILITY 250 mg, 500 mg tablets; 500 mg injection; 250 mg/5 mL
oral suspension

ACTION & *THERAPEUTIC EFFECT*
Thiazide diuretic whose primary action is production of diuresis by direct action on the distal convoluted
tubules. Inhibits reabsorption of sodium, potassium, and chloride ions.
*Promotes renal excretion of sodium
(and water), bicarbonate, and potassium. Antihypertensive mechanism
is due to decreased peripheral resistance and reduced blood pressure.*

USES Edema associated with CHF,
renal dysfunction ascites, hypertension.
UNLABELED USES Hypercalciuria.

CONTRAINDICATIONS Hypersensitivity to thiazide or sulfonamides;
anuria; hypokalemia; renal failure;
jaundiced neonates; SLE.
CAUTIOUS USE History of sulfa allergy; impaired renal or hepatic
function or gout; hypercalcemia, diabetes mellitus, older adult or debilitated patients, pancreatitis, sympathectomy; pregnancy (category C).

ROUTE & DOSAGE

Hypertension, Edema

Adult: **PO** 500 mg–1 g/day in 1–2
divided doses **IV** 500 mg–1 g/
day in 1–2 divided doses

Child (over 6 mo): **PO** 10–20 mg/
kg/day in 1–2 doses (max: 375
mg)
Infant (less than 6 mo): **PO** 10–20
mg/kg/day divided in 2 doses

ADMINISTRATION

Oral
- Give with or after food to prevent
  gastric irritation. Extent of absorption appears to be increased by
  taking it with food.
- Schedule daily doses to avoid
  nocturia and interrupted sleep.

Intravenous
- Reserve for emergency or when
  patient unable to take oral medication. ▪ IV administration to infants and children: Verify correct
  IV concentration and rate of infusion with prescriber.

PREPARE: Intermittent Reconstitute the 500-mg vial with at least
18 mL sterile water for injection.
▪ May be further diluted with
D5W or NS.
ADMINISTER: Intermittent: Give at
a rate of 500 mg over 5 min. ▪ Thiazide preparations are extremely
irritating to the tissues, and great
care **must be** taken to avoid extravasation. ▪ If infiltration occurs,
stop medication, remove needle,
and apply ice if area is small.

**INCOMPATIBILITIES Solution/addi-
tive: Amikacin, chlorpromazine, fluorouracil, hydralazine, insulin, levorphanol,
methadone, morphine, norepinephrine, pentobarbital, polymyxin B, procaine, prochlorperazine, promazine,
promethazine, streptomycin,
triflupromazine, vancomycin,
vitamin B complex with C,
warfarin. Y-site: Codeine, TPN.**

- Store tablets, PO solutions, and parenteral dosage forms at 15°–30° C (59°–86° F) unless otherwise directed by manufacturer. ▪ Unused reconstituted IV solutions may be stored at room temperature up to 24 h. Use only clear solutions.

ADVERSE EFFECTS (≥1%) **Body as a Whole:** Fever, respiratory distress, anaphylactic reaction. **CV:** Irregular heart beat, weak pulse, orthostatic hypotension. **GI:** Vomiting, acute pancreatitis, diarrhea. **Hematologic:** Agranulocytosis (rare), aplastic anemia (rare), asymptomatic hyperuricemia, hyperglycemia, glycosuria, SIADH secretion. **Metabolic:** Hypokalemia, hypercalcemia, hyponatremia, hypochloremic alkalosis, elevated cholesterol and triglyceride levels. **CNS:** Unusual fatigue, dizziness, mental changes, vertigo, headache. **Skin:** Urticaria, photosensitivity, skin rash.

DIAGNOSTIC TEST INTERFERENCE Chlorothiazide (thiazides) may cause: marked increases in *serum amylase* values, decrease in *PBI* determinations; increase in excretion of *PSP;* increase in *BSP retention;* false-negative *phentolamine* and *tyramine* tests; interference with *urine steroid* determinations, and possibly the *histamine test* for pheochromocytoma. Thiazides should be discontinued at least 3 days before *bentiromide test* (thiazides can invalidate test) and before *parathyroid function tests* because they tend to decrease calcium excretion.

INTERACTIONS Drug: Amphotericin B, CORTICOSTEROIDS increase hypokalemic effects of chlorothiazide; the hypoglycemic effects of SULFONYLUREAS and **insulin** may be antagonized; **cholestyramine, colestipol** decrease thiazide absorption; intensifies hypoglycemic and hypotensive effects of **diazoxide;** increased potassium and magnesium loss may cause **digoxin** toxicity; decreases **lithium** excretion, increasing its toxicity; increases risk of NSAID-induced renal failure and may attenuate diuresis.

PHARMACOKINETICS Absorption: Incompletely absorbed PO. **Onset:** 2 h PO; 15 min IV. **Peak:** 3–6 h PO; 30 min IV. **Duration:** 6–12 h PO; 2 h IV. **Distribution:** Throughout extracellular tissue; concentrates in kidney; crosses placenta. **Metabolism:** Does not appear to be metabolized. **Elimination:** In urine and breast milk. **Half-Life:** 45–120 min.

NURSING IMPLICATIONS

Assessment & Drug Effects

- Monitor for therapeutic effect. Antihypertensive action of a thiazide diuretic requires several days before effects are observed; usually optimum therapeutic effect is not established for 3–4 wk.
- Lab tests: Baseline and periodic blood count, serum electrolytes, CO_2, BUN, creatinine, uric acid, and blood glucose.
- Monitor for hyperglycemia. Thiazide therapy can cause hyperglycemia (see Appendix F) and glycosuria in diabetic and diabetic-prone individuals.
- Monitor patients with gout. Asymptomatic hyperuricemia can be produced because of interference with uric acid excretion.
- Establish baseline weight before initiation of therapy. Weigh patient at the same time each a.m. under standard conditions. A gain of more than 1 kg (2.2) within 2 or 3 days and a gradual weight gain over the week's period is reportable.
- Monitor BP closely during early drug therapy.

Common adverse effects in *italic*, life-threatening effects <u>underlined</u>; generic names in **bold;** classifications in SMALL CAPS; ♣ Canadian drug name; ⊙ Prototype drug

- Inspect skin and mucous membranes daily for evidence of petechiae in patients receiving large doses and those on prolonged therapy.
- Monitor I&O rates and patterns: Excessive diuresis may cause electrolyte imbalance and necessitate prompt dosage adjustment.
- Monitor patients on digitalis therapy for S&S of hypokalemia (see Appendix G), which can precipitate digitalis intoxication.

Patient & Family Education
- Urination will occur in greater amounts and with more frequency than usual, and there will be an unusual sense of tiredness. With continued therapy, diuretic action decreases; BP lowering effects usually are maintained, and sense of tiredness diminishes.
- Make position changes slowly to minimize risks associated with orthostatic hypotension.
- Report to prescriber any illness accompanied by prolonged vomiting or diarrhea.
- Avoid drinking large quantities of coffee or other caffeine drinks. Caffeine has a diuretic effect.
- Report S&S of hypokalemia, hypercalcemia, or hyperglycemia (see Appendix F).
- Hypokalemia may be prevented if the daily diet contains potassium-rich foods. Eat a banana and drink at least 6 oz orange juice every day.
- Report photosensitivity reaction to prescriber. Photosensitivity may occur 1½–2 wk after initial sun exposure.

CHLORPHENIRAMINE MALEATE
(klor-fen-eer′a-meen)

Aller-Chlor, Chlo-Amine, Chlor-Trimeton, Chlor-Tripolon ♦, Novopheniram ♦, Phenetron, Telachlor, Teldrin, Trymegan
Classification: ANTIHISTAMINE (H_1-RECEPTOR ANTAGONIST)
Therapeutic: ANTIHISTAMINE
Prototype: Diphenhydramine
Pregnancy Category: C

AVAILABILITY 2 mg, 4 mg tablets; 8 mg, 12 mg sustained release tablets; 2 mg/5 mL syrup

ACTION & *THERAPEUTIC EFFECT*
Antihistamine that competes with histamine for H_1-receptor sites on effector cells; thus it promotes capillary permeability and edema formation and constrictive action on respiratory, gastrointestinal, and vascular smooth muscles. *Has effective antihistamine reaction resulting in decreasing allergic symptomatology.*

USES Symptomatic relief of various uncomplicated allergic conditions; to prevent transfusion and drug reactions in susceptible patients, and as adjunct to epinephrine and other standard measures in anaphylactic reactions.

CONTRAINDICATIONS Hypersensitivity to antihistamines of similar structure; lower respiratory tract symptoms, narrow-angle glaucoma, obstructive prostatic hypertrophy or other bladder neck obstruction, GI obstruction or stenosis; premature and newborn infants; during or within 14 days of MAO INHIBITOR therapy.
CAUTIOUS USE Convulsive disorders, increased intraocular pressure, hyperthyroidism, cardiovascular disease, hepatic disease; BPH; GI obstruction; hypertension, diabetes mellitus, history of bronchial asthma, COPD, older adult

Common adverse effects in *italic*, life-threatening effects underlined; generic names in **bold**; classifications in SMALL CAPS; ♦ Canadian drug name; ❷ Prototype drug

305

patients, patients with G6PD deficiency, pregnancy (category C), lactation.

ROUTE & DOSAGE

Symptomatic Allergy Relief

Adult: **PO** 2–4 mg t.i.d. or q.i.d. or 8–12 mg b.i.d. or t.i.d. (max: 24 mg/day)
Geriatric: **PO** 4 mg daily or b.i.d. or 8 mg sustained release at bedtime
Child: **PO** 6–12 y, 2 mg q4–6h (max: 12 mg/day); 2–6 y, 1 mg q4–6h

ADMINISTRATION

Oral

- Give on an empty stomach for fastest response.
- Sustained release tablets should be swallowed whole and not crushed or chewed.
- Ensure that chewable tablets are chewed or crushed before being swallowed with a liquid.

ADVERSE EFFECTS (≥1%) **Body as a Whole:** Sensation of chest tightness. **CV:** Palpitation, tachycardia, mild hypotension or hypertension. **GI:** Epigastric distress, anorexia, nausea, vomiting, constipation, or diarrhea. **CNS:** *Drowsiness,* sedation, headache, dizziness, vertigo, fatigue, disturbed coordination, tremors, euphoria, nervousness, restlessness, insomnia. **Special Senses:** *Dryness of mouth,* nose, and throat, tinnitus, vertigo, acute labyrinthitis, thickened bronchial secretions, blurred vision, diplopia. **Urogenital:** Urinary frequency or retention, dysuria.

DIAGNOSTIC TEST INTERFERENCE

Antihistamines should be discontinued 4 days before **skin testing** procedures for allergy because they may obscure otherwise positive reactions.

INTERACTIONS Drug: Alcohol (ethanol) and other CNS DEPRESSANTS produce additive sedation and CNS depression.

PHARMACOKINETICS Absorption: Well absorbed from GI tract; about 45% of dose reaches systemic circulation intact. **Onset:** Within 6 h. **Peak:** 2–6 h. **Distribution:** Highest concentrations in lung, heart, kidney, brain, small intestine, and spleen. **Metabolism:** By CYP3A4. **Half-Life:** 12–43 h.

NURSING IMPLICATIONS

Assessment & Drug Effects

- Monitor for CNS depression and sedation, especially when chlorpheniramine is given in combination with other CNS depressants.
- Monitor BP in hypertensive patients since chlorpheniramine may elevate BP.

Patient & Family Education

- Avoid driving a car and other potentially hazardous activities until drug response has been determined.
- Avoid or minimize alcohol intake. Antihistamines have additive effects with alcohol.
- Report any of the following: Tinnitus or palpitations.
- Consult prescriber before taking additional OTC drugs for allergy relief.

CHLORPROMAZINE ⊕

(klor-proe′ma-zeen)

CHLORPROMAZINE HYDROCHLORIDE

Sonazine, Thorazine
Classifications: ANTIPSYCHOTIC, PHENOTHIAZINE; ANTIEMETIC
Therapeutic: ANTIPSYCHOTIC; ANTIEMETIC
Pregnancy Category: C

AVAILABILITY 10 mg, 25 mg, 50 mg, 100 mg, 200 mg tablets; 30 mg, 75 mg, 150 mg, 200 mg, 300 mg sustained release capsules; 10 mg/5 mL syrup; 30 mg/mL, 100 mg/mL oral concentrate; 25 mg/mL injection

ACTION & *THERAPEUTIC EFFECT*
Phenothiazine derivative with actions at all levels of CNS with a mechanism that produces strong antipsychotic effects. Antiemetic effect due to suppression of the chemoreceptor trigger zone (CTZ). Mechanism thought to be related to blockade of postsynaptic dopamine receptors in the brain. *Effective in decreasing psychotic symptoms. Also has antiemetic effects due to its action on the CTZ.*

USES To control manic phase of manic-depressive illness, for symptomatic management of psychotic disorders, including schizophrenia, in management of severe nausea and vomiting, to control excessive anxiety and agitation before surgery, and for treatment of severe behavior problems in children (e.g., attention deficit disorder). Also used for treatment of acute intermittent porphyria, intractable hiccups, and as adjunct in treatment of tetanus.

CONTRAINDICATIONS Hypersensitivity to phenothiazine derivatives, sulfite, or benzyl alcohol; withdrawal states from alcohol; CNS depression; comatose states, brain damage, bone marrow depression, Reye's syndrome; lactation.
CAUTIOUS USE Agitated states accompanied by depression, seizure disorders, respiratory impairment due to infection or COPD; glaucoma, diabetes, hypertensive disease, peptic ulcer, prostatic hypertrophy; thyroid, cardiovascular, and hepatic disorders; patients exposed to extreme heat or organophosphate insecticides; previously detected breast cancer; pregnancy (category C); children younger than 6 mo.

ROUTE & DOSAGE

Psychotic Disorders, Agitation

Adult: **PO** 25–100 mg t.i.d. or q.i.d., may need up to 1000 mg/day **IM/IV** 25–50 mg up to 600 mg q4–6h
Child: **PO** *Older than 6 mo,* 0.55 mg/kg q4–6h prn up to 500 mg/day **IM/IV** *Older than 6 mo,* 0.5–1 mg/kg q6–8h

Nausea and Vomiting

Adult: **PO** 10–25 mg q4–6h prn **IM/IV** 25–50 mg q4–6h prn
Child: **PO** *Older than 6 mo,* 0.55 mg/kg q4–6h prn up to 500 mg/day **IM/IV** *Older than 6 mo,* 0.55 mg/kg q6–8h

Dementia

Geriatric: **PO** Initial 10–25 mg 1–2 times/day, may increase q4–7days by 10–25 mg/day (max: 800 mg/day)

Intractable Hiccups

Adult: **PO/IM** 25–50 mg t.i.d. or q.i.d.

Tetanus

Adult: **IM/IV** 25–50 mg q6–8h
Child: **IM/IV** 0.5 mg/kg q6–8h

Nausea and Vomiting during Surgery

Adult/Adolescent: **IV** 2 mg q2min prn (max total: 25 mg)
Child/Infant (older than 6 mo): **IV** 1 mg q2min prn (max total: 0.25 mg/kg)

Common adverse effects in *italic*, life-threatening effects underlined; generic names in **bold**; classifications in SMALL CAPS; ♣ Canadian drug name; ⊙ Prototype drug

307

Intractable Hiccups

Adult/Adolescent: **IV** 25–50 mg in 500–1000 mL NS, not to exceed 1 mg/min

ADMINISTRATION

Oral

- Give with food or a full glass of fluid to minimize GI distress.
- Mix chlorpromazine concentrate just before administration in at least ½ glass juice, milk, water, coffee, tea, carbonated beverage, or with semisolid food.
- Ensure that sustained release form of drug is not chewed or crushed. It **must be** swallowed whole.

Intramuscular/Intravenous

- Avoid parenteral drug contact with skin, eyes, and clothing because of its potential for causing contact dermatitis.
- Keep patient recumbent for at least 30 min after parenteral administration. Observe closely. Report hypotensive reactions.

Intramuscular

- Inject IM preparations slowly and deep into upper outer quadrant of buttock. If irritation is a problem, consult prescriber about diluting medication with normal saline or 2% procaine. Rotate injection sites.

Intravenous

PREPARE: **Direct:** Dilute 25 mg with 24 mL of NS to yield 1 mg/mL. **Continuous:** May be further diluted in up to 1000 mL of NS.
ADMINISTER: **Direct:** Administer 1 mg or fraction thereof over 1 min for adults and over 2 min for children. **Continuous:** Give slowly at a rate not to exceed 1 mg/min.

- Lemon yellow color of parenteral preparation does not alter potency; if otherwise colored or markedly discolored, solution should be discarded.

INCOMPATIBILITIES **Solution/additive: Aminophylline, amphotericin B, ampicillin, chloramphenicol, chlorothiazide, methohexital, penicillin G, pentobarbital, phenobarbital. Y-site: Allopurinol, amifostine, amphotericin B cholesteryl complex, aztreonam, bivalirudin, cefepime, etoposide, fludarabine, furosemide, lansoprazole, linezolid, melphalan, methotrexate, paclitaxel, piperacillin/tazobactam, remifentanil, sargramostim.**

- All forms are stored preferably between 15°–30° C (59°–86° F) protected from light, unless otherwise specified by the manufacturer. Avoid freezing.

ADVERSE EFFECTS (≥1%) **Body as a Whole:** Idiopathic edema, muscle necrosis (following IM), SLE-like syndrome, sudden unexplained death. **CV:** Orthostatic hypotension, palpitation, tachycardia, ECG changes (usually reversible): Prolonged QT and PR intervals, blunting of T waves, ST depression. **GI:** Dry mouth; constipation, adynamic ileus, cholestatic jaundice, aggravation of peptic ulcer, dyspepsia, increased appetite. **Hematologic:** Agranulocytosis, thrombocytopenic purpura, pancytopenia (rare). **Metabolic:** Weight gain, hypoglycemia, hyperglycemia, glycosuria (high doses), enlargement of parotid glands. **CNS:** *Sedation, drowsiness,* dizziness, restlessness, neuroleptic malignant syndrome, tardive dyskinesias, tumor, syncope, headache, weakness, insomnia, reduced REM sleep, bizarre dreams, cerebral edema, convulsive seizures, hypothermia, inability to sweat, depressed cough

reflex, *extrapyramidal symptoms,* EEG changes. **Respiratory:** Laryngospasm. **Skin:** Fixed-drug eruption, urticaria, reduced perspiration, contact dermatitis, exfoliative dermatitis, photosensitivity, eczema, anaphylactoid reactions, hypersensitivity vasculitis; hirsutism (long-term therapy). **Special Senses:** Blurred vision, lenticular opacities, mydriasis, photophobia. **Urogenital:** Anovulation, infertility, pseudopregnancy, menstrual irregularity, gynecomastia, galactorrhea, priapism, inhibition of ejaculation, reduced libido, urinary retention and frequency.

DIAGNOSTIC TEST INTERFERENCE Chlorpromazine (phenothiazines) may increase *cephalin flocculation,* and possibly other *liver function tests;* also may increase *PBI.* False-positive result may occur for *amylase, 5-hydroxyindole acetic acid, phenylketonuria, porphobilinogens, urobilinogen (Ehrlich's reagent),* and *urine bilirubin (Bili-Labstix).* False-positive or false-negative *pregnancy test* results possibly caused by a metabolite of phenothiazines, which discolors urine depending on test used.

INTERACTIONS Drug: Alcohol, CNS DEPRESSANTS increase CNS depression; ANTACIDS, ANTIDIARRHEALS decrease absorption—space administration 2 h before or after administration of chlorpromazine; **phenobarbital** increases metabolism of phenothiazine; GENERAL ANESTHETICS increase excitation and hypotension; antagonizes antihypertensive action of **guanethidine; phenylpropanolamine** poses possibility of sudden death; TRICYCLIC ANTIDEPRESSANTS intensify hypotensive and anticholinergic effects; ANTICONVULSANTS decrease

seizure threshold—may need to increase anticonvulsant dose. **Herbal: Kava** increases risk and severity of dystonic reaction.

PHARMACOKINETICS Absorption: Rapid absorption with considerable first pass metabolism in liver; rapid absorption after IM. **Onset:** 30–60 min. **Peak:** 2–4 h PO; 15–20 min IM. **Duration:** 4–6 h. **Distribution:** Widely distributed; accumulates in brain; crosses placenta. **Metabolism:** In liver by CYP2D6. **Elimination:** In urine as metabolites; excreted in breast milk. **Half-Life:** Biphasic 2 and 30 h.

NURSING IMPLICATIONS
Assessment & Drug Effects
- Establish baseline BP (in standing and recumbent positions), and pulse, before initiating treatment.
- Monitor BP frequently. Hypotensive reactions, dizziness, and sedation are common during early therapy, particularly in patients on high doses and in the older adult receiving parenteral doses.
- Lab tests: Periodic CBC with differential, LFTs, and blood glucose.
- Monitor cardiac status with baseline ECG in patients with preexisting cardiovascular disease.
- Be alert for signs of neuroleptic malignant syndrome (see Appendix G). Report immediately.
- Report extrapyramidal symptoms that occur most often in patients on high dosage, the pediatric patient with severe dehydration and acute infection, the older adult, and women. Reduce smoking, if possible.
- Monitor I&O ratio and pattern: Urinary retention due to mental depression and compromised renal function may occur.
- Be alert to complaints of diminished visual acuity, reduced night

vision, photophobia, and a perceived brownish discoloration of objects. Patient may be more comfortable with dark glasses.

- Monitor diabetics or prediabetics on long-term, high-dose therapy for reduced glucose tolerance and loss of diabetes control.

- Ocular examinations, and EEG (in patients older than 50 y) are recommended before and periodically during prolonged therapy.

Patient & Family Education

- Take medication as prescribed and keep appointments for follow-up evaluation of dosage regimen. Improvement may not be experienced until 7 or 8 wk into therapy.

- May cause pink to red-brown discoloration of urine.

- Wear protective clothing and sunscreen lotion with SPF above 12 when outdoors, even on dark days. Photosensitivity causes exposed skin areas to have appearance of an exaggerated sunburn. If reaction occurs, report to prescriber.

- Practice meticulous oral hygiene. Oral candidiasis occurs frequently in patients receiving phenothiazines.

- Avoid driving a car or undertaking activities requiring precision and mental alertness until drug response is known.

- Do not abruptly stop this drug. Abrupt withdrawal of drug or deliberate dose skipping, especially after prolonged therapy with large doses, can cause onset of extrapyramidal symptoms (see Appendix F) and severe GI disturbances. When drug is to be discontinued, dosage **must be** tapered off gradually over a period of several weeks.

CHLORPROPAMIDE
(klor-proe′pa-mide)
Novopropamide ✦
Classifications: ANTIDIABETIC; SULFONYLUREA
Therapeutic: ANTIDIABETIC
Prototype: Glyburide
Pregnancy Category: C

AVAILABILITY 100 mg, 250 mg tablets

ACTION & *THERAPEUTIC EFFECT*
Longest-acting first-generation sulfonylurea compound. Lowers blood glucose by stimulating beta cells in pancreas to synthesize and release endogenous insulin. *Antidiabetic effect is due to the ability of the drug to stimulate beta cells of the pancreas to manufacture and release insulin. Therapeutic effectiveness is indicated by HbA1C level in normal range.*

USES Type 2 diabetes mellitus.
UNLABELED USES Neurogenic diabetes insipidus.

CONTRAINDICATIONS Known hypersensitivity to sulfonylureas and sulfonamides; diabetes complicated by severe infection; acidosis; severe renal, hepatic, or thyroid insufficiency; lactation.
CAUTIOUS USE Older adult patients, Addison's disease, CHF, and hepatic porphyria; pregnancy (category C). Safe use in children not established.

ROUTE & DOSAGE

Antidiabetic
Adult: **PO** Initial: 250 mg/day with breakfast, adjust by 50–125 mg/day q3–5days until glycemic control is achieved, up to 750 mg/day

Common adverse effects in *italic*, life-threatening effects underlined; generic names in **bold**; classifications in SMALL CAPS; ✦ Canadian drug name; ⊗ Prototype drug

ADMINISTRATION

Oral

- Give as a single morning dose with breakfast or 3 doses and taken with meals.
- Store at 15°–30° C (59°–86° F) in a tightly closed container, unless otherwise directed.

ADVERSE EFFECTS (≥1%) **Body as a Whole:** Flushing, photosensitivity, alcohol intolerance. **GI:** GI distress, anorexia, nausea, diarrhea, constipation, cholestatic jaundice. **Hematologic:** Leukopenia, thrombocytopenia, agranulocytosis. **Metabolic:** Hypoglycemia, antidiuretic effect (SIADH), dilutional hyponatremia, water intoxication. **CNS:** Drowsiness, muscle cramps, weakness, paresthesias. **Skin:** Rash, pruritus.

INTERACTIONS Drug: Adverse effects of ORAL ANTICOAGULANTS, **phenytoin,** SALICYLATES, NSAIDS may be increased along with those of chlorpropamide; THIAZIDE DIURETICS may increase blood sugar; **alcohol** produces disulfiram reaction; **probenecid,** MAO INHIBITORS may increase hypoglycemic effects. **Herbal: Garlic, ginseng** may increase hypoglycemic effects.

PHARMACOKINETICS Absorption: Readily from GI tract. **Onset:** 1 h. **Peak:** 3–6 h. **Distribution:** Highly protein bound; distributed into breast milk. **Metabolism:** In liver. **Elimination:** 80–90% in urine in 96 h. **Half-Life:** 36 h.

NURSING IMPLICATIONS

Assessment & Drug Effects

- Lab tests: Periodic fasting and postprandial blood glucose; HbA1C every 3 mo; baseline and periodic hematologic and hepatic studies are advisable, particularly in patients receiving high doses.

A CBC should be performed if symptoms of anemia appear.

- Report dizziness, shortness of breath, malaise, fatigue.
- Monitor for S&S of hypoglycemia (see Appendix F).
- Monitor I&O ratio and pattern: Infrequently, chlorpropamide produces an antidiuretic effect, with resulting severe hyponatremia, edema, and water intoxication. If fluid intake far exceeds output and edema develops (weight gain), report to the prescriber.

Patient & Family Education

- Report hypoglycemic episodes to prescriber. Because chlorpropamide has a long half-life, hypoglycemia can be severe.
- Report any of the following immediately to prescriber: Skin eruptions, malaise, fever, or photosensitivity. A change to another hypoglycemic agent may be indicated.

CHLORTHALIDONE
(klor-thal′i-done)

Thalitone
Classifications: ELECTROLYTE & WATER BALANCE AGENT; DIURETIC; ANTIHYPERTENSIVE
Therapeutic: DIURETIC; ANTIHYPERTENSIVE
Prototype: Hydrochlorothiazide
Pregnancy Category: B

AVAILABILITY 15 mg, 25 mg, 50 mg, 100 mg tablets

ACTION & *THERAPEUTIC EFFECT*
Sulfonamide derivative that increases excretion of sodium and chloride by inhibiting their reabsorption in the cortical diluting segment of the ascending loop of Henle. *Antihypertensive effect is correlated to*

Common adverse effects in *italic*, life-threatening effects underlined; generic names in **bold;** classifications in SMALL CAPS; ♣ Canadian drug name; ☺ Prototype drug

311

the decrease in extracellular and intracellular volumes. Decreased volume results in reduced cardiac output with subsequent decrease in peripheral resistance.

USES Edema associated with CHF, renal decompensation, hepatic cirrhosis, corticosteroid and estrogen therapy; as sole agent or with other antihypertensives to treat hypertension.

CONTRAINDICATIONS Hypersensitivity to sulfonamide or thiazide derivatives; anuria, hypokalemia; toxemia; hyperparathroidism; lactation; neonates with jaundice.

CAUTIOUS USE History of renal and hepatic disease, hyponatremia, hypochloremia; gout, SLE, diabetes mellitus; history of allergy or bronchial asthma; pregnancy (category B).

ROUTE & DOSAGE

Hypertension
Adult: **PO** 12.5–25 mg/day, may be increased to 100 mg/day if needed
Child: **PO** 2 mg/kg 3 times/wk

Edema
Adult: **PO** 50–100 mg/day, may be increased to 200 mg/day if needed

ADMINISTRATION

Oral
- Administer as single dose in a.m. to reduce potential for interrupted sleep because of diuresis.
- Consult prescriber when chlorthalidone is used as a diuretic; an intermittent dose schedule may reduce incidence of adverse reactions.
- Store tablets in tightly closed container at 15°–30° C (59°–86° F) unless otherwise advised.

ADVERSE EFFECTS (≥1%) **CV:** Orthostatic hypotension. **GI:** Anorexia, nausea, vomiting, diarrhea, constipation, cramping, jaundice. **Hematologic:** Agranulocytosis, thrombocytopenia, aplastic anemia. **CNS:** Dizziness, vertigo, paresthesias, headache. **Metabolic:** *Hypokalemia,* hyponatremia, hypochloremia, hypercalcemia, glycosuria, hyperglycemia, exacerbation of gout. **Skin:** Rash, urticaria, photosensitivity, vasculitis. **Urogenital:** Impotence.

INTERACTIONS Drug: Increased risk of **digoxin** toxicity because of hypokalemia; CORTICOSTEROIDS, **amphotericin B** increases hypokalemia; decreases **lithium** elimination; may antagonize the hypoglycemic effects of SULFONYL-UREAS; NSAIDs may attenuate diuretic effects; **cholestyramine** decreases thiazide absorption.

PHARMACOKINETICS Absorption: Readily from GI tract. **Onset:** 2 h. **Peak:** 3–6 h. **Duration:** 24–72 h. **Distribution:** Crosses placenta; appears in breast milk. **Elimination:** 30–60% in urine in 24 h. **Half-Life:** 54 h.

NURSING IMPLICATIONS

Assessment & Drug Effects
- Establish baseline BP measurements and check at regular intervals during period of dosage adjustment when chlorthalidone is used for hypertension.
- Be alert to signs of hypokalemia (see Appendix F). Older adult patients are more sensitive to adverse effects of drug-induced diuresis because of age-related changes in the cardiovascular and renal systems.
- Lab tests: Baseline and periodic: Serum electrolytes, serum uric acid, creatinine, BUN, and uric acid and blood glucose (especially in patients with diabetes).

Common adverse effects in *italic*, life-threatening effects <u>underlined</u>; generic names in **bold**; classifications in SMALL CAPS; ✦ Canadian drug name; ⊙ Prototype drug

- Monitor lithium and digoxin levels closely when either of these drugs is used concurrently.

Patient & Family Education

- Maintain adequate potassium intake, monitor weight, and make a daily estimate of I&O ratio.

CHLORZOXAZONE
(klor-zox′a-zone)

Classification: CENTRALLY ACTING SKELETAL MUSCLE RELAXANT
Therapeutic: SKELETAL MUSCLE RELAXER; ANTISPASMODIC
Prototype: Cyclobenzaprine
Pregnancy Category: C

AVAILABILITY 250 mg, 500 mg tablets

ACTION & *THERAPEUTIC EFFECT*
Centrally acting skeletal muscle relaxant that acts indirectly by depressing nerve transmission through polysynaptic pathways in spinal cord, subcortical centers, and brain stem; also possibly has a sedative effect. *Effectively controls muscle spasms and pain associated with musculoskeletal conditions.*

USES Symptomatic treatment of muscle spasm and pain associated with various musculoskeletal conditions.

CONTRAINDICATIONS Impaired liver function; alcoholism; hepatic disease, jaundice; lactation.
CAUTIOUS USE Patients with known allergies or history of drug allergies; renal impairment or failure; CNS depression; older adult patients; pregnancy (category C).

ROUTE & DOSAGE

Skeletal Muscle Relaxant

Adult: **PO** 250–500 mg t.i.d. or q.i.d. (max: 3 g/day)
Child: **PO** 20 mg/kg/day in 3–4 divided doses

ADMINISTRATION
Oral

- Give with food or meals to prevent gastric distress. If necessary, tablet may be crushed and mixed with food or liquid (e.g., milk, fruit juice).
- Store in tight container at 15°–30° C (59°–86° F) unless otherwise directed.

ADVERSE EFFECTS (≥1%) **GI:** Anorexia, heartburn, nausea, vomiting, constipation, diarrhea, abdominal pain, hepatotoxicity: Jaundice, liver damage. **CNS:** *Drowsiness, dizziness,* light-headedness, headache, malaise, overstimulation. **Skin:** Erythema, rash, pruritus, urticaria, petechiae, ecchymoses.

INTERACTIONS Drug: Alcohol, CNS DEPRESSANTS add to CNS depression.

PHARMACOKINETICS Absorption: Readily absorbed from GI tract. **Onset:** 1 h. **Peak:** 1–4 h. **Duration:** 3–4 h. **Distribution:** Not known if crosses placenta or distributed into breast milk. **Metabolism:** In liver. **Elimination:** In urine. **Half-Life:** 66 min.

NURSING IMPLICATIONS

Assessment & Drug Effects

- Monitor ambulation during early drug therapy; some patients may require supervision.
- Lab tests: Periodic LFTs are advised in patients receiving long-term therapy even if sporadic.
- Note: Since chlorzoxazone metabolite may discolor urine, dark urine cannot be a reliable sign of a hepatotoxic reaction.

Patient & Family Education

- Avoid activities requiring mental alertness, judgment, and physical coordination until reaction to drug is known, since sedation, drowsiness, and dizziness may occur.

Common adverse effects in *italic*, life-threatening effects underlined; generic names in **bold**; classifications in SMALL CAPS; ♦ Canadian drug name;⊘ Prototype drug

313

- Drug may discolor urine orange to purplish red, but this is of no clinical significance.
- Discontinue drug and notify prescriber if signs of hypersensitivity (see Appendix F) or of liver dysfunction appear (abdominal discomfort, yellow sclerae or skin, pruritus, malaise, nausea, vomiting).
- Check with prescriber before taking an OTC depressant (e.g., antihistamine, sedative, alcohol) since effects may be additive.

CHOLESTYRAMINE RESIN Ⓟ

(koe-less-tear′a-meen)

LoCHOLEST, Questran, Questran Light, Prevalite

Classifications: ANTILIPEMIC; BILE ACID SEQUESTRANT

Therapeutic: CHOLESTEROL-LOWERING

Pregnancy Category: C

AVAILABILITY 4 g powder for suspension; 1 g tablet

ACTION & *THERAPEUTIC EFFECT*
Anion-exchange resin used for its cholesterol-lowering effect. Adsorbs and combines with intestinal bile acids in exchange for chloride ions to form an insoluble, nonabsorbable complex that is excreted in the feces. As a result, bile salts are continually (but not entirely) prevented from reentry into the enterohepatic circulation, thus increasing fecal loss of bile acids. This leads to lowered serum total cholesterol by decreasing low-density lipoprotein (LDL) cholesterol. *The resin anion-exchange agent increases fecal loss of bile acids, which leads to lowered serum total cholesterol by decreasing (LDL) cholesterol, and reducing bile acid deposit in dermal tissues (decreasing pruritus). Serum triglyceride levels may increase or remain unchanged.*

USES As adjunct to diet therapy in management of patients with primary hypercholesterolemia (type IIa hyperlipidemia) with a significant risk of atherosclerotic heart disease and MI; for relief of pruritus secondary to partial biliary stasis.

UNLABELED USES To control diarrhea caused by excess bile acids in colon; for hyperoxaluria.

CONTRAINDICATIONS Complete biliary obstruction or biliary cirrhosis, cholelithiasis; GI obstruction; hypersensitivity to bile acid sequestrants; coagulopathy; lactation.

CAUTIOUS USE Bleeding disorders; hemorrhoids; impaired GI function, decreased GI motility; peptic ulcer, malabsorption states (e.g., steatorrhea); phenylketonuria (**Questran Light** only); renal disease; pregnancy (category C). Safe use in children 6 y or younger not established.

ROUTE & DOSAGE

Hypercholesterolemia

Adult: **PO** 4 g b.i.d. to q.i.d. a.c. and at bedtime, may need up to 24 g/day

Child: **PO** 240 mg/kg/day in 3 divided doses

Hyperlipoproteinemia

Adult: **PO** 4–8 g b.i.d. to q.i.d. a.c. and at bedtime (32 g/day or less)

Pruritus

Adult: **PO** 4 g b.i.d. to q.i.d. a.c. and at bedtime (16 g/day or less)

ADMINISTRATION

Oral

- Place contents of one packet or one level scoopful on surface of at least 120 to 180 mL (4–6 oz) of water or other preferred liquid. Permit drug to hydrate by standing

Common adverse effects in *italic*, life-threatening effects underlined; generic names in **bold**; classifications in SMALL CAPS; ♥ Canadian drug name; Ⓟ Prototype drug

without stirring 1–2 min, twirling glass occasionally, then stir until suspension is uniform. Rinse glass with small amount of liquid and have patient drink remainder to ensure entire dose is taken. Administer before meals.

- Store in tightly closed container at 15°–30° C (59°–86° F) unless otherwise specified.

ADVERSE EFFECTS (≥1%) **GI:** *Constipation,* fecal impaction, hemorrhoids, abdominal pain and distension, flatulence, bloating sensation, belching, nausea, vomiting, heartburn, anorexia, diarrhea, steatorrhea. **Endocrine:** Increased libido. **Metabolic:** Weight loss or gain, iron, calcium, vitamin A, D, and K deficiencies (from poor absorption); hypoprothrombinemia, hyperchloremic acidosis, decreased erythrocyte folate levels. **Skin:** Rash, irritations of skin, tongue, and perianal areas. **Special Senses:** Arcus juvenilis, uveitis.

DIAGNOSTIC TEST INTERFERENCE Cholestyramine therapy may be accompanied by increased *serum AST, phosphorus, chloride,* and *alkaline phosphatase* levels; decreased *serum calcium, sodium,* and *potassium* levels.

INTERACTIONS Drug: Decreases the absorption of ORAL ANTICOAGULANTS, **digoxin,** TETRACYCLINES, **penicillins,** **phenobarbital,** THYROID HORMONES, THIAZIDE DIURETICS, IRON SALTS, FAT-SOLUBLE VITAMINS (A, D, E, K) from the GI tract—administer cholestyramine 4 h before or 2 h after these drugs. Can bind to and affect absorption of any drug.

PHARMACOKINETICS Absorption: Not absorbed from GI tract. **Elimination:** Excreted in feces as insoluble complex.

NURSING IMPLICATIONS

Assessment & Drug Effects

- Be alert to early symptoms of hypoprothrombinemia (petechiae, ecchymoses, abnormal bleeding from mucous membranes, tarry stools) and report their occurrence promptly. Long-term use of cholestyramine resin can increase bleeding tendency.

- Monitor bowel function. Preexisting constipation may be worsened in the older adult and women.

- Consult prescriber regarding supplemental vitamins A and D and folic acid that may be required by patient on long-term therapy.

- Lab tests: Periodic CBC, platelet count, serum electrolytes, and lipid profile.

Patient & Family Education

- Report constipation to prescriber. High-bulk diet with adequate fluid intake is an essential adjunct to cholestyramine treatment and generally resolves the problems of constipation and bloating sensation.

- Do not omit doses. Sudden withdrawal can promote uninhibited absorption of other drugs taken concomitantly, leading to toxicity or overdosage.

- GI adverse effects usually subside after the first month of drug therapy.

- The following symptoms may be drug-induced and should be reported promptly: Severe gastric distress with nausea and vomiting, unusual weight loss, black stools, severe hemorrhoids (GI bleeding), sudden back pain.

CHOLINE MAGNESIUM TRISALICYLATE

(cho′leen mag-ne′si-um tri-sal′i-ci-late)

Common adverse effects in *italic,* life-threatening effects <u>underlined</u>; generic names in **bold;** classifications in SMALL CAPS; ♦ Canadian drug name; ⊘ Prototype drug

315

Classifications: ANALGESIC (SALICYLATE), NONSTEROIDAL ANTI-INFLAMMATORY DRUG (NSAID); ANTIPYRETIC
Therapeutic: ANALGESIC, NSAID
Prototype: Aspirin
Pregnancy Category: C first and second trimester; D third trimester

AVAILABILITY 500 mg, 750 mg, 1000 mg tablets; 500 mg/5 mL liquid

ACTION & *THERAPEUTIC EFFECT*

Choline magnesium trisalicylate, a nonsteroidal, anti-inflammatory preparation, acts by inhibiting prostaglandin synthesis by reversibly inhibiting cyclooxygenase (both COX-1 and COX-2), resulting in its anti-inflammatory properties as well as its analgesic property. *Has anti-inflammatory, analgesic, and antipyretic action.*

USES Osteoarthritis, rheumatoid arthritis, and other arthrides.

CONTRAINDICATIONS Hypersensitivity to nonacetylated salicylates; children younger than 6 y; children and teenagers with chickenpox, influenza, or flu symptoms because of the potential for Reye's syndrome; coagulopathy, anticoagulant therapy, G6PD deficiency; pregnancy (category D third trimester); contraindicated in late pregnancy, near term, or in labor and delivery.

CAUTIOUS USE Chronic renal and hepatic failure, history of GI disease, peptic ulcer; patients on coumadin or heparin, anemia; hypovolemic states; older adults; pregnancy (category C first and second trimester); lactation.

ROUTE & DOSAGE

Arthritis
Adult: **PO** 1–2.5 g/day in 1–3 divided doses (max: 4.5 g/day)

ADMINISTRATION

Oral
- Give with food to reduce gastric upset. Do not give with antacids.
- Store at 15°–30° C (59°–86° F).

ADVERSE EFFECTS (≥1%) GI: Vomiting, diarrhea. **CNS:** Headache, vertigo, confusion, drowsiness. **Special Senses:** Tinnitus.

INTERACTIONS Drug: Aminosalicylic acid increases risk of salicylate toxicity; **ammonium chloride** and other **acidifying agents** decrease its renal elimination, increasing risk of salicylate toxicity; ANTICOAGULANTS increase risk of bleeding; CARBONIC ANHYDRASE INHIBITORS enhance salicylate toxicity; CORTICOSTEROIDS compound ulcerogenic effects; increases **methotrexate** toxicity; low doses of salicylates may antagonize uricosuric effects of **probenecid, sulfinpyrazone.**

PHARMACOKINETICS Absorption: Readily absorbed from small intestine. **Onset:** 30 min. **Peak:** 1–3 h. **Metabolism:** In liver. **Elimination:** In urine. **Half-Life:** 2–3 h.

NURSING IMPLICATIONS

Assessment & Drug Effects

- As with other NSAIDS, the antipyretic and anti-inflammatory effects may mask usual S&S of infection or other diseases.
- Assess for GI discomfort; nausea, gastric irritation, indigestion, diarrhea, and constipation are frequent complaints.
- Monitor for S&S of bleeding. Closely monitor PT if used concurrently with warfarin.

Patient & Family Education

- Avoid taking aspirin, NSAIDS, or acetaminophen concurrently with drug.

Common adverse effects in *italic*, life-threatening effects underlined; generic names in **bold**; classifications in SMALL CAPS; ♥ Canadian drug name;☺ Prototype drug

- Avoid dangerous activities until reaction to drug is determined, due to possible CNS effects (e.g., vertigo, drowsiness).
- Report tinnitus or persistent gastric irritation and epigastric pain.
- Report any unexplained bruising or bleeding to prescriber.
- Hypoglycemic effects may be enhanced for those with type 2 diabetes taking an oral hypoglycemic agent (OHA).
- Do not give to children or teenagers with chickenpox, influenza, or flu symptoms because of association with Reye's syndrome.

CHORIONIC GONADOTROPIN
(go-nad'oh-troe-pin)
Pregnyl
Classification: HUMAN CHORIONIC GONADOTROPIN (HCG) HORMONE
Therapeutic: HCG HORMONE
Pregnancy Category: X

AVAILABILITY 10,000 unit vial

ACTION & *THERAPEUTIC EFFECT*
Human chorionic gonadotropin (HCG) is a polypeptide hormone produced by the placenta and extracted from urine during first trimester of pregnancy. Actions nearly identical to those of pituitary luteinizing hormone (LH). Promotes production of gonadal steroid hormones by stimulating interstitial cells of the testes to produce androgen, and the corpus luteum of the ovary to produce progesterone. *Administration of HCG to women of childbearing age with normal functioning ovaries causes maturation of the ovarian follicle and triggers ovulation.*

USES Prepubertal cryptorchidism not due to anatomic obstruction and male hypogonadism secondary to pituitary deficiency. Also used in conjunction with menotropins to induce ovulation and pregnancy in infertile women in whom the cause of anovulation is secondary; ovulation usually occurs within 18 h. To stimulate spermatogenesis in males with hypogonadism.
UNLABELED USES Corpus luteum dysfunction.

CONTRAINDICATIONS Known hypersensitivity to HCG, hypogonadism of testicular origin; hamster protein hypersensitivity; hypertrophy or tumor of pituitary, prostatic carcinoma or other androgen-dependent neoplasms, precocious puberty; ovarian failure; dysfunctional uterine bleeding; adrenal insufficiency; uncontrolled thyroid disease; children younger than 4 y; neonates; pregnancy (category X).
CAUTIOUS USE Epilepsy, migraine, asthma, cardiac or renal disease; endometriosis; thrombophlebitis; lactation.

ROUTE & DOSAGE

Prepubertal Cryptorchidism

Child: IM 4000 units 3 times/wk for 3 wk, *or* 5000 units every other day for 4 doses, *or* 500–1000 units 3 times/wk for 4–6 wk

Hypogonadotropic Hypogonadism

Adult: IM 500–1000 units 3 times/wk for 3 wk, then 2 times/wk for 3 wk *or* 4000 units 3 times/wk for 6–9 mo followed by 2000 units 3 times/wk for 3 mo

Stimulation of Spermatogenesis

Adult: IM 5000 units 3 times/wk until normal testosterone levels are achieved (4–6 mo), then 2000 units 2 times/wk with menotropins for 4 mo

---

Common adverse effects in *italic*, life-threatening effects <u>underlined</u>; generic names in **bold**; classifications in SMALL CAPS; ♣ Canadian drug name; ⊙ Prototype drug

C

Induction of Ovulation
Adult: **IM** 500–1000 units 1 day following last dose of menotropins

ADMINISTRATION
- Reconstitute only with diluent supplied by manufacturer.
- Following reconstitution solution is stable for 30–90 day, depending on manufacturer, when refrigerated; thereafter potency decreases.
- Store powder for injection at 15°–30° C (59°–86° F) unless otherwise directed.

ADVERSE EFFECTS (≥1%) **Body as a Whole:** Edema, pain at injection site, <u>arterial thromboembolism</u>. **Endocrine:** Gynecomastia, precocious puberty, increased urinary steroid excretion, ectopic pregnancy (incidence low). When used with menotropins (human menopausal gonadotropin): Ovarian hyperstimulation (ascites with or without pain, pleural effusion, ruptured ovarian cysts with resultant hemoperitoneum, multiple births). **CNS:** Headache, irritability, restlessness, depression, fatigue.

DIAGNOSTIC TEST INTERFERENCE *Pregnancy tests:* Possibility of false results.

INTERACTIONS Drug: No clinically significant drug interactions established. **Herbal: Black cohosh** may antagonize fertility effects.

PHARMACOKINETICS Onset: 2 h. **Peak:** 6 h. **Distribution:** Testes in males, ovaries in females. **Elimination:** 10–12% in urine within 24 h. **Half-Life:** 23 h.

NURSING IMPLICATIONS
Assessment & Drug Effects
- Assess prepubescent males for development of secondary sex characteristics.
- Assess females for and report excessive menstrual bleeding, irregular menstrual cycles, and abdominal/pelvic distention or pain.

Patient & Family Education
- Report promptly onset of abdominal pain and distension (ovarian hyperstimulation syndrome).
- Report to prescriber if the following appear: Axillary, facial, pubic hair; penile growth; acne; deepening of voice. Induction of androgen secretion by HCG may induce precocious puberty in patient treated for cryptorchidism.
- Observe for signs of fluid retention. A weight chart should be maintained for a biweekly record. Report to prescriber if weight gain is associated with edema.

CICLESONIDE
(ci-cle-so′nide)
Alvesco, Omnaris
Pregnancy Category: C
See Appendix A-3.

CICLOPIROXOLAMINE
(sye-kloe-peer′ox)
Loprox, Penlac Nail Lacquer
Classification: ANTIFUNGAL ANTIBIOTIC
Therapeutic: ANTIFUNGAL ANTIBIOTIC
Pregnancy Category: B

AVAILABILITY 1% cream, ointment; 8% nail lacquer; 1% shampoo

ACTION & *THERAPEUTIC EFFECT* Synthetic broad-spectrum antifungal with activity against pathogenic fungi. Inhibits transport of amino acids within fungal cell, thereby interfering with synthesis of fungal protein, RNA, and DNA. *Effective against*

Common adverse effects in *italic*, life-threatening effects <u>underlined</u>; generic names in **bold**; classifications in SMALL CAPS; ✤ Canadian drug name; ❂ Prototype drug

the following organisms: Dermato-phytes, yeasts, some species of Myco-plasma *and* Trichomonas vaginalis, *and certain strains of gram-positive and gram-negative bacteria.*

USES Topically for treatment of tinea cruris and tinea corporis (ringworm) due to *Trichophyton rubrum, Trichophyton mentagro-phytes, Epidermophyton floccosum,* and *Microsporum canis,* and for tinea (pityriasis) versicolor due to *M. furfur;* also cutaneous candidiasis (moniliasis) caused by *Candida albicans.* Nail lacquer indicated for onychomycosis of fingernails and toenails due to *T. rubrum;* seborrheic dermatitis of the scalp.

CONTRAINDICATIONS Hypersensitivity to ciclopiroxolamine or to any component in the formulation. Safe use in children younger than 10 y not established.

CAUTIOUS USE Type 1 diabetic patient; history of seizure disorder; immunosuppression; pregnancy (category B); lactation.

ROUTE & DOSAGE

Tinea

Adult: **Topical** Massage cream into affected area and surrounding skin twice daily, morning and evening

Onychomycosis

Adult: **Topical** Paint affected nail(s) under the surface of the nail and on the nail bed once daily at bedtime (at least 8 h before washing). After 7 days, remove lacquer with alcohol and remove or trim away unattached nail. Continue up to 48 wk.

Seborrheic Dermatitis

Adult: **Topical** Wet hair and apply approximately 1 tsp (5 mL) to the scalp (may use up to 10 mL for long hair), leave on scalp for 3 min, then rinse. Repeat treatment twice/wk × 4 wk, with a minimum of 3 days between applications.

ADMINISTRATION

- Wash hands thoroughly before and after treatments.
- Consult with prescriber about specific procedure for cleansing the skin before medication is applied. Regardless of method used, dry skin thoroughly before drug application.
- Avoid occlusive dressing, wrapping, or clothing over site where cream is applied.
- Store at 15°–30° C (59°–86° F) unless otherwise directed.

ADVERSE EFFECTS (≥1%) **Skin:** Irritation, pruritus, burning, worsening of clinical condition.

PHARMACOKINETICS Absorption: 1.3% absorbed through intact skin. **Distribution:** Distributed to epidermis, corium (dermis), including hair and hair follicles and sebaceous glands; not known if crosses placenta or is distributed into breast milk. **Elimination:** Excreted primarily by kidneys. **Half-Life:** 1.7 h.

NURSING IMPLICATIONS

Assessment & Drug Effects

- Monitor for therapeutic effectiveness. Tinea versicolor generally responds to drug treatment in about 2 wk. Tinea pedis ("athlete's foot"), tinea corporis (ringworm), tinea cruris ("jock itch"), and candidiasis (moniliasis) require about 4 wk of therapy.

Patient & Family Education

- Use medication for the prescribed time even though symptoms improve.
- Report skin irritation or other possible signs of sensitization. A reaction

Common adverse effects in *italic*, life-threatening effects <u>underlined</u>; generic names in **bold**; classifications in SMALL CAPS; ✦ Canadian drug name; ⊘ Prototype drug

319

C

suggestive of sensitization warrants drug discontinuation.
- Do not use occlusive dressings or wrappings.
- Avoid contact of drug in or near the eyes.
- Wear light clothing and footwear that will allow ventilation. Loose-fitting cotton underwear or socks are preferred.

CIDOFOVIR
(cye-do'fo-ver)
Vistide
Classification: ANTIVIRAL
Therapeutic: ANTIVIRAL
Prototype: Acyclovir
Pregnancy Category: C

AVAILABILITY 75 mg/mL injection

ACTION & *THERAPEUTIC EFFECT*
Cidofovir, a nucleotide analog, suppresses cytomegalovirus (CMV) replication by inhibiting CMV DNA polymerase. Cidofovir reduces the rate of viral DNA synthesis of CMV. *It is limited for use in treating CMV retinitis in patients with AIDS. Also effective against herpes viruses and other viruses.*

USES Treatment/prophylaxis of CMV retinitis in patients with AIDS.

CONTRAINDICATIONS Hypersensitivity to cidofovir, history of severe hypersensitivity to probenecid or other sulfa-containing medications; childbearing women and men without barrier contraception; severe renal dysfunction; lactation.
CAUTIOUS USE Renal function impairment, history of diabetes, myelosuppression, previous hypersensitivity to other nucleoside analogs; older adults; pregnancy (category C). Safety and effectiveness in children not established.

ROUTE & DOSAGE

CMV Retinitis: Induction & Maintenance
Adult: **IV** 5 mg/kg once weekly for 2 wk. Also give 2 g probenecid 3 h prior to infusion and 1 g 8 h after infusion (4 g total). Continue every 2 wk.

Renal Impairment Dosage Adjusment
If serum Cr increases by 0.3–0.4, lower dose to 3 mg/kg

ADMINISTRATION
- Pretreatment: Prehydrate with IV of 1 L NS infused over 1–2 h immediately before cidofovir infusion. If able to tolerate fluid load, infuse second liter over 1–3 h starting at beginning (or end) of cidofovir infusion.

Intravenous
PREPARE: **IV Infusion:** Dilute the calculated dose in 100 mL of NS.
ADMINISTER: **IV Infusion:** Give over 1 h at constant rate. ▪Do not coadminister with other agents with significant nephrotoxic potential.

- Store vials at 20°–25° C (68°–77° F); may store diluted IV solution at 2°–8° C (36°–46° F) for up to 24 h.

ADVERSE EFFECTS (≥1%) **Body as a Whole:** Infection, allergic reactions. **GI:** Nausea, vomiting, diarrhea. **Metabolic:** Metabolic acidosis. **Hematologic:** Neutropenia. **CNS:** *Fever, headache,* asthenia. **Respiratory:** Dyspnea, pneumonia. **Special Senses:** Ocular hypotony. **Urogenital:** *Nephrotoxicity, proteinuria.*

INTERACTIONS Drug: AMINOGLYCOSIDES, **amphotericin B, foscarnet, pentamidine** can increase risk of nephrotoxicity.

Common adverse effects in *italic*, life-threatening effects <u>underlined</u>; generic names in **bold**; classifications in SMALL CAPS; ♣ Canadian drug name; ⊕Prototype drug

PHARMACOKINETICS Duration:
Probenecid increases serum levels
and area under concentration–time
curve. **Elimination:** 80–100% in urine;
probenecid delays urinary excretion.

NURSING IMPLICATIONS

Assessment & Drug Effects

- Lab tests: Baseline and periodic
  serum creatinine, urine protein;
  periodic and WBC count with differential prior to each dose. Dose
  adjustments or discontinuation
  may be required.
- Periodic visual acuity tests and
  measurement of intraocular pressure are recommended.
- Monitor for S&S of hypersensitivity (see Appendix F). Report
  their appearance promptly.

Patient & Family Education

- Initiate or continue regular ophthalmologic exams.
- Be alert to potential adverse reactions caused by probenecid (e.g.,
  headache, nausea, vomiting, hypersensitivity reactions) and
  cidofovir.

CILOSTAZOL

(sil-os'tah-zol)
Pletal
Classifications: ANTIPLATELET;
PHOSPHODIESTERASE INHIBITOR
Therapeutic: PERIPHERAL VASO-
DILATOR; PLATELET AGGREGA-
TION INHIBITOR
Pregnancy Category: C

AVAILABILITY 50 mg, 100 mg tablets

ACTION & *THERAPEUTIC EFFECT*
Inhibition of an isoenzyme which
results in vasodilatation and inhibition of platelet aggregation induced by collagen or arachidonic
acid. *Increases the skin temperature of the extremities and improves*

*claudication. Effectiveness is indi-
cated by increased ability to walk
further without claudication.*

USES Intermittent claudication.

CONTRAINDICATIONS CHF of any
severity; hypersensitivity to cilostazol; acute MI; hemostatic disorders
or pathologic bleeding; lactation.
CAUTIOUS USE Cardiac arrhythmias,
MI within 6 mo; valvular heart disease; peptic ulcer disease; renal failure; pregnancy (category C). Safety
and efficacy in children younger
than 18 y are not established.

ROUTE & DOSAGE

Intermittent Claudication

Adult: **PO** 100 mg b.i.d., may
need to reduce to 50 mg b.i.d.
with concomitant CYP3A4 or
CYP2C19 inhibitors

ADMINISTRATION

Oral

- Give at least 30 min before or 2 h
  after a meal. Do not give with
  grapefruit juice.
- Store at 20°–25° C (68°–77° F).

ADVERSE EFFECTS (≥1%) **Body as
a Whole:** Back pain, *headache,* infection, myalgia. **CNS:** Dizziness,
vertigo. **CV:** Palpitations, tachycardia. **GI:** Abdominal pain, *abnormal
stools, diarrhea,* dyspepsia, flatulence, nausea. **Respiratory:** Cough,
pharyngitis, rhinitis.

**INTERACTIONS Drug: Diltiazem,
erythromycin, fluconazole,
fluvoxamine, fluoxetine, ketoconazole, itraconazole,** MAC-
ROLIDE ANTIBIOTICS, **nefazodone,
omeprazole, sertraline,** PROTEASE
INHIBITORS may increase cilostazol
levels and adverse effects. **Herbal:
Evening primrose oil** may increase bleeding risk. **Food:** High

---

Common adverse effects in *italic*, life-threatening effects underlined; generic names
in **bold**; classifications in SMALL CAPS; ♣ Canadian drug name; ⊘ Prototype drug

321

fat meals may increase peak concentrations. **Grapefruit juice** may increase concentration.

PHARMACOKINETICS Absorption: Well absorbed from GI tract. **Onset:** 2–4 wk. **Distribution:** 95–98% protein bound. Smoking may decrease serum levels. May be excreted in breast milk. **Metabolism:** Metabolized by CYP3A4 to active metabolites. **Elimination:** Metabolites primarily excreted in urine and feces. **Half-Life:** 11–13 h.

NURSING IMPLICATIONS

Assessment & Drug Effects

- Monitor therapeutic effectiveness indicated by ability to walk farther without leg pain.
- Monitor for S&S of CHF. Do not give cilostazol to patients with preexisting CHF.

Patient & Family Education

- Avoid grapefruit or grapefruit juice while taking cilostazol.
- Allow 2–12 wk for therapeutic response.

CIMETIDINE ⊘

(sye-met'i-deen)

Tagamet, Tagamet HB
Classification: ANTISECRETORY (H₂-RECEPTOR ANTAGONIST)
Therapeutic: ANTISECRETORY
Pregnancy Category: B

AVAILABILITY 100 mg, 200 mg, 400 mg, 800 mg tablets; 300 mg/5 mL liquid; 150 mg/mL injection

ACTION & *THERAPEUTIC EFFECT*
Has high selectivity for inhibition of histamine H₂-receptors on parietal cells of the stomach, thus suppressing all phases of daytime and nocturnal basal gastric acid secretion in the stomach. Indirectly reduces pepsin secretion. *Blocks the H₂-receptors on the parietal cells of the stomach, thus decreasing gastric acid secretion; raises the pH of the stomach and thereby reduces pepsin secretion.*

USES Short-term treatment of active duodenal ulcer and prevention of ulcer recurrence (at reduced dosage) after it is healed. Also used for short-term treatment of active benign gastric ulcer, pathologic hypersecretory conditions such as Zollinger-Ellison syndrome, and heartburn.
UNLABELED USES Prophylaxis of stress-induced ulcers, upper GI bleeding, and aspiration pneumonitis; gastroesophageal reflux; chronic urticaria; acetaminophen toxicity.

CONTRAINDICATIONS Known hypersensitivity to cimetidine or other H₂-receptor antagonists.
CAUTIOUS USE Older adults or critically ill patients; impaired renal or hepatic function; organic brain syndrome; gastric ulcers; immunocompromised patients, pregnancy (category B).

ROUTE & DOSAGE

Duodenal Ulcer
Adult: **PO** 300 mg q.i.d. *or* 400 mg b.i.d. *or* 800 mg at bedtime **IM/IV** 300 mg q6–8h
Child: **PO/IM/IV** 20–40 mg/kg/day in 4 divided doses
Neonate: **PO/IM/IV** 5–10 mg/kg/day divided q8–12h
Infant: **PO/IM/IV** 10–20 mg/kg/day divided q6–12h
Duodenal Ulcer, Maintenance Therapy
Adult: **PO** 400 mg at bedtime
Gastric Ulcer
Adult: **PO** 300 mg q.i.d. with meals and at bedtime **IM/IV** 300 mg q6–8h

Upper GI Bleed
Adult: **IV** 37.5 mg/h

Heartburn
Adult: **PO** 200 mg 2–4 times/day

Pathologic Hypersecretory Disease
Adult: **PO** 300 mg q.i.d. with meals and at bedtime, may increase up to 2400 mg/day **IM/ IV** 300 mg q6–8h, may increase up to 2400 mg/day

Renal Impairment Dosage Adjustment
CrCl less than 20 mL/min: Dose q12h

Hemodialysis Dosage Adjustment
Give scheduled dose after dialysis

ADMINISTRATION
Oral
■ Give 1 h before or 2 h after an antacid.

Intramuscular
■ Give IM injection undiluted into a large muscle.

Intravenous
IV administration to neonates, infants and children: Verify correct IV concentration and rate of infusion/injection with prescriber.

PREPARE: **Direct:** Dilute 300 mg in 18 mL D5W or NS to yield 300 mg/20 mL. **Intermittent:** Dilute 300 mg in 50 mL D5W or NS. **Continuous:** Further dilute in up to 1000 mL of selected IV solution.

ADMINISTER: **Direct:** Give 300 mg or fraction thereof over at least 5 min. **Intermittent:** Give over 15–20 min. **Continuous:** Give a loading dose of 150 mg at the intermittent infusion rate; then give continuous infusion equally spaced over 24 h.

INCOMPATIBILITIES **Solution/ additive:** Amphotericin B, cefazolin, chlorpromazine, pentobarbital, phenobarbital, secobarbital. **Y-site:** Allopurinol, amphotericin B cholesteryl complex, amsacrine, cefepime, indomethacin, lansoprazole, phenytoin, warfarin.

■ Parenteral solutions are stable for 48 h at room temperature when added to commonly used IV solutions for dilution. Follow manufacturer's directions. ■ Store all forms of cimetidine at 15°–30° C (59°–86° F) protected from light unless otherwise directed by manufacturer.

ADVERSE EFFECTS (≥1%) **Body as a Whole:** Fever. **CV (rare):** <u>Cardiac arrhythmias and cardiac arrest</u> after rapid IV bolus dose. **GI:** Mild transient diarrhea; severe diarrhea, constipation, abdominal discomfort. **Hematologic:** Increased prothrombin time; neutropenia (rare), <u>thrombocytopenia</u> (rare), <u>aplastic anemia</u>. **Metabolic:** Slight increase in serum uric acid, BUN, creatinine; transient pain at IM site; hypospermia. **Musculoskeletal:** Exacerbation of joint symptoms in patients with preexisting arthritis. **CNS:** Drowsiness, dizziness, light-headedness, depression, headache, reversible confusional states, paranoid psychosis. **Skin:** Rash, Stevens-Johnson syndrome, reversible alopecia. **Urogenital:** Gynecomastia and breast soreness, galactorrhea, reversible impotence.

DIAGNOSTIC TEST INTERFERENCE Cimetidine may cause false-positive *Hemoccult test for gastric bleeding* if test is performed within 15 min of oral cimetidine administration.

INTERACTIONS Drug: Cimetidine decreases the hepatic metabolism of **warfarin, phenobarbital,**

Common adverse effects in *italic*, life-threatening effects <u>underlined</u>; generic names in **bold**; classifications in SMALL CAPS; ✦ Canadian drug name; ◐ Prototype drug

323

phenytoin, diazepam, propranolol, lidocaine, theophylline, thus increasing their activity and toxicity; ANTACIDS may decrease absorption of cimetidine.

PHARMACOKINETICS Absorption: 70% of PO dose absorbed from GI tract. **Peak:** 1–1.5 h. **Distribution:** Widely distributed; crosses blood–brain barrier and placenta. **Metabolism:** In liver by CYP1A2 and 3A4. **Elimination:** Most of drug excreted in urine in 24 h; excreted in breast milk. **Half-Life:** 2 h.

NURSING IMPLICATIONS

Assessment & Drug Effects

- Monitor pulse of patient during first few days of drug regimen. Bradycardia should be reported. Pulse usually returns to normal within 24 h after drug discontinuation.
- Monitor I&O ratio and pattern: Particularly in the older adult, severely ill, and in patients with impaired renal function.
- Lab tests: Periodic evaluations of blood count and renal and hepatic function are advised during therapy.
- Be alert to onset of confusional states, particularly in the older adult or severely ill patient. Symptoms occur within 2–3 days after first dose; report immediately. Symptoms usually resolve within 3–4 days after therapy is discontinued.
- Check BP and report an elevation to the prescriber, if patient complains of severe headache.

Patient & Family Education

- Seek advice about self-medication with any OTC drug.
- Report breast tenderness or enlargement. Mild bilateral gynecomastia and breast soreness may occur after 1 mo or more of therapy. It may disappear spontaneously or remain throughout therapy.
- Report recurrence of gastric pain or bleeding (black, tarry stools or "coffee ground" vomitus) immediately, and notify prescriber if diarrhea continues more than 1 day.
- Avoid driving and other potentially hazardous activities until reaction to drug is known.
- Duodenal or gastric ulcer is a chronic, recurrent condition that requires long-term maintenance drug therapy.

CINACALCET HYDROCHLORIDE

(sin-a-kal′set)
Sensipar
Classifications: PARATHYROID HORMONE; CALCIUM RECEPTOR AGONIST
Therapeutic: PARATHYROID HORMONE
Pregnancy Category: C

AVAILABILITY 30 mg, 60 mg, 90 mg tablets

ACTION & *THERAPEUTIC EFFECT*
Directly lowers parathyroid hormone (PTH) levels by increasing sensitivity of calcium-sensing receptors on parathyroid gland to extracellular calcium. This causes decreased calcium and phosphate adsorption from bone, and thus decreased serum calcium and phosphate levels. *Lowers PTH production; this also decreases rate of bone turnover and bone fibrosis in chronic renal failure disease (CRFD).*

USES Treatment of secondary hyperparathyroidism in patients with chronic kidney disease on dialysis; hypercalcemia in patients with parathyroid cancer.

Common adverse effects in *italic*, life-threatening effects <u>underlined</u>; generic names in **bold**; classifications in SMALL CAPS; ♣ Canadian drug name; ⊕ Prototype drug

CONTRAINDICATIONS Hypersensitivity to cinacalcet; hypocalcemia or lower limit of normal serum calcium; chronic kidney disease patients not on dialysis; lactation; children younger than 18 y.

CAUTIOUS USE Moderate and severe hepatic impairment, history of seizures; hypotension; heart failure; history of arrhythmias; pregnancy (category C).

ROUTE & DOSAGE

Secondary Hyperparathyroidism

Adult: **PO** Start with 30 mg once daily; may increase q2–4wk until target iPTH of 150–300 pg/mL (max: 300 mg/day)

Hypercalcemia

Adult: **PO** 30 mg twice daily; titrate q2–4wk as 60 mg b.i.d., 90 mg b.i.d., then 90 mg 3–4 times daily as needed to normalize calcium concentrations

ADMINISTRATION

Oral

- Give with food or shortly after a meal.
- Tablets should be swallowed whole and not divided, crushed, or chewed.
- Do not give to patient with hypocalcemia.
- Store at 15°–30° C (59°–86° F).

ADVERSE EFFECTS (≥1%) **Body as a Whole:** Dizziness, asthenia, noncardiac chest pain, dialysis access infection. **CV:** Hypertension. **GI:** *Nausea, vomiting, diarrhea,* anorexia. **Metabolic:** Hypocalcemia. **Musculoskeletal:** *Myalgia,* adynamic bone disease (renal osteodystrophy).

INTERACTIONS Drug: May increase **amoxapine, atomoxetine, carvedilol, clozapine, codeine,** cyclobenzaprine, dexfenfluramine, dextromethorphan, donepezil, fenfluramine, flecainide, fluoxetine, haloperidol, hydrocodone, maprotiline, meperidine, methadone, methamphetamine, metoprolol, mexiletine, morphine, oxycodone, paroxetine, perphenazine, propafenone, propranolol, risperidone, thioridazine (use may be contraindicated with cinacalcet), **timolol, tramadol, trazodone,** TRICYCLIC ANTIDEPRESSANTS, **venlafaxine, zolpidem** levels; cinacalcet levels may be increased by strong CYP3A4 inhibitors such as **amiodarone, aprepitant, clarithromycin, dalfopristin, diltiazem, erythromycin, fluconazole, fluvoxamine, itraconazole, ketoconazole, miconazole, nefazodone, quinupristin, troleandomycin, verapamil, voriconazole. Food: Grapefruit juice** may increase cinacalcet levels.

PHARMACOKINETICS Peak: 2–6 h. **Distribution:** 93–97% protein bound. **Metabolism:** In liver by CYP3A4. **Elimination:** 80% by kidneys, 15% in feces. **Half-Life:** 30–40 h.

NURSING IMPLICATIONS

Assessment & Drug Effects

- Monitor for S&S of hypocalcemia (e.g., paresthesias, myalgias, cramping, tetany, convulsions).
- Withhold drug and notify prescriber for serum calcium less than 7.5 mg/dL or symptoms of hypocalcemia. Drug should not be resumed until serum calcium levels reach 8 mg/dL, and/or symptoms of hypocalcemia resolve.
- Lab tests: Baseline serum calcium; serum calcium and phosphorus within 1 wk after and iPTH 1–4 wk after initiation of drug or dose

Common adverse effects in *italic*, life-threatening effects underlined; generic names in **bold**; classifications in SMALL CAPS; ✚ Canadian drug name; ⊙ Prototype drug

325

adjustment; thereafter, monthly serum calcium and phosphorus (more often with a history of a seizure disorder), and iPTH every 1–3 mo.
- Closely monitor iPTH and serum calcium with concurrent administration of a strong CYP3A4 (e.g., ketoconazole, erythromycin, itraconazole).

Patient & Family Education
- Report promptly any of the following: Seizure or convulsion; muscle spasms or cramping of the abdomen, back, legs, face; burning, numbness, pricking, tickling, or tingling of the face, lips, tongue, hands, or feet; changes in mental status.

CIPROFLOXACIN HYDROCHLORIDE ⓟ

(ci-pro-flox'a-cin)

Cetraxal, Cipro, Cipro IV, Cipro XR, Proquin XR

CIPROFLOXACIN OPHTHALMIC

Ciloxan
Classification: QUINOLONE ANTIBIOTIC
Therapeutic: ANTIBIOTIC
Pregnancy Category: C

AVAILABILITY 100 mg, 250 mg, 500 mg, 750 mg tablets; 500 mg extended release tablets; 50 mg/mL, 100 mg/mL suspension; 200 mg, 400 mg injection; 3.5 mg/mL ophth solution; 0.2% otic drops

ACTION & _THERAPEUTIC EFFECT_
Synthetic quinolone that is a broad-spectrum bactericidal agent. Inhibits DNA-gyrase, an enzyme necessary for bacterial DNA replication and some aspects of transcription, repair, recombination, and transposition. _Effective against many gram-positive and aerobic gram-negative organisms._

USES UTIs, lower respiratory tract infections, skin and skin structure infections, bone and joint infections, GI infection or infectious diarrhea, chronic bacterial prostatitis, nosocomial pneumonia, acute sinusitis. Post-exposure prophylaxis for anthrax. **Ophthalmic:** Corneal ulcers, bacterial conjunctivitis caused by _Staphylococci, Streptococci,_ and _Pseudomonas aeruginosa._ **Otic:** Otitis externa.

CONTRAINDICATIONS Known hypersensitivity to ciprofloxacin or other quinolones, syphilis, viral infection; history of myasthenia gravis; peripheral neuropathy; tendon inflammation or tendon pain; lactation.

CAUTIOUS USE Known or suspected CNS disorders (i.e., severe cerebral arteriosclerosis or seizure disorders); myocardial ischemia, atrial fibrillation, QT prolongation, CHF; GI disease, colitis; CVA; uncorrected hypokalemia; severe renal impairment and crystalluria during ciprofloxacin therapy; pregnancy (category C); children.

ROUTE & DOSAGE

Uncomplicated UTI
Adult: **PO** 250 mg q12h or 500 mg XR daily × 3 days **IV** 200 mg q12h × 7–14 days

Complicated UTI
Adult: **PO** 1000 mg XR daily × 7–14 days **IV** 400 mg q12h × 7–14 days

Acute Sinusitis
Adult: **PO** 500 mg b.i.d. × 10 days

Moderate to Severe Systemic Infection
Adult: **PO** 500–750 mg q12h **IV** 200–400 mg q8–12h

Renal Impairment Dosage Adjustment

CrCl 30–50 mL/min: **PO** 250–500 mg q12h, **IV** No change in dose; *less than 30 mL/ min:* **PO** 250–500 mg q18h, **IV** 200–400 mg q18–24h

Bacterial Conjunctivitis

Adult: **Ophthalmic** 1–2 drops in conjunctival sac q2h while awake for 2 days, then 1–2 drops q4h while awake for the next 5 days **Ointment** ½-inch ribbon into conjunctival sac t.i.d. × 2 days, then b.i.d. × 5 days

Corneal Ulcers

Adult: **Ophthalmic** 2 drops q15min for 6 h, 2 drops q30min for the next 18 h, then 2 drops q1h for 24 h, then 2 drops q4h for 14 days

Otitis Externa

Adult/Adolescent/Child (older than 1 y): 0.25 mL into affected ears q12h × 7 days

ADMINISTRATION

▪ For patients with renal impairment, oral and IV doses are lowered according to creatinine clearance.

Oral

▪ Do not give an antacid within 4 h of the oral ciprofloxacin dose.

Intravenous

PREPARE: **Intermittent:** Dilute in NS or D5W to a final concentration of 0.5–2 mg/mL. ▪ Typical dilutions are 200 mg in 100–250 mL and 400 mg in 250–500 mL. *ADMINISTER:* **Intermittent:** Give slowly over 60 min. Avoid rapid infusion and use of a small vein. *INCOMPATIBILITIES* **Solution/ additive:** Aminophylline, amoxicillin, amoxicillin/clavulanate potassium, amphotericin B, ampicillin/sulbactam, ceftazidime, cefuroxime, clindamycin, heparin, metronidazole, piperacillin, sodium bicarbonate, ticarcillin. **Y-site:** Aminophylline, ampicillin, ampicillin/sulbactam, azithromycin, cefepime, dexamethasone, drotrecogin alfa, furosemide, heparin, hydrocortisone, lansoprazole, phenytoin, propofol, sodium bicarbonate, theophylline, TPN, warfarin.

▪ Discontinue other IV infusion while infusing ciprofloxacin or infuse through another site.
▪ Reconstituted IV solution is stable for 14 days refrigerated.

ADVERSE EFFECTS (≥1%) **GI:** Nausea, vomiting, diarrhea, cramps, gas, pseudomembranous colitis. **Metabolic:** Transient increases in liver transaminases, alkaline phosphatase, lactic dehydrogenase, and eosinophilia count. **Musculoskeletal:** Tendon rupture, cartilage erosion. **CNS:** Headache, vertigo, malaise, peripheral neuropathy, seizures (especially with rapid IV infusion). **Skin:** Rash, phlebitis, pain, burning, pruritus, and erythema at infusion site; photosensitivity. **Special Senses:** *Local burning and discomfort, crystalline precipitate on superficial portion of cornea,* lid margin crusting, scales, foreign body sensation, itching, and conjunctival hyperemia.

DIAGNOSTIC TEST INTERFERENCE Ciprofloxacin does not interfere with **urinary glucose** determinations using **cupric sulfate solution** or with **glucose oxidase tests;** may cause false positive on **opiate screening tests.**

INTERACTIONS Drug: May increase **theophylline** levels 15–30%;

Common adverse effects in *italic*, life-threatening effects <u>underlined</u>; generic names in **bold;** classifications in SMALL CAPS; ♣ Canadian drug name; ◐ Prototype drug

327

ANTACIDS, **sulcralfate, iron** decrease absorption of ciprofloxacin; may increase PT for patients on **warfarin. Food:** Calcium decreases the levels of ciprofloxacin.

PHARMACOKINETICS Absorption: 60–80% from GI tract. **Ophthalmic:** Minimal absorption through cornea or conjunctiva. **Onset:** Topical 0.5–2 h. **Duration:** Topical 12 h. **Peak:** Immediate release: 0.5–2 h; Cipro XR: 1–2.5 h; Proquin XR: 3.5–8.7 h. **Distribution:** Widely distributed including prostate, lung, and bone; crosses placenta; distributed into breast milk. **Elimination:** Primarily in urine with some biliary excretion. **Half-Life:** 3.5–4 h.

NURSING IMPLICATIONS

Assessment & Drug Effects

- Report tendon inflammation or pain. Drug should be discontinued.
- Lab tests: Culture and sensitivity tests should be done prior to initial dose. Treatment may be implemented pending results.
- Monitor urine pH; it should be less than 6.8, especially in the older adult and patients receiving high dosages of ciprofloxacin, to reduce the risk of crystalluria.
- Monitor I&O ratio and patterns: Patients should be well hydrated; assess for S&S of crystalluria.
- Monitor plasma theophylline concentrations, since drug may interfere with half-life.
- Administration with theophylline derivatives or caffeine can cause CNS stimulation.
- Assess for S&S of GI irritation (e.g., nausea, diarrhea, vomiting, abdominal discomfort) in clients receiving high dosages and in older adults.
- Monitor PT and INR in patients receiving coumarin therapy.
- Assess for S&S of superinfections (see Appendix F).

Patient & Family Education

- Immediately report tendon inflammation or pain. Drug should be discontinued.
- Fluid intake of 2–3 L/day is advised, if not contraindicated.
- Report sudden, unexplained joint pain.
- Restrict caffeine due to the following effects: Nervousness, insomnia, anxiety, tachycardia.
- Use sunscreen and avoid overexposure to sunlight.
- Report nausea, diarrhea, vomiting, and abdominal pain or discomfort.
- Use caution with hazardous activities until reaction to drug is known. Drug may cause light-headedness.

CISATRACURIUM BESYLATE

(cis-a-tra-kyoo-ri′um)

Nimbex

Classifications: NONDEPOLARIZING SKELETAL MUSCLE RELAXANT; NEUROMUSCULAR BLOCKER

Therapeutic: SKELETAL MUSCLE RELAXANT

Prototype: Atracurium

Pregnancy Category: B

AVAILABILITY 2 mg/mL, 10 mg/mL injection

ACTION & *THERAPEUTIC EFFECT*

Cisatracurium is a neuromuscular blocking agent with intermediate onset and duration of action compared with similar agents. It binds competitively to cholinergic receptors on the motor endplate of neurons, antagonizing the action of acetylcholine. *Blocks neuromuscular transmission of nerve impulses.*

USES Adjunct to general anesthesia to facilitate tracheal intubation and provide skeletal muscle relaxation during surgery or mechanical ventilation.

CONTRAINDICATIONS Hypersensitivity to cisatracurium or other related agents; rapid-sequence endotracheal intubation. Not studied in infants younger than 1 mo.

CAUTIOUS USE History of hemiparesis, electrolyte imbalances, burn patients, pulmonary disease, COPD; neuromuscular diseases (e.g., myasthenia gravis), older adults, renal function impairment, pregnancy (category B), lactation.

ROUTE & DOSAGE

Intubation

Adult: **IV** 0.15 or 0.20 mg/kg
Child (2–12 y): **IV** 0.1–0.15 mg/kg
Infant (older than 1 mo): **IV** 0.15 mg/kg

Maintenance

Adult: **IV** 0.03 mg/kg q20min prn or 1–2 mcg/kg/min
Child (2 y or older): **IV** 1–2 mcg/kg/min

Mechanical Ventilation in ICU

Adult: **IV** 3 mcg/kg/min (can range from 0.5 to 10.2 mcg/kg/min)

ADMINISTRATION

- Administer carefully adjusted, individualized doses using a peripheral nerve stimulator to evaluate neuromuscular function.
- Given only by or under supervision of expert clinician familiar with the drug's actions and potential complications.
- Have immediately available personnel and facilities for resuscitation and life support and an antagonist of cisatracurium.
- Note that 10-mL multiple-dose vials contain benzyl alcohol and should not be used with neonates.

Intravenous

PREPARE: Direct: Give undiluted. **IV Infusion:** Dilute 10 mg in 95 mL or 40 mg in 80 mL of compatible IV fluid to prepare 0.1 mg/mL or 0.4 mg/mL, respectively, IV solution. ▪ Compatible IV fluids include D5W, NS, D5/NS, D5/LR. **ICU IV Infusion (Mechanical Ventilation):** Dilute the contents of the 200 mg vial (i.e., 10 mg/mL) in 1000 mL or 500 mL of compatible IV fluid to prepare 0.2 mg/mL or 0.4 g/mL, respectively, IV solutions.

ADMINISTER: Direct: Give a single dose over 5–10 sec. **IV Infusion:** Adjust the rate based on patient's weight.

INCOMPATIBILITIES Solution/additive: Ketorolac, propofol (dose dependent). Y-site: Amphotericin B, amphotericin B cholesteryl complex, ampicillin, cefazolin, cefotaxime, cefotetan, cefuroxime, diazepam, furosemide, ganciclovir, heparin, methylprednisolone, sodium bicarbonate, thiopental, trimethoprim/sulfamethoxazole.

- Refrigerate vials at 2°–8° C (36°–46° F). Protect from light. Diluted solutions may be stored refrigerated or at room temperature for 24 h.

ADVERSE EFFECTS (≥1%) CV: Bradycardia, hypotension, flushing. **Respiratory:** Bronchospasm. **Skin:** Rash.

PHARMACOKINETICS Onset: Varies from 1.5 to 3.3 min (higher dose has faster onset). **Peak:** Varies from 1.5 to 3.3 min (higher dose has faster peak). **Duration:** Varies with dose from 46 to 121 min (higher dose, longer recovery time). **Metabolism:** Undergoes Hoffman elimination (pH- and temperature-dependent degradation) and hydrolysis by

Common adverse effects in *italic*, life-threatening effects underlined; generic names in **bold**; classifications in SMALL CAPS; ♣ Canadian drug name; ⊘ Prototype drug

329

plasma esterases. **Elimination:** In urine. **Half-Life:** 22 min.

NURSING IMPLICATIONS

Assessment & Drug Effects

- Time-to-maximum neuromuscular block is ≈1 min slower in the older adult.
- Monitor for bradycardia, hypotension, and bronchospasms; monitor ICU patients for spontaneous seizures.

CISPLATIN (cis-DDP, cis-PLATINUM II)

(sis'pla-tin)

Abiplatin ♣, Platinol
Classifications: ANTINEOPLASTIC; ALKYLATING AGENT
Therapeutic: ANTINEOPLASTIC
Prototype: Cyclophosphamide
Pregnancy Category: D

AVAILABILITY 1 mg/mL injection

ACTION & *THERAPEUTIC EFFECT*

A heavy metal complex that produces cross linkage in DNA of rapidly dividing cells, thus preventing DNA, RNA, and protein synthesis. *Cell cycle-nonspecific (i.e., effective throughout the entire cell life cycle).*

USES Established combination therapy (cisplatin, vinblastine, bleomycin) in patient with metastatic testicular tumors and with doxorubicin for metastatic ovarian tumors following appropriate surgical or radiation therapy.
UNLABELED USES Carcinoma of endometrium, bladder, head, and neck.

CONTRAINDICATIONS History of hypersensitivity to cisplatin or other platinum-containing compounds; impaired renal function of CrCl below 30 ml/min; severe myelosuppression; impaired hearing; active infection; history of gout and urate renal stones, renal failure; hypomagnesia; Raynaud syndrome; pregnancy (category D).
CAUTIOUS USE Previous cytotoxic drug or radiation therapy with other ototoxic and nephrotoxic drugs; peripheral neuropathy; hyperuricemia; electrolyte imbalances, moderate renal impairment; hepatic impairment; history of circulatory disorders. Safe use in children not established.

ROUTE & DOSAGE

Testicular Neoplasms
Adult: **IV** 20 mg/m^2/day for 5 days q3–4wk for 3 courses

Ovarian Neoplasms
Adult: **IV** *With cyclophosphamide:* 75–100 mg/m^2 once q4wk; *single agent:* 100 mg/m^2 once q4wk

Advanced Bladder Cancer
Adult: **IV** 50–75 mg/m^2 q3–4 wk

ADMINISTRATION

- Usually a parenteral antiemetic agent is administered 30 min before cisplatin therapy is instituted and given on a scheduled basis throughout day and night as long as necessary.
- Before the initial dose is given, hydration is started with 1–2 L IV infusion fluid to reduce risk of nephrotoxicity and ototoxicity.

Intravenous

PREPARE: IV Infusion: Use disposable gloves when preparing cisplatin solutions. If drug accidentally contacts skin or mucosa, wash immediately and thoroughly with soap and water. ▪ Do not

Common adverse effects in *italic*, life-threatening effects <u>underlined</u>; generic names in **bold**; classifications in SMALL CAPS; ♣ Canadian drug name; ☻ Prototype drug

use any equipment containing aluminum. ▪ Withdraw required dose and dilute in 2 L D5W 5% dextrose in ½ or ⅓ normal saline containing 37.5 g mannitol.

ADMINISTER: **IV Infusion:** Give 2 L over 6–8 h.

INCOMPATIBILITIES Solution/additive: 5% dextrose, fluorouracil, mesna, metoclopramide, sodium bicarbonate, thiotepa. Y-site: Amifostine, amphotericin B, cholesteryl, cefepime, lansoprazole, piperacillin/tazobactam, thiotepa, TPN.

▪ Hydration and forced diuresis are continued for at least 24 h after drug administration to ensure adequate urinary output.

▪ Store at 15°–30° C (59°–86° F). Do not refrigerate. Protect from light. Once vial is opened, solution is stable for 28 days protected from light or 7 days in fluorescent light.

ADVERSE EFFECTS (≥1%) **Body as a Whole:** <u>Anaphylactic-like reactions</u>. **CV:** Cardiac abnormalities. **GI:** *Marked nausea, vomiting,* anorexia, stomatitis, xerostomia, diarrhea, constipation. **Hematologic:** Myelosuppression (25–30% patients): <u>Leukopenia, thrombocytopenia</u>; hemolytic anemia, hemolysis. **Metabolic:** Hypocalcemia, *hypomagnesemia,* hyperuricemia, elevated AST, SIADH. **CNS:** Seizures, headache; peripheral neuropathies (may be irreversible): Paresthesia, unsteady gait, clumsiness of hands and feet, exacerbation of neuropathy with exercise, loss of taste. **Special Senses:** Ototoxicity (may be irreversible): Tinnitus, hearing loss, deafness, vertigo, blurred vision, changes in ability to see colors (optic neuritis, papilledema). **Urogenital:** Nephrotoxicity.

INTERACTIONS Drug: AMINOGLYCOSIDES, **amphotericin B, vancomycin,** other **nephrotoxic drugs** increase nephrotoxicity and acute renal failure—try to separate by at least 1–2 wk; AMINOGLYCOSIDES, **furosemide** increase risk of ototoxicity.

PHARMACOKINETICS Peak: Immediately after infusion. **Distribution:** Widely distributed in body fluids and tissues; concentrated in kidneys, liver, and prostate; accumulated in tissues. **Metabolism:** Not known. **Elimination:** 15–50% in urine within 24–48 h. **Half-Life:** 73–290 h.

NURSING IMPLICATIONS

Assessment & Drug Effects

▪ Obtain baseline ECG and cardiac monitoring during induction therapy because of possible myocarditis or focal irritability.

▪ Lab tests: The following tests should be done *before* initiating every course of therapy and repeated each week during treatment period: Serum uric acid, serum creatinine, BUN, urinary creatinine clearance. CBC and platelet counts are done weekly for 2 wk after each course of treatment. Monitor periodically serum electrolytes and LFTs.

▪ A repeat course of therapy should not be given until (1) serum creatinine is below 1.5 mg/dL; (2) BUN is below 25 mg/dL; (3) platelets 100,000/mm³ or more; (4) WBC 4000/mm³ or more; (5) audiometric test is within normal limits.

▪ Monitor urine output and specific gravity for 4 consecutive hours before treatment and for 24 h after therapy. A urine output of less than 75 mL/h necessitates medical intervention to avert a renal emergency.

▪ Audiometric testing should be performed before the first dose and before each subsequent

---

Common adverse effects in *italic*, life-threatening effects <u>underlined</u>; generic names in **bold**; classifications in SMALL CAPS; ✦ Canadian drug name; ⊙ Prototype drug

331

dose. Ototoxicity (reported in 31% of patients) may occur after a single dose of 50 mg/m². Children who receive repeated doses are especially susceptible.

- Monitor for anaphylactoid reactions (particularly in patient previously exposed to cisplatin), which may occur within minutes of drug administration.
- Monitor closely for dose-related adverse reactions. Drug action is cumulative; therefore severity of most adverse effects increases with repeated doses.
- Suspect ototoxicity if patient manifests tinnitus or difficulty hearing in the high frequency range.
- Monitor results of blood studies. The nadirs in platelet and leukocyte counts occur between day 18 and 23 (range: 7.5–45) with most patients recovering in 13–62 days.
- Monitor and report abnormal bowel elimination; diarrhea is a possible response to GI irritation.
- Inspect oral membranes for xerostomia (white patches and ulcerations) and tongue for signs of fungal overgrowth (black, furry appearance).
- Weigh the patient under standard conditions every day. A gradual ascending weight profile occurring over a period of several days should be reported.

Patient & Family Education

- Continue maintenance of adequate hydration (at least 3000 mL/24 h oral fluid if prescriber agrees) and report promptly: Reduced urinary output, flank pain, anorexia, nausea, vomiting, dry mucosae, itching skin, urine odor on breath, fluid retention, and weight gain.
- Avoid rapid changes in position to minimize risk of dizziness or falling.
- Tingling, numbness, and tremors of extremities, loss of vision, sense, and taste, and constipation

are early signs of neurotoxicity. Report their occurrence promptly to prevent irreversibility.

- Report tinnitus or any hearing impairment.
- Report promptly evidence of unexplained bleeding and easy bruising.
- Report unusual fatigue, fever, sore mouth and throat, abnormal body discharges.

CITALOPRAM HYDROBROMIDE

(cit-a-lo'pram)

Celexa

Classification: SELECTIVE SEROTONIN-REUPTAKE INHIBITOR (SSRI)
Therapeutic: ANTIDEPRESSANT
Prototype: Fluoxetine
Pregnancy Category: C

AVAILABILITY 20 mg, 40 mg tablets; 10 mg/5 mL oral solution

ACTION & THERAPEUTIC EFFECT
Selective serotonin reuptake inhibitor (SSRI) with an antidepressant effect presumed to be linked to its inhibition of CNS presynaptic neuronal uptake of serotonin. *Selective serotonin reuptake inhibition mechanism results in the antidepressant activity of citalopram.*

USES Depression.
UNLABELED USES Anxiety, hot flashes, obsessive-compulsive disorder, post-traumatic stress disorder, panic disorder.

CONTRAINDICATIONS Hypersensitivity to citalopram; unstable heart disease, recent MI; concurrent use of MAOIs or use within 14 days of discontinuing MAOIs; mania; volume depleted, hyponatremia; suicidal ideation.

CAUTIOUS USE Hypersensitivity to other SSRIs; hepatic insufficiency; history of potential suicide; older adults; dehydration; severe renal impairment or renal failure; cardiovascular disease (e.g., dysrhythmias, conduction defects, myocardial ischemia); history of drug abuse; history of seizure disorders or suicidal tendencies; bipolar disorder, history of mania; ECT treatments; pregnancy (category C), lactation; children younger than 18 y.

ROUTE & DOSAGE

Depression

Adult: **PO** Start at 20 mg daily, may increase to 40 mg daily if needed
Geriatric: **PO** 20 mg daily

ADMINISTRATION

Oral

- Do not begin this drug within 14 days of stopping an MAOI.
- Reduced doses are advised for the older adult and those with hepatic or renal impairment.
- Dose increments should be separated by at least 1 wk.
- Store at 15°–30° C (59°–86° F) in tightly closed container and protect from light.

ADVERSE EFFECTS (≥1%) **Body as a Whole:** Asthenia, fatigue, fever, arthralgia, myalgia. **CV:** Tachycardia, postural hypotension, hypotension. **GI:** *Nausea,* vomiting, diarrhea, dyspepsia, abdominal pain, *dry mouth,* anorexia, flatulence. **CNS:** Dizziness, *insomnia, somnolence,* agitation, tremor, anxiety, paresthesia, migraine, neuromalignant syndrome. **Respiratory:** URI, rhinitis, sinusitis. **Skin:** Increased sweating. **Urogenital:** Dysmenorrhea, decreased libido, ejaculation disorder, impotence.

INTERACTIONS Drug: Combination with MAOIS could result in hypertensive crisis, hyperthermia, rigidity, myoclonus, autonomic instability; **cimetidine** may increase citalopram levels; **linezolid** may cause serotonin syndrome. **Herbal:** **St. John's wort** may cause serotonin syndrome.

PHARMACOKINETICS Absorption: Rapidly absorbed from GI tract; approximately 80% reaches systemic circulation. **Peak:** Steady-state serum concentrations in 1 wk; peak blood levels at 4 h. **Distribution:** 80% protein bound; crosses placenta; distributed into breast milk. **Metabolism:** In liver by CYP3A4 and CYP2C9 enzymes. **Elimination:** 20% in urine, 80% in bile. **Half-Life:** 35 h.

NURSING IMPLICATIONS

Assessment & Drug Effects

- Watch closely for worsening of depression or emergence of suicidal ideations.
- Monitor for therapeutic effectiveness: Indicated by elevation of mood; 1–4 wk may be needed before improvement is noted.
- Lab tests: Monitor periodically hepatic functions, CBC, serum sodium, and lithium levels when the two drugs are given concurrently.
- Monitor periodically HR and BP, and carefully monitor complete cardiac status in person with known or suspected cardiac disease.
- Monitor closely older adult patients for adverse effects especially with doses greater than 20 mg/day.

Patient & Family Education

- Do not engage in hazardous activities until reaction to this drug is known.
- Avoid using alcohol while taking citalopram.

Common adverse effects in *italic*, life-threatening effects underlined; generic names in **bold**; classifications in SMALL CAPS; ♣ Canadian drug name; ⊘ Prototype drug

- Report immediately worsening of clinical condition, including suicidal ideation or other unusual changes in behavior.
- Report distressing adverse effects including any changes in sexual functioning or response.
- Periodic ophthalmology exams are advised with long-term treatment.

CLADRIBINE
(cla'dri-been)

Leustatin
Classifications: ANTINEOPLASTIC; ANTIMETABOLITE, PURINE ANTAGONIST
Therapeutic: ANTINEOPLASTIC; ANTIMETABOLITE
Prototype: 6-Mercaptopurine
Pregnancy Category: D

AVAILABILITY 1 mg/mL injection

ACTION & *THERAPEUTIC EFFECT*
Cladribine is a synthetic antineoplastic agent with selective toxicity toward certain normal and malignant lymphocytes and monocytes. It accumulates intracellularly, preventing repair of single-stranded DNA breaks and ultimately interfering with cellular metabolism and DNA synthesis. *Cladribine is cytotoxic to both actively dividing and quiescent lymphocytes and monocytes, inhibiting both DNA synthesis and repair.*

USES Treatment of hairy cell leukemia.
UNLABELED USES Advanced cutaneous T-cell lymphomas, acute myeloid leukemia, autoimmune hemolytic anemia, mycosis fungoides, chronic lymphotic leukemia, non-Hodgkin's lymphomas.

CONTRAINDICATIONS Hypersensitivity to cladribine; severe bone marrow suppression; pregnancy (category D).

CAUTIOUS USE Hepatic or renal impairment; previous radiation therapy or chemotherapy. Safety and efficacy in children not established.

ROUTE & DOSAGE

Hairy Cell Leukemia
Adult: **IV** 0.09–0.1 mg/kg/day by 7 days continuous infusion

ADMINISTRATION

- Use disposable gloves and protective clothing when handling the drug.
- Wash immediately if skin contact occurs.

Intravenous

PREPARE: **IV Infusion reservoir usually prepared by pharmacists as follows:** Use a 0.22 micron disposable hydrophilic syringe filter to add the required dose of cladribine to an infusion reservoir. Next, use the same method to add the required amount of bacteriostatic NS injection to bring the total volume of the solution to 100 mL. Aseptically aspirate air bubbles from the reservoir using a syringe and a dry second sterile filter or a sterile vent filter assembly.

ADMINISTER: **IV infusion (7-day dose):** Give through a central line and control by a pump device (e.g., Deltec pump) to deliver 100 mL evenly over 7 days.

INCOMPATIBILITIES **Solution/ additive:** Do not mix with any other diluents or drugs.

- Diluted solutions of cladribine may be stored refrigerated for up to 8 h prior to administration. ▪ Store unopened vials in refrigerator [2°–8° C (36°–46° F)], and protect from light.

ADVERSE EFFECTS (≥1%) **CNS:** Headache, dizziness. **GI:** Nausea, diarrhea. **Hematologic:** _Myelosuppression (neutropenia)_, _anemia_, thrombocytopenia. **Metabolic:** _Fever_. **CNS:** Headache, dizziness. **Urogenital:** Elevated serum creatinine.

INTERACTIONS Drug: Additive risk of bleeding with ANTICOAGULANTS, NSAIDS, PLATELET INHIBITORS, SALICYLATES.

PHARMACOKINETICS **Onset:** Therapeutic effect 10 days to 4 mo. **Duration:** 7–25+ mo. **Distribution:** Crosses placenta; distributed into breast milk. **Metabolism:** In malignant leukocytes, cladribine is phosphorylated to active forms, which are subsequently incorporated into cellular DNA. **Half-Life:** Initial 35 min, terminal 6.7 h.

NURSING IMPLICATIONS

Assessment & Drug Effects

- Monitor vital signs during and after drug infusion. Fever (above 100° F) is common during the 5th to 7th day in patients with hairy cell leukemia, and severe fever (above 104° F) may develop within the first month of therapy.
- Lab tests: Frequent hematologic studies; periodic serum creatinine and LFTs.
- Closely monitor hematologic status; myelosuppression is common during the first month after starting therapy.
- Monitor for and report S&S of infection. Note that within the first month, fever may occur in the absence of infection.
- With high doses of cladribine, monitor for neurologic toxicity and acute nephrotoxicity.

Patient & Family Education

- Be fully informed regarding adverse responses to the drug.
- Understand the need for close follow-up during and after treatment with the drug.

CLARITHROMYCIN
(clar'i-thro-my-sin)

Biaxin, Biaxin XL
Classification: MACROLIDE ANTIBIOTIC
Therapeutic: ANTIBIOTIC
Prototype: Erythromycin
Pregnancy Category: C

AVAILABILITY 250 mg, 500 mg tablets; 500 mg sustained release tablets; 125 mg/5 mL, 250 mg/5 mL suspension

ACTION & _THERAPEUTIC EFFECT_
A semisynthetic macrolide antibiotic that binds to the 50S ribosomal subunit of susceptible bacterial organisms and, thereby, blocks RNA-mediated bacterial protein synthesis of bacteria. _It is active against both aerobic and anaerobic gram-positive and gram-negative organisms._

USES Treatment of upper respiratory, lower respiratory infections; community-acquired pneumonia; acute maxillary sinusitis; otitis media; and skin and soft tissue infections. Prevention and treatment of _Mycobacterium avium_ complex (MAC) infections in patients with HIV. Used in combination for _Helicobacter pylori_.

CONTRAINDICATIONS Hypersensitivity to clarithromycin, erythromycin, or any other macrolide antibiotics; history of cholestatic jaundice/hepatic dysfunction with previous use of clarithromycin; acute porphyria; congenital QT prolongation, torsades de pointes; viral infections. Safety and efficacy in infants younger than 6 mo not established.
CAUTIOUS USE Renal impairment, older adults, GI disease, colitis; pregnancy (category C), lactation.

Common adverse effects in _italic_, life-threatening effects <u>underlined</u>; generic names in **bold;** classifications in SMALL CAPS; ♣ Canadian drug name; ☻ Prototype drug

335

ROUTE & DOSAGE

Mild to Moderate Infections
Adult: **PO** 250–500 mg b.i.d. × 7–14 days or 1000 mg XL daily for 7–14 days
Child: **PO** 7.5 mg/kg q12h

MAC Infections
Adult: **PO** 500 mg q12h
Child /Adolescent/Infant (over 6 mo): **PO** 7.5 mg/kg q12h

H. pylori Infections (with other medications)
Adult: **PO** 500 mg b.i.d. to t.i.d.

Renal Impairment Dosage Adjustment
CrCl less than 30 mL/min: Decrease dose by ½ or double the dosing interval

ADMINISTRATION

Oral
▪ Ensure that sustained release form of drug is not chewed or crushed. It **must be** swallowed whole.
▪ Shake suspension well before use.
▪ Store at 15°–30° C (59°–86° F).

ADVERSE EFFECTS (≥1%) **GI:** Diarrhea, abdominal discomfort, nausea, abnormal taste, dyspepsia. **Hematologic:** Eosinophilia. **CNS:** Headache. **Skin:** Rash, urticaria.

DIAGNOSTIC TEST INTERFERENCE May increase *serum AST* and *ALT levels.*

INTERACTIONS Drug: May increase **theophylline** levels; drugs known to interact with **erythromycin** (i.e., **digoxin, carbamazepine, triazolam, warfarin, ergotamine, dihydroergotamine**) should be monitored carefully for increased levels and toxicity; **pimozide** may increase risk of arrhythmias. **Food:** **Grapefruit juice** increases risk of adverse effects.

PHARMACOKINETICS Absorption: Readily from GI tract; 50% reaches the systemic circulation. **Peak:** 2–4 h. **Distribution:** Into most body tissue (excluding CNS); high pulmonary tissue concentrations. **Metabolism:** Partially in the liver; active 14-OH metabolite acts synergistically with the parent compound against *H. influenzae.* **Elimination:** 20% unchanged in urine; 10–15% of 14-OH metabolite excreted in urine. **Half-Life:** 3–5 h.

NURSING IMPLICATIONS

Assessment & Drug Effects
▪ Inquire about previous hypersensitivity to other macrolides (e.g., erythromycin) before treatment.
▪ Withhold drug and notify prescriber, if hypersensitivity occurs (e.g., rash, urticaria).
▪ Monitor for and report loose stools or diarrhea, since pseudomembranous colitis **must be** ruled out.
▪ When clarithromycin is given concurrently with anticoagulants, digoxin, or theophylline, blood levels of these drugs may be elevated. Monitor appropriate serum levels and assess for S&S of drug toxicity.

Patient & Family Education
▪ Complete prescribed course of therapy.
▪ Report rash or other signs of hypersensitivity immediately.
▪ Report loose stools or diarrhea even after completion of drug therapy.

CLEMASTINE FUMARATE
(klem′as-teen)
Tavist-1

Common adverse effects in *italic*, life-threatening effects underlined; generic names in **bold**; classifications in SMALL CAPS; ♣ Canadian drug name; ⊙ Prototype drug

Classification: ANTIHISTAMINE
(H₁-RECEPTOR ANTAGONIST)
Therapeutic: ANTIHISTAMINE
Prototype: Diphenhydramine
Pregnancy Category: B

AVAILABILITY 1.34 mg, 2.68 mg tablets; 0.67 mg/5 mL syrup

ACTION & *THERAPEUTIC EFFECT*
An antihistamine (H₁-receptor antagonist) that competes for H₁-receptor sites on cells, thus blocking histamine effectiveness. Has greater selectivity for peripheral H₁-receptors and, consequently, it produces little sedation. Has prominent antipruritic activity and low incidence of unpleasant adverse effects. *Effective in controlling various allergic reactions (e.g., nasal congestion, sneezing, itching).*

USES Symptomatic relief of allergic rhinitis and mild uncomplicated allergic skin manifestations such as urticaria and angioedema.

CONTRAINDICATIONS Hypersensitivity to clemastine or to other antihistamines of similar chemical structure; lower respiratory tract symptoms, including acute asthma; concomitant MAOI therapy; closed-angle glaucoma; children younger than 6 y; lactation.
CAUTIOUS USE History of bronchial asthma, COPD; increased intraocular pressure; GI or GU obstruction; hyperthyroidism; hepatic disease; cardiovascular disease, hypertension, older adults; children, pregnancy (category B).

ROUTE & DOSAGE

Allergic Rhinitis

Adult: **PO** 1.34 mg b.i.d., may increase up to 8.04 mg/day
Child: **PO** *Older than 6 y,* 0.67 mg b.i.d., may increase up to 4.02 mg/ day; *younger than 6 y,* 0.335–0.67 mg/kg/day in 2 divided doses (max: 1.34 mg/day)

Allergic Urticaria

Adult: **PO** 2.68 mg b.i.d. or t.i.d., may increase up to 8.04 mg/day
Child: **PO** 1.34 mg b.i.d., may increase up to 4.02 mg/day

ADMINISTRATION

Oral
- Drug may be administered with food, water, or milk to reduce possibility of gastric irritation.
- Older adult patients usually require less than average adult dose.
- Store at 15°–30° C (59°–86° F) unless otherwise directed.

ADVERSE EFFECTS (≥1%) **Body as a Whole:** <u>Anaphylaxis</u>, excess perspiration, chills. **CV:** Hypotension, palpitation, tachycardia, extrasystoles. **GI:** *Dry mouth,* epigastric distress, anorexia, nausea, vomiting, diarrhea, constipation. **Hematologic:** Hemolytic anemia, <u>thrombocytopenia</u>, <u>agranulocytosis</u>. **CNS:** Sedation, *transient drowsiness,* dry nose and throat, headache, dizziness, weakness, fatigue, disturbed coordination; confusion, restlessness, nervousness, hysteria, convulsions, tremors, irritability, euphoria, insomnia, paresthesias, neuritis. **Respiratory:** Dry nose and throat, thickening of bronchial secretions, tightness of chest, wheezing, nasal stuffiness. **Skin:** Urticaria, rash, photosensitivity. **Special Senses:** Vertigo, tinnitus, acute labyrinthitis, blurred vision, diplopia. **Urogenital:** Difficult urination, urinary retention, early menses.

INTERACTIONS Drug: Alcohol and other CNS DEPRESSANTS increase sedation; MAO INHIBITORS may prolong and intensify anticholinergic effects.

Common adverse effects in *italic*, life-threatening effects <u>underlined</u>; generic names in **bold**; classifications in SMALL CAPS; ✤ Canadian drug name; ⊘ Prototype drug

337

PHARMACOKINETICS Absorption:
Readily from GI tract. **Peak:** 5–7 h.
Duration: 10–12 h. **Distribution:**
Into breast milk. **Metabolism:** In
liver. **Elimination:** In urine.

NURSING IMPLICATIONS

Assessment & Drug Effects

- Monitor for drowsiness, poor coordination, or dizziness, especially in the older adult or debilitated.
  Supervision of ambulation may
  be warranted.
- Assess for symptomatic relief with
  use of the medication.
- Lab tests: Periodic hematologic
  studies with long-term use.

Patient & Family Education

- Check with prescriber before taking alcohol or other CNS depressants, since effects may be additive.
- Clemastine may cause lethargy
  and drowsiness; therefore, necessary safety precautions should be
  taken.
- Older adults should make position changes slowly and in stages,
  particularly from recumbent to
  upright posture, as dizziness and
  hypotension occur more frequently than in younger patients.
- Avoid driving and other potentially hazardous activities until response to the drug has been established.

CLEVIDIPINE BUTYRATE
(cle-vi-di′peen bu-ti′rate)

Cleviprex
Classifications: CALCIUM CHAN-
NEL BLOCKER; ANTIHYPERTENSIVE
Therapeutic: ANTIHYPERTENSIVE
Prototype: Nifedipine
Pregnancy Category: C

AVAILABILITY 0.5 mg/mL emulsion
for injection

ACTION & *THERAPEUTIC EFFECT*
A calcium channel blocker that
interferes with the influx of calcium during depolarization of arterial smooth muscle. Decreases
systemic vascular resistance, thus
lowering mean arterial pressure.
Decreases blood pressure.

USES Treatment of hypertension,
hypertensive emergency/urgency
when oral administration is neither
feasible nor desired.

CONTRAINDICATIONS Hypersensitivity to soybeans, soy products, eggs/egg products; defective
lipid metabolism (e.g., pathologic
hyperlipidemia, lipid nephrosis,
acute pancreatitis); severe aortic
stenosis.

CAUTIOUS USE Reflex tachycardia,
hypotension, heart failure; lipid
intake restriction; rebound hypertension following drug discontinuation; elderly; pregnancy (category
C); lactation. Safety and efficacy in
children not established.

ROUTE & DOSAGE

Hypertension

Adult: **IV** Initial dose of 1–2 mg/h.
Titrate dose to desired BP: May
initially double dose every 90 sec;
as BP approaches goal, decrease
dose increments to less than
double the previous dose and
lengthen time intervals between
doses to q5–10min.

ADMINISTRATION

Intravenous

PREPARE: **IV Infusion:** Supplied
premixed, ready to use. Invert
vial gently to produce a uniform
emulsion.
ADMINISTER: **IV Infusion:** Use infusion device that permits calibrated

rates. ▪ May infuse through a central or peripheral line using NS, D5W, D5W/NS, D5W/LR, LR, or 10% amino acid solution. ▪ Complete infusion within 4 h of entering vial.

INCOMPATIBILITIES **Solution/additive:** Do not dilute in any IV solution. **Y-site:** Unknown; do not mix.

▪ Store refrigerated at 2°–8° C (36°–46° F). Do not return unopened vials to refrigeration once they have reached room temperature.

ADVERSE EFFECTS (≥1%) **CNS:** Headache. **CV:** Hypotension, reflex tachycardia. **GI:** Nausea, vomiting.

PHARMACOKINETICS Distribution: 99.5% plasma protein bound. **Metabolism:** In the plasma. **Elimination:** Renal (63–74%) and fecal (7–22%). **Half-Life:** 15 min.

NURSING IMPLICATIONS

Assessment & Drug Effects

▪ Monitor HR and BP continuously during infusion. Increases in HR is a normal response to vasodilation and rebound hypertension may occur for at least 8 h after infusion is stopped.
▪ Monitor cardiac status continuously during infusion, especially with preexisting HF. Clevidipine may have a negative inotropic effect and exacerbate HF.

Patient & Family Education

▪ Report promptly any of the following: Signs of heart failure; visual changes, weakness, or other signs of neurologic impairment.

CLINDAMYCIN HYDROCHLORIDE 🅿️

(klin-da-mye'sin)

Cleocin, Dalacin C ✤

CLINDAMYCIN PALMITATE HYDROCHLORIDE

Cleocin Pediatric

CLINDAMYCIN PHOSPHATE

Cleocin Phosphate, Cleocin T, Dalacin C, Evoclin, Cleocin Vaginal Ovules or Cream
Classification: LINCOSAMIDE ANTIBIOTIC
Therapeutic: ANTIBIOTIC
Pregnancy Category: B

AVAILABILITY 75 mg, 150 mg, 300 mg capsules; 75 mg/5 mL oral suspension; 150 mg/mL injection; 2% vaginal cream; 100 mg suppositories; 10 mg gel, lotion; 1% foam

ACTION & *THERAPEUTIC EFFECT*
Semisynthetic derivative of lincomycin that suppresses protein synthesis by binding to 50 S subunits of bacterial ribosomes, and, therefore, inhibits other antibiotics (e.g., erythromycin) that act at this site. *Particularly effective against susceptible strains of anaerobic streptococci as well as aerobic gram-positive cocci.*

USES Serious infections when less toxic alternatives are inappropriate. Topical applications are used in treatment of acne vulgaris. Vaginal applications are used in treatment of bacterial vaginosis in nonpregnant women.
UNLABELED USES In combination with pyrimethamine for toxoplasmosis in patients with AIDS.

CONTRAINDICATIONS History of hypersensitivity to clindamycin or lincomycin; meningitis; history of ulcerative colitis, or antibiotic-associated colitis; viral infection.

Common adverse effects in *italic*, life-threatening effects <u>underlined</u>; generic names in **bold**; classifications in SMALL CAPS; ✤ Canadian drug name; 🅿️ Prototype drug

339

CAUTIOUS USE History of GI disease, severe hepatic disease; atopic individuals (history of eczema, asthma, hay fever); older adults; pregnancy (category B); lactation; children younger than 16 y.

ROUTE & DOSAGE

Moderate to Severe Infections

Adult: **PO** 150–450 mg q6h **IM/IV** 600–1200 mg/day in divided doses (max: 2700 mg/day)
Child: **PO** 10–30 mg/kg/day q6–8h **IM/IV** 20–40 mg/kg/day in divided doses
Neonate: **IM/IV** *7 days or younger, weight 2000 g or less,* 10 mg/kg/day q12h; *7 days or younger, weight greater than 2000 g,* 15 mg/kg/day q8h; *older than 7 days, weight less than 1200 g,* 10 mg/kg/day q12h; *older than 7 days, weight 1200 g–2000 g,* 15 mg/kg/day q8h; *older than 7 days, weight greater than 2000 g,* 20 mg/kg/day q6–8h

Acne Vulgaris

Adult: **Topical** Apply to affected areas b.i.d.; 1% foam daily application

Bacterial Vaginosis

Adult: **Topical** Insert 1 suppository intravaginally at bedtime × 3 days, or insert 1 applicator full of cream intravaginally at bedtime × 7 days

ADMINISTRATION

Oral

- Administer clindamycin capsules with a full [240 mL (8 oz)] glass of water to prevent esophagitis.
- Note expiration date of oral solution; retains potency for 14 days at room temperature. Do not refrigerate, as chilling causes thickening and thus makes pouring it difficult.

Intramuscular

- Deep IM injection is recommended. Rotate injection sites and observe daily for evidence of inflammatory reaction. Single IM doses should not exceed 600 mg.

Intravenous

IV administration to neonates, infants, and children: Verify correct IV concentration and rate of infusion with prescriber.

PREPARE: Intermittent: Each 18 mg **must be** diluted with at least 1 mL of D5W, NS, D5/.45% NaCl, or other compatible solution. ▪ Final concentration should never exceed 18 mg/mL.

ADMINISTER: Intermittent: Never give a bolus dose. ▪ Do not give more than 1200 mg in a single 1-h infusion. ▪ Infusion rate should not exceed 30 mg/min.

INCOMPATIBILITIES Solution/additive: Aminophylline, BARBITURATES, **calcium gluconate, ceftriaxone, ciprofloxacin, gentamicin, magnesium sulfate, ranitidine. Y-site: Allopurinol, azithromycin, doxapram, filgrastim, fluconazole, idarubicin, lansoprazole.**

- Store in tight containers at 15°–30° C (59°–86° F) unless otherwise directed.

ADVERSE EFFECTS (≥1%) **Body as a Whole:** Fever, serum sickness, sensitization, swelling of face (following topical use), generalized myalgia, superinfections, proctitis, vaginitis, pain, induration, sterile abscess (following IM injections); thrombophlebitis (IV infusion). **CV:** Hypotension (following IM), cardiac arrest (rapid IV). **GI:** *Diarrhea,* abdominal pain, flatulence, bloating, *nausea, vomiting,* pseudomembranous colitis; esophageal irritation, loss of taste, medicinal taste (high IV doses), jaundice, abnormal liver function tests. **Hematologic:** Leukopenia, eosinophilia, agranulocytosis, thrombocytopenia. **Skin:** *Skin rashes,* urticaria, pruritus, dryness, contact dermatitis, gram-negative folliculitis, irritation, oily skin.

DIAGNOSTIC TEST INTERFERENCE Clindamycin may cause increases in *serum alkaline phosphatase, bilirubin, creatine phosphokinase (CPK)* from muscle irritation following IM injection; *AST, ALT.*

INTERACTIONS Drug: Chloramphenicol, erythromycin possibly are mutually antagonistic to clindamycin; neuromuscular blocking action enhanced by NEUROMUSCULAR BLOCKING AGENTS **(atracurium, tubocurarine, pancuronium).**

PHARMACOKINETICS Absorption: Approximately 90% absorbed from GI tract; 10% of topical application is absorbed through skin. **Peak:** 45–60 min PO; 3 h IM. **Duration:** 6 h PO; 8–12 h IM. **Distribution:** Widely distributed except for CNS; crosses placenta; distributed into breast milk. **Metabolism:** In liver. **Elimination:** In urine and feces. **Half-Life:** 2–3 h.

NURSING IMPLICATIONS

Assessment & Drug Effects

- Lab tests: Culture and susceptibility testing should be performed initially. Periodic CBC with differential, liver, and kidney function tests.
- Monitor BP and pulse in patients receiving drug parenterally. Hypotension has occurred following IM injection. Advise patient to remain recumbent following drug administration until BP has stabilized.

- Severe diarrhea and colitis, including pseudomembranous colitis, have been associated with oral (highest incidence), parenteral, and topical clindamycin. Report immediately the onset of watery diarrhea, with or without fever. Symptoms may appear within a few days to 2 wk after therapy is begun or up to several weeks following cessation of therapy.
- Be alert to signs of superinfection (see Appendix F).
- Be alert for signs of anaphylactoid reactions (see Appendix F), that require immediate attention.

Patient & Family Education

- Report loose stools or diarrhea promptly.
- Stop drug therapy if significant diarrhea develops (more than 5 loose stools daily) and notify prescriber.
- Do not self-medicate with antidiarrheal preparations. Antiperistaltic agents may prolong and worsen diarrhea by delaying removal of toxins from colon.

CLOBETASOL PROPIONATE

(cloe-bay'ta-sol)

Clobex, Temovate, Embeline gel; Olux Foam
Pregnancy Category: C
See Appendix A-4.

CLOBAZAM

(kloe' ba zam)

Classifications: BENZODIAZEPINE; ANTICONVULSANT
Therapeutic: ANTICONVULSANT
Prototype: Diazepam
Pregnancy Category: C
Controlled Substance: Scheduled IV

AVAILABILITY 5 mg, 10 mg; 20 mg tablets

Common adverse effects in *italic,* life-threatening effects underlined; generic names in **bold;** classifications in SMALL CAPS; ♣ Canadian drug name; ☼ Prototype drug

C

ACTION & *THERAPEUTIC EFFECT*

Mechanism of action believed to involve potentiation of GABAergic neurotransmission resulting from binding at the benzodiazepine site of the $GABA_A$ receptor. *Helps inhibit development of seizures associated with Lennox-Gastaut syndrome.*

USES Adjunctive treatment of seizures associated with Lennox-Gastaut syndrome (LGS) in patients 2 y or older

CONTRAINDICATIONS Severe hepatic impairment; lactation.

CAUTIOUS USE Hepatic impairment; severe renal impairment; history of substance abuse; history of suicidal thoughts or behaviors; pregnancy (category C); depression; drug withdrawal; older adults. Safety and effectiveness in patients younger than 2 y not established.

ROUTE & DOSAGE

Adjunctive Treatment of Seizures

Adult and Children (2 y and older): **PO** *If over 30 kg, 5 mg b.i.d., increase to 10 mg b.i.d. on day 7, then increase to 20 mg b.i.d. on day 14.; If 30 kg or less, 5 mg once daily, increase to 5 mg b.i.d., on day 7, then increase to 10 mg b.i.d. on day 14*

Geriatric Patients: If over 30 kg, use normal dose regimen for 30 kg or less; may increase to 20 mg b.i.d. on day 21. If 30 kg or less, initial dose of 5 mg once daily, increase to 5 mg b.i.d. on day 14, then increase to 10 mg b.i.d.

Poor CYP2C19 metabolizers: Starting dose should be 5 mg/ day; dose titration should proceed slowly according to weight, but to half the normal dose regimen. If necessary, additional titration to the 20 mg/ day or 40 mg/day (depending on weight) may be started on day 21.

Hepatic Impairment Dosage Adjustment

Mild to moderate impairment (Child-Pugh score 5 to 9): Use geriatric dosing regimen

Severe impairment: Not recommended

Renal Impairment Dosage Adjustment

CrCl less than 30 ml/min: No dosage recommendations are available

ADMINISTRATION

Oral

- May be given whole or crushed and mixed in applesauce.
- Give without regard to timing of meals.
- Abrupt discontinuation of this drug should be avoided. Drug should be tapered off by decreasing the daily dose every week by 5 to 10 mg.
- Store at 20° –25° C (68° –77° F).

ADVERSE EFFECTS (≥5%) **Body as a Whole:** Fatigue, irritability, *pyrexia.* **CNS:** Aggresion, ataxia, drooling, dysarthria, insomnia, *lethargy,* psychomotor hyperactivity, sedation, *somnolence.* **GI:** Constipation, dysphagia, *vomiting.* **Metabolic:** Alterations in appetite. **Respiratory:** Bronchitis, cough, pneumonia, upper respiratory tract infection. **Urogenital:** Urinary tract infection.

INTERACTIONS Drug: Clobazam may decrease the effectiveness of

ORAL CONTRACEPTIVES and other drugs requiring CYP2D6 (e.g., **metoprolol, propranolol, aripiprazole, clozapine**). Strong and moderate inhibitors of CYP2C19 (e.g., **fluconazole, fluvoxamine, ticlopidine, omeprazole**) may result in increased levels of the active metabolite of clobazam. **Ethanol** increases the maximum plasma exposure of clobazam by 50%.

PHARMACOKINETICS Absorption: Approximately 100% bioavailable. **Peak:** 0.5–4 h. **Distribution:** 80–90% Plasma protein bound. **Metabolism:** Extensive hepatic oxidation to active and inactive metabolites. **Elimination:** Renal (82%) and fecal (11%). **Half-Life:** 36–42 h.

NURSING IMPLICATIONS

Assessment & Drug Effects

- Monitor for adverse CNS effects (e.g., somnolence, sedation, new or worsening depression, impaired judgment, impaired motor skills), especially when used concurrently with other CNS depressants.
- Monitor for and report promptly suicidal thoughts or behaviors, or thoughts of self-harm.
- Lab test: Baseline and periodic LFTs.

Patient & Family Education

- Promptly notify prescriber of suicidal ideation or thoughts of self-harm.
- Avoid potentially hazardous tasks, such as driving, until response to the drug is known.
- Do not drink alcohol while taking colbazam.
- Abruptly stopping colbazam may increase the risk of seizure activity.
- Women of childbearing age who use hormonal contraceptives should use alternative non-hormonal methods during and for 28 days after discontinuing colbazam.

- Notify prescriber immediately if you become pregnant or intend to become pregnant, or if you are breast feeding or intend to breast feed.

CLOCORTOLONE PIVALATE
(kloe-kor′toe-lone)
Cloderm
Pregnancy Category: C
See Appendix A-4.

CLOFARABINE
(clo-fa-ra′been)
Clolar
Classifications: ANTINEOPLASTIC; PURINE ANTIMETABOLITE
Therapeutic: ANTINEOPLASTIC
Prototype: 6-Mercaptopurine
Pregnancy Category: D

AVAILABILITY 1 mg/mL injection

ACTION & *THERAPEUTIC EFFECT* Clofarabine inhibits DNA repair within cancer cells, thus interfering with mitosis; it also disrupts the mitochondrial membrane, leading to cancer cell death. *Cytotoxic to rapidly proliferating and quiescent cancer cells.*

USES Relapsed or refractory acute lymphocytic leukemia (ALL) after at least 2 prior regimens.

CONTRAINDICATIONS Severe bone marrow suppression; active infection; pregnancy (category D); lactation.
CAUTIOUS USE Renal or hepatic function impairment; thrombocytopenia; neutropenia; previous chemotherapy or radiation therapy; females of childbearing age; history of viral infections such as herpes; history of cardiac disease or hypotension.

Common adverse effects in *italic*, life-threatening effects <u>underlined</u>; generic names in **bold**; classifications in SMALL CAPS; ♦ Canadian drug name; ☺ Prototype drug

343

ROUTE & DOSAGE

Acute Lymphocytic Leukemia
Adolescent/Child: **IV** 52 mg/m²/ day for 5 days

ADMINISTRATION

- Do not give drugs with known renal toxicity during the 5 days of clofarabine administration.

Intravenous

PREPARE: **IV Infusion:** Withdraw required dose from vial using a 0.2 micron filter syringe. • Further dilute in 100 mL or more of D5W or NS prior to infusion.
ADMINISTER: **IV Infusion:** Give over 2 h. • Note: It is recommended that IV fluids be given continuously throughout the 5 days of clofarabine administration.

- Store diluted solution at room temperature. Use within 24 h of mixing.

ADVERSE EFFECTS (≥1%) **CNS:** Anxiety, depression, dizziness, headache, irritability, somnolence. **CV:** *Tachycardia,* pericardial infusion, left ventricular systolic dysfunction (LSVT). **GI:** *Vomiting, nausea,* and *diarrhea,* abdominal pain, constipation. **Hematologic/ Lymphatic:** Anemia, <u>leukopenia</u>, <u>thrombocytopenia</u>, neutropenia, *febrile neutropenia.* **Hepatic:** Jaundice, hepatomegaly. **Metabolic:** Anorexia, decreased appetite, edema, decreased weight. **Musculoskeletal:** Arthralgia, back pain, myalgia. **Respiratory:** Cough, dyspnea, epistaxis, pleural effusion, respiratory distress. **Skin:** Dermatitis, contusion, dry skin, erythema, palmar-plantar erythrodysesthesia syndrome, pruritus. **Body as a Whole:** Increased risk of infection.

PHARMACOKINETICS Distribution: 47% protein bound. **Metabolism:** Negligible. **Elimination:** Primarily unchanged in the urine. **Half-Life:** 5.2 h.

NURSING IMPLICATIONS

Assessment & Drug Effects

- Monitor vital signs frequently during infusion of clofarabine.
- Monitor closely for S&S of capillary leak syndrome or systemic inflammatory response syndrome (e.g., tachypnea, tachycardia, hypotension, pulmonary edema). If either is suspected, immediately DC IV, institute supportive measures and notify prescriber.
- Monitor I&O rates and pattern and watch for S&S of dehydration, including dizziness, lightheadedness, fainting spells, or decreased urine output.
- Withhold drug and notify prescriber if hypotension develops for any reason during 5-day period of drug administration.
- Lab tests: Baseline and periodic CBC and platelet counts (more frequent with cytopenias); frequent LFTs and kidney function test during the 5 days of clofarabine therapy.

Patient & Family Education

- Report any distressing adverse effect of therapy to prescriber.
- Use effective measures to avoid pregnancy while taking this drug.

CLOMIPHENE CITRATE
(kloe′mi-feen)

Clomid, Milophene, Serophene
Classifications: OVULATION STIMULANT; NONSTEROID SELECTIVE ESTROGEN RECEPTOR MODULATOR (SERM)
Therapeutic: OVULATION STIMULANT; ANTIESTROGENIC
Pregnancy Category: X

AVAILABILITY 50 mg tablets

ACTION & *THERAPEUTIC EFFECT* Oral nonsteroidal selective estrogen receptor modulator (SERM) that induces ovulation in selected infrequently ovulating or anovulatory women. Clomiphene blocks the normal negative feedback of circulating estradiol on the hypothalamus, preventing estrogen from lowering the output of gonadotropin releasing hormone (GnRH). It acts by binding to hypothalamic estrogen receptors, decreasing their numbers, and thereby inhibiting receptor replenishment. *Inhibition of receptor replenishment results in a false hypoestrogenic state which stimulates pituitary release of luteinizing hormone (LH), follicle-stimulating hormone (FSH), and gonadotropins, leading to ovarian stimulation.*

USES Infertility in appropriately selected women desiring pregnancy whose partners are fertile and potent.

UNLABELED USES Male infertility, menstrual abnormalities, gynecomastia, fibrocystic breast disease, regulation of cycles in patients using rhythm method of contraception, endometrial hyperplasia, persistent lactation.

CONTRAINDICATIONS Neoplastic lesions, ovarian cyst; hepatic disease or dysfunction; abnormal uterine bleeding; endometriosis; primary ovarian failure; men with testicular failure; untreated thyroid disease; visual abnormalities; major depression or psychosis; thrombophlebitis; pregnancy (category X); lactation.

CAUTIOUS USE Polycystic ovarian enlargement, pelvic discomfort, sensitivity to pituitary gonadotropins.

ROUTE & DOSAGE

Infertility

Adult: **PO First course:** 50 mg/day for 5 days; start on 5th day of cycle following start of spontaneous or induced bleeding (with progestin) or at any time in the patient who has had no recent uterine bleeding **Second course if ovulation:** Repeat first course until conception or for 3 cycles **Second course if no ovulation:** 100 mg/day for 5 days as above (max: 100 mg/day)

ADMINISTRATION

Oral

- Each course of therapy should start on or about the 5th cycle day once ovulation has been established.
- Store at 15°–30° C (59°–86° F) in tightly capped, light-resistant container.

ADVERSE EFFECTS (≥1%) **Body as a Whole:** *Vasomotor flushes,* breast discomfort, abdominal pain, heavy menses, exacerbation of endometriosis; mental depression, headache, fatigue, insomnia, dizziness, vertigo. **GI:** Nausea, vomiting, increased appetite with weight gain, constipation, bloating. **Endocrine:** Spontaneous abortion, multiple ovulations, ovarian failure, *ovarian hyperstimulation syndrome, enlarged ovaries with multiple follicular cysts.* **Special Senses:** Transient blurring, diplopia, scotomas, photophobia, floaters, prolonged after-images. **Urogenital:** Urinary frequency, polyuria.

DIAGNOSTIC TEST INTERFERENCE Clomiphene may increase BSP retention; *plasma transcortin, thyroxine* and *sex hormone binding globulin* levels. Also increases *follicle-stimulating* and *luteinizing hormone* secretion in most patients.

Common adverse effects in *italic*, life-threatening effects underlined; generic names in **bold;** classifications in SMALL CAPS; ♣ Canadian drug name; ⊘ Prototype drug

345

INTERACTIONS Drug: No clinically significant drug interactions established. **Herbal: Black cohosh** may antagonize infertility treatments.

PHARMACOKINETICS Absorption: Readily absorbed from GI tract. **Metabolism:** In liver. **Elimination:** Primarily in feces in 5 days; the remainder is excreted slowly from enterohepatic pool or is stored in body fat for later release. **Half-Life:** 5 days.

NURSING IMPLICATIONS

Assessment & Drug Effects

- Monitor for abnormal bleeding. Report it immediately.
- Monitor for visual disturbances. Their occurrence indicates the need for a complete ophthalmologic evaluation. Drug will be stopped until symptoms subside.
- Pelvic pain indicates the need for immediate pelvic examination for diagnostic purposes.

Patient & Family Education

- Take the medicine at same time every day to maintain drug levels and prevent forgetting a dose.
- Missed dose: Take drug as soon as possible. If not remembered until time for next dose, double the dose, then resume regular dosing schedule. If more than one dose is missed, check with prescriber.
- Report these symptoms: Hot flushes resembling those associated with menopause; nausea, vomiting, headache.
- Report promptly yellowing of eyes, light-colored stools, yellow, itchy skin, and fever symptomatic of jaundice.
- Stop taking clomiphene if pregnancy is suspected.
- Because of the possibility of lightheadedness, dizziness, and visual disturbances, do not perform hazardous tasks requiring skill and coordination in an environment with variable lighting.
- Report promptly excessive weight gain, signs of edema, bloating, decreased urinary output.
- If clomiphene is continued more than 1 y, patient should have an ophthalmologic examination at regular intervals.

CLOMIPRAMINE HYDROCHLORIDE

(clo-mi´pra-meen)

Anafranil

Classification: TRICYCLIC ANTIDEPRESSANT

Therapeutic: ANTIPSYCHOTIC

Prototype: Imipramine

Pregnancy Category: C

AVAILABILITY 25 mg, 50 mg, 75 mg capsules

ACTION & *THERAPEUTIC EFFECT*
Inhibits reuptake of norepinephrine and serotonin at the presynaptic neuron. Of the tricyclic antidepressants (TCAs), it is the most selective and potent inhibitor of serotonin (5-HT) reuptake. *The basis of its antidepressant effects is thought to be due to the elevated serum levels of norepinephrine and serotonin.*

USES Obsessive-compulsive disorder (OCD).

UNLABELED USES Panic disorder, autism, agoraphobia, depression.

CONTRAINDICATIONS Hypersensitivity to other tricyclic compounds; acute recovery period after MI, QT elongation, cardiac arrhythmias (AV block, bundle-branch block); suicidal ideation.

CAUTIOUS USE History of convulsive disorders, prostatic hypertrophy, urinary retention, cardiovascular, hepatic, GI, or blood disorders; history of seizure disorder; respiratory depression; older adults; diabetes mellitus; GERD; Parkinson's disease; closed-angle glaucoma; asthma; bipolar disorder; history of suicidal tendencies; pregnancy (category C), lactation; children younger than 10 y.

ROUTE & DOSAGE

Obsessive-Compulsive Disorder

Adult: **PO** 25 mg daily, gradually increase to 100 mg daily as tolerated over 2 wk, then up to 250 mg daily
Adolescent/Child (older than 10 y): **PO** 25 mg daily, gradually increase to 100 mg daily or 3 mg/kg (whichever is less) as tolerated over 2 wk, then up to 200 mg or 3 mg/kg daily (whichever is less)

Pharmacogenetic Dosage Adjustment

Poor CYP2D6 metabolizers should receive 60% of normal dose

ADMINISTRATION

Oral

- Give with meals to reduce GI adverse effects.
- Following titration to the full dose, drug may be given as a single dose at bedtime to reduce daytime sedation.
- Store at 15°–30° C (59°–86° F).

ADVERSE EFFECTS (≥1%) **Body as a Whole:** Diaphoresis. **CV:** Hypotension, tachycardia. **GI:** Constipation, *dry mouth.* **Endocrine:** Galactorrhea, hyperprolactinemia, amenorrhea, *weight gain.* **Hematologic:** Leukopenia, agranulocytosis, thrombocytopenia, anemia. **CNS:** Mania, *tremor,* dizziness, hyperthermia, neuroleptic malignant syndrome, seizures (especially with abrupt withdrawal). **Urogenital:** Delayed ejaculation, anorgasmia.

DIAGNOSTIC TEST INTERFERENCE Increased glucose; may interfere with urine detection of **methadone.**

INTERACTIONS Drug: MAO INHIBITORS may precipitate hyperpyrexic crisis, tachycardia, or seizures; ANTIHYPERTENSIVE AGENTS potentiate orthostatic hypotension; CNS DEPRESSANTS, **alcohol** add to CNS depression; **norepinephrine** and other SYMPATHOMIMETICS may increase cardiac toxicity; **cimetidine** decreases hepatic metabolism, thus increasing imipramine levels; **methylphenidate** inhibits metabolism of **imipramine** and thus may increase its toxicity. **Herbal: Ginkgo** may decrease seizure threshold; **St. John's wort** may cause serotonin syndrome.

PHARMACOKINETICS Absorption: Rapidly from GI tract; 20–78% reaches systemic circulation. **Onset:** Approx 4–10 wk. **Peak:** 2–6 h. **Distribution:** Widely distributed including the CSF; crosses placenta. **Metabolism:** Extensive first-pass metabolism in the liver; active metabolite is desmethylclomipramine. **Elimination:** 50–60% in urine, 24–32% in feces. **Half-Life:** 20–30 h.

NURSING IMPLICATIONS

Assessment & Drug Effects

- Monitor for seizures, especially in those with predisposing factors or concurrent therapy with other drugs that lower seizure threshold.
- Lab tests: Periodic CBC with differential, platelet count, and Hct and Hgb. Monitor liver functions, especially with long-term therapy.
- Monitor for and report signs of neuroleptic malignant syndrome (see Appendix F).
- Monitor for sedation and vertigo, especially at the beginning of

Common adverse effects in *italic*, life-threatening effects underlined; generic names in **bold**; classifications in SMALL CAPS; ✦ Canadian drug name; ❷ Prototype drug

347

therapy and following dosage increases. Supervision of ambulation may be indicated.

- Notify prescriber of fever and complaints of sore throat since these may indicate need to rule out adverse hematologic changes.

Patient & Family Education

- Do not take nonprescribed drugs or discontinue therapy without consent of prescriber. Abrupt discontinuation may cause nausea, headache, malaise, or seizures.
- Men should understand that the drug may cause impotence or ejaculation failure.
- Report promptly a sore throat accompanied by fever.
- Use caution with ambulation until response to drug is known.
- Moderate alcohol intake since it may potentiate adverse drug effects.

CLONAZEPAM

(kloe-na′zi-pam)

Klonopin, Klonopin Wafers, Rivotril♦

Classifications: ANTICONVULSANT; BENZODIAZEPINE

Therapeutic: ANTICONVULSANT

Prototype: Diazepam

Pregnancy Category: D

Controlled Substance: Schedule IV

AVAILABILITY 0.5 mg, 1 mg, 2 mg tablets; 0.125 mg, 0.25 mg, 0.5 mg, 1 mg, and 2 mg orally disintegrating wafers

ACTION & THERAPEUTIC EFFECT
Benzodiazepine derivative with strong anticonvulsant activity that prevents seizures by potentiating the effects of GABA, an inhibitory neurotransmitter. Suppresses spread of seizure activity in the cortex, thalamus, and limbic regions of the brain. *Suppresses spike and wave discharge in absence seizures (petit mal) and decreases amplitude, frequency, duration, and spread of discharge in minor motor seizures.*

USES Alone or with other drugs in absence, myoclonic, and akinetic seizures, Lennox-Gastaut syndrome, absence seizures refractory to succinimides or valproic acid, and for infantile spasms and restless legs.

UNLABELED USES Panic disorder, complex partial seizure pattern, and generalized tonic-clonic convulsions.

CONTRAINDICATIONS Hypersensitivity to benzodiazepines; liver disease; acute narrow-angle glaucoma; pulmonary disease; coma or CNS depression; pregnancy (category D), lactation.

CAUTIOUS USE Renal or hepatic disease; COPD; drug-controlled open-angle glaucoma; bipolar disorder, preexisting depression; addiction-prone individuals; neuromuscular disease; mixed seizure disorders; children younger than 10 y.

ROUTE & DOSAGE

Seizures

Adult: **PO** 1.5 mg/day in 3 divided doses, increased by 0.5–1 mg q3days until seizures are controlled or until intolerable adverse effects (max recommended dose: 20 mg/day)

Child (younger than 10 y): **PO** 0.01–0.03 mg/kg/day (not to exceed 0.05 mg/kg/day) in 3 divided doses; may increase by 0.25–0.5 mg q3days until seizures are controlled or until intolerable adverse effects (max recommended dose: 0.2 mg/kg/day)

Panic Disorders

Adult: **PO** 1–2 mg/day in divided doses (max: 4 mg/day)

ADMINISTRATION

Oral

- Give largest dose at bedtime if daily dose cannot be equally divided.
- Place wafer form on tongue to dissolve.
- Store in tightly closed container protected from light at 15°–30° C (59°–86° F) unless otherwise specified.

ADVERSE EFFECTS (≥1%) **CV:** Palpitations. **GI:** Dry mouth, sore gums, anorexia, coated tongue, increased salivation, increased appetite, nausea, constipation, diarrhea. **Hematologic:** Anemia, leukopenia, thrombocytopenia, eosinophilia. **CNS:** *Drowsiness, sedation, ataxia,* insomnia, aphonia, choreiform movements, coma, dysarthria, "glassy-eyed" appearance, headache, hemiparesis, hypotonia, slurred speech, tremor, vertigo, confusion, depression, hallucinations, aggressive behavior problems, hysteria, suicide attempt. **Respiratory:** Chest congestion, respiratory depression, rhinorrhea, dyspnea, hypersecretion in upper respiratory passages. **Skin:** Hirsutism, hair loss, skin rash, ankle and facial edema. **Special Senses:** Diplopia, nystagmus, abnormal eye movements. **Urogenital:** Increased libido, dysuria, enuresis, nocturia, urinary retention.

DIAGNOSTIC TEST INTERFERENCE

Clonazepam causes transient elevations of ***serum transaminase*** and ***alkaline phosphatase.***

INTERACTIONS **Drug:** Alcohol

and other CNS DEPRESSANTS increase sedation and CNS depression; may increase **phenytoin** levels. **Herbal: Kava, valerian** may potentiate sedation.

PHARMACOKINETICS **Absorption:**

Readily absorbed from GI tract.

Onset: 60 min. **Peak:** 1–2 h. **Duration:** Up to 12 h in adults; 6–8 h in children. **Distribution:** Crosses placenta; distributed into breast milk. **Metabolism:** In liver. **Elimination:** In urine primarily as metabolites. **Half-Life:** 18–40 h.

NURSING IMPLICATIONS

Assessment & Drug Effects

- Monitor for signs of suicidal ideation in depressive individuals.
- Lab tests: Periodic LFTs, platelet counts, blood counts, and renal function tests.
- Both psychological and physical dependence may occur in the patient on long-term, high-dose therapy.
- Monitor for S&S of overdose, including somnolence, confusion, irritability, sweating, muscle and abdominal cramps, diminished reflexes, coma.

Patient & Family Education

- Report loss of seizure control promptly. Anticonvulsant activity is often lost after 3 mo of therapy; dosage adjustment may reestablish efficacy.
- Do not abruptly discontinue this drug. Abrupt withdrawal can precipitate seizures. Other withdrawal symptoms include convulsion, tremor, abdominal and muscle cramps, vomiting, sweating.
- Do not drive a car or engage in other activities requiring mental alertness and physical coordination until reaction to the drug is known. Drowsiness occurs in approximately 50% of patients.

✓CLONIDINE HYDROCHLORIDE

(kloe′ni-deen)

Catapres, Catapres-TTS, Dixaril✦, Duraclon, Kapvay

Common adverse effects in *italic*, life-threatening effects underlined; generic names in **bold**; classifications in SMALL CAPS; ✦ Canadian drug name; ⊘Prototype drug

349

Classifications: CENTRAL-ACTING ANTIHYPERTENSIVE; ANALGESIC
Therapeutic: ANTIHYPERTENSIVE; ANALGESIC
Prototype: Methyldopa
Pregnancy Category: C

AVAILABILITY 0.1 mg, 0.2 mg, 0.3 mg tablets; 0.1 mg/24 h, 0.2 mg/24 h, 0.3 mg/24 h transdermal patch; 100 mcg/mL, 500 mcg/mL injection; 0.1 mg extended release tablet

ACTION & *THERAPEUTIC EFFECT*
Centrally acting receptor agonist that stimulates alpha$_2$-adrenergic receptors in CNS to inhibit sympathetic cardioaccelerator and vasomotor centers. Central actions reduce plasma concentrations of norepinephrine. It decreases systolic and diastolic BP and heart rate. *Decreases systolic and diastolic BP and heart rate. Reportedly minimizes or eliminates many of the common clinical S&S associated with withdrawal of heroin, methadone, or other opiates.*

USES Hypertension, treatment of severe pain, ADHD.
UNLABELED USES Prophylaxis for migraine; treatment of dysmenorrhea, menopausal flushing, diarrhea, paroxysmal localized hyperhidroses, neuropathic pain; alcohol, smoking, opiate, and benzodiazepine withdrawal; Tourette's syndrome.

CONTRAINDICATIONS Hypersensitivity to clonidine; coagulopathy; lactation. **Clonidine ER:** Children younger than 6 y. **Patch:** Polyarteritis nodosa, scleroderma, SLE on affected areas. **Epidural:** Severe cardiovascular disease, or those who are hemodynamically unstable; infection at injection site; obstetric, postpartum, perioperative pain management; use above the C_4 dermatome. May be a rare case when use outweighs possible serious risk.

CAUTIOUS USE Severe coronary insufficiency, recent MI, sinus node dysfunction, cerebrovascular disease; diabetes mellitus; chronic renal failure; Raynaud's disease, thromboangiitis obliterans; history of hypotension, heart block, bradycardia, or CVD; history of syncope; history of depression; older adults; pregnancy (category C); children younger than 12 y.

ROUTE & DOSAGE

Hypertension

Adult: **PO** 0.1 mg b.i.d. or t.i.d., may increase by 0.1–0.2 mg/day until desired response is achieved (max: 2.4 mg/day)
Transdermal 0.1 mg patch once q7days, may increase by 0.1 mg q1–2wk
Geriatric: **PO** Start with 0.1 mg once daily
Child: **PO** 5–10 mcg/kg/day divided q8–12h, may increase to 5–25 mcg/kg/day divided q6h (max: 0.9 mg/day)

Severe Pain

Adult: **Epidural** Start infusion at 30 mcg/h and titrate to response. Use rates greater than 40 mcg/h with caution.
Child: **Epidural** Start infusion at 0.5 mcg/kg/h and titrate to response.

ADHD (extended release tablets)

Adolescent/Child (over 6 y): **PO** 0.1 mg qhs

ADMINISTRATION

Oral

- Ensure that extended release tablets are swallowed whole. They should not be crushed or chewed.

- Give last PO dose immediately before patient retires to ensure overnight BP control and to minimize daytime drowsiness.
- Oral dosage is increased gradually over a period of weeks so as not to lower BP abruptly (especially important in the older adult).
- During change from PO clonidine to transdermal system, PO clonidine should be maintained for at least 24 h after patch is applied. Consult prescriber.
- Do not abruptly discontinue drug. It should be withdrawn over a period of 2–4 days. Abrupt withdrawal may result in a hypertensive crisis within 8–18 h.
- Store in tightly closed container at 15°–30° C (59°–86° F) unless otherwise directed.

Transdermal

- Apply transdermal patch to dry skin, free of hair and rash. Avoid irritated, abraded, or scarred skin. Recommended areas for applying transdermal patch are upper outer arm and anterior chest. Rotate application sites and keep a record.

ADVERSE EFFECTS (≥1%) **CV:** *Hypotension (epidural),* postural hypotension (mild), peripheral edema, ECG changes, tachycardia, bradycardia, flushing, rapid increase in BP with abrupt withdrawal. **GI:** *Dry mouth, constipation,* abdominal pain, pseudo-obstruction of large bowel, altered taste, nausea, vomiting, hepatitis, hyperbilirubinemia, weight gain (sodium retention). **CNS:** *Drowsiness, sedation,* dizziness, headache, fatigue, weakness, sluggishness, dyspnea, vivid dreams, nightmares, insomnia, behavior changes, agitation, hallucination, nervousness, restlessness, anxiety, mental depression. **Skin:** Rash, pruritus, thinning of hair, exacerbation of psoriasis;

with transdermal patch: Hyperpigmentation, recurrent herpes simplex, skin irritation, contact dermatitis, mild erythema. **Special Senses:** Dry eyes. **Urogenital:** Impotence, loss of libido.

DIAGNOSTIC TEST INTERFERENCE Avoid use of transdermal patch during MRI. Possibility of decreased urinary excretion of *aldosterone, catecholamines,* and *VMA* (however, sudden withdrawal of clonidine may cause increases in these values); transient increases in *blood glucose;* weakly positive *direct antiglobulin (Coombs') tests.*

INTERACTIONS Drug: Alcohol and other CNS DEPRESSANTS add to CNS depression; TRICYCLIC ANTIDEPRESSANTS may reduce antihypertensive effects. OPIATE ANALGESICS increase hypotension with epidural clonidine. Increased risk of bradycardia or AV block when epidural clonidine is used with **digoxin,** CALCIUM CHANNEL BLOCKERS, or BETA BLOCKERS. Use with other ANTIHYPERTENSIVES can have added effect.

PHARMACOKINETICS Absorption: Readily absorbed from GI tract. **Onset:** 30–60 min PO; 1–3 days transdermal. **Peak:** 2–4 h PO; 2–3 days transdermal. **Duration:** 8 h PO; 7 days transdermal. **Distribution:** Widely distributed; crosses blood–brain barrier; not known if crosses placenta or distributed into breast milk. **Metabolism:** In liver. **Elimination:** 80% in urine, 20% in feces. **Half-Life:** 6–20 h.

NURSING IMPLICATIONS

Assessment & Drug Effects

- Monitor BR closely. Determine positional changes (supine, sitting, standing).
- With epidural administration, frequently monitor BP and HR.

Common adverse effects in *italic*, life-threatening effects underlined; generic names in **bold;** classifications in SMALL CAPS; ♣ Canadian drug name; ⊘ Prototype drug

351

Hypotension is a common side effect that may require intervention.

- Monitor BP closely whenever a drug is added to or withdrawn from therapeutic regimen.
- Monitor I&O during period of dosage adjustment. Report change in I&O ratio or change in voiding pattern.
- Determine weight daily. Patients not receiving a concomitant diuretic agent may gain weight, particularly during first 3 or 4 days of therapy, because of marked sodium and water retention.
- Supervise closely patients with history of mental depression, as they may be subject to further depressive episodes.

Patient & Family Education

- Although postural hypotension occurs infrequently, make position changes slowly, and in stages, particularly from recumbent to upright position, and dangle and move legs a few minutes before standing. Lie down immediately if faintness or dizziness occurs.
- Avoid potentially hazardous activities until reaction to drug has been determined due to possible sedative effects.
- Do not omit doses or stop the drug without consulting the prescriber.
- Do not take OTC medications, alcohol, or other CNS depressants without prior discussion with prescriber.
- Examine site when transdermal patch is removed and report to prescriber if erythema, rash, irritation, or hyperpigmentation occurs.
- If transdermal patch loosens, tape it in place with adhesive. The patch should never be cut or trimmed.

CLOPIDOGREL BISULFATE ℗

(clo-pi'do-grel)

Plavix

Classification: ANTIPLATELET

Therapeutic: PLATELET AGGREGATION INHIBITOR; ANTITHROMBOTIC

Pregnancy Category: B

AVAILABILITY 75 mg, 300 mg tablets

ACTION & *THERAPEUTIC EFFECT*
Inhibits platelet aggregation by selectively preventing the binding of adenosine diphosphate to its platelet receptor. The drug's effect on the adenosine diphosphate receptor of a platelet is irreversible. *Clopidogrel prolongs bleeding time, thereby reducing atherosclerotic events in high-risk patients.*

USES Acute coronary syndrome (ST or non-ST elevations). Secondary prevention of MI, stroke, and vascular death.

UNLABELED USES Reduction of restenosis after stent placement.

CONTRAINDICATIONS Hypersensitivity to clopidogrel; intracranial hemorrhage, peptic ulcer, or any other active pathologic bleeding. Discontinue clopidogrel 7 days before surgery and during lactation.

CAUTIOUS USE GI bleeding, peptic ulcer disease; hepatic impairment (moderate to severe); severe renal impairment; patients at risk for increased bleeding; pregnancy (category B). Safety and efficacy not established in children.

ROUTE & DOSAGE

Secondary Prevention

Adult: **PO** 75 mg daily

Secondary Prevention in Patients with STEMI (ST segment Elevated MI)

Adult: **PO** 75 mg daily with aspirin

Common adverse effects in *italic*, life-threatening effects underlined; generic names in **bold**; classifications in SMALL CAPS; ♦ Canadian drug name; ℗ Prototype drug

Acute Coronary Syndrome (Non-ST Elevation)

Adult: **PO** 300 mg loading dose then 75 mg daily (use with aspirin)

Pharmacogenetic Adjustment
Poor CYP2C19 metabolizers may need a higher initial dose.

ADMINISTRATION

Oral

- Do not administer to persons with active pathologic bleeding.
- Discontinue drug 7 days prior to surgery.
- Store at 15°–30° C (59°–86° F) in tightly closed container and protect from light.

ADVERSE EFFECTS (≥1%) **Body as a Whole:** Flu-like syndrome, fatigue, pain, arthralgia, back pain. **CV:** Chest pain, edema, hypertension, thrombocytopenic purpura. **GI:** Abdominal pain, dyspepsia, diarrhea, nausea, hypercholesterolemia. **Hematologic:** Thrombotic thrombocytopenic purpura, epistaxis. **CNS:** Headache, dizziness, depression. **Respiratory:** URI, dyspnea, rhinitis, bronchitis, cough. **Skin:** Rash, pruritus.

INTERACTIONS Drug: NSAIDS may increase risk of bleeding events. PROTON PUMP INHIBITORS may decrease effectiveness. **Fluoxetine, citalopram,** or **fluvosamine** may decrease effectiveness. **Herbal: Garlic, ginger, ginkgo, evening primrose oil** may increase risk of bleeding.

PHARMACOKINETICS Absorption: Rapidly from GI tract. **Onset:** 2 h; reaches steady state in 3–7 days. **Distribution:** 94–98% protein bound. **Metabolism:** via CYP2C19. **Elimination:** 50% in urine and 50% in feces. **Half-Life:** 8 h.

NURSING IMPLICATIONS

Assessment & Drug Effects

- Carefully monitor for and immediately report S&S of GI bleeding, especially when coadministered with NSAIDs, aspirin, heparin, or warfarin.
- Lab tests: Baseline test for CYP2C19 genotype; periodic platelet count and lipid profile.
- Evaluate patients with unexplained fever or infection for myelotoxicity.

Patient & Family Education

- Report promptly any unusual bleeding (e.g., black, tarry stools).
- Avoid chronic aspirin or NSAID use unless approved by prescriber.

CLORAZEPATE DIPOTASSIUM
(klor-az′e-pate)

Novoclopate ♦, **Tranxene, Tranxene-SD**
Classifications: ANXIOLYTIC; SEDATIVE-HYPNOTIC; ANTICONVULSANT; BENZODIAZEPINE
Therapeutic: ANTIANXIETY; SEDATIVE-HYPNOTIC; ANTICONVULSANT
Prototype: Lorazepam
Pregnancy Category: D
Controlled Substance: Schedule IV

AVAILABILITY 3.75 mg, 7.5 mg, 15 mg capsules and tablets; 11.25 mg, 22.5 mg long acting tablets

ACTION & *THERAPEUTIC EFFECT*

Anxiolytic benzodiazepine exerts its effects through enhancement of GABA-benzodiazepine receptor complex, an inhibitory neurotransmitter. Clorazepate has depressant effects on the CNS, thus controlling anxiety associated with stress and also resulting in sedative effects. *Effective in controlling anxiety and withdrawal symptoms of alcohol.*

Common adverse effects in *italic*, life-threatening effects <u>underlined</u>; generic names in **bold**; classifications in SMALL CAPS; ♦ Canadian drug name; ⊙ Prototype drug

USES Management of anxiety disorders, short-term relief of anxiety symptoms, as adjunct in management of partial seizures, and symptomatic relief of acute alcohol withdrawal.

CONTRAINDICATIONS Hypersensitivity to clorazepate and other benzodiazepines; acute narrow-angle glaucoma; depressive neuroses; pulmonary disease, COPD; psychotic reactions, drug abusers. Safe use during pregnancy (category D), lactation.

CAUTIOUS USE Older adults; debilitated patients; hepatic disease; kidney disease; Parkinson's disease; neuromuscular disease; seizure disorders; bipolar disorder, mania, history of suicidal tendencies. Safe use in children younger than 9 y not established.

ROUTE & DOSAGE

Anxiety
Adult: **PO** 15 mg/day at bedtime, may increase to 15–60 mg/day in divided doses (max: 60 mg/day)

Acute Alcohol Withdrawal
Adult: **PO** 30 mg followed by 30–60 mg in divided doses (max: 90 mg/day), taper by 15 mg/day over 4 days to 7.5–15 mg/day until patient is stable

Partial Seizures
Adult: **PO** 7.5 mg t.i.d.
Child (9–12 y): **PO** 3.75–7.5 mg b.i.d., may increase by no more than 3.75 mg/wk (max: 60 mg/day)

ADMINISTRATION
Oral
- Give with food to minimize gastric distress.
- Ensure that sustained-release form of drug is not chewed or crushed. It **must be** swallowed whole.
- Taper drug dose gradually over several days when drug is to be discontinued.
- Store in light-resistant container at 15°–30° C (59°–86° F) unless otherwise specified.

ADVERSE EFFECTS (≥1%) **Body as a Whole:** Allergic reactions. **CV:** Hypotension. **GI:** GI disturbances, abnormal liver function tests, xerostomia. **Hematologic:** Decreased Hct, blood dyscrasias. **CNS:** *Drowsiness,* ataxia, dizziness, headache, paradoxical excitement, mental confusion, insomnia. **Special Senses:** Diplopia, blurred vision.

INTERACTIONS Drug: Alcohol and other CNS DEPRESSANTS compound CNS depression; clorazepate increases effects of **cimetidine, disulfiram,** causing excessive sedation. **Herbal: Ginkgo** may decrease anticonvulsant effectiveness.

PHARMACOKINETICS Absorption: Decarboxylated in stomach; absorbed as active metabolite, desmethyldiazepam. **Peak:** 1 h. **Duration:** 24 h. **Distribution:** Crosses placenta; distributed into breast milk. **Metabolism:** In liver to oxazepam. **Elimination:** Primarily in urine. **Half-Life:** 30–200 h.

NURSING IMPLICATIONS
Assessment & Drug Effects
- Drowsiness, a common side effect, is more likely to occur at initiation of therapy and with dose increments on successive days.
- Lab tests: Periodic blood counts and tests of liver function should be performed throughout therapy.
- Monitor patient with history of cardiovascular disease in early therapy for drug-induced responses. If systolic BP drops more than

20 mm Hg or if there is a sudden increase in pulse rate, withhold drug and notify prescriber.

Patient & Family Education

- Take drug as prescribed and do not change dose or abruptly stop taking the drug without prescriber's approval.
- Do not self-dose with OTC drugs (cold remedies, sleep medications, antacids) without consulting prescriber.
- Avoid driving and other potentially hazardous activities until reaction to drug is known.
- Do not use alcohol and other CNS depressants while on clorazepate therapy.
- If a woman becomes pregnant during therapy or intends to become pregnant, communicate with prescriber about the desirability of discontinuing the drug.

CLOTRIMAZOLE
(kloe-trim′a-zole)

Canesten ♣, Gyne-Lotrimin, Gyne-Lotrimin-3, Lotrimin, Mycelex, Mycelex-G
Classifications: ANTIBIOTIC; AZOLE ANTIFUNGAL
Therapeutic: ANTIFUNGAL
Prototype: Fluconazole
Pregnancy Category: B (topical); C (oral)

AVAILABILITY 1% cream, solution, lotion; 10 mg troches; 100 mg, 200 mg, 500 mg vaginal tablets; 1% vaginal cream

ACTION & *THERAPEUTIC EFFECT*
Acts by altering fungal cell membrane permeability, permitting loss of phosphorous compounds, potassium, and other essential intracellular constituents with consequent loss of ability to replicate. *Has broad-spectrum fungicidal activity. Active against a wide variety of fungi, yeast, dermatophytes and certain gram-positive bacteria.*

USES Dermal infections including tinea pedis, tinea cruris, tinea corporis, tinea versicolor; also vulvovaginal and oropharyngeal candidiasis.
UNLABELED USES Trichomoniasis.

CONTRAINDICATIONS Ophthalmic uses; systemic mycoses.
CAUTIOUS USE Hyersensitivity to other azole antifungals; hepatic impairment, diabetes mellitus; HIV; pregnancy (category C for oral troches; category B for topical use); lactation. Safe use in children younger than 3 y not established.

ROUTE & DOSAGE

Dermal Infections

Adult: **Topical** Apply small amount onto affected areas b.i.d. a.m. and p.m.

Vulvovaginal Infections

Adult: **Intravaginal** Insert 1 applicator full or one 100 mg vaginal tablet into vagina at bedtime for 7 days, or one 500 mg vaginal tablet at bedtime for 1 dose

Oropharyngeal Candidiasis

Adult/Child: **PO** 1 troche (lozenge) 5 times/day q3h for 14 days

ADMINISTRATION

- Instruct patient taking the oral lozenge to allow it to dissolve slowly in mouth over 15–30 min for maximum effectiveness.
- Apply skin cream and solution preparations sparingly. Protect hands with latex gloves when applying medication.
- Avoid contact of clotrimazole preparations with the eyes.

Common adverse effects in *italic*, life-threatening effects underlined; generic names in **bold**; classifications in SMALL CAPS; ♣ Canadian drug name; ❂ Prototype drug

355

- Do not use occlusive dressings unless directed by prescriber to do so.
- Consult prescriber about skin cleansing procedure before applying medication. Regardless of procedure used, dry skin thoroughly.
- Store cream and solution formulations at 15°–30° C (59°–86° F); do not store troches or vaginal tablets above 35° C (95° F) unless otherwise directed.

ADVERSE EFFECTS (≥1%) **GI:** Abnormal liver function tests; occasional nausea and vomiting (with oral troche). **Skin:** Stinging, erythema, edema, vesication, desquamation, pruritus, urticaria, skin fissures. **Urogenital:** Mild burning sensation, lower abdominal cramps, bloating, cystitis, urethritis, mild urinary frequency, vulval erythema and itching, pain and vaginal soreness during intercourse.

INTERACTIONS Drug: Intravaginal preparations may inactivate SPERMICIDES.

PHARMACOKINETICS Absorption: Minimal systemic absorption; minimally absorbed topically. **Peak:** High saliva concentrations less than 3 h; high vaginal concentrations in 8–24 h. **Metabolism:** In liver. **Elimination:** Eliminated as metabolite in bile.

NURSING IMPLICATIONS

Assessment & Drug Effects

- Evaluate effectiveness of treatment. Report any signs of skin irritation with dermal preparations.
- Anticipate signs of clinical improvement within the first week of drug use.

Patient & Family Education

- Use clotrimazole as directed and for the length of time prescribed by prescriber.
- Generally, clinical improvement is apparent during first week of

therapy. Report to prescriber if condition worsens or if signs of irritation or sensitivity develop, or if no improvement is noted after 4 wk of therapy.

- If receiving the drug vaginally, your sexual partner may experience burning and irritation of penis or urethritis; refrain from sexual intercourse during therapy or have sexual partner wear a condom.

CLOZAPINE Ⓟ

(clo'za-pin)

Clozaril, Fazaclo
Classification: ATYPICAL ANTIPSYCHOTIC
Therapeutic: ANTIPSYCHOTIC
Pregnancy Category: B

AVAILABILITY 25 mg, 50 mg, 100 mg, 200 mg tablets and 12.5 mg, 25 mg, 100 mg orally disintegrating tablets

ACTION & THERAPEUTIC EFFECT Interferes with binding of dopamine to D_1 and D_2 receptors in the limbic region of brain. It binds primarily to nondopaminergic sites (e.g., alpha-adrenergic, serotonergic, and cholinergic receptors). *Limited to treatment of schizophrenia uncontrolled by other agents.*

USES Management of schizophrenia and schizoaffective disorder.
UNLABELED USES Bipolar disorder, dementia-related behavioral disorders, tremor.

CONTRAINDICATIONS Severe CNS depression, blood dyscrasia, history of bone marrow depression; patients with myeloproliferative disorders, uncontrolled epilepsy; clozapine-induced agranulocytosis, severe granulocytosis, chemotherapy, coma, leukemia, leukopenia, neutropenia, myocarditis; renal failure, dialysis, hepatitis, jaundice; infants, lactation.

CAUTIOUS USE Arrhythmias, GI disorders, narrow-angle glaucoma, hepatic and renal impairment, prostatic hypertrophy, history of seizures; cardiovascular and/or pulmonary disease; cerebrovascular disease, cardiac arrhythmias, tachycardia, dehydration, neurologic disease, tardive dyskinesia, previous history of agranulocytosis; surgery, glaucoma, infection, older adults; pregnancy (category B). Safety and efficacy in children younger than 16 y have not been established.

ROUTE & DOSAGE

Schizophrenia

Adult (older than 16 y): **PO** Initiate at 12.5 mg daily or b.i.d. then increase by 25–50 mg/day and titrate to a target dose of 350–450 mg/day in 3 divided doses, further increases (not more than twice weekly) can be made if necessary (max: 900 mg/day)

ADMINISTRATION

Oral

- Drug is usually withdrawn gradually over 1–2 wk if therapy must be discontinued.
- Store the drug away from heat or light.

ADVERSE EFFECTS (≥1%) **CV:** Orthostatic hypotension, *tachycardia,* ECG changes, increased risk of myocarditis especially during first month of therapy, pericarditis, pericardial effusion, cardiomyopathy, heart failure, MI, mitral insufficiency. **GI:** Nausea, dry mouth, constipation, hypersalivation. **Hematologic:** Agranulocytosis. **CNS:** Seizures, *transient fever,* sedation, neuroleptic malignant syndrome (rare), dystonic

reactions (rare). **Metabolic:** Hyperglycemia, diabetes mellitus. **Urogenital:** Urinary retention. **Other:** Increased mortality from severe hematologic, cardiovascular, and respiratory adverse effects.

INTERACTIONS Drug: Alcohol and other CNS DEPRESSANTS compound depressant effects; ANTICHOLINERGIC AGENTS potentiate anticholinergic effects; ANTIHYPERTENSIVE AGENTS may potentiate hypotension; ANTINEOPLASTIC AGENTS may potentiate bone marrow suppression. **Herbal: St. John's wort** and **kava** may increase sedation.

PHARMACOKINETICS Absorption: Readily absorbed from GI tract. **Onset:** 2–4 wk. **Peak:** 2.5 h. **Distribution:** Possibly distributed into breast milk. **Metabolism:** In liver. **Elimination:** 50% in urine, 30% in feces. **Half-Life:** 8–12 h.

NURSING IMPLICATIONS

Assessment & Drug Effects

- Lab tests: Baseline WBC and absolute neutrophil count **must be** made before initial treatment, every week for first 6 mo, then every 2 wk for next 6 mo, then every 4 wk, and weekly for 4 wk after the drug is discontinued; periodic blood glucose.
- Monitor diabetics for loss of glycemic control.
- Monitor for seizure activity; seizure potential increases at the higher dose level.
- Closely monitor for recurrence of psychotic symptoms if the drug is being discontinued.
- Monitor cardiovascular and respiratory status, especially during the first month of therapy. Report promptly S&S of potential cardiac problems.

Common adverse effects in *italic*, life-threatening effects underlined; generic names in **bold**; classifications in SMALL CAPS; ♦ Canadian drug name; ⊙ Prototype drug

357

- Monitor for development of tachycardia or hypotension, which may pose a serious risk for patients with compromised cardiovascular function.

Patient & Family Education
- Carefully monitor blood glucose levels if diabetic.
- Do not engage in any hazardous activity until response to the drug is known. Drowsiness and sedation are common adverse effects.
- Due to the risk of agranulocytosis (see Appendix F) it is important to comply with blood test regimen. Report flu-like symptoms, fever, sore throat, lethargy, malaise, or other signs of infection.
- Rise slowly to avoid orthostatic hypotension.
- Report immediately any of the following: Unexplained fatigue, especially with activity; shortness of breath, sudden weight gain or edema of the lower extremities.
- Take drug exactly as ordered.
- Do not use OTC drugs or alcohol without permission of prescriber.

COCAINE
(koe-kane′)

COCAINE HYDROCHLORIDE
Classification: LOCAL ANESTHETIC
Therapeutic: TOPICAL ANESTHETIC
Prototype: Procaine
Pregnancy Category: C
Controlled Substance: Schedule II

AVAILABILITY 4%, 10% topical solution

ACTION & *THERAPEUTIC EFFECT*
Alkaloid obtained from leaves of *Erythroxylon coca.* Topical application blocks nerve conduction and produces surface anesthesia accompanied by local vasoconstriction. Exerts adrenergic effect by potentiating action of endogenous epinephrine and norepinephrine, possibly by inhibiting reuptake of catecholamines into sympathetic nerve terminals. *Topical form of cocaine is a local anesthetic. Systemic absorption produces descending CNS stimulation, with intense, short-lived euphoria accompanied by indifference to pain or hunger and with illusions of great strength, endurance, and mental capacity, all the basis for drug abuse.*

USES Surface anesthesia of ear, nose, throat, rectum, and vagina.

CONTRAINDICATIONS Hypersensitivity to local anesthetics; sepsis in region of proposed application; acute MI, history of cardiac arrhythmias, cardiac disease; seizures or seizure disorders; thyrotoxicosis; cerebrovascular disease; Tourette's syndrome; MAOI therapy; lactation.
CAUTIOUS USE History of drug sensitivities, history of drug abuse; pregnancy (category C).

ROUTE & DOSAGE

Surface Anesthesia
Adult: **Topical** 1–10% solution (use greater than 4% solution with caution) (max single dose: 1 mg/kg)

ADMINISTRATION

Topical
- Exercise caution to ensure that drug is taken as prescribed.
- Preserve in tightly closed, light-resistant containers.

ADVERSE EFFECTS (≥1%) **Body as a Whole:** Formication ("cocaine bugs"), hypersensitivity reactions. **CV:** Tachycardia, <u>ventricular fibrillation, MI</u>, angina pectoris. **GI:** Nausea, vomiting, anorexia, abdominal pain. **CNS:** *CNS stimulation* and <u>CNS</u>

Common adverse effects in *italic*, life-threatening effects <u>underlined</u>; generic names in **bold**; classifications in SMALL CAPS; ♣ Canadian drug name; ☯ Prototype drug

depression (respiratory and circulatory failure). **Respiratory:** Pneumonia, lung damage (chronic cocaine smoking). **Special Senses:** Runny nose, perforated nasal septum; clouding, pitting, and ulceration of cornea.

INTERACTIONS Drug: Epinephrine entails risk of severe hypertension and arrhythmias; MAO INHIBITORS potentiate pharmacologic effects of cocaine.

PHARMACOKINETICS Absorption: Readily absorbed from mucous membranes; absorption limited by vasoconstriction. **Onset:** 1 min. **Peak:** 15–120 min. **Duration:** 30 min–2 h. **Distribution:** Crosses placenta; distributed into breast milk. **Metabolism:** Hydrolyzed in serum. **Elimination:** In urine; detectable for up to 30 h. **Half-Life:** 1–2.5 h.

NURSING IMPLICATIONS

Assessment & Drug Effects
▪ When used for anesthesia of throat, cocaine causes temporary paralysis of cilia of respiratory tract cells, reducing protection against aspiration. It also may interfere with pharyngeal stage of swallowing. Give nothing by mouth until sensation returns.
▪ Monitor cardiovascular status, especially in patients with known cardiac disease. Report promptly cardiac arrhythmias.

Patient & Family Education
▪ Promptly report angina or chest pain or respiratory distress.

CODEINE
(koe'deen)

CODEINE PHOSPHATE
Paveral ✤

CODEINE SULFATE
Classifications: NARCOTIC (OPIATE AGONIST) ANALGESIC; ANTITUSSIVE
Therapeutic: NARCOTIC ANALGESIC; ANTITUSSIVE
Prototype: Morphine
Pregnancy Category: C
Controlled Substance: Schedule II

AVAILABILITY 15 mg, 30 mg, 60 mg tablets; 15 mg/5 mL oral solution; 30 mg, 60 mg injection

ACTION & THERAPEUTIC EFFECT Opium agonist in the CNS. Analgesia is mediated through changes in the perception of pain at the spinal cord and higher levels in the CNS. The antitussive effects are mediated through direct action on receptors in the cough center of the medulla. *Analgesic potency is about one-sixth that of morphine; antitussive activity is also a little less than that of morphine.*

USES Symptomatic relief of mild to moderately severe pain when control cannot be obtained by nonnarcotic analgesics and to suppress hyperactive or nonproductive cough.

CONTRAINDICATIONS Hypersensitivity to codeine or other morphine derivatives; increased intracranial pressure, head injury, acute alcoholism; use during labor.
CAUTIOUS USE Prostatic hypertrophy, G6PD deficiency; GI disease; COPD, acute asthma; hepatic or renal disease; hepatitis; immunosuppression; hypothyroidism; debilitated patients, very young and very old patients; history of drug abuse; pregnancy (category C), lactation. Safe use in children not established.

ROUTE & DOSAGE

Analgesic
Adult: **PO/IM/Subcutaneous** 15–60 mg q.i.d.

Common adverse effects in *italic*, life-threatening effects underlined; generic names in **bold;** classifications in SMALL CAPS; ✤ Canadian drug name; ❷ Prototype drug

Child: **PO/IM/Subcutaneous** 0.5–1 mg/kg q4–6h prn (max: 60 mg/dose)

Antitussive

Adult: **PO** 10–20 mg q4–6h prn (max: 120 mg/24 h)
Child: **PO** 6–12 y, 5–10 mg q4–6h (max: 60 mg/24 h); 2–6 y, 2.5–5 mg q4–6h (max: 30 mg/24 h)

ADMINISTRATION

Oral
- Administer PO codeine with milk or other food to reduce possibility of GI distress.

Subcutaneous/Intramuscular
- Give parenterally to achieve greatest effectiveness. An oral dose is about 60% as effective as an equal parenteral dose.
- Preserve in tight, light-resistant containers at 15°–30° C (59°–86° F) unless otherwise directed.

ADVERSE EFFECTS (≥1%) **Body as a Whole:** Shortness of breath, anaphylactoid reaction. **CV:** Palpitation, hypotension, orthostatic hypotension, bradycardia, tachycardia, circulatory collapse. **GI:** *Nausea,* vomiting, *constipation.* **CNS:** *Dizziness,* light-headedness, *drowsiness,* sedation, lethargy, euphoria, agitation; restlessness, exhilaration, convulsions, narcosis, respiratory depression. **Skin:** Diffuse erythema, rash, urticaria, *pruritus,* excessive perspiration, facial flushing, fixed-drug eruption. **Special Senses:** Miosis. **Urogenital:** Urinary retention.

INTERACTIONS Drug: Alcohol and other CNS DEPRESSANTS augment CNS depressant effects. **Herbal:**

St. John's wort may cause increased sedation.

PHARMACOKINETICS Absorption: Readily from GI tract. **Onset:** 15–30 min. **Peak:** 1–1.5 h. **Duration:** 4–6 h. **Distribution:** Crosses placenta; distributed into breast milk. **Metabolism:** In liver. **Elimination:** In urine. **Half-Life:** 2.5–4 h.

NURSING IMPLICATIONS

Assessment & Drug Effects
- Record relief of pain and duration of analgesia.
- Evaluate effectiveness as cough suppressant. Treatment of cough is directed toward decreasing frequency and intensity of cough without abolishing cough reflex, need to remove bronchial secretions.
- Supervise ambulation and use other safety precautions as warranted since drug may cause dizziness and light-headedness.
- Monitor for nausea, a common side effect. Report nausea accompanied by vomiting. Change to another analgesic may be warranted.

Patient & Family Education
- Make position changes slowly and in stages, particularly from recumbent to upright posture. Lie down immediately if light-headedness or dizziness occurs.
- Lie down when feeling nauseated and notify prescriber if this symptom persists. Nausea appears to be aggravated by ambulation.
- Avoid driving and other potentially hazardous activities until reaction to drug is known. Codeine may impair ability to perform tasks requiring mental alertness.
- Do not take alcohol or other CNS depressants unless approved by prescriber.

Common adverse effects in *italic*, life-threatening effects <u>underlined</u>; generic names in **bold**; classifications in SMALL CAPS; ✤ Canadian drug name; ⊘ Prototype drug

COLCHICINE
(kol'chi-seen)

Colcyrus, Novocolchine ✤
Classification: ANTIGOUT
Therapeutic: ANTIGOUT
Pregnancy Category: C

AVAILABILITY 0.6 mg tablets

ACTION & THERAPEUTIC EF-FECT Has antimitotic and indirect anti-inflammatory properties by inhibiting the migration of neutrophils into the area of inflammation. It does appear to prevent the release of an inflammatory glycoprotein from phagocytes in the inflammatory process. *Inhibition of inflammation and reduction of pain and swelling, which occurs in gouty arthritis. Colchicine is nonanalgesic and nonuricosuric.*

USES Prophylactically for recurrent gouty arthritis or for acute gout, treatment or prevention of Mediterranean fever.
UNLABELED USES Sarcoid arthritis, chondrocalcinosis (pseudogout), arthritis associated with erythema nodosum, leukemia, adenocarcinoma, acute calcific tendonitis, multiple sclerosis, primary biliary cirrhosis, mycosis fungoides, and Paget's disease.

CONTRAINDICATIONS Hypersensitivity to the drug; blood dyscrasias; severe GI, renal, hepatic, or cardiac disease. Safe use in children not established.
CAUTIOUS USE Early manifestations of GI, renal, hepatic, or cardiac disease; hemotologic disorders; debilitated patients and older adults; pregnancy (category C).

ROUTE & DOSAGE

Acute Gouty Flare
Adult: **PO** 1.2 mg followed by 0.6 mg one hour later (max: 1.8 mg in one hour).

Prophylaxis
Adult: **PO** 0.6 mg once or twice daily. Adult with concurrent CYP3A4 inhibitor use **PO** 0.3 mg every day or every other day

Familial Mediterranean Fever
Adult: **PO** 1.2–2.4 mg in 1 or 2 doses.

Renal Impairment Dosage Adjustment
CrCl 30 mL/min: Use 50% of normal dose

ADMINISTRATION

Oral
- Administer oral drug with milk or food to reduce possibility of GI upset.
- Preserve in tight, light-resistant containers preferably between 15°–30° C (59°–86° F), unless otherwise directed by manufacturer.

ADVERSE EFFECTS (≥1%) **GI:** *Nausea, vomiting, diarrhea, abdominal pain,* anorexia, hemorrhagic gastroenteritis, steatorrhea, hepatotoxicity, pancreatitis. **Hematologic:** Neutropenia, <u>bone marrow depression</u>, <u>thrombocytopenia</u>, <u>agranulocytosis</u>, <u>aplastic anemia</u>. **CNS:** Mental confusion, peripheral neuritis, syndrome of muscle weakness (accompanied by elevated serum creatine kinase). **Skin:** Severe irritation and tissue damage if IV administration leaks around injection site. **Urogenital:** Azotemia, proteinuria, hematuria, oliguria.

Common adverse effects in *italic*, life-threatening effects <u>underlined</u>; generic names in **bold**; classifications in SMALL CAPS; ✤ Canadian drug name; ⊙ Prototype drug

361

DIAGNOSTIC TEST INTERFERENCE
False-positive *urine tests for RBCs and Hgb* reported.

INTERACTIONS Drug: May decrease intestinal absorption of vitamin B_{12}. Do not use with PROTEASE INHIBITORS. Avoid CYP3A4 inhibitors. **Food:** Grapefruit juice may increase adverse effects.

PHARMACOKINETICS Absorption: Rapidly from GI tract. **Peak:** 0.5–2 h; multiple peaks because of enterohepatic cycling. **Distribution:** Widely distributed; concentrates in leukocytes, kidney, liver, spleen, and intestinal tract. **Metabolism:** by P-glycoprotein and CYP3A4. **Elimination:** Primarily in feces.

NURSING IMPLICATIONS

Assessment & Drug Effects
- Lab tests: Baseline and periodic serum uric acid and creatinine, CBC, platelet count, serum electrolytes, and urinalysis.
- Monitor for dose-related adverse effects; they are most likely to occur during the initial course of treatment.
- Monitor for early signs of colchicine toxicity including weakness, abdominal discomfort, anorexia, nausea, vomiting, and diarrhea. Report to prescriber. To avoid more serious toxicity, drug should be discontinued promptly until symptoms subside.
- Monitor I&O ratio and pattern (during acute gouty attack): High fluid intake promotes excretion and reduces danger of crystal formation in kidneys and ureters.

Patient & Family Education
- Withhold drug and report to the prescriber the onset of GI symptoms or signs of bone marrow depression (nausea, sore throat, bleeding gums, sore mouth, fever, fatigue, malaise, unusual bleeding or bruising).
- Avoid fermented beverages such as beer, ale, and wine as they may precipitate gouty attack.

COLESEVELAM HYDROCHLORIDE
(co-less'e-ve-lam)
Welchol
Classifications: ANTIHYPERLIPIDEMIC; BILE ACID SEQUESTRANT
Therapeutic: CHOLESTEROL-LOWERING; BILE ACID SEQUESTRANT
Prototype: Cholestyramine resin
Pregnancy Category: B

AVAILABILITY 625 mg tablets; 3.75 mg powder for suspension

ACTION & *THERAPEUTIC EFFECT*
Anion exchange resin used for its cholesterol-lowering effect. Binds with bile salts in the intestinal tract to form an insoluble complex that is excreted in the feces, thus reducing circulating cholesterol and increasing serum LDL removal rate. Serum triglyceride levels may increase slightly. *Decreases serum LDL and total cholesterol level. Removes bile salts from the intestine.*

USES Hypercholesterolemia, hyperlipoproteinemia, type 2 diabetes.

CONTRAINDICATIONS Hypersensitivity to colesevelam; complete biliary obstruction; history of hypertriglyceridemia-induced pancreatitis; serum triglyceride concentrations greater than 500 mg/dL; bowel obstruction.
CAUTIOUS USE Preexisting GI disorders or bowel disease, primary biliary cirrhosis, partial biliary obstruction, biliary atresia; diabetes mellitus; hypertriglyceridemia; older adults,

Common adverse effects in *italic*, life-threatening effects <u>underlined</u>; generic names in **bold**; classifications in SMALL CAPS; ♣ Canadian drug name; Ⓟ Prototype drug

malabsorption states; bleeding disorders; pregnancy (category B).

ROUTE & DOSAGE

Hypercholesterolemia
Adult: **PO** 3 tablets b.i.d. with meals or 6 tablets daily with a meal, may be increased to 7 tablets/day or 1.875 gram packet b.i.d.

Type 2 Diabetes
Adult: **PO** 3 tablets b.i.d. or 6 tablets daily or 1.875 gram packet b.i.d.

ADMINISTRATION

Oral
- Give with meals (mandatory) and adequate liquid (e.g., 8 oz).
- Administer concurrently ordered drugs at least 4 h prior to colesevelam.
- Store at 15°–30° C (59°–86° F) with occasional fluctuations to 40° C (90° F); protect from moisture.

ADVERSE EFFECTS (≥1%) **Body as a Whole:** Infection, pain, flu-like syndrome, asthenia, myalgia. **CNS:** Headache. **GI:** Abdominal pain, *flatulence, constipation,* diarrhea, nausea, dyspepsia. **Respiratory:** Sinusitis, rhinitis, cough, pharyngitis.

INTERACTIONS Drug: May decrease absorption of **verapamil.** Can bind and affect absorption of any drug.

PHARMACOKINETICS Absorption: Not absorbed. **Metabolism:** Not metabolized. **Elimination:** 0.05% in urine.

NURSING IMPLICATIONS

Assessment & Drug Effects
- Lab tests: Periodic total cholesterol, LDL-C, HDL-C, and triglycerides; periodic blood glucose and HbA1C.

- Withhold drug and notify prescriber for triglycerides greater than 300 mg/dL.

Patient & Family Education
- Report S&S of GI distress (see Appendix F), especially constipation.

COLESTIPOL HYDROCHLORIDE
(koe-les'ti-pole)

Cholestabyl ✦, Colestid, Lestid ✦
Classifications: ANTIHYPERLIPIDEMIC; BILE ACID SEQUESTRANT
Therapeutic: CHOLESTEROL-LOWERING AGENT; BILE ACID SEQUESTRANT
Prototype: Cholestyramine
Pregnancy Category: C

AVAILABILITY 1 g tablets; 5 g powder for suspension

ACTION & *THERAPEUTIC EFFECT*
Insoluble chloride salt of a basic anion exchange resin with high molecular weight, which adsorbs and combines with intestinal bile acids in exchange for chloride ions to form an insoluble, nonabsorbable complex that is excreted in the feces. *Reduces circulating cholesterol and increases serum LDL removal rate. Serum triglycerides are not affected or are minimally increased.*

USES Pruritus associated with partial biliary obstruction; biliary cirrhosis; also as adjunct to diet therapy of patient with primary hypercholesterolemia (type IIa hyperlipoproteinemia) or with coronary artery disease unresponsive to diet or other measures alone.

UNLABELED USES Digitoxin overdose and hyperoxaluria and to control postoperative diarrhea caused by excess bile acids in colon.

CONTRAINDICATIONS Complete biliary obstruction, biliary cirrhosis;

Common adverse effects in *italic*, life-threatening effects underlined; generic names in **bold;** classifications in SMALL CAPS; ✦ Canadian drug name; ☻ Prototype drug

363

hypersensitivity to bile acid sequestrants; renal disease.

CAUTIOUS USE Hemorrhoids; bleeding disorders; malabsorption states; GI motility disorders, dysphagia; older adult; pregnancy (category C). Safe use in children not established.

ROUTE & DOSAGE

Hypercholesterolemia
Adult: **PO** 15–30 g/day in 2–4 doses a.c. and at bedtime, or 1–2 tabs 1–2 times/day
Digitalis Toxicity
Adult: **PO** 10 g followed by 5 g q6–8h as needed

ADMINISTRATION

Oral

- Give 30 min before a meal when ordered a.c.
- Ensure that tablets are not chewed or crushed. They **must be** swallowed whole.
- Always mix granule form with liquids, juices, soups, cereals, or pulpy fruits. Add powder to at least 90 mL fluid. When carbonated drink is used, slowly stir in a large glass because excess foaming may occur. Rinse glass with small amount extra fluid to be sure all the drug is taken.
- Drugs given concomitantly should be scheduled at least 1 h before or 4 h after ingestion of colestipol to reduce interference with their absorption (see drug interactions).
- Store at 15°–30° C (59°–86° F) in tightly closed container unless otherwise instructed.

ADVERSE EFFECTS (≥1%) **Body as a Whole:** Joint and muscle pain, arthritis, shortness of breath. **GI:** *Constipation,* abdominal pain or distention, belching, flatulence, nausea, vomiting, diarrhea. **Metabolic:** Transient increases in liver enzyme tests, serum phosphorus and chloride; decreases in serum sodium and potassium. **Skin:** Dermatitis, urticaria.

INTERACTIONS Drug: Because it decreases the absorption from the GI tract of ORAL ANTICOAGULANTS, **digoxin,** TETRACYCLINES, PENICILLINS, **phenobarbital,** THYROID HORMONES, THIAZIDE DIURETICS, IRON SALTS, FAT-SOLUBLE VITAMINS (A, D, E, K), administer cholestyramine 4 h before or 2 h after these drugs. Can bind and affect absorption of any drug.

PHARMACOKINETICS Absorption: Not absorbed from GI tract. **Elimination:** In feces as insoluble complex.

NURSING IMPLICATIONS

Assessment & Drug Effects

- Watch for changes in bowel elimination pattern. Constipation should not be allowed to persist without medical attention.
- Lab tests: Periodic total cholesterol, LDL-C, HDL-C, and triglycerides; periodic serum electrolytes. Report S&S of hyponatremia and hypokalemia (see Appendix F).

Patient & Family Education

- To prevent drug interactions, it is important to keep to established schedule for taking colestipol and other drugs. See Drug Interactions.
- If receiving prolonged therapy, report unusual bleeding (vitamin K deficiency). Colestipol prevents absorption of fat-soluble vitamins (A, D, E, K).

COLISTIMETHATE SODIUM
(koe-lis-ti-meth'ate)

Coly-Mycin M

Classifications: URINARY TRACT ANTIINFECTIVE; ANTIBIOTIC

Therapeutic: URINARY TRACT ANTIINFECTIVE
Prototype: Trimethoprim
Pregnancy Category: B

AVAILABILITY 150 mg injection

ACTION & *THERAPEUTIC EFFECT*
Acts by affecting phospholipid component in bacterial cytoplasmic membranes with resulting damage and leakage of essential intracellular components. *Bactericidal against most gram-negative organisms, but not effective against* Proteus *or* Neisseria *species.*

USES Severe, acute and chronic UTIs caused by organisms resistant to other antibiotics.

CONTRAINDICATIONS Hypersensitivity to polypeptide antibiotics; concurrent use of nephrotoxic and ototoxic drugs.

CAUTIOUS USE Impaired renal function; myasthenia gravis; older adult patients, pregnancy (category B), lactation; infants.

ROUTE & DOSAGE

Urinary Tract Infections
Adult/Child: **IM/IV** 2.5–5 mg/kg/day divided in 2–4 doses (max: 5 mg/kg/day)

Renal Impairment Dosage Adjustment
Serum CrCl 1.3–1.5 mg/dL: 2.5–3.8 mg/kg/day in 2 divided doses; *1.6–2.5 mg/dL:* 2.5 mg/kg/day in a single dose or 2 divided doses; *2.6–4 mg/dL:* 1.5 mg/kg q36h

ADMINISTRATION

Intramuscular
- Reconstitute each 150-mg vial with 2 mL of sterile water for injection to yield a concentration of 75 mg/mL. Swirl vial gently during reconstitution to avoid bubble formation.
- IM injection should be made deep into upper outer quadrant of buttock.
- Patients commonly experience pain at injection site. Rotate sites.

Intravenous

***PREPARE:* Direct/Intermittent:** Prepare first half of total daily dose as directed for IM then further dilute with 20 mL sterile water for injection. ▪ Prepare second half of total daily dose by diluting further in 50 mL or more of D5W, NS, D5/NS, LR or other compatible solution. ▪ IV infusion solution should be freshly prepared and used within 24 h.

***ADMINISTER:* Direct/Intermittent:** First half of total daily dose: Give slowly over 3–5 min. ▪ Second half of total daily dose: Starting 1–2 h after the first half dose has been given, infuse the second half dose over the next 22–23 h.

INCOMPATIBILITIES **Solution/additive: Cefazolin, cephapirin, erythromycin, hydrocortisone, hydroxyzine, kanamycin.**

- Reconstituted solution may be stored in refrigerator at 2°–8° C (36°–46° F) or at controlled room temperature of 15°–30° C (59°–86° F). Use within 7 days. ▪ Store unopened vials at controlled room temperature.

ADVERSE EFFECTS (≥1%) **Body as a Whole:** Drug fever, pain at IM site. **GI:** GI disturbances. **CNS:** Circumoral, lingual, and peripheral paresthesias; visual and speech disturbances, <u>neuromuscular blockade</u> (generalized muscle weakness, dyspnea, <u>respiratory depression or paralysis</u>), seizures, psychosis. **Respiratory:** <u>Respiratory arrest</u> after IM injection.

Common adverse effects in *italic*, life-threatening effects <u>underlined</u>; generic names in **bold**; classifications in SMALL CAPS; ♣ Canadian drug name; ⊙ Prototype drug

365

Skin: Pruritus, urticaria, dermatoses. **Special Senses:** Ototoxicity. **Urogenital:** <u>Nephrotoxicity</u>.

INTERACTIONS Drug: Tubocurarine, pancuronium, atracurium, AMINOGLYCOSIDES may compound and prolong respiratory depression; AMINOGLYCOSIDES, **amphotericin B, vancomycin** augment nephrotoxicity.

PHARMACOKINETICS Peak: 1–2 h IM. **Duration:** 8–12 h. **Distribution:** Widely distributed in most tissues except CNS; crosses placenta; distributed into breast milk in low concentrations. **Metabolism:** In liver. **Elimination:** 66–75% in urine within 24 h. **Half-Life:** 2–3 h.

NURSING IMPLICATIONS

Assessment & Drug Effects

- Lab tests: Baseline C&S and renal function tests prior to therapy; frequent renal function tests and urine drug levels during therapy.
- Report restlessness or dyspnea promptly. Respiratory arrest has been reported after IM administration.
- Monitor I&O ratio and patterns: Decrease in urine output or change in I&O ratio and rising BUN, serum creatinine, and serum drug levels (without dosage increase) are indications of renal toxicity. If they occur, withhold drug and report to prescriber.
- Be alert to neurologic symptoms: Changes in speech and hearing, visual changes, drowsiness, dizziness, ataxia, and transient paresthesias, and keep prescriber informed.
- Monitor closely postoperative patients who have received curariform muscle relaxants, ether, or sodium citrate for signs of neuromuscular blockade (delayed recovery, muscle weakness, depressed respiration).

Patient & Family Education

- Avoid operating a vehicle or other potentially hazardous activities while on drug therapy because of the possibility of transient neurologic disturbances.

CONIVAPTAN HYDROCHLORIDE ⊙

(con-i-vap'tin)

Vaprisol
Classifications: ELECTROLYTIC & WATER BALANCE AGENT; DIURETIC; VASOPRESSIN ANTAGONIST
Therapeutic: VASOPRESSIN ANTAGONIST; DIURETIC
Pregnancy Category: C

AVAILABILITY 5 mg/mL solution for injection

ACTION & *THERAPEUTIC EFFECT*
Conivaptan is a vasopressin receptor (V2) antagonist that reduces the effect of vasopressin in the kidney, thus increasing the excretion of free water into the renal collecting ducts. *Conivaptan increases urine output and decreases urine osmolality in patients with euvolemic hyponatremia, thus restoring serum sodium balance.*

USES Treatment of euvolemic hyponatremia (e.g., syndrome of inappropriate secretion of antidiuretic hormone, or SIADH) in hospitalized patients.

CONTRAINDICATIONS Hypersensitivity to conivaptan; CHF; hyponatremia associated with hypovolemia; hypotension, syncope; lactation.
CAUTIOUS USE Renal or hepatic function impairment; pregnancy

(category C). Safety and efficacy in children not established.

ROUTE & DOSAGE

Euvolemic Hyponatremia

Adult: **IV** 20 mg loading dose followed by 20 mg IV over 24 h. May repeat 20 mg/day dose for 1–3 days, or may titrate up to 40 mg/day based on response. Total duration of infusion should not exceed 4 days.

ADMINISTRATION

Intravenous

PREPARE: **IV Infusion:** Use a filter needle when withdrawing a drug from an ampule. *Loading dose infusion:* Withdraw 4 mL (20 mg) from one ampule and add to 100 mL of D5W. Gently invert the bag several times to mix. *Initial maintenance infusion:* Withdraw 4 mL (20 mg) from one ampule and add to 250 mL of D5W. Gently invert the bag several times to mix. *Maximum maintenance dose infusion:* Withdraw 8 mL (40 mg) from two ampules and add to 250 mL of D5W. Gently invert the bag several times to mix.

ADMINISTER: **IV Infusion:** Give via a large vein and change infusion site every 24 h. *Loading dose:* Give over 30 min. *Maintenance dose:* Give over 24 h. ▪ Frequently monitor the serum sodium level. A reduction in dose or discontinuation of infusion may be required if the serum sodium rises too rapidly. Discontinue infusion immediately and notify prescriber of a rise in serum sodium greater than 12 mEq/L/24 h. DO NOT resume infusion if serum sodium continues to rise. ▪ Infusion may

be resumed ONLY if hyponatremia persists or reoccurs and patient demonstrates no indication of neurologic impairment. If the serum sodium rises too slowly, the dose may be titrated up to 40 mg over 24 h.

INCOMPATIBILITIES **Solution/additive: Lactated Ringer's solution, sodium chloride 0.9%.**

▪ Store vials at 25° C (77° F). Ampules should be stored in the original container and protected from light until ready for use. ▪ After diluting with D5W, the solution should be used immediately, with infusion completed within 24 h of mixing.

ADVERSE EFFECTS (≥1%) **Body as a Whole:** Cannula-site reaction, *infusion-site reaction*, pain, peripheral edema, pyrexia, *thirst*. **CNS:** Confusional state, *headache*, insomnia. **CV:** <u>Atrial fibrillation</u>, hypertension, hypotension, orthostatic hypotension, phlebitis. **GI:** Constipation, diarrhea, dry mouth, nausea, vomiting. **Hematologic:** Anemia. **Metabolic:** Dehydration, hyperglycemia, hypoglycemia, *hypokalemia*, hypomagnesemia, hyponatremia. **Respiratory:** Pneumonia. **Skin:** Erythema. **Special Senses:** Oral candidiasis.

INTERACTIONS Drug: Compounds that inhibit CYP3A4 (e.g., **ketoconazole, itraconazole, clarithromycin, ritonavir, indinavir**) can increase conivaptan levels. Conivaptan can increase the levels of **digoxin** and drugs that require CYP3A4 for metabolism (e.g., **midazolam,** HMG COA REDUCTASE INHIBITORS, **amlodipine**). **Food: Grapefruit juice** may increase the level of conivaptan. **Herbal: St. John's wort** may decrease the level of conivaptan.

Common adverse effects in *italic*, life-threatening effects <u>underlined</u>; generic names in **bold**; classifications in SMALL CAPS; ♣ Canadian drug name; ⊘ Prototype drug

PHARMACOKINETICS Distribution: 99% protein bound. **Metabolism:** Extensive hepatic metabolism. **Elimination:** Primarily fecal elimination (83%) with minor renal elimination. **Half-Life:** 5 h.

NURSING IMPLICATIONS

Assessment & Drug Effects

- Monitor infusion site for erythema, phlebitis, or other site reaction.
- Monitor vital signs and neurologic status frequently; report immediately S&S of hypernatremia (see Appendix F).
- Lab tests: Baseline and frequent serum sodium, serum potassium, and urine osmolality.
- Monitor digoxin blood levels with concurrent therapy and assess for S&S of digoxin toxicity.
- Monitor I&O closely. Effective treatment is accompanied by increased urine output, whereas decreasing urine output and oliguria may indicate developing hypernatremia.

Patient & Family Education

- Report any of the following to a health care provider: Pain at the infusion site, dizziness, confusion, palpitations, swelling of hands or feet.

CORTISONE ACETATE

(kor'ti-sone)

Cortistan, Cortone
Classifications: ADRENOCORTICAL STEROID; ANTI-INFLAMMATORY
Therapeutic: ANTI-INFLAMMATORY; GLUCOCORTICOID REPLACEMENT; IMMUNOSUPPRESSANT
Prototype: Prednisone
Pregnancy Category: D

AVAILABILITY 5 mg, 10 mg, 25 mg tablets; 50 mg/mL injection

ACTION & *THERAPEUTIC EFFECT*

Short-acting synthetic steroid with prominent glucocorticoid activity and minimal mineralocorticoid effects. Cortisone is converted in the body to cortisol, resulting in metabolic effects including promotion of protein, carbohydrate, and fat metabolism and interference with linear growth in children. *Has anti-inflammatory and immunosuppressive actions. Suppresses inflammation caused by radiant, mechanical, chemical, and infectious stimuli. Also suppress immune responses in diseases, such as in asthma, urticaria, or renal allograft.*

USES Replacement therapy for primary or secondary adrenocortical insufficiency and inflammatory and allergic disorders.

CONTRAINDICATIONS Hypersensitivity to glucocorticoids; psychoses; viral, fungal, or bacterial diseases of skin; Cushing's syndrome, immunologic procedures; pregnancy (category D), lactation.
CAUTIOUS USE Diabetes mellitus; hypertension, CHF; older adults; active or arrested tuberculosis; coagulopathy; hepatic disease; psychosis, emotional instability; renal disease, seizure disorders; active or latent peptic ulcer.

ROUTE & DOSAGE

Replacement or Inflammatory Disorders
Adult: **PO/IM** 20–300 mg/day in 1 or more divided doses, try to reduce periodically by 10–25 mg/day to lowest effective dose *Child:* **PO** 2.5–10 mg/kg/day divided q6–8h **IM** 1–5 mg/kg/day divided q12–24h

Common adverse effects in *italic*, life-threatening effects underlined; generic names in **bold**; classifications in SMALL CAPS; ♣ Canadian drug name; ● Prototype drug

ADMINISTRATION

Oral

- Administer cortisone (usually in a.m.) with food or fluid of patient's choice to reduce gastric irritation.
- Sodium chloride and a mineralo-corticoid are usually given with cortisone as part of replacement therapy.

Intramuscular

- Shake bottle well before withdrawing dose.
- Give deep IM into a large muscle.
- Drug **must be** gradually tapered rather than withdrawn abruptly.
- Store at 15°–30° C (59°–86° F) in tightly closed container unless otherwise directed by manufacturer. Protect from heat and freezing.

ADVERSE EFFECTS (≥1%) **CV:** CHF, hypertension, *edema.* **GI:** *Nausea,* peptic ulcer, pancreatitis. **Endocrine:** Hyperglycemia. **Hematologic:** Thrombocytopenia. **Musculoskeletal:** *Compression fracture,* osteoporosis, muscle weakness. **CNS:** Euphoria, insomnia, vertigo, nystagmus. **Skin:** Impaired wound healing, petechiae, ecchymosis, acne. **Special Senses:** *Cataracts,* glaucoma, blurred vision.

INTERACTIONS Drug: BARBITURATES, **phenytoin, rifampin** decrease effects of cortisone.

PHARMACOKINETICS Absorption: Readily absorbed from GI tract. **Onset:** Rapid PO; 24–48 h IM. **Peak:** 2 h PO; 24–48 h IM. **Duration:** 1.25–1.5 days. **Distribution:** Concentrated in many tissues; crosses placenta; distributed into breast milk. **Metabolism:** In liver. **Elimination:** In urine. **Half-Life:** 0.5 h; HPA suppression: 8–12 h.

NURSING IMPLICATIONS

Assessment & Drug Effects

- Monitor for S&S of Cushing's syndrome (see Appendix F), especially in patients on long-term therapy.
- Lab tests: Periodic blood glucose and CBC with platelet count.
- Cortisone may mask some signs of infection, and new infections may appear.
- Be alert to clinical indications of infection: Malaise, anorexia, depression, and evidence of delayed healing. (Classic signs of inflammation are suppressed by cortisone.)
- Report ecchymotic areas, unexplained bleeding, and easy bruising.

Patient & Family Education

- Take drug exactly as prescribed. Do not alter dose intervals or stop therapy abruptly.
- Monitor weight and report a steady gain especially if it is accompanied by signs of fluid retention (e.g., edema of ankles or hands).
- Report changes in visual acuity, including blurring, promptly.
- Inform prescriber or dentist that cortisone is being taken.

CROMOLYN SODIUM ⊙

(kroe'moe-lin)

Crolom, Fivent ✦, Gastrocom, Intal, Opticrom, Rynacrom ✦, Vistacrom ✦

Classifications: RESPIRATORY AGENT; MAST CELL STABILIZER; ANTI-INFLAMMATORY; ANTIASTHMATIC

Therapeutic: ANTIASTHMATIC; ANTI-INFLAMMATORY

Pregnancy Category: B

AVAILABILITY 20 mg/2 mL solution for nebulization; 800 mcg

Common adverse effects in *italic*, life-threatening effects underlined; generic names in **bold**; classifications in SMALL CAPS; ✦ Canadian drug name; ⊙ Prototype drug

369

C

spray; 40 mg/mL nasal solution; 4% ophth solution; 100 mg/5 mL oral concentrate

ACTION & *THERAPEUTIC EFFECT*

Synthetic asthma prophylactic agent with unique action. Inhibits release of bronchoconstrictors—histamine and SRS-A (slow-reacting substance of anaphylaxis) from sensitized pulmonary mast cells, thereby suppressing an allergic response. Additionally, cromolyn may also reduce the release of inflammatory leukotrienes.

Particularly effective for IgE-mediated or "extrinsic asthma" precipitated by exposure to specific allergen (e.g., pollens, dust, animal dander), by inhibiting the release of bronchoconstrictors.

USES Primarily for prophylaxis of mild to moderate seasonal and perennial bronchial asthma and allergic rhinitis. Also used for prevention of exercise-related bronchospasm, prevention of acute bronchospasm induced by known pollutants or antigens, and for prevention and treatment of allergic rhinitis. Orally for systemic mastocytosis. **Ophthalmic use:** Allergic ocular disorders, conjunctivitis, vernal keratoconjunctivitis.

UNLABELED USES Orally for prophylaxis of GI and systemic reactions to food allergy.

CONTRAINDICATIONS Use of aerosol (because of fluorocarbon propellants) in patients with coronary artery disease or history of arrhythmias; dyspnea, acute asthma, status asthmaticus, or acute bronchospasm; patients unable to coordinate actions or follow instructions.

CAUTIOUS USE Renal or hepatic dysfunction; pregnancy (category B),

lactation. Safe use in children younger than 6 y not established.

ROUTE & DOSAGE

Allergies

Adult: **Inhalation** Metered dose inhaler or capsule: 1 spray or 1 capsule inhaled q.i.d.; nasal solution: 1 spray in each nostril 3–6 times/day at regular intervals
Child: **Inhalation** *Older than 6 y,* Metered dose inhaler or capsule: Same as for adult; *older than 6 y,* nasal solution: Same as for adult

Conjunctivitis

Adult: **PO** 2 ampules q.i.d. 30 min a.c. and at bedtime
Child (2–12 y): **PO** 1 ampule q.i.d. 30 min a.c. and at bedtime

Mastocytosis

See Appendix A-1.

ADMINISTRATION

Oral
- Give at least 30 min before meals.

Inhalation
- Patients should receive detailed instructions for each inhalation device. See manufacturer's instructions. Therapeutic effect is dependent on proper inhalation technique.
- Advise patient to clear as much mucus as possible before inhalation treatments.
- Instruct patient to exhale as completely as possible before placing inhaler mouthpiece between lips, tilt head backward and inhale rapidly and deeply with steady, even breaths. Remove inhaler from mouth, hold breath for a few seconds, then exhale into the air. Repeat until entire dose is taken.

Common adverse effects in *italic*, life-threatening effects underlined; generic names in **bold**; classifications in SMALL CAPS; ♣ Canadian drug name; Ⓟ Prototype drug

• Protect cromolyn from moisture and heat. Store in tightly closed, light-resistant container at 15°–30° C (59°–86° F) unless otherwise directed.

ADVERSE EFFECTS (≥1%) **Body as a Whole:** Peripheral eosinophilia, angioedema, bronchospasm, anaphylaxis (rare). **GI:** Swelling of parotid glands, dry mouth, slightly bitter aftertaste, *nausea,* vomiting, esophagitis. **CNS:** Headache, dizziness, peripheral neuritis. **Skin:** Erythema, urticaria, rash, contact dermatitis. **Special Senses:** *Sneezing, nasal stinging and burning,* dryness and *irritation of throat and trachea; cough;* nasal congestion, itchy, puffy eyes, lacrimation, *transient ocular burning, stinging.*

PHARMACOKINETICS Absorption: Approximately 8% of dose absorbed from lungs. **Onset:** 1 wk with regular use. **Peak:** 15 min. **Duration:** 4–6 h; may last as long as 2–3 wk. **Elimination:** In bile and urine in equal amounts. **Half-Life:** 80 min.

NURSING IMPLICATIONS

Assessment & Drug Effects

• Withhold drug and notify prescriber if any of the following occur; angioedema or bronchospasm.
• Monitor for exacerbation of asthmatic symptoms including breathlessness and cough that may occur in patients receiving cromolyn during corticosteroid withdrawal.
• For patients with asthma, therapeutic effects may be noted within a few days but generally not until after 1–2 wk of therapy.

Patient & Family Education

• Throat irritation, cough, and hoarseness can be minimized by gargling with water, drinking a few swallows of water, or by sucking on a lozenge after each treatment.
• Talk to your prescriber about what to do in the event of an acute asthmatic attack. Cromolyn is of no value in acute asthma.
• Cromolyn does not eliminate the continued need for therapy with bronchodilators, expectorants, antibiotics, or corticosteroids, but the amount and frequency of use of these medications may be appreciably reduced.
• Report any unusual signs or symptoms. Hypersensitivity reactions (see Signs & Symptoms, Appendix F) can be severe and life-threatening. Drug should be discontinued if an allergic reaction occurs.

CROTAMITON
(kroe-tam'i-ton)
Eurax
Classifications: SCABICIDE; ANTIPRURITIC
Therapeutic: SCABICIDE; ANTIPRURITIC
Prototype: Lindane
Pregnancy Category: C

AVAILABILITY 10% cream, lotion

ACTION & THERAPEUTIC EFFECT By unknown mechanisms, drug eradicates *Sarcoptes scabiei* and effectively relieves itching. *Scabicidal and antipruritic agent.*

USES Treatment of scabies and for symptomatic treatment of pruritus.

CONTRAINDICATIONS Application to acutely inflamed skin, raw or weeping surfaces, eyes, or mouth; history of previous sensitivity to crotamiton.

Common adverse effects in *italic*, life-threatening effects underlined; generic names in **bold**; classifications in SMALL CAPS; ♣ Canadian drug name; ⊘ Prototype drug

371

CAUTIOUS USE Pregnancy (category C); children.

ROUTE & DOSAGE

Scabies/Pruritus

Adult/Child: **Topical** Apply a thin layer of cream from neck to toes; apply a second layer 24 h later. Bathe 48 h after last application to remove drug.

ADMINISTRATION

Topical

- Shake container well before use of solution.
- The skin **must be** thoroughly dry before applying medication.
- If drug accidentally contacts eyes, thoroughly flush out medication with water.
- Pruritus treatment: Massage medication gently into affected areas until it is completely absorbed. Repeat as needed (usually effective for 6–10 h).
- Store in tightly closed containers at 15°–30° C (59°–86° F). Do not freeze.

ADVERSE EFFECTS (≥1%) **Skin:** Skin irritation (particularly with prolonged use), rash, erythema, sensation of warmth, allergic sensitization.

NURSING IMPLICATIONS

Assessment & Drug Effects

- Monitor for and report significant skin irritation or allergic sensitization.

Patient & Family Education

- Review package insert before treatment begins.
- Discontinue medication and report to prescriber if irritation or sensitization develops.

CYANOCOBALAMIN
(sye-an-oh-koe-bal′a-min)

Anacobin ♦, Bedoz ♦, Nascobal, Rubion ♦
Classification: VITAMIN B₁₂
Therapeutic: VITAMIN B₁₂
Pregnancy Category: A (PO or nasal spray); C (parenteral)

AVAILABILITY 25 mcg, 50 mcg, 100 mcg, 250 mcg tablets; 400 mcg/unit, 500 mcg/0.1 mL nasal gel; 500 mcg/0.1 mL nasal spray

ACTION & *THERAPEUTIC EFFECT*

Vitamin B_{12} is a cobalt-containing B complex vitamin essential for normal growth, cell reproduction, maturation of RBCs, nucleoprotein synthesis, maintenance of nervous system (myelin synthesis), and believed to be involved in protein and carbohydrate metabolism. Vitamin B_{12} deficiency results in megaloblastic anemia, dysfunction of spinal cord with paralysis, GI lesions. *Therapeutically effective for treatment of vitamin B_{12} deficiency and pernicious anemia.*

USES Vitamin B_{12} deficiency due to malabsorption syndrome as in pernicious (Addison's) anemia, sprue; GI pathology, dysfunction, or surgery; fish tapeworm infestation, and gluten enteropathy. Also used in B_{12} deficiency caused by increased physiologic requirements or inadequate dietary intake, and in vitamin B_{12} absorption (Schilling) test.

UNLABELED USES To prevent and treat toxicity associated with sodium nitroprusside.

CONTRAINDICATIONS History of sensitivity to vitamin B_{12}, other cobalamins, or cobalt; early Leber's

Common adverse effects in *italic*, life-threatening effects <u>underlined</u>; generic names in **bold**; classifications in SMALL CAPS; ♦ Canadian drug name; 🕪 Prototype drug

disease (hereditary optic nerve atrophy), indiscriminate use in folic acid deficiency.

CAUTIOUS USE Heart disease, anemia, pulmonary disease; pregnancy (category A for PO or nasal route, and category C for parenteral).

ROUTE & DOSAGE

Vitamin B$_{12}$ Deficiency

Adult: **IM/Deep Subcutaneous** 30 mcg/day for 5–10 days, then 100–200 mcg/mo
Child: **IM/Deep Subcutaneous** 100 mcg doses to a total of 1–5 mg over 2 wk, then 60 mcg/mo

Pernicious Anemia

Adult: **IM/Deep Subcutaneous** 100–1000 mcg/day for 2–3 wk, then 100–1000 mcg q2–4wk
Intranasal one pump in one nostril once weekly
Child: **IM** 30–50 mcg/day × 2 wk to total of 1000 mcg, then 100 mcg/mo
Infant: **IM** 1000 mcg/day × at least 2 wk, then 50 mcg/mo

Diagnosis of Megaloblastic Anemia

Adult: **IM/Deep Subcutaneous** 1 mcg/day for 10 days while maintaining a low folate and vitamin B$_{12}$ diet

Schilling Test

Adult: **IM/Deep Subcutaneous** 1000 mcg × 1 dose

Nutritional Supplement

Adult: **PO** 1–25 mcg/day
Child: **PO** *Younger than 1 y,* 0.3 mcg/day; *1 y or older,* 1 mcg/day

ADMINISTRATION

Oral

- PO preparations may be mixed with fruit juices. However, administer promptly since ascorbic acid affects the stability of vitamin B$_{12}$.
- Administration of oral vitamin B$_{12}$ with meals increases its absorption.

Subcutaneous/Intramuscular

- Give deep subcutaneous by slightly tenting the skin at the injection site.
- IM may be given into any normal IM injection site.
- Preserved in light-resistant containers at room temperature preferably at 15°–30° C (59°–86° F) unless otherwise directed by manufacturer.

ADVERSE EFFECTS (≥1%) **Body as a Whole:** Feeling of swelling of body, <u>anaphylactic shock, sudden death</u>. **CV:** Peripheral vascular thrombosis, pulmonary edema, CHF. **GI:** Mild transient diarrhea. **Hematologic:** Unmasking of polycythemia vera (with correction of vitamin B$_{12}$ deficiency). **Metabolic:** Hypokalemia. **Skin:** Itching, rash, flushing. **Special Senses:** Severe optic nerve atrophy (patients with Leber's disease).

DIAGNOSTIC TEST INTERFERENCE Most antibiotics, methotrexate, and pyrimethamine may produce invalid diagnostic *blood assays for vitamin B$_{12}$.* Possibility of false-positive test for *intrinsic factor antibodies.*

INTERACTIONS Drug: Alcohol, aminosalicylic acid, neomycin, colchicine may decrease absorption of oral cyanocobalamin; **chloramphenicol** may

Common adverse effects in *italic*, life-threatening effects <u>underlined</u>; generic names in **bold**; classifications in SMALL CAPS; ✚ Canadian drug name; ⊙ Prototype drug

373

interfere with therapeutic response to cyanocobalamin.

PHARMACOKINETICS Absorption:
Intestinal absorption requires presence of intrinsic factor in terminal ileum. **Distribution:** Widely distributed; principally stored in liver, kidneys, and adrenals; crosses placenta, excreted in breast milk. **Metabolism:** Converted in tissues to active coenzymes; enterohepatically cycled. **Elimination:** 50–95% of doses 100 mcg or more are excreted in urine in 48 h. **Half-Life:** 6 days (400 days in liver).

NURSING IMPLICATIONS

Assessment & Drug Effects

- Obtain a careful history of sensitivities. Sensitization to cyanocobalamin can take as long as 8 y to develop.
- Monitor vital signs in patients with cardiac disease and in those receiving parenteral cyanocobalamin, and be alert to symptoms of pulmonary edema, which generally occur early in therapy.
- Lab tests: Baseline reticulocyte and erythrocyte counts, Hgb, Hct, vitamin B_{12}, and serum folate levels; then repeated between 5 and 7 days after start of therapy and at regular intervals during therapy. Monitor potassium levels during the first 48 h.
- Characteristically, reticulocyte concentration rises in 3–4 days, peaks in 5–8 days, and then gradually declines as erythrocyte count and Hgb rise to normal levels (in 4–6 wk).
- Obtain a complete diet and drug history and inquire into alcohol drinking patterns for all patients receiving cyanocobalamin to identify and correct poor habits.

Patient & Family Education

- Notify prescriber of any intercurrent disease or infection. Increased dosage may be required.
- To prevent irreversible neurologic damage resulting from pernicious anemia, drug therapy **must be** continued throughout life.
- Rich food sources of B_{12} are nutrient-added breakfast cereals, vitamin B_{12}-fortified soy milk, organ meats, clams, oysters, egg yolk, crab, salmon, sardines, muscle meat, milk, and dairy products.

CYCLIZINE HYDROCHLORIDE
(sye'kli-zeen)

Marezine, Marzine ♥
Classifications: ANTIHISTAMINE (H_1-RECEPTOR ANTAGONIST); ANTIVERTIGO; ANTIEMETIC
Therapeutic: ANTIVERTIGO; ANTIEMETIC
Prototype: Meclizine
Pregnancy Category: B

AVAILABILITY 50 mg tablets

ACTION & THERAPEUTIC EFFECT Piperazine antihistamine (H_1-receptor antagonist) that exhibits CNS depression and anticholinergic, antispasmodic, local anesthetic, and antihistaminic activity. Has prominent depressant action on labyrinthine excitability and on conduction in vestibular-cerebellar pathways. *Produces marked antimotion and antiemetic effects.*

USES Motion sickness.

CONTRAINDICATIONS Increased intraocular pressure; asthma, closed-angle glaucoma.

CAUTIOUS USE Narrow-angle glaucoma; prostatic hypertrophy; elderly; obstructive disease of GU or GI tracts; postoperative patients; pregnancy (category B), lactation; children younger than 6 y.

ROUTE & DOSAGE

Motion Sickness

Adult/Adolescent: **PO** 50 mg q4–6 h (max: 200 mg/day)
Child: **PO** 6–12 y, 25 mg q6–8 h prn (max: 75 mg/day)

ADMINISTRATION

Oral
▪ Give dose 30 min prior to any activity likely to cause motion sickness.

ADVERSE EFFECTS (≥1%) **CV:** Hypotension, palpitation, tachycardia. **GI:** Anorexia, nausea, vomiting, diarrhea, or constipation, cholestatic jaundice. **CNS:** *Drowsiness,* excitement, euphoria, auditory and visual hallucinations, hyperexcitability alternating with drowsiness, convulsions, respiratory paralysis (rare). **Skin:** Urticaria, rash. **Special Senses:** *Dry mouth,* nose, and throat; blurred vision, diplopia; tinnitus. **Other:** Pain at IM injection site.

INTERACTIONS Drug: Alcohol, BARBITURATES, CNS DEPRESSANTS (e.g., HYPNOTICS, SEDATIVES, and ANXIOLYTICS) may compound effects of cyclizine.

PHARMACOKINETICS Onset: Rapid. **Duration:** 4–6 h. **Metabolism:** Unknown.

NURSING IMPLICATIONS

Assessment & Drug Effects
▪ Monitor postoperative patient's vital signs closely, as cyclizine can cause hypotension.

▪ Monitor for and report signs of CNS stimulation (e.g., hyperexcitability, euphoria). Dose reduction or discontinuation of drug may be indicated.

Patient & Family Education
▪ Take cyclizine with food or a glass of milk or water to minimize GI irritation.
▪ Do not drive a car or engage in other potentially hazardous activities until reaction to the drug is known. Adverse effects include drowsiness and dizziness.
▪ Alcohol, barbiturates, narcotic analgesic, and other CNS depressants may compound sedative action.

CYCLOBENZAPRINE HYDROCHLORIDE ⓟ

(sye-kloe-ben′za-preen)

Amrix, Flexeril
Classification: CENTRAL ACTING SKELETAL MUSCLE RELAXANT
Therapeutic: SKELETAL MUSCLE RELAXANT; ANTISPASMODIC
Pregnancy Category: B

AVAILABILITY 5 mg, 10 mg tablets; 15 mg, 30 mg extended release capsules

ACTION & *THERAPEUTIC EFFECT*
Relieves skeletal muscle spasm of local origin without interfering with muscle function. Believed to act primarily within CNS at brain stem. Depresses tonic somatic motor activity, although both gamma and alpha motor neurons are affected. *Relieves muscle spasm associated with acute, painful musculoskeletal conditions.*

USES Short-term adjunct to rest and physical therapy for relief

Common adverse effects in *italic*, life-threatening effects underlined; generic names in **bold**; classifications in SMALL CAPS; ♦ Canadian drug name; ⓟ Prototype drug

375

of muscle spasm associated with acute musculoskeletal conditions. Not effective in treatment of spasticity associated with cerebral palsy or cerebral or cord disease.
UNLABELED USES Fibromyalgia.

CONTRAINDICATIONS Acute recovery phase of MI, cardiac arrhythmias, heart block or conduction disturbances, QT prolongation; CHF, hyperthyroidism; closed-angle glaucoma, increased intraocular pressure; moderate or severe hepatic impairment; MAOI therapy; cerebral palsy.

CAUTIOUS USE Prostatic hypertrophy, history of urinary retention, seizures; cardiovascular disease; mild hepatic impairment; older adults, debilitated patients; history of psychiatric illness; pregnancy (category B), lactation. Safe use in children younger than 15 y not established.

ROUTE & DOSAGE

Muscle Spasm

Adult/Adolescent (at least 15 y old): **PO** 5–10 mg t.i.d. (max: 30 mg/day) or 15–30 mg daily extended release
Geriatric: Start with 5 mg, adjust dose slowly

Hepatic Impairment Dosage Adjustment

Mild: Start with 5 mg
Moderate to Severe: Not recommended

ADMINISTRATION

Oral

- Do not administer drug if patient is receiving an MAO inhibitor (e.g., furazolidone, isocarboxazid, pargyline, tranylcypromine).
- Do not open extended release capsules. They **must be** swallowed whole.
- Cyclobenzaprine is intended for short-term (2 or 3 wk) use.
- Store in tightly closed container, preferably at 15°–30° C (59°–86° F) unless otherwise directed by manufacturer.

ADVERSE EFFECTS (≥1%) **Body as a Whole:** <u>Edema of tongue</u> and face, sweating, myalgia, hepatitis, alopecia. Shares toxic potential of tricyclic antidepressants. **CV:** Tachycardia, syncope, palpitation, vasodilation, chest pain, orthostatic hypotension, dyspnea; with high doses, possibility of severe arrhythmias. **GI:** *Dry mouth,* indigestion, unpleasant taste, coated tongue, tongue discoloration, vomiting, anorexia, abdominal pain, flatulence, diarrhea, paralytic ileus. **CNS:** *Drowsiness, dizziness,* weakness, fatigue, asthenia, paresthesias, tremors, muscle twitching, insomnia, euphoria, disorientation, mania, ataxia. **Skin:** Pruritus, urticaria, skin rash. **Urogenital:** Increased or decreased libido, impotence.

INTERACTIONS Drug: Alcohol, BARBITURATES, other CNS DEPRESSANTS enhance CNS depression; potentiates anticholinergic effects of **phenothiazine** and other ANTICHOLINERGICS; MAO INHIBITORS may precipitate hypertensive crisis—use with extreme caution.

PHARMACOKINETICS Absorption: Well absorbed from GI tract with some first-pass elimination in liver. **Onset:** 1 h. **Peak:** 3–8 h. **Duration:** 12–24 h. **Distribution:** 93% protein bound. **Metabolism:** In liver to inactive metabolites. **Elimination:** Slowly in urine with some elimination in feces; may be excreted in breast milk. **Half-Life:** 1–3 days.

Common adverse effects in *italic*, life-threatening effects <u>underlined</u>; generic names in **bold**; classifications in SMALL CAPS; ✤ Canadian drug name; ❂ Prototype drug

NURSING IMPLICATIONS

Assessment & Drug Effects

- Supervision of ambulation may be indicated, especially in the older adult because of risk of drowsiness and dizziness.
- Withhold drug and notify prescriber if signs of hypersensitivity (e.g., pruritus, urticaria, rash) appear.

Patient & Family Education

- Avoid driving and other potentially hazardous activities until reaction to drug is known. Adverse effects include drowsiness and dizziness.
- Avoid alcohol and other CNS depressants (unless otherwise directed by prescriber) because cyclobenzaprine enhances their effects.
- Dry mouth may be relieved by increasing total fluid intake (if not contraindicated).
- Keep prescriber informed of therapeutic effectiveness. Spasmolytic effect usually begins within 1 or 2 days and may be manifested by lessening of pain and tenderness, increase in range of motion, and ability to perform ADL.

CYCLOPHOSPHAMIDE ⊙

(sye-kloe-foss'fa-mide)

Cytoxan, Neosar, Procytox ♦
Classifications: ANTINEOPLASTIC; NITROGEN MUSTARD; ALKYLATING AGENT
Therapeutic: ANTINEOPLASTIC
Pregnancy Category: D

AVAILABILITY 25 mg, 50 mg tablets; 100 mg, 200 mg, 500 mg, 1 g, 2 g vials

ACTION & *THERAPEUTIC EFFECT*
Cell-cycle–nonspecific alkylating agent that causes cross-linkage of DNA strands, thereby blocking synthesis of DNA, RNA, and protein. *Has pronounced antineoplastic effects and immunosuppressive activity.*

USES In treatment of malignant lymphoma, multiple myeloma, leukemias, mycosis fungoides (advanced disease), neuroblastoma, adenocarcinoma of ovary, carcinoma of breast, or malignant neoplasms of lung.
UNLABELED USES To prevent rejection in homotransplantation; to treat severe rheumatoid arthritis, multiple sclerosis, systemic lupus erythematosus, Wegener's granulomatosis, nephrotic syndrome.

CONTRAINDICATIONS Serious infections (including chickenpox, herpes zoster); live virus vaccines; severe myelosuppression; dehydration; severe hemorrhagic cystitis; pregnancy (category D) lactation.
CAUTIOUS USE History of radiation or cytotoxic drug therapy; hepatic and renal impairment, elderly; recent history of steroid therapy; bone marrow infiltration with tumor cells; history of urate calculi and gout; patients with leukopenia, thrombocytopenia; men and women of child-bearing age.

ROUTE & DOSAGE

Neoplasm

Adult: **PO Initial** 1–5 mg/kg/day; **Maintenance** 1–5 mg/kg q7–10 days **IV Initial** 40–50 mg/kg in divided doses over 2–5 days up to 100 mg/kg; **Maintenance** 10–15 mg/kg q7–10 days or 3–5 mg twice weekly
Child: **PO Initial** 2–8 mg/kg or 60–250 mg/m² ; **Maintenance** 2–5 mg/kg or 50–150 mg/m²

Common adverse effects in *italic*, life-threatening effects underlined; generic names in **bold**; classifications in SMALL CAPS; ♦ Canadian drug name; ⊙ Prototype drug

377

twice weekly **IV Initial** 2–8 mg/kg *or* 60–250 mg/m²

Renal Impairment Dosage Adjustment

CrCl less than 10 mL/min: Give 50% of dose; administer post-dialysis, give supplemental dose of 35%

ADMINISTRATION

Oral

- Administer PO drug on empty stomach. If nausea and vomiting are severe, however, it may be taken with food. An antiemetic medication may be prescribed to be given before the drug.
- Store cyclophosphamide PO solution in refrigerator at 2°–8° C (36°–46° F) and use within 14 days.

Intravenous

PREPARE: Direct: Add 5 mL bacteriostatic water for injection (paraben-preserved only) to each 100 mg and shake gently to dissolve. **Intermittent:** May be further diluted with 100–250 mL D5W, NS, D5/NS, LR, or other compatible solution.

ADMINISTER: Direct/Intermittent: Give each 100 mg or fraction thereof over 10–15 min.

INCOMPATIBILITIES Y-site: Amphotericin B cholesteryl complex, amphotericin B colloidal, diazepam, gemtuzumab, lansoprazole, phenytoin.

- Store at temperature between 2° and 30° C (36° and 86° F) unless otherwise recommended by the manufacturer.

ADVERSE EFFECTS (≥1%) **Body as a Whole:** Transient dizziness, fatigue, facial flushing, diaphoresis, drug fever, <u>anaphylaxis</u>, secondary neoplasia. **GI:** *Nausea, vomiting,* mucositis, *anorexia,* hepatotoxicity, diarrhea. **Hematologic:** <u>Leukopenia,</u> *neutropenia,* acute myeloid leukemia, anemia, thrombophlebitis, interference with normal healing. **Metabolic:** Severe hyperkalemia, SIADH, hyponatremia, weight gain (but without edema) or weight loss, hyperuricemia. **Respiratory:** <u>Pulmonary emboli</u> and edema, pneumonitis, <u>interstitial pulmonary fibrosis.</u> **Skin:** *Alopecia* (reversible), transverse ridging of nails, pigmentation of nail beds and skin (reversible), nonspecific dermatitis, <u>toxic epidermal necrolysis,</u> <u>Stevens-Johnson syndrome.</u> **Urogenital:** <u>Sterile hemorrhagic and</u> <u>nonhemorrhagic cystitis,</u> bladder fibrosis, nephrotoxicity.

DIAGNOSTIC TEST INTERFERENCE Cyclophosphamide suppresses positive reactions to **Candida, *mumps, trichophytons,*** and ***tuberculin PPD skin tests. Papanicolaou (PAP)*** smear may be falsely positive.

INTERACTIONS Drug: Succinylcholine, prolonged neuromuscular blocking activity; **doxorubicin** may increase cardiac toxicity.

PHARMACOKINETICS Absorption: Readily from GI tract. **Peak:** 1 h PO. **Distribution:** Widely distributed, including brain, breast milk; crosses placenta. **Metabolism:** In liver by CYP3A4. **Elimination:** In urine as active metabolites and unchanged drug. **Half-Life:** 4–6 h.

NURSING IMPLICATIONS

Assessment & Drug Effects

- Lab tests: Baseline total and differential leukocyte count, platelet count, and Hct, and repeat at least 2 times per week during maintenance period. Baseline and periodic LFTs, kidney function and serum electrolytes. Microscopic urine examinations are recommended after large IV doses.

- Thrombocytopenia is rare, but if it occurs (count of 100,000/mm³ or lower), assess for signs of unexplained bleeding or easy bruising. If platelet count indicates thrombocytopenia (100,000/mm³ or less), drug will be discontinued.

- Marked leukopenia is the most serious side effect. It can be fatal. Nadir may occur in 2–8 days after first dose but may be as late as 1 mo after a series of several daily doses. Leukopenia usually reverses 7–10 days after therapy is discontinued.

- During severe leukopenic period, protect patient from infection and trauma and from visitors and medical personnel who have colds or other infections.

- Report onset of unexplained chills, sore throat, tachycardia. Monitor temperature carefully and report an elevation immediately. The development of fever in a neutropenic patient (granulocyte count less than 1000) is a medical emergency because sepsis can develop quickly in these patients.

- Observe and report character of wound drainage. During period of neutropenia, purulent drainage may become serosanguineous because there are not enough WBC to create pus. Because of suppressed immune mechanisms, wound healing may be prolonged or incomplete.

- Monitor I&O ratio and patterns: Since the drug is a chemical irritant, PO and IV fluid intake is generally increased to help prevent renal irritation and hemorrhagic cystitis. Have patient void frequently, especially after each dose and just before retiring to bed.

- Watch for symptoms of water intoxication or dilutional hyponatremia;

patients are usually well hydrated as part of the therapy.

- Promptly report hematuria or dysuria. Drug schedule is usually interrupted and fluids are forced.

- Record body weight at least twice weekly (basis for dose determination). Alert prescriber to sudden change or slow, steady weight gain or loss over a period of time that appears inconsistent with caloric intake.

- Diarrhea may signal onset of hyperkalemia, particularly if accompanied by colicky pain, nausea, bradycardia, and skeletal muscle weakness. These symptoms warrant prompt reporting to prescriber.

- Monitor for hyperuricemia, which occurs commonly during early treatment period in patients with leukemias or lymphoma. Report edema of lower legs and feet; joint, flank, or stomach pain.

- Protect patient from potential sources of infection. Cyclophosphamide makes the patient particularly susceptible to varicella-zoster infections (chickenpox, herpes zoster).

- Report any sign of overgrowth with opportunistic organisms, especially in patient receiving corticosteroids or who has recently been on steroid therapy.

- Report fever, dyspnea, and nonproductive cough. Pulmonary toxicity is not common, but the already debilitated patient is particularly susceptible.

Patient & Family Education

- Adhere to dosage regimen and do not omit, increase, decrease, or delay doses. If for any reason drug cannot be taken, notify prescriber.

- Alopecia occurs in about 33% of patients on cyclophosphamide therapy. Hair loss may be noted 3 wk after therapy begins; regrowth (often differs in texture and color) usually starts 5–6 wk after drug is withdrawn and may occur while on maintenance doses.

- Use adequate means of contraception during and for at least 4 mo after termination of drug treatment. Breast-feeding should be discontinued before cyclophosphamide therapy is initiated.

- Amenorrhea may last up to 1 y after cessation of therapy in 10–30% of women.

CYCLOSERINE

(sye-kloe-ser'een)

Seromycin
Classification: ANTITUBERCULOSIS
Therapeutic: ANTITUBERCULOSIS
Pregnancy Category: C

AVAILABILITY 250 mg capsules

ACTION & *THERAPEUTIC EFFECT*
Broad-spectrum anti-infective that inhibits cell wall synthesis in susceptible strains of bacteria. It competitively interferes with the incorporation of D-alanine into the bacterial cell wall, resulting in cell death. *Effective against gram-positive and gram-negative bacteria and* Mycobacterium tuberculosis.

USES Treatment of tuberculosis, urinary tract infections.
UNLABELED USES Treatment of MAC.

CONTRAINDICATIONS Uncontrolled epilepsy; depression, severe anxiety, history of psychoses; severe renal insufficiency.
CAUTIOUS USE Renal impairment, anemia; chronic alcoholism; pregnancy (category C); lactation. Safe use in children not established.

ROUTE & DOSAGE

Tuberculosis
Adult: **PO** 250 mg q12h for 2 wk, may increase to 250

q6–8h OR 15–20 mg/kg/day (max: 1 g/day)
Urinary Tract Infection
Adult: **PO** 250 mg q12h for 2 wk

ADMINISTRATION

Oral
- Pyridoxine 200–300 mg/day may be ordered concurrently to prevent neurotoxic effects of cycloserine.
- Store in tightly closed container at 15°–30° C (59°–86° F) unless otherwise directed.

ADVERSE EFFECTS (≥1%) **CV:** Arrhythmias, CHF. **Hematologic:** Vitamin B_{12} and folic acid deficiency, megaloblastic or sideroblastic anemia. **CNS:** *Drowsiness,* anxiety, *headache,* tremors, myoclonic jerking, convulsions, vertigo, visual disturbances, speech difficulties (dysarthria), lethargy, depression, disorientation with loss of memory, confusion, nervousness, psychoses, tic episodes, character changes, hyperirritability, aggression, hyperreflexia, peripheral neuropathy, paresthesias, paresis, dyskinesias. **Skin:** Dermatitis, photosensitivity. **Special Senses:** Eye pain (optic neuritis), photophobia.

INTERACTIONS Drug: Alcohol increases risk of seizures; **ethionamide, isoniazid** potentiate neurotoxic effects; may inhibit **phenytoin** metabolism, increasing its toxicity.

PHARMACOKINETICS Absorption: 70–90% from GI tract. **Peak:** 3–4 h. **Distribution:** Distributed to lung, ascitic, pleural and synovial fluids, and CSF; crosses placenta; distributed into breast milk. **Metabolism:** Not metabolized. **Elimination:** 60–70% in urine within 72 h; small amount in feces. **Half-Life:** 10 h.

NURSING IMPLICATIONS

Assessment & Drug Effects

- Lab tests: Baseline C&S before initiation of therapy and periodically thereafter; weekly plasma drug levels; hematologic, renal function and LFTs at regular intervals.
- Maintenance of blood-drug level below 30 mg/mL considerably reduces incidence of neurotoxicity. Possibility of neurotoxicity increases when dose is 500 mg or more or when renal clearance is inadequate.
- Observe patient carefully for signs of hypersensitivity and neurologic effects. Neurotoxicity generally appears within first 2 wk of therapy and disappears after drug is discontinued.
- Drug should be withheld and prescriber notified or dosage reduced if symptoms of CNS toxicity or hypersensitivity reaction (see Appendix F) develop.

Patient & Family Education

- Take cycloserine after meals to prevent GI irritation.
- Notify prescriber immediately of the onset of skin rash and early signs of CNS toxicity (see Appendix F).
- Avoid potentially hazardous tasks such as driving until reaction to cycloserine has been determined.
- Take drug precisely as prescribed and to keep follow-up appointments. Continuous therapy may extend into months or years.

CYCLOSPORINE ⊃

(sye'kloe-spor-een)

Gengraf, Neoral, Sandimmune, Restasis
Classifications: BIOLOGICAL RESPONSE MODIFIER; IMMUNOSUPPRESSANT

Therapeutic: IMMUNOSUPPRESSANT; ANTIRHEUMATIC; ANTIPSORIATIC
Pregnancy Category: C

AVAILABILITY Sandimmune: 25 mg, 50 mg, 100 mg capsules; 100 mg/mL oral solution; **Gengraf, Neoral:** (Microemulsion) 25 mg, 100 mg capsules; 100 mg/mL oral solution; 50 mg/mL injection; **Restasis:** 0.05% ophth emulsion

ACTION & *THERAPEUTIC EFFECT*

Has immunosuppressant action by reducing transplant rejection due to selective and reversible inhibition of first phase of T-cell activation with T-lymphocytes (which normally stimulate antibody production). *Prevents allograft rejection in transplant patients. Additionally, it is a disease-modifying antirheumatic drug (DMARD) in RA patients that have not responded on methothrexate alone.*

USES In conjunction with adrenal corticosteroids to prevent organ rejection after kidney, liver, and heart transplants (allografts). Has had limited use in pancreas, bone marrow, and heart/lung transplantations. Also used for treatment of chronic transplant rejection in patients previously treated with other immunosuppressants; rheumatoid arthritis, severe psoriasis. Ophthalmic emulsion for the treatment of keratoconjunctivitis sicca.

UNLABELED USES Sjögren's syndrome, to prevent rejection of heart-lung and pancreatic transplants, ulcerative colitis.

CONTRAINDICATIONS Hypersensitivity to cyclosporine; recent contact with or bout of chickenpox, herpes zoster; administration of live virus vaccines to patient or family members; **Gengraf** and **Neoral** in psoriasis or RA patients

---

Common adverse effects in *italic*, life-threatening effects <u>underlined</u>; generic names in **bold**; classifications in SMALL CAPS; ♣ Canadian drug name; ⊃ Prototype drug

381

with abnormal renal function, uncontrolled hypertension, or malignancies; ocular infection, **PO form:** Lactation.

CAUTIOUS USE Renal, hepatic, pancreatic, or bowel dysfunction; biliary tract disease, jaundice, hyperkalemia; electrolyte imbalance, hyperuricemia; hypertension; infection; radiation therapy, older adults, encephalopathy, females of childbearing age, fungal or viral infection, gout, herpes infection, lymphoma; neoplastic disease, malabsorption problems (e.g., liver transplant patients); older adults; pregnancy (category C).

ROUTE & DOSAGE

Prevention of Organ Rejection

Adult/Child: **PO** 14–18 mg/kg beginning 4–12 h before transplantation and continued for 1–2 wk after surgery, then gradual reduction by 5%/wk, max dose of microemulsion: 10 mg/kg/day; **Maintenance** 5–10 mg/kg/day **IV** 5–6 mg/kg beginning 4–12 h before transplantation and continued after surgery until patient can take orally

Rheumatoid Arthritis (Neoral)

Adult: **PO** 2.5 mg/kg/day divided into 2 doses. May increase by 0.5–0.75 mg/kg/day q4wk to a max of 4 mg/kg/day.

Severe Psoriasis (Neoral)

Adult: **PO** 1.25 mg/kg b.i.d. If significant improvement has not occurred after 4 wk, may increase dose by 0.5 mg/kg/day every 2 wk to max of 4 mg/kg/day.

Keratoconjunctivitis Sicca

Adult: **Ophthalmic** 1 drop in affected eye(s) twice daily approximately 12 h apart

ADMINISTRATION

Oral

- Do not dilute oral solution with grapefruit juice. Dilute with orange or apple juice, stir well, then administer immediately.
- The various product brands may not be bioequivalent on a mg for mg basis. Do not interchange without prescriber supervision.

Intravenous

PREPARE: **IV Infusion:** Dilute each 1 mL immediately before administration in 20–100 mL of D5W or NS.

ADMINISTER: **IV Infusion:** Give by slow infusion over approximately 2–6 h. ▪ Rapid IV can result in nephrotoxicity.

INCOMPATIBILITIES **Solution/ additive: Magnesium sulfate. Y-site: Amphotericin B cholesteryl complex, TPN.**

- Store preferably at 15°–30° C (59°–86° F) in well-closed containers. Do not refrigerate. ▪ Protect ampules from light.

ADVERSE EFFECTS (≥1%) **Body as a Whole:** Lymphoma, gynecomastia, chest pain, leg cramps, edema, fever, chills, weight loss, increased risk of skin malignancies in psoriasis patients previously treated with methotrexate, psoralens, or UV light therapy. **CV:** *Hypertension,* <u>MI</u> (rare). **GI:** Gingival hyperplasia, diarrhea, nausea, *vomiting,* abdominal discomfort, anorexia, gastritis, constipation. **Hematologic:** <u>Leukopenia</u>, anemia, <u>thrombocytopenia</u>, *hypermagnesemia, hyperkalemia,* hyperuricemia, *decreased serum bicarbonate,* hyperglycemia. **CNS:** *Tremor,* convulsions, headache, paresthesias, hyperesthesia, flushing, night sweats, insomnia, visual hallucinations, confusion, anxiety,

Common adverse effects in *italic*, life-threatening effects <u>underlined</u>; generic names in **bold;** classifications in SMALL CAPS; ♣ Canadian drug name; ❷ Prototype drug

flat affect, depression, lethargy, weakness, paraparesis, ataxia, amnesia. **Skin:** *Hirsutism,* acne, oily skin, flushing. **Special Senses:** Sinusitis, tinnitus, hearing loss, sore throat. **Urogenital:** Urinary retention, frequency, *nephrotoxicity (oliguria).*

DIAGNOSTIC TEST INTERFERENCE *Hyperlipidemia* and abnormalities in *electrophoresis* reported; believed to be due to polyoxyl 35 castor oil (Cremophor) in IV cyclosporine.

INTERACTIONS Drug: AMINOGLYCO-SIDES, **danazol, diltiazem, doxycycline, erythromycin, ketoconazole, methylprednisolone, metoclopramide, nicardipine,** NSAIDS, **prednisolone, verapamil** may increase cyclosporine levels; **carbamazepine, isoniazid, octreotide, phenobarbital, phenytoin, rifampin** may decrease cyclosporine levels; **acyclovir,** AMINOGLYCO-SIDES, **amphotericin B, cimetidine, erythromycin, ketoconazole, melphalan, ranitidine, cotrimoxazole, trimethoprim** may increase risk of nephrotoxicity; POTASSIUM-SPARING DIURETICS, ACE INHIBITORS **(captopril, enalapril)** may potentiate hyperkalemia. **Food: Grapefruit juice** may increase concentration. **Herbal: St. John's wort** may decrease cyclosporine levels; **berberine** may increase toxicities.

PHARMACOKINETICS Absorption: Variably and incompletely absorbed (30%). Microemulsion formulation (**Neoral**) has less variability and may produce significantly higher serum levels compared with the standard formulation. **Peak:** 3–4 h. **Distribution:** Widely distributed; 33–47% distributed to plasma; 41–50% to RBCs; crosses placenta; distributed into breast milk.

Metabolism: In liver by CYP3A4, including significant first pass metabolism; considerable enterohepatic circulation. **Elimination:** Primarily in bile and feces; 6% in urine. **Half-Life:** 19–27 h.

NURSING IMPLICATIONS

Assessment & Drug Effects

- Observe patients receiving the drug parenterally for at least 30 min continuously after start of IV infusion, and at frequent intervals thereafter to detect allergic or other adverse reactions.

- Monitor I&O ratio and pattern: Nephrotoxicity has been reported in about one third of transplant patients. It has occurred in mild forms as late as 2–3 mo after transplantation. In severe form, it can be irreversible, and therefore early recognition is critical.

- Monitor vital signs. Be alert to indicators of local or systemic infection that can be fungal, viral, or bacterial. Also report significant rise in BP.

- Lab tests: Baseline and periodic renal function, liver function (AST, ALT, serum amylase, bilirubin, and alkaline phosphatase), and serum potassium.

- Lab tests: In psoriasis patients, CBC, BUN, uric acid, potassium, lipids, and magnesium should be monitored biweekly during first 3 mo.

- Periodic tests should be made of neurologic function. Neurotoxic effects generally occur over 13–195 days after initiation of cyclosporine therapy. Signs and symptoms are reportedly fully reversible with dosage reduction or discontinuation of drug.

- Monitor blood or plasma drug concentrations at regular intervals, particularly in patients receiving the drug orally for prolonged periods, as drug absorption is erratic.

Common adverse effects in *italic,* life-threatening effects underlined; generic names in **bold;** classifications in SMALL CAPS; ✤ Canadian drug name; ➋ Prototype drug

383

Patient & Family Education

- Take medication with meals to reduce nausea or GI irritation.
- Enhance palatability of oral solution by mixing it with milk, chocolate milk, or orange juice, preferably at room temperature. Mix in a glass rather than a plastic container. Stir well, drink immediately, and rinse glass with small quantity of diluent to assure getting entire dose.
- Take medication at same time each day to maintain therapeutic blood levels.
- Practice good oral hygiene. Inspect mouth daily for white patches, sores, swollen gums.
- Hirsutism is reversible with discontinuation of drug.

CYPROHEPTADINE HYDROCHLORIDE

(si-proe-hep′ta-deen)

Periactin, Vimicon ✦

Classifications: ANTIHISTAMINE; ANTIPRURITIC

Therapeutic: ANTIHISTAMINE

Prototype: Diphenhydramine

Pregnancy Category: B

AVAILABILITY 4 mg tablets

ACTION & *THERAPEUTIC EFFECT*
Potent piperidine antihistamine that acts by competing with histamine for serotonin and H_1-receptor sites, thus preventing histamine-mediated responses. *Has significant antipruritic, local anesthetic, and antiserotonin activity.*

USES Symptomatic relief of various allergic conditions.

UNLABELED USES Cushing's disease, carcinoid syndrome, vascular headaches, appetite stimulant.

CONTRAINDICATIONS Hypersensitivity to cyproheptadine; MAOI therapy within 14 days; angle-closure glaucoma; stenosing peptic ulcer, symptomatic BPH, bladder neck obstruction, pyloroduodernal obstruction; older adults, debilitated patients; acute asthma attack; newborns, premature infants, lactation. Safe use in children younger than 2 y not established.

CAUTIOUS USE Patients predisposed to urinary retention; glaucoma; asthma; COPD; increased intraocular pressure; hyperthyroidism; cardiovascular or hepatic disease; hypertension; children with a family history of SIDS; pregnancy (category B), children.

ROUTE & DOSAGE

Allergies

Adult: **PO** 4 mg t.i.d. or q.i.d. (4–20 mg/day), max: 0.5 mg/kg/day
Geriatric: **PO** Start with 4 mg b.i.d.
Child: **PO** 0.25 mg/kg/day in 3–4 divided doses (max: 12 mg/day for 2–6 y, 16 mg/day for 6–12 y)

ADMINISTRATION

Oral

- GI adverse effects may be minimized by administering drug with food or milk.
- Store in tightly covered container at 15°–30° C (59°–86° F) unless otherwise directed.

ADVERSE EFFECTS (≥1%) **GI:** *Dry mouth,* nausea, vomiting, epigastric distress, appetite stimulation, weight gain, transient decrease in fasting blood sugar level, increased serum amylase level, cholestatic jaundice. **CNS:** *Drowsiness,* dizziness, faintness, headache, tremulousness, fatigue, disturbed coordination. **Respiratory:**

Thickened bronchial secretions. **Skin:** Skin rash. **Special Senses:** Dry nose and throat. **Urogenital:** Urinary frequency, retention, and difficult urination.

INTERACTIONS Drug: Alcohol and CNS DEPRESSANTS add to CNS depression; TRICYCLIC ANTIDEPRESSANTS and other ANTICHOLINERGICS have additive anticholinergic effects; may inhibit pressor effects of **epinephrine.**

PHARMACOKINETICS Absorption: Readily absorbed from GI tract. **Duration:** 6–9 h. **Distribution:** Distribution into breast milk not known. **Metabolism:** In liver. **Elimination:** In urine.

NURSING IMPLICATIONS

Assessment & Drug Effects

- Monitor level of alertness. In some patients, the sedative effect disappears spontaneously after 3–4 days of drug administration.
- Since drug may cause dizziness, supervision of ambulation and other safety precautions may be warranted.

Patient & Family Education

- Avoid activities requiring mental alertness and physical coordination, such as driving a car, until reaction to the drug is known.
- Drug causes sedation, dizziness, and hypotension in older adults. Report these symptoms. Children are more apt to manifest CNS stimulation (e.g., confusion, agitation, tremors, hallucinations). Reduction in dosage may be indicated.
- Cyproheptadine may increase and prolong the effects of alcohol, barbiturates, narcotic analgesics, and other CNS depressants.
- Maintain sufficient fluid intake to help to relieve dry mouth and also reduce risk of cholestatic jaundice.

CYTARABINE

(sye-tare'a-been)

Classifications: ANTINEOPLASTIC; ANTIMETABOLITE, PURINE ANTAGONIST
Therapeutic: ANTINEOPLASTIC
Prototype: 6-Mercaptopurine
Pregnancy Category: D

AVAILABILITY 10 mg/mL liposomal, 20 mg/mL, 100 mg, 500 mg, 1 g, 2 g powder for injection

ACTION & *THERAPEUTIC EFFECT*
Pyrimidine analog with cell phase specificity affecting rapidly dividing cells in S phase (DNA synthesis). In certain conditions prevents development of cell from G_1 to S phase. Interferes with DNA and RNA synthesis in rapidly growing cells. *Antineoplastic agent which has strong myelosuppressant activity. Immunosuppressant properties are exhibited by obliterated cell-mediated immune responses, such as delayed hypersensitivity skin reactions.*

USES To induce and maintain remission in acute myelocytic leukemia, acute lymphocytic leukemia, and meningeal leukemia and for treatment of lymphomas. Used in combination with other antineoplastics in established chemotherapeutic protocols.

CONTRAINDICATIONS History of drug-induced myelosuppression; immunization procedures; active meningeal infection (**liposomal cytarabine**); pregnancy (category D) particularly during first trimester, lactation.
CAUTIOUS USE Impaired renal or hepatic function, elderly; neurologic disease; gout, drug-induced myelosuppression. Safe use in infants not established.

Common adverse effects in *italic*, life-threatening effects <u>underlined</u>; generic names in **bold**; classifications in SMALL CAPS; ♣ Canadian drug name; ⊅ Prototype drug

385

ROUTE & DOSAGE

Leukemias

Adult/Child: **IV** 100–200 mg/m² by continuous infusion over 24 h **Subcutaneous** 1 mg/kg 1–2 times/ wk **Intrathecal** 5–75 mg once q4days or once/day for 4 days

Renal Impairment Dosage Adjustment

Serum Cr of 1.5–1.9 mg/dL (or from baseline of 0.5–1.2 mg/dL): Reduce to 1 g/m²/dose. *Serum Cr of 2 or more (or greater than 1.2 mg/dL change):* Do not exceed 100 mg/m²/day.

ADMINISTRATION

Intrathecal

- For intrathecal injection, reconstitute with an isotonic, buffered diluent without preservatives. Follow manufacturer's recommendations.

Intravenous

PREPARE: **Direct:** Reconstitute with bacteriostatic water for injection (without benzyl alcohol for neonates) as follows: Add 5 mL to the 100-mg vial to yield 20 mg/mL; add 10 mL to the 500 mg vial to yield 50 mg/mL. **IV Infusion:** May be further diluted with 100 mL or more of D5W or NS.

ADMINISTER: **Direct:** Give at a rate of 100 mg or a fraction thereof over 3 min. **IV Infusion:** Give over 30 min or longer depending on the total volume of IV solution to be infused.

INCOMPATIBILITIES **Solution/additive: Fluorouracil, gentamicin, heparin, hydrocortisone, insulin, nafcillin, oxacillin, penicillin G. Y-site: Allopurinol, amphotericin B cholesteryl sulfate complex, gallium, ganciclovir, lansoprazole, TPN.**

- Store cytarabine in refrigerator until reconstituted. ▪ Reconstituted solutions may be stored at 15°–30° C (59°–86° F) for 48 h. Discard solutions with a slight haze.

ADVERSE EFFECTS (≥1%) **Body as a Whole:** Weight loss, sore throat, fever, thrombophlebitis and pain at injection site; pericarditis, bleeding (any site), pneumonia. Potentially carcinogenic and mutagenic. **GI:** *Nausea, vomiting,* diarrhea, stomatitis, oral or anal inflammation or ulceration, esophagitis, anorexia, hemorrhage, hepatotoxicity, jaundice. **Hematologic:** <u>*Leukopenia,* *thrombocytopenia,*</u> anemia, megaloblastosis, myelosuppression (reversible); transient hyperuricemia. **CNS:** Headache, <u>neurotoxicity</u>; peripheral neuropathy, brachial plexus neuropathy, personality change, neuritis, vertigo, lethargy, somnolence, confusion. **Skin:** Rash, erythema, freckling, cellulitis, skin ulcerations, pruritus, urticaria, bulla formation, desquamation. **Special Senses:** Conjunctivitis, keratitis, photophobia. **Urogenital:** Renal dysfunction, urinary retention.

INTERACTIONS Drug: GI toxicity may decrease **digoxin** absorption; decreases AMINOGLYCOSIDES activity against *Klebsiella pneumoniae.*

PHARMACOKINETICS Peak: 20–60 min subcutaneous. **Distribution:** Crosses blood–brain barrier and placenta. **Metabolism:** In liver. **Elimination:** 80% in urine in 24 h. **Half-Life:** 1–3 h.

NURSING IMPLICATIONS

Assessment & Drug Effects

- Inspect patient's mouth before the administration of each dose. Toxicity necessitating dosage alterations almost always occurs. Report adverse reactions immediately.

- Lab tests: Daily Hct and platelet counts and total and differential leukocyte counts during initial therapy. Serum uric acid and hepatic function tests should be performed at regular intervals throughout treatment period.
- Hyperuricemia due to rapid destruction of neoplastic cells may accompany cytarabine therapy. A regimen that includes a uricosuric agent such as allopurinol, urine alkalinization, and adequate hydration may be started. To reduce potential for urate stone formation, fluids are forced in excess of 2 L, if tolerated. Consult prescriber.
- Monitor I&O ratio and pattern.
- Monitor body temperature. Be alert to the most subtle signs of infection, especially low-grade fever, and report promptly.
- When platelet count falls below 50,000/mm³ and polymorphonuclear leukocytes to below 1000/mm³, therapy may be suspended. WBC nadir is usually reached in 5–7 days after therapy has been stopped. Therapy is restarted with appearance of bone marrow recovery and when preceding cell counts are reached.
- Provide good oral hygiene to diminish adverse effects and chance of superinfection. Stomatitis and cheilosis usually appear 5–10 days into the therapy.

Patient & Family Education

- Report promptly protracted vomiting or signs of nephrotoxicity (see Appendix F).
- Flu-like syndrome occurs usually within 6–12 wk after drug administration and may recur with successive therapy. Report chills, fever, achy joints and muscles.
- Practice good oral hygiene to minimize discomfort from stomatitis.
- Report any S&S of superinfection (see Appendix F).

CYTOMEGALOVIRUS IMMUNE GLOBULIN (CMVIG, CMV-IVIG)

(cy-to-meg'a-lo-vi-rus)

CytoGam

Classifications: BIOLOGICAL RESPONSE MODIFIER; IMMUNOGLOBULIN
Therapeutic: IMMUNOGLOBULIN
Prototype: Peginterferon alfa-2a
Pregnancy Category: C

AVAILABILITY 50 mg/mL injection

ACTION & THERAPEUTIC EFFECT
Cytomegalovirus immune globulin (CMVIG) is a preparation of immunoglobulin G (IgG) antibodies with high concentrations of antibodies directed against cytomegalovirus (CMV). *The CMV antibodies attenuate or reduce the incidence of serious CMV disease, such as CMV-associated pneumonia, CMV-associated hepatitis, and concomitant fungi and parasitic superinfections.*

USES Attenuation of primary cytomegalovirus (CMV) disease associated with kidney transplantation.
UNLABELED USES Prevention of CMV disease in other organ transplants (especially heart) when the recipient is seronegative for CMV and the donor is seropositive.

CONTRAINDICATIONS History of previous severe reactions associated with CMVIG or other human immunoglobulin preparations, selective immunoglobulin A (IgA) deficiency.
CAUTIOUS USE Myelosuppression, maltose or sucrose hypersensitivity; renal insufficiency, diabetes mellitus, age older than 65, volume depletion, sepsis, paraproteinemia; cardiac disease; pregnancy (category C), lactation.

Common adverse effects in *italic*, life-threatening effects underlined; generic names in **bold**; classifications in SMALL CAPS; ◆ Canadian drug name; ➋ Prototype drug

387

ROUTE & DOSAGE

Prevention of CMV Disease

Adult: **IV** 150 mg/kg within 72 h of transplantation, then 100 mg/kg 2, 4, 6, and 8 wk post-transplant, then 50 mg/kg 12 and 16 wk post-transplant

ADMINISTRATION

Intravenous

CMVIG should be administered through a separate IV line using an infusion pump. See manufacturer's directions if this is not possible.

***PREPARE:* IV Infusion:** Use a double-ended transfer needle or large syringe to reconstitute with 50 mL sterile water. ▪ Gently rotate vial to dissolve; do not shake. Allow 30 min to dissolve powder. Reconstituted solution contains 50 mg/mL. ▪ **Must be** completely infused within 12 h since solution contains no preservative.

***ADMINISTER:* IV Infusion:** Use a constant infusion pump and give at rate of 15, 30, 60 mg/kg/h over first 30 min, second 30 min, third 30 min, respectively. Monitor closely during and after each rate change. ▪ If flushing, nausea, back pain, fever, or chills develops, slow or temporarily discontinue infusion. ▪ If BP begins to decrease, stop infusion and institute emergency measures. **Infusion of Subsequent IV Doses:** The intervals for increasing the dose from 15 to 30 to 60 mg may be shortened from 30 to 15 min.

▪ Never infuse more than 75 mL/h of CMVIG.

▪ Reconstituted solution should be started within 6 h and completed within 12 h of preparation. Discard solution if cloudy.

ADVERSE EFFECTS (≥1%) **Body as a Whole:** Muscle aches, back pain, <u>anaphylaxis</u> (rare), fever and chills during infusion. **CV:** Hypotension, palpitations. **GI:** Nausea, vomiting, metallic taste. **CNS:** Headache, anxiety. **Respiratory:** Shortness of breath, wheezing. **Skin:** Flushing.

INTERACTIONS **Drug:** May interfere with the immune response to LIVE VIRUS VACCINES **(BCG, measles/ mumps/rubella, live polio),** defer vaccination with live viral vaccines for approximately 3 mo after administration of CMVIG; revaccination may be necessary if these vaccines were given shortly after CMVIG.

NURSING IMPLICATIONS

Assessment & Drug Effects

▪ Monitor vital signs preinfusion, before increases in infusion rate, periodically during infusion, and postinfusion.

▪ Notify prescriber immediately if any of the following occur: Flushing, nausea, back pain, fall in BP, other signs of anaphylaxis.

▪ Emergency drugs should be available for treatment of acute anaphylactic reactions.

▪ Monitor for CMV-associated syndromes (e.g., leukopenia, thrombocytopenia, hepatitis, pneumonia) and for superinfections.

Patient & Family Education

▪ Familiarize yourself with potential adverse effects and know which to report to prescriber.

▪ Defer vaccination with live viral vaccines for 3 mo after administration of CMVIG.

DABIGATRAN ETEXILATE
(dab-i-ga′tran e-tex′i-late)
Pradaxa

Common adverse effects in *italic*, life-threatening effects <u>underlined</u>; generic names in **bold**; classifications in SMALL CAPS; ✤ Canadian drug name; ⊘ Prototype drug

Classification: ANTICOAGULANT; DIRECT THROMBIN INHIBITOR
Therapeutic: ANTITHROMBOTIC; THROMBIN INHIBITOR
Prototype: Lepirudin
Pregnancy Category: C

AVAILABILITY 75 mg and 100 mg capsules

ACTION & *THERAPEUTIC EFFECT* A direct inhibitor of thrombin which prevents thrombin-induced platelet aggregation and conversion of fibrinogen into fibrin during the coagulation cascade. *It prolongs the aPTT and TT, and it prevents development of a thrombus.*

USES Reduction of the risk of stroke and systemic embolism in patients with nonvalvular atrial fibrillation.

CONTRAINDICATIONS History of serious hypersensitivity to dabigatran etexilate; active pathological bleeding.
CAUTIOUS USE Medications or conditions (e.g., labor and delivery, chronic NSAID use, use of antiplatelet agents) that predispose to bleeding; severe renal impairment; pregnancy (category C); lactation. Safety and efficacy in children not established.

ROUTE & DOSAGE

Risk of Stroke or Systemic Embolism Reduction
Adult: **PO** 150 mg b.i.d.

Renal Impairment Dosage Adjustment
CrCl greater than or equal to 15 to less than or equal to 30 mL/min: 75 mg b.i.d.; CrCl less than 15 mL/min: Not recommended

Conversion from Warfarin to Dabigatran
Discontinue warfarin and begin dabigatran when INR is less than 2.0

Conversion from Dabigatran to Warfarin
CrCl greater than 50 mL/min: Start warfarin 3 d before discontinuing dabigatran; CrCl 31–50 mL/min: Start warfarin 2 d before discontinuing dabigatran; CrCl 15–30 mL/min: Start warfarin 1 d before discontinuing dabigatran; CrCl less than 15 mL/min: Not recommended

ADMINISTRATION
Oral
- Ensure that capsule is swallowed whole. It should not be opened or chewed.
- Converting from a parenteral anticoagulant: Start dabigatran up to 2 h before the next dose of parenteral drug was due or at time of discontinuation of an IV anticoagulant.
- Withhold drug and report to prescriber if active bleeding is suspected.
- Store at 15–30° C (59–86° F). Store in original package to protect from moisture. Once bottle is opened, use contents within 30 days.

ADVERSE EFFECTS (≥1%) **GI:** Abdominal pain or discomfort, diarrhea, epigastric discomfort, erosive gastritis, esophagitis, gastric hemorrhage, gastroesophogeal reflux disorder, gastrointestinal ulcer, hemorrhagic gastritis, hemorrhagic erosive gastritis, nausea. **Hematological:** *Increased risk of bleeding.*

INTERACTIONS Drug: Concomitant use with P-GLYCOPROTEIN INDUCERS

Common adverse effects in *italic*, life-threatening effects underlined; generic names in **bold**; classifications in SMALL CAPS; ♥ Canadian drug name; ☺ Prototype drug

389

(**rifampin**) reduces the levels of dabigatran.

PHARMACOKINETICS Absorption:
3–7% bioavailable. **Peak:** 1 h. **Distribution:** 35% plasma protein bound. **Metabolism:** Esterase hydrolysis and glucuronide conjugation to active metabolites. **Elimination:** Primarily renal; unabsorbed drug excreted in feces. **Half-Life:** 12–17 h.

NURSING IMPLICATIONS

Assessment & Drug Effects
▪ Monitor for and promptly report S&S of active bleeding.
▪ Monitor for and report promptly adverse GI effects including: Epigastric pain or discomfort, GERD, or abdominal pain or discomfort.
▪ Lab test: Baseline serum creatinine; periodic aPTT and TT.

Patient & Family Education
▪ Seek emergency care for any of the following: Unusual bruising; pink or brown urine; red or black, tarry stools; coughing up blood; vomiting blood, or vomit that looks like coffee grounds.
▪ Report promptly to prescriber any of the following: Pain, swelling or discomfort in a joint; nose bleeds or bleeding from gums; headaches, dizziness, or weakness; menstrual bleeding or vaginal bleeding that is heavier than normal; indigestion, gastric reflux, or nausea.
▪ Alert all health care providers that you are taking dabigatran before any invasive procedure, including dental procedures.

DACARBAZINE
(da-kar′ba-zeen)
DTIC-Dome
Classifications: ANTINEOPLASTIC; ALKYLATING AGENT
Therapeutic: ANTINEOPLASTIC

Prototype: Cyclophosphamide
Pregnancy Category: C

AVAILABILITY 10 mg/mL injection

ACTION & *THERAPEUTIC EFFECT*
Cytotoxic agent that may have alkylating properties or inhibiting DNA synthesis by acting as a purine analog. It is cell-cycle nonspecific. Either mechanism would interfere with DNA replication, RNA transcription, and protein synthesis in rapidly proliferating cells. *Has carcinogenic, mutagenic, and teratogenic effects.*

USES As single agent or in combination with other antineoplastics in treatment of metastatic malignant melanoma, refractory Hodgkin's disease.
UNLABELED USES Various sarcomas and malignant glucagonoma.

CONTRAINDICATIONS Hypersensitivity to dacarbazine; severe bone marrow suppression; active infection; live vaccine; lactation.
CAUTIOUS USE Hepatic or renal impairment; previous radiation or chemotherapy; pregnancy (category C).

ROUTE & DOSAGE

Neoplasms
Adult: **IV** 2–4.5 mg/kg/day for 10 days repeated at 4-wk intervals or 250 mg/m²/day for 5 days repeated at 3-wk intervals

Hodgkin's Disease
Adult: **IV** 150 mg/m²/day × 5 days; repeat at 4-wk intervals

ADMINISTRATION

Intravenous
IV administration to infants and children: Verify correct IV concentration and rate of infusion with prescriber.

- Wear gloves when handling this drug. If solution gets into the eyes, wash them with soap and water immediately, then irrigate with water or isotonic saline.

PREPARE: Direct: Reconstitute drug with sterile water for injection to make a solution containing 10 mg/mL dacarbazine (pH 3.0–4.0) by adding 9.9 mL to 100 mg or 19.7 mL to 200 mg. **IV Infusion:** Dilute further reconstituted solution in 50–250 mL of D5W or NS.

ADMINISTER: Direct: Give by direct IV over at least 15 min. **IV Infusion (*preferred*):** Infuse IV over 30–60 min. ▪ If possible, avoid using antecubital vein or veins on dorsum of hand or wrist where extravasation could lead to loss of mobility of entire limb. ▪ Avoid veins in extremity with compromised venous and lymphatic drainage and veins near joint spaces.

INCOMPATIBILITIES Solution/additive: Allopurinol, heparin, hydrocortisone. Y-site: Allopurinol, cefepime, heparin, piperacillin/tazobactam.

▪ Administer dacarbazine only to patients under close supervision because close observation and frequent laboratory studies are required during and after therapy. ▪ *IV extravasation:* Monitor injection site frequently (instruct patient to do so, if able). Give prompt attention to patient's complaint of swelling, stinging, and burning sensation around injection site. ▪ Extravasation can occur painlessly and without visual signs. Danger areas for extravasation are dorsum of hand or ankle (especially if peripheral arteriosclerosis is present), joint spaces, and previously irradiated areas. ▪ If extravasation is suspected, infusion should be stopped

immediately and restarted in another vein. Report to the prescriber. Prompt institution of local treatment is IMPERATIVE.

▪ Store reconstituted solution up to 72 h at 4° C (39° F) or at room temperature 15°–30° C (59°–86° F) for up to 8 h. ▪ Store diluted reconstituted solution for 24 h at 4° C (39° F) or at room temperature for up to 8 h. Protect from light.

ADVERSE EFFECTS (≥1%) **Body as a Whole:** Hypersensitivity (erythematosus, urticarial rashes, hepatotoxicity, photosensitivity); facial paresthesia and flushing, flu-like syndrome, myalgia, malaise, anaphylaxis. **CNS:** Confusion, headache, seizures, blurred vision. **GI:** *Anorexia, nausea, vomiting.* **Hematologic:** Severe leukopenia and thrombocytopenia, mild anemia. **Skin:** Alopecia. **Other:** *Pain along injected vein.*

PHARMACOKINETICS Distribution: Localizes primarily in liver. **Metabolism:** In liver by CYP1A2. **Elimination:** 35–50% in urine in 6 h. **Half-Life:** 5 h.

NURSING IMPLICATIONS

Assessment & Drug Effects

- Monitor IV site carefully for extravasation; if suspected, discontinue IV immediately and notify prescriber.
- Note: Skin damage by dacarbazine can lead to deep necrosis requiring surgical debridement, skin grafting, and even amputation. Older adults, the very young, comatose, and debilitated patients are especially at risk. Other risk factors include establishing an IV line in a vein previously punctured several times and the use of nonplastic catheters.
- Lab tests: Monitor for hematopoietic toxicity that usually appears about 4 wk after first dose. Generally, a leukocyte count of less than

---

Common adverse effects in *italic*, life-threatening effects <u>underlined</u>; generic names in **bold**; classifications in SMALL CAPS; ◆ Canadian drug name; ☺ Prototype drug

391

D

3000/mm³ and a platelet count of less than 100,000/mm³ require suspension or cessation of therapy.

- Avoid, if possible, all tests and treatments during platelet nadir requiring needle punctures (e.g., IM). Observe carefully and report evidence of unexplained bleeding.
- Monitor for severe nausea and vomiting (greater than 90% of patients) that begin within 1 h after drug administration and may last for as long as 12 h.
- Check patient's mouth for ulcerative stomatitis prior to the administration of each dose.
- Monitor I&O ratio and pattern and daily temperature. Renal impairment extends the half-life and increases danger of toxicity. Report symptoms of renal dysfunction and even a slight elevation of temperature.

Patient & Family Education

- Learn about all potential adverse drug effects.
- Report flu-like syndrome that may occur during or even a week after treatment is terminated and last 7–21 days. Symptoms frequently recur with successive treatments.
- Avoid prolonged exposure to sunlight or to ultraviolet light during treatment period and for at least 2 wk after last dose. Protect exposed skin with sunscreen lotion (SPF 15) and avoid exposure in midday.
- Report promptly the onset of blurred vision or paresthesia.

DACTINOMYCIN
(dak-ti-noe-mye'sin)
Cosmegen
Classifications: ANTINEOPLASTIC; ANTHRACYCLINE (ANTIBIOTIC)
Therapeutic: ANTINEOPLASTIC
Prototype: Doxorubicin
Pregnancy Category: D

AVAILABILITY 0.5 mg injection

ACTION & *THERAPEUTIC EFFECT*
Potent cytotoxic cell cycle nonspecific antibiotic. Complexes with DNA, thereby inhibiting DNA, RNA, and protein synthesis in actively proliferating cells. Potentiates effects of x-ray therapy and the converse also appears likely. *Has antineoplastic properties that result from inhibiting DNA and RNA synthesis.*

USES To treat Wilms' tumor, rhabdomyosarcoma, carcinoma of testes and uterus, Ewing's sarcoma, solid malignancies, gestational trophoblastic neoplasia, and sarcoma botryoides.
UNLABELED USES Malignant melanoma, Kaposi's sarcoma, osteogenic sarcoma, among others.

CONTRAINDICATIONS Acute infection; pregnancy (category D), lactation.
CAUTIOUS USE Previous therapy with antineoplastics or radiation within 3–6 wk, bone marrow depression; infections; history of gout; impairment of kidney or liver function; obesity; chickenpox, herpes zoster, and other viral infections. Safe use in infants younger than 6 mo is not known.

ROUTE & DOSAGE

Neoplasms
Adult/Adolescent/Child/Infant (older than 6 mo): IV 500 mcg/day for 5 days max, may repeat at 2–4 wk intervals if tolerated (if patient is obese or edematous, give 400–600 mcg/m²/day to relate dosage to lean body mass); monitor for symptoms of toxicity from overdosage

Wilms' Tumor, Childhood Rhabdomyosarcoma, Ewing's Sarcoma, Nephroblastoma
Adult/Child: IV 15 mcg/kg/day × 5 days with other agents

Common adverse effects in *italic*, life-threatening effects underlined; generic names in **bold**; classifications in SMALL CAPS; ♣ Canadian drug name; ☻ Prototype drug

Gestational Trophoblastic Neoplasia

Adult: **IV** 12 mcg/kg/day × 5 days or 500 mcg × 2 days with other agents

Solid Tumor

Adult/Adolescent/Child/Infant (older than 6 mo): **IV** 50 mcg/kg (lower extremity) or 35 mcg/kg (upper extremity)

ADMINISTRATION

Intravenous

Use gloves and eye shield when preparing solution. If skin is contaminated, rinse with running water for 10 min; then rinse with buffered phosphate solution. ▪ If solution gets into the eyes, wash with water immediately; then irrigate with water or isotonic saline for 10 min.

PREPARE: **Direct:** Reconstitute 0.5 mg vial by adding 1.1 mL sterile water (without preservative) for injection; the resulting solution will contain approximately 0.5 mg/mL. **IV Infusion:** Further dilute reconstituted solution in 50 mL of D5W or NS for infusion.

ADMINISTER: **Direct:** Use two-needle technique for direct IV: Withdraw calculated dose from vial with one needle, change to new needle to give directly into vein without using an infusion. Give over 2–3 min. ▪ Or give directly into an infusing solution of D5W or NS, or into tubing or side arm of a running IV infusion. **IV Infusion:** Give diluted solution as a single dose over 15–30 min.

INCOMPATIBILITIES **Y-site:** Filgrastim.

▪ Store drug at 15°–30° C (59°–86° F) unless otherwise directed. Protect from heat and light.

ADVERSE EFFECTS (≥1%) **GI:** *Nausea, vomiting,* anorexia, abdominal pain, diarrhea, proctitis, GI ulceration, *stomatitis,* cheilitis, glossitis, dysphagia, hepatitis. **Hematologic:** Anemia (including aplastic anemia), agranulocytosis, *leukopenia, thrombocytopenia,* pancytopenia, reticulopenia. **Skin:** Acne, desquamation, hyperpigmentation and reactivation of erythema especially over previously irradiated areas, *alopecia* (reversible). **Other:** Malaise, fatigue, lethargy, fever, myalgia, anaphylaxis, gonadal suppression, hypocalcemia, hyperuricemia, thrombophlebitis; *necrosis, sloughing, and contractures at site of extravasation;* hepatitis, hepatomegaly.

INTERACTIONS Drug: Elevated **uric acid** level produced by dactinomycin may necessitate dose adjustment of ANTIGOUT AGENTS; effects of both dactinomycin and other MYELOSUPPRESSANTS are potentiated; effects of both **radiation** and dactinomycin are potentiated, and dactinomycin may reactivate erythema from previous radiation therapy; **vitamin K** effects (antihemorrhagic) decreased, leading to prolonged clotting time and potential hemorrhage.

PHARMACOKINETICS Distribution: Concentrated in liver, spleen, kidneys, and bone marrow; does not cross blood–brain barrier; crosses placenta. **Elimination:** 50% unchanged in bile and 10% in urine; only 30% in urine over 9 days. **Half-Life:** 36 h.

NURSING IMPLICATIONS

Assessment & Drug Effects

▪ Observe injection site frequently; if extravasation occurs, stop infusion immediately. Restart infusion in another vein. Report to prescriber. Institute prompt local treatment to

Common adverse effects in *italic*, life-threatening effects underlined; generic names in **bold**; classifications in SMALL CAPS; ✚ Canadian drug name; ⊙ Prototype drug

393

prevent thrombophlebitis and necrosis.

- Monitor for severe toxic effects that occur with high frequency. Effects usually appear 2–4 days after a course of therapy is stopped and may reach maximal severity 1–2 wk following discontinuation of therapy.

- Use antiemetic drugs to control nausea and vomiting, which often occur a few hours after drug administration. Vomiting may be severe enough to require intermittent therapy. Observe patient daily for signs of drug toxicity.

- Lab tests: Frequent renal, hepatic, and bone marrow function tests are advised; daily. WBC counts, and platelet counts every 3 days.

- Monitor temperature and inspect oral membranes daily for stomatitis.

- Monitor for stomatitis, diarrhea, and severe hematopoietic depression. These may require prompt interruption of therapy until drug toxicity subsides.

- Report onset of unexplained bleeding, jaundice, and wheezing. Also, be alert to signs of agranulocytosis (see Appendix F). Report to prescriber. Antibiotic therapy, protective isolation, and discontinuation of the antineoplastic are indicated.

- Observe and report symptoms of hyperuricemia (see Appendix F). Urge patient to increase fluid intake up to 3000 mL/day if allowed.

Patient & Family Education

- Note: Infertility is a possible, irreversible adverse effect of this drug.

- Learn preventative measures to minimize nausea and vomiting.

- Note: Alopecia (hair loss) is an anticipated reversible adverse effect of this drug. Seek appropriate supportive guidance.

DALFAMPRIDINE
(dal-fam′pri-deen)

Ampyra

Classification: CENTRAL NERVOUS SYSTEM AGENT; NEUROLOGIC AGENT; POTASSIUM CHANNEL BLOCKER

Therapeutic: CNS AGENT; NEUROLOGIC AGENT

Pregnancy Category: C

AVAILABILITY 10 mg extended release tablets

ACTION & THERAPEUTIC EFFECT Dalfampridine is a broad spectrum potassium channel blocker. The mechanism by which it exerts its therapeutic effect is not fully understood. *Improves walking speed in persons with multiple sclerosis.*

USES To improve walking speed in patients with multiple sclerosis.

CONTRAINDICATIONS History of seizures; moderate or severe renal impairment; concurrent use of other forms of 4-aminopyridine (4-AP, fampridine); lactation.

CAUTIOUS USE Urinary tract infections. Safe use in children younger than 18 y is not established.

ROUTE & DOSAGE

Multiple Sclerosis

Adult: **PO** 10 mg q12h

Renal Impairment Dosage Adjustment

CrCl 50 mL/min or less: Dalfampridine is contraindicated.

ADMINISTRATION

Oral

- Ensure that extended release tablet is swallowed whole. It should not be crushed or chewed.
- Store at 15–30C (59–86F).

ADVERSE EFFECTS (≥2%) **CNS:** Asthenia, balance disorder, dizziness, headache, *insomnia*, MS relapse, paresthesia. **GI:** Constipation, dyspepsia, nausea. **Musculoskeletal:** Back pain. **Respiratory:** Nasopharyngitis, pharyngolaryngeal pain. **Urogenital:** *Urinary tract infection.*

PHARMACOKINETICS Absorption: 96% bioavailability. **Peak:** 3–4 h. **Metabolism:** Minimal. **Elimination:** Renal as unchanged drug. **Half-Life:** 5.2–6.5 h.

NURSING IMPLICATIONS

Assessment & Drug Effects

- Monitor closely for signs of seizure activity. Withhold drug and notify prescriber if a seizure occurs.
- Monitor for and report signs of urinary tract infection.
- Lab tests: Baseline and periodic kidney function tests.
- Monitor I&O and report significant changes in output.

Patient & Family Education

- Stop taking drug and notify prescriber immediately if a seizure occurs.
- Tablets must be taken whole. Breaking the tablet may allow too much medication to be released too quickly increasing the risk of a seizure.
- Do not take more than 2 tablets in a 24 h period and maintain an approximate 12 h interval between doses.
- If you take too much dalfampridine, immediately call your prescriber or go to the nearest emergency room.

DALTEPARIN SODIUM
(dal-tep-a'rin)

Fragmin
Pregnancy Category: B
See Appendix A-2.

D

DANAZOL
(da'na-zole)

Cyclomen ✦
Classification: ANDROGEN/ANABOLIC STEROID
Therapeutic: ANABOLIC STEROID
Prototype: Testosterone
Pregnancy Category: X

AVAILABILITY 50 mg, 100 mg, 200 mg capsules

ACTION & *THERAPEUTIC EFFECT*
Synthetic androgen steroid; derivative of testosterone with dose-related mild androgenic effects. Suppresses pituitary output of FSH and LH, resulting in anovulation and associated amenorrhea. *Interrupts progress and pain of endometriosis by causing atrophy and involution of both normal and ectopic endometrial tissue.*

USES Palliative treatment of endometriosis when alternative hormonal therapy is ineffective, contraindicated, or intolerable. Also used to treat fibrocystic breast disease and hereditary angioedema.
UNLABELED USES To treat precocious puberty, gynecomastia, menorrhagia, premenstrual syndrome (PMS), chronic immune thrombocytopenic purpura (ITP), autoimmune hemolytic anemia, hemophilia A and B.

CONTRAINDICATIONS Undiagnosed abnormal genital bleeding; porphyria; peripheral neuropathy; vaginal bleeding; pregnancy (category X), lactation.
CAUTIOUS USE Migraine headache, epilepsy; seizure disorders; renal

---

Common adverse effects in *italic*, life-threatening effects <u>underlined</u>; generic names in **bold**; classifications in SMALL CAPS; ✦ Canadian drug name; ☉ Prototype drug

395

D

impairment; history of strokes; older adults.

ROUTE & DOSAGE

Endometriosis

Adult: **PO** 400 mg b.i.d. for 3–6 mo, start during menstruation or if pregnancy test is negative, may extend to 9 mo if necessary. Do not repeat regimen

Fibrocystic Breast Disease

Adult: **PO** 100–400 mg in 2 divided doses, start during menstruation or if pregnancy test is negative

Hereditary Angioedema

Adult: **PO** 200 mg b.i.d. or t.i.d., may decrease by 50% at intervals of 1–3 mo or longer, start during menstruation or if pregnancy test is negative

ADMINISTRATION

Oral

- Start therapy during menstruation, or after a negative pregnancy test.
- Store capsules at 15°–30° C (59°–86° F) in tightly closed container.

ADVERSE EFFECTS (≥1%) **Body as a Whole:** Hypersensitivity (skin rashes, nasal congestion). **Endocrine:** Androgenic effects (acne, mild hirsutism, deepening of voice, oily skin and hair, hair loss, edema, weight gain, pitch breaks, voice weakness, decrease in breast size); hypoestrogenic effects (*hot flashes;* sweating; emotional lability; nervousness; vaginitis with itching, drying, burning, or bleeding; *amenorrhea, irregular menstrual patterns*); impairment in glucose tolerance. **CNS:** Dizziness, sleep disorders, fatigue, tremor, irritability. **Special Senses:** Conjunctival edema. **CV:** Elevated BP. **GI:** Gastroenteritis, <u>hepatic damage</u> (rare), increased LDL, decreased HDL. **Urogenital:** Decreased libido. **Musculoskeletal:** Joint lock-up, joint swelling.

INTERACTIONS Herbal: Echinacea possibility of increased hepatotoxicity.

PHARMACOKINETICS Elimination: Other pharmacokinetic information is not known. **Half-Life:** 4.5 h.

NURSING IMPLICATIONS

Assessment & Drug Effects

- Routine breast examinations should be carried out during therapy. Carcinoma of the breast should be ruled out prior to start of therapy for fibrocystic breast disease. Advise patient to report to prescriber if any nodule enlarges or becomes tender or hard during therapy.
- Because danazol may cause fluid retention, patients with cardiac or renal dysfunction, epilepsy, or migraine should be observed closely during therapy, as these problems could worsen. Monitor weight.
- Drug-induced edema may compress the median nerve, producing symptoms of carpal tunnel syndrome. If patient complains of wrist pain that worsens at night, paresthesias in radial palmar aspect of the hand and fingers, consult prescriber.
- Lab tests: Baseline and periodic liver function tests LFTs. Patients with diabetes (or history of) should have blood glucose tests.

Patient & Family Education

- Note: Pain and discomfort are usually relieved in 2 or 3 mo; the nodularity in 4–6 mo. Menses may be regular or irregular in pattern during therapy.

- Note: Drug-induced amenorrhea is reversible. Ovulation and cyclic bleeding usually return within 60–90 days after therapeutic regimen is discontinued as well as the potential for conception.
- Use a nonhormonal contraceptive during treatment because ovulation may not be suppressed until 6–8 wk after therapy is begun. If pregnancy occurs while taking this drug, contact prescriber immediately.
- Report voice changes promptly. Virilizing adverse effects may persist even after drug therapy is terminated.

DANTROLENE SODIUM
(dan'troe-leen)
Dantrium
Classification: DIRECT-ACTING SKELETAL MUSCLE RELAXANT
Therapeutic: SKELETAL MUSCLE RELAXANT; ANTISPASMODIC
Pregnancy Category: C

AVAILABILITY 25 mg, 50 mg, 100 mg capsules; 20 mg vial

ACTION & *THERAPEUTIC EFFECT* Hydantoin derivative with peripheral skeletal muscle relaxant action. Directly relaxes spastic muscle by interfering with calcium ion release from sarcoplasmic reticulum within skeletal muscle. *Relief of muscle spasticity, however, may be accompanied by muscle weakness sufficient to affect overall functional capacity of the patient.*

USES Orally for the symptomatic treatment of skeletal muscle spasms secondary to spinal cord injury, stroke, cerebral palsy, multiple sclerosis. Used intravenously for the management of malignant hyperthermia. Oral dantrolene can be used prophylactically for patients with a history of malignant hyperthermia or with a family history of the disorder.

UNLABELED USES Neuroleptic malignant syndrome, exercise-induced muscle pain, and flexor spasms.

CONTRAINDICATIONS Active hepatic disease; when spasticity is necessary to sustain upright posture and balance in locomotion or to maintain increased body function; spasticity due to rheumatic disorders; active liver disease; lactation.

CAUTIOUS USE Impaired cardiac or pulmonary function, muscular sclerosis; neuromuscular disease; myopathy; patients older than 35 y, especially women; pregnancy (category C). Safe use in children younger than 5 y is not established.

ROUTE & DOSAGE

Relief of Spasticity

Adult: **PO** 25 mg once/day, increase to 25 mg b.i.d. to q.i.d., may increase q4–7days up to 100 mg b.i.d. to q.i.d.
Child: **PO** 0.5 mg/kg b.i.d., increase to 0.5 mg/kg t.i.d. or q.i.d., may increase by 0.5 mg/kg up to 3 mg/kg b.i.d. to q.i.d. (max: 100 mg q.i.d.)

Malignant Hyperthermia Treatment

Adult/Child: **IV** 1 mg/kg rapid direct IV push repeated prn up to a total of 10 mg/kg **PO** May be necessary to continue orally with 1–2 mg/kg q.i.d. for 1–3 days to prevent recurrence

Malignant Hyperthermia Prophylaxis

Adult: **IV** 1.5 mg/kg infusion over 1 h may be repeated

Common adverse effects in *italic*, life-threatening effects underlined; generic names in **bold**; classifications in SMALL CAPS; ♣ Canadian drug name; ⊙ Prototype drug

397

D

Hepatic Impairment Dosage Adjustment

Do not use in active liver disease

ADMINISTRATION

Oral

- Prepare oral suspension for a single dose, when necessary, by emptying contents of capsule(s) into fruit juice or other liquid. Shake suspension well before pouring. • Avoid contamination, keep refrigerated, and use within several days, since it does not contain a preservative.

Intravenous ─────────

PREPARE: **Direct:** Dilute each 20 mg with 60 mL sterile water without preservatives. Shake until clear. **Infusion:** Large volume used for prophylaxis may be transferred to plastic (not glass) infusion bags.
ADMINISTER: **Direct:** Give by rapid direct IV push. Avoid extravasation; solution has a high pH and therefore is extremely irritating to tissue. • Ensure IV patency prior to giving drug direct IV. **Infusion:** Give over 1 h.

- Store capsules in tightly closed, light-resistant container. Contents of vial (for IV use) **must be** protected from direct light and used within 6 h after reconstitution, since it does not contain a preservative. • Store both PO and parenteral forms at 15°–30° C (59°–86° F) unless otherwise directed.

ADVERSE EFFECTS (≥1%) **Body as a Whole:** Hypersensitivity (pruritus, urticaria, eczematoid skin eruption, photosensitivity, eosinophilic pleural effusion). **CNS:** Drowsiness, *muscle weakness,* dizziness, light-headedness, unusual fatigue, speech disturbances, headache, confusion, nervousness, mental depression, insomnia, euphoria, seizures. **CV:** Tachycardia, erratic BP. **Special Senses:** Blurred vision, diplopia, photophobia. **GI:** *Diarrhea,* constipation, nausea, vomiting, anorexia, swallowing difficulty, alterations of taste, gastric irritation, abdominal cramps, GI bleeding; hepatitis, jaundice, hepatomegaly, hepatic necrosis (all related to prolonged use of high doses). **Urogenital:** Crystalluria with pain or burning with urination, urinary frequency, urinary retention, nocturia, enuresis, difficult erection.

INTERACTIONS Drug: **Alcohol** and other CNS DEPRESSANTS compound CNS depression; **estrogens** increase risk of hepatotoxicity in women older than 35 y; **verapamil** and other CALCIUM CHANNEL BLOCKERS increase risk of ventricular fibrillation and cardiovascular collapse with IV dantrolene.

PHARMACOKINETICS **Absorption:** Incompletely absorbed from GI tract. **Peak:** 5 h. **Distribution:** Crosses placenta. **Metabolism:** In liver. **Elimination:** In urine chiefly as metabolites. **Half-Life:** 8.7 h.

NURSING IMPLICATIONS

Assessment & Drug Effects

- Monitor for therapeutic effectiveness. Improvement may not be apparent until 1 wk or more of drug therapy.
- Monitor vital signs during IV infusion. Also monitor ECG, CVP, and serum potassium.
- Supervise ambulation until patient's reaction to drug is known. Relief of spasticity may

be accompanied by some loss of strength.

- Note: Most common adverse effects are generally transient, lasting up to 14 days after initiation of therapy. Keep prescriber informed.
- Monitor patients with impaired cardiac or pulmonary function closely for cardiovascular or respiratory symptoms such as tachycardia, BP changes, feeling of suffocation.
- Monitor for and report symptoms of allergy and allergic pleural effusion: Shortness of breath, pleuritic pain, dry cough.
- Alert prescriber if improvement is not evident within 45 days. Drug may be discontinued because of the possibility of hepatotoxicity (see Appendix F).
- Lab tests: Baseline and periodic LFTs, blood cell counts, and renal function tests.
- Monitor bowel function. Persistent diarrhea may necessitate drug withdrawal. Severe constipation with abdominal distention and signs of intestinal obstruction have been reported.

Patient & Family Education

- Report promptly the onset of jaundice: Yellow skin or sclerae; dark urine, clay-colored stools, itching, abdominal discomfort. Hepatotoxicity frequently occurs between 3rd and 12th mo of therapy.
- Do not drive or engage in other potentially hazardous activities until response to drug is known.
- Do not use OTC medications, alcoholic beverages, or other CNS depressants unless otherwise advised by prescriber. Liver toxicity occurs more commonly when other drugs are taken concurrently.

DAPSONE ⊘

(dap'sone)

Aczone, Avlosulfon ✦
Classification: ANTILEPROSY (SULFONE)
Therapeutic: ANTILEPROSY
Pregnancy Category: C

AVAILABILITY 25 mg, 100 mg tablets; 5% gel

ACTION & THERAPEUTIC EFFECT
Sulfone derivative chemically related to sulfonamides, with bacteriostatic and bactericidal activity similar to that group. Interferes with bacterial cell growth by competitive inhibition of folic acid synthesis by susceptible organisms. It also interferes with alternative pathways of complement system. *Effective against dapsone-sensitive multibacillary (borderline, borderline lepromatous, or lepromatous) leprosy, and dapsone-sensitive paucibacillary (indeterminate, tuberculoid, or borderline tuberculoid) leprosy. Gel form is effective against acne vulgaris.*

USES All forms of leprosy. Used in dapsone-sensitive multibacillary leprosy (with clofazimine and rifampin) and in dapsone-sensitive paucibacillary leprosy (with rifampin, clofazimine, or ethionamide). Also used prophylactically in contacts of patients with all forms of leprosy except tuberculoid and indeterminate leprosy. Used for treatment of dermatitis herpetiformis. Gel used for acne vulgaris.
UNLABELED USES Chemoprophylaxis of malaria (with pyrimethamine), systemic and discoid lupus erythematosus, pemphigus vulgaris, dermatosis (especially those associated with bullous eruptions, mucocutaneous lesions, inflammation or pustules); rheumatoid

Common adverse effects in *italic*, life-threatening effects underlined; generic names in **bold**; classifications in SMALL CAPS; ✦ Canadian drug name; ⊘ Prototype drug

399

arthritis, allergic vasculitis; treatment of initial episodes of *P. carinii* pneumonia (with trimethoprim) in limited number of adults with AIDS.

CONTRAINDICATIONS Hypersensitivity to sulfones or its derivatives; advanced renal amyloidosis, anemia, methemoglobin reductase deficiency. **CAUTIOUS USE** Sulfonamide hypersensitivity; chronic renal, hepatic, pulmonary, or cardiovascular disease, refractory anemias, albuminuria, G6PD deficiency; pregnancy (category C), lactation.

ROUTE & DOSAGE

Tuberculoid and Indeterminate-Type Leprosy

Adult: **PO** 100 mg/day (with 6 mo of rifampin 600 mg/day) for a minimum of 3 y

Lepromatous and Borderline Lepromatous Leprosy

Adult: **PO** 100 mg/day for 10 y or more
Child: **PO** 1–2 mg/kg/day once daily in combination therapy (max: 100 mg/day)

Dermatitis Herpetiformis

Adult: **PO** 50 mg/day, may be increased to 300 mg/day if necessary (max: 500 mg/day)

Prophylaxis for Close Contacts of Patient with Multibacillary Leprosy

Adult: **PO** 50 mg/day
Child: **PO** *Younger than 6 mo,* 6 mg 3 times/wk; *6–23 mo,* 12 mg 3 times/wk; *2–5 y,* 25 mg 3 times/wk; *6–12 y,* 25 mg/day

P. carinii Pneumonia Prophylaxis

Adult: **PO** 50 mg b.i.d. or 100 mg daily
Child: **PO** 2 mg/kg once daily (max: 100 mg/day)

Acne

Apply pea-sized amount of gel to affected area b.i.d.

ADMINISTRATION

Oral

- Give with food to reduce possibility of GI distress.
- Store in tightly covered, light-resistant containers at 15°–30° C (59°–86° F). Drug discoloration apparently does not indicate a chemical change.

Topical

- Clean skin with soap and water before application.

ADVERSE EFFECTS (≥1%) **Body as a Whole:** Hypersensitivity (cutaneous reactions); erythema multiforme, exfoliative dermatitis, <u>toxic epidermal necrolysis</u> (rare), allergic rhinitis, urticaria, fever, infectious mononucleosis-like syndrome. **CNS:** Headache, nervousness, insomnia, vertigo; paresthesia, *muscle weakness.* **CV:** Tachycardia. **GI:** Anorexia, nausea, vomiting, abdominal pain; <u>toxic hepatitis</u>, cholestatic jaundice (reversible with discontinuation of drug therapy); increased ALT, AST, LDH; hyperbilirubinemia. **Hematologic:** In patient with or without G6PD deficiency; *dose-related hemolysis,* Heinz body formation, *methemoglobinemia with cyanosis,* hemolytic anemia; <u>aplastic anemia</u> (rare), <u>agranulocytosis</u>. **Skin:** Drug-induced lupus erythematosus, phototoxicity. **Special Senses:** Blurred vision, tinnitus. **Other:** Male infertility; sulfone syndrome (fever, malaise, exfoliative dermatitis, hepatic necrosis with jaundice, lymphadenopathy, methemoglobinemia, anemia).

INTERACTIONS Drug: Activated charcoal decreases dapsone absorption and enterohepatic circulation;

pyrimethamine, trimethoprim increase risk of adverse hematologic reactions; **rifampin** decreases dapsone levels 7–10 fold.

PHARMACOKINETICS Absorption: Rapidly and nearly completely absorbed from GI tract. **Peak:** 2–8 h. **Distribution:** Distributed to all body tissues; high concentrations in kidney, liver, muscle, and skin; crosses placenta; distributed into breast milk. **Metabolism:** In liver by CYP3A4. **Elimination:** 70–85% in urine; remainder in feces; traces of drug may be found in body for 3 wk after repeated doses. **Half-Life:** 20–30 h.

NURSING IMPLICATIONS

Assessment & Drug Effects

- Monitor for therapeutic effectiveness that may not appear for leprosy until after 3–6 mo of therapy.
- Lab tests: Baseline then weekly CBC during the first month of therapy, at monthly intervals for at least 6 mo, and semiannually thereafter.
- Determine periodic dapsone blood levels.
- Perform liver function tests in patients who complain of malaise, fever, chills, anorexia, nausea, vomiting, and have jaundice.
- Monitor severity of anemia. Nearly all patients demonstrate hemolysis.
- Monitor temperature during first few weeks of therapy. If fever is frequent or severe, leprosy reactional state should be ruled out.

Patient & Family Education

- Report symptoms of leprosy that do not improve within 3 mo or get worse to prescriber.
- Report the appearance of a rash with bullous lesions around elbows and other joints promptly. Drug-induced or worsening skin lesions require withdrawal of dapsone.
- Report symptoms of peripheral neuropathy with motor loss (muscle weakness) promptly.

DAPTOMYCIN

(dap-to-my'sin)

Cubicin

Classifications: ANTIBIOTIC; LIPOPEPTIDE
Therapeutic: ANTIBIOTIC
Pregnancy Category: B

AVAILABILITY 500 mg vial

ACTION & *THERAPEUTIC EFFECT*
Daptomycin is cyclic lipopeptide antibiotic. It binds to bacterial membranes of gram-positive bacteria causing rapid depolarization of the membrane potential leading to inhibition of protein, DNA, and RNA synthesis and bacterial cell death. *Daptomycin is effective against a broad spectrum of gram-positive organisms, including both susceptible and resistant strains of S. aureus.*

USES Complicated skin and skin structure infections, bacteremia.
UNLABELED USES Vancomycin-resistant enterococci.

CONTRAINDICATIONS Pseudomembranous colitis; myopathy; eosinophilic pneumonia.
CAUTIOUS USE Severe renal or hepatic impairment, end-stage renal failure; peripheral neuropathy; GI disease; history of rhabdomyolysis, or myopathy; older adults; pregnancy (category B), lactation. Safe use in infants or children younger than 18 y is not established.

ROUTE & DOSAGE

Skin Infections
Adult: **IV** 4 mg/kg q24h × 7–14 days

Common adverse effects in *italic*, life-threatening effects underlined; generic names in **bold**; classifications in SMALL CAPS; ♦ Canadian drug name; ⊘ Prototype drug

401

D

Bacteremia (S. aureus)

Adult: IV 6 mg/kg × 2–6 wk

Renal Impairment Dosage Adjustment

CrCl less than 30 mL/min: administer q48h

Hemodialyis Dosage Adjustment
Dose by CrCl, administer after dialysis

ADMINISTRATION

Intravenous

PREPARE: **IV Direct:** Reconstitute the 250 mg vial or the 500 mg vial with 5 mL or 10 mL, respectively, of NS to yield 50 mg/mL. **IV Infusion:** Further dilute the 50 mg/mL solution in 50–100 mL of NS.

ADMINISTER: **Direct:** Inject over 2 min. **IV Infusion:** Infuse over 30 min; if same IV line is used for infusion of other drugs, flush line before/after with NS.

INCOMPATIBILITIES **Solution/ additive:** **Dextrose**-containing solutions. **Y-site: Acyclovir, alemtuzumab, allopurinol, amphotericin B, amphotericin B liposomal, cytarabine, dantrolene, gemcitabine, gemtuzumab, ifosfamide, imipenem/cilastin, methotrexate, metronidazole, minocycline, mitomycin, nesiritide, nitroglycerin, pantoprazole, pentazocine, pentobarbital, phenytoin, quinidine, remifentanil, streptozocin, sulfentanil, thiopental, vancomycin.**

- Store unopened vials in 2°–8° C (36°–46° F). Avoid excessive heat.
- May store reconstituted, single-use vials or IV solution for 12 h at room temperature or 48 h if refrigerated.

ADVERSE EFFECTS (≥1%) **Body as a Whole:** Injection site reactions, fever, fungal infections. **CNS:** Headache, insomnia, dizziness. **CV:** Hypotension, hypertension. **GI:** Constipation, nausea, vomiting, diarrhea, abnormal liver function tests. **Hematologic:** Anemia, eosinophilia, leukocytosis, thrombocytopenia. **Metabolic:** Elevated CPK. **Musculoskeletal:** Limb pain, arthralgia. **Respiratory:** Dyspnea. **Skin:** Rash, pruritus. **Urogenital:** UTIs, renal failure.

INTERACTIONS Drug: Significant reactions not identified.

PHARMACOKINETICS Elimination: Primarily renal. **Half-Life:** 8 h.

NURSING IMPLICATIONS

Assessment & Drug Effects

- Monitor for and report: Muscle pain or weakness, especially with concurrent therapy with HMG-CoA reductase inhibitors (statin drugs); S&S of peripheral neuropathy, superinfection such as candidiasis.
- Lab tests: Baseline C&S before treatment is begun; baseline renal function tests; weekly CPK levels; PT/INR during first few days of daptomycin therapy with concurrent warfarin use; daily blood glucose monitoring in diabetics; serum electrolytes if S&S of hypokalemia or hypomagnesemia (see Appendix F) appear.
- Withhold drug and notify prescriber if S&S of myopathy develop with CPK elevation greater than 1000 units/L (~5 × ULN), or if CPK level is 10 × ULN or greater.

Patient & Family Education

- Report any of the following to the prescriber: Muscle pain, weakness or unusual tiredness; numbness or tingling; difficulty breathing or shortness of breath; severe diarrhea or vomiting; skin rash or itching.

Common adverse effects in *italic*, life-threatening effects underlined; generic names in **bold**, classifications in SMALL CAPS; ✤ Canadian drug name; ⊘ Prototype drug

DARBEPOETIN ALFA
(dar-be-po-e'tin)

ARANESP

Classifications: BLOOD FORMER; ERYTHROPOIESIS-STIMULATING AGENT

Therapeutic: ANTIANEMIC

Prototype: Epoetin alfa

Pregnancy Category: C

AVAILABILITY 25 mcg/mL, 40 mcg/mL, 60 mcg/mL, 100 mcg/mL, 150 mcg/mL, 200 mcg/mL, 300 mcg/mL, 500 mcg/mL vials; 40 mcg/0.4 mL, 60 mcg/0.3 mL, 100 mcg/0.5 mL, 150 mcg/0.3 mL, 200 mcg/0.4 mL, 300 mcg/0.6 mL, 500 mcg/mL syringe

ACTION & *THERAPEUTIC EFFECT*
An erythropoiesis-stimulating protein that stimulates red blood cell production in the bone marrow in response to hypoxia. Production of endogenous erythropoietin is impaired in patients with chronic renal failure (CRF) resulting in anemia. *Darbepoetin stimulates release of reticulocytes from the bone marrow into the blood stream where they mature into RBCs.*

USES Treatment of anemia in patients with chronic renal failure or chemotherapy-associated anemia, treatment of chemotherapy-induced anemia in nonmyeloid malignancies.

CONTRAINDICATIONS Patients with uncontrolled hypertension; serious hypersensitivity to darbepoetin or human albumin; antibody-mediated anemia due to anti-erythropoietin antibodies; pure red cell aplasia that begins after treatment with darbepoetin alfa or other related drugs.

CAUTIOUS USE Controlled hypertension, elevated hemoglobin, folic acid or vitamin B_{12} deficiencies, hematologic diseases; infections, inflammatory or malignant processes, osteofibrosis, occult blood loss, hemolysis, severe aluminum toxicity, bone marrow fibrosis, chronic renal failure patients not on dialysis; pregnancy (category C), lactation.

ROUTE & DOSAGE

Anemia

Adult: **IV/Subcutaneous** Initially, 0.45 mcg/kg once/wk. Reduce dose by 25% if there is a rapid increase (i.e., more than 1 g/dL in any 2-wk period) in Hgb or if the Hgb is approaching 12 g/dL. If the Hgb does not increase by 1 g/dL after 4 wk of therapy and iron stores are adequate, increase the dose by 25%. Maintenance dose is 0.26–0.65 mcg/kg once/wk.

Converting Epoetin Alfa to Darbepoetin

Adults: **IV/Subcutaneous** Estimate the starting dose of darbepoetin alfa based on the total weekly dose of epoetin alfa at the time of conversion. If the patient was receiving epoetin alfa 2–3 times/wk, administer darbepoetin alfa once/week; if the patient was receiving epoetin alfa once/week, administer darbepoetin alfa once every 2 wk. The route of administration (i.e., subcutaneous or IV) should be maintained.
Note: The following darbepoetin alfa dosage recommendations are estimates based on total amount of epoetin alfa administered per week. Because of individual variability, titrate doses to maintain the target Hgb.

Estimated Starting Dose (titrate to maintain target Hgb)

Common adverse effects in *italic*, life-threatening effects <u>underlined</u>; generic names in **bold**; classifications in SMALL CAPS; ♣ Canadian drug name; ⊙ Prototype drug

403

D

Previous weekly dose of epoetin alfa: 1500–2499 units/wk:
• Darbepoetin dose: 6.25 mcg/wk

Previous weekly dose of epoetin alfa: 2500–4999 units/wk:
• Darbepoetin dose: 10–12.5 mcg/wk

Previous weekly dose of epoetin alfa: 5000–10,999 units/wk:
• Darbepoetin dose: 20–25 mcg/wk

Previous weekly dose of epoetin alfa: 11,000–17,999 units/wk:
• Darbepoetin dose: 40 mcg/wk

Previous weekly dose of epoetin alfa: 18,000–33,999 units/wk:
• Darbepoetin dose: 60 mcg/wk

Previous weekly dose of epoetin alfa: 34,000–89,999 units/wk:
• Darbepoetin dose: 100 mcg/wk

Previous weekly dose of epoetin alfa: Greater than 90,000 units/wk:
• Darbepoetin dose: 200 mcg/wk

ADMINISTRATION

All Routes

- Correct deficiencies of folic acid or vitamin B_{12} prior to initiation of therapy.

Subcutaneous

- Do not shake solution. Shaking may denature the darbepoetin, rendering it biologically inactive.
- Inspect solution for particulate matter prior to use. Do not use if solution is discolored or if it contains particulate matter.
- Use only one dose per vial, and do not reenter vial.
- Do not give with any other drug solution.

Intravenous

PREPARE: **Direct:** Without shaking vial, withdraw the desired dose. Do not dilute. ▪ Discard the unused portion.

ADMINISTER: **Direct:** Give direct IV as a bolus dose over 1 min. ▪ Discard any unused portion of the vial. It contains no preservatives.

ADVERSE EFFECTS (≥1%) Body as a Whole: Injection site pain, *peripheral edema,* fatigue, fever, <u>death</u>, chest pain, fluid overload, access infection, access hemorrhage, flu-like symptoms, asthenia, *infection.* **CNS:** *Headache,* dizziness. **CV:** *Hypertension, hypotension, arrhythmias,* <u>cardiac arrest</u>, angina, chest pain, vascular access thrombosis, CHF, red cell aplasia. **GI:** *Nausea, vomiting, diarrhea,* constipation. **Musculoskeletal:** *Myalgia, arthralgia,* limb pain, back pain. **Respiratory:** *Upper respiratory infection, dyspnea, cough,* bronchitis. **Skin:** Pruritus. **Other:** Increased risk of thrombotic events and mortality in cancer patients.

INTERACTIONS Drug: No clinically significant reactions reported.

PHARMACOKINETICS Absorption: 37% absorbed from **subcutaneous** site. **Peak:** 24–72 h **Subcutaneous**. **Distribution:** Distribution confined primarily to intravascular space. **Elimination:** 10% in urine. **Half-Life:** 21 h **IV**, 49 h **Subcutaneous**.

NURSING IMPLICATIONS

Assessment & Drug Effects

- Control BP adequately prior to initiation of therapy and closely monitor and control during therapy. Report immediately S&S of CHF, cardiac arrhythmias, or sepsis. Note that hypertension is an adverse effect that **must be** controlled.
- Notify prescriber of a rapid rise in Hgb as dosage will need to be reduced because of risk of serious

hypertension. Note that BP may rise during early therapy as Hgb increases.

- Monitor for premonitory neurologic symptoms (i.e., aura, and report their appearance promptly). The potential for seizures exists during periods of rapid Hgb increase (e.g., greater than 1.0 g/dL in any 2-wk period).
- Monitor closely and report immediately S&S of thrombotic events (e.g., MI, CVA, TIA), especially for patients with CRF.
- Lab tests: Baseline and periodic iron stores, including transferrin and serum ferritin; Hgb twice weekly until stabilized and maintenance dose is established, then weekly for at least 4 wk, and at regular intervals thereafter; periodic CBC with differential and platelet count; periodic BUN, creatinine, serum phosphorus, and serum potassium.

Patient & Family Education

- Adhere closely to antihypertensive drug regimen and dietary restrictions.
- Monitor BP as directed by prescriber.
- Do not drive or engage in other potentially hazardous activity during the first 90 days of therapy because of possible seizure activity.
- Report any of the following to the prescriber: Chest pain, difficulty breathing, shortness of breath, severe or persistent headache, fever, muscle aches and pains, or nausea.

DARIFENACIN HYDROBROMIDE

(dar-i-fen′a-sin)

Enablex

Classifications: ANTICHOLINERGIC; MUSCARINIC RECEPTOR ANTAGONIST; BLADDER ANTISPASMODIC
Therapeutic: BLADDER ANTISPASMODIC
Prototype: Oxybutin
Pregnancy Category: C

AVAILABILITY 7.5 mg and 15 mg extended release tablets

ACTION & THERAPEUTIC EFFECT
Darifenacin is a selective M_3 muscarinic receptor antagonist. Muscarinic M_3 receptors play an important role in contraction of the urinary bladder smooth muscle and stimulation of salivary secretion. *Control of urinary incontinence due to urgency and frequency.*

USES Treatment of overactive bladder with symptoms of urge urinary incontinence, urgency, and frequency.

CONTRAINDICATIONS Hypersensitivity to darifenacin; severe hepatic impairment (Child-Pugh C class); urinary retention; gastric retention; pyloric stenosis; ileus; urinary retention; uncontrolled narrow-angle glaucoma.
CAUTIOUS USE Risk of urinary retention, clinically significant bladder outflow obstruction, renal disease; mild to moderate hepatic impairment; decreased GI motility, GERD, severe constipation, ulcerative colitis; myasthenia gravis; controlled narrow-angle glaucoma; pregnancy (category C), lactation.

ROUTE & DOSAGE

Overactive Bladder
Adult: **PO** 7.5–15 mg daily

Common adverse effects in *italic*, life-threatening effects underlined; generic names in **bold**; classifications in SMALL CAPS; ♣ Canadian drug name; ☻ Prototype drug

405

Moderate Hepatic Impairment (Child-Pugh B Class) Dosage Adjustment

Max dose: 7.5 mg daily

ADMINISTRATION

Oral

- Ensure that the drug is not chewed or crushed. It **must be** swallowed whole.
- Note: Dosage should not exceed 7.5 mg daily with moderate hepatic impairment (i.e., Child-Pugh B class) or concurrent therapy with potent inhibitors of CYP3A4 (e.g., itraconazole, clarithromycin, nefazodone, nelfinavir, ritonavir).
- Store 15°–30° C (59°–86° F). Protect from light.

ADVERSE EFFECTS (≥1%) **Body as a Whole:** Flu-like symptoms, urinary tract infection, angioedema **CNS:** *Headache,* asthenia, dizziness. **GI:** *Constipation, dry mouth, dyspepsia, nausea,* abdominal pain, diarrhea.

INTERACTIONS Drug: Potent inhibitors of CYP3A4 (e.g., **clarithromycin, erythromycin, itraconazole, ketoconazole, nefazodone, nelfinavir,** and **ritonavir**) increase darifenacin levels. Darifenacin will cause additive anticholinergic effects with other ANTICHOLINERGIC drugs. Darifenacin can increase **digoxin** concentrations. **Food: Grapefruit juice** may increase darifenacin levels.

PHARMACOKINETICS Absorption: 15–19% bioavailability. **Peak:** 7 h. **Distribution:** 98% protein bound. **Metabolism:** Extensive hepatic metabolism. **Elimination:** Renal and fecal. **Half-Life:** 13–19 h.

NURSING IMPLICATIONS

Assessment & Drug Effects

- Monitor for adverse effects of concurrently used drugs that have a narrow therapeutic window and are metabolized by CYP26D (e.g., flecainide, thioridazine, or TRICYCLIC ANTIDEPRESSANTS).
- Lab tests: Monitor digoxin levels with concurrent therapy and assess for S&S of digoxin toxicity.

Patient & Family Education

- Do not drive or engage in potentially hazardous activities until response to drug is known.
- Use caution in hot environments to minimize the risk of heat prostration.
- Report any of the following to a health care provider: Difficulty passing urine, unexplained nausea, or persistent constipation.

DARUNAVIR

(da-run′a-ver)

Prezista

Classifications: ANTIRETROVIRAL; PROTEASE INHIBITOR

Therapeutic: HIV PROTEASE INHIBITOR

Prototype: Saquinavir

Pregnancy Category: C

AVAILABILITY 75 mg, 150 mg, 400 mg, 600 mg tablets

ACTION & *THERAPEUTIC EFFECT*

Darunavir is an inhibitor of HIV-1 protease that selectively inhibits the cleavage of HIV polyproteins in infected cells, thereby preventing the maturation of virus particles. *Darunavir reduces viral load (decreases the number of RNA copies) and increases the number of T helper CD4 cells.*

USES Treatment of HIV infection with other antiretroviral agents.

CONTRAINDICATIONS Hypersensitivity to darunavir or protease inhibitors, ritonavir; severe hepatic impairment; pancreatitis; new or worsening liver dysfunction; lactation.

CAUTIOUS USE Hepatic function impairment, hepatitis; severe renal impairment, chronic renal failure; hemophilia A or B; diabetes mellitus; diabetes ketoacidosis; hyperglycemia; older adults; pregnancy (category C).

ROUTE & DOSAGE

HIV Infection, Treatment Naive

Adult: **PO** 800 mg daily with 100 mg ritonavir PO

HIV Infection, Treatment Experienced

Adult: **PO** 600 mg b.i.d. OR 800 mg daily with 100 mg ritonavir PO
Adolescent/Child (weight greater than 40 kg): **PO** 600 mg b.i.d. with 100 mg ritonavir PO
Adolescent/Child (weight 30–40 kg): **PO** 450 mg b.i.d. with 60 mg ritonavir PO
Child (older than 6 y, weight 20–30 kg): **PO** 375 mg b.i.d. with 50 mg ritonavir PO

ADMINISTRATION

Oral

- Give with food and coadminister with 100 mg ritonavir.
- Tablets **must be** swallowed whole.
- Store at 15°–30° C (59°–86° F). Protect from light and moisture.

ADVERSE EFFECTS (≥1%) **CNS:** *Headache.* **GI:** Abdominal pain, constipation, *diarrhea, vomiting.* **Skin:** Stevens-Johnson syndrome. **Hepatic:** Acute hepatitis.

INTERACTIONS Drug: AZOLE ANTI-FUNGALS and **indinavir** increase the levels of darunavir. Coadministration of other inhibitors of CYP3A4 may also increase darunavir. ANTICON-VULSANTS (e.g., **carbamazepine, phenobarbital, phenytoin**), CORTICOSTEROIDS (e.g., **dexamethasone**), **efavirenz,** RIFAMYCINS (e.g., **rifampin, rifabutin**), and **saquinavir** may decrease darunavir levels. Darunavir may increase the levels of AZOLE ANTIFUNGALS, CORTICOSTEROIDS, **efavirenz, indinavir,** RIFAMYCINS, **amiodarone, bepridil, lidocaine, quinidine,** CALCIUM CHANNEL BLOCKERS (e.g., **nifedipine, nicardipine, felodipine**), **clarithromycin,** IMMUNOSUPPRESSANTS (e.g., **cyclosporine, sirolimus, tacrolimus**), PHOSPHODIESTE-RASE TYPE 5 INHIBITORS (e.g., **sildenafil, tadalafil, vardenafil**), and trazodone, due in part to its ability to inhibit CYP3A4. Darunavir decreases the levels of the **lopinavir/ritonavir** combination, ORAL CONTRACEPTIVES (e.g., **ethinyl estradiol, norethindrone**), **methadone,** SELECTIVE SEROTONIN REUPTAKE INHIBITORS [SSRIs (e.g., **paroxetine, sertraline**)], and **warfarin.** Use of BENZODIAZEPINES (e.g., **midazolam, triazolam**) increases the risk of prolonged or increased sedation or respiratory depression. Use of ERGOT ALKALOIDS may increase ergot toxicity. Coadministration with HMG COA REDUCTASE INHIBITORS increases the risk of myopathy. Combination use with **pimozide** increases the risk of cardiac arrhythmias. **Food:** Food enhances the bioavailability of darunavir. **Herbal:** St. John's wort decreases the level of darunavir.

PHARMACOKINETICS Absorption: 82% absorbed (in combination with ritonavir). **Peak:** 2.5–4 h. **Distribution:** 95% protein bound. **Metabolism:** In the liver. **Elimination:** Primarily fecal (80%) with minor elimination in urine. **Half-Life:** 15 h.

Common adverse effects in *italic*, life-threatening effects underlined; generic names in **bold**; classifications in SMALL CAPS; ✦ Canadian drug name; ⊘ Prototype drug

407

NURSING IMPLICATIONS

Assessment & Drug Effects

- Monitor for and report S&S of pancreatitis, as this may be an indication for discontinuation of darunavir.
- Monitor for S&S of skin rash. Notify prescriber immediately if a severe rash appears.
- Monitor diabetics for loss of glycemic control.
- Lab tests: Periodic CD4+ cell count, plasma HIV-RNA, lipid profile, LFTs; and plasma glucose.
- Increase monitoring of INR with concurrent warfarin therapy.
- Monitor for adverse effects or loss of efficacy of concurrent medications, as many drug interactions occur with darunavir.

Patient & Family Education

- Follow directions for taking the drug (see Administration). If a dose is missed by more than 6 h, wait until the next regularly scheduled dose. If a dose is missed by less than 6 h, take a dose and continue with the next regularly scheduled dose.
- Ensure that you know which medicines should NOT be taken with darunavir, as serious consequences could occur.
- Report any of the following to a health care provider: Blistering, redness, or peeling skin or mucus membranes; severe skin rash.
- Use or add a barrier contraceptive if using an estrogen-containing oral contraceptive if you wish to prevent pregnancy.

DASATINIB

(das-a′ti-nib)

Sprycel

Classifications: ANTINEOPLASTIC BIOLOGIC RESPONSE MODIFIER; TYROSINE KINASE INHIBITOR (TKI)

Therapeutic: ANTINEOPLASTIC; TKI
Prototype: Gefitinib
Pregnancy Category: D

AVAILABILITY 20 mg, 50 mg, 70 mg, 80 mg, 100 mg, 140 mg tablets

ACTION & THERAPEUTIC EFFECT Dasatinib is a BCR-ABL tyrosine kinase inhibitor. BCR-ABL tyrosine kinase is an enzyme produced by a chromosomal translocation associated with chronic myeloid leukemia (CML) and certain types of acute lymphocytic leukemias (Ph+ ALL). *Dasatinib inhibits the growth of CML and ALL cell lines overexpressing BCR-ABL kinase.*

USES Treatment of chronic, accelerated, or myeloid or lymphoid blast phase chronic myelogenous leukemia (CML). Treatment of Philadelphia chromosome–positive (Ph+) acute lymphocytic leukemia (ALL) in adults.

CONTRAINDICATIONS Hypersensitivity to dasatinib; active bleeding. hypokalemia; hypomagnesemia; pregnancy (category D); lactation.
CAUTIOUS USE Hepatic impairment; bacterial or viral infection; history of GI bleeding; interstitial pneumonia; pleural effusion; QT prolongation; cardiac dysfunction. Safe use in children younger than 18 y not established.

ROUTE & DOSAGE

CML and Philadelphia Chromosome-Positive ALL
Adult: **PO** Starting dose 70 mg b.i.d. May increase/decrease dose by 20 mg based on response.

Chronic Phase CML
Adult: **PO** 100 mg daily

Dosage Adjustments for Neutropenia and Thrombocytopenia

Chronic phase CML where absolute neutrophil count (ANC) less than 500/m³ and/or platelets less than 50,000/mm³

Step 1: DC dasatinib until the ANC 1000/mm³ or more and platelets 50,000/mm³ or more
Step 2: If cell recovery occurs in 7 days or less, restart at original dose
Step 3: If platelets less than 25,000 mm³ and/or recurrence of ANC less than 500/mm³, resume at dose of 80 mg daily
Accelerated phase CML, blast phase CML, and Ph+ ALL where ANC less than 500/mm³ and/or platelets less than 10,000/mm³

Step 1: Assure that cytopenia is unrelated to the underlying leukemia. If so, DC until ANC 1000/mm³ or more and platelets 20,000/mm³ or more.
Step 2: Resume at starting dose
Step 3: If cytopenia recurs, repeat step 1 and resume at 50 mg b.i.d. (second episode) or 40 mg b.i.d. (third episode)
Step 4: If cytopenia is related to the underlying leukemia, may increase to 180 mg daily

ADMINISTRATION

Oral

- Do not crush or break tablets. They should be swallowed whole.
- Ensure that hypokalemia and hypomagnesemia are corrected prior to administering dasatinib.
- Store at 15°–30° C (59°–86° F).

ADVERSE EFFECTS (≥1%) **Body as a Whole:** Chills, contusion, febrile neutropenia, *hemorrhage, infection,* malaise, *pain, pyrexia,* tumor lysis syndrome, weight gain or weight loss. **CNS:** *Asthenia,* anxiety, confusional state, CNS bleeding, depression, dizziness, dysgeusia, *fatigue, headache,* insomnia, neuropathy, somnolence, syncope, tremor, vertigo. **CV:** Arrhythmia, chest pain, angina, <u>congestive heart failure</u>, pericardial effusion, cardiomegaly, hypertension, hypotension, <u>myocardial infarction</u>, palpitations. **GI:** *Abdominal distention and pain,* anal fissure, *anorexia,* ascites, colitis, constipation, *diarrhea,* dysphagia, gastritis, *GI bleeding, nausea,* oral soft tissue disorder, *vomiting, mucosal inflammation.* **Hematologic:** Anemia, neutropenia, pancytopenia, <u>*thrombocytopenia*</u>, elevated ALT and AST, hypocalcemia, hypophosphatemia. **Metabolic:** Appetite disturbances, *fluid retention, edema,* hyperuricemia. **Musculoskeletal:** *Arthralgia, musculoskeletal pain,* muscle inflammation, myalgia, musculoskeletal stiffness. **Respiratory:** Asthma, *cough, dyspnea,* lung infiltration, *plural effusion,* pneumonia, pulmonary edema, pulmonary hypertension, *upper respiratory tract infection.* **Skin:** Acne, alopecia, dermatitis, dry skin, hyperhidrosis, nail disorder, photosensitivity reaction, pigmentation disorder, pruritus, *skin rash.* **Special Senses:** Conjunctivitis, dry eye, tinnitus. **Urogenital:** Gynecomastia, renal failure, urinary frequency.

INTERACTIONS Drug: Aluminum- and **magnesium**-based ANTACIDS decrease dasatinib absorption. AZOLE ANTIFUNGAL AGENTS (e.g., **ketoconazole, itraconazole**), MACROLIDE ANTIBIOTICS (e.g., **clarithromycin, erythromycin,**

Common adverse effects in *italic*, life-threatening effects <u>underlined</u>; generic names in **bold**; classifications in SMALL CAPS; ♣ Canadian drug name; ⊘ Prototype drug

409

telithromycin), HIV PROTEASE IN-HIBITORS (e.g., **indinavir, nelfinavir, ritonavir, saquinavir**), **nefazodone,** and other inhibitors of CYP3A4 may increase dasatinib levels. Compounds that induce CYP3A4 (e.g., **carbamazepine, dexamethasone, phenobarbital, phenytoin, rifampin**) may decrease dasatinib levels. PROTON PUMP INHIBITORS and H₂ ANTAGO-NISTS may reduce the absorption of dasatinib due to long-term suppression of gastric acid secretion. Dasatinib may alter the plasma concentrations of other drugs that require CYP3A4 and have a narrow therapeutic window (e.g., **cyclosporine,** ERGOT ALKALOIDS). Dasatinib increases the levels of **simvastatin**. **Food:** Food enhances the bioavailability of dasatinib. **Herbal:** St. **John's wort** may decrease the level of dasatinib.

PHARMACOKINETICS Peak: 0.5–6 h. **Distribution:** 93–96% protein bound. **Metabolism:** Extensive hepatic metabolism. **Elimination:** Fecal. **Half-Life:** 3–5 h.

NURSING IMPLICATIONS

Assessment & Drug Effects

- Monitor for and report S&S of fluid retention (e.g., pleural or pericardial effusion, peripheral or pulmonary edema, ascites).
- Monitor for S&S of cardiac dysfunction (e.g., heart failure, arrhythmias). ECG monitoring may be needed to evaluate potential QT interval prolongation.
- Monitor for numerous adverse side effects of dasatinib. Immediately report suspected bleeding or infection.
- Lab tests: Baseline and periodic serum potassium and magnesium; baseline CBC with differential (including ANC and platelet

count), weekly for first 2 mo, then monthly; periodic LFTs.

Patient & Family Education

- Take antacids (if needed for GI distress) 2 h before or after dasatinib.
- Do not use OTC medications for heartburn (other than antacids) without consulting prescriber.
- Inform your prescriber if you are pregnant or planning to become pregnant, as dasatinib may harm the fetus.
- Report immediately to your health care provider any of the following: Bleeding (including wine- or coke-colored urine, or black tarry stools) or easy bruising, fever or other signs of an infection, severe lethargy or weakness.

DAUNORUBICIN HYDROCHLORIDE

(daw-noe-roo'bi-sin)
Cerubidine

DAUNORUBICIN CITRATED LIPOSOMAL

DaunoXome
Classifications: ANTINEOPLASTIC; ANTHRACYCLINE; (ANTIBIOTIC)
Therapeutic: ANTINEOPLASTIC
Prototype: Doxorubicin HCl
Pregnancy Category: D

AVAILABILITY Daunorubicin HCl: 10 mg, 20 mg, 50 mg, 100 mg lyophilized vials; **Daunorubicin Citrated Liposomal:** 2 mg/mL (equivalent to 50 mg daunorubicin base) injection

ACTION & *THERAPEUTIC EFFECT*
Cytotoxic and antimitotic anthracycline antibiotic that is cell-cycle specific for S-phase of cell division. Has rapid interaction with the DNA molecule changing its shape, thus

resulting in inhibition of DNA, RNA, and protein synthesis. *Antineoplastic effects against acute leukemias with decreased incidence of cardiotoxicity than doxorubicin.*

USES To induce remission in acute nonlymphocytic/lymphocytic leukemia, advanced HIV-associated Kaposi's sarcoma.
UNLABELED USES Non-Hodgkin's lymphoma.

CONTRAINDICATIONS Severe myelosuppression; immunizations (patient, family), and preexisting cardiac disease unless risk-benefit is evaluated; uncontrolled systemic infection; pregnancy (category D), lactation.
CAUTIOUS USE History of gout, urate calculi, hepatic or renal function impairment; older adults with inadequate bone reserve due to age or previous cytotoxic drug therapy, tumor cell infiltration of bone marrow, patient who has received potentially cardiotoxic drugs or related antineoplastics.

ROUTE & DOSAGE

Neoplasms

Adult: **IV** *Younger than 60 y,* 45 mg/m²/day on days 1, 2, and 3 of first course then days 1 and 2 of subsequent courses (max total cumulative dose: 500–600 mg/m²); *60 y or older,* 30 mg/m²/day on days 1, 2, and 3 of first course then days 1 and 2 of subsequent courses
Child: **IV** As combination therapy, *2 y or older,* 25 mg/m² weekly; *younger than 2 y,* 1 mg/kg

Kaposi's Sarcoma (DaunoXome)

Adult: **IV** 40 mg/m² over 1 h, repeat q2wk (withhold therapy if granulocyte count less than 750 cells/mm³)

Renal Impairment Dosage Adjustment

If serum Cr greater than 3 mg/dL, give 50% of dose

Hepatic Impairment Dosage Adjustment

Total bilirubin 1.2–3 mg/dL: Give 50% of dose
Greater than 3–5 mg/dL: Give 25% of dose
Greater than 5 mg/dL: Omit dose

ADMINISTRATION

Intravenous

Use gloves during preparation for infusion to prevent skin contact with this drug. If contact occurs, decontaminate skin with copious amounts of water with soap.

Daunorubicin HCl

PREPARE: **Direct:** Reconstitute 20 mg vial with 4 mL sterile water for injection. The concentration of the solution will be 5 mg/mL. • Withdraw dose into syringe containing 10–15 mL normal saline. **IV Infusion:** Dilute further in 100 mL NS as required.
ADMINISTER: **Direct:** Inject over approximately 3 min into the tubing or side arm of a rapidly flowing IV infusion of D5W or NS. **Infusion:** Give a single dose over 30–45 min.

Specific to DaunoXome

PREPARE: **IV Infusion:** Each vial of DaunoXome contains the equivalent of 50 mg daunorubicin base. Dilute with enough D5W to produce a concentration of 1 mg/1 mL.
ADMINISTER: **IV Infusion:** Give DaunoXome over 60 min. Do not use a filter with DaunoXome.

Common adverse effects in *italic*, life-threatening effects underlined; generic names in **bold**; classifications in SMALL CAPS; ✦ Canadian drug name; ❂ Prototype drug

411

INCOMPATIBILITIES **Solution/ additive:** **Dexamethasone, heparin. Y-site: Allopurinol, aztreonam, cefepime, fludarabine, lansoprazole, piperacillin/ tazobactam.**

- Avoid extravasation because it can cause severe tissue necrosis.

- Store reconstituted solution at room temperature (15°–30° C; 59°–86° F) for 24 h and under refrigeration at 2°–8° C (36°–46° F) for 48 h. Protect from light.

ADVERSE EFFECTS (≥1%) **Body as a Whole:** Fever. **CNS:** Amnesia, anxiety, ataxia, confusion, hallucinations, emotional lability, tremors. **CV:** Pericarditis, myocarditis, arrhythmias, peripheral edema, CHF, hypertension, tachycardia. **GI:** *Acute nausea and vomiting* (mild), anorexia, *stomatitis,* mucositis, diarrhea (occasionally) hemorrhage. **Urogenital:** Dysuria, nocturia, polyuria, dry skin. **Hematologic:** <u>Bone marrow depression</u> <u>*thrombocytopenia,*</u> <u>*leukopenia,*</u> anemia, **Skin:** Generalized *alopecia* (reversible), transverse pigmentation of nails, severe cellulitis or tissue necrosis at site of drug extravasation. **Endocrine:** Hyperuricemia, gonadal suppression.

PHARMACOKINETICS Distribution: Highest concentrations in spleen, kidneys, liver, lungs, and heart; does not cross blood–brain barrier; crosses placenta. **Metabolism:** In liver to active metabolite. **Elimination:** 25% in urine, 40% in bile. **Half-Life:** 18.5–26.7 h.

NURSING IMPLICATIONS

Assessment & Drug Effects

- Monitor for therapeutic effectiveness. A profound suppression of bone marrow is required to induce a complete remission. Nadirs for thrombocytes and leukocytes are usually reached in 10–14 days.

- Monitor serum bilirubin; drug dose needs to be reduced when bilirubin is greater than 1.2 mg/dL.

- Lab tests: Baseline and periodic Hct, platelet count, total and differential leukocyte count, serum uric acid, LFTs, and renal function tests.

- Monitor BP, temperature, pulse, and respiratory function during treatment.

- Monitor for S&S of acute CHF. It can occur suddenly, especially when total dosage exceeds 550 mg/m², or in patients with compromised heart function because of previous radiation therapy to heart area.

- Report immediately: Breathlessness, orthopnea, change in pulse and BP parameters. Early clinical diagnosis of drug-induced CHF is essential for successful treatment.

- Report promptly S&S of superinfections including elevation of temperature, chills, upper respiratory tract infection, tachycardia, overgrowth with opportunistic organisms because myelosuppression imposes risk of superimposed infection (see Appendix F).

- Protect patient from contact with persons with infections. The most hazardous period is during nadirs of thrombocytes and leukocytes.

- Control nausea and vomiting (usually mild) by antiemetic therapy.

- Inspect oral membranes daily. Mucositis may occur 3–7 days after drug is administered.

Patient & Family Education

- Note: Loss of hair is probable; recovery is usual in 6–10 wk.

- Use barrier contraceptives during treatment because this drug is teratogenic. Tell your prescriber immediately if you become pregnant during therapy.

Common adverse effects in *italic*, life-threatening effects <u>underlined</u>; generic names in **bold**; classifications in SMALL CAPS; ✿ Canadian drug name; ⊙ Prototype drug

• Note: A transient effect of the drug is to turn urine red on the day of infusion.

DECITABINE

(de-sit′a-bine)

Dacogen

Classifications: ANTINEOPLASTIC; ANTIMETABOLITE; PYRIMIDINE

Therapeutic: ANTINEOPLASTIC

Prototype: 5-Fluorouracil

Pregnancy Category: D

AVAILABILITY 50 mg lyophilized powder for injection

ACTION & *THERAPEUTIC EFFECT*
Decitabine is an antimetabolite that exerts antineoplastic effects after its direct incorporation into DNA and inhibition of DNA transferase, causing loss of cell differentiation and cell death. Nonproliferating cells are resistant to the effects of decitabine. *Decitabine-induced changes in neoplastic cells may restore normal function to genes that are critical for control of cellular differentiation and proliferation.*

USES Treatment of patients with myelodysplastic syndrome (MDS).

UNLABELED USES Treatment of chronic myelogenous leukemia (CML).

CONTRAINDICATIONS Hypersensitivity to decitabine; conception within 2 mo of drug use; renal failure patients with CrCl less than 2 mg/mL; liver dysfunction with transaminase greater than 2 × upper limit of normal (ULN), or serum bilirubin greater than 1.5 mg/dL; active infection; pregnancy (category D), lactation.

CAUTIOUS USE Moderate to severe renal failure; hepatic impairment. Safety and effectiveness in children not established.

ROUTE & DOSAGE

Myelodysplastic Syndrome

Adult: **IV** 15 mg/m² q8h × 3 days; repeat q6w for at least 4 cycles.

ADMINISTRATION

Intravenous

***PREPARE:* IV Infusion:** Caution should be exercised when handling and preparing decitabine. Procedures for proper handling and disposal of antineoplastic drugs should be applied. • Reconstitute each vial with 10 mL sterile water for injection to yield approximately 5 mg/mL at pH 6.7–7.3. Immediately after reconstitution, further dilute with NS, D5W, or LR to a final drug concentration of 0.1–1 mg/mL. • Use within 15 min of reconstitution (see Storage).

***ADMINISTER:* IV infusion:** Premedicate with standard antiemetic therapy. Give decitabine over 3 h. • **NOTE:** Withhold dose and notify prescriber of any of the following: Absolute neutrophil count (ANC) less than 1000/mcL; platelet count less than 50,000/mcL; serum creatinine at 2 mg/dL or higher; ALT, total bilirubin 2 × ULN or more; or an active or uncontrolled infection.

• Store vials at 15°–30° C (59°–86° F). Unless used within 15 min of reconstitution, the diluted solution **must be** prepared using cold (2°–8° C) infusion fluids and stored at 2°–8° C (36°–46° F) for up to a maximum of 7 h until administration.

ADVERSE EFFECTS (≥5 %) Body as a Whole: *Fatigue, pyrexia, Mycobacterium avium* complex infection, peripheral edema,

Common adverse effects in *italic*, life-threatening effects underlined; generic names in **bold**; classifications in SMALL CAPS; ✤ Canadian drug name; ❷ Prototype drug

413

bacteremia, candidal infection, cellulitis, injection site reactions, rigors, tenderness, transfusion reaction, sinusitis, staphylococcal infection. **CNS:** <u>Intracranial hemorrhage</u>, anxiety, confusion, dizziness, headache, hypesthesia, insomnia, pyrexia. **CV:** <u>Cardiorespiratory arrest</u>, cardiac murmur, hypotension. **GI:** *Nausea, vomiting, constipation, diarrhea,* abdominal distention and discomfort, anorexia, dyspepsia, gastroesophageal reflux disease, glossodynia, gingival bleeding, hemorrhoids, lip ulceration, stomatitis, tongue ulceration. **Hematologic:** *Anemia, neutropenia,* <u>*thrombocytopenia,*</u> hematoma, <u>leukopenia</u>, lymphadenopathy, thrombocythemia. **Metabolic:** *Hyperglycemia,* increased AST, decreased blood albumin, increased blood alkaline phosphatase, altered blood bicarbonate, decreased blood bilirubin, decreased blood chloride, increased blood lactate dehydrogenase, increased blood urea, decreased total protein, dehydration, hyperbilirubinemia, altered potassium levels, hypoalbuminemia, hypomagnesemia, hyponatremia. **Musculoskeletal:** Arthralgia, back pain, chest wall pain, musculoskeletal discomfort, myalgia, pain in limb. **Respiratory:** *Cough,* lung crackles, hypoxia, pharyngitis, pneumonia, pulmonary edema, rales. **Skin:** Alopecia, ecchymosis, erythema, pallor, *petechiae,* pruritus, rash, skin lesion, swelling face, urticaria. **Special Senses:** Blurred vision. **Urogenital:** Dysuria, urinary frequency, urinary tract infection.

PHARMACOKINETICS Distribution: Negligible plasma protein binding. **Half-Life:** 0.2–0.8 h.

NURSING IMPLICATIONS

Assessment & Drug Effects

- Monitor for and report S&S of pulmonary or peripheral edema, cardiac arrhythmias, new-onset depression, or infection.
- Lab tests: CBC with differentials and platelet count prior to each chemotherapy cycle; baseline and periodic LFTs and serum creatinine.
- Avoid IM injections with platelet counts less than 50,000/mcL.
- Monitor diabetics for loss of glycemic control.

Patient & Family Education

- Do not accept vaccinations during treatment with decitabine.
- Avoid contact with anyone who recently received the oral poliovirus vaccine.
- Women of childbearing age should avoid becoming pregnant while receiving decitabine.
- Men should not father a child while receiving decitabine and for 2 mo after the end of therapy.
- Report any of the following to a health care provider: Signs of infection such as fever, chills, sore throat; signs of bleeding such as easy bruising, black, tarry stools, blood in the urine; irregular heart rate; significant tiredness or weakness.

DEFEROXAMINE MESYLATE
(de-fer-ox′a-meen)
Desferal
Classifications: CHELATING AGENT; ANTIDOTE
Therapeutic: ANTIDOTE
Pregnancy Category: C

AVAILABILITY 500 mg, 2 g powder for injection

Common adverse effects in *italic*, life-threatening effects <u>underlined</u>; generic names in **bold**; classifications in SMALL CAPS; ♣ Canadian drug name; ☻ Prototype drug

ACTION & *THERAPEUTIC EFFECT*

Chelating agent with specific affinity for ferric ion and low affinity for calcium. Binds ferric ions to form a stable water soluble chelate readily excreted by kidneys. *Main effect is removal of iron from ferritin, hemosiderin, and transferrin in iron toxicity.*

USES Adjunct in treatment of acute iron intoxication or iron overload.
UNLABELED USES To promote aluminum excretion.

CONTRAINDICATIONS Severe renal disease, anuria, pyelonephritis; primary hemochromatosis; acute infection.
CAUTIOUS USE History of pyelonephritis; infants and children younger than 3 y; elderly, cardiac dysfunction; pregnancy (category C), lactation.

ROUTE & DOSAGE

Acute Iron Intoxication

Adult: **IM/IV** 1 g followed by 500 mg at 4 h intervals for 2 doses, subsequent doses of 500 mg q4–12h may be given if necessary (max: 6 g/24 h), infuse at 15 mg/kg/h or less
Child (older than 3 y): **IV** 15 mg/kg/h (max: 6 g/24 h) **IM** 40–90 mg/kg (up to 1 g) q4–8h

Chronic Iron Overload

Adult: **IM** 500 mg–1 g/day
Subcutaneous 1–2 g/day (20–40 mg/kg/day) infused over 8–24 h
Child (older than 3 y): **IM** 500 mg–1 g/day **Subcutaneous** 20–40 mg/kg/day over 8–12 h

ADMINISTRATION

Subcutaneous

- Reconstitute by adding 5 mL sterile water for injection to each 500

mg vial or 20 mL to each 2 gram vial to yield 100 mg/mL. Dissolve completely.
- Give subcutaneously over 8–24 h using a portable minipump device.

Intramuscular

- Reconstitute by adding 2 mL sterile water for injection to 500 mg vial or 8 mL to the 2 g vial to yield 250 mg/mL. Dissolve completely.
- Use IM route for all patients not in shock; preferred route for acute intoxication.

Intravenous

For infants and children: Verify correct IV concentration and rate with prescriber.

PREPARE: **IV Infusion:** Reconstitute by adding 5 mL sterile water for injection to 500 mg vial to yield 100 mg/mL. ▪ After drug is completely dissolved, withdraw prescribed amount from vial and add to NS, D5W, or LR solution.
ADMINISTER: **IV Infusion:** *Adult:* Give initial dose at a rate not to exceed 15 mg/kg/h; give two subsequent 500 mg dose at 125 mg/h; give any additional doses over 4–12 h. *Child:* Give at 15 mg/kg/h. ▪ Do not infuse IV rapidly; such infusion is associated with the occurrence of more adverse effects.
INCOMPATIBILITIES **Solution/admixture: Iron dextran.**

- Store at room temperature 15°–30° C (59°–86° F) for not longer than 1 wk. Protect from light.

ADVERSE EFFECTS (≥1%) **Body as a Whole:** Hypersensitivity (generalized itching, cutaneous wheal formation, rash, fever, <u>anaphylactoid reaction</u>). **CV:** Hypotension, tachycardia. **Special Senses:** Decreased hearing; blurred vision, decreased visual acuity and visual fields, color vision abnormalities, night blindness,

Common adverse effects in *italic*, life-threatening effects <u>underlined</u>; generic names in **bold**; classifications in SMALL CAPS; ✤ Canadian drug name; ⊘ Prototype drug

415

retinal pigmentary degeneration. **GI:** Abdominal discomfort, diarrhea. **Urogenital:** Dysuria, exacerbation of pyelonephritis, orange-rose discoloration of urine. **Other:** *Pain and induration at injection site.*

INTERACTIONS Drug: Use with **ascorbic acid** increases cardiac risk, **prochlorperazine** may cause loss of consciousness.

PHARMACOKINETICS Distribution: Widely distributed in body tissues. **Metabolism:** Forms nontoxic complex with iron. **Elimination:** Primarily in urine; some in feces.

NURSING IMPLICATIONS

Assessment & Drug Effects

- Lab tests: Baseline kidney function tests prior to drug administration.
- Monitor injection site. If pain and induration occur, move infusion to another site.
- Monitor I&O ratio and pattern. Report any change. Observe stools for blood (iron intoxication frequently causes necrosis of GI tract).
- Note: Periodic ophthalmoscopic (slit lamp) examinations and audiometry are advised for patients on prolonged or high-dose therapy for chronic iron overload.

Patient & Family Education

- Deferoxamine chelate makes urine turn a reddish color.
- Report blurred vision or any other visual abnormality.

DEGARELIX ACETATE

(de-ga're-lix)

Firmagon
Classification: GONADOTROPIN-RELEASING HORMONE (GNRH) ANTAGONIST
Therapeutic: GNRH ANTAGONIST

Prototype: Ganirelix acetate
Pregnancy Category: X

AVAILABILITY 80 mg, 240 mg powder for injection

ACTION & *THERAPEUTIC EFFECT*
A gonadotropin-releasing hormone (GnRH) receptor antagonist. It binds reversibly to pituitary GnRH receptors, reducing release of gonadotropins and, consequently, testosterone. *Testosterone suppression slows growth of androgensensitive prostate cancer cells as indicated by decrease in PSA values.*

USES Treatment of advanced prostate cancer.

CONTRAINDICATIONS Hypersensitivity to degarelix; women who are or may become pregnant (category X); lactation.
CAUTIOUS USE Congenital long QT syndrome; electrolyte abnormalities; CHF; elderly. Safety and effectiveness in children have not been established.

ROUTE & DOSAGE

Prostate Cancer

Adult: **Subcutaneous** Initial dose of 240 mg in two 120 mg injections, followed by 80 mg q28days

ADMINISTRATION

Subcutaneous

- Glove should be worn for reconstitution. Follow carefully manufacturer's guidelines for reconstitution of the powder vial.
- Administer into the abdominal area within 1 h of reconstitution.
- Initial dose: Give as 2 subcutaneous injections of 120 mg each (at a concentration of 40 mg/mL).

- Maintenance dose: Give 1 subcutaneous injection of 80 mg (at a concentration of 20 mg/mL).
- Store at 25° C (77° F), excursions permitted between 15°–30° C (59°–86° F).

ADVERSE EFFECTS (≥1%) **Body as a Whole:** *Injection-site reactions,* night sweats. **CNS:** Asthenia, chills, dizziness, fatigue, fever, headache, insomnia. **CV:** *Hot flashes,* hypertension. **GI:** Constipation, nausea, diarrhea. **Metabolic:** Increases in ALT, AST, and GGT, *weight gain.* **Musculoskeletal:** Arthralgia, back pain. **Urogenital:** Erectile dysfunction, gynecomastia, testicular atrophy, urinary tract infections.

INTERACTIONS Drug: Degarelix may cause an additive effect with other drugs that prolong the QT interval prolongation (e.g., ANTIARRHYTHMIC AGENTS, **chlorpromazine, dolasetron, droperidol, mefloquine, mesoridazine, moxifloxacin, pentamidine, pimozide, tacrolimus, thioridazine, ziprasidone**).

PHARMACOKINETICS Peak: 2 days. **Distribution:** 90% Plasma protein bound. **Metabolism:** Hepatobiliary. **Elimination:** Biliary excretion 70–80%; urinary excretion 20–30%. **Half-Life:** 53 days.

NURSING IMPLICATIONS

Assessment & Drug Effects
- Monitor ECG and QT interval, especially with electrolyte imbalances, a history of CHF, and concurrent Class IA (e.g., quinidine) or III (e.g., amiodarone) antiarrhythmics.
- Lab tests: Baseline and periodic PSA and serum testosterone, periodic LFTs.

Patient & Family Education
- Hot flashes are a common side effect that usually subside spontaneously.
- Degarelix can decrease bone density leading to osteoporosis. With long-term therapy, bone density tests are advisable.

D

DELAVIRDINE MESYLATE
(del-a-vir′deen)

Rescriptor
Classifications: ANTIVIRAL; NONNUCLEOSIDE REVERSE TRANSCRIPTASE INHIBITOR (NNRTI)
Therapeutic: ANTIVIRAL; NNRTI
Prototype: Efavirenz
Pregnancy Category: C

AVAILABILITY 100 mg tablets

ACTION & THERAPEUTIC EFFECT Nonnucleoside reverse transcriptase inhibitor (NNRTI) of HIV-1 binds directly to reverse transcriptase (RT) and disrupts RNA- and DNA-dependent DNA polymerase activities. *Prevents replication of the HIV-1 virus; resistant strains appear rapidly.*

USES Treatment of HIV infection in combination with other antiretroviral agents.

CONTRAINDICATIONS Hypersensitivity to delavirdine; lactation.
CAUTIOUS USE Impaired liver function; elderly; achlorhydria; pregnancy (category C).

ROUTE & DOSAGE

HIV Infection
Adult/Adolescent: **PO** 400 mg t.i.d.

Common adverse effects in *italic,* life-threatening effects <u>underlined</u>; generic names in **bold**; classifications in SMALL CAPS; ◆ Canadian drug name; ⊘ Prototype drug

417

D

ADMINISTRATION

Oral
- Disperse in water by adding a single dose to at least 3 oz of water, let stand for a few minutes, then stir to create a uniform suspension just prior to administration.
- Give drug to patients with achlorhydria with an acid beverage such as orange or cranberry juice.
- Store at 20°–25° C (68°–77° F) and protect from high humidity in a tightly closed container.

ADVERSE EFFECTS (≥1%) Body as a Whole: Headache, fatigue, allergic reaction, chills, edema, arthralgia. **CNS:** Abnormal coordination, agitation, amnesia, anxiety, confusion, dizziness. **CV:** Chest pain, bradycardia, palpitations, postural hypotension, tachycardia. **GI:** Nausea, vomiting, diarrhea, increased LFTs, abdominal cramps, anorexia, aphthous stomatitis. **Hematologic:** Neutropenia. **Respiratory:** Bronchitis, cough, dyspnea. **Skin:** *Rash,* pruritus.

INTERACTIONS Drug: ANTACIDS, H₂-RECEPTOR ANTAGONISTS decrease absorption; **didanosine** and **delavirdine** should be taken 1 h apart to avoid decreased delavirdine levels; **clarithromycin, fluoxetine, ketoconazole** may increase delavirdine levels; **carbamazepine, phenobarbital, phenytoin, rifabutin, rifampin** may decrease delavirdine levels; delavirdine may increase levels of **clarithromycin, indinavir, saquinavir, dapsone, rifabutin, alprazolam, midazolam, triazolam,** DIHYDROPYRIDINE, CALCIUM CHANNEL BLOCKERS (e.g., **nifedipine, nicardipine,** etc.), **quinidine, warfarin.** Use with HMG-COA REDUCTASE INHIBITORS may increase the risk of rhabdomyolysis. Use with **pimozide** may cause cardiac arrhythmias. Use with **trazodone** may increase **trazodone** levels. Use with inhaled **fluticasone** may increase

fluticasone levels. Use with HYPNOTICS, **alprazolam, midazolam, triazolam** can cause respiratory depression. **Herbal: St. John's wort** may decrease antiretroviral activity.

PHARMACOKINETICS Absorption: Rapidly from GI tract, 80% reaches systemic circulation. **Peak:** 1 h. **Distribution:** 98% protein bound. **Metabolism:** In the liver by CYP3A4. **Elimination:** Half in urine, 44% in feces. **Half-Life:** 2–11 h.

NURSING IMPLICATIONS

Assessment & Drug Effects
- Therapeutic effectiveness: Indicated by decreased viral load.
- Monitor for and immediately report appearance of a rash, generally within 1–3 wk of starting therapy; rash is usually diffuse, maculopapular, erythematous, and pruritic.

Patient & Family Education
- Take this drug exactly as prescribed. Missed doses increase risk of drug resistance.
- Do not take antacids and delavirdine at the same time; separate by at least 1 h.
- Report all prescription and nonprescription drugs used to prescriber because of multiple drug interactions.
- Discontinue medication and notify prescriber if rash is accompanied by any of the following: Fever, blistering, oral lesions, conjunctivitis, swelling, muscle or joint pain.

DEMECLOCYCLINE HYDROCHLORIDE
(dem-e-kloe-sye′kleen)

Declomycin
Classifications: ANTIBIOTIC; TETRACYCLINE
Therapeutic: ANTIBIOTIC

Common adverse effects in *italic*, life-threatening effects underlined; generic names in **bold**; classifications in SMALL CAPS; ✦ Canadian drug name; ⊙ Prototype drug

Prototype: Tetracycline
Pregnancy Category: D

AVAILABILITY 150 mg capsules; 150 mg, 300 mg tablets

ACTION & THERAPEUTIC EFFECT
Demeclocycline is a broad-spectrum, tetracycline antibiotic. It is pumped through the inner cytoplasmic membrane of bacteria. Demeclocycline blocks the binding of transfer RNA (tRNA) to messenger RNA (mRNA) of bacteria. Therefore, bacterial protein synthesis is inhibited and bacterial cells are destroyed. *Effective against both gram-positive and gram-negative bacteria.*

USES Similar to those of tetracycline.
UNLABELED USES Treatment of chronic SIADH (syndrome of inappropriate antidiuretic hormone) secretion.

CONTRAINDICATIONS Hypersensitivity to any of the tetracyclines; severe renal or hepatic disease; cirrhosis, common bile duct obstruction; period of tooth development in fetus; pregnancy (category D), lactation, children younger than 8 y (causes permanent yellow discoloration of teeth, enamel hypoplasia, and retarded bone growth).
CAUTIOUS USE Mild or moderate impaired renal or hepatic function; nephrogenic diabetes insipidus.

ROUTE & DOSAGE

Anti-Infective

Adult: **PO** 150 mg q6h or 300 mg q12h (max: 2.4 g/day)
Child (older than 8 y): **PO** 8–12 mg/kg/day divided q8–12h

Gonorrhea

Adult: **PO** 600 mg followed by 300 mg q12h for 4 days

SIADH

Adult: **PO** 600–1200 mg/day in 3–4 divided doses

ADMINISTRATION
Oral
- Give not less than 1 h before or 2 h after meals. Foods rich in iron (e.g., red meat or dark green vegetables) or calcium (e.g., milk products) impair absorption.
- Concomitant therapy: Do not give antacids with tetracyclines.
- Check expiration date before giving drug. Renal damage and death have resulted from use of outdated tetracyclines.
- Store in tight, light-resistant containers, preferably at 15°–30° C (59°–86° F) unless otherwise directed. Tetracyclines form toxic products when outdated or exposed to light, heat, or humidity.

ADVERSE EFFECTS (≥1%) **Body as a Whole:** Hypersensitivity [*photosensitivity,* pericarditis, anaphylaxis (rare)]. **GI:** *Nausea,* vomiting, *diarrhea,* esophageal irritation or ulceration, enterocolitis, abdominal cramps, anorexia. **Urogenital:** Diabetes insipidus, azotemia, hyperphosphatemia. **Skin:** Pruritus, erythematous eruptions, exfoliative dermatitis.

DIAGNOSTIC TEST INTERFERENCE
Like other tetracyclines, demeclocycline may cause false increases in *urine catecholamines (fluorometric* methods); false decreases in *urine urobilinogen;* and false-negative *urine glucose* with *glucose oxidase* methods (e.g., *Clinistix, TesTape*).

INTERACTIONS Drug: ANTACIDS, IRON PREPARATION, **calcium, magnesium, zinc, kaolin-pectin, sodium bicarbonate** can significantly decrease demeclocycline absorption;

Common adverse effects in *italic*, life-threatening effects underlined; generic names in **bold**; classifications in SMALL CAPS; ✦ Canadian drug name; ⊘ Prototype drug

419

effects of **desmopressin** and demeclocycline antagonized; increases **digoxin** absorption, increasing risk of **digoxin** toxicity; **methoxyflurane** increases risk of renal failure. **Food:** Dairy products significantly decrease demeclocycline absorption; food may decrease drug absorption also.

PHARMACOKINETICS Absorption: 60–80% absorbed from GI tract. **Peak:** 3–4 h. **Distribution:** Concentrated in liver; crosses placenta; distributed into breast milk. **Metabolism:** In liver; enterohepatic circulation. **Elimination:** 40–50% excreted in urine and 31% in feces in 48 h. **Half-Life:** 10–17 h.

NURSING IMPLICATIONS

Assessment & Drug Effects

- Lab tests: C&S prior to initial therapy and periodically during prolonged therapy. With prolonged therapy, periodic evaluations of electrolytes, and renal and hepatic function.
- Monitor I&O ratio and pattern and record weights in patients with impaired kidney or liver function, or on prolonged or high dose therapy.

Patient & Family Education

- Do not use antacids while taking this drug.
- Take drug on an empty stomach to enhance absorption. Because esophageal irritation and ulceration have been reported, take each dose with a full glass (240 mL) of water; remain upright for at least 90 sec after taking medication; and avoid taking drug within 1 h of lying down or bedtime.
- Notify prescriber if gastric distress is a problem; a snack or light meal free of dairy products may be added to the regimen.
- Report symptoms of superinfections (see Appendix F).

- Demeclocycline-induced phototoxic reaction can be unusually severe. Avoid sunlight as much as possible and use sunscreen.

DENILEUKIN DIFTITOX
(den-i-leu'kin dif'ti-tox)

Ontak

Classifications: ANTINEOPLASTIC; IMMUNOMODULATOR; INTERLEUKIN-2 (IL-2) RECEPTOR INHIBITOR

Therapeutic: ANTINEOPLASTIC; IL-2 RECEPTOR INHIBITOR

Pregnancy Category: C

AVAILABILITY 150 mcg/mL vial

ACTION & *THERAPEUTIC EFFECT*
A recombinant DNA cytotoxic protein that is an interleukin-2 (IL-2) receptor-specific protein that acts as an antineoplastic agent. It acts against malignant cells that express particular high-affinity IL-2 receptors on the cell surface, thus inhibiting cellular protein synthesis and causing cell death in malignant cells. *Effectiveness is indicated by reduced tumor burden. Interacts with high affinity to IL-2 receptors in particular leukemias and lymphomas.*

USES Persistent or recurrent T-cell lymphoma, mycosis fungoides.
UNLABELED USES Non-Hodgkin's lymphoma, psoriasis.

CONTRAINDICATIONS Hypersensitivity to denileukin, diphtheria toxin, or interleukin-2; serum albumin levels below 3 g/dL; lactation.
CAUTIOUS USE Cardiovascular disease; peripheral vascular disease, coronary artery disease; hepatic and renal impairment; elderly; preexisting lowering of serum albumin levels; pregnancy (category C). Safety and efficacy in children younger than 18 y are unknown.

ROUTE & DOSAGE

T-Cell Lymphoma
Adult: **IV** 9 or 18 mcg/kg/day for 5 days every 21 days

ADMINISTRATION
Intravenous

PREPARE: **IV Infusion:** Bring vials to room temperature (solution will be clear when room temperature is reached). Swirl to mix, but do not shake. ▪ Use only plastic syringe and plastic IV bag. Withdraw the calculated dose and inject it into an <u>empty</u> IV bag. Add NO MORE THAN 9 mL sterile saline without preservative to IV bag for each 1 mL of drug. ▪ Use within 6 h of preparation.
ADMINISTER: **IV Infusion:** Infuse over <u>at least</u> 15 min without an in-line filter. ▪ Stop infusion and notify prescriber if S&S of hypersensitivity occur.
INCOMPATIBILITIES **Solution/additive:** Do not physically mix with any other drug.

ADVERSE EFFECTS (≥1%) **Body as a Whole:** *Chills, fever, asthenia, infection, pain, headache, chest pain,* flu-like syndrome; injection site reaction; *acute hypersensitivity reaction (hypotension, back pain, dyspnea, vasodilation, rash, chest pain or tightness, tachycardia, dysphagia, syncope, <u>anaphylaxis</u>, myalgia, arthralgia.* **CNS:** *Dizziness, paresthesia, nervousness,* confusion, insomnia. **CV:** *Vascular leak syndrome (hypotension, edema, hypoalbuminemia); hypotension, vasodilation, tachycardia,* thrombotic events, hypertension, arrhythmia. **GI:** *Nausea, vomiting, anorexia, diarrhea,* constipation, dyspepsia, dysphagia. **Hematologic:** *Anemia,* <u>thrombocytopenia</u>, <u>leukopenia</u>. **Metabolic:**

Hypoalbuminemia; transaminase increase; edema; hypocalcemia; weight loss; dehydration; hypokalemia. **Respiratory:** *Dyspnea, cough, pharyngitis, rhinitis,* lung disorder. **Skin:** *Rash, pruritus, sweating.* **Urogenital:** *Hematuria, albuminuria, pyuria, increased creatinine.* **Special Senses:** Loss of visual acuity.

INTERACTIONS No clinically significant interactions established.

PHARMACOKINETICS Distribution: Primarily distributed to liver and kidneys. **Metabolism:** By proteolytic degradation. **Half-Life:** 70–80 min.

NURSING IMPLICATIONS
Assessment & Drug Effects
▪ Monitor and notify prescriber immediately for S&S of hypersensitivity or anaphylaxis that occur during/within 24 h of infusion.
▪ Monitor and notify prescriber immediately for S&S of flu-like syndrome that occur within several hours to days following infusion.
▪ Monitor outpatients for weight gain, developing edema, or declining blood pressure. Notify prescriber immediately for S&S of vascular leak syndrome (e.g., edema PLUS hypotension or hypoalbuminemia) that may occur within 2 wk of infusion.
▪ Lab tests: Baseline and weekly CBC with differential, platelet count, blood chemistry panel (including serum electrolytes, serum albumin, renal and LFTs).

Patient & Family Education
▪ Report S&S of infection promptly to prescriber.
▪ Check weight daily and report rapid weight gain or swelling of extremities promptly.
▪ Report bothersome adverse effects or S&S of infection or flu-like symptoms (e.g., fever, nausea, vomiting, diarrhea, rash).

Common adverse effects in *italic*, life-threatening effects <u>underlined</u>; generic names in **bold**; classifications in SMALL CAPS; ♣ Canadian drug name; ⊙ Prototype drug

421

D

DENOSUMAB

(den-o'-su-mab)

Prolia

Classification: MONOCLONAL ANTIBODY; RANK LIGAND INHIBITOR; BONE RESORPTION INHIBITOR

Therapeutic: MONOCLONAL ANTIBODY; BONE RESORPTION INHIBITOR

Pregnancy Category: C

AVAILABILITY 60 mg/mL solution for injection

ACTION & *THERAPEUTIC EFFECT*
Denosumab prevents RANKL (receptor activator of nuclear factor κB ligand, a protein essential for survival of osteoclasts) from activating its receptor, RANK, on the surface of osteoclasts. Prevention of the RANKL/RANK interaction inhibits osteoclast formation, function, and survival. *Decreases bone resorption and increases bone mass and strength.*

USES Treatment of osteoporosis in postmenopausal women who are at a high risk for fractures; prevention of bone metastases.

CONTRAINDICATIONS Preexisting hypocalcemia; lactation.

CAUTIOUS USE Predisposition to hypocalcemia and electrolyte imbalance (e.g., history of hypoparathyroidism, thyroid or parathyroid surgery, malabsorption syndromes, removal of small intestine, severe renal impairment [(creatinine clearance less than 30 mL/min) or on dialysis]. Safety and efficacy in children younger than 18 y are not established.

ROUTE & DOSAGE

Osteoporosis

Adult: **Subcutaneous** 60 mg every 6 m

Prevention of Bone metastases

Adult: **Subcutaneous** 120 mg every 4 wk

ADMINISTRATION

Subcutaneous

- Ensure that hypocalcemia is corrected prior to initiation of therapy.
- Bring syringe to room temperature by removing from refrigeration 15–30 min prior to injection. Do not warm any other way.
- Inject into upper arm, upper thigh, or abdomen.
- Store refrigerated.

ADVERSE EFFECTS (≥2%) **Body as a Whole:** Cystitis, herpes zoster infection, hypercholesterolemia, peripheral edema, pharyngitis. **CNS:** Asthenia, insomnia, sciatica, vertigo. **CV:** Anginal pectoris, atrial fibrillation. **GI:** Flatuence, gastroesophageal reflux disease, upper abdominal pain. **Hematological:** Anemia. **Musculoskeletal:** *Back pain*, bone pain, musculoskeletal pain, myalgia, *pain in extremety*, spinal osteoarthritis. **Respiratory:** Pneumonia, upper respiratory tract infection. **Skin:** Pruritis, rash.

PHARMACOKINETICS Peak: 10 d. **Half-Life:** 25.4 d.

NURSING IMPLICATIONS

Assessment & Drug Effects

- Monitor for S&S of hypocalcemia (see Appendix F) and report immediately if any of these appear.
- Lab tests: Baseline and periodic serum calcium, phosphorus, and magnesium.
- Monitor for and report promptly any of the following: S&S of infection; skin reaction such as dermatitis, eczema, rash; or jaw pain.

Patient & Family Education

- Ensure that calcium and vitamin D supplements (usually 1000 mg calcium and at least 400 IU vitamin D daily) are taken exactly as ordered.

Common adverse effects in *italic*, life-threatening effects <u>underlined</u>; generic names in **bold**; classifications in SMALL CAPS; ♣ Canadian drug name; ⊕ Prototype drug

- Notify your prescriber immediately if you develop any of the following: S&S of low calcium (e.g., spasms, twitches, cramps; numbness or tingling in your fingers, toes, or around your mouth); S&S of infection (e.g., fever or chills; skin that is red, hot, or swollen; abdominal pain; frequency, urgency, or burning with urination); signs of skin irritation (e.g., rash, itching, peeling); jaw pain.
- Practice meticulous oral hygiene while you are taking this drug.
- Inform prescriber if you become pregnant while taking this drug.

DESIPRAMINE HYDROCHLORIDE

(dess-ip′ra-meen)

Norpramin
Classification: TRICYCLIC ANTIDEPRESSANT
Therapeutic: TRICYCLIC ANTIDEPRESSANT
Prototype: Imipramine
Pregnancy Category: C

AVAILABILITY 10 mg, 25 mg, 50 mg, 75 mg, 100 mg, 150 mg tablets

ACTION & *THERAPEUTIC EFFECT*
Desipramine is a tricyclic antidepressant (TCA) and the active metabolite of imipramine. Antidepressant activity appears to be related to blocking reuptake of norepinephrine and serotonin in the CNS, thus increasing their levels. *Has antidepressant activity.*

USES Endogenous depression.
UNLABELED USES Attention deficit disorder, bulimia, diabetic neuropathy, panic disorder, postherpetic neuralgia.

CONTRAINDICATIONS Hypersensitivity to tricyclic compounds; recent MI, QT prolongation, cardiac arrhythmias, AV block, bundle branch block; concurrent use of MAOI therapy; suicidal ideation; lactation.
CAUTIOUS USE Urinary retention, prostatic hypertrophy; narrow-angle glaucoma; epilepsy; alcoholism; adolescents, older adults; bipolar disease; thyroid; cardiovascular, renal, and hepatic disease; suicidal tendency; ECT; elective surgery; pregnancy (category C). Safe use in children younger than 6 y is not established.

ROUTE & DOSAGE

Antidepressant

Adult: **PO** 75–100 mg/day at bedtime or in divided doses, may gradually increase to 150–300 mg/day (use lower doses in older adult patients)
Adolescent: **PO** 25–50 mg/day in divided doses (max: 100 mg/day)
Child (6–12 y): **PO** 1–3 mg/kg/day in divided doses (max: 5 mg/kg/day)

Pharmacogenetic Dosage Adjustment

CYP2D6 poor metabolizers: Start with 40% of normal dose

ADMINISTRATION

Oral

- Give drug with or immediately after food to reduce possibility of gastric irritation.
- Give maintenance dose at bedtime to minimize daytime sedation.
- Store drug in tightly closed container at 15°–30° C (59°–86° F) unless otherwise specified.

ADVERSE EFFECTS (≥1%) Body as a Whole: Hypersensitivity (rash, urticaria, photosensitivity). **CNS:** *Drowsiness,* dizziness, weakness,

---

Common adverse effects in *italic*, life-threatening effects underlined; generic names in **bold**; classifications in SMALL CAPS; ♣ Canadian drug name; ⊘ Prototype drug

423

fatigue, headache, insomnia, confusional states, depressive reaction, paresthesias, ataxia. **CV:** *Postural hypotension,* hypotension, palpitation, tachycardia, ECG changes, flushing, <u>heart block</u>. **Special Senses:** Tinnitus, parotid swelling; blurred vision, disturbances in accommodation, mydriasis, increased IOP. **GI:** *Dry mouth, constipation,* bad taste, diarrhea, nausea. **Urogenital:** *Urinary retention,* frequency, delayed micturition, nocturia; impaired sexual function, galactorrhea. **Hematologic:** <u>Bone marrow depression</u> and <u>agranulocytosis</u> (rare). **Other:** Sweating, craving for sweets, weight gain or loss, SIADH secretion, hyperpyrexia, eosinophilic pneumonia.

INTERACTIONS Drug: May somewhat decrease response to ANTIHYPERTENSIVES; CNS DEPRESSANTS, **alcohol,** HYPNOTICS, BARBITURATES, SEDATIVES potentiate CNS depression; may increase hypoprothrombinemic effect of ORAL ANTICOAGULANTS; **ethchlorvynol** may cause transient delirium; **levodopa,** SYMPATHOMIMETICS (e.g., **epinephrine, norepinephrine**) pose possibility of sympathetic hyperactivity with hypertension and hyperpyrexia; MAO INHIBITORS pose possibility of severe reactions, toxic psychosis, cardiovascular instability; **methylphenidate** increases plasma TCA levels; THYROID AGENTS may increase possibility of arrhythmias; **cimetidine** may increase plasma TCA levels. **Herbal: Ginkgo** may decrease seizure threshold; **St. John's wort** may cause **serotonin** syndrome.

PHARMACOKINETICS Absorption: Rapidly from GI tract and injection sites. **Peak:** 4–6 h. **Distribution:** Crosses placenta. **Metabolism:** In liver. **Elimination:** Primarily in urine. **Half-Life:** 7–60 h.

NURSING IMPLICATIONS

Assessment & Drug Effects

- Monitor children and adolescents for signs of suicidal ideation.
- Monitor for therapeutic effectiveness: Usually not realized until after at least 2 wk of therapy.
- Monitor BP and pulse rate during early phase of therapy, particularly in older adult, debilitated, or cardiovascular patients. If BP rises or falls more than 20 mm Hg or if there is a sudden increase in pulse rate or change in rhythm, withhold drug and inform prescriber.
- Note: Drowsiness, dizziness, and orthostatic hypotension are signs of impending toxicity in patient on long-term, high dosage therapy. Prolonged QT or QRS intervals indicate possible toxicity. Report to prescriber.
- Observe patient with history of glaucoma. Report symptoms that may signal acute attack: Severe headache, eye pain, dilated pupils, halos of light, nausea, vomiting.
- Monitor bowel elimination pattern and I&O ratio. Severe constipation and urinary retention are potential problems of TCA therapy.
- Note: Norpramin tablets may contain tartrazine, which can cause allergic-type reactions including bronchial asthma in susceptible individuals. Such individuals are frequently also sensitive to aspirin.

Patient & Family Education

- Make all position changes slowly and in stages, particularly from recumbent to standing position.
- Do not drive or engage in other potentially hazardous activities until reaction to drug is known.
- Take medication exactly as prescribed; do not change dose or dose intervals.

- Note: Patients who receive high doses for prolonged periods may experience withdrawal symptoms including headache, nausea, musculoskeletal pain, and weakness if drug is discontinued abruptly.
- Do not take OTC drugs unless prescriber has approved their use.
- Stop, or at least limit, smoking because it may increase the metabolism of desipramine, thereby diminishing its therapeutic action.

DESLORATADINE

(des-lor-a-ta′deen)

Clarinex, Clarinex Reditabs
Classifications: NONSEDATING ANTIHISTAMINE, H$_1$-RECEPTOR ANTAGONIST
Therapeutic: ANTIHISTAMINE; ANTIALLERGIC
Prototype: Loratadine
Pregnancy Category: C

AVAILABILITY 5 mg tablets; 2.5 mg, 5 mg orally dissolving tablets; 0.5 mg/mL syrup

ACTION & *THERAPEUTIC EFFECT* A long-acting, nonsedating antihistamine with selective H$_1$-receptor antagonist properties. Reduces human mast cell release of inflammatory cytokines. Therefore, it also exhibits antiallergic effects. *Desloratadine is effective in controlling allergic rhinitis and inhibiting histamine-induced wheals and flare (hives).*

USES Treatment of seasonal or perennial allergic rhinitis and idiopathic urticaria.

CONTRAINDICATIONS Hypersensitivity to desloratadine or loratadine; lactation; neonates; infants.
CAUTIOUS USE Renal and hepatic insufficiencies; bladder neck obstruction or urinary retention; prostatic hypertrophy; asthma; glaucoma;

pregnancy (category C). Safety and efficacy in children younger than 12 y not known.

ROUTE & DOSAGE

Allergic Rhinitis, Idiopathic Urticaria
Adult: **PO** 5 mg daily

Renal Impairment Dosage Adjustment
CrCl less than 50 mL/min: 5 mg every other day

Hepatic Impairment Dosage Adjustment
5 mg every other day

ADMINISTRATION

Oral
- Note that drug should be given every other day to patients with significant renal or hepatic impairment.
- Store between 2°–25° C (36°–77° F).

ADVERSE EFFECTS (≥1%) **Body as a Whole:** Pharyngitis, fatigue, flu-like symptoms, myalgia. **CNS:** Somnolence, dizziness. **GI:** Dry mouth, nausea, dry throat. **Urogenital:** Dysmenorrhea.

INTERACTIONS Drug: No clinically significant interactions established.

PHARMACOKINETICS Absorption: Well absorbed. **Peak:** 3 h. **Distribution:** 85–89% protein bound. **Metabolism:** Extensively metabolized in liver to 3-hydroxydesloratadine, an active metabolite. **Elimination:** Equally in urine and feces. **Half-Life:** 27 h.

NURSING IMPLICATIONS

Assessment & Drug Effects
- Monitor cardiovascular status and report significant changes in BP and palpitations or tachycardia.

Common adverse effects in *italic*, life-threatening effects underlined; generic names in **bold**; classifications in SMALL CAPS; ♣ Canadian drug name; ⊕ Prototype drug

425

- Lab tests: Monitor periodically renal function and LFTs.
- Concurrent drugs: Monitor ECG when used in combination with any other drug that can produce an additive effect causing QT interval prolongation.

Patient & Family Education
- Drug may cause significant drowsiness in older adult patients and those with liver or kidney impairment.
- Note: Concurrent use of alcohol and other CNS depressants may have an additive effect.
- Do not take this drug more often than every other day if you have renal impairment.

DESMOPRESSIN ACETATE

(des-moe-pres'sin)

DDAVP, Stimate
Classification: POSTERIOR PITUITARY HORMONE
Therapeutic: ANTIDIURETIC HORMONE (ADH)
Prototype: Vasopressin
Pregnancy Category: B

AVAILABILITY 0.1 mg, 0.2 mg tablets; 0.1% nasal solution; 0.15 mg/spray nasal spray; 4 mcg/mL, 15 mcg/mL injection

ACTION & *THERAPEUTIC EFFECT*
Synthetic analog of natural human posterior pituitary (antidiuretic) hormone, vasopressin. Reduces urine volume and osmolality of serum in patients with central diabetes insipidus by increasing reabsorption of water by kidney collecting tubules. Produces a dose-related increase in factor VIII (antihemophilic factor) and von Willebrand's factor. *Desmopressin is an effective replacement for antidiuretic hormone. It also*

can shorten or normalize bleeding time, and correct platelet adhesion abnormalities in certain patients with bleeding disorders.

USES To control and prevent symptoms and complications of central (neurohypophyseal) diabetes insipidus, and to relieve temporary polyuria and polydipsia associated with trauma or surgery in the pituitary region.

UNLABELED USES To increase factor VIII activity in selected patients with mild to moderate hemophilia A and in type I von Willebrand's disease or uremia, and to control enuresis in children.

CONTRAINDICATIONS Nephrogenic diabetes insipidus, type II B von Willebrand's disease; renal failure. **PO form:** Patients with fluid and electrolyte imbalance; **intranasal form:** Children with primary nocturnal enuresis (PNE).

CAUTIOUS USE Coronary artery insufficiency, hypertensive cardiovascular disease; history of hyponatremia; water intoxication with hyponatremia; severe CHF; older adults; renal impairment; history of thromboembolic disease; pregnancy (category B).

ROUTE & DOSAGE

Diabetes Insipidus

Adult: **Intranasal** 0.1–0.4 mL (10–40 mcg) in 1–3 divided doses **IV/Subcutaneous** 2–4 mcg in 2 divided doses **PO** 0.2–0.4 mg/day
Child: **Intranasal** *3 mo–12 y,* 0.05–0.3 mL in 1–2 divided doses **IV/Subcutaneous** 0.3 mcg/kg infused over 15–30 min **PO** 0.05 mg titrated to response

Enuresis

Adult: **Intranasal** 5–40 mcg at bedtime

Child (6 y or older): **PO** 0.2 mg at bedtime, may titrate up to 0.6 mg at bedtime

Von Willebrand's Disease

Adult/Child (older than 3 mo): **IV/Subcutaneous** 0.3 mcg/kg 30 min preoperative, may repeat in 48 h if needed

Renal Impairment Dosage Adjustment

CrCl less than 50 mL/min: Do not use

ADMINISTRATION

Oral

▪ Note that 0.2 mg PO is equivalent to 10 mcg (0.1 mL) intranasal.

Intranasal

▪ Follow manufacturer's instructions for proper technique with nasal spray.

▪ Give initial dose in the evening, and observe antidiuretic effect. Dose is increased each evening until uninterrupted sleep is obtained. If daily urine volume is more than 2 L after nocturia is controlled, morning dose is started and adjusted daily until urine volume does not exceed 1.5–2 L/24 h.

Subcutaneous

▪ Give undiluted.

Intravenous

PREPARE: **Direct:** Give undiluted for diabetes insipidus. **IV Infusion:** Dilute 0.3 mcg/kg in 10 mL of NS (children weighing 10 kg or less) or 50 mL of NS (children greater than 10 kg and adults) for von Willebrand's disease (type I).

ADMINISTER: **Direct:** Give direct IV over 30 s for diabetes insipidus. **IV Infusion:** Give over 15–30 min for von Willebrand's disease (type I).

▪ Store parenteral and nasal solution in refrigerator preferably at 4° C (39.2° F) unless otherwise directed. Avoid freezing. ▪ Nasal spray can be stored at room temperature. ▪ Discard solutions that are discolored or contain particulate matter.

ADVERSE EFFECTS (≥1%) **All:** Dose related. **CNS:** Transient headache, drowsiness, listlessness. **Special Senses:** Nasal congestion, rhinitis, nasal irritation. **GI:** Nausea, heartburn, mild abdominal cramps. **Other:** Vulval pain, shortness of breath, slight rise in BP, facial flushing, pain and swelling at injection site, hyponatremia.

INTERACTIONS Drug: Demeclocycline, lithium, other VASOPRESSORS may decrease antidiuretic response; **carbamazepine, chlorpropamide, clofibrate** may prolong antidiuretic response.

PHARMACOKINETICS Absorption: 10–20% through nasal mucosa. **Onset:** 15–60 min. **Peak:** 1–5 h. **Duration:** 5–21 h. **Distribution:** Small amount crosses blood–brain barrier; distributed into breast milk. **Half-Life:** 76 min.

NURSING IMPLICATIONS

Assessment & Drug Effects

▪ Monitor I&O ratio and pattern (intervals). Fluid intake **must be** carefully controlled, particularly in older adults and the very young to avoid water retention and sodium depletion.

▪ Weigh patient daily and observe for edema. Severe water retention may require reduction in dosage and use of a diuretic.

Common adverse effects in *italic*, life-threatening effects underlined; generic names in **bold**; classifications in SMALL CAPS; ✦ Canadian drug name; ⊙ Prototype drug

- Monitor BP during dosage-regulating period and whenever drug is administered parenterally.
- Lab tests: Monitor urine and plasma osmolality. An increase in urine osmolality and a decrease in plasma osmolality indicate effectiveness of treatment in diabetes insipidus.

Patient & Family Education
- Report upper respiratory tract infection or nasal congestion.
- Follow manufacturer's instructions for insertion to ensure delivery of drug high into nasal cavity and not down throat. A flexible calibrated plastic tube is provided.

DESONIDE
(dess'oh-nide)

DesOwen, Tridesilon
Pregnancy Category: C
See Appendix A-4.

DESOXIMETASONE
(des-ox-i-met'a-sone)

Topicort, Topicort-LP
Pregnancy Category: C
See Appendix A-4.

DEXAMETHASONE
(dex-a-meth'a-sone)

Baycadron, Decadron, Dexamethasone Intensol, Maxidex, Mymethasone

DEXAMETHASONE SODIUM PHOSPHATE

Classifications: ADRENAL CORTICOSTEROID; GLUCOCORTICOID
Therapeutic: ADRENAL CORTICOSTEROID
Prototype: Prednisone
Pregnancy Category: C

AVAILABILITY Dexamethasone: 0.5 mg, 0.75 mg, 1 mg, 1.5 mg, 2 mg, 4 mg, 6 mg tablets; 0.5 mg/5 mL, 1 mg/mL oral solution; **Dexamethasone sodium phosphate:** 4 mg/mL, 10 mg/mL injection, 0.1% cream; 0.1% ophth solution, suspension.

ACTION & *THERAPEUTIC EFFECT*
Long-acting synthetic adrenocorticoid with intense anti-inflammatory (glucocorticoid) activity and minimal mineralocorticoid activity. ***Anti-inflammatory action:*** Prevents accumulation of inflammatory cells at sites of infection; inhibits phagocytosis, lysosomal enzyme release, and synthesis of potent mediators of inflammation, prostaglandins, and leukotrienes; reduces capillary dilation and permeability. ***Immunosuppression:*** Probably due to prevention or suppression of delayed hypersensitivity immune reaction. *Has anti-inflammatory and immunosuppression properties.*

USES Adrenal insufficiency concomitantly with a mineralocorticoid; inflammatory conditions, allergic states, collagen diseases, hematologic disorders, cerebral edema, and addisonian shock. Also palliative treatment of neoplastic disease, as adjunctive short-term therapy in acute rheumatic disorders and GI diseases, and as a diagnostic test for Cushing's syndrome and for differential diagnosis of adrenal hyperplasia and adrenal adenoma.

UNLABELED USES As an antiemetic in cancer chemotherapy; as a diagnostic test for endogenous depression; and to prevent hyaline membrane disease in premature infants.

CONTRAINDICATIONS Systemic fungal infection, acute infections, active or resting tuberculosis, vaccinia, varicella, administration of live virus vaccines (to patient, family members), latent or active amebiasis; Cushing's syndrome; neonates or infants weighing less than 1300 g; lactation. *Topical use:* Rosacea, perioral dermatitis; venous stasis ulcers. *Ophthalmic use:* Primary open-angle glaucoma, eye infections, superficial ocular herpes simplex, keratitis, and tuberculosis of eye.

CAUTIOUS USE Stromal herpes simplex, keratitis, GI ulceration, renal disease, diabetes mellitus, hypothyroidism, myasthenia gravis, CHF, cirrhosis, psychic disorders, seizures; coagulopathy; pregnancy (category C); children.

ROUTE & DOSAGE

Allergies, Inflammation, Neoplasias

Adult: **PO** 0.25–4 mg b.i.d. to q.i.d. **IM** 8–16 mg q1–3wk or 0.8–1.6 mg intralesional q1–3wk **IV** 0.75–0.9 mg/kg/day divided q6–12h
Child: **PO/IV/IM** 0.08–0.3 mg/kg/day divided q6–12h

Adrenocorticol Function Abnormalities

Adult: **PO/IV** 0.75–9 mg/day in divided doses, adjust to patient response
Child: **PO/IV** 0.03–0.3 mg/kg/day in divided doses, adjust to patient response

Cerebral Edema

Adult: **IV** 10 mg followed by 4 mg q6h, reduce dose after 2–4 days then taper over 5–7 days

Child: **PO/IV/IM** 1–2 mg/kg loading dose, then 1–1.5 mg/kg/day divided q4–6h × 5 days (max: 16 mg/day)

Shock

Adult: **IV** 1–6 mg/kg as a single dose or 40 mg repeated q2–6h if needed or 20 mg bolus then 3 mg/kg/day

Dexamethasone Suppression Test

Adult: **PO** 0.5 mg q6h for 48 h

Cushing's Syndrome Diagnosis

Adult: **PO** 2 mg q6h × 48h

Inflammation

Adult/Child: **Ophthalmic/ Topical/Inhalation/Intranasal** See Appendix A.

ADMINISTRATION

Oral

- Give the once-daily dose in the a.m. with food or liquid of patient's choice.
- Taper dosage over a period of time before discontinuing because adrenal suppression can occur with prolonged use.
- Do not store or expose aerosol to temperature above 48.9° C (120° F); do not puncture or discard into a fire or an incinerator.

Intramuscular

- Give IM injection deep into a large muscle mass (e.g., gluteus maximus). Avoid subcutaneous injection: Atrophy and sterile abscesses may occur.
- Use repository form, dexamethasone acetate, for IM or local injection only. The white suspension settles on standing; mild shaking will resuspend drug.

Common adverse effects in *italic*, life-threatening effects <u>underlined</u>; generic names in **bold**; classifications in SMALL CAPS; ✤ Canadian drug name; ☻ Prototype drug

Intravenous

PREPARE: Direct: Give undiluted. **Intermittent:** Dilute in D5W or NS for infusion.

ADMINISTER: Direct: Give direct IV push over 30 sec or less. **Intermittent:** Set rate as prescribed or according to amount of solution to infuse.

INCOMPATIBILITIES Solution/additive: Daunorubicin, diphenhydramine, doxorubicin, doxapram, glycopyrrolate, metaraminol, phenobarbital, vancomycin. Y-site: Ciprofloxacin, fenoldopam, idarubicin, midazolam, topotecan.

▪ Store at 15°–30° C (59°–86° F) unless otherwise directed.

ADVERSE EFFECTS (≥1%) Aerosol Therapy: *Nasal irritation,* dryness, epistaxis, rebound congestion, bronchial asthma, anosmia, perforation of nasal septum. *Systemic Absorption*—**CNS:** Euphoria, insomnia, convulsions, increased ICP, vertigo, headache, psychic disturbances. **CV:** CHF, hypertension, *edema.* **Endocrine:** Menstrual irregularities, *hyperglycemia;* cushingoid state; growth suppression in children; hirsutism. **Special Senses:** *Posterior subcapsular cataract,* increased IOP, glaucoma, exophthalmos. **GI:** Peptic ulcer with possible perforation, abdominal distension, nausea, increased appetite, heartburn, dyspepsia, pancreatitis, bowel perforation, *oral candidiasis.* **Musculoskeletal:** Muscle weakness, loss of muscle mass, vertebral compression fracture, pathologic fracture of long bones, tendon rupture. **Skin:** Acne, *impaired wound healing,* petechiae, ecchymoses, diaphoresis, allergic dermatitis, hypo- or hyperpigmentation, subcutaneous and cutaneous atrophy, burning and tingling in perineal area (following IV injection).

DIAGNOSTIC TEST INTERFERENCE *Dexamethasone suppression test for endogenous depression:* False positive results may be caused by **alcohol, glutethimide, meprobamate;** false-negative results may be caused by high doses of benzodiazepines (e.g., **chlordiazepoxide** and **cyproheptadine**), long-term glucocorticoid treatment, **indomethacin, ephedrine,** estrogens or hepatic enzyme-inducing agents **(phenytoin)** may also cause false-positive results in *test for Cushing's syndrome.*

INTERACTIONS Drug: BARBITURATES, **phenytoin, rifampin** increase steroid metabolism—dosage of dexamethasone may need to be increased; **amphotericin B,** DIURETICS compound potassium loss; **ambenonium, neostigmine, pyridostigmine** may cause severe muscle weakness in patients with myasthenia gravis; may inhibit antibody response to VACCINES, TOXOIDS.

PHARMACOKINETICS Absorption: Readily from GI tract. **Onset:** Rapid. **Peak:** 1–2 h PO; 8 h IM. **Duration:** 2.75 days PO; 6 days IM; 1–3 wk intra lesional, intra-articular. **Distribution:** Crosses placenta; distributed into breast milk. **Elimination:** Hypothalamus-pituitary axis suppression: 36–54 h. **Half-Life:** 3–4.5 h.

NURSING IMPLICATIONS

Assessment & Drug Effects

- Monitor and report S&S of Cushing's syndrome (see Appendix F) or other systemic adverse effects.
- Monitor neonates born to a mother who has been receiving a corticosteroid during pregnancy for symptoms of hypoadrenocorticism.
- Monitor for S&S of a hypersensitivity reaction (see Appendix F). The acetate and sodium phosphate formulations may contain bisulfites, parabens, or both; these inactive ingredients are allergenic to some individuals.

Patient & Family Education

- Take drug exactly as prescribed.
- Report lack of response to medication or malaise, orthostatic hypotension, muscular weakness and pain, nausea, vomiting, anorexia, hypoglycemic reactions (see Appendix F), or mental depression to prescriber.
- Report changes in appearance and easy bruising to prescriber.
- Add potassium-rich foods to diet; report signs of hypokalemia (see Appendix F). Concomitant potassium-depleting diuretic can enhance dexamethasone-induced potassium loss.
- Note: Dexamethasone dose regimen may need to be altered during stress (e.g., surgery, infections, emotional stress, illness, acute bronchial attacks, trauma). Consult prescriber if change in living or working environment is anticipated.
- Discontinue drug gradually under the guidance of the prescriber.
- Note: It is important to prevent exposure to infection, trauma, and sudden changes in environmental factors, as much as possible, because drug is an immunosuppressor.

DEXCHLORPHENIRAMINE MALEATE

(dex-klor-fen-eer′a-meen)

Classifications: ANTIHISTAMINE; H₁-RECEPTOR ANTAGONIST
Therapeutic: ANTIHISTAMINE
Prototype: Diphenhydramine
Pregnancy Category: B

AVAILABILITY 4 mg sustained release tablets; 2 mg/5 mL syrup

ACTION & *THERAPEUTIC EFFECT*
H₁-receptor antagonist that competes for H₁-receptor sites on cells, thus blocking histamine release. *Has high antihistamine effects and moderate anticholinergic effects.*

USES Perennial and seasonal allergic rhinitis, other manifestations of allergy, and vasomotor rhinitis. Also as adjunct to epinephrine in treatment of anaphylactic reactions.

CONTRAINDICATIONS Hypersensitivity to antihistamines of similar class; closed-angle glaucoma; acute asthmatic attack, lower respiratory tract symptoms, newborns, premature infants, children younger than 2 y (**syrup** and **tablets**), **extended release tablets:** children younger than 6 y for 4 mg and 12 y for 6 mg.
CAUTIOUS USE Increased intraocular pressure; prostatic hypertrophy; hyperthyroidism; asthma, COPD; renal, hepatic, and cardiovascular disease; serious GI disorders; older adults; pregnancy (category B), lactation.

ROUTE & DOSAGE

Allergic Rhinitis

Adult: **PO** 4–6 mg at bedtime or q8–10h during the day
Child: **PO** 2–5 y, 0.5 mg q4–6h (max: 3 mg/24 h); 6–11 y, 1 mg q4–6h (max: 6 mg/24 h) or 4 mg at bedtime

---

Common adverse effects in *italic*, life-threatening effects underlined; generic names in **bold**; classifications in SMALL CAPS; ♣ Canadian drug name; ⊘ Prototype drug

D

ADMINISTRATION

Oral

- Ensure that sustained release form of drug is not chewed or crushed. It **must be** swallowed whole.
- Give medication with food, water, or milk to lessen GI distress.
- Store at 15°–30° C (59°–86° F) unless otherwise directed.

ADVERSE EFFECTS (≥1%) **CNS:** *Drowsiness,* dizziness, weakness, headache, excitation, neuritis, disturbed coordination, insomnia, euphoria, paresthesias. **Special Senses:** Vertigo, tinnitus, acute labyrinthitis; blurred vision. **CV:** Palpitations, tachycardia, hypotension, extrasystoles. **GI:** Nausea, vomiting, anorexia, *dry mouth,* constipation, diarrhea. **Urogenital:** Difficulty in urinating, *urinary retention,* urinary frequency, early menses. **Hematologic:** Agranulocytosis (rare), hemolytic or hypoplastic anemia. **Skin:** Skin eruptions, photosensitivity.

INTERACTIONS Drug: Alcohol and other CNS DEPRESSANTS, MAO INHIBITORS compound CNS depression.

PHARMACOKINETICS Absorption: Readily from GI tract. **Onset:** 15–30 min. **Peak:** 3 h. **Distribution:** Small amounts into breast milk. **Metabolism:** In liver. **Elimination:** In urine within 24 h.

NURSING IMPLICATIONS

Assessment & Drug Effects

- Supervise ambulation and take safety precautions, especially with older adult patients.
- Monitor I&O and assess for difficulty voiding (e.g., frequency or retention).

Patient & Family Education

- Swallow timed or sustained release tablet whole. Do not break, crush, or chew.

- Do not drive or engage in other potentially hazardous activities until reaction to drug is known.
- Ask prescriber about the use of alcohol, tranquilizers, sedatives, or other CNS depressants because the effects of dexchlorpheniramine will be additive.

DEXMEDETOMIDINE ⊕ HYDROCHLORIDE
(dex-med-e-to′mi-deen)

Precedex
Classifications: ALPHA₂-ADRENERGIC AGONIST; NONBARBITUATE SEDATIVE-HYPNOTIC
Therapeutic: SEDATIVE-HYPNOTIC
Pregnancy Category: C

AVAILABILITY 100 mcg/mL injection

ACTION & *THERAPEUTIC EFFECT* Stimulates alpha₂-adrenergic receptors in the CNS (primarily in the medulla oblongata) causing inhibition of the sympathetic vasomotor center of the brain resulting in sedative effects. *Sedative properties utilized in intubating patients and for initially maintaining them on a mechanical ventilator.*

USES Sedation of initially intubated or mechanically ventilated patients.

CONTRAINDICATIONS Hypersensitivity to dexmedetomidine; labor and delivery, including cesarean section.
CAUTIOUS USE Cardiac arrhythmias or cardiovascular disease, uncontrolled hypertension; hypotension; cerebrovascular disease; renal or hepatic insufficiency; signs of light anesthesia; pregnancy (category C), lactation; older adults over 65 y. Safety and efficacy in children younger than 18 y are unknown.

Common adverse effects in *italic*, life-threatening effects underlined; generic names in **bold**; classifications in SMALL CAPS; ◆ Canadian drug name; ⊕ Prototype drug

ROUTE & DOSAGE

Sedation

Adult: **IV** 1 mcg/kg loading dose infused over 10 min, then continue with infusion of 0.2–0.7 mcg/kg/h for up to 24 h adjusted to maintain sedation

Hepatic Impairment Dosage Adjustment

Reduce initial dosage

Renal Impairment Dosage Adjustment

CrCl less than 30 mL/min: Reduce initial dose

ADMINISTRATION

Intravenous

PREPARE: **Continuous:** Withdraw 2 mL of dexmedetomidine and add to 48 mL of NS to yield 4 mcg/mL. Shake gently to mix.
ADMINISTER: **Continuous:** Administer using a controlled infusion device. ▪ A loading dose of 1 mcg/kg is infused over 10 min followed by the ordered maintenance dose. Do **NOT** use administration set containing natural rubber. Do **NOT** infuse longer than 24 h.
INCOMPATIBILITIES **Y-site: Amphotericin B, diazepam, gemtuzumab, irinotecan, pantoprazole, phenytoin.**

▪ Store at 15°–30° C (59°–86° F).

ADVERSE EFFECTS (≥1%) **Body as a Whole:** Pain, infection. **CV:** *Hypotension,* bradycardia, atrial fibrillation. **GI:** *Nausea,* thirst. **Respiratory:** Hypoxia, pleural effusion, pulmonary edema. **Hematologic:** Anemia, leukocytosis. **Urogenital:** Oliguria.

INTERACTIONS **Drug:** BARBITURATES, BENZODIAZEPINES, GENERAL ANESTHETICS, OPIATE AGONISTS, ANXIOLYTICS, SEDATIVES/HYPNOTICS, **ethanol,** TRICYCLIC ANTIDEPRESSANTS, **tramadol,** PHENOTHIAZINES, SKELETAL MUSCLE RELAXANTS, **azatadine, brompheniramine, carbinoxamine, chlorpheniramine, clemastine, cyproheptadine, dexchlorpheniramine, dimenhydrinate, diphenhydramine, doxylamine, hydroxyzine, methdilazine, phenindamine, promethazine, tripelennamine** enhance CNS depression possibly prolong recovery from anesthesia.

PHARMACOKINETICS **Metabolism:** Extensively in liver (CYP2A6). **Elimination:** Primarily in urine. **Half-Life:** 2 h.

NURSING IMPLICATIONS

Assessment & Drug Effects

▪ Monitor for hypertension during loading dose; reduction of loading dose may be required.
▪ Monitor cardiovascular status continuously; notify prescriber immediately if hypotension or bradycardia occur.

DEXMETHYLPHENIDATE
(dex-meth-ill-fen'i-date)

Focalin, Focalin XR
Classification: CEREBRAL STIMULANT
Therapeutic: CEREBRAL STIMULANT
Prototype: Amphetamine
Pregnancy Category: C
Controlled Substance: Schedule II

AVAILABILITY 2.5 mg, 5 mg, 10 mg tablets; 5 mg, 10 mg, 20 mg extended release capsules

ACTION & *THERAPEUTIC EFFECT*
Thought to block reuptake of norepinephrine and dopamine into presynaptic neurons and, thereby increases release of these substances

Common adverse effects in *italic*, life-threatening effects underlined; generic names in **bold**; classifications in SMALL CAPS; ♣ Canadian drug name; ⊘ Prototype drug

433

into the synapse. *Is effective in controlling ADHD syndrome in conjunction with other measures (psychological, educational, and social).*

USES Attention deficit hyperactivity disorder (ADHD).

CONTRAINDICATIONS Hypersensitivity to dexmethylphenidate or methylphenidate; known structural cardiac abnormalities in children or adults, cardiomyopathy, congenital heart disease; coronary heart disease; severe agitation, anxiety, or tension; psychotic symptomatology; substance abuse; glaucoma; motor tics other than Tourette's syndrome; concurrent MAOI therapy or within 14 days of discontinuation; occurrence of seizures without a history; lactation.

CAUTIOUS USE Moderate to severe hepatic insufficiency; Tourette's syndrome; depression; emotional instability; bipolar disorder, history of suicides; alcoholism or drug dependence; history of seizure disorders; hypertension, CHF, cardiac arrhythmias; hyperthyroidism; older adults; pregnancy (category C). Safe use in children younger than 6 y is not established.

ROUTE & DOSAGE

Attention Deficit Hyperactivity Disorder

Adult: **PO** 2.5 mg b.i.d., may increase by 2.5–5 mg/day at weekly intervals to max of 20 mg/day. If converting from methylphenidate, start with ½ of methylphenidate dose. **Extended release:** 10 mg daily, may increase by 5 mg at weekly intervals to max of 20 mg/day.
Child (older than 6 y): **PO** 2.5 mg b.i.d., may increase by 2.5–5 mg/day at weekly intervals to max of 20 mg/day. If converting from methylphenidate, start with ½ of methylphenidate dose. **Extended release:** 5 mg daily, may increase by 5 mg at weekly intervals to max of 20 mg/day.

ADMINISTRATION

Oral

- Do not administer with or within 14 days following discontinuation of an MAO inhibitor.
- Give sustained release capsules whole. They should not be crushed or chewed.
- Give b.i.d. doses at least 4 h apart.
- Store at 15°–30° C (59°–86° F).

ADVERSE EFFECTS (≥1%) **Body as a Whole:** Fever, allergic reactions. **CNS:** Dizziness, insomnia, nervousness, tics, abnormal thinking, hallucinations, emotional lability, CNS overstimulation or sympathomimetic effects [angina, anxiety, agitation, biting, blurred vision, delirium, diaphoresis, flushing or pallor, hallucinations, hyperthermia, labile blood pressure and heart rate (hypotension or hypertension), mydriasis, palpitations, paranoia, purposeless movements, psychosis, sinus tachycardia, tachypnea, or tremor]. **CV:** Hypertension, tachycardia. **GI:** *Abdominal pain,* decreased appetite, nausea, vomiting.

INTERACTIONS Drug: Additive stimulant effects with other STIMULANTS (including **amphetamine, caffeine**); increased vasopressor effects with **dopamine, epinephrine, norepinephrine, phenylpropanolamine, pseudoephedrine;** MAO INHIBITORS may cause hypertensive crisis; antagonizes hypotensive effects of **guanethidine,** may inhibit metabolism and increase serum levels of **fosphenytoin, phenytoin,**

Common adverse effects in *italic*, life-threatening effects <u>underlined</u>; generic names in **bold**; classifications in SMALL CAPS; ♣ Canadian drug name; ❶ Prototype drug

phenobarbital, and **primidone, warfarin,** TRICYCLIC ANTIDEPRESSANTS.

PHARMACOKINETICS Absorption: Well absorbed. **Peak:** 1–1.5 h. **Metabolism:** De-esterified in liver. No interaction with CYP450 system. **Elimination:** Primarily in urine. **Half-Life:** 2.2 h.

NURSING IMPLICATIONS

Assessment & Drug Effects

- Withhold drug and notify prescriber if patient has a seizure. Monitor closely for loss of seizure control with a prior history of seizures.
- Monitor BP in all patients receiving this drug. Monitor cardiac status and report palpitations or other signs of arrhythmias.
- Monitor for potential abuse and dependence on this drug. Careful supervision is needed during drug withdrawal since severe depression may occur.
- Monitor for signs of aggression or psychotic behavior in adolescents and children.
- Lab tests: Periodic CBC, differential, platelet counts, and LFTs during prolonged therapy.
- Concurrent drugs: Monitor patients on BP-lowering drugs for loss of BP control. Monitor plasma levels of oral anticoagulants and anticonvulsants; doses of these drugs may need to be decreased.

Patient & Family Education

- Withhold drug and report immediately any of the following signs of overdose: Vomiting, agitation, tremors, muscle twitching, convulsions, confusion, hallucinations, delirium, sweating, flushing, headache, or high temperature.
- Note that drug is usually discontinued if improvement is not observed after appropriate dosage adjustment over 1 mo.

DEXRAZOXANE
(dex-ra-zox′ane)
Zinecard
Classifications: ANTINEOPLASTIC; CYTOPROTECTIVE AGENT; CARDIOPROTECTIVE
Therapeutic: CARDIOPROTECTIVE FOR DOXORUBICIN
Pregnancy Category: C

AVAILABILITY 250 mg, 500 mg vials for injection

ACTION & *THERAPEUTIC EFFECT*
A derivative of EDTA that readily penetrates cell membranes. Dexrazoxane is converted intracellularly to a chelating agent that interferes with iron-mediated free radical generation thought to be partially responsible for one form of cardiomyopathy. *Cardioprotective effect is related to its chelating activity.*

USES Reduction of cardiomyopathy associated with a cumulative doxorubicin dose of 300 mg/m².

CONTRAINDICATIONS Chemotherapy regimens that do not contain anthracycline; lactation.
CAUTIOUS USE Myelosuppression, prior radiation or chemotherapy; renal failure or impairment; older adults; pregnancy (category C). Safety and efficacy in children have not been established.

ROUTE & DOSAGE

Cardiomyopathy
Adult: **IV** 10 parts dexrazoxane to 1 part doxorubicin or 500 mg/m² for every 50 mg/m² of doxorubicin

Renal Impairment Dosage Adjustment
CrCl less than 40 mL/min: Use a 5:1 ratio of dexrazoxane to doxorubicin

D

ADMINISTRATION

Intravenous

Wear gloves when handling dexrazoxane. Immediately wash with soap and water if drug contacts skin or mucosa.

- Doxorubicin dose **must be** started within 30 min of beginning dexrazoxane.

PREPARE: **Direct:** Reconstitute by adding 25 or 50 mL of 0.167 M sodium lactate injection (provided by manufacturer) to the 250- or 500-mg vial, respectively, to produce a 10-mg/mL solution. **IV Infusion:** Further dilute reconstituted solution with NS or D5W to a concentration of 1.3–5 mg/mL for infusion.

ADMINISTER: **Direct:** Give bolus dose slowly. **IV Infusion:** Give over 10–15 min.

- Store reconstituted solutions for 6 h at 15°–30° C (59°–86° F).

ADVERSE EFFECTS (≥1%) **All:** Adverse effects of dexrazoxane are difficult to distinguish from those of the chemotherapeutic agents. Pain at injection site, <u>leukopenia</u>, <u>granulocytopenia</u>, and <u>thrombocytopenia</u> appear to occur more frequently with the addition of dexrazoxane than with placebo.

PHARMACOKINETICS Distribution: Not protein bound. **Metabolism:** In liver. **Elimination:** 42% in urine. **Half-Life:** 2–2.5 h.

NURSING IMPLICATIONS

Assessment & Drug Effects

- Monitor cardiac function. Drug does not eliminate risk of doxorubicin cardiotoxicity.
- Lab tests: Monitor LFTs, renal function, and hematopoietic status throughout course of therapy.

- Note: Adverse effects are likely due to concurrent cytotoxic drugs rather than dexrazoxane.

Patient & Family Education

- Report any of the following to prescriber: Worsening shortness of breath, swelling extremities, or chest pains.

DEXTRAN 40
(dex′tran)

Gentran 40, 10% LMD, Rheomacrodex
Classification: PLASMA VOLUME EXPANDER
Therapeutic: PLASMA VOLUME EXPANDER
Prototype: Albumin
Pregnancy Category: C

AVAILABILITY 10% solution in D5W or NS

ACTION & *THERAPEUTIC EFFECT*
Low-molecular-weight polysaccharide. As a hypertonic colloidal solution, produces immediate and short-lived expansion of plasma volume by increasing colloidal osmotic pressure and drawing fluid from interstitial to intravascular space. *Cardiovascular response to volume expansion includes increased BP, pulse pressure, CVP, cardiac output, venous return to heart, and urinary output.*

USES Adjunctively to expand plasma volume and provide fluid replacement in treatment of shock or impending shock. Also used in prophylaxis and therapy of venous thrombosis and pulmonary embolism. Used as priming fluid or as additive to other primers during extracorporeal circulation.

CONTRAINDICATIONS Hypersensitivity to dextrans, severe renal

failure, hypervolemic conditions, severe CHF, significant anemia, hypofibrinogenemia or other marked hemostatic defects including those caused by drugs, (e.g., heparin, warfarin); lactation.
CAUTIOUS USE Active hemorrhage; severe dehydration; chronic liver disease; impaired renal function; thrombocytopenia; patients susceptible to pulmonary edema or CHF; pregnancy (category C).

ROUTE & DOSAGE

Shock

Adult/Adolescent/Child: **IV** Up to 20 mL/kg in the first 24 h (doses up to 10 mL/kg/day may be given for a maximum of 4 additional days if needed)
Prophylaxis for Thromboembolic Complications

Adult: **IV** 500–1000 mL (10 mL/kg) on the day of operation followed by 500 mL/day for 2–3 days, may continue with 500 mL q2–3days for up to 2 wk if necessary
Priming for Extracorporeal Circulation

Adult: **IV** 10–20 mL/kg added to perfusion circuit

ADMINISTRATION

Intravenous
If blood is to be administered, draw a cross-match specimen before dextran infusion.

PREPARE: **IV Infusion:** Use only if seal is intact, vacuum is detectable, and solution is absolutely clear. ▪ No dilution required.

ADMINISTER: **IV Infusion:** Specific flow rate should be prescribed

by prescriber. ▪ For emergency treatment of shock in adults give first 500 mL rapidly (e.g., 20–40 mL/min); give remaining portion of the daily dose over 8–24 h or at the rate prescribed.

INCOMPATIBILITIES **Solution/additive: Amoxicillin, ampicillin, oxacillin, penicillin.**

▪ Store at a constant temperature, preferably 25° C (77° F). Once opened, discard unused portion because dextran contains no preservative.

ADVERSE EFFECTS (≥1%) **Body as a Whole:** Hypersensitivity (mild to generalized urticaria, pruritus, anaphylactic shock (rare), angioedema, dyspnea). **Other:** Renal tubular vacuolization (osmotic nephrosis), stasis, and blocking; oliguria, renal failure; increased AST and ALT, interference with platelet function, prolonged bleeding and coagulation times.

DIAGNOSTIC TEST INTERFERENCE When blood samples are drawn for study, notify laboratory that patient has received dextran. *Blood glucose:* False increases (utilizing *ortho-toluidine methods* or *sulfuric* or *acetic acid* hydrolysis). *Urinary protein:* False increases (utilizing *Lowry method*). *Bilirubin assays:* False increases when alcohol is used. *Total protein assays:* False increases using *biuret reagent. Rh testing, blood typing* and *cross-matching* procedures: Dextran may interfere with results (by inducing rouleaux formation) when *proteolytic enzyme techniques* are used (*saline agglutination* and *indirect antiglobulin methods* reportedly not affected).

Common adverse effects in *italic*, life-threatening effects underlined; generic names in **bold**; classifications in SMALL CAPS; ♣ Canadian drug name; ⊙ Prototype drug

437

INTERACTIONS Drug: May potentiate **abciximab** anticoagulant effects.

PHARMACOKINETICS Onset: Volume expansion within minutes of infusion. **Duration:** 12 h. **Metabolism:** Degraded to glucose and metabolized to CO_2 and water over a period of a few weeks. **Elimination:** 75% excreted in urine within 24 h; small amount excreted in feces.

NURSING IMPLICATIONS

Assessment & Drug Effects

- Evaluate patient's state of hydration before dextran therapy begins. Administration to severely dehydrated patients can result in renal failure.
- Lab tests: Baseline Hct prior to and after initiation of dextran (dextran usually lowers Hct). Notify prescriber if Hct is depressed below 30% by volume.
- Monitor vital signs and observe patient closely for at least the first 30 min of infusion. Hypersensitivity reaction is most likely to occur during the first few minutes of administration. Terminate therapy at the first sign of a hypersensitivity reaction (see Appendix F).
- Monitor CVP as an estimate of blood volume status and a guide for determining dosage. Normal CVP: 5–10 cm H_2O.
- Observe for S&S of circulatory overload (see Appendix F).
- Note: When sodium restriction is indicated, know that 500 mL of dextran 40 in 0.9% normal saline contains 77 mEq of both sodium and chloride.
- Monitor I&O ratio and check urine specific gravity at regular intervals. Low urine specific gravity may signify failure of renal dextran clearance and is an indication to discontinue therapy.
- Report oliguria, anuria, or lack of improvement in urinary output

(dextran usually causes an increase in urinary output). Discontinue dextran at first sign of renal dysfunction.
- High doses are associated with transient prolongation of bleeding time and interference with normal blood coagulation.

Patient & Family Education

- Report immediately S&S of bleeding: Easy bruising, blood in urine or dark tarry stool.

DEXTROAMPHETAMINE SULFATE

(dex-troe-am-fet′a-meen)

Dexampex, Dexedrine, Oxydess II ♣, Spancap No. 1
Classifications: RESPIRATORY AND CEREBRAL STIMULANT; AMPHETAMINE; ANOREXIANT
Therapeutic: AMPHETAMINE; ANOREXIANT
Prototype: Amphetamine
Pregnancy Category: C
Controlled Substance: Schedule II

AVAILABILITY 5 mg, 10 mg tablets; 5 mg, 10 mg, 15 mg sustained release capsules

ACTION & *THERAPEUTIC EFFECT*
Has anorexigenic action thought to result from CNS stimulation (twice that of amphetamine) and possibly from loss of acuity of smell and taste. *Is a more potent appetite suppressant than amphetamine. In hyperkinetic children, amphetamines reduce motor restlessness by an unknown mechanism.*

USES Adjunct in short-term treatment of exogenous obesity, narcolepsy, and attention deficit disorder with hyperactivity in children (also called minimal brain dysfunction or hyperkinetic syndrome).

UNLABELED USES Adjunct in epilepsy to control ataxia and drowsiness induced by barbiturates; to combat sedative effects of trimethadione in absence seizures.

CONTRAINDICATIONS Hypersensitivity to sympathomimetic amines, closed-angle glaucoma, agitated states, psychoses (especially in children), structural cardiac abnormalities, valvular heart disease; congenital heart disease, coronary heart disease, advanced arteriosclerosis, symptomatic heart disease, moderate to severe hypertension, hyperthyroidism, history of drug abuse, during or within 14 days of MAOI therapy; lactation.

CAUTIOUS USE Bipolar disease; salicylate hypersensitivity; seizure disorders; suicidal ideation, depression; salicylate hypersensitivity; pregnancy (category C). Safety and efficacy in children younger than 6 y for Narcolepsy and younger than 3 y for Attention Deficit Disorder have not been established.

ROUTE & DOSAGE

Narcolepsy

Adult: **PO** 5–20 mg 1–3 times/ day at 4–6 h intervals
Child: **PO** 6–12 y, 5 mg/day, may increase by 5 mg at weekly intervals; *older than 12 y,* 10 mg/ day, may increase by 10 mg at weekly intervals

Attention Deficit Disorder

Child: **PO** 3–5 y, 2.5 mg 1–2 times/day, may increase by 2.5 mg at weekly intervals; *6 y or older,* 5 mg 1–2 times/day, may increase by 5 mg at weekly intervals (max: 40 mg/day)

Obesity

Adult: **PO** 5–10 mg 1–3 times/ day or 10–15 mg of sustained release once/day 30–60 min a.c.

D

ADMINISTRATION

Oral

- Ensure that sustained release capsule is not chewed or crushed. It **must be** swallowed whole.
- Give 30–60 min before meals for treatment of obesity. Give long-acting form in the morning.
- Give last dose no later than 6 h before patient retires (10–14 h before bedtime for sustained release form) to avoid insomnia.
- Store in tightly closed containers at 15°–30° C (59°–86° F) unless otherwise directed.

ADVERSE EFFECTS (≥1%) **CNS:** Nervousness, *restlessness,* hyperactivity, *insomnia,* euphoria, dizziness, headache; *with prolonged use:* Severe depression, psychotic reactions. **CV:** Palpitations, tachycardia, elevated BP. **GI:** Dry mouth, unpleasant taste, anorexia, weight loss, diarrhea, constipation, abdominal pain. **Other:** Impotence, changes in libido, unusual fatigue, increased intraocular pressure, marked dystonia of head, neck, and extremities; sweating.

DIAGNOSTIC TEST INTERFERENCE Dextroamphetamine may cause significant elevations in *plasma corticosteroids* (evening levels are highest) and increases in *urinary epinephrine* excretion (during first 3 h after drug administration).

INTERACTIONS Drug: Acetazolamide, sodium bicarbonate decrease dextroamphetamine elimination; **ammonium chloride, ascorbic acid** increase dextroamphetamine elimination; effects of both BARBITURATES and

Common adverse effects in *italic,* life-threatening effects <u>underlined</u>; generic names in **bold;** classifications in SMALL CAPS; ✦ Canadian drug name; ❂ Prototype drug

439

dextroamphetamine may be antagonized; **furazolidone** may increase BP effects of AMPHETAMINES—interaction may persist for several weeks after discontinuing **furazolidone;** antagonizes antihypertensive effects of **guanethidine;** MAO INHIBITORS, **selegiline** can cause—hypertensive crisis (fatalities reported)—do not administer AMPHETAMINES during or within 14 days of these drugs; PHENO-THIAZINES may inhibit mood elevating effects of AMPHETAMINES; TRICYCLIC ANTIDEPRESSANTS enhance dextroamphetamine effects because of increased **norepinephrine** release; BETA-ADRENERGIC AGONISTS increase cardiovascular adverse effects.

PHARMACOKINETICS Absorption: Rapid. **Peak:** 1–5 h. **Duration:** Up to 10 h. **Distribution:** All tissues, especially the CNS. **Metabolism:** In liver. **Elimination:** Renal elimination; excreted in breast milk. **Half-Life:** 10–30 h.

NURSING IMPLICATIONS

Assessment & Drug Effects
- Monitor children, adolescents, and adults for signs and symptoms of adverse cardiac reactions (e.g., arrhythmias).
- Monitor growth rate closely in children.
- Monitor children and adolescents for development of aggressive or abnormal behaviors.
- Note: Tolerance to anorexiant effects may develop after a few weeks; however, tolerance does not appear to develop when dextroamphetamine is used to treat narcolepsy.

Patient & Family Education
- Swallow sustained release capsule whole with a liquid; do not chew or crush.
- Do not drive or engage in other potentially hazardous activities until response to drug is known.

- Drug is usually tapered off gradually following long-term use to avoid extreme fatigue, mental depression, and prolonged sleep pattern.

DEXTROMETHORPHAN HYDROBROMIDE
(dex-troe-meth-or′fan)

Balminil DM ♣, Benylin DM, Cremacoat 1, Delsym, DM Cough, Hold, Koffex ♣, Mediquell, Neo-DM ♣, Ornex DM ♣, Pedia Care, Pertussin 8 Hour Cough Formula, Robidex ♣, Robitussin DM, Romilar CF, Romilar Children's Cough, Sedatuss ♣, Sucrets Cough Control
Classification: ANTITUSSIVE
Therapeutic: ANTITUSSIVE
Prototype: Benzonatate
Pregnancy Category: C

AVAILABILITY 30 mg capsules; 2.5 mg, 5 mg, 7.5 mg, 10 mg/15 mL, 3.5 mg/5 mL, 7.5 mg/5 mL, 15 mg/5 mL liquid; 15 mg/15 mL, 10 mg/5 mL syrup

ACTION & *THERAPEUTIC EFFECT* Nonnarcotic derivative of levorphanol. Chemically related to morphine but without central hypnotic or analgesic effect. Controls cough spasms by depressing the cough center in medulla. Antitussive activity comparable to that of codeine. *Temporarily relieves coughing spasm.*

USES Temporary relief of cough spasms in nonproductive coughs due to colds, pertussis, and influenza.

CONTRAINDICATIONS Asthma, COPD, productive cough, persistent or chronic cough; severe hepatic function impairment.
CAUTIOUS USE Chronic pulmonary disease; enlarged prostate; concurrent use of MAOI; mild or moderate hepatic impairment; pregnancy

(category C); lactation. Safe use in children younger than 2 y not established.

ROUTE & DOSAGE

Cough

Adult: **PO** 10–20 mg q4h or 30 mg q6–8h (max: 120 mg/day) or 60 mg of sustained action liquid b.i.d.
Child: **PO** 2–6 y, 2.5–5 mg q4h or 7.5 mg q6–8h (max: 30 mg/day) or 15 mg sustained action liquid b.i.d.; 6–12 y, 5–10 mg q4h or 15 mg q6–8h (max: 60 mg/day) or 30 mg sustained action liquid b.i.d.

ADMINISTRATION

Oral

- Do not give lozenges to children younger than 6 y.
- Ensure that extended release form of drug is not chewed or crushed. It **must be** swallowed whole.
- Note: Although soothing local effect of the syrup may be enhanced if given undiluted, depression of cough center depends only on systemic absorption of drug.

ADVERSE EFFECTS (≥1%) **CNS:** Dizziness, drowsiness, CNS depression with very large doses; excitability, especially in children. **GI:** GI upset, constipation, abdominal discomfort.

INTERACTIONS Drug: High risk of excitation, hypotension, and hyperpyrexia with MAO INHIBITORS.

PHARMACOKINETICS Absorption: Readily from GI tract. **Onset:** 15–30 min. **Duration:** 3–6 h. **Metabolism:** In liver. **Elimination:** In urine.

NURSING IMPLICATIONS

Assessment & Drug Effects

- Monitor for dizziness and drowsiness, especially when concurrent

therapy with CNS depressant is used.

Patient & Family Education

- Note: Treatment aims to decrease the frequency and intensity of cough without completely eliminating protective cough reflex.
- While dextromethorphan is available OTC, any cough persisting longer than 1 wk–10 days needs to be medically diagnosed.

DIAZEPAM ⊘

(dye-az′e-pam)

Diastat, Diazemuls ♦, Valium
Classifications: BENZODIAZEPINE ANTICONVULSANT; ANXIOLYTIC
Therapeutic: ANTICONVULSANT; ANTIANXIETY
Pregnancy Category: D
Controlled Substance: Schedule IV

AVAILABILITY 2 mg, 5 mg, 10 mg tablets; 1 mg/mL, 5 mg/mL, 5 mg/5 mL oral solution; 5 mg/mL injection; 2.5 mg, 5 mg, 10 mg, 15 mg, 20 mg rectal gel

ACTION & *THERAPEUTIC EFFECT*
Long-acting benzodiazepine psychotherapeutic agent. Benzodiazepines act at the limbic, thalamic, and hypothalamic regions of the CNS and produce CNS depression resulting in sedation, hypnosis, skeletal muscle relaxation, and anticonvulsant activity dependent on the dosage. *Has antianxiety, anticonvulsant, and skeletal muscle relaxation properties.*

USES Drug of choice for status epilepticus. Management of anxiety disorders, for short-term relief of anxiety symptoms, to allay anxiety and tension prior to surgery, cardioversion and endoscopic procedures, as an amnesic, and treatment for restless legs. Also used to alleviate acute

Common adverse effects in *italic*, life-threatening effects underlined; generic names in **bold**; classifications in SMALL CAPS; ♦ Canadian drug name; ⊘ Prototype drug

441

withdrawal symptoms of alcoholism, voiding problems in older adults, and adjunctively for relief of skeletal muscle spasm associated with cerebral palsy, paraplegia, athetosis, stiffman syndrome, tetanus.

CONTRAINDICATIONS Acute narrow-angle glaucoma, untreated open-angle glaucoma; during or within 14 days of MAOI therapy; pregnancy (category D), lactation. **Injectable form:** Shock, coma, acute alcohol intoxication, depressed vital signs, obstetric patients. **CAUTIOUS USE** Epilepsy, psychoses, mental depression; myasthenia gravis; impaired hepatic or renal function; neuromuscular disease; bipolar disorder, dementia, Parkinson's disease; organic brain syndrome, psychosis, suicidal ideation; drug abuse, addiction-prone individuals. **Injectable form:** Extreme caution in older adults, the very ill, and patients with COPD, or asthma. Safe use in children younger than 30 days old not known.

ROUTE & DOSAGE

Status Epilepticus

Adult: **IV/IM** 5–10 mg, repeat if needed at 10–15 min intervals up to 30 mg, then repeat if needed q2–4h
Child (5 y or older): **IV** 1 mg/kg q2–5min (max: 10 mg), may repeat in 2–4 h
Child/Infant (1 mo–5 y): **IV** 0.2–0.5 mg slowly q2–5min up to 5 mg
Neonate: **IV** 0.1–0.3 mg/kg q15–30min (max total dose: 2 mg)

Muscle Spasm

Adult/Adolescent/Child (5 y or older): **IV** 5–10 mg q3–4h prn (larger dose for tetanus)

Child/Infant (1 mo–5 y): **IV** 1–2 mg q3–4h prn

Anxiety

Adult/Adolescent: **IV** 2–10 mg, repeat if needed in 3–4 h
Child/Infant (6 mo or older): **IV** 0.04–0.3 mg q2–4h (max: 0.6 mg/kg/8 h)

Alcohol Withdrawal

Adult: **IV** 10 mg then 5–10 mg in 3–4 h

Preoperative

Adult: **IV** 5–15 mg 5–10 min before procedure

ADMINISTRATION

Oral

- Ensure that sustained release form is not chewed or crushed. It **must be** swallowed whole. Give other tablets crushed with fluid or mixed with food if necessary.
- Supervise oral ingestion to ensure drug is swallowed.
- Avoid abrupt discontinuation of diazepam. Taper doses to termination.

Intramuscular

- Give deep into large muscle mass. Inject slowly. Rotate injection sites.

Intravenous

PREPARE: Direct: Do not dilute or mix with any other drug.
ADMINISTER: Direct: Give direct IV by injecting drug slowly, taking at least 1 min for each 5 mg (1 mL) given to adults and taking at least 3 min to inject 0.25 mg/kg body weight of children. ▪ If injection cannot be made directly into vein, inject slowly through infusion tubing as close as possible to vein insertion. ▪ The emulsion form is incompatible with

Common adverse effects in *italic*, life-threatening effects <u>underlined</u>; generic names in **bold**; classifications in SMALL CAPS; ♣ Canadian drug name; ☻ Prototype drug

PVC infusion sets. ▪ Avoid small veins and take extreme care to avoid intra-arterial administration or extravasation.

INCOMPATIBILITIES Solution/additive: Bleomycin, dobutamine, doxorubicin, epinephrine, fluorouracil, furosemide, glycopyrrolate, nalbuphine, sodium bicarbonate. Emulsion also incompatible with **morphine. Y-site:** Amphotericin B cholesteryl complex, atracurium, bivalirudin, cefepime, dexmedetomidine, diltiazem, fenoldopam, fluconazole, foscarnet, furosemide, heparin, hetastarch, lansoprazole, linezolid, meropenem, oxaliplatin, pancuronium, potassium chloride, propofol, remifentanil, tirofiban, vecuronium, vitamin B complex with C. Do not mix emulsion with any other drugs. Do not administer through **polyvinyl chloride (PVC)** infusion sets.

▪ Store in tight, light-resistant containers at 15°–30° C (59°–86° F), unless otherwise specified by manufacturer.

ADVERSE EFFECTS (≥1%) **Body as a Whole:** Throat and chest pain. **CNS:** *Drowsiness,* fatigue, ataxia, confusion, paradoxic rage, dizziness, vertigo, amnesia, vivid dreams, headache, slurred speech, tremor; EEG changes, tardive dyskinesia. **CV:** Hypotension, tachycardia, edema, <u>cardiovascular collapse</u>. **Special Senses:** Blurred vision, diplopia, nystagmus. **GI:** Xerostomia, nausea, constipation, hepatic dysfunction. **Urogenital:** Incontinence, urinary retention, gynecomastia (prolonged use), menstrual irregularities, ovulation failure. **Respiratory:** Hiccups, coughing, <u>laryngospasm</u>. **Other:**

Pain, venous thrombosis, phlebitis at injection site.

INTERACTIONS Drug: Alcohol, CNS DEPRESSANTS, ANTICONVULSANTS potentiate CNS depression; **cimetidine** increases diazepam plasma levels, increases toxicity; may decrease antiparkinson effects of **levodopa;** may increase **phenytoin** levels; smoking decreases sedative and antianxiety effects. **Herbal: Kava, valerian** may potentiate sedation.

PHARMACOKINETICS Absorption: Readily from GI tract; erratic IM absorption. **Onset:** 30–60 min PO; 15–30 min IM; 1–5 min IV. **Peak:** 1–2 h PO. **Duration:** 15 min–1 h IV; up to 3 h PO. **Distribution:** Crosses blood–brain barrier and placenta; distributed into breast milk. **Metabolism:** In liver to active metabolites. **Elimination:** Primarily in urine. **Half-Life:** 20–50 h.

NURSING IMPLICATIONS

Assessment & Drug Effects

▪ Monitor for adverse reactions. Most are dose related.

▪ Monitor for therapeutic effectiveness. Maximum effect may require 1–2 wk; patient tolerance to therapeutic effects may develop after 4 wk of treatment.

▪ Observe necessary preventive precautions for suicidal tendencies that may be present in anxiety states accompanied by depression.

▪ Observe patient closely and monitor vital signs when diazepam is given parenterally; hypotension, muscular weakness, tachycardia, and respiratory depression may occur.

▪ Lab tests: Periodic CBC and LFTs during prolonged therapy.

▪ Supervise ambulation. Adverse reactions such as drowsiness and

D

ataxia are more likely to occur in older adults and debilitated or those receiving larger doses. Dosage adjustment may be necessary.

- Monitor I&O ratio, including urinary and bowel elimination.
- Note: Psychic and physical dependence may occur in patients on long-term high dosage therapy, in those with histories of alcohol or drug addiction, or in those who self-medicate.

Patient & Family Education

- Avoid alcohol and other CNS depressants during therapy unless otherwise advised by prescriber. Concomitant use of these agents can cause severe drowsiness, respiratory depression, and apnea.
- Do not drive or engage in other potentially hazardous activities or those requiring mental precision until reaction to drug is known.
- Tell prescriber if you become or intend to become pregnant during therapy; drug may need to be discontinued.
- Take drug as prescribed; do not change dose or dose intervals.

DIAZOXIDE

(dye-az-ox′ide)

Proglycem
Classifications: VASODILATOR; ANTIHYPERTENSIVE; SULFONYLUREA; ANTIDIABETIC
Therapeutic: ANTIHYPERTENSIVE; HYPOGLYCEMIC AGENT
Prototype: Hydralazine
Pregnancy Category: C

AVAILABILITY 50 mg/mL oral suspension.

ACTION & *THERAPEUTIC EFFECT*
Rapid-acting thiazide nondiuretic hypotensive and hyperglycemic agent. Causes sodium and water

retention, thus decreasing urinary output. This is probably due to its increase of proximal tubular reabsorption of sodium. Has a dose related increase in blood glucose level caused by inhibition of insulin release from the pancreas, and to an extra pancreatic effect. *Reduces peripheral vascular resistance and BP by direct vasodilatory effect on peripheral arteriolar smooth muscles. Increases blood glucose level.*

USES Orally in treatment of various diagnosed hypoglycemic states due to hyperinsulinism when other medical treatment or surgical management has been unsuccessful or is not feasible.

CONTRAINDICATIONS Hypersensitivity to diazoxide; cerebral bleeding, eclampsia; aortic coarctation; AV shunt, significant coronary artery disease; pheochromocytoma; lactation. Use of oral diazoxide for functional hypoglycemia or in presence of increased bilirubin in newborns.
CAUTIOUS USE Diabetes mellitus; impaired cerebral or cardiac circulation; impaired renal function; patients taking corticosteroids or estrogen–progestogen combinations; hyperuricemia, history of gout, uremia; thiazide diuretic hypersensitivity; pregnancy (category C).

ROUTE & DOSAGE

Hypoglycemia
Adult/Child: **PO** 1 mg/kg PO q8h
Neonate/Infant: **PO** 8–10 mg/kg/day divided q8–12h

ADMINISTRATION
- Do not give darkened solutions. Store at 2°–30° C (36°–86° F) unless otherwise directed. Protect from light, heat, and freezing.

ADVERSE EFFECTS (≥1%) **CNS:** Headache, weakness, malaise, *dizziness*, polyneuritis, sleepiness, insomnia, euphoria, anxiety, extrapyramidal signs. **CV:** Palpitations, atrial and ventricular arrhythmias, flushing, shock; *orthostatic hypotension,* CHF, transient hypertension. **Special Senses:** Tinnitus, momentary hearing loss; blurred vision, transient cataracts, subconjunctival hemorrhage, ring scotoma, diplopia, lacrimation, papilledema. **GI:** *Nausea, vomiting,* abdominal discomfort, diarrhea, constipation, ileus, anorexia, transient loss of taste, impaired hepatic function. **Hematologic:** Transient neutropenia, eosinophilia, decreased Hgb/Hct, decreased IgG. **Body as a Whole:** Hypersensitivity (rash, fever, leukopenia); chest and back pain, muscle cramps. **Urogenital:** Decreased urinary output, nephrotic syndrome (reversible), hematuria, increased nocturia, proteinuria, azotemia; inhibition of labor. **Skin:** Monilial dermatitis, herpes, hirsutism; loss of scalp hair, sweating, sensation of warmth, burning, or itching. **Endocrine:** Advance in bone age (children), *hyperglycemia, sodium and water retention, edema,* hyperuricemia, glycosuria, enlargement of breast lump, galactorrhea; decreased immunoglobulinemia, hirsutism.

DIAGNOSTIC TEST INTERFERENCE Diazoxide can cause false-negative response to ***glucagon.***

INTERACTIONS Drug: SULFONYLUREAS antagonize effects; THIAZIDE DIURETICS may intensify hyperglycemia and antihypertensive effects; **phenytoin** increases risk of hyperglycemia, and diazoxide may increase **phenytoin** metabolism, causing loss of seizure control.

PHARMACOKINETICS Onset: 1 h. **Duration:** 8 h. **Distribution:** Crosses blood–brain barrier and placenta. **Metabolism:** Partially metabolized in the liver. **Elimination:** In urine. **Half-Life:** 21–45 h.

NURSING IMPLICATIONS

Assessment & Drug Effects
- Monitor closely for S&S of CHF (e.g., development of edema, weight gain). Sodium and fluid retention may precipitate CHF in those with preexisting cardiac disease.
- Lab tests: Baseline and periodic blood glucose, urine for glucose and ketones, serum electrolytes, CBC with differential, Hct, platelet count, AST, and serum uric acid, and at regular intervals in patients receiving multiple doses.
- Report promptly any change in I&O ratio.
- Oral administration usually does not produce marked effects on BP. However, do make periodic measurements of BP and vital signs.

Patient & Family Education
- Note: Drug may cause hyperglycemia and glycosuria. Closely monitor blood and urine glucose; report any abnormalities to prescriber.
- Report palpitations, chest pain, dizziness, fainting, or severe headache.

DIBUCAINE
(dye′byoo-kane)

Nupercainal
Classification: ANESTHETIC, LOCAL (AMIDE-TYPE)
Therapeutic: LOCAL ANESTHETIC
Prototype: Procaine
Pregnancy Category: C

Common adverse effects in *italic,* life-threatening effects underlined; generic names in **bold;** classifications in SMALL CAPS; ♣ Canadian drug name; ⊙ Prototype drug

445

AVAILABILITY 1% ointment

ACTION & *THERAPEUTIC EFFECT*
Long-acting anesthetic of the amide type that appears to inhibit initiation and conduction of nerve impulses by reducing permeability of nerve cell membrane to sodium ions. *Relief of pain and itching due to inhibiting conduction of nerve impulses.*

USES Fast, temporary relief of pain and itching due to hemorrhoids and other anorectal disorders, nonpoisonous insect bites, sunburn, minor burns, cuts, and scratches.

CONTRAINDICATIONS Hypersensitivity to amide-type anesthetics, children younger than 1 y.
CAUTIOUS USE Pregnancy (category C), lactation, children younger than 12 y.

ROUTE & DOSAGE

Itching Due to Insect Bites or Hemorrhoids
Adult: **Topical** Apply skin cream or ointment to affected area as needed [max: 1 oz (28 g)/24 h]; insert rectal ointment morning and evening and after each bowel movement
Child: **Topical** Apply skin cream or ointment to affected area as needed [max: ¼ oz (7 g)/24 h]

ADMINISTRATION
Topical
- Apply cream preparation after bathing or swimming (water soluble).
- Store at 15°–30° C (59°–86° F) in tight, light-resistant containers.

ADVERSE EFFECTS (≥1%) **Skin:** *Irritation, contact dermatitis; rectal bleeding (suppository).*

PHARMACOKINETICS Absorption: Poorly absorbed from intact skin;

readily absorbed from mucous membranes or abraded skin. **Onset:** 15 min. **Duration:** 2–4 h.

NURSING IMPLICATIONS
Patient & Family Education
- Discontinue if irritation or rectal bleeding (following use of rectal preparations) develops and consult prescriber.
- Prescriber may prescribe sitz baths 3–4 times/day to reduce the swelling and pain of hemorrhoids.
- Note: Medication is intended for temporary relief of mild to moderate itching or pain. Seek medical advice for continuing discomfort, pain, bleeding, or sensation of rectal pressure.

DICLOFENAC SODIUM
(di-klo′fen-ak)
PENNSAID, Solaraze, Voltaren, Voltaren-XR

DICLOFENAC POTASSIUM
Cambia, Cataflam, Zipsor

DICLOFENAC EPOLAMINE
Flector
Classifications: NONSTEROIDAL ANALGESIC, ANTI-INFLAMMATORY DRUG (NSAID)
Therapeutic: ANALGESIC, NSAID; ANTIPYRETIC
Prototype: Ibuprofen
Pregnancy Category: C

AVAILABILITY Diclofenac Sodium: 25 mg, 50 mg, 75 mg delayed release tablets; 100 mg sustained release tablets; 0.1% ophth solution; 1%, 3% gel; 1.5% topical solution. **Diclofenac Potassium:** 50 mg tablets, 50 mg powder for solution. **Diclofenac Epolamine:** 1.3% transdermal patch

ACTION & *THERAPEUTIC EFFECT*

Diclofenac competitively inhibits both cyclooxygenase (COX) isoenzymes, COX-1 and COX-2, by blocking arachidonic acid conversion to other chemicals, thus leading to its analgesic, antipyretic, and anti-inflammatory effects. It appears to be a potent inhibitor of cyclooxygenase, thereby decreasing the synthesis of prostaglandins. *Nonsteroidal anti-inflammatory drug (NSAID) with analgesic and antipyretic activity.*

USES Analgesic and antipyretic effects in symptomatic treatment of rheumatoid arthritis, osteoarthritis, and ankylosing spondylitis. Also acute gout; juvenile rheumatoid arthritis; various rheumatic conditions. **Ophthalmic:** Cataract surgery; photophobia associated with refractive surgery. **Topical:** Treatment of actinic keratosis. **Transdermal:** Acute pain.

CONTRAINDICATIONS Hypersensitivity to diclofenac, NSAIDS, or salicylate; patients in whom asthma, urticaria, angioedema, bronchospasm, severe rhinitis, history of GI bleeding; hepatic porphyria; shock, or other sensitivity reaction is precipitated by aspirin or other NSAIDS; perioperative CABG pain.

CAUTIOUS USE Patients receiving anticoagulant therapy; diabetes mellitus; history of GI disease or bleeding; hepatic disease; GU tract problems such as dysuria, cystitis, hematuria, nephritis, nephrotic syndrome, patients who must restrict their sodium intake; impaired hepatic function; SLE; heart failure, cardiac disease; hypertension; older adults, children; pregnancy (category C), lactation.

ROUTE & DOSAGE

Rheumatoid Arthritis

Adult: **PO** 150–200 mg/day in 3–4 divided doses or 75 mg delayed release daily or 100 mg sustained release daily
Child: **PO** 25 mg b.i.d. or t.i.d.

Osteoarthritis

Adult: **PO** 100–150 mg/day in 3–4 divided doses; 75 mg delayed release daily; 100 mg sustained release daily **Topical (gel)** 4 g for each knee, ankle or foot q.i.d. **(solution)** 40 drops to each affected knee q.i.d.

Ankylosing Spondylitis

Adult: **PO** 25 mg q.i.d. and 25 mg at bedtime

Cataract Surgery

Adult: **Ophthalmic** 1 drop of 0.1% solution in affected eye q.i.d. beginning 24 h after surgery and continuing for 2 wk

Actinic Keratosis

Adult: **Topical** Apply to affected area b.i.d. for 60–90 days

Acute Pain (Flector)

Adult: **Transdermal** Apply one patch to most painful area b.i.d.

ADMINISTRATION

Oral

- Ensure that sustained release forms of drug are not chewed or crushed. **Must be** swallowed whole.
- Minimize gastric irritation by administering it with a full glass of milk or food.
- Store at 15°–30° C (59°–86° F) away from heat and direct light.

Common adverse effects in *italic*, life-threatening effects <u>underlined</u>; generic names in **bold**; classifications in SMALL CAPS; ✤ Canadian drug name; ❶ Prototype drug

447

D

Topical/Transdermal

- Do not apply gel or patch to areas of skin irritation.
- Massage gel into skin of entire affected area. Do not wash area within 1 hr of application.
- Avoid application of any other topical products to treated area.
- Do not apply external heat or occlusive dressing to treated area.

ADVERSE EFFECTS (≥1%) **CNS:** Dizziness, headache, drowsiness. **Special Senses:** Tinnitus. **Skin:** Rash, pruritus. **GI:** *Dyspepsia,* nausea, vomiting, abdominal pain, cramps, constipation, diarrhea, indigestion, abdominal distension, flatulence, peptic ulcer; liver enzymes, transaminases increased, liver test abnormalities. **CV:** Fluid retention, hypertension, CHF. **Respiratory:** Asthma. **Body as a Whole:** Back, leg, or joint pain. **Endocrine:** Hyperglycemia. **Hematologic:** Prolonged bleeding time; inhibits platelet aggregation.

DIAGNOSTIC TEST INTERFERENCE *Liver function test* values may be increased. *Liver function test* abnormalities may return to normal despite continued use; however, if significant abnormalities occur, clinical signs and symptoms consistent with liver disease develop, or systemic manifestations such as eosinophilia or rash occur, the medication should be discontinued. *Serum uric acid* concentrations may be decreased because of increased *renal clearance.*

INTERACTIONS Drug: Increases **cyclosporine**-induced nephrotoxicity; increases **methotrexate** levels (increases toxicity); may decrease BP-lowering effects of DIURETICS; may increase levels and toxicity of **lithium;** may increase **digoxin** levels. **Herbal: Feverfew, garlic,** **ginger, ginkgo** may increase risk of bleeding.

PHARMACOKINETICS Absorption: Readily absorbed from GI tract; 50–60% reaches systemic circulation. **Peak:** 2–3 h. **Distribution:** Widely distributed including synovial fluid and into breast milk; 99% protein bound. **Metabolism:** Extensively metabolized in liver. **Elimination:** 50–70% in urine, 30–35% in feces. **Half-Life:** 1.2–2 h (PO); 12 h (transdermal).

NURSING IMPLICATIONS

Assessment & Drug Effects

- Lab tests: Periodic LFTs, serum uric acid concentrations, Hct, PT/INR, and blood glucose.
- Observe and report signs of bleeding (e.g., petechiae, ecchymoses, bleeding gums, bloody or black stools, cloudy or bloody urine).
- Monitor BP for hypertension and blood sugar for hyperglycemia.
- Monitor diabetics closely for loss of diabetic control.
- Monitor for increased serum sodium and potassium in patients receiving potassium-sparing diuretics.
- Monitor for S&S of CHF, including weight gains greater than 1 kg (2 lb)/24 h.
- Monitor for signs and symptoms of GI irritation and ulceration.

Patient & Family Education

Oral Form

- Do not lie down for 15–30 min after taking medicine to decrease esophageal irritation.
- Discontinue use with onset of ringing or buzzing in the ears, impaired hearing, dizziness, GI discomfort, or bleeding and notify prescriber.

- Do not take aspirin or other OTC analgesics without permission of the prescriber.
- Avoid alcohol or other CNS depressants.
- Do not drive or engage in other potentially hazardous activities until reaction to drug is known.
- Note: Diabetics need to monitor blood glucose carefully for loss of glycemic control.

DICLOXACILLIN SODIUM

(dye-klox-a-sill'in)

Classification: PENICILLIN ANTIBIOTIC
Therapeutic: PENICILLIN ANTIBIOTIC
Prototype: Penicillin G potassium
Pregnancy Category: B

AVAILABILITY 125 mg, 250 mg, 500 mg capsules

ACTION & *THERAPEUTIC EFFECT*
Semisynthetic, acid-stable, penicillinase-resistant penicillin. It inhibits the final stage of bacterial cell wall synthesis by preferentially binding to specific penicillin-binding proteins (PBPs) that are located inside bacterial cell wall; this leads to cell death. *Effective against penicillinase-producing staphylococci.*

USES Primarily in systemic infections caused by penicillinase-producing staphylococci and penicillin-resistant staphylococci.

CONTRAINDICATIONS Hypersensitivity to penicillins.
CAUTIOUS USE History of or suspected atopy or allergy (asthma, eczema, hives, hay fever); history of hypersensitivity to cephalosporins or carbapenem; GI disease, colitis; renal or hepatic impairment; pregnancy (category B); lactation.

ROUTE & DOSAGE

Mild to Moderate Infections
Adult: **PO** 125–500 mg q6h
Child (weight less than 40 kg): **PO** 12.5–25 mg/kg q6h (max: 4 g/day)

ADMINISTRATION
Oral
- Give on an empty stomach at least 1 h before or 2 h after meals. Food reduces drug absorption.
- Store capsules at room temperature in tight containers unless otherwise directed.

ADVERSE EFFECTS (≥1%) **Body as a Whole:** Hypersensitivity (pruritus, urticaria, rash, wheezing, sneezing, <u>anaphylaxis</u>; eosinophilia). **GI:** Nausea, vomiting, flatulence, *diarrhea*, abdominal pain. **Other:** Transient elevations of ALT, superinfections.

INTERACTIONS Drug: Probenecid decreases dicloxacillin elimination.

PHARMACOKINETICS Absorption: 35–76% absorbed from GI tract. **Peak:** 0.5–2 h. **Duration:** 4–6 h. **Distribution:** Distributed throughout body with highest concentrations in liver and kidney; low CSF penetration; crosses placenta; distributed into breast milk. **Metabolism:** In liver. **Elimination:** Primarily in urine with some elimination through bile. **Half-Life:** 30–60 min.

NURSING IMPLICATIONS
Assessment & Drug Effects
- Note: Take care to establish previous exposure and sensitivity to penicillins and cephalosporins as well as other allergic reactions of any kind before initiating therapy.

Common adverse effects in *italic*, life-threatening effects <u>underlined</u>; generic names in **bold;** classifications in SMALL CAPS; ♣ Canadian drug name; ⊘ Prototype drug

449

- Obtain C&S prior to initiation of therapy to determine susceptibility of causative organism. Therapy may begin pending test results.
- Lab tests: Baseline blood culture; weekly WBC with differential during prolonged therapy. Periodic LFTs, urinalysis, BUN, and creatinine are also advised for these patients.

Patient & Family Education

- Take medication around the clock. Do not miss a dose and continue taking medication until it is all gone, unless otherwise directed by prescriber.
- Check with prescriber if GI side effects appear.
- Watch for and report the signs of hypersensitivity reactions and superinfections (see Appendix F).

DICYCLOMINE HYDROCHLORIDE

(dye-sye'kloe-meen)

Bentyl, Bentylol ✦, Formulex ✦, Lomine ✦
Classifications: ANTICHOLINERGIC; ANTISPASMODIC
Therapeutic: GI ANTISPASMODIC
Prototype: Atropine
Pregnancy Category: B

AVAILABILITY 10 mg capsules; 20 mg tablets; 10 mg/5 mL syrup; 10 mg/mL injection

ACTION & THERAPEUTIC EFFECT Synthetic tertiary amine that relieves smooth muscle spasm by direct effect on the muscles as well as by antagonism of bradykinin and histamine-induced spasm in GI tract. *Exerts antispasmodic effect on the GI tract.*

USES Irritable bowel syndrome.

CONTRAINDICATIONS Hypersensitivity to anticholinergic drugs; obstructive diseases of GU and GI tracts, paralytic ileus, intestinal atony, biliary tract disease; closed-angle glaucoma; unstable cardiovascular status; severe ulcerative colitis, toxic megacolon, esophagitis; myasthenia gravis; peripheral neuropathy; lactation.

CAUTIOUS USE Prostatic hypertrophy; autonomic neuropathy; hyperthyroidism; coronary heart disease, CHF, arrhythmias, hypertension; hepatic or renal disease; GERD, hiatal hernia associated with esophageal reflux; pregnancy (category B). Safety and efficacy in children are not established.

ROUTE & DOSAGE

Irritable Bowel Disorders
Adult/Adolescent: **PO** 20–40 mg q.i.d. **IM** 20 mg q4–6h.

ADMINISTRATION

Oral

- Give 30 min before meals and at bedtime.

Intramuscular

- Give deep IM into a large muscle. Do NOT give IV.
- Store below 30° C (86° F) unless otherwise directed.

ADVERSE EFFECTS (≥1%) **All:** Dose related. **Body as a Whole:** Allergic reactions; curare-like effect (cyanosis, apnea, respiratory arrest); decreased sweating; suppression of lactation; urticaria. **CNS:** Lightheadedness, drowsiness, headache, insomnia, brief euphoria, fever, restlessness, irritability, coma, seizures. **CV:** Fluctuations in heart rate, palpitation, tachycardia. **GI:** *Dry mouth,* nausea, *constipation,* paralytic ileus, vomiting, diminished sense of taste, bloated feeling. **Urogenital:** Urinary hesitancy, *urinary retention,* impotence. **Special Senses:** Blurred vision.

PHARMACOKINETICS Absorption: Readily from GI tract. **Onset:** 1–2 h. **Duration:** 4 h. **Metabolism:** In liver. **Elimination:** 80% in urine, 10% in feces. **Half-Life:** 9–10 h.

NURSING IMPLICATIONS

Assessment & Drug Effects

- Monitor for adverse effects especially in infants. Treatment of infant colic with dicyclomine includes some risk, especially in infants younger than 2 mo of age. Infants younger than 6 wk have developed respiratory symptoms as well as seizures, fluctuations in heart rate, weakness, and coma within minutes after taking syrup formulation. Symptoms generally last 20–30 min and are believed to be due to local irritation.
- Monitor I&O to assess for urinary retention.
- If drug produces drowsiness and light-headedness, supervision of ambulation and other safety precautions are warranted.

Patient & Family Education

- Exercise caution in hot weather. Dicyclomine may increase risk of heatstroke by decreasing sweating, especially in older adults.
- Do not drive or engage in other potentially hazardous activities until reaction to drug is known.
- Report changes in urine volume, voiding pattern.

DIDANOSINE (DDI)

(di-dan′o-sine)

Videx, Videx EC

Classifications: ANTIRETROVIRAL; NUCLEOSIDE REVERSE TRANSCRIPTASE INHIBITOR (NRTI)

Therapeutic: ANTIRETROVIRAL (NRTI)
Prototype: Lamivudine
Pregnancy Category: B

D

AVAILABILITY 125 mg, 200 mg, 250 mg, 400 mg delayed release capsules; powder for oral solution

ACTION & *THERAPEUTIC EFFECT*

DDI interferes with the HIV RNA-dependent DNA polymerase (reverse transcriptase), thus preventing replication of the virus. *Synthetic purine nucleotide that inhibits replication of HIV.*

USES Advanced HIV infection in patients who are intolerant to zidovudine (AZT) or who demonstrate significant clinical or immunologic deterioration during zidovudine therapy.

CONTRAINDICATIONS Hypersensitivity to any of the components in the formulation; pancreatitis; PKU; lactation.

CAUTIOUS USE Individuals with peripheral vascular disease, history of neuropathy, chronic pancreatitis, renal impairment, or any liver impairment; patients on sodium restriction; renal failure, renal impairment; alcoholism; elderly; gout; concurrent use with stavudine in pregnancy; pregnancy (category B).

ROUTE & DOSAGE

HIV Infection

Adult/Adolescent/Child: **PO** Weight 60 kg or more, tablets, 400 mg daily or 200 mg b.i.d.; weight 25–60 kg, tablets, 250 mg daily or 125 mg b.i.d.; weight 20–25 kg, tablets 200 mg daily
Child/Infant (older than 8 mo): **PO (solution)** 120 mg/m² b.i.d.
Neonate/Infant (2 wk–8 mo): **PO** 100 mg/m² q8h

Common adverse effects in *italic*, life-threatening effects underlined; generic names in **bold**; classifications in SMALL CAPS; ◆ Canadian drug name; ❷ Prototype drug

451

D

Renal Impairment Dosage Adjustment

Varies based on patient weight and dosage form used; see package insert

ADMINISTRATION

Oral

- Give drug on an empty stomach. Food should not be consumed within 15–30 min of drug administration.
- Give with water. Do NOT give with fruit juice or any other acid-containing liquid.
- Ensure that delayed release forms are swallowed whole. They must not be crushed or chewed.
- Mix powder for oral solution (buffered) with at least 120 mL (4 oz) of water, stir until dissolved (requires 2–3 min), and immediately swallowed.
- Dosage reduction may be indicated in those with renal impairment.
- Store reconstituted liquid in a tightly closed container in refrigerator for up to 30 days.

ADVERSE EFFECTS (≥1%) **CV:** Palpitations, thrombophlebitis, arrhythmias, *vasodilation.* **CNS:** *Headache, dizziness, nervousness, insomnia, peripheral neuropathy,* lethargy, poor coordination, seizures. **Special Senses:** Retinal depigmentation, photophobia, blurred vision, optic neuritis, diplopia, blindness. **GI:** *Abdominal pain, nausea, vomiting, diarrhea,* constipation, stomatitis, dry mouth, pancreatitis, increased liver enzymes. **Hematologic:** Increased WBC, neutrophil, lymphocyte, and platelet counts; increased Hgb, thrombocytopenia, ecchymosis, hemorrhage, petechiae. **Metabolic:** Hypocalcemia, hypokalemia, hypomagnesemia, hyperuricemia (asymptomatic), *hypertriglyceridemia.* **Musculoskeletal:** Muscle atrophy, myalgia, arthritis, decreased strength. **Respiratory:** *Asthma, cough, dyspnea, epistaxis, rhinitis, rhinorrhea,* hypoventilation, pharyngitis, rhonchi or rales, sinusitis, congestion. **Skin:** Rash, impetigo, eczema, *pruritus, sweating,* erythema.

INTERACTIONS Drug: ALUMINUM- and MAGNESIUM-CONTAINING ANTACIDS may increase the aluminum- and magnesium-associated adverse effects of tablets. The effectiveness of **dapsone** in prophylaxis of *Pneumocystis carinii* pneumonia may be reduced by concomitant didanosine. May cause additive neuropathy with **zalcitabine** (ddC). **Food:** Absorption is significantly decreased by food. Take on an empty stomach.

PHARMACOKINETICS Absorption: Rapidly absorbed from GI tract when administered to fasting patient with antacids; 23–40% reaches systemic circulation. **Peak:** 0.6–1 h. **Distribution:** Distributed primarily to body water; 21% reaches CSF; crosses placenta. **Elimination:** 36% in urine. **Half-Life:** 0.8–1.5 h.

NURSING IMPLICATIONS

Assessment & Drug Effects

- Monitor for S&S of pancreatitis (e.g., abdominal pain, nausea, vomiting, elevated serum amylase). Report immediately to prescriber and withhold drug until ruled out.
- Monitor for S&S of peripheral neuropathy (e.g., numbness, tingling, burning, pain in hands or feet). Report to prescriber; dose reduction may be indicated.

- Monitor patients with renal impairment for drug toxicity and hypermagnesemia manifested by muscle weakness and confusion.
- Lab tests: Periodic CBC with differential, serum electrolytes including magnesium, uric acid, and lipid profile.

Patient & Family Education
- Report immediately to prescriber any of the following: Abdominal pain, nausea, or vomiting.

DIETHYLPROPION HYDROCHLORIDE ⊕

(dye-eth-il-proe'pee-on)

Nobesine ♣, Radtue
Classification: ANOREXIANT
Therapeutic: ANOREXIANT
Pregnancy Category: B
Controlled Substance: Schedule IV

AVAILABILITY 25 mg tablets; 75 mg sustained release tablets

ACTION & *THERAPEUTIC EFFECT*
Sympathomimetic amine chemically related to amphetamine. Anorexigenic action probably secondary to direct (CNS) stimulation of appetite control center in hypothalamus and limbic regions. *Suppresses appetite as a result of drug action on CNS appetite control center.*

USES As short-term (a few weeks) adjunct in a regimen of weight reduction based on caloric restriction in obesity management.

CONTRAINDICATIONS Known hypersensitivity or idiosyncrasy to sympathomimetic amines; severe hypertension, advanced arteriosclerosis, valvular heart disease; hyperthyroidism; glaucoma; history of drug abuse; anorexia nervosa; symptomatic cardiovascular disease, arrhythmias; MAOI therapy; pulmonary hypertension.

CAUTIOUS USE Hypertension, psychosis, mania, agitated states, epilepsy; diabetes mellitus; elderly, renal failure or impairment; seizure disorder; pregnancy (category B), lactation. Safe use in children younger than 6 y is not known.

ROUTE & DOSAGE

Obesity
Adult: **PO** 25 mg t.i.d. 30–60 min a.c. or 75 mg sustained release daily midmorning

ADMINISTRATION

Oral
- Give on an empty stomach, 30 min–1 h before meals.
- Note: Additional dose sometimes prescribed in midevening to control nighttime hunger. Rarely causes insomnia except in high doses.
- Store between 15°–30° C (59°–86° F) in well-closed container unless otherwise specified.

ADVERSE EFFECTS (≥1%) **Body as a Whole:** Hypersensitivity (urticaria, rash, erythema); muscle pain, dyspnea, hair loss, blurred vision, severe dermatoses (chronic intoxication), increased sweating. **CNS:** Mild euphoria, restlessness, *nervousness,* dizziness, headache, irritability, hyperactivity, insomnia, drowsiness, mood changes, lethargy, increase in convulsive episodes in patients with epilepsy. **CV:** Palpitation, tachycardia, precordial pain, rise in BP. **GI:** Nausea, vomiting, diarrhea, constipation, dry mouth, unpleasant taste. **Urogenital:** Impotence, changes in libido, gynecomastia, menstrual irregularities; polyuria, dysuria.

INTERACTIONS Drug: Acetazolamide, sodium bicarbonate decreases diethylpropion elimination;

Common adverse effects in *italic*, life-threatening effects underlined; generic names in **bold**; classifications in SMALL CAPS; ♣ Canadian drug name; ⊕ Prototype drug

453

ammonium chloride, ascorbic acid increases diethylpropion elimination; a BARBITURATE and diethylpropion taken together may antagonize the effects of both drugs; **furazolidone** may increase blood pressure effects of AMPHETAMINES, and interaction may persist for several weeks after discontinuation of **furazolidone; guanethidine** antagonizes antihypertensive effects; MAO INHIBITORS, **selegiline** can cause hypertensive crisis (fatalities reported)—AMPHETAMINES should not be administered at the same time as or within 14 days of these drugs; PHENOTHIAZINES may inhibit mood elevating effects of AMPHETAMINES; TRICYCLIC ANTIDEPRESSANTS enhance AMPHETAMINES' effects by increasing **norepinephrine** release; BETA AGONISTS increase cardiovascular adverse effects.

PHARMACOKINETICS Absorption: Readily from GI tract. **Duration:** 4 h, regular tablets; 10–14 h, sustained release. **Elimination:** In urine. **Half-Life:** 4–6 h.

NURSING IMPLICATIONS

Assessment & Drug Effects
- Observe patients with epilepsy closely for reduction in seizure control.
- Monitor diabetics for loss of glycemic control.
- Note: Varying degrees of psychologic and rarely physical dependence can occur.

Patient & Family Education
- Swallow sustained release tablets whole; do NOT chew.
- Do not drive or engage in other potentially hazardous activities until reaction to drug is known.
- If diabetic, closely monitor blood glucose values.

DIFLORASONE DIACETATE
(dye-flor'a-sone)

Florone, Florone E, Maxiflor, Psorcon
Pregnancy Category: C
See Appendix A-4.

DIFLUNISAL
(dye-floo'ni-sal)

Dolobid
Classifications: ANALGESIC, NONSTEROIDAL ANTI-INFLAMMATORY DRUG (NSAID)
Therapeutic: ANALGESIC, NSAID; ANTIPYRETIC; ANTIRHEUMATIC
Prototype: Ibuprofen
Pregnancy Category: C; D third trimester

AVAILABILITY 500 mg tablets

ACTION & *THERAPEUTIC EFFECT*
A long-acting nonsteroidal anti-inflammatory drug (NSAID) is a nonnarcotic analgesic agent. This NSAID has peripheral analgesic properties due to interfering with prostaglandin synthesis by inhibiting cyclooxygenase (COX) isoenzymes, COX-1 and COX-2. *Has analgesic and anti-inflammatory properties.*

USES Treatment of osteoarthritis and rheumatoid arthritis.
UNLABELED USES Other pain disorders.

CONTRAINDICATIONS Patients in whom aspirin or other NSAIDs precipitate an acute asthmatic attack (bronchospasm), urticaria, angioedema, severe rhinitis, or shock; active peptic ulcer, GI bleeding; severe salicylate hypersensitivity; treatment of perioperative pain in CABG care; pregnancy (category D third trimester).

CAUTIOUS USE History of upper GI disease; preexisting renal disease; impaired renal or hepatic function; alcoholics; compromised cardiac function, and other conditions associated with fluid retention; bone marrow suppression; geriatric patients; hypertension; patients who may be adversely affected by prolonged bleeding time; elderly; pregnancy (category C first and second trimester), lactation. Safe use in children younger than 12 y not established.

ROUTE & DOSAGE

Arthritis

Adult: PO 250–500 b.i.d. (max: 1500 mg/day)

ADMINISTRATION

Oral

- Give with water, milk, or food to reduce GI irritation. Food causes slight reduction in absorption rate, but does not affect total amount absorbed.
- Store at 15°–30° C (59°–86° F) in tightly closed containers unless otherwise directed.

ADVERSE EFFECTS (≥1%) **Body as a Whole:** Hypersensitivity syndrome (fever, chills, rash, eosinophilia, changes in renal and hepatic function, anaphylactic reactions with bronchospasm). **CNS:** Headache, drowsiness, insomnia, dizziness, vertigo, light-headedness, fatigue, weakness, nervousness, confusion, disorientation. **CV:** Palpitation, tachycardia, *peripheral edema.* **Special Senses:** Tinnitus, hearing loss; blurred vision, reduced visual acuity, changes in color vision, scotomas, corneal deposits, retinal disturbances. **GI:** *Nausea,* GI pain, flatulence, GI bleeding, peptic ulcer, anorexia, eructation, cholestatic jaundice. **Urogenital:** Hematuria, proteinuria, interstitial nephritis, renal failure. **Hematologic:** Prolonged PT, anemia, decreased serum uric acid, transient elevations of liver function tests. **Skin:** Rash, toxic epidermal necrolysis, exfoliative dermatitis, urticaria. **Other:** Weight gain, hyperventilation, dyspnea, photosensitivity.

DIAGNOSTIC TEST INTERFERENCE False elevation of **serum salicylate levels**.

INTERACTIONS Drug: ANTACIDS decrease diflunisal absorption; **aspirin** and other NSAIDS increase risk of GI bleeding; increases risk of **warfarin**-induced hypoprothrombinemia; increases **methotrexate** levels and toxicity.

PHARMACOKINETICS Absorption: Readily from GI tract. **Onset:** 1 h. **Peak:** 2–3 h. **Duration:** 12 h. **Distribution:** Probably crosses placenta; distributed into breast milk. **Metabolism:** In liver. **Elimination:** In urine. **Half-Life:** 8–12 h.

NURSING IMPLICATIONS

Assessment & Drug Effects

- Monitor for therapeutic effectiveness: Full anti-inflammatory effect for arthritis may not occur until 8 days to several weeks into therapy.
- Lab test: With prolonged use, periodic Hgb and Hct, PT/INR, and renal function tests.
- Note: Although the antipyretic effect is mild, chronic or high doses may mask fever in some patients.

Patient & Family Education

- Swallow tablet whole; do not crush or chew.
- Report onset of visual or auditory problems immediately to prescriber.

Common adverse effects in *italic*, life-threatening effects underlined; generic names in **bold**; classifications in SMALL CAPS; ♣ Canadian drug name; ✪ Prototype drug

455

- Be aware of I&O ratio and pattern and check for and report peripheral edema and unusual weight gain.
- Report promptly to prescriber the onset of melena (i.e., dark, tarry stools) or hematemesis (i.e., bloody or dark brown vomitus) or severe stomach pain.
- Do not drive or engage in other potentially hazardous activities until reaction to drug is known.

DIGOXIN ℗

(di-jox'in)

Lanoxicaps, Lanoxin
Classifications: CARDIAC GLYCOSIDE; INOTROPIC
Therapeutic: CARDIAC GLYCOSIDE; ANTIARRHYTHMIC
Pregnancy Category: C

AVAILABILITY 0.05 mg, 0.1 mg, 0.2 mg capsules; 0.125 mg, 0.25 mg, 0.5 mg tablets; 0.05 mg/mL elixir; 0.25 mg/mL, 0.1 mg/mL injection

ACTION & *THERAPEUTIC EFFECT*
Widely used cardiac glycoside that acts by increasing the force and velocity of myocardial systolic contraction (positive inotropic effect). It also decreases conduction velocity through the AV node. *Increases contractility of heart muscle (positive inotropic effect). Has antiarrhythmic properties that result from its effects on the AV node.*

USES Rapid digitalization and for maintenance therapy in CHF, atrial fibrillation, atrial flutter, paroxysmal atrial tachycardia.

CONTRAINDICATIONS Digitalis hypersensitivity, sick sinus syndrome, Wolff-Parkinson-White syndrome; ventricular fibrillation, ventricular tachycardia unless due to CHF. Full digitalizing dose not given if patient has received digoxin during previous week or if slowly excreted cardiotonic glycoside has been given during previous 2 wk.

CAUTIOUS USE Renal insufficiency, hypokalemia, advanced heart disease, cardiomyopathy, acute MI, incomplete AV block, cor pulmonale; hypothyroidism; lung disease; premature and immature infants, children, older adults, or debilitated patients; pregnancy (category C).

ROUTE & DOSAGE

Digitalizing Dose (Give ½ dose initially followed by ¼ at 8–12 h intervals)
Adult: **PO** 0.75–1.5 mg
IV 0.5–1 mg
Child: **IV** 2–10 y, 20–35 mcg/kg; *older than 10 y,* 8–12 mcg/kg **PO** 2–10 y, 30–40 mcg/kg; *older than 10 y,* 10–15 mcg/kg
Infant: **IV** 30–50 mcg/kg **PO** 35–60 mcg/kg
Neonate: **IV** *Preterm,* 15–25 mcg/kg; *full-term,* 20–30 mcg/kg
Maintenance Dose
Adult: **PO/IV** 0.1–0.375 mg/day
Child: **PO/IV** *Younger than 2 y,* 7.5–9 mcg/kg/day; *2–10 y,* 6–7.5 mcg/kg/day; *older than 10 y,* 0.125–0.25 mg/day
Neonate: **IV** 4–8 mcg/kg/day

ADMINISTRATION

Oral

- Give without regard to food. Administration after food may slightly delay rate of absorption, but total amount absorbed is not affected.
- Crush and mix with fluid or food if patient cannot swallow it whole.

Intravenous

PREPARE: **Direct:** Give undiluted or diluted in 4 mL of sterile water, D5W, or NS (less diluent may cause precipitation).
ADMINISTER: **Direct:** Give each dose over at least 5 min. ▪ Monitor IV site frequently. Infiltration of parenteral drug into subcutaneous tissue can cause local irritation and sloughing.
INCOMPATIBILITIES **Solution/additive: Dobutamine. Y-site: Amiodarone, amphotericin B cholesteryl complex, fluconazole, foscarnet, propofol.**

▪ Store tablets, elixir, and injection solution at 25° C (77° F) or at 15°–30° C (59°–86° F).

ADVERSE EFFECTS (≥1%) **CNS:** Fatigue, muscle weakness, headache, facial neuralgia, mental depression, paresthesias, hallucinations, confusion, drowsiness, agitation, dizziness. **CV:** Arrhythmias, hypotension, AV block. **Special Senses:** Visual disturbances. **GI:** Anorexia, *nausea,* vomiting, diarrhea. **Other:** Diaphoresis, recurrent malaise, dysphagia.

INTERACTIONS Drug: ANTACIDS, **cholestyramine, colestipol** decrease digoxin absorption; DIURETICS, CORTICOSTEROIDS, **amphotericin B,** LAXATIVES, **sodium polystyrene sulfonate** may cause hypokalemia, increasing the risk of digoxin toxicity; **calcium IV** may increase risk of arrhythmias if administered together with digoxin; **quinidine, verapamil, amiodarone, flecainide** significantly increase digoxin levels, and digoxin dose should be decreased by 50%; **erythromycin** may increase digoxin levels; **succinylcholine** may potentiate arrhythmogenic effects; **nefazodone** may increase digoxin levels. **Food:** High fiber intake may decrease absorption. **Herbal: Ginseng** increase digoxin toxicity; **ma huang, ephedra** may induce arrhythmias; **St. John's wort** decreases plasma concentration. **Lab Test: Panax ginseng** can falsely elevate concentrations with fluorescence polarization immunoassay (FPIA) or falsely lower concentrations when microparticle enzyme immunoassay (MEIA).

PHARMACOKINETICS Absorption: 70% PO tablets; 90% PO liquid and capsules. **Onset:** 1–2 h PO; 5–30 min IV. **Peak:** 6–8 h PO; 1–5 h IV. **Duration:** 3–4 days in fully digitalized patient. **Distribution:** Widely distributed; tissue levels significantly higher than plasma levels; crosses placenta. **Metabolism:** 14% in liver. **Elimination:** 80–90% by kidneys; may appear in breast milk. **Half-Life:** 34–44 h.

NURSING IMPLICATIONS

Assessment & Drug Effects

▪ Take apical pulse for 1 full min, noting rate, rhythm, and quality before administering drug.
▪ Withhold medication and notify prescriber if apical pulse falls below ordered parameters (e.g., less than 50 or 60/min in adults and less than 60 or 70/min in children).
▪ Be familiar with patient's baseline data (e.g., quality of peripheral pulses, blood pressure, clinical symptoms, serum electrolytes, creatinine clearance) as a foundation for making assessments.
▪ Lab tests: Baseline and periodic serum digoxin, potassium, magnesium, and calcium. Draw blood samples for determining plasma digoxin levels at least 6 h after daily dose and preferably just before next scheduled daily dose.
▪ Monitor for S&S of drug toxicity: In children, cardiac arrhythmias

Common adverse effects in *italic*, life-threatening effects <u>underlined</u>; generic names in **bold**; classifications in SMALL CAPS; ♣ Canadian drug name; ❷ Prototype drug

457

D

are usually reliable signs of early toxicity. Early indicators in adults (anorexia, nausea, vomiting, diarrhea, visual disturbances) are rarely initial signs in children.

- Monitor I&O ratio during digitalization, particularly in patients with impaired renal function. Also monitor for edema daily and auscultate chest for rales.

- Monitor serum digoxin levels closely during concurrent antibiotic–digoxin therapy, which can precipitate toxicity because of altered intestinal flora.

- Observe patients closely when being transferred from one preparation (tablet, elixir, or parenteral) to another; when tablet is replaced by elixir potential for toxicity increases since 30% or more of drug is absorbed.

Patient & Family Education

- Report to prescriber if pulse falls below 60 or rises above 110 or if you detect skipped beats or other changes in rhythm, when digoxin is prescribed for atrial fibrillation.

- Suspect toxicity and report to prescriber if any of the following occur: Anorexia, nausea, vomiting, diarrhea, or visual disturbances.

- Weigh each day under standard conditions. Report weight gain greater than 1 kg (2 lb)/day.

- Take digoxin PRECISELY as prescribed. Do not skip or double a dose or change dose intervals, and take it at same time each day.

- Do not to take OTC medications, especially those for coughs, colds, allergy, GI upset, or obesity, without prior approval of prescriber.

- Continue with brand originally prescribed unless otherwise directed by prescriber.

DIGOXIN IMMUNE FAB (OVINE)

(di-jox′in)

Digibind, DigiFab
Classification: ANTIDOTE
Therapeutic: ANTIDOTE
Pregnancy Category: C

AVAILABILITY 38 mg, 40 mg vial

ACTION & *THERAPEUTIC EFFECT*
Fragments of antidigoxin antibodies (Fab) used instead of whole antibody molecules permits more extensive and faster distribution to serum and toxic cellular sites. Fab acts by selectively complexing with circulating digoxin or digitoxin, thereby preventing drug from binding at receptor sites; the complex is then eliminated in urine. *Used as an antidote for digitalis toxicity.*

USES Treatment of potentially life-threatening digoxin or digitoxin intoxication in carefully selected patients.

CONTRAINDICATIONS Hypersensitivity to sheep products; renal or cardiac failure.
CAUTIOUS USE Prior treatment with sheep antibodies or ovine Fab fragments; mannitol hypersensitivity; history of allergies; impaired renal function or renal failure; elderly; pregnancy (category C), lactation.

ROUTE & DOSAGE

Serious Digoxin Toxicity Secondary to Overdose

Adult/Child: **IV** Dosages vary according to amount of digoxin to be neutralized; dosages are based on total body load or steady state serum digoxin concentrations (see package insert); some patients may require a second dose after several hours

Common adverse effects in *italic*, life-threatening effects <u>underlined</u>; generic names in **bold**; classifications in SMALL CAPS; ♣ Canadian drug name; ⊘ Prototype drug

ADMINISTRATION

Intravenous

PREPARE: **Direct:** Dilute each vial with 4 mL of sterile water for injection to yield 9.5 mg/mL for Digibind and 10 mg/mL for DigiFab; mix gently. **IV Infusion:** Dilute further with any volume of NS compatible with cardiac status. ▪ For those receiving less than 3 mg, further dilute to a concentration of 1 mg/mL by adding an additional 34 mL of NS to Digibind or 36 mL of NS to DigiFab. ▪ For very small doses for infants, reconstitute to a concentration of 10 mg/mL.

ADMINISTER: **Direct:** Give undiluted bolus only if cardiac arrest is imminent. **IV Infusion:** Give IV infusion over 30 min, preferably through a 0.22-micron membrane filter. ▪ For administration to infants: Reconstitute as for direct IV and administer with a tuberculin syringe. ▪ For small doses (e.g., 2 mg or less), dilute the reconstituted 40 mg vial with 36 mL of NS to yield 1 mg/mL. ▪ Closely monitor for fluid overload.

▪ Use reconstituted solutions promptly or refrigerated at 2°–8° C (36°–46° F) for up to 4 h.

ADVERSE EFFECTS (≥1%) Adverse reactions associated with use of digoxin immune Fab are related primarily to the effects of **digitalis** withdrawal on the heart (see Nursing Implications). Allergic reactions have been reported rarely. Hypokalemia.

DIAGNOSTIC TEST INTERFERENCE
Digoxin immune Fab may interfere with *serum digoxin* determinations by immunoassay tests.

PHARMACOKINETICS Onset: Less than 1 min after IV administration. **Elimination:** In urine over 5–7 days. **Half-Life:** 14–20 h.

NURSING IMPLICATIONS

Assessment & Drug Effects

▪ Perform skin testing for allergy prior to administration of immune Fab, particularly in patients with history of allergy or who have had previous therapy with immune Fab.

▪ Keep emergency equipment and drugs immediately available before skin testing is done or first dose is given and until patient is out of danger.

▪ Monitor for therapeutic effectiveness: Reflected in improvement in cardiac rhythm abnormalities, mental orientation and other neurologic symptoms, and GI and visual disturbances. S&S of reversal of digitalis toxicity occurs in 15–60 min in adults and usually within minutes in children.

▪ Baseline and frequent vital signs and EGG during administration.

▪ Lab tests: Baseline and periodic serum potassium and serum digoxin; serum digoxin or digitoxin concentration (this measurement will not be accurate for at least 5–7 days after therapy begins because of test interference by immune Fab).

▪ Note: Serum potassium is particularly critical during first several hours following administration of immune Fab. Monitor closely.

▪ Monitor closely: Cardiac status may deteriorate as inotropic action of digitalis is withdrawn by action of immune Fab. CHF, arrhythmias, increase in heart rate, and hypokalemia can occur.

▪ Make sure serum digoxin levels and ECG readings are obtained for at least 2–3 wk.

Patient & Family Education

▪ Tell your prescriber or health care professional about all other medications you are taking, including

Common adverse effects in *italic*, life-threatening effects underlined; generic names in **bold;** classifications in SMALL CAPS; ✤ Canadian drug name; ⊘ Prototype drug

459

non-prescription medications, nutritional supplements, or herbal products.

- Check with your prescriber before stopping or starting any of your medicines.

DIHYDROERGOTAMINE MESYLATE

(dye-hye-droe-er-got'a-meen)

D.H.E. 45, Migranal

Classifications: ALPHA-ADRENERGIC ANTAGONIST; ERGOT ALKALOID
Therapeutic: ANTIMIGRAINE
Prototype: Ergotamine
Pregnancy Category: X

AVAILABILITY 4 mg/mL nasal spray; 1 mg/mL injection

ACTION & *THERAPEUTIC EFFECT*
Alpha-adrenergic blocking agent and ergot alkaloid with direct constricting effect on smooth muscle of peripheral and cranial blood vessels. Additionally, its ergot properties act as selective serotonin agonists at the 5-HT$_1$ receptors located on intracranial blood vessels, which may also cause vasoconstriction of large intracranial conductance arteries; this correlates with relief of migraine headaches. *Reduces rate of serotonin-induced platelet aggregation. Has somewhat weaker vasoconstrictor action than ergotamine but greater adrenergic blocking activity, resulting in relief from migraine headaches.*

USES To prevent or abort headache or migraine.
UNLABELED USES To treat postural hypotension.

CONTRAINDICATIONS History of hypersensitivity to ergot preparations; peripheral vascular disease, coronary heart disease, MI, hypertension; peptic ulcer; severely impaired hepatic or renal function; sepsis; within 48 h of surgery; pregnancy (category X), lactation. Safe use in children younger than 6 y is not established.
CAUTIOUS USE Moderate or mild renal or hepatic impairment; obesity; diabetes mellitus; postmenopausal women; males older than 40 y; pulmonary heart disease; valvular heart disease; smokers.

ROUTE & DOSAGE

Migraine Headache

Adult: **IV/IM/Subcutaneous** 1 mg, may be repeated at 1 h intervals to a total of 3 mg IM or 2 mg IV/subcutaneous (max: 6 mg/wk) **Intranasal** 1 spray (0.5 mg) in each nostril, may repeat with additional spray in 15 min if no relief (max: 4 sprays/attack); wait 6–8 h before treating another attack (max: 8 sprays/24 h, 24 sprays/wk)

ADMINISTRATION

Intranasal

- Give at first warning of migraine headache.

Intramuscular/Subcutaneous

- Give at first warning of migraine headache.
- Withdraw IM or subcutaneous dose directly from ampule. Do not dilute.
- Note: Onset of action is about 20 min; when rapid relief is required, the IV route is prescribed.

Intravenous

PREPARE: Direct: Give undiluted.
ADMINISTER: Direct: Give at a rate of 1 mg/60 sec.

- Store at 15°–30° C (59°–86° F) unless otherwise directed. ■ Protect ampules from heat and light; do not freeze. ▪ Discard ampule if solution appears discolored.

ADVERSE EFFECTS (≥1%) **CV** Vasospasm: Coldness, numbness and tingling in fingers and toes, muscle pains and weakness of legs, precordial distress and pain, transient tachycardia or bradycardia, hypertension (large doses). **GI:** *Nausea, vomiting.* **Body as a Whole:** Dizziness, dysphoria, *localized edema and itching;* ergotism (excessive doses).

INTERACTIONS Drug: BETA-BLOCKERS, **erythromycin** increase peripheral vasoconstriction with risk of ischemia; increased **ergotamine** toxicity with drugs that inhibit CYP3A4 (e.g., PROTEASE INHIBITORS, **amprenavir, ritonavir, nelfinavir, indinavir, saquinavir**), MACROLIDE ANTIBIOTICS **(erythromycin), azithromycin, clarithromycin),** AZOLE ANTIFUNGALS **(ketoconazole, itraconazole, fluconazole, clotrimazole), nefazodone, fluoxetine, fluvoxamine. Food: Grapefruit juice** may increase toxicity.

PHARMACOKINETICS Onset: 15–30 min IM; less than 5 min IV. **Duration:** 3–4 h. **Distribution:** Probably distributed into breast milk. **Metabolism:** In liver by CYP3A4. **Elimination:** Primarily in urine; some in feces. **Half-Life:** 21–32 h.

NURSING IMPLICATIONS

Assessment & Drug Effects
▪ Monitor cardiac status, especially when large doses are given.
▪ Monitor for and report numbness and tingling of fingers and toes, extremity weakness, muscle pain, or intermittent claudication.

Patient & Family Education
▪ Take at first warning of migraine headache.

▪ Lie down in a quiet, darkened room for several hours after drug administration for best results.
▪ Report immediately if any of the following S&S develop: Chest pain, nausea, vomiting, change in heartbeat, numbness, tingling, pain or weakness of extremities, edema, or itching.
▪ Women should use effective means of contraception while using this drug. Notify prescriber if you become pregnant.

DILTIAZEM
(dil-tye′a-zem)
Cardizem, Cardizem CD, Cardizem LA, Cartia XT, Dilacor XR, Dilt-CD, Tiazac, Taztia XT

DILTIAZEM IV
Cardizem IV
Classifications: CALCIUM CHANNEL BLOCKING AGENT; ANTIANGINAL; ANTIHYPERTENSIVE
Therapeutic: ANTIANGINAL; ANTIARRHYTHMIC; ANTIHYPERSENSITIVE
Prototype: Verapamil
Pregnancy Category: C

AVAILABILITY 30 mg, 60 mg, 90 mg, 120 mg tablets; 120 mg, 180 mg, 240 mg sustained release tablets; 60 mg, 90 mg, 120 mg, 180 mg, 240 mg, 300 mg, 360 mg sustained release capsules; 120 mg, 180 mg, 240 mg, 300 mg, 360 mg, 420 mg extended release tablets; 25 mg, 50 mg vials

ACTION & *THERAPEUTIC EFFECT*
Inhibits calcium ion influx through slow channels into cell of myocardial and arterial smooth muscle. Improves myocardial perfusion, and reduces left ventricular workload. *Slows SA and AV node conduction (antiarrhythmic effect). Dilates coronary arteries and arterioles*

Common adverse effects in *italic*, life-threatening effects underlined; generic names in **bold**; classifications in SMALL CAPS; ♣ Canadian drug name; ⊕ Prototype drug

461

DILTIAZEM

and inhibits coronary artery spasm; thus myocardial oxygen delivery is increased (antianginal effect). By vasodilation of peripheral arterioles, drug decreases total peripheral vascular resistance and reduces arterial BP at rest (antihypertensive effect).

USES Vasospastic angina (Prinzmetal's variant or at rest angina), chronic stable (classic effort-associated) angina, essential hypertension. **IV form:** Atrial fibrillation, atrial flutter, supraventricular tachycardia.
UNLABELED USES Prevention of re-infarction in non-Q-wave MI.

CONTRAINDICATIONS Known hypersensitivity to drug; sick sinus syndrome (unless pacemaker is in place and functioning); acute MI; severe hypotension (systolic less than 90 mm Hg or diastolic less than 60 mm Hg); patients undergoing intracranial surgery; bleeding aneurysms.
CAUTIOUS USE Sinoatrial nodal dysfunction, sick sinus syndrome with functioning pacemaker; right ventricular dysfunction, CHF, severe bradycardia; conduction abnormalities; renal or hepatic impairment; older adults; pregnancy (category C). Safe use in children not established.

ROUTE & DOSAGE

Angina
Adult: **PO** 30 mg q.i.d., may increase q1–2 days as required (usual range: 180–360 mg/day in divided doses) **PO** (extended release) 120–180 mg daily

Hypertension
Adult/Adolescent: **PO** 120–240 mg daily or 20–120 mg b.i.d

Atrial Fibrillation/Flutter
Adult: **IV** 0.25 mg/kg IV bolus over 2 min, if inadequate

response, may repeat in 15 min with 0.35 mg/kg, followed by a continuous infusion of 5–10 mg/h (max: 15 mg/h for 24 h)

ADMINISTRATION

Oral
- Do not crush sustained release capsules or tablets. They **must be** swallowed whole.
- Withhold if systolic BP is less than 90 mm Hg or diastolic is less than 60 mm Hg.
- Give before meals and at bedtime.
- Store at 15°–30° C (59°–86° F).

Intravenous

PREPARE: Direct: Give undiluted. **Continuous:** For IV infusion, add to a volume of D5W, NS, or D5/0.45% NaCl that can be administered in 24 h or less.
ADMINISTER: Direct: Give as a bolus dose over 2 min. A second bolus may be given after 15 min. **Continuous:** Give at a rate 5–15 mg/h. Infusion duration longer than 24 h and infusion rate greater than 15 mg/h are not recommended.
INCOMPATIBILITIES Solution/additive: Furosemide. **Y-site:** Acetazolamide, acyclovir, allopurinol, aminophylline, amphotericin B, ampicillin, ampicillin/sulbactam, cefepime, chloramphenicol, dantrolene, diazepam, fluorouracil, furosemide, ganciclovir, hydrocortisone, insulin, ketorolac, lansoprazole, methotrexate, methylprednisolone, mezlocillin, micafungin, nafcillin, pantoprazole, pentobarbital, phenobarbital, phenytoin, piperacillin/tazobactam, rifampin, sodium bicarbonate, thiopental.

462

Common adverse effects in *italic*, life-threatening effects underlined; generic names in **bold**; classifications in SMALL CAPS; ♣ Canadian drug name; ✪ Prototype drug

ADVERSE EFFECTS (≥1%) **CNS:** *Headache,* fatigue, dizziness, asthenia, drowsiness, nervousness, insomnia, confusion, tremor, gait abnormality. **CV:** Edema, arrhythmias, angina, second- or third-degree AV block, bradycardia, CHF, flushing, hypotension, syncope, palpitations. **GI:** Nausea, constipation, anorexia, vomiting, diarrhea, impaired taste, weight increase. **Skin:** Rash.

INTERACTIONS Drug: BETA-BLOCK-ERS, **digoxin** may have additive effects on av node conduction prolongation; may increase **digoxin** or **quinidine** levels; **cimetidine** may increase diltiazem levels, thus increasing effects; may increase **cyclosporine** levels. May increase STATIN levels, monitor closely. Do not use more than 10 mg of **simvastatin** concurrently. **Herbal:** Monitor carefully if used with hawthorn.

PHARMACOKINETICS Absorption: Approximately 80% from GI tract, with 40% reaching systemic circulation. **Peak:** 2–3 h; 6–11 h sustained release; 11–18 h Cardizem LA. **Distribution:** Into breast milk. **Metabolism:** In liver (CYP3A4). **Elimination:** Primarily in urine with some elimination in feces. **Half-Life:** Oral 3.5–9 h, IV 2 h.

NURSING IMPLICATIONS

Assessment & Drug Effects
- Check BP and ECG before initiation of therapy and monitor particularly during dosage adjustment period.
- Lab tests: With concurrent therapy, monitor digoxin levels when initiating, adjusting, and discontinuing diltiazem.
- Monitor for and report S&S of CHF.
- Monitor for headache. An analgesic may be required.
- Supervise ambulation as indicated.

Patient & Family Education
- Make position changes slowly and in stages; light-headedness and dizziness (hypotension) are possible.
- Do not drive or engage in other potentially hazardous activities until reaction to drug is known.

DIMENHYDRINATE
(dye-men-hye′dri-nate)
Calm-X, Dimenhydrinate Injection, Dramamine
Classifications: ANTIHISTAMINE (H₁-RECEPTOR ANTAGONIST); ANTIVERTIGO
Therapeutic: ANTIVERTIGO; ANTIEMETIC
Prototype: Diphenhydramine
Pregnancy Category: B

AVAILABILITY 50 mg tablets; 50 mg chewable; 50 mg/mL injection; 15.62 mg/5 mL, 12.5 mg/4 mL, 12.5 mg/5 mL liquid

ACTION & THERAPEUTIC EFFECT H₁-receptor antagonist with antiemetic action thought to involve ability to inhibit cholinergic stimulation in vestibular and associated neural pathways. It has been reported to inhibit labyrinthine stimulation for up to 3 h. *Has antiemetic and antivertigo activity.*

USES Chiefly in prevention and treatment of motion sickness. Also has been used in management of vertigo, nausea, and vomiting associated with radiation sickness, labyrinthitis, Ménière's syndrome, stapedectomy, anesthesia, and various medications.

CONTRAINDICATIONS Narrowangle glaucoma, BPH; GI obstruction; urinary tract obstruction; CNS depression; lactation, neonates.
CAUTIOUS USE Convulsive disorders; asthma, COPD; severe hepatic

Common adverse effects in *italic*, life-threatening effects underlined; generic names in **bold**; classifications in SMALL CAPS; ◆ Canadian drug name; ⊘ Prototype drug

463

disease; PKU; history of porphyria; closed-angle glaucoma; older adults; pregnancy (category B). Safe use in children younger than 2 y not established.

ROUTE & DOSAGE

Motion Sickness

Adult: **PO** 50–100 mg q4–6h (max: 400 mg/24 h) **IV/IM** 50 mg as needed
Child: **PO** 2–6 y, up to 25 mg q6–8h (max: 75 mg/24 h); 6–12 y, 25–50 mg q6–8h (max: 150 mg/24 h) **IM** 1.25 mg/kg q.i.d. up to 300 mg/day

ADMINISTRATION

- First dose should be given 30–60 min before starting activity.

Oral

- Ensure that chewable tablets are chewed and not swallowed whole.

Intramuscular

- Give undiluted and inject deep IM into a large muscle.

Intravenous

PREPARE: **Direct:** Dilute each 50 mg in 10 mL of NS.
ADMINISTER: **Direct:** Give each 50 mg or fraction thereof over 2 min.
INCOMPATIBILITIES **Solution/additive: Aminophylline, amobarbital, chlorpromazine, glycopyrrolate, hydrocortisone, hydroxyzine, pentobarbital, phenobarbital, phenytoin, prochlorperazine, promazine, promethazine, thiopental.**

- Store preferably at 15°–30° C (59°–86° F), unless otherwise directed by manufacturer. ▪ Examine parenteral preparation for particulate matter and discoloration. Do not use unless absolutely clear.

ADVERSE EFFECTS (≥1%) CNS:

Drowsiness, headache, incoordination, dizziness, blurred vision, nervousness, restlessness, *insomnia (especially children).* **CV:** Hypotension, palpitation. **GI:** Dry mouth, nose, throat; anorexia, constipation or diarrhea. **Urogenital:** Urinary frequency, dysuria.

DIAGNOSTIC TEST INTERFERENCE

Skin testing procedures should not be performed within 72 h after use of an antihistamine.

INTERACTIONS Drug: Alcohol and

other CNS DEPRESSANTS enhance CNS depression, drowsiness; TRICYCLIC ANTIDEPRESSANTS compound anticholinergic effects.

PHARMACOKINETICS Absorption:

Readily absorbed from GI tract. **Onset:** 15–30 min PO; immediate IV; 20–30 min IM. **Duration:** 3–6 h. **Distribution:** Distributed into breast milk. **Elimination:** In urine.

NURSING IMPLICATIONS

Assessment & Drug Effects

- Use falls precautions and supervise ambulation; drug produces high incidence of drowsiness.
- Note: Tolerance to CNS depressant effects usually occurs after a few days of drug therapy; some decrease in antiemetic action may result with prolonged use.
- Monitor for dizziness, nausea, and vomiting; these may indicate drug toxicity.

Patient & Family Education

- Do not drive or engage in other potentially hazardous activities until response to drug is known.
- Take 30–60 min before departure to prevent motion sickness; repeat before meals and upon retiring.

DIMERCAPROL

(dye-mer-kap′role)

BAL in Oil

Classifications: CHELATING AGENT; ANTIDOTE

Therapeutic: ANTIDOTE

Pregnancy Category: C

AVAILABILITY 100 mg/mL injection

ACTION & *THERAPEUTIC EFFECT*

Dithiol compound that combines with ions of various heavy metals to form relatively stable, nontoxic, soluble complexes called chelates, which can be excreted; inhibition of enzymes by toxic metals is thus prevented. *Neutralizes the effects of various heavy metals.*

USES Acute poisoning by arsenic, gold, and mercury; as adjunct to edetate calcium disodium (EDTA) in treatment of lead encephalopathy.

UNLABELED USES Chromium dermatitis; ocular and dermatologic manifestations of arsenic poisoning, as adjunct to penicillamine to increase rate of copper excretion in Wilson's disease, and for poisoning with heavy metals.

CONTRAINDICATIONS Hepatic insufficiency (with exception of postarsenical jaundice); history of peanut oil hypersensitivity; severe renal insufficiency; poisoning due to cadmium, iron, selenium, or uranium; lactation.

CAUTIOUS USE Hypertension; oliguria; patients with G6PD deficiency; preexisting renal disease; rheumatoid arthritis; pregnancy (category C).

ROUTE & DOSAGE

Arsenic or Gold Poisoning

Adult/Child: **IM** 2.5–3 mg/kg q4h for first 2 days, then q.i.d. on third day, then b.i.d. for 10 days

Mercury Poisoning

Adult/Child: **IM** 5 mg/kg initially, followed by 2.5 mg/kg 1–2 times/day for 10 days

Acute Lead Encephalopathy

Adult/Child: **IM** 4 mg/kg initially, then 3–4 mg/kg q4h with EDTA for 2–7 days depending on response

ADMINISTRATION

Intramuscular

- Initiate therapy ASAP (within 1–2 h) after ingestion of the poison because irreversible tissue damage occurs quickly, particularly in mercury poisoning.
- Give by deep IM injection only. Local pain, gluteal abscess, and skin sensitization possible. Rotate injection sites and observe daily.
- Determine if a local anesthetic may be given with the injection to decrease injection site pain.
- Handle with caution; contact of drug with skin may produce erythema, edema, dermatitis.

ADVERSE EFFECTS (≥1%) **CNS:** Headache, anxiety, muscle pain or weakness, restlessness, paresthesias, tremors, *convulsions,* shock. **CV:** *Elevated BP,* tachycardia. **Special Senses:** Rhinorrhea; burning sensation, feeling of pain and constriction in throat. **GI:** Nausea, *vomiting;* burning sensation in lips and mouth, halitosis, salivation; abdominal pain, metabolic acidosis. **Urogenital:** Burning sensation in penis, renal damage. **Other:** Pains in chest or hands, pain and sterile abscess at injection site, sweating, reduction in polymorphonuclear leukocytes, dental pain.

DIAGNOSTIC TEST INTERFERENCE I^{131} *thyroid uptake* values may be decreased if test is done during or immediately following dimercaprol therapy.

Common adverse effects in *italic,* life-threatening effects underlined; generic names in **bold;** classifications in SMALL CAPS; ✦ Canadian drug name; ⊘ Prototype drug

465

INTERACTIONS Drug: Iron, cadmium, selenium, uranium form toxic complexes with dimercaprol.

PHARMACOKINETICS Peak: 30–60 min. **Distribution:** Distributed mainly in intracellular spaces, including brain; highest concentrations in liver and kidneys. **Elimination:** Completely excreted in urine and bile within 4 h. **Half-Life:** Short.

NURSING IMPLICATIONS

Assessment & Drug Effects

- Monitor vital signs. Elevations of systolic and diastolic BPs accompanied by tachycardia frequently occur within a few minutes following injection and may remain elevated up to 2 h.
- Note: Fever occurs in approximately 30% of children receiving treatment and may persist throughout therapy.
- Monitor I&O. Drug is potentially nephrotoxic. Report oliguria or change in I&O ratio to prescriber.
- Check urine daily for albumin, blood, casts, and pH. Blood and urinary levels of the metal serve as guides for dosage adjustments.
- Minor adverse reactions generally reach maximum 15–20 min after drug administration and subside in 30–90 min.

Patient & Family Education

- Drink as much fluid as the prescriber will permit.

DIMETHYL SULFOXIDE

(dye-meth′il sul-fox′ide)

DMSO, Rimso-50

Classifications: GENITOURINARY; LOCAL ANTI-INFLAMMATORY

Therapeutic: INTERSTITIAL CYSTITIS AGENT

Pregnancy Category: C

AVAILABILITY 50% solution

ACTION & *THERAPEUTIC EFFECT*

Reported effects include anti-inflammatory effects, membrane penetration, collagen dissolution, vasodilation, muscle relaxation, diuresis, initiation of histamine release at administration site, cholinesterase inhibition. *Has symptomatic relief of interstitial cystitis with local anti-inflammatory properties.*

USES Symptomatic treatment of interstitial cystitis.

UNLABELED USES Topical treatment of a variety of musculoskeletal disorders, arthritis, scleroderma, tendinitis, breast and prostate malignancies, retinitis pigmentosa, herpesvirus infections, head and spinal cord injuries, shock, and as a carrier to enhance penetration and absorption of other drugs. Also used to protect living cells and tissues during cold storage (cryo-protection). Widely used as an industrial solvent and in veterinary medicine for treatment of musculoskeletal injuries.

CONTRAINDICATIONS Urinary tract malignancy; lactation. Safe use in children is not established.

CAUTIOUS USE Hepatic or renal dysfunction; pregnancy (category C).

ROUTE & DOSAGE

Interstitial Cystitis

Adult: **Instillation** 50 mL of 50% solution instilled slowly into urinary bladder and retained for 15 min; may repeat q2wk until maximum relief obtained, then increase intervals between treatments

ADMINISTRATION

Instillation

- Apply analgesic lubricant such as lidocaine jelly to urethra to facilitate insertion of catheter.

- Instruct patient to retain instillation for 15 min and then expel it by spontaneous voiding.
- Note: Discomfort associated with instillation usually lessens with repeated administration. Prescriber may prescribe an oral analgesic or suppository containing belladonna and an opiate prior to instillation to reduce bladder spasm.
- Store at 15°–30° C (59°–86° F) unless otherwise directed. Protect from strong light. Avoid contact with plastics.

ADVERSE EFFECTS (≥1%) **Special Senses:** Transient disturbances in color vision, photophobia. **GI:** Nausea, diarrhea. Hypersensitivity: Local or generalized rash, erythema, pruritus, urticaria, swelling of face, dyspnea (<u>anaphylactoid reaction</u>). **Other:** Nasal congestion, headache, sedation, drowsiness. **Following instillation:** *Garlic-like odor on breath and skin; garlic-like taste; discomfort during administration; transient cystitis.* **Following topical application:** Vesicle formation.

INTERACTIONS Drug: Decreases effectiveness of **sulindac**, possibly causing severe peripheral neuropathy.

PHARMACOKINETICS Absorption: Readily absorbed systemically. **Peak:** 4–8 h. **Distribution:** Widely distributed in tissues and body fluids; penetrates blood–brain barrier; distributed into breast milk. **Metabolism:** Metabolized to dimethyl sulfide (garlic breath) and dimethyl sulfone. **Elimination:** Dimethyl sulfide excreted through lungs and skin; dimethyl sulfone may remain in serum longer than 2 wk and is excreted in urine and feces.

NURSING IMPLICATIONS

Assessment & Drug Effects

- Monitor and report level of bladder discomfort. In cases of severe discomfort prescriber may elect to do instillation under anesthesia.
- Monitor for visual disturbances. Complete eye evaluation, including slit-lamp examination, is recommended prior to and at regular intervals during therapy.

Patient & Family Education

- Note: Garlic-like taste may be experienced within minutes after drug instillation and may last for several hours. Garlic-like odor on breath and skin may last as long as 72 h.

DINOPROSTONE (PGE₂, PROSTAGLANDIN E₂)

(dye-noe-prost′one)

Cervidil, Prostin E₂, Prepidil
Classification: OXYTOCIC
Therapeutic: PROSTAGLANDIN; OXYTOCIC
Prototype: Oxytocin
Pregnancy Category: C

AVAILABILITY 20 mg suppository; **Prepidil:** 0.5 mg gel; **Cervidil:** 10 mg vaginal insert

ACTION & *THERAPEUTIC EFFECT*
Synthetic prostaglandin E₂ that appears to act directly on myometrium and vascular smooth muscle. Stimulation of gravid uterus in early weeks of gestation is more potent than that of oxytocin. *Contractions are qualitatively similar to those that occur during term labor. Has high success rate when used as abortifacient before twentieth week and for stimulation of labor in cases of intrauterine fetal death.*

Common adverse effects in *italic*, life-threatening effects <u>underlined</u>; generic names in **bold**; classifications in SMALL CAPS; ♣ Canadian drug name; ⊘ Prototype drug

467

DINOPROSTONE (PGE₂, PROSTAGLANDIN E₂)

USES To terminate pregnancy from twelfth week through second trimester as calculated from first day of last regular menstrual period; to evacuate uterine contents in management of missed abortion or intrauterine fetal death up to 28 wk gestational age; to manage benign hydatidiform mole; cervical ripening prior to labor induction.

CONTRAINDICATIONS Acute pelvic inflammatory disease; abnormal fetal position; history of pelvic surgery, uterine fibroids, cervical stenosis, active cardiac, pulmonary, renal, or hepatic disease.

CAUTIOUS USE History of hypertension, hypotension, asthma, epilepsy, anemia, diabetes mellitus; jaundice, history of hepatic, renal, or cardiovascular disease; cervicitis, acute vaginitis, infected endocervical lesion; previous history of caesarean section; pregnancy (category C).

ROUTE & DOSAGE

Induction of Labor

Adult: **Endocervical** Place *Prepidil* 0.5 mg endocervically, may repeat q6h (max: 1.5 mg); place *Cervidil* insert 10 mg transversely in the posterior fornix of the vagina, remove on onset of active labor or 12 h after insertion

Evacuation of Uterus

Adult: **Intravaginal** Insert suppository high in vagina, repeat q2–5h until abortion occurs or membranes rupture (max total dose: 240 mg)

ADMINISTRATION

Endocervical & Intravaginal

- Antiemetic and antidiarrheal medication may be prescribed to be given before dinoprostone to minimize GI side effects.
- Place vaginal insert in the vagina immediately after removal from the foil package. DO NOT use without retrieval system.
- Keep patient in supine position for 10 min after administration of suppository to prevent expulsion and enhance absorption.
- Store suppositories in freezer at temperature not exceeding −20° C (−4° F) unless otherwise specified.

ADVERSE EFFECTS (≥1%) **CNS:** Headache, tremor, tension. **CV:** Transient hypotension, flushing, cardiac arrhythmias. **GI:** *Nausea, vomiting, diarrhea.* **Urogenital:** Vaginal pain, endometritis, <u>uterine rupture</u>. **Respiratory:** Dyspnea, cough, hiccups. **Body as a Whole:** Chills, *fever,* dehydration, diaphoresis, rash.

INTERACTIONS Drug: OXYTOCICS used with extreme caution.

PHARMACOKINETICS Absorption: Slowly absorbed from vagina; Cervidil insert releases approximately 0.3 mg/h. **Onset:** 10 min. **Duration:** 2–3 h. **Distribution:** Widely distributed in body. **Metabolism:** Rapidly metabolized in lungs, kidneys, spleen, and other tissues. **Elimination:** Mainly in urine; some in feces.

NURSING IMPLICATIONS

Assessment & Drug Effects

- Observe patient carefully, after insertion of the drug. Rupture of the membranes is not a contraindication to drug, but be aware that profuse bleeding may result in expulsion of the suppository. Report wheezing, chest pain, dyspnea, and significant changes in BP and pulse to the prescriber.
- Monitor uterine contractions and observe for and report excessive

vaginal bleeding and cramping pain.

- Monitor vital signs. Fever is a physiologic response of the hypothalamus to use of dinoprostone and occurs within 15–45 min after insertion of suppository. Temperature returns to normal within 2–6 h after discontinuation of medication.

Patient & Family Education

- Continue taking your temperature (late afternoon) for a few days after discharge. Contact prescriber with onset of fever, bleeding, abdominal cramps, abnormal or foul-smelling vaginal discharge.
- Avoid douches, tampons, intercourse, and tub baths for at least 2 wk. Clarify with prescriber.

DIPHENHYDRAMINE HYDROCHLORIDE 🄳

(dye-fen-hye′dra-meen)

Allerdryl ♣, Benadryl, Benadryl Dye-Free, Sleep-Eze 3, Sominex Formula 2, Tusstat, Twilite, Valdrene

Classifications: CENTRALLY ACTING CHOLINERGIC ANTAGONIST; ANTIHISTAMINE; H₁-RECEPTOR ANTAGONIST

Therapeutic: ANTIHISTAMINE; SEDATIVE-HYPNOTIC; ANTIPARKINSON; ANTIDYSKINETIC; NONNARCOTIC ANTITUSSIVE

Pregnancy Category: C

AVAILABILITY 25 mg, 50 mg capsules, tablets; 6.25 mg/5 mL, 12.5 mg/5 mL syrup; 50 mg/mL injection

ACTION & *THERAPEUTIC EFFECT*

Diphenhydramine competes for H₁-receptor sites on effector cells, thus blocking histamine release. Effects in parkinsonism and drug-induced extrapyramidal symptoms are apparently related to its ability to suppress central cholinergic activity and to prolong action of dopamine by inhibiting its reuptake and storage. *Has antihistamine, antivertigo, antiemetic, antianaphylactic, antitussive, antidyskinetic, and sedative-hypnotic effects.*

USES Temporary symptomatic relief of various allergic conditions and to treat or prevent motion sickness, vertigo, and reactions to blood or plasma in susceptible patients. Also used in anaphylaxis as adjunct to epinephrine and other standard measures after acute symptoms have been controlled; in treatment of parkinsonism and drug-induced extrapyramidal reactions; as a nonnarcotic cough suppressant; as a sedative-hypnotic; and for treatment of intractable insomnia.

CONTRAINDICATIONS Hypersensitivity to antihistamines of similar structure; lower respiratory tract symptoms (including acute asthma); narrow-angle glaucoma; prostatic hypertrophy, bladder neck obstruction; GI obstruction or stenosis; lactation, premature neonates, and neonates.

CAUTIOUS USE History of asthma; COPD; convulsive disorders; increased IOP; hyperthyroidism; hypertension, cardiovascular disease; hepatic disease; diabetes mellitus; older adults, infants, and young children; pregnancy (category C). Use as a nighttime sleep aid in children younger than 2 y not established.

ROUTE & DOSAGE

Allergy Symptoms, Antiparkinsonism, Motion Sickness, Nighttime Sedation

Adult: PO 25–50 mg t.i.d. or q.i.d. (max: 300 mg/day) IV/IM 10–50 mg q4–6h (max: 400 mg/day)
Child: PO 2–6 y, 6.25 mg q4–6h (max: 300 mg/24 h); 6–12 y,

Common adverse effects in *italic*, life-threatening effects <u>underlined</u>; generic names in **bold**; classifications in SMALL CAPS; ♣ Canadian drug name; 🄳 Prototype drug

469

12.5–25 mg q4–6h (max: 300 mg/24 h) **IV/IM** 5 mg/kg/day divided into 4 doses (max: 300 mg/day)

Nonproductive Cough

Adult: **PO** 25 mg q4–6h (max: 100 mg/day)
Child: **PO** 2–6 y, 6.25 mg q4–6h (max: 25 mg/24 h); 6–12 y, 12.5 mg q4–6h (max: 50 mg/24 h)

ADMINISTRATION

Oral
- Give with food or milk to lessen GI adverse effects.
- For motion sickness: Give the first dose 30 min before exposure to motion; give remaining doses before meals and at bedtime.

Intramuscular
- Give IM injection deep into large muscle mass; alternate injection sites. Avoid perivascular or subcutaneous injections because of its irritating effects.
- Note: Hypersensitivity reactions (including anaphylactic shock) are more likely to occur with parenteral than PO administration.

Intravenous
PREPARE: **Direct:** Give undiluted.
ADMINISTER: **Direct:** Give at a rate of 25 mg or a fraction thereof over 1 min.
INCOMPATIBILITIES **Solution/additive: Amphotericin B, dexamethasone, iodipamide, lorazepam, methylprednisolone, metoclopramide, pentobarbital, phenobarbital, phenytoin, thiopental. Y-site: Allopurinol, aminophylline, amphotericin B cholesteryl complex, ampicillin, azathioprine, cefmandole, cefazolin, cefepime, cefmetazole, cefonicid,** cefperazone, cefotaxime, cefotetan, cefoxitin, ceftazidime, ceftobiprole, ceftriaxone, cefuroxime, cephalothin, chloramphenicol, dantrolene, diazepam, diazoxide, fluorouracil, foscarnet, furosemide, ganciclovir, indomethacin, insulin, ketorolac, lansoprazole, methylprednisolone, mezlocillin, nitroprusside, pantoprazole, pentobarbital, phenobarbital, phenytoin, sulfamethoxazole/trimethoprim.

- Store in tightly covered containers at 15°–30° C (59°–86° F) unless otherwise directed by manufacturer. Keep injection and elixir formulations in light-resistant containers.

ADVERSE EFFECTS (≥1%) **CNS:** *Drowsiness*, dizziness, headache, fatigue, disturbed coordination, tingling, heaviness and weakness of hands, tremors, euphoria, nervousness, restlessness, insomnia; confusion; (especially in children): Excitement, fever. **CV:** Palpitation, *tachycardia*, mild hypotension or hypertension, cardiovascular collapse. **Special Senses:** Tinnitus, vertigo, dry nose, throat, nasal stuffiness; blurred vision, diplopia, photosensitivity, dry eyes. **GI:** *Dry mouth*, nausea, epigastric distress, anorexia, vomiting, constipation, or diarrhea. **Urogenital:** Urinary frequency or retention, dysuria. **Body as a Whole:** Hypersensitivity (skin rash, urticaria, photosensitivity, anaphylactic shock). **Respiratory:** Thickened bronchial secretions, wheezing, sensation of chest tightness.

DIAGNOSTIC TEST INTERFERENCE Diphenhydramine should be discontinued 4 days prior to *skin testing* procedures for allergy because it may obscure otherwise positive reactions.

INTERACTIONS Drug: Alcohol and other CNS DEPRESSANTS, MAO INHIBITORS compound CNS depression.

Common adverse effects in *italic*; life-threatening effects underlined; generic names in **bold;** classifications in SMALL CAPS; ♣ Canadian drug name; ⊘ Prototype drug

PHARMACOKINETICS Absorption: Readily absorbed from GI tract but only 40–60% reaches systemic circulation. **Onset:** 15–30 min. **Peak:** 1–4 h. **Duration:** 4–7 h. **Distribution:** Crosses placenta; distributed into breast milk. **Metabolism:** In liver; some degradation in lung and kidney. **Elimination:** Mostly in urine within 24 h.

NURSING IMPLICATIONS

Assessment & Drug Effects

- Monitor cardiovascular status especially with preexisting cardiovascular disease.
- Monitor for adverse effects especially in children and the older adult.
- Supervise ambulation and institute falls precautions as necessary. Drowsiness is most prominent during the first few days of therapy and often disappears with continued therapy. Older adults are especially likely to manifest dizziness, sedation, and hypotension.

Patient & Family Education

- Do not use alcohol and other CNS depressants because of the possible additive CNS depressant effects with concurrent use.
- Do not drive or engage in other potentially hazardous activities until the response to drug is known.
- Increase fluid intake, if not contraindicated; drug has an atropine-like drying effect (thickens bronchial secretions) that may make expectoration difficult.

DIPHENOXYLATE HYDROCHLORIDE WITH ATROPINE SULFATE

(dye-fen-ox′i-late)

Diphenatol, Lofene, Lomanate, Lomotil, Lonox, Lo-Trol, Low-Quel, Nor-Mil

Classification: ANTIDIARRHEAL
Therapeutic: ANTIDIARRHEAL
Pregnancy Category: C
Controlled Substance: Schedule V

AVAILABILITY 2.5 mg tablets; 2.5 mg/5 mL liquid

ACTION & *THERAPEUTIC EFFECT* Diphenoxylate is a synthetic narcotic opiate agonist. Commercially available only with atropine sulfate to discourage deliberate overdosage. Inhibits mucosal receptors responsible for peristaltic reflex, thereby reducing GI motility. *Reduces GI motility.*

USES Adjunct in symptomatic management of diarrhea.

CONTRAINDICATIONS Hypersensitivity to diphenoxylate or atropine; severe dehydration or electrolyte imbalance, advanced liver disease, obstructive jaundice, diarrhea caused by pseudomembranous enterocolitis; diarrhea induced by poisons; glaucoma; lactation.

CAUTIOUS USE Advanced hepatic disease, abnormal liver function tests; renal function impairment, MAOI therapy; addiction-prone individuals, or those whose history suggests drug abuse; ulcerative colitis; children; pregnancy (category C). Safe use in children younger than 2 y not established.

ROUTE & DOSAGE

Diarrhea

Adult: **PO** 1–2 tablets or 1–2 teaspoons full (5 mL) 3–4 times/day (each tablet or 5 mL contains 2.5 mg diphenoxylate HCl and 0.025 mg atropine sulfate)
Child (2–12 y): **PO** 0.3–0.4 mg/kg/day of liquid in divided doses

---

Common adverse effects in *italic*, life-threatening effects underlined; generic names in **bold**; classifications in SMALL CAPS; ♣ Canadian drug name; ⊙ Prototype drug

471

D

ADMINISTRATION

Oral

- Crush tablet if necessary and give with fluid of patient's choice.
- Reduce dosage as soon as initial control of symptoms occurs.
- Withhold drug in presence of severe dehydration or electrolyte imbalance until appropriate corrective therapy has been initiated.
- Note: Treatment is generally continued for 24–36 h before it is considered ineffective.
- Store in tightly covered, light-resistant container, preferably 15°–30° C (59°–86° F), unless otherwise directed by manufacturer.

ADVERSE EFFECTS (≥1%) **Body as a Whole:** Hypersensitivity (pruritus, angioneurotic edema, giant urticaria, rash). **CNS:** Headache, sedation, drowsiness, dizziness, lethargy, numbness of extremities; restlessness, euphoria, mental depression, weakness, general malaise. **CV:** Flushing, palpitation, tachycardia. **Special Senses:** Nystagmus, mydriasis, blurred vision, miosis (toxicity). **GI:** Nausea, vomiting, anorexia, dry mouth, abdominal discomfort or distension, paralytic ileus, toxic megacolon. **Other:** Urinary retention, swelling of gums.

INTERACTIONS Drug: MAO INHIBITORS may precipitate hypertensive crisis; **alcohol** and other CNS DEPRESSANTS may enhance CNS effects; also see **atropine.**

PHARMACOKINETICS Absorption: Readily absorbed from GI tract. **Onset:** 45–60 min. **Peak:** 2 h. **Duration:** 3–4 h. **Distribution:** Distributed into breast milk. **Metabolism:** Rapidly metabolized to active and inactive metabolites in liver. **Elimination:** Slowly through bile into feces; small amount in urine. **Half-Life:** 4.4 h.

NURSING IMPLICATIONS

Assessment & Drug Effects

- Assess GI function; report abdominal distention and signs of decreased peristalsis.
- Monitor for S&S of dehydration (see Appendix F). It is essential to monitor young children closely; dehydration occurs more rapidly in this age group and may influence variability of response to diphenoxylate and predispose patient to delayed toxic effects.
- Monitor frequency and consistency of stools.

Patient & Family Education

- Take medication only as directed by prescriber.
- Notify prescriber if diarrhea persists or if fever, bloody stools, palpitation, or other adverse reactions occur.
- Do not drive or engage in other potentially hazardous activities until response to drug is known.

DIPIVEFRIN HYDROCHLORIDE

(dye-pi've-frin)

Propine
Pregnancy Category: B
See Appendix A-1.

DIPYRIDAMOLE

(dye-peer-id'a-mole)

Apo-Dipyridamole ✦, **Persantine**
Classifications: ANTIPLATELET; PLATELET AGGREGATE INHIBITOR

Common adverse effects in *italic*, life-threatening effects underlined; generic names in **bold;** classifications in SMALL CAPS; ✦ Canadian drug name; ⊘ Prototype drug

Therapeutic: PLATELET AGGREGATE INHIBITOR
Pregnancy Category: B

AVAILABILITY 25 mg, 50 mg, 75 mg tablets; 10 mg injection

ACTION & *THERAPEUTIC EFFECT*

Nonnitrate coronary vasodilator that increases coronary blood flow by selectively dilating coronary arteries, thereby increasing myocardial oxygen supply. Additionally, it exhibits antiplatelet aggregation activity. *Has antiplatelet, and coronary vasodilator effects.*

USES To prevent postoperative thromboembolic complications associated with prosthetic heart valves and as adjunct for thallium stress testing.

UNLABELED USES To reduce rate of reinfarction following MI; to prevent TIAs (transient ischemic attacks) and coronary bypass graft occlusion.

CAUTIOUS USE Hypotension, anticoagulant therapy; aspirin sensitivity; elderly; severe hepatic dysfunction; syncope; pregnancy (category B), lactation. Safety and efficacy in children younger than 12 y are not established.

ROUTE & DOSAGE

Prevention of Thromboembolism in Cardiac Valve Replacement
Adult: PO 75–100 mg q.i.d.
Child: PO 1–2 mg t.i.d.

Thromboembolic Disorders
Adult: PO 150–400 mg/day in divided doses

Thallium Stress Test
Adult: IV 0.142 mg/kg/min for 4 min

ADMINISTRATION

Oral

- Give on an empty stomach at least 1 h before or 2 h after meals, with a full glass of water. Prescriber may prescribe with food if gastric distress persists.

Intravenous

PREPARE: **Direct:** Dilute to at least a 1:2 ratio with 0.45% NaCl, NS, or D5W to yield a final volume of 20–50 mL.
ADMINISTER: **Direct:** Give a single dose over 4 min (0.142 mg/kg/min).

- Store in tightly closed container at 15°–30° C (59°–86° F) unless otherwise directed. Protect injection from direct light.

ADVERSE EFFECTS (≥1%) Usually dose related, minimal, and transient. **CNS:** Headache, dizziness, faintness, syncope, weakness. **CV:** Peripheral vasodilation, flushing. **GI:** Nausea, vomiting, diarrhea, abdominal distress. **Skin:** Skin rash, pruritus.

INTERACTIONS Drug: Other ANTICOAGULANTS can increase bleeding risk. **Herbal: Evening primrose oil, ginseng** can increase bleeding risk.

PHARMACOKINETICS Absorption: Readily absorbed from GI tract. **Peak:** 45–150 min. **Distribution:** Small amount crosses placenta. **Metabolism:** In liver. **Elimination:** Mainly in feces. **Half-Life:** 10–12 h.

NURSING IMPLICATIONS

Assessment & Drug Effects

- Monitor therapeutic effectiveness. Clinical response may not be evident before second or third month of continuous therapy. Effects include reduced frequency or elimination of anginal episodes,

Common adverse effects in *italic*, life-threatening effects underlined; generic names in **bold**; classifications in SMALL CAPS; ◆ Canadian drug name; ● Prototype drug

473

D

improved exercise tolerance, reduced requirement for nitrates.

Patient & Family Education

- Notify prescriber of any adverse effects.
- Make all position changes slowly and in stages, especially from recumbent to upright posture, if postural hypotension or dizziness is a problem.

DISOPYRAMIDE PHOSPHATE

(dye-soe-peer′a-mide)

Norpace, Norpace CR, Rythmodan ♦, Rythmodan-LA ♦

Classification: CLASS IA ANTIARRHYTHMIC
Therapeutic: CLASS IA ANTIARRHYTHMIC

Prototype: Procainamide
Pregnancy Category: C

AVAILABILITY 100 mg, 150 mg regular and sustained release capsules

ACTION & *THERAPEUTIC EFFECT*
Class IA antiarrhythmic agent that decreases myocardial conduction velocity and excitability in the atria, ventricles, and accessory pathways. It prolongs the QRS and QT intervals in normal sinus rhythm and atrial arrhythmias. *Acts as myocardial depressant by reducing rate of spontaneous diastolic depolarization in pacemaker cells, thereby suppressing ectopic focal activity.*

USES Treatment of ventricular tachycardia.

UNLABELED USES To treat or prevent serious refractory arrhythmias. To convert atrial fibrillation, atrial flutter, and paroxysmal atrial tachycardia to normal sinus rhythm.

CONTRAINDICATIONS Cardiogenic shock, preexisting 2nd or 3rd degree AV block (if no pacemaker is present); sick sinus syndrome (bradycardia-tachycardia); Wolff-Parkinson-White (WPW) syndrome or bundle branch block, history of torsades de pointes; cardiogenic shock; QT prolongation; uncompensated or inadequately compensated CHF, hypotension (unless secondary to cardiac arrhythmia), hypokalemia.

CAUTIOUS USE Myocarditis or other cardiomyopathy, underlying cardiac conduction abnormalities; hepatic or renal impairment; urinary tract disease (especially prostatic hypertrophy); diabetes mellitus; myasthenia gravis; older adults; narrow-angle glaucoma; family history of glaucoma; pregnancy (category C), lactation.

ROUTE & DOSAGE

Arrhythmias

Adult: **PO** *Weight greater than 50 kg,* 100–200 mg q6h or 300 mg loading dose; *weight less than 50 kg,* 100 mg q6h
Adolescent: **PO** 6–15 mg/kg/day in divided doses q6h
Child: **PO** *Younger than 1 y,* 10–30 mg/kg/day in divided doses q6h; *1–4 y,* 10–20 mg/kg/day in divided doses q6h; *4–12 y,* 10–15 mg/kg/day in divided doses q6h

ADMINISTRATION

Oral

- Start drug 6–12 h after last quinidine dose and 3–6 h after last procainamide dose for patients who have been receiving either quinidine or procainamide.
- Give sustained release capsules whole.

Common adverse effects in *italic*, life-threatening effects underlined; generic names in **bold**; classifications in SMALL CAPS; ♦ Canadian drug name; ❂ Prototype drug

- Do not use sustained release capsules in loading doses when rapid control is required or in patients with creatinine clearance of 40 mL/min or less.
- Start sustained release capsules 6 h after last dose of conventional capsule if change in drug form is made.
- Store at 15°–30° C (59°–86° F) unless otherwise directed.

ADVERSE EFFECTS (≥1%) **Body as a Whole:** Hypersensitivity (pruritus, urticaria, rash, photosensitivity, <u>laryngospasm</u>). **CNS:** Dizziness, headache, fatigue, muscle weakness, convulsions, paresthesias, nervousness, acute psychosis, peripheral neuropathy. **CV:** *Hypotension,* chest pain, edema, dyspnea, syncope, bradycardia, tachycardia; worsening of CHF or cardiac arrhythmia; <u>cardiogenic shock, heart block</u>; edema with weight gain. **Special Senses:** *Blurred vision,* dry eyes, increased IOP, precipitation of acute angle-closure glaucoma. **GI:** *Dry mouth, constipation,* epigastric or abdominal pain, cholestatic jaundice. **Urogenital:** *Hesitancy* and *retention,* urinary frequency, urgency, renal insufficiency. **Other:** Dry nose and throat, drying of bronchial secretions, initiation of uterine contractions (pregnant patient); muscle aches, precipitation of myasthenia gravis, <u>agranulocytosis</u> (rare), <u>thrombocytopenia</u>.

INTERACTIONS Drug: ANTICHOLINERGIC DRUGS (e.g., TRICYCLIC ANTIDEPRESSANTS, ANTIHISTAMINES) compound anticholinergic effects; other ANTIARRHYTHMICS compound toxicities; **phenytoin, rifampin** may increase disopyramide metabolism and decrease levels; may increase **warfarin**-induced hypoprothrombinemia.

PHARMACOKINETICS **Absorption:** Readily from GI tract; 60–83% reaches systemic circulation. **Onset:** 30 min–3.5 h. **Peak:** 1–2 h. **Duration:** 1.5–8.5 h. **Distribution:** Distributed in extracellular fluid; crosses placenta; distributed into breast milk. **Metabolism:** In liver. **Elimination:** 80% in urine, 10% in feces. **Half-Life:** 4–10 h.

NURSING IMPLICATIONS

Assessment & Drug Effects

- Check apical pulse before administering drug. Withhold drug and notify prescriber if pulse rate is slower than 60 bpm, faster than 120 bpm, or if there is any unusual change in rate, rhythm, or quality.
- Monitor ECG closely. The following signs are indications for drug withdrawal: Prolongation of QT interval and worsening of arrhythmia interval, QRS widening (greater than 25%).
- Monitor for rapid weight gain or other signs of fluid retention.
- Lab tests: Baseline and periodic blood glucose, and serum potassium. Correct hypokalemia or other imbalances before initiation of therapy.
- Monitor BP closely in all patients during periods of dosage adjustment and in those receiving high dosages.
- Monitor I&O, particularly in older adults and patients with impaired renal function or prostatic hypertrophy. Persistent urinary hesitancy or retention may necessitate lower dosage or discontinuation of drug.
- Report S&S of hyperkalemia (see Appendix F); it enhances drug's toxic effects.
- Monitor for S&S of CHF.

Common adverse effects in *italic*, life-threatening effects <u>underlined</u>; generic names in **bold**; classifications in SMALL CAPS; ♣ Canadian drug name; ✪ Prototype drug

475

D

Patient & Family Education

- Take drug precisely as prescribed to maintain regularity of heart-beat.
- Weigh daily under standard conditions and check ankles for edema. Report to prescriber a weekly weight gain of 1–2 kg (2–4 lb) or more.
- Make position changes slowly, particularly when getting up from lying down because of the possibility of hypotension; dangle legs for a few minutes before walking, and do not stand still for prolonged periods.
- Do not drive or engage in other potentially hazardous activities until response to drug is known.

DISULFIRAM

(dye-sul′fi-ram)

Antabuse, Cronetal, Ro-sulfiram
Classifications: ENZYME INHIBITOR; ANTIALCOHOL AGENT
Therapeutic: ALCOHOL ABUSE DETERRENT
Pregnancy Category: B

AVAILABILITY 250 mg, 500 mg tablets

ACTION & *THERAPEUTIC EFFECT*

Acts as a deterrent to alcohol ingestion by inhibiting the enzyme acetaldehyde dehydrogenase, which normally metabolizes alcohol in the body. *When a small amount of alcohol is ingested, a complex of highly unpleasant symptoms known as the disulfiram reaction occurs, which serves as a deterrent to further drinking.*

USES Adjunct in treatment of the patient with chronic alcoholism who sincerely wants to maintain sobriety.

CONTRAINDICATIONS Severe myocardial disease; cardiac disease; psychosis; patients who have recently ingested alcohol, metronidazole, paraldehyde; multiple drug dependence.
CAUTIOUS USE Diabetes mellitus; epilepsy; seizure disorders; hypothyroidism; coronary artery disease; cerebral damage; chronic and acute nephritis; renal disease; hepatic cirrhosis or insufficiency; abnormal EEG; pregnancy (category B), lactation.

ROUTE & DOSAGE

Alcoholism
Adult: **PO** 500 mg/day for 1–2 wk, then 125–500 mg/day (max: 500 mg/day)

ADMINISTRATION

Oral

- Daily does may be given at bedtime to minimize sedative effect of the drug. Decrease in dose may also reduce sedative effect.
- Make sure patient has abstained from alcohol and alcohol-containing preparations for at least 12 h and preferably 48 h before initiating therapy.
- Store at 15°–30° C (59°–86° F) unless otherwise directed. Protect tablets from light.

ADVERSE EFFECTS (≥1%) **Reaction with alcohol ingestion:** Flushing of face, chest, arms, pulsating headache, nausea, violent vomiting, thirst, sweating, marked uneasiness, confusion, weakness, vertigo, blurred vision, pruritic skin rash, hyperventilation, abnormal gait, slurred speech, disorientation, confusion, personality changes, bizarre behavior, psychoses, tachycardia, palpitation, chest pain, <u>hypotension to shock level arrhythmias, acute congestive failure.</u> **Severe**

reactions: <u>Marked respiratory depression, unconsciousness, convulsions, sudden death.</u> **Body as a Whole:** Hypersensitivity (allergic or acneiform dermatitis; urticaria, fixed-drug eruption). **CNS:** Drowsiness, fatigue, restlessness, headache, tremor, psychoses (usually with high doses), polyneuritis, peripheral neuropathy, optic neuritis. **GI:** Mild GI disturbances, garlic-like or metallic taste, <u>hepatotoxicity</u>, hypersensitivity hepatitis.

DIAGNOSTIC TEST INTERFERENCE
Disulfiram can reduce *uptake of* I^{131}; or decreases *PBI* test results (rare).

INTERACTIONS Drug: Alcohol (including in liquid OTC drugs, **IV nitroglycerin, IV cotrimoxazole**), **metronidazole, paraldehyde** will produce disulfiram reaction; **isoniazid** can produce neurologic symptoms; may increase blood levels and toxicity of **warfarin, paraldehyde,** BARBITURATES, **phenytoin.**

PHARMACOKINETICS Absorption: Readily absorbed from GI tract. **Onset:** Up to 12 h. **Duration:** Up to 2 wk. **Distribution:** Initially deposited in fat. **Metabolism:** Metabolized slowly in liver. **Elimination:** 5–20% excreted in feces; 20% remains in body for 1–2 wk; some may be excreted in breath as carbon disulfide.

NURSING IMPLICATIONS
Assessment & Drug Effects
- Lab tests: Baseline and follow-up transaminase studies every 10–14 days to detect hepatic dysfunction.
- Note: Disulfiram reaction occurs within 5–10 min following ingestion of alcohol and may last 30 min to several hours. Intensity of reaction varies with each individual, but is generally proportional to the amount of alcohol ingested.
- Treat patient with severe disulfiram reaction as though in shock. Monitor potassium levels, especially if patient has diabetes mellitus.

Patient & Family Education
- Understand fully the possible dangers if alcohol is ingested during disulfiram treatment before consenting to therapy.
- Report promptly to prescriber the onset of nausea with right upper quadrant pain or discomfort, itching, jaundiced sclerae or skin, dark urine, clay-colored stools. Withhold drug pending liver function tests.
- Note: Ingestion of even small amounts of alcohol or use of external applications that contain alcohol may be sufficient to produce a reaction. Read all labels and avoid use of anything containing alcohol.
- Alcohol sensitivity may last as long as 2 wk after disulfiram has been discontinued.
- Note: Adverse effects of drug are often experienced during first 2 wk of therapy; symptoms usually disappear with continued therapy or with dose reduction.
- Do not drive or engage in other potentially hazardous activities until response to drug is known.

DOBUTAMINE HYDROCHLORIDE
(doe-byoo'ta-meen)
Dobutrex
Classifications: ADRENERGIC AGONIST; VASOPRESSOR
Therapeutic: CARDIAC STIMULANT; IONOTROPIC
Prototype: Isoproterenol
Pregnancy Category: C

Common adverse effects in *italic*, life-threatening effects <u>underlined</u>; generic names in **bold**; classifications in SMALL CAPS; ♣ Canadian drug name; ❷ Prototype drug

477

AVAILABILITY 12.5 mg/mL injection

ACTION & *THERAPEUTIC EFFECT*
Produces inotropic effect by acting on beta-receptors and primarily on myocardial alpha-adrenergic receptors. Increases cardiac output and decreases pulmonary wedge pressure and total systemic vascular resistance. Also increases conduction through AV node, and has lower potential for precipitating arrhythmias than dopamine. *In CHF or cardiogenic shock it increases cardiac output, enhances renal perfusion, increases renal output, and renal sodium excretion.*

USES Inotropic support in short-term treatment of adults with cardiac decompensation due to depressed myocardial contractility (cardiogenic shock) resulting from either organic heart disease or from cardiac surgery.
UNLABELED USES To augment cardiovascular function in children undergoing cardiac catheterization, stress thallium testing.

CONTRAINDICATIONS History of hypersensitivity to other sympathomimetic amines or sulfites, ventricular tachycardia, idiopathic hypertrophic subaortic stenosis; hypovolemia.
CAUTIOUS USE Preexisting hypertension, hypotension; atrial fibrillation; acute MI; unstable angina, severe coronary artery disease; pregnancy (category C). Safe use in children younger than 2 y is not established.

ROUTE & DOSAGE

Cardiac Decompensation
Adult: **IV** 0.5–1 mcg/kg/min then titrate up to 2.5–15 mcg/kg/min (max: 40 mcg/kg/min)
Adolescent/Child: **IV** 2–20 mcg/kg/min

ADMINISTRATION
Intravenous

PREPARE: Continuous: Reconstitute by adding 10 mL sterile water for injection or D5W to 250-mg vial; if not completely dissolved, add an additional 10 mL of diluent. ▪ Further dilution is typical (e.g., 250 mg in 1000 mL yields 250 mcg/mL; 250 mg in 500 mL yields 500 mcg/mL; 250 mg in 250 mL yields 1000 mcg/mL). ▪ Use IV solutions within 24 h.
ADMINISTER: Continuous: Rate of infusion is determined by body weight and controlled by an infusion pump (preferred) or a microdrip IV infusion set. ▪ IV infusion rate and duration of therapy are determined by heart rate, blood pressure, ectopic activity, urine output, and whenever possible, by measurements of cardiac output and central venous or pulmonary wedge pressures.
INCOMPATIBILITIES Solution/additive: Acyclovir, alteplase, aminophylline, bumetanide, calcium chloride, calcium gluconate, diazepam, digoxin, furosemide, heparin, insulin, magnesium sulfate, phenytoin, potassium chloride, potassium phosphate, sodium bicarbonate. Y-site: Acyclovir, alteplase, aminophylline, amphotericin B cholesteryl sulfate, cefepime, foscarnet, furosemide, heparin, indomethacin, lansoprazole, pantoprazole, pemetrexed, phytonadione, piperacillin/tazobactam, thiopental, warfarin.

▪ Refrigerate reconstituted solution at 2°–15° C (36°–59° F) for 48 h or store for 6 h at room temperature.

ADVERSE EFFECTS (≥1%) **All:** Generally dose related. **CNS:** Headache, tremors, paresthesias, mild leg cramps, nervousness, fatigue (with overdosage). **CV:** *Increased heart rate and BP,* premature ventricular beats, palpitation, *anginal pain.* **GI:** Nausea, vomiting. **Other:** Nonspecific chest pain, shortness of breath.

INTERACTIONS Drug: GENERAL ANESTHETICS (especially **cyclopropane** and **halothane**) may sensitize myocardium to effects of CATECHOLAMINES such as dobutamine and lead to serious arrhythmias—use with extreme caution; BETA-ADRENERGIC BLOCKING AGENTS (e.g., **metoprolol, propranolol**) may make dobutamine ineffective in increasing cardiac output, but total peripheral resistance may increase—concomitant use generally avoided; MAO INHIBITORS, TRICYCLIC ANTIDEPRESSANTS potentiate pressor effects—use with extreme caution.

PHARMACOKINETICS Onset: 2–10 min. **Peak:** 10–20 min. **Metabolism:** Metabolized in liver and other tissues by COMT. **Elimination:** In urine. **Half-Life:** 2 min.

NURSING IMPLICATIONS

Assessment & Drug Effects
- Correct hypovolemia by administration of appropriate volume expanders prior to initiation of therapy.
- Monitor therapeutic effectiveness. At any given dosage level, drug takes 10–20 min to produce peak effects.
- Monitor ECG and BP continuously during administration.
- Note: Marked increases in blood pressure (systolic pressure is the most likely to be affected) and

heart rate, or the appearance of arrhythmias or other adverse cardiac effects are usually reversed promptly by reduction in dosage.
- Observe patients with preexisting hypertension closely for exaggerated pressor response.
- Monitor I&O ratio and pattern. Urine output and sodium excretion generally increase because of improved cardiac output and renal perfusion.

Patient & Family Education
- Report anginal pain to prescriber promptly.

DOCETAXEL
(doc-e-tax′el)
Taxotere
Classifications: ANTINEOPLASTIC; TAXANE
Therapeutic: ANTINEOPLASTIC
Prototype: Paclitaxel
Pregnancy Category: D

AVAILABILITY 20 mg, 80 mg injection

ACTION & *THERAPEUTIC EFFECT* Docetaxel is a semisynthetic analog of paclitaxel. Docetaxel, like paclitaxel, binds to the microtubule network essential for interphase and mitosis of the cell cycle. *Docetaxel stabilizes the microtubules involved in cell division and prevents their normal functioning; this results in inhibiting mitosis in cancer cells.*

USES Breast cancer, gastric cancer, prostate cancer, head/neck cancer, non-small cell lung cancer.

CONTRAINDICATIONS Hypersensitivity to docetaxel or other drugs formulated with polysorbate 80, paclitaxel; neutrophil count less than 1500 cells/mm^3, biliary tract disease, hepatic disease, jaundice,

Common adverse effects in *italic,* life-threatening effects <u>underlined;</u> generic names in **bold;** classifications in SMALL CAPS; ♣ Canadian drug name; ⊗ Prototype drug

479

D

intramuscular injections, thrombocytopenia, acute infection, pregnancy (category D), lactation. **CAUTIOUS USE** Bone marrow suppression, bone marrow transplant patients; CHF, ascites, peripheral edema, pleural effusion; radiation therapy; pulmonary disorders, acute bronchospasm; cardiac tamponade; dental disease, dental work, herpes infection; hypotension, elderly; infection. Safety and effectiveness in children younger than 16 y are not established.

ROUTE & DOSAGE

Breast Cancer

Adult: **IV** 60–100 mg/m² q3wk (premedicate patients with dexamethasone 8 mg b.i.d. × 5 days, starting 1 day prior to docetaxel)

Prostate Cancer

Adult: **IV** 75 mg/m² q21days plus prednisone (5 mg PO twice daily) for 10 cycles (premedicate patients with dexamethasone 8 mg 12h, 3h, and 1h prior to starting docetaxel infusion)

Non–Small-Cell Lung Cancer

Adult: **IV** 75 mg/m² q3wk

Head/Neck Cancer

Adult: **IV** 75 mg/m² q21 days × 4 cycles

ADMINISTRATION

▪ Note: If drug contacts skin during preparation, wash immediately with soap and water.

Intravenous

PREPARE: **IV Infusion:** Bring vials to room temperature for 5 min; add provided diluent to yield 10 mg/mL, gently rotate for 45 sec; let stand until surface foam dissipates. ▪Inject required amount of diluted solution into a 250-mL, or larger, bag of NS or D5W; the final concentration should be between 0.3–0.74 mg/mL. Mix completely by manual rotation. ▪Use glass or polypropylene bottles or polypropylene or polyolefin plastic bags and administer through polyethylene-lined administration sets. Do not use PVC administration sets or containers. ▪Mix completely by manual rotation. ▪Use within 4 h (including the 1 h infusion time). *ADMINISTER:* **IV Infusion** Give at a constant rate over 1 h. ▪Administer ONLY after premedication with corticosteroids to prevent hypersensitivity. ▪Reduce dose by 25% following severe neutropenia (less than 500 cells/mm³) for 7 days or longer for febrile neutropenia, severe cutaneous reactions, or severe peripheral neuropathy.

INCOMPATIBILITIES **Y-site: Amphotericin B, doxorubicin, methylprednisolone, nalbuphine.**

▪Refrigerate vials at 2°–8° C (36°–46° F). Protect from light. Do not store in PVC bags. ▪Store diluted solutions in refrigerator or at room temperature for 8 h.

ADVERSE EFFECTS (≥1%) **CNS:** Paresthesia, pain, burning sensation, weakness, confusion. **CV:** Hypotension, *fluid retention (peripheral edema, weight gain),* pleural effusion. **GI:** *Nausea, vomiting, diarrhea, stomatitis,* abdominal pain; increased liver function tests (AST or ALT). **Hematologic:** <u>Neutropenia,</u> leukopenia, <u>thrombocytopenia, anemia,</u> febrile neutropenia. **Skin:** Rash, localized eruptions, desquamation, *alopecia*, nail changes (hyper/hypopigmentation, onycholysis). **Body as a Whole:** *Hypersensitivity reactions,* infusion site reactions (hyperpigmentation, inflammation, redness, dryness, phlebitis, extravasation).

INTERACTIONS Drug: Possible interaction with other drugs metabolized by CYP3A4.

PHARMACOKINETICS Distribution: 97% protein bound. **Metabolism:** In liver by CYP3A4. **Elimination:** 80% in feces, 20% renally. **Half-Life:** 11.1 h.

NURSING IMPLICATIONS

Assessment & Drug Effects

- Lab tests: Monitor LFTs prior to each drug cycle. Monitor frequently CBCs with differential. Withhold drug and notify prescriber if platelets less than 100,000 or neutrophils less than 1500 cells/mm^3.
- Monitor for S&S of hypersensitivity (see Appendix F), which may develop within a few minutes of initiation of infusion. It is usually not necessary to discontinue infusion for minor reactions (i.e., flushing or local skin reaction).
- Assess throughout therapy and report cardiovascular dysfunction, respiratory distress; fluid retention; development of neurosensory symptoms; severe, cutaneous eruptions on feet, hands, arms, face, or thorax; and S&S of infection.

Patient & Family Education

- Learn common adverse effects and measures to control or minimize them when possible. Report immediately any distressing adverse effects.
- Note: It is extremely important to comply with corticosteroid therapy and monitoring of lab values.
- Avoid pregnancy during therapy.

DOCOSANOL
(doc'os-a-nol)

Abreva
Classification: ANTIVIRAL
Therapeutic: ANTIVIRAL
Pregnancy Category: C

AVAILABILITY 10% cream

ACTION & *THERAPEUTIC EFFECT*
Docosanol inhibits viral replication by interfering with the early intracellular events surrounding viral entry into target cells. *Believed to exert its antiviral effect by inhibiting fusion of the HSV (herpes virus) envelope with the human cell plasma membrane, therefore making it difficult for the virus to enter the cell and replicate.*

USES Treatment of herpes simplex infections of the face and lips (i.e., cold sores).

CONTRAINDICATIONS Hypersensitivity to docosanol or any of the inactive ingredients in the cream; immunosuppressant patients; lactation.

CAUTIOUS USE Pregnancy (category C). Safety and efficacy in children are not established.

ROUTE & DOSAGE

Herpes Simplex Infections
Adult: **Topical** Apply to lesions 5 times/day for up to 10 days, starting at onset of symptoms

ADMINISTRATION

Topical

- Apply cream only to the affected areas using a gloved finger. Rub in gently but completely.
- Do not apply near or in the eyes.
- Avoid application to the mucous membranes inside of the mouth.
- Store at 20°–25° C (68°–77° F).

Common adverse effects in *italic*, life-threatening effects <u>underlined</u>; generic names in **bold**; classifications in SMALL CAPS; ♣ Canadian drug name; ⊘ Prototype drug

481

D

ADVERSE EFFECTS (≥1%) **CNS:** Headache. **Skin:** Skin irritation, burning.

INTERACTIONS Drug: No clinically significant interactions established.

NURSING IMPLICATIONS

Assessment & Drug Effects
- Monitor severity and extent of infection.
- Notify prescriber if improvement is not seen within 10 days of initiating treatment

Patient & Family Education
- Wash hands before and after applying cream.
- Do not share this cream with any other individual as this may spread the herpes virus.
- Report to prescriber if your condition worsens or does not improve within 10 days of beginning treatment.

DOCUSATE CALCIUM (DIOCTYL CALCIUM SULFOSUCCINATE) ⊕

(dok'yoo-sate)

DCS, PMS-Docusate Calcium, Pro-Cal-Sof, Surfak

DOCUSATE POTASSIUM

Dialose, Diocto-K, Kasof

DOCUSATE SODIUM

Colace, Colace Enema, Dio-Sul, Disonate, DGSS, D-S-S, Duosol, Lax-gel, Laxinate 100, Modane, Soft, Pro-Sof, Regulax ◆, Regutol, Therevac-Plus, Therevac-SB
Classification: STOOL SOFTENER
Therapeutic: STOOL SOFTENER
Pregnancy Category: C

AVAILABILITY Docusate Calcium: 50 mg, 240 mg capsules; **Docusate Potassium:** 100 mg tablets; 240 mg capsules; **Docusate Sodium:** 100 mg tablets; 50 mg, 100 mg, 240 mg, 250 mg capsules; 50 mg/15 mL 60 mg/15 mL, 150 mg/15 mL syrup

ACTION & *THERAPEUTIC EFFECT* Anionic surface-active agent with emulsifying and wetting properties. *Detergent action lowers surface tension, permitting water and fats to penetrate and soften stools for easier passage.*

USES Prophylactically in patients who should avoid straining during defecation and for treatment of constipation associated with hard, dry stools (e.g., following anorectal surgery, MI).

CONTRAINDICATIONS Atonic constipation, nausea, vomiting, abdominal pain, fecal impaction, structural anomalies of colon and rectum, intestinal obstruction or perforation; use of docusate sodium in patients on sodium restriction; use of docusate potassium in patients with renal dysfunction. **CAUTIOUS USE** History of CHF, edema, diabetes mellitus; pregnancy (category C).

ROUTE & DOSAGE

Stool Softener
Adult: **PO** 50–500 mg/day **PR** 50–100 mg added to enema fluid *Child:* **PO** *Younger than 3 y,* 10–40 mg/day; *3–6 y,* 20–60 mg/day; *6–12 y,* 40–120 mg/day

ADMINISTRATION

Oral
- Give with a full glass of water if allowed.
- Store syrup formulations in tight, light-resistant containers at 15°–30° C (59°–86° F) unless directed otherwise.

Rectal

- Microenema: Insert full length of nozzle (half length for children) into the rectum. Squeeze entire contents of tube and remove completely before releasing grip on tube.
- Store in tightly covered containers.

ADVERSE EFFECTS (≥1%) **GI:** Occasional mild abdominal cramps, *diarrhea*, nausea, bitter taste. **Other:** Throat irritation (liquid preparation), rash.

INTERACTIONS Drug: Docusate will increase systemic absorption of **mineral oil.**

NURSING IMPLICATIONS

Assessment & Drug Effects

- Withhold drug if diarrhea develops and notify prescriber.
- Therapeutic effectiveness: Usually apparent 1–3 days after first dose.

Patient & Family Education

- Take sufficient liquid with each dose and increase fluid intake during the day, if allowed. Oral liquid (NOT syrup) may be administered in milk, fruit juice, or infant formula to mask bitter taste.
- Do not take concomitantly with mineral oil.
- Do not take for prolonged periods in lieu of proper dietary management or treatment of underlying causes of constipation.

DOFETILIDE

(do-fe-ti'lyde)

Tikosyn
Classifications: CLASS III ANTI-ARRHYTHMIC, POTASSIUM CHANNEL BLOCKER
Therapeutic: CLASS III ANTI-ARRHYTHMIC

Prototype: Amiodarone HCl
Pregnancy Category: C

AVAILABILITY 125 mcg, 250 mcg, 500 mcg capsules

ACTION & *THERAPEUTIC EFFECT*
Class III antiarrhythmic agent that prolongs the cardiac action potential by blocking the potassium channels and thus one or more of the potassium currents. Action results in suppression of arrhythmias dependent upon reentry of potassium ions. *Effectiveness indicated by correction of atrial arrhythmias.*

USES Symptomatic atrial fibrillation and flutter.

CONTRAINDICATIONS Hypersensitivity to dofetilide; QT prolongation; ventricular arrhythmias; history of torsades de pointes; electrolyte imbalances (e.g., hypokalemia, hypomagnesemia, etc.); renal failure; lactation.
CAUTIOUS USE Atrioventricular block, bradycardia, CHF, concurrent administration of potassium depleting diuretics, hepatic or renal impairment; history of moderate QT_c interval prolongation; moderate to severe hypertension; recent MI or unstable angina; vascular heart disease; older adults; pregnancy (category C). Safety and efficacy in children younger than 18 y are unknown.

ROUTE & DOSAGE

Atrial Fibrillation/Flutter

Adult: **PO** Based on creatinine clearance (CrCl) and QT_c interval, if QT_c increases by more than 15% from baseline or is greater than 500 milliseconds 2–3 h after initial dose. Decrease subsequent doses by 50%.

Common adverse effects in *italic*, life-threatening effects <u>underlined</u>; generic names in **bold;** classifications in SMALL CAPS; ♣ Canadian drug name; ◯ Prototype drug

483

D

Renal Impairment Dosage Adjustment

CrCl greater than 60 mL/min: 500 mcg b.i.d.; *40–60 mL/min:* 250 mcg b.i.d.; *20–39 mL/min:* 125 mcg b.i.d.

ADMINISTRATION

Oral
- Do not give dofetilide if QT/QT$_c$ interval greater than 420 milliseconds (or greater than 500 milliseconds with ventricular conduction abnormalities).
- Administer only with continuous ECG monitoring.
- Do not initiate therapy until 3 mo after withdrawal of previous antiarrhythmic therapy.
- Do not initiate therapy until 3 mo after amiodarone has been withdrawn or plasma level is less than 0.3 mcg/mL.
- Store at 15°–30° C (59°–86° F); protect from moisture and humidity.

ADVERSE EFFECTS (≥1%) **Body as a Whole:** Flu-like syndrome, back pain. **CNS:** *Headache,* dizziness, insomnia. **CV:** <u>*Torsades de pointes arrhythmia, ventricular arrhythmias,*</u> AV block, *chest pain.* **GI:** Nausea, diarrhea, abdominal pain. **Respiratory:** Respiratory infection, dyspnea. **Skin:** Rash.

INTERACTIONS Drug: Dofetilide levels increased by **verapamil, cimetidine, trimethoprim, ketoconazole, prochlorperazine, megestrol;** do not give with drugs known to increase the QT$_c$ interval such as **bepridil,** PHENOTHIAZINES, TRICYCLIC ANTIDEPRESSANTS, ORAL MACROLIDES, other ANTIARRHYTHMICS.

PHARMACOKINETICS Absorption: Greater than 90% bioavailable. **Peak:** 2–3 h. **Distribution:** 60–70%

protein bound. **Metabolism:** In liver. **Elimination:** Primarily excreted unchanged in urine. **Half-Life:** 10 h.

NURSING IMPLICATIONS

Assessment & Drug Effects
- Monitor ECG continuously during first 3 mo of therapy; then periodically.
- Do not discharge patient until 12 h after conversion to normal sinus rhythm.
- Lab tests: Baseline and periodic serum electrolytes (including magnesium), periodic CBC, and routine blood chemistry. Serum potassium **must be** within normal limits prior to and throughout therapy with dofetilide.
- Notify prescriber immediately of electrolyte imbalances, especially hypokalemia and hypomagnesemia.

Patient & Family Education
- Report immediately conditions that cause potassium loss (e.g., prolonged vomiting, diarrhea, excessive sweating).
- Do **NOT** take concurrently cimetidine, verapamil, ketoconazole, trimethoprim.

DOLASETRON MESYLATE
(dol-a-se′tron)

Anzemet
Classifications: SELECTIVE SEROTONIN (5-HT$_3$) RECEPTOR ANTAGONIST; ANTIEMETIC
Therapeutic: ANTIEMETIC
Prototype: Ondansetron
Pregnancy Category: B

AVAILABILITY 50 mg, 100 mg tablets; 20 mg/mL injection

ACTION & *THERAPEUTIC EFFECT*
Dolasetron is a selective serotonin

(5-HT$_3$) receptor antagonist used for control of nausea and vomiting. Serotonin receptors affected are located in the chemoreceptor trigger zone (CTZ) of the brain and peripherally on the vagal nerve terminal. Serotonin, released from the cells of the small intestine, activate 5-HT$_3$ receptors located on vagal efferent neurons, thus initiating the vomiting reflex. *Has effective antiemetic properties.*

USES Prevention of nausea and vomiting from emetogenic chemotherapy, prevention and treatment of postoperative nausea and vomiting.

CONTRAINDICATIONS Hypersensitivity to dolasetron; **IV route:** Nausea and vomiting associated with cancer chemotherapy; congenital QT prolongation syndrome; uncorrected hypokalemia or hypomagnesemia.

CAUTIOUS USE Patients who have or may develop prolongation of cardiac conduction intervals, particularly QT$_c$ (i.e., patients with potential hypokalemia, hypomagnesemia, diuretics; patients taking antiarrhythmic drugs and high-dose anthracycline therapy, etc.), pregnancy (category B), lactation. Safety and efficacy in children younger than 2 y are not established.

ROUTE & DOSAGE

Prevention of Chemotherapy-Induced Nausea and Vomiting
Adult/Child (older than 2 y): **PO** 100 mg 1 h prior to chemotherapy

Pre-/Postoperative Nausea and Vomiting
Adult: **IV** 12.5 mg 15 min before cessation of anesthesia or when postoperative nausea and vomiting occurs **PO** 100 mg within 2 h prior to surgery
Child (older than 2 y): **IV** 0.35 mg/kg up to 12.5 mg 15 min before cessation of anesthesia or when postoperative nausea and vomiting occurs **PO** 1.2 mg/kg up to 100 mg starting 2 h prior to surgery (may also mix IV formulation in apple or apple-grape juice and administer orally)

ADMINISTRATION

Oral
- Give within 2 h before surgery, when used for postoperative nausea.

Intravenous

PREPARE: **Direct:** Give undiluted. **IV Infusion:** Dilute in 50 mL of any of the following: NS, D5W, D5/0.45% NaCl, LR.

ADMINISTER: **Direct:** Inject undiluted drug over 30 sec. **IV Infusion:** Infuse diluted drug over 15 min.

INCOMPATIBILITIES Solution/additive **Potassium chloride.**

- Store at 20°–25° C (66°–77° F) and protect from light. - Diluted IV solution may be stored refrigerated up to 48 h.

ADVERSE EFFECTS (≥1%) **Body as a Whole:** Fever, fatigue, pain, chills or shivering. **CNS:** *Headache,* dizziness, drowsiness. **CV:** Hypertension. **GI:** *Diarrhea,* increased LFTs, abdominal pain. **Genitourinary:** Urinary retention.

INTERACTIONS Drugs: Avoid use with **apomorphine** due to hypotension; **ziprasidone** may prolong QT interval.

Common adverse effects in *italic*, life-threatening effects underlined; generic names in **bold**; classifications in SMALL CAPS; ✤ Canadian drug name; ⊘ Prototype drug

485

PHARMACOKINETICS Absorption: Rapidly absorbed from GI tract. **Peak:** 0.6 h IV, 1 h PO. **Distribution:** Crosses placenta, distributed into breast milk. **Metabolism:** Metabolized to hydrodolasetron by carbonyl reductase. Hydrodolasetron is metabolized in the liver by CYP2D6. **Elimination:** Primarily in urine as hydrodolasetron. **Half-Life:** 10 min dolasetron, 7.3 h hydrodolasetron.

NURSING IMPLICATIONS

Assessment & Drug Effects

- Lab tests: Baseline serum electrolytes before initiating drug. Hypokalemia and hypomagnesemia should be correct before initiating therapy. With prolonged therapy, periodically monitor LFTs, PTT, CBC with platelet count, and alkaline phosphatase.
- Monitor closely cardiac status especially with vomiting, excess diuresis, or other conditions that may result in electrolyte imbalances.
- Monitor ECG, especially in those taking concurrent antiarrhythmic or other drugs that may cause QT prolongation.
- Monitor for and report signs of bleeding (e.g., hematuria, epistaxis, purpura, hematoma).

Patient & Family Education

- Headache requiring analgesic for relief is a common adverse effect.

DONEPEZIL HYDROCHLORIDE ⓟ

(don-e'pe-zil)

Aricept, Aricept ODT

Classifications: CENTRAL ACTING CHOLINERGIC; CHOLINESTERASE INHIBITOR
Therapeutic: ANTIDEMENTIA; ALZHEIMER'S AGENT
Pregnancy Category: C

AVAILABILITY 5 mg, 10 mg tablets and orally disintegrating tablets; 23 mg tablet

ACTION & THERAPEUTIC EFFECT In early stages of Alzheimer's disease, pathologic changes in neurons result in deficiency of acetylcholine. A cholinesterase inhibitor, presumably elevates acetylcholine concentration in the cerebral cortex by slowing degrading acetylcholine released by remaining intact neurons. *Improves global function, cognition, and behavior of patients with mild to moderate Alzheimer's.*

USES Mild, moderate, or severe dementia of Alzheimer's type.
UNLABELED USES Vascular dementia, poststroke aphasia, memory improvement in multiple sclerosis patients.

CONTRAINDICATIONS Hypersensitivity to donepezil, tacrine, or piperidine derivatives; GI bleeding, jaundice; lactation; children.
CAUTIOUS USE Anesthesia, sick sinus rhythm, AV block, bradycardia, cardiac arrhythmias, cardiac disease, hypotension; hyperthyroidism, history of ulcers, abnormal liver function; history of asthma or obstructive pulmonary disease, history of seizures, urinary tract obstruction, intestinal obstruction; diarrhea, emesis, GI disease, renal failure, renal impairment, surgery; pregnancy (category C).

ROUTE & DOSAGE

Alzheimer's Disease Related Dementia
Adult: **PO** 5–10 mg at bedtime

ADMINISTRATION

Oral

- Give immediately before going to bed.

- Dosage increase from 5 to 10 mg is usually made only after 4–6 wk of therapy.
- Store at 15°–30° C (59°–86° F).

ADVERSE EFFECTS (≥1%) Body as a Whole: *Headache,* fatigue. CNS: *Insomnia,* dizziness, depression, tremor, irritability, vertigo, ataxia. CV: Syncope, hypertension, atrial fibrillation, hot flashes, hypotension. GI: *Nausea, diarrhea, vomiting, muscle cramps, anorexia,* GI bleeding, bloating, fecal incontinence, epigastric pain. Respiratory: Dyspnea. Skin: Pruritus, sweating, urticaria. Other: Ecchymoses, muscle cramps, dehydration, blurred vision, urinary incontinence, nocturia.

INTERACTIONS Drug: Keto-conazole, quinidine may inhibit donepezil metabolism; carbamazepine, dexamethasone, phenobarbital, phenytoin, rifampin may increase donepezil elimination; donepezil may interfere with the action of ANTICHOLIN-ERGIC AGENTS.

PHARMACOKINETICS Absorption: Rapidly from GI tract. Peak: 3–4 h. Distribution: 96% protein bound. Metabolism: In liver by CYP2D6 and CYP3A4. Elimination: Primarily in urine. Half-Life: 70 h.

NURSING IMPLICATIONS

Assessment & Drug Effects
- Monitor closely for S&S of GI ulceration and bleeding, especially with concurrent use of NSAIDs.
- Monitor carefully patients with a history of asthma or obstructive pulmonary disease.
- Monitor cardiovascular status; drug may have vagotonic effect on the heart, causing bradycardia, especially in presence of conduction abnormalities.

Patient & Family Education
- Exercise caution. Fainting episodes related to slowing the heart rate may occur.
- Report immediately to prescriber any S&S of GI ulceration or bleeding (e.g., "coffee-grounds" emesis, tarry stools, epigastric pain).

DOPAMINE HYDROCHLORIDE
(doe′pa-meen)

Classifications: ALPHA- AND BETA-ADRENERGIC AGONIST; INOTROPIC
Therapeutic: CARDIAC STIMULANT; VASOPRESSOR
Prototype: Epinephrine
Pregnancy Category: C

AVAILABILITY 40 mg/mL, 80 mg/mL, 160 mg/mL injection

ACTION & *THERAPEUTIC EFFECT*
Major cardiovascular effects produced by direct action on alpha- and beta-adrenergic receptors and on specific dopaminergic receptors in mesenteric and renal vascular beds. Positive inotropic effect on myocardium increases cardiac output with increase in systolic and pulse pressure and little or no effect on diastolic pressure. Improves circulation to renal vascular bed by decreasing renal vascular resistance with resulting increase in glomerular filtration rate and urinary output. *Due to its potential for inotropic, chronotropic, and vasopressor effects, dopamine has several clinical uses, including decreased cardiac output as well as correction of hypotension associated with cardiogenic and septic shock.*

USES To correct hemodynamic imbalance in shock syndrome due to MI (cardiogenic shock), trauma, endotoxic septicemia (septic shock), open heart surgery, and CHF.

Common adverse effects in *italic*, life-threatening effects <u>underlined</u>; generic names in **bold**; classifications in SMALL CAPS; ♣ Canadian drug name; ⊘ Prototype drug

487

D

UNLABELED USES Acute renal failure; cirrhosis; hepatorenal syndrome; barbiturate intoxication.

CONTRAINDICATIONS Pheochromocytoma; uncorrected tachyarrhythmias or ventricular fibrillation; persistent hypotension; children younger than 2 y.

CAUTIOUS USE Patients with history of occlusive vascular disease (e.g., Buerger's or Raynaud's disease); CAD; cold injury; acute MI; diabetic endarteritis, arterial embolism; pregnancy (category C), lactation, neonates.

ROUTE & DOSAGE

Shock/Surgery
Adult: IV 2–5 mcg/kg/min increased gradually up to 20–50 mcg/kg/min if necessary
Adolescent/Child: IV 1–5 mcg/kg/min increased gradually up to 20 mcg/kg/min

CHF
Adult: IV 3–10 mcg/kg/min

ADMINISTRATION

Intravenous

PREPARE: Continuous: Dilute just prior to administration. ▪ Dilute each ampule in one of the following: D5W, D5/NS, D5/LR, D5/0.45% NaCl, NS. ▪ Dilute 200 mg ampule in 250 mL, 500 mL, or 1000 mL IV solution to yield 800 mcg/mL, 400 mcg/mL, or 200 mcg/mL, respectively. Dilute 400 mg ampule in 250 mL, 500 mL, or 1000 mL IV solution to yield 1600 mcg/mL, 800 mcg/mL or 400 mcg/mL, respectively. ▪ Dilute 400 mg ampule in 250 mL, 500 mL, or 1000 mL IV solution to yield 1600 mcg/mL, 800 mcg/mL, or 400 mcg/mL, respectively. ▪ Dilute 800 mg ampule in 250 mL, 500 mL, or 1000 mL IV solution to yield 3200 mcg/mL, 1600 mcg/mL or 800 mcg/mL, respectively. ▪ Consult package information for other dilutions.

ADMINISTER: Continuous: Infusion rate is based on body weight. ▪ Infusion rate and guidelines for adjusting rate relative changes in blood pressure are prescribed by prescriber. ▪ Microdrip and other reliable metering device should be used for accuracy of flow rate.

INCOMPATIBILITIES **Solution/additive: Acyclovir, alteplase, amphotericin B, ampicillin, metronidazole, penicillin G, sodium bicarbonate. Y-site: Acyclovir, alteplase, amphotericin B cholesteryl complex, cefepime, doxycycline, furosemide, indomethacin, insulin, lansoprazole, sodium bicarbonate, thiopental.**

▪ Correct hypovolemia, if possible, with either whole blood or plasma before initiation of dopamine therapy. ▪ Monitor infusion continuously for free flow, and take care to avoid extravasation, which can result in tissue sloughing and gangrene. Use a large vein of the antecubital fossa. ▪ Antidote for extravasation: Stop infusion promptly and remove needle. Immediately infiltrate the ischemic area with 5–10 mg phentolamine mesylate in 10–15 mL of NS, using syringe and fine needle. Pediatric dosage of phentolamine should be 0.1–0.2 mg/kg (max: 10 mg per dose). ▪ Protect dopamine from light. Discolored solutions should not be used.

▪ Store reconstituted solution for 24 h at 2°–15° C (36°–59° F) or 6 h at room temperature 15°–30° C.

Common adverse effects in *italic*, life-threatening effects underlined; generic names in **bold**; classifications in SMALL CAPS; ✦ Canadian drug name; ⊙ Prototype drug

ADVERSE EFFECTS (≥1%) **CV:** *Hypotension,* ectopic beats, *tachycardia,* anginal pain, palpitation, vasoconstriction (indicated by disproportionate rise in diastolic pressure), cold extremities; less frequent: <u>Aberrant conduction</u>, bradycardia, widening of QRS complex, elevated blood pressure. **GI:** Nausea, vomiting. **CNS:** Headache. **Skin:** Necrosis, tissue sloughing with extravasation, <u>gangrene</u>, piloerection. **Other:** Azotemia, dyspnea, dilated pupils (high doses).

DIAGNOSTIC TEST INTERFERENCE
Dopamine may modify test response when histamine is used as a control for *intradermal skin tests.*

INTERACTIONS Drug: MAO IN-HIBITORS, ERGOT ALKALOIDS, increase alpha-adrenergic effects (headache, hyperpyrexia, hypertension); **guanethidine, phenytoin** may decrease dopamine action; BETA-BLOCKERS antagonize cardiac effects; ALPHA BLOCKERS antagonize peripheral vasoconstriction; **halothane, cyclopropane** increase risk of hypertension and ventricular arrhythmias.

PHARMACOKINETICS Onset: Less than 5 min. **Duration:** Less than 10 min. **Distribution:** Widely distributed; does not cross blood–brain barrier. **Metabolism:** Inactive in the liver, kidney, and plasma by monoamine oxidase and COMT. **Elimination:** In urine. **Half-Life:** 2 min.

NURSING IMPLICATIONS

Assessment & Drug Effects
▪ Monitor blood pressure, pulse, peripheral pulses, and urinary output at intervals prescribed by prescriber. Precise measurements are essential for accurate titration of dosage.

▪ Report the following indicators promptly to prescriber for use in decreasing or temporarily suspending dose: Reduced urine flow rate in absence of hypotension; ascending tachycardia; dysrhythmias; disproportionate rise in diastolic pressure (marked decrease in pulse pressure); signs of peripheral ischemia (pallor, cyanosis, mottling, coldness, complaints of tenderness, pain, numbness, or burning sensation).
▪ Monitor therapeutic effectiveness. In addition to improvement in vital signs and urine flow, other indices of adequate dosage and perfusion of vital organs include loss of pallor, increase in toe temperature, adequacy of nail bed capillary filling, and reversal of confusion or comatose state.

DORIPENEM
(dor-i-pen′em)
Doribax
Classifications: BETA-LACTAM ANTIBIOTIC; CARBAPENEM ANTIBIOTIC
Therapeutic: ANTIBIOTIC
Prototype: Imipenem-cilastatin
Pregnancy Category: B

AVAILABILITY 500 mg single-use vials

ACTION & *THERAPEUTIC EFFECT*
Inhibits essential penicillin-binding proteins resulting in inhibition of bacterial cell wall synthesis, resulting in bacterial cell death. *Bactericidal against aerobic and anaerobic gram-negative and gram-positive bacteria, and effectively resolves infection.*

USES Single-agent treatment of complicated intra-abdominal infections and urinary tract infections,

Common adverse effects in *italic*, life-threatening effects <u>underlined</u>; generic names in **bold**; classifications in SMALL CAPS; ♣ Canadian drug name; ⊘ Prototype drug

489

including pyelonephritis caused by susceptible organisms.

UNLABELED USES Hospital acquired pneumonia.

CONTRAINDICATIONS Hypersensitivity to doripenem, or beta-lactam antibiotics; multiple allergies; inhalation route.

CAUTIOUS USE Hypersensitivity to cephalosporins, penicillins; moderate to severe renal impairment; GI disease, colitis, IBD; history of a seizure disorder; bacterial meningitis; older adults; pregnancy (category B); lactation. Safe use in children and adolescents is not established.

ROUTE & DOSAGE

Complicated Intra-Abdominal Infection
Adult: **IV** 500 mg q8h × 5–14 days

Complicated UTI, Including Pyelonephritis
Adult: **IV** 500 mg q8h × 10 days

Renal Impairment Dosage Adjustment
CrCl 30–50 mL/min: 250 mg q8h; *greater than 10 mL/min but less than 30 mL/min:* 250 mg q12h

ADMINISTRATION

Intravenous

PREPARE: Intermittent: Add 10 mL of sterile water for injection or NS to the 500 mg or 250 mg vial, gently shake to form suspension; yields 50 mg/mL (500 mg vial) or 25 mg/mL (250 mg vial). ▪ *Preparation of 500 mg dose:* Withdraw contents of 500 mg vial with a 21-gauge needle and add to infusion bag of 100 mL of NS or D5W, gently shake until clear. Final concentration is approximately 4.5 mg/mL. ▪ *Preparation of 250 mg dose:* Withdraw contents of 250 mg vial with a 21-gauge needle and add to infusion bag of 50 or 100 mL of NS or D5W, gently shake until clear. Final concentration is approximately 4.2 mg/mL (50 mL bag) or 2.3 mg/mL (100 mL bag).

ADMINISTER: Intermittent: Infuse over 15–30 min.

INCOMPATIBILITIES Solution/additive: Do not combine with any other drug. **Y-site:** Do not add to Y-site.

ADVERSE EFFECTS (≥1%) **Body as a Whole:** Anaphylaxis, hypersensitivity reactions. **CNS:** *Headache.* **CV:** Phlebitis. **GI:** Diarrhea, nausea, oral candidiasis. **Hematologic:** Anemia. **Metabolic:** Elevated hepatic enzymes. **Skin:** Rash. **Urogenital:** Vulvomycotic infection

INTERACTIONS Drug: Doripenem decreases plasma levels of **valproic acid. Probenecid** increases doripenem plasma levels.

PHARMACOKINETICS Distribution: Minimal protein binding. **Metabolism:** In liver (18%) to inactive metabolite. **Elimination:** Urine (primarily unchanged). **Half-Life:** 1 h.

NURSING IMPLICATIONS

Assessment & Drug Effects

- Lab tests: Baseline C&S prior to therapy. Monitor periodically LFTs, Hct and Hgb.
- Determine history of hypersensitivity reactions to other beta-lactams, cephalosporins, penicillins, or other drugs.
- Discontinue drug and immediately report S&S of hypersensitivity (see Appendix F).
- Report S&S of superinfection or pseudomembranous colitis (see Appendix F).

Patient & Family Education

- Learn S&S of hypersensitivity, superinfection, and pseudomembranous colitis; report any of these to prescriber promptly.

DORNASE ALFA
(dor′naze)

Pulmozyme
Classifications: RESPIRATORY ENZYME; MUCOLYTIC
Therapeutic: MUCOLYTIC
Pregnancy Category: B

AVAILABILITY 1 mg/mL solution for inhalation

ACTION & THERAPEUTIC EFFECT

Dornase is a solution of recombinant human deoxyribonuclease (DNAse), an enzyme that selectively cleaves DNA. In cystic fibrosis (CF) viscous, purulent secretions in the airway reduce pulmonary function and lead to exacerbations of infection. Purulent pulmonary secretions contain very high concentrations of DNA released by degenerating leukocytes that are present in response to infection. *Dornase hydrolyzes the DNA in sputum of CF patients and reduces sputum viscosity, thus reducing incidence of respiratory tract infection.*

USES In combination with standard therapies to reduce the frequency of respiratory infections in patients with CF and to improve pulmonary function.
UNLABELED USES Chronic bronchitis.

CONTRAINDICATIONS Hypersensitivity to dornase or hamster protein.
CAUTIOUS USE Pregnancy (category B), lactation. Safe use in children younger than 3 mo not established.

ROUTE & DOSAGE

Cystic Fibrosis

Adult/Child (older than 3 mo):
Inhalation 2.5 mg (1 ampule) inhaled once daily using a recommended nebulizer, may increase to twice daily (do not mix with other agents in nebulizer)

ADMINISTRATION

Inhalation

- Do not dilute or mix with any other drugs or solutions in the nebulizer.
- Use only with nebulizer systems recommended by the drug manufacturer.
- Do not shake ampules; do not use ampules that have been at room temperature longer than 24 h or have become cloudy or discolored.
- Store refrigerated at 2°–8° C (36°–46° F) in protective foil pouch.

ADVERSE EFFECTS (≥1%) **Respiratory:** Hoarseness, sore throat, voice alterations, pharyngitis, laryngitis, cough, rhinitis. **Other:** Conjunctivitis, chest pain, rash.

PHARMACOKINETICS Absorption: Minimal systemic absorption. **Onset:** 3–8 days. **Duration:** Benefit lasts up to 4 days after discontinuing treatment.

NURSING IMPLICATIONS

Assessment & Drug Effects

- Monitor for improvement in dyspnea and sputum clearance.
- Monitor for S&S of hypersensitivity (see Appendix F). Patients with a history of hypersensitivity to bovine pancreatic dornase are at high risk.
- Monitor for adverse effects; rarely, dosage adjustments may be required.

Common adverse effects in *italic*, life-threatening effects <u>underlined</u>; generic names in **bold**; classifications in SMALL CAPS; ◆ Canadian drug name; ⊙ Prototype drug

Patient & Family Education

- Report rash, hives, itching, or other S&S of hypersensitivity to prescriber immediately.
- Know potential adverse effects and report those that are bothersome or do not disappear.
- Take a missed dose as soon as possible; if it is almost time for the next dose, skip the missed dose.

DORZOLAMIDE HYDROCHLORIDE
(dor-zol′a-mide)

Trusopt
Classifications: EYE PREPARATION; CARBONIC ANHYDRASE INHIBITOR
Therapeutic: ANTIGLAUCOMA; OCULAR ANTIHYPERTENSIVE
Prototype: Acetazolamide
Pregnancy Category: C

AVAILABILITY 2% ophth solution

ACTION & *THERAPEUTIC EFFECT*
Dorzolamide is a sulfonamide that inhibits carbonic anhydrase in the eye, thus reducing the rate of aqueous humor formation with subsequent lowering of IOP. Elevated IOP is a major risk factor in the pathogenesis of optic nerve damage and visual field loss due to glaucoma. *Lowers IOP in glaucoma or ocular hypertension.*

USES Ocular hypertension, open-angle glaucoma.

CONTRAINDICATIONS Previous hypersensitivity to dorzolamide.
CAUTIOUS USE History of hypersensitivity to other carbonic anhydrase inhibitors, sulfonamides, or thiazide diuretics; ocular infection or inflammation; recent ocular surgery; moderate to severe renal or hepatic insufficiency; angle-closure glaucoma; corneal abrasion; older adults, pregnancy (category C).

ROUTE & DOSAGE

Glaucoma, Ocular Hypertension

Adult/Adolescent/Child/Infant:
Ophthalmic 1 drop in affected eye t.i.d.

ADMINISTRATION

Instillation

- Apply gentle pressure to lacrimal sac during and immediately following drug instillation for about 1 min to lessen degree of systemic absorption.
- Administer at least 10 min apart, if another ophthalmic drug is being used concurrently.
- Store at 15°–30° C (59°–86° F).

ADVERSE EFFECTS (≥1%) CNS: Headache. **GI:** Bitter taste, nausea. **Special Senses:** *Transient burning or stinging, transient blurred vision,* superficial punctate keratitis, tearing, dryness, photophobia, ocular allergic reaction. **Skin:** Rash.

PHARMACOKINETICS Absorption: Some systemic absorption. **Onset:** 2 h. **Duration:** 8–12 h. **Distribution:** Into red blood cells. **Elimination:** In urine. **Half-Life:** RBC elimination about 4 mo.

NURSING IMPLICATIONS

Assessment & Drug Effects

- Inquire about previous hypersensitivity to sulfonamides prior to therapy.
- Withhold drug and notify prescriber if S&S of local or systemic hypersensitivity occur (see Appendix F).
- Withhold the drug and notify the prescriber if ocular irritation occurs.

Patient & Family Education

- Learn proper technique for applying eyedrops.

- Do not allow tip of drug dispenser to come in contact with the eye.
- Discontinue drug and report to prescriber: Ocular irritation, infection, or S&S of systemic hypersensitivity occur (see Appendix F).

DOXAPRAM HYDROCHLORIDE
(dox'a-pram)

Dopram
Classifications: CEREBRAL STIMULANT; RESPIRATORY STIMULANT; ANALEPTIC
Therapeutic: CEREBRAL STIMULANT; RESPIRATORY STIMULANT
Prototype: Caffeine
Pregnancy Category: B

AVAILABILITY 20 mg/mL injection

ACTION & *THERAPEUTIC EFFECT*
Short-acting analeptic capable of stimulating all levels of the cerebrospinal axis. Respiratory stimulation by direct medullary action or possibly by indirect activation of peripheral chemoreceptors increases tidal volume and slightly increases respiratory rate. *Decreases Pco_2 and increases Po_2 by increasing alveolar ventilation; may elevate BP and pulse rate by stimulation of brainstem vasomotor area. It is used to stimulate respiration postanesthesia, for drug-induced CNS depression, and for chronic pulmonary disease associated with acute hypercapnia.*

USES Short-term adjunctive therapy to alleviate postanesthesia and drug-induced respiratory depression. Also as a temporary measure (approximately 2 h) in hospitalized patients with COPD associated with acute respiratory insufficiency as an aid to prevent elevation of $Paco_2$ during administration of oxygen. (Not used with mechanical ventilation.)

UNLABELED USES Neonatal apnea refractory to xanthine therapy.

CONTRAINDICATIONS Epilepsy and other convulsive disorders; head injury, cerebral edema; ventilatory disorders, pulmonary fibrosis, flail chest, pneumothorax, airway obstruction, extreme dyspnea, or acute bronchial asthma; severe hypertension, severe coronary artery disease, uncompensated heart failure, CVA; MAOI; lactation.
CAUTIOUS USE History of bronchial asthma, COPD; cardiac disease, severe tachycardia, arrhythmias, hypertension; hyperthyroidism; pheochromocytoma; increased intracranial pressure; peptic ulcer, patients undergoing gastric surgery; acute agitation; pregnancy (category B). Safe use in children younger than 12 y not established.

ROUTE & DOSAGE

Postanesthesia

Adult: **IV** 0.5–1 mg/kg single injection (not more than 1.5 mg/kg), may repeat q5min up to 2 mg/kg total dose; infusion of 0.5–1 mg/kg (max total dose: 4 mg/kg)

Drug-Induced CNS Depression

Adult: **IV** 1–2 mg/kg repeat in 5 min, then q1–2h until patient awakens [if relapse occurs, resume q1–2h injections (max total dose: 3 g), if no response after priming dose, may give 1–3 mg/min for up to 2 h until patient awakens]

Chronic Obstructive Pulmonary Disease

Adult: **IV** 0.5–2 mg/kg OR 1–2 mg/min for a max of 2 h (max rate: 3 mg/min)

Common adverse effects in *italic*, life-threatening effects underlined; generic names in **bold**; classifications in SMALL CAPS; ♦ Canadian drug name; ⊕ Prototype drug

493

D

ADMINISTRATION

- IV administration to neonates: Verify correct IV concentration and rate of infusion with prescriber. Generally do not use in newborns because doxapram contains benzyl alcohol.
- Ensure adequacy of airway and oxygenation before initiation of doxapram therapy.

Intravenous

PREPARE: **Direct:** Give undiluted. **IV Infusion for CNS Depression:** Dilute 250 mg (12.5 mL) in 250 mL of D5W or NS. **IV Infusion for COPD:** Add 400 mg doxapram to 180 mL of D5W, D10W, or NS to yield 2 mg/mL.

ADMINISTER: **Direct for CNS Depression:** Give undiluted over 5 min. **IV Infusion for CNS Depression:** Give at a rate of 1–3 mg/min, depending on patient response. Never exceed 3 mg/min.
- Infusion should not be administered for longer than 2 h. **IV Infusion for COPD:** Infuse at 0.5–1.5 mL/min.

INCOMPATIBILITIES **Solution/additive: Aminophylline, ascorbic acid, CEPHALOSPORINS, dexamethasone, diazepam, digoxin, dobutamine, folic acid, furosemide, hydrocortisone, ketamine, methylprednisolone, minocycline, thiopental, ticarcillin. Y-site: Clindamycin.**

- Store at 15°–30° C (59°–86° F) unless otherwise directed.

ADVERSE EFFECTS (≥1%) CNS:
Dizziness, sneezing, apprehension, confusion, *involuntary movements,* hyperactivity, paresthesias; feeling of warmth and burning, especially of genitalia and perineum; flushing, sweating, hyperpyrexia, headache, pilomotor erection, pruritus, muscle tremor, rigidity, convulsions, *increased deep-tendon reflexes,* bilateral Babinski sign, *carpopedal spasm,* pupillary dilation, mild delayed narcosis. **CV:** *Mild to moderate increase in BP, sinus tachycardia,* bradycardia, extrasystoles, lowered T waves, PVCs, chest pains, tightness in chest. **GI:** Nausea, vomiting, diarrhea, salivation, sour taste. **Urogenital:** Urinary retention, frequency, incontinence. **Respiratory:** Dyspnea, tachypnea, cough, <u>laryngospasm, bronchospasm</u>, hiccups, rebound hypoventilation, hypocapnia with tetany. **Other:** Local skin irritation, thrombophlebitis with extravasation; decreased Hgb, Hct, and RBC count; elevated BUN; albuminuria.

INTERACTIONS Drug: MAO INHIBITORS, SYMPATHOMIMETIC AGENTS add to pressor effects.

PHARMACOKINETICS Onset: 20–40 s. Peak: 1–2 min. Duration: 5–12 min. Metabolism: Rapidly metabolized. Elimination: In urine as metabolites.

NURSING IMPLICATIONS

Assessment & Drug Effects

- Monitor IV site frequently. Extravasation or use of same IV site for prolonged periods can cause thrombophlebitis (see Appendix F) or tissue irritation.
- Monitor carefully and observe accurately: BP, pulse, deep tendon reflexes, airway, and arterial blood gases. All are essential guides for determining minimum effective dosage and preventing overdosage. Make baseline determinations for comparison.
- Lab tests: Draw arterial Po_2 and Pco_2 and O_2 saturation prior to both initiation of doxapram infusion and oxygen administration in

patients with COPD, and then at least every 30 min during infusion.

- Discontinue doxapram if arterial blood gases show evidence of deterioration and when mechanical ventilation is initiated.
- Observe patient continuously during therapy and maintain vigilance until patient is fully alert (usually about 1 h) and protective pharyngeal and laryngeal reflexes are completely restored.
- Notify prescriber immediately of any adverse effects. Be alert for early signs of toxicity: Tachycardia, muscle tremor, spasticity, hyperactive reflexes.
- Note: A mild to moderate increase in BP commonly occurs.
- Discontinue if sudden hypotension or dyspnea develops.

DOXAZOSIN MESYLATE

(dox-a′zo-sin)

Cardura
Classification: ALPHA₁-ADRENERGIC ANTAGONIST
Therapeutic: ANTIHYPERTENSIVE
Prototype: Prazosin
Pregnancy Category: B

AVAILABILITY 1 mg, 2 mg, 4 mg, 8 mg tablets

ACTION & *THERAPEUTIC EFFECT*
By selective competitive inhibition of alpha₁-adrenoreceptors, it produces vasodilation in both arterioles and veinous vessels resulting in both peripheral vascular resistance and reduced blood pressure. *Long-acting effect of lowering blood pressure in supine or standing individuals with most pronounced effect on diastolic pressure.*

USES Mild to moderate hypertension, benign prostatic hypertrophy.
UNLABELED USES CHF.

CONTRAINDICATIONS Hypersensitivity to doxazosin, prazosin, and terazosin; hypotension, syncope.
CAUTIOUS USE Hepatic impairment or disease; renal disease, impairment, or failure; pregnancy (category B); lactation. Safe use in children not established.

ROUTE & DOSAGE

Hypertension
Adult: **PO** Start with 1 mg at bedtime and titrate up to maximum of 16 mg/day in 1–2 divided doses
Geriatric: **PO** Start with 0.5 mg at bedtime

ADMINISTRATION

Oral
- Give initial dose at bedtime to minimize problems with postural hypotension and syncope.
- Individualize maintenance dose according to the standing BP response.
- Store at 15°–30° C (59°–86° F).

ADVERSE EFFECTS (≥1%) **CV:** *Orthostatic hypotension,* edema. **CNS:** Vertigo, *headache,* dizziness, somnolence, fatigue, nervousness, anxiety. **GI:** Nausea, abdominal pain. **Hematologic:** Leukopenia. **Skin:** Pruritus, eczema.

INTERACTIONS Drug: sildenafil, vardenafil, and **tadalafil** may enhance hypotensive effects.

PHARMACOKINETICS Absorption: Readily absorbed from GI tract; 62–69% of dose reaches systemic circulation. **Peak:** 2–6 h. **Duration:** Up to 24 h. **Distribution:** Highly protein bound (98–99%). **Metabolism:** Approximately 35% of dose is metabolized in liver. **Elimination:** 9% in urine, 63% in feces. **Half-Life:** 9–12 h.

Common adverse effects in *italic*, life-threatening effects underlined; generic names in **bold**; classifications in SMALL CAPS; ♣ Canadian drug name; ⊘ Prototype drug

D

NURSING IMPLICATIONS

Assessment & Drug Effects

- Monitor BP with patient lying down and standing; doses above 4 mg increase the risk of postural hypotension.
- Monitor BP 2–6 h after initial dose or any dose increase. This is when postural hypotension is most likely to occur.

Patient & Family Education

- Do not drive or engage in other potentially hazardous activities for 12–24 h after the first dose or an increase in dosage or when medication is restarted after an interruption in dosage.
- Use caution when rising from a sitting or supine position in order to avoid orthostatic hypotension and syncope; make position and directional changes slowly and in stages.
- Report to the prescriber episodes of dizziness or palpitations. These will require a dosage adjustment.

DOXEPIN HYDROCHLORIDE

(dox′e-pin)

Prudoxin, Silenor, Triadapin ✦, Zonalon
Classifications: TRICYCLIC ANTIDEPRESSANT; ANXIOLYTIC
Therapeutic: ANTIDEPRESSANT; ANTIANXIETY
Prototype: Imipramine
Pregnancy Category: C

AVAILABILITY 10 mg, 25 mg, 50 mg, 75 mg, 100 mg, 150 mg capsules; 10 mg/mL oral concentrate; 3 mg, 6 mg tablet

ACTION & THERAPEUTIC EFFECT
Dibenzoxepin is a tricyclic antidepressant (TCA) that inhibits serotonin reuptake from the synaptic gap; also inhibits norepinephrine reuptake to a moderate degree. *Effective for treatment of both depression and anxiety.*

USES Anxiety; depression; insomnia; atopic dermatitis; eczema; lichen simplex.
UNLABELED USES Migraine prophylaxis, neuralgia.

CONTRAINDICATIONS Hypersensitivity to doxepin; during acute recovery phase following MI; bundle branch block, cardiac arrhythmias, QT prolongation; ileus; glaucoma; increased intraocular pressure; prostatic hypertrophy; tendency for urinary retention; suicidal ideation within 14 days of using MAOIs; lactation.
CAUTIOUS USE Patients receiving electroconvulsive therapy, history of suicidal tendency, bipolar disorder, schizophrenia, psychosis; diabetes mellitus; GI disease; GERD; Parkinson's disease; seizure disorders; renal, cardiovascular, or hepatic dysfunction; older adults; pregnancy (category C). Safe use in children younger than 12 y not established.

ROUTE & DOSAGE

Depression/Anxiety

Adult: **PO** 25–150 mg/day in divided doses, may increase up to 300 mg/day (use lower doses in older adult patients)
Geriatric: **PO** 10–50 mg/day may increase to 150 mg/day

Insomnia (Silenor)

Adult: **PO** 3–6 mg at bedtime
Geriatric: **PO** 3 mg at bedtime

Dermatitis

Adult: **Topical** Apply a thin film q.i.d. with at least 3–4 h between

Common adverse effects in *italic*, life-threatening effects underlined; generic names in **bold**; classifications in SMALL CAPS; ✦ Canadian drug name; ⊘ Prototype drug

applications, may use up to 8 days

Pharmacogenetic Dosage Adjustment

CYP2D6 Poor metabolizers: Give 40% of normal starting oral dose

ADMINISTRATION

Oral

- Give oral concentrate diluted with approximately 120 mL water, milk, or fruit juice.
- Empty capsule and swallow contents with fluid or mix with food as necessary if it cannot be swallowed whole.
- Inform prescriber if daytime sedation is pronounced. Entire daily dose (up to 150 mg) may be prescribed for bedtime administration.

Topical

- Apply a thin film to affected areas; allow 3–4 h between applications.
- Store all forms at 15°–30° C (59°–86° F) in tightly closed, light-resistant container.

ADVERSE EFFECTS (≥1%) **All:** Anticholinergic. **CNS:** *Drowsiness,* dizziness, weakness, fatigue, headache, hypomania, confusion, tremors, paresthesias. **CV:** *Orthostatic hypotension,* palpitation, hypertension, tachycardia, ECG changes. **Special Senses:** Mydriasis, blurred vision, photophobia. **GI:** *Dry mouth,* sour or metallic taste, epigastric distress, constipation. **Urogenital:** Urinary retention, delayed micturition, urinary frequency. **Other:** Increased perspiration, tinnitus, weight gain, photosensitivity reaction, skin rash, agranulocytosis, *burning or stinging at application site,* edema.

INTERACTIONS **Drug:** May decrease some antihypertensive response to ANTIHYPERTENSIVES; CNS DEPRESSANTS, **alcohol,** HYPNOTICS, BARBITURATES, SEDATIVES potentiate CNS depression; may increase hypoprothrombinemic effect of ORAL ANTICOAGULANTS; **levodopa,** SYMPATHOMIMETICS (e.g., **epinephrine, norepinephrine**) introduce possibility of sympathetic hyperactivity with hypertension and hyperpyrexia; MAO INHIBITORS introduce possibility of severe reactions, toxic psychosis, cardiovascular instability; **methylphenidate** or **cimetidine** increases plasma levels; THYROID AGENTS may increase possibility of arrhythmias; be cautious with other drugs that prolong the QT interval (e.g., ANTIARRYTHMICS). **Herbal:** Ginkgo may decrease seizure threshold; **St. John's wort** may cause **serotonin** syndrome.

PHARMACOKINETICS **Absorption:** Rapidly from GI sites and through intact skin. **Peak:** 2 h. **Distribution:** Crosses placenta; distributed into breast milk. **Metabolism:** In liver. **Elimination:** Primarily in urine. **Half-Life:** 6–8 h.

NURSING IMPLICATIONS

Assessment & Drug Effects

- Monitor use of other CNS depressants, including alcohol. Danger of overdosage or suicide attempt is increased when patient uses excessive amounts of alcohol.
- Be alert to changes in voiding and evaluate patient for constipation and abdominal distention; drug has moderate to strong anticholinergic effects.

Patient & Family Education

- Maintain established dosage regimen and avoid change of intervals, doubling, reducing, or skipping doses.
- Consult prescriber about safe amount of alcohol, if any, that can

---

Common adverse effects in *italic*, life-threatening effects underlined; generic names in **bold**; classifications in SMALL CAPS; ✚ Canadian drug name; ⊙ Prototype drug

497

be taken. The actions of both alcohol and doxepin are potentiated when used together and for up to 2 wk after doxepin is discontinued.

- Do not drive or engage in other potentially hazardous activities until response to drug is known.

DOXERCALCIFEROL

(dox-er-kal′si-fe-rol)

Hectorol

Classification: VITAMIN D ANALOG

Therapeutic: ANTIHYPERPARATHYROID; VITAMIN D ANALOG

Prototype: Calcitriol

Pregnancy Category: B

AVAILABILITY 0.5 mcg, 1 mcg, 2.5 mcg capsule; 2 mcg/mL injection

ACTION & *THERAPEUTIC EFFECT* Vitamin D_2 analog that is activated by the liver. Activated vitamin D is needed for absorption of dietary calcium in the intestine, and the parathyroid hormone (PTH), which mobilizes calcium from the bone tissue. *Regulates the blood calcium level.*

USES Hyperparathyroidism.

CONTRAINDICATIONS Hypersensitivity to doxercalciferol or other vitamin D analogs; recent hypercalcemia, recent hyperphosphatemia, hypervitaminosis D.

CAUTIOUS USE Renal or hepatic insufficiency; renal osteodystrophy with hyperphosphatemia, prolonged hypercalcemia; pregnancy (category B), lactation. Safety and efficacy in children are not established.

ROUTE & DOSAGE

Secondary Hyperparathyroidism

Adult: **PO** 10 mcg 3 times/wk at dialysis, adjust dose as needed to lower iPTH into the range of 150–300 pg/mL by increasing the dose in 2.5 mcg increments every 8 wk (max: 60 mcg/wk) **IV** 4 mcg 3 times/wk at end of dialysis (max: 18 mcg/wk)

ADMINISTRATION

Oral

- Give at time of dialysis.
- Withhold drug and notify prescriber if any of the following occurs: iPTH less than 100 pg/mL, hypercalcemia, hyperphosphatemia, or product of serum calcium times serum phosphorus greater than 70.
- Store at 20°–25° C (66°–77° F); excursions to 15°–30° C (59°–86° F) are permitted.

Intravenous

***PREPARE:* Direct:** No dilution is needed. ▪ Withdraw appropriate dose from ampule using a filter needle. ▪ Change needles before injecting as an IV bolus. Discard any unused portion.

***ADMINISTER:* Direct:** Give a bolus injection at the end of dialysis sessions.

- Store at 15°–20° C (59°–77° F).

ADVERSE EFFECTS (≥1%) **Body as a Whole:** Abscess, *headache, malaise,* arthralgia. **CNS:** *Dizziness,* sleep disorder. **CV:** Bradycardia, *edema.* **GI:** Anorexia, constipation, dyspepsia, *nausea, vomiting.* **Respiratory:** *Dyspnea.* **Skin:** Pruritus. **Other:** Weight gain.

INTERACTIONS Drug: Cholestyramine, mineral oil may decrease absorption; MAGNESIUM-CONTAINING ANTACIDS may cause hypermagnesemia; other VITAMIN D ANALOGS may increase toxicity and hypercalcemia.

PHARMACOKINETICS Absorption: Absorbed from GI tract and is activated in the liver. **Peak:** 11–12 h. **Metabolism:** Activated by CYP 27 to form 1alpha, 25-$(OH)_2D_2$ (major metabolite) and 1alpha, 24-dihydroxy vitamin D_2 (minor metabolite). **Half-Life:** 32–37 h.

NURSING IMPLICATIONS

Assessment & Drug Effects
- Lab tests: Baseline and periodic iPTH, serum calcium, serum phosphorus. Monitor levels weekly during dose titration.
- Monitor for S&S of hypercalcemia (see Appendix F).

Patient & Family Education
- Do not take antacids without consulting the prescriber.
- Notify the prescriber if you become pregnant while taking this drug.
- Do not use mineral oil on the days doxercalciferol is taken. Mineral oil may decrease absorption of drug.
- Do not take nonprescription drugs containing magnesium while taking doxercalciferol.
- Report S&S of hypercalcemia immediately: Bone or muscle pain, dry mouth with metallic taste, rhinorrhea, itching, photophobia, conjunctivitis, frequent urination, anorexia and weight loss.

DOXORUBICIN HYDROCHLORIDE ℗

(dox-oh-roo′bi-sin)

Adriamycin, Rubex

DOXORUBICIN LIPOSOME

Doxil

Classifications: ANTINEOPLASTIC; ANTHRACYCLINE; (ANTIBIOTIC)
Therapeutic: ANTINEOPLASTIC
Pregnancy Category: D

AVAILABILITY 10 mg, 20 mg, 50 mg, 100 mg, 150 mg powder for injection; 2 mg/mL injection; 20 mg liposomal injection

ACTION & *THERAPEUTIC EFFECT* Cytotoxic agent with wide spectrum of antitumor activity. Intercalates with preformed DNA residues, blocking effective DNA and RNA transcription. A potent radiosensitizer capable of enhancing radiation reactions. *Highly destructive to rapidly proliferating cells and slowly developing carcinomas; selectively toxic to cardiac tissue.*

USES Doxorubicin: To produce regression in neoplastic conditions, including acute lymphoblastic and myeloblastic leukemias, transitional cell bladder cancer, breast cancer, Hodgkin's disease, ovarian cancer, small-cell lung cancer, non-Hodgkin's lymphoma, thyroid cancer, Wilms' tumor, soft tissue and bone sarcomas. **Doxorubicin Liposome:** Kaposi's sarcoma, progressive/refractory ovarian cancer, relapsed/refractory multiple myeloma.

CONTRAINDICATIONS History of hypersensitive reactions to conventional or liposomal doxorubicin; myelosuppression, thrombocytopenia; impaired cardiac function, obstructive jaundice, previous treatment with complete cumulative doses of doxorubicin or daunorubicin; pregnancy (category D), lactation.

CAUTIOUS USE Impaired hepatic or renal function; patients who have received cyclophosphamide or pelvic irradiation or radiotherapy to areas surrounding heart; preexisting heart disease; history of atopic dermatitis; children.

ROUTE & DOSAGE

CONVENTIONAL DOXORUBICIN

Acute Lymphatic Leukemia

Adult/Child: **IV** 30 mg/m² weekly × 4 wk

Acute Myelogenous Leukemia

Adult/Child: **IV** 30 mg/m² × 3 days (with cytarabine)

Transitional Bladder Cell Cancer

Adult: **IV** 30 mg/m²/dose once monthly

Hodgkin's Disease

Adult/Child: **IV** 25 mg/m² days 1 and 15, repeat q28days

Thyroid Cancer

Adult/Child: **IV** 60–75 mg/m² q3wk

Other Neoplasms

Adult: **IV** 40–50 mg/m² usually in combination with other agents (max total cumulative lifetime dose: 500–550 mg/m²)

Child: **IV** 35–75 mg/m² as single dose, repeat at 21-day interval, or 20–30 mg/m² once weekly (max total cumulative lifetime dose: 500–550 mg/m²)

Hepatic Impairment Dosage Adjustment

Bilirubin 1.2–3 mg/dL: Reduce dose by 50%; bilirubin 3–5 mg/dL: Reduce dose by 75%
Bilirubin greater than 5 mg/dL: Stop therapy

DOXORUBICIN LIPOSOME

Kaposi's Sarcoma

Adult: **IV** 20 mg/m² q3wk. Infuse over 30 min (do not use in-line filters).

Progressive/Refractory Ovarian Cancer

Adult: **IV** 50 mg/m² q4wk, minimum of 4 courses

Relapsed/Refractory Multiple Myeloma

Adult: **IV** 45 mg/m² q4wk, up to 6 cycles

Hepatic Impairment Dosage Adjustment

Bilirubin 1.2–3 mg/dL: Reduce dose 50%; bilirubin 3–5 mg/dL: Reduce dose by 75%

ADMINISTRATION

Intravenous

▪ IV administration to children: Verify correct IV concentration and rate of infusion with prescriber. ▪ Wear gloves and use caution when preparing drug solution. If powder or solution contacts skin or mucosa, wash copiously with soap and water.

Conventional Doxorubicin

PREPARE: **Direct:** *Vial reconstitution:* Dilute with 1 mL of nonbacteriostatic NS for each 2 mg of doxorubicin to yield a final concentraion of 2 mg/mL. ▪ For each mL of NS added, withdraw an equal volume of air from vial to minimize pressure buildup. Shake to dissolve. ▪ *Doxorubicin solutions:* Solutions of 2 mg/mL are available that can be further diluted in 50 mL or more of NS or D5W. *ADMINISTER:* **Direct:** Give bolus dose slowly into Y-site of freely running IV infusion of NS or D5W. ▪ If possible, use IV tubing attached to a needle inserted into a larger vein with a butterfly needle. ▪ Usually infused over 3–5 min.

■ Monitor for red streaking along vein or facial flushing which indicates need to slow infusion rate.

Lyophilized Doxorubicin

PREPARE: **IV Infusion:** Dilute doses up to 90 mg in 250 mL of D5W and doses greater than 90 mg in 500 mL D5W. Solution will be translucent but not clear, and will be red in color. ■ DO NOT use filters during preparation or administration.

ADMINISTER: **IV Infusion:** DO NOT give bolus injection or undiluted solution. ■ Infuse at 1 mg/min initially; may increase rate to complete infusion in 1 h if no adverse reactions occur. Slow infusion rate as warranted if an adverse reaction occurs. ■ Do not use a filter.

INCOMPATIBILITIES **Solution/additive:** *Conventional doxorubicin:* **Aminophylline, diazepam, fluorouracil. Y-site:** *Conventional doxorubicin:* **Allopurinol, amphotericin B cholesteryl sulfate, cefepime, gallium ganciclovir, lansoprazole, pemetrexed, prochlorperazine, propofol, TPN.** *Doxorubicin liposome:* **Amphotericin B, amphotericin B cholesteryl complex, hydroxyzine, mannitol, meperidine, metoclopramide, mitoxantrone, morphine, paclitaxel, piperacillin/tazobactam, promethazine, sodium bicarbonate.**

■ Facial flushing and local red streaking along the vein may occur if drug is administered too rapidly. ■ Avoid using antecubital vein or veins on dorsum of hand or wrist, if possible, where extravasation could damage underlying tendons and nerves. ■ Also avoid veins in extremity with compromised venous or lymphatic drainage.

■ Store reconstituted solution for 24 h at room temperature; refrigerated at 4°–10° C (39°–50° F) for 48 h. Protect from sunlight; discard unused solution.

ADVERSE EFFECTS (≥1%) **Body as a Whole:** Hypersensitivity (red flare around injection site, erythema, skin rash, pruritus, angioedema, urticaria, eosinophilia, fever, chills, <u>anaphylactoid reaction</u>). **CV:** <u>Serious, irreversible myocardial toxicity with delayed CHF, ventricular arrhythmias, acute left ventricular failure,</u> hypertension, hypotension, cardiomyopathy. **GI:** *Stomatitis,* esophagitis with ulcerations; nausea, vomiting, anorexia, inanition, diarrhea. **Hematologic:** *Severe myelosuppression* (60–85% of patients); <u>leukopenia (principally granulocytes),</u> <u>thrombocytopenia,</u> anemia. **Skin:** Hyperpigmentation of nail beds, tongue, and buccal mucosa (especially in blacks); *complete alopecia* (reversible), hyperpigmentation of dermal creases (especially in children), rash, *recall phenomenon (skin reaction due to prior radiotherapy).* **Other:** Lacrimation, drowsiness, fever, facial flush with too rapid IV infusion rate, microscopic hematuria, hyperuricemia, *hand-foot syndrome. With extravasation: severe cellulitis, vesication, tissue necrosis, lymphangitis, phlebosclerosis.*

INTERACTIONS Drug: BARBITURATES may decrease effects by increasing its hepatic metabolism; **streptozocin** may prolong doxorubicin half-life; agents affecting QT interval (e.g., **Bepridil, droperidol, erythromycin, haloperidol, methadone,** PHENOTHIAZINES, etc.)

Common adverse effects in *italic*, life-threatening effects underlined; generic names in **bold**; classifications in SMALL CAPS; ✦ Canadian drug name; ⦿ Prototype drug

501

D

may increase risk of cardiac side effects. Conventional formulation: Avoid use with **zidovudine,** monitor **warfarin** carefully.

PHARMACOKINETICS Distribution: Widely distributed; does not cross blood–brain barrier; 75% protein binding; does not cross placenta; passes into breast milk. **Metabolism:** In liver to active metabolite. **Elimination:** Primarily in bile. **Half-Life:** 30–50 h. *Doxorubicin Liposome:* **Distribution:** Vascular fluid. **Metabolism:** In plasma and liver. **Elimination:** In urine. **Half-Life:** 44–55 h.

NURSING IMPLICATIONS

Assessment & Drug Effects

- Care should be taken to avoid extravasation. Stop infusion, remove IV needle, and notify prescriber promptly if patient complains of stinging or burning sensation at the injection site.
- Monitor any area of extravasation closely for 3–4 wk. If ulceration begins (usually 1–4 wk after extravasation), a plastic surgeon should be consulted.
- Establish baseline data. Include temperature, pulse, respiration, BP, body weight, laboratory values, and I&O ratio and pattern.
- Lab tests: Baseline and periodic LFTs, renal function, CBC with differential throughout therapy.
- Note: The nadir of leukopenia (an expected 1000/mm^3) typically occurs 10–14 days after single dose, with recovery occurring within 21 days.
- Cardiac function must be evaluated prior to initiation of therapy, at regular intervals, and at end of therapy.
- Be alert to and report early signs of cardiotoxicity (see Appendix F). Acute life-threatening arrhythmias may occur within a few hours of drug administration.

- Report promptly objective signs of hepatic dysfunction (jaundice, dark urine, pruritus) or kidney dysfunction (altered I&O ratio and pattern, local discomfort with voiding).
- Report signs of superinfection (see Appendix F) promptly; these may result from antibiotic therapy during leukopenic period.
- Avoid rectal medications and use of rectal thermometer; rectal trauma is associated with bloody diarrhea resulting from an antiblastic effect on rapidly growing intestinal mucosal cells.

Patient & Family Education

- Note: Complete loss of hair (reversible) is an expected adverse effect. It may also involve eyelashes and eyebrows, beard and mustache, pubic and axillary hair. Regrowth of hair usually begins 2–3 mo after drug is discontinued.
- Drug turns urine red for 1–2 days after administration.
- Keep hands away from eyes to prevent conjunctivitis. Increased tearing for 5–10 days after a single dose is possible.
- Maintain fastidious oral hygiene, especially before and after meals. Stomatitis, generally maximal in second week of therapy, frequently begins with a burning sensation accompanied by erythema of oral mucosa that may progress to ulceration and dysphagia in 2 or 3 days.
- Exposure to doxorubicin during the first trimester of pregnancy can result in fetal abnormalities or fetal loss.

DOXYCYCLINE HYCLATE
(dox-i-sye′kleen)

Apo-Doxy ♥, **Doryx, Doxy, Doxycin** ♥, **Monodox, Novo-doxylin** ♥, **Vibramycin, Vibra-Tabs**

Common adverse effects in *italic*, life-threatening effects underlined; generic names in **bold;** classifications in SMALL CAPS; ♥ Canadian drug name; ✪ Prototype drug

Classifications: ANTIBIOTIC;
TETRACYCLINE
Therapeutic: ANTIBIOTIC
Prototype: Tetracycline
Pregnancy Category: D

AVAILABILITY 50 mg, 75 mg, 100
mg capsules, tablets; 200 mg injection

ACTION & *THERAPEUTIC EFFECT*
Semisynthetic broad-spectrum
long-acting tetracycline antibiotic
that is more lipophilic than the other tetracyclines allowing it to pass
through the lipid layer of bacteria
where reversible binding to the 30
S ribosomal subunits of bacteria
occurs. This blocks the binding of
transfer RNA (tRNA) to the messenger RNA (mRNA) of bacteria,
resulting in inhibition of bacterial
protein synthesis. *Primarily bacteriostatic against both gram-positive
and gram-negative bacteria. Similar in use to tetracycline.*

USES Similar to those of tetracycline
(e.g., chlamydial and mycoplasmal
infections); gonorrhea, syphilis in
penicillin-allergic patients; rickettsial diseases; acute exacerbations
of chronic bronchitis.
UNLABELED USES Treatment of
acute PID, leptospirosis, prophylaxis for rape victims, suppression
and chemoprophylaxis of chloroquine-resistant *Plasmodium falciparum* malaria, short-term prophylaxis and treatment of travelers'
diarrhea caused by enterotoxigenic
strains of *Escherichia coli.* Intrapleural administration for malignant
pleural effusions, post-exposure
anthrax treatment and prophylaxis.

CONTRAINDICATIONS Sensitivity to
any of the tetracyclines; use during
period of tooth development including last half of pregnancy; pregnancy

(category D), lactation, infants, and
children younger than 8 y except
for use in anthrax exposure (causes
permanent yellow discoloration of
teeth, enamel hypoplasia, and retardation of bone growth).
CAUTIOUS USE Alcoholism; hepatic
disease; GI disease; sulfite hypersensitivity; sunlight (UV) exposure.

ROUTE & DOSAGE

Anti-Infective
Adult: **PO/IV** 100 mg q12h on
day 1, then 100 mg/day as single
dose (max: 100 mg q12h)
Child (older than 8 y): **PO/IV** 4.4
mg/kg in 1–2 doses on day 1,
then 2.2–4.4 mg/kg/day in 1–2
divided doses

Gonorrhea
Adult: **PO** 200 mg immediately,
followed by 100 mg at bedtime,
then 100 mg b.i.d. for 3 days

Primary and Secondary Syphilis
Adult: **PO** 300 mg/day in divided
doses for at least 10 days

Travelers' Diarrhea
Adult: **PO** 100 mg/day during
risk period (up to 2 wk) beginning
day 1 of travel

Acute Pelvic Inflammatory Disease
Adult: **IV** 100 mg q12h until
improved, then 100 mg **PO** b.i.d.
to complete 14 days

Acne
Adult: **PO** 100 mg q12h on day 1,
then 100 mg daily
Child: **PO** Older than 8 y, weight
greater than 45 kg: 100 mg q12h
on day 1, then 100 mg daily
Older than 8 y, weight less than
45 kg: 2.2 mg/kg q12h on day
1, then 2.2 mg/kg/daily

Common adverse effects in *italic*, life-threatening effects underlined; generic names
in **bold**; classifications in SMALL CAPS; ✚ Canadian drug name; ⊘ Prototype drug

503

D

Anthrax Post-Exposure

Adult/Adolescent/Child (older than 8 y, weight greater than 45 kg): IV 100 mg q12h, then switch to PO for a total of 60

Child (weight 45 kg or less or 8 y or less): IV 2.2 mg/kg q12h, then switch to PO for a total of 60

ADMINISTRATION

Oral

- Check expiration date. Degradation products of tetracycline are toxic to the kidneys.
- Give with food or a full glass of milk to minimize nausea without significantly affecting bioavailability of drug (UNLIKE MOST TETRACYCLINES).
- Consult prescriber about ordering the oral suspension for patients who are bedridden or have difficulty swallowing.

Intravenous

PREPARE: **Intermittent:** Reconstitute by adding 10 mL sterile water for injection, or D5W, NS, LR, D5/LR, or other diluent recommended by manufacturer, to each 100 mg of drug. • Further dilute with 100–1000 mL (per 100 mg of drug) of compatible infusion solution to produce concentrations ranging from 0.1 to 1 mg/mL.

ADMINISTER: **Intermittent:** IV infusion rate will usually be prescribed by prescriber. Duration of infusion varies with dose but is usually 1–4 h. • Recommended minimum infusion time for 100 mg of 0.5 mg/mL solution is 1 h. Infusion should be completed within 12 h of dilution. • When diluted with LR or D5/LR, infusion **must be** completed within 6 h to ensure adequate stability.

- Protect all solutions from direct sunlight during infusion.

INCOMPATIBILITIES **Solution/additive: Potassium phosphate. Y-site: Allopurinol, heparin, meropenem, piperacillin/tazobactam, TPN.**

- Store oral and parenteral forms (prior to reconstitution) in tightly covered, light-resistant containers at 15°–30° C (59°–86° F) unless otherwise directed. • Refrigerate reconstituted solutions for up to 72 h. After this time, infusion **must be** completed within 12 h.

ADVERSE EFFECTS (≥1%) **Special Senses:** Interference with color vision. **GI:** Anorexia, *nausea,* vomiting, diarrhea, enterocolitis; esophageal irritation (oral capsule and tablet). **Skin:** Rashes, photosensitivity reaction. **Other:** Thrombophlebitis (IV use), superinfections.

DIAGNOSTIC TEST INTERFERENCE

Like other *tetracyclines,* doxycycline may cause false increases in *urinary catecholamines* (fluorometric methods); false decreases in *urinary urobilinogen;* false-negative *urine glucose* with *glucose oxidase methods* (e.g., *Clinistix, TesTape*); parenteral doxycycline (containing ascorbic acid) may cause false-positive determinations using *Benedict's reagent* or *Clinitest.*

INTERACTIONS **Drug:** ANTACIDS, **iron** preparation, **calcium, magnesium, zinc, kaolin-pectin, sodium bicarbonate** can significantly decrease absorption; effects of both doxycycline and **desmopressin** antagonized; increases **digoxin** absorption, thus increasing risk of **digoxin** toxicity; **methoxyflurane** increases risk of renal failure.

PHARMACOKINETICS **Absorption:** Completely absorbed from GI tract.

Common adverse effects in *italic*, life-threatening effects underlined; generic names in **bold**; classifications in SMALL CAPS; ✦ Canadian drug name; ⊙ Prototype drug

Peak: 1.5–4 h. **Distribution:** Penetrates eye, prostate, and CSF; crosses placenta; distributed into breast milk. **Metabolism:** In GI tract. **Elimination:** 20–30% in urine and 20–40% in feces in 48 h. **Half-Life:** 14–24 h.

NURSING IMPLICATIONS

Assessment & Drug Effects

- Report sudden onset of painful or difficult swallowing promptly to prescriber. Doxycycline (capsule and tablet forms) is associated with a comparatively high incidence of esophagitis, especially in patients older than 40 y.
- Report evidence of superinfections (see Appendix F).

Patient & Family Education

- Take capsule or tablet forms with a full glass (240 mL) of water to ensure passage into stomach and prevent esophageal ulceration. Avoid taking capsule or tablet within 1 h of lying down or retiring.
- Avoid exposure to direct sunlight and ultraviolet light during and for 4 or 5 days after therapy is terminated to reduce risk of phototoxic reaction. Phototoxic reaction appears like an exaggerated sunburn. Sunscreens provide little protection.

DRONABINOL
(droe-nab'i-nol)

Marinol, THC
Classifications: CANNABINOID; ANTIEMETIC
Therapeutic: ANTIEMETIC; APPETITE STIMULANT
Pregnancy Category: C
Controlled Substance: Schedule III

AVAILABILITY 2.5 mg, 5 mg, 10 mg capsules

ACTION & _THERAPEUTIC EFFECT_
Synthetic derivative of tetrahydro-cannabinol (THC), the principal psychoactive constituent of marijuana *(Cannabis sativa)*. Inhibits vomiting through the control mechanism in the medulla oblongata, producing potent antiemetic effect. Risk of drug abuse is high. *Produces potent antiemetic effect and is used to treat chemotherapy-induced nausea and vomiting.*

USES To treat chemotherapy-induced nausea and vomiting in cancer patients. Appetite stimulant for AIDS patients.
UNLABELED USES Glaucoma.

CONTRAINDICATIONS Nausea and vomiting caused by other than chemotherapeutic agents; hypersensitivity to dronabinol or sesame oil; lactation.
CAUTIOUS USE First exposure, especially in the older adult or cardiac patient; hypertension, cardiovascular disorders; epilepsy; psychiatric illness, patient receiving other psychoactive drugs; severe hepatic dysfunction; pregnancy (category C).

ROUTE & DOSAGE

Chemotherapy-Induced Nausea
Adult/Child: **PO** 5 mg/m² 1–3 h before administration of chemotherapy, then q2–4h after chemotherapy for a total of 4–6 doses, dose may be increased by 2.5 mg/m² (max: 15 mg/m² if necessary)
Appetite Stimulant
Adult: **PO** 2.5 mg b.i.d., before lunch and dinner

ADMINISTRATION

Oral

- Do not repeat dose following a CNS adverse reaction until patient's mental state has returned

---

Common adverse effects in *italic*, life-threatening effects <u>underlined</u>; generic names in **bold;** classifications in SMALL CAPS; ♣ Canadian drug name; ⊘ Prototype drug

505

to normal and the circumstances have been evaluated.
- Store at 8°–15° C (46°–59° F).

ADVERSE EFFECTS (≥1%) **CNS:** *Drowsiness,* psychologic high, dizziness, anxiety, confusion, euphoria, sensory or perceptual difficulties, impaired coordination, depression, irritability, headache, ataxia, memory lapse, paresthesias, paranoia, depersonalization, disorientation, tinnitus, nightmares, speech difficulty, facial flush, diaphoresis. **CV:** Tachycardia, orthostatic hypotension, hypertension, syncope. **GI:** Dry mouth, diarrhea, fecal incontinence. **Other:** Muscular pains.

INTERACTIONS Drug: Alcohol and other CNS DEPRESSANTS may exaggerate psychoactive effects of dronabinol; TRICYCLIC ANTIDEPRESSANTS, **atropine** may cause tachycardia.

PHARMACOKINETICS Absorption: Rapidly absorbed from GI tract, with bioavailability of 10–20%. **Peak:** 2–3 h. **Distribution:** Fat soluble; distributed to many organs; distributed into breast milk. **Metabolism:** In liver; extensive first-pass metabolism. **Elimination:** Principally in bile; 50% in feces within 72 h; 10–15% in urine. **Half-Life:** 25–36 h.

NURSING IMPLICATIONS

Assessment & Drug Effects
- Monitor patients with hypertension or heart disease for BP and cardiac status.
- Response to dronabinol is varied, and previous uneventful use does not guarantee that adverse reactions will not occur. Effects of drug may persist an unpredictably long time (days). Extended use at therapeutic dosage may cause accumulation of toxic amounts of dronabinol and its metabolites.

- Watch for disturbing psychiatric symptoms if dose is increased: Altered mental state, loss of coordination, evidence of a psychologic high (easy laughing, elation and heightened awareness), or depression.
- Note: Abrupt withdrawal is associated with symptoms (within 12 h) of irritability, insomnia, restlessness. Peak intensity occurs at about 24 h: Hot flashes, diaphoresis, rhinorrhea, watery diarrhea, hiccups, anorexia. Usually, syndrome is over in 96 h.

Patient & Family Education
- Do not drive or engage in other potentially hazardous activities that require alertness and judgment because of high incidence of dizziness and drowsiness.
- Understand potential (reversible) for drug-induced mood or behavior changes that may occur during dronabinol use.
- Do not ingest alcohol during period of systemic dronabinol effect. Effect on blood ethanol levels is complex and unpredictable.

DRONEDARONE
(dro-ne′da-rone)

Multaq
Classification: CLASS III ANTIARRHYTHMIC
Therapeutic: CLASS III ANTIARRHYTHMIC
Prototype: Amiodarone
Pregnancy Category: X

AVAILABILITY 400 mg tablets

ACTION & *THERAPEUTIC EFFECT* Has antiarrhythmic properties of all four classes of antiarrhythmic drugs. Known to inhibit potassium currents, sodium channels, and slow-L type calcium channels.

Also has antiadrenergic properties. *Reduces risk of hospitalization in patients with recent paroxysmal or persistent atrial fibrillation (AF) or atrial flutter (AFL).*

USES Recent episode of paroxysmal or persistent atrial fibrillation or atrial flutter in patients with associated CV risk factors and who are in sinus rhythm or who will be cardioverted.

CONTRAINDICATIONS NYHA Class IV HF or NYHA Class II-III HF with a recent decompensation requiring hospitalization or referral to a specialized HF clinic; second or third degree AV block or sick sinus syndrome (except with used in conjunction with a functioning pacemaker); bradycardia less than 40 bpm; QT_c interval elongation; severe hepatic impairment; pregnancy (category X), lactation.

CAUTIOUS USE HF; prolonged QT interval; hypokalemia, hypomagnesium; potassium-depleting diuretics; moderate liver impairment; women of child-bearing age. Safety and efficacy of children younger than 18 y have not been established.

ROUTE & DOSAGE

Atrial Fibrillation or Atrial Flutter
Adult: **PO** 400 mg b.i.d. with meals

ADMINISTRATION

Oral
- Give with morning and evening meal. Do NOT give with grapefruit juice.
- Store at 15°–3° C (56°–89° F).

ADVERSE EFFECTS (≥1%) **Body as a Whole:** Asthenia. **CV:** Bradycardia, *QT_c prolongation.* **GI:** Abdominal pain, diarrhea, dyspepsia, nausea, vomitting. **Metabolic:** *Increased serum creatinine,* hepatic injury. **Skin:** Dermatitis, eczema, erythematous, macula-papular rash, pruritus.

INTERACTIONS Drug: Concomitant use of CYP3A4 inducers (e.g., **rifampin, phenobarbital, carbamazepine, phenytoin**) can increase the levels of dronedarone. **Ketoconazole, itraconazole, clarithromycin**, and other inhibitors of CYP3A4 can increase the levels of dronedarone. Dronedarone can increase the levels of **digoxin** and other compounds requiring P-glycoprotein (P-gp) transport. BETA-BLOCKERS may provoke excessive bradycardia. **Verapamil** and **diltiazem** can potentiate dronedarone's effects on conduction. Use cautiously with **dabigatran**. **Food: Grapefruit juice** can increase the levels of dronedarone. **Herbal: St. John's wort** can decrease the levels of dronedarone.

PHARMACOKINETICS Peak: 3–6 h. **Distribution:** 98% Plasma protein bound. **Metabolism:** Extensive hepatic metabolism to active and inactive compounds. **Elimination:** 84% in the feces; 6% in the urine. **Half-Life:** 13–19 h.

NURSING IMPLICATIONS

Assessment & Drug Effects
- Monitor vital signs and ECG. Report promptly prolongation of the QT_c interval.
- Monitor for S&S of hepatic toxicity (see Appendix F).
- Report promptly S&S of worsening HF (e.g., rapid weight gain, dependent edema, increasing shortness of breath).
- Lab tests: Baseline and periodic potassium and magnesium levels; periodic serum creatinine; periodic digoxin levels with concurrent therapy.

Common adverse effects in *italic*, life-threatening effects underlined; generic names in **bold**; classifications in SMALL CAPS; ✦ Canadian drug name; ❼ Prototype drug

507

- Withhold drug and notify prescriber if hypokalemia or hypomagnesemia develops.

Patient & Family Education
- Report immediately any of the following: Shortness of breath, wheezing, chest tightness, coughing up frothy sputum, rapid weight gain, requiring more pillows to sleep at night.
- Women of childbearing age should use effective contraception while on this drug.
- Avoid grapefruit and grapefruit juice while taking this drug.

DROPERIDOL

(droe-per′i-dole)

Classifications: BUTYROPHENONE; ANTIEMETIC; ANXIOLYTIC
Therapeutic: ANTIEMETIC; ANTIANXIETY
Prototype: Haloperidol
Pregnancy Category: C

AVAILABILITY 2.5 mg/mL injection

ACTION & *THERAPEUTIC EFFECT*
Antagonizes emetic effects of morphine-like analgesics and other drugs that act on chemoreceptor trigger zone. *Sedative property reduces anxiety and motor activity without necessarily inducing sleep; patient remains responsive. Has antiemetic properties.*

USES To reduce nausea/vomiting association with surgery/diagnostic procedures.
UNLABELED USES Chemotherapy induced nausea-vomiting.

CONTRAINDICATIONS Known or suspected QT elongation; history of torsades de pointes; known intolerance to droperidol; hypokalemia, hypomagnesia, lactation. Safe use in children younger than 2 y is not established.
CAUTIOUS USE Older adult, debilitated, alcoholism, and other poor-risk patients; MAOI therapy; Parkinson's disease; cardiac disease; cardiac bradyarrhythmias, cardiac arrhythmias, CHF, hypotension; liver and kidney impairment or disease; pregnancy (category C).

ROUTE & DOSAGE

Postoperative Nausea and Vomiting Prevention Using Continual ECG Monitoring

Adult: **IV/IM** 2.5 mg; additional doses of 1.25 mg may be given
Child: **IV/IM** 0.1 mg/kg (max: 2.5 mg)

Renal Impairment Dosage Adjustment

Due to increased risk of QT prolongation and torsades de points continuous monitoring is required

ADMINISTRATION

Intramuscular
- Give undiluted.
- Give deep IM into a large muscle.

Intravenous
IV administration to infants and children: Verify correct rate of IV injection with prescriber.

***PREPARE:* Direct:** Give undiluted.
***ADMINISTER:* Direct:** *Adult:* Give at a rate of 2.5 mg or fraction thereof over 1–2 min. *Child:* Give a single dose over at least 2 min.
***INCOMPATIBILITIES* Solution/additive:** BARBITURATES. **Y-site: Allopurinol, amphotericin B cholesteryl complex, cefepime, cefotetan, fluorouracil, foscarnet, furosemide, heparin, leucovorin, methotrexate, nafcillin, piperacillin/tazobactam.**

- Store at 15°–30° C (59°–86° F), unless otherwise directed by manufacturer. Protect from light.

ADVERSE EFFECTS (≥1%) **CNS:** *Postoperative drowsiness, extrapyramidal symptoms:* dystonia, akathisia, oculogyric crisis; dizziness, restlessness, anxiety, hallucinations, mental depression. **CV:** *Hypotension, tachycardia,* irregular heartbeats *(prolonged QT_c interval even at low doses)*. **Other:** Chills, shivering, laryngospasm, bronchospasm.

INTERACTIONS Drugs: Additive effect with CNS depressants, **metoclopramide** may increase extrapyramidal symptoms, closely monitor other drugs affecting QT interval.

PHARMACOKINETICS Onset: 3–10 min. **Peak:** 30 min. **Duration:** 2–4 h; may persist up to 12 h. **Distribution:** Crosses placenta. **Metabolism:** In liver. **Elimination:** In urine and feces.

NURSING IMPLICATIONS

Assessment & Drug Effects

- Monitor ECG throughout therapy. Report immediately prolongation of QT_c interval.
- Monitor vital signs closely. Hypotension and tachycardia are common adverse effects.
- Exercise care in moving medicated patients because of possibility of severe orthostatic hypotension. Avoid abrupt changes in position.
- Observe patients for signs of impending respiratory depression carefully when receiving a concurrent narcotic analgesic.
- Note: EEG patterns are slow to return to normal during the postoperative period.
- Observe carefully and report promptly to prescriber early signs of acute dystonia: Facial grimacing, restlessness, tremors, torticol-

lis, oculogyric crisis. Extrapyramidal symptoms may occur within 24–48 h postoperatively.
- Note: Droperidol may aggravate symptoms of acute depression.

DULOXETINE HYDROCHLORIDE
(du-lox′e-teen)

Cymbalta
Classifications: ANTIDEPRESSANT; SEROTONIN NOREPINEPHRINE REUPTAKE INHIBITOR (SNRI)
Therapeutic: ANTIDEPRESSANT; SNRI; ANTIANXIETY; NEUROPATHIC PAIN RELIEVER
Prototype: Venlafaxine
Pregnancy Category: C in first and second trimester and D in third trimester

AVAILABILITY 20 mg, 30 mg, 60 mg delayed-release capsules

ACTION & *THERAPEUTIC EFFECT*
As a selective serotonin and norepinephrine reuptake inhibitor (SSNRI), duloxetine causes potentiation of serotonergic and noradrenergic activity in the CNS. Antidepressant and antianxiety effects are presumed to be due to its dual inhibition of CNS presynaptic neuronal uptake of serotonin and norepinephrine, thus increasing the serum levels of both substances. *Effective as an antidepressant, antianxiety, and neuropathic pain reliever.*

USES Treatment of major depression, generalized anxiety, fibromyalgia, diabetic peripheral neuropathy, musculoskeletal pain, osteoarthritis.
UNLABELED USES Stress urinary incontience.

CONTRAINDICATIONS Concurrent administration of MAOI therapy;

Common adverse effects in *italic*, life-threatening effects underlined; generic names in **bold**; classifications in SMALL CAPS; ♣ Canadian drug name; ⊘ Prototype drug

509

suicidal ideation; uncontrolled narrow-angle glaucoma; alcoholism; end-stage renal disease; hepatitis; jaundice; abrupt discontinuation; pregnancy (category D in third trimester), lactation.

CAUTIOUS USE Anorexia nervosa, bipolar disease; history of mania, history of suicidal tendencies; cardiac disease; renal impairment or renal failure; hypertension; pregnancy (category C in first and second trimester). Safe use in children younger than 18 y not established.

ROUTE & DOSAGE

Depression

Adult: PO 40–60 mg/day in one or two divided doses

Generalized Anxiety/Diabetic Neuropathy/Musculoskeletal Pain

Adult: PO 60 mg once daily

Fibromyalgia

Adult: PO 30 mg/day × 1 wk then 60 mg/day

ADMINISTRATION

Oral

- Do not initiate therapy within 14 days of the last dose of an MAOI.
- **Must be** swallowed whole. Do not cut, chew, or crush. Do not sprinkle on food or mix with liquids.
- Store at 15°–30° C (59°–86° F).

ADVERSE EFFECTS (≥1%) **Body as a Whole:** Fatigue, hot flashes. **CNS:** Dizziness, somnolence, tremor, *insomnia*. **GI:** *Nausea, dry mouth, constipation,* diarrhea, vomiting. **Metabolic:** Decreased appetite, weight loss. **Skin:** Increased sweating. **Special Senses:** Blurred vision. **Urogenital:** Decreased libido, abnormal orgasm, erectile dysfunction, ejaculatory dysfunction. Cholestatic jaundice and hepatitis.

INTERACTIONS Drug: Alcohol may result in increased liver function tests; MAOIS may result in hyperthermia, rigidity, mental status changes, myoclonus, autonomic instability, features resembling neuroleptic malignant syndrome; **cimetidine, fluoxetine, fluvoxamine, paroxetine, quinidine,** QUINOLONES may increase levels and half-life of dulox-etine; may increase levels and toxicity of **thioridazine,** TRICYCLIC ANTIDEPRESSANTS. **Amphetamine, dextroamphetamine, buspirone, cocaine, dexfenfluramine, fenfluramine, lithium, phentermine, sibutramine, nefazodone,** SSRIS, TRIPTANS, **tramadol, trazodone** may cause serotonin syndrome. **Herbal: St. John's wort, tryptophan** may cause serotonin syndrome.

PHARMACOKINETICS Peak: 6 h. **Metabolism:** In the liver by CYP2D6 and CYP1A2. **Elimination:** 70% in urine, 20% in feces. **Half-Life:** 12 h (8–17 h).

NURSING IMPLICATIONS

Assessment & Drug Effects

- Ensure that a complete list of all concurrent medications is obtained.
- Monitor for S&S of numerous drug-drug interactions (see Interaction section).
- Lab test: LFTs for unexplained abdominal pain or enlarged liver.
- Monitor closely for and report suicide ideation, especially when drug is initiated or dosage changed.
- Report emergence of any of the following: Anxiety, agitation, panic attacks, insomnia, irritability, hostility, psychomotor restlessness, hypomania, and mania.

■ Monitor BP, especially in those being treated for hypertension.

Patient & Family Education

■ The beneficial effects of this drug may not be felt for approximately 4 wk.

■ Report any of the following: Suicidal ideation (especially early in treatment or when dosage is changed), palpitations, anxiety, hyperactivity, agitation, panic attacks, insomnia, irritability, hostility, restlessness.

■ Do not abruptly discontinue taking this drug. Notify prescriber if side effects are bothersome.

■ Avoid or minimize use of alcohol while taking this drug.

■ Do not self-treat for coughs, colds, or allergies. Consult prescriber.

DUTASTERIDE

(du-tas'ter-ide)

Avodart

Classifications: ANTIANDROGEN; 5-ALPHA REDUCTASE INHIBITOR
Therapeutic: BENIGN PROSTATIC HYPERPLASIA (BPH) AGENT
Prototype: Finasteride
Pregnancy Category: X

AVAILABILITY 0.5 mg capsules

ACTION & _THERAPEUTIC EFFECT_
Specific inhibitor of the steroid 5-alpha-reductase, an enzyme necessary to convert testosterone into the potent androgen 5-alpha-dihydrotestosterone (DHT) in the prostate gland. _Decreases the production of testosterone in the prostate gland._

USES Treatment of benign prostatic hypertrophy (BPH).
UNLABELED USE Alopecia.

CONTRAINDICATIONS Hypersensitivity to dutasteride or finasteride; pregnancy (category X), lactation.

CAUTIOUS USE Hepatic impairment, obstructive uropathy. Safe use in children younger than 18 y is not established.

ROUTE & DOSAGE

BPH

Adult: **PO** 0.5 mg once daily

ADMINISTRATION

Oral

■ Do not handle capsules if you are or may become pregnant because of the potential for absorption of dutasteride and the subsequent risk to a developing male fetus.

■ Do not open or crush capsules. They **must be** swallowed whole.

■ Store at 15°–30° C (59°–86° F).

ADVERSE EFFECTS (≥1%) **Endocrine:** Gynecomastia. **Urogenital:** Ejaculation dysfunction, impotence, decreased libido.

DIAGNOSTIC TEST INTERFERENCE Lab Test: Dutasteride affects the serum PSA levels, so levels should be established after 3–6 mo of therapy.

INTERACTIONS Drug: Diltiazem, verapamil may decrease clearance of dutasteride. **Herbal:** May see exaggerated effects with **saw palmetto.**

PHARMACOKINETICS Absorption: Rapidly; 60% bioavailability. **Peak:** 2–3 h. **Distribution:** 99% protein bound. **Metabolism:** In liver by CYP3A4. **Elimination:** Primarily in feces. **Half-Life:** 5 wk.

NURSING IMPLICATIONS

Assessment & Drug Effects

■ Monitor voiding patterns, assessing for ease of starting a stream, frequency, and urgency.

■ Lab tests: Monitor baseline PSA and again at 3–6 mo to establish new

Common adverse effects in _italic_, life-threatening effects underlined; generic names in **bold;** classifications in SMALL CAPS; ♣ Canadian drug name; ☻ Prototype drug

511

baseline to use to assess potentially cancer-related changes in PSA.

Patient & Family Education

- Do not donate blood until at least 6 mo following last dose to prevent administration of dutasteride to a pregnant female transfusion recipient.
- Ejaculate volume might be decreased during treatment but this does not seem to interfere with normal sexual function.
- Note that the incidence of most drug-related sexual adverse events (impotence, decreased libido, and ejaculation disorder) typically decrease with duration of treatment.

DYPHYLLINE

(dye'fi-lin)

Dylix, Lufyllin, Protophylline ♦

Classifications: RESPIRATORY SMOOTH MUSCLE RELAXANT, XANTHINE; BRONCHODILATOR

Therapeutic: BRONCHODILATOR; ANTIASTHMATIC

Prototype: Theophylline

Pregnancy Category: C

AVAILABILITY 200 mg, 400 mg tablets, 100 mg/15 mL elixir

ACTION & THERAPEUTIC EFFECT Xanthine derivative of theophylline results in bronchodilation, myocardial stimulation, and smooth muscle relaxation. *Effective bronchodilator and antiasthmatic agent.*

USES Acute bronchial asthma and reversible bronchospasm associated with chronic bronchitis and emphysema.

CONTRAINDICATIONS Hypersensitivity to xanthine compounds; apnea in newborns.

CAUTIOUS USE Severe cardiac disease, hypertension, acute myocardial injury; renal or hepatic dysfunction; glaucoma; seizure disorders; hyperthyroidism; peptic ulcer; older adults; children; pregnancy (category C), lactation. Safe use in children under 6 y is not established.

ROUTE & DOSAGE

Asthma

Adult: **PO** Up to 15 mg/kg q.i.d.

ADMINISTRATION

Oral

- Give oral preparation with a full glass of water on an empty stomach (e.g., 1 h before or 2 h after meals) to enhance absorption. However, administration after meals may help to relieve gastric discomfort.
- Exercise care in the amount of elixir given to children because it has a high alcohol content (18–20%).

ADVERSE EFFECTS (≥1%) **CNS:** Headache, irritability, restlessness, dizziness, insomnia, light-headedness, muscle twitching, <u>convulsions</u>. **CV:** Palpitation, *tachycardia,* extrasystoles, flushing, hypotension. **GI:** *Nausea,* vomiting, diarrhea, anorexia, epigastric distress. **Respiratory:** Tachypnea. **Other:** Albuminuria, fever, dehydration.

INTERACTIONS Drug: BETA-BLOCKERS may antagonize bronchodilating effects of dyphylline; **halothane** increases risk of cardiac arrhythmias; **probenecid** may decrease dyphylline elimination.

PHARMACOKINETICS Absorption: Readily from GI tract. **Peak:** 1 h. **Metabolism:** In liver (but not to theophylline). **Elimination:** In urine. **Half-Life:** 2 h.

NURSING IMPLICATIONS

Assessment & Drug Effects

- Lab tests: Baseline and periodic pulmonary function tests.
- Monitor therapeutic effectiveness; usually occurs at a blood level of at least 12 mcg/mL.
- Note: Toxic dyphylline plasma levels, although rare with normal dosage, are a risk in patients with a diminished capacity for dyphylline clearance (e.g., those with CHF or hepatic impairment or who are older than 55 y or younger than 1 y).

Patient & Family Education

- Take medication consistently with or without food at the same time each day.
- Notify prescriber of adverse effects: Nausea, vomiting, insomnia, jitteriness, headache, rash, severe GI pain, restlessness, convulsions, or irregular heartbeat.
- Avoid alcohol and also large amounts of coffee and other xanthine-containing beverages (e.g., tea, cocoa, cola) during therapy.
- Consult prescriber before taking OTC preparations. Many OTC drugs for coughs, colds, and allergies contain nervous system stimulants.

ECHOTHIOPHATE IODIDE

(ek-oh-thye'oh-fate)

Phospholine Iodide
Pregnancy Category: C
See Appendix A-1.

ECONAZOLE NITRATE

(e-kone'a-zole)

Ecostatin ✦, Spectazole
Classifications: ANTIBIOTIC; AZOLE ANTIFUNGAL

Therapeutic: ANTIFUNGAL
Prototype: Fluconazole
Pregnancy Category: C

E

AVAILABILITY 1% cream

ACTION & *THERAPEUTIC EFFECT*
Azole antifungal antibiotic with broad spectrum of activity that disrupts normal fungal cell membrane permeability. *Active against dermatophytes, yeasts, and many other fungi.*

USES Topically for treatment of tinea pedis (athlete's foot or ringworm of foot), tinea cruris ("jock itch" or ringworm of groin), tinea corporis (ringworm of body), tinea versicolor, and cutaneous candidiasis (moniliasis).
UNLABELED USES Has been used for topical treatment of erythrasma and with corticosteroids for fungal or bacterial dermatoses associated with inflammation.

CONTRAINDICATIONS Infants younger than 3 mo.
CAUTIOUS USE Pregnancy (category C), lactation.

ROUTE & DOSAGE

Tinea Cruris, Tinea Corporis, Tinea Pedis, Cutaneous Candidiasis
Adult/Child: **Topical** Apply sufficient amount to affected areas twice daily, morning and evening
Tinea Versicolor
Adult: **Topical** Apply sufficient amount to affected areas once daily

ADMINISTRATION
Topical
- Cleanse skin with soap and water and dry thoroughly before applying medication (unless otherwise directed by prescriber). Wash

---

Common adverse effects in *italic*, life-threatening effects <u>underlined</u>; generic names in **bold**; classifications in SMALL CAPS; ✦ Canadian drug name; ❶ Prototype drug

513

E

hands thoroughly before and after treatments.

- Do not use occlusive dressings unless prescribed by prescriber.
- Store at less than 30° C (86° F) unless otherwise directed.

ADVERSE EFFECTS (≥1%) **Skin:** Burning, stinging sensation, pruritus, erythema.

INTERACTIONS Drug: No clinically significant interactions established.

PHARMACOKINETICS Absorption: Minimal percutaneous absorption through intact skin; increased absorption from denuded skin. **Peak:** 0.5–5 h. **Elimination:** Less than 1% of applied dose is eliminated in urine and feces.

NURSING IMPLICATIONS

Patient & Family Education

- Use medication for the prescribed time even if symptoms improve and report to prescriber skin reactions suggestive of irritation or sensitization.
- Notify prescriber if full course of therapy does not result in improvement. Diagnosis should be reevaluated.
- Do not to apply the topical cream in or near the eyes or intravaginally.

ECULIZUMAB

(e-cul-i-zu'mab)

Soliris

Classifications: BIOLOGICAL RESPONSE MODIFIER; MONOCLONAL ANTIBODY; IMMUNOGLOBULIN

Therapeutic: IMMUNOGLOBULIN

Prototype: Immune globulin

Pregnancy Category: C

AVAILABILITY 10 mg/mL injection

ACTION & *THERAPEUTIC EFFECT*
A monoclonal antibody (IgG) immunoglobulin molecule that binds with high affinity to complement C5 inhibiting formation of the terminal complement complex, C5b-9. *Inhibition of C5b-9 complement complex prevents complement-mediated hemolysis in those with RBC deficiency in patients with paroxysmal nocturnal hemoglobinuria (PNH), resulting from genetic mutation.*

USES Reduction of hemolysis in patients with paroxysmal nocturnal hemoglobinuria.

CONTRAINDICATIONS Serious meningococcal infections.

CAUTIOUS USE History of hypersensitivity to protein components; older adults; systemic infection; pregnancy (category C), lactation. Safe use in children younger than 18 y has not established.

ROUTE & DOSAGE

Paroxysmal Nocturnal Hemoglobinuria

Adult: **IV** 600 mg IV infusion every 7 days × 4 wk (a total of 4 doses); then 900 mg IV on day 7 after the 4th dose, and then 900 mg IV every 14 days thereafter

ADMINISTRATION

Intravenous

Note: Patients **must be** vaccinated against *Neisseria meningitidis* **at least 2 wk prior to** the first dose of eculizumab. Prior to initiating treatment, patients and prescribers **must be** enrolled in the Soliris™ Safety Registry.

***PREPARE:* IV Infusion:** Dilute to a final concentration of 5 mg/mL in NS, D5/0.45% NaCl, or LR by adding the required volume of eculizumab to an EQUAL volume of IV fluid. Invert bag to mix.
- Final infusion volumes will be

600 mg in 120 mL or 900 mg in 180 mL. ▪ Allow to come to room temperature prior to infusion. *ADMINISTER:* **IV Infusion:** Do NOT give direct IV. ▪ Give over 35 min via infusion pump or syringe pump. ▪ If infusion is slowed for an infusion reaction, the total infusion time should not exceed 2 h.

INCOMPATIBILITIES **Solution/ additive/Y-site:** Do not mix with any other drugs or solutions.

▪ Store infusion bags for 24 h at 2°–8° C (36°–46° F).

ADVERSE EFFECTS (≥1%) **Body as a Whole:** Herpes simplex infections, influenza-like illness, pain in extremity. **CNS:** *Fatigue, headache.* **GI:** Constipation, *nausea.* **Musculoskeletal:** *Back pain,* myalgia. **Respiratory:** *Cough, nasopharyngitis,* respiratory tract infection, sinusitis.

PHARMACOKINETICS **Half-Life:** 272 ± 82 h.

NURSING IMPLICATIONS

Assessment & Drug Effects
▪ Monitor for a hypersensitivity reaction throughout infusion and for at least 1 h after completion of the infusion.
▪ Monitor for early signs of meningococcal infection. Report immediately if an infection is suspected.
▪ Lab tests: Baseline and periodic RBC blood studies.

Patient & Family Education
▪ Although patient must be vaccinated against *N. meningitidis* prior to therapy with eculizumab, vaccination may not prevent meningitis. Report immediately any of the following: Moderate to severe headache with nausea or vomiting, stiff neck or stiff back, fever, rash, confusion, severe muscle aches with flu-like symptoms, and sensitivity to light.

EDETATE CALCIUM DISODIUM
(ed′e-tate)

Calcium Disodium Versenate
Classification: CHELATING AGENT
Therapeutic: CHELATING AGENT; ANTIPOISON
Pregnancy Category: B

E

AVAILABILITY 200 mg/mL injection

ACTION & *THERAPEUTIC EFFECT*
Chelating agent that combines with divalent and trivalent metals to form stable, nonionizing soluble complexes that can be readily excreted by kidneys. Action is dependent on ability of heavy metal to displace the less strongly bound calcium in drug molecules. *Chelating agent that binds with heavy metals such as lead to form a soluble complex that can be excreted through the kidney, thereby ridding the body of the poisonous substance.*

USES As adjunct in treatment of acute and chronic lead poisoning (plumbism). Generally used in combination with dimercaprol (BAL) in treatment of lead encephalopathy or when blood lead level exceeds 100 mcg/dL. Also used to diagnose suspected lead poisoning.
UNLABELED USES Treatment of poisoning from other heavy metals such as chromium, manganese, nickel, zinc, and possibly vanadium; removal of radioactive and nuclear fission products such as plutonium, yttrium, uranium. Not effective in poisoning from arsenic, gold, or mercury.

CONTRAINDICATIONS Severe kidney disease, active renal disease, anuria, oliguria; hepatitis; IV use in patients with lead encephalopathy not generally recommended (because of possible increase in intracranial pressure).

Common adverse effects in *italic*, life-threatening effects <u>underlined</u>; generic names in **bold**; classifications in SMALL CAPS; ✦ Canadian drug name; ☻ Prototype drug

515

CAUTIOUS USE Kidney dysfunction; active tubercular lesions; history of gout; cardiac arrhythmias; pregnancy (category B), lactation.

ROUTE & DOSAGE

Diagnosis of Lead Poisoning

Adult: IV/IM 500 mg/m² (max: 1 g) over 1 h, then collect urine for 24 h (if mcg lead:mg EDTA ratio in urine is greater than 1, the test is positive)

Child: IM 50 mg/kg (max: 1 g), then collect urine for 6–8 h, (if mcg lead:mg EDTA ratio in urine is greater than 0.5, the test is positive)

Treatment of Lead Poisoning

Adult/Child: IV 1–1.5 g/m²/day infused over 8–24 h for up to 5 days IM 1–1.5 g/m²/day divided q8–12h

Asymptomatic Lead Poisoning

Adult/Child: IV 1 g/m²/day infused over 8–24 h for up to 5 days

Lead Nephropathy/Renal Impairment Dosage Adjustment

Adult: IV *Based on serum CrCl less than 2 mg/dL:* 1 g/m²/day × 5 days; *2–3 mg/dL:* 500 mg/m²/day × 5 days; *3.1–4 × 3 doses mg/dL:* 500 mg/m² q48h; *greater than 4 mg/dL:* 500 mg/m² once/wk. Infuse over 8–24 h, may repeat monthly.

ADMINISTRATION

- Note: Calcium disodium edetate can produce potentially fatal effects when higher than recommended doses are used or when it is continued after toxic effects appear.

Intramuscular

- IM route preferred for symptomatic children and recommended

for patients with incipient or overt lead-induced encephalopathy.
- Add Procaine HCl to minimize pain at injection site (usually 1 mL of procaine 1% to each 1 mL of concentrated drug). Consult prescriber.

Intravenous

***PREPARE:* IV Infusion:** Dilute the 5 mL ampule with 250–500 mL of NS or D5W.

***ADMINISTER:* IV Infusion:** Warning: Rapid IV infusion may be LETHAL by suddenly increasing intracranial pressure in patients who already have cerebral edema. • Manufacturer recommends total daily dose over 8–12 h. Consult prescriber for specific rate.

INCOMPATIBILITIES **Solution/additive: Amphotericin B, D10W hydralazine, lactated Ringer's.**

ADVERSE EFFECTS (≥1%) **CV:** Hypotension, thrombophlebitis. **GI:** Anorexia, nausea, vomiting, diarrhea, abdominal cramps, cheilosis. **Hematologic:** Transient bone marrow depression, depletion of blood metals. **Urogenital:** <u>Nephrotoxicity</u> (renal tubular necrosis), proteinuria, hematuria. **Body as a Whole:** *Febrile reaction* (excessive thirst, fever, chills, severe myalgia, arthralgia, GI distress), *histamine-like reactions* (flushing, throbbing headache, sweating, sneezing, nasal congestion, lacrimation, postural hypotension, tachycardia).

DIAGNOSTIC TEST INTERFERENCE Edetate calcium disodium may decrease **serum cholesterol, plasma lipid** levels (if elevated), and **serum potassium** values. *Glycosuria* may occur with toxic doses.

INTERACTIONS Drug: May affect **insulin** requirements.

PHARMACOKINETICS Absorption: Well absorbed IM. **Onset:**

Common adverse effects in *italic*, life-threatening effects <u>underlined</u>; generic names in **bold;** classifications in SMALL CAPS; ✦ Canadian drug name; ✪ Prototype drug

1 h. **Peak:** Peak chelation 24–48 h. **Distribution:** Distributed to extracellular fluid; does not enter CSF. **Metabolism:** Not metabolized. **Elimination:** Chelated lead excreted in urine; 50% excreted in 1 h. **Half-Life:** 20–60 min IV, 90 min IM.

NURSING IMPLICATIONS

Assessment & Drug Effects

- Determine adequacy of urinary output prior to therapy. This may be done by administering IV fluids before giving first dose.
- Increase fluid intake to enhance urinary excretion of chelates. Avoid excess fluid intake, however, in patients with lead encephalopathy because of the danger of further increasing intracranial pressure. Consult prescriber regarding allowable intake.
- Monitor I&O. Since drug is excreted almost exclusively via kidneys, toxicity may develop if output is inadequate. Stop therapy if urine flow is markedly diminished or absent. Report any change in output or I&O ratio to prescriber.
- Lab tests: Obtain serum creatinine, calcium, and phosphorus before and during each course of therapy. Monitor baseline and frequent BUN levels and ECG during therapy. With prolonged therapy determine periodic determinations of blood trace element metals (e.g., copper, zinc, magnesium).
- Be alert for occurrence of febrile reaction that may appear 4–8 h after drug infusion (see ADVERSE EFFECTS).

EDROPHONIUM CHLORIDE

(ed-roe-foe′nee-um)

Enlon

Classifications: CHOLINERGIC MUSCLE STIMULANT; CHOLINESTERASE INHIBITOR

Therapeutic: ANTICHOLINESTERASE MUSCLE STIMULANT
Prototype: Neostigmine
Pregnancy Category: C

AVAILABILITY 10 mg/mL injection

ACTION & *THERAPEUTIC EFFECT*
Facilitates transmission of impulses across the myoneural junction by inhibiting the destruction of acetylcholine by cholinesteratse. *Acts as antidote to curariform drugs by displacing them from muscle cell receptor sites, thus permitting resumption of normal transmission of neuromuscular impulses.*

USES Differential diagnosis and as adjunct in evaluation of treatment requirements of myasthenia gravis; curare antagonist.

CONTRAINDICATIONS Hypersensitivity to anticholinesterase agents; cholinesterase inhibitor toxicity; intestinal and urinary obstruction; lactation.

CAUTIOUS USE Sulfite hypersensitivity; bronchial asthma; cardiac arrhythmias; bradycardia; peptic ulcer disease; hypotension; patients receiving digitalis; pregnancy (category C).

ROUTE & DOSAGE

Myasthenia Gravis Diagnosis

Adult: **IV** Prepare 10 mg in a syringe; inject 2 mg over 15–30 sec, if no reaction after 45 sec, inject the remaining 8 mg, may repeat test after 30 min **IM** Inject 10 mg; if cholinergic reaction occurs, retest after 30 min with 2 mg to rule out false-negative reaction *Child:* **IV** Weight 34 kg or less, 1 mg, if no response after 45 sec, dose may be titrated up to 5 mg

Common adverse effects in *italic*, life-threatening effects underlined; generic names in **bold**; classifications in SMALL CAPS; ✦ Canadian drug name; ❂ Prototype drug

517

IM 2 mg **IV** *Weight greater than 34 kg*, 2 mg, if no response after 45 sec, dose may be titrated up to 10 mg **IM** 5 mg

Evaluation of Myasthenia Treatment

Adult: **IV** 1–2 mg administered 1 h after last PO dose of anticholinesterase medication

Curare Antagonist

Adult: **IV** 1 mL (10 mg) over 30–45 sec, repeat as necessary. Max dose: 4 mL (40 mg).

ADMINISTRATION

- Note: Have antidote (atropine sulfate) immediately available and facilities for endotracheal intubation, tracheostomy, suction, assisted respiration, and cardiac monitoring for treatment of cholinergic reaction.

Intramuscular

- Give undiluted. IM route used if IV route not accessible.

Intravenous

PREPARE: **Direct/Infusion:** May be given undiluted or diluted in D5W or NS for infusion.

ADMINISTER: **Direct:** Inject 2 mg (adult and child weighing more than 34 kg) or 1 mg (child weighing 34 kg or less) over 15–30 sec; if no reaction after 45 sec, inject additional 8 mg (adult) or titrate up to a total of 8 mg additional (child weighing more than 34 kg) or titrate in 1 mg increments up to a total of 4 mg additional (child weighing 34 kg or less), may repeat test after 30 min. ▪ If cholinergic reaction (increased muscle weakness) is obtained after initial 1 or 2 mg, discontinue test and give atropine IV (as ordered). **IV Infusion:** Infuse over 1 h.

ADVERSE EFFECTS (≥1%) **Body as a Whole:** Severe adverse effects uncommon with usual doses. **CNS:** Weakness, muscle cramps, dysphoria, fasciculations, incoordination, dysarthria, dysphagia, convulsions, respiratory paralysis. **CV:** Bradycardia, irregular pulse, hypotension, pulmonary edema. **Special Senses:** Miosis, blurred vision, diplopia, lacrimation. **GI:** Diarrhea, abdominal cramps, nausea, vomiting, excessive salivation. **Respiratory:** Increased bronchial secretions, bronchospasm, laryngospasm, pulmonary edema. **Other:** Excessive sweating, urinary frequency, incontinence.

INTERACTIONS **Drug: Procainamide, quinidine** may antagonize the effects of edrophonium; DIGITALIS GLYCOSIDES increase the sensitivity of the heart to edrophonium; **succinylcholine, decamethonium** may prolong neuromuscular blockade.

PHARMACOKINETICS **Onset:** 30–60 sec IV; 2–10 min IM. **Duration:** 5–10 min IV; 5–30 min IM.

NURSING IMPLICATIONS

Assessment & Drug Effects

- Monitor vital signs. Observe for signs of respiratory distress. Patients older than 50 y are particularly likely to develop bradycardia, hypotension, and cardiac arrest.
- Edrophonium test for myasthenia gravis: All cholinesterase inhibitors (anticholinesterases) should be discontinued for at least 8 h before test. Positive response to edrophonium test consists of brief improvement in muscle strength unaccompanied by lingual or skeletal muscle fasciculations.
- Evaluation of myasthenic treatment: *Myasthenic response* (immediate subjective improvement with increased muscle strength,

Common adverse effects in *italic*; life-threatening effects <u>underlined</u>; generic names in **bold**; classifications in SMALL CAPS; ✦ Canadian drug name; ❶ Prototype drug

absence of fasciculations; generally indicates that patient requires larger dose of anticholinesterase agent or longer-acting drug); *Cholinergic response* [muscarinic adverse effects (lacrimation, diaphoresis, salivation, abdominal cramps, diarrhea, nausea, vomiting; accompanied by decrease in muscle strength; usually indicates over-treatment with cholinesterase inhibitor)]; *Adequate response* [no change in muscle strength; fasciculations may be present or absent; minimal cholinergic adverse effects (observed in patients at or near optimal dosage level)].

EFAVIRENZ ℗

(e-fa′vi-renz)

Sustiva

Classifications: ANTIRETROVIRAL; NONNUCLEOSIDE REVERSE TRANSCRIPTASE INHIBITOR (NNRTI)

Therapeutic: ANTIRETROVIRAL; NNRTI

Pregnancy Category: D (including first trimester)

AVAILABILITY 50 mg, 200 mg capsules; 600 mg tablets

ACTION & THERAPEUTIC EFFECT Nonnucleoside reverse transcriptase inhibitor (NNRTI) of HIV-1. Binds directly to reverse transcriptase and blocks RNA polymerase activities of the HIV-1 virus, thus preventing replication of the virus. *Prevents replication of the HIV-1 virus. Resistant strains appear rapidly. Effectiveness is indicated by reduction in viral load (plasma level HIV RNA).*

USES HIV-1 infection in combination with other antiretroviral agents.

CONTRAINDICATIONS Hypersensitivity to efavirenz; suicidal ideation; pregnancy (category D, including fetal harm in first trimester), lactation.

CAUTIOUS USE Liver disease, alcoholism, hepatitis B or C, hypertriglyceridemia, hypercholesterolemia, substance abuse, antimicrobial resistance, bipolar disorder, depression, suicidal tendencies, exfoliative dermatitis; females of childbearing age, CNS disorders; history of seizures; older adults. Safety and efficacy in children younger than 3 y or who weigh less than 13 kg (29 lb) are not known.

ROUTE & DOSAGE

HIV Infection

Adult/Adolescent: **PO** 600 mg daily
Child: **PO** *3 y or older*
Weight 10–15 kg: 200 mg daily
Weight 15–20 kg: 250 mg daily
Weight 20–25 kg: 300 mg daily
Weight 25–32.5 kg: 350 mg daily
Weight 32.5–40 kg: 400 mg daily
Weight greater than 40 kg: 600 mg daily

ADMINISTRATION

Oral

- Use bedtime dosing to increase tolerability of CNS adverse effects.
- Give exactly as ordered. Do not skip a dose or discontinue therapy without consulting the prescriber.
- Do not give efavirenz following a high fat meal.
- Store at 15°–30° C (59°–86° F) in a tightly closed container and protect from light.

ADVERSE EFFECTS (≥1%) **Body as a Whole:** Fatigue, fever. **CNS:** Dizziness, headache, hypoesthesia,

Common adverse effects in *italic*, life-threatening effects underlined; generic names in **bold**; classifications in SMALL CAPS; ♦ Canadian drug name; ℗ Prototype drug

519

impaired concentration, insomnia, abnormal dreams, somnolence, depression, nervousness, adverse psychiatric experiences. **CV:** Hypercholesterolemia. **GI:** *Nausea,* vomiting, *diarrhea,* dyspepsia, abdominal pain, flatulence, anorexia, increased liver function tests (ALT, AST). **Respiratory:** Cough. **Skin:** *Rash* (erythematous rash, pruritus, *maculopapular rash,* erythema multiforme, Stevens-Johnson syndrome, toxic epidermal necrolysis), increased sweating. **Urogenital:** Renal calculus, hematuria.

DIAGNOSTIC TEST INTERFERENCE False-positive urine tests for **marijuana.**

INTERACTIONS Drug: Decreased concentrations of **clarithromycin, indinavir, nelfinavir, saquinavir, voriconazole;** increased concentrations of **ritonavir, azithromycin, ethinyl estradiol.** Efavirenz levels are increased by **ritonavir, fluconazole** and decreased by **saquinavir, rifampin.** Additional drugs not recommended for administration with efavirenz include **midazolam, triazolam,** ERGOT DERIVATIVES, **warfarin. Herbal:** St. John's wort may decrease antiretroviral activity.

PHARMACOKINETICS Peak: 5 h; steady-state 6–10 days. **Distribution:** 99% protein bound. **Metabolism:** In liver by cytochrome P450 3A4 and 2B6; can induce (increase) its own metabolism. **Elimination:** 14–34% in urine, 16–61% in feces. **Half-Life:** 52–76 h after single dose, 40–55 h after multiple doses.

NURSING IMPLICATIONS

Assessment & Drug Effects

- Monitor for suicidal ideation in patients who are depressed, or who have a history of depression.

- Monitor GI status and evaluate ability to maintain a normal diet.
- Lab tests: Periodic LFTs and lipid profile.

Patient & Family Education

- Contact prescriber promptly if any of the following occurs: Skin rash, delusions, inappropriate behavior, suicidal ideation.
- Avoid pregnancy because of fetal harm in first trimester of pregnancy.
- Use or add barrier contraception if using hormonal contraceptive.
- Notify prescriber immediately if you become pregnant.
- Do not drive or engage in potentially hazardous activities until response to the drug is known. Dizziness, impaired concentration, and drowsiness usually improve with continued therapy.

EFLORNITHINE HYDROCHLORIDE

(e-flor'ni-theen)

Vaniqa

Classification: DERMATOLOGIC
Therapeutic: FACIAL HIRSUTISM AGENT
Pregnancy Category: C

AVAILABILITY 13.9% cream

ACTION & *THERAPEUTIC EFFECT* Inhibits enzyme activity in the skin that is required for hair growth. *Results in retarding the rate of hair growth.*

USES Reduction of unwanted facial hair in women.

CONTRAINDICATIONS Hypersensitivity to eflornithine or its components.

CAUTIOUS USE Bone marrow suppression; HIV; pregnancy·(category C), lactation. Safe use in children younger than 12 y is not established.

Common adverse effects in *italic*, life-threatening effects underlined; generic names in **bold;** classifications in SMALL CAPS; ✦ Canadian drug name; ❍ Prototype drug

ROUTE & DOSAGE

Hair Removal

Adult: **Topical** Apply thin layer to affected areas of the face and adjacent involved areas under the chin and rub in thoroughly b.i.d. at least 8 h apart

ADMINISTRATION

Topical

- Apply thin layer to affected skin areas on face and under chin and rub in thoroughly.
- Do not wash treated areas for at least 4 h after application.
- Store at 15°–30° C (59°–86° F).

ADVERSE EFFECTS (≥1%) **Body as a Whole** Facial edema. **CNS** Dizziness. **GI:** Dyspepsia, anorexia. **Skin:** *Acne, pseudofolliculitis barbae,* stinging, burning, pruritus, erythema, tingling, irritation, rash, alopecia, folliculitis, ingrown hair.

PHARMACOKINETICS **Absorption:** Less than 1% absorbed through intact skin. **Metabolism:** Not metabolized. **Elimination:** Primarily in urine. **Half-Life:** 8 h.

NURSING IMPLICATIONS

Assessment & Drug Effects

- Monitor for and report skin irritation.
- Note: Drug slows growth of facial hair, but is not a depilatory.

Patient & Family Education

- Note: Effect of drug is usually not apparent for 4–8 wk.
- Reduce frequency of drug application to once daily if skin irritation occurs. If irritation continues, contact prescriber.

ELETRIPTAN HYDROBROMIDE
(e-le-trip′tan)

Relpax
Classification: SEROTONIN 5-HT$_1$ RECEPTOR AGONIST
Therapeutic: ANTIMIGRAINE
Prototype: Sumatriptan
Pregnancy Category: C

E

AVAILABILITY 20 mg, 40 mg tablets

ACTION & *THERAPEUTIC EFFECT*
Eletriptan is a potent agonist at central serotonin 5-HT$_{1B}$, 5-HT$_{1D}$, and 5-HT$_{1F}$ receptors. Eletriptan stimulates presynaptic 5-HT$_{1D}$ receptors inhibiting dural vasodilation and agonizes vascular 5-HT$_{1B}$ receptors causing vasoconstriction of intracranial extracerebral vessels. *Inhibits dural vasodilation and inflammation, and causes vasoconstriction of painfully dilated intracranial extra-cerebral vessels, thus relieving the migraine headache. Also relieves photophobia, phonophobia, and nausea and vomiting associated with migraine attacks.*

USES Treatment of acute migraine attacks with or without aura.

CONTRAINDICATIONS Hypersensitivity to eletriptan; history of CAD; ischemic or vasospastic CAD, arteriosclerosis, history of MI; ischemic colitis, Raynaud's disease uncontrolled hypertension; CVA or TIA; within 24 h of administering of another ergotamine; lactation within 24 h after dose; severe hepatic insufficiency; hemiplegic or basilar migraine; peripheral vascular disease; concurrent MAOI therapy.

CAUTIOUS USE Hypotension in the elderly; older adults over 65 y; mild to moderate hepatic impairment; diabetes, obesity, smoking, high

---

Common adverse effects in *italic*, life-threatening effects <u>underlined</u>; generic names in **bold**; classifications in SMALL CAPS; ♦ Canadian drug name; ◐ Prototype drug

521

E

cholesterol; men older than 40 y; postmenopausal women. Use within 72 h of potent CYP3A4 metabolizing drugs; pregnancy (category C), lactation. Safe use in children younger than 18 y has not been established.

ROUTE & DOSAGE

Acute Migraine

Adult: **PO** 20 mg or 40 mg at onset of migraine (max: 40 mg/dose and 80 mg/day), may repeat dose in 2 h if partial response

Hepatic Impairment Dosage Adjustment

Severe Hepatic Impairment: Not recommended

ADMINISTRATION

Oral

- Give one tablet as soon as the migraine begins.
- May give 2nd tablet if headache improves but returns after 2 h.
- If 1st tablet is ineffective, do not give a 2nd without consulting prescriber.
- Do not give within 72 h of potent CYP3A4 inhibitors (see INTERACTIONS).
- Store at 15°–30° C (59°–86° F). Protect from light and moisture.

ADVERSE EFFECTS (≥1%) **Body as a Whole:** *Asthenia,* paresthesia, flushing, back pain, chills. **CNS:** Dizziness, drowsiness, headache, somnolence, hypertonia, hypesthesia. **CV:** Chest tightness/pressure, palpitation, hypertension. The following are rare, usually seen in patients with cardiovascular disease risk factors: Coronary vasospasm, transient myocardial ischemia, <u>MI</u>, ventricular tachycardia, atrial fibrillation, ven-

tricular fibrillation. **GI:** Abdominal pain, dyspepsia, dysphagia, nausea, vomiting, dry mouth. **Respiratory:** Pharyngitis. **Skin:** Sweating.

INTERACTIONS Drug: Drugs that inhibit CYP3A4 may increase eletriptan levels and toxicity, do not administer eletriptan within 72 h of AZOLE ANTIFUNGALS (especially **itraconazole, ketoconazole, voriconazole**), **amiodarone, cimetidine, dalfopristin, quinupristin, diltiazem, metronidazole, nicardipine, norfloxacin, quinine, verapamil, zafirlukast, zileuton,** MACROLIDE ANTIBIOTICS, NONNUCLEOTIDE REVERSE TRANSCRIPTASE INHIBITORS, PROTEASE INHIBITORS, SELECTIVE SEROTONIN REUPTAKE INHIBITORS, **sibutramine;** ERGOT ALKALOIDS may prolong vasospastic adverse reactions (do not use within 24 h of ergot-containing drugs); do not administer within 24 h of other 5-HT₁ AGONISTS (increases adverse effects). **Food: Grapefruit juice** may increase eletriptan levels and toxicity. **Herbal: echinacea, St. John's wort** may increase triptan toxicity.

PHARMACOKINETICS Absorption: Rapid with 50% reaching systemic circulation. **Onset:** 1–2 h. **Peak:** 1.5 h. **Distribution:** 85% protein bound. **Metabolism:** In liver by CYP3A4. **Elimination:** Nonrenal routes. **Half-Life:** 4–5 h.

NURSING IMPLICATIONS

Assessment & Drug Effects

- Monitor CV status carefully following first dose in patients at risk for coronary artery disease (e.g., history of hypertension, postmenopausal women, men older than 40 y, persons with known CAD risk factors) or who have coronary artery vasospasms.

- Report immediately chest pain, tightness in chest or throat that is severe or does not quickly resolve following a dose of eletriptan.
- Monitor therapeutic effectiveness. Pain relief is usually achieved within 1 h.

Patient & Family Education
- Note: If first dose is ineffective, do not take a second dose as it will not work for the same attack.
- Inform prescriber of all prescription, nonprescription, and herbal drugs you are taking. Do not add additional drugs without informing prescriber as many drugs interact with eletriptan.
- Report promptly any of the following: Headache more severe than usual, migraine; dizziness, faintness, blurred vision; chest, neck, or throat pain; irregular heart beat, palpitations; shortness of breath, wheezing, difficulty breathing; tingling, pain, or numbness in the face, hands, or feet; seizures; severe stomach pain, cramping, or bloody diarrhea.
- Do not drive or engage in any potentially hazardous task until reaction to drug is known.

EMEDASTINE DIFUMARATE ⊘
(em-e-das′teen di-foom′a-rate)
Emadine
Classifications: OCULAR; ANTIHISTAMINE; H_1-RECEPTOR ANTAGONIST
Therapeutic: OCULAR ANTIHISTAMINE
Pregnancy Category: B

AVAILABILITY 0.05% ophth solution

ACTION & *THERAPEUTIC EFFECT*
Emedastine is a selective antagonist at H_1-receptors. It blocks H_1-receptors and inhibits histamine-stimulated vascular permeability in the conjunctiva. *Relieves ocular pruritus related to allergic response to histamine.*

USES Temporary relief of seasonal allergic conjunctivitis.

CONTRAINDICATIONS Hypersensitivity to emedastine.
CAUTIOUS USE Hypersensitivity to other antihistamines, soft contact lenses; pregnancy (category B), lactation. Safety and efficacy in children younger than 3 y are not established.

ROUTE & DOSAGE

Allergic Conjunctivitis
Adult: **Ophthalmic** 1 drop in affected eye q.i.d.

ADMINISTRATION
Instillation
- Wash hands before and after use.
- Shake well before using. Apply drops in the center of the lower conjunctival sac. Do not touch eyelids with dropper.
- Gently close eyes for 1–2 min after installation of drops.
- Wait 10 min after installation of drug before inserting soft lenses into eyes.
- Store in a tightly closed bottle. Protect the solution from light.
- Do not use if discolored.

ADVERSE EFFECTS (≥1%) **CNS:** Headache. **Special Senses:** *Ocular irritation, mild transient stinging and burning,* conjunctival congestion, eyelid edema, eye pain, photophobia, abnormal lacrimation.

INTERACTIONS Drug: No clinically significant interactions established.

Common adverse effects in *italic*, life-threatening effects <u>underlined</u>; generic names in **bold**; classifications in SMALL CAPS; ♦ Canadian drug name; ⊘ Prototype drug

523

PHARMACOKINETICS Absorption: Minimal. **Half-Life:** 3–4 h.

NURSING IMPLICATIONS

Assessment & Drug Effects

- Monitor for S&S of hypersensitivity to the drug (see Appendix F).
- Evaluate safety of engaging in hazardous activities since drowsiness is a potential adverse effect.

Patient & Family Education

- Learn potential adverse responses to emedastine.
- Eye drops contain benzalkonium chloride, which may damage soft contact lenses. After instillation of drops, wait 10 min before inserting these contact lenses into the eye.
- Contact your prescriber if symptoms do not start to improve in 2 or 3 days.

EMLA (EUTECTIC MIXTURE OF LIDOCAINE AND PRILOCAINE)

EMLA Cream
Classification: LOCAL ANESTHETIC
Therapeutic: LOCAL ANESTHETIC
Prototype: Procaine
Pregnancy Category: B

AVAILABILITY 2.5% lidocaine/2.5% prilocaine cream

ACTION & *THERAPEUTIC EFFECT* EMLA cream is a mixture of lidocaine and prilocaine. The mixture forms a liquid at room temperature. *EMLA is a topical analgesic.*

USES Topical anesthetic on normal intact skin for local anesthesia.
UNLABELED USES Topical anesthetic prior to leg ulcer debridement; treatment of postherpetic neuralgia.

CONTRAINDICATIONS Patients with known sensitivity to local anesthetics; patients with congenital or idiopathic methemoglobinemia; tympanic membrane perforation.
CAUTIOUS USE Acutely ill, debilitated, or older adult patients; severe liver disease; pregnancy (category B), lactation. Safe use in children younger than 1 mo not established.

ROUTE & DOSAGE

Topical Anesthetic

Adult/Child (older than 1 mo):
Topical Apply 2.5 g of cream (½ of 5-g tube) over 20–25 cm² of skin, cover with occlusive dressing and wait at least 1 h, then remove dressing and wipe off cream, cleanse area with an antiseptic solution and prepare patient for the procedure.

ADMINISTRATION

Topical

- Apply a thick layer to skin (approximately ½ of 5-g tube per 20–25 cm² or 2 × 2 in) at site of procedure. Apply an occlusive dressing. Do not spread out cream. Seal edges of dressing well to avoid leakage.
- Apply EMLA cream 1 h before routine procedure and 2 h before painful procedure.
- Remove EMLA cream prior to skin puncture and clean area with an aseptic solution.
- Store at room temperature 15°–30° C (59°–86° F).

ADVERSE EFFECTS (≥1%) **Hematologic:** Methemoglobinemia, especially in infants, small children, and patients with G6PD deficiency. **Skin:** *Blanching and redness,* itching, heat sensation. **Body as a**

Whole: Edema, soreness, aching, numbness, heaviness. **Other:** The adverse effects of lidocaine could occur with large doses or if there is significant systemic absorption.

INTERACTIONS Drug: May cause additive toxicity with CLASS I ANTIARRHYTHMICS; may increase risk of developing methemoglobin when used with **acetaminophen, chloroquine, dapsone, fosphenytoin,** NITRATES and NITRITES, **nitric oxide, nitrofurantoin, nitroprusside, pamaquine, phenobarbital, phenytoin, primaquine, quinine,** or SULFONAMIDES.

PHARMACOKINETICS Absorption: Penetrates intact skin. **Onset:** 15–60 min. **Peak:** 2–3 h. **Duration:** 1–2 h after removal of cream. **Distribution:** Crosses blood–brain barrier and placenta, distributed into breast milk. **Metabolism:** In liver. **Elimination:** 98% of absorbed dose is excreted in urine. **Half-Life:** 60–150 min.

NURSING IMPLICATIONS

Assessment & Drug Effects

- Monitor for local skin reactions including erythema, edema, itching, abnormal temperature sensations, and rash. These reactions are very common and usually disappear in 1–2 h.
- Note: Patients taking Class I antiarrhythmic drugs may experience toxic effects on the cardiovascular system. EMLA should be used with caution in these patients.
- Wash immediately with water or saline if contact with the eye occurs; protect the eye until sensation returns.

Patient & Family Education

- Skin analgesia lasts for 1 h following removal of the occlusive dressing. Analgesia may be accompanied by temporary loss of all sensation in the treated skin. Advise caution until sensation returns.

EMTRICITABINE

(em-tri′ci-ta-been)

Emtriva

Classifications: ANTIRETROVIRAL; NUCLEOSIDE REVERSE TRANSCRIPTASE INHIBITOR (NRTI)

Therapeutic: ANTIRETROVIRAL, NRTI

Prototype: Zidovudine

Pregnancy Category: D

AVAILABILITY 200 mg capsules; 10 mg/mL oral solution.

ACTION & *THERAPEUTIC EFFECT*
Emtricitabine is a synthetic nucleoside reverse transcriptase analogue with inhibitory activity against HIV-1. It inhibits HIV-1 reverse transcriptase (RT), both by competing with the natural DNA nucleoside and by incorporation into viral DNA, which terminates the formation of the viral DNA chain. *The viral load is decreased as measured by an increase in CD4 leukocyte count and suppression of viral RNA.*

USES Treatment of HIV in combination with other antiretroviral agents. **UNLABELED USES** Treatment of chronic hepatitis B in HIV-positive patients, HIV prophylaxis.

CONTRAINDICATIONS Suicidal ideation; HBV infection; pregnancy (category D); lactation. **CAUTIOUS USE** Renal impairment, and with end-stage renal disease; hepatic impairment; history of mental illness including bipolar disorder, psychosis; history of suicidal tendencies; alcoholism; substance abuse; seizure disorders; hypercholesterolemia, hypertriglyceridemia; children less than 3 mo.

Common adverse effects in *italic*, life-threatening effects underlined; generic names in **bold**; classifications in SMALL CAPS; ✚ Canadian drug name; ✪ Prototype drug

525

E

ROUTE & DOSAGE

HIV

Adult/Adolescent/Child (over 33 kg): PO 200 mg once/day
Child (3 mo–17 y): PO 6 mg/kg days (max: 240 mg/day)
Neonate/Infant (less than 3 mo): 3 mg/kg daily

Renal Impairment Dosage Adjustment

CrCl 30–49 mL/min: 200 mg q48h; 15–29 mL/min: 200 mg q72h; less than 15 mL/min: 200 mg q96h

ADMINISTRATION

Oral

- Give at the same time daily.
- Store between 15°–30° C (59°–86° F) in a tightly closed container.

ADVERSE EFFECTS (≥1%) Body as a Whole: Asthenia, neuropathy, peripheral neuritis. **CNS:** *Headache,* depression, dizziness, insomnia. **GI:** *Diarrhea, nausea,* dyspepsia, abdominal pain, hepatomegaly. **Metabolic:** Lactic acidosis. **Musculoskeletal:** Arthralgia, myalgia, paresthesias. **Respiratory:** Cough, rhinitis. **Skin:** *Rash,* hyperpigmentation of palms and soles of feet.

PHARMACOKINETICS Absorption: 93% reaches systemic circulation. **Peak:** 1–2 h. **Distribution:** 4% protein bound. **Metabolism:** In liver. **Elimination:** Urine. **Half-Life:** 10 h (active metabolite has intracellular half-life of 39 h).

NURSING IMPLICATIONS

Assessment & Drug Effects

- Monitor individuals with a history of depression for S&S of suicidal ideation.

- Monitor closely for S&S of lactic acidosis, especially in persons with known risk factors such as female gender, obesity, alcoholism, or hepatic disease.
- Withhold drug and notify prescriber if S&S suggestive of lactic acidosis or hepatotoxicity occur.
- Lab tests: Baseline renal function tests; frequent LFTs and serum electrolytes during the last trimester of pregnancy; complete blood chemistry if lactic acidosis is suspected; and periodic lipid profile; serum cholesterol and triglycerides; bone density monitoring for history of osteoporosis.
- Monitor closely for severe exacerbation of hepatitis B in coinfected patients if this drug is discontinued.

Patient & Family Education

- May cause serious CNS effects. Avoid driving or operating machinery until individual reaction to the drug is known.
- Report any of the following to the prescriber: Difficulty breathing, shortness of breath, fast or irregular heartbeat; weight gain with fullness around waist and/or face; vomiting or diarrhea; unexplained muscle aches, pains, weakness, or fatigue; yellow eyes or skin.
- Avoid alcoholic drinks while taking this drug.
- Do not self-treat nausea, vomiting, or stomach pain. Contact prescriber for guidance.

ENALAPRIL MALEATE ◯

(e-nal′a-pril)

Vasotec

ENALAPRILAT

Vasotec I.V.

Common adverse effects in *italic*, life-threatening effects <u>underlined</u>; generic names in **bold**; classifications in SMALL CAPS; ◆ Canadian drug name; ◯ Prototype drug

Classifications: ANGIOTENSIN-CONVERTING ENZYME (ACE) INHIBITOR; ANTIHYPERTENSIVE
Therapeutic: ANTIHYPERTENSIVE
Pregnancy Category: C first trimester; D second and third trimester

AVAILABILITY 2.5 mg, 5 mg, 10 mg, 20 mg tablets; 1.25 mg/mL injection

ACTION & *THERAPEUTIC EFFECT*
Angiotensin-converting enzyme (ACE) inhibitor that catalyzes the conversion of angiotensin I to angiotensin II, a vasoconstrictor substance. Therefore, inhibition of ACE decreases angiotensin II levels, thus decreasing vasopressor activity and aldosterone secretion. Both actions achieve an antihypertensive effect by suppression of the renin–angiotensin–aldosterone system. ACE inhibitors also reduce peripheral arterial resistance (afterload), pulmonary capillary wedge pressure (PCWP), a measure of preload, pulmonary vascular resistance, and improve cardiac output as well as exercise tolerance. *Antihypertensive effect lowers blood pressure. Improvement in cardiac output results in increased exercise tolerance.*

USES Management of mild to moderate hypertension. Malignant, refractory, accelerated, and renovascular hypertension (except in bilateral renal artery stenosis or renal artery stenosis in a solitary kidney), CHF.
UNLABELED USES Hypertension or renal crisis in scleroderma.

CONTRAINDICATIONS Hypersensitivity to enalapril or captopril; hypotension. There has been evidence of fetotoxicity and kidney damage in newborns exposed to ACE inhibitors during pregnancy (category D second and third trimester).
CAUTIOUS USE Renal impairment, renal artery stenosis; patients with hypovolemia, receiving diuretics; undergoing dialysis; hepatic disease; bone marrow suppression; patients in whom excessive hypotension would present a hazard (e.g., cerebrovascular insufficiency); CHF; aortic stenosis, cardiomyopathy hepatic impairment; diabetes mellitus; pregnancy (category C first trimester); infants and children with CrCl less than 30 mL/min/1.73 m²; lactation.

ROUTE & DOSAGE

Hypertension
Adult: **PO** 2.5–5 mg/day, may increase to 10–40 mg/day in 1–2 divided doses **IV** 0.625–1.25 mg q6h, may give up to 5 mg q6h in hypertensive emergencies.
Neonate: **PO** 0.1 mg/kg q24h
Child: **PO** 0.08 mg/kg/day in 1–2 divided doses, may increase (max: 5 mg/kg/day) **IV** 5–10 mcg/kg/dose q8–24h.

Congestive Heart Failure
Adult: **PO** 2.5 mg b.i.d., may increase up to 5–20 mg/day in 1–2 divided doses (max: 40 mg/day).

Renal Impairment Dosage Adjustment
Enalapril: CrCl less than 30 mL/min: Start with 2.5 mg dose then titrate
Enalaprilat: CrCl less than 30 mL/min: Start with dose of 0.625 mg q6h then titrate

Hemodialysis Dosage Adjustment
Administer post-dialysis or give 20–25% supplemental dose

Common adverse effects in *italic*, life-threatening effects <u>underlined</u>; generic names in **bold**; classifications in SMALL CAPS; ♣ Canadian drug name; ⊘ Prototype drug

527

E

ADMINISTRATION

Oral

- Discontinue diuretics, if possible, for 2–3 days prior to initial oral dose to reduce incidence of hypotension. If the diuretic were discontinued, give an initial dose of 2.5 mg. Keep patient under medical supervision for at least 2 h and until BP has stabilized for at least an additional hour.
- Give with food or drink of patient's choice.
- Protect from heat and light. Expiration date: 30 mo following date of manufacture if stored at less than 30° C.
- Store tablets at 30° C (86° F); protect from heat and light.

Intravenous

Note: Verify correct IV concentration and rate of infusion/injection with prescriber for neonates, infants, children.

PREPARE: **Direct:** Give undiluted. **Intermittent:** Dilute in 50 mL of D5W, NS, D5/NS, D5/LR.
ADMINISTER: **Direct/Intermittent:** Give direct IV slowly over at least 5 min through a port of a free flowing infusion of D5W or NS or as an infusion over 5 min.
- Longer infusion time decreases risk of severe hypotension.
INCOMPATIBILITIES **Y-site:** Amphotericin B, amphotericin B cholesteryl, caspofungin, cefepime, dantrolene, diazepam, lansoprazole, phenytoin.

ADVERSE EFFECTS (≥1%) CNS:
Headache, dizziness, fatigue, nervousness, paresthesias, asthenia, insomnia, somnolence. **CV:** *Hypotension including postural hypotension;* syncope, palpitations, chest pain. **GI:** Diarrhea, nausea, abdominal pain, loss of taste, dyspepsia. **Hematologic:** Decreased Hgb and Hct. **Urogenital:** <u>Acute kidney failure</u>, deterioration in kidney function. **Skin:** Pruritus with and without *rash,* angioedema, erythema. **Metabolic:** Hyperkalemia. **Respiratory:** Cough. **Whole Body:** <u>Angioedema</u>.

INTERACTIONS Drug: Indomethacin and other NSAIDS may decrease antihypertensive activity; POTASSIUM SUPPLEMENTS, POTASSIUM-SPARING DIURETICS may cause hyperkalemia; may increase **lithium** levels and toxicity.

PHARMACOKINETICS Absorption: 70% from GI tract. **Onset:** 1 h PO; 15 min IV. **Peak:** 4–8 h PO; 4 h IV. **Duration:** 12–24 h PO; 6 h IV. **Distribution:** Limited amount crosses blood–brain barrier; crosses placenta. **Metabolism:** PO dose undergoes first-pass metabolism in liver to active form, enalaprilat. **Elimination:** 60% in urine, 33% in feces within 24 h. **Half-Life:** 2 h.

NURSING IMPLICATIONS

Assessment & Drug Effects

- Monitor for therapeutic effectiveness. Peak effects after the first IV dose may not occur for up to 4 h; peak effects of subsequent doses may exceed those of the first.
- Maintain bedrest and monitor BP for the first 3 h after the initial IV dose. First-dose phenomenon (i.e., a sudden exaggerated hypotensive response) may occur within 1–3 h of first IV dose, especially in the patient with very high blood pressure or one on a diuretic and controlled salt intake regimen. An IV infusion of normal saline for volume expansion may be ordered to counteract the hypotensive response. This initial

response is not an indicator to stop therapy.

- Monitor BP for first several days of therapy. If antihypertensive effect is diminished before 24 h, the total dose may be given as 2 divided doses.
- Report transient hypotension with lightheadedness. Older adults are particularly sensitive to drug-induced hypotension. Supervise ambulation until BP has stabilized.
- Lab tests: Monitor serum potassium and be alert to symptoms of hyperkalemia (K^+ greater than 5.7 mEq/L). Patients who have diabetes, impaired kidney function, or CHF are at risk of developing hyperkalemia during enalapril treatment. Monitor kidney function closely during first few weeks of therapy.

Patient & Family Education

- Full antihypertensive effect may not be experienced until several weeks after enalapril therapy starts.
- When drug is discontinued due to severe hypotension, the hypotensive effect may persist a week or longer after termination because of long duration of drug action.
- Do not follow a low-sodium diet (e.g., low-sodium foods or low-sodium milk) without approval from prescriber.
- Avoid use of salt substitute (principal ingredient: potassium salt) and potassium supplements because of the potential for hyperkalemia.
- Notify prescriber of a persistent nonproductive cough, especially at night, accompanied by nasal congestion.
- Report to prescriber promptly if swelling of face, eyelids, tongue, lips, or extremities occurs. Angioedema is a rare adverse effect

and, if accompanied by laryngeal edema, may be fatal.

- Do not drive or engage in other potentially hazardous activities until response to drug is known.

ENFUVIRTIDE
(en-fu-vir′tide)
Fuzeon
Classifications:ANTIRETROVIRAL; FUSION INHIBITOR
Therapeutic: ANTIRETROVIRAL
Pregnancy Category: C

AVAILABILITY 90 mg/mL injection

ACTION & *THERAPEUTIC EFFECT*
Enfuvirtide interferes with entry of HIV-1 into host cells by inhibiting fusion of the virus with the host cell membranes. In order for HIV-1 to enter and infect a human cell, the viral surface glycoprotein (gp41) must bind to the host CD4+ cells. Then, the viral glycoprotein undergoes a change in shape facilitating the fusion of viral membranes with the host cell membrane. Prevents entry of the HIV-1 virus into host cells. *Effectiveness is measured in reduction of viral load as measured by an increase in CD4 leucocyte count and suppression of viral RNA.*

USES Treatment of advanced HIV disease with evidence of resistance to other therapies.

CONTRAINDICATIONS Hypersensitivity to enfuvirtide or any of its components; HIV/HBV co-infected patients; severe hepatomegaly; lactation.
CAUTIOUS USE Hypersensitivity to mannitol; renal and hepatic impairment; renal clearance of less than 35 mL/min; bacterial pneumonia,

Common adverse effects in *italic*, life-threatening effects underlined; generic names in **bold**; classifications in SMALL CAPS; ♦ Canadian drug name; ⊘ Prototype drug

529

E

low initial CD4 count, past history of lung disease, high initial viral load, IV drug use; history of pulmonary disease; pregnancy (category C).

ROUTE & DOSAGE

Advanced HIV Disease

Adult/Adolescent (16 y or older or weight 42.6 kg or more):
Subcutaneous 90 mg b.i.d.
Child/Adolescent (6–16 y or weight less than 42.6 kg):
Subcutaneous 2 mg/kg (up to 90 mg) b.i.d.

ADMINISTRATION

Subcutaneous

- Reconstitute by adding 1.1 mL sterile water for injection into vial. Mix by gently tapping vial for 10 sec, then gently rolling in palms of hands. Ensure that no drug is remaining on vial wall. Allow vial to stand until powder completely dissolves (up to 45 min). Solution should be clear, colorless, and without bubbles or particulate matter.
- Bring refrigerated reconstituted solution to room temperature before injection. Ensure that powder is fully dissolved and solution is clear, colorless, and without bubbles or particulate matter.
- Inject into upper arm, abdomen, or anterior thigh.
- Rotate injection sites and inject in an area with no current injection site reaction.
- Store unreconstituted vials at 15°–30° C (59°–86° F) or refrigerated at 2°–6° C (3°–46° F); do not freeze.

ADVERSE EFFECTS (≥1%) **Body as a Whole:** Injection site reactions (pain, induration, erythema, nodules, cysts, pruritus, ecchymoses), infection at injection site, fatigue, systemic hypersensitivity reactions,

Guillain-Barré syndrome, asthenia, herpes simplex infections, influenza, lymphadenopathy, myalgia, peripheral neuropathy. **CNS:** Anxiety, depression, insomnia. **GI:** Diarrhea, nausea, abdominal pain, anorexia, constipation, dysgeusia, pancreatitis, weight loss. **Hematologic:** Eosinophilia, anemia. **Metabolic:** Increased amylase, increased lipase, increased ALT and AST, hypertriglyceridemia. **Respiratory:** Bacterial pneumonia, acute respiratory distress syndrome, cough, sinusitis. **Skin:** Pruritus, skin papilloma. **Special Senses:** Conjunctivitis. **Urogenital:** Glomerulonephritis.

INTERACTIONS Increases levels of **tripranavir** (dose adjustment not needed).

PHARMACOKINETICS Absorption: 84.3% absorbed from subcutaneous site. **Peak:** Average 4–8 h. **Distribution:** 92% protein bound. **Metabolism:** Catabolized into constituent amino acids. **Half-Life:** 4 h.

NURSING IMPLICATIONS

Assessment & Drug Effects

- Inspect subcutaneous sites for S&S of site reactions (e.g., itching, swelling, redness, pain, tenderness, or hardened skin) that usually last for less than 7 days postinjection.
- Monitor closely for S&S of pneumonia, especially with low initial CD4 count, high initial viral load, IV drug use, smoking, or prior history of lung disease.
- Lab tests: Periodic LFTs, serum lipase and amylase, lipid profile, and CBC with differential.

Patient & Family Education

- Report promptly S&S of infection at subcutaneous injection sites: Increased heat, redness, pain, or oozing.

Common adverse effects in *italic*, life-threatening effects underlined; generic names in **bold**; classifications in SMALL CAPS; ✦ Canadian drug name; Ⓟ Prototype drug

- Report promptly S&S of pneumonia: Cough with fever, rapid breathing, shortness of breath.

✓ENOXAPARIN ⊙
(e-nox′a-pa-rin)

Lovenox
Classifications: ANTICOAGULANT; LOW MOLECULAR WEIGHT HEPARIN
Therapeutic: ANTICOAGULANT; ANTITHROMBOTIC
Pregnancy Category: B

AVAILABILITY 30 mg/0.3 mL, 40 mg/0.4 mL, 60 mg/0.6 mL, 80 mg/0.8 mL, 100 mg/1 mL injection

ACTION & *THERAPEUTIC EFFECT* Low molecular weight heparin with antithrombotic properties. Does affect thrombin time (TT) and activated thromboplastin time (aPTT) up to 1.8 times the control value. Antithrombotic properties are due to its antifactor Xa and antithrombin (antifactor IIa) in the coagulation activities. *An effective anticoagulation agent, it is used for prophylactic treatment as an antithrombotic agent following certain types of surgery.*

USES Prevention of deep vein thrombosis (DVT) after hip, knee, or abdominal surgery, treatment of DVT and pulmonary embolism, management of acute ST elevation myocardial infarction (STEMI), non-Q wave MI.

CONTRAINDICATIONS Hypersensitivity to enoxaprin, porcine protein hypersensitivity, active major bleeding, GI bleeding, hemophilia, heparin hypersensitivity, heparin-induced thrombocytopenia (HIT), thrombocytopenia associated with an antiplatelet antibody in the presence of enoxaparin, bleeding disorders, idiopathic thrombocytopenic purpura (ITP).

CAUTIOUS USE Uncontrolled arterial hypertension, recent history of GI disease, conditions or surgery with increased risk of bleeding, hepatic disease, hypertension, coagulopathy, thrombocytopenia, dental disease, diabetic retinopathy, dialysis, diverticulitis, inflammatory bowel disease, peptic ulcer disease, older adults, endocarditis, renal disease, renal impairment, stroke, surgery, pregnancy (category B), lactation. Safe use in neonates, infants, and children has not been established.

ROUTE & DOSAGE

Prevention of DVT after Hip or Knee Surgery
Adult: **Subcutaneous** 30 mg b.i.d. for 10–14 days starting 12–24 h post-surgery

Prevention of DVT after Abdominal Surgery
Adult: **Subcutaneous** 40 mg daily starting 2 h before surgery and continuing for 7–10 days (max: 12 days)

Treatment of DVT and Pulmonary Embolus
Adult: **Subcutaneous** 1 mg/kg q12h or 1.5 mg/kg/day; monitor anti-Xa activity to determine appropriate dose

Non-Q Wave MI
Adult: **Subcutaneous** 1 mg/kg q12h for 2–8 days, give concurrently with aspirin 100–325 mg/day

Acute STEMI
Adult: **IV** 30 mg bolus plus 1 mg/kg subcutaneously, then 1 mg/kg q12h subcutaneously

Common adverse effects in *italic*, life-threatening effects underlined; generic names in **bold;** classifications in SMALL CAPS; ♦ Canadian drug name; ⊙ Prototype drug

E

Renal Impairment Dosage Adjustment

CrCl less than 30 mL/min: 30 mg or 1 mg/kg q24h

ADMINISTRATION

Subcutaneous

- Use a TB syringe or prefilled syringe to ensure accurate dosage.
- Do not expel the air bubble from the 30 or 40 mg prefilled syringe before injection.
- Place patient in a supine position for injection of the drug.
- Alternate injections between left and right anterolateral and posterolateral abdominal wall.
- Hold the skin fold between the thumb and forefinger and insert the whole length of the needle into the skin fold. Hold skin fold throughout the injection. Do not rub site post-injection.
- Store at 15°–30° C (59°–86° F).

Intravenous

PREPARE: **Direct:** Give undiluted.
ADMINISTER: **Direct:** Give bolus dose direct IV through an IV line. Flush before and after with NS or D5W to ensure that the IV line has been cleared. Do not mix with any other drugs or solutions.

ADVERSE EFFECTS (≥1%) **Body as a Whole:** Allergic reactions (rash, urticaria), fever, angioedema, arthralgia, pain and inflammation at injection site, peripheral edema, fever. **Digestive:** Abnormal liver function tests. **Hematologic:** Hemorrhage, thrombocytopenia, ecchymoses, anemia. **Respiratory:** Dyspnea. **Skin:** Rash, pruritus.

INTERACTIONS Drug: Aspirin, NSAIDS, **warfarin** can increase risk of hemorrhage. **Herbal: Garlic, ginger, ginkgo, feverfew** may increase risk of bleeding.

PHARMACOKINETICS Absorption: 91% from subcutaneous injection site. **Peak:** 3 h. **Duration:** 4.6 h. **Distribution:** Accumulates in liver, kidneys, and spleen. Does not cross placenta. **Elimination:** Primarily in urine. **Half-Life:** 4.6 h.

NURSING IMPLICATIONS

Assessment & Drug Effects

- Lab tests: Baseline coagulation studies; periodic CBC, platelet count, urine and stool for occult blood.
- Monitor platelet count closely. Withhold drug and notify prescriber if platelet count less than 100,000/mm³.
- Monitor closely patients with renal insufficiency and older adults who are at higher risk for thrombocytopenia.
- Monitor for and report immediately any sign or symptom of unexplained bleeding.

Patient & Family Education

- Report to prescriber promptly signs of unexplained bleeding such as: Pink, red, or dark brown urine; red or dark brown vomitus; bleeding gums or bloody sputum; dark, tarry stools.
- Do not take any OTC drugs without first consulting prescriber.

ENTACAPONE

(en-ta′ca-pone)

Comtan

Classifications: CATECHOLAMINE O-METHYLTRANSFERASE (COMT) INHIBITOR; ANTIPARKINSON
Therapeutic: ANTIPARKINSON
Prototype: Tolcapone
Pregnancy Category: C

AVAILABILITY 200 mg tablets

ACTION & *THERAPEUTIC EFFECT*
Selective inhibitor of catechola-mine O-methyltransferase (COMT). COMT is responsible for metabolizing levodopa to an intermediate compound 3-O-methyldopa, a chemical which interferes with the availability of levodopa to the brain. Therefore, it increases availability of levodopa in CNS. *Taken with levodopa, it decreases formation of 3-O-methyldopa, thus increasing the duration of the motor response of the brain to levodopa in Parkinson's disease, diminishing its manifestations.*

USES Adjunct to levodopa/carbidopa to treat Parkinson's disease.

CONTRAINDICATIONS Hypersensitivity to entacapone; concurrent MAO inhibitors; children.
CAUTIOUS USE Hepatic impairment; biliary obstruction; history of hypotension or syncope; pregnancy (category C), lactation.

ROUTE & DOSAGE

Parkinson's Disease
Adult: PO 200 mg administered with each dose of levodopa/ carbidopa up to 8 times/day

ADMINISTRATION
Oral
- Give simultaneously with each levodopa/carbidopa dose.
- **Must be** tapered if discontinued. Never discontinue abruptly.
- Do not administer to patients receiving nonselective MAO inhibitors.
- Store at 15°–30° C (59°–86° F).

ADVERSE EFFECTS (≥1%) **Body as a Whole:** Back pain, fatigue, asthenia. **CNS:** *Dyskinesia, hyperkinesia,* hypokinesia, dizziness, anxiety, somnolence, agitation. **GI:** Taste perversion, *nausea, diarrhea,* abdominal pain, constipation, vomiting, dry mouth, dyspepsia, flatulence, gastritis. **Respiratory:** Dyspnea. **Skin:** Increased sweating. **Other:** *Urine discoloration,* purpura.

INTERACTIONS Drug: Extreme caution **must be** used if administered with a nonselective MAOI; **bitol-terol, dobutamine, dopamine, epinephrine, isoetharine, iso-proterenol, methyldopa, nor-epinephrine** may increase heart rates, possibly cause arrhythmias, excessive changes in BP.

PHARMACOKINETICS Absorption: Rapidly absorbed, 35% bioavailable. **Peak:** 1 h. **Distribution:** Highly protein bound. **Metabolism:** Extensively metabolized in plasma and erythrocytes. **Elimination:** Primarily in feces. **Half-Life:** 2.4 h (terminal).

NURSING IMPLICATIONS
Assessment & Drug Effects
- Monitor carefully for hyperpyrexia, confusion, or emergence of Parkinson's S&S during drug withdrawal.
- Monitor for orthostatic hypotension and worsening of dyskinesia or hyperkinesia.
- Lab tests: Hgb and serum ferritin levels with prolonged therapy.

Patient & Family Education
- Take with levodopa/carbidopa; not effective alone.
- Do not discontinue abruptly; gradually reduce dosage.
- Exercise caution when rising from a sitting or lying position because faintness/dizziness can occur.
- Exercise caution with hazardous activities until reaction to the drug is known.
- Harmless brownish-orange discoloration of urine is possible.

E

Common adverse effects in *italic*, life-threatening effects <u>underlined</u>; generic names in **bold**; classifications in SMALL CAPS; ✤ Canadian drug name; ◎ Prototype drug

533

- Report unusual adverse effects (e.g., hallucinations/unexplained diarrhea).

ENTECAVIR

(en-te'ca-vir)

Baraclude

Classifications: ANTIRETROVIRAL; NUCLEOSIDE REVERSE TRANSCRIPTASE INHIBITOR (NRTI)

Therapeutic: ANTIRETROVIRAL; NRTI

Prototype: Lamivudine

Pregnancy Category: C

AVAILABILITY 0.5, 1 mg tablets; 0.05 mg/mL oral solution

ACTION & *THERAPEUTIC EFFECT*
Inhibits hepatitis B viral (HBV) DNA polymerase by inhibiting viral reverse transcriptase of messenger RNA that ultimately results in inhibiting the synthesis of HBV DNA. *The antiviral activity of entecavir inhibits HBV DNA synthesis.*

USES Hepatitis B infection.

CONTRAINDICATIONS Hypersensitivity to entecavir; lactic acidosis; severe hepatomegaly; HIV/HVB co-infected patients, if HIV is not being treated with highly active antiretroviral therapy; lactation.

CAUTIOUS USE Liver transplant patients; HIV patients; renal impairment, ESRF, dialysis; older adults; labor and delivery; pregnancy (category C). Safety and effectiveness in children younger than 16 y have not been established.

ROUTE & DOSAGE

Chronic Hepatitis B (nucleoside-treatment–naïve patients)

Adult/Adolescent (16 y or older):
PO 0.5 mg daily

Chronic Hepatitis B (lamivudine or telbivudine resistant patients/ chronic hepatitis B with decompensated liver disease)

Adult/Adolescent (16 y or older):
PO 1 mg daily

Renal Impairment Dosage Adjustment

CrCl 30–49 mL/min: Decrease dose by 50%; *10–29 mL/min:* Decrease dose by 70%; *less than 10 mL/min:* Decrease dose by 90%

ADMINISTRATION

Oral

- Give on an empty stomach (at least 2 h before/after a meal).
- Administer after hemodialysis.
- Store in a tightly closed container at 15°–30° C (59°–86° F)

ADVERSE EFFECTS (≥1%) CNS: Dizziness, fatigue, headache, insomnia, somnolence. **GI:** Diarrhea, dyspepsia, nausea, vomiting. **Metabolic:** Elevated liver enzymes (ALT, AST), hyperamylasemia, elevated lipase concentration, hyperbilirubinemia, fasting hyperglycemia, glycosuria, hematuria, lactic acidosis.

INTERACTIONS Drug: Use of entecavir with drugs that reduce renal function or compete for active tubular secretion may increase serum concentrations of either drug. **Food:** High-fat meal reduces oral absorption.

PHARMACOKINETICS Peak: 0.5–1 h. **Distribution:** 13% protein bound. **Metabolism:** Minimal. **Elimination:** Primarily in the urine. **Half-Life:** 128–149 h.

NURSING IMPLICATIONS

Assessment & Drug Effects

- Monitor closely for adverse reactions when drugs that are known

Common adverse effects in *italic*, life-threatening effects underlined; generic names in **bold**; classifications in SMALL CAPS; ◆ Canadian drug name; ⊘ Prototype drug

to affect renal function are taken concurrently.
- Lab tests: Periodic LFTs during treatment and for several months after drug is discontinued; periodic fasting plasma glucose.
- Monitor for S&S of lactic acidosis, including respiratory distress, tachycardia, and irregular HR.

Patient & Family Education
- Follow directions for taking the drug (see ADMINISTRATION).
- Do not discontinue medication without consent of prescriber.
- Do not drive or engage in potentially hazardous activities until response to drug is known.
- Inform prescriber if you are or plan to become pregnant.
- Report any of the following to a health care provider: Unexplained tiredness or weakness, unusual muscle pain, difficulty breathing, cold extremities, dizziness or lightheadedness, irregular heartbeat, loss of appetite, stomach pain, nausea, vomiting, clay-colored stool, dark urine, or jaundice.

EPHEDRINE SULFATE
(e-fed'rin sul-fate)

Classifications: ALPHA- AND BETA-ADRENERGIC AGONIST; BRON-CHODILATOR
Therapeutic: BRONCHODILATOR
Prototype: Epinephrine HCl
Pregnancy Category: C

AVAILABILITY 25 mg capsules; 50 mg/mL injection

ACTION & THERAPEUTIC EFFECT
Both indirect- and direct-acting sympathomimetic amine thought to act indirectly by releasing tissue stores of norepinephrine and directly by stimulation of alpha-, beta$_1$-, and beta$_2$-adrenergic receptors. Like epinephrine, contracts dilated arterioles of nasal mucosa, thus reducing engorgement and edema and facilitating ventilation and drainage. *Ephedrine relaxes bronchial smooth muscle, relieving mild bronchospasm, improving air exchange and increasing vital capacity.*

USES In treatment of mild cases of acute asthma and in patients with chronic asthma requiring continuing treatment and for temporary support of ventricular rate in Adams-Stokes syndrome.

CONTRAINDICATIONS History of hypersensitivity to ephedrine or other sympathomimetics; narrow-angle glaucoma; angina pectoris, coronary insufficiency, chronic heart disease, uncontrolled hypertension, cardiac arrhythmias, cardiomyopathy; hypovolemia; concurrent MAOI therapy; lactation. **CAUTIOUS USE** Hypertension, arteriosclerosis, closed-angle glaucoma; diabetes mellitus; hyperthyroidism; prostatic hypertrophy; pregnancy (category C).

ROUTE & DOSAGE

Bronchodilator, Nasal Decongestant
Adult: **PO** 25–50 mg q3–4h prn (max: 150 mg/24 h) **IM/IV/Subcutaneous** 12.5–25 mg
Child (older than 2 y): **PO** 2–3 mg/kg/day in 4–6 divided doses

Hypotension
Adult: **IM/Subcutaneous/IV** 5–25 mg slow IV, may repeat in 5–10 min if necessary (max: 150 mg/24 h)

Common adverse effects in *italic*, life-threatening effects underlined; generic names in **bold**; classifications in SMALL CAPS; ♣ Canadian drug name; ⊕ Prototype drug

535

E

Child: **IM/SC** 0.5 mg/kg or 16.7 mg/m² q4–6h

Narcolepsy

Adult: **PO** 25–50 mg q3–4h

Enuresis

Adult: **PO** 25 mg at bedtime

ADMINISTRATION

Oral

- Administer last dose a few hours before bedtime, if possible, to minimize insomnia.
- Store at 15°–30° C (59°–86° F) in tightly closed, light-resistant containers unless otherwise directed by the manufacturer.

Subcutaneous/Intramuscular

- Give undiluted.

Intravenous

PREPARE: **Direct:** Give undiluted.
ADMINISTER: **Direct:** Direct IV at a rate of 10 mg or fraction thereof over 30–60 sec.
INCOMPATIBILITIES **Solution/additive: Hydrocortisone, pentobarbital, phenobarbital, secobarbital, thiopental. Y-site: Amphotericin B liposome, azathioprine, caspofungin, dantrolene, diazepam, diazoxide, ganciclovir, pantoprazole, pentamidine, SMZ/TMP, thiopental.**

- Store in tightly closed, light-resistant containers. Do not use liquid formulation unless it is absolutely clear.

ADVERSE EFFECTS (≥1%) **CNS:** Headache, insomnia, *nervousness,* anxiety, tremulousness, giddiness. **CV:** Palpitation, tachycardia, precordial pain, cardiac arrhythmias. **GU:** Difficult or painful urination, acute urinary retention (especially older men with prostatism). **GI:** Nausea, vomiting, anorexia. **Body as a**

Whole: Sweating, thirst, overdosage: Euphoria, confusion, delirium, convulsions, pyrexia, hypertension, rebound hypotension, respiratory difficulty. **Skin:** Fixed-drug eruption. **Topical Use:** *Burning, stinging,* dryness of nasal mucosa, sneezing, rebound congestion.

DIAGNOSTIC TEST INTERFERENCE Ephedrine is generally withdrawn at least 12 h before *sensitivity tests* are made to prevent false-positive reactions.

INTERACTIONS Drug: MAO INHIBITORS, TRICYCLIC ANTIDEPRESSANTS, **furazolidone, guanethidine** may increase alpha-adrenergic effects (headache, hyperpyrexia, hypertension); **sodium bicarbonate** decreases renal elimination of ephedrine, increasing its CNS effects; **epinephrine, norepinephrine** compound sympathomimetic effects; effects of ALPHA and BETA-BLOCKERS and ephedrine antagonized.

PHARMACOKINETICS Absorption: Readily absorbed from GI tract. **Peak:** 15 min–1 h. **Duration:** Bronchodilation 2–4 h; cardiac and pressor effects up to 4 h PO and 1 h IV. **Distribution:** Widely distributed; crosses blood–brain barrier and placenta; distributed into breast milk. **Metabolism:** Small amounts metabolized in liver. **Elimination:** In urine. **Half-Life:** 3–6 h.

NURSING IMPLICATIONS

Assessment & Drug Effects

- Supervise continuously patients receiving ephedrine IV. Take baseline BP and other vital signs. Check BP repeatedly during first 5 min, then q3–5min until stabilized.
- Monitor I&O ratio and pattern, especially in older male patients. Encourage patient to void before

Common adverse effects in *italic*, life-threatening effects <u>underlined</u>; generic names in **bold;** classifications in SMALL CAPS; ✚ Canadian drug name; ❂ Prototype drug

taking medication (see ADVERSE EFFECTS).

Patient & Family Education
- Note: Ephedrine is a commonly abused drug. Learn adverse effects and dangers; take medication ONLY as prescribed.
- Do not take OTC medications for coughs, colds, allergies, or asthma unless approved by prescriber. Ephedrine is a common ingredient in these preparations.

EPINASTINE HYDROCHLORIDE

(e-pi-nas′teen)
Elestat
Pregnancy Category: C
See Appendix A-1.

EPINEPHRINE ⊕

(ep-i-ne′frin)
Epinephrine Pediatric, EpiPen Auto-Injector

EPINEPHRINE BITARTRATE

AsthmaHaler, Bronkaid Mist Suspension, Bronitin Mist Suspension, Epitrate

EPINEPHRINE HYDROCHLORIDE

Adrenalin Chloride, Bronkaid Mistometer, Dysne-Inhal, Epifrin, SusPhrine ♦

EPINEPHRINE, RACEMIC

Vaponefrin ♦
Classifications: ALPHA- AND BETA-ADRENERGIC AGONIST; CARDIAC STIMULANT; VASOPRESSOR
Therapeutic: ANTI-ANAPHYLACTIC; VASOPRESSOR
Pregnancy Category: C

AVAILABILITY 1:100, 1:1000, 2.25% solution for inhalation; 0.35 mg, 0.2 mg spray; 1:1000, 1:2000, 1:10,000, 1:100,000 injection; 1:200 suspension; 0.1%, 0.5%, 1%, 2% ophth solution; 0.1% nasal solution

ACTION & THERAPEUTIC EFFECT A catecholamine that acts directly on both alpha and beta receptors; it is the most potent activator of alpha receptors. Strengthens myocardial contraction; increases systolic but may decrease diastolic blood pressure; increases cardiac rate and cardiac output. Constricts bronchial arterioles and inhibits histamine release, thus reducing congestion and edema and increasing tidal volume and vital capacity. Relaxes uterine smooth musculature and inhibits uterine contractions. *Reverses anaphylatic reactions and provides temporary relief from acute asthmatic attack. Restores normal cardiac rhythm.*

USES Temporary relief of bronchospasm, acute asthmatic attack, mucosal congestion, hypersensitivity and anaphylactic reactions, syncope due to heart block or carotid sinus hypersensitivity, and to restore cardiac rhythm in cardiac arrest. Relaxes myometrium and inhibits uterine contractions; prolongs action and delays systemic absorption of local and intraspinal anesthetics. Used topically to control superficial bleeding. Ophthalmic preparation is used in management of simple (open-angle) glaucoma, generally as an adjunct to topical miotics and oral carbonic anhydrase inhibitors; also used as ophthalmic decongestant.

CONTRAINDICATIONS Hypersensitivity to sympathomimetic amines; narrow-angle glaucoma; hemorrhagic, traumatic, or cardiogenic shock; cardiac dilatation, cerebral arteriosclerosis, coronary insufficiency,

Common adverse effects in *italic*, life-threatening effects underlined; generic names in **bold**; classifications in SMALL CAPS; ♦ Canadian drug name; ⊕ Prototype drug

537

arrhythmias, organic heart or brain disease; during second stage of labor; for local anesthesia of fingers, toes, ears, nose, genitalia.

CAUTIOUS USE Older adults or debilitated patients; prostatic hypertrophy; hypertension; diabetes mellitus; hyperthyroidism; Parkinson's disease; tuberculosis; psychoneurosis; in patients with long-standing bronchial asthma and emphysema with degenerative heart disease; pregnancy (category C), lactation.

ROUTE & DOSAGE

Anaphylaxis
Adult: **Subcutaneous** 0.1–0.5 mg q10–15min prn **IV** 0.1–0.25 mg q5–15min
Child: **Subcutaneous** 0.01 mL/kg of, 1:1000 q10–15min prn **IV** 0.01 mL/kg of 1:1000 q10–15min
Neonate: **IV Intratracheal** 0.01–0.05 mg/kg q3–5min prn

Cardiac Arrest
Adult: **IV** 1 mg q3–5min as needed **Intracardiac** 0.1–1 mg
Child: **IV** 0.01 mg/kg q3–5min as needed (max: 1 mg) **Intracardiac** 0.05–0.1 mg/kg

Asthma
Adult: **Subcutaneous** 0.1–0.5 mg q20min–4h **Inhalation** 1 inhalation q4h prn
Child: **Subcutaneous** 0.01 mL/kg of 1:1000 q20min–4h **Inhalation** 1 inhalation q4h prn

Glaucoma
Adult/Child: **Instillation** 1–2 drops 0.25–2% solution 1/day or b.i.d.

Ocular Mydriasis
See Appendix A-1.

Nasal Hemostasis
Adult/Child: **Instillation** 1–2 drops 0.1% ophthalmic or 0.1% nasal solution

Topical Hemostatic
Adult/Child: **Topical** 1:50,000–1:1000 applied topically or 1:500,000–1:50,000 mixed with a local anesthetic

ADMINISTRATION

Inhalation
- Have patient in an upright position when aerosol preparation is used. The reclining position can result in overdosage by producing large droplets instead of fine spray.
- Instruct patient to rinse mouth and throat with water immediately after inhalation to avoid swallowing residual drug (may cause epigastric pain and systemic effects from the propellant in the aerosol preparation) and to prevent dryness of oropharyngeal membranes.

Instillation
- Instill nose drops with head in lateral, head-low position to prevent entry of drug into throat.
- Instruct patient to rinse nose dropper or spray tip with hot water after each use to prevent contamination of solution with nasal secretions.

Ophthalmic
- Remove soft contact lenses before instilling eye drops.
- Instruct patient to apply gentle finger pressure against nasolacrimal duct immediately after drug is instilled for at least 1 or 2 min following instillation to prevent excessive systemic absorption.

Subcutaneous
- Use tuberculin syringe to ensure greater accuracy in measurement of parenteral doses.

Common adverse effects in *italic*; life-threatening effects underlined; generic names in **bold**; classifications in SMALL CAPS; ♣ Canadian drug name; ⊘ Prototype drug

- Protect epinephrine injection from exposure to light at all times. Do not remove ampule or vial from carton until ready to use.
- Shake vial or ampule thoroughly to disperse particles before withdrawing epinephrine suspension into syringe; then inject promptly.
- Aspirate carefully before injecting epinephrine. Inadvertent IV injection of usual subcutaneous doses can result in sudden hypertension and possibly cerebral hemorrhage.
- Rotate injection sites and observe for signs of blanching. Vascular constriction from repeated injections may cause tissue necrosis.

Intravenous

Note: Verify correct rate of IV injection to neonates, infants, children with prescriber.

- Note: 1:1000 solution contains 1 mg/1 mL. 1:10,000 solution contains 0.1 mg/1 mL.

PREPARE: **Direct:** Dilute each 1 mg of 1:1000 solution with 10 mL of NS to yield 1:10,000 solution.
- The 1:10,000 solution may be given undiluted. **IV Infusion:** Dilute required dose in 250–500 mL of D5W.

ADMINISTER: **Direct:** Give each 1 mg over 1 min or longer; may give more rapidly in cardiac arrest. **IV Infusion:** 1–10 mcg/min titrated according to patient's condition.

INCOMPATIBILITIES **Solution/additive: Aminophylline, cephapirin, hyaluronidase, mephentermine, sodium bicarbonate, warfarin. Y-site: Ampicillin, thiopental, sodium bicarbonate.**

ADVERSE EFFECTS (≥1%) **Special Senses:** *Nasal burning or stinging,* dryness of nasal mucosa, sneezing, rebound congestion. *Transient stinging or burning of eyes,* lacrimation, brow ache, headache, rebound conjunctival hyperemia, allergy, iritis; with prolonged use: Melanin-like deposits on lids, conjunctiva, and cornea; corneal edema; loss of lashes (reversible); maculopathy with central scotoma in aphakic patients (reversible). **Body as a Whole:** *Nervousness,* restlessness, sleeplessness, fear, anxiety, *tremors,* severe headache, cerebrovascular accident, weakness, dizziness, syncope, pallor, sweating, dyspnea. **GI:** Nausea, vomiting. **CV:** Precordial pain, *palpitations,* hypertension, <u>MI</u>, tachyarrhythmias including <u>ventricular fibrillation</u>. **Respiratory:** Bronchial and <u>pulmonary edema</u>. **Urogenital:** Urinary retention. **Skin:** Tissue necrosis with repeated injections. **Metabolic:** Metabolic acidoses, elevated serum lactic acid, transient elevations of blood glucose. **CNS:** Altered state of perception and thought, psychosis.

INTERACTIONS Drug: May increase hypotension in circulatory collapse or hypotension caused by PHENOTHIAZINES, **oxytocin, entacapone.** Additive toxicities with other SYMPATHOMIMETICS **(albuterol, dobutamine, dopamine, isoproterenol, meta-proterenol, norepinephrine, phenylephrine, phenylpropanolamine, pseudoephedrine, ritodrine, salmeterol, terbutaline),** MAO INHIBITORS, TRYCYCLIC ANTIDEPRESSANTS. ALPHA- AND BETA-ADRENERGIC BLOCKING AGENTS (e.g., **ergotamine, propranolol**) antagonize effects of epinephrine. GENERAL ANESTHETICS increase cardiac irritability.

PHARMACOKINETICS Absorption: Inactivated in GI tract. **Onset:** 3–5 min, 1 h on conjunctiva. **Peak:** 20 min, 4–8 h on conjunctiva. **Duration:** 12–24 h topically. **Distribution:** Widely

Common adverse effects in *italic*, life-threatening effects <u>underlined</u>; generic names in **bold**; classifications in SMALL CAPS; ♣ Canadian drug name; ⊙ Prototype drug

539

distributed; does not cross blood–brain barrier; crosses placenta. **Metabolism:** In tissue and liver by monoamine oxidase (MAO) and catecholamine-methyltransferase (COMT). **Elimination:** Small amount unchanged in urine; excreted in breast milk.

NURSING IMPLICATIONS

Assessment & Drug Effects

- Check BP repeatedly when epinephrine is administered IV during first 5 min, then q3–5min until stabilized.
- Monitor BP, pulse, respirations, and urinary output and observe patient closely following IV administration. Continuous cardiac monitoring is recommended during IV infusion. If disturbances in cardiac rhythm occur, withhold epinephrine and notify prescriber immediately.
- Keep prescriber informed of any changes in intake-output ratio.
- Advise patient to report bronchial irritation, nervousness, or sleeplessness. Dosage should be reduced.
- Monitor blood glucose and HbA1C for loss of glycemic control if diabetic.

Patient & Family Education

- Report to prescriber if symptoms of asthma are not relieved in 20 min or if they become worse following inhalation.
- Be aware intranasal application may sting slightly.
- Administer ophthalmic drug at bedtime or following prescribed miotic to minimize mydriasis, with blurred vision and sensitivity to light (possible in some patients being treated for glaucoma).
- Transitory stinging may follow initial ophthalmic administration and that headache and browache occur

frequently at first but usually subside with continued use. Notify prescriber if symptoms persist.
- Discontinue epinephrine eye drops and consult a prescriber if signs of hypersensitivity develop (edema of lids, itching, discharge, crusting eyelids).
- Learn how to administer epinephrine subcutaneously. Keep medication and equipment available for home emergency. Confer with prescriber.
- Advise patient to report bronchial irritation, nervousness, or sleeplessness. Dosage should be reduced.
- Report tolerance to prescriber; may occur with repeated or prolonged use. Continued use of epinephrine in the presence of tolerance can be dangerous.
- Take medication only as prescribed and immediately notify prescriber of onset of systemic effects of epinephrine.
- Discard discolored or precipitated solutions.

EPIRUBICIN HYDROCHLORIDE

(e-pi-roo′bi-sin)

Ellence

Classifications: ANTINEOPLASTIC; ANTRACYCLINE; (ANTIBIOTIC)
Therapeutic: ANTINEOPLASTIC
Prototype: Doxorubicin HCl
Pregnancy Category: D

AVAILABILITY 2 mg/mL

ACTION & THERAPEUTIC EFFECT Cytotoxic antibiotic with wide spectrum of antitumor activity and strong immunosuppressive properties. Complexes with DNA causing the DNA helix to change shape, thus blocking effective DNA and RNA transcription. *Highly* *destructive to rapidly proliferating cells. Effectiveness indicated by tumor regression.*

USES Adjunctive therapy for axillary node-positive breast cancer.

CONTRAINDICATIONS Hypersensitivity to epirubicin and other related drugs; marked myelosuppression, severely impaired cardiac function, severe cardiac arrhythmias, recent MI; severe hepatic disease, jaundice; previous treatment with maximum doses of epirubicin, doxorubicin, or daunorubicin; pregnancy (category D), lactation.
CAUTIOUS USE Arrhythmias; mild or moderate liver dysfunction; severe renal insufficiency or renal failure.

ROUTE & DOSAGE

Breast Cancer

Adult: IV 100–120 mg/m² infused on day 1 of a 3–4 wk cycle or 50–60 mg/m² on day 1 and 8 of a 3–4 wk cycle (max cumulative dose: 900 mg/m²)

Hepatic Impairment Dosage Adjustment

Bilirubin 1.2–3 mg/dL: Give 50% of dose; *bilirubin over 3 mg/dL:* Give 25% of dose; *bilirubin greater than 5 mg/dL:* Skip dose

Toxicity Dosage Adjustment

Reduce dose by 25% if platelets less than 50,000/mm³, ANC less than 250/mm³, neutropenic fever, or grade 3 or 4 hematologic toxicity

ADMINISTRATION

Intravenous

Note: Pregnant women should **NOT** prepare or administer this drug. Wear protective goggles, gowns and disposable gloves and masks when handling this drug. Discard **ALL** equipment used in preparation of this drug in high-risk, waste-disposal bags for incineration. Treat accidental contact with skin or eyes by rinsing with copious amounts of water followed by prompt medical attention.

- Note: Reduce dosages when serum creatinine greater than 5 mg/dL or AST 2–4 times the upper limit of normal.

PREPARE: IV Infusion: Epirubicin is manufactured as a preservative free ready-to-use solution. The contents of a vial **must be** used within 24 h of first penetrating the rubber stopper. Discard unused solution.
ADMINISTER: IV Infusion: Measure ordered dose and inject into a port of a freely flowing IV solution of D5W or NS over 3–20 min. ▪ **DO NOT** give by direct IV push into a vein. ▪ Avoid IV sites that enter small veins or repeated injections into the same vein. ▪ Monitor IV site closely for S&S of extravasation and if suspected, notify prescriber immediately.
INCOMPATIBILITIES Solution/additive: ALKALINE SOLUTIONS (including **sodium bicarbonate**), **fluorouracil, heparin.**

- Store between 2°–8° C (36°–46° F). Protect from light.

ADVERSE EFFECTS (≥1%) **Body as a Whole:** *Lethargy,* fever. **CV:** Asymptomatic decrease in LVEF, CHF. **GI:** *Nausea, vomiting, mucositis, diarrhea,* anorexia. **Hematologic:** Leukopenia, neutropenia, anemia, thrombocytopenia, AML. **Skin:** *Alopecia, injection site reaction,*

Common adverse effects in *italic*, life-threatening effects <u>underlined</u>; generic names in **bold;** classifications in SMALL CAPS; ♣ Canadian drug name; ❷ Prototype drug

541

rash, itching, skin changes. **Other:** *Amenorrhea, hot flashes, infection, conjunctivitis/keratitis,* <u>secondary acute myelogenous leukemia</u> (related to cumulative dose).

INTERACTIONS Drug: **Cimetidine** increases epirubicin levels; concomitant use with cardioactive drugs (e.g., CALCIUM CHANNEL BLOCKERS) may affect cardiac function.

PHARMACOKINETICS Distribution: Widely distributed, 77% protein bound, concentrated in red blood cells. **Metabolism:** Extensively in liver, blood and other organs. Clearance is reduced in patients with hepatic impairment. **Elimination:** Primarily in bile, some urinary excretion; clearance decreases in older adult female patients. **Half-Life:** 33 h.

NURSING IMPLICATIONS

Assessment & Drug Effects

- Withhold drug and notify prescriber of any of the following: Neutrophil count less than 1500 cells/mm³, recent MI, suspicion of severe myocardial insufficiency.
- Obtain baseline and periodic (before each cycle of therapy) cardiac evaluation: Left ventricular ejection fraction, ECG and ECHO (tests are recommended especially in the presence of risk factors of cardiac toxicity).
- Monitor cardiac status closely throughout therapy as the risk of developing severe CHF increases rapidly when cumulative doses approach 900 mg/m². Report significant ECG changes immediately. Report immediately S&S of the following: Tachycardia, gallop rhythm, pleural effusion, pulmonary edema, dependent edema, ascites, or hepatomegaly.

- Lab tests: Baseline and periodic (before each cycle of therapy) CBC with differential, platelet count, serum total bilirubin, AST, serum creatinine.

Patient & Family Education

- Review all literature regarding the adverse effects of epirubicin therapy carefully.
- Report any of the following to prescriber immediately: Pain at the site of IV infusion, chest pain, palpitations, shortness of breath or difficulty breathing, sudden weight gain, swelling of hands, feet or legs, or any unexplained bleeding.
- Be aware that your urine may turn red for 1–2 days after receiving this drug. This change is expected and harmless.
- Do not take OTC cimetidine or any other OTC drug without consulting prescriber.
- Use effective means of contraception (both men and women) while on epirubicin therapy.

EPLERENONE
(e-ple're-none)

Inspra

Classifications: ELECTROLYE & WATER BALANCE AGENT; SELECTIVE ALDOSTERONE RECEPTOR ANTAGONIST (SARA); ANTIHYPERTENSIVE

Therapeutic: ANTIHYPERTENSIVE; DIURETIC; SARA

Prototype: Spironolactone

Pregnancy Category: B

AVAILABILITY 25 mg, 50 mg tablets

ACTION & THERAPEUTIC EFFECT Binds to mineralocorticoid receptors and blocks the binding of aldosterone, a component of the renin-angiotensin-aldosterone system (RAAS). Thus eplerenone blocks the primary effect of aldosterone which is sodium reabsorption. *Lowers blood pressure by inhibiting sodium and water retention, thus reducing total plasma volume.*

USES Treatment of hypertension. Adjunctive therapy for post MI heart failure.

CONTRAINDICATIONS Serum potassium greater than 5.5 mEq/L; type 2 diabetes with microalbuminuria; serum creatinine greater than 2 mg/dL in males or greater than 1.8 mg/dL in females; CrCl less than 50 mL/min; lactation.

CAUTIOUS USE Hepatic impairment; severe hepatic disease; pregnancy (category B). Safety and efficacy in children, infants, or neonates are not established.

ROUTE & DOSAGE

Hypertension

Adult: PO 50 mg once daily, may be increased to 50 mg b.i.d. or 100 mg daily, if inadequate response after 4 wk

Heart Failure/Post MI

Adult: PO 25 mg then titrate to 50 mg daily

Renal Impairment Dosage Adjustment

CrCl less than 50 mL/min (hypertension patient): Do not administer
CrCl less than 30 mL/min (heart failure patients): Do not administer

ADMINISTRATION

Oral

- Do not administer in combination with potassium supplements or potassium-sparing diuretics.

- Manufacturer recommends dosage reduction to 25 mg once daily with concurrent administration of erythromycin, saquinavir, verapamil, or fluconazole.
- Store at 15°–30° C (59°–86° F).

ADVERSE EFFECTS (≥1%) **Body as a Whole:** Fatigue, flu-like syndrome. **CNS:** Headache, dizziness. **CV:** Angina, MI. **GI:** Diarrhea, abdominal pain. **Endocrine:** Gynecomastia. **Metabolic:** *Hyperkalemia,* increased GGT, hypercholesterolemia, hypertriglyceridemia, decreased sodium levels. **Respiratory:** Cough. **Urogenital:** Albuminuria, abnormal vaginal bleeding.

INTERACTIONS Drug: ACE INHIBITORS, ANGIOTENSIN II RECEPTOR BLOCKERS, AZOLE ANTIFUNGALS (e.g., **fluconazole, itraconazole, ketoconazole**), **erythromycin, saquinavir, verapamil** may increase risk of hyperkalemia. **Food: Potassium**-containing SALT SUBSTITUTES may increase risk of hyperkalemia.

PHARMACOKINETICS Absorption: Rapidly absorbed. **Peak:** 1.5 h. **Distribution:** 50% protein bound, primarily to alpha$_1$-acid glycoproteins. **Metabolism:** In liver by CYP3A4. **Elimination:** 32% in feces, 67% in urine. **Half-Life:** 4–6 h.

NURSING IMPLICATIONS

Assessment & Drug Effects

- Monitor cardiovascular status with frequent BP determinations. Note that BP lowering usually occurs within 2 wk with maximal antihypertensive effects achieved within 4 wk.
- Lab tests: Monitor baseline and periodic serum potassium, serum sodium, renal function tests, lipid profile, and LFTs. Monitor type 2 diabetics for microalbuminuria.

Common adverse effects in *italic*, life-threatening effects underlined; generic names in **bold**; classifications in SMALL CAPS; ✚ Canadian drug name; ⊙ Prototype drug

543

■ Concurrent drugs: Monitor serum potassium levels more frequently when patient also receiving an ACE inhibitor or an angiontensin II receptor antagonist. Monitor frequently for lithium toxicity with concurrent use.

■ Withhold drug and notify prescriber for any of the following: Serum potassium greater than 5.5 mEq/L, serum creatinine greater than 2 mg/dL in males or greater than 1.8 mg/dL in females, creatinine clearance less than 50 mL/min, microalbuminuria in type 2 diabetics.

Patient & Family Education

■ Do not use potassium supplements, salt substitutes containing potassium, or contraindicated drugs (e.g., ketoconazole, itraconazole) without consulting prescriber.

■ Do not use OTC nonsteroidal anti-inflammatory drugs without consulting prescriber.

■ Do not drive or operate machinery until reaction to drug is known. It may cause dizziness.

EPOETIN ALFA (HUMAN RECOMBINANT ERYTHROPOIETIN) ℗

(e-po-e-tin)

Epogen, Eprex ♦, Procrit
Classification: HEMATOPOIETIC GROWTH FACTOR
Therapeutic: ANTIANEMIC; HUMAN ERYTHROPOIETIN
Pregnancy Category: C

AVAILABILITY 2000 units/mL, 3000 units/mL, 4000 units/mL, 10,000 units/mL, 20,000 units/mL

ACTION & *THERAPEUTIC EFFECT*
Human erythropoietin is produced in the kidney and stimulates bone marrow production of RBCs (erythropoiesis). Hypoxia and anemia generally increase the production of erythropoietin. Epoetin alpha stimulates RBC production. *Stimulates the production of RBCs in the bone marrow of severely anemic patients.*

USES Treatment of anemia.

CONTRAINDICATIONS Uncontrolled hypertension and known hypersensitivity to mammalian cell–derived products and albumin (human); hamster protein hypersensitivity; iron-deficiency anemia; pure red cell aplasia associated with erythropoietin protein drugs; neonates; infants.

CAUTIOUS USE Leukemia, sickle cell disease; coagulopathy; seizure disorders; pregnancy (category C); lactation.

ROUTE & DOSAGE

Anemia of CKD

Adult: **Subcutaneous/IV** Start with 50–100 units/kg/dose until target Hct range of 30–33% (max: 36%) is reached; if Hgb increases more than 1 g/dL and approaches 12 g/dL, reduce dose by 25%. If after 4 wk there is less than 1 g/dL, increase dose by 25%.
Child/Infant: **IV/SC** 50 units/kg 3 ×/wk (adjust as above)

Anemia Related to Chemotherapy

Adult: **SC** 150 units/kg 3 ×/wk or 40,000 units once/wk when Hgb below 10 g/dL
Adolescent/Child (over 5): **IV** 600 units/kg/wk when Hgb below 10 g/dL

Common adverse effects in *italic*, life-threatening effects <u>underlined</u>; generic names in **bold**; classifications in SMALL CAPS; ♦ Canadian drug name; ℗ Prototype drug

ADMINISTRATION

Subcutaneous

- Do not shake solution. Shaking may denature the glycoprotein, rendering it biologically inactive.
- Inspect solution for particulate matter prior to use. Do not use if solution is discolored or if it contains particulate matter.
- Use only one dose per vial, and do not reenter vial.
- Do not give with any other drug solution.

Intravenous

PREPARE: Direct: Give undiluted.
ADMINISTER: Direct: Give direct IV as a bolus dose over 1 min.
INCOMPATIBILITIES Solution/additive: D10W, normal saline.
- Discard any unused portion of the vial. It contains no preservatives.

- Store at 2°–8° C (36°–46° F). Do not freeze or shake.

ADVERSE EFFECTS (≥1%) CNS: Seizures, *headache*. **CV:** *Hypertension*. **GI:** Nausea, diarrhea. **Hematologic:** *Iron deficiency,* <u>thrombocytosis,</u> pure red cell aplasia, *clotting of AV fistula*. **Other:** Sweating, bone pain, arthralgias.

INTERACTIONS Drug: Do not give concurrently with **darbepoetin alfa.**

PHARMACOKINETICS Onset: 7–14 days. **Metabolism:** In serum. **Elimination:** Minimal recovery in urine. **Half-Life:** 4–13 h.

NURSING IMPLICATIONS

Assessment & Drug Effects

- Control BP adequately prior to initiation of therapy and closely monitor and control during therapy. Hypertension is an adverse effect that **must be** controlled.
- Be aware that BP may rise during early therapy as the Hct increases. Notify prescriber of a rapid rise in Hct (more than 4 points in 2 wk). Dosage will need to be reduced because of risk of serious hypertension.
- Monitor for hypertensive encephalopathy in patients with CRF during period of increasing Hct.
- Monitor for premonitory neurologic symptoms (i.e., aura, and report their appearance promptly). The potential for seizures exists during periods of rapid Hct increase (more than 4 points in 2 wk).
- Monitor closely for thrombotic events (e.g., MI, CVA, TIA), especially for patients with CRF.
- Lab tests: Baseline transferrin and serum ferritin. Determine Hct twice weekly until it is stabilized in the target range (30–33%) and the maintenance dose of epoetin alfa has been determined; then monitor at regular intervals. Perform CBC with differential and platelet count regularly. Monitor BUN, creatinine, and serum electrolytes regularly.

Patient & Family Education

- Important to comply with antihypertensive medication and dietary restrictions.
- Do not drive or engage in other potentially hazardous activity during the first 90 days of therapy because of possible seizure activity.
- Note: As Hct increases, there is an improved sense of well-being and quality of life. It is important to continue compliance with dietary and dialysis prescriptions.
- Understand that headache is a common adverse effect. Report if severe or persistent, may indicate developing hypertension.
- Keep all follow-up appointments.

Common adverse effects in *italic*, life-threatening effects <u>underlined</u>; generic names in **bold**; classifications in SMALL CAPS; ✢ Canadian drug name; ◉ Prototype drug

545

EPOPROSTENOL SODIUM ℗

(e-po-pros'te-nol)

Flolan

Classifications: PROSTAGLANDIN; PULMONARY ANTIHYPERTENSIVE

Therapeutic: PULMONARY ANTIHYPERTENSIVE

Pregnancy Category: B

AVAILABILITY 0.5 mg, 1.5 mg powder for injection

ACTION & *THERAPEUTIC EFFECT*

Naturally occurring prostaglandin that reduces right and left ventricular afterload, increases cardiac output, and increases stroke volume through its vasodilation effect. Potent pulmonary vasodilator that reduces pulmonary hypertension. *Potent vasodilator of pulmonary and systemic arterial vascular beds and an inhibitor of platelet aggregation.*

USES Long-term treatment of primary pulmonary hypertension in NYHA Class III and IV patients.

CONTRAINDICATIONS Chronic use with left ventricular systolic dysfunction in CHF patients, hypersensitivity to epoprostenol or related compounds.

CAUTIOUS USE Older adults, pregnancy (category B). Safety and efficacy in children are not established.

ROUTE & DOSAGE

Primary Pulmonary Hypertension

Adult: **IV** *Acute dose:* Initiate with 2 ng/kg/min, increase by 2 ng/kg/min q15min until dose-limiting effects occur (e.g., nausea, vomiting, headache, hypotension, flushing); *Chronic administration:*
Start infusion at 4 ng/kg/min less than the maximum tolerated infusion; if maximum tolerated infusion is 5 ng/kg/min or less, start *maintenance infusion* at 50% of maximum tolerated dose

ADMINISTRATION

▪ Note: Anticoagulation therapy is generally initiated along with epoprostenol to reduce the risk of developing thromboembolic disease.

Intravenous

PREPARE: **Continuous** Note: **Must be** reconstituted using sterile diluent for epoprostenol; must not be mixed with any other medications or solution prior to or during administration. ▪ To make 100 mL of 3000 ng/mL, add 5 mL of the supplied diluent to one 0.5 mg vial; withdraw 3 mL and add to enough diluent to make a total of 100 mL. ▪ To make 100 mL of 5000 ng/mL, add 5 mL of diluent to one 0.5 mg vial; withdraw contents of vial and add to enough diluent to make a total of 100 mL. ▪ To make 100 mL of 10,000 ng/mL, add 5 mL of diluent to each of two 0.5 mg vials; withdraw contents of each vial and add to enough diluent to make a total of 100 mL. ▪ To make 100 mL of 15,000 ng/mL, add 5 mL of diluent to a 1.5 mg vial; withdraw contents of vial and add to enough diluent to make a total of 100 mL.

ADMINISTER: **Continuous:** Give at ordered rate using an infusion control device. Avoid abrupt infusion interruption or large dosage reduction.

INCOMPATIBILITIES **Solution/additive:** Do not mix or infuse with any other parenteral drugs or solutions prior to or during administration.

Common adverse effects in *italic*, life-threatening effects underlined; generic names in **bold**; classifications in SMALL CAPS; ♣ Canadian drug name; ℗ Prototype drug

- Store unopened vials at 15°–25° C (59°–77° F). Protect from light. ■ See manufacturer's directions for stability or storage of reconstituted solutions.

ADVERSE EFFECTS (≥1%) **CNS:** *Chills, fever, flu-like syndrome, dizziness, syncope, headache, anxiety/nervousness,* hyperesthesia, paresthesia, dizziness. **CV:** *Tachycardia, hypotension, flushing, chest pain,* bradycardia. **GI:** *Diarrhea, nausea, vomiting,* abdominal pain. **Musculoskeletal:** *Jaw pain, myalgia, nonspecific musculoskeletal pain.* **Respiratory:** Dyspnea. **Other:** Dose-limiting effects.

INTERACTIONS Drug: Hypotension if administered with other VASODILATORS or ANTIHYPERTENSIVES.

PHARMACOKINETICS Peak: Approximately 15 min. **Metabolism:** Rapidly hydrolyzed at neutral pH in blood; also subject to enzyme degradation. **Elimination:** 82% in urine. **Half-Life:** Approximately 6 min.

NURSING IMPLICATIONS

Assessment & Drug Effects
- Assess carefully for development of pulmonary edema during dose ranging.
- Monitor respiratory and cardiovascular status frequently during entire period of chronic use of epoprostenol.
- Monitor for and report recurrence or worsening of symptoms associated with primary pulmonary hypertension (e.g., dyspnea, dizziness, exercise intolerance) or adverse effects of drug; dosage adjustments may be needed.

Patient & Family Education
- Learn correct techniques for storage, reconstitution, and administration of drug, and maintenance of catheter site (see ADMINISTRATION).

- Notify prescriber immediately of S&S of worsening primary pulmonary hypertension, adverse drug reactions, and S&S of infection at catheter site or sepsis.

E

EPROSARTAN MESYLATE

(e-pro-sar'tan)

Teveten

Classifications: RENIN ANGIOTENSIN SYSTEM ANTAGONIST; ANGIOTENSIN II RECEPTOR ANTAGONIST, ANTIHYPERTENSIVE

Therapeutic: ANTIHYPERTENSIVE

Prototype: Losartan potassium

Pregnancy Category: C first trimester; D second and third trimester

AVAILABILITY 400 mg, 600 mg tablets

ACTION & *THERAPEUTIC EFFECT*
Selectively blocks the binding of angiotensin II to the AT_1 receptors found in many tissues. This blocks vasoconstricting and aldosterone-secreting effects of angiotensin II, thus resulting in an antihypertensive effect. *It decreases both the systolic and diastolic BP.*

USES Treatment of hypertension.

CONTRAINDICATIONS Hypersensitivity to eprosartan, losartan, or other angiotensin II receptor antagonists; pregnancy (category D second and third trimester), lactation.

CAUTIOUS USE Angioedema, aortic or mitral value stenosis, coronary artery disease, cardiomyopathy, hypotension, CHF; biliary obstruction; older adults; severe hepatic dysfunction, renal artery stenosis, renal disease, renal impairment; pregnancy (category C first trimester). Safe use in children younger than 18 y not established.

Common adverse effects in *italic*, life-threatening effects underlined; generic names in **bold**; classifications in SMALL CAPS; ♣ Canadian drug name; ⊕ Prototype drug

547

ROUTE & DOSAGE

Hypertension

Adult: **PO** 600 mg daily (range 400–800 mg) (max: 800 mg/day)

ADMINISTRATION

Oral

- Correct volume depletion prior to therapy to prevent hypotension.
- Store at 15°–30° C (59°–86° F); protect from moisture and direct light.

ADVERSE EFFECTS (≥1%) **Body as a Whole:** Viral infection, fatigue, arthralgia. **CNS:** Depression. **GI:** Abdominal pain, hypertriglyceridemia. **Respiratory:** Upper respiratory infection, rhinitis, pharyngitis, cough.

PHARMACOKINETICS Absorption: Only 13% of oral dose reaches systemic circulation. **Peak:** 1–2 h. **Metabolism:** Minimal metabolism. **Elimination:** 61% in feces and 37% in urine. **Half-Life:** 5–9 h.

NURSING IMPLICATIONS

Assessment & Drug Effects

- Monitor BP periodically; do trough readings just before scheduled dose when possible.
- Monitor for S&S of angioedema (may occur within 30 min or as long as 30 days after initial dose).
- Lab tests: Periodic monitor LFTs, BUN and creatinine, serum potassium, CBC with differential periodically.

Patient & Family Education

- Inform prescriber immediately of pregnancy.
- Report episodes of dizziness especially associated with position changes.

- Report swelling of lips, tongue, face, or feeling of obstruction in neck immediately.

EPTIFIBATIDE

(ep-ti-fib′a-tide)

Integrilin

Classifications: ANTIPLATELET; PLATELET GLYCOPROTEIN (GP IIb/IIIa) INHIBITOR

Therapeutic: ANTIPLATELET

Prototype: Abciximab

Pregnancy Category: B

AVAILABILITY 0.75 mg/mL, 2 mg/mL injection

ACTION & *THERAPEUTIC EFFECT* Binds to the glycoprotein IIb/IIIa (GPIIb/IIIa) receptor sites of platelets. *Inhibits platelet aggregation by preventing fibrinogen, von Willebrand's factor, and other molecules from adhering to GPIIb/IIIa receptor sites on platelets.*

USES Treatment of acute coronary syndromes (unstable angina, non-Q-wave MI) and patients undergoing percutaneous coronary interventions (PCIs).

CONTRAINDICATIONS Hypersensitivity to eptifibatide; active bleeding; GI or GU bleeding within 6 wk; thrombocytopenia; renal failure requiring dialysis; coagulopathy; recent major surgery or trauma; intracranial neoplasm, intracranial bleeding within 6 mo; severe hypertension (systolic blood pressure greater than 200 mm Hg or diastolic blood pressure greater than 110 mm Hg), aneurysm.

CAUTIOUS USE Hypersensitivity to related compounds (e.g., abciximab, tirofiban, lamifiban); pregnancy (category B), lactation. Safety and effectiveness in children are not established.

ROUTE & DOSAGE

Acute Coronary Syndromes (ACS)

Adult: **IV** 180 mcg/kg initial bolus (max: 22.6 mg) followed by 2 mcg/kg/min until hospital discharge or up to 72 h

Percutaneous Coronary Interventions (PCI)

Adult: **IV** 180 mcg/kg initial bolus followed by 2 mcg/kg/min; after 10 min, a second 180 mcg/kg bolus should be given; the infusion should continue up to 24 h after the end of the procedure

Renal Impairment Dosage Adjustment

CrCl 10–49 mL/min: Give 1 mcg/kg/min continuous infusion

ADMINISTRATION

▪ Note: Review contraindications to administration prior to giving this drug.

Intravenous

PREPARE: **Direct:** Give undiluted.
ADMINISTER: **Direct:** Give bolus doses IV push over 1–2 min. **Continuous** Start continuous infusion immediately following bolus dose. ▪ Give undiluted directly from the 100-mL vial (at a rate based on patient's weight) using a vented infusion set. ▪ May be given in the same IV line with NS or D5/NS (either solution may contain up to 60 mEq KCl).

▪ Store unopened vials at 2°–8° C (36°–46° F) and protect from light.

Discard any unused portion in opened vial.

ADVERSE EFFECTS (≥1%) **CNS:** Intracranial bleed (rare). **GI:** GI bleeding. **Hematologic:** *Bleeding* (major bleeding 4.4–11%), anemia, <u>thrombocytopenia</u>.

INTERACTIONS Drug: ORAL ANTICOAGULANTS, NSAIDS, **dipyridamole, ticlopidine, dextran** may increase risk of bleeding.

PHARMACOKINETICS Duration: 6–8 h after stopping infusion. **Metabolism:** Minimally metabolized. **Elimination:** 50% in urine. **Half-Life:** 2.5 h.

NURSING IMPLICATIONS

Assessment & Drug Effects

▪ Lab tests: Prior to infusion determine PT/aPTT and INR, activated clotting time (ACT) for those undergoing percutaneous coronary intervention (PCI); Hct or Hgb; platelet count; and serum creatinine.
▪ Lab tests: Monitor aPTT and INR (target aPPT, 50–70 sec); during PCI (target ACT, 300–350 sec).
▪ Minimize all vascular and other trauma during treatment. When obtaining IV access, avoid using a noncompressible site such as the subclavian vein.
▪ Monitor carefully for and immediately report S&S of bleeding (e.g., femoral artery access site bleeding, intracerebral hemorrhage, GI bleeding).
▪ Immediately stop infusion of eptifibatide and heparin if bleeding at the arterial access site cannot be controlled by pressure.
▪ Achieve hemostasis at the arterial access site by standard compression for a minimum of 4 h prior to hospital discharge following discontinuation of eptifibatide and heparin.

Common adverse effects in *italic*, life-threatening effects <u>underlined</u>; generic names in **bold**; classifications in SMALL CAPS; ♣ Canadian drug name; ⊘ Prototype drug

549

ERGOCALCIFEROL

(er-goe-kal-si'fe-role)

Activated Ergosterol, Calcidol, Drisdol, D-ViSol, Ostoforte ✦, Radiostol ✦, Radiostol Forte ✦, Vitamin D₂

Classification: VITAMIN D ANALOG
Therapeutic: VITAMIN D ANALOG
Prototype: Calcitriol
Pregnancy Category: C

AVAILABILITY 8000 international units/mL oral liquid; 50,000 units capsules, tablets; 500,000 international units/mL injection

ACTION & *THERAPEUTIC EFFECT* The name vitamin D encompasses two related fat-soluble substances. Vitamin D acts like a hormone in that it is distributed through the circulation and plays a major regulatory role. Reponsible for regulation of serum calcium level. *Maintains normal blood calcium and phosphate ion levels by enhancing their intestinal absorption and by promoting mobilization of calcium from bone and renal tubular resorption of phosphate.*

USES Familial hypophosphatemia (vitamin D–resistant rickets), osteomalacia (adult rickets), anticonvulsant-induced rickets and osteomalacia, osteoporosis, renal osteodystrophy, hypocalcemia associated with hypoparathyroidism; prophylaxis and treatment of nutritional rickets. Also hypophosphatemia in Fanconi's syndrome.

UNLABELED USES With varying clinical results in lupus vulgaris, psoriasis, and rheumatoid arthritis.

CONTRAINDICATIONS Hypersensitivity to vitamin D, hypervitaminosis D, hypercalcemia, hyperphosphatemia, renal osteodystrophy with hyperphosphatemia, malabsorption syndrome, decreased kidney function. Safe use of amounts in excess of 400 international units (10 mcg) daily during pregnancy (category C) is not established.

CAUTIOUS USE Coronary disease; arteriosclerosis (especially in older adults); history of kidney stones; biliary tract disease; lactation.

ROUTE & DOSAGE

Nutritional Rickets, Osteomalacia

Adult: **PO** 25–125 mcg/day for 6–12 wk, may need up to 7.5 mg/day in patients with malabsorption
Child: **PO** 50–125 mcg/day, may need up to 250–625 mcg/day in patients with malabsorption

Vitamin D–Dependent Rickets

Adult: **PO** 250 mcg–1.5 mg/day, may need up to 12.5 mg/day (prolonged therapy with greater than 2.5 mg/day increases risk of toxicity)
Child: **PO** 75–125 mcg/day, may need up to 1.5 mg/day

Hypoparathyroidism, Pseudohypoparathyroidism

Adult: **PO** 625 mcg–5 mg/day, may need up to 10 mg/day (prolonged therapy with greater than 2.5 mg/day increases risk of toxicity)
Child: **PO** 1.25–5 mg/day, (prolonged therapy with greater than 2.5 mg/day increases risk of toxicity)

ADMINISTRATION

Oral

▪ Preserve in tightly covered, light-resistant containers. Drug decomposes when exposed to light and air.

Common adverse effects in *italic*, life-threatening effects underlined; generic names in **bold**; classifications in SMALL CAPS; ✦ Canadian drug name; ☻ Prototype drug

ADVERSE EFFECTS (≥1%) **Body as a Whole:** Fatigue, weakness, vertigo, tinnitus, ataxia, muscle and joint pain, hypotonia (infants), exanthema, rhinorrhea; pruritus; mild acidosis. **CNS:** Headache, drowsiness, convulsions. **GI:** Metallic taste, dry mouth, anorexia, nausea, vomiting, diarrhea, constipation, abdominal cramps. **Hematologic:** Anemia. **Musculoskeletal:** Calcification of soft tissues (kidneys, blood vessels, myocardium, lungs, skin). **Urogenital:** Nephrotoxicity (polyuria, hyposthenuria, polydipsia, nocturia, casts, albuminuria, hematuria), kidney failure. **CV:** Hypertension, cardiac arrhythmias. **Special Senses:** Conjunctivitis (calcific); photophobia. **Metabolic:** Osteoporosis (adults); weight loss, chronic hypervitaminosis D in children (mental and physical retardation, suppression of linear growth).

DIAGNOSTIC TEST INTERFERENCE
Vitamin D may cause false increase in *serum cholesterol* measurements *(Zlatkis-Zak reaction).*

INTERACTIONS Drug: Cholestyramine, colestipol, mineral oil may decrease absorption of vitamin D.

PHARMACOKINETICS Absorption: Readily from GI tract. **Peak:** After 4 wk. **Duration:** 2 mo or more. **Distribution:** Most of drug first appears in lymph, stored chiefly in liver and in skin, brain, spleen, and bones. **Metabolism:** In liver and kidney to active metabolites. **Elimination:** About 50% of PO dose in bile; may be stored in tissues for months. **Half-Life:** 12–24 h.

NURSING IMPLICATIONS

Assessment & Drug Effects
- Monitor closely patients receiving therapeutic doses of vitamin D;

must remain under close medical supervision.
- Lab tests: When high therapeutic doses are used, progress is followed by frequent (q2wk or more often) serum calcium, phosphorus, magnesium, alkaline phosphatase, BUN; periodic urine calcium, casts, albumin, and RBC. Blood calcium concentration is generally kept between 9 and 10 mg/dL.
- Monitor for hypercalcemia; in patients with osteomalacia a decrease in serum alkaline phosphatase may signal the onset of hypercalcemia.

Patient & Family Education
- Avoid magnesium-containing antacids and laxatives with chronic kidney failure when receiving vitamin D preparations since vitamin D increases the risk of magnesium intoxication than other patients.
- Do not use OTC medications unless approved by prescriber.

ERGOLOID MESYLATE
(er′goe-loid mess′i-late)
Gerimal
Classifications: ALPHA-ADRENERGIC ANTAGONIST; ERGOT ALKALOID
Therapeutic: ANTIDEMENTIA; ALZHEIMER'S AGENT
Prototype: Ergotamine tartrate
Pregnancy Category: X

AVAILABILITY 1 mg tablets

ACTION & THERAPEUTIC EFFECT
Produces peripheral vasodilation primarily by central action and may cause slight reduction in BP and heart rate. Relieves symptoms of cerebral arteriosclerosis. *Some improvements in Alzheimer's dementia symptoms, possibly by*

Common adverse effects in *italic*, life-threatening effects underlined; generic names in **bold**; classifications in SMALL CAPS; ✤ Canadian drug name; ⊙ Prototype drug

551

increasing cerebral metabolism with consequent increase in blood flow. Improvement may not be apparent until after 3–4 wk of therapy.

USES Vascular or Alzheimer's related dementia.

CONTRAINDICATIONS Acute or chronic psychosis; hypersensitivity to ergoloid; pregnancy (category X), lactation.

CAUTIOUS USE Acute intermittent porphyria; older adult; hepatic disease; hypotension; bradycardia.

ROUTE & DOSAGE

Senile Dementia of Alzheimer Type

Adult: **PO/SL** 1 mg t.i.d.; doses up to 4.5–12 mg/day have been used

ADMINISTRATION

Oral
▪ Store in tightly closed container.

ADVERSE EFFECTS (≥1%) **Body as a Whole:** Mostly dose related. **CV:** Orthostatic hypotension, dizziness or light-headedness, flushing, sinus bradycardia. **Special Senses:** Blurred vision, nasal stuffiness. **GI:** Anorexia, stomach cramps, transient nausea and vomiting, heartburn. **Skin:** Skin rash.

INTERACTIONS Drug: Use with other ERGOT ALKALOIDS may increase toxicities.

PHARMACOKINETICS Absorption: Incompletely from GI tract; 50% reaches systemic circulation. **Peak:** 1.5–3 h. **Metabolism:** Undergoes rapid first-pass metabolism in liver. **Elimination:** Primarily in feces. **Half-Life:** 2–12 h.

NURSING IMPLICATIONS

Assessment & Drug Effects
▪ Establish baseline values of BP and pulse; check at regular intervals throughout therapy.
▪ Report to prescriber sinus bradycardia (40 bpm); has been reported in patients receiving 1.5 mg doses. Pulse rate usually returns to normal within 2 days after drug is discontinued.
▪ Withdraw drug permanently if marked bradycardia or hypotension occurs.

Patient & Family Education
▪ Make position changes slowly, particularly from recumbent to upright posture, and move ankles and feet for a few minutes before walking.

ERGOTAMINE TARTRATE ℗

(er-got′a-meen)

Ergomar
Classifications: ALPHA-ADRENERGIC ANTAGONIST; ERGOT ALKALOID
Therapeutic: ANTIMIGRAINE
Pregnancy Category: X

AVAILABILITY 2 mg sublingual tablets

ACTION & *THERAPEUTIC EFFECT*
Natural amino acid alkaloid of ergot. Alpha-adrenergic blocking agent with direct stimulating action on cranial and peripheral vascular smooth muscles and depressant effect on central vasomotor centers. *In vascular headache, exerts vasoconstrictive action on previously dilated cerebral vessels, reduces amplitude of arterial pulsations, and antagonizes effects of serotonin.*

USES As single agent or in combination with caffeine to prevent or abort migraine, cluster headache (histamine cephalalgia), and other vascular headaches.

Common adverse effects in *italic*, life-threatening effects <u>underlined</u>; generic names in **bold**; classifications in SMALL CAPS; ♣ Canadian drug name; ℗ Prototype drug

CONTRAINDICATIONS Hypersensitivity to ergotamine; sepsis, obliterative vascular disease, thromboembolic disease, prolonged use of excessive dosage, liver and kidney disease, severe pruritus, diabetes mellitus; marked arteriosclerosis, history of MI, peripheral vascular disease; coronary artery disease, angina; basilar/hemiplegic migraine; hepatic disease; biliary tract disease; cholestasis; hypertension; infectious states, anemia, malnutrition; concurrent administration of potent CYP 3A4 inhibitors (e.g., protease inhibitors and macrolide antibiotics); pregnancy (category X).

CAUTIOUS USE Older adult patients; lactation. Safe use in children not established.

ROUTE & DOSAGE

Vascular Headaches

Adult: **SL** 1–2 mg followed by 1–2 mg q30min until headache abates or until max of 6 mg/24 h or 10 mg/wk

ADMINISTRATION

Sublingual
- Instruct patient to allow sublingual (SL) tablet to dissolve under tongue and not to drink, eat, or smoke while tablet is in place. Do not crush SL tablets.

ADVERSE EFFECTS (≥1%) **Body as a Whole:** Paresthesias; pain (spasms) of facial muscles, tongue, limbs, and lumbar region with difficulty in walking; muscle pains, *weakness,* numbness, coldness and cyanosis of digits (Raynaud's phenomenon). **CNS:** Delirium; convulsive seizures; confusion; depression; drowsiness. **GI:** *Nausea; vomiting;* diarrhea; abdominal pain; unquenchable thirst; partial necrosis of tongue, disagreeable aftertaste. **CV:** Rapid, weak, or irregular pulse; intermittent claudication, complete absence of medium- and large-vessel pulsations in extremities; precordial distress and pain; angina pectoris, transient bradycardia or tachycardia; elevated or lowered BP. **Skin:** Itching and cold skin; gangrene of nose, digits, ears. **Urogenital:** Kidney failure. **Other:** Symptoms of ergotism.

INTERACTIONS Drug: With high doses of BETA-ADRENERGIC BLOCKERS, SYMPATHOMIMETICS, possibility of additive vasoconstrictor effects; **erythromycin, troleandomycin** may cause severe peripheral vasospasm. **Eletriptan, naratriptan, rizatriptan, sumatriptan, or zolmitriptan** may increase risk of coronary ischemia, separate drugs by 24 h; AZOLE ANTIFUNGALS **(ketoconazole, itraconazole, fluconazole, clotrimazole), nefazodone, fluoxetine, fluvoxamine, amprenavir, delavirdine, efavirenz, indinavir, nelfinavir, ritonavir, and saquinavir,** may inhibit ergot metabolism and increase toxicity; **sibutramine, dexfenfluramine, nefazodone, fluvoxamine** may increase risk of serotonin syndrome. **Food: Grapefruit juice** may increase toxicity.

PHARMACOKINETICS Absorption: Variable. **Peak:** 0.5–3 h. **Distribution:** Crosses blood–brain barrier. **Metabolism:** Extensive first-pass metabolism in liver. **Elimination:** 96% in feces; excreted in breast milk. **Half-Life:** 2.7 h initial phase, 21 h terminal phase.

NURSING IMPLICATIONS

Assessment & Drug Effects
- Monitor adverse GI effects. Nausea and vomiting are adverse reactions that occur in about 10% of patients after they take ergotamine. Patient may need an antiemetic. Consult with prescriber.

Common adverse effects in *italic*, life-threatening effects <u>underlined</u>; generic names in **bold;** classifications in SMALL CAPS; ♣ Canadian drug name; ☻ Prototype drug

553

■ Monitor patients with PVD carefully for development of peripheral ischemia.

■ Monitor long-term effectiveness. Patients receiving high ergotamine doses for prolonged periods may experience increased frequency of headaches, fatigue, and depression. Discontinuation of the drug in these patients results in severe withdrawal headache that may last a few days.

■ Overdose symptoms: Nausea, vomiting, weakness, and pain in legs, numbness and tingling in fingers and toes, tachycardia or bradycardia, hypertension or hypotension, and localized edema.

Patient & Family Education

■ Begin drug therapy as soon after onset of migraine attack as possible, preferably during migraine prodrome (scintillating scotomas, visual field defects, nausea, paresthesias usually on side opposite to that of the migraine).

■ Notify prescriber if migraine attacks occur more frequently or are not relieved.

■ Lie down in a quiet, dark room for 2–3 h after drug administration.

■ Report muscle pain or weakness of extremities, cold or numb digits, irregular heartbeat, nausea, or vomiting. Carefully protect extremities from exposure to cold temperatures; provide warmth, but not heat, to ischemic areas.

■ Do NOT increase dosage without consulting prescriber; overdosage is the chief cause of adverse effects from the drug.

ERIBULIN MESYLATE

(er-e-bu′ lin)

Halaven
Classifications: ANTINEOPLASTIC; MITOTIC INHIBITOR

Therapeutic: ANTINEOPLASTIC
Pregnancy Category: D

AVAILABILITY 1 mg/2 mL solution for injection

ACTION & *THERAPEUTIC EFFECT*

A microtubule inhibitor that blocks completion of the cell cycle and prevents cell replication resulting in apoptotic cell death. *Interferes with mitosis and cell replication thus reducing growth and metastatic spread of cancer cells.*

USES Treatment of metastatic breast cancer in patients who have previously received at least two previous chemotherapeutic regimens that included an anthracycline and a taxane in either the adjuvant or metastatic setting

CONTRAINDICATIONS Congenital long QT syndrome; ANC less than 1000 mm³; platelets less than 75,000 mm³; Grade 3 or 4 nonhematologic toxicities.

CAUTIOUS USE Neutropenia; peripheral neuropathy; QT prolongation; hepatic impairment; preganacy; lactation. Safety and effectiveness in children have not been established.

ROUTE & DOSAGE

Metastatic Breast Cancer

Adult: **IV** 1.4 mg/m² days 1 and 8 of a 21-day cycle. Repeat cycle as needed or until unacceptable toxicities arise

Delay Doses for Any of the Following

If ANC less than 1000/mm³, platelets less than 75,000/mm³, or Grade 3 or 4 nonhematological toxicities: Do not administer eribulin; the day 8 dose may be delayed a max of 1 wk

Common adverse effects in *italic*, life-threatening effects underlined; generic names in **bold**; classifications in SMALL CAPS; ♣ Canadian drug name; ⊘ Prototype drug

If toxicities do not resolve to Grade 2 or better by day 15: **Omit the dose**

If toxicities resolve or improve to Grade 2 or better by day 15: **Administer eribulin at a reduced dose and initiate the next cycle no sooner than 2 wk later**

Hematologic Toxicity Dosage Adjustment

If ANC less than 500/mm³ for more than 7 days, ANC less than 1000/mm³ with fever or infection, platelets less than 25,000/mm³ or less than 50,000/mm³ requiring transfusion or day 8 of previous cycle omitted or delayed: **Permanently reduce dose to 1.1 mg/m²**

If while receiving 1.1 mg/m², recurrence of hematologic event occurs, or if day 8 of previous cycle omitted or delayed: **Permanently reduce dose to 0.7 mg/m²**

If while receiving 0.7 mg/m², recurrence of hematologic event occurs, or if day 8 of previous cycle omitted or delayed: **Discontinue eribulin**

Nonhematologic Toxicity Dosage Adjustment

If Grade 3 or 4 nonhematologic toxicity or if day 8 of previous cycle omitted or delayed: **Permanently reduce dose to 1.1 mg/m²**

While receiving 1.1 mg/m², if recurrence of Grade 3 or 4 nonhematologic toxicity occurs, or if day 8 of previous cycle omitted or delayed: **Permanently reduce dose to 0.7 mg/m²**

While receiving 0.7 mg/m², if recurrence of Grade 3 or 4 nonhematologic toxicity occurs, or if day 8 of previous cycle omitted or delayed: **Discontinue eribulin**

Hepatic Impairment Dosage Adjustment

Mild hepatic impairment (Child-Pugh class A): **Reduce dose to 1.1 mg/m²**

Moderate hepatic impairment (Child-Pugh class B): **Reduce dose to 0.7 mg/m²**

Renal Impairment Dosage Adjustment

CrCl 30–50 ml/min: **Reduce dose to 1.1 mg/m²**

ADMINISTRATION

Intravenous

- Correct hypokalemia or hypomagnesemia prior to initiating eribulin.

PREPARE: IV Infusion: May be given undiluted or diluted in 100 mL of NS. Do not dilute with dextrose.

ADMINISTER: IV Infusion: Give over 2–5 min. Do not administer in same IV line as dextrose or any other required medication.

INCOMPATIBILITIES Solution/additive: Incompatible with D5W for dilution. **Y-site:** None listed.

- Store unopened at 15°–30° C (59°–86° F). Store undiluted in a syringe or diluted in NL for up to 4 h at room temperature or for up to 24 h under refrigeration.

ADVERSE EFFECTS (≥1%) **Body as a Whole:** Mucosal inflammation, pyrexia. **CNS:** *Asthenia/fatigue,*

Common adverse effects in *italic*, life-threatening effects <u>underlined</u>; generic names in **bold**; classifications in SMALL CAPS; ♣ Canadian drug name; ☺ Prototype drug

555

depression, dizziness, *headache*, insomnia, *peripheral neuropathy*. **GI:** Abdominal pain, *constipation*, *diarrhea*, dry mouth, dyspepsia, *nausea*, stomatitis, vomiting. **Hematological:** *Anemia*, febrile neutropenia, *neutropenia*, thrombocytopenia. **Metabolic:** *Anorexia*, *decreased weight*, hypokalemia, *increased ALT*, peripheral edema. **Musculoskeletal:** *Arthralgia*, *back pain*, bone pain, muscle spasms, muscle weakness, *myalgia*, pain in extremitiy. **Respiratory:** Cough, *dyspnea*, upper respiratory tract infection. **Skin:** *Alopecia*, rash. **Special Senses:** Dysgeusia, increased lacrimation. **Urogenital:** Urinary tract infection.

INTERACTIONS Drug: Eribulin has been associated with QT prolongation. Coadministration of another drug that prolongs the QT interval (e.g., **disopyramide, procainamide, amiodarone, bretylium, clarithromycin, levofloxacin**) may cause additive effects.

PHARMACOKINETICS Distribution: 49–65% plasma protein bound. **Metabolism:** Minimal. **Elimination:** Fecal (82%) and renal (9%). **Half-Life:** 40 h.

NURSING IMPLICATIONS

Assessment & Drug Effects
- Monitor ECG in those with CHF, bradyarrhythmias, concurrent use of Class Ia and III antiarrhythmics, and electrolyte imbalances.
- Monitor closely for S&S of peripheral motor and sensory neuropathy.
- Monitor temperature. Report fever and/or S&S of infection.
- Withhold drug and notify prescriber of the following: Grade 3 or 4 peripheral neuropathy; hypokalemia or hypomagnesemia; prolonged QT interval.

- Lab tests: Monitor baseline and periodic serum electrolytes; monitor CBC with differential prior to each dose and more often with Grade 3 or 4 cytopenia; periodic LFTs.

Patient & Family Education
- Report promptly fever (100.5° F or greater) or other S&S of infection.
- Report S&S of peripheral neuropathy, including: Tingling, numbness, deep pain in the feet, legs, or arms; weakness or problems with balance; or difficulty with fine motor skills.
- Effective methods of birth control are recommended during therapy. Notify prescriber immediately if a pregnancy develops.
- Do not breast feed while taking this drug without consulting prescriber.

ERLOTINIB
(er-lo'ti-nib)
Tarceva
Classifications: ANTINEOPLASTIC; TYROSINE KINASE INHIBITOR-EPIDERMAL GROWTH FACTOR RECEPTOR (TKI-EGFR) INHIBITOR
Therapeutic: ANTINEOPLASTIC
Prototype: Gefitinib
Pregnancy Category: D

AVAILABILITY 25 mg, 100 mg, 150 mg tablets

ACTION & *THERAPEUTIC EFFECT* Erlotinib is a human epidermal growth factor receptor type 1 (HER1/EGFR) inhibitor. Antitumor action is believed to be due to inhibition of phosphorylation of tyrosine kinase associate with the EGFR present on the cell surface of both normal and cancer cells. *Inhibition of EGFR in cancer cells diminishes their capacity for cell proliferation, cell survival, and decreases metastases.*

USES Treatment of patients with locally advanced or metastatic non–small cell lung cancer (NSCLC), pancreatic cancer.

CONTRAINDICATIONS Hypersensitivity to erlotinib; severe hepatic impairment; pregnancy (category D); lactation.

CAUTIOUS USE Mild or moderate hepatic impairment; interstitial pulmonary disease (interstitial pneumonia, pneumonitis, alveolitis); myelosuppression; ocular toxicities (corneal ulcer, eye pain).

ROUTE & DOSAGE

Metastatic Non–Small-Cell Lung Cancer

Adult: **PO** 150 mg once daily

Pancreatic Cancer (with Gemcitabine)

Adult: **PO** 100 mg daily

Hepatic Impairment Dosage Adjustment

Discontinue use in patient with severe change in liver function

ADMINISTRATION

Oral

- Give at least 1 h before or 2 h after eating.
- Store at 15°–30° C (59°–86° F). Keep container tightly closed. Protect from light.

ADVERSE EFFECTS (≥1%) **Body as a Whole:** Infection. **GI:** *Diarrhea, anorexia, fatigue,* nausea, vomiting, stomatitis, abdominal pain. **Metabolic:** Increased LFTs. **Respiratory:** *Dyspnea,* cough, interstitial lung disease (sometimes fatal). **Skin:** *Acneiform rash,* pruritus, dry skin. **Special Senses:** Conjunctivitis, dry eyes.

INTERACTIONS Drug: Dose will need to be adjusted with concurrent strong **CYP3A4 inducers (e.g., carbamazepine, nevirapine, phenobarbital, phenytoin). Atazanavir, clarithromycin, indinavir, itraconazole, ketoconazole, nefazodone, nelfinavir, ritonavir, saquinavir, telithromycin, troleandomycin, voriconazole** may increase erlotinib levels and toxicity; increased bleeding with **warfarin. Herbal: St. John's wort** may decrease erlotinib levels.

PHARMACOKINETICS Absorption: 60% absorbed; food can increase to 100%. **Peak:** 4 h. **Metabolism:** In liver by CYP3A4. **Elimination:** In feces (83%). **Half-Life:** 36.2 h.

NURSING IMPLICATIONS

Assessment & Drug Effects

- Monitor closely changes in pulmonary function.
- Withhold drug and notify prescriber for acute onset of new or progressive pulmonary symptoms (e.g., dyspnea, cough, or fever) or significant changes in liver functions as indicated by elevated transaminases, bilirubin, and alkaline phosphatase.
- Lab tests: Periodic LFTs.

Patient & Family Education

- Report promptly any of the following: Severe or persistent diarrhea, nausea, anorexia, or vomiting; onset or worsening of unexplained shortness of breath or cough; eye irritation.
- Monitor closely PT/INR values with concurrent warfarin therapy.
- Women should use effective means to avoid pregnancy while taking this drug.

E

Common adverse effects in *italic,* life-threatening effects underlined; generic names in **bold;** classifications in SMALL CAPS; ✤ Canadian drug name; ⊘ Prototype drug

557

E

ERTAPENEM SODIUM

(er-ta-pen′em)

Invanz

Classification: BETA-LACTAM ANTIBIOTIC

Therapeutic: ANTIBIOTIC

Prototype: Imipenem-cilastatin

Pregnancy Category: B

AVAILABILITY 1 g for injection

ACTION & *THERAPEUTIC EFFECT*
Broad-spectrum carbapenem antibiotic that inhibits the cell wall synthesis of gram-positive and gram-negative bacteria by its strong affinity for penicillin-binding proteins (PBPs) of the bacterial cell wall. *Effective against both gram-positive and gram-negative bacteria. Highly resistant to most bacterial beta-lactamases.*

USES Complicated intra-abdominal infections, complicated skin and skin structure infections, community-acquired pneumonia, complicated UTI (including pyelonephritis), and acute pelvic infections due to susceptible bacteria.

CONTRAINDICATIONS Hypersensitivity to ertapenem, penicillins, or carbapenem antibiotics; hypersensitivity to amide-type local anesthetics such as lidocaine; hypersensitivity to meropenem or imipenem.

CAUTIOUS USE Renal impairment; history of CNS disorders; history of seizures; hypersensitivity to other beta-lactam antibiotics (penicillins, cephalosporins); hypersensitivity to other allergens; meningitis; pregnancy (category B); lactation (bottle feed during and for 5 days after therapy ends). Safe use in infants younger than 3 mo not established.

ROUTE & DOSAGE

Community-Acquired Pneumonia; Complicated UTI

Adult/Adolescent: **IV/IM** 1 g daily × 10–14 days; may switch to appropriate PO antibiotic after 3 days if responding
Child/Infant (older than 3 mo): **IV/IM** 15 mg/kg q12h × 10–14 days (max: 1 g/day)

Intra-Abdominal Infection

Adult/Adolescent: **IV/IM** 1 g daily × 5–14 days
Child/Infant (older than 3 mo): **IV/IM** 15 mg/kg b.i.d. × 5–14 days (max: 1 g/day)

Skin and Skin Structure Infections

Adult/Adolescent: **IV/IM** 1 g daily × 7–14 days
Child/Infant (older than 3 mo): **IV/IM** 15 mg/kg b.i.d. × 7–14 days (max: 1 g/day)

Acute Pelvic Infections

Adult: **IV/IM** 1 g daily × 3–10 days

Renal Impairment Dosage Adjustment

CrCl less than 30 mL/min: Reduce dose to 500 mg daily

ADMINISTRATION

Intramuscular

- Reconstitute 1 g vial with 3.2 mL of 1% lidocaine HCl injection (without epinephrine). Shake vial thoroughly to form solution. Use immediately.
- Inject deep IM into a large muscle mass (such as the gluteal muscles or lateral part of the thigh).
- The reconstituted IM solution should be used within 1 h after preparation. Note: DO NOT use this solution for IV administration.

Intravenous

PREPARE: **Intermittent for Adult/ Child:** Reconstitute 1 g vial with 10 mL of sterile water for injection, NS, or bacteriostatic water for injection. Shake well to dissolve. **Intermittent for Adult/Child (13 y or older):** Immediately after reconstitution, transfer contents to 50 mL of NS injection solution. **Intermittent for Child (3 mo–12 y):** Immediately after reconstition, transfer required dose to enough NS injection solution to yield a final concentration of 20 mg/mL or less. *ADMINISTER:* **Intermittent:** Infuse over 30 min. Note: Infusion should be completed within 6 h of reconstitution.

INCOMPATIBILITIES **Solution/additive:** Mannitol, sodium bicarbonate. **Y-site:** Allopurinol, amiodarone, amphotericin B, anidulafungin, caspofungin, chlorpromazine, dantrolene, danorubicin, diazepam, dobutamine, doxorubicin, droperidol, epirubicin, hydralazine, hydroxyzine, idarubicin, midazolam, minocycline, mitoxantrone, nicardipine, ondansetron, pentamidine, phenytoin, prochlorperazine, promethazine, quinidine, quinupristin/dalfopristin, thiopental, topotecan, verapamil.

▪ Store lyophilized powder above 25° C (77° F). ▪ Must use reconstituted solution stored at room temperature (not above 25° C/77° F) within 6 h. ▪ May store for 24 h under refrigeration. Use within 4 h of removal from refrigeration. ▪ Do not freeze.

ADVERSE EFFECTS (≥1%) **Body as a Whole:** Phlebitis or thrombosis at injection site, asthenia, fatigue, <u>death</u>, fever, leg pain. **CNS:** Anxiety, altered mental status, dizziness, headache, insomnia. **CV:** Chest pain, hypertension, hypotension, tachycardia, edema. **GI:** Abdominal pain, *diarrhea*, acid regurgitation, constipation, dyspepsia, nausea, vomiting, increased AST and ALT. **Respiratory:** Cough, dyspnea, pharyngitis, rales/rhonchi, and respiratory distress. **Skin:** Erythema, pruritus, rash. **Urogenital:** Vaginitis.

INTERACTIONS Drug: Probenecid decreases renal excretion. **Valproic acid** levels may be decreased.

PHARMACOKINETICS Absorption: 90% from IM site. **Peak:** 2.3 h. **Distribution:** 95% protein bound, distributes into breast milk. **Metabolism:** Hydrolysis of beta-lactam ring. **Elimination:** 80% in urine, 10% in feces. **Half-Life:** 4.5 h.

NURSING IMPLICATIONS

Assessment & Drug Effects

▪ Lab tests: Baseline C&S tests prior to therapy. Monitor periodically LFTs and kidney function.
▪ Determine history of hypersensitivity reactions to other beta-lactams, cephalosporins, penicillins, or other drugs.
▪ Discontinue drug and immediately report S&S of hypersensitivity (see Appendix F).
▪ Report S&S of superinfection or pseudomembranous colitis (see Appendix F).
▪ Monitor for seizures especially in older adults and those with renal insufficiency.
▪ Lab tests: Monitor AST, ALT, alkaline phosphatase, CBC, platelet count, and routine blood chemistry during prolonged therapy.

Patient & Family Education

▪ Learn S&S of hypersensitivity, superinfection, and pseudomembranous colitis (see Appendix F); report any of these to prescriber promptly.

Common adverse effects in *italic*, life-threatening effects <u>underlined</u>; generic names in **bold**; classifications in SMALL CAPS; ♣ Canadian drug name; ☯ Prototype drug

ERYTHROMYCIN ⊕
(er-ith-roe-mye'sin)
Apo-Erythro Base ♣, A/T/S, E-Mycin, Eryc, EryDerm, EryTab, Erythrocin, Erythromid ♣, Erythromycin Base, Novo-Rythro ♣, PCE, Ro-Mycin ♣

ERYTHROMYCIN STEARATE
Apo-Erythro-S ♣, Erythrocin Stearate, SK-Erythromycin

Classification: MACROLIDE ANTIBIOTIC
Therapeutic: ANTIBIOTIC
Pregnancy Category: C

AVAILABILITY **Erythromycin:** 250 mg, 333 mg, 500 mg tablets, capsules; 2% topical solution 2% gel; 2% ointment 2% pledgets; 5% ophth ointment; **Erythromycin Estolate:** 125 mg, 250 mg capsules 125 mg/mL, 250 mg/mL suspension; **Erythromycin Stearate:** 250 mg, 500 mg tablets

ACTION & *THERAPEUTIC EFFECT* Macrolide antibiotic that binds to the 50S ribosomal subunit, thus inhibiting bacterial protein synthesis. *More active against gram-positive organisms than against gram-negative organisms due to its superior penetration into gram-positive organisms.*

USES Pneumococcal pneumonia, *Mycoplasma pneumoniae* (primary atypical pneumonia), acute pelvic inflammatory disease caused by *Neisseria gonorrhoeae* in females sensitive to penicillin, infections caused by susceptible strains of staphylococci, streptococci, and certain strains of *Haemophilus influenzae.* Also used in intestinal amebiasis, Legionnaires' disease, uncomplicated urethral, endocervical, and rectal infections caused by *Chlamydia trachomatis,* for prophylaxis of ophthalmia neonatorum caused by *N. gonorrhoeae, C. trachomatis,* and for chlamydial conjunctivitis in neonates. Considered an acceptable alternative to penicillin for treatment of streptococcal pharyngitis, for prophylaxis of rheumatic fever and bacterial endocarditis, for treatment of diphtheria as adjunct to antitoxin and for carrier state, and as alternate choice in treatment of primary syphilis in patients allergic to penicillins. **Topical applications:** Pyodermas, acne vulgaris, and external ocular infections, including neonatal chlamydial conjunctivitis and gonococcal ophthalmia.

CONTRAINDICATIONS Hypersensitivity to erythromycins or other macrolide antibiotics; congenital QT prolongation; electrolyte imbalances. **Estolate:** History of hepatotoxicity in patients with hepatic disease. **CAUTIOUS USE** Impaired liver function; seizure disorders; history of GI disorders; pregnancy (category C).

ROUTE & DOSAGE

Moderate to Severe Infections
Adult: **PO** 250–500 mg q6h; 333 mg q8h
Child: **PO** 30–50 mg/kg/day divided q6h **Topical** Apply ointment to infected eye 1 or more times/day
Neonate: **PO** *7 days or younger,* 10 mg/kg q12h; *older than 7 days,* 10 mg/kg q8–12h **Topical** 0.5–1 cm in conjunctival sac once
Chlamydia Trachomatis Infections
Adult: **PO** 500 mg q.i.d. or 666 mg q8h
Child: **Topical** Apply 0.5–1 cm ribbon in lower conjunctival sacs shortly after birth

Common adverse effects in *italic*, life-threatening effects underlined; generic names in **bold**; classifications in SMALL CAPS; ♣ Canadian drug name; ⊕ Prototype drug

ADMINISTRATION

Oral

- Erythromycin base or stearate should be given on an empty stomach 1 h before or 3 h after meals. Do not give with, or immediately before or after, fruit juices.
- Erythromycin estolate and enteric-coated tablets may be given without regard to meals.
- Ensure that capsules and tablets are not chewed or crushed. They **must be** swallowed whole.

Topical

- Prophylaxis for neonatal eye infection: Ribbon of ointment approximately 0.5–1 cm long is placed into lower conjunctival sac of neonate shortly after birth. Use a new tube of erythromycin for each neonate.
- Store all forms at 15°–30° C (59°–86° F) in tightly capped containers unless otherwise directed by manufacturer.

ADVERSE EFFECTS (≥1%) **GI:** *Nausea, vomiting, abdominal cramping,* diarrhea, heartburn, anorexia. **Body as a Whole:** Fever, eosinophilia, urticaria, skin eruptions, fixed drug eruption, anaphylaxis. Superinfections by nonsusceptible bacteria, yeasts, or fungi. **Special Senses:** Ototoxicity: Reversible bilateral hearing loss, tinnitus, vertigo. **Digestive:** (Estolate) Cholestatic hepatitis syndrome. **Skin:** (Topical use) Erythema, desquamation, burning, tenderness, dryness or oiliness, pruritus.

DIAGNOSTIC TEST INTERFERENCE False elevations of *urinary catecholamines, urinary steroids,* and *AST, ALT* (by *colorimetric methods*).

INTERACTIONS Drug: Serum levels and toxicities of **alfentanil, bexarotene, carbamazepine, cevimeline, cilostazol, clozapine,** cyclosporine, disopyramide, estazolam, fentanyl, midazolam, methadone, modafinil, quinidine, sirolimus, digoxin, theophylline, triazolam, warfarin are increased. **Ergotamine, dihydroergotamine** may increase peripheral vasospasm. **Food: Grapefruit juice** may increase side effects.

PHARMACOKINETICS Absorption: Most erythromycins are absorbed in small intestine. **Peak:** 1–4 h PO. **Distribution:** Widely distributed to most body tissues; low concentrations in CSF; concentrates in liver and bile; crosses placenta. **Metabolism:** Partially in liver. **Elimination:** Primarily in bile; excreted in breast milk. **Half-Life:** 1.5–2 h.

NURSING IMPLICATIONS

Assessment & Drug Effects

- Report onset of GI symptoms after PO administration. These are dose related; if symptoms persist after dosage reduction, prescriber may prescribe drug to be given with meals in spite of impaired absorption.
- Monitor for adverse GI effects. Pseudomembranous enterocolitis (see Appendix F), a potentially life-threatening condition, may occur during or after antibiotic therapy.
- Observe for S&S of superinfection by overgrowth of nonsusceptible bacteria or fungi. Emergence of resistant staphylococcal strains is highly predictable during prolonged therapy.
- Lab tests: Periodic LFTs during prolonged therapy.
- Monitor for S&S of hepatotoxicity. Premonitory S&S include: Abdominal pain, nausea, vomiting, fever, leukocytosis, and eosinophilia; jaundice may or may not be present. Symptoms may appear a few days after initiation of drug

Common adverse effects in *italic*, life-threatening effects underlined; generic names in **bold**; classifications in SMALL CAPS; ✦ Canadian drug name;⊙ Prototype drug

561

but usually occur after 1–2 wk of continuous therapy. Symptoms are reversible with prompt discontinuation of erythromycin.

- Monitor for ototoxicity that appears to develop most frequently in patients receiving 4 g/day or more, older adults, female patients, and patients with kidney or liver dysfunction. It is reversible with prompt discontinuation of drug.

Patient & Family Education

- Notify prescriber for S&S of superinfection (see Appendix F).
- Notify prescriber immediately for S&S of pseudomembranous enterocolitis (see Appendix F), which may occur even after the drug is discontinued.
- Report any ototoxic effects including dizziness, vertigo, nausea, tinnitus, roaring noises, hearing impairment (see Appendix F).

ERYTHROMYCIN ETHYLSUCCINATE

(er-ith-roe-mye′sin)

Apo-Erythro-ES ♦, E.E.S., E.E.S.-200, E.E.S.-400, EryPed, Pediamycin
Classification: MACROLIDE ANTIBIOTIC
Therapeutic: ANTIBIOTIC
Prototype: Erythromycin
Pregnancy Category: B

AVAILABILITY 200 mg chewable tablet, 400 mg tablets; 100 mg/2.5 mL, 200 mg/5 mL, 400 mg/5 mL suspension

ACTION & _THERAPEUTIC EFFECT_
Macrolide antibiotic that binds to the 50S ribosomal subunit of bacteria, thus inhibiting bacterial protein synthesis. _More active against grampositive than gram-negative bacteria._

USES See ERYTHROMYCIN.

CONTRAINDICATIONS Hypersensitivity to erythromycins or any macrolide antibiotic; history of erythromycin-associated hepatitis; preexisting liver disease; congenital QT prolongation; electrolyte imbalances.
CAUTIOUS USE Myasthenia gravis; history of GI disease; seizure disorders; pregnancy (category B).

ROUTE & DOSAGE

Infection

Adult: **PO** 400 mg q6h up to 4 g/day according to severity of infection
Child: **PO** 30–50 mg/kg/day in 4 divided doses (max: 100 mg/kg/day) for severe infections

ADMINISTRATION

- Note: 400 mg erythromycin ethylsuccinate is approximately equal to 250 mg erythromycin base.

Oral

- Chewable tablets should be chewed and not swallowed whole.
- Suspensions are stable for 14 days at room temperature unless otherwise stated by manufacturer. Note expiration date.
- Store tablets in tight containers unless otherwise directed.

ADVERSE EFFECTS (≥1%) **GI:** Diarrhea, _nausea,_ vomiting, stomatitis, _abdominal cramps,_ anorexia, hepatotoxicity. **Skin:** Skin eruptions. **Special Senses:** Ototoxicity. **Body as a Whole:** Potential for superinfections.

INTERACTIONS Drug: Serum levels and toxicities of **alfentanil, bexarotene, carbamazepine, cevimeline, cilostazol, clozapine, cyclosporine, disopyramide, estazolam, fentanyl, midazolam, methadone, modafinil, quini-**

Common adverse effects in _italic_, life-threatening effects underlined; generic names in **bold**; classifications in SMALL CAPS; ♦ Canadian drug name; ⊘ Prototype drug

dine, sirolimus, digoxin, theophylline, triazolam, warfarin
are increased. **Ergotamine** may
increase peripheral vasospasm and
may increase risk of arrhythmias.

PHARMACOKINETICS Absorption:
Readily absorbed from GI tract.
Peak: 2 h. **Distribution:** Concentrates in liver; crosses placenta; distributed into breast milk. **Metabolism:** In liver. **Elimination:** Primarily
in bile and feces. **Half-Life:** 2–5 h.

NURSING IMPLICATIONS

Assessment & Drug Effects

▪ Lab tests: Determine C&S prior to
treatment. Periodic LFTs and
blood cell counts if therapy is
prolonged.

▪ Cholestatic hepatitis syndrome is
most likely to occur in adults who
have received erythromycin estolate for more than 10 days or
who have had repeated courses
of therapy. The condition generally clears within 3–5 days after
cessation of therapy.

Patient & Family Education

▪ Advise patient to report immediately the onset of adverse reactions and to be on the alert for
signs and symptoms associated
with jaundice (see Appendix F).

▪ Ototoxicity is most likely to occur
in patients receiving high dosage
or who have impaired kidney
function. Report immediately the
onset of tinnitus, vertigo, or hearing impairment.

ERYTHROMYCIN LACTOBIONATE

(er-ith-roe-mye' sin lak'-toe-bye'-
oh-nate)

Erythrocin Lactobionate-I.V.
Classification: MACROLIDE ANTI-
BIOTIC

Therapeutic: ANTIBIOTIC
Prototype: Erythromycin
Pregnancy Category: B

E

AVAILABILITY 500 mg, 1 g injection

ACTION & *THERAPEUTIC EFFECT*
Soluble salt of erythromycin that
binds to the 50S ribosome subunits
of susceptible bacteria, resulting in
the suppression of protein synthesis of bacteria. *More active against
gram-positive than gram-negative
bacteria.*

USES When oral administration is
not possible or the severity of infection requires immediate high serum levels. See erythromycin.

CONTRAINDICATIONS Hypersensitivity to erythromycin or macrolide
antibiotics; congenital QT prolongation; electrolyte imbalances.
CAUTIOUS USE Impaired liver function; seizure disorders; pregnancy
(category B).

ROUTE & DOSAGE

Infections
Adult/Child: **IV** 15–20 mg/kg/
day in 4 divided doses
Legionnaires' Disease
Adult: **IV** 0.5–1 g q6h × 21 days
Pelvic Inflammatory Disease
Adult: **IV** 500 mg q6h × 3d, then
convert to PO

ADMINISTRATION

Intravenous

PREPARE: **Intermittent/Continu-
ous:** Initial solution is prepared
by adding 10 mL sterile water for
injection without preservatives to
each 500 mg or fraction thereof.

Common adverse effects in *italic*, life-threatening effects underlined; generic names
in **bold**; classifications in SMALL CAPS; ♣ Canadian drug name; ⊙ Prototype drug

563

E

▪ Shake vial until drug is completely dissolved. **Intermittent:** Further dilute each 1 g dose in 100–250 mL of LR or NS. **Continuous (preferred):** Further dilute each 1 g in 1000 mL LR or NS. ▪ Give within 4 h.

ADMINISTER: **Intermittent:** Give 1 g or fraction thereof over 20–60 min. ▪ Slow rate if pain develops along course of vein. **Continuous (preferred):** Continuous infusion is administered slowly, usually over 6–24 h.

INCOMPATIBILITIES **Solution/additive:** **Dextrose**-containing solutions, **ascorbic acid, colistimethate, clindamycin, furosemide, heparin, linezolid, metaraminol, metoclopramide, tetracycline, vitamin B complex with C. Y-site: Cefepime, ceftazidime, fluconazole.**

▪ Store: **Gluceptate,** reconstituted solution is stable up to 7 days if refrigerated at 2°–8° C (36°–46° F); use solution diluted for infusion within 4 h. **Lactobionate,** reconstituted solution is stable up to 14 days if refrigerated at 2°–8° C (36°–46° F); use solution diluted for infusion within 8 h.

ADVERSE EFFECTS (≥1%) **Body as a Whole:** *Pain and venous irritation after IV injection;* allergic reactions, <u>anaphylaxis</u> (rare); superinfections. **GI:** *Nausea,* vomiting, diarrhea, *abdominal cramps,* variations in liver function tests following prolonged or repeated therapy. (See also ERYTHROMYCIN.)

INTERACTIONS Drug: Serum levels and toxicities of **alfentanil, bexarotene, carbamazepine, cevimeline, cilostazol, clozapine, cyclosporine, disopyramide, estazolam, fentanyl, midazolam, methadone, modafinil, quinidine,** **sirolimus, digoxin, theophylline, triazolam, warfarin** are increased. **Ergotamine** may increase peripheral vasospasm, and may increase risk of arrhythmias.

PHARMACOKINETICS Peak: 1 h. **Distribution:** Concentrates in liver; crosses placenta; distributed into breast milk. **Metabolism:** In liver. **Elimination:** Primarily in bile and feces; 12–15% in urine. **Half-Life:** 3–5 h.

NURSING IMPLICATIONS

Assessment & Drug Effects

▪ Lab tests: Determine C&S prior to initiation of therapy. Periodic LFTs with daily high doses or prolonged or repeated therapy.
▪ Monitor hearing impairment may occur with large doses of this drug. It may occur as early as the second day and as late as the third week of therapy.
▪ Monitor for S&S of thrombophlebitis (see Appendix F). IV infusion of large doses is reported to increase risk.

Patient & Family Education

▪ Notify prescriber immediately of tinnitus, dizziness, or hearing impairment.

ESCITALOPRAM OXALATE

(es-ci-tal′o-pram)
Lexapro
Classifications: ANTIDEPRESSANT; SELECTIVE SEROTONIN REUPTAKE INHIBITOR (SSRI)
Therapeutic: ANTIDEPRESSANT; SSRI
Prototype: Fluoxetine
Pregnancy Category: C

AVAILABILITY 5 mg, 10 mg, 20 mg tablets; 5 mg/5 mL liquid

ACTION & *THERAPEUTIC EFFECT* Selective serotonin reuptake

Common adverse effects in *italic*, life-threatening effects <u>underlined</u>; generic names in **bold**; classifications in SMALL CAPS; ♣ Canadian drug name;◉ Prototype drug

inhibitor (SSRI) in the CNS. Antidepressant effect is presumed to be linked to its inhibition of CNS presynaptic neuronal uptake of serotonin. *Selective serotonin reuptake inhibition mechanism results in the antidepressant activity with or without anxiety symptoms.*

USES Depression, generalized anxiety disorder.
UNLABELED USES Treatment of panic disorders, social anxiety disorders.

CONTRAINDICATIONS Hypersensitivity to citalopram; concurrent use of MAOIS or use within 14 days of discontinuing MAOIS; abrupt discontinuation; suicidal ideations; mania; bipolar depression; volume depleted.
CAUTIOUS USE Hypersensitivity to other SSRIs; suicidal tendencies; bipolar disorder; obsessive-compulsive disorder, major depressive disorder, all major psychiatric disorders especially pediatric patients; depression, history of mania, hypomania; hyponatremia, ethanol intoxication, ECT, dehydration, severe renal impairment, hepatic disease; older adults; history of seizure disorders; elderly; pregnancy (category C), lactation (not within 4 h of drug ingestion). Safety and efficacy in children younger than 12 y are unknown.

ROUTE & DOSAGE

Depression, Generalized Anxiety
Adult/Adolescent: **PO** 10 daily, may increase to 20 mg daily if needed after 1 wk
Geriatric: **PO** 10 mg daily

Panic Disorder
Adult: **PO** 5 daily, may increase to 20 mg daily if needed after 1 wk

Hepatic Impairment Dosage Adjustment
Adult: **PO** 10 daily

ADMINISTRATION

Oral
- Do not begin this drug within 14 days of stopping an MAOI.
- Dose increments should be separated by at least 1 wk.
- Store at 15°–30° C (59°–86° F) in tightly closed container and protect from light.

ADVERSE EFFECTS (≥1%) **Body as a Whole:** Fatigue, fever, arthralgia, myalgia. **CV:** Palpitation, hypertension. **GI:** *Nausea*, diarrhea, dyspepsia, abdominal pain, dry mouth, vomiting, flatulence, reflux. **CNS:** Dizziness, *insomnia, somnolence*, paresthesia, migraine, tremor, vertigo. **Metabolic:** Increased or decreased weight, hyponatremia. **Respiratory:** URI, rhinitis, sinusitis. **Skin:** Increased sweating. **Urogenital:** Dysmenorrhea, decreased libido, ejaculation disorder, impotence, menstrual cramps.

INTERACTIONS Drug: Combination with MAOI could result in hypertensive crisis, hyperthermia, rigidity, myoclonus, autonomic instability; **cimetidine** may increase escitalopram levels; **linezolid** may cause serotonin syndrome. Use with drugs affecting hemostasis (**aspirin, warfarin**) increases bleeding risk. **Herbal: St. John's wort** may cause serotonin syndrome.

PHARMACOKINETICS Absorption: Rapidly absorbed from GI tract. **Onset:** Approximately 1 wk. **Peak:** 3 h. **Distribution:** 80% protein bound; crosses placenta; distributed into breast milk. **Metabolism:** In liver by CYP3A4, 2C19, and 2D6 enzymes. **Elimination:** 20% in urine, 80% in bile. **Half-Life:** 25 h.

NURSING IMPLICATIONS

Assessment & Drug Effects

- Closely observe for worsening of depression or emergence of suicidality especially in adolescents or children.
- Lab tests: Monitor periodically LFTs, CBC, serum sodium, and lithium levels when the two drugs are given concurrently.
- Monitor periodically HR and BP, and carefully monitor complete cardiac status in person with known or suspected cardiac disease.
- Monitor closely older adult patients for adverse effects, especially with doses greater than 20 mg/day.

Patient & Family Education

- Report promptly changes in behavior such as anxiety, agitation, depression, panic attacks, aggressiveness, and suicidal ideation.
- Do not engage in hazardous activities until reaction to this drug is known.
- Avoid using alcohol while taking escitalopram.
- Inform prescriber of commonly used OTC drugs as there is potential for drug interactions. The use of aspirin and NSAIDs can affect coagulation and cause increased risk of bleeding.
- Report distressing adverse effects including any changes in sexual functioning or response.
- Periodic ophthalmology exams are advised with long-term treatment.

ESMOLOL HYDROCHLORIDE

(ess′moe-lol)

Brevibloc

Classification: BETA-ADRENERGIC ANTAGONIST; ANTIARRHYTHMIC

Therapeutic: ANTIARRHYTHMIC
Prototype: Propranolol
Pregnancy Category: C

AVAILABILITY 10 mg/mL, 250 mg/mL injection

ACTION & *THERAPEUTIC EFFECT*

Ultrashort-acting beta$_1$-adrenergic blocking agent with cardioselective properties. Inhibits the agonist effect of catecholamines by competitive binding at beta-adrenergic receptors. Antiarrhythmic properties occur at the AV node. *Effective as an antiarrhythmic agent on the AV-nodal conduction system. Blocks sympathetically mediated increases in cardiac rate and BP since it binds predominantly to beta$_1$-receptors in cardiac tissue.*

USES Supraventricular tachyarrhythmias (SVT) in perioperative and postoperative periods or in other critical situations. Also short-term treatment of noncompensating sinus tachycardia and in the control of heart rate for patients with MI.

UNLABELED USES Moderate postoperative hypertension; treatment of intense transient adrenergic response to surgical stress in cardiac as well as noncardiac surgery.

CONTRAINDICATIONS Hypersensitivity to esmolol; heart block greater than first degree; sinus bradycardia, cardiogenic shock; decompensated CHF; acute bronchospasm.

CAUTIOUS USE History of allergy; CHF; pulmonary disease such as bronchial asthma, COPD, or pulmonary edema; diabetes mellitus; kidney function impairment; pregnancy (category C); lactation. Safe use in children younger than 18 y not established.

Common adverse effects in *italic*, life-threatening effects <u>underlined</u>; generic names in **bold**; classifications in SMALL CAPS; ♣ Canadian drug name; ☻ Prototype drug

ROUTE & DOSAGE

Supraventricular Tachyarrhythmias

Adult: **IV** 500 mcg/kg loading dose followed by 50 mcg/kg/min, may increase dose q5–10min prn if response inadequate, may repeat loading dose followed by 100 mcg/kg/min × 4 min; may continue repeating loading dose and increasing 4-min dose by 50 mcg/kg/min prn (max: 200 mcg/kg/min)

Intraoperative/Postoperative Tachycardia

Adult: **IV** 80 mg bolus followed by 150 mcg/kg/min; increase if needed (max: 300 mcg/kg/min)

ADMINISTRATION

Intravenous

Note: Do not use the 2500 mg ampule for direct IV injection.

***PREPARE:* Direct:** Use the 10 mg/mL vial undiluted for the loading dose. **IV Infusion:** Use the concentrate (250 mg/mL) for infusion. ▪ Prepare maintenance infusion by adding 2.5 g to 250 mL or 5 g to 500 mL of IV solution to yield 10 mg/mL. Compatible diluents include D5W, D5/LR, D5/NS, D5/.45NS, LR.

***ADMINISTER:* Direct:** Give loading dose over 1 min. **IV Infusion:** ▪ Give maintenance infusion over 4 min. ▪ If adequate response is noted, continue maintenance infusion with periodic adjustments as needed. ▪ If an adequate response has not occurred, repeat loading dose and follow with an increased maintenance infusion of 100 mcg/kg/min. ▪ May continue titration

cycle with same loading dose while increasing maintenance infusion by 50 mcg/kg/min until desired end point is near. ▪ Then omit loading dose and titrate maintenance dose up or down by 25 to 50 mcg/kg/min until desired heart rate is reached.

INCOMPATIBILITIES Solution/additive: **Diazepam, procainamide, thiopental.** Y-site: **Amphotericin B cholesteryl, furosemide, warfarin.**

▪ Diluted infusion solution is stable for at least 24 h at room temperature.

ADVERSE EFFECTS (≥1%) **CNS:** Headache, *dizziness,* somnolence, confusion, agitation. **CV:** *Hypotension* (dose related), cold hands and feet, bradyarrhythmias, flushing, myocardial depression. **GI:** Nausea, vomiting. **Respiratory:** Dyspnea, chest pain, rhonchi, <u>bronchospasm</u>. **Skin:** *Infusion site inflammation* (redness, swelling, induration).

INTERACTIONS Drug: May increase **digoxin** IV levels 10–20%; **morphine** IV may increase esmolol levels by 45%; **succinylcholine** may prolong neuromuscular blockade.

PHARMACOKINETICS Onset: Less than 5 min. **Peak:** 10–20 min. **Duration:** 10–30 min. **Metabolism:** Hydrolyzed by RBC esterases. **Elimination:** In urine. **Half-Life:** 9 min.

NURSING IMPLICATIONS

Assessment & Drug Effects

▪ Monitor BP, pulse, ECG, during esmolol infusion. Hypotension may have its onset during the initial titration phase; thereafter the risk increases with increasing doses. Usually the hypotension experienced during esmolol infusion is resolved within 30 min after infusion is reduced or discontinued.

Common adverse effects in *italic*, life-threatening effects <u>underlined</u>; generic names in **bold**; classifications in SMALL CAPS; ♣ Canadian drug name; ◐ Prototype drug

567

- Change injection site if local reaction occurs. IV site reactions (burning, erythema) or diaphoresis may develop during infusion. Both reactions are temporary. Blood chemistry abnormalities have not been reported.

- Overdose symptoms: Discontinue administration if the following symptoms occur: Bradycardia, severe dizziness or drowsiness, dyspnea, bluish-colored fingernails or palms of hands, seizures.

ESOMEPRAZOLE MAGNESIUM

(e-so-me'pra-zole)

Nexium

Classification: PROTON PUMP INHIBITOR

Therapeutic: ANTIULCER

Prototype: Omeprazole

Pregnancy Category: B

AVAILABILITY 20 mg, 40 mg capsules; 20 mg, 40 mg powder for injection; 10 mg, 20 mg, 40 mg oral suspensions

ACTION & THERAPEUTIC EFFECT Isomer of omeprazole, a weak base that is converted to the active form in the highly acidic environment of the gastric parietal cells. Inhibits the enzyme H^+K^+-ATPase (the acid pump), thus suppressing gastric acid secretion. *Due to inhibition of the H^+K^+-ATPase, esomeprazole substantially decreases both basal and stimulated acid secretion through inhibition of the acid pump in parietal cells.*

USES Erosive esophagitis, gastrointestinal reflux disease (GERD), hypersecretory diseases, duodenal ulcer associated with *H. pylori* in combination with antibiotics.

CONTRAINDICATIONS Hypersensitivity to esomeprazole, magnesium, omeprazole, or other proton pump inhibitors; gastric malignancy; lactation.

CAUTIOUS USE Severe renal insufficiency; severe hepatic impairment; treatment for more than a year; gastric ulcers; elderly; IBD, GI disease; pregnancy (category B). Safe use in children younger than 1 y not established.

ROUTE & DOSAGE

Healing of Erosive Esophagitis

Adult/Adolescent: **PO/IV** 20–40 mg daily at least 1 h before meals times 4–8 wk

GERD, Erosive Esophagitis Maintenance

Adult/Adolescent: **PO/IV** 20–40 mg daily at least 1 h before meals times 4–8 wk

Child/Infant (older than 1 y): **PO** *20 kg or more:* 10–20 mg daily at least 1 h before meals up to 8 wk; *weight less than 20 kg:* 10 mg daily at least 1 h before meals up to 8 wk

Duodenal Ulcer

Adult: **PO** 40 mg daily times 10 days

Hypersecretory Disease (Zollinger-Ellison)

Adult: **PO** 40 mg b.i.d.

NSAID Ulcer Prophylaxis

Adult: **PO** 20–40 mg daily

GERD Associated with History or Erosive Esophagitis

Adult: **IV** 20–40 daily × 10 days
Adolescent/Child (over 55 kg): **IV** 20 mg daily
Adolescent/Child (less than 55 kg): 10 mg daily
Infant (over 1 mo): 0.5 mg/kg daily

Hepatic Impairment Dosage Adjustment
Child-Pugh class C: Do not exceed 20 mg/day

ADMINISTRATION

Oral

- Give at least 1 h before eating.
- Do not crush or chew capsule. **Must be** swallowed whole.
- Open capsule and mix pellets with applesauce (cold or room temperature) if patient cannot swallow capsules. Do NOT crush pellets. Applesauce should be swallowed immediately after mixing without chewing.
- May take with antacids.
- Store in the original blister package 15°–30° C (59°–86° F).

Intravenous

PREPARE: **Direct:** Reconstitute powder with 5 mL of NS. **IV Infusion:** Further dilute reconstituted solution in 50 mL of NS, LR, or D5W.
ADMINISTER: **Direct:** Withdraw required dose from reconstituted solution and give over no less than 3 min. **IV Infusion:** Give IV solution over 10–30 min.
INCOMPATIBILITIES Do not give simultaneously with any other medication through the same IV site or line.

- Flush IV line with NS, LR, or D5W before/after infusion.

- Store reconstituted solution at room temperature up to 30° C (86° F); give within 12 h of reconstitution with NS or LR and within 6 h of reconstitution with D5W.

ADVERSE EFFECTS (≥1%) **CNS:** Headache. **GI:** Nausea, vomiting, diarrhea, constipation, abdominal pain, flatulence, dry mouth.

INTERACTIONS Drug: May increase **diazepam, phenytoin, warfarin** levels. May decrease levels of **atazanavir** and **nelfinavir**. **Food:** Food decreases absorption by up to 35%.

PHARMACOKINETICS Absorption: Destroyed in acidic environment, therefore capsules are designed for delayed absorption in the small intestine. 70% reaches systemic circulation. **Metabolism:** In liver by CYP2C19. **Elimination:** Inactive metabolites excreted in both urine and feces. **Half-Life:** 1.5 h.

NURSING IMPLICATIONS

Assessment & Drug Effects

- Monitor for S&S of adverse CNS effects (vertigo, agitation, depression) especially in severely ill patients.
- Monitor phenytoin levels with concurrent use.
- Monitor INR/PT with concurrent warfarin use.
- Lab tests: Periodic LFTs, CBC, Hct and Hbg, urinalysis for hematuria and proteinuria.

Patient & Family Education

- Report any changes in urinary elimination such as pain or discomfort associated with urination to prescriber.
- Report severe diarrhea. Drug may need to be discontinued.

ESTAZOLAM
(es-ta-zo'lam)

Classifications: ANXIOLYTIC; SEDATIVE-HYPNOTIC; BENZODIAZEPINE **Therapeutic:** ANTIANXIETY; SEDATIVE-HYPNOTIC
Prototype: Lorazepam
Pregnancy Category: X
Controlled Substances: Schedule IV

AVAILABILITY 1 mg, 2 mg tablets

Common adverse effects in *italic*, life-threatening effects underlined; generic names in **bold**; classifications in SMALL CAPS; ♥ Canadian drug name; ✪ Prototype drug

569

E

ACTION & *THERAPEUTIC EFFECT*

Benzodiazepine whose effects (anxiolytic, sedative, hypnotic, skeletal muscle relaxant) are mediated by the inhibitory neurotransmitter gamma-aminobutyric acid (GABA). GABA acts at the thalamic, hypothalamic, and limbic levels of CNS. *Benzodiazepines generally decrease the number of awakenings from sleep. Stage 2 sleep is increased with all benzodiazepines. Estazolam shortens stages 3 and 4 (slow-wave sleep), and REM sleep is shortened. The total sleep time, however, is increased.*

USES Short-term management of insomnia.

CONTRAINDICATIONS Known sensitivity to benzodiazepines; acute closed-angle glaucoma, primary depressive disorders or psychosis; abrupt discontinuation; coma, shock, acute alcohol intoxication; pregnancy (category X), lactation.

CAUTIOUS USE Renal and hepatic impairment, renal failure; organic brain syndrome, alcoholism, benzodiazepine dependence, suicidal ideations, CNS depression, seizure disorder, status epilepticus; substance abuse; shock, coma; dementia, mania, psychosis; myasthenia gravis, Parkinson's disease; sleep apnea; open-angle glaucoma, GI disorders, older adult and debilitated patients; limited pulmonary reserve, pulmonary disease, COPD. Safe use in children younger than 18 y not established.

ROUTE & DOSAGE

Insomnia

Adult: **PO** 1 mg at bedtime, may increase up to 2 mg if necessary (older adult patients may start with 0.5 mg at bedtime)

ADMINISTRATION

Oral

- For older adult patients in good health, a 1 mg dose is indicated; reduce initial dose to 0.5 mg for debilitated or small older adult patients.
- Dosage reduction also may be needed in the presence of hepatic impairment.

ADVERSE EFFECTS (≥1%) **CNS:** Headache, dizziness, impaired coordination, hypokinesia, *somnolence,* hangover, weakness. **CV:** Palpitations, arrhythmias, syncope (all rare). **Hematologic:** Leukopenia, agranulocytosis. **GI:** Constipation, xerostomia, anorexia, flatulence, vomiting. **Musculoskeletal:** Arthritis, arthralgia, myalgia, muscle spasm.

INTERACTIONS Drug: Cimetidine may decrease metabolism of estazolam and increase its effects; **alcohol** and other CNS DEPRESSANTS may increase drowsiness; CYP3A4 inhibitors **(ketoconazole, itraconazole, nefazodone, diltiazem, fluvoxamine, cimetidine, isoniazid, erythromycin)** can increase concentrations and toxicity of estazolam; **carbamazepine, phenytoin, rifampin,** BARBITURATES may decrease estazolam concentrations. **Food: Grapefruit juice** greater than 1 quart may increase toxicity. **Herbal: Kava, valerian** may potentiate sedation.

PHARMACOKINETICS Absorption: Rapidly absorbed from GI tract. **Onset:** 20–30 min. **Peak:** 2 h. **Distribution:** Crosses rapidly into brain; crosses placenta; distributed into breast milk. **Metabolism:** Extensively in liver. **Elimination:** In urine. **Half-Life:** 10–24 h.

NURSING IMPLICATIONS

Assessment & Drug Effects

- Monitor for improvement in S&S of insomnia.

- Assess for excess CNS depression or daytime sedation.
- Assess for safety, especially with older adult or debilitated patients, as dizziness and impaired coordination are known adverse effects.

Patient & Family Education
- Learn adverse effects and report those experienced to the prescriber.
- Avoid using this drug in combination with other CNS depressant drugs or alcohol.
- Do not drive or engage in other potentially hazardous activities until response to drug is known.

ESTRADIOL 🅟
(ess-tra-dye′ole)
Alora, Climara, Divigel, Estrace, Estraderm, Estring, EstroGel, Evamist, Menostar, Vivelle, Vivelle DOT, Vagifem

ESTRADIOL ACETATE
Femring, Femtrace

ESTRADIOL CYPIONATE
Depo-Estradiol   Cypionate, Estro-Cyp

ESTRADIOL VALERATE
Delestrogen, Femogex ◆

ESTRADIOL HEMIHYDRATE
Estrasorb
Classification: ESTROGEN
Therapeutic: ESTROGEN
REPLACEMENT
Pregnancy Category: X

AVAILABILITY Estradiol: 0.025 mg, 0.0375 mg, 0.05 mg, 0.06 mg, 0.075 mg, 0.1 mg patch; 14 mcg/24 h transdermal patch; 0.5 mg, 1 mg, 2 mg tablets; 25 mcg vaginal tablets; 2 mg vaginal ring; 0.1 mg vaginal cream; 2.5 mg/g topical emulsion; 0.06%, 0.1% topical gel; 1.53 mg/ actuation transdermal spray; **Cypionate:** 5 mg/mL injection; **Valerate:** 10 mg/mL, 20 mg/mL, 40 mg/mL injection; **Acetate:** 0.45 mg, 0.9 mg, 1.8 mg tablets; 0.05 mg/day, 0.1 mg/day vaginal insert; **Hemihydrate:** 0.25% topical emulsion

ACTION & *THERAPEUTIC EFFECT*
Natural or synthetic steroid hormone secreted principally by the ovarian follicles, and also by the adrenals, corpus luteum, placenta, and testes. Estrogen binds to a specific intracellular receptor, forming a complex that stimulates synthesis of proteins responsible for estrogenic effects. Promotes endometrial lining development, but long-time use leads to abnormal endometrial hyperplasia, and abnormal bleeding. Conversely, estrogen-stimulated endometrium suddenly deprived of estrogen may bleed within 48–72 h. *Estradiol (estrogens) effects simulate those produced by the endogenous hormone. May mask onset of climacteric.*

USES Natural or surgical menopausal symptoms, kraurosis vulvae, atrophic vaginitis, primary ovarian failure, female hypogonadism, castration. Used adjunctively with diet, calcium, and physical therapy to prevent and treat postmenopausal osteoporosis; also for palliation in advanced prostatic carcinoma and inoperable metastatic breast cancer in women at least 5 y after menopause. Combined with progestins in many oral contraceptive formulations.

CONTRAINDICATIONS Estrogenic-dependent neoplasms, breast cancer (except in selected patients being treated for metastatic disease). History of thromboembolic disorders; active arterial thrombosis or thrombophlebitis;   undiagnosed abnormal genital bleeding; uterine fibroids; endometriosis; history of cholestatic disease; hepatic disease; thyroid dysfunction; blood dyscrasias; hypercalcemia; lupus (SLE); known or suspected pregnancy (category X).

CAUTIOUS USE Adolescents with incomplete bone growth; endometriosis; hypertension, cardiac insufficiency; diseases of calcium and phosphate metabolism (metabolic bone disease); cerebrovascular disease; mental depression; benign breast disease, family history of breast or genital tract neoplasm; diabetes mellitus; gallbladder disease; preexisting leiomyoma, abnormal mammogram, history of idiopathic jaundice of pregnancy; varicosities; asthma; epilepsy; migraine headaches; liver or kidney dysfunction; jaundice, acute intermittent porphyria, pyridoxine deficiency.

ROUTE & DOSAGE

Menopause, Atrophic Vaginitis, Kraurosis Vulvae, Female Hypogonadism, Female Castration, Primary Ovarian Failure

Adult: **PO** 0.45–2 mg/day in a cyclic regimen **Topical** 2–4 g vaginal cream intravaginally once/day for 1–2 wk, then 1–2 g/day for 1–2 wk, then 1 g 1–3 times/wk; Transdermal patch **Estraderm** twice weekly; **Climara, FemPatch, Menostar** qwk in a cyclic regimen; **Estrasorb** Apply 1 packet to the left thigh and calf and 1 packet to the right thigh and calf once daily in the morning; **EstroGel** Apply 1.25 g (one-half applicatorful) to one arm every day (usually in the morning). **IM Cypionate** 1–5 mg once q3–4wk; **Valerate** 10–25 mg once q4wk; **Acetate** Insert 1 vaginal ring into the upper third of the vaginal vault. Keep in place continuously for 3 mo, then remove. **Divigel** Apply one packet to upper thigh daily (alternate legs). **Evamist** Apply one spray to inner forearm daily, dose may be increased to 2–3 sprays daily

Metastatic Breast Cancer

Adult: **PO** 10 mg t.i.d.

Prostatic Cancer

Adult: **PO** 1–2 mg t.i.d. **IM Valerate** 30 mg once q1–2wk

Postpartum Breast Engorgement

Adult: **IM Valerate** 10–25 mg at end of first stage of labor

ADMINISTRATION

Oral

- Give with or immediately after solid food to reduce nausea.
- Protect tablets from light and moisture in well-closed container. Protect from freezing, unless otherwise directed by manufacturer.

Intravaginal

- Insert calibrated dosage applicator approximately 5 cm (2 in.) into vagina, directing it slightly back toward sacrum. Instill medication by pushing plunger. Patient should remain in recumbent position about 30 min to prevent losing the medication. Observe perineal area before each administration: If mucosa is red, swollen, or excoriated or if there is a change in vaginal discharge, report to prescriber.

Topical

- Cleanse and dry selected skin area. Apply as directed under Route & Dosage.

Transdermal

- Cleanse and dry selected skin area on trunk of body, preferably the abdomen. Avoid application to the breasts, to an irritated, abraded, oily area, or to the waistline. If system falls off, it may be reapplied, or if necessary, a new one can be applied. Return to original treatment schedule. Rotate application site with an interval of at least 1 wk between applications to a particular site.

Intramuscular

- Give deep into a large muscle.
- Store at 15°–30° C (59°–86° F); protect from light and freezing.

ADVERSE EFFECTS (≥1%) **CNS:** Headache, migraine, dizziness, mental depression, chorea, convulsions, increased risk of dementia. **CV:** Thromboembolic disorders, stroke, CAD, hypertension. **Special Senses:** Intolerance to contact lenses, worsening of myopia or astigmatism, scotomas. **GI:** Nausea, vomiting, anorexia, increased appetite, diarrhea, abdominal cramps or pain, constipation, bloating, colitis, acute pancreatitis, cholestatic jaundice, benign hepatoadenoma. **Urogenital:** Mastodynia, breast secretion, spotting, changes in menstrual flow, dysmenorrhea, amenorrhea, cervical erosion, altered cervical secretions, premenstrual-like syndrome, vaginal candidiasis, endometrial cystic hyperplasia, reactivation of endometriosis, increased size of preexisting fibromyomas, cystitis-like syndrome, hemolytic uremic syndrome, change in libido; in men: Gynecomastia, testicular atrophy, feminization, impotence (reversible). **Metabolic:** Reduced carbohydrate tolerance, hyperglycemia, hypercalcemia, folic acid deficiency, fluid retention. **Skin:** Dermatitis, pruritus, seborrhea, oily skin, acne; photosensitivity, chloasma, loss of scalp hair, hirsutism. **Body as a Whole:** Pain and postinjection flare at injection site; sterile abscess; leg cramps, weight changes. **Hematologic:** Acute intermittent porphyria.

DIAGNOSTIC TEST INTERFERENCE Estradiol reduces response of *metyrapone* test and excretion of *pregnanediol. Increases: BSP* retention, norepinephrine-induced *platelet aggregability, hydrocortisone, PBI, T₄, sodium, thyroxine-binding globulin (TBG), prothrombin and factors VII, VIII, IX,* and *X; serum triglyceride,* and *phospholipid* concentrations, *renin* substrate. *Decreases: Antithrombin III, pyridoxine* and *serum folate* concentrations, serum *cholesterol,* values for the *T₃ resin uptake* test, *glucose tolerance.* May cause false-positive test for *LE cells* or *antinuclear antibodies (ANA).*

INTERACTIONS Drug: BARBITURATES, **phenytoin, rifampin** decrease estrogen effect by increasing its metabolism; ORAL ANTICOAGULANTS may decrease hypoprothrombinemic effects; interfere with effects of **bromocriptine;** may increase levels and toxicity of **cyclosporine,** TRICYCLIC ANTIDEPRESSANTS, **theophylline;** decrease effectiveness of **clofibrate.**

PHARMACOKINETICS Absorption: Rapid from GI tract; readily through skin and mucous membranes; slow from IM injections. **Distribution:** Throughout body

Common adverse effects in *italic*, life-threatening effects underlined; generic names in **bold**; classifications in SMALL CAPS; ♣ Canadian drug name; ☯ Prototype drug

573

tissues, especially in adipose tissue; crosses placenta. **Metabolism:** Primarily in liver. **Elimination:** In urine; in breast milk.

NURSING IMPLICATIONS

Assessment & Drug Effects

- Monitor adverse GI effects. Nausea, frequently at breakfast time, usually disappears after 1 or 2 wk of drug use.
- Check BP on a regular basis in patients with cardiac or kidney dysfunction or hypertension; monitored carefully.
- Note: Severe hypercalcemia (greater than 15 mg/dL) may be caused by estradiol therapy in patients with breast cancer and bone metastasis.

Patient & Family Education

- Comply with established dosage schedule. Do not alter unless prescriber prescribes a change.
- Notify prescriber of intermittent breakthrough bleeding, spotting, bleeding, or unexplained and sudden pain.
- Determine weight under standard conditions 1 or 2 times/wk; report sudden weight gain or other signs of fluid retention.
- Notify prescriber of calf pain upon flexing foot and the following symptoms of thromboembolic disorders: Tenderness, swelling, and redness in extremity; sudden, severe headache or chest pain; slurring of speech; change in vision; tenderness, pain, sudden shortness of breath.
- Monitor blood glucose for loss of glycemic control if diabetic.
- Decrease caffeine intake, since estrogen depresses caffeine metabolism.
- Learn self-examination of breasts and follow a monthly schedule.

- Estrogen-induced feminization and impotence in male patients are reversible with termination of therapy.

ESTRAMUSTINE PHOSPHATE SODIUM

(ess-tra-muss'teen)

Emcyt

Classifications: ANTINEOPLASTIC; ALKYLATING AGENT; NITROGEN MUSTARD

Therapeutic: ANTINEOPLASTIC
Prototype: Cyclophosphamide
Pregnancy Category: D

AVAILABILITY 140 mg capsules

ACTION & *THERAPEUTIC EFFECT*
Conjugate of estradiol and the carbamate of nitrogen mustard. Incorporation of estramustine in tumor tissues is probably due to the presence of estramustine-binding protein (EMBP), which is found in prostate carcinoma, glioma, melanoma, and breast carcinoma. Binds to proteins and microtubulin resulting in microtubule changes in the cell division cycle, thus arresting cell division in the G2/M phase of the cell cycle. *Major effectiveness reported to be in patients who have been refractory to estrogen therapy alone.*

USES Prostate cancer.

CONTRAINDICATIONS Hypersensitivity to either estradiol or nitrogen mustard; active thrombophlebitis or thromboembolic disorders; pregnancy (category D), lactation.

CAUTIOUS USE History of thrombophlebitis, thromboses, or thromboembolic disorders; cerebrovascular or coronary artery disease; gallstones or peptic ulcer; impaired liver function; metabolic bone diseases associated with hypercalcemia;

diabetes mellitus; hypertension, conditions that might be aggravated by fluid retention (e.g., epilepsy, migraine, kidney dysfunction); older adults.

ROUTE & DOSAGE

Prostate Cancer
Adult: **PO** 14 mg/kg/day or 600 mg/m²/day in 3–4 divided doses

ADMINISTRATION

Oral
- Give with meals to reduce incidence of GI adverse effects. Some patients require drug withdrawal because of intolerable GI effects.
- Store at 2°–8° C (38°–46° F) in tight, light-resistant containers, unless otherwise directed by manufacturer.

ADVERSE EFFECTS (≥1%) **CNS:** Lethargy, emotional lability, insomnia, headache, anxiety. **CV:** CVA, <u>MI</u>, *thrombophlebitis,* CHF, *peripheral edema.* **GI:** *Nausea,* diarrhea, anorexia, flatulence, vomiting, thirst, GI bleeding. **Hematologic:** <u>Leukopenia</u>, <u>thrombocytopenia</u>, *abnormalities in liver function tests,* hypercalcemia, <u>bone marrow depression</u> (rare). **Respiratory:** Hoarseness, burning sensation in throat, dyspnea, upper respiratory discharge, <u>pulmonary emboli</u>. **Skin:** Rash, pruritus, urticaria, dry skin, easy bruising, flushing, peeling skin and fingertips, thinning hair. **Special Senses:** Tearing of eyes. **Urogenital:** Gynecomastia, breast tenderness, impotence. **Endocrine:** Decrease in glucose tolerance. **Musculoskeletal:** Leg cramps.

INTERACTIONS Food: Milk, dairy products, calcium supplements may decrease estramustine absorption.

PHARMACOKINETICS Absorption: Readily absorbed from GI tract. **Peak:** 2–3 h. **Metabolism:** Dephosphorylated in intestines to estramustine, estradiol, estrone, and nitrogen mustard; further metabolized in liver. **Elimination:** In feces via bile. **Half-Life:** 20 h.

NURSING IMPLICATIONS

Assessment & Drug Effects
- Monitor weight and examine for peripheral edema. Be mindful that drug can cause CHF.
- Monitor I&O ratio and pattern to prevent dehydration and electrolyte imbalance, especially with vomiting or diarrhea.
- Observe diabetics closely because of possibility of estramustine-induced reduction in glucose tolerance. Monitor baseline and periodic glucose tolerance tests.
- Lab tests: Perform baseline and periodic LFTs and bilirubin; repeat after drug has been discontinued for 2 mo.

Patient & Family Education
- Eat small meals at frequent intervals to reduce drug-induced nausea, eat slowly, and try cold food if food odors are offensive.
- Drink liquids 1 h before or 1 h after rather than with meals; clear liquids may be more palatable.

ESTROGEN-PROGESTIN COMBINATIONS (CONTRACEPTIVES)
Oral

Monophasic: Apri, Alesse, Aviane, Balziva, Brevicon, Cryselle, Demulen, Desogen, Gencept, Junel, Lessina, Levlite, Levora, Loestrin, Lo/Ovral, Low-Ogestrel, Microgestin, Modicon, Nordette, Norethin,

E

Common adverse effects in *italic*, life-threatening effects <u>underlined</u>; generic names in **bold;** classifications in SMALL CAPS; ♣ Canadian drug name; ◑ Prototype drug

575

E

Norinyl, Nortrel, Ogestrel, Ortho-Cept, Ortho-Cyclen, Ortho-Novum, Ovcon, Portia, Previfem, Seasonale, Sprintec, Yasmin, Yaz, Zovia
Biphasic: LoSeasonique, Kariva, Mircette, Ortho-Novum 10/11
Triphasic: Aranelle, Cyclessa, Enpresse, Estrostep, Estrostep Fe, Lybrel, Ortho-Novum 7/7/7, Ortho Tri-Cyclen, Ortho Tri-Cyclen Lo, Tri-Norinyl, Tri-Previfem, Tri-Sprintec, Triphasil, Trivora, Velivet
Postcoital Contraceptives: Plan B, Preven

Transdermal
Ortho Evra
Intravaginal
NuvaRing
Classification: ESTROGEN-PROGESTIN COMBINATIONS
Therapeutic: CONTRACEPTIVE
Prototype: Estradiol, Norethindrone
Pregnancy Category: X

AVAILABILITY Combination oral contraceptives contain one of the following estrogens and one of the following progestins. **Estrogen:** Ethinyl estradiol 10 mcg, 20 mcg, 25 mcg, 30 mcg, 35 mcg, 40 mcg, 50 mcg; mestranol 50 mcg; **Progestin:** Desogestrel 0.15 mg; drospirenone 3 mg; ethynodiol diacetate 1 mg; levonorgestrel 0.05 mg, 0.075 mg, 0.1 mg, 0.125 mg, 0.15 mg, 0.25 mg, 0.75 mg; norethindrone 0.4 mg, 0.5 mg, 0.75 mg, 1 mg; norethindrone acetate 1 mg, 1.5 mg; norgestimate 0.18 mg, 0.215 mg, 0.25 mg; norgestrel 0.3 mg, 0.5 mg; **Transdermal:** Norelgestromin 6 mg/0.75 mg ethinyl estradiol patch; **Vaginal:** Etonogestrel 11.7 mg/2.7 mg ethinyl estradiol vaginal insert

ACTION & *THERAPEUTIC EFFECT*
Three types of estrogen-progestin combinations are available:

(1) monophasic, fixed dosage of estrogen-progestin throughout the cycle; (2) biphasic, amount of estrogen remains the same throughout cycle, less progestin in first half of cycle and increased progestin in second half; (3) triphasic, estrogen amount is the same or varies throughout cycle, progestin amount varies. *Fixed combination of estrogen and progestin produces contraception by preventing ovulation and rendering reproductive tract structures hostile to sperm penetration and zygote implantation.*

USES To prevent conception and to treat hypermenorrhea and endometriosis; postcoital contraceptive or "morning after pill"; moderate acne in females 15 y or older (Tri-Cyclen).

CONTRAINDICATIONS Familial or personal history of or existence of breast or other estrogen-dependent neoplasm, recurrent chronic cystic mastitis, history of or existence of thrombophlebitis or thromboembolic disorders, cerebral vascular or coronary artery disease, MI, serious hepatic dysfunction, hepatic neoplasm, family history of hepatic porphyria, undiagnosed abnormal vaginal bleeding, women age 40 and over, adolescents with incomplete epiphyseal closure; pregnancy (category X), lactation.
CAUTIOUS USE History of depression, preexisting hypertension, or cardiac or renal disease; impaired liver function, history of migraine, convulsive disorders, or asthma; multiparous women with grossly irregular menses, diabetes, or familial history of diabetes; gallbladder disease, lupus erythematosus, rheumatic disease, varicosities, smokers.

ROUTE & DOSAGE

Contraception

Adult: **PO** 1 active tablet daily for 21 days, then placebo tablet or no tablets for 7 days, repeat cycle **Continuous regimen** (Seasonale) 1 tablet daily × 84 consecutive days. Wait 7 days for withdrawal bleeding before starting next cycle **Topical** Apply one patch once weekly for 3 wk, then have 1 wk patch-free before repeating the cycle **Intravaginal** Insert 1 ring on or before day 5 of the cycle. Remove ring after 3 wk, followed by a 1 wk rest. Then insert new ring.

Postcoital Contraception (Plan B, Preven, Ovral)

Adult: **PO Ovral,** 2 tablets within 72 h of intercourse, then 2 tablets 12 h later; 1 (Plan B) or 2 **(Preven)** tablets within 72 h of unprotected intercourse, take second dose of 1 (Plan B) or 2 **(Preven)** tablets 12 h later

ADMINISTRATION

- Give without regard to meals.
- Do not exceed 24-h intervals between the daily doses; taking with a meal or at bedtime is a helpful reminder.

ADVERSE EFFECTS (≥1%) **Body as a Whole:** Paresthesias. **CV:** Malignant hypertension, thrombotic and thromboembolic disorders, *mild to moderate increase in BP,* increase in size of varicosities, edema. **Endocrine:** Estrogen excess (*nausea,* bloating, menstrual tension, cervical mucorrhea, polyposis, *chloasma, hypertension,* migraine headache, breast fullness or tenderness, edema); estrogen deficiency (hypomenorrhea, *early or mid-cycle breakthrough bleeding,* increased spotting); progestin excess (hypomenorrhea, breast regression, *vaginal candidiasis,* depression, fatigue, weight gain, increased appetite, acne, oily scalp, hair loss); progestin deficiency (late-cycle breakthrough bleeding, amenorrhea). **GI:** *Nausea,* cholelithiasis, gallbladder disease, cholestatic jaundice, benign hepatic adenomas; diarrhea, constipation, abdominal cramps. **Metabolic:** *Decreased glucose tolerance,* pyridoxine deficiency (see also diagnostic test interferences), acute intermittent porphyria. **Skin:** Rash (allergic), photosensitivity (photoallergy or phototoxicity), irritation from patch. **Special Senses:** Unexplained loss of vision, optic neuritis, proptosis, diplopia, change in corneal curvature (steepening), intolerance to contact lenses, retinal thrombosis, papilledema. **Urogenital:** Ureteral dilation, increased incidence of urinary tract infection, hemolytic uremia syndrome, renal failure, increased risk of congenital anomalies, decreased quality and quantity of breast milk, dysmenorrhea, increased size of preexisting uterine fibroids, *menstrual disorders.* Foreign body sensation, coital problems, device expulsion, vaginal discomfort, vaginitis, leukorrhea from ring.

DIAGNOSTIC TEST INTERFERENCE

ORAL CONTRACEPTIVES (OCS) increase **BSP** retention, **prothrombin** and **coagulation factors II, VII, VIII, IX, X; platelet agregability, thyroid-binding globulin, PBI, T_4; transcortin; corticosteroid, triglyceride** and **phospholipid** levels; **ceruloplasmin, aldosterone, amylase, transferrin; renin** activity, **vitamin A.** OCS decrease

E

Common adverse effects in *italic;* life-threatening effects <u>underlined;</u> generic names in **bold;** classifications in SMALL CAPS; ♣ Canadian drug name; ⊙ Prototype drug

577

E

antithrombin III, T₃ resin uptake, serum folate, glucose tolerance, albumin, vitamin B₁₂ and reduce the ***metyrapone*** test response.

INTERACTIONS Drug: Aminoca-proic acid may increase clotting factors, leading to hypercoagula-ble state; BARBITURATES, ANTICONVUL-SANTS, ANTIBIOTICS, **rifampin,** ANTI-FUNGALS reduce efficacy of OCs and increase incidence of breakthrough bleeding and risk of pregnancy. May decrease efficacy of **lamotri-gine. Herbal: St. John's wort** may decrease efficacy of OCs.

PHARMACOKINETICS Absorption: Oral: Readily from GI tract; or from transdermal patch placed on ab-domen, buttock, upper outer arm and upper torso (excluding breast). Vaginal insert: Norgestrel 100% ab-sorbed, ethinyl estradiol 56% ab-sorbed. **Peak:** Patch: 48 h. **Duration:** Patch: 1 wk. **Distribution:** Widely distributed; crosses placenta; small amount distributed into breast milk. **Metabolism:** In liver. **Elimination:** In urine and feces. **Half-Life:** 6–45 h oral. Following removal of the patch: Norelgestromin 28 h, ethinyl estradiol 17 h; vaginal ring: Nor-gestrel 29 h; ethinyl estradiol 45 h.

NURSING IMPLICATIONS

Assessment & Drug Effects

- Check BP periodically. In some women, changes in BP occur within each cycle; in others, slow increase of pressure, particularly diastolic, over several months is significant. Drug-induced BP ele-vation is usually reversible with cessation of OCs.
- Nausea with or without vomiting occurs in approximately 10% of pa-tients during the first cycle and is reportedly one of the major rea-sons for voluntary discontinuation

of therapy. Most adverse effects tend to disappear in third or fourth cycle of use. Instruct patient to re-port symptoms that persist after fourth cycle. Dose adjustment or a different product may be indicated.

- Hirsutism and loss of hair are re-versible with discontinuation of OCs or by change of selected combination.
- Acne may improve, worsen, or develop for first time. In women on OCs for at least 1 y, postcon-traceptive acne sometimes occurs 3–4 mo after stopping drug and may continue for 6–12 mo.
- Anovulation or amenorrhea fol-lowing termination of OC regi-men may persist more than 6 mo. The user with pretreatment oli-gomenorrhea or secondary amen-orrhea is most apt to have over-suppression syndrome.

Patient & Family Education

- Use an additional method of birth control during the first week of the initial cycle.
- Consult patient information sup-plied with drug for management of missed doses.
- Ovulation is unlikely with omis-sion of 1 daily dose; however, the possibility of escaped ovulation, spotting, or breakthrough bleeding increases with each missed dose.
- Discontinue medication if intra-cycle bleeding resembling men-struation occurs. Begin taking tab-lets from a new compact on day 5. If bleeding persists, see prescriber.
- Transdermal patches: Apply only one patch at a time and never cut or otherwise alter a patch prior to application.
- See prescriber to rule out pregnan-cy if 2 consecutive periods are missed, before continuing on OCs.
- Learn breast self-examination and do every month.

Common adverse effects in *italic*, life-threatening effects underlined; generic names in **bold**; classifications in SMALL CAPS; ♣ Canadian drug name; ⊘ Prototype drug

- Record frequent weight checks to permit early recognition of fluid retention.
- Understand the increased risk of thromboembolic and cardiovascular problems and increased incidence of gallbladder disease with OC use. Be alert to manifestations of thrombotic or thromboembolic disorders: Severe headache (especially if persistent and recurrent), dizziness, blurred vision, leg or chest pain, respiratory distress, unexplained cough. Discontinue drug if any of these symptoms appear and report them promptly to prescriber.
- Report sudden abdominal pain immediately to prescriber in order to rule out ectopic pregnancy.
- Stop drug and contact prescriber if unexplained partial or complete, sudden or gradual loss of vision, protrusion of eyeballs, or blurred vision occurs.
- If OC use is accompanied by vaginal itching and irritation, report to prescriber promptly to rule out candidiasis.
- Monitor blood glucose closely if diabetic. Adjustment of antidiabetic medication may be necessary.
- Use alternate method of birth control when breast feeding until infant is weaned.

ESTROGENS, CONJUGATED

(ess'tro-jenz)

C.E.S. ✦, Cenestin, Enjuvia, Premarin, Progens, SCE-A Vaginal Cream
Classification: ESTROGENS
Therapeutic: FEMALE HORMONE REPLACEMENT THERAPY (HRT)
Prototype: Estradiol
Pregnancy Category: X

AVAILABILITY 0.3 mg, 0.45 mg, 0.625 mg, 0.9 mg, 1.25 mg, 2.5 mg tablets; 25 mg injection; 0.625 mg/g vaginal cream

ACTION & *THERAPEUTIC EFFECT*
Circulating estrogens modulate the pituitary secretion of the gonadotropins luteinizing hormone (LH) and follicle stimulating hormone (FSH) through a negative feedback mechanism. Estrogens act to reduce the elevated levels of these gonadotropins seen in postmenopausal women. *Binds to intracellular receptors that stimulate DNA and RNA to synthesize proteins responsible for effects of estrogen.*

USES Atrophic vaginitis, kraurosis vulvae, and abnormal bleeding (hormonal imbalance); also female hypogonadism, primary ovarian failure, vasomotor symptoms associated with menopause; to retard progression of osteoporosis and as palliative therapy of breast and prostatic carcinomas.
UNLABELED USES Infertility, hyperparathyroidism.

CONTRAINDICATIONS Breast cancer, except for palliative therapy; vaginal and cervical cancers; endometrial cancer; endometrial hyperplasia; abnormal vaginal bleeding; hepatic disease or cancer; hypercalcemia; ovarian cancer; history of thromboembolic disease; known or suspected pregnancy (category X).
CAUTIOUS USE Hypertension; gallbladder disease; diabetes mellitus; heart failure; kidney dysfunction.

ROUTE & DOSAGE

Menopause, Osteoporosis, Atrophic Vaginitis, Kraurosis Vulvae
Adult: **PO** 0.3–1.25 mg/day for 21 days each month, adjust to

Common adverse effects in *italic*, life-threatening effects <u>underlined</u>; generic names in **bold**; classifications in SMALL CAPS; ✦ Canadian drug name; ⊘ Prototype drug

579

lowest level that gives symptom control (0.625 mg/day or less) **IV/IM** 25 mg, repeated in 6–12 h if needed **Topical** 2–4 g of cream/day

Female Hypogonadism

Adult: **PO** 2.5–7.5 mg/day in 1–3 divided doses for 20 days, followed by a 10-day rest period

Breast Cancer

Adult: **PO** 10 mg t.i.d. for at least 3 mo

Prostatic Cancer Palliation

Adult: **PO** 1.25–2.5 mg t.i.d.

ADMINISTRATION

Oral

- Give at the same time each day.

Topical

- Use calibrated dosage applicator dispensed with the cream.

Intramuscular

- Reconstitute by first removing approximately 5 mL of air from the dry-powder vial, then slowly inject the supplied diluent to the vial by aiming it at the side of the vial. Gently agitate to dissolve but DO NOT SHAKE.
- Use within a few hours of reconstitution.

Intravenous

PREPARE: **Direct:** Reconstitute as for IM injection.
ADMINISTER: **Direct:** Give slowly at a rate of 5 mg/min. ▪ Estrogen solution is compatible with D5W and NS and may be added to IV tubing just distal to the needle if necessary.
INCOMPATIBILITIES **Solution/additive: Ascorbic acid. Y-site: Pantoprazole.**

▪ Store ampule and reconstituted solution at 2°–8° C (38°–46° F) and protected from light; stable for 60 days. ▪ Discard precipitated or discolored solution.

ADVERSE EFFECTS (≥1%) **CNS:** Headache, dizziness, depression, *libido changes.* **CV:** <u>Thromboembolic disorders,</u> hypertension. **GI:** *Nausea,* vomiting, diarrhea, bloating, cholestatic jaundice. **Urogenital:** Mastodynia, spotting, changes in menstrual flow, dysmenorrhea, amenorrhea. **Metabolic:** Reduced carbohydrate tolerance, fluid retention. **Other:** Leg cramps, intolerance to contact lenses.

INTERACTIONS Drug: BARBITURATES, **carbamazepine, phenytoin, rifampin** decrease estrogen effect by increasing its metabolism; ORAL ANTICOAGULANTS may decrease hypoprothrombinemic effects; interfere with effects of **bromocriptine;** may increase levels and toxicity of **cyclosporine,** TRICYCLIC ANTIDEPRESSANTS, **theophylline;** decrease effectiveness of **clofibrate.**

PHARMACOKINETICS Absorption: Rapid absorption from GI tract; readily absorbed through skin and mucous membranes (including vaginal mucosa); slow absorption from IM injections. **Distribution:** Distributed throughout body tissues, especially in adipose tissue; crosses placenta, excreted in breast milk. Conjugated estrogens are bound primarily to albumin. **Metabolism:** Metabolized primarily in liver to glucuronide and sulfate conjugates of estradiol, estrone, and estriol. **Elimination:** In urine. **Half-Life:** 4–18 h.

NURSING IMPLICATIONS

Assessment & Drug Effects

- See additional implications under estradiol.

- Monitor for and report break-through vaginal bleeding.
- Assess for relief of menopausal symptoms.
- Lab tests: Monitor serum phosphatase levels with prostate cancer.
- Monitor bone density annually when used for osteoporosis prophylaxis.

Patient & Family Education
- Be aware of importance of taking drug exactly as prescribed: Specifically, do not omit, increase, or decrease doses without advice of prescriber.
- Intravaginal administration: For self-administration, wash hands well before and after application, and avoid contact of denuded areas with the cream. Do not use tampons while on vaginal cream therapy.
- Notify prescriber promptly of adverse symptoms.
- Know signs of thrombophlebitis (see Appendix F) and report promptly if suspected.
- Review package insert to ensure understanding of estrogen therapy.

ESTROGENS, ESTERIFIED
(ess'tro-jenz)

Estratab, Menest, Menrium, Neo-Estrone ✚
Classification: ESTROGEN
Therapeutic: ESTROGEN; FEMALE HORMONE REPLACEMENT THERAPY (HRT)
Prototype: Estradiol
Pregnancy Category: X

AVAILABILITY 0.3 mg, 0.625 mg, 1.25 mg, 2.5 mg tablets

ACTION & *THERAPEUTIC EFFECT*
At the cellular level, estrogens increase cervical secretions, result in proliferation of the endometrium, and increase uterine tone. Estrogens also can affect bone calcium deposition and accelerate epiphyseal closure. Estrogens appear to prevent osteoporosis associated with the onset of menopause; they generally do not reverse bone density loss that has already developed. *Binds to intracellular receptors that stimulate DNA and RNA to synthesize proteins responsible for effects of estrogen.*

USES Atrophic vaginitis, kraurosis vulvae and abnormal bleeding (hormonal imbalance), female hypogonadism, castration, primary ovarian failure, vasomotor symptoms associated with menopause, palliative therapy of breast and prostatic carcinomas; prevention of osteoporosis.

CONTRAINDICATIONS Breast cancer; cervical cancer; endometrial cancer; endometrial hyperplasia; prostate cancer; hepatic disease or cancer; hypercalcemia; lupus (SLE); history of thromboembolic disease; known or suspected pregnancy (category X); lactation.
CAUTIOUS USE Hypertension; gallbladder disease; diabetes mellitus; heart failure; kidney dysfunction; migraine headaches; seizure disorders.

ROUTE & DOSAGE

Menopause
Adult: **PO** 0.3–1.25 mg/day for 21 days each month, adjust to lowest level that gives symptom control (0.625 mg/day or less)
Female Hypogonadism, Primary Ovarian Failure, Female Castration
Adult: **PO** 2.5–7.5 mg/day in 1–3 divided doses for 20 days followed by a 10-day rest period,

Common adverse effects in *italic*, life-threatening effects <u>underlined</u>; generic names in **bold**; classifications in SMALL CAPS; ✚ Canadian drug name; ⊕ Prototype drug

581

during last 5 days of estrogen, give a PO progestin

Breast Cancer

Adult: **PO** 10 mg t.i.d. for 2–3 mo

Prostatic Cancer (palliation)

Adult: **PO** 1.25–2.5 mg t.i.d. for several weeks

Prevention of Osteoporosis

Adult: **PO** 0.3 mg daily

ADMINISTRATION

Oral

- Give with food or fluid of patient's choice.
- Store tablets at 15°–30° C (59°–86° F) in a tightly closed container.

ADVERSE EFFECTS (≥1%) **CNS:** Headache, dizziness, depression, *libido changes.* **CV:** Thromboembolic disorders, hypertension. **GI:** *Nausea*, vomiting, diarrhea, bloating, cholestatic jaundice. **Urogenital:** Mastodynia, spotting, changes in menstrual flow, dysmenorrhea, amenorrhea. **Metabolic:** Reduced carbohydrate tolerance, fluid retention. **Other:** Leg cramps, intolerance to contact lenses.

INTERACTIONS Drug: BARBITURATES, **phenytoin, rifampin** decrease estrogen effect by increasing its metabolism; ORAL ANTICOAGULANTS may decrease hypoprothrombinemic effects; interfere with effects of **bromocriptine;** may increase levels and toxicity of **cyclosporine,** TCAS, **theophylline;** decrease effectiveness of **clofibrate.**

PHARMACOKINETICS Absorption: Well absorbed with first pass metabolism. **Metabolism:** Metabolized in GI mucosa and liver to estrone, further metabolized to inactive metabolites. **Elimination:** In urine and bile. **Half-Life:** 4–18.5 h.

NURSING IMPLICATIONS

Assessment & Drug Effects

- See nursing implications under estradiol.
- Monitor for and report breakthrough vaginal bleeding.
- Assess for relief of menopausal symptoms.
- Lab tests: Monitor serum phosphatase levels with prostate cancer.
- Monitor bone density annually when used for osteoporosis prophylaxis.

Patient & Family Education

- Be aware of importance of taking drug exactly as prescribed: Specifically, do not omit, increase, or decrease doses without advice of prescriber. Know what to do when a dose is missed.
- Review package insert to ensure understanding of estrogen therapy.

ESTRONE

(ess'trone)

Classification: ESTROGEN
Therapeutic: ESTROGEN; FEMALE HORMONE REPLACEMENT THERAPY (HRT)
Prototype: Estradiol
Pregnancy Category: X

AVAILABILITY 5 mg/mL injection

ACTION & *THERAPEUTIC EFFECT* Due to increased risk of serious complications from extended use, estrogen HRT or estrogen-progestin HRT should be prescribed for the shortest duration possible consistent with the treatment goals of post menopausal symptoms. *Replaces estrogen in postmenopausal women, relieving symptoms of menopause.*

USES Atrophic vaginitis, kraurosis vulvae, and abnormal bleeding (hormonal imbalance); also female

Common adverse effects in *italic*, life-threatening effects <u>underlined</u>; generic names in **bold**; classifications in SMALL CAPS; ◆ Canadian drug name; ⊘ Prototype drug

hypogonadism, primary ovarian failure, vasomotor symptoms associated with menopause, and as palliative therapy of prostatic carcinoma.

CONTRAINDICATIONS Breast cancer; liver dysfunction; history of thromboembolic disease; known or suspected pregnancy (category X), lactation.

CAUTIOUS USE Hypertension; gallbladder disease; diabetes mellitus; heart failure; kidney dysfunction; seizure disorders.

ROUTE & DOSAGE

Menopause
Adult: **IM** 0.1–0.5 mg 2–3 times/wk

Female Hypogonadism, Primary Ovarian Failure
Adult: **IM** 0.1–1 mg/wk in single or divided doses

Inoperable Prostatic Cancer Palliation
Adult: **IM** 2–4 mg/day 2–3 times/wk

ADMINISTRATION

Intramuscular
- Shake vial and syringe well to suspend medication before withdrawing and injecting medication.
- Give deep into a large muscle.
- Store at 15°–30° C (59°–86° F); protect from light and do not freeze.

ADVERSE EFFECTS (≥1%) **CNS:** Headache, dizziness, depression, *libido changes.* **CV:** <u>Thromboembolic disorders</u>, hypertension. **GI:** *Nausea,* vomiting, diarrhea, bloating, cholestatic jaundice. **Urogenital:** Mastodynia, spotting, changes in menstrual flow, dysmenorrhea, amenorrhea. **Metabolic:** Reduced carbohydrate tolerance, fluid retention. **Other:** Leg cramps, intolerance to contact lenses.

INTERACTIONS Drug: Carbamazepine, phenytoin, rifampin decrease estrogen levels because they increase metabolism; may enhance steroid effects of CORTICOSTEROIDS; may decrease anticoagulant effects of ORAL ANTICOAGULANTS.

PHARMACOKINETICS Absorption: Occurs over several days. **Metabolism:** Converts to estradiol in GI mucosa. **Half-Life:** 4–18.5 h.

NURSING IMPLICATIONS

Assessment & Drug Effects
- See nursing implications under estradiol.
- Monitor for and report breakthrough vaginal bleeding.
- Assess for relief of menopausal symptoms.
- Lab tests: Monitor serum phosphatase levels with prostate cancer.
- Monitor patients with conditions that may be influenced by fluid retention carefully (e.g., migraine, cardiac or kidney dysfunction, asthma, epilepsy, hypertension). Check BP on a regular basis.

Patient & Family Education
- Review package insert to assure understanding of estrogen therapy.
- Determine weight under standard conditions 1 or 2 times/wk and report sudden weight gain or other signs of fluid retention.
- Notify prescriber of the following symptoms of thromboembolic disorders immediately: Tenderness, swelling, and redness in extremity; calf pain upon flexing foot; sudden, severe headache or chest pain; slurring of speech; change in vision; sudden shortness of breath.

Common adverse effects in *italic*, life-threatening effects <u>underlined</u>; generic names in **bold**; classifications in SMALL CAPS; ♣ Canadian drug name; ⊙ Prototype drug

E

- Report symptoms of vaginal candidiasis (thick, white, curd-like secretions and inflamed, congested introitus) to permit appropriate treatment.

ESTROPIPATE
(es-troe-pi'pate)

Ogen, Ortho-Est
Classification: ESTROGEN
Therapeutic: ESTROGEN; FEMALE HORMONE REPLACEMENT THERAPY (HRT)
Prototype: Estradiol
Pregnancy Category: X

AVAILABILITY 0.625 mg, 1.25 mg, 2.5 mg, 5 mg tablets; 1.5 mg/g cream

ACTION & *THERAPEUTIC EFFECT*
Water-soluble preparation of pure crystalline estrone. Due to increased risk of serious complications from extended use, estrogen HRT or estrogen-progestin HRT should be prescribed for the shortest duration possible consistent with the treatment goals of postmenopausal symptoms. *Replaces estrogen in postmenopausal women relieving symptoms of menopause.*

USES Atrophic vaginitis, kraurosis vulvae, and abnormal bleeding (hormonal imbalance); also female hypogonadism, primary ovarian failure, vasomotor symptoms associated with menopause, and as palliative therapy of prostatic carcinoma.

CONTRAINDICATIONS Estrogen hypersensitivity; breast cancer; vaginal cancer; endometrial hyperplasia; history of thromboembolic disease; known or suspected pregnancy (category X); lactation.

CAUTIOUS USE Hypertension; gallbladder disease; diabetes mellitus; heart failure; kidney dysfunction; seizure disorders.

ROUTE & DOSAGE

Menopause, Atrophic Vaginitis, Kraurosis Vulvae
Adult: **PO** 0.75–6 mg/day for 21 days each month; adjust to lowest level that gives symptom control **Intravaginal** 2–4 g of cream once/day in a cyclic regimen
Female Hypogonadism, Primary Ovarian Failure, Female Castration
Adult: **PO** 1.5–9 mg/day in 1–3 divided doses for 21 days, followed by an 8–10-day drug-free period

ADMINISTRATION

Oral
- Give with food or fluid of patient's choice.

Intravaginal
- Apply vaginal cream using calibrated dosage applicator dispensed with drug. Squeeze tube of cream to force sufficient amount into applicator so that number on plunger indicating prescribed dose is level with top of barrel.
- Store at 15°–30° C (59°–86° F) in tightly closed containers unless otherwise directed.

ADVERSE EFFECTS (≥1%) **CNS:** Headache, dizziness, depression, *libido changes.* **CV:** <u>Thromboembolic disorders</u>, edema, hypertension. **GI:** *Nausea,* vomiting, diarrhea, bloating, cholestatic jaundice. **Urogenital:** Mastodynia, spotting, changes in menstrual

Common adverse effects in *italic*, life-threatening effects <u>underlined</u>; generic names in **bold**; classifications in SMALL CAPS; ♣ Canadian drug name; ⦿ Prototype drug

flow, dysmenorrhea, amenorrhea. **Metabolic:** Reduced carbohydrate tolerance, fluid retention. **Other:** Leg cramps, intolerance to contact lenses.

INTERACTIONS Drug: Carbamazepine, phenytoin, rifampin decrease estrogen levels because they increase its metabolism; may enhance steroid effects of CORTICOSTEROIDS; may decrease anticoagulant effects of ORAL ANTICOAGULANTS. **Herbal: St. John's wort** may decrease blood levels. **Dong quai, red clover, black cohosh,** and **saw palmetto** may have additive hormonal effects.

PHARMACOKINETICS Absorption: Absorbed with some metabolism occuring in GI tract. Some systemic absorption from vaginal administration. **Metabolism:** In GI tract and liver. **Half-Life:** 4–18.5 h.

NURSING IMPLICATIONS

Assessment & Drug Effects
- See nursing implications under estradiol.
- Monitor for and report breakthrough vaginal bleeding.
- Assess for relief of menopausal symptoms.
- Lab tests: Monitor serum phosphatase levels with prostate cancer.

Patient & Family Education
- Do not use tampons while on vaginal cream therapy.
- Intravaginal administration: For self-administration, wash hands well before and after application.
- Pull plunger out of barrel and wash applicator in warm soapy water after use.
- Note: Sudden discontinuation of vaginal cream after high dosage or prolonged use may evoke withdrawal bleeding.

ESZOPICLONE
(es-zo′pi-clone)
Lunesta
Classifications: SEDATIVE-HYPNOTIC
Therapeutic: SEDATIVE-HYPNOTIC
Pregnancy Category: C
Controlled Substance: Schedule IV

AVAILABILITY 1 mg, 2 mg, 3 mg tablets

ACTION & *THERAPEUTIC EFFECT*
Eszopiclone is a nonbenzodiazepine sedative-hypnotic agent. Mechanism of action believed to result from its interaction with GABA-receptor complexes at binding sites close to or coupled to benzodiazepine receptors in the brain. *Improves sleep maintenance in transient insomnia.*

USES Treatment of insomnia.

CONTRAINDICATIONS Hypersensitivity to eszopiclone; concurrent administration with CYP3A4 inhibitor drugs; alcohol intoxication; alcoholism; eszopiclone induced angioedema; suicidal tendencies or ideation.
CAUTIOUS USE Hepatic impairment; elderly or debilitated patients; signs and symptoms of depression; COPD; pregnancy (category C), lactation. Safe use in children younger than 18 y is not established.

ROUTE & DOSAGE

Insomnia
Adult: **PO** 2–3 mg at bedtime Geriatric: **PO** 1–2 mg at bedtime
Severe Hepatic Impairment Dosage Adjustment
Start with 2 mg or less

Common adverse effects in *italic*, life-threatening effects <u>underlined</u>; generic names in **bold**; classifications in SMALL CAPS; ♣ Canadian drug name; ⊙ Prototype drug

585

ADMINISTRATION

Oral
- Give immediately prior to bedtime.
- Store at 15°–30° C (59°–8°6 F).

ADVERSE EFFECTS (≥1%) CNS:
Anxiety, confusion, depression, dizziness, hallucinations, *headache*, irritability, decreased libido, nervousness, *somnolence*. **CV:** *Tachycardia,* pericardial infusion, left ventricular systolic dysfunction (LVSD). **GI:** Dry mouth, dyspepsia, nausea, vomiting. **GU:** Dysmenorrhea, gynecomastia. **Respiratory:** Infection. **Skin:** Rash, pruritus. **Special Senses:** *Unpleasant taste.*

INTERACTIONS Drug:
Inhibitors of CYP3A4, including (but not limited to) **amiodarone,** ANTIRETROVIRAL PROTEASE INHIBITORS, **aprepitant, clarithromycin, dalfopristin/quinupristin, delavirdine, diltiazem, efavirenz** (inducer or inhibitor), **erythromycin, fluconazole, fluoxetine, fluvoxamine, itraconazole, ketoconazole, mifepristone, nefazodone, norfloxacin,** other systemic AZOLE ANTIFUNGALS (**miconazole** and **voriconazole**), **troleandomycin,** and **zafirlukast** increase eszopiclone levels. **Ethanol** and other CNS DEPRESSANT agents can produce additive effects in combination with eszopiclone. **Herbal: St. John's wort** can increase eszopiclone levels.

PHARMACOKINETICS Absorption:
Rapidly absorbed from GI tract. **Distribution:** 52–59% protein bound. **Peak:** 1 h. **Metabolism:** Extensive hepatic metabolism. **Elimination:** Primarily in the urine. **Half-Life:** 5–6 h.

NURSING IMPLICATIONS

Assessment & Drug Effects
- Monitor for and report worsening insomnia and cognitive or behavioral changes.
- Monitor for suicidal ideation in depressive patients.
- Monitor for S&S of CNS depression when other CNS depressants are used concurrently.
- Supervise ambulation if patient is out of bed after taking eszopiclone.

Patient & Family Education
- Follow directions for taking the drug (see Administration).
- Do not take this drug unless you can get at least 8 h of sleep.
- Do not consume alcohol while taking this drug.
- Do not drive or engage in potentially hazardous activities until response to drug is known.
- Report any of the following to a health care provider: Worsening insomnia, cognitive or behavioral changes, problem with reproductive function.

ETANERCEPT ⊚
(e-tan'er-cept)

Enbrel

Classifications: BIOLOGICAL RESPONSE MODIFIER; IMMUNOMODULATOR; TUMOR NECROSIS FACTOR (TNF) MODIFIER

Therapeutic: DISEASE-MODIFYING ANTIRHEUMATIC (DMARD); ANTIPSORIATIC

Pregnancy Category: B

AVAILABILITY 25 mg, 50 mg injection; 50 mg/mL prefilled syringe

ACTION & THERAPEUTIC EFFECT
Fusion protein produced by recombinant DNA technology that binds specifically to tumor necrosis factor

(TNF) and blocks it from attaching to cell surface TNF receptors. This naturally occurring cytokine (e.g., TNF) is part of the normal immune and inflammatory response. TNF mediates inflammation and modulates cellular immune responses. Elevated levels of TNF are found in the synovial fluids of rheumatoid arthritis (RA) patients. *Effectiveness is indicated by improved RA symptomatology and/ or decreased inflammation in other inflammatory disorders.*

USES Reduction of the signs and symptoms of RA and psoriatic RA in adults, and polyarticular juvenile RA (JRA) in children with inadequate response to other disease-modifying antirheumatic drugs. Treatment of ankylosing spondylitis, moderate-severe chronic plaque psoriasis.

CONTRAINDICATIONS Hypersensitivity to etanercept; malignancy; benzyl alcohol; benzyl alcohol hypersensitivity; patients with sepsis or other active infection; agranulocytosis; malignancy; bleeding, hematologic disease, intramuscular administration, intravenous administration; latex hypersensitivity; sepsis; varicella; lactation.

CAUTIOUS USE Immunosuppression; autoimmune disease, bone marrow suppression; diabetes mellitus; hamster protein hypersensitivity; heart failure; multiple sclerosis, neoplastic disease, neurologic disease, seizure disorder, seizures; vaccination, varicella, vasculitis; pregnancy (category B). Safety and efficacy in children younger than 2 y are not established.

ROUTE & DOSAGE

Rheumatoid Arthritis, Psoriatic Arthritis, Ankylosing Spondylitis

Adult: **Subcutaneous** 25 mg twice weekly; or 50 mg once weekly

Juvenile RA

Adolescent/Child (2–17 y): **Subcutaneous** 0.4 mg/kg (max: 25 mg/dose) twice weekly or 0.8 mg/kg weekly

Plaque Psoriasis

Adult: **Subcutaneous** 50 twice weekly (3–4 days apart) for 3 mo, then 50 mg weekly

ADMINISTRATION

Subcutaneous
- Reconstitute by slowly injecting the supplied diluent into the vial. Swirl gently to dissolve and do not shake. Reconstituted solution should be clear and colorless. Use within 6 h.
- Inject into thigh, abdomen, upper arm; rotate injection sites and never inject into an old injection site or where skin is tender, bruised, red, or hard.
- Store reconstituted solution up to 6 h refrigerated at 2°–8° C (36°–46° F). Store unopened dose tray refrigerated at 2°–8° C (36°–46° F).

ADVERSE EFFECTS (≥1%) **Body as a Whole:** Asthenia, serious *infections,* sepsis, monitor for reactivation of tuberculosis, increased malignancy risk. **CNS:** Headache, dizziness, cerebral ischemia, depression, demyelinating disorders (multiple sclerosis, myelitis, optic neuritis). **CV:** Heart failure, MI, myocardial ischemia, hypertension, hypotension. **GI:** Abdominal pain, dyspepsia, cholecystitis, pancreatitis, GI hemorrhage. **Respiratory:** Rhinitis, URI, pharyngitis, cough, respiratory disorder, sinusitis, dyspnea may reactivate latent tuberculosis (TB). **Skin:** Rash; injection site reactions (*erythema, itching, pain, swelling*).

Common adverse effects in *italic*, life-threatening effects underlined; generic names in **bold**; classifications in SMALL CAPS; ✦ Canadian drug name; ⊙ Prototype drug

587

Musculoskeletal: Bursitis. **Hematologic:** <u>Pancytopenia</u>.

INTERACTIONS Drug: Concurrent or recent use with **azathioprine, cyclophosphamide, leflunomide, methotrexate** has been associated with pancytopenia.

PHARMACOKINETICS Onset: 1–2 wk. **Peak:** 72 h. **Half-Life:** 115 h.

NURSING IMPLICATIONS

Assessment & Drug Effects
▪ Monitor carefully for and immediately report S&S of infection.

Patient & Family Education
▪ A PPD test is recommended before starting therapy to check for TB.
▪ Discard all needles and syringes after use; do not reuse.
▪ Withhold etanercept and notify prescriber before resuming drug if you develop an infection or are exposed to varicella virus.
▪ Avoid vaccinations, in general, and live vaccines, in particular, while on etanercept.
▪ Note: Injection site reactions (e.g., redness, pain, swelling) are common in the first month of therapy but generally decrease over time.

ETHACRYNIC ACID
(eth-a-krin′ik)
Edecrin

ETHACRYNATE SODIUM
Classifications: ELECTROLYTIC AND WATER BALANCE AGENT; LOOP DIURETIC
Therapeutic: LOOP DIURETIC
Prototype: Furosemide
Pregnancy Category: B

AVAILABILITY 25 mg, 50 mg tablet; 50 mg injection

ACTION & *THERAPEUTIC EFFECT*
Inhibits sodium and chloride reabsorption in proximal tubule and most segments of loop of Henle. Promotes significant fluid excretion by inhibiting reabsorption of a large proportion of filtered sodium. *Rapid and potent diuretic effect resulting in hypotensive effect. Fluid and electrolyte loss may exceed that caused by thiazides.*

USES Severe edema associated with CHF, hepatic cirrhosis, ascites of malignancy, kidney disease, nephrotic syndrome, lymphedema.
UNLABELED USES Treatment of nephrogenic diabetes insipidus, hypercalcemia, mild to moderate hypertension, and as adjunct in therapy of hypertensive crisis complicated by pulmonary edema.

CONTRAINDICATIONS History of hypersensitivity to ethacrynic acid; increasing azotemia, anuria; hepatic coma; severe diarrhea, dehydration, electrolyte imbalance, hypotension; lactation.
CAUTIOUS USE Hepatic cirrhosis, history of hepatic encephalopathy; severe myocardial disease; older adults, cardiac patients; diabetes mellitus; history of gout; pulmonary edema associated with acute MI; diabetic mellitus; hyperaldosteronism; nephrotic syndrome; history of pancreatitis; pregnancy (category B). Safe use in infants and neonates not established. Safe use of IV form in children not established.

ROUTE & DOSAGE

Edema
Adult: **PO** 50–100 mg 1–2 times/ day, may increase by 25–50 mg prn up to 400 mg/day **IV** 0.5–1

Common adverse effects in *italic*, life-threatening effects <u>underlined</u>; generic names in **bold**; classifications in SMALL CAPS; ✦ Canadian drug name; ⊘ Prototype drug

mg/kg or 50 mg up to 100 mg,
may repeat if necessary
Child: **PO** 1 mg/kg daily, may
increase to 3 mg/kg/day

ADMINISTRATION

Oral
- Give after a meal or food to prevent gastric irritation.
- Schedule doses to avoid nocturia and thus sleep interference. Avoid administration within at least 4 h of bedtime, if possible. This recommendation may not apply to the patient who accumulates fluid and develops respiratory symptoms during sleep.

Intravenous

PREPARE: **Direct:** Reconstitute by adding 50 mL of D5W or NS to vial. ▪Use solution within 24 h. ▪Vials reconstituted with D5W may turn cloudy; if so, discard the vial.
ADMINISTER: **Direct:** Give at a rate of 10 mg/min. May give through tubing of a freely flowing, compatible infusion. ▪If a second IV dose is required, a new site should be selected to prevent thrombophlebitis.
INCOMPATIBILITIES Solution/additive: **Hydralazine, procainamide, ranitidine, tolazoline, triflupromazine.**

- Store oral and parenteral form at 15°–30° C (59°–86° F) unless otherwise directed.

ADVERSE EFFECTS (≥1%) CNS:
Headache, fatigue, apprehension, confusion. **CV:** *Postural hypotension* (dizziness, light-headedness). **Metabolic:** Hyponatremia, *hypokalemia,* hypochloremic alkalosis, hypomagnesemia, hypocalcemia, hypercalciuria, hyperuricemia, hypovolemia, hematuria, glycosuria, hyperglycemia, gynecomastia, elevated BUN, creatinine, and urate levels. **Special Senses:** Vertigo, tinnitus, sense of fullness in ears, temporary or permanent deafness. **GI:** Anorexia, diarrhea, nausea, vomiting, dysphagia, abdominal discomfort or pain, GI bleeding (IV use), abnormal liver function tests. **Hematologic:** <u>Thrombocytopenia, agranulocytosis</u> (rare), <u>severe neutropenia</u> (rare). **Skin:** Skin rash, pruritus. **Body as a Whole:** Fever, chills, acute gout; local irritation and thrombophlebitis with IV injection.

INTERACTIONS Drug: THIAZIDE
DIURETICS increase potassium loss; increased risk of **digoxin** toxicity from hypokalemia; CORTICOSTEROIDS, **amphotericin B** increase risk of hypokalemia; decreased **lithium** clearance, so increased risk of lithium toxicity; SULFONYLUREA effect may be blunted, causing hyperglycemia; ANTIHYPERTENSIVE AGENTS increase risk of orthostatic hypotension; AMINOGLYCOSIDES may increase risk of ototoxicity; **warfarin** potentiates hypoprothrombinemia.

PHARMACOKINETICS Absorption:
Rapidly absorbed from GI tract. **Onset:** 30 min PO; 5 min IV. **Peak:** 5 min PO; 15–30 min IV. **Duration:** 6–8 h PO; 2 h IV. **Distribution:** Does not cross CSF. **Metabolism:** Metabolized to cysteine conjugate. **Elimination:** 30–65% in urine; 35–40% in bile. **Half-Life:** 30–70 min.

NURSING IMPLICATIONS

Assessment & Drug Effects
- Observe closely following IV infusion. Rapid, copious diuresis following IV administration can produce hypotension.
- Monitor IV site closely. Extravasation of IV drug causes local pain and tissue irritation from dehydration and blood volume depletion.

Common adverse effects in *italic,* life-threatening effects <u>underlined</u>; generic names in **bold;** classifications in SMALL CAPS; ♣ Canadian drug name; ⊙ Prototype drug

589

E

- Monitor BP during initial therapy. Because orthostatic hypotension can occur, supervise ambulation.
- Monitor BP and pulse throughout therapy in patients with impaired cardiac function. Diuretic-induced hypovolemia may reduce cardiac output, and electrolyte loss promotes cardiotoxicity in those receiving digitalis (cardiac) glycosides.
- Establish baseline weight prior to start of therapy; weigh patient under standard conditions. Keep prescriber informed of weight loss or gain in excess of 1 kg (2 lb)/day.
- Monitor I&O ratio. Report promptly excessive diuresis, oliguria, hematuria, or sudden profuse diarrhea. Report signs to prescriber.
- Lab tests: Determine baseline and periodic blood count, serum electrolytes, CO_2, BUN, creatinine, blood glucose, uric acid, and LFTs.
- Observe for and report S&S of electrolyte imbalance: Anorexia, nausea, vomiting, thirst, dry mouth, polyuria, oliguria, weakness, fatigue, dizziness, faintness, headache, muscle cramps, paresthesias, drowsiness, mental confusion.
- Report immediately possible signs of thromboembolic complications (see Appendix F).
- Impaired glucose tolerance with hyperglycemia and glycosuria has occurred in patients receiving doses in excess of 200 mg/day.

Patient & Family Education

- Learn S&S of hypokalemia and hyponatremia (see Appendix F), and report any of these promptly to prescriber.
- Make position changes slowly, particularly from lying to upright posture.
- Notify prescriber immediately of any evidence of impaired hearing.

Hearing loss may be preceded by vertigo, tinnitus, or fullness in ears; it may be transient, lasting 1–24 h, or it may be permanent.

ETHAMBUTOL HYDROCHLORIDE

(e-tham′byoo-tole)

Etibi ♣, Myambutol
Classification: ANTITUBERCULOSIS
Therapeutic: ANTITUBERCULAR
Prototype: Isoniazid
Pregnancy Category: B

AVAILABILITY 100 mg, 400 mg tablets

ACTION & *THERAPEUTIC EFFECT*
Ethambutol appears to inhibit RNA synthesis and thus arrests multiplication of tubercle bacilli. The emergence of resistant strains is delayed by administering ethambutol in combination with other antituberculosis drugs. *Synthetic antituberculosis agent that is also effective against atypical mycobacterial infections.*

USES In conjunction with other antituberculosis agents in treatment of pulmonary tuberculosis.
UNLABELED USES Atypical mycobacterial infections.

CONTRAINDICATIONS Hypersensitivity to ethambutol; optic neuritis, patients unable to report changes in vision (young children, or unconscious patients).
CAUTIOUS USE Renal impairment, hepatic disease; gout; ocular defects (e.g., cataract, recurrent ocular inflammatory conditions, diabetic retinopathy); pregnancy (category B). Safe use in children younger than 6 y not established.

Common adverse effects in *italic*; life-threatening effects underlined; generic names in **bold**; classifications in SMALL CAPS; ♣ Canadian drug name; ⑫ Prototype drug

ROUTE & DOSAGE

Tuberculosis

Adult: **PO** 15 mg/kg q24h; for retreatment start with 25 mg/kg/day for 60 days, then decrease to 15 mg/kg/day
Child (6–12 y): **PO** 10–15 mg/kg/day

ADMINISTRATION

Oral

- Give with food if GI irritation occurs.
- Protect ethambutol from light, moisture, and excessive heat. Store at 15°–30° C (59°–86° F) in tightly closed container unless otherwise directed.

ADVERSE EFFECTS (≥1%) CNS:
Headache, dizziness, confusion, hallucinations, paresthesias, joint pains. **Special Senses:** Ocular toxicity: *Retrobulbar optic neuritis;* possibility of anterior optic neuritis with decrease in visual acuity, temporary loss of vision, constriction of visual fields, red–green color blindness, central and peripheral scotomas, eye pain, photophobia; retinal hemorrhage and edema. **GI:** Anorexia, nausea, vomiting, abdominal pain. **Body as a Whole:** Hypersensitivity (pruritus, dermatitis, <u>anaphylaxis</u>).

INTERACTIONS Drug: Aluminum-containing antacids can decrease absorption.

PHARMACOKINETICS Absorption:
70–80% from GI tract. **Peak:** 2–4 h. **Distribution:** Distributes to most body tissues; highest concentrations in erythrocytes, kidney, lungs, saliva; crosses placenta; distributed into breast milk. **Metabolism:** In liver. **Elimination:** 50% in urine within 24 h; 20–22% in feces. **Half-Life:** 3–4 h.

NURSING IMPLICATIONS

Assessment & Drug Effects

- Perform C&S prior to and periodically throughout therapy.
- Ophthalmoscopic examination is recommended prior to and at monthly intervals during therapy.
- Monitor I&O ratio in patients with renal impairment. Report oliguria or any significant changes in ratio or in laboratory reports of kidney function. Systemic accumulation with toxicity can result from delayed drug excretion.
- Lab tests: Monitor LFTs and kidney function tests, CBC, and serum uric acid levels at regular intervals throughout therapy.

Patient & Family Education

- Adhere to drug regimen exactly and keep follow-up appointments.
- Notify prescriber promptly of the onset of blurred vision, changes in color perception, constriction of visual fields, or any other visual symptoms. Have eyes checked periodically. Ethambutol can cause irreversible blindness due to optic neuritis.

ETHINYL ESTRADIOL
(eth′in-il ess-tra-dye′ole)

Estinyl, Feminone
Classification: ESTROGEN
Therapeutic: ESTROGEN; FEMALE HORMONE REPLACEMENT THERAPY (HRT)
Prototype: Estradiol
Pregnancy Category: X

AVAILABILITY 0.02 mg, 0.05 mg, 0.5 mg tablets

ACTION & *THERAPEUTIC EFFECT*
Potent oral estrogen given cyclically for short-term use. Ethinyl estradiol is not commonly used as

Common adverse effects in *italic*, life-threatening effects <u>underlined</u>; generic names in **bold**; classifications in SMALL CAPS; ♣ Canadian drug name; ⊗ Prototype drug

591

E

a single agent, but most commonly found in combination oral contraceptives. *May be used to prevent osteoporosis and relieve symptoms associated with menopause.*

USES Moderate to severe vasomotor symptoms associated with menopause; also postmenopausal osteoporosis, female hypogonadism, and as palliation for inoperable, metastatic cancer of female breast (at least 5 y postmenopause) and of the prostate.
UNLABELED USES Postcoital contraceptive.

CONTRAINDICATIONS Breast, ovarian, cervical, or endometrial cancer; endometrial hyperplasia; uterine or vaginal cancer; abnormal vaginal bleeding; hepatic disease or cancer; jaundice; MI; history of thromboembolic disease; heart failure; coagulopathies; lupus; known or suspected pregnancy (category X), lactation.
CAUTIOUS USE Hypertension; gallbladder disease; diabetes mellitus; kidney dysfunction.

ROUTE & DOSAGE

Menopause, Postmenopausal Osteoporosis
Adult: **PO** 0.02–0.05 mg/day for 21 days each month, adjust to lowest level that gives symptom control

Female Hypogonadism
Adult: **PO** 0.05 mg 1–3 times/day for 2 wk, followed by 2 wk of progestin, continue this regimen for 3–6 mo

Breast Cancer
Adult: **PO** 1 mg t.i.d. for 2–3 mo

Prostatic Cancer Palliation
Adult: **PO** 0.15–2 mg/day

Postcoital Contraceptive
Adult: **PO** 5 mg/day for 5 consecutive days beginning within 72 h of coitus

ADMINISTRATION

Oral
- Morning-after pill: Start drug within 24 h and not later than 72 h after sexual exposure when used as an emergency postcoital contraceptive. Perform a pregnancy test prior to dosing.
- Store at 15°–30° C (59°–86° F) in tight, light-resistant container.

ADVERSE EFFECTS (≥1%) **CNS:** Headache, dizziness, depression, *libido changes.* **CV:** <u>Thromboembolic disorders</u>, hypertension. **GI:** *Nausea,* vomiting, diarrhea, anorexia, weight changes, bloating, cholestatic jaundice. **Urogenital:** Mastodynia, breakthrough bleeding, changes in menstrual flow, dysmenorrhea, amenorrhea; in men: Impotence, gynecomastia, testicular atrophy. **Metabolic:** Reduced carbohydrate tolerance, fluid retention. **Body as a Whole:** Leg cramps, edema, intolerance to contact lenses.

INTERACTIONS Drug: **Carbamazepine, phenytoin, rifampin** decrease estrogen levels because they increase its metabolism; may enhance steroid effects of CORTICOSTEROIDS; may decrease anticoagulant effects of ORAL ANTICOAGULANTS.

PHARMACOKINETICS Absorption: 83% absorbed. **Metabolism:** Extensively metabolized in liver. **Elimination:** In urine and feces. **Half-Life:** 3–27 h.

NURSING IMPLICATIONS

Assessment & Drug Effects
- Check BP on a regular basis in patients with conditions that

E

may be influenced by fluid retention (migraine, cardiac or kidney dysfunction, asthma, epilepsy, hypertension).

Patient & Family Education
- Be aware that risk of blood clot formation is high. Notify prescriber immediately of calf pain upon foot flexion and the following symptoms of thromboembolic disorders: Tenderness, pain, swelling, and redness in extremity; sudden, severe headache or chest pain; slurring of speech; change in vision; sudden shortness of breath.
- Determine weight under standard conditions 1 or 2 times/wk and report sudden weight gain or other signs of fluid retention.
- Notify prescriber of yellow skin and sclera, pruritus, dark urine, and light-colored stools; history of jaundice in pregnancy increases the possibility of estrogen-induced jaundice.
- Report symptoms of vaginal candidiasis (thick, white, curd-like secretions and inflamed congested introitus) to permit appropriate treatment.
- Note: Estrogen-induced feminization and impotence in male patients are reversible with termination of therapy.
- Decrease caffeine intake from sources such as tea, coffee, and cola; estrogenic depression of caffeine metabolism may cause caffeinism.

ETHIONAMIDE

(e-thye-on-am′ide)

Trecator
Classifications: ANTITUBERCU-LOSIS; ANTILEPROSY (SULFONE)
Therapeutic: ANTITUBERCULAR; ANTI-LEPROSY

Prototype: Isoniazid
Pregnancy Category: C

AVAILABILITY 250 mg tablets

ACTION & THERAPEUTIC EFFECT
Ethionamide appears to inhibit mycolic acid synthesis, which disrupts the formation of the mycobacterial cell wall. *Effective against human and bovine strains of* Mycobacterium tuberculosis *and* M. kansasii *and some strains of* Mycobacterium avium-intracellulare *complex. Also active against* M. leprae.

USES Active tuberculosis infection (with other agents).
UNLABELED USES Atypical mycobacterial infections and tuberculous meningitis.

CONTRAINDICATIONS Hypersensitivity to ethionamide and chemically related drugs [e.g., isoniazid, niacin (nicotinamide)]; severe liver damage; hepatic encephalopathy.
CAUTIOUS USE Diabetes mellitus, liver dysfunction, history of psychiatric illnesses including depression; history of thyroid disease; pregnancy (category C), lactation, children younger than 12 y.

ROUTE & DOSAGE

Tuberculosis

Adult: **PO** 15–20 mg/kg/day
Child: **PO** 15–20 mg/kg/day in 2–3 equally divided doses (max: 1 g/day)

ADMINISTRATION

Oral
- Give with or after meals to minimize GI adverse effects. Some patients tolerate ethionamide best when it is taken as a single dose

Common adverse effects in *italic*; life-threatening effects underlined; generic names in **bold**; classifications in SMALL CAPS; ♣ Canadian drug name; ☻Prototype drug

593

after the evening meal or as a single dose at bedtime.

- About 50% of patients cannot tolerate a single dose larger than 500 mg because of GI adverse effects.
- Store in a cool, dry place at 8°–15° C (46°–59° F) in a tightly closed container unless otherwise directed.

ADVERSE EFFECTS (≥1%) **CNS:** Headache, restlessness, mental depression, drowsiness, dizziness, ataxia, hallucinations, paresthesias, convulsions. **GI:** Dose related and frequent; symptoms may be due to CNS stimulation rather than to GI irritation: Anorexia, *epigastric distress, nausea, vomiting,* metallic taste, *diarrhea,* stomatitis, sialorrhea. **Metabolic:** Elevated ALT, AST; hepatitis (with jaundice), hypothyroidism. **Urogenital:** Menorrhagia, impotence. **Body as a Whole:** Postural hypotension.

INTERACTIONS Drug: Cycloserine, isoniazid may increase neurotoxic effects.

PHARMACOKINETICS Absorption: 80% absorbed from GI tract. **Peak:** 3 h. **Duration:** 9 h. **Distribution:** Widely distributed including CSF; crosses placenta; distribution into breast milk unknown. **Metabolism:** In liver. **Elimination:** In urine. **Half-Life:** 3 h.

NURSING IMPLICATIONS

Assessment & Drug Effects

- Lab tests: Perform C&S prior to start of therapy. Baseline LFTs, CBC, and kidney function tests including urinalysis and every 2–4 wk during therapy.
- Report onset of skin rash. Progression to exfoliative dermatitis can occur if drug is not promptly discontinued.

- Monitor blood glucose closely in the diabetic until response to drug is established. Diabetics appear to be especially prone to hepatotoxicity (see Appendix F).

Patient & Family Education

- Avoid alcohol or use in moderation because ethionamide may increase potential for liver dysfunction.
- Notify prescriber of S&S of hepatotoxicity (see Appendix F); generally reversible if drug is promptly withdrawn.
- Make position changes slowly and in stages, particularly from lying to upright posture if experiencing hypotension.

ETHOSUXIMIDE ⓟ

(eth-oh-sux′i-mide)

Zarontin

Classification: SUCCINIMIDE ANTICONVULSANT

Therapeutic: ANTICONVULSANT

Pregnancy Category: C

AVAILABILITY 250 mg capsules; 250 mg/5 mL syrup

ACTION & THERAPEUTIC EFFECT Succinimide anticonvulsant that reduces the current in T-type calcium channel found on primary afferent neurons. Activation of the T channel causes low-threshold calcium spikes in neurons, believed to play a role in the spike-and-wave pattern observed during absence (petit mal) seizures. *Reduces frequency of epileptiform attacks, apparently by depressing motor cortex and elevating CNS threshold to stimuli.*

USES Management of absence (petit mal) seizures, myoclonic

Common adverse effects in *italic*, life-threatening effects underlined; generic names in **bold;** classifications in SMALL CAPS; ♣ Canadian drug name; ⓟ Prototype drug

seizures, and akinetic epilepsy. May be administered with other anticonvulsants when other forms of epilepsy coexist with petit mal.

CONTRAINDICATIONS Hypersensitivity to succinimides; severe liver or kidney disease; bone marrow suppression; use alone in mixed types of epilepsy (may increase frequency of grand mal seizures).

CAUTIOUS USE Hematologic disease; preexisting hepatic disease; intermittent porphyria; renal disease; pregnancy (category C). Safe use in children younger than 3 y not established.

ROUTE & DOSAGE

Absence Seizures

Adult/Child (6–12 y): **PO** 250 mg b.i.d., may increase q4–7days prn (max: 1.5 g/day)
Child (3–6 y): **PO** 250 mg/day, may increase q4–7days prn (max: 1.5 g/day)

ADMINISTRATION

Oral

- Give with food if GI distress occurs.
- Store all forms at 15°–30° C (59°–86° F); capsules in tight containers, and syrup in light-resistant containers; avoid freezing.

ADVERSE EFFECTS (≥1%) **CNS:** Drowsiness, hiccups, ataxia, dizziness, headache, euphoria, restlessness, irritability, anxiety, hyperactivity, aggressiveness, inability to concentrate, lethargy, confusion, sleep disturbances, night terrors, hypochondriacal behavior, muscle weakness, fatigue. **Special Senses:** Myopia. **GI:** Nausea, vomiting, *anorexia, epigastric distress,* abdominal pain, *weight loss,* diarrhea, constipation, gingival hyperplasia. **Uro-**

genital: Vaginal bleeding. **Hematologic:** Eosinophilia, leukopenia, thrombocytopenia, agranulocytosis, pancytopenia, aplastic anemia, positive direct Coombs' test. **Skin:** Hirsutism, pruritic erythematous skin eruptions, urticaria, alopecia, erythema multiforme, exfoliative dermatitis.

INTERACTIONS Drug: Carbamazepine decreases ethosuximide levels; **isoniazid** significantly increases ethosuximide levels; levels of both **phenobarbital** and ethosuximide may be altered with increased seizure frequency. **Herbal: Ginkgo** may decrease anticonvulsant effectiveness.

PHARMACOKINETICS Absorption: Readily from GI tract. **Peak:** 4 h; steady state: 4–7 days. **Metabolism:** In liver. **Elimination:** In urine; small amounts in bile and feces. **Half-Life:** 30 h (child), 60 h (adult).

NURSING IMPLICATIONS

Assessment & Drug Effects

- Lab tests: Perform baseline and periodic hematologic studies, LFTs and kidney function tests.
- Monitor adverse drug effects. GI symptoms, drowsiness, ataxia, dizziness, and other neurologic adverse effects occur frequently and indicate the need for dosage adjustment.
- Observe closely during period of dosage adjustment and whenever other medications are added or eliminated from the drug regimen. Therapeutic serum levels: 40–80 mcg/mL.
- Observe patients with prior history of psychiatric disturbances for behavioral changes. Close supervision is indicated. Drug should be withdrawn slowly if these symptoms appear.

Common adverse effects in *italic*, life-threatening effects <u>underlined</u>; generic names in **bold**; classifications in SMALL CAPS; ✦ Canadian drug name; ⓟ Prototype drug

595

Patient & Family Education

- Discontinue drug only under prescriber supervision; abrupt withdrawal of ethosuximide (whether used alone or in combination therapy) may precipitate seizures or petit mal status.
- Do not drive or engage in other potentially hazardous activities until response to drug is known.
- Monitor weight on a weekly basis. Report anorexia and weight loss to prescriber; may indicate need to reduce dosage.

ETIDRONATE DISODIUM ⓟ

(e-ti-droe'nate)

Didronel, EHDP
Classifications: BISPHOSPHONATE; BONE METABOLISM REGULATOR
Therapeutic: BONE METABOLISM REGULATOR
Pregnancy Category: C

AVAILABILITY 200 mg, 400 mg tablets

ACTION & THERAPEUTIC EFFECT
Diphosphate preparation with primary action on bone. Reduces elevated cardiac output associated with Paget's disease by decreasing vascularity of bone. Induces reversible hyperphosphatemia without adverse effects. *Slows rate of bone resorption and new bone formation in pagetic bone lesions and in normal remodeling process. Response of Paget's disease may be slow (1–3 mo) and may continue for months after treatment is discontinued.*

USES Symptomatic Paget's disease and heterotopic ossification due to spinal cord injury or after total hip replacement.

UNLABELED USES Prevention and treatment of corticosteroid-induced osteoporosis.

CONTRAINDICATIONS Enterocolitis; pathologic fractures; renal failure; lactation.

CAUTIOUS USE Renal impairment; asthma; colitis; dysphagia; esophagitis; gastritis; patients on restricted calcium and vitamin D intake; pregnancy (category C). Safe use in children younger than 18 y not established.

ROUTE & DOSAGE

Paget's Disease
Adult: **PO** 5–10 mg/kg/day for up to 6 mo or 11–20 mg/kg/day for up to 3 mo, may repeat after 3–6 mo off the drug if necessary
Heterotopic Ossification Due to Spinal Cord Injury
Adult: **PO** 20 mg/kg/day for 2 wk, then 10 mg/kg/day for an additional 10 wk
Heterotopic Ossification Due to Total Hip Arthroplasty
Adult: **PO** 20 mg/kg/day starting 1 mo before the procedure and continuing for 3 mo after

ADMINISTRATION

Oral

- Give as single dose on empty stomach 2 h before meals with full glass of water or juice to reduce gastric irritation.
- Relieve GI adverse effects by dividing total oral daily dose.

ADVERSE EFFECTS (≥1%) **GI:** Nausea, diarrhea, *loose bowel movements,* metallic or altered taste. **Musculoskeletal:** Increased or recurrent bone pain in pagetic sites, onset of bone pain in previously asymptomatic sites, increased risk of fractures in patient with Paget's disease. **Metabolic:** Hypocalcemia,

Common adverse effects in *italic*, life-threatening effects underlined; generic names in **bold**; classifications in SMALL CAPS; ♣ Canadian drug name; ⓟ Prototype drug

hyperphosphatemia, elevated serum phosphatase, suppressed mineralization of uninvolved skeleton (focal osteomalacia). **Urogenital:** Renal insufficiency (high doses).

INTERACTIONS Drug: CALCIUM SUPPLEMENTS, ANTACIDS, IRON AND OTHER MINERAL SUPPLEMENTS may decrease absorption of etidronate (give etidronate 2 h before other drugs). **Food:** Food, especially milk and dairy products, will decrease absorption of etidronate (give 2 h before meals).

PHARMACOKINETICS Absorption: Variably from GI tract. **Distribution:** 50% distributed to bone. **Metabolism:** Not metabolized. **Elimination:** 50% in urine. **Half-Life:** 6 h.

NURSING IMPLICATIONS

Assessment & Drug Effects

- Report persistent nausea or diarrhea; GI adverse effects may interfere with adequate nutritional status and need to be treated promptly.
- Monitor I&O ratio, serum creatinine, or BUN of patient with impaired kidney function.
- Lab tests: Periodic serum calcium and phosphate.
- Monitor for signs of hypocalcemia. Latent tetany (hypocalcemia) may be detected by Chvostek's and Trousseau's signs and a serum calcium value of 7–8 mg/dL.
- Note: Serum phosphate levels generally return to normal 2–4 wk after medication is discontinued.

Patient & Family Education

- Avoid eating 2 h before or after taking etidronate. Drug absorption is decreased by food, especially milk, milk products, and other foods high in calcium, mineral supplements, and antacids.

- Notify prescriber promptly of sudden onset of unexplained pain. Risk of pathological fractures increases when daily dose of 20 mg/kg is taken longer than 3 mo.
- Report promptly if bone pain, restricted mobility, heat over involved bone site occur.

ETODOLAC
(e-to'do-lac)

Classifications: ANALGESIC, NONSTEROIDAL ANTI-INFLAMMATORY AGENT (NSAID); DISEASE-MODIFYING ANTIRHEUMATIC DRUG (DMARD)
Therapeutic: ANALGESIC, NSAID; DMARD; ANTIPYRETIC
Prototype: Ibuprofen
Pregnancy Category: C first and second trimester; D third trimester

AVAILABILITY 400 mg, 500 mg tablets; 200 mg, 300 mg capsules; 400 mg, 500 mg, 600 mg sustained release tablets

ACTION & *THERAPEUTIC EFFECT* Inhibits cyclooxygenase (COX-1 and COX-2) enzyme activity and prostaglandin synthesis. NSAIDs may also suppress production of rheumatoid factor. *Produces analgesic and anti-inflammatory effects of an NSAID.*

USES Osteoarthritis and acute pain, rheumatoid arthritis.

CONTRAINDICATIONS Hypersensitivity to NSAIDs, salicylates; ulceration or inflammation; perioperative CABG pain; asthma, urticaria, or other allergic reactions to aspirin or other NSAIDs; S&S of developing liver disease; pregnancy (category D third trimester).
CAUTIOUS USE Renal impairment, liver function impairment,

Common adverse effects in *italic*, life-threatening effects underlined; generic names in **bold;** classifications in SMALL CAPS; ♦ Canadian drug name; ⊙ Prototype drug

597

GI disorders, history of GI ulceration, GI bleeding; cardiac disorders including fluid retention, hypertension, heart failure; dehydration; preexisting hematologic diseases (e.g., coagulopathy and hemophilia) or thrombocytopenia; IM injections; dental work; diabetes mellitus; surgery when hemostasis is required; immunosuppression, neutropenia; patients over 65 y, pregnancy (category C first and second trimester), lactation. Safe use in children younger than 6 y not established.

ROUTE & DOSAGE

Acute Pain

Adult: PO 200–400 mg q6–8h prn

Osteoarthritis

Adult: PO 600–1200 mg/day in divided doses, (max: 1200 mg/day; extended release form: 400–1000 mg daily)

Juvenile Rheumatoid Arthritis

Adolescent/Child (over 6): PO Dose based on weight
20–30 kg: 400 mg
31–45 kg: 600 mg
46–60 kg: 800 mg
over 60 kg: 1000 mg

ADMINISTRATION

Oral

- Give with food or antacid to reduce risk of GI ulceration.
- Ensure that sustained release form of drug is not chewed or crushed. It **must be** swallowed whole.
- Store at 15°–25° C (59°–77° F); tablets and capsules in bottles; sustained release capsules in unit-dose packages. Protect all forms from moisture.

ADVERSE EFFECTS (≥1%) **CV:** Fluid retention, edema. **CNS:** Dizziness, headache, drowsiness, insomnia. **GI:** *Dyspepsia, nausea, vomiting, diarrhea,* indigestion, heartburn, abdominal pain, constipation, flatulence, gastritis, melena, peptic ulcer, GI bleeding. **Hematologic:** Thrombocytopenia, increased bleeding time. **Skin:** Rash, pruritus. **Urogenital:** Urinary frequency. **Metabolic:** Hepatotoxicity. **Special Senses:** Blurred vision; tinnitus. **Respiratory:** Asthma.

DIAGNOSTIC TEST INTERFERENCE May cause a false-positive **urinary bilirubin** test and a false-positive **ketone** test done with the dipstick method. May cause a small decrease (1 to 2 mg/dL) in **serum uric acid** levels.

INTERACTIONS Drug: May reduce effects of **diuretics** and antihypertensive effects of **beta-blockers** and other ANTIHYPERTENSIVE MEDICATIONS. May increase **digoxin** and **lithium** levels and nephrotoxicity due to **cyclosporine. Herbal:** Feverfew, garlic, ginger, ginkgo may increase bleeding.

PHARMACOKINETICS Absorption: Readily from GI tract. **Onset:** 30 min. **Peak:** 1–2 h. **Duration:** 4–12 h. **Distribution:** Widely distributed; 99% protein bound; not known if crosses placenta or if distributed into breast milk. **Metabolism:** Extensively in liver. **Elimination:** 72% in urine, 16% in feces. **Half-Life:** 6–7 h.

NURSING IMPLICATIONS

Assessment & Drug Effects

- Assess for signs of GI ulceration and bleeding. Risk factors include high doses of etodolac, history of peptic ulcer disease, alcohol use, smoking, and concomitant use of aspirin.

- Assess carefully for fluid retention by monitoring weight and observing for edema in patients with a history of CHF.
- Monitor for decreased BP control in hypertensive patients.
- Lab tests: Periodic CBC, kidney function tests and LFTs.
- Monitor for drug toxicity when used concurrently with either digoxin or lithium.
- Monitor for rhinitis, urticaria, or other signs of allergic reactions.
- Monitor carefully increases in etodolac dosage with older adult patients; adverse effects are more pronounced.

Patient & Family Education
- Learn S&S of GI ulceration. Stop medication in presence of bleeding and contact the prescriber immediately.
- Do not take aspirin, which may potentiate ulcerogenic effects.

ETOPOSIDE
(e-toe-po'side)

Etopophos, Toposar
Classifications: ANTINEOPLASTIC; MITOTIC INHIBITOR
Therapeutic: ANTINEOPLASTIC, CELL-CYCLE SPECIFIC
Prototype: Vincristine
Pregnancy Category: D

AVAILABILITY 50 mg capsules; 20 mg/mL injection; 100 mg lyophilized powder for injection

ACTION & *THERAPEUTIC EFFECT*
Produces cytotoxic action by arresting G_2 (resting or premitotic) phase of cell cycle; also acts on S phase of DNA synthesis. High doses cause lysis of cells entering mitotic phase, and lower doses inhibit cells from entering prophase. *Antineoplastic effect is due to its ability to arrest mitosis (cell division).*

USES Lung cancer, testicular cancer, small cell lung cancer.
UNLABELED USES Hodgkin's and non-Hodgkin's lymphomas, acute myelogenous (nonlymphocytic) leukemia, ovarian cancer, thyoma, trophoblastic disease.

CONTRAINDICATIONS Severe bone marrow depression; severe hepatic or renal impairment; existing or recent viral infection, bacterial infection; intraperitoneal, intrapleural, or intrathecal administration; pregnancy (category D), lactation.
CAUTIOUS USE Impaired kidney or liver function; gout; radiation therapy. Safe use in children not established.

ROUTE & DOSAGE

Testicular Carcinoma
Adult: **IV** 100 mg/m²/day for 5 consecutive days or 100 mg/m² on days 1, 3, and 5 q3–4wk for 3–4 courses **PO** Twice the IV dose rounded to the nearest 50 mg
Small Cell Lung Carcinoma
Adult: **IV** 35 mg/m²/day for 4 consecutive days to 50 mg/m²/day for 5 consecutive days q3–4wk **PO** Twice the IV dose rounded to the nearest 50 mg
Hepatic Impairment Dosage Adjustment
Total bilirubin 1.5–3 mg/dL: Decrease by 50%, 3.1–5 mg/dL: Decrease by 75%, over 5 mg/dL: hold dose
Renal Impairment Dosage Adjustment
CrCl 45–60 mL/min: Reduce dose 15%; 30–44 mL/min: Reduce dose 20%; less than 30 mL/min: Reduce dose 25%

Common adverse effects in *italic*, life-threatening effects underlined; generic names in **bold**; classifications in SMALL CAPS; ♣ Canadian drug name; ⊙ Prototype drug

E

ADMINISTRATION

Oral

- Oral dose is usually in the range of 70–100 mg/m^2 daily, rounded to nearest 50 mg.
- Refrigerate capsules at 2°–8° C (36°–46° F) unless otherwise directed. Do not freeze.

Intravenous

Note: Wear disposable surgical gloves when preparing or disposing of etoposide. Wash immediately with soap and water if skin comes in contact with drug.

PREPARE: **IV Infusion:** *Etoposide concentration for injection:* Each 100 mg **must be** diluted with 250–500 mL of D5W or NS to produce final concentrations of 0.2–0.4 mg/mL. *Etoposide phosphate:* Add 5 or 10 mL of sterile water for injection, D5W, NS, bacteriostatic water for injection, or bacteriostatic NS for injection to yield 20 or 10 mg/mL etoposide, respectively. ■ May be given as prepared or further diluted to as low as 0.1 mg/mL in either D5W or NS.

ADMINISTER: **IV Infusion:** Give by slow IV infusion over 30–60 min to reduce risk of hypotension and bronchospasm. ■ Before administration, inspect solution for particulate matter and discoloration. Solution should be clear and yellow. If crystals are present, discard.

INCOMPATIBILITIES **Y-site:** **Cefepime, dantrolene, diazepam, filgrastim, gallium, gemtuzumab, idarubicin, indomethacin, lansoprazole, pantoprazole, phenytoin, thiopental.**

■ Diluted solutions with concentration of 0.2 mg/mL are stable for 96 h, and the 0.4 mg/mL solutions are stable for 24 h under normal room fluorescent light in glass or plastic (PVC) containers. ■ Phosphate solution is stable for 24 h at room temperature or refrigerated.

ADVERSE EFFECTS (≥1%) Body as a Whole:

Hypersensitivity (sweating, chills, fever, coryza, tachycardia; throat, back and general body pain; abdominal cramps, flushing, substernal chest pain, dyspnea, bronchospasm, pulmonary edema, anaphylactoid reaction). **CNS:** Peripheral neuropathy, paresthesias, weakness, somnolence, unusual tiredness, transient confusion. **CV:** Transient hypotension; thrombophlebitis with extravasation. **GI:** *Nausea, vomiting,* dyspepsia, anorexia, diarrhea, constipation, stomatitis. **Hematologic:** Leukopenia *(principally granulocytopenia),* thrombocytopenia, severe myelosuppression, *anemia, pancytopenia, neutropenia.* **Respiratory:** Pleural effusion, bronchospasm. **Skin:** *Reversible alopecia* (can progress to total baldness); radiation recall dermatitis; necrosis, *pain at IV site.*

INTERACTIONS Drug:

ANTICOAGULANTS, ANTIPLATELET AGENTS, NSAIDS, **aspirin** may increase risk of bleeding. Avoid concurrent use of LIVE VACCINES. **Food: Grapefruit juice** may decrease effect.

PHARMACOKINETICS Absorption:

Approximately 50% from GI tract. **Peak:** 1–1.5 h. **Distribution:** Variable penetration into CSF. **Metabolism:** In liver. **Elimination:** 44–60% in urine, 2–16% in feces over 3 days. **Half-Life:** 5–10 h.

NURSING IMPLICATIONS

Assessment & Drug Effects

■ Check IV site during and after infusion. Extravasation can cause thrombophlebitis and necrosis.

- Be prepared to treat an anaphylactoid reaction (see Appendix F). Stop infusion immediately if the reaction occurs.
- Monitor vital signs during and after infusion. Stop infusion immediately if and kidney function tests.
- Lab Tests: Prior to each cycle and at frequent intervals CBC with platelet count; periodic renal function and LFTs.
- Withhold therapy when an absolute neutrophil count is below 500/mm³ or a platelet count below 50,000/mm³.
- Be alert to evidence of patient complaints that might suggest development of leukopenia (see Appendix F), infection (immunosuppression), and bleeding.
- Protect patient from any trauma that might precipitate bleeding during period of platelet nadir particularly. Withhold invasive procedures if possible.

Patient & Family Education
- Learn possible adverse effects of etoposide, such as blood dyscrasias, alopecia, carcinogenesis, before treatment begins.
- Make position changes slowly, particularly from lying to upright position because transient hypotension after therapy is possible.
- Inspect mouth daily for ulcerations and bleeding. Avoid obvious irritants such as hot or spicy foods, smoking, alcohol.

ETRAVIRINE
(e-tra′vi-reen)

Intelence
Classifications: ANTIRETROVIRAL; NONNUCLEOSIDE REVERSE TRANSCRIPTASE INHIBITOR (NNRTI)
Therapeutic: ANTIRETROVIRAL; NNRTI

Prototype: Efavirenz
Pregnancy Category: B

AVAILABILITY 100 mg tablets

ACTION & THERAPEUTIC EFFECT Prevents replication of HIV-1 viruses by binding directly to reverse transcriptase, thus blocking RNA- and DNA-dependent polymerase activities. *Effectiveness is indicated by reduction in viral load (plasma level HIV RNA).*

USES HIV-1 infection in combination with other antiretroviral agents in treatment-experienced adult patients resistant to other antiretroviral agents (including other NNRTIs).

CONTRAINDICATIONS Severe skin reactions; lactation.
CAUTIOUS USE Severe hepatic impairment (Child-Pugh class C); concurrent hepatitis B or C, dyslipidemia; older adults; pregnancy (category B). Safety and efficacy in children not established.

ROUTE & DOSAGE

HIV Infection
Adult: **PO** 200 mg b.i.d. p.c.

ADMINISTRATION

Oral
- Give after a meal. Ensure that tablets are not chewed.
- May dissolve in water if patient cannot swallow tablets. Once dissolved, should be swallowed immediately. Rinse glass several times and instruct to swallow each time to ensure entire dose has been administered.
- Store at 15°–30° C (59°–86° F). Keep bottles closed tightly to

Common adverse effects in *italic*, life-threatening effects underlined; generic names in **bold;** classifications in SMALL CAPS; ♣ Canadian drug name; ⊙ Prototype drug

601

protect from moisture. Do not remove desiccant pouches from bottle.

ADVERSE EFFECTS (≥1%) Body as a Whole: Peripheral neuropathy, rhabdomyolysis, **CNS:** Fatigue, headache. **CV:** Hypertension. **GI:** Abdominal pain, *diarrhea, nausea,* vomiting. **Metabolic:** Elevated creatinine, elevated LDL, elevated total cholesterol, elevated triglycerides, elevated glucose, elevated ALT. **Skin:** *Rash.*

INTERACTIONS Drug: Compounds that inhibit CYP3A4, CYP2C9, and/ or CYP2C19 (e.g., **itraconazole, ketoconazole**) may increase plasma levels of etravirine. Compounds that induce CYP3A4, CYP2C9, and/ or CYP2C19 (e.g., **carbamazepine, phenobarbital, phenytoin**) may decrease plasma levels of etravirine. Etravirine may decrease the plasma levels of other compounds that require CYP3A4 for metabolism (e.g., **HIV protease inhibitors**) and may increase the plasma levels of other compounds that require CYP2C9 and/or CYP2C19 for metabolism (e.g., **diazepam, warfarin**). **Herbal: St. John's wort** may decrease etravirine levels.

PHARMACOKINETICS Peak: 2.5 to 4 h. **Distribution:** 99.9% protein bound. **Metabolism:** In liver by CYP2C9, CYP2C19, and CYP3A4. **Elimination:** 93.7% in feces, 1.2% in urine. **Half-Life:** 41 h.

NURSING IMPLICATIONS

Assessment & Drug Effects
- Monitor for and report promptly potentially serious adverse reactions, including skin hypersensitivity reactions, muscle pain indicative of rhabdomyolysis, and S&S of hepatic dysfunction.

- Monitor for and report S&S of opportunistic infections.
- Lab tests: Periodic CD4+ T cell count, plasma HIV RNA, CBC with platelet count, serum amylase, LFTs, renal function tests, and lipid profile.

Patient & Family Education
- Do not take on an empty stomach.
- Do not remove drying-agent pouches from medication bottle.
- Report promptly any of the following: Rash, S&S of infection, or unexplained muscle pain.
- Report use of all prescription and nonprescription medications, as well as herbs, to prescriber.

EVEROLIMUS
(e-ver-o-li'mus)
Afinitor, Zortress
Classifications: BIOLOGIC RESPONSE MODIFIER; IMMUNOMODULATOR; ANTINEOPLASTIC; IMMUNOSUPPRESSANT
Therapeutic: ANTINEOPLASTIC
Pregnancy Category: D

AVAILABILITY 0.25 mg, 0.5 mg, 0.75 mg, 2.5 mg, 5 mg, 10 mg tablets

ACTION & THERAPEUTIC EFFECT
Binds to an intracellular protein of cancer cells of the kidney that inhibits a major dysfunctional kinase pathway in renal cancer development. It also reduces the expression of vascular endothelial growth factor (VEGF) in these cells. *Reduces cell proliferation, angiogenesis, and glucose uptake in renal carcinoma cells.*

USES Renal cell cancer or prophylaxis of kidney transplant rejection, astrocytoma, pancreatic neuroendocrine tumor (PNET).

CONTRAINDICATIONS Hypersensitivity to everolimus, or other rapamycin derivatives; hypersensitivity to sirolimus (**Zortess**) only; severe hepatic impairment (Child-Pugh class C) severe non-infection pneumonitis; fungal infection; live vaccine; severe hereditary problems of galactose intolerance; pregnancy (category D), lactation.

CAUTIOUS USE Child-Pugh class B hepatic impairment; renal impairment; severe refractory hyperlipidemia; DM; older adults.

ROUTE & DOSAGE

Renal Cell Cancer (Afinitor)

Adult: **PO** 10 mg once daily. 15–20 mg daily for patients taking a strong CYP3A4 inducer.

PNET

Adult: **PO** 10 mg daily

Astrocytomia (Affinitor only)

Adult/Adolescent/Child (over 3): **PO**
Pt BSA over 2.2m²: **7.5 mg daily**
Pt BSA of 1.3–2.1m²: **5 mg daily**
Pt BSA of 0.5–1.2 m²: **2.5 mg daily**

Kidney Transplant Rejection Prophylaxis (Zortress only)

Adult: **PO** 0.75 mg q12h

Hepatic Impairment Dosage Adjustment

Moderate impairment (Child-Pugh class B): Reduce to 5 mg once daily.

Kidney Transplant Rejection

Adult: **PO** 0.75 mg q12h.

ADMINISTRATION

Oral

- Give at the same time each day with/without food.

- Ensure that the tablet is swallowed whole. It should not be crushed or chewed.
- Store at 15°–30° C (59°–86° F). Protect from light and moisture.

ADVERSE EFFECTS (≥3%) Body as a Whole: *Asthenia,* chest pain, chills, epistaxis, *fatigue, infection,* peripheral edema, pyrexia. **CNS:** Dizziness, dysgeusia, headache, insomnia, paresthesia. **CV:** Hypertension, <u>tachycardia</u>. **GI:** Abdominal pain, *diarrhea,* dry mouth, dysphagia, hemorrhoids, *mucosal inflammation,* nausea, *stomatitis,* vomiting. **Hematologic:** *Anemia, decreased hemoglobin,* decreased neutrophils, decreased platelets, *lymphopenia,* hemorrhage. **Metabolic:** Decreased weight, elevated AST and ALT, elevated bilirubin, elevated creatinine, *hypercholesterolemia, hypertriglyceridemia, hyperglycemia, hypophosphatemia.* **Musculoskeletal:** Jaw pain, pain in extremity. **Respiratory:** Alveolitis, bronchitis, *cough, dyspnea,* interstitial lung disease, lung infiltration, nasopharyngitis, pneumonia, *pneumonitis,* pulmonary alveolar hemorrhage, pulmonary effusion, pharyngolaryngeal pain, rhinorrhea, sinusitis. **Skin:** Acneiform dermatitis, dry skin, erythema, hand-foot syndrome, nail disorder, onychoclasis, pruritus, *rash,* skin lesion. **Special Senses:** Eyelid edema. **Urogenital:** <u>Renal failure</u>

INTERACTIONS Drug: Strong inhibitors of CYP3A4 and P-glycoprotein (e.g., **ketoconazole, erythromymin, veramapil**) increase everolimus levels. Strong inducers (e.g., **rifampin**) decrease everolimus levels.

PHARMACOKINETICS Peak: 1–2 h. **Distribution:** 74% plasma protein bound. **Metabolism:** In liver to inactive metabolites. **Elimination:**

Common adverse effects in *italic*, life-threatening effects <u>underlined</u>; generic names in **bold**; classifications in SMALL CAPS; ✚ Canadian drug name; ⊘ Prototype drug

603

Fecal (80%) and renal (5%). **Half-Life:** 30 h.

NURSING IMPLICATIONS

Assessment & Drug Effects

- Monitor for and promptly report S&S of a hypersensitivity reaction (e.g., anaphylaxis, dyspnea, flushing, chest pain, angioedema).
- Monitor pulmonary status and report promptly unexplained cough, shortness of breath, pain on inspiration, or diminished breath sounds.
- Monitor for and promptly report S&S of infection.
- Lab tests: Baseline and periodic CBC with differential, renal function tests, blood glucose, and lipid profile.

Patient & Family Education

- Report promptly any signs of infections, including sore throat, fever, and flu-like symptoms.
- Avoid live vaccinations and close contact with those who have received live vaccines.
- Practice meticulous oral hygiene. Do not use mouthwashes that contain alcohol or peroxide.
- Women should use adequate means of contraception to avoid pregnancy while on this drug and for 8 wk after ending treatment.

EXEMESTANE
(ex-e-mes'tain)

Aromasin
Classifications: ANTINEOPLASTIC; AROMATASE INHIBITOR
Therapeutic: ANTINEOPLASTIC
Prototype: Anastrozole
Pregnancy Category: D

AVAILABILITY 25 mg tablet

ACTION & *THERAPEUTIC EFFECT*

Steroidal aromatase inhibitor that suppresses plasma estrogens estradiol and estrone. The enzyme, aromatase converts estrone to estradiol. *Tumor regression is possible in postmenopausal women with estrogen dependent breast cancer. Effectiveness is indicated by evidence of reduction in tumor size.*

USES Estrogen-receptor positive early breast cancer following treatment with tamoxifen, treatment of advanced breast cancer in postmenopausal women whose disease has progressed following tamoxifen therapy.

CONTRAINDICATIONS Hypersensitivity to exemestane; coadministration of estrogen-containing drugs; pregnancy (category D).
CAUTIOUS USE Hepatic or renal insufficiency; GI disorders; cardiovascular disease; hyperlipidemia; lactation. Safe use in children not established.

ROUTE & DOSAGE

Early and Advanced Breast Cancer
Adult: **PO** 25 mg daily

ADMINISTRATION

Oral
- Give following a meal.
- Store at 15°–30° C (59°–86° F).

ADVERSE EFFECTS (≥1%) **Body as a Whole:** *Fatigue, hot flashes, pain,* flu-like symptoms; edema; fever; paresthesia. **CNS:** *Depression, insomnia, anxiety;* dizziness; headache. **CV:** Hypertension. **GI:** *Nausea,* vomiting, abdominal pain, anorexia, constipation, diarrhea, increased appetite. **Respiratory:**

Dyspnea, cough, bronchitis, sinusitis. **Skin:** Increased sweating, rash, itching. **Other:** UTI; lymphedema.

PHARMACOKINETICS **Absorption:**
Rapidly, approximately 42% reaches systemic circulation. **Distribution:** Extensive tissue distribution, 90% protein bound. **Metabolism:** Extensively in liver (CYP3A4). **Elimination:** Equally in urine and feces. **Half-Life:** 24 h.

NURSING IMPLICATIONS

Assessment & Drug Effects
- Lab tests: Baseline LFTs, BUN and creatinine; periodic WBC with differential, lipid profile, routine blood chemistry.

Patient & Family Education
- Review manufacturer's patient literature thoroughly to reinforce understanding of likely adverse effects.
- Report bothersome adverse effects to prescriber.

EXENATIDE ⊘
(e-xe′na-tide)

Byetta
Classification: ANTIDIABETIC, INCRETIN MIMETIC
Therapeutic: ANTIDIABETIC
Pregnancy Category: C

AVAILABILITY 5 mcg, 10 mcg pen injection

ACTION & *THERAPEUTIC EFFECT*
Improves glycemic control in type 2 diabetes mellitus by mimicking the functions of incretin, a glucagon-like peptide-1 (GLP-1). Exenatide enhances glucose-dependent insulin secretion by pancreatic beta-cells, suppresses inappropriately elevated glucagon secretion, and slows gastric emptying. These actions decrease glucagon stimulation of hepatic glucose output and decrease insulin demand. *Improves glycemic control by reducing fasting and postprandial glucose concentrations in patients with type 2 diabetes.*

USES Type 2 diabetes mellitus in combination with diet/exercise.

CONTRAINDICATIONS Hypersensitivity to exenatide, or cresol; type I diabetes; severe GI disease, diabetic ketoacidosis; gastroparesis; end-stage renal disease, severe renal impairment (CrCl less than 30 mL/min).

CAUTIOUS USE Renal impairment; renal disease; thyroid disease; older adults; pregnancy (category C); lactation. Safety and efficacy are not established in children.

ROUTE & DOSAGE

Type 2 Diabetes Mellitus
Adult: **Subcutaneous** Initial dose of 5 mcg b.i.d., within 60 min prior to the morning and evening meal. After 1 mo, may increase to 10 mcg b.i.d., within 60 min prior to the morning and evening meal.

ADMINISTRATION
- Give subcutaneously into thigh, abdomen, or upper arm within 60 min before the morning and evening meals. Do not administer after a meal.
- Do not give within 1 h of oral antibiotics, an oral contraceptive, or acetaminophen.
- Store at 36°–46° F (2°–8° C) and protect from light. Discard pen 30 days after first use. Do not use if pen has been frozen. After first use, pen may be kept at or below 77° F (25° C).

Common adverse effects in *italic*, life-threatening effects <u>underlined</u>; generic names in **bold**; classifications in SMALL CAPS; ♣ Canadian drug name; ⊘ Prototype drug

605

ADVERSE EFFECTS (≥1%) **CNS:** Asthenia, dizziness, restlessness, jittery feeling. **GI:** Nausea, vomiting, diarrhea, dyspepsia, anorexia, gastroesophageal reflux. **Metabolic:** Hypoglycemia, excessive sweating (hyperhidrosis or diaphoresis).

INTERACTIONS Drug: Due to its ability to slow gastric emptying, exenatide can decrease the rate and/or serum levels of oral medications that require GI absorption.

PHARMACOKINETICS Peak: 2 h. **Elimination:** Primarily in urine. **Half-Life:** 2.4 h.

NURSING IMPLICATIONS

Assessment & Drug Effects
- Monitor for and report S&S of significant GI distress, including NV&D.
- Monitor for S&S of hypoglycemia and S&S of acute pancreatitis (acute abdominal pain with/without vomiting). If pancreatitis is suspected, withhold drug and notify prescriber immediately.
- Lab tests: Periodic fasting and postprandial plasma glucose and periodic HbA1C; baseline and periodic renal function tests.

Patient & Family Education
- If a dose is missed, wait for the next scheduled dose.
- Discard any pen that has been in use for greater than 30 days.
- Exenatide may cause decreased appetite and some weight loss.
- Report significant GI distress to prescriber. Report promptly persistent, severe abdominal pain that may be accompanied by vomiting.

EZETIMIBE
(e-ze-ti′mibe)
Zetia, Ezetrol ♣
Classifications: ANTILIPEMIC; CHOLESTEROL ABSORPTION INHIBITOR
Therapeutic: CHOLESTEROL LOWERING
Pregnancy Category: C

AVAILABILITY 10 mg tablets

ACTION & *THERAPEUTIC EFFECT*
Selectively blocks the lining of the small intestine by inhibiting the absorption of cholesterol, but does not inhibit cholesterol synthesis in the liver or increase bile acid excretion. Thus it decreases the amount of intestinal cholesterol available to the liver. *Lowers both total cholesterol and low-density lipid (LDL) cholesterol, apo B, triglycerides, and increases HDL-C; its mechanism of action is complementary to statins.*

USES Treatment of primary hypercholesterolemia alone or with an HMG-CoA reductase inhibitor (statin); treatment of homozygous sitosterolemia as an adjunct to diet.

CONTRAINDICATIONS Hypersensitivity to ezetimibe; concurrent use with HMG-CoA reductase inhibitor in patients with active liver disease or elevated serum transaminases; moderate to severe hepatic disease; lactation.
CAUTIOUS USE Mild hepatic insufficiency; elderly; pregnancy (category C). Safe use in children younger than 10 y not established.

ROUTE & DOSAGE

Hypercholesterolemia
Adult: **PO** 10 mg daily

Common adverse effects in *italic;* life-threatening effects underlined; generic names in **bold;** classifications in SMALL CAPS; ♣ Canadian drug name; ☯ Prototype drug

ADMINISTRATION

Oral

- Give no sooner than 2 h before or 4 h after administration of a bile acid sequestrant such as cholestyramine.
- Store at 15°–30° C (59°–86° F). Protect from moisture.

ADVERSE EFFECTS (≥1%) **Body as a Whole:** Fatigue, arthralgia, back pain, myalgia, angioedema, myopathy. **CNS:** Dizziness, headache. **GI:** Abdominal pain, diarrhea. **Respiratory:** Pharyngitis, sinusitis, cough. **Hematologic:** <u>Thrombocytopenia</u>. **Skin:** Rash. **Other:** Hepatitis, pancreatitis, rhabdomyolysis.

INTERACTIONS Drug: BILE ACID SEQUESTRANTS (e.g., **cholestyramine**) may decrease absorption (give ezetimibe 2 h before or 4 h after these drugs); **cyclosporine** or FIBRIC ACID DERIVATIVES can significantly increase ezetimibe levels.

PHARMACOKINETICS Absorption: Well absorbed from the small intestine. **Peak:** 4–12 h. **Distribution:** Ezetimibe-glucuronide is 99% protein bound. **Metabolism:** Extensively conjugated to an active glucuronide compound (ezetimibe-glucuronide). Metabolized in small intestine and liver. **Elimination:** Primarily in feces. **Half-Life:** 22 h.

NURSING IMPLICATIONS

Assessment & Drug Effects

- Lab tests: Monitor baseline and periodic lipid profile; periodic Hgb and Hct and platelet count. Monitor baseline LFTs and when used with a statin, monitor periodic LFTs in accordance with the monitoring schedule for that statin.
- Assess for and report unexplained muscle pain, especially when used in combination with a statin drug.
- Monitor closely patients who take both ezetimibe and cyclosporine.

Patient & Family Education

- Report unexplained muscle pain, tenderness, or weakness.
- Females should use effective methods of contraception to prevent pregnancy while taking this drug in combination with a statin.

EZOGABINE

(e-zog' a-been)

Potiga

Classifications: ANTICONVULSANT; POTASSIUM CHANNEL OPENER

Therapeutic: ANTICONVULSANT

Pregnancy Category: C

AVAILABILITY 50 mg, 200 mg, 300 mg, 400 mg tablets

ACTION & THERAPEUTIC EFFECT Mechanism of action thought to be related to enhancement of transmembrane potassium currents resulting in stabilization of the resting membrane and reduced brain excitability. May also augment GABA-mediated currents. *Reduces frequency of partial onset seizures.*

USES Adjunctive therapy in the treatment of partial onset seizures.

CAUTIOUS USE Urinary hesitation or retention; history of hepatic or renal impairment; confusional states, psychotic symptoms (including suicidal ideation or behavior), hallucinations; dizziness and somnolence; prolonged QT interval or concurrent drugs that increase QT interval; CHF; ventricular hypertrophy; hypokalemia or hypomagnesemia;

Common adverse effects in *italic*, life-threatening effects <u>underlined</u>; generic names in **bold**; classifications in SMALL CAPS; ♣ Canadian drug name; ⊘ Prototype drug

607

pregnancy (category C) and lactation. Safety and effectiveness in children under 18 y not established.

ROUTE & DOSAGE

Partial Seizures

Adult 65 y or younger: **PO** 100 mg t.i.d. initially, may increase by 50 mg/wk to 200–400 mg t.i.d. *Adult over 65 y:* **PO** 50 mg t.i.d. initially, may increase by 50 mg/wk to 250 mg t.i.d.

Hepatic Impairment Dosage Adjustment

Moderate impairment (Child-Pugh 7 or higher up to 9): **Same as adult over 65 y**

Severe impairment (Child-Pugh greater than 9): 50 mg t.i.d. initially, may increase by 50 mg/wk to 200 mg t.i.d.

Renal Impairment Dosage Adjustment

CrCl less than 50 mL/min: 50 mg t.i.d. initially, may increase by 50 mg/wk to 200 mg t.i.d.

ADMINISTRATION

- May be given with or without food.
- Tablets should be swallowed whole and not crushed or chewed.
- Ezogabine is ordinarily discontinued over a period of three weeks.
- Store at (15°–30° C) 59°– 86° F.

ADVERSE EFFECTS (≥2%) **Body as a Whole:** Increased weight, influenza. **CNS:** Abnormal coordination, amnesia, anxiety, aphasia, asthenia, attention disturbance, balance disorder, *confusion*, disorientation, *dizziness*, dysrthria, dysphasia, *fatigue*, gait disturbances, *memory impairment*, paresthesia, *somnolence*, tremor, vertigo. **GI:** Constipation, dyspepsia, nausea. **Special Senses:** Blurred vision, diplopia. **Urogenital:** Chromaturia, dysuria, hematuria, urinary hesitation.

DIAGNOSTIC TEST INTERFERENCE

False elevations of serum and urine bilirubin.

INTERACTIONS Drug: Carbamazepine and **phenytoin** decrease the plasma levels of ezogabine. Ezogabine may increase the plasma levels of **digoxin** and decrease the plasma levels of **lamotrigine.** Ezogabine may cause an additive effect with other drugs that prolong the QT interval (e.g., **amiodarone, procainamide, droperidol, mesoridazine, moxifloxacin, pimozide, thioridazine**). **Food: Alcohol** may increase the levels of ezogabine.

PHARMACOKINETICS Peak: 0.5–2 h. **Distribution:** 80% plasma protein bound. **Metabolism:** In the liver to inactive metabolites. **Elimination:** Renal (86%) and fecal (14%). **Half-Life:** 7–11 h.

NURSING IMPLICATIONS

Assessment & Drug Effects

- Monitor for and report immediately: Suicidal ideation or any other psychiatric symptom, confusional states, dizziness, or somnolence.
- Monitor urinary output especially in those with BPH, cognitive impairment, or concurrent anticholinergic drugs. Report promptly signs of urinary retention.
- Lab tests: Baseline and periodic LFTs and renal function tests; monitor serum digoxin with concurrent therapy.
- Monitor for S&S of digoxin toxicity with concurrent therapy.

Patient & Family Education

- Report immediately: Suicidal thoughts, altered mental status, confusion, excessive drowsiness or dizziness, or decreased ability to urinate.
- Do not abruptly stop taking ezogabine unless told to do so by prescriber.
- Avoid alcohol while taking this drug.
- Do not drive or engage in potentially hazardous activities until response to drug is known.
- Do not breast feed while taking this drug without consulting prescriber.

FAMCICLOVIR

(fam-ci'clo-vir)

Famvir
Classification: ANTIVIRAL
Therapeutic: ANTIVIRAL
Prototype: Acyclovir
Pregnancy Category: B

AVAILABILITY 125 mg, 250 mg, 500 mg tablets

ACTION & *THERAPEUTIC EFFECT*
Prodrug of the antiviral agent penciclovir that prevents viral replication by inhibition of DNA synthesis in herpes virus–infected cells. *Effectiveness is indicated by decreasing pain and crusting of lesions followed by loss of vesicles, ulcers, and crusts. Interferes with DNA synthesis of herpes simplex virus type 1 and 2 (HSV-1 and HSV-2) infections, varicella-zoster virus, and cytomegalovirus.*

USES Management of acute herpes zoster, genital herpes, recurrent episodes of genital herpes in immunocompromised adults.

UNLABELED USES Bell's palsy, post herpetic neuralgia prophylaxis.

CONTRAINDICATIONS Hypersensitivity to famciclovir, lactation.
CAUTIOUS USE Renal or hepatic impairment, carcinoma, older adults, pregnancy (category B). Safety in children younger than 18 y is not established.

ROUTE & DOSAGE

Herpes Zoster, Treatment
Adult: **PO** 500 mg q8h for 7 days, start within 48–72 h of onset of rash

Renal Impairment Dosage Adjustment
CrCl 40–59 mL/min: 500 mg q12h; *20–39 mL/min:* 500 mg q24h

Treatment of Recurrent Genital Herpes
Adult: **PO** 125 mg b.i.d. × 5 days OR 1000 mg b.i.d.

Suppression of Recurrent Genital Herpes
Adult: **PO** 250 mg b.i.d. for up to 1 y

ADMINISTRATION

Oral
- Most effective when given within 72 h of appearance of a rash or within 6 h of onset of a genital lesion.
- Store at room temperature, 15°–30° C (59°–86° F).

ADVERSE EFFECTS (≥1%) **CNS:** *Headache,* somnolence, dizziness, paresthesias, fatigue, fever, rigors. **Hematologic:** Purpura. **GI:** Nausea, diarrhea, vomiting, constipation,

---

Common adverse effects in *italic*, life-threatening effects <u>underlined</u>; generic names in **bold**; classifications in SMALL CAPS; ♣ Canadian drug name; ♦ Prototype drug

609

anorexia, abdominal pain. **Body as a Whole:** Pharyngitis, sinusitis, pruritus.

INTERACTIONS Drug: Probenecid may decrease elimination; famciclovir may increase **digoxin** levels.

PHARMACOKINETICS Absorption: Readily absorbed from GI tract and rapidly converted to penciclovir in intestinal and liver tissue. **Onset:** Median times to full crusting of lesions, loss of vesicles, loss of ulcers, and loss of crusts were 6, 5, 7, and 19 days, respectively; median time to loss of acute pain was 21 days. **Peak:** 1 h. **Distribution:** Distributes into breast milk of animals. **Metabolism:** Metabolized in liver and intestinal tissue to penciclovir, which is the active antiviral agent. **Elimination:** Approximately 60% recovered in urine as penciclovir. **Half-Life:** Penciclovir 2–3 h.

NURSING IMPLICATIONS

Assessment & Drug Effects

▪ Lab tests: Baseline CBC and routine blood chemistry studies prior to and after short courses of therapy; periodically during prolonged treatment.

Patient & Family Education

▪ Learn potential adverse effects and report those that are bothersome to prescriber.

▪ Be aware that a full therapeutic response may take several weeks.

FAMOTIDINE

(fa-moe'ti-deen)

Pepcid, Pepcid AC
Classification: ANTISECRETORY (H₂-RECEPTOR ANTAGONIST)
Therapeutic: ANTIULCER
Prototype: Cimetidine
Pregnancy Category: B

AVAILABILITY 10 mg, 20 mg, 40 mg tablets; 40 mg/5 mL suspension; 10 mg/mL, 20 mg/50 mL injection

ACTION & *THERAPEUTIC EFFECT* A potent competitive inhibitor of histamine at its H_2 receptor sites in gastric parietal cells. Inhibits basal, nocturnal, meal-stimulated, and pentagastrin-stimulated gastric secretion as well as pepsin secretion. *Reduces parietal cell output of hydrochloric acid; thus, detrimental effects of acid on gastric mucosa are diminished.*

USES Short-term treatment of active duodenal ulcer. Maintenance therapy for duodenal ulcer patients on reduced dosage after healing of an active ulcer. Treatment of pathologic hypersecretory conditions (e.g., Zollinger-Ellison syndrome), benign gastric ulcer, gastroesophageal reflux disease (GERD), gastritis.
UNLABELED USES Stress ulcer prophylaxis.

CONTRAINDICATIONS Hypersensitivity to famotidine or other H_2-receptor antagonists; sudden GI bleeding; lactation.
CAUTIOUS USE Renal insufficiency; renal failure; PKU; hepatic disease; elderly; pregnancy (category B).

ROUTE & DOSAGE

Duodenal Ulcer
Adult: **PO** 40 mg at bedtime or 20 mg b.i.d. **PO Maintenance Therapy** 20 mg at bedtime **IV** 20 mg q12h
Child: **PO/IV** 0.25–0.5 mg/kg q12h (max: 40 mg/day)
Pathological Hypersecretory Conditions
Adult: **PO** 20–160 mg q6h **IV** 20 mg q6h

GERD, Gastritis

Adult: **PO** 10 mg b.i.d.
Child: **PO** 1 mg/kg/day in 2
divided doses (max: 40 mg b.i.d.)

**Renal Impairment Dosage
Adjustment**

CrCl less than 50 mL/min: 50% of
usual dose or usual dose q36–48h

ADMINISTRATION

Oral

- Give with liquid or food of patient's choice; an antacid may also
be given if patient is also on antacid therapy.
- Store at 15°–30° C (59°–86° F).
Protect from moisture and strong
light; do not freeze.

Intravenous

Note: Verify correct IV concentration and rate of infusion/injection
with prescriber before administration to infants or children.

***PREPARE:* Direct:** Dilute each 20
mg (2 mL) famotidine IV solution
(containing 10 mg/mL) with
D5W, NS, or other compatible IV
diluent (see manufacturer's directions) to a total volume of 5 or
10 mL. **Intermittent:** Dilute required dose with 100 mL compatible IV solution.
***ADMINISTER:* Direct:** Give over
not less than 2 min. **Intermittent:**
Infuse over 15–30 min.
***INCOMPATIBILITIES* Y-site: Am-
photericin B cholesteryl complex, azithromycin, cefepime,
piperacillin/tazobactam.**

- Store IV solution at 2°–8° C (36°–
46° F); reconstituted IV solution is
stable for 48 h at room temperature
15°–30° C (59°–86° F).

ADVERSE EFFECTS (≥1%) **CNS:**
Dizziness, headache, confusion, depression. **GI:** Constipation, diarrhea.

Skin: Rash, acne, pruritus, dry skin,
flushing. **Hematologic:** <u>Thrombo-
cytopenia</u>. **Urogenital:** Increases in
BUN and serum creatinine.

INTERACTIONS Drug: May inhibit
absorption of **itraconazole** or **ketoconazole.**

PHARMACOKINETICS Absorption:
Incompletely from GI tract (40–
50% reaches systemic circulation).
Onset: 1 h. **Peak:** 1–3 h PO; 0.5–3 h
IV. **Duration:** 10–12 h. **Metabolism:**
In liver. **Elimination:** In urine. **Half-
Life:** 2.5–4 h.

NURSING IMPLICATIONS

Assessment & Drug Effects

- Monitor for improvement in GI
distress.
- Monitor for signs of GI bleeding.

Patient & Family Education

- Be aware that pain relief may not
be experienced for several days
after starting therapy.

FAT EMULSION, INTRAVENOUS
(fat e-mul'sion)

Intralipid, Liposyn II, Soyacal
Classifications: CALORIC AGENT;
LIPID EMULSION
Therapeutic: NUTRITIONAL SUPPLE-
MENT; LIPID
Pregnancy Category: C

AVAILABILITY 10%, 20%, 30%
emulsion

ACTION & *THERAPEUTIC EFFECT*
Soybean oil in water emulsion containing egg yolk phospholipids and
glycerin. Liposyn 10% is safflower
oil in water emulsion containing
egg phosphatides and glycerin.
Used as a nutritional supplement.

USES Fatty acid deficiency. Also to
supply fatty acids and calories in

Common adverse effects in *italic*, life-threatening effects <u>underlined</u>; generic names
in **bold**; classifications in SMALL CAPS; ♣ Canadian drug name; ⊘ Prototype drug

611

high-density form to patients receiving prolonged TPN therapy who cannot tolerate high dextrose concentrations or when fluid intake **must be** restricted as in renal failure, CHF, ascites.

CONTRAINDICATIONS Hyperlipemia; bone marrow dyscrasias; impaired fat metabolism as in pathological hyperlipemia, lipoid nephrosis, acute pancreatitis accompanied by hyperlipemia.

CAUTIOUS USE Severe hepatic or pulmonary disease; coagulation disorders; anemia; when danger of fat embolism exists; history of gastric ulcers; diabetes mellitus; thrombocytopenia; newborns, premature neonates, infants with hyperbilirubinemia; pregnancy (category C).

ROUTE & DOSAGE

Prevention of Essential Fatty Acid Deficiency

Adult: **IV** 500 mL of 10% or 250 mL of 20% solution twice/wk (max: rate of 100 mL/h)
Child: **IV** 5–10 mL/kg/day twice/wk (max: 3–4 g/kg/day; max: rate of 100 mL/h)

Calorie Source in Fluid-Restricted Patients

Adult: **IV** Up to 2.5 g/kg or 60% of nonprotein calories daily (max: rate of 100 mL/h)
Child: **IV** Up to 4 g/kg or 60% of nonprotein calories daily (max: rate of 100 mL/h)
Premature Neonate: **IV** 0.25–0.5 g/kg/day, increase by 0.25–0.5 g/kg/day (max: 3–4 g/kg/day; max: infusion 0.15 g/kg/h)

ADMINISTRATION

Intravenous
Do not use if oil appears to be separating out of the emulsion.

PREPARE: **IV Infusion** Allow preparations that have been refrigerated to stand at room temperature for about 30 min before using whenever possible. ▪ Check with a pharmacist before mixing fat emulsions with electrolytes, vitamins, drugs, or other nutrient solutions.
ADMINISTER: **IV Infusion for Adult:** *10% emulsion:* Infuse at 1 mL/min for first 15–30 min; increase to 2 mL/min if no adverse reactions. ▪ *20% emulsion:* Infuse at 0.5 mL/min for first 15–30 min; increase to 2 mL/min if no adverse reactions occur. **IV Infusion for Child:** *10% emulsion:* Infuse at 0.1 mL/min for first 10–15 min; increase to 1 g/kg in 4 h if no adverse reactions occur. ▪ Do not exceed 100 mL/h. ▪ *20% emulsion:* Infuse at 0.05 mL/min for first 10–15 min; increase to 1 g/kg in 4 h if no adverse reactions occur. ▪ Do not exceed 50 mL/h. **IV Infusion for Premature Neonate:** Infuse at rate not to exceed 0.15 g/kg/h. **IV Infusion for All Patients:** Give fat emulsions via a separate peripheral site or by piggyback into same vein receiving amino acid injection and dextrose mixtures or give by piggyback through a Y-connector near infusion site so that the two solutions mix only in a short piece of tubing proximal to needle. ▪ Must hang fat emulsions higher than hyperalimentation solution bottle to prevent backup of fat emulsion into primary line. ▪ Do not use an in-line filter because size of fat particles is larger than pore size. ▪ Control flow rate of each solution by separate infusion pumps. ▪ Use a constant rate over 20–24 h to reduce risk of hyperlipemia in neonates and

prematures because they tend to metabolize fat slowly.

INCOMPATIBILITIES **Solution/ additive:** **Aminophylline, amphotericin B, ampicillin, calcium chloride, calcium gluconate, dextrose 10%, gentamicin, hetastarch, magnesium chloride, penicillin G, phenytoin, ranitidine, vitamin B complex.** **Y-site:** **Acyclovir, albumin, amphotericin B, cyclosporine, doxorubicin, doxycycline, droperidol, ganciclovir, haloperidol, heparin, hetastarch, hydromorphone, levorphanol, lorazepam, midazolam, minocycline, nalbuphine, ondansetron, pentobarbital, phenobarbital, potassium phosphate, sodium phosphate.**

▪ Discard contents of partly used containers. ▪ Store, unless otherwise directed by manufacturer, Intralipid 10% and Liposyn 10% at room temperature [25° C (77° F) or below]; refrigerate Intralipid 20%. Do not freeze.

ADVERSE EFFECTS (≥1%) **Body as a Whole:** Hypersensitivity reactions (to egg protein), irritation at infusion site. **Hematologic:** Hypercoagulability, thrombocytopenia in neonates. **GI:** *Transient increases in liver function tests, hyperlipemia.* **[Long-Term Administration]** Sepsis, jaundice (cholestasis), hepatomegaly, kernicterus (infants with hyperbilirubinemia), shock (rare).

DIAGNOSTIC TEST INTERFERENCE Blood samples drawn during or shortly after fat emulsion infusion may produce abnormally high *hemoglobin MCH and MCHC* values. Fat emulsions may cause transient abnormalities in *liver function tests* and may interfere with estimations of *serum bilirubin* (especially in infants).

INTERACTIONS Drug: No clinically significant interactions established.

NURSING IMPLICATIONS

Assessment & Drug Effects

▪ Observe patient closely. Acute reactions tend to occur within the first 2½ h of therapy.

▪ Lab tests: Determine baseline values for hemoglobin, platelet count, blood coagulation, LFTs lipid profile (especially serum triglycerides and cholesterol, free fatty acids). Repeat 1 or 2 times weekly during therapy in adults; more frequently in children. Report significant deviations promptly.

▪ Lab tests: Obtain daily platelet counts in neonates during first week of therapy, then every other day during second week, and 3 times a week thereafter because newborns are prone to develop thrombocytopenia.

▪ Note: Lipemia must clear after each daily infusion. Degree of lipemia is measured by serum triglycerides and cholesterol levels 4–6 h after infusion has ceased.

Patient & Family Education

▪ Report difficulty breathing, nausea, vomiting, or headache to prescriber.

FEBUXOSTAT
(fee-bux′o-stat)
Uloric
Classifications: ANTIGOUT; XANTHINE OXIDASE INHIBITOR
Therapeutic: ANTIGOUT
Prototype: Allopurinol
Pregnancy Category: C

Common adverse effects in *italic*, life-threatening effects underlined; generic names in **bold**; classifications in SMALL CAPS; ♣ Canadian drug name; ❷ Prototype drug

613

AVAILABILITY 40 mg, 80 mg tablets

ACTION & *THERAPEUTIC EFFECT*
Febuxostat decreases serum uric acid by inhibiting the enzyme needed to convert xanthine to uric acid (end product of protein catabolism). *Effectiveness is measured by decreasing serum uric acid level to less than 6 mg/dL.*

USES Management of hyperuricemia in patients with chronic gout.

CONTRAINDICATIONS Asymptomatic hyperuricemia.

CAUTIOUS USE History of MI or stroke; severe renal impairment (CrCl less than 30 mL/min); severe hepatic dysfunction (Child-Pugh class C); pregnancy (category C); lactation. Safety and effectiveness in children younger than 18 y have not been established.

ROUTE & DOSAGE

Gout
Adult: **PO** 40 mg once daily; can be increased to 80 mg once daily

ADMINISTRATION

Oral
- Concurrent therapy with an NSAID or colchicine is recommended to prevent gout flares during the first 6 mo of therapy.
- Store at 15°–30° C (59°–86° F) away from light.

ADVERSE EFFECTS (≥1%) **CV:** Atrial filbrilation, AV block, thromboembolic events. **GI:** Nausea. **Metabolic:** Elevated AST and ALT levels. **Musculoskeletal:** Arthralgia. **Skin:** Rash. **Whole Body:** Rhabdomyolysis.

INTERACTIONS Drug: Febuxostat will increase the levels of drugs requiring xanthine oxidase for normal metabolism (e.g., **6-mercaptopurine, azathioprine**)

PHARMACOKINETICS Absorption: 49%. **Peak:** 1–1.5 h. **Distribution:** Greater than 99% plasma protein bound. **Metabolism:** Extensive hepatic metabolism via oxidation and glucuronide conjugation. **Elimination:** Renal (49%) and fecal (45%). **Half-Life:** 5–8 h.

NURSING IMPLICATIONS

Assessment & Drug Effects
- Monitor for and report gout flares.
- Monitor CV status throughout therapy.
- Lab tests: Baseline serum uric acid, again at 2 wk, and periodically thereafter. LFTs at 2 and 4 mo from start of therapy and periodically thereafter.

Patient & Family Education
- Notify prescriber if you experience a gout flare, but do not stop taking this drug.
- NSAIDs are typically used to control gout flares. Consult prescriber.

FELBAMATE
(fel'ba-mate)
Felbatol
Classification: ANTICONVULSANT; HYDANTOIN
Therapeutic: ANTICONVULSANT
Prototype: Phenytoin
Pregnancy Category: C

AVAILABILITY 400 mg, 600 mg tablets; 600 mg/5 mL suspension

ACTION & *THERAPEUTIC EFFECT*
Anticonvulsant that blocks repetitive firing of neurons and increases seizure threshold; prevents seizure spread. *Increases seizure threshold and prevents seizure spread.*

USES Treatment of Lennox–Gastaut syndrome and partial seizures.

UNLABELED USES Treatment of generalized tonic/clonic seizures.

CONTRAINDICATIONS Hypersensitivity to felbamate, history of blood dyscrasia or hepatic dysfunction; bone marrow depression; active liver disease; elevated serum AST or ALT.

CAUTIOUS USE Hypersensitivity to other carbamates; renal impairment, renal failure; thrombocytopenia; iron-deficiency anemia; older adults; pregnancy (category C), lactation. Safety and effectiveness in children other than those with Lennox–Gastaut syndrome are not established.

ROUTE & DOSAGE

Partial Seizures

Adult: **PO** Initiate with 1200 mg/day in 3–4 divided doses, may increase by 600 mg/day q2wk (max: 3600 mg/day); when converting to monotherapy, reduce dose of concomitant anticonvulsants by 1/3 when initiating felbamate, then continue to decrease other anticonvulsants by 1/3 with each increase in felbamate q2wk; when using as adjunctive therapy, decrease other anticonvulsants by 20% when initiating felbamate and note that further reductions in other anticonvulsants may be required to minimize side effects and drug interactions

Lennox–Gastaut Syndrome

Child: **PO** Start at 15 mg/kg/day in 3 or 4 divided doses, reduce concurrent antiepileptic drugs by 20%, further reductions may be required to minimize side effects due to drug interactions, may increase felbamate by 15 mg/kg/day at weekly intervals (max: 45 mg/kg/day)

ADMINISTRATION

Oral

- Titrate dose under close clinical supervision.
- Shake suspension well before giving a dose.
- Store in airtight container at room temperature, 15°–30° C (59°–86° F).

ADVERSE EFFECTS (≥1%) **CNS:** Mild tremors, headache, dizziness, ataxia, diplopia, blurred vision; agitation, aggression, hallucinations, fatigue, psychological disturbances. **Endocrine:** Slight elevation of serum cholesterol, hyponatremia, hypokalemia, weight gain and loss. **GI:** *Nausea and vomiting,* anorexia, constipation, hiccup, taste disturbance, indigestion, esophagitis, increased appetite, <u>acute liver failure</u>. **Hematologic:** <u>*Aplastic anemia*</u>.

INTERACTIONS Drug: Felbamate reduces serum **carbamazepine** levels by a mean of 25%, but increases levels of its active metabolite, increases serum **phenytoin** levels approximately 20%, and increases **valproic acid** levels. **Herbal: Gingko** may decrease anticonvulsant effectiveness.

PHARMACOKINETICS Absorption: 90% from GI tract. Absorption of tablet not affected by food. **Onset:** Therapeutic effect approximately 14 days. **Peak:** Peak plasma levels at 1–6 h. **Distribution:** 20–25% protein bound, readily crosses the blood–brain barrier. **Metabolism:** In the liver via the cytochrome P450 system. **Elimination:** 40–50% excreted unchanged in urine, rest excreted in urine as metabolites. **Half-Life:** 20–23 h.

NURSING IMPLICATIONS

Assessment & Drug Effects

- Lab tests: Obtain baseline LFTs and complete hematologic studies

Common adverse effects in *italic*, life-threatening effects <u>underlined</u>; generic names in **bold**; classifications in SMALL CAPS; ♣ Canadian drug name; ⓟ Prototype drug

615

before initiating therapy, repeat frequently during therapy, and for a lengthy period after discontinuation of felbamate. Monitor serum sodium and potassium levels periodically.

- Report immediately any hematologic abnormalities.
- Note: When used concomitantly with either phenytoin or carbamazepine, carefully monitor serum levels of these drugs when felbamate is added, when adjustments in felbamate dosing are made, or when felbamate is discontinued.
- Monitor weight, because both weight gain and loss have been reported.
- Monitor for S&S of drug toxicity including GI distress and CNS toxicity.

Patient & Family Education
- Report promptly signs of liver dysfunction including jaundice, anorexia, GI discomfort, and fatigue.
- Report promptly signs of bone marrow suppression including infection, bleeding, easy bruising or signs of anemia.

FELODIPINE
(fel-o′di-peen)
Classifications: CALCIUM CHANNEL BLOCKER; ANTIHYPERTENSIVE
Therapeutic: ANTIHYPERTENSIVE
Prototype: Nifedipine
Pregnancy Category: C

AVAILABILITY 2.5 mg, 5 mg, 10 mg sustained release tablets

ACTION & *THERAPEUTIC EFFECT*
Calcium channel antagonist with high vascular selectivity that reduces systolic, diastolic, and mean arterial pressure at rest and during exercise. Felodipine inhibits influx of extracellular calcium across myocardial and vascular smooth muscle cell membranes. Resultant decrease in intracellular calcium inhibits contractility of smooth muscle, resulting in dilation of coronary and systemic arteries. *BP reduction is due to reduction in peripheral vascular resistance (afterload) against which the heart works. This reduces oxygen demand by the heart and may account for its effectiveness in chronic stable angina.*

USES Treatment of hypertension.
UNLABELED USES Angina.

CONTRAINDICATIONS Hypersensitivity to felodipine; sick sinus rhythm or second- or third-degree heart block except with the use of a pacemaker; abnormal aortic stenosis; hypotension; bradycardia; cardiogenic shock; acute MI; left ventricular dysfunction.
CAUTIOUS USE Hypotension, CHF, angina; aortic stenosis, cardiomyopathy; older adults; GERD; hiatal hernia; hepatic impairment; pregnancy (category C), lactation. Safety and efficacy in children are not established.

ROUTE & DOSAGE

Hypertension
Adult: **PO** 5–10 mg once/day (max: 10 mg/day)

Hepatic Impairment Dosage Adjustment
Start older adults and patients with impaired liver function at 2.5 mg daily

ADMINISTRATION
Oral
- Give tablet whole. Do not crush or chew tablets.
- Store at or below 30° C (86° F) in a tightly closed, light-resistant container.

ADVERSE EFFECTS (≥1%) **Body as a Whole:** Most adverse effects appear to be dose dependent. **CV:** Tachycardia, *palpitations, flushing, peripheral edema.* **CNS:** *Dizziness, fatigue,* headache. **GI:** Nausea, flatulence, diarrhea, dyspepsia. **Hematologic:** Small but significant decreases in Hct, Hgb, and RBC count.

DIAGNOSTIC TEST INTERFERENCE Serum *alkaline phosphatase* may be slightly but significantly increased. Plasma total and ionized *calcium* levels rise significantly. Serum *gamma-glutamyl transferase* may increase.

INTERACTIONS Drug: Adenosine may cause prolonged bradycardia if it is used to treat patients with toxic concentrations of CALCIUM CHANNEL BLOCKERS. **Carbamazepine, phenobarbital, phenytoin** may decrease felodipine effect. **Cimetidine** may increase felodipine bioavailability and adverse effect risk. Concomitant felodipine and **digoxin** administration produces only transient increases in plasma **digoxin** concentrations (35–40% increase), which are not sustained with continued administration. **Food: Grapefruit juice** may increase adverse effects.

PHARMACOKINETICS Absorption: Completely from GI tract; it undergoes extensive first-pass metabolism with only about 15% of dose reaching systemic circulation. **Onset:** Less than 1 h. **Peak:** 2–4 h. **Duration:** 20–24 h (sustained release formulation). **Distribution:** Greater than 99% bound to plasma proteins. **Metabolism:** Metabolized via hepatic cytochrome P-450 mixed function oxidase system. **Elimination:** 60–70% of metabolites are excreted in urine within 72 h. **Half-Life:** 10 h.

NURSING IMPLICATIONS

Assessment & Drug Effects

- Monitor BP carefully, especially at initiation of drug therapy, in patients older than 64 y, and in those with impaired liver function.
- Anticipate BP reduction with possible reflex heart rate increase (5–10 bpm) 2–5 h after dosing.
- Report sustained hypotension promptly; more common with concurrent beta-blocker therapy.
- Assess for and report reflex tachycardia; may precipitate angina.
- Monitor patients for possible digoxin toxicity when taking concurrent digoxin.

Patient Education

- Report peripheral edema, headache, or flushing to prescriber. These may necessitate discontinuation of drug.
- Get up from lying down slowly and in stages; there is potential for dizziness and hypotension.

FENOFIBRATE ℗

(fen-o-fi′brate)

Antara, Lofibra, Tricor, Triglide, TriLipix

Classifications: ANTILIPEMIC; FIBRATE

Therapeutic: CHOLESTEROL-LOWERING

Pregnancy Category: C

AVAILABILITY 48 mg, 50 mg, 154 mg, 160 mg tablets; 43 mg, 67 mg, 87 mg, 134 mg, 200 mg capsules or 50 mg, 100 mg, 150 mg, 160 mg capsules; 45 mg, 135 mg delayed release capsules

ACTION & THERAPEUTIC EFFECT Fibric acid derivative with lipid-regulating properties. Lowers plasma triglycerides by inhibiting triglyceride synthesis and, as a result,

Common adverse effects in *italic*, life-threatening effects <u>underlined</u>; generic names in **bold**; classifications in SMALL CAPS; ♣ Canadian drug name; ℗ Prototype drug

617

lowers VLDL production as well as stimulates the catabolism of triglyceride-rich lipoprotein (e.g., VLDL). Produces a moderate increase in HDL cholesterol levels in most patients. *Effectiveness indicated by reduction in the level of serum triglycerides and VLDL production.*

USES Adjunctive therapy to diet for patients with high triglycerides.

CONTRAINDICATIONS Hypersensitivity to fenofibrate or other fibric acid derivatives (e.g., clofibrate, benzofibrate); liver or severe kidney dysfunction; unexplained liver function abnormality; preexisting hepatic disease; primary biliary cirrhosis; preexisting gallbladder disease; thrombocytopenia; lactation. **CAUTIOUS USE** Renal impairment, older adults; history of bleeding disorders; myelosuppression; pregnancy (category C). Safety and efficacy in children younger than 10 y **(capsules)**, younger than 18 y **(tablets)** are not established.

ROUTE & DOSAGE

Hypertriglyceridemia
Adult: **PO** 43–200 mg/day depending on product

ADMINISTRATION

Oral

- Drug is usually discontinued after 2 mo if adequate lipid reduction is not achieved with the maximum recommended dose.
- Give at least 1 h before or 4–6 h after cholestyramine.
- Store at 15°–30° C (59°–86° F) in a tightly closed container and protect from light.

ADVERSE EFFECTS (≥1%) **Body as a Whole:** Asthenia, fatigue, infections, flu-like syndrome, localized pain, arthralgia. **CNS:** Headache, paresthesia, dizziness, insomnia. **CV:** Arrhythmia. **GI:** Dyspepsia, eructation, flatulence, nausea, vomiting, abdominal pain, constipation, diarrhea, increased appetite. **Respiratory:** Cough, rhinitis, sinusitis. **Skin:** Pruritus, rash. **Special Senses:** Earache, eye floaters, blurred vision, conjunctivitis, eye irritation. **Urogenital:** Decreased libido, polyuria, vaginitis.

INTERACTIONS Drug: May potentiate anticoagulant effects of **warfarin;** combination with an HMG-COA REDUCTASE INHIBITOR (STATIN) may result in rhabdomyolysis; **cholestyramine, colestipol** may decrease absorption (give fenofibrate 1 h before or 4–6 h after BILE ACID SEQUESTRANTS); may increase risk of nephrotoxicity of **cyclosporine.**

PHARMACOKINETICS Absorption: Well absorbed from the GI tract; increased with food. **Peak:** 6–8 h. **Distribution:** 99% protein bound; excreted in breast milk. **Metabolism:** Rapidly hydrolyzed by esterases to active metabolite, fenofibric acid. **Elimination:** 60% in urine, 25% in feces. **Half-Life:** 20 h.

NURSING IMPLICATIONS

Assessment & Drug Effects

- Lab tests: Periodically monitor lipid levels, LFTs, and CBC with differential.
- Assess for muscle pain, tenderness, or weakness and, if present, monitor CPK level. Withdraw drug with marked elevations of CPK or if myopathy is suspected.
- Monitor patients on coumarin-type drugs closely for prolongation of PT/INR.

Patient & Family Education

- Contact prescriber immediately if any of the following develops:

Unexplained muscle pain, tenderness, or weakness, especially with fever or malaise; yellowing of skin or eyes; nausea or loss of appetite; skin rash or hives.
- Inform prescriber regarding concurrent use of cholestyramine, oral anticoagulants, or cyclosporine.

FENOLDOPAM MESYLATE

(fen-ol'do-pam mes'y-late)

Corlopam

Classifications: NON-NITRATE VASODILATOR; DOPAMINE AGONIST; ANTIHYPERTENSIVE

Therapeutic: ANTIHYPERTENSIVE

Pregnancy Category: B

AVAILABILITY 10 mg/mL injection

ACTION & *THERAPEUTIC EFFECT*
Rapid-acting vasodilator that is a dopamine D_1-like receptor agonist. Exerts hypotensive effects by decreasing peripheral vascular resistance while increasing renal blood flow, diuresis, and natriuresis. Effectiveness *indicated by rapid reduction in BP. Decreases both systolic and diastolic pressures.*

USES Short-term (up to 48 h) management of severe hypertension.

CONTRAINDICATIONS Hypersensitivity to fenoldopam.

CAUTIOUS USE Asthmatic patients; hepatic cirrhosis, portal hypertension, or variceal bleeding; arrhythmias, tachycardia, or angina, particularly unstable angina; elevated IOP; angular-closure glaucoma; hypotension; hypokalemia; acute cerebral infarct or hemorrhage; pregnancy (category B), lactation.

ROUTE & DOSAGE

Severe Hypertension

Adult: **IV** 0.1–0.3 mcg/kg/min by continuous infusion for up to 48 h, may increase by 0.05–0.1 mcg/kg/min q15min (dosage range: 0.01–1.6 mcg/kg/min)

Child: **IV** 0.2 mcg/kg/min, may increase to 0.3–0.5 mcg/kg/min

ADMINISTRATION

Intravenous

PREPARE: Continuous for Adult: Dilute to a final concentration of 40 mcg/mL by adding 1 mL (10 mg), 2 mL (20 mg), or 3 mL (30 mg) of fenoldopam to 250, 500, or 1000 mL, respectively, of NS or D5W. **Continuous for Child:** Dilute to a final concentration of 60 mcg/mL by adding 0.6 mL (6 mg), 1.5 mL (15 mg) or 3 mL (3 mg) of fenoldopam to 100, 250, or 500 mL, respectively, of NS or D5W.

ADMINISTER: Continuous for Adult/Child: Give only by continuous infusion; never give a direct or bolus dose. ▪ Titrate initial dose up or down no more frequently than q15min.

INCOMPATIBILITIES **Y-site: Acyclovir, aminophylline, amphotericin B, ampicillin, bumetanide, cefoxitin, dantrolene, dexamethasone, diazepam, fosphenytoin, furosemide, ganciclovir, gemtuzumab, hetastarch, ketorolac, meropenem, mesna, methohexital, methylprednisolone, mitomycin, pantoprazole, pentobarbital, phenytoin, prochlorperazine, sodium bicarbonate, thiopental.**

▪ Note: Diluted solution is stable under normal room temperature and

Common adverse effects in *italic*, life-threatening effects underlined; generic names in **bold**; classifications in SMALL CAPS; ♦ Canadian drug name; ⊙ Prototype drug

619

light for 24 h. Discard any unused solution after 24 h. ▪ Store at 15°–30° C (59°–86° F) in a tightly closed container and protect from light.

ADVERSE EFFECTS (≥1%) Body as a Whole: Injection site reaction, pyrexia, nonspecific chest pain. **CNS:** Headache, nervousness, anxiety, insomnia, dizziness. **CV:** *Hypotension, tachycardia,* T-wave inversion, flushing, postural hypotension, extrasystoles, palpitations, bradycardia, heart failure, ischemic heart disease, <u>MI</u>, angina. **GI:** Nausea, vomiting, abdominal pain or fullness, constipation, diarrhea. **Metabolic:** Increased creatinine, BUN, glucose, transaminases, LDH; hypokalemia. **Respiratory:** Nasal congestion, dyspnea, upper respiratory disorder. **Skin:** Sweating. **Other:** UTI, leukocytosis, bleeding.

INTERACTIONS Use with BETA-BLOCKERS increases risk of hypotension.

PHARMACOKINETICS Onset: 5 min. **Peak:** 15 min. **Duration:** 15–30 min. **Distribution:** Crosses placenta. **Metabolism:** Conjugated in liver. **Elimination:** 90% in urine, 10% in feces. **Half-Life:** 5 min.

NURSING IMPLICATIONS

Assessment & Drug Effects

▪ Monitor BP and HR carefully at least q15min or more often as warranted; expect dose-related tachycardia.
▪ Lab tests: Carefully monitor serum electrolytes (especially serum potassium), BUN and creatinine, LFTs, and blood glucose and HbA1C.

FENOPROFEN CALCIUM

(fen-oh-proe′fen)

Nalfon

Classifications: ANALGESIC, NONSTEROIDAL ANTI-INFLAMMATORY DRUG (NSAID)

Therapeutic: ANALGESIC; NSAID; ANTIARTHRITIC; ANTIPYRETIC
Prototype: Ibuprofen
Pregnancy Category: B first and second trimester; D third trimester

AVAILABILITY 200 mg, 400 mg capsules; 600 mg tablets

ACTION & *THERAPEUTIC EFFECT*
Exhibits anti-inflammatory, analgesic, and antipyretic properties of an NSAID. Fenoprofen competitively inhibits both cyclooxygenase COX-1 and COX-2 enzymes by blocking arachidonate binding to prostaglandin G_2 resulting in its pharmacologic effects. *Has nonsteroidal, anti-inflammatory, antipyretic, antiarthritic properties that provide relief from mild to severe pain.*

USES Rheumatoid arthritis and osteoarthritis; relief of mild to moderate pain.
UNLABELED USES Juvenile rheumatoid arthritis, acute gouty arthritis, ankylosing spondylitis.

CONTRAINDICATION Hypersensitivity to fenoprofen or other NSAIDs; salicylate; history of nephrotic syndrome associated with aspirin or other NSAIDs; patient in whom urticaria, severe rhinitis, bronchospasm, angioedema, nasal polyps are precipitated by aspirin or other NSAIDs; severe renal or hepatic dysfunction; perioperative pain associated in CABG; pregnancy (category D third trimester).

CAUTIOUS USE History of upper GI tract disorders; lupus; older adults; renal failure; renal impairment; hemophilia or other bleeding tendencies; compromised cardiac function; hypertension; impaired hearing; pregnancy (category B first and second trimester); lactation. Safety in children not established.

ROUTE & DOSAGE

Inflammatory Disease
Adult: PO 400–600 mg t.i.d. or q.i.d. (max: 3200 mg/day)

Mild to Moderate Pain
Adult: PO 200 mg q4–6h prn

ADMINISTRATION

Oral

▪ Give on an empty stomach 30–60 min before or 2 h after meals. Give with meals, milk, or antacid (prescribed) if patient experiences GI disturbances.

▪ May crush tablets or empty capsule and mix with fluid or mix with food.

▪ Store capsules and tablets in tightly closed containers at 15°–30° C (59°–86° F); avoid freezing.

ADVERSE EFFECTS (≥1%) **CNS:** *Headache, drowsiness,* dizziness, fatigue, lassitude, tremor, confusion, insomnia, nervousness, depression. **Special Senses:** Tinnitus, decreased hearing, deafness; blurred vision. **GI:** *Indigestion, nausea, vomiting,* anorexia, *constipation,* diarrhea, flatulence, abdominal pain, dry mouth; infrequent: Gastritis, peptic ulcer, <u>GI bleeding</u>. **Urogenital:** Dysuria, cystitis, hematuria, oliguria, azotemia, anuria, allergic nephritis, papillary necrosis, nephrotoxicity (rare). **Hematologic:** (Infrequent) <u>Thrombocytopenia, hemolytic anemia, agranulocytosis, pancytopenia</u>. **Skin:** (May or may not be hypersensitivity reaction) Pruritus, rash, purpura, increased sweating, urticaria. **Body as a Whole:** Dyspnea, malaise, <u>anaphylaxis</u>, edema.

INTERACTIONS Drug: Fenoprofen may prolong bleeding time; should not be given with ORAL ANTICOAGULANTS, **heparin;** action and side effects of **phenytoin,** SULFONYLUREAS, SULFONAMIDES, and fenoprofen may be potentiated. **Herbal: Feverfew, garlic, ginger, gingko** may increase bleeding potential.

PHARMACOKINETICS Absorption: 80% from GI tract. **Onset:** 2 h. **Peak:** 2 h. **Duration:** 4–6 h. **Distribution:** Small amounts distributed into breast milk. **Metabolism:** In liver. **Elimination:** Primarily in urine; some biliary excretion. **Half-Life:** 3 h.

NURSING IMPLICATIONS

Assessment & Drug Effects

▪ Lab tests: Baseline Hct and Hgb, kidney function tests and LFTs.

▪ Baseline and periodic auditory and ophthalmic examinations are recommended in patients receiving prolonged or high-dose therapy.

▪ Monitor for S&S of GI bleeding.

Patient & Family Education

▪ Do not drive or engage in potentially hazardous activities until response to drug is known; fenoprofen may cause dizziness and drowsiness.

▪ Report immediately the onset of unexplained fever, rash, arthralgia, oliguria, edema, weight gain to prescriber. Possible symptoms of nephrotic syndrome are rapidly reversible if drug is promptly withdrawn.

▪ Understand that alcohol and aspirin may increase risk of GI ulceration and bleeding tendencies; avoid both unless otherwise advised by prescriber.

FENTANYL CITRATE
(fen′ta-nil)
Abstral, Actiq, Duragesic, Fentora, Ionsys, Onsolis, Lazanda, Sublimaze

F

Common adverse effects in *italic;* life-threatening effects <u>underlined;</u> generic names in **bold;** classifications in SMALL CAPS; ♣ Canadian drug name; ⊘ Prototype drug

621

Classifications: ANALGESIC; NAR-
COTIC (OPIATE AGONIST)
Therapeutic: NARCOTIC ANALGESIC
Prototype: Morphine
Pregnancy Category: C (B for
fentanyl injection)
Controlled Substance: Schedule II

AVAILABILITY 0.05 mg/mL injection;
100 mcg, 200 mcg, 300 mcg, 400
mcg lozenges; 200 mcg, 400 mcg,
600 mcg, 800 mcg, 1200 mcg, 1600
mcg lozenges on a stick; 12 mcg/h,
25 mcg/h, 50 mcg/h, 75 mcg/h, 100
mcg/h transdermal patch; 100 mcg,
200 mcg, 300 mcg, 400 mcg, 600
mcg, 800 mcg buccal tablet; 0.2 mg,
0.4 mg. 0.6 mg, 0.8 mg, 1.2 mg buccal film; 100 mcg, 200 mcg, 300 mcg,
400 mcg, 600 mcg, 800 mcg sublingual tablet; 100 mcg/actuation, 400
mcg/actuation nasal spray

ACTION & *THERAPEUTIC EFFECT*
Synthetic, potent narcotic agonist
analgesic that causes analgesia and
sedation. Its alterations in respiratory rate and alveolar ventilation may
persist beyond the analgesic effect.
*Provides analgesia for moderate to
severe pain as well as sedation.*

USES Short-acting analgesic during
operative and perioperative periods,
moderate or severe pain.

CONTRAINDICATIONS Patients
who have received MAO inhibitors
within 14 days; substance abuse;
myasthenia gravis; labor and delivery. **Transdermal patch:** Patients
not opioid tolerant, acute pain or
short term use; postoperative pain;
mild pain; intermittent pain.
CAUTIOUS USE Head injuries,
increased intracranial pressure;
older adults, debilitated, poor-risk
patients; cardiac diseases, angina,
hypotension, or cardiac arrhythmias; COPD, other respiratory

problems; liver and kidney dysfunction; bradyarrhythmias; children; pregnancy (category C) and
(category B for fentanyl injection).

ROUTE & DOSAGE

Premedication

Adult: **IV** 50–100 mcg 30–60 min
before surgery **PO** Suck on 400
mcg lozenge until sedated
Child: **PO** Suck on lozenge until
sedated, *weight 10–25 kg,* 200
mcg lozenge; *weight 25–35 kg,*
300 mcg lozenge; *weight 35–40
kg,* 400 mcg lozenge

Adjunct for Regional Anesthesia

Adult: **IM/IV** 50–100 mcg

General Anesthesia

Adult: **IV** 2–20 mcg/kg,
additional doses of 25–100 mcg
as required
Child: **IV** 2–3 mcg/kg as needed

Postoperative Pain

Adult: **IM/IV** 50–100 mcg
q1–2h prn
Child: **IM** 1.7–3.3 mcg/kg
q1–2h prn

Pain

Adult: **Transdermal** Individualize
and regularly reassess doses of
transdermal fentanyl; for patient
not already receiving an opioid,
the initial dose is 25 mcg/h patch
q3days; for patients already on
opioids, see package insert for
conversions **Stick lozenge (Actiq)**
200 mcg **(Actiq** only) as breakthrough agent **IV/IM** 50–110 mcg

ADMINISTRATION

Oral

- *Buccal tablet:* Do not push tablet
  through blister, as this may cause

Common adverse effects in *italic*, life-threatening effects <u>underlined</u>; generic names
in **bold**; classifications in SMALL CAPS; ✦ Canadian drug name; ☢ Prototype drug

damage to tablet. *Lozenge:* Place unit between cheek and lower gum, moving it from one side to the other using the handle. Instruct the patient to suck, not chew, the lozenge. Should be consumed over a 15-min period.

- *Buccal film:* Place film on inside of cheek with the pink side against the mucous membrane. Hold film against cheek for 5 sec. Film should be left in place until it dissolves (15–30 min). Liquids may be consumed after 5 min but solids should not be eaten until film dissolves. *Sublingual tablet:* Note that Abstral tablets are not equivalent on a mcg per mcg basis with other fentanyl products.

Intranasal
- *Nasal spray:* Lazanda is not equivalent to other fentanyl products on a mcg per mcg basis.

Intramuscular
- Inject undiluted into a large muscle.

Transdermal
- Place on nonirritated flat surface (e.g., chest, back, upper arm). The upper back is preferred to minimize unintended patch removal. Clip (not shave) hair at application site prior to system application. If needed, clean site prior to application only with clear water. Press patch in place for 30 sec. If gel from patch leaks out and contacts skin of patient or caregiver, wash thoroughly with water.

Intravascular

PREPARE: **Direct:** Give parenteral doses undiluted or diluted in 5 mL sterile water or NS.
ADMINISTER: **Direct:** Infuse over 3–5 min.
INCOMPATIBILITIES **Solution/additive: Fluorouracil, lidocaine.**

Y-site: Azithromycin, phenytoin.

- Store at 15°–30° C (59°–86° F) unless otherwise directed. Protect drug from light.

ADVERSE EFFECTS (≥1%) **CNS:** *Sedation,* euphoria, dizziness, diaphoresis, delirium, convulsions with high doses. **CV:** Hypotension, bradycardia, <u>circulatory depression, cardiac arrest</u>. **Special Senses:** Miosis, blurred vision. **GI:** *Nausea,* vomiting, constipation, ileus. **Respiratory:** Laryngospasm, bronchoconstriction, <u>respiratory depression or arrest</u>. **Body as a Whole:** Muscle rigidity, especially muscles of respiration after rapid IV infusion, urinary retention. **Skin:** Rash, contact dermatitis from patch.

INTERACTIONS Drug: Alcohol and other CNS DEPRESSANTS potentiate effects; MAO INHIBITORS may precipitate hypertensive crisis.

PHARMACOKINETICS Absorption: Absorbed through the skin, leveling off between 12–24 h. **Onset:** Immediate IV; 7–15 min IM; 12–24 h transdermal. **Peak:** 3–5 min IV; 24–72 h transdermal. **Duration:** 30–60 min IV; 1–2 h IM; 72 h transdermal. **Metabolism:** In liver by CYP3A4. **Elimination:** In urine. **Half-Life:** 17 h transdermal.

NURSING IMPLICATIONS

Assessment & Drug Effects
- Monitor vital signs and observe patient for signs of skeletal and thoracic muscle (depressed respirations) rigidity and weakness.
- Watch carefully for respiratory depression and for movements of various groups of skeletal muscle in extremities, external eye, and neck during postoperative period. These movements may present

Common adverse effects in *italic*, life-threatening effects <u>underlined</u>; generic names in **bold**; classifications in SMALL CAPS; ♣ Canadian drug name; ⊙ Prototype drug

623

patient management problems; report promptly.

- Note: Duration of respiratory depressant effect may be considerably longer than narcotic analgesic effect. Have immediately available oxygen, resuscitative and intubation equipment, and an opioid antagonist such as naloxone.

Patient & Family Education
- Follow exactly instructions for taking fentanyl and for disposal of unit provided in patient information.
- Exercise caution when engaging in hazardous activities until reaction to drug is known.
- Children exposed to buccal tablets are at high risk for respiratory depression. Keep out of reach of children.

✓FERROUS SULFATE ℗
(fer'rous sul'fate)

Feosol, Fer-In-Sol, Fer-Iron, Ferospace, Ferralyn, Fesofor, Hematinic, Mol-Iron, Novoferrosulfa ◆, Slow-Fe

FERROUS FUMARATE
(fer'rous foo'ma-rate)

Feco-T, Femiron, Feostat, Fersamal, Fumasorb, Fumerin, Hemocyte, Ircon-FA, Neo-Fer-50 ◆, Novofumar ◆, Palafer ◆, Palmiron

FERROUS GLUCONATE
(fer'rous gloo'koe-nate)

Fergon, Fertinic ◆, Novoferrogluc ◆, Simron
Classification: IRON PREPARATION
Therapeutic: ANTIANEMIC; IRON SUPPLEMENT
Pregnancy Category: A

AVAILABILITY Ferrous Sulfate: 167 mg, 200 mg, 324 mg, 325 mg tablets; 160 mg sustained release tablets, capsules; 90 mg/5 mL syrup; 220 mg/5 mL elixir; 75 mg/0.6 mL drops; **Ferrous Fumarate:** 63 mg, 100 mg, 200 mg, 324 mg, 325 mg, 350 mg tablets; 100 mg/5 mL suspension; 45 mg/0.6 mL drops; **Ferrous Gluconate:** 240 mg, 325 mg tablets

ACTION & *THERAPEUTIC EFFECT*
Ferrous sulfate: Standard iron preparation that corrects erythropoietic abnormalities induced by iron deficiency but does not stimulate erythropoiesis. ***Ferrous gluconate:*** Claimed to cause less gastric irritation and be better tolerated than ferrous sulfate. *Effectiveness is experienced within 48 h as a sense of well-being, increased vigor, improved appetite, and decreased irritability (in children). Reticulocyte response begins in about 4 days; it usually peaks in 7–10 days and returns to normal after 2 or 3 wk.*

USES To correct simple iron deficiency and to treat iron deficiency anemias. May be used prophylactically during periods of increased iron needs, as in infancy, childhood, and pregnancy.

CONTRAINDICATIONS Peptic ulcer, regional enteritis, ulcerative colitis; hemolytic anemias (in absence of iron deficiency), hemochromatosis, hemosiderosis, patients receiving repeated transfusions, pyridoxine-responsive anemia; cirrhosis of liver.
CAUTIOUS USE Hepatic disease; GI diseases; sulfite hypersensitivity; pregnancy (category A).

ROUTE & DOSAGE

Iron Deficiency
Adult: **PO Sulfate (30% elemental iron)** 750–1500 mg/day in 1–3

Common adverse effects in *italic*, life-threatening effects <u>underlined</u>; generic names in **bold**; classifications in SMALL CAPS; ◆ Canadian drug name; ℗ Prototype drug

divided doses; **Fumarate (33% elemental iron)** 200 mg t.i.d. or q.i.d.; **Gluconate (12% elemental iron)** 325–600 mg q.i.d., may be gradually increased to 650 mg q.i.d. as needed and tolerated
Child: **PO Sulfate (30% elemental iron)** *Younger than 6 y,* 75–225 mg/day in divided doses; *6–12 y,* 600 mg/day in divided doses; **Fumarate (33% elemental iron)** 3 mg/kg t.i.d.; **Gluconate (12% elemental iron)** *Younger than 6 y,* 100–300 mg/day in divided doses; *6–12 y,* 100–300 mg t.i.d.

Iron Supplement

Adult: **PO Sulfate** *Pregnancy,* 300–600 mg/day in divided doses; **Fumarate** 200 mg once/day; **Gluconate** 325–600 mg once/day
Child: **PO Fumarate** 3 mg/kg once/day; **Gluconate** *Younger than 6 y,* 100–300 mg/day in divided doses; *6–12 y,* 100–300 mg once/day
Infant: **PO Fumarate** *Low birth weight,* 2 mg/kg/day up to 15 mg/day; *3 y or younger,* 1 mg/kg/day (max: 15 mg/day)

ADMINISTRATION

Oral

- Give on an empty stomach if possible because oral iron preparations are best absorbed then (i.e., between meals). Minimize gastric distress if needed by giving with or immediately after meals with adequate liquid.
- Do not crush tablet or empty contents of capsule when administering.
- Do not give tablets or capsules within 1 h of bedtime.
- Consult prescriber about prescribing a liquid formulation or a less corrosive form, such as ferrous gluconate, if the patient experiences difficulty in swallowing tablet or capsule.
- Dilute liquid preparations well and give through a straw or placed on the back of tongue with a dropper to prevent staining of teeth and to mask taste. Instruct the patient to rinse mouth with clear water immediately after ingestion.
- Mix ferosol elixir with water; not compatible with milk or fruit juice. Fer-In-Sol (drops) may be given in water or in fruit or vegetable juice, according to manufacturer.
- Do not use discolored tablets.
- Store in tightly closed containers and protect from moisture. Store at 15°–30° C (59°–86° F).

ADVERSE EFFECTS (≥1%) GI: *Nausea, heartburn,* anorexia, *constipation,* diarrhea, epigastric pain, abdominal distress, *black stools.* **Special Senses:** Yellow-brown discoloration of eyes and teeth (liquid forms). **Large Chronic Doses in Infants** Rickets (due to interference with phosphorus absorption). **Massive Overdosage** Lethargy, drowsiness, nausea, vomiting, abdominal pain, diarrhea, local corrosion of stomach and small intestines, pallor or cyanosis, metabolic acidosis, <u>shock, cardiovascular collapse</u>, convulsions, <u>liver necrosis</u>, coma, renal failure, <u>death</u>.

DIAGNOSTIC TEST INTERFERENCE By coloring feces black, large iron doses may cause false-positive tests for ***occult blood with orthotoluidine (Hematest, Occultist, Labstix); guaiac reagent benzidine test*** is reportedly not affected.

Common adverse effects in *italic*, life-threatening effects <u>underlined</u>; generic names in **bold**; classifications in SMALL CAPS; ♣ Canadian drug name; ♥ Prototype drug

625

INTERACTIONS Drug: ANTACIDS decrease iron absorption; iron decreases absorption of TETRACYCLINES, **ciprofloxacin, ofloxacin; chloramphenicol** may delay iron's effects; iron may decrease absorption of **penicillamine. Food:** Food decreases absorption of iron; **ascorbic acid (vitamin C)** may increase iron absorption.

PHARMACOKINETICS Absorption: 5–10% absorbed in healthy individuals; 10–30% absorbed in iron-deficiency; food decreases amount absorbed. **Distribution:** Transported by transferrin to bone marrow, where it is incorporated into hemoglobin; crosses placenta. **Elimination:** Most of iron released from hemoglobin is reused in body; small amounts are lost in desquamation of skin, GI mucosa, nails, and hair; 12–30 mg/mo lost through menstruation.

NURSING IMPLICATIONS

Assessment & Drug Effects

- Lab tests: Monitor Hgb and reticulocyte values during therapy. Investigate the absence of satisfactory response after 3 wk of drug treatment.
- Continue iron therapy for 2–3 mo after the hemoglobin level has returned to normal (roughly twice the period required to normalize hemoglobin concentration).
- Monitor bowel movements as constipation is a common adverse effect.

Patient & Family Education

- Note: Ascorbic acid increases absorption of iron. Consuming citrus fruit or tomato juice with iron preparation (except the elixir) may increase its absorption.
- Be aware that milk, eggs, or caffeine beverages when taken with the iron preparation may inhibit absorption.
- Be aware that iron preparations cause dark green or black stools.
- Report constipation or diarrhea to prescriber; symptoms may be relieved by adjustments in dosage or diet or by change to another iron preparation.

FESOTERODINE
(fes-o-ter-o-deen)
Toviaz
Classifications: ANTICHOLINERGIC; MUSCARINIC RECEPTOR ANTAGONIST; BLADDER ANTISPASMODIC
Therapeutic: BLADDER ANTISPASMODIC
Prototype: Oxybutynin
Pregnancy Category: C

AVAILABILITY 4 mg, 8 mg extended release tablets

ACTION & *THERAPEUTIC EFFECT* A muscarinic receptor antagonist that reduces urinary incontinence, urgency, and frequency. It helps regulate the involuntary contractions of the bladder associated with sudden urges to urinate. *It controls urinary incontinence or overactive bladder (OAB).*

USES Treatment of overactive bladder in patients with urinary incontinence, urgency.

CONTRAINDICATIONS Severe hepatic impairment (Child-Pugh class C); gastric obstruction, paralytic ileus, uncontrolled narrow-angle glaucoma; urinary retention; severe BPH; severe renal insufficiency (CrCl less than 30 mL/min) use should be avoided or requires a dosage reduction; lactation.
CAUTIOUS USE Cross-sensitivity to tolterodine; mild to moderate

Common adverse effects in *italic*, life-threatening effects <u>underlined</u>; generic names in **bold**; classifications in SMALL CAPS; ♣ Canadian drug name; ⊘ Prototype drug

hepatic impairment (Child-Pugh class A and B); mild to moderate renal insufficiency; history of constipation; bladder outlet obstruction; history of decreased GI motility; severe constipation; palpitations; narrow-angle glaucoma; myasthenia gravis; pregnancy (category C). Safe use in children not established.

ROUTE & DOSAGE

Overactive Bladder

Adult: **PO** 4–8 mg daily. Do not exceed 4 mg daily with concurrent potent CYP3A4 inhibitors.

Hepatic Impairment Dosage Adjustment

Severe hepatic impairment (Child-Pugh class C): Not recommended

Renal Impairment Dosage Adjustment

CrCl less than 30 mL/min: Not recommended

ADMINISTRATION

Oral

- Give with water without regard to food.
- Do not break or crush extended release tablet. Ensure that it is swallowed whole.
- Store at 15°–30° C (59°–86° F).

ADVERSE EFFECTS (≥1%) **Body as a Whole:** Peripheral edema, upper respiratory tract infection, urinary tract infection, angioedema. **CNS:** Insomnia. **GI:** Constipation, *dry mouth*, dyspepsia, nausea. **Metabolic:** Elevated ALT levels, elevated GGT levels. **Musculoskeletal:** Back pain. **Respiratory:** Cough, dry throat. **Skin:** Rash. **Special Senses:** Dry eyes. **Urogenital:** Dysuria, urinary retention.

INTERACTIONS Drug: Potent CYP-3A4 INHIBITORS (e.g., **clarithromycin, itraconazole, ketoconazole, nefazodone, nelfinavir,** and **ritonavir**) increase fesoterodine levels. INDUCERS OF CYP3A4 (e.g., **rifampin**) can decrease fesoterodine levels.

PHARMACOKINETICS Absorption: 52% bioavailable. **Peak:** 5 h. **Distribution:** 50% plasma protein bound. **Metabolism:** Hepatic metabolism to active and inactive metabolites. **Elimination:** Renal (70%) and fecal (7%). **Half-Life:** 7 h.

NURSING IMPLICATIONS

Assessment & Drug Effects

- Monitor bowel and bladder function as urinary retention and constipation are potential adverse effects. Older adults are at greater risk for adverse effects, especially with the 8 mg dose.
- Monitor for and report bothersome anticholinergic effects (see Appendix F).

Patient & Family Education

- Use caution in hot environments to avoid heat prostration as fesoterodine causes decreased sweating and reduces body cooling in excessive heat.
- Exercise caution with hazardous activities until response to drug is known.
- Moderate alcohol consumption as it may enhance drowsiness caused by fesoterodine.

FEXOFENADINE

(fex-o-fen′a-deen)

Allegra

Classifications: NONSEDATING; ANTIHISTAMINE; H_1-RECEPTOR ANTAGONIST

Common adverse effects in *italic*, life-threatening effects underlined; generic names in **bold**; classifications in SMALL CAPS; ♣ Canadian drug name; ⊘ Prototype drug

627

Therapeutic: NONSEDATING ANTI-HISTAMINE
Prototype: Loratadine
Pregnancy Category: C

AVAILABILITY 30 mg, 60 mg, 180 mg tablets; 60 mg capsules; 30 mg orally disintegrating tablet; 30 mg/5 mL oral suspension

ACTION & *THERAPEUTIC EFFECT*
Competes with histamine for binding at the H_1-receptor. This blocks effects of histamine on H_1-receptors in bronchial smooth muscle resulting in decreased formation of edema, flare, and pruritus. *Inhibits antigen-induced bronchospasm and histamine release from mast cells. Efficacy is indicated by reduction of the following: Nasal congestion and sneezing; watery or red eyes; itching nose, palate, or eyes.*

USES Relief of symptoms associated with seasonal allergic rhinitis, and chronic urticaria.

CONTRAINDICATIONS Hypersensitivity to fexofenadine or terfenadine; neonates.
CAUTIOUS USE Mild to severe renal and hepatic insufficiency, hypertension, diabetes mellitus, ischemic heart disease, increased ocular pressure, hyperthyroidism, renal impairment, prostatic hypertrophy; elderly, pregnancy (category C), lactation; young children.

ROUTE & DOSAGE

Allergic Rhinitis
Adult/Adolescent: **PO** 60 mg b.i.d. OR 180 mg daily
Child (2–11 y): **PO** 30 mg b.i.d.
Child (6–11 y): **PO (oral disintegrating tablet)** 30 mg b.i.d.

Chronic Urticaria
Adult: **PO** 60 mg b.i.d. OR 180 mg daily
Child (2–11 y): **PO** 30 mg b.i.d.
Child (6–11 y): **PO (oral disintegrating tablet)** 30 mg b.i.d.
Child (younger than 2 y)/Infant (older than 6 mo): **PO** 15 mg b.i.d.

Renal Impairment Dosage Adjustment
CrCl less than 80 mL/min: Give normal dose only once per day

ADMINISTRATION
Oral
- Reduce starting dose for those with decreased kidney function.
- Do not give within 15 min of an aluminum- or magnesium-containing antacid.
- Store at 20°–25° C (68°–77° F). Protect from excess moisture.

ADVERSE EFFECTS (≥1%) **CNS:** *Headache,* drowsiness, fatigue. **GI:** Nausea, dyspepsia, throat irritation.

INTERACTIONS Drug: ANTACIDS will decrease serum level of fexofenadine. **Herbal: St. John's wort** will decrease serum level of fexofenadine. **Food: Grapefruit juice** or **apple juice** may decrease efficacy.

PHARMACOKINETICS Absorption: Rapidly from GI tract, 33% reaches systemic circulation. **Onset:** 1 h. **Peak:** 2–3 h. **Duration:** At least 12 h. **Distribution:** 60–70% bound to plasma proteins. **Metabolism:** Only 5% of dose metabolized in liver. **Elimination:** 80% in urine, 11% in feces. **Half-Life:** 14.4 h.

NURSING IMPLICATIONS
Assessment & Drug Effects
- Monitor therapeutic effectiveness, which is indicated by decreased

nasal congestion, sneezing, watery or red eyes, and itching nose, palate, or eyes.

Patient & Family Education
▪ Note: Drug is well tolerated and causes minimal adverse effects.

FIDAXOMICIN
(fye dax′ oh mye′ sin)
Dificid
Classifications: MACROLIDE ANTI-BIOTIC
Therapeutic: ANTIBIOTIC
Prototype: Erythromycin
Pregnancy Category: B

AVAILABILITY 200 mg tablets

ACTION & *THERAPEUTIC EFFECT*
Macrolide antibiotic that inhibits RNA-synthesis by RNA polymerases. *Bactericidal action on Clostridium difficile in the GI tract.*

USES Treatment of pseudomembranous colitis or *Clostridium difficile*-associated diarrhea (CDAD).

CAUTIOUS USE Not for use in systemic infections; hyperglycemia; metabolic acidosis; pregnancy; lactation. Safety and effectiveness in children younger than 18 y have not been established.

ROUTE & DOSAGE

Clostridium Difficile–Associated Diarrhea
Adult: **PO** 200 mg b.i.d. for 10 d

ADMINISTRATION

Oral
▪ May give without regard to food.
▪ Store between 15°–30° C (59°–86° F).

ADVERSE EFFECTS (≥2%) **GI:** Abdominal pain, GI hemorrhage, *nausea, vomiting.* **Hematological:** Anemia, neutropenia.

PHARMACOKINETICS Absorption: Minimal systemic absorption. **Peak:** 2 h. **Distribution:** Primarily confined to GI tract. **Metabolism:** Hydrolysis in GI tract to active metabolite. **Elimination:** Primarily fecal (92%). **Half-Life:** 11.7 h.

NURSING IMPLICATIONS

Assessment & Drug Effects
▪ Note that fidaxomicin is not effective for treatment of systemic infections.
▪ Monitor GI symptoms of CDAD (C. difficile-associated disease) and report increasing GI distress.
▪ Effectiveness is indicated by decreasing diarrhea and improvement in other S&S of GI distress.
▪ Lab tests: CBC with differential as needed.

Patient & Family Education
▪ It is common to feel better soon after beginning treatment, however, do not stop taking the drug until the full course of treatment is completed.

FILGRASTIM 🅟
(fil-gras′tim)
Neupogen
Classification: HEMATOPOIETIC GROWTH FACTOR
Therapeutic: ANTINEUTROPENIC; GRANULOCYTE COLONY-STIMULATING FACTOR (G-CSF)
Pregnancy Category: C

AVAILABILITY 300 mcg/mL injection

ACTION & *THERAPEUTIC EFFECT*
Human granulocyte colony-stimulating factor (G-CSF) produced by recombinant DNA

Common adverse effects in *italic*, life-threatening effects underlined; generic names in **bold**; classifications in SMALL CAPS; ♣ Canadian drug name; 🅟 Prototype drug

629

technology. Endogenous G-CSF regulates production of neutrophils within the bone marrow; primarily affects neutrophil proliferation, differentiation, and selected end-cell functional activity (including enhanced phagocytic activity and antibody-dependent killing). *Increases neutrophil proliferation and differentiation within the bone marrow.*

USES To decrease the incidence of infection, as manifested by febrile neutropenia, in patients with nonmyeloid malignancies receiving myelosuppressive anticancer drugs associated with a significant incidence of severe neutropenia with fever; to decrease neutropenia associated with bone marrow transplant; to treat chronic neutropenia; to mobilize peripheral blood stem cells (PBSCs) for autologous transplantation.

CONTRAINDICATIONS Hypersensitivity to *Escherichia coli*–derived proteins, concurrent administration with chemotherapy, radiation; ARDS.
CAUTIOUS USE Sickle cell disease; respiratory insufficiency; pregnancy (category C); lactation.

ROUTE & DOSAGE

Neutropenia

Adult/Child: **IV** 5 mcg/kg/day by 30 min infusion, may increase by 5 mcg/kg/day (max: 30 mcg/kg/day) **Subcutaneous** 5 mcg/kg/day as single dose, may increase by 5 mcg/kg/day (max: 20 mcg/kg/day)

Bone Marrow Transplant

Adult: **IV** 10 mcg/kg/day given 24 h after cytotoxic therapy and 24 h after bone marrow transfusion

ADMINISTRATION

Subcutaneous & Intravenous

- Do not administer filgrastim 24 h before or after cytotoxic chemotherapy. ▪ Use only one dose per vial; do not reenter the vial.
- Prior to injection, filgrastim may be allowed to reach room temperature for a maximum of 6 h.
- Discard any vial left at room temperature for longer than 6 h.

PREPARE: **Intermittent/Continuous:** May dilute with 10–50 mL D5W to yield 15 mcg/mL or greater. ▪ If more diluent is used to yield concentrations of 5–15 mcg/mL, 2 mL of 5% human albumin **must be** added for each 50 mL D5W (prior to adding filgrastim) to prevent adsorption to plastic IV infusion materials.
ADMINISTER: **Intermittent:** Give a single dose over 15–30 min. ▪ Flush line before/after with D5W. **Continuous:** Give a single dose over 4–24 h. ▪ Flush line before/after with D5W.
INCOMPATIBILITIES **Y-site: Amphotericin B, cefepime, cefotaxime, cefoxitin, ceftizoxime, ceftriaxone, cefuroxime, clindamycin, dactinomycin, etoposide, fluorouracil, furosemide, gentamicin, heparin, imipenem, mannitol, methylprednisolone, metronidazole, mitomycin, piperacillin, prochlorperazine, thiotepa.**

- Store refrigerated at 2°–8° C (36°–46° F). Do not freeze. Avoid shaking.

ADVERSE EFFECTS (≥1%) **CV:** Abnormal ST segment depression. **Hematologic:** Anemia. **GI:** Nausea, anorexia. **Body as a Whole:** *Bone pain,* hyperuricemia, *fever.*

DIAGNOSTIC TEST INTERFERENCE Elevations in **leukocyte alkaline**

Common adverse effects in *italic*, life-threatening effects <u>underlined</u>; generic names in **bold**; classifications in SMALL CAPS; ♣ Canadian drug name; ⊕ Prototype drug

phosphatase, **serum alkaline phosphatase, lactate dehydrogenase,** and **uric acid** have been reported. These elevations appear to be related to increased bone marrow activity.

INTERACTIONS Drug: Can interfere with activity of CYTOTOXIC AGENTS, do not use 24 h before or after CYTOTOXIC AGENTS.

PHARMACOKINETICS Absorption: Readily from subcutaneous site. **Onset:** 4 h. **Peak:** 1 h. **Elimination:** Probably in urine. **Half-Life:** 1.4–7.2 h.

NURSING IMPLICATIONS

Assessment & Drug Effects
- Lab tests: Obtain a baseline CBC with differential and platelet count prior to administering drug. Obtain CBC twice weekly during therapy to monitor neutrophil count and leukocytosis. Monitor Hct and platelet count regularly.
- Discontinue filgrastim if absolute neutrophil count exceeds 10,000/mm³ after the chemotherapy-induced nadir. Neutrophil counts should then return to normal.
- Assess degree of bone pain if present. Consult prescriber if nonnarcotic analgesics do not provide relief.

Patient & Family Education
- Report bone pain and, if necessary, to request analgesics to control pain.
- Note: Proper drug administration and disposal are important. A puncture-resistant container for the disposal of used syringes and needles should be available to the patient.

FINASTERIDE ⊕
(fin-as′te-ride)

Propecia, Proscar
Classifications: ANTIANDROGEN; 5-ALPHA REDUCTASE INHIBITOR

Therapeutic: ANTIANDROGEN
Pregnancy Category: X

AVAILABILITY 1 mg, 5 mg tablets

ACTION & *THERAPEUTIC EFFECT*
Specific inhibitor of the steroid 5-alpha-reductase, an enzyme necessary to convert testosterone into potent androgen 5-alpha-dihydrotestosterone (DHT) in the prostate gland. *Decreases the production of testosterone in the prostate gland.*

USES Benign prostatic hypertrophy, male pattern hair loss (androgenetic alopecia).

CONTRAINDICATIONS Hypersensitivity to finasteride; females, pregnancy (category X), lactation, children.
CAUTIOUS USE Hepatic impairment, obstructive uropathy.

ROUTE & DOSAGE

Benign Prostatic Hypertrophy
Adult: **PO** 5 mg/day
Male Pattern Hair Loss
Adult: **PO** 1 mg daily

ADMINISTRATION

Oral
- Crush tablets if necessary. Pregnant women should not handle the crushed drug; if absorbed through the skin it may be harmful to a male fetus.
- Store at 15°–30° C (59°–86° F) unless otherwise directed.

ADVERSE EFFECTS (≥1%) **Urogenital:** Impotence, decreased libido, decreased volume of ejaculate. **CV:** Postural hypotension. **Body as a Whole:** Asthenia.

Common adverse effects in *italic*, life-threatening effects underlined; generic names in **bold**; classifications in SMALL CAPS; ✤ Canadian drug name; ⊕ Prototype drug

DIAGNOSTIC TEST INTERFERENCE
Depresses levels of **DHT** and **prostate-specific antigen (PSA)**. **Testosterone** levels usually are increased.

INTERACTIONS Drug: No clinically significant interactions established. **Herbal: Saw palmetto** may potentiate effects of finasteride.

PHARMACOKINETICS Absorption: Readily from GI tract. **Onset:** 3–6 mo. **Duration:** 5–7 days. **Elimination:** 39% in urine, 57% in feces. **Half-Life:** 5–7 h.

NURSING IMPLICATIONS

Assessment & Drug Effects
- Evaluate carefully any sustained increase in serum PSA levels while patient is taking finasteride. It may indicate the presence of prostate cancer or noncompliance with the therapy.
- Monitor patients with a large residual urinary volume or decreased urinary flow. These patients may not be candidates for this therapy.

Patient & Family Education
- Use a barrier contraceptive to prevent pregnancy in a sexual partner.
- Be aware that impotence and decreased libido may occur with treatment.
- Report promptly any of the following: breast tenderness or enlargement; testicular pain; rash, itching, or swelling about the face and lips.

FINGOLIMOD
(fin-go'-li-mod)
Gilenya
Classification: BIOLOGIC RESPONSE MODIFIER; IMMUNOMODULATOR; SPHINGOSINE 1-PHOSPHATE RECEPTOR MODULATOR
Therapeutic: IMMUNOMODULATOR
Pregnancy Category: C

AVAILABILITY 0.5 mg capsules

ACTION & *THERAPEUTIC EFFECT*
Binds to sphingosine 1-phosphate receptors 1, 3, 4, and 5 thus reducing the number of lymphocytes leaving lymph nodes and entering the peripheral blood stream. Mechanism of action in MS is unknown but may involve reduction in the migration of lymphocytes into the CNS. *Reduces the frequency of exacerbations in relapsing MS and slows development of physical disability.*

USES Treatment of relapsing forms of multiple sclerosis to reduce the frequency of clinical exacerbations and to delay the accumulation of physical disability.

CAUTIOUS USE Bradyarrhythmia, second degree or higher AV block, sick sinus syndrome, prolonged QT interval, ischemic cardiac disease, congestive heart failure, or hypertension; concurrent immunosuppressive or immune modulating therapies; infection; varicella zoster antibody testing/vaccination; macula edema; decreased respiratory function; hepatic dysfunction; renal impairment; age 65 y or older; pregnancy (category C); lactation. Safety and efficacy in children with MS younger than 18 y have not been established.

ROUTE & DOSAGE

Multiple Sclerosis
Adult: **PO** 0.5 mg once daily
Hepatic Impairment
Mild or moderate hepatic impairment: No dosage adustment needed
Severe hepatic impairment (Child-Pugh class C, total score greater than 10): Fingolimod exposure can be doubled.

Common adverse effects in *italic*, life-threatening effects <u>underlined</u>; generic names in **bold**; classifications in SMALL CAPS; ♣ Canadian drug name; ⊕ Prototype drug

ADMINISTRATION

Oral

- May be given with or without food.
- Monitor closely for 6 h after first dose for S&S of bradycardia. If fingolimod is discontinued for more than 2 wk then reinitiated, observe for bradycardia for 6 h following reinitiated dose.
- Store at 15°–30°C (59°–86°F).

ADVERSE EFFECTS (≥1%) **Body as a Whole:** Asthenia, influenza, herpes infections. **CNS:** Headache, depression, dizziness, migraine, paresthesia. **CV:** Bradycardia, hypertension. **GI:** Diarrhea, gastroenteritis. **Hematologic:** <u>Leukopenia</u>, lymphopenia. **Metabolic:** Increased ALT and AST, increased GGT, increased triglycerides, weight loss. **Musculoskeletal:** Back pain. **Respiratory:** Bronchitis, cough, dyspnea, sinusitis. **Skin:** Alopecia, eczema, pruritis, tinea infections. **Special Senses:** Blurred vision, eye pain.

INTERACTIONS Drug: Ketoconazole increases levels of fingolimod; ANTINEOPLASTIC, IMMUNOSUPPRESSIVE or IMMUNOMODULATING agents may increase the risk of immunosuppresion; fingolimod causes an additional reduction of heart rate when used with **atenolol;** CLASS IA and CLASS III ANTIARRHYTMICS may increase risk of serious rhythm disturbances and BETA BLOCKERS may cause increased bradycardia during fingolimod initiation.

PHARMACOKINETICS Absorption: 93% bioavailable. **Peak:** 12–16 h. **Distribution:** 99.7% plasma protein bound. **Metabolism:** Phosphorylated to active metabolite; oxidized by multiple CYP450 enzymes.

Elimination: Renal (81%) and fecal (2.5%). **Half-Life:** 6–9 d.

NURSING IMPLICATIONS

Assessment & Drug Effects

- Baseline ECG is recommended to identify risk factors for bradycardia and AV block especially in those with known cardiac risk factors or concurrent antiarrhythmic drugs including beta-blockers and calcium channel blockers.
- Monitor HR and BP at baseline and for 6 h after the first dose. HR declines within 1 h and is maximal at approximately 6 h. Monitor HR during first month of therapy during which time a return to baseline should be seen.
- Lab tests: Baseline and periodic LFTs. Withhold drug and consult prescriber for liver enzyme values 3 × ULN.
- Monitor for and report promptly S&S of the following: Infection, hepatic dysfunction (e.g., unexplained nausea, vomiting, abdominal pain, fatigue, anorexia, or jaundice and/or dark urine), or dyspnea.
- Baseline ophthalmologic exam is recommended and should be performed if patient reports visual disturbances at any time while taking this drug.

Patient & Family Education

- Those who have not had chickenpox or received the vaccination should consider receiving the VZV vaccine prior to starting treatment with fingolimod.
- Promptly report any of the following: S&S of infection; visual disturbances; new onset or worsening dyspnea; unexplained nausea, vomiting, abdominal pain, fatigue, loss of appetite, jaundice and/or dark urine.
- Women of childbearing age should use effective contraception

Common adverse effects in *italic*, life-threatening effects <u>underlined</u>; generic names in **bold**; classifications in SMALL CAPS; ♣ Canadian drug name; ⊘ Prototype drug

633

during and for 2 mo following discontinuation of fingolimod.
- Notify prescriber immediately if a pregnancy occurs.

FLAVOXATE HYDROCHLORIDE
(fla-vox'ate)

Classifications: ANTICHOLINERGIC; SMOOTH MUSCLE RELAXANT
Therapeutic: URINARY TRACT ANTISPASMODIC
Prototype: Oxybutynin
Pregnancy Category: B

AVAILABILITY 100 mg tablets

ACTION & THERAPEUTIC EFFECT
Exerts spasmolytic action on smooth muscle. Increases urinary bladder capacity in patients with spastic bladder, possibly by direct action on detrusor muscle. Also demonstrates local anesthetic and analgesic action. *Has antispasmodic action on the urinary bladder.*

USES Symptomatic relief of dysuria, overactive bladder, urinary urgency/incontinence.

CONTRAINDICATIONS Pyloric or duodenal obstruction, obstructive intestinal lesions, ileus, achalasia, GI hemorrhage; obstructive uropathies of lower urinary tract. Safety in children younger than 12 y is not established.
CAUTIOUS USE Suspected or closed-angle glaucoma; myasthenia gravis; autonomic neuropathy; dehydration; older adults; pregnancy (category B).

ROUTE & DOSAGE

Dysuria, Nocturia, Incontinence/OAB
Adult: **PO** 100–200 mg t.i.d. or q.i.d.

ADMINISTRATION
Oral
- Give without regard to meals.
- Store at 15°–30° C (59°–86° F) unless otherwise directed.

ADVERSE EFFECTS (≥1%) **CNS:** Headache, vertigo, drowsiness, mental confusion (especially in older adults). **CV:** Palpitation, tachycardia. **Special Senses:** Blurred vision, increased intraocular tension, disturbances of eye accommodation. **GI:** Nausea, vomiting, dry mouth (and throat), constipation (with high doses). **Skin:** Dermatosis, urticaria. **Other:** Dysuria, hyperpyrexia, eosinophilia, leukopenia (rare).

INTERACTIONS Drug: May antagonize the GI motility effects of **metoclopramide,** may add to GI slowing caused by ANTIDIARRHEALS

PHARMACOKINETICS Elimination: 10–30% in urine within 6 h.

NURSING IMPLICATIONS
Assessment & Drug Effects
- Monitor heart rate. Report tachycardia.
- Those with suspected glaucoma should be closely monitored for increased intraocular tension.

Patient & Family Education
- Do not drive or engage in potentially hazardous activities until response to drug is known.
- Report adverse reactions to prescriber as well as clinical improvement or the lack of a favorable response.

FLECAINIDE ☉
(fle-kay'nide)
Tambocor
Classification: CLASS IC ANTIARRHYTHMIC

Therapeutic: CLASS IC ANTI-ARRHYTHMIC
Pregnancy Category: C

AVAILABILITY 50 mg, 100 mg, 150 mg tablets

ACTION & *THERAPEUTIC EFFECT*
Local (membrane) anesthetic and antiarrhythmic with electrophysiologic properties similar to other class IC antiarrhythmic drugs. Slows conduction velocity throughout myocardial conduction system, increases ventricular refractoriness. *Is an effective suppressant of PVCs and a variety of atrial and ventricular arrhythmias.*

USES Life-threatening ventricular arrhythmias.

UNLABELED USES Atrial tachycardia and other arrhythmias unresponsive to standard agents (e.g., quinidine), Wolff-Parkinson-White syndrome, and recurrent ventricular tachycardias.

CONTRAINDICATIONS Hypersensitivity to flecainide; preexisting second- or third-degree AV block, right bundle branch block when associated with a left hemiblock unless a pacemaker is present; cardiogenic shock, left ventricular dysfunction; recent acute MI; QT prolongation syndromes; electrolyte imbalances.

CAUTIOUS USE Hypersensitivity to amide local anesthetics; atrial fibrillation; cardiac arrhythmias; cardiac disease; sick sinus syndrome; severe or moderate hepatic or renal impairment; older adults; pregnancy (category C), children and infants.

ROUTE & DOSAGE

Life-Threatening Ventricular Arrhythmias
Adult: **PO** 100 mg q12h, may increase by 50 mg b.i.d. q4days (max: 400 mg/day)

Child: **PO** 1–3 mg/kg/day in 3 divided doses (max: 8 mg/kg/day)

ADMINISTRATION
Oral
- Do not increase dosage more frequently than every 4 days.
- Store in tightly covered, light-resistant containers at 15°–30° C (59°–86° F) unless otherwise directed.

ADVERSE EFFECTS (≥1%) **CNS:** *Dizziness,* headache, light-headedness, unsteadiness, paresthesias, fatigue. **CV:** Arrhythmias, chest pain, worsening of CHF. **Special Senses:** *Blurred vision, difficulty in focusing,* spots before eyes. **GI:** *Nausea,* constipation, change in taste perception. **Body as a Whole:** Dyspnea, fever, edema.

INTERACTIONS Drug: Cimetidine may increase flecainide levels; may increase **digoxin** levels 15–25%; BETA-BLOCKERS may have additive negative inotropic effects.

PHARMACOKINETICS Absorption: Readily from GI tract. **Peak:** 2–3 h. **Distribution:** Crosses placenta; distributed into breast milk. **Metabolism:** In liver. **Elimination:** Mainly in urine. **Half-Life:** 7–22 h.

NURSING IMPLICATIONS

Assessment & Drug Effects
- Correct preexisting hypokalemia or hyperkalemia before treatment is initiated.
- Note: ECG monitoring, including Holter monitor for ambulating patients, is recommended because of the possibility of drug-induced arrhythmias.
- Lab tests: Monitor plasma level, especially in patients with severe CHF or renal failure because drug elimination may be delayed in these patients.

Common adverse effects in *italic*, life-threatening effects underlined; generic names in **bold;** classifications in SMALL CAPS; ♣ Canadian drug name; ◑ Prototype drug

635

- Note: Effective trough plasma levels are between 0.7–1 mcg/mL. The probability of adverse reactions increases when trough levels exceed 1 mcg/mL.
- Monitor carefully during period of dose adjustment.

Patient & Family Education
- Note: It is VERY important to take this drug at the prescribed times.
- Report visual disturbances to prescriber.

FLOXURIDINE
(flox-yoor'i-deen)
FUDR
Classifications: ANTINEOPLASTIC; ANTIMETABOLITE, PYRIMIDINE
Therapeutic: ANTINEOPLASTIC
Prototype: Fluorouracil
Pregnancy Category: D

AVAILABILITY 500 mg powder for injection

ACTION & *THERAPEUTIC EFFECT*
Pyrimidine antagonist and cell-cycle specific that is catabolized to fluorouracil in the body; highly toxic because it blocks an enzyme essential to normal DNA and RNA synthesis. *Proliferative cells of neoplasms are affected more than healthy tissue cells.*

USES Palliative agent in management of selected patients with GI metastasis to liver.
UNLABELED USES Carcinoma of breast, ovary, cervix, urinary bladder, and prostate not responsive to other antimetabolites.

CONTRAINDICATIONS Existing or recent viral infections; pregnancy (category D); lactation.
CAUTIOUS USE Poor nutritional status, bone marrow depression, serious infections; high-risk patients: prior high-dose pelvic irradiation, impaired kidney or liver function.

ROUTE & DOSAGE

Carcinoma
Adult: **Intra-Arterial** 0.1–0.6 mg/kg/day by continuous intra-arterial infusion

ADMINISTRATION
Intra-arterial Infusion
PREPARE: **Direct:** Reconstitute with 5 mL sterile distilled water for injection; further dilute with D5W or NS injection to a volume appropriate for the infusion apparatus to be used.
ADMINISTER: **Direct:** It is administered by pump to overcome pressure in large arteries and to ensure a uniform rate. ▪Examine infusion site frequently for signs of extravasation. If this occurs, stop infusion and restart in another vessel.
INCOMPATIBILITIES Y-site: **Allopurinol, cefepime.**

- Keep reconstituted solutions, which are stable at 2°–8° C (36°–46° F), for no more than 2 wk.
- Store at 15°–30° C (59°–86° F) unless otherwise directed.

ADVERSE EFFECTS (≥1%) **CNS:** Vertigo, convulsions, depression, hemiplegia. **CV:** Myocardial ischemia, angina. **GI:** *Nausea, vomiting, stomatitis,* diarrhea, cramps, anorexia, enteritis, gastritis, esophagopharyngitis. **Hematologic:** Leukopenia, thrombocytopenia. **Skin:** Dermatitis, alopecia (usually reversible), *erythema* or increased skin pigmentation (photosensitivity), dry skin, pruritic ulcerations, rash. **Body as a Whole:** Hiccups, fever, epistaxis, decreased resistance to disease. **Urogenital:** Renal insufficiency.

INTERACTIONS Drug: Metronidazole may increase general floxuridine toxicity; may increase or decrease serum levels of **phenytoin, fosphenytoin; hydroxyurea** can decrease conversion to active metabolite.

PHARMACOKINETICS Distribution: Distributed to tumor, intestinal mucosa, bone marrow, liver, and CSF; probably crosses placenta. **Metabolism:** Rapidly metabolized in liver to fluorouracil. **Elimination:** 15% in urine, 60–80% through lungs as carbon dioxide. **Half-Life:** 16 min.

NURSING IMPLICATIONS

Assessment & Drug Effects

- Discontinue therapy promptly with onset of any of the following: Stomatitis, esophagopharyngitis, intractable vomiting, diarrhea, leukopenia (WBC less than 3500/mm^3), or rapidly falling WBC count, thrombocytopenia (platelets 100,000/mm^3), GI bleeding, hemorrhage from any site.
- Lab tests: Obtain baseline and periodic total and differential leukocyte counts, Hct, platelet count, serum uric acid, creatinine, and LFTs.

Patient & Family Education

- Be aware that floxuridine sometimes causes temporary thinning of hair.

FLUCONAZOLE ⊘

(flu-con′a-zole)

Diflucan
Classification: AZOLE ANTIFUNGAL
Therapeutic: ANTIFUNGAL
Pregnancy Category: C

AVAILABILITY 50 mg, 100 mg, 150 mg, 200 mg tablets; 10 mg/mL, 40 mg/mL suspension; 2 mg/mL injection

ACTION & THERAPEUTIC EFFECT Interferes with formation of ergosterol, the principal sterol in the fungal cell membrane leading to cell death. *Antifungal properties are related to the drug effect on the functioning of fungal cell membrane.*

USES Cryptococcal meningitis and oropharyngeal and systemic candidiasis, both commonly found in AIDS and other immunocompromised patients; vaginal candidiasis.

CONTRAINDICATIONS Hypersensitivity to fluconazole or other azole antifungals; lacatation.
CAUTIOUS USE AIDS or malignancy; hepatic impairment; structural cardiac disease; history of torsades de pointes or QT prolongation; renal impairment or failure; pregnancy (category C).

ROUTE & DOSAGE

Oropharyngeal Candidiasis

Adult: **PO/IV** 200 mg day 1, then 100 mg/day × 14 days
Child: **PO/IV** 3–6 mg/kg/day × 14 days

Esophageal Candidiasis

Adult: **PO/IV** 200 mg day 1, then 100 mg daily × 3 wk
Child/Infant: **PO/IV** 3–6 mg/kg/day × 21 days

Systemic Candidemia

Adult: **PO/IV** 400 mg day 1, then 200 mg daily × 4 wk
Child/Infant/Nenonate (older than 14 days): **PO/IV** 6 mg/kg q12h × 28 days

Common adverse effects in *italic*, life-threatening effects underlined; generic names in **bold**; classifications in SMALL CAPS; ♣ Canadian drug name; ⊘ Prototype drug

637

Neonate (0–14 days): **IV** 6 mg/kg q72h

Vaginal Candidiasis
Adult: **PO** 150 mg × 1 dose

Cryptococcal Meningitis
Adult: **PO/IV** 400 mg day 1, then 200 mg daily × 10–12 wk
Child/Infant/Neonate (older than 14 days): **PO/IV** 12 mg/kg day 1, then 6–12 mg/kg/day × 10–12 wk
Neonate (0–14 days): **IV** 6–12 mg/kg day 1, then 6–12 mg/kg q48h
Premature Neonates (0–14 days): **IV** 5–6 mg/kg q72h

Renal Impairment Dosage Adjustment
CrCl 50 mL/min or less (without concurrent dialysis): Give 50% of maintenance dose

Hemodialysis Dosage Adjustment
Administer full dose post-dialysis

ADMINISTRATION
Oral
- Take this medication for the full course of therapy, which may take weeks or months.
- Take next dose as soon as possible if you miss a dose; however, do not take a dose if it is almost time for next dose. Do not double dose.

Intravenous

PREPARE: **Continuous:** Packaged ready for use as a 2 mg/mL solution. Remove wrapper just prior to use.
ADMINISTER: **Continuous:** Give at a maximum rate of approximately 200 mg/h. Give after hemodialysis is completed.

- Do not use IV admixtures of fluconazole and other medications.
INCOMPATIBILITIES **Solution/additive:** Trimethoprim-sulfamethoxazole. **Y-site:** Amphotericin B, amphotericin B cholesteryl, ampicillin, calcium gluconate, cefotaxime, ceftazidime, ceftriaxone, cefuroxime, chloramphenicol, clindamycin, diazepam, digoxin, erythromycin, furosemide, haloperidol, hydroxyzine, imipenemcilastatin, pentamidine, piperacillin, ticarcillin, trimethoprimsulfamethoxazole.

ADVERSE EFFECTS (≥1%) **CNS:** Headache. **GI:** Nausea, vomiting, abdominal pain, diarrhea, increase in AST in patients with cryptococcal meningitis and AIDS. **Skin:** Rash.

INTERACTIONS Drug: Increased PT in patients on **warfarin;** may increase **alosetron, bexarotene, phenytoin, cevimeline, cilostazol, cyclosporine, dihydroergotamine, ergotamine, dofetilide, haloperidol, levobupivacaine, modafinil, zonisamide** levels and toxicity; hypoglycemic reactions with ORAL SULFONYLUREAS; decreased fluconazole levels with **rifampin, cimetidine;** may prolong the effects of **fentanyl, alfentanil, methadone.**

PHARMACOKINETICS Absorption: 90% from GI tract. **Peak:** 1–2 h. **Distribution:** Widely distributed, including CSF. **Metabolism:** 11% of dose metabolized in liver. **Elimination:** In urine. **Half-Life:** 20–50 h.

NURSING IMPLICATIONS
Assessment & Drug Effects
- Monitor for allergic response. Patients allergic to other azole antifungals may be allergic to fluconazole.

■ Lab tests: Monitor BUN, serum creatinine, and LFTs.
■ Note: Drug may cause elevations of the following laboratory serum values: ALT, AST, alkaline phosphatase, bilirubin.
■ Monitor for S&S of hepatotoxicity.

Patient & Family Education
■ Monitor carefully for loss of glycemic control if diabetic.
■ Inform prescriber of all medications being taken.

FLUCYTOSINE

(floo-sye'toe-seen)

Ancobon, Ancotil ♣
Classification: ANTIFUNGAL
Therapeutic: ANTIFUNGAL
Prototype: Fluconazole
Pregnancy Category: C

AVAILABILITY 250 mg, 500 mg capsules

ACTION & *THERAPEUTIC EFFECT*
Selectively penetrates fungal cell and is converted to fluorouracil, an antimetabolite believed to be responsible for antifungal activity. *Has antifungal activity against* Cryptococcus *and* Candida *as well as chromomycosis.*

USES Alone or in combination with amphotericin B for serious systemic infections caused by susceptible strains of *Cryptococcus* and *Candida* species.
UNLABELED USES *Chromomycosis.*

CONTRAINDICATIONS Lactation.
CAUTIOUS USE Hepatic disease; electrolyte imbalance; bone marrow depression, hematologic disorders, patients being treated with or having received radiation or bone marrow depressant drugs; dental disease; extreme caution in impaired kidney function; pregnancy (category C).

ROUTE & DOSAGE

Fungal Infection
Adult: **PO** 50–150 mg/kg/day divided q6h
Child: **PO** *Weight less than 50 kg,* 1.5–4.5 g/m²/day divided q6h; *weight greater than 50 kg,* 50–150 mg/kg/day divided q6h
Neonate: **PO** 50–100 mg/kg/ day in 1–2 divided doses

ADMINISTRATION
Oral
■ Lower dosages with longer dosage intervals are recommended in patients with serum creatinine of 1.7 mg/dL or higher. Check with prescriber.
■ Give capsules a few at a time over 15 min to decrease incidence and severity of nausea and vomiting.
■ Store in light-resistant containers at 15°–30° C (59°–86° F).

ADVERSE EFFECTS (≥1%) **CNS:** Confusion, hallucinations, headache, sedation, vertigo. **GI:** Nausea, vomiting, diarrhea, abdominal bloating, enterocolitis. **Hematologic:** Hypoplasia of bone marrow: Anemia, leukopenia, thrombocytopenia, agranulocytosis, eosinophilia. **Skin:** Rash. **Metabolic:** Elevated levels of serum alkaline phosphatase, AST, ALT, BUN, serum creatinine. **GI:** Hepatomegaly, hepatitis.

DIAGNOSTIC TEST INTERFERENCE False elevations of *serum creatinine* can occur with *Ektachem analyzer.*

INTERACTIONS Drug: Amphotericin B produces additive or synergistic effects and can increase flucytosine toxicity by inhibiting its renal clearance.

PHARMACOKINETICS Absorption: Readily from GI tract. **Peak:** 2 h.

Common adverse effects in *italic*, life-threatening effects underlined; generic names in **bold;** classifications in SMALL CAPS; ♣ Canadian drug name; ⊘ Prototype drug

639

Distribution: Widely distributed in body tissues including aqueous humor and CSF; crosses placenta. **Metabolism:** Minimal. **Elimination:** 75–90% in urine unchanged. **Half-Life:** 3–6 h.

NURSING IMPLICATIONS

Assessment & Drug Effects

- C&S tests should be performed before initiation of therapy and at weekly intervals during therapy. Organism resistance has been reported.
- Lab tests: Obtain baseline hematology, kidney function tests and LFTs on all patients before and at frequent intervals during therapy. Twice weekly leukocyte and differential counts with WBC with differential and platelet counts are recommended.
- Frequent assays of blood drug level are recommended, especially in patients with impaired kidney function to determine adequacy of drug excretion (therapeutic range: 25–120 mg/mL).
- Monitor I&O. Report change in I&O ratio or pattern. Because most of drug is eliminated unchanged by kidneys, compromised function can lead to drug accumulation.

Patient & Family Education

- Report fever, sore mouth or throat, and unusual bleeding or bruising tendency to prescriber.
- Be aware that the general duration of therapy is 4–6 wk, but it may continue for several months.

FLUDARABINE

(flu-dar'a-bine)

Fludara
Classifications: ANTINEOPLASTIC; ANTIMETABOLITE, PURINE ANALOG

Therapeutic: ANTINEOPLASTIC; IMMUNOSUPPRESSANT
Prototype: Mercaptopurine
Pregnancy Category: D

AVAILABILITY 50 mg powder or 25 mg/mL solution for injection

ACTION & *THERAPEUTIC EFFECT*

Inhibits DNA polymerase alpha, ribonucleotide reductase, and DNA primase, thus inhibiting DNA synthesis in tumor-sensitive cells. *Fludarabine has cytotoxic effects on lymphocytic leukemia and lymphoma as well as immunosuppressant properties.*

USES Treatment of B-cell chronic lymphocytic leukemia (CLL).
UNLABELED USES In combination therapy for the treatment of primary resistant or relapsing acute myelogenous leukemia (AML), acute lymphoblastic leukemia (ALL), and secondary AML; cutaneous T-cell lymphoma; macroglobulinemia; myelodysplastic syndrome; prolymphocytic leukemia (PLL); stem-cell transplant preparation; non Hodgkins lymphoma.

CONTRAINDICATIONS Hypersensitivity to fludarabine; live vaccines; severe myelosuppression; pregnancy (category D); lactation.
CAUTIOUS USE Moderate or severe renal impairment; patients at risk for tumor lysis syndrome; history of herpes or viral infection. Safe use in children not established.

ROUTE & DOSAGE

Treatment of Unresponsive B-cell Chronic Lymphocytic Leukemia

Adult: IV 25 mg/m² daily × 5 days; repeat q28days

Renal Impairment Dosage Adjustment

CrCl 50–70 mL/min: 20 mg/m²;

CrCL 30–49 mL/min: 15 mg/m²; *less than 30 mL/min:* Should not receive IV fludarabine

ADMINISTRATION

Intravenous

PREPARE: **IV Infusion** Exercise caution in the preparation and handling of fludarabine. Avoid exposure by inhalation or direct contact with skin or mucous membranes. ▪ Reconstitute each 50 mg vial by adding 2 mL of sterile water for injection to yield 25 mg/mL. The solution should dissolve within 15 sec and have a pH of 7.2–8.2. ▪ Further dilute in 100–125 mL of D5W or NS. ▪ Administer within 8 h of reconstitution.
ADMINISTER: **IV Infusion:** Give over 30 min.
INCOMPATIBILITIES **Y-site: Acyclovir, amiodarone, amphotericin B, chlorpromazine, dantrolene, daunorubicin, diazepam, ganciclovir, gemtuzumab, hydroxyzine, idarubicin, pantoprazole, phenytoin, prochlorperazine, quinupristin-dalfopristin, trastuzumab.**

▪ Store unreconstituted vials at 2°–8° C (36°–46° F). ▪ Discard any unused reconstituted product.

ADVERSE EFFECTS (≥1%) Body as a Whole: *Fever, chills, fatigue, infection, pain,* malaise, diaphoresis, anaphylaxis, hyperglycemia, dehydration. **CNS:** *Weakness,* paresthesia. **CV:** *Edema.* **GI:** *Nausea, vomiting, diarrhea, anorexia, stomatitis,* GI bleeding, esophagitis, mucositis. **Hematologic:** *Neutropenia,* thrombocytopenia, hemolytic anemia. **Musculoskeletal:** Myalgia. **Respiratory:** *Cough, pneumonia, dyspnea,* sinusitis, pharyngitis, upper respiratory tract infection. **Skin:** *Rash,* pruritus. **Special Senses:** Visual disturbance, hearing loss. **Urogenital:** Dysuria, urinary infection, hematuria.

INTERACTIONS Drug: Use with **pentostatin** increases risk of severe pulmonary toxicity. Do not give LIVE VACCINES due to decreased immune response. Use with **natalizumab** increases immunosuppression.

PHARMACOKINETICS Metabolism: Rapid conversion to active metabolite (2-fluoro-ara-A). **Elimination:** Renal. **Half-Life:** 7–12 h.

NURSING IMPLICATIONS

Assessment & Drug Effects

▪ Monitor for and report S&S of hemolysis, infection, tumor lysis syndrome (e.g., flank pain, hematuria), peripheral neuropathy, or respiratory distress.
▪ Lab tests: Baseline creatinine clearance and CBC with differential and platelet count, repeat prior to each treatment cycle, and more often as indicated; periodic serum electrolytes, serum uric acid, and kidney function tests.

Patient & Family Education

▪ Report any of the following to a health care provider: Fever, chills, cough, sore throat, or other signs of infection; pain or difficulty passing urine; signs of bleeding such as easy bruising, black, tarry stools, nosebleeds; signs of anemia such as excessive weakness, lightheadedness, or confusion; difficulty breathing or shortness of breath; decreased vision; mouth sores or skin rash.
▪ Avoid activities that could cause physical injury and predispose to severe bleeding.
▪ Women of childbearing age should avoid becoming pregnant while receiving fludarabine.

Common adverse effects in *italic*, life-threatening effects underlined; generic names in **bold**; classifications in SMALL CAPS; ✦ Canadian drug name; ⊕ Prototype drug

F

FLUDROCORTISONE ACETATE ⊙

(floo-droe-kor′ti-sone)

Florinef Acetate

Classifications: ADRENOCORTICAL STEROID; MINERALOCORTICOID

Therapeutic: MINERALOCORTICOID; ANTI-INFLAMMATORY

Pregnancy Category: C

AVAILABILITY 0.1 mg tablets

ACTION & *THERAPEUTIC EFFECT*
Long-acting synthetic steroid with potent mineralocorticoid activity. Small doses produce marked sodium retention, increased urinary potassium excretion, and elevated BP. *Synthetic corticosteroid replacement product for adrenocortical insufficiency.*

USES Partial replacement therapy for adrenocortical insufficiency and for treatment of salt-losing forms of congenital adrenogenital syndrome.

UNLABELED USES To increase systolic and diastolic blood pressure in patients with severe hypotension secondary to diabetes mellitus or to levodopa therapy.

CONTRAINDICATIONS Hypersensitivity to glucocorticoids, idiopathic thrombocytopenic purpura, psychoses, acute glomerulonephritis, viral or bacterial diseases of skin, systemic fungal infections; infections not controlled by antibiotics, active or latent amebiasis, hypercorticism, smallpox vaccination or other immunologic procedures.

CAUTIOUS USE Diabetes mellitus; chronic, active hepatitis positive for hepatitis B surface antigen; hyperlipidemia; cirrhosis; stromal herpes simplex; glaucoma, tuberculosis of eye; osteoporosis; convulsive disorders; hypothyroidism; diverticulitis; nonspecific ulcerative colitis; fresh intestinal anastomoses; active or latent peptic ulcer; gastritis; esophagitis; thromboembolic disorders; CHF; metastatic carcinoma; hypertension; renal insufficiency; history of allergies; active or arrested tuberculosis; myasthenia gravis; history of psychosis; pregnancy (category C), lactation; children.

ROUTE & DOSAGE

Adrenocortical Insufficiency

Adult: **PO** 0.1 mg/day, may range from 0.1 mg 3 times/wk to 0.2 mg/day

Child: **PO** 0.05–0.1 mg/day

Salt-Losing Adrenogenital Syndrome

Adult: **PO** 0.1–0.2 mg/day

Child: **PO** 0.05–0.1 mg/day

ADMINISTRATION

Oral

- Note: Concomitant oral cortisone or hydrocortisone therapy may be advisable to provide substitute therapy approximating normal adrenal activity.
- Store in airtight containers at 15°–30° C (59°–86° F). Protect from light.

ADVERSE EFFECTS (≥1%) **CNS:** Vertigo, headache, nystagmus, increased intracranial pressure with papilledema (usually after discontinuation of medication), mental disturbances, aggravation of preexisting psychiatric conditions, insomnia, ataxia (rare). **CV:** CHF, hypertension, thromboembolism (rare), tachycardia. **Endocrine:** Suppressed linear growth in children, decreased glucose tolerance; hyperglycemia,

Common adverse effects in *italic*, life-threatening effects underlined; generic names in **bold**; classifications in SMALL CAPS; ✦ Canadian drug name; ⊙ Prototype drug

manifestations of latent diabetes mellitus; hypocorticism; amenorrhea and other menstrual difficulties. **Special Senses:** Posterior subcapsular cataracts (especially in children), glaucoma, exophthalmos, increased intraocular pressure with optic nerve damage, perforation of the globe. **Metabolic:** Hypocalcemia; *sodium and fluid retention;* hypokalemia and hypokalemic alkalosis, negative nitrogen balance, decreased serum concentration of vitamins A and C. **GI:** *Nausea,* increased appetite, ulcerative esophagitis, pancreatitis, abdominal distension, peptic ulcer with perforation and hemorrhage, melena. **Hematologic:** Thrombocytopenia. **Musculoskeletal:** (Long-term use) Osteoporosis, compression fractures, muscle wasting and weakness, tendon rupture, aseptic necrosis of femoral and humeral heads. **Skin:** Skin thinning and atrophy, *acne, impaired wound healing;* petechiae, ecchymosis, easy bruising; suppression of skin test reaction; hypopigmentation or hyperpigmentation, hirsutism, acneiform eruptions, subcutaneous fat atrophy; allergic dermatitis, urticaria, angioneurotic edema, increased sweating. **Body as a Whole:** Anaphylactoid reactions (rare), aggravation or masking of infections; malaise, weight gain, obesity. **Urogenital:** Increased or decreased motility and number of sperm.

INTERACTIONS Drug: The antidiabetic effects of **insulin** and SULFONYLUREAS may be diminished; **amphotericin B,** DIURETICS may increase **potassium** loss; **warfarin** may decrease prothrombin time; **indomethacin, ibuprofen** can potentiate the pressor effect of fludrocortisone; ANABOLIC STEROIDS increase risk of edema and acne; **rifampin** may increase the hepatic metabolism of fludrocortisone.

PHARMACOKINETICS Absorption: Readily from GI tract. **Peak:** 1.7 h. **Metabolism:** In liver. **Half-Life:** 3.5 h.

NURSING IMPLICATIONS

Assessment & Drug Effects

- Monitor weight and I&O ratio to observe onset of fluid accumulation, especially if patient is on unrestricted salt intake and without potassium supplement. Report weight gain of 2 kg (5 lb)/wk.
- Monitor and record BP daily. If hypertension develops as a consequence of therapy, report to prescriber. Usually, the dose will be reduced to 0.05 mg/day.
- Check BP q4–6h and weight at least every other day during periods of dosage adjustment.
- Lab tests: Periodic serum electrolytes and ABGs during prolonged therapy.
- Monitor for S&S of hypokalemia and hyperkalemic metabolic alkalosis (see Appendix F).

Patient & Family Education

- Report signs of hypokalemia (see Appendix F).
- Be aware of signs of potassium depletion associated with high sodium intake: Muscle weakness, paresthesias, circumoral numbness; fatigue, anorexia, nausea, mental depression, polyuria, delirium, diminished reflexes, arrhythmias, cardiac failure, ileus, ECG changes.
- Eat foods with high potassium content.
- Signs of edema should be reported immediately. Sodium intake may or may not require regulation, depending on individual needs and clinical situation.
- Weigh daily under standard conditions and report steady weight gain.
- Report intercurrent infection, trauma, or unexpected stress of

Common adverse effects in *italic*, life-threatening effects underlined; generic names in **bold;** classifications in SMALL CAPS; ✙ Canadian drug name; ⊘ Prototype drug

643

any kind promptly when taking maintenance therapy.
- Carry medical identification at all times.

FLUMAZENIL ◐
(flu-ma′ze-nil)

Mazicon ✦, Romazicon
Classification: BENZODIAZEPINE ANTAGONIST
Therapeutic: BENZODIAZEPINE ANTIDOTE
Pregnancy Category: C

AVAILABILITY 0.1 mg/mL injection

ACTION & *THERAPEUTIC EFFECT*
Antagonizes the effects of benzodiazepine on the CNS, including sedation, impairment of recall, and psychomotor impairment. *Reverses the action of a benzodiazepine.*

USES Reversal of sedation induced by benzodiazepine for anesthesia or diagnostic or therapeutic procedures as well as through overdose.
UNLABELED USES Seizure disorders, alcohol intoxication, hepatic encephalopathy, facilitation of weaning from mechanical ventilation.

CONTRAINDICATIONS Hypersensitivity to flumazenil or to benzodiazepines; patients given a benzodiazepine for control of a life-threatening condition; patients showing signs of cyclic antidepressant overdose; seizure-prone individuals; during labor and delivery.

CAUTIOUS USE Hepatic function impairment, older adults, intensive care patients, head injury, anxiety or pain disorder; drug- and alcohol-dependent patients, and physical dependence upon

benzodiazepines; pregnancy (category C); lactation; children.

ROUTE & DOSAGE

Reversal of Sedation
Adult: **IV** 0.2 mg over 15 sec, may repeat 0.2 mg each min for 4 additional doses or a cumulative dose of 1 mg
Child: **IV** 0.01 mg/kg may repeat each min (max: 1 mg)

Benzodiazepine Overdose
Adult: **IV** 0.2 mg over 30 sec, if no response after 30 sec, then 0.3 mg over 30 sec, may repeat with 0.5 mg each min (max cumulative dose: 3 mg)

ADMINISTRATION
Intravenous

PREPARE: **Direct:** May give undiluted or diluted. If diluted use D5W, lactated Ringer's, NS.
ADMINISTER: **Direct for Reversal of Anesthesia:** Ensure patency of IV before administration of flumazenil, since extravasation will cause local irritation. ▪ Do not give as bolus dose. Give through an IV that is freely flowing into a large vein. **Direct for Reversal of Anesthesia or Sedation:** Give each dose slowly over 15 sec. ▪ In high-risk patients, slow the rate to provide the smallest effective dose. **Direct for Benzodiazepine Overdose:** Give each dose slowly over 30 sec.
▪ Use all diluted solutions within 24 h of dilution.

ADVERSE EFFECTS (≥1%) **CNS:** Emotional lability, headache, *dizziness,* agitation, *resedation,* seizures, blurred vision. **GI:** *Nausea, vomiting,* hiccups. **Other:** Shivering, pain at injection site, hypoventilation.

INTERACTIONS Drug: May antagonize effects of **zaleplon, zolpidem;** may cause convulsions or arrhythmias with TRICYCLIC ANTIDEPRESSANTS.

PHARMACOKINETICS Onset: 1–5 min. **Peak:** 6–10 min. **Duration:** 2–4 h. **Metabolism:** In the liver to inactive metabolites. **Elimination:** 90–95% in urine, 5–10% in feces within 72 h. **Half-Life:** 54 min.

NURSING IMPLICATIONS

Assessment & Drug Effects

- Monitor respiratory status carefully until risk of resedation is unlikely (up to 120 min). Drug may not fully reverse benzodiazepine-induced ventilatory insufficiency.
- Monitor carefully for seizures and take appropriate precautions.

Patient & Family Education

- Do not drive or engage in potentially hazardous activities until at least 18–24 h after discharge following a procedure.
- Do not ingest alcohol or nonprescription drugs for 18–24 h after flumazenil is administered or if the effects of the benzodiazepine persist.

FLUNISOLIDE

(floo-niss′oh-lide)

AeroBid, Nasalide, Nasarel
Pregnancy Category: C
See Appendix A-3.

FLUOCINOLONE ACETONIDE

(floo-oh-sin′oh-lone)

Fluoderm ✦, Synalar
Prototype: Hydrocortisone
Pregnancy Category: C
See Appendix A-4.

FLUOCINONIDE

(floo-oh-sin′oh-nide)

Lidemol, Lidex, Lidex-E, Lyderm, Topsyn, Vanos
Pregnancy Category: C
See Appendix A-4.

FLUORESCEIN SODIUM

(flure′e-seen)

Fluorescite
Classification: OPHTHALMIC DIAGNOSTIC AGENT
Therapeutic: OPHTHALMIC DIAGNOSTIC AGENT
Pregnancy Category: X

AVAILABILITY 100 mg/mL injection

ACTION & *THERAPEUTIC EFFECT* Mildly antiseptic fluorescent dye that demonstrates defects of the corneal epithelium. *Any break in the epithelial tissue allows the dye to enter the tissue of the eye. Epithelial damage will appear as a bright green area.*

USES Used intravenously as a diagnostic aid in retinal angiography. Also used as an antidote for aniline dye.

CONTRAINDICATIONS Intra-arterial administration; intrathecal administration; pregnancy (category X). **CAUTIOUS USE** History of hypersensitivity, allergies, bronchial asthma.

ROUTE & DOSAGE

Retinal Angiography

Adult: **IV** 5 mL of 10% solution or 3 mL of 25% solution injected rapidly in antecubital vein
Child: **IV** 7.5 mg/kg injected rapidly in antecubital vein

Common adverse effects in *italic*, life-threatening effects underlined; generic names in **bold;** classifications in SMALL CAPS; ✦ Canadian drug name; ⦿ Prototype drug

645

ADMINISTRATION

Intravenous

ADMINISTER: **IV Direct for Adult:** 5 mL of 10% solution or 3 mL of 25% solution injected rapidly in antecubital vein. **IV Direct for Child:** 7.5 mg/kg injected rapidly in antecubital vein.

ADVERSE EFFECTS (≥1%) **CNS:** Headache, paresthesias, pyrexia, convulsions. **CV:** Hypotension, transient dyspnea, acute pulmonary edema, basilar artery ischemia, syncope, <u>severe shock,</u> <u>cardiac arrest.</u> **GI:** Nausea, vomiting, strong metallic taste following high dosage. **Body as a Whole:** Hypersensitivity (urticaria, pruritus, angioneurotic edema, <u>anaphylactic reaction</u>). **Skin:** Thrombophlebitis at injection site, temporary discoloration of skin and urine.

NURSING IMPLICATIONS

Assessment & Drug Effects

- Have facilities for treatment of anaphylactic reaction immediately available (e.g., epinephrine 1:1000 for IV or IM use, an antihistamine, and oxygen).
- Discontinue fluorescein immediately if S&S of sensitivity develop.

Patient & Family Education

- Note: IV administration may impart a yellowish orange discoloration to skin and to urine. Skin discoloration usually fades in 6–12 h; urine clears in 24–36 h.

FLUOROMETHOLONE

(flure-oh-meth′oh-lone)

Flarex, FML Forte, FML Liqui-film

Pregnancy Category: C

See Appendix A-1.

FLUOROURACIL [5-FLUOROURACIL (5-FU)] 📑

(flure-oh-yoor′a-sil)

Carac, Efudex, Fluoroplex

Classifications: ANTINEOPLASTIC; ANTIMETABOLITE, PYRIMIDINE

Therapeutic: ANTINEOPLASTIC

Pregnancy Category: D

AVAILABILITY 50 mg/mL injection; 1%, 2%, 5% topical solution; 0.5%, 1%, 5% topical cream

ACTION & *THERAPEUTIC EFFECT* Pyrimidine antagonist and cell-cycle specific agent that blocks action of enzymes essential to normal DNA and RNA synthesis and may become incorporated in RNA to form a fraudulent molecule; unbalanced growth and death of cell follow. Exhibits higher affinity for tumor tissue than healthy tissue. *Highly toxic, especially to proliferative cells in neoplasms, bone marrow, and intestinal mucosa.*

USES Systemically as single agent or in combination with other antineoplastics for treatment of patients with inoperable neoplasms of breast, colon or rectum, stomach, pancreas, urinary bladder, ovary, cervix, liver. Also topically for solar or actinic keratoses and superficial basal cell carcinoma.

UNLABELED USES To induce repigmentation in vitiligo; actinic cheilitis; malignant effusions; mucosal leukoplakia.

CONTRAINDICATIONS Poor nutritional status; myelosuppression; pregnancy (category D), lactation.

CAUTIOUS USE Major surgery during previous month; history of high-dose pelvic irradiation, metastatic cell infiltration of bone marrow, previous use of alkylating agents; cardiac disease, CAD, angina; men

Common adverse effects in *italic*, life-threatening effects <u>underlined</u>; generic names in **bold;** classifications in SMALL CAPS; ♣ Canadian drug name; 📑 Prototype drug

and women in childbearing ages; hepatic or renal impairment.

ROUTE & DOSAGE

Carcinoma

Adult: **IV** 12 mg/kg/day for 4 consecutive days up to 800 mg or until toxicity develops or 12 days therapy, may repeat at 1-mo intervals; if toxicity occurs, 15 mg/kg once weekly can be given until toxicity subsides

Actinic and Solar Keratosis

Adult: **Topical** Apply cream or solution b.i.d. for 2–4 wk; apply Carac once daily

Superficial Basal Cell Carcinoma

Adult: **Topical** Apply 5% cream b.i.d. for 3–6 wk

Obesity Dosage Adjustment

Dose patient based on lean body mass

ADMINISTRATION

Topical

- Use gloved fingers to apply topical drug.
- Do not use occlusive dressings with topical drug. Use a porous gauze dressing for cosmetic purposes.
- Store at 15°–30° C (59°–86° F) unless otherwise directed. Protect from light and freezing.

Intravenous

Note: Parenteral dose is determined by actual weight unless patient is obese, in which case ideal weight is used.

- Safe handling: Double-glove with latex gloves, and change the double set after every 30 min of exposure. If a drug spill occurs, change gloves immediately after it is cleaned up.

PREPARE: **Direct/Infusion:** This drug may be given undiluted or further diluted in D5W or NS for infusion. ▪ If a precipitate forms, redissolve drug by heating to 60° C (140° F) and shake vigorously. Allow to cool to body temperature before administration.

ADMINISTER: **Direct/Infusion:** Give by direct IV injection over 1–2 min. ▪ Infuse over 2–24 h as ordered. **IV Extravasation:** Inspect injection site frequently; avoid extravasation. If it occurs, stop infusion and restart in another vein. ▪ Ice compresses may reduce danger of local tissue damage from infiltrated solution.

INCOMPATIBILITIES **Solution/additive: Carboplatin, chlorpromazine, cisplatin, cytarabine, diazepam, doxorubicin, epirubicin, fentanyl, leucovorin calcium, metoclopramide, morphine. Y-site: Aldesleukin, amphotericin B cholesteryl, droperidol, filgrastim, gallium, lansoprazole, ondansetron, TPN, topotecan, vinorelbine.**

- Fluorouracil solution is normally colorless to faint yellow. Slight discoloration during storage does not appear to affect potency or safety.
- Discard dark yellow solution.

ADVERSE EFFECTS (≥1%) **CNS:** Euphoria, acute cerebellar syndrome (dysmetria, nystagmus, ataxia, severe mental deterioration); pustular contact hypersensitivity. **CV:** Cardiotoxicity (rare), angina. **GI:** Anorexia, *nausea, vomiting, stomatitis,* esophagopharyngitis, medicinal taste, *diarrhea,* proctitis. **Hematologic:** Anemia, leukopenia, thrombocytopenia, eosinophilia. **Body as**

Common adverse effects in *italic,* life-threatening effects underlined; generic names in **bold;** classifications in SMALL CAPS; ♣ Canadian drug name; ⊘ Prototype drug

647

a Whole: Hypersensitivity: Pustular contact eruption, edema of face, eyes, tongue, legs. **Skin:** SLE-like dermatitis, *alopecia*, photosensitivity, erythema, increased pigmentation, skin dryness and fissuring, pruritic maculopapular rash. **[Topical]** Local pain, pruritus, hyperpigmentation, burning at site of application, dermatitis, suppuration, swelling, scarring, toxic granulation.

DIAGNOSTIC TEST INTERFERENCE
Fluorouracil may decrease *plasma albumin* (because of drug-induced protein malabsorption).

INTERACTIONS Drug: Metronidazole
may increase general floxuridine toxicity; may increase or decrease serum levels of **phenytoin, fosphenytoin; hydroxyurea** can decrease conversion to active metabolite.

PHARMACOKINETICS Distribution:
Distributed to tumor, intestinal mucosa, bone marrow, liver, and CSF; probably crosses placenta. **Metabolism:** In liver. **Elimination:** 15% in urine, 60–80% through lungs as carbon dioxide. **Half-Life:** 16 min.

NURSING IMPLICATIONS

Assessment & Drug Effects
- Lab tests: Obtain total and differential leukocyte counts before each dose is administered. Discontinue drug if leukopenia occurs (WBC less than 3500/mm³) or if patient develops thrombocytopenia (platelet count less than 100,000/mm³). Baseline and periodic checks of Hct, LFTs, kidney function tests are also advised.
- Use protective isolation of patient during leukopenic period (WBC less than 3500/mm³).

- Watch for and report signs of abnormal bleeding from any source during thrombocytopenic period (day 7–17); inspect skin for ecchymotic and petechial areas. Protect patient from trauma.
- Report disorientation or confusion; drug should be withdrawn immediately.
- Indications to discontinue drug: Severe stomatitis, leukopenia (WBC less than 3500/mm³ or rapidly decreasing count), intractable vomiting, diarrhea, thrombocytopenia (platelets less than 100,000/mm³), and hemorrhage from any site.
- Inspect patient's mouth daily. Promptly report cracked lips, xerostomia, white patches, and erythema of buccal membranes.
- Report development of maculopapular rash; it usually responds to symptomatic treatment and is reversible.
- Be aware of expected response of lesion to topical 5-FU: Erythema followed in sequence by vesiculation, erosion, ulceration, necrosis, epithelialization. Applications of drug are continued until ulcerative stage is reached (2–6 wk after initial applications) and then discontinued.

Patient & Family Education
- Understand that it is very important to report the first signs of toxicity: Anorexia, vomiting, nausea, stomatitis, diarrhea, GI bleeding.
- Avoid exposure to sunlight or ultraviolet lamp treatments. Protect exposed skin. Photosensitivity usually subsides 2–3 mo after last dose.
- Report promptly to prescriber any difficulty in maintaining balance while ambulating.
- Use contraception during 5-FU treatment. If you suspect you are pregnant, tell your prescriber.

√FLUOXETINE HYDROCHLORIDE ℗⁺

(flu'ox-e-tine)

Prozac, Prozac Weekly, Sarafem

Classifications: SELECTIVE SEROTON-IN REUPTAKE INHIBITOR (SSRI); ANTIDE-PRESSANT

Therapeutic: ANTIDEPRESSANT; SSRI

Pregnancy Category: C

AVAILABILITY 10 mg, 15 mg tablets; 10 mg, 20 mg, 40 mg capsules; 20 mg/5 mL solution; 90 mg sustained release capsules (Prozac Weekly)

ACTION & *THERAPEUTIC EFFECT*

A selective serotonin reuptake inhibitor (SSRI). Antidepressant effect is presumed to be linked to its inhibition of CNS neuronal uptake of serotonin, a neurotransmitter. *Effectiveness may take from several days to 5 wk to develop fully. Drug has antidepressant, antiobsessive-compulsive, and antibulimic actions.*

USES Depression, geriatric depression, obsessive-compulsive disorder (OCD), bulimia nervosa, premenstrual dysphoric disorder, panic disorder.

UNLABELED USES Obesity, fibromyalgia, hot flashes.

CONTRAINDICATIONS Hypersensitivity to fluoxetine or other SSRI drugs; concurrent administration with MAOIs, or thioridazine; children younger than 7 y for OCD, children younger than 8 y for depression.

CAUTIOUS USE Hepatic and renal impairment, renal failure, abrupt discontinuation, anorexia nervosa, mania, bleeding; hyponatremia, cardiac disease, dehydration, diabetes mellitus, patients with history of suicidal ideations or current suicidal tendencies; seizure disorders, ECT, hepatic disease. Older adults may require dose adjustments; pregnancy (category C), lactation.

ROUTE & DOSAGE

Depression, Obsessive-Compulsive Disorder

Adult: **PO** 20 mg/day in a.m., may increase by 20 mg/day at weekly intervals (max: 80 mg/day); 20 mg/day in a.m.; when stable may switch to 90 mg sustained release capsule qwk (max: 90 mg/wk)

Child (older than 7 y): **PO** 10–20 mg/day in a.m. (max: 60 mg/day for OCD)

Geriatric: **PO** Start with 10 mg/day

Premenstrual Dysphoric Disorder

Adult: **PO** 10–20 mg daily (max: 60 mg/day)

Bulimia Nervosa

Adult: **PO** 60 mg daily

Panic Disorder

Adult: **PO** 10 mg daily may increase to 20 mg daily

Pharmacogenetic Dosage Adjustment

CYP2D6 poor metabolizers: Start at 80% of normal dose

ADMINISTRATION

Oral

▪ Give as a single dose in morning. Give in two divided doses; one in a.m. and one at noon to prevent insomnia, when more than 20 mg/day prescribed.

▪ Provide suicidal or potentially suicidal patient with small quantities of prescription medication.

Common adverse effects in *italic*, life-threatening effects underlined; generic names in **bold;** classifications in SMALL CAPS; ♣ Canadian drug name; ℗ Prototype drug

649

F

- Monitor for worsening of depression or expression of suicidal ideations.
- Store at 15°–25° C (59°–77° F).

ADVERSE EFFECTS (≥1%) **CNS:** *Headache, nervousness, anxiety, insomnia,* drowsiness, fatigue, tremor, dizziness. **CV:** Palpitations, hot flushes, chest pain. **GI:** *Nausea, diarrhea,* anorexia, dyspepsia, increased appetite, dry mouth. **Skin:** Rash, pruritus, sweating, hypersensitivity reactions. **Special Senses:** Blurred vision. **Body as a Whole:** Myalgias, arthralgias, flu-like syndrome, hyponatremia. **Urogenital:** Sexual dysfunction, menstrual irregularities.

INTERACTIONS Drug: Concurrent use of **tryptophan** may cause agitation, restlessness, and GI distress; MAO INHIBITORS, **selegiline** may increase risk of severe hypertensive reaction and death; increases half-life of **diazepam;** may increase toxicity of TRICYCLIC ANTIDEPRESSANTS; AMPHETAMINES, **cilostazol, nefazodone, pentazocine, propafenone, sibutramine, tramadol, venlafaxine** may increase risk of serotonin syndrome; may inhibit metabolism of **carbamazepine, phenytoin, ritonavir;** increased ergotamine toxicity with **dihydroergotamine, ergotamine.** ANTIPSYCHOTICS like **pimozide** can cause QT prolongation. **Herbal: St. John's wort** may cause serotonin syndrome.

PHARMACOKINETICS Absorption: 60–80% from GI tract. **Onset:** 1–3 wk. **Peak:** 4–8 h. **Distribution:** Widely distributed, including CNS. **Metabolism:** In liver to active metabolite, norfluoxetine. **Elimination:** Greater than 80% in urine; 12% in feces. **Half-Life:**

Fluoxetine 2–3 days, norfluoxetine 7–9 days.

NURSING IMPLICATIONS

Assessment & Drug Effects

- Monitor children and adolescents for changes in behavior and suicidal ideation.
- Use with caution in the older adult patient or patient with impaired renal or hepatic function (may need lower dose).
- Supervise patients closely who are high suicide risks; especially during initial therapy.
- Monitor for S&S of anaphylactoid reaction (see Appendix F).
- Lab tests: Periodic serum electrolytes; monitor closely plasma glucose in diabetes.
- Monitor serum sodium level for development of hyponatremia, especially in patients who are taking diuretics or are otherwise hypovolemic.
- Monitor diabetics for loss of glycemic control; hypoglycemia has occurred during initiation of therapy, and hyperglycemia during drug withdrawal.
- Weigh weekly to monitor weight loss, particularly in the older adult or nutritionally compromised patient. Report significant weight loss to prescriber.
- Observe for and promptly report rash or urticaria and S&S of fever, leukocytosis, arthralgias, carpal tunnel syndrome, edema, respiratory distress, and proteinuria.
- Observe for dizziness and drowsiness and employ safety measures as indicated.
- Monitor for and report increased anxiety, nervousness, or insomnia; may need modification of drug dose.
- Monitor for seizures in patients with a history of seizures. Use appropriate safety precautions.

Patient & Family Education

- Notify prescriber of any rash; possible sign of a serious group of adverse effects.
- Do not drive or engage in potentially hazardous activities until response to drug is known; especially if dizziness is noted.
- Monitor blood glucose for loss of glycemic control if diabetic.
- Note: Drug may increase seizure activity in those with history of seizure.

FLUOXYMESTERONE
(floo-ox-ee-mess′te-rone)

Androxy, Ora Testryl ♦
Classification: ANDROGEN/ANABOLIC STEROID
Therapeutic: ANABOLIC STEROID; MALE HORMONE REPLACEMENT
Prototype: Testosterone
Pregnancy Category: X
Controlled Substance: Schedule III

AVAILABILITY 10 mg tablets

ACTION & *THERAPEUTIC EFFECT*
Short-acting, orally effective derivative of testosterone with hypercholesterolemic effect. *Replacement therapy for endogenous testosterone. Promotes recalcification of osseous metastases and regression of soft tissue lesions.*

USES In men as replacement therapy in conditions associated with testicular hormone deficiency; in women to antagonize effects of estrogen in androgen-responsive inoperable breast cancer.

CONTRAINDICATIONS Breast cancer in men, prostatic cancer, benign obstructive prostatic hypertrophy; hypercalcemia; diabetes mellitus; severe cardiorenal disease or liver damage; nephrosis or nephrotic phase of nephritis; history of MI; athletes; women with inoperable mammary cancer less than 1 y or greater than 5 y after menopause; pregnancy (category X), lactation; infants.

CAUTIOUS USE Older males, history of MI, or coronary disease, hepatic, renal or congestive heart failure, women; children.

ROUTE & DOSAGE

Male Hypogonadism
Adult: **PO** 5 mg 1–4 times per day
Metastatic Carcinoma of Female Breast
Adult: **PO** 10–40 mg/day in divided doses

ADMINISTRATION

Oral
- Give immediately before or with meals to diminish GI distress.

ADVERSE EFFECTS (≥1%) **Endocrine:** Virilization (women), gynecomastia (men). **Urogenital:** Priapism, impotence. **Metabolic:** Jaundice (reversible), hypoglycemia, hypercalcemia. **GI:** <u>Hepatocellular carcinoma</u>, peliosis hepatitis, nausea, vomiting, diarrhea, symptoms resembling peptic ulcer. **Body as a Whole:** <u>Anaphylactic reactions</u> (rare), *edema, acne.*

INTERACTIONS Drug: ORAL ANTICOAGULANTS increase risk of bleeding. Possibly increases risk of **cyclosporine** toxicity. **Insulin** and ORAL HYPOGLYCEMIC AGENTS may decrease **glucose** level; dose will need to be adjusted. **Herbal: Echinacea** may increase hepatotoxicity.

Common adverse effects in *italic*, life-threatening effects <u>underlined</u>; generic names in **bold**; classifications in SMALL CAPS; ♦ Canadian drug name; ⊘ Prototype drug

651

PHARMACOKINETICS Absorption: Readily from GI tract. **Metabolism:** In liver. **Half-Life:** 9.5 h.

NURSING IMPLICATIONS

Assessment & Drug Effects

- Lab test: Obtain baseline and periodic liver function and serum electrolytes, Hgb, Hct, and serum and urine calcium; also serial serum cholesterol in patients with history of MI or coronary artery disease.
- Monitor I&O ratio and pattern and weight, and check for edema; report significant changes.
- Monitor for S&S of hypercalcemia (see Appendix F); particularly likely in patients with metastatic breast carcinoma.
- Watch for symptoms of hypoglycemia (see Appendix F) and report to prescriber. Drug may reduce blood glucose in diabetic patients.
- Observe patient on concomitant anticoagulant therapy for ecchymotic areas, petechiae, or abnormal bleeding from any site. Close monitoring of PT and INR is essential.

Patient & Family Education

- Good personal hygiene, including meticulous skin care is very important (females and prepubertal males are especially likely to develop acne).
- Note and report symptoms of jaundice (see Appendix F) to prescriber. Dose adjustment may reverse the condition.
- Report menstrual irregularities.
- Report priapism (prolonged erection) to prescriber promptly, it is a symptom of overdosage. A temporary interruption of regimen may be indicated. Also report persistent GI distress, diarrhea, or the onset of jaundice.

- Be aware that virilization usually occurs. Report to prescriber any voice change (hoarseness or deepening), increased libido (associated with clitoral enlargement), hirsutism immediately. Usually, stopping therapy will end further development of symptoms but will not reverse hirsutism or voice change.

FLUPHENAZINE DECANOATE
(floo-fen'a-zeen)
Modecate Decanoate ♦

FLUPHENAZINE HYDROCHLORIDE
Moditen HCl ♦
Classifications: ANTIPSYCHOTIC; PHENOTHIAZINE
Therapeutic: ANTIPSYCHOTIC
Prototype: Chlorpromazine
Pregnancy Category: C

AVAILABILITY 1 mg, 2.5 mg, 5 mg, 10 mg tablets; 2.5 mg/5 mL elixir; 5 mg/mL oral concentrate; **Decanoate:** 25 mg/mL injection

ACTION & *THERAPEUTIC EFFECT* Potent phenothiazine, antipsychotic agent that blocks postsynaptic dopamine receptors in the brain. Similar to other phenothiazines with the following exceptions: More potent per weight, higher incidence of extrapyramidal complications, and lower frequency of sedative, hypotensive, and antiemetic effects. *Effective for treatment of antipsychotic symptoms including schizophrenia.*

USES Management of manifestations of psychotic disorders.

UNLABELED USES As antineuralgia adjunct.

CONTRAINDICATIONS Known hypersensitivity to phenothiazines; subcortical brain damage, comatose or severely depressed states, blood dyscrasias, renal or hepatic disease; lactation. **Parenteral form** not recommended for children younger than 12 y.

CAUTIOUS USE Older adults, previously diagnosed breast cancer; closed-angle glaucoma; GI disorders; significant pulmonary disease; renal failure; seizure disorders; history of suicidal ideation or high risk for suicide attempt; cardiovascular diseases; pheochromocytoma; history of convulsive disorders; patients exposed to extreme heat or phosphorous insecticides; peptic ulcer; respiratory impairment; pregnancy (category C).

ROUTE & DOSAGE

Psychosis

Adult: **PO** 0.5–10 mg/day in 1–4 divided doses (max: of 20 mg/day) **IM/Subcutaneous HCl** 2.5–10 mg/day divided q6–8h (max: 10 mg/day); **Decanoate** 12.5–25 mg q1–4wk; **Enanthate** 25 mg q2wk

Dementia Behavior

Geriatric: **PO** 1–2.5 mg/day, may increase every 4–7 days by 1–2.5 mg/day (max: 20 mg/day in 2–3 divided doses)

ADMINISTRATION

Oral

- Dilute oral concentrate in fruit juice, water, carbonated beverage, milk, soup. Avoid caffeine-containing beverages (cola, coffee) as a diluent, also tannic acid (tea) or pectinates (apple juice).

- Be careful not to contact skin or clothing with drug when preparing oral concentrate or liquid preparations for injection. Warn patient to avoid spilling drug. If drug contacts skin, rinse/flush skin promptly with warm water.

- Protect all preparations from light and freezing. Solutions may safely vary in color from almost colorless to light amber. Discard dark or otherwise discolored solutions.

- Store in tightly closed container at 15°–30° C (59°–86° F) unless otherwise specified by manufacturer. Protect all forms from light.

Intramuscular/Subcutaneous

- Fluphenazine hydrochloride (HCl) is given IM and fluphenazine decanoate may be given IM or subcutaneously.

ADVERSE EFFECTS (≥1%) **CNS:** *Extrapyramidal symptoms* (resembling Parkinson's disease), tardive dyskinesia, sedation, drowsiness, dizziness, headache, mental depression, catatonic-like state, impaired thermoregulation, grand mal seizures. **CV:** Tachycardia, hypertension, hypotension. **GI:** Dry mouth, nausea, epigastric pain, constipation, fecal impaction, cholecystic jaundice. **Urogenital:** Urinary retention, polyuria, inhibition of ejaculation. **Hematologic:** Transient leukopenia, agranulocytosis. **Skin:** Contact dermatitis. **Body as a Whole:** Peripheral edema. **Special Senses:** Nasal congestion, blurred vision, increased intraocular pressure, *photosensitivity*. **Endocrine:** Hyperprolactinemia.

INTERACTIONS Drug: Alcohol and other CNS DEPRESSANTS may potentiate depressive effects; decreases seizure threshold, may need to adjust dosage of ANTICONVULSANTS. **Herbal: Kava** may increase risk and severity of dystonic reactions.

Common adverse effects in *italic*, life-threatening effects underlined; generic names in **bold**; classifications in SMALL CAPS; ✦ Canadian drug name; ⊘ Prototype drug

653

PHARMACOKINETICS Absorption:
HCl is readily absorbed PO and IM;
decanoate, enanthate have delayed
IM absorption. **Onset:** 1 h HCl; 24–
72 h decanoate, enanthate. **Peak:**
0.5 h PO; 1.5–2 h IM HCl. **Duration:**
6–8 h HCl; 1–6 wk decanoate; 2–4
wk enanthate. **Distribution:** Crosses
blood–brain barrier and placenta.
Metabolism: In liver. **Half-Life:** 15 h
HCl; 3.6 days enanthate; 7–10 days
decanoate.

NURSING IMPLICATIONS

Assessment & Drug Effects

- Report immediately onset of mental depression and extrapyramidal
symptoms.
- Be alert for appearance of acute
dystonia (see Appendix F). Symptoms can be controlled by reducing dosage or by adding an antiparkinsonism drug such as
benztropine.
- Lab tests: Monitor kidney function
in patients on long-term treatment. Withhold drug and notify
prescriber if BUN is elevated. Also
perform WBC with differential,
and LFTs, periodically.
- Monitor BP during early therapy.
If systolic drop is more than 20
mm Hg, inform prescriber.
- Monitor I&O ratio and bowel elimination pattern. Check for abdominal distension and pain. Monitor
for xerostomia and constipation.

Patient & Family Education

- Do not drive or engage in potentially hazardous activities until response to drug is known.
- Do not alter dosage regimen or
stop taking drug abruptly.
- Be alert for adverse effects, early
detection is critical because drug
has a long duration of action. Inform prescriber promptly if following symptoms appear: Light-
colored stools, changes in vision,

sore throat, fever, cellulitis, rash,
any interference with movement.
- Be aware that it may be difficult
for you to adjust to extremes in
temperature. Extended exposure
to high environmental temperature, to sun's rays, or to a high fever places the patient taking this
drug at risk for heat stroke.
- Avoid exposure to sun; wear protective clothing and cover exposed skin surfaces with sun
screen lotion (SPF above 12).
- Avoid alcohol while on fluphenazine therapy.
- Note: Fluphenazine may discolor
urine pink to red or reddish brown.
- Periodic ophthalmologic exams
are recommended.

FLURANDRENOLIDE

(flure-an-dren'oh-lide)
Cordran, Cordran SP, Drenison ✦
Pregnancy Category: C
See Appendix A-4.

FLURAZEPAM HYDROCHLORIDE

(flure-az'e-pam)
**Apo-Flurazepam ✦, Dalmane,
Novoflupam ✦**
Classifications: SEDATIVE-HYPNO-
TIC; ANXIOLYTIC; BENZODIAZEPINE
Therapeutic: SEDATIVE-HYPNOTIC;
ANTIANXIETY
Prototype: Lorazepam
Pregnancy Category: X
Controlled Substance: Schedule IV

AVAILABILITY 15 mg, 30 mg capsules

ACTION & *THERAPEUTIC EFFECT*
Benzodiazepine derivative that enhances the GABA-benzodiazepine

receptor complex. GABA is an inhibitory neurotransmitter involved in anxiolytic and sedative effects. Flurazepam appears to act at the limbic and subcortical levels of CNS to produce sedation. *Reduces sleep induction time; produces marked reduction of stage 4 sleep (deepest sleep stage) while at the same time increasing duration of total sleep time.*

USES Hypnotic in management of all kinds of insomnia (e.g., difficulty in falling asleep, frequent nocturnal awakening or early morning awakening or both). Also for treatment of poor sleeping habits.

CONTRAINDICATIONS Prolonged administration; benzodiazepine hypersensitivity; ethanol intoxication; COPD, sleep apnea; respiratory depression; shock; coma; major depression or psychosis; intermittent porphyria; children younger than 15 y; pregnancy (category X), lactation.
CAUTIOUS USE Impaired renal or hepatic function; glaucoma; mental depression; psychoses, history of suicidal tendencies, bipolar disorder; intermittent porphyria; addiction-prone individuals; older adult or debilitated patients; COPD.

ROUTE & DOSAGE

Sedative, Hypnotic

Adult (15 y or older): **PO** 15–30 mg at bedtime
Geriatric: **PO** 15 mg at bedtime

ADMINISTRATION

Oral
- Give once patient is in bed and ready to fall asleep.
- Store in light-resistant container with childproof cap at 15°–30° C (59°–86° F) unless otherwise specified.

ADVERSE EFFECTS (≥1%) **CNS:** *Residual sedation, drowsiness,* light-headedness, dizziness, ataxia, headache, nervousness, apprehension, talkativeness, irritability, depression, hallucinations, nightmares, confusion, paradoxic reactions: Excitement, euphoria, hyperactivity, disorientation, coma (overdosage). **Special Senses:** Blurred vision, burning eyes. **GI:** Heartburn, nausea, vomiting, diarrhea, abdominal pain. **Body as a Whole:** Immediate allergic reaction, hypotension, granulocytopenia (rare), jaundice (rare).

DIAGNOSTIC TEST INTERFERENCE Flurazepam may increase serum levels of *total and direct bilirubin, alkaline phosphatase, AST,* and *ALT.* False-negative *urine glucose* reactions may occur with *Clinistix* and *Diastix;* no effect with *TesTape.*

INTERACTIONS Drug: Alcohol, CNS DEPRESSANTS, ANTICONVULSANTS potentiate CNS depression; **cimetidine, disulfiram** may increase flurazepam levels, thus increasing its toxicity. **Herbal: Kava, valerian** may potentiate sedation.

PHARMACOKINETICS Absorption: Readily from GI tract. **Onset:** 15–45 min. **Duration:** 7–8 h. **Distribution:** Crosses blood–brain barrier and placenta; distributed into breast milk. **Metabolism:** In liver to active metabolites. **Elimination:** Primarily in urine. **Half-Life:** 47–100 h.

NURSING IMPLICATIONS

Assessment & Drug Effects
- Monitor effectiveness. Hypnotic effect is apparent on second or third night of consecutive use and continues 1–2 nights after drug is stopped (drug has a long half-life).

Common adverse effects in *italic*, life-threatening effects underlined; generic names in **bold**; classifications in SMALL CAPS; ♣ Canadian drug name; ♦ Prototype drug

655

- Supervise ambulation. Residual sedation and drowsiness are relatively common. Excessive drowsiness, ataxia, vertigo, and falling occur more frequently in older adults or debilitated patients.
- Lab tests: Monitor blood counts, LFTs, and kidney function tests with repeated use.
- Be aware that withdrawal symptoms have occurred 3 days after abrupt discontinuation after prolonged use and include worsening of insomnia, dizziness, blurred vision, anorexia, GI upset, nasal congestion, paresthesias.

Patient & Family Education
- Avoid potentially hazardous activities until response to drug is known.
- Avoid alcohol. Concurrent ingestion with flurazepam intensifies CNS depressant effects; symptoms may occur even when alcohol is ingested as long as 10 h after last flurazepam dose.
- Be aware of the possible additive depressant effects when drug is combined with barbiturates, tranquilizers, or other CNS depressants.

FLURBIPROFEN SODIUM

(flure-bi´proe-fen)

Ansaid, Ocufen

Classifications: ANALGESIC, NONSTEROIDAL ANTI-INFLAMMATORY DRUG (NSAID); COX-1 AND COX-2 INHIBITOR; ANTIPYRETIC

Therapeutic: ANALGESIC, NSAID

Prototype: Ibuprofen

Pregnancy Category: B first or second trimester; D third trimester

AVAILABILITY 50 mg, 100 mg tablets; 0.03% ophth solution

ACTION & *THERAPEUTIC EFFECT*
Inhibits prostaglandin synthesis including in the conjunctiva and uvea by inhibiting the COX-1 or COX-2 enzymes. Ocular flurbiprofen reduces miosis, permitting maintenance of drug-induced mydriasis during surgical procedures. *An anti-inflammatory, nonsteroidal analgesic. Additionally, it inhibits migration of leukocytes into inflamed tissues, depresses monocyte function, and may inhibit platelet aggregation. Reduces miosis intraoperatively.*

USES Inhibition of intraoperative miosis; arthritis and other inflammatory diseases; mild to moderate pain.

UNLABELED USES Management of postoperative ocular inflammation, prevention of postcystoid macular edema.

CONTRAINDICATIONS Hypersensitivity to NSAIDs, or salicylates; epithelial herpes simplex; keratitis; perioperative pain from CABG; pregnancy (category D third trimester), lactation.

CAUTIOUS USE Patient who may be adversely affected by prolonged bleeding time; patient in whom asthma, rhinitis, or urticaria is precipitated by aspirin or other NSAIDs; pregnancy (category B first and second trimester). Safe use in children not established.

ROUTE & DOSAGE

Inflammatory Disease

Adult: **PO** 200–300 mg/day in 2–4 divided doses (max: 300 mg/day)

Mild to Moderate Pain

Adult: **PO** 50–100 mg q6–8h

Inhibition of Intraoperative Miosis

Adult: **Topical** 1 drop in eye approximately q30min beginning

Common adverse effects in *italic*, life-threatening effects underlined; generic names in **bold**; classifications in SMALL CAPS; ♣ Canadian drug name; ❂ Prototype drug

2 h before surgery for a total of 4 drops per affected eye

ADMINISTRATION

Topical

- Instill ophthalmic preparation with great care to avoid contamination of solution. Do not touch eye surface with dropper.

Oral

- Use the 300 mg dose for initiation of therapy or for acute exacerbations of disease.
- Store at 15°–30° C (59°–86° F) in tight, light-resistant container.

ADVERSE EFFECTS (≥1%) Special Senses: *Mild ocular stinging,* burning, itching, or foreign body sensation (transient). **Other:** Slowed corneal healing; increased bleeding time. **For adverse effects to oral preparations, see ibuprofen.**

INTERACTIONS Drug: ORAL ANTICOAGULANTS, **heparin** may prolong bleeding time; actions and side effects of both flurbiprofen and **phenytoin,** SULFONYLUREAS, or SULFONAMIDES may be potentiated. **Herbal:** Feverfew, garlic, ginger, gingko may increase bleeding potential.

PHARMACOKINETICS Absorption: 80% absorbed from GI tract. **Onset:** 2 h. **Peak:** 2 h. **Duration:** 6–8 h. **Distribution:** Small amounts distributed into breast milk. **Metabolism:** In liver. **Elimination:** Primarily in urine; some biliary excretion. **Half-Life:** 5 h.

NURSING IMPLICATIONS

Assessment & Drug Effects

- Observe patients with history of cardiac decompensation closely for evidence of fluid retention and edema.
- Lab tests: Baseline and periodic evaluations of Hgb, renal function tests, LFTs.

- Auditory and ophthalmologic examinations are recommended with prolonged or high-dose therapy.
- Monitor for GI distress and S&S of GI bleeding.

Patient & Family Education

- Report ocular irritation that persists after flurbiprofen use during surgery (tearing, dry eye sensation, dull eye pain, photophobia) to prescriber.
- Be alert for bleeding tendency and report unexplained bleeding, prolongation of bleeding time, or bruises.
- Notify prescriber immediately of passage of dark tarry stools, "coffee ground" emesis, frankly bloody emesis, or other GI distress, as well as blood or protein in urine, and onset of skin rash, pruritus, jaundice.
- Do not drive or engage in potentially hazardous activities until response to the drug is known.
- Avoid alcohol. Concurrent use may increase risk of GI ulceration and bleeding tendencies.

FLUTAMIDE ⊘

(flu′ta-mide)

Eulexin

Classifications: ANTINEOPLASTIC; ANTIANDROGEN

Therapeutic: ANTINEOPLASTIC; ANTIANDROGEN

Pregnancy Category: D

AVAILABILITY 125 mg capsules

ACTION & *THERAPEUTIC EFFECT*
Nonsteroidal, nonhormonal, antiandrogenic drug that inhibits androgen uptake or binding of androgen to target tissues (i.e., prostatic cancer cells). *Interferes with the binding of both testosterone and*

Common adverse effects in *italic*, life-threatening effects <u>underlined</u>; generic names in **bold**; classifications in SMALL CAPS; ♣ Canadian drug name; ⊘ Prototype drug

657

dihydrotestosterone to target tissue (i.e., prostate cancer cells).

USES In combination with luteinizing hormone-releasing hormone agonists (i.e., leuprolide) or castration for early stage and metastatic prostate cancer.

CONTRAINDICATIONS Hypersensitivity to flutamide; severe liver impairment if ALT is equal to twice the normal value; females; pregnancy (category D), lactation.
CAUTIOUS USE Lactase deficiency.

ROUTE & DOSAGE

Prostate Cancer
Adult: **PO** 250 mg (2 caps) q8h

ADMINISTRATION
Oral
- Use with caution in patients with severe hepatic impairment.
- Store at 2°–30° C (36°–86° F) in a tightly closed, light-resistant container.

ADVERSE EFFECTS (≥1%) **CNS:** Drowsiness, confusion, depression, anxiety, nervousness. **GI:** Diarrhea, nausea, vomiting, anorexia, hepatitis, cholestatic jaundice, encephalopathy, hepatic necrosis, <u>acute hepatic failure</u>, may increase ALT, AST, bilirubin. **Urogenital:** *Hot flashes, loss of libido, impotence.* **Hematologic:** Anemia, <u>leukopenia</u>, <u>thrombocytopenia</u>. **Skin:** Rash. **Body as a Whole:** Edema. **Endocrine:** Gynecomastia, galactorrhea.

INTERACTIONS Drug: May increase INR in patients on **warfarin.**

PHARMACOKINETICS Absorption: Readily absorbed from GI tract. **Onset:** Antiandrogenic activity 2.2 h; symptomatic relief 2–4 wk.

Duration: 3 mo–2.5 y, with an average of 10.5 mo. **Metabolism:** Metabolized in liver to at least 10 different metabolites; the major metabolite, 2-hydroxyflutamide (SCH-16423), is an alpha-hydroxylated derivative that is biologically active. **Elimination:** 98% in urine. **Half-Life:** 5–6 h.

NURSING IMPLICATIONS
Assessment & Drug Effects
- Monitor for symptomatic relief of bone pain.
- Assess for development of gynecomastia and galactorrhea; if these become bothersome, dosage reduction may be warranted.
- Lab tests: Monitor LFTs and serum bilirubin periodically.
- Monitor for and report development of a lupus-like syndrome.

Patient & Family Education
- Be aware of potential adverse effects of therapy.
- Notify prescriber immediately of the following: Pain in upper abdomen, yellowing of skin and eyes, dark urine, respiratory problems, rashes on face, difficulty urinating, sore throat, fever, chills.

FLUTICASONE
(flu-ti-ca'sone)

Advair, Flonase, Flovent, Flovent HFA, Cutivate, Veramyst
Pregnancy Category: C
See Appendixes A-3, A-4.

FLUVASTATIN
(flu-vah-stat'in)

Lescol, Lescol XL
Classifications: HMG-COA REDUCTASE INHIBITOR (STATIN); ANTIHYPERLIPEMIC

Therapeutic: CHOLESTEROL-LOWER-ING (STATIN)
Prototype: Lovastatin
Pregnancy Category: X

AVAILABILITY 20 mg, 40 mg capsules; 80 mg extended release tablet

ACTION & *THERAPEUTIC EFFECT*
Inhibits reductase 3-hydroxy-3-methylglutaryl coenzyme A (HMG-CoA) that is essential to hepatic production of cholesterol. Cholesterol-lowering effect triggers induction of LDL receptors, which promotes removal of LDL and VLDL remnants (precursors of LDL) from plasma. *Results in an increase in plasma HDL concentration. HDLs collect excess cholesterol from body cells and transport it to the liver for excretion.*

USES Adjunct to diet for the reduction of elevated total LDL cholesterol in patients with primary hypercholesterolemia (types IIa and IIb).
UNLABELED USES Other types of hyperlipidemias.

CONTRAINDICATIONS Hypersensitivity to fluvastatin, lovastatin, pravastatin, or simvastatin; active liver disease or unexplained persistent elevated liver function tests; pregnancy (category X), lactation.
CAUTIOUS USE Patients who consume substantial quantities of alcohol; history of liver disease; renal impairment. Safe use in children 10 y or younger not established.

ROUTE & DOSAGE

Hypercholesterolemia
Adult: **PO** 20 mg at bedtime, may increase up to 80 mg/day in 1–2 doses

ADMINISTRATION

Oral
- Ensure the extended release tablet is not chewed or crushed. It **must be** swallowed whole.
- Separate doses of this drug and bile-acid resin (e.g., cholestyramine) by at least 2 h when given concomitantly.
- Note: Dosage adjustments may be required in patients with significant renal or hepatic impairment.
- Store at room temperature, 15°–30° C (59°–86° F).

ADVERSE EFFECTS (≥1%) **CNS:** Headache, fatigue. **Body as a Whole:** Myalgia. **GI:** Dyspepsia, diarrhea, abdominal pain. **Skin:** Rash.

INTERACTIONS Drug: May increase risk of bleeding with **warfarin; cholestyramine** decreases fluvastatin absorption; **rifampin** increases metabolism of fluvastatin; may increase risk of myopathy and rhabdomyolysis with **gemfibrozil, fenofibrate, clofibrate.**

PHARMACOKINETICS Absorption: Readily from GI tract; about 24% reaches systemic circulation after first-pass metabolism. **Onset:** 3–6 wk. **Peak:** Serum level 0.5–1 h. **Distribution:** 98% protein bound; distributed into breast milk. **Metabolism:** In liver. **Elimination:** 95% in bile; 5% in urine. **Half-Life:** 0.5–1 h.

NURSING IMPLICATIONS
Assessment & Drug Effects
- Lab tests: Monitor lipoprotein levels; maximal lipid-lowering effect occurs in 4–6 wk. Monitor serum transaminase and CPK levels every 3–4 mo for the first year and periodically thereafter.
- Monitor PT and INR in patients on concurrent warfarin therapy; PT & INR may be prolonged.

Common adverse effects in *italic*, life-threatening effects underlined; generic names in **bold**; classifications in SMALL CAPS; ✚ Canadian drug name; ☻ Prototype drug

659

Patient & Family Education
- Take fluvastatin at bedtime.
- Be alert and report signs of bleeding immediately when also taking warfarin.
- Notify prescriber immediately of the following: Fever; rash; muscle pain, weakness, tenderness, or cramping.
- Reduce or eliminate alcohol consumption while taking fluvastatin.

FLUVOXAMINE

(flu-vox′a-meen)

Luvox, Luvox CR
Classifications: SELECTIVE SEROTONIN REUPTAKE INHIBITOR (SSRI); ANTIDEPRESSANT
Therapeutic: ANTIDEPRESSANT; SSRI
Prototype: Fluoxetine
Pregnancy Category: C

AVAILABILITY 25 mg, 50 mg, 100 mg tablets; 100 mg, 150 mg extended release capsules

ACTION & *THERAPEUTIC EFFECT*
Antidepressant with potent, selective, inhibitory activity on neuronal (5-HT) serotonin reuptake (SSRI). *Effective as an antidepressant and for control of obsessive-compulsive disorder and social anxiety.*

USES Treatment of obsessive-compulsive disorders, social anxiety disorder.
UNLABELED USES Post-traumatic stress disorder, depression, panic attacks.

CONTRAINDICATIONS Hypersensitivity to fluvoxamine or fluoxetine; suicidal ideation; concurrent MAOI therapy; bipolar disorder.
CAUTIOUS USE Liver disease, renal impairment, abrupt discontinuation; cardiac disease, dehydration, hyponatremia, older adults, ECT, seizure disorders, history of suicidal ideation, tobacco smoking; pregnancy (category C), lactation. Safety and efficacy in children younger than 8 y for obsessive compulsive disorder is not established.

ROUTE & DOSAGE

Obsessive-Compulsive Disorder

Adult: **PO** Start with 50 mg daily, may increase slowly up to 300 mg/day given every night or divided b.i.d. OR 100 mg extended release every night, may increase up (max: 300 mg/day)
Adolescent: **PO** Start with 25 mg daily, may increase up to 300 mg/day in divided doses
Child (8–11 y): **PO** Start with 25 mg every night, may increase by 25 mg q4–7days (max: 200 mg/day in divided doses)

Social Anxiety Disorder

Adult: **PO** 100 mg extended release caps every night, may increase as needed up to 300 mg/day

Pharmacogenetic Dosage Adjustment

Poor CYP2D6 metabolizers: Start with 70% of dose

ADMINISTRATION

Oral
- Do not open extended release capsules. They **must be** swallowed whole.
- Give starting doses at bedtime to improve tolerance to nausea and vomiting; both are common early in therapy.
- Store at room temperature, 15°–30° C (59°–86° F), away from moisture and light.

ADVERSE EFFECTS (≥1%) **CNS:** *Somnolence, headache, agitation, insomnia, dizziness,* seizures. **CV:**

Common adverse effects in *italic*, life-threatening effects <u>underlined</u>; generic names in **bold**; classifications in SMALL CAPS; ✤ Canadian drug name; ⊙ Prototype drug

Orthostatic hypotension, slight bradycardia. **GI:** *Nausea, vomiting, dry mouth, constipation, anorexia.* **Urogenital:** Sexual dysfunction. **Skin:** <u>Stevens-Johnson syndrome</u>, <u>toxic epidermal necrolysis</u> (rare).

DIAGNOSTIC TEST INTERFERENCE *Gamma-glutamyl transferase* increased by more than 3-fold following 3 wk of therapy.

INTERACTIONS Drug: Fluvoxamine has been shown to significantly increase plasma levels of **amitriptyline, clomipramine,** and other TRICYCLIC ANTIDEPRESSANTS to mildly increase levels of their metabolites. May antagonize the blood pressure-lowering effects of **atenolol** and other BETA-BLOCKERS. May increase levels and toxicity of **carbamazepine, mexiletine.** May increase **lithium** levels causing neurotoxicity, **serotonin** syndrome, somnolence, and mania. One report of increased **theophylline** levels with toxicity. Increases prothrombin time in patients on **warfarin;** increased ergotamine toxicity with **dihydroergotamine, ergotamine.** Use with CYP1A2 INHIBITORS **(thioridazine, pimozide, alosetron, tizanidine)** increases **fluvoxamine** levels and toxicity. **Food:** Grapefruit juice may increase risk of side effects. **Herbal:** Melatonin may increase and prolong drowsiness; **St. John's wort** may cause **serotonin** syndrome.

PHARMACOKINETICS Absorption: Almost completely absorbed from GI tract. **Onset:** 4–7 days. **Distribution:** Approximately 77% bound to plasma proteins; excreted in human breast milk but in an amount that poses little risk to the nursing infant. **Metabolism:** In liver. **Elimination:** Completely in urine. **Half-Life:** 16–24 h.

NURSING IMPLICATIONS

Assessment & Drug Effects
- Monitor for significant nausea and vomiting, especially during initial therapy.
- Monitor for worsening of depression or emergence of suicidal ideations especially in adolescents and children.
- Assess safety; drowsiness and dizziness are common adverse effects.
- Monitor PT and INR carefully with concurrent warfarin therapy; adjust warfarin as needed.

Patient & Family Education
- Note: Nausea and vomiting are common in early therapy. Notify prescriber if these adverse effects last more than a few days.
- Exercise caution with hazardous activity until response to the drug is known.

FOLIC ACID (VITAMIN B₉, PTEROYLGLUTAMIC ACID)
(fol'ic)

Apo-Folic ✦, Folacin, Novofolacid ✦

FOLATE SODIUM
Folvite Sodium
Classification: VITAMIN B₉
Therapeutic: VITAMIN SUPPLEMENT
Pregnancy Category: A

AVAILABILITY 0.4 mg, 0.8 mg, 1 mg tablets; 5 mg/mL injection

ACTION & *THERAPEUTIC EFFECT* Vitamin B₉ essential for nucleoprotein synthesis and maintenance of normal erythropoiesis. Acts against folic acid deficiency that results in production of defective DNA that leads to megaloblast formation and arrest of bone marrow maturation.

Common adverse effects in *italic*, life-threatening effects <u>underlined</u>; generic names in **bold**; classifications in SMALL CAPS; ✦ Canadian drug name; ⊘ Prototype drug

661

Stimulates production of RBCs, WBCs, and platelets in patients with megaloblastic anemias.

USES Folate deficiency, macrocytic anemia, and megaloblastic anemias associated with malabsorption syndromes, alcoholism, primary liver disease, inadequate dietary intake, pregnancy, infancy, and childhood.

CONTRAINDICATIONS Folic acid alone for pernicious anemia or other vitamin B$_{12}$ deficiency states; normocytic, refractory, aplastic, or undiagnosed anemia; neonates.

CAUTIOUS USE Pregnancy (category A).

ROUTE & DOSAGE

Therapeutic

Adult: **PO/IM/Subcutaneous/IV** 1 mg/day or less
Child: **PO/IM/Subcutaneous/IV** 1 mg/day or less

Maintenance

Adult: **PO/IM/Subcutaneous/IV** 0.4 mg/day or less
Child: **PO/IM/Subcutaneous/IV** 4 y or younger, up to 0.3 mg/day; older than 4 y, up to 0.1 mg/day
Infant: **PO/IM/Subcutaneous/IV** 0.1 mg/day

ADMINISTRATION

Oral
- Oral route is preferred to other routes.

Intramuscular/Subcutaneous
- Give undiluted. Use caution not to inject intradermally.

Intravenous

PREPARE: **Direct/Continuous:** Given undiluted.
ADMINISTER: **Direct/Continuous:** Give over 30–60 sec. • May also add to a continuous infusion.

INCOMPATIBILITIES Solution/additive: Calcium gluconate, chlorpromazine, dextrose 40% in water, doxapram.

- Store at 15°–30° C (59°–86° F) in tightly closed containers protected from light, unless otherwise directed.

ADVERSE EFFECTS (≥1%) Reportedly nontoxic. Slight flushing and feeling of warmth following IV administration.

DIAGNOSTIC TEST INTERFERENCE Falsely low serum *folate levels* may occur with *Lactobacillus casei assay* in patients receiving antibiotics such as TETRACYCLINES.

INTERACTIONS Drug: Chloramphenicol may antagonize effects of **folate** therapy; **phenytoin** metabolism may be increased, thus decreasing its levels.

PHARMACOKINETICS Absorption: Readily from proximal small intestine. **Peak:** 30–60 min PO. **Distribution:** Distributed to all body tissues; high concentrations in CSF; crosses placenta; distributed into breast milk. **Metabolism:** In liver to active metabolites. **Elimination:** Small amounts in urine in folate-deficient patients; large amounts excreted in urine with high doses.

NURSING IMPLICATIONS

Assessment & Drug Effects
- Obtain a careful history of dietary intake and drug and alcohol usage prior to start of therapy.
- Keep prescriber informed of patient's response to therapy.
- Monitor patients on phenytoin for subtherapeutic plasma levels.

Patient & Family Education
- Remain under close medical supervision while taking folic acid therapy. Adjustment of

maintenance dose should be made if there is threat of relapse.

FONDAPARINUX SODIUM
(fon-da-par'i-nux)

Arixtra
Classification: ANTICOAGULANT, SELECTIVE FACTOR XA INHIBITOR
Therapeutic: ANTICOAGULANT; ANTITHROMBOTIC
Pregnancy Category: B

AVAILABILITY 2.5 mg/0.5 mL, 5 mg/0.4 mL, 7.5 mg/0.6 mL, 10 mg/0.8 mL syringe

ACTION & *THERAPEUTIC EFFECT*
Fondaparinux sodium causes antithrombin III (ATIII)-mediated selective inhibition of Factor Xa. It potentiates the innate neutralization of Factor Xa by ATIII. This interrupts the blood coagulation cascade, inhibiting thrombin formation and, thus, thrombus development. *Effective in the prevention and treatment of deep-vein thrombosis measured by the laboratory value of the amount of anti-Xa assay expressed in mg.*

USES Prophylaxis for DVT or pulmonary embolism (PE) in patients undergoing hip or knee replacement surgery or abdominal surgery; treatment of acute DVT without PE with warfarin, treatment of PE with warfarin.

CONTRAINDICATIONS Hypersensitivity to fondaparinux; active bleeding; GI bleeding; severe renal impairment with a creatinine clearance of less than 30 mL/min; weight less than 50 kg; active major bleeding; bacterial endocarditis; intramuscular administration; thrombocytopenia associated with fondaparinux.

CAUTIOUS USE Renal impairment or disease; older adult; indwelling epidural catheter; dental disease, dental work; diabetic retinopathy; diverticulitis; endocarditis, epidural anesthesia; hemophilia, heparin-induced thrombocytopenia (HIT), hepatic disease, hypertension, idiopathic thrombocytopenia purpura (ITP); inflammatory bowel disease, lumbar puncture, spinal anesthesia; stroke; surgery; thrombocytopenia, thrombolytic therapy; vaginal bleeding, menstruation; peptic ulcer disease; bleeding disorders including a history of GI ulceration, etc., history of heparin-induced thrombocytopenia; pregnancy (category B), lactation. Safety and effectiveness in children have not been established.

ROUTE & DOSAGE

DVT, Pulmonary Embolism Prophylaxis

Adult: **Subcutaneous** *Weight greater than 50 kg,* 2.5 mg daily starting at least 6 h postsurgery × 5–9 days; *for hip fracture patients,* up to 24 days additional use

Treatment of DVT, Pulmonary Embolism

Adult: **Subcutaneous** *Weight less than 50 kg,* 5 mg; *50–100 kg,* 7.5 mg; *weight greater than 100 kg,* 10 mg once daily × 5–9 days

Renal Impairment Dosage Adjustment

CrCl 30–50 mL/min: Use with caution; *less than 30 mL/min:* Use is contraindicated

ADMINISTRATION

Subcutaneous
- Give no sooner than 6 h after surgery.

Common adverse effects in *italic*, life-threatening effects underlined; generic names in **bold;** classifications in SMALL CAPS; ♣ Canadian drug name; ◉ Prototype drug

663

F

- Inspect visually for particulate matter and discoloration prior to administration.
- Do not expel the air bubble from the syringe before the injection.
- Use prefilled syringe to inject into fatty tissue, alternating injection sites (e.g., between L and R abdominal wall).
- Store at 25° C (77° F); excursions permitted to 15°–30° C (59°–86° F).

ADVERSE EFFECTS (≥1%) **Body as a Whole:** Fever, edema. **CNS:** Insomnia, dizziness, confusion, headache. **CV:** Hypotension. **GI:** Nausea, constipation, vomiting, diarrhea, dyspepsia, elevated LFTs. **Endocrine:** Hypokalemia. **Hematologic:** Hemorrhage, *anemia,* hematoma. **Skin:** Irritation at injection site, rash, purpura, bullous eruption. **Urogenital:** UTI, urinary retention.

INTERACTIONS Drug: ANTICOAGULANTS, ANTIPLATELETS, NSAIDS, **aspirin** may increase risk of bleeding. **Herbal: Feverfew, ginkgo, ginger, evening primrose oil** may potentiate bleeding.

PHARMACOKINETICS Absorption: Rapidly and completely absorbed from subcutaneous injection site. **Peak:** 2–3 h. **Distribution:** Primarily in blood. **Metabolism:** Negligible metabolism. **Elimination:** In urine. **Half-Life:** 18 h.

NURSING IMPLICATIONS

Assessment & Drug Effects
- Monitor for S&S of bleeding or hemorrhage. If noted, withhold fondaparinux and notify prescriber immediately.
- Withhold fondaparinux and notify prescriber if platelet count falls below 100,000/mm³.
- Lab tests: Monitor baseline and periodic renal function ests; periodic CBC including platelet count, serum creatinine level, and stool occult blood tests. Lab test for measuring drug effectiveness is amount of anti-Xa assay expressed in mg.

Patient & Family Education
- Report any of the following to a health care provider: Signs of unexplained bleeding such as: Pink, red, or dark brown urine; red or dark brown vomitus; bleeding gums or bloody sputum; dark, tarry stools.
- Learn proper injection technique if you are to self-administer this drug.
- Do not take any OTC drugs without first consulting prescriber.

FORMOTEROL FUMARATE
(for-mo-ter'ol)

Foradil Aerolizer, Perforomist
Classifications: BETA-ADRENERGIC AGONIST; BRONCHODILATOR
Therapeutic: BRONCHODILATOR
Prototype: Albuterol
Pregnancy Category: C

AVAILABILITY 12 mcg inhalation capsules; 20 mcg/2mL solution for inhalation

ACTION & *THERAPEUTIC EFFECT*
Long-acting selective beta$_2$-adrenergic receptor agonist that stimulates production of intracellular cyclic AMP, which causes relaxation of bronchial smooth muscle. It also inhibits release of mediators of immediate hypersensitivity (e.g., histamine and leukotrienes) from mast cells in the lung. *Acts locally in lung as a bronchodilator; prevents bronchoconstriction that occurs during an asthma attack.*

USES Treatment of asthma, bronchitis, prevention of exercise induced asthma, prevention of bronchospasm in COPD.

Common adverse effects in *italic,* life-threatening effects <u>underlined</u>; generic names in **bold**; classifications in SMALL CAPS; ♣ Canadian drug name; ⊘ Prototype drug

CONTRAINDICATIONS Hypersensitivity to formoterol; significantly worsening or acutely deteriorating asthma; severe asthmatic attacks; paradoxical bronchospasm.

CAUTIOUS USE Cardiovascular disorders (especially coronary insufficiency, cardiac arrhythmias, and hypertension), QT prolongation; convulsive disorders; thyrotoxicosis; heightened responsiveness to sympathomimetic amines; diabetes mellitus; pregnancy (category C), lactation. Safe use in children younger than 5 y has not been established.

ROUTE & DOSAGE

Treatment of Asthma, COPD, Bronchitis/Emphysema

Adult/Child (5 y or older):
Inhaled Inhale contents of 1 capsule q12h

Prevention of Exercise-Induced Asthma

Adult/Child (12 y or older): **Inhaled** Inhale contents of 1 capsule at least 15 min before exercise, do not repeat for at least 12 h

---

ADMINISTRATION

Oral Inhalation

- Remove capsule from blister IMMEDIATELY before use.
- Avoid exposing capsules to moisture.
- Give capsules only by the oral inhalation route and only by using the Aerolizer Inhaler™. Review use of the Aerolizer Inhaler in *Patient Instructions for Use* provided by manufacturer. Do not use a spacer with the Aerolizer.
- Instruct patient not to swallow capsule and not to exhale into the Aerolizer.
- Store capsules in the blister at 20°–25° C (86°–77° F).

ADVERSE EFFECTS (≥1%) Body as a Whole: *Viral infections,* chest infection, chest pain, fatigue. **CNS:** Headache, tremor, dizziness, insomnia. **GI:** Abdominal pain, dyspepsia, nausea. **Respiratory:** Pharyngitis, bronchitis, dyspnea, tonsillitis, dysphonia, fatal exacerbation of asthma. **Skin:** Rash.

INTERACTIONS Drug: Effects may be antagonized by NONSELECTIVE BETA-BLOCKERS; XANTHINES, STEROIDS; DIURETICS may potentiate hypokalemia. Avoid use with MAOIs. Use with **dronderone** is contraindicated.

PHARMACOKINETICS Absorption: Rapidly absorbed. **Onset:** 1–3 min. **Peak:** 1–3 h. **Metabolism:** Metabolized by glucuronidation in the liver. **Elimination:** 60% in urine, 33% in feces. **Half-Life:** 10 h.

NURSING IMPLICATIONS

Assessment & Drug Effects

- Monitor cardiovascular status with periodic ECG, BP, and HR determinations.
- Withhold drug and notify prescriber immediately of S&S of bronchospasm.
- Lab tests: Monitor serum potassium and blood glucose periodically.
- Monitor diabetics closely for loss of glycemic control.

Patient & Family Education

- Do not take this drug more frequently than every 12 h.
- Use a short-acting inhaler if symptoms develop between doses of formoterol.
- Seek medical care immediately if a previously effective dosage regimen fails to provide the usual response, or if swelling about the face and neck and difficulty breathing develop.
- Report any of the following immediately to the prescriber: Rash,

---

Common adverse effects in *italic*, life-threatening effects underlined; generic names in **bold**; classifications in SMALL CAPS; ♣ Canadian drug name; ⊘ Prototype drug

665

hives, palpitations, chest pain, rapid heart rate, tremor or nervousness.
- Note to diabetics: Monitor blood glucose levels carefully since hyperglycemia is a possible adverse reaction.

F

FOSAMPRENAVIR CALCIUM
(fos-am-pre'na-vir)

Lexiva
Classifications: ANTIRETROVIRAL; PROTEASE INHIBITOR
Therapeutic: PROTEASE INHIBITOR
Prototype: Saquinavir
Pregnancy Category: C

AVAILABILITY 700 mg tablet; 50 mg/mL oral suspension

ACTION & *THERAPEUTIC EFFECT*
Fosamprenavir is a prodrug rapidly converted to amprenavir. Amprenavir is an HIV-1 protease inhibitor that binds to the active site of HIV-1 protease. Binding prevents processing of viral Gag and Gag-Pol polyprotein precursors, resulting in formation of immature noninfectious viral particles. *Inhibits normal replication of the HIV virus rending the virus noninfectious.*

USES Treatment of HIV infection in combination with other antiretroviral agents.
UNLABELED USES HIV prophylaxis (occupational exposure).

CONTRAINDICATIONS Hypersensitivity to amprenavir or sulfonamide; ergot derivatives, pimozide, midazolam, triazolam; coadministration of ritonavir, flecainide, and propafenone; severe hepatic impairment; hypercholesterolemia, hypertriglyceridemia; lactation.
CAUTIOUS USE Sulfonamide allergy; mild to moderate hepatic impairment; diabetes mellitus; diabetic ketoacidosis; elderly; hemophilia; pregnancy (category C). Safe use in children less than 6 y has not been established.

ROUTE & DOSAGE

HIV Infection

Adult/Adolescent: **PO** 700 mg b.i.d. in combination with 100 mg ritonavir b.i.d. (preferred if previously on a protease inhibitor); or 1400 mg b.i.d.; or 1400 mg daily in combination with 200 mg ritonavir daily
Child (at least 6 y): **PO** 18 mg/kg with ritonavir bid. (max: 700 mg/dose)

Hepatic Impairment Dosage Adjustment

Mild to moderate impairment: Reduce dose to 700 mg b.i.d. without ritonavir; *Severe hepatic impairment:* Not recommended.

ADMINISTRATION

Oral
- Ensure that patient is not receiving drugs contraindicated with fosamprenavir.
- Store at 15°–30° C (59°–86° F) in a tightly closed container.

ADVERSE EFFECTS (≥1%) **Body as a Whole:** Fatigue. **CNS:** *Oral/perioral paresthesia,* peripheral paresthesia, depression, mood disorders. **GI:** *Nausea, vomiting, diarrhea,* abdominal pain, taste disorders, increased triglycerides, and hyperglycemia. **Skin:** *Rash,* pruritus, <u>Stevens-Johnson syndrome</u>.

INTERACTIONS Note: Interaction profile can be significantly affected by coadministration with ritonavir. Metabolite is a strong inhibitor of CYP3A4. **Drug:** Administration with amiodarone, bepridil, dihydro-ergotamine, ergotamine, flecainide, itraconazole, ketoconazole, lidocaine, midazolam, pimozide, propafenone, quinidine, triazolam, and TRICYCLIC ANTIDEPRESSANTS may cause life-threatening reactions; **rifampin, rifabutin,** ORAL CONTRACEPTIVES, **phenobarbital, phenytoin, carbamazepine** decrease **amprenavir** concentrations; **amprenavir** may increase **dihydroergotamine, ergotamine, sildenafil** concentrations and toxicity; **amprenavir** may decrease **methadone** levels; monitor INR with **warfarin;** increased risk of myopathy and rhabdomyolysis with **lovastatin, simvastatin;** may decrease antiviral effectiveness of **delavirdine** or **lopinavir/ ritonavir. Herbal:** St. John's wort may decrease antiretroviral activity.

PHARMACOKINETICS Absorption: Rapidly hydrolyzed to amprenavir (active component) by gut enzymes. **Peak:** 2.5 h. **Metabolism:** In liver by CYP3A4. **Elimination:** 14% in urine, 75% in feces. **Half-Life:** 7.7 h.

NURSING IMPLICATIONS
Assessment & Drug Effects
- Ensure that patient has provided a complete list of all prescription, nonprescription, or herbal drugs being used.
- Monitor closely diabetics for loss of glycemic control.
- Monitor males taking PDE5 inhibitors for erectile dysfunction for adverse events including hypotension, visual changes, and priapism. Report promptly.
- Lab test: Baseline and periodic LFTs; periodic lipid profile; periodic blood glucose.

Patient & Family Education
- If you miss a dose by more than 4 h, wait and take the next dose at the regular time.
- Do not take other prescription, nonprescription, or herbal drugs without consulting prescriber.
- Monitor blood glucose more often than usual if diabetic.
- To prevent pregnancy, use a barrier contraceptive in addition to hormonal contraception.

FOSCARNET
(fos′car-net)
Classification: ANTIVIRAL
Therapeutic: ANTIVIRAL
Pregnancy Category: C

AVAILABILITY 24 mg/mL injection

ACTION & *THERAPEUTIC EFFECT* Selectively inhibits the viral-specific DNA polymerases and reverse transcriptases of susceptible viruses, thus preventing elongation of the viral DNA chain. *Effective against cytomegalovirus (CMV), herpes simplex virus types 1 and 2 (HSV-1, HSV-2), human herpesvirus 6 (HHV-6), Epstein-Barr virus (EBV), and varicella-zoster virus (VZV).*

USES CMV retinitis, mucocutaneous HSV, acyclovir-resistant HSV in immunocompromised patients.
UNLABELED USES Other CMV infections, herpes zoster infections in AIDS patients.

CONTRAINDICATIONS Hypersensitivity to foscarnet; lactation.
CAUTIOUS USE Renal impairment; cardiac disease; mineral and electrolyte imbalances, seizures, older adults; pregnancy (category C). Safety and efficacy in children are not established.

Common adverse effects in *italic*, life-threatening effects underlined; generic names in **bold;** classifications in SMALL CAPS; ♣ Canadian drug name; ☯ Prototype drug

667

ROUTE & DOSAGE

CMV Retinitis

Adult: **IV Induction** 60 mg/kg q8h for 2–3 wk OR 90 mg/kg q12h for 2–3 wk; induction may be repeated if relapse occurs

Recurrent CMV Retinitis

Adult: **IV** 90–120 mg/kg/day

Acyclovir-Resistant HSV in Immunocompromised Patients

Adult: **IV** 40 mg/kg q8–12h for up to 3 wk or until lesions heal

Renal Impairment Dosage Adjustment

See package insert.

ADMINISTRATION

▪ Note: Dose **must be** adjusted for renal insufficiency. See package insert for specific dosing adjustment.

Intravenous

PREPARE: **Direct:** Given undiluted (24 mg/mL) through a central line. ▪ For peripheral infusion, dilute to 12 mg/mL with D5W or NS. ▪ Do not give other IV solution or drug through the same catheter with foscarnet.

ADMINISTER: **Direct:** Give at a constant rate not to exceed 1 mg/kg/min over the specified period of infusion with an infusion pump. ▪ Do not increase the rate of infusion or shorten the specified interval between doses. ▪ Use prepared IV solutions within 24 h.

INCOMPATIBILITIES **Solution/additive: Lactated Ringer's, acyclovir, amphotericin B, diazepam, digoxin, diphenhydramine, dobutamine, droperidol, ganciclovir, haloperidol, leucovorin, lorazepam, midazolam, pentamidine, phenytoin, prochlorperazine, promethazine, sulfamethoxazole/trimethoprim, TPN, trimetrexate, vancomycin. Y-site: Acyclovir, amphotericin B, diazepam, digoxin, diphenhydramine, dobutamine, droperidol, ganciclovir, haloperidol, leucovorin, lorazepam, midazolam, pentamidine, phenytoin, prochlorperazine, promethazine, sulfamethoxazole/ trimethoprim, trimetrexate, vancomycin.**

▪ Prehydrate and continue daily hydration with 2.5 L of NS to reduce nephrotoxicity. ▪ Store according to manufacturer's directions.

ADVERSE EFFECTS (≥1%) CV:

Thrombophlebitis if infused through a peripheral vein. **CNS:** Tremor, muscle twitching, headache, weakness, fatigue, confusion, anxiety. **Endocrine:** *Hyperphosphatemia,* hypophosphatemia, hypocalcemia. **GI:** Nausea, vomiting, diarrhea. **Urogenital:** Penile ulceration. **Hematologic:** *Anemia,* leukopenia, thrombocytopenia. **Renal:** *Nephrotoxicity* (acute renal failure, tubular necrosis). **Skin:** Fixed drug eruption, rash.

DIAGNOSTIC TEST INTERFERENCE

May cause increase or decrease in serum *calcium, phosphorus,* and *magnesium.* Decreases *Hct* and *Hgb.* Increased serum *creatinine.*

INTERACTIONS Drug: AMINOGLY-

COSIDES, **amphotericin B, vancomycin** may increase risk of nephrotoxicity. **Etidronate, pamidronate, pentamidine (IV)** may exacerbate hypocalcemia.

PHARMACOKINETICS Onset: 3–7

days. **Duration:** Relapse usually occurs 3–4 wk after end of therapy.

Common adverse effects in *italic;* life-threatening effects <u>underlined;</u> generic names in **bold;** classifications in SMALL CAPS; ♣ Canadian drug name; ⊘ Prototype drug

Distribution: 3–28% of dose may be deposited in bone; variable penetration into CSF; crosses placenta; distributed into breast milk. **Metabolism:** Not metabolized. **Elimination:** 73–94% in urine. **Half-Life:** 3–4 h.

NURSING IMPLICATIONS

Assessment & Drug Effects

- Report serum creatinine and creatinine clearance values. Drug dose will be decreased in response to decreased clearance.
- Lab tests: Periodic CBC, serum electrolytes, serum creatinine, and creatinine clearance throughout therapy.
- Monitor for electrolyte imbalances.
- Monitor for seizures and take appropriate precautions.
- Question patients regarding local irritation of the penile or vulvovaginal epithelium. If either occurs, increase hydration and better personal hygiene.

Patient & Family Education

- Report perioral tingling, numbness, and paresthesia to prescriber immediately.
- Understand that drug is not a cure for CMV retinitis; regular ophthalmologic exams are necessary.
- Note: Good hydration is important to maintain adequate output of urine.

FOSFOMYCIN TROMETHAMINE
(fos-fo-my′sin)

Monurol
Classification: ANTIBIOTIC
Therapeutic: URINARY TRACT ANTI-INFECTIVE
Prototype: Nitrofurantoin
Pregnancy Category: B

AVAILABILITY 3 g packets

ACTION & *THERAPEUTIC EFFECT*
Synthetic, broad-spectrum, bactericidal agent that blocks the first steps in bacterial cell wall synthesis. *Acts as a bactericidal agent against* Enterococcus faecalis, E. faecium, *and* Escherichia coli. *In addition, it is effective against* Klebsiella, Proteus, *and* Serratia. *Effectiveness is indicated by improvement in cystitis symptoms within 2–3 days.*

USES Treatment of uncomplicated UTIs in women.

CONTRAINDICATIONS Hypersensitivity to fosfomycin.
CAUTIOUS USE Pregnancy (category B); lactation. Safety and efficiency in children younger than 12 y are not established.

ROUTE & DOSAGE

UTI
Adult: **PO** 3 g sachet dissolved in 3–4 oz of water as a single dose given once

ADMINISTRATION

Oral
- Pour entire contents of a single dose into 3–4 oz water (not hot), stir to dissolve completely, and give immediately. Drug must not be taken in the dry form.
- Store at 15°–30° C (59°–86° F).

ADVERSE EFFECTS (≥1%) **Body as a Whole:** Pain. **CNS:** *Headache,* dizziness. **GI:** *Diarrhea,* nausea, abdominal pain, dyspepsia. **Respiratory:** Rhinitis, pharyngitis. **Urogenital:** Vaginitis, dysmenorrhea.

INTERACTIONS Drug: Metoclopramide may decrease urinary excretion of fosfomycin.

Common adverse effects in *italic*, life-threatening effects underlined; generic names in **bold;** classifications in SMALL CAPS; ♣ Canadian drug name; ⊕ Prototype drug

669

PHARMACOKINETICS Absorption: Rapidly from GI tract, 37% of dose reaches systemic circulation as free acid. **Peak Urine Concentration:** 2–4 h. **Distribution:** Not protein bound, distributed to kidneys, bladder wall, prostate, and seminal vesicles. **Elimination:** Primarily in urine. **Half-Life:** 5.7 h.

NURSING IMPLICATIONS

Assessment & Drug Effects

▪ Lab tests: Obtain urine C&S before and after therapy.

Patient & Family Education

▪ Notify prescriber if symptoms do not improve in 2–3 days.

FOSINOPRIL

(fos-in'o-pril)

Monopril

Classifications: ANGIOTENSIN-CONVERTING ENZYME (ACE) INHIBITOR; ANTIHYPERTENSIVE

Therapeutic: ANTIHYPERTENSIVE; ACE INHIBITOR

Prototype: Enalapril

Pregnancy Category: D

AVAILABILITY 10 mg, 20 mg, 40 mg tablets

ACTION & *THERAPEUTIC EFFECT* Lowers BP by interrupting conversion sequences initiated by renin that leads to formation of angiotensin II, a potent vasoconstrictor. Inhibition of ACE also leads to decreased circulating aldosterone, a secretory response to angiotensin II stimulation. *Lowers blood pressure and reduces peripheral arterial resistance (afterload) and improves cardiac output as well as activity tolerance.*

USES Mild to moderate hypertension, CHF.

CONTRAINDICATIONS Hypersensitivity to fosinopril or any other ACE inhibitor(s); history of angioedema; renal artery stenosis; pregnancy (category D), lactation.

CAUTIOUS USE Impaired kidney function, autoimmune disease; collagen-vascular disease; hepatic disease; hyperkalemia, or surgery and anesthesia; black patients; aortic stenosis or cardiomyopathy; dialysis; older adult. Safety in children is not established.

ROUTE & DOSAGE

Hypertension, CHF
Adult: **PO** 5–40 mg once/day (max: 80 mg/day)

ADMINISTRATION

Oral

▪ An initial 5 mg dose is preferred in HF patients with moderate to severe renal failure or in those who have been recently diuresed.

▪ Store at 15°–30° C (59°–86° F) and protect from moisture.

ADVERSE EFFECTS (≥1%) **CV:** Hypotension. **CNS:** Headache, fatigue, dizziness. **Endocrine:** Hyperkalemia. **GI:** Nausea, vomiting, diarrhea. **Urogenital:** Proteinuria. **Respiratory:** Cough. **Skin:** Rash.

INTERACTIONS Drug: NSAIDS may decrease antihypertensive effects of fosinopril. POTASSIUM SUPPLEMENTS, POTASSIUM-SPARING DIURETICS increase risk of hyperkalemia. ACE inhibitors may increase **lithium** levels and toxicity.

PHARMACOKINETICS Absorption: Readily absorbed from GI tract; converted to its active form, fosinoprilat, in the liver. **Peak:** 3 h. **Duration:** 24 h. **Distribution:** Approximately 90% protein bound; crosses placenta. **Metabolism:** Hydrolyzed

---

by intestinal and hepatic esterases to its active form, fosinoprilat. **Elimination:** 44% in urine, 46% in feces. **Half-Life:** 3–4 h (fosinoprilat).

NURSING IMPLICATIONS

Assessment & Drug Effects

- Monitor for at least 2 h after initial dose for first-dose hypotension, especially in salt- or volume-depleted patients.
- Monitor BP at the time of peak effectiveness, 2–6 h after dosing and at the end of the dosing interval just before next dose.
- Report diminished antihypertensive effect toward the end of the dosing interval. An inadequate trough response may be an indication for dividing the daily dose.
- Lab tests: Monitor BUN and serum creatinine. Increases may necessitate dose reduction or discontinuation of the drug. Monitor serum potassium.
- Observe for S&S of hyperkalemia (see Appendix F).

Patient & Family Education

- Discontinue fosinopril and report to prescriber any of the following: S&S of angioedema (e.g., swelling of face or extremities, difficulty breathing or swallowing); syncope; chronic, nonproductive cough.
- Maintain adequate fluid intake and avoid potassium supplements or salt substitutes unless specifically prescribed by the prescriber.
- Report vomiting or diarrhea to prescriber immediately.

FOSPHENYTOIN SODIUM
(fos-phen'i-toin)

Cerebyx
Classification: ANTI-CONVULSANT; HYDANTOIN

Therapeutic: ANTICONVULSANT
Prototype: Phenytoin
Pregnancy Category: D

AVAILABILITY 150 mg, 750 mg vials

ACTION & *THERAPEUTIC EFFECT* Prodrug of phenytoin that converts to the anticonvulsant phenytoin that modulates the sodium channels of neurons, calcium flux across neuronal membranes, and enhances the sodium–potassium ATPase activity of neurons and glial cells. *Effective as an anticonvulsant agent by preventing seizure activity.*

USES Control of generalized convulsive status epilepticus and the prevention and treatment of seizures during neurosurgery, or as a parenteral short-term substitute for oral phenytoin.
UNLABELED USES Antiarrhythmic agent especially in treatment of digitalis-induced arrhythmia; treatment of trigeminal neuralgia (tic douloureux).

CONTRAINDICATIONS Hypersensitivity to hydantoin products, rash, seizures due to hypoglycemia, sinus bradycardia, complete or incomplete heart block; Adams-Stokes syndrome; pregnancy (category D).
CAUTIOUS USE Impaired liver or kidney function, alcoholism, hypotension, heart block, bradycardia, severe CAD, diabetes mellitus, hyperglycemia, respiratory depression, acute intermittent porphyria; lactation.

ROUTE & DOSAGE

Status Epilepticus

Adult: **IV Loading Dose** 15–20 mg PE/kg (PE = phenytoin sodium equivalents) administered at 100–150 mg PE/min

---

Common adverse effects in *italic*, life-threatening effects underlined; generic names in **bold**; classifications in SMALL CAPS; ♣ Canadian drug name; ❂ Prototype drug

IV Maintenance Dose 4–6 mg PE/kg/day

Substitution for Oral Phenytoin Therapy

Adult: **IV/IM** Substitute fosphenytoin at the same total daily dose in mg PE as the oral dose at a rate of infusion not greater than 150 mg PE/min

ADMINISTRATION

▪ Note: All dosing is expressed in phenytoin sodium equivalents (PE) to avoid the need to calculate molecular weight adjustments between fosphenytoin and phenytoin sodium doses. **Always** prescribe and fill fosphenytoin in PE units.

Intramuscular

▪ Follow institutional policy regarding maximum volume to inject into one IM site.

Intravenous

PREPARE: *Direct:* Dilute in D5W or NS to a concentration of 1.5–25 mg PE/mL.
ADMINISTER: *Direct:* Give 100–150 mg PE/min. Do not administer at a rate greater than 150 mg PE/min.
INCOMPATIBILITIES Y-site: **Fenoldopam, midazolam.**

▪ Store at 2°–8° C (36°–46° F); may store at room temperature not to exceed 48 h.

ADVERSE EFFECTS (≥1%) CNS:
Usually dose related. Paresthesia, tinnitus, *nystagmus, dizziness, somnolence, drowsiness,* ataxia, mental confusion, tremors, insomnia, headache, seizures, increased reflexes, dysarthria, intracranial hypertension. **CV:** Bradycardia, tachycardia, asystole, hypotension, hypertension, <u>cardiovascular collapse, cardiac arrest,</u> heart block, ventricular fibrillation, phlebitis. **Special Senses:** Photophobia, conjunctivitis, diplopia, blurred vision. **GI:** *Gingival hyperplasia,* nausea, vomiting, constipation, epigastric pain, dysphagia, loss of taste, weight loss, hepatitis, liver necrosis. **Hematologic:** <u>Thrombocytopenia</u>, <u>leukopenia</u>, leukocytosis, <u>agranulocytosis</u>, pancytopenia, eosinophilia; megaloblastic, hemolytic, or <u>aplastic anemias</u>. **Metabolic:** Fever, hyperglycemia, glycosuria, weight gain, edema, transient increase in serum thyrotropic (TSH) level, hyperkalemia, osteomalacia or rickets associated with hypocalcemia and elevated alkaline phosphatase activity. **Skin:** Alopecia, hirsutism (especially in young female); rash: Scarlatiniform, maculopapular, urticarial, morbilliform (may be fatal); bullous, exfoliative, or purpuric dermatitis; Stevens-Johnson syndrome, <u>toxic epidermal necrolysis</u>, keratosis, neonatal hemorrhage, *pruritus.* **Urogenital:** Acute renal failure, Peyronie's disease. **Respiratory:** Acute pneumonitis, pulmonary fibrosis. **Musculoskeletal:** Periarteritis nodosum, acute systemic lupus erythematosus, craniofacial abnormalities (with enlargement of lips). **Other:** Lymphadenopathy, injection site pain, chills.

DIAGNOSTIC TEST INTERFERENCE
Fosphenytoin may produce lower than normal values for ***dexamethasone*** or ***metyrapone*** tests; may increase serum levels of ***glucose, BSP,*** and ***alkaline phosphatase*** and may decrease ***PBI*** and ***urinary steroid*** levels.

INTERACTIONS Drug: Alcohol
decreases effects; OTHER ANTICONVULSANTS may increase or decrease fosphenytoin levels; fosphenytoin increases metabolism of CORTICOSTEROIDS, ORAL ANTICOAGULANTS, and

ORAL CONTRACEPTIVES, decreasing their effectiveness; **amiodarone, chloramphenicol, omeprazole** increase fosphenytoin levels; antituberculosis agents, **voriconazole** decrease fosphenytoin levels. **Food:** Folic acid, calcium, vitamin D absorption may be decreased by fosphenytoin; fosphenytoin absorption may be decreased by enteral nutrition supplements. **Herbal: Ginkgo** may decrease anticonvulsant effectiveness.

PHARMACOKINETICS Absorption: Completely absorbed after IM administration. **Peak:** 30 min IM. **Distribution:** 95–99% bound to plasma proteins, displaces phenytoin from protein binding sites; crosses placenta, small amount in breast milk. **Metabolism:** Converted to phenytoin by phosphatases; phenytoin is oxidized in liver to inactive metabolites. **Elimination:** Half-life 15 min to convert fosphenytoin to phenytoin, 22 h phenytoin; phenytoin metabolites excreted in urine.

NURSING IMPLICATIONS

Note: See **phenytoin** for additional nursing implications.

Assessment & Drug Effects
- Monitor ECG, BP, and respiratory function continuously during and for 10–20 min after infusion.
- Discontinue infusion and notify prescriber if rash appears. Be prepared to substitute alternative therapy rapidly to prevent withdrawal-precipitated seizures.
- Lab tests: Monitor CBC with differential, platelet count, serum electrolytes, and blood glucose.
- Allow at least 2 h after IV infusion and 4 h after IM injection before monitoring total plasma phenytoin concentration.

- Monitor diabetics for loss of glycemic control.
- Monitor carefully for adverse effects, especially in patients with renal or hepatic disease or hypoalbuminemia.

Patient & Family Education
- Be aware of potential adverse effects. Itching, burning, tingling, or paresthesia are common during and for some time following IV infusion.

FROVATRIPTAN
(fro-va-trip'tan)
Frova
Classification: SEROTONIN 5-HT$_1$ RECEPTOR AGONIST
Therapeutic: ANTIMIGRAINE
Prototype: Sumatriptan
Pregnancy Category: C

AVAILABILITY 2.5 mg tablets

ACTION & *THERAPEUTIC EFFECT* Selective agonist that binds with high affinity to 5-HT$_{1D}$, 5-HT$_{1B}$, 5-HT$_{1F}$ serotonin receptors, which are found on extracerebral and intracranial blood vessels, and on other structures in the CNS. This results in vasoconstriction and agonist effects on nerve terminals in the trigeminal system. *Activation of 5-HT$_1$ receptors results in constriction of cranial vessels that become dilated during a migraine attack, and reduced signal transmission in the pain pathways.*

USES Treatment of migraine headache with or without aura.

CONTRAINDICATIONS Hypersensitivity to frovatriptan; significant cardiovascular disease such as ischemic heart disease, coronary artery vasospasms, peripheral vascular disease, history of cerebrovascular events, or

---

Common adverse effects in *italic*, life-threatening effects <u>underlined</u>; generic names in **bold**; classifications in SMALL CAPS; ♣ Canadian drug name; ⊘ Prototype drug

673

F

uncontrolled hypertension; within 24 h of receiving another 5-HT$_1$ agonist or an ergotamine-containing or ergot-type drug; basilar or hemiplegic migraine.

CAUTIOUS USE Significant risk factors for coronary artery disease unless a cardiac evaluation has been done; hypertension; risk factors for cerebrovascular accident; impaired liver or kidney function; pregnancy (category C), lactation. Safe use in children younger than 18 y has not been established.

ROUTE & DOSAGE

Migraine Headache
Adult: **PO** 2.5 mg. If headache returns, may repeat after at least 2 h (max: 7.5 mg/24 h).

ADMINISTRATION

Oral

- Do not give within 24 h of an ergot-containing drug.
- Administer any time after symptoms of migraine appear.
- Do not administer a second dose without consulting the prescriber for any attack during which the FIRST dose did NOT work.
- Give a second dose if headache was relieved by first dose but symptoms return; however, wait at least 2 h after the first dose before giving a second dose.
- Do not give more than two doses in 24 h.
- Store at 15°–30° C (59°–86° F).

ADVERSE EFFECTS (≥1%) **Body as a Whole:** Fatigue, hot or cold sensation, flushing. **CNS:** Dizziness, headache, paresthesia, somnolence, insomnia, anxiety. **CV:** Chest pain, palpitation. **GI:** Dyspepsia, nausea, vomiting, diarrhea, dry mouth. **Musculoskeletal:** Skeletal pain. **Special Senses:** Abnormal vision. **Skin:** Sweating.

INTERACTIONS Drug: Dihydroergotamine, methysergide, other 5-HT$_1$ AGONISTS may cause prolonged vasospastic reactions; SSRIS, **sibutramine** have rarely caused weakness, hyperreflexia, and incoordination; MAOIS should not be used with 5-HT$_1$ AGONISTS. **Herbal: Gingko, ginseng, echinacea, St. John's wort** may increase triptan toxicity.

PHARMACOKINETICS Absorption: 20–30% bioavailability. **Peak:** 2–4 h. **Distribution:** 15% protein bound. **Metabolism:** In liver by CYP1A2. **Elimination:** 30% renally, 60% in feces. **Half-Life:** 26 h.

NURSING IMPLICATIONS

Assessment & Drug Effects

- Monitor cardiovascular status carefully following first dose in patients at relatively high risk for coronary artery disease (e.g., post-menopausal women, men older than 40 y, persons with known CAD risk factors), or who have coronary artery vasospasms.
- Report to prescriber immediately chest pain or tightness in chest or throat that is severe, or does not quickly resolve following a dose of frovatriptan.
- Pain relief usually begins within 10 min of ingestion, with complete relief in approximately 65% of all patients within 2 h.
- Monitor BP, especially in those being treated for hypertension.

Patient & Family Education

- Review patient information leaflet provided by the manufacturer carefully.
- Notify prescriber immediately if symptoms of severe angina (e.g., severe or persistent pain or tightness in chest, back, neck, or throat) or hypersensitivity (e.g., wheezing, facial swelling, skin rash, itching, or hives) occur.   .

Common adverse effects in *italic*, life-threatening effects <u>underlined</u>; generic names in **bold**; classifications in SMALL CAPS; ✚ Canadian drug name; ◉ Prototype drug

- Do not take any other serotonin receptor agonist (e.g., Imitrex, Maxalt, Zomig, Amerge) within 24 h of taking frovatriptan.
- Report any other adverse effects (e.g., tingling, flushing, dizziness) at next prescriber visit.

FULVESTRANT
(ful-ves'trant)

Faslodex

Classifications: ANTINEOPLASTIC; ANTIESTROGEN

Therapeutic: ANTINEOPLASTIC; ANTI-ESTROGEN

Prototype: Tamoxifen citrate
Pregnancy Category: D

AVAILABILITY 50 mg/mL

ACTION & *THERAPEUTIC EFFECT*
Fulvestrant is an estrogen receptor antagonist that selectively binds to the estrogen receptors (ER) of breast cancer cells. Estrogen stimulates the tumor growth of estrogen-sensitive breast tissue cancer cells in postmenopausal women. *In postmenopausal women, many breast cancers have postitive estrogen receptors (ERs), and the growth of these tumors is stimulated by estrogen. Therefore, fulvestrant decreases estrogen-sensitive breast tissue tumor growth.*

USES Treatment of hormone receptor-positive metastatic breast cancer in postmenopausal women with disease progression following antiestrogen therapy.

CONTRAINDICATIONS Hypersensitivity to fulvestrant; pregnancy (category D); lactation.
CAUTIOUS USE Moderate to severe liver impairment; biliary disease; coagulopathy; anticoagulant therapy. Safety and effectiveness in children not established.

ROUTE & DOSAGE

Metastatic Breast Cancer
Adult: **IM** 500 mg once/mo

ADMINISTRATION

Intramuscular

- Break the seal of the white plastic cover on the syringe luer connector to remove the cover with the attached rubber tip cap. Twist to lock the needle to the luer connector. Remove excess gas from the syringe (a small gas bubble may remain).
- Administer slowly in the buttock.
- Immediately activate needle protection device upon withdrawal from patient by pushing lever arm completely forward until needle tip is fully covered.
- Store in a refrigerator, 2°–8° C (36°–46° F) in original container.

ADVERSE EFFECTS (≥1%) **Body as a Whole:** *Asthenia, pain, injection site pain,* flu-like syndrome, fever, peripheral edema. **CNS:** Dizziness, insomnia, paresthesia, depression, anxiety. **CV:** *Vasodilation.* **GI:** *Nausea, vomiting, constipation, diarrhea,* anorexia. **Hematologic:** Anemia. **Musculoskeletal:** *Bone pain,* arthritis. **Respiratory:** *Pharyngitis, dyspnea, cough.* **Skin:** Rash, sweating.

PHARMACOKINETICS Peak: 7 days. **Duration:** 1 mo. **Distribution:** 99% protein bound. **Metabolism:** In liver via CYP3A4. **Elimination:** 90% in feces. **Half-Life:** 40 days.

NURSING IMPLICATIONS

Assessment & Drug Effects

- Monitor for S&S of tumor progression.
- Lab tests: Monitor periodic CBC with differential.

Common adverse effects in *italic*, life-threatening effects <u>underlined</u>; generic names in **bold**; classifications in SMALL CAPS; ♣ Canadian drug name; ⊘ Prototype drug

675

F

Patient & Family Education

- Use two methods of contraception while taking this drug. Immediately notify prescriber if you think you are pregnant.
- Report vaginal bleeding to prescriber. Understand the possibility of drug-induced menstrual irregularities before starting treatment.

√FUROSEMIDE ⊕

(fur-oh'se-mide)

Fumide ♣, Furomide ♣, Lasix, Luramide ♣

Classifications: ELECTROLYTIC AND WATER BALANCE AGENT; LOOP DIURETIC; ANTIHYPERTENSIVE
Therapeutic: LOOP DIURETIC; ANTIHYPERTENSIVE
Pregnancy Category: C

AVAILABILITY 20 mg, 40 mg, 80 mg tablets; 10 mg/mL, 40 mg/5 mL oral solution; 10 mg/mL injection

ACTION & *THERAPEUTIC EFFECT*
Rapid-acting potent sulfonamide "loop" diuretic and antihypertensive. Inhibits reabsorption of sodium and chloride primarily in loop of Henle and also in proximal and distal renal tubules. *An antihypertensive that decreases edema and intravascular volume, which lowers blood pressure.*

USES Treatment of edema associated with CHF, cirrhosis of liver, and kidney disease, including nephrotic syndrome. May be used for management of hypertension, alone or in combination with other antihypertensive agents, and for treatment of hypercalcemia. Has been used concomitantly with mannitol for treatment of severe cerebral edema, particularly in meningitis.

CONTRAINDICATIONS History of hypersensitivity to furosemide or sulfonamides; increasing oliguria, anuria, fluid and electrolyte depletion states; hepatic coma; preeclampsia, eclampsia.

CAUTIOUS USE Infants, older adults; hepatic disease; hepatic cirrhosis; renal disease, nephrotic syndrome; cardiogenic shock associated with acute MI; ventricular arrhythmias, CHF, diarrhea; history of SLE, history of gout; diabetes mellitus; pregnancy (category C), lactation.

ROUTE & DOSAGE

Edema

Adult: **PO** 20–80 mg in 1 or more divided doses up to 600 mg/day if needed **IV/IM** 20–40 mg in 1 or more divided doses up to 600 mg/day
Child: **PO** 2 mg/kg, may be increased by 1–2 mg/kg q6–8h (max: 6 mg/kg/dose) **IV/IM** 1–2 mg/kg, may be increased by 1 mg/kg q2h if needed (max: 6 mg/kg/dose)
Neonate: **PO** 1–4 mg/kg q12–24h **IV/IM** 1 mg/kg q12–24h

Hypertension

Adult: **PO** 10–40 mg b.i.d. (max: 480 mg/day)

ADMINISTRATION

Oral

- Give with food or milk to reduce possibility of gastric irritation.
- Schedule doses to avoid sleep disturbance (e.g., a single dose is generally given in the morning; twice-a-day doses at 8 a.m. and 2 p.m.).
- Store tablets at controlled room temperature, preferably at 15°–30° C (59°–86° F) unless otherwise directed. Protect from light.

- Store oral solution in refrigerator, preferably at 2°–8° C (36°–46° F). Protect from light and freezing.

Intramuscular
- Protect syringes from light once they are removed from package.
- Discard yellow or otherwise discolored injection solutions.

Intravenous
Note: Verify correct IV concentration and rate of infusion/injection with prescriber before administration to infants or children.

PREPARE: **Direct:** Give undiluted.
ADMINISTER: **Direct:** Give undiluted at a rate of 20 mg or a fraction thereof over 1 min. ▪ With high doses a rate of 4 mg/min is recommended to decrease risk of ototoxicity.
INCOMPATIBILITIES **Solution/additive: Amiodarone, buprenorphine, chlorpromazine, diazepam, dobutamine, erythromycin, fructose, gentamicin, isoproterenol, meperidine, metoclopramide, milrinone, netilmicin, pancuronium, papaveretum, prochlorperazine, promethazine, quinidine, thiamine. Y-site: Amrinone, amsacrine, azithromycin, chlorpromazine, ciprofloxacin, clarithromycin, diltiazem, dobutamine, dopamine, doxorubicin, droperidol, esmolol, fenoldopam, filgrastim, fluconazole, gemcitabine, gentamicin, hydralazine, idarubicin, labetalol, lansoprazole, levofloxacin, meperidine, methocarbamol, metoclopramide, midazolam, milrinone, morphine, netilmicin, nicardipine, ondansetron, quinidine, tetracycline, thiopental, tobramycin, vecuronium, vinblastine, vincristine, vinorelbine, TPN.**

- Use infusion solutions within 24 h.
- Store parenteral solution at controlled room temperature, preferably at 15°–30° C (59°–86° F) unless otherwise directed. Protect from light.

ADVERSE EFFECTS (≥1%) **CV:** Postural hypotension, dizziness with excessive diuresis, acute hypotensive episodes, <u>circulatory collapse</u>. **Metabolic:** Hypovolemia, dehydration, hyponatremia, *hypokalemia,* hypochloremia, metabolic alkalosis, hypomagnesemia, hypocalcemia (tetany), hyperglycemia, glycosuria, elevated BUN, hyperuricemia. **GI:** Nausea, vomiting, oral and gastric burning, anorexia, diarrhea, constipation, abdominal cramping, acute pancreatitis, jaundice. **Urogenital:** Allergic interstitial nephritis, irreversible renal failure, urinary frequency. **Hematologic:** Anemia, <u>leukopenia</u>, thrombocytopenic purpura; <u>aplastic anemia,</u> <u>agranulocytosis</u> (rare). **Special Senses:** Tinnitus, vertigo, feeling of fullness in ears, hearing loss (rarely permanent), blurred vision. **Skin:** Pruritus, urticaria, exfoliative dermatitis, purpura, photosensitivity, porphyria cutanea tarda, necrotizing angiitis (vasculitis). **Body as a Whole:** Increased perspiration; paresthesias; activation of SLE, muscle spasms, weakness; thrombophlebitis, pain at IM injection site.

DIAGNOSTIC TEST INTERFERENCE Furosemide may cause elevations in *BUN, serum amylase, cholesterol, triglycerides, uric acid* and *blood glucose* levels, and may decrease *serum calcium, magnesium, potassium,* and *sodium* levels.

INTERACTIONS Drug: OTHER DIURETICS enhance diuretic effects; with **digoxin** increased risk of toxicity because of hypokalemia;

Common adverse effects in *italic,* life-threatening effects <u>underlined</u>; generic names in **bold;** classifications in SMALL CAPS; ✦ Canadian drug name; ❂ Prototype drug

677

F

NON-DEPOLARIZING NEUROMUSCULAR BLOCKING AGENTS (e.g., **tubocurarine**) prolong neuromuscular blockage; CORTICOSTEROIDS, **amphotericin B** potentiate hypokalemia; decreased **lithium** elimination and increased toxicity; SULFONYLUREAS, **insulin** blunt hypoglycemic effects; NSAIDs may attenuate diuretic effects.

PHARMACOKINETICS Absorption: 60% PO dose from GI tract. **Peak:** 60–70 min PO; 20–60 min IV. **Onset:** 30–60 min PO; 5 min IV. **Duration:** 2 h. **Distribution:** Crosses placenta. **Metabolism:** Small amount in liver. **Elimination:** Rapidly in urine; 50% of oral dose and 80% of IV dose excreted within 24 h; excreted in breast milk. **Half-Life:** 30 min.

NURSING IMPLICATIONS

Assessment & Drug Effects

- Observe patients receiving parenteral drug carefully; closely monitor BP and vital signs. Sudden death from cardiac arrest has been reported.
- Monitor for S&S of hypokalemia (see Appendix F).
- Monitor BP during periods of diuresis and through period of dosage adjustment.
- Observe older adults closely during period of brisk diuresis. Sudden alteration in fluid and electrolyte balance may precipitate significant adverse reactions. Report symptoms to prescriber.
- Lab tests: Obtain frequent blood count, serum and urine electrolytes, CO_2, BUN, blood sugar, and uric acid values during first few months of therapy and periodically thereafter.
- Monitor urine and blood glucose and HbA1C closely in diabetics and patients with decompensated hepatic cirrhosis. Drug may cause hyperglycemia.

- Monitor I&O ratio and pattern. Report decrease or unusual increase in output. Excessive diuresis can result in dehydration and hypovolemia, circulatory collapse, and hypotension. Weigh patient daily under standard conditions.

Patient & Family Education

- Consult prescriber regarding allowable salt and fluid intake.
- Ingest potassium-rich foods daily (e.g., bananas, oranges, peaches, dried dates) to reduce or prevent potassium depletion.
- Learn S&S of hypokalemia (see Appendix F). Report muscle cramps or weakness to prescriber.
- Make position changes slowly because high doses of antihypertensive drugs taken concurrently may produce episodes of dizziness or imbalance.
- Avoid replacing fluid losses with large amounts of water.
- Avoid prolonged exposure to direct sun.

┃GABAPENTIN ⊙

(gab-a-pen'tin)
Gralise, Horizant, Neurontin
Classifications: ANTICONVULSANT; GABA ANALOG
Therapeutic: ANTICONVULSANT
Pregnancy Category: C

AVAILABILITY 100 mg, 300 mg, 400 mg capsules; 100 mg, 300 mg, 400 mg, 600 mg, 800 mg tablets; 300 mg, 600 mg extended release tablet

ACTION & *THERAPEUTIC EFFECT*

Gabapentin is a GABA neurotransmitter analog; however, it does not inhibit GABA uptake or degradation. It appears to interact with GABA cortical neurons, but its relationship to functional activity as an anticonvulsant is unknown. *Used in conjunction*

Common adverse effects in *italic*, life-threatening effects underlined; generic names in **bold**; classifications in SMALL CAPS; ♣ Canadian drug name; ⊙ Prototype drug

with other anticonvulsants to control certain types of seizures in patients with epilepsy. Effective in controlling painful neuropathies.

USES Adjunctive therapy for partial seizures with or without secondary generalization in adults, post-herpetic neuralgia, restless leg syndrome.

UNLABELED USES Add-on therapy for generalized seizures, peripheral neuropathy, migraine prophylaxis.

CONTRAINDICATIONS Hypersensitivity to gabapentin; suicidal ideations; lactation. Safety and efficacy in infants and children younger than 3 y are not established.

CAUTIOUS USE Status epilepticus, renal impairment, history of suicidal tendencies; psychiatric disorders; older adults; pregnancy (category C); children.

ROUTE & DOSAGE

Adjunctive Therapy for Seizure Disorder

Adult/Child (older than 12 y):
PO Start 300 mg on day 1, 300 mg b.i.d. on day 2, 300 mg t.i.d. on day 3, and continue to increase over a week to an initial total dose of 400 mg t.i.d. (1200 mg/day); may increase to 1800–2400 mg/ day depending on response (most patients receive 900–1800 mg/ day in 3 divided doses) 400 mg t.i.d. (1200 mg/day)
Child (3–12 y): **PO** Start 10–15 mg/kg/day in 3 divided doses, titrate q3days to target dose of 40 mg/kg/day in pts 3–4 y or 25–35 mg/kg/day in pts 5 y or older in 3 divided doses

Post-Herpetic Neuralgia

Adult: **PO** Start 300 mg day 1, 300 mg b.i.d. day 2, and 300 mg t.i.d. day 3; may increase up to 600 mg t.i.d. if needed
Gralise only: 300 mg day 1, 600 mg day 2, 900 mg days 3–6, 1200 mg days 7–10, 1500 mg days 11–14, 1800 mg day 15 and after

Restless Leg Syndrome (Horizant only)

Adult: **PO** 600 mg qd at 5 p.m.

Renal Impairment Dosage Adjustment

CrCl greater than 60 mL/min: 400 mg t.i.d.; *30–60 mL/min:* 300 mg b.i.d.; *15–30 mL/min:* 300 mg daily; *less than 15 mL/ min:* 300 mg every other day

Hemodialysis Dosage Adjustment
200–300 mg following dialysis

ADMINISTRATION

Oral
- Ensure that extend release tablet is swallowed whole, and not crushed or chewed.
- Separate doses of gabapentin and antacids by 2 h.
- Withdraw drug gradually over 1 wk; abrupt discontinuation may cause status epilepticus.
- Store at 15°–30° C (59°–86° F); protect from heat, moisture, and direct light.

ADVERSE EFFECTS (≥1%) **CNS:** *Drowsiness, fatigue,* dizziness, tremor, slurred speech, impaired concentration, headache, increased frequency of partial seizures. **Endocrine:** Weight gain. **GI:** Nausea, gastric upset, vomiting. **Special Senses:** Blurred vision, nystagmus. **Skin:** Rash, eczema.

INTERACTIONS Drug: Increase in **phenytoin** levels at higher doses (300–600 mg/day gabapentin). Does

Common adverse effects in *italic*, life-threatening effects underlined; generic names in **bold**; classifications in SMALL CAPS; ♣ Canadian drug name; ⊘ Prototype drug

679

not affect serum levels of other AN-
TICONVULSANTS. ANTACIDS reduce ab-
sorption of gabapentin. **Herbal:**
Ginkgo may decrease effectiveness.

PHARMACOKINETICS Absorption:
50–60% from GI tract. **Peak:** Peak
level 1–3 h; peak effect 2–4 wk.
Distribution: Crosses the blood–
brain barrier; readily passes into
cerebrospinal fluid; not bound to
plasma proteins; highest concentrations found in pancreas and kidneys. **Metabolism:** Does not appear
to be metabolized. **Elimination:** 76–
81% unchanged in 96 h; 10–23%
recovered in feces. **Half-Life:** 5–6 h.

NURSING IMPLICATIONS

Assessment & Drug Effects

- Monitor for therapeutic effectiveness;
  may not occur until several weeks
  following initiation of therapy.
- In those treated for seizure disorders, assess frequency of seizures:
  In rare cases, the drug has increased the frequency of partial
  seizures.
- Monitor for and report dizziness,
  somnolence, or other signs of
  CNS depression. Assess safety: Vision, concentration, and coordination impairment increase the
  risk for injury.
- Monitor for changes in behavior
  that may be indicative of suicidal
  ideation.

Patient & Family Education

- Learn potential adverse effects of
  drug.
- Notify prescriber immediately if
  any of the following occur: Increased seizure frequency, visual
  changes, unusual bruising or
  bleeding.
- Do not drive or engage in other
  potentially hazardous activities
  until response to drug is known.

- Do not abruptly discontinue use
  of drug; do not take drug within
  2 h of an antacid.

GALANTAMINE HYDROBROMIDE

(ga-lan'ta-meen)

Razadyne, Razadyne ER
Classifications: CENTRALLY ACTING
CHOLINERGIC; CHOLINESTERASE IN-
HIBITOR; ANTIDEMENTIA
Therapeutic: ANTIALZHEIMER'S; AN-
TIDEMENTIA
Prototype: Donezepril HCl
Pregnancy Category: B

AVAILABILITY 4 mg, 8 mg, 12 mg
tablets; 8 mg, 16 mg, 24 mg extended release capsules; 4 mg/mL oral
solution

ACTION & THERAPEUTIC EFFECT
Competitive and reversible inhibitor of acetylcholinesterase, which
is the enzyme responsible for the
hydrolysis (breakdown) of the
neurotransmitter, acetylcholine.
The cholinergic system is used in
processing needed for attention,
memory as well as modulation of
excitatory neurotransmission. *In
Alzheimer's disease cholinesterase
inhibitors are designed to offset loss
of presynaptic cholinergic function, slowing decline of memory
and maintaining ability to perform
functions of daily living.*

USES Treatment of mild to moderate dementia of Alzheimer's type.
UNLABELED USES Vascular dementia.

CONTRAINDICATIONS Hypersensitivity to galantamine; lactation, or
in children.
CAUTIOUS USE Bradycardia, heart
block or other cardiac conduction
disorders; asthma, COPD; potential

bladder outflow obstruction; a history of seizures or GI bleeding; pregnancy (category B).

ROUTE & DOSAGE

Alzheimer's Disease

Adult: **PO** Initiate with 4 mg b.i.d. times at least 4 wks, if tolerated may increase by 4 mg b.i.d. q4wk to target dose of 12 mg b.i.d. (8–16 mg b.i.d.)

Hepatic Impairment Dosage Adjustment

Not recommended with severe hepatic impairment

Renal Impairment Dosage Adjustment

CrCl less than 9 mL/min: Not recommended

ADMINISTRATION

Oral

- Give with meals (breakfast and dinner) to reduce the risk of nausea.
- Extended release capsules should be swallowed whole and not crushed or chewed.
- Make increases in dosage increments at 4-wk intervals.
- If drug is interrupted for several days or more, restart at the lowest dose and gradually increase to the current dose.
- Store at 15°–30° C (59°–86° F).

ADVERSE EFFECTS (≥1%) **Body as a Whole:** Weight loss, fatigue, rhinitis, syncope, malaise, asthenia, fever. **CNS:** Dizziness, headache, depression, insomnia, somnolence, tremor. **CV:** Bradycardia, chest pain. **GI:** *Nausea, vomiting,* diarrhea, anorexia, abdominal pain, dyspepsia, flatulence. **Hematologic:** Anemia. **Urogenital:** UTI, hematuria, incontinence. **Nervous System:** Tinnitus, leg cramps. **Other:** Increased mortality in patients with mild cognitive impairment.

INTERACTIONS Drug: Additive effects with other CHOLINESTERASE INHIBITORS (e.g., **succinylcholine, bethanecol**); **cimetidine, erythromycin, ketoconazole, paroxetine** may increase levels and toxicity.

PHARMACOKINETICS Absorption: Rapidly and completely. **Peak:** 1 h. **Distribution:** Mainly distributes to red blood cells. **Metabolism:** In liver by CYP2D6 and CYP3A4. **Elimination:** 95% in urine. **Half-Life:** 7 h (4.4–10 h).

NURSING IMPLICATIONS

Assessment & Drug Effects

- Monitor cardiovascular status including baseline and periodic EKG and BP readings. Assess for postural hypotension.
- Monitor I&O rates and pattern for urinary incontinence or urinary retention.
- Monitor appetite and food intake. Weigh weekly and report significant weight loss.
- Lab tests: Baseline ALT/AST, BUN and creatinine; periodic blood glucose, alkaline phosphatase, urinalysis, stool for occult blood.

Patient & Family Education

- Report any of the following to a health care provider immediately: Loss of weight, urinary retention, chest pain, palpitations, difficulty breathing, fainting, dark stools, blood in the urine.

GALLIUM NITRATE

(gal′li-um)

Ganite

Classification: BONE RESORPTION INHIBITOR

Common adverse effects in *italic*, life-threatening effects <u>underlined</u>; generic names in **bold**; classifications in SMALL CAPS; ♣ Canadian drug name; ⊕ Prototype drug

681

Therapeutic: BONE RESORPTION IN-
HIBITOR; CALCIUM REGULATOR
Pregnancy Category: C

AVAILABILITY 25 mg/mL injection

ACTION & *THERAPEUTIC EFFECT*

Exerts a hypocalcemic effect by inhibition of calcium resorption from
bone, possibly by reducing the rate
of bone metabolism. *Lowers calcium serum levels by inhibiting calcium resorption from bone.*

USES Hypercalcemia of malignancy.
UNLABELED USES Paget's disease,
painful bone metastases, adjuvant
therapy for bladder cancer and
lymphomas.

CONTRAINDICATIONS Severe renal impairment (serum creatinine
greater than 2.5 mg/dL); hypovolemia; hypocalcemia; lactation.
CAUTIOUS USE Renal function impairment; pregnancy (category C).
Safety and efficacy in children are
not established.

ROUTE & DOSAGE

Hypercalcemia
Adult: **IV** 100–200 mg/m²/day ×
5 days
Bone Metastases
Adult: **IV** 200 mg/m²/day × 7
days

ADMINISTRATION

Intravenous
Hydrate patient with oral or IV
NS to produce a urine output of
2 L/day; maintain adequate hydration throughout treatment.

PREPARE: **Continuous:** Dilute
each daily dose with 1000 mL NS
(preferred if not contraindicated)
or D5W.

ADMINISTER: **Continuous:** Infuse
over 24 h taking care to avoid rapid infusion. ▪ Control rate with infusion pump or microdrip device.
INCOMPATIBILITIES **Y-site: Cisplatin, cytarabine, doxorubicin,
haloperidol, hydromor-
phone.**

▪ Do not administer concurrently
with potentially nephrotoxic
drugs.

▪ Store IV solutions at 15°–30° C
(59°–86° F) for 48 h or refrigerated
at 2°–8° C (36°–46° F) for 7 days.
Discard unused portions.

ADVERSE EFFECTS (≥1%) **CNS:**
Fatigue, paresthesia, hyperthermia.
CV: Hypotension. **GI:** *Nausea, vomiting, diarrhea,* anorexia, stomatitis, dysgeusia, mucositis, metallic
taste. **Hematologic:** Anemia, granulocytopenia, thrombocytopenia.
Metabolic: Hypercalcemia, hypophosphatemia, hypomagnesemia.
Urogenital: Nephrotoxicity, acute
renal failure. **Other:** Optic neuritis,
maculopapular rash.

INTERACTIONS Drug: AMINOGLYCO-
SIDES, **amphotericin B, vancomycin**
increase the risk of nephrotoxicity.

PHARMACOKINETICS Onset: 48 h.
Duration: 4–14 days after discontinuation of therapy. **Distribution:**
Concentrates in tumors; distributed
to lung, skin, muscle, and heart with
high concentrations in liver and
kidney; not known if crosses placenta or is distributed into breast
milk. **Metabolism:** Not metabolized.
Elimination: 35–71% via kidneys
within first 24 h. **Half-Life:** 25–111 h.

NURSING IMPLICATIONS
Assessment & Drug Effects
▪ Ensure that patient is well hydrated. A urine output of 2 L/day is
desirable throughout treatment.

G

Common adverse effects in *italic*, life-threatening effects underlined; generic names
in **bold**; classifications in SMALL CAPS; ♣ Canadian drug name; ⊙ Prototype drug

- Lab tests: Monitor BUN and serum creatinine throughout therapy. Notify prescriber if serum creatinine exceeds 2.5 mg/dL; discontinue drug if this occurs. Also, check baseline serum calcium and serum phosphorus; follow with assessments daily and twice weekly, respectively.
- Note: If hypocalcemia occurs, withhold gallium nitrate and notify prescriber.

Patient & Family Education
- Learn S&S of hypocalcemia (see Appendix F). Notify prescriber immediately if any occur.

GANCICLOVIR

(gan-ci'clo-vir)

Cytovene, Zirgan
Classification: ANTIVIRAL; PURINE NUCLEOSIDE
Therapeutic: ANTIVIRAL
Prototype: Acyclovir
Pregnancy Category: C

AVAILABILITY 250 mg, 500 mg capsules; 500 mg powder for injection; 0.15% opthalmic gel

ACTION & *THERAPEUTIC EFFECT*
Ganciclovir is a synthetic purine nucleoside analog that is an antiviral drug active against cytomegalovirus (CMV). It inhibits the replication of CMV DNA. *Sensitive human viruses include CMV, herpes simplex virus-1 and -2 (HSV-1, HSV-2), Epstein-Barr virus, and varicella-zoster virus.*

USES CMV retinitis, prophylaxis and treatment of systemic CMV infections in immunocompromised patients including HIV-positive and transplant patients; dendritic keratitis.

CONTRAINDICATIONS Hypersensitivity to ganciclovir or acyclovir; infection; severe thrombocytopenia.
CAUTIOUS USE Valacyclovir or penciclovir hypersensitivity; renal impairment, bone marrow suppression; chemotherapy; radiation therapy; dehydration; secondary malignancy; older adults; pregnancy (category C), lactation; children younger than 3 mo.

ROUTE & DOSAGE

Induction Therapy
Adult/Child (older than 3 mo): **IV** 5 mg/kg q12h for 14–21 days (doses may range from 2.5–5 mg/kg q8–12h for 10–35 days)

Maintenance Therapy
Adult/Child: **IV** 5 mg/kg daily 7 days/wk or 6 mg/kg daily 5 days/wk **PO** 1000 mg t.i.d. or 500 mg 6 times/day q3h while awake

Prevention of CMV Disease in Transplant Recipients
Adult/Child: **IV** 5 mg/kg q12h 7–14 days, then 5 mg/kg daily or 6 mg/kg/day 5 days/wk

Dendritic Keratitis
Adult/Adolescent/Child (over 2): **Opthalmic** 1 drop in affected eye 5 times daily until healed then 1 drop in affected eye t.i.d. x 7 days

Renal Impairment Dosage Adjustment
CrCl 50–70 mL/min: Use 50% of dose; 25–50 mL/min: Use 50% of dose and q24h interval; 10–25 mL/min: Use 25% of dose and q24h interval

Hemodialysis Dosage Adjustment
Give dose post-dialysis

G

Common adverse effects in *italic*, life-threatening effects <u>underlined</u>; generic names in **bold**; classifications in SMALL CAPS; ♣ Canadian drug name; ⓓ Prototype drug

ADMINISTRATION

- Note: Do not administer if neutrophil count falls below 500/mm³ or platelet count falls below 25,000/mm³.
- Avoid direct contact of powder in capsules or solution with skin and mucous membranes. Wash thoroughly with soap and water if contact occurs.

Oral
- Give with food.

Intravenous

IV administration to infants and children: Verify correct IV concentration and rate of infusion with prescriber.

PREPARE: **Intermittent** Reconstitute the 500-mg vial using only 10 mL of sterile water (supplied) for injection immediately before use to yield 50 mg/mL. ▪ Shake well to dissolve. ▪ Withdraw the ordered amount and add to 100 mL of NS, D5W, or LR (volume less than 100 mL may be used, but the final concentration should be less than 10 mg/mL).

ADMINISTER: **Intermittent:** Give at a constant rate over 1 h. ▪ Avoid rapid infusion or bolus injection.

INCOMPATIBILITIES **Solution/additive:** Amino acid solutions (TPN), bacteriostatic water for injection, **foscarnet. Y-site: Aldesleukin, amifostine, amikacin, aminocaproic acid, aminophylline, amphotericin B colloidal, ampicillin, ampicillin/sulbactam, amsacrine, ascorbic acid, atracurium, azathioprine, aztreonam, benztropine, bumetanide, butorphanol, cefamandole, cefazolin, cefepime, cefmetazole, cefonicid, cefoperazone, cefotaxime, cefotetan, cefoxitin, ceftazidime, ceftizoxime, ceftriaxone, cefuroxime, cefphalothin, cephapirin, chloramphenicol, chlorpromazine, cimetidine, clindamycin, codeine, cytarabine, dantrolene, diazepam, diltiazem, diphenhydramine, dobutamine, dopamine, doxorubicin, doxycycline, ephedrine, epinephrine, epirubicin, erythromycin, esmolol, famotidine, fenoldopam, fludarabine, foscarnet, gemcitabine, gemtuzumab, gentamicin, haloperidol, hydralazine, hydrocortisone, hydroxyzine, idarubicin, imipenem/cilastin, irinotecan, isoproterenol, ketorolac, lidocaine, meperidine, metaraminol, methoxamine, methyldopate, methylprednisolone, metoclopramide, metronidazole, mezlocillin, midazolam, minocycline, morphine, mycophenolate, nalbuphine, netilmicin, norepinephrine, ondansetron, palonosetron, papaverine, penicillin G, pentazocine, phentolamine, phenylephrine, phenytoin, piperacillin/tazobactam, procainamide, prochlorperazine, promethazine, pyridoxine, quinidine, quinupristin/dalfopristin, sargramostim, sodium bicarbonate, streptokinase, succinylcholine, SMZ/TMP, tacrolimus, thiamine, ticarcillin, TPN, tobramycin, vancomycin, vercuronium, verapamil, vinorelbine.**

- Store reconstituted solutions refrigerated at 4° C; use within 12 h.
- Store infusion solution refrigerated up to 24 h of preparation.

ADVERSE EFFECTS (≥1%) **CNS:** *Fever,* headache, disorientation, mental status changes, ataxia, <u>coma</u>, confusion, dizziness, paresthesia, nervousness, somnolence, tremor.

CV: Edema, phlebitis. **GI:** *Nausea, diarrhea,* anorexia, elevated liver enzymes. **Hematologic:** <u>*Bone marrow suppression,*</u> thrombocytopenia, *granulocytopenia, eosinophilia,* <u>leukopenia,</u> hyperbilirubinemia. **Metabolic:** Hyperthermia, hypoglycemia. **Urogenital:** Infertility. **Skin:** Rash.

INTERACTIONS Drug: ANTINEOPLASTIC AGENTS, **amphotericin B, didanosine, trimethoprim-sulfamethoxazole (TMP-SMZ), dapsone, pentamidine, probenecid, zidovudine** may increase bone marrow suppression and other toxic effects of ganciclovir; may increase risk of nephrotoxicity from **cyclosporine;** may increase risk of seizures due to **imipenem-cilastatin.** Oral product increases **didanosine** levels.

PHARMACOKINETICS Onset: 3–8 days. **Duration:** Clinical relapse can occur 14 days to 3.5 mo after stopping therapy; positive blood and urine cultures recur 12–60 days after therapy. **Distribution:** Distributes throughout body including CSF, eye, lungs, liver, and kidneys; crosses placenta in animals; not known if distributed into breast milk. **Metabolism:** Not metabolized. **Elimination:** Unchanged in urine. **Half-Life:** 2.5–4.2 h.

NURSING IMPLICATIONS

Assessment & Drug Effects

- Lab tests: Neutrophil and platelet counts at least every other day during twice-daily dosing and weekly thereafter; more frequent monitoring may be indicated in certain patients. Monitor serum creatinine or creatinine clearance at least q2wk. Closely monitor renal function in the older adult.

- Inspect IV insertion site throughout infusion for signs and symptoms of phlebitis.

Patient & Family Education

- Note: Drug is not a cure for CMV retinitis; follow regular ophthalmologic examination schedule.
- Drink lots of fluids during therapy.
- Use barrier contraception throughout therapy and for at least 90 days afterwards.
- Maintain frequent hematologic monitoring.

GANIRELIX ACETATE 🅟

(gan-i-rel′ix)

Antagon

Classification: GONADOTROPIN-RELEASING HORMONE (GnRH) ANTAGONIST
Therapeutic: GnRH ANTAGONIST
Prototype: Ganirelix
Pregnancy Category: X

AVAILABILITY 250 mcg/0.5 mL syringe

ACTION & *THERAPEUTIC EFFECT*
Ganirelix is a gonadotropin-releasing hormone (GnRH) antagonist that suppresses pituitary gonadotropins and sex hormones. *It prevents LH surges in reproductive protocols, and causes shrinkage of uterine fibroids.*

USES Infertility treatment.

CONTRAINDICATIONS Prior hypersensitivity to ganirelix, LHRH, or other LHRH analogs, mannitol hypersensitivity; ovarian cyst; primary ovarian failure; pregnancy (category X), lactation.

CAUTIOUS USE History of current allergic disorders (e.g., asthma, hay fever, urticaria, eczema) or a history of allergic reactions to

Common adverse effects in *italic*, life-threatening effects <u>underlined</u>; generic names in **bold**; classifications in SMALL CAPS; ♣ Canadian drug name; 🅟 Prototype drug

685

medications; renal/hepatic dysfunction; endocrine disorders; alcohol consumption.

ROUTE & DOSAGE

Infertility

Adult: **Subcutaneous** After initiating follicle-stimulating hormone (FSH) therapy on day 2 or 3 of the cycle, give 250 mcg once daily during the early-to-mid-follicular phase

ADMINISTRATION

- Note: The packaging of the product, Antagon, contains natural rubber latex which may cause allergic reactions.

Subcutaneous

- Inject into subcutaneous tissue in the abdomen around the umbilicus or into the upper thigh.
- Rotate injection sites.
- Store at 5°–30° C (59°–86° F) and protect from light.

ADVERSE EFFECTS (≥1%) CNS:
Headache. **GI:** Abdominal pain, nausea. **Endocrine:** Ovarian hyperstimulation syndrome. **Skin:** Injection site reaction. **Urogenital:** Vaginal bleeding.

PHARMACOKINETICS Absorption:
91% from subcutaneous site. **Peak:** 1 h. **Distribution:** 81% protein bound. **Elimination:** 75% in feces; 22% in urine. **Half-Life:** 13–16 h.

NURSING IMPLICATIONS

Assessment & Drug Effects

- Exercise caution with patients with hypersensitivity to GnRH or with known allergic disorders (e.g., asthma, hay fever). These patients

should be carefully monitored after the first injection for S&S of an anaphylactic reaction.
- Lab tests: Monitor baseline and periodic CBC with differential, and periodic total bilirubin.

Patient & Family Education

- Report menstrual disorders (e.g., spotting, frank vaginal bleeding) to prescriber.
- Notify prescriber immediately if you think you are pregnant.

GATIFLOXACIN
(gat-i-flox′a-sin)
Zymer
Classifications: QUINOLONE ANTIBIOTIC
Therapeutic: ANTIBIOTIC
Prototype: Ciprofloxacin
Pregnancy Category: C

AVAILABILITY 200 mg, 400 mg tablets; 0.3% ophth solution

ACTION & *THERAPEUTIC EFFECT*
Synthetic quinolone that is a broad spectrum bactericidal agent. Inhibits topoisomerase II (DNA-gyrase), an enzyme necessary for bacterial replication, transcription, repair, and recombination. *Effective against gram-positive and gram-negative bacteria.*

USES Acute bacterial exacerbation of chronic bronchitis; acute sinusitis; community-acquired pneumonia; uncomplicated or complicated UTI; pyelonephritis; gonorrhea due to susceptible organisms.

CONTRAINDICATIONS Hypersensitivity to gatifloxacin or other quinolone antibiotics; diabetes mellitus; viral infections; lactation. **Ophthalmic** use in infants younger than 1 mo.
CAUTIOUS USE Patients with CNS disorders including seizures or

epilepsy; myasthenia gravis; GI disorders, renal dysfunction; hypersensitivity to other medications; pregnancy (category C). Safe use in children younger than 18 y not established.

ROUTE & DOSAGE

Acute Bacterial Exacerbation of Chronic Bronchitis, Complicated

Adult: **PO** 400 mg daily × 5 days

Complicated UTI, Acute Pyelonephritis

Adult: **PO** 400 mg daily × 7–10 days

Acute Sinusitis

Adult: **PO** 400 mg daily × 10 days

Community-Acquired Pneumonia

Adult: **PO** 400 mg daily × 7–14 days

Uncomplicated UTI

Adult: **PO** 400 mg as a single dose or 200 mg daily × 3 days

Uncomplicated Gonorrhea

Adult: **PO** 400 mg as a single dose

Renal Impairment Dosage Adjustment

CrCl less than 40 mL/min or on dialysis: 400 mg × 1 day, then 200 mg daily

ADMINISTRATION

Oral

- Give at least 4 h before or after an aluminum- or magnesium-containing antacid, or iron-containing products.
- Store at 15°–30° C (59°–86° F).

ADVERSE EFFECTS (≥1%) **Body as a Whole:** Headache, allergic reactions, chills, fever; back pain, chest pain. **CNS:** Dizziness, abnormal dreams, insomnia, paresthesia,

tremor, vasodilatation, vertigo. **CV:** Palpitation; peripheral edema. **GI:** Nausea, diarrhea, abdominal pain, constipation, dyspepsia, glossitis, oral moniliasis, stomatitis, vomiting. **Respiratory:** Dyspnea, pharyngitis. **Skin:** Rash, sweating. **Urogenital:** Vaginitis, dysuria, hematuria. **Special Senses:** Abnormal vision, taste perversion, tinnitus. **Metabolic:** Hyperglycemia, hypoglycemia. **Other:** Cartilage erosion.

DIAGNOSTIC TEST INTERFERENCE May cause false positive on *opiate screening tests.*

INTERACTIONS Drug: Probenecid decreases elimination of gatifloxacin; **ferrous sulfate,** ALUMINUM- or MAGNESIUM-CONTAINING ANTACIDS reduce absorption of gatifloxacin; gatifloxacin may cause slight increase in **digoxin** levels.

PHARMACOKINETICS Absorption: 96% from GI tract. **Peak:** 1–2 h PO. **Distribution:** 20% protein bound. **Metabolism:** Minimal metabolism (less than 1%). **Elimination:** Primarily in urine. **Half-Life:** 7–14 h (up to 35–40 h in severe renal failure).

NURSING IMPLICATIONS

Assessment & Drug Effects

- Monitor for S&S of CNS disturbance especially with history of cerebrovascular disease or seizures.
- Lab tests: C&S prior to initiation of therapy; baseline and periodic WBC with differential.
- Monitor diabetics for loss of glycemic control.
- Monitor for changes in digoxin blood levels with coadministered drugs.

Patient & Family Education

- Be aware that increased risk of seizures are associated with drug use in patient with history of seizures.

Common adverse effects in *italic*, life-threatening effects <u>underlined</u>; generic names in **bold**; classifications in SMALL CAPS; ♣ Canadian drug name; ⊕ Prototype drug

687

- Report unexplained dizziness or problems with balance, tendon pain, severe diarrhea, skin rash, mental status changes.

GEFITINIB ⓟ

(ge-fi′ti-nib)
Iressa
Classifications: ANTINEOPLASTIC; EPIDERMAL GROWTH FACTOR RECEPTOR-TYROSINE KINASE INHIBITOR (EGFR-TKI)
Therapeutic: ANTINEOPLASTIC
Pregnancy Category: D

AVAILABILITY 250 mg tablets

ACTION & *THERAPEUTIC EFFECT*
Gefitinib is a selective epidermal growth factor receptor-tyrosine kinase inhibitor (EGFR-TKI). EGFR is expressed or overexpressed in many cancers. EGFR expression is associated with poor prognosis for cancer, development of metastasis, and resistance to chemotherapy, hormonal therapy, and radiation therapy. *Inhibits up-regulation or overexpression of EGRF in cancer cells, thus diminishing their capacity for cell proliferation, cell survival, and decreasing their invasive capacity and metastases.*

USES Treatment of locally advanced or metastatic non–small-cell lung cancer after failure of both platinum and docetaxel therapy in patients who have previously used gefitinib.
UNLABELED USES Treatment of head and neck and other solid tumors.

CONTRAINDICATIONS Hypersensitivity to gefitinib; pregnancy (category D), lactation.
CAUTIOUS USE Severe renal impairment; hepatic impairment; bacterial/viral infection; dermatologic toxicities; GI disorders; interstitial lung disease (interstitial pneumonia, pneumonitis, and alveolitis), pulmonary fibrosis, respiratory insufficiency; myelosuppression; females of childbearing age; prior chemotherapy, radiation therapy; ocular toxicities (corneal ulcer, eye pain); children younger than 18 y.

ROUTE & DOSAGE

Non–Small-Cell Lung Cancer
Adult: **PO** 250 mg daily, may increase to 500 mg daily if on enzyme-inducing drugs
Head and Neck Cancers
Adult: **PO** 500 mg/day

ADMINISTRATION

Oral
- Give without regard to meals.
- Store tablets at 15°–30° C (59°–86° F).

ADVERSE EFFECTS (≥1%) **Body as a Whole:** Asthenia, peripheral edema. **GI:** *Diarrhea, nausea, vomiting,* anorexia, weight loss, stomatitis. **Respiratory:** Dyspnea, interstitial lung disease. **Skin:** *Acne/acneiform rash, dry skin,* pruritus, vesicular/bullous rash. **Special Senses:** Amblyopia, conjunctivitis, aberrant eyelash growth.

INTERACTIONS Drug: BARBITURATES, **bosentan, carbamazepine, dexamethasone, nevirapine, oxcarbazepine, phenytoin** or **fosphenytoin, rifampin, rifabutin, rifapentine** may increase metabolism and decrease levels of gefitinib; **amiodarone,** PROTEASE INHIBITORS, **cimetidine, clarithromycin, dalfopristin; quinupristin, delavirdine, efavirenz, erythromycin, fluconazole, fluvoxamine, fluoxetine, imatinib,**

Common adverse effects in *italic*, life-threatening effects underlined; generic names in **bold**; classifications in SMALL CAPS; ◆ Canadian drug name; ⓟ Prototype drug

itraconazole, ketoconazole, mifepristone, nefazodone, and voriconazole may increase levels and toxicity of gefitinib; may increase INR with **warfarin;** H₂-RECEPTOR ANTAGONISTS, PROTON PUMP INHIBITORS may decrease absorption of gefitinib. **Food: Grapefruit juice** may increase levels and toxicity of gefitinib. **Herbal: St. John's wort** may decrease levels of gefitinib.

PHARMACOKINETICS Absorption: Slowly absorbed, 60% reaches systemic circulation. **Peak:** 3–7 h. **Metabolism:** In liver primarily by CYP3A4. **Elimination:** 86% in feces. **Half-Life:** 48 h.

NURSING IMPLICATIONS

Assessment & Drug Effects

- Monitor pulmonary status and report promptly dyspnea, cough, and fever.
- Withhold drug and notify prescriber for significant elevations of transaminases, bilirubin, or alkaline phosphatase.
- Monitor for adverse effects, especially with concurrent use of drugs that may inhibit CYP3A4 (e.g., amiodarone, cimetidine, erythromycin, fluconazole, grapefruit juice, etc.). See INTERACTIONS.
- Lab tests: Periodic LFTs; frequent PT/INR with concurrent warfarin.

Patient & Family Education

- Report promptly any of the following: Eye pain or irritation; fever; breathing difficulty or shortness of breath; mouth sores.
- Inform prescriber of all prescription, nonprescription, or herbal drugs you are taking.
- Females should use reliable contraceptives while taking this drug.
- Minimize or avoid intake of grapefruit juice while taking this drug.

GEMCITABINE HYDROCHLORIDE
(gem-ci'ta-been)
Gemzar
Classifications: ANTINEOPLASTIC; ANTIMETABOLITE, PYRIMIDINE
Therapeutic: ANTINEOPLASTIC
Prototype: Fluorouracil
Pregnancy Category: D

AVAILABILITY 20 mg/mL injection

ACTION & *THERAPEUTIC EFFECT* Pyrimidine analog with cell phase specificity by affecting rapidly dividing cells in S phase (DNA synthesis). It also blocks the progression of cells from G_1 phase to S phase of cell cycle. Gemcitabine interferes with DNA synthesis by inhibiting ribonucleotide. In addition, if gemcitabine is incorporated into the DNA strand, it inhibits further growth of the strand. *Gemcitabine induces DNA fragmentation in dividing cells, resulting in the cell death of tumor cells.*

USES Locally advanced or metastatic adenocarcinoma of the pancreas, non–small-cell lung cancer, breast cancer.

CONTRAINDICATIONS Hypersensitivity to gemcitabine, severe thrombocytopenia, acute infection, pregnancy (category D), lactation.
CAUTIOUS USE Myelosuppression, neutropenia; renal or hepatic dysfunction; history of bleeding disorders, infection, previous cytotoxic or radiation treatment; older adults. Safety and effectiveness in children are not established.

ROUTE & DOSAGE

Pancreatic Cancer
Adult: **IV** 1000 mg/m² once weekly for up to 7 wk, followed by 1 wk rest

Common adverse effects in *italic*, life-threatening effects underlined; generic names in **bold**; classifications in SMALL CAPS; ◆ Canadian drug name; ● Prototype drug

689

from treatment; may repeat once weekly for 3 of every 4 wk

Non–Small-Cell Lung Cancer

Adult: **IV** 1000 mg/m² on days 1, 8, 15 of 28-day cycle OR 1250 mg/m² on days 1 and 8 of 21-day cycle. Given with cisplatin.

Breast Cancer

Adult: **IV** 1250 mg/m² on days 1 and 8 of 21-day cycle. Given with paclitaxel.

ADMINISTRATION

Intravenous

PREPARE: **IV Infusion:** Dilute with NS without preservatives by adding 5 mL or 25 mL to the 200 mg or 1 g vial, respectively, to yield 38 mg/mL. ▪ Shake to dissolve. ▪ Dilute further if necessary with NS to concentrations as low as 0.1 mg/mL.

ADMINISTER: **IV Infusion:** Infuse over 30 min. Infusion time greater than 60 min is associated with increased toxicity.

INCOMPATIBILITIES **Y-site:** Acyclovir, amphotericin B, cefopime, cefoperazone, cefotaxime, chloramphenicol, dantrolene, daptomycin, diazepam, furosemide, ganciclovir, imipenem/cilastatin, irinotecan, ketorolac, lansoprazole, methotrexate, methylprednisolone, mitomycin, nafcillin, pantoprazole, pemetrexed, phenytoin, piperacillin/tazobactam, prochlorperazine, thiopental.

▪ Store reconstituted solutions unrefrigerated at 20°–25° C (68°–77° F). Use within 24 h of reconstitution.

ADVERSE EFFECTS (≥1%) **CNS:** *Fever, flu-like syndrome (anorexia, headache, cough, chills, myalgia),*

paresthesias. **GI:** *Nausea, vomiting, diarrhea,* stomatitis, *transient elevations of liver transaminases.* **Hematologic:** <u>Myelosuppression (anemia, leukopenia, neutropenia, thrombocytopenia).</u> **Skin:** Bullous skin eruption, desquamation. **Urogenital:** Mild proteinuria and hematuria. **Other:** *Dyspnea, edema, peripheral edema, infection,* elevated liver function tests.

INTERACTIONS Drug: May increase effect of **warfarin** or ORAL ANTICOAGULANTS or NSAIDS. Do not use with **sargramostin** or **filgrastim.**

PHARMACOKINETICS Peak: Peak concentrations reached 30 min after infusion; lower clearance in women and older adult results in higher concentrations at any given dose. **Distribution:** Crosses placenta, distributed into breast milk. **Metabolism:** Intracellularly by nucleoside kinases to active diphosphate and triphosphate nucleosides. **Elimination:** 92–98% recovered in urine within 1 wk. **Half-Life:** 32–94 min.

NURSING IMPLICATIONS

Assessment & Drug Effects

▪ Lab tests: Monitor CBC with differential and platelet count prior to each dose. Monitor baseline and periodic kidney function tests and LFTs.

Patient & Family Education

▪ Learn about common adverse effects and measures to control or minimize when possible. Notify prescriber immediately of any distressing adverse effects.
▪ Note: Fever with flu-like symptoms, rash, and GI distress are very common.
▪ Females should use reliable contraception while taking this drug.

Common adverse effects in *italic*, life-threatening effects <u>underlined</u>; generic names in **bold**; classifications in SMALL CAPS; ♣ Canadian drug name; ◯ Prototype drug

GEMFIBROZIL
(gem-fi'broe-zil)
Lopid
Classifications: ANTILIPEMIC; FI-BRATE
Therapeutic: CHOLESTEROL-LOWERING
Prototype: Fenofibrate
Pregnancy Category: C

AVAILABILITY 600 mg tablets

ACTION & *THERAPEUTIC EFFECT*
Fibric acid derivative with lipid regulating properties. Blocks lipolysis of stored triglycerides in adipose tissue and inhibits hepatic uptake of fatty acids. *Decreases VLDL and therefore triglyceride synthesis. Produces a moderate increase in HDL cholesterol levels and reduces levels of total and LDL cholesterol and triglycerides.*

USES Patients with very high serum triglyceride levels (above 750 mg/dL) (type IV and V hyperlipidemia) who have not responded to intensive diet restriction and are at risk of pancreatitis and abdominal pain. Also severe familial hypercholesterolemia (type IIa or IIb) that developed in childhood and has failed to respond to dietary control or to other cholesterol-lowering drugs.

CONTRAINDICATIONS Gallbladder disease, biliary cirrhosis, hepatic or severe renal dysfunction. Safety and efficacy in children younger than 18 y are not established.
CAUTIOUS USE Diabetes mellitus, hypothyroidism; renal impairment; cholelithiasis; pregnancy (category C), lactation.

ROUTE & DOSAGE

Hypertriglyceridemia
Adult: **PO** 600 mg b.i.d. 30 min before morning and evening meal, may increase up to 1500 mg/day

ADMINISTRATION
Oral
- Give 30 min before breakfast and evening meal.
- Store at 15°–30° C (59°–86° F) unless otherwise directed.

ADVERSE EFFECTS (≥1%) **CNS:** Headache, dizziness, blurred vision. **GI:** *Abdominal* or *epigastric pain,* diarrhea, nausea, vomiting, flatulence. **Hematologic:** Eosinophilia, mild decreases in Hct, Hgb. **Musculoskeletal:** Painful extremities, back pain, muscle cramps, myalgia, arthralgia, swollen joints. **Skin:** Rash, dermatitis, pruritus, urticaria. **Endocrine:** Hypokalemia, moderate hyperglycemia.

INTERACTIONS Drug: May potentiate hypoprothrombinemic effects of ORAL ANTICOAGULANTS; **lovastatin** increases risk of myopathy and rhabdomyolysis; may increase hypoglycemic effects of ANTIDIABETIC MEDICATIONS.

PHARMACOKINETICS Absorption: Readily from GI tract. **Peak:** 1–2 h. **Metabolism:** Undergoes enterohepatic circulation. **Elimination:** In urine; 6% in feces. **Half-Life:** 1.3–1.5 h.

NURSING IMPLICATIONS
Assessment & Drug Effects
- Lab tests: Monitor baseline and at regular intervals during first year of therapy for serum LDL and VLDL, triglycerides, total cholesterol, CBC, blood glucose, liver function tests.
- Note: Mild decreases in WBC, Hgb, Hct may occur during early stage of treatment but generally stabilize with continued therapy.

Common adverse effects in *italic*, life-threatening effects underlined; generic names in **bold**; classifications in SMALL CAPS; ♣ Canadian drug name; ☻ Prototype drug

691

- Note: Drug is usually withdrawn if lipid response is inadequate after 3 mo of therapy.
- Notify prescriber if patient presents S&S suggestive of cholelithiasis or cholecystitis; gallbladder studies may be indicated. Symptoms often occur during the night or early morning; jaundice may or may not be present.

Patient & Family Education

- Do not drive or engage in other potentially hazardous activities until response to drug is known.
- Report promptly if you develop jaundice, pruritis, or unexplained, upper abdominal discomfort.

GEMIFLOXACIN

(gem-i-flox'a-cin)
Factive
Classifications: QUINOLONE ANTIBIOTIC
Therapeutic: ANTIBIOTIC
Prototype: Ciprofloxacin HCl
Pregnancy Category: C

AVAILABILITY 320 mg tablet

ACTION & *THERAPEUTIC EFFECT*
Gemifloxacin inhibits bacterial DNA gyrases (topoisomerase II), enzymes essential in replication, transcription, and repair of bacterial DNA. *Gemifloxacin is active against a wide range of gram-positive and gram-negative bacteria. It has greater activity against gram-positive cocci and penicillin- and ciprofloxacin-resistant Streptococcus pneumoniae than other fluoroquinolones.*

USES Treatment of acute exacerbations of chronic bronchitis, mild to moderate community-acquired pneumonia.

UNLABELED USES Acute sinusitis, UTI, acute pyelonephritis.

CONTRAINDICATIONS Hypersensitivity to gemifloxacin or other fluoroquinolone antibiotics; known QT prolongation; tendon pain; viral disease; lactation.
CAUTIOUS USE Hypokalemia, hypomagnesemia, or concurrent use of Class IA or III antiarrhythmic agents; bradycardia, acute myocardial ischemia; renal disease or impairment; hepatic disease; central nervous system disorders such as epilepsy; glucose 6-phosphate dehydrogenase deficiency; tendonitis; older adults; pregnancy (category C). Safe use in children 18 y or younger not established.

ROUTE & DOSAGE

Acute Exacerbation of Chronic Bronchitis
Adult: **PO** 320 mg daily × 5 days
Community-Acquired Pneumonia
Adult: **PO** 320 mg daily × 7 days
Sinusitis
Adult: **PO** 320 mg daily × 10 days
UTI
Adult: **PO** 320 mg daily × 3 days
Renal Impairment Dosage Adjustment
CrCl 40 mL/min or less: 160 mg daily

ADMINISTRATION

Oral

- Give 2 h before/3 h after drugs containing aluminum, magnesium, iron, zinc, or buffered tablets of any type.
- Give at least 2 h before sucralfate.
- Store at 15°–30° C (59°–86° F) and protect from light.

ADVERSE EFFECTS (≥1%) **CNS:** Headache. **GI:** Nausea, vomiting, diarrhea, elevated liver enzymes. **Skin:** Rash.

INTERACTIONS Drug: ANTACIDS, **didanosine (tablets and powder), iron, sevelamer, sucralfate** decrease absorption; may prolong the QT interval with **amiodarone, disopyramide, dofetilide, ibutilide, quinidine, procainamide, sotalol** leading to arrhythmias; may augment phototoxicity of RETINOIDS.

PHARMACOKINETICS Absorption: 71% absorbed. **Peak:** 0.5–2 h. **Metabolism:** Minimally in liver. **Elimination:** Primarily renal. **Half-Life:** 7 h.

NURSING IMPLICATIONS

Assessment & Drug Effects

- Monitor cardiac status with concurrent use of drugs that may prolong the QT interval. Report immediately bradycardia or S&S of heart failure.
- Withhold drug and report to prescriber any of the following: Tremors, restlessness, lightheadedness, confusion, hallucinations, paranoia, depression, nightmares, and insomnia.
- Lab tests: C&S prior to initiation of therapy; baseline and periodic serum electrolytes; frequent blood glucose levels in diabetics; CBC with differential and platelet count with prolonged treatment.

Patient & Family Education

- Use sunscreen and protective clothing outdoors. Avoid sun lamps.
- Stop gemifloxacin and notify prescriber for pain or swelling of a tendon or around a joint.
- Drink fluid liberally (unless contraindicated) while taking this drug.

- Do not drive or engage in other hazardous activities until reaction to drug is known.

GENTAMICIN SULFATE ⊕
(jen-ta-mye′sin)
Garamycin Ophthalmic, Genoptic
Classification: AMINOGLYCOSIDE ANTIBIOTIC
Therapeutic: ANTIBIOTIC
Pregnancy Category: D

AVAILABILITY 10 mg/mL, 40 mg/mL; 0.1% ointment, cream; 3 mg/mL ophth solution; 3 mg/g ophth ointment

ACTION & THERAPEUTIC EFFECT Broad-spectrum aminoglycoside antibiotic that binds irreversibly to 30S subunit of bacterial ribosomes, blocking a vital step in protein synthesis, and attachment of RNA molecules to bacterial ribosomes resulting in cell death. *Active against a wide variety of aerobic gram-negative but not anaerobic gram-negative bacteria. Also effective against certain gram-positive organisms, particularly penicillin-sensitive bacteria.*

USES Parenteral use restricted to treatment of serious infections of GI, respiratory, and urinary tracts, CNS, bone, skin, and soft tissue (including burns) when other less toxic antimicrobial agents are ineffective or are contraindicated. Has been used in combination with other antibiotics. Also used topically for primary and secondary skin infections and for superficial infections of external eye and its adnexa.
UNLABELED USES Prophylaxis of bacterial endocarditis in patients

---

undergoing operative procedures or instrumentation.

CONTRAINDICATIONS History of hypersensitivity to, or toxic reaction with any aminoglycoside antibiotic; pregnancy (category D), lactation.

CAUTIOUS USE Impaired renal function; history of eighth cranial (acoustic) nerve impairment; preexisting vertigo or dizziness or tinnitus; dehydration, fever; renal impairment; dehydration; Fabry disease; older adults, premature infants, neonates, infants; obesity, neuromuscular disorders: Myasthenia gravis, parkinsonian syndrome. Hypocalcemia, heart failure. **Topical:** Applied to widespread areas.

ROUTE & DOSAGE

Moderate to Severe Infection

Adult: **IV/IM** 1–2 mg/kg loading dose followed by 3–5 mg/kg/day in 3 divided doses **Intrathecal** 4–8 mg preservative free daily **Topical** 1–2 drops of solution in eye q4h up to 2 drops q1h or small amount of ointment b.i.d. or t.i.d.
Child: **IV/IM** 6–7.5 mg/kg/day in 3 divided doses **Intrathecal** *Older than 3 mo,* 1–2 mg preservative free daily
Neonate: **IV/IM** 2.5 mg/kg/day

Prophylaxis of Bacterial Endocarditis

Adult: **IV/IM** 1.5 mg/kg 30 min before procedure, may repeat in 8 h
Child (weight less than 27 kg): **IV/IM** 2 mg/kg 30 min before procedure, may repeat in 8 h

Obesity Dosage Adjustment

Dose based on IBW, in morbid obesity use IBW +0.4 (TBW–IBW)

Renal Impairment Dosage Adjustment

Reduce dose or extend dosing interval

ADMINISTRATION

Ophthalmic

- Apply pressure to inner canthus for 1 min immediately after instillation of drops.
- Have patient keep eyes closed for 1–2 min after administration of ophthalmic ointment to assure medication contact. Caution patient that vision will be blurred for a few minutes.

Topical

- Wash affected area with mild soap and water, rinse, and dry thoroughly. Gently apply small amount of medication to lesions; cover with sterile gauze.
- Do not apply topical preparations, particularly cream, to large denuded body surfaces because systemic absorption and toxicity are possible.

Intramuscular

- Give deep into a large muscle.
- Do not use solutions that are discolored or that contain particulate matter; drug for IV or IM is clear and colorless or slightly yellow.

Intrathecal

- Note: Intrathecal formulation is a clear and colorless solution.
- Use promptly after opening; contains no preservatives and any unused portion should be discarded.

Intravenous

PREPARE: Intermittent: Dilute a single dose with 50–200 mL of D5W or NS. ▪ For pediatric patients, amount of infusion fluid

Common adverse effects in *italic*, life-threatening effects underlined; generic names in **bold**; classifications in SMALL CAPS; ✦ Canadian drug name; ⬤ Prototype drug

may be proportionately smaller depending on patient's needs but should be sufficient to be infused over the same time period as for adults.

ADMINISTER: Intermittent: Give over 30 min–1 h. May extend infusion time to 2 h for a child.

INCOMPATIBILITIES Solution/additive: Fat emulsion, **TPN, amphotericin B, ampicillin,** CEPHALOSPORINS, **cytarabine, heparin, ticarcillin. Y-site:** Allopurinol, amphotericin B cholesteryl, azithromycin, furosemide, heparin, hetastarch, idarubicin, indomethacin, iodipamide, propofol, warfarin.

▪ Store all gentamicin solutions between 2°–30° C (36°–86° F) unless otherwise directed by manufacturer.

ADVERSE EFFECTS (≥1%) Special Senses: Ototoxicity (vestibular disturbances, impaired hearing), optic neuritis. **CNS:** Neuromuscular blockade: Skeletal muscle weakness, apnea, respiratory paralysis (high doses); arachnoiditis (intrathecal use). **CV:** Hypotension or hypertension. **GI:** Nausea, vomiting, transient increase in AST, ALT, and serum LDH and bilirubin; hepatomegaly, splenomegaly. **Hematologic:** Increased or decreased reticulocyte counts; granulocytopenia, thrombocytopenia (fever, bleeding tendency), thrombocytopenic purpura, anemia. **Body as a Whole:** Hypersensitivity (rash, pruritus, urticaria, exfoliative dermatitis, eosinophilia, burning sensation of skin, drug fever, joint pains, laryngeal edema, anaphylaxis). **Urogenital:** Nephrotoxicity: Proteinuria, tubular necrosis, cells or casts in urine, hematuria, rising BUN, nonprotein nitrogen, serum creatinine; *decreased creatinine clearance.*

Other: Local irritation and pain following IM use; thrombophlebitis, abscess, superinfections, syndrome of hypocalcemia (tetany, weakness, hypokalemia, hypomagnesemia). **Topical and Ophthalmic:** Photosensitivity, sensitization, erythema, pruritus; burning, stinging, and lacrimation (ophthalmic formulation).

INTERACTIONS Drug: Amphotericin B, capreomycin, cisplatin, methoxyflurane, polymyxin B, vancomycin, ethacrynic acid, and **furosemide** increase risk of nephrotoxicity. GENERAL ANESTHETICS and NEUROMUSCULAR BLOCKING AGENTS (e.g., **succinylcholine**) potentiate neuromuscular blockade. **Indomethacin** may increase gentamicin levels in neonates.

PHARMACOKINETICS Absorption: Well absorbed from IM site. **Peak:** 30–90 min IM. **Distribution:** Widely distributed in body fluids, including ascitic, peritoneal, pleural, synovial, and abscess fluids; poor CNS penetration; concentrates in kidney and inner ear; crosses placenta. **Metabolism:** Not metabolized. **Elimination:** Excreted unchanged in urine; small amounts accumulate in kidney and are eliminated over 10–20 days; small amount excreted in breast milk. **Half-Life:** 2–4 h.

NURSING IMPLICATIONS

Assessment & Drug Effects

▪ Lab tests: Perform C&S and renal function prior to first dose and periodically during therapy. Determine creatinine clearance and serum drug concentrations at frequent intervals.

▪ Note: Dosages are generally adjusted to maintain peak serum gentamicin concentrations of 4–10 mcg/mL, and trough concentrations of 1–2 mcg/mL. Prolonged

Common adverse effects in *italic*, life-threatening effects underlined; generic names in **bold**; classifications in SMALL CAPS; ♣ Canadian drug name; ⊘ Prototype drug

695

peak concentrations above 12 mcg/mL and trough concentrations above 2 mcg/mL are associated with toxicity.

- Draw blood specimens for peak serum gentamicin concentration 30 min–1h after IM administration, and 30 min after completion of a 30–60 min IV infusion. Draw blood specimens for trough levels just before the next IM or IV dose.
- Monitor vital signs and I&O. Keep patient well hydrated to prevent chemical irritation of renal tubules. Report oliguria, unusual appearance of urine, change in I&O ratio or pattern, and presence of edema (prolongs elimination time).
- Report promptly S&S of ototoxic effect (e.g., headache, dizziness or vertigo, nausea and vomiting with motion, ataxia, nystagmus, tinnitus, roaring noises, sensation of fullness in ears, hearing impairment).
- Watch for S&S of bacterial overgrowth (opportunistic infections) with resistant or nonsusceptible organisms (diarrhea, anogenital itching, vaginal discharge, stomatitis, glossitis).

Patient & Family Education

- Note: When using topical applications: Avoid excessive exposure to sunlight because of danger of photosensitivity; withhold medication and notify prescriber if condition fails to improve within 1 wk, worsens, or signs of irritation or sensitivity occur; and apply medication as directed and only for length of time prescribed (overuse can result in superinfections).

GLATIRAMER ACETATE

(gla-tir′a-mer)
Copaxone
Classifications: BIOLOGICAL RESPONSE MODIFIER; IMMUNOLOGIC

Therapeutic: IMMUNOSUPPRESSANT
Pregnancy Category: B

AVAILABILITY 20 mg injection

ACTION & *THERAPEUTIC EFFECT*
Glatiramer is a random synthetic copolymer of L-alanine, L-glutamic acid, L-lysine, and L-tyrosine. It modifies immune processes that are responsible for the pathogenesis of multiple sclerosis. *Its function is to reduce the relapse rate of multiple sclerosis (MS), a demyelinating disease of the CNS.*

USES Reduce frequency of relapses in patients with relapsing–remitting multiple sclerosis.

CONTRAINDICATIONS Hypersensitivity to glatiramer acetate or mannitol.
CAUTIOUS USE Immunosuppression, history of asthma or other respiratory disorders; pregnancy (category B), lactation. Safe use in children younger than 18 y has not been established.

ROUTE & DOSAGE

Multiple Sclerosis
Adult: **Subcutaneous** 20 mg daily

ADMINISTRATION
Subcutaneous

- Use recommended subcutaneous injection sites: Arms, abdomen, hips, and thighs.
- Reconstitute with supplied diluent, swirl gently, let stand at room temperature until completely dissolved, then use immediately.
- Do not store reconstituted drug. Before reconstitution, store vials at −20° to −10° C (−4° to −14° F).

ADVERSE EFFECTS (≥1%) **Body as a Whole:** *Asthenia, back pain,* chills, facial edema, fever, *flu-like syndrome, infection, pain, arthralgia.* **CNS:** Migraine, agitation, *anxiety, hypotonia.* **CV:** *Chest pain, palpitations,* syncope, tachycardia, *vasodilation.* **GI:** *Diarrhea, nausea,* anorexia, gastroenteritis, vomiting. **Respiratory:** *Dyspnea, rhinitis,* bronchitis. **Skin:** *Rash, pruritus, sweating.* **Other:** *Postinjection reaction (flushing, chest pain, palpitations, anxiety, dyspnea, constriction of throat, urticaria), injection site reactions (erythema, hemorrhage, pain, pruritus, urticaria, swelling),* ecchymoses, *lymphadenopathy,* ear pain, dysmenorrhea, urinary urgency.

NURSING IMPLICATIONS

Assessment & Drug Effects

- Monitor for therapeutic effectiveness: Indicated by longer remission periods and reduced frequency of attacks.
- Assess for systemic postinjection reactions (see PATIENT & FAMILY EDUCATION). Assure patient that reaction is self-limiting. Assess for local reactions at injection sites including erythema, itching, induration, and soreness.
- Monitor for S&S of compromised immune response (e.g., increasing frequency of infections).

Patient & Family Education

- Note: Systemic postinjection reaction with chest pain, palpitations, flushing, urticaria, anxiety, dyspnea, and laryngeal constriction may occur immediately after injection. These symptoms are transient (lasting from 30 sec–30 min), require no treatment, and resolve spontaneously.
- Report any distressing adverse drug effects.

GLIMEPIRIDE

(gli-me′pi-ride)

Amaryl

Classifications: ANTI-DIABETIC; SULFONYLUREA

Therapeutic: ANTIDIABETIC

Prototype: Glyburide

Pregnancy Category: C

AVAILABILITY 1 mg, 2 mg, 4 mg tablets

ACTION & *THERAPEUTIC EFFECT*

Second-generation sulfonylurea hypoglycemic agent that directly stimulates functioning pancreatic beta cells to secrete insulin, leading to a direct drop in blood glucose. Indirect action leads to increased sensitivity of peripheral insulin receptors, resulting in increased insulin binding in peripheral tissues. *Lowers blood sugar by increasing secretion of insulin from pancreatic beta cells. Glimepiride improves postprandial glycemic control.*

USES

Adjunct to diet and exercise in patients with type 2 diabetes.

CONTRAINDICATIONS

Hypersensitivity to glimepiride, diabetic ketoacidosis; nondiabetic patients with renal glycosuria; lactation.

CAUTIOUS USE Previous hypersensitivity to other sulfonylureas, sulfonamides, or thiazide diuretics; hypoglycemia or conditions predisposing to hypoglycemia (e.g., prolonged nausea and vomiting, alcohol ingestion, surgery; renal or hepatic function impairment, severe infections); pregnancy (category C). Safe use in children is not established.

ROUTE & DOSAGE

Type 2 Diabetes Mellitus

Adult: PO Start with 1–2 mg once daily with breakfast or first main

Common adverse effects in *italic*, life-threatening effects <u>underlined</u>; generic names in **bold**; classifications in SMALL CAPS; ♣ Canadian drug name; ◍ Prototype drug

697

meal, may increase to usual maintenance dose of 1–4 mg once daily (max: 8 mg/day)

ADMINISTRATION

Oral
- Give with breakfast or first main meal.
- Note: Maximum starting dose is 2 mg or less. With renal or hepatic insufficiency, initial recommended dose is 1 mg.
- Store in tightly closed container at 15°–30° C (59°–86° F).

ADVERSE EFFECTS (≥1%) **CNS:** Dizziness, asthenia, headache, blurred vision, changes in accommodation. **GI:** Nausea, vomiting, diarrhea, abdominal pain. **Hematologic:** Leukopenia, agranulocytosis (rare), thrombocytopenia. **Metabolic:** Hypoglycemia. **Skin:** Rash, pruritus, erythema, urticaria, maculopapular eruptions.

INTERACTIONS Drug: Hypoglycemic effects may be potentiated by other highly protein-bound drugs (e.g., ADRENERGIC ANTAGONISTS, **chloramphenicol,** MAO INHIBITORS, NSAIDS, **probenecid,** SALICYLATES, SULFONAMIDES, **warfarin**). CORTICOSTEROIDS, **phenytoin, isoniazid, nicotinic acid,** SYMPATHOMIMETIC AMINES, THIAZIDE DIURETICS may attenuate effects of glimepiride. **Herbal: Ginseng, garlic** may increase hypoglycemic effects.

PHARMACOKINETICS Absorption: Completely absorbed from GI tract. **Onset:** 1 h. **Peak:** 2–3 h. **Distribution:** Greater than 99.5% protein bound; probably secreted into breast milk. **Metabolism:** In liver by CYP2C9. **Elimination:** 60% in urine, 40% in feces. **Half-Life:** 5–9 h.

NURSING IMPLICATIONS

Assessment & Drug Effects
- Lab tests: Monitor fasting and postprandial blood glucose frequently. Monitor HgbA1C every 3–6 mo. Monitor periodically during long-term therapy: LFTs, serum osmolarity, serum sodium, and CBC with differential.
- Monitor for hypoglycemia especially with concurrent drugs which enhance hypoglycemic effects.

Patient & Family Education
- Take a missed dose as soon as possible unless it is almost time for next dose; NEVER take two doses at the same time.
- Avoid drinking alcohol or using OTC drugs without informing prescriber.
- Learn about adverse reactions and drug interactions.

GLIPIZIDE

(glip′i-zide)

Glucotrol, Glucotrol XL
Classifications: ANTIDIABETIC; SULFONYLUREA
Therapeutic: ANTIDIABETIC
Prototype: Glyburide
Pregnancy Category: C

AVAILABILITY 5 mg, 10 mg tablets; 5 mg, 10 mg sustained release tablets

ACTION & *THERAPEUTIC EFFECT* Second-generation sulfonylurea hypoglycemic agent that directly stimulates functioning pancreatic beta cells to secrete insulin, leading to an acute drop in blood glucose. Indirect action leads to altered numbers and sensitivity of peripheral insulin receptors, resulting in increased insulin binding. It also causes inhibition of hepatic glucose production and reduction in serum glucagon levels. *It lowers blood glucose level by stimulating pancreatic beta cells. Glypizide improves postprandial glycemic control.*

USES Adjunct to diet for control of hyperglycemia in patient with type 2 diabetes mellitus.

CONTRAINDICATIONS Hypersensitivity to sulfonylureas; diabetic ketoacidosis; lactation.

CAUTIOUS USE Impaired renal and hepatic function; thyroid disease; debilitated, malnourished patients; G6PD deficiency; trauma; surgery; patients with adrenal or pituitary insufficiency; older adults. **Extend release form:** Severe GI narrowing; pregnancy (category C). Safe use in children has not been established.

ROUTE & DOSAGE

Control of Hyperglycemia

Adult: **PO** 2.5–5 mg/day 30 min before breakfast, may increase by 2.5–5 mg q1–2wk; greater than 15 mg/day in divided doses 30 min before morning and evening meal (max: 40 mg/day); 5–10 mg sustained release tablets once/day

ADMINISTRATION

Oral

- Give once daily dosing 30 min before the first meal of the day.
- Ensure that sustained release form of drug is not chewed or crushed. It **must be** swallowed whole.
- Store in tightly closed, light-resistant container at 15°–30° C (59°–86° F).

ADVERSE EFFECTS (≥1%) **GI:** Nausea, diarrhea, constipation, gastralgia, cholestatic jaundice (rare). **Metabolic:** Hepatic porphyria, hypoglycemia. **Skin:** Erythema, morbilliform or maculopapular rash, pruritus, urticaria, eczema (transient). **Body as a Whole:** Hypersensitivity (fatigue, drowsiness, hunger, GI distress with heartburn, abdominal pain, anorexia). **CNS:** Transient drowsiness, headache, anxiety, ataxia, confusion; seizures, coma. **CV:** Tachycardia. **Special Senses:** Visual disturbances.

INTERACTIONS Drug: Alcohol produces **disulfiram**-like reaction in some patients; ORAL ANTICOAGULANTS, **chloramphenicol, clofibrate, phenylbutazone,** MAO INHIBITORS, SALICYLATES, **probenecid,** SULFONAMIDES may potentiate hypoglycemic actions; THIAZIDES may antagonize hypoglycemic effects; **cimetidine** may increase glipizide levels, causing hypoglycemia. **Herbal: Ginseng, garlic** may increase hypoglycemic effects.

PHARMACOKINETICS Absorption: Readily from GI tract. **Onset:** 15–30 min. **Peak:** 1–2 h. **Duration:** Up to 24 h. **Metabolism:** Metabolized extensively in liver. **Elimination:** Mainly in urine with some excretion via bile in feces. **Half-Life:** 3–5 h.

NURSING IMPLICATIONS

Assessment & Drug Effects

- Observe response to the initial dose, especially in older adult or debilitated patients; early signs of hypoglycemia are easily overlooked.
- Lab tests: Monitor fasting and postprandial blood glucose, and periodic HgbA1C. Monitor periodically during long-term therapy: LFTs, serum electrolytes, and serum osmolarity.

Common adverse effects in *italic*, life-threatening effects underlined; generic names in **bold**; classifications in SMALL CAPS; ♣ Canadian drug name; ⊘ Prototype drug

699

G

- Patients transferred from a sulfonyl-urea with a long half-life (e.g., chlorpropamide, half-life: 30–40 h) **must be** made aware of the potential for hypoglycemic responses (see Appendix F) for 1–2 wk because of potential overlapping of drug effect.
- Note: The first signs of hypoglycemia may be hard to detect in patients receiving concurrent beta-blockers or older adults.

Patient & Family Education

- Treat mild hypoglycemia (reaction without loss of consciousness or neurologic symptoms) with PO glucose and adjustment of dosage and meal pattern; monitor closely for at least 5–7 days to assure reestablishment of safe control. Severe hypoglycemia requires emergency hospitalization to permit treatment to maintain a blood glucose level above 100 mg/dL.
- Test fasting and postprandial blood glucose frequently.
- Keep all follow-up medical appointments and adhere to dietary instructions, regular exercise program, and scheduled and blood testing.
- When a drug that affects the hypoglycemic action of sulfonylureas (see DRUG INTERACTIONS) is withdrawn or added to the glipizide regimen, be alert to the added danger of loss of control. Urine and blood glucose tests and test for ketone bodies should be carefully monitored.
- Report promptly severe skin rash and pruritus as these may indicate a need for discontinuation of drug. Symptoms usually subside rapidly when drug is withdrawn.

GLUCAGON

(gloo´ka-gon)
GlucaGen
Classification: ANTIHYPOGLYCEMIC

Therapeutic: ANTIHYPOGLYCEMIC; DIAGNOSTIC TEST AID
Pregnancy Category: B

AVAILABILITY 1 mg powder for injection

ACTION & *THERAPEUTIC EFFECT* Recombinant glucagon identical to glucagon produced by alpha cells of islets of Langerhans. Stimulates uptake of amino acids and their conversion to glucose precursors. Promotes lipolysis in liver and adipose tissue with release of free fatty acid and glycerol, which further stimulates ketogenesis and hepatic gluconeogenesis. Action in hypoglycemia relies on presence of adequate liver glycogen stores. *Increases blood glucose secondary to gluconeogenesis, which is the breakdown of glycogen to glucose in the liver.*

USES Hypoglycemia, radiologic studies of GI tract.
UNLABELED USES GI disturbances associated with spasm, cardiovascular emergencies, and to overcome cardiotoxic effects of beta-blockers, quinidine, tricyclic antidepressants; as an aid in abdominal imaging; choking due to esophageal foreign body impaction.

CONTRAINDICATIONS Hypersensitivity to glucagon or protein compounds; depleted glycogen stores in liver; insulinemia; pheochromocytoma.
CAUTIOUS USE Cardiac disease, CAD; adrenal insufficiency; malnutrition; children; pregnancy (category B), lactation.

ROUTE & DOSAGE

Hypoglycemia
Adult/Adolescent/Child (over 25 kg): **IM/IV/Subcutaneous** 1 mg,

Common adverse effects in *italic*, life-threatening effects underlined; generic names in **bold**; classifications in SMALL CAPS; ◆ Canadian drug name; ☺ Prototype drug

may repeat q5–20min if no response for 1–2 more doses
Child: **IM/IV/Subcutaneous** *Weight less than 25 kg or younger than 8 y:* 0.5 mg (max dose: 1 mg)

Diagnostic Aid to Relax Stomach or Upper GI Tract
Adult: **IV** 0.2–0.5 mg **IM** 1 mg

Diagnostic Aid for Colon Exam
Adult: **IV** 0.5–0.75 mg **IM** 1–2 mg

ADMINISTRATION

Note: 1 mg = 1 unit

Subcutaneous/Intramuscular

- Dilute 1 unit (1 mg) of glucagon with 1 mL of diluent supplied by manufacturer.
- Use immediately after reconstitution of dry powder. Discard any unused portion.
- Note: Glucagon is incompatible in syringe with any other drug.

Intravenous

PREPARE: **Direct:** Prepare as noted for intramuscular injection. Do not use a concentration greater than 1 unit/mL.
ADMINISTER: **Direct:** Give 1 unit or fraction thereof over 1 min.
- May be given through a Y-site D5W (not NS) infusing.
INCOMPATIBILITIES Solution/additive: **Sodium chloride.**

- Store unreconstituted vials and diluent at 20°–25° C (68°–77° F).

ADVERSE EFFECTS (≥1%) **GI:** Nausea and vomiting. **Body as a Whole:** Hypersensitivity reactions. **Skin:** Stevens-Johnson syndrome (erythema multiforme). **Metabolic:** Hyperglycemia, hypokalemia.

INTERACTIONS Drug: May enhance effect of ORAL ANTICOAGULANTS.

PHARMACOKINETICS Onset: 5–20 min. **Peak:** 30 min. **Duration:** 1–1.5 h. **Metabolism:** In liver, plasma, and kidneys. **Half-Life:** 3–10 min.

NURSING IMPLICATIONS

Assessment & Drug Effects

- Be prepared to give IV glucose if patient fails to respond to glucagon. Notify prescriber immediately.
- Note: Patient usually awakens from (diabetic) hypoglycemic coma 5–20 min after glucagon injection. Give PO carbohydrate as soon as possible after patient regains consciousness.
- Note: After recovery from hypoglycemic reaction, symptoms such as headache, nausea, and weakness may persist.

Patient & Family Education

- Note: Prescriber may request that a responsible family member be taught how to administer glucagon subcutaneously or IM for patients with frequent or severe hypoglycemic reactions. Notify prescriber promptly whenever a hypoglycemic reaction occurs so the reason for the reaction can be determined.
- Review package insert and directions (see ADMINISTRATION).

GLYBURIDE ⊙

(glye′byoor-ide)
DiaBeta, Euglucon ♣, Glynase
Classifications: ANTIDIABETIC; SULFONYLUREA
Therapeutic: ANTIDIABETIC
Pregnancy Category: C

AVAILABILITY 1.25 mg, 2.5 mg, 5 mg tablets; 1.5 mg, 3 mg, 6 mg micronized tablets

Common adverse effects in *italic*, life-threatening effects underlined; generic names in **bold**; classifications in SMALL CAPS; ♣ Canadian drug name; ⊙ Prototype drug

701

ACTION & *THERAPEUTIC EFFECT*

One of the most potent of the second-generation sulfonylurea hypoglycemic agents. Appears to lower blood sugar concentration in both diabetic and nondiabetic individuals by sensitizing pancreatic beta cells to release insulin in the presence of elevated serum glucose levels. *Blood glucose-lowering effect persists during long-term glyburide treatment, but there is a gradual decline in meal-stimulated secretion of endogenous insulin toward pretreatment levels.*

USES Adjunct to diet and exercise to lower blood glucose in patients with type 2 diabetes mellitus.

CONTRAINDICATIONS Hypersensitivity to glyburide or sulfonylureas; diabetic ketoacidosis; type I diabetes mellitus; major surgery; severe trauma; severe infection; withhold 14 days before labor and delivery; lactation.

CAUTIOUS USE History of sulfonamide hypersensitivity; renal and hepatic impairment; cardiovascular disease; thyroid disease; mild renal impairment or hepatic disease; adrenal or pituitary insufficiency; older adults, debilitated, or malnourished patients; pregnancy (category C); children.

ROUTE & DOSAGE

Control of Hyperglycemia

Adult: **PO** 1.25–5 mg/day with breakfast, may increase by 2.5–5 mg q1–2wk; greater than 15 mg/day should be given in divided doses with morning and evening meal (max: 20 mg/day);

Micronized 1.5–3 mg/day (max: 12 mg/day)

ADMINISTRATION

Oral

- Give once daily in the morning with breakfast or with first main meal.
- Store in tightly closed, light-resistant container at 15°–30° C (59°–86° F).

ADVERSE EFFECTS (≥1%) **Metabolic:** Hypoglycemia. **GI:** Epigastric fullness, heartburn, nausea, vomiting. **Skin:** Pruritus, erythema, urticarial or cholestatic jaundice (rare) morbilliform eruptions. **Special Senses:** Blurred vision.

INTERACTIONS Drug: **Alcohol** causes disulfiram-like reaction in some patients; ORAL ANTICOAGULANTS, **chloramphenicol, clofibrate, phenylbutazone,** MAO INHIBITORS, SALICYLATES, **probenecid,** SULFONAMIDES, **clarithromycin** may potentiate hypoglycemic actions; THIAZIDES may antagonize hypoglycemic effects; **cimetidine** may increase glyburide levels, causing hypoglycemia. **Herbal: Ginseng, garlic** may increase hypoglycemic effects.

PHARMACOKINETICS Absorption: Readily absorbed from GI tract. **Onset:** 15–60 min. **Peak:** 1–2 h. **Duration:** Up to 24 h. **Distribution:** Distributed in highest concentrations in liver, kidneys, and intestines; crosses placenta. **Metabolism:** Extensively in liver. **Elimination:** Equally in urine and feces. **Half-Life:** 10 h.

NURSING IMPLICATIONS

Assessment & Drug Effects

- Monitor blood glucose levels carefully during the dangerous early treatment period when dosage is being individualized. Older

Common adverse effects in *italic*, life-threatening effects underlined; generic names in **bold**; classifications in SMALL CAPS; ♣ Canadian drug name; ☻ Prototype drug

adults are especially vulnerable to glyburide-induced hypoglycemia (see Appendix F) because the antidiabetic agent is long-acting.
- Note: The first signs of hypoglycemia may be hard to detect when the patient is also receiving a beta-blocker or is an older adult.
- Lab tests: Monitor at regular intervals: Fasting and postprandial blood glucose, HbA1C, and LFTs.

Patient & Family Education
- Eat or drink some form of sugar (e.g., corn syrup, orange juice with 2 or 3 tsp of table sugar) when symptoms of hypoglycemia occur. Report reaction to prescriber promptly.
- Remember that loss of control of diabetes may result from stress such as fever, surgery, trauma, or infection. Check blood glucose more frequently during stress periods.
- Keep all follow-up medical appointments and adhere to dietary instructions, regular exercise program, and scheduled blood testing.
- Report blurred vision to prescriber.

GLYCERIN
(gli′ser-in)

Fleet Babylax, Glycerol, Osmoglyn

GLYCERIN ANHYDROUS
Ophthalgan

Classifications: HYPEROSMOTIC LAXATIVE; ANTIGLAUCOMA
Therapeutic: HYPEROSMOTIC LAXATIVE; ANTIGLAUCOMA; OCULAR OSMOTIC DIURETIC
Pregnancy Category: C

AVAILABILITY 50% oral solution; suppositories; 4 mL/applicator, ophth solution

ACTION & *THERAPEUTIC EFFECT* Oral: Glycerin raises plasma osmotic pressure by withdrawing fluid from extravascular spaces; lowers ocular tension by decreasing volume of intraocular fluid. May also reduce CSF pressure. **Ocular topic application:** Reduces edema by hydroscopic effect. **Glycerin suppositories:** Apparently work by causing dehydration of exposed tissue, which produces an irritant effect, and by absorbing water from tissues, thus creating more bowel mass. Both actions stimulate peristalsis in the large bowel. *Reduces intraocular pressure by lowering intraocular fluid. Relieves constipation by absorption of water and stimulation of peristalsis.*

USES Orally to reduce elevated intraocular pressure (IOP) before or after surgery in patients with acute narrow-angle glaucoma, retinal detachment, or cataract and to reduce elevated CSF pressure. Sterile glycerin (anhydrous) is used topically to reduce superficial corneal edema resulting from trauma, surgery, or disease and to facilitate ophthalmoscopic examination. Used rectally (suppository or enema) to relieve constipation.

UNLABELED USES To reduce mortality due to strokes in older adults.

CONTRAINDICATIONS Diabetic ketoacidosis; moderate or severe renal impairment (CrCl less than 50 mL/min), renal failure.

CAUTIOUS USE Cardiac disease, mild renal impairment; hepatic disease; diabetes mellitus; thyroid disease; dehydrated or older adults; pregnancy (category C); lactation.

ROUTE & DOSAGE

Decrease IOP

Adult/Child: **PO** 1–1.8 g/kg 1–1.5 h before ocular surgery, may repeat q5h

Common adverse effects in *italic*, life-threatening effects underlined; generic names in **bold**; classifications in SMALL CAPS; ♣ Canadian drug name; ⊙ Prototype drug

703

Constipation

Adult/Child (6 y or older): **PR**
Insert 1 suppository or 5–15 mL of
enema high into rectum and retain
for 15 min

Child (younger than 6 y): **PR** Insert
1 infant suppository or 2–5 mL of
enema high into rectum and retain
for 15 min

Neonate: **PR** 0.5 mL of rectal
solution (enema)

Reduction of Corneal Edema

Adult: **Topical** 1–2 drops instilled
into eye q3–4h

ADMINISTRATION

Oral
- Pour oral solution over crushed
  ice and have patient sip through a
  straw.
- Prevent or relieve headache (from
  cerebral dehydration) by having
  patient lie down during and after
  administration of drug.

Rectal
- Ensure that suppository is inserted beyond rectal sphincter.

ADVERSE EFFECTS (≥1%) **CNS:**
Headache, dizziness, disorientation.
CV: Irregular heartbeat. **GI:** Nausea, vomiting, thirst, diarrhea, abdominal cramps, rectal discomfort,
hyperemia of rectal mucosa. **Metabolic:** Hyperglycemia, glycosuria,
dehydration, <u>hyperosmolar nonke-
totic coma</u>.

PHARMACOKINETICS Absorption:
Readily absorbed from GI tract after
oral administration; rectal preparations are poorly absorbed. **Onset:**
10 min PO. **Peak:** 30 min–2 h. **Duration:** 4–8 h. **Metabolism:** 80% metabolized in liver; 10–20% metabolized in kidneys to CO_2 and water

or utilized in glucose or glycogen
synthesis. **Elimination:** 7–14% excreted unchanged in urine. **Half-
Life:** 30–40 min.

NURSING IMPLICATIONS

Assessment & Drug Effects
- Consult prescriber regarding fluid
  intake in patients receiving drug
  for elevated IOP. Although hypotonic fluids will relieve thirst and
  headache caused by the dehydrating action of glycerin, these fluids
  may nullify its osmotic effect.
- Monitor glycemic control in diabetics. Drug may cause hyperglycemia (see Appendix F).

Patient & Family Education
- Evacuation usually comes 15–30
  min after administration of glycerin rectal suppository or enema.
- Note: Slight hyperglycemia and
  glycosuria may occur with PO use;
  adjustment in antidiabetic medication dosage may be required.

GLYCOPYRROLATE
(glye-koe-pye′roe-late)

Robinul, Robinul Forte
Classifications: ANTICHOLINERGIC;
ANTIMUSCARINIC; ANTISPASMODIC
Therapeutic: GI ANTISPASMODIC
Prototype: Atropine
Pregnancy Category: B

AVAILABILITY 1 mg, 2 mg tablets;
0.2 mg/mL injection

ACTION & *THERAPEUTIC EFFECT*
Synthetic anticholinergic (antimuscarinic) that inhibits muscarinic action of acetylcholine on autonomic
neuroeffector sites innervated by
postganglionic cholinergic nerves.
*Inhibits motility of GI and genitouri-
nary tract; it also decreases volume
of gastric and pancreatic secretions,
saliva, and perspiration.*

Common adverse effects in *italic*, life-threatening effects <u>underlined</u>; generic names
in **bold**; classifications in SMALL CAPS; ♣ Canadian drug name; ⊙ Prototype drug

USES Adjunctive management of peptic ulcer and other GI disorders associated with hyperacidity, hypermotility, and spasm. Also used parenterally as preanesthetic and intraoperative medication and to reverse neuromuscular blockade.

CONTRAINDICATIONS Glaucoma; asthma; prostatic hypertrophy, obstructive uropathy; obstructive lesions or atony of GI tract; achalasia; severe ulcerative colitis; myasthenia gravis; BPH; urinary tract obstruction; during cyclopropane anesthesia; neonates younger than 1 mo.

CAUTIOUS USE Autonomic neuropathy, hepatic or renal disease; cardiac arrhythmias; pregnancy (category B), lactation.

ROUTE & DOSAGE

Peptic Ulcer

Adult: **PO** 1 mg t.i.d or 2 mg b.i.d. or t.i.d. in equally divided intervals (max: 8 mg/day), then decrease to 1 mg b.i.d. **IM/IV** 0.1–0.2 mg 3–4 times/day

Reversal of Neuromuscular Blockade

Adult/Child: **IV** 0.2 mg administered with 1 mg of neostigmine or 5 mg pyridostigmine

Preanesthetic

Child: **PO** 40–100 mcg/kg t.i.d.–q.i.d. **IM** 4–10 mcg/kg q3–4h
Adult: **IM** 4 mcg/kg 30–60 min before procedure

ADMINISTRATION

Oral
▪ Give without regard to meals.

Intramuscular
▪ Give undiluted, deep into a large muscle.

Intravenous

PREPARE: **Direct:** Give undiluted. ▪ Inspect for cloudiness and discoloration. Discard if present.
ADMINISTER: **Direct:** Give 0.2 mg or fraction thereof over 1–2 min.
INCOMPATIBILITIES **Solution/additive: Chloramphenicol, dexamethasone, diazepam, dimenhydrinate, methohexital, methylprednisolone, pentazocine, phenobarbital, secobarbital, sodium bicarbonate, thiopental. Y-site: Propofol.**

▪ Store at 20°–25° C (68°–77° F).

ADVERSE EFFECTS (≥1%) **Body as a Whole:** *Decreased sweating,* weakness. **CNS:** Dizziness, drowsiness, overdosage (<u>neuromuscular blockade</u> with curare-like action leading to muscle weakness and <u>paralysis</u> is possible). **CV:** Palpitation, tachycardia. **GI:** *Xerostomia,* constipation. **GU:** *Urinary hesitancy or retention.* **Special Senses:** Blurred vision, mydriasis.

INTERACTIONS Drug: Amantadine, ANTIHISTAMINES, TRICYCLIC ANTIDEPRESSANTS, **quinidine, disopyramide, procainamide** compound anticholinergic effects; decreases **levodopa** effects; **methotrimeprazine** may precipitate extrapyramidal effects; decreases antipsychotic effects (decreased absorption) of PHENOTHIAZINES.

PHARMACOKINETICS Absorption: Poorly and incompletely absorbed from GI tract. **Onset:** 1 min IV; 15–30 min IM/Subcutaneous; 1 h PO. **Peak:** 30–45 min IM/Subcutaneous; 1 h PO. **Duration:** 2–7 h IM/Subcutaneous; 8–12 h PO. **Distribution:** Crosses placenta. **Metabolism:** Minimally in liver. **Elimination:** 85% in urine. **Half-Life:** 30–70 min (adult), 20–99 min (child), 20–120 min (infant).

G

Common adverse effects in *italic*, life-threatening effects <u>underlined</u>; generic names in **bold**; classifications in SMALL CAPS; ♣ Canadian drug name; ℞ Prototype drug

705

NURSING IMPLICATIONS

Assessment & Drug Effects

- Incidence and severity of adverse effects are generally dose related.
- Monitor I&O ratio and pattern particularly in older adults. Watch for urinary hesitancy and retention.
- Monitor vital signs, especially when drug is given parenterally. Report any changes in heart rate or rhythm.

Patient & Family Education

- Avoid high environmental temperatures (heat prostration can occur because of decreased sweating).
- Do not drive or engage in other potentially hazardous activities requiring mental alertness until response to drug is known.
- Use good oral hygiene, rinse mouth with water frequently and use a saliva substitute to lessen effects of dry mouth.

GOLD SODIUM THIOMALATE

(thye-oh-mah'late)

Myochrysine ♦

Classifications: BIOLOGICAL RESPONSE MODIFIER; IMMUNOMODULATOR; DISEASE-MODIFYING ANTIRHEUMATIC DRUG (DMARD)
Therapeutic: ANTIRHEUMATIC; GOLD COMPOUND
Prototype: Auranofin
Pregnancy Category: C

AVAILABILITY 50 mg/mL injection

ACTION & *THERAPEUTIC EFFECT*

Water-soluble gold compound. Appears to act by suppression of phagocytosis, altered immune responses, and possibly by inhibition of prostaglandin synthesis. *Has immunomodulatory and anti-inflammatory effects.*

USES Selected patients (adults and juveniles) with acute rheumatoid arthritis.
UNLABELED USES Psoriatic arthritis, Felty's syndrome.

CONTRAINDICATIONS History of severe toxicity from previous exposure to gold or other heavy metals; severe debilitation; SLE, Sjögren's syndrome in rheumatoid arthritis; renal disease; hepatic dysfunction, history of infectious hepatitis; uncontrolled diabetes or CHF.
CAUTIOUS USE History of drug allergies or hypersensitivity; marked hypertension; history of blood dyscrasias such as granulocytopenia or anemia caused by drug sensitivity; CHF; diabetes mellitus; previous kidney or liver disease; colitis, IBD; compromised cerebral or cardiovascular circulation; pregnancy (category C).

ROUTE & DOSAGE

Rheumatoid Arthritis

Adult: **IM** 10 mg wk 1, 25 mg wk 2, then 25–50 mg/wk to a cumulative dose of 1 g (if improvement occurs, continue at 25–50 mg q2wk for 2–20 wk, then q3–4wk indefinitely or until adverse effects occur)
Child: **IM** 10 mg test dose, then 1 mg/kg/wk or 2.5–5 mg for wk 1 and 2, followed by 1 mg/kg q1–4wk (max single dose: 50 mg)

ADMINISTRATION

Intramuscular

- Agitate vial before withdrawing dose to ensure uniform suspension.
- Give deep into upper outer quadrant of gluteus maximus with

patient lying down. Patient should remain recumbent for at least 30 min after injection because of the danger of "nitritoid reaction" (transient giddiness, vertigo, facial flushing, fainting).
- Observe for allergic reactions.
- Store in tight, light-resistant containers at 15°–30° C (59°–86° F). Do not use if any darker than pale yellow.

ADVERSE EFFECTS (≥1%) **CNS:** Dizziness, syncope, sweating, flushing. **CV:** Bradycardia. **GI:** Hepatitis, metallic taste, *stomatitis,* nausea, vomiting. **Hematologic:** Agranulocytosis, aplastic anemia, eosinophilia (all rare). **Urogenital:** Nephrotic syndrome, glomerulitis with hematuria, *proteinuria.* **Skin:** Transient pruritus, *erythema, dermatitis,* fixed drug eruption, alopecia, shedding of nails, gray to blue pigmentation of skin (chrysiasis). **Special Senses:** Gold deposits in ocular tissues, *photosensitivity.* **Body as a Whole:** Peripheral neuritis, angioneurotic edema, interstitial pneumonitis, anaphylaxis (rare). **Respiratory:** Pulmonary fibrosis.

INTERACTIONS Drug: ANTIMALARIALS, IMMUNOSUPPRESSANTS, **penicillamine, phenylbutazone** increase risk of blood dyscrasias.

PHARMACOKINETICS Absorption: Slowly and irregularly absorbed from IM site. **Peak:** 3–6 h. **Distribution:** Widely distributed, especially to synovial fluid, kidney, liver, and spleen; does not cross blood–brain barrier; crosses placenta. **Metabolism:** Not studied. **Elimination:** 60–90% of dose ultimately excreted in urine; also eliminated in feces; traces may be found in urine for 6 mo. **Half-Life:** 3–168 days.

NURSING IMPLICATIONS
Assessment & Drug Effects
- Lab tests: Prior to each injection, urinalysis for protein, blood, and sediment. Withhold drug and notify prescriber promptly if proteinuria or hematuria develops. Also do baseline Hgb and RBC, WBC count, differential count, platelet count before initiation of therapy and at regular intervals.
- Note: Rapid reduction in hemoglobin level, WBC count below 4000/mm³, eosinophil count above 5%, and platelet count below 100,000/mm³ signify possible toxicity.
- Interview and examine patient before each injection to detect occurrence of transient pruritus or dermatitis (both are common early indications of toxicity), stomatitis (sore tongue, palate, or throat), metallic taste, indigestion, or other signs and symptoms of possible toxicity. Interrupt treatment immediately and notify prescriber if any of these reactions occurs.
- Observe for allergic reaction, which may occur almost immediately after injection, 10 min after injection, or at any time during therapy. Withhold drug and notify prescriber if observed. Keep antidote dimercaprol (BAL) on hand during time of injection.

Patient & Family Education
- Therapeutic effects may not appear until after 2 mo of therapy.
- Notify prescriber of rapid improvement in joint swelling; this is indicative that you are closely approaching drug tolerance level.
- Use protective measures in sunlight. Exposure to sunlight may aggravate gold dermatitis.
- Notify prescriber at the appearance of unexplained skin bruising; this is always an indication for doing a platelet count.

Common adverse effects in *italic,* life-threatening effects underlined; generic names in **bold;** classifications in SMALL CAPS; ♣ Canadian drug name; ⊕ Prototype drug

707

- Know possible adverse reactions and report any symptom suggestive of toxicity immediately to prescriber.

GOLIMUMAB
(go-li-mu'mab)

Simponi

Classifications: BIOLOGIC RE-SPONSE MODIFIER; MONOCLONAL ANTIBODY; TUMOR NECROSIS FACTOR (TNF) MODIFIER; DISEASE-MODIFYING ANTIRHEUMATIC DRUG (DMARD)
Therapeutic: ANTIRHEUMATIC (DMARD); ANTIPSORIATIC
Prototype: Etanercept
Pregnancy Category: B

AVAILABILITY 50 mg/0.5 mL prefilled syringe solution; 50 mg/0.5 mL SmartJect auto injector solution

ACTION & *THERAPEUTIC EFFECT* A monoclonal antibody that binds to TNF-alpha, thus preventing it from binding to its receptors. TNF (a cytokine) is part of the immune and inflammatory response. However, elevated levels of TNF are found in the synovial fluids of rheumatoid arthritis (RA) patients. *Effectiveness is indicated by improved RA symtomatology and/or decreased inflammation in other inflammatory disorders.*

USES Treatment of moderately to severely active rheumatoid arthritis, active ankylosing spondylitis, and active psoriatic arthritis

CONTRAINDICATIONS Hypersensitivity to golimumab; serious infection or sepsis; fungal infection; history of TB or opportunistic infection; live vaccines; agranulocytosis; malignancy; lactation.

CAUTIOUS USE History of hepatitis B; rheumatoid arthritis; chronic or recurrent infections; history of HBV

infection; CHF; demyelization disorders, MS; cytopenias; pregnancy (category B). Safe use in children not established.

ROUTE & DOSAGE

Rheumatoid Arthritis
Adult: **Subcutaneous** 50 mg qmo
Ankylosing Spondylitis
Adult: **Subcutaneous** 50 mg qmo
Psoriatic Arthritis
Adult: **Subcutaneous** 50 mg qmo

ADMINISTRATION

Subcutaneous
- Allow prefilled syringe/autoinjector to come to room temperature for 30 min prior to injection. Do not warm any other way.
- Do not shake the autoinjector at any time. After injection, do not pull autoinjector away from skin until a second click sound (3–15 sec after the first sound) is heard.
- Rotate injection sites. Do not inject into areas that are tender, bruised, red, or hard.
- Do not initiate treatment in anyone with an active infection.
- Store refrigerated at 2°–8° C (36°–46° F) and protect from light by keeping in carton until use.

ADVERSE EFFECTS (≥1%) Body as a Whole: Influenza, injection site erythema. **CNS:** Dizziness, pyrexia. **CV:** Hypertension. **GI:** Oral herpes. **Metabolic:** Elevated ALT and AST levels. **Respiratory:** Bronchitis, *nasopharyngitis*, sinusitis, *upper respiratory tract infection*. **Skin:** Paraesthesia.

INTERACTIONS Drug: Abatacept, anakinra and other TNF-ALPHA BLOCKERS may increase the risk of serious infection.

PHARMACOKINETICS Peak: 2–6 days. **Half-Life:** 2 wk.

G

NURSING IMPLICATIONS

Assessment & Drug Effects

- Monitor closely for S&S of infection.
- Withhold drug and notify prescriber if symptoms of an infection develop.
- Monitor for and report new-onset and exacerbations of psoriasis.
- Lab tests: Baseline and periodic TB tests, periodic LFTs.

Patient & Family Education

- Contact prescriber immediately for any of the following: Symptoms of infection; jaundice; extreme fatigue; poor appetite or vomiting; or pain in the upper, right abdomen.
- If a case of pre-existing psoriasis worsens or if a new rash develops, contact prescriber.

GOSERELIN ACETATE

(gos-er'e-lin)

Zoladex
Classification: GONADOTROPIN-RELEASING HORMONE (GnRH) ANALOG
Therapeutic: GnRH ANALOG
Prototype: Leuprolide
Pregnancy Category: X for endometriosis, or endometrial thinning, and **Category D** for breast cancer

AVAILABILITY 3.6 mg, 10.8 mg subcutaneous implant

ACTION & *THERAPEUTIC EFFECT*

A synthetic form of luteinizing hormone-releasing hormone (LHRH or GnRH) that inhibits pituitary gonadotropin secretion. *With chronic administration, serum testosterone levels fall into the range normally seen with surgically castrated men.*

USES Prostate cancer, breast cancer. Endometrial thinning agent prior to endometrial ablation for dysfunctional uterine bleeding, endometriosis.
UNLABELED USES Benign prostatic hyperplasia, uterine leiomyomas.

CONTRAINDICATIONS Known hypersensitivity to an LHRH; hypercalcemia; pregnancy (category X for endometriosis, endometrial thinning and category D for breast cancer); lactation.
CAUTIOUS USE Renal impairment; family history of osteoporosis; osteoporosis; patients at risk for spinal cord compression; DM; CVD. Safety and efficacy in children are not established.

ROUTE & DOSAGE

Prostate Cancer, Breast Cancer, Endometriosis
Adult: **Subcutaneous** 3.6 mg q28days, 10.8 mg depot q12wk
Endometrial Thinning Prior to Endometrial Ablation
Adult: **Subcutaneous** 3.6 mg q28days

ADMINISTRATION

Subcutaneous

- Follow manufacturer's directions exactly for implanting the drug subcutaneously in the upper abdominal wall.
- Store at room temperature not to exceed 25° C (77° F).

ADVERSE EFFECTS (≥1%) **CNS:** Headache, tumor flare. **Endocrine:** Gynecomastia, breast swelling and tenderness, *postmenopausal symptoms* (*hot flashes*, vaginal dryness). **GI:** Nausea. **Urogenital:** Vaginal spotting, breakthrough bleeding, decreased libido, *impotence.* **Musculoskeletal:** Bone pain, bone loss **CV:** Increased risk of MI (men only).

G

Common adverse effects in *italic*, life-threatening effects <u>underlined</u>; generic names in **bold**; classifications in SMALL CAPS; ♣ Canadian drug name; ⊙ Prototype drug

709

DIAGNOSTIC TEST INTERFER-ENCE Interferes with pituitary gonadotropic and gonadal function tests during and for 12 wk after treatment.

PHARMACOKINETICS Absorption: Rapidly absorbed following subcutaneous administration. **Duration:** 29 days. **Elimination:** Excreted by kidneys. **Half-Life:** 4.9 h.

NURSING IMPLICATIONS

Assessment & Drug Effects

- Monitor carefully during the first month of therapy for S&S of spinal cord compression or ureteral obstruction in patients with prostate cancer. Report immediately to prescriber.
- Anticipate a transient worsening of symptoms (e.g., bone pain) during the first weeks of therapy in patients with prostate cancer.
- Lab tests: Periodic fasting blood glucose level.

Patient & Family Education

- Note: Sexual dysfunction in men and hot flashes may accompany drug use.
- Notify prescriber immediately of symptoms of spinal cord compression or urinary obstruction.

GRANISETRON

(gran'i-se-tron)

Granisol, Kytril, Sancuso
Classifications: ANTIEMETIC; 5-HT₃ ANTAGONIST
Therapeutic: ANTIEMETIC
Prototype: Ondansetron
Pregnancy Category: B

AVAILABILITY 1 mg tablets; 1 mg/mL injection; 2 mg/10 mL oral solution; 3.1 mg transdermal patch

ACTION & *THERAPEUTIC EFFECT*
Granisetron is a selective serotonin (5-HT₃) receptor antagonist. Serotonin receptors of the 5-HT₃ type are located centrally in the chemoreceptor trigger zone, and peripherally on the vagal nerve terminals. Serotonin released from the wall of the small intestine stimulates these vagal afferent neurons through the serotonin receptors, and initiates vomiting reflex. *Effective in preventing nausea and vomiting associated with cancer chemotherapy.*

USES Prevention of nausea and vomiting associated with initial and repeat courses of emetogenic cancer therapy, including high-dose cisplatin, postoperative nausea and vomiting.

CONTRAINDICATIONS Hypersensitivity to granisetron, or benzyl alcohol; GI obstruction; neonates.
CAUTIOUS USE Hypersensitivity to ondansetron or similar drugs; liver disease; pregnancy (category B), lactation; children 2 y or younger.

ROUTE & DOSAGE

Chemotherapy-Related Nausea and Vomiting

Adult/Child: **IV** *Older than 2 y,* 10 mcg/kg, beginning at least 30 min before initiation of chemotherapy (up to 40 mcg/kg per dose has been used) **PO** 1 mg b.i.d., start 1 mg up to 1 h prior to chemotherapy, then second tab 12 h later OR 2 mg daily
Adult: **Transdermal** Apply 1 patch q5days

Postoperative Nausea and Vomiting

Adult: **IV** 1 mg before anesthesia induction or before reversal of anesthesia

ADMINISTRATION

Oral
- Give only on the day of chemotherapy.

Transdermal
- Apply patch to upper outer arm 24–48 h before start of chemotherapy.
- Remove patch no sooner than 24 h after completion of chemotherapy.
- Patch may be left in place for up to 7 days.

Intravenous

PREPARE: **Direct:** Give undiluted. **IV Infusion:** Dilute in NS or D5W to a total volume of 20–50 mL.
- Prepare infusion at time of administration; do not mix in solution with other drugs.

ADMINISTER: **Direct:** Give a single dose over 30 sec. **IV Infusion:** Infuse diluted drug over 5 min or longer; complete infusion 20–30 min prior to initiation of chemotherapy.

INCOMPATIBILITIES **Y-site:** **Amphotericin B, doxorubicin.**

- Store at 15°–30° C (59°–86° F) for 24 h after dilution under normal lighting conditions.

ADVERSE EFFECTS (≥1%) **CNS:** *Headache,* dizziness, somnolence, insomnia, labile mood, anxiety, fatigue. **GI:** Constipation, diarrhea, elevated liver function tests.

INTERACTIONS Drug: Ketoconazole may inhibit metabolism.

PHARMACOKINETICS Onset: Several minutes. **Duration:** Approximately 24 h. **Distribution:** Widely distributed in body tissues. **Metabolism:** Appears to be metabolized in liver. **Elimination:** Excreted in urine as metabolites. **Half-Life:** 10–11 h in cancer patients, 4–5 h in healthy volunteers.

NURSING IMPLICATIONS

Assessment & Drug Effects
- Monitor the frequency and severity of nausea and vomiting.
- Lab tests: Monitor LFTs; elevated AST and ALT values usually normalize within 2 wk of last dose.
- Assess for headache, which usually responds to nonnarcotic analgesics.

Patient & Family Education
- Note: Headache requiring an analgesic for relief is a common adverse effect.
- Learn ways to manage constipation.

GRISEOFULVIN MICROSIZE
(gri-see-oh-ful′vin)

Fulvicin-U/F, Grifulvin V, Grisactin, Grisovin-FP ✦

GRISEOFULVIN ULTRAMICROSIZE

Fulvicin P/G, Grisactin Ultra, Gris-PEG

Classification: ANTIFUNGAL ANTIBIOTIC
Therapeutic: ANTIFUNGAL
Pregnancy Category: C second and third trimester

AVAILABILITY Griseofulvin Microsize: 250 mg, 500 mg tablets; 250 mg capsules; 125 mg/5 mL suspension; **Griseofulvin Ultramicrosize:** 125 mg, 165 mg, 250 mg, 330 mg tablets.

ACTION & *THERAPEUTIC EFFECT* Arrests metaphase of cell division by disrupting mitotic spindle structure in fungal cells. Deposits in keratin precursor cells and has special affinity for diseased tissue. It is tightly bound to new keratin of skin, hair, and nails that becomes highly resistant to fungal invasion. *Effective against various species of* Epidermophyton, Microsporum, *and* Trichophyton *(has no effect on other fungi, including* Candida, *bacteria, and yeasts).*

USES Mycotic disease of skin, hair, and nails not amenable to conventional topical measures. Concomitant use of appropriate topical agent may be required, particularly for tinea pedis.

UNLABELED USES Raynaud's disease, angina pectoris, and gout.

CONTRAINDICATIONS Hypersensitivity to griseofulvin; porphyria; hepatocellular failure; SLE; first trimester of pregnancy; lactation, prophylaxis against fungal infections.

CAUTIOUS USE Penicillin-sensitive patients (possibility if cross-sensitivity with penicillin exists; however, reportedly penicillin-sensitive patients have been treated without difficulty); hepatic impairment; pregnancy (category C second and third trimester); children 2 y or younger.

ROUTE & DOSAGE

Tinea Corporis, Tinea Cruris, Tinea Capitis

Adult: **PO** 500 mg microsize or 330–375 mg ultramicrosize daily in single or divided doses
Child: **PO** 10–20 mg/kg/day microsize or 5–10 mg/kg/day ultramicrosize in single or divided doses

Tinea Pedis, Tinea Unguium

Adult: **PO** 0.75–1 g microsize or 660–750 mg ultramicrosize daily in single or divided doses (decrease microsize dose to

500 mg/day after response is noted)
Child: **PO** 10–20 mg/kg/day microsize or 5–10 mg/kg/day ultramicrosize in single or divided doses

ADMINISTRATION

Oral
- Give with or after meals to allay GI disturbances.
- Give the microsize formulations with a high fat content meal (increases drug absorption rate) to enhance serum levels. Consult prescriber.
- Store at 15°–30° C (59°–86° F) in tightly covered containers unless otherwise directed.

ADVERSE EFFECTS (≥1%) Body as a Whole: Hypersensitivity (urticaria, photosensitivity, skin rashes, Steven-Johnson's syndrome, pruritus, fixed drug eruption, serum sickness syndromes, severe angioedema). **CNS:** *Severe headache,* insomnia, fatigue, mental confusion, impaired performance of routine functions, psychotic symptoms, vertigo. **GI:** Heartburn, nausea, vomiting, diarrhea, flatulence, dry mouth, thirst, decreased taste acuity, anorexia, unpleasant taste, furred tongue, oral thrush. **Hematologic:** <u>Leukopenia</u>, neutropenia, granulocytopenia, punctate basophilia, monocytosis. **Urogenital:** Nephrotoxicity (proteinuria); hepatotoxicity; estrogen-like effects (in children); aggravation of SLE. **Other:** Overgrowth of nonsusceptible organisms; candidal intertrigo.

INTERACTIONS Drug: Alcohol may cause flushing and tachycardia; BARBITURATES may decrease activity of griseofulvin; may decrease hypoprothrombinemic effects of

ORAL ANTICOAGULANTS; may increase **estrogen** metabolism, resulting in breakthrough bleeding, and decrease contraceptive efficacy of ORAL CONTRACEPTIVES.

PHARMACOKINETICS Absorption: Absorbed primarily from duodenum; microsize is variably and unpredictably absorbed; ultramicrosize is almost completely absorbed. **Peak:** 4–8 h. **Distribution:** Concentrates in skin, hair, nails, fat, and skeletal muscle; crosses placenta. **Metabolism:** In liver. **Elimination:** Mainly in urine; some excretion in perspiration. **Half-Life:** 9–24 h.

NURSING IMPLICATIONS

Assessment & Drug Effects

- Inquire about history of sensitivity to griseofulvin, penicillins, or other allergies prior to initiating treatment.
- Monitor food intake. Drug may alter taste sensations, and this may cause appetite suppression and inadequate nutrient intake.
- Lab tests: WBC with differential at least once weekly during first month of therapy or longer; periodic renal function tests and LFTs are also advised.

Patient & Family Education

- Continuing treatment as prescribed to prevent relapse, even if you experience symptomatic relief after 48–96 h of therapy.
- Note: Duration of treatment depends on time required to replace infected skin, hair, or nails, and thus varies with infection site. Average duration of treatment for tinea capitis (scalp ringworm), 4–6 wk; tinea corporis (body ringworm), 2–4 wk; tinea pedis (athlete's foot), 4–8 wk; tinea unguium (nail fungus), at least 4 mo for fingernails, depending on rate of growth, and 6 mo or more for toenails.
- Avoid exposure to intense natural or artificial sunlight, because photo-sensitivity-type reactions may occur.
- Note: Headaches often occur during early therapy but frequently disappear with continued drug administration.
- Avoid alcohol while taking this drug. Disulfiram-type reaction (see Appendix F) are possible with ingestion of alcohol during therapy.
- Pharmacologic effects of oral contraceptives may be reduced. Breakthrough bleeding and pregnancy may occur. Alternative forms of birth control should be used during therapy.

GUAIFENESIN ⓟ
(gwye-fen′e-sin)

Anti-Tuss, GG-Cen, Glyceryl Guaiacolate, Glycotuss, Glytuss, Guiatuss, Humibid, Hytuss, Malotuss, Mytussin, Mucinex, Resyl ♣, Robitussin
Classification: EXPECTORANT
Therapeutic: EXPECTORANT
Pregnancy Category: C

AVAILABILITY 100 mg/5 mL syrup; 100 mg/5 mL, 200 mg/5 mL liquid; 200 mg capsules; 300 mg sustained release capsules; 100 mg, 200 mg, 1200 mg tablets; 600 mg sustained release tablets

ACTION & *THERAPEUTIC EFFECT* Enhances reflex outflow of respiratory tract fluids by irritation of gastric mucosa. *Aids in expectoration by reducing adhesiveness and surface tension of secretions.*

USES To combat dry, nonproductive cough associated with colds

Common adverse effects in *italic*, life-threatening effects underlined; generic names in **bold**; classifications in SMALL CAPS; ♣ Canadian drug name; ⓟ Prototype drug

713

and bronchitis. A common ingredient in cough mixtures.

CONTRAINDICATIONS Hypersensitivity to guaifenesin; cough due to CHF, ACE inhibitor therapy, or tobacco smoking.
CAUTIOUS USE Chronic cough; asthma; pregnancy (category C), lactation.

ROUTE & DOSAGE

Cough
Adult: **PO** 200–400 mg q4h up to 2.4 g/day
Child: **PO** Younger than 2 y, 12 mg/kg/day in 6 divided doses; 2–5 y, 50–100 mg q4h up to 600 mg/day; 6–11 y, 100–200 mg q4h up to 1.2 g/day

ADMINISTRATION

Oral
- Ensure that sustained release form of drug is not chewed or crushed. It **must be** swallowed whole.
- Follow dose with a full glass of water if not contraindicated.
- Carefully observe maximum daily doses for adults and children.

ADVERSE EFFECTS (≥1%) **GI:** Low incidence of nausea. **CNS:** Drowsiness.

DIAGNOSTIC TEST INTERFERENCE May produce color interference with determinations of *urinary 5-hydroxyindoleacetic acid (5-HIAA)* and *vanillylmandelic acid (VMA).*

INTERACTIONS Drug: By inhibiting platelet function, guaifenesin may increase risk of hemorrhage in patients receiving **heparin** therapy.

NURSING IMPLICATIONS

Assessment & Drug Effects
- Monitor for therapeutic effectiveness. Persistent cough may indicate a serious condition requiring further diagnostic work.
- Notify prescriber if high fever, rash, or headaches develop.

Patient & Family Education
- Increase fluid intake to help loosen mucus; drink at least 8 glasses of fluid daily.
- Contact prescriber if cough persists beyond 1 wk.
- Contact prescriber if high fever, rash, or headache develops.

GUANFACINE HYDROCHLORIDE
(gwahn′fa-seen)

Intuniv, Tenex
Classifications: ALPHA-ADRENERGIC AGONIST; CENTRAL-ACTING ANTIHYPERTENSIVE
Therapeutic: ANTIHYPERTENSIVE
Prototype: Methyldopa
Pregnancy Category: B

AVAILABILITY 1 mg, 2 mg tablets; 1 mg, 2 mg, 3 mg, 4 mg extended release tablets

ACTION & *THERAPEUTIC EFFECT*
In cerebral cortex, stimulation of alpha$_2$-adrenoreceptors triggers inhibitory neurons to reduce central sympathetic outflow (i.e., impulses from vasomotor center to heart and blood vessels). **Extended release form:** Targets ADHD symptoms through central alpha$_2$-receptor activity in the prefrontal cortex. *Results in decreased peripheral vascular resistance, thus lowering blood pressure, and a slightly reduced (5 bpm) heart rate. Minimizes the signs and symptoms of ADHD in children.*

USES Management of mild-to-moderate hypertension; attention deficit hyperactivity disorder (ADHD) (**extended form** only).

UNLABELED USES Adjunct in heroin withdrawal; Tourette's syndrome.

CONTRAINDICATIONS Treatment of acute hypertension associated with toxemia of pregnancy; psychiatric disorders that mimic ADHD.

CAUTIOUS USE Severe coronary insufficiency, recent MI, cerebrovascular disease; chronic renal or hepatic failure; older adult; pregnancy (category B), lactation; children younger than 6 y (**extended release form**) and younger than 12 y (**tablet**).

ROUTE & DOSAGE

Hypertension

Adult: **PO** 1 mg/day at bedtime, may be gradually increased to 3 mg/day if needed

Attention Deficit Hyperactivity Disorder

Adult/Adolescent/Child (older than 6 y): **PO** 1 mg daily, titrate up (normal range: 1–4 mg daily)

ADMINISTRATION

Oral

- Ensure that extended release tablets are swallowed whole and not crushed or chewed.
- Usually given as a single dose at bedtime to reduce effect of somnolence.
- Discontinue treatment gradually with planned tapering of schedule.
- Store tablets at 15°–30° C (59°–86° F) in tightly closed container; protect from light.

ADVERSE EFFECTS (≥1%) **CNS:** Confusion, amnesia, mental depression, drowsiness, *dizziness, sedation,* headache, asthenia, *fatigue,* insomnia. **CV:** Bradycardia, palpitation, substernal pain. **Special Senses:** Rhinitis, tinnitus, taste change; vision disturbances, conjunctivitis, iritis. **GI:** *Dry mouth, constipation,* abdominal pain, diarrhea, dysphagia, nausea. **Urogenital:** *Impotence,* testicular disorder, urinary incontinence. **Musculoskeletal:** Leg cramps, hypokinesia. **Skin:** Dermatitis, pruritus, purpura, sweating. **Other:** Dyspnea.

INTERACTIONS Drug: **Alcohol** and other CNS DEPRESSANTS compound sedation and CNS depression. May increase **valproic acid** levels. Use cautiously with CYP3A4 INHIBITORS or INDUCERS.

PHARMACOKINETICS **Absorption:** Readily absorbed from GI tract; 70% protein bound. **Onset:** 2 h; 6 hr (extended release). **Peak:** 6 h. **Duration:** Up to 24 h. **Distribution:** Crosses placenta. **Metabolism:** In liver. **Elimination:** 80% ActHIB, Hiberix, Liquid PedvaxHIB in the urine in 24 h. **Half-Life:** 17 h.

NURSING IMPLICATIONS

Assessment & Drug Effects

- Do not discontinue abruptly; may cause plasma and urinary catecholamine increases leading symptoms of tachycardia, insomnia, anxiety, nervousness. Rebound hypertension (i.e., increases in BP to levels significantly greater than those before therapy) may occur 2–7 days after abrupt drug withdrawal, but serious effects rarely develop.
- Monitor BP until it is stabilized. Report a rise in pressure that occurs toward end of dose interval; a divided dose schedule may be ordered.
- Assess mental status and alertness. Adverse effects tend to be dose-dependent, increasing significantly with doses above 3 mg/day.

Common adverse effects in *italic,* life-threatening effects underlined; generic names in **bold;** classifications in SMALL CAPS; ♣ Canadian drug name; 🔟 Prototype drug

715

Patient & Family Education

- Continue drug even after you feel well. This is a maintenance dosage regimen (dose and dose intervals). If 2 or more doses are missed, consult prescriber about how to re-establish dosage regimen.
- Employ measures to keep mouth moist; saliva substitutes (e.g., Moi-Stir, Xero-Lube) are available OTC. If dry mouth persists longer than 2 wk, patient should check with dentist.
- Do not drive or engage in other potentially hazardous tasks requiring alertness until response to drug is known.
- Avoid alcohol and do not self-medicate with OTC drugs such as sleeping medications, or cough medications without consulting prescriber.

HAEMOPHILUS b CONJUGATE VACCINE (Hib)

(hee-mof'il-us)

ActHIB, Hiberix, Liquid Pedvax-HIB
Pregnancy Category: C
See Appendix J.

HALCINONIDE

(hal-sin'oh-nide)

Halog
Classifications: ANTI-INFLAMMATORY; FLUORINATED STEROID
Therapeutic: ANTI-INFLAMMATORY
Prototype: Hydrocortisone
Pregnancy Category: C

AVAILABILITY 0.1% ointment, cream, solution

ACTION & *THERAPEUTIC EFFECT*
Fluorinated steroid with substituted 17-hydroxyl group. Crosses cell membranes, complexes with nuclear DNA and stimulates synthesis of enzymes thought to be responsible for anti-inflammatory effects. *Exhibits anti-inflammatory, antipyretic, and vasoconstrictive properties.*

USES Relief of pruritic and inflammatory manifestations of corticosteroid-responsive dermatoses.

CONTRAINDICATIONS Use on large body surface area; long-term use; infection; acne vulgaris, acne rosacea, perioral dermatitis.
CAUTIOUS USE Hypersensitivity to corticosteroids; diabetes mellitus; older adults; skin abrasion; pregnancy (category C), lactation.

ROUTE & DOSAGE

Inflammation
Adult: **Topical** Apply thin layer b.i.d. or t.i.d.
Child: **Topical** Apply thin layer once/day

ADMINISTRATION

Topical
- Wash skin gently and dry thoroughly before each application.
- Note: Ointment is preferred for dry scaly lesions. Moist lesions are best treated with solution.
- Do not apply in or around the eyes.
- Do not apply occlusive dressings over areas covered with halcinonide unless specifically prescribed.
- Store at 15°–30° C (59°–86° F).

ADVERSE EFFECTS (≥1%) **Endocrine:** Reversible HPA axis suppression, hyperglycemia, glycosuria. **Skin:** Burning, itching, irritation, erythema, dryness, folliculitis, hypertrichosis, pruritus, acneiform

Common adverse effects in *italic*, life-threatening effects underlined; generic names in **bold**; classifications in SMALL CAPS; ♣ Canadian drug name; Prototype drug

eruptions, hypopigmentation, perioral dermatitis, allergic contact dermatitis, stinging cracking/tightness of skin, secondary infection, skin atrophy, striae, miliaria, telangiectasia.

PHARMACOKINETICS Absorption: Minimum through intact skin; increased from axilla, eyelid, face, scalp, scrotum, or with occlusive dressing.

NURSING IMPLICATIONS

Assessment & Drug Effects
- Discontinue if signs of infection or irritation occur.
- Monitor for systemic corticosteroid effects that may occur with occlusive dressings or topical applications over large areas of skin.

Patient & Family Education
- Do not use an occlusive dressing with this drug unless specifically directed to do so by prescriber.
- Wash your hands before and after applying this topical medicine.
- Do not get any of the medication in your eyes. If you do, rinse it out with plenty of cool tap water.

HALOPERIDOL ℗
(ha-loe-per′i-dole)
Haldol, Peridol✦

HALOPERIDOL DECANOATE
Haldol LA
Classification: ANTIPSYCHOTIC; BUTYROPHENONE
Therapeutic: ANTIPSYCHOTIC
Pregnancy Category: C

AVAILABILITY 0.5 mg, 1 mg, 2 mg, 5 mg, 10 mg, 20 mg tablets; 2 mg/mL oral solution; 5 mg/mL, 50 mg/mL, 100 mg/mL injection

ACTION & *THERAPEUTIC EFFECT*
Blocks postsynaptic dopamine (D_2) receptors in the limbic system of the brain. Decrease in dopamine neurotransmission has been correlated with its antipsychotic effects, and its higher instance of extrapyramidal effects. *Decreases psychotic manifestations and exerts strong antiemetic effect.*

USES Management of manifestations of psychotic disorders and for control of tics and vocal utterances of Gilles de la Tourette's syndrome; for treatment of agitated states in acute and chronic psychoses. Used for short-term treatment of hyperactive children and for severe behavior problems in children of combative, explosive hyperexcitability.
UNLABELED USES Cancer chemotherapy as an antiemetic in doses smaller than those required for antipsychotic effects; treatment of autism; alcohol dependence; chorea.

CONTRAINDICATIONS Parkinson's disease, seizure disorders, coma; severe neutropenia (ANC less than 1000/mm³); alcoholism; severe mental depression, CNS depression.
CAUTIOUS USE Older adult or debilitated patients, urinary retention, pulmonary disease; history of hypocalcemia; glaucoma, severe cardiovascular disorders, long QT syndrome, AV block, bundle-branch block, cardiac arrhythmias, uncompensated heart failure, recent acute MI; hematologic disease; thyrotoxicosis, or hyperthyroidism; pregnancy (category C), lactation. Safe use in children younger than 3 y is not established.

ROUTE & DOSAGE

Psychosis
Adult: **PO** 0.2–5 mg b.i.d. or t.i.d. **IM** 2–5 mg repeated q4h prn; **Decanoate:** 50–100 mg q4wk

---

Common adverse effects in *italic*, life-threatening effects underlined; generic names in **bold**; classifications in SMALL CAPS; ✦ Canadian drug name; ℗ Prototype drug

717

Child: **PO** 0.5 mg/day in 2–3 divided doses, may be increased by 0.5 mg q5–7days to 0.05–0.15 mg/kg/day

Severe Psychosis

Adult: **PO** 3–5 mg b.i.d. or t.i.d., may need up to 100 mg/day **IM** 2–5 mg, may repeat q.h. prn; **Decanoate:** 50–100 mg q4wk *Child:* **PO** 0.05–0.15 mg/kg/day in 2–3 divided doses

Dementia

Geriatric: **PO** 0.25–0.5 mg 1–2 times daily, may increase every 4–7 days (max: 4 mg/day in 2–3 divided doses)

Tourette's Disorder

Adult: **PO** 0.2–5 mg b.i.d. or t.i.d. *Child:* **PO** 0.05–0.075 mg/kg/day in 2–3 divided doses

Pharmacogenetic Dosage Adjustment

CYP3D6 poor metabolizers: Start with 75% of initial dose

ADMINISTRATION

Oral

- Give with a full glass (240 mL) of water or with food or milk.
- Taper dosing regimen when discontinuing therapy. Abrupt termination can initiate extrapyramidal symptoms.

Intramuscular

- Give by deep injection into a large muscle. Do not exceed 3 mL per injection site.
- Have patient recumbent at time of parenteral administration and for about 1 h after injection. Assess for orthostatic hypotension.
- Store in light-resistant container at 15°–30° C (59°–86° F), unless otherwise specified by manufacturer. Discard darkened solutions.

ADVERSE EFFECTS (≥1%) **CNS:** *Extrapyramidal reactions:* Parkinsonian symptoms, dystonia, akathisia, <u>tardive dyskinesia</u> (after long-term use); insomnia, restlessness, anxiety, euphoria, agitation, drowsiness, mental depression, lethargy, fatigue, weakness, tremor, ataxia, headache, confusion, vertigo; <u>neuroleptic malignant syndrome</u>, hyperthermia, grand mal seizures, exacerbation of psychotic symptoms. **CV:** Tachycardia, ECG changes, hypotension, hypertension (with overdosage). **Endocrine:** Menstrual irregularities, galactorrhea, lactation, gynecomastia, impotence, increased libido, hyponatremia, hyperglycemia, hypoglycemia. **Special Senses:** Blurred vision. **Hematologic:** Mild transient <u>leukopenia</u>, <u>agranulocytosis</u> (rare). **GI:** Dry mouth, anorexia, nausea, vomiting, constipation, diarrhea, hypersalivation. **Urogenital:** Urinary retention, priapism. **Respiratory:** <u>Laryngospasm</u>, bronchospasm, increased depth of respiration, bronchopneumonia, <u>respiratory depression</u>. **Skin:** Diaphoresis, maculopapular and acneiform rash, photosensitivity. **Other:** Cholestatic jaundice, variations in liver function tests, decreased serum cholesterol.

INTERACTIONS Drug: CNS DEPRESSANTS, OPIATES, **alcohol** increase CNS depression; may antagonize activity of ORAL ANTICOAGULANTS; ANTICHOLINERGICS may increase intraocular pressure; **methyldopa** may precipitate dementia.

PHARMACOKINETICS Absorption: Well absorbed from GI tract; 60% reaches systemic circulation. **Onset:** 30–45 min IM. **Peak:** 2–6 h PO; 10–20 min IM; 6–7 days decanoate. **Distribution:** Distributes mainly to liver with lower concentration in brain, lung, kidney, spleen, heart. **Metabolism:** In liver. **Elimination:** 40%

Common adverse effects in *italic*, life-threatening effects <u>underlined</u>; generic names in **bold**; classifications in SMALL CAPS; ♣ Canadian drug name; ⦿ Prototype drug

excreted in urine within 5 days; 15% eliminated in feces; excreted in breast milk. **Half-Life:** 13–35 h.

NURSING IMPLICATIONS

Assessment & Drug Effects

- Monitor for therapeutic effectiveness. Because of long half-life, therapeutic effects are slow to develop in early therapy or when established dosing regimen is changed. "Therapeutic window" effect (point at which increased dose or concentration actually decreases therapeutic response) may occur after long period of high doses. Close observation is imperative when doses are changed.
- Target symptoms expected to decrease with successful haloperidol treatment include hallucinations, insomnia, hostility, agitation, and delusions.
- Monitor patient's mental status daily.
- Monitor for neuroleptic malignant syndrome (NMS) (see Appendix F), especially in those with hypertension or taking lithium. Symptoms of NMS can appear suddenly after initiation of therapy or after months or years of taking neuroleptic (antipsychotic) medication. Immediately discontinue drug if NMS suspected.
- Monitor for parkinsonism and tardive dyskinesia (see Appendix F). Risk of tardive dyskinesia appears to be greater in women receiving high doses and in older adults. It can occur after long-term therapy and even after therapy is discontinued.
- Monitor for extrapyramidal (neuromuscular) reactions that occur frequently during first few days of treatment. Symptoms are usually dose related and are controlled by dosage reduction or concomitant administration of antiparkinson drugs.

- Be alert for behavioral changes in patients who are concurrently receiving antiparkinson drugs.
- Monitor for exacerbation of seizure activity.
- Observe patients closely for rapid mood shift to depression when haloperidol is used to control mania or cyclic disorders. Depression may represent a drug adverse effect or reversion from a manic state.
- Lab tests: Monitor WBC count with differential and liver function in patients on prolonged therapy.

Patient & Family Education

- Avoid use of alcohol during therapy.
- Do not drive or engage in other potentially hazardous activities until response to drug is known.
- Discuss oral hygiene with health care provider; dry mouth may promote dental problems. Drink adequate fluids.
- Avoid overexposure to sun or sunlamp and use a sunscreen; drug can cause a photosensitivity reaction.

HEMIN
(hee'min)

Panhematin

Classifications: HEMATOLOGIC; BLOOD DERIVATIVE; ENZYME INHIBITOR
Therapeutic: HEMATOLOGIC ENZYME INHIBITOR

Pregnancy Category: C

AVAILABILITY 313 mg powder for injection

ACTION & *THERAPEUTIC EFFECT*
Derived from processed red blood cells. Represses synthesis of porphyrin in liver or bone marrow by blocking production of delta-aminolevulinic acid (ALA) synthetase, an

---

essential enzyme in the porphyrin-heme biosynthetic pathway. *Effective in ameliorating recurrent attacks of acute intermittent porphyria (AIP).*

USES Acute intermittent porphyria (AIP).

CONTRAINDICATIONS History of hypersensitivity to hemin; anticoagulation therapy; porphyria cutanea tarda.

CAUTIOUS USE Pregnancy (category C), lactation. Safe use in children younger than 16 y is not established.

ROUTE & DOSAGE

Acute Intermittent Porphyria

Adult/Adolescent (over 16): **IV** 1–4 mg/kg/day for 3–14 days, do not repeat dose earlier than q12h (max: 6 mg/kg in 24 h)

ADMINISTRATION

Intravenous ──────────

PREPARE: **IV Infusion:** Reconstitute immediately before use by aseptically adding 43 mL sterile water for injection to vial to yield 7 mg/mL. ▪ Shake well for 2–3 min to dissolve all particles. ▪ Terminal filtration through a sterile 0.45 micron or smaller filter is recommended. ▪ Discard unused portions.

ADMINISTER: **IV Infusion:** Give a single dose over 10–15 min.

▪ Freeze and store lyophilized powder until time of use.

ADVERSE EFFECTS (≥1%) **Body as a Whole:** *Phlebitis* (when administered into small veins). **Hematologic:** Decreased Hct, anticoagulant effect (prolonged PT, thromboplastin time, <u>thrombocytopenia</u>, hypofibrinogenemia). **Urogenital:**

Reversible renal shutdown (with excessive doses).

INTERACTIONS Drug: Potentiates anticoagulant effects of ANTICOAGULANTS; BARBITURATES, ESTROGENS, CORTICOSTEROIDS may antagonize hemin effect.

PHARMACOKINETICS Duration: Can be detected in plasma up to 5 days. **Elimination:** Excess amounts eliminated in bile and urine.

NURSING IMPLICATIONS

Assessment & Drug Effects
▪ Monitor IV site for signs and symptoms of thrombophlebitis (see Appendix F).
▪ Monitor throughout therapy (decrease in these values indicates favorable clinical response): ALA, UPG (uroporphyrinogen), PBG (porphobilinogen or coproporphyrin).
▪ Monitor clinical effect of drug therapy by checking patient's symptoms and complaints associated with acute porphyria, which may include depression, insomnia, anxiety, disorientation, hallucinations, psychoses; dark urine, nausea, vomiting, abdominal pain, low back and leg pain, pareses (neuropathy), seizures.
▪ Monitor I&O and promptly report the onset of oliguria or anuria.

Patient & Family Education
▪ Notify prescriber of bruising, hematuria, tarry black stools, and nosebleeds.

HEPARIN SODIUM 🅟
(hep′a-rin)

Hepalean ♣, Heparin Sodium Lock Flush Solution, Hep-Lock
Classification: ANTICOAGULANT

Therapeutic: ANTICOAGULANT
Pregnancy Category: C

AVAILABILITY 10 units/mL, 100 units/mL, 1000 units/mL, 2000 units/mL, 5000 units/mL, 10,000 units/mL, 20,000 units/mL, 40,000 units/mL injection

ACTION & *THERAPEUTIC EFFECT*
Exerts direct effect on the cascade of blood coagulation by enhancing the inhibitory actions of antithrombin III (heparin cofactor) on several factors essential to normal blood clotting. This blocks the conversion of prothrombin to thrombin and fibrinogen to fibrin. *Inhibits formation of new clots. Has rapid anticoagulant effect. Does not lyse already existing thrombi but may prevent their extension and propagation.*

USES Prophylaxis and treatment of venous thrombosis and pulmonary embolism and to prevent thromboembolic complications arising from cardiac and vascular surgery, frostbite, and during acute stage of MI. Also used in treatment of disseminated intravascular coagulation (DIC), atrial fibrillation with embolization, and as anticoagulant in blood transfusions, extracorporeal circulation, and dialysis procedures.

UNLABELED USES Prophylaxis in hip and knee surgery. Heparin Sodium Lock Flush Solution is used to maintain potency of indwelling IV catheters in intermittent IV therapy or blood sampling. It is not intended for anticoagulant therapy.

CONTRAINDICATIONS History of hypersensitivity to heparin (white clot syndrome); active bleeding, bleeding tendencies; jaundice; ascorbic acid deficiency; inaccessible ulcerative lesions; visceral carcinoma; open wounds, extensive denudation of skin, suppurative thrombophlebitis; advanced kidney, liver, or biliary disease; active tuberculosis; bacterial endocarditis; continuous tube drainage of stomach or small intestines; threatened abortion; suspected intracranial hemorrhage, severe hypertension; recent surgery of eye, brain, or spinal cord; spinal tap; shock.

CAUTIOUS USE Alcoholism; history of allergy; during menstruation; immediate postpartum period; patients with indwelling catheters; older adults; use of acid-citrate-dextrose (ACD)-converted blood (may contain heparin); patients in hazardous occupations; cerebral embolism; pregnancy (category C), lactation.

ROUTE & DOSAGE

Treatment of Thromboembolism
Adult: **IV** 5000-unit bolus dose, then 20,000–40,000 units infused over 24 h, dose adjusted to maintain desired aPTT or 5000–10,000 units IV piggyback q4–6h **Subcutaneous** 10,000–20,000 units followed by 8000–20,000 units q8–12h
Child: **IV** 50 units/kg bolus, then 20,000 units/m²/24 h or 50–100 units/kg q4h
Open Heart Surgery
Adult: **IV** 150–400 units/kg
Prophylaxis of Embolism
Adult: **Subcutaneous** 5000 units q8–12h until patient is ambulatory

ADMINISTRATION
▪Note: Before administration, check coagulation test values; if results are not within therapeutic

Common adverse effects in *italic*, life-threatening effects underlined; generic names in **bold**; classifications in SMALL CAPS; ♣ Canadian drug name; ⊚Prototype drug

721

range, notify prescriber for dosage adjustment.

▪ Do not use solutions of heparin or heparin lock-flush that contain benzyl alcohol preservative in neonates.

Subcutaneous

▪ Use more concentrated heparin solutions for subcutaneous injection.

▪ Make injections into the fatty layer of the abdomen or just above the iliac crest. Avoid injecting within 5 cm (2 in.) of umbilicus or in a bruised area. Insert needle into tissue roll perpendicular to skin surface. Do not withdraw plunger to check entry into blood vessel.

▪ Systematically rotate injection sites and keep record.

▪ Exercise caution to avoid IM injection.

Intravenous

PREPARE: **Direct:** Give undiluted. **Intermittent/Continuous:** May add to any amount of NS, D5W, or LR for injection. ▪ Invert IV solution container at least 6 times to ensure adequate mixing.

ADMINISTER: **Direct:** Give a single dose over 60 sec. **Intermittent/Continuous (preferred):** Use infusion pump and give over 4–24 h.

INCOMPATIBILITIES **Solution/additive:** Alteplase, amikacin, atracurium, ciprofloxacin, codeine, cytarabine, dobutamine, doxorubicin, erythromycin, gentamicin, haloperidol, hyaluronidase, hydrocortisone, kanamycin, levorphanol, meperidine, methicillin, morphine, netilmicin, polymyxin B, promethazine, streptomycin, tetracycline, tobramycin, vancomycin. **Y-site:** Alteplase, amiodarone, amphotericin B cholesteryl, amsacrine, ciprofloxacin, clarithromycin,

dacarbazine, diazepam, dobutamine, doxorubicin, doxycycline, droperidol, ergotamine, filgrastim, gatifloxacin, gentamicin, haloperidol, idarubicin, isosorbide, levofloxacin, methotrimeprazine, mexiletine, nitroglycerin, phenytoin, polymyxin B, tobramycin, tramadol, triflupromazine, vancomycin, vinorelbine.

▪ Store at 15°–30° C (59°–86° F). Protect from freezing.

ADVERSE EFFECTS (≥1%) **Hematologic:** Spontaneous bleeding, *transient* thrombocytopenia, hypofibrinogenemia, "white clot syndrome." **Body as a Whole:** Fever, chills, urticaria, pruritus, skin rashes, itching and burning sensations of feet, numbness and tingling of hands and feet, elevated BP, headache, nasal congestion, lacrimation, conjunctivitis, chest pains, arthralgia, bronchospasm, anaphylactoid reactions. **Endocrine:** Osteoporosis, hypoaldosteronism, suppressed renal function, hyperkalemia; rebound hyperlipidemia (following termination of heparin therapy). **GI:** Increased AST, ALT. **Urogenital:** Priapism (rare). **Skin:** Injection site reactions: Pain, itching, ecchymoses, tissue irritation and sloughing; cyanosis and pains in arms or legs (vasospasm), reversible transient alopecia (usually around temporal area).

DIAGNOSTIC TEST INTERFERENCE Notify laboratory that patient is receiving heparin, when a test is to be performed. Possibility of false-positive rise in *BSP* test and in *serum thyroxine;* and increases in *resin T_3 uptake;* false-negative *^{125}I fibrinogen uptake.* Heparin prolongs *PT.* Valid readings may be obtained by drawing blood samples at least 4–6 h after an IV dose (but at any time during heparin infusion) and 12–24 h after a subcutaneous heparin dose.

INTERACTIONS Drug: May prolong PT, which is used to monitor therapy with ORAL ANTICOAGULANTS; **aspirin,** NSAIDs increase risk of bleeding; **nitroglycerin** IV may decrease anticoagulant activity; **protamine** antagonizes effects of heparin. **Herbal: Evening primrose oil, feverfew, ginkgo, ginger** may potentiate bleeding.

PHARMACOKINETICS Onset: 20–60 min subcutaneous. **Peak:** Within minutes. **Duration:** 2–6 h IV; 8–12 h subcutaneous. **Distribution:** Does not cross placenta; not distributed into breast milk. **Metabolism:** In liver and by reticuloendothelial system. **Elimination:** In urine. **Half-Life:** 90 min.

NURSING IMPLICATIONS

Assessment & Drug Effects

- Lab tests: Baseline blood coagulation tests, Hct, Hgb, RBC, and platelet counts prior to initiation of therapy and at regular intervals throughout therapy.
- Monitor aPTT levels closely.
- Note: In general, dosage is adjusted to keep aPTT between 1.5–2.5 times normal control level.
- Draw blood for coagulation test 30 min before each scheduled subcutaneous or intermittent IV dose and approximately q4h for patients receiving continuous IV heparin during dosage adjustment period. After dosage is established, tests may be done once daily.
- Patients vary widely in their reaction to heparin; risk of hemorrhage appears greatest in women, all patients older than 60 y, and patients with liver disease or renal insufficiency.

- Monitor vital signs. Report fever, drop in BP, rapid pulse, and other S&S of hemorrhage.
- Observe all needle sites daily for hematoma and signs of inflammation (swelling, heat, redness, pain).
- Antidote: Have on hand protamine sulfate (1% solution), specific heparin antagonist.

Patient & Family Education

- Protect from injury and notify prescriber of pink, red, dark brown, or cloudy urine; red or dark brown vomitus; red or black stools; bleeding gums or oral mucosa; ecchymoses, hematoma, epistaxis, bloody sputum; chest pain; abdominal or lumbar pain or swelling; unusual increase in menstrual flow; pelvic pain; severe or continuous headache, faintness, or dizziness.
- Note: Menstruation may be somewhat increased and prolonged; usually, this is not a contraindication to continued therapy if bleeding is not excessive.
- Learn correct technique for subcutaneous administration if discharged from hospital on heparin.
- Engage in normal activities such as shaving with a safety razor in the absence of a low platelet (thrombocyte) count. Usually, heparin does not affect bleeding time.
- Caution: Smoking and alcohol consumption may alter response to heparin and are not advised.
- Do not take aspirin or any other OTC medication without prescriber's approval.

HEPATITIS A VACCINE
(hep′a-ti-tis)

Havrix, Vaqta
Pregnancy Category: C
See Appendix J.

Common adverse effects in *italic*, life-threatening effects <u>underlined</u>; generic names in **bold**; classifications in SMALL CAPS; ♣ Canadian drug name; ⊙ Prototype drug

723

HEPATITIS B IMMUNE GLOBULIN

(hep'a-ti-tis)

HepaGam B, HyperHep, Nabi-HB
Pregnancy Category: C
See Appendix J.

HEPATITIS B VACCINE (RECOMBINANT) ⓟ

(hep'a-ti-tis)

Engerix-B, Recombivax HB
Pregnancy Category: C
See Appendix J.

HETASTARCH

(het'a-starch)

Hespan, Voluven
Classification: PLASMA EXPANDER
Therapeutic: PLASMA EXPANDER
Prototype: Albumin
Pregnancy Category: C

AVAILABILITY 6% injection

ACTION & *THERAPEUTIC EFFECT*
Synthetic starch closely resembling human glycogen. Acts much like albumin and dextran but is claimed to be less likely to produce anaphylaxis or to interfere with cross matching or blood typing procedures. *May prolong the aPTT and PT. In hypovolemic patients, it increases arterial and venous pressures, heart rate, cardiac output, urine output, as well as colloidal osmotic pressure.*

USES Treatment of hypovolemia, leukapheresis.
UNLABELED USES As a priming fluid in pump oxygenators for perfusion during extracorporeal circulation and as a cryoprotective agent for long-term storage of whole blood.

CONTRAINDICATIONS Severe bleeding disorders, CHF, treatment of shock not accompanied by hypovolemia, intracranial bleeding.
CAUTIOUS USE Hepatic or renal insufficiency, pulmonary edema in the very young or older adults, patients on sodium restriction; pregnancy (category C), lactation. Safe use in children is not established.

ROUTE & DOSAGE

Hypovolemia

Adult: **IV** 500–1000 mL or 20 mL/kg/day (max: 1500 mL/day)

Leukapheresis

Adult: **IV** 250–750 mL infused at a constant fixed ratio of 1:8 to venous whole blood

Renal Impairment Dosage Adjustment

CrCl less than 10 mL/min: Use original initial dose, then reduce doses by 25–50%

ADMINISTRATION

Intravenous

PREPARE: IV Infusion: Use undiluted as prepared by manufacturer.
ADMINISTER: IV Infusion: Specific flow rate is prescribed by prescriber. Rate may be as high as 20 mL/kg/h in acute hemorrhagic shock. ▪ Rate is usually reduced in patients with burns or septic shock.
INCOMPATIBILITIES **Y-site:** **Amikacin, amphotericin B, cefotaxime, cefoxitin, diazepam, gentamicin, ranitidine, sodium bicarbonate, theophylline, tobramycin.**

Common adverse effects in *italic*, life-threatening effects underlined; generic names in **bold;** classifications in SMALL CAPS; ✚ Canadian drug name; ⓟ Prototype drug

- Store at room temperature; avoid extremes of heat or cold. ■ Discard partially used bags.

ADVERSE EFFECTS (≥1%) **CV:** Peripheral edema, <u>circulatory overload, heart failure</u>. **Hematologic:** With large volumes, prolongation of PT, PTT, clotting time, and bleeding time; decreased Hct, Hgb, platelets, calcium, and fibrinogen; dilution of plasma proteins, hyperbilirubinemia, increased sedimentation rate. **Body as a Whole:** Pruritus, <u>anaphylactoid reactions</u> (periorbital edema, urticaria, wheezing), vomiting, mild fever, chills, influenza-like symptoms, headache, muscle pains, submaxillary and parotid glandular swelling.

PHARMACOKINETICS Duration: 24–36 h. **Distribution:** Remains in intravascular space. **Metabolism:** In reticuloendothelial system. **Elimination:** In urine with some biliary excretion.

NURSING IMPLICATIONS

Assessment & Drug Effects
- Monitor for S&S of hypersensitivity reaction (see Appendix F).
- Measure and record I&O. Report oliguria or significant changes in I&O ratio.
- Monitor BP and vital signs and observe patient for unusual bruising or bleeding.
- Lab tests: Monitor WBC count with differential, platelet count, and PT & PTT during leukapheresis.
- Observe for signs of circulatory overload (see Appendix F).
- Check laboratory reports of Hct values. Notify prescriber if there is an appreciable drop in Hct or if value approaches 30% by volume. Hct should not be allowed to drop below 30%.

Patient & Family Education
- Notify prescriber for any of the following: Difficulty breathing, nausea, chills, headache, itching.

HOMATROPINE HYDROBROMIDE ℞

(hoe-ma'troe-peen)

AK-Homatropine, Homatrine, Homatropine, Isopto Homatropine
Pregnancy Category: C
See Appendix A-1.

HUMAN PAPILLOMAVIRUS BIVALENT VACCINE ℞

(hu'man pap-ih-LO'ma VYE'rus)

Cervarix
Pregnancy Category: B
See Appendix J.

HUMAN PAPILLOMAVIRUS BIVALENT VACCINE ℞

(hu'man pap-ih LO'ma VYE'rus)

Zostavax
Pregnancy Category: C
See Appendix J.

HUMAN PAPILLOMAVIRUS QUADRIVALENT VACCINE ℞

(hu'man pap-ih-LO'ma VYE'rus)

Gardasil
Pregnancy Category: B
See Appendix J.

Common adverse effects in *italic*, life-threatening effects <u>underlined</u>; generic names in **bold;** classifications in SMALL CAPS; ♣ Canadian drug name; ℞ Prototype drug

725

HYALURONIDASE, OVINE

(hi-a-lu-ron'i-dase)

Amphadase, Hylenex, Vitrase
Classifications: HYALURONIC ACID
DERIVATIVE; ABSORPTION AND DIS-
PERSING ENHANCER
Therapeutic: ABSORPTION AND DIS-
PERSING ENHANCER
Pregnancy Category: C

AVAILABILITY 150 units/mL injection, 200 USP units/mL for injection

ACTION & *THERAPEUTIC EFFECT*
Hyaluronidase is a diffusing substance that modifies the permeability of connective tissue through the
hydrolysis of hyaluronic acid found
in the intercellular substance of
connective tissue. *It increases the*
absorption and dispersion of solu-
tions in the intercellular spaces.

USES Adjuvant to increase the absorption and dispersion of other injected
drugs; hypodermoclysis; adjunct in
subcutaneous urography for improving resorption of radiopaque agents.
UNLABELED USES Adjunct for ophthalmic anesthesia, treatment of
vitreous hemorrhage and diabetic
retinopathy.

CONTRAINDICATIONS Hypersensitivity to hyaluronidase or any other
ingredient in formulation; injection
into infected or acutely inflamed area,
area of swelling due to bites or stings;
corneal injection; injection by IV.
CAUTIOUS USE Pregnancy (category C), lactation.

ROUTE & DOSAGE

**Adjuvant to Increase the Absorption
and Dispersion of Other Drugs**
Adult/Adolescent/Child: 150
units (range: 50–300) added to
solution

Hypodermoclysis
Adult/Adolescent/Child: 150
units
Subcutaneous Urography
Adult/Adolescent/Child:
Subcutaneous 75 units prior to
contrast medium

ADMINISTRATION

Subcutaneous
- Give subcutaneously prior to
contrast media. Do not inject
near an infected or acutely inflamed area.
- Store unopened vial at 2°–8° C
(35°–46° F). After reconstitution,
store at 20°–25° C (59°–77° F), and
use within 6 h. Protect from light.

ADVERSE EFFECTS (≥1%) **CV:** Edema. **Other:** Injection site reaction
(e.g., erythema, irritation); enhanced adverse events associated
with coadministered drugs.

INTERACTIONS Drug: SALICYLATES,
CORTICOSTEROIDS, ESTROGENS, or H₁-
BLOCKERS may confer partial resistance to the action of hyaluronidase in some tissues.

NURSING IMPLICATIONS

Assessment & Drug Effects
- Monitor for S&S of hypersensitivity: Urticaria, erythema, chills,
nausea, vomiting, dizziness, tachycardia, and hypotension. Withhold and notify prescriber if hypersensitivity occurs.
- Note: Those receiving large doses
of salicylates, cortisone, ACTH,
estrogens, or antihistamines may
require larger amounts of hyaluronidase for equivalent dis-
persing effect.

Patient & Family Education
- Report immediately any of the following: Rash, itching, chills, nausea,
vomiting, dizziness, or palpitations.

Common adverse effects in *italic*, life-threatening effects <u>underlined</u>; generic names
in **bold**; classifications in SMALL CAPS; ✦ Canadian drug name; ⊙ Prototype drug

HYDRALAZINE HYDROCHLORIDE ⓟ

(hye-dral´a-zeen)

Classifications: NONNITRATE VA-SODILATOR; ANTIHYPERTENSIVE
Therapeutic: ANTIHYPERTENSIVE
Pregnancy Category: C

AVAILABILITY 10 mg, 25 mg, 50 mg, 100 mg tablets; 20 mg/mL vial

ACTION & *THERAPEUTIC EFFECT*

Reduces BP mainly by direct effect on vascular smooth muscles of arterial-resistance vessels, resulting in vasodilatation. *Reduces BP with diastolic response often being greater than systolic response. Vasodilation reduces peripheral resistance and substantially improves cardiac output, and renal and cerebral blood flow.*

USES Most commonly in stepped-care approach to treat moderate to severe hypertension. Also in early malignant hypertension and resistant hypertension that persists after sympathectomy.
UNLABELED USES Conjunctively with cardiac glycosides and other vasodilators in short-term treatment of acute CHF; unexplained pulmonary hypertension; eclampsia.

CONTRAINDICATIONS Monotherapy for CHF, mitral valvular rheumatic heart disease, MI, tachycardia.
CAUTIOUS USE Coronary heart disease; cerebrovascular accident, advanced renal impairment, coronary heart disease, renal disease; renal failure; SLE; use with MAO inhibitors; pregnancy (category C), lactation.

ROUTE & DOSAGE

Hypertension

Adult: **PO** 10–50 mg q.i.d. **IM** 10–50 mg q4–6h **IV** 10–20 mg q4–6h, may increase to 40 mg
Geriatric: **PO** Start with 10 mg 2–3 times/day
Child: **PO** 3–7.5 mg/kg/day in 4 divided doses **IV/IM** 0.1–0.2 mg/kg in divided doses (max: 20 mg)

Renal Impairment Dosage Adjustment

CrCl 10–50 mL/min: Dose q8h

ADMINISTRATION

Oral

- Give with food; bioavailability is increased by taking it with food.
- Discontinue gradually to avoid sudden rise in BP and acute heart failure.
- Inform patients of the dangers of abrupt withdrawal.

Intramuscular

- Give deep into a large muscle.

Intravenous

PREPARE: Direct: Give undiluted. Use immediately after being drawn into syringe. ▪ Do not add to IV solutions.
ADMINISTER: Direct: Give each 10 mg or fraction thereof over 1 min.
INCOMPATIBILITIES Solution/additive: Aminophylline, ampicillin, chlorothiazide, edetate calcium disodium, ethacrynate, hydrocortisone, mephentermine, methohexital, nitroglycerin, phenobarbital, verapamil, D5W. Y-site: Aminophylline, ampicillin, diazoxide, furosemide.

- Store at 15°–30° C (59°–86° F) in tight, light-resistant containers

Common adverse effects in *italic*, life-threatening effects <u>underlined</u>; generic names in **bold**; classifications in SMALL CAPS; ♣ Canadian drug name; ⓟ Prototype drug

727

unless otherwise directed. Avoid freezing.

ADVERSE EFFECTS (≥1%) **Body as a Whole:** Hypersensitivity (rash, urticaria, pruritus, fever, chills, arthralgia, eosinophilia, cholangitis, hepatitis, obstructive jaundice). **CNS:** *Headache,* dizziness, tremors. **CV:** *Palpitation,* angina, *tachycardia,* flushing, paradoxical pressor response. Overdose: Arrhythmia, shock. **Special Senses:** Lacrimation, conjunctivitis. **GI:** Anorexia, nausea, vomiting, diarrhea, constipation, abdominal pain, paralytic ileus. **Urogenital:** Difficulty in urination, glomerulonephritis. **Hematologic:** Decreased hematocrit and hemoglobin, anemia, agranulocytosis (rare). **Other:** Nasal congestion, muscle cramps, SLE-like syndrome, fixed drug eruption, edema.

DIAGNOSTIC TEST INTERFERENCE Positive *direct Coombs' tests* in patients with hydralazine-induced SLE. Hydralazine interferes with urinary *17-OHCS* determinations *(modified Glenn-Nelson technique).*

INTERACTIONS Drug: BETA-BLOCKERS and other ANTIHYPERTENSIVE AGENTS compound hypotensive effects.

PHARMACOKINETICS Absorption: Readily absorbed from GI tract. **Onset:** 20–30 min. **Peak:** 2 h. **Duration:** 2–6 h. **Distribution:** Crosses placenta; distributed into breast milk. **Metabolism:** In liver. **Elimination:** 90% in urine; 10% in feces. **Half-Life:** 2–8 h.

NURSING IMPLICATIONS

Assessment & Drug Effects
- Monitor BP and HR closely. Check every 5 min until it is stabilized at desired level, then every 15 min thereafter throughout hypertensive crisis.
- Lab tests: Baseline and periodic determinations of BUN, creatinine clearance, uric acid, serum potassium, blood glucose, and ECG. Baseline and periodic antinuclear antibody titer recommended with prolonged therapy.
- Monitor for S&S of SLE, especially with prolonged therapy.
- Monitor I&O when drug is given parenterally and in those with renal dysfunction.

Patient & Family Education
- Monitor weight, check for edema, and report weight gain to prescriber.
- Note: Some patients experience headache and palpitations within 2–4 h after first PO dose; symptoms usually subside spontaneously.
- Make position changes slowly and avoid standing still, hot baths/ showers, strenuous exercise, and excessive alcohol intake.
- Do not drive or engage in other potentially hazardous activities until response to drug is known.

✓HYDROCHLOROTHIAZIDE (HCTZ) Ⓟⓡ

(hye-droe-klor-oh-thye′a-zide)

Apo-Hydro ♣, Microzide, Oretic
Classifications: ELECTROLYTIC AND WATER BALANCE; THIAZIDE DIURETIC
Therapeutic: DIURETIC
Pregnancy Category: B

AVAILABILITY 12.5 mg capsules; 25 mg, 50 mg, 100 mg tablets; 50 mg/5 mL oral solution

ACTION & *THERAPEUTIC EFFECT*
Diuretic action is associated with

Common adverse effects in *italic,* life-threatening effects underlined; generic names in **bold;** classifications in SMALL CAPS; ♣ Canadian drug name; Ⓟ Prototype drug

drug interference with absorption of sodium ions across the distal renal tubular segment of the nephron. This enhances excretion of sodium, chloride, potassium, bicarbonates, and water. It also decreases cardiac output and reduces plasma and extracellular fluid volume. *Therapeutic effectiveness is measured by decrease in edema and lowering of blood pressure.*

USES Adjunct in treatment of edema associated with CHF, hepatic cirrhosis, renal failure, and in the management of hypertension.
UNLABELED USES Nephrogenic diabetes insipidus, hypercalciuria, and treatment of electrolyte disturbances associated with renal tubular acidosis.

CONTRAINDICATIONS Hypersensitivity to thiazides or other sulfonamides; anuria; electrolyte imbalance.
CAUTIOUS USE Bronchial asthma, allergy; hepatic cirrhosis; renal dysfunction; acid/base imbalance; CHF; stroke, CVA; history of gout, SLE; diabetes mellitus; older adults; excessive sunlight UV exposure; neonates with jaundice; pregnancy (category B), lactation.

ROUTE & DOSAGE

Edema

Adult: **PO** 25–100 mg/day in 1–3 divided doses (max: 200 mg/day)

Hypertension

Adult/Adolescent: **PO** 12.5–50 mg/day in 1–2 divided doses
Child/Infant (over 6 mo): **PO** 1–2 mg/kg/day in 2 divided doses
Neonate (younger than 6 mo): **PO** 2–4 mg/kg/day in 2 divided doses

ADMINISTRATION

Oral
- Give with food or milk to reduce GI upset.
- Schedule doses to avoid nocturia and interrupted sleep. If given in 2 doses, schedule second dose no later than 3 p.m.
- Store tablets in tightly closed container at 15°–30° C (59°–86° F) unless otherwise directed.

ADVERSE EFFECTS (≥1%) **CNS:** Mood changes, unusual tiredness or weakness, dizziness, light-headedness, paresthesias. **CV:** Irregular heartbeat, weak pulse, orthostatic hypotension. **GI:** Dry mouth, increased thirst, nausea, vomiting, anorexia, diarrhea, pancreatitis, jaundice. **Hematologic:** Agranulocytosis, thrombocytopenia, aplastic anemia, leukopenia. **Metabolic:** *Hyperglycemia,* glycosuria, *hyperuricemia, hypokalemia.* **Other:** Hypersensitivity reactions, photosensitivity, blurred vision, yellow vision (xanthopsia), muscle spasm.

DIAGNOSTIC TEST INTERFERENCE May interfere with ***parathyroid function tests, tyramine/phentolamine tests, histamine tests for pheochromocytoma.***

INTERACTIONS Drug: Amphotericin B, CORTICOSTEROIDS increase hypokalemic effects; SULFONYLUREAS, **insulin** may antagonize hypoglycemic effects; BILE ACID SEQUESTRANTS decrease THIAZIDE absorption; **diazoxide** intensifies hypoglycemic and hypotensive effects; increased **potassium** and **magnesium** loss may cause **digoxin** toxicity; decreases **lithium** excretion and increases toxicity; increases risk of NSAID-induced renal failure and may attenuate diuresis. Do not use with **dofetilide.** Withhold dose for

Common adverse effects in *italic*, life-threatening effects underlined; generic names in **bold**; classifications in SMALL CAPS; ✦ Canadian drug name; 🄿 Prototype drug

729

24 hr prior to **amifostine** usage. Monitor with **topiramate**.

PHARMACOKINETICS Absorption: Incompletely absorbed. **Onset:** 2 h. **Peak:** 4 h. **Duration:** 6–12 h. **Distribution:** Distributed throughout extracellular tissue; concentrates in kidney; crosses placenta; distributed in breast milk. **Metabolism:** Does not appear to be metabolized. **Elimination:** In urine. **Half-Life:** 45–120 min.

NURSING IMPLICATIONS

Assessment & Drug Effects

- Monitor for therapeutic effectiveness. Antihypertensive effects may be noted in 3–4 days; maximal effects may require 3–4 wk.
- Lab tests: Baseline and periodic determinations of serum electrolytes, blood counts, BUN, blood glucose, uric acid, CO_2, are recommended.
- Check BP at regular intervals.
- Monitor closely for hypokalemia; it increases the risk of digoxin toxicity.
- Monitor I&O and check for edema.
- Note: Drug may cause hyperglycemia and loss of glycemic control in diabetics.
- Note: Drug may cause orthostatic hypotension, dizziness.

Patient & Family Education

- Monitor weight daily.
- Note: Diabetic patients need to monitor blood glucose closely. This drug causes impaired glucose tolerance.
- Report signs of hypokalemia (see Appendix F) to prescriber.
- Change positions slowly; avoid hot baths or showers, extended exposure to sunlight, and sitting or standing still for long periods.
- Note: Photosensitivity reaction may occur 10–14 days after initial sun exposure.

HYDROCODONE BITARTRATE

(hye-droe-koe′done)

Dihydrocodeinone Bitartrate, Hycomet, Mycodone, Tussigon, Vicodin (with acetaminophen)
Classifications: NARCOTIC (OPIATE AGONIST) ANALGESIC; ANTITUSSIVE
Therapeutic: NARCOTIC ANALGESIC; ANTITUSSIVE
Prototype: Morphine
Pregnancy Category: C
Controlled Substance: Schedule II

AVAILABILITY 5 mg hydrocodone usually with 500 mg or more acetaminophen or 1.5 mg of homatropine

ACTION & *THERAPEUTIC EFFECT* CNS depressant with moderate to severe relief of pain. Suppresses cough reflex by direct action on cough center in medulla. *CNS depressant with moderate to severe relief of pain. Effective in cough suppression.*

USES Symptomatic relief of hyperactive or nonproductive cough and for relief of moderate to moderately severe pain. A common ingredient in a variety of proprietary mixtures.

CONTRAINDICATIONS Hypersensitivity to hydrocodone; acute or severe asthmatic bronchitis; COPD; upper airway obstruction.
CAUTIOUS USE Respiratory depression, asthma, emphysema; history of drug abuse or dependence; postoperative patients; hepatic or renal disease; renal impairment or failure; older adults, debilitated patients; children weighing less than 50 kg; G6PD deficiency; GI disease; patients with preexisting increased intracranial pressure; pregnancy (category C), lactation; children younger than 2 y.

ROUTE & DOSAGE

Mild to Moderate Pain, Cough

Adult: **PO** 5–10 mg q4–6h prn (max: 15 mg/dose)
Child (2–12 y): **PO** 1.25–5 mg q4–6h (max: 10 mg/dose)

ADMINISTRATION

Oral
- Give with food or milk to prevent GI irritation.
- Preserve in tight, light-resistant containers.

ADVERSE EFFECTS (≥1%) **GI:** Dry mouth, *constipation, nausea,* vomiting. **CNS:** Light-headedness, sedation, dizziness, *drowsiness,* euphoria, dysphoria. **Respiratory:** <u>Respiratory depression</u>. **Skin:** Urticaria, rash, pruritus.

INTERACTIONS Drug: Alcohol and other CNS DEPRESSANTS compound sedation and CNS depression. **Herbal: St. John's wort** increases sedation.

PHARMACOKINETICS Onset: 10–20 min. **Duration:** 3–6 h. **Distribution:** Crosses placenta; distributed into breast milk. **Metabolism:** In liver. **Elimination:** In urine. **Half-Life:** 3.8 h.

NURSING IMPLICATIONS

Assessment & Drug Effects
- Monitor for effectiveness of drug for pain relief.
- Monitor for nausea and vomiting, especially in ambulatory patients.
- Monitor respiratory status and bowel elimination.

Patient & Family Education
- Avoid hazardous activities until response to drug is determined.
- Do not use alcohol or other CNS depressants; may cause additive CNS depression.
- Drink plenty of liquids for adequate hydration.
- Do not take larger doses than prescribed since abuse potential is high.

✓HYDROCORTISONE ⓟ
(hye-droe-kor′ti-sone)
Aeroseb-HC, Cetacort, Cortaid, Cortenema, Dermolate, Hytone, Rectocort ♣, Synacort

HYDROCORTISONE ACETATE
Anusol HC, Carmol HC, Cortaid, Cort-Dome, Corticaine, Cortifoam, Cortiment ♣, Epifoam

HYDROCORTISONE CYPIONATE
Cortef

HYDROCORTISONE SODIUM SUCCINATE
A-Hydrocort, Solu-Cortef

HYDROCORTISONE VALERATE
Westcort

HYDROCORTISONE BUTYRATE
Locoid
Classification: ADRENOCORTICAL STEROID
Therapeutic: ANTI-INFLAMMATORY; IMMUNOSUPPRESSANT; ANTIPSORIATIC
Pregnancy Category: C

AVAILABILITY Hydrocortisone: 5 mg, 10 mg, 20 mg tablets; 0.5%, 1%, 2.5% cream, lotion, ointment, spray; **Hydrocortisone Acetate:** 25 mg/mL, 50 mg/mL suspension; 0.5%, 1% cream, ointment; **Hydrocortisone Cypionate:** 5 mg, 20 mg tablet; **Hydrocortisone Sodium Succinate:** 100 mg/2 mL, 250 mg/2 mL, 500 mg/4 mL, 1000 mg/8 mL

Common adverse effects in *italic*, life-threatening effects <u>underlined</u>; generic names in **bold**; classifications in SMALL CAPS; ♣ Canadian drug name; ⓟ Prototype drug

731

vials; **Hydrocortisone Valerate:** 0.2% cream, ointment; **Hydrocortisone Butyrate:** 0.1% cream, ointment, topical solution

ACTION & *THERAPEUTIC EFFECT*
Short-acting synthetic steroid with both glucocorticoid and mineralocorticoid properties that affect nearly all systems of the body. **Anti-inflammatory (glucocorticoid) action:** Stabilizes leukocyte lysosomal membranes; inhibits phagocytosis and release of allergic substances; suppresses fibroblast formation and collagen deposition; reduces capillary dilation and permeability; and increases responsiveness of cardiovascular system to circulating catecholamines. **Immunosuppressive action:** Modifies immune response to various stimuli; reduces antibody titers; and suppresses cell-mediated hypersensitivity reactions. **Mineralocorticoid action:** Promotes sodium retention, but under certain circumstances (e.g., sodium loading), enhances sodium excretion; promotes potassium excretion; and increases glomerular filtration rate (GFR). **Metabolic action:** Promotes hepatic gluconeogenesis, protein catabolism, redistribution of body fat, and lipolysis. *Has anti-inflammatory, immunosuppressive, and metabolic functions in the body.*

USES Replacement therapy in adrenocortical insufficiency; to reduce serum calcium in hypercalcemia, to suppress undesirable inflammatory or immune responses, to produce temporary remission in nonadrenal disease, and to block ACTH production in diagnostic tests. Use as anti-inflammatory or immunosuppressive agent largely replaced by synthetic glucocorticoids that have minimal mineralocorticoid activity. Topically for atopic dermatitis or inflammatory conditions.

CONTRAINDICATIONS Hypersensitivity to glucocorticoids, idiopathic thrombocytopenic purpura, psychoses, acute glomerulonephritis, viral or bacterial diseases of skin, infections not controlled by antibiotics, active or latent amebiasis, hypercorticism (Cushing's syndrome), smallpox vaccination or other immunologic procedures; acne; lactation (except for topical use). **Topical steroids:** Presence of varicella, vaccinia, on surfaces with compromised circulation.

CAUTIOUS USE Diabetes mellitus; chronic, active hepatitis positive for hepatitis B surface antigen; hyperlipidemia; cirrhosis; stromal herpes simplex; glaucoma, tuberculosis of eye; osteoporosis; convulsive disorders; hypothyroidism; diverticulitis; nonspecific ulcerative colitis; fresh intestinal anastomoses; active or latent peptic ulcer; gastritis; esophagitis; thromboembolic disorders; CHF; metastatic carcinoma; hypertension; renal insufficiency; history of allergies; active or arrested tuberculosis; systemic fungal infection; myasthenia gravis; pregnancy (category C); children; infants younger than 3 mo.

ROUTE & DOSAGE

Adrenal Insufficiency, Anti-Inflammatory
Adult: **PO** 10–320 mg/day in 3–4 divided doses **IV/IM** 15–800 mg/day in 3–4 divided doses (max: 2 g/day) **Subcutaneous** Sodium phosphate only, 15–240 mg/day
Child: **PO** 2.5–10 mg/kg/day in 3–4 divided doses **IV/IM** 1–5 mg/kg/day divided q12–24h

Common adverse effects in *italic*, life-threatening effects underlined; generic names in **bold**; classifications in SMALL CAPS; ♣ Canadian drug name; ⊙ Prototype drug

Intra-Articular, Intralesional (Acetate Salt)

Adult: **IM** 5–50 mg q3–5days for bursae; 5–50 mg once q1–4wk for joints

Anti-Inflammatory Agent

Adult: **Topical** Apply a small amount to the affected area 1–4 times/day **PR** Insert 1% cream, 10% foam, 10–25 mg suppository, or 100 mg enema nightly

Atopic Dermatitis

Adult/Adolescent/Child/Infant (older than 3 mo): **Topical** Apply sparingly b.i.d.

ADMINISTRATION

Note: Hydrocortisone phosphate may be given subcutaneously, IM, or IV. Hydrocortisone succinate may be given IM or IV.

Oral
- Give oral drug with food.

Rectal
- Administer retention enema preferably after a bowel movement; retain at least 1 h or all night if possible.

Topical
- Apply medication sparingly, rub until it disappears, and then reapply, leaving a thin coat over lesion. Cover area with transparent plastic or other occlusive device or vehicle only when so ordered.
- Avoid covering a weeping or exudative lesion.
- Note: Occlusive dressings usually are not applied to face, scalp, scrotum, axilla, and groin.
- Inspect skin carefully between applications for ecchymotic, petechial, and purpuric signs, maceration, secondary infection, skin atrophy, striae or miliaria; if

present, stop medication and notify prescriber.
- Store medication at 15°–30° C (59°–86° F) unless otherwise directed by manufacturer; protect from light and freezing.

Intramuscular
- Inject deep into gluteal muscle.

Intravenous
IV administration to infants, children: Verify correct IV concentration and rate of infusion/injection with prescriber.

PREPARE: **Direct (preferred):** Give undiluted. **Intermittent:** Dilute in 50–1000 mL of D5W, NS, or D5/NS.

ADMINISTER: **Direct:** Give each dose at a rate of 500 mg or fraction thereof over 1 min. **Intermittent:** Give over 10 min.

INCOMPATIBILITIES Solution/additive: **Amobarbital, ampicillin, bleomycin, colistimethate, dimenhydrinate, doxapram, doxorubicin, ephedrine, heparin, hydralazine, metaraminol, methicillin, nafcillin, pentobarbital, phenobarbital, prochlorperazine, promethazine, secobarbital, tetracycline.** Y-site: **Ergotamine, phenytoin.**
- Administer solutions that have been diluted for IV infusion within 24 h.

ADVERSE EFFECTS (≥1%) Body as a Whole: Hypersensitivity or <u>anaphylactoid reactions; aggravation or masking of infections;</u> malaise, weight gain, obesity; urogenital urinary frequency and urgency, enuresis increased or decreased motility and number of sperm. **CNS:** Vertigo, headache, nystagmus, ataxia (rare), increased intracranial pressure with papilledema (usually after discontinuation of

H

Common adverse effects in *italic*, life-threatening effects <u>underlined</u>; generic names in **bold**; classifications in SMALL CAPS; ♣ Canadian drug name; ⊙ Prototype drug

733

medication), mental disturbances, aggravation of preexisting psychiatric conditions, insomnia, anxiety, mental confusion, depression. **CV:** Syncopal episodes, thrombophlebitis, thromboembolism or fat embolism, palpitation, tachycardia, necrotizing angiitis, CHF, hypertension edema. **Endocrine:** Suppressed linear growth in children, decreased glucose tolerance; hyperglycemia, manifestations of latent diabetes mellitus; hypocorticism; amenorrhea and other menstrual difficulties; moon facies. **Special Senses:** Posterior subcapsular cataracts (especially in children), glaucoma, exophthalmos, increased intraocular pressure with optic nerve damage, perforation of the globe, fungal infection of the cornea, decreased or blurred vision. **Metabolic:** Hypocalcemia; *sodium* and *fluid retention;* hypokalemia and hypokalemic alkalosis decreased serum concentration of vitamins A and C; hyperglycemia, hypernatremia. **GI:** Cramping, bleeding, *nausea,* increased appetite, ulcerative esophagitis, pancreatitis, abdominal distention, peptic ulcer with perforation and hemorrhage, melena. **Hematologic:** <u>Thrombocytopenia</u>, polycythemia, ecchymoses. **Musculoskeletal:** Osteoporosis, compression fractures, muscle wasting and weakness, tendon rupture, aseptic necrosis of femoral and humeral heads. **Skin:** Skin thinning and atrophy, *acne, impaired wound healing;* petechiae, ecchymosis, easy bruising; suppression of skin test reaction; hypopigmentation or hyperpigmentation, hirsutism, acneiform eruptions, subcutaneous fat atrophy; allergic dermatitis, urticaria, angioneurotic edema, increased sweating. With parenteral therapy at IV site–pain, irritation, necrosis, atrophy, sterile abscess; Charcot-like arthropathy following intra-articular use; burning and tingling in perineal area (after IV injection).

DIAGNOSTIC TEST INTERFERENCE

Hydrocortisone (corticosteroids) may increase serum *cholesterol, blood glucose,* serum *sodium, uric acid* (in acute leukemia) and *calcium* (in bone metastasis). It may decrease serum *calcium, potassium, PBI, thyroxin (T_4), triiodothyronine (T_3) and reduce thyroid I 131* uptake. It increases *urine glucose* level and *calcium* excretion; decreases *urine 17-OHCS* and *17-KS* levels. May produce false-negative results with *nitroblue tetrazolium test* for systemic bacterial infection and may suppress reactions to skin tests.

INTERACTIONS **Drug:** BARBITURATES, **phenytoin, rifampin** may increase hepatic metabolism, thus decreasing cortisone levels; ESTROGENS potentiate the effects of hydrocortisone; NSAIDS compound ulcerogenic effects; **cholestyramine, colestipol** decrease hydrocortisone absorption; DIURETICS, **amphotericin B** exacerbate hypokalemia; ANTICHOLINESTERASE AGENTS (e.g., **neostigmine**) may produce severe weakness; immune response to VACCINES and TOXOIDS may be decreased.

PHARMACOKINETICS **Absorption:** Readily from GI tract and IM injection site. **Onset:** 1–2 h PO; immediately IV; 3–5 days PR. **Peak:** 1 h PO; 4–8 h IM. **Duration:** 1–1.5 days PO/IM; 0.5–4 wk intra-articular. **Distribution:** Distributed primarily to muscles, liver, skin, intestines, kidneys; crosses placenta. **Metabolism:** In liver. **Elimination:** HPA suppression 8–12 h; metabolites

excreted in urine; excreted in breast milk. **Half-Life:** 1.5–2 h.

NURSING IMPLICATIONS

Assessment & Drug Effects

- Establish baseline and continuing data on BP, weight, fluid and electrolyte balance, and blood glucose.
- Lab tests: Periodic serum electrolytes blood glucose, Hct and Hgb, platelet count, and WBC with differential.
- Monitor for adverse effects. Older adults and patients with low serum albumin are especially susceptible to adverse effects.
- Be alert to signs of hypocalcemia (see Appendix F).
- Ophthalmoscopic examinations are recommended every 2–3 mo, especially if patient is receiving ophthalmic steroid therapy.
- Monitor for persistent backache or chest pain; compression and spontaneous fractures of long bones and vertebrae present hazards.
- Monitor for and report changes in mood and behavior, emotional instability, or psychomotor activity, especially with long-term therapy.
- Be alert to possibility of masked infection and delayed healing (anti-inflammatory and immunosuppressive actions).
- Note: Dose adjustment may be required if patient is subjected to severe stress (serious infection, surgery, or injury).
- Note: Single doses of corticosteroids or use for a short period (less than 1 wk) do not produce withdrawal symptoms when discontinued, even with moderately large doses.

Patient & Family Education

- Expect a slight weight gain with improved appetite. After dosage is stabilized, notify prescriber of a sudden slow but steady weight increase [2 kg (5 lb)/wk].
- Avoid alcohol and caffeine; may contribute to steroid-ulcer development in long-term therapy.
- Do not ignore dyspepsia with hyperacidity. Report symptoms to prescriber and do NOT self-medicate to find relief.
- Do NOT use aspirin or other OTC drugs unless prescribed specifically by the prescriber.
- Note: A high protein, calcium, and vitamin D diet is advisable to reduce risk of corticosteroid-induced osteoporosis.
- Notify prescriber of slow healing, any vague feeling of being sick, or return to pretreatment symptoms.
- Do not abruptly discontinue drug; doses are gradually reduced to prevent withdrawal symptoms.
- Report exacerbation of disease during drug withdrawal.
- Apply topical preparations sparingly in small children. The hazard of systemic toxicity is higher because of the greater ratio of skin surface area to body weight.

HYDROMORPHONE HYDROCHLORIDE

(hye-droe-mor'fone)

Dilaudid, Dilaudid-HP, Exalgo

Classifications: NARCOTIC (OPIATE AGONIST); ANALGESIC

Therapeutic: NARCOTIC ANALGESIC; ANTITUSSIVE

Prototype: Morphine

Pregnancy Category: C; D in prolonged use or high doses at term

Controlled Substance: Schedule II

Common adverse effects in *italic*, life-threatening effects <u>underlined</u>; generic names in **bold**; classifications in SMALL CAPS; ◆ Canadian drug name; ☺ Prototype drug

735

AVAILABILITY 2 mg, 4 mg, 8 mg tablets; 5 mg/5 mL oral liquid; 1 mg/mL, 10 mg/mL injection; 8 mg, 12 mg, 16 mg extended release tablet.

ACTION & *THERAPEUTIC EFFECT* Potent opiate receptor agonist that does not alter pain threshold but changes the perception of pain in the CNS. *An effective narcotic analgesic that controls mild to moderate pain. Also has antitussive properties.*

USES Relief of moderate to severe pain and control of persistent nonproductive cough.

CONTRAINDICATIONS Intolerance to opiate agonists; opiate-naïve patients; acute or severe asthma; hypercapnia; bronchial asthma; upper airway obstruction; GI obstruction; ileus; severe respiratory depression; pregnancy (category D if used for prolonged periods or in high does at term); lactation.

CAUTIOUS USE Abrupt discontinuation, alcoholism; angina; biliary tract disease; older adults; epidural administration; GI disease; head trauma; heart failure; hepatic disease; hypotension, hypovolemia, oliguria, prostatic hypertrophy; pulmonary disease; COPD; respiratory depression; renal disease, renal impairment; paralytic ileus; increased intracranial pressure; inflammatory bowel disease, ulcerative colitis; latex hypersensitivity; labor; obstetric delivery; bladder obstruction; cardiac arrhythmias, cardiac disease; seizure disorder, seizures; substance abuse; surgery; urethral stricture, urinary retention; pregnancy (category C); children.

ROUTE & DOSAGE

Moderate to Severe Pain

Adult: **PO** 2.5–10 mg q4–6h prn in naïve patients **Subcutaneous/**

IM/IV 0.75–2 mg q4–6h depending on patient response
Child: **PO** 0.03–0.08 mg/kg q4–6h (max: 5 mg/dose) **IV** 0.015 mg/kg q4–6h prn

Antitussive

Adult: **PO** 1 mg q3–4h prn
Child (6–12 y): **PO** 0.5 mg q3–4h prn

ADMINISTRATION

Oral

- Ensure that extended release tablet is swallowed whole and not crushed or chewed.
- For chronic pain, around-the-clock dosing is recommended.

Subcutaneous/Intramuscular

- High-potency hydromorphone is highly concentrated, making delivery of exact small doses difficult. Use high-potency hydromorphone only if an accurate dose can be measured and delivered.
- Store at room 15°–30° C (59°–86° F) and protect from light.

Intravenous

IV administration to children: Verify correct IV concentration and rate of infusion with prescriber.

***PREPARE:* Direct:** May be given undiluted or diluted in 5 mL of sterile water or NS. **IV Infusion:** Solution typically diluted to 1 mg/mL (specific concentration is ordered by prescriber) in D5W, NS, or other compatible solution. ▪ *For Dilaudid-HP:* Reconstitute 250 mg dry powder vial immediately prior to use with 25 mL sterile water for injection to yield 10 mg/mL. ▪ Final dilution of Dilaudid-HP 250 and HP 500 (supplied 500 mg/50 mL) **must be** ordered by prescriber.

ADMINISTER: Direct: Give 2 mg or fraction thereof over 3–5 min. **IV Infusion:** Both final volume and rate of infusion **must be** ordered by prescriber.

INCOMPATIBILITIES Solution/additive: Prochlorperazine, sodium bicarbonate, thiopental. Y-site: Amphotericin B cholesteryl, ceftobiprole, dantrolene, gallium, lansoprazole, minocycline, phenytoin, sargramostim, thiopental.

▪ A slight discoloration in ampules or multidose vials causes no loss of potency. ▪ Store in tight, light-resistant containers at 15°–30° C (59°–86° F).

ADVERSE EFFECTS (≥1%) **GI:** Nausea, vomiting, constipation. **CNS:** Euphoria, dizziness, sedation, *drowsiness*. **CV:** Hypotension, bradycardia or tachycardia. **Respiratory:** <u>Respiratory depression</u>. **Special Senses:** Blurred vision.

INTERACTIONS Drug: Alcohol and other CNS DEPRESSANTS compound sedation and CNS depression. **Herbal: St. John's wort, kava** may increase sedation.

PHARMACOKINETICS Absorption: 60% absorbed from GI tract. **Onset:** 15 min IV, 30 min PO. **Peak:** 30–90 min. **Duration:** 3–4 h. **Distribution:** Crosses placenta; distributed into breast milk. **Metabolism:** In liver. **Elimination:** In urine. **Half-Life:** 2–3 h.

NURSING IMPLICATIONS

Assessment & Drug Effects

▪ Note baseline respiratory rate, rhythm, and depth and size of pupils before administration. Respirations of 12/min or less and miotis are signs of toxicity. Withhold drug and promptly notify prescriber.

▪ Monitor vital signs at regular intervals. Drug-induced respiratory depression may occur even with small doses and increases progressively with higher doses.

▪ Assess effectiveness of pain relief 30 min after medication administration.

▪ Monitor drug effects carefully in older adult or debilitated patients and those with impaired renal and hepatic function.

▪ Assess effectiveness of cough. Drug depresses cough and sigh reflexes and may induce atelectasis, especially in postoperative patients and those with pulmonary disease.

▪ Note: Nausea and orthostatic hypotension most often occur in ambulatory patients or when a supine patient assumes the head-up position.

▪ Monitor I&O ratio and pattern. Assess lower abdomen for bladder distension. Report oliguria or urinary retention.

▪ Monitor bowel pattern; drug-induced constipation may require treatment.

Patient & Family Education

▪ Request medication at the onset of pain and do not wait until pain is severe.

▪ Use caution with activities requiring alertness; drug may cause drowsiness, dizziness, and blurred vision.

▪ Avoid alcohol and other CNS depressants while taking this drug.

HYDROQUINONE

(hye'droe-kwin-one)

Eldopaque, Eldoquin, Esoterica Regular, Melanex, Porcelana, Solaquin

Classifications: PIGMENT AGENT; DEPIGMENTOR
Therapeutic: DEPIGMENTOR

Pregnancy Category: C

AVAILABILITY 1.5%, 2%, 3%, 4% cream, gel, solution

ACTION & *THERAPEUTIC EFFECT* Causes reversible bleaching of hyperpigmented skin due to increased melanin. Interferes with formation of new melanin but does not destroy existing pigment. Depresses melanin synthesis and melanocytic growth, possibly by increasing excretion of melanin from melanocytes. *Interferes with formation of new melanin but does not destroy existing pigment.*

USES Gradual bleaching of hyperpigmented skin conditions such as chloasma or melasma, severe freckling, senile lentigines (age spots or liver spots). Also as an antioxidant in topical preparations. Some formulations include a sunscreening agent (e.g., Porcelana with Sunscreen, Mercolized Cocrema, Pabaquinone, and Solaquin).

CONTRAINDICATIONS Hyersensitivity to hydroquinone, PABA, paraben, or sulfite; prickly heat, sunburn, irritated skin, depilatory usage.
CAUTIOUS USE Pregnancy (category C), lactation. Safe use in children younger than 12 y not established.

ROUTE & DOSAGE

Bleaching of Hyperpigmented Skin
Adult: **Topical** Apply thin layer and rub into hyperpigmented skin b.i.d., a.m. and p.m.

ADMINISTRATION
Topical
- Test skin for sensitivity before treatment is initiated. Apply small amount of drug (about 25 mm in diameter) to an unbroken patch of skin and check in 24 h. Do not use drug if vesicle formation, itching, or excessive inflammation occur. Minor redness is not a contraindication.
- Limit applications to an area no larger than that of face and neck.

ADVERSE EFFECTS (≥1%) **Skin:** Dryness and fissuring of paranasal and infraorbital areas, inflammatory reaction, erythema; stinging, tingling, burning sensations; irritation, sensitization, and contact dermatitis.

INTERACTIONS Drug: No clinically significant interactions established.

NURSING IMPLICATIONS
Assessment & Drug Effects
- Monitor for therapeutic effectiveness: In general, complete depigmentation occurs in 1–4 mo and lasts 2–6 mo after hydroquinone is discontinued. Once desired results are obtained, reduce amount and frequency of applications to the least that will maintain depigmentation.
- Discontinue if bleaching or skin lightening does not occur after 2 or 3 mo of therapy.

Patient & Family Education
- Use a sunscreen agent or a hydroquinone formulation containing a sunscreen for daytime applications.
- Wash drug off if rash or irritation develops and consult prescriber.
- Avoid contact with the eyes and not to use on open lesions, sunburned, irritated, or otherwise damaged skin.
- Continue use of protective clothing and sunscreening agent after treatment is terminated to reduce possibility of repigmentation.

HYDROXOCOBALAMIN (VITAMIN B$_{12 ALPHA}$)

(hye-drox-oh-koe-bal′a-min)

Hydrobexan, Hydroxo-12, LA-12
Classification: VITAMIN SUPPLEMENT
Therapeutic: VITAMIN B$_{12}$ REPLACEMENT

Prototype: Cyanocobalamin
Pregnancy Category: A (C if greater than RDA)

AVAILABILITY 1000 mcg/mL injection

ACTION & *THERAPEUTIC EFFECT*
Cobalamin derivative similar to cyanocobalamin (vitamin B$_{12}$). Essential for normal cell growth, cell reproduction maturation of RBCs, myelin synthesis, and believed to be involved in protein synthesis. *Effective in vitamin B$_{12}$ deficiency that results in megaloblastic anemia.*

USES Treatment of vitamin B$_{12}$ deficiency.
UNLABELED USES Cyanide poisoning and tobacco amblyopia.

CONTRAINDICATIONS History of sensitivity to vitamin B$_{12}$, other cobalamins, or cobalt; indiscriminate use in folic acid deficiency.
CAUTIOUS USE Pregnancy (category A; category C in greater than RDA), lactation, children.

ROUTE & DOSAGE

Vitamin B$_{12}$ Deficiency
Adult: **IM** 30 mcg/day for 5–10 days and then 100–200 mcg/mo or 1000 mcg every other day until remission and then 1000 mcg/mo
Child: **IM** 100 mcg doses to a total of 1–5 mg over 2 wk and then 30–50 mcg/mo

ADMINISTRATION

Intramuscular
- Give deep into a large muscle.

INTERACTIONS Drug: Chloramphenicol may interfere with therapeutic response to hydroxocobalamin.

PHARMACOKINETICS Distribution: Widely distributed; principally stored in liver, kidneys, and adrenals; crosses placenta. **Metabolism:** Converted in tissues to active coenzymes; enterohepatically cycled. **Elimination:** 50–95% of doses 100 mcg or greater are excreted in urine in 48 h; excreted in breast milk.

NURSING IMPLICATIONS

Assessment & Drug Effects
- Monitor for therapeutic effectiveness: Response to drug therapy is usually dramatic, occurring within 48 h. Effectiveness is measured by laboratory values and improvement in manifestations of vitamin B$_{12}$ deficiency.
- Lab tests: Prior to therapy determine reticulocyte and erythrocyte counts, Hgb, Hct, vitamin B$_{12}$, and serum folate levels; repeated 5–7 days after start of therapy and at regular intervals during therapy.
- Obtain a careful history of sensitivities. Sensitization can take as long as 8 y to develop.
- Monitor potassium levels during the first 48 h, particularly in patients with Addisonian pernicious anemia or megaloblastic anemia. Conversion to normal erythropoiesis can result in severe hypokalemia and sudden death.
- Monitor vital signs in patients with cardiac disease and be alert to symptoms of pulmonary

Common adverse effects in *italic*, life-threatening effects underlined; generic names in **bold**; classifications in SMALL CAPS; ♣ Canadian drug name; ❖ Prototype drug

739

edema; generally occur early in therapy.

- Monitor bowel function. Bowel regularity is essential for consistent absorption of oral preparations.

Patient & Family Education

- Notify prescriber of any intercurrent disease or infection. Increased dosage may be required.
- Note: It is imperative to understand that drug therapy **must be** continued throughout life for pernicious anemia to prevent irreversible neurologic damage.
- Neurologic damage is considered irreversible if there is no improvement after 1–1.5 y of adequate therapy.
- Dietary deficiency of vitamin B₁₂ has been observed in strict vegetarians (vegans) and their breast-fed infants as well as in the elderly.

HYDROXYCHLOROQUINE

(hye-drox-ee-klor'oh-kwin)
Plaquenil Sulfate
Classifications: BIOLOGICAL RESPONSE MODIFIER; ANTIMALARIAL; DISEASE MODIFYING RHEUMATIC DRUG (DMARD)
Therapeutic: ANTIMALARIAL; ANTIRHEUMATIC
Prototype: Chloroquine
Pregnancy Category: C

AVAILABILITY 200 mg tablets

ACTION & THERAPEUTIC EFFECT
Antimalarial activity results from forming complexes with DNA of parasite, thereby inhibiting replication and transcription to RNA and DNA synthesis of the parasite. *Effective against* Plasmodium vivax *and* Plasmodium malariae. *Also is effective as second line of defense for treatment of rheumatoid arthritis and SLE.*

USES Suppressive prophylaxis and treatment of acute malarial attacks due to all forms of susceptible malaria. Used adjunctively with primaquine for eradication of *Plasmodium vivax* and *Plasmodium malariae*. More commonly prescribed than chloroquine for treatment of rheumatoid arthritis and lupus erythematosus (usually in conjunction with salicylate or corticosteroid therapy).
UNLABELED USES Porphyria cutanea tarda.

CONTRAINDICATIONS Known hypersensitivity to retinal or visual field changes associated with quinoline compounds; psoriasis, porphyria, G6PD deficiency; long-term therapy in children.
CAUTIOUS USE Hepatic disease; alcoholism, use with hepatotoxic drugs; impaired renal function, porphoria; metabolic acidosis; patients with tendency to dermatitis; pregnancy (category C).

ROUTE & DOSAGE

Note: Doses are expressed in terms of hydroxychloroquine base:
400 mg tablet = 310 mg base;
800 mg tablet = 620 mg base

Acute Malaria
Adult: **PO** 620 mg base followed by 310 mg base at 6, 18, and 24 h
Child: **PO** 10 mg base/kg, then 5 mg base/kg at 6, 18, and 24 h

Malaria Suppression
Adult: **PO** 310 mg base the same day each week starting 2 wk before exposure and continuing for 4–6 wk after leaving the area of exposure

Common adverse effects in *italic*, life-threatening effects <u>underlined</u>; generic names in **bold**; classifications in SMALL CAPS; ♣ Canadian drug name; ⊙ Prototype drug

Child: **PO** 5 mg base/kg the same day each week starting 2 wk before exposure and continuing for 4–8 wk after leaving the area of exposure

Lupus Erythematosus

Adult: **PO** 310 mg base 1–2 times/day
Child: **PO** 3–5 mg/kg/day in 1–2 divided doses (max: 400 mg/day or 7 mg/kg/day)

Rheumatoid Arthritis

Adult: **PO** 400–600 mg/day until response, then decrease to lowest maintenance levels possible
Child: **PO** 3–5 mg/kg/day in 1–2 divided doses (max: 400 mg/day or 7 mg/kg/day)

ADMINISTRATION

Oral

- Give drug with meals or milk to reduce incidence of GI distress.
- Give antacids and laxatives at least 4 h before or after hydroxychloroquine.
- Store at 15°–30° C (59°–86° F) unless otherwise directed.

ADVERSE EFFECTS (≥1%) CNS: Fatigue, vertigo, headache, mood or mental changes, anxiety, *retinopathy,* blurred vision, difficulty focusing. **GI:** Anorexia, nausea, vomiting, diarrhea, abdominal cramps, weight loss. **Hematologic:** Hemolysis in patients with G6PD deficiency, agranulocytosis (rare), aplastic anemia (rare), thrombocytopenia. **Skin:** Bleaching or loss of hair, unusual pigmentation (blue-black) of skin or inside mouth, skin rash, itching.

INTERACTIONS Drug: Aluminum- and **magnesium-**containing ANTACIDS and LAXATIVES decrease hydroxychloroquine absorption; separate administrations by at least 4 h; hydroxychloroquine may interfere with response to **rabies vaccine.**

PHARMACOKINETICS Absorption: Rapidly and almost completely absorbed. **Peak:** 1–2 h. **Distribution:** Widely distributed; concentrates in lungs, liver, erythrocytes, eyes, skin, and kidneys; crosses placenta. **Metabolism:** Partially in liver to active metabolite. **Elimination:** In urine; excreted in breast milk. **Half-Life:** 70–120 h.

NURSING IMPLICATIONS

Assessment & Drug Effects

- Monitor for therapeutic effectiveness; may not appear for several weeks, and maximal benefit may not occur for 6 mo.
- Lab tests: Baseline and periodic blood cell counts on all patients on long-term therapy.
- Withhold drug and notify prescriber if weakness, visual symptoms, hearing loss, unusual bleeding, bruising, or skin eruptions occur.

Patient & Family Education

- Learn about adverse effects and their symptoms when taking prolonged therapy.
- Follow drug regimen exactly as prescribed by the prescriber.

HYDROXYPROGESTERONE CAPROATE

(hye-drox′-ee-proe-jes′-ter-one kap′-roe-ate)
Makena
Classifications: HORMONE; PROGESTIN
Therapeutic: PROGESTIN
Prototype: Progesterone
Pregnancy Category: B

AVAILABILITY 1250 mg/5 mL solution for injection

Common adverse effects in *italic,* life-threatening effects underlined; generic names in **bold;** classifications in SMALL CAPS; ✦ Canadian drug name; ❂ Prototype drug

741

ACTION & THERAPEUTIC EFFECT
The mechanism by which hydroxy-progesterone reduces the risk of preterm birth is unknown. *Decreases risk of recurrent preterm births.*

USES Decrease risk of premature birth in women with a singleton (i.e., one fetus) pregnancy who have a history of singleton spontaneous preterm birth.
UNLABELED USES Amenorrhea, dysfunctional uterine bleeding, endometrial cancer, test for endogenous estrogen production.

CONTRAINDICATIONS Current/history of thrombosis or thromboembolic disorders; current/history of breast cancer or other hormone-sensitive cancer; undiagnosed abnormal vaginal bleeding not related to pregnancy; cholestatic jaundice of pregnancy; benign or malignant liver tumors or other active liver disease; uncontrolled hypertension; postmenopausal status.
CAUTIOUS USE Hypersensitivity; diabetes or prediabetes; conditions exacerbated by fluid retention (e.g., preeclampsia, seizure disorder, migraine, asthma, cardiac or renal dysfunction); hypertension; jaundice; depression. Safety and effectiveness in children under 16 y not established.

ROUTE & DOSAGE

Prevention of Premature Birth

Adult: **IM** 250 mg weekly. Begin between 16 wks 0 d to 20 wk 6 d of pregnancy. Continue until wk 37.

ADMINISTRATION

Intramuscular

- Draw up 1 mL using an 18-guage needle. Change to a 21-gauge 1 and 1/2 inch needle.
- Slowly inject (over at least 1 min) into the upper outer quadrant of the gluteus maximus.
- Store at 15°–30° C (59°–86° F) upright and protect from light. Discard vial 5 wk after first use.

ADVERSE EFFECTS (≥1%) **Body as a Whole:** Hypersensitivity reactions, injection-site nodule, *injection-site pain*, injection-site pruritus, *injection-site swelling*, miscarriage, stillbirth. **CV:** Gestational hypertension or preeclampsia, thromboembolic disorders. **GI:** Diarrhea, nausea. **Metabolic:** Decreased glucose tolerance, fluid retention, gestational diabetes. **Skin:** Pruritis, *urticaria*.

INTERACTIONS Drug: Hydroxyprogesterone can decrease the plasma levels of drugs that are substrates for CYP1A2 (e.g., **clozapine, theophylline, tizanidine**), CYP2A6 (e.g., **acetaminophen, nicotine**), or CYP2B6 (i.e., **bupropion, efavirenz, methadone**).

PHARMACOKINETICS Peak: 3–7 d. **Distribution:** Extensively bound to plasma proteins. **Metabolism:** In liver. **Elimination:** Fecal (50%) and renal (30%). **Half-Life:** 7.8 d.

NURSING IMPLICATIONS

Assessment & Drug Effects

- Monitor periodically: BP, weight, and mental status.
- Monitor diabetics closely for loss of glycemic control.
- Promptly report development of any of the following: New onset hypertension; S&S of thromboembolic disorder; unexplained vaginal bleeding; sudden weight gain; jaundice; depression; S&S of hypersensitivity (see Appendix F).
- Lab tests: Baseline and periodic fasting blood sugar if diabetic or prediabetic.

Patient & Family Education

- Report to prescriber if injection site becomes inflamed or increasingly painful over time.
- Diabetics should frequently monitor blood sugar and report significant changes to the prescriber.

HYDROXYUREA

(hye-drox'ee-yoo-ree-ah)

Hydrea, Droxia

Classifications: ANTINEOPLASTIC; ANTIMETABOLITE

Therapeutic: ANTINEOPLASTIC

Pregnancy Category: D

AVAILABILITY 200 mg, 300 mg, 400 mg, 500 mg capsules

ACTION & *THERAPEUTIC EFFECT*

A cell-cycle-phase antineoplastic that causes an immediate inhibition of DNA synthesis by acting as an RNA reductase inhibitor, necessary for DNA synthesis but without interfering with the synthesis of RNA or protein. *Cytotoxic effect limited to tissues with high rates of cell proliferation.*

USES Palliative treatment of metastatic melanoma, chronic myelocytic leukemia; recurrent metastatic, or inoperable ovarian cancer. Also used as adjunct to x-ray therapy for treatment of advanced primary squamous cell (epidermoid) carcinoma of head (excluding lip), neck, lungs.

UNLABELED USES Psoriasis; combination therapy with radiation of lung carcinoma; sickle cell anemia.

CONTRAINDICATIONS Severe myelosuppression; severe anemia, thrombocytopenia; pregnancy (category D), lactation.

CAUTIOUS USE Recent use of other cytotoxic drugs or irradiation; bone marrow depression; renal dysfunction; HIV patients; older adults; history of gout. Safe use in children not established.

ROUTE & DOSAGE

Palliative Therapy

Adult: **PO** 80 mg/kg q3days or 20–30 mg/kg/day

Sickle Cell Disease

Adult: **PO** 15 mg/kg/day, may increase by 5 mg/kg/day (max: 35 mg/kg/day or until toxicity develops)

Renal Impairment Dosage Adjustment

CrCl 10–50 mL/min: Administer 50% of dose; *less than 10 mL/min:* Administer 20% of dose

Hemodialysis Dosage Adjustment

Administer dose after hemodialysis; no supplemental dose needed

ADMINISTRATION

Oral

- Open, mix with water, and give immediately when patient has difficulty swallowing capsule.
- Store in tightly covered container at 15°–30° C (59°–86° F) unless otherwise directed.

ADVERSE EFFECTS (≥1%) **CNS:** Rare: Headache, dizziness, hallucinations, convulsions. **GI:** Stomatitis, anorexia, nausea, vomiting, diarrhea, constipation. **Hematologic:** Bone marrow suppression (leukopenia, anemia, thrombocytopenia), megaloblastic erythropoiesis. **Skin:** Maculopapular rash, facial erythema, postirradiation erythema. **Urogenital:** Renal tubular dysfunction, elevated BUN,

Common adverse effects in *italic*, life-threatening effects underlined; generic names in **bold;** classifications in SMALL CAPS; ♣ Canadian drug name; ⊘ Prototype drug

743

serum, creatinine levels, hyperuricemia. **Body as a Whole:** Fever, chills, malaise.

INTERACTIONS Drug: No clinically significant interactions established.

PHARMACOKINETICS Absorption: Readily absorbed from GI tract. **Peak:** 2 h. **Distribution:** Crosses blood–brain barrier. **Metabolism:** In liver. **Elimination:** As respiratory CO_2 and as urea in urine.

NURSING IMPLICATIONS

Assessment & Drug Effects

- Lab tests: Determine status of kidney, liver, and bone marrow function before and periodically during therapy; monitor hemoglobin, WBC, platelet counts at least once weekly.
- Interrupt therapy if WBC drops to 2500/mm³ or platelets to 100,000/mm³.
- Monitor I&O. Advise patients with high serum uric acid levels to drink at least 10–12 240 mL (8 oz) glasses of fluid daily to prevent uric acid nephropathy.
- Note: Patients with marked renal dysfunction may rapidly develop visual and auditory hallucinations and hematologic toxicity.

Patient & Family Education

- Notify prescriber of fever, chills, sore throat, nausea, vomiting, diarrhea, loss of appetite, and unusual bruising or bleeding.
- Use barrier contraceptive during therapy. Drug is teratogenic.

HYDROXYZINE HYDROCHLORIDE ⊕

(hye-drox′i-zeen)

Atarax Syrup, Hyzine-50, Vistaril Intramuscular, Vistacon, Vistaject-25 & -50

HYDROXYZINE PAMOATE

Vistaril Oral

Classifications: ANTIHISTAMINE; H₁-RECEPTOR ANTAGONIST
Therapeutic: ANTIPRURITIC; ANTIANXIETY; ANTIEMETIC
Pregnancy Category: C

AVAILABILITY Hydroxyzine HCl: 10 mg, 25 mg, 50 mg tablets; 10 mg/5 mL syrup; 25 mg/5 mL oral suspension; 25 mg/mL, 50 mg/mL injection; **Hydroxyzine Pamoate:** 25 mg, 50 mg, 100 mg capsules; 25 mg/5 mL suspension

ACTION & *THERAPEUTIC EFFECT* H₁-receptor antagonist effective in treatment of histamine-mediated pruritus or other allergic reactions. Its tranquilizing effect is produced primarily by depression of hypothalamus and brain-stem reticular formation, rather than cortical areas. *Effective as an antianxiety agent and sedative. Additionally, it is an effective agent for pruritus and as an antiemetic agent.*

USES Emotional or psychoneurotic states characterized by anxiety, tension, or psychomotor agitation; to relieve anxiety, control nausea and emesis, and reduce narcotic requirements before or after surgery or delivery. Also used in management of pruritus due to allergic conditions (e.g., chronic urticaria), atopic and contact dermatoses, and in treatment of acute and chronic alcoholism with withdrawal symptoms or delirium tremens.

CONTRAINDICATIONS Known hypersensitivity to hydroxyzine; use as sole treatment in psychoses or depression; lactation.
CAUTIOUS USE History of allergies; GI disorders; cardiac disease;

·Common adverse effects in *italic*, life-threatening effects underlined; generic names in **bold**; classifications in SMALL CAPS; ♣ Canadian drug name; ⊕ Prototype drug

COPD; older adults; pregnancy (category C).

ROUTE & DOSAGE

Anxiety

Adult: **PO** 50–100 mg t.i.d. or q.i.d. **IM** 50–100 mg q4–6h
Child: **PO** *Younger than 6 y,* 50 mg/day in divided doses; *older than 6 y,* 50 mg/day in divided doses **IM** 1.1 mg/kg q4–6h

Pruritus

Adult: **PO** 25 mg t.i.d. or q.i.d. **IM** 25 mg q4–6h
Geriatric: **PO** 10 mg 3–4 times daily
Child: **PO** *Older than 6 y,* 50–100 mg/day in divided doses; *younger than 6 y,* 50 mg/day in divided doses **IM** 1.1 mg/kg q4–6h

Nausea

Adult: **IM** 25–100 mg q4–6h
Child: **IM** 1.1 mg/kg q4–6h

ADMINISTRATION

Oral

▪ Note: Tablets may be crushed and taken with fluid of patient's choice. Capsule may be emptied and contents swallowed with water or mixed with food. Liquid formulations are available.

Intramuscular

▪ Give deep into body of a relatively large muscle. The Z-track technique of injection is recommended to prevent subcutaneous infiltration.
▪ Recommended site: In adult, the gluteus maximus or vastus lateralis; in children, the vastus lateralis.
▪ Protect all forms from light. Store at 15°–30° C (59°–86° F) unless otherwise specified.

INCOMPATIBILITIES **Solution/additive: Aminophylline, amobarbital, chloramphenicol, dimenhydrinate, penicillin G, pentobarbital, phenobarbital.**

ADVERSE EFFECTS (≥1%) **CNS:** *Drowsiness* (usually transitory), sedation, dizziness, headache. **CV:** Hypotension. **GI:** *Dry mouth.* **Body as a Whole:** Urticaria, dyspnea, chest tightness, wheezing, involuntary motor activity (rare). **Hematologic:** Phlebitis, hemolysis, thrombosis. **Skin:** Erythematous macular eruptions, erythema multiforme, digital gangrene from inadvertent IV or intra-arterial injection, injection site reactions.

DIAGNOSTIC TEST INTERFERENCE Possible false positive *serum TCA screen.*

INTERACTIONS Drug: Alcohol and CNS DEPRESSANTS add to CNS depression; TRICYCLIC ANTIDEPRESSANTS and other ANTICHOLINERGICS have additive anticholinergic effects; may inhibit pressor effects of **epinephrine.**

PHARMACOKINETICS Absorption: Readily from GI tract. **Onset:** 15–30 min PO. **Duration:** 4–6 h. **Distribution:** Not known if it crosses placenta or is distributed into breast milk. **Metabolism:** In liver. **Elimination:** In bile.

NURSING IMPLICATIONS

Assessment & Drug Effects

▪ Evaluate alertness. Drowsiness may occur and usually disappears with continued therapy or following reduction of dosage.
▪ Monitor condition of oral membranes daily when patient is on high dosage of hydroxyzine.
▪ Reduce dosage of the depressant up to 50% when CNS depressants are prescribed concomitantly.

Common adverse effects in *italic*, life-threatening effects underlined; generic names in **bold;** classifications in SMALL CAPS; ♣ Canadian drug name; ⊙ Prototype drug

745

Patient & Family Education

- Do not drive or engage in other potentially hazardous activities until response to drug is known.
- Do NOT take alcohol and hydroxyzine at the same time.
- Relieve dry mouth by frequent warm water rinses, increasing fluid intake, and use of a salivary substitute (e.g., Moi-Stir, Xero-Lube).
- Give teeth scrupulous care. Avoid irritation or abrasion of gums and other oral tissues.

HYOSCYAMINE SULFATE

(hye-oh-sye′a-meen)

Anaspaz, Cystospaz, Levsin, Levsinex, NuLev

Classifications: ANTICHOLINERGIC; ANTIMUSCARINIC; ANTISPASMODIC
Therapeutic: GI ANTISPASMODIC
Prototype: Atropine
Pregnancy Category: C

AVAILABILITY 0.125 mg, 0.150 mg tablets; 0.125 mg sublingual tablets; 0.375 sustained release capsules; 0.125 mg orally disintegrating tablet 0.125 mg/mL oral solution; 0.125 mg/5 mL elixir; 0.5 mg/mL injection

ACTION & *THERAPEUTIC EFFECT*
Competitive inhibitor of acetylcholine at autonomic postganglionic cholinergic receptors. It decreases motility (smooth muscle tone) in GI, biliary, and urinary tracts. *Effective as a GI antispasmodic.*

USES GI tract disorders caused by spasm and hypermotility, as conjunct therapy with diet and antacids for peptic ulcer management, and as an aid in the control of gastric hypersecretion and intestinal hypermotility. Also symptomatic relief of biliary and renal colic, as a "drying agent" to relieve symptoms of acute rhinitis, to control preanesthesia salivation and respiratory tract secretions, to treat symptoms of parkinsonism, and to reduce pain and hypersecretion in pancreatitis.

CONTRAINDICATIONS Hypersensitivity to belladonna alkaloids, prostatic hypertrophy, obstructive diseases of GI or GU tract, ulcerative colitis, paralytic ileus or intestinal atony; myasthenia gravis.

CAUTIOUS USE Diabetes mellitus, cardiac disease, cardiac arrhythmias; autonomic neuropathy; closed-angle glaucoma; GERD, hiatal hernia; pulmonary disease; renal or hepatic disease; pregnancy (category C), lactation; children younger than 2 y.

ROUTE & DOSAGE

GI Spasms
Adult: **IV/IM/Subcutaneous** 0.25–0.5 mg q4h **PO/SL** 0.125–0.25 mg t.i.d. or q.i.d. prn *Child (2–12 y):* **PO** 0.0625–0.125 mg q4h prn (max: 0.75 mg/day)

ADMINISTRATION

- Note: Dose for older adults may need to be less than the standard adult dose. Observe patient carefully for signs of paradoxic reactions.

Oral

- Give preparations about 1 h before meals and at bedtime (at least 2 h after last meal).
- Ensure that sustained release form of drug is not chewed or crushed. It **must be** swallowed whole.

Intramuscular/Subcutaneous

- May be given undiluted.

Intravenous

PREPARE: Direct: Give undiluted.
ADMINISTER: Direct: Give a single dose over 60 sec.

- Store 15°–30° C (59°–86° F).

ADVERSE EFFECTS (≥1%) **CNS:** Headache, unusual tiredness or weakness, confusion, *drowsiness,* excitement in older adult patients. **CV:** Palpitations, tachycardia. **Special Senses:** *Blurred vision,* increased intraocular tension, cycloplegia, mydriasis. **GI:** *Dry mouth, constipation,* paralytic ileus. **Other:** *Urinary retention,* anhidrosis, suppression of lactation.

INTERACTIONS Drug: Aman-tadine, ANTIHISTAMINES, TRICYCLIC ANTI-DEPRESSANTS, **quinidine, di-sopyramide, procainamide** add anticholinergic effects; decreases **levodopa** effects; **methotrimep-razine** may precipitate extrapyramidal effects; decreases anti-psychotic effects of PHENOTHIAZINES (decreased absorption).

PHARMACOKINETICS Absorption: Well absorbed from all administration sites. **Onset:** 2–3 min IV; 20–30 min PO. **Peak:** 15–30 min IV; 30–60 min PO. **Duration:** 4–6 h (up to 12 h with sustained release form). **Distribution:** Distributed in most body tissues; crosses blood–brain barrier and placenta; distributed in breast milk. **Metabolism:** In liver. **Elimination:** In urine. **Half-Life:** 3.5–13 h.

NURSING IMPLICATIONS

Assessment & Drug Effects

- Monitor bowel elimination; may cause constipation.
- Monitor urinary output.
- Lessen risk of urinary retention by having patient void prior to each dose.

- Assess for dry mouth and recommend good practices of oral hygiene.

Patient & Family Education

- Avoid excessive exposure to high temperatures; drug-induced heat-stroke can develop.
- Do not drive or engage in other potentially hazardous activities until response to drug is known.
- Use dark glasses if experiencing blurred vision, but if this adverse effect persists, notify prescriber for dose adjustment or possible drug change.

IBANDRONATE SODIUM
Boniva

Classifications: BISPHOSPHONATE; BONE METABOLISM REGULATOR
Therapeutic: BONE METABOLISM REGULATOR
Prototype: Etidronate
Pregnancy Category: C

AVAILABILITY 2.5 mg and 150 mg tablets

ACTION & THERAPEUTIC EFFECT Ibandronate is a potent third-generation bisphosphonate. It inhibits activity of osteoclasts and reduces bone resorption and turnover in the matrix of the bone. *In postmenopausal women, it reduces the rate of bone turnover, resulting in a net gain in bone mass.*

USES Prevention and treatment of osteoporosis in postmenopausal women.

UNLABELED USES Treatment of metastatic bone disease in breast cancer.

CONTRAINDICATIONS Hypersensitivity to ibandronate; severe renal impairment; hypocalcemia, vitamin D

Common adverse effects in *italic;* life-threatening effects <u>underlined;</u> generic names in **bold;** classifications in SMALL CAPS; ♣ Canadian drug name; ⓟ Prototype drug

747

deficiency; inability to stand or sit up straight for 60 min; achalasia, esophageal stricture, dysphagia.

CAUTIOUS USE Mild or moderate renal impairment; history of GI bleeding or disease, esophagitis, esophageal or gastric ulcers; older adults; pregnancy (category C), lactation. Safe use in children younger than 18 y is not established.

ROUTE & DOSAGE

Postmenopausal Osteoporosis

Adult: **PO** 2.5 mg daily or 150 mg once monthly on the same day each month

Renal Impairment Dosage Adjustment

CrCl less than 30 mL/min: Use not recommended

ADMINISTRATION

- Correct hypocalcemia before administering ibandronate.
- Give at least 60 min before food, beverage, or other medications (including vitamins).
- Instruct to swallowed whole with a full glass of plain water (180–240 mL; 6–8 oz) while standing or sitting in an upright position.
- Keep patient sitting up or ambulating for 60 min after taking drug.
- Store 15°–30° C (59°–86° F).

ADVERSE EFFECTS (≥1%) **CNS:**
Dizziness, headache, nerve root lesion, vertigo. **GI:** Dyspepsia, constipation, diarrhea, esophagitis, gastritis, pharyngitis, nausea, vomiting. **Respiratory:** Upper respiratory infection, pharyngitis. **Skin:** Rash. **Body as a Whole:** Back pain. **Other:** Tooth disorder.

DIAGNOSTIC TEST INTERFERENCE
Interferes with the use of bone-imaging agents.

INTERACTIONS Drug: Concurrent administration of **calcium, magnesium,** or **iron** reduces ibandronate adsorption. **Food:** Food reduces ibandronate absorption (ibandronate should be taken in a fasting state).

PHARMACOKINETICS Absorption: Bioavailability poor (0.6%). **Peak:** 0.5–2 h. **Distribution:** 86–99% protein bound. **Metabolism:** None. **Elimination:** Renal. **Half-Life:** 10–60 h.

NURSING IMPLICATIONS

Assessment & Drug Effects

- Lab tests: Monitor albumin-adjusted serum calcium, serum phosphate, serum alkaline phosphatase, fasting and 24 h urinary calcium, and serum electrolytes; baseline and periodic renal function.
- Withhold drug and notify prescriber if the CrCl less than 30 mL/min.
- Diagnostic test: Bone density scan every 12–18 mo.
- Monitor for S&S of upper GI distress, especially with concurrent use of NSAIDs or aspirin.

Patient & Family Education

- Take the monthly dose (150 mg) on the same day each month. Carefully follow directions for taking the drug (see ADMINISTRATION).
- If a monthly dose is missed, and the next scheduled dose is more than 7 days away, take one 150 mg tablet the next morning then resume the original monthly schedule. Do not take two 150 mg tablets in the same week.
- Report to prescriber any of the following: Severe bone, joint, or muscle pain; heartburn, pain behind the sternum, difficulty or pain with swallowing.

⅋IBUPROFEN 🅟

(eye-byoo′proe-fen)

Advil, Amersol ♣, Caldolor, Children's Motrin, Ibuprin, Junior Strength Motrin Caplets, Motrin, Nuprin, Pediaprofen, Pamprin-IB, Rufen, Trendar

Classifications: ANALGESIC, NONSTEROIDAL ANTI-INFLAMMATORY DRUG (NSAID) (COX-1 AND COX-2 INHIBITOR); ANTIPYRETIC

Therapeutic: ANALGESIC, NSAID; ANTI-INFLAMMATORY; ANTIPYRETIC

Pregnancy Category: C

AVAILABILITY 100 mg, 200 mg, 400 mg, 600 mg, 800 mg tablets; 50 mg, 100 mg chewable tablets; 100 mg/5 mL, 100 mg/2.5 mL suspension; 40 mg/mL drops; 100 mg/mL injection

ACTION & _THERAPEUTIC EFFECT_
(COX-1 and COX-2) NSAID inhibitor with nonsteroidal anti-inflammatory activity that blocks prostaglandin synthesis. Its activity also includes modulation of T-cell function, inhibition of inflammatory cell chemotaxis, decreased release of superoxide radicals, or increased scavenging of these compounds at inflammatory sites. _Has nonsteroidal anti-inflammatory, analgesic, and antipyretic effects. Inhibits platelet aggregation and prolongs bleeding time._

USES Chronic, symptomatic rheumatoid arthritis and osteoarthritis; relief of mild to moderate pain; primary dysmenorrhea; reduction of fever.

UNLABELED USES Gout, juvenile rheumatoid arthritis, psoriatic arthritis, ankylosing spondylitis, vascular headache.

CONTRAINDICATIONS Patient in whom urticaria, severe rhinitis, bronchospasm, angioedema, nasal polyps are precipitated by aspirin or other NSAIDs; active peptic ulcer, bleeding abnormalities; perioperative pain related to CABG surgery.

CAUTIOUS USE History of GI ulceration; diabetes mellitus, impaired hepatic or renal function, chronic renal failure; hypertension, history of coronary artery disease, angina, MI; cardiac decompensation; patients with SLE; older adults; pregnancy (category C). Safe use in children younger than 6 mo not established.

ROUTE & DOSAGE

Inflammatory Disease
Adult: **PO** 400–800 mg t.i.d. or q.i.d. (max: 3200 mg/day)
Child: **PO** _Weight less than 20 kg,_ up to 400 mg/day in divided doses; _weight 20–30 kg,_ up to 600 mg/day in divided doses; _weight 30–40 kg,_ up to 800 mg/day in divided doses
Dysmenorrhea
Adult: **PO** 400 mg q4–6h up to 1200 mg/day
Mild to Moderate Pain
Adult: **PO** 400 mg q4–6h up to 1200 mg/day **IV** 400 mg q4–6h prn or 100–200 mg q4h prn
Fever
Adult: **PO** 200–400 mg t.i.d. or q.i.d. (max: 1200 mg/day)
Child (6 mo–12 y): **PO** 5–10 mg/kg q4–6h up to 40 mg/kg/day

ADMINISTRATION

Oral

- Give on an empty stomach, 1 h before or 2 h after meals. May be taken with meals or milk if GI intolerance occurs.

- Ensure that chewable tablets are chewed or crushed before being swallowed.
- Note: Tablet may be crushed if patient is unable to swallow it whole and mixed with food or liquid before swallowing.
- Store in tightly closed, light-resistant container unless otherwise directed by manufacturer.

Intravenous

Patients should be well hydrated before IV infusion to prevent renal damage.

PREPARE: Infusion: Dilute required dose with NS, D5W or LR to a final concentration of 4 mg/mL or less.

ADMINISTER: Infusion: Infuse over at least 30 min.

ADVERSE EFFECTS (≥1%) CNS:
Headache, dizziness, light-headedness, anxiety, emotional lability, fatigue, malaise, drowsiness, confusion, depression, aseptic meningitis. **CV:** Hypertension, palpitation, congestive heart failure (patient with marginal cardiac function); thrombotic events (MI, stroke); peripheral edema. **Special Senses:** Amblyopia (blurred vision, decreased visual acuity, scotomas, changes in color vision); nystagmus, visual-field defects; tinnitus, impaired hearing. **GI:** Dry mouth, gingival ulcerations, dyspepsia, *heartburn, nausea,* vomiting, anorexia, diarrhea, constipation, bloating, flatulence, epigastric or abdominal discomfort or pain, GI ulceration, *occult blood loss.* **Hematologic:** Thrombocytopenia, neutropenia, hemolytic or aplastic anemia, leukopenia; decreased Hgb, Hct; transitory rise in AST, ALT, serum alkaline phosphatase; rise in (Ivy) bleeding time. **GU:** Acute renal failure, polyuria, azotemia, cystitis, hematuria, nephrotoxicity, decreased creatinine clearance. **Skin:** Maculopapular and vesicobullous skin eruptions, erythema multiforme, pruritus, rectal itching, acne. **Body as a Whole:** Fluid retention with edema, Stevens-Johnson syndrome, toxic hepatitis, hypersensitivity reactions, anaphylaxis, bronchospasm, serum sickness, SLE, angioedema.

INTERACTIONS Drug: ORAL ANTICOAGULANTS, **heparin** may prolong bleeding time; may increase **lithium** and **methotrexate** toxicity. **Herbal: Feverfew, garlic, ginger, ginkgo** may increase bleeding potential.

PHARMACOKINETICS Absorption: 80% from GI tract (oral product). **Onset:** 1 h (antipyretic). **Peak:** 1–2 h. **Duration:** 6–8 h. **Metabolism:** In liver. **Elimination:** Primarily in urine; some biliary excretion. **Half-Life:** 2–4 h.

NURSING IMPLICATIONS

Assessment & Drug Effects

- Monitor for therapeutic effectiveness. Optimum response generally occurs within 2 wk to anti-inflammatory and/or analgesic effect (e.g., relief of pain, stiffness, or swelling; or improved joint flexion and strength).
- Observe patients with history of cardiac decompensation closely for evidence of fluid retention and edema.
- Lab tests: Baseline and periodic Hgb, renal function tests, LFTs.
- Auditory and ophthalmologic examinations are recommended in patients receiving prolonged or high-dose therapy.
- Monitor for GI distress and S&S of GI bleeding.

Common adverse effects in *italic*, life-threatening effects underlined; generic names in **bold**; classifications in SMALL CAPS; ✿ Canadian drug name; ⊘ Prototype drug

- Note: Symptoms of acute toxicity in children include apnea, cyanosis, response only to painful stimuli, dizziness, and nystagmus.

Patient & Family Education
- Notify prescriber immediately of passage of dark tarry stools, "coffee ground" emesis, frankly bloody emesis, or other GI distress, as well as blood or protein in urine, and onset of skin rash, pruritus, jaundice.
- Do not drive or engage in other potentially hazardous activities until response to the drug is known.
- Do not self-medicate with ibuprofen if taking prescribed drugs or being treated for a serious condition without consulting prescriber.
- Do not give to children younger than 3 mo or for longer than 2 days without consulting prescriber.
- Do not take aspirin concurrently with ibuprofen.
- Avoid alcohol and NSAIDs unless otherwise advised by prescriber. Concurrent use may increase risk of GI ulceration and bleeding tendencies.

IBUTILIDE FUMARATE
(i-bu'ti-lide)

Corvert
Classification: CLASS III ANTIARRHYTHMIC
Therapeutic: CLASS III ANTIARRHYTHMIC
Prototype: Amiodarone HCl
Pregnancy Category: C

AVAILABILITY 0.1 mg/mL injection

ACTION & *THERAPEUTIC EFFECT*
Ibutilide is a Class III antiarrhythmic that prolongs cardiac action potential and increases both atrial and ventricular refractoriness without affecting conduction. *Effective in treating recently occurring atrial arrhythmias. It may produce proarrhythmic effects that can be life threatening.*

USES Rapid conversion of atrial fibrillation or atrial flutter of recent onset.

CONTRAINDICATIONS Hypersensitivity to ibutilide, hypokalemia, hypomagnesemia.
CAUTIOUS USE History of CHF, cardiac ejection fraction of 35% or less, recent MI, prolonged QT intervals, ventricular arrhythmias; renal or liver disease, cardiovascular disorder other than atrial arrhythmias; pregnancy (category C), lactation. Safe use in children younger than 18 y not established.

ROUTE & DOSAGE

Atrial Fibrillation or Flutter
Adult: **IV** *Weight less than 60 kg, 0.01 mg/kg, may repeat in 10 min if inadequate response; weight 60 kg or greater, 1 mg, may repeat in 10 min if inadequate response*

ADMINISTRATION
- Hypokalemia and hypomagnesemia should be corrected prior to treatment with ibutilide.

Intravenous

***PREPARE:* Direct:** Give undiluted. **IV Infusion:** Contents of 1 mg vial may be diluted in 50 mL of D5W or NS to yield 0.017 mg/mL. ***ADMINISTER:* Direct/IV Infusion:** Give a single dose by direct injection or infusion over 10 min.

Common adverse effects in *italic*, life-threatening effects underlined; generic names in **bold**; classifications in SMALL CAPS; ◆ Canadian drug name; ⊙ Prototype drug

751

■ Stop injection/infusion as soon as presenting arrhythmia is terminated or with appearance of ventricular tachycardia or marked prolongation of QT or QT$_c$.

■ Store diluted solution up to 24 h at 15°–30° C (59°–86° F) or 48 h refrigerated at 2°–8° C (36°–46° F).

ADVERSE EFFECTS (≥1%) **CNS:** Headache. **CV:** Proarrhythmic effects (sustained and nonsustained polymorphic ventricular tachycardia), AV block, bundle branch block, ventricular extrasystoles, hypotension, postural hypotension, bradycardia, tachycardia, palpitations, prolonged QT segment. **GI:** Nausea.

INTERACTIONS Drug: Increased potential for proarrhythmic effects when administered with PHENO-THIAZINES, TRICYCLIC ANTIDEPRESSANTS, **amiodarone, disopyramide, quinidine, procainamide, sotalol** may cause prolonged refractoriness if given within 4 h of ibutilide.

PHARMACOKINETICS Onset: 30 min. **Metabolism:** In liver. **Elimination:** 82% in urine, 19% in feces. **Half-Life:** 6 h (range 2–21 h).

NURSING IMPLICATIONS

Assessment & Drug Effects

■ Observe with continuous ECG, BP, and HR monitoring during and for at least 4 h after infusion or until QT$_c$ has returned to baseline. Monitor for longer periods with liver dysfunction or if proarrhythmic activity is observed.

■ Lab tests: Baseline serum potassium and magnesium are recommended, as hypokalemia and hypomagnesemia should be corrected prior to beginning treatment with ibutilide.

■ Monitor for therapeutic effectiveness. Conversion to normal sinus rhythm normally occurs within 30 min of initiation of infusion.

Patient & Family Education

■ Consult prescriber and understand the potential risks of ibutilide therapy.

IDARUBICIN
(i-da-roo'bi-cin)
Idamycin PFS
Classifications: ANTINEOPLASTIC; ANTHRACYCLINE (ANTIBIOTIC)
Therapeutic: ANTINEOPLASTIC
Prototype: Doxorubicin
Pregnancy Category: D

AVAILABILITY 5 mg, 10 mg, 20 mg vials; 1 mg/mL injection

ACTION & *THERAPEUTIC EFFECT* Cytotoxic anthracycline that exhibits inhibitory effects on DNA topoisomerase II, an enzyme responsible for repairing faulty sections of DNA. It results in breaks in the helix of the DNA, and thus it affects RNA and protein synthesis in rapidly dividing cells. *Has antineoplastic and cytotoxic action on cancer cells that results in cell death.*

USES In combination with other antineoplastic drugs for treatment of AML.
UNLABELED USES Breast cancer, other solid tumors.

CONTRAINDICATIONS Myelosuppression, hypersensitivity to idarubicin or doxorubicin, pregnancy (category D), lactation.

CAUTIOUS USE Impaired renal or hepatic function; patients who have received irradiation or radiotherapy to areas surrounding heart; cardiac disease. Safe use in children younger than 2 y not established.

ROUTE & DOSAGE

Acute Myelogenous Leukemia (AML)
Adult: IV 12 mg/m² daily for 3 days injected slowly over 10–15 min

Acute Nonlymphocytic Leukemia, Acute Lymphocytic Leukemia
Child: IV 10–12 mg/m²/day for 3 days

Renal Impairment Dosage Adjustment
Creatinine greater than 2 mg/dL: Give 75% of dose

Hepatic Impairment Dosage Adjustment
Bilirubin 1.5–5 mg/dL: Give 50% of dose; *if greater than 5 mg/dL:* Do not use drug

ADMINISTRATION

Intravenous

IV administration to infants, children: Verify correct IV concentration and rate of infusion with prescriber.

PREPARE: IV Infusion: Reconstitute by adding 1 mL of nonbacteriostatic NS for each 1 mg of idarubicin to yield 1 mg/mL. ▪ Vials are under negative pressure, therefore, carefully insert needle into vial to reconstitute. ▪ Wash skin accidentally exposed with soap and water.

ADMINISTER: IV Infusion: Give slowly over 10–15 min into tubing of free flowing IV of NS or D5W. ▪ If extravasation is suspected, immediately stop infusion, elevate the arm, and apply ice pack for 30 min then q.i.d. for 30 min × 3 days.

INCOMPATIBILITIES Solution/additive: ALKALINE SOLUTIONS (i.e., **sodium bicarbonate**), **heparin. Y-site:** **Acyclovir, allopurinol, ampicillin/sulbactam, cefazolin, cefepime, ceftazidime, clindamycin, dexamethasone, etoposide, furosemide, gentamicin, heparin, hydrocortisone, imipenem/cilastatin, lorazepam, meperidine, methotrexate, mezlocillin, piperacillin/tazobactam, sargramostim, sodium bicarbonate, teniposide, vancomycin, vincristine.**

▪ Store reconstituted solutions up to 7 days refrigerated at 2°–8° C (36°–46° F) and 72 h at room temperature 15°–30° C (59°–86° F).

ADVERSE EFFECTS (≥1%) **CV:** CHF, atrial fibrillation, chest pain, <u>MI</u>. **GI:** *Nausea, vomiting, diarrhea, abdominal pain,* mucositis. **Hematologic:** <u>*Anemia, leukopenia,*</u> thrombocytopenia. **Other:** Nephrotoxicity, hepatotoxicity, *alopecia,* rash.

INTERACTIONS Drug: IMMUNOSUPPRESSANTS cause additive bone marrow suppression; ANTICOAGULANTS, NSAIDS, SALICYLATES, **aspirin,** THROMBOLYTIC AGENTS increase risk of bleeding; idarubicin may blunt the effects of **filgrastim, sargramostim.**

PHARMACOKINETICS Onset: Median time to remission 28 days. **Peak:** Serum level 4 h. **Duration:** Serum levels 120 h. **Distribution:** Concentrates in nucleated blood and bone marrow cells. **Metabolism:** In liver to idarubicinol, which

---

Common adverse effects in *italic*, life-threatening effects <u>underlined</u>; generic names in **bold;** classifications in SMALL CAPS; ❤ Canadian drug name; ❷ Prototype drug

753

may be as active as idarubicin. **Elimination:** 16% in urine; 17% in bile. **Half-Life:** Idarubicin 15–45 h, idarubicinol 45 h.

NURSING IMPLICATIONS

Assessment & Drug Effects

- Monitor infusion site closely, as extravasation can cause severe local tissue necrosis. Notify prescriber if pain, erythema, or edema develops at insertion site.
- Lab tests: Monitor hepatic and renal function, CBC with differential and coagulation studies periodically.
- Monitor cardiac status closely, especially in older adult patients or those with preexisting cardiac disease.
- Monitor hematologic status carefully; during the period of myelosuppression, patients are at high risk for bleeding and infection.
- Monitor for development of hyperuricemia secondary to lysis of leukemic cells.

Patient & Family Education

- Learn all potential adverse reactions to idarubicin.
- Anticipate possible hair loss.
- Discuss interventions to minimize nausea, vomiting, diarrhea, and stomatitis with health care providers.

IFOSFAMIDE
(i-fos′fa-mide)
Classifications: ANTINEOPLASTIC; ALKYLATING AGENT
Therapeutic: ANTINEOPLASTIC
Prototype: Cyclophosphamide
Pregnancy Category: D

AVAILABILITY 1 g, 3 g vials

ACTION & *THERAPEUTIC EFFECT*
Alkylated metabolite of ifosfamide that interacts with DNA as a cell cycle nonspecific agent. Antineoplastic action is primarily due to cross-linking of strands of DNA and RNA as well as inhibition of protein synthesis. *It has antineoplastic and cytotoxic action on cancer cells that results in cell death.*

USES In combination with other agents in various regimens for germ cell testicular cancer, soft tissue sarcomas, Ewing's sarcoma, and non-Hodgkin's lymphoma. Also used for lung and pancreatic sarcoma.

CONTRAINDICATIONS Previous hypersensitivity to ifosfamide; severe bone marrow depression; dehydration; pregnancy (category D), lactation.
CAUTIOUS USE Impaired renal function, renal failure; hepatic disease; prior radiation or prior therapy with other cytotoxic agents.

ROUTE & DOSAGE

Antineoplastic

Adult: **IV** 1.2 g/m²/day for 5 consecutive days; administer over at least 30 min, repeat q3wk or after recovery from hematologic toxicity (platelets 100,000/mm³ or greater; WBC 4000/mm³ or greater)

ADMINISTRATION

Intravenous

PREPARE: **IV Infusion:** Dilute each 1 g in 20 mL of sterile water or bacteriostatic water to yield 50 mg/mL. ▪ Shake well to dissolve. ▪ May be further diluted with D5W, NS, or LR to achieve concentrations of 0.6–20 mg/mL. ▪ Use solution prepared with sterile water within 6 h.

ADMINISTER: **IV Infusion:** Give slowly over 30 min. ▪ Note: Mesna is always given concurrently with ifosfamide; never give ifosfamide alone.

INCOMPATIBILITIES Y-site: **Cefepime, methotrexate.**

▪ Store reconstituted solution prepared with bacteriostatic solution up to a week at 30° C (86° F) or 6 wk at 5° C (41° F).

ADVERSE EFFECTS (≥1%) **CNS:** *Somnolence, confusion, hallucinations,* coma, dizziness, seizures, cranial nerve dysfunction. **GI:** *Nausea, vomiting,* anorexia, diarrhea, metabolic acidosis, hepatic dysfunction. **Hematologic:** <u>Neutropenia, thrombocytopenia</u>. **Urogenital:** <u>Hemorrhagic cystitis, nephrotoxicity</u>. **Skin:** *Alopecia,* skin necrosis with extravasation.

INTERACTIONS Drug: HEPATIC ENZYME INDUCERS (BARBITURATES, **phenytoin, chloral hydrate**) may increase hepatic conversion of ifosfamide to active metabolite; CORTICOSTEROIDS may inhibit conversion to active metabolites.

PHARMACOKINETICS **Distribution:** Distributed into breast milk. **Metabolism:** In liver via CYP3A4. **Elimination:** 70–86% in urine. **Half-Life:** 7–15 h.

NURSING IMPLICATIONS

Assessment & Drug Effects

▪ Lab tests: Monitor CBC with differential prior to each dose and at regular intervals; urinalysis prior to each dose for microscopic hematuria.
▪ Hold drug and notify prescriber if WBC count is below 2000/mm³ or platelet count is below 50,000/mm³.
▪ Reduce risk of hemorrhagic cystitis by hydrating with 3000 mL of fluid daily prior to therapy and for at least 72 h following treatment to ensure ample urine output.
▪ Monitor for and repost promptly any of the following CNS symptoms: Somnolence, confusion, depressive psychosis, and hallucinations.

Patient & Family Education

▪ Void frequently to lessen contact of irritating chemical with bladder mucosa by keeping well hydrated.
▪ Note: Susceptibility to infection may increase. Avoid people with infection. Notify prescriber of any infection, fever or chills, cough or hoarseness, lower back or side pain, painful or difficult urination.
▪ Check with prescriber immediately if there is any unusual bleeding or bruising, black tarry stools, or blood in urine or if pinpoint red spots develop on skin.
▪ Discuss possible adverse effects (e.g., alopecia, nausea, and vomiting) and measures that can minimize them with health care provider.

ILOPERIDONE
(i-lo-per′i-done)
Fanapt
Classification: ATYPICAL ANTIPSYCHOTIC
Therapeutic: ANTIPSYCHOTIC
Prototype: Clozapine
Pregnancy Category: C

AVAILABILITY 1 mg, 2 mg, 4 mg, 6 mg, 8 mg, 10 mg, 12 mg tablets

ACTION & *THERAPEUTIC EFFECT*
Is both a dopamine (D_2) and serotonin (5-HT_2) antagonist. *Effective in treating acute schizophrenia uncontrolled by other agents.*

Common adverse effects in *italic*, life-threatening effects <u>underlined</u>; generic names in **bold;** classifications in SMALL CAPS; ♣ Canadian drug name; ⊙ Prototype drug

755

USES Acute treatment of schizophrenia in patients who have not responded to other antipsychotic agents.

CONTRAINDICATIONS Hypersensitivity to iloperidone; elderly with dementia-related psychosis; suicidal ideation; recent acute MI; ANC less than 100 mm³; concurrent use of drugs that prolong QTc; lactation. **CAUTIOUS USE** Congenital long QT syndrome; history of cardiac arrhythmias; history of suicidal tendencies; cardiovascular disease; CVA; CHF; cerebrovascular disease; tardive dyskinesia; DM; history of seizures; history of leukopenia/neutropenia; hepatic impairment; patients at risk for aspiration pneumonia; older adult; pregnancy (category C). Safety and efficacy in children have not been established.

ROUTE & DOSAGE

Schizophrenia
Adult: **PO** Initial 1 mg b.i.d., then titrated to 6–12 mg b.i.d. Recommended titration schedule: Increase each b.i.d. dose by 2 mg a day from day 2 through 7 or until desired dose reached (max: 12 mg b.i.d.). Note: Reduce dose by 50% with concurrent use of a strong CYP2D6 inhibitor (e.g., fluoxetine or paroxetine) or strong CYP3A4 inhibitor (e.g., ketoconazole or clarithromycin).

ADMINISTRATION

Oral
- Note that gradual dose titration is recommended initially and whenever patient has been off drug for more than 3 days.
- Store at 15°–30° C (59°–86° F) and protect from light.

ADVERSE EFFECTS (≥1%) **Body as a Whole:** Fatigue. **CNS:** *Dizziness,* extrapyramidal disorder, lethargy, *somnolence,* tremor. **CV:** Hypotension, orthostatic hypotension, *tachycardia.* **GI:** Abdominal discomfort, diarrhea, *dry mouth, nausea.* **Metabolic:** Increased weight. **Musculoskeletal:** Arthralgia, musculoskeletal stiffness. **Respiratory:** Dyspnea, nasal congestion, nasopharyngitis, upper respiratory infection. **Skin:** Rash. **Special Senses:** Blurred vision. **Urogenital:** Ejaculation failure.

INTERACTIONS Drug: Potential additive QT prolongation if used in combination with drugs with similar effects (e.g., **disopyramide, procainamide, amiodarone**). Inhibitors of CYP3A4 (e.g., **ketoconazole, itraconazole, clarithromycin**) or CYP2D6 (e.g., **fluoxetine, paroxetine**) can increase iloperidone levels. **Food: Grapefruit juice** may increase iloperidone levels.

PHARMACOKINETICS Peak: 2–4 h. **Distribution:** 95% plasma protein bound. **Metabolism:** Extensive hepatic metabolism to active and inactive metabolites. **Elimination:** Renal (major) and fecal. **Half-Life:** 18–33 h.

NURSING IMPLICATIONS

Assessment & Drug Effects
- Monitor for suicidal ideation and report promptly if suspected.
- Monitor BP, HR, and weight. Monitor orthostatic vital signs with concurrent antihypertensive therapy or any condition that predisposes to hypotension (e.g., advanced age, dehydration).
- Monitor for orthostatic hypotension and syncope, especially early in therapy.

- Monitor ECG for prolongation of the QT_c interval.
- Monitor diabetics and those at risk for diabetes for loss of glycemic control.
- Lab tests: Baseline and periodic CBC with differential.

Patient & Family Education

- Be alert for and report worsening of condition, including ideas of suicide.
- Make position changes slowly, especially from a lying or sitting position to a standing position.
- If diabetic, monitor blood sugar closely for loss of control.
- Stop taking the drug and report immediately any of the following: Feeling faint or fainting, high fever, muscle rigidity, altered mental status, or palpitations.
- Do not drink alcohol while taking this drug.
- Avoid engaging in hazardous activities until response to drug is known.

ILOPROST
(i'lo-prost)
Classifications: PROSTAGLANDIN; PULMONARY ANTIHYPERTENSIVE
Therapeutic: PULMONARY ANTIHYPERTENSIVE
Prototype: Epoprostenol
Pregnancy Category: C

AVAILABILITY 20 mcg/mL, 10 mcg/mL solution

ACTION & *THERAPEUTIC EFFECT*
Iloprost is a synthetic analog of prostaglandin. It dilates systemic and pulmonary arterial vascular beds. *Dilation of the pulmonary arterial vessels reduces pulmonary hypertension.*

USES Treatment of pulmonary arterial hypertension in patients with New York Heart Association (NYHA) Class III or IV symptoms.

CONTRAINDICATIONS Systolic blood pressure less than 85 mm Hg; lactation.
CAUTIOUS USE Impaired hepatic function with at least Child-Pugh class B impairment, renal impairment with inhaled iloprost; older adults; asthma, acute respiratory infection, COPD; dialysis; syncope pregnancy (category C); children.

ROUTE & DOSAGE

Pulmonary Hypertension
Adult: **Inhaled** 2.5–5 mcg 6–9 times daily, but no more than q2h during waking hours

ADMINISTRATION
Inhalation

- Transfer the contents of one ampule to the drug delivery system medication chamber immediately before use. Follow instructions provided by manufacturer for the delivery system. Do not allow contact with skin or eyes.
- Do not administer if systolic BP is less than 85 mm Hg.
- Do not administer any sooner than 2 h after the previous dose.
- Discard any solution remaining in the medication chamber after the inhalation session.
- Store at 20°–25° C (68°–77° F).

ADVERSE EFFECTS (≥1%) **CNS:** *Headache*, insomnia. **CV:** *Hypotension, vasodilation (flushing)*, palpitations, syncope, chest pain, tachycardia, congestive heart failure. **GI:** *Nausea, vomiting.* **Hepatic:** Increased alkaline phosphatase, increased gamma-glutamyltransferase (GGT). **Musculoskeletal:** Back pain, muscle cramps, *trismus*. **Renal:** Kidney failure. **Respiratory:** *Cough*, dyspnea, hemoptysis, pneumonia, peripheral edema. **Body as a Whole:** *Flu-like syndrome*, tongue pain.

INTERACTIONS Drug: Enhanced hypotension when given with other VASODILATORS or ANTIHYPERTENSIVE agents; increased risk of bleeding when given with other ANTICOAGULANTS or ANTITHROMBOTIC agents.

PHARMACOKINETICS Distribution: 60% protein bound. **Metabolism:** Completely metabolized to inactive products. **Elimination:** Urine (major) and feces. **Half-Life:** 20–30 min.

NURSING IMPLICATIONS

Assessment & Drug Effects

- Supervise ambulation, especially with concurrent use of other drugs known to cause dizziness or syncope.
- Monitor vital signs closely during initiation of drug therapy.
- Monitor for and report S&S of heart failure.
- Withhold drug and notify prescriber if S&S of pulmonary edema appear.
- Lab tests: Monitor blood levels of anticoagulants when used concurrently.

Patient & Family Education

- Follow directions for taking the drug (see Administration). Iloprost inhalation should be used with the Prodose AAD system. Do not use it with other types of nebulizers.
- Make position changes slowly, especially when arising from a chair or bed.

- Do not drive or engage in potentially hazardous activities until response to drug is known.
- Report any of the following to a health care provider: Dizziness or fainting, especially upon exertion, or increased difficulty breathing.

IMATINIB MESYLATE

(i-ma'ti-nib)

Gleevec

Classifications: ANTINEOPLASTIC; MONOCLONAL ANTIBODY; EPIDERMAL GROWTH FACTOR RECEPTOR-TYROSINE KINASE INHIBITOR (EGFR-TKI)
Therapeutic: ANTINEOPLASTIC
Prototype: Gefitinib
Pregnancy Category: D

AVAILABILITY 100 mg, 400 mg tablet

ACTION & *THERAPEUTIC EFFECT*
Epidermal growth factor receptor-tyrosine kinase inhibitor (EGFR-TKI) that interferes with intracellular signaling pathways that are involved in the development of malignancies. Imatinib inhibits abnormal Bcr-Abl tyrosine kinase created by the Philadelphia chromosome abnormality in chronic myeloid leukemia (CLM). Tyrosine kinase is required for activation of a wide variety of intracellular activities vital to cell functioning and intracellular metabolic pathways. *Inhibits WBC cell proliferation and induces cell death in Bcr-Abl tyrosine kinase positive cells as well as in newly formed leukemic cells. Thus, it interferes with progression of chronic myeloid leukemia (CML). Additionally, imatinib inhibits proliferation and induces cell death in gastrointestinal stomal tumor (GIST) that express a mutation of an activated cKit tyrosine kinase.*

Common adverse effects in *italic*, life-threatening effects underlined; generic names in **bold**; classifications in SMALL CAPS; ♣ Canadian drug name; ⊘ Prototype drug

USES Treatment of CML in blast crisis, or in chronic phase after failure of interferon-alpha therapy; unresectable and/or metastatic malignant gastrointestinal stromal tumors (GISTs), acute lymphoblastic leukemia.

UNLABELED USES Acute lymphocytic leukemia (ALL), soft tissue sarcoma, recurrence of stomach and intestinal tumors.

CONTRAINDICATIONS Hypersensitivity to imatinib or any of its components; viral infections; intramuscular injections with concurrent thrombocytopenia; pregnancy (category D), lactation.

CAUTIOUS USE History of hypersensitivity to other monoclonal antibodies; hepatic or renal impairment; bleeding, bone marrow suppression; cardiac disease; dental disease; dental work; older adults, females of childbearing age; fungal infections; GI bleeding; heart failure; hepatic disease; infection; jaundice; peripheral edema, renal disease; vaccination, history of viral infection. Safe use in children younger than 3 y not established.

ROUTE & DOSAGE

CML Chronic Phase

Adult: **PO** 400–800 mg daily with a meal and large glass of water
Child (older than 3 y): **PO** 260 or 340 mg/m²/day in 1 or 2 divided dose(s)

CML Accelerated Phase or Blast Crisis

Adult: **PO** 600 mg daily with a meal and large glass of water

Acute Lymphoblastic Leukemia

Adult: **PO** 600 mg/day

GISTs

Adult: **PO** 400 mg daily

Hepatic Impairment Dosage Adjustment

Reduce dose to 300–400 mg daily

ADMINISTRATION

Oral

- Give with meal and large glass of water (at least 8 oz).
- Store at 15°–30° C (59°–86° F).

ADVERSE EFFECTS (≥1%) Body as a Whole: *Fluid retention, edema, fatigue,* weight gain, *fever,* night sweats, weakness. **CNS:** CNS hemorrhage, *headache.* **GI:** *Nausea, vomiting, diarrhea,* GI hemorrhage, dyspepsia, *abdominal pain, constipation, anorexia,* increased AST, ALT, and bilirubin. **Hematologic:** Hemorrhage, *neutropenia,* petechiae, epistaxis, pancytopenia (rare), thrombocytopenia (rare). **Metabolic:** Hypokalemia. **Musculoskeletal:** *Muscle cramps, pain, arthralgia,* myalgia. **Respiratory:** *Cough, dyspnea,* pharyngitis, pneumonia. **Skin:** *Rash,* pruritus.

INTERACTIONS Drug: **Clarithromycin, erythromycin, ketoconazole, itraconazole** may increase imatinib levels and toxicity; **carbamazepine, dexamethasone, phenobarbital, phenytoin, rifampin** may decrease imatinib levels; may increase levels of BENZODIAZEPINES, DIHYDROPYRIDINE, CALCIUM CHANNEL BLOCKERS (e.g., **nifedipine**), **warfarin. Herbal: St. John's wort** may decrease imatinib levels.

PHARMACOKINETICS Absorption: Well absorbed, 98% reaches systemic circulation. **Peak:** 2–4 h. **Metabolism:** Primarily by CYP3A4 in

Common adverse effects in *italic,* life-threatening effects underlined; generic names in **bold;** classifications in SMALL CAPS; ♣ Canadian drug name; ⊘ Prototype drug

759

liver. **Elimination:** Primarily in feces. **Half-Life:** 18 h imatinib, 40 h active metabolite.

NURSING IMPLICATIONS

Assessment & Drug Effects

- Monitor for S&S of fluid retention. Weigh daily and report rapid weight gain immediately.
- Lab tests: CBC with platelet count and differential weekly times 1 mo, biweekly for the 2nd mo, periodically thereafter as clinically indicated; baseline and monthly LFTs (AST, ALT, alkaline phosphatase, bilirubin); periodic serum creatinine and electrolytes.
- Withhold drug and notify prescriber for any of the following: Bilirubin greater than 3 × ULN, AST/ALT greater than 5 × ULN; treatment may be reinstituted when bilirubin less than 1.5 × ULN and AST/ALT less than 2.5 × ULN.

Patient & Family Education

- Do not take any OTC drugs (e.g., acetaminophen, St. John's wort) without consulting prescriber.
- Report any S&S of bleeding immediately to prescriber (e.g., black tarry stool, bright red or cola-colored urine, bleeding from gums).
- Report immediately to prescriber any unexplained change in mental status.
- Use effective means of contraception while taking this drug. Women of childbearing age should avoid becoming pregnant.

IMIPENEM-CILASTATIN SODIUM ⊘

(i-mi-pen'em sye-la-stat'in)
Primaxin
Classification: BETA-LACTAM ANTIBIOTIC

Therapeutic: ANTIBIOTIC
Pregnancy Category: C

AVAILABILITY 250 mg, 500 mg vials

ACTION & *THERAPEUTIC EFFECT*

Fixed combination of imipenem, a beta-lactam antibiotic, and cilastatin. Action of imipenem: Inhibition of mucopeptide synthesis in bacterial cell walls leading to cell death. Cilastatin increases the serum half-life of imipenem. *Effectively used for severe or resistant infections. Acts synergistically with aminoglycoside antibiotics against some isolates of* Pseudomonas aeruginosa.

USES Treatment of serious infections caused by susceptible organisms in the urinary tract, lower respiratory tract, bones and joints, skin and skin structures; also intraabdominal, gynecologic, and mixed infections; bacterial septicemia and endocarditis.

CONTRAINDICATIONS Hypersensitivity to any component of product, multiple allergens; carbapenem hypersensitivity; penicillin hypersensitivity.

CAUTIOUS USE Patients with CNS disorders (e.g., seizures, brain lesions, history of recent head injury); seizures; renal failure, renal impairment, renal disease; patients with history of cephalosporin allergies; pregnancy (category C), lactation.

ROUTE & DOSAGE

Serious Infections

Adult: **IV:** 250–500 mg infused over 20–30 min q6–8h (max dose: 4 g/day) **IM:** 500 or 750 mg q12h (max dose: 4 g/day)

Common adverse effects in *italic*, life-threatening effects underlined; generic names in **bold**; classifications in SMALL CAPS; ◆ Canadian drug name; ⊘ Prototype drug

Child: **IV** Older than 3 mo, 25 mg/kg q6h; 1–3 mo, 100 mg/kg/day in divided doses **IM** 15–25 mg/kg q12h
Neonate: **IV** Weight greater than 1500 g, 25 mg/kg q8–12h

Renal Impairment Dosage Adjustment

Make adjustments per package insert (based on CrCl)

ADMINISTRATION Caution: IM and IV solutions are NOT interchangeable; do NOT give IM solution by IV, and do NOT give IV solution as IM.

Intramuscular

- Reconstitute powder for IM injection as follows: Add 2 mL or 3 mL of 1% lidocaine HCl solution without epinephrine, respectively, to the 500 mg vial or the 750 mg vial. Agitate to form a suspension then withdraw and inject entire contents of the vial IM.
- Give IM suspension by deep injection into the gluteal muscle or lateral thigh.
- Use reconstituted IM injection within 1 h after preparation.

Intravenous

PREPARE: Intermittent: Reconstitute each dose with 10 mL of D5W, NS, or other compatible infusion solution. ▪ Agitate the solution until clear. Color should range from colorless to yellow. ▪ Further dilute with 100 mL of same solution used for initial dilution.
ADMINISTER: Intermittent: Give each 500 mg or fraction thereof over 20–30 min. Infuse larger doses over 40–60 min. ▪ DO NOT give as a bolus dose. ▪ Nausea appears to be related to infusion rate, and if it presents during infusion, slow the rate (occurs most frequently with 1-g doses).

INCOMPATIBILITIES **Solution/ additive: Lactated Ringer's,** some **dextrose**-containing solutions, **sodium bicarbonate, TPN. Y-site: Allopurinol, amiodarone, amphotericin B cholesteryl, azithromycin, etoposide, fluconazole, gemcitabine, lorazepam, meperidine, midzolam, milrinone, sargramostim, sodium bicarbonate.**

- Store according to manufacturer's recommendations; stability of IV solutions depends on diluent used for reconstitution. ▪ Most IV solutions retain potency for 4 h at 15°–30° C (59°–86° F) or for 24 h if refrigerated at 4° C (39° F). Avoid freezing.

ADVERSE EFFECTS (≥1%) **Body as a Whole:** Hypersensitivity (rash, fever, chills, dyspnea, pruritus), weakness, oliguria/anuria, polyuria, polyarthralgia; *phlebitis and pain at injection site,* superinfections. **CNS:** Seizures, dizziness, confusion, somnolence, encephalopathy, myoclonus, tremors, paresthesia, headache. **GI:** *Nausea, vomiting,* diarrhea, <u>pseudomembranous colitis</u>, hemorrhagic colitis, gastroenteritis, abdominal pain, glossitis, heartburn. **Respiratory:** Chest discomfort, hyperventilation, dyspnea. **Skin:** Rash, pruritus, urticaria, candidiasis, flushing, increased sweating, skin texture change, facial edema. **Metabolic:** Hyponatremia, hyperkalemia. **Special Senses:** Transient hearing loss; increased WBC, AST, ALT, alkaline phosphatase, BUN, LDH, creatinine; decreased Hgb, Hct, eosinophilia.

INTERACTIONS Drug: Aztreonam, cephalosporins, penicillins may antagonize the antibacterial effects. May affect **cyclosporine** levels.

PHARMACOKINETICS Distribution: Widely distributed; limited concentrations in CSF; crosses placenta; in breast milk. **Elimination:** 70% in urine within 10 h. **Half-Life:** 1 h.

NURSING IMPLICATIONS

Assessment & Drug Effects

- Determine previous hypersensitivity reaction to beta-lactam antibiotics (penicillins and cephalosporins) or to other allergens.
- Monitor for S&S of hypersensitivity (see Appendix F). Discontinue drug and notify prescriber if S&S occur.
- Monitor closely patients vulnerable to CNS adverse effects.
- Notify prescriber if focal tremors, myoclonus, or seizures occur; dosage adjustment may be needed.
- Monitor for S&S of superinfection (see Appendix F).
- Notify prescriber promptly to rule out pseudomembranous enterocolitis if severe diarrhea accompanied by abdominal pain and fever occurs (see Appendix F).
- Note: Sodium content derived from drug is high; consider in patient on restricted sodium intake.
- Monitor renal, hematologic, and liver function periodically.

Patient & Family Education

- Notify prescriber immediately to report pruritus or symptoms of respiratory distress.
- Report pain or discomfort at IV infusion site.
- Report loose stools or diarrhea promptly.

IMIPRAMINE HYDROCHLORIDE 📷

(im-ip′ra-meen)

Impril ♦, Novopramine ♦, Tofranil

IMIPRAMINE PAMOATE

Tofranil-PM

Classification: TRICYCLIC ANTIDEPRESSANT (TCA)
Therapeutic: ANTIDEPRESSANT
Pregnancy Category: D

AVAILABILITY 10 mg, 25 mg, 50 mg tablets; 25 mg/mL oral product; **Imipramine pamoate:** 75 mg, 100 mg, 125 mg, 150 mg capsules

ACTION & *THERAPEUTIC EFFECT* TCAs potentiate both norepinephrine and serotonin in the CNS by blocking their reuptake by presynaptic neurons. Imipramine decreases number of awakenings from sleep, markedly reduces time in REM sleep, and increases stage 4 sleep. Relief of nocturnal enuresis is due to anticholinergic activity and to nervous system stimulation, resulting in earlier arousal to sensation of full bladder. *Effective as an antidepressant. Relieves nocturnal enuresis in children.*

USES Depression, enuresis.
UNLABELED USES ADHD, bulimia nervosa, neuropathic pain, panic disorders, postherpetic neuralgia, overactive bladder, urinary incontinence.

CONTRAINDICATIONS Hypersensitivity to tricyclic drugs; concomitant use of MAOIs within 14 days; suicidal ideation; acute recovery period after MI, defects in bundle branch conduction, QT prolongation; severe renal or hepatic impairment; use of imipramine pamoate in children of any age; pregnancy (category D), lactation.
CAUTIOUS USE History of hypersensitivity to dibenzazepine compounds; respiratory difficulties;

cardiovascular, hepatic, or GI diseases; blood disorders; increased intraocular pressure, narrow-angle glaucoma; schizophrenia, hypomania or manic episodes, patient with suicidal tendencies, seizure disorders; prostatic hypertrophy, urinary retention; alcoholism, hyperthyroidism; electroshock therapy; older adults; adolescents, children.

ROUTE & DOSAGE

Depression
Adult: PO 75–100 mg/day (max: 300 mg/day) in 1 or more divided doses
Adolescent: PO 30–40 mg/day

Enuresis in Childhood
Adolescent: 10–25 mg at bedtime, may titrate up (max: 75 mg)
Child (6–12): 10–25 mg at bedtime, may titrate up (max: 50 mg)

Pharmacogenetic Dosage Adjustment
Poor CYP2D6 metabolizers: Start at 30% of normal dose

ADMINISTRATION
Oral
- Give with or immediately after food.
- Note: Single doses can be given at bedtime or q.a.m., respectively, if drowsiness or insomnia results.

ADVERSE EFFECTS (≥1%) Body as a Whole: Hypersensitivity (skin rash, erythema, petechiae, urticaria, pruritus, photosensitivity, angioedema of face, tongue, or generalized; drug fever). CNS: Sedation, drowsiness, dizziness, headache, fatigue, numbness, tingling (paresthesias) of extremities; incoordination, ataxia, tremors, peripheral neuropathy, extrapyramidal symptoms (including parkinsonism effects and tardive dyskinesia); lowered seizure threshold, altered EEG patterns, delirium, disturbed concentration, confusion, hallucinations, anxiety, nervousness, insomnia, vivid dreams, restlessness, agitation, shift to hypomania, mania; exacerbation of psychoses; hyperpyrexia. CV: Orthostatic hypotension, mild sinus tachycardia; arrhythmias, hypertension or hypotension, palpitation, MI, CHF, heart block, ECG changes, stroke, flushing, cold cyanotic hands and feet (peripheral vasospasm). Endocrine: Testicular swelling, gynecomastia (men), galactorrhea and breast enlargement (women), increased or decreased libido, ejaculatory and erectile disturbances, delayed or absent orgasm (male and female); elevation or depression of blood glucose levels. Special Senses: Nasal congestion, tinnitus; blurred vision, disturbances of accommodation, slight mydriasis, nystagmus. GI: Dry mouth, constipation, heartburn, excessive appetite, weight gain, nausea, vomiting, diarrhea, slowed gastric emptying time, flatulence, abdominal cramps, esophageal reflux, anorexia, stomatitis, increased salivation, black tongue, peculiar taste, paralytic ileus. Urogenital: Urinary retention, delayed micturition, nocturia, paradoxic urinary frequency. Hematologic: Bone marrow depression; agranulocytosis, eosinophilia, thrombocytopenia. Other: Excessive perspiration, cholestatic jaundice, precipitation of acute intermittent porphyria; dyspnea, changes in heat and cold tolerance, hair loss, syndrome of inappropriate anti-diuretic hormone secretion (SIADH).

Common adverse effects in italic, life-threatening effects underlined; generic names in bold; classifications in SMALL CAPS; ♣ Canadian drug name; ● Prototype drug

763

INTERACTIONS Drug: MAO IN-HIBITORS may precipitate hyperpyrexic crisis, tachycardia, or seizures; ANTIHYPERTENSIVE AGENTS potentiate orthostatic hypotension; CNS DEPRESSANTS, **alcohol** add to CNS depression; **norepinephrine** and other SYMPATHOMIMETICS may increase cardiac toxicity; **cimetidine** decreases hepatic metabolism, thus increasing imipramine levels; **methylphenidate** inhibits metabolism of imipramine and thus may increase its toxicity. **Herbal: Ginkgo** may decrease seizure threshold; **St. John's wort** may cause serotonin syndrome.

PHARMACOKINETICS Absorption: Completely absorbed from GI tract. **Peak:** 1–2 h. **Metabolism:** Metabolized to the active metabolite desipramine in liver. **Elimination:** Primarily in urine, small amount in feces; crosses placenta; may be secreted in breast milk. **Half-Life:** 8–16 h.

NURSING IMPLICATIONS

Assessment & Drug Effects

- Monitor for therapeutic effectiveness: May not occur for 2 wk or more.
- Monitor children and adolescents for increase in suicidality.
- Note: Dose sensitivity and adverse effects are most likely to occur in adolescents and older adults; lower initial doses are recommended for these patients.
- Lab tests: Monitor periodically LFTs, renal function, CBC with differential, and fluid and electrolyte balance.
- Monitor HR and BP frequently. Orthostatic hypotension may be marked in pretreatment hypertensive or cardiac patients.

- Note: During the first 2 wk of therapy, older adults may develop confusion, restlessness, disturbed sleep, forgetfulness. Symptoms last 3–20 days. Report to prescriber.
- Weigh patient under standard conditions biweekly: Report a gain of 0.5–1 kg (1.5–2 lb) within 2–3 days and frank edema.
- Monitor urinary and bowel elimination, at least until maintenance dosage is stabilized, to detect urinary retention or frequency, constipation, or paralytic ileus.
- Notify prescriber of extrapyramidal symptoms (tremors, twitching, ataxia, incoordination, hyperreflexia, drooling) in patients receiving large doses and especially in older adults.
- Monitor diabetic patients for loss of glycemic control. Hyperglycemia or hypoglycemia (see Appendix F) occur in some patients.

Patient & Family Education

- Report promptly signs of a worsening condition, suicidal ideation, or unusual changes in behavior, especially in children and adolescents.
- Change position slowly and in stages, especially from lying down to upright posture and dangle legs over bed for a few minutes before walking.
- Note: Effectiveness can decrease with continued drug administration in some patients. Inform prescriber if this occurs.
- Do NOT use OTC drugs while on a TCA without prescriber approval.
- Do not drive or engage in other potentially hazardous activities until response to drug is known.
- Avoid exposure to strong sunlight because of potential photosensitivity. Use sunscreen with at least SPF of 12–15 if allowed.

IMIQUIMOD
(i-mi'qui-mod)
Aldara, Zyclara
Classifications: KERATOLYTIC; IMMUNOMODULATOR
Therapeutic: IMMUNE RESPONSE MODIFIER; KERATOLYTIC
Pregnancy Category: C

AVAILABILITY 3.75%, 5% cream

ACTION & *THERAPEUTIC EFFECT*
An immune response modifier thought to induce cytokine production, including interferon-alfa, which may interfere with viral replication. *Despite destruction of HPV warts, latent or subclinical HPV infection can persist, and recurrence of visible warts is common.*

USES Treatment of external genital and perianal warts *(Condylomata acuminata),* actinic keratosis on the face and scalp of immunocompetent adults, and superficial basal cell carcinoma.
UNLABELED USES Treatment of common warts, herpes simplex virus.

CONTRAINDICATIONS Occlusive dressing; ocular exposure; excessive sun exposure or sunburn; UV exposure; surgery or drug treatment on affected area.
CAUTIOUS USE Hypersensitivity to benzyl alcohol or paraben; HIV infection; pregnancy (category C), lactation. Safe use in children younger than 12 y not established.

ROUTE & DOSAGE

Genital and Perianal Warts
Adult/Adolescent (older than 12 y): **Topical** Apply a thin layer to the affected areas once daily 3 times/wk just before bedtime.

Wash off cream after 6–10 h (max: 16 wk therapy).

Actinic Keratosis
Adult: **Topical** Apply a thin layer to the affected areas once daily 2 times/wk just before sleep for 16 wk. Wash off cream after 8 h.
Actinic keratosis on face/balding scalp (3.75% cream).
Adult: **Topical** Apply a thin layer once daily before bedtime for 2 wk.

Superficial Basal Cell Carcinoma
Adult: **Topical** Apply a thin layer to the affected areas once daily 5 times/wk just before sleep for 6 wk. Wash off cream after 8 h.

ADMINISTRATION
Topical
- Hand washing before and after application is recommended.
- Wash treatment area with soap and water and allow to dry thoroughly (at least 10 min).
- Single-use packets contain sufficient cream to cover an area of up to 20 cm² (approx. 8 in. by 8 in.).
- Instruct patient to apply a thin layer of cream (avoid using excessive cream), and work into area until no longer visible. Do not occlude the application site.
- After each treatment period, remove the cream by washing the treated area with soap and water.
- Store below 25° C (77° F).

ADVERSE EFFECTS (≥1%) **Body as a Whole:** Fungal infections, flu-like symptoms, myalgia. **CNS:** Headache. **Skin:** *Application site reactions, pruritus,* burning, bleeding, stinging, redness, tenderness, irritation, *erythema,* edema, weeping/exudates, dry skin, scabbing/crusting, hyperkeratosis, inflammatory reaction of female external genitalia.

Common adverse effects in *italic*, life-threatening effects underlined; generic names in **bold**; classifications in SMALL CAPS; ✦ Canadian drug name; ❂ Prototype drug

765

PHARMACOKINETICS Absorption: Minimal through intact skin.

NURSING IMPLICATIONS

Assessment & Drug Effects

- Monitor for and report promptly severe local inflammatory reactions on female external genitalia.

Patient & Family Education

- Uncircumcised males with warts under the foreskin: Pull back the foreskin and clean the area daily to help avoid penile skin reactions.
- Females should not apply cream directly into the vagina. Application to the labia may cause pain or swelling and may cause difficulty in passing urine.
- When being treated for actinic keratosis, avoid or minimize UV light exposure (artificial and sunlight). Wear protective clothing. If sunburn develops, avoid using imiquimod cream until fully recovered.

IMMUNE GLOBULIN INTRAMUSCULAR [IGIM, GAMMA GLOBULIN, IMMUNE SERUM GLOBULIN (ISG)] 🅿️

(im'mune glob'u-lin)

BayGam

IMMUNE GLOBULIN INTRAVENOUS (IGIV)

Flebogamma, Gammagard, Gammar-P IV, Gamunex, IGIV, Iveegam, Octagam, Sandoglobulin

IMMUNE GLOBULIN SUBCUTANEOUS (IGSC, SCIG)

Hizentra, Vivaglobin

Classifications: BIOLOGIC RESPONSE MODIFIER; IMMUNOGLOBULIN

Therapeutic: IMMUNOGLOBULIN
Pregnancy Category: C

AVAILABILITY IGIM: 2 mL, 10 mL vials **IGIV:** 5%, 10% solution; 50 mg/mL powder for injection **SC:** 20% solution

ACTION & *THERAPEUTIC EFFECT*

Sterile concentrated solution containing globulin (primarily IgG) prepared from human plasma of either venous or placental origin and processed by a special fractionating technique. *Like hepatitis B immune globulin (H-BIG), contains antibodies specific to hepatitis B surface antigen but in lower concentrations. Therefore, not considered treatment of first choice for postexposure prophylaxis against hepatitis B but usually an acceptable alternative when H-BIG is not available.*

USES IGIM: Provides passive immunity or to modify severity of certain infectious diseases [e.g., rubeola (measles), rubella (German measles), varicella-zoster (chickenpox), type A (infectious) hepatitis], and as replacement therapy in congenital agammaglobulinemia or IgG deficiency diseases. May be used as an alternative to H-BIG to provide passive immunity in hepatitis B infection. Also for postexposure prophylaxis of hepatitis non-A, non-B, and nonspecific hepatitis. **IGIV:** Principally as maintenance therapy in patients unable to manufacture sufficient quantities of IgG antibodies, in patients requiring an immediate increase in immunoglobulin levels, and when IM injections are contraindicated as in patients with bleeding disorders or who have small muscle mass. Also in chronic autoimmune thrombocytopenia and idiopathic thrombocytopenic purpura (ITP). Treatment of primary immunodeficiency disorders associated with defects in humoral immunity. **IGSC:** Primary immune deficiency. **UNLABELED USES** Kawasaki syndrome, chronic lymphocytic leukemia, AIDS, premature and low-birth-weight neonates, autoimmune neutropenia, HIV associated thrombocytopenia, or hemolytic anemia.

CONTRAINDICATIONS History of anaphylaxis or severe reaction to human immune serum globulin (IG) or to any ingredient in the formulations; persons with clinical hepatitis A; IGIV for patients with class-specific anti-IgA deficiencies; IGIM in severe thrombocytopenia or other bleeding disorders. **CAUTIOUS USE** Dehydration, diabetes mellitus, children, older adults, hypovolemia, IgA deficiency, infection; renal disease, renal impairment; sepsis; sucrose hypersensitivity; vaccination; viral infection; pregnancy (category C), lactation.

ROUTE & DOSAGE

Hepatitis A Exposure

Adult/Child: **IM** 0.02 mL/kg as soon as possible after exposure; if period of exposure will be 3 mo or longer, give 0.05–0.06 mL/kg once q4–6mo

Hepatitis B Exposure

Adult/Child: **IM** 0.02–0.06 mL/kg as soon as possible after exposure if H-BIG is unavailable

Rubella Exposure

Adult: **IM** 20 mL as single dose in susceptible pregnant women

Rubeola Exposure

Adult/Child: **IM** 0.25 mL/kg within 6 days of exposure

Varicella-Zoster Exposure

Adult/Child: **IM** 0.6–1.2 mL/kg promptly

Immunoglobulin Deficiency

*Dosages may vary between brands and formulations (see package insert)
Adult/Child: **IV** 200–400 mg/kg monthly **IM** 1.2 mL/kg followed by 0.6 mL/kg q2–4wk

Idiopathic Thrombocytopenia Purpura

Adult/Child: **IV** 400 mg/kg/day for 5 consecutive days or 1 g/kg × 1–2 days

Obesity Dosage Adjustment

Dose based on IBW or adjusted IBW

ADMINISTRATION

■Note: In hepatitis A (infectious hepatitis), immune globulin is most effective when given before or as soon as possible after exposure but not more than 2 wk after (incubation period for hepatitis A is 15–50 days). ■Do not give immune globulin to those presenting clinical manifestations of hepatitis A. ■For hepatitis B (serum hepatitis), give immune globulin within 24 h and not more than 7 days after exposure. ■Note: IGIM and IGIV formulations are NOT interchangeable.

Intramuscular

■ Give adults and older children injections into deltoid or anterolateral aspect of thigh; neonates and small children, into anterolateral aspect of thigh.
■ Avoid gluteal injections; however, when large volumes of immune globulin are prescribed or when large doses **must be** divided into several injections, the upper outer

quadrant of the gluteus has been used in adults.

Intravenous

PREPARE: IV Infusion: Refer to manufacturer's directions for information on reconstitution and dilution of the specific product.
ADMINISTER: IV Infusion: Flow rates vary with product being infused. Refer to manufacturer's directions for the specific product.
- Do not mix with other drugs.
- Discard partially used vial.
INCOMPATIBILITIES Do not mix other drugs with immunoglobulin.

- Store as directed by manufacturer for specific product. Avoid freezing.
- Do not use if turbidity has occurred or if product has been frozen.

ADVERSE EFFECTS (≥1%) Body as a Whole: *Pain, tenderness, muscle stiffness at IM site;* local inflammatory reaction, erythema, urticaria, angioedema, headache, malaise, fever, arthralgia, nephrotic syndrome, hypersensitivity (fever, chills, anaphylactic shock), infusion reactions (*nausea, flushing, chills,* headache, chest tightness, wheezing, skeletal pain, back pain, abdominal cramps, anaphylaxis), renal dysfunction, renal failure.

INTERACTIONS Drug: May interfere with antibody response to LIVE VIRUS VACCINES (measles/mumps/rubella); give VACCINES 14 days before or 3 mo after IMMUNEGLOBULINS.

PHARMACOKINETICS Peak: 2 days. **Distribution:** Rapidly and evenly distributed to intravascular and extravascular fluid compartments. **Half-Life:** 21–23 days.

NURSING IMPLICATIONS

Assessment & Drug Effects
- Make sure emergency drugs and appropriate emergency facilities are immediately available for treatment of anaphylaxis or sensitization.
- Note: Hypersensitivity reactions (see Appendix F) are most likely in patients receiving large IM doses, repeated injections, or rapid IV infusion.
- Monitor vital signs and infusion rate closely when patient is receiving IGIV.
- Note: IGIV has a mild diuretic effect in some patients due to presence of maltose.

Patient & Family Education
- Report immediately S&S of hypersensitivity (see Appendix F).
- Report immediately infusion symptoms of nausea, chills, headache, and chest tightness; these are indications to slow rate of infusion.
- Note: Passive immunity to measles (rubeola) lasts about 3–4 wk after immune globulin. In general, children 15 mo or younger need active immunization with measles virus vaccine 3 mo after IGIM.

INDACATEROL MALEATE
(in'-da-ka'-ter-ol mal'-ee-ate)

Arcapta
Classifications: BRONCHODILATOR; RESPIRATORY SMOOTH MUSCLE RELAXANT; BETA-ADRENERGIC AGONIST
Therapeutic: BRONCHODILATOR; RESPIRATORY SMOOTH MUSCLE RELAXANT
Prototype: Albuterol
Pregnancy Category: C

AVAILABILITY 75 mcg capsules containing powder for inhalation

ACTION & *THERAPEUTIC EFFECT*

Acts primarily on beta$_2$-adrenergic receptors in bronchial smooth muscle and, possibly on the small percentage of beta$_2$-adrenergic receptors found in cardiac muscle. *Causes relaxation of bronchial smooth muscle and bronchodilation.*

USES

Prophylactic, maintenance treatment of chronic obstructive pulmonary disorder (COPD), including chronic bronchitis and emphysema.

CONTRAINDICATIONS

Asthma (without concurrent use of a long-term asthma control drug); acutely deteriorating COPD; acute episodes of bronchospasm.

CAUTIOUS USE

Paradoxical bronchospasms; cardiovascular disease including hypertension; seizure disorder; thyrotoxicosis; sensitivity to sympathomimetic drugs; concurrent use of other drugs containing long-acting beta$_2$-agonist; pregnancy (category C); lactation. Safety and effectiveness in children not established.

ROUTE & DOSAGE

Chronic Obstructive Pulmonary Disorder

Adult: **PO Inhalation** 75 mcg once daily

ADMINISTRATION

Inhalation

- Capsules should be removed from blister immediately before use.
- Capsules should be used only with the neohaler device. Capsules must not be swallowed.
- The Arcapta neohaler should be administered at the same time each day.

- Store at 15°–30° C (59°–86° F) in foil package until ready to use.

ADVERSE EFFECTS (≥2%)

Body as a Whole: Peripheral edema. **CNS:** Headache. **GI:** Nausea. **Metabolic:** Hyperglycemia. **Musculoskeletal:** Muscle spasms, musculoskeletal pain. **Respiratory:** Cough, nasopharyngitis, oropharyngeal pain, sinusitis, upper respiratory tract infection.

INTERACTIONS

Drug: ADRENERGIC AGONISTS and MONOAMINE OXIDASE INHIBITORS may potentiate indacaterol. BETA-BLOCKERS may interfere with the actions of indacaterol. CORTICOSTEROIDS (e.g., **prednisone, dexamethasone**), **theophylline,** THIAZIDE DIURETICS OR LOOP DIURETICS may increase risk of hypokalemia. CYP3A4 inhibitors and/or P-gp efflux transporters (e.g., **ketoconazole,** HIV PROTEASE INHIBITORS, **erythromycin**) may increase the levels of indacaterol. Indacaterol has been associated with QT prolongation. Drugs that prolong the QT interval (e.g., **disopyramide, procainamide, amiodarone, bretylium, clarithromycin, levofloxacin**) may cause additive effects. TRICYCLIC ANTIDEPRESSANTS (e.g., **amitriptyline**) may potentiate cardiovascular effects of indacaterol.

PHARMACOKINETICS

Absorption: 43–45% bioavailable. **Peak:** 15 min. **Distribution:** 94–96% plasma protein bound. **Metabolism:** In the liver. **Elimination:** Primarily fecal. **Half-Life:** 45.5–126 h.

NURSING IMPLICATIONS

Assessment & Drug Effects

- Monitor HR, BP, and respiratory status. Report immediately deterioration in respiratory condition or

Common adverse effects in *italic*, life-threatening effects underlined; generic names in **bold**; classifications in SMALL CAPS; ♣ Canadian drug name; ⦾ Prototype drug

769

development of paradoxical bronchospasms following inhalation.

- Periodic ECG monitoring with concurrent use of other drugs associated with QT prolongation.
- Lab tests: Periodic serum potassium and more often with concurrent non-potassium diuretics; blood glucose in diabetics or prediabetics.

Patient & Family Education

- Do not use for relief of acute symptoms.
- Discontinue drug use and immediately notify prescriber of any of the following: Symptoms of COPD are worsening; indacaterol no longer controls the symptoms of COPD; breathing is worsened following inhalation of indacaterol; the concurrently prescribed short-acting beta$_2$-agonist drug becomes less effective, or more inhalations of the short-acting drug are required.
- Diabetics should monitor blood glucose level more frequently for loss of glycemic control.

INDAPAMIDE
(in-dap′a-mide)

Lozide ♦, Lozol
Classifications: ELECTROLYTIC AND WATER BALANCE AGENT; DIURETIC
Therapeutic: THIAZIDE-LIKE DIURETIC; ANTIHYPERTENSIVE
Prototype: Hydrochlorothiazide
Pregnancy Category: B

AVAILABILITY 1.25, 2.5 mg tablets

ACTION & *THERAPEUTIC EFFECT*
Sulfonamide derivative that has both diuretic and direct vascular effects. Acts on the proximal portion of the distal renal tubules. Enhances excretion of sodium, potassium, and water by interfering with sodium transfer across renal epithelium of tubules. *Hypotensive activity appears to result from a decrease in plasma and extracellular fluid volume, decreased peripheral vascular resistance, direct arteriolar dilation, and calcium channel blockade.*

USES Alone or with other antihypertensives in the management of hypertension in patients who have failed to respond to diet, exercise, or weight reduction.
UNLABELED USES Edema associated with CHF.

CONTRAINDICATIONS Hypersensitivity to indapamide or other sulfonamide derivatives, anuria, renal failure; lactation.
CAUTIOUS USE Electrolyte imbalance, hypokalemia, severe renal disease; impaired hepatic function or progressive liver disease; prediabetic and type II diabetic patient, hyperparathyroidism, thyroid disorders; SLE; sympathectomized patient; history of gout; pregnancy (category B). Safe use in children is not established.

ROUTE & DOSAGE

Hypertension, Edema
Adult: **PO** 2.5 mg once/day, may increase to 5 mg/day if needed

ADMINISTRATION

Oral
- Give with food or milk to reduce GI irritation.
- Administer in a.m. to prevent nocturia. Urge patient to take at least 240 mL (8 oz) of fluid (if allowed) with the medication.
- Store in tight, light-resistant container unless otherwise directed.

ADVERSE EFFECTS (≥1%) **CNS:** Headache, dizziness, fatigue, weakness,

Common adverse effects in *italic*, life-threatening effects underlined; generic names in **bold**; classifications in SMALL CAPS; ♦ Canadian drug name; ⊘ Prototype drug

muscle cramps or spasm, paresthe-sia, tension, anxiety, nervousness, agitation, vertigo, insomnia, mental depression, blurred vision, drowsiness. **CV:** Orthostatic hypotension, PVCs, dysrhythmias, flushing, palpitation. **GI:** Dry mouth, anorexia, nausea, vomiting, diarrhea, constipation, abdominal cramps or pain. **Urogenital:** Urinary frequency, nocturia, polyuria, glycosuria, impotence or reduced libido. **Skin:** Rash, hives, pruritus, vasculitis, photosensitivity. **Metabolic:** Dilutional hyponatremia, *hyperuricemia,* exacerbation of gout; *hypokalemia,* hyperglycemia, hypochloremia, hypercalcemia, increased BUN or creatinine, weight loss, exacerbation of SLE; increased cholesterol.

DIAGNOSTIC TEST INTERFERENCE Since indapamide may cause hypercalcemia (and hypophosphatemia), it is generally withheld before tests for *parathyroid function* are performed.

INTERACTIONS Drug: Effects of **diazoxide** and indapamide intensified; increased risk of **digoxin** toxicity with hypokalemia; decreased renal **lithium** clearance may increase risk of **lithium** toxicity.

PHARMACOKINETICS Absorption: Readily from GI tract. **Peak:** 2–2.5 h. **Duration:** Up to 36 h. **Metabolism:** In liver. **Elimination:** 60% in urine; 16–23% in feces. **Half-Life:** 14–18 h.

NURSING IMPLICATIONS

Assessment & Drug Effects
- Monitor BP periodically throughout therapy.
- Lab tests: Obtain baseline and periodic BUN, serum creatinine, uric acid, blood glucose, serum electrolytes, and fluid balance.
- Monitor for digitalis toxicity with concurrent therapy.

- Note: Electrolyte imbalances may be clinically serious with protracted vomiting and diarrhea, excessive sweating, GI drainage, and paracentesis.
- Report promptly signs of hyponatremia or hypokalemia (see Appendix F).
- Monitor diabetics for loss of glycemic control.

Patient & Family Education
- Notify prescriber of decreased urine output, dizziness, weakness or muscle cramps, nausea, jaundice, or blurred vision.
- Take precautions from sun exposure because of risk of photosensitivity.
- Record weight at least every other day; inspect ankles and legs for edema. Report unexplained, progressive weight gain [e.g., 1–1.5 kg (2–3 lb) in 2–3 days].

INDINAVIR SULFATE
(in-din′a-vir)
Crixivan
Classifications: ANTIRETROVIRAL; PROTEASE INHIBITOR
Therapeutic: PROTEASE INHIBITOR
Prototype: Saquinavir
Pregnancy Category: C

AVAILABILITY 100 mg, 200 mg, 333 mg, 400 mg capsules

ACTION & *THERAPEUTIC EFFECT*
Indinavir is an HIV protease inhibitor. HIV protease is an enzyme required to produce the polyprotein precursors of the functional proteins in infectious HIV. Indinavir binds to the protease active site and thus inhibits its activity. *Protease inhibitors*

Common adverse effects in *italic*, life-threatening effects underlined; generic names in **bold;** classifications in SMALL CAPS; ✦ Canadian drug name; ⊙ Prototype drug

771

prevent cleavage of HIV viral poly-proteins, resulting in formation of immature noninfectious virus particles.

USES Treatment of HIV infection, in combination with other agents.

CONTRAINDICATIONS Hypersensitivity to indinavir; severe leukocyturia of greater than 100 cells/high power field; lactation.

CAUTIOUS USE Hepatic dysfunction, hepatitis; renal impairment, history of nephrolithiasis, diabetes mellitus; hyperglycemia; concurrent HBV infection; history of adverse responses to other protease inhibitors; pregnancy (category C). Optimal dosing regimen for use in children has not been established.

ROUTE & DOSAGE

HIV

Adult: **PO** 800 mg (2 × 400 mg) q8h 1 h before or 2 h after meal
With ritonavir: 800 mg b.i.d. plus 100–200 mg ritonavir b.i.d.

ADMINISTRATION

Oral

- Give with water on an empty stomach 1 h before or 2 h after meal; if needed, may be given with a very light meal or beverage.
- Note: When didanosine and indinavir are ordered concurrently, give each on empty stomach at least 1 h apart.
- Do not administer concurrently with midazolam or triazolam.
- Store tightly closed with desiccant in original bottle.

ADVERSE EFFECTS (≥1%) **CNS:** Fatigue, headache, insomnia, dizziness, somnolence, nervousness, agitation, anxiety, paresthesia, peripheral neuropathy, tremor, vertigo. **CV:** Palpitations. **Hematologic:** Anemia, splenomegaly, lymphadenopathy. **GI:** *Nausea,* diarrhea, abdominal discomfort, dyspepsia, stomatitis, anorexia, dry mouth, cholecystitis, cholestasis, constipation, flatulence. **Skin:** Body odor, rash, pruritus, seborrhea, skin ulceration, dry skin, sweating, urticaria. **Other:** Myalgia, allergic reaction, bronchitis, cough, rhinitis, taste alterations, visual disturbances, hyperglycemia, diabetes, kidney stones.

INTERACTIONS Drug: Rifabutin, rifampin significantly decrease indinavir levels. **Ketoconazole** significantly increases indinavir levels. Indinavir could inhibit the metabolism and increase the toxicity of **amiodarone, midazolam, sildenafil, tadalafil, trazodone, triazolam, vardenafil.** Indinavir and **didanosine** should be administered at least 1 h apart on empty stomach to permit full absorption of each; increased **ergotamine** toxicity with indinavir. **Rosuvastatin** should not be used concurrently. **Herbal: St. John's wort,** garlic decreases ANTIRETROVIRAL activity of indinavir.

PHARMACOKINETICS Absorption: Rapidly from GI tract; a meal high in calories, fat, and protein significantly reduces absorption. **Distribution:** 60% protein bound. **Metabolism:** In liver by CYP3A4. **Elimination:** Primarily in feces (greater than 80%), 20% in urine.

NURSING IMPLICATIONS

Assessment & Drug Effects

- Lab tests: Monitor CBC with differential and platelet count, liver function tests, CPK, urinalysis, and serum amylase periodically.
- Assess for S&S of renal dysfunction, respiratory dysfunction, GI

distress, and other common adverse effects.

Patient & Family Education

- Learn drug interactions and potential adverse reactions. Drink plenty of liquids to minimize risk of renal stones.
- Notify prescriber of flank pain, hematuria, S&S of jaundice, or other distressing adverse effects.

INDOMETHACIN

(in-doe-meth'a-sin)

Indocid ♦, Indocin, Indocin SR
Classification: ANALGESIC, NONSTEROIDAL ANTI-INFLAMMATORY (NSAID)
Therapeutic: ANALGESIC, NSAID; ANTIRHEUMATIC
Prototype: Ibuprofen
Pregnancy Category: B first and second trimester; D third trimester

AVAILABILITY 25 mg, 50 mg capsules; 75 mg sustained release capsules; 25 mg/5 mL oral suspension; 50 mg suppositories; 1 mg injection

ACTION & *THERAPEUTIC EFFECT* Potent nonsteroidal compound that competes with COX-1 and COX-2 enzymes, thus interfering with formation of prostaglandin. Appears to reduce motility of polymorphonuclear leukocytes, development of cellular exudates, and vascular permeability in injured tissue resulting in its anti-inflammatory effects. Inhibition of prostaglandins is thought to promote closure of the patency of the ductus arterious. Antipyretic and anti-inflammatory actions may be related to its ability to inhibit prostaglandin biosynthesis. *It is a potent analgesic, anti-inflammatory, and antipyretic agent. Promotes closure of persistent patent ductus arteriosus.*

USES Palliative treatment in active stages of moderate to severe rheumatoid arthritis, ankylosing rheumatoid spondylitis, acute gouty arthritis, and osteoarthritis of hip in patients intolerant to or unresponsive to adequate trials with salicylates and other therapy. Also used IV to close patent ductus arteriosus in the premature infant.
UNLABELED USES To relieve biliary pain and dysmenorrhea, Paget's disease, athletic injuries, juvenile arthritis, idiopathic pericarditis.

CONTRAINDICATIONS Allergy to indomethacin, aspirin, or other NSAID; nasal polyps associated with angioedema; history of recurrent GI lesions; perioperative pain with CABG; pregnancy (D third trimester), lactation.
CAUTIOUS USE History of psychiatric illness, epilepsy, parkinsonism; impaired renal or hepatic function; hypertension; history of ulcer disease or GI bleeding; infection; coagulation disorders; uncontrolled infections; coagulation defects, CHF, fluid retention; older adults, persons in hazardous occupations; pregnancy (category B first and second trimester); children less than 12 y except for use in premature infants for patent ductus arteriosus.

ROUTE & DOSAGE

Rheumatoid Arthritis

Adult: **PO** 25–50 mg b.i.d or t.i.d. (max: 200 mg/day) or 75 mg sustained release 1–2 times/day

Pediatric Arthritis

Child: **PO** 1–2 mg/kg/day in 2–4 divided doses (max: 4 mg/kg/day) or 150–200 mg/day

Acute Gouty Arthritis

Adult: **PO/PR** 50 mg t.i.d. until pain is tolerable, then rapidly taper

Common adverse effects in *italic*, life-threatening effects underlined; generic names in **bold**; classifications in SMALL CAPS; ♦ Canadian drug name; ⊙ Prototype drug

773

Bursitis

Adult: **PO** 25–50 mg t.i.d. or q.i.d. (max: 200 mg/day) or 75 mg sustained release 1–2 times/day

Close Patent Ductus Arteriosus

Premature neonate: **IV** *Younger than 48 h,* 0.2 mg/kg followed by 2 doses of 0.1 mg/kg q12–24h; *2–7 days,* 0.2 mg/kg followed by 2 doses of 0.2 mg/kg q12–24h; *younger than 7 days,* 0.2 mg/kg followed by 2 doses of 0.25 mg/kg q12–24h

ADMINISTRATION

Oral

- Give immediately after meals, or with food, milk, or antacid (if prescribed) to minimize GI side effects.

Rectal

- Indomethacin rectal suppository use is contraindicated with history of proctitis or recent bleeding.

Intravenous

PREPARE: **Direct:** Dilute 1 mg with 1 mL of NS or sterile water for injection without preservatives. Resulting concentration (1 mg/mL) may be further diluted with an additional 1 mL for each 1 mg to yield 0.5 mg/mL.

ADMINISTER: **Direct:** Give by direct IV with a single dose given over 20–30 min.

INCOMPATIBILITIES **Y-site: Amino acid, calcium gluconate, cimetidine, dobutamine, dopamine, gentamicin, levofloxacin, tobramycin, tolazoline.**

- Avoid extravasation or leakage; drug can be irritating to tissue. - Discard any unused drug, since it contains no preservative.

- Store oral and rectal forms in tight, light-resistant containers unless otherwise directed. Do not freeze.

ADVERSE EFFECTS (≥1%) **Body as a Whole:** Hypersensitivity (rash, purpura, pruritus, urticaria, angioedema, angiitis, rapid fall in blood pressure, dyspnea, asthma syndrome in aspirin-sensitive patients), edema, weight gain, flushing, sweating. **CNS:** Headache, *dizziness,* vertigo, lightheadedness, syncope, fatigue, muscle weakness, ataxia, insomnia, nightmares, drowsiness, confusion, coma, convulsions, peripheral neuropathy, psychic disturbances (hallucinations, depersonalization, depression), aggravation of epilepsy, parkinsonism. **CV:** Elevated BP, palpitation, chest pains, tachycardia, bradycardia, CHF. **Special Senses:** Blurred vision, lacrimation, eye pain, visual field changes, corneal deposits, retinal disturbances including macula, *tinnitus,* hearing disturbances, epistaxis. **GI:** *Nausea, vomiting,* diarrhea, anorexia, bloating, abdominal distention, ulcerative stomatitis, proctitis, rectal bleeding, GI ulceration, hemorrhage, perforation, toxic hepatitis. **Hematologic:** Hemolytic anemia, aplastic anemia (sometimes fatal), agranulocytosis, leukopenia, thrombocytopenic purpura, inhibited platelet aggregation. **Urogenital:** Renal function impairment, hematuria, urinary frequency; vaginal bleeding, breast changes. **Skin:** Hair loss, exfoliative dermatitis, erythema nodosum, tissue irritation with extravasation. **Metabolic:** Hyponatremia, hypokalemia, hyperkalemia, hypoglycemia or hyperglycemia, glycosuria (rare).

DIAGNOSTIC TEST INTERFERENCE Increased *AST, ALT, bilirubin, BUN;* positive direct *Coombs' test.*

INTERACTIONS Drug: ORAL ANTICOAGULANTS, **heparin, alcohol** may prolong bleeding time; may increase

lithium toxicity; effects of ORAL ANTI-COAGULANTS, **phenytoin,** SALICYLATES, SULFONAMIDES, SULFONYLUREAS increased because of protein-binding displacement; increased toxicity including GI bleeding with SALICYLATES, NSAIDS; may blunt effects of ANTIHYPERTENSIVES and DIURETICS. **Herbal: Feverfew, garlic, ginger, ginkgo** may increase bleeding potential.

PHARMACOKINETICS Absorption: Completely absorbed from GI tract. **Onset:** 1–2 h. **Peak:** 3 h. **Duration:** 4–6 h. **Metabolism:** In liver. **Elimination:** Primarily in urine. **Half-Life:** 2.5–124 h.

NURSING IMPLICATIONS

Assessment & Drug Effects

- Monitor for therapeutic effectiveness: In acute gouty attack, relief of joint tenderness and pain is usually apparent in 24–36 h; swelling generally disappears in 3–5 days. In rheumatoid arthritis: Reduced fever, increased strength, reduced stiffness, and relief of pain, swelling, and tenderness.
- Question patient carefully regarding aspirin sensitivity before initiation of therapy.
- Observe patients carefully; instruct to report adverse reactions promptly to prevent serious and sometimes irreversible or fatal effects.
- Lab tests: Monitor renal function, LFTs, CBC with differential, BP and HR, visual and hearing acuity periodically.
- Monitor weight and observe dependent areas for signs of edema in patients with underlying cardiovascular disease.
- Monitor I&O closely and keep prescriber informed during IV administration for patent ductus arteriosus. Significant impairment of renal function is possible; urine output

may decrease by 50% or more. Also monitor BUN, serum creatinine, glomerular filtration rate, creatinine clearance, and serum electrolytes.

Patient & Family Education

- Notify prescriber of S&S of GI bleeding, visual disturbance, tinnitus, weight gain, or edema.
- Do not take aspirin or other NSAIDs; they increase possibility of ulcers.
- Note: Frontal headache is the most frequent CNS adverse effect; if it persists, dosage reduction or drug withdrawal may be indicated. Take drug at bedtime with milk to reduce the incidence of morning headache.
- Do not drive or engage in other potentially hazardous activities until response to drug is known.

INFLIXIMAB

(in-flix′i-mab)

Remicade

Classifications: BIOLOGICAL RESPONSE MODIFIER; MONOCLONAL ANTIBODY (IgG); TUMOR NECROSIS FACTOR (TNF) MODIFIER

Therapeutic: IMMUNOMODULATOR; ANTI-INFLAMMATORY; DISEASE-MODIFYING ANTIRHEUMATIC DRUG (DMARD)

Pregnancy Category: B

AVAILABILITY 100 mg powder for injection

ACTION & *THERAPEUTIC EFFECT*
An IgG$_1$-K is a monoclonal antibody that binds specifically to tumor necrosis factor-alpha (TNF-alpha), a cytokine. Thus, it prevents TNF-alpha from binding to its receptors. TNF-alpha induces proinflammatory cytokines. Infliximab reduces infiltration of inflammatory cells and TNF-alpha production in inflamed

Common adverse effects in *italic*, life-threatening effects <u>underlined</u>; generic names in **bold;** classifications in SMALL CAPS; ♣ Canadian drug name; ⊘ Prototype drug

775

areas of the intestine as seen in Crohn's disease. *Decreases GI inflammation in Crohn's and related diseases. It is also effectively used as a disease-modifying antirheumatic drug (DMARD).*

USES Moderately to severely active Crohn's disease, including fistulizing Crohn's disease, rheumatoid arthritis, ankylosing spondylitis, ulcerative colitis.

CONTRAINDICATIONS Severe hypersensitivity to infliximab; serious infection, sepsis; murine protein hypersensitivity; lactation.

CAUTIOUS USE History of allergic phenomena or untoward responses to monoclonal antibody preparation; chronic infections; previous history of TB infection; history of HBV infection; renal or hepatic impairment; multiple sclerosis (potential exacerbation); fungal infection; heart failure; human antichimeric antibody (HACA); leukopenia, thrombocytopenia; immunosuppressed patients; neoplastic disease; vasculitis; neurologic disease; neutropenia; seizure disorder; preexisting CNS demyelinating disorders; moderate to severe COPD; history of malignancy; older adults; pregnancy (category B).

ROUTE & DOSAGE

Crohn's Disease

Adult: **IV** 5 mg/kg infused over at least 2 h, repeat at 2 and 6 wk for fistulizing disease, then q8wk
Child: **IV** 5 mg/kg at weeks 0, 2, and 6, then 5 mg/kg q8wk

Rheumatoid Arthritis

Adult: **IV** 3 mg/kg at weeks 0, 2, and 6, then q8wk

Ulcerative Colitis

Adult: **IV** 5 mg/kg at weeks 0, 2, and 6, then 5 mg/kg q8wk

Ankylosing Spondylitis

Adult: **IV** 5 mg/kg at weeks 0, 2, and 6, then 5 mg/kg q6wk

ADMINISTRATION

- Note: Do not administer to a patient who has known or suspected sepsis.

Intravenous

PREPARE: **IV Infusion:** Reconstitute each 100 mg vial with 10 mL of sterile water for injection using a 21-gauge or smaller syringe. Inject sterile water against wall of vial, then gently swirl to dissolve but do not shake. • Let stand for 5 min. • Solution should be colorless to light yellow with a few translucent particles. Discard if particles are opaque. • Further dilute by first removing from a 250-mL IV bag of NS a volume of NS equal to the volume of reconstituted infliximab to be added to the IV bag. Slowly add the total volume of reconstituted infliximab solution to the 250-mL infusion bag and gently mix. • Infusion concentration should be 0.4 to 4 mg/mL. • Begin infusion within 3 h of preparation.

ADMINISTER: **IV Infusion:** Give over at least 2 h using a polyethylene-lined infusion set with an in-line, low-protein-binding filter (pore size 1.2 micron or less). • Flush infusion set before and after with NS to ensure delivery of total drug dose. • Discard unused infusion solution.

INCOMPATIBILITIES **Y-site:** Do not infuse with any other drugs.

- Store unopened vials at 2°–8° C (36°–46° F).

ADVERSE EFFECTS (≥1%) **Body as a Whole:** Fatigue, fever, pain, myalgia,

Common adverse effects in *italic*, life-threatening effects <u>underlined</u>; generic names in **bold**; classifications in SMALL CAPS; ♣ Canadian drug name; ⊙ Prototype drug

back pain, chills, hot flashes, arthralgia; infusion-related reactions (fever, chills, pruritus, urticaria, chest pain, hypotension, hypertension, dyspnea). Increased risk of lymphoma and opportunistic infections, including tuberculosis. **CNS:** Headache, dizziness, involuntary muscle contractions, paresthesias, vertigo, anxiety, depression, insomnia. **CV:** Chest pain, peripheral edema, hypotension, hypertension, tachycardia, anemia, CHF, pericardial effusion, systemic and cutaneous vasculitis. **GI:** Nausea, diarrhea, abdominal pain, vomiting, constipation, dyspepsia, flatulence, intestinal obstruction, ulcerative stomatitis, increased hepatic enzymes. **Hematologic:** Leukopenia, neutropenia, thrombocytopenia, pancytopenia. **Respiratory:** URI, pharyngitis, bronchitis, rhinitis, coughing, sinusitis, dyspnea. **Skin:** Rash, pruritus, acne, alopecia, fungal dermatitis, eczema, dry skin, increased sweating, urticaria. **Other:** Infections, development of autoantibodies, lupus-like syndrome, conjunctivitis, dysuria, urinary frequency.

INTERACTIONS Drug: May blunt effectiveness of VACCINES given concurrently.

PHARMACOKINETICS Distribution: Distributed primarily to the vascular compartment. **Half-Life:** 9.5 days.

NURSING IMPLICATIONS

Assessment & Drug Effects
- Discontinue IV infusion and notify prescriber for fever, chills, pruritus, urticaria, chest pain, dyspnea, hypo/hypertension.
- Monitor for up to 2 h post-infusion for an acute infusion reaction (e.g., chest pain, hypotension, hypertension, dyspnea).
- Monitor for and immediately report S&S of generalized infection.

Patient & Family Education
- Seek medical evaluation immediately if you suspect an infection.

INSULIN ASPART
(in′su-lyn)

NovoLog, NovoLog 50/50, NovoLog 70/30
Classifications: ANTIDIABETIC; RAPID-ACTING INSULIN
Therapeutic: RAPID-ACTING INSULIN
Prototype: Insulin injection
Pregnancy Category: B (**NovoLog**); C (**NovoLog 70/30**)

AVAILABILITY 100 units/mL injection; insulin aspart/insulin aspart protamine injection

ACTION & *THERAPEUTIC EFFECT* A recombinant insulin analog that is more rapidly absorbed than human insulin, with a more rapid onset and shorter duration than regular human insulin. *Provides better blood glucose control than regular human insulin when given before a meal.*

USES Treatment of diabetes mellitus.

CONTRAINDICATIONS Systemic allergic reactions; history of allergic reactions to insulin; hypoglycemia. **CAUTIOUS USE** Fever, hyperthyroidism, surgery or trauma; decreased insulin requirements due to diarrhea, nausea, or vomiting, malabsorption; renal or hepatic impairment, hypokalemia; pregnancy [category B (**NovoLog**) and category C (**NovoLog 70/30**)], lactation.

ROUTE & DOSAGE

Diabetes

Adult: **Subcutaneous** 0.25–0.7 units/kg/day injected 5–10 min before each meal **IV** Use only under close medical supervision in a clinical setting

Common adverse effects in *italic*, life-threatening effects underlined; generic names in **bold**; classifications in SMALL CAPS; ♣ Canadian drug name; ⊙ Prototype drug

777

ADMINISTRATION
- Use only if solution is absolutely clear.

Subcutaneous
- **Must be** given **no sooner** than 5–10 min before a meal.
- Draw up insulin aspart first when mixing with NPH insulin. Give injection immediately after it is mixed.
- Store refrigerated at 2°–8° C (36°–46° F); may be stored at room temperature, 15°–30° C (59°–86° F) for up to 28 days. Do not expose to excessive heat or sunlight, and do not freeze.

Intravenous

PREPARE: **Infusion:** Dilute with NS or D5W in a polypropylene infusion bag to a final concentration of 0.05–1 unit/mL.
ADMINISTER: **Infusion:** Give at rate ordered by prescriber.

ADVERSE EFFECTS (≥1%) **Body as a Whole:** Allergic reactions. **Endocrine:** Hypoglycemia, hypokalemia. **Skin:** Injection site reaction, lipodystrophy, pruritus, rash.

INTERACTIONS Drug: ORAL ANTIDIABETIC AGENTS, ACE INHIBITORS, **disopyramide, fluoxetine,** MAO INHIBITORS, **propoxyphene,** SALICYLATES, SULFONAMIDE ANTIBIOTICS, **octreotide** may enhance hypoglycemia; CORTICOSTEROIDS, **niacin, danazol,** DIURETICS, SYMPATHOMIMETIC AGENTS, PHENOTHIAZINES, THYROID HORMONES, ESTROGENS, PROGESTOGENS, **isoniazid, somatropin** may decrease hypoglycemic effects; BETA-BLOCKERS, **clonidine, lithium, alcohol** may either potentiate or weaken effects of insulin; **pentamidine** may cause hypoglycemia followed by hyperglycemia. **Herbal: Garlic, ginseng** may potentiate hypoglycemic effects.

PHARMACOKINETICS Absorption: Rapidly absorbed from subcutaneous injection site. **Onset:** 15 min. **Peak:** 1–3 h. **Duration:** 3–5 h. **Distribution:** Low protein binding. **Metabolism:** In liver with some metabolism in the kidneys. **Half-Life:** 81 min.

NURSING IMPLICATIONS

Assessment & Drug Effects
- Monitor for S&S of hypoglycemia (see Appendix F). Initial hypoglycemic response begins within 15 min and peaks 45–90 min after injection.
- Lab tests: Periodic postprandial blood glucose and HbA1C.
- Withhold drug and notify prescriber if patient is hypokalemic.

Patient & Family Education
- Eat immediately after injecting insulin aspart because it has a fast onset and short duration of action.
- Do not inject into areas with redness, swelling, itching, or dimpling.
- Ingest some form of sugar (e.g., orange juice, dissolved table sugar, honey) if symptoms of hypoglycemia develop, and seek medical assistance.
- Check blood sugar as prescribed, especially postprandial values; make note of and notify prescriber of fasting blood glucose less than 80 and greater than 120 mg/dL.
- Notify the prescriber of any of the following: Fever, infection, trauma, diarrhea, nausea or vomiting. Dosage adjustment may be needed.
- Do not take any other medication unless approved by the prescriber.

INSULIN DETEMIR
(in'su-lyn det'e-mir)

Levemir
Classifications: ANTI-DIABETIC; LONG-ACTING INSULIN

Common adverse effects in *italic*, life-threatening effects <u>underlined</u>; generic names in **bold**; classifications in SMALL CAPS; ♣ Canadian drug name; 🅟 Prototype drug

Therapeutic: LONG-ACTING INSULIN
Prototype: Insulin injection
Pregnancy Category: C

AVAILABILITY 100 units/mL available in 10 mL multidose vials and 3 mL prefilled syringes

ACTION & *THERAPEUTIC EFFECT*
Insulin detemir, a long-acting human insulin, exerts its action by binding to insulin receptors. Receptor-bound insulin lowers blood glucose by facilitating cellular uptake of glucose into skeletal muscle and fat, and inhibiting the output of glucose from the liver. *Insulin detemir is effective as a glucose-lowering agent, with glycemic control equivalent to that of NPH insulin.*

USES Treatment of type 1 and type 2 diabetes mellitus.

CONTRAINDICATIONS Hypersensitivity to insulin detemir, or cresol; use in insulin infusion pumps; diabetic ketoacidosis, coma, hyperosmolar hyperglycemic state, hypoglycemia.
CAUTIOUS USE Renal and hepatic impairment; older adults; cardiac disease, CHF, illness, stress; pregnancy (category C), lactation. Safe and effective use in children with type 2 diabetes has not been established.

ROUTE & DOSAGE

Diabetes

Adult/Child: **Subcutaneous**
Insulin-naïve patients: 0.1–0.2 units/kg daily in evening or 10 units daily or b.i.d. in evenly spaced doses. For those taking a basal insulin product (i.e., NPH insulin, insulin glargine), a unit-to-unit dose conversion can be used.

ADMINISTRATION

Subcutaneous
- Once-daily injections should be given with the evening meal or at bedtime. With twice-daily dosing, the evening dose may be given with the evening meal, at bedtime, or 12 h after the morning dose.
- Do not administer IV or IM. With thin patients, inject at a 45-degree angle into a pinched fold of skin to avoid IM injection.
- Do not mix with any other type of insulin. Do not use with an insulin infusion pump.
- Store unopened vials under refrigeration at 2°–8° C (36°–46° F). Once removed from refrigeration, pens, cartridges, and other delivery devices **must be** kept at room temperature (not to exceed 30° C or 85° F) and either used within 42 days or discarded.

INCOMPATIBILITIES **Solution/additive:** Insulin detemir should not be mixed with any other insulin preparations.

ADVERSE EFFECTS (≥1%) **[See INSULIN (REGULAR)] Body as a Whole:** Allergic reactions. **Metabolic:** Hypoglycemia, weight gain. **Skin:** Lipodystrophy, pruritus, rash.

DIAGNOSTIC TEST INTERFERENCE See INSULIN INJECTION (REGULAR).

INTERACTIONS Drug: See INSULIN INJECTION (REGULAR). **Herbal: Garlic** and **green tea** may potentiate hypoglycemic effects.

PHARMACOKINETICS Absorption: Slow, prolonged absorption over 24 h. **Peak:** 6–8 h. **Distribution:** 98–99% protein bound. **Half-Life:** 5–7 h.

NURSING IMPLICATIONS

Assessment & Drug Effects
- Monitor for S&S of hypoglycemia (see Appendix F), especially after

Common adverse effects in *italic*, life-threatening effects underlined; generic names in **bold;** classifications in SMALL CAPS; ♣ Canadian drug name; ⊘ Prototype drug

779

changes in insulin dose or type.

- Lab tests: Periodic fasting blood glucose and HbA1C; periodic serum potassium with concurrent potassium-lowering drugs.
- Monitor weight periodically.

Patient & Family Education

- Follow directions for taking the drug (see Administration). Rotate injection sites and never inject into an area with redness, swelling, itching, or dimpling.
- Know parameters for withholding drug. Check blood sugar as prescribed; notify prescriber of fasting blood glucose below 80 or above 120 mg/dL.
- Ingest some form of sugar (e.g., orange juice, dissolved table sugar, honey) if symptoms of hypoglycemia develop; and seek medical assistance.
- Notify the prescriber of any of the following: Fever, infection, trauma, diarrhea, nausea, or vomiting.
- Do not take any other medication unless approved by prescriber.

INSULIN GLARGINE

(in'su-lyn glar'geen)

Lantus, Lantus SoloStar
Classifications: ANTIDIABETIC; LONG-ACTING INSULIN
Therapeutic: LONG-ACTING INSULIN
Prototype: Insulin injection
Pregnancy Category: C

AVAILABILITY 100 units/mL injection; 3 mL cartridge; prefilled, disposable pen

ACTION & *THERAPEUTIC EFFECT*
A recombinant human insulin analog with a long duration of action. Lowers blood glucose levels over an extended period of time by stimulating peripheral glucose uptake especially in muscle and fat tissue. In addition, insulin inhibits hepatic glucose production. *Lowers blood glucose levels over an extended period of time. It also prevents the conversion of glucagon to glucose in the liver.*

USES Bedtime dosing of adults and children with type 1 diabetes, or adults with type 2 diabetes.

CONTRAINDICATIONS Prior hypersensitivity to insulin glargine; hypoglycemia.

CAUTIOUS USE Renal and hepatic impairment; older adults; illness; emotional disturbances; stress; pregnancy (category C), lactation. Safety and efficacy in children younger than 6 y of age in type 1 diabetes. Safety and efficacy in children with type 2 diabetes are unknown.

ROUTE & DOSAGE

Type 1 Diabetes

Adult/Child: **Subcutaneous** If not taking insulin, give 10 units at same time each day (usually at bedtime) once daily; if taking NPH or ultralente insulin once daily, give same dose at same time each day (usually at bedtime); if taking NPH insulin twice daily, give 80% of total daily dose at same time each day (usually at bedtime)

Type 2 Diabetes

Adult: **Subcutaneous** If already taking oral hypoglycemic drugs, start with 10 units at same time each day (usually at bedtime) once daily and adjust according to patient's needs

ADMINISTRATION

Subcutaneous

- Do not give this product IV.
- Give at same time each day (usually at bedtime) and do not mix with any other insulin product.

Common adverse effects in *italic*, life-threatening effects underlined; generic names in **bold**; classifications in SMALL CAPS; ♣ Canadian drug name; ⊘ Prototype drug

- Store in refrigerator at 2°–8° C (36°–46° F), may store at room temperature, 15°–30° C (59°–86° F). Discard opened refrigerated vials after 28 days and unrefrigerated vials after 14 days. Do not expose to excessive heat or sunlight, and do not freeze.

ADVERSE EFFECTS (≥1%) **Body as a Whole:** Allergic reactions. **Endocrine:** Hypoglycemia, hypokalemia. **Skin:** Injection site reaction, lipodystrophy, pruritus, rash.

INTERACTIONS Drug: ORAL ANTIDIABETIC AGENTS, ACE INHIBITORS, **disopyramide, fluoxetine,** MAO INHIBITORS, **propoxyphene,** SALICYLATES, SULFONAMIDE ANTIBIOTICS, **octreotide** may enhance hypoglycemia; CORTICOSTEROIDS, **niacin, danazol,** DIURETICS, SYMPATHOMIMETIC AGENTS, PHENOTHIAZINES, THYROID HORMONES, ESTROGENS, PROGESTOGENS, **isoniazid, somatropin** may decrease hypoglycemic effects; BETA-BLOCKERS, **clonidine, lithium, alcohol** may either potentiate or weaken effects of insulin; **pentamidine** may cause hypoglycemia followed by hyperglycemia. **Herbal: Garlic, ginseng** may potentiate hypoglycemic effects.

PHARMACOKINETICS Absorption: Slowly absorbed from subcutaneous injection site. **Onset:** 3–4 h. **Duration:** 10.4–24 h. **Metabolism:** In liver to active metabolites.

NURSING IMPLICATIONS

Assessment & Drug Effects

- Monitor for S&S of hypoglycemia (see Appendix F), especially after changes in insulin dose or type.
- Lab tests: Monitor fasting blood glucose and HbA1C periodically.
- Withhold drug and notify prescriber if patient is hypokalemic.

Patient & Family Education

- Do not inject into areas with redness, swelling, itching, or dimpling.
- Absorption patterns for this drug are not dependent on the injection site.
- Ingest some form of sugar (e.g., orange juice, dissolved table sugar, honey) if symptoms of hypoglycemia develop and seek medical assistance.
- Check blood sugar as prescribed; notify prescriber of fasting blood glucose less than 80 and greater than 120 mg/dL.
- Notify the prescriber of any of the following: Fever, infection, trauma, diarrhea, nausea, or vomiting. Dosage adjustment may be needed.
- Do not take any other medication unless approved by prescriber.

INSULIN GLULISINE
(in'su-lyn glu-li'seen)

Apidra, Apidra SoloSTAR
Classifications: ANTIDIABETIC; RAPID-ACTING INSULIN
Therapeutic: RAPID-ACTING INSULIN
Prototype: Insulin injection (Regular)
Pregnancy Category: C

AVAILABILITY 100 units/mL multidose (10 mL) vials; 3 mL cartridge system

ACTION & *THERAPEUTIC EFFECT*
Insulin glulisine, formed by recombinant DNA, is a rapid-acting insulin. Insulin lowers blood glucose by stimulating peripheral glucose uptake by skeletal muscle and fat and by inhibiting hepatic glucose production. Insulin causes lipolysis in the adipocytes, inhibits proteolysis, and enhances protein synthesis. *Insulin glulisine has a more rapid onset of*

Common adverse effects in *italic*, life-threatening effects underlined; generic names in **bold**; classifications in SMALL CAPS; ♣ Canadian drug name; ☻ Prototype drug

781

action and a shorter duration of action than regular human insulin; thus, it provides good postprandial blood glucose control.

USES Treatment of diabetes mellitus; type I diabetes mellitus in children.

CONTRAINDICATIONS Hypoglycemia; systemic allergy to insulin.

CAUTIOUS USE Renal impairment, hepatic dysfunction; thyroid disease; fever; older adults; children older than 4 y; pregnancy (category C), lactation. Safe use in children 4 y or younger not established.

ROUTE & DOSAGE

Diabetes

Adult/Adolescent/Child (4 y and older): **Subcutaneous** 5–10 units within 15 min before starting a meal or within 20 min after starting a meal. Dose should be individualized.

Adult/Adolescent/Child: **IV** 0.05–1 unit/mL via infusion; use under close supervision

ADMINISTRATION

Subcutaneous

▪ Give within 15 min before or up to 20 min after a meal.
▪ Store refrigerated at 36°–46° F (2°–8° C). Discard vial if frozen. Protect from light.

Intravenous

PREPARE: **Infusion:** Dilute with NS in a PVC bag to a final concentration of 0.05–1 unit/mL.
ADMINISTER: **Infusion:** Give at rate ordered by prescriber.

ADVERSE EFFECTS (≥1%) **[See INSULIN (REGULAR)] Body as a Whole:** Allergic reactions. **Metabolic:** Hypo-

glycemia. **Skin:** Injection site reactions, lipodystrophy, pruritus, rash.

DIAGNOSTIC TEST INTERFERENCE
See INSULIN INJECTION (REGULAR).

PHARMACOKINETICS Absorption: 70% bioavailable from injection sites. **Onset:** 15–30 min. **Peak:** 55 min. **Duration:** 3–4 h. **Metabolism:** In liver with some metabolism in the kidney. **Half-Life:** 42 min subcutaneous.

NURSING IMPLICATIONS

Assessment & Drug Effects

▪ Monitor for S&S of hypoglycemia (see Appendix F). Initial hypoglycemic response begins within 15 min and peaks, on average, 40–60 min after injection.
▪ Lab tests: Periodically monitor fasting and postprandial blood glucose and HbA1C.

Patient & Family Education

▪ Follow exactly directions for timing injection in relation to each meal.
▪ Do not inject into areas with redness, swelling, itching, or dimpling.
▪ If mixing with NPH human insulin, draw up insulin glulisine first. Inject immediately after mixing.
▪ Ingest some form of sugar (e.g., orange juice, dissolved table sugar, honey) if symptoms of hypoglycemia develop, and seek medical assistance.
▪ Check blood sugar as prescribed, especially postprandial values; notify prescriber of fasting blood glucose less than 80 and greater than 140 mg/dL.
▪ Notify the prescriber of any of the following: Fever, infection, trauma, diarrhea, nausea, or vomiting. Dosage adjustment may be needed.
▪ Do not take any other medication unless approved by the prescriber.

✓INSULIN (REGULAR) ⊕

(in'su-lyn)

Humulin R, Novolin R, Regular Insulin, Velosulin BR
Classifications: ANTIDIABETIC; SHORT-ACTING INSULIN
Therapeutic: SHORT-ACTING INSULIN
Pregnancy Category: B

AVAILABILITY 100 units/mL

ACTION & *THERAPEUTIC EFFECT*

Short-acting, clear, colorless solution of exogenous unmodified insulin extracted from beta cells in pork pancreas or synthesized by recombinant DNA technology (human). Enhances transmembrane passage of glucose across cell membranes in muscle and adipose tissue. Promotes conversion of glucose to glycogen in the liver. *It lowers blood glucose levels by increasing peripheral glucose uptake and by inhibiting the liver from changing glycogen to glucose.*

USES Emergency treatment of diabetic ketoacidosis or coma, to initiate therapy in patient with insulin-dependent diabetes mellitus, and in combination with intermediate-acting or long-acting insulin to provide better control of blood glucose concentrations in the diabetic patient. Used IV to stimulate growth hormone secretion (glucose counter-regulatory hormone) to evaluate pituitary growth hormone reserve in patient with known or suspected growth hormone deficiency. Other uses include promotion of intracellular shift of potassium in treatment of hyperkalemia (IV) and induction of hypoglycemic shock as therapy in psychiatry.

CONTRAINDICATIONS Hypersensitivity to insulin.
CAUTIOUS USE Renal impairment, renal failure; hepatic impairment, fever, thyroid disease; older adults; pregnancy (category B), children and infants.

ROUTE & DOSAGE

Diabetes Mellitus
Adult: **Subcutaneous** 5–10 units 30–60 min a.c. and at bedtime (dose adjustments based on blood glucose determinations)
Child: **Subcutaneous** 2–4 units 30–60 min a.c. and at bedtime (dose adjustments based on blood glucose determinations)
Ketoacidosis
Adult: **IV** 2.4–7.2 units loading dose, followed by 2.4–7.2 units/h continuous infusion
Child: **IV** 0.1 units/kg loading dose, followed by 0.1 units/h continuous infusion

ADMINISTRATION

▪ Note: Insulins should not be mixed unless prescribed by prescriber. In general, regular insulin is drawn up into syringe first. ▪Any change in the strength (e.g., U-40, U-100), brand (manufacturer), purity, type (regular, etc.), species (pork, human), or sequence of mixing two kinds of insulin is made by the prescriber only, since a simultaneous change in dosage may be necessary.

Subcutaneous
▪ Use an insulin syringe.
▪ Give regular insulin 30–60 min before a meal.
▪ Avoid injection of cold insulin; it can lead to lipodystrophy, reduced rate of absorption, and local reactions.
▪ Common injection sites: Upper arms, thighs, abdomen [avoid area over urinary bladder and 2 in.

(5 cm) around navel], buttocks, and upper back (if fat is loose enough to pick up). Rotate sites.

Intravenous

PREPARE: **Direct:** Give undiluted. **Continuous:** Typically diluted in NS or 0.45% NaCl. 100 units added to 1000 mL yields 0.1 units/mL. *ADMINISTER:* **Direct:** Give 50 units or a fraction thereof over 1 min. **Continuous:** Rate **must be** ordered by prescriber.

INCOMPATIBILITIES **Solution/ additive: Aminophylline, amobarbital, chlorothiazide, cytarabine, dobutamine, pentobarbital, phenobarbital, phenytoin, secobarbital, sodium bicarbonate, thiopental. Y-site: Dobutamine.**

▪ Regular insulin may be adsorbed into the container or tubing when added to an IV infusion solution. ▪ Amount lost is variable and depends on concentration of insulin, infusion system, contact duration, and flow rate. ▪ Monitor patient response closely.

▪ Insulin is stable at room temperature up to 1 mo. Avoid exposure to direct sunlight or to temperature extremes [safe range is wide: 5°–38° C (40°–100° F)]. Refrigerate but do not freeze stock supply. Insulin tolerates temperatures above 38° C with less harm than freezing.

ADVERSE EFFECTS (≥1%) **Body as a Whole:** Most adverse effects are related to hypoglycemia; <u>anaphylaxis</u> (rare), hyperinsulinemia (*profuse sweating,* hunger, headache, *nausea, tremulousness,* tremors, *palpitation,* tachycardia, weakness, fatigue, nystagmus, circumoral pallor); numb mouth, tongue, and other paresthesias; visual disturbances (diplopia, blurred vision, mydriasis), staring expression, confusion, personality changes, ataxia, incoherent speech, apprehension, irritability, inability to concentrate, personality changes, uncontrolled yawning, loss of consciousness, delirium, hypothermia, convulsions, Babinski reflex, <u>coma</u>. (Urine glucose tests will be negatives.) **CNS:** With overdose, psychic disturbances (i.e., aphasia, personality changes, maniacal behavior). **Metabolic:** Posthypoglycemia or rebound hyperglycemia (Somogyi effect), lipoatrophy and lipohypertrophy of injection sites; insulin resistance. **Skin:** Localized allergic reactions at injection site; generalized urticaria or bullae, lymphadenopathy.

DIAGNOSTIC TEST INTERFERENCE

Large doses of insulin may increase urinary excretion of *VMA.* Insulin can cause alterations in *thyroid function tests* and *liver function test* and may decrease *serum potassium* and *serum calcium.*

INTERACTIONS Drug: Alcohol, ANABOLIC STEROIDS, MAO INHIBITORS, **guanethidine,** SALICYLATES may potentiate hypoglycemic effects; **dextrothyroxine,** CORTICOSTEROIDS, **epinephrine** may antagonize hypoglycemic effects; **furosemide,** THIAZIDE DIURETICS increase **serum glucose** levels; **propranolol** and other BETA-BLOCKERS may mask symptoms of hypoglycemic reaction. **Herbal: Garlic, ginseng** may potentiate hypoglycemic effects.

PHARMACOKINETICS Absorption: Rapidly absorbed from IM and subcutaneous injections. **Onset:** 0.5–1 h. **Peak:** 2–4 h. **Duration:** 5–7 h. **Distribution:** Throughout extracellular fluids. **Metabolism:** In liver with some metabolism in kidneys. **Elimination:** Less than 2% excreted in urine. **Half-Life:** Biological, up to 13 h.

NURSING IMPLICATIONS

Assessment & Drug Effects

- Note: Frequency of blood glucose monitoring is determined by the insulin regimen and health status of the patient.
- Lab tests: Periodic fasting and postprandial blood glucose and HbA1C. Test urine for ketones in new, unstable, and type 1 diabetes; if patient has lost weight, exercises vigorously, or has an illness; whenever blood glucose is substantially elevated.
- Notify prescriber promptly for markedly elevated blood sugar or presence of acetone with sugar in the urine; may indicate onset of ketoacidosis.
- Monitor for hypoglycemia (see Appendix F) at time of peak action of insulin. Onset of hypoglycemia (blood sugar: 50–40 mg/dL) may be rapid and sudden.
- Check BP, I&O ratio, and blood glucose and ketones every hour during treatment for ketoacidosis with IV insulin.
- Patients with severe hypoglycemia are usually treated with glucagon, epinephrine, or IV glucose 10–50%. As soon as patient is fully conscious, oral carbohydrate (e.g., orange juice with sugar, Gatorade, or Pedialyte) to prevent secondary hypoglycemia may be used.

Patient & Family Education

- Learn correct injection technique.
- Inject insulin into the abdomen rather than a near muscle that will be heavily taxed, if engaged in active sports.
- Notify prescriber of local reactions at injection site; may develop 1–3 wk after therapy starts and last several hours to days, usually disappear with continued use.
- Do not change prescription lenses during early period of dosage regulation; vision stabilizes, usually 3–6 wk.
- Check your blood glucose often as directed by the prescriber. Hypoglycemia can result from excess insulin, insufficient food intake, vomiting, diarrhea, unaccustomed exercise, infection, illness, nervous or emotional tension, or overindulgence in alcohol.
- Respond promptly to beginning symptoms of hypoglycemia. Severe hypoglycemia is an emergency situation. Take 4 oz (120 mL) of any fruit juice or regular carbonated beverage [1.5–3 oz (45–90 mL) for child] followed by a meal of longer-acting carbohydrate or protein food. Failure to show signs of recovery within 30 min indicates need for emergency treatment.
- Carry some form of fast-acting carbohydrate (e.g., lump sugar, Life-Savers, or other candy) at all times to treat hypoglycemia.
- Check blood glucose regularly during menstrual period; loss of diabetes control (hyperglycemia or hypoglycemia) is common; adjust insulin dosage accordingly, as prescribed by prescriber.
- Notify prescriber immediately of S&S of diabetic ketoacidosis.
- Continue taking insulin during an illness, go to bed, and drink noncaloric liquids liberally (every hour if possible). Consult prescriber for insulin regulation if unable to eat prescribed diet.
- Avoid OTC medications unless approved by prescriber.

INSULIN, ISOPHANE (NPH)
(in'su-lyn)

Humulin N, Novolin N, ReliOn N
Classifications: ANTIDIABETIC; INTERMEDIATE-ACTING INSULIN

Common adverse effects in *italic*, life-threatening effects underlined; generic names in **bold**; classifications in SMALL CAPS; ♣ Canadian drug name; ⊕ Prototype drug

785

INSULIN, ISOPHANE (NPH)

Therapeutic: INTERMEDIATE ACT-
ING INSULIN
Prototype: Insulin
Pregnancy Category: B

AVAILABILITY 100 units/mL

ACTION & *THERAPEUTIC EFFECT*
Intermediate-acting, cloudy suspension of zinc insulin crystals
modified by protamine in a neutral
buffer. NPH Iletin II (pork), and In-
sulatard NPH are "purified" or "single component" insulins that have
been purified and are less likely to
cause allergic reactions than nonpu-
rified preparations. Lowers blood
glucose levels by increasing peripheral glucose uptake, especially by
skeletal muscle and fat tissue, and
by inhibiting the liver from changing glycogen to glucose. *Controls
postprandial hyperglycemia, usually without supplemental doses of
insulin injection.*

USES Used to control hyperglycemia in the diabetic patient. Mix-
tard and Novolin 70/30 are fixed
combinations of purified regular
insulin 30% and NPH 70%.

CONTRAINDICATIONS During episodes of hypoglycemia or in patients sensitive to any ingredient in
the formulation; intravenous route;
diabetic ketoacidosis; hyperosmolar hyperglycemic state.
CAUTIOUS USE In insulin-resistant
patients, hyperthyroidism or hypothyroidism; fever; older adults,
renal or hepatic impairment; pregnancy (category B); children.

ROUTE & DOSAGE

Diabetes Mellitus
Adult: **Subcutaneous** Individualized doses (see INSULIN, REGULAR)

ADMINISTRATION

Subcutaneous
▪ Give isophane insulin 30 min before first meal of the day. If necessary, a second smaller dose may
be prescribed 30 min before sup-
per or at bedtime.
▪ Ensure complete dispersion by
mixing thoroughly by gently rotating vial between palms and inverting it end to end several
times. Do not shake.
▪ Do NOT mix insulins unless prescribed by prescriber. In general,
when insulin injection (regular
insulin) is to be combined, it is
drawn first.
▪ Note: Isophane insulin may be
mixed with insulin injection without altering either solution.
▪ Store unopened vial at 2°–8° C
(36°–46° F). Avoid freezing and
exposure to extremes in temperature or to direct sunlight.

ADVERSE EFFECTS (≥1%) (see INSU-
LIN, REGULAR).

INTERACTIONS (see INSULIN, REGU-
LAR).

PHARMACOKINETICS Onset: 1–2 h.
Peak: 4–12 h NPH. **Duration:** 18–
24 h NPH. **Metabolism:** In liver and
kidney. **Elimination:** Less than 2%
excreted unchanged in urine. **Half-
Life:** Up to 13 h.

NURSING IMPLICATIONS

(see INSULIN, REGULAR)

Assessment & Drug Effects
▪ Suspect hypoglycemia if fatigue,
weakness, sweating, tremor, or
nervousness occur.

Patient & Family Education
▪ If insulin was given before breakfast, a hypoglycemic episode is
most likely to occur between

Common adverse effects in *italic*, life-threatening effects underlined; generic names
in **bold**; classifications in SMALL CAPS; ✦ Canadian drug name; ⊙ Prototype drug

mid-afternoon and dinnertime, when insulin effect is peaking. Advise to eat a snack in mid-afternoon and to carry sugar or candy to treat a reaction. A snack at bedtime will prevent insulin reaction during the night.
- Learn the S&S of hypoglycemia and hyperglycemia (see Appendix F).

INSULIN LISPRO
(in'su-lyn lis'pro)

Humalog
Classifications: ANTIDIABETIC; RAPID-ACTING INSULIN
Therapeutic: RAPID-ACTING INSULIN
Prototype: Insulin injection
Pregnancy Category: B

AVAILABILITY 100 units/mL

ACTION & *THERAPEUTIC EFFECT*
Insulin lispro of recombinant DNA origin is a human insulin that is a rapid-acting, glucose-lowering agent of shorter duration than human regular insulin. It lowers blood glucose levels by increasing peripheral glucose uptake, especially by skeletal muscle and fat tissue, and by inhibiting the liver from changing glycogen to glucose. *It lowers blood glucose levels and inhibits liver from changing glycogen to glucose.*

USES Treatment of diabetes mellitus.

CONTRAINDICATIONS During episodes of hypoglycemia or in patients sensitive to any ingredient in the formulation; intravenous administration.
CAUTIOUS USE In insulin-resistant patients, hyperthyroidism or hypothyroidism; older adults, renal or hepatic impairment; pregnancy (category B), lactation; children.

ROUTE & DOSAGE

Diabetes Mellitus (type 1)
Adult: **Subcutaneous** 5–10 units 0–15 min a.c. (dose adjustments based on blood glucose determinations)

ADMINISTRATION

Subcutaneous
- Give within 15 min before or immediately after a meal.
- Note: May be given in same syringe with longer-acting insulins but absorption may be delayed.

ADVERSE EFFECTS (≥1%) See INSULIN INJECTION, REGULAR.

INTERACTIONS See INSULIN INJECTION, REGULAR.

PHARMACOKINETICS Absorption: Rapidly absorbed from IM and subcutaneous injection sites. **Onset:** Less than 15 min. **Peak:** 0.5–1 h. **Duration:** 3–4 h. **Distribution:** Throughout extracellular fluids. **Metabolism:** Metabolized in liver with some metabolism in kidneys. **Elimination:** Less than 2% excreted in urine. **Half-Life:** Biological, up to 13 h.

NURSING IMPLICATIONS
(SEE INSULIN INJECTION, REGULAR)

Assessment & Drug Effects
- Assess for hypoglycemia from 1 to 3 h after injection.
- Assess highly insulin-dependent patients for need for increases in intermediate/long-acting insulins.

Patient & Family Education
- Note: Risk of hypoglycemia is greatest 1–3 h after injection.

Common adverse effects in *italic*, life-threatening effects underlined; generic names in **bold**; classifications in SMALL CAPS; ✦ Canadian drug name; ⊙ Prototype drug

787

INTERFERON ALFA-2b

(in-ter-feer'on)

Intron A

Classifications: BIOLOGICAL RESPONSE MODIFIER; IMMUNOMODULATOR; INTERFERON; ANTINEOPLASTIC
Therapeutic: ANTINEOPLASTIC; IMMUNOMODULATOR; INTERFERON; ANTIVIRAL
Prototype: Peg-interferon alfa-2a
Pregnancy Category: C

AVAILABILITY 5 million international units, 10 million international units, 18 million international units, 25 million international units, 50 million international units vials

ACTION & *THERAPEUTIC EFFECT* Interferon (IFN) alfa-2b, one of 4 types of alpha interferons, is a highly purified protein and natural product of human leukocytes within 4–6 h after viral stimulation. Produced by recombinant DNA technology. **Antiviral action:** Reprograms virus-infected cells to inhibit various stages of virus replication. **Antitumor action:** Suppresses cell proliferation. **Immunomodulating action:** Enhances phagocytic activity of macrophages and augments specific cytotoxicity of lymphocytes for target cells. The immune system and the interferon system of defense are complementary. *Has a broad spectrum of antiviral, cytotoxic, and immunomodulating activity (i.e., favorably adjusts immune system to better combat foreign invasion of antigens, cancers, and viruses).*

USES Hairy cell leukemia in splenectomized and non-splenectomized patients 18 y or older, chronic hepatitis B or C, malignant melanoma, condylomata acuminata, AIDS-related Kaposi's sarcoma.

UNLABELED USES Multiple sclerosis, condylomata acuminata.

CONTRAINDICATIONS Hypersensitivity to interferon alfa-2b or to any components of the product; colitis; severe cytopenia; pancreatitis; suicidal ideation; neonates.

CAUTIOUS USE Severe, preexisting cardiac, renal, or hepatic disease; pulmonary disease (e.g., COPD); diabetes mellitus, patients prone to ketoacidosis; coagulation disorders; severe myelosuppression; recent MI; previous dysrhythmias; history of depression or suicidal tendencies; pregnancy (category C), lactation.

ROUTE & DOSAGE

Hairy Cell Leukemia

Adult: **IM/Subcutaneous** 2 million units/m^2 3 times/wk

Kaposi's Sarcoma

Adult: **IM/Subcutaneous** 30 million units/m^2 3 times/wk

Condylomata Acuminata

Adult: **IM/Subcutaneous** 1 million units/m^2 3 times/wk

Chronic Hepatitis B or C

Adult: **Subcutaneous** 3 million units 3 times/wk × 18–24 mo

Malignant Melanoma

Adult: **IV** 20 million international units/m^2 daily for 5 days/wk × 4 wk; maintenance dose is 10 million international units/m^2 given subcutaneously weekly × 48 wk

Renal Impairment Dosage Adjustment

Not removed by dialysis

ADMINISTRATION
Subcutaneous Intramuscular
- Reconstitution: The final concentration with the amount of required diluent is determined by the condition being treated (see manufacturer's directions). Inject diluent (bacteriostatic water for injection) into interferon alfa-2b vial; gently agitate solution before withdrawing dose with a sterile syringe.
- Make sure reconstituted solution is clear and colorless to light yellow and free of particulate material; discard if there are particles or solution is discolored.
- Store vials and reconstituted solutions at 2°–8° C (36°–46° F); remains stable for 1 mo. Discard any remaining drug in reconstituted vials.

Intravenous

PREPARE: **IV Infusion:** Prepare **immediately** before use. Select the appropriate number of vials (i.e., 10, 18, or 50 million international units) of recombinant powder for injection and add to each the 1 mL of supplied diluent. Swirl gently to dissolve but do not shake. ▪ Further dilute by adding the required dose to 100 mL of NS. ▪ The final concentration should be less than 10 million international units/100 mL.
ADMINISTER: **IV Infusion:** Infuse over 20 min.
INCOMPATIBILITIES **Solution/ additive:** **Dextrose**-containing solutions.

ADVERSE EFFECTS (≥1%) **Body as a Whole:** *Flu-like syndrome (fever, chills) associated with myalgia and arthralgia,* leg cramps. **CNS:** Depression, nervousness, anxiety, confusion, *dizziness, fatigue,* somnolence, insomnia, altered mental states, ataxia, tremor, paresthesias, *headache.* **CV:** Hypertension, dyspnea, *hot flushes.* **Special Senses:** Epistaxis, pharyngitis, sneezing; abnormal vision. **GI:** Taste alteration, *anorexia,* weight loss, *nausea,* vomiting, stomatitis, *diarrhea,* flatulence. **Hematologic:** Mild <u>thrombocytopenia</u>, transient granulocytopenia, anemia, <u>*neutropenia,*</u> <u>leukemia</u>. **Skin:** Mild pruritus, mild alopecia, rash, dry skin, herpetic eruptions, nonherpetic cold sores, urticaria.

INTERACTIONS **Drug:** May increase **theophylline** levels; additive myelosuppression with ANTINEOPLASTICS, **zidovudine** may increase hematologic toxicity, increase **doxorubicin** toxicity, increase neurotoxicity with **vinblastine.** Use with **ribavirin** increases risk of hemolytic anemia; do not use in combination with **ribavirin** if CrCl less than 50 mL/min.

PHARMACOKINETICS **Peak:** 6–8 h. **Metabolism:** In kidneys. **Half-Life:** 6–7 h.

NURSING IMPLICATIONS
(see INTERFERON ALFA-2A)

Assessment & Drug Effects
- Assess hydration status; patient should be well hydrated, especially during initial stage of treatment and if vomiting or diarrhea occurs.
- Lab tests: Closely monitor CBC with differential and platelet counts.
- Monitor for and promptly report any of the following: Chest pain, dyspnea, ecchymoses, petechiae, fever, severe abdominal pain, or psychic disturbances.
- Assess for flu-like symptoms, which may be relieved by acetaminophen (if prescribed).
- Monitor level of GI distress and ability to consume fluids and food.

- Monitor mental status and alertness; implement safety precautions if needed.

Patient & Family Education
- Learn techniques for reconstitution and administration of drug.
- Seek medical attention promptly for any of the following: Chest pain, shortness of breath, easy bruising, persistent fever, decrease or loss of vision, severe abdominal pain, depression or suicidal ideation.
- Note: If flu-like symptoms develop, take acetaminophen as advised by prescriber and take interferon at bedtime.
- Use caution with hazardous activities until response to drug is known.
- Learn about adverse effects and notify prescriber about those that cause significant discomfort.

INTERFERON ALFACON-1
(in-ter-fer′on al′fa-con)

Infergen
Classifications: BIOLOGICAL RESPONSE MODIFIER; IMMUNOMODULATOR; INTERFERON; ANTIVIRAL
Therapeutic: ANTIVIRAL; INTERFERON
Prototype: Peg-interferon alfa-2a
Pregnancy Category: C

AVAILABILITY 9 mcg, 15 mcg injection

ACTION & *THERAPEUTIC EFFECT*
DNA recombinant Type 1 interferon. Its antiviral, antiproliferative, and natural killer (NK) cell activity is five times greater than interferon alpha-2b. *Effectiveness is measured by normalization of ALT level and serum HCV RNA less than 100 copies/mL. Type 1 interferons bind to the cell surface receptors inducing biological responses including antiviral, antiproliferative, and immunomodulatory activities.*

USES Treatment of chronic hepatitis C.

CONTRAINDICATIONS Hypersensitivity to alpha interferons or *E. coli* products; decompensated liver disease such as jaundice, ascites, etc.; suicidal ideations or severe psychiatric disorder; lactation.

CAUTIOUS USE History of severe psychiatric disorder, depression, or suicidal tendencies; preexisting cardiac disease, elderly, myelosuppression, previous hypersensitivity to interferon therapy; history of endocrine disorders; ophthalmic disorders or autoimmune disorders; pregnancy (category C). Safe use in children younger than 18 y not established.

ROUTE & DOSAGE

Chronic Hepatitis C
Adult: **Subcutaneous** 9 mcg 3 times/wk × 24 wk

ADMINISTRATION

Subcutaneous
- Allow at least 24 h to elapse between doses of interferon alfacon-1.
- Give only one dose per vial or per prefilled syringe. Enter each vial only once. Discard unused portion of a vial or prefilled syringe immediately.
- Initiate treatment only if acceptable baseline lab values are obtained: Platelet count 75 × 10⁹/L or greater, Hgb 100 g/L or greater, ANC 1500 × 10⁶/L or greater, serum creatinine less than 2 mg/dL, serum albumin 25 g/L or greater, bilirubin WNL, TSH, and T₄ WNL.
- Store vials and syringes at 2°–8° C (36°–46° F). Avoid direct sunlight and vigorous shaking.

ADVERSE EFFECTS (≥1%) **Body as a Whole:** *Asthenia, headache, fatigue, fever, chills, injection site reaction (pain, edema, hemorrhage, inflammation), pain, myalgia, arthralgia,* increased sweating. **CNS:** *Insomnia, depression, dizziness, paresthesia, nervousness, depression, anxiety,* agitation. **CV:** Hypertension, palpitation. **GI:** *Nausea, diarrhea, abdominal pain, anorexia, vomiting, dyspepsia,* constipation, flatulence, toothache, hemorrhoids, weight loss, hepatotoxicity. **Hematologic:** *Granulocytopenia,* <u>thrombocytopenia</u>, <u>leukopenia</u>, ecchymosis, lymphadenopathy, lymphocytosis. **Respiratory:** *Cough, bronchitis, dyspnea, pneumonia, rhinitis,* pharyngitis. **Skin:** *Alopecia, rash, dry skin, pruritus,* erythema. **Urogenital:** Dysmenorrhea, vaginitis, menstrual disorder.

INTERACTIONS No clinically significant interactions established.

PHARMACOKINETICS Peak: 24–36 h.

NURSING IMPLICATIONS

Assessment & Drug Effects
- Monitor for and report any of the following S&S immediately: Depression, suicidal ideation, suicide attempt, or other indications of psychiatric disturbance.
- Withhold drug and notify prescriber if symptoms of hepatic-decompensation such as jaundice or ascites develop. Withhold drug and notify prescriber if any other severe adverse reaction occurs.
- Lab tests: Baseline, 2 wk after initiation of therapy, and periodically thereafter: Platelet count, Hgb and Hct, WBC and ANC, serum creatinine, serum albumin, bilirubin, thyroid function, and triglyceride; periodic ALT to determine liver functions.

Patient & Family Education
- Report immediately any signs of psychiatric disturbance including depression, thoughts of suicide, nervousness, anxiety, agitation, apathy, or significant mood swings to prescriber.

INTERFERON BETA-1α
(in-ter-fer′on)
Avonex, Rebif
Classifications: BIOLOGICAL RESPONSE MODIFIER; IMMUNOMODULATOR; INTERFERON
Therapeutic: IMMUNOMODULATOR; INTERFERON
Prototype: Peg-interferon alfa-2a
Pregnancy Category: C

AVAILABILITY Avonex: 33 mcg vial; 30 mcg/5 mL prefilled syringe; **Rebif:** 22 mcg, 44 mcg vial

ACTION & THERAPEUTIC EFFECT
Interferon beta-1a is produced by recombinant DNA technology. Interferon beta-1a inhibits expression of pro-inflammatory cytokines including INF-G, thought to be a major factor in triggering the autoimmune reaction that leads to multiple sclerosis. It is believed that INF-G stimulates cytotoxic T-cells and causes degradation by macrophages' enzymes on the myelin sheath of neurons in the spinal cord. *Effective in improving time of onset of progression in disability; it was significantly longer in patients treated with interferon beta-1a.*

USES Relapsing-remitting multiple sclerosis.

CONTRAINDICATIONS Previous hypersensitivity to interferon-beta or human albumin, albumin hypersensitivity, hamster protein hypersensitivity; lactation (using **Rebif**).

CAUTIOUS USE Suicidal tendencies, depression, preexisting psychiatric disorders; bone marrow depression; cardiac disease; seizure disorders; thyroid disease; hepatic impairment; pregnancy (category C), lactation (using **Avonex**). Safety and efficacy in children younger than 18 y are not established.

ROUTE & DOSAGE

Multiple Sclerosis

Adult: **IM Avonex** 30 mcg qwk
Subcutaneous Rebif 44 mcg
3 times/wk

ADMINISTRATION

Intramuscular

- **Avonex:** Reconstitute single use Avonex vial (33 mcg of lyophilized powder) with 1.1 mL of supplied diluent and swirl gently to dissolve.
- Withdraw 1 mL for administration.
- Discard any residual drug as the product contains no preservatives.
- Use within 6 h of reconstitution.

Subcutaneous

- **Rebif:** Give at the same time each day (preferably in the late afternoon or evening) on the same three days of the week at least 48 h apart each week.
- Dose is usually titrated up from 8.8 mcg to 44 mcg three times a week over a 4-wk period.
- Inject subcutaneously using either a 22 or 44 mcg prefilled syringe. Discard any residual drug as the product contains no preservatives.
- Store unreconstituted vials or prefilled syringes at 2°–8° C (36°–46° F).
- May store for up to 30 days at room temperature up to 25° C (77° F). Do not use beyond expiration date.

ADVERSE EFFECTS (≥1%) **Body as a Whole:** Alopecia, myalgias, *flu-like syndrome,* anaphylaxis. **CNS:** Headache, *fever,* fatigue, lethargy, depression, somnolence, weakness, agitation, malaise, confusion or reduced ability to concentrate, anxiety, dementia, emotional lability, depersonalization, suicide attempts, worsening of psychiatric disorders. **CV:** Tachycardia, CHF (rare). **GI:** Nausea, vomiting, *diarrhea, hepatic injury.* **Hematologic:** Leukopenia, anemia, pancytopenia (rare), thrombocytopenia (rare). **Metabolic:** Hypocalcemia, elevated serum creatinine, elevated liver transaminases. **Skin:** Local skin necrosis at injection site, *pain at injection site.*

PHARMACOKINETICS Peak: Avonex 7.8–9.8 h; **Rebif** 16 h. **Metabolism:** Rapidly inactivated in body fluids and tissue. **Half-Life:** Avonex 8.6–10 h; **Rebif** 69 h.

NURSING IMPLICATIONS

Assessment & Drug Effects

- Withhold drug and notify prescriber if depression or suicidal ideation develops or if there is a worsening of psychiatric symptoms.
- Monitor patients with cardiac disease carefully for worsening cardiac function.
- Lab tests: Monitor periodically LFTs, renal function tests, routine blood chemistry, and CBC with differential, and platelet count. Monitor thyroid function tests q6mo with preexisting thyroid dysfunction or when clinically indicated.

Patient & Family Education

- Take a missed dose as soon as possible but not within 48 h of next scheduled dose.

Common adverse effects in *italic*, life-threatening effects underlined; generic names in **bold**; classifications in SMALL CAPS; ✦ Canadian drug name; ⊘ Prototype drug

- Learn about common adverse effects, especially flu-like syndrome (headache, fatigue, fever, rigors, chest pain, back pain, myalgia).
- Withhold drug and notify prescriber of depression or suicidal ideation or exacerbation of a pre-existing seizure disorder.
- Women who become pregnant should notify prescriber promptly.

INTERFERON BETA-1b

(in-ter-fer'on)

Betaseron, Extavia
Classifications: BIOLOGICAL RESPONSE MODIFIER; IMMUNOMODULATOR; INTERFERON; ANTINEOPLASTIC
Therapeutic: IMMUNOMODULATOR; ANTIVIRAL
Prototype: Peg-interferon alfa-2a
Pregnancy Category: C

AVAILABILITY 0.3 mg vial

ACTION & *THERAPEUTIC EFFECT*

Interferon beta-1b is a glycoprotein produced by recombinant DNA technique. It is thought to inhibit expression of pro-inflammatory cytokines including INF-G, thought to be a major factor in triggering the autoimmune reaction. It is believed that INF-G stimulates cytotoxic T-cells and causes degradation by macrophages' enzymes on the myelin sheath of neurons in the spinal cord. *Possess antiviral, antiproliferative, antitumor, and immunomodulatory activity. The effectiveness of interferon beta-1b for multiple sclerosis (MS) is based on the assumption that MS is an immunologically mediated illness.*

USES Relapsing and relapsing-remitting multiple sclerosis.

UNLABELED USES AIDS, AIDS-related Kaposi's sarcoma, metastatic renal cell carcinoma, malignant melanoma, cutaneous T-cell lymphoma, acute hepatitis C.

CONTRAINDICATIONS Previous hypersensitivity to interferon beta-1b or human albumin, mannitol hypersensitivity; history of depression or suicidal tendencies; lactation.
CAUTIOUS USE Suicidal/mental disorders especially chronic depression; seizures; cardiac disease; pregnancy (category C) but may cause a spontaneous abortion. Safety and efficacy in children younger than 18 y are not established.

ROUTE & DOSAGE

Multiple Sclerosis
Adult: **Subcutaneous** 0.25 mg (8 million international units) every other day

ADMINISTRATION

Subcutaneous
- Reconstitute by adding 1.2 mL of the supplied diluent (0.54% NaCl) to vial and gently swirl. Do NOT shake. The resultant solution contains 0.25 mg (8 million units)/mL.
- Discard reconstituted solution if it contains particulate matter or is discolored. Also discard unused solution.
- Rotate injection sites; use 27-gauge needle to administer drug.
- Store vials under refrigeration, 2°–8° C (36°–46° F) or at room temperature.

ADVERSE EFFECTS (≥1%) **CNS:** Headache, *fever,* fatigue, dizziness, lethargy, depression, somnolence, weakness, agitation, malaise, confusion or reduced ability to concentrate, anxiety, dementia, emotional

Common adverse effects in *italic*, life-threatening effects underlined; generic names in **bold**; classifications in SMALL CAPS; ♣ Canadian drug name; ⊙ Prototype drug

793

lability, depersonalization, suicide attempts. **CV:** Tachycardia, peripheral edema, CHF (rare). **GI:** Nausea, vomiting, *diarrhea.* **Hematologic:** Leukopenia, thrombocytopenia, anemia. **Metabolic:** Hypocalcemia, elevated serum creatinine, elevated liver transaminases, autoimmune hepatitis, hepatic failure. **Skin:** Local skin necrosis at injection site, rash, *pain at injection site.* **Body as a Whole:** Alopecia, myalgias, *flu-like syndrome.*

INTERACTIONS Drug: Zidovudine (AZT) levels are increased, resulting in toxicity.

PHARMACOKINETICS Absorption: About 50% absorbed from subcutaneous sites. **Distribution:** Penetrates intact blood–brain barrier poorly; crosses placenta; distributed into breast milk. **Metabolism:** Rapidly inactivated in body fluids and tissue.

NURSING IMPLICATIONS

Assessment & Drug Effects

- Monitor vital signs, neurologic status, and neuropsychiatric status frequently during therapy.
- Lab tests: Monitor LFTs at 1, 3, and 6 mo after initiation of therapy and as clinically warranted thereafter; monitor renal function, complete blood counts, and serum electrolytes periodically.
- Assess for and promptly treat flu-like symptom complex (fever, chills, myalgia, etc.).
- Assess injection sites; pain and redness are common reactions. Report tissue ulceration promptly.

Patient & Family Education

- Learn and understand potential adverse drug reactions.
- Learn proper technique for solution preparation and injection.
- Self-medicate with acetaminophen (if not contraindicated) if flu-like symptom complex develops.

- Avoid prolonged exposure to sunlight.
- Use caution when performing hazardous activities until response to drug is known.

INTERFERON GAMMA-1b
(in-ter-feer′on)
Actimmune
Classifications: BIOLOGICAL RESPONSE MODIFIER; IMMUNOMODULATOR; INTERFERON; ANTINEOPLASTIC
Therapeutic: IMMUNOMODULATOR; ANTINEOPLASTIC; ANTIVIRAL
Prototype: Peg-interferon alfa-2a
Pregnancy Category: C

AVAILABILITY 100 mcg (2 million IU)/ 0.5 mL vial

ACTION & *THERAPEUTIC EFFECT*
Immunomodulatory: Interferon gamma is produced by T-cells and natural killer (NK) cells after activation with immune or inflammatory stimuli. Interferon gamma stimulates macrophages to increase IL-12 and TNF-alpha production, which enhances interferon gamma synthesis. Interleukin-10 down-regulates interferon gamma production by NK and T-cells by preventing macrophage secretion of IL-12 and TNF-alpha. **Antineoplastic:** It also exerts antitumor effects by increasing expression of tumor suppressor genes and activating macrophages to lyse tumor cells. **Antiviral:** Has potent phagocyte-activating effects that include stimulating macrophages and generation of toxic oxygen metabolites (i.e., free radicals) capable of destroying virally infected cells. *Is a naturally occurring cytokine with antiviral, immunomodulatory, and antiproliferative activity. It enhances*

Common adverse effects in *italic*, life-threatening effects underlined; generic names in **bold;** classifications in SMALL CAPS; ♣ Canadian drug name; ⊘ Prototype drug

phagocyte function in chronic granulomatous disease also enhances osteoclast function in malignant osteopetrosis.

USES Chronic granulomatous disease, severe malignant osteopetrosis.
UNLABELED USES Idiopathic pulmonary fibrosis, refractory mycobacterium infection, ovarian cancer.

CONTRAINDICATIONS Hypersensitivity to interferon gamma or products derived from *E. coli;* suicidal ideation; pre-existing severe psychiatric disorder or the development of one; lactation.
CAUTIOUS USE Preexisting cardiac disease, CHF, cardiac arrhythmias; seizure disorders and compromised CNS function; myelosuppression; suicidal tendencies; older adults; pregnancy (category C). Safety and efficacy in infants younger than 1 y are not established.

ROUTE & DOSAGE

Chronic Granulomatous Disease, Osteopetrosis

Adult/Child: **Subcutaneous** BSA *0.5 m² or greater,* 50 mcg/m² 3 times weekly
Adult/Child: **Subcutaneous** BSA *0.5 m² or less,* 1.5 mcg/kg 3 times weekly

Idiopathic Pulmonary Fibrosis

Adult: **Subcutaneous** 180–200 mcg 3 times weekly

ADMINISTRATION

- Note: Pretreatment (4 h before) with acetaminophen is recommended to reduce headache, myalgia, and fever. Treatment should be continued 24 h postinjection.

Subcutaneous
- Do not shake vial. Inject subcutaneously undiluted into right or left deltoid area or anterior thigh area.
- Avoid intradermal or IV injection. Rotate injection sites.
- Store 2°–8° C (36°–46° F); do not freeze. Discard any unused portions or any vial left at room temperature for more than 12 h.

ADVERSE EFFECTS (≥1%) **Body as a Whole:** *Fever, fatigue, chills,* myalgia, arthralgia, night sweats. **CNS:** *Headache,* altered mental status, ataxia, confusion, dizziness, Parkinsonian symptoms, disorientation, seizures, hallucinations. **CV:** Heart block, heart failure, DVT, hypotension, <u>MI</u>, syncope, tachyarrhythmia. **GI:** *Nausea, vomiting, diarrhea.* **Hematologic:** <u>Leukopenia</u>, <u>thrombocytopenia</u>. **Respiratory:** Bronchospasm, interstitial pneumonitis, pulmonary embolism, tachypnea. **Skin:** Local skin necrosis at injection site, *pain at injection site, rash.* **Urogenital:** Reversible renal insufficiency.

INTERACTIONS Drug: Use cautiously with **aminophylline, fosphenytoin, phenytoin, theophylline, warfarin.**

PHARMACOKINETICS Absorption: 90% absorbed from subcutaneous site. **Peak:** 7 h. **Half-Life:** 5.9 h.

NURSING IMPLICATIONS

Assessment & Drug Effects

- Monitor CV status frequently. Report promptly severe hypotension and/or syncope.
- Monitor for and report S&S of infection.
- Lab tests: Baseline and at 3 mo CBC with differential and platelet counts; periodic LFTs, renal

Common adverse effects in *italic*, life-threatening effects <u>underlined</u>; generic names in **bold**; classifications in SMALL CAPS; ✦ Canadian drug name; ⊙ Prototype drug

795

function tests, and complete blood chemistry.

Patient & Family Education

- Report promptly: Skin rash, itching, unusual weakness or tiredness, chest pain or palpitations, or signs of an infection.
- Do not accept vaccination with a live vaccine during or for 3 mo following the end of therapy.

IODOQUINOL
(eye-oh-do-kwin'ole)

Yodoxin
Classification: AMEBICIDE
Therapeutic: AMEBICIDE
Prototype: Emetine
Pregnancy Category: C

AVAILABILITY 210 mg, 650 mg tablets

ACTION & THERAPEUTIC EFFECT Direct-acting (contact) amebicide. *Effective against both trophozoites and cyst forms of* Entamoeba histolytica *in intestinal lumen.*

USES Intestinal amebiasis and for asymptomatic passers of cysts. Commonly used either concurrently or in alternating courses with another intestinal amebicide.
UNLABELED USES Balantidiasis and *Acrodermatitis enteropathica;* traveler's diarrhea; shampoo preparation (Sebaquin) used for control of seborrheic dermatitis of scalp.

CONTRAINDICATIONS Hypersensitivity to any 8-hydroxyquinoline or to iodine-containing preparations or foods; hepatic or renal damage; lactation.
CAUTIOUS USE Severe thyroid disease; minor self-limiting problems; prolonged high-dosage therapy; preexisting optic neuropathy; pregnancy (category C).

ROUTE & DOSAGE

Intestinal Amebiasis

Adult: **PO** 650 mg t.i.d. for 20 days (max: 2 g/day); may repeat after a 2–3 wk drug-free interval
Child: **PO** 30–40 mg/kg/day in 2–3 divided doses for 20 days (max: 1.95 g/day); may repeat after a 2–3 wk drug-free interval

ADMINISTRATION

Oral

- Give drug after meals. If patient has difficulty swallowing tablet, crush and mix with applesauce.

ADVERSE EFFECTS (≥1%) **Body as a Whole:** Hypersensitivity (urticaria, pruritus). **CNS:** Headache, agitation, retrograde amnesia, vertigo, ataxia, peripheral neuropathy (especially in children); muscle pain, weakness usually below T12 vertebrae, dysesthesias especially of lower limbs, paresthesias, increased sense of warmth. **Special Senses:** Blurred vision, optic atrophy, optic neuritis, permanent loss of vision. **GI:** Nausea, vomiting, anorexia, abdominal cramps, diarrhea, constipation, rectal irritation and itching. **Skin:** Discoloration of hair and nails, acne, hair loss, urticaria, various forms of skin eruptions. **Hematologic:** Agranulocytosis (rare). **Endocrine:** Thyroid hypertrophy, iodism [generalized furunculosis (iodine toxiderma), skin eruptions, fever, chills, weakness].

DIAGNOSTIC TEST INTERFERENCE Iodoquinol can cause elevations of **PBI** and decrease of **I-131 uptake** (effects may last for several weeks to 6 mo even after discontinuation of therapy). *Ferric chloride test for PKU* (phenylketonuria) may yield false-positive results if iodo-quinol is present in urine.

PHARMACOKINETICS Absorption: Small amount from GI tract. **Elimination:** In feces.

NURSING IMPLICATIONS

Assessment & Drug Effects

- Monitor I&O ratio. Record characteristics of stools: Color, consistency, frequency, presence of blood, mucus, or other material.
- Note: Ophthalmologic examinations are recommended at regular intervals during prolonged therapy.
- Monitor and report immediately the onset of blurred or decreased vision or eye pain. Also report symptoms of peripheral neuropathy: Pain, numbness, tingling, or weakness of extremities.

Patient & Family Education

- Report any of the following: Skin rash, blurred vision, fever or other signs of infection.
- Complete full course of treatment. Stool needs to be examined again 1, 3, and 6 mo after termination of treatment.
- Note: Intestinal amebiasis is spread mainly by contaminated water, raw fruits or vegetables, flies, roaches, and hand-to-mouth transfer of infected feces. It is very important to wash hands after defecation and before eating.

IPECAC SYRUP

(ip'e-kak)

Ipecac Syrup
Classifications: ANTIDOTE; EMETIC
Therapeutic: EMETIC
Pregnancy Category: C

AVAILABILITY 15 mL, 30 mL doses

ACTION & THERAPEUTIC EFFECT Contains cephaeline (produces emesis) and emetine, a toxic alkaloid that is excreted slowly from the body. Emetine can cause potentially fatal cumulative toxicity with repeated use. It appears to inhibit protein synthesis and energy production in muscle tissue with resultant skeletal and cardiac muscle toxicity. *Acts locally on gastric mucosa and centrally on chemoreceptor trigger zone (CTZ) in the medulla to induce vomiting.*

USES Emergency emetic to remove unabsorbed ingested poisons.

CONTRAINDICATIONS Comatose, semicomatose, inebriated, deeply sedated patients; shock; depressed gag reflex; seizures, active or impending; treatment of ingested strong alkali, acids, or other corrosives, strychnine, petroleum distillates, volatile oils, or rapid-acting CNS depressants.

CAUTIOUS USE Impaired cardiac function; arteriosclerosis; cerebrovascular disease; head trauma; older adults; pregnancy (category C), lactation.

ROUTE & DOSAGE

Emergency Emesis

Adult: **PO** 30 mL followed by 1–2 240 mL (8 oz) glasses of water, may repeat once in 20 min if necessary
Child (older than 1y): **PO** 15 mL followed by 1–2 240 mL (8 oz) glasses of water, may repeat once in 20 min if necessary
Younger than 1y: 5–10 mL followed by 120–240 mL (4–8 oz) of water, may repeat once in 20 min if necessary

ADMINISTRATION

Oral

- Do not confuse with ipecac fluid extract, which is 14 times stronger

Common adverse effects in *italic*, life-threatening effects underlined; generic names in **bold**; classifications in SMALL CAPS; ♣ Canadian drug name; ⓟ Prototype drug

797

and has caused deaths when mistakenly given at the same dosage as ipecac syrup.

- Do not induce vomiting if victim is unconscious, semiconscious, or convulsing.
- Store in tight containers at temperature not exceeding 25° C (77° F).

ADVERSE EFFECTS (≥1%) **Body as a Whole:** Achy, stiff muscles, severe myopathy, convulsions, coma. **CV:** Cardiomyopathy, cardiotoxicity, cardiac arrhythmias, atrial fibrillation, tachycardia, chest pain, dyspnea, hypotension, fatal myocarditis. **GI:** Diarrhea, mild GI upset. If drug is not vomited but absorbed or if ipecac overdosage: persistent vomiting, gastroenteritis, bloody diarrhea, sensory disturbances, stomach cramps, tremor.

PHARMACOKINETICS Onset: 15–30 min. **Duration:** 25 min. **Elimination:** Metabolite can be detected in urine up to 60 days after excessive doses.

NURSING IMPLICATIONS

Assessment & Drug Effects

- Note: Emetic effect occurs in 15–30 min and continues for 20–25 min. If vomiting does not occur in 20–30 min, repeat dose once.
- Contact prescriber immediately if vomiting does not occur within 15–20 min after a second dose. Dosage should be recovered by gastric lavage and activated charcoal if necessary.
- Note: Ipecac syrup can be cardiotoxic if not vomited and allowed to be absorbed.
- Report immediately to prescriber if vomiting persists longer than 2–3 h after ipecac syrup is given.

Patient & Family Education

- Call an emergency room, poison control center, or prescriber before using ipecac syrup.

IPILIMUMAB
(ip'-i-lim'-ue-mab)

Yervoy
Classifications: MONOCLONAL ANTIBODY; ANTINEOPLASTIC
Therapeutic: ANTINEOPLASTIC
Prototype: Basilixumab
Pregnancy Category: C

AVAILABILITY 50 mg/10 mL and 200 mg/40 mL solution for injection

ACTION & THERAPEUTIC EFFECT
A recombinant, human monoclonal antibody thought to augment T-cell activation and proliferation thus enhancing T-cell mediated antitumor immune responses. Enhances the immune system's ability to seek out and destroy metastatic melanoma cells.

USES Treatment of unresectable or metastatic malignant melanoma.
CAUTIOUS USE Immune-mediated hepatitis, dermatitis, neuropathies, endocrinopathies, nephritis, pneumonitis, meningitis, pericarditis, uveitis, iritis, and hemolytic anemia; pregnancy (category C); lactation. Safety and effectiveness in children not established.

ROUTE & DOSAGE

Malignant Melanoma
Adult: **IV** 3 mg/kg q3wks for a total of 4 doses
Adverse Reaction Dosage Adjustment
Moderate immune-mediated adverse reaction or for symptomatic endocrinopathy: Withhold dose If the adverse reaction completely or partial resolves (Grade 0–1) and if patient is receiving less than 7.5 mg prednisone or equivalent per day: Resume at 3 mg/kg

Common adverse effects in *italic*, life-threatening effects underlined; generic names in **bold**; classifications in SMALL CAPS; ♣ Canadian drug name; ⊙ Prototype drug

q3wks until all 4 doses are given or 16 wks from first dose, whichever comes first.

If treatment course not completed within 16 wks: Permanently discontinue ipilimumab

If moderate adverse reactions are persistent: Permanently discontinue ipilimumab

If corticosteroid dose cannot be reduced to 7.5 mg prednisone; or if severe or life-threatening adverse reactions occur: Permanently discontinue ipilimumab

ADMINISTRATION

Intravenous

PREPARE: IV Infusion: Place vials at room temperature for 5 min. Do not shake. Withdraw required volume and add to an IV bag with enough NS or D5W to yield a final concentration of 1–2 mg/mL. Gently invert IV bag to mix.

ADMINISTER: IV Infusion: Give over 90 min through an IV line containing a low-protein-binding in-line filter. Flush line with NS or D5W after each dose.

INCOMPATIBILITIES Solution/additive: Do not mix ipilimumab, or administer as an infusion with other drugs or compounds.

- Store between 2°–8°C (36°–46° F). Protect from light. May store diluted solution for up to 24 h under refrigeration or at 20°–25°C (68°–77°F). Discard partially used vials.

ADVERSE EFFECTS (≥1%) Body as a Whole: *Fatigue.* **Hepatic:** AST or ALT greater 5 × the ULN or total bilirubin greater than 3 × the ULN. **GI:** *Diarrhea;* enterocolitis; GI

hemorrhage; GI perforation. **Skin:** *Pruritis, rash;* Stevens-Johnson syndrome, toxic epidermal necrolysis.

PHARMACOKINETICS Half-Life: 14.7 d.

NURSING IMPLICATIONS

Assessment & Drug Effects

- Use the *Nursing Immune-Mediated Adverse Reaction Symptom Checklist* (found on the product web site) to assess the patient.
- Monitor closely for and report immediately severe adverse reactions that may occur during or after (weeks to months) infusing. These include but are not limited to: Severe motor or sensory neuropathy; severe skin reactions; colitis with abdominal pain, fever, ileus, peritoneal signs, increased stool frequency (7 or more over baseline) or stool incontinence, need for intravenous hydration for more than 24 h, and GI hemorrhage.
- Lab tests, baseline and before each dose: LFTs and thyroid function tests; periodic renal function tests.
- Report immediately ALT elevations of more than 5 × the ULN or total bilirubin elevations more than 3 × the ULN.

Patient & Family Education

- Read the *Medication Guide* for ipilimumab prior to each infusion.
- Report immediately to prescribe any adverse reaction experienced during or after (weeks to months) infusion.
- Women should inform prescriber if they become pregnant.

IPRATROPIUM BROMIDE
(i-pra-troe′pee-um)
Atrovent, Atrovent HFA
Classifications: ANTICHOLINERGIC; ANTIMUSCARINIC; BRONCHODILATOR

Common adverse effects in *italic*, life-threatening effects underlined; generic names in **bold**; classifications in SMALL CAPS; ♣ Canadian drug name; ⊙ Prototype drug

799

Therapeutic: BRONCHODILATOR
Prototype: Atropine
Pregnancy Category: B

AVAILABILITY 0.02% solution for inhalation; 18 mcg inhaler; 0.03%, 0.06% nasal spray

ACTION & *THERAPEUTIC EFFECT*
Results in bronchodilation by inhibiting acetylcholine at its receptor sites, thereby blocking cholinergic bronchomotor tone (bronchoconstriction); also abolishes vagally mediated reflex bronchospasm triggered by such nonspecific agents as cigarette smoke, inert dusts, cold air, and a range of inflammatory mediators (e.g., histamine). *Produces local, site-specific effects on the larger central airways including bronchodilation and prevention of bronchospasms.*

USES Maintenance therapy in COPD including chronic bronchitis and emphysema; nasal spray for perennial rhinitis and symptomatic relief of rhinorrhea associated with the common cold.
UNLABELED USES Perennial nonallergic rhinitis.

CONTRAINDICATIONS Use as primary treatment for acute episodes; hypersensitivity to atropine, bromides, peanut oils, soy lecithin.
CAUTIOUS USE Narrow-angle glaucoma; prostatic hypertrophy, bladder neck obstruction; pregnancy (category B). Safe use in children 3 y or younger **(inhalation)** or 5 y or younger **(intranasal)** is not established.

ROUTE & DOSAGE

COPD
Adult: **Inhalation** 2 inhalations of MDI q.i.d. at no less than 4 h intervals (max: 12 inhalations in 24 h) **Nebulizer** 500 mcg (1 unit dose vial) q6–8h
Child (3–12 y): **Inhalation** 1–2 inhalations t.i.d. (max: 6/day) **Nebulizer** 125–250 mcg t.i.d.

Rhinitis
Adult (5 y or older): **Intranasal** 2 sprays of 0.03% each nostril b.i.d. or t.i.d.

Common Cold
Adult: **Intranasal** 2 sprays of 0.06% each nostril t.i.d. or q.i.d. up to 4 days

ADMINISTRATION
Intranasal/Inhalation/Nebulizer
- Demonstrate aerosol use and check return demonstration.
- Wait 3 min between inhalations if more than one inhalation per dose is ordered.
- Avoid contact with eyes.

ADVERSE EFFECTS (≥1%) **Special Senses:** Blurred vision (especially if sprayed into eye), difficulty in accommodation, acute eye pain, worsening of narrow-angle glaucoma. **GI:** Bitter taste, dry oropharyngeal membranes. With higher doses: Nausea, constipation. **Respiratory:** *Cough,* hoarseness, exacerbation of symptoms, drying of bronchial secretions, mucosal ulcers, epistaxis, nasal dryness. **Skin:** Rash, hives. **Urogenital:** Urinary retention. **CNS:** Headache.

PHARMACOKINETICS Absorption: 10% of inhaled dose reaches lower airway; approximately 0.5% of dose is systemically absorbed. **Peak:** 1.5–2 h. **Duration:** 4–6 h. **Elimination:** 48% of dose excreted in feces; less than 5% excreted in urine. **Half-Life:** 1.5–2 h.

Common adverse effects in *italic*, life-threatening effects underlined; generic names in **bold**; classifications in SMALL CAPS; ✦ Canadian drug name; ✪ Prototype drug

NURSING IMPLICATIONS

Assessment & Drug Effects

- Monitor respiratory status; auscultate lungs before and after inhalation.
- Report treatment failure (exacerbation of respiratory symptoms) to prescriber.

Patient & Family Education

- Note: This medication is not an emergency agent because of its delayed onset and the time required to reach peak bronchodilation.
- Review patient information sheet on proper use of nasal spray.
- Allow 30–60 sec between puffs for optimum results. Do not let medication contact your eyes.
- Wait 5 min between this and other inhaled medications. Check with prescriber about sequence of administration.
- Take medication only as directed, noting some leniency in number of puffs within 24 h. Supervise child's administration until certain all of dose is being administered.
- Rinse mouth after medication puffs to reduce bitter taste.
- Discuss changes in normal urinary pattern with the prescriber (more common in older adults).
- Call prescriber if you note changes in sputum color or amount, ankle edema, or significant weight gain.

IRBESARTAN

(ir-be-sar′tan)

Avapro

Classifications: ANGIOTENSIN II RECEPTOR ANTAGONIST; ANTIHYPERTENSIVE

Therapeutic: ANTIHYPERTENSIVE

Prototype: Losartan

Pregnancy Category: C first trimester; D second and third trimester

AVAILABILITY 75 mg, 150 mg, 300 mg tablets

ACTION & *THERAPEUTIC EFFECT*

Irbesartan is an angiotensin II receptor (type AT_1) antagonist. Irbesartan selectively blocks the binding of angiotensin II to the AT_1 receptors found in many tissues (e.g., vascular smooth muscle, adrenal glands), resulting in vasodilation of vascular smooth muscle. *This blocks vasoconstricting and aldosterone-secreting effects of angiotensin II, thus resulting in an antihypertensive effect.*

USES Hypertension, treatment of diabetic nephropathy in patients with hypertension and type 2 diabetes.

UNLABELED USES CHF.

CONTRAINDICATIONS Hypersensitivity to irbesartan, losartan, or valsartan; hypovolemia; pregnancy (category D second and third trimester), lactation.

CAUTIOUS USE Arterial stenosis of the renal artery, hepatic disease; severe CHF, African American patients; pregnancy (category C first trimester). Safe use in children not established.

ROUTE & DOSAGE

Hypertension

Adult: **PO** Start with 150 mg once daily, may increase to 300 mg/day

ADMINISTRATION

Oral

- Correct volume depletion prior to initiation of therapy to prevent hypotension. Titrate daily dose up to 300 mg; larger doses, however, are not likely to provide additional benefit.

ADVERSE EFFECTS (≥1%) **Body as a Whole:** Edema, fatigue, pain. **CNS:** Dizziness, headache, anxiety, nervousness. **CV:** Tachycardia, chest pain. **GI:** Diarrhea, dyspepsia,

---

Common adverse effects in *italic*, life-threatening effects <u>underlined</u>; generic names in **bold**; classifications in SMALL CAPS; ♣ Canadian drug name; ⊕ Prototype drug

801

nausea, vomiting, abdominal pain. **Respiratory:** Upper respiratory infection, cough, sinus disorder, pharyngitis, rhinitis. **Skin:** Rash. **Other:** UTI, hepatitis.

PHARMACOKINETICS Absorption: Rapidly absorbed from GI tract, 60–80% bioavailability. **Distribution:** 90% protein bound. **Metabolism:** In the liver primarily by CYP2C9. **Elimination:** Primarily in feces. **Half-Life:** 11–15 h.

NURSING IMPLICATIONS

Assessment & Drug Effects

- Monitor for therapeutic effectiveness: Maximum pressure lowering effect may not be evident for 6–12 wk; indicated by decreases in systolic and diastolic BP.
- Monitor BP periodically; trough readings, just prior to the next scheduled dose, should be made when possible.
- Lab tests: Monitor periodically BUN and creatinine, serum potassium, and CBC with differential.

Patient & Family Education

- Inform prescriber immediately if you become pregnant.
- Notify prescriber of episodes of dizziness, especially when making position changes.

IRINOTECAN HYDROCHLORIDE

(eye-ri-no′te-can)

Camptosar

Classifications: ANTINEOPLASTIC; CAMPTOTHECIN ANALOG
Therapeutic: ANTINEOPLASTIC
Prototype: Topotecan
Pregnancy Category: D

AVAILABILITY 20 mg/mL injection

ACTION & *THERAPEUTIC EFFECT*
Irinotecan is a camptothecin analog that displays antitumor activity by inhibiting the intranuclear enzyme topoisomerase I (DNA-gyrase). By inhibiting topoisomerase I, irinotecan and its active metabolite SN-38 cause double-stranded DNA damage during the synthesis (S) phase of DNA synthesis. *Irinotecan inhibits both DNA and RNA synthesis.*

USES Metastatic carcinoma of colon or rectum.

CONTRAINDICATIONS Previous hypersensitivity to irinotecan, topotecan, or other camptothecin analogs; acute infection, diarrhea, pregnancy (category D), lactation.

CAUTIOUS USE Gastrointestinal disorders, myelosuppression, renal or hepatic function impairment, history of bleeding disorders, previous cytotoxic or radiation therapy. Safe use in children not established.

ROUTE & DOSAGE

Metastatic Carcinoma

Adult: **IV** 125 mg/m² once weekly for 4 wk, then a 2-wk rest period (future courses may be adjusted to range from 50 to 150 mg/m² depending on tolerance; see complete prescribing information for specific dosage adjustment recommendations based on toxic effects)

Pharmacogenetic Dosage Adjustment

Patients with UGT1A1*28 allele have increased risk of side effects, start with decreased dose

ADMINISTRATION

Intravenous
Administer only after premedication (at least 30 min prior) with an antiemetic.

• Wash immediately with soap and water if skin contacts drug during preparation.

PREPARE: **IV Infusion:** Dilute the ordered dose in enough D5W (preferred) or NS to yield a concentration of 0.12–2.8 mg/mL. • Typical amount of diluent used is 250–500 mL.
ADMINISTER: **IV Infusion:** Infuse over 90 min. • Closely monitor IV site; if extravasation occurs, immediately flush with sterile water and apply ice.
INCOMPATIBILITIES **Y-site: Gemcitabine.**

• Store undiluted at 15°–30° C (59°–86° F) and protect from light. Use reconstituted solutions within 24 h.

ADVERSE EFFECTS (≥1%) **Body as a Whole:** *Asthenia, fever, pain,* chills, edema, abdominal enlargement, back pain. **CNS:** Headache, *insomnia, dizziness.* **CV:** Vasodilation/flushing. **GI:** *Diarrhea (early and late onset), dehydration, nausea, vomiting, anorexia, weight loss, constipation, abdominal cramping and pain,* flatulence, stomatitis, dyspepsia, increased alkaline phosphatase and AST. **Hematologic:** Leukopenia, neutropenia, *anemia.* **Respiratory:** *Dyspnea,* cough, rhinitis. **Skin:** *Alopecia,* sweating, rash.

INTERACTIONS Drug: ANTICOAGULANTS, ANTIPLATELET AGENTS, NSAIDS may increase risk of bleeding; **carbamazepine, phenytoin, phenobarbital** may decrease irinotecan levels. **Herbal:** St. John's wort may decrease irinotecan levels.

PHARMACOKINETICS Peak: 1 h. **Distribution:** Irinotecan is 30% protein bound; active metabolite SN-38 is 95% protein bound. **Metabolism:** In liver by carboxylesterase enzyme to active metabolite SN-38. **Elimination:** 10 h for SN-38; 20% excreted in urine. **Half-Life:** 10–20 h.

NURSING IMPLICATIONS

Assessment & Drug Effects

• Lab tests: Monitor WBC with differential, Hgb, and platelet count before each dose; monitor closely coagulation parameters especially with concurrent use of other drugs which affect these parameters.

• Lab tests: Monitor fluid and electrolyte balance closely during and after periods of diarrhea. Monitor LFTs, renal function tests and blood glucose periodically.

• Monitor for acute GI distress, especially early diarrhea (within 24 h of infusion), which may be preceded by diaphoresis and cramping, and late diarrhea (more than 24 h after infusion).

Patient & Family Education

• Learn about common adverse effects and measures to control or minimize when possible.

• Notify prescriber immediately when you experience diarrhea, vomiting, and S&S of infection. Diarrhea requires prompt treatment to prevent serious fluid and electrolyte imbalances.

IRON DEXTRAN
(i'ern dek'stran)

DexFerrum, INFeD, Proferdex
Classifications: BLOOD FORMER; IRON SUPPLEMENT; ANTIANEMIC
Therapeutic: ANTIANEMIC; IRON SUPPLEMENT
Prototype: Ferrous sulfate
Pregnancy Category: C

AVAILABILITY 50 mg elemental iron/mL

ACTION & *THERAPEUTIC EFFECT*
A complex of ferric hydroxide with dextran in solution for injection. Reticuloendothelial cells of liver, spleen, and bone marrow dissociate iron (ferric ion)

Common adverse effects in *italic*, life-threatening effects underlined; generic names in **bold**; classifications in SMALL CAPS; ♣ Canadian drug name; ☺ Prototype drug

803

from iron dextran complex. The released ferric ion combines with transferrin and is transported to bone marrow, where it is incorporated into hemoglobin. *Effective in replacement of iron needed in iron deficiency anemia, thus replenishing hemoglobin and depleted iron stores.*

USES Only in patients with clearly established iron deficiency anemia when oral administration of iron is unsatisfactory or impossible.

CONTRAINDICATIONS Hypersensitivity to the product; all anemias except iron-deficiency anemia; acute phase of infectious renal disease.

CAUTIOUS USE Rheumatoid arthritis, ankylosing spondylitis; renal disease; SLE; impaired hepatic function; cardiac disease; history of allergies or asthma; pregnancy (category C), lactation. Use not recommended in infants younger than 4 mo old.

ROUTE & DOSAGE

Iron Deficiency

Adult: **IM/IV** Dose is individualized and determined based on patient's weight and hemoglobin (see package insert); do not administer more than 100 mg (2 mL) of iron dextran within 24 h

Child: **IM/IV**
Weight less than 5 kg: No more than 0.5 mL (25 mg)/day;
Weight 5–10 kg: No more than 1 mL (50 mg)/day;
Weight greater than 10 kg: No more than 2 mL (100 mg)/day

ADMINISTRATION

Note: The multiple-dose vial is used ONLY for IM injections. It is not suitable for IV use because it contains a preservative (phenol).

Test Dose

- Give a test dose of 0.5 mL over a 5 min period before the first IM or IV therapeutic dose to observe patient's response to the drug. Have epinephrine (0.5 mL of a 1:1000 solution) immediately available for hypersensitivity emergency.
- Note: Although anaphylactic reactions (see Appendix F) usually occur within a few minutes after injection, it is recommended that 1 h or more elapse before remainder of initial dose is given following test dose.

Intramuscular

- Use the multiple-dose vial ONLY for IM injections. It is not suitable for IV use because it contains a preservative (phenol).
- Give injection only into the muscle mass in upper outer quadrant of buttock (never in the upper arm). In small child, use the lateral thigh. Use a 2- or 3-inch, 19- or 20-gauge needle. The Z-track technique is recommended. Use one needle to withdraw drug from container and another needle for injection.
- Note: If patient is receiving IM in standing position, patient should be bearing weight on the leg opposite the injection site; if in bed, patient should be in the lateral position with injection site uppermost.

Intravenous

Ensure that ONLY the vial for IV use is selected.

PREPARE: **Direct:** Give undiluted. **IV Infusion:** Dilute in 250–1000 mL of NS.

ADMINISTER: **Direct** *Test Dose:* A test dose is given before the first IV therapeutic dose. ▪*DexFerrum:* Give test dose of 25 mg (0.5 mL) slowly over 5 min. ▪*INFeD:* Give test dose over 30 sec. Wait 1–2 h and if no adverse reaction occurs, give the remainder of the first dose by IV infusion.

IV Infusion: Infuse at a rate not to exceed 50 mg (1 mL) or fraction thereof over 60 sec. Avoid rapid infusion.
INCOMPATIBILITIES **Solution/additive:** TPN.
▪ After infusion is completed, flush vein with 10 mL of NS.
▪ Have patient remain in bed for at least 30 min after IV administration to prevent orthostatic hypotension. Monitor BP and pulse.
▪ Store below 30° C (86° F) unless otherwise directed.

ADVERSE EFFECTS (≥1%) **Body as a Whole:** Hypersensitivity (urticaria, skin rash, allergic purpura, pruritus, fever, chills, dyspnea, arthralgia, myalgia; <u>anaphylaxis</u>). **CNS:** Headache, shivering, transient paresthesias, syncope, dizziness, <u>coma</u>, seizures. **CV:** *Peripheral vascular flushing (rapid IV), hypotension,* precordial pain or pressure sensation, tachycardia, <u>fatal cardiac arrhythmias, circulatory collapse</u>. **GI:** Nausea, vomiting, transient loss of taste perception, metallic taste, diarrhea, melena, abdominal pain, hemorrhagic gastritis, intestinal necrosis, hepatic damage. **Skin:** Sterile abscess and brown skin discoloration (IM site), local phlebitis (IV site), lymphadenopathy, *pain at IM injection site.* **Metabolic:** Hemosiderosis, metabolic acidosis, hyperglycemia, reactivation of quiescent rheumatoid arthritis, exogenous hemosiderosis. **Hematologic:** Bleeding disorder with severe toxicity.

DIAGNOSTIC TEST INTERFERENCE Falsely elevated ***serum bilirubin*** and falsely decreased ***serum calcium*** values may occur. Large doses of iron dextran may impart a brown color to serum drawn 4 h after iron administration. ***Bone scans*** involving Tc-99m diphosphonate have shown dense areas of activity along contour of iliac

crest 1–6 days after IM injections of iron dextran.

INTERACTIONS May decrease absorption of oral **iron, chloramphenicol** may decrease effectiveness of iron, a toxic complex may form with **dimercaprol.**

PHARMACOKINETICS Absorption: 60% from IM site by 3 days; 90% absorbed by 1–3 wk. **Distribution:** Crosses placenta; distributed into breast milk. **Metabolism:** In reticuloendothelial system. **Half-Life:** 6 h.

NURSING IMPLICATIONS

Assessment & Drug Effects
▪ Monitor for therapeutic effectiveness: Anticipated response to parenteral iron therapy is an average weekly hemoglobin rise of about 1 g/day. Peak levels are generally reached in about 4–8 wk.
▪ Note: Systemic reactions may occur over 24 h after parenteral iron has been administered. Large IV doses are associated with increased frequency of adverse effects.
▪ Lab tests: Periodic determinations of Hgb, Hct, and reticulocyte count should be made.

Patient & Family Education
▪ Do not take oral iron preparations when receiving iron injections.
▪ Eat foods high in iron and vitamin C.
▪ Notify prescriber of any of the following: Backache or muscle ache, chills, dizziness, fever, headache, nausea or vomiting, paresthesias, pain or redness at injection site, skin rash or hives, or difficulty breathing.

IRON SUCROSE INJECTION
(i′ron su′crose)
Venofer
Classifications: BLOOD FORMER; IRON REPLACEMENT; ANTIANEMIC

Common adverse effects in *italic*, life-threatening effects <u>underlined</u>; generic names in **bold**; classifications in SMALL CAPS; ♣ -Canadian drug name; ⊘ Prototype drug

805

Therapeutic: ANTIANEMIC; IRON DE-FICIENCY REPLACEMENT
Prototype: Ferrous sulfate
Pregnancy Category: B

AVAILABILITY 20 mg elemental iron/mL

ACTION & *THERAPEUTIC EFFECT* A complex of iron (ferric) (III) hydroxide in sucrose. It is dissociated by the reticuloendothelial system (RES) into iron and sucrose. Normal erythropoiesis depends on the concentration of iron and erythropoietin available in the plasma; both are decreased in renal failure. *Increases serum iron level in chronic renal failure patients, and results in increased hemoglobin level.*

USES Treatment of iron deficiency anemia in patients with chronic renal failure (with or without concurrent administration of erythropoietin).

CONTRAINDICATIONS Patients with iron overload, hypersensitivity to Venofer, or for anemia not caused by iron deficiency; hemochromatosis.
CAUTIOUS USE Patients with a history of hypotension; older adults, decreased renal, hepatic, or cardiac function; pregnancy (category B), lactation. Safety and effectiveness in infants or children are not established.

ROUTE & DOSAGE

Iron Deficiency Anemia

Adult: **IV Hemodialysis dependent (HDD-CKD):** 100 mg given at least 15 min per hemodialysis session (cumulative dose: 1000 mg). **Non-hemodialysis dependent (NDD-CKD):** 200 mg on 5 different occasions within the 14-day period. **Peritoneal dialysis dependent (PDD-CKD):** 300 mg on days 1 and 15, then 400 mg 14 days later

ADMINISTRATION

Intravenous

PREPARE: **Direct/Infusion: HDD-CKD:** Give direct IV undiluted or diluted immediately prior to infusion in a maxiumum of 100 mL NS. ▪ **NDD-CKD:** Give direct IV undiluted. ▪ **PDD-CKD:** Dilute 300–400 mg in a maxiumum of 250 mL of NS for infusion.
ADMINISTER: **Direct:** Give the undiluted solution slowly by direct IV over 2–5 min. **IV Infusion** ▪ Infusion diluted solution for **HDD-CKD patient** over at least 15 min and for **PDD-CKD patient** over 90 min. ▪ Avoid rapid infusion.
INCOMPATIBILITIES **Solution/additive:** Do not mix with other medications or parenteral nutrition solutions.

▪ Store unopened vials preferably at 25° C (77° F), but room temperature permitted. Discard unused portion in opened vial.

ADVERSE EFFECTS (≥1%) **Body as a Whole:** Fever, pain, asthenia, malaise, <u>anaphylactoid reactions</u>. **Cardiovascular:** *Hypotension,* chest pain, hypertension, hypervolemia. **Digestive:** Nausea, vomiting, diarrhea, abdominal pain, elevated liver function tests. **Musculoskeletal:** *Leg cramps,* muscle pain. **CNS:** Headache, dizziness. **Respiratory:** Dyspnea, pneumonia, cough. **Skin:** Pruritus, injection site reaction.

INTERACTIONS Drug: May reduce absorption of ORAL IRON PREPARATIONS.

PHARMACOKINETICS Peak: 4 wk. **Distribution:** Primarily to blood with some distribution to liver, spleen, bone marrow. **Metabolism:** Dissociated to iron and sucrose in reticuloendothelial system. **Elimination:** Sucrose is eliminated in urine, 5% of iron excreted in urine. **Half-Life:** 6 h.

NURSING IMPLICATIONS

Assessment & Drug Effects

- Withhold drug and notify prescriber when serum ferritin level equals or exceeds established guidelines.
- Stop infusion and notify prescriber for S&S overdosage or infusing too rapidly: Hypotension, edema; headache, dizziness, nausea, vomiting, abdominal pain, joint or muscle pain, and paresthesia.
- Lab tests: Periodic serum ferritin, transferrin saturation, Hct, and Hgb.
- Monitor patient carefully during the first 30 min after initiation of IV therapy for signs of hypersensitivity and anaphylactoid reaction (see Appendix F).

Patient & Family Education

- Report any of the following promptly: Itching, rash, chest pain, headache, dizziness, nausea, vomiting, abdominal pain, joint or muscle pain, and numbness and tingling.

ISOCARBOXAZID

(eye-soe-kar-box'a-zid)

Marplan

Classifications: ANTIDEPRESSANT; MONOAMINE OXIDASE INHIBITOR (MAOI)

Therapeutic: ANTIDEPRESSANT
Prototype: Phenelzine
Pregnancy Category: C

AVAILABILITY 10 mg tablets

ACTION & *THERAPEUTIC EFFECT*
Inhibits monoamine oxidase, the enzyme involved in the catabolism of catecholamine neurotransmitters and serotonin. *Effectiveness as an antidepressant is due to its inhibition of MAO.*

USES Symptomatic treatment of depressed patients refractory to or intolerant of TCAs or electroconvulsive therapy.

CONTRAINDICATIONS Hypersensitivity to MAO inhibitors; pheochromocytoma; children (younger than 16 y); older adults (over 60 y) or debilitated patients; cardiac arrhythmias, hypertension, CVA; severe renal or hepatic impairment; history of headache; increased intracranial pressure, surgery; stroke, head trauma; suicidal ideation; lactation.
CAUTIOUS USE Hyperthyroidism, parkinsonism, epilepsy, schizophrenia; bipolar disorder; psychosis; suicidal risks; dental work; pregnancy (category C).

ROUTE & DOSAGE

Refractory Depression
Adult: **PO** 10–30 mg/day in 1–3 divided doses (max: 30 mg/day)

ADMINISTRATION

Oral

- Note: Dosage is individualized on the basis of patient response. Lowest effective dosage should be used.
- Store in a tight, light-resistant container.

ADVERSE EFFECTS (≥1%) CNS: Dizziness, light-headedness, tiredness, weakness, *drowsiness,* vertigo, headache, *overactivity,* hyperreflexia, muscle twitching, tremors, mania hypomania, *insomnia,* confusion, memory impairment. **CV:** *Orthostatic hypotension,* paradoxical hypertension, palpitation, tachycardia, other arrhythmias. **Special Senses:** *Blurred vision,* nystagmus, glaucoma. **GI:** Increased appetite, weight gain, *nausea,* diarrhea, *constipation, anorexia,* black tongue, *dry mouth,* abdominal pain. **Urogenital:** Dysuria, *urinary retention,* incontinence, sexual disturbances. **Body as a Whole:** Peripheral edema, excessive sweating, chills, skin rash, hepatitis, jaundice.

Common adverse effects in *italic*, life-threatening effects underlined; generic names in **bold**; classifications in SMALL CAPS; ♦ Canadian drug name; ⊘ Prototype drug

807

INTERACTIONS Drug: TRICYCLIC ANTIDEPRESSANTS, **fluoxetine,** AMPHETAMINES, **ephedrine, reserpine, guanethidine, buspirone, methyldopa, dopamine, levodopa, tryptophan** may precipitate hypertensive crisis, headache, or hyperexcitability; **alcohol** and other CNS DEPRESSANTS compound CNS depressant effects; **meperidine** can cause fatal cardiovascular collapse; ANESTHETICS exaggerate hypotensive and CNS depressant effects; **metrizamide** increases risk of seizures; compounds hypotensive effects of DIURETICS and other ANTIHYPERTENSIVE AGENTS. **Food:** All **tyramine**-containing foods (aged cheeses, processed cheeses, sour cream, wine, champagne, beer, pickled herring, anchovies, caviar, shrimp, liver, dry sausage, figs, raisins, overripe bananas or avocados, chocolate, soy sauce, bean curd, yeast extracts, yogurt, papaya products, meat tenderizers, broad beans) may precipitate hypertensive crisis. **Herbal: Ginseng, ephedra, ma huang, St. John's wort** may precipitate hypertensive crisis.

PHARMACOKINETICS Duration: Up to 2 wk. **Metabolism:** In liver.

NURSING IMPLICATIONS

Assessment & Drug Effects

- Monitor for and report promptly signs of clinical deterioration or suicidal ideation. Children and adolescents are at particular risk.
- Monitor for therapeutic effectiveness: May be apparent within 1 wk or less, but in some patients there may be a time lag of 3–4 wk before improvement occurs.
- Monitor BP. Monitor for orthostatic hypotension by evaluating BP with patient recumbent and standing.
- Check for peripheral edema daily and monitor weight several times weekly.
- Note: Toxic symptoms from overdosage or from ingestion of contraindicated substances (e.g., foods high in tyramine) may occur within hours.
- Monitor I&O and bowel elimination patterns.

Patient & Family Education

- Monitor closely behavior of children and adolescents; report immediately unusual changes in behavior or suicidal ideation.
- Make position changes slowly and in stages; lie down or sit down if faintness occurs.
- Use caution when performing potentially hazardous activities.
- Consult prescriber before self-medicating with OTC agents (e.g., cough, cold, hay fever, or diet medications).
- Avoid alcohol and excessive caffeine-containing beverages and tryptophan and tyramine-containing foods including cheeses, yeast, meat extracts, smoked or pickled meat, poultry, or fish, fermented sausages, and overripe fruit.

ISOMETHEPTENE/ DICHLORALPHENAZONE/ ACETAMINOPHEN

(i-so-meth′ep-tene/di-chlor-al-phen′a-zone/a-cet′a-min-o-phen)
Isopap, Duradrin, Midrin, Migratine
Classifications: SYMPATHOMIMETIC; NONNARCOTIC ANALGESIC
Therapeutic: ANTIMIGRAINE; NONNARCOTIC ANALGESIC
Pregnancy Category: C
Controlled Substance: Schedule C-IV

AVAILABILITY 65 mg; **isomeptene mucate,** 100 mg; **dichloralphenazone,** 325 mg; **APAP** capsules

ACTION & *THERAPEUTIC EFFECT*
Isometheptene is a sympathomimetic amine that acts by constricting cranial and cerebral arterioles. Isometheptene relieves vascular headaches. Dichloralphenazone is a mild sedative that helps reduce headache pain. Acetaminophen is a mild analgesic. *Effective as a mild sedative, reduces headache pain as well as being a mild analgesic.*

USES Relief for tension, vascular, and migraine headaches.

CONTRAINDICATIONS Patients with glaucoma; severe renal disease, organic heart disease; hepatic disease; concurrent MAO inhibitors.
CAUTIOUS USE Hypertension; peripheral vascular disease, and recent cardiovascular attacks; older adults; pulmonary disease; pregnancy (category C), lactation.

ROUTE & DOSAGE

Tension Headache
Adult: **PO** 1–2 capsules q4h up to 8 capsules/24 h
Migraine Headache
Adult: **PO** 2 capsules at onset, then 1 capsule qh until relief (max: 5 capsules/12 h)

ADMINISTRATION
Oral
▪ Do not give this drug to anyone who is concurrently using an MAOI. Allow 14 days to elapse between discontinuation of the MAOI and administration of this drug.

▪ Do not give more than 8 capsules in a 24 h period.
▪ Store at 15°–30° C (59°–86° F) in a dry place.

ADVERSE EFFECTS (≥1%) **CNS:** Transient dizziness. **GI:** Acetaminophen hepatotoxicity. **Skin:** Rash.

INTERACTIONS Drug: MAOIS may cause hypertensive crisis; other **acetaminophen**-containing drugs (including OTC) may increase risk of hepatotoxicity.

PHARMACOKINETICS Absorption: Rapidly absorbed. **Metabolism:** Dichloralphenazone is metabolized to chloral hydrate and antipyrine. See ACETAMINOPHEN and CHLORAL HYDRATE for more detail. **Elimination:** Renal and hepatic. **Half-Life:** 12 h.

NURSING IMPLICATIONS
Assessment & Drug Effects
▪ Monitor BP closely with preexisting hypertension.
▪ Monitor lower extremity perfusion with a history of PVD.

Patient & Family Education
▪ Avoid, or moderate, alcohol use while taking this drug.
▪ Do not drive or engage in other potentially hazardous activities until response to drug is known.
▪ Report any decrease in tolerance to walking if you have a history of PVD.

ISONIAZID (ISONICOTINIC ACID HYDRAZIDE) 🅿

(eye-soe-nye'a-zid)
INH, Isotamine ✦**, Laniazid, Nydrazid, PMS Isoniazid** ✦
Classifications: ANTI-INFECTIVE; ANTITUBERCULOSIS
Therapeutic: ANTITUBERCULOSIS
Pregnancy Category: C

Common adverse effects in *italic*, life-threatening effects underlined; generic names in **bold**; classifications in SMALL CAPS; ✦ Canadian drug name; 🅿 Prototype drug

809

AVAILABILITY 50 mg, 100 mg, 300 mg tablets; 50 mg/5 mL syrup; 100 mg/mL injection

ACTION & *THERAPEUTIC EFFECT* Hydrazide of isonicotinic acid with highly specific action against *Mycobacterium tuberculosis.* Postulated to act by interfering with biosynthesis of bacterial proteins, nucleic acid, and lipids. *Exerts bacteriostatic action against actively growing tubercle bacilli; may be bactericidal in higher concentrations.*

USES Treatment of all forms of active tuberculosis caused by susceptible organisms and as preventive in high-risk persons (e.g., household members, persons with positive tuberculin skin test reactions). May be used alone or with other tuberculostatic agents.

UNLABELED USES Treatment of atypical mycobacterial infections; tuberculous meningitis; action tremor in multiple sclerosis.

CONTRAINDICATIONS History of isoniazid-associated hypersensitivity reactions, including hepatic injury; acute liver damage of any etiology; lactation.

CAUTIOUS USE Chronic liver disease; HIV infection; hepatitis; severe renal dysfunction; history of convulsive disorders; chronic alcoholism; persons older than 50 y; pregnancy (category C).

ROUTE & DOSAGE

Treatment of Active Tuberculosis
Adult: **PO/IM** 5 mg/kg (max: 300 mg/day)
Child: **PO/IM** 10–20 mg/kg (max: 300–500 mg/day)

Preventive Therapy
Adult: **PO** 300 mg/day

Child: **PO** 10 mg/kg up to 300 mg/day or 15 mg/kg 3 times/wk

ADMINISTRATION

Oral
- Give on an empty stomach at least 1 h before or 2 h after meals. If GI irritation occurs, drug may be taken with meals.

Intramuscular
- Note: Isoniazid solution for IM injection tends to crystallize at low temperatures; if this occurs, solution should be allowed to warm to room temperature to redissolve crystals before use.
- Give deep into a large muscle and rotate injection sites; local transient pain may follow IM injections.
- Store in tightly closed, light-resistant containers.

ADVERSE EFFECTS (≥1%) **Body as a Whole:** Drug-related fever, rheumatic and lupus erythematosus-like syndromes, irritation at injection site; hypersensitivity (fever, chills, skin eruption, vasculitis). **CNS:** *Paresthesias, peripheral neuropathy,* headache, unusual tiredness or weakness, tinnitus, dizziness, hallucinations. **Special Senses:** Blurred vision, visual disturbances, optic neuritis, atrophy. **GI:** Nausea, vomiting, epigastric distress, dry mouth, constipation; hepatotoxicity (*elevated AST, ALT;* bilirubinemia, jaundice, <u>hepatitis</u>). **Hematologic:** <u>Agranulocytosis</u>, hemolytic or <u>aplastic anemia</u>, <u>thrombocytopenia</u>, eosinophilia, methemoglobinemia. **Metabolic:** Decreased vitamin B_{12} absorption, pyridoxine (vitamin B_6) deficiency, pellagra, gynecomastia, hyperglycemia, glycosuria, hyperkalemia, hypophosphatemia, hypocalcemia, acetonuria, metabolic acidosis, proteinuria. **Other:** Dyspnea, urinary retention (males).

Common adverse effects in *italic*, life-threatening effects <u>underlined</u>; generic names in **bold**; classifications in SMALL CAPS; ♣ Canadian drug name; ⊕ Prototype drug

DIAGNOSTIC TEST INTERFERENCE Isoniazid may produce false-positive results using *copper sulfate tests* (e.g., *Benedict's solution, Clinitest*) but not with glucose oxidase methods (e.g., *Clinistix, Dextrostix, TesTape*).

INTERACTIONS Drug: Cycloserine, ethionamide enhance CNS toxicity; may increase **phenytoin** levels, resulting in toxicity; ALUMINUM-CONTAINING ANTACIDS decrease GI absorption; **disulfiram** may cause coordination difficulties or psychotic reactions; **alcohol** increases risk of hepatotoxicity. **Food:** Food decreases rate and extent of isoniazid absorption; should be taken 1 h before meals.

PHARMACOKINETICS Absorption: Readily from GI tract; food may reduce rate and extent of absorption. **Peak:** 1–2 h. **Distribution:** Distributed to all body tissues and fluids including the CNS; crosses placenta. **Metabolism:** Inactivated by acetylation in liver. **Elimination:** 75–96% in urine in 24 h; excreted in breast milk. **Half-Life:** 1–4 h.

NURSING IMPLICATIONS

Assessment & Drug Effects

- Monitor for therapeutic effectiveness: Evident within the first 2–3 wk of therapy. Over 90% of patients receiving optimal therapy have negative sputum by the sixth month.
- Withhold drug and notify prescriber immediately of a hypersensitivity reaction; generally occurs within 3–7 wk after initiation of therapy.
- Perform appropriate susceptibility tests before initiation of therapy and periodically thereafter to detect possible bacterial resistance.

- Lab tests: Monitor LFTs periodically. Isoniazid hepatitis (sometimes fatal) usually develops during the first 3–6 mo of treatment, but may occur at any time during therapy; much more frequent in patients 35 y or older, especially in those who ingest alcohol daily.
- Monitor for and report promptly signs of hepatic toxicity (see Appendix F).
- Monitor for and report promptly signs of peripheral neuritis (e.g., paresthesias of feet and hands with numbness, tingling, burning).
- Monitor BP during period of dosage adjustment. Some experience orthostatic hypotension; therefore, caution against rapid positional changes.
- Monitor diabetics for loss of glycemic control.
- Check weight at least twice weekly under standard conditions.

Patient & Family Education

- Report promptly any of the following signs of liver toxicity: Unexplained weakness, nausea/vomiting, loss of appetite, dark urine, jaundice, clay-colored stools.
- Avoid or at least reduce alcohol intake while on isoniazid therapy because of increased risk of hepatotoxicity.

ISOPROTERENOL HYDROCHLORIDE ⓟ

(eye-soe-proe-ter'e-nole)

Isuprel

Classifications: BETA-ADRENERGIC AGONIST; BRONCHODILATOR; CARDIAC STIMULATOR

Therapeutic: BRONCHODILATOR; ANTIARRHYTHMIC; CARDIAC STIMULATOR

Pregnancy Category: C

Common adverse effects in *italic*, life-threatening effects underlined; generic names in **bold**; classifications in SMALL CAPS; ♣ Canadian drug name; ⓟ Prototype drug

811

ISOPROTERENOL HYDROCHLORIDE

AVAILABILITY 0.2 mg/mL, 0.02 mg/mL injection

ACTION & *THERAPEUTIC EFFECT*
Synthetic sympathomimetic amine that acts directly on beta$_1$- and beta$_2$-adrenergic receptors that relaxes bronchospasm, and, by increasing ciliary motion, facilitates expectoration of pulmonary secretions. Induces stimulation of beta$_1$-adrenergic receptors and results in increased cardiac output and cardiac workload by increasing strength of contraction through positive inotropic and chronotropic effects on the heart. It also shortens AV conduction time and its refractory period in patients with heart block. *Reverses bronchospasm and facilitates removal of bronchial secretion. Increases cardiac output and cardiac workload. Also has antiarrhythmic properties by affecting AV node conduction.*

USES Reversible bronchospasm induced by anesthesia. As cardiac stimulant in cardiac arrest, carotid sinus hypersensitivity, cardiogenic and bacteremic shock, Adams-Stokes syndrome, or ventricular arrhythmias. Used in treatment of shock that persists after replacement of blood volume.

UNLABELED USES Treatment of status asthmaticus in children.

CONTRAINDICATIONS Preexisting cardiac arrhythmias associated with tachycardia; tachycardia caused by digitalis intoxication, central hyperexcitability, cardiogenic shock secondary to coronary artery occlusion and MI; ventricular fibrillation.

CAUTIOUS USE Sensitivity to sympathomimetic amines; older adult and debilitated patients, hypertension, coronary insufficiency and other cardiovascular disorders, angina; renal dysfunction, hyperthyroidism, diabetes, prostatic hypertrophy, glaucoma, tuberculosis; pregnancy (category C), lactation.

ROUTE & DOSAGE

Bronchospasm
Adult: **IV** 0.01–0.02 mg prn
Cardiac Arrhythmias/Cardiac Resuscitation
Adult: **IV** 0.02–0.06 mg bolus, followed by 5 mcg/min infusion
Child: **IV** 0.1 mcg/kg/min by continuous infusion
Shock/Hypoperfusion
Adult: **IV** 0.5–5 mcg/min

ADMINISTRATION
Intravenous
Note: Maximum concentration on IV solution for both adults and children: 20 mcg/mL (0.02 mg/mL)

PREPARE: **Direct IV Injection for Adult with AV Block/Arrhythmia/Bradycardia/Cardiac Arrest:** Dilute 1 mL (0.2) of 1:5000 solution with 9 mL NS or D5W to produce a 1:50,000 (0.02 mg/mL) solution or use 1:50,000 solution undiluted. **Continuous Infusion for Adult with AV Block/Arrhythmia/Bradycardia/Cardiac Arrest:** Dilute 10 mL (2 mg) of 1:5000 solution in 500 mL D5W to produce a 1:250,000 (4 mcg/mL) solution. **IV Infusion for Adult with Shock Hypoperfusion:** Dilute 5 mL (1 mg) of 1:5000 solution in 500 mL D5W to produce a 1:500,000 (2 mcg/mL) solution. **Direct IV Injection for Adult with Bronchospasm:** Dilute

812

Common adverse effects in *italic*, life-threatening effects underlined; generic names in **bold**; classifications in SMALL CAPS; ♣ Canadian drug name; ◉ Prototype drug

1 mL (0.2 mg) of 1:5000 solution with 9 mL NS or D5W to produce a 1:50,000 solution undiluted. **Continuous Infusion for Child with AV Block/Bradycardia:** Dilute to a range of 4–12 mcg/mL in 100 mL of D5W or NS.

ADMINISTER: **Direct IV for Adult/Child:** Give at a rate of 0.2 mg or fraction thereof over 1 min. ▪ Flush with 15–20 mL NS. **Continuous IV Infusion for Adult/Child:** Rate is adjusted according to patient response. Infusion rate is generally decreased or infusion may be temporarily discontinued if heart rate exceeds 110 bpm, because of the danger of precipitating arrhythmias. ▪Microdrip or constant-infusion pump is recommended to prevent sudden influx of large amounts of drug. ▪ IV administration is regulated by continuous ECG monitoring. ▪ Patient **must be** observed and response to therapy **must be** monitored continuously.

INCOMPATIBILITIES **Solution/additive: Aminophylline, diazepam, furosemide, sodium bicarbonate.**

▪Isoproterenol solutions lose potency with standing. ▪ Discard if precipitate or discoloration is present.

ADVERSE EFFECTS (≥1%) **CNS:** Headache, mild tremors, nervousness, anxiety, insomnia, excitement, fatigue. **CV:** Flushing, palpitations, tachycardia, unstable BP, anginal pain, <u>ventricular arrhythmias</u>. **GI:** Swelling of parotids (prolonged use), bad taste, buccal ulcerations (sublingual administration), nausea. **Other:** Severe prolonged asthma attack, sweating, bronchial irritation and edema. **Acute Poisoning:** Overdosage, especially after excessive use of aerosols (*tachycardia*, palpitations, nervousness, nausea, vomiting).

INTERACTIONS Drug: Epinephrine and other SYMPATHOMIMETIC AMINES, TRICYCLIC ANTIDEPRESSANTS increase effects and cause cardiac toxicity. HALOGENATED GENERAL ANESTHETICS exacerbate arrhythmias; while BETA-BLOCKERS antagonize effects.

PHARMACOKINETICS Absorption: Rapidly from parenteral administration. **Onset:** Immediate. **Metabolism:** Metabolized by COMT in liver, lungs, and other tissues. **Elimination:** 40–50% unchanged in urine.

NURSING IMPLICATIONS

Assessment & Drug Effects

▪ Check pulse before and during IV administration. Rate greater than 110 usually indicates need to slow infusion rate or discontinue infusion. Consult prescriber for guidelines.
▪ Incidence of arrhythmias is high, particularly when drug is administered IV to patients with cardiogenic shock or ischemic heart disease, digitalized patients, or to those with electrolyte imbalance.
▪ Note: Tolerance to bronchodilating effect and cardiac stimulant effect may develop with prolonged use.
▪ Note: Once tolerance has developed, continued use can result in serious adverse effects including rebound bronchospasm.

ISOSORBIDE DINITRATE
(eye-soe-sor′bide)

Coronex ♦, Dilatrate-SR, Iso-Bid, Isordil, Novosorbide ♦
Classification: NITRATE VASODILATOR
Therapeutic: VASODILATOR; ANTIANGINAL
Prototype: Nitroglycerin
Pregnancy Category: C

Common adverse effects in *italic*, life-threatening effects <u>underlined</u>; generic names in **bold**; classifications in SMALL CAPS; ♦ Canadian drug name; ⊘ Prototype drug

813

AVAILABILITY 2.5 mg, 5 mg, 10 mg sublingual tablets; 5 mg, 10 mg chewable tablets; 5 mg, 10 mg, 20 mg, 30 mg, 40 mg tablets; 40 mg sustained release tablets, capsules

ACTION & *THERAPEUTIC EFFECT*
Relaxes vascular smooth muscle with resulting vasodilation. Dilation of peripheral blood vessels tends to cause peripheral pooling of blood, decreased venous return to heart, and decreased left ventricular end-diastolic pressure, with consequent reduction in myocardial oxygen consumption. *Has an antianginal effect as a result of vasodilation of the coronary arteries.*

USES Relief of acute anginal attacks and for management of long-term angina pectoris.
UNLABELED USES Alone or in combination with a cardiac glycoside or with other vasodilators (e.g., hydralazine, prazosin, for refractory CHF; diffuse esophageal spasm without gastroesophageal reflux and heart failure).

CONTRAINDICATIONS Hypersensitivity to nitrates or nitrites; severe anemia; hyperthyroidism; head trauma; increased intracranial pressure; recent MI; GI disease; children.
CAUTIOUS USE Glaucoma, hypotension, hypovolemia; hyperthyroidism; hepatic disease; elderly; pregnancy (category C), lacation.
Extended release form: GI disease such as hypermotility or malabsorption syndrome.

ROUTE & DOSAGE

Angina Prophylaxis
Adult: **PO** 2.5–30 mg q.i.d. a.c. and at bedtime; Sublingual tablet 2.5–10 mg q4–6h; Chewable tablet 5–30 mg chewed q2–3h; Sustained release tablets 40 mg q6–12h
Acute Anginal Attack
Adult: **PO** Sublingual tablet 2.5–10 mg q2–3h prn; Chewable tablet 5–30 mg chewed prn for relief

ADMINISTRATION
Oral
- Do not confuse with isosorbide, an oral osmotic diuretic.
- Give regular oral forms on an empty stomach (1 h a.c. or 2 h p.c.). If patient complains of vascular headache, however, it may be taken with meals.
- Advise patient not to eat, drink, talk, or smoke while sublingual tablet is under tongue.
- Instruct patient to place sublingual tablet under tongue at first sign of an anginal attack. If pain is not relieved, repeat dose at 5–10 min intervals to a maximum of 3 doses. If pain continues, notify prescriber or go to nearest hospital emergency room.
- Chewable tablet **must be** thoroughly chewed before swallowing.
- Do not crush sustained release form. It **must be** swallowed whole.
- Have patient sit when taking rapid-acting forms of isosorbide dinitrate (sublingual and chewable tablets) because of the possibility of faintness.
- Store in tightly closed container in a cool, dry place. Do not expose to extremes of temperature.

ADVERSE EFFECTS (≥1%) **Body as a Whole:** Hypersensitivity reaction, paradoxical increase in anginal pain, methemoglobinemia (overdose). **CNS:** Headache, dizziness, weakness, *light-headedness,* restlessness. **CV:** Palpitation, postural

Common adverse effects in *italic*, life-threatening effects underlined; generic names in **bold;** classifications in SMALL CAPS; ♣ Canadian drug name; ⊕ Prototype drug

hypotension, tachycardia. **GI:** Nausea, vomiting. **Skin:** *Flushing,* pallor, perspiration, rash, exfoliative dermatitis.

INTERACTIONS Drug: Alcohol may enhance hypotensive effects and lead to cardiovascular collapse; ANTIHYPERTENSIVE AGENTS, PHENOTHIAZINES add to hypotensive effects.

PHARMACOKINETICS Absorption: Significant first pass metabolism with PO absorption, with 10–90% reaching systemic circulation. **Onset:** 2–5 min SL; within 1 h regular tabs; within 3 min chewable tabs; 30 min sustained release tabs. **Duration:** 1–2 h SL; 4–6 h regular tabs; 0.5–2 h chewable tabs; 6–8 h sustained release tabs. **Metabolism:** In liver. **Elimination:** 80–100% in urine within 24 h.

NURSING IMPLICATIONS

Assessment & Drug Effects

- Monitor effectiveness of drug in relieving angina.
- Note: Headaches tend to decrease in intensity and frequency with continued therapy but may require administration of analgesic and reduction in dosage.
- Note: Chronic administration of large doses may produce tolerance and thus decrease effectiveness of nitrate preparations.

Patient & Family Education

- Make position changes slowly, particularly from recumbent to upright posture, and dangle feet and ankles before walking.
- Lie down at the first indication of light-headedness or faintness.
- Keep a record of anginal attacks and the number of sublingual tablets required to provide relief.
- Do not drink alcohol because it may increase possibility of light-headedness and faintness.

ISOSORBIDE MONONITRATE
(eye-soe-sor'bide)

Ismo, Imdur, Monoket
Classification: NITRATE VASODILATOR
Therapeutic: ANTIANGINAL
Prototype: Nitroglycerin
Pregnancy Category: C

AVAILABILITY 10 mg, 20 mg tablets; 30 mg, 60 mg, 120 mg sustained release tablets

ACTION & *THERAPEUTIC EFFECT*
Isosorbide mononitrate is a long-acting metabolite of the coronary vasodilator isosorbide dinitrate. It decreases preload as measured by pulmonary capillary wedge pressure (PCWP), and left ventricular end volume and diastolic pressure (LVEDV), with a consequent reduction in myocardial oxygen consumption. *It is equally or more effective than isosorbide dinitrate in the treatment of chronic, stable angina. It is a potent vasodilator with antianginal and antiischemic effects.*

USES Prevention of angina. Not indicated for acute attacks.

CONTRAINDICATIONS Hypersensitivity to nitrates; severe anemia; closed-angle glaucoma; recent MI; postural hypotension, head trauma, cerebral hemorrhage (increases intracranial pressure).

CAUTIOUS USE Older adults, hypotension; pregnancy (category C), lactation. **Extend form** should not be used in patients with GI disease (e.g. GI motility, malabsorption syndrome).

ROUTE & DOSAGE

Prevention of Angina

Adult: **PO** Regular release (ISMO, Monoket) 20 mg b.i.d. 7 h apart; Sustained release (Imdur) 30–60 mg

Common adverse effects in *italic*, life-threatening effects <u>underlined</u>; generic names in **bold**; classifications in SMALL CAPS; ♣ Canadian drug name; ⊘ Prototype drug

815

every morning, may increase up to 120 mg once daily after several days if needed (max dose: 240 mg)

ADMINISTRATION

Oral

- Give first dose in morning on arising and second dose 7 h later with twice daily dosing regimen. Give in morning on arising with once daily dosing.
- Store sustained release tablets in a tight container.

ADVERSE EFFECTS (≥1%) **CNS:** Headache, agitation, anxiety, confusion, loss of coordination, hypoesthesia, hypokinesia, insomnia or somnolence, nervousness, migraine headache, paresthesia, vertigo, ptosis, tremor. **CV:** Aggravation of angina, abnormal heart sounds, murmurs, MI, transient hypotension, palpitations. **Hematologic:** Hypochromic anemia, purpura, thrombocytopenia, methemoglobinemia (high doses). **GI:** Nausea, vomiting, dry mouth, abdominal pain, constipation, diarrhea, dyspepsia, flatulence, tenesmus, gastric ulcer, hemorrhoids, gastritis, glossitis. **Metabolic:** Hyperuricemia, hypokalemia. **GU:** Renal calculus, UTI, atrophic vaginitis, dysuria, polyuria, urinary frequency, decreased libido, impotence. **Respiratory:** Bronchitis, pneumonia, upper respiratory tract infection, nasal congestion, bronchospasm, coughing, dyspnea, rales, rhinitis. **Skin:** Rash, pruritus, hot flashes, acne, abnormal texture. **Special Senses:** Diplopia, blurred vision, photophobia, conjunctivitis.

INTERACTIONS Drug: Alcohol may cause severe hypotension and cardiovascular collapse. **Aspirin** may increase nitrate serum levels. CALCIUM CHANNEL BLOCKERS may cause orthostatic hypotension.

PHARMACOKINETICS Absorption:

Completely and rapidly absorbed from GI tract; 93% reaches systemic circulation. **Onset:** 1 h. **Peak:** Regular release 30–60 min; sustained release 3–4 h. **Duration:** Regular release 5–12 h; sustained release 12 h. **Metabolism:** In liver by denitration and conjugation to inactive metabolites. **Elimination:** Primarily by kidneys. **Half-Life:** 4–5 h.

NURSING IMPLICATIONS

Assessment & Drug Effects

- Monitor cardiac status, frequency and severity of angina, and BP.
- Assess for and report possible S&S of toxicity, including orthostatic hypotension, syncope, dizziness, palpitations, light-headedness, severe headache, blurred vision, and difficulty breathing.
- Lab tests: Monitor serum electrolytes periodically.

Patient & Family Education

- Do not crush or chew sustained release tablets. May break tablets in two and take with adequate fluid (4–8 oz).
- Do not withdraw drug abruptly; doing so may precipitate acute angina.
- Maintain correct dosing interval with twice daily dosing.
- Note: Geriatric patients are more susceptible to the possibility of developing postural hypotension.
- Avoid alcohol ingestion and aspirin unless specifically permitted by prescriber.

ISOTRETINOIN (13-*cis*-RETINOIC ACID) ℗

(eye-soe-tret'i-noyn)

Amnesteem, Claravis, Sotret
Classification: ANTIACNE (RETINOID)

Therapeutic: ANTIACNE; ANTINE-OPLASTIC
Pregnancy Category: X

AVAILABILITY 10 mg, 20 mg, 40 mg capsules

ACTION & *THERAPEUTIC EFFECT*
Highly toxic metabolite of retinol (vitamin A). Principal actions: Regulation of cell (e.g., epithelial) differentiation and proliferation and of altered lipid composition on skin surface. Decreases sebum secretion by reducing sebaceous gland size; inhibits gland cell differentiation; blocks follicular keratinization. *Has antiacne properties and may be used as a chemotherapeutic agent for epithelial carcinomas.*

USES Treatment of severe recalcitrant cystic or conglobate acne in patient unresponsive to conventional treatment, including systemic antibiotics.
UNLABELED USES Lamellar ichthyosis, oral leukoplakia, hyperkeratosis, acne rosacea, scarring gram-negative folliculitis; adjuvant therapy of basal cell carcinoma of lung and cutaneous T-cell lymphoma (mycosis fungoides); psoriasis; chemoprevention for prostate cancer.

CONTRAINDICATIONS Tinnitus; hypersensitivity to parabens (preservatives in the formulation), retinoid hypersensitivity, leukopenia, neutropenia; UV exposure; pregnancy (category X), females of childbearing age, lactation.
CAUTIOUS USE Coronary artery disease; major depression, psychosis, history of suicides, alcoholism; hepatitis, hepatic disease; visual disturbance; rheumatologic disorders, osteoporosis; history of pancreatitis, inflammatory bowel disease; diabetes mellitus; obesity; retinal disease; elevated triglycerides, hyperlipidemia.

ROUTE & DOSAGE

Cystic Acne
Adult: **PO** 0.5–1 mg/kg/day in 2 divided doses (max recommended dose: 2 mg/kg/day) for 15–20 wk
Disorders of Keratinization
Adult: **PO** Up to 4 mg/kg/day in divided doses

ADMINISTRATION
Oral
- Give with or shortly after meals.
- Note: A single course of therapy provides adequate control in many patients. If a second course is necessary, it is delayed at least 8 wk because improvement may continue without the drug.
- Store in tight, light-resistant container. Capsules remain stable for 2 y.

ADVERSE EFFECTS (≥1%) **Body as a Whole:** Most are dose-related (i.e., occurring at doses greater than 1 mg/kg/day), reversible with termination of therapy. **CNS:** Lethargy, headache, fatigue, visual disturbances, pseudotumor cerebri, paresthesias, dizziness, depression, psychosis, suicide (rare). **Special Senses:** Reduced night vision, dry eyes, papilledema, eye irritation, *conjunctivitis,* corneal opacities. **GI:** *Dry mouth,* anorexia, nausea, vomiting, abdominal pain, nonspecific GI symptoms, acute hepatotoxic reactions (rare), inflammation and bleeding of gums, increased AST, ALT, acute pancreatitis. **Hematologic:** Decreased Hct, Hgb, elevated sedimentation rate. **Musculoskeletal:** Arthralgia; bone, joint, and muscle pain and stiffness;

Common adverse effects in *italic*, life-threatening effects underlined; generic names in **bold**; classifications in SMALL CAPS; ♣ Canadian drug name; ⊘ Prototype drug

817

chest pain, skeletal hyperostosis (especially in athletic people and with prolonged therapy), mild bruising, decreased bone mineral density. **Skin:** *Cheilitis,* skin fragility, dry skin, pruritus, peeling of face, palms, and soles; photosensitivity (photoallergic and phototoxic), erythema, skin infections, petechiae, rash, urticaria, exaggerated healing response (painful exuberant granulation tissue with crusting), brittle nails, alopecia **Respiratory:** Epistaxis, *dry nose.* **Metabolic:** Hyperuricemia, *increased serum concentrations of triglycerides by 50–70%,* serum cholesterol by 15–20%, VLDL cholesterol by 50–60%, LDL cholesterol by 15–20%.

INTERACTIONS Drug: VITAMIN A SUPPLEMENTS increase toxicity; decreases effectiveness of ESTROGEN hormonal contraceptives in oral form as well as topical/injectable/implantable/insertable ESTROGEN hormonal birth control. Use with systemic CORTICOSTEROIDS or **phenytoin** may increase bone loss.

PHARMACOKINETICS Absorption: Rapid absorption after slow dissolution in GI tract; 25% of administered drug reaches systemic circulation. **Peak:** 3.2 h. **Distribution:** Not fully understood; appears in liver, ureters, adrenals, ovaries, and lacrimal glands. **Metabolism:** In liver; enterohepatically cycled. **Elimination:** In urine and feces in equal amounts. **Half-Life:** 10–20 h.

NURSING IMPLICATIONS

Assessment & Drug Effects

- Lab tests: Determine baseline blood lipids at outset of treatment, then at 2 wk, 1 mo, and every month thereafter throughout course of therapy; LFTs at 2- or 3-wk intervals for 6 mo and once a month thereafter during treatment.
- Report signs of liver dysfunction (jaundice, pruritus, dark urine) promptly.
- Monitor closely for loss of glycemic control in diabetic and diabetic-prone patients.
- Monitor for development of depression and suicidal ideation.
- Note: Persistence of hypertriglyceridemia (levels above 500–800 mg/dL) despite a reduced dose indicates necessity to stop drug to prevent onset of acute pancreatitis.

Patient & Family Education

- Maintain drug regimen even if during the first few weeks transient exacerbations of acne occur. Recurring symptoms may signify response of deep unseen lesions.
- Discontinue medication at once and notify prescriber if visual disturbances occur along with nausea, vomiting, and headache.
- Rule out pregnancy within 2 wk of starting treatment. Use a reliable contraceptive 1 mo before, throughout, and 1 mo after therapy is discontinued.
- Do not self-medicate with multivitamins, which usually contain vitamin A. Toxicity of isotretinoin is enhanced by vitamin A supplements.
- Avoid or minimize exposure of the treated skin to sun or sunlamps. Photosensitivity (photoallergic and phototoxic) potential is high.
- Notify prescriber immediately of abdominal pain, rectal bleeding, or severe diarrhea, which are possible symptoms of drug-induced inflammatory bowel disease.

Common adverse effects in *italic,* life-threatening effects underlined; generic names in **bold;** classifications in SMALL CAPS; ♣ Canadian drug name; ● Prototype drug

- Keep lips moist and softened (use thin layer of lubricant such as petroleum jelly); dry mouth and cheilitis (inflamed, chapped lips), frequent adverse effects of isotretinoin.
- Notify prescriber of joint pain, such as pain in the great toe (symptom of gout and hyperuricemia).

ISOXSUPRINE HYDROCHLORIDE

(eye-sox′syoo-preen)

Tri Soxuprine

Classifications: BETA-ADRENERGIC AGONIST; ALPHA-ADRENERGIC RECEPTOR INHIBITOR; VASODILATOR
Therapeutic: VASODILATOR
Prototype: Isoprotenerol
Pregnancy Category: C

AVAILABILITY 10 mg, 20 mg tablets

ACTION & *THERAPEUTIC EFFECT* Sympathomimetic with beta-adrenergic stimulant activity and with an inhibitory effect on alpha receptors. Vasodilating action on arteries within skeletal muscles is greater than on cutaneous vessels. *Has both cerebral and peripheral vasodilatory properties.*

USES Adjunctive therapy in treatment of cerebral vascular insufficiency and peripheral vascular disease, such as Raynaud's disease.

CONTRAINDICATIONS Immediately postpartum; presence of arterial bleeding; parenteral use in presence of hypotension, fetal distress; intrauterine fetal death; vaginal bleeding; tachycardia.
CAUTIOUS USE Bleeding disorders; severe cerebrovascular disease, severe obliterative coronary artery disease, recent MI; pregnancy (category C), lactation.

ROUTE & DOSAGE

Cerebral Vascular Insufficiency, Peripheral Vascular Disease
Adult: **PO** 10–20 mg t.i.d. or q.i.d.

ADMINISTRATION

Oral
- May give without regard to meals.

ADVERSE EFFECTS (≥1%) **CV:** Flushing, orthostatic hypotension with light-headedness, faintness; palpitation, tachycardia. **CNS:** Dizziness, nervousness, trembling, weakness. **GI:** Nausea, vomiting, abdominal distress, abdominal distention.

PHARMACOKINETICS Absorption: Readily from GI tract. **Peak:** 1 h. **Duration:** 3 h. **Distribution:** Crosses placenta. **Metabolism:** In blood. **Elimination:** In urine. **Half-Life:** 1.25 h.

NURSING IMPLICATIONS

Assessment & Drug Effects
- Monitor for therapeutic effectiveness: Response to treatment of peripheral vascular disorders may take several weeks. Evaluate clinical manifestations of arterial insufficiency.
- Monitor BP and pulse; may cause hypotension and tachycardia. Supervise ambulation.
- Observe both mother and baby for hypotension and irregular and rapid heartbeat if isoxsuprine is used to delay premature labor. Hypocalcemia, hypoglycemia, and ileus have been observed in babies born of mothers taking isoxsuprine.

Patient & Family Education
- Notify prescriber of adverse reactions (skin rash, palpitation, flushing) promptly; symptoms are

Common adverse effects in *italic*, life-threatening effects underlined; generic names in **bold**; classifications in SMALL CAPS; ♣ Canadian drug name; ⊙ Prototype drug

819

usually effectively controlled by dosage reduction or discontinuation of drug.

- Prevent orthostatic hypotension by making position changes slowly and in stages, particularly from lying down to sitting upright and avoid standing still.
- Note: For treatment of menstrual cramps, isoxsuprine is usually started 1–3 days before onset of menstruation and continued until pain is relieved or menstrual flow stops.

ISRADIPINE

(is-ra′di-peen)

DynaCirc CR

Classifications: CALCIUM CHANNEL ANTAGONIST; ANTIHYPERTENSIVE

Therapeutic: ANTIHYPERTENSIVE; ANTIANGINAL

Prototype: Nifedipine

Pregnancy Category: C

AVAILABILITY 2.5 mg, 5 mg capsules; 5 mg, 10 mg sustained release tablets

ACTION & *THERAPEUTIC EFFECT*

Inhibits calcium ion influx into cardiac muscle and smooth muscle without changing calcium concentrations, thus affecting contractility. Isradipine relaxes coronary vascular smooth muscle. It significantly decreases systemic vascular resistance and reduces BP at rest and during isometric and dynamic exercise. *Reduces BP and has an antianginal effect.*

USES Mild to moderate hypertension.

UNLABELED USES Angina.

CONTRAINDICATIONS Hypersensitivity to isradipine or other calcium channel drugs.

CAUTIOUS USE CHF, acute MI, severe bradycardia, cardiogenic shock, ventricular dysfunction; older adult; mild renal impairment, hepatic impairment; GERD, hiatal hernia with esophageal reflux; pregnancy (category C), lactation.

ROUTE & DOSAGE

Hypertension

Adult: **PO** 2.5 mg b.i.d. (max: 20 mg/day); DynaCirc CR 5 mg daily

ADMINISTRATION

Oral

- Do not crush sustained release form. It **must be** swallowed whole.
- Note: After the first 2–4 wk of therapy, dose may be increased for improved BP control in increments of 5 mg/day at 2–4 wk intervals up to a maximum dose of 20 mg/day.
- Store in tight, light-resistant container.

ADVERSE EFFECTS (≥1%) **CNS:** Headache, dizziness, fainting, fatigue, sleep disturbances, vertigo. **CV:** Flushing, ankle edema, palpitations, tachycardia, hypotension, chest pain, CHF. **GI:** Nausea, vomiting, abdominal discomfort, constipation, increased liver enzymes. **Respiratory:** Dyspnea. **Skin:** Rash, decreased skin sensation.

INTERACTIONS Drug: Adenosine may prolong bradycardia. May increase **cyclosporine** levels and toxicity.

PHARMACOKINETICS Absorption: Rapidly and completely absorbed from GI tract, but only 15–24% reaches systemic circulation because of first-pass metabolism. **Onset:** 1 h. **Peak:** 2–3 h. **Duration:** 12 h. **Distribution:** Not known if crosses placenta or is distributed into breast milk. **Metabolism:** Extensive first-pass metabolism in liver. **Elimination:** 70% in urine as inactive metabolites; 30% in feces. **Half-Life:** 5–11 h.

NURSING IMPLICATIONS

Assessment & Drug Effects

- Monitor BP throughout course of therapy.
- Monitor patients with a history of CHF carefully, especially with concurrent beta-blocker use. Promptly report S&S of worsening heart failure.
- Monitor ambulation, especially with older adult patients, until response to drug is known.

Patient & Family Education

- Notify prescriber promptly of shortness of breath, palpitations, or other signs of adverse cardiovascular effects.
- Do not drive or engage in other potentially hazardous activities until response to drug is known.

ITRACONAZOLE

(i-tra-con′a-zole)

Sporanox

Classifications: ANTIBIOTIC; AZOLE ANTIFUNGAL

Therapeutic: AZOLE ANTIFUNGAL

Prototype: Fluconazole

Pregnancy Category: C

AVAILABILITY 100 mg capsules; 10 mg/mL oral solution; 250 mg/ 25 mL solution for injection

ACTION & *THERAPEUTIC EFFECT*
Interferes with formation of ergosterol, the principal sterol in the fungal cell membrane that,

when depleted, interrupts fungal membrane functioning. *Antifungal properties affect the fungal cell membrane functioning.*

USES Treatment of systemic fungal infections caused by blastomycosis, histoplasmosis, aspergillosis, onychomycosis due to dermatophytes of the toenail with or without fingernail involvement; oropharyngeal and esophageal candidiasis; orally to treat superficial mycoses (*Candida,* pityriasis versicolor).

UNLABELED USES Systemic and vaginal candidiasis.

CONTRAINDICATIONS Hypersensitivity to itraconazole; hypotension; CrCl less than 30 mL/min; ventricular dysfunction as in CHF; or history of CHF when treating onychomycosis; neuropathy; systemic candidiasis; lactation.

CAUTIOUS USE Hypersensitivity to other azole antifungals, achlorhydria; GERD; COPD, cystic fibrosis; dialysis; older adults, females of childbearing age; hepatic disease, hepatitis, HIV infection; hypochlorhydria; pulmonary disease; renal disease, renal impairment; valvular heart disease, ventricular dysfunction; angina, cardiac disease; pregnancy (category C). Safe use in children has not been established.

ROUTE & DOSAGE

Blastomycosis, Nonmeningeal Histoplasmosis, Aspergillosis

Adult: **IV** 200 mg b.i.d. x 2 days then 200 mg once daily **PO** 200–400 mg daily mg once daily. Continue for at least 3 mo

Oropharyngeal Candidiasis (solution)

Adult: **PO** 200 mg daily for 1–2 wk

Common adverse effects in *italic,* life-threatening effects underlined; generic names in **bold;** classifications in SMALL CAPS; ♣ Canadian drug name; ☻ Prototype drug

821

Esophageal Candidiasis

Adult: **PO** 100 mg daily for at least 3 wk (max: 200 mg/day)

Life-Threatening Infections

Adult: **PO** 200 mg t.i.d. × 3 days

Onychomycosis

Adult: **PO** 200 mg daily × 3 mo

ADMINISTRATION

Oral

- Give capsules with a full meal.
- Give oral solution without food. Liquid should be vigorously swished for several seconds and swallowed.
- Do not interchange oral solution and capsules.
- Divide dosages greater than 200 mg/day into two doses.
- Store liquid at or below 25° C (77° F).

ADVERSE EFFECTS (≥1%) **CV:** Hypertension with higher doses. **CNS:** Headache, dizziness, fatigue, somnolence, hearing loss (euphoria, drowsiness less than 1%). **Endocrine:** Gynecomastia, hypokalemia (especially with higher doses), hypertriglyceridemia. **GI:** *Nausea, vomiting, dyspepsia, abdominal pain, diarrhea, anorexia, flatulence, gastritis;* elevations of serum transaminases, alkaline phosphatase, and bilirubin. **Urogenital:** Decreased libido, impotence. **Skin:** Rash, pruritus. **Acute Poisoning:** Severe toxicity (doses exceeding 400 mg daily have been associated with higher risk of hypokalemia, hypertension, adrenal insufficiency).

INTERACTIONS Drug: Use with oral **midazolam, nisoldipine, pimozide, quinidine, dofetilide, triazolam** is contraindicated. Itraconazole may increase levels and toxicity of **ergotamine, dihydroergotamine,** ORAL HYPOGLYCEMIC AGENTS **warfarin, ritonavir, indinavir, vinca alkaloids, busulfan, ergonovine, methylergonovine, midazolam, triazolam, diazepam, nifedipine, nicardipine, amlodipine, felodipine, lova-statin, simvastatin, cyclosporine, tacrolimus, methylprednisolone, digoxin.** Combination with **dofetilide, levomethadyl, oral midazolam, pimozide, quinidine, triazolam** may cause severe cardiac events including cardiac arrest or sudden death. Itraconazole levels are decreased by **carbamazepine, phenytoin, phenobarbital, isoniazid, rifabutin, rifampin. Herbal: St. John's wort** and **garlic** may decrease itraconazole levels.

PHARMACOKINETICS Absorption: Best when taken with food. **Onset:** 2 wk–3 mo. **Peak:** Peak levels at 1.5–5 h. Steady-state concentrations reached in 10–14 days. **Distribution:** Highly protein bound, minimal concentrations in CSF. Higher concentrations in tissues than in plasma. **Metabolism:** Extensively in liver by CYP3A4, may undergo enterohepatic recirculation. **Elimination:** 35% in urine, 55% in feces. **Half-Life:** 34–42 h.

NURSING IMPLICATIONS

Assessment & Drug Effects

- Lab tests: C&S tests should be done before initiation of therapy. Drug may be started pending results. Monitor LFTs especially in those with preexisting hepatic abnormalities.
- Monitor for digoxin toxicity when given concurrently with digoxin.
- Monitor PT and INR carefully when given concurrently with warfarin.

Common adverse effects in *italic*; life-threatening effects underlined; generic names in **bold**; classifications in SMALL CAPS; ♣ Canadian drug name; ⊙ Prototype drug

- Monitor for S&S of hypersensitivity (see Appendix F); discontinue drug and notify prescriber if noted.

Patient & Family Education
- Take capsules, but NOT oral solution, with food.
- Notify prescriber promptly for S&S of liver dysfunction, including anorexia, nausea, and vomiting; weakness and fatigue; dark urine and clay-colored stool.
- Note: Risk of hypoglycemia may increase in diabetics on oral hypoglycemic agents.

IVERMECTIN

(i-ver-mec′tin)
Stromectol
Classifications: ANTHELMINTIC
Therapeutic: ANTHELMINTIC; ANTIPARASITIC
Prototype: Mebendazole
Pregnancy Category: C

AVAILABILITY 3 mg tablets

ACTION & THERAPEUTIC EFFECT
A semisynthetic anthelmintic agent which is a broad-spectrum antiparasitic agent that causes an increase in permeability to chloride ions of the parasitic cell membrane, resulting in hyperpolarization of nerve or muscle cells of parasites. This results in their paralysis and cell death. *Causes cell death of parasites.*

USES Treatment of strongyloidiasis of the intestinal tract, onchocerciasis.

CONTRAINDICATIONS Hypersensitivity to ivermectin.
CAUTIOUS USE Asthma; older adults; moderate or severe hepatic

disease; hyperreactive onchodermatitis; pregnancy (category C); lactation; children less than 15 kg of weight.

ROUTE & DOSAGE

Strongyloides

Adult/Child (weight 15 kg or greater): **PO** 200 mcg/kg × 1 dose

Onchocerciasis

Adult/Child (weight 15 kg or greater): **PO** 150 mcg/kg × 1 dose, may repeat q3–12mo prn

ADMINISTRATION

Oral
- Give tablets with water rather than any other type of liquid.
- Store below 30° C (86° F).

ADVERSE EFFECTS (≥1%) **Body as a Whole:** *Fever,* peripheral edema. **CNS:** Dizziness. **CV:** Tachycardia. **GI:** Diarrhea, nausea. **Skin:** *Pruritus, rash.* **Other:** Arthralgia/synovitis, lymphadenopathy.

INTERACTIONS Drug: May increase effect of **warfarin.**

PHARMACOKINETICS Peak: 4 h. **Distribution:** Distributed into breast milk. **Metabolism:** In the liver. **Elimination:** In feces over 12 days. **Half-Life:** 16 h.

NURSING IMPLICATIONS
Assessment & Drug Effects
- Monitor for therapeutic effectiveness: Indicated by negative stool samples.
- Monitor for cardiovascular effects such as orthostatic hypotension and tachycardia.
- Monitor for and report inflammatory conditions of the eyes.

Patient & Family Education
- Get a follow-up stool examination to determine effectiveness of

Common adverse effects in *italic*, life-threatening effects underlined; generic names in **bold**; classifications in SMALL CAPS; ♣ Canadian drug name; ⊕ Prototype drug

823

treatment. Treatment for worms does not kill adult parasites; repeated follow-up and retreatment are usually needed.

- Notify prescriber if eye discomfort develops.

IXABEPILONE
(ix-a-be-pi'lone)
Ixempra
Classifications: ANTINEOPLASTIC AGENT; EPOTHILONE
Therapeutic: ANTINEOPLASTIC AGENT; ANTIMITOTIC
Pregnancy Category: D

AVAILABILITY 15 mg and 45 mg lyophilized powder, single-use vials

ACTION & *THERAPEUTIC EFFECT* Binds directly to microtubules needed to form the spindles required in mitosis of dividing cells. *Blocks new cell formation during the mitotic phase of their cell division cycle, thus leading to cancer cell death.*

USES Metastatic or locally advanced breast cancer alone or in combination with capecitabine in patients who have failed therapy with an anthracycline and a taxane. Monotherapy is indicated only if a patient has also failed with capecitabine therapy.

CONTRAINDICATIONS Hepatic impairment in patients with AST or ALT greater than 10 × ULN, and/or bilirubin greater than 3 × ULN; concomitant use of capecitabine and bilirubin greater than 1 × ULN, or AST or ALT greater than 2.5 × ULN; grade 4 neuropathy or any other grade 4 toxicity; pregnancy (category D); lactation.

CAUTIOUS USE Hypersensitivity to ixabepilone; monotherapy of patients with hepatic impairment baseline values of AST or ALT greater than 5 × ULN.

ROUTE & DOSAGE

Breast Cancer
Adult: **IV** 40 mg/m² over 3 h q3wk

Obesity Dosage Adjustment
BSA greater than 2.2 m2: Dosage should be calculated based on 2.2 m² instead of actual m²

Dosage Adjustments
- **Grade 2 neuropathy 7 days or more, or grade 3 neuropathy less than 7 days, or grade 3 toxicity other than neuropathy:** 32 mg/m²
- **Neutrophil less than 500 cells/mm³ 7 days or more, or febrile neutropenia, or platelets less than 25,000/mm³ or platelets less than 50,000/mm³ with bleeding:** Decrease dose by 20%
- **Grade 3 neuropathy 7 days or more or disabling neuropathy, or any grade 4 toxicity:** Do not administer
- **Regimen with a strong CYP3A4 inhibitor:** 20 mg/m²

Hepatic Impairment Dosage Adjustment in Monotherapy
- **AST and ALT 10 × ULN or less and bilirubin 1.5 × ULN or less:** 32 mg/m²
- **AST and ALT 10 × ULN or less and bilirubin greater than 1.5 × ULN but 3 × ULN or less:** 20–30 mg/m²
- **AST and ALT greater than 10 × ULN or bilirubin greater than 3 × ULN:** Do not administer

ADMINISTRATION

Intravenous

Use gloves when handling vials containing ixabepilone.

PREPARE: **IV Infusion:** Supplied in kit containing a powder vial and diluent vial. Allow kit to come to room temperature for 30 min before reconstitution. ▪ Slowly inject diluent into the powder vial to yield 2 mg/mL. Swirl gently and invert vial to dissolve. ▪ Further dilute in LR solution in DEHP-free bags. Select a volume of LR to produce a final concentration of 0.2–0.6 mg/mL. Mix thoroughly.

ADMINISTER: **IV Infusion:** Use DEHP-free infusion line with a 0.2–1.2 micron in-line filter. ▪ Infuse at a rate appropriate to the total volume of solution. Complete infusion within 6 h of preparation.

INCOMPATIBILITIES **Solution/additive:** Diluents other than **lactated Ringer's injection** should not be combined with ixabepilone. **Y-site:** Do not use a Y-site connection with this drug.

▪ Store drug kit refrigerated at 2°–8° C (36°–46° F) in original packaging. ▪ Reconstituted solution may be stored in the vial for a maximum of only 1 h at room temperature/light. ▪ Once further diluted with lactated Ringer's injection, solution is stable at room temperature/light for 6 h.

ADVERSE EFFECTS (≥1%) **Body as a Whole:** Chest pain, dehydration, edema, *fatigue,* hypersensitivity reactions, pain, peripheral neuropathy, pyrexia. **CNS:** Dizziness, *headache,* insomnia. **CV:** Flushing. **GI:** *Abdominal pain, anorexia, constipation, diarrhea,* gastroesophageal reflux disease (GERD), *mucositis, nausea,* *stomatitis, vomiting,* taste disorder. **Hematologic:** Anemia, leukopenia, *neutropenia,* thrombocytopenia. **Metabolic:** Weight loss. **Musculoskeletal:** *Arthralgia, myalgia, musculoskeletal pain.* **Respiratory:** Cough, dyspnea, upper respiratory tract infection. **Skin:** *Alopecia,* exfoliation, hyperpigmentation, nail disorder, palmar-plantar erythrodysesthesia syndrome, pruritus, rash. **Special Senses:** Lacrimation.

INTERACTIONS **Drug:** Inhibitors of CYP3A4 (e.g., HIV PROTEASE INHIBITORS, MACROLIDE ANTIBIOTICS, AZOLE ANTIFUNGAL AGENTS) increase the plasma level of ixabepilone. Strong CYP3A4 inducers (e.g., **dexamethasone, phenytoin, carbamazepine, rifampin, rifabutin, phenobarbital**) decrease the plasma level of ixabepilone. **Food: Grapefruit** and **grapefruit juice** increase the plasma level of ixabepilone. **Herbal: St. John's wort** decreases the plasma level of ixabepilone.

PHARMACOKINETICS **Distribution:** 67–77% protein bound. **Metabolism:** In liver. **Elimination:** Stool (major) and urine (minor). **Half-Life:** 52 h.

NURSING IMPLICATIONS

Actions & Drug Effects

▪ Monitor for signs of an infusion-related hypersensitivity reaction.
▪ Monitor for and promptly report signs of neuropathy.
▪ Lab tests: Baseline and periodic WBC count with differential, platelet count, LFTs; periodic serum electrolytes.

Patient & Family Education

▪ Report promptly any of the following: Numbness and tingling of the hands or feet, S&S of infection (e.g., fever of 100.5° F or greater, chills, cough, burning or pain on

Common adverse effects in *italic,* life-threatening effects underlined; generic names in **bold;** classifications in SMALL CAPS; ♣ Canadian drug name; ⊙ Prototype drug

825

urination), hives, itching, rash, flushing, swelling, shortness of breath, difficulty breathing, chest tightness or pain, palpitations or unusual weight gain.
- Use effective contraceptive measures to prevent pregnancy.

KANAMYCIN
(kan-a-mye′sin)

Kantrex
Classifications: ANTIBIOTIC; AMINOGLYCOSIDE
Therapeutic: ANTIBIOTIC
Prototype: Gentamicin
Pregnancy Category: D

AVAILABILITY 1 g vials

ACTION & *THERAPEUTIC EFFECT*
Broad-spectrum, aminoglycoside antibiotic that binds irreversibly to aminoglycoside-binding sites on 30 S ribosomal subunit of bacteria, subsequently inhibiting bacterial protein synthesis. *Usually bactericidal in action. Active against many aerobic gram-negative microorganisms, as well as some gram-positive bacteria. It is not effective against anaerobic gram-negative bacteria.*

USES Parenterally for short-term treatment of serious infections; intraperitoneally after fecal spill during surgery; as irrigation solution; and as aerosol treatment. In conjunction with other drugs to treat tuberculosis in patients resistant to conventional therapy.

CONTRAINDICATIONS History of hypersensitivity to kanamycin or other aminoglycosides; history of drug-induced ototoxicity, preexisting hearing loss, vertigo, or tinnitus; long-term therapy;

intraperitoneally to patients under effects of inhalation anesthetics or skeletal muscle relaxants; pregnancy (category D).
CAUTIOUS USE Impaired renal function; older adults, neonates, and infants (immature renal systems); myasthenia gravis; parkinsonian syndrome; lactation.

ROUTE & DOSAGE

Serious Infection
Adult/Child: **IV/IM** 15 mg/kg/ day in equally divided doses q8–12h
Adult: **Intraperitoneal** 500 mg diluted in 20 mL sterile water instilled through wound catheter **Inhalation** 250 mg diluted in 3 mL NS administered per nebulizer q6–12h **Irrigation** 0.25% solution prn

Renal Impairment Dosage Adjustment
CrCl 50–80 mL/min: Give 60–90% of dose; *10–50 mL/ min:* Give 30–70% of dose or q12h; *less than 10 mL/min:* Give 20–30% of dose or q24–48h

ADMINISTRATION

Intramuscular
- Administer IM injection deep into upper outer quadrant of gluteal muscle (often painful).
- Observe sites daily for signs of irritation; rotate injection sites.

Intravenous

PREPARE: **Intermittent for Adult:** Dilute each 500 mg with at least 100 mL NS, D5W, D5/NS, or other compatible solution. **Intermittent for Child:** Dilute

K

each 2.5–5 mg in 1 mL of NS, D5W, D5/SW, or other compatible solution.
ADMINISTER: **Intermittent for Adult/Child:** Over 30–60 min. **Intermittent for Child:** Use a constant-rate volumetric infusion device for administration.
INCOMPATIBILITIES **Solution/additive: Amphotericin B, cefazolin, cefoxitin, ceftazidime, cefuroxime, cephalothin, cephapirin, chlorpheniramine, colistimethate, heparin, hyaluronidase, hydrocortisone, lincomycin, methicillin, methohexital, nitrofurantoin, pentobarbital, phenobarbital, prochlorperazine, sodium bicarbonate, thiopental, warfarin.**

▪Store vials at 15°–30° C (59°–86° F) unless otherwise directed. Some vials may darken with time, but this does not affect potency. ▪Discard partially used vials within 48 h.

ADVERSE EFFECTS (≥1%) **All:** Dose related. **Body as a Whole:** Eosinophilia, maculopapular rash, pruritus, urticaria, drug fever, anaphylaxis. **CNS:** Dizziness, circumoral and other paresthesias, optic neuritis, peripheral neuritis, headache, restlessness, tremors, lethargy, convulsions; neuromuscular paralysis, respiratory depression (rarely). **Special Senses:** Deafness (can be irreversible), *tinnitus, vertigo* or *dizziness,* ataxia, nystagmus. **GI:** Nausea, vomiting, diarrhea, appetite changes, abdominal discomfort, stomatitis, proctitis, malabsorption syndrome (with prolonged oral administration). **Hematologic:** Anemia, increased or decreased reticulocytes, granulocytopenia, agranulocytosis, thrombocytopenia, purpura. **Urogenital:** Nephrotoxicity; hematuria, urine casts and cells, proteinuria; elevated serum creatinine and BUN. **Other:** Superinfections; local pain; nodular formation at injection site.

INTERACTIONS Drug: Amphotericin B, cisplatin, methoxyflurane, vancomycin add to nephrotoxicity; GENERAL ANESTHETICS, SKELETAL MUSCLE RELAXANTS add to neuromuscular blocking effects; **capreomycin** compounds ototoxicity and nephrotoxicity; LOOP AND THIAZIDE DIURETICS, carboplatin may increase risk of ototoxicity.

PHARMACOKINETICS Absorption: Readily absorbed from peritoneal cavity, bronchial tree, and wounds. **Peak:** 1–2 h. **Distribution:** Crosses placenta; distributed into breast milk. **Elimination:** 80–90% in urine within 24 h. **Half-Life:** 2–4 h.

NURSING IMPLICATIONS

Assessment & Drug Effects

▪ Lab tests: Monitor baseline C&S, urinalysis, and kidney function prior to initiation of therapy and periodically thereafter. Monitor serum sodium, potassium, calcium, and magnesium.

▪ Notify prescriber immediately of signs of renal irritation: Albuminuria, casts, red and white cells in urine, increasing BUN, and serum creatinine, decreasing creatinine clearance, oliguria, and edema.

▪ Monitor peak and trough serum kanamycin concentrations: Assess peak specimen 30–60 min after IM administration; 30 min after completion of a 30–60 min IV infusion. Assess trough levels just before the next IM or IV dose.

K

Common adverse effects in *italic,* life-threatening effects underlined; generic names in **bold;** classifications in SMALL CAPS; ♣ Canadian drug name; ⊙ Prototype drug

827

- Keep patient well hydrated to prevent chemical irritation of renal tubules.
- Monitor I&O. Report decrease in urine output or change in I&O ratio.
- Determine baseline weight and vital signs and monitor at regular intervals during therapy.
- Report signs of superinfection (see Appendix F).
- Monitor for hearing and balance problems; stop drug if ototoxicity occurs. Tinnitus is not a reliable index of ototoxicity in the very elderly. Risk of ototoxicity is high in patients with impaired renal function, older adults, poorly hydrated patients, and with therapy 5 days or longer.
- Note: Deafness has occurred 2–7 days or more after termination of therapy in patients with impaired renal function.

Patient & Family Education

- Report ototoxic symptoms such as dizziness, hearing loss, weakness, or loss of balance; drug may need to be discontinued.

KAOLIN AND PECTIN

(kay'oh-lin and pek'tin)
Kaolin w/Pectin
Classification: ANTIDIARRHEAL
Therapeutic: ANTIDIARRHEAL
Prototype: Loperamide
Pregnancy Category: C

AVAILABILITY 5.2 g kaolin/260 mg pectin/30 mL, 90 g kaolin/2 g pectin/30 mL

ACTION & *THERAPEUTIC EFFECT*
Kaolin is hydrated aluminum silicate adsorbant. Kaolin is reported to have adsorbent, protectant, and demulcent properties. Mechanism of action of pectin may help consolidate stool. *Effective as an antidiarrheal agent.*

USES Adjunct in symptomatic treatment of mild to moderately severe acute diarrhea. Commonly used in antidiarrheal combination products.

CONTRAINDICATIONS Suspected obstructive bowel lesion, pseudomembranous colitis, diarrhea associated with bacterial toxins; presence of fever; use for more than 48 h without medical direction; lactation.
CAUTIOUS USE Older adults; infants or children 3 y or younger, pregnancy (category C).

ROUTE & DOSAGE

Diarrhea

Adult: **PO** 60–120 mL of regular suspension or 45–90 mL of concentrated suspension after each loose bowel movement
Child: **PO** 3–5 y, 15–30 mL regular suspension or 15 mL concentrated suspension after each loose bowel movement; 6–11 y, 30–60 mL regular suspension or 30 mL concentrated suspension after each loose bowel movement; 12 y or older, 60 mL regular suspension or 45 mL concentrated suspension after each loose bowel movement

ADMINISTRATION

Oral

- Administer at least 2–4 h before other oral medications.
- Shake suspension well before pouring.
- Store in tightly closed container at 15°–30° C (59°–86° F) unless

otherwise directed. Protect from freezing.

ADVERSE EFFECTS (≥1%) **GI:** Constipation usually mild and transient.

INTERACTIONS Drug: Chloroquine, digoxin, penicillamine, tetracycline, ciprofloxacin, and most other drugs.

PHARMACOKINETICS Absorption: Not absorbed from GI tract.

NURSING IMPLICATIONS

Assessment & Drug Effects

- Assess for abdominal distention and number of stools per day.
- Note: Fecal impaction may result from taking kaolin and pectin, especially in older adults.
- Note: Drug may decrease absorption of any orally administered medication.

Patient & Family Education

- Do not exceed prescribed dosage.
- Notify prescriber if diarrhea is not controlled within 48 h or if fever develops.

KETOCONAZOLE

(ke-to-con′a-zol)

Extina, Kuric, Nizoral, Nizoral A-D, Xolegel
Classification: AZOLE ANTIFUNGAL
Therapeutic: AZOLE ANTIFUNGAL
Prototype: Fluconazole
Pregnancy Category: C

AVAILABILITY 200 mg tablets; 2% cream; 2% shampoo; 2% foam; 2% gel

ACTION & THERAPEUTIC EFFECT Interferes with formation of ergosterol, the principal sterol in the fungal cell membrane that, when depleted, interrupts membrane function by increasing its permeability. *Antifungal properties are related to the drug effect on the fungal cell membrane functioning.*

USES Oral—Severe systemic fungal infections including candidiasis (e.g., oral thrush, candiduria), chronic mucocutaneous candidiasis, pulmonary and disseminated coccidioidomycosis, histoplasmosis, paracoccidioidomycosis, blastomycosis, and chromomycosis. **Topical**—Tinea corporis and tinea cruris (caused by *Epidermophyton floccosum, Trichophyton mentagrophytes,* and *Trichophyton rubrum*) and in treatment of tinea versicolor (pityriasis) caused by *Malassezia furfur (Pityrosporum obiculare),* seborrheic dermatitis. **UNLABELED USES Oral**—Onychomycosis, vaginal candidiasis, Cushing's syndrome associated with adrenal or pituitary adenoma; precocious puberty, dysfunctional hirsutism, and as swish and swallow preparation for prophylaxis against fungal infections in patients with neutropenia induced by cancer chemotherapy and in patients with AIDS.

CONTRAINDICATIONS Hypersensitivity to ketoconazole or any component in the formulation; chronic alcoholism, fungal meningitis; onychomycosis; ocular exposure, ophthalmic administration. **CAUTIOUS USE** Azole antifungal hypersensitivity; achlorhydria, hypochlorhydria; asthma; severe hepatic impairment; alcoholism; older adult; HIV infection; hyperactive onchodermatitis; pregnancy (category C), lactation. Safe use in children younger than 2 y is not established.

Common adverse effects in *italic*, life-threatening effects <u>underlined</u>; generic names in **bold;** classifications in SMALL CAPS; ♣ Canadian drug name; ☺ Prototype drug

829

ROUTE & DOSAGE

Fungal Infections

Adult: **PO** 200–400 mg once/day
Topical Apply 1–2 times/day to affected area and surrounding skin
Child (older than 2 y): **PO** 3.3–6.6 mg/kg/day as single dose

Dandruff

Adult/Child: **Topical** Shampoo twice a week for 4 wk with at least 3 days between shampoos **Topical (Extina)** Apply b.i.d. × 4 wk; **(Xolegel)** Apply daily × 2 wk

ADMINISTRATION

Oral

- Give with water, fruit juice, coffee, or tea; drug requires an acid medium for dissolution and absorption.
- Relieve nausea and vomiting during early therapy by taking drug with food and dividing into 2 daily doses.
- Do not give with antacids.
- Store in tightly covered container at 15°–30° C (59°–86° F) unless otherwise directed.

Topical

- Apply sufficient shampoo to produce lather to wash scalp and hair and gently massage over entire scalp area for 1 min, rinse hair thoroughly and repeat, leaving shampoo on scalp for 3 min. Rinse thoroughly.

ADVERSE EFFECTS (≥1%) **Oral— Body as a Whole:** Skin rash, erythema, urticaria, pruritus, angioedema, anaphylaxis. **GI:** *Nausea, vomiting,* anorexia, epigastric or abdominal pain, constipation, diarrhea, transient elevation in serum liver enzymes, fatal hepatic necrosis (rare). **Hematologic:** With high doses, lowers serum testosterone and ACTH-induced corticosteroid serum levels, transient decreases in serum cholesterol and triglycerides; hyponatremia (rare). **Urogenital:** Gynecomastia (males), breast pain; uterine bleeding, loss of libido, impotence, oligospermia, hair loss. **Other:** Acute hypoadrenalism (reduction of adrenal stress syndrome), renal hypofunction. **Topical—Skin:** Mild transient erythema, severe irritation, pruritus, stinging.

INTERACTIONS **Drug: Alcohol** may cause sunburn-like reaction; ANTACIDS, ANTICHOLINERGICS, H$_2$-RECEPTOR ANTAGONISTS decrease ketoconazole absorption; **isoniazid, rifampin** increase ketoconazole metabolism, thus decreasing its activity; levels of **phenytoin** and ketoconazole decreased; may increase levels of **cyclosporine** or **carbamazepine,** increasing the risk of toxicity; **warfarin** may potentiate hypoprothrombinemia; may increase ergotamine toxicity of **dihydroergotamine, ergotamine;** may increase concentration and toxicity of **trazodone. Herbal: Echinacea** may increase risk of hepatotoxicity.

PHARMACOKINETICS **Absorption:** Erratically from GI tract (needs an acid pH); minimal absorption topically. **Peak:** 1–2 h. **Distribution:** Distributed to saliva, urine, sebum, and cerumen; CSF levels unpredictable; distributed into breast milk. **Metabolism:** In liver (CYP3A4). **Elimination:** Primarily in feces, 13% in urine. **Half-Life:** 8 h.

NURSING IMPLICATIONS
Assessment & Drug Effects
- Lab tests: Monitor baseline LFTs (AST, ALT, alkaline phosphatase, and bilirubin) and repeat at least monthly throughout therapy.
- Monitor for S&S of hepatotoxicity (see Appendix F). Discontinue drug immediately to prevent irreversible liver damage and report to prescriber.

Patient & Family Education
- Report S&S of hepatotoxicity promptly to prescriber (see Appendix F).
- Note: Drowsiness and dizziness are early and time-limited adverse effects.
- Do not drive or engage in potentially hazardous activities until response to drug is known.
- Avoid OTC drugs for gastric distress, such as Rolaids, Tums, Alka-Seltzer and check with prescriber before taking any other nonprescription medicines.
- Do not alter dose or dose interval and do not stop taking ketoconazole before consulting the prescriber.
- Notify prescriber if skin condition fails to respond to topical therapy or worsens or if signs of irritation or sensitivity occur.

KETOPROFEN
(kee-toe-proe'fen)

Classifications: ANALGESIC, NONSTEROIDAL ANTI-INFLAMMATORY DRUG (NSAID); ANTIPYRETIC
Therapeutic: ANALGESIC, NSAID
Prototype: Ibuprofen
Pregnancy Category: B first and second trimester D in third trimester

AVAILABILITY 50 mg, 75 mg capsules; 200 mg sustained release capsules

ACTION & THERAPEUTIC EFFECT Nonsteroidal anti-inflammatory drug (NSAID) that inhibits both COX-1 and COX-2 enzymes; thus it also inhibits prostaglandin synthesis, and therefore interferes with the inflammatory process. It inhibits platelet aggregation and prolongs bleeding time. *Has analgesic, anti-inflammatory, antiarthritic, and antiplatelet properties.*

USES Acute or long-term treatment of rheumatoid arthritis and osteoarthritis; primary dysmenorrhea; headache; symptomatic relief of postoperative, dental, and postpartum pain; visceral pain associated with cancer.
UNLABELED USES Reiter's syndrome, juvenile arthritis, acute gouty arthritis, biliary pain, renal colic.

CONTRAINDICATIONS Patient in whom aspirin, salicylate, or another NSAID induces asthma, urticaria, bronchospasm, severe rhinitis, shock; perioperatively for CABG surgery; renal nephritis, nephritic syndrome; pregnancy (category D third trimester); lactation.
CAUTIOUS USE History of GI disease, GI bleeding, active ulcer; renal or hepatic impairment; patient who may be adversely affected by prolongation of bleeding time; heart failure, fluid retention; hypertension; patient receiving diuretics; geriatric patient; anemia; dental work; myasthenia gravis; pregnancy (category B first and second trimester). Safe use in

children younger than 16 y not established.

ROUTE & DOSAGE

Arthritis

Adult: **PO** 75 mg t.i.d. or 50 mg q.i.d. (max: 300 mg/day) or 200 mg sustained release daily
Geriatric: **PO** Start with 25 mg q.i.d., may also start with 50 mg t.i.d.

Mild to Moderate Pain, Dysmenorrhea

Adult: **PO** 12.5–50 mg q6–8h

ADMINISTRATION

Oral

- Ensure that extended-release capsule is swallowed whole. It should not be crushed or chewed.
- Give with food, milk, or prescribed antacid to reduce GI irritation.
- Store drug at 15°–30° C (59°–86° F) in tightly closed, light-resistant container unless otherwise directed.

ADVERSE EFFECTS (≥1%) CNS:
Trouble in sleeping, nervousness, *headache,* dizziness; depression, drowsiness, confusion, migraine, vertigo. **CV:** Peripheral edema, palpitations, hypertension, tachycardia. **Special Senses:** Visual disturbances, conjunctivitis, eye pain, retinal hemorrhage, pigmentation changes; dry nose or throat, tinnitus, hearing impairment. **GI:** *Dyspepsia,* <u>drug-induced peptic ulcer, GI bleeding</u>, nausea, vomiting, diarrhea, constipation, flatulence, stomach pain, anorexia, dry mouth, gingivitis, rectal burning and hemorrhage, melena, jaundice, elevated ALT, AST. **Hematologic:**
Prolonged bleeding time, anemia, purpura, <u>agranulocytosis</u>, thrombocytosis. **Urogenital:** Gynecomastia, changes in libido, urinary tract irritation (dysuria, frequency/urgency), renal impairment. **Respiratory:** <u>Laryngospasm, bronchospasm, laryngeal edema</u>, pharyngitis. **Skin:** Rash, pruritus, urticaria, erythema, photosensitivity. **Endocrine:** Aggravation of diabetes mellitus.

INTERACTIONS Drug: ORAL ANTICOAGULANTS, **heparin** may prolong bleeding time; may increase **lithium** toxicity; may increase **methotrexate** toxicity. **Herbal:** **Feverfew, garlic, ginger, ginkgo** increases bleeding potential.

PHARMACOKINETICS Absorption: Readily from GI tract. **Onset:** 1–2 h. **Peak:** 1–2 h. **Duration:** 4–6 h. **Metabolism:** In liver. **Elimination:** Primarily in urine, some biliary excretion. **Half-Life:** 1.1–4 h.

NURSING IMPLICATIONS

Assessment & Drug Effects

- Lab tests: Monitor baseline and periodic evaluations of hemoglobin, renal and hepatic function.
- Monitor for and report tinnitus, hearing impairment, and visual disturbance, especially during prolonged or high-dose therapy.
- Monitor for S&S of GI ulceration (e.g., stool for occult blood, persistent indigestion).

Patient & Family Education

- Report promptly signs of jaundice (see Appendix F) as well as the following: Blurred vision, tinnitus, urinary urgency or frequency, unexplained bleeding, weight gain with edema.

Common adverse effects in *italic*, life-threatening effects <u>underlined</u>; generic names in **bold**; classifications in SMALL CAPS; ♣ Canadian drug name; ⊕ Prototype drug

- Note: Possible CNS adverse effects (e.g., light-headedness, dizziness, drowsiness).
- Do not drive or engage in potentially hazardous activities until response to drug is known.
- Note: Alcohol, aspirin, or other NSAIDs may increase risk of GI ulceration and bleeding tendencies and therefore should be avoided.
- Tell dentist or surgeon that you are taking ketoprofen.

KETOROLAC TROMETHAMINE
(ke-tor′o-lac)

Acular, Acular LS, Acuvail, SPIRX
Classifications: ANALGESIC, NON-STEROIDAL ANTI-INFLAMMATORY DRUG (NSAID); ANTIPYRETIC
Therapeutic: NONNARCOTIC ANALGESIC; NSAID
Prototype: Ibuprofen
Pregnancy Category: C first and second trimester; D third trimester

AVAILABILITY 10 mg tablets; 15 mg/mL, 30 mg/mL injection; 0.4%, 0.5% ophth solution; nasal spray

ACTION & THERAPEUTIC EFFECT It inhibits synthesis of prostaglandins by inhibiting both COX-1 and COX-2 enzymes. Is a peripherally acting analgesic. It inhibits platelet aggregation and prolongs bleeding time. *Exhibits analgesic, anti-inflammatory, and antipyretic activity. Effective in controlling acute postoperative pain.*

USES *Short-term* management of pain; ocular itching due to seasonal allergic conjunctivitis, reduction of postoperative pain and photophobia after refractive surgery.

CONTRAINDICATIONS Hypersensitivity to ketorolac; hypersensitivity reaction to aspirin, salicylates, or other NSAIDs; during labor and delivery; surgery; patients with severe renal impairment or at risk for renal failure due to volume depletion; perioperative use in CABG surgery for 10–14 days after; patients with risk of bleeding; history of GI bleeding or peptic ulcer; pre- or intraoperatively; intrathecal or epidural administration; in combination with other NSAIDs; pregnancy (category D third trimester).
CAUTIOUS USE Impaired renal or hepatic function; Crohn's disease or IBD; bleeding disorders; corticosteroid therapy; myelosuppressive chemotherapy; older adults; debilitated patients; diabetes mellitus; lactase deficiency; coagulation disorders; SLE; CHF; history of hypertension, MI or stroke; pregnancy (category C first and second trimester); children younger than 3 y with ophthalmic solution. Safe use in children younger than 2 y for all other forms not established.

ROUTE & DOSAGE

Pain

Adult: **IV Loading Dose** 30 mg (15 mg if less than 50 kg) **IM** 30–60 mg loading dose, then 15–30 mg q6h [max: 150 mg/day on first day, then 120 mg subsequent days (30 mg load, then 15 mg q6h if less than 50 kg)] **PO** 10 mg q6h prn (max: 40 mg/day) max duration all routes 5 days
Nasal Spray 1 spray q6–8h (max: 8 per day)
Geriatric: **IV Loading Dose** 15 mg **IM** 30 mg loading dose, then 15 mg q6h **PO** 5–10 mg q6h prn

---

Common adverse effects in *italic*, life-threatening effects underlined; generic names in **bold;** classifications in SMALL CAPS; ♣ Canadian drug name; ⓟ Prototype drug

833

(max: 40 mg/day) max duration all routes 5 days

Pain after Refractive Surgery

Adult: **Ophthalmic** *Acular LS only* 1 drop in operative eye q.i.d. up to 4 days

Allergic Conjunctivitis

Adult: **Ophthalmic** 1 drop 0.5% solution q.i.d.

ADMINISTRATION

WARNING: Do not administer IV, IM, or PO ketorolac longer than 5 days.

Oral

- Give with food to reduce GI effects.

Instillation

- Do not touch container to the eye when applying ophthalmic drops.

Intramuscular

- Inject IM drug slowly and deeply into a large muscle.
- Rotate injection sites to avoid injection site pain in patients receiving multiple doses.

Intravenous ──────

PREPARE: **Direct:** Give undiluted. *ADMINISTER:* **Direct:** Give IV bolus dose over at least 15 sec. Preferred method is to give through a Y-tube in a free-flowing IV.
INCOMPATIBILITIES **Solution/additive:** Haloperidol, hydroxyzine, meperidine, morphine, prochlorperazine, promethazine. **Y-site:** Acyclovir, amphotericin B, azathioprine, azithromycin, calcium chloride, capsofungin, chlorpromazine, dantrolene, diazepam, diazoxide, diltiazem, diphenhydramine, dobutamine, doxycycline, epirubicin, erythromycin, esmolol, fenoldopam, gemcitabine, gemtuzumab, haloperidol, hydroxyzine, idarubicin, labetalol, levofloxacin, metaraminol, midazolam, milrinone acetate, minocycline, nalbuphine, pantoprazole, papaverine, pentamidine, pentazocine, phentolamine, phenytoin, prochlorperazine, promethazine, protamine, pyridoxine, quinidine, quinupristin/dalfopristin, rocuronium, sulfamethoxazole/trimethoprim, tolazoline, vancomycin, vecuronium, vinorelbine.

- Store all forms at 15°–30° C (59°–86° F).

ADVERSE EFFECTS (≥1%) CNS:

Drowsiness, dizziness, headache. **GI:** *Nausea,* dyspepsia, GI pain, hemorrhage. **Other:** Edema, sweating, pain at injection site.

INTERACTIONS Drug:

May increase **methotrexate** levels and toxicity; do not use with **pentoxifylline**. **Herbal:** Feverfew, garlic, ginger, ginkgo increased bleeding potential.

PHARMACOKINETICS Peak: 45–60 min. **Distribution:** Into breast milk. **Metabolism:** In liver. **Elimination:** In urine. **Half-Life:** 4–6 h.

NURSING IMPLICATIONS

Assessment & Drug Effects

- Correct hypovolemia prior to administration of ketorolac.
- Lab tests: Periodic serum electrolytes and LFTs; urinalysis (for hematuria and proteinuria) with long-term use.

- Monitor urine output in older adults and patients with a history of cardiac decompensation, renal impairment, heart failure, or liver dysfunction as well as those taking diuretics. Discontinuation of drug will return urine output to pretreatment level.
- Monitor for S&S of GI distress or bleeding including nausea, GI pain, diarrhea, melena, or hematemesis. GI ulceration with perforation can occur anytime during treatment. Drug decreases platelet aggregation and thus may prolong bleeding time.
- Monitor for fluid retention and edema in patients with a history of CHF.

Patient & Family Education

- Watch for S&S of GI ulceration and bleeding (e.g., bloody emesis, black tarry stools) during long-term therapy.
- Note: Possible CNS adverse effects (e.g., light-headedness, dizziness, drowsiness).
- Do not drive or engage in potentially hazardous activities until response to drug is known.
- Do not use other NSAIDs while taking this drug.

KETOTIFEN FUMARATE

(kee-toe-tye′fen)
Zaditor
Pregnancy Category: C
See Appendix A-1.

LABETALOL HYDROCHLORIDE

(la-bet′a-lole)
Trandate
Classifications: ALPHA- & BETA-ADRENERGIC ANTAGONIST; ANTIHYPERTENSIVE

Therapeutic: ANTIHYPERTENSIVE
Prototype: Propranolol
Pregnancy Category: B first and second trimester; D third trimester

AVAILABILITY 100 mg, 200 mg, 300 mg tablet; 5 mg/mL injection

ACTION & *THERAPEUTIC EFFECT* Acts as an adrenergic receptor blocking agent that combines selective alpha activity and nonselective beta-adrenergic blocking actions. The alpha blockade results in vasodilation, decreased peripheral resistance, and orthostatic hypotension. It has beta-blocking effects on the sinus node, AV node, and ventricular muscle, which lead to bradycardia, delay in AV conduction, and depression of cardiac contractility. *Effective in reducing blood pressure by vasodilation as well as depression of cardiac contractility.*

USES Mild, moderate, and severe hypertension. May be used alone or in combination with other antihypertensive agents, especially thiazide diuretics.

CONTRAINDICATIONS NSAID or salicylate hypersensitivity; bronchial asthma; uncontrolled cardiac failure, heart block (greater than first degree), cardiogenic shock, severe bradycardia; systolic blood pressure less than 100 mm Hg; perioperative CABG pain; pregnancy (category D third trimester).
CAUTIOUS USE Nonallergic bronchospastic disease (e.g., COPD); renal disease, renal failure, hepatic disease; well-compensated patients with history of heart

L

Common adverse effects in *italic*, life-threatening effects <u>underlined</u>; generic names in **bold;** classifications in SMALL CAPS; ✦ Canadian drug name; ⊘ Prototype drug

835

failure; acute MI; coronary artery disease; pheochromocytoma; liver dysfunction, jaundice; diabetes mellitus; SLE; myasthenia gravis; PVD; older adults; pregnancy (category B first and second trimester); lactation. Safety and efficacy in children not established.

ROUTE & DOSAGE

Hypertension

Adult: **PO** 100 mg b.i.d., may gradually increase to 200–400 mg b.i.d. (max: 1200–2400 mg/day). **IV** 20 mg slowly over 2 min, with 40–80 mg q10min if needed up to 300 mg total or 2 mg/min continuous infusion (max: 300 mg total dose)
Geriatric: **PO** Start with 100 mg daily **IV** 20 mg slowly over 2 min, with 40–80 mg q10min if needed up to 300 mg total or 2 mg/min continuous infusion (max: 300 mg total dose)

ADMINISTRATION

Oral

▪ Give with or immediately after food consistently. Food increases drug bioavailability.

Intravenous

Note: Amount of IV solution may be changed depending on patient status.

PREPARE: **Direct:** Give undiluted. **Continuous:** Dilute 300 mg in 240 mL of D5W, NS, D5/NS, LR, or other compatible IV solution to yield 1 mg/mL.

ADMINISTER: **Direct:** Give a 20-mg dose slowly over 2 min. ▪ Maximum hypotensive effect occurs 5–15 min after each administration. **Continuous:** Normal rate is 2 mg/min. ▪ Controlled infusion pump device is recommended for maintaining accurate flow rate during IV infusion. ▪ Keep patient supine when receiving labetalol IV. ▪ Take BP immediately before administration. Rate is adjusted according to BP response. ▪ Discontinue drug once the desired BP is attained.

INCOMPATIBILITIES **Solution/additive: Sodium bicarbonate, ceftriaxone, tenecteplase. Y-site: Amphotericin B cholesteryl, furosemide, heparin, nafcillin, thiopental, warfarin.**

▪ Store at 2°–30° C (36°–86° F) unless otherwise advised. Do not freeze. ▪ Protect tablets from moisture.

ADVERSE EFFECTS (≥1%) **CNS:** Dizziness, fatigue/malaise, headache, tremors, transient paresthesias (especially scalp tingling), hypoesthesia (numbness) following IV, mental depression, drowsiness, sleep disturbances, nightmares. **CV:** *Postural hypotension,* angina pectoris, palpitation, bradycardia, syncope, pedal or peripheral edema, pulmonary edema, CHF, flushing, cold extremities, arrhythmias (following IV), paradoxical hypertension (patients with pheochromocytoma). **Special Senses:** Dry eyes, vision disturbances, nasal stuffiness, rhinorrhea. **GI:** Nausea, vomiting, dyspepsia, constipation, diarrhea, taste disturbances, cholestasis with or without jaundice, increases in serum transaminases, dry mouth.

Common adverse effects in *italic*, life-threatening effects underlined; generic names in **bold**; classifications in SMALL CAPS; ♣ Canadian drug name; ⓟ Prototype drug

Urogenital: Acute urinary retention, difficult micturition, impotence, ejaculation failure, loss of libido, Peyronie's disease. **Respiratory:** Dyspnea, bronchospasm. **Skin:** Rashes of various types, increased sweating, pruritus. **Body as a Whole:** Myalgia, muscle cramps, toxic myopathy, antimitochondrial antibodies, positive antinuclear antibodies (ANA), SLE syndrome, pain at IV injection site.

DIAGNOSTIC TEST INTERFERENCE False increases in *urinary catecholamines* when measured by *nonspecific trihydroxyindole (THI) reaction* (due to labetalol metabolites) but not with specific *radioenzymatic* or *high-performance liquid chromatography assay techniques.*

INTERACTIONS Drug: Cimetidine may increase effects of labetalol; **glutethimide** decreases effects of labetalol; **halothane** adds to hypotensive effects; may mask symptoms of hypoglycemia caused by ORAL SULFONYLUREAS, **insulin;** BETA AGONISTS antagonize effects of labetalol.

PHARMACOKINETICS Absorption: Readily from GI tract, only 25% reaches systemic circulation due to first pass metabolism. **Onset:** 20 min–2 h PO; 2–5 min IV. **Peak:** 1–4 h PO; 5–15 min IV. **Duration:** 8–24 h PO; 2–4 h IV. **Distribution:** Crosses placenta; distributed into breast milk. **Metabolism:** In liver (CYP-2D6). **Elimination:** 60% in urine, 40% in bile. **Half-Life:** 3–8 h.

NURSING IMPLICATIONS
Assessment & Drug Effects
- Monitor BP and pulse during dosage adjustment period. Use standing BP as indicator for making dosage adjustments for oral drugs and assessing patient's tolerance of dosage increases. Take after patient stands for 10 min. Clarify with prescriber.
- Monitor BP at 5 min intervals for 30 min after IV administration; then at 30 min intervals for 2 h; then hourly for about 6 h; and as indicated thereafter.
- Monitor diabetic patients closely; drug may mask usual cardiovascular response to acute hypoglycemia (e.g., tachycardia).
- Maintain patient in supine position for at least 3 h after IV administration. Then determine patient's ability to tolerate elevated and upright positions before allowing ambulation. Manage this slowly.
- Lab tests: Periodic LFTs.

Patient & Family Education
- Note: Postural hypotension is most likely to occur during peak plasma levels (i.e., 2–4 h after drug administration).
- Make all position changes slowly and in stages, particularly from lying to upright position. Older adult patients are especially sensitive to hypotensive effects.
- Do not drive or engage in other potentially hazardous activities until response to drug is known.
- Diabetics should closely monitor blood sugar for loss of glycemic control.
- Do not abruptly stop taking this drug. It is usually discontinued gradually.

L

LACOSAMIDE
(lac-os′a-mide)
Vimpat
Classification: ANTICONVULSANT

Common adverse effects in *italic,* life-threatening effects <u>underlined</u>; generic names in **bold;** classifications in SMALL CAPS; ♣ Canadian drug name; ⊘ Prototype drug

837

Therapeutic: ANTICONVULSANT
Pregnancy Category: C
Controlled Substance: Schedule V

AVAILABILITY 50 mg, 100 mg, 150 mg, 200 mg tablets; 200 mg/20 mL single dose vial, 10 mg/mL oral solution.

ACTION & THERAPEUTIC EFFECT

Lacosamide selectively enhances slow inactivation of voltage-gated sodium channels, thus stabilizing hyper-excitable membranes and inhibiting repetitive neuronal firing. *Decreases frequency of partial-onset seizures in those treated with multiple antiepileptic drugs.*

USES Adjunctive therapy in the treatment of partial-onset seizures.
UNLABELED USES Neuropathic pain.

CONTRAINDICATIONS Severe hepatic impairment; suicidal ideation; lactation.
CAUTIOUS USE History of multiorgan hypersensitivity reactions; CrCl 30 mL/min or less; renal disease; mild to moderate hepatic impairment; history of suicidal tendencies, history of chronic depression; cardiovascular disease, cardiac conduction problems (e.g. second-degree AV block); myocardial ischemia, heart failure; seizure disorders; diabetic neuropathy; older adults; pregnancy (category C). Safety and effectiveness in children less than 17 y have not been established.

ROUTE & DOSAGE

Partial-Onset Seizures

Adult: **PO or IV** 50 mg b.i.d.; may increase by 100 mg/day qwk to 100–200 mg b.i.d. Note: Patients may be switched from PO to IV (or vice versa) using equivalent doses and administration frequency.

Hepatic Impairment Dosage Adjustment

Mild to moderate impairment: Maximum daily dose of 300 mg
Severe impairment: Not recommended

Renal Impairment Dosage Adjustment

CrCl less than 30 mL/min: Maximum daily dose of 300 mg

ADMINISTRATION

Oral

- Do not abruptly stop medication; it should be withdrawn gradually, over a minimum of 1 wk.

- Store tablets at 15°–30° C (59°–86° F).

Intravenous

PREPARE: IV Infusion: May give undiluted or diluted in NS, D5W, LR.
ADMINISTER: IV Infusion: Give over 30–60 min.
INCOMPATIBILITIES Solution/additive: Do not mix with other drugs.

- May be stored diluted for up to 26 h at 15°–30° C (59°–86° F).

ADVERSE EFFECTS (≥1%) Body as a Whole: Asthenia, contusion, *fatigue*, gait disturbance, injection site pain and irritation, skin laceration. **CNS:** *Ataxia*, balance disorder, depression; *dizziness, headache*, memory impairment, nystagmus, somnolence, tremor, vertigo. **GI:** Di-arrhea, *nausea*, vomiting. **Skin:** Pruritus. **Special Senses:** *Blurred vision, diplopia.*

Common adverse effects in *italic*, life-threatening effects underlined; generic names in **bold**; classifications in SMALL CAPS; ✦ Canadian drug name; ⊕ Prototype drug

INTERACTIONS Drug: May prolong QT interval when given with other drugs known to affect QT interval (e.g., ANTIARRYTHMICS, **astemizole, bepridil, droperidol**).

PHARMACOKINETICS Absorption: Approximately 100% oral absorption. **Peak:** 1–4 h. **Distribution:** Less than 15% protein bound. **Metabolism:** Hepatic oxidation to less active metabolite. **Elimination:** Primarily fecal (95%). **Half-Life:** 13 h.

NURSING IMPLICATIONS

Assessment & Drug Effects
- Monitor for and record all seizure activity.
- Monitor for and report promptly suicidal behavior and ideation.
- Monitor closely patients with known cardiac conduction abnormalities. Baseline and periodic ECGs are recommended
- Lab test: Periodic CBC with differential, especially with long-term therapy.
- Monitor older adults for adverse reactions after each upward dose titration. Institute safety precautions as needed to avoid injury from falls.

Patient & Family Education
- Exercise caution with hazardous activities until reaction to drug is known.
- Report promptly feelings of depression, unusual changes in mood or behavior, or thoughts of inflicting harm to self.
- Report bothersome neurologic symptoms such as dizziness, loss of balance, and blurred vision.
- Do not abruptly stop taking this drug. It **must be** tapered off.

LACTULOSE
(lak'tyoo-lose)

Cephulac, Chronulac
Classifications: HYPEROSMOTIC LAXATIVE; NEUROLOGIC
Therapeutic: LAXATIVE; AMMONIUM DETOXICANT
Pregnancy Category: B

AVAILABILITY 10 g/15 mL solution, syrup

ACTION & THERAPEUTIC EFFECT Reduces blood ammonia by acidifying colon contents, thus retarding diffusion of nonionic ammonia (NH_3) from colon to blood while promoting its migration from blood to colon. In the acidic colon, NH_3 is converted to nonabsorbable ammonium ions (NH_4^+) and is then expelled in feces. *Osmotic effect of lactulose moves water from plasma to intestines, softening stools, and stimulates peristalsis by pressure from water content of stool. Decreases blood ammonia in a patient with hepatic encephalopathy. Effectiveness is marked by improved EEG patterns and mental state (clearing of confusion, apathy, and irritation).*

USES Prevention and treatment of portal-systemic encephalopathy (PSE), including stages of hepatic precoma and coma, and by prescription for relief of chronic constipation.
UNLABELED USES To restore regular bowel habit posthemorrhoidectomy; to evacuate bowel in older adult patients with severe constipation after barium studies; and for treatment of chronic constipation in children.

CONTRAINDICATIONS Low galactose diet.
CAUTIOUS USE Diabetes mellitus; concomitant use with electrocautery

Common adverse effects in *italic*, life-threatening effects underlined; generic names in **bold**; classifications in SMALL CAPS; ✦ Canadian drug name; ☺ Prototype drug

839

procedures (proctoscopy, colonoscopy); older adult and debilitated patients; pediatric use; pregnancy (category B), lactation.

ROUTE & DOSAGE

Prevention and Treatment of Portal-Systemic Encephalopathy

Adult: **PO** 30–45 mL t.i.d. or q.i.d. adjusted to produce 2–3 soft stools/day

Adolescent/Child: **PO** 40–90 mL/day in divided doses adjusted to produce 2–3 soft stools/day

Infant: **PO** 2.5–10 mL/day in 3–4 divided doses adjusted to produce 2–3 soft stools/day

Management of Acute Portal-Systemic Encephalopathy

Adult: **PO** 30–45 mL q1–2h until laxation is achieved, then adjusted to produce 2–3 soft stools/day **Rectal** 300 mL diluted with 700 mL water given via rectal balloon catheter, and retained for 30–60 min, may repeat in 4–6 h if necessary or until patient can take PO

Chronic Constipation

Adult: **PO** 30–60 mL/day prn
Child: **PO** 7.5 mL/day after breakfast

ADMINISTRATION

Oral

- Give with fruit juice, water, or milk (if not contraindicated) to increase palatability. Laxative effect is enhanced by taking with ample liquids. Avoid meal times.

Rectal

- Administer as a retention enema via a rectal balloon catheter. If solution is evacuated too soon, instillation may be promptly repeated.

- Do not freeze. Avoid prolonged exposure to temperatures above 30° C (86° F) or to direct light. Normal darkening does not affect action, but discard solution that is very dark or cloudy.

ADVERSE EFFECTS (≥1%) **GI:** Flatulence, borborygmi, belching, abdominal cramps, pain, and distention (initial dose); *diarrhea* (excessive dose); nausea, vomiting, colon accumulation of hydrogen gas; hypernatremia.

INTERACTIONS **Drug:** LAXATIVES may incorrectly suggest therapeutic action of lactulose.

PHARMACOKINETICS **Absorption:** Poorly absorbed from GI tract. **Metabolism:** In gut by intestinal bacteria.

NURSING IMPLICATIONS

Assessment & Drug Effects

- In children if the initial dose causes diarrhea, dosage is reduced immediately. Discontinue if diarrhea persists.
- Promote fluid intake (1500–2000 mL/day or greater) during drug therapy for constipation; older adults often self-limit liquids. Lactulose-induced osmotic changes in the bowel support intestinal water loss and potential hypernatremia. Discuss strategy with prescriber.

Patient & Family Education

- Laxative action is not instituted until drug reaches the colon; therefore, about 24–48 h is needed.
- Do not self-medicate with another laxative due to slow onset of drug action.
- Notify prescriber if diarrhea (i.e., more than 2 or 3 soft stools/day) persists more than 24–48 h. Diarrhea is a sign of overdosage. Dose adjustment may be indicated.

LAMIVUDINE ⓟ

(lam-i-vu'deen)

Epivir, Epivir-HBV, Heptovir ♥
Classifications: ANTIRETROVIRAL;
NUCLEOSIDE REVERSE TRANSCRIPTASE
INHIBITOR (NRTI)
Therapeutic: ANTIRETROVIRAL; NRTI
Pregnancy Category: C

AVAILABILITY 100 mg, 150 tablets;
5 mg/mL, 10 mg/mL oral solution

ACTION & *THERAPEUTIC EFFECT*
Lamivudine is a synthetic nucleoside analog reverse transcriptase inhibitor. It inhibits the transcription of the HIV viral RNA chain as well as the hepatitis B viral RNA chain. *Antiviral action is effective against HIV viruses and hepatitis B (HBV) viral infections.*

USES HIV infection in combination with zidovudine; treatment of chronic hepatitis B.

CONTRAINDICATIONS Hypersensitivity to lamivudine, lactation.
CAUTIOUS USE Renal impairment, renal failure; diabetes mellitus; obesity; pregnancy (category C).

ROUTE & DOSAGE

HIV Infection

Adult: Epivir **PO** 150 mg b.i.d.
Child (3 mo–16 y): Epivir **PO**
4 mg/kg b.i.d. (max: 150 mg b.i.d.)

Renal Impairment Dosage Adjustment

CrCl 30–49 mL/min: 150 mg daily; *15–29 mL/min:* 150 mg first dose, then 100 mg daily; *5–14 mL/min:* 150 mg first dose, then 50 mg daily; *less than 5 mL/min:* 50 mg first dose, then 25 mg daily

Chronic Hepatitis B

Adult: Epivir-HBV **PO** 100 mg daily

Renal Impairment Dosage Adjustment

CrCl 30–49 mL/min: 100 mg first dose, then 50 mg daily; *15–29 mL/min:* 100 mg first dose, then 25 mg daily; *5–14 mL/min:* 35 mg first dose, then 15 mg daily; *less than 5 mL/min:* 35 mg first dose, then 10 mg daily

ADMINISTRATION

Oral
- Give Epivir b.i.d. in combination with AZT. The recommended dose for adults who weigh less than 50 kg (110 lb) is 2 mg/kg. Give Epivir-HBV daily; do NOT give in combination with AZT.
- Store solution at 2°–25° C (36°–77° F) tightly closed.

ADVERSE EFFECTS (≥1%) **CNS:** *Neuropathy, insomnia,* sleep disorders, *dizziness,* depression, *headache,* fatigue, *fever, chills.* **GI:** *Nausea, diarrhea,* vomiting, anorexia, abdominal pain, cramps, dyspepsia, increased LFTs (ALT, amylase), hepatomegaly with steatosis. **Hematologic:** Neutropenia, anemia, thrombocytopenia. **Musculoskeletal:** Myalgia, arthralgia, malaise, pain. **Skin:** Rash. **Respiratory:** Nasal symptoms, cough. **Metabolic:** Lactic acidosis.

INTERACTIONS Drug: Increases the C_{max} of **zidovudine. Trimethoprim-sulfamethoxazole** increases serum levels of lamivudine. Increased risk of lactic acidosis in combination with other REVERSE TRANSCRIPTASE INHIBITORS and ANTIRETROVIRAL AGENTS.

PHARMACOKINETICS Absorption: Rapidly absorbed from GI tract

L

(86% reaches systemic circulation). **Distribution:** Low binding to plasma proteins. **Metabolism:** Minimal. **Elimination:** Excreted primarily unchanged in urine. **Half-Life:** 2–4 h.

NURSING IMPLICATIONS

Assessment & Drug Effects

- Monitor children closely for S&S of pancreatitis; if they occur, immediately stop drug and notify prescriber.
- Lab tests: Monitor CBC with differential, kidney & liver function, and serum amylase throughout therapy.
- Monitor for and report all significant adverse reactions.

Patient & Family Education

- Report any of the following immediately: Nausea, vomiting, anorexia, abdominal pain, jaundice.
- Note: The long-term effects of lamivudine are unknown.

LAMOTRIGINE

(la-mo′tri-geen)

Lamictal, Lamictal CD, Lamictal XR
Classification: ANTICONVULSANT
Therapeutic: ANTICONVULSANT
Pregnancy Category: C

AVAILABILITY 25 mg, 100 mg, 150 mg, 200 mg tablets; 2 mg, 5 mg, 25 mg chewable tablets; 25 mg, 50 mg, 100 mg, 200 mg orally disintegrating tablet; 25 mg, 50 mg, 100 mg, 200 mg extended release tablet

ACTION & *THERAPEUTIC EFFECT*
May act by inhibiting the release of glutamate and aspartate, excitatory neurotransmitters at voltage-sensitive sodium channels, resulting in decreased seizure activity in the brain. This stabilizes neuronal membranes. *Effectiveness is measured by decreasing seizure activity.*

USES Adjunctive therapy for partial seizures. Generalized tonic–clonic, grand mal, or myoclonic seizures in adults, treatment of bipolar disorder (immediate release only).
UNLABELED USES Absence seizures, prevention of migraines.

CONTRAINDICATIONS Hypersensitivity to lamotrigine, development of any skin rash unless it is clearly not related to the drug; suicidal ideation especially if severe or abrupt in onset; lactation. Lamotrigine ER is not approved for children younger than 13 y.
CAUTIOUS USE Renal insufficiency, bipolar disorder, history of suicidal tendencies; CHF, cardiac or liver function impairment; severe renal impairment; moderate to severe hepatic impairment; older adults; pregnancy (category C). Safety and efficacy in acute treatment of a mood episode have not been established. Safety and efficacy have not been established in children with a mood disorder who are younger than 16 y. Safety and efficacy for treatment of seizures in children 2 y or younger are not established.

ROUTE & DOSAGE

Partial Seizures, Patients Receiving Anticonvulsants Other Than Valproic Acid

Adult/Adolescent: **PO** Start with 50 mg daily for 2 wk, then 50 mg b.i.d. for 2 wk, may titrate up to 300–500 mg/day in 2 divided doses (max: 700 mg/day)
Child (2–12 y): **PO** 0.3 mg/kg/day in divided doses × 2 wk, then 0.6 mg/kg/day in divided doses × 2 wk (max: 15 mg/kg/day or 400 mg/day)

Common adverse effects in *italic*, life-threatening effects underlined; generic names in **bold**; classifications in SMALL CAPS; ♣ Canadian drug name; ⊘ Prototype drug

Partial Seizures, Patients Receiving Valproic Acid

Adult: **PO** Start with 25 mg every other day for 2 wk, then 25 mg daily for 2 wk, may titrate up to 150 mg/day in 2 divided doses (max: 200 mg/day)

Child (2–16 y): **PO** 0.15 mg/kg/day in divided doses × 2 wk, then increase to 0.3 mg/kg/day in divided doses × 2 wk (max: 5 mg/kg/day or 250 mg/day)

Bipolar Disorder, Patients Not Receiving Valproate or Carbamazepine

Adult: **PO** Start with 25 mg daily for 2 wk, then 50 mg daily for 2 wk, then 100 mg/day for 1 wk, then 200 mg daily

Bipolar Disorder, Patients Receiving Valproic Acid

Adult/Adolescent (older than 16 y): **PO** Start with 25 mg every other day for 2 wk, then 25 mg daily for 2 wk, then 50 mg daily for 1 wk, then 100 mg daily

Bipolar Disorder, Patients Receiving Carbamazepine

Adult/Adolescent (older than 16 y): **PO** Start with 50 mg daily for 2 wk, then 50 mg b.i.d. for 2 wk, then 100 b.i.d. for 1 wk, then 150 mg b.i.d. for 1 wk, then 200 mg b.i.d.

Hepatic Impairment Dosage Adjustment

Reduce dose by 25%

Severe Hepatic Impairment and Ascites Dosage Adjustment

Reduce dose by 50%

ADMINISTRATION

Oral

- *Orally disintegrating tablet (ODT)* should be placed on the tongue and moved around the mouth. It will rapidly disintegrate. It may be swallowed with/without water or food.
- Ensure that *chewable tablets* are chewed or crushed before being swallowed with a liquid.
- *Chewable dispersible tablets* may be swallowed whole, chewed, or mixed in water or diluted juice.
- When discontinued, drug should be tapered off gradually over a 2-wk period, unless patient safety is at risk.

ADVERSE EFFECTS (≥1%) **CNS:** *Dizziness, ataxia, somnolence, headache,* aphasia, vertigo, confusion, slurred speech, irritability, depression, incoordination, hostility. **GI:** *Nausea,* vomiting, anorexia, abdominal pain, diarrhea, dyspepsia, constipation. **Urogenital:** Hematuria, dysmenorrhea, vaginitis. **Special Senses:** *Diplopia, blurred vision.* **Musculoskeletal:** Peripheral neuropathy, chills, tremor, arthralgia. **Skin:** Rash (including <u>Stevens-Johnson syndrome, toxic epidermal necrolysis</u>), urticaria, pruritus, alopecia, acne. **Respiratory:** *Rhinitis,* pharyngitis, cough.

INTERACTIONS **Drug:** Car**bamazepine, phenobarbital, primidone, phenytoin, fosphenytoin, rifampin,** ORAL CONTRACEPTIVES may decrease lamotrigine levels. **Valproic acid** may increase lamotrigine levels. Lamotrigine may decrease serum levels of **valproic acid.** May affect efficacy of ORAL CONTRACEPTIVES. Chronic **acetaminophen** use may affect serum concentrations of lamotrigine. **Herbal:** **Ginkgo** may decrease anticonvulsant effectiveness. **Evening primrose** oil may affect seizure threshold.

Common adverse effects in *italic*, life-threatening effects <u>underlined</u>; generic names in **bold**; classifications in SMALL CAPS; ♣ Canadian drug name; ☯ Prototype drug

843

PHARMACOKINETICS Absorption: Readily from GI tract; 98% reaches systemic circulation. **Onset:** 12 wk. **Peak:** 1–4 h. **Distribution:** 55% protein bound; crosses placenta; distributed into breast milk. **Metabolism:** In liver to inactive metabolite. **Elimination:** Can induce own metabolism; excreted in urine. **Half-Life:** 25–30 h.

NURSING IMPLICATIONS

Assessment & Drug Effects

- Withhold drug if rash develops and immediately report to prescriber.
- Monitor the plasma levels of lamotrigine and other anticonvulsants when given concomitantly.
- Monitor patients with bipolar disorder for worsening of their symptoms and suicidal ideation. Withhold the drug and immediately report to prescriber.
- Monitor for adverse reactions when lamotrigine is used with other anticonvulsants, especially valproic acid.
- Be aware of drug interactions and closely monitor when interacting drugs are added or discontinued.

Patient & Family Education

- Do not take drug if a skin rash develops. Contact your prescriber immediately.
- Notify prescriber for any of the following: Worsening seizure control, skin rash, ataxia, blurred vision or diplopia, fever or flu-like symptoms.
- Do not drive or engage in other potentially hazardous activities until response to the drug is known.
- Use protection from sunlight or ultraviolet light until tolerance is known; drug increases photosensitivity.
- Women using oral contraceptives to avoid pregnancy should add a barrier contraceptive.
- Schedule periodic ophthalmologic exams with long-term use.
- Do not discontinue abruptly.

LANREOTIDE ACETATE
(lan-re′o-tide)

Somatuline Depot
Classifications: SOMATOSTATIN ANALOG; ANTIGROWTH HORMONE
Therapeutic: ACROMEGALY AGENT; SOMATOSTATIN HORMONE REPLACEMENT
Prototype: Octreotide
Pregnancy Category: C

AVAILABILITY 60 mg, 90 mg, 120 mg single-dose syringe

ACTION & *THERAPEUTIC EFFECT*

An analog of natural somatostatin produced by the hypothalamus. Somatostatin inhibits the secretion of growth hormone (GH) that in turn inhibits secretion of insulin-like growth factor-1 (IGF-1). *Inhibition of IGF-1 suppresses proliferation of chondrocytes and stops bone growth.*

USES Long-term treatment of acromegaly.

UNLABELED USES Treatment of diarrhea and cutaneous flushing, with carcinoid neuroendocrine tumors; hyperthyroidism.

CONTRAINDICATIONS Hypersensitivity to lanreotide; lactation.

CAUTIOUS USE Cholelithiasis, gallbladder disease; gastroparesis; hypothyroidism; diabetes mellitus; cardiac disease, bradycardia; moderate to severe hepatic or renal impairment; pregnancy (category C). Safety and efficacy in children have not been established.

Common adverse effects in *italic*, life-threatening effects underlined; generic names in **bold**; classifications in SMALL CAPS; ♣ Canadian drug name; ⊕ Prototype drug

ROUTE & DOSAGE

Acromegaly

Adult: **Subcutaneous** 90 mg q4wk × 3 mo; then 60–120 mg q4wk depending on response

Hepatic/Renal Impairment Dosage Adjustment

Moderate/severe impairment: Initial dose of 60 mg then adjust

ADMINISTRATION

Subcutaneous

- Remove from refrigeration 30 min before injection to allow to come to room temperature, but keep in sealed pouch.
- Inject deeply into subcutaneous tissue in the upper, outer quadrant of the buttock. Do not fold skin. Insert needle rapidly and to its full length at a right angle to skin. Alternate right and left sides.
- Store refrigerated and protect from light in original package.

ADVERSE EFFECTS (≥1%) **Body as a Whole:** Injection site pain, pancreatitis. **CNS:** Headache. **CV:** Bradycardia. **GI:** *Abdominal pain, cholelithiasis,* constipation, *diarrhea,* flatulence, loose stools, *nausea,* vomiting. **Hematologic:** Anemia. **Metabolic:** Decreased weight. **Musculoskeletal:** Arthralgia. **Skin:** Pruritus.

INTERACTIONS Drug: BETA-BLOCKERS have an additive effect on the reduction of heart rate. Lanreotide may increase the bioavailability of **bromocriptine.** Coadministration may decrease the bioavailability of **cyclosporine.** Lanreotide may increase the plasma level of drugs metabolized by CYP 450 enzymes.

PHARMACOKINETICS Absorption: 69–79% depending on dose. **Peak:** Within 24 h. **Metabolism:** Extensive, only 5% excreted unchanged. **Elimination:** Stool (95%) and urine (5%). **Half-Life:** 23–30 days.

NURSING IMPLICATIONS

Assessment & Drug Effects

- Monitor cardiovascular status as bradycardia and hypertension are potential adverse effects.
- Lab tests: Baseline and periodic GH, IGF-1, blood glucose; baseline LFTs and renal function tests; periodic thyroid function tests, HCT and Hgb.
- Periodic gallbladder motility tests are recommended.
- Monitor diabetics for loss of glycemic control.

Patient & Family Education

- Ensure that proper technique for subcutaneous drug administration is utilized.
- Diabetics should monitor blood glucose values often as this drug may elevate or lower blood glucose.
- Report to prescriber severe pain in the upper right area of the stomach/abdomen, along with nausea and vomiting. These symptoms could indicate the presence of gallstones.

LANSOPRAZOLE

(lan′so-pra-zole)

Prevacid, Prevacid 24 HR

DEXLANSOPRAZOLE

DexiLant

Classifications: ANTISECRETORY; PROTON PUMP INHIBITOR

Therapeutic: ANTIULCER; ANTISECRETORY

Prototype: Omeprazole
Pregnancy Category: B

L

Common adverse effects in *italic*, life-threatening effects underlined; generic names in **bold**; classifications in SMALL CAPS; ♣ Canadian drug name; ⊘ Prototype drug

845

AVAILABILITY 15 mg, 30 mg sustained release capsules; 15 mg, 30 mg orally disintegrating tablets; **Dexlansoprazole (delayed release):** 30 mg, 60 mg capsules

ACTION & *THERAPEUTIC EFFECT* Suppresses gastric acid secretion by inhibiting the H+, K+-ATPase enzyme [the acid (proton H+) pump] in the parietal cells. *Suppresses gastric acid formation in the stomach.*

USES Short-term treatment of duodenal ulcer (up to 4 wk) and erosive esophagitis (up to 8 wk), pathologic hypersecretory disorders, gastric ulcers; in combination with antibiotics for *Helicobacter pylori.* Gastroesophageal reflux disease (GERD).
UNLABELED USES Stress gastritis prophylaxis.

CONTRAINDICATIONS Hypersensitivity to lansoprazole or dexlansoprazole; proton pump inhibitors (PPIs) hypersensitivity, lactation.
CAUTIOUS USE Hepatic impairment; gastric malignancy; pregnancy (category B), infants.

ROUTE & DOSAGE

Duodenal Ulcer
Adult: **PO** 15 mg once daily × 4 wk (treatment) or longer (maintenance)

Erosive Esophagitis
Adult: **PO** 30 mg once daily × 8 wk, then decrease to 15 mg once daily

GERD
Adult: **PO** 15 mg once daily for up to 8 wk
Child (1–11 y): **PO** 15–30 mg daily for up to 12 wk (max: 30 mg/day)

Hypersecretory Disorder
Adult: **PO** 60 mg once daily (max: 120 mg/day in divided doses), may need to be adjusted for hepatic impairment

NSAID-Associated Gastric Ulcer
Adult: **PO** 30 mg daily for up to 8 wk

H. pylori
Adult: **PO** 30 mg b.i.d. × 2 wk, in combination with antibiotics

Hepatic Impairment Dosage Adjustment
Severe hepatic disease (Child-Pugh class C): Reduce dose (max: 30 mg/day) **(Dexlansoprazole)**

Healing Erosive Esophagitis (Dexlansoprazole)
Adult: **PO** 60 mg daily for up to 8 wk

Maintenance of Healed Erosive Esophagitis (Dexlansoprazole)
Adult: **PO** 30 mg daily

Non-Erosive GERD (Dexlansoprazole)
Adult: **PO** 30 mg daily for 4 wk

ADMINISTRATION

Oral
- *All forms:* Administer dosage 30 min a.c. Give once daily dose before breakfast.
- Give at least 30 min prior to any concurrent sucralfate therapy.
- Do not crush or chew capsules. Capsules can be opened and granules sprinkled on food or mixed with 40 mL of apple juice and administered through an NG tube. Do not crush or chew granules.
- Note: Disintegrating tablets contain phenylalanine and should not be used for patients with PKU. Capsule and syrup formulations do not contain phenylalanine.

Common adverse effects in *italic*, life-threatening effects <u>underlined</u>; generic names in **bold**; classifications in SMALL CAPS; ♣ Canadian drug name; ⊕ Prototype drug

ADVERSE EFFECTS (≥1%) **CNS:** Fatigue, dizziness, headache. **GI:** Nausea, *diarrhea,* constipation, anorexia, increased appetite, thirst, elevated serum transaminases (AST, ALT). **Skin:** Rash.

INTERACTIONS Drug: May decrease **theophylline** levels. **Sucralfate** decreases lansoprazole bioavailability. May interfere with absorption of **ketoconazole, digoxin, ampicillin,** or IRON SALTS. Use with **warfarin** may increase INR. May alter **tacrolimus** concentration. **Food:** Food reduces peak lansoprazole levels by 50%.

PHARMACOKINETICS Absorption: Rapidly from GI tract after leaving stomach. **Onset:** Acid reduction within 2 h; ulcer relief within 1 wk. **Peak:** 1.5–3 h. Dexlansoprazole has 2 peaks, at 1–2 h then at 4–5 h. **Duration:** 24 h. **Distribution:** 97% bound to plasma proteins. **Metabolism:** In liver via CYP2C19 and 3A4. **Elimination:** 14–25% in urine; part of dose eliminated in bile and feces. **Half-Life:** 1.5 h.

NURSING IMPLICATIONS

Assessment & Drug Effects
- Lab tests: Monitor CBC, kidney and liver function tests, and serum gastric levels periodically.
- Monitor for therapeutic effectiveness of concurrently used drugs that require an acid medium for absorption (e.g., digoxin, ampicillin, ketoconazole).

Patient & Family Education
- Inform prescriber of significant diarrhea.

LANTHANUM CARBONATE
(lan-tha′num)

Fosrenol
Classifications: ELECTROLYTE AND WATER BALANCE AGENT; PHOSPHATE BINDER

Therapeutic: PHOSPHATE BINDER
Prototype: Sevelamer hydrochloride
Pregnancy Category: C

AVAILABILITY 500 mg, 750 mg, 1 g chewable tablets

ACTION & *THERAPEUTIC EFFECT*
Lanthanum is used for the management of hyperphosphatemia in end-stage renal disease; it is a calcium/aluminum-free phosphate binding agent. It has a higher affinity for binding to phosphate than calcium or aluminum. Low systemic absorption minimizes the risk of aluminum intoxication and hypercalcemia. Lanthanum decreases phosphate absorption from the diet. Dietary phosphate bound to lanthanum carbonate is excreted in the feces. *Lowers serum phosphate.*

USES Reduce serum phosphate levels in patients with end-stage renal disease.

CONTRAINDICATIONS Prior hypersensitivity to lanthanum carbonate, lactation.
CAUTIOUS USE Bowel obstruction, Crohn's disease, acute peptic ulcer, ulcerative colitis; pregnancy (category B); children younger than 18 y.

ROUTE & DOSAGE

Hyperphosphatemia
Adult: **PO** 250–500 mg t.i.d. with or immediately after meals; may titrate up every 2–3 wk in increments of 750 mg/day to achieve acceptable serum phosphate levels (max: 3750 mg/day)

ADMINISTRATION

Oral
- Give with or immediately after a meal.

L

Common adverse effects in *italic,* life-threatening effects underlined; generic names in **bold**; classifications in SMALL CAPS; ✤ Canadian drug name; ◉ Prototype drug

847

- Tablets **must be** chewed completely before swallowing. Whole tablets should not be swallowed.
- Store at 15°–30° C (59°–86° F).

ADVERSE EFFECTS (≥1%) **CNS:** Headache. **CV:** Hypotension. **GI:** Nausea, vomiting, diarrhea, abdominal pain, constipation. **Respiratory:** Bronchitis, rhinitis. **Other:** Dialysis graft occlusion.

PHARMACOKINETICS Absorption: Minimal from GI tract. **Metabolism:** Not metabolized. **Elimination:** In feces. **Half-Life:** 53 h.

NURSING IMPLICATIONS

Assessment & Drug Effects
- Monitor for dialysis graft occlusion, as lanthanum therapy may increase occlusion risk.
- Lab tests: Serum phosphate levels during dosage titration and regularly throughout treatment; periodic serum calcium, bicarbonate, and chloride.

Patient & Family Education
- Chew chewable tablets completely, then swallow.
- Report promptly any of the following: Headache, drowsiness, dizziness, fainting, confusion, irritability, nausea, vomiting, or loss of appetite.

LAPATINIB DITOSYLATE

(la-pa′ti-nib di-toe′si-late)

Tykerb

Classifications: BIOLOGICAL RESPONSE MODIFIER; ANTINEOPLASTIC; EPIDERMAL GROWTH FACTOR RECEPTOR-TYROSINE KINASE INHIBITOR (EGFR-TKI)

Therapeutic: ANTINEOPLASTIC
Prototype: Gefitinib
Pregnancy Category: D

AVAILABILITY 250 mg tablets

ACTION & THERAPEUTIC EFFECT
An inhibitor of intracellular tyrosine kinase domains of both epidermal growth factor receptors [EGFR (ErbB1) and HER2 (ErbB2)] required for cell proliferation of certain breast cancers. Inhibits ErbB-driven tumor cell growth in those who are positive for the HER2 receptor.

USES Treatment of advanced or metastatic breast cancer in patients whose tumor overexpresses the human epidermal receptor type 2 (HER2) protein and who have received prior therapy.

CONTRAINDICATIONS Hypersensitivity to lapatinib, capecitabine, doxifluridine, 5-FU; decreased left ventricular ejection fraction (LVEF) of grade 1, that is below lower limits of normal (LLN); myelosuppression; dihydropyrimidine dehydrogenase (DPD) deficiency; active infection; jaundice; severe renal failure or impairment; hypokalemia, hypomagnesemia; females of childbearing age; pregnancy (category D), lactation.

CAUTIOUS USE Moderate to severe hepatic impairment; coronary artery disease; angina, cardiac arrhythmias, congenital QT_c prolongation syndrome. Safe use in children younger than 18 y not established.

ROUTE & DOSAGE

Breast Cancer
Adult: **PO** 1250 mg daily on days 1–21 with capecitabine 2000 mg/m²/day q12h on days 1–14; repeat in 21-day cycle.
Postmenopausal women: **PO** 1500 mg/day with letrozole 2.5 mg/day
Hepatic Impairment Dosage Adjustment
Severe hepatic impairment (Child-Pugh class C): Reduce dose

ADMINISTRATION

Oral

- Give lapatinib at least 1 h before/after a meal.
- Give capecitabine with food or within 30 min after food.
- Note that concurrent use with strong CYP3A4 inhibitors/inducers should be avoided (see Drug Interactions). If concurrent use is necessary, dosage adjustments are required.
- Store at 15°–30° C (59°–86° F) in a tightly closed container.

ADVERSE EFFECTS (≥1%) **CNS:** Insomnia, *fatigue*. **CV:** QT prolongation. **GI:** *Diarrhea,* dyspepsia, mucosal inflammation, *nausea,* stomatitis, *vomiting*. **Hematologic:** Neutropenia, thrombocytopenia. **Metabolic:** Elevated ALT and AST levels, hyperbilirubinemia. **Musculoskeletal:** Back pain, pain in extremities. **Respiratory:** Dyspnea, pneumonitis. **Skin:** Dry skin, *palmar-plantar erythrodysesthesia, rash*.

INTERACTIONS Drug: INHIBITORS OF CYP3A4 (**ketoconazole, clarithromycin, atazanavir, indinavir, nefazodone, nelfinavir, ritonavir, saquinavir, telithromycin, voriconazole**) will increase lapatinib plasma level. INDUCERS OF CYP3A4 (**dexamethasone, phenytoin, carbamazepine, rifampin, rifabutin, rifapentine, phenobarbital**) will decrease lapatinib plasma level. Lapatinib can increase plasma levels of **theophylline** and **warfarin**. **Food: Grapefruit juice** may increase the plasma level of lapatinib; co-administration with food increases lapatinib plasma levels. **Herbal: St. John's wort** will decrease plasma levels of lapatinib.

PHARMACOKINETICS Peak: 4 h. **Distribution:** 9% Plasma protein bound. **Metabolism:** Extensive hepatic metabolism. **Elimination:** Fecal (major) and renal (minor). **Half-Life:** 24 h.

NURSING IMPLICATIONS

Assessment & Drug Effects

- Prior to initiating therapy, hypokalemia and hypomagnesemia should be corrected.
- Monitor cardiac status (i.e., LV ejection fraction, ECG with QT measurement) throughout therapy.
- Monitor for and report severe diarrhea as it may cause dehydration and serious electrolyte imbalances.
- Lab tests: Baseline and periodic serum electrolytes; periodic CBC with differential and platelet count, Hgb, Hct, LFTs.
- Monitor for theophylline and warfarin toxicity with concurrent use.

Patient & Family Education

- Adhere to directions regarding medication and food.
- Report promptly any of the following: Palpitations, shortness of breath, severe diarrhea.
- Do not eat grapefruit or drink grapefruit juice while taking lapatinib.
- Do not take St. John's wort or OTC medications for stomach ulcers while taking lapatinib unless approved by the prescriber.
- Women are advised to use effective means of contraception while taking lapatinib.

LATANOPROST ⊕

(la-tan'o-prost)

Xalatan
Classifications: EYE PREPARATION; PROSTAGLANDIN
Therapeutic: PROSTAGLANDIN
Pregnancy Category: B

AVAILABILITY 0.005% solution

ACTION & *THERAPEUTIC EF-FECT* Prostaglandin analog that is thought to reduce intraocular pressure (IOP) by increasing the outflow of aqueous humor. *Reduces elevated intraocular pressure in patients with open-angle glaucoma.*

USES Treatment of open-angle glaucoma, ocular hypertension, and elevated intraocular pressure (IOP).

CONTRAINDICATIONS Hypersensitivity to latanoprost or another component in the solution; intraocular infection; conjunctivitis.

CAUTIOUS USE Active intraocular inflammation such as iritis or uveitis; patients at risk for macular edema; hepatic or renal impairment; pregnancy (category B), lactation. Safety and effectiveness in children are not established.

ROUTE & DOSAGE

Glaucoma

Adult: **Ophthalmic** 1 drop in affected eye(s) daily in evening

ADMINISTRATION

Installation

- Ensure that contact lenses are removed prior to installation and not reinserted for 15 min after installation.
- Apply only to affected eye(s). Ensure that only one drop is instilled.
- Do not allow tip of dropper to touch eye.
- Wait at least 5 min before/after instillation of other eyedrops.
- Refrigerate at 2°–8° C (36°–46° F). Protect from light.

ADVERSE EFFECTS (≥1%) **Body as a Whole:** Headaches, asthenia, flu-like symptoms. **GI:** Abnormal liver function tests. **Skin:** Rash. **Special Senses:** *Conjunctival hyperemia, growth of eyelashes, ocular pruritus,* ocular dryness, visual disturbance, ocular burning, foreign body sensation, eye pain, pigmentation of the periocular skin, blepharitis, cataract, superficial punctate keratitis, eyelid erythema, ocular irritation, and eyelash darkening, eye discharge, tearing, photophobia, allergic conjunctivitis, increases in iris pigmentation (brown pigment), conjunctival edema.

INTERACTIONS Drug: Precipitation may occur if mixed with eyedrops containing **thimerosal;** space other EYE PREPARATIONS at least 5 min apart.

PHARMACOKINETICS Absorption: Absorbed through the cornea. **Onset:** 3–4 h. **Peak IOP Reduction:** 8–12 h. **Distribution:** Minimal systemic distribution. **Metabolism:** Hydrolyzed in aqueous humor to active form. **Elimination:** Renally excreted. **Half-Life:** 17 min.

NURSING IMPLICATIONS

Assessment & Drug Effects

- Withhold eyedrops and notify prescriber if acute intraocular inflammation (iritis or uveitis) or external eye inflammation are noted.
- Note that increased pigmentation of the iris and eyelid, and additional growth of eyelashes on the treated eye are adverse effects that may develop gradually over months to years.

Patient & Family Education

- Contact prescriber immediately if any ocular reaction occurs, especially conjunctivitis and lid reactions.

- Note: Increased pigmentation of the iris and eyelid, and additional growth of eyelashes on the treated eye, are possible adverse effects of this drug. Persons with light colored eyes receiving treatment to one eye may develop a darker eye.

LEFLUNOMIDE
(le-flu'no-mide)

Arava
Classifications: BIOLOGICAL RESPONSE MODIFIER; IMMUNOMODULATOR; ANTI-INFLAMMATORY
Therapeutic: DISEASE-MODIFYING ANTIRHEUMATIC DRUG (DMARD)
Pregnancy Category: X

AVAILABILITY 10 mg, 20 mg, 100 mg tablets

ACTION & *THERAPEUTIC EFFECT*
An immunomodulator that demonstrates anti-inflammatory effects. Suppression of pyrimidine synthesis in T and B lymphocytes interferes with RNA and protein synthesis in cells that are involved in the inflammatory process within affected joints. Reduction in activity of these lymphocytes leads to reduced cytokine and antibody-mediated destruction of the synovial joints as well as attenuation of the inflammatory process. *Reduces the S&S of rheumatoid arthritis (RA), retards structural joint damage, and improves physical function.*

USES Active RA.

CONTRAINDICATIONS Hepatic disease; jaundice; lactase deficiency; hypersensitivity to leflunomide; patients with positive hepatitis B or C serology; malignancy, particularly lymphoproliferative disorders; severe immunosuppression; vaccination, infants; uncontrolled infection; pregnancy (category X), lactation.

CAUTIOUS USE Renal insufficiency; renal failure; alcoholism; immunosuppression; lactase deficiency; infection. Use in children younger than 18 y has not been fully studied.

ROUTE & DOSAGE

Rheumatoid Arthritis
Adult: **PO** Initiate with a loading dose of 100 mg/day × 3 days, then maintenance dose of 20 mg daily, may decrease to 10 mg/day if higher dose is not tolerated

ADMINISTRATION

Oral
- Initiate with a 3-day loading dose followed by a lower maintenance dose.

ADVERSE EFFECTS (≥1%) **Body as a Whole:** Allergic reaction, asthenia, flu-like syndrome, infection, pain, back pain, arthralgia, leg cramps, synovitis, tenosynovitis. **CNS:** Dizziness, headache, paresthesias, peripheral neuropathy. **CV:** Hypertension, chest pain. **GI:** *Diarrhea,* increased LFTs (ALT and AST), abdominal pain, anorexia, dyspepsia, gastroenteritis, nausea, mouth ulcer, vomiting, weight loss, hepatotoxicity. **Metabolic:** Hypokalemia. **Respiratory:** Bronchitis, cough, respiratory infection, pharyngitis, pneumonia, rhinitis, sinusitis. **Skin:** Rash, alopecia, eczema, pruritus, dry skin, Stevens-Johnson syndrome, toxic epidermal necrolysis (rare). **Urogenital:** UTI.

L

Common adverse effects in *italic*, life-threatening effects underlined; generic names in **bold;** classifications in SMALL CAPS; ♣ Canadian drug name; ⊘ Prototype drug

851

INTERACTIONS Drug: Rifampin may significantly increase leflunomide levels; **cholestyramine, charcoal** decrease absorption; caution should be used with other hepatotoxic drugs.

PHARMACOKINETICS Absorption: Approximately 80% reaches systemic circulation. **Peak:** 6–12 h for active metabolite. **Distribution:** Greater than 99% protein bound. **Metabolism:** Metabolized primarily to M1 (active metabolite). **Elimination:** 43% in urine, 48% in feces. **Half-Life:** 19 days for active metabolite.

NURSING IMPLICATIONS

Assessment & Drug Effects

- Lab tests: Baseline screening to rule out hepatitis B or C; baseline and monthly LFTs × 12 mo, then every 6 mo thereafter.
- Monitor carefully for and report immediately S&S of infection; withhold leflunomide if infection is suspected.
- Monitor BP and weight periodically. Doses greater than 25 mg/day are associated with a greater incidence of side effects such as alopecia, weight loss, and elevated liver enzymes.

Patient & Family Education

- Use reliable contraception while taking leflunomide.
- Note: Both women and men need to discontinue leflunomide and undergo a drug elimination procedure prescribed by the prescriber BEFORE conception.
- Withhold drug if you develop an infection and notify the prescriber before resuming the drug.
- Notify prescriber about any of the following: Hair loss, weight loss, GI distress, rash, or itching.

LEPIRUDIN 🅟

(le-pir'u-din)

Refludan
Classifications: ANTICOAGULANT; DIRECT THROMBIN INHIBITOR
Therapeutic: ANTITHROMBOTIC; THROMBIN INHIBITOR
Pregnancy Category: B

AVAILABILITY 50 mg powder for injection

ACTION & *THERAPEUTIC EFFECT* Highly specific direct inhibitor of thrombin, including thrombin entrapped within established clots. One molecule of lepirudin binds to one molecule of thrombin and thereby blocks the thrombogenic activity of thrombin. Increases PT/INR and aPTT values in relation to the dose given. *Has antithrombotic activity. Its effectiveness is indicated by aPTT value in target range of 1.5 to 2.5.*

USES Anticoagulation in patients with heparin-induced thrombocytopenia (HIT).

CONTRAINDICATIONS Hypersensitivity to lepirudin; intracranial bleeding; patients with increased risk of bleeding (e.g., recent surgery, CVA, advanced kidney impairment); lactation.

CAUTIOUS USE Serious liver injury (e.g., cirrhosis); renal impairment; pregnancy (category B); children.

ROUTE & DOSAGE

Anticoagulation

Adult: **IV** 0.4 mg/kg initial bolus (max: 44 mg) followed by 0.15 mg/kg/h (max: 16.5 mg/h) for 2–10 days, adjust rate to maintain aPTT of 1.5–2.5

Common adverse effects in *italic*, life-threatening effects <u>underlined</u>; generic names in **bold**; classifications in SMALL CAPS; ♣ Canadian drug name; 🅟 Prototype drug

Renal Impairment Dosage Adjustment

CrCl 45–60 mL/min: Initial dose 0.2 mg/kg, then 0.075 mg/kg/h; 30–44 mL/min: Initial dose 0.2 mg/kg then 0.045 mg/kg/h; 15–29 mL/min: Initial dose 0.2 mg/kg, then 0.0225 mg/kg/h; less than 15 mL/min: Do not use

ADMINISTRATION

Intravenous

PREPARE: **Direct:** Reconstitute by adding 1 mL of sterile water for injection or NS to the 50-mg vial. ▪ To prepare bolus dose, withdraw reconstituted solution into a 10-mL syringe and dilute to 10 mL with sterile water for injection, NS or D5W to yield 5 mg/mL. **Continuous:** Transfer the contents of two reconstituted vials into 250 or 500 mL of D5W or NS to yield 0.4 or 0.2 mg/mL, respectively.

ADMINISTER: **Direct:** Give over 15–20 sec. **Continuous:** Give at a rate determined by body weight. If aPTT ratio is above the target range (1.5–2.5), stop infusion for 2 h, then restart at 50% of original rate (no additional IV bolus should be given). The aPTT ratio should be determined again in 4 h. ▪ If aPTT ratio is below the target range (1.5–2.5), the infusion rate should be stepped up in 20% increments. The aPTT ratio should be determined again 4 h later. ▪ Note: An infusion rate greater than 0.21 mg/kg/h is not advised without checking for coagulation abnormalities that might impair an appropriate aPTT response.

▪ Diluted solution is stable for 24 h during infusion. Store unopened vials at 2°–25° C (36°–77° F).

ADVERSE EFFECTS (≥1%) **CNS:** Intracranial bleeding. **CV:** Heart failure, ventricular fibrillation, pericardial effusion, MI. **GI:** Abnormal LFTs. **Hematologic:** Bleeding from injection site, anemia, hematoma, bleeding, hematuria, GI and rectal bleeding, epistaxis, hemothorax, vaginal bleeding. **Respiratory:** Pneumonia, cough, bronchospasm, stridor, dyspnea. **Skin:** Allergic skin reactions. **Body as a Whole:** Sepsis, abnormal kidney function, multiorgan failure.

INTERACTIONS **Drug: Warfarin,** NSAIDS, SALICYLATES, ANTIPLATELET AGENTS increases risk of bleeding. **Herbal: Feverfew, ginkgo, ginger, valerian** may potentiate bleeding.

PHARMACOKINETICS **Distribution:** Distributed primarily to extracellular compartment. **Metabolism:** By catabolic hydrolysis in serum. **Elimination:** 48% in urine. **Half-Life:** 1.3 h.

NURSING IMPLICATIONS

Assessment & Drug Effects

▪ Lab tests: Baseline aPTT prior to initiation of therapy (withhold therapy and notify prescriber if baseline aPTT ratio 2.5 or greater); aPTT 4 h after start of therapy and at least once daily (more often with renal or hepatic impairment) thereafter.
▪ Give with extreme caution to those at increased risk for bleeding.
▪ Monitor carefully for bleeding events (e.g., from puncture wounds, hematoma, hematuria); and report immediately.
▪ Do not give oral anticoagulants until lepirudin dose has been reduced and aPTT ratio lowered to just above 1.5.

L

Common adverse effects in *italic*, life-threatening effects underlined; generic names in **bold**; classifications in SMALL CAPS; ♣ Canadian drug name; ⓟ Prototype drug

853

LETROZOLE
(le'tro-zole)

Femara
Classifications: ANTINEOPLASTIC;
AROMATASE INHIBITOR
Therapeutic: ANTINEOPLASTIC
Prototype: Anastrozole
Pregnancy Category: D

AVAILABILITY 2.5 mg tablets

ACTION & *THERAPEUTIC EFFECT*
Nonsteroid competitive inhibitor of aromatase, the enzyme that converts androgens to estrogens. It does not inhibit adrenal steroid synthesis. *Results in the regression of estrogen-dependent tumors.*

USES Treatment of breast cancer in postmenopausal women.
UNLABELED USES Infertility, delayed puberty.

CONTRAINDICATIONS Hypersensitivity to letrozole; pregnant women, women of childbearing age, premenopausal females, hormone replacement therapy (HRT); pregnancy (category D).
CAUTIOUS USE Moderate to severe hepatic impairment; older adults, lactation. Safety and efficacy in children are not established.

ROUTE & DOSAGE

Breast Cancer
Adult: **PO** 2.5 mg daily

Hepatic Impairment Dosage Adjustment

Severe hepatic impairment (Child-Pugh class C): Reduce the dose by 50%

ADMINISTRATION
Oral
▪ Give without regard to food.

ADVERSE EFFECTS (≥1%) **Body as a Whole:** Fatigue, peripheral edema, asthenia, weight increase, *musculoskeletal pain,* arthralgia, angioedema. **CNS:** Headache, somnolence, dizziness. **CV:** Chest pain, hypertension, hypercholesterolemia. **GI:** Nausea, vomiting, constipation, diarrhea, abdominal pain, anorexia, dyspepsia. **Respiratory:** Dyspnea, cough. **Skin:** Hot flushes, rash, pruritus.

INTERACTIONS Drug: ESTROGENS, ORAL CONTRACEPTIVES could interfere with the pharmacologic action of letrozole.

PHARMACOKINETICS Absorption: Rapidly from GI tract. **Metabolism:** In liver (CYP3A4 and 2A6). **Elimination:** 90% in urine. **Half-Life:** 2 days.

NURSING IMPLICATIONS
Assessment & Drug Effects
▪ Lab tests: Periodically monitor serum calcium and CBC with differential.
▪ Monitor carefully for S&S of thrombophlebitis or thromboembolism; report immediately.

Patient & Family Education
▪ Notify prescriber immediately if S&S of thrombophlebitis develop (see Appendix F).

LEUCOVORIN CALCIUM
(loo-koe-vor'in)

LEVOLEUCOVORIN
(levo-loo-koe-vor'in)

Fusilev
Classifications: BLOOD FORMER;
ANTIANEMIC

Common adverse effects in *italic*, life-threatening effects <u>underlined</u>; generic names in **bold**; classifications in SMALL CAPS; ♣ Canadian drug name; ⊘ Prototype drug

Therapeutic: ANTIANEMIC; CHEMO-
THERAPEUTIC PROTECTANT
Pregnancy Category: C

AVAILABILITY 5 mg, 10 mg, 15 mg,
25 mg tablets; 50 mg, 100 mg, 350
mg vials

ACTION & *THERAPEUTIC EFFECT*
Both leucovorin and levoleucovorin are reduced forms of folic
acid. Unlike folic acid, they do not
require enzymatic reduction by dihydrofolate reductase. Thus, they
are readily available as an essential
cell growth factor. During antineoplastic therapy, both forms of the
drug prevent serious toxicity by
protecting cells from the action
of folic acid antagonists. *Antidote
against folic acid antagonists such
as methotrexate.*

USES Folate-deficient megaloblastic anemias due to sprue, pregnancy, and nutritional deficiency
when oral therapy is not feasible.
Also to prevent or diminish toxicity of antineoplastic folic acid
antagonists, particularly methotrexate. Also to treat advanced
colorectal cancer when given
concurrently with 5-fluorouracil
(5-FU).

CONTRAINDICATIONS Hypersensitivity to folic acid or folinic acid;
undiagnosed anemia, pernicious
anemia, or other megaloblastic
anemias secondary to vitamin
B_{12} deficiency; oral form with
stomatitis.
CAUTIOUS USE Renal dysfunction, third space fluid collection
(i.e., ascites, pleural effusion),
renal function impairment, or inadequate hydration; older adults;
seizure disorders; pregnancy (category C), lactation.

ROUTE & DOSAGE

***Fusilev is dosed at half the
normal dose of racemic
leucovorin***

Megaloblastic Anemia
Adult/Child: **IV/IM** Up to 1 mg/
day

**Leucovorin Rescue for
Methotrexate Toxicity**
Adult/Child: **PO/IM/IV** 10 mg/
m^2 q6h until serum methotrexate
levels are reduced

Levoleucovorin Rescue
*Adult/Adolescent/Child (older
than 6 y):* **IV** 7.5 mg q6h × 10
doses

**Inadvertent Overdose of
Methotrexate (Levoleucovorin)**
*Adult/Adolescent/Child (older
than 6 y):* **IV** 7.5 mg q6h until
serum methotrexate levels are
below 0.01 micromolar

**Leucovorin Rescue for Other
Folate Antagonist Toxicity**
Adult/Child: **PO/IM/IV** 5–15
mg/day

Advanced Colorectal Cancer
Adult: **IV** 200 mg/m^2 followed by
fluorouracil 370 mg/m^2

ADMINISTRATION

Oral
▪ Note: Oral route is NOT recommended for doses higher than 25 mg
or if patient is likely to vomit.

Intramuscular
▪ Give deep into a large muscle.

Intravenous

PREPARE: **Leucovorin Direct:**
Give 1 mL (3 mg) ampules,
which contain benzyl alcohol,

L

Common adverse effects in *italic*, life-threatening effects underlined; generic names
in **bold**; classifications in SMALL CAPS; ♣ Canadian drug name; ◑ Prototype drug

855

undiluted. **Leucovorin IV Infusion:** For doses less than 10 mg/m², reconstitute each 50 mg in 5 mL (10 mg/1 mL in 10 mL) of bacteriostatic water for injection with benzyl alcohol as a preservative. ▪For doses greater than 10 mg/m² reconstitute, as above, but with sterile water for injection without a preservative. Final concentration is 10 mg/mL. ▪Further dilute in 100–500 mL of IV solutions (e.g., D5W, NS, LR) to yield a concentration of 10–20 mg/mL of IV solution. **Levoleucovorin Direct:** Reconstitute the 50-mg vial with 5.3 mL NS injection to yield 10 mg/mL. ▪ May further dilute, immediately, in NS or D5W to 0.5–5 mg/mL. ▪ Do not mix with other solutions or additives.

ADMINISTER: **Leucovorin/Levoleucovorin Direct:** Give 160 mg or fraction thereof over 1 min. **Leucovorin/Levoleucovorin IV Infusion:** Do not exceed direct IV rate. ▪Give more slowly if the volume of IV solution to be infused is large.

INCOMPATIBILITIES -leucovorin only: **Solution/additive: Fluorouracil. Y-site: Amphotericin B cholesteryl complex, carboplatin, droperidol, epirubicin, foscarnet, gemtuzumab, lansoprazole, pamidronate, pantoprazole, quinpristin/dalfopristin, sodium bicarbonate.**

▪**Leucovorin:** Use solution reconstituted with bacteriostatic water within 7 days. ▪Use solution reconstituted with sterile water for injection immediately. **Levoleucovorin:** Solutions with NS may be held at 15°–30° C (59°–86° F) for up to 12 h. Solutions with D5W may be held at 15°–30° C (59°–86° F) for up to 4 h. ▪Protect from light.

ADVERSE EFFECTS (≥1%) **Body as a Whole:** Allergic sensitization (urticaria, pruritus, rash, wheezing). **Hematologic:** Thrombocytosis.

INTERACTIONS Drug: May enhance adverse effects of **fluorouracil;** may reverse therapeutic effects of **trimethoprim-sulfamethoxazole.**

PHARMACOKINETICS Onset: Within 30 min. **Peak:** 0.9 h (levoleucovorin). **Duration:** 3–6 h. **Distribution:** Crosses placenta; distributed into breast milk. **Metabolism:** In liver and intestinal mucosa to tetrahydrofolic acid derivatives. **Elimination:** 80–90% in urine, 5–8% in feces. **Half-Life:** 6 h; 0.77 h (levoleucovorin).

NURSING IMPLICATIONS

Assessment & Drug Effects
▪ Monitor neurologic status. Use of leucovorin alone in treatment of pernicious anemia or other megaloblastic anemias associated with vitamin B_{12} deficiency can result in an apparent hematological remission while allowing already present neurologic damage to progress.
▪ Lab tests: Baseline and daily serum creatinine and methotrexate levels; frequent urine pH to ensure urinary alkalinization (pH at or above 7).

Patient & Family Education
▪ Notify prescriber of S&S of a hypersensitivity reaction immediately (see Appendix F).

LEUPROLIDE ACETATE ℗
(loo-proe′lide)
Eligard, Lupron, Lupron Depot, Lupron Depot-Ped, Viadur

Classification: GONADOTROPIN-RELEASING HORMONE (GnRH) ANALOG
Therapeutic: GnRH ANALOG
Pregnancy Category: X

AVAILABILITY 5 mg/mL injection; 3.75 mg, 7.5 mg, 11.25 mg, 15 mg, 22.5 mg, 30 mg microspheres for injection (depot formulations); implant

ACTION & THERAPEUTIC EFFECT
Occupies and desensitizes pituitary GnRH receptors, resulting initially in release of gonadotropins LH and FSH and stimulation of ovarian and testicular steroidogenesis. *Antitumor effect: May inhibit growth of hormone-dependent tumors as indicated by reduction in concentrations of PSA and serum testosterone to levels equal to or less than pretreatment levels. Contraceptive effect: By inhibiting gonadotropin release, ovulation or spermatogenesis is suppressed. Has antitumor effect in males and contraceptive effects in both males and females.*

USES Palliative treatment of advanced prostatic carcinoma; endometriosis; anemia caused by leiomyomata; percocious puberty.
UNLABELED USES Breast cancer; BPH, PMS.

CONTRAINDICATIONS Known hypersensitivity to benzyl alcohol, GnRH analog hypersensitivity; following orchiectomy or estrogen therapy; metastatic cerebral lesions; menstruation, abnormal vaginal bleeding, pregnancy (category X), lactation.
CAUTIOUS USE Life-threatening carcinoma in which rapid symptomatic relief is necessary; osteoporosis; older adults.

ROUTE & DOSAGE

Palliative Treatment for Prostate Cancer
Adult: **Subcutaneous** 1 mg/day **IM** 7.5 mg/mo or 22.5 mg q3mo or 30 mg q4mo or 45 mg q6mo (depot preparation)

Endometriosis
Adult: **IM** 3.75 mg qmo or 11.25 mg q3mo (max: 6 mo)

Precocious Puberty
Child (over 37.5 kg): **IM** 15 mg q4wk
(25–37.5 kg): Give 11.25 mg q4wk
(less than 25 kg): Give 7.5 mg q4wk

Percocious Puberty (Depot Form)
Child (2–11 y): **IM** 11.25 or 30 mg q3mo
Child (1–12 y): **SC** 50 mcg/kg daily may titrate up

ADMINISTRATION

Subcutaneous
- Do not use Depot form for subcutaneous injection.
- Rotate injection sites.

Intramuscular
- Prepare solution for Depot-Ped injection using a 22-gauge needle (or syringe provided by manufacturer), withdraw 1.5 mL of diluent from the supplied ampule and inject it into the vial. Shake well to form a uniform suspension. Withdraw entire contents and administer immediately.
- Do not administer parenteral drug formulation if particulate matter or discoloration is present.
- Refrigerate unopened vials. Store vial in use at room temperature

Common adverse effects in *italic*, life-threatening effects underlined; generic names in **bold**; classifications in SMALL CAPS; ♣ Canadian drug name; ⊙ Prototype drug

857

for several months with minimal loss of potency. Protect from light and freezing.

ADVERSE EFFECTS (≥1%) Body as a Whole: *Disease flare (worsening of S&S of carcinoma),* injection site irritation, asthenia, fatigue, fever, facial swelling. **CNS:** Dizziness, pain, headache, paresthesia. **CV:** *Peripheral edema,* cardiac arrhythmias, <u>MI</u>. **Endocrine:** *Hot flushes, impotence, decreased libido,* gynecomastia, breast tenderness, amenorrhea, vaginal bleeding, thyroid enlargement, hypoglycemia. **GI:** Nausea, vomiting, constipation, anorexia, sour taste, GI bleeding, diarrhea. **Musculoskeletal:** Increased bone pain, myalgia. **Renal:** Increased hematuria, dysuria, flank pain. **Respiratory:** Pleural rub, pulmonary fibrosis flare. **Hematologic:** Decreased Hct, Hgb. **Skin:** Pruritus, rash, hair loss.

INTERACTIONS Drug: ANDROGENS, ESTROGENS counteract therapeutic effects.

PHARMACOKINETICS Absorption: Readily absorbed from subcutaneous or IM sites. **Metabolism:** By enzymes in hypothalamus and anterior pituitary. **Half-Life:** 3 h.

NURSING IMPLICATIONS

Assessment & Drug Effects
- Monitor PSA and testosterone levels in males with prostate cancer. A gradual rise in values after their decrease may signify treatment failure.
- Inspect injection site. If local hypersensitivity reactions occur (erythema, induration), suspect sensitivity to benzyl alcohol. Report to prescriber.
- Monitor I&O ratio and pattern. Report hematuria and decreased output. Carefully monitor voiding problems.

Patient & Family Education
- When used for prostate cancer, bone pain and voiding problems (i.e., symptoms of tumor obstruction) usually increase during first several weeks of continuous treatment but are transient. Hot flushes also may be experienced.
- Notify prescriber of neurologic S&S (paresthesia and weakness in lower limbs). Exercise caution when walking without assistance.
- When used for endometriosis, continuous treatment may cause amenorrhea and other menstrual irregularities.

LEVALBUTEROL HYDROCHLORIDE

(lev-al-bu′ter-ole)

Xopenex, Xopenex HFA
Classifications: SHORT-ACTING BETA-ADRENERGIC AGONIST; BRONCHODILATOR
Therapeutic: BRONCHODILATOR
Prototype: Albuterol
Pregnancy Category: C

AVAILABILITY 0.63 mg/3 mL, 1.25 mg/3 mL inhalation solution

ACTION & *THERAPEUTIC EFFECT*
An isomer of albuterol with beta₂-adrenergic agonist properties; acts on the $beta_2$ receptors of the smooth muscles of the bronchial tree, thus resulting in bronchodilation. *Effective bronchodilator that decreases airway resistance, facilitates mucous drainage, and increases vital capacity.*

USES Treatment or prevention of bronchospasm in patients with reversible obstructive airway disease.

CONTRAINDICATIONS Hypersensitivity to levalbuterol or albuterol; angioedema, lactation.

CAUTIOUS USE Cardiovascular disorders especially coronary insufficiency, cardiac arrhythmias, hypertension, QT elongation, convulsive disorders; diabetes mellitus, diabetic ketoacidosis; older adults; seizures; status asthmaticus, tachycardia; hypersensitivity to sympathetic amines; hyperthyroidism, thyrotoxicosis; pregnancy (category C); children younger than 6 y.

ROUTE & DOSAGE

> **Bronchospasm**
>
> *Adult:* **Inhalation** 0.63 mg by nebulization t.i.d. q6–8h, may increase to 1.25 mg t.i.d. if needed
> *Child (6–11 y):* **Inhalation** 0.31 mg by nebulization t.i.d. q6–8h (max: 0.63 mg t.i.d.)

ADMINISTRATION

Inhalation
- Use vials within 2 wk of opening pouch. Protect vial from light and use within 1 wk after removal from pouch. Use only if solution in vial is colorless.

***INCOMPATIBILITIES* Solution/additive:** Compatibility when mixed with other drugs in a nebulizer has not been established.

- Store at 15°–25° C (59°–77° F) in protective foil pouch.

ADVERSE EFFECTS (≥1%) **Body as a Whole:** Allergic reactions, flu syndrome, pain. **CNS:** Migraine, dizziness, nervousness, tremor, anxiety. **CV:** Tachycardia. **GI:** Dyspepsia. **Respiratory:** Increased cough, viral infection, rhinitis, sinusitis, turbinate edema, paradoxical bronchospasm. **Endocrine:** Increase in serum glucose.

INTERACTIONS Drug: BETA-ADRENERGIC BLOCKERS may antagonize levalbuterol effects; MAOI, TRICYCLIC ANTIDEPRESSANTS may potentiate levalbuterol effects on vascular system; ECG changes or hypokalemia may be exacerbated by LOOP or THIAZIDE DIURETICS.

PHARMACOKINETICS Onset: 5–15 min. **Duration:** 3–6 h. **Half-Life:** 3.3 h.

NURSING IMPLICATIONS

Assessment & Drug Effects
- Monitor for S&S of CNS or cardiovascular stimulation (e.g., BP, HR, respiratory status).
- Lab tests: Periodic serum potassium levels especially with co-administered loop or thiazide diuretics.
- Monitor diabetics for loss of glycemic control.

Patient & Family Education
- Seek medical advice immediately if a previously effective dose becomes ineffective.
- Report immediately to prescriber: Chest pains or palpitations, swelling of the eyelids, tongue, lips, or face; increased wheezing or difficulty breathing.
- Do not use drug more frequently than prescribed.
- Exercise caution with hazardous activities; dizziness and vertigo are possible side effects.
- Check with prescriber before taking OTC cold medication.

✓LEVETIRACETAM
(lev-e-tir′a-ce-tam)

Keppra, Keppra XR
Classification: ANTICONVULSANT
Therapeutic: ANTICONVULSANT
Pregnancy Category: C

AVAILABILITY 250 mg, 500 mg, 750 mg, 1000 mg tablets; 500 mg, 700 mg extended release tablets; 100 mg/mL oral solution; 100 mg/mL injection

---

Common adverse effects in *italic*, life-threatening effects <u>underlined</u>; generic names in **bold**; classifications in SMALL CAPS; ♣ Canadian drug name; ✪ Prototype drug

ACTION & *THERAPEUTIC EFFECT*

The precise mechanism of anti-epileptic effects is unknown. It is a broad spectrum antiepileptic agent that does not involve GABA inhibition. It prevents epileptiform burst firing and propagation of seizure activity. *Inhibits complex partial seizures and prevents epileptic and seizure activity.*

USES Adjunctive therapy for partial onset, myoclonic, tonic clonic seizures.

CONTRAINDICATIONS Hypersensitivity to levetiracetam; suicidal ideation; labor; lactation; **Extended release tablets:** children younger than 12.

CAUTIOUS USE Renal impairment; renal disease; renal failure; older adults; history of psychosis or depression, suicidal tendencies; pregnancy (category C); children younger than 4 y.

ROUTE & DOSAGE

Partial Onset Seizures

Adult/Adolescent (older than 16 y): **PO** 500 mg b.i.d., may increase by 500 mg b.i.d. q2wk (max: 3000 mg/day) **IV** 500 mg b.i.d., may increase q2wk (max: 3000 mg/day)
Child (4–15 y): **PO** 10 mg/kg b.i.d.; may increase by 20 mg/kg q2wk up to 60 mg/kg/day

Tonic Clonic Seizures

Adult/Adolescent (older than 16 y): **PO/IV** 500 mg b.i.d., increase by 1000 mg q2wk to dose of 3000 mg/day
Child (older than 6 y): **PO** 10 mg/kg b.i.d., increase by 20 mg/kg

q2wk to dose of 60 mg/kg/day in 2 doses

Myoclonic Seizures

Adult/Adolescent/Child (at least 12 y): **PO** 500 mg b.i.d., increase by 1000 mg/day q2wk to recommended dose of 3000 mg/day in divided doses

Renal Impairment Dosage Adjustment

CrCl 50–80 mL/min: 500–1000 mg q12h (IR or IV form), 1000–2000 mg q24 (XR form); *30–49 mL/min:* 250–750 mg q12h (IR or IV form), 500–1500 mg q24h (XR form); *less than 30 mL/min:* 250–500 mg q12h (IR or IV form), 500–1000 mg q24h (XR form)

Hemodialysis Dosage Adjustment

500–1000 mg q24h; 250–500 mg supplemental following hemodialysis

Hepatic Impairment Dosage Adjustment

Child-Pugh class C: Reduce to 1/2 dose

ADMINISTRATION

Oral

- Dose increment changes should be made no more often than at 2-wk intervals.
- Taper dose if discontinued.
- Give supplemental doses to dialysis patients after dialysis.
- Store at 15°–30° C (59°–86° F).

ADVERSE EFFECTS (≥1%) **Body as a Whole:** *Asthenia, headache, infection,* pain. **CNS:** *Somnolence,* amnesia, anxiety, ataxia, depression, dizziness, emotional lability, hostility, nervousness, vertigo, paradoxical increase in seizures

Common adverse effects in *italic*, life-threatening effects <u>underlined</u>; generic names in **bold**; classifications in SMALL CAPS; ✤ Canadian drug name; ⊘ Prototype drug

(as add-on therapy). **GI:** Anorexia. **Respiratory:** Cough, pharyngitis, rhinitis, sinusitis. **Special Senses:** Diplopia. **Other:** Increased symptoms of depression; suicidal ideation.

INTERACTIONS Drug: Levetiracetam does not affect **estrogen, warfarin,** or **digoxin** levels or affect levels of other antiepileptic drugs. **Sevelamer, colesevelam** may decrease effectiveness.

PHARMACOKINETICS Absorption: Rapidly and almost completely absorbed. **Peak:** 1 h; steady-state 2 days. **Distribution:** Less than 10% protein bound. **Metabolism:** Minimal hepatic metabolism. **Elimination:** Renally eliminated. **Half-Life:** 7.1 h (9.6 h in older adults).

NURSING IMPLICATIONS

Assessment & Drug Effects
- Monitor individuals with a history of psychosis or depression for signs and symptoms of suicidal tendencies, suicidal ideation, and suicidality. Report any of these symptoms to the prescriber.
- Monitor and notify prescriber of difficulty with gait or coordination.
- Lab tests: Periodic CBC with differential, Hct and Hgb, LFTs.

Patient & Family Education
- Monitor for signs and symptoms of suicidality, especially in children with a history of depression or psychosis.
- Do not drive or engage in potentially hazardous activities until response to drug is known.
- Do not abruptly discontinue drug. MUST use gradual dose reduction/taper.

LEVOBUNOLOL
(lee-voe-byoo'noe-lole)

Betagan
Pregnancy Category: C
See Appendix A-1.

LEVOCETIRIZINE
(lev-o-ce-tir'i-zeen)

Xyzal
Pregnancy Category: B
See Cetirizine.

LEVODOPA (L-DOPA) 🔵
(lee-voe-doe'pa)

Classifications: DOPAMINE RECEPTOR AGONIST; ANTIPARKINSON
Therapeutic: ANTIPARKINSON
Pregnancy Category: C

AVAILABILITY 100 mg, 250 mg, 500 mg tablets and capsules

ACTION & *THERAPEUTIC EFFECT* A metabolic precursor of dopamine, a catecholamine neurotransmitter. Levodopa readily crosses the blood–brain barrier; it is believed that the dopamine level is severely reduced in parkinsonism. Levodopa restores dopamine levels in extra pyramidal centers of the brain. *Effective in controlling the involuntary muscle movement such as tremors and rigidity associated with Parkinson's disease.*

USES Idiopathic Parkinson's disease, postencephalitic and arteriosclerotic parkinsonism, and parkinsonism symptoms associated with manganese and carbon monoxide poisoning.
UNLABELED USES To relieve pain of herpes zoster (shingles), liver

Common adverse effects in *italic*, life-threatening effects underlined; generic names in **bold**; classifications in SMALL CAPS; ✤ Canadian drug name; 🔵 Prototype drug

861

coma (caused by cirrhosis or fulminating hepatitis), bone pain in metastatic breast carcinoma, adjunctive therapy in CHF.

CONTRAINDICATIONS Known hypersensitivity to levodopa; narrow-angle glaucoma patients with suspicious pigmented lesion or history of melanoma; acute psychoses, within 2 wk of use of MAOIs, suicidal ideation; lactation.

CAUTIOUS USE Cardiovascular, kidney, liver, or endocrine disease, history of MI with residual arrhythmias; peptic ulcer; convulsions; history of suicidal tendencies; depression; bipolar disorder; psychiatric disorders; chronic wide-angle glaucoma; diabetes; pulmonary diseases, bronchial asthma; pregnancy (category C). Safe use in children not established.

ROUTE & DOSAGE

Parkinson's Disease

Adult: **PO** 500 mg to 1 g daily in 2 or more equally divided doses, may be increased by 100–750 mg q3–7days (max: 8 g/day); used in combination with carbidopa, decrease levodopa dose by 75–80%

ADMINISTRATION

Oral

- Give with food to reduce nausea. Absorption is decreased with high-protein meals.
- Crush tablets or empty capsule content into fruit juice as needed.
- Store in tight, light-resistant containers.

ADVERSE EFFECTS (≥1%) **CNS:** *Choreiform and involuntary movements,* increased hand tremor, bradykinetic episodes (on–off phenomena), trismus, grinding of teeth (bruxism), ataxia, muscle twitching, numbness, weakness, fatigue, headache, opisthotonos, confusion, agitation, anxiety, euphoria, insomnia, nightmares; psychotic episodes with paranoid delusions or hallucinations, severe depression, including suicidal tendencies, hypomania. **CV:** *Orthostatic hypotension;* palpitations, tachycardia, hypertension. **Special Senses:** *Blepharospasm,* diplopia, blurred vision, dilated pupils. **GI:** *Anorexia, nausea, vomiting,* abdominal distress, flatulence, dry mouth, dysphagia, sialorrhea; burning sensation of tongue, bitter taste, diarrhea or constipation; GI bleeding, hepatotoxicity. **Body as a Whole:** Flushing, increased sweating, weight gain or loss, edema, dark sweat or urine. **Urogenital:** Urinary retention or incontinence, increased sexual drive, priapism, postmenopausal bleeding. **Skin:** Skin rashes, loss of hair. **Respiratory:** Rhinorrhea, bizarre breathing patterns.

DIAGNOSTIC TEST INTERFERENCE Elevated *BUN, AST, ALT, alkaline phosphatase, LDH, bilirubin, protein-bound iodine,* serum level of *growth hormone;* decreased *glucose tolerance; hypokalemia,* decreased *WBC, Hgb, Hct. Urine glucose:* False-negative tests may result with use of *glucose oxidase methods* (e.g., *Clinistix, TesTape*) and false-positive results with the *copper reduction method* (e.g., *Clinitest*), especially in patients receiving large doses. It is reported that *Clinistix* and *TesTape* may be used if reading is taken at margin of wet and dry tape. *Urinary ketones:* There is possibility of false-positive tests by dipsticks [e.g., **Acetest** (equivocal), **Keto-stix, Labstix**]; **Serum and urinary uric acid:** False elevations by **colorimetric methods,** but not with **uricase; Urinary protein:** False increases by **Lowry method; Urinary VMA:** False decreases by **Pisano method; Urinary cat-echolamine:** False increases by **Hingerty method. PKU urine test:** Interference.

INTERACTIONS Drug: MAO INHIBI-TORS may precipitate hypertensive crisis; TRICYCLIC ANTIDEPRESSANTS augment postural hypotension; PHENOTHIAZINES, **haloperidol** may antagonize the therapeutic effects of levodopa; **pyridoxine** can reverse effects of levodopa; ANTICHO-LINERGICS may exacerbate abnormal involuntary movements; **methyl-dopa** may increase toxic CNS effects; HALOGENATED GENERAL AN-ESTHETICS increase risk of arrhythmi-as. **Food:** Food decreases the rate and extent of levodopa absorption. **Herbal: Kava** may worsen parkin-sonian symptoms.

PHARMACOKINETICS Absorption: Rapidly and well absorbed from GI tract; lower absorption if taken with food. **Peak:** 1–3 h. **Distribution:** Widely distributed in body. **Metabolism:** Most of drug is decarboxylated to dopamine in lumen of GI tract, liver, and serum. **Elimination:** 80–85% of dose excreted in urine in 24 h. **Half-Life:** 1 h.

NURSING IMPLICATIONS

Assessment & Drug Effects

- Monitor vital signs, particularly during period of dosage adjustment. Report alterations in BP, pulse, and respiratory rate and rhythm.

- Supervise ambulation as indicated. Orthostatic hypotension is usually asymptomatic, but some patients experience dizziness and syncope. Tolerance to this effect usually develops within a few months of therapy.

- Make accurate observations and report adverse reactions and therapeutic effects promptly. Rate of dosage increase is determined primarily by patient's tolerance and response to drug.

- Monitor all patients closely for behavior changes.

- Lab tests: Monitor blood glucose and HbA1C, CBC, Hgb and Hct, serum potassium, and liver and kidney function periodically.

- Report promptly muscle twitching and spasmodic winking (blepharospasm); these are early signs of overdosage. Patients on full therapeutic doses for longer than 1 y may develop such abnormal involuntary movements as well as jerky arm and leg movements. Symptoms tend to increase if dosage is not reduced.

- Report to prescriber any S&S of the on–off phenomenon sometimes associated with chronic management: Rapid unpredictable swings in intensity of motor symptoms of parkinsonism evidenced by increase in bradykinesia (attacks of "leg freezing" or slow body movement).

- Monitor mental status for S&S of drug-induced neuropsychiatric adverse reactions.

Patient & Family Education

- Do not take with high-protein foods. Also avoid high consumption of food sources of pyridoxine, including wheat germ, green vegetables, bananas, whole-grain cereals, muscular and glandular meats (especially liver), legumes.

Common adverse effects in *italic*, life-threatening effects underlined; generic names in **bold**; classifications in SMALL CAPS; ♣ Canadian drug name; ⚙ Prototype drug

863

- Do not take OTC preparations or fortified cereals unless approved by prescriber. Multivitamins, anti-nauseants, and fortified cereals usually contain vitamin B_6.
- Make positional changes slowly, particularly from lying to upright position, and dangle legs a few minutes before standing.
- Resume activities gradually, observing safety precautions to avoid injury. Elevation of mood and sense of well-being may precede objective improvement. Significant improvement usually occurs during second or third wk of therapy, but may not occur for 6 mo or more in some patients.
- Follow prescribed drug regimen. Sudden withdrawal of medication can lead to parkinsonism crisis (with return of marked rigidity, akinesia, tremor, hyperpyrexia) or neuroleptic malignant syndrome (NMS).
- A metabolite of levodopa may cause urine to darken and sweat to be dark-colored.

LEVOFLOXACIN

(lev-o-flox′a-sin)

Levaquin, Iquix, Quixin

Classifications: QUINOLONE ANTIBIOTIC

Therapeutic: ANTIBIOTIC

Prototype: Ciprofloxacin

Pregnancy Category: C

AVAILABILITY 250 mg, 500 mg, 750 mg tablets; 25 mg/mL solution; 25 mg/mL injection; 0.5%, 1.5% ophth solution

ACTION & *THERAPEUTIC EFFECT*
A broad-spectrum fluoroquinolone antibiotic that inhibits DNA-gyrase, an enzyme necessary for bacterial replication, transcription, repair, and recombination. *Effective against many aerobic gram-positive and aerobic gram-negative organisms.*

USES Treatment of maxillary sinusitis, acute exacerbations of bacterial bronchitis, community-acquired pneumonia, uncomplicated skin/skin structure infections, UTI, acute pyelonephritis caused by susceptible bacteria; acute bacterial sinusitis; chronic bacterial prostatitis; bacterial conjunctivitis.

CONTRAINDICATIONS Hypersensitivity to levofloxacin and quinolone antibiotics; tendon pain, inflammation or rupture; syphilis; viral infections; phototoxicity; suicidal ideation; psychotic manifestations; manifestations of peripheral neuropathy; hypoglycemic reaction to drug; QT prolongation, hypokalemia; lactation.

CAUTIOUS USE History of suicidal ideation; psychosis; anxiety, confusion, depression; known or suspected CNS disorders predisposed to seizure activity; risk factors associated with potential seizures (e.g., some drug therapy, renal insufficiency), dehydration, renal impairment (CrCl less than 50 mL/min); colitis; cardiac arrhythmias; older adults; pregnancy (category C). Safe use in children younger than 6 mo not established.

ROUTE & DOSAGE

Infections

Adult: **PO** 500 mg q24h × 10 days **IV** 500 mg infused over 60 min q24h × 7–14 days

Community-Acquired Pneumonia

Adult: **PO/IV** 750 mg q24h × 5 days

Uncomplicated UTI

Adult: **PO/IV** 250 mg q24h × 14 days

Complicated UTI, Pyelonephritis

Adult: **PO/IV** 250 mg q24h × 10 days

Acute Bacterial Sinusitis

Adult: **PO/IV** 750 mg daily × 5 days

Chronic Bacterial Prostatitis

Adult: **PO/IV** 500 mg q24h × 28 days

Skin & Skin Structure Infections

Adult: **PO** 750 mg q24h × 14 days

Inhaled Anthrax

Adult/Adolescent/Child (weight at least 50 kg): **IV** 500 mg daily × 60 days
Infant/Child (older than 6 mo and weight less than 50 kg): **IV** 8 mg/kg q12h (no more than 250 mg/dose)

Renal Impairment Dosage Adjustment

For initial dose of 500 mg, adjust as follows: *CrCl 20–50 mL/min:* 250 mg q24h; *less than 20 mL/min:* 250 mg q48h
For initial dose of 750 mg, adjust as follows: *CrCl 20–50 mL/min:* 750 mg q48h; *10–19 mL/min:* 500 mg q48h; *less than 20 mL/min:* 250 mg q48h

Bacterial Conjunctivitis

Adult: **Ophthalmic** Days 1–2, 1–2 drops in affected eye(s) q2h while awake (max: 8 times/day), days 3–7, 1–2 drops in affected eye(s) q4h while awake (max: 4 times/day)

ADMINISTRATION

Oral

- Do not give oral drug within 2 h of drugs containing aluminum or magnesium (antacids), iron, zinc, or sucralfate.

Intravenous

PREPARE: **Intermittent:** Withdraw the desired dose from 500 or 750 mg (25 mg/mL) single-use vial. ▪ Add to enough D5W, NS, D5/NS, D5/LR, or other compatible solutions to produce a concentration of 5 mg/mL [e.g., 500 mg (or 20 mL) added to 80 mL]. ▪ Discard any unused drug remaining in the vial.

ADMINISTER: **Intermittent:** Infuse 500 mg or less over 60 min. ▪ Infuse 750 mg over at least 90 min. ▪ Do NOT give a bolus dose or infuse too rapidly.

***INCOMPATIBILITIES* Y-site:** Do not add any drugs to levofloxacin solution or infuse simultaneously through the same line (manufacturer recommendation). **Azithromycin, furosemide, heparin, indomethacin, insulin, nitroglycerin, nitroprusside, propofol.**

- Store tablets in a tightly closed container. IV solution is stable for 72 h at 25° C (77° F).

ADVERSE EFFECTS (≥1%) **CNS:** Headache, insomnia, dizziness. **GI:** Nausea, diarrhea, constipation, vomiting, abdominal pain, dyspepsia. **Skin:** Rash, pruritus. **Special Senses:** Decreased vision, foreign body sensation, transient ocular burning, ocular pain, photophobia. **Urogenital:** Vaginitis. **Body as a Whole:** Injection site pain or inflammation, chest or back pain, fever, pharyngitis. **Other:** Cartilage erosion.

Common adverse effects in *italic*, life-threatening effects underlined; generic names in **bold**; classifications in SMALL CAPS; ♣ Canadian drug name; ☯ Prototype drug

865

DIAGNOSTIC TEST INTERFERENCE
May cause false positive on *opiate screening tests.*

INTERACTIONS Drug: Magnesium or **aluminum**-containing antacids, **sucralfate, iron, zinc** may decrease levofloxacin absorption; NSAIDs may increase risk of CNS reactions, including seizures; may cause hyper- or hypoglycemia in patients on ORAL HYPOGLYCEMIC AGENTS.

PHARMACOKINETICS Absorption: Rapidly from GI tract. **Peak:** PO 1–2 h. **Distribution:** Penetrates lung tissue, 24–38% protein bound. **Metabolism:** Minimally in the liver. **Elimination:** Primarily unchanged in urine. **Half-Life:** 6–8 h.

NURSING IMPLICATIONS
Assessment & Drug Effects
- Lab tests: Do C&S test prior to beginning therapy.
- Withhold therapy and report to prescriber immediately any of the following: Skin rash or other signs of a hypersensitivity reaction (see Appendix F); CNS symptoms such as seizures, restlessness, confusion, hallucinations, depression; skin eruption following sun exposure; symptoms of colitis such as persistent diarrhea; joint pain, inflammation, or rupture of a tendon.
- Monitor diabetics on oral hypoglycemic agents for loss of glycemic control.

Patient & Family Education
- Learn important indications for discontinuing drug and immediately notifying prescriber.
- If tendon pain occurs, discontinue the drug and notify the prescriber.
- Consume fluids liberally while taking levofloxacin.

- Allow a minimum of 2 h between drug dosage and taking any of the following: Aluminum or magnesium antacids, iron supplements, multivitamins with zinc, or sucralfate.
- Avoid exposure to excess sunlight or artificial UV light.
- Closely monitor blood glucose if taking oral hypoglycemic agents for diabetic control.

LEVOLEUCOVORIN
(levo-loo-koe-vor′in)
Fusilev
Pregnancy Category: C
See Leucovorin.

LEVONORGESTREL-RELEASING INTRAUTERINE SYSTEM
(lee′vo-nor-jes-trel)
Mirena
Classification: PROGESTIN HORMONE
Therapeutic: PROGESTIN
Prototype: Norgestrel
Pregnancy Category: X

AVAILABILITY 52 mg unit

ACTION & THERAPEUTIC EFFECT
A progestogen that induces morphological changes in the endometrium including glandular atrophy, leukocytic infiltration, and decrease in glandular and stromal mitoses. Contraceptive effect may result by preventing follicular maturation and ovulation, thickening of the cervical mucus of the uterus, thus preventing passage of sperm into the uterus, or decreasing ability of sperm to survive in an environment of altered endometrium. *Effective contraceptive.*

866

Common adverse effects in *italic*, life-threatening effects underlined; generic names in **bold**; classifications in SMALL CAPS; ♣ Canadian drug name; ⊕ Prototype drug

USES Hormonal contraception.

CONTRAINDICATIONS Hypersensitivity to any component of the product; previously inserted IUD which has not been removed; suspicion of pregnancy; within 6 wk of giving birth or prior to complete involution of the uterus; history of ectopic pregnancy or any condition which predisposes to ectopic pregnancy; history of uterine anomalies which distort the uterine cavity; acute PID or history of PID unless there has been a subsequent intrauterine pregnancy; cervicitis or vaginitis or other lower genital tract infection; genital actinomycosis; woman or partner has multiple sex partners; vaginal bleeding of unknown etiology; postpartum endometriosis or septic abortion in past 3 mo; abnormal Pap or suspected/known cervical neoplasm; known or suspected carcinoma of the breast; acute liver disease or liver tumor; immune deficiency states; pregnancy (category X).

CAUTIOUS USE Women at risk for venereal disease; anemia; diabetes mellitus; history of psychic depression; intermittent porphyria; fluid retention; history of migraines; impaired liver function; presence or history of salpingitis; venereal disease; genital bleeding of unknown etiology; coagulopathy; previous pelvic surgery.

ROUTE & DOSAGE

Contraception

Adult: **Intrauterine** Insert device on 7th day of menstrual cycle; may leave in place up to 5 y

ADMINISTRATION

Intrauterine

- Inserted only by prescriber or other person qualified by spe-

cial training in the intrauterine system.

ADVERSE EFFECTS (≥1%) **CV:** Hypertension. **GI:** Abdominal pain, nausea. **Endocrine:** Breast tenderness/pain. **Hematologic:** Anemia. **Metabolic:** Weight gain. **CNS:** Depression, emotional lability, headache (including migraine), nervousness. **Skin:** Acne, alopecia, eczema. **Urogenital:** Amenorrhea, dysmenorrhea, leukorrhea, decreased libido, vaginal moniliasis, vulvovaginal disorders, cervicitis, dyspareunia.

INTERACTIONS Drug: No clinically significant interactions established.

PHARMACOKINETICS Peak: Few weeks. **Duration:** 5 y. **Distribution:** 86% protein bound. **Metabolism:** In liver. **Elimination:** In both urine and feces. **Half-Life:** 37 h.

NURSING IMPLICATIONS

Assessment & Drug Effects

- Monitor for decreased pulse, perspiration, or pallor during insertion. Keep patient supine until these signs have disappeared.
- Monitor BP especially with preexisting hypertension.

Patient & Family Education

- Report S&S of PID immediately: (e.g., prolonged or heavy bleeding, unusual vaginal discharge, abdominal or pelvic pain or tenderness, painful sexual intercourse, chills, fever, and flu-like symptoms).
- Report any of the following to prescriber immediately: Migraine (if not experienced before) or exceptionally severe headache, or jaundice.

Common adverse effects in *italic*, life-threatening effects <u>underlined</u>; generic names in **bold;** classifications in SMALL CAPS; ✚ Canadian drug name; Ⓟ Prototype drug

867

LEVORPHANOL TARTRATE

(lee-vor'fa-nole)

Levo-Dromoran

Classifications: ANALGESIC; NARCOTIC (OPIATE AGONIST)

Therapeutic: NARCOTIC ANALGESIC

Prototype: Morphine sulfate

Pregnancy Category: B; D with long-time use or high doses

Controlled Substance: Schedule II

AVAILABILITY 2 mg tablets; 2 mg/mL injection

ACTION & *THERAPEUTIC EFFECT* A potent synthetic morphine derivative with agonist activity only. Reported to cause less nausea, vomiting, and constipation than equivalent doses of morphine but may produce more sedation, smooth-muscle relaxation, and respiratory depression. *More potent as an analgesic and has somewhat longer duration of action than morphine.*

USES To relieve moderate to severe pain.

CONTRAINDICATIONS Hypersensitivity to levorphanol; labor and delivery, pregnancy (category D with long time use or high doses); lactation.

CAUTIOUS USE Patients with impaired respiratory reserve, or depressed respirations from another cause (e.g., severe infection, obstructive respiratory conditions, chronic bronchial asthma); head injury or increased intracranial pressure; acute MI; cardiac dysfunction; liver disease, biliary surgery, alcohol or delirium tremens; liver or kidney dysfunction, hypothyroidism, Addison's disease, toxic psychosis, prostatic hypertrophy, or urethral stricture; older adults, other vulnerable populations; pregnancy (category B short term use of low doses).

ROUTE & DOSAGE

Moderate to Severe Pain

Adult: **PO** 2 mg q6–8h prn **IV** 1 mg q3–6h prn **Subcutaneous/IV** 1–2 mg q6–8h

ADMINISTRATION

Oral/Intramuscular/Subcutaneous

- Give in the smallest effective dose to minimize the possibility of tolerance and physical dependence.
- Rotate injection sites.
- Store tablets at 15°–30° C (59°–86° F) unless otherwise directed. Store in tightly covered, light-resistant containers.

Intravenous

PREPARE: **Direct:** May be given undiluted or diluted in 5 mL of NS or sterile water.

ADMINISTER: **Direct:** Give at a rate of 2 mg or fraction thereof over 5 min. ▪ AVOID rapid injection. ▪ May inject into Y-site of compatible infusion solution.

INCOMPATIBILITIES **Solution/additive: Aminophylline, ammonium chloride, amobarbital, chlorothiazide, heparin, methicillin, nitrofurantoin, novobiocin, pentobarbital, perphenazine, phenobarbital, phenytoin, secobarbital, sodium bicarbonate, sodium iodide, sulfadiazine, sulfisoxazole diethanolamine, thiopental. Y-site: Pantoprazole, trastuzumab.**

ADVERSE EFFECTS (≥1%) CNS: Euphoria, *sedation, drowsiness,* nervousness, confusion. **CV:** Hypotension, arrhythmias. **GI:** *Nausea,* vomiting, dry mouth, cramps, *constipation.* **Urogenital:** Urinary frequency, urinary retention, sedation. **Special Senses:** Blurred vision. **Respiratory:** Respiratory depression. **Body as a Whole:** Physical dependence.

INTERACTIONS Drug: Alcohol and other CNS DEPRESSANTS compound sedation and CNS depression. **Herbal: St. John's wort** may increase sedation.

PHARMACOKINETICS Peak: 60–90 min (PO); 15–30 min (IM). **Duration:** 6–8 h. **Distribution:** Crosses placenta; distributed into breast milk. **Metabolism:** In liver. **Elimination:** In urine. **Half-Life:** 11–16 h.

NURSING IMPLICATIONS

Assessment & Drug Effects

- Assess degree of pain relief. Drug is most effective when peaks and valleys of pain relief are avoided.
- Monitor bowel function.
- Monitor ambulation, especially in older adult patients.

Patient & Family Education

- Do not drive or engage in other potentially hazardous activities.
- Avoid alcohol and other CNS depressants unless approved by prescriber.
- Note: Ambulation may increase frequency of nausea and vomiting.
- Increase fluid and fiber intake to offset constipating effects of the drug.

LEVOTHYROXINE SODIUM (T₄) ⊕

(lee-voe-thye-rox′een)

Eltroxin ♣, Levoxyl, Levolet, Novothyrox, Synthroid, Unithroid

Classification: THYROID HORMONE REPLACEMENT
Therapeutic: THYROID HORMONE REPLACEMENT
Pregnancy Category: A

AVAILABILITY 25 mcg, 50 mcg, 75 mcg, 88 mcg, 100 mcg, 112 mcg, 125 mcg, 137 mcg, 150 mcg, 175 mcg, 200 mcg, 300 mcg tablets; 200 mcg, 500 mcg injection

ACTION & *THERAPEUTIC EFFECT* Synthetically prepared levo-isomer of thyroxine (T₄, principal component of thyroid gland secretions, determines normal thyroid function). Principal effects include diuresis, loss of weight and puffiness, increased sense of well-being and activity tolerance, plus rise of T₃ and T₄ serum levels toward normal. *By replacing decreased or absent thyroid hormone, it restores metabolic rate of a hypothyroid individual.*

USES Specific replacement therapy for diminished or absent thyroid function resulting from primary or secondary atrophy of gland, surgery, excessive radiation or antithyroid drugs, congenital defect. Administered orally for hypothyroid state; administered IV for myxedematous coma or other thyroid dysfunctions demanding rapid replacement, as well as in failure to respond to oral therapy.

CONTRAINDICATIONS Hypersensitivity to levothyroxine;

Common adverse effects in *italic*, life-threatening effects underlined; generic names in **bold**; classifications in SMALL CAPS; ♣ Canadian drug name; ⊕ Prototype drug

869

LEVOTHYROXINE SODIUM (T₄)

thyrotoxicosis; severe cardio-vascular conditions, acute MI; obesity treatment; adrenal insufficiency.
CAUTIOUS USE Cardiac disease, angina pectoris, cardiac arrhythmias, hypertension; diabetes mellitus; older adult, impaired kidney function, pregnancy (category A).

ROUTE & DOSAGE

Thyroid Replacement

Adult: **PO** 25–50 mcg/day, gradually increased by 50–100 mcg q1–4wk to usual dose of 100–400 mcg/day **IV/IM** ½ established oral dose (usually 50–100 mcg daily)
Child: **PO** 0–6 mo, 8–10 mcg/kg/day or 25–50 mcg/day; *6–12 mo,* 6–8 mcg/kg/day or 50–75 mcg/day; *1–5 y,* 5–6 mcg/kg/day or 75–100 mcg/day; *6–12 y,* 4–5 mcg/kg/day or 100–150 mcg/day; *older than 12 y,* 2–3 mcg/kg/day or greater than 150 mcg/day

Myxedematous Coma

Adult: **IV** 200–500 mcg day 1, additional 100–300 mcg on day 2 if needed

ADMINISTRATION
Oral
- Give as a single dose, preferably 1 h before or 2 h after breakfast. Give consistently with respect to meals.
- Maintenance dosage for older adults may be 25% lower than for heavier and younger adults.
- Store in tight, light-resistant container.

Intravenous

PREPARE: Direct: Reconstitute vial by adding 5 mL of NS for injection to each 100 mcg. Shake well to dissolve. Use immediately.
ADMINISTER: Direct: Give bolus dose over 1 min.
INCOMPATIBILITIES Do not mix with other medications.

- Store dry powder at 15°–30° C (59°–86° F).

ADVERSE EFFECTS (≥1%) CNS: Irritability, nervousness, *insomnia,* headache (pseudotumor cerebri in children), tremors, craniosynostosis (excessive doses in children). **CV:** Palpitations, tachycardia, arrhythmias, angina pectoris, hypertension. **GI:** Nausea, diarrhea, change in appetite. **Urogenital:** Menstrual irregularities. **Body as a Whole:** Weight loss, heat intolerance, sweating, fever, leg cramps, temporary hair loss (children).

INTERACTIONS Drug: Cholestyramine, colestipol decrease absorption of levothyroxine; **epinephrine, norepinephrine** increase risk of cardiac insufficiency; ORAL ANTICOAGULANTS may potentiate hypoprothrombinemia.

PHARMACOKINETICS Absorption: Variable and incompletely absorbed from GI tract (50–80%). **Peak:** 3–4 wk. **Duration:** 1–3 wk. **Distribution:** Gradually released into tissue cells. **Half-Life:** 6–7 days.

NURSING IMPLICATIONS
Assessment & Drug Effects
- Monitor HR and BP. Report promptly tachycardia or suspected arrhythmias.
- Monitor for adverse effects during early adjustment. If metabolism

increases too rapidly, especially in older adults and heart disease patients, symptoms of angina or cardiac failure may appear.

- Lab tests: Baseline and periodic tests of thyroid function. Closely monitor PT/INR and assess for evidence of bleeding if patient is receiving concurrent anticoagulant therapy. A decrease in anticoagulant dosage may be needed 1–4 wk after concurrent levothyroxine is started.

- Monitor bone age, growth, and psychomotor function in children.

- Some children have partial hair loss after a few months; it returns even with continued therapy.

- Synthroid 100 and 300 mcg tablets contain tartrazine, which may cause an allergic-type reaction in certain patients; particularly those who are hypersensitive to aspirin.

Patient & Family Education

- Thyroid replacement therapy is usually lifelong.
- Notify prescriber immediately of signs of toxicity (e.g., chest pain, palpitations, nervousness).
- Avoid OTC medications unless approved by prescriber.

LIDOCAINE HYDROCHLORIDE ⊕

(lye′doe-kane)

Anestacon, Dilocaine, L-Caine, Lidoderm, Lida-Mantle, Lidoject-1, LidoPen Auto Injector, Nervocaine, Octocaine, Xylocaine, Xylocard ♣
Classifications: CLASS IB ANTIARRHYTHMIC; LOCAL ANESTHETIC (AMIDE TYPE)

Therapeutic: CLASS IB ANTIARRHYTHMIC; LOCAL ANESTHETIC; ANTICONVULSANT
Pregnancy Category: B

AVAILABILITY **Antidysrhythmic:** 300 mg/3 mL autoinjector; 0.2%, 0.4%, 0.8%, 1%, 2%, 4%, 10%, 20% injections; **Local Anesthetic:** 0.5%, 1%, 1.5%, 2%, 4% injection; **Topical:** 2%, 2.5%, 4%, 5% solution; 2.5%, 5% ointment; 0.5%, 4% cream; 0.5%, 2.5% gel; 0.5%, 10% spray; 2% jelly; 0.5% patch; 0.5 mg intradermal patch

ACTION & *THERAPEUTIC EFFECT* Exerts antiarrhythmic action (Class IB) by suppressing automaticity in His-Purkinje system. Combines with fast sodium channels in myocardial cell membranes, thus inhibiting sodium influx into myocardial cells. Thus it decreases ventricular depolarization, automaticity, and excitability during diastole. As a local anesthetic, it decreases pain through a reversible nerve conduction blockade. *Suppresses automaticity in His-Purkinje system of the heart and elevates electrical stimulation threshold of ventricle during diastole. Prompt, intense, and longer-lasting local anesthetic than procaine.*

USES Rapid control of ventricular arrhythmias occurring during acute MI, cardiac surgery, and cardiac catheterization and those caused by digitalis intoxication. Also as surface and infiltration anesthesia and for nerve block, including caudal and spinal block anesthesia and to relieve local discomfort of skin and mucous membranes. **Patch** for relief of pain associated with post-herpetic neuralgia.

Common adverse effects in *italic*, life-threatening effects underlined; generic names in **bold**; classifications in SMALL CAPS; ♣ Canadian drug name; ⊕ Prototype drug

UNLABELED USES Refractory status epilepticus.

CONTRAINDICATIONS History of hypersensitivity to amide-type local anesthetics; application or injection of lidocaine anesthetic in presence of severe trauma or sepsis, blood dyscrasias, post-MI; supraventricular arrhythmias, Stokes-Adams syndrome, untreated sinus bradycardia, severe degrees of sinoatrial, atrioventricular, and intraventaricular heart block.

CAUTIOUS USE Liver or kidney disease, CHF, marked hypoxia, respiratory depression, hypovolemia, shock; myasthenia gravis; debilitated patients, older adults; family history of malignant hyperthermia (fulminant hypermetabolism); pregnancy (category B), lactation. **Topical use:** In eyes, over large body areas, over prolonged periods, in severe or extensive trauma or skin disorders.

ROUTE & DOSAGE

Ventricular Arrhythmias

Adult: **IV** 50–100 mg bolus at a rate of 20–50 mg/min, may repeat in 5 min, then start infusion of 1–4 mg/min immediately after first bolus, not more than 300 mg/h **IM** 200–300 mg, may repeat once after 60–90 min
Child: **IV** 1 mg/kg bolus dose, then 20–50 mcg/kg/min infusion

Anesthetic Uses

Adult: **Infiltration** 0.5–1% solution **Nerve Block** 1–2% solution **Epidural** 1–2% solution **Caudal** 1–1.5% solution **Spinal** 5% with glucose **Saddle Block** 1.5% with dextrose **Topical** 2.5–5% jelly, ointment, cream, or solution

Post-Herpetic Neuralgia

Adult: **Topical** Apply up to 3 patches over intact skin in most painful areas once for up to 12 h per 24 h period

ADMINISTRATION

Intramuscular

- Give in deltoid muscle as preferred site.

Topical

- Do not apply topical lidocaine to large areas of skin or to broken or abraded surfaces. Consult prescriber about covering area with a dressing.
- Avoid topical preparation contact with eyes.

Intravenous

- Note: Do not use lidocaine solutions containing preservatives for spinal or epidural (including caudal) block. Use ONLY lidocaine HCl injection without preservatives or epinephrine that is specifically labeled for IV injection or infusion.

PREPARE: Direct: Give undiluted. **IV Infusion:** Use D5W for infusion. For adults, add 1 g to 250 or 500 mL to yield 2 or 4 mg/mL, respectively; for children, add 120 mg to 100 m to yield 1.2 mg/mL. ▪ Do not use solutions with particulate matter or discoloration.

ADMINISTER: Direct: Give at a rate of 50 mg or fraction thereof over 1 min. **IV Infusion:** Use microdrip and infusion pump. *Adult:* Rate of flow is usually 4 mg/min or less. *Child:* Infuse at 30 mcg/kg/min.

INCOMPATIBILITIES Solution/additive: Ampicillin, cefazolin, methohexital, phenytoin.

Common adverse effects in *italic*, life-threatening effects underlined; generic names in **bold;** classifications in SMALL CAPS; ♣ Canadian drug name; ⊘ Prototype drug

Y-site: Amphotericin B choles- teryl complex, phenytoin, thiopental.

▪ Discard partially used solutions of lidocaine without preservatives.

ADVERSE EFFECTS (≥1%) **CNS:** Drowsiness, dizziness, light-head- edness, restlessness, confusion, disorientation, irritability, appre- hension, euphoria, wild excite- ment, numbness of lips or tongue and other paresthesias including sensations of heat and cold, chest heaviness, difficulty in speaking, difficulty in breathing or swallow- ing, muscular twitching, tremors, psychosis. With high doses: Con- vulsions, respiratory depression and arrest. **CV:** With high doses: hy- potension, bradycardia, conduction disorders including heart block, cardiovascular collapse, cardiac ar- rest. **Special Senses:** Tinnitus, de- creased hearing; blurred or double vision, impaired color perception. **Skin:** Site of topical application may develop erythema, edema. **GI:** An- orexia, nausea, vomiting. **Body as a Whole:** Excessive perspiration, soreness at IM site, local thrombo- phlebitis (with prolonged IV infu- sion), hypersensitivity reactions (urticaria, rash, edema, anaphylac- toid reactions).

DIAGNOSTIC TEST INTERFERENCE Increases in *creatine phosphoki- nase (CPK)* level may occur for 48 h after IM dose and may inter- fere with test for presence of MI.

INTERACTIONS Drug: Lidocaine patch may increase toxic effects of **tocainide, mexiletine;** BARBI- TURATES decrease lidocaine activity; **cimetidine,** BETA-BLOCKERS, **quini- dine** increase pharmacologic ef- fects of lidocaine; **phenytoin** in- creases cardiac depressant effects;

procainamide compounds neuro- logic and cardiac effects.

PHARMACOKINETICS Absorp- tion: Topical application is 3% ab- sorbed through intact skin. **Onset:** 45–90 sec IV; 5–15 min IM; 2–5 min topical. **Duration:** 10–20 min IV; 60–90 min IM; 30–60 min topi- cal; greater than 100 min injected for anesthesia. **Distribution:** Cross- es blood–brain barrier and pla- centa; distributed into breast milk. **Metabolism:** In liver via CYP3A4 and 2D6. **Elimination:** In urine. **Half-Life:** 1.5–2 h.

NURSING IMPLICATIONS

Assessment & Drug Effects

▪ Stop infusion immediately if ECG indicates excessive cardiac de- pression (e.g., prolongation of PR interval or QRS complex and the appearance or aggravation of arrhythmias).

▪ Monitor BP and ECG constantly; assess respiratory and neurologic status frequently to avoid poten- tial overdosage and toxicity.

▪ Auscultate lungs for basilar rales, especially in patients who tend to metabolize the drug slowly (e.g., CHF, cardiogenic shock, hepatic dysfunction).

▪ Watch for neurotoxic effects (e.g., drowsiness, dizziness, con- fusion, paresthesias, visual distur- bances, excitement, behavioral changes) in patients receiving IV infusions or with high lidocaine blood levels.

Patient & Family Education

▪ Swish and spit out when using li- docaine solution for relief of mouth discomfort; gargle for use in pharynx, may be swallowed (as prescribed).

▪ Oral topical anesthetics (e.g., Xy- locaine Viscous) may interfere

Common adverse effects in *italic*, life-threatening effects underlined; generic names in **bold**; classifications in SMALL CAPS; ♣ Canadian drug name; ◑ Prototype drug

873

with swallowing reflex. Do NOT ingest food within 60 min after drug application; especially pediatric, geriatric, or debilitated patients.

LINAGLIPTIN

(lin' a glip' tin)

Tradjenta

Classifications: HORMONE MODIFIER; ANTIDIABETIC; INCRETIN MODIFIER; DIPEPTIDYL PEPTIDASE-4 (DPP-4) INHIBITOR

Therapeutic: ANTIDIABETIC; DDP-4 INHIBITOR

Prototype: Sitagliptin

Pregnancy Category: B

AVAILABILITY 5 mg tablets

ACTION & *THERAPEUTIC EFFECT*
Slows inactivation of incretin hormones released by the intestine. When the blood glucose begins to rise, incretin hormones stimulate insulin secretion and reduce glucagon secretion, resulting in decreased hepatic glucose production. *Sitagliptin lowers both fasting and postprandial plasma glucose levels.*

USES Treatment of type 2 diabetes mellitus in combination with diet and exercise.

CONTRAINDICATIONS History of hypersensitivity to linagliptin (e.g., urticaria, angioedema, or bronchial hyperreactivity).

CAUTIOUS USE Concurrent use with an insulin secretagogue (e.g., sulfonylurea); pregnancy (category C), lactation.

ROUTE & DOSAGE

Type 2 Diabetes Mellitus

Adult: **PO** 5mg once daily

ADMINISTRATION

Oral
- May be given with or without food.
- Store at 15°–30° C (59°–86° F).

ADVERSE EFFECTS (≥2%) **Body as a Whole:** Hypersensitivity reactions. **CNS:** Headache. **Metabolic:** Hypoglycemia, hyperlipidemia, hypertriglyceridemia, increased uric acid levels, weight gain. **Musculoskeletal:** Arthralgia, back pain, myalgia. **Respiratory:** Cough, nasopharyngitis.

INTERACTIONS Drug: Strong inducers of CYP3A4 (e.g., **rifampin, dexamethasone, phenytoin, phenobarbital**) or P-glycoprotein (e.g., **rifampin**) may decrease the therapeutic effect of linagliptin. Combination use with a SULFONYLUREA (e.g., **glyburide**) may increase the risk of hypoglycemia.

PHARMACOKINETICS Absorption: 30% bioavailable. **Peak:** 1.5 h. **Metabolism:** Primarily excreted unchanged. **Elimination:** Enterohepatic system (80%) and renal elimination (5%). **Half-Life:** 12 h.

NURSING IMPLICATIONS

Assessment & Drug Effects

- Monitor for S&S of hypoglycemia when used in combination with a sulfonylurea drug or insulin.
- Lab tests: Baseline and periodic HbA1C.

Patient & Family Education

- Seek medical attention during periods of stress or illness as dosage adjustments may be required.
- Monitor both fasting and postprandial blood glucose levels as directed.
- Note that when taken alone to control diabetes, linagliptin is

unlikely to cause hypoglycemia because it only works when the blood sugar is rising.

LINCOMYCIN HYDROCHLORIDE

(lin-koe-mye′sin)

Lincocin

Classification: LINCOSAMIDE ANTIBIOTIC
Therapeutic: ANTIBIOTIC
Prototype: Clindamycin
Pregnancy Category: B

AVAILABILITY 300 mg injection

ACTION & *THERAPEUTIC EFFECT*

Derived from *Streptomyces lincolnensis* and binds to the 50S ribosomal subunits of the bacteria inhibiting protein synthesis, eventually resulting in inhibition of bacterial cell growth or bacterial cell death. *Effective against most common gram-positive pathogens. Also effective against many anaerobic bacteria.*

USES Reserved for treatment of serious infections caused by susceptible bacteria in penicillin-allergic patients or patients for whom penicillin is inappropriate.

CONTRAINDICATIONS Previous hypersensitivity to lincomycin and clindamycin; impaired liver function, known monilial infections (unless treated concurrently); lactation.
CAUTIOUS USE Impaired kidney function; history of GI disease, particularly colitis; history of liver, endocrine, or metabolic diseases; history of asthma, hay fever, eczema, drug or other allergies; older adult patients, pregnancy (category B); infants younger than 1 mo.

ROUTE & DOSAGE

Infections

Adult: **IM** 600 mg q12–24 h **IV** 600 mg–1 g q8–12h (max: 8 g/day)
Adolescent/Child/Infant (over 1 mo): **IM** 10 mg/kg q12–24h **IV** 10–20 mg/kg/day divided q8–12h

ADMINISTRATION

Intramuscular

- Give injection deep into large muscle mass; inject slowly to minimize pain. Rotate injection sites.

Intravenous

PREPARE: **Intermittent:** Dilute each 1 g of lincomycin in at least 100 mL of D5W, NS, or other compatible solution.
ADMINISTER: **Intermittent:** Give at a rate of 1 g/h or less.
INCOMPATIBILITIES **Solution/additive: Colistimethate, kanamycin, methicillin, penicillin G, phenytoin.**

- Follow manufacturer's directions for further information on reconstitution, storage time, compatible IV fluids, and IV administration rates.

ADVERSE EFFECTS (≥1%) Body as a Whole: Hypersensitivity [pruritus, urticaria, skin rashes, exfoliative and vesiculobullous dermatitis, erythema multiforme (rare), angioedema, photosensitivity, anaphylactoid reaction, serum sickness]; superinfections (proctitis, pruritus ani, vaginitis); vertigo, dizziness, headache, generalized myalgia, thrombophlebitis following IV use; pain at IM injection site. **CV:** Hypotension, syncope, cardiopulmonary arrest (particularly after

L

Common adverse effects in *italic*, life-threatening effects underlined; generic names in **bold**; classifications in SMALL CAPS; ♣ Canadian drug name; ● Prototype drug

875

rapid IV). **GI:** Glossitis, stomatitis, *nausea, vomiting,* anorexia, decreased taste acuity, unpleasant or altered taste, abdominal cramps, *diarrhea,* acute enterocolitis, <u>pseudomembranous colitis (potentially fatal)</u>. **Hematologic:** <u>Neutropenia, leukopenia, agranulocytosis</u>, thrombocytopenic purpura, <u>aplastic anemia</u>. **Special Senses:** Tinnitus.

INTERACTIONS Drug: Kaolin and pectin decrease lincomycin absorption; **tubocurarine, pancuronium** may enhance neuromuscular blockade.

PHARMACOKINETICS Peak: 30 min IM. **Duration:** 12–14 h IM; 14 h IV. **Distribution:** High concentrations in bone, aqueous humor, bile, and peritoneal, pleural, and synovial fluids; crosses placenta; distributed into breast milk. **Metabolism:** Partially in liver. **Elimination:** In urine and feces. **Half-Life:** 5 h.

NURSING IMPLICATIONS

Assessment & Drug Effects

- Lab tests: Perform C&S initially and during therapy to determine continued microbial susceptibility. Periodic liver and kidney function tests and CBC are indicated during prolonged drug therapy.
- Take a careful history of previous sensitivities to drugs or other allergens.
- Monitor BP and pulse. Have patient remain recumbent following drug administration until BP stabilizes.
- Monitor closely and report changes in bowel frequency. Discontinue drug if significant diarrhea occurs.
- Diarrhea, acute colitis, or pseudomembranous colitis (see Appendix F) may occur up to several weeks after cessation of therapy.
- Examine IM/IV injection sites daily for signs of inflammation.
- Monitor serum drug levels closely in patients with severe impairment of kidney function.
- Monitor for S&S of superinfections that are most likely to occur when therapy exceeds 10 days (see Appendix F).

Patient & Family Education

- Notify prescriber immediately of symptoms of hypersensitivity (see Appendix F). Drug should be discontinued.
- Notify prescriber promptly of the onset of perianal irritation, diarrhea, or blood and mucus in stools.

LINDANE ⊕
(lin′dane)

Gamma Benzene, Kwell, Scabene

Classifications: SCABICIDE; PEDICULICIDE

Therapeutic: ANTIPARASITIC; PEDICULICIDE

Pregnancy Category: C

AVAILABILITY 1% lotion, shampoo

ACTION & *THERAPEUTIC EFFECT*

Action related to its direct absorption by parasites and ova (nits). Drug absorption through the parasite exoskeleton results in death of parasites and their ova. *Has ectoparasitic and ovicidal activity against the two variants of* Pediculus humanus, Pediculus capitis *(head louse) and* Pediculus pubis *(crab louse), and the arthropod* Sarcoptes scabiei *(scabies).*

USES To treat head and crab lice and scabies infestations and to eradicate their ova.

CONTRAINDICATIONS Premature neonates, patient with known seizure disorders; application to eyes, face, mucous membranes, urethral meatus, open cuts or raw, weeping surfaces; prolonged or excessive applications or simultaneous application of creams, ointments, oils; extensive dermatitis; uncontrolled seizures; lactation.

CAUTIOUS USE History of seizures; HIV infection; alcoholism; pregnancy (category C); infants, children younger than 10 y, or individuals weighing less than 110 lb.

ROUTE & DOSAGE

Lice and Scabies Infestation

Adult/Child: **Topical** Apply to all body areas except the face, leave lotion on 8–12 h, then rinse off; leave shampoo on 5 min, then rinse thoroughly; do NOT repeat in less than 1 wk

ADMINISTRATION

Note: Caregiver needs to wear plastic disposable or rubber gloves when applying lindane, especially if pregnant or applying medication to more than one patient, to avoid prolonged skin contact.

Topical

- Remove all skin lotions, creams, and oil-based hair dressings completely and allow skin to dry and cool before applying lindane; this will reduce percutaneous absorption.
- Shake cream or lotion container well. *Scabies:* Apply thin film

from neck down over entire body surface including soles of feet. Avoid face and urethral meatus. Pay particular attention to intertriginous areas (finger webs and other body creases and folds), wrists, elbows, and belt line. Rub drug in; allow skin to dry and cool after application. After 8–12 h, remove medication by bath or shower. *Crab lice:* Apply thin film of drug to hair and skin of pubic area and, if infected, to thighs, trunk, axillary areas. Leave in place 8–12 h and follow with bath or shower. Observation of living lice after 7 days indicates the need for reapplication.

- Shampoo *(head lice):* Apply sufficient quantity to wet hair and skin. Work drug thoroughly onto hair shafts and scalp and allow to remain in place 4 min. Add small amounts of water sufficient to make a thick lather; then rinse well with water. Pay particular attention to areas above and behind ears and occipital region. Use fine-tooth comb or tweezers to remove remaining nit shells. If necessary, treatment may be repeated after 7 days but not more than twice in 1 wk. *Crab lice:* See above. Repeat treatment after 7 days only if live lice can be demonstrated.
- Store in tight container away from direct light and heat. Protect from freezing.

ADVERSE EFFECTS (≥1%) **CNS:** CNS stimulation (usually after accidental ingestion or misuse of product): Restlessness, dizziness, tremors, convulsions; seizures; <u>death</u>. **Body as a Whole:** Inhalation (headache, nausea, vomiting, irritation of ENT). **Skin:** Eczematous eruptions.

Common adverse effects in *italic*, life-threatening effects <u>underlined</u>; generic names in **bold**; classifications in SMALL CAPS; ♣ Canadian drug name; ⊘ Prototype drug

877

INTERACTIONS Drug: No clinically significant interactions established.

PHARMACOKINETICS Absorption: Slowly and incompletely absorbed through intact skin; maximum absorption from face, scalp, axillae. **Distribution:** Stored in body fat. **Metabolism:** In liver. **Elimination:** In urine and feces.

NURSING IMPLICATIONS

Assessment & Drug Effects

- Monitor for seizure activity in individuals with a history of seizures. Withhold drug and report to prescriber immediately.
- Burrows made by scabies mites (may or may not be visible) appear as grayish black straight or S-shaped lines with a papule containing the mite at one end and surrounded by a mild erythematous area.

Patient & Family Education

- Lindane is highly toxic drug if topical applications are excessive or if swallowed or inhaled. Keep out of reach of children.
- Note: Lindane shampoo is an effective disinfectant for personal items such as combs, brushes.
- Skin penetration with scabies mites causes an intolerable itching that may persist 2–3 wk after they have been killed.
- Discontinue medication and notify prescriber if skin eruptions appear.
- Do not apply medication to face, mouth, open skin lesions, or to eyelashes; avoid contact with eyes. If accidental eye contact occurs, flush with water.

LINEZOLID
(lin-e-zo'lid)
Zyvox, Zyvoxam ♣

Classification: OXAZOLIDINONE ANTIBIOTIC
Therapeutic: ANTIBIOTIC
Pregnancy Category: C

AVAILABILITY 600 mg tablets; 100 mg/5 mL suspension; 200 mg/100 mL, 600 mg/300 mL injection

ACTION & *THERAPEUTIC EFFECT*
Synthetic antibiotic that binds to a site on the 23S ribosomal RNA of bacteria, which prevents the bacterial RNA translation process, thus preventing further growth. *Is bactericidal against gram-positive, gram-negative, and anaerobic bacteria. Bacteriostatic against enterococci and staphylococci, and bactericidal against streptococci.*

USES Treatment of vancomycin-resistant *Enterococcus faecium* (VREF), nosocomial pneumonia, bactertemia, complicated and uncomplicated skin and skin structure infections, community-acquired pneumonia.

CONTRAINDICATIONS Hypersensitivity to linezolid; concurrent MAOI therapy.
CAUTIOUS USE History of thrombocytopenia, thrombocytopenia; patients on serotonin reuptake inhibitors, or adrenergic agents, active alcoholism, anemia, bleeding, bone marrow suppression, cardiac arrhythmias, cardiac disease, cerebrovascular disease, chemotherapy, coagulopathy, colitis, diarrhea, hypertension, hyperthyroidism, leukopenia, MI, radiographic contrast administration, spinal anesthesia, surgery, hypertension; phenylketonuria; carcinoid syndrome; pregnancy (category C), lactation.

Common adverse effects in *italic*, life-threatening effects underlined; generic names in **bold**; classifications in SMALL CAPS; ♣ Canadian drug name; ⊙ Prototype drug

ROUTE & DOSAGE

Vancomycin-Resistant *Enterococcus faecium*

Adult/Adolescent (older than 12 y): **PO/IV** 600 mg q12h × 14–28 days
Neonate/Infant/Child: **PO/IV** 10 mg/kg q8h × 14–28 days

Nosocomial or Community-Acquired Pneumonia, Complicated Skin Infections

Adult/Adolescent (older than 12 y): **PO/IV** 600 mg q12h × 10–14 days
Infant/Child: **PO/IV** 10 mg/kg q8h × 10–14 days

Uncomplicated Skin Infections

Adult: **PO** 400 mg q12h × 10–14 days
Adolescent: **PO** 600 mg q12h × 10–14 days
Infant/Child: **PO** 10 mg/kg q12h × 10–14 days

ADMINISTRATION Note: No dosage adjustment is necessary when switching from IV to oral administration.

Oral

- Reconstitute suspension by adding 123 mL distilled water in two portions; after adding first half, shake to wet all of the powder, then add second half of water and shake vigorously to produce a uniform suspension with a concentration of 100 mg/5 mL.
- Before each use, mix suspension by inverting bottle 3–5 times, but DO NOT SHAKE. Discard unused suspension after 21 days.

Intravenous

PREPARE: Intermittent: IV solution is supplied in a single-use, ready-to-use infusion bag. Remove from protective wrap immediately prior to use. ▪ Check for minute leaks by firmly squeezing bag. Discard if leaks are detected.

ADMINISTER: Intermittent: Do not use infusion bag in a series connection. ▪ Give over 30–120 min. If IV line is used to infuse other drugs, flush before and after with D5W, NS, or LR.

INCOMPATIBILITIES **Solution/additive: Ceftriaxone, erythromycin, trimethoprim-sulfamethoxazole. Y-site: Amphotericin B, chlorpromazine, dantrolene, diazepam, pantoprazole, pentamidine, phenytoin, thiopental.**

- Store at 25° C (77° F) preferred; 15°–30° C (59°–86° F) permitted. Protect from light and keep bottles tightly closed.

ADVERSE EFFECTS (≥1%) **Body as a Whole:** Fever. **GI:** Diarrhea, nausea, vomiting, constipation, taste alteration, abnormal LFTs, tongue discoloration. **Hematologic:** Thrombocytopenia, leukopenia. **CNS:** Headache, insomnia, dizziness. **Skin:** Rash. **Urogenital:** Vaginal moniliasis.

INTERACTIONS Drug: MAO INHIBITORS may cause hypertensive crisis; **pseudoephedrine** may cause elevated BP; may cause **serotonin** syndrome with SELECTIVE SEROTONIN REUPTAKE INHIBITORS. **Food:** Tyramine-containing food may cause elevated BP. **Herbal: Ginseng, ephedra, ma huang** may lead to elevated BP, headache, nervousness.

PHARMACOKINETICS Absorption: Rapidly absorbed, 100% bioavailable. **Peak:** 1–2 h PO. **Distribution:** 31% protein bound. **Metabolism:**

Common adverse effects in *italic*, life-threatening effects underlined; generic names in **bold;** classifications in SMALL CAPS; ♣ Canadian drug name; ⊙ Prototype drug

879

By oxidation. **Elimination:** Primarily in urine. **Half-Life:** 6–7 h.

NURSING IMPLICATIONS
Assessment & Drug Effects
- Monitor for S&S of: Bleeding; hypertension; or pseudomembranous colitis that begins with diarrhea.
- Lab tests: C&S before initiating therapy and during therapy as indicated; drug may be started pending results. Monitor complete blood count, including platelet count and Hgb and Hct, in those at risk for bleeding or with longer than 2 wk of linezolid therapy.

Patient & Family Education
- Report any of the following to prescriber promptly: Onset of diarrhea; easy bruising or bleeding of any type; or S&S of superinfection (see Appendix F), S&S of seizure activity.
- Avoid foods and beverages high in tyramine (e.g., aged, fermented, pickled, or smoked foods, and beverages). Limit tyramine intake to less than 100 mg per meal (see *Information for Patients* provided by the manufacturer).
- Do not take OTC cold remedies or decongestants without consulting prescriber.
- Note for phenylketonurics: Each 5 mL oral suspension contains 20 mg phenylalanine.

LIOTHYRONINE SODIUM (T₃)
(lye-oh-thye'roe-neen)
Cytomel, Triostat
Classification: THYROID HORMONE
Therapeutic: THYROID HORMONE REPLACEMENT

Prototype: Levothyroxine sodium
Pregnancy Category: A

AVAILABILITY 5 mcg, 25 mcg, 50 mcg tablets; 10 mcg/mL injection

ACTION & *THERAPEUTIC EFFECT*
Synthetic form of natural thyroid hormone (T₃). Shares actions and uses of thyroid but has more rapid action and more rapid disappearance of effect, permitting quick dosage adjustment, if necessary. *Replacement therapy for absent or decreased thyroid hormone. Principal effect is an increase in the metabolic rate of all body tissues.*

USES Replacement or supplemental therapy for cretinism, myxedema, goiter, secondary (pituitary) or tertiary (hypothalamic) hypothyroidism, and T₃ suppression test.

CONTRAINDICATIONS Hypersensitivity to liothyronine; thyrotoxicosis; obesity treatment; severe cardiovascular conditions, acute MI, uncontrolled hypertension; adrenal insufficiency.
CAUTIOUS USE Angina pectoris, hypertension; diabetes mellitus; impaired kidney function, renal failure; older adult; pregnancy (category A), lactation.

ROUTE & DOSAGE

Thyroid Replacement
Adult: **PO** 25–75 mcg/day
Geriatric: **PO** 5 mcg/day, increase by 5 mcg/day every 1–2 wk
Child: **PO** 5 mcg/day gradually increased by 5 mcg/day q3–4days until desired response

Myxedema
Adult: **PO** 5–100 mcg/day **IV** 25–50 mcg, may repeat between

4 and 12 h after previous dose. Target dose greater than 65 mcg/day (max: 100 mcg/day). *Geriatric:* **PO** Start at 5 mcg/day

Goiter

Adult: **PO** 5–75 mcg/day
Geriatric: **PO** Start at 5 mcg/day
Child: **PO** 5 mcg/day, increase by 5 mcg q1–2 wk (usual maintenance dose 15–20 mcg/day)

T₃ Suppression Test

Adult: **PO** 75–100 mcg/day × 7 days

ADMINISTRATION

Oral
- Give daily before breakfast.

Intravenous

PREPARE: Direct: Give undiluted.
ADMINISTER: Direct: Give each 10 mcg or fraction thereof over 1 min.

- Store tablets in heat-, light-, and moisture-proof container.

ADVERSE EFFECTS (≥1%) **Endocrine:** Result from overdosage evidenced as S&S of hyperthyroidism (see Appendix F). **Musculoskeletal:** Accelerated rate of bone maturation in children.

INTERACTIONS Drug: Cholestyramine, colestipol decrease absorption; **epinephrine, norepinephrine** increase risk of cardiac insufficiency; ORAL ANTICOAGULANTS may potentiate hypoprothrombinemia.

PHARMACOKINETICS Absorption: Completely absorbed from GI tract. **Peak:** 24–72 h. **Duration:** Up to 72 h. **Distribution:** Gradually released into tissue cells. **Half-Life:** 6–7 days.

NURSING IMPLICATIONS

Assessment & Drug Effects
- Watch for possible additive effects during the early period of liothyronine substitution for another preparation, particularly in older adults, children, and patients with cardiovascular disease. Residual actions of other thyroid preparations may persist for weeks.
- Metabolic effects of liothyronine persist a few days after drug withdrawal.
- Withhold drug and notify prescriber at onset of overdosage symptoms (hyperthyroidism, see Appendix F); usually therapy can be resumed with lower dosage.

Patient & Family Education
- Take medication exactly as ordered.
- Learn S&S of hyperthyroidism (see Appendix F); notify prescriber promptly if they appear.

LIOTRIX (T₃-T₄)
(lye′oh-trix)

Thyrolar
Classification: THYROID HORMONE
Therapeutic: THYROID HORMONE REPLACEMENT
Prototype: Levothyroxine sodium
Pregnancy Category: A

AVAILABILITY 0.0125 mcg, 3.1 mcg, 6.25 mcg, 12.5 mcg, 25 mcg, 37.5 mcg

ACTION & *THERAPEUTIC EFFECT*
Synthetic levothyroxine (T₄) and liothyronine (T₃) that influence growth and maturation of tissues, increase energy expenditure, and affect turnover of essentially all substrates. These hormones play an integral role in metabolic processes, and are important to development of the CNS in newborns.

Common adverse effects in *italic*, life-threatening effects underlined; generic names in **bold**; classifications in SMALL CAPS; ♣ Canadian drug name; 🔵 Prototype drug

881

Increases metabolic rate of all body tissues.

USES Replacement or supplemental therapy for cretinism, myxedema, goiter, and secondary (pituitary) or tertiary (hypothalamic) hypothyroidism. Also with antithyroid agents in thyrotoxicosis and to prevent goitrogenesis and hypothyroidism.

CONTRAINDICATIONS Untreated thyrotoxicosis, acute MI, morphologic hypogonadism, nephrosis, adrenal deficiency due to hypopituitarism; tartrazine dye hypersensitivity, obesity treatment.
CAUTIOUS USE Myxedema; hypertension, angina, cardiac arrhythmias, cardiac disease, coronary artery disease; older adults; hypertension; arteriosclerosis; kidney dysfunction, pregnancy (category A), lactation; neonates, infants, children.

ROUTE & DOSAGE

Thyroid Replacement
Adult/Child: **PO** 12.5–30 mcg/day, gradually increase to desired response

ADMINISTRATION
Oral
- Give as a single daily dose, preferably before breakfast.
- Make dose increases at 1- to 2-wk intervals.
- Store in heat-, light-, and moisture-proof container. Shelf-life: 2 y.

ADVERSE EFFECTS (≥1%) **CNS:** Nervousness, headache, tremors, insomnia. **CV:** Palpitation, tachycardia, angina pectoris, cardiac arrhythmias, hypertension, CHF. **GI:** Nausea, abdominal cramps, diarrhea. **Body as a Whole:** Weight loss, heat intolerance, fever, sweating, menstrual irregularities. **Musculoskeletal:** Accelerated rate of bone maturation in infants and children.

INTERACTIONS Drug: Cholestyramine, colestipol decrease absorption; **epinephrine, norepinephrine** increase risk of cardiac insufficiency; ORAL ANTICOAGULANTS may potentiate hypoprothrombinemia.

NURSING IMPLICATIONS
Assessment & Drug Effects
- Watch for possible additive effects during the early period of liothyronine substitution for another preparation, particularly in older adults, children, and patients with cardiovascular disease. Residual actions of other thyroid preparations may persist for weeks.
- Note: Metabolic effects of liotrix persist a few days after drug withdrawal.
- Withhold drug and notify prescriber at onset of overdosage symptoms (hyperthyroidism, see Appendix F); usually therapy can be resumed with lower dosage.
- Monitor diabetics for glycemic control; an increase in insulin or oral hypoglycemic may be required.

Patient & Family Education
- Notify prescriber of headache (euthyroid patients); may indicate need for dosage adjustment or change to another thyroid preparation.
- Take medication exactly as ordered.
- Learn S&S of hyperthyroidism (see Appendix F); notify prescriber if they appear.

Common adverse effects in *italic*, life-threatening effects <u>underlined</u>; generic names in **bold**; classifications in SMALL CAPS; ♣ Canadian drug name; ⊙ Prototype drug

LIRAGLUTIDE
(lir-a-glu'tide)
Victoza
Classification: ANTIDIABETIC; GLUCAGON-LIKE PEPTIDE-1 RECEPTOR AGONIST; INCRETIN MIMETICS
Therapeutic: ANTIDIABETIC
Prototype: Exenatide
Pregnancy Category: C

AVAILABILITY 18 mg/3 mL pre-filled pen solution for injection

ACTION & THERAPEUTIC EFFECT
Liraglutide is a glucagon-like peptide-1 (GLP-1) receptor agonist that causes increased insulin release and decreased glucagon release in the presence of elevated blood glucose, and delays the rate of gastric emptying. *Liraglutide lowers postprandial blood glucose levels and helps normalize HbA1C.*

USES Treatment of type 2 diabetes mellitus in combination with diet and exercise.

CONTRAINDICATIONS Family or personal history of medullary thyroid carcinoma; history of multiple endocrine neoplasia syndrome type 2 (MEN 2); Type 1 DM, diabetic ketoacidosis; lactation.
CAUTIOUS USE History of pancreatitis; history of severe hypoglycemia; gastroparesis; renal or hepatic impairment; concurrent use with insulin secretagogues (e.g., sulfonylurea). Safe use in children 18 y or younger not established.

ROUTE & DOSAGE

Type 2 Diabetes Mellitus
Adult: **Subcutaneous** Initial dose of 0.6 mg once daily for 1 wk. After 1 wk, increase dose to 1.2–1.8 mg once daily to achieve glycemic control.

ADMINISTRATION
Subcutaneous
- Inject into abdomen, thigh, or upper arm without regard to meals.
- Injection timing can be changed without dose adjustment (i.e., injection may be given any time of day).
- Store refrigerated until first use, then may be stored refrigerated at 15°–30°C (59°–86°F) for up to 30 days.

ADVERSE EFFECTS (≥1%) Body as a Whole: Influenza. **CNS:** Dizziness, *headache*. **CV:** Increased blood pressure. **GI:** *Constipation, diarrhea, nausea, vomiting.* **Musculoskeletal:** Back pain. **Respiratory:** Nasopharyngitis, sinusitis, *upper respiratory tract infection.* **Urogenital:** Urinary tract infection.

INTERACTIONS Drug: Due to its ability to slow gastric emptying, liraglutide can decrease absorption rate and plasma levels of oral medications.

PHARMACOKINETICS Peak: 8–12 h. **Distribution:** 98% plasma protein bound. **Metabolism:** Peptide hydrolysis/degradation. **Elimination:** Renal and fecal as inactive metabolites. **Half-Life:** 12–13 h.

NURSING IMPLICATIONS
Assessment & Drug Effects
- Monitor for S&S of hypoglycemia. Note that the initial week of dosing (0.6 mg/d) is not effective for glycemic control but is designed to reduce GI distress.
- Monitor for and report S&S of significant GI distress, including nausea, vomiting, and diarrhea.
- Monitor for and promptly report S&S of acute pancreatitis (acute abdominal pain with/without vomiting). If pancreatitis is suspected, withhold drug and notify prescriber immediately.

Common adverse effects in *italic*, life-threatening effects underlined; generic names in **bold**; classifications in SMALL CAPS; ♣ Canadian drug name; ⊘ Prototype drug

883

- Lab tests: Frequent fasting and postprandial plasma glucose and periodic HbA1C; periodic renal function tests and LFTs.

Patient & Family Education
- Monitor blood glucose daily as directed. Report to prescriber significant hypoglycemia.
- Report promptly any of the following: A lump in the neck; hoarseness; difficulty swallowing or difficulty breathing; significant GI distress such as persistent, severe abdominal pain that may be accompanied by vomiting.
- Discard any pen that has been in use for greater than 30 days.
- Liraglutide may cause decreased appetite and some weight loss.

LISDEXAMFETAMINE DIMESYLATE

(lis-dex-am-fet′a-meen)
Vyvanse
Classifications: CEREBRAL STIMULANT; AMPHETAMINE; ANOREXIGENIC
Therapeutic: STIMULANT; ANOREXIGENIC
Prototype: Amphetamine
Pregnancy Category: C
Controlled Substance: Schedule II

AVAILABILITY 20 mg, 30 mg, 40 mg, 50 mg, 60 mg, and 70 mg capsules

ACTION & *THERAPEUTIC EFFECT*
An isomer of amphetamine that has anorexigenic action; this is thought to result from CNS stimulation and possibly from loss of acuity of smell and taste. *In hyperkinetic children, amphetamines reduce motor restlessness by an unknown mechanism.*

USES Treatment of attention-deficit hyperactivity disorder (ADHD).

CONTRAINDICATIONS Hypersensitivity to sympathomimetic amines, dextroamphetamine, or amphetamine; advanced arteriosclerosis; structural cardiac abnormalities, cardiomyopathy, cardiac arrhythmias, or symptomatic cardiovascular disease; moderate to severe hypertension; glaucoma; agitated states; patients with history of drug abuse; during or within 14 days of administering MAOIs; hyperthyroidism; seizure disorders; tics or Tourette syndrome; substance abuse; emergence of new psychotic or manic symptoms caused by amphetamine use; lactation.
CAUTIOUS USE Controlled hypertension, heart failure, recent MI, or recent ventricular arrhythmia; preexisting psychotic disorder; suicidal tendencies; bipolar disorder, depression; history of aggressive or hostile behavior; alcoholism; pregnancy (category C); children younger than 6 y.

ROUTE & DOSAGE

Attention-Deficit Hyperactivity Disorder
Adult/Child (6–12 y): **PO** 30 mg daily in a.m.; may increase to 50–70 mg daily at weekly intervals (max: 70 mg daily)

ADMINISTRATION

Oral
- Give daily dose in the morning.
- Capsule may be taken whole or opened and dissolved in a glass of water.
- Store at 15°–30° C (59°–86° F) and protect from light.

ADVERSE EFFECTS (≥1%) **Body as a Whole:** Pyrexia. **CNS:** Affect lability, dizziness, *headache, insomnia, irritability,* somnolence, tic. **GI:** *Abdominal pain,* dry mouth, nausea, vomiting. **Metabolic:** Decreased appetite, weight loss. **Skin:** Rash.

DIAGNOSTIC TEST INTERFERENCE
Can cause a significant elevation in plasma CORTICOSTEROID levels and may interfere with *urinary steroid determinations.*

INTERACTIONS Drug: **Chlorpromazine** and **haloperidol** inhibit the CNS stimulant effects of amphetamines. **Furazolidone** and MAO INHIBITORS can increase adverse effects. **Lithium** may inhibit the effects of lisdexamfetamine. **Propoxyphene** can potentiate the CNS stimulation of lisdexamfetamine. Compounds that acidify the urine lower the plasma levels of lisdexamfetamine. Lisdexamfetamine inhibits the actions of **adrenergic blockers.** Co-administration of ANTIHISTAMINES with lisdexamfetamine can counteract desired sedative effects. Lisdexamfetamine may antagonize the hypotensive effects of ANTIHYPERTENSIVE AGENTS. Lisdexamfetamine may delay the absorption of **ethosuximide** and **phenytoin.** Lisdexamfetamine may potentiate the actions of TRICYCLIC ANTIDEPRESSANTS, **meperidine** and **norepinephrine.**

PHARMACOKINETICS Absorption: Rapidly from GI tract. **Peak:** 1 h. **Distribution:** Extensive throughout body. **Metabolism:** Prodrug converted in liver to dextroamphetamine. **Elimination:** Urine (96%). **Half-Life:** 1 h (lisdexamfetamine) 6–8 h (dextroamphetamine).

NURSING IMPLICATIONS

Assessment & Drug Effects

- Monitor children, adolescents, and adults for signs and symptoms of adverse cardiac reactions (e.g., hypertension, arrhythmias). Report promptly exertional chest pain or syncope.
- Monitor closely growth rate in children.

- Typically therapy is interrupted or dosage reduced periodically to assess effectiveness in behavior disorders.
- Monitor children and adolescents for development of aggressive or abnormal behaviors.

Patient & Family Education

- Do not drive or engage in other potentially hazardous activities until response to drug is known.
- Report promptly any of the following: Chest pain with activity, new or worse behavior or thought problems, psychotic symptoms (e.g., hearing voices, believing things that are not true).
- Taper drug gradually following long-term use to avoid extreme fatigue, mental depression, and prolonged abnormal sleep pattern.

L

LISINOPRIL
(ly-sin′o-pril)
Prinivil, Zestril
Classifications: ANTIHYPERTENSIVE; ANGIOTENSIN-CONVERTING ENZYME (ACE) INHIBITOR
Therapeutic: ANTIHYPERTENSIVE
Prototype: Enalapril
Pregnancy Category: D

AVAILABILITY 2.5 mg, 5 mg, 10 mg, 20 mg, 30 mg, 40 mg tablets

ACTION & THERAPEUTIC EFFECT
Lowers BP by specific inhibition of the angiotensin-converting enzyme (ACE). This interrupts conversion sequences initiated by renin that form angiotensin II, a potent vasoconstrictor. ACE inhibition alters hemodynamics without compensatory reflex tachycardia or changes in cardiac output (except in patients with CHF). *Improves cardiac output and exercise tolerance. Aldosterone is also reduced, thus permitting a potassium-sparing effect.*

Common adverse effects in *italic*, life-threatening effects underlined; generic names in **bold**; classifications in SMALL CAPS; ♣ Canadian drug name; ⊘ Prototype drug

885

USES Hypertension, alone or concomitantly with other classes of antihypertensive agents; CHF; to improve MI survival.

CONTRAINDICATIONS History of angioedema related to treatment with an ACE inhibitor, ACE inhibitor hypersensitivity; pregnancy (category D), lactation.

CAUTIOUS USE Impaired kidney function, renal artery stenosis, renal disease, renal failure, hyperkalemia, aortic stenosis, cardiomyopathy; cerebrovascular disease; collagen vascular disease; CAD; dialysis; heart failure, hyperkalemia, hypotension, hypovolemia; African Americans; autoimmune diseases, especially systemic lupus erythematosus (SLE); older adults; children younger than 6 y.

ROUTE & DOSAGE

Hypertension

Adult: **PO** 10 mg once/day, may increase up to 20–40 mg 1–2 times/day (max: 80 mg/day)
Child (6–16 y): **PO** Start at 0.07 mg/kg (max: 5 mg) once/day (max: 40 mg/day)
Geriatric: **PO** Initial 2.5–5 mg/day, may increase by 2.5–5 mg/day every 1–2 wk (max: 40 mg/day)

Heart Failure

Adult: **PO** 5–40 mg/day

ADMINISTRATION

Oral

- Monitor drug effect for several hours or until the BP is stabilized for at least 1 additional hour. Concurrent administration with a diuretic may compound hypotensive effect.
- Store away from both moisture and heat.

ADVERSE EFFECTS (≥1%) **CNS:** Headache, dizziness, fatigue. **CV:** Hypotension, chest pain. **GI:** Nausea, vomiting, diarrhea, anorexia, constipation, intestinal angioedema. **Hematologic:** Neutropenia. **Respiratory:** Dyspnea, cough. **Skin:** Rash. **Metabolic:** Azotemia, hyperkalemia, increased BUN, and creatinine levels.

INTERACTIONS Drug: Indomethacin and other NSAIDs may decrease antihypertensive activity; POTASSIUM SUPPLEMENTS, POTASSIUM-SPARING DIURETICS may cause hyperkalemia; may increase **lithium** levels and toxicity.

PHARMACOKINETICS Absorption: 25% absorbed from GI tract. **Onset:** 1 h. **Peak:** 6–8 h. **Duration:** 24 h. **Distribution:** Limited amount crosses blood–brain barrier; crosses placenta; small amount distributed in breast milk. **Metabolism:** Is not metabolized. **Elimination:** Primarily in urine. **Half-Life:** 12 h.

NURSING IMPLICATIONS

Assessment & Drug Effects

- Place patient in supine position and notify prescriber if sudden and severe hypotension occurs within the first 1–5 h after initial drug dose; greatest risk for hypotension is in patients who are sodium- or volume-depleted because of diuretic therapy.
- Measure BP just prior to dosing to determine whether satisfactory control is being maintained for 24 h. If the antihypertensive effect is diminished in less than 24 h, an increase in dosage may be necessary.
- Monitor closely for angioedema of extremities, face, lips, tongue, glottis, and larynx. Discontinue drug promptly and notify prescriber if such symptoms appear;

carefully monitor for airway obstruction until swelling is relieved.

- Monitor serum sodium and serum potassium levels for hyponatremia and hyperkalemia.
- Lab tests: Determine WBC count prior to initiation of treatment, every month for the first 3–6 mo of therapy, and at periodic intervals for 1 y. Withhold therapy and notify prescriber if neutropenia (neutrophil count less than 1000/mm^3) develops; kidney function tests at periodic intervals, especially in patients with severe volume or sodium replacement or those with severe CHF.

Patient & Family Education

- Discontinue drug and contact prescriber immediately for severe hypersensitivity reaction (e.g., hoarseness, swelling of the face, mouth, hands, or feet, or sudden trouble breathing).
- Be aware of importance of proper diet, including sodium and potassium restrictions. Do NOT use salt substitute containing potassium.
- Continued compliance with high BP medication is very important. If a dose is missed, take it as soon as possible but not too close to next dose.
- Do not drive or engage in other potentially hazardous activities until response to the drug is known.
- With concomitant therapy, lisinopril increases the risk of lithium toxicity.
- Notify prescriber promptly of any indication of infection (e.g., sore throat, fever).
- Do not store drug in a moist area. Heat and moisture may cause the medicine to break down.

LITHIUM CARBONATE ⊕
(li'thee-um)

Eskalith, Eskalith CR, Lithane, Lithobid, Lithonate, Lithotabs

LITHIUM CITRATE
Cibalith-S

Classification: ANTIPSYCHOTIC; MOOD STABILIZER
Therapeutic: ANTIPSYCHOTIC; ANTIMANIC; ANTIDEPRESSANT
Pregnancy Category: D

AVAILABILITY Lithium Carbonate: 150 mg, 300 mg, 600 mg capsules; 300 mg, 450 mg sustained release tablets; Lithium Citrate: 300 mg/5 mL syrup

ACTION & *THERAPEUTIC EFFECT*
The lithium ion behaves in the body much like the sodium ion; but its exact mechanism of action is unclear. Competes with various physiologically important cations: Na$^+$, K$^+$, Ca^{2+}, Mg^{2+}; therefore, it affects cell membranes, body water, and neurotransmitters. At the synapse, it accelerates catecholamine destruction, inhibits the release of neurotransmitters and decreases sensitivity of postsynaptic receptors. Decreases overactivity of receptors involved in stimulating manic states. *Effective response evidenced by changed facial affect, improved posture, assumption of self-care, improved ability to concentrate, improved sleep pattern.*

USES Control and prophylaxis of acute mania and the acute manic phase of mixed bipolar disorder.
UNLABELED USES Acute and recurrent depression (unipolar affective disorder), schizophrenic disorders, disorders of impulse control, alcohol dependence, antineoplastic drug-induced neutropenia, aplastic anemia, SIADH, cyclic neutropenia.

Common adverse effects in *italic*, life-threatening effects underlined; generic names in **bold**; classifications in SMALL CAPS; ◆ Canadian drug name; ⊕ Prototype drug

887

CONTRAINDICATIONS History of ACE inhibitor induced angio-edema; significant cardiovascular or kidney disease, brain damage, severe debilitation, dehydration or sodium depletion; patients on low-salt diet or receiving diuretics; pregnancy (category D), lactation. **CAUTIOUS USE** Thyroid disease; epilepsy; cardiac disease, cardiac arrhythmias, dehydration, diarrhea; fever, hyponatremia, hypothyroidism, concurrent infection; leukemia; mental status changes, organic brain syndrome, parkinsonism; psoriasis; renal disease, renal impairment; seizure disorder, sick sinus syndrome, sodium restriction, risk of suicidal ideation, thyroid disease, urinary retention; diabetes mellitus; severe infections; urinary retention; older adults; children younger than 12 y.

ROUTE & DOSAGE

Mania

Adult: **PO Loading Dose** 600 mg t.i.d. or 900 mg sustained release b.i.d. or 30 mL (48 mEq) of solution t.i.d. **PO Maintenance Dose** 300 mg t.i.d. or q.i.d. or 15–20 mL (24–32 mEq) solution in 2–4 divided doses (max: 2.4 g/day)
Child: **PO** 15–60 mg/kg/day in divided doses

ADMINISTRATION

Oral
- Give with meals.
- Ensure that sustained release tablets are not chewed or crushed; **must be** swallowed whole.
- Protect from light and moisture.

ADVERSE EFFECTS (≥1%) **CNS:** Dizziness, *headache, lethargy,* drowsiness, *fatigue,* slurred speech, psychomotor retardation, giddiness, incontinence, restlessness, seizures, confusion, blackout spells, disorientation, *recent memory loss,* stupor, coma, EEG changes. **CV:** Arrhythmias, hypotension, vasculitis, <u>peripheral circulatory collapse</u>, ECG changes. **Special Senses:** Impaired vision, transient scotomas, tinnitus. **Endocrine:** Diffuse thyroid enlargement, hypothyroidism, *nephrogenic diabetes insipidus,* transient hyperglycemia, glycosuria, hyponatremia. **GI:** *Nausea, vomiting, anorexia, abdominal pain, diarrhea, dry mouth,* metallic taste. **Musculoskeletal:** *Fine hand tremors,* coarse tremors, choreoathetotic movements; fasciculations, clonic movements, incoordination including ataxia, *muscle weakness,* hyperreflexia, encephalopathic syndrome (weakness, lethargy, fever, tremors, confusion, extrapyramidal symptoms). **Skin:** Thought to be toxicity rather than allergy: Pruritus, maculopapular rash, hyperkeratosis, chronic folliculitis, transient acneiform papules (face, neck, intertriginous areas), anesthesia of skin, cutaneous ulcers, drying and thinning of hair, allergic vasculitis. **Hematologic:** *Reversible leukocytosis* (14,000 to 18,000/mm³). **Urogenital:** Albuminuria, oliguria, urinary incontinence, polyuria, polydipsia, increased uric acid excretion. **Body as a Whole:** Edema, weight gain (common) or loss, exacerbation of psoriasis; flu-like symptoms.

INTERACTIONS Drug: Carbamazepine, haloperidol, PHENOTHIAZINES increase risk of neurotoxicity, extrapyramidal effects, and tardive dyskinesias; DIURETICS, NSAIDS, **methyldopa, probenecid,** TETRACYCLINES decrease renal clearance of lithium, increasing pharmacologic and toxic effects; THEOPHYLLINES, **urea, sodium bicarbonate,**

sodium or potassium citrate increase renal clearance of lithium, decreasing its pharmacologic effects. BETA-BLOCKERS may mask signs of toxicity.

PHARMACOKINETICS Absorption: Readily absorbed from GI tract. **Peak:** 0.5–3 h carbonate; 15–60 min citrate. **Distribution:** Crosses blood–brain barrier and placenta; distributed into breast milk. **Metabolism:** Not metabolized. **Elimination:** 95% in urine, 1% in feces, 4–5% in sweat. **Half-Life:** 20–27 h.

NURSING IMPLICATIONS

Assessment & Drug Effects

- Monitor response to drug. Usual lag of 1–2 wk precedes response to lithium therapy. Keep prescriber informed of progress.
- Lab test: Periodic lithium levels (draw blood sample prior to next dose or 8–12 h after last dose); periodic thyroid function tests.
- Monitor for S&S of lithium toxicity (e.g., vomiting, diarrhea, lack of coordination, drowsiness, muscular weakness, slurred speech when level is 1.5–2 mEq/L; ataxia, blurred vision, giddiness, tinnitus, muscle twitching, coarse tremors, polyuria when greater than 2 mEq/L). Withhold one dose and call prescriber. Drug should not be stopped abruptly. Note that lithium-induced tremors may be masked by concurrent use of beta-blockers.
- Monitor older adults carefully to prevent toxicity, which may occur at serum levels ordinarily tolerated by other patients.
- Be alert to and report symptoms of hypothyroidism (see Appendix F).
- Weigh patient daily; check ankles, tibiae, and wrists for edema. Report changes in I&O ratio, sudden weight gain, or edema.

- Report early signs of extra-pyramidal reactions promptly to prescriber.

Patient & Family Education

- Be alert to increased output of dilute urine and persistent thirst. Dose reduction may be indicated.
- Contact prescriber if diarrhea or fever develops. Avoid practices that may encourage dehydration: Hot environment, excessive caffeine beverages (diuresis).
- Drink plenty of liquids (2–3 L/day) during stabilization period and at least 1–1.5 L/day during ongoing therapy.
- Avoid self-prescribed low-salt regimen, self-dosing with antacids containing sodium, and high-sodium foods (e.g., prepared meats and diet soda).
- Do not drive or engage in other potentially hazardous activities until response to drug is known. Lithium may impair both physical and mental ability.
- Use effective contraceptive measures during lithium therapy. If therapy is continued during pregnancy, serum lithium levels **must be** closely monitored to prevent toxicity.

LODOXAMIDE

(lo-dox′a-mide)

Alomide
Pregnancy Category: B
See Appendix A-1.

LOMUSTINE

(loe-mus′teen)

CeeNU, CCNU
Classifications: ANTINEOPLASTIC; ALKYLATING AGENT; NITROSOUREA
Therapeutic: ANTINEOPLASTIC
Prototype: Cyclophosphamide
Pregnancy Category: D

Common adverse effects in *italic*, life-threatening effects underlined; generic names in **bold**; classifications in SMALL CAPS; ♣ Canadian drug name; ⊘ Prototype drug

889

AVAILABILITY 10 mg, 40 mg, 100 mg capsules

ACTION & *THERAPEUTIC EFFECT*
Lipid-soluble alkylating nitrosourea with actions like those of carmustine (e.g., cell-cycle-nonspecific activity against rapidly proliferating cell populations). Inhibits synthesis of both DNA and RNA. *Has antineoplastic and myelosuppressive effect.*

USES Palliative therapy in addition to other modalities or with other chemotherapeutic agents in malignant glioma and as secondary therapy in Hodgkin's disease.

UNLABELED USES GI, lung, and renal carcinomas, non-Hodgkin's lymphomas, malignant melanoma, and multiple myeloma.

CONTRAINDICATIONS Immunization with live virus vaccines, viral infections; severe bone marrow suppression; active infection; pregnancy (category D), lactation.

CAUTIOUS USE Patients with decreased circulating platelets, leukocytes, or erythrocytes; kidney or liver function impairment; previous cytotoxic or radiation therapy; pulmonary disease.

ROUTE & DOSAGE

Palliative Therapy

Adult/Adolescent: **PO** 130 mg/m² as single dose, repeated in 6 wk; subsequent doses based on hematologic response (WBC greater than 4000/mm³, platelets greater than 100,000/mm³)
Child: **PO** 75–150 mg/m² q6wk

ADMINISTRATION

Oral
- Give on an empty stomach to reduce possibility of nausea, may also give an antiemetic before drug to prevent nausea.

- Store capsules away from excessive heat (over 40° C).

ADVERSE EFFECTS (≥1%) **CNS:** Lethargy, ataxia, disorientation. **GI:** Anorexia, *nausea, vomiting,* stomatitis, transient elevations of LFTs. **Hematologic:** Delayed (cumulative) myelosuppression: (Thrombocytopenia, leukopenia); anemia. **Skin:** Alopecia, skin rash, itching. **Urogenital:** Nephrotoxicity. **Respiratory:** Pulmonary toxicity (rare).

INTERACTIONS Drug: Cimetidine can increase bone marrow toxicity; ANTICOAGULANTS, NSAIDS, SALICYLATES increase risk of bleeding.

PHARMACOKINETICS Absorption: Readily absorbed from GI tract. **Peak:** 1–6 h. **Distribution:** Readily crosses blood–brain barrier; crosses placenta; distributed into breast milk. **Metabolism:** In liver to several active metabolites. **Elimination:** In urine. **Half-Life:** 16–48 h.

NURSING IMPLICATIONS

Assessment & Drug Effects
- Lab tests: Monitor blood counts weekly for at least 6 wk after last dose. Liver and kidney function tests should be performed periodically.
- A repeat course is not given until platelets have returned to above 100,000/mm³ and leukocytes to above 4000/mm³.
- Avoid invasive procedures during nadir of platelets.
- Thrombocytopenia occurs about 4 wk and leukopenia about 6 wk after a dose, persisting 1–2 wk.
- Inspect oral cavity daily for S&S of superinfections (see Appendix F) and stomatitis or xerostomia.

Patient & Family Education
- Nausea and vomiting may occur 3–5 h after drug administration, usually lasting less than 24 h.

- Anorexia may persist for 2 or 3 days after a dose.
- Notify prescriber of signs of sore throat, cough, fever. Also report unexplained bleeding or easy bruising.
- Use reliable contraceptive measures during therapy.
- Be aware of the possibility of hair loss while taking this drug.
- A given dose may include capsules of different colors; the pharmacist prepares prescribed dose by combining various capsule strengths.

LOPERAMIDE 🄟
(loe-per'a-mide)

Imodium, Imodium AD, Kaopectate III, Maalox Antidiarrheal, Pepto Diarrhea Control
Classification: ANTIDIARRHEAL
Therapeutic: ANTIDIARRHEAL
Pregnancy Category: C

AVAILABILITY 2 mg tablets, capsules; 1 mg/mL, 1 mg/5 mL liquid

ACTION & *THERAPEUTIC EFFECT*
Inhibits GI peristaltic activity by direct action on circular and longitudinal intestinal muscles. Prolongs transit time of intestinal contents, increases consistency of stools, and reduces fluid and electrolyte loss. *Effectiveness as an antidiarrheal agent is due to prolonging transit time in the colon.*

USES Acute nonspecific diarrhea, chronic diarrhea associated with inflammatory bowel disease, and to reduce fecal volume from ileostomies.

CONTRAINDICATIONS Conditions in which constipation should be avoided, ileus, severe colitis, bacterial gastroenteritis; acute diarrhea caused by broad-spectrum antibiotics (pseudomembranous colitis) or associated with microorganisms that penetrate intestinal mucosa (e.g., toxigenic *Escherichia coli, Salmonella,* or *Shigella*); GI bleeding; lactation.

CAUTIOUS USE Dehydration; diarrhea caused by invasive bacteria; ulcerative colitis; impaired liver function; prostatic hypertrophy; history of narcotic dependence; pregnancy (category C); children younger than 2 y.

ROUTE & DOSAGE

Acute Diarrhea
Adult: **PO** 4 mg followed by 2 mg after each unformed stool (max: 16 mg/day)
Child: **PO** 2–6 y, 1 mg t.i.d.; 6–8 y, 2 mg b.i.d.; 8–12 y, 2 mg t.i.d.

Chronic Diarrhea
Adult: **PO** 4 mg followed by 2 mg after each unformed stool until diarrhea is controlled (max: 16 mg/day)
Child: **PO** 0.1 mg/kg after each unformed stool (usually 1 mg)

ADMINISTRATION
Oral
- Do not give prn doses to a child with acute diarrhea.

ADVERSE EFFECTS (≥1%) **Body as a Whole:** Hypersensitivity (skin rash); fever. **CNS:** Drowsiness, fatigue, dizziness, CNS depression (overdosage). **GI:** Abdominal discomfort or pain, abdominal distention, bloating, constipation, nausea, vomiting, anorexia, dry mouth; <u>toxic megacolon</u> (patients with ulcerative colitis).

INTERACTIONS Drug: No clinically significant interactions established.

Common adverse effects in *italic*, life-threatening effects <u>underlined</u>; generic names in **bold**; classifications in SMALL CAPS; ♣ Canadian drug name; 🄟 Prototype drug

891

PHARMACOKINETICS Absorption: Poorly absorbed from GI tract. **Onset:** 30–60 min. **Peak:** 2.5 h solution; 4–5 h capsules. **Duration:** 4–5 h. **Metabolism:** In liver. **Elimination:** Primarily in feces, less than 2% in urine. **Half-Life:** 11 h.

NURSING IMPLICATIONS

Assessment & Drug Effects

- Monitor therapeutic effectiveness. Chronic diarrhea usually responds within 10 days. If improvement does not occur within this time, it is unlikely that symptoms will be controlled by further administration.
- Discontinue if there is no improvement after 48 h of therapy for acute diarrhea.
- Monitor fluid and electrolyte balance.
- Notify prescriber promptly if the patient with ulcerative colitis develops abdominal distention or other GI symptoms (possible signs of potentially fatal toxic megacolon).

Patient & Family Education

- Notify prescriber if diarrhea does not stop in a few days or if abdominal pain, distention, or fever develops.
- Record number and consistency of stools.
- Do not drive or engage in other potentially hazardous activities until response to drug is known.
- Do not take alcohol and other CNS depressants concomitantly unless otherwise advised by prescriber; may enhance drowsiness.
- Learn measures to relieve dry mouth; rinse mouth frequently with water, suck hard candy.

LOPINAVIR/RITONAVIR
(lop-i-na′ver/rit-o-na′ver)
Kaletra

Classifications: ANTIRETROVIRAL; PROTEASE INHIBITOR
Therapeutic: PROTEASE INHIBITOR
Prototype: Saquinavir mesylate
Pregnancy Category: C

AVAILABILITY 200 mg lopinavir/ 50 mg ritonavir, 100 mg lopinavir/ 25 mg ritonavir tablets; 400 mg lopinavir/100 mg ritonavir/5 mL suspension

ACTION & *THERAPEUTIC EFFECT* Lopinavir, an HIV protease inhibitor that inhibits the activity of HIV protease and prevents the cleavage of viral polyproteins essential for the maturation of HIV. Ritonavir inhibits the CYP3A metabolism of lopinavir, thereby, increasing the blood level of lopinavir. *Decreases plasma HIV RNA level; reduces viral load as a result of the combined therapy of the two drugs in HIV infected patients.*

USES Treatment of HIV infection in combination with other antiretroviral agents.

CONTRAINDICATIONS Hypersensitivity to lopinavir or ritonavir; lactation.
CAUTIOUS USE Hepatic impairment, patients with hepatitis B or C, older adults; DM; history of pancreatitis; cardiac disease; potential for PR prolongation and QT prolongation; conduction abnormalities, ischemic heart disease, and cardiomyopathy; pregnancy (category C); infants less than 14 days old.

ROUTE & DOSAGE

HIV Infection—Treatment Naïve
Adult: **PO** 800/200 mg or 400/100 mg b.i.d. daily

Common adverse effects in *italic*, life-threatening effects <u>underlined</u>; generic names in **bold**; classifications in SMALL CAPS; ♣ Canadian drug name; ⦿ Prototype drug

HIV Infection—Treatment Experienced

Adult: **PO** 400/100 mg (3 tablets or 5 mL suspension) b.i.d., increase dose to 533/133 mg (4 tablets or 6.5 mL) b.i.d., with concurrent efavirenz or nevirapine

Child: **PO** with concurrent efavirenz or nevirapine *6 mo–12 y, weight 7–15 kg,* 12/3 mg/kg b.i.d.; *weight 15–40 kg,* 10/2.5 mg/kg; *weight greater than 40 kg,* 400/100 mg b.i.d., increase dose *weight 7–15 kg,* 13/3.25 mg/kg; *weight 15–40 kg,* 11/2.75 mg/kg; *weight greater than 40 kg,* 533/133 mg b.i.d. *Adolescent/Child/Infant (6 mo to 18 y):* **PO** Without concurrent efavirenz, nevirapine, fosamprenavir, or nelfinavir 230/57.5 mg/m² b.i.d. *Infant (14 days to 6 mo):* **PO** 16/4 mg/kg or 300/75 mg/m² b.i.d.

ADMINISTRATION

Note: Take with food.

Oral

- Give with a meal or light snack.
- Note: If didanosine is concurrently ordered, give didanosine 1 h before or 2 h after lopinavir/ritonavir.
- Store refrigerated at 2°–8° C (36°–46° F). If stored at room temperature 25° C (77° F) or below, discard after 2 mo.

ADVERSE EFFECTS (≥1%) **Body as a Whole:** Asthenia, pain. **GI:** Abdominal pain, abnormal stools, *diarrhea, nausea,* vomiting. **CNS:** Headache, insomnia, abnormal taste. **Skin:** Rash. **Metabolic:** Hypercholesterolemia, increased triglycerides, ALT increased, weakness. **Hematologic:** Platelets decreased, amenorrhea.

INTERACTIONS **Drug:** Flecainide, **propafenone, pimozide** may lead to life-threatening arrhythmias; **rifampin** may decrease antiretroviral response; **dihydroergotamine, ergonovine, ergotamine, methylergonovine** may lead to acute ergot toxicity; HMG-COA REDUCTASE INHIBITORS may increase risk of myopathy and rhabdomyolysis; BENZODIAZEPINES may have prolonged sedation or respiratory depression; **efavirenz, nevirapine,** ANTICONVULSANTS, STEROIDS may decrease lopinavir levels; **delavirdine, ritonavir** may increase lopinavir levels; may increase levels of **amprenavir, indinavir, saquinavir, ketoconazole, itraconazole, midazolam, triazolam, rifabutin, sildenafil, atorvastatin, cerivastatin,** IMMUNOSUPPRESSANTS; may decrease levels of **atovaquone, methadone, ethinyl estradiol;** may increase trazodone toxicity; decrease efficacy of hormonal contraceptives, increases midazolam concentration and toxicity. Also see INTERACTIONS in **ritonavir** monograph. **Herbal: St. John's wort, garlic** may decrease effect.

PHARMACOKINETICS **Absorption:** Increased absorption when taken with food. **Peak:** 4 h. **Distribution:** 98–99% protein bound. **Metabolism:** Extensively metabolized by CYP3A. **Elimination:** Primarily in feces. **Half-Life:** 5–6 h lopinavir.

NURSING IMPLICATIONS

Assessment & Drug Effects

- Monitor for S&S of: Pancreatitis, especially with marked triglyceride elevations; new onset diabetes or loss of glycemic control; hypothyroidism or Cushing's syndrome.
- Lab test: Periodically monitor fasting blood glucose, AST and ALT,

---

Common adverse effects in *italic*, life-threatening effects <u>underlined</u>; generic names in **bold**; classifications in SMALL CAPS; ♣ Canadian drug name; ⊘ Prototype drug

893

total cholesterol and triglycerides, serum amylase, inorganic phosphorus, CBC with differential, and thyroid functions.

Patient & Family Education

- Report all prescription and non-prescription drugs being taken. Do not use herbal products, especially St. John's wort, without first consulting the prescriber.
- Become familiar with the potential adverse effects of this drug; report those that are bothersome to prescriber.
- Concurrent use of sildenafil (Viagra) increases risk for adverse effects such as hypotension, changes in vision, and sustained erection; promptly report any of these to the prescriber.
- Use additional or alternative contraceptive measures if estrogen-based hormonal contraceptives are being used.

LORATADINE ⊕

(lor'a-ta-deen)

Alavert, Claritin, Claritin Reditabs

Classifications: NONSEDATING ANTIHISTAMINE; H_1-RECEPTOR ANTAGONIST
Therapeutic: NONSEDATING ANTIHISTAMINE

Pregnancy Category: B

AVAILABILITY 10 mg tablets; 1 mg/mL syrup

ACTION & *THERAPEUTIC EFFECT*
Long-acting nonsedating antihistamine with selective peripheral H_1-receptor sites, thus blocking histamine release. Loratadine is a long-acting H_1-receptor antagonist of histamine that disrupts capillary permeability, edema formation, and constriction of respiratory, GI, and vascular smooth muscle. *Effective in*

relieving allergic reactions related to histamine release.

USES Relief of symptoms of seasonal allergic rhinitis; idiopathic chronic urticaria.

CONTRAINDICATIONS Hypersensitivity to loratadine, lactation.
CAUTIOUS USE Hepatic and renal impairment, renal disease, renal failure; asthma; pregnancy (category B).

ROUTE & DOSAGE

Allergic Rhinitis
Adult: **PO** 10 mg once/day on an empty stomach; start patients with liver disease with 10 mg every other day
Child: **PO** *Weight less than 30 kg,* 5 mg daily; *weight greater than 30 kg,* 10 mg daily

ADMINISTRATION

Oral

- Give on an empty stomach, 1 h before or 2 h after a meal.
- Store in a tightly closed container.

ADVERSE EFFECTS (≥1%) **CNS:** Dizziness, dry mouth, fatigue, headache, somnolence, altered salivation and lacrimation, thirst, flushing, anxiety, depression, impaired concentration. **CV:** Hypotension, hypertension, palpitations, syncope, tachycardia. **GI:** Nausea, vomiting, flatulence, abdominal distress, constipation, diarrhea, weight gain, dyspepsia. **Body as a Whole:** Arthralgia, myalgia. **Special Senses:** Blurred vision, earache, eye pain, tinnitus. **Skin:** Rash, pruritus, photosensitivity.

PHARMACOKINETICS Absorption: Readily from GI tract. **Onset:** 1–3 h. **Peak:** 8–12 h; reaches steady state levels in 3–5 days. **Duration:** 24 h. **Distribution:** Distributed into breast milk. **Metabolism:** In liver to active metabolite, descarboethoxyloratidine. **Elimination:** In urine and feces. **Half-Life:** 12–15 h.

NURSING IMPLICATIONS

Assessment & Drug Effects

- Assess carefully for and report distressing or dangerous S&S that occur after initiation of the drug. A variety of adverse effects, although not common, are possible. Some are an indication to discontinue the drug.
- Monitor cardiovascular status and report significant changes in BP and palpitations or tachycardia.

Patient & Family Education

- Drug may cause significant drowsiness in older adult patients and those with liver or kidney impairment.
- Note: Concurrent use of alcohol and other CNS depressants may have an additive effect.

LORAZEPAM ℗

(lor-a′ze-pam)

Ativan
Classifications: ANXIOLYTIC; SEDATIVE-HYPNOTIC; BENZODIAZEPINE
Therapeutic: ANTIANXIETY; SEDATIVE-HYPNOTIC
Pregnancy Category: D
Controlled Substance: Schedule IV

AVAILABILITY 0.5 mg, 1 mg, 2 mg tablets; 2 mg/mL oral solution; 2 mg/mL, 4 mg/mL injection

ACTION & *THERAPEUTIC EFFECT*
Most potent of the available benzodiazepines. Effects (antianxiety, sedative, hypnotic, and skeletal muscle relaxant) are mediated by the inhibitory neurotransmitter GABA. Action sites are thalamic, hypothalamic, and limbic levels of CNS. *Antianxiety agent that also causes mild suppression of REM sleep, while increasing total sleep time.*

USES Management of anxiety disorders and for short-term relief of symptoms of anxiety. Also used for preanesthetic medication to produce sedation and to reduce anxiety and recall of events related to day of surgery; for management of status epilepticus.

UNLABELED USES Chemotherapy-induced nausea and vomiting.

CONTRAINDICATIONS Known sensitivity to benzodiazepines; acute narrow-angle glaucoma; primary depressive disorders or psychosis; COPD; coma, shock, sleep apnea; acute alcohol intoxication; dementia; pregnancy (category D), and lactation.

CAUTIOUS USE Renal or hepatic impairment; renal failure; organic brain syndrome; myasthenia gravis; narrow-angle glaucoma; pulmonary disease; mania; psychosis; suicidal tendency; history of seizure disorders; GI disorders; older adult and debilitated patients; *PO:* children younger than 12 y.

ROUTE & DOSAGE

Antianxiety

Adult: **PO** 2–6 mg/day in divided doses (max: 10 mg/day)
Geriatric: **PO** 0.5–1 mg/day (max: 2 mg/day)
Child: **PO/IV** 0.05 mg/kg q4–8h (max: 2 mg/dose)

Insomnia

Adult: **PO** 2–4 mg at bedtime
Geriatric: **PO** 0.5–1 mg at bedtime

Premedication

Adult: **IM** 2–4 mg (0.05 mg/kg) at least 2 h before surgery

L

Common adverse effects in *italic*, life-threatening effects <u>underlined</u>; generic names in **bold**; classifications in SMALL CAPS; ♣ Canadian drug name; ℗ Prototype drug

IV 0.044 mg/kg up to 2 mg 15–20 min before surgery
Child: **PO/IV/IM** 0.05 mg/kg (range: 0.02–0.09 mg/kg)

Status Epilepticus

Adult: **IV** 4 mg injected slowly at 2 mg/min, may repeat dose once if inadequate response after 10 min

ADMINISTRATION

Oral

▪ Increase the evening dose when higher oral dosage is required, before increasing daytime doses.

Intramuscular

▪ Injected undiluted, deep into a large muscle mass.

Intravenous

▪ IV administration to neonates, infants, children: Verify correct IV concentration and rate of infusion with prescriber. ▪ Patients older than 50 y may have more profound and prolonged sedation with IV lorazepam (usual max initial dose: 2 mg).

PREPARE: **Direct:** Prepare lorazepam immediately before use. Dilute with an equal volume of sterile water, D5W, or NS.

ADMINISTER: **Direct:** Inject directly into vein or into IV infusion tubing at rate not to exceed 2 mg/min and with repeated aspiration to confirm IV entry. ▪ Take extreme precautions to PREVENT intra-arterial injection and perivascular extravasation.

INCOMPATIBILITIES **Solution/additive: Dexamethasone. Y-site: Aldesleukin, aztreonam, fluconazole, foscarnet, gallium, idarubicin, imipenem/cilastatin, omeprazole, ondansetron, sargramostim, sufentanil, thiopental, TPN with albumin.**

▪ Keep parenteral preparation in refrigerator; do not freeze. ▪ Do not use a discolored solution or one with a precipitate.

ADVERSE EFFECTS (≥1%) **Body as a Whole:** Usually disappear with continued medication or with reduced dosage. **CNS:** Anterograde amnesia, *drowsiness, sedation,* dizziness, weakness, unsteadiness, disorientation, depression, sleep disturbance, restlessness, confusion, hallucinations. **CV:** Hypertension or hypotension. **Special Senses:** Blurred vision, diplopia; depressed hearing. **GI:** Nausea, vomiting, abdominal discomfort, anorexia.

INTERACTIONS Drug: Alcohol, CNS DEPRESSANTS, ANTICONVULSANTS potentiate CNS depression; **cimetidine** increases lorazepam plasma levels, increases toxicity; lorazepam may decrease antiparkinsonism effects of **levodopa;** may increase **phenytoin** levels; smoking decreases sedative and antianxiety effects. **Herbal: Kava, valerian** may potentiate sedation.

PHARMACOKINETICS Absorption: Readily absorbed from GI tract. **Onset:** 1–5 min IV; 15–30 min IM. **Peak:** 60–90 min IM; 2 h PO. **Duration:** 12–24 h. **Distribution:** Crosses placenta; distributed into breast milk. **Metabolism:** Not metabolized in liver. **Elimination:** In urine. **Half-Life:** 10–20 h.

NURSING IMPLICATIONS

Assessment & Drug Effects

▪ IM or IV lorazepam injection of 2–4 mg is usually followed by a depth of drowsiness or sleepiness that permits patient to respond to simple instructions whether patient appears to be asleep or awake.

- Supervise ambulation of older adult patients for at least 8 h after lorazepam injection to prevent falling and injury.
- Lab tests: Assess CBC and LFTs periodically for patients on long-term therapy.
- Supervise patient who exhibits depression with anxiety closely; the possibility of suicide exists, particularly when there is apparent improvement in mood.

Patient & Family Education
- Do not drive or engage in other hazardous activities for at least 24–48 h after receiving IM injection of lorazepam.
- Do not consume alcoholic beverages for at least 24–48 h after an injection and avoid when taking an oral regimen.
- Notify prescriber if daytime psychomotor function is impaired; a change in regimen or drug may be needed.
- Terminate regimen gradually over a period of several days. Do not stop long-term therapy abruptly; withdrawal may be induced with feelings of panic, tonic–clonic seizures, tremors, abdominal and muscle cramps, sweating, vomiting.
- Discuss discontinuation of drug with prescriber if you wish to become pregnant.

LOSARTAN POTASSIUM ℗

(lo-sar'tan)

Cozaar
Classifications: ANGIOTENSIN II RECEPTOR ANTAGONIST; ANTIHYPERTENSIVE
Therapeutic: ANTIHYPERTENSIVE
Pregnancy Category: C first trimester; D second and third trimester

AVAILABILITY 25 mg, 50 mg, 100 mg tablets

ACTION & *THERAPEUTIC EFFECT*
Angiotensin II receptor (type AT_1) antagonist acts as a potent vasoconstrictor and primary vasoactive hormone of the renin–angiotensin–aldosterone system. Selectively blocks the binding of angiotensin II to the AT_1 receptors found in many tissues (e.g., vascular smooth muscle, adrenal glands). *Antihypertensive effect is due to vasodilation and inhibition of aldosterone effects on sodium and water retention.*

USES Hypertension.

CONTRAINDICATIONS Hypersensitivity to losartan, pregnancy (category D second and third trimester), lactation.

CAUTIOUS USE Patients on diuretics, heart failure; hyperkalemia; hypovolemia; renal or hepatic impairment, pregnancy (category C first trimester); children younger than 6 y.

ROUTE & DOSAGE

Hypertension
Adult: **PO** 25–50 mg in 1–2 divided doses (max: 100 mg/day); start with 25 mg/day if volume depleted (i.e., on diuretics)

ADMINISTRATION
Oral
- Note: Starting dose is reduced 50% in patients with possible volume depletion or a history of liver disease.

ADVERSE EFFECTS (≥1%) **CNS:** Dizziness, insomnia, headache. **GI:** Diarrhea, dyspepsia. **Musculoskeletal:** Muscle cramps, myalgia, back or leg pain. **Respiratory:** Nasal congestion, cough, upper respiratory infection, sinusitis.

INTERACTIONS Drug: Phenobarbital decreases serum levels of losartan and its metabolite.

Common adverse effects in *italic*, life-threatening effects underlined; generic names in **bold**; classifications in SMALL CAPS; ♣ Canadian drug name; ℗ Prototype drug

PHARMACOKINETICS Absorption: Rapidly absorbed from GI tract; approximately 25–33% reaches systemic circulation. **Peak:** 6 h. **Duration:** 24 h. **Distribution:** Highly bound to plasma proteins; does not appear to cross blood–brain barrier. **Metabolism:** Extensively metabolized in liver by cytochrome P450 enzymes to an active metabolite. **Elimination:** 35% in urine, 60% in feces. **Half-Life:** Losartan 1.5–2 h; metabolite 6–9 h.

NURSING IMPLICATIONS

Assessment & Drug Effects
▪ Monitor BP at drug trough (prior to a scheduled dose).
▪ Inadequate response may be improved by splitting the daily dose into twice-daily dose.
▪ Lab tests: Monitor CBC, electrolytes, liver & kidney function with long-term therapy.

Patient & Family Education
▪ Do not use potassium supplements or salt substitutes without consulting prescriber.
▪ Notify prescriber of symptoms of hypotension (e.g., dizziness, fainting).
▪ Notify prescriber immediately of pregnancy.

LOTEPREDNOL ETABONATE
(lo-te′pred-nol e-ta-bo′nate)

Alrex, Lotemax
Pregnancy Category: C
See Appendix A-1.

LOVASTATIN 🅟
(loe-vah-stat′in)

Altoprev, Mevacor
Classifications: ANTILIPEMIC; LIPID-LOWERING; HMG-COA REDUCTASE INHIBITOR (STATIN)

Therapeutic: LIPID-LOWERING; STATIN
Pregnancy Category: X

AVAILABILITY 10 mg, 20 mg, 40 mg tablets; 10 mg, 20 mg, 40 mg, 60 mg extended release tablets

ACTION & *THERAPEUTIC EFFECT* Reduces plasma cholesterol levels by interfering with body's ability to produce its own cholesterol. This cholesterol-lowering effect triggers induction of LDL receptors, which promote removal of LDL and VLDL remnants (precursors of LDL) from plasma. Also results in an increase in plasma HDL concentrations (HDL collects excess cholesterol from body cells and transports it to liver for excretion). *Reduces plasma cholesterol levels by interfering with body's ability to produce its own cholesterol, and it also lowers LDL and VLDL cholesterol.*

USES Adjunct to diet for treatment of primary moderate hypercholesterolemia (types IIa and IIb) when diet and other nonpharmacologic measures have failed to reduce elevated total LDL cholesterol levels. Lovastatin is less effective in treatment of homozygous familial hypercholesterolemia than primary hypercholesterolemia, possibly because in these persons LDL receptors are not functional.

CONTRAINDICATIONS Active liver disease, unexplained elevations of serum transaminases; cholestasis, hepatic encephalopathy, hepatic disease, hepatitis, jaundice; rhabdomyolysis; surgery, trauma; hypotension, renal failure; pregnancy (category X), lactation.

CAUTIOUS USE Patient who consumes substantial quantities of alcohol; history of liver disease; electrolyte imbalance, endocrine disease; infection, myopathy, renal disease,

---

renal impairment, seizure disorder. Patient with risk factors predisposing to development of kidney failure secondary to rhabdomyolysis; females of child-bearing age; children younger than 10 y.

ROUTE & DOSAGE

Hypercholesterolemia
Adult: **PO** 20–40 mg 1–2 times/ day

ADMINISTRATION

Oral
- Give with the evening meal if daily. Give the first of 2 daily doses with breakfast.
- Ensure that extended release tablets are not crushed or chewed. They **must be** swallowed whole.
- Store tablets at 5°–30° C (41°–86° F) in light-resistant, tightly closed container.

ADVERSE EFFECTS (≥1%) **Body as a Whole:** Generally well tolerated. **CNS:** Dizziness, mild transient headache, insomnia, fatigue. **Special Senses:** Blurred vision. **GI:** Dyspepsia, dysgeusia, heartburn, nausea, constipation, diarrhea, flatus, abdominal pain, and cramps. **Metabolic:** Increases in serum transaminases, elevated creatine phosphokinase (CPK). **Skin:** Rash, pruritus.

INTERACTIONS Drug: Clarithromycin, clofibrate, cyclosporine, danazol, erythromycin, fenofibrate, fluconazole, gemfibrozil, itraconazole, ketoconazole, miconazole, niacin, and PROTEASE INHIBITORS increase risk of myopathy and rhabdomyolysis; potentiate hypoprothrombinemia with **warfarin. Food: Grapefruit juice** (greater than 1 qt/day) may increase risk of myopathy and rhabdomyolysis.

PHARMACOKINETICS Absorption: 30% from GI tract; extensive first-pass metabolism. **Onset:** 2 wk. **Peak:** 4–6 wk. **Distribution:** Crosses blood–brain barrier and placenta; distributed into breast milk. **Metabolism:** In liver to active metabolites. **Elimination:** 83% in feces; 10% in urine. **Half-Life:** 1.1–1.7 h.

NURSING IMPLICATIONS

Assessment & Drug Effects
- Lab tests: Monitor LFTs q4–6wk during first 15 mo of therapy. Monitor blood cholesterol levels and lipid profile periodically.
- Drug-induced increases in serum transaminases, usually not associated with jaundice or other clinical S&S, return to normal when drug is discontinued. If these values rise and remain at 3 times upper level of normal, drug will be discontinued and liver biopsy considered.

Patient & Family Education
- Notify prescriber promptly of muscle tenderness or pain, especially if accompanied by fever or malaise. If CPK is elevated or if myositis is diagnosed, drug will be discontinued.
- Avoid or at least reduce alcohol consumption.
- Understand that lovastatin is not a substitute for, but an addition to, diet therapy.

LOXAPINE HYDROCHLORIDE
(lox'a-peen)
Loxitane IM ✤

LOXAPINE SUCCINATE
Loxapac ✤
Classification: ANTIPSYCHOTIC
Therapeutic: ANTIPSYCHOTIC
Prototype: Chlorpromazine
Pregnancy Category: C

Common adverse effects in *italic*, life-threatening effects <u>underlined</u>; generic names in **bold**; classifications in SMALL CAPS; ✤ Canadian drug name; ⊕ Prototype drug

AVAILABILITY 5 mg, 10 mg, 25 mg, 50 mg capsules

ACTION & *THERAPEUTIC EFFECT*
This antipsychotic blocks postsynaptic dopamine receptors in limbic system and increases dopamine turnover by blockade of D_2-receptors in that region. After approximately 12 wk of chronic therapy, depolarization blockade of dopamine occurs, decreasing dopamine neurotransmission, correlating with its antipsychotic effects. *Stabilizes emotional component of schizophrenia by acting on subcortical level of CNS.*

USES Schizophrenia.
UNLABELED USES Agitation.

CONTRAINDICATIONS Severe drug-induced CNS depression; Parkinson's disease; comatose states, dementia-related psychosis in older adults; lactation.
CAUTIOUS USE Glaucoma, prostatic hypertrophy, urinary retention, history of convulsive disorders, cardiovascular disease; alcoholism; brain tumor; older adults; hematologic disease; hepatic disease; peptic ulcer disease; renal impairment; thyroid disease; pregnancy (category C); children younger than 16 y.

ROUTE & DOSAGE

Schizophrenia

Adult: **PO** Start with 10 mg b.i.d. and rapidly increase to maintenance dose of 60–100 mg/day in divided doses (max: 250 mg/day)

ADMINISTRATION
Oral
- Give with food, milk, or water to reduce possibility of stomach irritation.

ADVERSE EFFECTS (≥1%) **CNS:** *Drowsiness,* sedation, dizziness, syncope, EEG changes, paresthesias, staggering gait, muscle weakness, *extrapyramidal effects,* akathisia, <u>tardive dyskinesia, neuroleptic malignant syndrome</u>. **CV:** *Orthostatic hypotension,* hypertension, tachycardia. **Special Senses:** Nasal congestion, tinnitus; blurred vision, ptosis. **GI:** Constipation, dry mouth. **Skin:** Dermatitis, facial edema, pruritus, photosensitivity. **Urogenital:** Urinary retention, menstrual irregularities. **Body as a Whole:** Polydipsia, weight gain or loss, hyperpyrexia, transient <u>leukopenia</u>.

INTERACTIONS Drug: Alcohol and other CNS DEPRESSANTS potentiate CNS depression; will inhibit vasopressor effects of **epinephrine**.

PHARMACOKINETICS Absorption: Readily absorbed from GI tract. **Onset:** 20–30 min. **Peak:** 1.5–3 h. **Duration:** 12 h. **Distribution:** Widely distributed; crosses placenta; distributed into breast milk. **Metabolism:** In liver. **Elimination:** 50% in urine, 50% in feces. **Half-Life:** 19 h.

NURSING IMPLICATIONS
Assessment & Drug Effects
- Monitor baseline BP pattern prior and during therapy; both hypotension and hypertension have been reported as adverse reactions.
- Observe carefully for extrapyramidal effects such as acute dystonia (see Appendix F) during early therapy. Most symptoms disappear with dose adjustment or with antiparkinsonism drug therapy.

Common adverse effects in *italic*, life-threatening effects <u>underlined</u>; generic names in **bold**; classifications in SMALL CAPS; ♣ Canadian drug name; ⊙ Prototype drug

- Discontinue therapy and report promptly to prescriber the first signs of impending tardive dyskinesia (fine vermicular movements of the tongue) when patient is on long-term treatment.
- Monitor I&O and bowel elimination patterns and check for bladder distention. Depressed patients often fail to report urinary retention or constipation.
- Risk of seizures is increased in those with history of convulsive disorders.

Patient & Family Education

- Do NOT change dosage regimen in any way without prescriber approval.
- Avoid self-dosing with OTC drugs unless approved by the prescriber.
- Drowsiness usually decreases with continued therapy. If it persists and interferes with daily activities, consult prescriber. A change in time of administration or dose may help.
- Avoid potentially hazardous activity until response to drug is known.
- Learn measures to relieve dry mouth; rinse mouth frequently with water, suck hard candy. Avoid commercial products that may contain alcohol and enhance drying and irritation.
- Notify prescriber of blurred or colored vision.
- Do not take drug dose and notify prescriber of following: Light-colored stools, bruising, unexplained bleeding, prolonged constipation, tremor, restlessness and excitement, sore throat and fever, rash.
- Stay out of bright sun; cover exposed skin with sunscreen.

LUBIPROSTONE
(lu-bi-pros'tone)
Amitiza

Classifications: LAXATIVE AND STOOL SOFTENER; CHLORIDE CHANNEL ACTIVATOR
Therapeutic: LAXATIVE AND STOOL SOFTENER
Pregnancy Category: C

AVAILABILITY 24 mcg capsule

ACTION & THERAPEUTIC EFFECT
Lubiprostone activates chloride channels in the intestine that enhance chloride-rich intestinal fluid secretion without changing serum sodium and potassium concentrations. *The increase in intestinal fluid secretion enhances intestinal motility, thereby increasing the passage of stool and alleviating symptoms associated with chronic idiopathic constipation.*

USES Treatment of chronic idiopathic constipation; constipation predominant irritable bowel syndrome (IBS-C).

CONTRAINDICATIONS Hypersensitivity to lubiprostone; history of mechanical GI obstruction: Crohn's disease, volvulus, diverticulitis, etc.; severe diarrhea; lactation.
CAUTIOUS USE GI disease, hepatic impairment; pregnancy (category C); children.

ROUTE & DOSAGE

Chronic Idiopathic Constipation
Adult: **PO** 24 mcg b.i.d.
IBS-C
Adult: **PO** 8 mcg b.i.d.

Hepatic Dose Adjustment (for Chronic Constipation)
Child-Pugh class B: Start with 16 mcg b.i.d.
Child-Pugh class C: Start with 8 mcg b.i.d.

Common adverse effects in *italic*, life-threatening effects underlined; generic names in **bold**; classifications in SMALL CAPS; ◆ Canadian drug name; ⊙ Prototype drug

901

(for IBS)
Child-Pugh class C: Start with 8 mcg daily

ADMINISTRATION

Oral
- Administer with food to minimize nausea.
- Capsule should be swallowed whole.
- Do not administer to a patient with severe diarrhea or suspected bowel obstruction.
- Store at 15°–30° C (59°–86° F).

ADVERSE EFFECTS (≥1%) Body as a Whole: Chest pain, peripheral edema, pyrexia. **CNS:** Anxiety, depression, dizziness, fatigue, *headache,* insomnia. **CV:** Hypertension. **GI:** *Abdominal pain and discomfort,* constipation, *diarrhea,* dry mouth, dyspepsia, *flatulence,* viral gastroenteritis, gastroesophageal reflux disease, loose stools, *nausea,* vomiting. **Musculoskeletal:** Arthralgia, back pain, pain in extremities. **Respiratory:** Bronchitis, cough, dyspnea, nasopharyngitis, sinusitis, upper respiratory infection. **Urogenital:** Urinary tract infection.

PHARMACOKINETICS Absorption: Very low. **Peak:** 1.1 h (M3). **Distribution:** 94% protein bound. **Metabolism:** Extensive nonhepatic metabolism. **Elimination:** Urine (major) and feces. **Half-Life:** 0.9–1.4 h.

NURSING IMPLICATIONS

Assessment & Drug Effects
- Monitor for and report S&S of bowel obstruction.
- Lab tests: Baseline and periodic LFTs.

Patient & Family Education
- Report to prescriber if you experience severe or prolonged diarrhea, or new or worsening abdominal pain or dyspnea following dosing.
- Do not drive or engage in potentially hazardous activities until response to drug is known.

LURASIDONE
(lu-ra´-si-done)
Latuda
Classification: ATYPICAL ANTIPSYCHOTIC
Therapeutic: ANTIPSYCHOTIC
Prototype: Clozapine
Pregnancy Category: B

AVAILABILITY 40 mg, 80 mg tablets

ACTION & *THERAPEUTIC EFFECT* The mechanism of action is unknown but the efficacy of lurasidone in schizophrenia is thought to be mediated through central dopamine Type 2 (D_2) and serotonin Type 2 ($5HT_{2A}$) receptor antagonism. *Controls schizophrenic ideation and behavior.*

USES Indicated for the treatment of patients with schizophrenia.

CONTRAINDICATIONS Hypersensitivity to lurasidone; older adults with dementia-related psychosis.

CAUTIOUS USE Moderate and severe renal impairment; Child-Pugh class B and C hepatic impairment; history of seizures or Neuroleptic Malignant Syndrome (NMS); pregnancy (category C); lactation. Safety and efficacy in children not established.

ROUTE & DOSAGE

Schizophrenia
Adult: **PO** 40–80 mg once daily

Hepatic Impairment Dosage Adjustment

Moderate and severe hepatic impairment (Child-Pugh class B and C): Reduce dose (max: 40 mg once daily)

Renal Impairment Dosage Adjustment

Moderate to severe renal impairment: Reduce dose (max: 40 mg once daily)

ADMINISTRATION

Oral

- Give with food (at least 350 calories).
- Store at 15°–30°C (59°–86°F).

ADVERSE EFFECTS (≥1%) **Body as a Whole:** Agitation, anxiety, fatigue, insomnia, restlessness. **CNS:** *Akathisia*, dizziness, dystonia, *Parkinsonism, somnolence.* **CV:** Tachycardia. **GI:** Abdominal pain, diarrhea, dyspepsia, *nausea*, salivary hypersecretion, vomiting. **Metabolic:** Elevated ALT and AST levels, elevated serum creatinine. **Musculoskeletal:** Back pain, decreased appetite. **Skin:** Pruritis, rash. **Special Senses:** Blurred vision.

INTERACTIONS Drug: Inhibitors of CYP3A4 (**ketoconazole, diltiazem**) increase the levels of lurasidone; inducers of CYP3A4 (**rifampin**) decrease the levels of lurasidone. **Food: Grapefruit juice** can increase the levels of lurasidone. **Herbal: St. John's wort** can decrease the levels of lurasidone.

PHARMACOKINETICS Absorption: Approximately 9–19% is orally absorbed. **Peak:** 1–3 h.

Distribution: 99% plasma protein bound. **Metabolism:** Hepatic oxidation to active and inactive metabolites. **Elimination:** Fecal (80%) and renal (9%). **Half-Life:** 18 h.

NURSING IMPLICATIONS

Assessment & Drug Effects

- Monitor for and report promptly suicidal ideation.
- Monitor orthostatic vital signs, especially in those at risk for hypotension. Risk is highest at time of dose initiation or escalation.
- Lab tests: Periodic blood glucose, lipid profile, LFTs, and CBC with differential.
- Monitor closely patients with neutropenia. Report promptly development of fever or other signs of infection.
- Monitor weight and report significant weight gain.
- Monitor diabetics closely for loss of glycemic control.
- Monitor for and report promptly seizure activity, S&S of Neuroleptic Malignant Syndrome (NMS) or tardive dyskinesia (see Appendix F).

Patient & Family Education

- Do not engage in hazardous activities until response to drug is known.
- Rise slowly from a lying or sitting position to avoid dizziness and fainting.
- Report immediately any of the following: Suicidal thoughts, fever, muscle rigidity, palpitations, signs of infection, dizziness or fainting upon standing.
- Avoid situations that may cause overheating or dehydration.
- Avoid alcohol while taking lurasidone.
- Monitor blood glucose levels frequently if diabetic.

Common adverse effects in *italic*, life-threatening effects underlined; generic names in **bold**; classifications in SMALL CAPS; ♣ Canadian drug name; ⊙ Prototype drug

903

LYMPHOCYTE IMMUNE GLOBULIN
(lim'fo-site)

Antithymocyte Globulin, ATG, Atgam

Classifications: BIOLOGICAL RESPONSE MODIFIER; IMMUNOGLOBULIN **Therapeutic:** LYMPHOCYTE IMMUNOGLOBULIN; ANTITHYMOCYTE GLOBULIN; IMMUNOSUPPRESSANT
Prototype: Immune globulin
Pregnancy Category: C

AVAILABILITY 50 mg/mL injection

ACTION & *THERAPEUTIC EFFECT*
An immunoglobulin (IgG) and lymphocyte-selective immunosuppressant derived from human thymus lymphocytes. During rejection of allografts, human leukocyte antigens (HLAs) bind to peptides and form complexes. Helper T-lymphocytes activate these complexes and produce interleukins, cytotoxic T-cells, and natural killer cells, resulting in destruction of transplanted tissue. *Alters the formation of T-lymphocytes, thus reversing acute allograft rejection.*

USES Primarily to prevent or delay onset or to reverse acute renal allograft rejection.
UNLABELED USES Moderate and severe aplastic anemia in patients unsuitable for bone marrow transplantation, T-cell malignancy, acute and chronic graft-vs-host disease, and to prevent rejection of skin allografts.

CONTRAINDICATIONS Hypersensitivity to lymphocyte immune globulin or other equine gamma globulin preparations; history of previous systemic reaction to drug, hemorrhagic diatheses; leporine protein hypersensitivity; use in kidney transplant patient not receiving a concomitant immunosuppressant; fungal or viral infections.
CAUTIOUS USE Hypotension, infection, leukopenia, lymphoma, neoplastic disease, thrombocytopenia, vaccination, varicella; pregnancy (category C), lactation; children (experience limited).

ROUTE & DOSAGE

Renal Allotransplantation
Adult: **IV** 10–30 mg/kg/day by slow IV infusion
Child: **IV** 5–25 mg/kg/day by slow IV infusion

Prevention of Allograft Rejection
Adult: **IV** 15 mg/kg/day for 14 days followed by 15 mg/kg every other day for 14 days

Treatment of Allograft Rejection
Adult: **IV** 10–15 mg/kg/day for 14 days followed by 15 mg/kg every other day for 14 days if needed

Aplastic Anemia
Adult/Child: **IV** 10–20 mg/kg/day × 8–14 days followed by 10–20 mg/kg every other day for 7 doses

ADMINISTRATION

Intravenous
Note: Administer lymphocyte immune globulin (ATG) ONLY if experienced with immunosuppressant therapy and management of kidney transplant patients.

▪Do an intradermal skin test to rule out allergy to the drug before

Common adverse effects in *italic*, life-threatening effects underlined; generic names in **bold**; classifications in SMALL CAPS; ♣ Canadian drug name; ⊙ Prototype drug

first dose. ▪ Inject 0.1 mL of a 1:1000 dilution (5 mcg equine IgG in normal saline) and a saline control. ▪ If local reaction occurs (wheal or erythema more than 10 mm) or if there is pseudopod formation, itching, or local swelling, use caution during infusion. ▪ Discontinue infusion if systemic reaction develops (generalized rash, tachycardia, dyspnea, hypotension, anaphylaxis).

PREPARE: **IV Infusion:** Withdraw required dose of ATG concentrate and inject into IV solution container of 0.45% NaCl or NS. Invert IV container during injection of ATG to prevent its contact with air inside container. Use enough IV solution to create a concentration 4 mg/mL or less. ▪ Inspect concentrate and diluted solution for particulate matter (may develop during storage) and discoloration; discard if present.
ADMINISTER: **IV Infusion:** Give through an in-line 0.2–1 micron filter into a high-flow vein to decrease potential for phlebitis and thrombosis. Give over 4 h or longer (usually 4–8 h). ▪ Must finish infusion within 12 h of preparation.

▪ Total storage time for diluted solutions: NO MORE than 12 h (including storage time and actual infusion time). ▪ Refrigerate ampules and diluted solutions (if prepared before time of infusion) at 2°–8° C (35°–46° F). Do not freeze.

ADVERSE EFFECTS (≥1%) **CNS:** Headache, paresthesia, seizures. **CV:** Peripheral thrombophlebitis, hypotension, hypertension. **GI:** Nausea, vomiting, diarrhea, stomatitis, hiccups, epigastric pain, abdominal distention. **Hematologic:** *Leukopenia, thrombocytopenia.* **Musculoskeletal:** Arthralgia, myalgias, chest or back pain. **Respiratory:** Dyspnea, laryngospasm, pulmonary edema. **Skin:** *Rash, pruritus,* urticaria, wheal and flare. **Body as a Whole:** *Chills, fever,* night sweats, pain at infusion site, hyperglycemia, systemic infection, wound dehiscence; anaphylaxis, serum sickness, herpes simplex virus reactivation.

INTERACTIONS Drug: Azathioprine, CORTICOSTEROIDS, other IMMUNOSUPPRESSANTS increase degree of immunosuppression.

PHARMACOKINETICS Distribution: Poorly distributed into lymphoid tissues (spleen, lymph nodes); probably crosses placenta and into breast milk. **Elimination:** About 1% of dose is excreted in urine. **Half-Life:** Approximately 6 days.

NURSING IMPLICATIONS
Assessment & Drug Effects
▪ Discontinue infusion and initiate appropriate therapy promptly with onset of anaphylactic response (respiratory distress; pain in chest, flank, back; hypotension, anxiety).
▪ Monitor closely BP, vital signs, and patient's complaints during entire administration period. Prompt treatment is indicated for symptoms of anaphylaxis (incidence: 1%), serum sickness, or allergic response. Always have equipment for assisted respiration, epinephrine, antihistamines, corticosteroid, and vasopressor available at bedside.
▪ Watch closely for S&S of serum sickness: Fever, malaise, arthralgia, nausea, vomiting, lymphadenopathy and morbilliform eruptions on trunk and extremities. Rash begins as asymptomatic pale pink macules in

L

Common adverse effects in *italic*, life-threatening effects <u>underlined</u>; generic names in **bold**; classifications in SMALL CAPS; ♣ Canadian drug name; ⓟ Prototype drug

905

periumbilical region, axilla, and groin, then rapidly becomes generalized. Serum sickness usually occurs 6–18 days after initiation of therapy; may occur during drug administration or when treatment is stopped.

- Monitor carefully for S&S of thrombocytopenia, concurrent infection, and leukopenia; patient usually receives concomitant corticosteroids and antimetabolites.
- Monitor patient's temperature and attend to complaints of sore throat or rhinorrhea. Report to prescriber.

Patient & Family Education

- Notify prescriber immediately of pain in chest, flank, or back; chills; pruritus; night sweats; sore throat.

MAFENIDE ACETATE

(ma′fe-nide)

Sulfamylon
Classification: SULFONAMIDE ANTIBIOTIC
Therapeutic: ANTIBIOTIC
Prototype: Sulfisoxazole
Pregnancy Category: C

AVAILABILITY 5% powder; 11.2% cream

ACTION & THERAPEUTIC EFFECT
Produces marked reduction of bacterial growth in vascular tissue. Active in presence of purulent matter and serum. *Bacteriostatic against many gram-positive and gram-negative organisms, including Pseudomonas aeruginosa, and certain strains of anaerobes.*

USES Treatment of second- and third-degree burns.

CONTRAINDICATIONS History of hypersensitivity to mafenide; respiratory (inhalation) injury, pulmonary infection; lactation.

CAUTIOUS USE Impaired kidney or pulmonary function, burn patients with acute kidney failure; pregnancy (category C); children younger than 3 mo.

ROUTE & DOSAGE

Burns

Adult: **Topical** Apply aseptically to burn areas to a thickness of approximately 15 mm ($^1/_{16}$ in) once or twice daily

ADMINISTRATION

Topical

- Apply cream or solution aseptically to cleansed, debrided burn areas with sterile gloved hand.
- Cover burn areas with cream at all times. Make reapplications to areas from which cream has been removed as necessary.
- Store in tight, light-resistant containers. Avoid extremes of temperature.

ADVERSE EFFECTS (≥1%) **Hypersensitivity:** Pruritus, rash, urticaria, blisters, eosinophilia. **Skin:** *Intense pain, burning, or stinging at application sites,* bleeding of skin, excessive body water loss, excoriation of new skin, superinfections. **Other:** Metabolic acidosis.

PHARMACOKINETICS Absorption: Rapidly from burn surface. **Peak:** 2–4 h. **Metabolism:** Rapidly inactivated in blood to a weak carbonic anhydrase inhibitor. **Elimination:** Via kidneys.

NURSING IMPLICATIONS

Assessment & Drug Effects

- Monitor vital signs. Report immediately changes in BP, pulse, and respiratory rate and volume.

Common adverse effects in *italic*, life-threatening effects underlined; generic names in **bold**; classifications in SMALL CAPS; ♣ Canadian drug name; ⊘ Prototype drug

- Monitor I&O. Report oliguria or changes in I&O ratio and pattern.
- Lab tests: Monitor fluid and electrolyte status throughout therapy; acid–base balance should be monitored in patients with extensive burns and in those with pulmonary or kidney dysfunction.
- Be alert to S&S of metabolic acidosis (see Appendix F).
- Be alert to evidence of superinfections (see Appendix F), particularly in and below burn eschar.
- Observe carefully; accuracy is critical. It is frequently difficult to distinguish between adverse reactions to mafenide and the effects of severe burns.
- Note: Allergic reactions have reportedly occurred 10–14 days after initiation of mafenide therapy. Temporary discontinuation of drug may be necessary.
- Report intense local pain to prescriber; pain caused by drug may require administration of analgesic.

Patient & Family Education

- Apply only a thin dressing over burns unless otherwise directed.
- Therapy is usually continued until healing is progressing well (usually 60 days) or site is ready for grafting (after about 35–40 days). It is not withdrawn while there is a possibility of infection unless adverse reactions intervene.
- Report any of the following to the prescriber immediately: Foul-smelling drainage from wounds, bleeding at wound site, unexplained fever.

MAGNESIUM CITRATE

(mag-nes′i-um)

Citrate of Magnesia, Citroma, Citro-Nesia

Classification: SALINE CATHARTIC
Therapeutic: LAXATIVE

Prototype: Magnesium hydroxide
Pregnancy Category: A

AVAILABILITY 1.75 g/30 mL solution

ACTION & *THERAPEUTIC EFFECT*
Hyperosmotic laxative that promotes bowel evacuation by causing osmotic retention of fluid; this distends colon and stimulates peristaltic activity. *Evacuates bowels.*

USES To evacuate bowel prior to certain surgical and diagnostic procedures and to help eliminate parasites and toxic materials after treatment with a vermifuge.

CONTRAINDICATIONS Severe renal impairment, renal failure; nausea, vomiting, diarrhea, abdominal pain, acute surgical abdomen; intestinal impaction, obstruction or perforation; rectal bleeding; use of solutions containing sodium bicarbonate in patients on sodium-restricted diets.

CAUTIOUS USE Mild or moderate renal impairment; cardiac disease; older adults; pregnancy (category A), lactation; children younger than 2 y.

ROUTE & DOSAGE

Bowel Evacuation

Adult: **PO** 240 mL once
Child: **PO** 2–6 y, 4–12 mL; 6–12 y, 50–100 mL

ADMINISTRATION

Oral
- Give on an empty stomach with a full (240 mL) glass of water. Time dosing so that it does not interfere with sleep. Drug produces a watery or semifluid evacuation in 2–6 h.
- Chill solution by pouring it over ice or refrigerate it until ready to use to increase palatability.

Common adverse effects in *italic*, life-threatening effects underlined; generic names in **bold**; classifications in SMALL CAPS; ♣ Canadian drug name; 🅟 Prototype drug

907

- Be aware that once container is opened, effervescence will decrease. This does not effect the quality of preparation.
- Store at 2°–30° C (36°–86° F) in tightly covered containers.

ADVERSE EFFECTS (≥1%) GI: Abdominal cramps, nausea, fluid and electrolyte imbalance, hypermagnesemia (prolonged use).

INTERACTIONS Drug: May decrease effectiveness of **digoxin**, ORAL ANTICOAGULANTS, PHENOTHIAZINES; will decrease absorption of **ciprofloxacin**, TETRACYCLINES; **sodium polystyrene sulfonate** will bind magnesium, decreasing its effectiveness.

PHARMACOKINETICS Onset: 0.5–2 h.

NURSING IMPLICATIONS

Assessment & Drug Effects
- Monitor for dehydration, hypokalemia, and hyponatremia (see Appendix F) since drug may cause intense bowel evacuation.

Patient & Family Education
- Do not use for routine treatment of constipation (especially in older adult).
- Expect some degree of abdominal cramping.

MAGNESIUM HYDROXIDE ℗
(mag-nes′i-um)

Magnesia, Magnesia Magma, Milk of Magnesia, M.O.M.
Classifications: SALINE CATHARTIC; ANTACID
Therapeutic: LAXATIVE; ANTACID
Pregnancy Category: A

AVAILABILITY 311 mg tablets; 400 mg/5 mL, 800 mg/5 mL suspension

ACTION & *THERAPEUTIC EFFECT*
Aqueous suspension of magnesium hydroxide with rapid and long-acting neutralizing action. Causes osmotic retention of fluid, which distends colon, resulting in mechanical stimulation of peristaltic activity. *Acts as antacid in low doses and as mild saline laxative at higher doses.*

USES Short-term treatment of occasional constipation, for relief of GI symptoms associated with hyperacidity, and as adjunct in treatment of peptic ulcer. Also has been used in treatment of poisoning by mineral acids and arsenic, and as mouthwash to neutralize acidity.

CONTRAINDICATIONS Abdominal pain, nausea, vomiting, chronic diarrhea, severe kidney dysfunction, fecal impaction, intestinal obstruction or perforation, rectal bleeding, colostomy, ileostomy.
CAUTIOUS USE Renal impairment, renal disease; older adults; pregnancy (category A), lactation; children younger than 2 y.

ROUTE & DOSAGE

Laxative
Adult: **PO** 2.4–4.8 g (30–60 mL)/ day in 1 or more divided doses
Child: **PO** 2–5 y, 0.4–1.2 g (5–15 mL)/day in 1 or more divided doses; 6–11 y, 1.2–2.4 g (15–30 mL)/day in 1 or more divided doses

ADMINISTRATION

Oral
- Shake bottle well before pouring to assure mixing of suspension.
- Follow drug with at least a full glass of water to enhance drug action for laxative effect. Administer in the morning or at bedtime. Most

Common adverse effects in *italic*, life-threatening effects <u>underlined</u>; generic names in **bold**; classifications in SMALL CAPS; ♣ Canadian drug name; ℗ Prototype drug

effective when taken on an empty stomach.

- Store at 15°–30° C (59°–86° F) in tightly covered container. Slowly absorbs carbon dioxide on exposure to air. Avoid freezing.

ADVERSE EFFECTS (≥1%) **GI:** Nausea, vomiting, abdominal cramps, *diarrhea.* **Urogenital:** Alkalinization of urine. **Body as a Whole:** Weakness, lethargy, mental depression, hyporeflexia, dehydration, coma. **Metabolic:** Electrolyte imbalance with prolonged use. **CV:** Hypotension, bradycardia, complete heart block and other ECG abnormalities. **Respiratory:** Respiratory depression.

INTERACTIONS Drug: Milk of Magnesia decreases absorption of **chlordiazepoxide, dicumarol, digoxin, isoniazid,** QUINOLONES, TETRACYCLINES.

PHARMACOKINETICS Absorption: 15–30% of magnesium is absorbed. **Onset:** 3–6 h. **Distribution:** Small amounts distributed in saliva and breast milk. **Elimination:** In feces; some renal excretion.

NURSING IMPLICATIONS

Assessment & Drug Effects
- Evaluate the patient's continued need for drug. Prolonged and frequent use of laxative doses may lead to dependence. Additionally, even therapeutic doses can raise urinary pH and thereby predispose susceptible patients to urinary infection and urolithiasis.
- Lab tests: Monitor serum magnesium with signs of hypermagnesemia such as bradycardia (see Appendix F), especially with frequent use or any degree of renal impairment.

Patient & Family Education
- Investigate the cause of persistent or recurrent constipation or gastric distress with prescriber.

MAGNESIUM OXIDE
(mag-nes′i-um)

Mag-Ox, Maox, Par-Mag, Uro-Mag
Classifications: ANTACID; SALINE CATHARTIC
Therapeutic: ANTACID; MAGNESIUM SUPPLEMENT; LAXATIVE
Prototype: Magnesium hydroxide
Pregnancy Category: A

AVAILABILITY 400 mg, 420 mg, 500 mg tablets; 140 mg capsules

ACTION & THERAPEUTIC EFFECT Nonsystemic antacid with high neutralizing capacity and relatively long duration of action. *Acts as an antacid in low doses and a mild saline laxative at higher doses. Also effective as a magnesium supplement.*

USES Essentially the same as magnesium hydroxide. May also be used as magnesium supplement.

CONTRAINDICATIONS Abdominal pain, nausea, vomiting, diarrhea, severe kidney dysfunction, fecal impaction, intestinal obstruction or perforation, ileus; rectal bleeding, colostomy, ileostomy; AV block; hypermagnesia.
CAUTIOUS USE Cardiac disease, renal disease, renal impairment; electrolyte imbalance; pregnancy (category A), lactation; children younger than 2 y.

ROUTE & DOSAGE

Antacid
Adult: **PO** 280–1500 mg with water or milk q.i.d., p.c. and at bedtime
Laxative
Adult: **PO** 2–4 g with water or milk at bedtime

Common adverse effects in *italic*, life-threatening effects underlined; generic names in **bold**; classifications in SMALL CAPS; ♣ Canadian drug name; ☯ Prototype drug

909

Magnesium Supplement

Adult: **PO** 400–1200 mg/day in divided doses

ADMINISTRATION

Oral
- Separate administration of this drug from other oral drugs by 1–2 h.
- Store at 15°–30° C (59°–86° F) in airtight containers. On exposure to air, magnesium oxide rapidly absorbs moisture and carbon dioxide.

ADVERSE EFFECTS (≥1%) **GI:** *Diarrhea,* abdominal cramps, nausea; hypermagnesemia, kidney stones (chronic use).

INTERACTIONS Drug: See magnesium hydroxide.

PHARMACOKINETICS Absorption: 30–50% from GI tract. **Elimination:** In urine.

NURSING IMPLICATIONS

Assessment & Drug Effects
- Monitor for dehydration, hypokalemia, and hyponatremia (see Appendix F) since drug may cause intense bowel evacuation.
- Lab tests: Check patients on prolonged therapy periodically for electrolyte imbalance (i.e., hypermagnesemia).

Patient & Family Education
- Liquid preparation is reportedly more effective than the tablet form, as with other antacids.

MAGNESIUM SALICYLATE

(mag-nes'i-um)

Doan's Pills, Magan, Mobidin
Classification: ANALGESIC; NONSTEROIDAL ANTI-INFLAMMATORY DRUG (NSAID)

Therapeutic: NONNARCOTIC ANALGESIC, NSAID; ANTIPYRETIC
Prototype: Aspirin
Pregnancy Category: C first and second trimester; D third trimester

AVAILABILITY 467 mg, 500 mg, 580 mg caplets; 545 mg, 600 mg tablets

ACTION & *THERAPEUTIC EFFECT*
Sodium-free salicylate derivative that is a nonsteroidal anti-inflammatory drug (NSAID). It inhibits prostaglandin synthesis. *In equal doses, less potent than aspirin as an analgesic and antipyretic. Has anti-inflammatory effects.*

USES Relief of pain and inflammation in rheumatoid arthritis, osteoarthritis, bursitis, and other musculoskeletal disorders.

CONTRAINDICATIONS Hypersensitivity to salicylates; erosive gastritis, peptic ulcer; advanced renal insufficiency, liver damage; thrombolytic therapy; bleeding disorders; before surgery; pregnancy (category D third trimester).

CAUTIOUS USE Serious acid-base imbalances; renal disease, history of GI bleeding, or peptic ulcers; SLE; history of acute bronchospasm; pregnancy (category C first and second trimester), lactation; children younger than 12 y.

ROUTE & DOSAGE

Analgesic/Antipyretic
Adult: **PO** 650 mg t.i.d. or q.i.d.
Arthritic Conditions
Adult: **PO** Up to 9.6 g/day in divided doses

ADMINISTRATION

Oral

- Give with a full glass of water, food, or milk to minimize gastric irritation.

ADVERSE EFFECTS (≥1%) **Body as a Whole:** Salicylism [dizziness, drowsiness, tinnitus, hearing loss, nausea, vomiting, hypermagnesemia (with high doses in patients with renal insufficiency)].

INTERACTIONS Drug: Aminosalicylic acid increases risk of SALICYLATE toxicity; **ammonium chloride** and other ACIDIFYING AGENTS decrease renal elimination and increase risk of SALICYLATE toxicity; anticoagulants—added risk of bleeding with ANTICOAGULANTS; CARBONIC ANHYDRASE INHIBITORS enhance SALICYLATE toxicity; CORTICOSTEROIDS compound ulcerogenic effects; increases **methotrexate** toxicity; low doses of SALICYLATES may antagonize uricosuric effects of **probenecid, sulfinpyrazone.**

PHARMACOKINETICS Absorption: Well absorbed from the GI tract. **Peak:** 20 min. **Distribution:** Widely distributed with high levels of salicylic acid in liver and kidney, crosses placenta, excreted in breast milk. **Metabolism:** Salicylic acid is metabolized in liver. **Elimination:** In kidneys. **Half-Life:** 2–3 h with single dose, 15–30 h with chronic dosing.

NURSING IMPLICATIONS

Assessment & Drug Effects

- Lab tests: Monitor serum magnesium levels for hypermagnesemia if used in high dosages or in patients with any degree of renal impairment.

Patient & Family Education

- Report to prescriber promptly tinnitus, hearing loss, or dizziness.

- Do not to take aspirin-containing drugs without consent of prescriber.
- Check ingredients. Doan's pills may contain acetaminophen plus salicylamide.

MAGNESIUM SULFATE
(mag-nes′i-um)

Epsom Salt
Classifications: SALINE CATHARTIC; ELECTROLYTE REPLACEMENT; ANTICONVULSANT
Therapeutic: LAXATIVE; ELECTROLYTE REPLACEMENT; ANTICONVULSANT
Prototype: Magnesium hydroxide
Pregnancy Category: A

AVAILABILITY 0.8 mEq/mL, 1 mEq/mL, 4 mEq/mL injection

ACTION & THERAPEUTIC EFFECT Orally: Acts as a laxative by osmotic retention of fluid, which distends colon, increases water content of feces, and causes mechanical stimulation of bowel activity. **Parenterally:** Acts as a CNS depressant and also as a depressant of smooth, skeletal, and cardiac muscle function. Anticonvulsant properties thought to be produced by CNS depression by decreasing the amount of acetylcholine liberated from motor nerve terminals, producing peripheral neuromuscular blockade. *Effective parenterally as a CNS depressant, smooth muscle relaxant and anticonvulsant in labor and delivery, and cardiac disorders. When taken orally, it is a laxative.*

USES Orally to relieve acute constipation and to evacuate bowel in preparation for x-ray of intestines. Parenterally to control seizures in toxemia of pregnancy, epilepsy, and acute nephritis and

Common adverse effects in *italic*, life-threatening effects underlined; generic names in **bold**; classifications in SMALL CAPS; ◆ Canadian drug name; ⊙ Prototype drug

911

for prophylaxis and treatment of hypomagnesemia. Topically to reduce edema, inflammation, and itching.

UNLABELED USES To inhibit premature labor (tocolytic action) and as adjunct in hyperalimentation, to alleviate bronchospasm of acute asthma, to reduce mortality post-MI.

CONTRAINDICATIONS Myocardial damage; AV heart block; cardiac arrest except for certain arrhythmias; hypermagnesemia; GI obstruction; **IV:** during the 2 h preceding delivery; **Orally:** in patients with abdominal pain, nausea, vomiting, fecal impaction, or intestinal irritation, obstruction, or perforation.

CAUTIOUS USE Renal disease; renal failure; renal impairment; acute MI; digitalized patients; neuromuscular blocking agents, or cardiac glycosides; pregnancy (category A), children.

ROUTE & DOSAGE

Laxative

Adult: PO 10–15 g once/day

Seizures

Adult: IV 1 g, may need to repeat dose

Preeclampsia, Eclampsia

Adult: IM/IV 4–5 g slowly; simultaneously, 5 g **IM** in alternate buttocks q4h

Hypomagnesemia

Adult: IM/IV *Mild,* 1 g q6h for 4 doses; *Severe,* 5 g infused over 3 h
Child: IV 25–50 mg/kg q4–6h prn (max single dose: 2000 mg)

Total Parenteral Nutrition

Adult: IV 0.5–3 g/day

ADMINISTRATION

Oral

- Give in the morning or mid-afternoon in a glass of water for laxative action. Disguise bitter, salty taste by chilling or flavoring with lemon or orange juice.

Intramuscular

- Give deep using the 50% concentration for adults and the 20% concentration for children.

Intravenous

Note: Verify correct IV concentration and rate of infusion for administration to infants, children with prescriber.

PREPARE: **Direct/IV Infusion:** Give solutions with concentrations of 20% or less undiluted. ▪ Dilute more concentrated solutions to 20% (200 mg/mL) or less with D5W or NS.

ADMINISTER: **Direct:** Give at a rate of 150 mg over at least 1 min. Note: 20% solution contains 200 mg/mL, 10% solution contains 100 mg/mL. **IV Infusion:** Give required dose over 4 h. Do not exceed the direct rate.

INCOMPATIBILITIES **Solution/additive:** 10% fat emulsion, amphotericin B, calcium, chlorpromazine, clindamycin, cyclosporine, dobutamine, hydralazine, polymyxin B, procaine, prochlorperazine, sodium bicarbonate. **Y-site:** Amiodarone, amphotericin B, cholesteryl, cefepime, ciprofloxacin, haloperidol, lansoprazole.

ADVERSE EFFECTS (≥1%) **Body as a Whole:** Flushing, sweating, extreme thirst, sedation, confusion, depressed reflexes or no reflexes, muscle weakness, <u>flaccid paralysis</u>, hypothermia. **CV:** Hypotension, depressed cardiac function, <u>complete heart block, circulatory collapse</u>. **Respiratory:** <u>Respiratory paralysis</u>. **Metabolic:** Hypermagnesemia, hypocalcemia, dehydration, electrolyte imbalance including hypocalcemia with repeated laxative use.

INTERACTIONS Drug: NEUROMUSCULAR BLOCKING AGENTS add to respiratory depression and apnea.

PHARMACOKINETICS Onset: 1–2 h PO; 1 h IM. **Duration:** 30 min IV; 3–4 h PO. **Distribution:** Crosses placenta; distributed into breast milk. **Elimination:** In kidneys.

NURSING IMPLICATIONS
Assessment & Drug Effects
- Observe constantly when given IV. Check BP and pulse q10–15 min or more often if indicated.
- Lab tests: Monitor plasma magnesium levels in patients receiving drug parenterally (normal: 1.8–3 mEq/L). Plasma levels in excess of 4 mEq/L are reflected in depressed deep tendon reflexes and other symptoms of magnesium intoxication (see ADVERSE EFFECTS). Cardiac arrest occurs at levels in excess of 25 mEq/L. Monitor calcium and phosphorus levels also.
- Early indicators of magnesium toxicity (hypermagnesemia) include cathartic effect, profound thirst, feeling of warmth, sedation, confusion, depressed deep tendon reflexes, and muscle weakness.
- Monitor respiratory rate closely. Report immediately if rate falls below 12.
- Check urinary output, especially in patients with impaired kidney function. Therapy is generally not continued if urinary output is less than 100 mL during the 4 h preceding each dose.

- Observe newborns of mothers who received parenteral magnesium sulfate within a few hours of delivery for signs of toxicity, including respiratory and neuromuscular depression.
- Observe patients receiving drug for hypomagnesemia for improvement in the following signs of deficiency: Irritability, choreiform movements, tremors, tetany, twitching, muscle cramps, tachycardia, hypertension, psychotic behavior.
- Have calcium gluconate readily available in case of magnesium sulfate toxicity.

Patient & Family Education
- Drink sufficient water during the day when drug is administered orally to prevent net loss of body water.

M

MANNITOL ⊘
(man'ni-tole)

Osmitrol
Classifications: ELECTROLYTIC AND WATER BALANCE AGENT; OSMOTIC DIURETIC
Therapeutic: OSMOTIC DIURETIC
Pregnancy Category: B

AVAILABILITY 5%, 10%, 15%, 20%, 25% injection

ACTION & *THERAPEUTIC EFFECT*
Increases rate of electrolyte excretion by the kidney, particularly sodium, chloride, and potassium. Induces diuresis by raising osmotic pressure of glomerular filtrate, thereby inhibiting tubular reabsorption of water and solutes. Reduces elevated intraocular and cerebrospinal pressures by increasing plasma osmolality, thus inducing diffusion of water from these fluids back into plasma and extravascular

Common adverse effects in *italic*, life-threatening effects <u>underlined</u>; generic names in **bold**; classifications in SMALL CAPS; ♣ Canadian drug name; ⊘ Prototype drug

913

space. *Osmotic diuretic that reduces intracranial pressure, cerebral edema, intraocular pressure, and promotes diuresis, thus preventing or treating oliguria.*

USES To promote diuresis in prevention and treatment of oliguric phase of acute kidney failure following cardiovascular surgery, severe traumatic injury, surgery in presence of severe jaundice, hemolytic transfusion reaction. Also used to reduce elevated intraocular (IOP) and intracranial pressure (ICP), to measure glomerular filtration rate (GFR), to promote excretion of toxic substances, to relieve symptoms of pulmonary edema, and as irrigating solution in transurethral prostatic reaction to minimize hemolytic effects of water. Commercially available in combination with sorbitol for urogenital irrigation.

CONTRAINDICATIONS Anuria; severe renal failure with azotemia or increasing oliguria; marked pulmonary congestion or edema; severe CHF; metabolic edema; hypovolemia; organic CNS disease, intracranial bleeding; shock, severe dehydration; concomitantly with blood.

CAUTIOUS USE Electrolyte imbalance; older adult; pregnancy (category B), lactation; children.

ROUTE & DOSAGE

Acute Kidney Failure

Adult: **IV Test Dose** 0.2 g/kg over 3–5 min if partial response may repeat test dose 1 time. If still negative, do not use. **Treatment** 50–100 g as 15–20% solution over 90 min to several hours
Child: **IV Test Dose** 0.2 g/kg (max: 12.5 g) over 3–5 min
Positive Response Urine flow of 1 mL/kg/h for 1–2 h **Maintenance** 0.25–0.5 g/kg q4–6h

Edema, Ascites

Adult: **IV** 100 g as a 10–20% solution over 2–6 h

Elevated IOP or ICP

Adult: **IV** 1.5–2 g/kg as a 15–25% solution over 30–60 min

Acute Chemical Toxicity

Adult: **IV** 100–200 g depending on urine output

Measurement of GFR

Adult: **IV** 100 mL of 20% solution diluted with 180 mL NaCl injection infused at a rate of 20 mL/min

ADMINISTRATION

Intravenous

Note: Verify correct IV concentration and rate of infusion for administration to infants, children with prescriber.

PREPARE: **IV Infusion:** Give undiluted.

ADMINISTER: **IV Infusion:** Give a single dose over 30–90 min. *Oliguria:* A test dose is given to patients with marked oliguria to check adequacy of kidney function. Response is considered satisfactory if urine flow of at least 30–50 mL/h is produced over 2–3 h after drug administration; then rate is adjusted to maintain urine flow at 30–50 mL/h with a single dose usually being infused over 90 min or longer. ▪ Concentrations higher than 15% have a greater tendency to crystallize. ▪ Use an administration set with a 5 micron in-line IV filter when infusing concentrations of 15% or above.

INCOMPATIBILITIES Solution/additive: Furosemide, imipenem-cilastatin, meropenem, potassium chloride, sodium chloride, whole blood. Y-site: **Cefepime, doxorubicin liposome, filgrastim, pantoprazole.**

▪ Store at 15°–30° C (59°–86° F) unless otherwise directed. Avoid freezing.

ADVERSE EFFECTS (≥1%) **CNS:** Headache, tremor, convulsions, dizziness, transient muscle rigidity. **CV:** Edema, CHF, angina-like pain, hypotension, hypertension, thrombophlebitis. **Eye:** Blurred vision. **GI:** Dry mouth, nausea, vomiting. **Urogenital:** Marked diuresis, urinary retention, nephrosis, uricosuria. **Metabolic:** *Fluid and electrolyte imbalance,* especially hyponatremia; dehydration, acidosis. **Other:** With extravasation (local edema, skin necrosis; chills, fever, allergic reactions).

INTERACTIONS Drug: Increases urinary excretion of **lithium,** SALICYLATES, BARBITURATES, **imipramine, potassium.**

PHARMACOKINETICS Onset: 1–3 h diuresis; 30–60 min IOP; 15 min ICP. **Duration:** 4–6 h IOP; 3–8 h ICP. **Distribution:** Confined to extracellular space; does not cross blood–brain barrier except with very high plasma levels in the presence of acidosis. **Metabolism:** Small quantity metabolized to glycogen in liver. **Elimination:** Rapidly excreted by kidneys. **Half-Life:** 100 min.

NURSING IMPLICATIONS

Assessment & Drug Effects

▪ Take care to avoid extravasation. Observe injection site for signs of inflammation or edema.
▪ Lab tests: Monitor closely serum and urine electrolytes and kidney function during therapy.
▪ Measure I&O accurately and record to achieve proper fluid balance.

▪ Monitor vital signs closely. Report significant changes in BP and signs of CHF.
▪ Monitor for possible indications of fluid and electrolyte imbalance (e.g., thirst, muscle cramps or weakness, paresthesias, and signs of CHF).
▪ Be alert to the possibility that a rebound increase in ICP sometimes occurs about 12 h after drug administration. Patient may complain of headache or confusion.
▪ Take accurate daily weight.

Patient & Family Education

▪ Report any of the following: Thirst, muscle cramps or weakness, paresthesia, dyspnea, or headache.
▪ Family members should immediately report any evidence of confusion.

MAPROTILINE HYDROCHLORIDE
(ma-proe′ti-leen)
Classification: TETRACYCLIC ANTIDEPRESSANT
Therapeutic: ANTIDEPRESSANT
Prototype: Mirtazapine
Pregnancy Category: B

AVAILABILITY 25 mg, 50 mg, 75 mg tablets

ACTION & *THERAPEUTIC EFFECT* It selectively inhibits reuptake of norepinephrine at CNS adrenergic synapses; this appears to produce antidepressant as well as antianxiety effects of maprotiline. *Useful in depression associated with anxiety and sleep disturbances.*

USES Treatment of depressive neurosis (dysthymic disorder) and manic-depressive illness, depressed type (major depressive disorder).

Common adverse effects in *italic,* life-threatening effects <u>underlined</u>; generic names in **bold;** classifications in SMALL CAPS; ◆ Canadian drug name; ❷ Prototype drug

915

UNLABELED USES Bulimia, pain, panic attack, enuresis.

CONTRAINDICATIONS Acute MI, AV block, cardiac arrhythmias, QT prolongation; MAOI therapy within 14 days; tricyclic antidepressant therapy; history of alcoholism; suicidal ideation.

CAUTIOUS USE History of seizure activity; psychotic disorders; history of suicidal tendencies; diabetes mellitus; hepatic disease; GI disease; GERD; BPH; respiratory depression; labor and delivery; pregnancy (category B), lactation; children younger than 18 y.

ROUTE & DOSAGE

Mild to Moderate Depression

Adult: **PO** Start at 75 mg/day and may increase q2wk up to 150 mg/day in single or divided doses
Geriatric: **PO** Start with 25 mg at bedtime may increase to 50–75 mg/day

Severe Depression

Adult: **PO** Start at 100–150 mg/day may increase up to 300 mg/day in single or divided doses if needed

Pharmacogenetic Dosage Adjustment

Poor CYP2D6 metabolizers: Start with 40% of dose

ADMINISTRATION

Oral

- Give as single dose or in divided doses. Initiate therapy with low dosages to reduce risk of seizures.
- Store at 15°–30° C (59°–86° F) unless otherwise specified.

ADVERSE EFFECTS (≥1%) **CNS:** Seizures, exacerbation of psychosis, hallucinations, tremors, excitement, confusion, dizziness, *drowsiness*. **CV:** *Orthostatic hypotension*, hypertension, tachycardia. **Special Senses:** Accommodation disturbances, blurred vision, mydriasis. **GI:** Nausea, vomiting, epigastric distress, *constipation, dry mouth*. **Urogenital:** *Urinary retention*, frequency. **Skin:** Hypersensitivity reactions (skin rash, urticaria, photosensitivity).

INTERACTIONS Drug: May decrease some response to ANTIHYPERTENSIVES; CNS DEPRESSANTS, **alcohol,** HYPNOTICS, BARBITURATES, SEDATIVES potentiate CNS depression; may increase hypoprothrombinemic effect of ORAL ANTICOAGULANTS; with **levodopa,** SYMPATHOMIMETICS (e.g., **epinephrine, norepinephrine**) there is possibility of sympathetic hyperactivity with hypertension and **hyperpyrexia;** with MAO INHIBITORS or **linezolid** there is possibility of severe reactions, toxic psychosis, cardiovascular instability; **methylphenidate** increases plasma TCA levels; THYROID DRUGS increase possibility of arrhythmias; **cimetidine** may increase plasma TCA levels.

PHARMACOKINETICS Absorption: Slowly absorbed from GI tract. **Peak:** 12 h. **Distribution:** Distributed chiefly to brain, lungs, liver, and kidneys. **Metabolism:** In liver. **Elimination:** 70% in urine, 30% in feces. **Half-Life:** 51 h.

NURSING IMPLICATIONS

Assessment & Drug Effects

- Monitor for therapeutic effectiveness; 2–3 wk are usually necessary for full effect.
- Monitor for increased suicidality, unusual changes in behavior, or suicide attempt. Inform the prescriber immediately.
- Assess level of sedative effect. If recovering patient becomes too lethargic to care for personal

hygiene or to maintain food intake and interactions with others, report to prescriber.

- Monitor bowel elimination pattern and I&O ratio. Severe constipation and urinary retention are potential problems, especially in the older adult. Advise increased fluid intake (at least 1500 mL/day).
- Observe seizure precautions; risk of seizures appears to be high in heavy drinkers.
- Bear in mind that if patient uses excessive amounts of alcohol, potentiated effects of maprotiline may increase the danger of overdosage or suicide attempt.

Patient & Family Education

- Report symptoms of stomatitis and dry mouth when taking high doses. Sore or dry mouth can lead to lack of compliance.
- Use caution with tasks that require alertness and skill; ability may be impaired during early therapy.
- Do not change dose or dose schedule without consulting prescriber.
- Do not use OTC drugs unless approved by prescriber.
- Avoid alcohol; the effects of maprotiline are potentiated when both are used together and for 2 wk after maprotiline is discontinued.

MARAVIROC

(mar-a-vir'ok)

Selzentry

Classifications: ANTIRETROVIRAL; FUSION INHBITOR; CELLULAR CHEMOKINE RECEPTOR (CCR5) ANTAGONIST

Therapeutic: ANTIRETROVIRAL

Pregnancy Category: B

AVAILABILITY 150 mg, 300 mg tablets

ACTION & *THERAPEUTIC EFFECT*

Selectively binds to human chemokine coreceptor-5 (CCR-5) on cell membranes of helper T cell lymphocytes preventing interaction with the HIV-1 envelope protein necessary for the HIV virus to enter helper T cells. *Prevents infection of helper T cells by HIV-1 viruses with CCR-5 tropism.*

USES Treatment of human immunodeficiency virus (HIV-1) infection in combination with other antiretroviral agents.

CONTRAINDICATIONS Patients with dual/mixed, or chemokine-related receptor (CCR-4)-tropic HIV-1 virus; treatment naïve adults or children with HIV-1; S&S of hepatitis, allergic reaction to drug, or hepatoxicity; lactation.

CAUTIOUS USE Hepatic impairment, hepatitis B or C; renal impairment; CrCl less than 50 mL/min; cardiac disease or increased risk for cardiovascular events; older adults; pregnancy (category B); children younger than 16 y.

ROUTE & DOSAGE

Regimen without CYP3A Inducers or Inhibitors

Adult/Adolescent: **PO** 300 mg b.i.d.

Regimen with CYP3A Inhibitor with/without CYP3A Inducer

Adult/Adolescent: **PO** 150 mg b.i.d.

Regimen with CYP3A Inducers without a Strong CYP3A Inhibitor

Adult/Adolescent: **PO** 600 mg b.i.d.

Renal Impairment

See package insert.

Common adverse effects in *italic*, life-threatening effects underlined; generic names in **bold**; classifications in SMALL CAPS; ♣ Canadian drug name; ⓟ Prototype drug

917

ADMINISTRATION

- **Must be** given in combination with other antiretroviral drugs.
- Store at 15°–30° C (59°–86° F).

ADVERSE EFFECTS (≥1%)

Body as a Whole: Appetite disorders, herpes infection, pain and discomfort, *pyrexia.* **CNS:** Depression, disturbances in consciousness, *sleep disorders, dizziness,* paresthesias and dysesthesias, peripheral neuropathies, sensory abnormalities. **CV:** Vascular hypertension disorders. **GI:** *Abdominal pain,* constipation, dyspepsia, stomatitis, ulceration. **Hematologic:** Neutropenia. **Metabolic:** Elevated AST levels. **Musculoskeletal:** Joint-related signs and symptoms, muscle pains, *musculoskeletal symptoms.* **Respiratory:** Breathing abnormalities, bronchitis, bronchospasm, *cough,* influenza, paranasal sinus disorder, pneumonia, respiratory tract disorders, sinusitis, *upper respiratory tract infection.* **Skin:** Apocrine and eccrine gland disorders, benign neoplasms, dermatitis, eczema, folliculitis, lipodystrophies, pruritis, *rash.* **Urogenital:** Bladder and urethral symptoms, condyloma acuminatum, urinary tract signs and symptoms.

INTERACTIONS

Drug: STRONG CYP3A4 INHIBITORS (HIV PROTEASE INHIBITORS with the exception of **tipranavir/ritonavir, delavirdine, ketoconazole, itrazonazole, clarithromycin**) increase maraviroc plasma level. CYP3A4 INDUCERS (**efavirenz, rifampin, carbamazepine, phenobarbital, phenytoin**) decrease maraviroc plasma level. **Food:** Coadministration with a high-fat meal decreases the plasma levels of maraviroc. **Herbal:** St. John's wort may decrease the plasma levels of maraviroc.

PHARMACOKINETICS

Absorption: Bioavailability is 23–33%. **Peak:** 0.5–4 h (dose-dependent). **Distribution:** 75% protein bound. **Metabolism:** In liver via CYP3A4. **Elimination:** Primarily in stool. **Half-Life:** 14–18 h.

NURSING IMPLICATIONS

Assessment & Drug Effects

- Monitor for and report promptly S&S of hepatotoxicity, hepatitis or infection.
- Monitor BP especially in those on antihypertensive drugs and with a history of postural hypotension.
- Monitor CV status especially in those with preexisting conditions that cause myocardial ischemia.
- Lab tests: Baseline and periodic CD4+ cell count and HIV RNA viral load; periodic LFTs, WBC with differential.

Patient & Family Education

- Exercise caution when arising from a lying or sitting position. Dizziness is a common adverse effect.
- Do not engage in dangerous activities until response to drug is known.
- Report promptly any of the following: Itchy rash, yellow skin or eyes, nausea or vomiting, upper abdominal pain, flu-like symptoms, unexplained fatigue.

MEBENDAZOLE ⊙

(me-ben'da-zole)

Classification: ANTHELMINTIC
Therapeutic: ANTHELMINTIC
Pregnancy Category: C

AVAILABILITY 100 mg tablets

ACTION & *THERAPEUTIC EFFECT*

Carbamate with unusually broad-spectrum of anthelmintic activity. Inhibits formation of worm's microtubules and inhibits glucose and other nutrient uptake by susceptible helminths. *Effective against susceptible helminths (nematodes) by interfering with their survival.*

USES Treatment of *Trichuris trichiura* (whipworm), *Enterobius vermicularis* (pinworm), *Ascaris lumbricoides* (roundworm), *Ancylostoma duodenale* (common hookworm), *Necator americanus* (American hookworm) in single or mixed infections.
UNLABELED USES Beef, dwarf, and pork tapeworm and threadworm infections.

CONTRAINDICATIONS Hypersensitivity to mebendazole.
CAUTIOUS USE Inflammatory bowel disease, ulcerative colitis, Crohn's disease; hepatic disease; pregnancy (category C), lactation; children younger than 2y.

ROUTE & DOSAGE

Enterobiasis
Adult: **PO** 100 mg as single dose
Child: **PO** 100 mg as single dose
Other Infestations
Adult: **PO** 100 mg b.i.d. × 3 days
Child: **PO** 100 mg b.i.d. × 3 days

ADMINISTRATION
Oral
▪ Allow tablets to be chewed and swallowed, or crushed and mixed with food if needed.

ADVERSE EFFECTS (≥1%) **GI:** Transient abdominal pain, diarrhea. **Body as a Whole:** Dizziness, fever (possibly due to tissue necrosis in cysts).

INTERACTIONS Drug: Carbamazepine, phenytoin can increase metabolism of mebendazole.

PHARMACOKINETICS Absorption: Minimal from GI tract (2–10% of dose). **Metabolism:** Metabolized to inactive metabolite. **Elimination:** Primarily in feces. **Half-Life:** 3–9 h.

NURSING IMPLICATIONS
Assessment & Drug Effects
▪ Initiate second course of treatment if cure does not occur within 3 wk.
▪ Examine and treat all family members simultaneously because pinworms are readily transmitted from person to person.

Patient & Family Education
▪ Practice thorough hand washing after touching any potentially contaminated item.
▪ Change underclothing, bedclothes, towels, and facecloths daily; bathe frequently, preferably by showering. Infected person should sleep alone.

MECHLORETHAMINE HYDROCHLORIDE
(me-klor-eth′a-meen)
Mustargen
Classifications: ANTINEOPLASTIC; ALKYLATING AGENT; NITROGEN MUSTARD
Therapeutic: ANTINEOPLASTIC
Prototype: Cyclophosphamide
Pregnancy Category: D

AVAILABILITY 10 mg powder for injection

ACTION & *THERAPEUTIC EFFECT*
Analog of mustard gas that forms highly reactive carbonium ion that causes cross-linking and abnormal base-pairing in DNA, thereby

---

Common adverse effects in *italic*; life-threatening effects underlined; generic names in **bold**; classifications in SMALL CAPS; ♣ Canadian drug name; ☻ Prototype drug

919

interfering with DNA replication and RNA and protein synthesis. Cell-cycle nonspecific inhibitor of DNA and RNA synthesis. *Antineoplastic agent that simulates actions of x-ray therapy, but nitrogen mustards produce more acute tissue damage and more rapid recovery.*

USES Generally confined to nonterminal stages of neoplastic disease, as single agent or in combination with other agents in palliative treatment of Hodgkin's disease (stages III and IV), lymphosarcoma, mycosis fungoides, polycythemia vera, bronchogenic carcinoma, chronic myelocytic or chronic lymphocytic leukemia. Also for intrapleural, intrapericardial, and intraperitoneal palliative treatment of metastatic carcinoma resulting in effusion.

CONTRAINDICATIONS Myelosuppression; infectious granuloma; known infectious diseases, acute herpes zoster; intracavitary use with other systemic bone marrow suppressants; pregnancy (category D), lactation.
CAUTIOUS USE Bone marrow infiltration with malignant cells, chronic lymphocytic leukemia; men or women in childbearing age.

ROUTE & DOSAGE

Advanced Hodgkin's Disease

Adult: **IV** 6 mg/m² on day 1 and 8 of a 28-day cycle

Other Neoplasms

Adult: **IV** 0.4 mg/kg given as a single dose or in divided doses of 0.1–0.2 mg/kg/day, may repeat course in 3–6 wk

Obesity Dosage Adjustment

Dose based on IBW

ADMINISTRATION

Intravenous

Wear surgical gloves during preparation and administration of solution. ▪ Avoid inhalation of vapors and dust and contact of drug with eyes and skin. ▪ Flush contaminated area immediately if drug contacts the skin. Use copious amounts of water for at least 15 min, followed by 2% sodium thiosulfate solution. Irritation may appear after a latent period. ▪ Irrigate immediately if eye contact occurs. Use copious amounts of NS followed by ophthalmologic examination as soon as possible.

PREPARE: **Direct:** Reconstitute immediately before use by adding 10 mL sterile water for injection or NS injection to vial to yield 1 mg/mL. With needle still in stopper, shake vial several times to dissolve. ▪ Discard colored solution or contents of any vial with drops of moisture.

ADMINISTER: **Direct:** To reduce risk of severe infections from extravasation or high concentration of the drug, inject slowly over 3–5 min into tubing or sidearm of freely flowing IV infusion. ▪ Flush vein with running IV solution for 2–5 min to clear tubing of any remaining drug. ▪ Be alert for extravasation. Treat promptly with subcutaneous or intradermal injection with isotonic sodium thiosulfate solution (1/6 molar) and application of ice compresses intermittently for a 6–12 h period to reduce local tissue damage and discomfort. ▪ Tissue induration and tenderness may persist 4–6 wk, and tissue may slough.

INCOMPATIBILITIES **Solution/additive: D5W, methohexital, normal saline. Y-site: Allopurinol, cefepime.**

Common adverse effects in *italic*, life-threatening effects <u>underlined</u>; generic names in **bold**; classifications in SMALL CAPS; ✚ Canadian drug name; ⊙ Prototype drug

ADVERSE EFFECTS (≥1%) CNS:
Neurotoxicity: Vertigo, tinnitus, headache, drowsiness, peripheral neuropathy, light-headedness, paresthesias, cerebral deterioration, coma. **GI:** Stomatitis, xerostomia, anorexia, *nausea, vomiting,* diarrhea. **Hematologic:** Leukopenia, _thrombocytopenia,_ lymphocytopenia, _agranulocytosis, anemia,_ hyperheparinemia. **Skin:** Pruritus, hyperpigmentation, herpes zoster, alopecia. **Urogenital:** Amenorrhea, azoospermia, chromosomal abnormalities, hyperuricemia. **Body as a Whole:** Weakness, hypersensitivity reactions. *With extravasation: Painful inflammatory reaction, tissue sloughing, thrombosis, thrombophlebitis.*

INTERACTIONS Drug: May reduce effectiveness of ANTIGOUT AGENTS by raising serum **uric acid** levels; dosage adjustments may be necessary; may prolong neuromuscular blocking effects of **succinylcholine;** may potentiate bleeding effects of ANTICOAGULANTS, SALICYLATES, NSAIDS, PLATELET INHIBITORS.

PHARMACOKINETICS Metabolism: Rapid hydrolysis and demethylation. **Elimination:** In urine. **Half-Life:** Less than 1 min.

NURSING IMPLICATIONS

Assessment & Drug Effects

- Establish baseline data for body weight, I&O ratio and pattern, and blood labs as reference for design of drug and care regimens.
- Lab tests: Monitor CBC with differential and platelet count. Periodic serum uric acid levels.
- Record daily weight. Alert prescriber to sudden or slow, steady weight gain.
- Monitor and record patient's fluid losses carefully. Prolonged

vomiting and diarrhea can produce volume depletion.

- Report immediately petechiae, ecchymoses, or abnormal bleeding from intestinal and buccal membranes. Keep injections and other invasive procedures to a minimum during period of thrombocytopenia.
- Report symptoms of agranulocytosis (e.g., unexplained fever, chills, sore throat, tachycardia, and mucosal ulceration).
- Prevent exposure to people with infection, especially upper respiratory tract infections.
- Note and record state of hydration of oral mucosa, condition of gingiva, teeth, tongue, mucosa, and lips.

Patient & Family Education

- Report any signs of bleeding immediately.
- Use caution to prevent falls or other traumatic injuries, especially during periods of low platelet counts.
- Increase fluid intake up to 3000 mL/day if allowed to minimize risk of kidney stones. Report promptly all symptoms, including flank or joint pain, swelling of lower legs and feet, changes in voiding pattern.

MECLIZINE HYDROCHLORIDE ℗

(mek´li-zeen)

Antivert, Antrizine, Bonamine ♣, Bonine, Dizmiss, RuVert-M
Classification: ANTIHISTAMINE; H₁-RECEPTOR ANTAGONIST; ANTIVERTIGO
Therapeutic: ANTIHISTAMINE, ANTIVERTIGO
Pregnancy Category: B

AVAILABILITY 12.5 mg, 25 mg, 50 mg tablets; 25 mg, 30 mg capsules

Common adverse effects in *italic*, life-threatening effects underlined; generic names in **bold**; classifications in SMALL CAPS; ♣ Canadian drug name; ℗ Prototype drug

921

M

ACTION & *THERAPEUTIC EFFECT*
Long-acting antihistamine, with marked effect in blocking histamine-induced vasopressive response but only slight anticholinergic action. Marked depressant action on labyrinthine excitability and on conduction in vestibular-cerebellar pathways. *Exhibits antivertigo, and antiemetic effects.*

USES Management of nausea, vomiting, and dizziness associated with motion sickness and in vertigo associated with diseases affecting vestibular system.

CONTRAINDICATIONS Hypersensitivity to meclizine; GI obstruction, ileus.
CAUTIOUS USE Angle-closure glaucoma, older adults, asthma, prostatic hypertrophy, pregnancy (category B), lactation. Safety in children younger than 12 y is not established.

ROUTE & DOSAGE

Motion Sickness
Adult: **PO** 25–50 mg 1 h before travel, may repeat q24h if necessary for duration of journey
Vertigo
Adult: **PO** 25–100 mg/day in divided doses

ADMINISTRATION
Oral
- Give without regard to meals.
- Ensure that chewable tablets are chewed or crushed before being swallowed with a liquid.

ADVERSE EFFECTS (≥1%) CNS: *Drowsiness.* **GI:** Dry mouth. **Special Senses:** Blurred vision. **Body as a Whole:** Fatigue.

INTERACTIONS Drug: Alcohol, CNS DEPRESSANTS may potentiate sedative effects of meclizine.

PHARMACOKINETICS Absorption: Readily absorbed from GI tract. **Onset:** 1 h. **Duration:** 8–24 h. **Distribution:** Crosses placenta. **Elimination:** Primarily in feces. **Half-Life:** 6 h.

NURSING IMPLICATIONS
Assessment & Drug Effects
- Supervision of ambulation, particularly with the older adult, since drug may cause drowsiness.
- Assess effectiveness of drug and inform prescriber when prescribed for vertigo; dosage adjustment may be required.

Patient & Family Education
- Do not drive or engage in potentially hazardous activities until response to drug is known.
- Be aware that sedative action may add to that of alcohol, barbiturates, narcotic analgesics, or other CNS depressants.
- Take 1 h before departure when prescribed for motion sickness.

MECLOFENAMATE SODIUM
(me-kloe-fen-am′ate)

Classification: ANALGESIC, NON-STEROIDAL ANTI-INFLAMMATORY DRUG (NSAID)
Therapeutic: ANALGESIC, NSAID; ANTI-PYRETIC; ANTIRHEUMATIC
Prototype: Ibuprofen
Pregnancy Category: C first and second trimester; D third trimester

AVAILABILITY 50 mg, 100 mg capsules

ACTION & *THERAPEUTIC EFFECT*
Inhibits prostaglandin synthesis by inhibiting both the COX-1 and COX-2 enzymes necessary for its synthesis and competes for binding at prostaglandin receptor sites.

Does not appear to alter course of arthritis. *Palliative anti-inflammatory and analgesic activity.*

USES Symptomatic treatment of acute or chronic rheumatoid arthritis and osteoarthritis. Also in combination with gold salts or corticosteroids in treatment of rheumatoid arthritis.

UNLABELED USES Management of psoriatic arthritis, mild to moderate postoperative pain, dysmenorrhea.

CONTRAINDICATIONS Hypersensitivity to aspirin or other NSAIDS; active peptic ulcer, ulcerative colitis; perioperative pain related to CABG surgery; renal disease; pregnancy (category D third trimester), patient designated as functional class IV rheumatoid arthritis (incapacitated, bedridden, etc).

CAUTIOUS USE History of upper GI tract disease; coronary artery disease; acute MI, cardiac arrhythmias; CVA; diabetes mellitus; SLE; compromised cardiac and kidney function, or other conditions predisposing to fluid retention; pregnancy (category C first and second trimester), lactation; children younger than 14 y.

ROUTE & DOSAGE

Inflammatory Disease

Adult: PO 200–400 mg/day in 3–4 divided doses (max: 400 mg/day)

ADMINISTRATION

Oral

- Give with food or milk if patient complains of GI distress. An aluminum and magnesium hydroxide antacid (Maalox) also may be prescribed. Consult prescriber if symptoms persist.

- Withhold dose and report to prescriber if significant diarrhea occurs.
- Store at 15°–30° C (59°–86° F) in airtight, light-resistant container.

ADVERSE EFFECTS (≥1%) **CNS:** *Dizziness,* vertigo, lack of concentration, confusion, *headache,* tinnitus. **CV:** Edema. **GI:** *Severe diarrhea (dose-related),* peptic ulceration, GI bleeding, dyspepsia, abdominal pain, *nausea,* vomiting (may be severe), flatulence, eructation, pyrosis, anorexia, constipation. **GI:** *Abnormal liver function tests,* cholestatic jaundice. **Special Senses:** Blurred vision. **Urogenital:** Elevated BUN and creatinine, kidney failure. **Skin:** Rash, pruritus, urticaria.

INTERACTIONS Drug: ORAL ANTICOAGULANTS, **heparin** may prolong bleeding time; may increase **lithium** toxicity; increases pharmacologic and toxic activity of **phenytoin,** SULFONYLUREAS, SULFONAMIDES, **warfarin** through protein-binding displacement. **Herbal: Feverfew, garlic, ginger, ginkgo** increase bleeding potential.

PHARMACOKINETICS Absorption: Rapidly and completely from GI tract. **Peak:** 1–2 h. **Duration:** 2–4 h. **Distribution:** Crosses placenta. **Metabolism:** In liver. **Elimination:** 60% in urine, 30% in feces. **Half-Life:** 2–3.3 h.

NURSING IMPLICATIONS

Assessment & Drug Effects

- Report diarrhea promptly. It is the most frequent adverse effect and usually dose related.
- Lab tests: Monitor kidney function where incidence of adverse reactions is potentially high because drug is excreted primarily by the

M

Common adverse effects in *italic,* life-threatening effects underlined; generic names in **bold;** classifications in SMALL CAPS; ♣ Canadian drug name; ⊕ Prototype drug

923

kidneys. Monitor PT, PTT, and INR frequently with concurrent anticoagulant therapy.

- Monitor I&O ratio. Encourage fluid intake of at least 8 glasses of liquid a day.
- Consider sodium content of meclofenamate tablets if patient is on restricted sodium intake.

Patient & Family Education

- Stop taking drug and promptly notify the prescriber if nausea, vomiting, severe diarrhea, and abdominal pain occur. Generally dose reduction or temporary withdrawal will control symptoms.
- Report to prescriber without delay: Blurred vision, tinnitus, or taste disturbances.
- Schedule ophthalmic examinations before and periodically during treatment and whenever you experience visual disturbances.
- Notify prescriber if you become pregnant.
- Weigh under standard conditions (similar clothing, same time of day) twice weekly. Report weight gain of more than 2.5 to 3.5 kg (3–4 lb)/wk as well as signs of edema: Swollen ankles, tibiae, hands, feet.
- Dizziness, a troublesome early side effect, frequently disappears in time. Avoid driving a car or potentially hazardous activities until response to drug is known.
- Report immediately to prescriber any sign of bleeding (e.g., melena, epistaxis, ecchymosis) when taking concomitant oral anticoagulant.

MEDROXYPROGESTERONE ACETATE

(me-drox'ee-proe-jess'te-rone)

Depo-Provera, Depo-subQ Provera 104, Provera
Classification: PROGESTIN

Therapeutic: PROGESTIN
Prototype: Progesterone
Pregnancy Category: X

AVAILABILITY 2.5 mg, 5 mg, 10 mg tablets; 104 mg/0.65 mL, 150 mg/mL, 400 mg/mL injection

ACTION & *THERAPEUTIC EFFECT*
Induces and maintains endometrium, preventing uterine bleeding; inhibits production of pituitary gonadotropin, thus preventing ovulation and producing thick cervical mucus resistant to passage of sperm. *Slows release of luteinizing hormone (LH) preventing follicular maturation and ovulation.*

USES Dysfunctional uterine bleeding; secondary amenorrhea; parenteral form **(Depo-Provera)** used in adjunctive, palliative treatment of inoperable, recurrent, and metastatic endometrial or renal carcinoma; contraception; endometriosis-associated pain.
UNLABELED USES Obstructive sleep apnea.

CONTRAINDICATIONS History of thromboembolic disorders; breast cancer, cervical cancer, uterine cancer, vaginal cancer; hepatic disease; abnormal vaginal bleeding, incomplete abortion; pregnancy (category X); lactation.
CAUTIOUS USE Asthma, seizure disorders, CVA; migraine, cardiac or kidney dysfunction, liver disease.
IM: Due to potential for irreversible bone loss, long-term birth control use should be avoided or use in adolescents and early adulthood.

ROUTE & DOSAGE

Secondary Amenorrhea
Adult: **PO** 5–10 mg/day for 5–10 days beginning any time if

endometrium is adequately estrogen primed (withdrawal bleeding occurs in 3–7 days after discontinuing therapy)

Abnormal Bleeding Due to Hormonal Imbalance

Adult: **PO** 5–10 mg/day for 5–10 days beginning on the assumed or calculated 16th or 21st day of menstrual cycle; if bleeding is controlled, administer 2 subsequent cycles

Carcinoma

Adult: **IM** 400–1000 mg/wk; continue at 400 mg/mo if improvement occurs and disease stabilizes

Contraceptive

Adult: **IM** 100 mg q3mo

Sleep Apnea

Adult: **PO** 20 mg t.i.d.

ADMINISTRATION

Oral

- Oral drug may be given with food to minimize GI distress.

Intramuscular

- Administer IM deep into a large muscle.
- Store both formulations at 15°–30° C (59°–86° F); protect from freezing.

ADVERSE EFFECTS (≥1%) **CNS:** <u>Cerebral thrombosis or hemorrhage</u>, depression. **CV:** Hypertension, pulmonary embolism, edema. **GI:** Vomiting, nausea, cholestatic jaundice, abdominal cramps. **Urogenital:** *Breakthrough bleeding,* changes in menstrual flow, dysmenorrhea, vaginal candidiasis. **Skin:** Angioneurotic edema. **Body as a Whole:** Weight changes; *breast tenderness,* enlargement or secretion. **Musculoskeletal:** Loss of bone mineral density.

INTERACTIONS Drug: Aminoglutethimide decreases serum concentrations of medroxyprogesterone; BARBITURATES, **carbamazepine, oxcarbazepine, phenytoin, primidone, rifampin, modafinil, rifabutin, topiramate** can increase metabolism and decrease serum levels of medroxyprogesterone. **Herbal:** Intermenstrual bleeding and loss of contraceptive efficacy may occur with **St. John's wort.**

PHARMACOKINETICS Peak: 2–4 h PO, 3 wk IM. **Distribution:** Greater than 90% protein bound. **Metabolism:** In liver. **Elimination:** Primarily in feces. **Half-Life:** 30 days PO, 50 days IM.

NURSING IMPLICATIONS

Assessment & Drug Effects

- See progesterone for numerous additional nursing implications.
- Be aware that IM injection may be painful. Monitor sites for evidence of sterile abscess. A residual lump and discoloration of tissue may develop.
- Monitor for S&S of thrombophlebitis (see Appendix F).
- Note: Planned menstrual cycling with medroxyprogesterone may benefit the patient with a history of recurrent episodes of abnormal uterine bleeding.

Patient & Family Education

- Be aware that after repeated IM injections, infertility and amenorrhea may persist as long as 18 mo.
- Learn breast self-examination.
- Review package insert to ensure complete understanding of progestin therapy.

MEFENAMIC ACID

(me-fe-nam'ik)

Ponstel
Classification: ANALGESIC, NONSTEROIDAL ANTI-INFLAMMATORY DRUG (NSAID)

M

Common adverse effects in *italic*, life-threatening effects <u>underlined</u>; generic names in **bold**; classifications in SMALL CAPS; ♣ Canadian drug name; ⊕ Prototype drug

925

Therapeutic: ANALGESIC, NSAID; ANTIPYRETIC
Prototype: Ibuprofen
Pregnancy Category: C

AVAILABILITY 250 mg tablets

ACTION & *THERAPEUTIC EFFECT*
NSAID that inhibits COX-1 and COX-2 enzymes necessary for prostaglandin synthesis. It affects platelet function. *Analgesic and anti-inflammatory actions.*

USES Short-term relief of mild to moderate pain including primary dysmenorrhea.

CONTRAINDICATIONS Hypersensitivity to mefenamic; GI inflammation, or ulceration.
CAUTIOUS USE Hypersensitivity to aspirin, history of kidney or liver disease; blood dyscrasias; asthma; cardiac arrhythmias; CHF; edema; diabetes mellitus; SLE; pregnancy (category C), lactation; children younger than 14 y. Long-term use increases risk of serious adverse events (see DRUG INTERACTIONS).

ROUTE & DOSAGE

Mild to Moderate Pain
Adult: **PO Loading Dose** 500 mg **PO Maintenance Dose** 250 mg q6h prn

ADMINISTRATION
Oral
- Give with meals, food, or milk to minimize GI adverse effects.
- Duration of therapy should not exceed 1 wk (manufacturer's warning).

ADVERSE EFFECTS (≥1%) **CNS:** Drowsiness, insomnia, dizziness, nervousness, confusion, headache. **GI:** *Severe diarrhea,* ulceration, and bleeding; *nausea, vomiting,* abdominal cramps, flatus, constipation, hepatic toxicity. **Hematologic:** Prolonged prothrombin time, severe autoimmune hemolytic anemia (long-term use), leukopenia, eosinophilia, agranulocytosis, thrombocytopenic purpura, megaloblastic anemia, pancytopenia, bone marrow hypoplasia. **Urogenital:** Nephrotoxicity, dysuria, albuminuria, hematuria, elevation of BUN. **Skin:** Urticaria, rash, facial edema. **Special Senses:** Eye irritation, loss of color vision (reversible), blurred vision, ear pain. **Body as a Whole:** Perspiration. **CV:** Palpitation. **Respiratory:** Dyspnea; acute exacerbation of asthma; bronchoconstriction (in patients sensitive to aspirin).

DIAGNOSTIC TEST INTERFERENCE False-positive reactions for *urinary bilirubin* (using *diazo tablet test*).

INTERACTIONS Drug: Mefenamic acid may prolong bleeding time with ORAL ANTICOAGULANTS, **heparin;** may increase **lithium** toxicity; increases pharmacologic and toxic activity of **phenytoin,** SULFONYLUREAS, SULFONAMIDES, **warfarin** because of protein binding displacement. **Herbal:** Feverfew, garlic, ginger, ginkgo increase bleeding potential.

PHARMACOKINETICS Absorption: Rapidly and completely from GI tract. **Peak:** 2–4 h. **Duration:** 6 h. **Distribution:** Distributed in breast milk. **Metabolism:** Partially in liver. **Elimination:** 50% in urine, 50% in feces. **Half-Life:** 2 h.

NURSING IMPLICATIONS
Assessment & Drug Effects
- Assess patients who develop severe diarrhea and vomiting for

Common adverse effects in *italic*, life-threatening effects underlined; generic names in **bold**; classifications in SMALL CAPS; ♣ Canadian drug name; ⊙ Prototype drug

dehydration and electrolyte imbalance.

- Lab tests: With long-term therapy (not recommended) obtain periodic complete blood counts, Hct and Hgb, and kidney function tests.

Patient & Family Education

- Discontinue drug promptly if diarrhea, dark stools, hematemesis, ecchymoses, epistaxis, or rash occur and do not use again. Contact prescriber.
- Notify prescriber if persistent GI discomfort, sore throat, fever, or malaise occur.
- Do not drive or engage in potentially hazardous activities until response to drug is known. It may cause dizziness and drowsiness.
- Monitor blood glucose for loss of glycemic control if diabetic.

MEFLOQUINE HYDROCHLORIDE

(me-flo′quine)
Classifications: ANTIPROTOZOAL; ANTIMALARIAL
Therapeutic: ANTIMALARIAL
Prototype: Chloroquine
Pregnancy Category: C

AVAILABILITY 250 mg tablets

ACTION & *THERAPEUTIC EFFECT*
Antimalarial agent that inhibits replication of parasites. *Effective against all types of malaria, including chloroquine-resistant malaria.*

USES Treatment of mild to moderate acute malarial infections, prevention of chloroquine-resistant malaria caused by *Plasmodium falciparum* and *P. vivax.*

CONTRAINDICATIONS Hypersensitivity to mefloquine or a related compound; with a calcium channel

blocking agent, severe heart arrhythmias, history of QT_c prolongation; aggressive behavior; active depression, or history of depression, suicidal ideation; generalized anxiety disorder, psychosis, schizophrenia, or other major psychiatric disorders; seizure disorders; lactation.
CAUTIOUS USE Persons piloting aircraft or operating heavy machinery; pregnancy (category C).

ROUTE & DOSAGE

Note: The U.S. Public Health Service does NOT recommend its use in children less than 15 kg or in pregnant women

Treatment of Malaria

Adult: **PO** 1250 mg (5 tablets) as single oral dose taken with at least 8 oz of water
Child: **PO** 20–25 mg/kg as single dose

Prophylaxis for Malaria

Adult: **PO** 250 mg once/wk × 4 wk (beginning 1 wk before travel), then 250 mg every other wk for duration of exposure and for 2 doses after leaving endemic area
Child: **PO** Given weekly: *Weight 15–19 kg,* ¼ tablet; *weight 20–30 kg,* ½ tablet; *weight 31–45 kg,* ¾ tablet

ADMINISTRATION

Oral

- Give with food and at least 8 oz water.
- Do not give concurrently with quinine or quinidine; wait at least 12 h beyond last dose of either drug before administering mefloquine.
- Store at 15°–30° C (59°–86° F).

ADVERSE EFFECTS (≥1%) **Body as a Whole:** Arthralgia, chills, fatigue,

Common adverse effects in *italic*, life-threatening effects underlined; generic names in **bold**; classifications in SMALL CAPS; ✦ Canadian drug name; ⊘ Prototype drug

927

fever. **CNS:** Dizziness, nightmares, visual disturbances, headache, syncope, confusion, psychosis, aggression, suicide ideation (rare). **CV:** Bradycardia, ECG changes (including QT$_c$ prolongation), first-degree AV block. **GI:** Nausea, vomiting, abdominal pain, anorexia, diarrhea. **Skin:** Rash, itching.

DIAGNOSTIC TEST INTERFERENCE Transient increase in *liver transaminases.*

INTERACTIONS Drug: Mefloquine can prolong cardiac conduction in patients taking BETA-BLOCKERS, CALCIUM CHANNEL BLOCKERS, and possibly **digoxin. Quinine** may decrease plasma mefloquine concentrations. Mefloquine may decrease **valproic acid** serum concentrations by increasing its hepatic metabolism. Administration with **chloroquine** may increase risk of seizures. Increased risk of cardiac arrest and seizures with **quinidine.**

PHARMACOKINETICS Absorption: 85% absorbed, concentrates in red blood cells. **Onset:** 59 and 28 h for parasite and fever clearance times in patients with *P. vivax* infections, respectively; 166 and 93 h in patients with *P. malariae* infections. **Distribution:** Concentrated in red blood cells due to high-affinity binding to red blood cell membranes; 98% protein bound; distributed minimally into breast milk. **Metabolism:** In liver. **Elimination:** Primarily in bile and feces. **Half-Life:** 10–21 days (shorter in patients with acute malaria).

NURSING IMPLICATIONS

Assessment & Drug Effects

▪ Monitor carefully during prophylactic use for development of unexplained anxiety, depression, restlessness, or confusion; such manifestations may indicate a need to discontinue the drug.
▪ Evaluate cardiac and liver functions periodically with prolonged use.
▪ Lab tests: Monitor periodically CBC with differential during prolonged use.
▪ Monitor blood levels of anticonvulsants with concomitant therapy closely.

Patient & Family Education

▪ Take drug on the same day each week when used for malaria prophylaxis.
▪ Do not perform potentially hazardous activities until response to drug is known.
▪ Report any of the following immediately: Fever, sore throat, muscle aches, visual problems, anxiety, confusion, mental depression, hallucinations.

MEGESTROL ACETATE
(me-jess′trole)
Megace, Megace ES
Classifications: ANTINEOPLASTIC; PROGESTIN
Therapeutic: ANTINEOPLASTIC; APPETITE ENHANCER
Prototype: Progesterone
Pregnancy Category: X (oral suspension); D (tablets)

AVAILABILITY 40 mg/mL, 125 mg/mL suspension; 20 mg, 40 mg tablets

ACTION & THERAPEUTIC EFFECT
Progestational hormone with antineoplastic properties for which an antiluteinizing effect mediated via the pituitary has been postulated. *Effective for treating breast, renal cell, or endometrial carcinoma. Also effective as an appetite enhancer. Has a local*

Common adverse effects in *italic*, life-threatening effects underlined; generic names in **bold**; classifications in SMALL CAPS; ♦ Canadian drug name; ⊘ Prototype drug

effect when instilled directly into the endometrial cavity.

USES Palliative agent for treatment of advanced carcinoma of breast or endometrium, AIDS-related wasting or cachexia.

CONTRAINDICATIONS Diagnostic test for pregnancy; pregnancy (category X oral suspension, category D tablet); lactation.
CAUTIOUS USE Older adults; severe hepatic disease; diabetes mellitus; renal impairment; thromboembolic disease.

ROUTE & DOSAGE

Palliative Treatment for Advanced Breast Cancer
Adult: **PO** 40 mg q.i.d.

Palliative Treatment for Advanced Endometrial Cancer
Adult: **PO** 40–320 mg/day in divided doses

Appetite Stimulation
Adult: **PO** 200 mg q6h

HIV-Related Cachexia
Adult: **PO (suspension)** 800 mg daily or 625 mg of **Megace ES**

ADMINISTRATION
Oral
- Give with meals or food if GI distress occurs.
- Shake oral suspension well before use.
- Store at 15°–30° C (59°–86° F) in tightly closed container.

ADVERSE EFFECTS (≥1%) **Urogenital:** Vaginal bleeding. **Body as a Whole:** Breast tenderness, headache, increased appetite, weight gain, allergic-type reactions (including bronchial asthma). **GI:**
Abdominal pain, nausea, vomiting. **Hematologic:** DVT.

INTERACTIONS Drug: May increase levels of **warfarin;** may decrease renal clearance of **dofetilide.**

PHARMACOKINETICS Absorption: Appears to be well absorbed from GI tract. **Onset:** Onset of objective response in breast cancer in 6–8 wk. **Peak:** 1–3 h. **Duration:** 3–12 mo. **Metabolism:** Completely metabolized in liver. **Elimination:** 57–78% of dose excreted in urine within 10 days.

NURSING IMPLICATIONS
Assessment & Drug Effects
- Monitor weight periodically.
- Notify prescriber if abdominal pain, headache, nausea, vomiting, or breast tenderness become pronounced.
- Monitor for allergic reactions, including breathing distress characteristic of asthma, rash, urticaria, anaphylaxis, tachypnea, anxiety. Stop medication if they appear and notify prescriber.

Patient & Family Education
- Use contraception measures during therapy for carcinoma.
- Learn breast self-examination.
- Learn S&S of thrombophlebitis (see Appendix F).
- Review package insert to ensure understanding of megestrol therapy.

MELOXICAM
(mel-ox′-i-cam)
Mobic
Classification: ANALGESIC, NON-STEROIDAL ANTI-INFLAMMATORY DRUG (NSAID)

Common adverse effects in *italic*, life-threatening effects <u>underlined</u>; generic names in **bold;** classifications in SMALL CAPS; ♣ Canadian drug name; ❿ Prototype drug

929

Therapeutic: ANALGESIC, NSAID; ANTIPYRETIC; ANTIRHEUMATIC
Prototype: Ibuprofen
Pregnancy Category: C

AVAILABILITY 7.5 mg tablets

ACTION & *THERAPEUTIC EFFECT*

A nonsteroidal anti-inflammatory drug (NSAID) that inhibits both COX-1 and COX-2 enzymes that are necessary for synthesis of prostaglandin, which is part of the inflammatory response. *Exhibits analgesic and anti-inflammatory actions. It improves the S&S of RA.*

USES Relief of the signs and symptoms of osteoarthritis, rheumatoid arthritis.

CONTRAINDICATIONS Hypersensitivity to meloxicam, aspirin, salicylates, or NSAIDs; GI bleeding; severe hepatic disease; perioperative pain with CABG surgery; lactation. **CAUTIOUS USE** History of coagulation defects, liver dysfunction, gastrointestinal disease, anemia; asthma; bone marrow suppression; dehydration, edema, GI diseases; heart failure; jaundice; hepatic or renal impairment; hypertension, hypovolemia, immunosuppression; asthma; lactase deficiency, advanced renal dysfunction; hypertension or cardiac conditions aggravated by fluid retention and edema; older adults; females of childbearing age; pregnancy (category C); children.

ROUTE & DOSAGE

Osteoarthritis
Adult: **PO** 7.5–15 mg once daily
Rheumatoid Arthritis
Adult: **PO** 15 mg once daily

ADMINISTRATION

Oral

- Do not exceed the maximum recommended daily dose of 15 mg.
- Use the lowest effective dose for the shortest duration to minimize risk of serious adverse effects.
- Store at 15°–30° C (59°–86° F).

ADVERSE EFFECTS (≥1%) Body as a Whole: Edema, fall, flu-like syndrome, pain. **GI:** Abdominal pain, diarrhea, dyspepsia, flatulence, nausea, constipation, <u>ulceration, GI bleed</u>. **Hematologic:** Anemia. **Musculoskeletal:** Arthralgia. **CNS:** Dizziness, headache, insomnia. **Respiratory:** Pharyngitis, upper respiratory tract infection, cough. **Skin:** Rash, pruritus. **Urogenital:** Micturition frequency, urinary tract infection.

INTERACTIONS Drug: May decrease effectiveness of ACE INHIBITORS, DIURETICS; **aspirin, warfarin** may increase risk of bleed; may increase **lithium** levels and toxicity. **Herbal: Feverfew, garlic, ginger, ginkgo** may increase bleeding potential.

PHARMACOKINETICS Absorption: 89% bioavailable. **Peak:** 4–5 h. **Distribution:** Greater than 99% protein bound, distributes into synovial fluid. **Metabolism:** In liver (CYP2C9). **Elimination:** Equally in urine and feces. **Half-Life:** 15–20 h.

NURSING IMPLICATIONS

Assessment & Drug Effects

- Monitor for and immediately report S&S of GI ulceration or bleeding, including black, tarry stool, abdominal or stomach pain; hepatotoxicity, including fatigue, lethargy, pruritus, jaundice, flu-like symptoms; skin rash; weight gain and edema.

- Withhold drug and notify prescriber if hepatotoxicity or GI bleeding is suspected.
- Monitor carefully patients with a history of CHF, HTN, or edema for fluid retention.
- Lab tests: Monitor Hgb and Hct, CBC with differential, LFTs, serum electrolytes, BUN, and creatinine within 3 mo of initiating therapy and every 6–12 mo thereafter; with high-risk patients (e.g., older than 60 y, history of peptic ulcer disease, prolonged or high-dose NSAID therapy, concurrent use of corticosteroids or anticoagulants) monitor within first 3–4 wk and every 3–6 mo thereafter.
- Coadministered drugs: With warfarin, closely monitor INR when meloxicam is initiated or dose changed; monitor for lithium toxicity, especially during addition, withdrawal, or change in dose of meloxicam.

Patient & Family Education
- Report any of the following to the prescriber immediately: Nausea, black tarry stool, abdominal or stomach pain, unexplained fatigue or lethargy, itching, jaundice, flu-like symptoms, skin rash, weight gain, or edema.
- Discontinue drug if hepatotoxicity or GI bleeding is suspected. Note that GI bleeding may occur without forewarning and is more likely in older adults, in those with a history of ulcers or GI bleeding, and with alcohol consumption and cigarette smoking.
- Do not take aspirin or other NSAIDs while on this medication.

MELPHALAN
(mel'fa-lan)
Alkeran
Classifications: ANTINEOPLASTIC; ALKYLATING AGENT

Therapeutic: ANTINEOPLASTIC
Prototype: Cyclophosphamide
Pregnancy Category: D

AVAILABILITY 2 mg tablets; 50 mg/vial injection

ACTION & *THERAPEUTIC EFFECT*
Forms a highly reactive carbonium ion that causes cross-linking and abnormal base-pairing in DNA, thereby interfering with DNA and RNA replication as well as protein synthesis. *Antineoplastic effects result from its activity against both resting and rapidly dividing tumor cells.*

USES Palliative treatment of multiple myeloma and other neoplasms, including Hodgkin's disease and carcinomas of breast and ovary.
UNLABELED USES Polycythemia vera.

CONTRAINDICATIONS Severe bone marrow suppression; pregnancy (category D), lactation.
CAUTIOUS USE Recent treatment with other chemotherapeutic agents; moderate to severe anemia, neutropenia, or thrombocytopenia; renal or hepatic impairment; men and women of child bearing age; older adults.

ROUTE & DOSAGE

Multiple Myeloma
Adult: **PO** 6 mg/day for 2–3 wk, drug then withdrawn for 4–5 wk, restart at 2 mg/day when WBC and platelet counts start to rise **IV** 16 mg/m² over 15 min q2wk for 4 doses

Epithelial Ovarian Cancer
Adult: **PO** 0.2 mg/kg/day for 5 days as single course, may repeat course q4–5wk

M

Common adverse effects in *italic*, life-threatening effects underlined; generic names in **bold**; classifications in SMALL CAPS; ♣ Canadian drug name; ☻ Prototype drug

ADMINISTRATION

Oral
▪ Give with meals to reduce nausea and vomiting. An antiemetic may be ordered.

Intravenous

PREPARE: **IV Infusion:** Reconstitute melphalan powder by **RAPIDLY** injecting 10 mL of the provided diluent into the vial to yield 5 mg/mL. Shake vigorously until clear. ▪ Immediately dilute further with NS to a concentration of 0.45 mg/mL or less. ▪ Note: 45 mg in 100 mL yields 0.45 mg/mL. ▪ Do not refrigerate reconstituted solution prior to infusion.

ADMINISTER: **IV Infusion:** Give over 15 min or longer. Administration **must be** completed within 60 min of reconstitution of drug because both reconstituted and diluted solutions are unstable. ▪ Ensure patency of IV site prior to infusion.

INCOMPATIBILITIES **Solution/Additive: D5W, lactated Ringer's. Y-site: Amphotericin B, chlorpromazine.**

▪ Store at 15°–30° C (59°–86° F) in light-resistant, airtight containers.

ADVERSE EFFECTS (≥1%) **Hematologic:** <u>Leukopenia, agranulocytosis, thrombocytopenia,</u> anemia, acute nonlymphatic leukemia. **Body as a Whole:** Uremia, angioneurotic peripheral edema. **GI:** Nausea, vomiting, stomatitis. **Skin:** Temporary alopecia. **Respiratory:** Pulmonary fibrosis.

INTERACTIONS **Drug:** Increases risk of nephrotoxicity with **cyclosporine, cimetidine** may decrease efficacy. **Food:** Food decreases absorption.

PHARMACOKINETICS **Absorption:** Incompletely and variably absorbed from GI tract. **Peak:** 2 h. **Distribution:** Widely distributed to all tissues. **Metabolism:** By spontaneous hydrolysis in plasma. **Elimination:** 25–50% in feces; 25–30% in urine. **Half-Life:** 1.5 h.

NURSING IMPLICATIONS

Assessment & Drug Effects
▪ Lab tests: Monitor WBC and platelet counts 2–3 times/wk during dosage adjustment period; determine WBC each week for 6–8 wk during maintenance therapy. Monitor serum uric acid levels.
▪ Monitor laboratory reports to anticipate leukopenic and thrombocytopenic periods.
▪ A degree of myelosuppression is maintained during therapy so as to keep leukocyte count in range of 3000–3500/mm³.
▪ Assess for flank and joint pains that may signal onset of hyperuricemia.

Patient & Family Education
▪ Be alert to onset of fever, profound weakness, chills, tachycardia, cough, sore throat, changes in kidney function, or prolonged infections and report to prescriber.
▪ Understand that reversible hair loss is an expected adverse effect.

MEMANTINE
(me-man'teen)
Namenda
Classifications: N-METHYL-D-ASPARTATE (NMDA) RECEPTOR ANTAGONIST; ANTIDEMENTIA
Therapeutic: ANTIDEMENTIA; ANTI-ALZHEIMER'S
Pregnancy Category: B

AVAILABILITY 5 mg, 10 mg tablets; 10 mg/5 mL solution, 7 mg, 14 mg, 21 mg, 28 mg extended release capsules

ACTION & *THERAPEUTIC EFFECT*
Excess glutamate may play a role in Alzheimer's disease by overstimulating NMDA receptors. Blockade of NMDA receptors may slow intracellular calcium accumulation, preventing nerve damage without interfering with actions of glutamate that are required for memory and learning. *Improves cognitive functioning in moderate to severe Alzheimer's disease (AD) and in mild to moderate vascular dementia.*

USES Treatment of symptoms of moderate to severe Alzheimer's disease.

UNLABELED USES Treatment of moderate to severe vascular dementia, nystagmus.

CONTRAINDICATIONS Known memantine hypersensitivity; renal failure.

CAUTIOUS USE Moderate to severe renal impairment; severe hepatic impairment; history of seizure disorder; older adults; pregnancy (category B), lactation; children.

ROUTE & DOSAGE

Alzheimer's Disease
Adult: **PO** (immediate release) Initiate with 5 mg once daily, increase dose by 5 mg/wk over a 3-wk period to target dose of 10 mg b.i.d. **PO** (extended release) 7 mg daily, increase at weekly intervals to 28 mg daily

Severe Renal Impairment Dosage Adjustment
Decrease to 5 mg b.i.d.

ADMINISTRATION
Oral
- Ensure that extended-release capsule is swallowed whole. It should not be opened or chewed.
- Note: The recommended interval between dose increases is 1 wk.
- Dose reductions should be considered with moderate renal impairment.
- Store between 15°–30° C (59°–86° F).

ADVERSE EFFECTS (≥1%) **Body as a Whole:** Fatigue, pain, flu-like symptoms, peripheral edema. **CNS:** Dizziness, headache, confusion, somnolence, hallucinations, agitation, insomnia, abnormal gait, depression, anxiety, syncope, TIA, vertigo, ataxia, hypokinesia, aggressive reaction. **CV:** Hypertension, cardiac failure. **GI:** Constipation, vomiting, diarrhea, nausea, anorexia. **Hematologic:** Anemia. **Metabolic:** Weight loss, increased alkaline phosphatase. **Musculoskeletal:** Back pain, arthralgia. **Respiratory:** Coughing, dyspnea, bronchitis, upper respiratory infections, pneumonia. **Skin:** Rash. **Special Senses:** Conjunctivitis. **Urogenital:** Urinary incontinence, UTI, frequent micturition.

INTERACTIONS Drug: Drugs that increase the pH of the urine (CARBONIC ANHYDRASE INIBITORS, **sodium bicarbonate**) may increase levels of memantine; may enhance the effects of **amantadine, dextromethorphan, ketamine, bromocriptine, pergolide, pramipexole,** and **ropinirole;** may enhance the adverse effects of **levodopa**-containing drugs.

PHARMACOKINETICS Absorption: 100% from GI tract. **Duration:** 4–6 h. **Distribution:** Easily crosses the blood–brain barrier. **Metabolism:**

M

Common adverse effects in *italic*, life-threatening effects underlined; generic names in **bold**; classifications in SMALL CAPS; ♣ Canadian drug name; ☻ Prototype drug

933

Minimal. **Elimination:** Primarily excreted unchanged in urine. Increases in urinary pH can decrease elimination of drug. **Half-Life:** 60–80 h.

NURSING IMPLICATIONS

Assessment & Drug Effects

- Monitor respiratory and CV status, especially with preexisting heart disease.
- Assess for and report S&S of focal neurologic deficits (e.g., TIA, ataxia, vertigo).
- Lab tests: Periodic Hct and Hgb, serum sodium, alkaline phosphatase, and blood glucose.
- Monitor diabetics for loss of glycemic control.

Patient & Family Education

- Report any of the following to the prescriber: Problems with vision, skin rash, shortness of breath, swelling in throat or tongue, agitation or restlessness, confusion, dizziness, or incontinence.
- Do not drive or engage in other hazardous activities until reaction to drug is known.

MENINGOCOCCAL DIPHTHERIA TOXOID CONJUGATE

(me-nin'joe-kok-al)

Menactra, Menveo
Pregnancy Category: C
See Appendix J.

MEPERIDINE HYDROCHLORIDE

(me-per'i-deen)

Demerol, Pethadol ✦, Pethidine Hydrochloride ✦
Classification: NARCOTIC (OPIATE AGONIST) ANALGESIC
Therapeutic: NARCOTIC ANALGESIC

Prototype: Morphine
Pregnancy Category: B (D at term)
Controlled Substance: Schedule II

AVAILABILITY 50 mg, 100 mg tablets; 50 mg/5 mL syrup; 10 mg/mL, 25 mg/mL, 50 mg/mL, 75 mg/mL, 100 mg/mL injection

ACTION & *THERAPEUTIC EFFECT*
Analgesia is mediated through changes in the perception of pain at the spinal cord (mu_2, delta, kappa receptors) and higher levels in the CNS (mu_1 and $kappa_3$ receptors). *Control of moderate to severe pain. Does not alter pain threshold.*

USES Relief of moderate to severe acute pain, for preoperative medication, for support of anesthesia, and for obstetric analgesia.

CONTRAINDICATIONS Hypersensitivity to meperidine; convulsive disorders; acute abdominal conditions prior to diagnosis; chronic pain; pregnancy (category D at term).
CAUTIOUS USE Head injuries, increased intracranial pressure; asthma and other respiratory conditions; supraventricular tachycardias; prostatic hypertrophy; urethral stricture; glaucoma; older adult or debilitated patients; impaired kidney or liver function, hypothyroidism, Addison's disease; pregnancy (category B).

ROUTE & DOSAGE

Moderate to Severe Pain

NOTE: Should be titrated to pain response
Adult: **PO/Subcutaneous/IM/IV** 50–150 mg q3–4h prn
Child: **PO/Subcutaneous/IM/IV** 1–1.8 mg/kg q3–4h (max: 100 mg q4h) prn

Preoperative

Adult: **IM/Subcutaneous** 50–100 mg 30–90 min before surgery
Child: **IM/Subcutaneous** 1.1–2.2 mg/kg 30–90 min before surgery

Obstetric Analgesia

Adult: **IM/Subcutaneous** 50–100 mg when pains become regular, may be repeated q1–3h

Hepatic/Renal Impairment Dosage Adjustment

Adjust based on patient response

ADMINISTRATION

Oral

- Give syrup formulation in half a glass of water. Undiluted syrup may cause topical anesthesia of mucous membranes.

Subcutaneous and Intramuscular Injections

- Be aware that subcutaneous route is painful and can cause local irritation. IM route is generally preferred when repeated doses are required.
- Aspirate carefully before giving IM injection to avoid inadvertent IV administration. IV injection of undiluted drug can cause a marked increase in heart rate and syncope.

Intravenous

Note: Verify correct IV concentration and rate of infusion/injection for administration to infants or children with prescriber.

PREPARE: Direct: Dilute 50 mg in at least 5 mL of NS or sterile water to yield 10 mg/mL. **IV Infusion:** Dilute to a concentration of 1–10 mg/mL in NS, D5W, or other compatible solution.

ADMINISTER: Direct: Give slowly over 3–5 min at a rate not to exceed 25 mg/min. Slower injection

preferred. **IV Infusion:** Usually given through a controlled infusion device at a rate not to exceed 25 mg/min.

INCOMPATIBILITIES **Solution/additive: Aminophylline,** BARBITURATES, **furosemide, heparin, methicillin, morphine, phenytoin, sodium bicarbonate. Y-site: Allopurinol, amphotericin B cholesteryl complex, cefepime, doxorubicin liposome, furosemide, idarubicin, imipenem/cilastatin, lansoprazole, mezlocillin, minocycline, tetracycline.**

- Store at 15°–30° C (59°–86° F) in tightly closed, light-resistant containers unless otherwise directed by manufacturer.

ADVERSE EFFECTS (≥1%) **Body as a Whole:** Allergic (*Pruritus,* urticaria, skin rashes, wheal and flare over IV site), profuse perspiration. **CNS:** *Dizziness,* weakness, euphoria, dysphoria, *sedation,* headache, uncoordinated muscle movements, disorientation, decreased cough reflex, miosis, corneal anesthesia, <u>respiratory depression</u>. Toxic doses: Muscle twitching, tremors, hyperactive reflexes, excitement, hypersensitivity to external stimuli, agitation, confusion, hallucinations, dilated pupils, <u>convulsions</u>. **CV:** Facial flushing, light-headedness, hypotension, syncope, palpitation, bradycardia, tachycardia, <u>cardiovascular collapse, cardiac arrest (toxic doses)</u>. **GI:** Dry mouth, *nausea,* vomiting, *constipation,* biliary tract spasm. **Urogenital:** Oliguria, urinary retention. **Respiratory:** <u>Respiratory depression in newborn, bronchoconstriction</u> (large doses). **Skin:** Phlebitis (following IV use), pain, tissue irritation and induration, particularly following subcutaneous injection. **Metabolic:**

Common adverse effects in *italic,* life-threatening effects <u>underlined</u>; generic names in **bold;** classifications in SMALL CAPS; ♣ Canadian drug name; ☯ Prototype drug

935

Increased levels of serum amylase, BSP retention, bilirubin, AST, ALT.

DIAGNOSTIC TEST INTERFERENCE
High doses of meperidine may interfere with **gastric emptying studies** by causing delay in gastric emptying.

INTERACTIONS Drug: Alcohol and other CNS DEPRESSANTS, **cimetidine** cause additive sedation and CNS depression; AMPHETAMINES may potentiate CNS stimulation; MAO INHIBITORS, **selegiline** may cause excessive and prolonged CNS depression, convulsions, cardiovascular collapse; **phenytoin** may increase toxic meperidine metabolites. **Herbal: St. John's wort** may increase sedation.

PHARMACOKINETICS Absorption: 50–60% from GI tract. **Onset:** 15 min PO; 10 min IM, Subcutaneous; 5 min IV. **Peak:** 1 h PO, IM, Subcutaneous. **Duration:** 2–4 h PO, IM, Subcutaneous; 2 h IV. **Distribution:** Crosses placenta; distributed into breast milk. **Metabolism:** In liver. **Elimination:** In urine. **Half-Life:** 3–5 h.

NURSING IMPLICATIONS

Assessment & Drug Effects

- Give narcotic analgesics in the smallest effective dose and for the least period of time compatible with patient's needs.
- Assess patient's need for prn medication. Record time of onset, duration, and quality of pain.
- Note respiratory rate, depth, and rhythm and size of pupils in patients receiving repeated doses. If respirations are 12/min or below and pupils are constricted or dilated (see ACTION and USES) or breathing is shallow, or if signs of CNS hyperactivity are present, consult prescriber before administering drug.

- Monitor vital signs closely. Heart rate may increase markedly, and hypotension may occur. Meperidine may cause severe hypotension in postoperative patients and those with depleted blood volume.
- Schedule deep breathing, coughing (unless contraindicated), and changes in position at intervals to help to overcome respiratory depressant effects.
- Chart patient's response to drug and evaluate continued need.
- Repeated use can lead to tolerance as well as psychic and physical dependence of the morphine type.
- Be aware that abrupt discontinuation following repeated use results in morphine-like withdrawal symptoms. Symptoms develop more rapidly (within 3 h, peaking in 8–12 h) and are of shorter duration than with morphine. Nausea, vomiting, diarrhea, and pupillary dilatation are less prominent, but muscle twitching, restlessness, and nervousness are greater than produced by morphine.

Patient & Family Education

- Exercise caution with ambulation and moving from a lying/sitting position to a standing position.
- Be aware nausea, vomiting, dizziness, and faintness associated with fall in BP are more pronounced when walking than when lying down (these symptoms may also occur in patients without pain who are given meperidine). Symptoms are aggravated by the head-up position.
- Do not drive or engage in potentially hazardous activities until any drowsiness and dizziness have passed.
- Do not take other CNS depressants or drink alcohol because of their additive effects.

Common adverse effects in *italic*, life-threatening effects underlined; generic names in **bold**; classifications in SMALL CAPS; ♣ Canadian drug name; ⚈ Prototype drug

MEPHOBARBITAL

(me-foe-bar′bi-tal)

Mebaral

Classifications: ANTICONVULSANT; BARBITURATE; SEDATIVE-HYPNOTIC

Therapeutic: ANTICONVULSANT; SEDATIVE-HYPNOTIC

Prototype: Phenobarbital
Pregnancy Category: D
Controlled Substance: Schedule IV

AVAILABILITY 32 mg, 50 mg, 100 mg tablets

ACTION & *THERAPEUTIC EFFECT*

Long-acting barbiturate that limits the spread of seizure activity by increasing the threshold for motor cortex stimuli. Exerts strong sedative effect and mild hypnotic effect. *Reduces seizure activity by decreasing excitability in nerve cells. Depresses CNS producing drowsiness, hypnosis, and sedation.*

USES To control absence or tonic-clonic seizures, anxiety, sedation induction, and maintenance.

CONTRAINDICATIONS Hypersensitivity to barbiturates; coma; ethanol intoxication, hepatic encephalopathy; porphyria; status epilepticus; pregnancy (category D).

CAUTIOUS USE Fever, hyperthyroidism, alcoholism; respiratory disorders, COPD, sleep apnea; mental status changes, major depression; suicidal ideation; liver, kidney, or cardiac dysfunction; lactation; children.

ROUTE & DOSAGE

Anticonvulsant

Adult: **PO** 200–600 mg/day in divided doses
Child: **PO** 6–12 mg/kg/day in divided doses

Sedative/anxiety

Adult: **PO** 32–100 mg t.i.d. or q.i.d.
Child: **PO** *Younger than 5 y,* 16–32 mg t.i.d. or q.i.d.; *5 y or older,* 32–64 mg t.i.d. or q.i.d.

ADMINISTRATION

Oral

- Change from other anticonvulsant by gradually tapering off the former as mephobarbital doses are increased to maintain seizure control.
- When prescribed concurrently with phenobarbital, dose should be about one-half the amount of each used alone. When prescribed concurrently with phenytoin, the dose of phenytoin is usually reduced.
- Reduce discontinued drug dosage gradually over 4 or 5 days to avoid precipitating seizures of status epilepticus.

ADVERSE EFFECTS (≥1%) **CNS:** *Drowsiness,* dizziness, unsteadiness, hangover, paradoxical excitement. **GI:** Nausea, vomiting, constipation. **Body as a Whole:** Hypersensitivity reactions, respiratory depression.

INTERACTIONS Drug: Alcohol, CNS DEPRESSANTS compound CNS depression; may decrease absorption and increase metabolism of ORAL ANTICOAGULANTS; increases metabolism of CORTICOSTEROIDS, ORAL CONTRACEPTIVES, ANTICONVULSANTS, **digitoxin,** possibly decreasing their effects; ANTIDEPRESSANTS potentiate adverse effects; **griseofulvin** decreases absorption of mephobarbital. **Herbal: Kava, valerian** may potentiate sedation.

PHARMACOKINETICS Absorption: 50% from GI tract. **Onset:** 60 min. **Duration:** 10–12 h. **Metabolism:** In liver to phenobarbital. **Elimination:** In urine. **Half-Life:** 34 h.

M

Common adverse effects in *italic*, life-threatening effects underlined; generic names in **bold;** classifications in SMALL CAPS; ♣ Canadian drug name; ⊙ Prototype drug

937

NURSING IMPLICATIONS

Assessment & Drug Effects

- Monitor respiratory status, especially with concurrent CNS therapy with other drugs.
- Be prepared for paradoxical response to barbiturate therapy (i.e., irritability, marked excitement, aggression in children, depression, confusion) in older adults, debilitated patients, or children.

Patient & Family Education

- Be aware that abrupt cessation after prolonged therapy may result in withdrawal symptoms (tremulousness, weakness, insomnia, delirium, convulsions).
- Avoid driving and potentially hazardous activities until response to drug has stabilized.
- Do not take alcohol in any amount with a barbiturate.

MEPROBAMATE ℗
(me-proe-ba′mate)

Classifications: CARBAMATE; ANXIOLYTIC; SEDATIVE-HYPNOTIC
Therapeutic: ANTIANXIETY; SEDATIVE-HYPNOTIC
Pregnancy Category: D
Controlled Substance: Schedule IV

AVAILABILITY 200 mg, 400 mg tablets

ACTION & *THERAPEUTIC EFFECT* Carbamate derivative and CNS depressant. Acts on multiple sites in CNS and appears to block corticothalamic impulses. *Antianxiety agent. Hypnotic doses suppress REM sleep.*

USES To relieve anxiety and tension of psychoneurotic states and as adjunct in disease states associated with anxiety and tension. Also used to promote sleep in anxious, tense patients.

CONTRAINDICATIONS History of hypersensitivity to meprobamate or related carbamates; history of acute intermittent porphyria; pregnancy (category D), lactation.

CAUTIOUS USE Impaired kidney or liver function; convulsive disorders; history of alcoholism or drug abuse; patients with suicidal tendencies; children younger than 6 y.

ROUTE & DOSAGE

Sedative
Adult: **PO** 1.2–1.6 g/day in 3–4 divided doses (max: 2.4 g/day) *Child (6 y or older):* **PO** 100–200 mg b.i.d. or t.i.d.
Hypnotic
Adult: **PO** 400–800 mg *Geriatric:* **PO** 200 mg *Child (6 y or older):* **PO** 200 mg

ADMINISTRATION

Oral

- Give with food to minimize gastric distress.
- Treat physical dependence by gradual drug withdrawal over 1–2 wk to prevent onset of withdrawal symptoms.
- Store at 15°–30° C (59°–86° F) unless otherwise specified by manufacturer.

ADVERSE EFFECTS (≥1%) **Body as a Whole:** Allergy or idiosyncratic reactions (itchy, urticarial, or erythematous maculopapular rash; <u>exfoliative dermatitis</u>, petechiae, purpura, ecchymoses, eosinophilia, peripheral edema, angioneurotic edema, adenopathy, fever, chills, proctitis, bronchospasm, oliguria, anuria, <u>Stevens-Johnson syndrome</u>); <u>anaphylaxis</u>. **CNS:** *Drowsiness* and *ataxia,* dizziness, vertigo, slurred speech, headache,

M

weakness, paresthesias, impaired visual accommodation, paradoxic euphoria and rage reactions, seizures in epileptics, panic reaction, rapid EEG activity. **CV:** Hypotensive crisis, syncope, palpitation, tachycardia, arrhythmias, transient ECG changes, circulatory collapse (toxic doses). **GI:** Anorexia, nausea, vomiting, diarrhea. **Hematologic:** Aplastic anemia (rare): Leukopenia, agranulocytosis, thrombocytopenia, exacerbation of acute intermittent porphyria. **Respiratory:** Respiratory depression.

DIAGNOSTIC TEST INTERFERENCE
Meprobamate may cause falsely high *urinary steroid* determinations. *Phentolamine* tests may be falsely positive; meprobamate should be withdrawn at least 24 h and preferably 48–72 h before the test.

INTERACTIONS Drug: **Alcohol, entacapone,** TRICYCLIC ANTIDEPRESSANTS, ANTIPSYCHOTICS, OPIATES, SEDATING ANTIHISTAMINES, **pentazocine, tramadol,** MAOIS, SEDATIVE-HYPNOTICS, ANXIOLYTICS may potentiate CNS depression. **Herbal: Kava, valerian** may potentiate sedation.

PHARMACOKINETICS Absorption:
Well absorbed from GI tract. **Peak:** 1–3 h. **Onset:** 1 h. **Distribution:** Uniformly throughout body; crosses placenta. **Metabolism:** Rapidly in liver. **Elimination:** Renally excreted; excreted in breast milk. **Half-Life:** 10–11 h.

NURSING IMPLICATIONS
Assessment & Drug Effects
- Supervise ambulation, if necessary. Older adults and debilitated patients are prone to oversedation and to the hypotensive effects, especially during early therapy.
- Utilize safety precautions for hospitalized patients. Hypnotic doses

may cause increased motor activity during sleep.
- Consult prescriber if daytime psychomotor function is impaired. A change in regimen or drug may be indicated.
- Withdraw gradually in physically dependent patients to prevent preexisting symptoms or withdrawal reactions within 12–48 h: Vomiting, ataxia, muscle twitching, mental confusion, hallucinations, convulsions, trembling, sleep disturbances, increased dreaming, nightmares, insomnia. Symptoms usually subside within 12–48 h.

Patient & Family Education
- Take drug as prescribed. Psychic or physical dependence may occur with long-term use of high doses.
- Be aware that tolerance to alcohol will be lowered.
- Make position changes slowly, especially from lying down to upright; dangle legs for a few minutes before standing.
- Avoid driving or engaging in hazardous activities until response to drug is known.
- Report immediately onset of skin rash, sore throat, fever, bruising, unexplained bleeding.

MEQUINOL/TRETINOIN
(me-qui'nol/tre-ti'noyn)
Solagé
Classification: RETINOID
Therapeutic: DEPIGMENTING AGENT; RETINOID
Prototype: Isotretinoin
Pregnancy Category: X

AVAILABILITY 2%/0.01% solution

ACTION & THERAPEUTIC EFFECT
Mequinol is a depigmenting agent and tretinoin is a retinoid used to

Common adverse effects in *italic*, life-threatening effects underlined; generic names in **bold**; classifications in SMALL CAPS; ❖ Canadian drug name; ⊙ Prototype drug

939

M

improve dermatologic changes (e.g., fine wrinkling, mottled hyperpigmentation, roughness) associated with photo-damage and aging. Mequinol's mechanism of depigmentation is probably due to oxidation by tyrosine to cytotoxic products in melanocytes, and/or inhibition of melanin formation. Tretinoin, a retinoid, is used to improve photo-damage to the skin by acting via retinoic acid receptors (RARs). *Mequinol has depigmenting properties; tretinoin improves sun-damaged skin.*

USES Treatment of solar lentigines (age spots).
UNLABELED USES Facial wrinkles.

CONTRAINDICATIONS Hypersensitivity to mequinol or tretinoin; pregnancy (category X), lactation.
CAUTIOUS USE History of hypersensitivity to acitretin, isotretinoin, etretinate, or other vitamin A derivatives, or hydroquinone; patients with eczema, moderate to severe skin pigmentation, vitiligo; cold weather; children.

ROUTE & DOSAGE

Solar Lentigines
Adult: **Topical** Apply to solar lentigines b.i.d. at least 8 h apart

ADMINISTRATION

Topical
- Apply doses at least 8 h apart; avoid application to unaffected areas.
- Avoid contact with eyes, lips, mucous membranes, or paranasal creases.
- Protect from light.

ADVERSE EFFECTS (≥1%) **Skin:** *Erythema, burning, stinging, tingling, desquamation, pruritus,* skin irritation, temporary hypopig-

mentation, rash, dry skin, crusting, application site reaction.

INTERACTIONS Drug: THIAZIDE DIURETICS, TETRACYCLINES, FLUOROQUINOLONES, PHENOTHIAZINES, SULFONAMIDES may augment phototoxicity.

PHARMACOKINETICS Absorption: 4.4% through skin. **Peak:** 1–2 h.

NURSING IMPLICATIONS

Assessment & Drug Effects
- Monitor for and report peeling, erythema, or hypopigmentation.
- Monitor for signs of tretinoin toxicity: Headache, fever, weakness, and fatigue.

Patient & Family Education
- Do not apply larger than recommended amounts.
- Do not wash affected area for at least 6 h after drug application; do not apply cosmetics to affected area for at least 30 min after drug application.
- Minimize exposure to sunlight or sunlamps. Use extra caution if also taking concurrently other drugs that are photosensitizing (e.g., thiazide diuretics, phenothiazines).
- Notify prescriber if vitiligo (hypopigmentation of skin) or S&S of tretinoin toxicity develop (see ASSESSMENT & DRUG EFFECTS).

MERCAPTOPURINE (6-MP, 6-MERCAPTOPURINE) ℗

(mer-kap-toe-pyoor′een)
Purinethol
Classifications: ANTINEOPLASTIC; ANTIMETABOLITE, PURINE ANTAGONIST
Therapeutic: ANTINEOPLASTIC; IMMUNOSUPPRESSANT
Pregnancy Category: D

AVAILABILITY 50 mg tablets

ACTION & *THERAPEUTIC EFFECT*
Antimetabolite and purine antagonist that inhibits purine metabolism. Blocks conversion of inosinic acid to adenine and xanthine ribotides within sensitive tumor cells. Also inhibits adenine-containing coenzymes, suggesting an influence over multiple cellular reactions. *Has delayed immunosuppressive properties and carcinogenic potential.*

USES Primarily for acute lymphocytic and myelogenous leukemia. Response in adults is less than in children, but mercaptopurine is initial drug of choice. In chronic granulocytic leukemia, produces temporary remission.
UNLABELED USES Prevention of transplant graft rejection; SLE; rheumatoid arthritis; Crohn's disease.

CONTRAINDICATIONS Prior resistance to mercaptopurine; infections; pregnancy (category D); lactation.
CAUTIOUS USE Impaired kidney or liver function.

ROUTE & DOSAGE

Leukemias
Adult/Child: **PO Loading Dose**
2.5 mg/kg/day, may increase up to 5 mg/kg/day after 4 wk if needed **PO Maintenance Dose** 1.25–2.5 mg/kg/day

ADMINISTRATION
Oral
- Give total daily dose at one time.
- Reduce dose of mercaptopurine usually by 1/3–1/4 when given concurrently with allopurinol.
- Store tablets in light- and air-resistant container.

ADVERSE EFFECTS (≥1%) **GI:** Stomatitis, esophagitis, anorexia, nausea, vomiting, diarrhea, intestinal ulcerations, impaired liver function, hepatic necrosis. **Hematologic:** Leukopenia, anemia, eosinophilia, pancytopenia, thrombocytopenia, abnormal bleeding, bone marrow hypoplasia. **Urogenital:** Hyperuricemia, oliguria, renal impairment. **Skin:** Rash. **Body as a Whole:** Drug fever.

INTERACTIONS Drug: Allopurinol may inhibit metabolism and thus increase toxicity of mercaptopurine; may potentiate or antagonize anticoagulant effects of **warfarin.**

PHARMACOKINETICS Absorption: Approximately 50% absorbed from GI tract. **Peak:** 2 h. **Distribution:** Distributes into total body water. **Metabolism:** Rapidly by xanthine oxidase in liver. **Elimination:** 11% in urine within 6 h. **Half-Life:** 20–50 min.

NURSING IMPLICATIONS
Assessment & Drug Effects
- Lab tests: Monitor closely CBC with differential, platelet count, Hgb, Hct, and LFTs.
- Monitor for S&S of liver damage. Hepatic toxicity occurs most often when dose exceeds 2.5 mg/kg/day. Jaundice signals onset of hepatic toxicity and may necessitate terminating use.
- Withhold drug and notify prescriber at the first sign of an abnormally large or rapid fall in platelet and leukocyte counts.
- Record baseline data related to I&O ratio and pattern and body weight.
- Check vital signs daily. Report febrile states promptly.
- Protect patient from exposure to trauma, infections, or other stresses

Common adverse effects in *italic*, life-threatening effects underlined; generic names in **bold**; classifications in SMALL CAPS; ♣ Canadian drug name; ⊙ Prototype drug

941

(restrict visitors and personnel who have colds) during periods of leukopenia.
- Report nausea, vomiting, or diarrhea. These may signal excessive dosage, especially in adults.
- Watch for signs of abnormal bleeding (ecchymoses, petechiae, melena, bleeding gums) if thrombocytopenia develops; report immediately.

Patient & Family Education
- Report any signs of bleeding (e.g., hematuria, bruising, bleeding gums).
- Report signs of hepatic toxicity (see Appendix F).
- Increase hydration (10–12 glasses of fluid daily) to reduce risk of hyperuricemia. Consult prescriber about desirable volume.
- Notify prescriber of onset of chills, nausea, vomiting, flank or joint pain, swelling of legs or feet, or symptoms of anemia.

MEROPENEM
(mer-o'pe-nem)

Merrem
Classification: CARBAPENEM ANTIBIOTIC
Therapeutic: ANTIBIOTIC
Prototype: Imipenem
Pregnancy Category: B

AVAILABILITY 500 mg, 1 g injection

ACTION & *THERAPEUTIC EFFECT*
Broad-spectrum carbapenem antibiotic that inhibits cell wall synthesis of bacteria by its strong affinity for penicillin-binding proteins of bacterial cell wall. *Effective against both gram-positive and gram-negative bacteria.*

USES Complicated appendicitis and peritonitis, bacterial meningitis caused by susceptible bacteria, complicated skin infections, intra-abdominal infections, skin/soft tissue infections.

UNLABELED USES Febrile neutropenia.

CONTRAINDICATIONS Hypersensitivity to meropenem, other carbapenem antibiotics or history of anaphylactic reactions to beta-lactams.

CAUTIOUS USE History of asthma or allergies, renal impairment, renal disease; epileptics, history of neurologic disorders, older adults, pregnancy (category B), lacatation; children younger than 3 mo.

ROUTE & DOSAGE

Intra-Abdominal Infections

Adult/Child (weight greater than 50 kg): **IV** 1 g q8h
Child (3 mo or older, weight less than 50 kg): **IV** 20 mg/kg q8h (max: 1 g q8h)

Bacterial Meningitis

Adult/Child (weight greater than 50 kg): **IV** 2 g q8h
Child (3 mo or older, weight less than 50 kg): **IV** 40 mg/kg q8h (max: 2 g q8h)

Complicated Skin Infection

Adult/Child (weight greater than 50 kg): **IV** 500 mg q8h
Child (older than 3 mo, weight less than 50 kg): **IV** 10 mg/kg q8h (max: 500 mg q8h)

Renal Impairment Dosage Adjustment

CrCl 26–50 mL/min: q12h;
10–25 mL/min: ½ dose q12h; *less than 10 mL/min:* ½ dose q24h

ADMINISTRATION

Intravenous
Note: Dosage reduction is recommended for older adults.

M

PREPARE: **Direct:** Reconstitute the 500-mg or 1-g vial, respectively, by adding 10 or 20 mL sterile water for injection to yield approximately 50 mg/mL. ▪ Shake to dissolve and let stand until clear. **IV Infusion:** Further dilute reconstituted solution in 50–250 mL of D5W, NS, or D5/NS.

ADMINISTER: **Direct:** Give doses of 5–20 mL over 3–5 min. **IV Infusion:** Give over 15–30 min.

INCOMPATIBILITIES **Solution/additive: D5W, lactated Ringer's, mannitol, amphotericin B, metronidazole, multivitamins, sodium bicarbonate. Y-site: Amphotericin B, diazepam, doxycycline, metronidazole, ondansetron, zidovudine.**

▪ Store undiluted at 15°–30° C (59°–86° F), diluted IV solutions should generally be used within 1 h of preparation.

ADVERSE EFFECTS (≥1%) **GI:** Diarrhea, nausea, vomiting, constipation. **Other:** Inflammation at injection site, phlebitis, thrombophlebitis. **CNS:** Headache. **Skin:** Rash, pruritus, diaper rash. **Body as a Whole:** Apnea, oral moniliasis, sepsis, shock. **Hematologic:** Anemia.

INTERACTIONS Drug: Probenecid delays meropenem excretion; may decrease **valproic acid** serum levels.

PHARMACOKINETICS Distribution: Attains high concentrations in bile, bronchial secretions, cerebrospinal fluid. **Metabolism:** Renal and extrarenal metabolism via dipeptidases or nonspecific degradation. **Elimination:** In urine. **Half-Life:** 0.8–1 h.

NURSING IMPLICATIONS

Assessment & Drug Effects

▪ Lab tests: Perform C&S tests prior to therapy. Monitor periodically LFTs and kidney function.

▪ Determine history of hypersensitivity reactions to other beta-lactams, cephalosporins, penicillins, or other drugs.

▪ Discontinue drug and immediately report S&S of hypersensitivity (see Appendix F).

▪ Report S&S of superinfection or pseudomembranous colitis (see Appendix F).

▪ Monitor for seizures especially in older adults and those with renal insufficiency.

Patient & Family Education

▪ Learn S&S of hypersensitivity, superinfection, and pseudomembranous colitis; report any of these to prescriber promptly.

MESALAMINE ℗

(me-sal'a-meen)

Asacol, Canasa, Lialda, Pentasa, Rowasa, Salofalk ♦

Classifications: ANTI-INFLAMMATORY; PROSTAGLANDIN INHIBITOR

Therapeutic: GI; ANTI-INFLAMMATORY

Pregnancy Category: B

AVAILABILITY 250 mg controlled release capsule (Pentasa); 400 mg delayed release tablet (Asacol); 1.2 g delayed release tablet (Lialda); 500 mg suppository, 4 g/60 mL rectal suspension (Rowasa); 500 mg suppositories (Canasa)

ACTION & *THERAPEUTIC EFFECT* Thought to diminish inflammation by blocking cyclooxygenase and inhibiting prostaglandin synthesis in the colon. *Provides topical anti-inflammatory action in the colon of patients with ulcerative colitis.*

USES Indicated in active mild to moderate distal ulcerative colitis,

Common adverse effects in *italic*, life-threatening effects <u>underlined</u>; generic names in **bold**; classifications in SMALL CAPS; ♦ Canadian drug name; ℗ Prototype drug

943

proctosigmoiditis, or proctitis; maintenance of remission of ulcerative colitis.
UNLABELED USES Crohn's disease.

CONTRAINDICATIONS Hypersensitivity to mesalamine.
CAUTIOUS USE Sulfite hypersensitivity; sensitivity to sulfasalazine; renal disease, renal impairment; pregnancy (category B), lactation.

ROUTE & DOSAGE

Ulcerative Colitis

Adult: **Rectal (Rowasa)** 4 g once/day at bedtime, enema should be retained for about 8 h if possible or 1 suppository (500 mg) b.i.d.; **(Canasa)** 500 mg b.i.d., may increase up to 500 mg t.i.d. **PO (Asacol)** 800 mg t.i.d. × 6 wk; **(Pentasa)** 500 mg t.i.d. × 6 wk; **(Lialda)** 2.4 g daily or 4.8 mg daily **Maintenance Dose (Asacol)** 800 mg b.i.d. or 400 mg q.i.d.
Child: **PO** 50 mg/kg/day divided q6–12h

ADMINISTRATION

Oral

- Ensure that controlled-release and enteric forms of the drug are not crushed or chewed.
- Shake the bottle well to make sure the suspension is mixed.

Rectal

- Use rectal suspension at bedtime with the objective of retaining it all night.
- Store at 15°–30° C (59°–86° F) away from heat and light.

ADVERSE EFFECTS (≥1%) **CNS:** *Headache,* fatigue, asthenia, malaise, weakness, dizziness. **GI:** *Abdominal pain, cramps,* or *discomfort,* flatulence, nausea, diarrhea, constipation, hemorrhoids, rectal pain, hepatitis (rare). **Skin:** Sensitivity reactions, rash, pruritus, alopecia. **Body as a Whole:** Fever. **Hematologic:** <u>Thrombocytopenia</u> (rare), eosinophilia. **Urogenital:** Interstitial nephritis.

INTERACTIONS Drug: May decrease the absorption of **digoxin.**

PHARMACOKINETICS Absorption: Rectal 5–35% absorbed from colon depending on retention time of enema or suppository. PO Asacol, approximately 28% absorbed; 80% of drug is released in colon 12 h after ingestion. PO Pentasa, 50% of drug is released in colon at a pH less than 6. **Peak:** 3–6 h. **Distribution:** Rectal administration may reach as high as the ascending colon. Asacol is released in the ileum and colon; Pentasa is released in the jejunum, ileum, and colon. Low concentrations of mesalamine and higher concentrations of its metabolites are excreted in breast milk. **Metabolism:** Rapidly acetylated in the liver and colon wall. **Elimination:** Primarily in feces; absorbed drug excreted in urine. **Half-Life:** 2–15 h (depending on formulation).

NURSING IMPLICATIONS

Assessment & Drug Effects

- Lab tests: Monitor carefully urinalysis, BUN, and creatinine, especially in patients with preexisting kidney disease. The kidney is the major target organ for toxicity.
- Assess for S&S of allergic-type reactions (e.g., hives, itching, wheezing, anaphylaxis). Suspension contains a sulfite that may cause reactions in asthmatics and some nonasthmatic persons.
- Expect response to therapy within 3–21 days; however, the usual course of therapy is from 3–6 wk depending on symptoms and sigmoidoscopic examinations.

Patient & Family Education

- Report to prescriber promptly: Cramping, abdominal pain, bloody diarrhea, or other signs of rectal irritation.
- Check with prescriber before using any new medicine (prescription or OTC).
- Continue medication for full time of treatment even if you are feeling better.

MESNA
(mes'na)

Mesnex
Classifications: CHEMOPROTECTANT; DETOXIFYING AGENT
Therapeutic: DETOXIFYING AGENT
Pregnancy Category: B

AVAILABILITY 100 mg/mL injection; 400 mg tablet

ACTION & *THERAPEUTIC EFFECT*
Detoxifying agent used to inhibit hemorrhagic cystitis induced by ifosfamide. *Reacts chemically with urotoxic ifosfamide metabolites, resulting in their detoxification, and thus significantly decreases the incidence of hematuria.*

USES Prophylaxis for ifosfamide-induced hemorrhagic cystitis. Not effective in preventing hematuria due to other pathologic conditions such as thrombocytopenia.
UNLABELED USES Reduces the incidence of cyclophosphamide-induced hemorrhagic cystitis.

CONTRAINDICATIONS Hypersensitivity to mesna or other thiol compounds.
CAUTIOUS USE Autoimmune diseases; infants (injection); pregnancy (category B), lactation; neonates.

ROUTE & DOSAGE

Ifosfamide-Induced Hemorrhagic Cystitis
Adult: **IV** Dose = 20% of ifosfamide dose given 15 min before ifosfamide administration and 4 and 8 h after ifosfamide dose **PO** 40% of ifosfamide dose 2 and 6 h after each ifosfamide dose

ADMINISTRATION

- Note: To be effective, mesna **must be** administered with each dose of ifosfamide.

Intravenous

PREPARE: Direct: Add 4 mL of D5W, NS, or LR for each 100 mg of mesna to yield 20 mg/mL.
ADMINISTER: Direct: Give a single dose by direct IV over 60 sec.
INCOMPATIBILITIES Solution/additive: Carboplatin, cisplatin, ifosfamide with epirubicin. Y-site: Amphotericin B cholesteryl complex, lansoprazole.
- Inspect parenteral drug products visually for particulate matter and discoloration prior to administration. ▪ Discard any unused portion of the ampul because drug oxidizes on contact with air.

- Refrigerate diluted solutions or use within 6 h of mixing even though diluted solutions are chemically and physically stable for 24 h at 25° C (77° F). ▪ Store unopened ampul at 15°–30° C (59°–86° F) unless otherwise specified.

ADVERSE EFFECTS (≥1%) **GI:** *Bad taste in mouth, soft stools,* nausea, vomiting.

DIAGNOSTIC TEST INTERFERENCE
May produce a false-positive result in test for ***urinary ketones.***

M

Common adverse effects in *italic*, life-threatening effects underlined; generic names in **bold;** classifications in SMALL CAPS; ♣ Canadian drug name; ❶ Prototype drug

945

INTERACTIONS Drug: May decrease the effect of **warfarin.**

PHARMACOKINETICS Bioavailability: 45%–79% **Metabolism:** Rapidly oxidized in liver to active metabolite dimesna; dimesna is further metabolized in kidney. **Elimination:** 65% in urine within 24 h. **Half-Life:** Mesna 0.36 h, dimesna 1.17 h.

NURSING IMPLICATIONS

Assessment & Drug Effects

- Monitor urine for hematuria.
- About 6% of patients treated with mesna along with ifosfamide still develop hematuria.

Patient & Family Education

- Mesna prevents ifosfamide-induced hemorrhagic cystitis; it will not prevent or alleviate other adverse reactions or toxicities associated with ifosfamide therapy.
- Report any unusual or allergic reactions to prescriber.
- Check with prescriber before using any new prescription or OTC medicine.

METAPROTERENOL SULFATE

(met-a-proe-ter'e-nole)

Classifications: BETA-ADRENERGIC AGONIST; BRONCHODILATOR
Therapeutic: BRONCHODILATOR
Prototype: Albuterol
Pregnancy Category: C

AVAILABILITY 10 mg, 20 mg tablets; 10 mg/5 mL syrup; 75 mg, 150 mg metered dose inhaler; 0.4%, 0.6%, 5% solution for inhalation

ACTION & *THERAPEUTIC EFFECT*
Potent synthetic beta-adrenergic agonist that acts selectively on beta$_2$-adrenergic receptors to relax smooth muscle of bronchi, uterus, and blood vessels supplying skeletal muscles. *Effective as a bronchodilator; additionally, it controls bronchospasm in asthmatics.*

USES Bronchodilator in symptomatic relief of asthma and reversible bronchospasm associated with bronchitis and emphysema.
UNLABELED USES Treatment and prophylaxis of heart block and to avert progress of premature labor (tocolytic action).

CONTRAINDICATIONS Sensitivity to metaproterenol or other sympathomimetic agents; seizure disorders; diabetes mellitus; hyperthyroidism.
CAUTIOUS USE Older adults; hypertension, cardiovascular disorders including coronary artery disease, cardiac arrhythmias, QT prolongation; MAOI therapy; pregnancy (category C), lactation. Safety in children younger than 12 y **(aerosol)** and children younger than 6 y **(tablets)** is not established.

ROUTE & DOSAGE

Bronchospasm
Adult: **PO** 20 mg q6–8h **Metered Dose Inhaler** 2–3 inhalations q3–4h (max: 12 inhalations/day) **Nebulizer** 5–10 inhalations of undiluted 5% solution **IPPB** 2.5 mL of 0.4–0.6% solution q4–6h
Geriatric: **PO** 10 mg 3–4 times/day, may increase to 20 mg 3–4 times/day
Child: **PO** *Younger than 2 y,* 0.4 mg/kg t.i.d.–q.i.d.; *2–6 y,* 1.2–2.6 mg/kg/day in 3–4 divided doses; *6–9 y,* 10 mg q6–8h; *older than 9 y,* 20 mg q6–8h

ADMINISTRATION

- Note: Patient may use tablets and aerosol concomitantly.

M

Oral
- Give with food to reduce GI distress.

Inhalation
- Instruct patient to shake metered dose aerosol container, exhale through nose as completely as possible, administer aerosol while inhaling deeply through mouth, and to hold breath about 10 sec before exhaling slowly. Administer second inhalation 10 min after first.
- Store all forms at 15°–30° C (59°–86° F); protect from light and heat.

ADVERSE EFFECTS (≥1%) **CNS:** Nervousness, weakness, drowsiness, *tremor (particularly after PO administration),* headache, fatigue. **CV:** *Tachycardia,* hypertension, cardiac arrest, palpitation. **GI:** Nausea, vomiting, bad taste. **Urogenital:** Occasional difficulty in micturition and muscle cramps. **Respiratory:** Throat irritation, cough, exacerbation of asthma.

INTERACTIONS Drug: Epinephrine, other SYMPATHOMIMETIC BRONCHODILATORS may compound effects of metaproterenol; MAO INHIBITORS, TRICYCLIC ANTIDEPRESSANTS potentiate action of metaproterenol on vascular system; the effects of both metaproterenol and BETA-ADRENERGIC BLOCKERS are antagonized.

PHARMACOKINETICS Absorption: 40% of PO doses reach systemic circulation. **Onset:** Inhaled: 1 min; PO 15 min. **Peak:** 1 h all routes. **Duration:** Inhaled: 1–5 h; PO 4 h. **Metabolism:** In liver. **Elimination:** In urine.

NURSING IMPLICATIONS
Assessment & Drug Effects
- Monitor respiratory status. Auscultate lungs before and after inhalation to determine efficacy of drug in decreasing airway resistance.
- Monitor cardiac status. Report tachycardia and hypotension.

Patient & Family Education
- Report failure to respond to usual dose. Drug may have shorter duration of action after long-term use.
- Do not increase dose or frequency unless ordered by prescriber; there is the possibility of serious adverse effects.
- Anticipate tremor as a possible adverse effect.

METFORMIN 🅿
(met-for'min)
Fortamet, Glucophage, Glucophage XR, Glumetza, Riomet
Classifications: ANTIDIABETIC; BIGUANIDE
Therapeutic: ANTIHYPERGLYCEMIC
Pregnancy Category: B

AVAILABILITY 500 mg, 850 mg, 1000 mg tablets; 500 mg, 750 mg, 1000 mg sustained release tablets; 100 mg/mL oral solution

ACTION & *THERAPEUTIC EFFECT*
Biguanide oral hypoglycemic agent thought to both increase the binding of insulin to its receptors and potentiate insulin action. Improves tissue sensitivity to insulin, increases glucose transport into skeletal muscles and fat, and suppresses gluconeogenesis and hepatic production of glucose. *Effective in lowering serum glucose level and, ultimately, the HbA1C value.*

USES Treatment of type 2 diabetes mellitus as adjunct to diet and exercise.

CONTRAINDICATIONS Hypersensitivity to metformin; hepatic or cardiopulmonary insufficiency;

Common adverse effects in *italic*, life-threatening effects underlined; generic names in **bold**; classifications in SMALL CAPS; ♣ Canadian drug name; 🅿 Prototype drug

947

alcoholism; concurrent infection; acute MI, cardiogenic shock; diabetic ketoacidosis; hypoxemia, lactic acidosis; radiographic contrast administration; renal disease, renal failure, renal impairment; sepsis; surgery.

CAUTIOUS USE Previous hypersensitivity to phenformin or buformin; anemia; coma; dehydration, diarrhea; ethanol intoxication; fever; gastroparesis, GI obstruction; heart failure; hyperthyroidism, pituitary insufficiency; polycystic ovary syndrome; trauma, emesis; older adults; pregnancy (category B), lactation; children younger than 10 y.

ROUTE & DOSAGE

Type 2 Diabetes Mellitus

Adult: **PO** Start with 500 mg daily to t.i.d. or 850 mg daily to b.i.d. with meals, may increase by 500–850 mg/day q1–3wk (max: 2550 mg/day); or start with 500 mg sustained release with p.m. meal, may increase by 500 mg/day at p.m. meal qwk (max: 2000 mg/day)
Adolescent/Child (older than 10 y): **PO** Glucophage only; 500 mg b.i.d., may increase by 500 mg/day qwk (max: 2000 mg/day)

ADMINISTRATION

Oral

- Ensure that extended release tablets are not crushed or chewed. They **must be** swallowed whole.
- Use a calibrated oral syringe or container to measure the oral solution for accurate dosing.
- Give with or shortly after main meals.
- Withhold metformin 48 h before and 48 h after receiving IV contrast dye.

- Dose increments are usually made at 2- to 3-wk intervals.
- Store at 15°–30° C (59°–86° F).

ADVERSE EFFECTS (≥1%) **CNS:** Headache, dizziness, agitation, fatigue. **Metabolic:** Lactic acidosis. **GI:** *Nausea, vomiting, abdominal pain, bitter or metallic taste, diarrhea, bloatedness, anorexia;* malabsorption of amino acids, vitamin B_{12}, and folic acid possible.

INTERACTIONS Drug: **Captopril, furosemide, nifedipine** may increase risk of hypoglycemia. **Cimetidine** reduces clearance of metformin. Concomitant therapy with AZOLE ANTIFUNGAL AGENTS (**fluconazole, ketoconazole, itraconazole**) and ORAL HYPOGLYCEMIC DRUGS has been reported in severe hypoglycemia. IODINATED RADIOCONTRAST DYES can cause lactic acidosis and acute kidney failure. **Amiloride, cimetidine digoxin, dofetilide, midodrine, morphine, procainamide, quinidine, quinine, ranitidine, triamterene, trimethoprim,** or **vancomycin** may decrease metformin elimination by competing for common renal tubular transport systems. **Acarbose** may decrease metformin levels. **Iodinated contrast dyes** may cause lactic acidosis or acute kidney failure. **Herbal:** Garlic, ginseng, glucomannan may increase hypoglycemic effects. **Guar gum** decreases absorption.

PHARMACOKINETICS Absorption: 50–60% of dose reaches systemic circulation. **Peak:** 1–3 h. **Distribution:** Not bound to plasma proteins. **Metabolism:** Not metabolized. **Elimination:** In urine. **Half-Life:** 6.2–17.6 h.

NURSING IMPLICATIONS

Assessment & Drug Effects

- Monitor vital signs and fasting and postprandial blood glucose values.

- Report promptly any of the following signs of lactic acidosis: Malaise, myalgia, somnolence, respiratory depression, abdominal distress.
- Lab tests: Monitor baseline and periodic LFTs and kidney function tests; drug contraindicated in the presence of renal or hepatic insufficiency. Monitor blood glucose and HbA1C, and lipid profile periodically.
- Monitor known or suspected alcoholics carefully for decreased liver function.
- Monitor cardiopulmonary status throughout course of therapy; cardiopulmonary insufficiency may predispose to lactic acidosis.

Patient & Family Education

- Be aware that hypoglycemia is not a risk when drug is taken in recommended therapeutic doses unless combined with other drugs which lower blood glucose.
- Report to prescriber immediately S&S of infection, which increase the risk of lactic acidosis (e.g., abdominal pains, nausea, and vomiting, anorexia).
- Report promptly severe vomiting, diarrhea, fever, or any illness that causes limited fluid intake.
- Avoid drinking alcohol while taking this drug.

METHADONE HYDROCHLORIDE

(meth'a-done)

Dolophine, Methadose

Classifications: NARCOTIC (OPIATE AGONIST); ANALGESIC

Therapeutic: NARCOTIC ANALGESIC; TOXICOLOGY AGENT

Prototype: Morphine

Pregnancy Category: C

Controlled Substance: Schedule II

AVAILABILITY 5 mg, 10 mg, 40 mg tablets; 1 mg/mL, 2 mg/mL, 10 mg/mL oral solution; 10 mg/mL injection

ACTION & *THERAPEUTIC EFFECT*
Synthetic narcotic that is a CNS depressant, which causes sedation and respiratory depression. Highly addictive, with abuse potential; abstinence syndrome develops more slowly, and withdrawal symptoms are less intense but more prolonged. *Relieves severe pain and manages withdrawal therapy from narcotics, especially heroin.*

USES To relieve severe pain; for detoxification and temporary maintenance treatment in hospital and in federally controlled maintenance programs for ambulatory patients with narcotic abstinence syndrome.

CONTRAINDICATIONS Severe pulmonary disease; COPD; obstetric analgesia; lactation.

CAUTIOUS USE History of QT prolongation; liver, kidney, or cardiac dysfunction; pregnancy (category C).

ROUTE & DOSAGE

Pain

Adult: **PO/Subcutaneous/IM** 2.5–10 mg q3–4h prn **IV** 2.5–10 mg q8–12h prn (opiate naïve patient)

Child: **PO/IV/Subcutaneous/IM** 0.1–0.2 mg/kg q4h × 2–3 doses, then q6–12h prn (max: 5–10 mg/dose)

Detoxification Treatment

Adult: **PO/Subcutaneous/IM** 15–40 mg once/day, usually maintained at 20–120 mg/day

M

Common adverse effects in *italic*, life-threatening effects underlined; generic names in **bold**; classifications in SMALL CAPS; ♣ Canadian drug name; ✪ Prototype drug

949

Renal Impairment Dosage Adjustment

CrCl less than 10 mL/min: Use 50–75% of dose

ADMINISTRATION

Oral

- Give for analgesic effect in the smallest effective dose to minimize the possible tolerance and physical and psychic dependence.
- Dilute dispersible tablets in 120 mL of water or fruit juice and allow at least 1 min for dispersion.

Subcutaneous/Intramuscular

- Note: IM route is preferred over subcutaneous when repeated parenteral administration is required (subcutaneous injections may cause local irritation and induration). Rotate injection sites.

Intravenous

PREPARE: **Direct/IV Infusion:** May be given undiluted or diluted with 1–5 mL of NS.
ADMINISTER: **Direct/IV Infusion:** Give over 5 or more minutes.
INCOMPATIBILITIES **Y-site: Phenytoin.**

- Store at 15°–30° C (59°–86° F) in tight, light-resistant containers.

ADVERSE EFFECTS (≥1%) **CNS:** *Drowsiness,* light-headedness, dizziness, hallucinations. **GI:** Nausea, vomiting, dry mouth, *constipation.* **Body as a Whole:** Transient fall in BP, bone and muscle pain. **Urogenital:** Impotence. **Respiratory:** Respiratory depression.

INTERACTIONS Drug: Alcohol and other CNS DEPRESSANTS, **cimetidine** add to sedation and CNS depression; AMPHETAMINES may potentiate CNS stimulation; with MAO INHIBITORS, **selegiline, furazolidone**

causes excessive and prolonged CNS depression, convulsions, cardiovascular collapse. **Food: Grapefruit juice** may increase serum levels and adverse effects. **Herbal: St. John's wort** decreases plasma levels.

PHARMACOKINETICS Absorption: Well absorbed from GI tract, variable IM absorption. **Onset:** 30–60 min PO; 10–20 min IM/Subcutaneous. **Peak:** 1–2 h. **Duration:** 6–8 h PO, IM, Subcutaneous; may last 22–48 h with chronic dosing. **Distribution:** Crosses placenta; distributed into breast milk. **Metabolism:** In liver (CYP3A4). **Elimination:** In urine. **Half-Life:** 15–25 h.

NURSING IMPLICATIONS

Assessment & Drug Effects

- Evaluate patient's continued need for methadone for pain. Adjustment of dosage and lengthening of between-dose intervals may be possible.
- Monitor respiratory status. Principal danger of overdosage, as with morphine, is extreme respiratory depression.
- Be aware that because of the cumulative effects of methadone, abstinence symptoms may not appear for 36–72 h after last dose and may last 10–14 days. Symptoms are usually of mild intensity (e.g., anorexia, insomnia, anxiety, abdominal discomfort, weakness, headache, sweating, hot and cold flashes).
- Observe closely for recurrence of respiratory depression during use of narcotic antagonists such as naloxone.

Patient & Family Education

- Be aware that orthostatic hypotension, sweating, constipation, drowsiness, GI symptoms, and other transient adverse effects of therapeutic doses appear to be more prominent in ambulatory patients. Most adverse effects disappear over a period of several weeks.

- Make position changes slowly, particularly from lying down to upright position; sit or lie down if you feel dizzy or faint.
- Do not drive or engage in potentially hazardous activities until response to drug is known.

METHAMPHETAMINE HYDROCHLORIDE

(meth-am-fet′a-meen)

Desoxyephedrine, Desoxyn
Classifications: ADRENERGIC AGONIST; CEREBRAL STIMULANT; AMPHETAMINE
Therapeutic: CEREBRAL STIMULANT; ANOREXIANT
Prototype: Amphetamine sulfate
Pregnancy Category: C
Controlled Substance: Schedule II

AVAILABILITY 5 mg tablets; 5 mg, 10 mg, 15 mg long-acting tablets

ACTION & THERAPEUTIC EFFECT
CNS stimulant actions approximately equal to those of amphetamine, but accompanied by less peripheral activity. *CNS stimulation results in increased motor activity, diminished sense of fatigue, alertness, increased focus, and mood elevation. Anorexigenic effect is due to direct inhibition of hypothalamic appetite center.*

USES Short-term adjunct in management of exogenous obesity, as adjunctive therapy in attention deficit disorder (ADD), narcolepsy, epilepsy, and postencephalitic parkinsonism, and in treatment of certain depressive reactions, especially when characterized by apathy and psychomotor retardation.

CONTRAINDICATIONS Hypersensitivity or idiosyncracy to sympathomimetic amines; children with structural cardiac abnormalities; glaucoma; advanced arteriosclerosis; symptomatic cardiovascular disease; moderate to severe hypertension; hyperthyroidism; patients in agitated state or history of drug abuse; lactation.

CAUTIOUS USE Mild hypertension; psychopathic personalities; hyperexcitability states; history of suicide attempts; older adult or debilitated patients; pregnancy (category C); ADHD treatment in children younger than 6 y or for obesity treatment in children younger than 12 y; longer term use in children.

ROUTE & DOSAGE

Attention Deficit Disorder

Child (6 y or older): **PO** 2.5–5 mg 1–2 times/day, may increase by 5 mg at weekly intervals up to 20–25 mg/day

Obesity

Adult: **PO** 2.5–5 mg 1–3 times/day 30 min before meals or 5–15 mg of long-acting form once/day

ADMINISTRATION

Oral
- Give early in the day, if possible, to avoid insomnia.
- Ensure that long-acting tablets are not chewed or crushed; these need to be swallowed whole.
- Give 30 min before each meal when used for treatment of obesity. If insomnia results, advise patient to inform prescriber.
- Preserve in tight, light-resistant containers.

ADVERSE EFFECTS (≥1%) **CNS:** Restlessness, tremor, hyperreflexia, insomnia, headache, nervousness, anxiety, dizziness, euphoria or

Common adverse effects in *italic*; life-threatening effects underlined; generic names in **bold**; classifications in SMALL CAPS; ◆ Canadian drug name; ☢ Prototype drug

951

dysphoria. **CV:** Palpitation, arrhythmias, hypertension, hypotension, circulatory collapse. **GI:** Dry mouth, unpleasant taste, nausea, vomiting, diarrhea, constipation. **Special Senses:** Increased intraocular pressure.

INTERACTIONS Drug: Acetazolamide, sodium bicarbonate decreases methamphetamine elimination; **ammonium chloride, ascorbic acid** increases methamphetamine elimination; effects of both methamphetamine and BARBITURATES may be antagonized; **furazolidone** may increase BP effects of AMPHETAMINES—interaction may persist for several weeks after discontinuing **furazolidone;** antagonizes antihypertensive effects of **guanethidine;** MAO INHIBITORS, **selegiline** can cause hypertensive crisis (fatalities reported)—do not administer AMPHETAMINES during or within 14 days of administration of these drugs; PHENOTHIAZINES may inhibit mood elevating effects of AMPHETAMINES; TRICYCLIC ANTIDEPRESSANTS enhance methamphetamine effects because they increase norepinephrine release; BETA-ADRENERGIC AGONISTS increase adverse cardiovascular effects of AMPHETAMINES.

PHARMACOKINETICS Absorption: Readily absorbed from the GI tract. **Duration:** 6–12 h. **Distribution:** All tissues especially the CNS; excreted in breast milk. **Metabolism:** In liver. **Elimination:** Renal elimination.

NURSING IMPLICATIONS
Assessment & Drug Effects
- Monitor weight throughout period of therapy.
- Be alert for a paradoxical increase in depression or agitation in depressed patients. Report

immediately; drug should be withdrawn.
- Do not exceed duration of a few weeks for treatment of obesity.

Patient & Family Education
- Be alert for development of tolerance; happens readily, and prolonged use may lead to drug dependence. Abuse potential is high.
- Withdrawal after prolonged use is frequently followed by lethargy that may persist for several weeks.
- Weigh every other day under standard conditions and maintain a record of weight loss.

METHAZOLAMIDE
(meth-a-zoe′la-mide)
Classifications: EYE PREPARATION; CARBONIC ANHYDRASE INHIBITOR; ANTIGLAUCOMA
Therapeutic: ANTIGLAUCOMA
Prototype: Acetazolamide
Pregnancy Category: C

AVAILABILITY 25 mg, 50 mg tablets

ACTION & *THERAPEUTIC EFFECT*
Inhibits carbonic anhydrase activity in eye by reducing rate of aqueous humor formation with consequent lowering of intraocular pressure. *Effective in lowering intraocular pressure in glaucoma patients.*

USES Adjunctive treatment in chronic simple (open-angle) glaucoma and secondary glaucoma and preoperatively in acute angle-closure glaucoma when delay of surgery is desired in order to lower intraocular pressure. May be used concomitantly with miotic and osmotic agents.

CONTRAINDICATIONS Glaucoma due to severe peripheral anterior synechiae, severe or absolute glaucoma, hemorrhagic glaucoma; hypokalemia, hyponatremia; dialysis; hepatic disease; renal disease, anuria, renal failure.

CAUTIOUS USE Pulmonary disease, COPD; diabetes mellitus; renal impairment; pregnancy (category C), lactation.

ROUTE & DOSAGE

Glaucoma
Adult: PO 50–100 mg b.i.d. or t.i.d.

ADMINISTRATION

Oral

- Give with meals to minimize GI distress.

ADVERSE EFFECTS (≥1%) Body as a Whole: Malaise, drowsiness, fatigue, lethargy. **GI:** Mild GI disturbance, anorexia. **CNS:** Headache, vertigo, paresthesias, mental confusion, depression.

INTERACTIONS Drug: Renal excretion of AMPHETAMINES, **ephedrine, flecainide, quinidine, procainamide,** TRICYCLIC ANTIDEPRESSANTS may be decreased, thereby enhancing or prolonging their effects; increases renal excretion of **lithium;** excretion of **phenobarbital** may be increased; **amphotericin B,** CORTICOSTEROIDS may add to potassium loss; hypokalemia caused by methazolamide may predispose patients on DIGITALIS GLYCOSIDES to **digitalis** toxicity; patients on high doses of SALICYLATES are at higher risk for SALICYLATE toxicity.

PHARMACOKINETICS Absorption: Slowly from GI tract. **Onset:** 2–4 h. **Peak:** 6–8 h. **Duration:** 10–18 h. **Distribution:** Throughout body, concentrating in RBCs, plasma, and kidneys; crosses placenta. **Metabolism:** Partially in liver. **Elimination:** Primarily in urine.

NURSING IMPLICATIONS

Assessment & Drug Effects

- Supervise ambulation in older adult, since drug may cause vertigo.
- Assess patient's ability to perform ADL since drug may cause fatigue and lethargy.
- Lab tests: Obtain periodic serum electrolytes, especially in older adults. Monitor lithium levels with concurrent administration of lithium and methazolamide.

Patient & Family Education

- Be aware that drug may cause drowsiness. Advise caution with hazardous activities until response to drug is known.

M

METHENAMINE HIPPURATE

(meth-en′a-meen hip′yoo-rate)
Hiprex, Urex

METHENAMINE MANDELATE

Classification: URINARY TRACT ANTI-INFECTIVE
Therapeutic: URINARY TRACT ANTI-INFECTIVE
Prototype: Trimethoprim
Pregnancy Category: C

AVAILABILITY Methenamine Hippurate: 1 g tablets; **Methenamine Mandelate:** 0.5 g, 1 g tablets; 0.5 g/5 mL suspension

ACTION & *THERAPEUTIC EFFECT*

Tertiary amine that liberates formaldehyde in an acid medium, which is a nonspecific antibiotic agent with bactericidal activity. *Currently used only for suppression and prophylaxis of frequently recurring urinary tract infections such as in patients with neurogenic bladder*

Common adverse effects in *italic*, life-threatening effects underlined; generic names in **bold;** classifications in SMALL CAPS; ♣ Canadian drug name; ⊙ Prototype drug

953

or in those who require intermittent catheterization routinely.

USES Prophylactic treatment of recurrent urinary tract infections (UTIs). Also long-term prophylaxis when residual urine is present (e.g., neurogenic bladder).

CONTRAINDICATIONS Renal insufficiency; liver disease; gout; severe dehydration; lactation.

CAUTIOUS USE Oral suspension for patients susceptible to lipoid pneumonia (e.g., older adults, debilitated patients); gout; pregnancy (category C).

ROUTE & DOSAGE

UTI Prophylaxis

Adult: **PO** *(Hippurate)* 1 g b.i.d.; (Mandelate) 1 g q.i.d.
Child: **PO** 6 y or younger, *(Mandelate)* 18.4 mg/kg q.i.d.; *6–12 y,* *(Hippurate)* 0.5–1 g b.i.d.; *(Mandelate)* 500 mg q.i.d. or 50 mg/kg/day in 3 divided doses

ADMINISTRATION

Oral

- Give after meals and at bedtime to minimize gastric distress.
- Give oral suspension with caution to older adult or debilitated patients because of the possibility of lipid (aspiration) pneumonia; it contains a vegetable oil base.
- Store at 15°–30° C (59°–86° F) in tightly closed container; protect from excessive heat.

ADVERSE EFFECTS (≥1%) **GI:** Nausea, vomiting, diarrhea, abdominal cramps, anorexia. **Renal:** Bladder irritation, dysuria, frequency, albuminuria, hematuria, crystalluria.

DIAGNOSTIC TEST INTERFERENCE Methenamine (formaldehyde) may produce falsely elevated values for **urinary catecholamines** and **urinary steroids (17-hydroxycorticosteroids)** (by **Reddy method**). Possibility of false **urine glucose determinations** with **Benedict's** test. Methenamine interferes with **urobilinogen** and possibly **urinary VMA** determinations.

INTERACTIONS Drug: Sulfamethoxazole forms insoluble precipitate in acid urine; **acetazolamide, sodium bicarbonate** may prevent hydrolysis to formaldehyde.

PHARMACOKINETICS Absorption: Readily from GI tract, although 10–30% of dose is hydrolyzed to formaldehyde in stomach. **Peak:** 2 h. **Duration:** Up to 6 h or until patient voids. **Distribution:** Crosses placenta; distributed into breast milk. **Metabolism:** Hydrolyzed in acid pH to formaldehyde. **Elimination:** In urine. **Half-Life:** 4 h.

NURSING IMPLICATIONS

Assessment & Drug Effects

- Monitor urine pH; value of 5.5 or less is required for optimum drug action.
- Monitor I&O ratio and pattern; drug most effective when fluid intake is maintained at 1500 or 2000 mL/day.
- Consult prescriber about changing to enteric-coated tablet if patient complains of gastric distress.
- Supplemental acidification to maintain pH of 5.5 or below required for drug action may be necessary. Accomplish by drugs (ascorbic acid, ammonium chloride) or by foods.

Patient & Family Education

- Do not self-medicate with OTC antacids containing sodium bicarbonate or sodium carbonate (to prevent raising urine pH).

- Achieve supplementary acidification by limiting intake of foods that can increase urine pH [e.g., vegetables, fruits, and fruit juice (except cranberry, plum, prune)] and increasing intake of foods that can decrease urine pH (e.g., proteins, cranberry juice, plums, prunes).

METHIMAZOLE

(meth-im′a-zole)

Tapazole

Classification: ANTITHYROID HORMONE

Therapeutic: ANTITHYROID
Prototype: Propylthiouracil
Pregnancy Category: D

AVAILABILITY 5 mg, 10 mg, 15 mg, 20 mg tablets

ACTION & *THERAPEUTIC EFFECT*
Inhibits synthesis of thyroid hormones as the drug accumulates in the thyroid gland. Does not affect existing T_3 or T_4 levels. *Corrects hyperthyroidism by inhibiting synthesis of the thyroid hormone.*

USES Hyperthyroidism and prior to surgery or radiotherapy of the thyroid; may be used cautiously to treat hyperthyroidism in pregnancy.

CONTRAINDICATIONS Pregnancy (category D).
CAUTIOUS USE Bone marrow suppression; older adults; hepatic disease.

ROUTE & DOSAGE

Hyperthyroidism

Adult: **PO** 5–15 mg q8h
Child: **PO** 0.2–0.4 mg/kg/day divided q8h

ADMINISTRATION

Oral

- Give at same time each day relative to meals.
- Store at 15°–30° C (59°–86° F) in light-resistant container.

ADVERSE EFFECTS (≥1%) **GI:** Hepatotoxicity (rare). **Endocrine:** Hypothyroidism. **Hematologic:** Leukopenia, agranulocytosis, granulocytopenia, thrombocytopenia, pancytopenia, and aplastic anemia. **Musculoskeletal:** Arthralgia. **CNS:** Peripheral neuropathy, drowsiness, neuritis, paresthesias, vertigo. **Skin:** Rash, alopecia, skin hyperpigmentation, urticaria, and pruritus. **Urogenital:** Nephrotic syndrome.

INTERACTIONS Drug: Can reduce anticoagulant effects of **warfarin;** may increase serum levels of **digoxin;** may alter **theophylline** levels; may need to decrease dose of BETA-BLOCKERS.

PHARMACOKINETICS Absorption: Readily absorbed from GI tract. **Onset:** 30–40 min. **Peak:** 1 h. **Duration:** 2–4 h. **Distribution:** Crosses placenta; distributed into breast milk. **Elimination:** 12% in urine within 24 h. **Half-Life:** 5–13 h.

NURSING IMPLICATIONS

Assessment & Drug Effects

- Lab tests: Baseline and periodic thyroid function tests; periodic prothrombin time and LFTs.
- Closely monitor PT and INR in patients on oral anticoagulants. Anticoagulant activity may be potentiated.

Patient & Family Education

- Be aware that skin rash or swelling of cervical lymph nodes may indicate need to discontinue drug and change to another antithyroid agent. Consult prescriber.
- Notify prescriber promptly if the following symptoms appear:

M

Common adverse effects in *italic*, life-threatening effects <u>underlined</u>; generic names in **bold**; classifications in SMALL CAPS; ✦ Canadian drug name; ❂ Prototype drug

955

Bruising, unexplained bleeding, sore throat, fever, jaundice.
- Methimazole does not induce hypothyroiditis.

METHOCARBAMOL

(meth-oh-kar'ba-mole)

Robaxin

Classifications: CENTRALLY ACTING SKELETAL MUSCLE RELAXANT; CARBAMATE
Therapeutic: SKELETAL MUSCLE RELAXANT
Prototype: Cyclobenzaprine
Pregnancy Category: C

AVAILABILITY 500 mg, 750 mg tablet; 100 mg/mL injection

ACTION & *THERAPEUTIC EFFECT*
Exerts skeletal muscle relaxant action by depressing multisynaptic pathways in the spinal cord and possibly by sedative effect. *Acts on multisynaptic pathways in spinal cord that control muscular spasms.*

USES Adjunct to physical therapy and other measures in management of discomfort associated with acute musculoskeletal disorders. Also used intravenously as adjunct in management of neuromuscular manifestations of tetanus.

CONTRAINDICATIONS Comatose states; CNS depression; acidosis, older adults; kidney dysfunction.
CAUTIOUS USE Epilepsy; renal disease, renal failure, renal impairment, seizure disorder; females of childbearing age; pregnancy (category C), lactation, children younger than 16 y.

ROUTE & DOSAGE

Acute Musculoskeletal Disorders

Adult: **PO** 1.5 g q.i.d. for 2–3 days, then 4–4.5 g/day in 3–6 divided doses **IV/IM** 1 g q8h

Tetanus

Adult: **IV** 1–3 g may be repeated q6h
Child: **PO** 15 mg/kg repeated q6h as needed up to 1.8 g/m²/day for 3 consecutive days if necessary

ADMINISTRATION

Oral
- Tablets may be crushed, suspended in water, and given through an NG tube.

Intramuscular
- Do not exceed IM dose of 5 mL (0.5 g) into each gluteal region. Insert needle deep and carefully aspirate. Inject drug slowly. Rotate injection sites and observe daily for evidence of irritation.

Intravenous

PREPARE: Direct: May be given undiluted or diluted in up to 250 mL of NS or D5W. **IV Infusion:** May dilute in up to 250 mL of NS or D5W.
ADMINISTER: Direct: Give at a rate of 300 mg or fraction thereof over 1 min or longer. **IV Infusion:** Infuse at a rate consistent with amount of fluid, but do not exceed direct rate. ▪Keep patient recumbent during and for at least 15 min after IV injection in order to reduce possibility of orthostatic hypotension and other adverse reactions. ▪ Monitor vital signs and IV flow rate. ▪Take care to avoid extravasation of IV solution, which may result in thrombophlebitis and sloughing.
***INCOMPATIBILITIES* Y-site: Furosemide.**

- Store at 15°–30° C (59°–86° F).

ADVERSE EFFECTS (≥1%) **Body as a Whole:** Fever, <u>anaphylactic reaction</u>, flushing, syncope, convulsions.

Skin: Urticaria, pruritus, rash, thrombophlebitis, pain, sloughing (with extravasation). **Special Senses:** Conjunctivitis, blurred vision, nasal congestion. **CNS:** *Drowsiness, dizziness, light-headedness,* headache. **CV:** Hypotension, bradycardia. **GI:** Nausea, metallic taste. **Hematologic:** Slight reduction of white cell count with prolonged therapy. **Renal:** Polyethylene glycol in the injection may increase preexisting acidosis and urea retention in patients with renal impairment.

DIAGNOSTIC TEST INTERFERENCE
Methocarbamol may cause false increases in **urinary 5-HIAA** (with **nitrosonaphthol reagent**) and **VMA (Gitlow method).**

INTERACTIONS Drug: Alcohol and other CNS DEPRESSANTS enhance CNS depression.

PHARMACOKINETICS Absorption: Readily absorbed from GI tract. **Onset:** 30 min. **Peak:** 1–2 h. **Metabolism:** In liver. **Elimination:** In urine. **Half-Life:** 1–2 h.

NURSING IMPLICATIONS

Assessment & Drug Effects
- Lab tests: Obtain periodic WBC counts during prolonged therapy.
- Monitor vital signs closely during IV infusion.
- Supervise ambulation following parenteral administration.

Patient & Family Education
- Make position changes slowly, particularly from lying down to upright position; dangle legs before standing.
- Be aware that adverse reactions after oral administration are usually mild and transient and subside with dosage reduction. Use caution regarding drowsiness and dizziness. Avoid activities requiring mental alertness and physical coordination until response to drug is known.
- Urine may darken to brown, black, or green on standing.

METHOTREXATE SODIUM ⊕
(meth-oh-trex′ate)

MTX

Classifications: ANTINEOPLASTIC; ANTIMETABOLITE; IMMUNOSUPPRESSANT; DISEASE-MODIFYING ANTIRHEUMATIC DRUG (DMARD)
Therapeutic: ANTINEOPLASTIC; ANTIFOLATE; ANTIRHEUMATIC; ANTIPSORIATIC
Pregnancy Category: X

AVAILABILITY 2.5 mg tablets; 2.5 mg/mL, 25 mg/mL injection

ACTION & THERAPEUTIC EFFECT
Antimetabolite and folic acid antagonist that blocks folic acid participation in nucleic acid synthesis, thereby interfering with mitotic cell process. Rapidly proliferating tissues (malignant cells, bone marrow, and psoriasis) are sensitive to interference of the mitotic process by this drug. *Induces remission slowly; use often preceded by other antineoplastic therapies. Additionally has immunosuppressant effects, antipsoriatic, and antirheumatic effects.*

USES Principally in combination regimens to maintain induced remissions in neoplastic diseases. Effective in treatment of gestational choriocarcinoma and hydatidiform mole and as immunosuppressant in kidney transplantation, for acute and subacute leukemias and leukemic meningitis, especially in children. Used in lymphosarcoma, in certain inoperable tumors of head, neck, and pelvis, and in mycosis fungoides. Also used to treat severe psoriasis

Common adverse effects in *italic*, life-threatening effects underlined; generic names in **bold;** classifications in SMALL CAPS; ♣ Canadian drug name; ⊕ Prototype drug

957

nonresponsive to other forms of therapy, rheumatoid arthritis.

UNLABELED USES Psoriatic arthritis, SLE, polymyositis.

CONTRAINDICATIONS Hepatic and renal insufficiency; alcohol; ultraviolet exposure to psoriatic lesions; preexisting blood dyscrasias; men and women in childbearing age; pregnancy (category X); lactation.

CAUTIOUS USE Infections; peptic ulcer, ulcerative colitis; very young or old patients; cancer patients with preexisting bone marrow impairment; poor nutritional status.

ROUTE & DOSAGE

Trophoblastic Neoplasm
Adult: **PO/IM** 15–30 mg/day for 5 days, repeat for 3–5 courses

Leukemia
Adult: **IM/IV Loading Dose** 3.3 mg/m^2/day **PO/IM/IV Maintenance Dose** 30 mg/m^2 weekly in 2 doses

Meningeal Leukemia
Child: **Intrathecal** 10–15 mg/m^2

Lymphoma
Adult: **PO** 10–25 mg/kg for 4–8 days

Osteosarcoma
Adult: **IV** 12 g/m^2, dose repeated at weeks 4, 5, 6, 7, 11, 12, 15, 16, 29, 39, 44, 45

Psoriasis/Rheumatoid Arthritis
Adult: **PO** 2.5 mg q12h for 3 doses each wk or 7.5 mg once/wk
Child: **PO/IM** 5–15 mg/m^2/wk as single dose or in 3 divided doses 12 h apart

Mycosis Fungoides
Adult: **PO/IM** 5–50 mg weekly

ADMINISTRATION

Oral
- May be taken without respect to meals.
- Avoid skin exposure and inhalation of drug particles.

Intramuscular
- Inject deeply into a large muscle.

Intravenous
Note: Verify correct IV concentration and rate of infusion for administration to children with prescriber.

PREPARE: **Direct:** Reconstitute powder vial by adding 2 mL of NS or D5W without preservatives to each 5 mg to yield 2.5 mg/mL. Reconstitute 1 g high-dose vial with 19.4 mL D5W or NS to yield 50 mg/mL. **IV Infusion:** Further dilute contents of the reconstituted 1 g high-dose vial in D5W or NS to a 25 mg/mL or less.

ADMINISTER: **Direct:** Give at rate of 10 mg or fraction thereof over 60 sec. **IV Infusion:** Give over 1–4 h or as prescribed.

INCOMPATIBILITIES **Solution/additive: Bleomycin, metoclopramide, prednisolone, ranitidine. Y-site: Chlorpromazine, droperidol, gemcitabine, idarubicin, ifosfamide, midazolam, nalbuphine, promethazine, propofol.**

- Preserve drug in tight, light-resistant container.

ADVERSE EFFECTS (≥1%) **CNS:** *Headache,* drowsiness, blurred vision, dizziness, aphasia, hemiparesis; arachnoiditis, convulsions (after intrathecal administration); mental confusion, tremors, ataxia, coma. **GI:** <u>Hepatotoxicity</u>, GI ulcerations and hemorrhage, *ulcerative stomatitis, glossitis, gingivitis,* pharyngitis, nausea, vomiting, diarrhea, <u>hepatic cirrhosis</u>. **Urogenital:** Defective oogenesis or spermatogenesis, nephropathy, hematuria, menstrual dysfunction, infertility, abortion, fetal defects. **Hematologic:** *Leukopenia, thrombocytopenia,* anemia, <u>marked myelosuppression, aplastic bone marrow</u>, telangiectasis, thrombophlebitis at intra-arterial catheter site, hypogammaglobulinemia, and hyperuricemia. **Skin:** Erythematous rashes, pruritus, urticaria, folliculitis, vasculitis, photosensitivity, depigmentation, hyperpigmentation, alopecia. **Body as a Whole:** Malaise, undue fatigue, systemic toxicity (after intrathecal and intra-arterial administration), chills, fever, decreased resistance to infection, septicemia, osteoporosis, metabolic changes precipitating diabetes and <u>sudden death, pneumonitis, pulmonary fibrosis</u>.

DIAGNOSTIC TEST INTERFERENCE Severe reactions may occur when *live vaccines* are administered because of immunosuppressive activity of methotrexate.

INTERACTIONS Drug: Acitretin, alcohol, azathioprine, sulfasalazine increase risk of hepatotoxicity; **chloramphenicol, etretinate,** SALICYLATES, NSAIDS, SULFONAMIDES, SULFONYLUREAS, **phenylbutazone, phenytoin,** TETRACYCLINES, **PABA, penicillin, probenecid** may increase methotrexate levels with increased toxicity; **folic acid** may alter response to methotrexate. May increase **theophylline** levels; **cholestyramine** enhances methotrexate clearance. **Herbal: Echinacea** may increase risk of hepatotoxicity. **Food: Caffeine** greater than 180 mg/day (3–4 cups) may decrease effectiveness for rheumatoid arthritis.

PHARMACOKINETICS Absorption: Readily absorbed from GI tract. **Peak:** 0.5–2 h IM/IV; 1–4 h PO.

Distribution: Widely distributed with highest concentrations in kidneys, gallbladder, spleen, liver, and skin; minimal passage across blood–brain barrier; crosses placenta; distributed into breast milk. **Metabolism:** In liver. **Elimination:** Primarily in urine. **Half-Life:** 2–4 h.

NURSING IMPLICATIONS

Assessment & Drug Effects

- Lab tests: Obtain baseline LFTs and kidney function, CBC with differential, platelet count, and chest x-rays. Repeat weekly during therapy. Monitor blood glucose and HbA1C periodically in diabetics.
- Prolonged treatment with small frequent doses may lead to hepatotoxicity, which is best diagnosed by liver biopsy.
- Monitor for and report ulcerative stomatitis with glossitis and gingivitis, often the first signs of toxicity. Inspect mouth daily; report patchy necrotic areas, bleeding and discomfort, or overgrowth (black, furry tongue).
- Monitor I&O ratio and pattern. Keep patient well hydrated (about 2000 mL/24 h).
- Prevent exposure to infections or colds during periods of leukopenia. Be alert to onset of agranulocytosis (cough, extreme fatigue, sore throat, chills, fever) and report symptoms promptly.
- Be alert for and report symptoms of thrombocytopenia (e.g., ecchymoses, petechiae, epistaxis, melena, hematuria, vaginal bleeding, slow and protracted oozing following trauma).

Patient & Family Education

- Report promptly any of the following: Diarrhea, mouth sores, fever, dehydration, cough, bleeding, shortness of breath, any signs of infection, or a skin rash.

M

Common adverse effects in *italic*, life-threatening effects <u>underlined</u>; generic names in **bold**; classifications in SMALL CAPS; ♣ Canadian drug name; ⊘ Prototype drug

959

- Avoid or moderate alcohol ingestion, which increases the incidence and severity of methotrexate hepatotoxicity.
- Practice fastidious mouth care to prevent infection, provide comfort, and maintain adequate nutritional status.
- Do not self-medicate with vitamins. Some OTC compounds may include folic acid (or its derivatives), which alters methotrexate response.
- Use contraceptive measures during and for at least 3 mo following therapy.
- Avoid exposure to sunlight and ultraviolet light. Wear sunglasses and sunscreen.

METHOXSALEN ℗
(meth-ox′a-len)

8-MOP, Oxsoralen, Uvadex
Classifications: PSORALEN; PIGMENTING AGENT
Therapeutic: PIGMENTING AGENT; ANTIPSORIATIC
Pregnancy Category: C

AVAILABILITY 10 mg capsules, 20 mcg/mL solution; 1% lotion

ACTION & *THERAPEUTIC EFFECT*
Plant derivative with strong photosensitizing effects: Used with ultraviolet-A light (UVA) in therapeutic regimens called PUVA (P-psoralen). After photoactivation by long wavelength, UVA, methoxsalen combines with epidermal cell DNA causing photo-damage (cytotoxic action). *Photo-damage inhibits rapid and uncontrolled epidermal cell turnover characteristic of psoriasis. Results in an inflammatory reaction with erythema. Strongly melanogenic.*

USES With controlled exposure to UVA to repigment vitiliginous skin and for symptomatic treatment of severe disabling psoriasis that is refractory to other forms of therapy.
UNLABELED USES (PUVA therapy) mycosis fungoides.

CONTRAINDICATIONS Sunburn, sensitivity (or its history) to psoralens, diseases associated with photosensitivity (e.g., SLE, albinism, melanoma or its history); invasive squamous cell cancer; cataract; aphakia; previous exposure to arsenic or ionizing radiation.
CAUTIOUS USE Hepatic insufficiency; GI disease; chronic infection; treatment with known photosensitizing agents; immunosuppressed patient; cardiovascular disease; pregnancy (category C), lactation. Safety **(lotion)** in children younger than 12 y is not established. Safety **(oral)** in children is not established.

ROUTE & DOSAGE

Idiopathic Vitiligo

Adult: **Topical** Apply lotion 1–2 h before exposure to UV light once/wk

Psoriasis

Adult: **PO** Give 1.5–2 h before exposure to UV light 2–3 times/wk: *weight less than 30 kg,* 10 mg; *weight 30–50 kg,* 20 mg; *weight 51–65 kg,* 30 mg; *weight 66–80 kg,* 40 mg; *weight 81–90 kg,* 50 mg; *weight 91–115 kg,* 60 mg; *weight greater than 115 kg,* 70 mg

ADMINISTRATION

- Note: Methoxsalen therapy with UV light (PUVA therapy) should be done under the complete control of a prescriber with special competence and experience in photochemotherapy.

Oral
- Give with milk or food to prevent GI distress.
- Maintain consistent time relationship between food–drug ingestion. Food digestion and absorption appear to affect drug serum levels.

Topical
- Only small (less than 10 cm²), well-defined areas are treated with lotion. Systemic treatment is used for large areas.
- Apply lotion with cotton swabs, allow to dry 1–2 min, then reapply. Protect borders of the lesion with petrolatum and sunscreen lotion to prevent hyperpigmentation.
- Use finger cots or gloves to apply lotion and prevent photosensitization and burned skin.
- Apply sunscreen lotion to the skin for about one third of the initial exposure time during PUVA therapy until there is sufficient tanning. Do not apply to psoriatic areas before treatment.
- Store lotion and capsules at 15°–30° C (59°–86° F) in light-resistant containers unless otherwise directed by manufacturer.

ADVERSE EFFECTS (≥1%) **CNS:** Nervousness, dizziness, headache, mental depression or excitation, vertigo, insomnia. **Special Senses:** Cataract formation, ocular damage. **GI:** Cheilitis, *nausea* and other GI disturbances, toxic hepatitis. **Skin:** Phototoxic effects: <u>Severe edema and erythema</u>, *pruritus*, painful blisters; <u>burning</u>, peeling, thinning, freckling, and accelerated aging of skin; hyper- or hypopigmentation; severe skin pain (lasting 1–2 mo), photoallergic contact dermatitis (with topical use), exacerbation of latent photosensitive dermatoses, <u>malignant melanoma</u>

(rare). **Body as a Whole:** Transient loss of muscular coordination, edema, leg cramps, systemic immune effects, drug fever.

INTERACTIONS Drug: Anthralin, coal tar, griseofulvin, PHENOTHIA-ZINES, **nalidixic acid,** SULFONAMIDES, BACTERIOSTATIC SOAPS, TETRACYCLINES, THIAZIDES compound photosensitizing effects. **Food:** Food will increase peak and extent of absorption.

PHARMACOKINETICS Absorption: Variably from GI tract. **Peak:** 2 h. **Duration:** 8–10 h. **Distribution:** Preferentially taken up by epidermal cells; distributes into lens of eye. **Elimination:** 80–90% in urine within 8 h. **Half-Life:** 0.75–2.4 h.

NURSING IMPLICATIONS

Assessment & Drug Effects
- Schedule a pretreatment ophthalmologic exam to rule out cataracts; repeat periodically during treatment and at yearly intervals thereafter.
- Lab tests: Monitor CBC, LFTs and kidney function, and antinuclear antibody tests during oral therapy.
- Fair-skinned patients appear to be at greatest risk for phototoxicity from PUVA therapy (see ADVERSE EFFECTS).
- Be aware that repigmentation is more rapid on fleshy areas (i.e., face, abdomen, buttocks) than on hands or feet.

Patient & Family Education
- Expect that effective repigmentation may require 6–9 mo of treatment; periodic treatment usually is necessary to retain pigmentation. If, after 3 mo of treatment, there is no apparent response, drug is discontinued.
- Avoid additional exposure to UV light (direct or indirect) for at least

---

Common adverse effects in *italic*, life-threatening effects <u>underlined</u>; generic names in **bold**; classifications in SMALL CAPS; ♣ Canadian drug name; ℗ Prototype drug

961

8 h after oral drug ingestion and UVA exposure.

- Understand intended treatment schedule: After topical application, the initial sunlight exposure is limited to 1 min, with subsequent gradual and incremental exposures by prescription.
- Avoid additional UV light for 24–48 h after topical application and UVA exposure.
- Wear sunscreen lotion (with SPF 15 or higher) and protective clothing (hat, gloves) to cover all exposed areas including lips, to prevent burning or blistering if sunlight cannot be avoided after the treatment.
- Do not sunbathe for at least 48 h after PUVA treatment. Sunburn and photochemotherapy are additive in the production of burning and erythema.
- Wear wraparound sunglasses with UVA-absorbing properties both indoors and outdoors during daylight hours for 24 h. Do not substitute prescription sunglasses or photosensitive darkening glasses; they may actually increase danger of cataract formation.
- Alert prescriber to appearance of new psoriatic areas, flares, or regressed cleared skin areas during treatment and maintenance periods.

METHSCOPOLAMINE BROMIDE
(meth-skoe-pol′a-meen)

Pamine

Classifications: ANTICHOLINERGIC; ANTIMUSCARINIC; ANTISPASMODIC

Therapeutic: ANTISPASMODIC; ANTISECRETORY

Prototype: Atropine

Pregnancy Category: C

AVAILABILITY 2.5 mg tablets

ACTION & *THERAPEUTIC EFFECT*
Decreases GI tone and amplitude as well as frequency of peristaltic contractions of the esophagus, stomach, duodenum, jejunum, ileum, and colon. *Its spasmolytic and antisecretory actions are quantitatively similar to those of atropine but last longer.*

USES Adjunct in treatment of peptic ulcer, irritable bowel syndrome, and a variety of other GI conditions. Also may be used to control excessive sweating and salivation, migraine headaches, and premenstrual cramps.

CONTRAINDICATIONS Hypersensitivity to any of the drug's constituents; prostatic hypertrophy; pyloric obstruction; intestinal atony; tachycardia, cardiac disease; MS; pyloric stenosis; lactation.

CAUTIOUS USE Older adult and debilitated patients; COPD; pregnancy (category C).

ROUTE & DOSAGE

Irritable Bowel Syndrome
Adult: **PO** 2.5–5 mg 30 min a.c. and at bedtime

ADMINISTRATION
Oral
- Give 30 min before meals and at bedtime.
- Preserve in tight, light-resistant containers.

ADVERSE EFFECTS (≥1%) **GI:** Dry mouth, constipation. **Special Senses:** Blurred vision. **CNS:** Dizziness, drowsiness, flushing of skin. **Urogenital:** Urinary hesitancy or retention.

INTERACTIONS Drug: Amantadine, TRICYCLIC ANTIDEPRESSANTS increase anticholinergic effects; may increase effects of **atenolol,**

digoxin; may decrease effectiveness of PHENOTHIAZINES.

PHARMACOKINETICS Absorption: Erratic after PO administration. **Onset:** Approximately 1 h. **Duration:** 4–6 h. **Elimination:** Primarily in urine and bile; some unchanged drug excreted in feces.

NURSING IMPLICATIONS

Assessment & Drug Effects

- Incidence and severity of adverse effects are generally dose related. Dosage is usually maintained at a level that produces slight dryness of mouth.
- Report urinary retention promptly; may indicate need to discontinue drug.

Patient & Family Education

- Do not drive or engage in potentially hazardous activities until response to drug is known.
- Make position changes slowly and in stages.
- Learn measures to relieve dry mouth; rinse mouth frequently with water, suck hard candy.

METHYCLOTHIAZIDE

(meth-i-kloe-thye′a-zide)

Duretic ♣

Classifications: THIAZIDE DIURETIC; ANTIHYPERTENSIVE

Therapeutic: THIAZIDE DIURETIC; ANTIHYPERTENSIVE

Prototype: Hydrochlorothiazide

Pregnancy Category: D first trimester; C second and third trimester

AVAILABILITY 2.5 mg, 5 mg tablets

ACTION & *THERAPEUTIC EFFECT*
Diuretic effect results from inhibition of the renal tubular reabsorption of electrolytes. Excretion of sodium and chloride is enhanced, along with a loss of potassium ions via the kidney. BP is lowered, probably by the loss of sodium, chloride and water, and, consequently, blood volume. Edema is also decreased in CHF patients by the same mechanism. *Antihypertensive effect as well as enhanced excretion of sodium and water.*

USES Antihypertensive treatment and adjunctively in the management of edema associated with CHF, renal pathology, and hepatic cirrhosis.

CONTRAINDICATIONS Hypersensitivity to thiazides, and sulfonamide derivatives; anuria, hypokalemia, pregnancy (category D first trimester), lactation.

CAUTIOUS USE Renal disease; impaired kidney or liver function; older adults; gout; SLE; hypercalcemia; diabetes mellitus; pregnancy (category C second and third trimester); children.

ROUTE & DOSAGE

Edema
Adult: **PO** 2.5–10 mg once/day or 3–5 times/wk
Hypertension
Adult: **PO** 2.5–10 mg/day
Child: **PO** 0.05–0.2 mg/kg/day

ADMINISTRATION

Oral

- Give early in a.m. after eating (reduces gastric irritation) to prevent sleep interruption because of diuresis. If 2 doses are ordered, administer second dose no later than 3 p.m.
- Store at 15°–30° C (59°–86° F) unless otherwise instructed.

Common adverse effects in *italic,* life-threatening effects <u>underlined;</u> generic names in **bold;** classifications in SMALL CAPS; ♣ Canadian drug name; 🔄 Prototype drug

963

ADVERSE EFFECTS (≥1%) **Body as a Whole:** Postural hypotension, sialadenitis, unusual fatigue, dizziness, paresthesias. **Skin:** Photosensitivity. **Special Senses:** Yellow vision. **Metabolic:** *Hypokalemia.* **Hematologic:** <u>Agranulocytosis.</u>

INTERACTIONS Drug: Amphotericin B, CORTICOSTEROIDS increase hypokalemic effects; may antagonize hypoglycemic effects of **insulin,** SULFONYLUREAS; **cholestyramine, colestipol** decrease thiazide absorption; intensifies hypoglycemic and hypotensive effects of **diazoxide;** increased potassium and magnesium loss may cause **digoxin** toxicity; decreases **lithium** excretion, increasing its toxicity; NSAIDS may attenuate diuresis, and risk of NSAID-induced kidney failure increased.

PHARMACOKINETICS Absorption: Incompletely absorbed. **Onset:** 2 h. **Peak:** 6 h. **Duration:** Greater than 24 h. **Distribution:** Distributed throughout extracellular tissue; concentrates in kidney; crosses placenta; distributed in breast milk. **Metabolism:** Does not appear to be metabolized. **Elimination:** In urine.

NURSING IMPLICATIONS

Assessment & Drug Effects

- Expect antihypertensive effects in 3–4 days; maximal effects may require 3–4 wk.
- Monitor BP and I&O ratio during first phase of antihypertensive therapy. Report a sudden fall in BP, which may initiate severe postural hypotension and potentially dangerous perfusion problems, especially in the extremities.
- Lab tests: Periodic serum electrolytes and CBC with differential.
- Monitor patient for S&S of hypokalemia (see Appendix F). Report promptly. Prescriber may change dose and institute replacement therapy.

Patient & Family Education

- Eat a balanced diet to protect against hypokalemia; generally not severe even with long-term therapy. Prevent onset by eating potassium-rich foods including a banana (about 370 mg potassium) and at least 180 mL (6 oz) orange juice (about 330 mg potassium) every day.
- Watch carefully for loss of glycemic control (diabetics) and early signs of hyperglycemia (see Appendix F). Symptoms are slow to develop.
- Avoid OTC drugs unless the prescriber approves them. Many preparations contain both potassium and sodium, and may induce electrolyte imbalance adverse effects.
- Older adults are more responsive to excessive diuresis; orthostatic hypotension may be a problem.
- Change positions slowly and in stages from lying down to upright positions; avoid hot baths or showers, extended exposure to sunlight, and standing still. Accept assistance as necessary to prevent falling.
- Do not drive or engage in potentially hazardous activities until adjustment to the hypotensive effects of drug has been made.

METHYLDOPA ℗

(meth-ill-doe′pa)

Apo-Methyldopa ♣, **Novomedopa** ♣

METHYLDOPATE HYDROCHLORIDE

(meth-ill-doe′pate)

Classifications: CENTRALY ACT-ING ANTIHYPERTENSIVE; ALPHA-ADRENERGIC AGONIST
Therapeutic: CENTRALLY ACTING ANTIHYPERTENSIVE
Pregnancy Category: B

AVAILABILITY 125 mg, 250 mg, 500 mg tablets; 50 mg/mL injection

ACTION & *THERAPEUTIC EFFECT*
Structurally related to catecholamines and their precursors. Inhibits decarboxylation of dopa, thereby reducing concentration of dopamine, a precursor of norepinephrine. It also inhibits the precursor of serotonin. Reduces renal vascular resistance; maintains cardiac output without acceleration, but may slow heart rate; tends to support sodium and water retention. *Lowers standing and supine BP.*

USES Treatment of sustained moderate to severe hypertension, particularly in patients with kidney dysfunction. Also used in selected patients with carcinoid disease. Parenteral form has been used for treatment of hypertensive crises but is not preferred because of its slow onset of action.

CONTRAINDICATIONS Active liver disease (hepatitis, cirrhosis); pheochromocytoma; blood dyscrasias.
CAUTIOUS USE History of impaired liver or kidney function or disease; renal failure; autoimmune disease; cardiac disease; angina pectoris; history of mental depression; Parkinson's disease; young or older adult patients; pregnancy (category B).

ROUTE & DOSAGE

Hypertension

Adult: **PO** 250 mg b.i.d. or t.i.d., may be increased up to 3 g/day in divided doses, usual range

250–1000 mg total per day **IV** 250–500 mg q6h, may be increased up to 1 g q6h
Geriatric: **PO** 125 mg b.i.d. or t.i.d., may increase gradually (max: 3 g/day)
Child: **PO** 10 mg/kg/day in 2–4 divided doses (max: 3 g/day) **IV** 2–4 mg/kg/day in divided doses (max: 3 g/day)

Renal Impairment Dosage Adjustment

CrCl greater than 50 mL/min: Dose q8h; *10–50 mL/min:* Dose q8–12h; *less than 10 mL/min:* Dose q12–24h

ADMINISTRATION

Oral
▪ Make dosage increases in evening to minimize daytime sedation.

Intravenous

PREPARE: **Intermittent:** Dilute in 100–200 mL of D5W, as needed, to yield 10 mg/mL.
ADMINISTER: **Intermittent:** Give over 30–60 min.
INCOMPATIBILITIES **Solution/additive: Amphotericin B, hydrocortisone, methohexital, tetracycline. Y-site: Fat emulsion.**

ADVERSE EFFECTS (≥1%) **Body as a Whole:** Hypersensitivity (*fever,* skin eruptions, ulcerations of soles of feet, flu-like symptoms, lymphadenopathy, eosinophilia). **CNS:** *Sedation, drowsiness,* sluggishness, headache, weakness, fatigue, dizziness, vertigo, *decrease in mental acuity,* inability to concentrate, amnesia-like syndrome, parkinsonism, mild psychoses, depression, nightmares. **CV:** Orthostatic hypotension, syncope, bradycardia, myocarditis, edema, weight gain (*sodium and*

Common adverse effects in *italic*, life-threatening effects <u>underlined</u>; generic names in **bold;** classifications in SMALL CAPS; ✚ Canadian drug name; ⊘ Prototype drug

965

water retention), paradoxic hypertensive reaction (especially with IV administration). **GI:** Diarrhea, constipation, abdominal distention, malabsorption syndrome, nausea, vomiting, dry mouth, sore or black tongue, sialadenitis, abnormal liver function tests, jaundice, hepatitis, <u>hepatic necrosis</u> (rare). **Hematologic:** *Positive direct Coombs' test* (common especially in African-Americans), <u>granulocytopenia</u>. **Special Senses:** *Nasal stuffiness.* **Endocrine:** Gynecomastia, lactation, *decreased libido, impotence,* hypothermia (large doses), positive tests for lupus and rheumatoid factors. **Skin:** Granulomatous skin lesions.

DIAGNOSTIC TEST INTERFERENCE
Methyldopa may interfere with *serum creatinine* measurements using *alkaline picrate method, AST* by *colorimetric methods,* and *uric acid* measurements by *phosphotungstate method* (with high methyldopa blood levels); it may produce false elevations of *urinary catecholamines* and increase in *serum amylase* in methyldopa-induced sialadenitis.

INTERACTIONS Drug: AMPHETA-
MINES, TRICYCLIC ANTIDEPRESSANTS, PHENOTHIAZINES, BARBITURATES may attenuate antihypertensive response; methyldopa may inhibit effectiveness of **ephedrine; haloperidol** may exacerbate psychiatric symptoms; with **levodopa** additive hypotension, increased CNS toxicity, especially psychosis; increases risk of **lithium** toxicity; **methotrimeprazine** causes excessive hypotension; MAO INHIBITORS may cause hallucinations; **phenoxybenzamine** may cause urinary incontinence. **Herbal: Licorice** may affect electrolyte levels; **ephedra, yohimbe, ginseng** may decrease efficacy.

PHARMACOKINETICS Absorption:
About 50% absorbed from GI tract. **Peak:** 4–6 h. **Duration:** 24 h PO; 10–16 h IV. **Distribution:** Crosses placenta, distributed into breast milk. **Metabolism:** In liver and GI tract. **Elimination:** Primarily in urine. **Half-Life:** 1.7 h.

NURSING IMPLICATIONS

Assessment & Drug Effects
- Check BP and pulse at least q30min until stabilized during IV infusion and observe for adequacy of urinary output.
- Take BP at regular intervals in lying, sitting, and standing positions during period of dosage adjustment.
- Supervision of ambulation in older adults and patients with impaired kidney function; both are particularly likely to manifest orthostatic hypotension with dizziness and light-headedness during period of dosage adjustment.
- Monitor fluid and electrolyte balance and I&O. Weigh patient daily, and check for edema because methyldopa favors sodium and water retention.
- Lab tests: Baseline and periodic blood counts and LFTs especially during first 6–12 wk of therapy or if patient develops unexplained fever; periodic serum electrolytes.
- Be alert that rising BP indicating tolerance to drug effect may occur during week 2 or 3 of therapy.

Patient & Family Education
- Exercise caution with hot baths and showers, prolonged standing in one position, and strenuous exercise that may enhance orthostatic

M

hypotension. Make position changes slowly, particularly from lying down to upright posture; dangle legs a few minutes before standing.

- Be aware that transient sedation, drowsiness, mental depression, weakness, and headache commonly occur during first 24–72 h of therapy or whenever dosage is increased. Symptoms tend to disappear with continuation of therapy or dosage reduction.
- Avoid potentially hazardous tasks such as driving until response to drug is known; drug may affect ability to perform activities requiring concentrated mental effort, especially during first few days of therapy or whenever dosage is increased.
- Do not to take OTC medications unless approved by prescriber.

METHYLERGONOVINE MALEATE

(meth-ill-er-goe-noe'veen)

Methergine
Classifications: ERGOT ALKALOID; OXYTOCIC
Therapeutic: OXYTOCIC
Prototype: OXYTOCIN
Pregnancy Category: C

AVAILABILITY 0.2 mg tablets; 0.2 mg/mL injections

ACTION & *THERAPEUTIC EFFECT*

Ergot alkaloid that induces rapid, sustained tetanic uterine contraction that shortens third stage of labor and reduces blood loss. *Administered after delivery of the placenta to minimize the risk of postpartal hemorrhage.*

USES Routine management after delivery of placenta and for postpartum atony, subinvolution, and hemorrhage. With full obstetric supervision, may be used during second stage of labor.

CONTRAINDICATIONS Hypersensitivity to ergot preparations; induction of labor; use prior to delivery of placenta; threatened spontaneous abortion; prolonged use; uterine sepsis; hypertension; toxemia; angina; arteriosclerosis; CAD; dysfunctional uterine bleeding; eclampsia; hypertension; MI; neonates; PVD; preeclampsia; Raynaud's disease; sepsis; stroke; thromboangiitis obliterans; thrombophlebitis.
CAUTIOUS USE Diabetes mellitus; hepatic disease; migraine headaches; renal failure, renal impairment; pulmonary disease; pregnancy (category C), lactation.

ROUTE & DOSAGE

M

Postpartum Hemorrhage
Adult: **PO** 0.2 q6–8h × 2–7 days
IM/IV 0.2 mg q2–4h (max: 5 doses)

ADMINISTRATION

- Use parenteral routes only in emergencies.

Oral
- Note: Dosing should not exceed 1 wk.

Intramuscular
- Inject undiluted deep into a large muscle.

Intravenous

PREPARE: **Direct:** Give undiluted or diluted in 5 mL of NS.
ADMINISTER: **Direct:** Give 0.2 mg or fraction thereof over 60 sec.
- Do not use ampules containing discolored solution or visible particles.

- Store at 15°–30° C (59°–86° F) unless otherwise directed. Protect from light.

Common adverse effects in *italic*, life-threatening effects underlined; generic names in **bold;** classifications in SMALL CAPS; ♣ Canadian drug name; ⊘ Prototype drug

ADVERSE EFFECTS (≥1%) **GI:** *Nausea, vomiting* (especially with IV doses). **CV:** Severe hypertensive episodes, bradycardia. **Body as a Whole:** Allergic phenomena including <u>shock</u>, ergotism.

INTERACTIONS Drug: PARENTERAL SYMPATHOMIMETICS, other ERGOT ALKALOIDS, TRIPTANS add to pressor effects and carry risk of hypertension; PROTEASE INHIBITORS, **itraconazole** may increase the risk of toxicity.

PHARMACOKINETICS Absorption: Readily from GI tract. **Onset:** 5–15 min PO; 2–5 min IM; immediate IV. **Duration:** 3 or more h PO; 3 h IM; 45 min IV. **Distribution:** Distributed into breast milk. **Metabolism:** Slowly in liver. **Elimination:** Mainly in feces, small amount in urine. **Half-Life:** 0.5–2 h.

NURSING IMPLICATIONS

Assessment & Drug Effects

- Monitor vital signs (particularly BP) and uterine response during and after parenteral administration of methylergonovine until partum period is stabilized (about 1–2 h).
- Notify prescriber if BP suddenly increases or if there are frequent periods of uterine relaxation.

Patient & Family Education

- Report severe cramping or increased bleeding.
- Report any of the following: Cold or numb fingers or toes, nausea or vomiting, chest or muscle pain.

METHYLNALTREXONE BROMIDE

(meth-yl-nal-trex'own bro'mide)

Relistor
Classification: NARCOTIC (OPIATE ANTAGONIST)

Therapeutic: NARCOTIC ANTAGONIST
Prototype: Naloxone
Pregnancy Category: B

AVAILABILITY 12 mg/0.6 mL solution for injection

ACTION & THERAPEUTIC EFFECT A selective, peripherally acting antagonist of opioid binding to muopioid receptors in tissues such as the GI tract. *Decreases constipating effects of opioids without interfering with analgesic effect of opioids in the CNS.*

USES Treatment of opioid-induced constipation in patients with advanced illness who are receiving palliative care when response to laxative therapy has not been sufficient.

UNLABELED USES Management of nausea and vomiting related to morphine. Treatment of pruritus related to morphine. Management of urinary retention caused by opioids.

CONTRAINDICATIONS Known or suspected mechanical GI obstruction; severe or persistent diarrhea. **CAUTIOUS USE** Severe renal impairment; pregnancy (category B); lactation. Safety and efficacy in children not established.

ROUTE & DOSAGE

Opioid-Induced Constipation

Adult: **Subcutaneous** Administer every other day based on weight: *weight less than 38 kg,* 0.15 mg/kg; *weight 38 to less than 62 kg,* 8 mg; *weight 62 to less than 114 kg,* 12 mg

Renal Impairment Dosage Adjustment

CrCl less than 30 mL/min: Reduce normal adult dose by 50%

Common adverse effects in *italic*, life-threatening effects <u>underlined</u>; generic names in **bold**; classifications in SMALL CAPS; ♣ Canadian drug name; ⊙ Prototype drug

ADMINISTRATION

Subcutaneous

- An 8 mg dose equals 0.4 mL and a 12 mg dose equals 0.6 mL.
- Insert needle at a 45-degree angle into a pinched fold of skin on the abdomen, thigh, or upper arm. Release skin and inject. Rotate injection sites.
- Store at 20°–25° C (68°–77° F) away from light. May store drawn up into syringe for 24 h at room temperature with ambient light.

ADVERSE EFFECTS (≥1%) **CNS:** Dizziness. **GI:** *Abdominal pain*, diarrhea, *flatulence, nausea*.

PHARMACOKINETICS Peak: 0.5 h. **Distribution:** 11–15% protein bound. **Metabolism:** Hepatic. **Elimination:** Primarily eliminated unchanged (85%) in urine and feces. **Half-Life:** 8 h.

NURSING IMPLICATIONS

Assessment & Drug Effects

- Monitor bowel pattern.
- Withhold drug and report promptly severe or persistent diarrhea.

Patient & Family Education

- Ensure that patient/caregiver knows how to correctly inject subcutaneous medication.
- Stop methylnaltrexone and notify prescriber if severe or persistent diarrhea develops.

METHYLPHENIDATE HYDROCHLORIDE

(meth-ill-fen'i-date)

Concerta, Daytrana, Focalin XR, Metadate CD, Metadate ER, Methylin, Methylin ER, Ritalin, Ritalin LA, Ritalin SR

Classification: CEREBRAL STIMULANT
Therapeutic: CEREBRAL STIMULANT
Prototype: Amphetamine
Pregnancy Category: C
Controlled Substance: Schedule II

AVAILABILITY 5 mg, 10 mg, 20 mg tablets; 2.5 mg, 5 mg, 10 mg chewable tablets; 5 mg/5 mL, 10 mg/5 mL oral solution; 10 mg, 20 mg, 30 mg, 40 mg, 50 mg, 60 mg sustained release capsules; 10 mg, 18 mg, 20 mg, 27 mg, 36 mg, 54 mg sustained release tablets; 10 mg, 15 mg, 20 mg, 30 mg transdermal patch

ACTION & *THERAPEUTIC EFFECT* Acts mainly on cerebral cortex exerting a stimulant effect. Results in mild CNS and respiratory stimulation with potency intermediate between amphetamine and caffeine. *Effects are more prominent on mental rather than on motor activities. Also believed to have an anorexiant effect.*

USES Adjunctive therapy in hyperkinetic syndromes characterized by attention deficit disorder, narcolepsy.
UNLABELED USES Depression.

CONTRAINDICATIONS Hypersensitivity to drug; history of marked anxiety, agitation; aortic stenosis; serious cardiac disorders including arrhythmias; valvular heart disease; structural cardiac abnormalities; ventricular dysfunction; motor tics; history or diagnosis of Tourette's disease; glaucoma; concurrent use of MAOIs or within 14 days of their use; substance abuse; severe anxiety, psychosis, major depression, suicidal ideation.
CAUTIOUS USE Alcoholism; emotionally unstable individual; abrupt discontinuation; recent MI, anxiety,

cardiac arrhythmias, cardiac disease, dysphagia, esophageal stricture, GI obstruction, heart failure, hepatic disease, hyperthyroidism, history of paralytic ileus, cystic fibrosis, mania, radiographic contrast administration, seizure disorder, hypertension; history of seizures; older adults; pregnancy (category C); lactation; children younger than 6 y of age.

ROUTE & DOSAGE

Narcolepsy

Adult: **PO** 10 mg b.i.d. or t.i.d. 30–45 min p.c. (range: 10–60 mg/day)
Adolescent/Child (older than 6 y): **PO** 5 mg b.i.d., may increase weekly (max dose: 60 mg/day)

Attention Deficit Disorder

Adult: **PO** Immediate release products: 20–30 mg daily in divided doses. Concerta extended release product: 18–36 mg daily
Adolescent/Child (older than 6 y): **PO** 5–10 mg before breakfast and lunch, with a gradual increase of 5–10 mg/wk as needed (max: 60 mg/day) or 20–40 mg sustained release daily before breakfast (max dose: 72 mg daily). Concerta extended release product: 18 mg daily (max: 54 mg/day) **Transdermal patch** 10 mg patch worn for 9 hours × 1 wk then taper as needed. Increase no more than once weekly. Apply 2 h before desired effect.

ADMINISTRATION

Oral

- Give 30–45 min before meals. To avoid insomnia, give last dose before 6 p.m.

- Ensure that sustained release form is not chewed or crushed. It **must be** swallowed whole.
- Can open Metadate CD capsules and sprinkle on food
- Store at 15°–30° C (59°–86° F).

Transdermal

- Apply patch to hip area 2 h before desired effect and remove not later than 9 h after application. Patch may be removed earlier than 9 h if a shorter duration of effect is desired.
- Alternate application site daily. Do not apply under tight clothing.

ADVERSE EFFECTS (≥1%) **CNS:** Dizziness, drowsiness, *nervousness, insomnia.* **CV:** Palpitations, changes in BP and pulse rate, angina, cardiac arrhythmias, exacerbation of underlying CV conditions. **Special Senses:** Difficulty with accommodation, blurred vision. **GI:** Dry throat, anorexia, nausea; hepatotoxicity; abdominal pain. **Body as a Whole:** Hypersensitivity reactions (rash, fever, arthralgia, urticaria, <u>exfoliative dermatitis</u>, erythema multiforme); long-term growth suppression.

INTERACTIONS Drug: MAO INHIBITORS may cause hypertensive crisis; antagonizes hypotensive effects of **guanethidine,** potentiates action of CNS STIMULANTS (e.g. **amphetamine, caffeine**); may inhibit metabolism and increase serum levels of **fosphenytoin, phenytoin, phenobarbital,** and **primidone, warfarin,** TRICYCLIC ANTIDEPRESSANTS.

PHARMACOKINETICS Absorption: Readily from GI tract. Transdermal absorption increased with heat or inflamed skin. **Peak:** 1.9 h; 4–7 h sustained release, 2 h transdermal. **Duration:** 3–6 h; 8 h sustained release. **Elimination:** In urine.

NURSING IMPLICATIONS

Assessment & Drug Effects

- Monitor BP and pulse at appropriate intervals.
- Lab tests: Obtain periodic CBC with differential and platelet counts during prolonged therapy.
- Monitor closely patient with a history of drug dependence or alcoholism. Chronic abusive use can lead to tolerance, psychic dependence, and psychoses.
- Assess patient's condition with periodic drug-free periods during prolonged therapy.
- Supervise drug withdrawal carefully following prolonged use. Abrupt withdrawal may result in severe depression and psychotic behavior.

Patient & Family Education

- Report adverse effects to prescriber, particularly nervousness and insomnia. These effects may diminish with time or require reduction of dosage or omission of afternoon or evening dose.
- Check weight at least 2 or 3 times weekly and report weight loss. Check height and weight in children; failure to gain in either should be reported to prescriber.
- Withhold patch from an ADHD child who exhibits anxiety, tension or agitation. Consult prescriber.
- Do not apply heat or heating pad over area where patch is located.

METHYLPREDNISOLONE

(meth-ill-pred-niss'oh-lone)

Medrol

METHYLPREDNISOLONE ACETATE

Depo-Medrol

METHYLPREDNISOLONE SODIUM SUCCINATE

A-Methapred, Solu-Medrol

Classifications: BIOLOGICAL RESPONSE MODIFIER; IMMUNOSUPPRESSANT; ADRENAL CORTICOSTEROID; GLUCOCORTICOID

Therapeutic: GLUCOCORTICOID; ANTI-INFLAMMATORY

Prototype: Prednisone
Pregnancy Category: C

AVAILABILITY Methylprednisolone: 2 mg, 4 mg, 8 mg, 16 mg, 24 mg, 32 mg tablets; **Methylprednisolone Acetate:** 20 mg/mL, 40 mg/mL, 80 mg/mL injection; **Methylprednisolone Sodium Succinate:** 40 mg, 125 mg, 500 mg, 1 g, 2 g powder for injection

M

ACTION & *THERAPEUTIC EFFECT*

Intermediate-acting synthetic adrenal corticosteroid with glucocorticoid activity. It inhibits phagocytosis, and release of allergic substances. It also modifies the immune response of the body to various stimuli. **Sodium succinate** form is characterized by rapid onset of action and is used for emergency therapy of short duration. *Has anti-inflammatory and immunosuppressive properties.*

USES An anti-inflammatory agent in the management of acute and chronic inflammatory diseases, for palliative management of neoplastic diseases, and for control of severe acute and chronic allergic processes. *High-dose, short-term therapy:* Management of acute bronchial asthma, prevention of fat embolism in patient with long-bone fracture. Short-term management of rheumatic disorders.

UNLABELED USES Acetate form used as a long-acting contraceptive

Common adverse effects in *italic*, life-threatening effects <u>underlined</u>; generic names in **bold**; classifications in SMALL CAPS; ◆ Canadian drug name; ◯ Prototype drug

971

and for spinal cord injury, lupus nephritis, multiple sclerosis.

CONTRAINDICATIONS Hypersensitivity to corticosteroid drugs; Kaposi sarcoma; systemic fungal infections; use of solutions with benzyl alcohol preservative for premature infants or neonates.

CAUTIOUS USE Cushing's syndrome; GI disease, GI ulceration; hepatic disease; renal disease; hypertension; varicella, vaccinia; CHF; diabetes mellitus; ocular herpes simplex; glaucoma; coagulopathy; emotional instability or psychotic tendencies; pregnancy (category C), lactation.

ROUTE & DOSAGE

Inflammation

Adult: **PO** 2–60 mg/day in 1 or more divided doses **IM** (Acetate) 10–80 mg/wk weekly or every other week; (Succinate) 10–80 mg daily **IV** 10–40 mg prn or 30 mg/kg q4–6h × 48 h
Child: **PO/IM/IV** 0.5–1.7 mg/kg/day divided q6–12h

Status Asthmaticus

Adult/Child: **IV** 2 mg/kg then 1–5 mg/kg qh

Acute Spinal Cord Injury

Adult/Child: **IV** 30 mg/kg over 15 min, followed in 45 min by 5.4 mg/kg/h × 23 h

Obesity Dosage Adjustment

Dose based on IBW if lower than actual weight

ADMINISTRATION

Oral

- Crush tablet before and give with fluid of patient's choice.
- Note: Preparation less irritating if given with food.

- Use alternate day therapy when given over long period.

Intramuscular

- Give injection deep into large muscle (not deltoid).

Intravenous

Note: Do NOT use methylprednisolone acetate for IV.

PREPARE: **Direct/Intermittent:** Available in ACT-O-Vial from which the desired dose may be withdrawn after initial dilution with supplied diluent. ▪ May be further diluted according to prescriber's orders. Recommended dilution is 0.25 mg/mL.

ADMINISTER: **Direct:** Give each 500 mg or fraction thereof over 2–3 min. **Intermittent:** Give over 15–30 min.

INCOMPATIBILITIES **Solution/additive:** **Dextrose 5%/sodium chloride 0.45%, aminophylline, calcium gluconate, glycopyrrolate, metaraminol, nafcillin, penicillin G sodium. Y-site: Allopurinol, amsacrine, ciprofloxacin, cisatracurium** (2 mg/mL or greater concentration), **diltiazem, docetaxel, etoposide, filgrastim, fenoldopam, gemcitabine, ondansetron, paclitaxel, potassium chloride, propofol, sargramostim, vinorelbine.**

- Store at 15°–30° C (59°–86° F). Do not freeze.

ADVERSE EFFECTS (≥1%) **CNS:** Euphoria, headache, insomnia, confusion, psychosis. **CV:** CHF, edema. **GI:** Nausea, vomiting, peptic ulcer. **Musculoskeletal:** Muscle weakness, delayed wound healing, muscle wasting, osteoporosis, aseptic necrosis of bone, spontaneous fractures. **Endocrine:** Cushingoid features, growth suppression in

Common adverse effects in *italic*, life-threatening effects underlined; generic names in **bold**; classifications in SMALL CAPS; ♣ Canadian drug name; ⊙ Prototype drug

children, carbohydrate intolerance, hyperglycemia. **Special Senses:** Cataracts. **Hematologic:** Leukocytosis. **Metabolic:** Hypokalemia.

INTERACTIONS Drug: Amphotericin B, furosemide, THIAZIDE DIURETICS increase potassium loss; with ATTENUATED VIRUS VACCINES, may enhance virus replication or increase vaccine adverse effects; **isoniazid, phenytoin, phenobarbital, rifampin** decrease effectiveness of methylprednisolone because they increase metabolism of STEROIDS.

PHARMACOKINETICS Absorption: Readily absorbed from GI tract. **Peak:** 1–2 h PO; 4–8 days IM. **Duration:** 1.25–1.5 days PO; 1–5 wk IM. **Metabolism:** In liver. **Half-Life:** Greater than 3.5 h; HPA suppression: 18–36 h.

NURSING IMPLICATIONS

Assessment & Drug Effects

▪ Lab tests: Monitor periodically LFTs, kidney function, thyroid function, CBC, serum electrolytes, weight, and total cholesterol.
▪ Monitor diabetics for loss of glycemic control.
▪ Monitor serum potassium and report S&S of hypokalemia (see Appendix F).
▪ Monitor for and report S&S of Cushing's syndrome (see Appendix F).

Patient & Family Education

▪ Consult prescriber for any of the following: Slow wound healing, significant insomnia or confusion, or unexplained bone pain.
▪ Do not alter established dosage regimen (i.e., not to increase, decrease, or omit doses or change dose intervals). Withdrawal symptoms (rebound inflammation, fever) can be induced with sudden discontinuation of therapy.

▪ Report onset of signs of hypocorticism adrenal insufficiency immediately: Fatigue, nausea, anorexia, joint pain, muscular weakness, dizziness, fever.

METHYLTESTOSTERONE
(meth-ill-tess-toss'te-rone)

Android, Metandren ♣, Testred, Virilon
Classification: ANDROGEN/ANABOLIC STEROID
Therapeutic: ANABOLIC STEROID
Prototype: Testosterone
Pregnancy Category: X
Controlled Substance: Schedule III

AVAILABILITY 10 mg, 25 mg tablets

ACTION & *THERAPEUTIC EFFECT* Short-acting steroid with androgen/anabolic activity ratio (1:1) similar to that of testosterone but less effective than its esters. *Androgen activity is similar to testosterone; used in replacement therapy, and palliative treatment of postmenopausal female breast cancer.*

USES Androgen replacement therapy, delayed puberty (male), palliation of female mammary cancer (1–5 y postmenopausal), postpartum breast engorgement.

CONTRAINDICATIONS Liver dysfunction; prostate cancer; severe cardiac, renal, or hepatic disease; pregnancy (category X), lactation.
CAUTIOUS USE Mild or moderate liver, kidney, or cardiac dysfunction; heart failure; diabetes mellitus; prostatic hypertrophy.

ROUTE & DOSAGE

Replacement
Adult: **PO** 10–50 mg/day in divided doses

Common adverse effects in *italic*, life-threatening effects underlined; generic names in **bold**; classifications in SMALL CAPS; ♣ Canadian drug name; ⦿ Prototype drug

973

Breast Cancer

Adult: **PO** 50–200 mg/day in divided doses for duration of therapeutic response or no longer than 3 mo if no remission

Postpartum Breast Engorgement

Adult: **PO** 80 mg/day for 3–5 days

ADMINISTRATION

Oral

- Place buccal tablets between cheek and gum. Ensure that tablet is absorbed, not chewed or swallowed; and eating or drinking avoided until absorption is complete.
- Store at 15°–30° C (59°–86° F). Avoid freezing.

ADVERSE EFFECTS (≥1%) **GI:** <u>Cholestatic hepatitis with jaundice</u>, irritation of oral mucosa with buccal administration. **Urogenital:** Renal calculi (especially in immobilized patient), priapism. **Endocrine:** *Acne, gynecomastia, edema,* oligospermia, menstrual irregularities.

INTERACTIONS Drug: Increases risk of bleeding associated with ORAL ANTICOAGULANTS; possibly increases risk of **cyclosporine** toxicity; may decrease glucose level, making adjustment of doses of **insulin,** SULFONYLUREAS necessary. **Herbal: Echinacea** may increase risk of hepatotoxicity.

PHARMACOKINETICS Absorption: Readily from GI tract. **Metabolism:** In liver. **Elimination:** In urine.

NURSING IMPLICATIONS

Assessment & Drug Effects

- Lab tests: Monitor LFTs periodically; report signs of hepatic toxicity (see Appendix F).
- Monitor for flank pain, abdominal pain radiating to groin, or other symptoms of renal calculi.

Patient & Family Education

- Be prepared for distressing and undesirable adverse effects of virilization (women) since dosage sufficient to produce remission in breast cancer is quantitatively similar to that used for androgen replacement in the male.
- Report signs of virilism promptly. Voice change and hirsutism may be irreversible, even after drug is withdrawn.
- Report priapism (men) or other signs of excess sexual stimulation. The prescriber will terminate therapy.
- Report symptoms of jaundice with or without pruritus to prescriber; appears to be dose related. If liver function tests are altered at the same time, this drug will be withdrawn.

METIPRANOLOL HYDROCHLORIDE

(me-ti-pran'ol-ol)

OptiPranolol
Pregnancy Category: C
See Appendix A-1.

METOCLOPRAMIDE HYDROCHLORIDE ⊙

(met-oh-kloe-pra'mide)

Emex ✦, Maxeran ✦, Metozolv ODT, Octamide PFS, Reglan
Classifications: GI STIMULANT; PROKINETIC AGENT
Therapeutic: GI STIMULANT; ANTIEMETIC
Pregnancy Category: B

AVAILABILITY 5 mg, 10 mg tablets; 5 mg/5 mL solution; 5 mg/mL injection; 5 mg, 10 mg orally disintegrating tablet

ACTION & THERAPEUTIC EFFECT
Potent central dopamine receptor

antagonist that increases resting tone of esophageal sphincter, and tone and amplitude of upper GI contractions. Thus gastric emptying and intestinal transit are accelerated. Antiemetic action results from drug-induced elevation of CTZ threshold and enhanced gastric emptying. *Effective as an antiemetic agent as part of a chemotherapy regimen. In diabetic gastroparesis, it relieves anorexia, nausea, vomiting, or persistent fullness after meals.*

USES Management of diabetic gastric stasis (gastroparesis); to prevent nausea and vomiting associated with emetogenic cancer chemotherapy (e.g., cisplatin, dacarbazine) or surgery; to facilitate intubation of small bowel; symptomatic treatment of gastroesophageal reflux.

UNLABELED USES Tourette's syndrome, hiccups.

CONTRAINDICATIONS Sensitivity or intolerance to metoclopramide; uncontrolled seizures; allergy to sulfiting agents; pheochromocytoma; mechanical GI obstruction or perforation; ileus; symptomatic control of tardive dyskinesia; lactation.

CAUTIOUS USE CHF, cardiac disease; sulfite hypersensitivity, asthma, hypokalemia, hypertension; history of depression; hepatic disease, infertility, methemoglobin reductase deficiency, Parkinson's disease, kidney dysfunction; GI hemorrhage; G6PD deficiency, procainamide hypersensitivity, seizure disorder, seizures, tardive dyskinesia; history of intermittent porphyria; older adults; pregnancy (category B); longer than 12 wk.

ROUTE & DOSAGE

Gastroesophageal Reflux

Adult: **PO** 10–15 mg q.i.d. a.c. and at bedtime

Child: **PO** 0.1–0.2 mg/kg q.i.d.

Diabetic Gastroparesis

Adult: **PO/IV/IM** 10 mg q.i.d. a.c. and at bedtime for 2–8 wk
Geriatric: **PO** 5 mg a.c and at bedtime

Small-Bowel Intubation, Radiologic Examination

Adult: **IM/IV** 10 mg administered over 1–2 min
Child: **IM/IV** Younger than 6 y, 0.1 mg/kg over 1–2 min; 6–14 y, 2.5–5 mg over 1–2 min

Chemotherapy-Induced Emesis

Adult: **PO** 20–40 mg q4–6h, may repeat **IM/IV** 2 mg/kg 30 min before antineoplastic administration, may repeat q2h for 2 doses, then q3h for 3 doses if needed

Postoperative Nausea/Vomiting

Adult: **IM/IV** 10–20 mg near end of surgery

ADMINISTRATION

Oral

- Give 30 min before meals and at bedtime.
- Remove orally disintegrating tablet (ODT) from blister immediately before use. Place on tongue. ODT will melt and should then be swallowed.

Intravenous

Note: Verify correct IV concentration and rate of infusion for administration to infants or children with prescriber.

PREPARE: **Direct:** Doses of 10 mg or less may be given undiluted. **IV Infusion:** Doses greater than 10 mg IV should be diluted in at least 50 mL of D5W, NS, D5/0.45% NaCl, LR or other compatible solution.

M

Common adverse effects in *italic*, life-threatening effects underlined; generic names in **bold**; classifications in SMALL CAPS; ✤ Canadian drug name; ⊘ Prototype drug

975

ADMINISTER: **Direct:** Give over 1–2 min (or longer in pediatric patients). **IV Infusion:** Give over not less than 15 min.

INCOMPATIBILITIES **Solution/additive:** Calcium gluconate, chloramphenicol, cisplatin, dexamethasone, diphenhydramine, erythromycin, floxacillin, fluorouracil, furosemide, lorazepam, methotrexate, penicillin G potassium, sodium bicarbonate, TETRACYCLINES. **Y-site:** Allopurinol, amphotericin B cholesteryl complex, amsacrine, carmustine, cefepime, ceftobiprole, dantrolene, diazepam, diazoxide, doxorubicin liposome, furosemide, ganciclovir, gemtuzumab, milrinone acetate, lansoprazole, phenytoin, propofol, sulfamethoxazole/trimethoprim, TPN.

▪ Discard open ampules; do not store for future use. ▪ Store at 15°–30° C (59°–86° F) in light-resistant bottle. Tablets are stable for 3 y; solutions and injections, for 5 y.

ADVERSE EFFECTS (≥1%) CNS:
Mild sedation, fatigue, restlessness, agitation, headache, insomnia, disorientation, *extrapyramidal symptoms* (acute dystonic type), tardive dyskinesia, neurologic malignant syndrome with injection. **GI:** Nausea, constipation, *diarrhea,* dry mouth, altered drug absorption. **Skin:** Urticarial or maculopapular rash. **Body as a Whole:** Glossal or periorbital edema. **Hematologic:** Methemoglobinemia. **Endocrine:** Galactorrhea, gynecomastia, amenorrhea, impotence. **CV:** <u>Hypertensive crisis</u> (rare).

DIAGNOSTIC TEST INTERFERENCE
Metoclopramide may interfere with gonadorelin test by increasing *serum prolactin* levels.

INTERACTIONS Drug: Alcohol
and other CNS DEPRESSANTS add to sedation; ANTICHOLINERGICS, OPIATE ANALGESICS may antagonize effect on GI motility; PHENOTHIAZINES may potentiate extrapyramidal symptoms; may decrease absorption of **acetaminophen, aspirin, atovaquone, diazepam, digoxin, lithium, tetracycline;** may antagonize the effects of **amantadine, bromocriptine, levodopa, pergolide, ropinirole, pramipexole;** may cause increase in extrapyramidal and dystonic reactions with PHENOTHIAZINES, THIOANTHENES, **droperidol, haloperidol, loxapine, metyrosine;** may prolong neuromuscular blocking effects of **succinylcholine.**

PHARMACOKINETICS Absorption:
Readily from GI tract. **Onset:** 30–60 min PO; 10–15 min IM; 1–3 min IV. **Peak:** 1–2 h. **Duration:** 1–3 h. **Distribution:** To most body tissues including CNS; crosses placenta; distributed into breast milk. **Metabolism:** Minimally in liver. **Elimination:** 95% in urine, 5% in feces. **Half-Life:** 2.5–6 h.

NURSING IMPLICATIONS
Assessment & Drug Effects
▪ Report immediately the onset of restlessness, involuntary movements, facial grimacing, rigidity, or tremors. Extrapyramidal symptoms are most likely to occur in children, young adults, and the older adult and with high-dose treatment of vomiting associated with cancer chemotherapy. Symptoms can take months to regress.

▪ Be aware that during early treatment period, serum aldosterone may be elevated; after prolonged administration periods, it returns to pretreatment level.

- Lab tests: Periodic serum electrolytes.
- Monitor for possible hypernatremia and hypokalemia (see Appendix F), especially if patient has HF or cirrhosis.
- Adverse reactions associated with increased serum prolactin concentration (galactorrhea, menstrual disorders, gynecomastia) usually disappear within a few weeks or months after drug treatment is stopped.

Patient & Family Education

- Avoid driving and other potentially hazardous activities for a few hours after drug administration.
- Avoid alcohol and other CNS depressants.
- Report S&S of acute dystonia, such as trembling hands and facial grimacing (see Appendix F), immediately.

METOLAZONE

(me-tole′a-zone)

Zaroxolyn ♦
Classifications: THIAZIDE DIURETIC; ANTIHYPERTENSIVE
Therapeutic: DIURETIC; ANTIHYPERTENSIVE
Prototype: Hydrochlorothiazide
Pregnancy Category: D

AVAILABILITY 2.5 mg, 5 mg, 10 mg tablets

ACTION & *THERAPEUTIC EFFECT*

Diuretic action is associated with interference with transport of sodium ions across renal tubular epithelium. This enhances excretion of sodium, chloride, potassium, bicarbonate, and water. *Produces a decrease in the systolic and diastolic BPs, and reduces edema in CHF and kidney failure patients.*

USES Management of hypertension as sole agent or to enhance effectiveness of other antihypertensives in severe form of hypertension; also edema associated with CHF and kidney disease.

CONTRAINDICATIONS Hypersensitivity to metolazone and sulfonamides; anuria, hypokalemia; hepatic coma or precoma; SLE; pregnancy (category D), lactation.
CAUTIOUS USE History of gout; allergies; kidney and liver dysfunction; older adult.

ROUTE & DOSAGE

Edema
Adult: **PO** 5–20 mg/day
Child: **PO** 0.2–0.4 mg/kg/day divided q12–24h
Hypertension
Adult: **PO** 2.5–5 mg/day

ADMINISTRATION

Oral

- Do not interchange slow availability tablets and rapid availability tablets. They are not equivalent.
- Schedule doses to avoid nocturia and interrupted sleep. Give early in a.m. after eating to prevent gastric irritation (if given in 2 doses, schedule second dose no later than 3 p.m.).
- Store at 15°–30° C (59°–86° F) in tightly closed container.

ADVERSE EFFECTS (≥1%) **GI:** Cholestatic jaundice. **Body as a Whole:** Vertigo, orthostatic hypotension. **Hematologic:** Venous thrombosis, leukopenia. **Metabolic:** Dehydration, *hypokalemia, hyperuricemia, hyperglycemia.*

INTERACTIONS Drug: Amphotericin B, CORTICOSTEROIDS increase hypokalemic effects; may

Common adverse effects in *italic*, life-threatening effects <u>underlined</u>; generic names in **bold**; classifications in SMALL CAPS; ♦ Canadian drug name; ⓟ Prototype drug

977

antagonize hypoglycemic effects of SULFONYLUREAS, **insulin; cholestyramine, colestipol** decrease thiazide absorption; intensifies hypoglycemic and hypotensive effects of **diazoxide;** because of increased potassium and magnesium loss, may cause **digoxin** toxicity; decreases **lithium** excretion, increasing its toxicity; NSAIDs may attenuate diuresis—increased risk of NSAID-induced kidney failure.

PHARMACOKINETICS Absorption: Incomplete. **Onset:** 1 h. **Peak:** 2–8 h. **Duration:** 12–24 h. **Distribution:** Distributed throughout extracellular tissue; concentrates in kidney; crosses placenta; distributed in breast milk. **Metabolism:** Does not appear to be metabolized. **Elimination:** In urine. **Half-Life:** 14 h.

NURSING IMPLICATIONS

Assessment & Drug Effects

▪ Anticipate overdosage and adverse reactions in geriatric patients; may be more sensitive to effects of usual adult dose.
▪ Terminate therapy when adverse reactions are moderate to severe.
▪ Expect possible antihypertensive effects in 3 or 4 days, but 3–4 wk are required for maximum effect.
▪ Lab tests: Determine serum potassium at regular intervals. Prolonged treatment and inadequate potassium intake increase potential for hypokalemia (see Appendix F). Periodic plasma glucose and urinalysis determinations.

Patient & Family Education

▪ Do not drink alcohol; it potentiates orthostatic hypotension.
▪ Antihypertensive therapy may require as adjunct a high-potassium, low-sodium, and low-calorie diet.
▪ Include potassium-rich foods in the diet.

▪ Be aware that if hypokalemia develops, dietary potassium supplement of 1000–2000 mg (25–50 mEq) is usually adequate treatment.

METOPROLOL TARTRATE

(me-toe′proe-lole)

Apo-Metoprolol ✦, Betaloc ✦, Lopressor, Toprol XL
Classifications: CARDIOSELECTIVE; BETA-ADRENERGIC ANTAGONIST; ANTIHYPERTENSIVE; ANTIANGINAL
Therapeutic: ANTIHYPERTENSIVE; ANTIANGINAL
Prototype: Propranolol
Pregnancy Category: C

AVAILABILITY 25 mg, 50 mg, 100 mg tablets; 25 mg, 50 mg, 100 mg, 200 mg sustained release tablets; 1 mg/mL injection

ACTION & *THERAPEUTIC EFFECT*
Beta-adrenergic antagonist with preferential effect on beta$_1$ receptors located primarily on cardiac muscle. Antihypertensive action may be due to competitive antagonism of catecholamines at cardiac adrenergic neuron sites, drug-induced reduction of sympathetic outflow to the periphery, and to suppression of renin activity. *Reduces heart rate and cardiac output at rest and during exercise; lowers both supine and standing BP, slows sinus rate, and decreases myocardial automaticity. Antianginal effect is like that of propranolol.*

USES Management of mild to severe hypertension (monotherapy or in combination with a thiazide or vasodilator or both); long-term treatment of angina pectoris and prophylactic management of stable angina pectoris reduce the risk of mortality after an MI.

UNLABELED USES CHF, migraine prophylaxis, arrythmias.

CONTRAINDICATIONS Cardiogenic shock, sinus bradycardia, advanced AV block without a pacemaker, bradycardia, sick sinus syndrome; pheochromocytoma, hypotension, moderate to severe cardiac failure, right ventricular failure secondary to pulmonary hypertension; abrupt discontinuation.

CAUTIOUS USE Impaired liver or kidney function; cardiomegaly, CHF controlled by digitalis and diuretics; major depression; bronchial asthma and other bronchospastic diseases; history of allergy; thyrotoxicosis; diabetes mellitus; PVD; myasthenia gravis; cerebrovascular insufficiency; pregnancy (category C), lactation; children younger than 6 y.

ROUTE & DOSAGE

Hypertension

Adult: **PO** 50–100 mg/day in 1–2 divided doses, may increase weekly up to 100–450 mg/day
Geriatric: **PO** 25 mg/day (range: 25–300 mg/day)
Child (older than 6 y): **PO** 1 mg/kg daily (max: 200 mg)

Angina Pectoris

Adult: **PO** 100 mg/day in 2 divided doses, may increase weekly up to 100–400 mg/day

Myocardial Infarction

Adult: **IV** 5 mg q2min for 3 doses, followed by PO therapy **PO** 50 mg q6h for 48 h, then 100 mg b.i.d.

ADMINISTRATION

Oral

- Ensure that sustained-release form is not chewed or crushed. It **must be** swallowed whole.

- Give with food to slightly enhance absorption; however, administration with food not essential. It is important to give with or without food consistently to minimize possible variations in bioavailability.

Intravenous

PREPARE: **Direct:** Give undiluted.
ADMINISTER: **Direct:** Give at a rate of 5 mg over 60 sec. ▪ Note conditions which are contraindications to drug administration.
INCOMPATIBILITIES **Y-site: Amphotericin B cholesteryl complex.**

- Store at 15°–30° C (59°–86° F). Protect from heat, light, and moisture.

ADVERSE EFFECTS (≥1%) **Body as a Whole:** Hypersensitivity (erythematous rash, fever, headache, muscle aches, sore throat, <u>laryngospasm</u>, respiratory distress). **CNS:** *Dizziness, fatigue, insomnia,* increased dreaming, mental depression. **CV:** *Bradycardia,* palpitation, cold extremities, Raynaud's phenomenon, intermittent claudication, angina pectoris, CHF, intensification of AV block, AV dissociation, <u>complete heart block, cardiac arrest</u>. **GI:** Nausea, *heartburn,* gastric pain, diarrhea or constipation, flatulence. **Hematologic:** Eosinophilia, thrombocytopenic and nonthrombocytopenic purpura, <u>agranulocytosis</u> (rare). **Skin:** Dry skin, pruritus, skin eruptions. **Special Senses:** Dry mouth and mucous membranes. **Metabolic:** Hypoglycemia. **Respiratory:** Bronchospasm (with high doses), *shortness of breath.*

DIAGNOSTIC TEST INTERFERENCE In common with other beta-blockers, metoprolol may cause elevated ***BUN*** and ***serum creatinine levels*** (patients with severe heart disease),

Common adverse effects in *italic*; life-threatening effects <u>underlined</u>; generic names in **bold**; classifications in SMALL CAPS; ♣ Canadian drug name; ⊕ Prototype drug

979

elevated *serum transaminase, alkaline phosphatase, lactate dehydrogenase,* and *serum uric acid.*

INTERACTIONS Drug: BARBITU-RATES, **rifampin** may decrease effects of metoprolol; **cimetidine, methimazole, propylthiouracil,** ORAL CONTRACEPTIVES may increase effects of metoprolol; additive bradycardia with **digoxin;** effects of both metoprolol and **hydralazine** may be increased; **indomethacin** may attenuate hypotensive response; BETA AGONISTS and metoprolol mutually antagonistic; **verapamil** may increase risk of heart block and bradycardia; increases **terbutaline** serum levels.

PHARMACOKINETICS Absorption: Readily from GI tract; 50% of dose reaches systemic circulation. **Onset:** 15 min. **Peak:** 1.5 h; 20 min (IV). **Duration:** 13–19 h. **Distribution:** Crosses blood–brain barrier and placenta; distributed into breast milk. **Metabolism:** Extensively in liver (CYP2D6). **Elimination:** In urine. **Half-Life:** 3–4 h.

NURSING IMPLICATIONS

Assessment & Drug Effects

- Take apical pulse and BP before administering drug. Report to prescriber significant changes in rate, rhythm, or quality of pulse or variations in BP prior to administration.
- Monitor BP, HR, and ECG carefully during IV administration.
- Expect maximal effect on BP after 1 wk of therapy.
- Take several BP readings close to the end of a 12 h dosing interval to evaluate adequacy of dosage for patients with hypertension, particularly in patients on twice daily doses. Some patients require doses 3 times a day to maintain satisfactory control.
- Observe hypertensive patients with CHF closely for impending heart failure: Dyspnea on exertion, orthopnea, night cough, edema, distended neck veins.
- Lab tests: Obtain baseline and periodic evaluations of blood cell counts, blood glucose, LFTs and kidney function.
- Monitor I&O, daily weight; auscultate daily for pulmonary rales.
- Monitor for and report signs of mental depression.

Patient & Family Education

- Learn how to take radial pulse before each dose. Report to prescriber if pulse is slower than base rate (e.g., 60 bpm) or becomes irregular. Consult prescriber for parameters.
- Reduce insomnia or increased dreaming by avoiding late evening doses.
- Monitor blood glucose (diabetics) for loss of glycemic control. Drug may mask some symptoms of hypoglycemia (e.g., BP and HR changes) and prolong hypoglycemia. Be alert to other possible signs of hypoglycemia not affected by metoprolol and report to prescriber if present: Sweating, fatigue, hunger, inability to concentrate.
- Protect extremities from cold and do not smoke. Report cold, painful, or tender feet or hands or other symptoms of Raynaud's disease (intermittent pallor, cyanosis or redness, paresthesias). Prescriber may prescribe a vasodilator.
- Report immediately to prescriber the onset of problems with vision.
- Learn measures to relieve dry mouth; rinse mouth frequently with water, increase noncalorie liquid intake if inadequate,

M

chew sugarless gum or suck hard candy.

- Relieve eye dryness by using sterile artificial tears available OTC.
- Do not drive or engage in potentially hazardous activities until response to drug is known.
- Reduce dosage gradually over a period of 1–2 wk when drug is discontinued. Sudden withdrawal can result in increase in anginal attacks and MI in patients with angina pectoris and thyroid storm in patients with hyperthyroidism.

METRONIDAZOLE ℗

(me-troe-ni′da-zole)

Flagyl, Flagyl ER, Flagyl IV, MetroCream, MetroGel, MetroGel Vaginal, MetroLotion, Noritate, Vandazole
Classifications: ANTITRICHOMONAL; AMEBICIDE
Therapeutic: AMEBICIDE
Pregnancy Category: B

AVAILABILITY 250 mg, 500 mg tablets; 375 mg capsules; 750 mg sustained release tablets; 500 mg vials; 0.75% lotion, emulsion; 0.75%, 1% cream; 0.75%, 1% gel

ACTION & THERAPEUTIC EFFECT Synthetic compound with trichomonacidal and amebicidal activity as well as antibacterial activity. *Has direct trichomonacidal and amebicidal activity; exhibits antibacterial activity against obligate anaerobic bacteria, gramnegative anaerobic bacilli, and Clostridia.*

USES Asymptomatic and symptomatic trichomoniasis in females and males; acute intestinal amebiasis and amebic liver abscess; preoperative prophylaxis in colorectal surgery, elective hysterectomy or vaginal repair, and emergency appendectomy. IV metronidazole is used for the treatment of serious infections caused by susceptible anaerobic bacteria in intra-abdominal infections, skin infections, gynecologic infections, septicemia, and for both pre- and postoperative prophylaxis, bacterial vaginosis. *Topical:* Rosacea.

UNLABELED USES Treatment of pseudomembranous colitis, Crohn's disease, *H. pylori* eradication, bacterial vaginosis prophylaxis, gastric ulcer, pelvic inflammatory disease.

CONTRAINDICATIONS Blood dyscrasias; active CNS disease; lactation. **CAUTIOUS USE** Coexistent candidiasis; seizure disorders; heart failure; severe hepatic disease; renal impairment/failure; alcoholism; liver disease; older adults; pregnancy (category B).

ROUTE & DOSAGE

Trichomoniasis

Adult: **PO** 2 g once or 250 mg t.i.d. × 7 days
Child/Infant: **PO** 15–30 mg/kg/ day q8h × 7 days

Giardiasis, *Gardnerella*

Adult: **PO** 500 mg b.i.d. × 7 days OR 750 ER tablet daily × 7 days
Vaginal Once or twice daily × 5 days

Amebiasis

Adult: **PO** 500–750 mg t.i.d. × 5–10 days
Child: **PO** 35–50 mg/kg/day in 3 divided doses × 10 days

Anaerobic Infections

Adult: **PO** 7.5 mg/kg q6h (max: 4 g/day) **IV Loading Dose** 15 mg/ kg **IV Maintenance Dose** 7.5 mg/ kg q6h (max: 4 g/day)

Common adverse effects in *italic*, life-threatening effects underlined; generic names in **bold**; classifications in SMALL CAPS; ♣ Canadian drug name; ℗ Prototype drug

981

Child: **PO** 15–35 mg/kg/day divided q8h (max: 4 g/day) **IV** 30 mg/kg/day divided q6h *Neonate:* **IV** *Weight less than 1.2 kg, 7.5 mg q48h; younger than 7 days/weight 1.2 kg–2 kg, 7.5 mg q24h; younger than 7 days/weight greater than 2 kg, 15 mg/kg q12h; older than 7 days/weight 1.2 kg–2 kg, 15 mg/kg q12h; older than 7 days/weight greater than 2 kg, 30 mg/kg q12h*

Pseudomembranous Colitis

Adult: **PO** 250–500 mg 3–4 times daily × 10 days

Rosacea

Adult: **Topical** Apply 0.75% gel as a thin film to affected area b.i.d.; apply 1% gel as a thin film to affected area daily

ADMINISTRATION

Oral

- Crush tablets before ingestion if patient cannot swallow whole.
- Ensure that Flagyl ER (extended release form) is not chewed or crushed. It **must be** swallowed whole. Give on an empty stomach, 1 h before or 2 h after meals.
- Give immediately before, with, or immediately after meals or with food or milk to reduce GI distress.
- Give lower than normal doses in presence of liver disease.

Topical

- Apply a thin film to affected area only.

Intravenous

Note: Verify correct IV concentration and rate of infusion for administration to neonates, infants, or children with prescriber.

PREPARE: **Intermittent:** Single-dose flexible containers (500 mg/100 mL) are ready for use without further dilution. ▪ *Flagyl IV powder vial:* Sequence for preparing solution (important) consists of (1) reconstitution with 4.4 mL sterile water or NS, (2) dilution in IV solution to yield 8 mg/mL in NS, D5W, or LR, (3) pH neutralization with approximately 5 mEq sodium bicarbonate injection for each 500 mg of Flagyl IV used. ▪ Avoid use of aluminum-containing equipment when manipulating IV product (including syringes equipped with aluminum needles or hubs).

ADMINISTER: **Intermittent:** Give IV solution slowly at a rate of one dose per hour.

INCOMPATIBILITIES **Solution/additive: TPN, amoxicillin/clavulanate, aztreonam, dopamine. Y-site: Amphotericin B cholesteryl complex, aztreonam, filgrastim, meropenem, warfarin.**

▪ Note: Precipitation occurs if neutralized solution is refrigerated. ▪ Note: Use diluted and neutralized solution within 24 h of preparation.

▪ Store at 15°–30° C (59°–86° F); protect from light. ▪ Reconstituted Flagyl IV is chemically stable for 96 h when stored below 30° C (86° F) in room light. ▪ Diluted and neutralized IV solutions containing Flagyl IV should be used within 24 h of mixing.

ADVERSE EFFECTS (≥1%) **Body as a Whole:** Hypersensitivity (rash, urticaria, pruritus, flushing), fever, fleeting joint pains, overgrowth of *Candida.* **CNS:** Vertigo, headache, ataxia, confusion, irritability, depression, restlessness, weakness, fatigue, drowsiness, insomnia, paresthesias, sensory neuropathy (rare). **GI:** *Nausea,* vomiting, anorexia, epigastric distress, abdominal cramps, diarrhea, constipation, dry mouth, metallic or bitter taste, proctitis. **Urogenital:** Polyuria, dysuria, pyuria, incontinence, cystitis, decreased libido, dyspareunia, dryness of vagina and vulva, sense of pelvic pressure. **Special Senses:** Nasal congestion. **CV:** ECG changes (flattening of T wave).

DIAGNOSTIC TEST INTERFERENCE
Metronidazole may interfere with certain chemical analyses for *AST,* resulting in decreased values.

INTERACTIONS Drug: ORAL ANTI-COAGULANTS potentiate hypoprothrombinemia; **alcohol** may elicit disulfiram reaction; oral solutions of **citalopram, ritonavir; lopinavir/ritonavir,** and IV formulations of **sulfamethoxazole; trimethoprim, nitroglycerin** may elicit disulfiram reaction due to the alcohol content of the dosage form; **disulfiram** causes acute psychosis; **phenobarbital** increases metronidazole metabolism; may increase **lithium** levels; **fluorouracil, azathioprine** may cause transient neutropenia.

PHARMACOKINETICS Absorption: 80% absorbed from GI tract. **Peak:** 1–3 h. **Distribution:** Widely distributed to most body tissues, including CSF, bone, cerebral and hepatic abscesses; crosses placenta; distributed in breast milk. **Metabolism:** 30–60% in liver. **Elimination:** 77% in urine; 14% in feces within 24 h. **Half-Life:** 6–8 h.

NURSING IMPLICATIONS
Assessment & Drug Effects
▪ Discontinue therapy immediately if symptoms of CNS toxicity (see Appendix F) develop. Monitor especially for seizures and peripheral neuropathy (e.g., numbness and paresthesia of extremities).

▪ Lab tests: Obtain total and differential WBC counts before, during, and after therapy, especially if a second course is necessary.

▪ Monitor for S&S of sodium retention, especially in patients on corticosteroid therapy or with a history of CHF.

▪ Monitor patients on lithium for elevated lithium levels.

▪ Report appearance of candidiasis or its becoming more prominent with therapy to prescriber promptly.

▪ Repeat feces examinations, usually up to 3 mo, to ensure that amebae have been eliminated.

Patient & Family Education
▪ Adhere closely to the established regimen without schedule interruption or changing the dose.

▪ Refrain from intercourse during therapy for trichomoniasis unless male partner wears a condom to prevent reinfection.

▪ Have sexual partners receive concurrent treatment. Asymptomatic trichomoniasis in the male is a frequent source of reinfection of the female.

▪ Do not drink alcohol during therapy; may induce a disulfiram-type reaction (see Appendix F). Avoid alcohol or alcohol-containing medications for at least 48 h after treatment is completed.

▪ Urine may appear dark or reddish brown (especially with higher than recommended doses). This appears to have no clinical significance.

▪ Report symptoms of candidal overgrowth: Furry tongue, color changes of tongue, glossitis, stomatitis; vaginitis, curd-like, milky

M

Common adverse effects in *italic,* life-threatening effects <u>underlined;</u> generic names in **bold;** classifications in SMALL CAPS; ✤ Canadian drug name; ◎ Prototype drug

983

vaginal discharge; proctitis. Treatment with a candidacidal agent may be indicated.

METYROSINE
(me-tye'roe-seen)
Demser
Classification: ENZYME INHIBITOR
Therapeutic: PHEOCHROMOCYTOMA AGENT
Pregnancy Category: C

AVAILABILITY 250 mg capsules

ACTION & *THERAPEUTIC EFFECT* Blocks the enzyme tyrosine hydroxylase, thus inhibiting the conversion of tyrosine to DOPA, which is the initial and rate-setting step in synthesis of catecholamines (dopamine, epinephrine, norepinephrine). *In patients with pheochromocytoma, reduces catecholamine synthesis as much as 80%, ameliorating hypertensive attacks and associated symptoms.*

USES Short-term management of pheochromocytoma until surgery is performed, in long-term control when surgery is contraindicated, and in patients with malignant pheochromocytoma.
UNLABELED USES Has been used in selected patients with schizophrenia to potentiate antipsychotic effects of phenothiazines.

CONTRAINDICATIONS Control of essential hypertension; dehydration.
CAUTIOUS USE Impaired liver or kidney function; Parkinson's disease; pregnancy (category C), lactation; children younger than 12 y.

ROUTE & DOSAGE

Pheochromocytoma
Adult: **PO** 250 mg q.i.d., may increase to 2–3 g/day in divided doses (max: 4 g/day)

ADMINISTRATION
Oral
- Give each dose with a full glass of water and be consistent about time medication is to be taken.
- Store at 15°–30° C (59°–86° F).

ADVERSE EFFECTS (≥1%) **CNS:** *Sedation,* fatigue; *extrapyramidal signs: Drooling, difficulty in speaking (dysarthria), tremors,* jaw stiffness (trismus); frank parkinsonism, psychic disturbances (anxiety, depression, hallucinations, disorientation, confusion), headache, muscle spasms. **GI:** *Diarrhea,* nausea, vomiting, abdominal pain, dry mouth. **Skin:** Rash, urticaria. **Urogenital:** Transient dysuria, crystalluria, hematuria, impotence, failure of ejaculation. **Endocrine:** Breast swelling, galactorrhea. **Body as a Whole:** Peripheral edema, nasal stuffiness, shortness of breath. **Hematologic:** Eosinophilia.

DIAGNOSTIC TEST INTERFERENCE False increases in *urinary catecholamines* may occur because of catechol metabolites of metyrosine.

INTERACTIONS Drug: Alcohol and other CNS DEPRESSANTS add to sedation and CNS depression; **droperidol, haloperidol,** PHENOTHIAZINES potentiate extrapyramidal effects.

PHARMACOKINETICS Absorption: Readily absorbed from GI tract. **Peak:** 2–3 days. **Duration:** 3–4 days. **Distribution:** Crosses blood–brain barrier. **Elimination:** In urine. **Half-Life:** 3.4–7.2 h.

NURSING IMPLICATIONS
Assessment & Drug Effects

- Monitor therapeutic effectiveness with frequent assessment of vital signs.
- Monitor I&O ratio and pattern. Fluid intake **must be** enough (e.g., 10–12 glasses or more) to maintain urinary output of 2000 mL or more to minimize risk of crystalluria.
- Perform routine urinalysis; if crystals occur, increase fluid intake further. If crystalluria persists, decrease drug dosage or discontinue.
- Lab tests: Obtain baseline and periodic measurements of urinary catecholamines and their metabolites (metanephrines and VMA). Other baseline and regular determinations include LFTs, kidney function tests, and blood and urine glucose tests.
- Supervise ambulation. Sedative effects occur commonly within the first 24 h after drug is started. Maximal sedative effects in 2 or 3 days.

Patient & Family Education

- Notify prescriber if following adverse effects occur: Diarrhea, particularly if it is severe or persists, painful urination, jaw stiffness, drooling, difficult speech, tremors, disorientation. Dosage reduction or discontinuation of drug may be indicated.
- Avoid driving and potentially hazardous activities until response to drug is known.
- Be aware that abrupt withdrawal of metyrosine may result in psychic stimulation, feeling of increased energy, temporary changes in sleep pattern (usually insomnia). Symptoms may last for 2 or 3 days.
- Carry medical identification at all times if on prolonged therapy and notify all prescribers and dentists involved in care about drug regimen.

MEXILETINE
(mex-il′e-teen)

Mexitil
Classification: CLASS IB ANTIARRHYTHMIC
Therapeutic: CLASS IB ANTIARRHYTHMIC
Prototype: Lidocaine
Pregnancy Category: C

AVAILABILITY 150 mg, 200 mg, 250 mg capsules

ACTION & *THERAPEUTIC EFFECT* Analog of lidocaine with class IB antiarrhythmic properties. Shortens action potential refractory period duration and improves resting potential. Produces modest suppression of sinus node automatically and AV nodal conduction. Prolongs the His-to-ventricular interval only if patient has preexisting conduction disturbance. *Has antiarrhythmic properties for ventricular disturbances.*

USES Acute and chronic ventricular arrhythmias; prevention of recurrent cardiac arrests; suppression of PVCs due to ventricular tachyarrhythmias.

UNLABELED USES Wolff-Parkinson-White syndrome and supraventricular arrhythmias.

CONTRAINDICATIONS Severe left ventricular failure, cardiogenic shock, severe bradyarrhythmias. Preexisting second- or third-degree heart block without pacemaker; cardiogenic shock; lactation.

CAUTIOUS USE Patients with sinus node conduction irregularities, intraventricular conduction abnormalities; hypotension; severe congestive heart

Common adverse effects in *italic*, life-threatening effects <u>underlined</u>; generic names in **bold**; classifications in SMALL CAPS; ♣ Canadian drug name; ⊚ Prototype drug

985

failure; renal failure; liver dysfunction; pregnancy (category C).

ROUTE & DOSAGE

Ventricular Arrhythmias
Adult: **PO** 200–300 mg q8h (max: 1200 mg/day)
Child: **PO** 1.4–5 mg/kg q8h

ADMINISTRATION

Oral
- Give with food or milk to reduce gastric distress.

ADVERSE EFFECTS (≥1%) **CNS:** *Dizziness, tremor, nervousness, incoordination,* headache, blurred vision, paresthesias, numbness. **CV:** <u>Exacerbated arrhythmias</u>, palpitations, chest pain, syncope, hypotension. **GI:** *Nausea, vomiting, heartburn,* diarrhea, constipation, dry mouth, abdominal pain. **Skin:** Rash. **Body as a Whole:** Dyspnea, edema, arthralgia, fever, malaise, hiccups. **Urogenital:** Impotence, urinary retention.

INTERACTIONS Drug: Phenytoin, phenobarbital, rifampin may decrease mexiletine levels; **cimetidine, fluvoxamine** may increase mexiletine levels; may increase **theophylline** levels; may increase proarrhythmic effects of **dofetilide** (separate administration by at least 1 wk).

PHARMACOKINETICS Absorption: Readily from GI tract. **Peak:** 2–3 h. **Distribution:** Distributed into breast milk. **Metabolism:** In liver. **Elimination:** In urine; renal elimination increases with urinary acidification. **Half-Life:** 10–12 h.

NURSING IMPLICATIONS

Assessment & Drug Effects
- Check pulse and BP before administration; make sure both are stabilized.

- Effective serum concentration range is 0.5–2 mcg/mL.
- Lab tests: Baseline and periodic LFTs.
- Supervise ambulation in the weak, debilitated patient or the older adult during drug stabilization period. CNS adverse reactions predominate (e.g., intention tremors, nystagmus, blurred vision, dizziness, ataxia, confusion, nausea).
- Encourage drug compliance; affected particularly by the distressing adverse effects of tremor, ataxia, and eye symptoms.
- Check frequently with patient about adherence to drug regimen. If adverse effects are increasing, consult prescriber. Dose adjustment or discontinuation may be needed.

Patient & Family Education
- Learn about pulse parameters to be reported: Changes in rhythm and rate (bradycardia = pulse below 60); symptomatic bradycardia (light-headedness, syncope, dizziness), and postural hypotension.

MICAFUNGIN
(my-ca-fun'gin)
Mycamine
Classification: ANTIFUNGAL; ECHINOCANDIN
Therapeutic: ANTIFUNGAL
Prototype: Caspofungin
Pregnancy Category: C

AVAILABILITY 50 mg vial

ACTION & *THERAPEUTIC EFFECT*
Micafungin is an antifungal agent that inhibits the synthesis of glucan, an essential component of fungal cell walls. Micafungin does not allow *Candida* fungi to replicate. *Has*

antifungal effects against various species of Candida.

USES Treatment of patients with esophageal candidiasis, and for prophylaxis of *Candida* infections in patients undergoing hematopoietic stem cell transplantation. Susceptible organisms include *C. albicans, C. glabrata, C. krusei, C. parapsilosis,* and *C. tropicalis.*
UNLABELED USES Treatment of pulmonary *Aspergillus* infection.

CONTRAINDICATIONS Hypersensitivity to any component in micafungin.
CAUTIOUS USE Hepatic and renal dysfunction; older adult; pregnancy (category C); lactation; children younger than 18 y.

ROUTE & DOSAGE

Esophageal Candidiasis
Adult: **IV** 150 mg/day over 1 h × 14 days
Candidiasis Prophylaxis in Hematopoietic Stem Cell Transplantation Patients
Adult: **IV** 50 mg/day over 1 h × 18 days

ADMINISTRATION

Intravenous

PREPARE: **IV Infusion:** Reconstitute the 50 or 100 mg vial with 5 mL NS (without a bacteriostatic agent) to yield 10 mg/mL or 20 mg/mL, respectively. ▪ Gently swirl, but do not shake, to dissolve. Solution should be clear. ▪ Add required dose to 100 mL NS.
ADMINISTER: **IV Infusion:** Give slowly over 1 h. ▪ Flush existing IV line with NS before/after infusion. ▪ Protect IV solution from light.
INCOMPATIBILITIES Do not mix or infuse with any other medications.

▪ Store reconstituted vial and IV solution for up to 24 h at 25° C (77° F).
▪ Protect from light.

ADVERSE EFFECTS (≥1%) **CNS:** *Headache,* dizziness, somnolence. **CV:** Flushing, hypertension, phlebitis. **GI:** *Nausea, vomiting, diarrhea,* abdominal pain. **Hematologic/ Lymphatic:** Anemia, hemolytic anemia, leukemia, neutropenia, <u>thrombocytopenia</u>. **Hepatic:** Elevated liver enzymes, jaundice. **Metabolic:** Hypocalcemia, hypokalemia, hypomagnesemia. **Skin:** Pruritus, rash. **Body as a Whole:** Injection site pain, pyrexia, rigors.

INTERACTIONS Drug: Micafungin increases levels of **sirolimus** and **nifedipine.**

PHARMACOKINETICS Distribution: 99% protein bound. **Metabolism:** Biotransformation primarily in the liver. **Elimination:** Fecal (major) and renal. **Half-Life:** 14–17 h.

NURSING IMPLICATIONS

Assessment & Drug Effects
▪ Monitor for S&S of hypersensitivity during IV infusion; frequently monitor IV site for thrombophlebitis.
▪ Monitor for S&S of hemolytic anemia (i.e., jaundice).
▪ Lab tests: Periodic LFTs, kidney function tests, serum electrolytes, and CBC.
▪ Monitor blood levels of sirolimus or nifedipine with concurrent therapy. If sirolimus or nifedipine toxicity occurs, dosages of these drugs should be reduced.

Patient & Family Education
▪ Report immediately any of the following: Facial swelling, wheezing, difficulty breathing or swallowing, tightness in chest, rash, hives, itching, or sensation of warmth.

Common adverse effects in *italic,* life-threatening effects <u>underlined</u>; generic names in **bold;** classifications in SMALL CAPS; ✦ Canadian drug name; ⊘ Prototype drug

987

MICONAZOLE NITRATE

(mi-kon'a-zole)

Desenex, Femizol-M, Fungoid, Lotrimin AF, Micatin, Monistat 3, Monistat 7, Monistat-Derm, M-Zole, Tetterine

Classification: AZOLE ANTIFUNGAL
Therapeutic: ANTIFUNGAL
Prototype: Fluconazole
Pregnancy Category: C

AVAILABILITY 100 mg, 200 mg vaginal suppositories; 2% cream; 2% ointment; 2% powder; 2% spray; 2% solution

ACTION & *THERAPEUTIC EFFECT*
Broad-spectrum agent with fungicidal activity. Appears to inhibit uptake of components essential for cell reproduction and growth as well as cell wall structure, thus promoting cell death of fungi. *Effective against* Candida albicans *and other species of this genus. Inhibits growth of common dermatophytes, and the organism responsible for tinea versicolor.*

USES Vulvovaginal candidiasis, tinea pedis (athlete's foot), tinea cruris, tinea corporis, and tinea versicolor caused by dermatophytes.

CONTRAINDICATIONS Hypersensitivity to miconazole.
CAUTIOUS USE Hypersensitivity to azole antifungals; diabetes mellitus; bone marrow suppression; pregnancy (category C), lactation, children younger than 2 y **(topical)** and children younger than 12 y **(vaginal).**

ROUTE & DOSAGE

Fungal Infection
Adult: **Topical** Apply cream sparingly to affected areas twice a day, and once daily for tinea versicolor, for 2 wk (improvement expected in 2–3 days, tinea pedis is treated for 1 mo to prevent recurrence) **Intravaginal** Insert suppository or vaginal cream each night × 7 days (100 mg) or 3 days (200 mg)

ADMINISTRATION

Oral
- Buccal tablet should be applied to the upper gum each morning after brushing teeth. Tablet should not be crushed, chewed, or swallowed. Tablet may be repositioned if detached within the first 6 h, but if tablets are swallowed, a new tablet should be applied.

Topical
- Apply cream sparingly to intertriginous areas (between skin folds) to avoid maceration of skin.
- Massage affected area gently until cream disappears.
- Store at 15°–30° C (59°–86° F) unless otherwise directed.

ADVERSE EFFECTS (≥1%) Urogenital: Vulvovaginal burning, itching, or irritation; maceration, allergic contact dermatitis.

INTERACTIONS Drug: May increase INR with **warfarin;** may inactivate **nonoxynol-9** spermicides.

PHARMACOKINETICS Absorption: Small amount absorbed from vagina. **Metabolism:** Rapidly metabolized in liver. **Elimination:** In urine and feces. **Half-Life:** 2.1–24 h.

NURSING IMPLICATIONS

Assessment & Drug Effects
- Expect clinical improvement from topical application in 1 or 2 wk. If no improvement in 4 wk, diagnosis

is reevaluated. Treat tinea pedis infection for 1 mo to assure permanent recovery.

Patient & Family Education
- Complete full course of treatment to ensure recovery.
- Do not interrupt vaginal application during menstrual period.
- Avoid contact of drug with eyes.

MIDAZOLAM HYDROCHLORIDE
(mid′az-zoe-lam)

Classifications: ANESTHETIC; BENZODIAZEPINE; ANXIOLYTIC; SEDATIVE-HYPNOTIC
Therapeutic: ANESTHETIC; ANTIANXIETY; SEDATIVE-HYPNOTIC
Prototype: Lorazepam
Pregnancy Category: D
Controlled Substance: Schedule IV

AVAILABILITY 2 mg/mL syrup; 1 mg/mL, 5 mg/mL injection

ACTION & *THERAPEUTIC EFFECT*
Short-acting benzodiazepine that intensifies activity of gamma-aminobenzoic acid (GABA), a major inhibitory neurotransmitter of the brain, interfering with its reuptake and promoting its accumulation at neuronal synapses. Calms the patient, relaxes skeletal muscles, and in high doses produces sleep. *Is a CNS depressant with muscle relaxant, sedative-hypnotic, anticonvulsant, and amnestic properties.*

USES Sedation before general anesthesia, induction of general anesthesia; to impair memory of perioperative events (anterograde amnesia); for conscious sedation prior to short diagnostic and endoscopic procedures; and as the hypnotic supplement to nitrous oxide and oxygen (balanced anesthesia) for short surgical procedures.

CONTRAINDICATIONS Intolerance to benzodiazepines; acute narrow-angle glaucoma; shock, coma; acute alcohol intoxication; intra-arterial injection; status asthmaticus; pregnancy (category D), obstetric delivery, lactation.

CAUTIOUS USE COPD; chronic kidney failure; cardiac disease; pulmonary insufficiency; dementia; electrolyte imbalance; neuromuscular disease; Parkinson's disease; psychosis; CHF; bipolar disorder; older adults. .

ROUTE & DOSAGE

Conscious Sedation

Adult: **IM** 0.07–0.08 mg/kg 30–60 min before procedure **IV** 1–2.5 mg, may repeat in 2 min prn; Intubated Patients, 0.05–0.2 mg/kg/h by continuous infusion
Child: **IM** 0.08 mg/kg × 1 dose **PR** 0.3 mg/kg × 1 dose; Intubated Patients, 2 mcg/kg/min by continuous infusion, may increase by 1 mcg/kg/min q30min until light sleep is induced
Neonate: **IV** 0.5–1 mcg/kg/min

IV Induction for General Anesthesia

Adult: **IV** Premedicated, 0.15–0.25 mg/kg over 20–30 sec, allow 2 min for effect **IV** Non-premedicated, 0.3–0.35 mg/kg over 20–30 sec, allow 2 min for effect
Child: **IV** 0.15 mg/kg followed by 0.05 mg/kg q2min × 1–3 doses

Status Epilepticus

Child: **IV Loading Dose** Older than 2 mo, 0.15 mg/kg **IV Maintenance Dose** 1 mcg/kg/min infusion, may titrate upward as needed q5min

M

Common adverse effects in *italic*; life-threatening effects underlined; generic names in **bold**; classifications in SMALL CAPS; ♣ Canadian drug name; ⊙ Prototype drug

989

Preoperative Sedation

Child: **PO** *Younger than 5 y,* 0.5 mg/kg; *older than 5 y,* 0.4–0.5 mg/kg

ADMINISTRATION

Intramuscular

- Inject IM drug deep into a large muscle mass.

Intravenous

PREPARE: **Direct:** Dilute in D5W or NS to a concentration of 0.25 mg/mL (e.g., 1 mg in 4 mL or 5 mg in 20 mL). **IV Infusion:** Add 5 mL of the 5 mg/mL concentration to 45 mL of D5W or NS to yield 0.5 mg/mL.

ADMINISTER: **Direct for Conscious Sedation:** Give over 2 min or longer. **Direct for Induction of Anesthesia:** Give over 20–30 sec. **Direct for Neonate:** DO NOT give bolus dose; give over at least 2 min. **IV Infusion:** Give at a rate based on weight.

INCOMPATIBILITIES **Solution/additive: Lactated Ringer's, pentobarbital, perphenazine, prochlorperazine. Y-site: Albumin, amoxicillin, amoxicillin/clavulanate, amphotericin B cholesteryl complex, ampicillin, bumetanide, butorphanol, ceftazidime, cefuroxime, clonidine, dexamethasone, foscarnet, fosphenytoin, furosemide, hydrocortisone, imipenem/cilastatin, methotrexate, nafcillin, omeprazole, sodium bicarbonate, thiopental,** TPN, **trimethoprim/sulfamethoxazole.**

- Store at 15°–30° C (59°–86° F), therapeutic activity is retained for 2 y from date of manufacture.

ADVERSE EFFECTS (≥1%) **CNS:** *Retrograde amnesia,* headache, euphoria, drowsiness, excessive sedation, confusion. **CV:** Hypotension. **Special Senses:** Blurred vision, diplopia, nystagmus, pinpoint pupils. **GI:** Nausea, vomiting. **Respiratory:** Coughing, laryngospasm (rare), respiratory arrest. **Skin:** Hives, swelling, burning, pain, induration at injection site, tachypnea. **Body as a Whole:** Hiccups, chills, weakness.

INTERACTIONS Drug: Alcohol, CNS DEPRESSANTS, ANTICONVULSANTS potentiate CNS depression; **cimetidine** increases midazolam plasma levels, increasing its toxicity; may decrease antiparkinsonism effects of **levodopa;** may increase **phenytoin** levels; **smoking** decreases sedative and antianxiety effects. **Food: Grapefruit juice** (greater than 1 qt/day) may increase risk of myopathy and rhabdomyolysis. **Herbal: Kava, valerian** may potentiate sedation. **Echinacea, St. John's wort** may reduce efficacy.

PHARMACOKINETICS Onset: 1–5 min IV; 5–15 min IM, 20–30 min PO. **Peak:** 20–60 min. **Duration:** Less than 2 h IV; 1–6 h IM. **Distribution:** Crosses blood–brain barrier and placenta. **Metabolism:** In liver (CYP3A4). **Elimination:** In urine. **Half-Life:** 1–4 h.

NURSING IMPLICATIONS

Assessment & Drug Effects

- Inspect insertion site for redness, pain, swelling, and other signs of extravasation during IV infusion.
- Monitor closely for indications of impending respiratory arrest. Resuscitative drugs and equipment should be immediately available.
- Monitor for hypotension, especially if the patient is premedicated with a narcotic agonist analgesic.

Common adverse effects in *italic*, life-threatening effects underlined; generic names in **bold**; classifications in SMALL CAPS; ✤ Canadian drug name; ⊘ Prototype drug

- Monitor vital signs for entire recovery period. In obese patient, half-life is prolonged during IV infusion; therefore, duration of effects is prolonged (i.e., amnesia, postoperative recovery).
- Be aware that overdose symptoms include somnolence, confusion, sedation, diminished reflexes, coma, and untoward effects on vital signs.

Patient & Family Education

- Do not drive or engage in potentially hazardous activities until response to drug is known. You may feel drowsy, weak, or tired for 1–2 days after drug has been given.

MIDODRINE HYDROCHLORIDE

(mid'o-dreen)

Classification: VASOPRESSOR
Therapeutic: ANTIHYPOTENSIVE
Prototype: Dexmedetomide
Pregnancy Category: C

AVAILABILITY 2.5 mg, 5 mg, 10 mg tablets

ACTION & *THERAPEUTIC EFFECT*
Vasopressor and alpha$_1$ agonist that activates the alpha-adrenergic receptors of the arteries and veins, resulting in increased vascular tone and elevation in blood pressure. *Affects standing, sitting, and supine systolic and diastolic blood pressures. Effectiveness indicated by an increase in 1-min standing systolic BP and subjective feelings of clinical improvement.*

USES Orthostatic hypotension.

CONTRAINDICATIONS Severe organic heart disease; heart failure; kidney disease, renal failure; urinary retention; pheochromocytoma; thyrotoxicosis; MAOI therapy;

persistent and excessive supine hypertension.
CAUTIOUS USE Renal impairment, hepatic impairment; history of visual problems; diabetes with hypotension or visual disorders; pregnancy (category C), lactation. Safety and efficacy in children are not established.

ROUTE & DOSAGE

Orthostatic Hypotension

Adult: **PO** 10 mg t.i.d. during the daytime hours, dosed not less than 3 h apart with last dose at least 4 h before bedtime (max: 20 mg/dose)

ADMINISTRATION

Oral

- Do not give at bedtime or before napping (within 4 h of lying supine for any length of time).
- Give with caution in persons with pretreatment, supine systolic BP 170 mm Hg or higher.
- Store at 15°–30° C (59°–86° F).

ADVERSE EFFECTS (≥1%) **Body as a Whole:** *Paresthesia,* chills, pain, facial flushing. **CNS:** Confusion, nervousness, anxiety. **CV:** *Hypertension.* **GI:** Dry mouth. **Skin:** *Pruritus, piloerection,* rash. **Urogenital:** *Dysuria, urinary retention, urinary frequency.*

INTERACTIONS Drug: May antagonize effects of **doxazosin, prazosin, terazosin;** may potentiate vasoconstrictive effects of **ephedrine, phenylephrine, pseudoephedrine;** may cause hypertensive crisis with MAOIS.

PHARMACOKINETICS Absorption: Rapidly from GI tract. **Peak:** Midodrine 0.5 h; desglymidodrine 1–2 h. **Metabolism:** Rapidly metabolized

M

Common adverse effects in *italic*, life-threatening effects underlined; generic names in **bold**; classifications in SMALL CAPS; ♣ Canadian drug name; ⊘ Prototype drug

991

to the active metabolite. **Elimination:** In urine. **Half-Life:** 25 min.

NURSING IMPLICATIONS

Assessment & Drug Effects

- Lab tests: Evaluate LFTs and kidney function prior to initiating therapy.
- Monitor supine and standing BP regularly. Withhold drug and notify prescriber if supine BP increases excessively; determine acceptable parameters.
- Monitor carefully effect of the drug in diabetics with orthostatic hypotension and those taking fludrocortisone acetate, which may increase intraocular pressure.

Patient & Family Education

- Take last daily dose 4 h before bedtime.
- Report immediately to prescriber sensations associated with supine hypertension (e.g., pounding in ears, headache, blurred vision, awareness of heart beating).
- Discontinue drug and report to prescriber if S&S of bradycardia develop (e.g., dizziness, pulse slowing, fainting).
- Do not take allergy drugs, cold preparations, or diet pills without consulting prescriber.

MIGLITOL
(mig′li-toll)
Glyset
Classifications: ANTIDIABETIC; ALPHA-GLUCOSIDASE INHIBITOR
Therapeutic: ANTIDIABETIC; GLYCEMIC CONTROL ENHANCER
Prototype: Acarbose
Pregnancy Category: B

AVAILABILITY 25 mg, 50 mg, 100 mg tablets

ACTION & _THERAPEUTIC EFFECT_
Enzyme inhibition of intestinal glucosidases delaying the formation of glucose from saccharides in the small intestine. Miglitol does not enhance insulin secretion. _It delays the digestion of carbohydrates, lowers the postprandial hyperglycemia, and reduces the levels of glysylated hemoglobin (HbA1C) in type 2 diabetics._

USES Adjunct to diet for control of type 2 diabetes.
UNLABELED USES Type 1 diabetes.

CONTRAINDICATIONS Hypersensitivity to miglitol; diabetic ketoacidosis; digestive or absorptive disorders; history of or partial intestinal obstruction, IBD; lactation.
CAUTIOUS USE Hypersensitivity to acarbose; creatinine clearance greater than 2 mg/dL; high stress conditions (i.e., surgery, trauma, etc.); pregnancy (category B). Safety and efficacy in children younger than 18 y unknown.

ROUTE & DOSAGE

Type 2 Diabetes Mellitis
Adult: **PO** 25 mg t.i.d. at the start of each meal, may increase after 4–8 wk to 50 mg t.i.d. (max: 100 mg t.i.d.)

ADMINISTRATION

Oral

- Give drug with first bite of each of the three main meals.
- Store at 15°–30° C (59°–86° F).

ADVERSE EFFECTS (≥1%) **GI:** _Abdominal pain, diarrhea, flatulence._ **Skin:** Rash. **Metabolic:** Hypoglycemia.

INTERACTIONS Drug: Miglitol may reduce bioavailability of **propranolol, ranitidine; charcoal, pancreatin, amylase, pancrelipase** may decrease effectiveness of miglitol.

Common adverse effects in _italic_, life-threatening effects <u>underlined</u>; generic names in **bold;** classifications in SMALL CAPS; ♣ Canadian drug name; ☻ Prototype drug

Herbal: Garlic, ginseng may potentiate hypoglycemic effects.

PHARMACOKINETICS Absorption: 25 mg dose is completely absorbed, amount absorbed decreases with increasing dose to where 100 mg dose is 50–70% absorbed. **Peak:** 2–3 h. **Distribution:** Minimal protein binding (less than 4%). **Metabolism:** Not metabolized. **Elimination:** Half-life 2 h; 95% excreted unchanged in urine, lower doses should be used in patients with renal impairment.

NURSING IMPLICATIONS

Assessment & Drug Effects

- Monitor for therapeutic effectiveness: Indicated by improved blood glucose levels and decreased HbA1C.
- Monitor for S&S of hypoglycemia when used in combination with sulfonylureas, insulin, other hypoglycemia agents.
- Lab tests: Monitor daily postprandial blood glucose values and HbA1C q3mo.
- Treat hypoglycemia with oral glucose (dextrose); miglitol interferes with the breakdown of sucrose (table sugar).

Patient & Family Education

- Keep a source of oral glucose available to treat low blood sugar; miglitol prevents digestive breakdown of table sugar.
- Abdominal discomfort, flatulence, and diarrhea tend to diminish with continued therapy.

MILNACIPRAN

(mil-na-see'pran)
Savella
Classifications: ANTIDEPRESSANT; SEROTONIN NOREPINEPHRINE REUPTAKE INHIBITOR (SNRI); ANALGESIC

Therapeutic: ANALGESIC; SNRI
Prototype: Venlafaxine
Pregnancy Category: C

AVAILABILITY 12.5 mg, 25 mg, 50 mg, 100 mg tablets

ACTION & *THERAPEUTIC EFFECT*

Exact mechanism of central pain inhibition is unknown. Is a potent inhibitor of both neuronal norepinephrine and serotonin reuptake without affecting uptake of other neurotransmitters. *Effective in reducing the pain associated with fibromyalgia.*

USES Management of fibromyalgia. **UNLABELED USES** Treatment of depression.

CONTRAINDICATIONS Within 14 days discontinued use of MAOIs; abrupt discontinuation of milnacipran; suicidal ideation; major depressive disorder; uncontrolled narrow-angle glaucoma; substantial alcohol use; chronic liver disease; ESRD; lactation.

CAUTIOUS USE Suicidal tendencies; history of seizures or depression; history of cardiac disease or pre-existing tachyarrhythmias; male obstructive uropathies; history of GI bleeding; moderate and severe renal impairment; hepatic impairment; older adults; pregnancy (category C); children younger than 17 y.

ROUTE & DOSAGE

Fibromyalgia

Adults/Adolescents (17 y or older): **PO** Initial dose of 12.5 mg once daily on day 1, increase to 12.5 mg b.i.d. on days 2 and 3, 25 mg b.i.d. on days 4–7, and then 50 mg b.i.d. Dose can be

M

Common adverse effects in *italic*, life-threatening effects underlined; generic names in **bold**; classifications in SMALL CAPS; ✤ Canadian drug name; ● Prototype drug

993

increased to 100 mg b.i.d. if needed.

Renal Impairment Dosage Adjustment

CrCl 5–29 mL/min: Decrease dose by 50% (i.e., 25–50 mg b.i.d.); less than 5 mL/min: Use not recommended

ADMINISTRATION

Oral

- Dose titration should occur over a period of 1 wk to the recommended dose.
- Give with food, if needed, to improve tolerability of drug.
- Do not give within 14 days of an MAOI.
- Store at 15°–30° C (59°–86° F).

ADVERSE EFFECTS (≥2%) **Body as a Whole:** Chest pain and discomfort, chills. **CNS:** Anxiety, *dizziness, headache,* hypoesthesia, *insomnia,* migraine, paresthesia, tension headache, tremor. **CV:** *Hot flush,* increased blood pressure, increased heart rate, palpitations, tachycardia. **GI:** Abdominal pain, *constipation,* dry mouth, *nausea,* vomiting. **Metabolic:** Decreased appetite. **Respiratory:** Dyspnea, upper respiratory tract infection. **Skin:** Hyperhidrosis, pruritus, rash. **Special Senses:** Blurred vision. **Urogenital:** Decreased urine flow, dysuria, ejaculation disorder, erectile dysfunction, libido decreased, prostatitis, scrotal pain, testicular pain and swelling, urinary hesitation and retention, urethral pain.

INTERACTIONS Drug: Lithium may increase risk of serotonin syndrome. Milnacipran may inhibit antihypertensive effect of **clonidine** and other ALPHA₂ AGONISTS. **Digoxin** may increase the risk of postural hypotension and tachycardia. Milnacipran may increase bleeding with **warfarin, aspirin**, and NSAIDS; concurrent use with **epinephrine** or **norepinephrine** may cause paroxysmal hypertension and arrhythmia; concurrent use with SELECTIVE SEROTONIN REUPTAKE INHIBITORS, SELECTIVE NOREPINEPHRINE REUPTAKE INHIBITORS, **tramadol**, OR 5-HT-2B/2D AGONISTS (TRIPTANS) may cause additive serotonergic effects, hypertension, and coronary vasoconstriction.

PHARMACOKINETICS Absorption: 85–90% bioavailability. **Distribution:** Minimal (13%) plasma protein binding. **Metabolism:** Less than 50% metabolized by liver. **Elimination:** Primarily renal. **Half-Life:** 6–8 h.

NURSING IMPLICATIONS

Assessment & Drug Effects

- Monitor for and report promptly unusual changes in behavior (e.g., depression, anxiety, panic attack, insomnia, aggressiveness, mania) or suicidal ideation. Monitor closely during initial few months of therapy and during periods of dosage adjustment.
- Monitor HR and BP closely and report promptly sustained BP elevations. Pre-existing hypertension should be controlled before initiating this drug.
- Monitor for orthostatic hypotension and tachycardia with concurrent digoxin use.
- Lab tests: Periodic serum sodium, especially with concurrent diuretic therapy.

Patient & Family Education

- Do not abruptly stop taking this drug. It should be tapered off gradually after extended use.
- Report prompt unusual changes in behavior or suicidal ideas.
- Do not engage in potentially hazardous activities until reaction to drug is known.

Common adverse effects in *italic*, life-threatening effects <u>underlined</u>; generic names in **bold**; classifications in SMALL CAPS; ♣ Canadian drug name; ❷ Prototype drug

- Exercise care to take prescribed BP medications exactly as ordered.
- Concurrent use of aspirin or NSAIDs is not recommended due to increased risk of bleeding. Consult prescriber.
- Avoid consuming alcohol while taking this drug.

MILRINONE LACTATE ⓟ

(mil'ri-none)

Primacor
Classifications: INOTROPIC AGENT; VASODILATOR
Therapeutic: INOTROPIC AGENT
Prototype: Milrinone acetate
Pregnancy Category: C

AVAILABILITY 200 mcg/mL, 1 mg/mL injection

ACTION & *THERAPEUTIC EFFECT*
Has a positive inotropic action and is a vasodilator with little chronotropic activity. Inhibitory action against cyclic-AMP phosphodiesterase in cardiac and smooth vascular muscle. Increases cardiac contractility and myocardial contractility. *Therefore, increases cardiac output and decreases pulmonary wedge pressure and vascular resistance, without increasing myocardial oxygen demand or significantly increasing heart rate.*

USES Short-term management of CHF.
UNLABELED USES Short-term use to increase the cardiac index in patients with low cardiac output after surgery. To increase cardiac function prior to heart transplantation.

CONTRAINDICATIONS Hypersensitivity to milrinone; valvular heart disease; acute MI.

CAUTIOUS USE Atrial fibrillation, atrial flutter; renal disease; renal impairment, renal failure; older adults; pregnancy (category C), lactation; children.

ROUTE & DOSAGE

Heart Failure
Adult: **IV Loading Dose** 50 mcg/kg IV over 10 min **IV Maintenance Dose** 0.375–0.75 mcg/kg/min

ADMINISTRATION

Intravenous
Note: Correct preexisting hypokalemia before administering milrinone. ▪ See manufacturer's information for dosage reduction in the presence of renal impairment.

PREPARE: **IV Infusion Loading Dose:** Give undiluted or dilute each 1 mg in 1 mL NS or 0.45% NaCl. **IV Infusion Maintenance Dose:** Dilute 20 mg of milrinone in D5W, NS, or 0.45% NaCl to yield: 100 mcg/mL with 180 mL diluent; 150 mcg/mL with 113 mL diluent; 200 mcg/mL with 80 mL diluent.
ADMINISTER: **IV Infusion Loading Dose:** Give 50 mcg/kg over 10 min. **IV Infusion Maintenance Dose:** Give at a rate based on weight. Use a microdrip set and infusion pump.
INCOMPATIBILITIES **Solution/additive: Furosemide, procainamide. Y-site: Furosemide, imipenem/cilastatin, procainamide.**

▪ Store according to manufacturer's directions.

ADVERSE EFFECTS (≥1%) **CV:** Increased ectopic activity, PVCs,

Common adverse effects in *italic*, life-threatening effects underlined; generic names in **bold**; classifications in SMALL CAPS; ♣ Canadian drug name; ⓟ Prototype drug

995

ventricular tachycardia, ventricular fibrillation, supraventricular arrhythmias; possible increase in angina symptoms, hypotension.

INTERACTIONS Drug: Disopyramide may cause excessive hypotension.

PHARMACOKINETICS Peak: 2 min. **Duration:** 2 h. **Distribution:** 70% protein bound. **Elimination:** 80–85% excreted unchanged in urine within 24 h. Active renal tubular secretion is primary elimination pathway. **Half-Life:** 1.7–2.7 h.

NURSING IMPLICATIONS

Assessment & Drug Effects

- Monitor cardiac status closely during and for several hours following infusion. Supraventricular and ventricular arrhythmias have occurred.
- Monitor BP and promptly slow or stop infusion in presence of significant hypotension. Closely monitor those with recent aggressive diuretic therapy for decreasing blood pressure.
- Monitor fluid and electrolyte status. Hypokalemia should be corrected whenever it occurs during administration.

Patient & Family Education

- Report immediately angina that occurs during infusion to prescriber.
- Be aware that drug may cause a headache, which can be treated with analgesics.

MINOCYCLINE HYDROCHLORIDE

(mi-noe-sye′kleen)

Arestin, Dynacin, Minocin, Solodyn

Classification: TETRACYCLINE ANTIBIOTIC
Therapeutic: ANTIBIOTIC
Prototype: Tetracycline
Pregnancy Category: D

AVAILABILITY 50 mg, 75 mg, 100 mg capsules; 50 mg, 75 mg, 100 mg tablets; 50 mg/5 mL suspension; 1 mg sustained release microspheres

ACTION & *THERAPEUTIC EFFECT*
Semisynthetic tetracycline derivative with bacteriostatic action that appears to be a result of reversible binding to ribosomal units of susceptible bacteria and inhibition of bacterial protein synthesis. *Effective against gram-positive and gram-negative bacteria, but usually used against gram-negative bacteria.*

USES Treatment of mucopurulent cervicitis, granuloma inguinale, lymphogranuloma venereum, proctitis, bronchitis, lower respiratory tract infections caused by *Mycoplasma pneumoniae,* Rickettsial infections, chlamydial infections, non-gonococcal urethritis, chlamydial conjunctivitis, plague, brucellosis, bartonellosis, tularemia, UTI, and prostatitis; acne vulgaris, gonorrhea, cholera, meningococcal carrier state.

CONTRAINDICATIONS Hypersensitivity to tetracyclines; oral administration in meningococcal infections; sunlight (UV) exposure; pregnancy (category D).
CAUTIOUS USE Renal and hepatic impairment; older adults; lactation; children younger than 8 y.

ROUTE & DOSAGE

Anti-Infective
Adult: **PO** 200 mg followed by 100 mg q12h

Common adverse effects in *italic*, life-threatening effects <u>underlined</u>; generic names in **bold**; classifications in SMALL CAPS; ♣ Canadian drug name; ⊘ Prototype drug

Child (older than 8 y): **PO** 4.4 mg/kg followed by 2 mg/kg q12h

Acne

Adult: **PO** 50 mg 1–3 times/day

Moderate to Severe Acne (Solodyn only)

Adult/Adolescent/Child (over 12): **PO** 1 mg/kg daily x 12 wk

Meningococcal Carrier State

Adult: **PO** 100 mg q12h x 5 days
Child (older than 8 y): **PO** 4 mg/kg followed by 2 mg/kg q12h x 5 days (max: 100 mg/dose)

ADMINISTRATION

Oral

- Shake suspension well before administration.
- Ensure that sustained release tablets are swallowed whole.
- Check expiration date. Outdated tetracycline can cause severe adverse effects.

ADVERSE EFFECTS (≥1%) **CNS:** *Weakness, light-headedness, ataxia, dizziness, or vertigo.* **GI:** Nausea, cramps, diarrhea, flatulence. **Hepatic:** Hepatitis, increased liver enzyme, <u>hepatotoxicity</u>.

INTERACTIONS Drug: ANTACIDS, **iron, calcium, magnesium, zinc, kaolin and pectin, sodium bicarbonate, bismuth subsalicylate** can significantly decrease minocycline absorption; effects of both **desmopressin** and minocycline antagonized; increases **digoxin** absorption, increasing risk of **digoxin** toxicity; **methoxyflurane** increases risk of kidney failure. **Food:** Dairy products significantly decrease minocycline absorption; food may also decrease its absorption.

PHARMACOKINETICS Absorption: 90–100% from GI tract. **Peak:** 2–3 h. **Distribution:** Tends to accumulate in adipose tissue; crosses placenta; distributed into breast milk. **Metabolism:** Partially metabolized. **Elimination:** 20–30% in feces; ~12% in urine. **Half-Life:** 11–26 h.

NURSING IMPLICATIONS

Assessment & Drug Effects

- Obtain history of hypersensitivity reactions prior to administration; drug is contraindicated with known tetracycline hypersensitivity.
- Lab: C&S should be drawn prior to initiation of therapy.
- Monitor carefully for signs of hypersensitivity response (see Appendix F), particularly in patients with history of allergies, especially to drugs.
- Monitor at-risk patients for S&S of superinfection (see Appendix F).
- Assess risk of toxic effects carefully; increases with renal and hepatic impairment.
- Supervise ambulation, since lightheadedness, dizziness, and vertigo occur frequently.

Patient & Family Education

- Avoid hazardous activities or those requiring alertness while taking minocycline.
- Use sunscreen when outdoors and otherwise protect yourself from direct sunlight since photosensitivity reaction may occur.
- Report vestibular adverse effects (e.g., dizziness), which usually occur during first week of therapy. Effects are reversible if drug is withdrawn.
- Report loose stools or diarrhea or other signs of superinfection promptly to prescriber.
- Use or add barrier contraceptive while taking this drug if using hormonal contraceptive.

Common adverse effects in *italic*, life-threatening effects <u>underlined</u>; generic names in **bold**; classifications in SMALL CAPS; ♣ Canadian drug name; ✪ Prototype drug

997

MINOXIDIL

(mi-nox′i-dill)

Rogaine
Classifications: NONNITRATE VASODILATOR; ANTIHYPERTENSIVE
Therapeutic: ANTIHYPERTENSIVE
Prototype: Hydralazine
Pregnancy Category: C

AVAILABILITY 2.5 mg, 10 mg tablets; 2% solution

ACTION & *THERAPEUTIC EFFECT*
Direct-acting vasodilator that appears to act by blocking calcium uptake through cell membranes. Reduces elevated systolic and diastolic blood pressures in supine and standing positions, by decreasing peripheral vascular resistance. *Effective as an antihypertensive. It increases heart rate and cardiac output. Topical minoxidil: reverses balding to some degree.*

USES Treat severe hypertension that is symptomatic or associated with damage to target organs and is not manageable with maximum therapeutic doses of a diuretic plus two other antihypertensive drugs. Used with a diuretic to prevent fluid retention and a beta adrenergic blocking agent (e.g., propranolol) or an alpha-adrenergic agonist (e.g., clonidine or methyldopa) to prevent tachycardia. **Topical:** to treat alopecia areata and male pattern alopecia.

CONTRAINDICATIONS Pheochromocytoma; mild hypertension; recent acute MI, dissecting aortic aneurysm, valvular dysfunction; pulmonary hypertension; lactation.
CAUTIOUS USE Severe renal impairment; recent MI (within preceding month); coronary artery disease, chronic CHF; pregnancy (category C); children younger than 12 y.

ROUTE & DOSAGE

Hypertension

Adult: **PO** 5 mg/day, increased q3–5 days up to 40 mg/day in single or divided doses as needed (max: 100 mg/day)
Child: **PO** 0.2 mg/kg/day (max: 5 mg/day) initially, gradually increased to 0.25–1 mg/kg/day in divided doses (max: 50 mg/day)

Alopecia

Adult: **Topical** Apply 1 mL of 2% solution to affected area b.i.d.

ADMINISTRATION

Oral

- Dose increments are usually made at 3–5 days intervals. If more rapid adjustment is necessary, adjustments can be made q6h with careful monitoring.

Topical

- Do not apply topical product to an irritated scalp (e.g., sunburn, psoriasis).
- Store at 15°–30° C (59°–86° F) in tightly covered container unless otherwise directed.

ADVERSE EFFECTS (≥1%) **CV:** *Tachycardia,* angina pectoris, *ECG changes,* pericardial effusion and tamponade, rebound hypertension (following drug withdrawal); *edema,* including pulmonary edema; *CHF (salt and water retention).* **Skin:** *Hypertrichosis,* transient pruritus, darkening of skin, hypersensitivity rash, <u>Stevens-Johnson syndrome</u>. With topical use: Itching, flushing, scaling, dermatitis, folliculitis. **Body as a Whole:** Fatigue.

INTERACTIONS Drug: Epineph-rine, norepinephrine cause excessive cardiac stimulation; **guanethidine** causes profound orthostatic hypotension.

PHARMACOKINETICS Absorption: Readily absorbed from GI tract. **Onset:** 30 min PO; at least 4 mo topical. **Peak:** 2–8 h PO. **Duration:** 2–5 days PO; new hair growth will remain 3–4 mo after withdrawal of topical. **Distribution:** Widely distributed including into breast milk. **Metabolism:** In liver. **Elimination:** 97% in urine and feces. **Half-Life:** 4.2 h.

NURSING IMPLICATIONS

Assessment & Drug Effects

- Take BP and apical pulse before administering medication and report significant changes. Consult prescriber for parameters.
- Lab tests: Periodic serum electrolytes.
- Do not stop drug abruptly. Abrupt reduction in BP can result in CVA and MI. Keep prescriber informed.
- Monitor fluid and electrolyte balance closely throughout therapy. Sodium and water retention commonly occur. Consult prescriber regarding sodium restriction. Monitor potassium intake and serum potassium levels in patient on diuretic therapy.
- Monitor I&O and daily weight. Report unusual changes in I&O ratio or daily weight gain, greater than 1 kg (2 lb).
- Observe patient daily for edema and auscultate lungs for rales. Be alert to signs and symptoms of CHF (see Appendix F).
- Observe for symptoms of pericardial effusion or tamponade. Symptoms are similar to those of CHF, but additionally patient may have paradoxical pulse (normal inspiratory reduction in systolic BP may fall as much as 10–20 mm Hg).

Patient & Family Education

- Learn about usual pulse rate and count radial pulse for one full minute before taking drug. Report an increase of 20 or more bpm.
- Notify prescriber promptly if the following S&S appear: Increase of 20 or more bpm in resting pulse; breathing difficulty; dizziness; light-headedness; fainting; edema (tight shoes or rings, puffiness, pitting); weight gain, chest pain, arm or shoulder pain; easy bruising or bleeding.
- Be aware of possibility of hypertrichosis: Elongation, thickening, and increased pigmentation of fine body hair, especially of face, arms, and back. Develops 3–9 wk after start of therapy and occurs in approximately 80% of patients; reversible within 1–6 mo after drug withdrawal.

MIRTAZAPINE ℗

(mir-taz′a-peen)
Remeron, Remeron SolTab
Classifications: TETRACYCLIC ANTIDEPRESSANT; ANXIOLYTIC
Therapeutic: ANTIDEPRESSANT; ANTIANXIETY
Pregnancy Category: C

AVAILABILITY 15 mg, 30 mg, 45 mg tablets and orally disintegrating tablets; 7.5 mg tablet

ACTION & THERAPEUTIC EFFECT
Tetracyclic antidepressant pharmacologically and therapeutically similar to tricyclic antidepressants. Tetracyclics enhance central nonadrenergic and serotonergic activity; mechanism of action thought to be due to normalization of neurotransmission efficacy.

Common adverse effects in *italic*, life-threatening effects underlined; generic names in **bold**; classifications in SMALL CAPS; ♦ Canadian drug name; ℗ Prototype drug

999

Mirtazapine is a potent antagonist of 5-HT_2 and 5-HT_3 serotonin receptors. *Acts as antidepressant. Effectiveness is indicated by mood elevation.*

USES Treatment of depression.
UNLABELED USES Pruritus, tremor.

CONTRAINDICATIONS Hypersensitivity to mirtazapine or mianserin; hypersensitivity to other antidepressants (e.g., tricyclic antidepressants and MAOI depressants), acute MI; fever, infection; agranulocytosis, suicidal ideation; jaundice; ethanol intoxication; lactation.

CAUTIOUS USE History of cardiovascular or GI disorders; BPH, urinary retention; preexisting hematological disease; thrombocytopenia; narrow-angle glaucoma, increased intraocular pressure; renal impairment, renal failure; moderate to severe hepatic impairment; hypercholesterolemia, hypertriglyceridemia, cardiac disease; angina, cardiac arrhythmias; bipolar disorder, mania, bone marrow suppression, PKU, history of MI; CVD; seizure disorder, seizures; depression; history of suicidal tendencies; hypovolemia, surgery; closed-angle glaucoma; ileus, GI obstruction, dehydration; diabetes mellitus, diabetic ketoacidosis; older adults; pregnancy (category C). Safety and effectiveness in children are not established.

ROUTE & DOSAGE

Depression

Adult: **PO** 15 mg/day in single dose at bedtime, may increase q1–2wk (max: 45 mg/day)
Geriatric: **PO** Use lower doses

Renal or Hepatic Impairment Dosage Adjustment

Use lower doses

ADMINISTRATION

Oral

- Give preferably prior to sleep to minimize injury potential.
- Begin drug no sooner than 14 days after discontinuation of an MAO inhibitor.
- Reduce dosage as warranted with severe renal or hepatic impairment and in older adults.
- Store at 20°–25° C (68°–77° F) in tight, light-resistant container.

ADVERSE EFFECTS (≥1%) **Body as a Whole:** Asthenia, flu syndrome, back pain, edema, malaise. **CNS:** *Somnolence,* dizziness, abnormal dreams, abnormal thinking, tremor, confusion, depression, agitation, vertigo, twitching. **CV:** Hypertension, vasodilation. **GI:** Nausea, vomiting, abdominal pain, *increased appetite/ weight gain, dry mouth, constipation,* anorexia, cholecystitis, stomatitis, colitis, abnormal liver function tests. **Respiratory:** Dyspnea, cough, sinusitis. **Skin:** Pruritus, rash. **Urogenital:** Urinary frequency.

INTERACTIONS Drug: Additive cognitive and motor impairment with **alcohol** or BENZODIAZEPINES; increase risk of hypertensive crisis with MAOIS. **Herbal: Kava, valerian** may potentiate sedative effects.

PHARMACOKINETICS Absorption: Rapidly absorbed from GI tract, 50% reaches systemic circulation. **Peak:** 2 h. **Distribution:** 85% protein bound. **Metabolism:** In liver by cytochrome P450 system (CYP2D6, CYP1A2, CYP3A4). **Elimination:** 75% in urine, 15% in feces. **Half-Life:** 20–40 h.

NURSING IMPLICATIONS

Assessment & Drug Effects

- Lab tests: Monitor WBC count with differential, lipid profile, and ALT/AST periodically.

- Monitor for worsening of depression or suicidal ideation.
- Assess for weight gain and excessive somnolence or dizziness.
- Monitor for orthostatic hypotension with a history of cardiovascular or cerebrovascular disease. Periodically monitor ECG especially in those with known cardiovascular disease.
- Monitor those with history of seizures for lowering of the seizure threshold.

Patient & Family Education
- Do not drive or engage in potentially hazardous activities until response to drug is known.
- Do not use alcohol while taking drug.
- Report immediately unexplained fever or S&S of infection, especially flu-like symptoms, to prescriber.
- Do not take other prescription or OTC drugs without consulting prescriber.
- Make position changes slowly especially from lying or sitting to standing. Report dizziness, palpitations, and fainting.
- Monitor weight periodically and report significant weight gains.

MISOPROSTOL
(my-so-prost'ole)
Cytotec
Classification: PROSTAGLANDIN
Therapeutic: PROSTAGLANDIN
Pregnancy Category: X

AVAILABILITY 100 mcg, 200 mcg tablets

ACTION & *THERAPEUTIC EFFECT*
Synthetic prostaglandin E_1 analog, with both antisecretory (inhibiting gastric acid secretion) and mucosal protective properties. Increases bicarbonate and mucosal protective properties. Inhibits basal and nocturnal gastric acid secretion and acid secretion in response to a variety of stimuli, including meals, histamine, pentagastrin, and coffee. Produces uterine contractions that may endanger pregnancy and cause a miscarriage. *Inhibits basal and nocturanal gastric acid secretion.*

USES Prevention of NSAID (including aspirin-induced) gastric ulcers in patients at high risk of complications from a gastric ulcer (e.g., the older adult and patients with a concomitant debilitating disease or a history of ulcers). Drug is taken for the duration of NSAID therapy and does not interfere with the efficacy of the NSAID.

UNLABELED USES Short-term treatment of duodenal ulcers; cervical ripening and induction of labor.

CONTRAINDICATIONS History of allergies to prostaglandins; **topical:** Abnormal fetal position, caesarean section, ectopic pregnancy; fetal disease, incomplete abortion; multiparity, placenta previa, vaginal bleeding; pregnancy (category X).

CAUTIOUS USE Renal impairment; IBD; lactation.

ROUTE & DOSAGE

Prevention of NSAID-Induced Ulcers
Adult: **PO** 100–200 mcg q.i.d. p.c. and at bedtime or 200 mcg b.i.d. or t.i.d.

ADMINISTRATION
Oral
- Give with food to minimize GI adverse effects (manufacturer recommendation).
- Store away from heat, light, and moisture.

M

Common adverse effects in *italic*, life-threatening effects underlined; generic names in **bold;** classifications in SMALL CAPS; ♣ Canadian drug name; ❷ Prototype drug

1001

ADVERSE EFFECTS (≥1%) **CNS:** Headache. **GI:** *Diarrhea, abdominal pain,* nausea, flatulence, dyspepsia, vomiting, constipation. **Urogenital:** Spotting, cramps, dysmenorrhea, uterine contractions.

INTERACTIONS Drug: MAGNESIUM-CONTAINING ANTACIDS may increase diarrhea.

PHARMACOKINETICS Absorption: Readily from GI tract; extensive first pass metabolism. **Onset:** 30 min. **Peak:** 60–90 min. **Duration:** At least 3 h. **Metabolism:** In liver. **Elimination:** Primarily in urine; small amount in feces. **Half-Life:** 20–40 min.

NURSING IMPLICATIONS

Assessment & Drug Effects

- Monitor for diarrhea; may be minimized by giving drug after meals and at bedtime. Diarrhea is a common adverse effect that is dose related and usually self-limiting (often resolving in 8 days).

Patient & Family Education

- Avoid using concurrent magnesium-containing antacids because of increased incidence of diarrhea.
- Report postmenopausal bleeding to prescriber; it may be drug related.
- Avoid pregnancy during misoprostol therapy; use an effective contraception method while taking drug.
- Drug has abortifacient property. Contact prescriber and immediately discontinue drug if you become pregnant.

MITOMYCIN

(mye-toe-mye′sin)

Mutamycin, Mytozytrex
Classifications: ANTINEOPLASTIC (ANTIBIOTIC); ANTHRACYCLINE

Therapeutic: ANTINEOPLASTIC
Prototype: Doxorubicin
Pregnancy Category: D

AVAILABILITY 5 mg, 20 mg, 40 mg injection

ACTION & *THERAPEUTIC EFFECT* Potent antibiotic antineoplastic effective in certain tumors unresponsive to surgery, radiation, or other agents. It selectively inhibits synthesis of DNA. At high concentrations, cellular and enzymatic RNA as well as protein synthesis are suppressed. *Highly destructive to rapidly proliferating cells and slowly developing carcinomas.*

USES In combination with other chemotherapeutic agents in palliative, adjunctive treatment of disseminated adenocarcinoma of breast, pancreas, or stomach, squamous cell carcinoma of head, neck, lung, and cervix. Not recommended to replace surgery or radiotherapy or as a single primary therapeutic agent.

CONTRAINDICATIONS Hypersensitivity or idiosyncratic reaction; severe bone marrow suppression; coagulation disorders or bleeding tendencies; over-hydration; pregnancy (category D), lactation. **CAUTIOUS USE** Renal impairment; myelosuppression; pulmonary disease or respiratory insufficiency; older adults.

ROUTE & DOSAGE

Cancer

Adult/Child: **IV** 10–20 mg/m²/day as a single dose q6–8wk, additional doses based on hematologic response

Common adverse effects in *italic*, life-threatening effects underlined; generic names in **bold**; classifications in SMALL CAPS; ✦ Canadian drug name; ◉ Prototype drug

Renal Impairment Dosage Adjustment

CrCl less than 10 mL/min: Use 75% of dose

ADMINISTRATION

Intravenous

Note: Verify correct IV concentration and rate of infusion/injection for administration to children with prescriber.

PREPARE: **Direct:** Reconstitute each 5 mg vial with 10 mL sterile water for injection. Shake to dissolve. If product does not clear immediately, allow to stand at room temperature until solution is obtained. Reconstituted solution is purple. **IV Infusion:** Reconstituted solution may be further diluted to concentrations of 20–40 mcg in D5W, NS, or LR.

ADMINISTER: **Direct:** Give reconstituted solution over 5–10 min or longer. **IV Infusion:** Give over 10 min or longer as determined by total volume of solution. ▪ D5W IV solutions **must be** infused within 3 h of preparation (see storage, below). ▪ Monitor IV site closely. Avoid extravasation to prevent extreme tissue reaction (cellulitis) to the toxic drug.

INCOMPATIBILITIES **Solution/additive:** DEXTROSE-CONTAINING SOLUTIONS, **bleomycin. Y-site:** Aztreonam, cefepime, etoposide, filgrastim, gemcitabine, piperacillin/tazobactam, sargramostim, topotecan, vinorelbine.

▪ Store drug reconstituted with sterile water for injection (0.5 mg/mL) for 14 days refrigerated or 7 days at room temperature. ▪ Drug diluted in D5W (20–40 mcg/mL) is stable at room temperature for 3 h.

ADVERSE EFFECTS (≥1%) CNS: Paresthesias. **GI:** Stomatitis, *nausea, vomiting,* anorexia, hematemesis, diarrhea. **Hematologic:** <u>Bone marrow toxicity</u> (*<u>thrombocytopenia, leukopenia</u>* occurring 4–8 wk after treatment onset), thrombophlebitis, anemia. **Respiratory:** <u>Acute bronchospasm</u>, hemoptysis, dyspnea, nonproductive cough, pneumonia, <u>interstitial pneumonitis</u>. **Skin:** Desquamation; induration, pain, necrosis, cellulitis at injection site; reversible alopecia, purple discoloration of nail beds. **Body as a Whole:** Pain, headache, fatigue, edema. **Urogenital:** <u>Hemolytic uremic syndrome</u>, renal toxicity.

PHARMACOKINETICS Metabolism: Metabolized rapidly in liver. **Elimination:** In urine. **Half-Life:** 23–78 min.

NURSING IMPLICATIONS

Assessment & Drug Effects

▪ Lab tests: Perform frequent WBC with differential, platelet count, PT, INR, aPTT, Hgb, Hct, and serum creatinine during and for at least 7 wk after treatment.

▪ Do not administer if serum creatinine is greater than 1.7 mg/dL or if platelet count falls below 150,000/mm³ and WBC is down to 4000/mm³ or if prothrombin or bleeding times are prolonged.

▪ Monitor I&O ratio and pattern. Report any sign of impaired kidney function: Change in ratio, dysuria, hematuria, oliguria, frequency, urgency. Keep patient well hydrated (at least 2000–2500 mL orally daily if tolerated). Drug is nephrotoxic.

▪ Observe closely for signs of infection. Monitor body temperature frequently.

▪ Inspect oral cavity daily for signs of stomatitis or superinfection (see Appendix F).

Common adverse effects in *italic*, life-threatening effects <u>underlined</u>; generic names in **bold**; classifications in SMALL CAPS; ✦ Canadian drug name; ❖ Prototype drug

1003

Patient & Family Education

- Report respiratory distress to prescriber immediately.
- Report signs of common cold to prescriber immediately.
- Understand that hair loss is reversible after cessation of treatment.

MITOTANE

(mye'toe-tane)

Lysodren

Classification: ANTINEOPLASTIC
Therapeutic: ANTINEOPLASTIC
Pregnancy Category: C

AVAILABILITY 500 mg tablets

ACTION & *THERAPEUTIC EFFECT*
Cytotoxic agent with suppressant action on the adrenal cortex. Modifies peripheral metabolism of steroids and reduces production of adrenal steroids. Extra-adrenal metabolism of cortisol is altered, leading to reduction in 17-hydroxycorticosteroids (17-OHCS); however, plasma levels of corticosteroids do not fall. *Cytotoxic agent with suppressant action on the adrenal cortex.*

USES Inoperable adrenal cortical carcinoma (functional and nonfunctional).
UNLABELED USES Cushing's syndrome.

CONTRAINDICATIONS Shock, severe trauma; lactation.
CAUTIOUS USE Liver disease; infection; preexisting neurologic disease; pregnancy (category C); children.

ROUTE & DOSAGE

Adrenocortical Carcinoma

Adult: **PO** Initially 1–6 g/day in divided doses t.i.d. or q.i.d. then increased to 9–10 g/day in divided doses (tolerated dose range: 2–16 g/day)

ADMINISTRATION

Oral

- Withhold temporarily and consult prescriber if shock or trauma occurs, since adrenal suppression is its prime action. Exogenous steroids may be required until the already depressed adrenal starts secreting steroids.
- Store at 15°–30° C (59°–86° F) in tight, light-resistant containers.

ADVERSE EFFECTS (≥1%) **CNS:** Vertigo, dizziness, drowsiness, tiredness, depression, *lethargy, sedation,* headache, confusion, tremors. **CV:** Hypertension, hypotension, flushing **GI:** *Anorexia, nausea, vomiting, diarrhea.* **Urogenital:** Hematuria, hemorrhagic cystitis, albuminuria. **Endocrine:** Adrenocortical insufficiency. **Special Senses:** Blurred vision, diplopia, lens opacity, toxic retinopathy. **Body as a Whole:** Generalized aching, fever, muscle twitching, hypersensitivity reactions, hyperpyrexia. **Skin:** *Rash,* cutaneous eruptions and pigmentation. **Metabolic:** *Hypouricemia, hypercholesterolemia.*

DIAGNOSTIC TEST INTERFERENCE Mitotane decreases ***protein-bound iodine (PBI)*** and ***urinary 17-OHCS levels***.

INTERACTIONS Drug: Potentiates sedative effects of **alcohol** and other CNS DEPRESSANTS; may increase the metabolism of **phenytoin, phenobarbital, warfarin,** decreasing their effectiveness. POTASSIUM SPARING DIURETICS may decrease the effect.

PHARMACOKINETICS Absorption: Approximately 40% absorbed from GI tract. **Onset:** 2–4 wk. **Peak:** 3–5 h. **Distribution:** Deposits in most body tissues, especially adipose tissue. **Metabolism:** In liver. **Elimination:** 10% in urine, 1–17% in feces. **Half-Life:** 18–159 days.

NURSING IMPLICATIONS

Assessment & Drug Effects

- Monitor pulse and BP for early signs of shock (adrenal insufficiency).
- Observe for symptoms of hepatotoxicity (see Appendix F). Report them promptly, since reduced hepatic capacity can increase toxicity of mitotane and because dose may have to be decreased.
- Notify prescriber if following persist and become more severe: Aching muscles, fever, flushing, and muscle twitching.
- Monitor obese patient for symptoms of adrenal hypofunction. Because a large portion of the drug deposits in fatty tissue, the obese are particularly susceptible to prolonged adverse effects.
- Make neurologic and behavioral assessments at regular intervals throughout therapy.

Patient & Family Education

- Be aware that mitotane does not cure but does reduce tumor mass, pain, weakness, anorexia, and steroid symptoms.
- Report symptoms of adrenal insufficiency (weakness, fatigue, orthostatic hypotension, pigmentation, weight loss, dehydration, anorexia, nausea, vomiting, and diarrhea) to prescriber.
- Exercise caution when driving or performing potentially hazardous tasks requiring alertness because of drug-induced drowsiness, tiredness, dizziness. Symptoms tend to recede with continuation in therapy.

MITOXANTRONE HYDROCHLORIDE

(mi-tox'an-trone)

Novantrone

Classification: ANTINEOPLASTIC

Therapeutic: ANTINEOPLASTIC; IMMUNOSUPPRESSANT
Prototype: Doxorubicin
Pregnancy Category: D

AVAILABILITY 2 mg/mL injection

ACTION & *THERAPEUTIC EFFECT*

Non-cell-cycle specific antitumor agent with less cardiotoxicity than doxorubicin. Interferes with DNA synthesis by intercalating with the DNA double helix, blocking effective DNA and RNA transcription. *Highly destructive to rapidly proliferating cells in all stages of cell division.*

USES In combination with other drugs for the treatment of acute nonlymphocytic leukemia (ANLL) in adults, bone pain in advanced prostate cancer. Reducing neurologic disability and/or frequency of clinical relapses in multiple sclerosis.

UNLABELED USES Breast cancer, non-Hodgkin's lymphomas, autologous bone marrow transplant.

CONTRAINDICATIONS Hypersensitivity to mitoxantrone; myelosuppression; baseline LVEF less than 50%; multiple sclerosis; pregnancy (category D), lactation.

CAUTIOUS USE Impaired cardiac function; impaired liver and kidney function; systemic infections; previous treatment with daunorubicin or doxorubicin due to increased possibility of decreased cardiac function; children.

ROUTE & DOSAGE

Combination Therapy (with Cytarabine) for ANLL

Adult: **IV Induction Therapy:** 12 mg/m²/day on days 1–3, may need to repeat induction course **IV Consolidation Therapy:** 12 mg/m²

M

Common adverse effects in *italic*, life-threatening effects <u>underlined</u>; generic names in **bold;** classifications in SMALL CAPS; ♣ Canadian drug name; ● Prototype drug

1005

on days 1 and 2 (max lifetime dose: 80–120 mg/m²)

Prostate Cancer

Adult: IV 12–14 mg/m² q21days

Multiple Sclerosis

Adult: IV 12 mg/m² q3mo (max lifetime dose: 140 mg/m²) Discontinue drug in MS patients if LVEF drops below 50% or if there is a clinically significant reduction in LVEF.

ADMINISTRATION

Intravenous

If mitoxantrone touches skin, wash immediately with copious amounts of warm water.

PREPARE: **IV Infusion: Must be** diluted prior to use. Withdraw contents of vial and add to at least 50 mL of D5W or NS. ▪ May be diluted to larger volumes to extend infusion time. ▪ Use goggles, gloves, and protective gown during drug preparation and administration.

ADMINISTER: **IV Infusion:** Administer into the tubing of a freely running IV of D5W or NS and infused over at least 3 min or longer (i.e., 30–60 min) depending on the total volume of IV solution. ▪ If extravasation occurs, stop infusion and immediately restart in another vein.

INCOMPATIBILITIES **Solution/additive: Heparin, hydrocortisone, paclitaxel. Y-site: Amphotericin B cholesteryl complex, ampicillin, ampicillin/sulbactam, atenolol, azithromycin, aztreonam, cefazolin, cefoperazone, ceftaxime, cefoxitin, ceftazidime, ceftriaxone, cefuroxime,** **clindamycin, dantrolene, dexamethasone, diazepam, digoxin, cefepime, doxorubicin liposome, ertapenem, foscarnet, fosphenytion, furosemide, gemtuzumab, heparin, idarubicin, lansoprazole, methylprednisolone, nafcillin, nitroprusside, paclitaxel, pantoprazole, pemetrexed, phenytoin, piperacillin/tazobactam, propofol, ticarcillin, voriconazole, TPN.**

▪ Discard unused portions of diluted solution. ▪ Once opened, multiple-use vials may be stored refrigerated at 2°–8° C (35°–46° F) for 14 days.

ADVERSE EFFECTS (≥1%) **CV:** Arrhythmias, decreased left ventricular function, *CHF*, tachycardia, ECG changes, MI (occurs with cumulative doses of greater than 80–100 mg/m²), edema, increased risk of cardiotoxicity. **GI:** *Nausea, vomiting,* constipation, diarrhea, hepatotoxicity. **Hematologic:** Leukopenia, thrombocytopenia. **Other:** Discolors urine and sclera a blue-green color. **Skin:** Mild phlebitis, blue skin discoloration, alopecia.

INTERACTIONS Drug: May impair immune response to VACCINES such as influenza and pneumococcal infections. May have increased risk of infection with **yellow fever vaccine.**

PHARMACOKINETICS Distribution: Rapidly taken up by tissues and slowly released into plasma, 95% protein bound. **Metabolism:** In liver. **Elimination:** Primarily in bile. **Half-Life:** 37 h.

NURSING IMPLICATIONS

Assessment & Drug Effects

▪ Monitor IV insertion site. Transient blue skin discoloration may occur at site if extravasation has occurred.

- Monitor cardiac functioning throughout course of therapy including LVEF; report signs and symptoms of CHF or cardiac arrhythmias.
- Lab tests: Perform LFTs prior to and during course of treatment. Monitor serum uric acid levels and initiate hypouricemic therapy before antileukemic therapy. Monitor carefully CBC with differential prior to and during therapy.

Patient & Family Education

- Understand potential adverse effects of mitoxantrone therapy.
- Expect urine to turn blue-green for 24 h after drug administration; sclera may also take on a bluish color.
- Be aware that stomatitis/mucositis may occur within 1 wk of therapy.
- Do not risk exposure to those with known infections during the periods of myelosuppression.

MODAFINIL
(mod-a′fi-nil)
Provigil, Alertec ✦

ARMODAFINIL
Nuvigil
Classification: CNS STIMULANT, ANALEPTIC
Therapeutic: CNS STIMULANT; ANTINARCOLEPTIC
Pregnancy Category: C
Controlled Substance: Schedule IV

AVAILABILITY 100 mg, 200 mg capsules; 50 mg, 150 mg, 200 mg tablets

ACTION & *THERAPEUTIC EFFECT*
Primary sites of CNS stimulant activity of modafinil appear to be in the hippocampus, the centrolateral nucleus of the thalamus, and the central nucleus of the amygdala. Modafinil may increase excitatory transmission in the thalamus and hippocampus. *Modafinil causes wakefulness, increased locomotor activity, and psychoactive and euphoric effects.*

USES Improve wakefulness in patients with narcolepsy or excessive sleepiness associated with shift work sleep disorder, obstructive sleep apnea/hypopnea syndrome.
UNLABELED USES Fatigue related to organic brain syndrome or multiple sclerosis.

CONTRAINDICATIONS Hypersensitivity to modafinil; acute MI, valvular heart disease.
CAUTIOUS USE Cardiovascular disease including left ventricular hypertrophy; cardiac disease, ischemic ECG changes, chest pain, arrhythmias, mitral valve prolapse, recent MI, unstable angina; history of drug or alcohol abuse; psychosis or emotional instability, depression, mania; leukopenia, MI, neurologic disease, hypertension, severe hepatic disease, severe renal impairment, renal failure, seizure disorder, sleep apnea; older adults; pregnancy (category C), lactation; children younger than 16 y.

ROUTE & DOSAGE

Narcolepsy, Fatigue
Adult: **PO (Provigil)** 200 mg/day as single dose in the morning; **(Nuvigil)** 150–250 mg qa.m.
Shift-Work Sleep Disorder
Adult: **PO (Nuvigil)** 150 mg 1 h prior to shift
Hepatic Impairment Dosage Adjustment
Reduce dose by 50%

Common adverse effects in *italic*, life-threatening effects underlined; generic names in **bold**; classifications in SMALL CAPS; ✦ Canadian drug name; ❷ Prototype drug

1007

ADMINISTRATION

Oral

- Give in the morning shortly after awakening.
- Store at 15°–30° C (59°–86° F).

ADVERSE EFFECTS (≥1%) **Body as a Whole:** Chest pain, neck pain, chills, eosinophilia. **CNS:** *Headache*, nervousness, dizziness, depression, anxiety, cataplexy, insomnia, paresthesia, dyskinesia, hypertonia. **CV:** Hypotension, hypertension, vasodilation, arrhythmia, syncope. **GI:** *Nausea*, diarrhea, dry mouth, anorexia, abnormal LFTs, vomiting, mouth ulcer, gingivitis, thirst. **Respiratory:** Rhinitis, pharyngitis, lung disorder, dyspnea. **Skin:** Dry skin. **Special Senses:** Amblyopia, abnormal vision.

INTERACTIONS Drug: Methylphenidate may delay absorption of modafinil; modafinil may decrease levels of **cyclosporine**, ORAL CONTRACEPTIVES; modafinil may increase levels of **clomipramine, phenytoin, warfarin,** TRICYCLIC ANTIDEPRESSANTS.

PHARMACOKINETICS Absorption: Rapidly absorbed. **Peak:** 2–4 h. **Distribution:** Approximately 60% protein bound. **Metabolism:** In liver to inactive metabolites. **Elimination:** In urine. **Half-Life:** 15 h.

NURSING IMPLICATIONS

Assessment & Drug Effects

- Therapeutic effectiveness: Indicated by improved daytime wakefulness.
- Monitor BP and cardiovascular status, especially with preexisting hypertension and mitral valve prolapse or other CV condition.
- Monitor for S&S of psychosis, especially when history of psychotic episodes exists.
- Lab tests: Periodic LFTs.

- Coadministered drugs: Monitor INR with warfarin for first several months and when dosage is changed; monitor for toxicity with phenytoin.

Patient & Family Education

- Use barrier contraceptive instead of/in addition to hormonal contraceptive.
- Inform prescriber of all prescription or OTC drugs in/added to your regimen.
- Notify prescriber if any S&S of an allergic reaction appear.

MOEXIPRIL HYDROCHLORIDE

(mo-ex'i-pril)

Univasc

Classifications: ANGIOTENSIN-CONVERTING ENZYME (ACE) INHIBITOR; ANTIHYPERTENSIVE

Therapeutic: ANTIHYPERTENSIVE

Prototype: Enalapril

Pregnancy Category: C first trimester; D in second and third trimester

AVAILABILITY 7.5 mg, 15 mg tablets

ACTION & *THERAPEUTIC EFFECT*
ACE inhibitor that results in decreased conversion of angiotensin I to angiotensin II. Results in decreased vasopressor activity and aldosterone secretion. Lowering angiotensin II plasma levels results in blood pressure decreases and plasma renin activity increases. *ACE inhibition and decreased aldosterone secretion are responsible for its antihypertensive effect.*

USES Hypertension.

UNLABELED USES CHF, left ventricular dysfunction.

CONTRAINDICATIONS Hypersensitivity to moexipril; history of angioedema related to an ACE inhibitor; pregnancy (category D second and third trimester).

CAUTIOUS USE Hypersensitivity to any other ACE inhibitor; renal impairment, renal artery stenosis, volume-depleted patients; hypertensive patient with CHF; history of autoimmune disease; severe liver dysfunction; immunosuppressed patients; hyperkalemia; patients undergoing surgery/anesthesia; preexisting neutropenia; pregnancy (category C first trimester), lactation. Safety and efficacy in children are not established.

ROUTE & DOSAGE

Hypertension

Adult: **PO** 7.5 mg once/day, may increase up to 30 mg/day in divided doses

Renal Impairment Dosage Adjustment

CrCl 40 mL/min or less: Start with 3.75 mg daily (also if patient is volume depleted or on diuretics)

ADMINISTRATION

Oral

- Give 1 h before or 2 h after meals. Food greatly reduces absorption of moexipril.
- May need to reduce starting dose 50% in patients with possible volume depletion or a history of renal insufficiency.
- Store at 15°–30° C (59°–86° F).

ADVERSE EFFECTS (≥1%) **CNS:** Headache, *dizziness,* drowsiness, sleep disturbances, nervousness, anxiety, mood changes. **CV:** Hypotension, chest pain, angina, peripheral edema, <u>MI</u>, palpitations, arrhythmias. **Endocrine:** Hyperkalemia. **GI:** Diarrhea, nausea, dyspepsia,

abdominal pain, taste disturbances, constipation, vomiting, dry mouth, pancreatitis. **Urogenital:** Urinary frequency, increased BUN and serum creatinine. **Hematologic:** Neutropenia, hemolytic anemia. **Respiratory:** Cough, pharyngitis, rhinitis, flu-like symptoms. **Skin:** <u>Angioedema</u> (rare), rash, flushing.

INTERACTIONS Drug: Capsaicin may exacerbate cough. NSAIDS may reduce antihypertensive effects. May increase **lithium** levels and toxicity. POTASSIUM SUPPLEMENTS and POTASSIUM-SPARING DIURETICS may increase risk of hyperkalemia. **Food:** Food greatly reduces absorption of moexipril.

PHARMACOKINETICS Absorption: Readily absorbed from GI tract; approximately 13% of active metabolite reaches systemic circulation; absorption greatly reduced by food. **Onset:** 1 h. **Duration:** 24 h. **Distribution:** Approximately 50% protein bound. **Metabolism:** In liver to moexiprilat (active metabolite). **Elimination:** 13% in urine, 53% in feces. **Half-Life:** 2–9 h.

NURSING IMPLICATIONS

Assessment & Drug Effects

- Monitor closely for systematic hypotension that may occur within 1–3 h of first dose, especially in those with high blood pressure, on a diuretic or restricted salt intake, or otherwise volume depleted.
- Monitor BP and HR frequently during initiation of therapy, whenever a diuretic is added, and periodically throughout therapy.
- Determine trough BP (just before next dose) before dose adjustments are made.
- Lab tests: Monitor serum electrolytes, WBC with differential, Hct

M

and Hgb, urinalysis, LFTs and kidney function tests periodically throughout therapy.

- Supervise therapeutic response closely in patients with CHF.

Patient & Family Education

- Report to prescriber immediately swelling around face or neck or in extremities.
- Report S&S of hypotension (e.g., dizziness, weakness, syncope); nonproductive cough; skin rash; flu-like symptoms; jaundice; irregular heartbeat or chest pains; and dehydration from vomiting, diarrhea, or diaphoresis.
- Consult prescriber before using potassium-containing salt substitutes.

MOMETASONE FUROATE
(mo-met'a-sone)

Asmanex Twisthaler, Elocon, Nasonex
Pregnancy Category: C
See Appendix A-3.

MONTELUKAST
(mon-te-lu'cast)

Singulair
Classifications: BRONCHODILATOR (RESPIRATORY SMOOTH MUSCLE RELAXANT); LEUKOTRIENE RECEPTOR ANTAGONIST
Therapeutic: BRONCHODILATOR
Prototype: Zafirlukast
Pregnancy Category: B

AVAILABILITY 5 mg, 10 mg tablets; 4 mg chewable tablets; 4 mg oral granules

ACTION & *THERAPEUTIC EFFECT* Selective receptor antagonist of leukotriene, thus inhibiting bronchoconstriction. Leukotrienes (inflammatory agents) induce bronchoconstriction and mucus production. Elevated sputum and blood levels of leukotrienes are present during acute asthma attacks. Montelukast controls asthmatic attacks by inhibiting leukotriene release as well as inflammatory action associated with the attack. *Effectiveness is indicated by improved pulmonary functions and better controlled asthmatic symptoms.*

USES Prophylaxis and chronic treatment of asthma or allergic rhinitis; exercise-induced bronchoconstriction (EIB).

CONTRAINDICATIONS Hypersensitivity to montelukast; acute asthma attacks; bronchoconstriction due to acute asthma; status asthmaticus; concurrent use of aspirin or NSAIDs with known allergy to either.
CAUTIOUS USE Hypersensitivity to other leukotriene receptor antagonists (e.g., zafirlukast, zileuton); severe liver disease; jaundice, PKU; severe asthma; pregnancy (category B), lactation; children younger than 6 mo.

ROUTE & DOSAGE

Asthma
Adult: **PO** 10 mg daily in evening *Child:* **PO** 12 mo–5 y, 4 mg daily in evening; 6–14 y, 5 mg chewable tablet daily in evening
EIB
Adult/Adolescent (older than 15 y): **PO** 10 mg 2 h before exercise (not more than 1 per day)

ADMINISTRATION

Oral

- Give in the evening for maximum effectiveness.
- Ensure chewable tablets for children are not swallowed whole.

- Store at 15°–30° C (59°–86° F) in a tightly closed container and protect from light.

ADVERSE EFFECTS (≥1%) **Body as a Whole:** Asthenia, fever, trauma. **CNS:** Dizziness, *headache.* **GI:** Abdominal pain, dyspepsia, gastroenteritis, dental pain, abnormal liver function tests (ALT, AST), diarrhea, nausea. **Respiratory:** Nasal congestion, cough, influenza, laryngitis, pharyngitis, sinusitis. **Skin:** Rash. **Urogenital:** Pyuria.

PHARMACOKINETICS Absorption: Rapidly absorbed from GI tract, bioavailability 64%. **Peak:** 3–4 h for oral tablet, 2–2.5 h for chewable tablet. **Distribution:** Greater than 99% protein bound. **Metabolism:** Extensively metabolized by CYP3A4 and 2C9. **Elimination:** In feces. **Half-Life:** 2.7–5.5 h.

NURSING IMPLICATIONS

Assessment & Drug Effects

- Monitor effectiveness carefully when used in combination with phenobarbital or other potent cytochrome P450 enzyme inducers.
- Lab test: Periodic liver function tests.

Patient & Family Education

- Do not use for reversal of an acute asthmatic attack.
- Inform prescriber if short-acting inhaled bronchodilators are needed more often than usual with montelukast.
- Use chewable tablets (contain phenylalanine) with caution with PKU.

MORPHINE SULFATE ℗

(mor'feen)

Astramorph PF, Avinza, Depo-Dur, Duramorph, Embeda, **Epimorph ◆, Infumorph, Kadian, MS Contin, MSIR, Oramorph SR, RMS, Roxanol, Statex ◆**

Classifications: ANALGESIC; NARCOTIC (OPIATE AGONIST)

Therapeutic: NARCOTIC ANALGESIC

Pregnancy Category: C (D in long-term use, high dose, or close to term)

Controlled Substance: Schedule II

AVAILABILITY 10 mg, 15 mg, 30 mg tablets/capsules; 10 mg, 15 mg, 20 mg, 30 mg, 50 mg, 60 mg, 90 mg, 100 mg, 120 mg, 200 mg controlled release tablets/capsules; 10 mg/2.5 mL, 10 mg/5 mL, 20 mg/mL, 20 mg/5 mL, 30 mg/1.5 mL, 100 mg/5 mL oral solution; 0.5 mg/mL, 1 mg/mL, 2 mg/mL, 4 mg/mL, 5 mg/mL, 8 mg/mL, 10 mg/mL, 15 mg/mL, 25 mg/mL, 50 mg/mL injection; 10 mg/mL extended release lysosomal injection; 5 mg, 10 mg, 20 mg, 30 mg suppositories

ACTION & *THERAPEUTIC EFFECT* Natural opium alkaloid with agonist activity that binds with the same receptors as endogenous opioid peptides. Narcotic agonist effects are identified with different locations of receptors: Analgesia at supraspinal level, euphoria, respiratory depression and physical dependence; analgesia at spinal level, sedation and miosis; and dysphoric, hallucinogenic, and cardiac stimulant effects. *Controls severe pain; also used as an adjunct to anesthesia.*

USES Symptomatic relief of severe acute and chronic pain after nonnarcotic analgesics have failed and as preanesthetic medication; also used to relieve dyspnea of acute left ventricular failure and pulmonary edema and pain of MI.

CONTRAINDICATIONS Hypersensitivity to opiate agonists; convulsive

Common adverse effects in *italic*, life-threatening effects underlined; generic names in **bold**; classifications in SMALL CAPS; ◆ Canadian drug name; ℗ Prototype drug

1011

disorders; acute bronchial asthma, respiratory depression; circulatory shock; chemical-irritant induced pulmonary edema; hypovolemia; undiagnosed acute abdominal conditions; following biliary tract surgery and surgical anastomosis; pancreatitis, GI ileus; severe liver insufficiency; Addison's disease; hypothyroidism; during labor for delivery of a premature infant, premature infants; pregnancy (category D in long-term use, when high dose is used, or close to term).

CAUTIOUS USE Head trauma; increased cranial pressure; toxic psychosis; mild or moderate hepatic or renal impairment; cardiac arrhythmias, CVD; ulcerative colitis; constipation; emphysema; kyphoscoliosis; cor pulmonale; severe obesity; reduced blood volume; BPH; renal disease; history of substance abuse or alcoholism; very old, very young, or debilitated patients; labor, pregnancy (category C for low doses, short-term use, and not close to term); lactation.

ROUTE & DOSAGE

Pain Relief

Adult: **PO** 10–30 mg q4h prn or 15–30 mg sustained release q8–12h; **(Kadian)** dose q12–24h, increase dose prn for pain relief; **(Avinza)** dose q24h **IV** 2.5–15 mg/70 kg q2–4h or 0.8–10 mg/h by continuous infusion, may increase prn to control pain or 5–10 mg given epidurally q24h

Epidural (DepoDur only) 10–15 mg as single dose 30 min before surgery (max: 20 mg) **IM/Subcutaneous** 5–20 mg q4h **PR** 10–20 mg q4h prn

Child: **IV** 0.05–0.1 mg/kg q4h or 0.025–2.6 mg/kg/h by continuous infusion (max: 10 mg/dose) **IM/Subcutaneous** 0.1–0.2 mg/kg q4h (max: 15 mg/dose) **PO** 0.2–0.5 mg/kg q4–6h; 0.3–0.6 mg/kg sustained release q12h

Neonate: **IV/IM/Subcutaneous** 0.05 mg/kg q4–8h (max: 0.1 mg/kg/dose) or 0.01–0.02 mg/kg/h

Renal Impairment Dosage Adjustment

CrCl 10–50 mL/min: Use 75% of dose, if lower use 50% of dose

ADMINISTRATION

Oral

- A fixed, individualized schedule is recommended when narcotic analgesic therapy is started to provide effective management; blood levels can be maintained and peaks of pain can be prevented (usually a 4-h interval is adequate).
- Lower dosages are recommended for older adult or debilitated patients.
- Do not break in half, crush, or allow sustained release tablet to be chewed.
- Do not give patient sustained release tablet within 24 h of surgery.
- Dilute oral solution in approximately 30 mL or more of fluid or semisolid food. A calibrated dropper comes with the bottle. Read labels carefully when using liquid preparation; available solutions: 20 mg/mL; 100 mg/mL.

Intramuscular/Subcutaneous

- Give undiluted.

Intravenous

Note: Verify correct IV concentration and rate of infusion/injection for administration to neonates, infants, or children with prescriber.

Common adverse effects in *italic*, life-threatening effects <u>underlined</u>; generic names in **bold**; classifications in SMALL CAPS; ✦ Canadian drug name; ☉ Prototype drug

PREPARE: **Direct:** Dilute 2–10 mg in at least 5 mL of sterile water for injection. **Continuous:** Typically diluted to a range of 0.1–1 mg/mL. ▪ More concentrated solutions may be required with fluid restriction.

ADMINISTER: **Direct:** Give a single dose over 4–5 min. Avoid rapid administration. **Continuous:** Infuse via a controlled infusion device at a rate determined by patient response as ordered.

INCOMPATIBILITIES **Solution/additive:** Alteplase, aminophylline, amobarbital, chlorothiazide, floxacillin, fluorouracil, haloperidol, heparin, meperidine, nitrofurantoin, pentobarbital, phenobarbital, perphenazine, phenytoin, sodium bicarbonate, thiopental. **Y-site:** Amphotericin B cholesteryl complex, azithromycin, cefepime, doxorubicin liposome, gallium, minocycline, phenytoin, sargramostim, tetracycline.

▪ Store at 15°–30° C (59°–86° F). Avoid freezing. Refrigerate suppositories. Protect all formulations from light.

ADVERSE EFFECTS (≥1%) **Body as a Whole:** Hypersensitivity [*pruritus,* rash, urticaria, edema, hemorrhagic urticaria (rare), <u>anaphylactoid reaction</u> (rare)], sweating, skeletal muscle flaccidity; cold, clammy skin, hypothermia. **CNS:** Euphoria, insomnia, disorientation, visual disturbances, dysphoria, paradoxic CNS stimulation (restlessness, tremor, delirium, insomnia), convulsions (infants and children); decreased cough reflex, drowsiness, dizziness, deep sleep, coma, continuous intrathecal infusion may cause granulomas leading to paralysis. **Special Senses:** Miosis. **CV:** Bradycardia, palpitations, syncope; flushing of face, neck, and upper thorax; orthostatic hypotension, <u>cardiac arrest</u>. **GI:** *Constipation,* anorexia, dry mouth, biliary colic, *nausea,* vomiting, elevated transaminase levels. **Urogenital:** Urinary retention or urgency, dysuria, oliguria, reduced libido or potency (prolonged use). **Other:** Prolonged labor and respiratory depression of newborn. **Hematologic:** Precipitation of porphyria. **Respiratory:** <u>Severe respiratory depression</u> (as low as 2–4/min) or <u>arrest</u>; pulmonary edema.

DIAGNOSTIC TEST INTERFERENCE False-positive *urine glucose* determinations may occur using *Benedict's solution. Plasma amylase* and *lipase* determinations may be falsely positive for 24 h after use of morphine; *transaminase levels* may be elevated.

INTERACTIONS Drug: CNS DEPRESSANTS, SEDATIVES, BARBITURATES, BENZODIAZEPINES, and TRICYCLIC ANTIDEPRESSANTS potentiate CNS depressant effects. Use MAO INHIBITORS cautiously; they may precipitate hypertensive crisis. PHENOTHIAZINES may antagonize analgesia. Use with **alcohol** may lead to potentially fatal overdoses. **Herbal: Kava, valerian, St. John's wort** may increase sedation.

PHARMACOKINETICS Absorption: Variably from GI tract. **Peak:** 60 min PO; 20–60 min PR; 50–90 min subcutaneous; 30–60 min IM; 20 min IV. **Duration:** Up to 7 h. **Distribution:** Crosses blood–brain barrier and placenta; distributed in breast milk. **Metabolism:** In liver. **Elimination:** 90% in urine in 24 h; 7–10% in bile.

M

NURSING IMPLICATIONS

Assessment & Drug Effects

- Obtain baseline respiratory rate, depth, and rhythm and size of pupils before administering the drug. Respirations of 12/min or below and miosis are signs of toxicity. Withhold drug and report to prescriber.
- Observe patient closely to be certain pain relief is achieved. Record relief of pain and duration of analgesia.
- Differentiate among restlessness as a sign of pain and the need for medication, restlessness associated with hypoxia, and restlessness caused by morphine-induced CNS stimulation (a paradoxic reaction that is particularly common in women and older adult patients).
- Monitor carefully those at risk for severe respiratory depression after epidural or intrathecal injection: Older adult or debilitated patients or those with decreased respiratory reserve (e.g., emphysema, severe obesity, kyphoscoliosis).
- Continue monitoring for respiratory depression for at least 24 h after each epidural or intrathecal dose.
- Assess vital signs at regular intervals. Morphine-induced respiratory depression may occur even with small doses, and it increases progressively with higher doses (generally max: 90 min after subcutaneous, 30 min after IM, and 7 min after IV).
- Encourage changes in position, deep breathing, and coughing (unless contraindicated) at regularly scheduled intervals. Narcotic analgesics also depress cough and sigh reflexes and thus may induce atelectasis, especially in postoperative patients.
- Be alert for nausea and orthostatic hypotension (with light-headedness and dizziness) in ambulatory patients or when a supine patient assumes the head-up position or in patients not experiencing severe pain.
- Monitor I&O ratio and pattern. Report oliguria or urinary retention. Morphine may dull perception of bladder stimuli; therefore, encourage the patient to void at least q4h. Palpate lower abdomen to detect bladder distention.

Patient & Family Education

- Avoid alcohol and other CNS depressants while receiving morphine.
- Do not use of any OTC drug unless approved by prescriber.
- Do not ambulate without assistance after receiving drug.
- Use caution or avoid tasks requiring alertness (e.g., driving a car) until response to drug is known since morphine may cause drowsiness, dizziness, or blurred vision.

MOXIFLOXACIN HYDROCHLORIDE

(mox-i-flox'a-sin)

Avelox, Moxeza, Vigamox
Classifications: QUINOLONE ANTIBIOTIC
Therapeutic: ANTIBIOTIC
Prototype: Ciprofloxacin
Pregnancy Category: C

AVAILABILITY 400 mg tablets; 0.5% ophth solution; 400 mg/250 mL infusion

ACTION & THERAPEUTIC EFFECT
Moxifloxacin is a synthetic broad-spectrum fluoroquinolone antibiotic. It inhibits DNA gyrase, an enzyme required for DNA replication,

transcription, repair, and recombination of bacterial DNA. *Broadspectrum antibiotic that is bactericidal against gram-positive and gram-negative organisms.*

USES Treatment of acute bacterial sinusitis, acute bacterial exacerbation of chronic bronchitis, community-acquired pneumonia, skin and skin structure infections, bacterial conjunctivitis, complicated skin infections.

CONTRAINDICATIONS Hypersensitivity to moxifloxacin or other quinolones; moderate to severe hepatic insufficiency; syphilis; patients with history of prolonged QT$_c$ interval on ECG, history of ventricular arrhythmias, atrial fibrillation, hypokalemia, bradycardia, acute myocardial ischemia, acute MI; tendon pain; viral infection; history of torsades de pointes; lactation.

CAUTIOUS USE CNS disorders; cerebrovascular disease, colitis, diarrhea, GI disease; diabetes mellitus; mild or moderate heart insufficiency; myasthenia gravis; seizure disorder; sunlight (UV) exposure; pregnancy (category C). **Ocular preparation:** Use in children younger than 1 y.

ROUTE & DOSAGE

Acute Bacterial Sinusitis, Acute Bacterial Exacerbation of Chronic Bronchitis, Community-Acquired Pneumonia, Skin Infections
Adult: **PO/IV** 400 mg daily × 5–14 days
Complicated Skin Infection
Adult: **PO/IV** 400 mg daily × 7–21 days
Bacterial Conjunctivitis
Adult/Child (older than 1 y): **Ophthalmic** 1 drop in affected eye(s) t.i.d. × 7 days

ADMINISTRATION

Oral

- Administer 4 h before or 8 h after multivitamins (containing iron or zinc), antacids (containing magnesium, calcium, or aluminum), sucralfate, or didanosine.

Intravenous

PREPARE: IV Infusion: Avelox (400 mg) is supplied in ready-to-use 250 mL IV bags. No further dilution is necessary.
ADMINISTER: IV Infusion: Give over 60 min. AVOID RAPID OR BOLUS DOSE.

- Store at 15°–30° C (59°–86° F); protect from high humidity.

ADVERSE EFFECTS (≥1%) **CNS:** Dizziness, headache, peripheral neuropathy. **GI:** Nausea, diarrhea, abdominal pain, vomiting, taste perversion, abnormal liver function tests, dyspepsia. **Musculoskeletal:** Tendon rupture, cartilage erosion. **CV:** Arrythmia, QT prolongation

DIAGNOSTIC TEST INTERFERENCE May cause false positive on *opiate screening tests.*

INTERACTIONS Drug: Iron, zinc, ANTACIDS, **aluminum, magnesium, calcium, sucralfate** decrease absorption; **atenolol, erythromycin,** ANTIPSYCHOTICS, TRICYCLIC ANTIDEPRESSANTS, **quinidine, procainamide, amiodarone, sotalol** may cause prolonged QT$_c$ interval.

PHARMACOKINETICS Absorption: 90% bioavailable. **Steady State:** 3 d. **Distribution:** 50% protein bound. **Metabolism:** In liver. **Elimination:** Unchanged drug: 20% in urine, 25% in feces; metabolites: 38% in feces, 14% in urine. **Half-Life:** 12 h.

Common adverse effects in *italic*, life-threatening effects underlined; generic names in **bold**; classifications in SMALL CAPS; ♣ Canadian drug name; ❶ Prototype drug

1015

NURSING IMPLICATIONS

Assessment & Drug Effects

- Monitor for and notify prescriber immediately of adverse CNS effects.
- Notify prescriber immediately for S&S of hypersensitivity (see Appendix F).
- Lab tests: C&S before initiation of therapy and baseline serum potassium with history of hypokalemia.

Patient & Family Education

- Exercise care in timing of consumption of vitamins and antacids (see ADMINISTRATION).
- Drink fluids liberally, unless directed otherwise.
- Increased seizure potential is possible, especially when history of seizure exists.
- Stop taking drug and notify prescriber if experiencing palpitations, fainting, skin rash, severe diarrhea, ankle/foot pain, agitation, insomnia.
- Avoid engaging in hazardous activities until reaction to drug is known.

MUPIROCIN

(mu-pi-ro'sin)

Bactroban, Bactroban Nasal
Classification: PSEUDOMONIC ACID ANTIBIOTIC
Therapeutic: ANTIBIOTIC
Pregnancy Category: B

AVAILABILITY 2% ointment; cream

ACTION & THERAPEUTIC EFFECT
Topical antibacterial produced by fermentation of *Pseudomonas fluorescens*. Inhibits bacterial protein synthesis by binding with the bacterial transfer RNA. *Susceptible bacteria are* Staphylococcus aureus *[including methicillin-resistant (MRSA)* and beta-lactamase-producing strains] and other *Staphylococcus and* Streptococcus pyogenes.

USES Impetigo due to *Staphylococcus aureus*, beta-hemolytic *Streptococci*, and *Streptococcus pyogenes;* nasal carriage of *S. aureus*.

UNLABELED USES Superficial skin infections; burns.

CONTRAINDICATIONS Hypersensitivity to any of its components and for ophthalmic use; lactation (do not apply to breast); moderate to severe renal impairment.

CAUTIOUS USE Pregnancy (category B), lactation; children younger than 12 y **(intranasal form)**.

ROUTE & DOSAGE

Impetigo

Adult/Child: **Topical** Apply to affected area t.i.d., if no response in 3–5 days, reevaluate (usually continue for 1–2 wk)

Elimination of Staphylococcal Nasal Carriage

Child: **Intranasal** Apply intranasally b.i.d. to q.i.d. for 5–14 days

ADMINISTRATION

Topical

- Apply thin layer of medication to affected area.
- Cover area being treated with a gauze dressing if desired.

ADVERSE EFFECTS (≥1%) **Skin:** Burning, stinging, pain, pruritus, rash, erythema, dry skin, tenderness, swelling. **Special Senses:** Intranasal, local stinging, soreness, dry skin, pruritus.

INTERACTIONS Drug: Incompatible with **salicylic acid 2%**; do not mix in HYDROPHILIC VEHICLES (e.g., **Aquaphor**) or COAL TAR SOLUTIONS; **chloramphenicol** may interfere with bactericidal action of mupirocin.

PHARMACOKINETICS Absorption: Not systemically absorbed.

NURSING IMPLICATIONS

Assessment & Drug Effects

- Watch for signs and symptoms of superinfection (see Appendix F). Prolonged or repeated therapy may result in superinfection by nonsusceptible organisms.
- Reevaluate drug use if patient does not show clinical response within 3–5 days.
- Discontinue the drug and notify prescriber if signs of contact dermatitis develop or if exudate production increases.

Patient & Family Education

- Discontinue drug and contact prescriber if a sensitivity reaction or chemical irritation occurs (e.g., increased redness, itching, burning).

MUROMONAB-CD3

(myoo-roe-moe′nab)

Orthoclone OKT3

Classifications: BIOLOGICAL RESPONSE MODIFIER; MONOCLONAL ANTIBODY; IMMUNOSUPPRESSANT
Therapeutic: IMMUNOSUPPRESSANT
Prototype: Cyclosporine
Pregnancy Category: C

AVAILABILITY 1 mg/mL injection

ACTION & *THERAPEUTIC EFFECT*
Murine monoclonal antibody (purified IgG_2). Specifically targets the T_3 (CD_3) antigen site of the human T-cell membrane. CD_3-positive T-cells are rapidly removed from circulation, and T-lymphocyte action leading to renal inflammation and destruction is blocked. This reverses graft rejection. *CD_3-positive T-lymphocyte immunosuppressive activity results in reversing graft rejection of a transplanted kidney.*

USES Acute allograft rejection in kidney transplant patients.
UNLABELED USES Acute allograft rejection in heart and liver transplant patients.

CONTRAINDICATIONS Intolerance to any product of murine origin; fluid overload; uncompensated heart failure; weight gain of more than 3% within week prior to treatment; infection: Chickenpox (existing, recent, including recent exposure), herpes zoster; seizure disorders; lactation.
CAUTIOUS USE Recent MI; ischemic cardiac disease, CAD; pulmonary edema; repeated courses; pregnancy (category C).

ROUTE & DOSAGE

Transplant Rejection
Adult: IV 5 mg/day for 10–14 days
Child: IV *Weight 30 kg or less,* 2.5 mg daily; *weight greater than 30 kg,* 5 mg daily

ADMINISTRATION
Note: Only persons experienced with immunosuppressive therapy and management of kidney transplant patients should administer muromonab-CD3 and only in an area equipped with staff and facilities to deal with cardiac resuscitation.

Common adverse effects in *italic*, life-threatening effects underlined; generic names in **bold**; classifications in SMALL CAPS; ♣ Canadian drug name; ◉ Prototype drug

1017

Intravenous

Note: Verify correct rate of IV injection for administration to infants or children with prescriber.

- Administer IV methylprednisolone sodium succinate before and IV hydrocortisone sodium succinate 30 min after muromonab-CD3 to decrease incidence of first dose reaction.
- Be aware that concomitant maintenance immunosuppressive therapy is reduced or discontinued during drug therapy with muromonab-CD3 and resumed about 3 days prior to end of therapy.

PREPARE: Direct: Give undiluted. Do not shake ampule. • Draw sterile solution into syringe through a low protein-binding 0.2- or 0.22-micron filter. • Discard filter; attach syringe to an appropriate needle for IV bolus injection.

ADMINISTER: Direct: Give by rapid (bolus) injection. • Do not give by IV infusion or in conjunction with other drug solutions.

- Store at 2°–8° C (36°–46° F) unless otherwise stipulated. Avoid freezing.

ADVERSE EFFECTS (≥1%) **All:** Especially during first 2 days of therapy. **GI:** *Nausea, vomiting, diarrhea.* **Respiratory:** <u>Severe pulmonary edema</u>, *dyspnea, chest pain, wheezing.* **Body as a Whole:** *Fever, chills,* malaise, *tremor, increased susceptibility to cytomegalovirus, herpes simplex,* Pneumocystis carinii, Legionella, Cryptococcus, Serratia *organisms, and gram-negative bacteria.* **CV:** Tachycardia.

PHARMACOKINETICS Onset: The number of circulating CD_3-positive T-cells decreases within minutes. **Peak:** 2–7 days. **Duration:** 7 days.

NURSING IMPLICATIONS

Assessment & Drug Effects

- Monitor patient's response closely for 48 h for first dose reaction (occurs within 45–60 min after first dose and lasts several hours). It may occur (less severe) after second dose; then usually does not occur with subsequent doses. Symptoms: Chills, dyspnea, malaise, high fever.
- Assess and monitor vital signs. If temperature rises above 37.8° C (100° F), suspect infection (commonly observed in first 45 days of therapy). Take temperature before treatment and several hours after drug administration to detect first signs of infection.
- Consult prescriber if patient has a fever exceeding 37.8° C (100° F) before treatment. Make immediate attempts to lower temperature to at least 37.8° C (100° F) with antipyretics before muromonab-CD3 is administered.
- Be alert to susceptibility of patient with pretreatment fluid overload to acute pulmonary edema (may be fatal). Be prepared for prompt intubation, oxygenation, and corticosteroid drug administration should it occur.
- Lab tests: Periodic WBC with differential, CD_3 T cells, LFTs, kidney function tests.

Patient & Family Education

- Report any of the following to prescriber: Chest pain, difficulty breathing, wheezing, nausea and vomiting, significant weight gain, an infection, or fever.
- Use an effective method of birth control for 12 wk following the end of therapy.

MYCOPHENOLATE MOFETIL
(my-co-phen'o-late mo'fe-till)
CellCept

MYCOPHENOLATE ACID
Myfortic
Classifications: BIOLOGICAL RESPONSE MODIFIER; IMMUNOSUPPRESSANT
Therapeutic: IMMUNOSUPPRESSANT
Prototype: Cyclosporine
Pregnancy Category: D

AVAILABILITY 250 mg capsules; 500 mg tablets; 180 mg, 360 mg delayed release tablets; 500 mg injection; 200 mg/mL oral solution

ACTION & *THERAPEUTIC EFFECT*
Prodrug of mycophenolic acid with immunosuppressant properties; inhibits T- and B-lymphocyte responses, thus it inhibits antibody formation and generation of cytotoxic T-cells. It may also inhibit recruitment of leukocytes into sites of inflammation and graft sites in transplant patients. *Antirejection effects attributed to decreased number of activated lymphocytes in the graft site. It is synergistic with cyclosporine.*

USES Prophylaxis of organ rejection in patients receiving allogenic kidney, liver, or heart transplants.
UNLABELED USES Treatment of rheumatoid arthritis and psoriasis.

CONTRAINDICATIONS Hypersensitivity to mycophenolate mofetil, mycophenolic acid, polysorbate 80; vaccination, varicella; severe neutropenia; pregnancy (category D), lactation.
CAUTIOUS USE Viral or bacterial infections; presence or history of carcinoma; bone marrow suppression; active peptic ulcer disease; cholestasis; gallbladder disease; GI disease, severe diarrhea; malabsorption syndromes; hepatic encephalopathy, hepatic or renal impairment; renal failure, uremia; herpes infection, infection; hypoalbuminemia; PKU; lactase deficiency; females of childbearing age, older adults. **PO form:** Infants younger than 3 mo. **IV form:** Safety and efficacy in children have not been established.

ROUTE & DOSAGE

Note: **CellCept** and **Myfortic** are not interchangeable.

Prophylaxis for Kidney Transplant Rejection

Adult: **PO/IV** Start within 24 h of transplant, 1 g (mofetil) or 720 mg (sodium) b.i.d. in combination with corticosteroids and cyclosporine
Child: **PO** 600 mg/m^2 (mofetil) or 400 mg/m^2 (sodium) b.i.d. (max: 2 g/day mofetil, 720 mg/day sodium)

Prophylaxis for Heart/Liver Transplant Rejection

Adult: **PO/IV** 1.5 g (mofetil) b.i.d. started within 24 h of transplant

Toxicity Dosage Adjusment

If neutropenia develops, stop or reduce dose

ADMINISTRATION

Oral
- Give oral drug on an empty stomach.
- Adjust dosage with severe chronic kidney failure.
- Do not open or crush capsules; avoid contact with powder in capsules, and wash thoroughly with soap and water if contact occurs.

Common adverse effects in *italic*, life-threatening effects <u>underlined</u>; generic names in **bold**; classifications in SMALL CAPS; ✦ Canadian drug name; ✪ Prototype drug

1019

M

Intravenous

PREPARE: **IV Infusion:** Reconstitute each vial with 14 mL D5W. Further dilute each 500 mg in an additional 70 mL of D5W to yield 6 mg/mL.

ADMINISTER: **IV Infusion:** Slowly infuse over 2 h or longer. Avoid rapid injection.

INCOMPATIBILITIES **Solution/additive & Y-site:** Do not mix or infuse with other medications.

- Begin IV mycophenolate mofetil within 24 h of transplant and continue for up to 14 days.
- Switch patient to oral drug as soon as possible.
- Store at 15°–30° C (59°–86° F).

ADVERSE EFFECTS (≥1%) **CNS:** *Headache, tremor,* insomnia, dizziness, weakness. **CV:** *Hypertension.* **Endocrine:** Hyperglycemia, hypercholesterolemia, hypophosphatemia, hypokalemia, hyperkalemia, *peripheral edema,* increased risk of miscarriage. **GI:** *Diarrhea, constipation, nausea,* anorexia, vomiting, *abdominal pain, dyspepsia.* **Urogenital:** *UTI, hematuria,* renal tubular necrosis, burning, frequency, vaginal burning or itching, vaginal bleeding, kidney stones. **Hematologic:** <u>Leukopenia, anemia, thrombocytopenia,</u> hypochromic anemia, leukocytosis. **Respiratory:** *Respiratory infection, dyspnea,* increased cough, pharyngitis. **Skin:** Rash. **Body as a Whole:** Leg or hand cramps, bone pain, myalgias, <u>sepsis (bacterial, fungal, viral), progressive multifocal leukoencephalopathy.</u>

INTERACTIONS Drug: Acyclovir, ganciclovir may increase mycophenolate serum levels. ANTACIDS, **cholestyramine** decreases. **Mycophenolate** absorption. **Mycophenolate** may decrease protein binding of **phenytoin** or **theophylline,** causing increased serum levels. **Azathioprine** increases risk of adverse effects

PHARMACOKINETICS Absorption: Rapidly from GI tract; 94% reaches systemic circulation; absorption decreased by food. **Onset:** 4 wk. **Metabolism:** In liver to active form, mycophenolic acid. **Elimination:** 87% in urine. **Half-Life:** 11 h.

NURSING IMPLICATIONS

Assessment & Drug Effects

- Prior to initiating therapy: Baseline CBC with differential.
- Withhold dose and notify prescriber if neutropenia develps (ANC less than 1.3×10^3/mcL).
- Lab tests: Monitor CBC weekly for first month, biweekly for second and third months, then once per month for first year; periodic kidney function tests, LFTs, serum electrolytes, lipase, amylase, blood glucose, and routine urinalysis.
- Monitor for and report any S&S of sepsis or infection.

Patient & Family Education

- Comply exactly with dosing regimen and scheduled laboratory tests.
- Report to prescriber immediately S&S of infection, such as UTI or respiratory infection, or signs of bleeding (e.g., black tarry stools, blood in urine, easy bruising).
- Report all troubling adverse reactions (e.g., blood in urine and swelling in arms and legs) to prescriber as soon as possible.
- Avoid taking OTC antacids simultaneously with mycophenolate mofetil. Separate the two drugs by 2 h.
- Women should use effective contraception during and for 6 wk after treatment is completed.

NABILONE

(nab'i-lone)

Cesamet

Classifications: SYNTHETIC CANNABINOID; ANTIEMETIC

Therapeutic: ANTIEMETIC; CANNABINOID

Pregnancy Category: C

Controlled Substance: Schedule II

AVAILABILITY 1 mg capsules

ACTION & *THERAPEUTIC EFFECT*

Nabilone is a synthetic cannabinoid with multiple effects on the CNS. It is thought that the antiemetic effect results from its interaction with the cannabinoid receptor system (CB1-receptor) in neural tissues. In therapeutic doses, it produces relaxation, drowsiness, and euphoria. *It effectively controls emesis in patients receiving chemotherapy when other drugs have failed.*

USES Prevention and treatment of nausea and vomiting in adult patients induced by cancer chemotherapy refractory to standard antiemetic therapy.

CONTRAINDICATIONS Hypersensitivity to any cannabinoid; hypovolemia; lactation.

CAUTIOUS USE History of psychosis; older adults; pregnancy (category C); children.

ROUTE & DOSAGE

Nausea and Vomiting

Adult: **PO** Initial dose of 1 or 2 mg b.i.d. 1–3 h before chemotherapy. May increase to max of 2 mg t.i.d. May continue for 48 h after last dose of chemotherapy.

ADMINISTRATION

Oral

- Give 1–3 h before chemotherapy is begun. A dose of 1–2 mg the night before chemotherapy may be helpful in relieving nausea.
- Store at 15°–30° C (59°–86° F).

ADVERSE EFFECTS (≥1%) **CNS:** Asthenia, *ataxia, confusion difficulties,* depersonalization, *depression,* disorientation, *drowsiness, dysphoria, euphoria,* headache, sedation, *sleep disturbance, vertigo.* **CV:** Hypotension. **GI:** Anorexia, *dry mouth,* increased appetite, nausea. **Special Senses:** *Visual disturbances.*

INTERACTIONS Drug: SEDATIVES, HYPNOTICS, and other psychoactive substances can potentiate the CNS effects of nabilone. Coadministration of cannabinoids with **amphetamine, cocaine,** TRICYCLIC ANTIDEPRESSANTS, and/or SYMPATHOMIMETIC AGENTS can produce additive hypertension and tachycardia. Coadministration of cannabinoids with ANTIHISTAMINES or ANTICHOLINERGIC AGENTS can produce additive tachycardia and drowsiness. Coadministration of cannabinoids with BARBITURATES, BENZODIAZEPINES, **buspirone, ethanol, lithium,** MUSCLE RELAXANTS, OPIOIDS, and other CNS DEPRESSANTS can produce additive drowsiness and CNS-depressant effects. **Food: Alcohol** can potentiate the CNS effects of nabilone.

PHARMACOKINETICS Absorption: Complete absorption from GI tract. **Peak:** 2 h. **Metabolism:** Extensive hepatic metabolism. **Elimination:** Fecal (major) and urine. **Half-Life:** 2 h.

NURSING IMPLICATIONS

Assessment & Drug Effects

- Monitor for and report S&S of adverse psychiatric reactions (e.g.,

Common adverse effects in *italic*, life-threatening effects underlined; generic names in **bold;** classifications in SMALL CAPS; ♣ Canadian drug name; ☻ Prototype drug

1021

disorientation, hallucinations, psychosis) for 48–72 h after last dose of nabilone.

- Monitor for S&S of tachycardia and postural hypotension, especially in the older adult and those with a history of heart disease or hypertension.
- Lab tests: Periodic CBC with Hgb and Hct.

Patient & Family Education

- Do not use alcohol or other CNS depressants while using this medication.
- Do not drive or engage in potentially hazardous activities until response to drug is known.
- Report any of the following to a health care provider: Confusion, disorientation, hallucinations, or other bizarre behavior.

NABUMETONE

(na-bu-me′tone)

Relafen

Classification: ANALGESIC, NONSTEROIDAL ANTI-INFLAMMATORY DRUG (NSAID)

Therapeutic: ANALGESIC, NSAID; ANTIRHEUMATIC; ANTIPYRETIC

Prototype: Ibuprofen

Pregnancy Category: C

AVAILABILITY 500 mg, 750 mg tablets

ACTION & *THERAPEUTIC EFFECT*
Blocks prostaglandin synthesis by inhibiting cyclooxygenase, an enzyme that converts arachidonic acid to precursors of prostaglandins. *Anti-inflammatory, analgesic, and antipyretic effects. Effective antirheumatic agent. Inhibits platelet aggregation and prolongs bleeding time.*

USES Rheumatoid arthritis and osteoarthritis.

CONTRAINDICATIONS Patients in whom urticaria, severe rhinitis, bronchospasm, angioedema, or nasal polyps are precipitated by aspirin or other NSAIDs; salicylate hypersensitivity; active peptic ulcer; bleeding abnormalities; CABG perioperative pain; lactation.

CAUTIOUS USE Hypertension, fluid retention, heart failure; history of GI ulceration, impaired liver or kidney function, chronic kidney failure, cardiac decompensation, bone marrow suppression; patients with SLE; elderly; pregnancy (category C); children younger than 6 mo.

ROUTE & DOSAGE

Rheumatoid & Osteoarthritis
Adult: **PO** 1000 mg/day as a single dose, may increase (max: 2000 mg/day in 1–2 divided doses)

ADMINISTRATION

Oral

- Give with food, milk, or antacid (if prescribed) to reduce the possibility of GI upset.
- Store at 15°–30° C (59°–86° F).

ADVERSE EFFECTS (≥1%) GI: Diarrhea, abdominal pain, nausea, dyspepsia, flatulence, melena, ulcers, constipation, dry mouth, gastritis. **CNS:** Tinnitus, dizziness, headache, insomnia, vertigo, fatigue, diaphoresis, nervousness, somnolence. **Skin:** Rash, pruritus.

INTERACTIONS Drug: May attenuate the antihypertensive response to DIURETICS. NSAIDs increase the risk of **methotrexate** toxicity. **Food:** Food may increase the peak but not the overall absorption of nabumetone. **Herbal: Feverfew, garlic, ginger, ginkgo** may increase bleeding potential.

PHARMACOKINETICS Absorption: Readily absorbed from GI tract; approximately 35% is converted to its active metabolite on first pass through the liver. **Onset:** 1–3 wk for antirheumatic action. **Peak:** 3–6 h. **Distribution:** 99% protein bound; distributes into synovial fluid. **Metabolism:** In liver to its active metabolite, 6-methoxy-2-naphthylacetic acid (6MNA). **Elimination:** 80% of dose is excreted in urine as 6MNA; 10% excreted in feces. **Half-Life:** 24 h (6MNA).

NURSING IMPLICATIONS

Assessment & Drug Effects

- Lab tests: Obtain baseline and periodic evaluation of Hgb and Hct levels with prolonged or high-dose therapy.
- Monitor for signs and symptoms of GI bleeding.

Patient & Family Education

- Use caution with hazardous activities since nabumetone may cause dizziness, drowsiness, and blurred vision.
- Report abdominal pain, nausea, dyspepsia, or black tarry stools.
- Be aware that alcohol and aspirin will increase the risk of GI ulceration and bleeding.
- Notify your prescriber if any of the following occur: Persistent headache, skin rash or itching, visual disturbances, weight gain, or edema.

NADOLOL
(nay-doe′lole)

Corgard
Classifications: BETA-ADRENERGIC ANTAGONIST; ANTIHYPERTENSIVE
Therapeutic: ANTIHYPERTENSIVE
Prototype: Propranolol
Pregnancy Category: C

AVAILABILITY 20 mg, 40 mg, 80 mg, 120 mg, 160 mg tablets

ACTION & *THERAPEUTIC EFFECT* Nonselective beta-adrenergic blocking agent that inhibits response to adrenergic stimuli by competitively blocking these receptors within the heart. Reduces heart rate and cardiac output at rest and during exercise, and also decreases conduction velocity through AV node and myocardial automaticity. *Decreases both systolic and diastolic BP at rest and during exercise.*

USES Hypertension, either alone or in combination with a diuretic. Also long-term prophylactic management of angina pectoris.

CONTRAINDICATIONS Bronchial asthma, severe COPD, inadequate myocardial function, sinus bradycardia, greater than first-degree conduction block, overt cardiac failure, cardiogenic shock; abrupt withdrawal; lactation.

CAUTIOUS USE CHF; diabetes mellitus; hyperthyroidism; renal failure, renal impairment; pregnancy (category C); children younger than 18 y.

ROUTE & DOSAGE

Hypertension, Angina
Adult: **PO** 40 mg once/day, may increase up to 240–320 mg/day in 1–2 divided doses

ADMINISTRATION
Note: Dose is usually titrated up in 40–80 mg increments until optimum dose is achieved.

Oral

- Do not discontinue abruptly; reduce dosage over a 1–2-wk period. Abrupt withdrawal can precipitate MI or thyroid storm in susceptible patients.

Common adverse effects in *italic*, life-threatening effects <u>underlined</u>; generic names in **bold**; classifications in SMALL CAPS; ✦ Canadian drug name; ⊘ Prototype drug

1023

- Store at 15°–30° C (59°–86° F); protect drug from light.

ADVERSE EFFECTS (≥1%) Body as a Whole: Hypersensitivity (rash, pruritus, <u>laryngospasm, respiratory disturbances</u>). **CV:** *Bradycardia, peripheral vascular insufficiency (Raynaud's type),* palpitation, postural hypotension, conduction or rhythm disturbances, CHF. **GI:** Dry mouth. **CNS:** *Dizziness, fatigue,* sedation, headache, paresthesias, behavioral changes. **Special Senses:** Blurred vision, dry eyes. **Skin:** Dry skin. **Urogenital:** Impotence.

INTERACTIONS Drug: NSAIDS may decrease hypotensive effects; may mask symptoms of a hypoglycemic reaction to **insulin,** SULFONYLUREAS; **prazosin, terazosin** may increase severe hypotensive response to first dose.

PHARMACOKINETICS Absorption: 30–40% of PO dose absorbed. **Peak:** 2–4 h. **Duration:** 17–24 h. **Distribution:** Widely distributed; crosses placenta; distributed in breast milk. **Elimination:** 70% in urine; also in feces. **Half-Life:** 10–24 h.

NURSING IMPLICATIONS
Assessment & Drug Effects

- Assess heart rate and BP before administration of each dose. Withhold drug and notify prescriber if apical pulse drops below 60 bpm or systolic BP below 90 mm Hg.
- Monitor weight. Advise patient to report weight gain of 1–1.5 kg (2–3 lb) in a day and any other possible signs of CHF (e.g., cough, fatigue, dyspnea, rapid pulse, edema).
- Evaluate effectiveness for patients with angina by reduction in frequency of anginal attacks and improved exercise tolerance. Improvement should coincide with steady state serum concentration

reached within 6–9 days. Keep prescriber informed of drug effect.

- Monitor patients with diabetes mellitus closely. Beta-adrenergic blockade produced by nadolol may prevent important clinical manifestations of hypoglycemia (e.g., tachycardia, BP changes).
- Monitor I&O ratio and creatinine clearance in patients with impaired kidney function or with cardiac problems. Dosage intervals will be lengthened with decreases in creatinine clearance.

Patient & Family Education

- Check pulse before taking each dose. Do not take your medication if pulse rate drops below 60 (or other parameter set by prescriber) or becomes irregular. Consult your prescriber right away.
- Do not stop taking your medication or alter dosage without consulting your prescriber.
- Do not drive or engage in potentially hazardous activities until response to drug is known.

NAFARELIN ACETATE
(na-fa're-lin)

Synarel
Classification: GONADOTROPIN-RELEASING HORMONE (GnRH) ANALOG
Therapeutic: GnRH ANALOG
Prototype: Leuprolide
Pregnancy Category: X

AVAILABILITY 0.2 mg/spray solution

ACTION & *THERAPEUTIC EFFECT*
Potent agonist analog of gonadotropin-releasing hormone (GnRH). Inhibits pituitary gonadotropin secretion of LH and FSH. *Decrease in serum estradiol or testosterone concentrations results in the quiescence of tissues and functions that depend on LH and FSH.*

USES Endometriosis and precocious puberty.
UNLABELED USES Uterine leiomyomas, benign prostatic hypertrophy.

CONTRAINDICATIONS Hypersensitivity to GnRH or GnRH agonist analog; undiagnosed abnormal vaginal bleeding; pregnancy (category X), lactation.
CAUTIOUS USE Polycystic ovarian disease; osteoporosis.

ROUTE & DOSAGE

Endometriosis

Adult: **Inhalation** 2 inhalations/day (200 mcg/inhalation), one in each nostril, begin between days 2 and 4 of menstrual cycle; in patients with persistent regular menstruation after 2 mo of therapy, may increase to 800 mcg/day as 2 inhalations (one in each nostril) b.i.d.; do not exceed 6 mo of treatment

Precocious Puberty

Child: **Inhalation** 800–1200 mcg/day divided q8–12h

ADMINISTRATION

Inhalation

- Withhold any topical nasal decongestant, if being used, until at least 30 min after nafarelin administration.
- Store at 15°–30° C (59°–86° F); protect from light.

ADVERSE EFFECTS (≥1%) **GI:** *Bloating, abdominal cramps,* weight gain, nausea. **Endocrine:** *Hot flashes, anovulation, amenorrhea, vaginal dryness,* galactorrhea. **Metabolic:** Decreased bone mineral content (reversible). **CNS:** Transient headache, inertia, mild depression, moodiness, fatigue. **Respiratory:** Nasal irritation. **Urogenital:** *Impotence, decreased libido,* dyspareunia.

DIAGNOSTIC TEST INTERFERENCE Increased **alkaline phosphatase;** marked increase in **estradiol** in first 2 wk, then decrease to below baseline; decreased **FSH** and **LH** levels; decreased **testosterone** levels.

PHARMACOKINETICS Absorption: 21% absorbed from nasal mucosa. **Onset:** 4 wk. **Peak:** 12 wk. **Duration:** 30–50 days after discontinuing drug. **Distribution:** 78–84% bound to plasma proteins; crosses placenta. **Metabolism:** Hydrolyzed in kidney. **Elimination:** 44–55% in urine over 7 days, 19–44% in feces. **Half-Life:** 2.7 h.

NURSING IMPLICATIONS

Assessment & Drug Effects

- Make appropriate inquiries about breakthrough bleeding, which may indicate that patient has missed successive drug doses.
- Monitor for and report immediately S&S of thromboembolism, including signs of TIA, CVA, or MI.

Patient & Family Education

- Read the information pamphlet provided with nafarelin.
- Inform prescriber if breakthrough bleeding occurs or menstruation persists.
- Use or add barrier contraceptive during treatment.

NAFCILLIN SODIUM
(naf-sill'in)

Classifications: BETA-LACTAM ANTIBIOTIC; PENICILLIN
Therapeutic: ANTIBIOTIC
Prototype: Penicillin G potassium
Pregnancy Category: B

AVAILABILITY 1 g, 2 g injection

ACTION & THERAPEUTIC EFFECT Semisynthetic, acid-stable, penicillinase-resistant penicillin that interferes

Common adverse effects in *italic*, life-threatening effects underlined; generic names in **bold;** classifications in SMALL CAPS; ♣ Canadian drug name; ⊙ Prototype drug

1025

with synthesis of mucopeptides essential to formation and integrity of bacterial cell wall leading to bacterial cell lysis. *Effective against both penicillin-sensitive and penicillin-resistant strains of Staphylococcus aureus. Also active against pneumococci and group A beta-hemolytic streptococci.*

USES Primarily, infections caused by penicillinase-producing staphylococci. Serum concentrations are considerably enhanced by concurrent use of probenecid.

CONTRAINDICATIONS Hypersensitivity to penicillins, cephalosporins, and other allergens; use of oral drug in severe infections, gastric dilatation, cardiospasm, or intestinal hypermotility.

CAUTIOUS USE History of or suspected atopy or allergy (eczema, hives, hay fever, asthma); GI disease; hepatic disease; pregnancy (category B); lactation.

ROUTE & DOSAGE

Staphylococcal Infections

Adult: **IV** 500 mg–1 g q4h (max: 12 g/day) **IM** 500 mg q4–6h
Child: **IV** 50–200 mg/kg/day divided q4–6h (max: 12 g/day) **IM** *Weight greater than 40 kg,* 500 mg q4–6h; *weight less than 40 kg,* 25 mg/kg b.i.d.
Neonate: **IV** 50–100 mg/kg/day divided q6–12h **IM** 25–50 mg/kg b.i.d.

ADMINISTRATION

Intramuscular

- Reconstitute each 500 mg with 1.7 mL of sterile water for injection or NaCl injection to yield 250 mg/mL. Shake vigorously to dissolve.
- In adults: Make certain solution is clear. Select site carefully. Inject deeply into gluteal muscle. Rotate injection sites.
- In children: The preferred IM site in children younger than 3 y is the midlateral or anterolateral thigh. Check agency policy.
- Label and date vials of reconstituted solution. Remains stable for 7 days under refrigeration and for 3 days at 15°–30° C (59°–86° F).

Intravenous
Note: Verify correct IV concentration and rate of infusion in neonates, infants, children with prescriber.

PREPARE: **Direct:** Reconstitute as for IM injection. ▪ Further dilute with 15–30 mL of D5W, NS, or 0.45% NaCl. **Intermittent:** Dilute the required dose of reconstituted solution in 100–150 mL of compatible IV solution. **Continuous:** Add the required dose of reconstituted solution to a volume of IV solution that maintains concentration of drug between 2–40 mg/mL.
ADMINISTER: **Direct:** Give over at least 10 min. **Intermittent:** Give over 30–90 min. **Continuous:** Give at ordered rate.
INCOMPATIBILITIES **Solution/additive: Aminophylline, ascorbic acid, aztreonam, bleomycin, cytarabine, gentamicin, hydrocortisone, methylprednisolone, promazine. Y-site: Diltiazem, droperidol, insulin regular, labetalol, midazolam, nalbuphine, pentazocine, vancomycin, verapamil.**

- Note: Usually, limit IV therapy to 24–48 h because of the possibility of thrombophlebitis (see Appendix F), particularly in older adults.
- Discard unused portions 24 h after reconstitution.

ADVERSE EFFECTS (≥1%) **Body as a Whole:** Drug fever, <u>anaphylaxis</u>

(particularly following parenteral therapy). **GI:** Nausea, vomiting, *diarrhea,* increase in serum transaminase activity (following IM). **Hematologic:** Eosinophilia, thrombophlebitis following IV; neutropenia (long-term therapy). **Metabolic:** Hypokalemia (with high IV doses). **Skin:** Urticaria, pruritus, rash, pain and tissue irritation. **Urogenital:** Allergic interstitial nephritis.

DIAGNOSTIC TEST INTERFERENCE
Nafcillin in large doses can cause false-positive *urine protein* tests using *sulfosalicylic acid method* or serum protein tests.

INTERACTIONS Drug: May antagonize hypoprothrombinemic effects of **warfarin. Probenecid** increases serum concentrations.

PHARMACOKINETICS Peak: 30–120 min IM; 15 min IV. **Duration:** 4–6 h IM. **Distribution:** Distributes into CNS with inflamed meninges; crosses placenta; distributed into breast milk, 90% protein bound. **Metabolism:** Enters enterohepatic circulation. **Elimination:** Primarily in bile; 10–30% in urine. **Half-Life:** 1 h.

NURSING IMPLICATIONS
Assessment & Drug Effects
- Lab tests: Perform C&S prior to initiation of therapy and periodically thereafter.
- Obtain a careful history before therapy to determine any prior allergic reactions to penicillins, cephalosporins, and other allergens.
- Inspect IV site for inflammatory reaction. Also check IV site for leakage; in the older adult patient especially, loss of tissue elasticity with aging may promote extravasation around the needle.
- Note: Allergic reactions, principally rash, occur most commonly.

- Lab tests: Baseline and periodic WBC with differential; periodic LFTs, and kidney function tests with nafcillin therapy longer than 2 wk.
- Monitor neutrophil count. Nafcillin-induced neutropenia (agranulocytosis) occurs commonly during third week of therapy. It may be associated with malaise, fever, sore mouth, or throat. Perform periodic assessments of liver and kidney functions during prolonged therapy.
- Be alert for signs of bacterial or fungal superinfections (see Appendix F) in patients on prolonged therapy.
- Determine IV sodium intake for patients with sodium restriction. Nafcillin sodium contains approximately 3 mEq of sodium per gram.

Patient & Family Education
- Report promptly S&S of neutropenia (see Assessment & Drug Effects), superinfection, or hypokalemia (see Appendix F).

N

NAFTIFINE
(naf′ti-feen)
Naftin
Classifications: ANTIBIOTIC; ANTIFUNGAL
Therapeutic: ANTIFUNGAL
Prototype: Terbinafine
Pregnancy Category: B

AVAILABILITY 1% cream, gel

ACTION & *THERAPEUTIC EFFECT*
Synthetic broad-spectrum antifungal agent that may be fungicidal depending on the organism. Interferes in the synthesis of ergosterol, the principal sterol in the fungus cell membrane. Ergosterol becomes depleted and membrane function is affected. *Effective against topical infections caused by fungal organisms.*

Common adverse effects in *italic*, life-threatening effects underlined; generic names in **bold**; classifications in SMALL CAPS; ♣ Canadian drug name; ❷ Prototype drug

1027

USES Tinea pedis, tinea cruris, and tinea corporis.

CONTRAINDICATIONS Hypersensitivity to naftifine; occlusive dressing. **CAUTIOUS USE** Pregnancy (category B), lactation. Safety and efficacy in children are not established.

ROUTE & DOSAGE

Tinea Infections

Adult: **Topical** Apply cream daily, or apply gel twice daily, up to 4 wk

ADMINISTRATION

Topical

- Gently massage into affected area and surrounding skin. Wash hands before and after application.
- Do not apply occlusive dressing unless specifically directed to do so.
- Store at 15°–30° C (59°–86° F).

ADVERSE EFFECTS (≥1%) **Skin:** Burning or stinging, dryness, erythema, itching, local irritation.

PHARMACOKINETICS Absorption: 2.5–6% absorbed through intact skin. **Onset:** 7 days. **Metabolism:** In liver. **Elimination:** In urine and feces. **Half-Life:** 2–3 days.

NURSING IMPLICATIONS

Assessment & Drug Effects

- Assess for irritation or sensitivity to cream; these are indications to discontinue use.
- Reevaluate use of drug if no improvement is noted after 4 wk.

Patient & Family Education

- Learn correct application technique.
- Avoid contact with eyes or mucous membranes.

NALBUPHINE HYDROCHLORIDE

(nal'byoo-feen)

Nubain

Classifications: ANALGESIC; NARCOTIC (OPIATE AGONIST-ANTAGONIST)

Therapeutic: NARCOTIC ANALGESIC

Prototype: Pentazocine

Pregnancy Category: B; D in prolonged use or in high dose at term

AVAILABILITY 10 mg/mL, 20 mg/mL injection

ACTION & *THERAPEUTIC EFFECT* Synthetic narcotic analgesic with agonist and weak antagonist properties that is a potent analgesic. *Analgesic action that relieves moderate to severe pain with apparently low potential for dependence.*

USES Symptomatic relief of moderate to severe pain. Also preoperative sedation analgesia and as a supplement to surgical anesthesia.

CONTRAINDICATIONS History of hypersensitivity to nalbuphine, opiate agonists; pregnancy (category D in prolonged use or in high doses at term). **CAUTIOUS USE** History of emotional instability or drug abuse; head injury, increased intracranial pressure; cardiac disease; impaired respirations, COPD; GI disorders; impaired kidney or liver function; MI; biliary tract surgery; pregnancy (category B; see CONTRAINDICATIONS), lactation.

ROUTE & DOSAGE

Moderate to Severe Pain

Adult: **IV/IM/Subcutaneous** 10 mg/70 kg q3–6h prn (max: 160 mg/day)

Common adverse effects in *italic*, life-threatening effects underlined; generic names in **bold**; classifications in SMALL CAPS; ✤ Canadian drug name; ⊙ Prototype drug

Surgery Anesthesia Supplement
Adult: **IV** 0.3–3 mg/kg, then 0.25–0.5 mg/kg as required

ADMINISTRATION

Intramuscular/Subcutaneous
▪ Inject undiluted.

Intravenous ———————

PREPARE: **Direct:** Give undiluted.
ADMINISTER: **Direct:** Give at a rate of 10 mg or fraction thereof over 3–5 min.
INCOMPATIBILITIES **Solution/additive: Diazepam, dimenhydrinate, ketorolac, pentobarbital, promethazine, thiethylperazine. Y-site: Allopurinol, amphotericin B cholesteryl, cefepime, docetaxel, methotrexate, nafcillin, pemetrexed, piperacillin/tazobactam, sargramostim, sodium bicarbonate.**

▪ Store at 15°–30° C (59°–86° F), avoid freezing.

ADVERSE EFFECTS (≥1%) **CV:** Hypertension, hypotension, bradycardia, tachycardia, flushing. **GI:** Abdominal cramps, bitter taste, *nausea, vomiting,* dry mouth. **CNS:** *Sedation, dizziness,* nervousness, depression, restlessness, crying, euphoria, dysphoria, distortion of body image, unusual dreams, confusion, hallucinations; numbness and tingling sensations, headache, vertigo. **Respiratory:** Dyspnea, asthma, <u>respiratory depression</u>. **Skin:** Pruritus, urticaria, burning sensation, *sweaty, clammy skin.* **Special Senses:** Miosis, blurred vision, speech difficulty. **Urogenital:** Urinary urgency.

INTERACTIONS Drug: Alcohol and other CNS DEPRESSANTS add to CNS depression.

PHARMACOKINETICS Onset: 2–3 min IV; 15 min IM. **Peak:** 30 min IV. **Duration:** 3–6 h. **Distribution:** Crosses placenta. **Metabolism:** In liver. **Elimination:** In urine. **Half-Life:** 5 h.

NURSING IMPLICATIONS

Assessment & Drug Effects
▪ Assess respiratory rate before drug administration. Withhold drug and notify prescriber if respiratory rate falls below 12.
▪ Watch for allergic response in persons with sulfite sensitivity.
▪ Administer with caution to patients with hepatic or renal impairment.
▪ Monitor ambulatory patients; nalbuphine may produce drowsiness.
▪ Watch for respiratory depression of newborn if drug is used during labor and delivery.
▪ Avoid abrupt termination of nalbuphine following prolonged use, which may result in symptoms similar to narcotic withdrawal: Nausea, vomiting, abdominal cramps, lacrimation, nasal congestion, piloerection, fever, restlessness, anxiety.

Patient & Family Education
▪ Do not drive or engage in potentially hazardous activities until response to drug is known.
▪ Avoid alcohol and other CNS depressants.

N

NALOXONE HYDROCHLORIDE ⓟ
(nal-ox'one)

Narcan
Classification: NARCOTIC (OPIATE ANTAGONIST)
Therapeutic: NARCOTIC ANTAGONIST
Pregnancy Category: C

AVAILABILITY 0.02 mg/mL, 0.4 mg/mL, 1 mg/mL injection

Common adverse effects in *italic*; life-threatening effects <u>underlined</u>; generic names in **bold**; classifications in SMALL CAPS; ♣ Canadian drug name; ⓟ Prototype drug

ACTION & *THERAPEUTIC EFFECT*

A potent narcotic antagonist, essentially free of agonistic (morphine-like) properties. *Reverses the effects of opiates, including respiratory depression, sedation, and hypotension.*

USES Narcotic overdosage; complete or partial reversal of narcotic depression. Drug of choice when nature of depressant drug is not known and for diagnosis of suspected acute opioid overdosage. Challenge for opioid dependence.

UNLABELED USES Shock and to reverse alcohol-induced or clonidine-induced coma or respiratory depression.

CONTRAINDICATIONS Hypersensitivity to naloxone, naltrexone, nalmefene; respiratory depression due to nonopioid drugs; substance abuse.

CAUTIOUS USE Known or suspected narcotic dependence; brain tumor, head trauma, increased ICP; history of substance abuse; cardiac irritability; seizure disorders; pregnancy (category C), lactation.

ROUTE & DOSAGE

Opiate Overdose

Adult: **IV** 0.4–2 mg, may repeat q2–3min up to 10 mg if necessary
Child (5 y or older and weight at least 20 kg): **IV** 2 mg, may repeat q2–3min if needed
Child/Infant (weight less than 20 kg): **IV** 0.01–0.1 mg/kg, may repeat q2–3min up to 10 mg if necessary
Neonate: **IV/Subcutaneous/IM** 0.01 mg/kg, may repeat q2–3min

Postoperative Opiate Depression

Adult: **IV** 0.1–0.2 mg, may repeat q2–3min for up to 3 doses if necessary

Child: **IV** 0.005–0.01 mg/kg, may repeat q2–3min up to 3 doses if necessary

Challenge for Opioid Dependence

Adult: **IM** 0.2 mg, observe for 30 sec for signs/symptoms of withdrawal, if no signs/symptoms then 0.6 mg and observe for 20 min

ADMINISTRATION

Intramuscular/Subcutaneous
- Inject undiluted.

Intravenous

PREPARE: **Direct:** May be given undiluted. **IV Infusion:** Dilute 2 mg in 500 mL of D5W or NS to yield 4 mcg/mL (0.004 mg/mL).
ADMINISTER: **Direct:** Give bolus dose over 10–15 sec. **IV Infusion:** Adjust rate according to patient response.
INCOMPATIBILITIES **Y-site: Amphotericin B cholesteryl complex, lansoprazole.**

- Use IV solutions within 24 h.
- Store at 15°–30° C (59°–86° F), protect from excessive light.

ADVERSE EFFECTS (≥1%) **Body as a Whole:** Reversal of analgesia, tremors, hyperventilation, slight drowsiness, sweating. **CV:** Increased BP, tachycardia. **GI:** Nausea, vomiting. **Hematologic:** Elevated partial thromboplastin time.

INTERACTIONS Drug: Reverses analgesic effects of NARCOTIC (OPIATE) AGONISTS and NARCOTIC (OPIATE) AGONIST-ANTAGONISTS.

PHARMACOKINETICS Onset: 2 min. **Duration:** 45 min. **Distribution:** Crosses placenta. **Metabolism:** In liver. **Elimination:** In urine. **Half-Life:** 60–90 min.

NURSING IMPLICATIONS

Assessment & Drug Effects

- Observe patient closely; duration of action of some narcotics may exceed that of naloxone. Keep prescriber informed; repeat naloxone dose may be necessary.
- May precipitate opiate withdrawal if administered to a patient who is opiate dependent.
- Note: Effects of naloxone generally start to diminish 20–40 min after administration and usually disappear within 90 min.
- Monitor respirations and other vital signs.
- Monitor surgical and obstetric patients closely for bleeding. Naloxone has been associated with abnormal coagulation test results. Also observe for reversal of analgesia, which may be manifested by nausea, vomiting, sweating, tachycardia.

Patient & Family Education

- Report postoperative pain that emerges after administration of this drug to prescriber.

NALTREXONE HYDROCHLORIDE

(nal-trex'one)

Vivitrol

METHYLNALTREXONE

Relistor
Classification: NARCOTIC (OPIATE ANTAGONIST)
Therapeutic: NARCOTIC ANTAGONIST
Prototype: Naloxone HCl
Pregnancy Category: C

AVAILABILITY 50 mg tablets; 380 mg injection; **Methylnaltrexone:** 12 mg/0.6 mL injection

ACTION & *THERAPEUTIC EFFECT*
Opioid antagonist with a mechanism of action that appears to result from competitive binding at opioid receptor sites, thus it reduces euphoria and drug craving without supporting addiction. *Weakens or completely and reversibly blocks the subjective effects (the "high") of IV opioids and analgesics possessing both agonist and antagonist activity.*

USES Alcoholism, opiate agonist dependence. Opioid-related constipation in patients nonresponsive to laxatives.
UNLABELED USES Pruritus.

CONTRAINDICATIONS Patients receiving opioid analgesics; opiate agonist use within 7–10 days; acute opioid agonist withdrawal; opioid-dependent patient; acute hepatitis, liver failure, hepatic encephalopathy; suicidal ideation; any individual who (1) fails naloxone challenge, (2) has a positive urine screen for opioids, or (3) has a history of sensitivity to naltrexone; lactation.
CAUTIOUS USE Mild to moderate hepatic impairment (Childl-Pugh class A or B); history of suicidal tendencies; renal impairment; pregnancy (category C); **IM form:** Special at-risk patients: Thrombocytopenia, coagulopathy (e.g., hemophilia), severe hepatic impairment; children younger than 18 y.

ROUTE & DOSAGE

Opioid Dependence

Adult: **PO** 25 mg followed by another 25 mg in 1 h if no withdrawal response; maintenance regimen of 50–150 mg/day is individualized (max: 800 mg/day)
Adult: **IM** 380 mg q4w

Alcohol Dependence

Adult: **PO** 50 mg once/day x 12 wk **IM** 380 mg q 4 wk

Opioid-Related Constipation (Relistor)

Adult: **Subcutaneous** *Weight less than 38 kg or greater than 114 kg,* 0.15 mg/kg every other day (max: 0.15 mg/kg in 24 h); *weight 38–62 kg,* 8 mg every other day (max: 8 mg/24 h); *weight 62–114 kg,* 12 mg every other day (max: 12 mg/24 h)

Renal Impairment Dosage Adjustment (Relistor)

CrCl less than 30 mL/min: Reduce dose by 50%

ADMINISTRATION

Oral
▪ Give without regard to food.

Intramuscular
▪ Give IM into the gluteal muscle, alternating buttocks per injection. Aspirate before injection to ensure that drug is not injected IV.

Subcutaneous (Methylnaltrexone Only)
▪ Give subcutaneously into upper arm, abdomen, or thigh.

ADVERSE EFFECTS (≥1%) **GI:** Dry mouth, anorexia, *nausea, vomiting,* constipation, *abdominal cramps/pain,* <u>hepatotoxicity</u>. **Musculoskeletal:** *Muscle and joint pains.* **CNS:** *Difficulty sleeping, anxiety, headache, nervousness,* reduced or increased energy, irritability, dizziness, depression. **Skin:** Skin rash. **Hematologic: IM extended release form:** Hematoma formation at injection site. **Body as a Whole:** Chills.

INTERACTIONS Drug: Increased somnolence and lethargy with PHENOTHIAZINES; reverses analgesic effects of NARCOTIC (OPIATE) AGONISTS and NARCOTIC (OPIATE) AGONIST-ANTAGONISTS.

PHARMACOKINETICS Absorption: Rapidly from GI tract; 20% reaches systemic circulation (first pass effect). **Onset:** 15–30 min. **Peak:** 1 h; 30 min (Relistor). **Duration:** 24–72 h PO; 4 wk IM. **Metabolism:** In liver to active metabolite. **Elimination:** In urine. **Half-Life:** 10–13 h PO, 5–10 days IM.

NURSING IMPLICATIONS

Assessment & Drug Effects
▪ Lab tests: Check LFTs before the treatment is started, at monthly intervals for 6 mo, and then periodically as indicated.

Patient & Family Education
▪ Note: Naltrexone therapy may put you in danger of overdosing if you use opiates. Small doses even at frequent intervals will give no desired effects; however, a dose large enough to produce a high is dangerous and may be fatal.
▪ It may be possible to transfer from methadone to naltrexone. This can be done after gradual withdrawal and final discontinuation of methadone.
▪ Report promptly onset of signs of hepatic toxicity (see Appendix F) to prescriber. The drug will be discontinued.
▪ Do not self-dose with OTC drugs for treatment of cough, colds, diarrhea, or analgesia. Many available preparations contain small doses of an opioid. Consult prescriber for safe drugs if they are needed.
▪ Tell a doctor or dentist before treatment that you are using naltrexone.
▪ Wear identification jewelry indicating naltrexone use.

NANDROLONE DECANOATE

(nan'droe-lone)

Classification: ANABOLIC/ANDRO-
GEN STEROID
Therapeutic: ANABOLIC STEROID
Prototype: Testosterone
Pregnancy Category: X
Controlled Substance: Schedule III

AVAILABILITY 100 mg/mL, 200 mg/mL injection

ACTION & *THERAPEUTIC EFFECT*
Synthetic steroid with high ratio of anabolic activity to androgenic activity. Actions last 3–4 wk. *Increases hemoglobin and red cell mass and increases lean body mass in patients with cachexia (muscle wasting).*

USES Control of metastatic breast cancer, management of anemia of renal insufficiency.

CONTRAINDICATIONS Males with prostate or breast cancer; severe cardiac disease; liver dysfunction, severe renal disease; nephrotic syndrome, hypercalcemia; pregnancy (category X), lactation.
CAUTIOUS USE BPH; history of MI; CAD; diabetes mellitus; heart failure; children younger than 2 y.

ROUTE & DOSAGE

Anemia (Decanoate)

Adult: **IM** 50–200 mg/wk
Child (2–13 y): **IM** 25–50 mg
q3–4wk

Metastatic Breast Cancer

Adult: **IM** 50–100 mg/wk

ADMINISTRATION

Intramuscular

- Inject drug deep IM, preferably into gluteal muscle in adult. Follow agency policy regarding IM site in small child.

- Intermittent therapy is usually recommended (4-mo course of treatment followed by 6–8-wk rest period).

ADVERSE EFFECTS (≥1%) **Body as a Whole:** Muscle cramps. **GI:** *Nausea, vomiting,* diarrhea, anorexia, abdominal fullness, cholestatic jaundice, <u>hepatic necrosis, hepatocellular neoplasms.</u> **Hematologic:** <u>Leukopenia.</u> **Metabolic:** Sodium, chloride, water, potassium, phosphate, and calcium retention, ankle edema, glucose intolerance, increased cholesterol. **CNS:** Excitation, insomnia, chills, toxic confusion. **Endocrine:** *Acne, virilization.*

INTERACTIONS Drug: May increase hypoprothrombinemic effects of **warfarin;** may decrease **insulin** and SULFONYLUREA requirements; CORTICOSTEROIDS may increase edema. **Herbal: Echinacea** may increase risk of hepatotoxicity.

PHARMACOKINETICS Absorption: Slowly absorbed from IM injection site over 4 days. **Peak:** 3–6 days. **Metabolism:** In liver to active metabolite. **Half-Life:** 6–8 days.

NURSING IMPLICATIONS

Assessment & Drug Effects

- Lab tests: Obtain baseline and periodic LFTs and electrolyte levels.
- Monitor for S&S of hepatic toxicity (see Appendix F) and electrolyte imbalance, especially hyperkalemia and hypercalcemia (see Appendix F).
- Monitor diabetics for loss of glycemic control.

Patient & Family Education

- Note: In women, the drug may cause virilization (e.g., increased facial and body hair, deepening of voice).

Common adverse effects in *italic,* life-threatening effects <u>underlined</u>; generic names in **bold;** classifications in SMALL CAPS; ♣ Canadian drug name; ⊙ Prototype drug

1033

NAPHAZOLINE HYDROCHLORIDE ⊙

(naf-az'oh-leen)

Ak-Con, Albalon, Allerest, Clear Eyes, Comfort, Degest-2, Nafaz-air, Naphcon, Privine, VasoClear, Vasocon

Classifications: EYE AND EAR PREPARATION; VASOCONSTRICTOR; ALPHA-ADRENERGIC AGONIST; DECONGESTANT

Therapeutic: DECONGESTANT
Pregnancy Category: C

AVAILABILITY 0.012%, 0.02%, 0.03%, 0.1% ophth solution; 0.05% nasal solution

ACTION & *THERAPEUTIC EFFECT*
Direct-acting alpha-adrenergic agonist that produces rapid and prolonged vasoconstriction of arterioles. *It decreases fluid exudation and mucosal engorgement.*

USES Nasal decongestant and ocular vasoconstrictor.

CONTRAINDICATIONS Narrow-angle glaucoma; concomitant use with MAO inhibitors or tricyclic antidepressants; lactation.

CAUTIOUS USE Hypertension, cardiac irregularities, advanced arteriosclerosis; diabetes; hyperthyroidism; older adults; pregnancy (category C); children younger than 6 y.

ROUTE & DOSAGE

Congestion

Adult: **Intranasal** 2 drops or sprays of 0.05% solution in each nostril q3–6h for no more than 3–5 days **Ophthalmic** See Appendix A.

Child: **Intranasal** 1–2 drops or sprays of 0.025% solution q3–6h for no more than 3–5 days

ADMINISTRATION

Instillation

- Instill nasal spray with patient in upright position. If administered in reclining position, a stream rather than a spray may be ejected, with possibility of systemic reaction.
- Minimize amount of drug swallowed by taking care not to direct the flow toward nasopharynx and by positioning patient properly with the head tilted slightly downward.
- Store at 15°–30° C (59°–86° F), protect from freezing.

ADVERSE EFFECTS (≥1%) Body as a Whole: Hypersensitivity reactions, headache, nausea, weakness, sweating, drowsiness, hypothermia, coma. **CV:** Hypertension, bradycardia, shock-like hypotension. **Special Senses:** Transient nasal stinging or burning, dryness of nasal mucosa, pupillary dilation, increased intraocular pressure, rebound redness of the eye.

INTERACTIONS Drug: TRICYCLIC ANTIDEPRESSANTS, **maprotiline** may potentiate pressor effects.

PHARMACOKINETICS Onset: Within 10 min. **Duration:** 2–6 h.

NURSING IMPLICATIONS

Assessment & Drug Effects

- Watch for rebound congestion and chemical rhinitis with frequent and continued use.
- Monitor BP periodically for development or worsening of hypertension, especially with ophthalmic route.
- Overdose: Bradycardia and hypotension can result. Report promptly.

Patient & Family Education

- Do not exceed prescribed regimen. Systemic effects can result from swallowing excessive medication.
- Discontinue medication and contact prescriber if nasal congestion is not relieved after 5 days.
- Prevent contamination of eye solution by taking care not to touch eyelid or surrounding area with dropper tip.

NAPROXEN

(na-prox'en)

Apo-Naproxen ✦, **EC-Naprosyn, Naprelan, Naprosyn, Naxen** ✦, **Novonaprox** ✦

NAPROXEN SODIUM

Aleve, Anaprox, Anaprox DS
Classifications: ANALGESIC, NON-STEROIDAL ANTI-INFLAMMATORY DRUG (NSAID); ANTIPYRETIC
Therapeutic: NONNARCOTIC ANALGESIC, NSAID
Prototype: Ibuprofen
Pregnancy Category: B first and second trimester; D third trimester

AVAILABILITY 200 mg, 250 mg, 375 mg, 500 mg tablets; 375 mg, 500 mg sustained release tablets

ACTION & THERAPEUTIC EFFECT Propionic acid derivative that is an NSAID. Mechanism of action is related to inhibition of prostaglandin synthesis by inhibiting COX-1 and COX-2 isoenzymes. *Analgesic, anti-inflammatory, and antipyretic effects; also inhibits platelet aggregation and prolongs bleeding time.*

USES Anti-inflammatory and analgesic effects in symptomatic treatment of acute and chronic rheumatoid arthritis, juvenile arthritis (naproxen only), and for treatment of primary dysmenorrhea. Also management of ankylosing spondylitis, osteoarthritis, and gout.

UNLABELED USES Paget's disease of bone, Bartter's syndrome.

CONTRAINDICATIONS Active peptic ulcer; patients in whom asthma, rhinitis, urticaria, bronchospasm, or shock is precipitated by aspirin or other NSAIDs; perioperative pain associated with CABG; hypersensitivity to any NSAID; cardiac disease; pregnancy (category D third trimester); lactation.

CAUTIOUS USE History of upper GI tract disorders; impaired kidney, liver, or cardiac function; patients on sodium restriction **(naproxen sodium);** low pretreatment Hgb concentration; fluid retention, hypertension, heart failure; coagulopathy; SLE; pregnancy (category B first and second trimester); children younger than 2 y.

ROUTE & DOSAGE

Note: 275 mg naproxen sodium = 250 mg naproxen

Inflammatory Disease

Adult: **PO** 250–500 mg b.i.d. (max: 1000 mg/day naproxen, 1100 mg/day naproxen sodium); Naprelan is dosed daily
Child (older than 2 y): **PO** 10–15 mg/kg/day in 2 divided doses (max: 1000 mg/day)

Mild to Moderate Pain, Dysmenorrhea

Adult: **PO** 500 mg followed by 200–250 mg q6–8h prn up to 1250 mg/day
Child (older than 2 y): **PO** 5–7 mg/kg q8–12h

N

Common adverse effects in *italic*, life-threatening effects <u>underlined</u>; generic names in **bold**; classifications in SMALL CAPS; ✦ Canadian drug name; ⊘ Prototype drug

1035

ADMINISTRATION

Oral

- Ensure that extended release or enteric-coated form is not chewed or crushed. It **must be** swallowed whole.
- Give with food or an antacid (if prescribed) to reduce incidence of GI upset.
- Store at 15°–30° C (59°–86° F) in tightly closed container; protect from freezing.

ADVERSE EFFECTS (≥1%) CNS:
Headache, drowsiness, dizziness, light-headedness, depression. **CV:** Palpitation, dyspnea, peripheral edema, CHF, tachycardia. **Special Senses:** Blurred vision, tinnitus, hearing loss. **GI:** *Anorexia, heartburn,* indigestion, *nausea,* vomiting, thirst, GI bleeding, elevated serum ALT, AST. **Hematologic:** Thrombocytopenia, leukopenia, eosinophilia, inhibited platelet aggregation, agranulocytosis (rare). **Skin:** Pruritus, rash, ecchymosis. **Urogenital:** Nephrotoxicity. **Respiratory:** Pulmonary edema.

DIAGNOSTIC TEST INTERFERENCE
Transient elevations in **BUN** and serum *alkaline phosphatase* may occur. Naproxen may interfere with some urinary assays of **5-HIAA** and may cause falsely high **urinary 17-KGS** levels (using **m-dinitrobenzene reagent**). Naproxen should be withdrawn 72 h before adrenal function tests.

INTERACTIONS Drug: Bleeding time
effects of ORAL ANTICOAGULANTS, **heparin** may be prolonged; may increase **lithium** toxicity. **Herbal:** Feverfew, garlic, ginger, ginkgo, evening primrose oil may increase bleeding potential.

PHARMACOKINETICS Absorption:
Almost completely from GI tract when taken on empty stomach. **Peak:** 2 h naproxen; 1 h naproxen sodium. **Duration:** 7 h. **Metabolism:** In liver. **Elimination:** Primarily in urine; some biliary excretion (less than 1%). **Half-Life:** 12–15 h.

NURSING IMPLICATIONS

Assessment & Drug Effects

- Take detailed drug history prior to initiation of therapy. Observe for signs of allergic response in those with aspirin or other NSAID sensitivity.
- Lab tests: Obtain baseline and periodic evaluations of Hgb, LFTs, and kidney function in patients receiving prolonged or high dose therapy.
- Baseline and periodic auditory and ophthalmic examinations are recommended in patients receiving prolonged or high dose therapy.
- Monitor therapeutic effectiveness. Patients with arthritis may experience symptomatic relief (reduction in joint pain, swelling, stiffness) within 24–48 h with naproxen sodium therapy and in 2–4 wk with naproxen.

Patient & Family Education

- Be aware that the therapeutic effect of naproxen may not be experienced for 3–4 wk.
- Do not drive or engage in potentially hazardous activities until response to drug is known.
- Avoid alcohol and aspirin (as well as other NSAIDs) unless otherwise advised by a prescriber. Potential to increase risk of GI ulceration and bleeding.
- Tell your dentist or surgeon if you are taking naproxen before any treatment; it may prolong bleeding time.

NARATRIPTAN

(nar-a-trip'tan)

Amerge

Classification: SEROTONIN 5-HT$_1$ RECEPTOR AGONIST

Therapeutic: ANTIMIGRAINE

Prototype: Sumatriptan

Pregnancy Category: C

AVAILABILITY 1 mg, 2.5 mg tablets

ACTION & *THERAPEUTIC EFFECT*

Binds to the serotonin receptors (5-HT$_{1D}$ and 5-HT$_{1B}$) on intracranial blood vessels, resulting in selective vasoconstriction of dilated vessels in the carotid circulation. It also inhibits the release of inflammatory neuropeptides associated with a migraine attack. *Inhibits vasoconstriction of dilated vessels selectively. This results in the relief of acute migraine headache attacks.*

USES Acute migraine headaches with or without aura.

CONTRAINDICATIONS Hypersensitivity to naratriptan; severe renal impairment (creatinine clearance less than 15 mL/min); severe hepatic impairment; history of ischemic heart disease (i.e., angina pectoris, MI), arteriosclerosis, cardiac arrhythmias; cardiac disease, CAD, peripheral vascular disease; cerebrovascular syndromes (i.e., strokes or TIA); uncontrolled hypertension; patients with hemiplegic or basilar migraine; older adults.

CAUTIOUS USE Cardiovascular disease; renal or hepatic insufficiency; elderly; pregnancy (category C), lactation. Safety and efficacy in children younger than 18 y are not established.

ROUTE & DOSAGE

Acute Migraine

Adult: **PO** 1–2.5 mg; may repeat in 4 h if necessary (max: 5 mg/24 h); patients with mild or moderate renal or hepatic impairment should not exceed 2.5 mg/24 h

ADMINISTRATION

Oral

- Give any time after symptoms of migraine appear. If the first tablet was effective but symptoms return, a second tablet may be given, but no sooner than 4 h after the first. Do not exceed 5 mg in 24 h.
- If there is no response to the first tablet, contact prescriber before administering a second tablet.
- Do not give within 24 h of an ergot-containing drug or other 5-HT$_1$ agonist.
- Store at 2°–25° C (36°–77° F); protect from light.

ADVERSE EFFECTS (≥1%) **Body as a Whole:** Asthenia, fatigue, malaise, pain, pressure sensation, paresthesias, throat pressure, warm/cold sensations, hot flushes. **CNS:** Somnolence, dizziness, drowsiness, headache, hypesthesia, decreased mental acuity, euphoria, tremor. **CV:** Coronary artery vasospasm, transient myocardial ischemia, MI, ventricular tachycardia, ventricular fibrillation, chest pain/tightness/heaviness, palpitations. **GI:** Dry mouth, nausea, vomiting, diarrhea. **Respiratory:** Dyspnea. **Skin:** Flushing.

INTERACTIONS Drug: Dihydroergotamine, methysergide, and other 5-HT$_1$ AGONISTS may cause prolonged vasospastic reactions; SSRIS

N

Common adverse effects in *italic*, life-threatening effects underlined; generic names in **bold**; classifications in SMALL CAPS; ✦ Canadian drug name; ⊙ Prototype drug

1037

have rarely caused weakness, hyperreflexia, and incoordination; MAOIS should not be used with 5-HT$_1$ AGONISTS. **Herbal: Gingko, ginseng, echinacea, St. John's wort** may increase triptan toxicity.

PHARMACOKINETICS Absorption: Rapidly absorbed, 70% bioavailability. **Peak:** 2–4 h. **Distribution:** 28–31% protein bound. **Metabolism:** In liver. **Elimination:** Primarily in urine. **Half-Life:** 6 h.

NURSING IMPLICATIONS

Assessment & Drug Effects
- Monitor carefully cardiovascular status following first dose in patients at risk for CAD (e.g., postmenopausal women, men older than 40 y, persons with known CAD risk factors) or coronary artery vasospasms.
- Be aware that ECG is recommended following first administration of naratriptan to someone with known CAD risk factors and periodically with long-term use.
- Report immediately to the prescriber: Chest pain, nausea, or tightness in chest or throat that is severe or does not quickly resolve.
- Obtain periodic cardiovascular evaluation with continued use.

Patient & Family Education
- Carefully review patient information leaflet and guidelines for administration.
- Contact prescriber immediately for any of the following: Symptoms of angina (e.g., severe and/or persistent pain or tightness in chest or throat, severe nausea); hypersensitivity (e.g., wheezing, facial swelling, skin rash, or hives); or abdominal pain.
- Report any other adverse effects (e.g., tingling, flushing, dizziness) at next prescriber visit.

NATALIZUMAB
(na-tal'-i-zu-mab)
Tysabri
Classifications: BIOLOGICAL RESPONSE MODIFIER; MONOCLONAL ANTIBODY; INTEGRIN INHIBITOR
Therapeutic: IMMUNOMODULATOR; MONOCLONAL ANTIBODY (IgG)
Prototype: Basiliximab
Pregnancy Category: C

AVAILABILITY 300 mg/15 mL injection

ACTION & *THERAPEUTIC EFFECT*
Natalizumab is a recombinant immunoglobulin-G4 (IgG4) monoclonal antibody thought to interfere with the migration of lymphocytes and monocytes into the CNS endothelium of patients with multiple sclerosis, thereby reducing inflammation and demyelination of CNS white matter. *Inhibition of T-cell infiltration into the brain is thought to impede the demyelinating process of MS. It reduces relapses and occurrence of brain lesions. Natalizumab is also thought to attenuate T-lymphocyte–mediated intestinal inflammation in Crohn's disease and possibly ulcerative colitis.*

USES Treatment of relapsing forms of multiple sclerosis, treatment of Crohn's disease.

CONTRAINDICATIONS Prior hypersensitivity to natalizumab; murine protein hypersensitivity; have or have had progressive multifocal leukoencephalopathy (PML); active infection; S&S of PML; females of childbearing age; lactation.
CAUTIOUS USE Diabetes mellitus, immunocompromised patients; exposure to infection or tuberculosis; hepatic dysfunction;

pregnancy (category C); children younger than 18 y.

ROUTE & DOSAGE

Multiple Sclerosis/Moderate to Severe Crohn's Disease

Adult: **IV** 300 mg infused over 1 h every 4 wk

ADMINISTRATION

Intravenous

PREPARE: **IV Infusion:** Before and after dilution, solution should be colorless and clear to slightly opaque. Do not use if the solution has visible particles, flakes, color, or is cloudy. ▪ Withdraw 300 mg (15 mL) from the vial and add to an IV bag with 100 mL of NS. Do not use with any other diluent. ▪ Gently invert the bag to mix; do not shake. ▪ The IV solution **must be** used within 8 h.

ADMINISTER: **IV Infusion:** Flush IV line before/after with NS. Infuse over 1 h. ▪ Do not give a bolus dose. ▪ Stop infusion immediately if S&S of hypersensitivity appear.

INCOMPATIBILITIES **Solution/ additive/Y-site:** Do not mix or infuse with other drugs.

▪ Store IV solution for up to 8 h at 2°–8° C (36°–46° F). ▪ Allow solution to warm to room temperature before administration.

ADVERSE EFFECTS (≥1%) **Body as a Whole:** Anaphylaxis (rare, usually within 2 h of infusion), infections, fatigue, rigors, risk of progressive multifocal leukoencephalopathy. **CNS:** Depression, headache, syncope, tremor. **CV:** Chest discomfort. **GI:** Abdominal discomfort, abnormal liver function tests. **Hematologic:** Local bleeding from infusion site. **Musculoskeletal:** Arthralgia. **Respiratory:** Pneumonia. **Skin:** Acute urticaria. **Urogenital:** Urinary frequency, irregular menstruation, amenorrhea, dysmenorrhea. **Other:** Infusion-related reactions (headache, dizziness, fatigue, hypersensitivity reactions, urticaria, pruritus, and rigors).

INTERACTIONS Drug: May reduce the effectiveness of VACCINES and TOXOIDS; may increase risk of infection with IMMUNOSUPPRESSANTS.

PHARMACOKINETICS Half-Life: 11 days.

NURSING IMPLICATIONS

Assessment & Drug Effects

▪ During IV infusion and for 1–2 h after, monitor closely for S&S of hypersensitivity (e.g., urticaria, dizziness, fever, rash, chills, pruritus, nausea, flushing, hypotension, dyspnea, and chest pain).
▪ Monitor neurologic status frequently. Report promptly any emerging S&S of dysfunction.
▪ Lab tests: Baseline and periodic CBC with differential.

Patient & Family Education

▪ Report immediately any of the following during/after IV infusion: Difficulty breathing, wheezing or shortness of breath, swelling or tightness about the neck and throat, chest pain, skin rash or hives.
▪ Report promptly S&S of infection (e.g., cough, fever, chills, or sore throat).

NATAMYCIN

(na-ta-mye'sin)

Natacyn
Classification: ANTIFUNGAL ANTIBIOTIC

N

Common adverse effects in *italic*, life-threatening effects underlined; generic names in **bold**; classifications in SMALL CAPS; ✦ Canadian drug name; ⊙ Prototype drug

1039

Therapeutic: ANTIFUNGAL
Prototype: Amphotericin B
Pregnancy Category: C

AVAILABILITY 5% suspension

ACTION & *THERAPEUTIC EFFECT*
Mechanism of action is by binding to sterols in the fungal cell membrane resulting in cell death of fungi. *Effective against many yeasts and filamentous fungi including* Candida, Aspergillus, Cephalosporium, Fusarium, *and* Penicillium.

USES Blepharitis, conjunctivitis, and keratitis caused by susceptible fungi. Drug of choice for *Fusarium solani* keratitis.
UNLABELED USES Oral, cutaneous, and vaginal candidiasis; intranasal treatment of pulmonary aspergillosis.

CONTRAINDICATIONS Hypersensitivity to natamycin or any of its components.
CAUTIOUS USE Pregnancy (category C), lactation. Safety and efficacy in children are not established.

ROUTE & DOSAGE

Fungal Keratitis

Adult: **Ophthalmic** 1 drop in conjunctival sac of infected eye q1–2h for 3–4 days, then decrease to 1 drop q6–8h, then gradually decrease to 1 drop q4–7days

ADMINISTRATION

Instillation
- Wash hands thoroughly before and after treatment. Infection is easily transferred from infected to noninfected eye and to other individuals.
- Shake well before using.
- Store at 2°–24° C (36°–75° F).

ADVERSE EFFECTS (≥1%) **Special Senses:** Blurred vision, photophobia, eye pain. Uneven adherence of suspension to epithelial ulcerations or in fornices.

PHARMACOKINETICS Absorption:
Drug adheres to ulcerated surface of the cornea and is retained in conjunctival fornices. Does not appear to be systemically absorbed.

NURSING IMPLICATIONS

Assessment & Drug Effects
- Inspect eye for response and tolerance at least twice weekly.
- Note: Lack of improvement in keratitis within 7–10 days suggests that causative organisms may not be susceptible to natamycin. Reevaluation is indicated and possibly a change in therapy.

Patient & Family Education
- Learn appropriate technique for application of eye drops.
- Expect temporary light sensitivity. Be prepared to wear sunglasses outdoors after drug administration and perhaps for a few hours indoors.
- Return to ophthalmologist for reevaluation of eye problem if you experience symptoms of conjunctivitis: Pain, discharge, itching, scratching "foreign body sensation," changes in vision.
- Do not share facecloths and hand towels; this will help prevent transmission of the fungal infection.

NATEGLINIDE
(nat-e-gli′nide)
Starlix
Classifications: ANTIDIABETIC; MEGLITINIDE

Common adverse effects in *italic*, life-threatening effects underlined; generic names in **bold**; classifications in SMALL CAPS; ♣ Canadian drug name; ⊕ Prototype drug

Therapeutic: ANTIDIABETIC
Prototype: Repaglinide
Pregnancy Category: C

AVAILABILITY 60 mg, 120 mg tablets

ACTION & *THERAPEUTIC EFFECT*
Lowers blood glucose levels by stimulating the release of insulin from the pancreatic cells of a type 2 diabetic. Significantly reduces postprandial blood glucose in type 2 diabetics and improves glycemic control when given before meals. *Effectiveness is indicated by preprandial blood glucose between 80 and 120 mg/dL and HbA1C 6.5% or less.*

USES Alone or in combination with metformin for the treatment of non-insulin-dependent diabetes mellitus.

CONTRAINDICATIONS Prior hypersensitivity to nateglinide. Type 1 (insulin-dependent) diabetes mellitus, diabetic ketoacidosis; hypoglycemia.

CAUTIOUS USE Renal impairment; liver dysfunction; adrenal or pituitary insufficiency; malnutrition; infection, trauma, surgery or unusual stress; surgery; trauma; pregnancy (category C), lactation.

ROUTE & DOSAGE

Diabetes Mellitus
Adult: **PO** 60–120 mg t.i.d. 1–30 min prior to meals

ADMINISTRATION

Oral
▪ Give, preferably, 1–30 min before meals. Omit the dose if the meal is skipped. Add a dose if an extra meal is eaten. Never double the dose.

▪ Store at 15°–30° C (59°–86° F).

ADVERSE EFFECTS (≥1%) **Body as a Whole:** Back pain, flu-like symptoms. **CV:** Dizziness. **GI:** Diarrhea. **Metabolic:** Hypoglycemia. **Musculoskeletal:** Arthropathy. **Respiratory:** Upper respiratory infection, bronchitis, cough.

INTERACTIONS Drug: NSAIDS, SALICYLATES, MAO INHIBITORS, BETA-ADRENERGIC BLOCKERS, may potentiate hypoglycemic effects; THIAZIDE DIURETICS, CORTICOSTEROIDS, THYROID PREPARATIONS, SYMPATHOMIMETIC AGENTS may attenuate hypoglycemic effects. **Herbal: Garlic, ginseng** may potentiate hypoglycemic effects.

PHARMACOKINETICS Absorption: Rapidly absorbed, 73% bioavailability. **Peak:** 1 h. **Distribution:** 98% protein bound. **Metabolism:** In liver by CYP2C9 (70%) and CYP3A4 (30%). **Elimination:** Primarily in urine. **Half-Life:** 1.5 h.

NURSING IMPLICATIONS

Assessment & Drug Effects
▪ Lab tests: Frequent 2 h postprandial BS monitoring, FBS monitoring and HbA1C q3mo.
▪ Monitor carefully for S&S of hypoglycemia especially during the 1-wk period following transfer from a longer acting sulfonylurea.

Patient & Family Education
▪ Take only before a meal to lessen the chance of hypoglycemia.
▪ When transferred to nateglinide from another oral hypoglycemia drug, start nateglinide the morning after the other agent is stopped, unless directed otherwise by prescriber.
▪ Watch for S&S of hyperglycemia or hypoglycemia (see Appendix F);

Common adverse effects in *italic*, life-threatening effects underlined; generic names in **bold;** classifications in SMALL CAPS; ♣ Canadian drug name; ⊙ Prototype drug

1041

report poor blood glucose control to prescriber.

▪ Report gastric upset or other bothersome GI symptoms to prescriber.

NEBIVOLOL HYDROCHLORIDE

(ne-bi-vol'ol)

Bystolic

Classifications: BETA-ADRENERGIC ANTAGONIST; ANTIHYPERTENSIVE

Therapeutic: ANTIHYPERTENSIVE

Prototype: Propranolol

Pregnancy Category: C

AVAILABILITY 2.5 mg, 5 mg, and 10 mg tablets

ACTION & THERAPEUTIC EFFECT A beta-adrenergic receptor blocker that is a beta-1 selective antagonist in majority of individuals and a nonselective beta-blocker in poor metabolizers. At higher doses nebivolol blocks both beta-1 and beta-2 receptors. *Effectiveness is measured by decreasing both systolic and diastolic pressures associated with hypertension.*

USES Management of hypertension either alone or in combination with other antihypertensive agents. **UNLABELED USES** Management of heart failure.

CONTRAINDICATIONS Severe bradycardia; greater than first degree heart block; sick sinus syndrome without pacemaker; severe hepatic impairment (Child-Pugh greater than class C); decompensated HF; bronchospastic disease; lactation.

CAUTIOUS USE Compensated CHF; history of angina or recent MI; peripheral vascular disease; moderate hepatic and moderate to severe renal impairment; spontaneous hypoglycemia or DM; pregnancy (category C). Safety and efficacy in children not established.

ROUTE & DOSAGE

Hypertension

Adult: **PO** 5 mg PO daily; can increase q2wk up to 40 mg daily

Hepatic Impairment Dosage Adjustment

Moderate hepatic impairment (Child-Pugh class B): Decrease to 2.5 mg daily and increase cautiously as needed; *severe hepatic impairment:* Use is contraindicated

Renal Impairment Dosage Adjustment

CrCl less than 30 mL/min: Decrease to 2.5 mg PO daily and cautiously increase as needed

ADMINISTRATION

Oral

▪ May give without regard to meals.
▪ Store at 20°–25° C (68°–77° F) in a tight, light-resistant container.

ADVERSE EFFECTS (≥1%) Body as a Whole: Chest pain, peripheral edema. **CNS:** Asthenia, dizziness, fatigue, *headache*, insomnia, paresthesia. **CV:** Bradycardia. **GI:** Abdominal pain, diarrhea, nausea. **Metabolic:** Decreased HDL levels, decreased platelet count, hypercholesterolemia, hypertriglyceridemia, hyperuricemia, increased blood urea nitrogen. **Respiratory:** Dyspnea. **Skin:** Rash.

INTERACTIONS Drug: Catecholamine-depleting agents (**reserpine, guanethidine**) may produce excessive reduction in sympathetic activity if used with nebivolol. Compounds that inhibit CYP2D6 (**fluoxetine, paroxetine, propafenone, quinidine**) may increase nebivolol levels. If used in combination with **clonidine**, simultaneous discontinuation of both drugs may cause life-threatening increases in blood pressure. Combination use with **digoxin** may increase the risk of bradycardia. **Cimetidine** increases the levels of nebivolol metabolites. **Verapamil** and **diltiazem** may increase the pharmacologic effects of nebivolol. Nebivolol may decrease the clearance of **disopyramide**. Nebivolol may decrease the AUC and C_{max} of **sildenafil.**

PHARMACOKINETICS Peak: 1.5–4 h. **Distribution:** 98% Plasma protein bound. **Metabolism:** Hepatic; extent depends on genetic profile. **Elimination:** In urine (38–67%) and feces (13–44%). **Half-Life:** 12–19 h depending on genetic differences in metabolism.

NURSING IMPLICATIONS

Assessment & Drug Effects
- Monitor closely BP and HR. Report promptly significant bradycardia or S&S of heart failure.
- Monitor closely during the perioperative period for depressed cardiac functioning.
- Monitor diabetics for loss of glycemia control.
- Monitor respiratory status in those at risk for bronchospasm.
- Lab tests: Periodic lipid profile and LFTs; frequent blood glucose monitoring with DM.

Patient & Family Education
- Use caution with dangerous activities until reaction to drug is known.
- Report promptly any of the following: Sudden weight gain, increasing shortness of breath, swelling in lower legs and feet; heart rate less than 60 beats per minute or other value established by prescriber.
- Diabetics may experience hypoglycemia without the usual signs and symptoms while on this drug.
- Do not abruptly stop taking this medication. It should be tapered off over 1–2 wk.

NEFAZODONE
(nef-a-zo'done)
Classifications: ANTIDEPRESSANT; SEROTONIN NOREPINEPHRINE RE-UPTAKE INHIBITOR (SNRI)
Therapeutic: ANTIDEPRESSANT; SNRI
Prototype: Fluoxetine HCl
Pregnancy Category: C

N

AVAILABILITY 50 mg, 100 mg, 150 mg, 200 mg, 250 mg tablets

ACTION & THERAPEUTIC EFFECT
Antidepressant with a dual mechanism of action. Inhibits neuronal serotonin ($5\text{-}HT_2$) and norepinephrine reuptake. *Effective in treating major depression without major cardiovascular adverse effects.*

USES Treatment of depression.

CONTRAINDICATIONS Hypersensitivity to nefazodone or alcohol; active hepatic disease, hepatitis, jaundice; elevated hepatic transaminase levels; MAOI therapy; mania; severe restlessness, suicidal ideation; surgery.
CAUTIOUS USE History of seizure disorders, seizures; renal impairment; recent MI, unstable cardiac disease; hypotension; angina, stroke, hypovolemia, dehydration,

Common adverse effects in *italic*, life-threatening effects <u>underlined</u>; generic names in **bold**; classifications in SMALL CAPS; ♣ Canadian drug name; ⊘ Prototype drug

1043

bipolar disorder; history of mania; ECT therapy; older adults, women of childbearing age; pregnancy (category C), lactation. Safety and efficacy in children younger than 18 y are not established.

ROUTE & DOSAGE

Depression

Adult: **PO** 50–100 mg b.i.d., may need to increase up to 300–600 mg/day in 2–3 divided doses
Geriatric: **PO** Start with 50 mg b.i.d.

ADMINISTRATION

Oral

▪ Do not give within 14 days of discontinuation of an MAO inhibitor.
▪ Store at 15°–30° C (59°–86° F).

ADVERSE EFFECTS (≥1%) Body as a Whole: Anaphylactic reactions, angioedema. **CNS:** *Headache, dizziness, drowsiness,* asthenia, tremor, insomnia, agitation, anxiety. **GI:** Dry mouth, constipation, nausea, liver toxicity, <u>liver failure</u>. **Special Senses:** Visual disturbances, blurred vision, scotomata. **Endocrine:** Galactorrhea, gynecomastia, serotonin syndrome. **Skin:** <u>Stevens-Johnson syndrome</u>.

INTERACTIONS Drug: May cause serotonin syndrome (see Appendix F) with MAOIS or SSRIS; may increase plasma levels of some BENZODI-AZEPINES, including **alprazolam** and **triazolam.** May decrease plasma levels and effects of **propranolol.** May increase levels and toxicity of **buspirone, carbamazepine, cilostazol, digoxin;** reports of QT_c prolongation and ventricular arrhythmias with **pimozide;** increased risk of rhabdomyolysis with **lovastatin, simvastatin;** increased risk of **ergotamine** toxicity with **dihydroergotamine, ergotamine. Herbal: St. John's wort** may cause **serotonin** syndrome.

PHARMACOKINETICS Onset: 1 wk. **Peak:** 3–5 wk. **Metabolism:** In liver to at least two active metabolites. **Half-Life:** Nefazodone 3.5 h, metabolites 2–33 h.

NURSING IMPLICATIONS

Assessment & Drug Effects

▪ Monitor for worsening of depression or emergence of suicidal ideation.
▪ Evaluate concurrent drugs for possible interactions.
▪ Monitor patients with a history of seizures for increased activity.
▪ Assess safety, as dizziness and drowsiness are common adverse effects.
▪ Lab tests: Monitor periodically LFTs and CBC during long-term therapy.

Patient & Family Education

▪ Be aware that significant improvement in mood may not occur for several weeks following initiation of therapy.
▪ Do not drive or engage in potentially hazardous activities until response to the drug is known.
▪ Report changes in visual acuity.
▪ Report signs of jaundice such as yellow coloration of the cornea of the eye or other S&S of liver dysfunction (anorexia, GI complaints, malaise, etc.).

NELARABINE

Arranon
Classifications: ANTINEOPLASTIC; PYRIMIDINE, ANTIMETABOLITE
Therapeutic: ANTINEOPLASTIC

Common adverse effects in *italic*, life-threatening effects <u>underlined</u>; generic names in **bold**; classifications in SMALL CAPS; ♣ Canadian drug name; ⊙ Prototype drug

Prototype: 5-Fluorouracil
Pregnancy Category: D

AVAILABILITY 5 mg/mL solution

ACTION & THERAPEUTIC EFFECT
Nelarabine inhibits DNA synthesis in lymphoblastic T-cells of acute leukemia and lymphoma. *The incorporation of a nelarabine metabolite in the leukemic blast cells halts DNA synthesis and causes cell death.*

USES Treatment of patients with T-cell acute lymphoblastic leukemia lymphoma.

CONTRAINDICATIONS Hepersensitivity to nelarabine; severe bone marrow suppression; older adults; pregnancy (category D); lactation.
CAUTIOUS USE Severe renal impairment, severe renal failure; hepatic impairment; risk of infection, bleeding.

ROUTE & DOSAGE

Adult T-Cell Leukemia/ Lymphoma

Adult: IV 1500 mg/m² on days 1, 3, and 5, repeated every 21 days
Child: IV 650 mg/m² over 1 h for 5 days, repeated every 21 days

Toxicity Dosage Adjustment

Grade 2 or higher neurologic toxicity: Discontinue therapy; *hematologic toxicities:* Delay therapy

ADMINISTRATION
- Standard IV hydration, urine alkalinization, and prophylaxis with allopurinol are advised to manage hyperuricemia in those at risk for tumor lysis syndrome.
- Use gloves and protective clothing to prevent skin contact.

Intravenous

PREPARE: **IV Infusion:** Do not dilute. Transfer the required dose to a PVC or glass container for infusion.
ADMINISTER: **IV Infusion for Adult:** Give over 2 h. **IV Infusion for Child:** Give over 1 h. ▪ Discontinue IV and notify prescriber for neurologic adverse events of NCI Common Toxicity Criteria grade 2 or greater.

- Store vials at 15°–30° C (59°–86° F). Nelarabine is stable in PVC bags or glass infusion containers for 8 h up to 30° C.

ADVERSE EFFECTS (≥1%) **Body as a Whole:** Abnormal gait, *fatigue, pyrexia,* rigors. **CNS:** *Asthenia,* ataxia, *dizziness, headache, hypoesthesia, neuropathy, paresthesia, somnolence,* tremor. **CV:** Chest pain, *edema,* hypotension, *petechiae,* sinus tachycardia. **GI:** Abdominal pain, *constipation, diarrhea, nausea, vomiting,* stomatitis. **Hematologic/Lymphatic:** *Anemia, neutropenia, thrombocytopenia,* increased risk of infection. **Hepatic:** AST levels increased. **Metabolic:** Anorexia, dehydration, hyperglycemia. **Musculoskeletal:** Arthralgia, back pain, muscular weakness, *myalgia,* pain in extremities. **Respiratory:** *Cough, dyspnea, pleural effusion,* epistaxis, wheezing.

PHARMACOKINETICS Distribution: Extensive. **Metabolism:** Bioactivation to ara-GTP, oxidized to uric acid. **Elimination:** Renal. **Half-Life:** 3 h (active metabolite).

NURSING IMPLICATIONS
Assessment & Drug Effects
- Monitor for and report immediately S&S of adverse CNS effects, including altered mental status

Common adverse effects in *italic*, life-threatening effects underlined; generic names in **bold;** classifications in SMALL CAPS; ♣ Canadian drug name; ⊘ Prototype drug

1045

(e.g., confusion, severe somnolence), seizures, and peripheral neuropathy (e.g., numbness, paresthesias, motor weakness, ataxia, paralysis). Note: Previous or concurrent treatment with intrathecal chemotherapy or previous craniospinal irradiation may increase risk of CNS toxicity.

- Monitor for S&S of bleeding, especially with platelet counts less than 50,000/mm³.
- Lab tests: Baseline and periodic CBC with differential and platelet count; periodic serum electrolytes, serum uric acid, LFTs, and renal function test.
- Monitor diabetics for loss of glycemic control.
- Note: Previous or concurrent treatment with intrathecal chemotherapy or previous craniospinal irradiation may increase risk of CNS toxicity.

Patient & Family Education
- Do not drive or engage in potentially hazardous activities until response to drug is known.
- Report any of the following to a health care provider: Seizures; tingling or numbness in hands and feet; problems with fine motor coordination; unsteady gait and increased weakness with ambulating; fever or other signs of infections; black tarry stools, blood tinged urine, or other signs of bleeding.
- Use effective contraceptive measures to avoid pregnancy while taking this drug.

NELFINAVIR MESYLATE
(nel-fin′a-vir)

Viracept
Classifications: ANTIRETROVIRAL; PROTEASE INHIBITOR

Therapeutic: PROTEASE INHIBITOR
Prototype: Saquinavir
Pregnancy Category: B

AVAILABILITY 250 mg, 625 mg tablets; 50 mg/g powder for oral suspension

ACTION & *THERAPEUTIC EFFECT*
Inhibits HIV-1 protease, which is responsible for the production of HIV-1 viral particles in an infected individual. This prevents the cleavage of viral polypeptide, resulting in the production of an immature, noninfectious virus. *Effectiveness is indicated by decreased viral load.*

USES Treatment of HIV infection in combination with other agents.

CONTRAINDICATIONS Hypersensitivity to nelfinavir; pancreatitis; lactation.

CAUTIOUS USE Liver function impairment, hemophilia; diabetes mellitus, hyperglycemia; pregnancy (category B); children younger than 2 y.

ROUTE & DOSAGE

HIV Infection

Adult/Adolescent: **PO** 750 mg t.i.d. or 1250 mg b.i.d. with food
Child (2–13 y): **PO** 20–30 mg/kg t.i.d. with food (max: 750 mg/dose)

ADMINISTRATION
Oral
- Give with a meal or light snack.
- Oral powder may be mixed with a small amount of water, milk, soy milk, or dietary supplements; liquid should be consumed immediately. Do not mix oral powder in original container nor with acid food or juice (e.g., orange or apple juice, or applesauce).
- Store at 15°–30° C (59°–86° F).

Common adverse effects in *italic*, life-threatening effects underlined; generic names in **bold**; classifications in SMALL CAPS; ♣ Canadian drug name; ⊙ Prototype drug

ADVERSE EFFECTS (≥1%) **Body as a Whole:** Allergic reactions, back pain, fever, malaise, pain, asthenia, myalgia, arthralgia. **CNS:** Headache, anxiety, depression, dizziness, insomnia, seizures. **GI:** Abdominal pain, *diarrhea,* nausea, flatulence, anorexia, dyspepsia, GI bleeding, hepatitis, vomiting, pancreatitis, increased liver function tests. **Hematologic:** Anemia, leukopenia, thrombocytopenia. **Respiratory:** Dyspnea, pharyngitis, rhinitis. **Skin:** Rash, pruritus, sweating, urticaria. **Endocrine:** Lipodystrophy, insulin resistance, dyslipidemia.

INTERACTIONS Drug: Other PROTEASE INHIBITORS, **ketoconazole** may increase nelfinavir levels; **rifabutin, rifampin,** PROTON PUMP INHIBITORS may decrease nelfinavir levels; nelfinavir will decrease ORAL CONTRACEPTIVE levels; may increase levels of **amiodarone, atorvastatin, simvastatin, slidenafil,** PDE 5 INHIBITORS; increase risk of **ergotamine** toxicity with **dihydroergotamine, ergotamine,** HMG-COA REDUCTASE INHIBITORS may have increased risk of rhabdomyolysis. BENZODIAZEPINES may increase risk of sedation. **Herbal: St. John's wort, garlic** may decrease antiretroviral activity. Use with **red yeast rice** may increase risk of rhabdomylosis.

PHARMACOKINETICS Absorption: Food increases the amount of drug absorbed. **Distribution:** Greater than 98% protein bound. **Metabolism:** In the liver (CYP3A). **Elimination:** Primarily in feces. **Half-Life:** 3.5–5 h.

NURSING IMPLICATIONS

Assessment & Drug Effects

- Monitor hemophiliacs (type A or B) closely for spontaneous bleeding.

- Monitor carefully patients with hepatic impairment for toxic drug effects.

Patient & Family Education

- Drug **must be** taken exactly as prescribed. Do not alter dose or discontinue drug without consulting prescriber.
- Use a barrier contraceptive even if using hormonal contraceptives.
- Be aware that diarrhea is a common adverse effect that can usually be controlled by OTC medications.

NEOMYCIN SULFATE

(nee-oh-mye′sin)

Mycifradin, Myciguent, Neo-Tabs, Neo-fradin

Classification: AMINOGLYCOSIDE ANTIBIOTIC
Therapeutic: ANTIBIOTIC
Prototype: Gentamicin
Pregnancy Category: C

N

AVAILABILITY 500 mg tablet; 125 mg/5 mL oral solution; 3.5 mg/g ointment, cream

ACTION & *THERAPEUTIC EFFECT*
Aminoglycoside antibiotic that inhibits bacterial protein synthesis through irreversible binding to the 30S ribosomal subunit within susceptible bacteria, thus causing bacteria not to replicate. *Active against a wide variety of gram-negative bacteria. Effective against certain gram-positive organisms, particularly penicillin-sensitive and some methicillin-resistant strains of* Staphylococcus aureus *(MRSA).*

USES Severe diarrhea caused by enteropathogenic *Escherichia coli;* preoperative intestinal antisepsis; to inhibit nitrogen-forming bacteria of GI tract in patients with cirrhosis

Common adverse effects in *italic*, life-threatening effects <u>underlined</u>; generic names in **bold;** classifications in SMALL CAPS; ♣ Canadian drug name; ⊙ Prototype drug

1047

or hepatic coma and for urinary tract infections caused by susceptible organisms. Also topically for short-term treatment of eye, ear, and skin infections.

CONTRAINDICATIONS Hypersensitivity to aminoglycosides; use of oral drug in patients with intestinal obstruction; ulcerative bowel lesions; IBD; topical applications over large skin areas; aminoglycosides.

CAUTIOUS USE Dehydration; renal disease, renal impairment; hearing impairment; myasthenia gravis, parkinsonism; pregnancy (category C), lactation, children. **Topical otic applications:** Patients with perforated eardrum.

ROUTE & DOSAGE

Intestinal Antisepsis

Adult: **PO** 1 g q1h × 4 doses, then 1 g q4h × 5 doses
Child: **PO** 10.3 mg/kg q4–6h for 3 days

Hepatic Coma

Adult: **PO** 4–12 g/day in 4 divided doses for 5–6 days
Child: **PO** 437.5–1225 mg/m² q6h for 5–6 days

Diarrhea

Adult: **PO** 50 mg/kg in 4 divided doses for 2–3 days
Child: **PO** 8.75 mg/kg q6h for 2–3 days

Cutaneous Infections

Adult: **Topical** Apply 1–3 times/day

ADMINISTRATION

Oral

- Preoperative bowel preparation: Saline laxative is generally given immediately before neomycin therapy is initiated.

Topical

- Consult prescriber about what to use for cleansing skin before each application.
- Make sure ear canal is clean and dry prior to instillation for topical therapy of external ear.

ADVERSE EFFECTS (≥1%) **Body as a Whole:** <u>Neuromuscular blockade</u> with muscular and <u>respiratory paralysis</u>; hypersensitivity reactions. **GI:** Mild laxative effect, diarrhea, nausea, vomiting; prolonged therapy: Malabsorption-like syndrome including cyanocobalamin (vitamin B_{12}) deficiency, low serum cholesterol. **Urogenital:** <u>Nephrotoxicity</u>. **Special Senses:** <u>Ototoxicity</u>. **Skin:** *Redness,* scaling, pruritus, dermatitis.

INTERACTIONS Drug: May decrease absorption of **cyanocobalamin.**

PHARMACOKINETICS Absorption: 3% absorbed from GI tract in adults; up to 10% absorbed in neonates. **Peak:** 1–4 h. **Elimination:** 97% excreted unchanged in feces. **Half-Life:** 3 h.

NURSING IMPLICATIONS

Assessment & Drug Effects

- Monitor closely for ototoxicity and nephrotoxicity. Risk is greatest in those with impaired renal function.
- Lab tests: Obtain baseline and daily urinalysis for albumin, casts, and cells; BUN every other day; baseline and daily creatinine clearance in elderly patients.
- Monitor I&O in patients receiving drug orally. Report oliguria or changes in I&O ratio. Inadequate neomycin excretion results in high serum drug levels and risk of nephrotoxicity.

Patient & Family Education

▪ Stop treatment and consult your prescriber if irritation occurs when you are using topical neomycin. Allergic dermatitis is common.

▪ Report any unusual symptom related to ears or hearing (e.g., tinnitus, roaring sounds, loss of hearing acuity, dizziness).

▪ Do not exceed prescribed dosage or duration of therapy.

NEOSTIGMINE METHYLSULFATE ℗

(nee-oh-stig′meen)

Prostigmin
Classifications: CHOLINERGIC AGENT; CHOLINESTERASE INHIBITOR
Therapeutic: CHOLINESTERASE INHIBITOR
Pregnancy Category: C

AVAILABILITY 1:1000, 1:2000, 1:4000 injection

ACTION & *THERAPEUTIC EFFECT*
Produces reversible cholinesterase inhibition or inactivation with direct stimulant action on voluntary muscle fibers and possibly on autonomic ganglia and CNS neurons. Allows intensified and prolonged effect of acetylcholine at cholinergic synapses (basis for use in myasthenia gravis). *Produces generalized cholinergic response including miosis, increased tonus of intestinal and skeletal muscles, constriction of bronchi and ureters, slower pulse rate, and stimulation of salivary and sweat glands.*

USES To prevent and treat postoperative abdominal distention and urinary retention; for symptomatic control of and sometimes for differential diagnosis of myasthenia gravis; and to reverse the effects of

nondepolarizing muscle relaxants (e.g., tubocurarine).

CONTRAINDICATIONS Hypersensitivity to neostigmine, cholinergics; bromides; cholinesterase inhibitor toxicity; GI obstruction; ileus; bradycardia, hypotension; mechanical obstruction of intestinal or urinary tract; peritonitis; lactation.

CAUTIOUS USE Recent ileorectal anastomoses; epilepsy; bronchial asthma; hepatic disease; bradycardia, recent coronary occlusion; vagotonia; cardiac arrhythmias; renal failure; renal impairment; renal disease; hyperthyroidism; peptic ulcer; seizure disorder; pregnancy (category C).

ROUTE & DOSAGE

Diagnosis of Myasthenia Gravis
Adult: **IM** 0.02 mg/kg
Child: **IM** 0.04 mg/kg
Treatment of Myasthenia Gravis
Adult: **IM/Subcutaneous** 0.5–2.5 mg q1–3h (max: 10 mg/day)
Child: **IM/Subcutaneous** 0.01–0.04 mg/kg q2–4h
Reversal of Nondepolarizing Neuromuscular Blockade
Adult: **IV** 0.5–2.5 mg slowly (max dose: 5 mg); may repeat
Child: **IV** 0.025–0.08 mg/kg
Infant: **IV** 0.025–0.1 mg/kg
Postoperative Abdominal Distention and Urinary Retention
Adult: **IM/Subcutaneous** 0.25 mg q4–6h for 2–3 days
Myasthenia Gravis
Adult: **IV** 0.5–2 mg q1–3h
Child: **IV** 0.01–0.04 mg/kg q2–4h

N

Common adverse effects in *italic*, life-threatening effects underlined; generic names in **bold**; classifications in SMALL CAPS; ♣ Canadian drug name; ℗ Prototype drug

1049

Renal Impairment Dosage Adjustment

CrCl 10–50 mL/min: Use 50% of dose; *less than 10 mL/min:* Use 25% of dose

ADMINISTRATION

Intramuscular/Subcutaneous

- Note: 1 mg = 1 mL of the 1:1000 solution; 0.5 mg = 1 mL of the 1:2000 solution; 0.25 mg = 1 mL of the 1:4000 solution.
- Give undiluted.

Intravenous

PREPARE: **Direct:** Give undiluted.
ADMINISTER: **Direct:** Give at a rate of 0.5 mg or a fraction thereof over 1 min.

ADVERSE EFFECTS (≥1%) **Body as a Whole:** Muscle cramps, *fasciculations,* twitching, pallor, fatigability, generalized weakness, paralysis, agitation, fear, death. **CV:** Tightness in chest, bradycardia, hypotension, elevated BP. **GI:** *Nausea,* vomiting, eructation, epigastric discomfort, abdominal cramps, diarrhea, involuntary or difficult defecation. **CNS:** CNS stimulation. **Respiratory:** *Increased salivation* and bronchial secretions, sneezing, cough, dyspnea, diaphoresis, respiratory depression. **Special Senses:** Lacrimation, miosis, blurred vision. **Urogenital:** Difficult micturition.

INTERACTIONS Drug: **Succinylcholine decamethonium** may prolong phase I block or reverse phase II block; neostigmine antagonizes effects of **tubocurarine; atracurium, vecuronium, pancuronium; procainamide, quinidine, atropine** antagonize effects of neostigmine.

PHARMACOKINETICS Onset: 10–30 min IM or IV. **Peak:** 20–30 min IM or IV. **Distribution:** Not reported to cross placenta or appear in breast milk. **Metabolism:** In liver. **Elimination:** 80% of drug and metabolites excreted in urine within 24 h. **Half-Life:** 50–90 min.

NURSING IMPLICATIONS

Assessment & Drug Effects

- Check pulse before giving drug to bradycardic patients. If below 60/min or other established parameter, consult prescriber. Atropine will be ordered to restore heart rate.
- Monitor respiration, maintain airway or assisted ventilation, and give oxygen as indicated when used as antidote for tubocurarine or other nondepolarizing neuromuscular blocking agents (usually preceded by atropine).
- Monitor pulse, respiration, and BP during period of dosage adjustment in treatment of myasthenia gravis.
- Report promptly and record accurately the onset of myasthenic symptoms and drug adverse effects in relation to last dose.
- Note carefully time of muscular weakness onset. It may indicate whether patient is in cholinergic or myasthenic crisis: Weakness that appears approximately 1 h after drug administration suggests cholinergic crisis (overdose) and is treated by prompt withdrawal of neostigmine and immediate administration of atropine. Weakness that occurs 3 h or more after drug administration is more likely due to myasthenic crisis (underdose or drug resistance) and is treated by more intensive anticholinesterase therapy.
- Record drug effect and duration of action. S&S of myasthenia gravis relieved by neostigmine

include lid ptosis; diplopia; drooping facies; difficulty in chewing, swallowing, breathing, or coughing; and weakness of neck, limbs, and trunk muscles.

- Manifestations of neostigmine overdosage often appear first in muscles of neck and those involved in chewing and swallowing, with muscles of shoulder girdle and upper extremities affected next.

- Report to prescriber if patient does not urinate within 1 h after first dose when used to relieve urinary retention.

Patient & Family Education

- Keep a diary of "peaks and valleys" of muscle strength.

- Keep an accurate record for prescriber of your response to drug. Learn how to recognize adverse effects, how to modify dosage regimen according to your changing needs, or how to administer atropine if necessary.

- Be aware that certain factors may require an increase in size or frequency of dose (e.g., physical or emotional stress, infection, menstruation, surgery), whereas remission requires a decrease in dosage.

NEPAFENAC
(nep′a-fe-nac)
Nevanac
Pregnancy Category: C
See Appendix A-1.

NESIRITIDE
(nes-ir′i-tide)
Natrecor
Classifications: CARDIOVASCULAR; ATRIAL NATRIURETIC PEPTIDE HORMONE

Therapeutic: ATRIAL NATRIURETIC HORMONE (ANH)
Pregnancy Category: C

AVAILABILITY 1.5 mg vial

ACTION & *THERAPEUTIC EFFECT*
Nesiritide is a human B-type natriuretic peptide (hBNP), produced by recombinant DNA, which mimics the actions of human atrial natriuretic hormone (ANH). ANH is secreted by the right atrium when atrial blood pressure increases. Nesiritide, like ANH, inhibits antidiuretic hormone (ADH) by increasing urine sodium loss by the kidney and triggering the formation of a large volume of dilute urine. Nesiritide binds to a cyclic nucleic acid, which results in smooth muscle cell relaxation. *Effective in causing smooth muscle relaxation. The drug also causes dilation of veins and arteries. It is effective in managing dyspnea at rest in patients with acute CHF.*

USES Acute treatment of decompensated CHF in patients who have dyspnea at rest or with minimal activity.

CONTRAINDICATIONS Hypersensitivity to nesiritide, *Escherichia coli* protein, patients with a systolic blood pressure less than 90 mm Hg, cardiogenic shock, patients with low cardiac filling pressures, patients who should not receive vasodilators, such as those with significant valvular stenosis, restrictive or obstructive cardiomyopathy, pericardial perfusion; constrictive pericarditis, pericardial tamponade.
CAUTIOUS USE Pregnancy (category C), lactation. Safety and efficacy in pediatric patients have not been established.

Common adverse effects in *italic*, life-threatening effects underlined; generic names in **bold**; classifications in SMALL CAPS; ♣ Canadian drug name; ⊙ Prototype drug

1051

ROUTE & DOSAGE

Acute Decompensated CHF

Adult: IV 2 mcg/kg bolus administered over 60 sec, followed by a continuous infusion of 0.01 mcg/kg/min (0.1 mL/kg/h) (max: 0.03 mcg/kg/min). Monitor blood pressure. If hypotension occurs, the dose should be reduced or discontinued. The infusion can subsequently be restarted at a dose that is reduced by 30% (with no bolus administration) after stabilization of hemodynamics.

ADMINISTRATION

Intravenous

PREPARE: **Direct and IV Infusion:** Reconstitute one 1.5 mg vial by adding 5 mL of IV solution removed from a 250 mL bag of selected diluent (i.e., D5W, NS, D5/0.45% NaCl, D5/0.2% NaCl). ■ Rock the vial gently so that all surfaces, including the stopper, contact the diluent ensuring complete reconstitution. Do not shake the vial. ■ Add the entire contents of the vial to the 250 mL IV bag to yield approximately 6 mcg/mL. Invert the bag several times to mix completely. ■ Use within 24 h. ■ Prime the IV tubing with 25 mL prior to connecting to the vascular access port.

ADMINISTER: **Direct:** Bolus dose **MUST BE** withdrawn from the prepared infusion bag. Determine dose as follows: Bolus volume (mL) = (0.33) × (patient weight in kg). ■ Give the bolus dose over 60 sec through an IV port in the tubing. **IV Infusion:** Infuse remainder of IV infusion immediately following the bolus dose. ■ Determine the infusion rate as follows: Flow rate (mL/h) = (0.1) × (patient weight in kg).

INCOMPATIBILITIES **Solution/additive: Promethazine. Y-site: Bumetanide, enalaprilat, ethacrynic acid, furosemide, heparin, hydralazine, regular insulin, micafungin.**

■ Store at controlled room temperature at 20°–25° C (68°–77° F) or refrigerated.

ADVERSE EFFECTS (≥1%) Body as a Whole: Headache, back pain, catheter pain, fever, injection site pain, leg cramps. **CNS:** Insomnia, dizziness, anxiety, confusion, paresthesia, somnolence, tremor. **CV:** *Hypotension,* ventricular tachycardia, ventricular extrasystoles, angina, bradycardia, tachycardia, atrial fibrillation, AV node conduction abnormalities. **GI:** Abdominal pain, nausea, vomiting. **Respiratory:** Cough, hemoptysis, apnea. **Skin:** Sweating, pruritus, rash. **Special Senses:** Amblyopia. **Renal:** Renal failure in acutely decompensated heart failure patients.

INTERACTIONS Drug: Additive effects with ANTIHYPERTENSIVES.

PHARMACOKINETICS Onset: 15 min. **Duration:** Greater than 60 min depending on dose. **Metabolism:** Proteolytic cleavage, proteolysis. **Half-Life:** 18 min.

NURSING IMPLICATIONS

Assessment & Drug Effects

■ Monitor hemodynamic parameters (e.g., BP, PCWP, HR, ECG) throughout therapy.
■ Establish hypotension parameters prior to initiating therapy. Notify prescriber immediately if systolic BP falls below 90 mm Hg.
■ Reduce the dose or withhold the drug if hypotension occurs during administration. Reinitiate therapy infusion only after hypotension is corrected. Subsequent doses following a hypotensive episode are

Common adverse effects in *italic*, life-threatening effects <u>underlined</u>; generic names in **bold**; classifications in SMALL CAPS; ✚ Canadian drug name; ⊕ Prototype drug

usually reduced by 30% and given without a prior bolus dose.
- Lab tests: Baseline and periodic serum creatinine.

NEVIRAPINE
(ne-vir'a-peen)
Viramune, Viramune XR
Classifications: ANTIRETROVIRAL; NONNUCLEOSIDE REVERSE TRANSCRIPTASE INHIBITOR (NNRTI)
Therapeutic: ANTIRETROVIRAL; NNRTI
Prototype: Efavirenz
Pregnancy Category: B

AVAILABILITY 200 mg tablets; 10 mg/mL oral suspension; 400 mg extended release tablet

ACTION & *THERAPEUTIC EFFECT*
Nonnucleoside reverse transcriptase inhibitor (NNRTI) of HIV-1. Binds directly to reverse transcriptase and blocks RNA- and DNA-dependent polymerase activities, thus preventing replication of the virus. *Prevents replication of the HIV-1 virus. Resistant strains appear rapidly.*

USES Treatment of HIV with other agents.
UNLABELED USES Prevention of maternal-fetal HIV transmission.

CONTRAINDICATIONS Hypersensitivity to nevirapine; development of rash; severe skin reactions to the drug; hepatitis B or C; or early possbile S&S of hepatitis; increased transaminases combined with rash or sign of hepatotoxicity; hepatic impairment of Child-Pugh class B or C use in post-exposure to HIV prophylaxis treatment; hormonal contraception; lactation.
CAUTIOUS USE Renal disease; hemodialysis; mild hepatic impairment, CNS disorders; pregnant women with CD4+ lymphocyte

counts greater than 250/mm³; pregnancy (category B); neonates.

ROUTE & DOSAGE

HIV
Adult/Adolescent: **PO** 200 mg once daily for 14 days, then increase to 200 mg b.i.d. (immediate release) or 400 mg daily (extended release)
Child/Infant (older than 15 days): **PO** 150 mg/m² daily × 14 days, then 150 mg/m² twice daily (max: 400 mg) (max: 200 mg/dose)
Hepatic Impairment Dosage Adjustment
Do not use in Child-Pugh class B or C

ADMINISTRATION
Oral
- Reinitiate with 200 mg/day for 14 days, then increase to b.i.d. dosing, when dosing is interrupted for more than 7 days.
- Ensure that extended release tablet is swallowed whole. It must not be crushed or chewed.
- Store at 15°–30° C (59°–86° F) in a tightly closed container.

ADVERSE EFFECTS (≥1%) **Body as a Whole:** Fever, paresthesia, myalgia. **CNS:** Headache. **GI:** Nausea, diarrhea, abdominal pain, hepatitis, increased liver function tests, hepatotoxicity (including fulminant and cholestatic hepatitis, hepatic necrosis, and hepatic failure, especially with long-term use). **Hematologic:** Anemia, neutropenia. **Skin:** *Rash,* Stevens-Johnson syndrome.

INTERACTIONS Drug: May decrease plasma concentrations of PROTEASE INHIBITORS, ORAL CONTRACEPTIVES; may decrease **methadone, dronedarone** levels. **Fluconazole** may increase adverse effects. **Herbal:**

Common adverse effects in *italic*, life-threatening effects underlined; generic names in **bold**; classifications in SMALL CAPS; ♣ Canadian drug name; ☻ Prototype drug

1053

St. John's wort, garlic may decrease antiretroviral activity.

PHARMACOKINETICS Absorption: Rapidly from GI tract. **Peak:** 4h. **Distribution:** 60% protein bound, crosses placenta, distributed into breast milk. **Metabolism:** In liver (CYP3A). **Elimination:** Primarily in urine. **Half-Life:** 25–40 h.

NURSING IMPLICATIONS

Assessment & Drug Effects

- Lab tests: Obtain baseline and periodic LFTs and kidney function tests, routine blood chemistry, and CBC.
- Monitor weight, temperature, respiratory status with chest x-ray throughout therapy.
- Monitor carefully, especially during first 6 wk of therapy, for severe rash (with or without fever, blistering, oral lesions, conjunctivitis, swelling, joint aches, or general malaise).
- Withhold drug and notify prescriber if rash develops or liver function tests are abnormal.

Patient & Family Education

- Learn about common adverse effects.
- Withhold drug and notify prescriber if severe rash appears.
- Do not drive or engage in potentially hazardous activities until response to drug is known. There is a high potential for drowsiness and fatigue.
- Use or add barrier contraceptive if using hormonal contraceptive.

NIACIN (VITAMIN B₃, NICOTINIC ACID)

(nye'a-sin)

Niacor, Niaspan, Nicobid, Nico-400, Nicotinex, Novoniacin ♣, Slo-Niacin, Tri-B3 ♣

NIACINAMIDE (NICOTINAMIDE)

Classifications: VITAMIN B₃; ANTI-LIPEMIC

Therapeutic: ANTILIPEMIC; LIPID-LOWERING AGENT

Pregnancy Category: C

AVAILABILITY 50 mg, 100 mg, 250 mg, 500 mg tablets; 125 mg, 250 mg, 400 mg, 500 mg, 750 mg, 1000 mg sustained release tablets, capsules

ACTION & THERAPEUTIC EFFECT Water-soluble, heat-stable, B-complex vitamin (B₃) that functions with riboflavin as a control agent in coenzyme system that converts protein, carbohydrate, and fat to energy through oxidation-reduction. Niacinamide, an amide of niacin, is used as an alternative in the prevention and treatment of pellagra. *Produces vasodilation by direct action on vascular smooth muscles. Inhibits hepatic synthesis of VLDL, cholesterol, and triglyceride, and, indirectly, LDL. Large doses effectively reduce elevated serum cholesterol and total lipid levels in hypercholesterolemia and hyperlipidemic states.*

USES In prophylaxis and treatment of pellagra, usually in combination with other B-complex vitamins, and in deficiency states accompanying carcinoid syndrome, isoniazid therapy, Hartnup's disease, and chronic alcoholism. Also in adjuvant treatment of hyperlipidemia (elevated cholesterol or triglycerides) in patients who do not respond adequately to diet or weight loss. Also as vasodilator in peripheral vascular disorders, Ménière's disease, and labyrinthine syndrome, as well as to counteract LSD toxicity and to distinguish between psychoses of dietary and nondietary origin.

CONTRAINDICATIONS Hypersensitivity to niacin; hepatic impairment; severe hypotension; hemorrhaging or arterial bleeding; active peptic ulcer; lactation.

CAUTIOUS USE History of gallbladder disease, liver disease, and peptic ulcer; severe renal impairment; glaucoma; angina; coronary artery disease; diabetes mellitus; predisposition to gout; allergy; thrombocytopenia; pregnancy (category C); children younger than 16 y **(extended release).**

ROUTE & DOSAGE

Niacin Deficiency
Adult: **PO** 10–20 mg/day

Pellagra
Adult: **PO** 50–100 mg 3–4 times/ day
Child: **PO** 50–100 mg t.i.d.

Hyperlipidemia
Adult: **PO** 1.5–3 g/day in divided doses, may increase up to 6 g/ day if necessary
Child: **PO** 100–250 mg/day in 3 divided doses, may increase by 250 mg/day q2–3wk as tolerated

ADMINISTRATION

Oral
- Give with meals to decrease GI distress. Give with cold water (not hot beverage) to facilitate swallowing.
- Ensure that sustained release form is not chewed or crushed. It **must be** swallowed whole.
- Store at 15°–30° C (59°–86° F) in a light and moisture proof container.

ADVERSE EFFECTS (≥1%) **CNS:** *Transient headache, tingling of extremities,* syncope. With chronic use: Nervousness, panic, toxic amblyopia, proptosis, blurred vision, loss of central vision. **CV:** *Generalized flushing with sensation of warmth,* postural hypotension, vasovagal attacks, arrhythmias (rare). **GI:** *Abnormalities of liver function tests; jaundice, bloating, flatulence, nausea,* vomiting, GI disorders, activation of peptic ulcer, xerostomia. **Skin:** *Increased sebaceous gland activity,* dry skin, skin rash, *pruritus,* keratitis nigricans. **Metabolic:** Hyperuricemia, hyperglycemia, glycosuria, hypoprothrombinemia, hypoalbuminemia.

DIAGNOSTIC TEST INTERFERENCE Niacin causes elevated serum *bilirubin, uric acid, alkaline phosphatase, AST, ALT, LDH* levels and may cause *glucose intolerance.* Decreases *serum cholesterol* 15–30% and may cause false elevations with certain *fluorometric methods* of determining *urinary catecholamines.* Niacin may cause false-positive *urine glucose* tests using *copper sulfate reagents* (e.g., *Benedict's* solution).

INTERACTIONS Drug: Potentiates hypotensive effects of ANTIHYPERTENSIVE AGENTS.

PHARMACOKINETICS Absorption: Readily from GI tract. **Peak:** 20–70 min. **Distribution:** Into breast milk. **Metabolism:** In liver. **Elimination:** Primarily in urine. **Half-Life:** 45 min.

NURSING IMPLICATIONS

Assessment & Drug Effects
- Monitor therapeutic effectiveness and record effect of therapy on clinical manifestations of deficiency (fiery red tongue,

Common adverse effects in *italic*, life-threatening effects underlined; generic names in **bold**; classifications in SMALL CAPS; ♣ Canadian drug name; ☻ Prototype drug

1055

excessive saliva secretion and infection of oral membranes, nausea, vomiting, diarrhea, confusion). Therapeutic response usually begins within 24 h.

- Lab tests: Obtain baseline and periodic blood glucose and LFTs in patients receiving prolonged high dose therapy.
- Monitor diabetics and patients on high doses for decreased glucose tolerance and loss of glycemic control.
- Observe patients closely for evidence of liver dysfunction (jaundice, dark urine, light-colored stools, pruritus) and hyperuricemia in patients predisposed to gout (flank, joint, or stomach pain; altered urine excretion pattern).

Patient & Family Education

- Be aware that you may feel warm and flushed in face, neck, and ears within first 2 h after oral ingestion and it may last several hours. Effects are usually transient and subside as therapy continues.
- Sit or lie down and avoid sudden posture changes if you feel weak or dizzy. Report these symptoms and persistent flushing to your prescriber.
- Be aware that alcohol and large doses of niacin cause increased flushing and sensation of warmth.
- Avoid exposure to direct sunlight until lesions have entirely cleared if you have skin manifestations.

NICARDIPINE HYDROCHLORIDE

(ni-car'di-peen)

Cardene, Cardene SR
Classifications: CALCIUM CHANNEL BLOCKER; ANTIHYPERTENSIVE
Therapeutic: ANTIHYPERTENSIVE; ANTIANGINAL

Prototype: Nifedipine
Pregnancy Category: C

AVAILABILITY 20 mg, 30 mg capsules; 30 mg, 45 mg, 60 mg sustained release capsules; 2.5 mg/mL injection

ACTION & *THERAPEUTIC EFFECT*
Calcium channel entry blocker that inhibits the transmembrane influx of calcium ions into cardiac muscle and smooth muscle, thus affecting contractility. Selectively affects vascular smooth muscle more than cardiac muscle. *Significantly decreases systemic vascular resistance. It reduces BP at rest and during isometric and dynamic exercise.*

USES Either alone or with betablockers for chronic, stable (effort-associated) angina; either alone or with other antihypertensives for essential hypertension.
UNLABELED USES CHF, cerebral ischemia, migraine.

CONTRAINDICATIONS Hypersensitivity to nicardipine; advanced aortic stenosis; cardiogenic shock; hypotension; lactation.
CAUTIOUS USE CHF; renal and hepatic impairment; severe bradycardia; older adult; GERD; hiatal hernia; renal disease; renal impairment; acute stroke; acute myocardial infarction; pregnancy (category C).

ROUTE & DOSAGE

Hypertension, Angina
Adult: **PO** 20–40 mg t.i.d. or 30–60 mg SR b.i.d. **IV** Initiation of therapy in a drug-free patient: 5 mg/h initially, increase dose by 2.5 mg/h q15min (or faster) (max: 15 mg/h); for severe hypertension: 4–7.5 mg/h; for

postop hypertension: 10–15 mg/h initially, then 1–3 mg/h

Substitute for Oral Nicardipine

Adult: **IV** 20 mg q8h **PO** 0.5 mg/h; 30 mg q8h **PO** 1.2 mg/h; 40 mg q8h **PO** 2.2 mg/h

ADMINISTRATION

Note: To prevent symptoms of withdrawal, do not abruptly discontinue drug.

Oral

- Give on empty stomach. High-fat meals may decrease blood levels.
- Ensure that sustained release form is not chewed or crushed. It **must be** swallowed whole.
- When converting from IV to oral dose, give first dose of t.i.d. regimen 1 h before discontinuing infusion.

Intravenous

PREPARE: **IV Infusion:** Dilute each 25 mg ampule with 240 mL of D5W or NS to yield 0.1 mg/mL.

ADMINISTER: **IV Infusion:** Usually initiated at 50 mL/h (5 mg/h) with rate increases of 25 mL/h (2.5 mg/h) q5–15min up to a maximum of 150 mL/h. • Infusion is usually slowed to 30 mL/h once the target BP is reached. *Substitute for oral doses:* Oral 20 mg q8h, IV equivalent is 0.5 mg/h; oral 30 mg q8h, IV equivalent is 1.2 mg/h; oral 40 mg q8h, IV equivalent is 2.2 mg/h.

INCOMPATIBILITIES **Solution/additive: Sodium bicarbonate. Y-site: Ampicillin, ampicillin/sulbactam, furosemide, heparin, thiopental.**

ADVERSE EFFECTS (≥1%) CNS:
Dizziness or headache, fatigue, anxiety, depression, paresthesias, insomnia, somnolence, nervousness. **CV:** Pedal edema, hypotension, flushing, palpitations, tachycardia, increased angina. **GI:** Anorexia, nausea, vomiting, dry mouth, constipation, dyspepsia. **Skin:** Rash, pruritus. **Body as a Whole:** Arthralgia or arthritis.

INTERACTIONS Drug: Adenosine
prolongs bradycardia. **Amiodarone** may cause sinus arrest and AV block. **Benazepril** blunts increase in heart rate and increase in plasma **norepinephrine** and **aldosterone** seen with nicardipine. BETA-BLOCKERS cause hypotension and bradycardia. **Cimetidine** increases levels of nicardipine, resulting in hypotension. Concomitant nicardipine and **cyclosporine** result in significant increase in **cyclosporine** serum concentrations 1–30 days after initiation of nicardipine therapy; following withdrawal of nicardipine, **cyclosporine** levels decrease. **Magnesium,** when used to retard premature labor, may cause severe hypotension and neuromuscular blockade. **Food: Grapefruit juice** (greater than 1 qt/day) may increase plasma concentrations and adverse effects.

PHARMACOKINETICS Absorption:
Immediately 35% of oral dose reaches systemic circulation. **Onset:** 1 min IV; 20 min PO. **Peak:** 0.5–2 h. **Duration:** 3 h IV. **Distribution:** 95% protein bound; distributed in breast milk. **Metabolism:** Rapidly and extensively in liver (CYP3A4); active metabolite has less than 1% activity of parent compound. **Elimination:** 35% in feces, 60% in urine; not affected by hemodialysis. **Half-Life:** 8.6 h.

NURSING IMPLICATIONS

Assessment & Drug Effects

- Establish baseline data before treatment is started including BP and pulse.

Common adverse effects in *italic*, life-threatening effects underlined; generic names in **bold**; classifications in SMALL CAPS; ✤ Canadian drug name; ⊙ Prototype drug

1057

- Monitor closely BP values during initiation and titration of dosage. Hypotension with or without an increase in heart rate may occur.
- Avoid too rapid reduction in either systolic or diastolic pressure during parenteral administration.
- Discontinue IV infusion if hypotension or tachycardia develop.
- Observe for large peak and trough differences in BP. Initially, measure BP at peak effect (1–2 h after dosing) and at trough effect (8 h after dosing).

Patient & Family Education

- Record and report any increase in frequency, duration, and severity of angina when initiating or increasing dosage. Keep a record of nitroglycerin use and promptly report any changes in previous anginal pattern.
- Do not change dosage regimen without consulting prescriber.
- Be aware that abrupt withdrawal may cause an increased frequency and duration of chest pain. This drug **must be** gradually tapered under medical supervision.
- Rise slowly from a recumbent position; avoid driving or operating potentially dangerous equipment until response to nicardipine is known.
- Notify prescriber if any of the following occur: Irregular heartbeat, shortness of breath, swelling of the feet, pronounced dizziness, nausea, or drop in BP.

√NICOTINE ☻

(nik'o-teen)

Nicotrol NS, Nicotrol Inhaler, Commit

NICOTINE POLACRILEX
Nicorette Gum, Nicorette DS

NICOTINE TRANSDERMAL SYSTEM
Habitrol, Nicoderm, Nicotrol, ProStep
Classifications: SMOKING DETERRENT; CHOLINERGIC RECEPTOR ANTAGONIST
Therapeutic: SMOKING DETERRENT
Pregnancy Category: D (nasal spray, transdermal system) C (gum)

AVAILABILITY 2 mg, 4 mg gum; 2 mg, 4 mg lozenges; 0.5 mg spray; 4 mg inhaler; 7 mg/day, 14 mg/day, 21 mg/day, 5 mg/day, 10 mg/day, 15 mg/day, 11 mg/day, 22 mg/day transdermal patch

ACTION & *THERAPEUTIC EFFECT*
Ganglionic cholinergic receptor antagonist that has both adrenergic and cholinergic effects. Includes stimulant and depressant effects on the peripheral nervous system and CNS; respiratory stimulation; peripheral vasoconstriction; increased heart rate, cardiac output, and stroke volume; increased tone and motor activity of GI smooth muscles; increased bronchial secretions (initially); antidiuretic activity. Heavy smokers are tolerant of these effects. *Rationale for use is to reduce withdrawal symptoms accompanying cessation of smoking. Success rate appears to be greatest in smokers with high "physical" type of nicotine dependence.*

USES In conjunction with a medically supervised behavior modification program, as a temporary and alternate source of nicotine by the nicotine-dependent smoker who is withdrawing from cigarette smoking.

CONTRAINDICATIONS Nonsmokers, immediate post-MI period; life-threatening arrhythmias; active

Common adverse effects in *italic*, life-threatening effects underlined; generic names in **bold**; classifications in SMALL CAPS; ♣ Canadian drug name; ☻ Prototype drug

temporomandibular joint disease; severe angina pectoris; women with childbearing potential (unless effective contraception is used). **Nicotine Transdermal, Inhaler System:** Pregnancy (category D).

CAUTIOUS USE Vasospastic disease (e.g., Buerger's disease, Prinzmetal's variant angina), cardiac arrhythmias, hyperthyroidism, type 1 diabetes, pheochromocytoma, esophagitis, oral and pharyngeal inflammation; denture use, denture caps, or partial bridges; hypertension and peptic ulcer disease (active or inactive); GERD. **Gum:** Pregnancy (category C). During lactation, only if benefit of a smoking cessation program outweighs risks; children younger than 18 y.

ROUTE & DOSAGE

Smoking Cessation

Adult: **PO** Chew 1 piece of gum whenever have urge to smoke, may be repeated as needed (max: 30 pieces of gum/day) **Intranasal** 1 dose = 2 sprays, 1 in each nostril, start with 1–2 doses (2–4 sprays) each hour (max: 5 doses/h, 40 doses/day), may continue for 3 mo
Topical Apply 1 transdermal patch q24h by the following schedule: *Habitrol, Nicoderm:* 21 mg/day × 6 wk, 14 mg/day × 2 wk, 7 mg/day × 2 wk; *weight less than 45 kg (100 lb), smoke less than ½ pack/day, or have cardiovascular disease,* 14 mg/day × 6 wk, 7 mg/day × 2–4 wk
ProStep: 22 mg/day × 4–8 wk, 11 mg/day × 2–4 wk; *weight less than 45 kg (100 lb), smoke less than ½ pack/day, or have cardiovascular disease,* 11 mg/day × 4–8 wk. *Nicotrol:* Apply 1 transdermal patch 16 h/day by the following schedule: 15 mg/day × 4–12 wk, 10 mg/day × 2–4 wk, 5 mg/day × 2–4 wk

ADMINISTRATION

Oral

- Note: Most adverse local effects (irritation of tongue, mouth, and throat, jaw-muscle aches, dislike of taste) are transient and subside in a few days. Modification of the chewing technique may help.

Transdermal

- Remove the old patch before applying the next new patch.
- Apply patch to nonhairy, clean, dry skin site; immediately remove from protective container.
- Store at or below 30° C (86° F); patches are sensitive to heat.

ADVERSE EFFECTS (≥1%) **CNS:** *Headache, dizziness, light-headedness,* insomnia, irritability, dependence on nicotine. **CV:** Arrhythmias, tachycardia, palpitations, hypertension. **GI:** Air swallowing, *jaw ache, nausea,* belching, salivation, anorexia, dry mouth, laxative effects, constipation, *indigestion,* diarrhea, dyspepsia, vomiting, sialorrhea, abdominal pain, diarrhea. **Respiratory:** *Sore mouth or throat, cough, hiccups,* hoarseness; injury to mouth, teeth, temporomandibular joint pain, *irritation/tingling of tongue.* **Skin:** *Erythema, pruritus, local edema, rash;* skin reactions may be delayed, occurring after 3 wk of patch use. **Special Senses:** *Runny nose, nasal irritation, throat irritation, watering eyes,* minor epistaxis, nasal ulceration. **Body as a Whole:** Acute overdose/nicotine intoxication (perspiration; severe headache; dizziness; disturbed hearing and

Common adverse effects in *italic*, life-threatening effects <u>underlined</u>; generic names in **bold**; classifications in SMALL CAPS; ♣ Canadian drug name; ❷ Prototype drug

1059

vision; mental confusion; severe weakness; fainting; hypotension; dyspnea; weak, rapid, irregular pulse; seizures); <u>death</u> (from <u>respiratory failure</u> secondary to drug-induced <u>respiratory muscle paralysis</u>).

INTERACTIONS Drug: May increase metabolism of **caffeine, theophylline, acetaminophen, insulin, oxazepam, pentazocine propranolol. Food:** Coffee, cola may decrease nicotine absorption from nicotine gum.

PHARMACOKINETICS Absorption: Approximately 90% of the nicotine in a piece of gum is released slowly over 15–30 min; rate of release is controlled by vigor and duration of chewing; readily absorbed from buccal mucosa; transdermal 75–90% absorbed through skin; 53–58% of nasal spray is absorbed. **Peak:** Transdermal 8–9 h; nasal spray 4–15 min. **Distribution:** Crosses placenta; distributed into breast milk. **Metabolism:** In liver, primarily to cotinine. **Elimination:** In urine. **Half-Life:** 30–120 min.

NURSING IMPLICATIONS

Assessment & Drug Effects

- Be aware that transient erythema, pruritus, or burning is common with transdermal patch and usually disappears 24 h after patch removal.
- Differentiate cutaneous hypersensitivity (contact sensitization) that does not resolve in 24 h from a transient local reaction. The former is an indication to discontinue the transdermal patch.

Patient & Family Education

- Chew a piece of gum for approximately 30 min to get the full dose of nicotine.

- Chew only one piece of gum at a time. Chewing gum too rapidly can cause excessive buccal absorption and lead to adverse effects: Nausea, hiccups, throat irritation.
- Gradually decrease number of pieces of gum chewed in 24 h. Usually, a period of 3 mo is allowed before tapering use of gum.
- Promptly discontinue use of transdermal patch and notify prescriber if a severe or persistent local or generalized skin reaction occurs.
- Smoking while using the transdermal nicotine patch increases the risk of adverse reactions.

NIFEDIPINE ⚫
(nye-fed′i-peen)

Adalat CC, Nifedical XL, Procardia, Procardia XL

Classifications: CALCIUM CHANNEL BLOCKER; ANTIANGINAL, ANTIHYPERTENSIVE

Therapeutic: ANTIHYPERTENSIVE, ANTIANGINAL

Pregnancy Category: C

AVAILABILITY 10 mg, 20 mg capsules; 30 mg, 60 mg, 90 mg sustained release tablets

ACTION & *THERAPEUTIC EFFECT* Blocks calcium ion influx across cell membranes of cardiac muscle and vascular smooth muscle. Reduces myocardial oxygen utilization and supply and relaxes and prevents coronary artery spasm. Decreases peripheral vascular resistance and increases cardiac output. *The rise in peripheral blood flow is the basis for use in treatment of Raynaud's phenomenon as well as hypertension. Effective antianginal agent.*

USES Vasospastic "variant" or Prinzmetal's angina and chronic stable angina without vasospasm. Mild to moderate hypertension.

UNLABELED USES Vascular headaches; Raynaud's phenomenon; asthma; cardiomyopathy; primary pulmonary hypertension.

CONTRAINDICATIONS Known hypersensitivity to nifedipine; unstable angina; acute MI; cardiogenic shock; aortic stenosis; GI obstruction.

CAUTIOUS USE GERD; CHF; pregnancy (category C), lactation; children.

ROUTE & DOSAGE

Angina
Adult: **PO** 10–20 mg t.i.d. up to 180 mg/day **Extended release:** 30–60 mg daily
Hypertension
Adult: **PO** 10–20 mg t.i.d. up to 180 mg/day or 30–60 mg extended release once/day

ADMINISTRATION

Oral

- Do not give within the first 1–2 wk following an MI.
- Use only the sustained release form to treat chronic hypertension. Ensure that sustained release form is not chewed or crushed. It **must be** swallowed whole.
- Discontinue drug gradually, with close medical supervision to prevent severe hypertension and other adverse effects.
- Store at 15°–25° C (59°–77° F); protect from light and moisture.

ADVERSE EFFECTS (≥1%) **Body as a Whole:** Sore throat, weakness, fever, sweating, chills, febrile reaction. **CNS:** *Dizziness, light-headedness,* nervousness, mood changes, weakness, jitteriness, sleep disturbances, blurred vision, retinal ischemia, difficulty in balance, *headache.* **CV:** Hypotension, *facial flushing, heat sensation,* palpitations, *peripheral edema,* MI (rare), prolonged systemic hypotension with overdose. **GI:** Nausea, heartburn, *diarrhea,* constipation, cramps, flatulence, gingival hyperplasia, hepatotoxicity. **Musculoskeletal:** Inflammation, joint stiffness, muscle cramps. **Respiratory:** Nasal congestion, dyspnea, cough, wheezing. **Skin:** Dermatitis, pruritus, urticaria. **Urogenital:** Sexual difficulties, possible male infertility.

DIAGNOSTIC TEST INTERFERENCE Nifedipine may cause mild to moderate increases of *alkaline phosphatase, CPK, LDH, AST, ALT.*

INTERACTIONS Drug: BETA-BLOCKERS may increase likelihood of CHF; may increase risk of **phenytoin** toxicity. Do not use with strong CYP3A4 inducers (e.g., **carbamazepine, phenobarbital, rifabutin, rifampin**). **Herbal: Melatonin** may increase blood pressure and heart rate. **Ginkgo, ginseng** may increase plasma concentrations. **St. John's wort** may decrease plasma concentrations. **Food: Grapefruit juice** (greater than 1 qt/day) may increase plasma concentrations and adverse effects.

PHARMACOKINETICS Absorption: Readily from GI tract; 45–75% reaches systemic circulation. **Onset:** 10–30 min. **Peak:** 30 min. **Distribution:** Distributed into breast milk. **Metabolism:** In liver. **Elimination:** 75–80% in urine, 15% in feces. **Half-Life:** 2–5 h.

Common adverse effects in *italic,* life-threatening effects underlined; generic names in **bold;** classifications in SMALL CAPS; ♣ Canadian drug name; ◑ Prototype drug

NURSING IMPLICATIONS

Assessment & Drug Effects

- Monitor BP carefully during titration period. Patient may become severely hypotensive, especially if also taking other drugs known to lower BP. Withhold drug and notify prescriber if systolic BP less than 90.
- Monitor blood sugar in diabetic patients. Nifedipine has diabetogenic properties.
- Monitor for gingival hyperplasia and report promptly. This is a rare but serious adverse effect (similar to phenytoin-induced hyperplasia).

Patient & Family Education

- Keep a record of nitroglycerin use and promptly report any changes in previous pattern. Occasionally, people develop increased frequency, duration, and severity of angina when they start treatment with this drug or when dosage is increased.
- Be aware that withdrawal symptoms may occur with abrupt discontinuation of the drug (chest pain, increase in anginal episodes, MI, dysrhythmias).
- Inspect gums visually every day. Changes in gingivae may be gradual, and bleeding may be exhibited only with probing.
- Seek prompt treatment for symptoms of gingival hyperplasia (easy bleeding of gingivae and gradual enlarging of gingival mass, especially on buccal side of lower anterior teeth). Drug will be discontinued if gingival hyperplasia occurs.
- Research shows that smoking decreases the efficacy of nifedipine and has direct and adverse effects on the heart in the patient on nifedipine treatment.

NILOTINIB HYDROCHLORIDE

(ni-lot'i-nib hy-dro-chlor'ide)

Tasigna

Classifications: ANTINEOPLASTIC; TYROSINE KINASE INHIBITOR (TKI)

Therapeutic: ANTINEOPLASTIC; TKI

Prototype: Gefitinib

Pregnancy Category: D

AVAILABILITY 150 mg, 200 mg capsules

ACTION & THERAPEUTIC EFFECT
Nilotinib is a tyrosine kinase inhibitor designed to selectively inhibit the BCR-ABL tyrosine kinase on the Philadelphia chromosome found in chronic myelogenous leukemia (CML). It prevents tyrosine kinase enzyme from converting to its active conformation. Thus, it prevents proliferation of BCR-ABL cells and ultimately induces cell death in CML. Nilotinib enhances binding site affinity by offering alternate binding pathways for the ABL tyrosine kinases. *Increased kinase selectivity and binding site affinity makes nilotinib more potent than other similar drugs (imatinib) in preventing the proliferation of CML.*

USES Treatment of Philadelphia chromosome positive (Ph+) chronic myelogenous leukemia (CML) in patients resistant to or intolerant to prior therapy that included imatinib.

CONTRAINDICATIONS Hypoglycemia; hypomagnesemia; hypokalemia; long QT syndrome; severe galactose or lactose intolerance; pregnancy (category D), lactation.

CAUTIOUS USE History of pancreatitis; history of cardiac disease, recent MI, CHF, unstable angina; myelosuppression; total gastrectomy; hepatic impairment. Safety and efficacy in children not established.

ROUTE & DOSAGE

Chronic Myelogenous Leukemia

Adult: **PO** 300–400 mg b.i.d.

QT Prolongation Dosage Adjustment

See package insert.

Myelosuppression Dosage Adjustment

• If ANC less than 1 × 10⁹/L or platelet count less than 50 × 10⁹/L, withhold nilotinib.

• If ANC greater than 1 × 10⁹/L and platelet count greater than 50 × 10⁹/L within 2 wk, resume previous dose.

• If ANC less than 1 × 10⁹/L or platelet count less than 50 × 10⁹/L for more than 2 wk, resume at 400 mg PO daily

Nonhematologic Abnormalities Dosage Adjustment

For grade 3 or higher serum lipase, amylase, bilirubin, or hepatic transaminases: Withhold nilotinib and monitor abnormal level(s).
Resume nilotinib at 400 mg PO daily if toxicity resolves to grade 1 or lower.
Other moderate/severe nonhematologic toxicities: Withhold nilotinib until toxicity resolves, then resume at 400 mg PO daily. May increase to 400 mg b.i.d.

ADMINISTRATION

Oral

▪ Give on an empty stomach, at least 2 h before or 1 h after eating.
▪ Ensure that capsules are swallowed whole with water.
▪ For those unable to swallow, contents of capsule may be added to 1 tsp of applesauce immediately before administration.

▪ Note: Hypokalemia and hypomagnesemia should be corrected prior to drug administration.
▪ Store at 15°–30° C (59°–86° F)

ADVERSE EFFECTS (≥1%) **Body as a Whole:** *Peripheral edema, pyrexia.* **CNS:** *Asthenia,* dizziness, *fatigue, headache,* insomnia, paresthesia. **CV:** Flushing, hypertension, palpitations, QT prolongation. **GI:** *Abdominal pain, constipation, diarrhea,* dyspepsia, flatulence, *nausea.* **Hematologic:** *Anemia, neutropenia,* pancytopenia, *thrombocytopenia.* **Metabolic:** Decreased albumin, elevated alkaline phosphatase, elevated ALT, elevated AST, elevated bilirubin, *elevated lipase, hyperglycemia,* hyperkalemia, hypocalcemia, hypokalemia, hyponatremia, *hypophosphatemia.* **Musculoskeletal:** *Arthralgia, back pain, bone pain,* chest pain, *muscle spasms, myalgia, pain in extremity.* **Respiratory:** *Cough,* dysphonia, *dyspnea, nasopharyngitis.* **Skin:** Alopecia, dry skin, eczema, hyperhidrosis, night sweats, *pruritus, rash,* urticaria. **Special Senses:** *Vertigo.*

INTERACTIONS Drug: CYP3A4 Inducers **carbamazepine, dexamethasone, phenobarbital, phenytoin, rifabutin, rifampin, rifapentine**) decrease nilotinib levels. Inhibitors of CYP3A4 (**clarithromycin, indinavir, itraconazole, ketoconazole, nefazodone, nelfinavir, ritonavir, saquinavir, telithromycin, voriconazole**) increase nilotinib levels. Nilotinib increases the levels of **midazolam, warfarin. Food:** Food increases bioavailability. **Grapefruit** juice may increase nilotinib levels. **Herbal: St. John's wort** decreases nilotinib levels.

PHARMACOKINETICS Peak: 3 h. **Distribution:** 98% plasma protein bound. **Metabolism:** Hepatic to inactive metabolites. **Elimination:** 93% fecal. **Half-Life:** 17 h.

Common adverse effects in *italic*, life-threatening effects underlined; generic names in **bold**; classifications in SMALL CAPS; ✦ Canadian drug name; ⊘ Prototype drug

1063

NURSING IMPLICATIONS

Assessment & Drug Effects

- Obtain a baseline ECG, then again 7 days after first drug dose, and periodically thereafter. Withhold drug and report immediately QT prolongation.
- Monitor closely patients with hepatic impairment for QT interval prolongation.
- Lab tests: Baseline and periodic serum electrolytes; LFTs; CBC q2wk first 2 mo then monthly thereafter; periodic serum lipase.
- Monitor diabetics for loss of glycemic control.

Patient & Family Education

- Nilotinib should not be taken with food (see Administration).
- Women of childbearing age should use reliable forms of contraception, including a barrier type.
- Avoid grapefruit products while on nilotinib.

NILUTAMIDE

(ni-lu′ta-mide)

Nilandron

Classifications: ANTINEOPLASTIC; ANTIANDROGEN

Therapeutic: ANTINEOPLASTIC; ANTI-ANDROGEN

Prototype: Flutamide

Pregnancy Category: C

AVAILABILITY 150 mg tablets

ACTION & *THERAPEUTIC EFFECT*
Nonsteroidal with antiandrogen activity. Blocks the effects of testosterone at the androgen receptor sites, thus preventing the normal androgenic response. *Effective in blocking testosterone in treatment of metastatic prostate carcinoma.*

USES Use with surgical castration for metastatic prostate cancer.

CONTRAINDICATIONS Severe hepatic impairment; severe respiratory insufficiency; hypersensitivity to nilutamide; lactation.

CAUTIOUS USE Asian patients relative to causing interstitial pneumonitis; alcoholics; pregnancy (category C). Safety and effectiveness in children are not established.

ROUTE & DOSAGE

Metastatic Prostate Cancer

Adult: **PO** 300 mg daily × 30 days, then 150 mg daily

ADMINISTRATION

Oral

- Give first dose on the day of or day after surgical castration.
- Store below 15°–30° C (59°–86° F) and protect from light.

ADVERSE EFFECTS (≥1%) Body as a Whole: *Hot flushes, impotence, decreased libido, malaise,* edema, weight loss, arthritis. **CNS:** Nervousness, paresthesias. **CV:** Angina, heart failure, syncope. **GI:** Diarrhea, GI hemorrhage, melena, dry mouth. **Respiratory:** Cough, interstitial lung disease, rhinitis. **Skin:** Pruritus. **Other:** Alcohol intolerance. **Special Senses:** Cataracts, photophobia.

INTERACTIONS Drug: Carbamazepine, rifampin, phenytoin may decrease level; **fluconazole, gemfibrozil, omeprazole** may increase levels. **Herbal: St. John's wort** may decrease levels.

PHARMACOKINETICS Absorption: Rapidly from GI tract. **Metabolism:** In the liver (CYP2C19). **Elimination:** In urine. **Half-Life:** 38–50 h.

NURSING IMPLICATIONS

Assessment & Drug Effects

- Obtain baseline chest x-ray before treatment and periodically thereafter.
- Closely monitor for S&S of pneumonitis; at the first sign of adverse pulmonary effects, withhold drug and notify prescriber. Abnormal ABGs may indicate need to discontinue drug.
- Lab tests: Monitor LFTs before beginning treatment and at 3-mo intervals; if serum transaminases increase greater than 2–3 × ULN, treatment is usually discontinued.
- Monitor patients taking phenytoin, theophylline, or warfarin closely for toxic levels of these drugs.

Patient & Family Education

- Report to prescriber immediately the following S&S of adverse effects on lungs: Development of chest pain, dyspnea, and cough with fever.
- Report S&S of liver injury to prescriber: Jaundice, dark urine, fatigue, or signs of GI distress including nausea, vomiting, abdominal pain.
- Use caution when moving from lighted to dark areas because the drug may slow visual adaptation to darkness. Tinted glasses may partially alleviate the problem.

NIMODIPINE

(ni-mo′di-peen)

Classifications: CALCIUM CHANNEL BLOCKER; CEREBRAL ANTISPASMODIC
Therapeutic: CEREBRAL ANTISPASMODIC
Prototype: Nifedipine
Pregnancy Category: C

AVAILABILITY 30 mg capsule

ACTION & *THERAPEUTIC EFFECT*

Calcium channel blocking agent that is relatively selective for cerebral arteries compared with arteries elsewhere in the body. *Reduces vascular spasms in cerebral arteries during a stroke.*

USES To improve neurologic deficits due to spasm following subarachnoid hemorrhage from ruptured congenital intracranial aneurysms in patients who are in good neurologic condition posticus (e.g., Hunt and Hess Grades I–III).

UNLABELED USES Migraine headaches, ischemic seizures.

CONTRAINDICATIONS Near-fatal reaction to intravenous administration; hypotension; cardiogenic shock; lactation.
CAUTIOUS USE Hepatic impairment; acute MI; bradycardia, heart failure, ventricular dysfunction; elderly; pregnancy (category C). Safety and effectiveness in children are not established.

ROUTE & DOSAGE

Subarachnoid Hemorrhage

Adult: **PO** 60 mg q4h for 21 days, start therapy within 96 h of subarachnoid hemorrhage

Hepatic Impairment Dosage Adjustment

Decrease dose to 30 mg q4h

ADMINISTRATION

Oral

- Make a hole in both ends of the capsule with an 18-gauge needle and extract the contents into a syringe if patient is unable to swallow. Empty the contents into an enteral (if in use) tube and wash down with 30 mL of NS.
- Store below 40° C (104° F); protect from light.

Common adverse effects in *italic*, life-threatening effects underlined; generic names in **bold**; classifications in SMALL CAPS; ♣ Canadian drug name; ✪ Prototype drug

ADVERSE EFFECTS (≥1%) **CNS:** Headache. **CV:** *Hypotension.* **GI:** Hemorrhage, mild, transient increase in liver function tests.

INTERACTIONS Drug: Hypotensive effects increased when combined with other CALCIUM CHANNEL BLOCKERS. **Food: Grapefruit juice** (greater than 1 qt/day) may increase plasma concentrations and adverse effects.

PHARMACOKINETICS Absorption: Readily from GI tract; approximately 13% reaches systemic circulation (first pass metabolism). **Peak:** 1 h. **Distribution:** Crosses blood–brain barrier; possibly crosses placenta; distributed into breast milk. **Metabolism:** 85% in liver; 15% in kidneys. **Elimination:** Greater than 50% in urine, 32% in feces. **Half-Life:** 8–9 h.

NURSING IMPLICATIONS

Assessment & Drug Effects

▪ Take apical pulse prior to administering drug and hold it if pulse is below 60. Notify the prescriber.

▪ Establish baseline data before treatment is started: BP, pulse, and laboratory evaluations of liver and kidney function.

▪ Monitor frequently for adverse drug effects, including hypotension, peripheral edema, tachycardia, or skin rash.

▪ Monitor frequently for dizziness or light-headedness in older adult patients; risk of hypotension is increased.

Patient & Family Education

▪ Report gradual weight gain and evidence of edema (e.g., tight rings on fingers, ankle swelling).

▪ Keep follow-up appointments for monitoring of progress during therapy.

NISOLDIPINE
(ni-sol′di-peen)

Sular

Classifications: CALCIUM CHANNEL BLOCKER; ANTIHYPERTENSIVE

Therapeutic: ANTIHYPERTENSIVE; ANTIANGINAL

Prototype: Nifedipine

Pregnancy Category: C

AVAILABILITY 8.5 mg, 17 mg, 20 mg, 25.5 mg, 30 mg, 34 mg, 40 mg extended release tablets

ACTION & *THERAPEUTIC EFFECT* Inhibits calcium ion influx across cell membranes of cardiac muscle and vascular smooth muscle, which results in vasodilation, inotropism, and negative chronotropism. Inhibits vasoconstriction in the peripheral vasculature. *Significantly reduces total peripheral resistance, decreases blood pressure, and increases cardiac output. It is also a potent coronary vasodilator.*

USES Hypertension.

UNLABELED USES CHF, angina.

CONTRAINDICATIONS Hypersensitivity to nisoldipine or other calcium blockers; systolic BP less than 90 mm Hg, advanced aortic stenosis, cardiogenic shock, severe hypotension, acute MI, sick sinus syndrome; lactation.

CAUTIOUS USE Liver dysfunction; severe obstructive coronary artery disease; class II to IV heart failure, especially with concurrent administration of a beta-blocker; paroxysmal atrial fibrillation; GERD; CHF; digital ischemia, ulceration, or gangrene; nonobstructive hypertrophic cardiomyopathy; Duchenne muscular dystrophy; older adults; pregnancy (category C); children.

ROUTE & DOSAGE

Hypertension

Adult: **PO** 17 mg daily may increase by 8.5 mg weekly as needed
Geriatric: **PO** 8.5 mg daily, may increase weekly as needed

ADMINISTRATION

Oral

- Give drug with food to decrease GI distress, but do not give with grapefruit juice or a high-fat meal.
- Ensure that extended release form is not chewed or crushed. It **must be** swallowed whole.
- Drug is usually discontinued gradually to prevent adverse effects.
- Store at 15°–30° C (59°–86° F).

ADVERSE EFFECTS (≥1%) **CNS:** Dizziness, anxiety, tremor, weakness, fatigue, *headache.* **CV:** Hypotension, peripheral edema, palpitations, orthostatic hypotension. **GI:** Abdominal pain, cramps, constipation, dry mouth, diarrhea, nausea. **Skin:** *Flushing,* rash, erythema, urticaria. **Urogenital:** Urinary frequency. **Respiratory:** Pulmonary edema (patients with CHF), wheezing, dyspnea. **Body as a Whole:** Myalgia.

INTERACTIONS Drug: May cause significant increase in **digoxin** level in patients with CHF. BETA-BLOCKERS may cause hypotension and bradycardia. **Phenytoin, carbamazepine, phenobarbital** may significantly decrease levels. Azole antifungals may affect metabolism; avoid combination. **Food:** High-fat food increases availability.

PHARMACOKINETICS Absorption: Rapidly from GI tract; 4–8% reaches systemic circulation. **Peak Effect:** 1–3 h. **Duration:** 8–12 h for hypertension, 7–8 h for angina. **Distribution:** 99% protein bound. **Metabolism:** Extensively in liver. **Elimination:** 70–75% in urine as metabolites. **Half-Life:** 2–14 h.

NURSING IMPLICATIONS

Assessment & Drug Effects

- Monitor blood pressure carefully during period of drug initiation and with dosage increments.
- Monitor cardiovascular status especially heart rate, frequency of angina attacks, or worsening heart failure.
- Assess for and report edematous weight gain.
- Monitor digoxin levels closely with concurrent use and watch for S&S of digoxin toxicity (see Appendix F).

Patient & Family Education

- Do not discontinue the drug abruptly.
- Report symptoms of orthostatic hypotension or other bothersome adverse effects to prescriber.
- Do not drive or engage in potentially hazardous activities until response to drug is known.

NITAZOXANIDE

(nit-a-zox′-a-nide)
Alinia
Classification: ANTIPROTOZOAL
Therapeutic: ANTIPROTOZOAL
Prototype: Metronidazole
Pregnancy Category: B

AVAILABILITY 100 mg/5 mL oral suspension; 500 mg tablets

ACTION & *THERAPEUTIC EFFECT*
Antiprotozoal activity believed to be due to interference with an essential enzyme needed for anaerobic energy metabolism in protozoa. *Inhibits*

Common adverse effects in *italic*, life-threatening effects underlined; generic names in **bold**; classifications in SMALL CAPS; ♣ Canadian drug name; ⊙ Prototype drug

1067

growth of sporozoites and oocysts of Cryptosporidium parvum *and* trophozoites *of* Giardia lamblia.

USES Diarrhea caused by *Cryptosporidium parvum* and *Giardia lamblia.*

CONTRAINDICATIONS Prior hypersensitivity to nitazoxanide.

CAUTIOUS USE Hepatic and biliary disease, renal disease, renal impairment, renal failure, and combined renal and hepatic disease; pregnancy (category B); lactation. Safety and efficacy in children younger than 1 y have not been studied.

ROUTE & DOSAGE

Diarrhea
Adult/Adolescent: **PO** 500 mg q12h × 3 days
Child (1–3 y): **PO** 100 mg q12h × 3 days; *4–11 y:* 200 mg q12h × 3 days

ADMINISTRATION

Oral
- Prepare suspension as follows: Tap bottle until powder loosens. Draw up 48 mL of water, add half to bottle, shake to suspend powder, then add remaining 24 mL of water and shake vigorously.
- Give required dose (5 or 10 mL) with food.
- Keep container tightly closed, and shake well before each administration.
- Suspension may be stored for 7 days at 15°–30° C (59°–86° F), after which any unused portion **must be** discarded.

ADVERSE EFFECTS (≥1%) **CNS:** Headache. **GI:** Abdominal pain, diarrhea, vomiting.

INTERACTIONS Food: Increases levels.

PHARMACOKINETICS Peak: 1–4 h. **Distribution:** 99% protein bound. **Metabolism:** Rapidly hydrolyzed in liver to an active metabolite, tizoxanide (desacetyl-nitazoxanide). **Elimination:** In urine, bile, and feces.

NURSING IMPLICATIONS

Assessment & Drug Effects
- Monitor for therapeutic effectiveness: No watery stools and 2 or less soft stools with no hematochezia within the past 24 h or no symptoms and no unformed stools within the past 48 h.
- Monitor closely patients with preexisting hepatic or biliary disease for adverse reactions.
- Assess appetite, level of abdominal discomfort and extent of bloating.
- Assess frequency and quantity of diarrhea and monitor total hydration status.
- Weigh daily to aid in assessment of possible fluid loss from diarrhea.

Patient & Family Education
- Note that 5 mL of the oral suspension contains approximately 1.5 g of sucrose.
- Report either no improvement in or worsening of diarrhea and abdominal discomfort.

NITROFURANTOIN
(nye-troe-fyoor′an-toyn)
Furadantin, Novo-Furan ✦

NITROFURANTOIN MACROCRYSTALS
Macrobid, Macrodantin

Classifications: URINARY TRACT ANTI-INFECTIVE; NITROFURAN
Therapeutic: URINARY TRACT ANTI-BIOTIC
Pregnancy Category: B

AVAILABILITY 25 mg/5 mL suspension; 25 mg, 50 mg, 100 mg capsules

ACTION & *THERAPEUTIC EFFECT*
Synthetic nitrofuran derivative presumed to act by interfering with several bacterial enzyme systems. Highly soluble in urine and reportedly most active in acid urine. Antimicrobial concentrations in urine exceed those in blood. *Active against wide variety of gram-negative and gram-positive microorganisms.*

USES Uncomplicated urinary tract infection, including cystitis.

CONTRAINDICATIONS Hypersensitivity to nitrofurantoin including hepatic dysfunction; anuria, oliguria, significant impairment of kidney function (CrCl less than 60 mL/min); G6PD deficiency; history of cholestatic jaundice; pregnancy at term (38–42 wk), labor, or obstetric delivery; lactation.

CAUTIOUS USE History of asthma, anemia, diabetes, vitamin B deficiency, hepatic disease; pulmonary disease; mild to moderate renal disease; electrolyte imbalance, debilitating disease; B$_{12}$ deficiency; pregnancy (category B); infants younger than 1 mo.

ROUTE & DOSAGE

UTI, Cystitis

Adult: **PO** 50–100 mg q.i.d. × 7 days, or 3 days after sterile urine sample
Child/Infant (1 mo–12 y): **PO** 1.25–1.75 mg/kg q6h (max: 400 mg/day)
Adult/Adolescent (Macrobid only): **PO** 100 mg q12h × 7 days

Chronic Suppressive Therapy for UTI

Adult: **PO** 50–100 mg at bedtime

Child (1 mo–12 y): **PO** 1–2 mg/kg/day in 1–2 divided doses (max: 100 mg/day)

Renal Impairment Dosage Adjustment

Avoid if CrCl less than 60 mL/min

ADMINISTRATION

Oral
- Give with food or milk to minimize gastric irritation.
- Avoid crushing tablets because of the possibility of tooth staining; dilute oral suspension in milk, water, or fruit juice, and rinse mouth thoroughly after taking drug.

ADVERSE EFFECTS (≥1%) **CNS:** Peripheral neuropathy, headache, nystagmus, drowsiness, vertigo. **GI:** *Anorexia, nausea, vomiting,* abdominal pain, diarrhea, cholestatic jaundice, hepatic necrosis. **Hematologic (rare):** Hemolytic or megaloblastic anemia (especially in patients with G6PD deficiency), granulocytosis, eosinophilia. **Body as a Whole:** Angioedema, anaphylaxis, drug fever, arthralgia. **Respiratory:** Allergic pneumonitis, asthmatic attack (patients with history of asthma), pulmonary sensitivity reactions (interstitial pneumonitis or fibrosis). **Skin:** Skin eruptions, pruritus, urticaria, exfoliative dermatitis, transient alopecia. **Urogenital:** Genitourinary superinfections (especially with *Pseudomonas*), crystalluria (older adult patients), dark yellow or brown urine. **Other:** Tooth staining from direct contact with oral suspension and crushed tablets (infants).

DIAGNOSTIC TEST INTERFERENCE
Nitrofurantoin metabolite may produce false-positive ***urine glucose*** test results with ***Benedict's reagent***.

Common adverse effects in *italic*, life-threatening effects underlined; generic names in **bold**; classifications in SMALL CAPS; ✤ Canadian drug name; ⊘ Prototype drug

1069

INTERACTIONS Drug: ANTACIDS may decrease absorption of nitrofurantoin; **nalidixic acid,** other QUINOLONES may antagonize antimicrobial effects; **probenecid, sulfinpyrazone** increase risk of nitrofurantoin toxicity.

PHARMACOKINETICS Absorption: Readily from GI tract. **Peak:** Urine: 30 min. **Distribution:** Crosses placenta; distributed into breast milk. **Metabolism:** Partially in liver. **Elimination:** Primarily in urine. **Half-Life:** 20 min.

NURSING IMPLICATIONS

Assessment & Drug Effects

- Lab tests: Perform C&S prior to therapy; recommended in patients with recurrent infections.
- Monitor I&O. Report oliguria and any change in I&O ratio.
- Be alert to signs of urinary tract superinfections (e.g., milky urine, foul-smelling urine, perineal irritation, dysuria).
- Assess for nausea (which occurs fairly frequently). May be relieved by using macrocrystalline preparation (Macrodantin).
- Watch for acute pulmonary sensitivity reaction, usually within first week of therapy and apparently more common in older adults. May be manifested by mild to severe flu-like syndrome.
- With prolonged therapy, monitor for subacute or chronic pulmonary sensitivity reaction, commonly manifested by insidious onset of malaise, cough, dyspnea on exertion, altered ABGs.
- Monitor for S&S of peripheral neuropathy, which can be severe and irreversible. Withhold drug and notify prescriber immediately.

Patient & Family Education

- Report promptly muscle weakness, tingling, numbness.

- Nitrofurantoin may impart a harmless brown color to urine.
- Consult prescriber regarding fluid intake. Generally, fluids are not forced; however, intake should be adequate.

NITROGLYCERIN ⊕
(nye-troe-gli′ser-in)

Minitran, Nitrocap, Nitrodisc, Nitro-Dur, Nitrogard, Nitrogard-SR, Nitrong SR, Nitrospan, Nitrostat, Nitrostat I.V., ProStakan
Classification: NITRATE VASODILATOR
Therapeutic: ANTIANGINAL; VASODILATOR
Pregnancy Category: C

AVAILABILITY 5 mg/mL injection; 0.3 mg, 0.4 mg, 0.6 mg sublingual tablets; 2.5 mg, 6.5 mg, 9 mg sustained release tablets, capsules; 0.1 mg/h, 0.2 mg/h, 0.3 mg/h, 0.4 mg/h, 0.6 mg/h, 0.8 mg/h transdermal patch; 0.4%, 2% ointment

ACTION & *THERAPEUTIC EFFECT*
Organic nitrate and potent vasodilator that relaxes vascular smooth muscle. After conversion to nitric oxide, it leads to dose-related dilation of both venous and arterial blood vessels. Promotes peripheral pooling of blood, reduction of peripheral resistance, and decreased venous return to the heart. Both left ventricular preload and afterload are reduced and myocardial oxygen consumption or demand is decreased. *Produces antianginal, anti-ischemic, and antihypertensive effects.*

USES Prophylaxis, treatment, and management of angina pectoris. IV nitroglycerin is used to control BP in perioperative hypertension, CHF associated with acute MI; to produce controlled hypotension during surgical procedures, and to treat angina pectoris in patients who have not re-sponded to nitrate or beta-blocker therapy. Treatment of pain associated with chronic anal fissure.

UNLABELED USES Sublingual and topical to reduce cardiac workload in patients with acute MI and in CHF. Ointment for adjunctive treatment of Raynaud's disease.

CONTRAINDICATIONS Hypersensitivity, idiosyncrasy, or tolerance to nitrates; severe anemia; head trauma, increased ICP; **Sublingual:** Early MI; severe anemia, increased ICP; hypersensitivity to nitroglycerin. **Sustained release form:** Glaucoma. **IV form:** Hypotension, uncorrected hypovolemia, constrictive pericarditis, pericardial tamponade; restrictive cardiomyopathy.

CAUTIOUS USE Severe liver or kidney disease, conditions that cause dry mouth, pregnancy (category C), lactation.

ROUTE & DOSAGE

Angina

Adult: **Sublingual** 1–2 sprays (0.4–0.8 mg) or a 0.3–0.6-mg tablet q3–5min as needed (max: 3 doses in 15 min) **PO** 1.3–9 mg q8–12h **IV** Start with 5 mcg/min and titrate q3–5min until desired response (up to 200 mcg/min) **Transdermal Unit** Apply once q24h or leave on for 10–12 h, then remove and have a 10–12 h nitrate free interval **Topical** Apply 1.5–5 cm (½–2 in) of ointment q4–6h
Child: **IV** 0.25–0.5 mcg/kg/min, titrate by 0.5–1 mcg/kg/min q3–5min (max: 5 mg/kg/min)

Anal Fissure

Adult: **Topical** 1 inch every 12 h for up to 3 wk

ADMINISTRATION

Sublingual Tablet

- Give 1 tablet and if pain is not relieved, give additional tablets at 5-min intervals, but not more than 3 tablets in a 15-min period.
- Typically available for self-administration in their original container. Instruct in correct use. Request patient to report all attacks.
- Instruct to sit or lie down upon first indication of oncoming anginal pain and to place tablet under tongue or in buccal pouch (hypotensive effect of drug is intensified in the upright position).

Sustained Release Tablet or Capsule

- Give on an empty stomach (1 h before or 2 h after meals), with a full glass of water. Ensure it is swallowed whole.
- Be aware that sustained release form helps to prevent anginal attacks; it is not intended for immediate relief of angina.

Transdermal Ointment

- Using dose-determining applicator (paper application patch) supplied with package, squeeze prescribed dose onto this applicator. Using applicator, spread ointment in a thin, uniform layer to premarked 5.5 by 9 cm (2¼ by 3½ in.) square. Place patch with ointment side down onto nonhairy skin surface (areas commonly used: Chest, abdomen, anterior thigh, forearm). Cover with transparent wrap and secure with tape. Avoid getting ointment on fingers.
- Rotate application sites to prevent dermal inflammation and sensitization. Remove ointment from previously used sites before reapplication.

N

Common adverse effects in *italic*, life-threatening effects <u>underlined</u>; generic names in **bold**; classifications in SMALL CAPS; ♣ Canadian drug name; ⊘ Prototype drug

1071

Transdermal Unit

- Apply transdermal unit (transdermal patch) at the same time each day, preferably to skin site free of hair and not subject to excessive movement. Avoid abraded, irritated, or scarred skin. Clip hair if necessary.
- Change application site each time to prevent skin irritation and sensitization.

Intravenous

- Check to see if patient has transdermal patch or ointment in place before starting IV infusion. The patch (or ointment) is usually removed to prevent overdosage.

PREPARE: **IV Infusion:** Nitroglycerin is available undiluted and premixed in D5W IV solutions of varying concentrations. ▪ *IV Infusion from Concentrate:* Use only non-PVC plastic or glass bottles and manufacturer-supplied IV tubing. ▪ Withdraw contents of one vial (25 or 50 mg) into syringe and inject immediately into 500 mL of IV solution to minimize contact with plastic; yields 50 mcg/mL or 100 mcg/mL. ▪ If less fluid is desired, add 5 mg to 100 mL to yield 50 mcg/mL. Other concentrations within the range of 25–400 mcg/mL may be used. ▪ Do not exceed 400 mcg/mL.

ADMINISTER: **IV Infusion:** Give by continuous infusion regulated exactly by an infusion pump. ▪ IV dosage titration requires careful and continuous hemodynamic monitoring.

INCOMPATIBILITIES **Solution/additive: Caffeine, hydralazine, phenytoin. Y-site: Alteplase, levofloxacin.**

- Use only glass containers for storage of reconstituted IV solution. Polyvinyl chloride (PVC) plastic can absorb nitroglycerin and therefore should not be used. ▪ Non-polyvinyl-chloride (non-PVC) sets are recommended or provided by manufacturer.

ADVERSE EFFECTS (≥1%) **CNS:** *Headache,* apprehension, blurred vision, weakness, vertigo, dizziness, faintness. **CV:** *Postural hypotension,* palpitations, tachycardia (sometimes with paradoxical bradycardia), increase in angina, syncope, and <u>circulatory collapse</u>. **GI:** Nausea, vomiting, involuntary passing of urine and feces, abdominal pain, dry mouth. **Hematologic:** Methemoglobinemia (high doses). **Skin:** Cutaneous vasodilation with flushing, rash, exfoliative dermatitis, contact dermatitis with transdermal patch; topical allergic reactions with ointment: Pruritic eczematous eruptions, <u>anaphylactoid reaction</u> characterized by oral mucosal and conjunctival edema. **Body as a Whole:** Muscle twitching, pallor, perspiration, cold sweat; local sensation in oral cavity at point of dissolution of sublingual forms.

DIAGNOSTIC TEST INTERFERENCE Nitroglycerin may cause increases in determinations of ***urinary catecholamines*** and ***VMA;*** may interfere with the ***Zlatkis-Zak color reaction,*** causing a false report of decreased ***serum cholesterol.***

INTERACTIONS Drug: Alcohol, antihypertensive agents compound hypotensive effects; IV nitroglycerin may antagonize **heparin** anticoagulation. Vasodilating effects may be enhanced by **sildenafil, vardenafil,** or **tadalafil,** so this combination should be avoided.

PHARMACOKINETICS Absorption: Significant loss to first pass metabolism after oral dosing. **Onset:** 2 min SL; 3 min PO; 30 min ointment.

Duration: 30 min SL; 3–5 h PO; 3–6 h ointment. **Distribution:** Widely distributed; not known if distributes to breast milk. **Metabolism:** Extensively in liver. **Elimination:** Inactive metabolites in urine. **Half-Life:** 1–4 min.

NURSING IMPLICATIONS

Assessment & Drug Effects

- Administer IV nitroglycerin with extreme caution to patients with hypotension or hypovolemia since the IV drug may precipitate a severe hypotensive state.
- Monitor patient closely for change in levels of consciousness and for dysrhythmias.
- Be aware that moisture on sublingual tissue is required for dissolution of sublingual tablet. However, because chest pain typically leads to dry mouth, a patient may be unresponsive to sublingual nitroglycerin.
- Assess for headaches. Approximately 50% of all patients experience mild to severe headaches following nitroglycerin. Transient headache usually lasts about 5 min after sublingual administration and seldom longer than 20 min. Assess degree of severity and consult with prescriber about analgesics and dosage adjustment.
- Supervise ambulation as needed, especially with older adult or debilitated patients. Postural hypotension may occur even with small doses of nitroglycerin. Patients may complain of dizziness or weakness due to postural hypotension.
- Take baseline BP and heart rate with patient in sitting position before initiation of treatment with transdermal preparations.
- One hour after transdermal (ointment or unit) medication has been applied, check BP and pulse again with patient in sitting position. Report measurements to prescriber.
- Assess for and report blurred vision or dry mouth.
- Assess for and report the following topical reactions: Contact dermatitis from the transdermal patch; pruritus and erythema from the ointment.

Patient & Family Education

- Store tablet form in its original container.
- Sit or lie down upon first indication of oncoming anginal pain.
- Relax for 15–20 min after taking tablet to prevent dizziness or faintness.
- Be aware that pain not relieved by 3 sublingual tablets over a 15-min period may indicate acute MI or severe coronary insufficiency. Contact prescriber immediately or go directly to emergency room.
- Note: Sublingual tablets may be taken prophylactically 5–10 min prior to exercise or other stimulus known to trigger angina (drug effect lasts 30–60 min).
- Keep record for prescriber of number of angina attacks, amount of medication required for relief of each attack, and possible precipitating factors.
- Remove transdermal unit or ointment immediately from skin and notify prescriber if faintness, dizziness, or flushing occurs following application.
- You can use a sublingual formulation while transdermal unit or ointment is in place.
- Report blurred vision or dry mouth. Both warrant withdrawal of drug.
- Change position slowly and avoid prolonged standing. Dizziness, light-headedness, and syncope (due to postural hypotension) occur most frequently in older adults.
- Report any increase in frequency, duration, or severity of anginal attack.

N

Common adverse effects in *italic*, life-threatening effects underlined; generic names in **bold;** classifications in SMALL CAPS; ✦ Canadian drug name; ⊙ Prototype drug

1073

NITROPRUSSIDE SODIUM

(nye-troe-pruss'ide)

Nitropress

Classifications: NONNITRATE VASO-DILATOR; ANTIHYPERTENSIVE

Therapeutic: ANTIHYPERTENSIVE; VASODILATOR

Prototype: Hydralazine

Pregnancy Category: C

AVAILABILITY 50 mg injection

ACTION & *THERAPEUTIC EFFECT*
Potent, rapid-acting hypotensive agent that acts directly on vascular smooth muscle to produce peripheral vasodilation, with consequently marked lowering of arterial BP, mild decrease in cardiac output, and moderate lowering of peripheral vascular resistance. *Effective antihypertensive agent used for rapid reduction of high blood pressure.*

USES Short-term, rapid reduction of BP in hypertensive crises and for producing controlled hypotension during anesthesia to reduce bleeding.

UNLABELED USES Refractory CHF or acute MI.

CONTRAINDICATIONS Compensatory hypertension, as in atriovenous shunt or coarctation of aorta, and for control of hypotension in patients with inadequate cerebral circulation; lactation.

CAUTIOUS USE Hepatic insufficiency, hypothyroidism, severe renal impairment, hyponatremia, older adult patients with low vitamin B_{12} plasma levels or with Leber's optic atrophy; pregnancy (category C).

ROUTE & DOSAGE

Hypertensive Crisis

Adult: **IV** 0.3–0.5 mcg/kg/min (average 3 mcg/kg/min)

Child: **IV** 1 mcg/kg/min (average 3 mcg/kg/min) (max: 5 mcg/kg/min)

ADMINISTRATION

Intravenous

PREPARE: **Continuous:** Dissolve each 50 mg in 2–3 mL of D5W. Further dilute in 250 mL D5W to yield 200 mcg/mL or 500 mL D5W to yield 100 mcg/mL. ▪ Lower concentrations may be desirable depending on patient weight. ▪ Following reconstitution, solutions usually have faint brownish tint; if solution is highly colored, do not use it. ▪ Promptly wrap container with aluminum foil or other opaque material to protect drug from light.

ADMINISTER: **Continuous:** Administer by infusion pump or similar device that will allow precise measurement of flow rate required to lower BP. ▪ Give at the rate required to lower BP, usually between 0.5–10 mcg/kg/min. ▪ DO NOT exceed the maximum dose of 10 mcg/kg/min nor give this dose for longer than 10 min.

INCOMPATIBILITIES **Solution/additive: Amiodarone, propafenone. Y-site: Cisatracurium, haloperidol, levofloxacin.**

▪ Store reconstituted solutions and IV solution at 15°–30° C (59°–86° F) protected from light; stable for 24 h.

ADVERSE EFFECTS (≥1%) **Body as a Whole:** Diaphoresis, apprehension, restlessness, muscle twitching, retrosternal discomfort. <u>Thiocyanate toxicity</u> (profound hypotension, tinnitus, blurred vision, fatigue, metabolic acidosis, pink skin color, absence of reflexes, faint heart sounds, loss of consciousness). **CV:** Profound hypotension, palpitation, increase or transient lowering of pulse rate, bradycardia, tachycardia, ECG changes. **GI:** Nausea, retching, abdominal pain. **Metabolic:** Increase in serum creatinine, fall or rise in total plasma cobalamins. **CNS:** Headache, dizziness. **Special Senses:** Nasal stuffiness. **Other:** Irritation at infusion site.

PHARMACOKINETICS Onset: Within 2 min. **Duration:** 1–10 min after infusion is terminated. **Metabolism:** Rapidly converted to cyanogen in erythrocytes and tissue, which is metabolized to thiocyanate in liver. **Elimination:** Excreted in urine primarily as thiocyanate. **Half-Life:** (Thiocyanate): 2.7–7 days.

NURSING IMPLICATIONS

Assessment & Drug Effects

- Monitor constantly to titrate IV infusion rate to BP response.
- Relieve adverse effects by slowing IV rate or by stopping drug; minimize them by keeping patient supine.
- Notify prescriber immediately if BP begins to rise after drug infusion rate is decreased or infusion is discontinued.
- Monitor I&O.
- Lab tests: Monitor blood thiocyanate level in patients receiving prolonged treatment or in patients with severe kidney dysfunction (levels usually are not allowed to exceed 10 mg/dL). Determine plasma cyanogen level following 1 or 2 days of therapy in patients with impaired liver function.

NIZATIDINE

(ni-za′ti-deen)

Axid, Axid AR

Classifications: H$_2$-RECEPTOR ANTAGONIST; ANTISECRETORY
Therapeutic: ANTIULCER; ANTISECRETORY
Prototype: Cimetidine
Pregnancy Category: B

AVAILABILITY 75 mg tablets; 150 mg, 300 mg capsules; 15 mg/mL oral solution

ACTION & THERAPEUTIC EFFECT Inhibits secretion of gastric acid by reversible, competitive blockage of histamine at the H$_2$ receptor, particularly those in the gastric parietal cells. *Significantly reduces nocturnal gastric acid secretion for up to 12 h.*

USES Active duodenal ulcers; maintenance therapy for duodenal ulcers.

CONTRAINDICATIONS Hypersensitivity to nizatidine; lactation.
CAUTIOUS USE Hypersensitivity to other H$_2$-receptor antagonists; renal impairment or renal failure; older adults; pregnancy (category B); children 12 y or younger.

ROUTE & DOSAGE

Active Duodenal Ulcer
Adult: **PO** 150 mg b.i.d. or 300 mg at bedtime

Maintenance Therapy
Adult: **PO** 150 mg at bedtime

ADMINISTRATION

Oral

- Give drug usually once daily at bedtime. Dose may be divided and given twice daily.
- Administer oral liquid drug using a calibrated measuring device.
- Be aware that antacids consisting of aluminum and magnesium

Common adverse effects in *italic*, life-threatening effects <u>underlined</u>; generic names in **bold**; classifications in SMALL CAPS; ♣ Canadian drug name; ◯ Prototype drug

1075

hydroxides with simethicone decrease nizatidine absorption by about 10%. Administer the antacid 2 h after nizatidine.

ADVERSE EFFECTS (≥1%) **CNS:** Somnolence, fatigue. **Skin:** Pruritus, sweating. **Metabolic:** Hyperuricemia.

INTERACTIONS Drug: May decrease absorption of **delavirdine, didanosine, itraconazole, ketoconazole;** ANTACIDS may decrease absorption of nizatidine. May increase **alcohol** levels.

PHARMACOKINETICS Absorption: Greater than 90% from GI tract. **Peak:** 0.5–3 h. **Metabolism:** In liver. **Elimination:** 60% in urine unchanged. **Half-Life:** 1–2 h.

NURSING IMPLICATIONS

Assessment & Drug Effects

- Monitor patient for alleviation of symptoms. Most ulcers should heal within 4 wk.
- Monitor for persistence of ulcer symptoms in patients who continue to smoke during therapy.
- Lab tests: Periodic LFTs and kidney function tests with long-term therapy.

Patient & Family Education

- Take medications for the full course of therapy even though symptoms may be relieved.
- Do not take other prescription or OTC medications without consulting prescriber.
- Stop smoking; smoking adversely affects healing of ulcers and effectiveness of the drug.

NONOXYNOL-9
(noe-nox′ee-nole)
Conceptrol, Delfen, Emko, Gynol II, Koromex

Classification: SPERMICIDE CONTRACEPTIVE
Therapeutic: SPERMICIDE CONTRACEPTIVE
Pregnancy Category: C

AVAILABILITY 1%, 2%, 2.2%, 3.5%, 4%, 5% gel; 8%, 12.5% foam; 2.27%, 100 mg, 150 mg suppositories

ACTION & *THERAPEUTIC EFFECT*
Nonionic surfactant spermicidal incorporated into foams, gels, jelly, or suppositories. Immobilizes sperm by cell membrane disruption. *Applied over the cervix, blocks entrance to uterus by sperm, traps and absorbs seminal fluid, then releases the immediately available spermicide.*

USES As barrier contraceptive alone or in conjunction with a vaginal diaphragm or with a condom.

CONTRAINDICATIONS Cystocele, prolapsed uterus, sensitivity or allergy to polyurethane or to nonoxynol-9; vaginitis; history of TSS; immediately after delivery or abortion.
CAUTIOUS USE HIV patients; menstruation; pregnancy (category C).

ROUTE & DOSAGE

Contraceptive

Adult: **Topical** Apply or insert 30–60 min before intercourse. Repeat before each intercourse.

ADMINISTRATION

Topical

- Apply foams, gels, jelly, cream: Fully load intravaginal applicator and insert about ⅔ of its length [7.5–10 cm (3–4 in.)] into vagina.
- Use with diaphragm: Place 1–3 tsp spermicide formulation in

Common adverse effects in *italic*, life-threatening effects underlined; generic names in **bold**; classifications in SMALL CAPS; ♣ Canadian drug name; ⊘ Prototype drug

dome prior to insertion. After diaphragm is in place, additional spermicide is recommended. Leave spermicide and diaphragm in place 6 h after intercourse.

ADVERSE EFFECTS (≥1%) **Urogenital:** *Candidiasis;* vaginal irritation and dryness; increase in vaginal infections; menstrual and nonmenstrual <u>toxic shock syndrome (TSS)</u>.

INTERACTIONS Drug: Intravaginal AZOLE ANTIFUNGALS may inactivate the spermicides.

PHARMACOKINETICS Onset: Spermicidal action is prompt upon contact with sperm; minimal systemic absorption.

NURSING IMPLICATIONS

Patient & Family Education

- Stop using nonoxynol-9 if pregnancy is suspected.
- Report symptoms of vaginal infection to prescriber: Burning, inflammation, intense vaginal and vulvar itching, cheesy, curd-like discharge, painful intercourse, dysuria. Nonoxynol-9 antifungal properties are weaker than its antibacterial potency, thus vulvovaginal candidiasis frequently occurs.
- Use spermicide before the first and every subsequent act of intercourse.

NOREPINEPHRINE BITARTRATE
(nor-ep-i-nef′rin)

Levarterenol, Levophed, Nor-adrenaline

Classifications: ADRENERGIC AGONIST; VASOCONSTRICTOR
Therapeutic: VASOPRESSOR; CARDIAC INOTROPIC

Prototype: Epinephrine
Pregnancy Category: C

AVAILABILITY 1 mg/mL injection

ACTION & *THERAPEUTIC EFFECT*
Direct-acting sympathomimetic amine identical to natural catecholamine norepinephrine. Acts directly and predominantly on alpha-adrenergic receptors; little action on beta receptors except in heart (beta₁ receptors). Causes vasoconstriction and cardiac stimulation; also produces powerful constrictor action on resistance and capacitance blood vessels. *Peripheral vasoconstriction and moderate inotropic stimulation of heart result in increased systolic and diastolic blood pressure, myocardial oxygenation, coronary artery blood flow, and workload of the heart.*

USES To restore BP in certain acute hypotensive states such as shock, sympathectomy, pheochromocytomectomy, spinal anesthesia, poliomyelitis, MI, septicemia, blood transfusion, and drug reactions. Also as adjunct in treatment of cardiac arrest.

CONTRAINDICATIONS Use as sole therapy in hypovolemic states, except as temporary emergency measure; mesenteric or peripheral vascular thrombosis; profound hypoxia or hypercarbia; use during cyclopropane or halothane anesthesia; hypertension; hyperthyroidism.

CAUTIOUS USE Severe heart disease; older adult patients; within 14 days of MAOI therapy; patients receiving tricyclic antidepressants; pregnancy (category C), lactation.

ROUTE & DOSAGE

Hypotension
Adult: **IV** Initial 0.5–1 mcg/min, titrate to response; usual range 8–30 mcg/min
Child: **IV** 0.05–0.1 mcg/kg/min; titrate to response (max: 1–2 mcg/kg/min)

Common adverse effects in *italic*, life-threatening effects <u>underlined</u>; generic names in **bold**; classifications in SMALL CAPS; ♣ Canadian drug name; ⊗ Prototype drug

ADMINISTRATION

Intravenous

PREPARE: **IV Infusion:** Dilute a 4 mL ampule in 1000 mL of D5W or D5/NS. ▪ More concentrated solutions (e.g., 4 mg in 500 mL to yield 8 mcg/mL) may be used based on fluid requirements. ▪ Do not use solution if discoloration or precipitate is present. Protect from light.

ADMINISTER: **IV Infusion:** Initial rate of infusion is 2–3 mL/min (8–12 mcg/min), then titrated to maintain BP, usually 0.5–1 mL/min (2–4 mcg/min). ▪ An infusion pump is used. Usually give at the slowest rate possible required to maintain BP. Constantly monitor flow rate. ▪ Check infusion site frequently and immediately report any evidence of extravasation: Blanching along course of infused vein (may occur without obvious extravasation), cold, hard swelling around injection site. ▪ Antidote for extravasation ischemia: Phentolamine, 5–10 mg in 10–15 mL NS injection, is infiltrated throughout affected area (using syringe with fine hypodermic needle) as soon as possible. ▪ If therapy is to be prolonged, change infusion sites at intervals to allow effect of local vasoconstriction to subside. ▪ Avoid abrupt withdrawal; when therapy is discontinued, infusion rate is slowed gradually.

INCOMPATIBILITIES **Solution/additive: Aminophylline, amobarbital, ampicillin, whole blood, cephapirin, chlorothiazide, chlorpheniramine, diazepam, pentobarbital, phenobarbital, phenytoin, secobarbital, sodium bicarbonate, sodium iodide, streptomycin, thiopental, warfarin. Y-site: Insulin, thiopental.**

ADVERSE EFFECTS (≥1%) **Body as a Whole:** Restlessness, anxiety, *tremors,* dizziness, weakness, insomnia, pallor, plasma volume depletion, edema, hemorrhage, intestinal, <u>hepatic</u>, or renal <u>necrosis</u>, retrosternal and pharyngeal pain, profuse sweating. **CV:** Palpitation, hypertension, reflex bradycardia, <u>fatal arrhythmias</u> (large doses), severe hypertension. **GI:** Vomiting. **Metabolic:** Hyperglycemia. **CNS:** Headache, violent headache, <u>cerebral hemorrhage</u>, convulsions. **Respiratory:** Respiratory difficulty. **Skin:** Tissue necrosis at injection site (with extravasation). **Special Senses:** Blurred vision, photophobia.

INTERACTIONS Drug: ALPHA- and BETA-BLOCKERS antagonize pressor effects; ERGOT ALKALOIDS, **furazolidone, guanethidine, methyldopa,** TRICYCLIC ANTIDEPRESSANTS may potentiate pressor effects; **halothane, cyclopropane** increase risk of arrhythmias.

PHARMACOKINETICS Onset: Very rapid. **Duration:** 1–2 min after infusion. **Distribution:** Localizes in sympathetic nerve endings; crosses placenta. **Metabolism:** In liver and other tissues by catecholamine O-methyltransferase and monoamine oxidase. **Elimination:** In urine.

NURSING IMPLICATIONS

Assessment & Drug Effects

▪ Monitor constantly while patient is receiving norepinephrine. Take baseline BP and pulse before start of therapy, then q2min from initiation of drug until stabilization occurs at desired level, then every 5 min during drug administration.
▪ Adjust flow rate to maintain BP at low normal (usually 80–100 mm Hg systolic) in normotensive patients. In previously hypertensive patients, systolic is generally maintained no

higher than 40 mm Hg below pre-existing systolic level.

- Observe carefully and record mental status (index of cerebral circulation), skin temperature of extremities, and color (especially of earlobes, lips, nail beds) in addition to vital signs.
- Monitor I&O. Urinary retention and kidney shutdown are possibilities, especially in hypovolemic patients. Urinary output is a sensitive indicator of the degree of renal perfusion. Report decrease in urinary output or change in I&O ratio.
- Be alert to patient's complaints of headache, vomiting, palpitation, arrhythmias, chest pain, photophobia, and blurred vision as possible symptoms of overdosage. Reflex bradycardia may occur as a result of rise in BP.
- Continue to monitor vital signs and observe patient closely after cessation of therapy for clinical sign of circulatory inadequacy.

NORETHINDRONE ⊕

(nor-eth-in′drone)

Micronor, Norlutin, Nor-Q.D.

NORETHINDRONE ACETATE

Aygestin ♣, Norlutate ♣
Classification: PROGESTIN
Therapeutic: PROGESTIN; CONTRA-CEPTIVE
Pregnancy Category: X

AVAILABILITY 0.35 mg, 5 mg tablets

ACTION & *THERAPEUTIC EFFECT*
Synthetic progestation hormone with androgenic, anabolic, and estrogenic properties. Progestin-only contraceptives alter cervical mucus, exert progestational effect on endometrium, interfere with implantation, and, in some cases, suppress

ovulation. *Contraceptive that suppresses the midcycle surge of luteinizing hormone (LH).*

USES Amenorrhea, abnormal uterine bleeding due to hormonal imbalance in absence of organic pathology; endometriosis. Also alone or in combination with an estrogen for birth control.

CONTRAINDICATIONS Thromboembolic disorders, cerebral vascular or coronary vascular disease; carcinoma of breast, endometrium, or liver; abnormal vaginal bleeding; known or suspected pregnancy (category X).
CAUTIOUS USE Cardiac disease; history of depression, seizure disorders, migraine; diabetes mellitus; CHF; history of thrombophlebitis or thromboembolic disease; lactation; children younger than 16 y.

ROUTE & DOSAGE

Amenorrhea

Adult: **PO Norethindrone** 5–20 mg on day 5 through day 25 of menstrual cycle; **Acetate** 2.5–10 mg on day 5 through day 25 of menstrual cycle

Endometriosis

Adult: **PO Norethindrone** 10 mg/day for 2 wk; increase by 5 mg/day q2wk up to 30 mg/day, dose may remain at this level for 6–9 mo or until breakthrough bleeding; **Acetate** 5 mg/day for 2 wk, increase by 2.5 mg/day q2wk up to 15 mg/day, dose may remain at this level for 6–9 mo or until breakthrough bleeding

Progestin-Only Contraception

Adult: **PO Norethindrone** 0.35 mg/day starting on day 1 of menstrual flow, then continuing indefinitely

Common adverse effects in *italic*, life-threatening effects underlined; generic names in **bold**; classifications in SMALL CAPS; ♣ Canadian drug name; ⊕ Prototype drug

1079

ADMINISTRATION

Oral

- Note: Dosing schedule is based on a 28-day menstrual cycle.
- Use or add a barrier contraceptive when starting the minipill regimen (progestin-only contraception) for the first cycle or for 3 wk to ensure full protection.
- Protect drug from light and from freezing.

ADVERSE EFFECTS (≥1%) **CNS:** <u>Cerebral thrombosis or hemorrhage</u>, depression. **CV:** Hypertension, <u>pulmonary embolism</u>, edema. **GI:** Nausea, vomiting, cholestatic jaundice, abdominal cramps. **Urogenital:** *Breakthrough bleeding,* cervical erosion, changes in menstrual flow, dysmenorrhea, vaginal candidiasis. **Other:** *Weight changes; breast tenderness,* enlargement or secretion.

INTERACTIONS Drug: BARBITURATES, **carbamazepine, fosphenytoin, modafinil, phenytoin, primidone, pioglitazone, rifampin rifabutin, rifapentine, topiramate, troglitazone** can decrease contraceptive effectiveness.

PHARMACOKINETICS Absorption: Readily absorbed from GI tract. **Metabolism:** In liver. **Elimination:** In urine and feces as metabolites.

NURSING IMPLICATIONS

Assessment & Drug Effects

- Monitor for S&S of thrombophlebitis (see Appendix F).
- Withhold drug and notify prescriber if any of the following occur: Sudden, complete, or partial loss of vision, proptosis, diplopia, or migraine headache.

Patient & Family Education

- Wait at least 3 mo before becoming pregnant after stopping the minipill to prevent birth defects.

Use a barrier or nonhormonal method of contraception until pregnancy is desired.

- If you have not taken all your pills and you miss a period, consider the possibility of pregnancy after 45 days from the last menstrual period; stop using this drug until pregnancy is ruled out.
- If you have taken all your pills and you miss 2 consecutive periods, rule out pregnancy and use a barrier or nonhormonal method of contraception before continuing the regimen.
- Promptly report prolonged vaginal bleeding or amenorrhea.
- Keep appointments for physical checkups (q6–12mo) while you are taking hormonal birth control.

NORFLOXACIN

(nor-flox'a-sin)

Noroxin

Classification: QUINOLONE ANTIBIOTIC

Therapeutic: ANTIBIOTIC

Prototype: Ciprofloxacin

Pregnancy Category: C

AVAILABILITY 400 mg tablets

ACTION & *THERAPEUTIC EFFECT*
Potent broad-spectrum antibiotic activity. Alters structure of bacterial DNA gyrase, thus promoting double-stranded DNA breakage, thus interfering with synthesis of bacterial protein and blocking bacterial survival. *Active against many bacterial pathogens of the urinary tract.*

USES Complicated and uncomplicated urinary tract infection (UTI); gonorrhea, prostatitis.

UNLABELED USES Gastroenteritis and prevention of travelers' diarrhea.

CONTRAINDICATIONS Use in individual with known factors that predispose to seizures; history of hypersensitivity to norfloxacin and other quinolone antibiotics; history of QT prolongation; tendon pain; lactation.

CAUTIOUS USE Impaired kidney function, adolescents if skeletal growth is complete; G6PD deficiency; GI disease; myasthenia gravis; pregnancy (category C); children younger than 18 y.

ROUTE & DOSAGE

Urinary Tract Infection/Prostatitis
Adult: **PO** 400 mg b.i.d. × 3–21 days (depending on causative agent)
Gonorrhea or Gonococcal Urethritis (Not CDC Recommended)
Adult: **PO** 800 mg once/day
Renal Impairment Dosage Adjustment
CrCl 30 mL/min or less: 400 mg once daily

ADMINISTRATION

Oral

- Give 1 h before or 2 h after meals with a full glass of water.
- Administer concomitant antacid at least 2 h after norfloxacin to prevent interference with absorption. Aluminum or magnesium ions in the antacid may bind to and form insoluble complexes with the quinolone in GI tract.
- Store at 40° C (104° F) or lower in tightly closed container. Do not freeze.

ADVERSE EFFECTS (≥1%) Musculo-skeletal: Joint swelling, cartilage erosion in weight-bearing joints, tendonitis. In immunosuppressed adult: Acute ankle and hip pain followed by acute pain, tenderness,

and swelling of tendon sheath of middle finger of both hands after 4 wk of therapy. **CNS:** *Headache,* dizziness, light-headedness, fatigue, drowsiness, somnolence, depression, insomnia, seizures, peripheral neuropathy. **GI:** *Nausea,* abdominal pain, diarrhea, vomiting, anorexia, dyspepsia, dysphagia, dry mouth, bitter taste, heartburn, flatulence, pruritus ani, increased serum AST, ALT, alkaline phosphatase. **Hematologic:** Leukopenia, neutropenia. **Urogenital:** With high doses: Crystalluria (not associated with renal toxicity), vulvar irritation.

DIAGNOSTIC TEST INTERFERENCE
May cause false positive on **opiate screening tests.**

INTERACTIONS Drug: ANTACIDS, iron, sucralfate, zinc decrease absorption; **nitrofurantoin** may antagonize antibacterial effects; may increase hypoprothrombinemic effects of **warfarin;** may cause slight increase in **theophylline** levels; concurrent administration with CLASS IA and CLASS III ANTIARRHYTHMICS may result in development of QT prolongation as well as torsades de pointes.

PHARMACOKINETICS Absorption: 30–40% from GI tract. **Peak:** 1–2 h. **Distribution:** Renal parenchyma, gallbladder, liver, prostate; crosses placenta; distributed into breast milk. **Metabolism:** In liver. **Elimination:** In urine and feces. **Half-Life:** 3–4 h.

NURSING IMPLICATIONS
Assessment & Drug Effects
- Collect urine specimens for testing before initiating antibiotic.
- Monitor patient for tendon pain. Norfloxacin should be discontinued and prescriber informed.
- Lab tests: Periodic WBC with differential, LFTs, including alkaline phosphatase, especially with prolonged use.

N

Common adverse effects in *italic*, life-threatening effects underlined; generic names in **bold;** classifications in SMALL CAPS; ♣ Canadian drug name; ⊘ Prototype drug

- Report to the prescriber if patient is adequately hydrated, yet I&O ratio and pattern changes are noted, or if condition does not improve within a few days. Dosage may need to be modified.

Patient & Family Education
- Take drug at same time each day.
- Take drug exactly as prescribed. Erratic dosing can encourage emergence of resistant bacteria; underdosing or premature discontinuation of treatment can cause return of UTI symptoms.
- Keep fluid intake high (at least 2500–3000 mL/day if tolerated) to provide adequate urine output and hydration, important in the prevention of crystalluria (rare side effect).

NORMAL SERUM ALBUMIN, HUMAN ℗

(al-byoo′min)

Albuminar, Albutein, Buminate, Plasbumin
Classifications: PLASMA DERIVATIVE; PLASMA VOLUME EXPANDER
Therapeutic: PLASMA VOLUME EXPANDER
Pregnancy Category: C

AVAILABILITY 5%, 20%, 25% injection

ACTION & *THERAPEUTIC EFFECT*
Obtained by fractionating pooled venous and placental human plasma, which is then sterilized by filtration and heated to minimize transmitting hepatitis B or HIV. Plasma volume expander that increases the osmotic pressure of plasma. *Expands volume of circulating blood by osmotically shifting tissue fluid into general circulation.*

USES To restore plasma volume and maintain cardiac output in hypovolemic shock; for prevention and treatment of cerebral edema; as adjunct in exchange transfusion for hyperbilirubinemia and erythroblastosis fetalis; to increase plasma protein level in treatment of hypoproteinemia; and to promote diuresis in refractory edema. Also used for blood dilution prior to or during cardiopulmonary bypass procedures. Has been used as adjunct in treatment of adult respiratory distress syndrome (ARDS).

CONTRAINDICATIONS Hypersensitivity to albumin; severe anemia; cardiac failure; within 24 h of severe burns; heart failure; patients with normal or increased intravascular volume.
CAUTIOUS USE Low cardiac reserve, pulmonary disease, absence of albumin deficiency; liver or kidney failure, dehydration, hypertension, hypernatremia; restricted sodium intake; pregnancy (category C).

ROUTE & DOSAGE

Emergency Volume Replacement
Adult: **IV** 25 g, may repeat in 15–30 min if necessary (max: 250 g)

Colloidal Volume Replacement (Nonemergency)
Child: **IV** 12.5 g, may repeat in 15–30 min if necessary

Hypoproteinemia
Adult: **IV** 50–75 g (max: 2 mL/min)
Child: **IV** 25 g (max: 2 mL/min)

ADMINISTRATION

Intravenous

***PREPARE:* IV Infusion:** Normal serum albumin, 5%, is infused without further dilution. ▪ Normal serum albumin, 20% and 25%, may be infused undiluted or diluted in NS or D5W (with sodium restriction).

Common adverse effects in *italic*, life-threatening effects <u>underlined</u>; generic names in **bold;** classifications in SMALL CAPS; ♦ Canadian drug name; ℗ Prototype drug

***ADMINISTER:* IV Infusion for Hypovolemic Shock**: Give initially as rapidly as necessary to restore blood volume. As blood volume approaches normal, rate should be reduced to avoid circulatory overload and pulmonary edema.
- Give 5% albumin at rate not exceeding 2–4 mL/min. Give 20% and 25% albumin at a rate not to exceed 1 mL/min. **IV Infusion with Normal Blood Volume:** Give 5% albumin human at a rate not to exceed 5–10 mL/min; give 20% and 25% albumin at a rate not to exceed 2 or 3 mL/min. **IV Infusion for Children:** Usual rate is 25%–50% of the adult rate.

***INCOMPATIBILITIES* Solution/additive: Amino acids, verapamil. Y-site: Fat emulsion, midazolam, vancomycin, verapamil.**

- Store at temperature not to exceed 37° C (98.6° F). ▪ Use solution within 4 h, once container is opened, because it contains no preservatives or antimicrobials. Discard unused portion.

ADVERSE EFFECTS (≥1%) **Body as a Whole:** Fever, chills, flushing, increased salivation, headache, back pain. **Skin:** Urticaria, rash. **CV:** Circulatory overload, pulmonary edema (with rapid infusion); hypotension, hypertension, dyspnea, tachycardia. **GI:** Nausea, vomiting.

DIAGNOSTIC TEST INTERFERENCE False rise in ***alkaline phosphatase*** when albumin is obtained partially from pooled placental plasma (levels reportedly decline over period of weeks).

NURSING IMPLICATIONS

Assessment & Drug Effects
- Monitor BP, pulse and respiration, and IV albumin flow rate. Adjust flow rate as needed to avoid too rapid a rise in BP.
- Lab tests: Monitor dosage of albumin using plasma albumin (normal): 3.5–5 g/dL; total serum protein (normal): 6–8.4 g/dL; Hgb; Hct; and serum electrolytes.
- Observe closely for S&S of circulatory overload and pulmonary edema (see Appendix F). If S&S appear, slow infusion rate just sufficiently to keep vein open, and report immediately to prescriber.
- Monitor I&O ratio and pattern. Report changes in urinary output. Increase in colloidal osmotic pressure usually causes diuresis, which may persist 3–20 h.
- Withhold fluids completely during succeeding 8 h, when albumin is given to patients with cerebral edema.

Patient & Family Education
- Report chills, nausea, headache, or back pain to prescriber immediately.

NORTRIPTYLINE HYDROCHLORIDE

(nor-trip′ti-leen)

Aventyl, Pamelor
Classification: TRICYCLIC ANTIDEPRESSANT
Therapeutic: ANTIDEPRESSANT
Prototype: Imipramine
Pregnancy Category: D

AVAILABILITY 10 mg, 25 mg, 50 mg, 75 mg capsules; 10 mg/5 mL solution

ACTION & *THERAPEUTIC EFFECT* Secondary amine derivative of amitriptyline that inhibits the action of many chemical agents including catecholamines. Mood elevation may be due to its inhibition of reuptake of serotonin or another neurotransmitter at the presynaptic

Common adverse effects in *italic*, life-threatening effects <u>underlined</u>; generic names in **bold**; classifications in SMALL CAPS; ♣ Canadian drug name; ⊘ Prototype drug

1083

membrane. *Effective in improving depressive moods.*

USES Endogenous depression. Similar in actions, uses, limitations, and interactions to imipramine.
UNLABELED USES Nocturnal enuresis in children.

CONTRAINDICATIONS Hypersensititivity to tricyclic antidepressants; acute recovery period after MI; AV block; history of QT prolongation; suicidal ideation; during or within 14 days of MAO inhibitor therapy; pregnancy (category D), lactation.
CAUTIOUS USE Narrow-angle glaucoma, cardiac disease; history of suicidal tendencies; hyperthyroidism, concurrent use with electroshock therapy; history of suicides; Parkinson's disease; asthma; bipolar disorder; older adults; children younger than 6 y.

ROUTE & DOSAGE

Antidepressant

Adult: **PO** 25 mg t.i.d. or q.i.d., gradually increased to 100–150 mg/day
Geriatric: **PO** Start with 10–25 mg at bedtime, increase by 25 mg q3days to 75 mg at bedtime (max: 150 mg/day)
Adolescent: **PO** 30–50 mg/day in divided doses
Child (6–12 y): **PO** 10–20 mg/day in 3–4 divided doses

Nocturnal Enuresis

Child: **PO** 6–7 y, 10 mg/day; 8–11 y, 10–20 mg/day; older than 11 y, 25–35 mg/day given 30 min before bedtime

Pharmacogenetic Dosage Adjustment

Poor CYP2D6 metabolizers: Start with 50% of dose

ADMINISTRATION

Oral

- Give with food to decrease gastric distress.
- In older adults, total daily dose may be given once a day at bedtime (preferred).
- Be aware that nortriptyline is a 4% alcohol solution.
- Supervise drug ingestion to be sure patient swallows medication.
- Store at 15°–30° C (59°–86° F) in tightly closed container.

ADVERSE EFFECTS (≥1%) **Body as a Whole:** Tremors, hyperhidrosis. **CV:** *Orthostatic hypotension.* **GI:** Paralytic ileus, *dry mouth.* **Hematologic:** Agranulocytosis (rare). **CNS:** Drowsiness, confusional state (especially in older adults and with high dosage). **Skin:** Photosensitivity reaction. **Special Senses:** Blurred vision. **Urogenital:** *Urinary retention.*

INTERACTIONS Drug: May decrease response to ANTIHYPERTENSIVES; CNS DEPRESSANTS, **alcohol,** HYPNOTICS, BARBITURATES, SEDATIVES potentiate CNS depression; may increase hypoprothrombinemic effect of ORAL ANTICOAGULANTS; **levodopa,** SYMPATHOMIMETICS (e.g., **epinephrine, norepinephrine**) pose possibility of sympathetic hyperactivity with hypertension and hyperpyrexia; MAO INHIBITORS pose possibility of severe reactions: Toxic psychosis, cardiovascular instability; **methylphenidate** increases plasma TCA levels; THYROID DRUGS may increase possibility of arrhythmias; **cimetidine** may increase plasma TCA levels. **Herbal: Ginkgo** may decrease seizure threshold. **St. John's wort** may cause serotonin syndrome (see Appendix F).

PHARMACOKINETICS Absorption: Rapidly from GI tract. **Peak:** 7–8.5 h.

Duration: Crosses placenta; distributed in breast milk. **Metabolism:** In liver (CYP2D6). **Elimination:** Primarily in urine. **Half-Life:** 16–90 h.

NURSING IMPLICATIONS
Assessment & Drug Effects

- Be aware that nortriptyline has a narrow therapeutic plasma level range, or "therapeutic window." Drug levels above or below the therapeutic window are associated with decreased rate of response.
- Therapeutic response may not occur for 2 wk or more.
- Monitor carefully for signs and symptoms of suicidality in children and adults.
- Monitor BP and pulse rate during adjustment period of TCA therapy. If systolic BP falls more than 20 mm Hg or if there is a sudden increase in pulse rate, withhold medication and notify the prescriber.
- Notify prescriber if psychotic signs increase.
- Inspect oral membranes daily if patient is on high doses of TCA. Urge outpatient to report stomatitis or dry mouth. Sore mouth can be a major cause of poor nutrition and noncompliance. Consult prescriber about use of a saliva substitute (e.g., VA-Oralube, Moi-Stir).
- Monitor bowel elimination pattern and I&O ratio. Urinary retention and severe constipation are potential problems, especially in older adults. Advise increased fluid intake; consult prescriber about stool softener.
- Observe patient with history of glaucoma. Symptoms that may signal acute attack (severe headache, eye pain, dilated pupils, halos of light, nausea, vomiting) should be reported promptly.

Patient & Family Education

- Be aware that your ability to perform tasks requiring alertness and skill may be impaired. Do not engage in hazardous activities until response to drug is known.
- Do not use OTC drugs unless prescriber approves.
- Consult prescriber about safe amount of alcohol, if any, that can be ingested. Alcohol and nortriptyline both have increased effects when used together and for up to 2 wk after the TCA is discontinued.
- Nortriptyline enhances the effects of barbiturates and other CNS depressants are enhanced.

NYSTATIN
(nye-stat′in)
Mycostatin, Nadostine ✦, Nilstat, Nyaderm ✦, Nystex, O-V Statin
Classification: ANTIFUNGAL ANTIBIOTIC
Therapeutic: ANTIFUNGAL
Prototype: Amphotericin B
Pregnancy Category: C

N

AVAILABILITY 500,000 unit tablets; 100,000 units/mL oral suspension; 200,000 troches; 100,000 units vaginal tablets; 100,000 units/g cream, ointment, powder

ACTION & *THERAPEUTIC EFFECT*
Nontoxic, nonsensitizing antifungal antibiotic that binds to sterols in fungal cell membrane, thereby changing membrane potential and allowing leakage of intracellular components that leads to fungi cell death. *Fungistatic and fungicidal activity against a variety of yeasts and fungi.*

USES Local infections of skin and mucous membranes caused by *Candida* sp. including *Candida*

---

albicans (e.g., paronychia; cutaneous, oropharyngeal, vulvovaginal, and intestinal candidiasis).

CONTRAINDICATIONS Vaginal infections caused by *Gardnerella vaginalis* or *Trichomonas* sp.
CAUTIOUS USE Diabetes mellitus; pregnancy (category C), lactation.

ROUTE & DOSAGE

Candida Infections

Adult: **PO** 500,000–1,000,000 units t.i.d.; 1–4 troches 4–5 times/day; Suspension: 400,000–600,000 units q.i.d. **Intravaginal** 1–2 tablets daily for 2 wk
Child: **PO** Suspension: 400,000–600,000 units q.i.d.
Infant: **PO** 100,000–200,000 units q.i.d.

ADMINISTRATION

Oral

- Give reconstituted powder for oral suspension immediately after mixing.
- Rinse mouth with 1–2 tsp oral suspension. Should be kept in mouth (swish) as long as possible (at least 2 min), then liquid should be spit out or swallowed (if "swish and swallow" is ordered).
- For children, infants: Apply drug with swab to each side of mouth. Avoid food or drink for 30 min after treatment.
- The troche dosage form should dissolve in mouth (about 30 min). Troches should not be chewed or swallowed. Food and drink should be avoided during period of dissolving and for 30 min after treatment.

Topical

- Do not apply occlusive dressings over topical applications unless specifically directed to do so.

Intravaginal

- Store vaginal tablets in refrigerator below 15° C (59° F).

ADVERSE EFFECTS (≥1%) **GI:** Nausea, vomiting, epigastric distress, diarrhea (especially with high oral doses).

PHARMACOKINETICS Absorption: Poorly absorbed from GI tract. **Elimination:** In feces.

NURSING IMPLICATIONS

Assessment & Drug Effects

- Monitor oral cavity, especially the tongue, for signs of improvement.

Patient & Family Education

- This drug may cause contact dermatitis. Stop using the drug and report to prescriber if redness, swelling, or irritation develops.
- Take for oral candidiasis (thrush) treatment after meals and at bedtime.
- Care of dentures: Remove dentures before each rinse with oral suspension and before use of troche. Remove dentures at night (infection occurs more frequently in person who wears dentures 24 h a day).
- Dust shoes and stockings, as well as feet, with nystatin dusting powder.
- Gently clean infected areas with tepid water before each application of topical preparation.
- Continue medication for vulvovaginal candidiasis during menstruation.
- Use vaginal tablets up to 6 wk before term to prevent thrush in the newborn.

OCTREOTIDE ACETATE ⓟ

(oc-tre′o-tide)

Sandostatin, Sandostatin LAR depot
Classification: SOMATOSTATIN ANALOG

Therapeutic: HORMONE SUPPRES-
SANT; ACROMEGALY AGENT; ANTIDI-
ARRHEAL
Pregnancy Category: B

AVAILABILITY 0.05 mg/mL, 0.1 mg/
mL, 0.2 mg/mL, 0.5 mg/mL, 1 mg/mL
injection; 10 mg/5 mL, 20 mg/5 mL,
30 mg/5 mL depot injection

ACTION & *THERAPEUTIC EFFECT* A
long-acting peptide that mimics
natural hormone somatostatin.
Suppresses secretion of serotonin,
pancreatic peptides, gastrin,
vasoactive intestinal peptide, insulin, glucagon, secretin, and motilin.
*Stimulates fluid and electrolyte ab-
sorption from the GI tract and prolongs intestinal transit time; also
inhibits the growth hormone.*

USES Symptomatic treatment of severe diarrhea and flushing episodes associated with metastatic
carcinoid tumors. Also watery diarrhea associated with vasoactive intestinal peptide (VIP) tumors, acromegaly.

UNLABELED USES Acromegaly associated with pituitary tumors, fistula drainage, variceal bleeding,
hepatorenal syndrome, orthostatic
hypotension.

CONTRAINDICATIONS Hypersensitivity to octreotide.

CAUTIOUS USE Cholelithiasis, renal
impairment; dialysis; hepatic disease, liver cirrhosis; cardiac disease,
CHF; diabetes, TPN administration;
hypothyroidism; older adults; pregnancy (category B); lactation.

ROUTE & DOSAGE

Carcinoid Syndrome

Adult: **Subcutaneous/IV** 100–
600 mcg/day in 2–4 divided
doses, titrate to response **IM**

May switch to depot injection
after 2 wk at 20 mg q4wk ×
3 mo

VIPoma

Adult: **Subcutaneous/IV** 200–
300 mcg/day in 2–4 divided
doses, titrate to response **IM** May
switch to depot injection after 2
wk at 20 mg q4wk × 2 mo

Acromegaly

Adult: **Subcutaneous** 50 mcg
t.i.d., titrate up to 100 mcg–500
mcg t.i.d. **IM** May switch to depot
injection after 2 wk at 20 mg
q4wk × at least 3 mo, then
reassess

Renal Impairment Dosage Adjustment

Dialysis: Reduce dose

ADMINISTRATION

Subcutaneous/Intramuscular

- **Sandostatin LAR Depot** should
  be given IM. Reconstitute according to manufacturer's directions.
- **Sandostatin** may be given subcutaneously or IV.
- Minimize GI side effects by giving
  injections between meals and at
  bedtime.
- Avoid multiple injections into
  the same site. Rotate subcutaneous sites on abdomen, hip,
  and thigh.
- Give deep IM into a large muscle.
  To reduce local irritation, allow solution to reach room temperature
  before injection and administer
  slowly.

Intravenous

PREPARE: Direct: Give **Sandostatin** undiluted. **Intermittent:** Dilute in 50–200 mL D5W.

0

Common adverse effects in *italic*, life-threatening effects <u>underlined</u>; generic names
in **bold**; classifications in SMALL CAPS; ✚ Canadian drug name; ⊘ Prototype drug

1087

ADMINISTER: **Direct:** Give a single dose over 3–5 min. In emergency (eg, carcinoid crisis), give rapid IV bolus over 60 sec. **Intermittent:** Give over 15–30 min. *INCOMPATIBILITIES* **Solution/additive: Fat emulsion, regular insulin. Y-site: Dantrolene, diazepam, micafungin, phenytoin, pantoprazole.**

ADVERSE EFFECTS (≥1%) **CNS:** Headache, fatigue, dizziness. **GI:** *Nausea, diarrhea, abdominal pain* and discomfort. **CV:** Bradycardia. **Metabolic:** Hypoglycemia, hyperglycemia, increased liver transaminases, hypothyroidism (after long-term use), cholelithiasis, pancreatitis. **Body as a Whole:** Flushing, edema, *injection site pain*, pruritus.

INTERACTIONS Drug: May decrease **cyclosporine** levels; may alter other drug and nutrient absorption because of alterations in GI motility.

PHARMACOKINETICS Absorption: Rapidly from subcutaneous injection. **Peak:** 0.4 h. **Duration:** Up to 12 h. **Metabolism:** In liver. **Elimination:** In urine. **Half-Life:** 1.5 h.

NURSING IMPLICATIONS

Assessment & Drug Effects
- Lab tests: Periodic blood glucose, LFTs, and serum electrolytes. As specific conditions indicate, periodic: Plasma serotonin levels with carcinoid tumors, plasma VIP levels with VIPoma, serum GH, serum IGF-1, and thyroid function tests.
- Monitor for hypoglycemia and hyperglycemia (see Appendix F), because octreotide may alter the balance between insulin, glucagon, and growth hormone.

- Monitor fluid and electrolyte balance, as octreotide stimulates fluid and electrolyte absorption from GI tract.
- Monitor vitals signs, especially BP.
- Monitor bowel function, including bowel sounds and stool consistency.

Patient & Family Education
- Learn proper technique for subcutaneous injection if self-medication is required.
- Note: Preferred sites for subcutaneous injections of octreotide are the hip, thigh, and abdomen. Multiple injections at the same subcutaneous injection site within short periods of time are not recommended. This is to avoid irritating the area.

OFATUMUMAB
(o-fa-tu'mu-mab)

Arzerra
Classifications: BIOLOGICAL RESPONSE MODIFIER; MONOCLONAL ANTIBODY; ANTINEOPLASTIC
Therapeutic: ANTINEOPLASTIC
Prototype: Basiliximab
Pregnancy Category: C

AVAILABILITY 100 mg/5 mL solution for injection

ACTION & *THERAPEUTIC EFFECT*
A CD20 cytolytic IgG1 kappa monoclonal antibody. The CD20 molecule is present on normal B lymphocytes, both mature and immature lymphocytes as well as on B-cell chronic lymphocytic leukemia (CLL). Ofatumumab causes B-cell lysis possibly by antibody-dependent, cell-mediated cytotoxicity. *Effectiveness in treatment of CLL refractory to the standard drug regimen is based*

on the clinical improvement in response to ofatumumab.

USES Chronic lymphocytic leukemia refractory to fludarabine and alemtuzumab.

CONTRAINDICATIONS Serious infusion reaction; moderate to severe COPD; leukoencephalopathy; viral hepatitis; live vaccines.
CAUTIOUS USE History of hepatitis B; elderly; pregnancy (category C); lactation; children.

ROUTE & DOSAGE

Chronic Lymphocytic Leukemia (CLL)

Adult: **IV** Initial dose of 300 mg followed 1 wk later by 2000 mg qwk for 7 doses, followed by 2000 mg q4wk for 4 doses

ADMINISTRATION

Intravenous

PREPARE: **Infusion:** Do not shake drug vials. ▪ Determine the volume of the required drug dose and withdraw an equal volume of NS from a 1000 mL polyolefin IV bag. Add ofatumumab to the IV bag and mix by gentle inversion.
ADMINISTER: **Infusion:** Infuse through an in-line filter and PVC administration set supplied with product. ▪ Do **NOT** give IV push or bolus. ▪ Do not mix or administer with any other drugs or solutions. *For doses 1 and 2:* Initiate infusion at 12 mL/h. If no infusion reaction occurs from 0–30 min, may increase rate q30min as follows: 31–60 min, 25 mL/h; 61–90 min, 50 mL/h; 91–120 min, 100 mL/h; after 120 min, 200 mL/h. ▪ *For doses 3–12:* Initiate infusion at 25 mL/h. If

no infusion reaction occurs from 0–30 min, may increase rate q30min as follows: 31–60 min, 50 mL/h; 61–90 min, 100 mL/h; 91–120 min, 200 mL/h; after 120 min, 400 mL/h.

▪ Store diluted solution between 2°–8° C (36°–46° F). ▪ Use within 12 h of preparation. Discard solution after 24 h.

ADVERSE EFFECTS (≥1%) **Body as a Whole:** Chills, *fatigue,* herpes zoster infection, *infusion reactions,* peripheral edema, *pyrexia,* sepsis. **CNS:** Headache, insomnia. **CV:** Hypertension, hypotension, tachycardia. **GI:** *Diarrhea, nausea.* **Hematologic:** Anemia, neutropenia. **Musculoskeletal:** Back pain, muscle spasms. **Respiratory:** *Bronchitis, cough, dyspnea,* nasopharyngitis, *pneumonia,* sinusitis, *upper respiratory tract infections.* **Skin:** Hyperhidrosis, *rash,* urticaria.

PHARMACOKINETICS Half-Life: 14 days.

NURSING IMPLICATIONS
Assessment & Drug Effects
▪ Monitor for infusion reactions and stop infusion for any of the following: Bronchospasm, dyspnea, angioedema, flushing, significant changes in BP, tachycardia, back or abdominal pain, fever, rash, or urticaria.
▪ Monitor for and report promptly S&S of changes in neurologic status or suspected intestinal obstruction.
▪ Lab tests: Baseline and periodic CBC with differential.

Patient & Family Education
▪ Report promptly any of the following: New or worsening abdominal pain or nausea, confusion,

dizziness, loss of balance, difficulty talking or problems with vision, sore throat, fever, or other signs of infections.

■ Avoid live vaccinations and close contact with those who have received live vaccines.

OFLOXACIN
(o-flox′a-cin)

Floxin, Ocuflox
Classification: QUINOLONE ANTIBIOTIC
Therapeutic: ANTIBIOTIC
Prototype: Ciprofloxacin
Pregnancy Category: C

AVAILABILITY 200 mg, 300 mg, 400 mg tablets; 0.3% ophth solution; 0.3% otic solution

ACTION & *THERAPEUTIC EFFECT* A fluoroquinolone antibiotic that inhibits DNA gyrase, an enzyme necessary for bacterial DNA replication and some aspects of its transcription, repair, recombination, and transposition. *Has a broad spectrum of activity against gram-positive and gram-negative bacteria. Most effective against aerobic and anaerobic gram-negative bacteria.*

USES Uncomplicated gonorrhea, prostatitis, respiratory tract infections, skin and skin structure infections, urinary tract infections, superficial ocular infections, pelvic inflammatory disease. Otic: Otitis externa, otitis media with perforated tympanic membranes.

UNLABELED USES EENT infections, *Helicobacter pylori* infections, *Salmonella* gastroenteritis, anthrax.

CONTRAINDICATIONS Hypersensitivity to ofloxacin or other quinolone antibacterial agents; tendon pain; sunlight (UV) exposure; QT prolongation; viral infection; lactation.

CAUTIOUS USE Renal disease; patients with a history of epilepsy, psychosis, or increased intracranial pressure, cerebrovascular disease, CNS disorders such as seizures, epilepsy, myasthenia gravis; GI disease, colitis, dehydration; syphilis; atrial fibrillation; acute MI; CVA; pregnancy (category C); children and adolescents (except for **opthalmic otic** preparation).

ROUTE & DOSAGE

Uncomplicated Gonorrhea (Not CDC Recommended)
Adult: **PO** 400 mg for 1 dose

Respiratory Tract and Skin and Skin Structure Infections
Adult: **PO** 200–400 mg q12h × 10 days

Urinary Tract Infection
Adult: **PO** 200 mg q12h × 3–10 days

Pelvic Inflammatory Disease
Adult: **PO** 400 mg b.i.d. × 14 days

Prostatitis
Adult: **PO** 300 mg b.i.d. × 6 wk

Superficial Ocular Infections
Adult/Adolescent/Child:
Ophthalmic Instill 1–2 drops q2–4h for first 2 days, then q.i.d. for up to 5 additional days

Otitis Media with Perforation
Adult: **Otic** 10 drops (0.5 mL) q12h for 14 days
Child (1 y or older): **Otic** 5 drops (0.25 mL) q12h for 14 days

Otitis Externa

Adult: **Otic** 10 drops (0.5 mL) q12h for 7 days
Child (6 mo–13 y): **Otic** 5 drops (0.25 mL) q12h for 7 days

Renal Impairment Dosage Adjustment

CrCl 20–50 mL/min: Dose should be given q24h; *less than 20 mL/min:* ½ the dose q24h

Hepatic Impairment Dosage Adjustment

Severe impairment: 400 mg daily

ADMINISTRATION

Oral

- Do not give with meals.
- Avoid administering mineral supplements or vitamins with iron or zinc within 2 h of drug.
- Do not give antacids with magnesium, aluminum, or sucralfate within 4 h before or 2 h after drug.

Instillation

- Do NOT allow tip of dropper for ocular preparation to contact any surface.

ADVERSE EFFECTS (≥1%) **CNS:** *Headache, dizziness, insomnia,* hallucinations. **GI:** Nausea, vomiting, diarrhea, GI discomfort. **Urogenital:** Pruritus, pain, irritation, burning, vaginitis, vaginal discharge, dysmenorrhea, menorrhagia, dysuria, urinary frequency. **Skin:** Pruritus, rash. **Other:** Cartilage erosion.

DIAGNOSTIC TEST INTERFERENCE May cause false positive on *opiate screening tests.*

INTERACTIONS Drug: Ofloxacin absorption decreased when it is administered with MAGNESIUM- or ALUMINUM-CONTAINING ANTACIDS. Other CATIONS, including **calcium, iron,** and **zinc,** also appear to interfere with ofloxacin absorption. May have additive effect with ANTIDIABETICS.

PHARMACOKINETICS Absorption: 90–98% from GI tract. **Peak:** 1–2 h. **Distribution:** Distributes to most tissues; 50% crosses into CSF with inflamed meninges; 20–32% protein bound; crosses placenta; distributed into breast milk. **Metabolism:** Slightly in liver. **Elimination:** 72–98% in urine within 48 h. **Half-Life:** 5–7.5 h.

NURSING IMPLICATIONS

Assessment & Drug Effects

- Lab tests: Do C&S tests prior to initial dose. Treatment may be implemented pending results.
- Determine history of hypersensitivity reactions to quinolones or other drugs before therapy is started.
- Withhold ofloxacin and notify prescriber at first sign of tendon pain, a skin rash, or other allergic reaction.
- Monitor for seizures, especially in patients with known or suspected CNS disorders. Discontinue ofloxacin and notify prescriber immediately if seizure occurs.
- Assess for signs and symptoms of superinfection (see Appendix F).

Patient & Family Education

- Drink fluids liberally unless contraindicated.
- Be aware that dizziness or lightheadedness may occur; use appropriate caution.
- Avoid excessive sunlight or artificial ultraviolet light because of the possibility of phototoxicity.

O

Common adverse effects in *italic,* life-threatening effects underlined; generic names in **bold;** classifications in SMALL CAPS; ♣ Canadian drug name; ◐ Prototype drug

1091

OLANZAPINE

(o-lan′za-peen)

Zyprexa, Zyprexa Relprevv Zyprexa Zydis
Classification: ATYPICAL ANTIPSYCHOTIC
Therapeutic: ANTIPSYCHOTIC, ANTIMANIC
Prototype: Clozapine
Pregnancy Category: C

AVAILABILITY 2.5 mg, 5 mg, 7.5 mg, 10 mg, 15 mg tablets; 10 mg, 15 mg, 20 mg orally disintegrating tablets; 10 mg powder for injection; 210 mg, 300 mg, 405 mg extended release powder for injection

ACTION & *THERAPEUTIC EFFECT*
Antipsychotic activity is thought to be due to antagonism for both serotonin 5-HT$_{2A/2C}$ and dopamine D$_{1-4}$ receptors. May inhibit the CNS presynaptic neuronal reuptake of serotonin and dopamine. *Has effective antipsychotic activity.*

USES Management of schizophrenia, treatment of bipolar disorder, acute agitation (IM).
UNLABELED USES Alzheimer's dementia, acute psychosis.

CONTRAINDICATIONS Hypersensitivity to olanzapine; abrupt discontinuation, coma, severe CNS depression; dementia-related psychosis in elderly; lactation. **IM form:** Tardive dyskinesia; infants.
CAUTIOUS USE Known cardiovascular disease, neurologic disease, stroke, cerebrovascular disease, Parkinson's disease; history of seizures, conditions that predispose to hypotension (i.e., dehydration, hypovolemia); history of syncope; history of breast cancer; Japanese; diabetes mellitus; BPH; closed-angle glaucoma; paralytic ileus; urinary retention; hepatic or renal impairment, jaundice; predisposition to aspiration pneumonia; history of or high risk for suicide; schizophrenia; elevated triglyceride levels; pregnancy (category C). Safety and effectiveness in children 6 y and younger are not established. **IM form:** Elderly, debilitated or at risk for hypotension.

ROUTE & DOSAGE

Psychotic Disorders

Adult: **PO** Start with 5–10 mg once/day, may increase by 2.5–5 mg qwk until desired response (usual range 10–15 mg/day, max: 20 mg/day) **IM** (extended release) 150–300 mg q2wk or 405 mg q4wk
Adolescent: 2.5–5 mg daily (may increase to 10 mg daily)
Geriatric: **PO** Start with 5 mg once/day **IM** (extended release) 150 mg q4wk

Bipolar Mania

Adult: **PO** Start with 10–15 mg once/day, may increase by 5 mg q24h if needed
Adolescent: 2.5–5 mg daily (may increase to 10 mg daily)

Acute Agitation

Adult: **IM** 10 mg, do not repeat more frequently than q2h (max: 30 mg/24h)
Geriatric: **IM** 2.5–5 mg once

ADMINISTRATION

Oral

- Do not push orally disintegrating tablet through blister foil. Peel foil back and remove tablet.

Tablet will disintegrate with/without liquid.

Intramuscular

- *Short-acting formulation:* Reconstitute with 2.1 mL of sterile water for injection to yield 5 mg/mL. Use within 1 h of reconstitution.
- *Extended-release formulation:* Reconstitute with supplied syringe and diluent. Use gloves and flush with water if skin contact is made. To produce a 150 mg/mL solution for injection add 1.3 mL of diluent to the 210 mg vial; or add 1.8 mL of diluent to the 300 mg vial; or add 2.3 mL of diluent to the 405 mg vial.
- Give deep IM into the gluteal muscle. Do not inject more than 5 mL into one site.

ADVERSE EFFECTS (≥1%) **Body as a Whole:** *Weight gain,* fever, back and chest pain, peripheral and lower extremity edema, joint pain, twitching, premenstrual syndrome. **CNS:** *Somnolence, dizziness, headache, agitation, insomnia, nervousness, hostility,* anxiety, personality disorder, akathisia, hypertonia, tremor amnesia, euphoria, stuttering, extrapyramidal symptoms (dystonic events, *parkinsonism, akathisia),* tardive dyskinesia. **CV:** Postural hypotension, hypotension, tachycardia. **Special Senses:** Amblyopia, blepharitis. **GI:** Abdominal pain, constipation, dry mouth, increased appetite, increased salivation, nausea, vomiting, elevated liver function tests. **Metabolic:** Hyperglycemia, diabetes mellitus. **Urogenital:** Premenstrual syndrome, hematuria, urinary incontinence, metrorrhagia. **Respiratory:** Rhinitis, cough, pharyngitis, dyspnea. **Skin:** Rash.

INTERACTIONS Drug: May enhance hypotensive effects of ANTIHYPERTENSIVES. May enhance effects of other CNS ACTIVE DRUGS, **alcohol. Carbamazepine, omeprazole, rifampin** may increase metabolism and clearance of olanzapine. **Fluvoxamine** may inhibit metabolism and clearance of olanzapine.

PHARMACOKINETICS Absorption: Rapidly from GI tract; 60% reaches systemic circulation. **Onset:** 15 min IM. **Peak:** 6 h. **Distribution:** 93% protein bound, secreted into breast milk of animals (human secretion unknown). **Metabolism:** In liver (CYP1A2). **Elimination:** Approximately 57% in urine, 30% in feces. **Half-Life:** 21–54 h.

NURSING IMPLICATIONS

Assessment & Drug Effects

- Monitor closely cerebrovascular status in elderly patients with dementia-related psychosis.
- Monitor diabetics for loss of glycemic control.
- Withhold drug and immediately report S&S of neuroleptic malignant syndrome (see Appendix F); assess for and report S&S of tardive dyskinesia (see Appendix F).
- Lab tests: Periodically monitor ALT, especially in those with hepatic dysfunction or being treated with other potentially hepatotoxic drugs. Periodic blood glucose monitoring.
- Monitor BP and HR periodically. Monitor temperature, especially under conditions such as strenuous exercise, extreme heat, or treatment with other anticholinergic drugs.
- Monitor for and report orthostatic hypotension, especially during the initial dose-titration period.

Common adverse effects in *italic*, life-threatening effects underlined; generic names in **bold**; classifications in SMALL CAPS; ♣ Canadian drug name; ⓟ Prototype drug

1093

- Monitor for seizures, especially in older adults and cognitively impaired persons.

Patient & Family Education
- Carefully monitor blood glucose levels if diabetic.
- Do not drive or engage in potentially hazardous activities until response to drug is known; drug increases risk of orthostatic hypotension and cognitive impairment.
- Learn common adverse effects and possible drug interactions.
- Avoid alcohol and do not take additional medications without informing prescriber.
- Do not become overheated; avoid conditions leading to dehydration.

OLMESARTAN MEDOXOMIL
(ol-me-sar′tan)

Benicar
Classification: ANGIOTENSIN II RECEPTOR (TYPE AT$_1$) ANTAGONIST, ANTIHYPERTENSIVE
Therapeutic: ANTIHYPERTENSIVE
Prototype: Losartan
Pregnancy Category: C first trimester; D second and third trimester

AVAILABILITY 5 mg, 20 mg, 40 mg tablets

ACTION & *THERAPEUTIC EFFECT*
Angiotensin II receptor (type AT$_1$) antagonist acts as a potent vasodilator and primary vasoactive hormone of the renin-angiotensin-aldosterone system. Selectively blocks the binding of angiotensin II to the AT$_1$ receptors found in many tissues (e.g., vascular smooth muscle, adrenal glands). *Antihypertensive effect is due to its potent vasodilation effect.*

USES Treatment of hypertension.

CONTRAINDICATIONS Hypersensitivity to pimecrolimus or components in the cream; Netherton's syndrome; application to active cutaneous viral infection; pregnancy (category D second and third trimester); lactation.
CAUTIOUS USE Renal artery stenosis; heart failure; severe renal impairment; hypovolemia; pregnancy (category C first trimester); children.

ROUTE & DOSAGE

Hypertension
Adult: **PO** 20 mg daily, may increase to 40 mg daily. Start with 5–10 mg daily if volume depleted.

ADMINISTRATION

Oral
- Determine if patient is volume depleted (e.g., patients treated with diuretics) prior to first administration of drug. If volume depletion is suspected, a lower starting dose is recommended.
- Store at 20°–25° C (68°–77° F).

ADVERSE EFFECTS (≥1%) **Body as a Whole:** Back pain, flu-like symptoms. **CNS:** Headache. **CV:** Hypotension (especially if dehydrated). **GI:** Diarrhea. **Metabolic:** Increased CPK, hyperglycemia, hypertriglyceridemia. **Respiratory:** Bronchitis, pharyngitis, rhinitis, sinusitis, upper respiratory infection. **Urogenital:** Hematuria.

INTERACTIONS Drug: May increase hypotensive effect of other ANTIHYPERTENSIVES; may cause hyperkalemia with POTASSIUM-SPARING DIURETICS, POTASSIUM SUPPLEMENTS; increase risk of lithium toxicity. **Herbal: Ephedra, ma huang** may antagonize antihypertensive effects.

PHARMACOKINETICS Absorption: Rapidly absorbed, 26% reaches systemic circulation. **Peak:** 1–2 h. **Distribution:** 99% protein bound. **Metabolism:** Not metabolized by CYP 450 system. **Elimination:** 50% in urine, 50% in feces. **Half-Life:** 13 h.

NURSING IMPLICATIONS

Assessment & Drug Effects

- Monitor closely any volume-depleted patient following initial drug doses. If serious hypotension occurs, place patient in supine position and notify prescriber immediately.
- Monitor BP and HR at drug trough (prior to a scheduled dose). Report hypotension or bradycardia.
- Monitor drug effectiveness, especially in African-Americans, when olmesartan is used as monotherapy.
- Lab tests: Monitor baseline and periodic renal functions; monitor CBC, electrolytes, and LFTs with long-term therapy.

Patient & Family Education

- Discontinue drug and notify prescriber if you experience swelling of the face, tongue, or throat, or if you believe you are pregnant.
- Notify prescriber of symptoms of hypotension (e.g., dizziness, fainting).

OLOPATADINE HYDROCHLORIDE

(o-lo-pa'ta-deen)

Patase, Pataday, Patanol
Pregnancy Category: C
See Appendix A-1.

OLSALAZINE SODIUM

(ol-sal'a-zeen)

Dipentum
Classification: ANTI-INFLAMMATORY
Therapeutic: GI ANTI-INFLAMMATORY
Prototype: Mesalamine
Pregnancy Category: C

AVAILABILITY 250 mg capsules

ACTION & *THERAPEUTIC EFFECT*
Converted to 5-aminosalicylic acid (5-ASA) by colonic bacteria. The 5-ASA is absorbed slowly, resulting in a very high local concentration in the colon. 5-ASA inhibits prostaglandin production in the colon, thus leading to its anti-inflammatory properties. *5-ASA has anti-inflammatory activity in ulcerative colitis.*

USES Maintenance therapy in patients with ulcerative colitis.

CONTRAINDICATIONS Hypersensitivity to salicylates or 5-ASA.

CAUTIOUS USE Patients with preexisting kidney disease; elderly; colitis; pregnancy (category C), lactation. Safety and effectiveness in children are not established.

ROUTE & DOSAGE

Ulcerative Colitis
Adult: **PO** 500 mg b.i.d., may increase up to 1.5–3 g/day in 2–4 divided doses

ADMINISTRATION

Oral
- Give with food in two evenly divided doses.

ADVERSE EFFECTS (≥1%) **CNS:** Headache. **GI:** *Diarrhea,* nausea, abdominal pain, indigestion, vomiting, bloating. **Skin:** Rash. **Body as a Whole:** Arthralgia.

Common adverse effects in *italic*, life-threatening effects <u>underlined</u>; generic names in **bold**; classifications in SMALL CAPS; ✤ Canadian drug name; ⊙ Prototype drug

1095

PHARMACOKINETICS Absorption: 1–3% from GI tract; high colonic concentrations are associated with efficacy. **Metabolism:** Prodrug metabolized to 2 molecules of 5-ASA; **Elimination:** Primarily in feces. **Half-Life:** At least 6 h.

NURSING IMPLICATIONS

Assessment & Drug Effects

- Monitor kidney function in patients with preexisting renal disease.
- Monitor for S&S of a hypersensitivity reaction (see Appendix F). Withhold olsalazine and notify prescriber at first sign of an allergic response.

Patient & Family Education

- Report diarrhea, a possible adverse effect, to the prescriber.

OMALIZUMAB

(o-mal-i-zoo′mab)

Xolair
Classification: BIOLOGICAL RESPONSE MODIFIER; MONOCLONAL ANTIBODY; RESPIRATORY ANTI-INFLAMMATORY
Therapeutic: ANTIALLERGIC; ANTIASTHMATIC; ANTI-INFLAMMATORY
Pregnancy Category: B

AVAILABILITY 75 mg, 150 mg vial

ACTION & THERAPEUTIC EFFECT DNA recombinant monoclonal antibody that selectively binds to human IgE. It inhibits binding of IgE to high-affinity IgE receptors on the surface of mast cells and basophils, limiting the release of inflammatory mediators. *Inhibits release of mediators of the allergic response and has an anti-inflammatory action on the respiratory system.*

USES Control of moderate to severe allergic asthma.
UNLABELED USES Seasonal allergic rhinitis, food allergies.

CONTRAINDICATIONS Hypersensitivity to omalizumab; severe infections, including chickenpox and other viral infections; acute bronchospasm, acute asthma, status asthmaticus; malignancies.
CAUTIOUS USE Live vaccines; pregnancy (category B), lactation; children younger than 12 y.

ROUTE & DOSAGE

Allergic Asthma

Adult/Adolescent: **Subcutaneous** 150–375 mg q2–4wk. Dose is based on baseline IgE serum levels.

ADMINISTRATION

Subcutaneous

- Reconstitute as follows: (1) Draw 1.4 mL of sterile water for injection into a 3-mL syringe with a 1-inch, 18-gauge needle. (2) Place vial upright on flat surface and inject sterile water. Keep vial upright and gently swirl for about 1 min to wet powder. Do not shake. (3) Gently swirl vial for 5–10 sec q5min to dissolve remaining solids. Some vials may take longer than 20 min to dissolve. Do not use if not completely dissolved by 40 min. (4) Once dissolved, invert vial for 15 sec to allow solution to drain toward stopper. (5) Using a new 3-mL syringe with a 1-inch, 18-gauge needle, insert needle into inverted vial with tip at the very bottom of solution, then withdraw solution. Before removing needle from vial, pull the plunger to end of syringe barrel to remove all solution from inverted vial. (6) Replace 18-gauge needle with a 25-gauge needle for subcutaneous injection. (7) Expel air, large bubbles, and any excess solution to obtain the required 1.2 mL dose. A thin layer of small bubbles may remain at top of the solution in syringe.

- Give subcutaneously and rotate injection sites. Solution is viscous and takes 5–10 sec to inject.
- Use within 8 h of reconstitution when stored in the vial at 2°–8° C (36°–46° F), or within 4 h of reconstitution when stored at room temperature.

ADVERSE EFFECTS (≥1%) **Body as a Whole:** <u>Anaphylaxis/anaphylactoid reactions</u>, *injection site reactions (bruising, erythema, warmth, burning, stinging, pruritus, hive formation, pain, induration, inflammation)*, fatigue, generalized pain. **CNS:** Headache, dizziness. **GI:** *Nausea, vomiting, diarrhea, abdominal pain.* **Hematologic:** Epistaxis, menorrhagia, hematoma, anemia. **Musculoskeletal:** Arthralgia. **Respiratory:** Upper respiratory tract infections, sinusitis, pharyngitis. **Skin:** Rash, pruritus, urticaria, dermatitis. **Special Senses:** Earache.

PHARMACOKINETICS Absorption: Slowly absorbed from subcutaneous site; 53–71% reaches systemic circulation. **Peak:** 7–8 days. **Half-Life:** 22 days.

NURSING IMPLICATIONS

Assessment & Drug Effects
- Monitor for injection site reactions including bruising, redness,

warmth, burning, stinging, itching, hive formation, pain, indurations, mass, and inflammation.
- Lab test: Platelet counts if signs of increased tendency to bleed appear.

Patient & Family Education
- Do not use this drug for relief of acute bronchospasm or status asthmaticus.
- Promptly report any of the following: Bleeding or unusual bruising, difficulty breathing or shortness of breath, skin rash or hives.
- Do not accept a live virus vaccine without consulting prescriber.

OMEGA-3 FATTY ACIDS (EICOSAPENTAENOIC ACID AND DOCOSAHEXAENOIC ACID) EPA & DHA

(o-me′ga-3)

Dr. Sears OmegaRx, Eskimo-3, Fish Oil, Omega-3 Fatty Acids, ICAR Prenatal Essential Omega-3, Mega Twin EPA, Natrol DHA Neuromins, Natrol Omega-3, Natural Fish Oil, Oleomed Heart, Omacor, Omega-3 Fish Oil Concentrate, Sea Omega, ZonePerfect Omega 3

Classifications: NUTRITIONAL SUPPLEMENT; OMEGA-3 FATTY ACIDS

Therapeutic: OMEGA-3 FATTY ACIDS; ANTILIPEMIC

Pregnancy Category: C

AVAILABILITY 100 mg, 200 mg, 300 mg, 360 mg, 375 mg, 500 mg, 517 mg, 840 mg, 1000 mg, 1760 mg capsules; 900 mg/5 mL and 1800 mg/5 mL oil for oral ingestion

ACTION & *THERAPEUTIC EFFECT* Mechanism of action of

Common adverse effects in *italic*, life-threatening effects <u>underlined</u>; generic names in **bold;** classifications in SMALL CAPS; ♣ Canadian drug name; ⊙ Prototype drug

1097

omega-3-acid ethyl esters is not completely understood. May include inhibition of acetyl-CoA and increased peroxisomal beta-oxidation in the liver. *Triglyceride lowering is the most consistent effect observed.*

USES Adjunct to diet to reduce hypertriglyceridemia.
UNLABELED USES Adjunct nutritional supplementation for hypertriglyceridemia, rheumatoid arthritis, or for the general purpose of maintaining a healthy heart.

CONTRAINDICATIONS Hypersensitivity to any component of the medication.
CAUTIOUS USE Known sensitivity or allergy to fish; pregnancy (category C), lactation; infants.

ROUTE & DOSAGE

Hypertriglyceridemia (Lovaza Rx form)
Adult: **PO** 4 g daily (as single or divided dose)

ADMINISTRATION

Oral
- The daily dose may be given as one dose or divided b.i.d.
- Store 15°–30° C (59°–86° F).

ADVERSE EFFECTS (≥1%) **Body as a Whole:** Back pain, flu syndrome, unspecified pain. **GI:** Diarrhea, dyspepsia, eructation, nausea, vomiting. **Metabolic/Nutritional:** Increased total cholesterol and/or LDL levels, weight gain. **Skin:** Rash. **Special Senses:** Halitosis, taste disturbances.

INTERACTIONS Drug: ANTICOAGULANTS and THROMBOLYTICS are affected by inhibition of platelet aggregation with omega-3 fatty acids.

PHARMACOKINETICS Metabolism: Extensive liver metabolism.

NURSING IMPLICATIONS

Assessment & Drug Effects
- Monitor for S&S of hypersensitivity in those with known allergy to fish.
- Monitor diabetics for loss of glycemic control.
- Lab tests: Baseline and periodic lipid profile.
- Note: Poor therapeutic response after 2 mo is an indication to discontinue drug.
- Monitor blood levels of anticoagulants with concurrent therapy.

Patient & Family Education
- Do not take omega-3 fatty acids without consulting prescriber if you have a chronic medical disorder.

OMEPRAZOLE 🅟
(o-me′pra-zole)

Losec ✦, Prilosec, Prilosec OTC, Zegerid
Classifications: PROTON PUMP INHIBITOR; ANTISECRETORY
Therapeutic: ANTIULCER
Pregnancy Category: C

AVAILABILITY 10 mg, 20 mg, 40 mg capsules; 2.5 mg, 10 mg powder for oral suspension; 20 mg delayed release tablet

ACTION & *THERAPEUTIC EFFECT*
An antisecretory compound that is a gastric acid pump inhibitor. Suppresses gastric acid secretion by inhibiting the H+, K+-ATPase enzyme system [the acid (proton H+) pump] in the parietal cells. *Suppresses gastric acid secretion relieving gastrointestinal distress and promoting ulcer healing.*

Common adverse effects in *italic*, life-threatening effects underlined; generic names in **bold;** classifications in SMALL CAPS; ✦ Canadian drug name; 🅟 Prototype drug

USES Duodenal and gastric ulcer. Gastroesophageal reflux disease including severe erosive esophagitis (4 to 8 wk treatment). Long-term treatment of pathologic hypersecretory conditions such as Zollinger-Ellison syndrome, multiple endocrine adenomas, and systemic mastocytosis. In combination with clarithromycin to treat duodenal ulcers associated with *Helicobacter pylori*. Dyspepsia occurring more than twice weekly.

UNLABELED USES Healing or prevention of NSAID-related ulcers; stress gastritis prophylaxis.

CONTRAINDICATIONS Long-term use for gastroesophageal reflux disease (GERD), duodenal ulcers; proton pump inhibitors (PPIs), hypersensitivity; lactation; use of **Zegerid** in metabolic alkalosis, hypocalcemia, vomiting, GI bleeding.

CAUTIOUS USE Dysphagia; metabolic or respiratory alkalosis; hepatic disease; pregnancy (category C); children younger than 1 y; **OTC form:** Children younger than 18 y.

ROUTE & DOSAGE

Gastroesophageal Reflux, Erosive Esophagitis, Duodenal Ulcer
Adult/Adolescent: **PO** 20–40 mg once/day for 4–8 wk
Gastric Ulcer
Adult: **PO** 20 mg b.i.d. for 4–8 wk
Hypersecretory Disease
Adult: **PO** 60 mg once/day up to 120 mg t.i.d.
Duodenal Ulcer Associated with *H. pylori*
Adult: **PO** 40 mg once/day for 14 days, then 20 mg/day for

14 days, in combination with clarithromycin 500 mg t.i.d. for 14 days
Dyspepsia
Adult: **PO** 20 mg daily × 14 days

ADMINISTRATION

Oral

- Give before food, preferably breakfast; capsules **must be** swallowed whole (do not open, chew, or crush).
- Note: Antacids may be administered with omeprazole.
- *For NG tube administration:* Into a catheter-tipped syringe, empty a 2.5 mg packet of omeprazole spheres into 5 mL of water or a 10 mg packet into 15 mL of water. Immediately shake syringe then allow to thicken for 2–3 min. Shake syringe again, then inject into NG tube.

ADVERSE EFFECTS (≥1%) **CNS:** Headache, dizziness, fatigue. **GI:** Diarrhea, abdominal pain, nausea, mild transient increases in liver function tests. **Urogenital:** Hematuria, proteinuria. **Skin:** Rash.

DIAGNOSTIC TEST INTERFERENCE Omeprazole has been reported to significantly impair peak cortisol response to exogenous ACTH. This finding is undergoing further investigation. May result in false-negative 13C-urea breath test.

INTERACTIONS Drug: May increase **diazepam, phenytoin, warfarin** levels. May affect levels of ANTIRETROVIRAL AGENTS. **Herbal: Ginkgo, St. John's wort** may decrease plasma concentrations. **Food:** Food decreases absorption by up to 35%.

PHARMACOKINETICS Absorption: Poorly from GI tract; 30–40%

Common adverse effects in *italic*, life-threatening effects underlined; generic names in **bold**; classifications in SMALL CAPS; ♣ Canadian drug name; ⊘ Prototype drug

1099

reaches systemic circulation. **Onset:** 0.5–3.5 h. **Peak:** Peak inhibition of gastric acid secretion: 5 days. **Metabolism:** In liver (CYP2C19). **Elimination:** 80% in urine, 20% in feces. **Half-Life:** 0.5–1.5 h.

NURSING IMPLICATIONS

Assessment & Drug Effects
- Lab tests: Monitor urinalysis for hematuria and proteinuria. Periodic LFTs with prolonged use.

Patient & Family Education
- Report any changes in urinary elimination such as pain or discomfort associated with urination, or blood in urine.
- Report severe diarrhea; drug may need to be discontinued.

√ONDANSETRON HYDROCHLORIDE ℗

(on-dan′si-tron)

Zofran, Zofran ODT, Zuplenz
Classifications: 5-HT₃ ANTAGONIST; ANTIEMETIC
Therapeutic: ANTIEMETIC
Pregnancy Category: B

AVAILABILITY 4 mg, 8 mg, 16 mg, 24 mg tablets; 4 mg, 8 mg orally disintegrating tablets; 4 mg/5 mL oral solution; 2 mg/mL, 8 mg/50 mL, 32 mg/50 mL injection; 4 mg, 8 mg oral soluble film

ACTION & *THERAPEUTIC EFFECT*
Selective serotonin (5-HT₃) receptor antagonist. Serotonin receptors are located centrally in the chemoreceptor trigger zone (CTZ) and peripherally on the vagal nerve terminals. Serotonin is released from the wall of the small intestine and stimulates the vagal efferent nerves through the serotonin receptors and initiates the vomiting reflex. *Prevents nausea and vomiting associated with cancer chemotherapy and anesthesia.*

USES Prevention of nausea and vomiting associated with initial and repeated courses of cancer chemotherapy, including high-dose cisplatin; postoperative nausea and vomiting.
UNLABELED USES Treatment of hyperemesis gravidarum

CONTRAINDICATIONS Hypersensitivity to ondansetron.
CAUTIOUS USE Hepatic disease; QT prolongation; PKU; pregnancy (category B), lactation; **PO:** children younger than 4 y.

ROUTE & DOSAGE

Prevention of Chemotherapy-Induced Nausea and Vomiting
Adult/Adolescent: **PO** 8–24 mg 30 min before chemotherapy, then q8h times 2 more doses **IV** 32 mg or three 0.15 mg/kg doses starting 30 min before chemotherapy, then 4 and 8 h after
Adult/Child/Infant (6 mo–18 y): **IV** 0.15 mg/kg infused over 15 min beginning 30 min before start of chemotherapy, then 4 and 8 h after first dose of ondansetron
Child (older than 4 y): **PO** 4 mg 30 min before chemotherapy, then q8h times 2 more doses

Nausea and Vomiting with Highly Emetogenic Chemotherapy
Adult: **PO** Single 24 mg dose 30 min before administration of

single-day highly emetogenic chemotherapy

Postoperative Nausea and Vomiting

Adult: **PO** 8–16 mg 1 h preoperatively **IM/IV** 4 mg injected immediately prior to anesthesia induction or once postoperatively if patient experiences nausea/vomiting shortly after surgery
Child/Infant: **IV** 1 mo–12 y, weight less than 40 kg, 0.1 mg/kg, 1 mo–12 y, weight greater than 40 kg, 4 mg dose

Hepatic Impairment Dosage Adjustment

Child-Pugh class C: Maximum dose 8 mg/day

ADMINISTRATION

Oral

- Give tablets 30 min prior to chemotherapy and 1–2 h prior to radiation therapy.
- Do NOT push orally disintegrating tablet through blister foil. Peel foil back and remove tablet. Tablets will disintegrate with/without liquid.

Intravenous

PREPARE: **Direct for Postoperative N&V:** May be given undiluted. **IV Infusion for Chemotherapy-Induced N&V:** Dilute a single dose in 50 mL of D5W or NS. ▪ May be further diluted in selected IV solution.
ADMINISTER: **Direct for Postoperative N&V:** Give over at least 30 sec, 2–5 min preferred. **IV Infusion for Chemotherapy-Induced N&V:** Give over 15 min.

- When three separate doses are administered, infuse each over 15 min.

INCOMPATIBILITIES **Solution/additive: Meropenem. Y-site: Acyclovir, allopurinol, aminophylline, amphotericin B, amphotericin B cholesteryl, ampicillin, ampicillin/sulbactam, amsacrine, cefepime, fluorouracil, furosemide, ganciclovir, lansoprazole, lorazepam, meropenem, methylprednisolone, pemetrexed, sargramostim, sodium bicarbonate, TPN.**

AVDVERSE EFFECTS (≥1%) **CNS:** Dizziness and light-headedness, *headache, sedation.* **GI:** *Diarrhea,* constipation, dry mouth, transient increases in liver aminotransferases and bilirubin. **Body as a Whole:** Hypersensitivity reactions.

INTERACTIONS Drug: Rifampin may decrease ondansetron levels.

PHARMACOKINETICS Peak: 1–1.5 h. **Metabolism:** In liver (CYP3A4). **Elimination:** 44–60% in urine within 24 h; ~25% in feces. **Half-Life:** 3 h.

NURSING IMPLICATIONS

Assessment & Drug Effects

- Monitor fluid and electrolyte status. Diarrhea, which may cause fluid and electrolyte imbalance, is a potential adverse effect of the drug.
- Monitor cardiovascular status, especially in patients with a history of coronary artery disease. Rare cases of tachycardia and angina have been reported.

Patient & Family Education

- Be aware that headache requiring an analgesic for relief is a common adverse effect.

Common adverse effects in *italic*, life-threatening effects underlined; generic names in **bold;** classifications in SMALL CAPS; ✦ Canadian drug name; ◉ Prototype drug

1101

OPIUM, POWDERED OPIUM TINCTURE (LAUDANUM)

(oh'pee-um)

Deodorized Opium Tincture
Classifications: NARCOTIC (OPIATE AGONIST) ANALGESIC; ANTI-DI-ARRHEAL
Therapeutic: NARCOTIC ANALGESIC; ANTIDIARRHEAL
Prototype: Morphine
Pregnancy Category: C
Controlled Substance: Schedule II

AVAILABILITY 10%, 2 mg/5 mL liquid

ACTION & *THERAPEUTIC EFFECT*
Contains several natural alkaloids including morphine, codeine, papaverine. Antidiarrheal due to inhibition of GI motility and propulsion; leads to prolonged transit of intestinal contents, desiccation of feces, and constipation. *Antidiarrheal activity due to inhibition of GI motility.*

USES Symptomatic treatment of acute diarrhea and to treat severe withdrawal symptoms in neonates born to women addicted to opiates.

CONTRAINDICATIONS Diarrhea caused by poisoning (until poison is completely eliminated).
CAUTIOUS USE History of opiate agonist dependence; asthma; severe prostatic hypertrophy; hepatic disease; pregnancy (category C), lactation.

ROUTE & DOSAGE

Acute Diarrhea
Adult: **PO** 0.6 mL q.i.d. up to 1 mL q.i.d. (max: 6 mL/day)

Child: **PO** 0.005–0.01 mL/kg q3–4h (max: 6 doses/24 h)
Neonatal Withdrawal
Child: **PO** Make a 1:25 aqueous dilution, then give 3–6 drops q3–6h as needed or 0.2 mL q3h, may increase by 0.05 mL q3h until withdrawal symptoms are controlled, then gradually decrease dose after withdrawal symptoms have stabilized

ADMINISTRATION

Oral

- Do not confuse this preparation with camphorated opium tincture (paregoric), which contains only 2 mg anhydrous morphine/5 mL, thus requiring a higher dose volume than that required for therapeutic dose of Deodorized Opium Tincture.
- Give drug diluted with about one third glass of water to ensure passage of entire dose into stomach.
- Store in tight, light-resistant containers.

ADVERSE EFFECTS (≥1%) GI: Nausea and other GI disturbances. **CNS:** Depression of CNS.

INTERACTIONS Drug: Alcohol and other CNS DEPRESSANTS add to CNS effects.

PHARMACOKINETICS Absorption: Variable absorption from GI tract. **Distribution:** Crosses placenta; distributed into breast milk. **Metabolism:** In liver. **Elimination:** In urine.

NURSING IMPLICATIONS

Assessment & Drug Effects
- Withhold medication and report to prescriber if respirations are

12/min or below or have changed in character and rate.

- Discontinue as soon as diarrhea is controlled; note character and frequency of stools.
- Offer small amounts of fluid frequently but attempt to maintain 3000–4000 mL fluid total in 24 h.
- Monitor body weight, I&O ratio and pattern, and temperature. If patient develops fever of 38.8° C (102° F) or above, electrolyte and hydration levels may need to be evaluated. Consult prescriber.

Patient & Family Education

- Be aware that constipation may be a consequence of antidiarrheal therapy but that normal habit pattern usually is reestablished with resumption of normal dietary intake.
- Note: Addiction is possible with prolonged use or with drug abuse.

OPRELVEKIN
(o-prel've-kin)

Neumega
Classifications: BLOOD FORMER; HEMATOPOIETIC GROWTH FACTOR
Therapeutic: HEMATOPOIETIC GROWTH FACTOR
Prototype: Epoetin alfa
Pregnancy Category: C

AVAILABILITY 5 mg injection

ACTION & THERAPEUTIC EFFECT
Hematopoietic growth factor (interleukin-11) that is produced by recombinant DNA. *Effectiveness indicated by return of postnadir platelet count toward normal (50,000 or higher). Increases platelet count in a dose-dependent manner.*

USES Prevention of severe thrombocytopenia following myelosuppressive chemotherapy (not effective after myeloablative chemotherapy).

CONTRAINDICATIONS Hypersensitivity to oprelvekin; myeloablative chemotherapy; myeloid malignancies; lactation.

CAUTIOUS USE Left ventricular dysfunction, cardiac disease, CHF, history of atrial arrhythmias, or other arrhythmias; electrolyte imbalance, hypokalemia; respiratory disease; papilledema; thromboembolic disorders; older adults; cerebrovascular disease, stroke, TIAs; pleural effusion, pericardial effusion, ascites; ICP, brain tumor, visual disturbances; hepatic or renal dysfunction; pregnancy (category C); children.

ROUTE & DOSAGE

Thrombocytopenia

Adult: **Subcutaneous** 50 mcg/kg once daily starting 6–24 h after completing chemotherapy and continuing until platelet count is 50,000 cells/mcL or higher or up to 21 days
Child/Infant (8 mo–17 y): **Subcutaneous** 75–100 mcg/kg once daily starting 6–24 h after completing chemotherapy and continuing until platelet count is 50,000 cells/mcL or higher or up to 21 days

ADMINISTRATION

- Note: Do not use if solution is discolored or if it contains particulate matter.

Subcutaneous

- Reconstitute solution by gently injecting 1 mL of sterile water for injection (without preservative) toward the sides of the vial. Keep needle in vial and gently swirl to dissolve but do not shake solution. Without removing needle, withdraw specified amount of oprelvekin for injection.

Common adverse effects in *italic*, life-threatening effects underlined; generic names in **bold**; classifications in SMALL CAPS; ♣ Canadian drug name; ⊙ Prototype drug

1103

- Give as single dose into the abdomen, thigh, hip, or upper arm.
- Discard any unused portion of the vial. It contains no preservatives.
- Use reconstituted solution within 3 h; store at 2°–8° C (36°–46° F) until used.
- Store unopened vials at 2°–8° C (36°–46° F). Do not freeze.

ADVERSE EFFECTS (≥1%) **Body as a Whole:** *Edema, neutropenic fever, fever,* asthenia, pain, chills, myalgia, bone pain, dehydration. **CNS:** *Headache, dizziness, insomnia,* nervousness. **CV:** *Tachycardia,* vasodilation, palpitations, syncope, atrial fibrillation/flutter, peripheral edema, capillary leak syndrome. **GI:** *Nausea, vomiting, mucositis, diarrhea,* oral moniliasis, anorexia, constipation, dyspepsia. **Hematologic:** Ecchymosis. **Respiratory:** *Dyspnea, rhinitis, cough, pharyngitis,* pleural effusion, pulmonary edema, exacerbation of preexisting pleural effusion. **Skin:** Alopecia, *rash,* skin discoloration, exfoliative dermatitis. **Special Senses:** Conjunctival injection, amblyopia.

INTERACTIONS Drug: No clinically significant interactions established.

PHARMACOKINETICS Absorption: 80% from subcutaneous injection site. **Onset:** Days 5–9. **Duration:** 7 days after last dose. **Distribution:** Distributes to highly perfused organs. **Elimination:** In urine. **Half-Life:** 6.9 h.

NURSING IMPLICATIONS

Assessment & Drug Effects
- Lab tests: Monitor platelet counts until adequate recovery; periodically monitor CBC with differential and serum electrolytes.
- Monitor carefully for and immediately report S&S of fluid overload, hypokalemia, and cardiac arrhythmias.

- Monitor persons with preexisting fluid retention carefully (e.g., CHF, pleural effusion, ascites) for worsening of symptoms.

Patient & Family Education
- Review patient information leaflet with special attention to administration directions.
- Report any of the following to the prescriber: Shortness of breath, edema of arms and/or legs, chest pain, unusual fatigue or weakness, irregular heartbeat, blurred vision.

ORLISTAT
(or'li-stat)

Alli, Xenical
Classifications: ANORECTANT; NON-SYSTEMIC LIPASE INHIBITOR
Therapeutic: ANORECTANT
Prototype: Diethylpropion
Pregnancy Category: B

AVAILABILITY 60 mg, 120 mg capsules

ACTION & *THERAPEUTIC EFFECT*
Nonsystemic inhibitor of gastrointestinal lipase. Reduces intestinal absorption of dietary fat by forming inactive enzymes with pancreatic and gastric lipase in the GI tract. *Indicated by weight loss/decreased body mass index (BMI). Reduces caloric intake in obese individuals.*

USES Weight loss and weight maintenance in conjunction with diet and exercise.

CONTRAINDICATIONS Hypersensitivity to orlistat; malabsorption syndrome; cholestasis; gallbladder disease; hypothyroidism; organic causes of obesity; anorexia nervosa, bulimia nervosa.
CAUTIOUS USE Gastrointestinal diseases including frequent diarrhea; known dietary deficiencies in fat

soluble vitamins (i.e., A, D, E); history of calcium oxalate nephrolithiasis or hyperoxaluria; older adults; pregnancy (category B), lactation; children younger than 12 y.

ROUTE & DOSAGE

Weight Loss

Adult/Adolescent (older than 12 y): **PO** 60–120 mg t.i.d. with each main meal containing fat

ADMINISTRATION

Oral

- Give during or up to 1 h after a meal containing fat.
- Omit dose with nonfat-containing meal or if meal is skipped.
- Store at 15°–30° C (59°–86° F). Keep bottle tightly closed; do **NOT** use after the printed expiration date.

ADVERSE EFFECTS (≥1%) **Body as a Whole:** Fatigue. **CNS:** *Headache, dizziness, anxiety.* **CV:** Hypertension, stroke. **GI:** *Oily spotting, flatus with discharge, fecal urgency, fatty/oily stool, oily evacuation, increased defecation,* fecal incontinence, *abdominal pain/discomfort,* nausea, infectious diarrhea, rectal pain/discomfort, tooth disorder, gingival disorder, vomiting. **Skin:** Rash. **Urogenital:** Menstrual irregularity.

DIAGNOSTIC TEST INTERFERENCE

Monitor PT/INR in patients on chronic stable doses of **warfarin.**

INTERACTIONS Drug: Orlistat may increase absorption of **pravastatin;** may decrease absorption of fat soluble VITAMINS (A, D, E, K), **amiodarone, cyclosporine.**

PHARMACOKINETICS Absorption: Minimal. **Metabolism:** In gastrointestinal wall. **Elimination:** In feces. **Half-Life:** 1–2 h.

NURSING IMPLICATIONS

Assessment & Drug Effects

- Monitor weight and BMI; closely monitor diabetics for hypoglycemia.
- Coadministered drugs: Monitor PT/INR with warfarin.
- Monitor BP frequently, especially with preexisting hypertension.

Patient & Family Education

- Take a daily multivitamin containing fat-soluble vitamins at least 2 h before/after orlistat.
- Remember common GI adverse effects typically resolve after 4 wk therapy.
- Avoid high-fat meals to minimize adverse GI effects. Distribute fat calories over three main meals daily.
- Monitor weight several times weekly. Diabetics: Monitor blood glucose carefully following any weight loss.

ORPHENADRINE CITRATE

(or-fen'a-dreen)

Norflex

Classification: CENTRAL ACTING SKELETAL MUSCLE RELAXANT

Therapeutic: SKELETAL MUSCLE RELAXANT

Prototype: Cyclobenzaprine

Pregnancy Category: C

AVAILABILITY 100 mg sustained release tablets; 30 mg/mL injection

ACTION & *THERAPEUTIC EFFECT*

Tertiary amine anticholinergic and central-acting skeletal muscle relaxant. Relaxes tense skeletal muscles indirectly, possibly by analgesic action or by atropine-like central action. *Relieves skeletal muscle spasm.*

USES To relieve muscle spasm discomfort associated with acute musculoskeletal conditions.

Common adverse effects in *italic*, life-threatening effects underlined; generic names in **bold**; classifications in SMALL CAPS; ♣ Canadian drug name; ⊚ Prototype drug

1105

CONTRAINDICATIONS Narrow-angle glaucoma; achalasia; pyloric or duodenal obstruction, stenosing peptic ulcers; prostatic hypertrophy or bladder neck obstruction, urinary tract obstruction; myasthenia gravis; cardiospasm; tachycardia.

CAUTIOUS USE History of cardiac disease, arrhythmias, coronary insufficiency; asthma; GERD; hepatic disease; renal disease; renal impairment; older adults; pregnancy (category C), lactation; children.

ROUTE & DOSAGE

Muscle Spasm

Adult: **PO** 100 mg b.i.d. **IM/IV** 60 mg q12h, convert to PO therapy as soon as possible

ADMINISTRATION

Oral
- Ensure that sustained release form is not chewed or crushed. It **must be** swallowed whole.

Intramuscular
- Give undiluted deep into a large muscle.

Intravenous

PREPARE: Direct: Give undiluted. Protect from light.
ADMINISTER: Direct: Give at a rate of 60 mg (2 mL) over 5 min with patient in supine position. ▪ Keep supine for 5–10 min post-injection.

ADVERSE EFFECTS (≥1%) **CNS:** *Drowsiness,* weakness, headache, dizziness; mild CNS stimulation (high doses: restlessness, anxiety, tremors, confusion, hallucinations, agitation, tachycardia, palpitation, syncope). **Special Senses:** Increased ocular tension, dilated pupils, blurred vision. **GI:** *Dry mouth,* nausea, vomiting, abdominal cramps, constipation. **Urogenital:** *Urinary hesitancy* or *retention.*

Body as a Whole: Hypersensitivity [pruritus, urticaria, rash, anaphylactic reaction (rare)].

INTERACTIONS Drug: Propoxyphene may cause increased confusion, anxiety, and tremors; may worsen schizophrenic symptoms, or increase risk of tardive dyskinesia with **haloperidol;** additive CNS depressant with ANXIOLYTICS, SEDATIVES, HYPNOTICS, **butorphanol, nalbuphine,** OPIATE AGONISTS, **pentazocine, tramadol, cyclobenzaprine** may increase anticholinergic effects. **Herbal: Valerian, kava** potentiate sedation.

PHARMACOKINETICS Absorption: Readily from GI tract. **Peak:** 2 h. **Duration:** 4–6 h. **Distribution:** Rapidly distributed in tissues; crosses placenta. **Metabolism:** In liver. **Elimination:** In urine. **Half-Life:** 14 h.

NURSING IMPLICATIONS

Assessment & Drug Effects
- Lab tests: Periodic blood, urine, and LFTs with prolonged therapy.
- Report complaints of mouth dryness, urinary hesitancy or retention, headache, tremors, GI problems, palpitation, or rapid pulse to prescriber. Dosage reduction or drug withdrawal is indicated.
- Monitor elimination patterns. Older adults are particularly sensitive to anticholinergic effects (urinary hesitancy, constipation); closely observe.
- Monitor therapeutic drug effect. In the patient with parkinsonism, orphenadrine reduces muscular rigidity but has little effect on tremors. Some reduction in excessive salivation and perspiration may occur, and patient may appear mildly euphoric.

Patient & Family Education

- Relieve mouth dryness by frequent rinsing with clear tepid water, increasing noncaloric fluid intake, sugarless gum, or lemon drops. If these measures fail, a saliva substitute may help.
- Do not drive or engage in potentially hazardous activities until response to drug is known.
- Avoid concomitant use of alcohol and other CNS depressants; these may potentiate depressant effects.

OSELTAMIVIR PHOSPHATE

(o-sel'tam-i-vir)

Tamiflu
Classification: ANTIVIRAL
Therapeutic: ANTIVIRAL
Pregnancy Category: C

AVAILABILITY 30 mg, 45 mg, 75 mg capsule; 6 mg/mL, 12 mg/mL suspension

ACTION & *THERAPEUTIC EFFECT*
Inhibits influenza A and B viral neuraminidase enzyme, preventing the release of newly formed virus from the surface of the infected cells. Inhibits replication of the influenza A and B virus. *Effectiveness indicated by relief of flu symptoms. Prevents viral spread across the mucous lining of the respiratory tract.*

USES Treatment of uncomplicated acute influenza in adults symptomatic for no more than 2 days; prophylaxis of influenza.
UNLABELED USES H1N1 influenza.

CONTRAINDICATIONS Hypersensitivity to oseltamivir; severe hepatic impairment or severe hepatic disease; viral infections other than flu. Safety in immunosuppression has not been established.

CAUTIOUS USE Hereditary fructose intolerance; cardiac disease; COPD, pediatric patients with asthma; mild or moderate hepatic impairment; psychiatric disorders; renal impairment; pregnancy (category C), lactation; infants younger than 1 y or neonates. Safety and efficacy in chronic cardiac/respiratory disease are not established.

ROUTE & DOSAGE

Influenza Treatment

Adult/Adolescent: **PO** 75 mg b.i.d. × 5 days
Child (1–12 y): **PO** Weight greater than 40 kg, 75 mg b.i.d. × 5 days; weight 23–40 kg, 60 mg b.i.d. × 5 days; weight 15–22 kg, 45 mg b.i.d. × 5 days; weight less than 15 kg, 30 mg b.i.d. × 5 days

Influenza Prevention

Adult/Adolescent/Child (older than 1 y): **PO** 75 mg daily × 10 days; begin within 2 days of contact with infected person

Renal Impairment Dosage Adjustment

CrCl less than 30 mL/min: (treatment) 75 mg daily (prophylaxis) 75 mg every other day or 30 mg daily

ADMINISTRATION

Oral

- Give with food to decrease the risk of GI upset.
- Start within 48 h of onset of flu symptoms.
- Take missed dose as soon as possible unless next dose is due within 2 h.
- Store at 15°–30° C (59°–86° F); protect from moisture, keep dry.

Common adverse effects in *italic*, life-threatening effects underlined; generic names in **bold;** classifications in SMALL CAPS; ♣ Canadian drug name; ⊙ Prototype drug

1107

ADVERSE EFFECTS (≥1%) **Body as a Whole:** Fatigue. **CNS:** Dizziness, headache, insomnia, vertigo. **GI:** Nausea, vomiting, diarrhea, abdominal pain. **Respiratory:** Bronchitis, cough.

PHARMACOKINETICS Absorption: Readily absorbed, 75% bioavailable. **Distribution:** 42% protein bound. **Metabolism:** Extensively metabolized to active metabolite oseltamivir carboxylate by liver esterases. **Elimination:** Primarily in urine. **Half-Life:** 1–2 h; oseltamivir carboxylate 6–10 h.

NURSING IMPLICATIONS

Assessment & Drug Effects

- Monitor ambulation in frail and older adult patients due to potential for dizziness and vertigo.
- Monitor children for abnormal behavior such as delirium or self-injury.

Patient & Family Education

- Contact your prescriber regarding the use of this drug in children.

OXACILLIN SODIUM

(ox-a-sill′in)

Bactocill
Classifications: PENICILLIN ANTIBIOTIC; PENICILLINASE-RESISTANT PENICILLIN
Therapeutic: ANTIBIOTIC
Prototype: Penicillin G potassium
Pregnancy Category: B

AVAILABILITY 1 g, 2 g, 10 g injection

ACTION & *THERAPEUTIC EFFECT* Semisynthetic, acid-stable, penicillinase-resistant isoxazolyl penicillin. Oxacillin inhibits final stage of bacterial cell wall synthesis by preferentially binding to specific penicillin-binding proteins (PBPs) located within the bacterial cell wall, leading to destruction of the cell wall of the organism. *It is highly active against most penicillinase-producing staphylococci, and is generally ineffective against gram-negative bacteria and methicillin-resistant staphylococci (MRSA).*

USES Infections caused by staphylococci.
UNLABELED USES Catheter-related infections.

CONTRAINDICATIONS Infections caused by staphylococci. Hypersensitivity to penicillins or cephalosporins.
CAUTIOUS USE History of or suspected atopy or allergy (hives, eczema, hay fever, asthma); history of GI disease; hepatic disease; renal disease; pregnancy (category B), lactation (may cause infant diarrhea), premature infants, neonates.

ROUTE & DOSAGE

Staphylococcal Infections

Adult/Adolescent/Child over 40 kg: **IV** 250 mg–1 g q4–6h (max: 12 g/day)
Child (under 40 kg): **IV** 50–100 mg/kg/day divided q4–6h (max: 12 g/day)
Neonate: **IV** 50–100 mg/kg/day divided q6–12h

ADMINISTRATION
Note: The total sodium content (including that contributed by buffer) in each gram of oxacillin is approximately 3.1 mEq or 71 mg.

Intravascular

Note: Verify correct IV concentration and rate of infusion/injection with prescriber before IV administration to neonates, infants, children.

PREPARE: **Direct:** Reconstitute each 500 mg or fraction thereof with 5 mL with sterile water for injection or NS to yield 250 mg/1.5 mL. **Intermittent:** Dilute required dose of reconstituted solution in 50–100 mL of D5W, NS, D5/NS, or LR. **Continuous:** Dilute required dose of reconstituted solution in up to 1000 mL of compatible IV solutions.

ADMINISTER: **Direct:** Give at a rate of 1 g or fraction thereof over 10 min. **Intermittent:** Give over 15–30 min. **Continuous:** Give over 6 h.

INCOMPATIBILITIES **Solution/additive: Cytarabine, verapamil. Y-site: Amphotericin B, caffeine citrate, calcium, dantrolene, diazepam, diazoxide, dobutamine, doxycycline, esmolol, haloperidol, hydralazine, milrinone acetate, minocycline, netilmicin, pentazocine, phenytoin, polymixin B, promethazine, protamine, pyridoxine, quinidine, sodium bicarbonate, succinylcholine, tobramycin, verapamil.**

ADVERSE EFFECTS (≥1%) **Body as a Whole:** Thrombophlebitis (IV therapy), superinfections, wheezing, sneezing, fever, <u>anaphylaxis</u>. **GI:** Nausea, vomiting, flatulence, *diarrhea*, hepatocellular dysfunction (elevated AST, ALT, hepatitis). **Hematologic:** Eosinophilia, <u>leukopenia</u>, <u>thrombocytopenia</u>, granulocytopenia, <u>agranulocytosis</u>; neutropenia (reported in children). **Skin:** Pruritus, rash, urticaria.

Urogenital: Interstitial nephritis, transient hematuria, albuminuria, azotemia (newborns and infants on high doses).

DIAGNOSTIC TEST INTERFERENCE Oxacillin in large doses can cause false-positive *urine protein tests* using sulfosalicylic acid methods.

PHARMACOKINETICS Peak: 30–120 min IM; 15 min IV. **Duration:** 4–6 h IM. **Distribution:** Distributes into CNS with inflamed meninges; crosses placenta; distributed into breast milk, 90% protein bound. **Metabolism:** Enters enterohepatic circulation. **Elimination:** Primarily in urine, some in bile. **Half-Life:** 0.5–1 h.

NURSING IMPLICATIONS

Assessment & Drug Effects

- Ask patient prior to first dose about hypersensitivity reactions to penicillins, cephalosporins, and other allergens.
- Lab tests: Periodic LFTs, CBC with differential, platelet count, and urinalysis.
- Hepatic dysfunction (possibly a hypersensitivity reaction) has been associated with IV oxacillin; it is reversible with discontinuation of drug. Symptoms may resemble viral hepatitis or general signs of hypersensitivity and should be reported promptly: Hives, rash, fever, nausea, vomiting, abdominal discomfort, anorexia, malaise, jaundice (with dark yellow to brown urine, light-colored or clay-colored stools, pruritus).
- Withhold next drug dose and report the onset of hypersensitivity reactions and superinfections (see Appendix F).

Common adverse effects in *italic*, life-threatening effects <u>underlined</u>; generic names in **bold;** classifications in SMALL CAPS; ♣ Canadian drug name; ⊙ Prototype drug

1109

Patient & Family Education

- Take oral medication around the clock; do not miss a dose. Take all of the medication prescribed even if you feel better, unless otherwise directed by prescriber.

OXALIPLATIN
(ox-a-li-pla′tin)
Eloxatin
Classifications: ANTINEOPLASTIC; ALKYLATING AGENT
Therapeutic: ANTINEOPLASTIC
Prototype: Cyclophosphamide
Pregnancy Category: D

AVAILABILITY 5 mg/mL injection

ACTION & *THERAPEUTIC EFFECT*
Oxaliplatin forms inter- and intra-strand DNA cross-links that inhibit DNA replication and transcription. The cytotoxicity of oxaliplatin is cell-cycle nonspecific. *Antitumor activity of oxaliplatin in combination with 5-fluorouracil (5-FU) has antiproliferative activity against colon carcinoma that is greater than either compound alone.*

USES Metastatic cancer of colon and rectum.
UNLABELED USES Non–small-cell lung cancer, non-Hodgkin's lymphoma, ovarian cancer.

CONTRAINDICATIONS History of known allergy to oxaliplatin or other platinum compounds; myelosuppression; pregnancy (category D); lactation.
CAUTIOUS USE Renal impairment, because clearance of ultrafilterable platinum is decreased in mild, moderate, and severe renal impairment; hepatic impairment; older adults; children.

ROUTE & DOSAGE

Metastatic Colon or Rectal Cancer
Adult: **IV** 85 mg/m² infused over 120 min once every 2 wk × 6 mo; adjust for toxicities

Renal Impairment Dosage Adjustment
CrCl less than 19 mL/min: Omit dose or change therapy

ADMINISTRATION

Intravenous
Premedication with an antiemetic is recommended.

PREPARE: **IV Infusion:** NEVER reconstitute with NS or any solution containing chloride. ▪ Reconstitute the 50 mg vial or the 100 mg vial by adding 10 mL or 20 mL, respectively, of sterile water for injection or D5W. ▪ MUST further dilute in 250–500 mL of D5W for infusion.

ADMINISTER: **IV Infusion:** Do NOT use needles or infusion sets containing aluminum parts. ▪ Flush infusion line with D5W before and after administration of any other concomitant medication. ▪ Give over 120 min with frequent monitoring of the IV insertion site. ▪ Discontinue at the first sign of extravasation and restart IV in a different site.

INCOMPATIBILITIES **Solution/additive:** CHLORIDE-CONTAINING SOLUTIONS, ALKALINE SOLUTIONS, including **sodium bicarbonate, 5-fluorouracil (5-FU). Y-site:** ALKALINE SOLUTIONS, including **sodium bicarbonate, diazepam, 5-fluorouracil (5-FU).**

▪ Store reconstituted solution up to 24 h under refrigeration at 2°–8° C (36°–46° F). ▪ After final dilution, the IV solution may be stored for 6 h at room temperature [20°–25° C (68°–77° F)] or up to 24 h under refrigeration.

ADVERSE EFFECTS (≥1%) **Body as a Whole:** *Fever, edema, pain,* allergic reaction, arthralgia, rigors. **CNS:** *Fatigue, neuropathy, headache,* dizziness, insomnia. **CV:** Chest pain. **GI:** *Diarrhea, nausea, vomiting, anorexia, stomatitis, constipation, abdominal pain,* reflux, dyspepsia, taste perversion, mucositis, flatulence. **Hematologic:** *Anemia,* leukopenia, thrombocytopenia, neutropenia, thromboembolism. **Metabolic:** Hypokalemia, dehydration. **Respiratory:** *Dyspnea, cough,* rhinitis, pharyngitis, epistaxis, hiccup. **Skin:** Flushing, rash, alopecia, injection site reaction. **Urogenital:** Dysuria.

INTERACTIONS Drug: AMINOGLYCOSIDES, **amphotericin B, vancomycin,** and other **nephrotoxic drugs** may increase risk of renal failure.

PHARMACOKINETICS Distribution: Greater than 90% protein bound. **Metabolism:** Rapid and extensive nonenzymatic biotransformation. **Elimination:** Primarily in urine. **Half-Life:** 391 h.

NURSING IMPLICATIONS

Assessment & Drug Effects

- Monitor for S&S of hypersensitivity (e.g., rash, urticaria, erythema, pruritus; rarely, bronchospasm and hypotension). Discontinue drug and notify prescriber if any of these occur.
- Monitor insertion site. Extravasation may cause local pain and inflammation that may be severe and lead to complications, including necrosis.
- Monitor for S&S of coagulation disorders including GI bleeding, hematuria, and epistaxis.

- Monitor for S&S of peripheral neuropathy (e.g., paresthesia, dysesthesia, hypoesthesia in the hands, feet, perioral area, or throat, jaw spasm, abnormal tongue sensation, dysarthria, eye pain, and chest pressure). Symptoms may be precipitated or exacerbated by exposure to cold temperature or cold objects.
- Lab tests: Before each administration cycle, monitor WBC count with differential, hemoglobin, platelet count, and blood chemistries (including ALT, AST, bilirubin, and creatinine). Monitor baseline and periodic renal functions.
- Do not apply ice to oral mucous membranes (e.g., mucositis prophylaxis) during the infusion of oxaliplatin as cold temperature can exacerbate acute neurological symptoms.

Patient & Family Education

- Use effective methods of contraception while receiving this drug.
- Avoid cold drinks, use of ice, and cover exposed skin prior to exposure to cold temperature or cold objects.
- Do not drive or engage in potentially hazardous activities until response to drug is known.
- Report any of the following to a health care provider: Difficulty writing, buttoning, swallowing, walking; numbness, tingling or other unusual sensations in extremities; non-productive cough or shortness of breath; fever, particularly if associated with persistent diarrhea or other evidence of infection.
- Report promptly S&S of a bleeding disorder such as black tarry stool, coke-colored or frankly bloody urine, bleeding from the nose or mucous membranes.

Common adverse effects in *italic*, life-threatening effects underlined; generic names in **bold**; classifications in SMALL CAPS; ✚ Canadian drug name; ⊙ Prototype drug

1111

OXANDROLONE

(ox-an'dro-lone)

Oxandrin
Classification: ANDROGEN/ANA-BOLIC STEROID
Therapeutic: ANABOLIC STEROID
Prototype: Testosterone
Pregnancy Category: X
Controlled Substance: Schedule III

AVAILABILITY 2.5 mg tablets

ACTION & *THERAPEUTIC EFFECT*
Synthetic steroid with anabolic and androgenic activity. *Androgenic activity:* Responsible for the growth spurt of the adolescent and for growth termination by epiphyseal closure. Increases erythropoiesis, possibly by stimulating production of erythropoietin, and promotes vascularization and darkening of skin. *Anabolic activity:* Increases protein metabolism and decreases its catabolism. Large doses suppress spermatogenesis, thereby causing testicular atrophy. Controls development and maintenance of secondary sexual characteristics. *Antagonizes effects of estrogen excess on female breast and endometrium. Responsible for the growth spurt of the adolescent male and onset of puberty.*

USES Adjunctive therapy to promote weight gain, offset protein catabolism associated with prolonged administration of corticosteroids, relieve bone pain accompanying osteoporosis.

CONTRAINDICATIONS Hypersensitivity or toxic reactions to androgens; severe cardiac, hepatic, or renal disease; possibility of virilization of external genitalia of female fetus; polycythemia; hypercalcemia; known or suspected prostatic or breast cancer in males; benign prostatic hypertrophy with obstruction; patients easily stimulated sexually; asthenic males who may react adversely to androgenic overstimulation; conditions aggravated by fluid retention; hypertension; pregnancy (category X), lactation.

CAUTIOUS USE Cardiac, hepatic, and mild to moderate renal disease, hypercholesterolemia, heart failure, peripheral edema, arteriosclerosis, coronary artery disease, MI; cholestasis; DM; BPH; prepubertal males, acute intermittent porphyria; older adults.

ROUTE & DOSAGE

Weight Gain
Adult: **PO** 2.5 mg b.i.d. to q.i.d. (max: 20 mg/day) for 2–4 wk
Child: **PO** 0.1 mg/kg/day

ADMINISTRATION
Oral
- Individualize doses; great variations in response exist.
- Store at 15°–30° C (59°–86° F).

ADVERSE EFFECTS (≥1%) **CNS:** Habituation, excitation, insomnia, depression, changes in libido. **Urogenital:** *Males:* Phallic enlargement, increased frequency or persistence of erections, inhibition of testicular function, testicular atrophy, oligospermia, impotence, chronic priapism, epididymitis, bladder irritability; *Females:* Clitoral enlargement, menstrual irregularities. **Hepatic:** Cholestatic jaundice with or without hepatic necrosis and death, hepatocellular neoplasms, peliosis hepatitis (long-term use). **Skin:** Hirsutism and male pattern baldness in females, acne. **Endocrine:** Gynecomastia, deepening of voice in females, premature closure of epiphyses in children, edema, decreased glucose tolerance.

DIAGNOSTIC TEST INTERFERENCE
May decrease levels of thyroxine-binding globulin (decreased total T_4 and increased T_3 RU and free T_4).

INTERACTIONS Drug:
May increase INR with **warfarin.** May inhibit metabolism of ORAL HYPOGLYCEMIC AGENTS. Concomitant STEROIDS may increase edema. **Herbal: Echinacea** may increase risk of hepatotoxicity.

PHARMACOKINETICS Half-Life:
10–13 h (increased in elderly patients).

NURSING IMPLICATIONS

Assessment & Drug Effects
- Monitor weight closely throughout therapy.
- Assess for and report development of edema or S&S of jaundice (see Appendix F).
- Lab tests: Monitor periodically LFTs, lipid profile, Hct and Hgb, PT and INR, serum electrolytes, and CPK.
- Withhold and notify prescriber if hypercalcemia develops in breast cancer patient.
- Monitor growth in children closely.

Patient & Family Education
- Women: Report signs of virilization, including acne and changes in menstrual periods.
- Men: Report too frequent or prolonged erections or appearance/worsening of acne.
- Report S&S of jaundice (see Appendix F) or edema.
- Monitor blood glucose for loss of glycemic control if diabetic.

OXAPROZIN
(ox-a-pro′zin)

Daypro
Classification: ANALGESIC, NONSTEROIDAL ANTI-INFLAMMATORY DRUG (NSAID)
Therapeutic: NONNARCOTIC ANALGESIC, NSAID; ANTIRHEUMATIC; ANTIPYRETIC
Prototype: Ibuprofen
Pregnancy Category: C first and second trimester; D third trimester

AVAILABILITY 600 mg tablets

ACTION & *THERAPEUTIC EFFECT*
Long-acting NSAID agent, which is an effective prostaglandin synthetase inhibitor. It inhibits COX-1 and COX-2 enzymes needed for prostaglandin synthesis at the site of inflammation. *Has anti-inflammatory, antipyretic, and analgesic properties.*

USES Treatment of osteoarthritis and rheumatoid arthritis.
UNLABELED USES Ankylosing spondylitis, chronic pain, gout, oral surgery pain, temporal arteritis, tendinitis.

CONTRAINDICATIONS Hypersensitivity to oxaprozin or any other NSAID; complete or partial syndrome of nasal polyps; angioedema; CABG perioperative pain; pregnancy (category D third trimester); lactation.
CAUTIOUS USE History of GI bleeding, alcoholism, smoking; history of severe hepatic dysfunction, renal insufficiency; cardiac disease; coagulopathy; photosensitivity; older adults; pregnancy (category C first and second trimester). Safety and effectiveness in children younger than 6 y are not established.

ROUTE & DOSAGE

Osteoarthritis, Rheumatoid Arthritis
Adult: **PO** 600–1200 mg daily (max: 1800 mg/day or 25 mg/kg, whichever is lower)

Common adverse effects in *italic,* life-threatening effects <u>underlined</u>; generic names in **bold;** classifications in SMALL CAPS; ♣ Canadian drug name; ⊘ Prototype drug

1113

ADMINISTRATION

Oral

- Give with meals or milk to decrease GI distress.
- Divide doses in those unable to tolerate once-daily dosing.
- Use lower starting doses for those with renal or hepatic dysfunction, advanced age, low body weight, or a predisposition to GI ulceration.

ADVERSE EFFECTS (≥1%) **CNS:** Tinnitus, headache, insomnia, somnolence. **GI:** Diarrhea, abdominal pain, nausea, dyspepsia, flatulence, melena, ulcers, constipation, dry mouth, gastritis. **Skin:** Rash, pruritus. **Urogenital:** Dysuria, urinary frequency.

DIAGNOSTIC TEST INTERFERENCE May cause false-positive reactions for BENZODIAZEPINES with *urine drug-screening* tests.

INTERACTIONS Drug: May attenuate the antihypertensive response to DIURETICS. NSAIDS increase the risk of **methotrexate** or **lithium** toxicity. May increase **aspirin** toxicity. **Herbal: Feverfew, garlic, ginger, ginkgo** may increase risk of bleeding.

PHARMACOKINETICS Absorption: Readily from GI tract. **Peak:** 125 min. **Onset:** 1–6 wk for maximum therapeutic effect. **Distribution:** 99% protein bound. Distributes into synovial fluid, crosses placenta. Distributed into breast milk. **Metabolism:** In the liver. **Elimination:** 60% in urine, 30–35% in feces. **Half-Life:** 40 h.

NURSING IMPLICATIONS

Assessment & Drug Effects

- Monitor for S&S of GI bleeding, especially in patients with a history of inflammation or ulceration of upper GI tract, or those treated chronically with NSAIDs.

- Monitor patients with CHF for increased fluid retention and edema. Report rapid weight increases accompanied by edema.
- Lab tests: Perform baseline and periodic evaluation of Hgb, LFTs and kidney function. Auditory and ophthalmologic exams are recommended with prolonged or high-dose therapy.

Patient & Family Education

- Report immediately dark tarry stools, "coffee ground" or bloody emesis, or other GI distress.
- Avoid aspirin or other NSAIDs without explicit permission of prescriber.
- Be aware of the possibility of photosensitivity, which results in a rash on sun-exposed skin.
- Report immediately to prescriber ringing in ears, decreased hearing, or blurred vision.
- Do not exceed ordered dose. The goal of therapy is lowest effective dose.

OXAZEPAM

(ox-a′ze-pam)

Ox-Pam ♣, Serax, Zapex ♣
Classifications: ANXIOLYTIC; SEDATIVE-HYPNOTIC; BENZODIAZEPINE
Therapeutic: ANTIANXIETY; SEDATIVE-HYPNOTIC
Prototype: Lorazepam
Pregnancy Category: D
Controlled Substance: Schedule IV

AVAILABILITY 10 mg, 15 mg, 30 mg capsules; 15 mg tablets

ACTION & *THERAPEUTIC EFFECT*

Benzodiazepine derivative related to lorazepam. Effects are mediated by the inhibitory neurotransmitter GABA, and acts on the thalamic, hypothalamic, and limbic levels of CNS. *Has anxiolytic,*

Common adverse effects in *italic*, life-threatening effects <u>underlined</u>; generic names in **bold**; classifications in SMALL CAPS; ♣ Canadian drug name; ⊙ Prototype drug

sedative, hypnotic, and skeletal muscle relaxant effects.

USES Management of anxiety and tension associated with a wide range of emotional disturbances. Also to control acute withdrawal symptoms in chronic alcoholism.

CONTRAINDICATIONS Hypersensitivity to oxazepam and other benzodiazepines; respiratory depression; psychoses, suicidal ideation; acute alcohol intoxication; acute-angle glaucoma; pregnancy (category D), lactation.
CAUTIOUS USE Impaired kidney and liver function; alcoholism; addiction-prone patients; COPD; history of seizures; history of suicide; mental depression; bipolar disorder; older adult and debilitated patients; children younger than 6 y.

ROUTE & DOSAGE

Anxiety
Adult: **PO** 10–30 mg t.i.d. or q.i.d.
Acute Alcohol Withdrawal
Adult: **PO** 15–30 mg t.i.d. or q.i.d.

ADMINISTRATION
Oral
- Give with food if GI upset occurs.
- Store in tightly closed container at 15°–30° C (59°–86° F) unless otherwise specified.

ADVERSE EFFECTS (≥1%) **CNS:** *Drowsiness,* dizziness, mental confusion, vertigo, ataxia, headache, lethargy, syncope, tremor, slurred speech, paradoxic reaction (euphoria, excitement). **GI:** Nausea, xerostomia, jaundice. **Skin:** Skin rash, edema. **CV:** Hypotension, edema. **Hematologic:** <u>Leukopenia</u>. **Urogenital:** Altered libido.

INTERACTIONS Drug: Alcohol, CNS DEPRESSANTS, ANTICONVULSANTS potentiate CNS depression; **cimetidine** increases oxazepam plasma levels, increasing its toxicity; may decrease antiparkinsonism effects of **levodopa;** may increase **phenytoin** levels; smoking decreases sedative and antianxiety effects. **Herbal: Kava, valerian** may potentiate sedation.

PHARMACOKINETICS Absorption: Readily absorbed from GI tract. **Peak:** 2–3 h. **Distribution:** Crosses placenta; distributed into breast milk. **Metabolism:** In liver. **Elimination:** Primarily in urine, some in feces. **Half-Life:** 2–8 h.

NURSING IMPLICATIONS
Assessment & Drug Effects
- Observe older adult patients closely for signs of overdosage. Report to prescriber if daytime psychomotor function is depressed.
- Monitor for increased signs and symptoms of suicidality.
- Lab tests: Perform LFTs and white blood cell counts on a regular planned basis.
- Note: Excessive and prolonged use may cause physical dependence.

Patient & Family Education
- Report promptly any mild paradoxic stimulation of affect and excitement with sleep disturbances that may occur within the first 2 wk of therapy. Dosage reduction is indicated.
- Consult prescriber before self-medicating with OTC drugs.
- Do not drive or engage in potentially hazardous activities until response to drug is known.
- Do not drink alcoholic beverages while taking oxazepam. The CNS depressant effects of each agent may be intensified.

Common adverse effects in *italic*, life-threatening effects <u>underlined</u>; generic names in **bold;** classifications in SMALL CAPS; ✚ Canadian drug name; ⊘ Prototype drug

1115

- Contact prescriber if you intend to or do become pregnant during therapy about discontinuing the drug.
- Withdraw drug slowly following prolonged therapy to avoid precipitating withdrawal symptoms (seizures, mental confusion, nausea, vomiting, muscle and abdominal cramps, tremulousness, sleep disturbances, unusual irritability, hyperhidrosis).

OXCARBAZEPINE

(ox-car′ba-ze-peen)

Trileptal
Classification: ANTICONVULSANT
Therapeutic: ANTICONVULSANT
Prototype: Carbamazepine
Pregnancy Category: C

AVAILABILITY 150 mg, 300 mg, 600 mg tablets; 300 mg/5 mL suspension

ACTION & *THERAPEUTIC EFFECT*
Anticonvulsant properties may result from blockage of voltage-sensitive sodium channels, which results in stabilization of hyperexcited neural membranes. *Inhibits repetitive neuronal firing, and decreased propagation of neuronal impulses.*

USES Monotherapy or adjunctive therapy in the treatment of partial seizures in adults and children age 4–16 y.

CONTRAINDICATIONS Hypersensitivity to oxcarbazepine.
CAUTIOUS USE Older adults; renal impairment; renal failure; infertility, hyponatremia, SIADH, and drugs associated with SIADH as an adverse effect; pregnancy (category C), lactation; children younger than 4 y.

ROUTE & DOSAGE

Partial Seizures

Adult: **PO** Start with 300 mg b.i.d. and increase by 600 mg/day qwk to 2400 mg/day in 2 divided doses for monotherapy or 1200 mg/day as adjunctive therapy
Child: **PO** *4–16 y,* initiate with 8–10 mg/kg/day divided b.i.d. (max: 600 mg/day), gradually increase weekly to target dose (divided b.i.d.) based on weight: *Weight 20–29 kg,* 900 mg/day; *weight 29.1–39 kg,* 1200 mg/day; *weight greater than 39 kg,* 1800 mg/day

Renal Impairment Dosage Adjustment

CrCl less than 30 mL/min: Initiate at ½ usual starting dose (300 mg b.i.d.)

ADMINISTRATION

Oral

- Initiate therapy at one-half the usual starting dose (300 mg/day) if creatinine clearance less than 30 mL/min.
- Do not abruptly stop this medication; withdraw drug gradually when discontinued to minimize seizure potential.
- Store preferably at 25° C (77° F), but room temperature permitted. Keep container tightly closed.

ADVERSE EFFECTS (≥1%) **Body as a Whole:** *Fatigue,* asthenia, peripheral edema, generalized edema, chest pain, weight gain. **CV:** Hypotension. **GI:** *Nausea, vomiting, abdominal pain,* diarrhea, dyspepsia, constipation, gastritis, anorexia, dry mouth. **Hematologic:** Lymphadenopathy. **Metabolic:**

Hyponatremia. **Musculoskeletal:** Muscle weakness. **CNS:** *Headache, dizziness, somnolence, ataxia, nystagmus, abnormal gait,* insomnia, tremor, nervousness, agitation, abnormal coordination, speech disorder, confusion, abnormal thinking, aggravate convulsions, emotional lability. **Respiratory:** Rhinitis, cough, bronchitis, pharyngitis. **Skin:** Acne, hot flushes, purpura, <u>Stevens-Johnson syndrome, toxic epidermal necrolysis</u>. **Special Senses:** *Diplopia, vertigo, abnormal vision,* abnormal accommodation, taste perversion, ear ache. **Urogenital:** Urinary tract infection, micturition frequency, vaginitis.

INTERACTIONS Drug: **Carbamazepine, phenobarbital, phenytoin, valproic acid, verapamil;** CALCIUM CHANNEL BLOCKERS may decrease oxcarbazepine levels; may increase levels of **phenobarbital, phenytoin;** may decrease levels of **felodipine,** ORAL CONTRACEPTIVES. **Herbal: Ginkgo** may decrease anticonvulsant effectiveness. **Evening primrose oil** may decrease the seizure threshold.

PHARMACOKINETICS Absorption: Rapidly and completely from GI tract. **Peak:** Steady-state levels reached in 2–3 days. **Distribution:** 40% protein bound. **Metabolism:** Extensively metabolized in liver to active 10-monohydroxy metabolite (MHD). **Elimination:** 95% in kidneys. **Half-Life:** 2 h, MHD 9 h.

NURSING IMPLICATIONS

Assessment & Drug Effects

- Monitor for and report S&S of: Hyponatremia (e.g., nausea, malaise, headache, lethargy, confusion); CNS impairment (e.g., somnolence, excessive fatigue, cognitive deficits, speech or language problems, incoordination, gait disturbances).
- Monitor phenytoin levels when administered concurrently.
- Lab tests: Periodic serum sodium, T_4 level; when oxcarbazepine is used as adjunctive therapy, closely monitor plasma level of the concomitant antiepileptic drug during titration of the oxcarbazepine dose.

Patient & Family Education

- Notify prescriber of the following: Dizziness, excess drowsiness, frequent headaches, malaise, double vision, lack of coordination, or persistent nausea.
- Exercise special caution with concurrent use of alcohol or CNS depressants.
- Use caution with potentially hazardous activities and driving until response to drug is known.
- Use or add barrier contraceptive since drug may render hormonal methods ineffective.

OXICONAZOLE NITRATE
(ox-i-con′a-zole)

Oxistat
Classification: AZOLE ANTIFUNGAL
Therapeutic: ANTIFUNGAL
Prototype: Fluconazole
Pregnancy Category: B

AVAILABILITY 1% cream, lotion

ACTION & *THERAPEUTIC EFFECT*
Topical synthetic antifungal agent that presumably works by altering the cellular membrane of the fungi, resulting in increased membrane permeability, secondary metabolic effects, and growth inhibition. *Effective against fungi.*

Common adverse effects in *italic*, life-threatening effects <u>underlined</u>; generic names in **bold;** classifications in SMALL CAPS; ♣ Canadian drug name; ℗ Prototype drug

1117

USES Topical treatment of tinea pedis, tinea cruris, and tinea corporis due to *Trichophyton rubrum* and *Trichophyton mentagrophytes;* also used for cutaneous candidiasis caused by *Candida albicans* and *Candida tropicalis.*

CONTRAINDICATIONS Hypersensitivity to oxiconazole.
CAUTIOUS USE Hypersensitivity to other azole antifungals; pregnancy (category B), lactation.

ROUTE & DOSAGE

Tinea and Other Dermal Infections
Adult: **Topical** Apply to affected area once daily in the evening

ADMINISTRATION
Topical
- Apply cream to cover the affected areas once daily (in the evening).
- Treat tinea corporis and tinea cruris for 2 wk; tinea pedis for 1 mo to reduce the possibility of recurrence.
- Store at 15°–30° C (59°–86° F).

ADVERSE EFFECTS (≥1%) **Skin:** Transient burning and stinging, dryness, erythema, pruritus, and local irritation.

PHARMACOKINETICS Absorption: Less than 0.3% is absorbed systemically.

NURSING IMPLICATIONS
Patient & Family Education
- Use only externally. Do not use intravaginally.
- Discontinue drug and contact prescriber if irritation or sensitivity develops.
- Avoid contact with eyes.
- Contact prescriber if no improvement is noted after the prescribed treatment period.

OXYBUTYNIN CHLORIDE ⊙
(ox-i-byoo′ti-nin)
Ditropan, Ditropan XL, Gelnique, Oxytrol
Classifications: ANTICHOLINERGIC; ANTIMUSCARINIC; GU ANTISPASMODIC
Therapeutic: GU ANTISPASMODIC
Pregnancy Category: B

AVAILABILITY 5 mg tablets; 5 mg, 10 mg sustained release tablets; 5 mg/5 mL syrup; 3.9 mg/day transdermal patch; 10% topical gel

ACTION & *THERAPEUTIC EFFECT* Synthetic tertiary amine that exerts direct antispasmodic action and inhibits muscarinic effects of acetylcholine on smooth muscle of the urinary muscle. *Prominent antispasmodic activity of the urinary muscle.*

USES To relieve symptoms associated with voiding in patients with uninhibited neurogenic bladder and reflex neurogenic bladder. Also has been used to relieve pain of bladder spasm following transurethral surgical procedures.

CONTRAINDICATIONS Hypersensitivity of oxybutynin; narrow-angle glaucoma, myasthenia gravis, partial or complete GI obstruction, gastric retention, paralytic ileus, intestinal atony (especially older adult or debilitated patients), megacolon, severe colitis, GU obstruction, urinary retention, unstable cardiovascular status; lactation. **Extended Release Form:** With renal impairment.
CAUTIOUS USE Older adults; autonomic neuropathy, hiatus hernia

with reflex esophagitis; hepatic or renal dysfunction; urinary infection; hyperthyroidism; CHF, coronary artery disease, hypertension; prostatic hypertrophy; pregnancy (category B); children.

ROUTE & DOSAGE

Overactive Bladder

Adult/Adolescent: **PO** 5 mg 2–4 times/day (max: 20 mg/day) or 5 mg sustained release daily, may increase up to 30 mg/day **Topical** Apply 1 patch twice weekly; or apply contents of 1 package once daily
Child (older than 6 y): **PO** 5 mg b.i.d., not more than three doses/day (immediate release) or 5 mg daily (sustained release)
Geriatric: **PO** 2.5–5 mg b.i.d. (max: 15 mg/day) or 5 mg sustained release daily, may increase up to 30 mg/day
Topical Apply 1 patch twice weekly
Child: **PO** 1–5 y, 0.2 mg/kg b.i.d.–q.i.d.; *older than 5 y,* 5 mg b.i.d. (max: 15 mg/day)

ADMINISTRATION

Oral

- Ensure that sustained release form is not chewed or crushed. It **must be** swallowed whole.

Topical

- Ensure that old patch is removed prior to application of new patch.
- Gel may be applied to abdomen, upper arms/shoulder, and thigh. Squeeze entire contents of the gel packet onto application site. Gently rub into skin until dried. Avoid contact with eyes, nose, open sores, recently shaved skin, and skin with rashes.

ADVERSE EFFECTS (≥1%) **Body as a Whole:** Severe allergic reactions including urticaria, skin rashes, suppression of lactation, decreased sweating, fever. **CNS:** *Drowsiness,* dizziness, weakness, insomnia, restlessness, psychotic behavior (overdosage). **CV:** Palpitations, tachycardia, flushing. **Special Senses:** Mydriasis, *blurred vision,* cycloplegia, increased ocular tension. **GI:** *Dry mouth,* nausea, vomiting, *constipation,* bloated feeling. **Skin:** *Pruritus at application site,* rash, application site reactions, erythema. **Urogenital:** Urinary hesitancy or retention, impotence.

PHARMACOKINETICS Absorption: Diffuses across intact skin. **Onset:** 0.5–1 h. **Peak:** 3–6 h. **Duration:** 6–10 h. **PO:** 96 h transdermal. **Metabolism:** In liver. **Elimination:** Primarily in urine. **Half-Life:** 2–5 h.

NURSING IMPLICATIONS

Assessment & Drug Effects

- Periodic interruptions of therapy are recommended to determine patient's need for continued treatment. Tolerance has occurred in some patients.
- Keep prescriber informed of expected responses to drug therapy (e.g., effect on urinary frequency, urgency, urge incontinence, nocturia, completeness of bladder emptying).
- Monitor patients with colostomy or ileostomy closely; abdominal distention and the onset of diarrhea in these patients may be early signs of intestinal obstruction or of toxic megacolon.

Patient & Family Education

- Do not drive or engage in potentially hazardous activities until response to drug is known.

Common adverse effects in *italic*, life-threatening effects underlined; generic names in **bold;** classifications in SMALL CAPS; ♣ Canadian drug name; ● Prototype drug

1119

- Exercise caution in hot environments. By suppressing sweating, oxybutynin can cause fever and heat stroke.

OXYCODONE HYDROCHLORIDE

(ox-i-koe'done)

OxyContin, Percolone, Endocodone, OxyFAST, Roxicodone, Oxceta

Classifications: NARCOTIC (OPIATE AGONIST); ANALGESIC
Therapeutic: NARCOTIC ANALGESIC
Prototype: Morphine
Pregnancy Category: B; D for prolonged use or use of high doses at term
Controlled Substance: Schedule II

AVAILABILITY 5 mg, 15 mg, 30 mg tablets; **OxyContin:** 10 mg, 20 mg, 40 mg, 80 mg, 160 mg sustained release tablets; 5 mg/5 mL, 20 mg/mL oral solution

ACTION & *THERAPEUTIC EFFECT* Semisynthetic derivative of an opium agonist that binds with stereospecific receptors in various sites of CNS to alter both perception of pain and emotional response to pain. *Active against moderate to moderately severe pain. Appears to be more effective in relief of acute than long-standing pain.*

USES Relief of moderate to moderately severe pain, neuralgia. Relieves postoperative, postextractional, postpartum pain.

CONTRAINDICATIONS Hypersensitivity to oxycodone and principle drugs with which it is combined; bronchial asthma; pregnancy (category D for prolonged use or high doses at term); lactation.

CAUTIOUS USE Alcoholism; renal or hepatic disease; viral infections; Addison's disease; cardiac arrhythmias; chronic ulcerative colitis; history of drug abuse or dependency; gallbladder disease, acute abdominal conditions; head injury, intracranial lesions; hypothyroidism; BPH; respiratory disease; urethral stricture; peptic ulcer or coagulation abnormalities (combination products containing aspirin); older adult or debilitated patients; pregnancy (category B except for prolonged use); children.

ROUTE & DOSAGE

Moderate to Severe Pain

Adult: **PO** 5–10 mg q6h prn; (may titrate up to 10–30 mg q4h prn)
Controlled release product: 10 mg q12h (titrate up)

ADMINISTRATION

Oral

- Ensure that sustained release form is not chewed or crushed. It **must be** swallowed whole.
- Store at 15°–30° C (59°–86° F). Protect from light.

ADVERSE EFFECTS (≥1%) **CNS:** Euphoria, dysphoria, lightheadedness, dizziness, *sedation.* **GI:** Anorexia, nausea, vomiting, *constipation,* jaundice, hepatotoxicity (combinations containing acetaminophen). **Respiratory:** Shortness of breath, respiratory depression. **Skin:** Pruritus, skin rash. **CV:** Bradycardia. **Body as a Whole:** Unusual bleeding or bruising. **Urogenital:** Dysuria, frequency of urination, urinary retention.

DIAGNOSTIC TEST INTERFERENCE *Serum amylase* levels may be elevated because oxycodone causes spasm of sphincter of Oddi. ***Blood***

glucose determinations: False decrease (measured by ***glucose oxidase-peroxidase method***). ***5-HIAA determination:*** False positive with use of ***nitroisonaphthol reagent*** (quantitative test is unaffected).

INTERACTIONS Drug: Alcohol and other CNS DEPRESSANTS add to CNS depressant activity. **Herbal: St. John's wort** may increase sedation.

PHARMACOKINETICS Absorption: Readily from GI tract. **Onset:** 10–15 min. **Peak:** 30–60 min. **Duration:** 4–5 h. **Distribution:** Crosses placenta; distributed into breast milk. **Metabolism:** In liver. **Elimination:** Primarily in urine. **Half-Life:** 3–5 h.

NURSING IMPLICATIONS

Assessment & Drug Effects

- Monitor patient's response closely, especially to sustained-release preparations.
- Consult prescriber if nausea continues after first few days of therapy.
- Note: Light-headedness, dizziness, sedation, or fainting appear to be more prominent in ambulatory than in nonambulatory patients and may be alleviated if patient lies down.
- Evaluate patient's continued need for oxycodone preparations. Psychic and physical dependence and tolerance may develop with repeated use. The potential for drug abuse is high.
- Lab tests: Monitor LFTs and hematologic status periodically in patients on high dosage.
- Be aware that serious overdosage of any oxycodone preparation presents problems associated with a narcotic overdose (respiratory depression, circulatory collapse, extreme somnolence progressing to stupor or coma).

Patient & Family Education

- Do not alter dosage regimen by increasing, decreasing, or shortening intervals between doses. Habit formation and liver damage may result.
- Avoid potentially hazardous activities such as driving a car or operating machinery while using oxycodone preparation.
- Do not drink large amounts of alcoholic beverages while using oxycodone preparations; risk of liver damage is increased.
- Check with prescriber before taking OTC drugs for colds, stomach distress, allergies, insomnia, or pain.
- Inform surgeon or dentist that you are taking an oxycodone preparation before any surgical procedure is undertaken.

OXYMETAZOLINE HYDROCHLORIDE

(ox-i-met-az'oh-leen)

Afrin, Dristan Long Lasting, Duramist Plus, Duration, Nafrine ♣, Neo-Synephrine 12 Hour, Nostrilla, Sinex Long Lasting
Classifications: NASAL PREPARATION; DECONGESTANT
Therapeutic: DECONGESTANT
Prototype: Naphazoline
Pregnancy Category: C

AVAILABILITY 0.025%, 0.05% solution

ACTION & *THERAPEUTIC EFFECT*

Sympathomimetic agent that acts directly on alpha receptors of sympathetic nervous system resulting in relief of nasal congestion. *Constricts smaller arterioles in nasal passages and has prolonged decongestant effect.*

Common adverse effects in *italic*, life-threatening effects underlined; generic names in **bold**; classifications in SMALL CAPS; ♣ Canadian drug name; ✪ Prototype drug

1121

USES Relief of nasal congestion in a variety of allergic and infectious disorders of the upper respiratory tract; used as nasal tampon to facilitate intranasal examination or before nasal surgery. Also used as adjunct in treatment and prevention of middle ear infection by decreasing congestion of eustachian ostia.

CONTRAINDICATIONS Closed-angle glaucoma; lactation.
CAUTIOUS USE Within 14 days of MAO inhibitors, CAD, hypertension, hyperthyroidism, DM: pregnancy (category C); children younger than 2 y.

ROUTE & DOSAGE

Nasal Congestion
Adult: **Intranasal** 2–3 drops or 2–3 sprays of 0.05% solution into each nostril b.i.d. for up to 3–5 days *Child:* **Intranasal** *2–5 y,* 2–3 drops or 2–3 sprays of 0.025% solution into each nostril b.i.d. for up to 3–5 days; *older than 6 y,* same as for adult

ADMINISTRATION

Instillation
- Place spray nozzle in nostril without occluding it and tilt head slightly forward prior to instillation of spray; instruct patient to sniff briskly during administration.
- Rinse dropper or spray tip in hot water after each use to prevent contamination of solution by nasal secretions.

ADVERSE EFFECTS (≥1%) **Special Senses:** *Burning,* stinging, dryness of nasal mucosa, *sneezing.* **Body as a Whole:** Headache, lightheadedness, drowsiness, insomnia, palpitations, *rebound congestion.*

PHARMACOKINETICS Onset: 5–10 min. **Duration:** 6–10 h.

NURSING IMPLICATIONS

Assessment & Drug Effects
- Monitor for S&S of excess use. If noted, discuss possibility of rebound congestion.

Patient & Family Education
- Wash hands carefully after handling oxymetazoline. Anisocoria (inequality of pupil size, blurred vision) can develop if eyes are rubbed with contaminated fingers.
- Do not to exceed recommended dosage. Rebound congestion (chemical rhinitis) may occur with prolonged or excessive use.
- Systemic effects can result from swallowing excessive medication.

OXYMETHOLONE
(ox-i-meth'oh-lone)
Anadrol-50
Classification: ANDROGEN/ANABOLIC STEROID
Therapeutic: ANABOLIC STEROID
Prototype: Testosterone
Pregnancy Category: X
Controlled Substance: Schedule III

AVAILABILITY 50 mg tablets

ACTION & *THERAPEUTIC EFFECT*
Mechanism of action in refractory anemias is unclear but may be due to direct stimulation of bone marrow, protein anabolic activity, or to androgenic stimulation of erythropoiesis. *Stimulates formation of red blood cells in the bone marrow. Stimulates bone growth, aids in bone matrix reconstitution.*

USES Aplastic anemia.

CONTRAINDICATIONS Hypersensitivity to oxymetholone; prostatic hypertrophy with obstruction; prostatic or male breast cancer; carcinoma of the breast in women with hypercalcemia; hepatic decompensation; hepatic carcinoma; nephrosis or nephrotic stage of nephritis; females of childbearing age; pregnancy (category X), lactation.
CAUTIOUS USE Prepubertal males; older males; DM; history of cardiac, renal, or hepatic impairment; CAD; artherosclerosis; children.

ROUTE & DOSAGE

Aplastic Anemia
Adult: **PO** 1–5 mg/kg/day

ADMINISTRATION
Oral
- For treatment of anemias, a minimum trial period of 3–6 mo is recommended, since response tends to be slow.
- Store at 15°–30° C (59°–86° F). Protect from heat and light.

ADVERSE EFFECTS (≥1%) **Endocrine:** Androgenic in women: Suppression of ovulation, lactation, or menstruation; *hoarseness or deepening of voice* (often irreversible); *hirsutism; oily skin; acne;* clitoral enlargement; regression of breasts; male-pattern baldness. Hypoestrogenic effects in women: Flushing, sweating; vaginitis with pruritus, drying, bleeding; menstrual irregularities. **Males:** Premature epiphyseal closure, phallic enlargement, priapism. Postpubertal: Testicular atrophy, decreased ejaculatory volume, azoospermia, oligospermia (after prolonged administration or excessive dosage), impotence, epididymitis, gynecomastia; increased risk of atherosclerosis. **CV:** *Edema,* skin flush. **GI:** *Nausea, vomiting, anorexia,* diarrhea, jaundice, hepatotoxicity. **Urogenital:** Bladder irritability. **Metabolic:** Hypercalcemia. **Hematologic:** Bleeding (with concurrent anticoagulant therapy), iron deficiency anemia, leukemia.

INTERACTIONS Drug: May enhance hypoprothrombinemic effects of **warfarin. Herbal: Echinacea** may increase risk of hepatotoxicity.

PHARMACOKINETICS Absorption: Readily from GI tract. **Metabolism:** In liver. **Elimination:** In urine. **Half-Life:** 9 h.

NURSING IMPLICATIONS
Assessment & Drug Effects
- Monitor patient with a history of seizures closely because an increase in their frequency may be noted.
- Monitor periodically for edema that may develop with or without CHF.
- Monitor for hypercalcemia (see Appendix F), especially in women with breast cancer.
- Lab tests: Periodic serum calcium; periodic LFTs are especially important for the older adult patient. Drug should be stopped with first sign of liver toxicity (jaundice).

Patient & Family Education
- Monitor blood glucose for loss of glycemic control if diabetic.
- Women: Notify prescriber of signs of virilization.

OXYMORPHONE HYDROCHLORIDE
(ox-i-mor'fone)
Numorphan, Opana, Opana ER
Classifications: NARCOTIC (OPIATE AGONIST); ANALGESIC
Therapeutic: NARCOTIC ANALGESIC

Common adverse effects in *italic*, life-threatening effects underlined; generic names in **bold**; classifications in SMALL CAPS; ♣ Canadian drug name; ◑ Prototype drug

1123

Prototype: Morphine
Pregnancy Category: C
Controlled Substance: Schedule II

AVAILABILITY 1 mg/mL, 1.5 mg/mL injection; 5 mg suppositories; 10 mg extended release tablets; 10 mg tablets

ACTION & *THERAPEUTIC EFFECT*
CNS opiate agonist that is effect for analgesic action for moderate to severe pain. Produces mild sedation. *Effective in relief of moderate to severe pain.*

USES Relief of moderate to severe pain, preoperative medication, obstetric analgesia, support of anesthesia, and relief of anxiety in patients with dyspnea associated with acute ventricular failure and pulmonary edema.

CONTRAINDICATIONS Hypersensitivity to oxymorphone; pulmonary edema resulting from chemical respiratory irritants; ileus; respiratory depression without appropriate monitoring; acute or severe bronchial asthma; upper airway obstruction; hypercapia; moderate and severe hepatic function impairment; status asthmaticus.

CAUTIOUS USE Alcoholism; biliary tract disease; bladder obstruction; severe pulmonary disease, respiratory insufficiency, COPD; head injury; circulatory shock; history of seizures; renal impairment; depression; older adults; pregnancy (category C), lactation; children younger than 12 y.

ROUTE & DOSAGE

Moderate to Severe Pain
Adult: **PO** 10–20 mg q4–6h prn; extended release 5–10 mg q12h **Subcutaneous/IM** 1–1.5 mg q4–6h prn **IV** 0.5 mg q4–6h then

switch to alternate route **PR** 5 mg q4–6h prn
Analgesia during Labor
Adult: **IM** 0.5–1 mg

ADMINISTRATION
Subcutaneous/Intramuscular
▪ Give undiluted.
Intravenous

***PREPARE:* Direct:** May be given undiluted or diluted in 5 mL of sterile water or NS.
***ADMINISTER:* Direct:** Give at a rate of 0.5 mg over 2–5 min.

▪ Protect drug from light. Store suppositories in refrigerator 2°–15° C (36°–59° F).

ADVERSE EFFECTS (≥1%) **GI:**
Nausea, vomiting, euphoria. **CNS:**
Dizziness, lightheadedness, sedation. **Respiratory:** <u>Respiratory depression</u> (see morphine), apnea, <u>respiratory arrest</u>. **Body as a Whole:** Sweating, coma, shock. **CV:** <u>Cardiac arrest</u>, circulatory depression.

INTERACTIONS Drug: Alcohol and other CNS DEPRESSANTS add to CNS depression; **propofol** increases risk of bradycardia.

PHARMACOKINETICS Onset: 5–10 min IV; 10–15 min IM; 15–30 min PR. **Peak:** 1–1.5 h. **Duration:** 3–6 h. **Distribution:** Crosses placenta. **Metabolism:** In liver. **Elimination:** In urine. **Half-Life:** PO 7–9 h; extended release 9–11 h.

NURSING IMPLICATIONS
Assessment & Drug Effects
▪ Monitor respiratory rate. Withhold drug and notify prescriber if rate falls below 12 breaths/ min.

Common adverse effects in *italic*, life-threatening effects <u>underlined</u>; generic names in **bold**; classifications in SMALL CAPS; ♣ Canadian drug name; ⊙ Prototype drug

- Supervise ambulation and advise patient of possible light-headedness. Older adult and debilitated patients are most susceptible to CNS depressant effects of drug.
- Evaluate patient's continued need for narcotic analgesic. Prolonged use can lead to dependence of morphine type.
- Medication contains sulfite and may precipitate a hypersensitivity reaction in susceptible patient.

Patient & Family Education
- Use caution when walking because of potential for injury from dizziness.
- Do not consume alcohol while taking oxymorphone.

OXYTOCIN INJECTION ⊕

(ox-i-toe'sin)

Pitocin
Classification: OXYTOCIC
Therapeutic: OXYTOCIC
Pregnancy Category: X

AVAILABILITY 10 units/mL injection

ACTION & *THERAPEUTIC EFFECT*
Synthetic polypeptide identical pharmacologically to natural oxytocin released by posterior pituitary. By direct action on myofibrils, produces phasic contractions characteristic of normal delivery. Uterine sensitivity to oxytocin increases during gestational period and peaks sharply before parturition. *Effective in initiating or improving uterine contractions at term.*

USES To initiate or improve uterine contraction at term, management of inevitable, incomplete, or missed abortion; stimulation of uterine contractions during third stage of labor; stimulation to overcome uterine inertia; control of postpartum hemorrhage and promotion of postpartum uterine involution. Also used to induce labor in cases of maternal diabetes, preeclampsia, eclampsia, and erythroblastosis fetalis.

CONTRAINDICATIONS Hypersensitivity to oxytocin; significant cephalopelvic disproportion, unfavorable fetal position or presentations that are undeliverable without conversion before delivery, obstetric emergencies that favor surgical intervention, fetal distress in which delivery is not imminent, prematurity, placenta previa, prolonged use in severe toxemia or uterine inertia, hypertonic uterine patterns, conditions predisposing to thromboplastin or amniotic fluid embolism (dead fetus, abruptio placentae), grand multiparity, invasive cervical carcinoma, primipara older than 35 y, past history of uterine sepsis or of traumatic delivery; pregnancy (category X); lactation.
CAUTIOUS USE Preeclampsia; history of seizures; history of mental hypertension.

ROUTE & DOSAGE

Labor Induction

Adult: **IV** 0.5–2 milliunits/min, may increase by 1–2 milliunits/min q15–60min (max: 20 milliunits/min); dose is decreased when labor is established. High dose regimen: 6 milliunits/min, may increase by 6 milliunits/min q15–60min until contraction pattern established.

Postpartum Bleeding

Adult: **IM** 10 units total dose **IV** Infuse a total of 10–40 units at a rate of 20–40 milliunits/min after delivery

---

Common adverse effects in *italic*, life-threatening effects <u>underlined</u>; generic names in **bold;** classifications in SMALL CAPS; ♣ Canadian drug name; ⊕ Prototype drug

1125

| **Incomplete Abortion** |
| *Adult:* **IV** 10–20 milliunits/min |

ADMINISTRATION

Intramuscular

- Give 10 units IM after delivery of the placenta.

Intravenous

PREPARE: **IV Infusion:** When diluting oxytocin for IV infusion, rotate bottle gently to distribute medicine throughout solution. **IV Infusion for Inducing Labor:** Add 10 units (1 mL) to 1 L of D5W, NS, LR, or D5NS to yield 10 milliunits/mL. **IV Infusion for Postpartum Bleeding/Incomplete Abortion:** Add 10–40 units (1–4 mL) to 1 L of D5W, NS, LR, or D5NS to yield 10–40 milliunits/mL.

ADMINISTER: **IV Infusion:** Use an infusion pump for accurate control of infusion rate. **IV Infusion for Inducing Labor:** Initially infuse 0.5–2 milliunits/min; increase by 1–2 milliunits/min at 30–60 min intervals. **IV Infusion for Postpartum Bleeding:** Initially infuse 10–40 milliunits/min, then adjust to control uterine atony. **IV Infusion for Incomplete Abortion:** Infuse 10–20 milliunits/min. Do not exceed 30 units in 12 h.

INCOMPATIBILITIES **Solution/additive: Fibrinolysin, norepinephrine, prochlorperazine, warfarin.**

ADVERSE EFFECTS (≥1%) **Body as a Whole:** Fetal trauma from too rapid propulsion through pelvis, fetal <u>death</u>, anaphylactic reactions, postpartum hemorrhage, precordial pain, edema, cyanosis or redness of skin. **CV:** Fetal bradycardia and arrhythmias, maternal cardiac arrhythmias, hypertensive episodes, subarachnoid <u>hemorrhage</u>, increased blood flow, <u>fatal afibrinogenemia</u>, ECG changes, PVCs, <u>cardiovascular spasm and collapse</u>. **GI:** Neonatal jaundice, maternal nausea, vomiting. **Endocrine:** ADH effects leading to severe water intoxication and hyponatremia, hypotension. **CNS:** Fetal <u>intracranial hemorrhage</u>, anxiety. **Respiratory:** Fetal hypoxia, maternal dyspnea. **Urogenital:** Uterine hypertonicity, tetanic contractions, <u>uterine rupture</u>, pelvic hematoma.

INTERACTIONS Drug: VASOCONSTRICTORS cause severe hypertension; **cyclopropane anesthesia** causes hypotension, maternal bradycardia, arrhythmias. **Herbal: Ephedra, ma huang** may cause hypertension.

PHARMACOKINETICS Duration: 1 h. **Distribution:** Distributed throughout extracellular fluid; small amount may cross placenta. **Metabolism:** Rapidly destroyed in liver and kidneys. **Elimination:** Small amounts excreted unchanged in urine. **Half-Life:** 3–5 min.

NURSING IMPLICATIONS

Assessment & Drug Effects

- Start flow charts to record maternal BP and other vital signs, I&O ratio, weight, strength, duration, and frequency of contractions, as well as fetal heart tone and rate, before instituting treatment.

- Monitor fetal heart rate and maternal BP and pulse at least q15min during infusion period; evaluate tonus of myometrium during and between contractions and record on flow chart. Report change in rate and rhythm immediately.

- Stop infusion to prevent fetal anoxia, turn patient on her side,

Common adverse effects in *italic*, life-threatening effects <u>underlined</u>; generic names in **bold**; classifications in SMALL CAPS; ♣ Canadian drug name; ⊘ Prototype drug

and notify prescriber if contractions are prolonged (occurring at less than 2-min intervals) and if monitor records contractions about 50 mm Hg or if contractions last 90 sec or longer. Stimulation will wane rapidly within 2–3 min. Oxygen administration may be necessary.

- If local or regional (caudal, spinal) anesthesia is being given to the patient receiving oxytocin, be alert to the possibility of hypertensive crisis (sudden intense occipital headache, palpitation, marked hypertension, stiff neck, nausea, vomiting, sweating, fever, photophobia, dilated pupils, bradycardia or tachycardia, constricting chest pain).
- Monitor I&O during labor. If patient is receiving drug by prolonged IV infusion, watch for symptoms of water intoxication (drowsiness, listlessness, headache, confusion, anuria, weight gain). Report changes in alertness and orientation and changes in I&O ratio (i.e., marked decrease in output with excessive intake).
- Check fundus frequently during the first few postpartum hours and several times daily thereafter.

Patient & Family Education
- Be aware of purpose and anticipated effect of oxytocin.
- Report sudden, severe headache immediately to health care providers.

PACLITAXEL ⊕
(pac-li-tax′el)

Abraxane, Taxol
Classifications: ANTINEOPLASTIC; TAXANE

Therapeutic: ANTINEOPLASTIC; ANTIMICROTUBULE
Pregnancy Category: D

AVAILABILITY 6 mg/mL injection; 100 mg powder for injection (with 900 mg human albumin)

ACTION & THERAPEUTIC EFFECT
During cell division paclitaxel is an antimicrotubular agent that interferes with the microtubule network essential for interphase and mitosis. This results in abnormal spindle formation and multiple asters during mitosis. *Interferes with growth of rapidly dividing cells including cancer cells, and eventually causes cell death. Additionally, the breakup of the cytoskeleton within nondividing cells interrupts intracellular transport and communications.*

USES Ovarian cancer, breast cancer, Kaposi's sarcoma, non–small-cell lung cancer (NSCLC).
UNLABELED USES Squamous cell head and neck cancer, small-cell lung cancer, endometrial cancer, esophageal cancer, gastric cancer, testicular cancer, germ cell tumors, and other solid tumors, leukemia, melanoma.

CONTRAINDICATIONS Taxol: Hypersensitivity to paclitaxel, or taxane; baseline neutrophil count less than 1500 cells/mm³; thrombocytopenia; with AIDS-related Kaposi's sarcoma baseline neutrophil count less than 1000 cells/mm³; pregnancy (category D), lactation. For **Abraxane:** Baseline neutrophil count less than 1500 cells/mm³.
CAUTIOUS USE Cardiac arrhythmias, cardiac disease; impaired liver function; alcoholism; older

Common adverse effects in *italic*, life-threatening effects underlined; generic names in **bold;** classifications in SMALL CAPS; ♦ Canadian drug name; ⊕ Prototype drug

1127

adults; peripheral neuropathy. Safety and efficacy in children are not established.

ROUTE & DOSAGE

Ovarian Cancer, NSCLC
Adult: **IV** 135 mg/m² 24 h infusion repeated q3wk

Breast Cancer
Adult: **IV** 175–250 mg/m² over 3 h q3wk
Abraxane: **IV** 260 mg/m² over 30 min q3wk

Kaposi's Sarcoma
Adult: **IV** 135 mg/m² infused over 3 h q3wk or 100 mg/m² infused over 3 h q2wk

ADMINISTRATION

Intravenous
Note: Premedication as follows (except with **Abraxane**) to avoid severe hypersensitivity: Dexamethasone 20 mg (10 mg with AIDS-related Kaposi's) PO/IV at 12 and 6 h prior to infusion; diphenhydramine 50 mg IV 30–60 min prior to infusion; and cimetidine 300 mg or ranitidine 50 mg IV 30 min before infusion.
- Do not administer to patients with AIDS-related Kaposi's unless neutrophil count is at least 1000/mm³; for all others, do not administer unless neutrophil count is at least 1500/mm³.
- Follow institutional or standard guidelines for preparation, handling, and disposal of cytotoxic agents.

PREPARE: **IV Infusion: Dilution of Conventional Paclitaxel:** Do not use equipment or devices containing polyvinyl chloride (PVC) in preparation of infusion.
- Dilute to a final concentration of 0.3–1.2 mg/mL in any of the following: D5W, NS, D5/NS, or D5W in Ringer's injection. The prepared solution may be hazy, but this does not indicate a loss of potency.

Abraxane Vial Reconstitution: Slowly inject 20 mL NS over at least 1 min onto the inside wall of the vial to yield 5 mg/mL.
- DO NOT inject directly into the cake powder. Allow vial to sit for at least 5 min, then gently swirl for at least 2 min to completely dissolve. If foaming occurs, let stand for at least 15 min until foam subsides. ▪ If particulates or settling are visible, gently invert vial to ensure complete resuspension prior to use. ▪ Remove the required dose and inject into an empty sterile, PVC or non-PVC type IV bag.

ADMINISTER: **IV Infusion:** Because tissue necrosis occurs with extravasation, frequently assess patency of a peripheral IV site.
- **Conventional Paclitaxel:** Infuse over 3 h through IV tubing containing an in-line (0.22 micron or less) filter. Do not use equipment containing PVC.
- **Abraxane:** DO NOT use an in-line filter. Infuse over 30 min.

INCOMPATIBILITIES **Solution/additive: PVC bags** and **infusion sets** should be avoided (except with Abraxane) due to leaching of DEHP (plasticizer). ▪ Do not mix with any other medications. **Y-site: Amphotericin B, amphotericin B cholesteryl sulfate complex, chlorpromazine, doxorubicin liposome, hydroxyzine, methylprednisolone, mitoxantrone.**

- **Conventional paclitaxel** solutions diluted for infusion are stable at room temperature (approximately 25° C/77° F) for up to 27 h.

- Reconstituted **Abraxane** should be used immediately but may be kept refrigerated for up to 8 h if needed.

ADVERSE EFFECTS (≥1%) CV: Ventricular tachycardia, ventricular ectopy, *transient bradycardia,* chest pain. **CNS:** Fatigue, headaches, *peripheral neuropathy,* weakness, seizures. **GI:** *Nausea, vomiting,* diarrhea, taste changes, *mucositis,* elevations in serum triglycerides. **Hematologic:** <u>Neutropenia, anemia, thrombocytopenia</u>. **Body as a Whole:** *Hypersensitivity reactions (hypotension, dyspnea with <u>bronchospasm</u>, urticaria, abdominal and extremity pain, diaphoresis, <u>angioedema</u>), myalgias, arthralgias, alopecia.* **Skin:** *Alopecia,* tissue necrosis with extravasation. **Urogenital:** Minor elevations in kidney and liver function tests.

INTERACTIONS Drug: Increased myelosuppression if **cisplatin, doxorubicin** is given before paclitaxel; **ketoconazole** can inhibit metabolism of paclitaxel; additive bradycardia with BETA-BLOCKERS, **digoxin, verapamil;** additive risk of bleeding with ANTICOAGULANTS, NSAIDS, PLATELET INHIBITORS (including **aspirin**), THROMBOLYTIC AGENTS.

PHARMACOKINETICS Distribution: Greater than 90% protein bound; does not cross CSF. **Metabolism:** In liver (CYP3A4, 2C8). **Elimination:** Feces 70%, urine 14%. **Half-Life:** 1–9 h.

NURSING IMPLICATIONS

Assessment & Drug Effects

- Monitor for hypersensitivity reactions, especially during first and second administrations of the paclitaxel. S&S requiring treatment, but not necessarily

discontinuation of the drug, include dyspnea, hypotension, and chest pain. Discontinue immediately and manage symptoms aggressively if angioedema and generalized urticaria develop.
- Monitor vital signs frequently, especially during the first hour of infusion. Bradycardia occurs in approximately 12% of patients, usually during infusion. It does not normally require treatment. Cardiac monitoring is indicated for those with severe conduction abnormalities.
- Lab tests: Monitor hematologic status throughout course of treatment. Severe neutropenia is common but usually of short duration (less than 500/mm³ for less than 7 days) with the nadir occurring about day 11. Thrombocytopenia occurs less often and is less severe with the nadir around day 8 or 9. The incidence and severity of anemia increase with exposure to paclitaxel.
- Monitor for peripheral neuropathy, the severity of which is dose dependent. Severe symptoms occur primarily with higher than recommended doses.

Patient & Family Education

- Immediately report to prescriber S&S of paclitaxel hypersensitivity: Difficulty breathing, chest pain, palpitations, angioedema (subcutaneous swelling usually around face and neck), and skin rashes or itching.
- Be sure to have periodic blood work as prescribed.
- Avoid aspirin, NSAIDS, and alcohol to minimize GI distress.
- Be aware of high probability of developing hair loss (greater than 80%).

P

Common adverse effects in *italic*, life-threatening effects <u>underlined</u>; generic names in **bold;** classifications in SMALL CAPS; ✚ Canadian drug name; ⊘ Prototype drug

1129

PALIFERMIN

(pal-i-fur'men)

Kepivance

Classifications: BIOLOGIC RE-SPONSE MODIFIER; KERATINOCYTE GROWTH FACTOR; CYTOKINE

Therapeutic: KERATINOCYTE GROWTH FACTOR (KGF)

Pregnancy Category: C

AVAILABILITY 6.25 mg powder for injection

ACTION & *THERAPEUTIC EFFECT*
Naturally occurring keratinocyte growth factor (KGF) is produced and regulated in response to epithelial tissue injury. Binding of KGF to its receptors in epithelial cells results in proliferation, differentiation, and repair of injury to epithelial cells. Palifermin is a synthetic form of KCG; thus it enhances replacement of injured cells. *Reduces the incidence of severe oral mucositis that interferes with food consumption in the cancer patient.*

USES Reduction of the incidence and duration of severe oral mucositis in patients with hematologic malignancies who are receiving myelotoxic therapy requiring hematopoietic stem cell support.

CONTRAINDICATIONS Hypersensitivity to *Escherichia coli*-derived protein, palifermin; nonhematologic malignancies; within 24 h of chemotherapy.

CAUTIOUS USE Use contraception for females of childbearing age; pregnancy (category C), lactation; children.

ROUTE & DOSAGE

Oral Mucositis

Adult: **IV** 60 mcg/kg/day for 3 days before and 3 days after

myelotoxic therapy. **Premyelotoxic therapy:** Final dose should be given 24–48 h before therapy. **Postmyelotoxic therapy:** First dose should be given after but on the same day of hematopoietic stem cell infusion, and at least 4 days after the most recent administration of palifermin.

ADMINISTRATION

Intravenous

Do not give within 24 h before/after or during myelotoxic chemotherapy.

PREPARE: **Direct:** Reconstitute powder with 1.2 mL sterile water to yield 5 mg/mL. Gently swirl to dissolve but do not shake. ▪ Powder will dissolve in about 3 min. Should be used immediately.

ADMINISTER: **Direct:** Give as a bolus dose. ▪ If heparin is used to maintain the IV line, flush before/after with NS. ▪ If diluted solution was refrigerated, may warm to room temperature for up to 1 h but protect from light.

INCOMPATIBILITIES **Y-site: Heparin.**

▪ Store powder vial at 2°–8° C (36°–46° F). Protect from light. ▪ If needed, may store reconstituted solution refrigerated for up to 24 h. ▪ Discard any reconstituted solution left at room temperature for longer than 1 h.

ADVERSE EFFECTS (≥1%) **Body as a Whole:** *Edema, fever, pain.* **CNS:** *Dysesthesia.* **GI:** *Mouth/tongue thickness or discoloration, taste alterations.* **Metabolic:** *Elevated serum amylase, elevated serum lipase.* **Musculoskeletal:** *Arthralgia.* **Skin:** *Erythema, pruritus, rash.* **Urogenital:** *Proteinuria.*

INTERACTIONS Drug: Administration of palifermin within 24 h of **myelotoxic chemotherapy** increases the severity and duration of oral mucositis.

PHARMACOKINETICS Distribution: Extravascular distribution. **Half-Life:** 4.5 h.

NURSING IMPLICATIONS

Assessment & Drug Effects

• Monitor for improvement in mucositis.

• Monitor for S&S of oral toxicities and skin toxicities.

Patient & Family Education

• Report any of the following to a health care provider: Alteration of taste, discoloration or enlargement of the tongue, lack of sensation around the mouth, skin rash, itching, or edema.

PALIPERIDONE

(pa-li'per-i-done)

Invega

PALIPERIDONE PALMITATE

Invega Sustenna

Classification: ATYPICAL ANTI-PSYCHOTIC

Therapeutic: ANTIPSYCHOTIC

Prototype: Clozapine

Pregnancy Category: C

AVAILABILITY 1.5 mg, 3 mg, 6 mg, 9 mg extended release tablets; extended release injection

ACTION & *THERAPEUTIC EFFECT*
Interferes with binding of dopamine to dopamine type 2 (D_2) receptors, serotonin (5-HT_{2A}) receptors, and alpha-adrenergic receptors. *Effective in controlling symptoms of schizophrenia as well as other psychotic symptoms.*

USES Treatment of schizophrenia, schizoaffective disorder.

CONTRAINDICATIONS Hypersensitivity to paliperidone, risperidone; concurrent administration with drugs that produce QT_c prolongation including Class IA or Class III antiarrhythmic medications, and antibiotics that prolong the QT interval as well as other antipsychotic medications; hyperglycemia, ketoacidosis; GI narrowing (pathologic or iatrogenic); older adults with dementia-related psychosis; lactation.

CAUTIOUS USE History of cerebrovascular events; hypovolemia, dehydration; cardiovascular disease; renal impairment; older adults; history of hypotension; CNS pathology; systemic infection; DM, obesity, family history of seizures; Parkinson's disease; disorders that may lead to aspiration pneumonia (e.g., severe Alzheimer's dementia); potential for suicidality; pregnancy (category C); children younger than 12 y.

P

ROUTE & DOSAGE

Schizophrenia
Adult: **PO** Initially 6 mg/day; may adjust up/down in 3 mg increments; at least 5 day intervals needed for dosage increments (max: 12 mg/day) **IM** 234 mg on day 1, then 156 mg 1 wk later, then monthly dose of 117 mg (dose may vary based on patient response) **Converting PO to IM** 3 mg/day PO = 39–78 mg/mo IM; 60 mg/day PO = 117 mg/mo IM; 12 mg/day PO = 234 mg/mo IM *Adolescent:* **PO** 3 mg daily

Common adverse effects in *italic*, life-threatening effects underlined; generic names in **bold**; classifications in SMALL CAPS; ♣ Canadian drug name; ⊘ Prototype drug

1131

Schizoaffective Disorder

Adult: **PO** 6 mg daily, dose adjusted at intervals of more than 4 days (max: 12 mg/day)

Renal Impairment Dosage Adjustment

CrCl 50–79 mL/min: Max 6 mg/day; *IM injection:* Use 156 on day 1 and 117 mg 1 wk later, then monthly 78 mg injections *CrCl 10–49 mL/min:* Max 3 mg/day; IM not recommended

ADMINISTRATION

- Extended release tablets **must be** swallowed whole. They should not be chewed or crushed.
- Give in the morning with or without food.
- Store at 15°–30° C (59°–86° F). Protect from moisture.

ADVERSE EFFECTS (≥1%) **Body as a Whole:** Back pain, cough, pain in extremity, pyrexia; injection site reaction. **CNS:** *Akathisia, anxiety,* asthenia, dizziness, dystonia, extrapyramidal disorder, fatigue, *headache,* hypertonia, parkinsonism, *somnolence,* tremor. **CV:** Atrioventricular block, bundle branch block, ECG T-wave abnormalities, hypertension, orthostatic hypotension, QT_c prolongation, sinus arrhythmia, tachycardia. **GI:** Abdominal pain, dry mouth, dyspepsia, nausea, salivary hypersecretion. **Metabolic:** Increased insulin levels. **Special Senses:** Blurred vision.

INTERACTIONS Drug: Enhanced CNS depression with **alcohol** or CNS DE-PRESSANTS. Paliperidone may enhance the effects of ANTIHYPERTENSIVE AGENTS. Paliperidone can diminish the effects of DOPAMINE AGONISTS (**levodopa, bromocriptine, cabergoline, pergolide, pramipexole, ropinirole**).

Carbamazepine decreases effectiveness. **Food:** High fat/high caloric meal increases paliperidone levels.

PHARMACOKINETICS Absorption: Bioavailability is 28%. **Peak:** 24 h. **Distribution:** 74% protein bound. **Metabolism:** In liver (26–41%). **Elimination:** Urine (major, 50–70% unchanged) and stool (minor). **Half-Life:** 23 h.

NURSING IMPLICATIONS

Assessment & Drug Effects

- Baseline ECG recommended to rule out congenital, long-QT syndrome.
- Prior to initiating therapy, hypokalemia and hypomagnesemia should be corrected.
- Monitor CV status and monitor BP especially in those prone to hypotension.
- Reassess patient periodically in order to maintain on the lowest effective drug dose.
- Monitor closely neurologic status of older adults.
- Supervise closely those with suicidal ideation.
- Monitor closely those at risk for seizures.
- Assess degree of cognitive and motor impairment, and assess for environmental hazards.
- Lab tests: Baseline and periodic serum electrolytes; periodic blood glucose, and complete blood counts.
- Monitor diabetics for loss of glycemic control.

Patient & Family Education

- Exercise caution with hazardous activities until response to drug is known.
- Carefully monitor blood glucose levels if diabetic.

Common adverse effects in *italic*, life-threatening effects underlined; generic names in **bold**; classifications in SMALL CAPS; ✦ Canadian drug name; ⊘ Prototype drug

- Be aware of the risk of orthostatic hypotension.
- The shell of the tablet may be eliminated in the stool whole, but this does not mean the drug was not absorbed.
- Monitor for signs and symptoms of suicidal ideation.
- Be aware of the possibility of seizure activity.

PALIVIZUMAB
(pal-i-viz′u-mab)
Synagis
Classifications: IMMUNOMODULATOR; MONOCLONAL ANTIBODY; IMMUNOGLOBULIN
Therapeutic: IMMUNOGLOBULIN (IGG)
Pregnancy Category: C

AVAILABILITY 100 mg vial

ACTION & *THERAPEUTIC EFFECT*
Monoclonal antibody ($IgG1_k$) produced by recombinant DNA technology to the respiratory syncytial virus (RSV). *Provides passive immunity against respiratory syncytial virus. Indicated by prevention of lower respiratory tract infection.*

USES Prevention of serious lower respiratory tract infections in children susceptible to RSV.

CONTRAINDICATIONS Hypersensitivity to palivizumab; lactation.
CAUTIOUS USE Hypersensitivity to other immunoglobulin preparations, blood products, or other medications; kidney or liver dysfunction; acute RSV infection; pregnancy (category C).

ROUTE & DOSAGE

RSV
Child: **IM** 15 mg/kg qmo during RSV season

ADMINISTRATION
Intramuscular
- Reconstitute solution by gently injecting 1 mL of sterile water for injection (without preservative) toward the sides of the vial. Gently swirl for 30 sec to dissolve (do not shake solution). Allow to stand at room temperature for at least 20 min until solution clears.
- Give IM only into the anterolateral aspect of the thigh. Volumes greater than 1 mL should be divided and given in different sites.
- Use reconstituted solution within 6 h. Discard any unused portion of the vial. It contains no preservatives.

ADVERSE EFFECTS (≥1%) **Body as a Whole:** *Otitis media,* pain, hernia. **GI:** Increased AST, diarrhea, nausea, vomiting, gastroenteritis. **Respiratory:** *URI, rhinitis,* pharyngitis, cough, wheeze, bronchiolitis, asthma, croup, dyspnea, sinusitis, apnea. **Skin:** *Rash.*

PHARMACOKINETICS Half-Life: 20 days.

NURSING IMPLICATIONS
Assessment & Drug Effects
- Lab tests: Periodic monitoring of LFTs may be warranted.
- Monitor carefully for and immediately report S&S of respiratory illness including fever, cough, wheezing, and chest retractions.
- Assess for and report erythema or indurations at injection site.

Patient & Family Education
- Contact prescriber for S&S of respiratory illness, vomiting, diarrhea, or if redness develops at injection site.

P

Common adverse effects in *italic*, life-threatening effects underlined; generic names in **bold**; classifications in SMALL CAPS; ♥ Canadian drug name; ◐ Prototype drug

1133

PALONOSETRON

(pal-o-no'si-tron)
Aloxi
Classifications: SEROTONIN 5-HT₃ RECEPTOR ANTAGONIST; ANTIEMETIC
Therapeutic: ANTIEMETIC
Prototype: Ondansetron
Pregnancy Category: B

AVAILABILITY 0.25 mg/5 mL injection

ACTION & *THERAPEUTIC EFFECT*
Selectively blocks serotonin 5-HT₃ receptors found centrally in the chemoreceptor trigger zone (CTZ) of the hypothalamus, and peripherally at vagal nerve endings in the intestines. *Prevents acute chemotherapy-induced nausea and vomiting associated with initial and repeat courses of moderately or highly emetogenic chemotherapy.*

USES Prevention of acute and delayed nausea and vomiting associated with highly emetogenic cancer chemotherapy.
UNLABELED USES Postoperative nausea/vomiting.

CONTRAINDICATIONS Hypersensitivity to palonosetron; lactation.
CAUTIOUS USE Dehydration; cardiac arrhythmias, QT prolongation; electrolyte imbalance; pregnancy (category B); children younger than 18 y.

ROUTE & DOSAGE

Prevention of Chemotherapy-Induced Nausea and Vomiting
Adult: **IV** 0.25 mg infused over 30 sec 30 min prior to chemotherapy; do not repeat for at least 7 days

Hepatic Impairment/Renal Impairment Dosage Adjustment
No adjustment necessary

ADMINISTRATION

Intravenous

PREPARE: **Direct:** Do not dilute and do not mix with other drugs.
ADMINISTER: **Direct:** Give over 30 sec. Flush IV line with NS before and after administration.
INCOMPATIBILITIES Do not mix with other drugs.

- Store at room temperature of 15°–30° C (59°–86° F). Protect from light.

ADVERSE EFFECTS (≥1%) **CNS:** Headache, anxiety, dizziness. **GI:** Constipation, diarrhea, abdominal pain. **Dermatologic:** Pruritus.

INTERACTIONS Drug: Can cause profound hypotension with **apomorphine.**

PHARMACOKINETICS Metabolism: In liver (CYP2D6, 1A2, 3A4). **Elimination:** Primarily renal. **Half-Life:** 40 h.

NURSING IMPLICATIONS

Assessment & Drug Effects
- Monitor closely cardiac status especially in those taking diuretics or otherwise at risk for hypokalemia or hypomagnesemia, with congenital QT syndrome, or patients taking antiarrhythmic or other drugs that lead to QT prolongation.

Patient & Family Education
- Report promptly any of the following: Difficulty breathing, wheezing, or shortness of breath; palpitations or chest tightness; skin rash or itching; swelling of the face, tongue, throat, hands, or feet.

PAMIDRONATE DISODIUM

(pa-mi′dro-nate)

Aredia
Classification: BISPHOSPHONATE (BONE METABOLISM REGULATORY)
Therapeutic: BONE METABOLISM REGULATORY
Prototype: Etidronate
Pregnancy Category: D

AVAILABILITY 30 mg, 60 mg, 90 mg powder for injection; 3 mg, 6 mg, 9 mg solution for injection

ACTION & *THERAPEUTIC EFFECT*
A bone-resorption inhibitor thought to absorb calcium phosphate crystals into bone. May also inhibit osteoclast activity, thus contributing to inhibition of bone resorption. *Reduces bone turnover and, when used in combination with adequate hydration, it increases renal excretion of calcium, thus reducing serum calcium concentrations.*

USES Hypercalcemia of malignancy and Paget's disease, bone metastases in multiple myeloma or breast cancer.
UNLABELED USES Primary hyperparathyroidism, osteoporosis prophylaxis.

CONTRAINDICATIONS Hypersensitivity to pamidronate; breast cancer, severe renal disease, hypercalcemia, hypercholesterolemia, polycythemia, prostatic cancer; pregnancy (category D).
CAUTIOUS USE Heart failure, nephrosis or nephrotic syndrome, moderate renal disease, chronic kidney failure; hepatic disease, cholestasis; peripheral edema, prostate hypertrophy; cancer patients with stomatitis; lactation. Safe use in children not established.

ROUTE & DOSAGE

Moderate Hypercalcemia of Malignancy (corrected calcium 12–13.5 mg/dL)
Adult: **IV** 60–90 mg infused over 4–24 h, may repeat in 7 days

Severe Hypercalcemia of Malignancy (corrected calcium greater than 13.5 mg/dL)
Adult: **IV** 90 mg infused over 4–24 h, may repeat in 7 days

Paget's Disease
Adult: **IV** 30 mg once daily for 3 days (90 mg total)

Osteolytic Metastases
Adult: **IV** 90 mg once/mo

ADMINISTRATION

Intravenous

PREPARE: **IV Infusion:** Add 10 mL sterile water for injection to reconstitute the 30 or 90 mg vial to yield 3 or 9 mg/mL, respectively. Allow to completely dissolve. **IV Infusion for Hypercalcemia of Malignancy:** Withdraw the required dose and dilute in D5W, NS, or 1/2NS as follows: Use 1000 mL. **IV Infusion for Paget's Disease and Multiple Myeloma:** Withdraw the required dose and dilute in D5W, NS, or 1/2NS as follows: Use 500 mL. **IV Infusion for Breast Cancer Bone Metastases:** Withdraw the required dose and dilute in D5W, NS, or 1/2NS as follows: Use 250 mL.
ADMINISTER: **IV Infusion:** Regulate infusion rate carefully. Rapid infusion may cause renal damage. **IV Infusion for Hypercalcemia of Malignancy:** Infuse over 2–24 h. **IV Infusion for Paget's disease and Multiple Myeloma:** Infuse over 4 h. **IV Infusion for**

P

Breast Cancer Bone Metastases: Infuse over 2 h.

INCOMPATIBILITIES **Solution/ additive:** CALCIUM-CONTAINING SOLUTIONS (including **lactated Ringer's**).

- Refrigerate reconstituted pamidronate solution at 2°–8° C (36°–46° F); the IV solution may be stored at room temperature. Both are stable for 24 h.

ADVERSE EFFECTS (≥1%) **Body as a Whole:** *Fever with or without rigors* generally occurs within 48 h and subsides within 48 h despite continued therapy; *thrombophlebitis at injection site;* general malaise lasting for several weeks; transient increase in bone pain; jaw osteonecrosis. **Metabolic:** *Hypocalcemia.* **GI:** Nausea, abdominal pain, *epigastric discomfort.* **CV:** Hypertension. **Skin:** Rash.

INTERACTIONS Drug: Concurrent use of **foscarnet** may further decrease serum levels of ionized calcium.

PHARMACOKINETICS Absorption: 50% of dose is retained in body. **Onset:** 24–48 h. **Peak:** 6 days. **Duration:** 2 wk–3 mo. **Distribution:** Accumulates in bone; once deposited, remains bound until bone is remodeled. **Metabolism:** Not metabolized. **Elimination:** 50% excreted in urine unchanged. **Half-Life:** 28 h.

NURSING IMPLICATIONS

Assessment & Drug Effects

- Assess IV injection site for thrombophlebitis.
- Lab tests: Monitor serum calcium, phosphate, magnesium, and potassium at frequent intervals; CBC with differential; Hct and Hgb; and kidney function tests throughout course of therapy.

- Monitor for S&S of hypocalcemia, hypokalemia, hypomagnesemia, and hypophosphatemia.
- Monitor for seizures especially in those with a preexisting seizure disorder.
- Monitor vital signs. Be aware that drug fever, which may occur with pamidronate use, is self-limiting, usually subsiding in 48 h even with continued therapy.
- Monitor I&O and hydration status. Patient should be adequately hydrated, without fluid overload.

Patient & Family Education

- Be aware that transient, self-limiting fever with/without chills may develop.
- Generalized malaise, which may last for several weeks following treatment, is an anticipated adverse effect.
- Report to prescriber immediately perioral tingling, numbness, and paresthesia. These are signs of hypocalcemia.

PANCRELIPASE ℗

(pan-kre-li′pase)

Cotazym, Cotazym-S, Festal II, Ilozyme, Ku-Zyme-Hp, Pancrease, Ultrase, Viokase, Zenpep

Classification: ENZYME REPLACEMENT THERAPY

Therapeutic: PANCREATIC ENZYME REPLACEMENT THERAPY

Pregnancy Category: C

AVAILABILITY Tablets or capsules containing lipase, protease, and amylase

ACTION & *THERAPEUTIC EFFECT*

Pancreatic enzyme concentrate of porcine origin standardized for lipase content. Similar to natural pancreatin but on a weight basis has 12 times the lipolytic activity and at least 4 times the trypsin and amylase content of pancreatin. *Facilitates hydrolysis of fats into glycerol and fatty acids, starches into dextrins and sugars, and proteins into peptides for easier absorption.*

USES Replacement therapy in symptomatic treatment of malabsorption syndrome due to cystic fibrosis and other conditions associated with exocrine pancreatic insufficiency.

CONTRAINDICATIONS History of allergy to porcine protein or enzymes; esophageal strictures; pancreatitis.

CAUTIOUS USE GI disease, Crohn's disease, short bowel syndrome; CF; pregnancy (category C), lactation.

ROUTE & DOSAGE

Pancreatic Insufficiency

Adult /Adolescent/Child (over 4 years): **PO** 500 lipase units/kg per meal
Child (1–4 years): **PO** 1000 lipase units/kg/meal
Infant (up to 12 mo): **PO** 2000–4000 lipase units/120 mL of formula or breast-feeding

ADMINISTRATION

Oral

- Ensure that enteric-coated preparations are not crushed or chewed.
- Note: For children, powder form may be sprinkled on food.
- Open capsule and sprinkle contents on soft food, which should be swallowed without chewing to prevent mucus membrane irritation. Follow with a full glass of water or juice. Cimetidine, ranitidine, or an antacid may be prescribed to be given before pancrelipase to prevent drug's destruction by gastric pepsin and acid pH.

- Determine dosage in relation to fat content in diet (suggested ratio: 300 mg pancrelipase for each 17 g dietary fat).

ADVERSE EFFECTS (≥1%) **GI:** Anorexia, nausea, vomiting, diarrhea. **Metabolic:** Hyperuricosuria.

INTERACTIONS Drug: Iron absorption may be decreased.

PHARMACOKINETICS Distribution: Acts locally in GI tract. **Elimination:** Feces.

NURSING IMPLICATIONS

Assessment & Drug Effects

- Monitor I&O and weight. Note appetite and quality of stools, weight loss, abdominal bloating, polyuria, thirst, hunger, itching. Pancreatic insufficiency is frequently associated with steatorrhea, bulky stools, and insulin-dependent diabetes.

Patient & Family Education

- Learn proper timing of medication in relation to meals.

P

PANCURONIUM BROMIDE

(pan-kyoo-roe'nee-um)
Classification: NONDEPOLARIZING SKELETAL MUSCLE RELAXANT
Therapeutic: SKELETAL MUSCLE RELAXANT
Prototype: Atracurium
Pregnancy Category: C

AVAILABILITY 1 mg/mL, 2 mg/mL injection

ACTION & *THERAPEUTIC EFFECT* Synthetic curariform nondepolarizing neuromuscular blocking agent that produces skeletal muscle relaxation or paralysis by competing with acetylcholine at cholinergic receptor sites on the

Common adverse effects in *italic*, life-threatening effects <u>underlined</u>; generic names in **bold;** classifications in SMALL CAPS; ♣ Canadian drug name; �Prototype drug

1137

skeletal muscle endplate and thus blocks nerve impulse transmission. *Induces skeletal muscle relaxation or paralysis.*

USES Adjunct to anesthesia to induce skeletal muscle relaxation. Also to facilitate management of patients undergoing mechanical ventilation.

CONTRAINDICATIONS Hypersensitivity to the drug or bromides; tachycardia.
CAUTIOUS USE Debilitated patients; dehydration; myasthenia gravis; neuromuscular disease; pulmonary, liver, or kidney disease; fluid or electrolyte imbalance; pregnancy (category C), lactation.

ROUTE & DOSAGE

Skeletal Muscle Relaxation

Adult/Child/Infant: **IV** 0.04–0.1 mg/kg initial dose, may give additional doses of 0.01 mg/kg at 30–60 min intervals
Neonate: **IV** 0.02 mg/kg test dose, then 0.03 mg/kg

Obesity Dosage Adjustment

Use IBW

Renal Impairment Dosage Adjustment

CrCl 10–50 mL/min: Use 50% of dose; *less than 10 mL/min:* Do not use

ADMINISTRATION

Intravenous
Plastic syringe may be used for administration, but drug may adsorb to plastic with prolonged storage.

PREPARE: Direct: Give undiluted.
ADMINISTER: Direct: Give over 30–90 sec.

INCOMPATIBILITIES Solution/additive: Furosemide.

- Refrigerate at 2°–8° C (36°–46° F). Do not freeze.

ADVERSE EFFECTS (≥1%) **CV:** *Increased pulse rate and BP,* ventricular extrasystoles. **Skin:** Transient acneiform rash, burning sensation along course of vein. **Body as a Whole:** Salivation, skeletal muscle weakness, <u>respiratory depression</u>.

DIAGNOSTIC TEST INTERFERENCE Pancuronium may decrease ***serum cholinesterase*** concentrations.

INTERACTIONS Drug: GENERAL ANESTHETICS increase neuromuscular blocking and duration of action; AMINOGLYCOSIDES, **bacitracin, polymyxin B, clindamycin, lidocaine,** parenteral **magnesium, quinidine, quinine, trimethaphan, verapamil** increase neuromuscular blockade; DIURETICS may increase or decrease neuromuscular blockade; **lithium** prolongs duration of neuromuscular blockade; NARCOTIC ANALGESICS possibly add to respiratory depression; **succinylcholine** increases onset and depth of neuromuscular blockade; **phenytoin** may cause resistance to or reversal of neuromuscular blockade.

PHARMACOKINETICS Onset: 30–45 sec. **Peak:** 2–3 min. **Duration:** 60 min. **Distribution:** Well distributed to tissues and extracellular fluids; crosses placenta in small amounts. **Metabolism:** Small amount in liver. **Elimination:** Primarily in urine. **Half-Life:** 2 h.

NURSING IMPLICATIONS

Assessment & Drug Effects

- Assess cardiovascular and respiratory status continuously.

- Observe patient closely for residual muscle weakness and signs of respiratory distress during recovery period. Monitor BP and vital signs. Peripheral nerve stimulator may be used to assess the effects of pancuronium and to monitor restoration of neuromuscular function.
- Note: Consciousness is not affected by pancuronium. Patient will be awake and alert but unable to speak.

PANITUMUMAB

(pan-i-tu-mu′mab)
Vectibix
Classifications: ANTINEOPLASTIC; BIOLOGIC RESPONSE MODIFIER; MONOCLONAL ANTIBODY; EPIDERMAL GROWTH FACTOR RECEPTOR (EGFR) INHIBITOR
Therapeutic: ANTINEOPLASTIC
Prototype: Gefitinib
Pregnancy Category: C

AVAILABILITY 20 mg/mL solution for injection in 5 mL, 10 mL, and 20 mL vials

ACTION & *THERAPEUTIC EFFECT*
Overexpression of epidermal growth factor receptors (EGFRs) occurs in many human cancers, including those of the colon and rectum. EGFRs control the activity of intracellular tyrosine kinases that regulate transcription of DNA molecules involved in cellular growth, survival, motility, proliferation, and transformation. *Panitumumab inhibits upregulation or overexpression of EGFR in cancer cells, decreasing their capacity for cell proliferation, cell survival, and decreasing their invasive capacity and metastases.*

USES Treatment of EGFR-expressing metastatic colorectal carcinoma in patients with disease progression on or following fluoropyrimidine-, oxaliplatin-, and irinotecan-containing chemotherapy regimens.

CONTRAINDICATIONS Pulmonary fibrosis; interstitial lung disease. Use contraception for females of childbearing age; lactation.
CAUTIOUS USE Photosensitivity with drug use; electrolyte imbalances, especially hypomagnesemia, and hypocalcemia; lung disorders; pregnancy (category C). Safe use in children not established.

ROUTE & DOSAGE

Metastatic Colorectal Carcinoma
Adult: **IV** 6 mg/kg q14days

Dosage Adjustments for Infusion Reactions and Dermatologic Reactions

Mild or moderate infusion reactions (Grade 1 or 2): Reduce infusion rate by 50%
Severe infusion reactions (Grade 3 or 4): Discontinue permanently
Intolerable or severe dermatologic toxicity (greater than Grade 3): Withhold drug. If toxicity does not improve to at least grade 2 within 1 mo, permanently discontinue. If toxicity improves to at least grade 2 and patient is symptomatically improved after withholding no more than 2 doses, resume at 50% of original dose. If toxicities recur, discontinue permanently. If toxicities do not recur, subsequent doses may be increased by increments of 25% of original dose until 6 mg/kg is reached.

ADMINISTRATION

Intravenous

PREPARE: IV Infusion: Dilute doses up to 1000 mg with NS to a

Common adverse effects in *italic*, life-threatening effects <u>underlined</u>; generic names in **bold**; classifications in SMALL CAPS; ♣ Canadian drug name; ⊘ Prototype drug

1139

total volume of 100 mL. ▪ Dilute higher doses with NS to a total volume of 150 mL. ▪ Final concentration should not exceed 10 mg/mL. ▪ Mix by gentle inversion and do not shake. Solution will contain small translucent particles that will be removed by filtration during infusion.

ADMINISTER: IV Infusion: Infuse doses less than 1000 mg over 60 min. ▪ Infuse doses greater than 1000 mg over 90 min. ▪ Use an infusion pump and a 0.2 or 0.22 micron in-line filter. ▪ Flush the line before/after infusion with NS. ▪ Discontinue infusion immediately if an anaphylactic reaction is suspected (i.e., bronchospasm, fever, chills, hypotension).

▪ Store unopened vials at 2°–8° C (36°–46° F). Protect vials from direct sunlight. ▪ Use diluted infusion solution within 6 h if stored at room temperature, or within 24 h if stored at 2°–8° C (36°–46° F).

ADVERSE EFFECTS (≥1%) **Body as a Whole:** *Fatigue,* <u>infectious sequelae and septic death,</u> infusion reactions, peripheral edema. **GI:** *Abdominal pain, constipation, diarrhea,* mucosal inflammation, *nausea,* stomatitis, *vomiting.* **Metabolic:** *Hypomagnesemia.* **Respiratory:** *Cough,* pulmonary fibrosis. **Skin:** *Acneiform dermatitis, dry skin, erythema, pruritus, skin exfoliation, skin fissures, nail disorders, paronychia.* **Special Senses:** Conjunctivitis, eye/eyelid irritation, increased lacrimation, ocular hyperemia.

PHARMACOKINETICS Half-Life: 7.5 days.

NURSING IMPLICATIONS

Assessment & Drug Effects

▪ Monitor for S&S of a severe infusion reaction; check vital signs q30min during infusion and 30 min post-infusion.

▪ Monitor for and report S&S of dermatologic toxicity such as acnelike dermatitis, pruritus, erythema, rash, skin exfoliation, dry skin, and skin fissures; inflammatory or infectious sequelae in those who experience severe dermatologic toxicities.

▪ Withhold drug and notify prescriber for any signs of drug toxicity.

▪ Lab tests: Periodic serum electrolytes during and for 8 wk following completion of therapy.

Patient & Family Education

▪ Immediately report any discomfort experienced during and shortly after drug infusion.

▪ Wear sunscreen and limit sun exposure while receiving panitumumab.

▪ Report any of the following to a health care provider: Any signs of irritation, inflammation, or infection of the skin, nails, or eyes; shortness of breath or any other breathing difficulty.

▪ Women of childbearing age should use reliable means of contraception during and for 6 mo after the last dose of panitumumab.

PANTOPRAZOLE SODIUM
(pan-to′pra-zole)
Protonix, Protonix IV
Classifications: GASTRIC PROTON PUMP INHIBITOR; ANTISECRETORY
Therapeutic: ANTIULCER
Prototype: Omeprazole
Pregnancy Category: B

AVAILABILITY 20 mg, 40 mg enteric coated tablets; 40 mg injection; 40 mg delayed release oral suspension

ACTION & *THERAPEUTIC EFFECT*

Gastric acid pump inhibitor that belongs to a class of antisecretory compounds. Gastric acid secretion is decreased by inhibiting the H^+, K^+-ATPase enzyme system responsible for acid production. *Suppresses gastric acid secretion by inhibiting the acid (proton H^+) pump in the parietal cells.*

USES
Short-term treatment of erosive esophagitis associated with gastroesophageal reflux disease (GERD), hypersecretory disease.
UNLABELED USES Peptic ulcer disease.

CONTRAINDICATIONS
Hypersensitivity to pantoprazole or other proton pump inhibitors (PPIs); lactation.
CAUTIOUS USE Mild to severe hepatic insufficiency, cirrhosis; concurrent administration of EDTA-containing products; pregnancy (category B). Safety and effectiveness in children younger than 18 y are not established.

ROUTE & DOSAGE

Erosive Esophagitis
Adult: **PO** 40 mg daily × 8–16 wks **IV** 40 mg daily × 7–10 days

Hypersecretory Disease
Adult: **PO** 40 mg b.i.d. (doses up to 240 mg/day have been used) **IV** 80 mg b.i.d.; adjust based on acid output

Renal Impairment/Hepatic Impairment Dosage Adjustment
Adjustment not needed

Hemodialysis Dosage Adjustment
Drug not removed

ADMINISTRATION

Oral
- Do not crush or break in half. **Must be** swallowed whole.
- Granules for oral suspension should be given 30 min before meals. Granules should be put into apple juice or applesauce, not water.
- *NG tube administration:* Add granules for suspension to a catheter tip syringe. Add 40 mL of apple juice in 10 mL increments to fully suspend granules and ensure that all of the drug is washed into the stomach.
- Store preferably at 20°–25° C (66°–77° F), but room temperature permitted.

Intravenous

PREPARE: IV Infusion: Reconstitute each 40 mg vial with 10 mL NS to yield 4 mg/mL. ▪ The required dose of 40 or 80 mg may be further diluted to a **total volume** of 100 mL in D5W, NS, or LR to yield 0.4 mg/mL or 0.8 mg/mL, respectively.

ADMINISTER: IV Infusion: Give through a dedicated line or flushed IV line before and after each dose with D5W, NS, or LR. ▪ Give the 4 mg/mL concentration over at least 2 min. ▪ Infuse the 0.4 or 0.8 mg/mL concentration over 15 min.

INCOMPATIBILITIES Solution/additive: Solutions containing **zinc.** **Y-site: Midazolam, zinc.**

- Reconstituted solution may be stored for up to 6 h at 15–30° C (59–86° F) before further dilution. ▪ The diluted 100 mL solution should be infused within 24 h or infused within 24 h of reconstitution.

ADVERSE EFFECTS (≥1%) **GI:** Diarrhea, *flatulence, abdominal pain.* **CNS:** Headache, insomnia. **Skin:** Rash.

INTERACTIONS Drug: May decrease absorption of **ampicillin,**

Common adverse effects in *italic*, life-threatening effects underlined; generic names in **bold**; classifications in SMALL CAPS; ♣ Canadian drug name; ● Prototype drug

1141

IRON SALTS, **itraconazole, ketoconazole;** increases INR with **warfarin. Herbal: Ginkgo** may decrease plasma levels. **Lab Test:** May cause false-positive urine tetrahydrocannabinol (THC) test.

PHARMACOKINETICS Absorption: Well absorbed with 77% bioavailability. **Peak:** 2.4 h. **Distribution:** 98% protein bound. **Metabolism:** In liver (CYP2C19). **Elimination:** 71% in urine, 18% in feces. **Half-Life:** 1 h.

NURSING IMPLICATIONS

Assessment & Drug Effects

▪ Monitor for and immediately report S&S of angioedema or a severe skin reaction.

▪ Lab tests: Urea breath test 4–6 wk after completion of therapy.

Patient & Family Education

▪ Contact prescriber promptly if any of the following occur: Peeling, blistering, or loosening of skin; skin rash, hives, or itching; swelling of the face, tongue, or lips; difficulty breathing or swallowing.

PAPAVERINE HYDROCHLORIDE

(pa-pav′er-een)
Para-Time
Classification: NONNITRATE VASODILATOR
Therapeutic: VASODILATOR; SMOOTH MUSCLE RELAXANT
Prototype: Hydralazine
Pregnancy Category: C

AVAILABILITY 30 mg/mL injection

ACTION & *THERAPEUTIC EFFECT*
Exerts nonspecific direct spasmolytic effect on smooth muscles unrelated to innervation. Acts directly on myocardium, depresses conduction and irritability, and prolongs refractory period. *Relaxes the smooth muscle of the heart as well as produces relaxation of vascular smooth muscles.*

USES Smooth muscle spasms.
UNLABELED USES Impotence, cardiac bypass surgery.

CONTRAINDICATIONS Parenteral use in complete AV block.
CAUTIOUS USE Glaucoma; myocardial depression; QT prolongation, angina pectoris; recent stroke; pregnancy (category C); lactation.

ROUTE & DOSAGE

Cerebral and Peripheral Ischemia
Adult: **IM/IV** 30–120 mg q3h as needed

ADMINISTRATION

Intramuscular
▪ Aspirate carefully before injecting IM to avoid inadvertent entry into blood vessel, and administer slowly.

Intravenous
▪ IV administration to children: Verify correct IV concentration and rate of infusion with prescriber.

PREPARE: Direct: Give undiluted or diluted in an equal volume of sterile water for injection.
ADMINISTER: Direct: Give slowly over 1–2 min. AVOID rapid injection.

INCOMPATIBILITIES Solution/additive: **Aminophylline, heparin, lactated Ringer's.**

ADVERSE EFFECTS (≥1%) **Body as a Whole:** General discomfort, facial flushing, sweating, weakness, coma. **CNS:** Dizziness, drowsiness, headache, sedation. **CV:** Slight rise in BP, paroxysmal tachycardia, transient ventricular ectopic rhythms, AV block, arrhythmias. **GI:** Nausea, anorexia, constipation, diarrhea, abdominal distress, dry mouth and throat, hepatotoxicity (jaundice, eosinophilia, abnormal liver function tests); with rapid IV administration. **Respiratory:** Increased depth of respiration, respiratory depression, fatal apnea. **Skin:** Pruritus, skin rash. **Special Senses:** Diplopia, nystagmus. **Urogenital:** Priapism.

INTERACTIONS Drug: May decrease **levodopa** effectiveness; **morphine** may antagonize smooth muscle relaxation effect of papaverine.

PHARMACOKINETICS Absorption: Readily from GI tract. **Peak:** 1–2 h. **Duration:** 12 h sustained release. **Metabolism:** In liver. **Elimination:** In urine chiefly as metabolites. **Half-Life:** 90 min.

NURSING IMPLICATIONS

Assessment & Drug Effects
- Monitor pulse, respiration, and BP in patients receiving drug parenterally. If significant changes are noted, withhold medication and report promptly to prescriber.
- Lab tests: Perform LFTs and blood tests periodically. Hepatotoxicity (thought to be a hypersensitivity reaction) is reversible with prompt drug withdrawal.

Patient & Family Education
- Notify prescriber if any adverse effect persists or if GI symptoms, jaundice, or skin rash appear. Liver function tests may be indicated.

- Do not drive or engage in potentially hazardous activities until response to drug is known. Alcohol may increase drowsiness and dizziness.

PAREGORIC (CAMPHORATED OPIUM TINCTURE)

(par-e-gor′ik)

Classifications: ANTIDIARRHEAL; NARCOTIC (OPIATE AGONIST) ANALGESIC
Therapeutic: ANTIDIARRHEAL
Prototype: Loperamide
Pregnancy Category: C
Controlled Substance: Schedule III

AVAILABILITY 2 mg/5 mL liquid

ACTION & *THERAPEUTIC EFFECT* Pharmacologic activity is due to morphine content, but is 25 times less potent. Decreases GI motility and effective propulsive peristalsis while diminishing digestive secretions. *Delayed transit of intestinal contents results in desiccation of feces and constipation.*

USES Short-term treatment for symptomatic relief of acute diarrhea and abdominal cramps.

CONTRAINDICATIONS Hypersensitivity to opium alkaloids; diarrhea caused by poisons (until eliminated); COPD.
CAUTIOUS USE Asthma; liver disease; GI disease; history of opiate agonist dependence; severe prostatic hypertrophy; pregnancy (category C), lactation, children.

ROUTE & DOSAGE

Acute Diarrhea

Adult: **PO** 5–10 mL after loose bowel movement, 1–4 times daily if needed
Child: **PO** 0.25–0.5 mL/kg 1–4 times/day

P

Common adverse effects in *italic*, life-threatening effects underlined; generic names in **bold**; classifications in SMALL CAPS; ♣ Canadian drug name; ⊙ Prototype drug

1143

ADMINISTRATION

Oral

- Give paregoric in sufficient water (2 or 3 swallows) to ensure its passage into the stomach (mixture will appear milky).

ADVERSE EFFECTS (≥1%) **GI:** Anorexia, nausea, vomiting, *constipation,* abdominal pain. **Body as a Whole:** Dizziness, faintness, drowsiness, facial flushing, sweating, physical dependence.

INTERACTIONS Drug: Alcohol and other CNS DEPRESSANTS add to CNS effects.

PHARMACOKINETICS Absorption: Readily from GI tract. **Duration:** 4–5 h. **Distribution:** Crosses placenta; distributed into breast milk. **Metabolism:** In liver. **Elimination:** In urine. **Half-Life:** 2–3 h.

NURSING IMPLICATIONS

Assessment & Drug Effects

- Paregoric may worsen the course of infection-associated diarrhea by delaying the elimination of pathogens.
- Be aware that adverse effects are primarily due to morphine content. Paregoric abuse results because of the narcotic content of the drug.
- Assess for fluid and electrolyte imbalance until diarrhea has stopped.

Patient & Family Education

- Adhere strictly to prescribed dosage schedule.
- Maintain bed rest if diarrhea is severe with a high level of fluid loss.
- Replace fluids and electrolytes as needed for diarrhea. Drink warm clear liquids and avoid dairy products, concentrated sweets, and cold drinks until diarrhea stops.

- Observe character and frequency of stools. Discontinue drug as soon as diarrhea is controlled. Report promptly to prescriber if diarrhea persists more than 3 days, if fever is higher than 38.8° C (102° F), abdominal pain develops, or if mucus or blood is passed.
- Understand that constipation is often a consequence of antidiarrheal treatment and a normal elimination pattern is usually established as dietary intake increases.

PARICALCITOL

(par-i-cal′ci-tol)
Zemplar
Classification: VITAMIN D ANALOG
Therapeutic: VITAMIN D ANALOG
Prototype: Calcitriol
Pregnancy Category: C

AVAILABILITY 2 mcg/mL, 5 mcg/mL vial; 1 mcg, 2 mcg, 4 mcg capsules

ACTION & *THERAPEUTIC EFFECT* Synthetic vitamin D analog that reduces parathyroid hormone (PTH) activity levels in chronic kidney failure (CRF) patients. Lowers serum levels of calcium and phosphate. Decreases the parathyroid hormone as well as bone resorption in some patients. *Effectiveness indicated by iPTH levels less than 1.5–3 times the upper limit of normal.*

USES Prevention and treatment of secondary hyperparathyroidism associated with CRF.

CONTRAINDICATIONS Hypersensitivity to paricalcitol; hypercalcemia; evidence of vitamin D toxicity; concurrent administration of phosphate preparations and vitamin D.
CAUTIOUS USE Severe liver disease; abnormally low levels of PTH;

Common adverse effects in *italic*, life-threatening effects underlined; generic names in **bold**; classifications in SMALL CAPS; ♣ Canadian drug name; ☻ Prototype drug

pregnancy (category C), lactation; children younger than 5 y.

ROUTE & DOSAGE

Secondary Hyperparathyroidism
Adult: **IV** 0.04 mcg/kg–0.1 mcg/kg (max: 0.24 mcg/kg), no more than every other day during dialysis **PO** iPTH less than 500 pg/mL, 1 mcg/day or 2 mcg 3 times/wk; iPTH greater than 500 pg/mL, 2 mcg/day or 4 mcg 3 times/wk

Renal Impairment Dosage Adjustment

Stage 5 Chronic Kidney Disease (CKD) (see package insert).

ADMINISTRATION
Oral
- Give no more frequently than every other day when dosing 3 times/wk.
- Store at 15–30° C (59–86° F).

Intravenous

PREPARE: Direct: Give undiluted.
ADMINISTER: Direct: Give IV bolus dose anytime during dialysis.

- Store at 25° C (77° F). Discard unused portion of a single dose vial.

ADVERSE EFFECTS (≥1%) **Body as a Whole:** Chills, feeling unwell, fever, flu-like symptoms, sepsis, edema. **CNS:** Light-headedness. **CV:** Palpitations. **GI:** Dry mouth, <u>GI bleeding</u>, *nausea,* vomiting. **Respiratory:** Pneumonia. **Metabolic:** Hypercalcemia.

INTERACTIONS Drug: Hypercalcemia may increase risk of **digoxin** toxicity; may increase **magnesium** absorption and toxicity in renal failure. Do not use with **vitamin D analogs**. **Herbal:** Be cautious of

vitamin D content in herbal and OTC products.

PHARMACOKINETICS Distribution: Greater than 99% protein bound. **Metabolism:** Via CYP3A4. **Elimination:** In feces. **Half-Life:** 15 h.

NURSING IMPLICATIONS
Assessment & Drug Effects
- Monitor for S&S of hypercalcemia (see Appendix F).
- Lab tests: Serum calcium and phosphate 2 times a wk during initiation of therapy; then monthly; serum PTH q3mo; periodic serum magnesium, alkaline phosphatase, 24-urinary calcium and phosphate. Increase frequency of lab tests during dosage adjustments.
- Withhold drug and notify prescriber if hypercalcemia occurs.
- Coadministered drugs: Monitor for digoxin toxicity if serum calcium level is elevated.

Patient & Family Education
- Report immediately any of the following to the prescriber: Weakness, anorexia, nausea, vomiting, abdominal cramps, diarrhea, muscle or bone pain, or excessive thirst.
- Adhere strictly to dietary regimen of calcium supplementation and phosphorus restriction to ensure successful therapy.
- Avoid excessive use of aluminum-containing compounds such as antacids/vitamins.

PAROMOMYCIN SULFATE ⦿
(par-oh-moe-mye´sin)
Classifications: AMINOGLYCOSIDE ANTIBIOTIC; AMEBICIDE
Therapeutic: AMEBICIDE
Pregnancy Category: C

Common adverse effects in *italic*, life-threatening effects <u>underlined</u>; generic names in **bold**; classifications in SMALL CAPS; ♣ Canadian drug name; ⦿ Prototype drug

1145

AVAILABILITY 250 mg capsules

ACTION & *THERAPEUTIC EFFECT*
Aminoglycoside antibiotic with broad-spectrum antibacterial activity. *Exerts direct bactericidal and amebicidal action, primarily in lumen of GI tract. Ineffective against extraintestinal amebiasis.*

USES Acute and chronic intestinal amebiasis; adjunctive therapy for hepatic coma.

CONTRAINDICATIONS Aminoglycoside hypersensitivity; intestinal obstruction; impaired kidney function.
CAUTIOUS USE GI ulceration; renal failure; renal impairment; older adults; myasthenia gravis; parkinsonism; pregnancy (category C).

ROUTE & DOSAGE

Intestinal Amebiasis
Adult/Child: **PO** 25–35 mg/kg divided in 3 doses for 5–10 days
Hepatic Coma
Adult: **PO** 4 g/day in divided doses for 5–6 days

ADMINISTRATION

Oral
- Give after meals to prevent gastric distress.

ADVERSE EFFECTS (≥1%) **CNS:** Headache, vertigo. **GI:** *Diarrhea, abdominal cramps,* steatorrhea, *nausea, vomiting, heartburn,* secondary enterocolitis. **Skin:** Exanthema, rash, pruritus. **Special Senses:** Ototoxicity. **Urogenital:** Nephrotoxicity (in patients with GI inflammation or ulcerations). **Body as a Whole:** Eosinophilia, overgrowth of nonsusceptible organisms.

DIAGNOSTIC TEST INTERFERENCE Prolonged use of paromomycin may cause reduction in ***serum cholesterol.***

INTERACTIONS Drug: May decrease absorption of **cyanocobalamin.**

PHARMACOKINETICS Absorption: Poorly from intact GI tract. **Elimination:** In feces.

NURSING IMPLICATIONS

Assessment & Drug Effects
- Monitor therapeutic effectiveness. Criterion of cure is absence of amoebae in stool specimens examined at weekly intervals for 6 wk after completion of treatment, and thereafter at monthly intervals for 2 y.
- Monitor for appearance of a superinfection during therapy (see Appendix F).
- Lab test: Baseline WBC with differential. Repeat if superinfection is suspected.
- Monitor closely patients with history of GI ulceration for nephrotoxicity and ototoxicity (see Appendix F). Drug absorption can take place through diseased mucosa.

Patient & Family Education
- Do not prepare, process, or serve food until treatment is complete when receiving drug for intestinal amebiasis. Isolation is not required.
- Practice strict personal hygiene, particularly hand washing after defecation and before eating food.

PAROXETINE
(par-ox′e-teen)

Pexeva, Paxil, Paxil CR
Classifications: ANTIDEPRESSANT; SELECTIVE SEROTONIN 5-HT REUPTAKE INHIBITOR (SSRI)
Therapeutic: ANTIDEPRESSANT; SSRI; ANTIANXIETY

Prototype: Fluoxetine
Pregnancy Category: D

AVAILABILITY 10 mg, 20 mg, 30 mg, 40 mg tablets; 12.5 mg, 25 mg, 37.5 mg sustained release tablets; 10 mg/5 mL suspension

ACTION & *THERAPEUTIC EFFECT*
Antidepressant that is a serotonin 5-HT reuptake inhibitor. It is highly potent and a highly selective inhibitor of serotonin reuptake by neurons in CNS. *Efficacious in depression resistant to other antidepressants and in depression complicated by anxiety.*

USES Depression, obsessive-compulsive disorders, panic attacks, excessive social anxiety, generalized anxiety, post-traumatic stress disorder (PTSD), premenstrual dysphoric disorder (PMDD).
UNLABELED USES Diabetic neuropathy, myoclonus, bipolar depression in conjunction with lithium, chronic headache, premature ejaculation, fibromyalgia.

CONTRAINDICATIONS Hypersensitivity to paroxetine; suicidal ideation; concomitant use of MAO inhibitors, alcohol; pregnancy (category D).
CAUTIOUS USE History of mania, suicidal tendencies; anorexia nervosa, ECT therapy; seizure disorder; renal/hepatic impairment, renal failure; history of metabolic disorders; volume-depleted patients, recent MI, unstable cardiac disease; lactation; children and adolescents.

ROUTE & DOSAGE

Depression
Adult: **PO** 10–50 mg/day (max: 80 mg/day); 25 mg sustained release daily in morning, may increase by 12.5 mg (max: 62.5 mg/day); use lower starting doses for patients with renal or hepatic insufficiency and geriatric patients
Geriatric: **PO** Start with 10 mg/day (12.5 mg/day sustained release), [max: 40 mg/day (50 mg/day sustained release)]

Obsessive-Compulsive Disorder
Adult: **PO** 20–60 mg/day
Panic Attacks
Adult: **PO** 40 mg/day
Social Anxiety Disorder
Adult: **PO** 20–60 mg/day
Generalized Anxiety, PTSD
Adult: **PO** Start with 10 mg once daily, may increase by 10 mg/day at weekly intervals if needed to target dose of 40 mg once daily (max: 60 mg/day)
Geriatric: **PO** Start with 10 mg PO once daily, may increase by 10 mg/day at weekly intervals if needed (max: 40 mg/day)

Premenstrual Dysphoric Disorder
Adult: **PO** 12.5 mg once daily (up to 25 mg once daily) throughout the month or daily for 2 wk before menstrual period

Pharmacogenetic Dosage Adjustment
Poor CYP2D6 metabolizers: Start with 65% of dose

ADMINISTRATION
Oral
- Ensure that sustained release form is not chewed or crushed. **Must be** swallowed whole.
- Be aware that at least 14 days should elapse when switching a patient from/to an MAO inhibitor to/from paroxetine.

ADVERSE EFFECTS (≥1%) **CV:** Postural hypotension. **CNS:** *Headache,*

Common adverse effects in *italic*, life-threatening effects underlined; generic names in **bold**; classifications in SMALL CAPS; ✦ Canadian drug name; ⊘ Prototype drug

1147

tremor, agitation or nervousness, anxiety, paresthesias, dizziness, insomnia, *sedation*. **GI:** *Nausea,* constipation, vomiting, anorexia, diarrhea, dyspepsia, flatulence, increased appetite, taste aversion, *dry mouth*. **Urogenital:** Urinary hesitancy or frequency, change in male fertility. **Hepatic:** Isolated reports of elevated liver enzymes. **Special Senses:** Blurred vision. **Skin:** Diaphoresis, rash, pruritus. **Metabolic:** Hyponatremia in older adult. **Body as a Whole:** Bone fracture (in older adults).

INTERACTIONS Drug: Activated charcoal reduces absorption of paroxetine. **Cimetidine** increases paroxetine levels. MAO INHIBITORS, **selegiline** may cause an increased vasopressor response leading to hypertensive crisis or death. **Phenytoin** can cause liver enzyme induction resulting in lower paroxetine levels and shorter half-life. **Warfarin** may increase risk of bleeding and **thioridazine** levels, and prolong QT_c interval leading to heart block; increase **ergotamine** toxicity with **dihydroergotamine, ergotamine.** May reduce the efficacy of **tamoxifen. Herbal: St. John's wort** may cause serotonin syndrome (headache, dizziness, sweating, agitation).

PHARMACOKINETICS Absorption: 99% from GI tract. **Onset:** 2 wk. **Peak:** 5–8 h. **Distribution:** Very lipophilic. 95% protein bound. Distributes into breast milk. **Metabolism:** Extensively in the liver to inactive metabolites via CYP2D6. **Elimination:** Less than 2% is excreted unchanged in urine. 65% of dose appears in urine as metabolites. Metabolites of paroxetine are also excreted in feces, presumably via bile. **Half-Life:** 24 h.

NURSING IMPLICATIONS

Assessment & Drug Effects

- Monitor for worsening of depression or emergence of suicidal ideation. Closely monitor those younger than 18 y for suicidal thinking and behavior.
- Monitor for adverse effects, which include headache, weakness, sedation, dizziness, insomnia; nausea, constipation, or diarrhea; dry mouth; sweating; male ejaculatory disturbance. These occur in more than 10% of all patients and may result in poor compliance with drug regimen.
- Monitor older adult for fluid and sodium imbalances.
- Monitor for significant weight loss.
- Monitor patients with history of mania for reactivation of condition.
- Monitor patients with preexisting cardiovascular disease carefully because paroxetine may adversely affect hemodynamic status.

Patient & Family Education

- Monitor children and adolescents for changes in behavior that may indicate suicidal ideation.
- Use caution when operating hazardous machinery or equipment until response to drug is known.
- Concurrent use of alcohol may increase risk of adverse CNS effects.
- Adaptation to some adverse effects (especially dizziness and nausea) may occur over a period of 4–6 wk.
- Do not stop drug therapy after improvement in emotional status occurs.
- Notify prescriber of any distressing adverse effects.

PAZOPANIB
(pas-o′pa-nib)
Votrient
Classifications: ANTINEOPLASTIC; TYROSINE KINASE INHIBITOR (TKI)

Common adverse effects in *italic*, life-threatening effects <u>underlined</u>; generic names in **bold**; classifications in SMALL CAPS; ✤ Canadian drug name; ✪ Prototype drug

Therapeutic: ANTINEOPLASTIC; TKI
Prototype: Gefitinib
Pregnancy Category: D

ALT over 3 × ULN and bilirubin 2 × ULN: Stop treatment

AVAILABILITY 200 mg tablets

ACTION & *THERAPEUTIC EFFECT*
A multi-tyrosine kinase inhibitor (TKI) of vascular endothelial growth factor receptor (VEGFR). Overexpression of VEGFR is present in many cancers. *Pazopanib inhibits growth of advanced renal cell carcinoma.*

USES Treatment of advanced renal cell carcinoma

CONTRAINDICATIONS Hepatotoxicity; severe hepatic impairment; ALT elevation greater than 3 × ULN concurrently with bilirubin elevation of greater than 2 × ULN; cerebral or GI bleeding within last 6 mo; uncontrolled hypertension; surgical procedures; pregnancy (category D); lactation.

CAUTIOUS USE Risk for QT prolongation; risk for or history of thrombotic event; risk for or history of GI perforation or fistula; moderate hepatic impairment; older adults; children younger than 18 y.

ROUTE & DOSAGE

Renal Cell Carcinoma
Adult: **PO** 800 mg once daily reduce to 400 mg once daily if larger dose isn't tolerated or if patient is taking a strong CYP3A4 inhibitor

Hepatic Impairment Dosage Adjustment
ALT 3–8 × ULN: Monitor liver function weekly until ALT returns to baseline
ALT over 8 × ULN: Stop therapy until ALT returns to baseline

ADMINISTRATION
Oral
▪ Give without food at least 1 h before or 2 h after a meal.
▪ Ensure that the tablets are swallowed whole. They should not be crushed or chewed.

ADVERSE EFFECTS (≥1%) **Body as a Whole:** Alopecia, *asthenia,* epistaxis, *fatigue, hair color changes.* **CNS:** *Headache.* **CV:** Chest pain, *hypertension,* myocardial infarction, QT elevation, transient ischemic attack. **GI:** *Abdominal pain, diarrhea, dyspepsia, nausea,* rectal hemorrhage, *vomiting.* **Hematologic:** Leukopenia, *lymphocytopenia,* neutropenia, thrombocytopenia. **Metabolic:** *Alterations in glucose, anorexia, AST and ALT elevation, decreased magnesium, decreased phosphorus, decreased sodium,* decreased weight, *elevated bilirubin,* hypothyroidism, lipase enzyme elevation, proteinuria. **Respiratory:** Hemoptysis. **Skin:** Facial edema, palmar-plantar erythrodysesthesia, rash, skin depigmentation. **Special Senses:** Dysgeusia. **Urogenital:** Hematuria.

INTERACTIONS Drug: Strong INHIBITORS OF CYP3A4 (e.g., **ketoconazole, ritonavir, clarithromycin**) may increase pazopanib levels. INDUCERS OF CYP3A4 (e.g., **rifampin**) may decrease pazopanib levels. Pazopanib may increase the levels of other drugs that require CYP3A4, CYP2D6, or CYP2C8 for their metabolism. **Food: Grapefruit juice** may increase pazopanib levels.

PHARMACOKINETICS Peak: 2–4 h. **Distribution:** Greater than 99%

Common adverse effects in *italic*, life-threatening effects underlined; generic names in **bold**; classifications in SMALL CAPS; ♣ Canadian drug name; ⊘ Prototype drug

1149

plasma protein bound. **Metabolism:** Hepatic oxidation by CYP3A4. **Elimination:** Primarily fecal. **Half-Life:** 30.9 h.

NURSING IMPLICATIONS

Assessment & Drug Effects

- Monitor BP closely. Consult prescriber for desired parameters and report promptly BP elevations above desired levels.
- Monitor cardiac status, especially in those at higher risk for QT interval prolongation. ECG monitoring as warranted.
- Lab tests: Baseline LFTs, then q4wk for 4 mo, and periodically thereafter; periodic thyroid function tests, urinalysis for proteinuria; baseline and periodic serum electrolytes.
- Withhold drug and notify prescriber immediately if ALT exceeds 3 × ULN and bilirubin exceeds 2 × ULN.

Patient & Family Education

- Report promptly any of the following: Unexplained signs of bleeding, jaundice, unusually dark urine, unusual tiredness, or pain in the right upper abdomen.
- Do not take OTC drugs, herbs, vitamins or dietary supplements without consulting prescriber.
- Women of childbearing age should use adequate means of contraception to avoid pregnancy while on this drug.

PEGFILGRASTIM

(peg-fil-gras′tim)

Neulasta

Classifications: HEMATOPOIETIC GROWTH FACTOR; GRANULOCYTE COLONY-STIMULATING FACTOR (G-CSF)
Therapeutic: HEMATOPOIETIC GROWTH FACTOR; G-CSF
Prototype: Filgrastim
Pregnancy Category: C

AVAILABILITY 10 mg/mL injection

ACTION & *THERAPEUTIC EFFECT*

Human granulocyte colony-stimulating factor (G-CSF) produced by recombinant DNA. Endogenous G-CSF regulates the production of neutrophils within the bone marrow; primarily affects neutrophil proliferation, differentiation, and selected end-cell functional activity (including enhanced phagocytic activity, antibody-dependent killing, and increased expression of some functions associated with cell-surface antigens). *Increases neutrophil proliferation and differentiation within the bone marrow.*

USES To decrease the incidence of infection, as manifested by febrile neutropenia, in patients with non-myeloid malignancies receiving myelosuppressive anticancer drugs associated with a significant incidence of severe neutropenia with fever; to decrease neutropenia associated with bone marrow transplant; to treat chronic neutropenia.

CONTRAINDICATIONS Hypersensitivity to *E. coli*-derived proteins, 14 days before or 24 h after administration of chemotherapy; myeloid cancers; splenomegaly; ARDS.
CAUTIOUS USE Sickle cell disorders. For use in peripheral blood stem cells (PBSC) mobilization; neutropenic patients with sepsis; leukemia; pregnancy (category C), lactation; children weighing less than 45 kg.

ROUTE & DOSAGE

Neutropenia

Adult (weight greater than 45 kg):
Subcutaneous 6 mg once per chemotherapy cycle at least 24 h after chemotherapy

ADMINISTRATION

Subcutaneous

- Do not administer pegfilgrastim in the period 14 days before or 24 h after cytotoxic chemotherapy.
- Use only one dose per vial; do not reenter the vial.
- Prior to injection, pegfilgrastim may be allowed to reach room temperature for a maximum of 6 h. Discard any vial left at room temperature for longer than 6 h.
- Aspirate prior to injection to avoid injection into a blood vessel. Inject subcutaneously; do not inject intradermally. Recommended injection sites include outer area of upper arms, abdomen (excluding 2-in. area around navel), front of middle thighs, and upper outer areas of the buttocks.
- Store refrigerated at 2°–8° C (36°–46° F). Do not freeze. Avoid shaking.

ADVERSE EFFECTS (≥1%) **Body as a Whole:** *Bone pain,* hyperuricemia, *fever.* **Hematologic:** Anemia. **GI:** Nausea, anorexia, increased LFTs. **Body as a Whole:** *Bone pain,* hyperuricemia, *fever.*

INTERACTIONS Drug: Can interfere with activity of CYTOTOXIC AGENTS; do not use 14 days before or less than 24 h after CYTOTOXIC AGENTS; **lithium** may increase release of neutrophils.

PHARMACOKINETICS Absorption: Readily absorbed from subcutaneous site. **Half-Life:** 15–80 h.

NURSING IMPLICATIONS

Assessment & Drug Effects

- Lab tests: Obtain a baseline CBC with differential and platelet count prior to administering drug. Obtain CBC twice weekly during therapy to monitor neutrophil count and leukocytosis. Monitor Hct and platelet count regularly.

- Discontinue pegfilgrastim if absolute neutrophil count exceeds 10,000/mm³ after the chemotherapy-induced nadir. Neutrophil counts should then return to normal.
- Monitor patients with preexisting cardiac conditions closely. MI and arrhythmias have been associated with a small percent of patients receiving pegfilgrastim.
- Monitor temperature q4h. Incidence of infection should be reduced after administration of pegfilgrastim.
- Assess degree of bone pain if present. Consult prescriber if non-narcotic analgesics do not provide relief.

Patient & Family Education

- Report bone pain and, if necessary, request analgesics to control pain.
- Note: Proper drug administration and disposal is important. A puncture-resistant container for the disposal of used syringes and needles should be utilized.

PEGINTERFERON ALFA-2A ⊘

(peg-in-ter-fer′on)

Pegasys

Classifications: BIOLOGIC RESPONSE MODIFIER; IMMUNOMODULATOR; ALPHA INTERFERON

Therapeutic: ANTIVIRAL; ANTIHEPATITIS

Pregnancy Category: C

AVAILABILITY 180 mcg/mL vials; 180 mcg/0.5 mL prefilled syringes

ACTION & *THERAPEUTIC EFFECT*

Interferon-stimulated genes modulate processes leading to inhibition of viral replication in infected cells, inhibition of cell proliferation, and immunomodulation. Stimulates production of effector proteins that raise body temperature, and causes

Common adverse effects in *italic*, life-threatening effects underlined; generic names in **bold**; classifications in SMALL CAPS; ♣ Canadian drug name; ⊘ Prototype drug

1151

reversible decreases in leukocyte and platelet counts. *Induces antiviral effects by activation of macrophages, natural killer cells, and T-cells, thus boosting cellular immunity and suppressing hepatic inflammation and replication of hepatitis C virus.*

USES Chronic hepatitis C or chronic hepatitis B.
UNLABELED USES Renal cell carcinoma, chronic myelogenous leukemia.

CONTRAINDICATIONS Hypersensitivity to peginterferon alfa-2a or any of its components; severe immunosuppression; autoimmune thyroid diseases (e.g., Graves' disease, thyroiditis); autoimmune hepatitis; dental work; sepsis; *E. coli* hypersensitivity, neonates and infants because it contains benzyl alcohol.
CAUTIOUS USE History of neuropsychiatric disorder; alcoholism, substance abuse, bipolar disorder, mania, psychosis; bone marrow suppression; cardiac arrhythmias, history of MI, cardiac disease, heart failure, uncontrolled hypertension; pulmonary disease, including COPD; thyroid dysfunction; DM; autoimmune disorders; ulcerative and hemorrhagic colitis; pancreatitis; pulmonary disorders; liver impairment; HBV or HIV coinfection; retinal disease; renal impairment with creatinine clearance less than 50 mL/min; organ transplant recipients; older adults; pregnancy (category C); lactation; children younger than 18 y.

ROUTE & DOSAGE

Chronic Hepatitis B or C

Adult: **Subcutaneous** 180 mcg once weekly × 48 wk, may be used with **ribavirin**

Renal Impairment Dosage Adjustment

End stage renal disease: Reduce dose to 135 mcg once weekly

Hepatic Impairment Dosage Adjustment

Reduce dose to 90 mcg once weekly if LFTs progressively increase over baseline

ADMINISTRATION

Subcutaneous

- Give dose on the same day of each week. Administer subcutaneously in the abdomen or thigh and rotate injection sites.
- Warm refrigerated vial by rolling in hands for about 1 min. Do not use if particulate matter is visible in the vial or product is discolored. Discard any unused portion.
- Withhold drug and notify prescriber for any of the following: ANC less than 750/mm³ or platelet count less than 50,000/mm³.
- Store in the refrigerator at 36°–46° F (2°–8° C), do not freeze or shake. Protect from light. Vials are for single use only.

ADVERSE EFFECTS (≥1%) **Body as a Whole:** *Musculoskeletal pain, myalgia, arthralgia, fatigue, inflammation at injection site, flu-like symptoms, rigors, fever,* pain, malaise, asthenia, exacerbation of autoimmune disease. **CNS:** *Headache, depression,* anxiety, *irritability, insomnia, dizziness,* impaired concentration, impaired memory, suicidal ideation/attempts. **GI:** *Nausea, diarrhea, abdominal pain, anorexia,* dry mouth. **Hematologic:** Thrombocytopenia, *neutropenia.* **Skin:** *Alopecia, pruritus,* dermatitis, sweating, rash.

INTERACTIONS Drug: May increase **theophylline** levels; increased risk of fetal defects with **ribavirin;** additive myelosuppression with ANTINEOPLASTICS. Use with NNRTIS or PROTEASE INHIBITORS increases risk of hepatic damage. May require dosage adjustments of NRTIS.

PHARMACOKINETICS Peak: 72–96 h. **Elimination:** 30% in urine. **Half-Life:** 80 h.

NURSING IMPLICATIONS

Assessment & Drug Effects

- Monitor for S&S of hypersensitivity (e.g., angioedema, bronchoconstriction) and, if noted, institute appropriate medical action immediately. Note that transient rashes are not an indication to discontinue treatment.
- Withhold drug and notify prescriber for any of the following: Severe neuropsychiatric events (e.g., psychosis, hallucinations, suicidal ideation, depression, bipolar disorders and mania), severe neutropenia or thrombocytopenia, abdominal pain accompanied by bloody diarrhea and fever, S&S of pancreatitis, new or worsening ophthalmologic disorders, or any other severe adverse event (SEE CAUTIOUS USE).
- Monitor respiratory and cardiovascular status; report dyspnea, chest pain, and hypotension immediately; perform baseline and periodic ECG and chest X-ray.
- Lab tests: Baseline and periodic creatinine clearance, uric acid, CBC with differential, platelet count, Hct and Hgb, TSH, ALT, AST, bilirubin, blood glucose; retest CBC with differential, platelet count, Hct and Hgb after 2 wk and other blood chemistries after 4 wk. Serum HCV RNA levels after 24 wk of treatment.

- Baseline and periodic ophthalmology exams are recommended.

Patient & Family Education

- If you miss a drug dose and remember within 2 days of the scheduled dose, take the dose and continue with your regular schedule. If more than 2 days have passed, contact prescriber for instructions.
- Notify prescriber immediately for any of the following: Severe depression or suicidal thoughts, severe chest pain, difficulty breathing, changes in vision, unusual bleeding or bruising, bloody diarrhea, high fever, severe stomach or lower back pain, severe chest pain, development of a new or worsening of a preexisting skin condition.
- Follow up with lab tests; compliance with lab testing is extremely important while taking this drug.
- Do not drive or engage in other potentially hazardous activities until reaction to drug is known.
- Women should use reliable means of contraception while taking this drug and notify prescriber immediately if they become pregnant.

PEGINTERFERON ALFA-2B

(peg-in-ter-fer′on)

PEG-Intron, Peg-Intron Redipen, Sylatron
Classifications: BIOLOGIC RESPONSE MODIFIER; IMMUNOMODULATOR; ALPHA INTERFERON
Therapeutic: ANTIVIRAL; ANTIHEPATITIS
Prototype: Peginterferon alfa-2a
Pregnancy Category: C

AVAILABILITY 50 mcg, 80 mcg, 120 mcg, 150 mcg powder for injection

ACTION & *THERAPEUTIC EFFECT*
Binds to specific membrane receptors

Common adverse effects in *italic*, life-threatening effects <u>underlined</u>; generic names in **bold**; classifications in SMALL CAPS; ♣ Canadian drug name; ⊙ Prototype drug

1153

on the cell surface, thereby initiating suppression of cell proliferation, enhanced phagocytic activity of macrophages, augmentation of specific cytotoxic lymphocytes for target cells, and inhibition of viral replication in virus-infected cells. *Induces antiviral effects by activation of macrophages, natural killer cells, and T-cells, thus boosting cellular immunity and suppressing hepatic inflammation and replication of hepatitis C virus.*

USES Chronic hepatitis C, melanoma.
UNLABELED USES Renal carcinoma, hepatitis B.

CONTRAINDICATIONS Hypersensitivity to peginterferon; autoimmune hepatitis; decompensated liver disease {Child Pugh (class B and C)} in cirrhotic chronic hepatitis C patients before or during treatment; **Sylatron:** Hepatic decompensation {Child-Pugh (class B and C)}; persistently severe or worsening S&S of life-threatening neuropsychiatric, autoimmune, ischemic, or infectious disorders.

CAUTIOUS USE History of neuropsychiatric disorder; suicidal tendencies; bone marrow suppression; ulcerative and hemorrhagic colitis; pulmonary disorders; HBV or HIV coinfection; thyroid dysfunction; DM; cardiovascular disease; autoimmune disorders; pulmonary disease, COPD; retinal disease; renal impairment with creatinine clearance less than 50 mL/min; older adults; pregnancy (category C), lactation. Safety and efficacy in children younger than 3 y are not established.

ROUTE & DOSAGE

Chronic Hepatitis C
Adult: **Subcutaneous** 1 mcg/kg/wk based on weight and injected once weekly × 1 y

Melanoma (Sylatron only)
Adult: **Subcutaneous** 6 mcg/kg/wk × 8 doses then 3 mcg/kg/wk for up to 5 y

ADMINISTRATION

Subcutaneous
- Give dose on the same day of each week.
- *Vial reconstitution:* Be aware that two Safety Lok™ syringes are provided in the drug package: One for reconstitution and one for injection. Reconstitute with only 0.7 mL of supplied diluent and discard remaining diluent. Enter the vial only once as it does not contain a preservative. Swirl gently to produce a clear, colorless solution. Use solution immediately.
- *Redipen use:* To reconstitute the lyophilized powder in the Redipen, hold the Redipen upright with dose button down and press the 2 halves of the pen together until there is an audible click. Gently invert the pen to mix but do not shake. Select the dose by pulling back on the dosing button until the dark bands are visible and turning the button until the dark band is aligned with the correct dose.
- Store dry vial at 15°–30° C (59°–86° F). If necessary, store reconstituted solution up to 24 h at 2°–8° C (36°–46° F).

ADVERSE EFFECTS (≥1%) **Body as a Whole:** *Musculoskeletal pain, fatigue, inflammation at injection site, flu-like symptoms, rigors, fever, weight loss, viral infection,* pain, malaise, hypertonia. **CNS:** *Headache, depression, anxiety, emotional lability, irritability, insomnia, dizziness.* **GI:** *Nausea, anorexia, diarrhea, abdominal pain,* vomiting, dyspepsia, hepatomegaly. **Endocrine:** Hypothyroidism. **Hematologic:** Thrombocytopenia, neutropenia. **Respiratory:** *Pharyngitis*, sinusitis, cough. **Skin:** *Alopecia, pruritus, dry skin,* sweating, rash, flushing.

INTERACTIONS Drug: May increase **theophylline** levels; additive myelosuppression with ANTINEOPLASTICS; **zidovudine** may increase hematologic toxicity; increase **doxorubicin** toxicity, increase neurotoxicity with **vinblastine; aldesleukin (IL-2)** may potentiate the risk of kidney failure. Use with NNRTIS or PROTEASE INHIBITORS may increase risk of hepatic damage. May require dosage adjustments of NRTIS.

PHARMACOKINETICS Peak: 15–44 h. **Duration:** 48–72 h. **Elimination:** 30% in urine. **Half-Life:** 40 h (22–60 h).

NURSING IMPLICATIONS

Assessment & Drug Effects

- Monitor for S&S of hypersensitivity (e.g., angioedema, bronchoconstriction) and, if noted, institute appropriate medical action immediately. Note that transient rashes are not an indication to discontinue treatment.
- Monitor for and report immediately S&S of neuropsychiatric disorders (e.g., psychosis, hallucinations, suicidal ideation, depression).
- Monitor respiratory and cardiovascular status; report dyspnea, chest pain, and hypotension immediately; baseline and periodic ECG and chest X-ray.
- Lab tests: Baseline and periodic creatinine clearance, CBC with differential, platelet count, Hct and Hgb, TSH, ALT, AST, bilirubin, blood glucose; with diabetics or hypertensives. Serum HCV RNA levels are assessed after 24 wk of treatment.
- Withhold drug and notify prescriber for any of the following: Severe neuropsychiatric events, severe

neutropenia or thrombocytopenia, abdominal pain accompanied by bloody diarrhea and fever, S&S of pancreatitis, or any other severe adverse event (see CAUTIOUS USE).

- Baseline and periodic ophthalmology exams are recommended.

Patient & Family Education

- Drink fluids liberally while taking this drug, especially during the initial stages of therapy.
- Learn reasons for withholding drug (see ASSESSMENT & DRUG EFFECTS).
- Use effective means of contraception while taking this drug. Women should not become pregnant.
- Follow up with lab tests; compliance with lab testing is extremely important while taking this drug.

PEGLOTICASE

(peg-lo'ti-case)

Krystexxa

Classification: ANTIGOUT; URIC ACID METABOLIZING ENZYME
Therapeutic: ANTIGOUT
Pregnancy Category: C

AVAILABILITY 8 mg/mL injection

ACTION & *THERAPEUTIC EFFECT*
Pegloticase is a recombinant uric acid specific enzyme that catalyzes the conversion of uric acid to a water soluble, inert metabolite readily eliminated by the kidney. *It lowers the serum uric acid, thus reducing uric acid-induced inflammation.*

USES Treatment of chronic gout.

CONTRAINDICATIONS Established G6PD deficiency; lactation.
CAUTIOUS USE African or Mediterranean ancestry; uric acid level above 6 mg/mL; retreatment with pegloticase following discontinuation of therapy for longer than

---

Common adverse effects in *italic*, life-threatening effects underlined; generic names in **bold;** classifications in SMALL CAPS; ♣ Canadian drug name; ⊘ Prototype drug

6 wk; history of CHF; pregnancy (category C); children younger than 18 y.

ROUTE & DOSAGE

Chronic Gout
Adult: IV 8 mg over 2 hrs q8wk

ADMINISTRATION

Intravenous
- Note: Premedication with antihistamines and corticosteroids is required to minimize the risk of anaphylaxis and infusion reactions.
- Pegloticase should only be administered in a health care setting and by health care providers trained to manage anaphylaxis and infusion reactions.

PREPARE: **IV Infusion:** Do not shake vial. ▪ Withdraw 1 mL of pegloticase from the 2 mL vial and inject into 250 mL of NS. Do not mix or dilute with other drugs. ▪ Invert IV bag to mix but do not shake. ▪ Discard unused pegloticase remaining in vial.
ADMINISTER: **IV Infusion:** If refrigerated, bring to room temperature prior to infusion. DO NOT give IV push or bolus. Infuse over NO LESS THAN 120 min. ▪ Monitor closely throughout infusion and for 2 h post-infusion for S&S of anaphylaxis or an infusion reaction (e.g., urticarial, erythema, dyspnea, flushing, chest discomfort, chest pain, and rash). ▪ If S&S of anaphylaxis or an infusion reaction occur, stop infusion and immediately notify prescriber. ▪ If anaphylaxis is ruled out and infusion is restarted, it should be run at a slower rate. ▪ Infusion bags may be stored refrigerated for 4 h after dilution.

INCOMPATIBILITIES **Solution/additive:** Do not mix with another drug. **Y-site:** Do not mix with another drug.

- Store unopened vials in refrigerator and protect from light. Do not shake or freeze.

ADVERSE EFFECTS (≥5%) **Body as a Whole:** Anaphylaxis, contusion, ecchymosis, gout flare, infusion reaction, pruritus. **CV:** Chest pain. **GI:** Nausea, vomiting. **Respiratory:** Dyspnea, nasopharyngitis. **Skin:** Erythema, urticaria.

INTERACTIONS Drug: Potential interaction with other pegylated products (**pegfilgrastim, peginterferon alfa-2a**).

NURSING IMPLICATIONS

Assessment & Drug Effects
- Monitor closely during infusion and for 2 h post-infusion for S&S of anaphylaxis (see Appendix F) or an infusion reaction (see IV ADMINISTRATION).
- Assess vital signs frequently during and for 2 h post-infusion.
- Monitor cardiac status, especially with preexisting congestive heart failure.
- Lab tests: Baseline and periodic serum uric acid levels.
- Notify prescriber if uric acid level is 6 mg/mL or above. Two consecutive readings above 6 mm/mL may indicate need to discontinue treatment.

Patient & Family Education
- Report promptly any discomfort (e.g., wheezing, facial swelling, skin rash, redness of the skin, difficulty breathing, flushing, chest discomfort, chest pain, rash) during or following IV infusion.
- Goat flares may increase during the first 3 mo of therapy and are

not a reason to discontinue pegloticase.

PEGVISOMANT
(peg-vis′o-mant)

Somavert

Classifications: GROWTH HORMONE MODIFIER; GROWTH HORMONE RECEPTOR ANTAGONIST

Therapeutic: GROWTH HORMONE RECEPTOR ANTAGONIST

Pregnancy Category: B (use only when clearly needed)

AVAILABILITY 10 mg, 15 mg, 20 mg powder for injection

ACTION & *THERAPEUTIC EFFECT*
A growth hormone (GH) receptor antagonist that binds to GH receptors on cell surfaces where it blocks its action and ability to stimulate production of insulin-like growth factor I (IGF-I). *Produces a significant decrease in the level of serum insulin-like growth factor I (IGF-I), the primary mediator of GH effects on body tissues.*

USES Treatment of acromegaly when other treatments have failed or are inappropriate.

CONTRAINDICATIONS Hypersensitivity to pegvisomant or latex.

CAUTIOUS USE Pituitary tumors; DM; hepatic and/or renal impairment; neoplastic disease; older adults; pregnancy (category B use only when clearly needed); lactation; children.

ROUTE & DOSAGE

Acromegaly

Adult: **Subcutaneous** 40 mg loading dose, then 10 mg once daily. Adjust dose in 5 mg increments, up to 30 mg/day, based on serum IGF-I concentrations.

ADMINISTRATION

Subcutaneous

- Allow vials to reach room temperature, then reconstitute by adding 1 mL of supplied diluent (sterile water for injection) to the vial. Direct diluent against the glass wall of vial, then mix by gently rolling between palms of hands to dissolve. DO NOT SHAKE. ▪ Solution should be clear and colorless. ▪ Use within 6 h of reconstitution.
- Inject subcutaneously and exercise caution not to inject IV.
- Rotate injection sites and do not use any site more than once every 1–2 mo.
- Store vials of powder at 2°–8° C (36°–46° F).

ADVERSE EFFECTS (≥1%) **Body as a Whole:** Asthenia, flu-like syndrome, infection, injection site reactions, back pain, paresthesias, peripheral edema. **CNS:** Dizziness. **CV:** Angina, chest pain, hypertension, MI. **GI:** Elevated liver function tests, diarrhea, nausea, vomiting. **Metabolic:** Hypercholesterolemia, hypoglycemia, and low titer nonneutralizing antigrowth hormone antibodies. **Musculoskeletal:** Arthralgia. **Respiratory:** Sinusitis.

DIAGNOSTIC TEST INTERFERENCE Similar to growth hormone and may cross-react with ***growth hormone assays.*** Do not use these assays to monitor pegvisomant therapy.

INTERACTIONS Drug: OPIATE AGONISTS may lead to higher pegvisomant

Common adverse effects in *italic*, life-threatening effects underlined; generic names in **bold;** classifications in SMALL CAPS; ♣ Canadian drug name; ⊙ Prototype drug

1157

dosing requirements; may need to decrease doses of **insulin,** ORAL ANTI-DIABETIC AGENTS; **octreotide** may affect response.

PHARMACOKINETICS Absorption: 57% from subcutaneous injection site. **Peak:** 33–77 h. **Half-Life:** 6 days.

NURSING IMPLICATIONS

Assessment & Drug Effects

- Montior CV status with baseline and periodic BP measurements.
- Monitor diabetics for loss of glycemic control.
- Withhold drug and notify prescriber for significant elevation in AST/ALT or S&S of hepatitis.
- Lab tests: IGF-I levels 4–6 wk after initiation of therapy or any dose adjustment, then q6mo after IGF-I levels have normalized; periodic LFTs and lipid profile; frequent blood glucose monitoring, especially if diabetic.

Patient & Family Education

- Report promptly any of the following: Chest pain or tightness, signs of infection (e.g., fever, chills, flu-like symptoms).
- Discontinue drug and notify prescriber immediately if jaundice appears.
- Do not drive or engage in other hazardous activities until reaction to drug is known.

PEMETREXED
(pe-me-trex′ed)
Alimta
Classifications: ANTINEOPLASTIC; ANTIMETABOLITE, ANTIFOLATE
Therapeutic: ANTINEOPLASTIC
Prototype: Methotrexate
Pregnancy Category: D

AVAILABILITY 500 mg powder for injection

ACTION & *THERAPEUTIC EFFECT* Suppresses tumor growth by inhibiting both DNA synthesis and folate metabolism at multiple target enzymes. *Appears to arrest the cell cycle, thus inhibiting tumor growth.*

USES Treatment of malignant pleural mesothelioma that is unresectable or in patients that are not surgery candidates in combination with cisplatin; treatment of locally advanced or metastatic non-small-cell lung cancer (NSCLC).
UNLABELED USES Solid tumors, including bladder, breast, colorectal, gastric, head and neck, pancreatic, and renal cell cancers.

CONTRAINDICATIONS Mannitol hypersensitivity; creatinine clearance is less than 45 mL/min, renal failure, active infection; vaccines; pregnancy (category D); lactation.
CAUTIOUS USE Anemia, thrombocytopenia, neutropenia, dental disease; older adults; hepatic disease, renal impairment; hypoalbuminemia, hypovolemia, dehydration, ascites, pleural effusion; children younger than 18 y.

ROUTE & DOSAGE

Malignant Mesothelioma, Non–Small-Cell Lung Cancer
Adult: **IV** 500 mg/m² on day 1 of each 21-day cycle
Renal Impairment Dosage Adjustment
CrCl less than 45 mL/min: Not recommended

ADMINISTRATION

Intravenous
Pre-/posttreatment with folic acid, vitamin B₁₂, and dexamethasone are needed to reduce hematologic and gastrointestinal

toxicity, and the possibility of severe cutaneous reactions from pemetrexed.

PREPARE: IV Infusion: Reconstitute each 500 mg vial with 20 mL of preservative-free NS. ▪ Do not use any other diluent. Swirl gently to dissolve. Each vial will contain 25 mg/mL. ▪ Withdraw the needed amount of reconstituted solution and add to 100 mL of preservative-free NS. ▪ Discard any unused portion.

ADMINISTER: IV Infusion: Do NOT give a bolus dose. ▪ Infuse over 10 min.

INCOMPATIBILITIES Solution/additive: Solutions containing **calcium, lactated Ringer's. Y-site: Amphotericin B, calcium, cefazolin, cefotaxime, cefotetan, cefoxitin, ceftazidime, chlorpromazine, ciprofloxacin, dobutamine, doxorubicin, doxycycline, droperidol, gemcitabine, gentamicin, irinotecan, metronidazole, minocycline, mitoxantrone, nalbuphine, ondansetron, prochlorperazine, tobramycin, topotecan.**

▪ Store unopened single-use vials at room temperature between 15°–30° C (59°–86° F). ▪ The reconstituted drug is stable for up to 24 h at 2°–8° C (36°–46° F) or at 25° C (77° F).

ADVERSE EFFECTS (≥1%) **Body as a Whole:** *Fatigue, fever,* hypersensitivity reaction, edema, myalgia, arthralgia. **CNS:** Neuropathy, *mood alteration, depression.* **CV:** Chest pain, thromboembolism. **GI:** *Nausea, vomiting, constipation, anorexia, stomatitis, diarrhea,* dehydration, dysphagia, esophagitis, odynophagia, increased LFTs. **Hematologic:** *Neutropenia,*

leukopenia, anemia, thrombocytopenia. **Respiratory:** *Dyspnea.* **Skin:** *Rash, desquamation,* alopecia. **Urogenital:** *Increases serum creatinine,* renal failure.

INTERACTIONS Drug: Increased risk of renal toxicity with other nephrotoxic drugs (**acyclovir, adefovir, amphotericin B,** AMINOGLYCOSIDES, **carboplatin, cidofovir, cisplatin, cyclosporine, foscarnet, ganciclovir, sirolimus, tacrolimus, vancomycin**); NSAIDs may increase risk of renal toxicity in patients with preexisting renal insufficiency; may cause additive risk of bleeding with ANTICOAGULANTS, PLATELET INHIBITORS, **aspirin,** THROMBOLYTIC AGENTS.

PHARMACOKINETICS Metabolism: Not extensively. **Elimination:** Primarily in urine. **Half-Life:** 3.5 h.

NURSING IMPLICATIONS

Assessment & Drug Effects

▪ Withhold drug and notify prescriber if the absolute neutrophil count (ANC) is less than 1500 cells/mm^3 or the platelet count is less than at least 100,000 cells/mm^3, or if the CrCl is less than 45 mL/min.

▪ Lab tests: Baseline and periodic CBC with differential; monitor for nadir and recovery before each dose (on day 8 and 15, respectively, of each cycle); periodic LFTs, serum creatinine and BUN.

▪ Notify prescriber for S&S of neuropathy (paresthesia) or thromboembolism.

Patient & Family Education

▪ Report promptly any of the following to prescriber: Symptoms of anemia (e.g., chest pain, unusual weakness or tiredness, fainting spells, light-headedness, shortness of breath); symptoms

P

Common adverse effects in *italic*, life-threatening effects underlined; generic names in **bold;** classifications in SMALL CAPS; ✚ Canadian drug name; ⊘ Prototype drug

1159

of poor blood clotting (e.g., bruising; red spots on skin; black, tarry stools; blood in urine); symptoms of infection (e.g., fever or chills, cough, sore throat, pain or difficulty passing urine); symptoms of liver problems (e.g., yellowing of skin).

- Do not take nonsteroidal anti-inflammatory drugs (NSAIDs) without first consulting the prescriber.

PEMIROLAST POTASSIUM

(pem-ir'o-last po-tass'i-um)

Alamast
Pregnancy Category: C
See Appendix A-1.

PENBUTOLOL

(pen-bu'tol-ol)

Levatol
Classifications: BETA-ADRENERGIC ANTAGONIST; ANTIHYPERTENSIVE
Therapeutic: ANTIHYPERTENSIVE
Prototype: Propranolol
Pregnancy Category: C

AVAILABILITY 20 mg tablets

ACTION & *THERAPEUTIC EFFECT*
Synthetic beta$_1$- and beta$_2$-adrenergic blocking agent that competes with epinephrine and norepinephrine for available beta receptor sites. Lowers both supine and standing BP in hypertensive patients. Hypotensive effect is associated with decreased cardiac output, suppressed renin activity as well as beta blockage. *Effective in lowering mild to moderate blood pressure.*

USES Mild to moderate hypertension alone or with other antihypertensive agents.

CONTRAINDICATIONS Hypersensitivity to penbutolol; cardiogenic shock, acute CHF, sinus bradycardia, second and third degree AV block; bronchial asthma, acute bronchospasm; Raynaund's disease; COPD.

CAUTIOUS USE Cardiac failure; PVD; chronic bronchitis; diabetes; mental depression; myasthenia gravis, cerebrovascular insufficiency, stroke; renal disease; pregnancy (category C), lactation. Safety and effectiveness in children is not established.

ROUTE & DOSAGE

Hypertension
Adult: **PO** 10–20 mg daily, may increase to 40–80 mg/day

ADMINISTRATION

Oral
- Discontinue by reducing the dose gradually over 1 to 2 wk.

ADVERSE EFFECTS (≥1%) **CNS:** Dizziness, fatigue, *headache*, insomnia. **CV:** AV block, bradycardia. **GI:** Nausea, diarrhea, dyspepsia. **Respiratory:** Cough, dyspnea. **Urogenital:** Impotence.

INTERACTIONS Drug: DIURETICS and other HYPOTENSIVE AGENTS increase hypotensive effect; effects of **albuterol, metaproterenol, terbutaline, pirbuterol,** and **penbutolol** are antagonized; NSAIDs blunt hypotensive effect; decreases hypoglycemic effect of **glyburide; amiodarone** increases risk of bradycardia and sinus arrest.

PHARMACOKINETICS Absorption: Readily from GI tract. **Peak:** 2–3 h. **Duration:** 20 h. **Metabolism:** In liver. **Elimination:** In urine. **Half-Life:** 5 h.

Common adverse effects in *italic*, life-threatening effects <u>underlined</u>; generic names in **bold;** classifications in SMALL CAPS; ◆ Canadian drug name; ⊘ Prototype drug

NURSING IMPLICATIONS

Assessment & Drug Effects

- Take apical pulse before administering drug. If pulse is below 60, or other established parameter, hold the drug and contact prescriber.
- Take a BP reading before giving drug, if BP is not stabilized. If systolic pressure is 90 mm Hg or less, hold drug and contact prescriber.
- Check BP near end of dosage interval or before administration of next dose to evaluate effectiveness.
- Monitor therapeutic effectiveness. Full effectiveness of the drug may not be seen for 4–6 wk.
- Watch for S&S of bronchial constriction. Report promptly and withhold drug.
- Monitor diabetics for loss of glycemic control. Drug suppresses clinical signs of hypoglycemia (e.g., BP changes, increased pulse rate) and may prolong hypoglycemic state.
- Monitor carefully for exacerbation of angina during drug withdrawal.

Patient & Family Education

- Do not discontinue the drug without prescriber's advice because of the possible exacerbation of ischemic heart disease.
- If diabetic, report persistent S&S of hypoglycemia (see Appendix F) to prescriber (diabetics).
- Avoid driving or other potentially hazardous activities until response to drug is known.
- Make position changes slowly and avoid prolonged standing. Notify prescriber if dizziness and light-headedness persist.
- Comply with and do not alter established regimen (i.e., do not omit, increase, or decrease dosage or change dosage interval).
- Avoid prolonged exposure of extremities to cold.

- Avoid excesses of alcohol. Heavy alcohol consumption [i.e., greater than 60 mL (2 oz)/day] may elevate arterial pressure; therefore, to maintain treatment effectiveness, either avoid alcohol or drink moderately (less than 60 mL/day). Consult prescriber.

PENCICLOVIR
(pen-cy′clo-vir)

Denavir
Classification: ANTIVIRAL
Therapeutic: TOPICAL ANTIVIRAL
Prototype: Acyclovir
Pregnancy Category: B

AVAILABILITY 10 mg/g cream

ACTION & *THERAPEUTIC EFFECT*

Antiviral agent active against herpes simplex virus type 1 (HSV-1) and type 2 (HSV-2). HSV-1 and HSV-2 infected cells phosphorylate penciclovir utilizing viral thymidine kinase. Competes with viral DNA, thus inhibiting both viral DNA synthesis and replication. *Effectiveness is measured in decreased viral load.*

USES Treatment of recurrent herpes labialis (cold sores).

CONTRAINDICATIONS Hypersensitivity to penciclovir or famciclovir, lactation.

CAUTIOUS USE Acyclovir, or related antiviral hypersensitivity; pregnancy (category B). Safety and efficacy in children younger than 12 y have not been established. Safety in immunocompromised patients is not established.

ROUTE & DOSAGE

Cold Sores

Adult: **Topical** Apply q2h while awake × 4 days

ADMINISTRATION

Topical

- Apply as soon as possible after developing lesion.
- Do not apply to mucous membranes or near the eyes.
- Store at or below 30° C (86° F). Do not freeze.

ADVERSE EFFECTS (≥1%) **CNS:** Headache. **Skin:** Erythema.

PHARMACOKINETICS Absorption: Minimally absorbed from cold sore.

NURSING IMPLICATIONS

Assessment & Drug Effects

- Monitor the extent of lesions and treatment effectiveness.

Patient & Family Education

- Wash hands before and after application. Avoid contact of drug with eyes.
- Apply sunscreen to lips; may minimize recurrence of lesions.

PENICILLAMINE

(pen-i-sill′a-meen)

Cuprimine, Depen
Classifications: CHELATING AGENT; DISEASE-MODIFIYING ANTIRHEUMATIC DRUG (DMARD)
Therapeutic: CHELATING AGENT; ANTIRHEUMATIC (DMARD)
Pregnancy Category: D

AVAILABILITY 250 mg capsules

ACTION & _THERAPEUTIC EFFECT_
Combines chemically with cystine to form a soluble disulfide complex that prevents stone formation and may even dissolve existing cystic stones. Forms stable soluble chelate with copper, zinc, iron, lead, mercury, and possibly other heavy metals and promotes their excretion in urine. Mechanism of action in rheumatoid arthritis appears to be related to inhibition of collagen formation. _With Wilson's disease, therapeutic effectiveness is indicated by improvement in psychiatric and neurologic symptoms, visual symptoms, and liver function. With rheumatoid arthritis, therapeutic effectiveness is indicated by improvement in grip strength, decrease in stiffness following immobility, reduction of pain, decrease in sedimentation rate and rheumatoid factor._

USES To promote renal excretion of excess copper in Wilson's disease (hepatolenticular degeneration); active rheumatoid arthritis in patients who have failed to respond to conventional therapy; cystinuria.
UNLABELED USES Scleroderma, primary biliary cirrhosis, porphyria cutanea tarda, lead poisoning.

CONTRAINDICATIONS Hypersensitivity to penicillamine or to any penicillin; history of penicillamine-related aplastic anemia or agranulocytosis; rheumatoid arthritis patients with renal insufficiency or who are pregnant; renal failure; concomitant administration with drugs that can cause severe hematologic or renal reactions (e.g., antimalarials, gold salts); pregnancy (category D), lactation.
CAUTIOUS USE Allergy-prone individuals; DM; renal disease, renal impairment; hepatic impairment, hepatic disease; history of hematologic disease.

ROUTE & DOSAGE

Wilson's Disease

Adult: **PO** 250 mg q.i.d., with 3 doses 1 h a.c. and the last dose at least 2 h after the last meal
Child: **PO** 20 mg/kg/day in 2–4 divided doses (max: 1 g/day)

Common adverse effects in _italic_, life-threatening effects <u>underlined</u>; generic names in **bold**; classifications in SMALL CAPS; ✤ Canadian drug name; ☺ Prototype drug

Cystinuria

Adult: **PO** 250–500 mg q.i.d., with doses adjusted to limit urinary excretion of cystine to 100–200 mg/day
Child: **PO** 30 mg/kg/day in 4 divided doses with doses adjusted to limit urinary excretion of cystine to 100–200 mg/day

Rheumatoid Arthritis (RA)

Adult: **PO** 125–250 mg/day; may increase at 1–3 mo intervals up to 1–1.5 g/day
Child: **PO** 3 mg/kg/day (up to 250 mg/day) × 3 mo, then 6 mg/kg/day (up to 500 mg/day) in 2 divided doses × 3 mo [max: 10 mg/kg/day (up to 1.5 g/day) in 3–4 divided doses]

Lead Poisoning

Child: **PO** 30–40 mg/kg/day in 3–4 divided doses (max: 1.5 g/day); initiate at 25% target dose, gradually increase to full dose over 2–3 wk

Renal Impairment Dosage Adjustment

CrCl less than 50 mL/min: Avoid use

Hemodialysis Dosage Adjustment

In RA patients dose may be decreased from 250 mg daily to 250 mg 3 times/wk

ADMINISTRATION

Oral

- Give on empty stomach (60 min before or 2 h after meals) to avoid absorption of metals in foods by penicillamine.
- Give contents in 15–30 mL of chilled fruit juice or pureed fruit (e.g., applesauce) if patient cannot swallow capsules or tablets.

ADVERSE EFFECTS (≥1%) **Body as a Whole:** Fever, arthralgia, lymphadenopathy, thyroiditis, SLE-like syndrome, thrombophlebitis, hyperpyrexia, myasthenia gravis syndrome, tingling of feet, weakness. **GI:** *Anorexia, nausea, vomiting,* epigastric pain, diarrhea, oral lesions, *reduction or loss of taste perception (particularly salt and sweet), metallic taste,* activation of peptic ulcer, pancreatitis. **Urogenital:** Membranous glomerulopathy, *proteinuria,* hematuria. **Hematologic:** Thrombocytopenia, leukopenia, agranulocytosis, thrombotic thrombocytopenic purpura, hemolytic anemia, aplastic anemia. **Metabolic:** Pyridoxine deficiency. **Skin:** *Generalized pruritus, urticaria,* mammary hyperplasia, alveolitis, skin friability, excessive skin wrinkling, *early and late occurring rashes,* pemphigus-like rash, alopecia. **Special Senses:** Tinnitus, optic neuritis, ptosis.

INTERACTIONS **Drug:** ANTIMALARIALS, CYTOTOXICS, **gold** therapy may potentiate hematologic and renal adverse effects; **iron** may decrease penicillamine absorption.

PHARMACOKINETICS **Absorption:** Readily from GI tract. **Peak:** 1 h. **Distribution:** Crosses placenta. **Metabolism:** In liver. **Elimination:** In urine. **Half-Life:** 1–7 h.

NURSING IMPLICATIONS

Assessment & Drug Effects

- Lab tests: Baseline WBC with differential, direct platelet counts, Hgb, and urinalyses prior to initiation of therapy and every 3 days during the first month of therapy, then every 2 wk thereafter. Perform LFTs before start of therapy and at least twice yearly thereafter.

Common adverse effects in *italic,* life-threatening effects underlined; generic names in **bold;** classifications in SMALL CAPS; ♣ Canadian drug name; ⊙ Prototype drug

1163

- Withhold drug and contact prescriber if the patient with rheumatoid arthritis develops proteinuria greater than 1 g (some clinicians accept greater than 2 g) or if platelet count drops to less than 100,000/mm³, or platelet count falls below 3500–4000/mm³, or neutropenia occurs.

Patient & Family Education

- Note: Clinical evidence of therapeutic effectiveness may not be apparent until 1–3 mo of drug therapy.
- Take exactly as prescribed. Allergic reactions occur in about one third of patients receiving penicillamine. Temporary interruptions of therapy increase possibility of sensitivity reactions.
- Take temperature nightly during first few months of therapy. Fever is a possible early sign of allergy.
- Observe skin over pressure sites: Knees, elbows, shoulder blades, toes, buttocks. Penicillamine increases risk of skin breakdown.
- Report unusual bruising or bleeding, sore mouth or throat, fever, skin rash, or any other unusual symptoms to prescriber.

PENICILLIN G BENZATHINE
(pen-i-sill'in)

Bicillin, Bicillin L-A, Permapen
Classifications: BETA-LACTAM ANTIBIOTIC; NATURAL PENICILLIN
Therapeutic: ANTIBIOTIC
Prototype: Penicillin G potassium
Pregnancy Category: B

AVAILABILITY 300,000 units/mL, 600,000 units/mL, 1,200,000 units/2 mL, 2,400,000 units/4 mL injection

ACTION & *THERAPEUTIC EFFECT*
Acid-stable, penicillinase-sensitive,

long-acting form of natural penicillin. Acts by interfering with synthesis of mucopeptides essential to formation and integrity of the bacterial cell wall. *Effective against many strains of* Staphylococcus aureus, *gram-positive cocci, gram-negative cocci. Also effective against gram-positive bacilli and gram-negative bacilli as well as some strains of* Salmonella, Shigella, *and spirochetes.*

USES Infections highly susceptible to penicillin G, such as streptococcal, pneumococcal, and staphylococcal infections, venereal disease such as syphilis (including early, late, and congenital forms), and nonvenereal diseases (e.g., yaws, bejel, and pinta). Also used in prophylaxis of rheumatic fever.

CONTRAINDICATIONS Hypersensitivity to penicillins; IV administration. **CAUTIOUS USE** History of or suspected allergy (eczema, hives, hay fever, asthma); hypersensitivity to cephalosporins or carbapenems; history of colitis; IBD; renal disease, renal impairment; GI disease; pregnancy (category B); lactation; infants, neonates.

ROUTE & DOSAGE

Mild to Moderate Infections
Adult: **IM** 1,200,000 units once/ day
Child: **IM** *Weight greater than 27 kg,* 900,000 units once/day; *weight less than 27 kg,* 300,000–600,000 units once/day

Syphilis
Adult: **IM** Less than 1 y duration: 2,400,000 units as single dose; greater than 1 y duration: 2,400,000 units/wk for 3 wk
Child: **IM** Congenital: 50,000 units/kg as single dose

Common adverse effects in *italic*, life-threatening effects underlined; generic names in **bold**; classifications in SMALL CAPS; ♣ Canadian drug name; ⊙ Prototype drug

Prophylaxis for Rheumatic Fever

Adult: **IM** 1,200,000 units q4wk
Child: **IM** 1,200,000 units
q3–4wk

ADMINISTRATION

Intramuscular

- Do not confuse penicillin G benzathine with preparations containing procaine penicillin G (e.g., Bicillin C-R).
- Make IM injection deep into upper outer quadrant of buttock. In infants and small children, the preferred site is the midlateral aspect of the thigh.
- Shake multiple-dose vial vigorously before withdrawing desired IM dose. Shake prepared cartridge unit vigorously before injecting drug.
- Select IM site with care. Injection into or near a major peripheral nerve can result in nerve damage.
- Inadvertent IV administration has resulted in arterial occlusion and cardiac arrest.
- Make injections at a slow steady rate to prevent needle blockage.
- Store at 15°–30° C (59°–86° F).

ADVERSE EFFECTS (≥1%) **Body as a Whole:** *Local pain,* tenderness, and fever associated with IM injection, chills, fever, wheezing, <u>anaphylaxis</u>, neuropathy, <u>nephrotoxicity</u>; superinfections, Jarisch-Herxheimer reaction in patients with syphilis. **Skin:** Pruritus, urticaria, and other skin eruptions. **Hematologic:** Eosinophilia, hemolytic anemia, and other blood abnormalities. Also see PENICILLIN G POTASSIUM.

INTERACTIONS Drug: **Probenecid** decreases renal elimination; may decrease efficacy of ORAL CONTRACEPTIVES.

PHARMACOKINETICS Absorption: Slowly absorbed from IM site.

Peak: 12–24 h. **Duration:** 26 days. **Distribution:** Crosses placenta; distributed into breast milk. **Metabolism:** Hydrolyzed to penicillin in body. **Elimination:** Excreted slowly by kidneys.

NURSING IMPLICATIONS

Note: See penicillin G potassium for numerous additional clinical implications.

Assessment & Drug Effects

- Determine history of hypersensitivity reactions to penicillins, cephalosporins, or other allergens prior to initiation of drug therapy.
- Lab tests: Perform C&S tests prior to initiation of therapy and periodically thereafter. Perform periodic renal function tests.

Patient & Family Education

- Report immediately to prescriber the onset of an allergic reaction. There is great risk of severe and prolonged reactions because drug is absorbed so slowly.

PENICILLIN G POTASSIUM ℗
(pen-i-sill'in)
Megacillin ✣, Pfizerphen-G

PENICILLIN G SODIUM

Classifications: BETA-LACTAM ANTIBIOTIC; NATURAL PENICILLIN
Therapeutic: ANTIBIOTIC
Pregnancy Category: B

AVAILABILITY 1,000,000 units, 5,000,000 units, 10,000,000 units, 20,000,000 unit vials; 1,000,000 units/50 mL, 2,000,000 units/50 mL 3,000,000 units/50 mL injection

ACTION & *THERAPEUTIC EFFECT*
Acid-labile, penicillinase-sensitive, natural penicillin that acts by interfering with synthesis of mucopeptides

Common adverse effects in *italic*, life-threatening effects <u>underlined</u>; generic names in **bold**; classifications in SMALL CAPS; ✣ Canadian drug name; ℗ Prototype drug

1165

essential to formation and integrity of bacterial cell wall. Antimicrobial spectrum is narrow compared to that of semisynthetic penicillins. *Highly active against gram-positive cocci (e.g., non-penicillinase-producing* Staphylococcus, Streptococcus *groups) and gram-negative cocci. Also effective against gram-positive bacilli and gram-negative bacilli as well as some strains of* Salmonella *and* Shigella *and spirochetes.*

USES Moderate to severe systemic infections caused by penicillin-sensitive microorganisms. Certain staphylococcal infections; streptococcal infections. Also used as prophylaxis in patients with rheumatic or congenital heart disease. Since oral preparations are absorbed erratically and thus **must be** given in comparatively high doses, this route is generally used only for mild or stabilized infections or long-term prophylaxis.

CONTRAINDICATIONS Hypersensitivity to any of the penicillins or corn; nausea, vomiting, hypermotility, gastric dilatation; cardiospasm; viral infections; patients on sodium restriction. **Oral:** Severe infections.
CAUTIOUS USE History of or suspected allergy (asthma, eczema, hay fever, hives); history of allergy to cephalosporins; GI disorders; kidney or liver dysfunction, electrolyte imbalance; renal disease or renal impairment; myasthenia gravis, epilepsy, pregnancy (category B); neonates, young infants. Use during lactation may lead to sensitization of infants.

ROUTE & DOSAGE

Moderate to Severe Infections
Adult: **IV/IM** 2–24 million units divided q4h
Child: **IV/IM** 250,000–400,000 units/kg divided q4h

ADMINISTRATION

Note: Check whether prescriber has prescribed penicillin G potassium or sodium.

Intramuscular
- Do not use the 20,000,000 unit dosage form for IM injection.
- Reconstitute for IM: Loosen powder by shaking bottle before adding diluent (sterile water for injection or sterile NS). Keep the total volume to be injected small. Solutions containing up to 100,000 units/mL cause the least discomfort. Adding 10 mL diluent to the 1,000,000 unit vial = 100,000 units/mL. Shake well to dissolve.
- Select IM site carefully. IM injection is made deep into a large muscle mass. Inject slowly. Rotate injection sites.

Intravenous ──────────

PREPARE: **Intermittent/Continuous:** Reconstitute as for IM injection then withdraw the required dose and add to 100–1000 mL of D5W or NS IV solution, depending on length of each infusion.
ADMINISTER: **Intermittent:** *Adults:* Give over at least 1 h; *Infants and Children:* Give over 15–30 min. **Continuous:** Give at a rate required to infuse the daily dose in 24 h. ▪ With high doses, IV penicillin G should be administered slowly (usually over 24 h) to prevent electrolyte imbalance from potassium or sodium content. ▪ Prescriber will often prescribe specific flow rate.
INCOMPATIBILITIES **Solution/additive: Dextran 40, fat emulsion, aminophylline, amphotericin B, cephalothin, chlorpromazine, dopamine, hydroxyzine, metaraminol,**

metoclopramide, pentobarbital, prochlorperazine, promazine, sodium bicarbonate, TETRACYCLINES, **thiopental.**

- Store dry powder (for parenteral use) at room temperature. After reconstitution (initial dilution), store solutions for 1 wk under refrigeration. - Intravenous infusion solutions containing penicillin G are stable at room temperature for at least 24 h.

ADVERSE EFFECTS (≥1%) Body as a Whole: Coughing, sneezing, feeling of uneasiness; systemic anaphylaxis, fever, widespread increase in capillary permeability and vasodilation with resulting edema (mouth, tongue, pharynx, larynx), laryngospasm, malaise, serum sickness (fever, malaise, pruritus, urticaria, lymphadenopathy, arthralgia, angioedema of face and extremities, neuritis prostration, eosinophilia), SLE-like syndrome, Injection site reactions (pain, inflammation, abscess, phlebitis), superinfections (especially with *Candida* and gram-negative bacteria), neuromuscular irritability (twitching, lethargy, confusion, stupor, hyperreflexia, multifocal myoclonus, localized or generalized seizures, coma). **CV:** Hypotension, circulatory collapse, cardiac arrhythmias, cardiac arrest. **GI:** Vomiting, diarrhea, severe abdominal cramps, nausea, epigastric distress, diarrhea, flatulence, dark discoloration of tongue, sore mouth or tongue. **Urogenital:** Interstitial nephritis, Loeffler's syndrome, vasculitis. **Hematologic:** Hemolytic anemia, thrombocytopenia. **Metabolic:** Hyperkalemia (penicillin G potassium); hypokalemia, alkalosis, hypernatremia, CHF (penicillin G sodium). **Respiratory:** Bronchospasm, asthma. **Skin:** Itchy palms or axilla, pruritus, *urticaria,* flushed skin, *delayed skin rashes* ranging from urticaria to exfoliative dermatitis, Stevens-Johnson syndrome, fixed-drug eruptions, contact dermatitis.

DIAGNOSTIC TEST INTERFERENCE
Blood grouping and compatibility tests: Possible interference associated with penicillin doses greater than 20 million units daily. ***Urine glucose:*** Massive doses of penicillin may cause false-positive test results with ***Benedict's solution*** and possibly ***Clinitest*** but not with ***glucose oxidase methods*** (e.g., ***Clinistix, Diastix, TesTape***). ***Urine protein:*** Massive doses of penicillin can produce false-positive results when turbidity measures are used (e.g., ***acetic acid*** and ***heat, sulfosalicylic acid***); ***Ames reagent*** reportedly not affected. ***Urinary PSP excretion tests:*** False decrease in urinary excretion of PSP. ***Urinary steroids:*** Large IV doses of penicillin may interfere with accurate measurement of ***urinary 17-OHCS*** (***Glenn-Nelson technique*** not affected).

INTERACTIONS Drug: Probenecid decreases renal elimination; penicillin G may decrease efficacy of ORAL CONTRACEPTIVES; **colestipol** decreases penicillin absorption; POTASSIUM-SPARING DIURETICS may cause hyperkalemia with penicillin G potassium. **Food:** Food increases breakdown in stomach.

PHARMACOKINETICS Peak: 15–30 min IM. **Distribution:** Widely distributed; good CSF concentrations with inflamed meninges; crosses placenta; distributed in breast milk. **Metabolism:** 16–30% metabolized. **Elimination:** 60% in urine within 6 h. **Half-Life:** 0.4–0.9 h.

NURSING IMPLICATIONS
Assessment & Drug Effects
- Obtain an exact history of patient's previous exposure and

Common adverse effects in *italic,* life-threatening effects underlined; generic names in **bold;** classifications in SMALL CAPS; ♣ Canadian drug name; ⊘ Prototype drug

1167

sensitivity to penicillins and cephalosporins and other allergic reactions of any kind prior to treatment with penicillin.

- Hypersensitivity reactions are more likely to occur with parenteral penicillin than with the oral drug. Skin rash is the most common type allergic reaction and should be reported promptly to prescriber.
- Lab tests: Perform C&S tests prior to initiation of therapy. Evaluate renal, hepatic, and hematologic systems at regular intervals in patients on high-dose therapy. Additionally, check electrolyte balance periodically in patients receiving high parenteral doses.
- Observe all patients closely for at least 30 min following administration of parenteral penicillin. The rapid appearance of a red flare or wheal at the IM or IV injection site is a possible sign of sensitivity. Also suspect an allergic reaction if patient becomes irritable, has nausea and vomiting, breathing difficulty, or sudden fever. Report any of the foregoing to prescriber immediately.
- Be aware that reactions to penicillin may be rapid in onset or may not appear for days or weeks. Symptoms usually disappear fairly quickly once drug is stopped, but in some patients may persist for 5 days or more.
- Allergy to penicillin is unpredictable. It has occurred in patients with a negative history of penicillin allergy and also in patients with no known prior contact with penicillin (sensitization may have occurred from penicillin used commercially in foods and beverages).
- Be alert for neuromuscular irritability in patients receiving parenteral penicillin in excess of 20 million units/day who have renal insufficiency, hyponatremia, or underlying CNS disease, notably myasthenia gravis or epilepsy. Seizure precautions are indicated. Symptoms usually begin with twitching, especially of face and extremities.

- Monitor I&O, particularly in patients receiving high parenteral doses. Report oliguria, hematuria, and changes in I&O ratio. Consult prescriber regarding optimum fluid intake. Dehydration increases the concentration of drug in kidneys and can cause renal irritation and damage.
- Observe closely for signs of toxicity, especially in neonates, young infants, the older adult, and patients with impaired kidney function receiving high-dose penicillin therapy. Urinary excretion of penicillin is significantly delayed in these patients.
- Observe patients on high-dose therapy closely for evidence of bleeding, and bleeding time should be monitored. (In high doses, penicillin interferes with platelet aggregation.)

Patient & Family Education

- Understand that hypersensitivity reaction may be delayed. Report skin rashes, itching, fever, malaise, and other signs of a delayed reaction to prescriber immediately (see ADVERSE EFFECTS).
- Notify prescriber if following symptoms appear when taking penicillin for treatment of syphilis: Headache, chills, fever, myalgia, arthralgia, malaise, and worsening of syphilitic skin lesions. Reaction is usually self-limiting. Check with prescriber if symptoms do not improve within a few days or get worse.
- Report S&S of superinfection (see Appendix F).

PENICILLIN G PROCAINE
(pen-i-sill'in)

Classifications: BETA-LACTAM ANTIBIOTIC; NATURAL PENICILLIN

Therapeutic: ANTIBIOTIC
Prototype: Penicillin G potassium
Pregnancy Category: B

AVAILABILITY 600,000 units/mL, 300,000 units/mL

ACTION & *THERAPEUTIC EFFECT*
Long-acting form of penicillin G. The procaine salt has low solubility and thus creates a tissue depot from which penicillin is slowly absorbed. Slower onset of action than penicillin G potassium, but longer duration of action. It inhibits the final stage of bacterial cell wall synthesis by binding to specific penicillin-binding proteins (PBPs) located in the bacterial cell wall. This results in cell death of bacteria. *Same actions and antibacterial activity as for penicillin G potassium and is similarly inactivated by penicillinase and gastric acid.*

USES Moderately severe infections due to penicillin G-sensitive microorganisms that are susceptible to low but prolonged serum penicillin concentrations. Commonly, uncomplicated pneumococcal pneumonia, uncomplicated gonorrheal infections, and all stages of syphilis. May be used concomitantly with penicillin G or probenecid when more rapid action and higher blood levels are indicated.

CONTRAINDICATIONS History of hypersensitivity to any of the penicillins, or to procaine or any other "caine-type" local anesthetic.
CAUTIOUS USE History of or suspected allergy, hypersensitivity to cephalosporins, carbapenem; asthmatics; GI disease, renal disease; renal impairment; pregnancy (category B); lactation; infants, neonates.

ROUTE & DOSAGE

Moderate to Severe Infections

Adult: **IM** 600,000–1,200,000 units once/day
Child: **IM** 300,000 units once/day

Pneumococcal Pneumonia

Adult: **IM** 600,000 units q12h

Uncomplicated Gonorrhea

Adult: **IM** 4,800,000 units divided between 2 different injection sites at one visit preceded by 1 g of probenecid 30 min before injections

Syphilis

Adult: **IM** Primary, secondary, latent: 600,000 units/day for 8 days; late latent, tertiary, neurosyphilis: 600,000 units/day for 10–15 days
Child: **IM** 500,000–1,000,000 units/m^2 once/day

ADMINISTRATION

Intramuscular

- Shake multiple-dose vial thoroughly before withdrawing medication to ensure uniform suspension of drug.
- Use 20-gauge needle to avoid clogging.
- Give IM deep into upper outer quadrant of gluteus muscle; in infants and small children midlateral aspect of thigh is generally preferred. Select IM site carefully. Accidental injection into or near major peripheral nerves and blood vessels can cause neurovascular damage.
- Aspirate carefully before injecting drug to avoid entry into a blood vessel. Inadvertent IV administration reportedly has resulted in pulmonary infarcts and death.

P

Common adverse effects in *italic*, life-threatening effects <u>underlined</u>; generic names in **bold**; classifications in SMALL CAPS; ♣ Canadian drug name; ⓟ Prototype drug

1169

• Inject drug at a slow, but steady rate to prevent needle blockage. Give in two sites if the dose is very large. Rotate injection sites.

ADVERSE EFFECTS (≥1%) Body as a Whole: Procaine toxicity [e.g., mental disturbances (anxiety, confusion, depression, combativeness, hallucinations), expressed fear of impending death, weakness, dizziness, headache, tinnitus, unusual tastes, palpitation, changes in pulse rate and BP, seizures]. Also see PENICILLIN G POTASSIUM.

INTERACTIONS Drug: Probenecid decreases renal elimination; may decrease efficacy of ORAL CONTRACEPTIVES.

PHARMACOKINETICS Absorption: Slowly from IM site. **Peak:** 1–3 h. **Duration:** 15–20 h. **Distribution:** Crosses placenta; distributed into breast milk. **Metabolism:** Hydrolyzed to penicillin in body. **Elimination:** By kidneys within 24–36 h.

NURSING IMPLICATIONS

Assessment & Drug Effects

• Obtain an exact history of patient's previous exposure and sensitivity to penicillins, cephalosporins, and to procaine, and other allergic reactions of any kind prior to treatment.
• Test patient by injecting 0.1 mL of 1–2% procaine hydrochloride intradermally if sensitivity is suspected. Appearance of a wheal, flare, or eruption indicates procaine sensitivity.
• Be alert to the possibility of a transient toxic reaction to procaine, particularly when large single doses are administered. The reaction manifested by mental disturbance and other symptoms (see ADVERSE EFFECTS) occurs almost immediately and usually subsides after 15–30 min.

Patient & Family Education
• Report any skin reaction at the site of injection.
• Report onset of rash, itching, fever, chills or other symptoms of an allergic reaction to prescriber.

PENICILLIN V
PENICILLIN V POTASSIUM
(pen-i-sill'in)

Apo-Pen-VK ✦, Beepen VK, Betapen-VK, Ledercillin VK, Nado-pen-V ✦, Novopen-VK ✦, Penicillin VK, Pen-V, Pen-Vee K, Robicillin VK, V-Cillin K, Veetids
Classifications: BETA-LACTAM ANTIBIOTIC; NATURAL PENICILLIN
Therapeutic: ANTIBIOTIC
Prototype: Penicillin G potassium
Pregnancy Category: B

AVAILABILITY 250 mg, 500 mg tablets; 125 mg/5 mL, 250 mg/5 mL suspension

ACTION & *THERAPEUTIC EFFECT* Acid-stable analog of penicillin G with which it shares actions. It binds with the necessary penicillin-binding proteins (PBP) in cell wall of bacteria interfering with cell wall synthesis and resulting in cell lysis. *Penicillin V is bactericidal and is inactivated by penicillinase. Less active than penicillin G against gonococci and other gram-negative microorganisms.*

USES Mild to moderate infections caused by susceptible *Streptococci, Pneumococci,* and *Staphylococci.* Also Vincent's infection and as prophylaxis in rheumatic fever.

CONTRAINDICATIONS Hypersensitivity to any penicillin.
CAUTIOUS USE History of or suspected allergy (hay fever, asthma,

hives, eczema) reactions; hypersensitivity to cephalosporins, beta-lactamase inhibitors, or carbapenem; GI disease; cystic fibrosis; renal impairment, hepatic impairment; pregnancy (category B); lactation.

ROUTE & DOSAGE

Mild to Moderate Infections

Adult: **PO** 125–500 mg q6h
Child (younger than 12 y): **PO** 15–50 mg/kg/day in 3–6 divided doses

Endocarditis Prophylaxis

Adult: **PO** 2 g 30–60 min before procedure, then 500 mg q6h for 8 doses
Child (weight less than 30 kg): **PO** 1 g 30–60 min before procedure, then 250 mg q6h for 8 doses

ADMINISTRATION

Oral

- Give after a meal rather than on an empty stomach; drug may be better absorbed and result in higher blood levels.
- Shake well before pouring. Following reconstitution, oral solution is stable for 14 days under refrigeration.

ADVERSE EFFECTS (≥1%) **Body as a Whole:** Nausea, vomiting, *diarrhea,* epigastric distress. *Hypersensitivity reactions* (e.g., flushing, pruritus, urticaria or other skin eruptions, eosinophilia, <u>anaphylaxis</u>; hemolytic anemia, <u>leukopenia, thrombocytopenia</u>, neuropathy, superinfections).

INTERACTIONS Drug: Probenecid decreases renal elimination; may decrease efficacy of ORAL CONTRACEPTIVES; **colestipol** decreases absorption. **Food:** Food increases breakdown in stomach.

PHARMACOKINETICS Absorption: 60–73% absorbed from GI tract. **Peak:** 30–60 min. **Duration:** 6 h. **Distribution:** Highest levels in kidneys; crosses placenta; distributed into breast milk. **Elimination:** In urine. **Half-Life:** 30 min.

NURSING IMPLICATIONS

Note: See penicillin G potassium for numerous additional nursing implications.

Assessment & Drug Effects

- Obtain careful history concerning hypersensitivity reactions to penicillins, cephalosporins, and other allergens before therapy begins.
- Lab tests: Perform C&S tests prior to initiation and at regular intervals throughout therapy. Evaluate renal, hepatic, and hematologic systems at regular intervals in patients receiving prolonged therapy.

Patient & Family Education

- Take penicillin V around the clock at specific intervals to maintain a constant blood level.
- Do not miss any doses and continue taking medication until it is all gone unless otherwise directed by the prescriber.
- Discontinue medication and promptly report to prescriber the onset of hypersensitivity reactions and superinfections (see Appendix F).
- Use specially marked measuring device to ensure accurate doses of oral liquid preparation.

PENTAMIDINE ISETHIONATE

(pen-tam′i-deen)

Nebupent, Pentacarinat ♦, Pentam 300
Classification: ANTIPROTOZOAL
Therapeutic: ANTIPROTOZOAL
Pregnancy Category: C

P

Common adverse effects in *italic,* life-threatening effects <u>underlined</u>; generic names in **bold;** classifications in SMALL CAPS; ♦ Canadian drug name; ❶ Prototype drug

1171

AVAILABILITY 300 mg injection; 300 mg aerosol

ACTION & *THERAPEUTIC EFFECT*
Aromatic diamide antiprotozoal drug that appears to block parasite reproduction by interfering with nucleotide (DNA, RNA), phospholipid, and protein synthesis. *Effective against the protozoan parasite* Pneumocystis carinii *in AIDS patients.*

USES *P. carinii* pneumonia (PCP).
UNLABELED USES African trypanosomiasis and visceral leishmaniasis. (Drug supplied for latter uses is through the Centers for Disease Control and Prevention, Atlanta, GA.)

CONTRAINDICATIONS QT prolongation, history of torsades de pointes; lactation.
CAUTIOUS USE Hypertension, hypotension; hyperglycemia; pancreatitis; hypoglycemia; hypocalcemia; blood dyscrasias; liver or kidney dysfunction; diabetes mellitus; pancreatitis; asthma; cardiac arrhythmias; pregnancy (category C).

ROUTE & DOSAGE

Treatment of *Pneumocystis carinii* Pneumonia

Adult/Child: **IM/IV** 4 mg/kg/day for 14–21 days; infuse IV over 60 min

Prophylaxis of *Pneumocystis carinii* Pneumonia

Adult: **Inhaled** 300 mg per nebulizer q3–4wk
Child: **IV/IM** 4 mg/kg monthly

ADMINISTRATION

Inhaled
- Reconstitute contents of one vial in 6 mL sterile water (not saline) and administer using nebulizer.

- Do not mix with any other drug.

Intramuscular
- Dissolve contents of 1 vial (300 mg) in 3 mL sterile water for injection.
- Give deep IM into a large muscle.
- The IM injection is painful and frequently causes local reactions (pain, indurations, swelling). Select alternate sites for daily doses and institute local treatment if indicated.

Intravenous
PREPARE: IV Infusion: Dissolve contents of 1 vial in 3–5 mL sterile water for injection or D5W.
- Further dilute in 50–250 mL of D5W.
ADMINISTER: IV Infusion: Give over 60 min.
INCOMPATIBILITIES Y-site: Aldesleukin, CEPHALOSPORINS, **fluconazole, foscarnet, linezolid.**

- Note: IV solutions are stable at room temperature for up to 24 h. Protect solution from light.

ADVERSE EFFECTS (≥1%) **CNS:** Confusion, hallucinations, neuralgia, dizziness, sweating. **CV:** <u>Sudden, severe hypotension</u>, cardiac arrhythmias, ventricular tachycardia, phlebitis. **GI:** Anorexia, nausea, vomiting, pancreatitis, unpleasant taste. **Urogenital:** <u>Acute kidney failure</u>. **Hematologic:** <u>Leukopenia, thrombocytopenia</u>, anemia. **Metabolic:** <u>Hypoglycemia, hypocalcemia, *hyperkalemia*</u>. **Respiratory:** *Cough, bronchospasm,* laryngitis, shortness of breath, chest pain, <u>pneumothorax</u>. **Skin:** Stevens-Johnson syndrome, facial flush (with IV injection), *local reactions at injection site.*

INTERACTIONS Drug: AMINOGLYCOSIDES, **amphotericin B, cidofovir, cisplatin, ganciclovir, cyclosporine, vancomycin,** other nephrotoxic drugs increase risk of nephrotoxicity.

PHARMACOKINETICS Absorption: Readily after IM injection. **Distribution:** Leaves bloodstream rapidly to bind extensively to body tissues. **Elimination:** 50–66% in urine within 6 h; small amounts found in urine for as long as 6–8 wk. **Half-Life:** 6.5–13.2 h.

NURSING IMPLICATIONS

Assessment & Drug Effects

- Monitor BP and HR continuously during the infusion, every half hour for 2 h thereafter, and then every 4 h until BP stablizes. Sudden severe hypotension may develop after a single dose. Place patient in supine position while receiving the drug.
- Lab tests: Monitor periodically serum electrolytes, renal function, CBC with differential, platelet count, and blood glucose.
- Measure and record I&O ratio and pattern.
- Be alert and report promptly S&S of impending kidney dysfunction (e.g., changed I&O ratio, oliguria, edema).
- Characteristics of pneumonia in the immunocompromised patient include constant fever, scanty (if any) sputum, dyspnea, tachypnea, and cyanosis.
- Monitor temperature changes and institute measures to lower the temperature as indicated. Fever is a constant symptom in *P. carinii* pneumonia, but may be rapidly elevated [as high as 40° C (104° F)] shortly after drug infusion.

Patient & Family Education

- Report promptly to prescriber increasing respiratory difficulty.

- Monitor blood glucose for loss of glycemic control if diabetic.
- Report any unusual bruising or bleeding. Avoid using aspirin or other NSAIDs.
- Increase fluid intake (if not contraindicated) to 2–3 qt (L) per day.

PENTAZOCINE HYDROCHLORIDE ⓟ
(pen-taz'oh-seen)

Talwin
Classifications: NARCOTIC (OPIATE AGONIST-ANTAGONIST); ANALGESIC
Therapeutic: NARCOTIC ANALGESIC
Pregnancy Category: C
Controlled Substance: Schedule IV

AVAILABILITY 30 mg/mL injection

ACTION & THERAPEUTIC EFFECT Synthetic analgesic with potency approximately one-third that of morphine. Opiates exert their effects by stimulating specific opiate receptors that produce analgesia, respiratory depression, and euphoria as well as physical dependence. *Effective for moderate to severe pain relief. Acts as weak narcotic antagonist and has sedative properties.*

USES Relief of moderate to severe pain; also used for preoperative analgesia or sedation, and as supplement to surgical anesthesia.

CONTRAINDICATIONS Hypersensitivity to sulfite; head injury, increased intracranial pressure; seizures; emotionally unstable patients, or history of drug abuse; pregnancy (other than labor).
CAUTIOUS USE Impaired kidney or liver function; cardiac disease; COPD, asthmas, respiratory depression; GI obstruction; biliary surgery; patients

Common adverse effects in *italic*, life-threatening effects underlined; generic names in **bold**; classifications in SMALL CAPS; ♣ Canadian drug name; ⓟ Prototype drug

1173

with MI who have nausea and vomiting; pregnancy (category C), lactation; children younger than 12 y.

ROUTE & DOSAGE

Moderate to Severe Pain (Excluding Patients in Labor)

Adult: IM/IV/Subcutaneous 30–60 mg q3–4h (max: 360 mg/day)
Child: IM 15–30 mg

Women in Labor

Adult: IM 20–30 mg; 20 mg may be repeated 1 or 2 times at 2–3 h intervals

Renal Impairment Dosage Adjustment

CrCl 10–50 mL/min: Give 75% of dose; less than 10 mL/min: Give 50% of dose

ADMINISTRATION

Subcutaneous/Intramuscular

- IM is preferred to subcutaneous route when frequent injections over an extended period are required.
- Observe injection sites daily for signs of irritation or inflammation.

Intravenous

PREPARE: **Direct:** Give undiluted or diluted with 1 mL sterile water for injection for each 5 mg.
ADMINISTER: **Direct:** Give slowly at a rate of 5 mg over 60 sec.
INCOMPATIBILITIES **Solution/additive: Aminophylline,** BARBITURATES, **sodium bicarbonate, glycopyrrolate, heparin, nafcillin. Y-site: Nafcillin.**

ADVERSE EFFECTS (≥1%) **Body as a Whole:** Flushing, allergic reactions, shock. **CNS:** *Drowsiness,* sweating, *dizziness, light-headedness, euphoria,* psychotomimetic effects, confusion, anxiety, hallucinations, disturbed dreams, bizarre thoughts, euphoria and other mood alterations. **CV:** Hypertension, palpitation, tachycardia. **GI:** *Nausea, vomiting,* constipation, dry mouth, alterations of taste. **Urogenital:** Urinary retention. **Respiratory:** Respiratory depression. **Skin:** Injection-site reactions (induration, nodule formation, sloughing, sclerosis, cutaneous depression), rash, pruritus. **Special Senses:** Visual disturbances.

INTERACTIONS Drug: Alcohol and other CNS DEPRESSANTS add to CNS depression; NARCOTIC ANALGESICS may precipitate narcotic withdrawal syndrome.

PHARMACOKINETICS Onset: 15 min IM, Subcutaneous; 2–3 min IV. **Peak:** 1 h IM, 15 min IV. **Duration:** 3 h IM, 1 h IV. **Distribution:** Crosses placenta. **Metabolism:** Extensively in liver. **Elimination:** Primarily in urine; small amount in feces. **Half-Life:** 2–3 h.

NURSING IMPLICATIONS

Assessment & Drug Effects

- Monitor therapeutic effect. Tolerance to analgesic effect sometimes occurs. Psychologic and physical dependence have been reported in patients with history of drug abuse, but rarely in patients without such history. Addiction liability matches that of codeine.
- Monitor vital signs and assess for respiratory depression. Keep supine to minimize adverse effects.
- Monitor drug-induced CNS depression.
- Be aware that pentazocine may produce acute withdrawal symptoms in some patients who have been receiving opioids on a regular basis.
- Monitor I&O as drug may cause urinary retention.

Common adverse effects in *italic*, life-threatening effects underlined; generic names in **bold**; classifications in SMALL CAPS; ♣ Canadian drug name; ⦿ Prototype drug

Patient & Family Education

- Avoid driving and other potentially hazardous activities until response to drug is known.
- Do not discontinue drug abruptly following extended use; may result in chills, abdominal and muscle cramps, yawning, runny nose, tearing, itching, restlessness, anxiety, drug-seeking behavior.

PENTOBARBITAL
(pen-toe-bar′bi-tal)

PENTOBARBITAL SODIUM
Nembutal Sodium, Novopentobarb ◆

Classifications: ANXIOLYTIC; SEDATIVE-HYPNOTIC; BARBITURATE; ANTICONVULSANT
Therapeutic: ANTIANXIETY; SEDATIVE-HYPNOTIC; ANTICONVULSANT
Prototype: Secobarbital
Pregnancy Category: D
Controlled Substance: Schedule II

AVAILABILITY 50 mg/mL injection

ACTION & *THERAPEUTIC EFFECT*
Short-acting barbiturate with anticonvulsant properties. Potent respiratory depressant. Initially, barbiturates suppress REM sleep, but with chronic therapy REM sleep returns to normal. *Effective as a sedative and hypnotic and anticonvulsant.*

USES Sedative or hypnotic for preanesthetic medication, induction of general anesthesia, adjunct in manipulative or diagnostic procedures, and emergency control of acute convulsions.

CONTRAINDICATIONS History of sensitivity to barbiturates; parturition, fetal immaturity, uncontrolled pain; ethanol intoxication; hepatic encephalopathy; porphyria; suicidal ideation; pregnancy (category D), lactation.

CAUTIOUS USE COPD, sleep apnea; heart failure; hypertension, hypotension, pulmonary disease; alcoholism; mental status changes, suicidality, major depression; neonates; renal impairment, renal failure; children.

ROUTE & DOSAGE

Preoperative Sedation
Adult: **IM** 150–200 mg in 2 divided doses
Child: **IV** 1–3 mg/kg (max: 100 mg)
Hypnotic
Adult: **IM** 150–200 mg **IV** 100 mg q1–3min up to 500 mg dose
Child: **IM** 2–6 mg/kg (max: 100 mg)
Status Epilepticus
Adult: **IV** 2–15 mg/kg loading, then 0.5–3 mg/kg/h
Child: **IM** 5–15 mg/kg loading, then 0.5–5 mg/kg/h

ADMINISTRATION
Note: Do not give within 14 days of starting/stopping an MAO inhibitor.

Intramuscular
- Do not use parenteral solutions that appear cloudy or in which a precipitate has formed.
- Make IM injections deep into large muscle mass, preferably upper outer quadrant of buttock. Aspirate needle carefully before injecting it to prevent inadvertent entry into blood vessel. Inject no more than 5 mL (250 mg) in any one site because of possible tissue irritation.

Intravenous

***PREPARE*: Direct:** Give undiluted or diluted (preferred) with sterile water, D5W, NS, or other compatible IV solutions.

P

Common adverse effects in *italic*, life-threatening effects underlined; generic names in **bold;** classifications in SMALL CAPS; ◆ Canadian drug name; ◑ Prototype drug

1175

ADMINISTER: **Direct:** Give slowly. Do not exceed rate of 50 mg/min.

INCOMPATIBILITIES **Solution/additive:** Atropine, butorphanol, chlorpheniramine, chlorpromazine, cimetidine, codeine, dimenhydrinate, diphenhydramine, droperidol, ephedrine, fentanyl, glycopyrrolate, hydrocortisone, hydroxyzine, inulin, levorphanol, meperidine, methadone, midazolam, morphine, nalbuphine, norepinephrine, TETRACYCLINES, penicillin G, pentazocine, perphenazine, phenytoin, promazine, prochlorperazine, promethazine, ranitidine, sodium bicarbonate, streptomycin, succinylcholine, triflupromazine, vancomycin. **Y-site:** Amphotericin B cholesteryl, fenoldopam, TPN.

▪ Take extreme care to avoid extravasation. Necrosis may result because parenteral solution is highly alkaline. ▪ Do not use cloudy or precipitated solution.

ADVERSE EFFECTS (≥1%) **Body as a Whole:** Drowsiness, lethargy, hangover, paradoxical excitement in the older adult patient. **CV:** Hypotension with rapid IV. **Respiratory:** With rapid IV (<u>respiratory depression, laryngospasm</u>, bronchospasm, <u>apnea</u>).

INTERACTIONS Drug: Phenmetrazine antagonizes effects of pentobarbital; CNS DEPRESSANTS, **alcohol,** SEDATIVES add to CNS depression; MAO INHIBITORS cause excessive CNS depression; **methoxyflurane** creates risk of nephrotoxicity. **Herbal: Kava, valerian** may potentiate sedation.

PHARMACOKINETICS Onset: 10–15 min IM; 1 min IV. **Duration:** 15 min IV. **Distribution:** Crosses placenta. **Metabolism:** Primarily in liver. **Elimination:** In urine. **Half-Life:** 4–50 h.

NURSING IMPLICATIONS

Assessment & Drug Effects

▪ Monitor BP, pulse, and respiration q3–5min during IV administration. Observe patient closely; maintain airway. Have equipment for artificial respiration immediately available.

▪ Observe patient closely for adverse effects for at least 30 min after IM administration of hypnotic dose.

▪ Monitor for hypersensitivity reactions (see Appendix F) especially with a history of asthma or angioedema.

▪ Monitor for adverse CNS effects including exacerbation of depression and suicidal ideation.

▪ Monitor those in acute pain, children, the elderly, and debilitated patients for paradoxical excitement restlessness.

▪ Lab tests: Periodic pentobarbital levels. Note: Plasma levels greater than 30 mcg/mL may be toxic and 65 mcg/mL and above may be lethal.

▪ Concurrent drug: Monitor warfarin and phenytoin levels frequently to ensure therapeutic range.

Patient & Family Education

▪ Exercise caution when driving or operating machinery for the remainder of day after taking drug.

▪ Avoid alcohol and other CNS depressants for 24 h after receiving this drug.

▪ Women using oral contraceptives should use an additional, alternative form of contraception.

PENTOXIFYLLINE
(pen-tox-i'fi-leen)

Pentoxil, Trental
Classifications: HEMATOLOGIC; RED BLOOD CELL MODIFIER; BLOOD VISCOSITY REDUCER

Therapeutic: RED BLOOD CELL MOD-IFIER; BLOOD VISCOSITY IMPROVER
Pregnancy Category: C

AVAILABILITY 400 mg sustained release tablets

ACTION & *THERAPEUTIC EFFECT*

Useful in restoration of blood flow through capillary microcirculation that has been compromised by structural and flow dynamic changes in cerebral and peripheral vascular disorders. Maintains the flexibility of RBCs, increasing erythrocyte cAMP activity, thus allowing erythrocyte membranes to maintain their integrity and become more resistant to deformity. Improvement in blood viscosity results in increased blood flow to the microcirculation and enhanced tissue oxygenation. *Results in increased blood flow to the extremities, reduced pain and paresthesia of intermittent claudication.*

USES Intermittent claudication associated with occlusive peripheral vascular disease; diabetic angiopathies.

UNLABELED USES To improve psychopathologic symptoms in patient with cerebrovascular insufficiency and to reduce incidence of stroke in the patient with recurrent TIAs.

CONTRAINDICATIONS Intolerance to pentoxifylline or to methylxanthines (caffeine and theophylline); intracranial bleeding; retinal bleeding; lactation.
CAUTIOUS USE Angina, hypotension, arrhythmias, cerebrovascular disease; peptic ulcer disease; renal failure; renal impairment; risk of bleeding; pregnancy (category C); children younger than 18 y.

ROUTE & DOSAGE

Intermittent Claudication
Adult: **PO** 400 mg t.i.d. with meals

ADMINISTRATION

Oral
- Give on an empty stomach or with food; be consistent with time of day and relationship to food in establishing the daily regimen.
- Store tablets at 15°–30° C (59°–86° F).

ADVERSE EFFECTS (≥1%) **Body as a Whole:** Fever, flushing, convulsions, somnolence, loss of consciousness. **CNS:** Agitation, nervousness, *dizziness,* drowsiness, headache, insomnia, tremor, confusion. **CV:** Angina, chest pain, dyspnea, arrhythmias, palpitations, hypotension, edema, flushing. **Eye:** Blurred vision, conjunctivitis, scotomas. **GI:** Abdominal discomfort, belching, flatus, bloating, diarrhea, *dyspepsia, nausea, vomiting.* **Skin:** Brittle fingernails, pruritus, rash, urticaria. **Other:** Earache, unpleasant taste, excessive salivation, leukopenia, malaise, sore throat, swollen neck glands, weight change.

INTERACTIONS Drug: Ciprofloxacin, cimetidine may increase levels and toxicity, warfarin may have additive effects. **Herbal: Evening primrose oil, ginseng** may increase bleeding risk.

PHARMACOKINETICS Absorption: Readily from GI tract; 10–50% reaches systemic circulation (first pass metabolism). **Peak:** 2–4 h. **Distribution:** Distributed into breast milk. **Metabolism:** In liver and erythrocytes. **Elimination:** Primarily in urine. **Half-Life:** 0.4–0.8 h.

NURSING IMPLICATIONS
Assessment & Drug Effects
- Monitor therapeutic effectiveness which is indicated by relief from pain and cramping in calf muscles, buttocks, thighs, and feet during exercise and improves

Common adverse effects in *italic*, life-threatening effects underlined; generic names in **bold**; classifications in SMALL CAPS; ✤ Canadian drug name; ⊙ Prototype drug

1177

walking performance (time and duration).

- Monitor BP if patient is also on antihypertensive treatment. Drug may slightly decrease an already stabilized BP, necessitating a reduced dose of the hypotensive drug.

Patient & Family Education

- Consult prescriber to determine CV status and capacity before reestablishing walking as exercise.
- Pay particular attention to care of the feet because of arterial insufficiency (diminished perfusion to feet).
- Be aware that bleeding and prolonged PT/INR associated with this treatment have been reported. Report promptly unexplained bleeding, easy bruising, nose bleed, pinpoint rash to prescriber.
- Avoid driving or working with hazardous machinery until drug response has stabilized because of potential for tiredness, blurred vision, dizziness.

P

PERINDOPRIL ERBUMINE

(per-in′do-pril)

Aceon

Classifications: ANGIOTENSIN-CONVERTING ENZYME (ACE) INHIBITOR; ANTIHYPERTENSIVE

Therapeutic: ANTIHYPERTENSIVE

Prototype: Captopril

Pregnancy Category: D

AVAILABILITY 2 mg, 4 mg, 8 mg tablets

ACTION & _THERAPEUTIC EFFECT_
Angiotensin-converting enzyme (ACE) inhibitor. ACE catalyzes the conversion of angiotensin I to angiotensin II, a potent vasoconstrictor substance. Lowers BP by inhibition of ACE. Reduced aldosterone is

associated with potassium-sparing effect. In addition, it decreases systemic vascular resistance (afterload) and pulmonary capillary wedge pressure (PCWP), a measure of preload, and improves cardiac output as well as activity tolerance. _Effective in lowering blood pressure by vasodilatation resulting from inhibition of ACE. Improves cardiac output as well as activity tolerance in CAD._

USES Hypertension, myocardial infarction prophylaxis.

UNLABELED USES Heart failure.

CONTRAINDICATIONS Hypersensitivity to perindopril or any other ACE inhibitor; history of angioedema induced by an ACE inhibitor, patients with hypertrophic cardiomyopathy, renal artery stenosis; pregnancy (category D).

CAUTIOUS USE Renal insufficiency, volume-depleted patients, severe liver dysfunction; autoimmune diseases, immunosuppressant drug therapy; hyperkalemia or potassium-sparing diuretics; surgery; neutropenia; febrile illness; older adults; lactation.

ROUTE & DOSAGE

Hypertension, Stable Coronary Artery Disease

Adult: **PO** 2–4 mg once daily, may be increased to 8 mg daily in 1 or 2 divided doses (max: 16 mg/day)

Myocardial Infarction Prophylaxis

Adult: **PO** 4 mg daily × 2 wk, then 8 mg daily

Renal Impairment Dosage Adjustment

CrCl 30–59 mL/min: Start 2 mg daily; _CrCl 16–29 mL/min:_ Start

2 mg every other day; *CrCl less than 16 mL/min:* Give 2 mg on dialysis days only

ADMINISTRATION

Oral

- Manufacturer recommends an initial dose of 2–4 mg in 1 or 2 divided doses if concurrently ordered diuretic cannot be discontinued 2–3 days before beginning perinodopril. Consult prescriber.
- Give on an empty stomach 1 h before meals.
- Dosage adjustments are generally made at intervals of at least 1 wk.
- Store at 20°–25° C (68°–77° F) and protect from moisture.

ADVERSE EFFECTS (≥1%) **CNS:** Dizziness, light-headedness (in the absence of postural hypotension), headache, mood and sleep disorders, fatigue. **CV:** Palpitations. **Endocrine:** Hyperkalemia. **GI:** Nausea, vomiting, epigastric pain, diarrhea, taste disturbances, dyspepsia. **Urogenital:** Proteinuria, impotence, sexual dysfunction. **Special Senses:** Dry eyes, blurred vision. **Body as a Whole:** *Cough,* angioedema, pruritus, muscle cramps, sinusitis, hypertonia, fever. **Skin:** Rash.

INTERACTIONS Drug: POTASSIUM-SPARING DIURETICS (**amiloride, spironolactone, triamterene**) may increase the risk of hyperkalemia. POTASSIUM SUPPLEMENTS increase the risk of hyperkalemia; lithium levels can be increased. Use with **azathioprine** may cause anemia and leukopenia. **Pregabalin** may increase risk of angioedema. **Food:** Food can decrease drug absorption 35%.

PHARMACOKINETICS Absorption: Readily from GI tract, absorption significantly decreased when taken with food. **Peak: Perindopril:** 1 h; **perindoprilat:** 3–7 h. **Duration:** 24 h. **Metabolism:** Hydrolyzed in the liver to its active form, perindoprilat. **Elimination:** Primarily in urine. **Half-Life: Perindopril:** 0.8–1 h, **perindoprilat:** 30–120 h.

NURSING IMPLICATIONS

Assessment & Drug Effects

- Monitor BR and HR carefully following initial dose for several hours until stable, especially in patients using concurrent diuretics, on salt restriction, or volume depleted.
- Place patient immediately in a supine position if excess hypotension develops.
- Lab tests: Periodic serum potassium, serum sodium, BUN and creatinine, ALT, blood glucose, lipid profile, and WBC with differential.
- Monitor closely kidney function in patients with CHF.
- Monitor serum lithium levels and assess for S&S of lithium toxicity when used concurrently; increased caution is needed when diuretic therapy is also used.

Patient & Family Education

- Discontinue drug and immediately report S&S of angioedema (i.e., swelling) of face or extremities to prescriber. Seek emergency help for swelling of the tongue or any other signs of potential airway obstruction.
- Be aware that light-headedness can occur, especially during early therapy; excess fluid loss of any kind (e.g., vomiting, diarrhea) will increase risk of hypotension and fainting.
- Avoid using potassium supplements unless specifically directed to do so by prescriber.

Common adverse effects in *italic*, life-threatening effects underlined; generic names in **bold**; classifications in SMALL CAPS; ♣ Canadian drug name; ⊘ Prototype drug

1179

- Report S&S of infection (e.g., sore throat, fever) promptly to prescriber.

PERMETHRIN ⊙

(per-meth'rin)

Nix, Elimite, Acticin
Classifications: SCABICIDE; PEDICULICIDE
Therapeutic: SCABICIDE; PEDICULICIDE
Pregnancy Category: B

AVAILABILITY 5% cream; 1% liquid

ACTION & *THERAPEUTIC EFFECT*
Pediculicidal and ovicidal activity against *Pediculus humanus* var. *capitis* (head louse). Inhibits sodium ion influx through nerve cell membrane channels, resulting in delayed repolarization of the action potential and paralysis of the pest. *It prevents burrowing into host's skin. Since lice are completely dependent on blood for survival, they die within 24–48 h. Also active against ticks, mites, and fleas.*

USES Pediculosis capitis.

CONTRAINDICATIONS Hypersensitivity to pyrethrins, chrysanthemums, sulfites, or other preservatives or dyes; acute inflammation of the scalp; lactation.
CAUTIOUS USE Children younger than 2 y **(liquid)**, and less than 2 mo **(lotion)**; asthma; pregnancy (category B).

ROUTE & DOSAGE

Head Lice

Adult/Child (older than 2 y):
Topical Apply sufficient volume to clean wet hair to saturate the hair and scalp; leave on 10 min, then rinse hair thoroughly

ADMINISTRATION
Topical
- Saturate scalp as well as hair with the lotion; this is not a shampoo. Shake lotion well before application.
- Hair should be washed with regular shampoo before treatment with permethrin, thoroughly rinsed and dried.
- Rinse hair and scalp thoroughly and dry with a clean towel following 10 min exposure to the medication. Head lice are usually eliminated with one treatment.
- Store drug away from heat at 15°–25° C (59°–77° F) and direct light. Avoid freezing.

ADVERSE EFFECTS (≥1%) **Skin:** *Pruritus, transient tingling,* burning, stinging, numbness; erythema, edema, rash.

PHARMACOKINETICS Absorption: Less than 2% of amount applied is absorbed through intact skin. **Metabolism:** Rapidly hydrolyzed to inactive metabolites. **Elimination:** Primarily in urine.

NURSING IMPLICATIONS
Assessment & Drug Effects
- Do not attempt therapy if patient is known to be sensitive to any pyrethrin or pyrethroid. Stop treatment if a reaction occurs.

Patient & Family Education
- When hair is dry, comb with a fine-tooth comb (furnished with medication) to remove dead lice and remaining nits or nit shells.
- Be aware that drug remains on hair shaft up to 14 days; therefore, recurrence of infestation rarely occurs (less than 1%).
- Inspect hair shafts daily for at least 1 wk to determine drug effectiveness.

Common adverse effects in *italic*, life-threatening effects underlined; generic names in **bold**; classifications in SMALL CAPS; ✤ Canadian drug name; ⊙ Prototype drug

Contact prescriber if live lice are observed after 7 days. Signs of inadequate treatment: Itching, redness of skin, skin abrasion, infected scalp areas.

- Resume regular shampooing after treatment; residual deposit of drug on hair is not reduced.
- Be aware that drug is usually irritating to the eyes and mucosa. Flush well with water if medicine accidentally gets into eyes.

PERPHENAZINE
(per-fen'a-zeen)
Classifications: PHENOTHIAZINE ANTIPSYCHOTIC; ANTIEMETIC
Therapeutic: ANTIPSYCHOTIC; ANTIEMETIC
Prototype: Chlorpromazine
Pregnancy Category: C

AVAILABILITY 2 mg, 4 mg, 6 mg, 8 mg, 16 mg tablets

ACTION & *THERAPEUTIC EFFECT*
Affects all parts of CNS, particularly the hypothalamus. Antipsychotic effect is due to its ability to antagonize neurotransmitter dopamine by acting on its receptors in the brain. Antiemetic action results from direct blockade of dopamine in the chemoreceptor trigger zone (CTZ) in the medulla. *Has antipsychotic and antiemetic properties.*

USES Psychotic disorders, symptomatic control of severe nausea and vomiting.

CONTRAINDICATIONS Hypersensitivity to perphenazine and other phenothiazines; preexisting liver damage; suspected or established subcortical brain damage, comatose states, CNS depression; dementia-related psychosis; hepatic encephalopathy; QT prolongation; bone marrow depression; lactation.

CAUTIOUS USE Previously diagnosed breast cancer; liver or kidney dysfunction; renal impairment; cardiovascular disorders; alcohol withdrawal; epilepsy; psychic depression, patients with suicidal tendency; cardiac and pulmonary disease; glaucoma; history of intestinal or GU obstruction; patients who will be exposed to extremes of heat or cold, or to phosphorous insecticides; older adults or debilitated patients; pregnancy (category C); children younger than 12 y.

ROUTE & DOSAGE

Psychotic Disorders
Adult/Adolescent: **PO** 8–16 mg b.i.d. to q.i.d. (max: 64 mg/day)
Nausea
Adult: **PO** 8–16 mg daily in divided doses (up to 24 mg/day)
Hemodialysis Dosage Adjustment Not dialyzable
Pharmacogenetic Adjustment Poor CYP2D6 metabolizers: Start with 30% of dose

ADMINISTRATION

Oral
- Ensure that sustained release form is not chewed or crushed. **Must be** swallowed whole.

ADVERSE EFFECTS (≥1%) **CNS:** *Extrapyramidal effects (dystonic reactions, akathisia, parkinsonian syndrome, tardive dyskinesia), sedation,* convulsions. **CV:** *Orthostatic hypotension,* tachycardia, bradycardia. **Special Senses:** Mydriasis, blurred vision, corneal and lenticular deposits. **GI:** Constipation, *dry mouth,* increased appetite, adynamic ileus, abnormal liver function tests, cholestatic jaundice. **Urogenital:** *Urinary retention,* gynecomastia, menstrual

P

Common adverse effects in *italic*, life-threatening effects underlined; generic names in **bold;** classifications in SMALL CAPS; ♣ Canadian drug name; ⊘ Prototype drug

1181

irregularities, inhibited ejaculation. **Hematologic:** Agranulocytosis, thrombocytopenic purpura, aplastic or hemolytic anemia. **Body as a Whole:** Photosensitivity, itching, erythema, urticaria, angioneurotic edema, drug fever, anaphylactoid reaction, sterile abscess. Nasal congestion, decreased sweating. **Metabolic:** Hyperprolactinemia, galactorrhea, weight gain.

DIAGNOSTIC TEST INTERFERENCE Perphenazine may cause falsely abnormal *thyroid function* tests because of elevations of *thyroid globulin.*

INTERACTIONS Drug: Alcohol and other CNS DEPRESSANTS enhance CNS depression; ANTACIDS, ANTIDIARRHEALS may decrease absorption of phenothiazines; ANTICHOLINERGIC AGENTS add to anticholinergic effects including fecal impaction and paralytic ileus; BARBITURATES, ANESTHETICS increase hypotension and excitation. **Herbal: Kava** increased risk and severity of dystonic reactions.

PHARMACOKINETICS Absorption: Poorly absorbed from GI tract; 20% reaches systemic circulation. **Peak:** 4–8 h PO. **Duration:** 6–12 h. **Distribution:** Crosses placenta. **Metabolism:** In liver (CYP2D6) with some metabolism in GI tract. **Elimination:** In urine and feces. **Half-Life:** 9.5 h.

NURSING IMPLICATIONS

Assessment & Drug Effects

- Establish baseline BP before initiation of drug therapy and check it at regular intervals, especially during early therapy.
- Report restlessness, weakness of extremities, dystonic reactions (spasms of neck and shoulder muscles, rigidity of back, difficulty swallowing or talking); motor restlessness (akathisia: inability to be still); and parkinsonian syndrome (tremors, shuffling gait, drooling, slow speech). A high incidence of extrapyramidal effects accompanies use of perphenazine, particularly with high doses.
- Withhold medication and report IMMEDIATELY to prescriber S&S of tardive dyskinesia (i.e., fine, wormlike movements or rapid protrusions of the tongue, chewing motions, lip smacking). Patients on long-term therapy are at high risk. Teach patients and responsible family members about symptoms because early reporting is essential.
- Lab tests: Obtain CBC with differential, LFTs and kidney function studies.
- ECG and ophthalmologic examination are recommended prior to initiation and periodically during therapy.
- Suspect hypersensitivity, withhold drug, and report to prescriber if jaundice appears between weeks 2 and 4.
- Monitor urine and bowel elimination pattern.

Patient & Family Education

- Make all position changes slowly and in stages, particularly from recumbent to upright posture, and to lie down or sit down if light-headedness or dizziness occurs.
- Do not drive or engage in potentially hazardous activities until response to drug is known.
- Discontinue drug and report to prescriber immediately if jaundice appears between weeks 2 and 4.
- Avoid long exposure to sunlight and to sunlamps. Photosensitivity results in skin color changes from brown to blue-gray.

- Adhere strictly to dosage regimen. Contact prescriber before changing it for any reason.
- Drug should be discontinued gradually over a period of several weeks following prolonged therapy.
- Avoid OTC drugs unless prescriber prescribes them.
- Be aware that perphenazine may discolor urine reddish brown.

PHENAZOPYRIDINE HYDROCHLORIDE

(fen-az-oh-peer′i-deen)

Azo-Standard, Baridium, Geridium, Phenazo ♦, Phenazodine, Pyridiate, Pyridium, Pyronium ♦, Urodine, Urogesic
Classification: URINARY TRACT ANALGESIC
Therapeutic: URINARY TRACT ANALGESIC
Pregnancy Category: B

AVAILABILITY 95 mg, 97.2 mg, 100 mg, 150 mg, 200 mg tablets

ACTION & THERAPEUTIC EFFECT
Azo dye that has local anesthetic action on urinary tract mucosa, which imparts little or no antibacterial activity. *Effective as a urinary tract analgesic.*

USES Symptomatic relief of pain, burning, frequency, and urgency arising from irritation of urinary tract mucosa, as from infection, trauma, surgery, or instrumentation.

CONTRAINDICATIONS Renal insufficiency, renal disease including glomerulonephritis, pyelonephritis, renal failure, uremia; hepatic disease; glucose-6-phosphate dehydrogenase deficiency, severe hepatitis.
CAUTIOUS USE GI disturbances; older adults; pregnancy (category B), lactation.

ROUTE & DOSAGE

Cystitis
Adult: **PO** 200 mg t.i.d.
Child: **PO** 12 mg/kg/day in 3 divided doses

ADMINISTRATION
Oral
- Give with or after meals.

ADVERSE EFFECTS (≥1%) **Body as a Whole:** Headache, vertigo. **GI:** Mild GI disturbances. **Urogenital:** Kidney stones, transient acute kidney failure. **Metabolic:** Methemoglobinemia, hemolytic anemia. **Skin:** Skin pigmentation. **Special Senses:** May stain soft contact lenses.

DIAGNOSTIC TEST INTERFERENCE
Phenazopyridine may interfere with any urinary test that is based on color reactions or spectrometry: *Bromsulphalein* and *phenolsulphonphthalein* excretion tests; urinary *glucose* test using *Clinistix* or *TesTape* (*copper-reduction methods* such as *Clinitest* and *Benedict's test* reportedly not affected); *bilirubin* using "foam test" or *Ictotest; ketones* using *nitroprusside* (e.g., *Acetest, Ketostix,* or *Gerhardt ferric chloride*); *urinary protein* using *Albustix, Albutest,* or *nitric acid ring test;* urinary *steroids; urobilinogen; assays* for *porphyrins.*

PHARMACOKINETICS **Absorption:**
Readily absorbed from GI tract. **Distribution:** Crosses placenta in trace amounts. **Metabolism:** In liver and other tissues. **Elimination:** Primarily in urine.

NURSING IMPLICATIONS
Assessment & Drug Effects
- Monitor for therapeutic effectiveness as indicated by relief from pain and burning upon urination.

Common adverse effects in *italic*, life-threatening effects underlined; generic names in **bold;** classifications in SMALL CAPS; ♦ Canadian drug name; ⊕ Prototype drug

1183

Patient & Family Education

▪ Drug will impart an orange to red color to urine and may stain clothing.

PHENELZINE SULFATE ⓟ
(fen'el-zeen)

Classifications: ANTIDEPRESSANT; MONOAMINE OXIDASE (MAO) INHIBITOR
Therapeutic: ANTIDEPRESSANT; MAO INHIBITOR
Pregnancy Category: C

AVAILABILITY 15 mg tablets

ACTION & *THERAPEUTIC EFFECT*
Potent hydrazine monoamine oxidase (MAO) inhibitor. Antidepressant action believed to be due to irreversible inhibition of MAO, thereby permitting increased concentrations of endogenous epinephrine, norepinephrine, serotonin, and dopamine within presynaptic neurons and at receptor sites. *Antidepressant utilization is limited to individuals who do not respond well to other classes of antidepressants.*

USES Atypical or nonendogenous depression.

CONTRAINDICATIONS Hypersensitivity to MAO inhibitors; suicidal ideation; pheochromocytoma; untreated hyperthyroidism; cardiac arrhythmias, uncontrolled hypertension; increased intracranial pressure; intracranial bleeding; atonic colitis; glaucoma; frequent headaches; bipolar depression; accompanying alcoholism or drug addiction; paranoid schizophrenia; older adults or debilitated patients; lactation.

CAUTIOUS USE Epilepsy; pyloric stenosis; DM; manic-depressive states; agitated patients; schizophrenia or psychosis; seizures; suicidal tendencies; chronic brain syndromes; pregnancy (category C); children and adolescents.

ROUTE & DOSAGE

Depression
Adult: **PO** 15 mg t.i.d., rapidly increase to at least 60 mg/day, may need up to 90 mg/day

ADMINISTRATION

Oral

▪ Avoid rapid discontinuation, particularly after high dosage, since a rebound effect may occur (e.g., headache, excitability, hallucinations, and possibly depression).
▪ Store in tightly covered containers away from heat and light.

ADVERSE EFFECTS (≥1%) Body as a Whole: Dizziness or vertigo, headache, *orthostatic hypotension,* drowsiness or *insomnia,* weakness, fatigue, edema, tremors, twitching, akathisia, ataxia, hyperreflexia, faintness, hyperactivity, marked agitation, anxiety, seizures, trismus, opisthotonos, <u>respiratory depression, coma.</u> **CNS:** Mania, hypomania, confusion, memory impairment, delirium, hallucinations, euphoria, acute anxiety reaction, toxic precipitation of schizophrenia, convulsions, peripheral neuropathy. **CV:** <u>Hypertensive crisis</u> (intense occipital headache, palpitation, marked hypertension, stiff neck, nausea, vomiting, sweating, fever, photophobia, dilated pupils, bradycardia or tachycardia, constricting chest pain, intracranial bleeding), hypotension or hypertension, <u>circulatory collapse.</u> **GI:** *Constipation, dry mouth, nausea,* vomiting, *anorexia,* weight gain. **Hematologic:** Normocytic and normochromic anemia, <u>leukopenia.</u> **Skin:** Hyperhidrosis, skin rash, photosensitivity. **Special Senses:** Blurred vision.

DIAGNOSTIC TEST INTERFERENCE
Phenelzine may cause a slight false increase in *serum bilirubin.*

INTERACTIONS Drug: TRICYCLIC ANTI-DEPRESSANTS may cause hyperpyrexia, seizures; **fluoxetine, sertraline, paroxetine** may cause serotonin syndrome (see Appendix F); SYMPATHOMIMETIC AGENTS (e.g., **amphetamine, phenylephrine, phenylpropanolamine**), **guanethidine** and **reserpine** may cause hypertensive crisis; CNS DEPRESSANTS have additive CNS depressive effects; OPIATE ANALGESICS (especially **meperidine**) may cause hypertensive crisis and circulatory collapse; **buspirone,** hypertension; GENERAL ANESTHETICS, prolonged hypotensive and CNS depressant effects; hypertension, headache, hyperexcitability reported with **dopamine, methyldopa, levodopa, tryptophan; metrizamide** may increase risk of seizures; HYPOTENSIVE AGENTS and DIURETICS have additive hypotensive effects. **Food:** Aged meats or aged cheeses, protein extracts, sour cream, alcohol, anchovies, liver, sausages, over-ripe figs, bananas, avocados, chocolate, soy sauce, bean curd, natural yogurt, fava beans—**tyramine**-containing foods—may precipitate hypertensive crisis. Avoid **chocolate** or **caffeine. Herbal:** Ginseng, **ephedra, ma huang, St. John's wort** may cause hypertensive crisis.

PHARMACOKINETICS Absorption: Readily absorbed from GI tract. **Onset:** 2 wk. **Metabolism:** Rapidly metabolized. **Elimination:** 79% of metabolites excreted in urine in 96 h.

NURSING IMPLICATIONS

Assessment & Drug Effects

▪ Prior to initiation of treatment, evaluate patient's BP in standing and recumbent positions.

▪ Monitor BP and pulse between doses when titrating initial dosages. Observe closely for evidence of adverse drug effects. Thereafter, monitor at regular intervals throughout therapy.

▪ Monitor children, adolescents, and adults for changes in behavior that may indicate suicidality.

▪ Lab tests: Baseline and periodic CBC and LFTs.

▪ Report immediately if hypomania (exaggeration of motility, feelings, and ideas) occurs as depression improves. This reaction may also appear at higher than recommended doses or with long-term therapy.

▪ Observe for and report therapeutic effectiveness of drug: Improvement in sleep pattern, appetite, physical activity, interest in self and surroundings, as well as lessening of anxiety and bodily complaints.

▪ Observe patient with diabetes closely for S&S of hypoglycemia (see Appendix F).

▪ Patients on prolonged therapy should be checked periodically for altered color perception, changes in fundi or visual fields. Changes in red-green vision may be the first indication of eye damage.

Patient & Family Education

▪ Maximum antidepressant effects generally appear in 2–6 wk and persist several weeks after drug withdrawal.

▪ Avoid self-medication. OTC preparations (e.g., cough, cold, and hay fever remedies, appetite suppressants) can precipitate severe hypertensive reactions if taken during therapy or within 2–3 wk after discontinuation of an MAO inhibitor.

▪ Report immediately to prescriber the onset of headache and palpitation, or any other unusual effects which may indicate need to discontinue therapy.

- Do not consume foods and beverages containing tyramine or tryptophan or drugs containing pressor agent. These can cause severe hypertensive reactions. Get a list from your care provider.
- Avoid drinking excessive caffeine and chocolate beverages (e.g., coffee, tea, cocoa, or cola).
- Make position changes slowly, especially from recumbent to upright posture, and dangle legs over bed a few minutes before rising to walk. Avoid standing still for prolonged periods. Also avoid hot showers and baths (resulting vasodilatation may potentiate hypotension); lie down immediately if feeling light-headed or faint.
- Check weight 2 or 3 times per wk and report unusual gain.
- Report jaundice. Hepatotoxicity is believed to be a hypersensitivity reaction unrelated to dosage or duration of therapy.

PHENOBARBITAL ⊕
(fee-noe-bar'bi-tal)
Solfoton

PHENOBARBITAL SODIUM
Luminal
Classifications: ANTICONVULSANT; SEDATIVE-HYPNOTIC; BARBITURATE
Therapeutic: ANTICONVULSANT; SEDATIVE-HYPNOTIC
Pregnancy Category: D
Controlled Substance: Schedule IV

AVAILABILITY 15 mg, 16 mg, 30 mg, 60 mg, 90 mg, 100 mg tablets; 16 mg capsules; 15 mg/5 mL, 20 mg/5 mL liquid; 30 mg/mL, 60 mg/mL, 65 mg/mL, 130 mg/mL injection

ACTION & *THERAPEUTIC EFFECT*
Long-acting barbiturate, while the sodium form of phenobarbital is short acting; both forms have anticonvulsant properties. Sedative and hypnotic effects appear to be due primarily to interference with impulse transmission of cerebral cortex by inhibition of reticular activating system. Limiting spread of seizure activity results by increasing the threshold for motor cortex stimulation. *Effective as a sedative, hypnotic, and an anticonvulsant with no analgesic effect.*

USES Long-term management of tonic-clonic (grand mal) seizures and partial seizures; status epilepticus, eclampsia, febrile convulsions in young children. Also used as a sedative in anxiety or tension states; in pediatrics as preoperative and postoperative sedation and to treat pylorospasm in infants.
UNLABELED USES Treatment and prevention of hyperbilirubinemia in neonates and in the management of chronic cholestasis; benzodiazepine withdrawal.

CONTRAINDICATIONS Hypersensitivity to barbiturates; manifest hepatic or familial history of porphyria; severe respiratory or kidney disease; history of previous addiction to sedative hypnotics; alcohol intoxication; uncontrolled pain; renal failure, anuria; pregnancy (category D).
CAUTIOUS USE Impaired liver, kidney, cardiac, or respiratory function; sleep apnea; COPD; history of allergies; patients with fever; hyperthyroidism; diabetes mellitus or severe anemia; seizure disorders; during labor and delivery; patient with borderline hypoadrenal function; older adult or debilitated patients; lactation, young children and neonates.

ROUTE & DOSAGE

Anticonvulsant

Adult: **PO/IV** 1–3 mg/kg/day in divided doses
Child: **PO/IV** 4–8 mg/kg/day in divided doses

Status Epilepticus

Adult: **IV** 300–800 mg, then 120–240 mg q20min (total max: 1–2 g)
Child: **IV** 10–20 mg/kg in single or divided doses, then 5 mg/kg/dose q15–30min (total max: 40 mg/kg)
Neonate: **IV** 15–20 mg/kg in single or divided doses

Sedative/Hypnotic

Adult: **PO** 30–120 mg/day **IV/IM** 100–320 mg/day
Child: **PO** 2 mg/kg/day in 3 divided doses **IV/IM** 3–5 mg/kg

Renal Impairment Dosage Adjustment

CrCl less than 10 mL/min: Dose q12–16h

Hemodialysis Dosage Adjustment

20–50% dialyzed

ADMINISTRATION

Oral

▪ Give crushed and mixed with a fluid or with food if patient cannot swallow pill. Do not permit patient to swallow dry crushed drug.

Intramuscular

▪ Give IM deep into large muscle mass; do not exceed 5 mL at any one site.

Intravenous

Note: Verify correct IV concentration and rate of infusion for neonates, infants, children with prescriber. Use IV route ONLY if other routes are not feasible.

PREPARE: Direct: May be given undiluted or diluted in 10 mL of sterile water for injection.

ADMINISTER: Direct: Give 60 mg or fraction thereof over at least 60 sec. Give within 30 min after preparation.

***INCOMPATIBILITIES* Solution/additive:** Ampicillin, cephalothin, chlorpromazine, cimetidine, clindamycin, codeine phosphate, dexamethasone, diphenhydramine, erythromycin, ephedrine, hydralazine, hydrocortisone sodium succinate, hydroxyzine, insulin, kanamycin, levorphanol, meperidine, methadone, methylphenidate, morphine, nitrofurantoin, norepinephrine, pentazocine, pentobarbital, phytonadione, procaine, prochlorperazine, promazine, promethazine, sodium bicarbonate, streptomycin,; TETRACYCLINES, vancomycin, warfarin. **Y-site:** Amphotericin B cholesteryl complex, hydromorphone, TPN with albumin.

▪ Be aware that extravasation of IV phenobarbital may cause necrotic tissue changes that necessitate skin grafting. Check injection site frequently.

ADVERSE EFFECTS (≥1%) **Body as a Whole:** Myalgia, neuralgia, <u>CNS depression, coma, and death.</u> **CNS:** *Somnolence,* nightmares, insomnia, "hangover," headache, anxiety, thinking abnormalities, dizziness, nystagmus, irritability, paradoxic excitement and exacerbation of hyperkinetic behavior (in children); confusion or depression or marked excitement (older adult or debilitated patients); ataxia. **CV:** Bradycardia, syncope, hypotension. **GI:** Nausea,

Common adverse effects in *italic,* life-threatening effects <u>underlined</u>; generic names in **bold**; classifications in SMALL CAPS; ♣ Canadian drug name; ◑ Prototype drug

1187

vomiting, constipation, diarrhea, epigastric pain, liver damage. **Hematologic:** Megaloblastic anemia, <u>agranulocytosis</u>, <u>thrombocytopenia</u>. **Metabolic:** Hypocalcemia, osteomalacia, rickets. **Musculoskeletal:** Folic acid deficiency, vitamin D deficiency. **Respiratory:** <u>Respiratory depression</u>. **Skin:** Mild maculopapular, morbilliform rash; erythema multiforme, <u>Stevens-Johnson syndrome, exfoliative dermatitis (rare)</u>.

DIAGNOSTIC TEST INTERFERENCE

BARBITURATES may affect *bromsulphalein* retention tests (by enhancing liver uptake and excretion of dye) and increase *serum phosphatase.*

INTERACTIONS Drug: Alcohol, CNS DEPRESSANTS compound CNS depression; phenobarbital may decrease absorption and increase metabolism of ORAL ANTICOAGULANTS; increases metabolism of CORTICOSTEROIDS, ORAL CONTRACEPTIVES, ANTICONVULSANTS, **digitoxin,** possibly decreasing their effects; ANTIDEPRESSANTS potentiate adverse effects of phenobarbital; **griseofulvin** decreases absorption of phenobarbital; **quinine** increases plasma levels. **Herbal: Kava, valerian** may potentiate sedation.

PHARMACOKINETICS Absorption:

70–90% slowly from GI tract. **Peak:** 8–12 h PO; 30 min IV. **Duration:** 4–6 h IV. **Distribution:** 20–45% protein bound; crosses placenta; enters breast milk. **Metabolism:** In liver (CYP2C19). **Elimination:** In urine. **Half-Life:** 2–6 days.

NURSING IMPLICATIONS

Assessment & Drug Effects

- Observe patients receiving large doses for at least 30 min to ensure that sedation is not excessive.
- Chronic use in children or infants requires continuous assessment

related to normal cognitive and behavioral functioning.

- Keep patient under constant observation when drug is administered IV, and record vital signs at least every hour or more often if indicated.
- Check IV injection site very frequently to prevent extravasation of phenobarbital. It could result in tissue damage requiring skin grafting.
- Lab tests: Periodic LFTs, CBC with differential, Hct and Hgb, serum folate, and vitamin D levels during prolonged therapy.
- Monitor serum drug levels. Serum concentrations greater than 50 mcg/mL may cause coma. Therapeutic serum concentrations of 15–40 mcg/mL produce anticonvulsant activity in most patients. These values are usually attained after 2 or 3 wk of therapy with a dose of 100–200 mg/day.
- Expect barbiturates to produce restlessness when given to patients in pain because these drugs do not have analgesic action.
- Be prepared for paradoxical responses and report promptly in older adult or debilitated patient and children [i.e., irritability, marked excitement (inappropriate tearfulness and aggression in children), depression, and confusion].
- Monitor for drug interactions. Barbiturates increase the metabolism of many drugs, leading to decreased pharmacologic effects of those drugs.
- Monitor for and report chronic toxicity symptoms (e.g., ataxia, slurred speech, irritability, poor judgment, slight dysarthria, nystagmus on vertical gaze, confusion, insomnia, somatic complaints).

Patient & Family Education

- Be aware that anticonvulsant therapy may cause drowsiness during first few weeks of treatment, but

Common adverse effects in *italic*, life-threatening effects <u>underlined</u>; generic names in **bold**; classifications in SMALL CAPS; ♣ Canadian drug name; ☻ Prototype drug

this usually diminishes with continued use.

- Avoid potentially hazardous activities requiring mental alertness until response to drug is known.
- Do not consume alcohol in any amount when taking a barbiturate; it may severely impair judgment and abilities.
- Increase vitamin D-fortified foods (e.g., milk products) because drug increases vitamin D metabolism. A vitamin D supplement may be prescribed.
- Maintain adequate dietary folate intake: Fresh vegetables (especially green leafy), fresh fruits, whole grains, liver. Long-term therapy may result in nutritional folate (B$_9$) deficiency. A supplement of folic acid may be prescribed.
- Adhere to drug regimen (i.e., do not change intervals between doses or increase or decrease doses) without contacting prescriber.
- Do not stop taking drug abruptly because of danger of withdrawal symptoms (8–12 h after last dose), which can be fatal.
- Report to prescriber the onset of fever, sore throat or mouth, malaise, easy bruising or bleeding, petechiae, jaundice, rash when on prolonged therapy.
- Avoid pregnancy when receiving barbiturates. Use or add barrier device to hormonal contraceptive when taking prolonged therapy.

PHENOXYBENZAMINE HYDROCHLORIDE

(fen-ox-ee-ben′za-meen)

Dibenzyline

Classifications: ALPHA-ADRENERGIC RECEPTOR ANTAGONIST

Therapeutic: AGENT FOR PHEOCHROMOCYTOMA; ANTIHYPERTENSIVE

Prototype: Prazosin
Pregnancy Category: C

AVAILABILITY 10 mg capsules

ACTION & *THERAPEUTIC EFFECT*
Long-acting alpha-adrenergic receptor antagonist that produces noncompetitive blockade of alpha-adrenergic receptor sites at postganglionic synapses. Alpha-receptor sites are thus unable to react to the endogenous or exogenous sympathomimetic agents epinephrine and norepinephrine. *Blocks excitatory effects of epinephrine, including vasoconstriction, but does not affect adrenergic cardiac inhibitory actions. It produces a "chemical sympathectomy" and maintains it.*

USES Management of pheochromocytoma.
UNLABELED USES Raynaud's phenomenon, benign prostatic hyperplasia, acrocyanosis.

CONTRAINDICATIONS Instances when fall in BP would be dangerous; lactation.
CAUTIOUS USE Marked cerebral or coronary arteriosclerosis, compensated CHF, coronary artery disease; renal insufficiency; respiratory infections; older adults; pregnancy (category C).

ROUTE & DOSAGE

Management of Pheochromocytoma
Adult: **PO** 10 mg b.i.d., may increase by 10 mg/day every other day (usual range: 20–40 mg/day in 2–3 divided doses)

ADMINISTRATION
Oral
- Give with milk or in divided doses to reduce gastric irritation.

Common adverse effects in *italic*, life-threatening effects underlined; generic names in **bold**; classifications in SMALL CAPS; ◆ Canadian drug name; ◯ Prototype drug

1189

- Preserve in airtight containers protected from light.

ADVERSE EFFECTS (≥1%) **Body as a Whole:** *Dizziness,* fainting, drowsiness, sedation, tiredness, weakness, lethargy, confusion, headache, <u>shock</u>. **CNS:** CNS stimulation (large doses). **CV:** *Postural hypotension, tachycardia,* palpitation. **GI:** Dry mouth. **Urogenital:** Inhibition of ejaculation. **Respiratory:** *Nasal congestion.* **Skin:** Allergic contact dermatitis. **Special Senses:** *Miosis,* drooping of eyelids.

INTERACTIONS Drug: Inhibits effects of **dexmedetomide, norepinephrine, phenylephrine;** additive hypotensive effects with ANTIHYPERTENSIVES.

PHARMACOKINETICS Absorption: Variably (approximately 30%) from GI tract. **Onset:** 2 h. **Peak:** 4–6 h. **Duration:** 3–4 days. **Distribution:** Accumulates in adipose tissue. **Elimination:** 80% in urine and bile within 24 h. **Half-Life:** 24 h.

NURSING IMPLICATIONS

Assessment & Drug Effects

- Monitor BP and note pulse quality, rate, and rhythm in recumbent and standing positions during period of dosage adjustment. Observe patient closely for at least 4 days from one dosage increment to the next; hypotension and tachycardia are most likely to occur in standing position.
- Drug has cumulative action, thus onset of therapeutic effects may not occur until after 2 wk of therapy, and full therapeutic effects may not be apparent for several more weeks.

Patient & Family Education

- Make position changes slowly, particularly from reclining to upright posture, and dangle legs and exercise ankles and feet for a few minutes before standing.
- Be aware that light-headedness, dizziness, and palpitations usually disappear with continued therapy but may reappear under conditions that promote vasodilation, such as strenuous exercise or ingestion of a large meal or alcohol.
- Pupil constriction, nasal stuffiness, and inhibition of ejaculation generally decrease with continued therapy.
- Do not take OTC medications for coughs, colds, or allergy without approval of prescriber. Many contain agents that cause BP elevation.

PHENTERMINE HYDROCHLORIDE
(phen-ter′meen)

Adi-pex-P
Classification: ANOREXIANT
Therapeutic: APPETITE SUPPRESSANT
Prototype: Diethylpropion
Pregnancy Category: C
Controlled Substance: Schedule IV

AVAILABILITY 37.5 mg tablets; 15 mg, 30 mg, 37.5 mg capsules

ACTION & *THERAPEUTIC EFFECT* Sympathetic amine with actions that include CNS stimulation and blood pressure elevation. *Appetite suppression or metabolic effects along with diet adjustment result in weight loss in obese individuals.*

USES Short-term (8–12 wk) adjunct for weight loss.

CONTRAINDICATIONS Known hypersensitivity to sympathetic amines; history of hypertension, moderate to severe hypertension,

advanced arteriosclerosis, cardiovascular disease; agitated states; psychosis; schizophrenia; history of drug abuse; during or within 14 days of administration of MAO inhibitor; glaucoma; lactation.
CAUTIOUS USE Mild hypertension, DM; older adults; pregnancy (category C); children younger than 16 y.

ROUTE & DOSAGE

Obesity
Adult: **PO** 15–30 mg 2 hr after breakfast or 1 tablet daily after breakfast

ADMINISTRATION

Oral
- Ensure that at least 14 days have elapsed between the first dose of phentermine and the last dose of an MAO inhibitor.
- Give 30 min before meals.
- Do not administer if an SSRI is currently prescribed.
- Store in a tight container.

ADVERSE EFFECTS (≥1%) **Body as a Whole:** Hypersensitivity (urticaria, rash, erythema, burning sensation), chest pain, excessive sweating, clamminess, chills, flushing, fever, myalgia. **CV:** Palpitations, tachycardia, arrhythmias, hypertension or hypotension, syncope, precordial pain, pulmonary hypertension. **GI:** Dry mouth, altered taste, nausea, vomiting, abdominal pain, diarrhea, constipation, stomach pain. **Endocrine:** Gynecomastia. **Hematologic:** Bone marrow suppression, agranulocytosis, leukopenia. **Musculoskeletal:** Muscle pain. **CNS:** Overstimulation, nervousness, restlessness, dizziness, insomnia, weakness, fatigue, malaise, anxiety, euphoria, drowsiness, depression, agitation, dysphoria, tremor, dyskinesia, dysarthria, confusion,

incoordination, headache, change in libido. **Skin:** Hair loss, ecchymosis. **Special Senses:** Mydriasis, blurred vision. **Urogenital:** Dysuria, polyuria, urinary frequency, impotence, menstrual upset.

INTERACTIONS Drug: MAO INHIBITORS, **furazolidone** may increase pressor response resulting in hypertensive crisis. TRICYCLIC ANTIDEPRESSANTS may decrease anorectic response. May decrease hypotensive effects of **guanethidine.**

PHARMACOKINETICS Absorption: Absorbed from the small intestine. **Duration:** 4–14 h. **Elimination:** Primarily in urine. **Half-Life:** 19–24 h.

NURSING IMPLICATIONS

Assessment & Drug Effects
- Assess for tolerance to the anorectic effect of the drug. Withhold drug and report to prescriber when this occurs.
- Lab tests: Periodic CBC with differential and blood glucose.
- Monitor periodic cardiovascular status, including BP, exercise tolerance, peripheral edema.
- Monitor weight at least 3 times/wk.

Patient & Family Education
- Do not take this drug late in the evening because it could cause insomnia.
- Report immediately any of the following: Shortness of breath, chest pains, dizziness or fainting, swelling of the extremities.
- Tolerance to the appetite suppression effects of the drug usually develops in a few weeks. Notify prescriber, but do not increase the drug dose.
- Weigh yourself at least 3 times/wk at the same time of day with the same amount of clothing.

Common adverse effects in *italic*, life-threatening effects underlined; generic names in **bold;** classifications in SMALL CAPS; ✦ Canadian drug name; ⊙ Prototype drug

PHENTOLAMINE MESYLATE

(fen-tole′a-meen)

Regitine, Rogitine ♣

Classifications: ALPHA-ADRENERGIC
RECEPTOR ANTAGONIST; VASODILATOR
Therapeutic: VASODILATOR
Prototype: Prazosin
Pregnancy Category: C

AVAILABILITY 5 mg injection

ACTION & *THERAPEUTIC EFFECT*
Alpha-adrenergic blocking agent
that competitively blocks alpha-
adrenergic receptors, but action is
transient and incomplete. Causes
vasodilation and decreases general vascular resistance as well as
pulmonary arterial pressure, primarily by direct action on vascular
smooth muscle. *Prevents hypertension resulting from elevated
levels of circulating epinephrine or
norepinephrine.*

USES Diagnosis of pheochromocytoma and to prevent or control hypertensive episodes prior to or during pheochromocytomectomy.

UNLABELED USES Prevention of
dermal necrosis and sloughing following IV administration or extravasation of norepinephrine.

CONTRAINDICATIONS Hypersensitivity to phentolamine; MI (previous or present), CAD; peptic ulcer
disease; lactation.
CAUTIOUS USE Gastritis; pregnancy (category C).

ROUTE & DOSAGE

To Prevent Hypertensive Episode during Surgery
Adult: **IV/IM** 5 mg 1–2 h before
surgery, repeat as needed
Child: **IV/IM** 0.05–0.1 mg/kg/
dose (max: 5 mg/dose)

To Test for Pheochromocytoma
Adult: **IV/IM** 5 mg
Child: **IV/IM** 0.05–0.1 mg/kg
(max: 5 mg)

To Treat Extravasation
Adult: **Intradermal** 5–10 mg
diluted in 10 mL of normal saline
injected into affected area within
12 h of extravasation
Child: **Intradermal** 0.1–0.2 mg/
kg diluted with normal saline
injected into affected area within
12 h of extravasation

ADMINISTRATION

Note: Place patient in supine position when receiving drug parenterally. ▪ Monitor BP and pulse q2min
until stabilized.

Intramuscular
▪ Reconstitute 5 mg vial with 1 mL
of sterile water for injection.

Intravenous

PREPARE: Direct: Reconstitute as
for IM. May be further diluted with
up to 10 mL of sterile water. ▪ Use
immediately.
ADMINISTER: Direct: Give a single
dose over 60 sec.

ADVERSE EFFECTS (≥1%) **Body as
a Whole:** Weakness, dizziness,
flushing, *orthostatic hypotension.*
GI: *Abdominal pain, nausea,
vomiting, diarrhea, exacerbation
of peptic ulcer.* **CV:** *Acute and prolonged hypotension, tachycardia,
anginal pain,* cardiac arrhythmias, MI, cerebrovascular spasm,
shock-like state. **Special Senses:**
Nasal stuffiness, conjunctival
infection.

INTERACTIONS Drug: May antagonize BP raising effects of **epinephrine, ephedrine.**

PHARMACOKINETICS Peak: 2 min IV; 15–20 min IM. **Duration:** 10–15 min IV; 3–4 h IM. **Elimination:** In urine. **Half-Life:** 19 min.

NURSING IMPLICATIONS

Assessment & Drug Effects

Test for pheochromocytoma:

- *IV administration:* Keep patient at rest in supine position throughout test, preferably in quiet darkened room. ▪ Prior to drug administration, take BP q10min for at least 30 min to establish that BP has stabilized before IV injection. ▪ Record BP immediately after injection and at 30-sec intervals for first 3 min; then at 1-min intervals for next 7 min.
- *IM administration:* Post-injection, BP determinations at 5-min intervals for 30–45 min.

Patient & Family Education

- Avoid sudden changes in position, particularly from reclining to upright posture and dangle legs and exercise ankles and toes for a few minutes before standing to walk.
- Lie down or sit down in head-low position immediately if light-headed or dizzy.

PHENYLEPHRINE HYDROCHLORIDE

(fen-ill-ef′rin)

AK-Dilate Ophthalmic, Alconefrin, Isopto Frin, Mydfrin, Neo-Synephrine, Nostril, Rhinall, Sinarest Nasal, Sinex

Classifications: EYE AND NOSE PREPARATION; ALPHA-ADRENERGIC AGONIST; MYDRIATIC; VASOPRESSOR; DECONGESTANT

Therapeutic: VASOCONSTRICTOR; DECONGESTANT; MYDRIATIC

Prototype: Dexmedetomide
Pregnancy Category: C

AVAILABILITY 10 mg chewable tablet; 0.125%, 0.16%, 0.5%, 1% nasal solution; 0.12%, 2.5%, 10% ophth solution; 10 mg/mL injection

ACTION & *THERAPEUTIC EFFECT*

Potent, synthetic, direct-acting sympathomimetic with strong alpha-adrenergic cardiac stimulant actions. Elevates systolic and diastolic pressures through arteriolar constriction. Reduces intraocular pressure by increasing outflow and decreasing rate of aqueous humor secretion. *Effective antihypotensive agent. Topical applications to eye produce vasoconstriction and prompt mydriasis of short duration, usually without causing cycloplegia. Nasal decongestant action qualitatively similar to that of epinephrine but more potent and has longer duration of action.*

USES Parenterally to maintain BP during anesthesia, to treat vascular failure in shock, and to overcome paroxysmal supraventricular tachycardia. Used topically for rhinitis of common cold, allergic rhinitis, and sinusitis; in selected patients with wide-angle glaucoma; as mydriatic for ophthalmoscopic examination or surgery, and for relief of uveitis.

CONTRAINDICATIONS Severe CAD, severe hypertension, atrial fibrillation, atrial flutter, cardiac arrhythmias; severe organic cardiac disease, cardiomyopathy; uncontrolled hypertension; ventricular fibrillation or tachycardia; acute MI, angina; cerebral arteriosclerosis, MAOI; labor, delivery. **Ophthalmic preparations:** Narrow-angle glaucoma.

CAUTIOUS USE Hyperthyroidism; DM; 21 days before or following termination of MAO inhibitor therapy; older adults; pregnancy (category C). **Ophthalmic:** Lactation.

ROUTE & DOSAGE

Hypotension

Adult: **IM/Subcutaneous** 1–10 mg (initial dose not to exceed 5 mg) q10–15min as needed **IV** 0.1–0.18 mg/min until BP stabilizes; then 0.04–0.06 mg/min for maintenance

Ophthalmoscopy

See Appendix A

Supraventricular Tachycardia

Adult: **IV** 0.25–0.5 mg bolus, then 0.1–0.2 mg doses (total max: 1 mg)

Vasoconstrictor

Adult: **Ophthalmic** See Appendix A-1 **Intranasal** Small amount of nasal jelly placed into each nostril q3–4h as needed or 2–3 drops or sprays of 0.25–0.5% solution q3–4h as needed
Child: **Intranasal** *Younger than 6 y,* 2–3 drops or sprays of 0.125% solution q3–4h as needed; *6–12 y,* 2–3 drops or sprays of 0.25% solution q3–4h as needed

ADMINISTRATION

Instillation

- Nasal preparations: Instruct patient to blow nose gently (with both nostrils open) to clear nasal passages before administration of medication.
- Instillation (drops): Tilt head back while sitting or standing up, or lie on bed and hang head over side. Stay in position a few minutes to permit medication to spread

through nose. (Spray): With head upright, squeeze bottle quickly and firmly to produce 1 or 2 sprays into each nostril; wait 3–5 min, blow nose, and repeat dose. (Jelly): Place in each nostril and sniff it well back into nose.

- Clean tips and droppers of nasal solution dispensers with hot water after use to prevent contamination of solution. Droppers of ophthalmic solution bottles should not touch any surface including the eye.
- Ophthalmic preparations: To avoid excessive systemic absorption, apply pressure to lacrimal sac during and for 1–2 min after instillation of drops.

Subcutaneous/Intramuscular

- Give undiluted.

Intravenous

Note: Ensure patency of IV site prior to administration.

PREPARE: Direct: Dilute each 10 mg (1 mL) of 1% solution in 9 mL of sterile water. **IV Infusion:** Dilute each 10 mg in 500 mL D5W or NS (concentration: 0.02 mg/mL).

ADMINISTER: Direct: Give a single dose over 60 sec. **IV Infusion:** Titrate to maintain BP.

INCOMPATIBILITIES Solution/additive: Phenytoin Y-site: Propofol, thiopental.

- Protect from exposure to air, strong light, or heat, any of which can cause solutions to change color to brown, form a precipitate, and lose potency.

ADVERSE EFFECTS (≥1%) **Special Senses:** *Transient stinging,* lacrimation, brow ache, headache, blurred vision, allergy (pigmentary deposits on lids, conjunctiva, and cornea with prolonged use), increased sensitivity to light. *Rebound nasal congestion* (hyperemia and edema of mucosa), *nasal burning*, stinging, dryness, *sneezing*. **CV:** Palpitation, tachycardia, bradycardia (overdosage), extrasystoles, hypertension. **Body as a Whole:** Trembling, sweating, pallor, sense of fullness in head, tingling of extremities, sleeplessness, dizziness, light-headedness, weakness, restlessness, anxiety, precordial pain, *tremor*, severe visceral or peripheral vasoconstriction, necrosis if IV infiltrates.

INTERACTIONS Drug: ERGOT ALKALOIDS, **guanethidine, reserpine**, TRICYCLIC ANTIDEPRESSANTS increase pressor effects of phenylephrine; **halothane, digoxin** increase risk of arrhythmias; MAO INHIBITORS cause hypertensive crisis; **oxytocin** causes persistent hypertension; ALPHA-BLOCKERS, BETA-BLOCKERS antagonize effects of phenylephrine.

PHARMACOKINETICS Onset: Immediate IV; 10–15 min IM/Subcutaneous. **Duration:** 15–20 min IV; 30–120 min IM/Subcutaneous; 3–6 h topical. **Metabolism:** In liver and tissues by monoamine oxidase.

NURSING IMPLICATIONS

Assessment & Drug Effects

- Monitor infusion site closely as extravasation may cause tissue necrosis and gangrene. If extravasation does occur, area should be immediately injected with 5–10 mg of phentolamine (Regitine) diluted in 10–15 mL of NS.
- Monitor pulse, BP, and central venous pressure (q2–5min) during IV administration.
- Control flow rate and dosage to prevent excessive dosage. IV overdoses can induce ventricular dysrhythmias.
- Observe for congestion or rebound miosis after topical administration to eye.

Patient & Family Education

- Be aware that instillation of 2.5–10% strength ophthalmic solution can cause burning and stinging.
- Do not exceed recommended dosage regardless of formulation.
- Inform the prescriber if no relief is experienced from preparation in 5 days.
- Wear sunglasses in bright light because after instillation of ophthalmic drops, pupils will be large and eyes may be more sensitive to light than usual. Stop medication and notify prescriber if sensitivity persists beyond 12 h after drug has been discontinued.
- Be aware that some ophthalmic solutions may stain contact lenses.

PHENYTOIN ⊘
(fen'i-toy-in)
Dilantin-125, Dilantin

PHENYTOIN SODIUM EXTENDED
Dilantin Kapseals, Phentek

PHENYTOIN SODIUM PROMPT
Dilantin
Classifications: ANTICONVULSANT; HYDANTOIN
Therapeutic: ANTICONVULSANT
Pregnancy Category: D

AVAILABILITY 100 mg capsule; 100 mg, 200 mg, 300 mg sustained release capsule; 50 mg chewable tablet; 125 mg/5 mL suspension; 50 mg/mL injection

ACTION & THERAPEUTIC EFFECT
Anticonvulsant action elevates the seizure threshold and/or limits the spread of seizure discharge. Phenytoin is accompanied by reduced voltage, frequency, and

Common adverse effects in *italic*, life-threatening effects underlined; generic names in **bold**; classifications in SMALL CAPS; ♣ Canadian drug name; ⊘ Prototype drug

1195

spread of electrical discharges within the motor cortex. *Inhibits seizure activity. Effective in treating arrhythmias associated with QT prolongation.*

USES To control tonic-clonic (grand mal) seizures, psychomotor and nonepileptic seizures (e.g., Reye's syndrome, after head trauma). Also used to prevent or treat seizures occurring during or after neurosurgery. Is not effective for absence seizures.

UNLABELED USES Antiarrhythmic agent (phenytoin IV) especially in treatment of digitalis-induced arrhythmias; treatment of trigeminal neuralgia (tic douloureux).

CONTRAINDICATIONS Hypersensitivity to hydantoin products; rash; seizures due to hypoglycemia; sinus bradycardia, complete or incomplete heart block; Adams-Stokes syndrome; pregnancy (category D).
CAUTIOUS USE Impaired liver or kidney function; alcoholism; blood dyscrasias; hypotension, severe myocardial insufficiency, impending or frank heart failure; pancreatic adenoma; DM, hyperglycemia; respiratory depression; acute intermittent porphyria; older adult, debilitated, or gravely ill patients.

ROUTE & DOSAGE

Anticonvulsant
Adult: **PO** 15–20 mg/kg loading dose, then 300 mg/day in 1–3 divided doses, may be gradually increased by 100 mg/wk until seizures are controlled **IV** 10–15 mg/kg then 300 mg/day in divided doses
Child: **PO/IV** 15–20 mg/kg loading dose, then 5 mg/kg in 2 divided doses

ADMINISTRATION

Oral

- Ensure that sustained release form is not chewed or crushed. **Must be** swallowed whole.
- Do not give within 2–3 h of antacid ingestion.
- Shake suspension vigorously before pouring to ensure uniform distribution of drug.
- Note: Prompt release capsules and chewable tablets are not intended for once-a-day dosage since drug is too quickly bioavailable and can therefore lead to toxic serum levels.
- Use sustained release capsules ONLY for once-a-day dosage regimens.

Intravenous

Note: Verify correct rate of IV injection for administration to infants or children with prescriber.

- Inspect solution prior to use. May use a slightly yellowed injectable solution safely. Precipitation may be caused by refrigeration, but slow warming to room temperature restores clarity.

PREPARE: **Direct:** Give undiluted. Use only when clear without precipitate.

ADMINISTER: **Direct for Adult** Give 50 mg or fraction thereof over 1 min (25 mg/min in older adult or when used as antiarrhythmic). • Follow with an injection of sterile saline through the same in-place catheter or needle. DO NOT use solutions containing dextrose.
Direct for Child/Neonate: Give 1 mg/kg/min. • Follow with an injection of sterile saline through the same in-place catheter or needle. DO NOT use solutions containing dextrose.

INCOMPATIBILITIES Solution/additive: 5% dextrose, lactated Ringer's, fat emulsion, sodium chloride, amikacin, aminophylline, bretylium, cephalothin, cephapirin, chloramphenicol, chlordiazepoxide, clindamycin, codeine phosphate, diphenhydramine, dobutamine, hydromorphone, insulin, levorphanol, lidocaine, lincomycin, meperidine, metaraminol, methadone, morphine, nitroglycerin, norepinephrine, penicillin G, pentobarbital, phenylephrine, phytonadione, procaine, prochlorperazine, secobarbital, streptomycin, warfarin. Y-site: Amikacin, amphotericin B cholesteryl complex, bretylium, cimetidine, ciprofloxacin, clarithromycin, clindamycin, diltiazem, dobutamine, enalaprilat, fenoldopam, gatifloxacin, heparin, hydromorphone, lidocaine, linezolid, methadone, morphine, ondansetron, potassium chloride, propofol, sufentanil, tacrolimus, theophylline, TPN, vitamin B complex with C.

- Observe injection site frequently during administration to prevent infiltration. Local soft tissue irritation may be serious, leading to erosion of tissues.

ADVERSE EFFECTS (≥1%) **CNS:** Usually dose-related: Nystagmus, *drowsiness,* ataxia, dizziness, mental confusion, tremors, insomnia, headache, seizures. **CV:** Bradycardia, hypotension, cardiovascular collapse, ventricular fibrillation, phlebitis. **Special Senses:** Photophobia, conjunctivitis, diplopia, blurred vision. **GI:** *Gingival hyperplasia,* nausea, vomiting, constipation, epigastric pain, dysphagia, loss of taste, weight loss, hepatitis, liver necrosis.

Hematologic: *Thrombocytopenia, leukopenia,* leukocytosis, agranulocytosis, pancytopenia, eosinophilia; megaloblastic, hemolytic, or aplastic anemias. **Metabolic:** Fever, hyperglycemia, glycosuria, weight gain, edema, transient increase in serum thyrotropic (TSH) level, osteomalacia or rickets associated with hypocalcemia and elevated alkaline phosphatase activity. **Skin:** Alopecia, hirsutism (especially in young female); rash: scarlatiniform, maculopapular, urticaria, morbilliform; bullous, exfoliative, or purpuric dermatitis; Stevens-Johnson syndrome, toxic epidermal necrolysis, keratosis, neonatal hemorrhage. **Urogenital:** Acute renal failure, Peyronie's disease. **Respiratory:** Acute pneumonitis, pulmonary fibrosis. **Body as a Whole:** Periarteritis nodosum, acute systemic lupus erythematosus, craniofacial abnormalities (with enlargement of lips); lymphadenopathy.

DIAGNOSTIC TEST INTERFERENCE Phenytoin (HYDANTOINS) may produce lower than normal values for *dexamethasone* or *metyrapone* tests; may increase serum levels of *glucose, BSP,* and *alkaline phosphatase* and may decrease *PBI* and *urinary steroid* levels.

INTERACTIONS Drug: Alcohol decreases phenytoin effects; OTHER ANTICONVULSANTS may increase or decrease phenytoin levels; phenytoin may decrease absorption and increase metabolism of ORAL ANTICOAGULANTS; phenytoin increases metabolism of CORTICOSTEROIDS, ORAL CONTRACEPTIVES, and **nisoldipine,** decreasing their effectiveness; **amiodarone, chloramphenicol, omeprazole,** and **ticlopidine** increase phenytoin levels; ANTITUBERCULOSIS AGENTS decrease phenytoin levels. **Food: Folic acid, calcium,**

P

Common adverse effects in *italic,* life-threatening effects <u>underlined;</u> generic names in **bold;** classifications in SMALL CAPS; ✚ Canadian drug name; ⊘ Prototype drug

1197

and **vitamin D** absorption may be decreased by phenytoin; phenytoin absorption may be decreased by enteral nutrition supplements. **Herbal: Ginkgo** may decrease anticonvulsant effectiveness.

PHARMACOKINETICS Absorption: Completely from GI tract. **Peak:** 1.5–3 h prompt release; 4–12 h sustained release. **Distribution:** 95% protein bound; crosses placenta; small amount in breast milk. **Metabolism:** Oxidized in liver to inactive metabolites. **Elimination:** By kidneys. **Half-Life:** 22 h.

NURSING IMPLICATIONS

Assessment & Drug Effects

- Monitor infusion site closely as extravasation may cause tissue necrosis.
- Continuously monitor vital signs and symptoms during IV infusion and for an hour afterward. Watch for respiratory depression. Constant observation and a cardiac monitor are necessary with older adults or patients with cardiac disease. Margin between toxic and therapeutic IV doses is relatively small.
- Be aware of therapeutic serum concentration: 10–20 mcg/mL; toxic level: 30–50 mcg/mL; lethal level: 100 mcg/mL. Steady-state therapeutic levels are not achieved for at least 7–10 days.
- Lab tests: Periodic serum phenytoin concentration; CBC with differential, platelet count, and Hct and Hgb; serum glucose, serum calcium, and serum magnesium; and LFTs.
- Observe patient closely for neurologic adverse effects following IV administration.
- Monitor for gingival hyperplasia, which appears most commonly in children and adolescents and never occurs in patients without teeth.

- Make sure patients on prolonged therapy have adequate intake of vitamin D-containing foods and sufficient exposure to sunlight.
- Monitor diabetics for loss of glycemic control.
- Monitor for S&S of hypocalcemia (see Appendix F), especially in patients receiving other anticonvulsants concurrently, as well as those who are inactive, have limited exposure to sun, or whose dietary intake is inadequate.
- Observe for symptoms of folic acid deficiency: Neuropathy, mental dysfunction.
- Be alert to symptoms of hypomagnesemia (see Appendix F); neuromuscular symptoms: Tetany, positive Chvostek's and Trousseau's signs, seizures, tremors, ataxia, vertigo, nystagmus, muscular fasciculations.

Patient & Family Education

- Be aware that drug may make urine pink or red to red-brown.
- Report symptoms of fatigue, dry skin, deepening voice when receiving long-term therapy because phenytoin can unmask a low thyroid reserve.
- Do not alter prescribed drug regimen. Stopping drug abruptly may precipitate seizures and status epilepticus.
- Do not request/accept change in drug brand when refilling prescription without consulting prescriber.
- Understand the effects of alcohol: Alcohol intake may increase phenytoin serum levels, leading to phenytoin toxicity.
- Discontinue drug immediately if a measles-like skin rash or jaundice appears and notify prescriber.
- Be aware that influenza vaccine during phenytoin treatment may increase seizure activity. Consult prescriber.

PHYSOSTIGMINE SALICYLATE

(fi-zoe-stig'meen)

Classification: CHOLINESTERASE INHIBITOR
Therapeutic: ANTICHOLINERGIC ANTIDOTE, CHOLINESTERASE INHIBITOR
Prototype: Neostigmine
Pregnancy Category: C

AVAILABILITY 1 mg/mL injection

ACTION & *THERAPEUTIC EFFECT*

Physostigmine competes with acetylcholine (ACE) for its binding site on acetylcholinesterase, thereby potentiating action of ACE on the skeletal muscle, GI tract and within the CNS. *Effective in reversing anticholingeric toxicity.*

USES To reverse anticholinergic toxicity.

CONTRAINDICATIONS Asthma; DM; gangrene, cardiovascular disease; mechanical obstruction of intestinal or urogenital tract; peptic ulcer disease; asthma; any vagotonic state; closed-angle glaucoma; secondary glaucoma; inflammatory disease of iris or ciliary body; lactation.
CAUTIOUS USE Epilepsy; parkinsonism; bradycardia; hyperthyroidism; seizure disorders; hypotension; pregnancy (category C).

ROUTE & DOSAGE

Reversal of Anticholinergic Effects

Adult: **IM/IV** 0.5–2 mg (IV not faster than 1 mg/min), repeat as needed
Child: **IV** 0.02 mg/kg/dose, may repeat q5–10min (max total dose: 2 mg)

ADMINISTRATION

- Use only clear, colorless solutions. Red-tinted solution indicates oxidation, and such solutions should be discarded.

Intramuscular
- Give undiluted.

Intravenous
Note: Verify correct rate of IV injection for infants or children with prescriber.

PREPARE: **Direct:** Give undiluted. *ADMINISTER:* **Direct for Adult:** Give slowly at a rate of no more than 1 mg/min. ▪ Rapid administration and overdosage can cause a cholinergic crisis. **Direct for Child:** Give 0.5 mg or fraction thereof over at least 1 min. ▪ Rapid administration and overdosage can cause a cholinergic crisis.
INCOMPATIBILITIES **Solution/additive:** Phenytoin, ranitidine. **Y-site:** Dobutamine.

ADVERSE EFFECTS (≥1%) **Body as a Whole:** *Sweating,* cholinergic crisis (acute toxicity), hyperactivity, respiratory distress, convulsions. **CNS:** Restlessness, hallucinations, twitching, tremors, *sweating,* weakness, ataxia, convulsions, collapse. **GI:** *Nausea, vomiting, epigastric pain, diarrhea, salivation.* **Urogenital:** Involuntary urination or defecation. **Special Senses:** Miosis, *lacrimation,* rhinorrhea. **Respiratory:** Dyspnea, bronchospasm, respiratory paralysis, pulmonary edema. **Cardiovascular:** Irregular pulse, palpitations, bradycardia, rise in BP.

INTERACTIONS Drug: Antagonizes effects of **echothiophate, isofluorphate.**

PHARMACOKINETICS Absorption: Readily from mucous membranes, muscle, subcutaneous tissue; 10–12% absorbed from GI tract. **Onset:** 3–8 min IM/IV. **Duration:** 0.5–5 h IM/IV.

Common adverse effects in *italic,* life-threatening effects underlined; generic names in **bold;** classifications in SMALL CAPS; ♣ Canadian drug name; ⊘ Prototype drug

1199

Distribution: Crosses blood–brain barrier. **Metabolism:** In plasma by cholinesterase. **Elimination:** Small amounts in urine. **Half-Life:** 15–40 min.

NURSING IMPLICATIONS

Assessment & Drug Effects

- Monitor vital signs and state of consciousness closely in patients receiving drug for atropine poisoning. Since physostigmine is usually rapidly eliminated, patient can lapse into delirium and coma within 1 to 2 h; repeat doses may be required.
- Monitor closely for adverse effects related to CNS and for signs of sensitivity to physostigmine. Have atropine sulfate readily available for clinical emergency.
- Discontinue parenteral or oral drug if following symptoms arise: Excessive salivation, emesis, frequent urination, or diarrhea.
- Eliminate excessive sweating or nausea with dose reduction.

PHYTONADIONE (VITAMIN K₁)

(fye-toe-na-dye'one)

Mephyton
Classifications: VITAMIN K; ANTIDOTE
Therapeutic: VITAMIN K; ANTIDOTE
Pregnancy Category: C

AVAILABILITY 5 mg tablets; 2 mg/mL, 10 mg/mL injection

ACTION & *THERAPEUTIC EFFECT*

Fat-soluble substance chemically identical to and with similar activity as naturally occurring vitamin K. Vitamin K is essential for hepatic biosynthesis of blood clotting Factors II, VII, IX, and X. *Promotes liver synthesis of clotting factors.*

USES Drug of choice as antidote for overdosage of coumarin and

indandione oral anticoagulants. Also reverses hypoprothrombinemia secondary to administration of oral antibiotics, quinidine, quinine, salicylates, sulfonamides, excessive vitamin A, and secondary to inadequate absorption and synthesis of vitamin K (as in obstructive jaundice, biliary fistula, ulcerative colitis, intestinal resection, prolonged hyperalimentation). Also prophylaxis of and therapy for neonatal hemorrhagic disease.

CONTRAINDICATIONS Hypersensitivity to phytonadione, benzyl alcohol or castor oil; severe liver disease.
CAUTIOUS USE Biliary tract disease, obstructive jaundice, pregnancy (category C). **IV:** Older adults.

ROUTE & DOSAGE

Anticoagulant Overdose

Adult: **PO/Subcutaneous/IM** 2.5–10 mg; rarely up to 50 mg/day, may repeat parenteral dose after 6–8 h if needed or PO dose after 12–24 h **IV** Emergency only: 10–15 mg at a rate of 1 mg/min or less, may be repeated in 4 h if bleeding continues

Hemorrhagic Disease of Newborns

Infant: **IM/Subcutaneous** 0.5–1 mg immediately after delivery, may repeat in 6–8 h if necessary

Other Prothrombin Deficiencies

Adult: **IM/Subcutaneous/IV** 2–25 mg
Child/Infant: **IM/Subcutaneous/IV** 0.5–5 mg

ADMINISTRATION

Oral

- Bile salts must be given with tablets if patient has deficient bile production.

Common adverse effects in *italic*; life-threatening effects <u>underlined</u>; generic names in **bold**; classifications in SMALL CAPS; ♣ Canadian drug name; ❂ Prototype drug

- Store in tightly closed container and protect from light. Vitamin K is rapidly degraded by light.

Intramuscular/Subcutaneous

- Subcutaneous route is preferred. IM route has been associated with severe reactions.
- Apply gentle pressure to site following injection. Swelling (internal bleeding) and pain sometimes occur with injection.

Intravenous

Note: Reserve IV route only for emergencies.

PREPARE: Direct: Dilute a single dose in 10 mL D5W, NS, or D5/NS. ▪ Protect infusion solution from light.

ADMINISTER: Direct: Give solution immediately after dilution at a rate not to exceed 1 mg/min.

INCOMPATIBILITIES Solution/additive: Ascorbic acid, cephalothin, dobutamine, doxycycline, magnesium sulfate, nitrofurantoin, phenobarbital, ranitidine, thiopental, vancomycin, warfarin. Y-site: Dobutamine.

▪ Protect infusion solution from light by wrapping container with aluminum foil or other opaque material. ▪ Discard unused solution and contents in open ampule.

ADVERSE EFFECTS (≥1%) **Body as a Whole:** Hypersensitivity or anaphylaxis-like reaction: Facial flushing, cramp-like pains, convulsive movements, chills, fever, diaphoresis, weakness, dizziness, shock, cardiac arrest. **CNS:** Headache (after oral dose), brain damage, death. **GI:** Gastric upset. **Hematologic:** Paradoxic hypoprothrombinemia (patients with severe liver disease), severe hemolytic anemia. **Metabolic:** Hyperbilirubinemia,

kernicterus. **Respiratory:** Bronchospasm, dyspnea, sensation of chest constriction, respiratory arrest. **Skin:** Pain at injection site, hematoma, and nodule formation, erythematous skin eruptions (with repeated injections). **Special Senses:** Peculiar taste sensation.

DIAGNOSTIC TEST INTERFERENCE Falsely elevated **urine steroids** (by modifications of **Reddy, Jenkins, Thorn procedure**).

INTERACTIONS Drug: Antagonizes effects of **warfarin; cholestyramine, colestipol, mineral oil** decrease absorption of oral phytonadione.

PHARMACOKINETICS Absorption: Readily from intestinal lymph if bile is present. **Onset:** 6–12 h PO; 1–2 h IM/Subcutaneous; 15 min IV. **Peak:** Hemorrhage usually controlled within 3–8 h; normal prothrombin time may be obtained in 12–14 h after administration. **Distribution:** Concentrates briefly in liver after absorption; crosses placenta; distributed into breast milk. **Metabolism:** Rapidly in liver. **Elimination:** In urine and bile.

NURSING IMPLICATIONS

Assessment & Drug Effects

- Monitor patient constantly. Severe reactions, including fatalities, have occurred during and immediately after IV and IM injection (see ADVERSE EFFECTS).
- Lab tests: Baseline and frequent PT/INR.
- Frequency, dose, and therapy duration are guided by PT/INR clinical response.
- Monitor therapeutic effectiveness which is indicated by shortened PT, INR, bleeding, and clotting times, as well as decreased hemorrhagic tendencies.

P

Common adverse effects in *italic*; life-threatening effects underlined; generic names in **bold**; classifications in SMALL CAPS; ♣ Canadian drug name; ◑ Prototype drug

1201

- Be aware that patients on large doses may develop temporary resistance to coumarin-type anticoagulants. If oral anticoagulant is reinstituted, larger than former doses may be needed. Some patients may require change to heparin.

Patient & Family Education

- Maintain consistency in diet and avoid significant increases in daily intake of vitamin K–rich foods when drug regimen is stabilized. Know sources rich in vitamin K: Asparagus, broccoli, cabbage, lettuce, turnip greens, pork or beef liver, green tea, spinach, watercress, and tomatoes.

PILOCARPINE HYDROCHLORIDE ℗

PILOCARPINE NITRATE

(pye-loe-kar′peen)

Adsorbocarpine, Isopto Carpine, Minims Pilocarpine ♣, Miocarpine ♣, Ocusert, Pilo, Pilocar, Salagen
Classifications: EYE PREPARATION; MIOTIC (ANTIGLAUCOMA); DIRECT-ACTING CHOLINERGIC
Therapeutic: ANTIGLAUCOMA
Pregnancy Category: C

AVAILABILITY 0.25%, 0.5%, 1%, 2%, 3%, 4%, 5%, 6%, 8%, 10% ophth solution; 4% ophth gel; 20 mcg/h, 40 mcg/h ocular insert; 5 mg tablets

ACTION & *THERAPEUTIC EFFECT*
In open-angle glaucoma, pilocarpine causes contraction of the ciliary muscle, increasing the outflow of aqueous humor, which reduces intraocular pressure (IOP). In closed-angle glaucoma, it induces miosis by opening the angle of the anterior chamber of the eye, through which aqueous humor exits. *Decrease in*

IOP results from stimulation of ciliary and papillary sphincter muscles, thus facilitating outflow of aqueous humor.

USES Open-angle and angle-closure glaucomas; to reduce IOP and to protect the lens during surgery and laser iridotomy; to counteract effects of mydriatics and cycloplegics following surgery or ophthalmoscopic examination; to treat xerostomia.

CONTRAINDICATIONS Secondary glaucoma, acute iritis, acute inflammatory disease of anterior segment of eye; uncontrolled asthma; lactation. **Ocular therapeutic system:** Not used in acute infectious conjunctivitis, keratitis, retinal detachment, or when intense miosis is required, or with contact lens use.
CAUTIOUS USE Bronchial asthma; biliary tract disease; COPD; hypertension; pregnancy (category C).

ROUTE & DOSAGE

Acute Glaucoma

Adult/Child: **Ophthalmic** 1 drop of 1–2% solution in affected eye q5–10min for 3–6 doses, then 1 drop q1–3h until IOP is reduced

Chronic Glaucoma

Adult/Child: **Ophthalmic** 1 drop of 0.5–4% solution in affected eye q4–12h or 1 ocular system **(Ocusert)** q7days

Miotic

Adult/Child: **Ophthalmic** 1 drop of 1% solution in affected eye

Xerostomia

Adult: **PO** 5 mg t.i.d., may increase up to 10 mg t.i.d.

ADMINISTRATION

Oral

- Give with a full glass of water, if not contraindicated.

Instillation

- Note: During acute phase, prescriber may prescribe instillation of drug into unaffected eye to prevent bilateral attack of acute glaucoma.
- Apply gentle digital pressure to periphery of nasolacrimal drainage system for 1–2 min immediately after instillation of drops to prevent delivery of drug to nasal mucosa and general circulation.

ADVERSE EFFECTS (≥1%) **CNS:** Oral (asthenia, headaches, dizziness, chills). **Special Senses:** Ciliary spasm with brow ache, twitching of eyelids, eye pain with change in eye focus, miosis, *diminished vision in poorly illuminated areas,* blurred vision, reduced visual acuity, sensitivity, contact allergy, lacrimation, follicular conjunctivitis, conjunctival irritation, cataract, <u>retinal detachment</u>. **GI:** *Nausea,* vomiting, abdominal cramps, diarrhea, epigastric distress, *salivation.* **Respiratory:** Bronchospasm, rhinitis. **CV:** Tachycardia. **Body as a Whole:** Tremors, *increased sweating,* urinary frequency.

INTERACTIONS Drug: The actions of pilocarpine and **carbachol** are additive when used concomitantly. Oral form may cause conduction disturbances with BETA-BLOCKERS. Antagonizes the effects of concurrent ANTICHOLINERGIC DRUGS (e.g., **atropine, ipratropium**). **Food:** High-fat meal decreases absorption of pilocarpine.

PHARMACOKINETICS Absorption: Topical penetrates cornea rapidly; readily absorbed from GI tract. **Onset:** Miosis 10–30 min; IOP reduction 60 min; salivary stimulation 20 min. **Peak:** Miosis 30 min; IOP reduction 75 min; salivary stimulation 60 min. **Duration:** Miosis 4–8 h; IOP reduction 4–14 h (7 days with Ocusert); salivary stimulation 3–5 h. **Metabolism:** Inactivated at neuronal synapses and in plasma. **Elimination:** In urine. **Half-Life:** 0.76–1.35 h.

NURSING IMPLICATIONS

Assessment & Drug Effects

- Be aware that hourly tonometric tests may be done during early treatment because drug may cause an initial transitory increase in IOP.
- Monitor changes in visual acuity.
- Monitor for adverse effects. Brow pain and myopia tend to be more prominent in younger patients and generally disappear with continued use of drug.

Patient & Family Education

- Understand that therapy for glaucoma is prolonged and that adherence to established regimen is crucial to prevent blindness.
- Do not drive or engage in potentially hazardous activities until vision clears. Drug causes blurred vision and difficulty in focusing.
- Discontinue medication if symptoms of irritation or sensitization persist and report to prescriber.

PIMECROLIMUS

(pim-e-cro-lim′us)

Elidel

Classifications: BIOLOGIC RESPONSE MODIFIER; IMMUNOMODULATOR

Therapeutic: IMMUNOSUPPRESSANT; ANTI-INFLAMMATORY

Prototype: Cyclosporine

Pregnancy Category: C

AVAILABILITY 1% cream

ACTION & *THERAPEUTIC EFFECT*

Pimecrolimus selectively inhibits inflammatory action of skin cells by blocking T-cell activation and

Common adverse effects in *italic*, life-threatening effects <u>underlined</u>; generic names in **bold;** classifications in SMALL CAPS; ♣ Canadian drug name; ☻ Prototype drug

1203

cytokine release. It appears to inhibit the production of IL-2, IL-4, IL-10, and interferon gamma in T-cells. *Produces significant anti-inflammatory activity without evidence of skin atrophy.*

USES Short-term intermittent treatment of mild to moderate atopic dermatitis.

CONTRAINDICATIONS Hypersensitivity to pimecrolimus or components in the cream; Netherton's syndrome; application to active cutaneous viral infection; occlusive dressing; artificial or natural sunlight (UV) exposure; continuous long-term use; lactation.

CAUTIOUS USE Infection at topical treatment sites; history of untoward effects with topical cyclosporine or tacrolimus; skin papillomas; immunocompromised patients; pregnancy (category C); children younger than 2 y.

ROUTE & DOSAGE

Atopic Dermatitis
Adult: **Topical** Apply thin layer to affected skin b.i.d. (avoid continuous long-term use)

ADMINISTRATION
Topical
▪ Do not apply to any skin surface that appears to be infected.

ADVERSE EFFECTS (≥1%) **Body as a Whole:** Flu-like symptoms, infections, fever, increased risk of cancer. **CNS:** Headache. **GI:** Gastroenteritis, abdominal pain, nausea, vomiting, diarrhea, constipation. **Respiratory:** Sore throat, *upper respiratory infection, cough,* nasal congestion, asthma exacerbation, rhinitis, epistaxis. **Skin:** *Burning,* irritation, pruritus, skin infection, impetigo, folliculitis, skin papilloma,

herpes simplex dermatitis, urticaria, acne. **Special Senses:** Ear infection, earache, conjunctivitis.

PHARMACOKINETICS Absorption: Minimal through intact skin. **Metabolism:** No evidence of skin-mediated metabolism, metabolized in liver by CYP3A4. **Elimination:** Primarily in feces.

NURSING IMPLICATIONS
Assessment & Drug Effects
▪ Assess for and report persistent skin irritation that develops following application of the cream and lasts for more than 1 wk.

Patient & Family Education
▪ Minimize exposure of treated area to natural or artificial sunlight.
▪ Immediately report a new or changed skin lesion to the prescriber.
▪ Stop topical application once signs of dermatitis have disappeared. Resume application at the first sign of recurrence.
▪ Wash hands thoroughly after application if hands are not the treatment sites.
▪ Report any significant skin irritation that results from application of the cream.

PIMOZIDE
(pi'moe-zide)
Orap
Classification: ANTIPSYCHOTIC
Therapeutic: ANTIPSYCHOTIC; CNS DOPAMINERGIC RECEPTOR ANTAGONIST
Prototype: Haloperidol
Pregnancy Category: C

AVAILABILITY 1 mg, 2 mg tablets

ACTION & *THERAPEUTIC EFFECT*
Potent central dopamine antagonist

P

that alters release and turnover of central dopamine stores. Blockade of CNS dopaminergic receptors results in suppression of the motor and phonic tics that characterize Tourette's disorder. *Effective in suppressing motor and phonic tics associated with Tourette's disorder.*

USES Tourette's syndrome resistant to standard therapy.
UNLABELED USES Schizophrenia, delusions of parasitosis.

CONTRAINDICATIONS Treatment of simple tics other than those associated with Tourette's disorder; drug-induced tics; history of cardiac dysrhythmias and conditions marked by prolonged QT syndrome, patient taking drugs that may prolong QT interval (e.g., quinidine); congenital heart defects; cardiac arrhythmias; electrolyte imbalance; Parkinson's disease; severe toxic CNS depression; lactation.
CAUTIOUS USE Kidney and liver dysfunction; cardiac disease; glaucoma; BPH; urinary retention; pregnancy (category C). Safe use in children younger than 12 y is not known.

ROUTE & DOSAGE

Tourette's Disorder
Adult: **PO** 1–2 mg/day in divided doses, gradually increase dose every other day up to 0.2 mg/kg/day or 10 mg/day in divided doses (max: 0.2 mg/kg/day or 10 mg/day)
Adolescent: **PO** 0.05 mg/kg/day at bedtime, gradually increase as needed (max: 0.2 mg/kg/day or 10 mg, whichever is less)

ADMINISTRATION
Oral
▪ Increase drug dose gradually, usually over 1–3 wk, until maintenance dose is reached.
▪ Follow regimen prescribed by prescriber for withdrawal: Usually slow, gradual changes over a period of days or weeks (drug has a long half-life). Sudden withdrawal may cause reemergence of original symptoms (motor and phonic tics) and of neuromuscular adverse effects of the drug.

ADVERSE EFFECTS (≥1%) **Body as a Whole:** *Akathisia,* speech disorder, *torticollis, tremor,* handwriting changes, *akinesia,* fainting, hyperpyrexia, tardive dyskinesia, *rigidity, oculogyric crisis,* hyperreflexia; seizures, neuroleptic malignant syndrome; *extrapyramidal dysfunction,* hyperthermia, autonomic dysfunction; diaphoresis, weight changes, asthenia, chest pain, periorbital edema. **CNS:** Headache, *sedation, drowsiness,* insomnia, seizures, stupor. **CV:** Prolongation of QT interval, inverted or flattened T wave, appearance of U wave, labile blood pressure. **Urogenital:** Loss of libido, impotence, nocturia, urinary frequency, amenorrhea, dysmenorrhea, mild galactorrhea, urinary retention, acute renal failure. **Respiratory:** Dyspnea, respiratory failure. **Skin:** Sweating, skin irritation. **Special Senses:** Visual disturbances, photosensitivity, decreased accommodation, blurred vision, cataracts. **GI:** Increased salivation, nausea, vomiting, diarrhea, anorexia, abdominal cramps, constipation.

INTERACTIONS Drug: Alcohol and other CNS DEPRESSANTS increase CNS depression; ANTICHOLINERGIC AGENTS (e.g., TRICYCLIC ANTIDEPRESSANTS,

Common adverse effects in *italic,* life-threatening effects underlined; generic names in **bold;** classifications in SMALL CAPS; ♣ Canadian drug name; ⊘ Prototype drug

1205

atropine) increase anticholinergic effects; PHENOTHIAZINES, TRICYCLIC ANTIDEPRESSANTS, ANTIARRHYTHMICS, MACROLIDE ANTIBIOTICS, AZOLE ANTIFUNGALS, PROTEASE INHIBITORS, **nefazodone, sertraline, zileuton** increase risk of arrhythmias and heart block; pimozide antagonizes effects of ANTICONVULSANTS—there is loss of seizure control. **Food: Grapefruit juice** (greater than 1 qt/day) may increase plasma concentrations and adverse effects.

PHARMACOKINETICS Absorption: Slowly and variably from GI tract (40–50% absorbed). **Peak:** 6–8 h. **Metabolism:** In liver (by CYP3A4). **Elimination:** 80–85% in urine, 15–20% in feces. **Half-Life:** 55 h.

NURSING IMPLICATIONS

Note: See haloperidol for additional nursing implications.

Assessment & Drug Effects
- Obtain ECG baseline data at beginning of therapy and check periodically, especially during dosage adjustments.
- Notify prescriber immediately for widening or prolongation of the QT interval, which suggests developing cardiotoxicity.
- Risk of tardive dyskinesia appears to be greatest in women, older adults, and those on high-dose therapy.
- Be aware that extrapyramidal reactions often appear within the first few days of therapy, are dose-related, and usually occur when dose is high.
- Be aware that anticholinergic effects (dry mouth, constipation) may increase as dose is increased.

Patient & Family Education
- Adhere to established drug regimen (i.e., do not change dose or intervals and discontinue only with prescriber's guidance).
- Use measures to relieve dry mouth (frequent rinsing with water, saliva substitute, increased fluid intake) and constipation (increased dietary fiber, drink 6–8 glasses of water daily).
- Do not drive or engage in potentially hazardous activities because drug-caused hand tremors, drowsiness, and blurred vision may impair alertness and abilities.
- Pseudoparkinsonism symptoms are usually mild and reversible with dose adjustment.
- Be alert to the earliest symptom of tardive dyskinesia ("flycatching"— an involuntary movement of the tongue), and report promptly to the prescriber.
- Return to prescriber for periodic assessments of therapy benefit and cardiac status.
- Understand dangers of ingesting alcohol to prevent augmenting CNS depressant effects of pimozide.

PINDOLOL
(pin′doe-lole)

Classifications: BETA-ADRENERGIC RECEPTOR ANTAGONIST; ANTIHYPERTENSIVE; ANTIANGINAL
Therapeutic: ANTIHYPERTENSIVE; ANTIANGINAL
Prototype: Propranolol
Pregnancy Category: B

AVAILABILITY 5 mg, 10 mg tablets

ACTION & *THERAPEUTIC EFFECT*
Hypotensive mechanism results from its competitively blocking beta-adrenergic receptors primarily in myocardium, and beta receptors within smooth muscle. *Lowers blood pressure by decreasing peripheral*

vascular resistance. Exerts vasodilation as well as hypotensive effects.

USES Management of hypertension.
UNLABELED USES Angina.

CONTRAINDICATIONS Bronchospastic diseases, asthma; severe bradycardia, cardiogenic shock, AV block, sick sinus syndrome; cardiac failure; pulmonary failure; lactation.
CAUTIOUS USE Nonallergic bronchospasm; COPD; CHF; diabetes mellitus; hyperthyroidism; myasthenia gravis; impaired liver and kidney function; pregnancy (category B). Safety and efficacy in children not established.

ROUTE & DOSAGE

Hypertension
Adult: **PO** 5 mg b.i.d., may increase by 10 mg/day q3–4wk if needed up (max: 60 mg/day in 2–3 divided doses)
Geriatric: **PO** Start with 5 mg daily

ADMINISTRATION
Oral
- Give drug at same time of day each day with respect to time of food intake for most predictable results.
- Withdraw or discontinue treatment gradually over a period of 1–2 wk.

ADVERSE EFFECTS (≥1%) **CNS:** *Fatigue,* dizziness, insomnia, drowsiness, confusion, fainting, decreased libido. **CV:** *Bradycardia,* hypotension, CHF. **GI:** Nausea, *diarrhea, constipation,* flatulence. **Respiratory:** Bronchospasm, pulmonary edema, dyspnea. **Body as a Whole:**

Back or joint pain. Sensitivity reactions seen as antinuclear antibodies (ANA) (10–30% of patients). **Hematologic:** Agranulocytosis. **Urogenital:** Impotence. **Metabolic:** Hypoglycemia (may mask symptoms of a hypoglycemic reaction).

INTERACTIONS Drug: DIURETICS and other HYPOTENSIVE AGENTS increase hypotensive effect; effects of **albuterol, metaproterenol, terbutaline, pirbuterol** and **pindolol** antagonized; NSAIDs blunt hypotensive effect; decreases hypoglycemic effect of **glyburide; amiodarone** increases risk of bradycardia and sinus arrest.

PHARMACOKINETICS Absorption: Rapidly from GI tract; 50–95% reaches systemic circulation (first pass metabolism). **Onset:** 3 h. **Peak:** 1–2 h. **Duration:** 24 h. **Distribution:** Distributed into breast milk. **Metabolism:** 40–60% in liver. **Elimination:** In urine. **Half-Life:** 3–4 h.

NURSING IMPLICATIONS
Assessment & Drug Effects
- Monitor HR and BP. Report bradycardia and hypotension. Dosage adjustment may be indicated.
- Note: Hypotensive effect may begin within 7 days but is not at maximum therapeutically until about 2 wk after beginning of treatment.
- Lab test: Periodic CBC with differential, kidney function tests, and blood glucose.

Patient & Family Education
- Pindolol masks the dizziness and sweating symptoms of hypoglycemia. Monitor blood glucose for loss of glycemic control.
- Adhere to the prescribed drug regimen; if a change is desired, consult prescriber first. Abrupt withdrawal of drug might precipitate a

Common adverse effects in *italic*, life-threatening effects underlined; generic names in **bold**; classifications in SMALL CAPS; ✦ Canadian drug name; ⊙ Prototype drug

1207

thyroid crisis in a patient with hyperthyroidism, and angina in the patient with ischemic heart disease, leading to an MI.

PIOGLITAZONE HYDROCHLORIDE

(pi-o-glit′a-zone)

Actos

Classifications: ANTIDIABETIC; THIAZOLIDINEDIONE

Therapeutic: ANTIDIABETIC; INSULIN SENSITIZER

Prototype: Rosiglitazone maleate

Pregnancy Category: C

AVAILABILITY 15 mg, 30 mg, 45 mg tablets

ACTION & *THERAPEUTIC EFFECT*
Decreases hepatic glucose output and increases insulin-dependent muscle glucose uptake in skeletal muscle and adipose tissue. Improves glycemic control in noninsulin-dependent diabetics (type 2) by enhancing insulin sensitivity of cells without stimulating pancreatic insulin secretion. *Improves glycemic control as indicated by improved blood glucose levels and decreased HbA1C to 6.5 or lower.*

USES Adjunct to diet in the treatment of type 2 diabetes mellitus.

CONTRAINDICATIONS Hypersensitivity to pioglitazone, troglitazone, rosiglitazone, englitazone; type 1 diabetes, or treatment of DKA; New York Heart Association (NYHA) Class III or IV heart failure; active liver disease or ALT levels greater than 2.5 times normal limit; jaundice; lactation.

CAUTIOUS USE Liver dysfunction; cardiovascular disease; hypertension, CHF, anemia, edema; renal impairment; older adults; pregnancy (category C). Safety and efficacy in children younger than 18 y are not recommended.

ROUTE & DOSAGE

Type 2 Diabetes Mellitus

Adult: **PO** 15–30 mg once daily (max: 45 mg daily)

Disease State Dosage Adjustment

Patients with NHYA Class I or II heart failure (max dose: 15 mg once daily)

Drug Interaction Dosage Adjustment

If used with gemfibrizol (max dose: 15 mg)

ADMINISTRATION

Oral

- Give without regard to food.
- Do not initiate therapy if baseline serum ALT greater than 2.5 times normal.
- Store at 15°–30° C (59°–86° F) in tightly closed container; protect from humidity and moisture.

ADVERSE EFFECTS (≥1%) **Body as a Whole:** Headache, myalgia, edema, rhabdomyolysis. **CV:** Edema, fluid retention, exacerbation of heart failure. **GI:** Tooth disorder. **Respiratory:** *Upper respiratory tract infection,* sinusitis, pharyngitis. **Metabolic:** Hypoglycemia, mild anemia.

INTERACTIONS Drug: Pioglitazone may decrease serum levels of ORAL CONTRACEPTIVES; **ketoconazole, gemfibrozil** may increase serum levels of **pioglitazone. Bosentan** may decrease effect. **Herbal: Garlic, ginseng** may potentiate hypoglycemic effects.

PHARMACOKINETICS Absorption: Rapidly absorbed. **Peak:** 2 h;

steady state concentrations within 7 days. **Duration:** 24 h. **Distribution:** Greater than 99% protein bound. **Metabolism:** In liver to active metabolites. **Elimination:** Primarily in bile and feces. **Half-Life:** 16–24 h.

NURSING IMPLICATIONS

Assessment & Drug Effects

- Monitor for S&S hypoglycemia (possible when insulin/sulfonylureas are coadministered).
- Monitor closely for S&S of CHF or exacerbation of symptoms with preexisting CHF.
- Lab tests: Baseline serum ALT, then q2mo for first year, then periodically (more often if elevated); periodic HbA1C, Hgb and Hct, and lipid profile.
- Withhold drug and notify prescriber if ALT greater than 3 × ULN or patient has jaundice.
- Monitor weight and notify prescriber of development of edema.

Patient & Family Education

- Be aware that resumed ovulation is possible in nonovulating premenopausal women.
- Use or add barrier contraceptive if using hormonal contraception.
- Report immediately to prescriber: Unexplained anorexia, nausea, vomiting, abdominal pain, fatigue, dark urine; or S&S of fluid retention such as weight gain, edema, or activity intolerance.
- Combination therapy: May need adjustment of other antidiabetic drugs to avoid hypoglycemia.
- Learn of and adhere strictly to guideliness for liver function tests. Be sure to have blood tests for liver function every 2 mo for first year; then periodically.

PIPERACILLIN/ TAZOBACTAM ℗

(pi-per'a-cil-lin/taz-o-bac'tam)
Zosyn
Classifications: ANTIBIOTIC; EXTENDED SPECTRUM PENICILLIN
Therapeutic: ANTIBIOTIC
Pregnancy Category: B

AVAILABILITY 2 g, 3 g, 4 g injection

ACTION & *THERAPEUTIC EFFECT* Antibacterial combination product consisting of the semisynthetic piperacillin and the beta-lactamase inhibitor tazobactam. Tazobactam has little antibacterial activity itself; however, in combination with piperacillin, it extends the spectrum of bacteria that are susceptible to piperacillin. *Two-drug combination has antibiotic activity against an extremely broad spectrum of gram-positive, gram-negative, and anaerobic bacteria.*

USES Treatment of moderate to severe appendicitis, uncomplicated and complicated skin and skin structure infections, endometritis, pelvic inflammatory disease, or nosocomial or community-acquired pneumonia caused by beta-lactamase-producing bacteria.

CONTRAINDICATIONS Hypersensitivity to piperacillin, tazobactam, penicillins; coagulopathy.
CAUTIOUS USE Hypersensitivity to cephalosporins, carbapenem or beta-lactamase inhibitors such as clavulanic acid and sulbactam; GI disease, colitis; CF; eczema; kidney failure; complicated urinary tract infections; pregnancy (category B), lactation.

ROUTE & DOSAGE

Moderate to Severe Infections

Adult: **IV** 3.375 g q6h, infused over 30 min, for 7–10 days

P

Common adverse effects in *italic*, life-threatening effects underlined; generic names in **bold;** classifications in SMALL CAPS; ♣ Canadian drug name; ℗ Prototype drug

1209

Child: **IV** *Younger than 6 mo,* 150–300 mg piperacillin/kg/day divided q6–8h; *6 mo or older,* 240 mg piperacillin component/kg/day divided q8h

Nosocomial Pneumonia

Adult: **IV** 4.5 g q6h, infused over 30 min, for 7–10 days

Renal Insufficiency Dosage Adjustment

CrCl 20–40 mL/min: 2.25 g q6h; *less than 20 mL/min:* 2.25 g q8h

Hemodialysis Dosage Adjustment

2.25 g q12h (for nosocomial pneumonia dose q8h); give additional 0.75 g after dialysis session

ADMINISTRATION

Note: Verify correct IV concentration and rate of infusion for administration to infants or children with prescriber.

Intravenous

PREPARE: **Intermittent:** Reconstitute powder with 5 mL of diluent (e.g., D5W, NS) for each 1 g or fraction thereof; shake well until dissolved. ▪ Further dilute to a total of 50 mL or less in selected diluent [e.g., NS, sterile water for injection, D5W, dextran 6% in NS, and LR only with solution containing EDTA].

ADMINISTER: **Intermittent:** Give over at least 30 min. ▪ DO NOT administer through a line with another infusion.

INCOMPATIBILITIES **Solution/additive: Aminoglycosides, lactated Ringer's, albumin, blood products, solutions containing sodium bicarbonate. Y-site: Acyclovir, aminoglycosides, amiodarone, amphotericin B, amphotericin B cholesteryl complex, azithromycin, chlorpromazine, cisatracurium, cisplatin, dacarbazine, daunorubicin, dobutamine, doxorubicin, doxorubicin liposome, doxycycline, droperidol, famotidine, ganciclovir, gatifloxacin, gemcitabine, haloperidol, hydroxyzine, idarubicin, miconazole, minocycline, mitomycin, mitoxantrone, nalbuphine, prochlorperazine, promethazine, streptozocin, vancomycin.**

ADVERSE EFFECTS (≥1%) **CNS:** Headache, insomnia, fever. **GI:** Diarrhea, constipation, nausea, vomiting, dyspepsia, pseudomembranous colitis. **Skin:** Rash, pruritus, hypersensitivity reactions.

INTERACTIONS Drug: May increase risk of bleeding with ANTICOAGULANTS; **probenecid** decreases elimination of piperacillin.

PHARMACOKINETICS Distribution: Distributes into many tissues, including lung, blister fluid, and bile; crosses placenta; distributed into breast milk. **Metabolism:** In liver. **Elimination:** In urine. **Half-Life:** 0.7–1.2 h.

NURSING IMPLICATIONS

Assessment & Drug Effects

▪ Obtain history of hypersensitivity to penicillins, cephalosporins, or other drugs prior to administration.

▪ Lab tests: C&S prior to first dose of the drug; start drug pending results. Monitor hematologic status with prolonged therapy (Hct and Hgb, CBC with differential and platelet count).

▪ Monitor patient carefully during the first 30 min after initiation of the infusion for signs of hypersensitivity (see Appendix F).

Patient & Family Education
- Report rash, itching, or other signs of hypersensitivity immediately.
- Report loose stools or diarrhea as these may indicate pseudomembranous colitis.

PIRBUTEROL ACETATE
(pir-bu'ter-ol)

Maxair
Classifications: BETA-ADRENERGIC AGONIST; BRONCHODILATOR
Therapeutic: BRONCHODILATOR
Prototype: Albuterol
Pregnancy Category: C

AVAILABILITY 0.2 mg aerosol

ACTION & *THERAPEUTIC EFFECT*
Selective agonist of beta$_2$-adrenergic receptors that relaxes bronchospasm and increases ciliary motion. Activates enzyme that catalyzes the conversion of ATP to cyclic adenosine monophosphate (cAMP). Increased cAMP is associated with relaxation of bronchial smooth muscle and inhibition of the release of histamine and other mediators of hypersensitivity from mast cells. *Effective bronchodilator and decreases the release of mediators within the mast cell that cause a hypersensitivity reaction.*

USES Prevention and reversal of bronchospasm associated with asthma.

CONTRAINDICATIONS Hypersensitivity to pirbuterol or any other adrenergic agent such as epinephrine, albuterol, or isoproterenol; lactation.
CAUTIOUS USE Heart disease, irregular heartbeat; QT prolongation, AV block; high blood pressure, history of stroke or seizures;

DM; Parkinson's disease; thyroid disease; prostate disease; glaucoma; pregnancy (category C); children younger than 12 y.

ROUTE & DOSAGE

Asthma

Adult/Child (older than 12 y):
Inhaled 2 inhalations (0.4 mg) q6h (max: 12 inhalations/day)

ADMINISTRATION

Inhalation
- Shake inhaler canister well immediately before using.
- Direct patient to exhale deeply, loosely close lips around mouthpiece, then inhale slowly and deeply through mouthpiece while pressing top of canister.
- Store at 15°–30° C (59°–86° F).

ADVERSE EFFECTS (≥1%) **CNS:** Nervousness, headache, dizziness, tremor. **CV:** Palpitations, tachycardia. **GI:** Dry mouth, nausea, glossitis, abdominal pain, cramps, anorexia, diarrhea, stomatitis. **Other:** Cough, tolerance.

INTERACTIONS Drug: Epinephrine and other SYMPATHOMIMETIC BRONCHODILATORS may have additive effects. BETA-BLOCKERS may antagonize the effects.

PHARMACOKINETICS Onset: 5 min. **Peak:** 30 min. **Duration:** 3–4 h. **Metabolism:** In liver. **Elimination:** By kidneys. **Half-Life:** 2–3 h.

NURSING IMPLICATIONS

Assessment & Drug Effects
- Monitor arterial blood gases and pulmonary functions periodically.
- Monitor vital signs. Report tachycardia, palpitations, and hypertension or hypotension.

P

Common adverse effects in *italic*, life-threatening effects underlined; generic names in **bold**; classifications in SMALL CAPS; ♣ Canadian drug name; ⊘ Prototype drug

1211

Patient & Family Education
- Learn proper technique for using the inhaler.
- Report palpitations, chest pain, nervousness, tremors, or other bothersome adverse effects promptly to prescriber.
- Contact prescriber immediately if symptoms of asthma worsen or you do not respond to the usual dose.
- Adhere rigidly to dosing directions and contact prescriber if breathing difficulty persists.

PIROXICAM
(peer-ox'i-kam)
Feldene
Classifications: NONNARCOTIC ANALGESIC; NONSTEROIDAL ANTI-INFLAMMATORY DRUG (NSAID)
Therapeutic: ANALGESIC; NSAID; ANTIPYRETIC
Prototype: Ibuprofen
Pregnancy Category: C

AVAILABILITY 10 mg, 20 mg capsules

ACTION & *THERAPEUTIC EFFECT*
Nonsteroidal anti-inflammatory agent that strongly inhibits enzyme cyclooxygenase, both COX1 and COX2, the catalyst of prostaglandin synthesis. Decreased prostaglandin results in anti-inflammatory properties, analgesic and antipyretic effects. *Decreases inflammatory processes in bone-joint disease as well as has analgesic and antipyretic effects.*

USES Acute and long-term relief of mild to moderate pain and for symptomatic treatment of osteoarthritis and rheumatoid arthritis.
UNLABELED USES Acute and chronic relief of mild to moderate pain.

CONTRAINDICATIONS Hypersensitivity to NSAIDs or salicylates; hemophilia; active peptic ulcer, GI bleeding; CABG perioperative pain; ST-elevated MI; lactation.
CAUTIOUS USE History of upper GI disease including ulcerative colitis; SLE; kidney dysfunction; hepatic disease; CHF; acute MI; compromised cardiac function; hypertension or other conditions predisposing to fluid retention; renal disease; alcoholism; coagulation disorders; older adults; pregnancy (category C). Safe use in children not established.

ROUTE & DOSAGE

Arthritis, Pain
Adult: **PO** 10–20 mg 1–2 times/day

ADMINISTRATION
Oral
- Give at the same time every day.
- Give capsule with food or fluid to help reduce GI irritation.
- Dose adjustments should be made on basis of clinical response at intervals of weeks rather than days in order to prevent over dosage.
- Store in tightly closed container at 15°–30° C (59°–86° F) unless otherwise directed.

ADVERSE EFFECTS (≥1%) **CNS:** Somnolence, dizziness, vertigo, depression, insomnia, nervousness. **CV:** Peripheral edema, hypertension, worsening of CHF, exacerbation of angina. **Special Senses:** Tinnitus, hearing loss, blurred vision, reduced visual acuity, changes in color vision, scotomas, corneal deposits, retinal disturbances. **GI:** *Nausea, vomiting, dyspepsia,* GI bleeding, diarrhea, constipation, flatulence, dry mouth, peptic ulceration, anorexia, jaundice, hepatitis. **Hematologic:** Anemia, decreases in Hgb, Hct; *leukopenia*, eosinophilia, <u>aplastic anemia</u>; *thrombocytopenia, prolonged bleeding time.* **Skin:** Urticaria, erythema multiforme, maculopapular, vesiculobullous rash; photosensitivity, sweating, Stevens-Johnson syndrome, bruising, dermatitis. **Body as a Whole:** Allergic rhinitis, <u>angioedema</u>, fever, palpitations, syncope, muscle cramps, fever, hypersensitivity reactions. **Metabolic:** Hypoglycemia, hyperglycemia, hyperkalemia, weight gain. **Urogenital:** Dysuria, <u>acute kidney failure</u>, papillary necrosis, hematuria, proteinuria, nephrotic syndrome. **Respiratory:** <u>Bronchospasm</u>, dyspnea.

INTERACTIONS Drug: ORAL ANTICOAGULANTS, **heparin** may prolong bleeding time; may increase **lithium** toxicity; **alcohol, aspirin** increase risk of GI hemorrhage. **Herbal: Feverfew, garlic, ginger, ginkgo** may increase bleeding potential.

PHARMACOKINETICS Absorption: Extensively from GI tract. **Onset:** 1 h analgesia; 7 days for rheumatoid arthritis. **Peak:** 3–5 h analgesia; 2–4 wk antirheumatic. **Duration:** 48–72 h analgesia. **Distribution:** Small amount distributed into breast milk. **Metabolism:** Extensively in liver. **Elimination:** Primarily in urine, some in bile (less than 5%). **Half-Life:** 30–86 h.

NURSING IMPLICATIONS

Assessment & Drug Effects

- Wait at least 7 days to evaluate antirheumatic effect.
- Clinical evidence of benefits from drug therapy include pain relief in motion and in rest, reduction in night pain, stiffness, and swelling; increased ROM (range of motion) in all joints.
- Be aware that adverse effects may not appear for 7–10 days after start of therapy (except for an allergic reaction).
- Lab tests: Periodic BUN, ALT, AST, CBC, Hgb and Hct in patient (especially the older adult) receiving drug for an extended period.

Patient & Family Education

- If a dose is missed, take drug when missed dose is discovered if it is 6–8 h before the next scheduled dose. Otherwise, omit dose and reestablish regimen at next scheduled hour.
- Do not self-dose with aspirin or other OTC drug without prescriber's advice.
- Do not increase dosage beyond prescribed regimen. Understand that long half-life of drug may cause delayed therapeutic effect. Higher than recommended doses are associated with increased incidence of GI irritation and peptic ulcer.
- Incidence of GI bleeding with this drug is relatively high. Report symptoms of GI bleeding (e.g., dark, tarry stools, coffee-colored emesis) or severe gastric pain promptly to prescriber.
- Be alert to symptoms of drug-induced anemia: Profound fatigue, skin and mucous membrane pallor, lethargy.
- Avoid alcohol since it may increase the risk of GI bleeding.
- Be alert to signs of hypoprothrombinemia including bruises, pinpoint rash, unexplained bleeding, nose bleed, blood in urine, when piroxicam is taken concomitantly with an anticoagulant.
- Do not drive or engage in potentially hazardous activities until response to drug is known.
- Drink at least 6–8 full glasses of water daily and report signs of renal insufficiency (see Appendix F) to prescriber because most of

P

Common adverse effects in *italic*, life-threatening effects <u>underlined</u>; generic names in **bold**; classifications in SMALL CAPS; ♣ Canadian drug name; ◉ Prototype drug

1213

drug is excreted by kidneys and impaired kidney function increases danger of toxicity.

PITAVASTATIN CALCIUM

(pit-a-vah-stat′in)

Livalo

Classifications: ANTIHYPERLIPIDEMIC; LIPID-LOWERING AGENT; HMG-COA REDUCTASE INHIBITOR (STATIN)

Therapeutic: ANTILIPEMIC; LIPID-LOWERING AGENT; STATIN

Prototype: Lovastatin

Pregnancy Category: X

AVAILABILITY 1 mg, 2 mg, 4 mg tablets

ACTION & THERAPEUTIC EFFECT
Pitavastatin is a HMG-CoA reductase inhibitor that reduces plasma cholesterol levels by interfering with the production of cholesterol in the liver. It also promotes removal of LDL and VLDL from plasma. *Effectiveness is measured by decrease in blood level of total cholesterol, LDL cholesterol and triglycerides and an increase in HDL cholesterol.*

USES Treatment of hypercholesterolemia, hyperlipoproteinemia, and/or hypertriglyceridemia as an adjunct to dietary control. Pitavastatin reduces elevated total cholesterol, LDL cholesterol, apolipoprotein B, and triglyceride concentrations, and increases HDL cholesterol in patients with primary hypercholesterolemia or mixed dyslipidemia.

UNLABELED USES Regression of coronary atherosclerosis in patients with acute coronary syndrome (ACS).

CONTRAINDICATIONS Hypersensitivity to pitavastatin; myopathy and rhabdomyolysis; acute renal failure; ESRD not on hemodialysis; active liver disease; pregnancy (category X); lactation.

CAUTIOUS USE Moderate or mild renal impairment; older adults. Safety and effectiveness in children have not been established.

ROUTE & DOSAGE

Hypercholesterolemia, Hyperlipoproteinemia, and/or Hypertriglyceridemia

Adult: **PO** Initially 2 mg once daily; can be adjusted to 1–4 mg once daily

Renal Impairment Dosage Adjustment

CrCl 30 to less than 60 mL/min: Initially, 1 mg once daily; do not exceed 2 mg once daily
CrCl less than 30 mL/min on hemodialysis: Initially, 1 mg once daily; do not exceed 2 mg once daily
CrCl less than 30 mL/min not on hemodialysis: Not recommended

ADMINISTRATION

Oral
- May give without regard to meals any time of day.
- Store at 15°–30° C (59°–86° F).

ADVERSE EFFECTS (≥1%) **Body as a Whole:** Pain in extremity, influenza. **CNS:** Headache. **GI:** Constipation, diarrhea. **Hematologic:** Elevated creatine phosphokinase (CPK), elevated transaminase (AST and ALT) levels. **Musculoskeletal:** Arthralgia, back pain, myalgia. **Respiratory:** Nasopharyngitis. **Skin:** Pruritus, rash, urticaria.

INTERACTIONS Drug: Cyclosporine, erythromycin, HIV PROTEASE

INHIBITORS, and **rifampin** increase pitavastatin levels. FIBRATES may increase the risk of myopathy and rhabdomyolysis. Combination use with **niacin** may increase the risk of skeletal muscle effects.

PHARMACOKINETICS Absorption: 51% bioavailable. **Peak:** 1 h. **Distribution:** Greater than 99% plasma protein bound. **Metabolism:** In liver. **Elimination:** Primarily fecal (79%) with minor renal (15%) elimination. **Half-Life:** 12 h.

NURSING IMPLICATIONS

Assessment & Drug Effects

- Monitor for and report muscle pain, tenderness, or weakness.
- Withhold drug and notify prescriber for ALT/AST greater than 3 × ULN.
- Lab tests: Lipid profile at baseline, at 4 wk, and periodically thereafter; LFTs at baseline, at 12 wk after initiation or elevation of dose, and periodically thereafter; creatine kinase levels if patient experiences muscle pain.

Patient & Family Education

- Report promptly unexplained muscle pain, tenderness, or weakness.
- Avoid or minimize alcohol consumption while on this drug.
- Use effective contraceptive measures to avoid pregnancy while taking this drug.

PLASMA PROTEIN FRACTION
(plas′ma)

Plasmanate, Plasma-Plex, Protenate
Classification: PLASMA VOLUME EXPANDER
Therapeutic: PLASMA VOLUME EXPANDER; ALBUMIN
Prototype: Normal serum albumin, human
Pregnancy Category: C

AVAILABILITY 5% injection

ACTION & *THERAPEUTIC EFFECT*
Provides plasma proteins that increase colloidal osmotic pressure within the intravascular compartment equal to human plasma. It shifts water from the extravascular tissues back into the intravascular space, thus expanding plasma volume. *Used to maintain cardiac output by expanding plasma volume in the treatment of shock due to various causes.*

USES Emergency treatment of hypovolemic shock due to burns, trauma, surgery, infections; temporary measure in treatment of blood loss when whole blood is not available; to replenish plasma protein in patients with hypoproteinemia (if sodium restriction is not a problem).

CONTRAINDICATIONS Hypersensitivity to albumin; severe anemia; cardiac failure; patients undergoing cardiopulmonary bypass surgery.
CAUTIOUS USE Patients with low cardiac reserve; absence of albumin deficiency; liver or kidney failure; pregnancy (category C).

ROUTE & DOSAGE

Plasma Volume Expansion
Adult: **IV** 250–500 mL at a maximum rate of 10 mL/min
Child: **IV** 6.6–30 mL/kg at a rate of 5–10 mL/min

Hypoproteinemia
Adult: **IV** 1–1.5 L/day infused at a rate not to exceed 5–8 mL/min

ADMINISTRATION

Intravenous
Do not use solutions that show a sediment or appear turbid.

- Do not use solutions that have been frozen.

Common adverse effects in *italic*, life-threatening effects underlined; generic names in **bold**; classifications in SMALL CAPS; ♣ Canadian drug name; ⊘ Prototype drug

1215

PREPARE: **IV Infusion:** Give undiluted. ▪ Once container is opened, solution should be used within 4 h because it contains no preservatives. ▪ Discard unused portions.
ADMINISTER: **IV Infusion:** Rate of infusion and volume of total dose will depend on patient's age, diagnosis, degree of venous and pulmonary congestion, Hct, and Hgb determinations. As with any oncotically active solution, infusion rate should be relatively slow. Range may vary from 1–10 mL/min.

INCOMPATIBILITIES PROTEIN HYDROLYSATES or solutions containing alcohol.

ADVERSE EFFECTS (≥1%) **GI:** Nausea, vomiting, hypersalivation, headache. **Body as a Whole:** Tingling, chills, fever, cyanosis, chest tightness, backache, urticaria, erythema, shock (systemic anaphylaxis), circulatory overload, pulmonary edema.

NURSING IMPLICATIONS

Assessment & Drug Effects

▪ Monitor vital signs (BP and pulse). Frequency depends on patient's condition. Flow rate adjustments are made according to clinical response and BP. Slow or stop infusion if patient suddenly becomes hypotensive.

▪ Report a widening pulse pressure (difference between systolic and diastolic); it correlates with increase in cardiac output.

▪ Report changes in I&O ratio and pattern.

▪ Observe patient closely during and after infusion for signs of hypervolemia or circulatory overload (see Appendix F). Report these symptoms immediately to prescriber.

▪ Make careful observations of patient who has had either injury or surgery in order to detect bleeding points that failed to bleed at lower BP.

PLERIXAFOR
(ple-rix′a-for)

Mozobil

Classification: BIOLOGICAL RESPONSE MODIFIER; INHIBITOR OF CXCR4 CHEMOKINE RECEPTOR

Therapeutic: CXCR4 CHEMOKINE RECEPTOR INHIBITOR; STEM CELL MOBILIZER

Pregnancy Category: D

AVAILABILITY 24 mg/1.2 mL solution for injection

ACTION & *THERAPEUTIC EFFECT*
Plerixafor is a hematopoietic stem cell mobilizer that induces leukocytosis and increases the number of circulating hematopoietic progenitor cells. *Plerixafor enables collection of peripheral blood stem cell (PBSC) for autologous transplantation.*

USES In combination with granulocyte colony-stimulating factor (G-CSF), used to mobilize peripheral blood stem cell (PBSC) for collection and subsequent autologous transplantation in patients with non-Hodgkin's lymphoma and multiple myeloma.

CONTRAINDICATIONS Leukemia; pregnancy (category D); lactation.
CAUTIOUS USE Neutrophil count greater than 50,000/mm³; thrombocytopenia; splenomegaly; moderate-to-severe renal impairment; concurrent drugs that reduce renal function. Safety and efficacy in children have not been established.

ROUTE & DOSAGE

Peripheral Blood Stem Cell Mobilization

Adult: **Subcutaneous** 0.24 mg/kg once daily (max: 0.40 mg/day), 11 h prior to initiation of apheresis;

Common adverse effects in *italic*, life-threatening effects <u>underlined</u>; generic names in **bold**; classifications in SMALL CAPS; ♣ Canadian drug name; ⊙ Prototype drug

administer with filgrastim for up to 4 consecutive days.

Renal Impairment Dosage Adjustment

CrCl less than or equal to 50 ml/min: Reduce dose to 0.16 mg/kg (max: 27 mg/day)

Obesity Dosage Adjustment
Dose based on actual body weight up to 175% of IBW

ADMINISTRATION

Subcutaneous
- Give undiluted approximately 11 h prior to apheresis.
- Calculate the required dose as follows: 0.012 × actual weight in kg = mL to administer. Note that each vial contains 1.2 mL of 20 mg/mL solution. Discard unused solution.
- Store single-use vials at 15°–30°C (59°–86°F).

ADVERSE EFFECTS (≥5%) **Body as a Whole:** *Arthralgia, injection-site reactions.* **CNS:** *Dizziness, fatigue, headache,* insomnia. **GI:** *Diarrhea,* flatulence, *nausea, vomiting.*

PHARMACOKINETICS **Peak:** 30–60 min. **Distribution:** 58% plasma protein bound. **Metabolism:** Not significantly metabolized. **Elimination:** Primarily in urine. **Half-Life:** 3–5 h.

NURSING IMPLICATIONS

Assessment & Drug Effects
- Monitor for and promptly report: Signs of a systemic reaction (e.g., urticaria, periorbital swelling, dyspnea, or hypoxia); signs of splenic enlargement (e.g., upper abdominal pain and/or scapular or shoulder pain).
- Monitor for orthostatic hypotension and syncope during or within

1 h of drug injection. Monitor ambulation or take other precautions as warranted.
- Lab tests: Baseline and periodic WBC with differential and platelet count, kidney function tests.
- Monitor injection sites for skin reactions.

Patient & Family Education
- Report promptly any discomfort experienced during or shortly after drug injection.
- Exercise caution when changing position from lying or sitting to standing as rapid movement shortly after injection may trigger fainting.
- Report skin reactions at the injection site (e.g., itching, rash swelling, or pain).
- Females should use effective contraception while taking this drug.

PNEUMOCOCCAL 13-VALENT VACCINE, DIPHTHERIA
(noo-moe′ĸoĸ-al)

Prevnar 13
Pregnancy Category: C
See Appendix J.

PODOPHYLLUM RESIN (PODOPHYLLIN)
(pode-oh-fill′um)

Podo-ben, Podofin

PODOFILOX
Condylox
Classification: KERATOLYTIC
Therapeutic: CYTOTOXIC; KERATOLYTIC
Pregnancy Category: C

AVAILABILITY **Podophyllum:** 25% liquid; **Podofilox:** 0.5% gel, solution

Common adverse effects in *italic*, life-threatening effects underlined; generic names in **bold**; classifications in SMALL CAPS; ♣ Canadian drug name; ⊕ Prototype drug

1217

ACTION & *THERAPEUTIC EFFECT*

Potent cytotoxic and keratolytic agent that directly affects epithelial cell metabolism, causing degeneration and arrest of mitosis. *Slow disruption of cells and tissue erosion as a result of its caustic action. Selectively affects embryonic and tumor cells more than adult cells.*

USES Benign growths including external genital and perianal warts, papillomas, fibroids.

CONTRAINDICATIONS Birthmarks, moles, or warts with hair growth from them; cervical, urethral, oral warts; normal skin and mucous membranes peripheral to treated areas; DM; patient with poor circulation; irritated, or bleeding skin; application of drug over large area. **CAUTIOUS USE** Pregnancy (category C), lactation. Safe use in children is not known.

ROUTE & DOSAGE

Condylomata Acuminata

Adult: **Topical** Use 10% solution and repeat 1–2 times/wk for up to 4 applications

Verruca Vulgaris (Common Wart)

Adult: **Topical** Apply 0.5% solution q12h for up to 4 wk

Multiple Superficial Epitheliomatosis, Keratoses

Adult: **Topical** Apply 0.5% solution or gel daily for several days

ADMINISTRATION

Note: Use 10–25% solution for areas less than 10 cm^2 or 5% solution for areas of 10–20 cm^2, anal, or genital warts; apply drug to dry surface, allowing area to dry between drops, wash off after 1–4 h.

Topical

- Avoid podophyllum resin contact with eyes or similar mucosal surfaces; if it occurs, flush thoroughly with lukewarm water for 15 min and remove film precipitated by the water.
- Avoid application of drug to normal tissue. If it occurs, remove with alcohol. Protect surfaces surrounding area to be treated with a layer of petrolatum or flexible collodion.
- Remove drug thoroughly with soap and water after each treatment of accessible tissue surface.
- Apply a protective coat of talcum powder after treatment and drying of anogenital area.
- Remove drug with alcohol, if application causes extreme pain, pruritus, or swelling.
- Store in a tight, light-resistant container; avoid exposure to excessive heat.

ADVERSE EFFECTS (≥1%) **Body as a Whole:** <u>Severe systemic toxicity</u> (sometimes fatal), sensorimotor neuropathy (reversible), symptomatic orthostatic hypotension, paresthesias and weakness of extremities, stocking-glove sensory loss, absent ankle reflexes, decreased response to painful stimuli. **CNS:** Lethargy, mental confusion, disorientation, delirium, agitation, seizures, progressive stupor, polyneuritis, pyrexia, coma, visual and auditory hallucinations, acute psychotic reaction, ataxia, hypotonia, areflexia, increased CSF protein, paralytic ileus. **CV:** Sinus tachycardia. **Hematologic:** <u>Bone marrow suppression</u> similar to that caused by antineoplastic drug toxicity, *leukopenia, thrombocytopenia*. **GI:** *Nausea, vomiting, diarrhea, abdominal pain,* hepatotoxicity, increased serum concentrations of LDH, AST, and alkaline phosphatase. **Urogenital:** <u>Renal failure</u>, urinary retention. **Respiratory:** Decreased respirations, <u>apnea</u>, hyperventilation.

NURSING IMPLICATIONS

Assessment & Drug Effects

- Warts become blanched, then necrotic within 24–48 h. Sloughing begins after about 72 h with no scarring. Frequently, a mild topical anti-infective agent, with or without a dressing, is applied until the healing is complete.
- Monitor neurologic status. Sensorimotor polyneuropathy, if it occurs, appears about 2 wk after application of drug, worsens for 3 mo, and may persist for up to 9 mo. Cerebral effects may persist for 7–10 days; ataxia, hypotonia, and areflexia improve more slowly than effects on sensorium.

Patient & Family Education

- Learn proper technique of treatment if self-administered as treatment of verruca vulgaris (common wart). Also be fully aware of the need to report treatment failure to prescriber.
- Be aware that as with any STD, the patient's sex partner should be examined.
- Systemic toxicity may be severe and serious and is associated with application of drug to large areas, to tissue that is friable, bleeding, or recently biopsied, or for prolonged time. Toxicity may occur within hours of application. There are significant dangers from overuse or misuse of this drug.
- Learn symptoms of toxicity and report any that appear promptly to prescriber (see ADVERSE EFFECTS).

POLYCARBOPHIL
(pol-i-kar'boe-fil)
Equalactin, FiberNorm
Classifications: BULK-PRODUCING LAXATIVE; ANTIDIARRHEAL
Therapeutic: BULK LAXATIVE; ANTIDIARRHEAL
Prototype: Psyllium
Pregnancy Category: C

AVAILABILITY 500 mg, 625 mg tablets; 500 mg, 625 mg chewable tablets

ACTION & THERAPEUTIC EFFECT
Hydrophilic agent that absorbs free water in intestinal tract and opposes dehydrating forces of bowel by forming a gelatinous mass. *Restores more normal moisture level and motility in the lower GI tract; produces well-formed stool and reduces diarrhea.*

USES Constipation or diarrhea associated with acute bowel syndrome, diverticulosis, irritable bowel and in patients who should not strain during defecation. Also choleretic diarrhea, diarrhea caused by small-bowel surgery or vagotomy, and disease of terminal ileum.

CONTRAINDICATIONS Partial or complete GI obstruction; fecal impaction; dysphagia; acute abdominal pain; rectal bleeding; undiagnosed abdominal pain, or other symptoms symptomatic of appendicitis; poisonings; before radiologic bowel examination; bowel surgery.
CAUTIOUS USE Renal failure, renal impairment; pregnancy (category C); children younger than 3 y.

ROUTE & DOSAGE

Constipation or Diarrhea
Adult: **PO** 1 g q.i.d. prn (max: 6 g/day)

Common adverse effects in *italic*, life-threatening effects <u>underlined</u>; generic names in **bold**; classifications in SMALL CAPS; ♣ Canadian drug name; ⦿ Prototype drug

1219

Child: **PO** *3–6 y,* 500 mg b.i.d. prn (max: 1.5 g/day); *6–12 y,* 500 mg t.i.d. prn (max: 3 g/day)

ADMINISTRATION

Oral

- Chewable tablets should be chewed well before swallowing.
- Give each dose with a full glass [240 mL (8 oz)] of water or other liquid.
- Repeat dose every 30 min up to the maximum dose in 24 h with severe diarrhea.
- Store at 15°–30° C (59°–86° F) in tightly closed container unless otherwise directed.

ADVERSE EFFECTS (≥1%) **GI:** Esophageal blockage, intestinal impaction, *abdominal fullness.* **Metabolic:** Low serum potassium, elevated blood glucose levels (with extended use). **Respiratory:** Asthma. **Skin:** Skin rash.

INTERACTIONS Drug: May decrease absorption and clinical effects of ANTIBIOTICS, **warfarin, digoxin, nitrofurantoin,** SALICYLATES.

PHARMACOKINETICS Absorption: Not absorbed from GI tract. **Onset:** 12–24 h. **Peak:** 1–3 days.

NURSING IMPLICATIONS

Assessment & Drug Effects

- Determine duration and severity of diarrhea in order to anticipate signs of fluid-electrolyte losses.
- Monitor and record number and consistency of stools per day, presence and location of abdominal discomfort (i.e., tenderness, distention), and bowel sounds.
- Monitor and record I&O ratio and pattern. Dehydration is indicated if output is less than 30 mL/h.

- Inspect oral cavity for dryness, and be alert to systemic signs of dehydration (e.g., thirst and fever). Dehydration from an episode of diarrhea appears rapidly in young children and older adults.

Patient & Family Education

- Consult prescriber if sudden changes in bowel habit persist more than 1 wk, action is minimal or ineffective for 1 wk, or if there is no antidiarrheal action within 2 days.
- Be aware that extended use of this drug may cause dependence for normal bowel function.
- Do not discontinue polycarbophil unless prescriber advises if also taking an oral anticoagulant, digoxin, salicylates, or nitrofurantoin.

POLYMYXIN B SULFATE

(pol-i-mix′in)

Classifications: ANTIBIOTIC; POLYMYXIN
Therapeutic: ANTIBIOTIC
Pregnancy Category: C (IV/IM form) B (topical form)

AVAILABILITY 500,000 unit injection

ACTION & THERAPEUTIC EFFECT Antibiotic that binds to lipid phosphates in bacterial membranes and changes permeability to permit leakage of cytoplasm from bacterial cells, resulting in cell death. *Bactericidal against susceptible gram-negative but not gram-positive organisms.*

USES Topically and in combination with other anti-infectives or corticosteroids for various superficial infections of eye, ear, mucous membrane, and skin. Concurrent systemic anti-infective therapy may be required for treatment of intraocular infection and severe progressive corneal ulcer. Used parenterally only in hospitalized patients for treatment of severe acute infections of urinary tract, bloodstream, and meninges; and in combination with Neosporin for continuous bladder irrigation to prevent bacteremia associated with use of indwelling catheter.

CONTRAINDICATIONS Hypersensitivity to polymyxin antibiotics; concurrent and sequential use of other nephrotoxic and neurotoxic drugs; respiratory insufficiency; concurrent use of products that inhibit peristalsis, skeletal muscle relaxants, ether, or sodium citrate.
CAUTIOUS USE Impaired kidney function, renal failure; inflammatory bowel disease; myasthenia gravis; pulmonary disease; **IV/IM form:** Pregnancy (category C), lactation; **topical form:** Pregnancy (category B). Safety and efficacy in children less than 1 mo not known.

ROUTE & DOSAGE

Infections

Adult/Child: **IV** 15,000–25,000 units/kg/day divided q12h **IM** 25,000–30,000 units/kg/day divided q4–6h **Intrathecal** 50,000 units × 3–4 days then every other day; *older than 2 y,* 20,000 units × 3–4 days, then 25,000 units every other day
Infant: **IV/IM** Up to 40,000 units/kg/day

Renal Impairment Dosage Adjustment

CrCl 5–20 mL/min: 7500–12,500 units/kg/day IV divided q12h; *less than 5 mL/min:* 2250–3750 units/kg/day IV divided q12h

ADMINISTRATION

Intramuscular

- Routine administration by IM routes not recommended because it causes intense discomfort, along the peripheral nerve distribution, 40–60 min after IM injection.
- Make IM injection in adults deep into upper outer quadrant of buttock. Select IM site carefully to avoid injection into nerves or blood vessels. Rotate injection sites. Follow agency policy for IM site used in children.

Intravenous

PREPARE: **Intermittent:** Reconstitute by dissolving 500,000 units in 5 mL sterile water for injection or NS to yield 100,000 units/mL. Withdraw a single dose and then further dilute in 300–500 mL of D5W.
ADMINISTER: **Intermittent:** Infuse over period of 60–90 min. ▪ Inspect injection site for signs of phlebitis and irritation.
INCOMPATIBILITIES **Solution/additive: Amphotericin B, cefazolin, cephalothin, cephapirin, chloramphenicol, heparin, nitrofurantoin, prednisolone, tetracycline.**

- Protect unreconstituted product and reconstituted solution from light and freezing. Store in refrigerator at 2°–8° C (36°–46° F). ▪ Parenteral solutions are stable for 1 wk when refrigerated. Discard unused portion after 72 h.

ADVERSE EFFECTS (≥1%) **Body as a Whole:** Irritability, facial flushing, ataxia, circumoral, lingual, and peripheral paresthesias (stocking-glove distribution); severe pain (IM site), thrombophlebitis (IV site), superinfections, electrolyte disturbances (prolonged use; also reported in

P

Common adverse effects in *italic*, life-threatening effects <u>underlined</u>; generic names in **bold**; classifications in SMALL CAPS; ♣ Canadian drug name; ⊘ Prototype drug

1221

patients with acute leukemia); local irritation and burning (topical use), anaphylactoid reactions (rare). **CNS:** Drowsiness, dizziness, vertigo, convulsions, coma; neuromuscular blockade (generalized muscle weakness, respiratory depression or arrest); meningeal irritation, increased protein and cell count in cerebrospinal fluid, fever, headache, stiff neck (intrathecal use). **Special Senses:** Blurred vision, nystagmus, slurred speech, dysphagia, ototoxicity (vestibular and auditory) with high doses. **GI:** GI disturbances. **Urogenital:** Albuminuria, cylindruria, azotemia, hematuria; nephrotoxicity.

INTERACTIONS Drug: ANESTHETICS and NEUROMUSCULAR BLOCKING AGENTS may prolong skeletal muscle relaxation. AMINOGLYCOSIDES and **amphotericin B** have additive nephrotoxic potential.

PHARMACOKINETICS Peak: 2 h IM. **Distribution:** Widely distributed except to CSF, synovial fluid, and eye; does not cross placenta. **Metabolism:** Unknown. **Elimination:** 60% excreted unchanged in urine. **Half-Life:** 4.3–6 h.

NURSING IMPLICATIONS

Assessment & Drug Effects

- Lab tests: Obtain C&S tests prior to first dose and periodically thereafter to determine continuing sensitivity of causative organisms. Perform baseline serum electrolytes and kidney function tests before parenteral therapy. Frequent monitoring of kidney function and serum drug levels is advised during therapy. Monitor electrolytes at regular intervals during prolonged therapy.
- Review electrolyte results. Patients with low serum calcium are particularly prone to develop neuromuscular blockade.

- Inspect tongue every day. Assess for S&S of superinfection (see Appendix F). Polymyxin therapy supports growth of opportunistic organisms. Report symptoms promptly.
- Monitor I&O. Maintain fluid intake sufficient to maintain daily urinary output of at least 1500 mL. Some degree of renal toxicity usually occurs within first 3 or 4 days of therapy even with therapeutic doses. Consult prescriber.
- Withhold drug and report findings to prescriber for any of the following: Decreases in urine output (change in I&O ratio), proteinuria, cellular casts, rising BUN, serum creatinine, or serum drug levels (not associated with dosage increase). All can be interpreted as signs of nephrotoxicity.
- Nephrotoxicity is generally reversible, but it may progress even after drug is discontinued. Therefore, close monitoring of kidney function is essential, even following termination of therapy.
- Be alert for respiratory arrest after the first dose and also as long as 45 days after initiation of therapy. It occurs most commonly in patients with kidney failure and high plasma drug levels and is often preceded by dyspnea and restlessness.

Patient & Family Education

- Report to prescriber immediately any muscle weakness, shortness of breath, dyspnea, depressed respiration. These symptoms are rapidly reversible if drug is withdrawn.
- Report promptly to prescriber transient neurologic disturbances (burning or prickling sensations, numbness, dizziness). All occur commonly and usually respond to dosage reduction.
- Report promptly to prescriber the onset of stiff neck and headache (possible symptoms of neurotoxic

reactions, including neuromuscular blockade). This response is usually associated with high serum drug levels or nephrotoxicity.

- Report promptly S&S of superinfection (see Appendix F).
- Report any S&S of colitis for up to 2 mo following discontinuation of drug.

PORACTANT ALPHA

(por-ac'tant)

Curosurf

Classification: LUNG SURFACTANT
Therapeutic: LUNG SURFACTANT

AVAILABILITY 80 mg/mL suspension

ACTION & *THERAPEUTIC EFFECT*
Endogenous pulmonary surfactant lowers the surface tension on alveoli surfaces during respiration, and stabilizes the alveoli against collapse at resting pressures. *Alleviates respiratory distress syndrome (RDS) in premature infants caused by deficiency of surfactant.*

USES Treatment (rescue) of respiratory distress syndrome in premature infants.

CONTRAINDICATIONS Hypersensitivity to porcine products or poractant alpha.
CAUTIOUS USE Infants born greater than 3 wk after ruptured membranes; intraventricular hemorrhage of grade III or IV; major congenital malformations; nosocomial infection; pretreatment of hypothermia or acidosis due to increased risk of intracranial hemorrhage; lactation.

ROUTE & DOSAGE

Respiratory Distress Syndrome
Neonate: **Intratracheal** 2.5 mL/kg birth weight, may repeat with

1.25 mL/kg q12h × 2 more doses if needed (max: 5 mL/kg)

ADMINISTRATION
Note: Correction of acidosis, hypotension, anemia, hypoglycemia, and hypothermia is recommended prior to administration of poractant alfa.

Intratracheal
- Warm vial slowly to room temperature; gently turn upside down to form uniform suspension, but do NOT shake.
- Withdraw slowly the entire contents of a vial (concentration equals 80 mg/mL) into a 3 or 5 mL syringe through a large gauge (greater than 20 gauge) needle.
- Attach a 5 French catheter, precut to 8 cm, to the syringe.
- Fill the catheter with poractant alfa and discard excess through the catheter so that only the total dose to be given remains in the syringe.
- Refer to specific instruction provided by manufacturer for proper dosing technique. Follow instructions carefully regarding installation of drug and ventilation of infant. Note that catheter tip should not extend beyond distal tip of endotracheal tube.
- Store refrigerated at 2°–8° C (36°–46° F) and protect from light. Do not shake vials. Do not warm to room temperature and return to refrigeration more than once.

ADVERSE EFFECTS (≥1%) **CV:** Bradycardia, hypotension. **Respiratory:** Intratracheal tube blockage, oxygen desaturation; pulmonary hemorrhage.

PHARMACOKINETICS Not studied.

NURSING IMPLICATIONS
Assessment & Drug Effects
- Stop administration of poractant alfa and take appropriate measures

Common adverse effects in *italic*, life-threatening effects underlined; generic names in **bold**; classifications in SMALL CAPS; ♣ Canadian drug name; ⊕ Prototype drug

1223

if any of the following occur: Transient episodes of bradycardia, decreased oxygen saturation, reflux of poractant alfa into endotracheal tube, or airway obstruction. Dosing may resume after stabilization.

- Do not suction airway for 1 h after poractant alfa instillation unless there is significant airway obstruction.

POSACONAZOLE
(pos-a-con′a-zole)

Noxafil
Classification: AZOLE ANTIFUNGAL
Therapeutic: ANTIFUNGAL
Prototype: Fluconazole
Pregnancy Category: C

AVAILABILITY 200 mg/5 mL oral suspension

ACTION & *THERAPEUTIC EFFECT*
Azole antifungals inhibit ergosterol synthesis, the principal sterol in the fungal cell membrane, thus interfering with the functions of fungal cell membrane. This results in increased membrane permeability causing leakage of cellular contents. *Has a broad spectrum of antifungal activity against common fungal pathogens.*

USES Prophylactic treatment of invasive *Aspergillus* and *Candida*, chemotherapy induced neutropenia, oropharyngeal candidiasis, fungal prophylaxis.

UNLABELED USES Treatment of febrile neutropenia or refractory invasive fungal infection; treatment of periorbital cellulitis due to *Rhizopus* sp.; treatment of refractory histoplasmosis; treatment of refractory coccidioidomycosis; treatment of fungal necrotizing fasciitis.

CONTRAINDICATIONS Hypersensitivity to posaconazole; coadministration with ergot alkaloids, or CYP3A4 substrates; history of QT prolongation; abnormal levels of potassium, magnesium, or calcium; lactation.

CAUTIOUS USE Hypersensitivity to other azole antifungal antibiotics; hepatic disease or hepatitis; cardiac arrhythmias; history of proarrhythmic conditions; CHF, myocardial ischemia, atrial fibrillation; AIDS; pregnancy (category C); children younger than 13 y.

ROUTE & DOSAGE

Prophylactic Treatment of Invasive *Aspergillus* and *Candida* Infections
Adult: **PO** 200 mg t.i.d.
Thrush
Adult/Adolescent: **PO** 100 mg b.i.d. × 1 day then 100 mg qd × 13 days

ADMINISTRATION

Oral
- Shake well before use. Give with a full meal or liquid nutritional supplement.
- Store at 15°–30° C (59°–86° F).

ADVERSE EFFECTS (≥1%) **Body as a Whole:** Anxiety, *bacteremia, dizziness, edema, fatigue, fever, headache, infection, insomnia, rigors,* weakness. **CNS:** QT/QT$_c$ prolongation, tremor. **CV:** *Hypertension, hypotension, tachycardia.* **GI:** *Abdominal pain, anorexia constipation, diarrhea, dyspepsia, mucositis, nausea,* vomiting. **Hematologic:** *Anemia, febrile neutropenia, neutropenia, petechiae,* <u>thrombocytopenia</u>. **Metabolic:** *Bilirubinemia,* creatinine levels increased, elevated liver enzymes, hypocalcemia, *hyperglycemia,* *hypokalemia, hypomagnesemia.* **Musculoskeletal:** *Arthralgia, back pain, musculoskeletal pain.* **Respiratory:** *Cough, dyspnea, epistaxis, pharyngitis,* upper respiratory tract infection. **Skin:** *Pruritus, rash.* **Special Senses:** Blurred vision, taste disturbances. **Urogenital:** *Vaginal hemorrhage.*

INTERACTIONS Drug: Rifabutin and **phenytoin** increase the metabolism of posaconazole resulting in decreased plasma levels. **Cimetidine** decreases the absorption of posaconazole. Posaconazole is known to increase the plasma levels of **cyclosporine, tacrolimus, rifabutin, midazolam,** and **phenytoin.** Coadministration with other drugs that cause QT prolongation (e.g., **quinidine**) can result in torsades de pointes. Posaconazole may increase the plasma levels of ERGOT ALKALOIDS, VINCA ALKALOIDS, HMG COA REDUCTASE INHIBITORS, and CALCIUM CHANNEL BLOCKERS. **Food:** Administration with food increases absorption of posaconazole.

PHARMACOKINETICS Peak: 3–5 h. **Distribution:** 98% protein bound. **Metabolism:** Conjugated to inactive metabolites. **Elimination:** Primarily fecal elimination (71%) with minor renal elimination. **Half-Life:** 35 h.

NURSING IMPLICATIONS

Assessment & Drug Effects

- Monitor for and report S&S of breakthrough fungal infections, especially in those with severe renal impairment, or experiencing vomiting and diarrhea, or who cannot tolerate a full meal or supplement along with posaconazole.
- Monitor and report degree of improvement of oropharyngeal candidiasis.

- Monitor those with proarrhythmic conditions for development of arrhythmias.
- Lab tests: Baseline and periodic LFTs; baseline serum electrolytes.
- Withhold drug and notify prescriber of abnormal serum potassium, magnesium, or calcium levels.
- Monitor blood levels of phenytoin, cyclosporine, tacrolimus, and sirolimus with concurrent therapy. Monitor for adverse effects of concurrently administered statins or calcium channel blockers.

Patient & Family Education

- Do not take any prescription or nonprescription drugs without informing your prescriber.
- Know parameters for withholding drug (i.e., inability to take with a full meal or nutritional supplement).
- Report immediately any of the following to your health care provider: Vomiting, diarrhea, inability to eat, jaundice of skin, yellowing of eyes, itching, or skin rash.

P

√**POTASSIUM CHLORIDE**
(poe-tass′ee-um)

Apo-K ♣, K-10, Kalium Durules ♣, Kaochlor, Kaochlor-20 Concentrate, Kaon-Cl, KCl 5% and 20%, K-Long ♣, Klor, Klor-10%, Klor-Con, Kloride, Klorvess, Klotrix, K-Dur, K-Lyte/Cl, K-tab, Micro-K Extentabs, SK-Potassium Chloride, Slo-Pot ♣, Slow-K

POTASSIUM GLUCONATE

Kaon, Kaylixir
Classification: ELECTROLYTIC REPLACEMENT SOLUTION
Therapeutic: ELECTROLYTE REPLACEMENT
Pregnancy Category: C

Common adverse effects in *italic*, life-threatening effects <u>underlined</u>; generic names in **bold**; classifications in SMALL CAPS; ♣ Canadian drug name; ⊚ Prototype drug

1225

AVAILABILITY Chloride: 6.7 mEq, 8 mEq, 10 mEq, 20 mEq sustained release tablets; 500 mg, 595 mg tablets; 20 mEq, 25 mEq, 50 mEq effervescent tablets; 20 mEq/15 mL, 40 mEq/15 mL, 45 mEq/15 mL liquid; 15 mEq, 20 mEq, 25 mEq powder; 2 mEq/mL injection; 10 mEq, 20 mEq, 30 mEq, 40 mEq, 60 mEq, 90 mEq vials; **Gluconate:** 20 mEq/15 mL liquid

ACTION & *THERAPEUTIC EFFECT*

Principal intracellular cation that is essential for maintenance of intracellular isotonicity, transmission of nerve impulses, contraction of cardiac, skeletal, and smooth muscles, maintenance of normal kidney function, and for enzyme activity. *Effectiveness in hypokalemia is measured by serum potassium concentration greater than 3.5 mEq/liter.*

USES To prevent and treat potassium deficit secondary to diuretic or corticosteroid therapy. Also indicated when potassium is depleted by severe vomiting, diarrhea; intestinal drainage, fistulas, or malabsorption; prolonged diuresis, diabetic acidosis. Effective in the treatment of hypokalemic alkalosis (chloride, not the gluconate).

CONTRAINDICATIONS Severe renal impairment; severe hemolytic reactions; untreated Addison's disease; crush syndrome; early postoperative oliguria (except during GI drainage); adynamic ileus; acute dehydration; heat cramps, hyperkalemia, patients receiving potassium-sparing diuretics, digitalis intoxication with AV conduction disturbance.

CAUTIOUS USE Cardiac or kidney disease; systemic acidosis; slow-release potassium preparations in presence of delayed GI transit or Meckel's diverticulum; extensive tissue breakdown (such as severe burns); pregnancy (category C), lactation.

ROUTE & DOSAGE

Hypokalemia

Adult: **PO** 10–100 mEq/day in divided doses **IV** 10–60 mEq/h diluted to at least 10–20 mEq/100 mL of solution (max: 200–400 mEq/day, monitor higher doses carefully)
Child: **PO** 1–3 mEq/kg/day in divided doses; sustained release tablets not recommended **IV** Up to 3 mEq/kg/24 h at a rate less than 0.02 mEq/kg/min

ADMINISTRATION

Oral

▪ Give while patient is sitting up or standing (never in recumbent position) to prevent drug–induced esophagitis. Some patients find it difficult to swallow the large sized KCl tablet.

▪ Do not crush or allow to chew any potassium salt tablets. Observe to make sure patient does not suck tablet (oral ulcerations have been reported if tablet is allowed to dissolve in mouth).

▪ Swallow whole tablet with a large glass of water or fruit juice (if allowed) to wash drug down and to start esophageal peristalsis.

▪ Follow exactly directions for diluting various liquid forms of KCl. In general, dilute each 20 mEq potassium in at least 90 mL water or juice and allow to dissolve completely before administration.

▪ Dilute liquid forms as directed before giving it through nasogastric tube.

Intravenous

PREPARE: IV Infusion: Add desired amount to 100–1000 mL IV solution (compatible with all standard solutions). ▪ Usual maximum is 80 mEq/1000 mL, however, 40 mEq/L is preferred to lessen irritation to veins. Note: **NEVER** add KCl to an IV bag/bottle which is hanging. ▪ After adding KCl invert bag/bottle several times to ensure even distribution.

ADMINISTER: **IV Infusion for Adult/Child:** KCl is **never** given direct IV or in concentrated amounts by any route. ▪ Too rapid infusion may cause fatal hyperkalemia. **IV Infusion for Adult:** Infuse at rate not to exceed 10 mEq/h; in emergency situations, may infuse very cautiously up to 40 mEq/h with continuous cardiac monitoring. **IV Infusion for Child:** Infuse at a rate not to exceed 0.5–1.0 mEq/kg/h.

INCOMPATIBILITIES **Solution/additive: Furosemide, pentobarbital, phenobarbital, succinylcholine. Y-site: Amphotericin B cholesteryl complex, azithromycin, chlordiazepoxide, chlorpromazine, diazepam, ergotamine, methylprednisolone, phenytoin.**

▪ Take extreme care to prevent extravasation and infiltration. At first sign, discontinue infusion and select another site.

ADVERSE EFFECTS (≥1%) **GI:** *Nausea, vomiting,* diarrhea, abdominal distention. **Body as a Whole:** Pain, mental confusion, irritability, listlessness, paresthesias of extremities, muscle weakness and heaviness of limbs, difficulty in swallowing, <u>flaccid paralysis</u>. **Urogenital:** Oliguria, anuria. **Hematologic:** Hyperkalemia. **Respiratory:** <u>Respiratory distress</u>. **CV:** Hypotension, bradycardia; <u>cardiac depression, arrhythmias, or arrest</u>; altered sensitivity to digitalis glycosides. *ECG changes in hyperkalemia:*

Tenting (peaking) of T wave (especially in right precordial leads), lowering of R with deepening of S waves and depression of RST; prolonged P-R interval, widened QRS complex, decreased amplitude and disappearance of P waves, prolonged QT interval, signs of right and left bundle block, <u>deterioration of QRS contour and finally ventricular fibrillation and death</u>.

INTERACTIONS Drug: POTASSIUM-SPARING DIURETICS, ANGIOTENSIN-CONVERTING ENZYME (ACE) INHIBITORS may cause hyperkalemia.

PHARMACOKINETICS Absorption: Readily from upper GI tract. **Elimination:** 90% in urine, 10% in feces.

NURSING IMPLICATIONS

Assessment & Drug Effects

▪ Monitor I&O ratio and pattern in patients receiving the parenteral drug. If oliguria occurs, stop infusion promptly and notify prescriber.
▪ Lab test: Frequent serum electrolytes are warranted.
▪ Monitor for and report signs of GI ulceration (esophageal or epigastric pain or hematemesis).
▪ Monitor cardiac status of patients receiving parenteral potassium. Irregular heartbeat is usually the earliest clinical indication of hyperkalemia.
▪ Be alert for potassium intoxication (hyperkalemia, see S&S, Appendix F); may result from any therapeutic dosage, and the patient may be asymptomatic.
▪ The risk of hyperkalemia with potassium supplement increases (1) in older adults because of decremental changes in kidney function associated with aging, (2) when dietary intake of potassium suddenly increases, and (3) when kidney function is significantly compromised.

P

Common adverse effects in *italic*, life-threatening effects <u>underlined</u>; generic names in **bold**; classifications in SMALL CAPS; ♣ Canadian drug name; ⊘ Prototype drug

1227

Patient & Family Education

- Do not be alarmed when the tablet carcass appears in your stool. The sustained release tablet (e.g., Slow-K) utilizes a wax matrix as carrier for KCl crystals that passes through the digestive system.
- Learn about sources of potassium with special reference to foods and OTC drugs.
- Do not use any salt substitute unless it is specifically ordered by the prescriber. These contain a substantial amount of potassium and electrolytes other than sodium.
- Do not self-prescribe laxatives. Chronic laxative use has been associated with diarrhea-induced potassium loss.
- Notify prescriber of persistent vomiting because losses of potassium can occur.
- Report continuing signs of potassium deficit to prescriber: Weakness, fatigue, polyuria, polydipsia.
- Advise dentist or new prescriber that a potassium drug has been prescribed as long-term maintenance therapy.
- Do not open foil-wrapped powders and tablets before use.

POTASSIUM IODIDE

(poe-tass'ee-um)

Pima, SSKI, Thyro-Block ♥
Classifications: ANTITHYROID; EXPECTORANT
Therapeutic: ANTITHYROID; EXPECTORANT
Prototype: Guaifenesin
Pregnancy Category: D

AVAILABILITY 325 mg/5 mL syrup; 1 g/mL solution

ACTION & *THERAPEUTIC EFFECT* Appears to increase secretion of respiratory fluids by direct action on bronchial tissue, thereby decreasing mucus viscosity. When the thyroid gland is hyperplastic, excess iodide ions temporarily inhibit secretion of thyroid hormone, foster accumulation in thyroid follicles, and decrease vascularity of gland. *Administration for hyperthyroidism is limited to short-term therapy. As an expectorant, the iodine ion increases mucous secretion formation in bronchi, and decreases viscosity of mucus.*

USES To facilitate bronchial drainage and cough in emphysema, asthma, chronic bronchitis, bronchiectasis, and respiratory tract allergies characterized by difficult-to-raise sputum. Also used alone for hyperthyroidism or in conjunction with antithyroid drugs and propranolol in treatment of thyrotoxic crisis; in immediate preoperative period for thyroidectomy to decrease vascularity, fragility, and size of thyroid gland and for treatment of persistent or recurring hyperthyroidism that occurs in Graves' disease patients. Used as a radiation protectant in patients receiving radioactive iodine and to shield the thyroid gland from radiation in the wake of a serious nuclear plant accident. (Use as an expectorant has been largely replaced by other agents.)

CONTRAINDICATIONS Hypersensitivity or idiosyncrasy to iodine; hyperthyroidism; hyperkalemia; acute bronchitis; pregnancy (category D), lactation.
CAUTIOUS USE Renal impairment; cardiac disease; pulmonary tuberculosis; Addison's disease.

ROUTE & DOSAGE

To Reduce Thyroid Vascularity
Adult/Child: **PO** 50–250 mg t.i.d. for 10–14 days before surgery

Common adverse effects in *italic*, life-threatening effects <u>underlined</u>; generic names in **bold**; classifications in SMALL CAPS; ♥ Canadian drug name; ⊚ Prototype drug

Expectorant

Adult: **PO** 300–650 mg p.c. b.i.d. or t.i.d.

Child: **PO** 60–250 mg p.c. b.i.d. or t.i.d.

Thyroid Blocking in Radiation Emergency

Adult: **PO** 130 mg/day for 10 days

Child: **PO** *Younger than 1 y,* 65 mg/day for 10 days; *older than 1 y,* 130 mg/day for 10 days

ADMINISTRATION

Oral

- Give with meals in a full glass (240 mL) of water or fruit juice and at bedtime with juice to disguise salty taste and minimize gastric distress.
- Avoid giving KI with milk; absorption of the drug may be decreased by dairy products.
- Adhere strictly to schedule and accurate dose measurements when iodide is administered to prepare thyroid gland for surgery, particularly at end of treatment period when possibility of "escape" (from iodide) effect on thyroid gland increases.
- Place container in warm water and gently agitate to dissolve if crystals are noted in the solution.
- Discard any solution that has turned a brownish yellow on standing, especially if exposed to light (caused by liberated trace of free iodine).
- Store in airtight, light-resistant container.

ADVERSE EFFECTS (≥1%) **GI:** Diarrhea, nausea, vomiting, stomach pain, nonspecific small bowel lesions (associated with enteric coated tablets). **Body as a Whole:** <u>Angioneurotic edema</u>, cutaneous and mucosal hemorrhage, fever, arthralgias, lymph node enlargement, eosinophilia, paresthesias, periorbital edema, weakness. *Iodine poisoning (iodism):* Metallic taste, stomatitis, salivation, coryza, sneezing; swollen and tender salivary glands (sialadenitis), frontal headache, vomiting (blue vomitus if stomach contained starches, otherwise yellow vomitus), bloody diarrhea. **Metabolic:** Hyperthyroid adenoma, goiter, hypothyroidism, collagen disease-like syndromes. **CV:** Irregular heartbeat. **CNS:** Mental confusion. **Skin:** Acneiform skin lesions (prolonged use), flare-up of adolescent acne. **Respiratory:** Productive cough, pulmonary edema.

DIAGNOSTIC TEST INTERFERENCE

Potassium iodide may alter *thyroid function* test results and may interfere with *urinary 17-OHCS* determinations.

INTERACTIONS Drug: ANTITHYROID DRUGS, **lithium** may potentiate hypothyroid and goitrogenic actions; POTASSIUM-SPARING DIURETICS, POTASSIUM SUPPLEMENTS, ACE INHIBITORS increase risk of hyperkalemia.

PHARMACOKINETICS Absorption: Adequately absorbed from GI tract. **Distribution:** Crosses placenta. **Elimination:** Cleared from plasma by renal excretion or thyroid uptake.

NURSING IMPLICATIONS

Assessment & Drug Effects

- Lab tests: Determine serum potassium levels before and periodically during therapy.
- Keep prescriber informed about characteristics of sputum: Quantity, consistency, color.

Patient & Family Education

- Report to prescriber promptly the occurrence of abdominal pain, distension, nausea, or vomiting.

Common adverse effects in *italic*, life-threatening effects <u>underlined</u>; generic names in **bold**; classifications in SMALL CAPS; ♣ Canadian drug name; ⊘ Prototype drug

1229

- Report clinical S&S of iodism (see ADVERSE EFFECTS). Usually, symptoms will subside with dose reduction and lengthening intervals between doses.
- Avoid foods rich in iodine if iodism develops: Seafood, fish liver oils, and iodized salt.
- Be aware that sudden withdrawal following prolonged use may precipitate thyroid storm.
- Do not use OTC drugs without consulting prescriber. Many preparations contain iodides and could augment prescribed dose [e.g., cough syrups, gargles, asthma medication, salt substitutes, cod liver oil, multiple vitamins (often suspended in iodide solutions)].
- Be aware that optimum hydration is the best expectorant when taking KI as an expectorant. Increase daily fluid intake.

PRALATREXATE

(pra-la-trex′ate)
Folotyn
Classifications: ANTINEOPLASTIC; ANTIMETABOLITE, ANTIFOLATE
Therapeutic: ANTINEOPLASTIC
Prototype: Methotrexate
Pregnancy Category: D

AVAILABILITY 20 mg/mL, 40 mg/2 mL solution for injection

ACTION & *THERAPEUTIC EFFECT*
Antimetabolite and folic acid antagonist. Blocks folic acid participation in nucleic acid synthesis, thereby interfering with cell division (mitosis). Rapidly dividing cells, including cancer cells, are sensitive to this interference in the mitotic process. *Effective in treatment of relapsed or refractory peripheral T-cell lymphoma (PTCL).*

USES Treatment of relapsed or refractory peripheral T-cell lymphoma (PTCL).

CONTRAINDICATIONS Concomitant administration of hepatotoxicity drugs; pregnancy (category D); lactation.
CAUTIOUS USE Thrombocytopenia, neutropenia, anemia; moderate to severe renal function impairment; liver function impairment; ulcerative colitis; poor nutritional status.

ROUTE & DOSAGE

Peripheral T-Cell Lymphoma
Adult: **IV** 30 mg/m^2 over 3–5 min. Repeat weekly for 6 wk in 7-wk cycles.

Adjustments Based on Drug Hematologic Toxicities
- Platelet count less than 50,000/mm^3 for 1 wk: Omit dose, resume dose when platelet count is at least 50,000/mm^3
- Platelet count less than 50,000/mm^3 for 2 wk: Omit dose, resume at 20 mg/m^2 when platelet count is at least 50,000/mm^3
- Platelet count less than 50,000/mm^3 for 3 wk: Discontinue therapy.
- ANC 500–1,000/mm^3 and no fever for 1 wk: Omit dose; resume dose when ANC is at least 1,000/mm^3.
- ANC 500–1,000/mm^3 and fever or ANC less than 500/mm^3 for 1 wk: Omit dose; give G-CSF or GM-CSF support; resume dose and continue G-CSF or GM-CSF support when ANC is at least 1,000/mm^3 and fever resolves.
- ANC 500–1,000/mm^3 and fever or ANC less than 500/mm^3 for 2 wks or

Common adverse effects in *italic*, life-threatening effects <u>underlined</u>; generic names in **bold**; classifications in SMALL CAPS; ✦ Canadian drug name; ❍ Prototype drug

recurrence of toxicity: Omit dose; give G-CSF or GM-CSF support; resume at 20 mg/m² and continue G-CSF or GM-CSF support when ANC is at least 1,000/mm³ and fever resolves.

- ANC 500–1,000/mm³ and fever or ANC less than 500/mm³ for 3 wk's duration or 2nd recurrence of toxicity: Discontinue therapy

Adjustments Based on Mucositis

- Grade 2 mucositis: Omit dose; resume dose when toxicity grade is 1 or less
- Grade 2 mucositis recurrence: Omit dose; resume at 20 mg/m² when toxicity grade is 1 or less
- Grade 3 mucositis: Omit dose; resume at 20 mg/m² when toxicity grade is 1 or less
- Grade 4 mucositis: Discontinue therapy

Adjustments for All Other Treatment-Related Toxicities

- Grade 3 toxicities: Omit dose and resume at 20 mg/m² when toxicity is grade 2 or less
- Grade 4 toxicities: Discontinue therapy

Hepatic Impairment Dosage Adjustment

Included above with general grade 3 and grade 4 treatment-related toxicities.

ADMINISTRATION

Intravenous

PREPARE: Direct: Withdraw from vial into syringe immediately before use. Do not dilute. Use gloves and other protective clothing during handling and preparation.

- Flush thoroughly if drug contacts skin or mucous membranes.

ADMINISTER: Direct: Give over 3–5 min via a side port of a free-flowing NS IV line. ▪ Withhold drug and notify prescriber for any of the following: Platelet count less than 100,000/mcL for first dose or less than 50,000/mcL for all subsequent doses, ANC less than 1000/mcL, or grade 2 or higher mucositis.

- Refrigerate at 2°–8° C (36°–46° F) until use and protect from light. ▪ Vials are stable at room temperature for up to 72 h if left in original carton.

ADVERSE EFFECTS (≥10%) Body as a Whole: Abdominal pain, asthenia, *edema, fatigue,* night sweats, pain in extremity, *pyrexia,* sepsis. **CV:** Tachycardia. **GI:** *Constipation, diarrhea, mucositis, nausea, vomiting.* **Hematologic:** Anemia, leukopenia, *neutropenia, thrombocytopenia.* **Metabolic:** Anorexia, elevated AST and ALT, hypokalemia. **Musculoskeletal:** Back pain. **Respiratory:** *Cough, dyspnea, epistaxis,* pharyngolaryngeal pain, upper respiratory tract infection. **Skin:** Pruritus, rash.

INTERACTIONS Drug: Probenecid may increase pralatrexate levels. Drugs that are subject to substantial renal clearance (e.g., NSAIDS, **trimethoprim/sulfamethoxazole**) may delay the clearance of pralatrexate.

PHARMACOKINETICS Distribution: 67% bound to plasma proteins. **Metabolism:** Not extensively metabolized. **Elimination:** Primarily excreted in the urine. **Half-Life:** 12–18 h.

NURSING IMPLICATIONS

Assessment & Drug Effects

- Monitor vitals signs. Report immediately S&S of infection, especially fever of 100.5° F or greater.

Common adverse effects in *italic,* life-threatening effects underlined; generic names in **bold;** classifications in SMALL CAPS; ✚ Canadian drug name; ⊘ Prototype drug

1231

- Monitor status of mucus membranes because mucositis is a dose-limiting toxicity.
- Lab tests: Prior to each dose, CBC with differential and platelet count; baseline and periodic LFTs and kidney function tests.

Patient & Family Education
- Practice meticulous oral hygiene.
- Monitor for S&S of an infection. Contact the prescriber immediately if an infection is suspected or if your temperature is elevated.
- Report promptly unexplained bleeding or symptoms of anemia (e.g., excessive weakness, fatigue, intolerance to activity, pale skin).
- Folic acid and vitamin B_{12} supplementation are recommended to reduce the risk of drug-related toxicities. Consult with prescriber.
- Use effective means of contraception to avoid pregnancy while taking this drug. If a pregnancy does occur, notify prescriber immediately.
- Do not take aspirin or NSAIDs without consulting prescriber.

PRALIDOXIME CHLORIDE
(pra-li-dox'eem)

2-PAM, Protopam Chloride
Classifications: CHOLINESTERASE RECEPTOR AGONIST; DETOXIFICATION AGENT
Therapeutic: ANTIDOTE; CHOLINESTERASE ENHANCER
Pregnancy Category: C

AVAILABILITY 1 g injection

ACTION & *THERAPEUTIC EFFECT*
Reactivates cholinesterase by displacing the enzyme from its receptor sites; the free enzyme then can resume its function of degrading accumulated acetylcholine, thereby restoring normal neuromuscular transmission. *More active against effects of anticholinesterases at skeletal neuromuscular junction than at autonomic effector sites or in CNS respiratory center; therefore, atropine **must be** given concomitantly to block effects of acetylcholine and its accumulation in these sites.*

USES Antidote in treatment of poisoning by organophosphate insecticides and pesticides with anticholinesterase activity (e.g., parathion, TEPP, sarin) and to control overdosage by anticholinesterase drugs used in treatment of myasthenia gravis (cholinergic crisis).

UNLABELED USES To reverse toxicity of echothiophate ophthalmic solution.

CONTRAINDICATIONS Hypersensitivity to pralidoxime; use in poisoning by insecticide of the carbonate class (Sevin), inorganic phosphates, or organophosphates having no anticholinesterase activity.

CAUTIOUS USE Myasthenia gravis; renal insufficiency; asthma; peptic ulcer; severe cardiac disease; pregnancy (category C), lactation.

ROUTE & DOSAGE

Organophosphate Poisoning

Adult/Adolescent (at least 17 years): **IV** 1–2 g in 100 mL NS infused over 15–30 min; or 1–2 g as 5% solution in sterile water over not less than 5 min, may repeat after 1 h if muscle weakness not relieved. **IM** 600 mg, may repeat in 15 min to total dose of 1800 mg if IV route is not feasible.
Adolescent (less than 17 years old)/Child/Infant: **IV** 20–50 mg/kg. May repeat in 1–2 h if needed (max dose: 2 g).

Anticholinesterase Overdose in Myasthenia Gravis

Adult: **IV** 1–2 g in 100 mL NS infused over 15–30 min, followed by increments of 250 mg q5min prn

ADMINISTRATION

Subcutaneous/Intramuscular

- Give only if unable to give IV; NOT preferred routes.
- Reconstitute as for direct IV injection (see below).

Intravenous

PREPARE: **Direct:** Reconstitute 1-g vial by adding 20 mL sterile water for injection to yield 50 mg/mL (a 5% solution). ▪ If pulmonary edema is present, give without further dilution. **IV Infusion (preferred):** Further dilute reconstituted solution in 100 mL NS.

ADMINISTER: **Direct:** In pulmonary edema, 1 g or fraction thereof over 5 min; do not exceed 200 mg/min. **IV Infusion (preferred):** Give over 15–30 min.

- Stop infusion or reduce rate if hypertension occurs.

ADVERSE EFFECTS (≥1%) **CNS:** Dizziness, headache, drowsiness. **GI:** Nausea. **Special Senses:** Blurred vision, diplopia, impaired accommodation. **CV:** Tachycardia, hypertension (dose-related). **Body as a Whole:** Hyperventilation, muscular weakness, laryngospasm, muscle rigidity.

INTERACTIONS Drug: May potentiate the effects of BARBITURATES.

PHARMACOKINETICS Peak: 5–15 min IV; 10–20 min IM. **Distribution:** Distributed throughout extracellular fluids; crosses blood–brain barrier slowly if at all. **Metabolism:** Probably in liver. **Elimination:** Rapidly in urine. **Half-Life:** 0.8–2.7 h.

NURSING IMPLICATIONS

Assessment & Drug Effects

- Monitor BP, vital signs, and I&O. Report oliguria or changes in I&O ratio.
- Monitor closely. It is difficult to differentiate toxic effects of organophosphates or atropine from toxic effects of pralidoxime.
- Be alert for and report immediately: Reduction in muscle strength, onset of muscle twitching, changes in respiratory pattern, altered level of consciousness, increases or changes in heart rate and rhythm.
- Observe necessary safety precautions with unconscious patient because excitement and manic behavior reportedly may occur following recovery of consciousness.
- Keep patient under close observation for 48–72 h, particularly when poison was ingested, because of likelihood of continued absorption of organophosphate from lower bowel.
- In patients with myasthenia gravis, overdosage with pralidoxime may convert cholinergic crisis into myasthenic crisis.

PRAMIPEXOLE DIHYDROCHLORIDE

(pra-mi-pex′ole)

Mirapex, Mirapex ER
Classifications: DOPAMINE RECEPTOR AGONIST; ANTIPARKINSON
Therapeutic: ANTIPARKINSON
Prototype: Levodopa
Pregnancy Category: C

AVAILABILITY 0.125 mg, 0.25 mg, 0.75 mg, 1 mg, 1.5 mg tablets; 0.375 mg, 0.75 mg, 1.5 mg, 2.25 mg, 3 mg, 3.75 mg, 4 mg extended release tablet

ACTION & THERAPEUTIC EFFECT
Nonergot dopamine receptor

P

Common adverse effects in *italic*, life-threatening effects underlined; generic names in **bold;** classifications in SMALL CAPS; ♣ Canadian drug name; ⊙ Prototype drug

1233

agonist for treatment of Parkinson's disease. Exhibits high affinity for the D_2 subfamily of dopamine receptors in the brain and higher binding affinity to D_3 than other dopamine receptor subtypes. *Effectiveness is indicated by improved control of neuromuscular functioning. Improves ADLs.*

USES Treatment of idiopathic Parkinson's disease and moderate to severe primary restless legs syndrome (RLS).

CONTRAINDICATIONS Hypersensitivity to pramipexole or ropinirole; lactation.

CAUTIOUS USE Renal impairment; impulse control/compulsive behavior symptoms; history of orthostatic hypotension; pregnancy (category C). Safety and efficacy in children are not established.

ROUTE & DOSAGE

Parkinson's Disease

Adult: **PO Immediate release:** Start with 0.125 mg t.i.d. gradually increase every 5–7 days to a maximum dose of 1.5 mg t.i.d. **Extended release:** 0.375 mg PO daily, may increase after 5 days up to a maximum of 4.5 mg/day.

Restless Legs Syndrome

Adult: **PO** 0.125 mg taken 2–3 h before bed; dose can be increased every 4–7 days

Renal Impairment Dosage Adjustment

CrCl 35–60 mL/min: Same titration schedule dosed b.i.d. (max: 1.5 mg b.i.d.); *15–35 mL/min:* Same titration schedule dosed daily (max: 1.5 mg daily)

ADMINISTRATION

Oral

- Ensure that extended release tablet is swallowed whole and not crushed or chewed.
- Give with food if nausea develops.

ADVERSE EFFECTS (≥1%) Body as a Whole: *Asthenia,* general edema, malaise, fever, decreased weight. **CNS:** *Dizziness, somnolence, sudden sleep attacks, insomnia, hallucinations, dyskinesia, extrapyramidal syndrome,* headache, confusion, amnesia, hypesthesia, dystonia, akathisia, myoclonus, peripheral edema. **CV:** *Postural hypotension,* chest pain. **GI:** *Nausea, constipation,* anorexia, dysphagia, dry mouth. **Respiratory:** Dyspnea, rhinitis. **Urogenital:** Decreased libido, impotence, urinary frequency or incontinence. **Special Senses:** Vision abnormalities.

INTERACTIONS Drug: Cimetidine decreases clearance; BUTYROPHENONES, metoclopramide, PHENOTHIAZINES, ANTIPSYCHOTICS may antagonize effects.

PHARMACOKINETICS Absorption: Rapidly from GI tract, greater than 90% bioavailability. **Peak:** 2 h. **Distribution:** 15% protein bound. **Metabolism:** Minimally in the liver. **Elimination:** Primarily in urine. **Half-Life:** 8–12 h.

NURSING IMPLICATIONS

Assessment & Drug Effects

- Monitor for S&S of orthostatic hypotension, especially when the dosage is increased.
- Monitor cardiac status, especially in those with significant orthostatic hypotension.
- Lab tests: Monitor BUN and creatinine periodically; monitor CPK with complaints of muscle pain.

- Monitor for and report signs of tardive dyskinesia (see Appendix F).

Patient & Family Education
- Hallucinations are an adverse effect of this drug and occur more often in older adults.
- Make position changes slowly especially from a lying or sitting to standing.
- Use caution with potentially dangerous activities until response to drug is known; drowsiness is a common adverse effect.
- Avoid alcohol and use extra caution if taking other prescribed CNS depressants; both may exaggerate drowsiness, dizziness, and orthostatic hypotension.
- Do not abruptly stop taking this drug. It should be discontinued over a period of 1 wk.

PRAMLINTIDE
Symlin
Classifications: ANTIDIABETIC; AMYLIN ANALOG
Therapeutic: ANTIHYPERGLYCEMIC
Pregnancy Category: C

AVAILABILITY 0.6 mg/mL injection

ACTION & *THERAPEUTIC EFFECT*
Pramlintide is a synthetic analog of human amylin, a hormone secreted by pancreatic beta cells. In type 2 diabetic patients using insulin and in type 1 diabetics, beta cells in the pancreas are either damaged or destroyed, resulting in reduced secretion of both insulin and amylin after meals. Amylin reduces postmeal glucagon levels, thus lowering serum glucose level. *Pramlintide is an antihyperglycemic drug that controls postprandial blood glucose levels.*

USES Adjunct treatment of diabetes mellitus type 1 and type 2 in patients who use mealtime insulin therapy and who have failed to achieve desired glucose control despite optimal insulin therapy.

CONTRAINDICATIONS Hypersensitivity to pramlintide, cresol; noncompliance with insulin regime or medical care; HbA1C greater than 9%; hypoglycemia; gastroparesis; concomitant use of drugs to stimulate GI motility; renal failure or dialysis.
CAUTIOUS USE Osteoporosis; alcohol; thyroid disease; pregnancy (category C), lactation. Safety and efficacy in children not established.

ROUTE & DOSAGE

Type 1 Diabetes Mellitus
Adult: **Subcutaneous** 15 mcg immediately before each major meal; may increase by 15 mcg increments if no clinically significant nausea for 3–7 days. If nausea or vomiting persists at 45 mcg or 60 mcg, reduce to 30 mcg.

Type 2 Diabetes Mellitus
Adult: **Subcutaneous** 60 mcg immediately before each major meal; may increase to 120 mcg if no clinically significant nausea for 3–7 days. If nausea or vomiting persists at 120 mcg, reduce to 60 mcg.

ADMINISTRATION
Subcutaneous
- Give subcutaneously into the abdomen or thigh (not the arm) immediately before each major meal. Rotate injection sites.
- Never mix pramlintide and insulin in the same syringe. Separate injection sites.
- Use a U100 insulin syringe to administer. One unit of pramlintide

P

Common adverse effects in *italic*, life-threatening effects <u>underlined</u>; generic names in **bold**; classifications in SMALL CAPS; ♣ Canadian drug name; ⊙ Prototype drug

1235

drawn from a 0.6 mg/mL vial contains 6 mcg of medication. Thus, a 30 mcg dose is equal to 5 units in a U100 syringe.

- Do not administer to patients with HbA1C greater than 9% or those taking drugs to stimulate GI motility.
- Note: When initiating pramlintide therapy, insulin dose reduction is required.
- Store at 2°–8° C (36°–46° F), and protect from light. Do not freeze. Discard vials that have been frozen or overheated. Discard open vials after 28 days.

ADVERSE EFFECTS (≥1%) **CNS:** Dizziness, fatigue, *headache.* **GI:** Abdominal pain, *anorexia, nausea, vomiting.* **Musculoskeletal:** Arthralgia. **Respiratory:** *Coughing,* pharyngitis. **Body as a Whole:** *Allergic reaction, inflicted injury.*

INTERACTIONS Drugs: Pramlintide can decrease rate and/or extent of GI absorption of other oral drugs. Significant slowing of gastric motility with ANTIMUSCARINICS.

PHARMACOKINETICS Absorption: 30–40% bioavailability. **Peak:** 20 min. **Distribution:** 40% protein bound. **Metabolism:** Extensive renal metabolism. **Half-Life:** 48 min.

NURSING IMPLICATIONS

Assessment & Drug Effects
- Monitor for severe hypoglycemia, which usually occurs within 3 h of injection. Hypoglycemia is worse in type 1 diabetics.
- Monitor diabetics for loss of glycemic control.
- Lab tests: Baseline and periodic HbA1C; frequent pre/postmeal plasma glucose levels.
- Withhold drug and notify prescriber for clinically significant

nausea or increased frequency or severity of hypoglycemia.

Patient & Family Education
- Note: Patients should reduce a.c. rapid-acting or short-acting insulin dosages by 50% when pramlintide is initiated. Check with prescriber.
- Do not drive or engage in potentially hazardous activities until response to drug is known.
- Report any of the following to prescriber: Persistent, significant nausea; episodes of hypoglycemia (e.g., hunger, headache, sweating, tremor, irritability, or difficulty concentrating).

PRAMOXINE HYDROCHLORIDE
(pra-mox'een)

Fleet Relief Anesthetic Hemorrhoidal, Prax, ProctoFoam, Tronolane, Tronothane ♣
Classifications: LOCAL ANESTHETIC (MUCOSAL); ANTIPRURITIC
Therapeutic: LOCAL ANESTHETIC; ANTIPRURITIC
Prototype: Procaine
Pregnancy Category: C

AVAILABILITY 1% cream, gel, lotion, spray

ACTION & *THERAPEUTIC EFFECT*
Produces anesthesia by blocking conduction and propagation of sensory nerve impulses in skin and mucous membranes. *Provides temporary relief from pain and itching on skin or mucous membrane.*

USES To relieve pain caused by minor burns and wounds; for temporary relief of pruritus secondary to dermatoses, hemorrhoids, and anal fissures; and to facilitate sigmoidoscopic examination.

CONTRAINDICATIONS Application to large areas of skin; prolonged use; preparation for laryngopharyngeal examination, bronchoscopy, or gastroscopy.

CAUTIOUS USE Extensive skin disorders; pregnancy (category C), lactation. Safety in children younger than 2 y is not established.

ROUTE & DOSAGE

Relief of Minor Pain and Itching

Adult/Child (older than 2 y):
Topical Apply t.i.d. or q.i.d.

ADMINISTRATION

Topical
- Clean thoroughly and dry rectal area before use for temporary relief of hemorrhoidal pain and itching.
- Administer rectal preparations in the morning and evening and after bowel movement or as directed by prescriber.
- Apply lotion or cream to affected surfaces with a gloved hand. Wash hands thoroughly before and after treatment.
- Do not apply to eyes or nasal membranes.

ADVERSE EFFECTS (≥1%) **Skin:** Burning, stinging, sensitization.

PHARMACOKINETICS Onset: 3–5 min. **Duration:** Up to 5 h.

NURSING IMPLICATIONS

Assessment & Drug Effects
- Monitor for and report promptly significant tissue irritation or sloughing.

Patient & Family Education
- Drug is usually discontinued if condition being treated does not improve within 2–3 wk or if it worsens, or if rash or condition not present before treatment appears, or if treated area becomes inflamed or infected.

- Discontinue and consult prescriber if rectal bleeding and pain occur during hemorrhoid treatment.

PRASUGREL
(pra-soo′grel)

Effient
Classifications: ANTIPLATELET; PLATELET INHIBITOR; ADP RECEPTOR ANTAGONIST
Therapeutic: PLATELET INHIBITOR
Prototype: Clopidogrel
Pregnancy Category: B

AVAILABILITY 5 mg, 10 mg tablets

ACTION & *THERAPEUTIC EFFECT*
Prasugrel is an inhibitor of platelet activation and aggregation through irreversible binding to adenosine diphosphate (ADP) receptors on platelets. *Prasugrel prolongs bleeding time, thereby reducing atherosclerotic events in selected high risk patient managed with percutaneous coronary intervention (PCI).*

USES Prophylaxis of arterial thromboembolism in patients with acute coronary syndrome, including unstable angina, non–ST-elevation myocardial infarction (NSTEMI), or ST-elevation acute myocardial infarction (STEMI), who are being managed with percutaneous coronary intervention (PCI).

CONTRAINDICATIONS Active pathologic bleeding disorder; active bleeding; history of TIA or stroke; within 7 days of CABG surgery or any surgery; concomitant use of NSAIDs or other drugs that increase risk of bleeding.

CAUTIOUS USE Severe hepatic impairment (Child-Pugh class C, total score greater than 10); older adults over 75 y; pregnancy (category B); lactation.

P

Common adverse effects in *italic*, life-threatening effects underlined; generic names in **bold;** classifications in SMALL CAPS; ♣ Canadian drug name; ⮀ Prototype drug

1237

ROUTE & DOSAGE

Thromboembolism Prophylaxis

Adult: **PO** Younger than 75 y, weight 60 kg or greater; 60 mg loading dose, then 10 mg once daily; younger than 75 y, weight less than 60 kg; 60 mg loading dose, then 5 mg once daily

ADMINISTRATION

Oral

- Give without regard to food.
- Daily aspirin is recommended with prasugrel.
- Do not administer to patient with active bleeding or who is likely to undergo urgent CABG.
- Store at 15°–30° C (59°–86° F).

ADVERSE EFFECTS (≥1%) **Body as a Whole:** Fatigue, non-cardiac chest pain, pain in extremity, peripheral edema, pyrexia. **CNS:** Dizziness, headache. **CV:** Atrial fibrillation, bradycardia, hypertension, hypotension. **GI:** Diarrhea, gastrointestinal hemorrhage, nausea. **Hematologic:** Epistaxis, increased bleeding tendency. **Metabolic:** Hypercholesterolemia, hyperlipidemia, leukopenia. **Musculoskeletal:** Back pain. **Respiratory:** Cough, dyspnea.

INTERACTIONS Drug: Warfarin or NSAIDs may increase the risk of bleeding.

PHARMACOKINETICS Absorption: 79% or higher. **Peak:** 30 min. **Metabolism:** Hydrolysis and oxidation to active metabolite. **Elimination:** Urine (68%) and feces (27%). **Half-Life:** 7 h.

NURSING IMPLICATIONS

Assessment & Drug Effects

- Monitor vital signs. Suspect bleeding if patient is hypotensive and has recently undergone an invasive or surgical procedure.
- Monitor for and report promptly any S&S of active bleeding.
- Lab tests: Periodic lipid profile.

Patient & Family Education

- Report promptly unexplained prolonged or excessive bleeding, or blood in urine or stool.
- Report immediately any of the following: Weakness, extremely pale skin, purple skin patches, jaundice, or fever.
- Inform all medical providers that you are taking prasugrel.
- Do not take OTC anti-inflammatory or pain medications without consulting prescriber.

PRAVASTATIN

(pra-vah-stat'in)

Pravachol
Classifications: ANTIHYPERLIPEMIC; HMG-COA REDUCTASE INHIBITOR (STATIN)
Therapeutic: ANTILIPEMIC; STATIN
Prototype: Lovastatin
Pregnancy Category: X

AVAILABILITY 10 mg, 20 mg, 40 mg, 80 mg tablets

ACTION & THERAPEUTIC EFFECT
Competitively inhibits 3-hydroxy-3-methylglutaryl-coenzyme A (HMG-CoA) reductase, the enzyme that catalyzes cholesterol biosynthesis. HMG-CoA reductase inhibitors (statins) increase serum HDL cholesterol, decrease serum LDL cholesterol, VLDL cholesterol, and plasma triglyceride levels. It is effective in reducing total and LDL cholesterol in various forms of hypercholesterolemia.

USES Hypercholesterolemia (alone or in combination with bile acid sequestrants) and familial hypercholesterolemia.

CONTRAINDICATIONS Hypersensitivity to pravastatin; active liver disease or unexplained elevated liver function test; hepatic encephalopathy, hepatitis, jaundice, rhabdomyolysis; pregnancy (category X), lactation.
CAUTIOUS USE Alcoholics, history of liver disease; renal impairment; renal disease; seizure disorders. Safe use in children younger than 8 y not established.

ROUTE & DOSAGE

Hyperlipidemia
Adult: **PO** 10–80 mg daily
Child (8–13 y): **PO** 20 mg daily

ADMINISTRATION

Oral
- Give without regard to meals.
- Give in the evening.

ADVERSE EFFECTS (≥1%) **GI:** Nausea, diarrhea, abdominal pain, vomiting, constipation, flatulence, heartburn, transient elevations in serum liver transaminase levels. **Other:** Fatigue, rhinitis, cough, transient elevations in CPK.

INTERACTIONS Drug: May increase PT when administered with **warfarin.**

PHARMACOKINETICS Absorption: Poorly from GI tract; 17% reaches systemic circulation. **Onset:** 2 wk. **Peak:** 4 wk. **Distribution:** 43–55% protein bound; does not cross blood–brain barrier; crosses placenta; distributed into breast milk. **Metabolism:** Extensive first-pass metabolism in liver; has no active metabolites. **Elimination:** 20% of dose excreted in urine, 71% in feces. **Half-Life:** 1.8–2.6 h.

NURSING IMPLICATIONS

Assessment & Drug Effects
- Lab tests: Baseline LFTs at start of therapy and then at 12 wk. If normal at 12 wk, may change to semiannual monitoring. Monitor cholesterol levels throughout therapy.
- Monitor coagulation studies with patients receiving concurrent warfarin therapy. PT may be prolonged.
- Monitor CPK levels if patient experiences unexplained muscle pain.

Patient & Family Education
- Report unexplained muscle pain, tenderness, or weakness, especially if accompanied by malaise or fever, to prescriber promptly.
- Report signs of bleeding to prescriber promptly when taking concomitant warfarin therapy.

PRAZIQUANTEL
(pray-zi-kwon'tel)
Biltricide
Classification: ANTHELMINTIC
Therapeutic: ANTHELMINTIC
Prototype: Mebendazole
Pregnancy Category: B

AVAILABILITY 600 mg tablets

ACTION & *THERAPEUTIC EFFECT*
Synthetic agent with broad-spectrum anthelmintic activity against all developmental stages of schistosomes and other trematodes (flukes) and against cestodes (tapeworm). Increases permeability of parasite cell membrane to calcium. Leads to immobilization of their suckers and dislodgment from their residence in blood vessel walls. *Active against all developmental stages of schistosomes, including cercaria (free-swimming larvae). Also active against other*

trematodes (flukes) and cestodes (tapeworms).

USES All stages of schistosomiasis (bilharziasis) caused by all *Schistosoma* species pathogenic to humans. Other trematode infections caused by Chinese liver fluke.

UNLABELED USES Lung, sheep liver, and intestinal flukes and tapeworm infections.

CONTRAINDICATIONS Hypersensitivity to praziquantel; ocular cysticercosis. Women should not breastfeed on day of praziquantel therapy or for 72 h after last dose of drug.

CAUTIOUS USE Hepatic disease; cardiac arrhythmias; pregnancy (category B); children younger than 4 y.

ROUTE & DOSAGE

Schistosomiasis

Adult/Child (older than 4 y): **PO** 60 mg/kg in 3 equally divided doses at 4–6 h intervals on the same day, may repeat in 2–3 mo after exposure

Other Trematodes

Adult/Child (older than 4 y): **PO** 75 mg/kg in 3 equally divided doses at 4–6 h intervals on the same day

Cestodiasis (Adult or Intestinal Stage)

Adult: **PO** 10–20 mg/kg as single dose

Cestodiasis (Larval or Tissue Stage)

Adult: **PO** 50 mg/kg in 3 divided doses/day for 14 days

ADMINISTRATION

Oral
- Give dose with food and fluids. Tablets can be broken into

quarters but should NOT be chewed.
- Advise patient to take sufficient fluid to wash down the medication. Tablets are soluble in water; gagging or vomiting because of bitter taste may result if tablets are retained in the mouth.
- Store tablets in tight containers at less than 30° C (86° F).

ADVERSE EFFECTS (≥1%) CNS: *Dizziness, headache, malaise,* drowsiness, lassitude, <u>CSF reaction syndrome</u> (exacerbation of neurologic signs and symptoms such as seizures, increased CSF protein concentration, increased anticysticercal IgG levels, hyperthermia, intracranial hypertension) in patient treated for cerebral cysticercosis. **GI:** *Abdominal pain or discomfort with or without nausea;* vomiting, anorexia, diarrhea. **Hepatic:** *Increased AST, ALT (slight).* **Skin:** Pruritus, urticaria. **Body as a Whole:** Fever, sweating, symptoms of host-mediated immunologic response to antigen release from worms (fever, eosinophilia).

DIAGNOSTIC TEST INTERFERENCE Be mindful that selected drugs may interfere with stool studies for ova and parasites: ***Iron, bismuth, oil (mineral*** or ***castor), Metamucil*** (if ingested within 1 wk of test), ***barium, antibiotics, antiamebic*** and ***antimalarial drugs,*** and ***gallbladder dye*** (if administered within 3 wk of test).

INTERACTIONS Drug: Phenytoin can lead to therapeutic failure. **Food: Grapefruit juice** (greater than 1 qt/day) may increase plasma concentrations and adverse effects.

PHARMACOKINETICS Absorption: Rapidly, 80% reaches systemic circulation. **Peak:** 1–3 h. **Distribution:** Enters cerebrospinal fluid. **Metabolism:** Extensively to inactive metabolites. **Elimination:** Primarily in urine. **Half-Life:** 0.8–1.5 h.

NURSING IMPLICATIONS

Assessment & Drug Effects
- Patient is reexamined in 2 or 3 mo to ensure complete eradication of the infections.

Patient & Family Education
- Do not drive or operate other hazardous machinery on day of treatment or the following day because of potential drug-induced dizziness and drowsiness.
- Usually, all schistosomal worms are dead 7 days following treatment.
- Contact prescriber if you develop a sustained headache or high fever.

PRAZOSIN HYDROCHLORIDE ⊕
(pra′zoe-sin)

Minipress
Classifications: ALPHA-ADRENERGIC RECEPTOR ANTAGONIST; ANTIHYPERTENSIVE; VASODILATOR
Therapeutic: ANTIHYPERTENSIVE
Pregnancy Category: C

AVAILABILITY 1 mg, 2 mg, 5 mg capsules

ACTION & *THERAPEUTIC EFFECT*
Selective inhibition of alpha$_1$-adrenoceptors that produces vasodilation in both resistance (arterioles) and capacitance (veins) vessels with the result that both peripheral vascular resistance and blood pressure are reduced. *Lowers blood pressure in supine and standing positions with most pronounced effect on diastolic pressure.*

USES Treatment of hypertension.
UNLABELED USES Severe refractory congestive heart failure, Raynaud's disease or phenomenon, ergotamine-induced peripheral ischemia, pheochromocytoma, benign prostatic hypertrophy.

CONTRAINDICATIONS Hypotension.
CAUTIOUS USE Renal impairment; chronic kidney failure; hypertensive patient with cerebral thrombosis; angina; men with sickle cell trait; older adults; pregnancy (category C); lactation.

ROUTE & DOSAGE

Hypertension

Adult: **PO** Start with 1 mg at bedtime, then 1 mg b.i.d. or t.i.d., may increase to 20 mg/day in divided doses
Child: **PO** Start with 5 mcg/kg q6h, gradually increase to 25 mcg/kg q6h (max: 15 mg or 0.4 mg/kg/day)

ADMINISTRATION

Oral
- Give initial dose at bedtime to reduce possibility of adverse effects such as postural hypotension and syncope. However, if first dose is taken during the day, advise patient not to drive a car for about 4 h after ingestion of drug.
- Give drug with food to reduce incidence of faintness and dizziness; food may delay absorption but does not affect extent of absorption.
- Store in tightly closed container away from strong light. Do not freeze.

ADVERSE EFFECTS (≥1%) **CNS:** *Dizziness, headache, drowsiness,* nervousness, vertigo, depression,

Common adverse effects in *italic*, life-threatening effects <u>underlined</u>; generic names in **bold**; classifications in SMALL CAPS; ♣ Canadian drug name; ⊕ Prototype drug

1241

paresthesia, insomnia. **CV:** Edema, dyspnea, syncope *first-dose phenomenon,* postural hypotension, *palpitations,* tachycardia, angina. **Special Senses:** Blurred vision, tinnitus, reddened sclerae. **GI:** Dry mouth, *nausea,* vomiting, diarrhea, constipation, abdominal discomfort, pain. **Urogenital:** Urinary frequency, incontinence, priapism (especially in men with sickle cell anemia), impotence. **Skin:** Rash, pruritus, alopecia, lichen planus. **Body as a Whole:** Diaphoresis, epistaxis, nasal congestion, arthralgia, transient <u>leukopenia</u>, increased serum uric acid, and BUN.

INTERACTIONS Drug: DIURETICS, HYPOTENSIVE AGENTS and alcohol increase hypotensive effects. **Sildenafil, vardenafil,** and **tadalafil** may enhance hypotensive effects.

PHARMACOKINETICS Absorption: Approximately 60% of oral dose reaches the systemic circulation. **Onset:** 2 h. **Peak:** 2–4 h. **Duration:** Less than 24 h. **Distribution:** Widely distributed, including into breast milk. **Metabolism:** Extensively in liver. **Elimination:** 6–10% in urine, rest in bile and feces. **Half-Life:** 2–4 h.

NURSING IMPLICATIONS

Assessment & Drug Effects

- Be alert for first-dose phenomenon (rare adverse effect: 0.15% of patients); characterized by a precipitous decline in BP, bradycardia, and consciousness disturbances (syncope) within 90–120 min after the initial dose of prazosin. Recovery is usually within several hours. Preexisting low plasma volume (from diuretic therapy or salt restriction), beta-adrenergic therapy, and recent stroke appear to increase the risk of this phenomenon.

- Monitor blood pressure. If it falls precipitously with first dose, notify prescriber promptly.

- Full therapeutic effect may not be achieved until 4–6 wk of therapy.

Patient & Family Education

- Avoid situations that would result in injury if you should faint, particularly during early phase of treatment. In most cases, effect does not recur after initial period of therapy; however, it may occur during acute febrile episodes, when drug dose is increased, or when another antihypertensive drug is added to the medication regimen.

- Make position and direction changes slowly and in stages. Dangle legs and move ankles a minute or so before standing when arising in the morning or after a nap.

- Lie down immediately if you experience light-headedness, dizziness, a sense of impending loss of consciousness, or blurred vision. Attempting to stand or walk may result in a fall.

- Do not drive or engage in other potentially hazardous activities until response to drug is known.

- Take drug at same time(s) each day.

- Report priapism or impotence. A change in the drug regimen usually reverses these difficulties.

- Do not take OTC medications, especially remedies for coughs, colds, and allergy, without consulting prescriber.

- Be aware that adverse effects usually disappear with continuation of therapy, but dosage reduction may be necessary.

PREDNISOLONE

(pred-niss′oh-lone)

Prelone

PREDNISOLONE ACETATE

Flo-Pred, Pred Forte, Pred Mild

PREDNISOLONE SODIUM PHOSPHATE

AK-Pred, Inflamase Forte, Inflamase Mild

Classification: ADRENAL CORTICOSTEROID
Therapeutic: ANTI-INFLAMMATORY; IMMUNOSUPPRESSANT
Prototype: Prednisone
Pregnancy Category: C

AVAILABILITY Prednisolone: 1 mg, 2.5 mg, 5 mg tablet; 5 mg/5 mL, 15 mg/5 mL syrup; **Acetate:** 1% ophth suspension; **Sodium Phosphate:** 5 mg/5 mL liquid; 0.125%, 1%, 0.9%, 0.11% ophth solution

ACTION & *THERAPEUTIC EFFECT*

Has glucocorticoid activity similar to naturally occurring hormone. It prevents or suppresses inflammation and immune responses. Its actions include inhibition of leukocyte infiltration at the site of inflammation, interference in the function of inflammatory mediators, and suppression of humoral immune responses. *Effective as an anti-inflammatory agent.*

USES Principally as an anti-inflammatory and immunosuppressant agent.

CONTRAINDICATIONS Fungal infections; GI bleeding.
CAUTIOUS USE Cataracts; coagulopathy; DM; seizure disorders; renal disease; psychosis; emotional instability; GI disorders; pregnancy (category C).

ROUTE & DOSAGE

Anti-Inflammatory

Adult: PO 5–60 mg/day in single or divided doses **Ophthalmic** See Appendix A-1
Child: PO 0.1–2 mg/kg/day in divided doses

ADMINISTRATION

Oral
- Give with meals to reduce gastric irritation. If distress continues, consult prescriber about possible adjunctive antacid therapy.

Alternate-Day Therapy (ADT) for Patient on Long-Term Therapy
- With ADT, the 48-h requirement for steroids is administered as a single dose every other morning.
- Be aware that ADT minimizes adverse effects associated with long-term treatment while maintaining the desired therapeutic effect.
- See PREDNISONE for numerous additional nursing implications.

ADVERSE EFFECTS (≥1%) **Endocrine:** Hirsutism (occasional), adverse effects on growth and development of the individual and on sperm. **Special Senses:** Perforation of cornea (with topical drug). **Body as a Whole:** Sensitivity to heat; fat embolism, hypotension and shock-like reactions. **CNS:** Insomnia. **GI:** Gastric irritation or ulceration. **Skin:** Ecchymotic skin lesions; vasomotor symptoms. Also see PREDNISONE.

INTERACTIONS Drug: BARBITURATES, **phenytoin, rifampin** increase steroid metabolism, therefore may need increased doses of prednisolone; **amphotericin B,** DIURETICS add to **potassium** loss; **ambenonium, neostigmine, pyridostigmine** may cause severe muscle weakness in patients with myasthenia gravis; VACCINES, TOXOIDS may inhibit antibody response. **Food: Licorice** may elevate plasma levels and adverse effects.

PHARMACOKINETICS Absorption: Readily from GI tract. **Peak:** 1–2 h. **Duration:** 1–1.5 days. **Distribution:** Crosses placenta; distributed into breast milk. **Metabolism:** In liver.

P

Common adverse effects in *italic*, life-threatening effects underlined; generic names in **bold**; classifications in SMALL CAPS; ✦ Canadian drug name; ⊘ Prototype drug

1243

Elimination: HPA suppression: 24–36 h; in urine. **Half-Life:** 3.5 h.

NURSING IMPLICATIONS

Assessment & Drug Effects

- Be alert to subclinical signs of lack of improvement such as continued drainage, low-grade fever, and interrupted healing. In diseases caused by microorganisms, infection may be masked, activated, or enhanced by corticosteroids. Observe and report exacerbation of symptoms after short period of therapeutic response.
- Be aware that temporary local discomfort may follow injection of prednisolone into bursa or joint.

Patient & Family Education

- Adhere to established dosage regimen (i.e., do not increase, decrease, or omit doses or change dose intervals).
- Report gastric distress or any sign of peptic ulcer.

USES May be used as a single agent or conjunctively with antineoplastics in cancer therapy; also used in treatment of myasthenia gravis and inflammatory conditions as an immunosuppressant; acute respiratory distress syndrome, Addison's disease, adrenal hyperplasic, gout, gouty arthritis, headache, hemolytic anemia, sarcoidosis, Stevens-Johnson syndrome.
UNLABELED USES Absence seizures.

CONTRAINDICATIONS Systemic fungal infections and known hypersensitivity; cataracts.
CAUTIOUS USE Patients with infections; nonspecific ulcerative colitis; diverticulitis; active or latent peptic ulcer; renal insufficiency; coagulopathy; psychosis; seizure disorders; thromboembolic disease; hypertension; osteoporosis; myasthenia gravis; pregnancy (category C).

PREDNISONE ⓟ
(pred′ni-sone)

Apo-Prednisone ♣, Deltasone, Meticorten, Orasone, Panasol, Prednicen-M, Sterapred, Winpred ♣
Classification: ADRENAL CORTICOSTEROID
Therapeutic: IMMUNOSUPPRESSANT; ANTI-INFLAMMATORY
Pregnancy Category: C

AVAILABILITY 1 mg, 2.5 mg, 5 mg, 10 mg, 20 mg, 50 mg tablets; 5 mg/5 mL, 5 mg/mL solution

ACTION & *THERAPEUTIC EFFECT*
Immediate-acting synthetic analog of hydrocortisone that is biotransformed in the liver into prednisolone. *Has anti-inflammatory and immunosuppressant properties.*

ROUTE & DOSAGE

Anti-Inflammatory
*Doses are highly individualized (ranges are provided)
Adult: **PO** 5–60 mg/day in single or divided doses
Child: **PO** 0.1–0.15 mg/kg/day in single or divided doses
Acute Asthma
Child: **PO** *Younger than 1 y,* 1–2 mg/kg/day × 3–5 days or 10 mg q12h; *1–4 y,* 20 mg q12h; *5–13 y,* 30 mg q12h; *older than 13 y,* 40 mg q12h × 3–5 days

ADMINISTRATION

Oral

- Crush tablet and give with fluid of patient's choice if unable to swallow whole.

P

- Give at mealtimes or with a snack to reduce gastric irritation.
- Dose adjustment may be required if patient is subjected to severe stress (serious infection, surgery, or injury) or if a remission or disease exacerbation occurs.
- Do not abruptly stop drug. Reduce dose gradually by scheduled decrements (various regimens) to prevent withdrawal symptoms and permit adrenals to recover from drug-induced partial atrophy.

Alternate-Day Therapy (ADT) for Patient on Long-Term Therapy

- With ADT, the 48-h requirement for steroids is administered as a single dose every other morning.
- Be aware that ADT minimizes adverse effects associated with long-term treatment while maintaining the desired therapeutic effect.

ADVERSE EFFECTS (≥1%) **CNS:** Euphoria, headache, insomnia, confusion, psychosis. **CV:** CHF, edema. **GI:** Nausea, vomiting, peptic ulcer. **Musculoskeletal:** Muscle weakness, delayed wound healing, muscle wasting, osteoporosis, aseptic necrosis of bone, spontaneous fractures. **Endocrine:** Cushingoid features, growth suppression in children, carbohydrate intolerance, hyperglycemia. **Special Senses:** Cataracts. **Hematologic:** Leukocytosis. **Metabolic:** Hypokalemia.

INTERACTIONS Drug: BARBITURATES, **phenytoin, rifampin** increase steroid metabolism—increased doses of prednisone may be needed; **amphotericin B,** DIURETICS increase **potassium** loss; **ambenonium, neostigmine, pyridostigmine** may cause severe muscle weakness in patients with myasthenia gravis; may inhibit antibody response to VACCINES, TOXOIDS.

PHARMACOKINETICS Absorption: Readily from GI tract. **Peak:** 1–2 h. **Duration:** 1–1.5 days. **Distribution:** Crosses placenta; distributed into breast milk. **Metabolism:** In liver. **Elimination:** Hypothalamus-pituitary axis suppression: 24–36 h; in urine. **Half-Life:** 3.5 h.

NURSING IMPLICATIONS

Assessment & Drug Effects

- Establish baseline and continuing data regarding BP, I&O ratio and pattern, weight, fasting blood glucose level, and sleep pattern. Start flow chart as reference for planning individualized pharmacotherapeutic patient care.
- Check and record BP during dose stabilization period at least 2 times daily. Report an ascending pattern.
- Monitor patient for evidence of HPA axis suppression during long-term therapy by determining plasma cortisol levels at weekly intervals.
- Lab tests: Obtain fasting blood glucose, serum electrolytes, and routine laboratory studies at regular intervals during long-term steroid therapy.
- Be aware that older adult patients and patients with low serum albumin are especially susceptible to adverse effects because of excess circulating free glucocorticoids.
- Be alert to signs of hypocalcemia (see Appendix F). Patients with hypocalcemia have increased requirements for pyridoxine (vitamin B_6), vitamins C and D, and folates.
- Be alert to possibility of masked infection and delayed healing (anti-inflammatory and immunosuppressive actions). Prednisone suppresses early classic signs of inflammation. When patient is on an extended therapy regimen,

P

Common adverse effects in *italic*, life-threatening effects underlined; generic names in **bold**; classifications in SMALL CAPS; ♣ Canadian drug name; ⊙ Prototype drug

1245

incidence of oral *Candida* infection is high. Inspect mouth daily for symptoms: White patches, black furry tongue, painful membranes and tongue.

- Monitor bone density. Compression and spontaneous fractures of long bones and vertebrae present hazards, particularly in long-term corticosteroid treatment of rheumatoid arthritis or diabetes, in immobilized patients, and older adults.

- Be aware of previous history of psychotic tendencies. Watch for changes in mood and behavior, emotional stability, sleep pattern, or psychomotor activity, especially with long-term therapy, that may signal onset of recurrence. Report symptoms to prescriber.

- Monitor for withdrawal syndrome (e.g., myalgia, fever, arthralgia, malaise) and hypocorticism (e.g., anorexia, vomiting, nausea, fatigue, dizziness, hypotension, hypoglycemia, myalgia, arthralgia) with abrupt discontinuation of corticosteroids after long-term therapy.

Patient & Family Education
- Take drug as prescribed and do not alter dosing regimen or stop medication without consulting prescriber.

- Be aware that a slight weight gain with improved appetite is expected, but after dosage is stabilized, a sudden slow but steady weight increase [2 kg (5 lb)/wk] should be reported to prescriber.

- Avoid or minimize alcohol, which may contribute to steroid-ulcer development in long-term therapy.

- Report symptoms of GI distress to prescriber and do not self-medicate to find relief.

- Do not use aspirin or other OTC drugs unless they are prescribed specifically by the prescriber.

- Be fastidious about personal hygiene; give special attention to foot care, and be particularly cautious about bruising or abrading the skin.

- Report persistent backache or chest pain (possible symptoms of vertebral or rib fracture) that may occur with long-term therapy.

- Tell dentist or new prescriber about prednisone therapy.

PREGABALIN
Lyrica
Classifications: ANTICONVULSANT; GABA-ANALOG; ANALGESIC/MISCELLANEOUS; ANXIOLYTIC
Therapeutic: ANTICONVULSANT; ANALGESIC; ANTI-ANXIETY
Prototype: Gabapentin
Pregnancy Category: C
Controlled Substance: Schedule V

AVAILABILITY 25 mg, 50 mg, 75 mg, 100 mg, 150 mg, 200 mg, 225 mg, 300 mg capsules

ACTION & *THERAPEUTIC EFFECT*
Pregabalin is an analog of gamma-aminobutyric acid (GABA) that increases neuronal GABA levels and reduces calcium currents in the calcium channels of neurons; this may account for its control of pain and anxiety. Its affinity for voltage-gated calcium channels may account for its antiseizure activity. *Has analgesic, anti-anxiety, and anticonvulsant properties.*

USES Management of neuropathic pain associated with diabetic peripheral neuropathy, adjunctive therapy for adult patients with partial-onset seizures, management of postherpetic neuralgia, fibromyalgia.
UNLABELED USES Treatment of generalized anxiety disorders, treatment of social anxiety disorder, treatment of moderate pain.

Common adverse effects in *italic*, life-threatening effects underlined; generic names in **bold**; classifications in SMALL CAPS; ♣ Canadian drug name; ◯ Prototype drug

CONTRAINDICATIONS Hypersensitivity to pregabalin or gabapentin; suicidal ideation or worsening of depression or unusual changes in mood or behavior; lactation.

CAUTIOUS USE Renal impairment or failure, hemodialysis; history of suicidal tendencies or depression; history of drug abuse or alcohol; history of angioedema; CHF, NYHA (Class III or IV) cardiac status; older adults; pregnancy (category C). Safe use in children younger than 18 y not established.

ROUTE & DOSAGE

Neuropathic Pain (Diabetic Peripheral Neuropathy)
Adult: **PO** 50–100 mg t.i.d.

Partial-Onset Seizures
Adult: **PO** Initial dose 75 mg or less b.i.d or 50 mg t.i.d.; may increase to 300 mg b.i.d. or 200 mg t.i.d.

Fibromyalgia
Adult: **PO** 75 mg b.i.d., may increase up to 150 b.i.d. within first week, then up to 225 b.i.d. (max dose: 450 mg/day)

Postherpetic Neuralgia
Adult: **PO** Initial dose 75 mg b.i.d. or 50 mg t.i.d.; may increase to 150–300 mg b.i.d. or 100–200 mg t.i.d.

Renal Impairment Dosage Adjustment
CrCl 30–60 mL/min: 75–300 mg/day given in 2 or 3 divided doses; *15–30 mL/min:* 25–150 mg/day given in 1 or 2 divided doses; *less than 15 mL/min:* 25–75 mg once daily

Hemodialysis Dosage Adjustment
Dose based on renal function, give supplemental dose

ADMINISTRATION

Oral
- Dosage reduction is required with renal dysfunction.
- Drug should not be abruptly stopped; discontinue by tapering over a minimum of 1 wk.
- Give a supplemental dose immediately following dialysis.
- Store at 15°–30° C (59°– 86° F).

ADVERSE EFFECTS (≥1%) **Body as a Whole:** *Accidental injury,* flu syndrome, pain. **CNS:** Abnormal gait, amnesia, *ataxia,* confusion, *dizziness,* euphoria, headache, incoordination, myoclonus, nervousness, neuropathy, *somnolence,* speech disorder, abnormal thinking, tremor, twitching, vertigo. **CV:** Chest pain. **GI:** Constipation, dry mouth, flatulence, increased appetite, vomiting. **GU:** Urinary incontinence. **Metabolic/Nutritional:** Edema, facial edema, hypoglycemia, *peripheral edema, weight gain.* **Musculoskeletal:** Back pain, myasthenia. **Respiratory:** Bronchitis, dyspnea. **Special Senses:** Abnormal vision, *blurry vision, diplopia.*

INTERACTIONS Drug: Concomitant use with THIAZOLIDINEDIONES may exacerbate weight gain and fluid retention.

PHARMACOKINETICS Absorption: 90% bioavailability. **Peak:** 1.5 h. **Metabolism:** Negligible. **Elimination:** Primarily in the urine. **Half-Life:** 6 h.

NURSING IMPLICATIONS

Assessment & Drug Effects
- Monitor for and report promptly mental status or behavior changes (e.g., anxiety, panic attacks, restlessness, irritability, depression, suicidal thoughts).
- Monitor for weight gain, peripheral edema, and S&S of heart failure,

Common adverse effects in *italic*, life-threatening effects underlined; generic names in **bold**; classifications in SMALL CAPS; ✚ Canadian drug name; ⊘ Prototype drug

1247

especially with concurrent thiazoli-dinedione (e.g., rosiglitazone) therapy.

- Lab tests: Baseline and periodic kidney function tests; periodic platelet counts.
- Monitor diabetics for increased incidences of hypoglycemia.
- Supervise ambulation especially when other CNS drugs are used concurrently.

Patient & Family Education

- Do not drive or engage in potentially hazardous activities until response to drug is known.
- Report any of the following to a health care provider: Changes in vision (i.e., blurred vision); dizziness and incoordination; weight gain and swelling of the extremities, behavior or mood changes, especially suicidal thoughts.
- Avoid alcohol consumption while taking this drug.
- Inform your prescriber if you plan to become pregnant or father a child.

PRIMAQUINE PHOSPHATE

(prim′a-kween)
Primaquine
Classification: ANTIMALARIAL
Therapeutic: ANTIMALARIAL
Prototype: Chloroquine
Pregnancy Category: C

AVAILABILITY 26.3 mg tablets; 5 g, 25 g, 100 g, 500 g powder

ACTION & *THERAPEUTIC EFFECT*
Acts on primary exoerythrocytic forms of *Plasmodium vivax* and *Plasmodium falciparum*. Destroys late tissue forms of *P. vivax* and thus effects radical cure (prevents relapse). *Gametocidal activity against all species of Plasmodia that infect humans; interrupts transmission of malaria.*

USES To prevent relapse ("radical" or "clinical" cure) of *P. vivax* and *P. ovale* malarias and to prevent attacks after departure from areas where *P. vivax* and *P. ovale* malarias are endemic. With clindamycin for the treatment of *Pneumocystis carinii* pneumonia (PCP) in AIDS.

CONTRAINDICATIONS Hypersensitivity to primaquine or iodoquinol; rheumatoid arthritis; lupus erythematosus (SLE); hemolytic drugs, concomitant or recent use of agents capable of bone marrow depression (e.g., quinacrine; patients with G6PD deficiency).

CAUTIOUS USE Bone marrow depression; hematologic disease; methemoglobin reductase deficiency; pregnancy (category C), lactation.

ROUTE & DOSAGE

Malaria Treatment
Adult: **PO** 30 mg daily for 14 days concomitantly or consecutively with chloroquine or hydroxychloroquine on first 3 days of acute attack
Child: **PO** 0.5 mg/kg daily for 14 days concomitantly or consecutively with chloroquine or hydroxychloroquine on first 3 days of acute attack

Malaria Prophylaxis
Adult: **PO** 15 mg daily for 14 days beginning immediately after leaving malarious area
Child: **PO** 0.3 mg/kg daily for 14 days beginning immediately after leaving malarious area

ADMINISTRATION
Oral

- Give drug at mealtime or with an antacid (prescribed); may prevent

or relieve gastric irritation. Notify prescriber if GI symptoms persist.
- Store in tight, light-resistant containers.

ADVERSE EFFECTS (≥1%) **Hematologic:** <u>Hematologic reactions including granulocytopenia and acute hemolytic anemia in patients with G6PD deficiency</u>, moderate leukocytosis or *leukopenia*, anemia, granulocytopenia, agranulocytosis. **GI:** Nausea, vomiting, epigastric distress, abdominal cramps. **Skin:** Pruritus. **Metabolic:** Methemoglobinemia (cyanosis). **Body as a Whole:** Headache, confusion, mental depression. **Special Senses:** Disturbances of visual accommodation. **CV:** Hypertension, arrhythmias (rare).

INTERACTIONS Drug: Toxicity of both **quinacrine** and primaquine increased.

PHARMACOKINETICS Absorption: Readily from GI tract. **Peak:** 6 h. **Metabolism:** Rapidly in liver to active metabolites. **Elimination:** In urine. **Half-Life:** 3.7–9.6 h.

NURSING IMPLICATIONS

Assessment & Drug Effects
- Be aware drug may precipitate acute hemolytic anemia in patients with G6PD deficiency, an inherited error of metabolism carried on the X chromosome, present in about 10% of American black males and certain white ethnic groups: Sardinians, Sephardic Jews, Greeks, and Iranians. Whites manifest more intense expression of hemolytic reaction than do blacks. Screen for prior to initiation of therapy.
- Lab tests: Perform repeated hematologic studies (particularly blood cell counts and Hgb) and urinalyses during therapy.

Patient & Family Education
- Examine urine after each voiding and to report to prescriber darkening of urine, red-tinged urine, and decrease in urine volume. Also report chills, fever, precordial pain, cyanosis (all suggest a hemolytic reaction). Sudden reductions in hemoglobin or erythrocyte count suggest an impending hemolytic reaction.

PRIMIDONE
(pri′mi-done)
Apo-Primidone ♦, Mysoline
Classifications: ANTICONVULSANT; BARBITURATE
Therapeutic: ANTICONVULSANT
Prototype: Phenobarbital
Pregnancy Category: D

AVAILABILITY 50 mg, 250 mg tablets; 250 mg/5 mL suspension

ACTION & *THERAPEUTIC EFFECT* Antiepileptic properties result from raising the seizure threshold and changing seizure patterns. *Effective as an anticonvulsant in all types of seizure disorders except absence seizures.*

USES Alone or concomitantly with other anticonvulsant agents in the prophylactic management of complex partial (psychomotor) and generalized tonic-clonic (grand mal) seizures.

UNLABELED USES Essential tremor.

CONTRAINDICATIONS Hypersensitivity to barbiturates, porphyria; ethanol intoxication, hepatic encephalopathy, suicidal ideation; pregnancy (category D).

CAUTIOUS USE Chronic lung disease, sleep apnea; liver or kidney disease, dialysis; hyperactive children; mental status changes, major depression, suicidal tendencies; lactation; children.

Common adverse effects in *italic*, life-threatening effects <u>underlined</u>; generic names in **bold**; classifications in SMALL CAPS; ♦ Canadian drug name; ❂ Prototype drug

1249

ROUTE & DOSAGE

Seizures

Adult/Child (8 y or older): **PO** 250 mg/day, increased by 250 mg/wk (max: 2 g in 2–4 divided doses)
Child (Younger than 8 y): **PO** 125 mg/day, increased by 125 mg/wk (max: 2 g/day in 2–4 divided doses)

ADMINISTRATION

Oral

- Give whole or crush with fluid of patient's choice.
- Give with food if drug causes GI distress.

ADVERSE EFFECTS (≥1%) CNS:

Drowsiness, sedation, vertigo, ataxia, headache, excitement (children), confusion, unusual fatigue, hyperirritability, emotional disturbances, acute psychoses (usually patients with psychomotor epilepsy). **Special Senses:** Diplopia, nystagmus, swelling of eyelids. **GI:** *Nausea, vomiting, anorexia.* **Hematologic:** <u>Leukopenia, thrombocytopenia,</u> eosinophilia, decreased serum folate levels, megaloblastic anemia (rare). **Skin:** Alopecia, maculopapular or morbilliform rash, edema, lupus erythematosus-like syndrome. **Urogenital:** Impotence. **Body as a Whole:** Lymphadenopathy, osteomalacia.

INTERACTIONS Drug: Alcohol,

CNS DEPRESSANTS compound CNS depression; **phenobarbital** may decrease absorption and increase metabolism of ORAL ANTICOAGULANTS; increases metabolism of CORTICOSTEROIDS, ORAL CONTRACEPTIVES, ANTICONVULSANTS, **digitoxin,** possibly decreasing their effects; ANTIDEPRESSANTS potentiate adverse effects of primidone; **griseofulvin** decreases absorption of primidone. **Herbal: Kava, valerian** may potentiate sedation.

PHARMACOKINETICS Absorption:

Approximately 60–80% from GI tract. **Peak:** 4 h. **Distribution:** Distributed into breast milk. **Metabolism:** In liver to phenobarbital and PEMA. **Elimination:** In urine. **Half-Life:** Primidone 3–24 h, PEMA 24–48 h; phenobarbital 72–144 h.

NURSING IMPLICATIONS

Assessment & Drug Effects

- Lab tests: Perform baseline and periodic CBC, complete blood chemistry (q6mo), and primidone blood levels. (Therapeutic blood level for primidone: 5–10 mcg/mL.)
- Monitor primidone plasma levels (concentrations of primidone greater than 10 mcg/mL are usually associated with significant ataxia and lethargy).
- Therapeutic response may not be evident for several weeks.
- Observe for S&S of folic acid deficiency: Mental dysfunction, psychiatric disorders, neuropathy, megaloblastic anemia. Determine serum folate levels if indicated.

Patient & Family Education

- Avoid driving and other potentially hazardous activities during beginning of treatment because drowsiness, dizziness, and ataxia may be severe. Symptoms tend to disappear with continued therapy; if they persist, dosage reduction or drug withdrawal may be necessary.
- Avoid alcohol and other CNS depressants unless otherwise directed by prescriber.
- Do not take OTC medications unless approved by prescriber.
- Pregnant women should receive prophylactic vitamin K therapy

P

for 1 mo prior to and during delivery to prevent neonatal hemorrhage.
- Withdraw primidone gradually to avoid precipitating status epilepticus.

PROBENECID

(proe-ben'e-sid)

Benemid, Benuryl ♣, Probalan, SK-Probenecid
Classifications: URICOSURIC; ANTIGOUT
Therapeutic: ANTIGOUT
Pregnancy Category: B

AVAILABILITY 0.5 g tablet

ACTION & *THERAPEUTIC EFFECT*

Competitively inhibits renal tubular reabsorption of uric acid, thereby promoting its excretion and reducing serum urate levels. *Prevents formation of new tophaceous deposits and uric acid buildup in the serum and tissues. As an additive to penicillin, it increases serum concentration of penicillins and prolongs their serum concentration.*

USES Hyperuricemia in chronic gouty arthritis and tophaceous gout.

UNLABELED USES Adjuvant to therapy with penicillin G and penicillin analogs to elevate and prolong plasma concentrations of these antibiotics; to promote uric acid excretion in hyperuricemia secondary to administration of thiazides and related diuretics, furosemide, ethacrynic acid, pyrazinamide.

CONTRAINDICATIONS Blood dyscrasias; uric acid kidney stones; during or within 2–3 wk of acute gouty attack; overexcretion of uric

acid (greater than 1000 mg/day); patients with creatinine clearance less than 50 mg/min; use with penicillin in presence of known renal impairment; use for hyperuricemia secondary to cancer chemotherapy. **CAUTIOUS USE** History of peptic ulcer; pregnancy (category B), lactation; children younger than 2 y.

ROUTE & DOSAGE

Gout
Adult: **PO** 250 mg b.i.d. for 1 wk, then 500 mg b.i.d. (max: 3 g/day)
Adjunct for Penicillin or Cephalosporin Therapy
Adult: **PO** 500 mg q.i.d. or 1 g with single dose therapy (e.g., gonorrhea)
Child (2–14 y or weight less than 50 kg): **PO** 25–40 mg/kg/day in 4 divided doses

ADMINISTRATION

Oral
- Therapy is usually not initiated during an acute gouty attack. Consult prescriber.
- Minimize GI adverse effects by giving after meals, with food, milk, or antacid (prescribed). If symptoms persist, dosage reduction may be required.
- Give with a full glass of water if not contraindicated.
- Be aware that prescriber may prescribe concurrent prophylactic doses of colchicine for first 3–6 mo of therapy because frequency of acute gouty attacks may increase during first 6–12 mo of therapy.

ADVERSE EFFECTS (≥1%) Body as a Whole: Flushing, dizziness, fever, anaphylaxis. **CNS:** *Headache.* **GI:** *Nausea, vomiting, anorexia,* sore gums, hepatic necrosis (rare).

Common adverse effects in *italic*, life-threatening effects underlined; generic names in **bold**; classifications in SMALL CAPS; ♣ Canadian drug name; Prototype drug

1251

Urogenital: Urinary frequency. **Hematologic:** Anemia, <u>hemolytic anemia (possibly related to G6PD deficiency), aplastic anemia (rare)</u>. **Musculoskeletal:** Exacerbations of gout, uric acid kidney stones. **Skin:** Dermatitis, pruritus. **Respiratory:** <u>Respiratory depression</u>.

DIAGNOSTIC TEST INTERFERENCE
False-positive ***urine glucose*** tests are possible with ***Benedict's solution*** or ***Clinitest*** [***glucose oxidase methods*** not affected (e.g., ***Clinistix, TesTape***)].

INTERACTIONS Drug: SALICYLATES may decrease uricosuric activity; may decrease **methotrexate** elimination, causing increased toxicity; decreases **nitrofurantoin** efficacy and increases its toxicity. Decreases clearance of PENICILLINS, CEPHALOSPORINS, and NSAIDS.

PHARMACOKINETICS Absorption: Readily from GI tract. **Onset:** 30 min. **Peak:** 2–4 h. **Duration:** 8 h. **Distribution:** Crosses placenta. **Metabolism:** In liver. **Elimination:** In urine. **Half-Life:** 4–17 h.

NURSING IMPLICATIONS
Assessment & Drug Effects
- Lab tests: Periodic serum urate levels, Hct and Hgb, and urinalysis. Determine acid-base balance periodically when urinary alkalinizers are used.
- Patients taking sulfonylureas may require dosage adjustment. Probenecid enhances hypoglycemic actions of these drugs (see DIAGNOSTIC TEST INTERFERENCES).
- Expect urate tophaceous deposits to decrease in size. Classic locations are in cartilage of ear pinna and big toe, but they can occur in bursae, tendons, skin, kidneys, and other tissues.

Patient & Family Education
- Drink fluid liberally (approximately 3000 mL/day) to maintain daily urinary output of at least 2000 mL or more. This is important because increased uric acid excretion promoted by drug predisposes to renal calculi.
- Prescriber may advise restriction of high-purine foods during early therapy until uric acid level stabilizes. Foods high in purine include organ meats (sweetbreads, liver, kidney), meat extracts, meat soups, gravy, anchovies, and sardines. Moderate amounts are present in other meats, fish, seafood, asparagus, spinach, peas, dried legumes, wild game.
- Avoid alcohol because it may increase serum urate levels.
- Do not stop taking drug without consulting prescriber. Irregular dosage schedule may sharply elevate serum urate level and precipitate acute gout.
- Report symptoms of hypersensitivity to prescriber. Discontinuation of drug is indicated.
- Do not take aspirin or other OTC medications without consulting prescriber. If a mild analgesic is required, acetaminophen is usually allowed.

PROCAINAMIDE HYDROCHLORIDE ℗
(proe-kane-a′mide)
Pronestyl, Pronestyl SR
Classification: CLASS IA ANTIARRHYTHMIC
Therapeutic: CLASS IA ANTIARRHYTHMIC
Pregnancy Category: C

AVAILABILITY 250 mg, 375 mg, 500 mg tablets, capsules; 250 mg, 500 mg, 750 mg, 1000 mg sustained release tablets; 100 mg/mL, 500 mg/mL injection

ACTION & *THERAPEUTIC EFFECT*

Potent class IA antiarrhythmic that depresses excitability of myocardium to electrical stimulation, reduces conduction velocity in atria, ventricles, and His-Purkinje system. Produces peripheral vasodilation and hypotension, especially with IV use. *Effectively used for atrial arrhythmias; suppresses automaticity of His-Purkinje ventricular muscle.*

USES Prophylactically to maintain normal sinus rhythm following conversion of atrial flutter or fibrillation by other methods; to prevent recurrence of paroxysmal atrial fibrillation and tachycardia, paroxysmal AV junctional rhythm, ventricular tachycardia, ventricular and atrial premature contractions. Also cardiac arrhythmias associated with surgery and anesthesia.

UNLABELED USES Malignant hyperthermia.

CONTRAINDICATIONS Hypersensitivity to procainamide or procaine, yellow dye 5 (tartrazine); blood dyscrasias; bundle branch block; complete AV block, second and third degree AV block unassisted by pacemaker; QT prolongation, torsades de pointes; non-life-threatening ventricular arrhythmias; leukopenia or agranulocytosis; SLE; concurrent use with other antiarrhythmic agents; myasthenia gravis.

CAUTIOUS USE Patient who has undergone electrical conversion to sinus rhythm; first-degree heart block; bone marrow suppression or cytopenia; hypotension, cardiac enlargement, CHF, MI, ischemic heart disease; coronary occlusion, ventricular dysrhythmia from digitalis intoxication, ventricular arrhythmias;

hepatic or renal insufficiency; electrolyte imbalance; bronchial asthma; aspirin hypersensitivity; cytopenia; pregnancy (category C), lactation.

ROUTE & DOSAGE

Arrhythmias

Adult: **PO** 50 mg/kg/day in divided doses (b.i.d. for Procanbid); max: 5 g/day **IM** 0.5–1 g q4–8h until able to take PO **IV** 100 mg q5min at a rate of 25–50 mg/min until arrhythmia is controlled or 1 g given, then 2–6 mg/min
Child: **PO** 15–50 mg/kg/day divided q3–6h **IM** 50 mg/kg/day divided q3–6h until PO tolerated **IV** 3–6 mg/kg q10–30min (max: 100 mg/dose), then 20–80 mcg/kg/min

Renal Impairment Dosage Adjustment

Oral doses CrCl 10–50 mL/min: Give q6–12h; *less than 10 mL/min:* Give q8–24h. **IV doses** Reduce loading dose to 12 mg/kg, then maintenance by $^1/_3$ to $^2/_3$.

Hemodialysis Dosage Adjustment

Give 200 mg supplemental dose post dialysis

ADMINISTRATION

Oral

- Give first PO dose at least 4 h after last IV dose.
- Give oral preparation on empty stomach, 1 h before or 2 h after meals, with a full glass of water to enhance absorption. If drug causes gastric distress, give with food.
- Crush immediate release (but NOT sustained release) tablet if patient is unable to swallow it whole.

Common adverse effects in *italic*, life-threatening effects <u>underlined</u>; generic names in **bold**; classifications in SMALL CAPS; ♣ Canadian drug name; ● Prototype drug

- Swallow sustained release tablet whole. It has a wax matrix that is not absorbed but appears in the stool.

Intramuscular

- IM route should be used only when IV route is not feasible.

Intravenous

Use IV route for emergency situations.

PREPARE: **Direct:** Dilute each 100 mg with 5–10 mL of D5W or sterile water for injection. **IV Infusion:** Add 1 g of procainamide to 250–500 mL of D5W solution to yield 4 mg/mL in 250 mL or 2 mg/mL in 500 mL.

ADMINISTER: **Direct:** Usual rate is 20 mg/min. Faster rates (up to 50 mg/min) should be used with caution. **IV Infusion for Adult:** 2–6 mg/min. **IV Infusion for Child:** 20–80 mcg/kg/min. ▪ Control IV administration over several hours by assessment of procainamide plasma levels. ▪ Use an infusion pump with constant monitoring. ▪ Keep patient in supine position. ▪ Be alert to signs of too rapid administration of drug (speed shock: Irregular pulse, tight feeling in chest, flushed face, headache, loss of consciousness, shock, cardiac arrest).

INCOMPATIBILITIES **Solution/additive: Bretylium, esmolol, ethacrynate, milrinone, phenytoin. Y-site: Milrinone acetate (amrinone), milrinone.**

- Store solution for up to 24 h at room temperature and for 7 days under refrigeration at 2°–8° C (36°–46° F). ▪ Slight yellowing does not alter drug potency, but discard solution if it is markedly discolored or precipitated.

ADVERSE EFFECTS (≥1%) **CNS:** Dizziness, psychosis. **CV:** Severe

hypotension, pericarditis, <u>ventricular fibrillation</u>, AV block, tachycardia, flushing. **GI:** Bitter taste, nausea, vomiting, diarrhea, anorexia, (all mostly PO). **Hematologic:** <u>Agranulocytosis with repeated use</u>; <u>thrombocytopenia</u>. **Body as a Whole:** Fever, muscle and joint pain, angioneurotic edema, myalgia, *SLE-like syndrome (50% of patients on large doses for 1 y):* Polyarthralgias, pleuritic pain, pleural effusion. **Skin:** Maculopapular rash, pruritus. erythema, skin rash.

DIAGNOSTIC TEST INTERFERENCE

It may alter results of the *edrophonium test.*

INTERACTIONS Drug: Other AN-TIARRHYTHMICS add to therapeutic and toxic effects; ANTICHOLINERGIC AGENTS compound anticholinergic effects; ANTIHYPERTENSIVES add to hypotensive effects; **cimetidine** may increase levels with increase in toxicity.

PHARMACOKINETICS Absorption: 75–95% from GI tract. **Peak:** 15–60 min IM; 30–60 min PO. **Duration:** 3 h; 8 h with sustained release. **Distribution:** Distributed to CSF, liver, spleen, kidney, brain, and heart; crosses placenta; distributed into breast milk. **Metabolism:** In liver to *N*-acetyl-procainamide (NAPA), an active metabolite (30–60% metabolized to NAPA). **Elimination:** In urine. **Half-Life:** 3 h procainamide, 6 h NAPA.

NURSING IMPLICATIONS

Assessment & Drug Effects

- Check apical radial pulses before each dose during period of adjustment to the oral route.
- Patients with severe heart, liver, or kidney disease and hypotension are at particular risk for adverse effects.
- Monitor the patient's ECG and BP continuously during IV drug administration.

Common adverse effects in *italic*, life-threatening effects <u>underlined</u>; generic names in **bold;** classifications in SMALL CAPS; ✤ Canadian drug name; ⊘ Prototype drug

- Discontinue IV drug temporarily when (1) arrhythmia is interrupted, (2) severe toxic effects are present, (3) QRS complex is excessively widened (greater than 50%), (4) PR interval is prolonged, or (5) BP drops 15 mm Hg or more. Obtain rhythm strip and notify prescriber.
- Ventricular dysrhythmias are usually abolished within a few minutes after IV dose and within an hour after PO or IM administration.
- Report promptly complaints of chest pain, dyspnea, and anxiety. Digitalization may have preceded procainamide in patients with atrial arrhythmias. Cardiotonic glycosides may induce sufficient increase in atrial contraction to dislodge atrial mural emboli, with subsequent pulmonary embolism.
- Therapeutic procainamide blood levels are reached in approximately 24 h if kidney function is normal but are delayed in presence of renal impairment.

Patient & Family Education

- Keep a record of weekly weight. Notify prescriber if weight gain of 1 kg (2 lb) or more is accompanied by local edema.
- Record and report date, time, and duration of fibrillation episodes when taking maintenance doses: Light-headedness, giddiness, weakness, or faintness.
- Keep a record of pulse rates. Report to prescriber changes in rate or quality.
- Report to prescriber signs of reduced procainamide control: Weakness, irregular pulse, unexplained fatigability, anxiety.
- Do not double dose or change an interval because a previous dose was missed. Take procainamide at evenly spaced intervals around the clock unless otherwise prescribed.

PROCAINE HYDROCHLORIDE ℗

(proe'kane)

Novocain

Classification: LOCAL ANESTHETIC (ESTER-TYPE)

Therapeutic: LOCAL ANESTHETIC

Pregnancy Category: C

AVAILABILITY 1%, 10% injection

ACTION & *THERAPEUTIC EFFECT* Decreases sodium flux into nerve cell, thus depressing initial depolarization and preventing propagation and conduction of the nerve impulse. *Local anesthetic action produces loss of sensation and motor activity in circumscribed areas that are treated.*

USES Spinal anesthesia and epidural and peripheral nerve block by injection and infiltration methods.

CONTRAINDICATIONS Known hypersensitivity to procaine or to other drugs of similar chemical structure, to PABA, and to parabens; generalized septicemia, inflammation, or sepsis at proposed injection site; cerebrospinal diseases (e.g., meningitis, syphilis); heart block, hypotension, hypertension; bowel pathology, GI hemorrhage; coagulopathy, anticoagulants, *thrombocytopenia.*

CAUTIOUS USE Debilitated, acutely ill patients; obstetric delivery; increased intra-abdominal pressure; known drug allergies and sensitivities; impaired cardiac function, dysrhythmias; shock; older adults; pregnancy (category C), lactation.

ROUTE & DOSAGE

Spinal Anesthesia

Adult: **Intrathecal** 10% solution diluted with NS at 1 mL/5 sec

Common adverse effects in *italic*, life-threatening effects <u>underlined</u>; generic names in **bold;** classifications in SMALL CAPS; ◆ Canadian drug name; ℗ Prototype drug

1255

Infiltration Anesthesia/ Peripheral Nerve Block
Adult: **Regional** 0.25–0.5% solution

ADMINISTRATION

Subcutaneous

- Reconstitute solution: To prepare 60 mL of a 0.5% solution (5 mg/mL), dilute 30 mL of 1% solution with 30 mL sterile distilled water. Add 0.5–1 mL epinephrine 1:1000/ 100 mL anesthetic solution for vasoconstrictive effect (1:200,000– 1:100,000).
- Do not use solutions that are cloudy, discolored, or that contain crystals. Discard unused portion of solutions not containing a preservative. Avoid use of solution with preservative for spinal, epidural, or caudal block.
- With subcutaneous administration, inject slowly with frequent aspirations to avoid inadvertent intravascular administration, which can lead to a systemic reaction.

INCOMPATIBILITIES **Solution/additive: Aminophylline, amobarbital, chlorothiazide, magnesium sulfate, phenobarbital, phenytoin, secobarbital, sodium bicarbonate.**

ADVERSE EFFECTS (≥1%) **CNS:** Anxiety, nervousness, dizziness, circumoral paresthesia, tremors, drowsiness, sedation, convulsions, respiratory arrest. With spinal anesthesia: postspinal headache, arachnoiditis, palsies, spinal nerve paralysis, meningism. **Special Senses:** Tinnitus, blurred vision. **CV:** Myocardial depression, arrhythmias including bradycardia (also fetal bradycardia); hypotension. **GI:** Nausea, vomiting. **Skin:** Cutaneous lesions of delayed onset, urticaria, pruritus, angioneurotic edema, sweating, syncope, anaphylactoid reaction. **Urogenital:** Urinary retention, fecal or urinary incontinence, loss of perineal sensation and sexual function, slowing of labor and increased incidence of forceps delivery (all with caudal or epidural anesthesia).

INTERACTIONS Drug: May antagonize effects of SULFONAMIDES; increased risk of hypotension with MAOIS, ANTIHYPERTENSIVES.

PHARMACOKINETICS Absorption: Rapidly from injection site. **Onset:** 2–5 min. **Duration:** 1 h. **Metabolism:** Hydrolyzed by plasma pseudocholinesterases. **Elimination:** 80% of metabolites excreted in urine. **Half-Life:** 7.7 min.

NURSING IMPLICATIONS

Assessment & Drug Effects

- Be aware that reactions during dental procedure are usually mild, transient, and produced by epinephrine added to local anesthetic (e.g., headache, palpitation, tachycardia, hypertension, dizziness).
- Use procaine with epinephrine with caution in body areas with limited blood supply (e.g., fingers, toes, ears, nose). If used, inspect particular area for evidence of reduced perfusion (vasospasm): Pale, cold, sensitive skin.
- Hypotension is the most important complication of spinal anesthesia. Risk period is during first 30 min after induction and is intensified by changes in position that promote decreased venous return, or by preexisting hypertension, pregnancy, old age, or hypovolemia.

Patient & Family Education

- Understand that there will be temporary loss of sensation in the area of the injection.
- Do not consume hot liquids or foods until sensation returns when drug used for dental procedure.

PROCARBAZINE HYDROCHLORIDE

(proe-kar'ba-zeen)

Matulane, Natulan ✦

Classifications: ANTINEOPLASTIC; ALKYLATING AGENT

Therapeutic: ANTINEOPLASTIC

Prototype: Cyclophosphamide

Pregnancy Category: D

AVAILABILITY 50 mg capsules

ACTION & *THERAPEUTIC EFFECT*

Hydrazine derivative with anti-metabolite properties that is cell cycle–specific for the S phase of cell division. Suppresses mitosis at interphase, and causes chromatin derangement. *Highly toxic to rapidly proliferating tissue.*

USES Adjunct in palliative treatment of Hodgkin's disease.

UNLABELED USES Solid tumors.

CONTRAINDICATIONS Severe myelosuppression; pheochromocytoma; alcohol ingestion; foods high in tyramine content; sympathomimetic drugs. MAO inhibitors should be discontinued 14 days prior to therapy; tricyclic antidepressants discontinued 7 days before therapy; pregnancy (category D), lactation.

CAUTIOUS USE Hepatic or renal impairment; cardiac disease; bipolar disorder, mania, paranoid schizophrenia; G6PD deficiency; parkinsonism; following radiation or chemotherapy before at least 1 mo has elapsed; alcoholism; infection; DM.

ROUTE & DOSAGE

Adjunct for Hodgkin's Disease

Adult: **PO** 2–4 mg/kg/day in single or divided doses for 1 wk, then 4–6 mg/kg/day until WBC less than 4000/mm³ or platelets are less than 100,000/mm³ or maximum response obtained; drug is then discontinued until bone marrow recovery is satisfactory; treatment is started again at 1–2 mg/kg/day

Child: **PO** 50 mg/m²/day in single or divided doses for 1 wk, then 100 mg/m²/day until WBC is less than 4000/mm³ or platelets are less than 100,000/mm³ or maximum response obtained; drug is then discontinued until bone marrow recovery is satisfactory; treatment is started again at 50 mg/m²/day

ADMINISTRATION

Oral

- Do not give if WBC count is less than 4000/mm³ or platelet count is less than 100,000/mm³. Consult prescriber.
- Store at 15°–30° C (59°–86° F). Protect from freezing, moisture, and light.

ADVERSE EFFECTS (≥1%) **CNS:** Myalgia, arthralgia, paresthesias, weakness, fatigue, lethargy, drowsiness, neuropathies, mental depression, acute psychosis, hallucinations, dizziness, headache, ataxia, nervousness, insomnia, coma, confusion, seizures. **GI:** *Severe nausea and vomiting,* anorexia, stomatitis, dry mouth, dysphagia, diarrhea, constipation, jaundice, ascites. **Hematologic:** Bone marrow suppression (leukopenia, anemia, thrombocytopenia),

Common adverse effects in *italic*, life-threatening effects underlined; generic names in **bold;** classifications in SMALL CAPS; ✦ Canadian drug name; ❂ Prototype drug

1257

hemolysis, bleeding tendencies. **Skin:** Dermatitis, pruritus, herpes, hyperpigmentation, flushing, alopecia. **Respiratory:** *Pleural effusion, cough,* hoarseness. **CV:** Hypotension, tachycardia. **Body as a Whole:** Chills, fever, sweating, photosensitivity; intercurrent infections. **Urogenital:** Gynecomastia, depressed spermatogenesis, atrophy of testes.

INTERACTIONS Drug: **Alcohol,** PHENOTHIAZINES, and other CNS DEPRESSANTS add to CNS depression; TRICYCLIC ANTIDEPRESSANTS, MAO INHIBITORS, SYMPATHOMIMETICS, **ephedrine, phenylpropanolamine** may precipitate hypertensive crisis, hyperpyrexia; seizures, or death. Procarbazine may enhance the effects of **CNS depressants.** A disulfiram-like reaction may occur following ingestion of **alcohol. Food: Tyramine**-containing foods may precipitate hypertensive crisis [see **phenelzine sulfate** (MAO INHIBITOR)].

PHARMACOKINETICS Absorption: Readily from GI tract. **Peak:** 1 h. **Distribution:** Widely distributed with high concentrations in liver, kidneys, intestinal wall, and skin. **Metabolism:** In liver. **Elimination:** In urine. **Half-Life:** 1 h.

NURSING IMPLICATIONS

Assessment & Drug Effects

- Monitor baseline and periodic BP, weight, temperature, pulse, and I&O ratio and pattern.
- Lab tests: Determine hematologic status (Hgb, Hct, WBC with differential, reticulocyte, and platelet counts) initially and at least q3–4days. Hepatic and renal studies (transaminase, alkaline phosphatase, BUN, urinalysis) are also indicated initially and at least weekly during therapy.

- Protect patient from exposure to infection and trauma when nadir of leukopenia is approached. Note and report changes in voiding pattern, hematuria, and dysuria (possible signs of urinary tract infection). Monitor I&O ratio and temperature closely.
- Withhold drug and notify prescriber of any of the following: CNS S&S (e.g., paresthesias, neuropathies, confusion); leukopenia [WBC count less than 4000/mm³; thrombocytopenia (platelet count less than 100,000/mm³)]; hypersensitivity reaction, the first small ulceration or persistent spot of soreness in oral cavity, diarrhea, and bleeding.
- Monitor for and report any of the following: Chills, fever, weakness, shortness of breath, productive cough. Drug will be discontinued.
- Assess for signs of liver dysfunction: Jaundice (yellow skin, sclerae, and soft palate), frothy or dark urine, clay-colored stools.
- Tolerance to nausea and vomiting (most common adverse effects) usually develops by end of first week of treatment. Doses are kept at a minimum during this time. If vomiting persists, therapy will be interrupted.

Patient & Family Education

- Avoid OTC nose drops, cough medicines, and antiobesity preparations containing ephedrine, amphetamine, epinephrine, and tricyclic antidepressants because they may cause hypertensive crises. Do not to use OTC preparations without prescriber's approval.
- Report to prescriber any sign of impending infection.
- Do not eat foods high in tyramine content (e.g., aged cheese, beer, wine).

Common adverse effects in *italic*, life-threatening effects underlined; generic names in **bold;** classifications in SMALL CAPS; ♣ Canadian drug name; ⊙ Prototype drug

- Avoid alcohol; ingestion of any form of alcohol may precipitate a disulfiram-type reaction (see Appendix F).
- Report to prescriber immediately signs of hemorrhagic tendencies: Bleeding into skin and mucosa, easy bruising, nose bleeds, or blood in stool or urine. Bone marrow depression often occurs 2–8 wk after start of therapy.
- Avoid excessive exposure to the sun because of potential photosensitivity reaction: Cover as much skin area as possible with clothing, and use sunscreen lotion (SPF higher than 12) on all exposed skin surfaces.
- Use caution while driving or performing hazardous tasks until response to drug is known since drowsiness, dizziness, and blurred vision are possible adverse effects.
- Use contraceptive measures during procarbazine therapy.

PROCHLORPERAZINE ⊕
(proe-klor-per′a-zeen)
Compazine, Compro

PROCHLORPERAZINE EDISYLATE
Compazine

PROCHLORPERAZINE MALEATE
Compazine, Stemetil ✦
Classifications: ANTIPSYCHOTIC; PHENOTHIAZINE; ANTIEMETIC
Therapeutic: ANTIPSYCHOTIC; ANTIEMETIC
Pregnancy Category: C

AVAILABILITY 5 mg, 10 mg, 25 mg tablets; 10 mg, 15 mg, 30 mg sustained release capsule; 2.5 mg, 5 mg, 25 mg suppositories; 5 mg/mL injection; **Edisylate:** 5 mg/mL injection

ACTION & *THERAPEUTIC EFFECT*
Strong antipsychotic effects thought to be due to blockade of postsynaptic dopamine receptors in the brain. Antiemetic effect is produced by suppression of the chemoreceptor trigger zone (CTZ). *Effective antipsychotic and antiemetic properties.*

USES Management of manifestations of psychotic disorders, of excessive anxiety, tension, and agitation, and to control severe nausea and vomiting.
UNLABELED USES Behavioral syndromes in dementia.

CONTRAINDICATIONS Hypersensitivity to phenothiazines; bone marrow depression; blood dyscrasias, jaundice; comatose or severely depressed states; dementia-related psychosis in elderly; children weighing less than 9 kg (20 lb) or younger than 2 y of age; pediatric surgery; short-term vomiting in children or vomiting of unknown etiology; Reye's syndrome or other encephalopathies; history of dyskinetic reactions or epilepsy; lactation. **CAUTIOUS USE** Patient with previously diagnosed breast cancer, children with acute illness or dehydration; Parkinson's disease; GI obstruction; hepatic disease; seizure disorders; urinary retention, BPH; pregnancy (category C).

ROUTE & DOSAGE

Severe Nausea, Vomiting
Adult: **PO** 5–10 mg 3–4 times/ day; sustained release: 10–15 mg q12h **IM** 5–10 mg q3–4h up to 40 mg/day **IV** 2.5–10 mg q3–4h (max: 40 mg/day) **PR** 25 mg b.i.d.

Common adverse effects in *italic*, life-threatening effects <u>underlined</u>; generic names in **bold;** classifications in SMALL CAPS; ✦ Canadian drug name; ⊕ Prototype drug

1259

Child (weight greater than 9 kg):
PO/PR 2.5 mg 1–3 times/day or
5 mg b.i.d. (max: 15 mg/day)
IM 0.13 mg/kg q3–4h

Psychotic Disorders

Adult: **PO** 5–10 mg 3–4 times/
day; titrate up q2–3days **IM** 10–20
mg; may repeat q1–4h to gain
control, then q4–6h
Child (2–12 y): **PO/PR** 2.5 mg
2–3 times/day (max: 20 mg daily
ages 2–5 and 25 mg daily ages
6–12)

ADMINISTRATION

Oral

- Dosages for older adults, emaciated patients and children should be increased slowly.
- Ensure that sustained release form is not chewed or crushed. **Must be** swallowed whole.
- Do not give oral concentrate to children.
- Avoid skin contact with oral concentrate or injection solution because of possibility of contact dermatitis.

Rectal

- Ensure that suppository is inserted beyond the anal sphincter.

Intramuscular

- Do not inject drug subcutaneously.
- Make injection deep into the upper outer quadrant of the buttock in adults. Follow agency policy regarding IM injection site for children.

Intravenous

PREPARE: Direct: May be given undiluted or diluted in small amounts of NS. **IV Infusion:** Dilute in 50–100 mL of D5W, NS, D5/0.45% NaCl, LR or other compatible solution.

ADMINISTER: Direct: DO NOT give a bolus dose. Give at a maximum rate of 5 mg/min. **IV Infusion:** Give over 15–30 min. Do not exceed direct IV rate.
INCOMPATIBILITIES Solution/additive: **Aminophylline, amphotericin B, ampicillin, calcium gluconate, cephalothin, chloramphenicol, chlorothiazide, dimenhydrinate, epinephrine, erythromycin, furosemide, hydrocortisone, hydromorphone, kanamycin, ketorolac, methohexital, midazolam, morphine, penicillin G sodium, pentobarbital, phenobarbital, tetracycline, thiopental, vancomycin, warfarin.** Y-site: **Aldesleukin, allopurinol, amifostine, amphotericin B cholesteryl complex, aztreonam, bivalirudin, cefepime, etoposide, fenoldopam, filgrastim, fludarabine, foscarnet, gemcitabine, piperacillin-tazobactam.**

- Discard markedly discolored solutions; slight yellowing does not appear to alter potency.

ADVERSE EFFECTS (≥1%) CNS:

Drowsiness, dizziness, *extrapyramidal reactions (akathisia, dystonia, or parkinsonism),* <u>persistent tardive dyskinesia</u>, acute catatonia. **CV:** Hypotension. **GI:** Cholestatic jaundice. **Skin:** Contact dermatitis, photosensitivity. **Endocrine:** Galactorrhea, amenorrhea. **Special Senses:** Blurred vision. **Hematologic:** <u>*Leukopenia,* agranulocytosis</u>.

INTERACTIONS Drug: Alcohol,

CNS DEPRESSANTS increase CNS depression; ANTACIDS, ANTIDIARRHEALS decrease absorption, therefore, administer 2 h apart; **phenobarbital** increases metabolism of prochlorperazine; GENERAL ANESTHETICS

Common adverse effects in *italic*; life-threatening effects <u>underlined</u>; generic names in **bold**; classifications in SMALL CAPS; ✤ Canadian drug name; ⦿ Prototype drug

increase excitation and hypotension; antagonizes antihypertensive action of **guanethidine; phenylpropanolamine** poses possibility of sudden death; TRICYCLIC ANTIDEPRESSANTS intensify hypotensive and anticholinergic effects; decreases seizure threshold—ANTICONVULSANT dosage may need to be increased. **Herbal: Kava** may increase risk and severity of dystonic reactions.

PHARMACOKINETICS Absorption: Readily from GI tract. **Onset:** 30–40 min PO; 60 min PR; 10–20 min IM. **Duration:** 3–4 h PO; 10–12 h sustained release PO; 3–4 h PR; up to 12 h IM. **Distribution:** Crosses placenta; distributed into breast milk. **Metabolism:** In liver. **Elimination:** In urine.

NURSING IMPLICATIONS

Assessment & Drug Effects

- Position carefully to prevent aspiration of vomitus; may have depressed cough reflex.
- Most older adult and emaciated patients and children, especially those with dehydration or acute illness, appear to be particularly susceptible to extrapyramidal effects. Be alert to onset of symptoms: Early in therapy watch for pseudoparkinson's and acute dyskinesia. After 1–2 mo, be alert to akathisia.
- Keep in mind that the antiemetic effect may mask toxicity of other drugs or make it difficult to diagnose conditions with a primary symptom of nausea.
- Lab tests: Periodic CBC with differential in long-term therapy.
- Be alert to signs of high core temperature: Red, dry, hot skin; full bounding pulse; dilated pupils; dyspnea; confusion; temperature over 40.6° C (105° F); elevated BP. Exposure to high environmental temperature places this patient at risk for heat stroke. Inform prescriber and institute measures to reduce body temperature rapidly.

Patient & Family Education

- Take drug only as prescribed and do not alter dose or schedule. Consult prescriber before stopping the medication.
- Avoid hazardous activities such as driving a car until response to drug is known because drug may impair mental and physical abilities, especially during first few days of therapy.
- Be aware that drug may color urine reddish brown. It also may cause the sun-exposed skin to turn gray-blue.
- Protect skin from direct sun's rays and use a sunscreen lotion (SPF higher than 12) to prevent photosensitivity reaction.
- Withhold dose and report to the prescriber if the following symptoms persist more than a few hours: Tremor, involuntary twitching, exaggerated restlessness. Other reportable symptoms include light-colored stools, changes in vision, sore throat, fever, rash.

P

PROGESTERONE ⊕

(proe-jess'ter-one)

Crinone Gel, Endrometrin, Gesterol 50, Progestaject, Prometrium

Classification: PROGESTIN
Therapeutic: PROGESTIN
Pregnancy Category: X; B (vaginal gel in first trimester)

AVAILABILITY 100 mg capsules; 50 mg/mL injection; 4%, 8% gel; 100 mg vaginal insert

Common adverse effects in *italic*; life-threatening effects underlined; generic names in **bold**; classifications in SMALL CAPS; ♣ Canadian drug name; ⊕ Prototype drug

1261

ACTION & *THERAPEUTIC EFFECT*

Has estrogenic, anabolic, and androgenic activity. Transforms endometrium from proliferative to secretory state; suppresses pituitary gonadotropin secretion, thereby blocking follicular maturation and ovulation. *Relaxes estrogen-primed myometrium and prohibits spontaneous contraction of uterus. Sudden drop in blood levels of progestin (and estradiol) causes "withdrawal bleeding" from endometrium. Intrauterine placement of progesterone hypothetically inhibits sperm survival, and suppresses endometrial proliferation (antiestrogenic effect).*

USES Secondary amenorrhea, functional uterine bleeding, endometriosis, and premenstrual syndrome. Largely supplanted by new progestins, which have longer action and oral effectiveness. Treatment of infertile women with progesterone deficiency.

CONTRAINDICATIONS Hypersensitivity to progestins, known or suspected breast or genital malignancy; use as a pregnancy test; thrombophlebitis, thromboembolic disorders; ectopic pregnancy; cerebral apoplexy (or its history); severely impaired liver function or disease; undiagnosed vaginal bleeding, incomplete abortion; use during first 4 mo of pregnancy (category X); other than vaginal gel use for assisted reproductive technology (ART) in early pregnancy.

CAUTIOUS USE Anemia; diabetes mellitus; history of psychic depression; persons susceptible to acute intermittent porphyria or with conditions that may be aggravated by fluid retention (asthma, seizure disorders, cardiac or kidney function, migraine); impaired liver function; previous ectopic pregnancy; presence or history of salpingitis; venereal disease; unresolved abnormal Pap smear; genital bleeding of unknown etiology; previous pelvic surgery; lactation. **Vaginal gel:** in early first trimester (category B).

ROUTE & DOSAGE

Amenorrhea

Adult: **IM** 5–10 mg for 6–8 consecutive days **PO** 400 mg at bedtime × 10 days **Vaginal gel** 45 mg every other day (up to 6 doses)

Uterine Bleeding

Adult: **IM** 5–10 mg/day for 6 days

Premenstrual Syndrome

Adult: **PR** 200–400 mg/day

Assisted Reproductive Technology

Adult: **Vaginally** 90 mg gel daily or b.i.d. until placental autonomy OR 10–12 wk; 100 mg insert 2–3 times daily up to 10 wk

ADMINISTRATION

Oral

- Give at bedtime and advise caution with ambulation because drug may cause drowsiness or dizziness.
- Do not give oral capsules, which contain peanut oil, to patients allergic to peanuts.

Intramuscular

- Immerse vial in warm water momentarily to redissolve crystals (if present) and to facilitate aspiration of drug into syringe.
- Inject deeply IM. Injection site may be irritated. Inspect IM sites carefully and rotate areas systematically.

Vaginal

- When given using the supplied applicators, a measured dose of 4% vaginal gel contains 45 mg and a measured dose of 8% gel contains 90 mg. Dosage increases from 4% to 8% cannot be achieved by an increase in volume of 4% gel; the 8% gel must be used to supply a 90 mg dose.

- Store drug at 15°–30° C (59°–86° F) unless otherwise specified by manufacturer. Protect from freezing and light.

ADVERSE EFFECTS (≥1%) **CNS:** Migraine headache, *dizziness,* lethargy, mental depression, somnolence, insomnia. **CV:** Thromboembolic disorder, pulmonary embolism. **Special Senses:** Change in vision, proptosis, diplopia, papilledema, retinal vascular lesions. **GI:** Hepatic disease, cholestatic jaundice; *nausea,* vomiting, *abdominal cramps.* **Urogenital:** Vaginal candidiasis, chloasma, cervical erosion and changes in secretions, *breakthrough bleeding,* dysmenorrhea, amenorrhea, pruritus vulvae. **Metabolic:** Hyperglycemia, decreased libido, transient increase in sodium and chloride excretion, pyrexia. **Skin:** *Acne,* pruritus, allergic rash, photosensitivity, urticaria, hirsutism, alopecia. **Body as a Whole:** *Edema, weight changes;* pain at injection site; fatigue. **Endocrine:** Gynecomastia, galactorrhea.

DIAGNOSTIC TEST INTERFERENCE PROGESTINS may decrease levels of *urinary pregnanediol* and increase levels of *serum alkaline phosphatase, plasma amino acids, urinary nitrogen,* and *coagulation factors VII, VIII, IX,* and *X.* They also decrease *glucose tolerance* (may cause false-positive *urine glucose tests*) and lower *HDL* (high-density lipoprotein) levels.

INTERACTIONS Drug: BARBITURATES, **carbamazepine, phenytoin,**
rifampin may alter contraceptive effectiveness; **ketoconazole** may inhibit progesterone metabolism; may antagonize effects of **bromocriptine.**

PHARMACOKINETICS Absorption: Rapid from IM site; PO peaks at 3 h. **Metabolism:** Extensively in liver. **Elimination:** Primarily in urine; excreted in breast milk. **Half-Life:** 5 min.

NURSING IMPLICATIONS

Assessment & Drug Effects
- Record baseline data for comparative value about patient's weight, BP, and pulse at onset of progestin therapy. Report deviations promptly.
- Lab tests: Periodic LFTs, blood glucose, and serum electrolytes.
- Monitor for and report immediately S&S of thrombophlebitis or thromboembolic disease.

Patient & Family Education
- Avoid exposure to UV light and prolonged periods of time in the sun. Photosensitivity severity is related to both time of exposure and dose. A phototoxic drug reaction usually looks like an exaggerated sunburn and occurs within 5–18 h after exposure to sun and is maximal by 36–72 h.
- Use sunscreen lotion (SPF higher than 12) on exposed skin surfaces whenever outdoors, even on dark days.
- Inform prescriber promptly if any of the following occur: Sudden severe headache or vomiting, dizziness or fainting, numbness in an arm or leg, pain in calves accompanied by swelling, warmth, and redness; acute chest pain or dyspnea.
- Report to prescriber promptly unexplained sudden or gradual, partial or complete loss of vision, ptosis, or diplopia.

P

Common adverse effects in *italic,* life-threatening effects underlined; generic names in **bold**; classifications in SMALL CAPS; ♣ Canadian drug name; ⊘ Prototype drug

1263

- Monitor for loss of glycemic control if diabetic.
- Notify prescriber if you become or suspect pregnancy. Learn the potential risk to the fetus from exposure to progestin.

✓PROMETHAZINE HYDROCHLORIDE

(proe-meth'a-zeen)

Histantil ✦, Phenergan
Classifications: ANTIHISTAMINE; ANTIEMETIC; ANTIVERTIGO
Therapeutic: ANTIHISTAMINE; ANTIEMETIC; ANTIVERTIGO
Prototype: Prochlorperazine
Pregnancy Category: C

AVAILABILITY 12.5 mg, 25 mg, 50 mg tablets; 6.25 mg/5 mL syrup; 12.5 mg, 25 mg, 50 mg suppositories; 25 mg/mL, 50 mg/mL injection

ACTION & *THERAPEUTIC EFFECT*
An antihistamine that exerts antiserotonin, anticholinergic, and local anesthetic action. Antiemetic action thought to be due to depression of CTZ in medulla. *Long-acting derivative of phenothiazine with marked antihistamine activity and prominent sedative, amnesic, antiemetic, and anti-motion sickness actions.*

USES Motion sickness, nausea/vomiting, induction of sedation, pruritus.
UNLABELED USES Allergic rhinitis, hyperemesis gravidarum, nystagmus.

CONTRAINDICATIONS Hypersensitivity to phenothiazines; acute MI; angina, atrial fibrillation, atrial flutter, cardiac arrhythmias, cardiomyopathy, uncontrolled hypertension; MAOI therapy; comatose or severely depressed states; hepatic encephalopathy, hepatic diseases; acutely ill or dehydrated children; children with Reye's syndrome; lactation.

CAUTIOUS USE Impaired liver function; epilepsy; bone marrow depression; cardiovascular disease; peripheral vascular disease; asthma; acute or chronic respiratory impairment (particularly in children); hypertension; narrow angle glaucoma; stenosing peptic ulcer, pyloroduodenal obstruction; prostatic hypertrophy; bladder neck obstruction; older adult or debilitated patients; pregnancy (category C); children younger than 2 y.

ROUTE & DOSAGE

Motion Sickness
Adult: **PO/PR** 25 mg q12h prn
Child (older than 2 y): **PO/PR** 12.5–25 mg q12h prn (max: 25 mg/dose)

Nausea
Adult: **PO/PR/IM/IV** 12.5–25 mg q4–6h prn
Child (older than 2 y): **PO/PR/IM/IV** 0.5 mg/pound q4–6h prn (max: 25 mg/dose)

Pruritus
Adult: **PO/PR** 12.5 mg q.i.d. or 25 mg at bedtime **IM/IV** 25 mg, repeat in 2 h if necessary, switch to PO
Child (older than 2 y): **PO** 6.25–12.5 mg q.i.d. **IV/IM** 0.5 mg/pound (max: 12.5 mg)

Sedation
Adult: **PO/IM/IV** 25–50 mg/dose
Child (older than 2 y): **PO/IM/IV** 0.5 mg/pound (max: 25 mg dose)

ADMINISTRATION
Oral
- Give with food, milk, or a full glass of water may minimize GI distress.

Common adverse effects in *italic*, life-threatening effects underlined; generic names in **bold**; classifications in SMALL CAPS; ✦ Canadian drug name; ⊘ Prototype drug

- Tablets may be crushed and mixed with water or food before swallowing.
- Oral doses for allergy are generally prescribed before meals and on retiring or as single dose at bedtime.

Rectal
- Ensure that suppository is inserted beyond the rectal sphincter.

Intramuscular
- Give IM injection deep into large muscle mass. Aspirate carefully before injecting drug. Intra-arterial injection can cause arterial or arteriolar spasm, with resultant gangrene.
- Subcutaneous injection (also contraindicated) can cause chemical irritation and necrosis. Rotate injection sites and observe daily.

Intravenous

PREPARE: **Direct:** The 25 mg/mL concentration may be given undiluted. ▪ Dilute the 50 mg/mL concentration in NS to yield no more than 25 mg/mL (e.g., diluting the 50 mg/mL concentration in 4 mL yields 10 mg/mL). ▪ Inspect parenteral drug before preparation. Discard if it is darkened or contains precipitate.
ADMINISTER: **Direct:** Give each 25 mg or fraction thereof over at least 1 min.
INCOMPATIBILITIES **Solution/additive:** Aminophylline, ampicillin, carbenicillin, cefazolin, cefotetan, ceftizoxime, chloramphenicol, chlordiazepoxide, chlorothiazide, dexamethasone, dimenhydrinate floxacillin, furosemide, heparin, hydrocortisone, ketorolac, methicillin, methohexital, nalbuphine, nitrofurantoin, penicillin G sodium, pentobarbital, phenobarbital, thiopental. **Y-site:** Acyclovir, aldesleukin, allopurinol, aminophylline,

amphotericin B cholesteryl complex, ampicillin, azathioprine, bretylium, carmustine, cefamandole, cefepime, cefmetazole, cefonicid, cefoperazone, cefotaxime, cefotetan, cefoxitin, ceftazidime, ceftobiprole, ceftriaxone, cefuroxime, cephalothin, cephapirin, chloramphenicol, clindamycin, dantrolene, dexamethasone, diazepam, diazoxide, doxorubicin liposome, ertapenem, fluorouracil, foscarnet, furosemide, ganciclovir, gemtuzumab, heparin, indomethacin, ketorolac, lansoprazole, methicillin, methotrexate, methylprednisolone, mezlocillin, milrinone acetate, minocycline, nafcillin, nitroprusside, oxacillin, pantoprazole, penicillin G, pentobarbital, phenobarbital, phenytoin, piperacillin/tazobactam, streptokinase, TPN, urokinase.

- Store at 15°–30° C (59°–86° F) in tight, light-resistant container unless otherwise directed.

ADVERSE EFFECTS (≥1%) **Body as a Whole:** Deep sleep, coma, convulsions, cardiorespiratory symptoms, extrapyramidal reactions, nightmares (in children), CNS stimulation, abnormal movements. **Respiratory:** Irregular respirations, respiratory depression, apnea. **CNS:** Sedation *drowsiness,* confusion, dizziness, disturbed coordination, restlessness, tremors. **CV:** Transient mild hypotension or hypertension. **GI:** Anorexia, nausea, vomiting, constipation. **Hematologic:** Leukopenia, agranulocytosis. **Special Senses:** *Blurred vision, dry mouth,* nose, or throat. **Skin:** Photosensitivity. **Urogenital:** Urinary retention.

DIAGNOSTIC TEST INTERFERENCE May produce false results

Common adverse effects in *italic*, life-threatening effects underlined; generic names in **bold**; classifications in SMALL CAPS; ♣ Canadian drug name; ❶ Prototype drug

1265

with **urinary pregnancy tests**. Promethazine can cause significant alterations of **flare response** in **intradermal allergen tests** if performed within 4 days of patient receiving promethazine.

INTERACTIONS Drug: Alcohol and other CNS DEPRESSANTS add to CNS depression and anticholinergic effects.

PHARMACOKINETICS Absorption: Readily from GI tract. **Onset:** 20 min PO/PR/IM; 5 min IV. **Duration:** 2–8 h. **Distribution:** Crosses placenta. **Metabolism:** In liver (CYP2D6, 2B6). **Elimination:** Slowly in urine and feces.

NURSING IMPLICATIONS

Assessment & Drug Effects

- Supervise ambulation. Promethazine sometimes produces marked sedation and dizziness.
- Be aware that antiemetic action may mask symptoms of unrecognized disease and signs of drug overdose as well as dizziness, vertigo, or tinnitus associated with toxic doses of aspirin or other ototoxic drugs.
- Patients in pain may develop involuntary (athetoid) movements of upper extremities following parenteral administration. These symptoms usually disappear after pain is controlled.
- Monitor respiratory function in patients with respiratory problems, particularly children. Drug may suppress cough reflex and cause thickening of bronchial secretions.

Patient & Family Education

- For motion sickness: Take initial dose 30–60 min before anticipated travel and repeat at 8–12 h intervals if necessary. For duration of journey, repeat dose on arising and again at evening meal.

- Do not drive or engage in other potentially hazardous activities requiring mental alertness and normal reaction time until response to drug is known.
- Avoid sunlamps or prolonged exposure to sunlight. Use sunscreen lotion during initial drug therapy.
- Do not take OTC medications without prescriber's approval.
- Avoid alcohol and other CNS depressants.
- Relieve dry mouth by frequent rinses with water or by increasing noncaloric fluid intake (if allowed), chewing sugarless gum, or sucking hard candy. If these measures fail, add a saliva substitute (e.g., Moi-Stir, Orex, Xero-Lube).

PROPAFENONE
(pro-pa′fen-one)

Rythmol, Rythmol SR

Classification: CLASS IC ANTIARRHYTHMIC

Therapeutic: CLASS IC ANTIARRHYTHMIC

Prototype: Flecainide

Pregnancy Category: C

AVAILABILITY 150 mg, 225 mg, 300 mg tablets; 225 mg, 325 mg, 425 mg sustained release capsule

ACTION & THERAPEUTIC EFFECT
Class IC antiarrhythmic drug with a direct stabilizing action on myocardial membranes. Reduces spontaneous automaticity. Exerts a negative inotropic effect on the myocardium. *Decreases rate of single and multiple PVCs and suppresses ventricular arrhythmias.*

USES Ventricular arrhythmias, atrial fibrillation, paroxysmal supraventricular tachycardia.

UNLABELED USES Wolff-Parkinson-White syndrome.

Common adverse effects in *italic*, life-threatening effects underlined; generic names in **bold**; classifications in SMALL CAPS; ♣ Canadian drug name; ⊕ Prototype drug

CONTRAINDICATIONS Hypersensitivity to propafenone; uncontrolled CHF, cardiogenic shock, sinoatrial, AV or intraventricular disorders (e.g., sick sinus node syndrome, AV block) without a pacemaker; cardiogenic shock; bradycardia, QT prolongation; marked hypotension; bronchospastic disorders; electrolyte imbalances; non–life-threatening arrhythmias.

CAUTIOUS USE CHF, COPD, chronic bronchitis; AV block; hepatic/renal impairment; older adult patients; pregnancy (category C), lactation. Safe use in children not established.

ROUTE & DOSAGE

Ventricular Arrhythmias (immediate release)

Adult: **PO** Initiate with 150 mg q8h, may be increased at 3–4 days intervals (max: 300 mg q8h)

Atrial Fibrillation

Adult: **PO** (immediate release) 150 mg q8h, titrate to response (usually 450–900 mg/day) **PO** (extended release) 225 mg q12h increase to response (max: 425 mg q12h).

ADMINISTRATION

- Dosage increments are usually made gradually with older adults or those with previous extensive myocardial damage.
- Significant dose reduction is warranted with severe liver dysfunction. Consult prescriber.
- Store at 15°–30° C (59°–86° F).

ADVERSE EFFECTS (≥1%) **CNS:**
Blurred vision, dizziness, paresthesias, fatigue, somnolence, vertigo, headache. **CV:** Arrhythmias, ventricular tachycardia, hypotension, bundle branch block, AV block, complete

heart block, sinus arrest, CHF. **Hematologic:** *Leukopenia,* granulocytopenia (both rare). **GI:** Nausea, abdominal discomfort, constipation, vomiting, dry mouth, *taste alterations,* cholestatic hepatitis. **Skin:** Rash.

INTERACTIONS Drug: **Amiodarone, quinidine** increases the levels and toxicity of propafenone. May increase levels and toxicity of TRICYCLIC ANTIDEPRESSANTS, **cyclosporine, digoxin,** BETA-BLOCKERS, **theophylline,** and **warfarin** may increase levels of both **propafenone** and **diltiazem. Phenobarbital** decreases levels of **propafenone.** Use with caution in drugs that prolong the QT interval.

PHARMACOKINETICS Absorption: Readily from GI tract. **Peak:** 3.5 h. **Distribution:** 97% protein bound, concentrates in the lung. Crosses placenta, distributed into breast milk. **Metabolism:** Extensively metabolized in the liver. **Elimination:** 18.5–38% of dose excreted in urine as metabolites. **Half-Life:** 5–8 h.

NURSING IMPLICATIONS

Assessment & Drug Effects

- Monitor cardiovascular status frequently (e.g., ECG, Holter monitor) to determine effectiveness of drug and development of new or worsened arrhythmias.
- Monitor closely patients with preexisting CHF for worsening of this condition. Monitor for digoxin toxicity with concurrent use, because drug may increase serum digoxin levels.
- Report development of second- or third-degree AV block or significant widening of the QRS complex. Dosage adjustment may be warranted.

Patient & Family Education

- Report to prescriber any of following: Chest pain, palpitations,

Common adverse effects in *italic*, life-threatening effects underlined; generic names in **bold**; classifications in SMALL CAPS; ♣ Canadian drug name; ⊘ Prototype drug

1267

blurred or abnormal vision, dyspnea, or signs and symptoms of infection.

- Be aware when taking concurrent warfarin of possible increase in plasma levels that increase bleeding risk. Report unusual bleeding or bruising.
- Monitor radial pulse daily and report decreased heart rate or development of an abnormal heartbeat.
- Be aware of possibility of dizziness and need for caution with walking, especially in older adult or debilitated patients.

PROPANTHELINE BROMIDE

(proe-pan'the-leen)

Pro-Banthine, Propanthel ◆

Classifications: ANTICHOLINERGIC; ANTIMUSCARINIC; ANTISPASMODIC
Therapeutic: ANTISPASMODIC
Prototype: Atropine
Pregnancy Category: C

AVAILABILITY 7.5 mg, 15 mg tablets

ACTION & THERAPEUTIC EFFECT Has potent antimuscarinic activity and postganglionic nicotinic receptor blocking action. *Decreases motility (smooth muscle tone) in the GI, biliary, and urinary tracts, resulting in antispasmodic action.*

USES Adjunct in treatment of peptic ulcer, irritable bowel syndrome, pancreatitis, ureteral and urinary bladder spasm. Also used prior to radiologic diagnostic procedures to reduce duodenal motility.

CONTRAINDICATIONS Narrow-angle glaucoma; tachycardia; MI; paralytic ileus, GI obstructive disease; hemorrhagic shock; myasthenia gravis.

CAUTIOUS USE CAD, CHF, cardiac arrhythmias; liver disease, ulcerative colitis, hiatus hernia, esophagitis; kidney disease; BPH; glaucoma; debilitated patients; hyperthyroidism; autonomic neuropathy; brain damage; Down's syndrome; spastic disorders; pregnancy (category C), lactation. Safe use in children not established.

ROUTE & DOSAGE

Irritable Bowel Syndrome

Adult: **PO** 15 mg 30 min a.c. and 30 mg at bedtime (max: 120 mg/day)
Geriatric: **PO** 7.5 mg 2–3 times/day a.c. (max: 90 mg/day)

ADMINISTRATION

Oral

- Give 30–60 min before meals and at bedtime. Advise not to chew tablet; drug is bitter.
- Give at least 1 h before or 1 h after an antacid (or antidiarrheal agent).
- Store dry powder and tablets at 15°–30° C (59°–86° F); protect from freezing and moisture.

ADVERSE EFFECTS (≥1%) **GI:** *Constipation, dry mouth.* **Special Senses:** Blurred vision, mydriasis, increased intraocular pressure. **CNS:** Drowsiness. **Urogenital:** Decreased sexual activity, difficult urination.

INTERACTIONS Drug: Decreased absorption of **ketoconazole;** ORAL POTASSIUM may increase risk of GI ulcers. **Food:** Food significantly decreases absorption.

PHARMACOKINETICS Absorption: Incompletely from GI tract. **Onset:** 30–45 min. **Duration:** 4–6 h. **Metabolism:** 50% in GI tract before absorption; 50% in liver. **Elimination:** Primarily in urine; some in bile. **Half-Life:** 9 h.

NURSING IMPLICATIONS

Assessment & Drug Effects

- Assess bowel sounds, especially in presence of ulcerative colitis, since paralytic ileus may develop, predisposing to toxic megacolon.
- Be aware that older adult or debilitated patients may respond to a usual dose with agitation, excitement, confusion, drowsiness. Stop drug and report to prescriber if these symptoms are observed.
- Check BP, heart sounds and rhythm periodically in patients with cardiac disease.

Patient & Family Education

- Void just prior to each dose to minimize risk of urinary hesitancy or retention. Record daily urinary volume and report problems to prescriber.
- Relieve dry mouth by rinsing with water frequently, chewing sugar-free gum or sucking hard candy.
- Maintain adequate fluid and high-fiber food intake to prevent constipation.
- Make all position changes slowly and lie down immediately if faintness, weakness, or palpitations occur. Report symptoms to prescriber.
- Do not drive or engage in potentially hazardous activities until response to drug is known.

PROPOFOL

(pro′po-fol)

Diprivan

FOSPROPOFOL

Lusendra

Classifications: GENERAL ANESTHESIA; SEDATIVE-HYPNOTIC
Therapeutic: SEDATIVE-HYPNOTIC; GENERAL ANESTHESIA
Pregnancy Category: B

AVAILABILITY 10 mg/mL, 35 mg/mL injection

ACTION & *THERAPEUTIC EFFECT*

Sedative-hypnotic used in the induction and maintenance of anesthesia or sedation. Rapid onset (40 sec) and minimal excitation during induction of anesthesia. *Effectively used for conscious sedation and maintenance of anesthesia.*

USES Induction or maintenance of anesthesia as part of a balanced anesthesia technique; conscious sedation in mechanically ventilated patients.

CONTRAINDICATIONS Hypersensitivity to propofol or propofol emulsion, which contain soybean oil and egg phosphatide; patients with increased intracranial pressure or impaired cerebral circulation; obstetrical procedures; lactation.
CAUTIOUS USE Patients with severe cardiac or respiratory disorders, respiratory depression, hypoxemia, hypertension; history of epilepsy or seizures; hypovolemia; older adults; pregnancy (category B). **Propofol:** Do not use for induction of anesthesia in children younger than 3 y and for maintenance of anesthesia in infants younger than 2 mo. **Fospropofol:** Children younger than 18 y.

ROUTE & DOSAGE

Induction of Anesthesia

Adult: **IV** 2–2.5 mg/kg q10sec until induction onset
Adult (older than 55 y): **IV** 1–1.5 mg/kg q10sec until induction onset
Adolescent/Child (3 y or older): **IV** 2.5–3.5 mg/kg over 20–30 sec

Maintenance of Anesthesia

Adult: **IV** 100–200 mcg/kg/min

Common adverse effects in *italic*, life-threatening effects underlined; generic names in **bold**; classifications in SMALL CAPS; ♣ Canadian drug name; ❶ Prototype drug

1269

Adult (older than 55 y): **IV** 50–100 mcg/kg/min

Child/Infant (older than 2 months): **IV** 125–150 mcg/kg/min

Conscious Sedation

Adult: **IV** 5 mcg/kg/min for at least 5 min, may increase by 5–10 mcg/kg/min q5–10min until desired level of sedation is achieved (may need maintenance rate of 5–80 mcg/kg/min)

Monitored Anesthesia Care or Sedation (Fospropofol)

Adult (healthy): **IV** *Weight greater than 90 kg:* 577.5 mg bolus, then may supplement no more than q4min to a max of 140 mg/dose; *weight 61–89 kg:* 6.5 mg/kg (max: 577.5 mg) bolus, then supplement doses up to 1.6 mg/kg no more than q4min; *weight less than 60 kg:* 385 mg bolus, supplement up to 105 mg/dose no more than q4min

Adult (severe systemic disease): **IV** *Weight greater than 90 kg:* 437.5 mg as bolus; give supplemental doses up to 105 mg per dose no more frequently than q4min; *weight 61–89 kg:* Initially, give 75% of standard dosing regimen; give supplemental doses of 75% of standard dosing regimen; *weight less than 60 kg:* 297.5 mg bolus; give supplemental doses to a max of 70 mg per dose no more frequently than q4min

Elderly: Use decreased starting dose

ADMINISTRATION

- Use strict aseptic technique to prepare propofol for injection; drug emulsion supports rapid growth of microorganisms.

- Inspect for particulate matter and discoloration. Discard if either is noted.

- Shake well before use. Inspect for separation of the emulsion. Do not use if there is evidence of separation of phases of the emulsion.

Intravenous

PREPARE: **IV Infusion:** Give undiluted or diluted in D5W to a concentration not less than 2 mg/mL. Begin drug administration immediately after preparation and complete within 6 h.

ADMINISTER: **IV Infusion:** Use syringe or volumetric pump to control rate. ▪ Rate is determined by patient weight in kg. Depending on the form of the drug, indication, patient's health status and age, drug my be given by variable rate infusion or intermittent IV bolus (usually over 3–5 min). ▪ Administer immediately after spiking the vial. Complete infusion within 6 h.

INCOMPATIBILITIES **Y-site: Amikacin, amphotericin B, ascorbic acid, atracurium, atropine, bretylium, calcium chloride, ciprofloxacin, cisatracurium, diazepam, digoxin, doripenem, doxorubicin, hydroxocobalamine, gentamicin, levofloxacin, methotrexate, methylprednisolone, metoclopramide, minocycline, mitoxantrone, netilmicin, nimodipine, phenytoin, remifentanil, tobramycin, verapamil.**

- Store unopened between 4° C (40° F) and 22° C (72° F). Refrigeration is not recommended. Protect from light.

ADVERSE EFFECTS (≥1%) CNS: Headache, dizziness, *twitching,* *bucking, jerking, thrashing, clonic/ myoclonic movements.* **Special Senses:** Decreased intraocular pressure. **CV:** Hypotension, <u>ventricular asystole</u> (rare). **GI:** Vomiting, abdominal cramping. **Respiratory:** Cough, hiccups, apnea. **Other:** Pain at injection site.

DIAGNOSTIC TEST INTERFERENCE
Propofol produces a temporary reduction in *serum cortisol levels.* However, propofol does not seem to inhibit adrenal responsiveness to *ACTH.*

INTERACTIONS Drug: Concurrent continuous infusions of propofol and **alfentanil** produce higher plasma levels of **alfentanil** than expected. CNS DEPRESSANTS cause additive CNS depression.

PHARMACOKINETICS Onset: 9–36 sec. **Duration:** 6–10 min. **Distribution:** Highly lipophilic, crosses placenta, excreted in breast milk. **Metabolism:** Extensively in the liver (CYP2B6, 2C9). **Elimination:** Approximately 88% of the dose is recovered in the urine as metabolites. **Half-Life:** 5–12 h.

NURSING IMPLICATIONS
Assessment & Drug Effects
- Monitor hemodynamic status and assess for dose-related hypotension.
- Take seizure precautions. Tonic-clonic seizures have occurred following general anesthesia with propofol.
- Be alert to the potential for drug-induced excitation (e.g., twitching, tremor, hyperclonus) and take appropriate safety measures.
- Provide comfort measures; pain at the injection site is quite common especially when small veins are used.

PROPRANOLOL HYDROCHLORIDE 🅟
(proe-pran'oh-lole)
Apo-Propranolol ✦, Inderal, Inderal LA, InnoPran XL, Novopranol ✦
Classifications: BETA-ADRENERGIC RECEPTOR ANTAGONIST; ANTIHYPERTENSIVE; CLASS II ANTIARRHYTHMIC
Therapeutic: ANTIHYPERTENSIVE; CLASS II ANTIARRHYTHMIC; ANTIANGINAL
Pregnancy Category: C

AVAILABILITY 10 mg, 20 mg, 40 mg, 60 mg, 80 mg, 90 mg tablets; 60 mg, 80 mg, 120 mg, 160 mg sustained release capsules; 4 mg/mL, 8 mg/mL, 80 mg/mL solution; 1 mg/mL injection

ACTION & *THERAPEUTIC EFFECT*
Nonselective beta-blocker of both cardiac and bronchial adrenoreceptors that competes with epinephrine and norepinephrine for available beta receptor sites. In higher doses, it depresses cardiac function including contractility and arrhythmias. Lowers both supine and standing blood pressures in hypertensive patients. *Reduces heart rate, myocardial irritability (Class II antiarrhythmic), and force of contraction, depresses automaticity of sinus node and ectopic pacemaker, and decreases AV and intraventricular conduction velocity. Hypotensive effect is associated with decreased cardiac output. Has migraine prophylactic effects.*

USES Management of cardiac arrhythmias, myocardial infarction, tachyarrhythmias associated with digitalis intoxication, anesthesia, and thyrotoxicosis, hypertrophic subaortic stenosis, angina pectoris due to coronary atherosclerosis,

Common adverse effects in *italic*, life-threatening effects <u>underlined</u>; generic names in **bold**; classifications in SMALL CAPS; ✦ Canadian drug name; 🅟 Prototype drug

1271

pheochromocytoma, hereditary essential tremor; also treatment of hypertension alone, but generally with a thiazide or other antihypertensives.

UNLABELED USES Anxiety states, migraine prophylaxis, essential tremors, schizophrenia, tardive dyskinesia, acute panic symptoms (e.g., stage fright), recurrent GI bleeding in cirrhotic patients, treatment of aggression and rage, drug induced tremors.

CONTRAINDICATIONS Hypersensitivity to propranolol; greater than first-degree heart block; right ventricular failure secondary to pulmonary hypertension; ventricular dysfunction; sinus bradycardia, cardiogenic shock, significant aortic or mitral valvular disease; bronchial asthma or bronchospasm, severe COPD, pulmonary edema; abrupt discontinuation; major depression; PVD, Raynaud's disease.

CAUTIOUS USE Peripheral arterial insufficiency; history of systemic insect sting reaction; patients prone to nonallergenic bronchospasm (e.g., chronic bronchitis, emphysema); major surgery; cerebrovascular disease, stroke; renal or hepatic disease; pheochromocytoma, vasospastic angina; diabetes mellitus; patients prone to hypoglycemia; hyperthyroidism, thyrotoxicosis; surgery; myasthenia gravis and other skeletal muscular diseases; Wolff-Parkinson-White syndrome; older adults; pregnancy (category C), lactation.

ROUTE & DOSAGE

Hypertension

Adult: **PO Immediate release:** 40 mg b.i.d., usually need 160–480 mg/day in divided doses;

InnoPran XL dose 80 mg each night, may increase to 120 mg at bedtime. **Other extended release forms:** 80 mg daily increase up to 120–160 mg

Angina

Adult: **PO Immediate release:** 10–20 mg/day in divided doses (increase up to 160–320 mg/day) **Extended release:** 80 mg QD (increase to 160–320 mg/day)

Arrhythmias

Adult: **PO** 10–30 mg q6–8h **IV** 1–3 mg q4h

Acute MI

Adult: **PO** 180–240 mg/day in divided doses

Migraine Prophylaxis

Adult: **PO** 80 mg/day in divided doses, may need 160–240 mg/day

ADMINISTRATION

- Take apical pulse and BP before administering drug. Withhold drug if heart rate less than 60 bpm or systolic BP less than 90 mm Hg. Consult prescriber for parameters.

Oral

- Do not give within 2 wk of an MAO inhibitor.
- Note that InnoPran XL should be given at bedtime.
- Be consistent with regard to giving with food or on an empty stomach to minimize variations in absorption.
- Ensure that sustained release form is not chewed or crushed. **Must be** swallowed whole.
- Reduce dosage gradually over a period of 1–2 wk and monitor patient closely when discontinued.

Common adverse effects in *italic*, life-threatening effects <u>underlined</u>; generic names in **bold**; classifications in SMALL CAPS; ♣ Canadian drug name; ⊘ Prototype drug

Intravenous

Note: Verify correct IV concentration and rate of infusion for neonates with prescriber.

- Take apical pulse and BP before administering drug. Withhold drug if heart rate less than 60 bpm or systolic BP less than 90 mm Hg. ▪Consult prescriber for parameters.

PREPARE: Direct: May be given undiluted or dilute each 1 mg in 10 mL of D5W. **Intermittent:** Dilute a single dose in 50 mL of NS.

ADMINISTER: Direct: Give each 1 mg or fraction thereof over 1 min. **Intermittent:** Give each dose over 15–20 min.

INCOMPATIBILITIES Y-site: Amphotericin B cholesteryl complex, dantrolene, diazepam, diazoxide, indomethacin, insulin, lansoprazole, paclitaxel, pantoprazole, phenytoin, piperacillin/tazobactam.

- Store at 15°–30° C (59°–86° F) in tightly closed, light-resistant containers.

ADVERSE EFFECTS (≥1%) **Body as a Whole:** Fever; pharyngitis; respiratory distress, weight gain, LE-like reaction, cold extremities, leg fatigue, arthralgia, anaphylactic/anaphylactoid reactions. **Urogenital:** Impotence or decreased libido. **Skin:** Erythematous, psoriasis-like eruptions; pruritus, Stevens-Johnson syndrome, toxic epidermal necrolysis, erythema multiforme, exfoliative dermatitis, urticaria. Reversible alopecia, hyperkeratoses of scalp, palms, feet; nail changes, dry skin. **CNS:** Drug-induced psychosis, sleep disturbances, depression, *confusion*, agitation, giddiness, light-headedness, *fatigue,* vertigo, syncope, weakness, *drowsiness,* insomnia, vivid dreams, visual hallucinations, delusions, reversible organic brain syndrome. **CV:** Palpitation, profound *bradycardia,* AV heart block, cardiac standstill, hypotension, angina pectoris, tachyarrhythmia, acute CHF, peripheral arterial insufficiency resembling Raynaud's disease, myotonia, paresthesia of *hands.* **Special Senses:** Dry eyes (gritty sensation), visual disturbances, conjunctivitis, tinnitus, hearing loss, nasal stuffiness. **GI:** Dry mouth, nausea, vomiting, heartburn, diarrhea, constipation, flatulence, abdominal cramps, mesenteric arterial thrombosis, ischemic colitis, pancreatitis. **Hematologic:** Transient eosinophilia, thrombocytopenic or nonthrombocytopenic purpura, agranulocytosis. **Metabolic:** Hypoglycemia, hyperglycemia, hypocalcemia (patients with hyperthyroidism). **Respiratory:** Dyspnea, laryngospasm, bronchospasm.

DIAGNOSTIC TEST INTERFERENCE BETA-ADRENERGIC BLOCKERS may produce false-negative test results in exercise tolerance ECG tests, and elevations in *serum potassium, peripheral platelet count, serum uric acid, serum transaminase, alkaline phosphatase, lactate dehydrogenase, serum creatinine, BUN,* and an increase or decrease in *blood glucose* levels in diabetic patients.

INTERACTIONS Drug: PHENOTHIAZINES have additive hypotensive effects. BETA-ADRENERGIC AGONISTS (e.g., **albuterol**) antagonize effects. **Atropine** and TRICYCLIC ANTIDEPRESSANTS block bradycardia. DIURETICS and other HYPOTENSIVE AGENTS increase hypotension. High doses of **tubocurarine** may potentiate neuromuscular blockade. **Cimetidine** decreases clearance, increases effects. ANTACIDS, **ascorbic acid** may decrease absorption. **Herbal: Black pepper** may increase plasma levels.

Common adverse effects in *italic*, life-threatening effects underlined; generic names in **bold**; classifications in SMALL CAPS; ✚ Canadian drug name; ⊘ Prototype drug

1273

PROPRANOLOL HYDROCHLORIDE

PHARMACOKINETICS Absorption:
Completely from GI tract; undergoes extensive first-pass metabolism. **Peak:** 60–90 min immediate release; 6 h sustained release; 5 min IV. **Distribution:** Widely distributed including CNS, placenta, and breast milk. **Metabolism:** Almost completely in liver (CYP1A2, 2D6). **Elimination:** 90–95% in urine as metabolites; 1–4% in feces. **Half-Life:** 2.3 h.

NURSING IMPLICATIONS

Assessment & Drug Effects

- Obtain careful medical history to rule out allergies, asthma, and obstructive pulmonary disease. Propranolol can cause bronchiolar constriction even in normal subjects.
- Monitor apical pulse, respiration, BP, and circulation to extremities closely throughout period of dosage adjustment. Consult prescriber for acceptable parameters.
- Evaluate adequate control or dosage interval for patients being treated for hypertension by checking blood pressure near end of dosage interval or before administration of next dose.
- Be aware that adverse reactions occur most frequently following IV administration soon after therapy is initiated; however, incidence is also high following oral use in the older adult and in patients with impaired kidney function. Reactions may or may not be dose related.
- Lab tests: Obtain periodic hematologic, kidney, liver, and cardiac functions when propranolol is given for prolonged periods.
- Monitor I&O ratio and daily weight as significant indexes for detecting fluid retention and developing heart failure.
- Consult prescriber regarding allowable salt intake. Drug plasma volume may increase with consequent risk of CHF if dietary sodium is not restricted in patients not receiving concomitant diuretic therapy.
- Fasting for more than 12 h may induce hypoglycemic effects fostered by propranolol.
- If patient complains of cold, painful, or tender feet or hands, examine carefully for evidence of impaired circulation. Peripheral pulses may still be present even though circulation is impaired. Caution patient to avoid prolonged exposure of extremities to cold.

Patient & Family Education

- Learn usual pulse rate and take radial pulse before each dose. Report to prescriber if pulse is below the established parameter or becomes irregular.
- Be aware that propranolol suppresses clinical signs of hypoglycemia (e.g., BP changes, increased pulse rate) and may prolong hypoglycemia.
- Understand importance of compliance. Do not alter established regimen (i.e., do not omit, increase, or decrease dosage or change dosage interval).
- Do not discontinue abruptly; can precipitate withdrawal syndrome (e.g., tremulousness, sweating, severe headache, malaise, palpitation, rebound hypertension, MI, and life-threatening arrhythmias in patients with angina pectoris).
- Be aware that drug may cause mild hypotension (experienced as dizziness or light-headedness) in normotensive patients on prolonged therapy. Make position changes slowly and avoid prolonged standing. Notify prescriber if symptoms persist.
- Do not drive or engage in potentially hazardous activities until response to drug is known.
- Consult prescriber before self-medicating with OTC drugs.

Common adverse effects in *italic*, life-threatening effects underlined; generic names in **bold**; classifications in SMALL CAPS; ♣ Canadian drug name; ⊘ Prototype drug

- Inform dentist, surgeon, or ophthalmologist that you are taking propranolol (drug lowers normal and elevated intraocular pressure).

PROPYLTHIOURACIL (PTU) ⓟ
(proe-pill-thye-oh-yoor'a-sill)

Propyl-Thyracil ✚
Classification: ANTITHYROID
Therapeutic: ANTITHYROID
Pregnancy Category: D

AVAILABILITY 50 mg tablets

ACTION & *THERAPEUTIC EFFECT*
Interferes with use of iodine and blocks synthesis of thyroxine (T_4) and triiodothyronine (T_3). Antithyroid action is delayed days and weeks until preformed T_3 and T_4 are degraded. *Effective as an antithyroid agent in various hyperthyroid conditions.*

USES Hyperthyroidism, iodine-induced thyrotoxicosis, and hyperthyroidism associated with thyroiditis; to establish euthyroidism prior to surgery or radioactive iodine treatment; palliative control of toxic nodular goiter.

CONTRAINDICATIONS Hypersensitivity to propylthiouracil; concurrent administration of sulfonamides or coal tar derivatives such as aminopyrine or antipyrine; pregnancy (category D), lactation.
CAUTIOUS USE Infection; bone marrow depression; impaired liver function; patients at risk for liver failure.

ROUTE & DOSAGE

Hyperthyroidism

Adult: **PO** 300–450 mg/day divided q8h, may need 600–1200 mg/day initially

Geriatric: **PO** 150–300 mg/day divided q8h
Child: **PO** 6–10 y, 50–150 mg/day; *older than 10 y,* 150–300 mg/day or 150 mg/m²/day
Neonates: **PO** 5–10 mg/kg/day

Thyrotoxic Crisis

Adult: **PO** 200 mg q4–6h until full control achieved

ADMINISTRATION
Oral
- Give at the same time each day with relation to meals. Food may alter drug response by changing absorption rate.
- If drug is being used to improve thyroid state before radioactive iodine (RAI) treatment, discontinue 3 or 4 days before treatment to prevent uptake interference. PTU therapy may be resumed if necessary 3–5 days after the RAI administration.
- Store drug at 15°–30° C (59°–86° F) in light-resistant container.

ADVERSE EFFECTS (≥1%) **CNS:** Paresthesias, headache, vertigo, drowsiness, neuritis. **GI:** Nausea, vomiting, diarrhea, dyspepsia, loss of taste, sialoadenitis, hepatitis. **Hematologic:** Myelosuppression, lymphadenopathy, periarteritis, hypoprothrombinemia, *thrombocytopenia*, *leukopenia*, agranulocytosis. **Metabolic:** Hypothyroidism (goitrogenic): Enlarged thyroid, reduced GI motility, periorbital edema, puffy hands and feet, bradycardia, cool and pale skin, worsening of ophthalmopathy, sleepiness, fatigue, mental depression, dizziness, vertigo, sensitivity to cold, paresthesias, nocturnal muscle cramps, changes in menstrual periods, unusual weight gain. **Skin:** Skin rash, urticaria, pruritus,

P

Common adverse effects in *italic*, life-threatening effects underlined; generic names in **bold**; classifications in SMALL CAPS; ✚ Canadian drug name; ⓟ Prototype drug

1275

hyperpigmentation, lightening of hair color, abnormal hair loss. **Body as a Whole:** Drug fever, lupus-like syndrome, arthralgia, myalgia, hypersensitivity vasculitis.

DIAGNOSTIC TEST INTERFERENCE
Propylthiouracil may elevate *prothrombin time* and serum *alkaline phosphatase, AST, ALT* levels.

INTERACTIONS Drug: Amiodarone, potassium iodide, sodium iodide, THYROID HORMONES can reverse efficacy.

PHARMACOKINETICS Absorption: Rapidly from GI tract. **Peak:** 1–1.5 h. **Distribution:** Appears to concentrate in thyroid gland; crosses placenta; some distribution into breast milk. **Metabolism:** Rapidly to inactive metabolites. **Elimination:** 35% in urine within 24 h. **Half-Life:** 1–2 h.

NURSING IMPLICATIONS
Assessment & Drug Effects
- Be aware that about 10% of patients with hyperthyroidism have leukopenia less than 4000 cells/mm^3 and relative granulocytopenia.
- Observe for signs of clinical response to PTU (usually within 2 or 3 wk): Significant weight gain, reduced pulse rate, reduced serum T$_4$.
- Lab tests: Baseline and periodic T$_3$ and T$_4$; periodic LFTs and CBC with differential and platelet count.
- Satisfactory euthyroid state may be delayed for several months when thyroid gland is greatly enlarged.
- Be alert to signs of hypoprothrombinemia: Ecchymoses, purpura, petechiae, unexplained bleeding. Warn ambulatory patients to report these signs promptly.
- Be alert for important diagnostic signs of excess dosage: Contraction of a muscle bundle when pricked, mental depression, hard and nonpitting

edema, and need for high thermostat setting and extra blankets in winter (cold intolerance).
- Monitor for urticaria (occurs in 3–7% of patients during weeks 2–8 of treatment). Report severe rash.

Patient & Family Education
- Note that PTU treatment may be reinstituted if surgery fails to produce normal thyroid gland function.
- Report severe skin rash or swelling of cervical lymph nodes. Therapy may be discontinued.
- Report to prescriber sore throat, fever, and rash immediately (most apt to occur in first few months of treatment). Drug will be discontinued and hematologic studies initiated.
- Avoid use of OTC drugs for asthma, or cough treatment without checking with the prescriber. Iodides sometimes included in such preparations are contraindicated.
- Learn how to take pulse accurately and check daily. Report to prescriber continued tachycardia.
- Report diarrhea, fever, irritability, listlessness, vomiting, weakness; these are signs of inadequate therapy or thyrotoxicosis.
- Chart weight 2 or 3 times weekly; clinical response is monitored through changes in weight and pulse.
- Do not alter drug regimen (e.g., increase, decrease, omit doses, change dosage intervals).
- Check with prescriber about use of iodized salt and inclusion of seafood in the diet.

PROTAMINE SULFATE
(proe′ta-meen)
Classifications: ANTIDOTE; HEPARIN ANTAGONIST
Therapeutic: ANTIHEMORRHAGIC
Pregnancy Category: C

AVAILABILITY 10 mg/mL injection

ACTION & *THERAPEUTIC EFFECT*
Combines with heparin to produce a stable complex; thus it neutralizes the anticoagulant effect of heparin. *Effective antidote to heparin overdose.*

USES Antidote for heparin overdosage (after heparin has been discontinued).
UNLABELED USES Antidote for heparin administration during extracorporeal circulation.

CONTRAINDICATIONS Hemorrhage not induced by heparin overdosage; lactation.
CAUTIOUS USE Cardiovascular disease; history of allergy to fish; vasectomized or infertile males; diabetes mellitus; patients who have received protamine-containing insulin; pregnancy (category C).

ROUTE & DOSAGE

> **Antidote for Heparin Overdose**
> *Adult/Child:* **IV** 1 mg for every 100 units of heparin to be neutralized (max: 100 mg in a 2 h period), give the first 25–50 mg by slow direct IV and the rest over 2–3 h

ADMINISTRATION

Note: Titrate dose carefully to prevent excess anticoagulation because protamine has a longer half-life than heparin and also has some anticoagulant effect of its own.

Intravenous
Note: Verify correct IV concentration and rate of infusion for infants or children with prescriber.

PREPARE: Direct: May be given as supplied direct IV. **Continuous:** Dilute in 50 mL or more of NS or D5W.

ADMINISTER: Direct Give each 50 mg or fraction thereof slowly over 10–15 min. ▪ NEVER give more than 50 mg in any 10 min period or 100 mg in any 2 h period.

Continuous: Do not exceed direct rate. Give over 2–3 h or longer as determined by coagulation studies.
INCOMPATIBILITIES Solution/additive: RADIOCONTRAST MATERIALS, **furosemide.**

▪ Store protamine sulfate injection at 15°–30° C (59°–86° F). ▪ Solutions do not contain preservatives and should not be stored.

ADVERSE EFFECTS (≥1%) **CV:** *Abrupt drop in BP* (with rapid IV infusion), bradycardia. **Body as a Whole:** Urticaria, angioedema, pulmonary edema, anaphylaxis, dyspnea, lassitude; transient flushing and feeling of warmth. **GI:** Nausea, vomiting. **Hematologic:** Protamine overdose or "heparin rebound" (hyperheparinemia).

INTERACTIONS No clinically significant interactions established.

PHARMACOKINETICS Onset: 5 min. **Duration:** 2 h.

NURSING IMPLICATIONS
Assessment & Drug Effects
▪ Do not use protamine if only minor bleeding occurs during heparin therapy because withdrawal of heparin will usually correct minor bleeding within a few hours.
▪ Monitor BP and pulse q15–30min, or more often if indicated. Continue for at least 2–3 h after each dose, or longer as dictated by patient's condition.
▪ Lab tests: Monitor aPTT or ACT values. Coagulation tests are usually performed 5–15 min after

Common adverse effects in *italic*, life-threatening effects underlined; generic names in **bold**; classifications in SMALL CAPS; ♣ Canadian drug name; ✪ Prototype drug

1277

administration of protamine, and again in 2–8 h if desirable.

- Observe closely patients undergoing extracorporeal dialysis or patients who have had cardiac surgery for bleeding (heparin rebound). Even with apparent adequate neutralization of heparin by protamine, bleeding may occur 30 min to 18 h after surgery. Monitor vital signs closely. Additional protamine may be required in these patients.

PROTEIN C CONCENTRATE (HUMAN)
(pro′teen)

Ceprotin
Classifications: HEMATOLOGIC; THROMBOLYTIC; PROTEIN INHIBITOR
Therapeutic: PROTEIN C REPLACEMENT THERAPY
Pregnancy Category: C

AVAILABILITY 500 unit, 1000 unit vials of lyophilized powder

ACTION & THERAPEUTIC EFFECT Protein C is a critical element in a pathway that provides a natural mechanism for control of the coagulation system. This prevents excess procoagulant responses to activating stimuli. *Protein C is necessary to decrease thrombin generation and intravascular clot formation.*

USES Treatment of patients with severe congenital protein C deficiency; protein C replacement therapy for the prevention and treatment of venous thrombosis and purpura fulminans in children and adults.

CONTRAINDICATIONS Hypersensitivity to human protein C; concurrent administration with tissue plasminogen activator (tPA); hypernatremia; lactation.

CAUTIOUS USE Concurrent administration of anticoagulants; heparin induced thrombocytopenia (HIT); renal impairment; hepatic impairment; older adults; pregnancy (category C).

ROUTE & DOSAGE

Acute Episodes of Venous Thrombosis and Purpura Fulminans and Short-Term Prophylaxis
Adult: **IV** Initial dose 100–120 units/kg; then 60–80 units/kg q6h × 3. Maintenance dose: 45–60 units/kg q6–12h.

ADMINISTRATION

Intravenous

PREPARE: **Direct/IV Infusion:** Bring powder and supplied diluent to room temperature. Insert supplied double-ended transfer needle into diluent vial, then invert and rapidly insert into protein C powder vial. (If vacuum does not draw diluent into vial, discard.) ▪Remove transfer needle and gently swirl to dissolve. ▪Resulting solution concentration is 100 units/mL and it should be colorless to slightly yellowish, clear to slightly opalescent and free from visible particles. ▪Withdraw required dose with the supplied filter needle.

ADMINISTER: **Direct/IV Infusion:** Infuse at 2 mL/min. **Direct/IV Infusion for Child weighing greater than 10 kg:** Infuse at 0.2 mL/kg/min.

- Store at room temperature for no more than 3 h after reconstitution.
- Prior to reconstitution, protect from light.

ADVERSE EFFECTS (≥1%) **Body as a Whole:** Fever, hyperhidrosis, hypersensitivity reactions (rash, pruritits), restlessness. **CNS:** Light-headedness. **CV:** Hemothorax, hypotension.

INTERACTIONS Drug: Protein C concentrate can increase bleeding caused by **alteplase, reteplase,** or **tenecteplase.**

PHARMACOKINETICS Peak: 0.5–1 h. **Half-Life:** 9.9 h.

NURSING IMPLICATIONS

Assessment & Drug Effects

- Monitor for and promptly report S&S of bleeding or hypersensitivity reactions (see Appendix F).
- Monitor vital signs including BP and temperature.
- Lab tests: Baseline and periodic protein C activity, protein C trough level with acute thrombotic events; platelet counts; frequent serum sodium with renal function impairment.

Patient & Family Education

- Report immediately early signs of hypersensitivity reactions including hives, generalized itching, tightness in chest, wheezing, difficulty breathing.
- Report immediately any signs of bleeding including black tarry stools, pink/red-tinged urine, unusual bruising.

PROTRIPTYLINE HYDROCHLORIDE
(proe-trip'te-leen)
Triptil ✦, Vivactil
Classification: TRICYCLIC ANTI-DEPRESSANT
Therapeutic: ANTIDEPRESSANT
Prototype: Imipramine
Pregnancy Category: C

AVAILABILITY 5 mg, 10 mg tablets

ACTION & *THERAPEUTIC EFFECT*
Tricyclic antidepressant (TCA) that is believed to enhance actions of nor-epinephrine and serotonin by blocking their reuptake at the neuronal membrane. *Effective in the treatment of depressed individuals, particularly those who are withdrawn.*

USES Treatment of depression (notably withdrawn/anergic patients).

CONTRAINDICATIONS Hypersensitivity to TCAs; concurrent use of MAOIs; acute recovery phase following MI or within 14 days of use; QT prolongation, bundle branch block; cardiac conduction defects; suicidal ideation.

CAUTIOUS USE Hepatic, cardiovascular, or kidney dysfunction; DM; hyperthyroidism; history of alcoholism; patients with insomnia; asthma; bipolar disorder; suicidal tendencies; children and adolescents; pregnancy (category C), lactation.

ROUTE & DOSAGE

Antidepressant
Adult: **PO** 15–40 mg/day in 3–4 divided doses (max: 60 mg/day)
Adolescent: **PO** 5 mg t.i.d.

ADMINISTRATION

Oral

- Give whole or crush and mix with fluid or food.
- Give dosage increases in the morning dose to prevent sleep interference and because this TCA has psychic energizing action.
- Give last dose of day no later than mid-afternoon; insomnia rather than drowsiness is a frequent adverse effect.
- Store at 15°–30° C (59°–86° F) in tightly closed container.

ADVERSE EFFECTS (≥1%) **Body as a Whole:** Photosensitivity, <u>edema</u>

Common adverse effects in *italic*, life-threatening effects <u>underlined</u>; generic names in **bold**; classifications in SMALL CAPS; ✦ Canadian drug name; ☻ Prototype drug

1279

(general or of face and <u>tongue</u>). **GI:** *Xerostomia, constipation,* paralytic ileus. **Special Senses:** Blurred vision. **Urogenital:** *Urinary retention.* **CNS:** Insomnia, headache, confusion. **CV:** Change in heat or cold tolerance; *orthostatic hypotension, tachycardia.*

INTERACTIONS Drug: May decrease some response to ANTIHYPERTENSIVES; CNS DEPRESSANTS, **alcohol,** HYPNOTICS, BARBITURATES, SEDATIVES potentiate CNS depression; ORAL ANTICOAGULANTS may increase hypoprothrombinemic effects; **ethchlorvynol** causes transient delirium; **levodopa** SYMPATHOMIMETICS (e.g., **epinephrine, norepinephrine**) increases possibility of sympathetic hyperactivity with hypertension and hyperpyrexia; MAO INHIBITORS present possibility of severe reactions—toxic psychosis, cardiovascular instability; **methylphenidate** increases plasma TCA levels; THYROID DRUGS may increase possibility of arrhythmias; **cimetidine** may increase plasma TCA levels. **Herbal: Ginkgo** may decrease seizure threshold; **St. John's wort** may cause serotonin syndrome (headache, dizziness, sweating, agitation).

PHARMACOKINETICS Absorption: Rapidly from GI tract. **Peak levels:** 24–30 h. **Distribution:** Crosses placenta; distributed into breast milk. **Metabolism:** In liver. **Elimination:** Primarily in urine. **Half-Life:** 54–98 h.

NURSING IMPLICATIONS

Assessment & Drug Effects

- Monitor therapeutic effectiveness. Onset of initial effect characterized by increased activity and energy is fairly rapid, usually within 1 wk after therapy is initiated.

Maximum effect may not occur for 2 wk or more.

- Monitor adolescents as well as adults for changes in behavior that may indicate suicidality. Suicide is an inherent risk with any depressed patient and may remain until there is significant improvement.

- Monitor vital signs closely and CV system responses during early therapy, particularly in patients with cardiovascular disorders and older adults receiving daily doses in excess of 20 mg. Withhold drug and inform prescriber if BP falls more than 20 mm Hg or if there is a sudden increase in pulse rate.

- Lab tests: Periodic LFTs.

- Monitor I&O ratio and bowel pattern during early therapy and when patient is on large doses.

- Assess and advise prescriber as indicated for prominent anticholinergic effects (xerostomia, blurred vision, constipation, paralytic ileus, urinary retention, delayed micturition).

- Assess condition of oral membranes frequently; institute symptomatic treatment if necessary. Xerostomia can interfere with appetite, fluid intake, and integrity of tooth surfaces.

Patient & Family Education

- Report promptly changes in mood or behavior indicative of suicidal thinking.

- Consult prescriber about safe amount of alcohol, if any, that can be taken. Actions of both alcohol and protriptyline are potentiated when used together for up to 2 wk after the TCA is discontinued.

- Consult prescriber before taking any OTC medications.

- Be aware that effects of barbiturates and other CNS depressants are enhanced by TCAs.

- Avoid potentially hazardous activities requiring alertness and skill until response to drug is known.

- Avoid exposure to the sun without protecting skin with sunscreen lotion (SPF higher than 12). Photosensitivity reactions may occur.

PSEUDOEPHEDRINE HYDROCHLORIDE
(soo-doe-e-fed′rin)

Cenafed, Decongestant Syrup, Dorcol Children's Decongestant, Eltor ◆, Eltor 120 ◆, Halofed, Novafed, PediaCare, Pseudofrin ◆, Robidrine ◆, Sudafed, Sudrin

Classifications: ALPHA- AND BETA-ADRENERGIC RECEPTOR AGONIST; DECONGESTANT
Therapeutic: NASAL DECONGESTANT
Prototype: Epinephrine
Pregnancy Category: C

AVAILABILITY 30 mg, 60 mg tablets; 120 mg, 240 mg sustained release tablets; 15 mg/5 mL, 30 mg/5 mL liquid; 7.5 mg/0.8 mL drops

ACTION & THERAPEUTIC EFFECT Sympathomimetic amine that produces decongestion of respiratory tract mucosa by stimulating the sympathetic nerve endings including alpha-, beta$_1$-, and beta$_2$-receptors. *Effect is caused by vasoconstriction and thus increased nasal airway patency.*

USES Symptomatic relief of nasal congestion associated with rhinitis, coryza, and sinusitis and for eustachian tube congestion.

CONTRAINDICATIONS Hypersensitivity to sympathomimetic amines; severe hypertension; severe coronary artery disease; use within 14 days of MAOIs; hyperthyroidism; prostatic hypertrophy.
CAUTIOUS USE Hypertension, heart disease, renal impairment; acute MI,

angina; closed-angle glaucoma; concurrent use of ACE INHIBITOR; Pregnancy (Category C), lactation. Safe use in children younger than 2 y **(PO form)**, children younger than 12 y **(sustained release form)** is not established.

ROUTE & DOSAGE

Nasal Congestion
Adult: **PO** 60 mg q4–6h or 120 mg sustained release q12h
Geriatric: **PO** 30–60 mg q6h prn
Child: **PO** 2–6 y, 15 mg q4–6h (max: 60 mg/day); 6–11 y, 30 mg q4–6h (max: 120 mg/day)

ADMINISTRATION
Oral
- Ensure that sustained release form is not chewed or crushed. **Must be** swallowed whole.

ADVERSE EFFECTS (≥1%) **Body as a Whole:** *Transient stimulation,* tremulousness, difficulty in voiding. **CV:** Arrhythmias, palpitation, *tachycardia.* **CNS:** *Nervousness,* dizziness, headache, sleeplessness, numbness of extremities. **GI:** Anorexia, dry mouth, nausea, vomiting.

INTERACTIONS Drug: Other SYMPATHOMIMETICS increase pressor effects and toxicity; MAO INHIBITORS may precipitate hypertensive crisis; BETA-BLOCKERS may increase pressor effects; may decrease antihypertensive effects of **guanethidine, methyldopa, reserpine.**

PHARMACOKINETICS Absorption: Readily from GI tract. **Onset:** 15–30 min. **Duration:** 4–6 h (8–12 h sustained release). **Distribution:** Crosses placenta; distributed into breast milk. **Metabolism:** Partially metabolized in liver. **Elimination:** In urine.

Common adverse effects in *italic*, life-threatening effects underlined; generic names in **bold**; classifications in SMALL CAPS; ◆ Canadian drug name; ⊚ Prototype drug

1281

NURSING IMPLICATIONS

Assessment & Drug Effects

- Monitor HR and BP, especially in those with a history of cardiac disease. Report tachycardia or hypertension.

Patient & Family Education

- Do not take drug within 2 h of bedtime because drug may act as a stimulant.
- Discontinue medication and consult prescriber if extreme restlessness or signs of sensitivity occur.
- Consult prescriber before concomitant use of OTC medications; many contain ephedrine or other sympathomimetic amines and might intensify action of pseudoephedrine.

PSYLLIUM HYDROPHILIC MUCILLOID 🅿

(sill'i-um)

Hydrocil Instant, Karasil ♦, Konsyl, Metamucil, Modane Bulk, Perdiem Plain, Reguloid, Serutan, Siblin, V-Lax
Classification: BULK-PRODUCING LAXATIVE
Therapeutic: BULK LAXATIVE
Pregnancy Category: C

AVAILABILITY 3.4 g/dose powder; 2.5 g, 3.4 g, 4.03 g/teaspoon granules

ACTION & *THERAPEUTIC EFFECT*
Bulk-producing laxative that absorbs liquid in the GI tract, facilitating peristalsis and bowel motility. *Bulk-producing laxative that promotes peristalsis and natural elimination.*

USES Chronic atonic or spastic constipation and constipation associated with rectal disorders or anorectal surgery.

CONTRAINDICATIONS Esophageal and intestinal obstruction, dysphagia; nausea, vomiting, fecal impaction, acute abdomen; undiagnosed abdominal pain, appendicitis.
CAUTIOUS USE Diabetics; pregnancy (category C); children younger than 6 y.

ROUTE & DOSAGE

Constipation or Diarrhea
Adult: **PO** 1–2 rounded tsp or 1 packet 1–3 times/day prn
Child (younger than 6 y): **PO** 1 tsp in water at bedtime

ADMINISTRATION

Oral

- Fill an 8-oz (240-mL) water glass with cool water, milk, fruit juice, or other liquid; sprinkle powder into liquid; stir briskly; and give immediately (if effervescent form is used, add liquid to powder). Granules should not be chewed.
- Follow each dose with an additional glass of liquid to obtain best results.
- Exercise caution with older adult patient who may aspirate the drug.

ADVERSE EFFECTS (≥1%) Hematologic: Eosinophilia. **GI:** Nausea and vomiting, diarrhea (with excessive use); GI tract strictures when drug used in dry form, abdominal cramps.

INTERACTIONS Drug: Psyllium may decrease absorption and clinical effects of ANTIBIOTICS, **warfarin, digoxin, nitrofurantoin,** SALICYLATES.

PHARMACOKINETICS Absorption: Not absorbed from GI tract. **Onset:** 12–24 h. **Peak:** 1–3 days.

NURSING IMPLICATIONS

Assessment & Drug Effects

- Report promptly to prescriber if patient complains of retrosternal

---

Common adverse effects in *italic*, life-threatening effects underlined; generic names in **bold**; classifications in SMALL CAPS; ♣ Canadian drug name; 🅿 Prototype drug

pain after taking the drug. Drug may be lodged as a gelatinous mass (because of poor mixing) in the esophagus.

- Monitor therapeutic effectiveness. When psyllium is used as either a bulk laxative or to treat diarrhea, the expected effect is formed stools. Laxative effect usually occurs within 12–24 h. Administration for 2 or 3 days may be needed to establish regularity.
- Assess for complaints of abdominal fullness. Smaller, more frequent doses spaced throughout the day may be indicated to relieve discomfort of abdominal fullness.
- Monitor warfarin and digoxin levels closely if either is given concurrently.

Patient & Family Education

- Note sugar and sodium content of preparation if on low-sodium or low-calorie diet. Some preparations contain natural sugars, whereas others contain artificial sweeteners.
- Understand that drug works to relieve both diarrhea and constipation by restoring a more normal moisture level to stool.
- Be aware that drug may reduce appetite if it is taken before meals.

PYRANTEL PAMOATE
(pi-ran'tel)

Pin-X
Classification: ANTHELMINTIC
Therapeutic: ANTHELMINTIC
Prototype: Mebendazole
Pregnancy Category: C

AVAILABILITY 180 mg, 250 mg, 720 mg capsules; 50 mg/mL suspension

ACTION & *THERAPEUTIC EFFECT*
Exerts selective depolarizing neuromuscular blocking action that results in spastic paralysis of worm. *Causes evacuation of worms from intestines.*

USES *Enterobius vermicularis* (pinworm) and *Ascaris lumbricoides* (roundworm) infestations.
UNLABELED USES Hookworm infestations; trichostrongylosis.

CONTRAINDICATIONS Hypersensitivity to pyrantel.
CAUTIOUS USE Liver dysfunction; malnutrition; dehydration; anemia; pregnancy (category C).

ROUTE & DOSAGE

Pinworm or Roundworm
Adult/Child: **PO** 11 mg/kg as a single dose (max: 1 g)

ADMINISTRATION

Oral

- Shake suspension well before pouring it to ensure accurate dosage.
- Give with milk or fruit juices and without regard to prior ingestion of food or time of day.
- Store below 30° C (86° F). Protect from light.

ADVERSE EFFECTS (≥1%) **CNS:** Dizziness, headache, drowsiness, insomnia. **GI:** Anorexia, *nausea,* vomiting, abdominal distention, diarrhea, *tenesmus,* transient elevation of AST. **Skin:** Skin rashes.

INTERACTIONS Drug: Piperazine and pyrantel may be mutually antagonistic.

PHARMACOKINETICS Absorption: Poorly from GI tract. **Peak:** 1–3 h. **Metabolism:** In liver. **Elimination:** Greater than 50% in feces, 7% in urine.

Common adverse effects in *italic*, life-threatening effects <u>underlined</u>; generic names in **bold**; classifications in SMALL CAPS; ♦ Canadian drug name; ⊘ Prototype drug

1283

NURSING IMPLICATIONS

Assessment & Drug Effects

- Lab tests: Monitor baseline and periodic LFTs in individuals with known liver dysfunction.

Patient & Family Education

- Do not drive or engage in other potentially hazardous activities until response to drug is known.

PYRAZINAMIDE

(peer-a-zin′a-mide)

PZA, Tebrazid ✦
Classifications: ANTIBIOTIC; ANTI-TUBERCULOSIS
Therapeutic: ANTITUBERCULOSIS
Pregnancy Category: C

AVAILABILITY 500 mg tablets

ACTION & *THERAPEUTIC EFFECT*
Pyrazinoic acid amide, analog of nicotinamide. *Bacteriostatic against* Mycobacterium tuberculosis. *Not used as sole agent against TB infection.*

USES Treatment of tuberculosis infection (with other agents).

CONTRAINDICATIONS Severe liver damage, acute gout.
CAUTIOUS USE History of gout or diabetes mellitus; impaired kidney function; alcoholism; history of peptic ulcer; acute intermittent porphyria; pregnancy (category C), lactation.

ROUTE & DOSAGE

Tuberculosis

Adult/Child: **PO** 15–30 mg/kg/day (max: 3 g/day)
Renal Impairment Dosage Adjustment
CrCl 10–50 mL/min: extend interval to q48–72 hours; *CrCl less than 10 mL/min:* Extend interval to q72h

ADMINISTRATION

Oral

- Discontinue drug if hepatic reactions (jaundice, pruritus, icteric sclerae, yellow skin) or hyperuricemia with acute gout (severe pain in great toe and other joints) occurs.
- Store at 15°–30° C (59°–86° F) in tightly closed container.

ADVERSE EFFECTS (≥1%) Body as a Whole: *Active gout,* arthralgia, lymphadenopathy. **Urogenital:** Difficulty in urination. **CNS:** Headache. **Skin:** Urticaria. **Hematologic:** Hemolytic anemia, decreased plasma prothrombin. **GI:** Splenomegaly, fatal hemoptysis, aggravation of peptic ulcer, *hepatotoxicity, abnormal liver function tests.* **Metabolic:** *Rise in serum uric acid.*

DIAGNOSTIC TEST INTERFERENCE
Pyrazinamide may produce a temporary decrease in ***17-ketosteroids*** and an increase in ***protein-bound iodine.***

INTERACTIONS Drug: Increase in liver toxicity (including fatal hepatoxicity in when treating latent TB) with **rifampin.**

PHARMACOKINETICS Absorption: Readily from GI tract. **Peak:** 2 h. **Distribution:** Crosses blood–brain barrier. **Metabolism:** In liver. **Elimination:** Slowly in urine. **Half-Life:** 9–10 h.

NURSING IMPLICATIONS

Assessment & Drug Effects

- Observe and supervise closely. Patients should receive at least one other effective antituberculosis agent concurrently.
- Examine patients at regular intervals and question about possible signs of toxicity: Liver enlargement or tenderness, jaundice, fever, anorexia, malaise, impaired vascular

Common adverse effects in *italic*, life-threatening effects underlined; generic names in **bold**; classifications in SMALL CAPS; ✦ Canadian drug name; ⊙ Prototype drug

integrity (ecchymoses, petechiae, abnormal bleeding).

- Hepatic reactions appear to occur more frequently in patients receiving high doses.
- Lab tests: Baseline LFTs (especially AST, ALT, serum bilirubin) and at 2–4 wk intervals during therapy. Blood uric acid determinations are advised before, during, and following therapy.

Patient & Family Education

- Report to prescriber onset of difficulty in voiding. Keep fluid intake at 2000 mL/day if possible.
- Monitor blood glucose (diabetics) for possible loss of glycemic control.

PYRETHRINS
(peer'e-thrins)

A-200 Pyrinate, Barc, Blue, Pyrinate, Pyrinyl, R & C, RID, TISIT, Triple X
Classifications: ANTIPARASITIC; PEDICULICIDE
Therapeutic: SCABICIDE
Prototype: Permethrin
Pregnancy Category: C

AVAILABILITY 0.18%, 0.2%, 0.3% liquid; 0.3% gel; 0.3% shampoo

ACTION & *THERAPEUTIC EFFECT*
Pediculicide solution that acts as a contact poison affecting the parasite's nervous system, causing paralysis and death. *Controls head lice, pubic (crab) lice, and body lice and their eggs (nits).*

USES External treatment of *Pediculus humanus* infestations.

CONTRAINDICATIONS Sensitivity to solution components; skin infections and abrasions.
CAUTIOUS USE Ragweed-sensitized patient; asthma; preg-

nancy (category C), lactation, infants, children.

ROUTE & DOSAGE

Pediculus humanus Infestations
Adult: **Topical** See ADMINISTRATION for appropriate application

ADMINISTRATION
Topical

- Apply enough solution to completely wet infested area, including hair. Allow to remain on area for 10 min.
- Wash and rinse with large amounts of warm water.
- Use fine-toothed comb to remove lice and eggs from hair.
- Shampoo hair to restore body and luster.
- Repeat treatment once in 24 h if necessary.
- Repeat treatment in 7–10 days to kill newly hatched lice.
- Do not apply to eyebrows or eyelashes without consulting prescriber.
- Flush eyes with copious amounts of warm water if accidental contact occurs.

ADVERSE EFFECTS (≥1%) **Body as a Whole:** Irritation with repeated use.

NURSING IMPLICATIONS
Patient & Family Education

- Do not swallow, inhale, or allow pyrethrins to contact mucosal surfaces or the eyes.
- Discontinue use and consult prescriber if treated area becomes irritated.
- Examine each family member carefully; if infested, treat immediately to prevent spread or reinfestation of previously treated patient.

P

Common adverse effects in *italic*, life-threatening effects underlined; generic names in **bold**; classifications in SMALL CAPS; ♣ Canadian drug name; ☻ Prototype drug

1285

- Dry clean, boil, or otherwise treat contaminated clothing. Sterilize (soak in pyrethrins) combs and brushes used by patient.
- Do not share combs, brushes, or other headgear with another person.

PYRIDOSTIGMINE BROMIDE
(peer-id-oh-stig'meen)

Mestinon, Regonol
Classifications: CHOLINERGIC MUS-CLE STIMULANT; ANTICHOLINESTERASE
Therapeutic: CHOLINESTERASE INHIB-ITOR
Prototype: Neostigmine
Pregnancy Category: C

AVAILABILITY 60 mg/5 mL syrup; 60 mg tablet; 180 mg extended release tablet; 5 mg/mL injection

ACTION & *THERAPEUTIC EFFECT*
Indirect-acting cholinergic that inhibits cholinesterase activity. Facilitates transmission of impulses across myoneural junctions by blocking destruction of acetylcholine. *Has direct stimulant action on voluntary muscle fibers and possibly on autonomic ganglia and CNS neurons. Produces increased tone in skeletal muscles.*

USES Myasthenia gravis and as an antagonist to nondepolarizing skeletal muscle relaxants (e.g., curariform drugs).

CONTRAINDICATIONS Hypersensitivity to anticholinesterase agents; mechanical obstruction of urinary or intestinal tract; hypotension; lactation.
CAUTIOUS USE Hypersensitivity to bromides; bronchial asthma; epilepsy; recent cardiac occlusion; renal impairment; vagotonia; hyperthyroidism; peptic ulcer; cardiac dysrhythmias; bradycardia; pregnancy (category C).

ROUTE & DOSAGE

Myasthenia Gravis
Adult: **PO** 60 mg–1.5 g/day spaced according to response of individual patient; sustained release: 180–540 mg 1–2 times/day at intervals of at least 6 h **IM/IV** Approximately ¹/₃₀ of PO dose
Child: **PO** 7 mg/kg/day divided into 5–6 doses
Neonates: **PO** 5 mg q4–6h **IM/IV** 0.05–0.15 mg/kg q4–6h

Reversal of Muscle Relaxants
Adult: **IV** 10–20 mg immediately preceded by IV atropine

ADMINISTRATION

Oral
- Give with food or fluid.
- Ensure that sustained release form is not chewed or crushed. **Must be** swallowed whole.
- Note: A syrup is available. Some patients may not like it because it is sweet; try to make it more palatable by giving it over ice chips. The syrup formulation contains 5% alcohol.

Intramuscular
- Note: Parenteral dose is about ¹/₃₀ the oral adult dose.
- Give deep IM into a large muscle.

Intravenous
PREPARE: Direct: Give undiluted. Do NOT add to IV solutions.
ADMINISTER: Direct: Give at a rate of 0.5 mg over 1 min for myasthenia gravis; 5 mg over 1 min for reversal of muscle relaxants.

- Store at 15°–30° C (59°–86° F). Protect from light and moisture.

ADVERSE EFFECTS (≥1%) **Skin:** Acneiform rash. **Hematologic:** Thrombophlebitis (following IV administration). **GI:** *Nausea, vomiting, diarrhea.* **Special Senses:** *Miosis.* **Body as a Whole:** *Excessive salivation and sweating,* weakness, fasciculation. **Respiratory:** Increased bronchial secretion, <u>bronchoconstriction</u>. **CV:** Bradycardia, hypotension.

INTERACTIONS Drug: Atropine NONDEPOLARIZING MUSCLE RELAXANTS antagonize effects of pyridostigmine.

PHARMACOKINETICS Absorption: Poorly from GI tract. **Onset:** 30–45 min PO; 15 min IM; 2–5 min IV. **Duration:** 3–6 h. **Distribution:** Crosses placenta. **Metabolism:** In liver and in serum and tissue by cholinesterases. **Elimination:** In urine.

NURSING IMPLICATIONS

Assessment & Drug Effects

- Report increasing muscular weakness, cramps, or fasciculations. Failure of patient to show improvement may reflect either underdosage or overdosage.
- Observe patient closely if atropine is used to abolish GI adverse effects or other muscarinic adverse effects because it may mask signs of overdosage (cholinergic crisis): Increasing muscle weakness, which through involvement of respiratory muscles can lead to death.
- Monitor vital signs frequently, especially respiratory rate.
- Observe for signs of cholinergic reactions (see Appendix F), particularly when drug is administered IV.
- Observe neonates of myasthenic mothers, who have received pyridostigmine, closely for difficulty in breathing, swallowing, or sucking.
- Observe patient continuously when used as muscle relaxant

antagonist. Airway and respiratory assistance **must be** maintained until full recovery of voluntary respiration and neuromuscular transmission is assured. Complete recovery usually occurs within 30 min.

Patient & Family Education

- Be aware that duration of drug action may vary with physical and emotional stress, as well as with severity of disease.
- Report onset of rash to prescriber. Drug may be discontinued.
- Sustained release tablets may become mottled in appearance; this does not affect their potency.

PYRIDOXINE HYDROCHLORIDE (VITAMIN B₆)
(peer-i-dox′een)
Classification: VITAMIN
Therapeutic: VITAMIN B₆ REPLACEMENT
Pregnancy Category: A (C if greater than RDA)

AVAILABILITY 25 mg, 50 mg, 100 mg, 250 mg, 500 mg tablets; 100 mg/mL injection

ACTION & *THERAPEUTIC EFFECT* Water-soluble complex of three closely related compounds with B₆ activity. Converted in body to pyridoxal, a coenzyme that functions in protein, fat, and carbohydrate metabolism and in facilitating release of glycogen from liver and muscle. In protein metabolism, participates in enzymatic transformations of amino acids and conversion of tryptophan to niacin and serotonin. Aids in energy transformation in brain and nerve cells, and is thought to stimulate heme production. *Effectiveness is evaluated by improvement of B₆ deficiency manifestations: Nausea, vomiting, skin lesions resembling those of riboflavin and niacin deficiency,*

Common adverse effects in *italic*, life-threatening effects <u>underlined</u>; generic names in **bold**; classifications in SMALL CAPS; ♣ Canadian drug name; ☻ Prototype drug

1287

edema, CNS symptoms, hypochromic microcytic anemia.

USES Prophylaxis and treatment of pyridoxine deficiency, as seen with inadequate dietary intake, drug-induced deficiency (e.g., isoniazid, oral contraceptives), and inborn errors of metabolism (vitamin B_6–dependent convulsions or anemia). Also to prevent chloramphenicol-induced optic neuritis, to treat acute toxicity caused by overdosage of cycloserine, hydralazine, isoniazid (INH); alcoholic polyneuritis; sideroblastic anemia associated with high serum iron concentration. Has been used for management of many other conditions ranging from nausea and vomiting in radiation sickness and pregnancy to suppression of postpartum lactation.

CONTRAINDICATIONS IV form: Cardiac disease.
CAUTIOUS USE Renal impairment; neonatal prematurity with renal impairment; cardiac disease; pregnancy [category A or (C if greater than RDA)].

ROUTE & DOSAGE

Dietary Deficiency
Adult: **PO/IM/IV** 10–20 mg/day × 2–3 wk
Child: **PO** 5–25 mg/day × 3 wk, then 1.5–2.5 mg/day

Pyridoxine Deficiency Syndrome
Adult: **PO/IM/IV** Initial dose up to 600 mg/day may be required; then up to 50 mg/day

Isoniazid-Induced Deficiency
Adult: **PO/IM/IV** 100 mg/day × 3 wk, then 30 mg/day
Child: **PO** 10–50 mg/day × 3 wk, then 1–2 mg/kg/day

Pyridoxine-Dependent Seizures
Neonate/Infant: **PO/IM/IV** 50–100 mg/day

ADMINISTRATION

Oral
- Ensure that sustained release and enteric forms are not chewed or crushed. **Must be** swallowed whole.

Intramuscular
- Give deep IM into a large muscle.

Intravenous

PREPARE: **Direct:** Give undiluted. **Continuous:** May be added to most standard IV solutions.
ADMINISTER: **Direct:** Give at a rate of 50 mg or fraction thereof over 60 seconds. **Continuous:** Give according to ordered rate for infusion.

- Store at 15°–30° C (59°–86° F) in tight, light-resistant containers. Avoid freezing.

ADVERSE EFFECTS (≥1%) **Body as a Whole:** Paresthesias, slight flushing or feeling of warmth, temporary burning or stinging pain in injection site. **CNS:** Somnolence seizures (particularly following large parenteral doses). **Metabolic:** Low folic acid levels.

INTERACTIONS Drug: Isoniazid, cycloserine, penicillamine, hydralazine, and ORAL CONTRACEPTIVES may increase pyridoxine requirements; may reverse or antagonize therapeutic effects of **levodopa.**

PHARMACOKINETICS Absorption: Readily from GI tract. **Distribution:** Stored in liver; crosses placenta. **Metabolism:** In liver. **Elimination:** In urine.

NURSING IMPLICATIONS

Assessment & Drug Effects

- Monitor neurologic status to determine therapeutic effect in deficiency states.
- Record a complete dietary history so poor eating habits can be identified and corrected (a single vitamin deficiency is rare; patient can be expected to have multiple vitamin deficiencies).
- Lab tests: Periodic Hct and Hgb, and serum iron.

Patient & Family Education

- Learn rich dietary sources of vitamin B₆: Yeast, wheat germ, whole grain cereals, muscle and glandular meats (especially liver), legumes, green vegetables, bananas.
- Do not self-medicate with vitamin combinations (OTC) without first consulting prescriber.

PYRIMETHAMINE

(peer-i-meth'a-meen)

Daraprim
Classification: FOLIC ACID ANTAGONIST
Therapeutic: ANTIMALARIAL
Prototype: Chloroquine
Pregnancy Category: C

AVAILABILITY 25 mg tablets

ACTION & *THERAPEUTIC EFFECT*
Long-acting folic acid antagonist. Selectively inhibits action of dehydrofolic reductase in parasites with resulting blockade of folic acid metabolism. *Prevents development of fertilized gametes in the mosquito and thus helps to prevent transmission of malaria.*

USES Prophylaxis and treatment of malaria; toxoplasmosis.

CONTRAINDICATIONS Chloroguanide-resistant malaria; hypersensitivity to sulfonamides; megaloblastic anemia caused by folate deficiency; lactation.

CAUTIOUS USE Convulsive disorders; asthma; bone marrow suppression; folate deficiency; hepatic disease; renal disease; seizure disorder; pregnancy (category C); children younger than 2 mo.

ROUTE & DOSAGE

Malaria Treatment
Adult/Adolescent/Child (over 4 y): **PO** 25 mg daily x 2 days

Malaria Chemoprophylaxis
Adult/Adolescent/Child (over 10 y): **PO** 25 mg once/wk
Child: **PO** *Younger than 4 y,* 6.25 mg once/wk; *4–10 y,* 12.5 mg once/wk

Toxoplasmosis
Adult: **PO** 50–75 mg/day with a sulfonamide for 1–3 wk, then decrease dose by half and continue for 1 mo
Child: **PO** 1 mg/kg/day divided into 2 doses with a sulfonamide for 1–3 wk, then decrease to 0.5 mg/kg/day for 1 mo (max: 25 mg/day)

ADMINISTRATION

Oral

- Minimize GI distress by giving with meals. If symptoms persist, dosage reduction may be necessary.
- Give on same day each week for malaria prophylaxis. Begin when individual enters malarious area and continue for 10 wk after leaving the area.

ADVERSE EFFECTS (≥1%) **GI:** *Anorexia, vomiting, atrophic glossitis, abdominal cramps, diarrhea.* **Skin:** Skin rashes. **Hematologic:** *Folic acid deficiency (megaloblastic*

Common adverse effects in *italic*, life-threatening effects <u>underlined</u>; generic names in **bold**; classifications in SMALL CAPS; ♣ Canadian drug name; ⊙ Prototype drug

1289

anemia, leukopenia, thrombocy-topenia, pancytopenia, diarrhea).
CNS: CNS stimulation including convulsions, respiratory failure.

INTERACTIONS Drug: Folic acid, *para*-aminobenzoic acid (PABA) may decrease effectiveness against toxoplasmosis.

PHARMACOKINETICS Absorption: Readily from GI tract. **Peak:** 2 h. **Distribution:** Concentrates in kidneys, lungs, liver, and spleen; distributed into breast milk. **Elimination:** Slowly in urine; excretion may extend over 30 days or longer. **Half-Life:** 54–148 h.

NURSING IMPLICATIONS

Assessment & Drug Effects

- Monitor patient response closely. Dosages required for treatment of toxoplasmosis approach toxic levels.
- Lab tests: Perform blood counts, including platelets, twice weekly during therapy.
- Withhold drug and notify prescriber if hematologic abnormalities appear.

Patient & Family Education

- Be aware that folic acid deficiency may occur with long-term use of pyrimethamine. Report to prescriber weakness, and pallor (from anemia), ulcerations of oral mucosa, superinfections, glossitis; GI disturbances such as diarrhea and poor fat absorption, fever. Folate (folinic acid) replacement may be prescribed. Increase food sources of folates (if allowed) in diet.

QUAZEPAM
(qua′ze-pam)
Doral
Classifications: BENZODIAZEPINE; ANXIOLYTIC; SEDATIVE-HYPNOTIC

Therapeutic: ANTIANXIETY; SEDATIVE-HYPNOTIC
Prototype: Lorazepam
Pregnancy Category: X
Controlled Substance:: Schedule IV

AVAILABILITY 15 mg tablets

ACTION & *THERAPEUTIC EFFECT*
Believed to potentiate gamma-aminobutyric acid (GABA) neuronal inhibition in the limbic, neocortical, and mesencephalic reticular systems. *Significantly decreases total wake time and significantly increases sleep time. REM sleep is essentially unchanged.*

USES Insomnia characterized by difficulty in falling asleep, frequent nocturnal awakenings, or early morning awakenings.

CONTRAINDICATIONS Hypersensitivity to quazepam or benzodiazepines; sleep apnea; pregnancy (category X), lactation.

CAUTIOUS USE Impaired liver and kidney function; compromised respiratory function; history of seizures; elderly; debilitated clients. Safety and effectiveness in children younger than 18 y are not established.

ROUTE & DOSAGE

Insomnia
Adult: **PO** 7.5–15 mg at bedtime

ADMINISTRATION

Oral

- Initial dose is usually 15 mg but can often be effectively reduced after several nights of therapy.
- Use lowest effective dose in older adults as soon as possible.

ADVERSE EFFECTS (≥1%) **CNS:** *Drowsiness, headache,* fatigue, dizziness, dry mouth. **GI:** Dyspepsia.

Common adverse effects in *italic*, life-threatening effects underlined; generic names in **bold**; classifications in SMALL CAPS; ♣ Canadian drug name; ⊘ Prototype drug

INTERACTIONS Drug: Alcohol, CNS DEPRESSANTS, ANTICONVULSANTS potentiate CNS depression; **cimetidine** increases quazepam plasma levels, increasing its toxicity; may decrease antiparkinsonism effects of **levodopa;** may increase **phenytoin** levels; **smoking** decreases sedative effects of quazepam. **Herbal: Kava, valerian** may potentiate sedation.

PHARMACOKINETICS Absorption: Readily from GI tract. **Onset:** 30 min. **Peak:** 2 h. **Distribution:** Crosses placenta; distributed into breast milk. **Metabolism:** In liver to active metabolites. **Elimination:** In urine and feces. **Half-Life:** 39 h.

NURSING IMPLICATIONS

Assessment & Drug Effects
- Monitor for respiratory depression in patients with chronic respiratory insufficiency.
- Monitor for suicidal tendencies in previously depressed clients.
- Daytime drowsiness is more likely to occur in older adult clients.

Patient & Family Education
- Inform prescriber about any alcohol consumption and prescription or nonprescription medication that you take. Avoid alcohol use since it potentiates CNS depressant effects.
- Inform prescriber immediately if you become pregnant. This drug causes birth defects.
- Do not drive or engage in potentially hazardous activities until response to drug is known.
- Do not increase the dose of this drug; inform prescriber if the drug no longer works.
- This drug may cause daytime sedation, even for several days after drug is discontinued.

QUETIAPINE FUMARATE
(ke-ti-a′peen)

Seroquel, Seroquel XR
Classification: ATYPICAL ANTIPSYCHOTIC
Therapeutic: ANTIPSYCHOTIC
Prototype: Clozapine
Pregnancy Category: C

AVAILABILITY 25 mg, 100 mg, 200 mg tablets; 50 mg, 200 mg, 300 mg, 400 mg extended release tablets

ACTION & THERAPEUTIC EFFECT Antagonizes multiple neurotransmitter receptors in the brain including serotonin ($5\text{-}HT_{1A}$ and $5\text{-}HT_2$) as well as dopamine D_1 and D_2 receptors. *Effectiveness indicated by a reduction in psychotic behavior.*

USES Management of schizophrenia, maintenance of acute bipolar disorder, and add-on therapy for major depressive disorder.
UNLABELED USES Management of agitation and dementia.

CONTRAINDICATIONS Hypersensitivity to quetiapine; alcohol use; suicidal ideation; lactation; dementia-related psychosis.
CAUTIOUS USE Liver function impairment, **immediate release tablets:** Older adults, or hepatic impairment; cardiovascular disease (history of MI or ischemic heart disease, heart failure, arrhythmias, CVA, hypotension, dehydration, treatment with antihypertensives); history of seizures or suicide; breast cancer; Alzheimer's, Parkinson's disease; patient at risk for aspiration pneumonia; debilitated patients; cerebrovascular disease; pregnancy (category C). Safe use in children younger than 10 y is not established. **Extended release tablets:** Safe use in children younger than 18 y is not established.

Common adverse effects in *italic*, life-threatening effects underlined; generic names in **bold**; classifications in SMALL CAPS; ♣ Canadian drug name; ⊘ Prototype drug

1291

ROUTE & DOSAGE

Bipolar Depression
Adult: **PO** Day 1: 50 mg at bedtime, Day 2: 100 mg at bedtime, Day 3: 200 mg at bedtime, then 300 mg daily at bedtime

Schizophrenia
Adult: **PO** *(Immediate release)* Start 25 mg b.i.d., may increase by 25–50 mg b.i.d. to t.i.d. on the second or third day as tolerated to a target dose of 300–400 mg/day divided b.i.d. to t.i.d., may adjust dose by 25–50 mg b.i.d. daily as needed (max: 800 mg/day); *(Extended release)* 300 mg daily at bedtime, titrate up to 400–800 mg daily (max: 800 mg/day)

Manic Episodes in Bipolar Disorder Monotherapy or with Lithium/Divalproex (Immediate Release Only)
Adult: **PO** Start with total of 100 mg (in two doses) day 1, increase to 400 mg/day (in two doses) by day 4 OR extended release 300 mg on day 1, then 600 mg on day 2, may adjust by 400–800 mg/day based on response
Geriatric: Titrate more slowly due to risk of orthostatic hypotension

Bipolar I Disorder Maintenance
Adult: **PO** 200–400 mg b.i.d.

Hepatic Impairment Dosage Adjustment
Immediate release: Start with 25 mg dose and increase by 25–50 mg/day; *Extended release:* Start with 50 mg PO on day 1, then increase by 50 mg/day to the lowest effective and tolerable dose

ADMINISTRATION

Oral
- Dose is usually retitrated over a period of several days when patient has been off the drug for longer than 1 wk.
- Follow recommended lower doses and slower titration for the older adults, the debilitated, and those with hepatic impairment or a predisposition to hypotension.
- Store at 15°–30° C (59°–86° F).

ADVERSE EFFECTS (≥1%) **Body as a Whole:** Asthenia, fever, hypertonia, dysarthria, flu syndrome, weight gain, peripheral edema, increased risk of suicidal thinking. **CNS:** *Dizziness, headache, somnolence.* **CV:** Postural hypotension, tachycardia, palpitations. **GI:** Dry mouth, dyspepsia, abdominal pain, constipation, anorexia. **Metabolic:** Hyperglycemia, diabetes mellitus. **Respiratory:** Rhinitis, pharyngitis, cough, dyspnea. **Skin:** Rash, sweating. **Hematologic:** Leukopenia.

INTERACTIONS Drug: BARBITURATES, **carbamazepine, phenytoin, rifampin, thioridazine** may increase clearance of quetiapine. Quetiapine may potentiate the cognitive and motor effects of **alcohol,** enhance the effects of ANTIHYPERTENSIVE AGENTS, antagonize the effects of **levodopa** and DOPAMINE AGONISTS. **Ketoconazole, itraconazole, fluconazole, erythromycin** may decrease clearance of quetiapine. Drugs that increase the QT interval (e.g., **amiodarone, clarithromycin,** ANTIARRHYTHMICS, **haloperidol**) increase risk of cardiac effects. Other ANTIPSYCHOTICS increase the risk of adverse effects. **Herbal:** St. John's wort may cause **serotonin** syndrome (see Appendix F).

Common adverse effects in *italic*, life-threatening effects underlined; generic names in **bold**; classifications in SMALL CAPS; ♣ Canadian drug name; ⊘ Prototype drug

PHARMACOKINETICS Absorption: Rapidly and completely absorbed from GI tract. **Peak:** 1.5 h. **Distribution:** 83% protein bound. **Metabolism:** In liver (CYP3A4). **Elimination:** 73% in urine, 20% in feces. **Half-Life:** 6 h.

NURSING IMPLICATIONS

Assessment & Drug Effects

- Monitor diabetics for loss of glycemic control.
- Monitor for changes in behavior that may indicate suicidality.
- Reassess need for continued treatment periodically.
- Withhold the drug and immediately report S&S of tardive dyskinesia or neuroleptic malignant syndrome (see Appendix F).
- Lab tests: Periodic LFTs, lipid profile, thyroid function, blood glucose, CBC with differential.
- Monitor ECG periodically, especially in those with known cardiovascular disease.
- Baseline cataract exam is recommended when therapy is started and at 6 mo intervals thereafter.
- Monitor patients with a history of seizures for lowering of the seizure threshold.

Patient & Family Education

- Carefully monitor blood glucose levels if diabetic.
- Exercise caution with potentially dangerous activities requiring alertness, especially during the first week of drug therapy or during dose increments.
- Make position changes slowly, especially when changing from lying or sitting to standing to avoid dizziness, palpitations, and fainting.
- Avoid alcohol consumption and activities that may cause overheating and dehydration.

QUINAPRIL HYDROCHLORIDE

(quin′a-pril)

Accupril

Classifications: ANGIOTENSIN-CONVERTING ENZYME (ACE) INHIBITOR; ANTIHYPERTENSIVE
Therapeutic: ANTIHYPERTENSIVE
Prototype: Enalapril
Pregnancy Category: D

AVAILABILITY 5 mg, 10 mg, 20 mg, 40 mg tablets

ACTION & *THERAPEUTIC EFFECT*
Potent, long-acting ACE inhibitor that lowers BP by interrupting the conversion sequences initiated by renin to form angiotensin II, a vasoconstrictor. Also decreases circulating aldosterone, a secretory response to angiotensin II stimulation. Reduces PCWP, systemic vascular resistance, and mean arterial pressure, with concurrent increases in cardiac output, cardiac index, and stroke volume. *Lowers BP by producing vasodilation. Effective in the treatment of CHF since it improves cardiac indicators.*

USES Mild to moderate hypertension, CHF.

CONTRAINDICATIONS Hypersensitivity to quinapril or other ACE inhibitors; history of angioedema; pregnancy (category D), lactation.
CAUTIOUS USE Renal insufficiency; autoimmune disease, volume-depleted patients, aortic stenosis, hypertrophic cardiomyopathy; renal artery stenosis, neutropenia; children.

ROUTE & DOSAGE

Hypertension, CHF
Adult: **PO** 10–20 mg daily, may increase up to 80 mg/day in 1–2 divided doses

Q

Common adverse effects in *italic*, life-threatening effects underlined; generic names in **bold**; classifications in SMALL CAPS; ♣ Canadian drug name; ◉ Prototype drug

1293

Geriatric: **PO** Start with 2.5–5 mg daily

Renal Impairment Dosage Adjustment

CrCl 30–60 mL/min: 5 mg daily initially; *less than 10–30 mL/min:* 2.5 mg/day initially

ADMINISTRATION

Oral

- When patient has been treated with a diuretic, the diuretic is usually discontinued 2–3 days before beginning quinapril. If the diuretic cannot be discontinued, initial quinapril dose is usually lowered to 5 mg.
- Store at 15°–30° C (59°–86° F) and protect from moisture.

ADVERSE EFFECTS (≥1%) **CV:** Edema, hypotension. **CNS:** Dizziness, fatigue, headache. **GI:** Nausea, vomiting, diarrhea. **Hematologic:** Eosinophilia, neutropenia. **Metabolic:** Hyperkalemia, proteinuria. **Respiratory:** Cough. **Body as a Whole:** <u>Angioedema</u>, myalgia.

DIAGNOSTIC TEST INTERFERENCE May increase ***BUN*** or ***serum creatinine.***

INTERACTIONS Drug: POTASSIUM-SPARING DIURETICS may increase risk of hyperkalemia. May elevate serum **lithium** levels, resulting in **lithium** toxicity.

PHARMACOKINETICS Absorption: Rapidly from GI tract. **Onset:** 1 h. **Peak:** 2–4 h. **Duration:** Up to 24 h. **Distribution:** 97% bound to plasma proteins; crosses placenta; not known if distributed into breast milk. **Metabolism:** Extensively metabolized in liver to its active metabolite, quinaprilat. **Elimination:** 50–60% in urine, primarily as quinaprilat; 30% in feces. **Half-Life:** 2 h.

NURSING IMPLICATIONS

Assessment & Drug Effects

- Following initial dose, monitor for several hours for first-dose hypotension, especially in salt- or volume-depleted patients (e.g., those pretreated with a diuretic).
- Monitor BP at time of peak effectiveness, 2–4 h after dosing, and at end of dosing interval just before next dose.
- Report diminished antihypertensive effect toward end of dosing interval. Inadequate trough response may indicate need to divide daily dose.
- Lab tests: Baseline and periodic kidney function tests; periodic serum potassium; periodic WBC in those with collagen vascular disease or renal impairment.
- Observe for S&S of hyperkalemia (see Appendix F).

Patient & Family Education

- Discontinue quinapril and report S&S of angioedema (e.g., swelling of face or extremities, difficulty breathing or swallowing) to prescriber.
- Maintain adequate fluid intake and avoid potassium supplements or salt substitutes unless specifically prescribed by prescriber.
- Light-headedness and dizziness may occur, especially during the initial days of therapy. If fainting occurs, stop taking quinapril until the prescriber has been consulted.

QUINIDINE SULFATE

(kwin'i-deen sul-fate)

Apo-Quinidine ♣, Novoquinidin ♣

QUINIDINE GLUCONATE

Classification: CLASS IA ANTI-ARRHYTHMIC

Common adverse effects in *italic*, life-threatening effects <u>underlined</u>; generic names in **bold**; classifications in SMALL CAPS; ♣ Canadian drug name; ◐ Prototype drug

Therapeutic: CLASS IA ANTIARRHYTH-MIC
Prototype: Procainamide
Pregnancy Category: C

AVAILABILITY Quinidine sulfate:
200 mg, 300 mg tablets; 300 mg sustained release tablets; **Quinidine gluconate:** 324 mg sustained release tablets; 80 mg/mL injection

ACTION & *THERAPEUTIC EFFECT*
Class IA antiarrhythmic that depresses myocardial excitability, contractility, automaticity, and conduction velocity as well as prolongs refractory period. *Depresses myocardial excitability, conduction velocity, and irregularity of nerve impulse conduction.*

USES Premature atrial, AV junctional, and ventricular contraction; paroxysmal atrial tachycardia, chronic ventricular tachycardia (when not associated with complete heart block); maintenance therapy after electrical conversion of atrial fibrillation or flutter; life-threatening malaria.

CONTRAINDICATIONS Hypersensitivity or idiosyncrasy to quinine or quinidine; thrombocytopenic purpura resulting from prior use of quinidine; intraventricular conduction defects, complete AV block, AV conduction disorders; left bundle branch block; marked QRS widening; thyrotoxicosis; extensive myocardial damage, frank CHF, hypotensive states; history of drug-induced torsades de pointes.
CAUTIOUS USE Incomplete heart block; impaired kidney or liver function; bronchial asthma or other respiratory disorders; myasthenia gravis; potassium imbalance; pregnancy (category C), lactation.

ROUTE & DOSAGE

Conversion to and/or Maintenance of Sinus Rhythm Sulfate Immediate Release
Adult: **PO** 200–300 mg q6–8h until sinus rhythm restored or toxicity occurs (max: 3–4 g)

Sulfate Extended Release
Adult: **PO** 324–648 mg q8–12h

Gluconate
Adult: **IM** 600 mg salt, then 400 mg salt; can repeat q2h if needed **IV** 800 mg salt, monitor closely

Malaria
Adult/Adolescent/Child: **IV** 24 mg/kg loading dose, then 12 mg/kg q8h OR 10 mg/kg loading dose, then 0.02 mg/kg/min for 24 h

Renal Impairment Dosage Adjustment
CrCl less than 10 mL/min: Give 75% of dose

Hemodialysis Dosage Adjustment
Give a 200 mg supplement dose post-dialysis

ADMINISTRATION

Oral
- Give with a full glass of water on an empty stomach for optimum absorption (i.e., 1 h before or 2 h after meals). Administer drug with food if GI symptoms occur (nausea, vomiting, diarrhea are most common). Do not administer with grapefruit juice.
- Ensure that extended release tablets are swallowed whole. They should not be crushed or chewed.
- Store in tight, light-resistant containers away from excessive heat.

Q

Common adverse effects in *italic*, life-threatening effects <u>underlined</u>; generic names in **bold**; classifications in SMALL CAPS; ♣ Canadian drug name; ☑ Prototype drug

1295

Intramuscular

- Aspirate carefully before injection to avoid inadvertent entry into blood vessel.

Intravenous

PREPARE: **IV Infusion:** Dilute 800 mg (10 mL) in at least 40 mL D5W to yield a maximum concentration of 16 mg/mL.

ADMINISTER: **IV Infusion:** Give via infusion pump at a rate not to exceed 16 mg (1 mL)/min.

INCOMPATIBILITIES Solution/additive: Amiodarone, atracurium, furosemide. Y-site: Acyclovir, aminophylline, amphotericin B, ampicillin, aztreonam, bivalirudin, CEPHALOSPORINS, clindamycin, dantrolene, daptomycin, dexamethasone, diazoxide, ertapenem, furosemide, ganciclovir, heparin in dextrose, hydrocortisone, indomethacin, insulin, ketorolac, methicillin, methylprednisolone, mezlocillin, minocycline, nafcillin, nitroprusside, oxacillin, pantoprazole, premetrexed, penicillin, pentobarbital, phenobarbital, phenytoin, piperacillin, sodium bicarbonate, SMP/TMX, ticarcillin.

- Use supine position during drug administration; severe hypotension is most likely to occur in patients receiving drug via IV.

- Protect IV solutions from light and heat to prevent brownish discoloration and possibly precipitation.

ADVERSE EFFECTS (≥1%) **CNS:** Headache, fever, tremors, apprehension, delirium, syncope with sudden loss of consciousness, seizures. **CV:** Hypotension, CHF, <u>widened QRS complex</u>, bradycardia, <u>heart block</u>, atrial flutter, <u>ventricular</u> <u>flutter, fibrillation</u> or tachycardia; quinidine syncope, <u>torsades de pointes</u>. **Special Senses:** Mydriasis, blurred vision, disturbed color perception, reduced visual field, photophobia, diplopia, night blindness, scotomas, optic neuritis, disturbed hearing (tinnitus, auditory acuity). **GI:** *Nausea, vomiting, diarrhea, abdominal pain,* hepatic dysfunction. **Hematologic:** <u>Acute hemolytic anemia</u> (especially in patients with G6PD deficiency), hypoprothrombinemia, <u>leukopenia</u>. <u>Thrombocytopenia</u>, <u>agranulocytosis</u> (both rare). **Body as a Whole:** Cinchonism (nausea, vomiting, headache, dizziness, fever, tremors, vertigo, tinnitus, visual disturbances), <u>angioedema</u>, acute asthma, <u>respiratory depression, vascular collapse</u>. **Skin:** Rash, urticaria, cutaneous flushing with intense pruritus, photosensitivity. **Metabolic:** SLE, hypokalemia.

INTERACTIONS Drug: May increase **digoxin** levels by 50%; **amiodarone** may increase quinidine levels, increasing its risk of heart block; other ANTIARRHYTHMICS, PHENOTHIAZINES, **reserpine** add to cardiac depressant effects; ANTICHOLINERGIC AGENTS add to vagolytic effects; CHOLINERGIC AGENTS may antagonize cardiac effects; ANTICONVULSANTS, BARBITURATES, **rifampin** increase the metabolism of quinidine, thus decreasing its efficacy; CARBONIC ANHYDRASE INHIBITORS, **sodium bicarbonate,** CHRONIC ANTACIDS decrease renal elimination of quinidine, thus increasing its toxicity; **verapamil** causes significant hypotension; may increase hypoprothrombinemic effects of **warfarin. Diltiazem** may increase levels and decrease elimination of quinidine. **Food: Grapefruit juice** (greater than 1 qt/day) may decrease absorption.

Q

PHARMACOKINETICS Absorption: Almost completely from GI tract. **Onset:** 1–3 h. **Peak:** 0.5–1 h. **Duration:** 6–8 h. **Distribution:** Widely distributed to most body tissues except the brain; crosses placenta; distributed into breast milk. **Metabolism:** In liver (CYP3A4). **Elimination:** Greater than 95% in urine, less than 5% in feces. **Half-Life:** 6–8 h.

NURSING IMPLICATIONS

Assessment & Drug Effects

- Observe cardiac monitor and report immediately the following indications for stopping quinidine: (1) Sinus rhythm, (2) widening QRS complex in excess of 25% (i.e., longer than 0.12 sec), (3) changes in QT interval or refractory period, (4) disappearance of P waves, (5) sudden onset of or increase in ectopic ventricular beats (extrasystoles, PVCs), (6) decrease in heart rate to 120 bpm. Also report immediately any worsening of minor side effects.
- Continuous monitoring of ECG and BP is required. Observe patient closely (check sensorium and be alert for any sign of toxicity); determine plasma quinidine concentrations frequently when large doses (more than 2 g/day) are used or when quinidine is given parenterally (i.e., quinidine gluconate).
- Observe patient closely following each parenteral dose. Amount of subsequent dose is gauged by response to preceding dose.
- Monitor vital signs q1–2h or more often as needed during acute treatment. Count apical pulse for a full minute. Report any change in pulse rate, rhythm, or quality or any fall in BP.
- Severe hypotension is most likely to occur in patients receiving high oral doses or parenteral quinidine (i.e., quinidine gluconate).

- Lab tests: Periodic blood counts, serum electrolytes, LFTs, and kidney during long-term therapy. Periodic serum quinidine (target range 2–6 micrograms/mL or higher).
- Monitor I&O. Diarrhea occurs commonly during early therapy; most patients become tolerant to this side effect. Evaluate serum electrolytes, acid-base, and fluid balance when symptoms become severe; dosage adjustment may be required.

Patient & Family Education

- Report feeling of faintness to prescriber. "Quinidine syncope" is caused by quinidine-induced changes in ventricular rhythm resulting in decreased cardiac output and syncope.
- Do not self-medicate with OTC drugs without advice from prescriber.
- Do not increase, decrease, skip, or discontinue doses without consulting prescriber.
- Notify prescriber immediately of disturbances in vision, ringing in ears, sense of breathlessness, onset of palpitations, and unpleasant sensation in chest. Be sure to note the time of occurrence and duration of chest symptoms.

Q

QUININE SULFATE

(kwye′nine)

Novoquinine ◆
Classification: ANTIMALARIAL
Therapeutic: ANTIMALARIAL
Prototype: Chloroquine
Pregnancy Category: X

AVAILABILITY 325 mg capsules

ACTION & *THERAPEUTIC EFFECT*
Inhibits protein synthesis and depresses many enzyme systems in malaria parasite. *Effective against* Plasmodium vivax *and* Plasmodium

Common adverse effects in *italic*; life-threatening effects underlined; generic names in **bold**; classifications in SMALL CAPS; ◆ Canadian drug name; ⊙ Prototype drug

1297

malariae *but not* Plasmodium falciparum. *Generally replaced by less toxic and more effective agents in treatment of malaria.*

USES Chloroquine-resistant falciparum malaria and in combination with other antimalarials for radical cure of relapsing vivax malaria; also relief of nocturnal recumbency leg cramps.
UNLABELED USES Restless leg syndrome.

CONTRAINDICATIONS Hypersensitivity to quinine; tinnitus, optic neuritis; myasthenia gravis; G6PD deficiency; pregnancy (category X).
CAUTIOUS USE Cardiac arrhythmias; restless leg syndrome. Same precautions as for quinidine sulfate when used in patients with cardiovascular conditions; lactation.

ROUTE & DOSAGE

Acute Malaria
Adult: **PO** 650 mg q8h for 3 days
Child: **PO** 25 mg/kg/day in three divided doses q8h for 3 days
Malaria Chemoprophylaxis
Adult: **PO** 325 mg b.i.d. for 6 wk
Nocturnal Leg Cramps
Adult: **PO** 260–300 mg at bedtime

ADMINISTRATION

Oral
- Give with or after meals or a snack to minimize gastric irritation. Quinine has potent local irritant effect on gastric mucosa. Do not crush capsule; drug is not only irritating but also extremely bitter.
- Store in tight, light-resistant containers.

ADVERSE EFFECTS (≥1%) **Body as a Whole:** Cinchonism (tinnitus,

decreased auditory acuity, dizziness, vertigo, headache, visual impairment, *nausea, vomiting, diarrhea,* fever); hypersensitivity (cutaneous flushing, visual impairment, pruritus, skin rash, fever, gastric distress, dyspnea, tinnitus); hypothermia, coma. **CNS:** Confusion, excitement, apprehension, syncope, delirium, convulsions, blackwater fever (extensive intravascular hemolysis with renal failure), death. **CV:** Angina, hypotension, tachycardia, cardiovascular collapse. **Hematologic:** Leukopenia, thrombocytopenia, agranulocytosis, hypoprothrombinemia, hemolytic anemia. **Respiratory:** Decreased respiration.

DIAGNOSTIC TEST INTERFERENCE Quinine may interfere with determinations of **urinary catecholamines** (**Sobel** and **Henry modification procedure**) and **urinary steroids** (**17-hydroxycorticosteroids**) (modification of **Reddy, Jenkins, Thorn** method). May cause false positive for OPIOIDS.

INTERACTIONS Drug: May increase **digoxin** levels; ANTICHOLINERGIC AGENTS add to vagolytic effects; CHOLINERGIC AGENTS may antagonize cardiac effects; ANTICONVULSANTS, BARBITURATES, **rifampin** increase the metabolism of quinine, thus decreasing its efficacy; CARBONIC ANHYDRASE INHIBITORS, **sodium bicarbonate,** CHRONIC ANTACIDS decrease renal elimination of quinine, thus increasing its toxicity; **warfarin** may increase hypoprothrombinemic effects. **Amantadine, carbamazepine, phenobarbital** levels may be increased. Avoid use with **ritonavir. Food: Grapefruit juice** (greater than 1 qt/day) may increase plasma concentrations and adverse effects.

PHARMACOKINETICS Absorption: Well from GI tract. **Peak:** 1–3 h.

Q

Duration: 6–8 h. **Distribution:** Widely distributed to most body tissues except the brain; crosses placenta; distributed into breast milk. **Metabolism:** In liver. **Elimination:** Greater than 95% in urine, less than 5% in feces. **Half-Life:** 8–21 h.

NURSING IMPLICATIONS

Assessment & Drug Effects
- Be alert for S&S of rising plasma concentration of quinine marked by tinnitus and hearing impairment, which usually do not occur until concentration is 10 mcg/mL or more.
- Follow the same precautions with quinine as are used with quinidine in patients with atrial fibrillation; quinine may produce cardiotoxicity in these patients.

Patient & Family Education
- Learn possible adverse reactions and report onset of any unusual symptom promptly to prescriber.

QUINUPRISTIN/DALFOPRISTIN
(quin-u-pris′tin/dal′fo-pris-tin)
Synercid
Classifications: STREPTOGRAMIN ANTIBIOTIC; CYCLIC MACROLIDE
Therapeutic: ANTIBIOTIC
Pregnancy Category: B

AVAILABILITY 500 mg vial (150 mg quinupristin/350 mg dalfopristin)

ACTION & THERAPEUTIC EFFECT
Dalfopristin inhibits the early phase of protein synthesis of bacteria, while quinupristin inhibits the late phase of protein synthesis of bacteria. Both actions lead to death of the bacteria organisms. *Effectiveness indicated by clinical improvement in S&S of life-threatening bacteremia. Active against gram-positive*

pathogens including vancomycin-resistant Enterococcus faecium (VREF), as well as some gram-negative anaerobes.

USES Serious or life-threatening infections associated with VREF bacteremia; complicated skin and skin structure infections caused by *Staphylococcus aureus* or *Streptococcus pyogenes*.

CONTRAINDICATIONS Hypersensitivity to quinupristin/dalfopristin, pristinamycin, other streptogramins. **CAUTIOUS USE** Renal or hepatic dysfunction; IBD; GI disease; pregnancy (category B); lactation. Safe use in children younger than 16 y not established.

ROUTE & DOSAGE

Vancomycin-Resistant *Enterococcus faecium*
Adult: **IV** 7.5 mg/kg infused over 60 min q8h
Complicated Skin and Skin Structure Infections
Adult: **IV** 7.5 mg/kg infused over 60 min q12h × 7 days

ADMINISTRATION

Intravenous

***PREPARE:* Intermittent:** Reconstitute a single 500 mg vial by adding 5 mL D5W or sterile water for injection to yield 100 mg/mL. ▪ Gently swirl to dissolve but do NOT shake. Allow solution to clear. ▪ Withdraw the required dose and further dilute by adding to 100 mL (central line) or 250–500 mL (peripheral site) of D5W.
***ADMINISTER:* Intermittent:** Flush line before and after with D5W. Do NOT use saline. ▪ Administer over 1 h.

Common adverse effects in *italic*, life-threatening effects <u>underlined</u>; generic names in **bold**; classifications in SMALL CAPS; ♣ Canadian drug name; ⊘ Prototype drug

INCOMPATIBILITIES **Solution/additive: Saline solutions** and **lactated Ringer's** solution (flush lines with **D5W** before infusing other drugs). **Y-site:** Any drugs diluted in **saline.**

- Refrigerate unopened vials. After reconstitution solution is stable for 5 h at room temperature and 54 h refrigerated.

ADVERSE EFFECTS (≥1%) **Body as a Whole:** Headache, pain, *myalgia, arthralgia.* **GI:** Nausea, diarrhea, vomiting. **Skin:** Rash, pruritus. **Other:** *Inflammation, pain, or edema at infusion site, other infusion site reactions,* thrombophlebitis.

INTERACTIONS Drug: Inhibits CYP3A4 metabolism of **cyclosporine, midazolam, nifedipine,** PROTEASE INHIBITORS, **vincristine, vinblastine, docetaxel, paclitaxel, diazepam, tacrolimus, carbamazepine, quinidine, lidocaine, disopyramide.**

PHARMACOKINETICS Distribution: Moderately protein bound. **Metabolism:** Metabolized to several active metabolites. **Elimination:** Primarily in feces (75–77%). **Half-Life:** 3 h quinupristin, 1 h dalfopristin.

NURSING IMPLICATIONS

Assessment & Drug Effects
- Monitor for S&S of infusion site irritation; change infusion site if irritation is apparent.
- Monitor for cutaneous reaction (e.g., pruritus/erythema of neck, face, upper body).
- Lab tests: C&S from site of infection prior to initiating therapy; WBC with differential; and LFTs (especially with preexisting hepatic insufficiency).

Patient & Family Education
- Report burning, itching, or pain at infusion site to prescriber.
- Report any sensation of swelling of face and tongue; difficulty swallowing.

RABEPRAZOLE SODIUM
(rab-e-pra'zole)

AcipHex
Classification: GASTRIC PROTON PUMP INHIBITOR
Therapeutic: ANTIULCER
Prototype: Omeprazole
Pregnancy Category: B

AVAILABILITY 20 mg tablets

ACTION & *THERAPEUTIC EFFECT*
Gastric proton pump inhibitor that specifically suppresses gastric acid secretion by inhibiting the H+, K+-ATPase enzyme system [the acid (proton H+) pump] in the parietal cells of the stomach. Produces an antisecretory effect on the hydrogen ion (H+) in the parietal cells. *Effectiveness indicated by a negative for* H. pylori *with preexisting gastric ulcer; also by elimination of S&S of GERD or peptic ulcers.*

USES Healing and maintenance of healing of erosive or ulcerative gastroesophageal reflux disease (GERD); healing of duodenal ulcers; treatment of hypersecretory conditions.

CONTRAINDICATIONS Hypersensitivity to rabeprazole, lansoprazole, or omeprazole or proton pump inhibitors (PPIs); lactation.
CAUTIOUS USE Severe hepatic impairment; mild to moderate hepatic disease; Japanese; older adults; pregnancy (category B). Safety and efficacy in children younger than 16 y are not established.

Common adverse effects in *italic*, life-threatening effects underlined; generic names in **bold**; classifications in SMALL CAPS; ♣ Canadian drug name; ⊙ Prototype drug

ROUTE & DOSAGE

Healing of Erosive GERD

Adult: **PO** 20 mg daily × 48 wk, may continue up to 16 wk if needed
Adolescent: **PO** 20 mg daily for up to 8 wk

Maintenance Therapy for GERD

Adult: **PO** 20 mg daily

Healing Duodenal Ulcer

Adult: **PO** 20 mg daily × 4 wk

Hypersecretory Disease

Adult: **PO** 60 mg daily in 1–2 divided doses (max: 100 mg daily or 60 mg b.i.d.)

ADMINISTRATION

Oral

- Ensure that the tablet is swallowed whole. It should not be crushed or chewed.
- Store at 15°–30° C (59°–86° F).

ADVERSE EFFECTS (≥1%) **Body as a Whole:** Headache. **Skin:** (Rare) <u>Stevens-Johnson syndrome, toxic epidermal necrolysis, erythema multiforme.</u>

INTERACTIONS Drug: May decrease absorption of **ketoconazole;** may increase **digoxin** levels; may decrease **nelfinavir** levels. **Herbal: Ginkgo** may decrease plasma levels.

PHARMACOKINETICS Absorption: 52% bioavailability. **Distribution:** 96% protein bound. **Metabolism:** In liver by (CYP3A4, 2C19). **Elimination:** Primarily in urine. **Half-Life:** 1–2 h.

NURSING IMPLICATIONS

Assessment & Drug Effects

- Lab tests: Periodic LFTs.
- Coadministered drugs: Monitor for changes in digoxin blood level.

Patient & Family Education

- Report diarrhea, skin rash, other bothersome adverse effects to prescriber.

RALOXIFENE HYDROCHLORIDE

(ra-lox'i-feen)
Evista
Classification: SELECTIVE ESTROGEN RECEPTOR ANTAGONIST/AGONIST
Therapeutic: OSTEOPOROSIS PROPHY-LACTIC
Prototype: Tamoxifen
Pregnancy Category: X

AVAILABILITY 60 mg tablets

ACTION & *THERAPEUTIC EFFECT*

Tamoxifen analog that exhibits selective estrogen receptor antagonist activity on uterus and breast tissue. Prevents tissue proliferation in both sites. Decreases bone resorption and increases bone density. *Effectiveness indicated by increased bone mineral density. Reduces the risk of invasive breast cancer in high risk postmenopausal women (e.g., breast cancer in situ, or atypical hyperplasia).*

USES Prevention and treatment of osteoporosis in postmenopausal women; breast cancer prophylaxis.

CONTRAINDICATIONS Active or past history of a thromboembolic event; hypersensitivity to raloxifene; severe hepatic impairment; pregnancy (category X), lactation.
CAUTIOUS USE Concurrent use of raloxifene and estrogen hormone replacement therapy and lipid-lowering agents; hyperlipidemia; hepatic impairment; moderate or severe renal impairment.

R

Common adverse effects in *italic*, life-threatening effects <u>underlined</u>; generic names in **bold**; classifications in SMALL CAPS; ♣ Canadian drug name; ⓟ Prototype drug

1301

ROUTE & DOSAGE

Prevention/Treatment of Osteoporosis
Adult: **PO** 60 mg daily

Breast Cancer Prophylaxis
Postmenopausal Adult: **PO** 60 mg daily

ADMINISTRATION

Oral

- Calcium and vitamin D supplementation are recommended with raloxifene: 1500 mg/day of elemental calcium and 400–800 units/day of vitamin D.
- Store at 15°–30° C (59°–86° F) in a tightly closed container and protect from light.

ADVERSE EFFECTS (≥1%) **Body as a Whole:** Infection, flu-like syndrome, leg cramps, fever, arthralgia, myalgia, arthritis. **CNS:** Migraine headache, depression, insomnia. **CV:** *Hot flashes*, chest pain, peripheral edema, decreased serum cholesterol. **GI:** Nausea, dyspepsia, vomiting, flatulence, GI disorder, gastroenteritis, weight gain. **Respiratory:** Sinusitis, pharyngitis, cough, pneumonia, laryngitis. **Skin:** Rash, sweating. **Urogenital:** Vaginitis, UTI, cystitis, leukorrhea, endometrial disorder, breast pain, vaginal bleeding.

INTERACTIONS Drug: Use of ESTROGENS not recommended; absorption reduced by **cholestyramine;** use with **warfarin** or other coumarin derivatives may result in changes in prothrombin time (PT). **Herbal: Soy isoflavones** should be used with caution.

PHARMACOKINETICS Absorption: 60% absorbed, absolute bioavailability 2%. **Metabolism:** Extensive first-pass metabolism in liver. **Elimination:** Primarily in feces. **Half-Life:** 27.7–32.5 h.

NURSING IMPLICATIONS

Assessment & Drug Effects

- Lab tests: Periodically monitor LFTs; with concurrent oral anticoagulants, carefully monitor PT and INR.
- Monitor carefully for and immediately report S&S of thromboembolic events.
- Do not give drug concurrently with cholestyramine; however, if unavoidable, space the two drugs as widely as possible.

Patient & Family Education

- Contact prescriber immediately if unexplained calf pain or tenderness occurs.
- Avoid prolonged restriction of movement during travel.
- Drug does not prevent and may induce hot flashes.
- Do not take drug with other estrogen-containing drugs.
- Raloxifene is normally discontinued 72 h prior to prolonged immobilization (e.g., post-surgical recovery, prolonged bedrest). Consult prescriber.

RALTEGRAVIR

(ral-te-gra′vir)

Isentress
Classifications: ANTIRETROVIRAL; INTEGRASE INHIBITOR
Therapeutic: ANTIRETROVIRAL
Pregnancy Category: C

AVAILABILITY 400 mg tablets

ACTION & *THERAPEUTIC EFFECT*
Inhibits HIV-1 integrase, an enzyme required for integration of proviral DNA into the helper T-cell genome, thus preventing formation of the HIV-1 provirus. *Inhibiting integration prevents replication and proliferation of the HIV-1 virus.*

USES In combination with other antiretroviral agents for the treatment of HIV-1 infection in treatment-experienced adult patients who have evidence of viral replication and HIV-1 strains resistant to multiple antiretroviral agents.

CONTRAINDICATIONS Treatment of naïve HIV-1 patients; lactation. **CAUTIOUS USE** Hepatitis; mild to moderate hepatic impairment; lactase deficiency; history of psychiatric disease or suicidal ideation; older adults; pregnancy (category C); children younger than 16 y. Safety and efficacy in patients with severe hepatic impairment are unknown.

ROUTE & DOSAGE

HIV-1 Infection
Adult/Adolescent: **PO** 400 mg b.i.d.

HIV Infection (with Concurrent Rifampin)
Adult/Adolescent: **PO** 800 mg b.i.d.

ADMINISTRATION
- May be given without regard to food.
- Give before dialysis.
- Store at 15–30° C (59–86° F).

ADVERSE EFFECTS (≥1%) **Body as a Whole:** Asthenia, fatigue, pyrexia. **CNS:** Dizziness, headache. **GI:** Abdominal pain, *diarrhea*, nausea, vomiting. **Skin:** Lipodystrophy. **Hematologic:** Anemia, neutropenia.

INTERACTIONS Drug: Atazanavir may increase plasma levels of raltegravir; **rifampin** and **tipranavir/ritonavir** may decrease plasma levels of raltegravir.

PHARMACOKINETICS Peak: 3 h. **Distribution:** 83% protein bound.

Metabolism: In the liver. **Elimination:** Stool and urine. **Half-Life:** 9 h.

NURSING IMPLICATIONS

Assessment & Drug Effects
- Monitor for and report S&S of immune reconstitution syndrome (inflammatory response to residual opportunistic infections such as MAC, CMV, PCP, or reactivation of varicella zoster).
- Lab tests: Baseline and periodic CD4+ cell count and HIV RNA viral load; periodic CBC with differential, LFTs.
- Monitor diabetics for loss of glycemic control.

Patient & Family Education
- Inform prescriber immediately if you plan to become or become pregnant during therapy.
- Report promptly unexplained leg pain or muscle cramping.

RAMELTEON
(ra-mel'tee-on)

Rozerem
Classifications: MELATONIN RECEPTOR AGONIST; SEDATIVE-HYPNOTIC
Therapeutic: SEDATIVE-HYPNOTIC
Pregnancy Category: C

AVAILABILITY 8 mg tablets

ACTION & *THERAPEUTIC EFFECT*
Ramelteon is a melatonin receptor agonist with high affinity for melatonin receptors in the brain. This activity is believed to promote sleep, as these receptors, in response to endogenous melatonin, are thought to be involved in maintaining the circadian rhythm underlying the normal sleep-wake cycle. *Effective in promoting onset of sleep.*

USES Treatment of insomnia characterized by difficulty with sleep onset.

Common adverse effects in *italic*, life-threatening effects underlined; generic names in **bold**; classifications in SMALL CAPS; ♣ Canadian drug name; ⊘ Prototype drug

1303

CONTRAINDICATIONS Hypersensitivity to ramelteon; severe hepatic function impairment (Child-Pugh class C); severe sleep apnea or severe COPD; severe depression; suicidal ideation; lactation.

CAUTIOUS USE Moderate hepatic function impairment (Child-Pugh class B); depression with suicidal tendencies; older adults; pregnancy (category C).

ROUTE & DOSAGE

Insomnia
Adult: **PO** 8 mg within 30 min of bedtime

ADMINISTRATION

Oral
- Give within 30 min of bedtime.
- Do not administer to anyone on concurrent fluvoxamine therapy without alerting prescriber. This combination causes a dramatic increase in rameleton blood level.
- Store at 15°–30° C (59°– 86° F).

ADVERSE EFFECTS (≥1%) **CNS:** Depression, dizziness, fatigue, headache, insomnia, somnolence. **GI:** Diarrhea, unpleasant taste, nausea. **Musculoskeletal:** Arthralgia, myalgia. **Respiratory:** Upper respiratory tract infection.

INTERACTIONS Drug: Concurrent use with **ethanol** produces additive CNS depressant effects; **ketoconazole, itraconazole,** and **fluvoxamine** increase rameleton levels; other CYP1A2 INHIBITORS (e.g., **ciprofloxacin, enoxacin, mexiletine, norfloxacin, tacrine**) may also increase rameleton levels; **rifampin** decreases rameleton levels. **Food:** High-fat meal, **grapefruit** or **grapefruit juice** increase rameleton levels.

PHARMACOKINETICS Absorption: 84%. **Peak:** 45 min. **Distribution:** 82% protein bound. **Metabolism:** Rapid

and extensive first pass hepatic metabolism; one metabolite, M-II, is active. **Elimination:** Primarily renal. **Half-Life:** 1–2.5 h.

NURSING IMPLICATIONS

Assessment & Drug Effects
- Monitor for and report worsening insomnia and cognitive or behavioral changes.
- Monitor for S&S of decreased testosterone levels (e.g., loss of libido) or increased prolactin levels (galactorrhea).
- Lab test: Baseline LFTs.

Patient & Family Education
- Do not take with or immediately after a high fat meal.
- Do not drive or engage in potentially hazardous activities until response to drug is known.
- Do not consume alcohol while taking this drug.
- Report any of the following to prescriber: Worsening insomnia, cognitive or behavioral changes, problem with reproductive function.

RAMIPRIL
(ram'i-pril)

Altace
Classifications: ANGIOTENSIN-CONVERTING ENZYME (ACE) INHIBITOR; ANTIHYPERTENSIVE
Therapeutic: ANTIHYPERTENSIVE
Prototype: Enalapril
Pregnancy Category: C first trimester; D second and third trimester

AVAILABILITY 1.25 mg, 2.5 mg, 5 mg, 10 mg capsules; 1.25 mg, 2.5 mg, 5 mg, 10 mg tablets

ACTION & *THERAPEUTIC EFFECT*
Reduces peripheral vascular resistance by inhibiting the formation of angiotensin II, a potent vasoconstrictor. This also decreases serum aldosterone levels and reduces peripheral

arterial resistance (afterload) as well as improves cardiac output and exercise tolerance. *Lowers BP, and improves cardiac output as well as exercise tolerance.*

USES Mild to moderate hypertension, CHF, stroke prophylaxis, myocardial infarction prophylaxis, post myocardial infarction.
UNLABELED USES Diabetic nephropathy, proteinuria.

CONTRAINDICATIONS Hypersensitivity to ramipril or any other ACE inhibitor, patients with history of angioneurotic edema; jaundice; hyperkalemia; pregnancy (category D second and third trimester), lactation.
CAUTIOUS USE Impaired kidney or liver function, surgery or anesthesia; CHF; pregnancy (category C first trimester). Safety and effectiveness in children are not established.

ROUTE & DOSAGE

Hypertension, CHF, Stroke Prophylaxis, Myocardial Infarction Prophylaxis
Adult: PO 2.5–5 mg daily, may increase up to 20 mg/day in 1–2 divided doses
Post Myocardial Infarction
Adult: PO 1.25–2.5 mg b.i.d. (may titrate up to 5 mg b.i.d.)

ADMINISTRATION

Oral
- When patient has been treated with a diuretic, the diuretic is usually discontinued 2–3 days before beginning ramipril. If the diuretic cannot be discontinued, initial ramipril dose is usually lowered to 1.25 mg.
- Store at 15°–30° C (59°–86° F) and protect from moisture.

ADVERSE EFFECTS (≥1%) **CNS:** Dizziness, fatigue, headache. **GI:** Nausea, vomiting, diarrhea, eructation. **Metabolic:** Hyperkalemia, hyponatremia. **Skin:** Erythema, pruritus. **Body as a Whole:** Angioedema. **Respiratory:** Cough.

INTERACTIONS Drug: POTASSIUM-SPARING DIURETICS may increase risk of hyperkalemia. May elevate serum **lithium** levels, resulting in lithium toxicity. NSAIDS may attenuate antihypertensive effects. Use with **azathioprine** increases risk of hematologic side effects. **Pregabalin** use increases the risk of angioedema.

PHARMACOKINETICS Absorption: 60% absorbed from GI tract. **Onset:** 2 h. **Peak:** 6–8 h. **Duration:** Up to 24 h. **Distribution:** Crosses placenta; not known if distributed into breast milk. **Metabolism:** Rapidly metabolized in liver to its active metabolite, ramiprilat. **Elimination:** 40–60% in urine, 40% in feces. **Half-Life:** 2–3 h.

NURSING IMPLICATIONS

Assessment & Drug Effects
- Monitor BP at time of peak effectiveness, 3–6 h after dosing and at end of dosing interval just before next dose.
- Report diminished antihypertensive effect.
- Monitor for first-dose hypotension, especially in salt- or volume-depleted persons.
- Lab tests: Monitor BUN and serum creatinine periodically. Increases may necessitate dose reduction or discontinuation of drug. Monitor serum potassium values.
- Observe for S&S of hyperkalemia (see Appendix F).

Patient & Family Education
- Discontinue drug and report S&S of angioedema to prescriber (e.g., swelling of face or extremities, difficulty breathing or swallowing).

R

Common adverse effects in *italic*, life-threatening effects underlined; generic names in **bold**; classifications in SMALL CAPS; ♣ Canadian drug name; ⊙ Prototype drug

1305

- Maintain adequate fluid intake and avoid potassium supplements or salt substitutes unless specifically prescribed by the prescriber.
- Light-headedness and dizziness may occur, especially during the initial days of therapy. If fainting occurs, stop taking ramipril until the prescriber has been consulted.

RANITIDINE HYDROCHLORIDE

(ra-nye'te-deen)

Zantac, Zantac-75
Classification: ANTISECRETORY (H$_2$-RECEPTOR ANTAGONIST)
Therapeutic: ANTIULCER
Prototype: Cimetidine
Pregnancy Category: B

AVAILABILITY 75 mg, 150 mg, 300 mg tablets; 25 mg, 150 mg effervescent tablets; 150 mg, 300 mg capsules; 15 mg/mL syrup; 1 mg/mL, 25 mg/mL injection

ACTION & *THERAPEUTIC EFFECT*
Potent anti-ulcer drug that competitively inhibits histamine action at H$_2$-receptor sites on parietal cells, blocking gastric acid secretion. Indirectly reduces pepsin secretion. *Blocks daytime and nocturnal basal gastric acid secretion stimulated by histamine and reduces gastric acid release in response to food, caffeine, pentagastrin, and insulin.*

USES Short-term treatment of active duodenal ulcer; maintenance therapy for duodenal ulcer patient after healing of acute ulcer; treatment of gastroesophageal reflux disease; short-term treatment of active, benign gastric ulcer; treatment of pathologic GI hypersecretory conditions (e.g., Zollinger-Ellison syndrome, systemic mastocytosis, and postoperative hypersecretion); heartburn.

CONTRAINDICATIONS Hypersensitivity to ranitidine; acute porphyria.
CAUTIOUS USE Hypersensitivity to H$_2$-blockers; hepatic and renal dysfunction; renal failure; elderly; PKU; pregnancy (category B), lactation, infants younger than 1 mo; **OTC form:** Children younger than 12 y.

ROUTE & DOSAGE

Duodenal Ulcer, Gastric Ulcer, Gastroesophageal Reflux
Adult: **PO** 150 mg b.i.d. or 300 mg at bedtime **IV** 50 mg q6–8h; 150–300 mg/24 h by continuous infusion
Child: **PO** 4–5 mg/kg/day divided q8–12h (max: 300 mg/day) **IM/IV** 2–4 mg/kg/day divided q6–8h (max: 200 mg/day)
Infant (younger than 2 wk): **PO** 1.5–2 mg/kg/day divided q12h **IV** 1.5 mg/kg/day divided q12h or 0.04 mg/kg/h by continuous infusion

Duodenal Ulcer, Maintenance Therapy
Adult: **PO** 150 mg at bedtime

Pathologic Hypersecretory Conditions
Adult: **PO** 150 mg b.i.d. up to 6 g/day **IV** 1 mg/kg/h, adjusted for gastric output

Heartburn
Adult: **PO** 75–150 mg b.i.d.

Renal Impairment Dosage Adjustment
CrCl less than 50 mL/min: Use PO dose q24h, use IV dose q18–24h
Hemodialysis Dosage Adjustment
Time dose to administer at the end of dialysis

Common adverse effects in *italic*, life-threatening effects underlined; generic names in **bold**; classifications in SMALL CAPS; ♣ Canadian drug name; ☯ Prototype drug

ADMINISTRATION

Oral

- May be given without regard to meals.
- Effervescent tablets should not be chewed, swallowed whole, or allowed to dissolve on tongue. Dissolve 25 mg tablet in at least 5 mL of water; dissolve 150 mg tablet in 6–8 oz water. Allow tablet to dissolve completely before administration.
- Store tablets in light-resistant, tightly capped container at 15°–30° C (59°–86° F) in a dry place.

Intramuscular

- Note: Does not need to be diluted.

Intravenous

Note: Verify correct IV concentration and rate of infusion for infants and children with prescriber.

PREPARE: **Direct:** Dilute 50 mg NS, D5W, LR, or other compatible IV solution to a total volume of 20 mL. **Intermittent:** Dilute 50 mg in 50–100 mL of NS, D5W, LR, or other compatible IV solution. **Continuous:** Dilute total daily dose in 250 mL of NS, D5W, LR, or other compatible IV solution. Final concentration should be 2.5 mg/mL or less.

ADMINISTER: **Direct:** Give at a rate of 4 mL/min or 20 mL over not less than 5 min. **Intermittent:** Give over 15–30 min. **Continuous:** Give over 24 h. Do not exceed 6.25 mg/h.

INCOMPATIBILITIES **Solution/additive:** Amphotericin B, atracurium, cefazolin, cefoxitin, ceftazidime, cefuroxime, clindamycin, chlorpromazine, diazepam, ethacrynic acid, hydroxyzine, methotrimeprazine, midazolam, pentobarbital, phenobarbital, phytonadione. **Y-site:** Amphotericin B cholesteryl complex, hetastarch in normal saline, insulin.

- Schedule dose to coincide with end of treatment if patient is having hemodialysis.

ADVERSE EFFECTS (≥1%) CNS: Headache, malaise, dizziness, somnolence, insomnia, vertigo, mental confusion, agitation, depression, hallucinations in older adults. **CV:** Bradycardia (with rapid IV push). **GI:** Constipation, nausea, abdominal pain, diarrhea. **Skin:** Rash. **Hematologic:** Reversible decrease in WBC count, thrombocytopenia. **Body as a Whole:** Hypersensitivity reactions, anaphylaxis (rare).

DIAGNOSTIC TEST INTERFERENCE Ranitidine may produce slight elevations in *serum creatinine* (without concurrent increase in *BUN*); (rare) increases in *AST, ALT, alkaline phosphatase, LDH,* and total *bilirubin.* Produces false-positive tests for *urine protein* with *Multistix* (use *sulfosalicylic acid* instead).

INTERACTIONS Drug: May reduce absorption of **cefpodoxime, cefuroxime, delavirdine, ketoconazole, itraconazole.**

PHARMACOKINETICS Absorption: Incompletely from GI tract (50% reaches systemic circulation). **Peak:** 2–3 h PO. **Duration:** 8–12 h. **Distribution:** Distributed into breast milk. **Metabolism:** In liver. **Elimination:** In urine, with some excreted in feces. **Half-Life:** 2–3 h.

NURSING IMPLICATIONS

Assessment & Drug Effects

- Potential toxicity results from decreased clearance (elimination), which causes prolonged action; greatest risk for toxicity is in older adult patients or those with hepatic or renal dysfunction.

R

- Lab tests: Periodic LFTs. Monitor creatinine clearance if renal dysfunction is present or suspected.
- Be alert for early signs of hepatotoxicity: Jaundice (dark urine, pruritus, yellow sclera and skin), elevated transaminases (especially ALT) and LDH.
- Long-term therapy may lead to vitamin B_{12} deficiency.

Patient & Family Education

- Note: Long duration of action provides ulcer pain relief that is maintained through the night as well as the day.
- Adhere to scheduled periodic laboratory checkups during ranitidine treatment.
- Do not supplement therapy with OTC remedies for gastric distress or pain without prescriber's advice.

RANOLAZINE
(ra-no'la-zeen)

Ranexa
Classifications: ANTIANGINAL; PARTIAL FATTY ACID OXIDATION (PFOX) INHIBITOR
Therapeutic: ANTIANGINAL
Pregnancy Category: C

AVAILABILITY 500 mg, 1000 mg extended release tablets

ACTION & THERAPEUTIC EFFECT
Ranolazine is a partial fatty-acid oxidation inhibitor that shifts myocardial metabolism away from fatty acids to glucose. This shift requires less oxygen for oxidation and results in decreased oxygen demand. *Improves exercise tolerance and angina symptoms.*

USES Treatment of chronic stable angina in combination with calcium channel blockers, beta-blockers, or nitrates.

CONTRAINDICATIONS Severe hepatic impairment; severe renal impairment, renal failure, hypokalemia, hypomagnesemia; history of acute MI; lactation.
CAUTIOUS USE History of QT prolongation or torsades de pointes; renal impairment; older adult; pregnancy (category C); children.

ROUTE & DOSAGE

Chronic Stable Angina
Adult: **PO** 500 mg b.i.d. (max: 1000 mg b.i.d.) (patients taking concurrent CYP3A4 inhibitors have a max dose of 500 mg b.i.d.)

ADMINISTRATION

Oral
- **Must be** swallowed whole. Should not be crushed, broken, or chewed.
- Store at 15°–30° C (59°–86° F).

ADVERSE EFFECTS (≥1%) **Body as a Whole:** Peripheral edema. **CNS:** *Dizziness,* headache. **CV:** Palpitations. **GI:** Abdominal pain, *constipation,* dry mouth, nausea, vomiting. **Respiratory:** Dyspnea. **Special Senses:** Tinnitus, vertigo.

DIAGNOSTIC TEST INTERFERENCE
Ranolazine is not known to interfere with any diagnostic laboratory test.

INTERACTIONS Drug: INHIBITORS OF P-GLYCOPROTEIN (e.g., **ritonavir, cyclosporine**) may increase ranolazine absorption. Ranolazine increases the plasma concentrations of **digoxin** and **simvastatin.** INHIBITORS OF CYP3A4 [e.g., **diltiazem, erythromycin, grapefruit juice,** HIV PROTEASE INHIBITORS, **ketoconazole,** MACROLIDE ANTIBIOTICS (especially **ketoconazole**), **verapamil**] can increase plasma levels and QT_c elevation. **Paroxetine,** a CYP2D6 INHIBITOR, increases the plasma levels of ranolazine. CLASS I OR

Common adverse effects in *italic*, life-threatening effects underlined; generic names in **bold**; classifications in SMALL CAPS; ◆ Canadian drug name; ⊙ Prototype drug

III ANTIARRHYTHMICS (e.g., **quinidine, dofetilide, sotalol**), **thioridazine,** and **ziprasidone** can cause additive increases in QT$_c$ elevation. **Food: Grapefruit juice.**

PHARMACOKINETICS Absorption: 73% of PO dose absorbed. **Peak:** 2–5 h. **Distribution:** 62% protein bound. **Metabolism:** Extensive hepatic metabolism. **Elimination:** 75% in urine; 25% in feces. **Half-Life:** 7 h.

NURSING IMPLICATIONS

Assessment & Drug Effects

- Monitor ECG at baseline and periodically for prolongation of the QT$_c$ interval.
- Lab tests: Baseline and periodic HbA1C in diabetics.
- Monitor blood levels of digoxin with concurrent therapy.
- When coadministered with simvastatin, monitor for and report unexplained muscle weakness or pain.

Patient & Family Education

- Do not engage in hazardous activities until response to drug is known.
- Contact prescriber if you experience fainting while taking ranolazine.
- Do not drink grapefruit juice or eat grapefruit while taking this drug.

RASAGILINE

(ras-a-gi′leen)

Azilect

Classifications: MONOAMINE OXIDASE-B (MAO-B) INHIBITOR; ANTIPARKINSON
Therapeutic: ANTIPARKINSON
Pregnancy Category: C

AVAILABILITY 0.5 mg, 1 mg tablets

ACTION & *THERAPEUTIC EFFECT*
Rasagiline is a potent monoamine oxidase B (MAO-B) inhibitor that prevents the enzyme monoamine oxidase B from breaking down dopamine in the brain. Rasagiline also interferes with dopamine reuptake at synapses in the brain. *Rasagiline helps to overcome dopaminergic motor dysfunction in Parkinson's disease.*

USES Treatment of Parkinson's disease, either as monotherapy or as an adjunct to levodopa.

CONTRAINDICATIONS Moderate to severe hepatic impairment; alcoholism; biliary cirrhosis; pheochromocytoma; elective surgery; increased intracranial pressure; cerebrovascular disease, intracranial bleeding, recent head trauma, stroke; controlled hypertension; concurrent use of MAOI therapy or antihypertensive drugs; elective surgery.

CAUTIOUS USE Mild hepatic dysfunction; cardiovascular disease; diabetes mellitus; asthma, bronchitis, hyperthyroidism; postural or orthostatic hypotension; migraine headaches; moderate to severe renal impairment, anuria; epilepsy or preexisting seizure disorders; pregnancy (category C), lactation; children.

ROUTE & DOSAGE

Parkinson's Disease
Adult: **PO** 1 mg/day as monotherapy; 0.5–1 mg/day if adjunctive therapy

Hepatic Impairment Dosage Adjustment
Mild Impairment: 0.5 mg/day

ADMINISTRATION

Oral

- May be given without regard to food.
- Store at 15°–30° C (59°–86° F).

Common adverse effects in *italic*, life-threatening effects <u>underlined</u>; generic names in **bold**; classifications in SMALL CAPS; ♣ Canadian drug name; ☻ Prototype drug

1309

ADVERSE EFFECTS (≥1%) **Body as a Whole:** Accidental injury, allergic reaction, alopecia, gingivitis, hernia, infection, neck pain, pruritus. **CNS:** Abnormal dreams, abnormal gait, amnesia, anxiety, asthenia, ataxia, confusion, depression, dizziness, *dyskinesia, dystonia, fall,* fever, flu-like syndrome, hallucinations, *headache,* hyperkinesias, hypertonia, malaise, neuropathy, neck pain, paresthesia, somnolence, syncope, tremor, vertigo. **CV:** Angina pectoris, bundle branch block, cerebrovascular accident, chest pain, postural hypotension. **GI:** Abdominal pain, anorexia, constipation, diarrhea, dry mouth, dyspepsia, dysphagia, gastroenteritis, GI hemorrhage, *nausea,* vomiting. **Hematologic:** Anemia, hemorrhage. **Metabolic:** Abnormal liver function tests, albuminuria, weight loss. **Musculoskeletal:** Arthralgia, arthritis, bursitis, leg cramps, myasthenia, tenosynovitis. **Respiratory:** Asthma, dyspnea, epistaxis, increased cough, rhinitis. **Skin:** Ecchymosis, eczema, skin carcinoma, skin ulcer, sweating, urticaria, vesiculobullous rash. **Special Senses:** Conjunctivitis. **Urogenital:** Decreased libido, hematuria, impotence, urinary incontinence.

INTERACTIONS Drug: INHIBITORS OF CYP1A2 (e.g., **atazanavir, ciprofloxacin, mexiletine, tacrine**) may increase rasagiline plasma levels. Rasagiline increases the plasma levels of ANESTHETICS; thus it **must be** discontinued 14 days prior to elective surgery. Rasagiline can cause severe CNS toxicity, including hyperpyrexia and death, with ANTIDEPRESSANTS, SELECTIVE SEROTONIN REUPTAKE INHIBITORS (SSRI), SEROTONIN-NOREPINEPHRINE REUPTAKE INHIBITORS (SNRI), NONSELECTIVE MAO INHIBITORS, or SELECTIVE MAO-B INHIBITORS. Rasagiline can increase the plasma levels of **cyclobenzaprine** and SYMPATHOMIMETIC AMINES. Rasagiline and **dextromethorphan** can cause brief episodes of psychosis and bizarre behavior. Rasagiline can potentiate the dopaminergic effects of **levodopa.** Rasagiline can increase the plasma levels of **meperidine, methadone, propoxyphene,** and **tramadol,** resulting in coma, severe hypertension or hypotension, severe respiratory depression, convulsions, and death. **Herbal:** Rasagiline increases the plasma levels of **St. John's wort.**

PHARMACOKINETICS Absorption: Rapidly absorbed with 36% bioavailability. **Peak:** 1 h. **Distribution:** 88–94% protein bound. **Metabolism:** Extensive hepatic metabolism. **Elimination:** Primarily renal (62%) with minor fecal elimination. **Half-Life:** 3 h.

NURSING IMPLICATIONS

Assessment & Drug Effects

▪ Monitor for and report S&S of dopaminergic side effects (e.g., dyskinesia, hallucinations, etc.) with concurrent levodopa.
▪ Monitor for and report suspicious skin changes suggestive of melanoma or other skin cancers.
▪ Lab tests: Baseline and periodic LTFs; periodic renal function tests, CBC with Hct and Hgb.
▪ Note all contraindicated drugs and drug groups and exercise caution not to administer a contraindicated substance.
▪ Note all drug interactions and monitor for the indicated effects.

Patient & Family Education

▪ Do not take any prescription or nonprescription drug without consulting prescriber.
▪ Periodic skin examinations should be scheduled with a dermatologist. If you notice changes in a skin mole or new skin lesion, contact the dermatologist.

- Avoid foods and beverages containing tyramine (e.g., aged cheeses and meats, tap beer, red wine, soybean products).
- Make position changes slowly, especially when standing from a lying or sitting position.
- Do not drive or engage in potentially hazardous activities until response to drug is known.
- Report immediately any of the following to a health care provider: Palpitations, severe headache, blurred vision, difficulty thinking, seizures, chest pain, unexplained nausea or vomiting, or any sudden weakness or paralysis.

RASBURICASE
(ras-bur′i-case)
Elitek, Fasturtec ♦
Classifications: ANTIGOUT; ANTI-METABOLITE
Therapeutic: ANTIGOUT
Pregnancy Category: C

AVAILABILITY 1.5 mg/vial powder for injection

ACTION & *THERAPEUTIC EFFECT* A recombinant urate-oxidase enzyme produced by DNA technology. In humans, uric acid is the final step in the catabolic pathway of purines. Rasburicase catalyzes enzymatic oxidation of uric acid; thus it is only active at the end of the purine catabolic pathway. *Used to manage plasma uric acid levels in pediatric patients with leukemia, lymphoma, and solid tumor malignancies who are receiving anticancer therapy that results in tumor lysis, and therefore elevates plasma uric acid.*

USES Initial management of increased uric acid levels secondary to tumor lysis.

CONTRAINDICATIONS Hypersensitivity to rasburicase; deficiency in glucose-6-phosphate dehydrogenase (G6PD); history of anaphylaxis; hemolytic reactions or methemoglobinemia reactions to rasburicase; lactation.

CAUTIOUS USE Patients at risk for G6PD deficiency (e.g., African or Mediterranean ancestry); asthma; bone marrow suppression, pregnancy (category C); children younger than 1 mo. Safety and efficacy in adults and elderly are unknown.

ROUTE & DOSAGE

Hyperuricemia
Child (older than 1 mo): **IV** 0.15–0.2 mg/kg/day × 5 days starting 4–24 h before chemotherapy

ADMINISTRATION

Intravenous

PREPARE: **IV Infusion:** Reconstitute each 1.5 mg vial with 1 mL of the provided diluent and mix by swirling very gently. **Do not shake.** Discard if particulate matter is visible or if product is discolored after reconstitution. ▪ Remove the predetermined dose from the reconstituted vials and inject into enough NS in an infusion bag to achieve a final total volume of 50 mL.
ADMINISTER: **IV Infusion:** Give over 30 min. **DO NOT GIVE BOLUS DOSE.** Infuse through an **unfiltered** line used for no other medications. ▪ If a separate line is not possible, flush the line with at least 15 mL of saline solution before/after infusion of rasburicase.
▪ Immediately discontinue IV infusion and institute emergency measures for S&S of anaphylaxis

Common adverse effects in *italic*, life-threatening effects <u>underlined</u>; generic names in **bold**; classifications in SMALL CAPS; ♦ Canadian drug name; 🅿 Prototype drug

1311

including chest pain, dyspnea, hypotension, and/or urticaria.
INCOMPATIBILITIES Do not mix or infuse with other drugs.

ADVERSE EFFECTS (≥1%) **Body as a Whole:** *Fever;* sepsis, severe hypersensitivity reactions including anaphylaxis at any time during treatment. **CNS:** *Headache.* **GI:** *Mucositis, vomiting, nausea, diarrhea, abdominal pain.* **Hematologic:** Neutropenia. **Skin:** *Rash.*

DIAGNOSTIC TEST INTERFERENCE May give false elevations for **uric acid** if blood sample is left at room temperature.

PHARMACOKINETICS Half-Life: 18 h.

NURSING IMPLICATIONS
Assessment & Drug Effects
- Lab tests: Patients at higher risk for G6PD deficiency (e.g., patients of African or Mediterranean ancestry) should be screened prior to starting therapy as this deficiency is a contraindication for this drug.
- Monitor closely for S&S of hypersensitivity and be prepared to institute emergency measures for anaphylaxis.
- Monitor cardiovascular, respiratory, neurologic, and renal status throughout therapy.

Patient & Family Education
- Report immediately any distressing S&S to prescriber.

REMIFENTANIL HYDROCHLORIDE
(rem-i-fent'a-nil)
Ultiva
Classifications: ANALGESIC, NARCOTIC (OPIATE AGONIST); GENERAL ANESTHESIA

Therapeutic: NARCOTIC ANALGESIC; GENERAL ANESTHESIA
Prototype: Morphine
Pregnancy Category: C
Controlled Substance: Schedule II

AVAILABILITY 1 mg/mL, 2 mg/mL, 5 mg/mL injection

ACTION & THERAPEUTIC EFFECT Synthetic, potent narcotic agonist analgesic that is rapidly metabolized; therefore respiratory depression is of shorter duration when discontinued. *Used as the analgesic component of an anesthesia regime.*

USES Analgesic during induction and maintenance of general anesthesia, as the analgesic component of monitored anesthesia care.

CONTRAINDICATIONS Hypersensitivity to fentanyl analogs, epidural or intrathecal administration.
CAUTIOUS USE Head injuries, increased intracranial pressure; debilitated, morbid obesity, poor-risk patients; COPD, other respiratory problems, bradyarrhythmia; older adults; pregnancy (category C), lactation. Safety in labor and delivery has not been demonstrated.

ROUTE & DOSAGE

Adjunct to Anesthesia
Adult: **IV** 0.5–1 mcg/kg/min or 1 mcg/kg bolus
Child: **IV** Birth–2 mo, 0.4–1 mcg/kg/min; 1–12 y, 0.5–1 mcg/kg/min or 1 mcg/kg bolus

Obesity Dosage Adjustment
Dose based on IBW

ADMINISTRATION
Intravenous
IV administration to infants and children: Verify correct IV

concentration and rate of infusion with prescriber.

PREPARE: **Direct/Continuous Infusion** Reconstitute by adding 1 mL of sterile water for injection, D5W, NS, D5NS, 1/2NS, or D5LR to each 1 mg of remifentanil to yield 1 mg/mL. Shake well to dissolve. ▪ Further dilute to a final concentration of 20, 25, 50, or 250 mcg/mL by adding the required dose to the appropriate amount of IV solution.

ADMINISTER: **Direct/Continuous Infusion** Give at the ordered rate according to patient's weight. ▪ Note that bolus doses should NOT be given during a continuous infusion of remifentanil. ▪ Flush IV tubing thoroughly following infusion.

INCOMPATIBILITIES **Solution/additive:** Unknown. **Y-site: Amphotericin B, amphotericin B cholesteryl, chlorpromazine, diazepam.**

▪ Clear IV tubing completely of the drug following discontinuation of remifentanil infusion to ensure that inadvertent administration of the drug will not occur at a later time. ▪ Reconstituted solution is stable for 24 h at room temperature. Store vials of powder at 2°–25° C (36°–77° F).

ADVERSE EFFECTS (≥1%) **Body as a Whole:** Muscle rigidity, shivering. **CNS:** Dizziness, headache. **CV:** Hypotension, hypertension, bradycardia. **GI:** *Nausea*, vomiting. **Respiratory:** Respiratory depression, apnea. **Skin:** Pruritus.

INTERACTIONS Drug: Alcohol and other CNS DEPRESSANTS potentiate effects; MAO INHIBITORS may precipitate hypertensive crisis.

PHARMACOKINETICS Duration: 12 min. **Distribution:** 70% protein bound. **Metabolism:** Hydrolyzed by nonspecific esterases in the blood and tissues. **Elimination:** In urine. **Half-Life:** 3–10 min.

NURSING IMPLICATIONS

Assessment & Drug Effects

▪ Monitor vital signs during postoperative period; observe for and immediately report any S&S of respiratory distress or respiratory depression, or skeletal and thoracic muscle rigidity and weakness.
▪ Monitor for adequate postoperative analgesia.

REPAGLINIDE ⊘

(rep-a-gli′nide)
Prandin, GlucoNorm ♦
Classifications: ANTIDIABETIC; MEGLITINIDE
Therapeutic: ANTIHYPERGLYCEMIC
Pregnancy Category: C

AVAILABILITY 0.5 mg, 1 mg, 2 mg tablets

ACTION & *THERAPEUTIC EFFECT* Hypoglycemic agent that lowers blood glucose levels by stimulating release of insulin from the pancreatic islets. *Significantly reduces postprandial blood glucose in type 2 diabetes [preprandial blood glucose between 80 and 120 mg/dL and HbA1C (glycosylated Hgb less than 6.5%)]. Minimal effects on fasting blood glucose.*

USES Adjunct to diet and exercise in type 2 diabetes. May also be used in combination with metformin.

CONTRAINDICATIONS Hypersensitivity to repaglinide; insulin-dependent diabetes, diabetic ketoacidosis, lactation. **CAUTIOUS USE** Hypoglycemia; loss of glycemic control due to

Common adverse effects in *italic*, life-threatening effects underlined; generic names in **bold**; classifications in SMALL CAPS; ♦ Canadian drug name; ⊘ Prototype drug

1313

R

secondary failure; hepatic impairment; severe renal impairment; older adults, surgery, fever, systemic infection, trauma; pregnancy (category C); children.

ROUTE & DOSAGE

Type 2 Diabetes

Adult: **PO** Initial dose: 0.5 mg 15–30 min a.c.; initial dose for patients previously using glucose-lowering agents: 1–2 mg 15–30 min a.c. (2–4 doses/day depending on meal pattern; max: 16 mg/day); dosage range: 0.5–4 mg 15–30 min a.c.

ADMINISTRATION

Oral

- Give within 30 min of beginning a meal.
- Store at 15°–30° C (59°–86° F) in a tightly closed container and protect from moisture.

ADVERSE EFFECTS (≥1%) **Body as a Whole:** Arthralgia, back pain, paresthesia, allergy. **CNS:** Headache. **CV:** Chest pain, angina. **GI:** Nausea, diarrhea, constipation, vomiting, dyspepsia. **Respiratory:** URI, sinusitis, rhinitis, bronchitis. **Metabolic:** *Hypoglycemia.*

INTERACTIONS Drug: Erythromycin, ketoconazole may inhibit metabolism and potentiate hypoglycemia; BARBITURATES, **carbamazepine, rifabutin, rifampin, rifapentine, pioglitazone** may induce metabolism and cause hyperglycemia; **gemfibrozil** may increase risk of hypoglycemia and duration of action. **Herbal: Ginseng, garlic** may increase hypoglycemic effects. **Food: Grapefruit juice** (greater than 1 qt/day) may increase plasma concentrations and adverse effects.

PHARMACOKINETICS Absorption: Rapidly from GI tract, 56% bioavailability. **Peak:** 1 h. **Distribution:** 98% protein bound. **Metabolism:** In liver (CYP3A4). **Elimination:** 90% in feces. **Half-Life:** 1 h.

NURSING IMPLICATIONS

Assessment & Drug Effects

- Lab tests: Frequent FBS and postprandial blood glucose monitoring and HbA1C q3mo to determine effective dose.
- Monitor carefully for S&S of hypoglycemia especially during the 1-wk period following transfer from a longer-acting sulfonylurea.

Patient & Family Education

- Take only with meals to lessen the chance of hypoglycemia. If a meal is skipped, skip a dose; if a meal is added, add a dose.
- Start repaglinide the morning after the other agent is stopped when changing from another oral hypoglycemia drug.
- Be alert for S&S of hyperglycemia or hypoglycemia (see Appendix F); report poor blood glucose control to prescriber.

RESERPINE ⊘

(re-ser′peen)

Serpalan, Sk-Reserpine
Classifications: RAUWOLFIA ALKALOID; ANTIHYPERTENSIVE
Therapeutic: ANTIHYPERTENSIVE
Pregnancy Category: D

AVAILABILITY 0.1 mg, 0.25 mg tablets

ACTION & *THERAPEUTIC EFFECT*
Interferes with binding of serotonin at receptor sites, decreases synthesis of norepinephrine by depleting dopamine (its precursor), and competitively inhibits their reuptake in storage granules.

Depletes norepinephrine and serotonin in CNS, peripheral nervous system, heart, and other organs and tissues. *Sympathetic inhibition seen in small but persistent decrease in BP, frequently associated with bradycardia, and reduced cardiac output.*

USES Mild essential hypertension and as adjunctive therapy with other antihypertensive agents in the more severe forms of hypertension. Also used in agitated psychotic states, primarily in patients intolerant to phenothiazine or patients who also require antihypertensive medication.

UNLABELED USES Reduce vasospastic attacks in Raynaud's phenomenon and other peripheral vascular disorders, and for short-term symptomatic treatment of thyrotoxicosis.

CONTRAINDICATIONS Hypersensitivity to rauwolfia alkaloids; history of mental depression; acute peptic ulcer, ulcerative colitis; patients receiving electroconvulsive therapy; within 7–14 days of MAO inhibitor therapy; pregnancy (category D), lactation.

CAUTIOUS USE Renal insufficiency; cardiac arrhythmias; cardiac damage; cerebrovascular accident; history of peptic ulcers; epilepsy; bronchitis; asthma; debilitated patients; gallstones; obesity; chronic sinusitis; parkinsonism; pheochromocytoma; older adults. Safe use in children not established.

ROUTE & DOSAGE

Hypertension

Adult: **PO** 0.5 mg/day initially, reduced to 0.1–0.25 mg/day
Geriatric: **PO** Start with 0.05 mg daily, increase by 0.05 mg/wk

ADMINISTRATION

Oral

- Give with meals or with milk or other food to minimize possibility of gastric irritation (drug increases gastric secretions).
- Store in tight, light-resistant containers, preferably at 15°–30° C (59°–86° F), unless otherwise directed by manufacturer.

ADVERSE EFFECTS (≥1%) **CNS:** *Drowsiness,* sedation, *lethargy,* mental depression, nervousness, anxiety, nightmares, increased dreaming, headache, dizziness, increased appetite, dull sensorium; prolonged use of large doses: CNS stimulation (parkinsonian syndrome): Tremors, muscle rigidity; <u>respiratory depression</u>, convulsions, hypothermia. **CV:** Bradycardia, *edema,* orthostatic hypotension, increased AV conduction time (prolonged therapy); angina-like symptoms, arrhythmias, CHF (rare). **Special Senses:** *Nasal congestion,* epistaxis, lacrimation, blurred vision; miosis, ptosis, conjunctival congestion (acute toxicity). **GI:** Dry mouth or excessive salivation, nausea, vomiting, abdominal cramps, diarrhea, reactivation of peptic ulcer (hypersecretion), heartburn, biliary colic. **Hematologic:** Thrombocytopenic purpura, anemia, prolonged BT. **Body as a Whole:** Hypersensitivity (pruritus, rash, asthma), muscle aches, dysuria, fixed-drug eruptions. **Urogenital:** Menstrual irregularities, breast engorgement, galactorrhea, gynecomastia, feminization (males), impaired sexual function, impotence.

DIAGNOSTIC TEST INTERFERENCE Possibility of elevated ***blood glucose*** values; however, it is also reported that reserpine may decrease thiazide-induced hyperglycemia. Increase in ***serum prolactin*** with

R

Common adverse effects in *italic,* life-threatening effects underlined; generic names in **bold**; classifications in SMALL CAPS; ✦ Canadian drug name; ⊙ Prototype drug

1315

chronic administration of **rauwolfia** alkaloids; overdoses may cause initial increase in excretion of **urinary catecholamines;** decreases with chronic administration. Large doses may cause initial rise in **urinary 5 HIAA** excretion. Initial IM doses may increase **urinary VMA** excretion followed by decrease by end of third day of therapy (with oral or parenteral administration). Possible interference with **urinary steroid** colorimetric determinations: **17-OHCS** and **17-KS.**

INTERACTIONS Drug: Diuretics, other HYPOTENSIVE AGENTS compound hypotensive effects; CARDIAC GLYCOSIDES (**digoxin**) may increase risk of arrhythmias; MAO INHIBITORS may cause excitation and hypertension; CNS DEPRESSANTS compound depression; may decrease response to **levodopa. Herbal:** St. John's wort may antagonize hypotensive effects.

PHARMACOKINETICS Peak: 2 h. **Distribution:** Widely distributed, especially to adipose tissue; crosses blood–brain barrier and placenta; distributed in breast milk. **Metabolism:** Extensively metabolized to inactive compounds. **Elimination:** Slowly excreted, 60% in feces within 96 h and 10% in urine. **Half-Life:** 4.5 and 11.3 h.

NURSING IMPLICATIONS

Assessment & Drug Effects

- Assess vital signs at frequent intervals. (Note: Drop in BP may be accompanied by bradycardia.)
- Lab tests: Periodic CBC with differential, platelet count, serum electrolytes, and plasma glucose.
- Supervise ambulation as indicated; postural hypotension occurs more frequently in elderly patients.
- Monitor I&O, especially in patients with impaired kidney function. Report changes in I&O ratio and pattern.

- Full therapeutic effect of oral drug for hypertension may not occur until 2–3 wk of therapy, and effects may persist for as long as 4–6 wk after drug is discontinued.
- Be aware that mental depression is a serious adverse effect and may be severe. It occurs most commonly in high dosage regimens (e.g., 0.5–1 mg/day or more) and may not appear until 2–8 mo of therapy and may last for several months after drug is withdrawn.

Patient & Family Education

- Take drug at the same time each day, do not skip or double doses, and do not stop therapy without advice of prescriber.
- Do not drive or engage in potentially hazardous activities until response to drug is known.
- Learn about possible adverse effects and report promptly to prescriber.
- Report the following possible beginning symptoms of depression: Early morning insomnia, anorexia, inability to concentrate, despondency, self-deprecation, attitude of detachment, mood swings, or impotence.
- Make position changes slowly, particularly from recumbent to upright posture, and lie down or sit down (head-low position) if patient feels faint. Do not take hot showers or hot tub baths, and do not to stand still for prolonged periods. Report symptoms of dizziness or lightheadedness to prescriber.
- Check for edema and record weight daily. Consult prescriber about gain of 1–2 kg (3–5 lb) in 1 wk.
- Do not take OTC medications without consulting prescriber or pharmacist (many preparations for coughs and colds contain agents that affect the actions of reserpine).

Common adverse effects in *italic*, life-threatening effects underlined; generic names in **bold**; classifications in SMALL CAPS; ✦ Canadian drug name; ⊙ Prototype drug

RESPIRATORY SYNCYTIAL VIRUS IMMUNE GLOBULIN (RSV-IVIG)

(res-pir'a-tory sin-cy'ti-al)

RespiGam

Classifications: BIOLOGICAL RESPONSE MODIFIER; IMMUNOGLOBULIN (IgG)

Therapeutic: IMMUNOGLOBULIN
Prototype: Immune globulin
Pregnancy Category: C

AVAILABILITY 2500 mg/50 mL vial

ACTION & *THERAPEUTIC EFFECT*
Contains IgG immune globulin antibodies from human plasma. *The preparation contains large amounts of RSV-neutralizing antibodies.*

USES Prevention of serious lower respiratory tract infection caused by RSV in children younger than 24 mo with bronchopulmonary dysplasia or history of premature birth; hypervolemia.

CONTRAINDICATIONS Previous severe reaction to virus immune globulin or other human immunoglobulin preparation, selective IgA deficiency; congenital heart disease, fluid overload; hepatic disease; lactation.

CAUTIOUS USE Immunodeficiency, AIDS, pulmonary disease; CHF; renal failure; pregnancy (category C).

ROUTE & DOSAGE

RSV

Child/Infant/Neonate: **IV** 750 mg/kg, may repeat monthly as needed

ADMINISTRATION

Intravenous

PREPARE: **IV Infusion:** Give undiluted. Do not shake vial.
ADMINISTER: **IV Infusion:** Infuse at 1.5 mL/kg/h for first 15 min,
then 3 mL/kg/h for next 15 min, then 6 mL/kg/h for rest of infusion. Infuse vial contents undiluted through a separate IV line if possible; if "piggyback" **must be** used, see manufacturer's directions. DO NOT EXCEED IV INFUSION RATES given in Route & Dosage table! ▪ Use a constant infusion pump.

INCOMPATIBILITIES **Solution/additive or Y-site:** Do not mix with other drugs.

▪ Store vials at 2°–8° C (35°–46° F). Begin infusion within 6 h after vial is entered and complete within 12 h.

ADVERSE EFFECTS (≥1%) **Body as a Whole:** Fever, pyrexia, fluid overload. **CV:** Tachycardia, hypertension. **GI:** Vomiting, diarrhea, gastroenteritis. **Respiratory:** Respiratory distress, wheezing, rales, hypoxia, hypoxemia, tachypnea. **Skin:** Injection site inflammation.

INTERACTIONS Drug: May interfere with immune response to LIVE VIRUS VACCINES (mumps, rubella, measles), may need to repeat vaccine if given within 10 mo of RespiGam.

PHARMACOKINETICS Half-Life: 22–28 days.

NURSING IMPLICATIONS

Assessment & Drug Effects
▪ Monitor closely during and after each IV rate change.
▪ Assess vital signs and respiratory status prior to infusion, during and after each rate change, and at 30-min intervals until 30 min after infusion is completed, and periodically thereafter for 24 h.
▪ Slow infusion immediately if S&S of fluid overload appear and report to prescriber.
▪ Lab tests: Monitor routine blood chemistry, serum electrolytes, blood gases, osmolality.

R

Common adverse effects in *italic*, life-threatening effects underlined; generic names in **bold**; classifications in SMALL CAPS; ♣ Canadian drug name; ☻ Prototype drug

1317

- Monitor for aseptic meningitis syndrome, which may begin up to 2 days after infusion.

Patient & Family Education
- Be aware of the possibility of aseptic meningitis syndrome; learn S&S to report (headache, drowsiness, fever, photophobia, painful eye movements, muscle rigidity, nausea, vomiting).

RETAPAMULIN
Altabax
Classifications: ANTIBIOTIC; PLEUROMUTILIN
Therapeutic: TOPICAL ANTIBIOTIC
Pregnancy Category: B

AVAILABILITY 1% topical ointment

ACTION & *THERAPEUTIC EFFECT*
Selectively inhibits bacterial protein synthesis at a site on the 50S subunit of the bacterial ribosome. *Effective against* Staphylococcus *(MRSA) and* Streptococcus *organisms.*

USES Topical treatment of impetigo due to susceptible stains of *Staphylococcus aureus* (methicillin-sensitive strains only) or *Streptococcus pyogenes* in patients 9 mo of age or older.

CONTRAINDICATIONS Hypersensitivity to retapamulin.
CAUTIOUS USE Pregnancy (category B); lactation; children less than 9 mo.

ROUTE & DOSAGE

Impetigo Infection
Adult/Child/Infant (9 mo or older): Apply in thin layer b.i.d. × 5 days

ADMINISTRATION
Topical
- Apply a thin layer to infected region. May cover with gauze dressing.
- Store at 15°–30° C (59°–86° F).

ADVERSE EFFECTS (≥1%) **Body as a Whole:** Application-site irritation, pyrexia. **CNS:** Headache. **GI:** Diarrhea, nausea. **Metabolic:** Creatinine phosphokinase increased. **Respiratory:** Nasopharyngitis. **Skin:** Eczema, pruritus.

PHARMACOKINETICS Absorption: Minimal systemic absorption. **Metabolism:** In liver.

NURSING IMPLICATIONS
Assessment & Drug Effects
- Monitor for excessive skin irritation. Report swelling, blistering, or oozing.

Patient & Family Education
- Report any of the following at application site: Redness, itching, burning, swelling, blistering, or oozing.

RETEPLASE RECOMBINANT
(re'te-plase)
Retavase
Classification: THROMBOLYTIC ENZYME, TISSUE PLASMINOGEN ACTIVATOR (T-PA)
Therapeutic: THROMBOLYTIC
Prototype: Alteplase
Pregnancy Category: C

AVAILABILITY 10.4 international unit vials

ACTION & *THERAPEUTIC EFFECT*
DNA recombinant human tissue-type plasminogen activator (t-PA) that acts as a catalyst in the cleavage of plasminogen to plasmin.

Responsible for degrading the fibrin matrix of a clot. *Has antithrombolytic properties.*

USES Thrombolysis management of acute MI to reduce the incidence of CHF and mortality.

CONTRAINDICATIONS Active internal bleeding, history of CVA, recent neurologic surgery or trauma, intercranial neoplasm, or aneurysm, bleeding disorders, severe uncontrolled hypertension.
CAUTIOUS USE Any condition in which bleeding constitutes a significant hazard (i.e., severe hepatic or renal disease, CVA, hypertension, acute pancreatitis, septic thrombophlebitis); pregnancy (category C), lactation. Safety and efficacy in children are not established.

ROUTE & DOSAGE

Thrombolysis during Acute MI
Adult: **IV** 10 units injected over 2 min. Repeat dose in 30 min (20 units total).

ADMINISTRATION

Intravenous ───────────

PREPARE: **Direct:** Reconstitute using only the diluent, syringe, needle, and dispensing pin provided with reteplase. ▪ Withdraw diluent with syringe provided. Remove needle from syringe, replace with dispensing pin and transfer diluent to vial of reteplase. Leave pin and syringe in place in vial and swirl to dissolve. Do NOT shake. ▪ When completely dissolved, remove 10 mL solution, replace dispensing pin with a 20-gauge needle.
ADMINISTER: **Direct:** Flush IV line before and after with 30 mL NS or D5W and do NOT give any other drug simultaneously through the same IV line. ▪ Give a single dose evenly over 2 min.
INCOMPATIBILITIES **Solution/additive: Heparin. Y-site: Bivalirudin, heparin.**

▪ Store drug kit unopened at 2°–25° C (36°–77° F).

ADVERSE EFFECTS (≥1%) **Hematologic:** *Hemorrhage* (including *intracranial*, GI, genitourinary), anemia. **CV:** Reperfusion arrhythmias.

DIAGNOSTIC TEST INTERFERENCE Causes decreases in plasminogen and fibrinogen, making ***coagulation*** and ***fibrinolytic*** tests unreliable.

INTERACTIONS Drug: Aspirin, abciximab, dipyridamole, heparin may increase risk of bleeding.

PHARMACOKINETICS Elimination: In urine. **Half-Life:** 13–16 min.

NURSING IMPLICATIONS

Assessment & Drug Effects
▪ Discontinue concomitant heparin immediately if serious bleeding not controllable by local pressure occurs and, if not already given, withhold the second reteplase bolus.
▪ Monitor carefully all potential bleeding sites; monitor for S&S of internal hemorrhage (e.g., GI, GU, intracranial, retroperitoneal, pulmonary).
▪ Monitor carefully cardiac status for arrhythmias associated with reperfusion.
▪ Avoid invasive procedures, arterial and venous punctures, IM injections, and nonessential handling of the patient during reteplase therapy.

Patient & Family Education
▪ Report changes in consciousness or signs of bleeding to prescriber immediately.

Common adverse effects in *italic*, life-threatening effects underlined; generic names in **bold**; classifications in SMALL CAPS; ♣ Canadian drug name; ☻ Prototype drug

1319

RH$_0$(D) IMMUNE GLOBULIN
(row)

RhoGAM, Rhophylac, WinRho SDF

RH$_0$(D) IMMUNE GLOBULIN MICRO-DOSE

BayRho-D Mini Dose, MICRho-GAM

Classifications: BIOLOGICAL RESPONSE MODIFIER; IMMUNOGLOBULIN (IgG)
Therapeutic: IMMUNOGLOBULIN
Prototype: Immune globulin
Pregnancy Category: C

AVAILABILITY RhoGAM, MICRhoGAM: 5% solution in prefilled syringes; **Rhophylac:** 300 mcg prefilled syringe; **WinRho SDF:** 120 mcg, 300 mcg, 1000 mcg vials

ACTION & *THERAPEUTIC EFFECT*
Sterile nonpyrogenic gamma globulin solution containing immunoglobulins (IgG), which provides passive immunity by suppressing active antibody response and formation of anti-Rh$_0$(D) in Rh-negative [Rh$_0$(D)-negative] individuals previously exposed to Rh-positive [Rh$_0$(D)-positive, D^u-positive] blood. *Effective for exposure in Rh-negative women when Rh-positive fetal RBCs enter maternal circulation during third stage of labor, fetal-maternal hemorrhage (as early as second trimester), amniocentesis, or other trauma during pregnancy, termination of pregnancy, and following transfusion with Rh-positive RBC, whole blood, or components (platelets, WBC) prepared from Rh-positive blood.*

USES To prevent isoimmunization in Rh-negative individuals exposed to Rh-positive RBC (see above). Rh$_0$(D) immune globulin microdose is for use only after spontaneous or induced abortion or termination of ectopic pregnancy up to and including 12 wk of gestation. Treatment of idiopathic thrombocytopenia purpura.

CONTRAINDICATIONS Rh$_0$(D)-positive patient; person previously immunized against Rh$_0$(D) factor, severe immune globulin hypersensitivity, bleeding disorders.
CAUTIOUS USE IgA deficiency; pregnancy (category C); neonates.

ROUTE & DOSAGE

Note: Only WinRho SDF and Rhophylac can be given IV. BayRho-D and RhoGAM are available in regular and mini-dose vials.

Antepartum Prophylaxis

Adult: **IM/IV** 300 mcg at approximately 28-wk gestation; followed by 1 vial of mini-dose or 120 mcg within 72 h of delivery if infant is Rh-positive

Postpartum Prophylaxis

Adult: **IM/IV** 300 mcg preferably within 72 h of delivery if infant is Rh-positive

Following Amniocentesis, Miscarriage, Abortion, Ectopic Pregnancy

Adult: **IM** If over 13-wk gestation, 300 mcg, preferably within 3 h but at least within 72 h; if less than 13 wk, give 50 mcg

Transfusion Accident

Adult: **IM/IV** 300 mcg for each volume of RBCs infused divided by 15, given within at least 72 h of accident

Common adverse effects in *italic*, life-threatening effects underlined; generic names in **bold**; classifications in SMALL CAPS; ♣ Canadian drug name; ⊘ Prototype drug

Child: **IV** Administer 600 mcg q8h until total dose given. Exposure to positive whole blood 9 mcg/mL, exposure to positive RBCs 18 mcg/mL. **IM** Administer 1200 mcg q12h until total dose given. Exposure to positive whole blood 12 mcg/mL, exposure to positive RBCs 24 mcg/mL.

Idiopathic Thrombocytopenia Purpura

Adult/Child: **IV** 50 mcg/kg, then 25–60 mcg/kg depending on response

ADMINISTRATION

- BayRho-D (HyperRHO S/D), MIC-RhoGam, and RhoGAM are administered by IM route only. NEVER give IV.
- WinRho SDF and Rhophylac may be given IM or IV depending on the indication.

Intramuscular

- Use the deltoid muscle. Give in divided doses at different sites, all at once or at intervals, as long as the entire dose is given within 72 h after delivery or termination of pregnancy.
- Reconstitute lyophilized powder with 1.25 mL of NS (using the same method to dissolve as for IV). Give immediately after reconstitution.
- Observe patient closely for at least 20 min after administration. Keep epinephrine immediately available; systemic allergic reactions sometimes occur.

Intravenous

PREPARE: **Direct:** No dilution is required for products supplied in liquid form. ▪ Reconstitute lyophilized powder vials according to specific manufacturer's directions.

▪ WinRho SDF: Remove entire contents of vial to obtain the labeled dosage. If partial vial is needed for dosage calculation, withdraw the entire contents to ensure accurate calculation of dosage requirement. ▪ Rhophylac: Bring to room temperature before use.

ADMINISTER: **Direct:** Rhophylac: Give at a rate of 2 mL per 15–60 sec for ITP. ▪ WinRho SDF: Give over 3–5 min.

▪ Refrigerate commercially prepared solutions, although it may remain stable up to 30 days at room temperature according to manufacturer. ▪ Discard solutions that have been frozen. ▪ Store powder at 2°–8° C (36°–46° F) unless otherwise directed; avoid freezing.

ADVERSE EFFECTS (≥1%) Body as a Whole: Injection site irritation, slight fever, myalgia, lethargy.

INTERACTIONS Drug: May interfere with immune response to LIVE VIRUS VACCINE; should delay use of LIVE VIRUS VACCINES for 3 mo after administration of Rh₀(D) immune globulin.

PHARMACOKINETICS Peak: 2 h IV, 5–10 days IM. Half-Life: 25 days.

NURSING IMPLICATIONS

Assessment & Drug Effects

- Obtain history of systemic allergic reactions to human immune globulin preparations prior to drug administration.
- Send sample of newborn's cord blood to laboratory for cross-match and typing immediately after delivery and before administration of Rh₀(D) immune globulin. Confirm that mother is Rh₀(D) and Dᵘ-negative. Infant **must be** Rh-positive.

Patient & Family Education

- Be aware that administration of Rh₀(D) immune globulin (antibody)

R

Common adverse effects in *italic*, life-threatening effects <u>underlined</u>; generic names in **bold**; classifications in SMALL CAPS; ♣ Canadian drug name; ❷ Prototype drug

1321

prevents hemolytic disease of the newborn in a subsequent pregnancy.

RIBAVIRIN

(rye-ba-vye′rin)

Copegus, Rebetol, RibaPak, Ribasphere, Virazole
Classification: ANTIVIRAL
Therapeutic: ANTIVIRAL
Prototype: Acyclovir
Pregnancy Category: X

AVAILABILITY 6 g vial for nebulizer; 200 mg, 400 mg, 600 mg tablets; 200 mg capsules; 40 mg/mL oral solution

ACTION & THERAPEUTIC EFFECT Synthetic nucleoside with broad-spectrum antiviral activity against DNA and RNA viruses. Mode is believed to involve multiple mechanisms including selective interference with viral ribonucleic protein synthesis. *Active against many RNA and DNA viruses, including respiratory syncytial virus (RSV), influenza A and B, parainfluenza, measles, mumps, Lassa fever, enterovirus 72 (formerly called hepatitis A), yellow fever, HIV, herpes simplex virus (HSV-1 and HSV-2), and vaccinia.*

USES Aerosol product used for selected infants and young children with respiratory syncytial virus (RSV). Oral product used in combination with interferon-alfa-2b to treat hepatitis C or in combination with peginterferon alpha for treatment of hepatitis C.

UNLABELED USES Prophylaxis and treatment of influenza A and B, pneumonia caused by adenovirus; Lassa fever, measles, HSV-1, HSV-2, enterovirus 72 (formerly hepatitis A), SARS, cytomegalovirus.

CONTRAINDICATIONS Mild RSV infections of lower respiratory tract; infants requiring simultaneous assisted ventilation; pancreatitis; autoimmune hepatitis; renal failure; hemoglobinopathy; pregnancy (category X), lactation.

CAUTIOUS USE COPD, asthma; anemia; history of MI, cardiac arrhythmias, cardiac disease; older adults, decreased renal, hepatic, or cardiac function; respiratory depression; history of depression or suicidal tendencies.

ROUTE & DOSAGE

RSV
Child: **Inhalation** 20 mg via SPAG-2 nebulizer administered over 12–18 h/day for 3–7 days

Hepatitis C
(in combination with interferon-alfa)
Adult: **PO** *Weight greater than 75 kg,* 600 mg b.i.d. for 24–48 wk; *weight less than 75 kg,* 400 mg in a.m., 600 mg in p.m. for 24–48 wk
Child/Adolescent: **PO** *Weight greater than 73 kg,* 15 mg/kg/day or 1200 mg/day in divided doses; *weight 60–73 kg,* 15 mg/kg/day or 1000 mg/day in divided doses; *weight 47–59 kg,* 15 mg/kg/day or 800 mg/day in divided doses; *older than 3 y, weight less than 47 kg,* 15 mg/kg/day in divided doses

Chronic Hepatitis C (with Peginterferon Alfa-2b)
Adult: **PO** 800–1400 mg daily in divided doses
Adolescent/Child (older than 3 y): **PO** *Weight greater than 73 kg,* 15 mg/kg/day or 1200 mg/day in divided doses; *weight 60–73 kg,* 15 mg/kg/day or 1000 mg/day in divided doses; *weight 47–59 kg,* 15 mg/kg/day or 800 mg/day

in divided doses; *weight less than 47 kg*, 15 mg/kg/day in divided doses

Renal Impairment Dosage Adjustment

CrCl less than 50 mL/min: Oral ribavirin should not be used

ADMINISTRATION

- Give tablets with food. Ensure that tablets and capsules are swallowed whole. They should not be opened, crushed, or chewed.

Inhalation

- Administer only by SPAG-2 aerosol generator, following manufacturer's directions.
- Caution: Ribavirin has demonstrated teratogenicity in animals. Advise pregnant health care personnel of the potential teratogenic risks associated with exposure during ribavirin administration to patients.
- Do not give other aerosol medication concomitantly with ribavirin.
- Discard solution in the SPAG-2 reservoir at least q24h and whenever liquid level is low before fresh reconstituted solution is added.
- Store unopened vial in a dry place at 15°–25° C (59°–78° F) unless otherwise directed.
- Following reconstitution, store solution at 20°–30° C (68°–86° F) for 24 h.

ADVERSE EFFECTS (≥1%) CV: Hypotension (faintness, light-headedness, unusual fatigue), MI, cardiac arrest. **Special Senses:** Conjunctivitis, erythema of eyelids. **Hematologic:** Reticulocytosis, hemolytic anemia (especially in combination with interferon alpha). **Respiratory:** Deterioration of respiratory function, dyspnea, apnea, chest soreness, bacterial pneumonia,

ventilator dependence. **GI:** Transient increases in AST, ALT, bilirubin; abdominal cramps, jaundice.

INTERACTIONS Drug: Ribavirin may antagonize the antiviral effects of **zidovudine** against HIV; increased risk of fetal defects with **peginterferon**. Use with **azathioprine** and **peginterferon alfa-2a** increases risk of pancytopenia.

PHARMACOKINETICS Absorption: Rapidly absorbed orally (44%) and systemically from lungs. **Peak:** Inhaled 60–90 min. PO 1.7–3 h. **Distribution:** Crosses placenta; distributed into breast milk. **Metabolism:** In cells to an active metabolite. **Elimination:** 85% in urine, 15% in feces. **Half-Life:** 24 h in plasma, 16–40 days in RBCs.

NURSING IMPLICATIONS

Assessment & Drug Effects

- Obtain specimens for rapid diagnosis of RSV infection before therapy is initiated or at least during the first 24 h of ribavirin therapy. Do not continue therapy without laboratory confirmation of RSV infection.
- Lab tests: Baseline CBC with differential and platelet count, repeat at 2 and 4 wk, and periodically thereafter; baseline and periodic serum electrolytes, LFTs, and TSH.
- Monitor respiratory function and fluid status closely during therapy. Note baseline rate and character of respirations and pulse. Observe for signs of labored breathing: Dyspnea, apnea; rapid, shallow respirations, intercostal and substernal retraction, nasal flaring, limited excursion of lungs, cyanosis. Auscultate lungs for abnormal breath sounds.
- Observe patients requiring simultaneous assisted ventilation closely for S&S of worsening pulmonary function. Check equipment carefully every 2 h, including endotracheal tube, for malfunction. Precipitation

R

Common adverse effects in *italic*, life-threatening effects underlined; generic names in **bold**; classifications in SMALL CAPS; ♥ Canadian drug name; ⊙ Prototype drug

1323

of ribavirin and accumulation of fluid in tubing can obstruct the apparatus and cause inadequate ventilation and gas exchange.

- Monitor cardiac status, including ECG, especially in those with pre-exisiting cardiac dysfunction.

Patient & Family Education
- Both male and female patients should take every precaution to prevent pregnancy during treatment and for 6 mo following the end of therapy.
- Inform prescriber immediately if a pregnancy occurs within 6 mo of completing therapy.
- Drink fluids liberally unless otherwise advised by prescriber.
- Use caution with hazardous activities until response to drug is known.

RIBOFLAVIN (VITAMIN B₂)

(rye'bo-flay-vin)
Riboflavin (Vitamin B₂)
Classification: VITAMIN
Therapeutic: VITAMIN REPLACEMENT
Pregnancy Category: A (C if greater than RDA)

AVAILABILITY 50 mg, 100 mg tablets

ACTION & *THERAPEUTIC EFFECT*
Water-soluble vitamin and component of the flavoprotein enzymes, which work together with a wide variety of proteins to catalyze many cellular respiratory reactions by which the body derives its energy. *Evaluated by improvement of clinical manifestations of deficiency: Digestive disturbances, headache, burning sensation of skin (especially "burning" feet), cracking at corners of mouth (cheilosis), glossitis, seborrheic dermatitis (and other skin lesions), mental depression, corneal vascularization (with photophobia, burning and itchy eyes,* *lacrimation, roughness of eyelids), anemia, neuropathy.*

USES To prevent riboflavin deficiency and to treat ariboflavinosis; also to treat microcytic anemia and as a supplement to other B vitamins in treatment of pellagra and beri-beri.

CAUTIOUS USE Pregnancy (category A; category C if greater than RDA).

ROUTE & DOSAGE

Nutritional Supplement
Adult: **PO** 5–10 mg/day
Child: **PO** 1–4 mg/day
Nutritional Deficiency
Adult: **PO** 5–30 mg/day in divided doses
Child: **PO** 3–10 mg/day

ADMINISTRATION
Oral
- Give with food to enhance absorption.
- Store in airtight containers protected from light.

ADVERSE EFFECTS (≥1%) **Urogenital:** May discolor urine bright yellow.

DIAGNOSTIC TEST INTERFERENCE
In large doses, riboflavin may produce yellow-green fluorescence in *urine* and thus cause false elevations in certain *fluorometric determinations* of *urinary catecholamines.*

INTERACTIONS Drug: No clinically significant interactions established.

PHARMACOKINETICS Absorption: Readily absorbed from GI tract. **Distribution:** Little is stored; excess amounts are excreted in urine. **Elimination:** In urine. **Half-Life:** 66–84 min.

NURSING IMPLICATIONS

Assessment & Drug Effects

▪ Collaborate with prescriber, dietitian, patient, and responsible family member in planning for diet. A complete dietary history is an essential part of vitamin replacement so that poor eating habits can be identified and corrected. Deficiency in one vitamin is usually associated with other vitamin deficiencies.

Patient & Family Education

▪ Be aware that large doses may cause an intense yellow discoloration of urine.

▪ Note: Rich dietary sources of riboflavin are found in liver, kidney, beef, pork, heart, eggs, milk and milk products, yeast, whole-grain cereals, vitamin B–enriched breakfast cereals, green vegetables, and mushrooms.

RIFABUTIN

(rif-a-bu'tin)

Mycobutin

Classifications: ANTIBIOTIC; ANTITUBERCULOSIS
Therapeutic: ANTITUBERCULOSIS
Prototype: Rifampin
Pregnancy Category: B

AVAILABILITY 150 mg capsules

ACTION & *THERAPEUTIC EFFECT*

Semisynthetic bacteriostatic antibiotic. Mode of action may be to inhibit DNA-dependent RNA polymerase in susceptible bacterial cells but not in human cells. *Effective against* Mycobacterium avium *complex (MAC) (or* M. avium-intracellulare*) and many strains of* M. tuberculosis.

USES

The prevention of disseminated *Mycobacterium avium* complex (MAC) disease in patients with advanced HIV infection.

CONTRAINDICATIONS

Hypersensitivity to rifabutin or any other rifamycins; lactation.

CAUTIOUS USE

Older adults, pregnancy (category B).

ROUTE & DOSAGE

Prevention of MAC

Adult: **PO** 300 mg daily, may give 150 mg b.i.d. if nausea is a problem
Child: **PO** 75 mg daily

ADMINISTRATION

Oral

▪ Give the usual dose of 300 mg/day or in two divided doses of 150 mg with food if needed to reduce GI upset.

▪ Store at room temperature, 15°–30° C (59°–86° F), unless otherwise directed.

ADVERSE EFFECTS (≥1%) **CNS:** *Headache.* **GI:** *Abdominal pain, dyspepsia, nausea, taste perversion, increased liver enzymes.* **Hematologic:** Thrombocytopenia, eosinophilia, leukopenia, neutropenia. **Skin:** Rash. **Other:** *Turns urine, feces, saliva, sputum, perspiration, and tears orange. Soft contact lenses may be permanently discolored.*

INTERACTIONS Drug: May decrease levels of BENZODIAZEPINES, BETA-BLOCKERS, **clofibrate, dapsone,** NARCOTICS, ANTICOAGULANTS, CORTICOSTEROIDS, **cyclosporine, quinidine,** ORAL CONTRACEPTIVES, PROGESTINS, SULFONYLUREAS, **ketoconazole, fluconazole,** BARBITURATES, **theophylline,** and ANTICONVULSANTS, resulting in therapeutic failure.

PHARMACOKINETICS Absorption: 12–20% of oral dose reaches the systemic circulation. **Peak:** 2–3 h. **Distribution:** 85% protein bound. Widely distributed, high

R

Common adverse effects in *italic*, life-threatening effects underlined; generic names in **bold**; classifications in SMALL CAPS; ♣ Canadian drug name; ⊘ Prototype drug

1325

concentrations in the lungs, liver, spleen, eyes, and kidney. Crosses placenta, distributed into breast milk. **Metabolism:** In the liver. Causes induction of hepatic enzymes. **Elimination:** Approximately 53% of dose is excreted in urine as metabolites, 30% is excreted in feces. **Half-Life:** 16–96 h (average 45 h).

NURSING IMPLICATIONS

Assessment & Drug Effects

- Monitor patients for S&S of active TB. Report immediately.
- Lab tests: Monitor periodic blood work for neutropenia and thrombocytopenia.
- Evaluate patients on concurrent oral hypoglycemic therapy for loss of glycemic control.
- Review patient's complete drug regimen because dosage adjustment of a significant number of drugs may be needed when rifabutin is added to regimen.

Patient & Family Education

- Learn S&S of TB and MAC (e.g., persistent fever, progressive weight loss, anorexia, night sweats, diarrhea) and notify prescriber if any of these develop.
- Notify prescriber of following: Muscle or joint pain, eye pain or other discomfort, chest pain with dyspnea, rash, or a flu-like syndrome.
- Be aware that urine, feces, saliva, sputum, perspiration, tears, and skin may be colored brown-orange. Soft contact lens may be permanently discolored.
- Rifabutin may reduce the activity of a wide variety of drugs. Provide a complete and accurate list of concurrent drugs to the prescriber for evaluation.

RIFAMPIN ⓟ

(rif'am-pin)

Rifadin, Rofact ♦

Classifications: ANTIBIOTIC; ANTITUBERCULOSIS

Therapeutic: ANTITUBERCULOSIS

Pregnancy Category: C

AVAILABILITY 150 mg, 300 mg capsules; 600 mg injection

ACTION & *THERAPEUTIC EFFECT*
Semisynthetic derivative of rifamycin B that inhibits DNA-dependent RNA polymerase activity in susceptible bacterial cells, thereby suppressing RNA synthesis. *Active against* Mycobacterium tuberculosis, M. leprae, Neisseria meningitidis, *and a wide range of gram-negative and gram-positive organisms.*

USES Initial treatment and retreatment of tuberculosis; as short-term therapy to prevent meningococcal infection.

UNLABELED USES Chemoprophylaxis in contacts of patients with *Haemophilus influenzae* type B infection; leprosy (especially dapsone-resistant leprosy); Legionnaire's disease, endocarditis, pruitus.

CONTRAINDICATIONS Hypersensitivity to rifampin; obstructive biliary disease; meningococcal disease; intermittent rifampin therapy.

CAUTIOUS USE Hepatic disease; history of alcoholism; IBD. Concomitant use of other hepatotoxic agents; pregnancy (category C).

ROUTE & DOSAGE

Pulmonary Tuberculosis

Adult: **PO/IV** 600 mg daily with other agents

Common adverse effects in *italic*, life-threatening effects <u>underlined</u>; generic names in **bold**; classifications in SMALL CAPS; ♦ Canadian drug name; ⓟ Prototype drug

Adults (with HIV): **PO/IV** 10 mg/kg (max: 600 mg) daily for 2 mo with other agents
Child: **PO/IV** 10–20 mg/kg/day (max: 600 mg/day) with other agents
Infant/Child (with HIV):
PO/IV 10–20 mg/kg (max: 600 mg) daily for 2 mo with other agents

Meningococcal Carriers

Adult: **PO/IV** 600 mg q12h for 2 consecutive days
Adolescent/Child/Infant: **PO/IV** 10–20 mg/kg q12h for 2 consecutive days (max: 600 mg/day)
Neonate: **PO/IV** 5 mg/kg b.i.d for 2 days

Hepatic Impairment Adjustment

Do not exceed 8 mg/kg/day

Renal Impairment Adjustment

CrCl less than 10 mL/min: Reduce dose by 50%

ADMINISTRATION

Oral

- Give 1 h before or 2 h after a meal. Peak serum levels are delayed and may be slightly lower when given with food; capsule contents may be emptied into fluid or mixed with food.
- Note: An oral suspension can be prepared from capsules for use with pediatric patients. Consult pharmacist for directions.
- Keep a desiccant in bottle containing capsules to prevent moisture causing instability.

Intravenous

PREPARE: **IV Infusion:** Reconstitute vial by adding 10 mL of sterile water for injection to each 600-mg to yield 60 mg/mL. Swirl to dissolve. ▪ Withdraw the ordered dose and further dilute in 500 mL of D5W (preferred) or NS. ▪ If absolutely necessary, 100 mL of D5W or NS may be used.

ADMINISTER: **IV Infusion:** Infuse 500 mL solution over 3 h and 100 mL solution over 30 min. ▪ Note: A less concentrated solution infused over a longer period is preferred.

INCOMPATIBILITIES **Solution/additive: Minocycline. Y-site: Amiodarone, diltiazem, tramadol.**

- Use NS solutions within 24 h and D5W solutions within 4 h of preparation.

ADVERSE EFFECTS (≥1%) **CNS:** Fatigue, drowsiness, headache, ataxia, confusion, dizziness, inability to concentrate, generalized numbness, pain in extremities, muscular weakness. **Special Senses:** Visual disturbances, transient low-frequency hearing loss, conjunctivitis. **GI:** *Heartburn, epigastric distress, nausea, vomiting, anorexia, flatulence, cramps, diarrhea,* <u>pseudomembranous colitis,</u> *transient elevations in liver function tests* (bilirubin, BSP, alkaline phosphatase, ALT, AST), pancreatitis. **Hematologic:** <u>Thrombocytopenia,</u> transient <u>leukopenia,</u> anemia, including hemolytic anemia. **Body as a Whole:** Hypersensitivity (fever, pruritus, urticaria, skin eruptions, soreness of mouth and tongue, eosinophilia, hemolysis), flu-like syndrome. **Urogenital:** Hemoglobinuria, hematuria, <u>acute renal failure,</u> light-chain proteinuria, menstrual disorders, <u>hepatorenal syndrome,</u> (with intermittent therapy). **Respiratory:** Hemoptysis. **Other:** Increasing lethargy, liver enlargement and tenderness, jaundice, brownish-red or orange discoloration of skin, sweat, saliva, tears, and feces; unconsciousness.

DIAGNOSTIC TEST INTERFERENCE Rifampin interferes with contrast

R

Common adverse effects in *italic*, life-threatening effects <u>underlined</u>; generic names in **bold**; classifications in SMALL CAPS; ✦ Canadian drug name; ⊙ Prototype drug

1327

media used for *gallbladder study;* therefore, test should precede daily dose of rifampin. May also cause retention of *BSP*. Inhibits standard assays for *serum folate* and *vitamin B$_{12}$*, may cause false positive opiate urine screen.

INTERACTIONS Drug: Do not use with PROTEASE INHIBITORS and **nevirapine** as it may increase treatment failure rate. **Alcohol, isoniazid, pyrazinamide, ritonavir,** saquinavir increase risk of drug-induced hepatotoxicity decreases concentrations of **alfentanil, alosetron, alprazolam, amprenavir,** BARBITURATES, BENZODIAZEPINES, **carbamazepine, atovaquone, cevimeline, chloramphenicol, clofibrate,** CORTICOSTEROIDS, **cyclosporine, dapsone, delavirdine, diazepam, digoxin, diltiazem, disopyramide, estazolam, estramustine, fentanyl, fosphenytoin, fluconazole galantamine, indinavir, itraconazole, ketoconazole, lamotrigine, levobupivacaine, lopinavir, methadone, metoprolol, mexiletine, midazolam, nelfinavir,** ORAL SULFONYLUREAS, ORAL CONTRACEPTIVES, **phenytoin,** PROGESTINS, **propafenone, propranolol, quinidine, quinine, ritonavir, sirolimus, theophylline,** THYROID HORMONES, **tocainide, tramadol, verapamil, warfarin, zaleplon,** and **zonisamide,** leading to potential therapeutic failure. Do not use with **simvastatin.**

PHARMACOKINETICS Absorption: Readily from GI tract. **Peak:** 2–4 h. **Distribution:** Widely distributed, including CSF; crosses placenta; distributed into breast milk. **Metabolism:** In liver to active and inactive metabolites; is enterohepatically cycled. **Elimination:** Up to 30% in urine, 60–65% in feces. **Half-Life:** 3 h.

NURSING IMPLICATIONS

Assessment & Drug Effects

- Lab tests: Periodic LFTs are advised. Closely monitor patients with hepatic disease.
- Check prothrombin time daily or as necessary to establish and maintain required anticoagulant activity when patient is also receiving an anticoagulant.

Patient & Family Education

- Do not interrupt prescribed dosage regimen. Hepatorenal reaction with flu-like syndrome has occurred when therapy has been resumed following interruption.
- Be aware that drug may impart a harmless red-orange color to urine, feces, sputum, sweat, and tears. Soft contact lenses may be permanently stained.
- Report onset of jaundice, hypersensitivity reactions, and persistence of GI adverse effects to prescriber.
- Use or add barrier contraceptive if using hormonal contraception. Concomitant use of rifampin and oral contraceptives leads to decreased effectiveness of the contraceptive and to menstrual disturbances (spotting, breakthrough bleeding).
- Keep drug out of reach of children.

RIFAPENTINE
(rif'a-pen-teen)
Priftin
Classifications: ANTIBIOTIC; ANTITUBERCULOSIS; MYCOBACTERIUM
Therapeutic: ANTITUBERCULOSIS
Prototype: Rifampin
Pregnancy Category: C

AVAILABILITY 150 mg tablets

ACTION & *THERAPEUTIC EFFECT*
Rifamycin derivative that inhibits

Common adverse effects in *italic*, life-threatening effects underlined; generic names in **bold**; classifications in SMALL CAPS; ♣ Canadian drug name; ⦾ Prototype drug

DNA-dependent RNA polymerase activity in susceptible bacterial cells, thereby suppressing RNA synthesis. *Effective against* Mycobacterium tuberculosis, *indicated by improvement in clinical S&S (e.g., fever, cough, pleuritic pain, fatigue) and on chest x-ray.*

USES Pulmonary tuberculosis in conjunction with at least one other antitubercular agent.

CONTRAINDICATIONS Hypersensitivity to any rifamycins (e.g., rifampin, rifabutin, rifapentine); porphyria; lactation.
CAUTIOUS USE Patients with abnormal liver function tests or hepatic disease; HIV disease; older adults; pregnancy (category C). Safe use in children younger than 12 y not established.

ROUTE & DOSAGE

> **Tuberculosis: Short-Course Therapy**
> *Adult/Adolescent:* PO 600 mg twice weekly (at least 72 h apart) × 2 mo, then 600 mg once weekly × 4 mo

ADMINISTRATION

Oral

- Give with an interval of NO LESS than 72 h between doses.
- Give with food to minimize GI upset.
- Store at 15°–30° C (59°–86° F) in a tightly closed container and protect from excess moisture.

ADVERSE EFFECTS (≥1%) **CNS:** Headache, dizziness. **CV:** Hypertension. **GI:** Increased liver function tests (ALT, AST), anorexia, nausea, vomiting, dyspepsia, diarrhea. **GU:** *Hyperuricemia,* pyuria, proteinuria, hematuria, urinary casts. **Hematologic:** Neutropenia, lymphopenia, anemia, leukopenia, thrombocytosis. **Respiratory:** Hemoptysis. **Skin:** Rash, pruritus, acne. **Body as a Whole:** Arthralgia, pain.

INTERACTIONS Drug: Decreased activity of ORAL CONTRACEPTIVES, **phenytoin, disopyramide, mexiletine, quinidine, tocainide, warfarin, fluconazole, itraconazole, ketoconazole, diazepam,** BETA-BLOCKERS, CALCIUM CHANNEL BLOCKERS, CORTICOSTEROIDS, **haloperidol,** SULFONYLUREAS, **cyclosporine, tacrolimus, levothyroxine,** NARCOTIC ANALGESICS, **quinine,** REVERSE TRANSCRIPTASE INHIBITORS, TRICYCLIC ANTIDEPRESSANTS, **sildenafil, theophylline.**

PHARMACOKINETICS Absorption: 70% absorbed. **Peak:** 5–6 h. **Distribution:** 97.7% protein bound. **Metabolism:** Hydrolyzed by esterase enzyme to active metabolite in liver; inducer of cytochromes P450 3A4 and 2C8/9. **Elimination:** 70% in feces, 17% in urine. **Half-Life:** 13.3 h.

NURSING IMPLICATIONS

Assessment & Drug Effects

- Lab tests: Sputum smear and culture, CBC, baseline LFTs (especially serum transaminases) to rule out preexisting hepatic disease and serum creatinine and BUN.
- Monitor carefully for S&S of toxicity with concurrent use of oral anticoagulants, digitalis preparations, or anticonvulsants.

Patient & Family Education

- Follow strict adherence to the prescribed dosing schedule to prevent emergence of resistant strains of tuberculosis.
- Be aware that food may be useful in preventing GI upset.
- Report immediately any of the following to the prescriber: Fever,

R

Common adverse effects in *italic*, life-threatening effects underlined; generic names in **bold**; classifications in SMALL CAPS; ♣ Canadian drug name; ⊘ Prototype drug

1329

weakness, nausea or vomiting, loss of appetite, dark urine or yellowing of eyes or skin, pain or swelling of the joints, severe or persistent diarrhea.
- Use or add barrier contraceptive if using hormonal contraception.

RIFAXIMIN
(ri-fax′i-min)

Xifaxan
Classifications: RIFAMYCIN ANTIBIOTIC; MYCOBACTERIUM
Therapeutic: MYCOBACTERIUM
Prototype: Rifampin
Pregnancy Category: C

AVAILABILITY 200 mg, 550 mg tablets

ACTION & *THERAPEUTIC EFFECT*
A rifamycin antibiotic that inhibits bacterial RNA synthesis by binding to DNA-dependent RNA polymerase, thereby blocking RNA transcription. *Its spectrum of activity includes gram-positive and gram-negative aerobes and anaerobes.*

USES Treatment of traveler's diarrhea, hepatic encephalopathy.
UNLABELED USES Crohn's disease, diverticulitis, irritable bowel syndrome.

CONTRAINDICATIONS Hypersensitivity to rifaximin, other rifamycin antimicrobial agents or to any of its components; dysentery; lactation.
CAUTIOUS USE Diarrhea with fever and/or blood in the stool, or diarrhea due to organisms other than *E. coli;* IBD, worsening diarrhea or diarrhea persisting for longer than 24–48 h; pregnancy (category C); children younger than 12 y.

ROUTE & DOSAGE

Traveler's Diarrhea
Adult/Adolescent: **PO** 200 mg t.i.d. for 3 days

Reduce Risk of Hepatic Encephalopathy Recurrence
Adult: **PO** 550 mg b.i.d.

ADMINISTRATION
Oral
- May be given without regard to food.
- Store at 15°–30° C (59°–86° F).

ADVERSE EFFECTS (≥1%) **Body as a Whole:** Fever. **CNS:** Headache. **GI:** *Flatulence,* abdominal pain, rectal tenesmus, defecation urgency, nausea, constipation, vomiting.

PHARMACOKINETICS Absorption: Less than 0.4% absorbed orally. **Peak:** 1.21 h. **Elimination:** In feces. **Half-Life:** 5.85 h.

NURSING IMPLICATIONS

Assessment & Drug Effects
- Withhold drug and notify prescriber if diarrhea worsens or lasts longer than 48 h after starting drug; an alternative treatment should be considered.
- Report promptly the appearance of blood in the stool.

Patient & Family Education
- Report promptly any of the following: Fever; difficulty breathing; skin rash, itching, or hives; worsening diarrhea during or after treatment or blood in the stool.

RILPIVIRINE
(ril-pi′vi-reen)

Edurant
Classifications: ANTIRETROVIRAL; NONNUCLEOSIDE REVERSE TRANSCRIPTASE INHIBITOR (NNRTI)
Therapeutic: ANTIRETROVIRAL; NNRTI
Prototype: Efavirenz
Pregnancy Category: B

AVAILABILITY 25 mg tablets

ACTION & *THERAPEUTIC EFFECT*
A nonnucleoside reverse transcriptase inhibitor (NNRTI) of human immunodeficiency virus type 1 (HIV-1). *Inhibits HIV-1 replication and slows/prevents disease progression.*

USES Treatment of HIV-1 infection in combination with other antiretroviral agents in adult patients who are treatment-naïve.

CONTRAINDICATIONS Concurrent drugs that significantly decrease rilpivirine plasma level such as anticonvulsants, antimycobacterials, proton pump inhibitors, systemic dexamethasone, St. John's wort (see Drug Interactions); lactation.
CAUTIOUS USE Concurrent drugs with a known risk of Torsade de Pointes; congenital prolonged QT interval or concurrent drugs that prolong the QT interval; depressive disorders or suicidal ideation; older adult; severe renal impairment; severe hepatic impairment (Child-Pugh class C); pregnancy (category B). Safety and effectiveness in children not established.

ROUTE & DOSAGE

HIV-1 Infection
Adult: **PO** 25 mg once daily

ADMINISTRATION
Oral
- Give with a full meal.
- If antacids are prescribed, they must be given at least 2 h before or 4 h after rilpivirine.
- Store at 15–30° C (59–86° F) and protect from light.

ADVERSE EFFECTS (≥1%) **Body as a Whole:** Cholecystitis, cholelithiasis. **CNS:** Abnormal dreams, anxiety, depressive disorders, dizziness, fatigue, headache, insomnia, sleep disorders, somnolence. **GI:** Abdominal pain, decreased appetite, diarrhea, nausea, vomiting. **Metabolic:** *Increase ALT and AST*, increased bilirubin, *increased cholesterol*, increased creatinine, increased triglycerides. **Skin:** Rash. **Urogenital:** Glomerulonephritis.

INTERACTIONS Drug: Strong CYP3A4 inhibitors (e.g., **atazanavir, clarithromycin, delaviridine, indinavir, itraconazole, ketoconazole, nefazodone, nelfinavir, ritonavir, saquinavir,** and **telithromycin**) **voriconazole,** can increase rilpivirine levels, while strong CYP3A4 inducers (e.g. **rifampin, dexamethasone, phenytoin, carbamazepine, efavirenz,** and **phenobarbital**) can decrease rilpivirine levels. Antacids, GLUCOCORTICOIDS, H₂ RECEPTOR ANTAGONISTS, **methadone** can decrease rilpivirine. Rilpivirine can decrease the levels of **methadone.** Rilpivirine has been associated with QT prolongation. Coadministration of other drug that prolongs the QT interval (e.g., **disopyramide, procainamide, amiodarone, bretylium, clarithromycin, levofloxacin**) may cause additive effects. **Food:** Administration with a meal enhances absorption. **Herbal: St. John's wort** may decrease the levels of rilpivirine.

PHARMACOKINETICS Peak: 4-5 h. **Distribution:** 99.7% plasma protein bound. **Metabolism:** In liver. **Elimination:** Fecal (85%) and renal (6%). **Half-Life:** 50 h.

NURSING IMPLICATIONS
Assessment & Drug Effects
- Monitor closely for adverse effects in those with severe renal impairment.

Common adverse effects in *italic*, life-threatening effects underlined; generic names in **bold**; classifications in SMALL CAPS; ♣ Canadian drug name; ⊘ Prototype drug

1331

- Monitor for and report promptly signs of depression or suicidal ideation.
- Lab tests: Periodic plasma HIV RNA; periodic lipid profile, LFTs, and renal functions.

Patient & Family Education
- Drug absorption is improved when taken with a full meal.
- Seek immediate medical assistance if you experience depression or thoughts of suicide.
- Do not self-treat depression with St. John's wort.
- Report to prescriber all prescription and nonprescription drugs and herbal products you use.
- Do not use over-the-counter stomach medications without consent of your health care provider.
- If you use an antacid, take it at least 2 h before or 4 h after taking rilpivirine.

RILUZOLE

(ri-lu′zole)
Rilutek
Classifications: AMYOTROPHIC LATERAL SCLEROSIS (ALS) AGENT; GLUTAMATE ANTAGONIST
Therapeutic: ALS AGENT
Pregnancy Category: C

AVAILABILITY 50 mg tablets

ACTION & THERAPEUTIC EFFECT Glutamate antagonist that inhibits the presynaptic release of glutamic acid in the CNS. Effectiveness based on theory that pathogenesis of amyotrophic lateral sclerosis (ALS) is related to injury of motor neurons by glutamate. *Believed to reduce the degeneration of neurons in ALS.*

USES Treatment of ALS.

CONTRAINDICATIONS Hypersensitivity to riluzole; ALT levels are 5 × ULN or if clinical jaundice develops; lactation.
CAUTIOUS USE Hepatic dysfunction, renal impairment; hypertension, history of other CNS disorders; older adults; pregnancy (category C). Safe use in children younger than 12 y is not established.

ROUTE & DOSAGE

ALS
Adult: **PO** 50 mg q12h at least 1 h before or 2 h after meals

ADMINISTRATION

Oral
- Give at same time daily and at least 1 h before or 2 h after a meal. Do not give before/after a high-fat meal.
- Store at room temperature; protect from bright light.

ADVERSE EFFECTS (≥1%) **Body as a Whole:** *Asthenia,* headache, back pain, malaise, arthralgia, weight loss, peripheral edema, flu-like syndrome. **CNS:** Hypertonia, depression, dizziness, dry mouth, insomnia, somnolence, circumoral paresthesia. **CV:** Hypertension, tachycardia, phlebitis, palpitation. **GI:** Abdominal pain, *nausea,* vomiting, dyspepsia, anorexia, diarrhea, flatulence, stomatitis. **Respiratory:** *Decreased lung function,* rhinitis, increased cough, apnea, bronchitis, dysphagia, dyspnea. **Skin:** Pruritus, eczema, alopecia, exfoliative dermatitis (rare). **Urogenital:** UTI.

INTERACTIONS Drug: BARBITURATES, **carbamazepine** may increase risk of hepatotoxicity.

PHARMACOKINETICS Absorption: Well absorbed from GI tract, 60% reaches systemic circulation. **Peak:** Steady-state levels by day 5. **Distribution:** 96% protein bound.

Metabolism: In liver by CYP1A2. **Elimination:** 90% in urine. **Half-Life:** 12 h.

NURSING IMPLICATIONS

Assessment & Drug Effects

- Lab tests: Monitor periodically Hct and Hgb, routine blood chemistries, and alkaline phosphatase. If febrile illness develops, monitor WBC count. Baseline LFTs, then monthly for 3 mo, then q3mo for remainder of first year, and periodically thereafter.
- Withhold drug and notify prescriber if liver enzymes are elevated.

Patient & Family Education

- Report any febrile illness to prescriber.
- Do not engage in potentially hazardous activities until response to drug is known.
- Learn common adverse effects and possible adverse interaction with alcohol.

RIMANTADINE

(ri-man'ta-deen)

Flumadine

Classifications: ANTIVIRAL; ADMANTANE

Therapeutic: ANTIVIRAL

Prototype: Amantadine

Pregnancy Category: C

AVAILABILITY 100 mg tablets

ACTION & *THERAPEUTIC EFFECT*

Antiviral agent thought to exert an inhibitory effect early in the viral replication cycle, probably by interfering with the viral uncoating procedure of the influenza A virus. Inhibits synthesis of both viral RNA and viral protein, thus causing viral destruction. *Prevents or interrupts influenza A infections.*

USES Prophylaxis and treatment of influenza A.

CONTRAINDICATIONS Hypersensitivity to rimantadine and amantadine; lactation, children younger than 1 y for prophylaxis treatment.

CAUTIOUS USE History of seizures; renal or hepatic impairment; older adults; pregnancy (category C). Safe use in children younger than 17 y for treatment of Influenza A is not known.

ROUTE & DOSAGE

Prophylaxis of Influenza A

Adult/Child (10 y or older): **PO** 100 mg b.i.d.
Child (1–9 y): **PO** 5 mg/kg daily (max: 150 mg/day) in divided doses
Geriatric: **PO** 100 mg daily

Treatment of Influenza A

Adult/Adolescent (older than 17 y): **PO** 100 mg b.i.d. started within 48 h of symptoms and continued for 5–7 days from initial symptoms
Geriatric: **PO** 100 mg daily started within 48 h of symptoms and continued for 5–7 days from initial symptoms

Hepatic Impairment Dosage Adjustment

100 mg daily with severe liver disease

Renal Impairment Dosage Adjustment

CrCl 10–30 mL/min: Extend dosing interval to 24 h

ADMINISTRATION

Oral

- Store at 15°–30° C (59°–86° F).

ADVERSE EFFECTS (≥1%) **CNS:** Nervousness, dizziness, headache, sleep disturbances, fatigue or malaise, drowsiness, anticholinergic effects. **GI:** Nausea, vomiting, diarrhea,

R

Common adverse effects in *italic*, life-threatening effects underlined; generic names in **bold**; classifications in SMALL CAPS; ♣ Canadian drug name; ❶ Prototype drug

1333

dyspepsia, dry mouth, anorexia, abdominal pain.

INTERACTIONS Drug: Intranasal influenza vaccine should not be used within 48 h.

PHARMACOKINETICS Absorption: Readily absorbed from GI tract. **Peak:** Serum levels 3.2–4.3 h. **Distribution:** Concentrates in respiratory secretions. **Metabolism:** Extensively in liver. **Elimination:** By kidneys. **Half-Life:** 20–36 h.

NURSING IMPLICATIONS

Assessment & Drug Effects

- Monitor carefully for seizure activity in patients with a history of seizures. Seizures are an indication to discontinue the drug.
- Monitor cardiac, respiratory, and neurologic status while on drug. Report palpitations, hypertension, dyspnea, or pedal edema.

Patient & Family Education

- Report bothersome adverse effects to prescriber; especially hallucinations, palpitations, difficulty breathing, and swelling of legs.
- Use caution with hazardous activities until reaction to drug is known.

RIMEXOLONE
(rim-ex′o-lone)
Vexol
Pregnancy Category: C
See Appendix A-1.

RISEDRONATE SODIUM
(ri-se-dron′ate)
Actonel
Classifications: BISPHOSPHONATE; BONE METABOLISM REGULATOR
Therapeutic: BONE RESORPTION INHIBITOR; OSTEOPOROSIS TREATMENT
Prototype: Etidronate disodium
Pregnancy Category: C

AVAILABILITY 5 mg, 30 mg, 35 mg, 150 mg tablets

ACTION & *THERAPEUTIC EFFECT* Diphosphate preparation with primary action on bone. Lowers serum alkaline phosphatase, presumably by decreasing release of phosphate from bone and increasing excretion of parathyroid hormone. Slows rate of bone resorption and new bone formation in pagetic bone lesions and in normal remodeling process. *Effectiveness indicated by decreased bone and joint pain and improved bone density.*

USES Paget's disease, prevention and treatment of osteoporosis.
UNLABELED USES Osteolytic metastases.

CONTRAINDICATIONS Hypersensitivity to risedronate or other bisphosphonates; hypocalcemia, vitamin D deficiency; severe renal impairment (CrCl less than 30 mL/min); lactation.
CAUTIOUS USE Renal impairment; CHF; hyperphosphatemia or vitamin D deficiency; hepatic disease; UGI disease; fever related to infection or other causes; pregnancy (category C). Safety and efficacy in children younger than 18 y are not established.

ROUTE & DOSAGE

Paget's Disease
Adult: **PO** 30 mg daily for 2 mo, may repeat after 2 mo rest if necessary

Prevention and Treatment of Osteoporosis
Adult (female): **PO** 5 mg daily OR 35 mg once weekly OR 75 mg daily for 2 consecutive days each month OR 150 mg once monthly
Adult (men): **PO** 35 mg weekly
Adult (with chronic systemic glucocorticoid): **PO** 5 mg daily

ADMINISTRATION

Oral

- Give on an empty stomach (at least 30 min before first food or drink of the day) with at least 6–8 oz plain water. Ensure that tablet is swallowed whole. It should not be crushed or chewed.
- Note: Patient should be upright. Maintain upright position and empty stomach for at least 30 min after administration.
- Space calcium supplements and antacids as far as possible from risedronate.
- Store at 15°–30° C (59°–86° F) in a tightly closed container and protect from light.

ADVERSE EFFECTS (≥1%) **Body as a Whole:** Flu-like syndrome, asthenia, arthralgia, bone pain, leg cramps, myasthenia. **CNS:** Headache, dizziness. **CV:** Chest pain, peripheral edema. **GI:** *Diarrhea,* abdominal pain, nausea, constipation, belching, colitis. **Respiratory:** Bronchitis, sinusitis. **Skin:** Rash. **Special Senses:** Amblyopia, tinnitus, dry eyes.

DIAGNOSTIC TEST INTERFERENCE May interfere with the use of *bone-imaging agents.*

INTERACTIONS Drug: Calcium, ANTACIDS significantly decrease absorption, use with NONSTEROIDAL ANTI-INFLAMMATORIES may increase risk of gastric ulcer.

PHARMACOKINETICS Absorption: Minimally absorbed from GI tract, bioavailability 0.63%. **Peak:** 1 h. **Distribution:** Approximately 60% of dose is distributed to bone. **Metabolism:** Not metabolized. **Elimination:** In urine; unabsorbed drug excreted in feces. **Half-Life:** 220 h.

NURSING IMPLICATIONS

Assessment & Drug Effects

- Lab tests: Baseline and periodic serum calcium, phosphorus, and alkaline phosphatase.
- Monitor carefully for and immediately report S&S of GI bleeding and hypocalcemia.

Patient & Family Education

- Administration guidelines regarding upright position, empty stomach, and spacing relative to calcium supplements and antacids **must be** strictly followed.
- Report any of the following to prescriber: Eye irritation, significant GI upset, or flu-like symptoms.

RISPERIDONE

(ris-per'i-done)

Risperdal, Risperdal M-TAB, Risperdal Consta
Classification: ATYPICAL ANTIPSYCHOTIC
Therapeutic: ANTIPSYCHOTIC
Prototype: Clozapine
Pregnancy Category: C

AVAILABILITY 0.25 mg, 0.5 mg, 1 mg, 2 mg, 3 mg, 4 mg tablets; 0.5 mg, 1 mg, 2 mg, 3 mg, 4 mg quick-dissolving tablets; 1 mg/mL solution; 12.5 mg, 25 mg, 37.5 mg, 50 mg injection

ACTION & *THERAPEUTIC EFFECT*

Interferes with binding of dopamine to D_2-interlimbic region of the brain, serotonin ($5\text{-}HT_2$) receptors, and alpha-adrenergic receptors in the occipital cortex. It has low to moderate affinity for the other serotonin ($5\text{-}HT$) receptors. *Effective in controlling symptoms of*

Common adverse effects in *italic*, life-threatening effects underlined; generic names in **bold**; classifications in SMALL CAPS; ♣ Canadian drug name; ⊙ Prototype drug

1335

schizophrenia as well as other psychotic symptoms.

USES Treatment of schizophrenia; treatment of bipolar disorder; irritability associated with autism.

UNLABELED USES Management of patients with dementia-related psychotic symptoms. Adjunctive treatment of behavioral disturbances in patients with mental retardation.

CONTRAINDICATIONS Hypersensitivity to risperidone; elderly with dementia-related psychosis; QT prolongation, Reye's syndrome, brain tumor, severe CNS depression, head trauma; tardive dyskinesia; sunlight (UV) exposure, tanning beds; lactation.

CAUTIOUS USE Older adults; arrhythmias, hypotension, breast cancer, blood dyscrasia, cardiac disorders, cerebrovascular disease, hypotension, dehydration, diabetes mellitus, diabetic ketoacidosis, hyperglycemia, hypokalemia, hypomagnesemia, hyponatremia, MI, obesity, orthostatic hypotension, mild or moderate CNS depression, coma; GI obstruction, dysphagia; electrolyte imbalance, ethanol intoxication, heart failure, renal or hepatic dysfunction; seizure disorder, seizures, suicidal ideation; stroke, Parkinson's disease; pregnancy (category C). Children younger than 13 y for schizophrenia; children younger than 10 y for bipolar disease; children younger than 5 y for autism.

ROUTE & DOSAGE

Schizophrenia

Adult/Adolescent (older than 13 y): **PO** 1–2 mg/day in 1 or 2 doses, then titrate up (max: 8 mg/day) **IM** 25 mg once q2wk (max: 50 mg)
Geriatric: **PO** Start 0.5 mg b.i.d. and increase by 0.5 mg b.i.d. daily to an initial target of 1.5 mg

b.i.d. (max: 4 mg/day) **IM** 25 mg once q2wk (max: 25 mg)

Bipolar Disorder

Adult/Adolescent (older than 10 y): **PO** 2–3 mg once daily for up to 3 wk (max: 6 mg/day)
Geriatric: **PO** Start with 0.5 mg b.i.d. and increase by 0.5 mg b.i.d. daily to an initial target of 1.5 mg b.i.d. (max: 4 mg/day). May convert to once daily dosing after stabilized in b.i.d. 2–3 days.

Irritability Associated with Autism

Adolescent/Child (5 y or older, weight 20 kg or greater): **PO** 0.5 mg daily; after 4 days, increase to 1 mg daily; can increase by 0.5 mg q2wk
Child (5 y or older, weight less than 20 kg): **PO** 0.25 mg daily; after 4 days, increase to 0.5 mg daily; can increase by 0.25 mg q2wk

Renal Impairment Dosage Adjustment

CrCl less than 30 mL/min: Start with 0.5 mg b.i.d., increase by 0.5 mg b.i.d. daily to an initial target of 1.5 mg b.i.d., may increase by 0.5 mg b.i.d. at weekly intervals (max: 6 mg/day); lower **IM** dose may be required

Hepatic Impairment Dosage Adjustment

Start with dose of 0.5 mg b.i.d.

ADMINISTRATION

Oral

- The oral solution may be mixed with water, orange juice, low-fat milk, or coffee. It is not compatible with cola or tea.
- Orally disintegrating tablets should not be removed from the blister

---

until immediately before administration. Tablets disintegrate immediately and may be swallowed with/without liquid.
- Store at 15°–30° C (59°–86° F).

Intramuscular
- Reconstitute the 25, 37.5, or 50 mg vial using the supplied 2 mL prefilled syringe. Shake vigorously for at least 10 sec to produce a uniform, thick, milky suspension. If 2 min or more pass before injection, shake vial again.
- Give deep IM into the upper-outer quadrant of the gluteal muscle with the supplied needle; do not substitute. Follow the manufacturer's instructions for use of the SmartSite Needle-Free Vial Access Device and Needle-Pro device.
- Store unopened vials at 2°–8° C (36°–46° F). Protect from light.

ADVERSE EFFECTS (≥1%) **Body as a Whole:** Orthostatic hypotension with initial doses, sweating, weakness, fatigue. **CNS:** *Sedation, drowsiness, headache,* transient blurred vision, *insomnia,* disinhibition, *agitation,* anxiety, increased dream activity, dizziness, catatonia, *extrapyramidal symptoms* (akathisia, dystonia, pseudoparkinsonism), especially with doses greater than 10 mg/day, <u>neuroleptic malignant syndrome</u> (rare), increased risk of stroke in elderly. **CV:** Prolonged QT_c interval, tachycardia. **GI:** Dry mouth, dyspepsia, nausea, vomiting, diarrhea, constipation, abdominal pain, elevated liver function tests (AST, ALT). **Endocrine:** Galactorrhea. **Metabolic:** Hyperglycemia, diabetes mellitus. **Respiratory:** Rhinitis, cough, dyspnea. **Skin:** Photosensitivity. **Urogenital:** Urinary retention, menorrhagia, decreased sexual desire, erectile dysfunction, sexual dysfunction male and female.

DIAGNOSTIC TEST INTERFERENCE *Liver function tests (AST, ALT)* are elevated.

INTERACTIONS Drug: Risperidone may enhance the effects of certain ANTIHYPERTENSIVE AGENTS. May antagonize the antiparkinson effects of **bromocriptine, cabergoline, levodopa, pergolide, pramipexole, ropinirole. Carbamazepine, phenytoin, phenobarbital, rifampin** may decrease risperidone levels. **Clozapine** may increase risperidone levels.

PHARMACOKINETICS Absorption: Rapidly; not affected by food. **Onset:** Therapeutic effect 1–2 wk. **Peak:** 1–2 h. **Distribution:** 0.7 L/kg; in animal studies, risperidone has been found in breast milk. **Metabolism:** Primarily in liver by cytochrome P450 with an active metabolite, 9-hydroxyrisperidone. **Elimination:** 70% in urine; 14% in feces. **Half-Life:** 20 h for slow metabolizers, 30 h for fast metabolizers.

NURSING IMPLICATIONS
Assessment & Drug Effects
- Monitor diabetics for loss of glycemic control.
- Reassess patients periodically and maintain on the lowest effective drug dose.
- Monitor closely neurologic status of older adults.
- Monitor cardiovascular status closely; assess for orthostatic hypotension, especially during initial dosage titration.
- Monitor closely those at risk for seizures.
- Assess degree of cognitive and motor impairment, and assess for environmental hazards.
- Lab tests: Monitor periodically blood glucose, serum electrolytes, LFTs and complete blood counts.

Common adverse effects in *italic*, life-threatening effects <u>underlined</u>; generic names in **bold**; classifications in SMALL CAPS; ✦ Canadian drug name; ⊘ Prototype drug

1337

Patient & Family Education

- Carefully monitor blood glucose levels if diabetic.
- Do not engage in potentially hazardous activities until the response to drug is known.
- Be aware of the risk of orthostatic hypotension.
- Avoid alcohol while taking this drug.

RITONAVIR

(ri-ton'a-vir)

Norvir

Classifications: ANTIRETROVIRAL; PROTEASE INHIBITOR

Therapeutic: PROTEASE INHIBITOR

Prototype: Saquinavir

Pregnancy Category: B

AVAILABILITY 100 mg capsules; 80 mg/mL solution

ACTION & *THERAPEUTIC EFFECT*

HIV protease is an enzyme required to produce the polyprotein procurers of functional proteins in infectious HIV. Prevent cleavage of the viral polyproteins, resulting in the formation of immature noninfectious virus particles. *Protease inhibitor of both HIV-1 and HIV-2 resulting in the formation of noninfectious viral particles.*

USES

Alone or in combination with other antiretroviral agents or protease inhibitors for treatment of HIV infection. Often used to increase the effect of other antiretrovirals.

CONTRAINDICATIONS

Hypersensitivity to ritonavir; concurrent use of saquinavir and ritonavir in patients with conditions that affect cardiac electrical activity; antimicrobial resistance to protease inhibitors; pancreatitis; lactation.

CAUTIOUS USE Hepatic diseases, liver enzyme abnormalities, or hepatitis, jaundice; diabetes mellitus, diabetic ketoacidosis, hyperlipidemia, hypertriglyceridemia; hemophilia A or B, renal insufficiency; pregnancy (category B). Safe use in children younger than 1 mo has not been established.

ROUTE & DOSAGE

HIV
Adult: **PO** 600 mg b.i.d. 1 h before or 2 h after meal (may take with a light snack)
Child (older than 1 mo): **PO** 300–400 mg/m² b.i.d. (max: 600 mg b.i.d.), start with 250 mg/m² b.i.d., increase by 50 mg/m² q2–3days

ADMINISTRATION

Oral

- Give preferably with food; oral solution may be mixed with chocolate milk or nutritional therapy liquids within 1 h of dosing to improve taste.
- Store refrigerated at 2°–8° C (36°–46° F). Protect from light in tightly closed container.

ADVERSE EFFECTS (≥1%) **Body as a Whole:** Myalgia, allergic reaction, bronchitis, cough, rhinitis, taste alterations, visual disturbances, dysuria, hyperglycemia, diabetes. **CNS:** *Asthenia,* fatigue, headache, fever, malaise, circumoral or peripheral paresthesia, insomnia, dizziness, somnolence, abnormal thinking, amnesia, agitation, anxiety, confusion, convulsions, aphasia, ataxia, diplopia, emotional lability, euphoria, hallucinations, decreased libido, nervousness, neuralgia, neuropathy, peripheral neuropathy, paralysis, tremor, vertigo. **CV:** Palpitations, vasodilation, hypotension, postural hypotension, syncope, tachycardia, <u>prolonged QT interval</u>. **Hematologic:** Anemia, <u>thrombocytopenia</u>, lymphadenopathy. **GI:** *Nausea, diarrhea, vomiting,* abdominal pain, dyspepsia, stomatitis, anorexia, dry mouth, constipation, flatulence, cholecystitis, cholestasis, abnormal liver function tests, hepatitis. **Skin:** Rash, sweating, acne, contact dermatitis, pruritus, urticaria, skin ulceration, dry skin.

INTERACTIONS Drug: Carbamazepine, dexamethasone, phenobarbital, phenytoin, rifabutin, rifampin, smoking can decrease ritonavir levels. **Ritonavir** may increase serum levels and toxicity of **clarithromycin,** especially in patients with renal insufficiency (reduce **clarithromycin** dose in patients with CrCl less than 60 mL/min); **alfuzosin, amiodarone, bepridil, bupropion, clozapine, desipramine; dihydroergotamine, flecainide, fluticasone, meperidine, pimozide, piroxicam, propoxyphene, quinidine, rifabutin, saquinavir, trazodone.** Ritonavir decreases levels of ORAL CONTRACEPTIVES, **theophylline;** may increase **ergotamine** toxicity with **dihydroergotamine, ergotamine;** may increase systemic steroid exposure with **fluticasone.** Liquid formulation may cause disulfiram-like reaction with **alcohol** or **metronidazole.** See the complete prescribing information for a comprehensive table of potential, but not studied, drug interactions. Use with **darunavir** may increase risk of hepatotoxicity. **Herbal: St. John's wort, garlic** may decrease antiretroviral activity.

PHARMACOKINETICS Absorption: Rapidly from GI tract. **Peak:** 2–4 h. **Distribution:** 98–99% protein bound. **Metabolism:** In liver (CYP3A4). **Elimination:** Primarily in feces (greater than 80%).

NURSING IMPLICATIONS

Assessment & Drug Effects

- Lab tests: Monitor periodically CBC with differential and platelet count, LFTs, kidney function, serum albumin, lipid profile, CPK, serum amylase, electrolytes, blood glucose HbA1C, and alkaline phosphatase.
- Withhold drug and notify prescriber in the presence of abnormal liver function.
- Assess for S&S of GI distress, peripheral neuropathy, and other potential adverse effects.

Patient & Family Education

- Learn potential adverse reactions and drug interactions; report to prescriber use of any OTC or prescription drugs.
- Take this drug exactly as prescribed. Do not skip doses. Take at same time each day.
- Do not take ritonavir with any of the following drugs as fatal reactions may occur: Amiodarone, astemizole, alfuzosin, bepridil, dihydroergotamine, ergotamine, ergonovine, flecainide, methylergonovine, midazolam, pimozide, propafenone, quinidine, triazolam, voriconazole.

RITUXIMAB

(rit-ux′i-mab)

Rituxan

Classifications: ANTINEOPLASTIC; BIOLOGICAL RESPONSE MODIFIER; MONOCLONAL ANTIBODY; IMMUNOGLOBULIN (IGG); DISEASE-MODIFYING ANTIRHEUMATIC DRUG (DMARD)

Therapeutic: ANTINEOPLASTIC; ANTIRHEUMATIC; DMARD

Pregnancy Category: C

AVAILABILITY 10 mg/mL injection

ACTION & *THERAPEUTIC EFFECT* Rituximab is an IgG kappa

Common adverse effects in *italic*, life-threatening effects <u>underlined</u>; generic names in **bold**; classifications in SMALL CAPS; ♣ Canadian drug name; ⊘ Prototype drug

1339

immunoglobulin. As a monoclonal antibody it binds with the CD20 antigen on the surface of normal and malignant B lymphocytes. B-cells are believed to play a role in the pathogenesis of rheumatoid arthritis and associated chronic synovitis. B-cells may be acting at multiple sites in the autoimmune/inflammatory process including rheumatoid factor and other autoantibody production, antigen presentation, T-cell activation, and/or proinflammatory cytokine production. Rituximab-induced depletion of peripheral B-lymphocytes in patients with rheumatoid arthritis (RA) and in B-cell non-Hodgkin's lymphoma results in a rapid and sustained depletion of circulating and tissue-based (e.g., thymus, spleen) B-lymphocytes. *Rituximab effectiveness in both rheumatoid arthritis and non-Hodgkin's lymphoma is measured by induced depletion of peripheral B-lymphocytes. In combination with fludrabine and cyclophosphamide effective in treating patients previously untreated or previously treated CD20 positive Chronic lymphocytic leukemia (CLL).*

USES Rheumatoid arthritis, CLL, non-Hodgkin's lymphoma, Wegener's granulomatosis.

UNLABELED USES Acute lymphocytic leukemia, idiopathic thrombocytopenic purpura, multiple sclerosis, hemolytic anemia, mantle cell lymphoma.

CONTRAINDICATIONS Hypersensitivity to murine proteins, rituximab, or abciximab; serious infection; life-threatening cardiac arrhythmias; severe mucocutaneous reaction to rituximab; oliguria, rising serum creatinine; lactation.

CAUTIOUS USE Prior exposure to murine-based monoclonal antibodies; history of allergies; asthma and other pulmonary disease (increased risk of bronchospasm);

respiratory insufficiency; viral hepatitis B (HBV); CAD; thrombocytopenia; history of cardiac arrhythmias; hypertension, renal impairment; older adults; pregnancy (category C). Safety and efficacy in children are not established.

ROUTE & DOSAGE

Non-Hodgkin's Lymphoma

Adult: **IV** 375 mg/m^2 repeat dose on days 8, 15, and 22 (total of 4 doses)

Rheumatoid Arthritis

Adult: **IV** 1000 mg on days 1 and 15 (with methotrexate)

CLL

Adult: **IV** 375 mg/m^2 on day 1 then 500 mg/m^2 (in combination with fludarabine/cyclophosphamide) on day 1 of cycles 2–6.

Wegener's Granulomatosis

Adult: **IV** 375 mg/m^2 weekly x 4 wk

ADMINISTRATION

Intravenous

***PREPARE:* IV Infusion:** Dilute ordered dose to 1–4 mg/mL by adding to an infusion bag of NS or D5W. ▪ Examples: 500 mg in 400 mL yields 1 mg/mL; 500 mg in 75 mL yields 4 mg/mL. ▪ Gently invert bag to mix. Discard unused portion left in vial.

***ADMINISTER:* IV Infusion:** Infuse first dose at a rate of 50 mg/h; may increase rate at 50 mg/h increments q30min to maximum rate of 400 mg/h. ▪ For subsequent doses, infuse at a rate of 100 mg/h and increase by 100 mg/h increments q30min up to maximum rate of 400 mg/h. ▪ Slow or stop infusion if S&S of hypersensitivity appear (see Appendix F).

INCOMPATIBILITES: **Y-site: aldesleukin, amphotericin B colloidal, ciprofloxacin, cyclosporine, daunorubicin, doxorubicin, furosemide, minocycline, ondansetron, quinupristin/dalfopristin, sodium bicarbonate, topotecan, vancomycin.**

- Store unopened vials at 2°–8° C (36°–46° F) and protect from light.

ADVERSE EFFECTS (≥1%) **Body as a Whole:** Angioedema, *fatigue,* asthenia, night sweats, *fever, chills,* myalgia. **CNS:** Headache, dizziness, depression. **CV:** Hypotension, tachycardia, peripheral edema. **GI:** *Nausea,* vomiting, throat irritation, anorexia, abdominal pain, hepatitis B reactivation with fulminant hepatitis, hepatic failure, and death. **Hematologic:** Leukopenia, thrombocytopenia, anemia, neutropenia. **Respiratory:** Bronchospasm, dyspnea, rhinitis. **Skin:** Pruritus, rash urticaria. **Other:** Infusion-related reactions: *Fever, chills, rigors, pruritus, urticaria, pain, flushing,* chest pain, hypotension, hypertension, dyspnea; fatal infusion-related reactions have been reported.

INTERACTIONS Drug: ANTIHYPERTENSIVE AGENTS should be stopped 12 h prior to avoid excessive hypotension; **cisplatin** may cause additive nephrotoxicity.

PHARMACOKINETICS Duration: 6–12 mo. **Half-Life:** 60–174 h (increases with multiple infusions).

NURSING IMPLICATIONS

Assessment & Drug Effects

- Lab tests: Baseline and periodic CBC with differential, LFTs, and renal function tests. CBC with differential, peripheral CD20+ B lymphocytes.
- Monitor carefully BP and ECG status during infusion and immediately report S&S of hypersensitivity (e.g., fever, chills, urticaria, pruritus, hypotension, bronchospasms; see Appendix F for others).

Patient & Family Education

- Do not take antihypertensive medication within 12 h of rituximab infusions.
- Note: Use effective contraception during and for up to 12 mo following rituximab therapy.
- Report any of the following experienced during infusion: Itching, difficulty breathing, tightness in throat, dizziness, headache, or nausea.

RIVAROXABAN

(riv'-a-rox'-a-ban)
Xarelto
Classifications: ANTICOAGULANT; ANTITHROMBOTIC; SELECTIVE FACTOR XA INHIBITOR
Therapeutic: ANTICOAGULANT; ANTITHROMBOTIC
Pregnancy Category: C

AVAILABILITY 10 mg tablets

ACTION & *THERAPEUTIC EFFECT* A factor Xa inhibitor that selectively blocks the active site of factor Xa thus blocking activation of factor X to factor Xa (FXa) via the intrinsic and extrinsic pathways. *Inhibits the coagulation cascade thus preventing thrombosis formation.*

USES Prophylactic treatment to prevent postoperative deep venous thrombosis (DVT) in patients undergoing knee or hip replacement surgery.

CONTRAINDICATIONS Hypersensitivity to rivaroxaban; major active bleeding; severe renal impairment (CrCl less than 30 mL/min); moderate-to-severe hepatic impairment (Child-Pugh class B or C).
CAUTIOUS USE Spinal/epidural anesthesia or spinal puncture; concurrent

Common adverse effects in *italic,* life-threatening effects underlined; generic names in **bold**; classifications in SMALL CAPS; ✦ Canadian drug name; ✪ Prototype drug

1341

platelet aggregation inhibitors, other antithrombotic agents, fibrinolytic therapy, thienopyridines and chronic use of NSAIDs; mild hepatic impairment (Child-Pugh class A); moderate renal impairment (CrCl 49–30 mL/min); pregnancy (category C). Safety and effectiveness in pediatric patients are not established.

ROUTE & DOSAGE

Deep Venous Thrombosis

Adult: **PO** 10 mg once daily beginning 6–10 h after surgery once hemostasis has been established; continue for 12 days following knee replacement surgery or 35 days following hip replacement surgery; may increase to 20 mg if coadministered with a combined P-glycoprotein and strong CYP3A4 inducer

Hepatic Impairment Dosage Adjustment

Mild impairment (Child-Pugh class A): Avoid use if coagulopathy is present
Moderate or severe impairment (Child-Pugh class B or C): Avoid use

Renal Impairment Dosage Adjustment

CrCl 30 to < 50 mL/min: Observe closely and evaluate any signs or symptoms of blood loss
CrCl < 30 mL/min: Avoid use. Discontinue use in patients who develop acute renal failure

ADMINISTRATION

Oral

- May be given without regard to food.
- *Following spinal/epidural anesthesia or spinal puncture:* Wait at least 18 h after the last dose of rivaroxaban to remove an epidural catheter. Do not give the next rivaroxaban dose earlier than 6 h after the removal of catheter. If traumatic puncture occurs, wait 24 h before giving rivaroxaban.
- Store at 15°–30° C (59°–86° F).

ADVERSE EFFECTS (≥1%) **Body as a Whole:** Pain in extremity, wound secretion. **CNS:** Syncope. **CV:** Thrombocytopenia. **Hematological:** *Bleeding complications.* **Metabolic:** Elevated ALT and AST, elevated bilirubin, *elevated gamma-glutamyl-transferase.* **Musculoskeletal:** Muscle spasm. **Skin:** Blister, pruritus.

INTERACTIONS Drug: Concomitant use with drugs that are combined P-gp and CYP3A4 inhibitors (**ketoconazole, ritonavir, clarithromycin, erythromycin, itraconazole, lopinavir, indinavir, conivaptan, azithromycin, diltiazem, verapamil, quinidine, ranolazine, dronedarone, amiodarone, felodipine**) may increase rivaroxaban exposure and increase bleeding risk. Concomitant use with drugs that are combined P-gp and CYP3A4 inducers (e.g., **rifampin, carbamazepine, phenytoin**) may decrease rivaroxaban levels. Concomitant use with **clopidogrel** increases bleeding time and tendencies. **Herbal: St. John's wort** may decrease rivaroxaban levels.

PHARMACOKINETICS Absorption: 80–100% bioavailable. **Peak:** 2–4 h. **Distribution:** 92–95% Plasma protein bound. **Metabolism:** Approximately 50% metabolized to inactive compounds. **Elimination:** Renal (66%) and fecal (28%). **Half-Life:** 5–9 h.

NURSING IMPLICATIONS

Assessment & Drug Effects

- Monitor vital signs closely and report immediately S&S of bleeding

R

and internal hemorrhage (e.g., intracranial and epidural hematoma, retinal hemorrhage, GI bleeding).

- Monitor frequently and report immediately S&S of neurologic impairment such as tingling, numbness (especially in the lower limbs), and muscular weakness.
- Report immediately neurologic impairment or an unexplained drop in BP or falling Hgb & Hct values.
- Lab tests: Baseline LFTs and kidney functions; baseline and periodic Hgb & Hct.

Patient & Family Education

- Report promptly any of the following: Unusual bleeding or bruising; blood in urine or tarry stools; tingling or numbness (especially in the lower limbs); and muscular weakness.
- Notify prescriber immediately if you become pregnant or intend to become pregnant, or if you are breast feeding or intend to breast feed.
- Confer with prescriber before using a nonsteroidal anti-inflammatory drug (NSAID), aspirin or herbal products as these may increase risk of bleeding.

RIVASTIGMINE TARTRATE

(ri-vas'tig-meen)

Exelon

Classifications: CHOLINESTERASE INHIBITOR; ANTIDEMENTIA
Therapeutic: ANTIALZHEIMER'S
Pregnancy Category: B

AVAILABILITY 2 mg/mL oral solution; 1.5 mg, 3 mg, 4.5 mg, 6 mg capsules; transdermal patch

ACTION & *THERAPEUTIC EFFECT*

Inhibits acetylcholinesterase G_1 form of this enzyme in the cerebral cortex and the hippocampus. The G_1 form of acetylcholinesterase is found in higher levels in the brains of patients with Alzheimer's disease. *Inhibits acetylcholinesterase more specifically in the brain (hippocampus and cortex) than in the heart or skeletal muscle in Alzheimer's disease.*

USES Treatment of mild to moderate dementia of the Alzheimer's type, Parkinson's disease related dementia.

CONTRAINDICATIONS Hypersensitivity to rivastigmine or carbamate derivatives; lactation.

CAUTIOUS USE History of toxicity to cholinesterase inhibitors (e.g., tacrine); DM; sick sinus syndrome or other supraventricular cardiac conduction conditions; asthma or obstructive pulmonary disease; GI disorders including intestinal obstruction; history of PUD or GI bleeding; urogenital tract obstruction; Parkinson's disease; history of seizures; hepatic or renal insufficiency; low body weight; pregnancy (category B).

ROUTE & DOSAGE

R

Alzheimer's/Parkinson's Related Dementia

Adult/Geriatric: **PO** Start with 1.5 mg b.i.d. with food, may increase by 1.5 mg b.i.d. q2wk if tolerated, target dose 3–6 mg b.i.d. (max: 12 mg b.i.d.) (if discontinued for a few doses, restart at last dose or lower; if treatment is interrupted for several days, reinitiate with 1.5 mg b.i.d. and titrate q2wk as above)
Transdermal

Adult: Apply 4.6 mg patch daily, after 4 wk, can increase to 9.5 mg patch if tolerated

---

Common adverse effects in *italic*, life-threatening effects <u>underlined</u>; generic names in **bold**; classifications in SMALL CAPS; ◆ Canadian drug name; ❷ Prototype drug

ADMINISTRATION

Oral

- Give both capsules and liquid with food.
- Give liquid form undiluted or mixed with water, juice, or soda (do not mix with other liquids). Stir completely to dissolve. Ensure that entire mixture is swallowed.
- Withhold drug and notify prescriber if significant anorexia, nausea, or vomiting occur.
- Store capsules and oral solution below 25° C (77° F). Ensure that bottle of liquid is in an UPRIGHT position.

Transdermal

- Apply patch at same time each day to back, upper arm, or chest.
- Rotate application sites and do not reapply to same site for at least 14 days.
- Do not tape or otherwise cover the patch.
- If a patch falls off, do not reapply. Replace with a new patch and remove it following the original schedule for replacement.
- Store patch at 15°–30° C (59°–86° F) in sealed pouch until ready to use.

ADVERSE EFFECTS (≥1%) **Body as a Whole:** Asthenia, increased sweating, syncope, fatigue, malaise, flu-like syndrome. **CV:** Hypertension. **GI:** *Nausea, vomiting, anorexia*, dyspepsia, *diarrhea, abdominal pain*, constipation, flatulence, eructation. **Metabolic:** Weight loss. **CNS:** *Dizziness, headache*, somnolence, tremor, insomnia, confusion, depression, anxiety, hallucination, aggressive reaction. **Respiratory:** Rhinitis.

INTERACTIONS Drug: May exaggerate muscle relations with **succinylcholine** and other NEUROMUSCULAR BLOCKING AGENTS, may attenuate effects of ANTICHOLINERGIC AGENTS.

PHARMACOKINETICS Absorption: Well absorbed. **Peak:** 1 h. **Duration:** 10 h. **Distribution:** Crosses blood–brain barrier with CSF peak concentrations in 1.4–2.6 h. **Metabolism:** By cholinesterase-mediated hydrolysis. **Elimination:** In urine. **Half-Life:** 1.5 h.

NURSING IMPLICATIONS

Assessment & Drug Effects

- Monitor cognitive function and ability to perform ADLs.
- Monitor for and report S&S of GI distress: Anorexia, weight loss, nausea and vomiting.
- Lab tests: Periodic serum electrolytes, Hgb and Hct, urinalysis, blood glucose HbA1C, especially with long-term therapy.
- Monitor ambulation as dizziness is a common adverse effect.
- Monitor diabetics for loss of glycemic control.

Patient & Family Education

- Review instruction sheet provided with liquid form of the drug.
- Monitor weight at least weekly.
- Report any of the following to the prescriber: Loss of appetite, weight loss, significant nausea and/or vomiting.
- Supervise activity since there is a high potential for dizziness.

RIZATRIPTAN BENZOATE

(ri-za-trip′tan ben′zo-ate)

Maxalt, Maxalt-MLT
Classification: SEROTONIN 5-HT₁ RECEPTOR AGONIST (TRIPTANS)
Therapeutic: ANTIMIGRAINE
Prototype: Sumatriptan
Pregnancy Category: C

AVAILABILITY 5 mg, 10 mg tablets; 5 mg, 10 mg disintegrating tablets

ACTION & *THERAPEUTIC EFFECT* Selective (5-HT₁B/₁D) receptor agonist that reverses the vasodilation of cranial blood vessels associated with a migraine. *Activation of the 5-HT$_{1B/1D}$ receptors reduces the pain pathways associated with the migraine headache as well as reversing vasodilation of cranial blood vessels.*

USES Acute migraine headaches with or without aura.

CONTRAINDICATIONS Hypersensitivity to rizatriptan; CAD; Prinzmetal's angina (potential for vasospasm); ischemic heart disease; risk factors for CAD such as hypertension, hypercholesterolemia, obesity, diabetes, smoking, and strong family history; hemiplegia; concurrent administration with ergotamine drugs or sumatriptan; concurrent administration with MAOIS; or within 14 days of use; basilar or hemiplegic migraine.

CAUTIOUS USE Hypersensitivity to sumatriptan; renal or hepatic impairment; hypertension; asthmatic patients; pregnancy (category C), lactation. Safety and effectiveness in children younger than 18 y are not established.

ROUTE & DOSAGE

Acute Migraine

Adult: **PO** 5–10 mg, may repeat in 2 h if necessary (max: 30 mg/24 h); 5 mg with concurrent propranolol (max: 15 mg/24 h)

ADMINISTRATION

Oral
- Give any time after symptoms of migraine appear. If symptoms return, a second tablet may be given but no sooner than 2 h after the first.
- Do not exceed 30 mg (three doses) in any 24 h period.
- Do not give within 24 h of an ergot-containing drug or another 5-HT$_1$ agonist.

- Store at 15°–30° C (59°–86° F) and protect from light and moisture.

ADVERSE EFFECTS (≥1%) **Body as a Whole:** Asthenia, fatigue, pain, pressure sensation, paresthesias, throat pressure, warm/cold sensations. **CNS:** Somnolence, dizziness, headache, hypesthesia, decreased mental acuity, euphoria, tremor. **CV:** Coronary artery vasospasm, transient myocardial ischemia, <u>MI</u>, ventricular tachycardia, ventricular fibrillation, chest pain/tightness/heaviness, palpitations. **GI:** Dry mouth, nausea, vomiting, diarrhea. **Respiratory:** Dyspnea. **Skin:** Flushing. **Endocrine:** Hot flashes.

INTERACTIONS Drug: Propranolol may increase concentrations of rizatriptan, use smaller rizatriptan doses; **dihydroergotamine, methysergide,** other 5-HT$_1$ AGONISTS may cause prolonged vasospastic reactions; SSRIS have rarely caused weakness, hyperreflexia, and incoordination; MAOIS should not be used with 5-HT$_1$ AGONISTS. **Herbal: St. John's wort** may increase triptan toxicity.

PHARMACOKINETICS Absorption: 45% of oral dose reaches systemic circulation. **Peak:** 1–1.5 h for oral tabs; 1.6–2.5 h for orally disintegrating tablets. **Metabolism:** Via oxidative deamination by monoamine oxidase A. **Elimination:** Primarily in urine (82%). **Half-Life:** 2–3 h.

NURSING IMPLICATIONS

Assessment & Drug Effects
- Monitor cardiovascular status carefully following first dose in patients at risk for CAD (e.g., postmenopausal women, men older than 40 y, persons with known CAD risk factors) or coronary artery vasospasms.

Common adverse effects in *italic*, life-threatening effects <u>underlined</u>; generic names in **bold**; classifications in SMALL CAPS; ♣ Canadian drug name; ⊙ Prototype drug

1345

- ECG is recommended following first administration of rizatriptan to someone with known CAD risk factors.
- Report immediately to prescriber: Chest pain or tightness in chest or throat that is severe or does not quickly resolve.
- Monitor periodically cardiovascular status with continued rizatriptan use.

Patient & Family Education
- Do not exceed 30 mg (three doses) in 24 h.
- Allow orally disintegrating tablets to dissolve on tongue; no liquid is needed.
- Contact prescriber immediately if any of the following develop following rizatriptan use: Symptoms of angina (e.g., severe and/or persistent pain or tightness in chest or throat), hypersensitivity (e.g., wheezing, facial swelling, skin rash, or hives), abdominal pain.
- Report any other adverse effects (e.g., tingling, flushing, dizziness) at next prescriber visit.

ROFLUMILAST
(ro-flu'mi-last)
Daliresp
Classifications: RESPIRATORY AGENT; ANTIINFLAMMATORY AGENT; PHOSPHODIESTERASE 4 (PDE4) INHIBITOR
Therapeutic: RESPIRATORY AGENT; ANTIINFLAMMATORY AGENT
Pregnancy Category: C

AVAILABILITY 500 mcg tablets

ACTION & *THERAPEUTIC EFFECT*
Selectively inhibits phosphodiesterase 4 (PDE4), a major enzyme found in inflammatory and immune cells, thus indirectly suppressing the release of cytokines and other products of inflammation. *Has an anti-inflammatory effect in pulmonary diseases.*

USES Reduction of the prevalence of chronic obstructive pulmonary disease (COPD) exacerbations in patients with severe COPD associated with chronic bronchitis and a history of exacerbations.

CONTRAINDICATIONS Moderate-to-severe liver impairment (Child-Pugh class B or C); lactation.
CAUTIOUS USE Mild hepatic impairment (Child-Pugh class A); history of psychiatric illness including anxiety, depression, or suicidal ideation; clinically significant weight loss; pregnancy (category C). The safety and effectiveness in children have not been established in children.

ROUTE & DOSAGE

Chronic Obstructive Pulmonary Disease
Adult: **PO** 500 mcg once daily

ADMINISTRATION
Oral
- May be given with or without food.
- Store at 15°–30° C (59°–86° F).

ADVERSE EFFECTS (≥1%) **Body as a Whole:** Influenza. **CNS:** Anxiety, depression, dizziness, headache, insomnia, tremor. **GI:** Abdominal pain, decreased appetite, *diarrhea*, dyspepsia, gastritis, nausea, vomiting, *weight loss*. **Musculoskeletal:** Back pain, muscle spasms. **Respiratory:** Rhinitis, sinusitis. **Urogenital:** Urinary tract infection.

INTERACTIONS Drug: Coadministration with strong CYP3A4 or CYP3A4/CYP1A2 inhibitors (e.g., **cimetidine, ketoconazole, ritonavir, erythromycin,**

Common adverse effects in *italic*, life-threatening effects underlined; generic names in **bold**; classifications in SMALL CAPS; ♣ Canadian drug name; ☯ Prototype drug

fluvoxamine) may increase the levels of roflumilast. Coadministration with strong CYP3A4 inducers (i.e., **carbamazepine, phenobarbital, phenytoin,** RIFAMYCINS) may decrease the levels of roflumilast. Coadministration with **ethinyl estradiol** and **gestodene** may increase the levels of roflumilast. **Food:** None listed. **Herbal: St. John's wort** may decrease the levels of roflumilast.

PHARMACOKINETICS Absorption: 80% Bioavailable. **Peak:** 1 h. **Distribution:** 99% Plasma protein bound. **Metabolism:** In liver to active and inactive metabolites. **Elimination:** Primarily renal. **Half-Life:** 17 h (30 h for active metabolite).

NURSING IMPLICATIONS

Assessment & Drug Effects
▪ Monitor for and report promptly new onset or worsening insomnia, anxiety, depression, suicidal thoughts, or other significant mood changes.
▪ Monitor weight at regular intervals. Report significant weight loss to prescriber.
▪ Lab tests: Baseline and periodic LFTs.

Patient & Family Education
▪ Roflumilast is not a bronchodilator and does not provide relief of acute bronchospasm.
▪ Report to prescriber any of the following: Disturbing mood changes such as anxiety or depression; difficulty sleeping; suicidal thoughts; persistent nausea and/or diarrhea.
▪ Monitor your weight and report unexplained weight loss.
▪ Do not self-treat depression with St. John's wort.

ROMIDEPSIN
(rom-i-dep′sin)
Istodax
Classifications: ANTINEOPLASTIC; HISTONE DEACETYLASE (HDAC) INHIBITOR
Therapeutic: ANTINEOPLASTIC
Pregnancy Category: D

AVAILABILITY 10 mg powder for reconstitution for injection

ACTION & *THERAPEUTIC EFFECT*
Romidepsin inhibits histone deacetylase, an enzyme required for completion of the cell cycle in certain cancers, thus inducing cell cycle arrest and apoptosis in these cancer cells. The mechanism of antineoplastic effect has not been fully characterized. *Effective as a cytotoxic agent against cutaneous T-cell lymphoma.*

USES Treatment of cutaneous or peripheral T-cell lymphoma.

CONTRAINDICATIONS Severe bone marrow depression; pregnancy (category D); lactation.
CAUTIOUS USE Risk factors for potassium or magnesium imbalance, thrombocytopenia, or leukopenia; congenital long QT syndrome, history of significant CV disease; moderate-to-severe hepatic impairment; end-stage renal disease. Safe use in children not established.

ROUTE & DOSAGE

Cutaneous or Peripheral T-Cell Lymphoma (CTCL)
Adult: **IV** 14 mg/m² over 4 h on days 1, 8, and 15 q28d. Repeat cycle until disease progression has been treated or unacceptable toxicity occurs.

R

Common adverse effects in *italic*, life-threatening effects underlined; generic names in **bold**; classifications in SMALL CAPS; ♣ Canadian drug name; ⊘ Prototype drug

1347

Toxicity Dosage Adjustment (National Cancer Institute Common Toxicity Criteria, CTC)

For CTC grade 3 or 4 neutropenia or thrombocytopenia: Hold romidepsin until ANC is 1,500/mm³ or higher and/or platelet count is 75,000/mm³ or higher, or parameters return to baseline values. Then, resume romidepsin 14 mg/m². For CTC grade 4 febrile neutropenia or thrombocytopenia requiring platelet transfusion: Hold romidepsin until cytopenia is no greater than grade 1 and permanently reduce dose to 10 mg/m². For CTC grade 2 or 3 nonhematological toxicity: Hold romidepsin until toxicity is no greater than grade 1 or returns to baseline. Resume romidepsin 14 mg/m². For recurrent CTC grade 3 nonhematological toxicity: Hold romidepsin until toxicity is no greater than grade 1 or returns to baseline and permanently reduce dose to 10 mg/m². For CTC grade 4 nonhematological toxicity: Hold romidepsin until toxicity is no greater than grade 1 or returns to baseline and permanently reduce dose to 10 mg/m². For recurrent CTC grade 3 or 4 nonhematological toxicity after dose reduction: Discontinue therapy.

ADMINISTRATION

Intravenous

PREPARE: **IV Infusion:** Avoid contact with medication during preparation. ▪ Withdraw 2 mL of supplied diluent and inject slowly into the 10 mL powder vial to yield 5 mg/mL. Swirl to dissolve. ▪ Withdraw the required dose of reconstituted solution and add to 500 mL NS and invert container to dissolve. ▪ Should be administered as soon as possible after dilution.

ADMINISTER: **IV Infusion:** Give over 4 h. Note: Ensure that potassium and magnesium are within normal range before administration.

▪ Reconstituted vials are stable for 8 h and IV solutions are stable for 24 h at room temperature. Store unopened vials at 15°–30° C (59°–86° F).

INCOMPATIBILITIES **Solution/additive:** Do not mix with another drug. **Y-site:** Do not mix with another drug.

ADVERSE EFFECTS (≥20%) **Body as a Whole:** *Asthenia,* fatigue, hyperglycemia, *infections,* pyrexia. **CV:** ECG ST-T wave changes, hypotension. **GI:** *Anorexia, constipation, diarrhea, dysgeusia, nausea, vomiting.* **Hematological:** *Anemia, neutropenia,* <u>leukopenia</u>, lymphopenia, *<u>thrombocytopenia</u>*. **Metabolic:** Elevated ALT, elevated AST, hypermagnesemia, hyperuricemia, hypoalbuminemia, hypocalcemia, hypokalemia, hypomagnesemia, hyponatremia, hypophosphatemia. **Skin:** Pruritis, dermatitis, exfoliative dermatitis.

INTERACTIONS Drug: Coadministration of CYP3A4 inducers (e.g., **carbamazepine, phenobarbital, phenytoin, rifampin**) can decrease the levels of romidepsin. Coadministration of strong CYP3A inhibitors (e.g., **atazanavir, indinavir, itraconazole, ketoconazole, nefazodone, nelfinavir, ritonavir, saquinavir, telithromycin, voriconazole**) can increase the levels of romidepsin. Coadministration of

Common adverse effects in *italic*, life-threatening effects <u>underlined</u>; generic names in **bold**; classifications in SMALL CAPS; ♣ Canadian drug name; ⊘ Prototype drug

drugs that prolong the QT interval (e.g., **amiodarone, bretylium, disopyramide, dofetilide, procainamide, quinidine, sotalol**), **arsenic trioxide, chlorpromazine, cisapride, dolasetron, droperidol, mefloquine, mesoridazine, moxifloxacin, pentamidine, pimozide, tacrolimus, thioridazine, and ziprasidone**) may produce life-threatening arrhythmias. **Food:** Grapefruit juice can increase the levels of romidepsin. **Herbal: St. John's wort** can decrease the levels of romidepsin.

PHARMACOKINETICS Distribution: 92–94% plasma protein bound. **Metabolism:** Extensive hepatic metabolism by CYP3A4. **Elimination:** Primarily biliary excretion. **Half-Life:** 3 h.

NURSING IMPLICATIONS

Assessment & Drug Effects
- Monitor ECG at baseline and periodically throughout treatment.
- Lab tests: Baseline and prior to each dose: Serum electrolytes, CBC with differential including platelet count; baseline and periodic LFTs and kidney function tests.
- Report immediately electrolyte imbalances or QT prolongation.

Patient & Family Education
- Report promptly any of the following: Excessive nausea, chest pain, shortness of breath, or abnormal heart beat.
- Seek medical attention if unusual bleeding occurs.
- Women of childbearing age who use an estrogen-containing contraceptive should add a barrier contraceptive to prevent pregnancy.

ROPINIROLE HYDROCHLORIDE

(ro-pi'ni-role)

Requip, Requip XL

Classifications: DOPAMINE RECEPTOR AGONIST; ANTIPARKINSON
Therapeutic: ANTIPARKINSON
Prototype: Levodopa
Pregnancy Category: C

AVAILABILITY 0.25 mg, 0.5 mg, 1 mg, 2 mg, 3 mg, 4 mg, 5 mg tablets; 2 mg, 4 mg, 6 mg, 8 mg, 12 mg extended release tablets

ACTION & *THERAPEUTIC EFFECT*
Nonergot dopamine receptor agonist that has high affinity for the D_2 and D_3 subfamily of dopamine receptors. *Effectiveness indicated by improvement in idiopathic Parkinson's disease.*

USES Idiopathic Parkinson's disease, restless legs syndrome.

CONTRAINDICATIONS Hypersensitivity to ropinirole or pramipexole.
CAUTIOUS USE Hepatic impairment; severe renal impairment; mental instability; concomitant use of CNS depressants; pregnancy (category C), lactation. Safety and efficacy in children are not established.

ROUTE & DOSAGE

Parkinson's Disease

Adult: **PO Immediate release:** Start with 0.25 mg t.i.d., titrate up by 0.25 mg/dose t.i.d. qwk to a target dose of 1 mg t.i.d.; if response is still not satisfactory, may continue to increase by 1.5 mg/day qwk to a dose of 9 mg/day, and then by 3 mg/day or less weekly (max: 24 mg/day) **Extended release:** 2 mg daily for 1–2 wk then increase by 2 mg/day at one week intervals (max: 24 mg/day)

Restless Legs Syndrome

Adult: **PO** Take 0.25 mg 1–3 h before bed × 2 days, increase to

R

Common adverse effects in *italic*, life-threatening effects underlined; generic names in **bold**; classifications in SMALL CAPS; ♣ Canadian drug name; ⬤ Prototype drug

1349

0.5 mg for the first wk, then increase by 0.5 mg qwk to a maximum of 4 mg

ADMINISTRATION

Oral

- Give with food to reduce occurrence of nausea.
- Do not crush extended release tablets. Ensure that they are swallowed whole.
- Drug should not be abruptly discontinued. Dose should be tapered over a period of days.
- Store at 15°–30° C (59°–86° F).

ADVERSE EFFECTS (≥ 1%) Body as a Whole: Increased sweating, dry mouth, flushing, asthenia, *fatigue,* pain, edema, malaise, *viral infection,* UTI, impotence. **CNS:** *Dizziness, somnolence, sudden sleep attacks,* hallucinations, confusion, amnesia, hypesthesia, yawning, hyperkinesia, impaired concentration, vertigo, hallucinations. **CV:** *Syncope,* chest pain, orthostatic symptoms, hypertension, palpitations, atrial fibrillation, extrasystoles, hypotension, tachycardia, peripheral edema, peripheral ischemia. **GI:** *Nausea, vomiting, dyspepsia,* abdominal pain, anorexia, flatulence. **Respiratory:** Pharyngitis, rhinitis, sinusitis, bronchitis, dyspnea. **Special Senses:** Abnormal vision, xerophthalmia, eye abnormality.

INTERACTIONS Drug: Ropinirole levels may be increased by ESTROGENS, QUINOLONE ANTIBIOTICS, **cimetidine, diltiazem, erythromycin, fluvoxamine, mexiletine, tacrine;** effects may be antagonized by PHENOTHIAZINES, BUTYROPHENONES, **metoclopramide zileuton** may increase ropinerole levels.

PHARMACOKINETICS Absorption: Rapidly from GI tract; 55% bioavailability. **Peak:** 1–2 h. **Distribution:** 30–40% protein bound. **Metabolism:** In liver (CYP1A2). **Elimination:** Primarily in urine. **Half-Life:** 6 h.

NURSING IMPLICATIONS

Assessment & Drug Effects

- Lab test: Periodically monitor BUN and creatinine, hepatic function.
- Monitor cardiac status. Report increases in BP and HR to prescriber.
- Monitor carefully for orthostatic hypotension, especially during dose escalation.
- Monitor level of alertness. Institute appropriate precautions to prevent injury due to dizziness or drowsiness.

Patient & Family Education

- Be aware that hallucinations are a possible adverse effect and occur more often in older adults.
- Make position changes slowly, especially after long periods of lying or sitting. Postural hypotension is common, especially during early treatment.
- Exercise caution with hazardous activities requiring alertness since drowsiness and sedation are common adverse effects. Effects are additive with alcohol or other CNS depressants.
- Report behavioral changes (e.g., impulsive behavior) to prescriber.

ROPIVACAINE HYDROCHLORIDE

(ro-piv′i-cane)

Naropin

Classification: LOCAL ANESTHETIC (ESTER-TYPE)

Therapeutic: LOCAL ANESTHETIC

Prototype: Procaine HCl

Pregnancy Category: B

Common adverse effects in *italic*, life-threatening effects underlined; generic names in **bold**; classifications in SMALL CAPS; ♣ Canadian drug name; ✪ Prototype drug

R

AVAILABILITY 2 mg/mL, 5 mg/mL, 7.5 mg/mL, 10 mg/mL injection

ACTION & *THERAPEUTIC EFFECT* Blocks the generation and conduction of nerve impulses, probably by increasing the threshold for electrical excitability. *Local anesthetic action produces loss of sensation and motor activity in areas of the body close to the injection site.*

USES Local and regional anesthesia, postoperative pain management, anesthesia/pain management for obstetric procedures.

CONTRAINDICATIONS Hypersensitivity to ropivacaine or any local anesthetic of the amide type; generalized septicemia, inflammation or sepsis at the proposed injection site; cerebral spinal diseases (e.g., meningitis); heart block, hypotension, hypertension, GI hemorrhage.

CAUTIOUS USE Debilitated, older adult, or acutely ill patients; arrhythmias, shock; pregnancy (category B), lactation.

ROUTE & DOSAGE

Surgical Anesthesia
Adult: **Epidural** 25–200 mg (0.5–1% solution) **Nerve block** 5–250 mg (0.5%, 0.75% solution)
Labor Pain
Adult: **Epidural** 20–40 mg (0.2% solution)
Postoperative Pain Management
Adult: **Epidural** 12–20 mg/h (0.2% solution) **Infiltration** 2–200 mg (0.2–0.5% solution)

ADMINISTRATION

Intrathecal

- Avoid rapid injection of large volumes of ropivacaine. Incremental doses should always be used to achieve the smallest effective dose and concentration.
- Use an infusion concentration of 2 mg/mL (0.2%) for postoperative analgesia.
- Do not use disinfecting agents containing heavy metal ions (e.g., mercury, copper, zinc, etc.) on skin insertion site or to clean the ropivacaine container top.
- Discard continuous infusions solution after 24 h; it contains no preservatives.
- Store unopened at 20°–25° C (68°–77° F).

ADVERSE EFFECTS (≥1%) **Body as a Whole:** Pain, fever, rigors, hypoesthesia. **CNS:** Paresthesia, headache, dizziness, anxiety. **CV:** *Hypotension,* bradycardia, hypertension, tachycardia, chest pain, fetal bradycardia. **GI:** Nausea. **Skin:** Pruritus. **Urogenital:** Urinary retention, oliguria. **Hematologic:** Anemia.

INTERACTIONS Drug: Additive adverse effects with other LOCAL ANESTHETICS.

PHARMACOKINETICS Onset: 1–30 min (average 10–20 min) depending on dose/route of administration. **Duration:** 0.5–8 h depending on dose/route of administration. **Distribution:** 94% protein bound. **Metabolism:** In the liver by CYP1A. **Elimination:** In urine. **Half-Life:** 1.8–4.2 h.

NURSING IMPLICATIONS

Assessment & Drug Effects

- Monitor carefully cardiovascular and respiratory status throughout treatment period. Assess for hypotension and bradycardia.
- Report immediately S&S of CNS stimulation or CNS depression.

R

Common adverse effects in *italic*, life-threatening effects underlined; generic names in **bold**; classifications in SMALL CAPS; ✤ Canadian drug name; ⊘ Prototype drug

1351

Patient & Family Education
- Report any of the following to prescriber immediately: Restlessness, anxiety, tinnitus, blurred vision, tremors.

ROSIGLITAZONE MALEATE ℗ℛ

(ros-i-glit′a-zone)

Avandia
Classifications: ANTIDIABETIC; THIAZOLIDINEDIONE
Therapeutic: ANTIHYPERGLYCEMIC
Pregnancy Category: C

AVAILABILITY 2 mg, 4 mg, 8 mg tablets

ACTION & *THERAPEUTIC EFFECT*
Antidiabetic agent that lowers blood sugar levels by improving target cell response to insulin in Type 2 diabetics. It reduces cellular insulin resistance and decreases hepatic glucose output (gluconeogenesis). *Reduces hyperglycemia and hyperlipidemia, thus improving hyperinsulinemia without stimulating pancreatic insulin secretion. Effectiveness indicated by decreased HbA1C.*

USES Type 2 diabetes.

CONTRAINDICATIONS Hypersensitivity to rosiglitazone; patients with New York Heart Association Class III and IV cardiac status (e.g., CHF); symptomatic heart failure; active hepatic disease; lactation.
CAUTIOUS USE As monotherapy in Type 1 diabetes mellitus or diabetic ketoacidosis; CHF or risk for CHF; patients with ongoing edema; hepatic impairment; older adults; pregnancy (category C). Safety and efficacy in children younger than 18 y are not established.

ROUTE & DOSAGE

Type 2 Diabetes Mellitus
Adult: **PO** Start at 4 mg daily or 2 mg b.i.d., may increase after 12 wk (max: 8 mg/day in 1–2 divided doses)
Hepatic Impairment Dosage Adjustment
Do not use if ALT greater than 2.5 × ULN

ADMINISTRATION

Oral
- May be given without regard to meals.
- Store at 15°–30° C (59°–86° F) in tight, light-resistant container.

ADVERSE EFFECTS (≥1%) **Body as a Whole:** Edema, anemia, headache, back pain, fatigue. **CV:** Edema, fluid retention, exacerbation of heart failure, <u>increased risk of heart attack</u>. **GI:** Diarrhea. **Respiratory:** Upper respiratory tract infection, sinusitis. **Special Senses:** Macular edema. **Other:** Hyperglycemia.

INTERACTIONS Drug: Insulin may increase risk of heart failure or edema; enhance hypoglycemia with ORAL ANTIDIABETIC AGENTS, **ketoconazole, gemfibrozil** may increase effect. **Bosentan** may reduce effect. **Herbal: Garlic, ginseng, green tea** may potentiate hypoglycemic effects.

PHARMACOKINETICS Absorption: 99% from GI tract. **Peak:** 1 h, food delays time to peak by 1.75 h. **Duration:** Greater than 24 h. **Distribution:** Greater than 99% protein bound. **Metabolism:** In liver (CYP2C8) to inactive metabolites. **Elimination:** 64% urine, 23% feces. **Half-Life:** 3–4 h. Liver disease increases serum concentrations and increases half-life by 2 h.

NURSING IMPLICATIONS

Assessment & Drug Effects

- Monitor for S&S of hypoglycemia (possible when insulin/sulfonylureas are coadministered).
- Monitor for S&S of CHF or exacerbation of symptoms with preexisting CHF.
- Lab tests: Baseline LFTs, then q2mo for first year; then periodically (more often when elevated); periodic HbA1C, Hgb and Hct, and lipid profile.
- Withhold drug and notify prescriber if ALT greater than 2.5 times normal or patient jaundiced.
- Monitor weight and notify prescriber of development of edema.

Patient & Family Education

- Report promptly any of the following: Rapid weight gain, edema, shortness of breath, or exercise intolerance.
- Be aware that resumed ovulation is possible in nonovulating premenopausal women.
- Use or add barrier contraceptive if using hormonal contraception.
- Report immediately to prescriber: S&S of liver dysfunction such as unexplained anorexia, nausea, vomiting, abdominal pain, fatigue, dark urine; or S&S of fluid retention such as weight gain, edema, or activity intolerance.
- Combination therapy: May need adjustment of other antidiabetic drugs to avoid hypoglycemia.

√ROSUVASTATIN
(ro-su-va-sta′ten)
Crestor
Classifications: ANTIHYPERLIPIMIC; HMG-COA REDUCTASE INHIBITOR (STATIN)

Therapeutic: ANTILIPEMIC; STATIN
Prototype: Lovastatin
Pregnancy Category: X

AVAILABILITY 5 mg, 10 mg, 20 mg, 40 mg tablets

ACTION & *THERAPEUTIC EFFECT*

Rosuvastatin is a potent inhibitor of HMG-CoA reductase, an enzyme that catalyzes the conversion of HMG-CoA to mevalonic acid, an early and rate-limiting step in cholesterol biosynthesis. This results in reducing the amount of mevalonic acid, a precursor of cholesterol. It increases the number of hepatic HDL receptors on the cell surface to enhance uptake and catabolism of LDL. It also inhibits hepatic synthesis of very low density lipoprotein (VLDL). *Reduces total cholesterol and LDL cholesterol; additionally, lowers plasma triglycerides and VLDL while increasing HDL.*

USES Adjunct to diet for the reduction of LDL cholesterol and triglycerides in patients with primary hypercholesterolemia, hypertriglyceridemia, and mixed dyslipidemia.

CONTRAINDICATIONS Hypersensitivity to any component of the product, active liver disease, pregnancy (category X), women of childbearing potential not using appropriate contraceptive measures, lactation.
CAUTIOUS USE Concomitant use of cyclosporine and gemfibrozil; excessive alcohol use or history of liver disease; renal impairment; advanced age; hypothyroidism; children younger than 10 y.

R

Common adverse effects in *italic*, life-threatening effects underlined; generic names in **bold**; classifications in SMALL CAPS; ♣ Canadian drug name; ⊕ Prototype drug

1353

ROUTE & DOSAGE

Hyperlipidemia

Adult/Adolescent/Child (older than 10 y): **PO** 10 mg once daily (5–40 mg/day), max dose 40 mg/day; If taking cyclosporine, gemfibrozil, lopinavir/ritonavir, start with 5 mg/day
Geriatric: Initial dose of 5 mg/day

Renal Impairment Dosage Adjustment

CrCl less than 30 mL/min: 5 mg once daily (max: 10 mg/day)

ADMINISTRATION

Oral

- Persons of Asian descent may be slow metabolizers and may require half the normal dose.
- May give any time of day without regard to food.
- Do not give within 2 h of an antacid.
- Store at or below 30° C (86° F).

ADVERSE EFFECTS (≥1%) Body as a Whole: Asthenia, back pain, flu-like syndrome, chest pain, infection, pain, peripheral edema. **CNS:** Headache, dizziness, insomnia, hypertonia, paresthesia, depression, anxiety, vertigo, neuralgia. **CV:** Hypertension, angina, vasodilatation, palpitations. **GI:** Diarrhea, dyspepsia, nausea, abdominal pain, constipation, gastroenteritis, vomiting, flatulence, gastritis. **Endocrine:** Diabetes. **Hematologic:** Anemia, ecchymosis. **Musculoskeletal:** Myalgia, arthritis, arthralgia, rhabdomyolysis (especially with dose greater than 40 mg). **Respiratory:** Pharyngitis, rhinitis, sinusitis, bronchitis, increased cough, dyspnea, pneumonia, asthma. **Skin:** Rash, pruritus. **Urogenital:** UTI.

INTERACTIONS Drug: Atazanavir, cyclosporine, gemfibrozil, niacin may increase risk of rhabdomyolysis; ANTACIDS may decrease rosuvastatin absorption; may cause increase in INR with **warfarin. Herbal: Red-yeast rice** increases rhabdomyolysis risk.

PHARMACOKINETICS Absorption: Well absorbed. **Peak:** 3–5 h. **Metabolism:** Limited metabolism in the liver (not CYP3A4). **Elimination:** Primarily in feces (90%). **Half-Life:** 20 h.

NURSING IMPLICATIONS

Assessment & Drug Effects

- Monitor for and report promptly S&S of myopathy (e.g., skeletal muscle pain, tenderness or weakness).
- Withhold drug and notify prescriber if CPK levels are markedly elevated (10 or more × ULN) or if myopathy is diagnosed or suspected.
- Lab tests: CPK levels for S&S of myopathy; periodic LFTs; more frequent INR values with concomitant warfarin therapy.
- Monitor CV status, especially with a known history of hypertension or heart disease.
- Monitor diabetics for loss of glycemic control.

Patient & Family Education

- Do not take antacids within 2 h of taking this drug.
- Women should use reliable means of contraception to prevent pregnancy while taking this drug.

SALMETEROL XINAFOATE

(sal-me'ter-ol xin'a-fo-ate)

Serevent

Classifications: BETA$_2$-ADRENERGIC AGONIST; RESPIRATORY SMOOTH MUSCLE RELAXANT; BRONCHODILATOR
Therapeutic: BRONCHODILATOR
Prototype: Albuterol
Pregnancy Category: C

AVAILABILITY 50 mcg powder diskus for inhalation

ACTION & *THERAPEUTIC EFFECT*

Long-acting beta$_2$-adrenoreceptor agonist that stimulates beta$_2$-adrenoreceptors, relaxes bronchospasm, and increases ciliary motility, thus facilitating expectoration. *Relaxes bronchospasm and increases ciliary motility, thus facilitating expectoration of pulmonary secretions.*

USES Maintenance therapy for asthma or bronchospasm associated with COPD. Prevention of exercise-induced bronchospasm. Do not use to treat acute bronchospasm.

CONTRAINDICATIONS Hypersensitivity to salmeterol; other long-acting beta$_2$-adrenergic agonists; primary treatment of status asthmaticus; acute bronchospasm; MAOI therapy.

CAUTIOUS USE Cardiovascular disorders, cardiac arrhythmias, hypertension, QT prolongation; history of seizures or thyrotoxicosis; hyperthyroidism; pheochromocytoma; liver and renal impairment, diabetes mellitus, sensitivity to other beta-adrenergic agonists; older adults; women in labor; pregnancy (category C), lactation; children younger than 4 y.

ROUTE & DOSAGE

Asthma or Bronchospasm (COPD associated)

Adult/Child (4 y or older):
Inhalation 1 powder diskus (50 mcg) b.i.d. approximately 12 h apart

Prevention of Exercise-Induced Bronchospasm

Adult/Child (4 y or older):
Inhalation 1 powder diskus (50 mcg) 30–60 min before exercise

ADMINISTRATION

Inhalation

- Do not use to relieve symptoms of acute asthma.
- Activate diskus by moving lever until it clicks. Patient should exhale fully (not into diskus), place diskus in mouth, and inhale quickly and deeply through the diskus. Diskus should be removed and breath held for 10 sec.
- Store at room temperature, 15°–30° C (59°–86° F).

ADVERSE EFFECTS (≥1%) **CNS:** Dizziness, headache, tremor. **CV:** Palpitations, sinus tachycardia. **Respiratory:** Respiratory arrest (rare). **Skin:** Rash. **Body as a Whole:** Tolerance (tachyphylaxis).

INTERACTIONS Drug: Effects antagonized by BETA-BLOCKERS, MAOIs.

PHARMACOKINETICS Onset: 10–20 min. **Peak:** Effect 2 h. **Duration:** Up to 12 h. **Distribution:** 94–95% protein bound. **Metabolism:** Salmeterol is extensively metabolized by hydroxylation. **Elimination:** Primarily in feces. **Half-Life:** 3–4 h.

S

Common adverse effects in *italic*, life-threatening effects underlined; generic names in **bold**; classifications in SMALL CAPS; ♣ Canadian drug name; ⊘ Prototype drug

1355

NURSING IMPLICATIONS

Assessment & Drug Effects

- Withhold drug and notify prescriber immediately if bronchospasms occur following its use.
- Monitor cardiovascular status; report tachycardia.
- Lab tests: Monitor LFTs periodically with long-term therapy.

Patient & Family Education

- Never use a spacer device with the drug.
- Do not use this drug to treat an acute asthma attack.
- Notify prescriber immediately of worsening asthma or failure to respond to the usual dose of salmeterol.
- Do not use an additional dose prior to exercise if taking twice-daily doses of salmeterol.
- Take the preexercise dose 30–60 min before exercise and wait 12 h before an additional dose.

SALSALATE
(sal'sa-late)

Mono-Gesic, Salsitab
Classifications: ANALGESIC (SALICYLATE); NONSTEROIDAL ANTI-INFLAMMATORY DRUG (NSAID)
Therapeutic: NONNARCOTIC ANALGESIC, NSAID; DISEASE-MODIFYING ANTIRHEUMATIC DRUG (DMARD)
Prototype: Aspirin
Pregnancy Category: C

AVAILABILITY 500 mg, 750 mg tablets

ACTION & *THERAPEUTIC EFFECT*
Anti-inflammatory and analgesic activity of salsalate may be mediated through inhibition of the prostaglandin synthetase enzyme complex. *Has analgesic, anti-inflammatory, and antirheumatic effects.*

USES Symptomatic treatment, rheumatoid arthritis, osteoarthritis, and related rheumatic disorders.

CONTRAINDICATIONS Hypersensitivity to salicylates or NSAIDs; chronic renal insufficiency; peptic ulcer; children younger than 12 y; hemophilia; chickenpox; influenza, tinnitus.

CAUTIOUS USE Liver function impairment; older adults; pregnancy (category C), lactation.

ROUTE & DOSAGE

Arthritis
Adult: **PO** 1500 mg b.i.d. or 1000 mg t.i.d. (max: 4 g/day)

ADMINISTRATION

Oral

- Give with a full glass of water or food or milk to reduce GI adverse effects.

ADVERSE EFFECTS (≥1%) GI: Nausea, dyspepsia, heartburn, vomiting, diarrhea, risk of GI bleed. **Special Senses:** Tinnitus, hearing loss (reversible). **Body as a Whole:** Vertigo, flushing, headache, confusion, hyperventilation, sweating. **CNS:** Drowsiness.

DIAGNOSTIC TEST INTERFERENCE False-negative results for ***Clinistix;*** false-positives for ***Clinitest.***

INTERACTIONS Drug: Aminosalicylic acid increases risk of salicylate toxicity. **Ammonium chloride** and other ACIDIFYING AGENTS decrease renal elimination and increase risk of salicylate toxicity. ANTICOAGULANTS increase risk of bleeding. ORAL HYPOGLYCEMIC AGENTS increase hypoglycemic activity with salsalate doses greater than 2 g/day. CARBONIC ANHYDRASE INHIBITORS enhance salicylate toxicity. CORTICOSTEROIDS add to ulcerogenic effects. **Methotrexate** toxicity is increased. Low doses of salicylates may antagonize uricosuric effects of **probenecid** and **sulfinpyrazone. Herbal: Feverfew, garlic, ginger, ginkgo** may increase bleeding potential.

PHARMACOKINETICS Absorption: Readily absorbed from small intestine. **Peak:** 1.5–4 h. **Metabolism:** Hydrolyzed in liver, GI mucosa, plasma, whole blood, and other tissues. **Elimination:** In urine. **Half-Life:** 1 h.

NURSING IMPLICATIONS

Assessment & Drug Effects
- Symptom relief is gradual (may require 3–4 days to establish steady-state salicylate level).
- Monitor for adverse GI effects, especially in patient with a history of peptic ulcer disease.

Patient & Family Education
- Do not take another salicylate (e.g., aspirin) while on salsalate therapy.
- Monitor blood glucose for loss of glycemic control in diabetes; drug may induce hypoglycemia when used with sulfonylureas.
- Report tinnitus, hearing loss, vertigo, rash, or nausea.

SAQUINAVIR MESYLATE ⓟ
(sa-quin′a-vir mes′y-late)
Invirase
Classifications: ANTIRETROVIRAL; PROTEASE INHIBITOR
Therapeutic: PROTEASE INHIBITOR ·
Pregnancy Category: B

AVAILABILITY 200 mg gelatin capsules; 500 mg tablets

ACTION & *THERAPEUTIC EFFECT*
Synthetic peptide that inhibits the activity of HIV protease and prevents the cleavage of viral polyproteins essential for the maturation of HIV. *Effectiveness indicated by reduced viral load (decreased number of RNA copies), and increased number of T helper CD4 cells.*

USES HIV infection (used with other antiretroviral agents).

CONTRAINDICATIONS Significant hypersensitivity to saquinavir; concurrent use of saquinavir and ritonavir in patients with conditions that affect cardiac electrical activity; severe hepatic impairment; antimicrobial resistance to other protease inhibitors, monotherapy, lactation.
CAUTIOUS USE Mild to moderate hepatic insufficiency; severe renal impairment; hepatitis B or C; diabetes mellitus, diabetic ketoacidosis; hemophilia A or B; older adults; pregnancy (category B). Safety and efficacy in HIV-infected children younger than 16 y are not established.

ROUTE & DOSAGE

HIV
Adult/Adolescent: **PO** 1000 mg b.i.d. with ritonavir 100 mg b.i.d.

ADMINISTRATION
Oral
- Give with or up to 2 h after a full meal to ensure adequate absorption and bioavailability. Give with ritonavir.
- Do not administer to anyone taking rifampin or rifabutin

S

Common adverse effects in *italic*, life-threatening effects underlined; generic names in **bold**; classifications in SMALL CAPS; ♣ Canadian drug name; ⓟ Prototype drug

1357

because these drugs significantly decrease the plasma level of saquinavir.

- Store at 15°–30° C (59°–86° F) in tightly closed bottle.

ADVERSE EFFECTS (≥1%) **CNS:** Headache, paresthesia, numbness, dizziness, peripheral neuropathy, ataxia, confusion, convulsions, hyperreflexia, hyporeflexia, tremor, agitation, amnesia, anxiety, depression, excessive dreaming, hallucinations, euphoria, irritability, lethargy, somnolence. **CV:** Chest pain, hypertension, hypotension, syncope. **Endocrine:** Dehydration, hyperglycemia, diabetes, weight changes. **Hematologic:** Anemia, splenomegaly, thrombocytopenia, pancytopenia. **GI:** *Nausea, diarrhea, abdominal discomfort,* dyspepsia, mucosal damage, change in appetite, dry mouth. **Skin:** Rash, pruritus, acne, erythema, seborrhea, hair changes, photosensitivity, skin ulceration, dry skin. **Body as a Whole:** Myalgia, allergic reaction. **Respiratory:** Bronchitis, cough, dyspnea, epistaxis, hemoptysis, laryngitis, rhinitis. **Special Senses:** Xerophthalmia, earache, taste alterations, tinnitus, visual disturbances.

INTERACTIONS Drug: Do not use with **sildenafil, alfuzosin,** or **salmeterol. Rifampin, rifabutin** significantly decrease saquinavir levels. **Phenobarbital, phenytoin, dexamethasone, carbamazepine** may also reduce saquinavir levels. Saquinavir levels may be increased by **delavirdine, ketoconazole, ritonavir, clarithromycin, indinavir.** May increase serum levels of **triazolam, midazolam,** ERGOT DERIVATIVES, **nelfinavir, simvastatin.**

May increase risk of **ergotamine** toxicity of **dihydroergotamine, ergotamine.** Dosage adjustments are needed when used with **bosentan, tadalafil,** or **colchicine. Herbal:** St. John's wort, garlic may decrease antiretroviral activity. **Food: Grapefruit juice** (greater than 1 qt/day) may increase plasma concentrations and adverse effects.

PHARMACOKINETICS Absorption: Rapidly from GI tract; food significantly increases bioavailability. **Distribution:** 98% protein bound. **Metabolism:** In liver (CYP3A4), first-pass metabolism. **Elimination:** Primarily in feces (greater than 80%). **Half-Life:** 13 h.

NURSING IMPLICATIONS

Assessment & Drug Effects

- Lab tests: Monitor serum electrolytes, CBC with differential, liver function, blood glucose and HbA1C, CPK, and serum amylase prior to initiating therapy and periodically thereafter.
- Monitor for and report S&S of peripheral neuropathy.
- Assess for buccal mucosa ulceration or other distressing GI S&S.
- Monitor weight periodically.
- Monitor for toxicity if any of the following drugs is used concomitantly: Calcium channel blockers, clindamycin, dapsone, quinidine, triazolam, or simvastatin.

Patient & Family Education

- Take drug within 2 h of a full meal.
- Be aware of all drugs which should not be taken concurrently with saquinavir.
- Be aware that saquinavir is not a cure for HIV infection and that its long-term effects are unknown.
- Report any distressing adverse effects to prescriber.

SARGRAMOSTIM (GM-CSF)

(sar-gra′mos-tim)

Leukine

Classifications: HEMATOPOIETIC GROWTH FACTOR; GRANULOCYTE MACROPAHGE COLONY STIMULATING FACTOR (GM-CSF)

Therapeutic: HEMATPOIETIC GROWTH FACTOR; GM-CSF

Prototype: Filgrastim

Pregnancy Category: C

AVAILABILITY 250 mcg, 500 mcg injection

ACTION & _THERAPEUTIC EFFECT_
Recombinant human granulocyte macrophage colony stimulating factor (GM-CSF) is produced by recombinant DNA technology. GM-CSF is a hematopoietic growth factor that stimulates proliferation and differentiation of progenitor cells in the granulocyte-macrophage pathways. _Effectiveness is measured by an increase in the number of mature white blood cells (i.e., neutrophil count)._

USES Febrile neutropenia, neutropenia caused by chemotherapy, peripheral blood stem cell (PBSC) mobilization.

UNLABELED USES Neutropenia secondary to other diseases; aplastic anemia, Crohn's disease, malignant melanoma.

CONTRAINDICATIONS Hypersensitivity to GM-CSF, yeast-derived products; excessive leukemic myeloid blasts in bone marrow or blood greater than or equal to 10%; within 24 h of chemotherapy or radiation treatment; if ANC exceeds 20,000 cells/mm³ discontinue drug or use half the dose; increased growth of tumor size; lactation.

CAUTIOUS USE Hypersensitivity to benzl alcohol; history of cardiac arrhythmias, preexisting cardiac disease, renal or hepatic dysfunction, CHF, hypoxia, myelodysplastic syndromes; pulmonary infiltrates; fluid retention; kidney and liver dysfunction; use in AML for adults younger than 55 y; pregnancy (category C). Safety and efficacy in children are not established.

ROUTE & DOSAGE

Neutropenia following Stem Cell Transplantation

Adult: **IV** 250 mcg/m²/day infused over 2 h for 21 days, begin 2–4 h after bone marrow transfusion and not less than 24 h after last dose of chemotherapy or 12 h after last radiation therapy

Following PBSC

Adult: **IV/Subcutaneous** 250 mg/m²/day

Neutropenia following Chemotherapy/Febrile Neutropenia

Adult: **IV/Subcutaneous** 250 mcg/m²/day starting ~day 11 or 4 days following induction chemotherapy

ADMINISTRATION

- Note: Do not give within 24 h preceding or following chemotherapy or within 12 h preceding or following radiotherapy.

Subcutaneous

- Reconstitute each 250 mcg vial with 1 mL of sterile water for injection (without preservative). Direct sterile water against side of vial and swirl gently. Avoid excessive or vigorous agitation. Do not shake. Use without further dilution for subcutaneous injection.

S

Intravenous

Note: Verify correct IV concentration and rate of infusion administration in infants and children with prescriber.

PREPARE: **IV Infusion:** Reconstitute as for subcutaneous, then further dilute reconstituted solution with NS. If the final concentration is less than 1 mcg/mL, add albumin (human) to NS before addition of sargramostim. ▪Use 1 mg albumin per 1 mL of NS to give a final concentration of 0.1% albumin. ▪Administer as soon as possible and within 6 h of reconstitution or dilution for IV infusion. ▪Discard after 6 h. ▪Sargramostim vials are single-dose vials, do not reenter or reuse. Discard unused portion.

ADMINISTER: **IV Infusion:** Give over 2, 4, or 24 h as ordered. ▪Do not use an in-line membrane filter. ▪ Interrupt administration and reduce the dose by 50% if absolute neutrophil count exceeds 20,000/mm³ or if platelet count exceeds 500,000/mm³. Notify prescriber. ▪Reduce the IV rate 50% if patient experiences dyspnea during administration. ▪Discontinue infusion if respiratory symptoms worsen. Notify prescriber.

INCOMPATIBILITIES **Y-site:** Acyclovir, amphotericin B, ampicillin, ampicillin/sulbactam, amsacrine, cefonicid, ceftazidime, chlorpromazine, ganciclovir, haloperidol, hydrocortisone, hydromorphone, hydroxyzine, imipenem/cilastatin, lorazepam, methylprednisolone, mitomycin, morphine, nalbuphine, ondansetron, piperacillin, sodium bicarbonate, tobramycin.

▪ Refrigerate the sterile powder, the reconstituted solution, and store diluted solution at 2°–8° C (36°–46° F). ▪ Do not freeze or shake.

ADVERSE EFFECTS (≥1%) **CNS:** Lethargy, malaise, headache, fatigue. **CV:** Abnormal ST segment depression, supraventricular arrhythmias, edema, *hypotension, tachycardia,* <u>pericardial effusion,</u> pericarditis. **Hematologic:** Anemia, <u>thrombocytopenia.</u> **GI:** Nausea, vomiting, diarrhea, anorexia. **Body as a Whole:** *Bone pain, myalgia, arthralgias,* weight gain, hyperuricemia, *fever.* **Respiratory:** Pleural effusion. **Skin:** *Rash, pruritus.* **Other:** *First-dose reaction* (some or all of the following symptoms: hypotension, tachycardia, fever, rigors, flushing, nausea, vomiting, diaphoresis, back pain, leg spasms, and dyspnea).

INTERACTIONS Drug: CORTICOSTEROIDS and **lithium** should be used with caution because it may potentiate the myeloproliferative effects.

PHARMACOKINETICS Absorption: Readily from subcutaneous site. **Onset:** 3–6 h. **Peak:** 1–2 h. **Duration:** 5–10 days Subcutaneous. **Elimination:** Probably in urine. **Half-Life:** 80–150 min.

NURSING IMPLICATIONS

Assessment & Drug Effects

▪ Lab tests: Baseline and biweekly CBC with differential and platelet count; biweekly LFTs and kidney function tests in patients with established kidney or liver dysfunction.

▪ Discontinue treatment if WBC 50,000/mm³. Notify the prescriber.

▪ Monitor cardiac status. Occasional transient supraventricular arrhythmias have occurred during administration, particularly in those with a history of cardiac

S

arrhythmias. Arrhythmias are reversed with discontinuation of drug.

- Give special attention to respiratory symptoms (dyspnea) during and immediately following IV infusion, especially in patients with preexisting pulmonary disease.
- Use drug with caution in patients with preexisting fluid retention, pulmonary infiltrates, or CHF. Peripheral edema, pleural or pericardial effusion has occurred after administration. It is reversible with dose reduction.
- Notify prescriber of any severe adverse reaction immediately.

Patient & Family Education
- Notify nurse or prescriber immediately of any adverse effect (e.g., dyspnea, palpitations, peripheral edema, bone or muscle pain) during or after drug administration.

SAXAGLIPTIN

(sax-a-glip′tin)

Onglyza

Classifications: ANTIDIABETIC; INCRETIN MODIFIER; DIPEPTIDYL PEPTIDASE-4 (DPP-4) INHIBITOR

Therapeutic: ANTIDIABETIC; HORMONE MODIFIER; DDP-4 INHIBITOR

Prototype: Sitagliptin
Pregnancy Category: B

AVAILABILITY 2.5 mg, 5 mg tablets

ACTION & *THERAPEUTIC EFFECT*
Saxagliptin slows inactivation of incretin hormones [e.g., glucagon-like peptide-1 (GLP-1) and glucose-dependent insulinotropic polypeptide (GIP)]. As plasma glucose rises, incretin hormones stimulate release of insulin from the pancreas and GLP-1 also lowers glucagon secretion, resulting in reduced hepatic glucose production. *In type 2 diabetics, saxagliptin elevates the level of incretin hormones, thus increasing insulin secretion and reducing glucagon secretion. It lowers both fasting and postprandial plasma glucose levels.*

USES Treatment of type 2 diabetes mellitus.

CONTRAINDICATIONS Hypersensitivity to saxagliptin; type 1 diabetes mellitus, ketoacidosis; concurrent administration with insulin.

CAUTIOUS USE Renal impairment; older adults; pregnancy (category B); lactation. Safe use in children younger than 18 y not established.

ROUTE & DOSAGE

Type 2 Diabetes Mellitus
Adult: **PO** 2.5–5 mg daily.
Limited to 2.5 mg once daily when used with a strong CYP3A4/5 inhibitor.

Renal Impairment Dosage Adjustment
CrCl less than or equal to 50 mL/min: 2.5 mg once daily

ADMINISTRATION

Oral
- May be taken without regard to meals.
- Dosing in the older adults should be based on creatinine clearance.
- Store at 15°–30° C (59°–86° F).

ADVERSE EFFECTS (≥1%) **Body as a Whole:** Peripheral edema. **CNS:** *Headache.* **GI:** Abdominal pain, gasteroenteritis, vomiting.

Common adverse effects in *italic*, life-threatening effects underlined; generic names in **bold**; classifications in SMALL CAPS; ♣ Canadian drug name; ⊙ Prototype drug

1361

Metabolic: Hypoglycemia. **Respiratory:** *Nasopharyngitis, upper respiratory tract infection.* **Skin:** Facial edema, urticaria. **Special Senses:** Sinusitis. **Urogenital:** *Urinary tract infection.*

DIAGNOSTIC TEST INTERFERENCE Dose-related decrease in *absolute lymphocyte* count.

INTERACTIONS Drug: Rifampin and other INDUCERS OF CYP3A4/5 enzymes decrease saxagliptin levels. Moderate (e.g., **amprenavir, aprepitant, erythromycin, fluconazole, fosamprenavir, verapamil**) and strong (e.g., **atazanavir, clarithromycin, indinavir, itraconazole, ketaconazole, nefazodone, nelfinavir, ritonavir, saquinavir, telithromycin**) INHIBITORS OF CYP3A4/5 increase saxagliptin levels. **Food: Grapefruit juice** increases saxagliptin levels.

PHARMACOKINETICS Peak: 2 h. **Distribution:** Negligible plasma protein binding. **Metabolism:** Hepatic metabolism to active and inactive compounds. **Elimination:** Renal (75%) and fecal (22%). **Half-Life:** 2.5–3.1 h.

NURSING IMPLICATIONS

Assessment & Drug Effects

- Monitor for and report S&S of significant GI distress including NV&D.
- Monitor for S&S of hypoglycemia when used in combination with a sulfonylurea or insulin.
- Lab tests: Baseline and periodic creatinine clearance; periodic fasting and postprandial plasma glucose and HbA1C; lymphocyte count during periods of infection.

Patient & Family Education

- Carry out blood glucose monitoring as directed by prescriber.

- Consult prescriber during periods of stress and illness as dosage adjustments may be required.
- When taken alone to control diabetes, saxagliptin is unlikely to cause hypoglycemia because it only works when the blood sugar is rising after food intake.

SCOPOLAMINE
(skoe-pol′a-meen)
Transderm Scōp, Transderm-V ♣

SCOPOLAMINE HYDROBROMIDE
Hyoscine, Isopto-Hyoscine, Scopace, Murocoll, Triptone
Classifications: ANTICHOLINERGIC; ANTIMUSCARINIC; ANTISPASMODIC; ANTIVERTIGO
Therapeutic: ANTISPASMODIC; ANTIEMETIC; ANTIVERTIGO
Prototype: Atropine
Pregnancy Category: C

AVAILABILITY Scopolamine: 1.5 mg transdermal patch; **Scopolamine HBr:** 0.4 mg tablets; 0.3 mg/mL, 0.4 mg/mL, 0.86 mg/mL, 1 mg/mL injection; 0.25% ophth solution

ACTION & *THERAPEUTIC EFFECT* Antimuscarinic agent that inhibits the action on acetylcholine (ACh) on postganglionic cholinergic nerves as well as on smooth muscles that lack cholinergic innervation. *Produces CNS depression with marked sedative and tranquilizing effects for use in anesthesia. Effective as a preanesthetic agent to control bronchial, nasal, pharyngeal, and salivary secretions. Additionally, it prevents nausea and vomiting associated with motion sickness.*

USES In obstetrics with morphine to produce amnesia and sedation ("twilight sleep") and as preanesthetic medication. To control spasticity (and drooling) in postencephalitic parkinsonism, paralysis agitans, and other spastic states, as prophylactic agent for motion sickness and as mydriatic and cycloplegic in ophthalmology. Therapeutic system: (**Transderm Scōp**) is used to prevent nausea and vomiting associated with motion sickness.

CONTRAINDICATIONS Hypersensitivity to anticholinergic drugs; narrow angle glaucoma; severe ulcerative colitis, GI obstruction; urinary tract obstruction diseases; toxemia of pregnancy.
CAUTIOUS USE Hypertension; patients older than 40 y, pyloric obstruction, autonomic neuropathy; thyrotoxicosis, liver disease; CAD; CHF; tachycardia, or other tachyarrhythmias; paralytic ileus; hiatal hernia, mild or moderate ulcerative colitis, gastric ulcer, GERD; renal impairment; parkinsonism; COPD, asthma or allergies; hyperthyroidism; brain damage; spastic paralysis; Down syndrome; older adults; pregnancy (category C); lactation; children, infants.

ROUTE & DOSAGE

Preanesthetic

Adult: **PO** 0.4–0.8 mg **IM/Subcutaneous/IV** 0.3–0.6 mg q4–6h
Child: **PO/IM/Subcutaneous/IV** 6 mcg/kg q6–8h (max: 0.3 mg/dose)

Motion Sickness

Adult: **Topical** 1 patch q72h starting 12 h before anticipated travel
Child: **PO** 6 mcg/kg 1 h before anticipated travel

Refraction

Adult: **Ophthalmic** 1–2 drops in eye 1 h before refraction

Uveitis

Adult: **Ophthalmic** 1–2 drops in eye up to q.i.d.

ADMINISTRATION

Instillation
- Minimize possibility of systemic absorption by applying pressure against lacrimal sac during and for 1 or 2 min following instillation of eye drops.

Transdermal
- Apply transdermal disc system (Transderm Scōp, a controlled-release system) to dry surface behind the ear.
- Replace with another disc on another site behind the ear if disc system becomes dislodged.

Subcutaneous or Intramuscular
- Give undiluted.

Intravenous

PREPARE: Direct: Dilute required dose with an equal volume of sterile water for injection.
ADMINISTER: Direct: Give a single dose slowly over 2–3 min.

- Preserve in tight, light-resistant containers.

ADVERSE EFFECTS (≥1%) **Body as a Whole:** Fatigue, dizziness, *drowsiness,* disorientation, restlessness, hallucinations, toxic psychosis. **GI:** *Dry mouth and throat, constipation.* **Urogenital:** Urinary retention. **CV:** Decreased heart rate. **Special Senses:** Dilated pupils, photophobia, blurred vision, *local irritation,* follicular conjunctivitis. **Respiratory:** <u>Depressed</u>

S

Common adverse effects in *italic*, life-threatening effects <u>underlined</u>; generic names in **bold**; classifications in SMALL CAPS; ✦ Canadian drug name; ⊘ Prototype drug

1363

respiration. **Skin:** Local irritation from patch adhesive, rash.

INTERACTIONS Drug: Amantadine, ANTIHISTAMINES, TRICYCLIC ANTIDEPRESSANTS, **quinidine, disopyramide, procainamide** add to anticholinergic effects; decreases **levodopa** effects; **methotrimeprazine** may precipitate extrapyramidal effects; decreases antipsychotic effects (decreased absorption) of PHENOTHIAZINES. **Food: Grapefruit juice** (greater than 1 qt/day) may increase plasma concentrations and adverse effects.

DIAGNOSTIC TEST INTERFERENCE Lab Test: Interferes with *gastric secretion test.*

PHARMACOKINETICS Absorption: Readily from GI tract and percutaneously. **Peak:** 20–60 min. **Duration:** 5–7 days. **Distribution:** Crosses placenta; distributed to CNS. **Metabolism:** In liver. **Elimination:** In urine.

NURSING IMPLICATIONS

Assessment & Drug Effects

- Observe patient closely; some patients manifest excitement, delirium, and disorientation shortly after drug is administered until sedative effect takes hold.
- Use of side rails is advisable, particularly for older adults, because of amnesic effect of scopolamine.
- In the presence of pain, scopolamine may cause delirium, restlessness, and excitement unless given with an analgesic.
- Be aware that tolerance may develop with prolonged use.
- Terminate ophthalmic use if local irritation, edema, or conjunctivitis occur.

Patient & Family Education

- Vision may blur when used as mydriatic or cycloplegic; do not drive or engage in potentially hazardous activities until vision clears.

- Place disc on skin site the night before an expected trip or anticipated motion for best therapeutic effect.
- Wash hands carefully after handling scopolamine. Anisocoria (unequal size of pupils, blurred vision can develop by rubbing eye with drug-contaminated finger).

SECOBARBITAL SODIUM ⊕

(see-koe-bar′bi-tal)

Seconal Sodium
Classifications: SEDATIVE-HYPNOTIC; BARBITURATE; ANXIOLYTIC
Therapeutic: SEDATIVE-HYPNOTIC
Pregnancy Category: D
Controlled Substance: Schedule II

AVAILABILITY 50 mg, 100 mg capsules

ACTION & *THERAPEUTIC EFFECT*
Short-acting barbiturate with CNS depressant effects as well as mood alteration from excitation to mild sedation, hypnosis, and deep coma. Depresses the sensory cortex, decreases motor activity, alters cerebellar function and produces drowsiness, sedation, and hypnosis. *Alters cerebellar function and produces drowsiness, sedation, and hypnosis.*

USES Hypnotic for simple insomnia and preoperatively to provide basal hypnosis for general, spinal, or regional anesthesia.

CONTRAINDICATIONS History of sensitivity to barbiturates; porphyria; hepatic coma; severe respiratory disease; parturition; fetal immaturity; uncontrolled pain. Use of sterile injection containing polyethylene glycol vehicle in patients with renal insufficiency; pregnancy (category D).
CAUTIOUS USE Pregnant women with toxemia or history of bleeding; labor and delivery; seizure disorders;

aspirin hypersensitivity; liver function impairment; renal impairment; hyperthyroidism; diabetes mellitus; depression; history of suicidal tendencies or drug abuse; acute or chronic pain; severe anemia; older adults (short term use only), debilitated individuals; lactation; children younger than 6 y.

ROUTE & DOSAGE

Sedative

Adult: **PO** 100–300 mg/day in 3 divided doses
Child: **PO** 4–6 mg/kg/day in 3 divided doses

Preoperative Sedative

Adult: **PO** 100–300 mg 1–2 h before surgery
Child: **PO** 50–100 mg 1–2 h before surgery

Hypnotic

Adult: **PO** 100–200 mg

ADMINISTRATION

Oral

- Give hypnotic dose only after patient retires for the evening.
- Crush and mix with a fluid or with food if patient cannot swallow pill.

ADVERSE EFFECTS (≥1%) **CNS:** Drowsiness, lethargy, hangover, paradoxical excitement in older adults. **Respiratory:** Respiratory depression, laryngospasm.

INTERACTIONS Drug: Phenmetrazine antagonizes effects of secobarbital; CNS DEPRESSANTS, **alcohol,** SEDATIVES compound CNS depression; MAO INHIBITORS cause excessive CNS depression; **methoxyflurane** increases risk of nephrotoxicity. **Herbal: Kava, valerian** may potentiate sedation.

PHARMACOKINETICS Absorption: 90% from GI tract. **Onset:** 15–30 min. **Duration:** 1–4 h. **Distribution:** Crosses placenta; distributed into breast milk. **Metabolism:** In liver. **Elimination:** In urine. **Half-Life:** 30 h.

NURSING IMPLICATIONS

Assessment & Drug Effects

- Be alert to unexpected responses and report promptly. Older adults or debilitated patients and children sometimes have paradoxical response to barbiturate therapy (i.e., irritability, marked excitement as inappropriate tearfulness and aggression in children, depression, and confusion).
- Be aware that barbiturates do not have analgesic action, and may produce restlessness when given to patients in pain.
- Lab tests: Periodic LFTs and hematology tests, serum folate and vitamin D levels during prolonged therapy.
- Be alert for acute toxicity (intoxication) characterized by profound CNS depression, respiratory depression, hypoventilation, cyanosis, cold clammy skin, hypothermia, constricted pupils (but may be dilated in severe intoxication), shock, oliguria, tachycardia, hypotension, respiration arrest, circulatory collapse, and death.

Patient & Family Education

- Do not drive or engage in potentially hazardous activities until response to drug is established.
- Store barbiturates in a safe place; not on the bedside table or other readily accessible places. It is possible to forget having taken the drug, and in half-wakened conditions take more and accidentally overdose.

Common adverse effects in *italic*, life-threatening effects underlined; generic names in **bold**; classifications in SMALL CAPS; ✥ Canadian drug name; ⊙ Prototype drug

1365

- Do not become pregnant. Use or add barrier contraception if using hormonal contraceptives.
- Report onset of fever, sore throat or mouth, malaise, easy bruising or bleeding, petechiae, jaundice, rash to prescriber during prolonged therapy.
- Do not consume alcohol in any amount when taking a barbiturate. It may severely impair judgment and abilities.

SELEGILINE HYDROCHLORIDE (L-DEPRENYL)
(se-leg′i-leen)

Carbex, Eldepryl, Emsam, Zelapar
Classifications: ANTIPARKINSON; ANTIDEPRESSANT (MAOI)
Therapeutic: ANTIPARKINSON; ANTIDEPRESSANT
Pregnancy Category: C

AVAILABILITY 5 mg tablets, capsules; 1.25 mg orally disintegrating tab; 6 mg, 9 mg, 12 mg transdermal patch

ACTION & *THERAPEUTIC EFFECT*
Increase in dopaminergic activity is thought to be primarily due to selective inhibition of MAO type B activity. Ability of selegiline to control parkinsonism is thought to be due to increased dopaminergic activity. It interferes with dopamine reuptake at the synapse of neurons as well as its inhibition of MAO type B dopaminergic activity in the brain. Interference with dopamine reuptake at the MAO type A dopaminergic receptors in the brain is thought to be the mechanism for antidepression. *Effectiveness is measured in decreased tremors, reduced akinesia, improved speech and motor abilities*

as well as improved walking. At slightly higher doses it is an effective antidepressant.

USES Adjunctive therapy of Parkinson's disease for patients being treated with levodopa and carbidopa who exhibit deterioration in the quality of their response to therapy, major depressive disorder.
UNLABELED USES Attention deficit/hyperactivity disorder, extrapyramidal symptoms.

CONTRAINDICATIONS Hypersensitivity to selegiline; uncontrolled hypertension; suicidal ideation; lactation.
CAUTIOUS USE Hypertension; history of suicide, bipolar disorder; psychosis; pregnancy (category C). Safety and efficacy in adolescents and children are not established.

ROUTE & DOSAGE

Parkinson's Disease
Adult: **PO** 5 mg b.i.d. with breakfast and lunch (doses greater than 10 mg/day are associated with increased risk of toxicity due to MAO inhibition) **PO (Zelapar)** 1.25 mg daily × 6 wk (max: 2.5 mg daily)
Geriatric: **PO** Start with 5 mg qa.m.

Depression
Adult: **Transdermal** 6 mg/day, may increase by 3 mg/day q2wk up to 12 mg/day

ADMINISTRATION
Oral
- Do not give daily doses exceeding 10 mg/day.
- Note: Concurrent levodopa and carbidopa doses are usually

reduced 10–30% after 2–3 days of selegiline therapy.
- Do not use concurrently with opioids (especially meperidine).
- Store at 15°–30° C (59°–86° F).

Transdermal
- Do not cut or trim patch.
- Before application wash the area with soap and warm water. Dry thoroughly.
- Apply to upper torso, upper thigh, or outer surface of upper arm. Do not apply to hairy, oily, irritated, broken, or calloused skin.
- Rotate sites.
- Wash hands after application.

ADVERSE EFFECTS (≥1%) **CNS:** Sleep disturbances, psychosis, agitation, confusion, dyskinesia, dizziness, hallucinations, dystonia, akathisia. **CV:** Hypotension. **GI:** Anorexia, *nausea,* vomiting, abdominal pain, constipation, diarrhea.

INTERACTIONS Drug: TRICYCLIC ANTIDEPRESSANTS may cause hyperpyrexia, seizures; **fluoxetine, sertraline, paroxetine** may cause hyperthermia, diaphoresis, tremors, seizures, delirium; SYMPATHOMIMETIC AGENTS (e.g., **amphetamine, phenylephrine, phenylpropanolamine**), **guanethidine,** and **reserpine** may cause hypertensive crisis; CNS DEPRESSANTS have additive CNS depressive effects; OPIATE ANALGESICS (especially **meperidine**) may cause hypertensive crisis and circulatory collapse; **buspirone,** hypertension; GENERAL ANESTHETICS: Prolonged hypotensive and CNS depressant-effects; hypertension, headache, hyperexcitability reported with **dopamine, methyldopa, levodopa, tryptophan; metrizamide** may increase risk of seizures; HYPOTENSIVE AGENTS and DIURETICS have additive hypotensive effects. **Food:** Aged meats

or aged cheeses, protein extracts, sour cream, alcohol, anchovies, liver, sausages, overripe figs, bananas, avocados, chocolate, soy sauce, bean curd, natural yogurt, fava beans—**tyramine**-containing foods—may precipitate hypertensive crisis (less frequent with usual doses of **selegiline** than with other MAOIS). **Herbal: Ginseng, ephedra, ma huang, St. John's wort** may cause hypertensive crisis.

PHARMACOKINETICS Absorption: Rapid; 73% reaches systemic circulation. **Onset:** 1 h. **Duration:** 1–3 days. **Distribution:** Crosses placenta; not known if distributed into breast milk. **Metabolism:** In liver to *N*-desmethyldeprenyl-amphetamine and methamphetamine. **Elimination:** In urine. **Half-Life:** 15 min (metabolites 2–20 h).

NURSING IMPLICATIONS

Assessment & Drug Effects
- Monitor vital signs, particularly during period of dosage adjustment. Report alterations in BP or pulse. Indications for discontinuation of the drug include orthostatic hypotension, hypertension, and arrhythmias.
- Monitor for changes in behavior that may indicate increase suicidality, especially in adolescents or children being treated for depression.
- Monitor all patients closely for behavior changes (e.g., hallucinations, confusion, depression, delusions).

Patient & Family Education
- Do not exceed the prescribed drug dose.
- Report symptoms of MAO inhibitor-induced hypertension (e.g., severe headache, palpitations,

---

Common adverse effects in *italic*, life-threatening effects underlined; generic names in **bold**; classifications in SMALL CAPS; ✦ Canadian drug name; ⊘ Prototype drug

1367

neck stiffness, nausea, vomiting) immediately to prescriber.

- Do not drive or engage in potentially hazardous activities until response to drug is known.

- Make positional changes slowly and in stages. Orthostatic hypotension is possible as well as dizziness, light-headedness, and fainting.

- If the transdermal patch falls off, apply a new patch to a new area, and resume previous schedule.

- Only one should be worn at a given time. Remove the old transdermal patch.

SELENIUM SULFIDE

(se-lee′nee-um)

Exsel, Selsun, Selsun Blue
Classifications: ANTIBIOTIC, TOPICAL; ANTIFUNGAL
Therapeutic: TOPICAL ANTIFUNGAL; ANTISEBORRHEIC
Pregnancy Category: C

AVAILABILITY 1% lotion, shampoo

ACTION & *THERAPEUTIC EFFECT*
Absorption of selenium sulfide into epithelial tissue cells is followed by degradation of compound to selenium and sulfide ions. Selenium ions block enzyme systems involved in epithelial cell growth. As a result, cell turnover rate is reduced. *Active against* Pityrosporum ovale, *a yeast-like fungus found in the normal flora of the scalp. Also decreases rate of growth of the epithelial cells of the scalp and other epithelial layers of cells in the body.*

USES Itching and flaking of the scalp associated with dandruff, seborrheic dermatitis of the scalp, and tinea versicolor.

CONTRAINDICATIONS Kidney failure or biliary tract obstruction, GI

malfunction; Wilson's disease; application to damaged or inflamed skin surfaces.

CAUTIOUS USE Prolonged skin contact; use in genital area or skin folds; pregnancy (category C); lactation; children younger than 2 y.

ROUTE & DOSAGE

Dandruff Control, Seborrheic Dermatitis

Adult/Child: **Topical** Massage 5–10 mL of a 1–2.5% solution into wet scalp and leave on for 2–3 min, rinse thoroughly, then repeat application and rinse well again (initially, shampoo 2 times/wk for 2 wk, then decrease to once q1–4wk prn)

Tinea Versicolor

Adult/Child: **Topical** Apply a 2.5% solution to affected area with a small amount of water to form a lather, leave on for 10 min, then rinse thoroughly, repeat once/day for 7 days

ADMINISTRATION

Topical

- Wash hands thoroughly after application of selenium sulfide to affected areas. Remove jewelry before treatment; drug will damage it.

- Rinse genital areas and skin folds well with water and dry thoroughly after treatment for tinea versicolor to prevent irritation.

- Store at 15°–30° C (59°–86° F) in tight container; protected from heat. Avoid freezing.

ADVERSE EFFECTS (≥1%) **Skin:** *Skin irritation (stinging),* rebound oiliness of scalp, hair discoloration, diffuse hair loss (reversible), systemic toxicity (if applied to abraded, infected skin).

PHARMACOKINETICS Absorption: No percutaneous absorption if skin is intact.

NURSING IMPLICATIONS

Assessment & Drug Effects
- Monitor therapeutic effectiveness.

Patient & Family Education
- Rinse thoroughly with water if lotion contacts eyes in order to prevent chemical conjunctivitis.
- Do not use drug more frequently than required to maintain control of dandruff.
- Hair loss is reversible, usually within 2–3 wk after treatment is discontinued.
- Discontinue use if skin is irritated or treatment fails. Systemic toxicity may result from application of lotion to damaged skin (percutaneous absorption) or from prolonged use (overdosage). Toxicity symptoms include tremors, anorexia, occasional vomiting, lethargy, weakness, severe perspiration, garlicky breath, lower abdominal pain. Symptoms disappear 10–12 days after treatment is stopped.

SENNA (SENNOSIDES)
(sen'na)

Black-Draught, Gentlax B, Senexon, Senokot, Senolax
Classification: STIMULANT LAXATIVE
Therapeutic: STIMULANT LAXATIVE
Prototype: Bisacodyl
Pregnancy Category: C

AVAILABILITY 8.6 mg, 15 mg, 25 mg tablets; 8.6 mg/5 mL, 15 mg/5 mL syrup

ACTION & *THERAPEUTIC EFFECT* Senna glycosides are converted in colon to active aglycone, which stimulates peristalsis. Concentrate is purified and standardized for uniform action and is claimed to produce less colic than crude form. *Peristalsis stimulated by conversion of drug to active chemical.*

USES Acute constipation and preoperative and preradiographic bowel evacuation.

CONTRAINDICATIONS Hypersensitivity; appendicitis, fecal impaction; fluid and electrolyte imbalances; irritable colon, nausea, vomiting, undiagnosed abdominal pain, intestinal obstruction.

CAUTIOUS USE Diabetes mellitus; fluid and electrolyte imbalances; pregnancy (category C); children younger than 6 y.

ROUTE & DOSAGE

Constipation
Adult: **PO Standard Senna Concentrate** 1–2 tablets or ½–1 tsp at bedtime (max: 4 tablets or 2 tsp b.i.d.); **Syrup, Liquid** 10–15 mL at bedtime
Child: **PO Standard Senna Concentrate** *Weight greater than 27 kg,* 1 tablet or ½ tsp at bedtime; **Syrup, Liquid** *1 mo–1 y,* 1.25–2.5 mL at bedtime; *1–5 y,* 2.5–5 mL at bedtime; *5–15 y,* 5–10 mL at bedtime

ADMINISTRATION

Oral
- Give at bedtime, generally.
- Avoid exposing drug to excessive heat; protect fluid extracts from light.

ADVERSE EFFECTS (≥1%) GI: Abdominal cramps, flatulence, nausea, watery diarrhea, excessive loss of water and electrolytes, weight loss, melanotic segmentation of colonic mucosa (reversible).

Common adverse effects in *italic*, life-threatening effects <u>underlined</u>; generic names in **bold**; classifications in SMALL CAPS; ♦ Canadian drug name; 🔟 Prototype drug

1369

PHARMACOKINETICS Onset: 6–10 h; may take up to 24 h. **Metabolism:** In liver. **Elimination:** In feces.

NURSING IMPLICATIONS

Assessment & Drug Effects

▪ Reduce dose in patients who experience considerable abdominal cramping.

Patient & Family Education

▪ Be aware that drug may alter urine and feces color; yellowish brown (acid), reddish brown (alkaline).
▪ Continued use may lead to dependence. Consult prescriber if constipation persists.
▪ See bisacodyl for additional nursing implications.

SERTACONAZOLE NITRATE

(ser-ta-con′a-zole)

Ertaczo
Classifications: ANTIBIOTIC; AZOLE ANTIFUNGAL
Therapeutic: ANTIFUNGAL
Prototype: Fluconazole
Pregnancy Category: C

AVAILABILITY 2% cream

ACTION & *THERAPEUTIC EFFECT* It is believed that azole antifungals act primarily by inhibiting cytochrome P450–dependent synthesis of ergosterol, a key component of the cell membrane of fungi resulting in fungal cell injury. *Has a broad spectrum of activity against common fungal pathogens.*

USES Treatment of tinea pedis in immunocompetent patients.

CONTRAINDICATIONS Onychomycosis.

CAUTIOUS USE History of hypersensitivity to azole antifungals; pregnancy (category C), lactation; children younger than 12 y.

ROUTE & DOSAGE

Tinea Pedis
Adult/Child (older than 12 y):
Topical Apply thin layer to affected area twice daily for 4 wk

ADMINISTRATION

Topical
▪ Cleanse the affected area and dry thoroughly before application.
▪ Apply a thin layer of the cream to affected area between the toes and the immediately surrounding healthy skin. Gently rub into the skin.
▪ Store at 15°–30° C (57°–86° F).

ADVERSE EFFECTS (≥1%) **Skin:** Contact dermatitis, dry skin, burning, application site reaction, skin tenderness.

PHARMACOKINETICS Absorption: Negligible through intact skin.

NURSING IMPLICATIONS

Assessment & Drug Effects
▪ Monitor for clinical improvement, which should be seen about 2 wk after initiating treatment.

Patient & Family Education
▪ Report any of the following: Severe skin irritation, redness, burning, blistering, or itching.
▪ Do not stop using this medication prematurely. Athlete's foot takes about 4 wk to clear completely.
▪ Nursing mothers should ensure that this topical cream does not accidentally get on the breast.

SERTRALINE HYDROCHLORIDE

(ser′tra-leen)

Zoloft
Classifications: ANTIDEPRESSANT; SELECTIVE SEROTONIN REUPTAKE INHIBITOR (SSRI)
Therapeutic: ANTIDEPRESSANT; SSRI

Prototype: Fluoxetine
Pregnancy Category: C

AVAILABILITY 25 mg, 50 mg, 100 mg tablets; 20 mg/mL liquid

ACTION & *THERAPEUTIC EFFECT*
Potent inhibitor of serotonin (5-HT) reuptake in the brain. Chronic administration results in downregulation of norepinephrine, a reaction found with other effective antidepressants. *Effective in controlling depression, obsessive-compulsive disorder, anxiety, and panic disorder.*

USES Major depression, obsessive-compulsive disorder, panic disorder, social anxiety disorder, premenstrual dysphoric disorder, generalized anxiety, post-traumatic stress disorder.
UNLABELED USES Eating disorders, generalized anxiety disorder.

CONTRAINDICATIONS Patients taking MAO inhibitors or within 14 days of discontinuing MAO inhibitor; concurrent use of Antabuse; suicidal ideation, hyponatremia; mania or hypomania.
CAUTIOUS USE Seizure disorders, major affective disorders, bipolar disorder, history of suicide; liver dysfunction, renal impairment; abrupt discontinuation; anorexia nervosa, recent history of MI or unstable cardiac disease, dehydration; diabetes mellitus; ECT therapy, older adults; pregnancy (category C), lactation. Safe use in children younger than 6 y is not established.

ROUTE & DOSAGE

Depression, Anxiety
Adult: **PO** Begin with 50 mg/day, gradually increase every few weeks according to response (range: 50–200 mg)

Geriatric: **PO** Start with 25 mg/day
Premenstrual Dysphoric Disorder
Adult: **PO** Begin with 50 mg/day for first cycle, may titrate up to 150 mg/day
Obsessive-Compulsive Disorder
Adult: **PO** Begin with 50 mg/day, may titrate at weekly intervals up to 200 mg/day
Child (6–12 y): **PO** Begin with 25 mg/day, may increase by 50 mg/wk, as tolerated and needed, up to 200 mg/day

ADMINISTRATION
Oral
- Give in the morning or evening.
- Do not give concurrently with an MAO inhibitor or within 14 days of discontinuing an MAO inhibitor.
- Dilute concentrate before use with 4 oz of water, ginger ale, lemon/lime soda, lemonade, or orange juice ONLY. Give immediately after mixing. Caution with latex sensitivity, as the dropper contains dry natural rubber.

ADVERSE EFFECTS (≥1%) **CV:** Palpitations, chest pain, hypertension, hypotension, edema, syncope, tachycardia. **CNS:** *Agitation, insomnia, headache, dizziness, somnolence, fatigue,* ataxia, incoordination, vertigo, abnormal dreams, aggressive behavior, delusions, hallucinations, emotional lability, paranoia, suicidal ideation, depersonalization. **Endocrine:** Gynecomastia, male sexual dysfunction. **GI:** Nausea, vomiting, diarrhea, constipation, indigestion, anorexia, flatulence, abdominal pain, dry mouth. **Special Senses:** Exophthalmos,

Common adverse effects in *italic*, life-threatening effects underlined; generic names in **bold**; classifications in SMALL CAPS; ♣ Canadian drug name; ☯ Prototype drug

1371

blurred vision, dry eyes, diplopia, photophobia, tearing, conjunctivitis, mydriasis. **Skin:** Rash, urticaria, acne, alopecia. **Respiratory:** Rhinitis, pharyngitis, cough, dyspnea, bronchospasm. **Body as a Whole:** Myalgia, arthralgia, muscle weakness, bone fracture (older adults). **Metabolic:** Hyponatremia in older adults.

DIAGNOSTIC TEST INTERFERENCE
May cause asymptomatic elevations in *liver function tests.* Slight decrease in *uric acid.*

INTERACTIONS Drug: MAOIS (e.g., **selegiline, phenelzine**) should be stopped 14 days before sertraline is started because of serious problems with other SEROTONIN REUPTAKE INHIBITORS (shivering, nausea, diplopia, confusion, anxiety). **Sertraline** may increase levels and toxicity of **diazepam, pimozide, tolbutamide.** Use cautiously with other centrally acting CNS drugs; increase risk of **ergotamine** toxicity with **dihydroergotamine, ergotamine.** Concentrate interacts with **disulfiram. Herbal: St. John's wort** may cause **serotonin** syndrome (headache, dizziness, sweating, agitation). **Food: Grapefruit juice** (greater than 1 qt/day) may increase plasma concentrations and adverse effects.

PHARMACOKINETICS Absorption: Slowly from GI tract. **Onset:** 2–4 wk. **Distribution:** 99% protein bound; distribution into breast milk unknown. **Metabolism:** Extensive first-pass metabolism in liver to inactive metabolites. **Elimination:** 40–45% in urine, 40–45% in feces. **Half-Life:** 24 h.

NURSING IMPLICATIONS

Assessment & Drug Effects
- Supervise patients at risk for suicide closely during initial therapy.
- Monitor for worsening of depression or emergence of suicidal ideation.
- Monitor older adults for fluid and sodium imbalances.
- Monitor patients with a history of a seizure disorder closely.
- Lab tests: Monitor PT and INR with patients receiving concurrent warfarin therapy.

Patient & Family Education
- Report diarrhea, nausea, dyspepsia, insomnia, drowsiness, dizziness, or persistent headache to prescriber.
- Report emergence of agitation, irritability, hostility or aggression, mania.
- Report signs of bleeding promptly to prescriber when taking concomitant warfarin.

SEVELAMER HYDROCHLORIDE ℗

(se-vel'a-mer)

Renagel, Renvela

Classifications: ELECTROLYTE AND WATER BALANCE AGENT; PHOSPHATE BINDER

Therapeutic: PHOSPHATE BINDER

Pregnancy Category: C

AVAILABILITY 400 mg, 800 mg tablets; oral powder for suspension

ACTION & *THERAPEUTIC EFFECT*
Polymer that binds intestinal phosphate; interacts with phosphate by way of ion-exchange and hydrogen binding. Advantageously, does not contain aluminum or calcium in treating hyperphosphatemia in

end-stage kidney failure. *Effectiveness indicated by a serum phosphate level 6.0 mg/dL or less.*

USES Hyperphosphatemia.

CONTRAINDICATIONS Hypersensitivity to sevelamer HCl; hypophosphatemia; fecal impaction; bowel obstruction; appendicitis; dysphagia, GI bleeding, major GI surgery; lactation.

CAUTIOUS USE GI motility disorders; vitamin deficiencies (especially vitamins D, E, and K and folic acid); pregnancy (category C); children younger than 18 y.

ROUTE & DOSAGE

Hyperphosphatemia
Adult: **PO** 800–1600 mg t.i.d. based on severity of hyperphosphatemia

ADMINISTRATION

Oral
- Give with meals.
- Give other oral medications 1 h before or 3 h after Renagel.
- Store at 15°–30° C (59°–86° F); protect from moisture.

ADVERSE EFFECTS (≥1%) **Body as a Whole:** Headache, infection, pain. **CV:** Hypertension, hypotension, thrombosis. **GI:** Diarrhea, dyspepsia, vomiting, nausea, constipation, flatulence. **Respiratory:** Increased cough.

NURSING IMPLICATIONS

Assessment & Drug Effects
- Lab tests: Obtain frequent serum phosphate levels.

Patient & Family Education
- Take daily multivitamin supplement approved by prescriber.

SILDENAFIL CITRATE ℗

(sil-den′a-fil ci′trate)

Revatio, Viagra

Classifications: PHOSPHODIESTERASE (PDE) INHIBITOR; IMPOTENCE; PULMONARY ANTIHYPERTENSIVE

Therapeutic: PULMONARY ANTIHYPERTENSIVE; IMPOTENCE

Pregnancy Category: B

AVAILABILITY 20 mg, 25 mg, 50 mg, 100 mg tablets; 10 mg/12.5 mL injection

ACTION & THERAPEUTIC EFFECT Enhances vasodilation effect of nitric oxide in the corpus cavernosus of the penis, thus sustaining an erection. PDE-5 inhibitors reduce pulmonary vasodilation by sustaining levels of cyclic guanosine monophosphate (cGMP). Additionally, sildenafil produces a reduction in the pulmonary to systemic vascular resistance ratio. *Effective for treatment of erectile dysfunction, whether organic or psychogenic in origin. Sildenafil produces a significant improvement in arterial oxygenation in pulmonary arterial hypertension (PAH).*

USES Erectile dysfunction, pulmonary arterial hypertension.

UNLABELED USES Altitude sickness, Raynaud's phenomenon, sexual dysfunction, anorgasmy.

CONTRAINDICATIONS Hypersensitivity to sildenafil.

CAUTIOUS USE CAD with unstable angina, heart failure, MI, cardiac arrhythmias, stroke within 6 mo of starting drug; nitrate therapy; hypotension and hypertension; risk factors for CVA; aortic stenosis; anatomic deformity of the penis; sickle cell anemia, polycythemia; multiple myeloma; leukemia; active bleeding

S

Common adverse effects in *italic*, life-threatening effects underlined; generic names in **bold**; classifications in SMALL CAPS; ♣ Canadian drug name; ℗ Prototype drug

1373

or a peptic ulcer, GERD, hiatal hernia; coagulopathy; retinitis pigmentosa; hepatic disease, hepatitis, cirrhosis; severe renal impairment; older adults; pregnancy (category B); lactation; children and infants.

ROUTE & DOSAGE

Erectile Dysfunction

Adult: **PO** 50 mg 0.5–4 h before sexual activity (dose range: 25 to 100 mg once/day); max dose: 25 mg/day with itraconazole or ketoconazole; max dose: 25 mg/48 h with ritonavir
Geriatric: **PO** 25 mg approximately 1 h before sexual activity

Pulmonary Arterial Hypertension

Adult: **PO** 20 mg t.i.d. (4–6 h apart) **IV** 10 mg t.i.d.

Hepatic Impairment Dosage Adjustment

Child-Pugh class A or B: Starting dose of 25 mg

Renal Impairment Dosage Adjustment

CrCl less than 30 mL/min: Starting dose of 25 mg

ADMINISTRATION

Oral

- For erectile dysfunction: Dose 1 h prior to sexual activity (effective range is 0.5–4 h).
- Do not give within 24 h of taking any medication with nitrates (i.e., nitroglycerin).

Intravenous

PREPARE: **Direct:** Give undiluted.
ADMINISTER: **Direct:** Give as a bolus dose.

- Store at 15°–30° C (59°–86° F) in a tightly closed container; protect from light.

ADVERSE EFFECTS (≥1%) **Body as a Whole:** Face edema, photosensitivity, shock, asthenia, pain, chills, fall, allergic reaction, arthritis, myalgia. **CNS:** *Headache,* dizziness, migraine, syncope, cerebral thrombosis, ataxia, neuralgia, paresthesias, tremor, vertigo, depression, insomnia, somnolence, abnormal dreams. **CV:** Flushing, chest pain, <u>MI</u>, angina, AV block, tachycardia, palpitation, hypotension, postural hypotension, <u>cardiac arrest</u>, <u>sudden cardiac death</u>, heart failure, cardiomyopathy, abnormal ECG, edema. **GI:** Dyspepsia, diarrhea, abdominal pain, vomiting, colitis, dysphagia, gastritis, gastroenteritis, esophagitis, stomatitis, dry mouth, abnormal liver function tests, thirst. **Respiratory:** Nasal congestion, asthma, dyspnea, laryngitis, pharyngitis, sinusitis, bronchitis, cough. **Skin:** Rash, urticaria, pruritus, sweating, <u>exfoliative dermatitis</u>. **Urogenital:** UTI. **Special Senses:** Abnormal vision (color changes, photosensitivity, blurred vision, sudden vision loss). **Hematologic:** Anemia, <u>leukopenia</u>. **Metabolic:** Gout, hyperglycemia, hyperuricemia, hypoglycemia, hypernatremia.

INTERACTIONS Drug: NITRATES increase risk of serious hypotension; if used within 4 h of **doxazosin, prazosin, terazosin, tamsulosin; cimetidine, erythromycin, ketoconazole, itraconazole,** PROTEASE INHIBITORS increase sildenafil levels; **rifampin** can decrease sildenafil levels. **Food: Grapefruit juice** (greater than 1 qt/day) may increase plasma concentrations and adverse effects.

PHARMACOKINETICS Absorption: Rapidly from GI tract. **Peak:** 30–120 min. **Distribution:** 96% protein bound. **Metabolism:** In liver (CYP3A4 and 2C9). **Elimination:** 80%

in feces, 12% in urine. **Half-Life:** 4 h.

NURSING IMPLICATIONS

Assessment & Drug Effects
- Monitor carefully for and immediately report S&S of cardiac distress.

Patient & Family Education
- Do not take sildenafil within 4 h of taking doxazosin, prazosin, terazosin, or tamsulosin.
- Consuming a high-fat meal before taking drug may cause delay in drug action.
- Report to prescriber: Headaches, flushing, chest pain, indigestion, blurred vision, sensitivity to light, changes in color vision.

SILODOSIN
(sil'o-do-sin)

Rapaflo
Classifications: ALPHA-1 ADRENERGIC RECEPTOR ANTAGONIST; GENITOURINARY SMOOTH MUSCLE RELAXANT
Therapeutic: GENITOURINARY SMOOTH MUSCLE RELAXANT
Prototype: Tamsulosin
Pregnancy Category: B

AVAILABILITY 4 mg, 8 mg capsules

ACTION & THERAPEUTIC EFFECT
Selective antagonist of post-synaptic alpha-1 adrenoreceptors located in the prostate, bladder base, bladder neck, prostatic capsule, and prostatic urethra. *Blockade of these alpha-1 adrenoreceptors causes the smooth muscle in these tissues to relax, resulting in improvement in urine flow and reduction in signs and symptoms of benign prostatic hyperplasia (BPH).*

USES
Treatment of the signs and symptoms of BPH

CONTRAINDICATIONS
Severe renal impairment (CrCl less than 30 mL/min); severe hepatic impairment (Child-Pugh score greater than or equal to 10).
CAUTIOUS USE Moderate renal impairment; history of hypotension; cataract surgery; older adults; pregnancy (category B); lactation. Safe use in children not established.

ROUTE & DOSAGE

Benign Prostatic Hyperplasia
Adult: **PO** 8 mg once daily with a meal

Renal Impairment Dosage Adjustment
CrCl 30–49 mL/min: Reduce dose to 4 mg once daily; *less than 30 mL/min:* Not recommended

ADMINISTRATION

Oral
- Give with meals.
- Store at 15°–30° C (59°–86° F). Protect from light and moisture.

ADVERSE EFFECTS (≥1%) **Body as a Whole:** Abdominal pain, asthenia. **CNS:** Dizziness, headache, insomnia. **CV:** Orthostatic hypotension. **GI:** Diarrhea. **Respiratory:** Nasal congestion, nasopharyngitis, rhinorrhea, sinusitis. **Urogenital:** *Retrograde ejaculation.*

DIAGNOSTIC TEST INTERFERENCE
Increased *prostate specific antigen (PSA).*

INTERACTIONS Drug: Strong CYP3A4 INHIBITORS (e.g., **itraconazole, ritonavir**) or strong P-GLYCOPROTEIN INHIBITORS (e.g., **ketoconazole**) greatly increases silodosin levels. Moderate CYP3A4 INHIBITORS (e.g., **diltiazem, erythromycin, verapamil**) may increase silodosin levels. Other ALPHA-BLOCKERS can cause additive

Common adverse effects in *italic*, life-threatening effects underlined; generic names in **bold**; classifications in SMALL CAPS; ♣ Canadian drug name; ⊘ Prototype drug

1375

pharmacodynamic effects. INHIBITORS OF UDP-GLUCURONOSYLTRANSFERASE 2B7 (e.g., **probenecid, valproic acid, fluconazole**) may increase silodosin levels.

PHARMACOKINETICS Absorption: 32% bioavailable. **Distribution:** Approximately 97% plasma protein bound. **Metabolism:** Hepatic oxidation and conjugation. **Elimination:** Renal and fecal. **Half-Life:** 13.3 h.

NURSING IMPLICATIONS

Assessment & Drug Effects
- Monitor I&O and ease of voiding.
- Monitor orthostatic vital signs (lying and then standing) at the beginning of therapy. Report a systolic pressure drop of 15 mm Hg or greater and HR increase of 15 beats or greater upon standing.
- Monitor for orthostatic hypotension, especially at the beginning of therapy and in those taking concurrent antihypertensive drugs.

Patient & Family Education
- Make position changes slowly and in stages to minimize risk of dizziness and fainting.
- Avoid hazardous activities until reaction to drug is known.
- Report unexplained skin eruptions or purple skin patches.
- If cataract surgery is planned, inform ophthalmologist that you are taking silodosin.

SILVER SULFADIAZINE ℞

(sul-fa-dye′a-zeen)

Silvadene
Classification: SULFONAMIDE
Therapeutic: TOPICAL ANTIINFECTIVE
Pregnancy Category: B

AVAILABILITY 1%/50 g cream

ACTION & *THERAPEUTIC EFFECT*

Silver salt is released slowly and exerts bactericidal effect only on bacterial cell membrane and wall, rather than by inhibiting folic acid synthesis. *Broad antimicrobial activity including many gram-negative and gram-positive bacteria and yeast.*

USES Prevention and treatment of sepsis in second- and third-degree burns.

CONTRAINDICATIONS Hypersensitivity to other sulfonamides; pregnant women at term.

CAUTIOUS USE Impaired kidney or liver function; porphyria; impaired respiratory function; G6PD deficiency; thrombocytopenia, leukopenia, hematological disease; pregnancy (category B), lactation; preterm infants; neonates younger than 2 mo.

ROUTE & DOSAGE

Burn Wound Treatment

Adult/Child: **Topical** Apply 1% cream 1–2 times/day to thickness of approximately 1.5 mm ($\frac{1}{16}$ in.)

ADMINISTRATION

Topical
- Do not use if cream darkens; it is water soluble and white.
- Apply with sterile, gloved hands to cleansed, debrided burned areas. Reapply cream to areas where it has been removed by patient activity; cover burn wounds with medication at all times.
- Bathe patient daily (in whirlpool or shower or in bed) as aid to debridement. Reapply drug.
- Note: Dressings are not required but may be used if necessary. Drug does not stain clothing.
- Store at room temperature away from heat.

ADVERSE EFFECTS (≥1%) **Body as a Whole:** Pain (occasionally), burning, itching, rash, reversible <u>leukopenia</u>. Potential for toxicity as for other sulfonamides if applied to extensive areas of the body surface.

INTERACTIONS Drug: PROTEOLYTIC ENZYMES are inactivated by silver in cream.

PHARMACOKINETICS Absorption: Not absorbed through intact skin, however, approximately 10% could be absorbed when applied to second- or third-degree burns. **Distribution:** Distributed into most body tissues. **Metabolism:** In the liver. **Elimination:** In urine.

NURSING IMPLICATIONS

Assessment & Drug Effects

- Observe for and report hypersensitivity reaction: Rash, itching, or burning sensation in unburned areas.
- Lab tests: Obtain serum sulfa concentrations, urinalysis, and kidney function tests when drug is applied to extensive areas. Significant quantities of drug may be absorbed.
- Observe patient for reactions attributed to sulfonamides.
- Note: Analgesic may be required. Occasionally, pain is experienced on application; intensity and duration depend on depth of burn.
- Continue treatment until satisfactory healing or burn site is ready for grafting, unless adverse reactions occur.

SIMVASTATIN

(sim-vah-sta′-tin)
Zocor
Classifications: HMG-COA REDUCTASE INHIBITOR (STATIN); ANTIHYPERLIPEMIC
Therapeutic: ANTILIPEMIC; STATIN
Prototype: Lovastatin
Pregnancy Category: X

AVAILABILITY 5 mg, 10 mg, 20 mg, 40 mg, 80 mg tablets

ACTION & *THERAPEUTIC EFFECT* Inhibitor of 3-hydroxy-3-methylglutaryl coenzyme A (HMG-CoA) reductase. HMG-CoA reductase inhibitors increase HDL cholesterol, and decrease LDL cholesterol, and total cholesterol synthesis. *Effectiveness indicated by decreased serum triglycerides, decreased LDL, cholesterol, and modest increases in HDL cholesterol.*

USES Hypercholesterolemia (alone or in combination with other agents), familial hypercholesterolemia. Reduces risk of CHD death and nonfatal MI and stroke.

CONTRAINDICATIONS Hypersensitivity to simvastatin; active liver disease or unexplained elevation of transaminase, hepatic encephalopathy, hepatitis, jaundice, AST or ALT of 3 × ULN; rhabdomyolysis, acute renal failure; cholestasis; myopathy; MS; pregnancy (category X), lactation. **CAUTIOUS USE** Homozygous familial hypercholesterolemia, history of liver disease, alcoholics; renal disease, renal impairment; DM; seizure disorder; children younger than 10 y.

ROUTE & DOSAGE

Hypercholesterolemia, Hyperlipidemia, CV Prevention
Adult: **PO** 10–20 mg each evening (max: 80 mg daily). *Adolescent/Child (older than 10 y and in females at least 1 y post-menarche):* **PO** 10 mg each night (may increase to 40 mg each night)
Renal Impairment Dosage Adjustment
CrCl less than 20 mL/min: Start with 5 mg each night

Common adverse effects in *italic*, life-threatening effects <u>underlined</u>; generic names in **bold**; classifications in SMALL CAPS; ♣ Canadian drug name; ☉ Prototype drug

1377

ADMINISTRATION

Oral

- Adjust dosage usually at 4-wk intervals.
- Give in the evening.
- Store at 15°–30° C (59°–86° F).

ADVERSE EFFECTS (≥1%) **CV:** Angina. **CNS:** Dizziness, headache, vertigo, asthenia, fatigue, insomnia. **GI:** Nausea, diarrhea, vomiting, abdominal pain, constipation, flatulence, heartburn, transient elevations in liver transaminases, transient elevations in CPK. **Body as a Whole:** Fatigue. **Respiratory:** Upper respiratory infection, rhinitis, cough.

INTERACTIONS Drug: Avoid use with **itraconazole, ketoconazole, erythromycin, clarithromycin, telithromycin,** PROTEASE INHIBITORS, **nefazodone.** Simvastatin dose will need to be decreased when given with **gemfibrozil, cyclosporine, danazol, amiodarone, verapamil, diltiazem.** May increase PT when administered with **warfarin; cyclosporine, gemfibrozil, fenofibrate, clofibrate,** antilipemic doses of **niacin, fluconazole, miconazole, sildenafil, tacrolimus,** may increase serum levels and increase risk of myopathy, rhabdomyolysis and acute kidney failure. Avoid use with **rifampin.** Use with **amiodarone** increases risk of rhabdomyolysis. Use with **amlodipine** increases risk of myopathy. **Food: Grapefruit juice** (greater than 1 qt/day) may increase risk of myopathy, rhabdomyolysis. **Herbal: St. John's wort** may decrease efficacy.

PHARMACOKINETICS Absorption: Rapidly from GI tract. **Onset:** 2 wk. **Peak:** 4–6 wk. **Distribution:** 95% protein bound; achieves high liver concentrations; crosses placenta. **Metabolism:** Extensive first-pass metabolism to its active metabolite. **Elimination:** 60% in bile and feces.

NURSING IMPLICATIONS

Assessment & Drug Effects

- Assess for and report unexplained muscle pain. Determine CPK level at onset of muscle pain.
- Lab tests: Obtain baseline and periodic (q6mo) LFTs during the first year and yearly thereafter. Monitor cholesterol levels throughout therapy.
- Monitor coagulation studies with patients receiving concurrent warfarin therapy. PT may be prolonged.

Patient & Family Education

- Report unexplained muscle pain, tenderness, or weakness, especially if accompanied by malaise or fever, to prescriber.
- Report signs of bleeding to prescriber promptly when taking concurrent warfarin.
- Moderate intake of grapefruit juice while taking this medication.

SIPULEUCEL-T

(sip-u-lew′cel)

Provenge

Classifications: BIOLOGIC RESPONSE MODIFIER; ACTIVE CELL IMMUNOTHERAPEUTIC; ANTINEOPLASTIC

Therapeutic: ANTINEOPLASTIC

Pregnancy Category: Undetermined. Not indicated for use in women.

AVAILABILITY 250 mL suspension containing 50 million autologous CD54+ cells activated with PAP-GM-CSF

ACTION & *THERAPEUTIC EFFECT*

While the mechanism of action is unknown, sipuleucel-T is designed to induce an immune response against prostatic acid phosphatase (PAP), an antigen expressed on most prostate cancer cells. *Stimulation of humoral and T-cell mediated responses are thought to slow tumor progression and improve survival.*

USES Prostate cancer.

CONTRAINDICATIONS Pregnancy category undetermined; not for use in women.

CAUTIOUS USE Concomitant use of chemotherapy; history of infusion reactions.

ROUTE & DOSAGE

Prostate Cancer
Adult: IV 250 mL infusion q2wk for a total of 3 doses

ADMINISTRATION

Intravenous
- Use universal precautions when handling sipuleucel-T.
- Note: Premedication is recommended with oral acetaminophen and an antihistamine (e.g., diphenhydramine) approximately 30 min prior to administration.

PREPARE: **IV Infusion:** ▪ Product is intended only for autologous use (i.e., patient donates to self). ▪ Open the outer cardboard shipping box to verify the product and patient-specific labels located on the top of the insulated container. ▪ Do NOT remove the insulated container from the shipping box, or open the lid of the insulated container, until the patient is ready for infusion. Inspect for leakage.

▪ Contents of bag will be slightly cloudy, with a cream-to-pink color. Gently tilt bag to resuspend contents; inspect for clumps and clots that should disperse with gentle manual mixing. ▪ Do not administer if the bag leaks during handling or if clumps remain in the bag.

ADMINISTER: **IV Infusion:** ▪ Infusion must be started prior to the expiration date and time on *Cell Product Disposition Form* and label. Give over 60 min. ▪ Administer through a dedicated line and do NOT use a cell filter. Ensure that entire contents of bag are infused. ▪ Monitor closely during and for 1 h after infusion for an infusion reaction (e.g., dyspnea, hypertension, tachycardia, chills, fever, nausea). Infusion may be stopped or slowed depending on severity of reaction. ▪ Do not reinitiate infusion with a bag held at room temperature for more than 3 h.

INCOMPATIBILITIES **Solution/ additive:** Do not mix with another drug. **Y-site:** Do not mix with another drug.

ADVERSE EFFECTS (≥ 5%) **Body as a Whole:** *Chills, citrate toxicity, fever,* hot flush, influenza-like illness, *pain,* peripheral edema. **CNS:** *Asthenia, dizziness, fatigue, headache,* insomnia, *paresthesia,* tremor. **GI:** Anorexia, *constipation,* diarrhea, *nausea, vomiting,* weight loss. **Hematologic:** *Anemia.* **Musculoskeletal:** *Back pain,* bone pain, *joint ache,* neck pain, *muscle ache,* muscle spasms, musculoskeletal chest pain, musculoskeletal pain, *pain in extremity.* **Respiratory:** Cough, dyspnea, upper respiratory infection. **Skin:** Rash, sweating. **Urogenital:** Hematuria, urinary tract infection.

S

Common adverse effects in *italic*, life-threatening effects underlined; generic names in **bold**; classifications in SMALL CAPS; ♣ Canadian drug name; ☺ Prototype drug

1379

INTERACTIONS Drug: Due to the ability of sipuleucel-T to stimulate the immune system, concomitant use of immunosuppressive agents (e.g., **corticosteroids**) may alter the efficacy and/or safety of sipuleucel-T.

NURSING IMPLICATIONS

Assessment & Drug Effects

▪ Observe closely during infusion and for at least 1 h following for an infusion reaction (see Administration). Monitor vital signs throughout observation period. Note that an acute infusion reaction is more likely after the second infusion.
▪ Report immediately to prescriber if an infusion reaction occurs.

Patient & Family Education

▪ Report immediately any of the following: Fever, chills, fatigue, breathing problems, dizziness, high blood pressure, palpitations, nausea, vomiting, headache, or muscle aches.

SIROLIMUS

(sir-o-li′mus)

Rapamune
Classifications: IMMUNOMODULATOR; IMMUNOSUPPRESSANT
Therapeutic: IMMUNOSUPPRESSANT
Prototype: Cyclosporine
Pregnancy Category: C

AVAILABILITY 0.5 mg, 1 mg, 2 mg tablets; 1 mg/mL oral solution

ACTION & *THERAPEUTIC EFFECT*

Immunomodulator structurally related to tacrolimus with immunosuppressive activity. Active in reducing a transplant rejection by inhibiting the response of helper T-lymphocytes and B-lymphocytes to cytokinesis [(interleukin) IL-2, IL-4, and IL-5]. *Inhibits antibody production and acute transplant rejection reaction in autoimmune disorders [e.g., systemic lupus erythematosus (SLE)]. Indicated by nonrejection of transplanted organ.*

USES Prophylaxis of kidney transplant rejection.

CONTRAINDICATIONS Hypersensitivity to sirolimus; lung or liver transplant patients; soya lecithin (soy fatty acids) hypersensitivity; lymphoma, neoplastic disease; females of childbearing age; lactation.
CAUTIOUS USE Hypersensitivity to or concurrent administration with tacrolimus; impaired renal function; renal transplant patients; dialysis patients, UV exposure, retransplant patients, multiorgan transplant recipients, African American transplant patients; viral or bacterial infection; hypertriglyceridemia, hyperlipidemia, DM, atrial fibrillation, CHF, hypervolemia, palpitations; hepatic disease; coronary artery disease; myelosuppression; liver disease; pregnancy (category C); children younger than 13 y.

ROUTE & DOSAGE

Kidney Transplant

Adult/Adolescent (over 40 kg):
PO 6 mg loading dose immediately after transplant, then 2 mg/day. Doses will need to be much higher if using cyclosporine or corticosteroids.
Adolescent (13 y or older, weight less than 40 kg): **PO** 3 mg/m^2 loading dose immediately after transplant, then 1 mg/m^2/day. Doses will need to be much higher if not on cyclosporine.

Hepatic Impairment Dosage Adjustment

Loading dose does not need to be modified. Reduce maintenance dose by 33% in moderate or mild impairment. In severe impairment, reduce by 50%.

ADMINISTRATION

Oral

- Give 4 h after oral cyclosporine.
- Tablets should be swallowed whole. They should not be crushed or chewed.
- Add prescribed amount of sirolimus oral solution to a glass containing 2 oz (60 mL) or more of water or orange juice (do not use any other type of liquid). Stir vigorously and administer immediately. Refill glass with 4 oz (120 mL) or more of water or orange juice. Stir vigorously and administer immediately.
- Give consistently with respect to amount and type of food.
- Refrigerate; protect from light; use multidose bottles within 1 mo of opening.

ADVERSE EFFECTS (≥1%) **Body as a Whole:** *Asthenia, back pain, chest pain, fever, pain, arthralgia;* flu-like syndrome; generalized edema; infection; lymphocele; malaise; <u>sepsis</u>, arthrosis, bone necrosis, leg cramps, myalgia, osteoporosis, tetany, abscess, ascites, cellulitis, chills, face edema, hernia, pelvic pain, peritonitis. **CNS:** *Insomnia, tremor, headache,* anxiety, confusion, depression, dizziness, emotional lability, hypertonia, hyperesthesia, hypotonia, neuropathy, paresthesia, somnolence. **CV:** *Hypertension,* atrial fibrillation, CHF, hypervolemia, hypotension, palpitation, peripheral vascular disorder, postural hypotension, syncope, tachycardia, thrombophlebitis, thrombosis, vasodilation.

GI: *Constipation, diarrhea, dyspepsia, nausea, vomiting, abdominal pain,* anorexia, dysphagia, eructation, esophagitis, flatulence, gastritis, gastroenteritis, gingivitis, gum hyperplasia, ileus, mouth ulceration, oral moniliasis, stomatitis, abnormal liver function tests. **Hematologic:** <u>Anemia</u>, <u>thrombocytopenia</u>, <u>leukopenia</u>, hemorrhage, ecchymosis, leukocytosis, lymphadenopathy, polycythemia, thrombotic, thrombocytopenic purpura. **Metabolic:** *Edema, hypercholesterolemia, hyperkalemia, hyperlipidemia, hypokalemia, hypophosphatemia, peripheral edema, weight gain,* Cushing's syndrome, diabetes, acidosis, hypercalcemia, hyperglycemia, hyperphosphatemia, hypocalcemia, hypoglycemia, hypomagnesemia, hyponatremia; increased LDH, alkaline phosphatase, BUN, creatine phosphokinase, ALT, or AST; weight loss. **Respiratory:** *Dyspnea, pharyngitis, upper respiratory tract infection,* asthma, atelectasis, bronchitis, cough, epistaxis, hypoxia, lung edema, pleural effusion, pneumonia, rhinitis, sinusitis. **Skin:** *Acne, rash,* fungal dermatitis, hirsutism, pruritus, skin hypertrophy, skin ulcer, sweating. **Urogenital:** *UTI,* albuminuria, bladder pain, dysuria, hematuria, hydronephrosis, impotence, kidney pain, nocturia, renal tubular necrosis, oliguria, pyuria, scrotal edema, incontinence, urinary retention, glycosuria. **Special Senses:** Abnormal vision, cataract, conjunctivitis, deafness, ear pain, otitis media, tinnitus.

INTERACTIONS Drug: Sirolimus concentrations increased by **clarithromycin, cyclosporine, diltiazem, erythromycin, ketoconazole, itraconazole, telithromycin;** sirolimus concentrations decreased by **rifabutin, rifampin;** VACCINES may be less effective with sirolimus; **tacrolimus** increases mortality, hepatic artery thrombosis, and

S

Common adverse effects in *italic*, life-threatening effects <u>underlined</u>; generic names in **bold**; classifications in SMALL CAPS; ♣ Canadian drug name; ⊙ Prototype drug

1381

graft loss. **Food: Grapefruit juice** significantly increases plasma levels. High fat meals increase levels. **Herbal: St. John's wort** decreases efficacy.

PHARMACOKINETICS Absorption: Rapidly with 14% bioavailability. **Peak:** 2 h. **Distribution:** 92% protein bound, distributes in high concentrations to heart, intestines, kidneys, liver, lungs, muscle, spleen, and testes. **Metabolism:** In liver (CYP3A4). **Elimination:** 91% in feces, 2.2% in urine. **Half-Life:** 62 h.

NURSING IMPLICATIONS

Assessment & Drug Effects
- Monitor for S&S of graft rejection.
- Control hyperlipidemia prior to initiating drug.
- Lab tests: Draw trough whole-blood sirolimus levels 1 h before a scheduled dose. Obtain periodic lipid profile, CBC with differential, fasting plasma glucose, blood chemistry, BUN, and creatinine (especially with other drugs known to cause renal impairment).

Patient & Family Education
- Avoid grapefruit juice within 2 h of taking sirolimus.
- Limit exposure to sunlight (UV exposure).
- Note: Decreased effectiveness possible for vaccines during therapy.
- Use or add barrier contraceptive before, during, and for 12 wk after discontinuing therapy.

SITAGLIPTIN ⊙

(sit-a-glip'tin)
Januvia
Classifications: ANTIDIABETIC; INCRETIN MODIFIER; DIPEPTIDYL PEPTIDASE-4 (DPP-4) INHIBITOR
Therapeutic: ANTIDIABETIC; HORMONE MODIFIER; DDP-4 INHIBITOR
Pregnancy Category: C

AVAILABILITY 25 mg, 50 mg, and 100 mg tablets

ACTION & *THERAPEUTIC EFFECT* Sitagliptin slows inactivation of incretin hormones [e.g., glucagon-like peptide-1 (GLP-1) and glucose-dependent insulinotropic polypeptide (GIP)] that are released by the intestine. As plasma glucose rises, incretin hormones stimulate release of insulin from the pancreas, and GLP-1 also lowers glucagon secretion, resulting in reduced hepatic glucose production. *Sitagliptin lowers both fasting and postprandial plasma glucose levels.*

USES Adjunct treatment of type 2 diabetes mellitus in combination with exercise and diet.

CONTRAINDICATIONS Type I diabetes mellitus, diabetic ketoacidosis.
CAUTIOUS USE Moderate to severe renal impairment, renal failure, hemodialysis; older adults; pregnancy (category C), lactation. Safe use in children younger than 18 y has not been established.

ROUTE & DOSAGE

Type 2 Diabetes Mellitus
Adult: **PO** 100 mg/day
Renal Impairment Dosage Adjustment
CrCl between 30 mL/min and 50 mL/min: 50 mg/day; *less than 30 mL/min:* 25 mg/day

ADMINISTRATION

Oral
- May be given without regard to meals.
- Note that dosage adjustment is recommended for moderate to severe renal impairment.
- Store at 20°–25° C (68°–77° F).

ADVERSE EFFECTS (≥1%) **CNS:** Headache. **Respiratory:** Nasopharyngitis, upper respiratory tract infection. **Endocrine:** Acute pancreatitis.

INTERACTIONS Drug: Sitagliptin may increase **digoxin** levels. QUINOLONES may increase blood glucose.

PHARMACOKINETICS Absorption: 87% absorbed. **Peak:** 1–4 h. **Distribution:** 38% protein bound. **Metabolism:** 20% metabolized in the liver. **Elimination:** Primarily renal (87%) with minor elimination in the kidneys. **Half-Life:** 12.4 h.

NURSING IMPLICATIONS

Assessment & Drug Effects

- Monitor for and report S&S of significant GI distress, including NV&D.
- Monitor for S&S of hypoglycemia when used in combination with a sulfonylurea drug or insulin.
- Lab tests: Baseline and periodic CrCl; periodic fasting and postprandial plasma glucose and HbA1C.
- Monitor blood levels of digoxin with concurrent therapy.

Patient & Family Education

- Note: When taken alone to control diabetes, sitagliptin is unlikely to cause hypoglycemia because it only works when your blood sugar is rising.

SODIUM BICARBONATE NA(HCO₃)

(sod′i-um bi-car′bon-ate)

Sodium Bicarbonate
Classifications: FLUID AND ELECTROLYTE BALANCE AGENT; ANTACID
Therapeutic: ANTACID
Pregnancy Category: C

AVAILABILITY 325 mg, 520 mg, 650 mg tablets; 4.2%, 5%, 7.5%, 8.4% injection

ACTION & *THERAPEUTIC EFFECT* Rapidly neutralizes gastric acid to form sodium chloride, carbon dioxide, and water. After absorption of sodium bicarbonate, plasma alkali reserve is increased and excess sodium and bicarbonate ions are excreted in urine, thus rendering urine less acid. *Short-acting, potent systemic antacid; rapidly neutralizes gastric acid or systemic acidosis.*

USES Systemic alkalinizer to correct metabolic acidosis (as occurs in diabetes mellitus, shock, cardiac arrest, or vascular collapse), to minimize uric acid crystallization associated with uricosuric agents, to increase the solubility of sulfonamides, and to enhance renal excretion of barbiturate and salicylate overdosage. Commonly used as home remedy for relief of occasional heartburn, indigestion, or sour stomach. Used topically as paste, bath, or soak to relieve itching and minor skin irritations such as sunburn, insect bites, prickly heat, poison ivy, sumac, or oak. Sterile solutions are used to buffer acidic parenteral solutions to prevent acidosis. Also as a buffering agent in many commercial products (e.g., mouthwashes, douches, enemas, ophthalmic solutions).

CONTRAINDICATIONS Prolonged therapy with sodium bicarbonate; patients losing chloride (as from vomiting, GI suction, diuresis); hypocalcemia; metabolic alkalosis; respiratory alkalosis; peptic ulcer.
CAUTIOUS USE Edema, sodium-retaining disorders; heart disease, hypertension; preexisting respiratory acidosis; renal disease, renal insufficiency; hyperkalemia, hypokalemia; older adults; pregnancy (category C).

Common adverse effects in *italic*, life-threatening effects underlined; generic names in **bold**; classifications in SMALL CAPS; ♦ Canadian drug name; ⊘ Prototype drug

1383

ROUTE & DOSAGE

Antacid

Adult: **PO** 0.3–2 g 1–4 times/day or ½ tsp of powder in glass of water

Urinary Alkalinizer

Adult: **PO** 4 g initially, then 1–2 g q4h
Child: **PO** 84–840 mg/kg/day in divided doses

Cardiac Arrest

Adult: **IV** 1 mEq/kg initially, then 0.5 mEq/kg q10min depending on arterial blood gas determinations (8.4% solutions contain 50 mEq/50 mL), give over 1–2 min
Child: **IV** 0.5–1 mEq/kg q10min depending on arterial blood gas determinations, give over 1–2 min

Metabolic Acidosis

Adult/Child: **IV** Dose adjusted according to pH, base deficit, $PaCO_2$, fluid limits, and patient response

ADMINISTRATION

Oral

- Do not add oral preparation to calcium-containing solutions.

Intravenous

PREPARE: **Direct/IV Infusion** May give 4.2% (0.5 mEq/mL) and 5% (0.595 mEq/mL) NaHCO₃ solutions undiluted. ▪ Dilute 7.5% (0.892 mEq/mL) and 8.4% (1 mEq/mL) solutions with compatible IV solutions to a maximum concentration of 0.5 mEq/mL. ▪ For infants and children, dilute to at least 4.2%.

ADMINISTER: **Direct:** Give a bolus dose over 1–2 min only in emergency situations. ▪ For neonates or infants younger than 2 y, use only 4.2% solution for direct IV injection. **IV Infusion:** Usual rate is 2–5 mEq/kg over 4–8 h; do not exceed 50 mEq/h. ▪ Flush line before/after with NS. ▪ Stop infusion immediately if extravasation occurs. Severe tissue damage has followed tissue infiltration.

INCOMPATIBILITIES **Solution/additive: Alcohol 5%, lactated Ringer's, amoxicillin, ascorbic acid, bupivacaine, carboplatin, carmustine, ciprofloxacin, cisplatin, codeine, corticotropin, dobutamine, dopamine, epinephrine, glycopyrrolate, hydromorphone, imipenem-cilastatin, insulin, isoproterenol, labetalol, levorphanol, magnesium sulfate, meperidine, meropenem, methadone, metoclopramide, morphine, norepinephrine, oxytetracycline, penicillin G, pentazocine, pentobarbital, phenobarbital, procaine, promazine, streptomycin, succinylcholine, tetracycline, thiopental, vancomycin, vitamin B complex with C. Y-site: Allopurinol, amiodarone, amphotericin B cholesteryl complex, calcium chloride, ciprofloxacin, cisatracurium, diltiazem, doxorubicin liposome, fenoldopam, hetastarch, idarubicin, imipenem/cilastatin, leucovorin, lidocaine, midazolam, milrinone acetate, nalbuphine, ondansetron, oxacillin, sargramostim, verapamil, vincristine, vindesine, vinorelbine.**

- Store in airtight containers. ▪ Note expiration date.

ADVERSE EFFECTS (≥1%) **GI:** *Belching, gastric distention,* flatulence. **Metabolic:** Metabolic alkalosis; electrolyte imbalance: Sodium overload (pulmonary edema), hypocalcemia (tetany), hypokalemia, milk-alkali syndrome, dehydration. **Other:** Rapid IV in neonates (hypernatremia, reduction in CSF pressure, <u>intracranial hemorrhage</u>). **Skin:** Severe tissue damage following extravasation of IV solution. **Urogenital:** Renal calculi or crystals, impaired kidney function.

DIAGNOSTIC TEST INTERFERENCE Small increase in *blood lactate* levels (following IV infusion of sodium bicarbonate); false-positive *urinary protein* determinations (using *ames reagent, sulfacetic acid,* heat and *acetic acid* or *nitric acid ring method*); elevated *urinary urobilinogen* levels (*urobilinogen* excretion increases in alkaline urine).

INTERACTIONS Drug: May decrease absorption of **ketoconazole;** may decrease elimination of **dextroamphetamine, ephedrine, pseudoephedrine, quinidine;** may increase elimination of **chlorpropamide, lithium,** SALICYLATES, TETRACYCLINES.

PHARMACOKINETICS Absorption: Readily from GI tract. **Onset:** 15 min. **Duration:** 1–2 h. **Elimination:** In urine within 3–4 h.

NURSING IMPLICATIONS

Assessment & Drug Effects

- Be aware that long-term use of oral preparation with milk or calcium can cause milk-alkali syndrome: Anorexia, nausea, vomiting, headache, mental confusion, hypercalcemia, hypophosphatemia, soft tissue calcification, renal and ureteral calculi, renal insufficiency, metabolic alkalosis.
- Lab tests: Urinary alkalinization: Monitor urinary pH as a guide to dosage (pH testing with nitrazine paper may be done at intervals throughout the day and dosage adjustments made accordingly).
- Lab tests: Metabolic acidosis: Monitor patient closely by observations of clinical condition; measurements of acid-base status (blood pH, Po_2, Pco_2, HCO_3^-, and other electrolytes, are usually made several times daily during acute period). Observe for signs of alkalosis (over treatment) (see Appendix F).
- Observe for and report S&S of improvement or reversal of metabolic acidosis (see Appendix F).

Patient & Family Education

- Do not use sodium bicarbonate as antacid. A nonabsorbable OTC alternative for repeated use is safer.
- Do not take antacids longer than 2 wk except upon advice and supervision of a prescriber. Self-medication with routine doses of sodium bicarbonate or soda mints may cause sodium retention and alkalosis, especially when kidney function is impaired.
- Be aware that commonly used OTC antacid products contain sodium bicarbonate: Alka-Seltzer, Bromo-Seltzer, Gaviscon.

SODIUM FERRIC GLUCONATE COMPLEX

(so′di-um fer′ric glu′co-nate)

Ferrlecit

Classifications: NUTRITIONAL SUPPLEMENT; IRON PREPARATION
Therapeutic: ANTIANEMIC; IRON REPLACEMENT
Prototype: Ferrous sulfate
Pregnancy Category: B

AVAILABILITY 62.5 mg elemental iron/5 mL ampule

ACTION & *THERAPEUTIC EFFECT* Stable iron complex used to restore iron loss in chronic kidney failure

Common adverse effects in *italic*, life-threatening effects <u>underlined</u>; generic names in **bold**; classifications in SMALL CAPS; ♣ Canadian drug name; ⊘ Prototype drug

1385

patients. The use of erythropoietin therapy and blood loss through hemodialysis require iron replacement. The ferric ion combines with transferrin and is transported to bone marrow where it is incorporated into hemoglobin. *Effectiveness indicated by improved Hgb and Hct, iron saturation, and serum ferritin levels.*

USES Treatment of iron deficiency in patients on chronic hemodialysis and receiving erythropoietin therapy.

CONTRAINDICATIONS Any anemia not related to iron deficiency; hypersensitivity to sodium ferric gluconate complex; hemochromatosis, hemosiderosis; hemolytic anemia; thalassemia; neonates.

CAUTIOUS USE Hypersensitivity to benzyl alcohol; active or suspected infection; cardiac disease; hepatic disease; older adults; pregnancy (category B), lactation. Safety and efficacy in children younger than 6 y are not established.

ROUTE & DOSAGE

Iron Deficiency in Dialysis Patients
Adult: **IV** 125 mg infused over 1 h *Child (older than 6 y):* **IV** 1.5 mg/kg infused over 1 h (max: 125 mg/dose)

ADMINISTRATION

Intravenous

PREPARE: **Direct for Adult:** May be given undiluted. **Direct for Child:** Dilute required doses in 25 mL NS. **IV Infusion for Adult/Child:** Dilute 125 mg in 100 mL of NS. ▪ Use immediately after dilution. *ADMINISTER:* **Direct for Adult:** Give no faster than 12.5 mg/min. **IV Infusion for Adult/Child:** Give over NOT less than 60 min.

INCOMPATIBILITIES **Solution/additive:** Do not mix with any other medications or add to parenteral nutrition solutions.

▪ Store unopened ampules at 20°–25° C (68°–77° F).

ADVERSE EFFECTS (≥1%) **Body as a Whole:** <u>Hypersensitivity reaction (cardiovascular collapse, cardiac arrest, bronchospasm, oral/pharyngeal edema,</u> dyspnea, <u>angioedema,</u> urticaria, pruritus). **CV:** Flushing, hypotension.

PHARMACOKINETICS Not studied.

NURSING IMPLICATIONS

Assessment & Drug Effects

▪ Monitor closely for S&S of severe hypersensitivity (see Appendix F) during IV administration.
▪ Monitor vital signs periodically during IV administration (transient hypotension possible especially during dialysis).
▪ Stop infusion immediately and notify prescriber if hypersensitivity is suspected.
▪ Lab tests: Periodic Hgb, Hct, Fe saturation, serum ferritin.

Patient & Family Education

▪ Report to prescriber immediately: Difficulty breathing, itching, flushing, rash, weakness, light-headedness, pain, or any other discomfort during infusion.

SODIUM FLUORIDE

(sod'i-um)

Fluorinse, Fluoritab, Flura-Drops, Karidium, Pediaflor, Point-Two, Thera-Flur-N
Classifications: ELECTROLYTE REPLACEMENT; DENTAL PROPHYLACTIC
Therapeutic: DENTAL PROPHYLACTIC
Pregnancy Category: B (topical); C (oral)

Common adverse effects in *italic*, life-threatening effects <u>underlined</u>; generic names in **bold**; classifications in SMALL CAPS; ♣ Canadian drug name; ⊕ Prototype drug

AVAILABILITY 0.25 mg, 0.5 mg, 1 mg tablets; 0.125 mg, 0.25 mg, 0.5 mg drops; 0.2 mg/mL solution; 0.02%, 0.04%, 0.09%, 2% rinse; 0.5%, 1.2% gel

ACTION & *THERAPEUTIC EFFECT*

Source of the fluorine ion, a trace element. Incorporates into developing tooth enamel, hardens surfaces, and increases resistance to cariogenic microbial processes. Topical application reduces acid production by bacteria in dental plaque and promotes remineralization of acid-damaged enamel. Application to exposed root surfaces supports formation of insoluble materials within dentinal tubules, thereby blocking transport of offending stimuli. *Oral form stimulates osteoblastic activity leading to increased bone mass. Topical application reduces acid production by bacteria in dental plaque and promotes remineralization of enamel.*

USES When fluoride ion concentration in drinking water is 0.7 ppm or less, to prevent periodontal disease and dental caries, to treat dental cervical hypersensitivity, and to control dental caries associated with xerostomia.

UNLABELED USES Adjunctive treatment of osteoporosis; management of bone lesions in multiple myeloma; to reduce bone pain in patient with metastatic prostatic carcinoma; to stabilize progression of hearing loss in a limited number of patients with otosclerosis.

CONTRAINDICATIONS When daily intake of fluoride from drinking water exceeds 0.7 ppm; low-sodium or sodium-free diets; rheumatoid arthritis; hypersensitivity to fluoride; GI disease.

CAUTIOUS USE Renal dysfunction; rheumatoid arthritis; arthralgia;

topical: pregnancy (category B); **oral:** pregnancy (category C), lactation; gels or dental rinses by children younger than 6 y, 1 mg tablet or rinse in children younger than 3 y, or 1 mg rinse in children younger than 6 y.

ROUTE & DOSAGE

Prevent Periodontal Disease (Drinking Water Concentration Less Than 0.3 ppm)

Child: **PO** *Birth–2 y,* 0.25 mg/day; *2–3 y,* 0.5 mg/day; *3–13 y,* 1 mg/day

Prevent Periodontal Disease (Drinking Water Concentration 0.3–0.7 ppm)

Child: **PO** *Birth–2 y,* 0.125 mg/day; *2–3 y,* 0.25 mg/day; *3–13 y,* 0.5 mg/day

Prevent Dental Caries

Child: **Topical** *6–12 y,* 5 mL of 0.2% solution daily; *older than 12 y,* 10 mL of 0.2% solution daily

Desensitization of Exposed Root Surfaces

Child: **Topical** 0.2% rinsing solution once nightly after brushing and flossing

ADMINISTRATION

Oral

- Avoid giving sodium fluoride with milk or dairy products. Calcium from these products combines with fluorine, decreasing its absorption.
- Give drops preferably after meals. Give undiluted or mixed with fluids or foods.
- Dissolve tablets in the mouth or chew before swallowing. Administer at bedtime (after brushing the teeth).

S

Common adverse effects in *italic*, life-threatening effects <u>underlined</u>; generic names in **bold**; classifications in SMALL CAPS; ✚ Canadian drug name; ⊙ Prototype drug

1387

Topical

- Apply all fluorine preparations after thoroughly brushing and flossing; preferably at bedtime.
- Do not swallow topical or rinse preparations.
- If patient's mouth is sore, the neutral preparation (Thera-Flur N) is better tolerated.
- Use as treatment for dental hypersensitivity: Thoroughly brush teeth; then swish PO solution around and between teeth for 1 min; expectorate. If gel is used, apply a few drops to toothbrush and brush gently onto affected surfaces.
- Apply Gel-drops with applicators supplied by the dentist. Spread gel on inner surfaces of applicators, which are placed over lower and upper teeth at the same time. User bites down lightly for 6 min, then removes applicators and rinses mouth thoroughly. Applicators are cleaned with cold water.
- Store all forms in tight plastic or paraffin-lined glass containers (sodium fluoride reacts with ordinary glass at a slow but appreciable rate) at 15°–30° C (59°–86° F). Avoid freezing.

ADVERSE EFFECTS (≥1%) **Skin:** Rash, atopic dermatitis, urticaria, stomatitis. **Body as a Whole:** GI and respiratory allergic reactions, salty or soapy taste, dehydration, thirst, excessive salivation, muscle weakness, tremors, <u>shock, death from cardiac and respiratory failure</u>. **Musculoskeletal:** Dental fluorosis (brown or white mottling of tooth enamel), osseous fluorosis (patchy mineralization and possible decrease in bone strength).

INTERACTIONS Drug: Aluminum, calcium, magnesium-containing products may decrease **fluoride** absorption.

PHARMACOKINETICS Absorption: Readily from GI tract. **Distribution:** Fluoride is stored in bones and teeth; crosses placenta; distributed into breast milk. **Elimination:** Rapidly excreted, primarily in urine with small amounts in feces.

NURSING IMPLICATIONS

Assessment & Drug Effect
- Monitor therapeutic effectiveness.

Patient & Family Education
- Do not eat, drink, or rinse mouth for at least 30 min after using the rinsing solution.
- Do not exceed recommended dosage. If mottling of teeth occurs, notify dentist.
- Apply sodium fluoride gel or solution used in orthodontic treatment regimen immediately before attachment or reattachment of the tooth-encircling bands.
- To be effective, fluoride supplementation **must be** consistent and continuous from infancy until 12–14 y.
- Consult dentist about continuing fluoride therapy if you move or there is a change in water supply (mottling may occur if drinking water has fluoride content greater than 1.5 ppm).

SODIUM OXYBATE (GHB)
(sod'i-um ox'y-bate)
Xyrem
Classification: CENTRAL NERVOUS SYSTEM (CNS) DEPRESSANT
Therapeutic: CNS DEPRESSANT
Pregnancy Category: C
Controlled Substance: Schedule III

AVAILABILITY 500 mg/mL solution

ACTION & *THERAPEUTIC EFFECT*
The precise mechanism by which sodium oxybate produces CNS depression including profound decreases

in level of consciousness, with instances of coma and death is not understood. Additionally the mechanism of anticataplexy in narcolepsy is also unknown. Sodium oxybate is GHB, a known drug of abuse. *Produces anticataplectic effects in narcolepsy and decreases the number of cataplexy events in individuals with narcolepsy. Also has sedative and amnestic properties.*

USES Treatment of cataplexy in patients with narcolepsy.

CONTRAINDICATIONS Alcohol or sedative-hypnotics or other CNS depressants; psychosis; coma; eclampsia; patients with succinic semialdehyde dehydrogenase deficiency; compromised respiratory drive, severe depression, or suicidal ideation.

CAUTIOUS USE Hepatic dysfunction; compromised respiratory function; sleep disorders; history of seizures; heart failure, hypertension, impaired hepatic or renal function; previous history of depressive illness or suicide attempt; sleep-walking; older adults; pregnancy (category C); children younger than 16 y.

ROUTE & DOSAGE

Cataplexy

Adult: **PO** Start with 2.25 g given at bedtime while in bed and repeated 2.5–4 h later. Dose may be increased by 1.5 g/day every 2 wk to a max of 9 g/day in 2 divided doses.

Hepatic Impairment Dosage Adjustment

Reduce dose by 50% in patients with hepatic impairment

ADMINISTRATION

Oral

- Give at bedtime at least 2–3 h after the evening meal.
- Dilute each dose with 2 oz (60 mL) of water in the dosing cups provided.
- Instruct patient to remain in bed after taking sodium oxybate.
- Discard any diluted dose that has not been used within 24 h.
- Store at 15°–30° C (59°–86° F).

ADVERSE EFFECTS (≥1%) **Body as a Whole:** *Pain, infection,* flu-like syndrome, asthenia, allergic reactions, chills. **CNS:** Confusion, depression, sleepwalking, *headache, dizziness, somnolence,* nervousness, abnormal dreams, insomnia, agitation, ataxia, convulsion, stupor, tremor. **CV:** Hypertension. **GI:** *Nausea,* diarrhea, vomiting, dyspepsia, abdominal pain, anorexia, constipation. **Metabolic:** Increased alkaline phosphatase, edema, hypercholesteremia, hypocalcemia, weight gain. **Respiratory:** *Pharyngitis,* rhinitis, sinusitis **Skin:** Increased sweating, acne, alopecia, rash. **Special Senses:** Amblyopia, tinnitus. **Urogenital:** Urinary incontinence, dysmenorrhea, albuminuria, cystitis, hematuria, metrorrhagia, urinary frequency.

INTERACTIONS Drug: Alcohol, SEDATIVE-HYPNOTICS, other CNS DEPRESSANTS may increase CNS depressant effects. **Food:** High-fat meal will significantly reduce absorption.

PHARMACOKINETICS Absorption: Incompletely absorbed, 25% reaches systemic circulation. **Peak:** 0.05–1.25 h. **Metabolism:** Oxidized in the Krebs' cycle to carbon dioxide and water. **Elimination:** Primarily eliminated as carbon dioxide in respiration. **Half-Life:** 0.5–1 h.

NURSING IMPLICATIONS

Assessment & Drug Effects

- Monitor for and report immediately any of the following: Seizure,

Common adverse effects in *italic*, life-threatening effects underlined; generic names in **bold**; classifications in SMALL CAPS; ✤ Canadian drug name; ☻ Prototype drug

1389

respiratory depression, or decreased level of consciousness.
- Monitor closely patients with hepatic insufficiency for adverse events.
- Monitor for and report excessive weight gain and development of edema.
- Lab tests: Perform baseline LFTs; monitor periodically serum electrolytes and lipid profile.

Patient & Family Education
- Do not take sodium oxybate at any time other than at night, immediately before bedtime.
- Be consistent with timing of the evening meal and take this drug at least 2–3 h after eating.
- Prepare both doses prior to bedtime. After ingesting each dose remain in bed.
- Do not consume alcohol or use other sedative hypnotic drugs with sodium oxybate.
- Do not drive or engage in potentially hazardous activities until reaction to drug is known.

SODIUM POLYSTYRENE SULFONATE

(pol-ee-stye'reen)

Kayexalate, SPS Suspension
Classifications: ELECTROLYTE AND WATER BALANCE; CATION EXCHANGE
Therapeutic: CATION EXCHANGE
Pregnancy Category: C

AVAILABILITY 15 g/60 mL suspension; 100 mg/g powder

ACTION & *THERAPEUTIC EFFECT*
Sulfonic cation-exchange resin that removes potassium by exchanging sodium ion for potassium, particularly in large intestine; potassium-containing resin is then excreted through the bowel. *Removes potassium by exchanging sodium ion for potassium through the large intestine.*

USES Hyperkalemia.

CONTRAINDICATIONS Hypersensitivity to Kayexalate; hypokalemia; GI obstruction; hypokalemia; lactation.
CAUTIOUS USE Acute or chronic kidney failure; low birth weight infants; neonates with reduced gut; patients receiving digitalis preparations; patients who cannot tolerate even a small increase in sodium load (e.g., CHF, severe hypertension, and marked edema); older adults; pregnancy (category C).

ROUTE & DOSAGE

Hyperkalemia
Adult: **PO** 15 g suspended in 70% sorbitol or 20–100 mL of other fluid 1–4 times/day **Rectal** 30–50 g/100 mL 70% sorbitol q6h as warm emulsion high into sigmoid colon
Child: **PO** Calculate appropriate amount on exchange rate of 1 mEq of potassium per gram of resin and suspend in 70% sorbitol or other appropriate solution (usual dose: 1 g/kg q6h) **Rectal** 1 g/kg q2–6h

ADMINISTRATION

Oral
- Give as a suspension in a small quantity of water or in syrup. Usual amount of fluid ranges from 20–100 mL or approximately 3–4 mL/g of drug.

Rectal
- Use warm fluid (as prescribed) to prepare the emulsion for enema.
- Administer at body temperature and introduce by gravity, keeping suspension particles in solution

S

by stirring. Flush suspension with 50–100 mL of fluid; then clamp tube and leave it in place.

- Urge patient to retain enema at least 30–60 min but as long as several hours if possible.
- Irrigate colon (after enema solution has been expelled) with 1 or 2 qt flushing solution (non-sodium containing). Drain returns constantly through a Y-tube connection.
- Store remainder of prepared solution for 24 h; then discard.

ADVERSE EFFECTS (≥1%) **GI:** *Constipation, fecal impaction (in older adults);* anorexia, gastric irritation, nausea, vomiting, diarrhea (with sorbitol emulsions). **Metabolic:** Sodium retention, hypocalcemia, hypokalemia, hypomagnesemia.

INTERACTIONS Drug: ANTACIDS, LAXATIVES containing **calcium** or **magnesium** may decrease potassium exchange capability of the resin.

PHARMACOKINETICS Absorption: Not absorbed systemically. **Onset:** Several hours to days. **Metabolism:** Not metabolized. **Elimination:** In feces.

NURSING IMPLICATIONS

Assessment & Drug Effects

- Lab tests: Determine serum potassium levels daily throughout therapy. Monitor acid–base balance, electrolytes, and minerals in patients receiving repeated doses.
- Serum potassium levels do not always reflect intracellular potassium deficiency. Observe patient closely for early clinical signs of severe hypokalemia (see Appendix F). ECGs are also recommended.
- Consult prescriber about restricting sodium content from dietary and other sources since drug contains approximately 100 mg

(4.1 mEq) of sodium per g (1 tsp, 15 mEq sodium).

Patient & Family Education

- Check bowel function daily. Usually, a mild laxative is prescribed to prevent constipation (common adverse effect). Older adult patients are particularly prone to fecal impaction.

SOLIFENACIN SUCCINATE
(sol-i-fen′a-sin)
VESIcare
Classifications: ANTICHOLINERGIC; ANTIMUSCARINIC; ANTISPASMODIC
Therapeutic: ANTISPASMODIC
Prototype: Oxybutynin
Pregnancy Category: C

AVAILABILITY 5 mg, 10 mg tablets

ACTION & *THERAPEUTIC EFFECT*
Solifenacin is a selective muscarinic antagonist that depresses both voluntary and involuntary bladder contractions caused by detrusor overactivity. *Solifenacin improves the volume of urine per void and reduces the frequency of incontinent and urgency episodes.*

USES Treatment of overactive bladder (OAB).

CONTRAINDICATIONS Hypersensitivity to solifenacin; severe hepatic impairment; gastric retention; uncontrolled narrow-angle glaucoma; urinary retention; toxic megacolon; GI obstruction; ileus; GERD; lactation.
CAUTIOUS USE Bladder outflow obstruction; concurrent use of ketoconazole or other potent CYP3A4 inhibitors; obstructive disorders; decreased GI motility; severe hepatic impairment; history of QT prolongation or concurrent use of medications known to prolong the QT

Common adverse effects in *italic*, life-threatening effects underlined; generic names in **bold**; classifications in SMALL CAPS; ✦ Canadian drug name; ⊘ Prototype drug

1391

interval; controlled narrow-angle glaucoma; renal impairment; renal disease; renal failure; mild to moderate hepatic impairment; older adults; pregnancy (category C).

ROUTE & DOSAGE

Overactive Bladder

Adult: PO 5 mg once daily; may be increased to 10 mg once daily if tolerated (max: 5 mg/day if taking drugs that inhibit CYP3A4—see INTERACTIONS, Drug)

Hepatic Impairment Dosage Adjustment

If moderate hepatic impairment, do not exceed 5 mg/day. If severe hepatic impairment, do not use.

Renal Impairment Dosage Adjustment

CrCl less than 30 mL/min: Max dose 5 mg/day

ADMINISTRATION

Oral

- Tablets should be swallowed whole.
- Store at 15°–30° C (59°–86° F).

ADVERSE EFFECTS (≥1%) **Body as a Whole:** Edema, fatigue. **CNS:** Dizziness, depression. **CV:** Hypertension. **GI:** *Dry mouth, constipation,* nausea, vomiting, dyspepsia, upper abdominal pain. **Respiratory:** Cough. **Special Senses:** Blurred vision, dry eyes. **Urogenital:** Urinary tract infection, urinary retention.

INTERACTIONS Drug: CYP3A4 IN-HIBITORS (e.g., **clarithromycin, delavirdine, diltiazem, efavirenz, erythromycin, fluconazole, fluvoxamine, itraconazole, nefazodone, norfloxacin, omeprazole,** PROTEASE INHIBITORS, **quinine, verapamil, troleandomycin, voricon-azole, zafirlukast**) may increase levels and toxicity (max dose: 5 mg/day); **amantadine, amoxapine, bupropion, clozapine, cyclobenzaprine, diphenhydramine, disopyramide, maprotiline, olanzapine, orphenadrine,** PHE-NOTHIAZINES, TRICYCLIC ANTIDEPRESSANTS have additive anticholinergic adverse effects. **Food: Grapefruit juice** may increase solifenacin levels and toxicity.

PHARMACOKINETICS Absorption: 90% absorbed from GI tract. **Peak:** 3–8 h. **Metabolism:** Extensively metabolized in the liver by CYP3A4. **Elimination:** Primarily in urine, 22% in feces. **Half-Life:** 45–68 h.

NURSING IMPLICATIONS

Assessment & Drug Effects

- Monitor bladder function and report promptly urinary retention.
- Monitor ECG in patients with a known history of QT prolongation or patients taking medications that prolong the QT interval.

Patient & Family Education

- Stop taking this drug and report to prescriber if urinary retention occurs.
- Report promptly any of the following: Blurred vision or difficulty focusing vision, palpitations, confusion, or severe dizziness.
- Report to prescriber problems with bowel elimination, especially constipation lasting 3 days or longer.
- Exercise caution in hot environments, as the risk of heat prostration increases with this drug.

SOMATROPIN ⊕

(soe-ma-troe'pin)

Accretropin, Genotropin, Genotropin Miniquick, Humatrope, Norditropin, Nutropin,

S

Nutropin AQ, NuSpin, Omnitrope, Saizen, Serostim, Tev-Tropin, Zorbtive
Classification: GROWTH HORMONE
Therapeutic: GROWTH HORMONE
Pregnancy Category: B or C (depending on the brand)

AVAILABILITY 0.2 mg, 0.4 mg, 0.6 mg, 0.8 mg, 1 mg, 1.2 mg, 1.4 mg, 1.5 mg, 1.6 mg, 4 mg, 5 mg, 6 mg, 8 mg, 10 mg, 12 mg, 24 mg injection; 5 mg/1.5 mL, 15 mg/1.5 mL prefilled syringe

ACTION & *THERAPEUTIC EFFECT*
Recombinant growth hormone with the natural sequence of 191 amino acids characteristic of endogenous growth hormone (GH), of pituitary origin. *Induces growth responses in children.*

USES Growth failure due to GH deficiency; replacement therapy prior to epiphyseal closure in patients with idiopathic GH deficiency; GH deficiency secondary to intracranial tumors or panhypopituitarism; inadequate GH secretion; short stature in girls with Turner's syndrome; AIDS wasting syndrome; short bowel syndrome; small for gestational age.

UNLABELED USES Growth deficiency in children with rheumatoid arthitis.

CONTRAINDICATIONS Hypersensitivity to somatropin, growth hormone or any ingredient; patient with closed epiphyses; underlying progressive intracranial tumor, acute critical illness including open heart surgery, abdominal surgery, respiratory failure; diabetic retinopathy; respiratory insufficiency; during chemotherapy, radiation therapy, active neoplastic disease; selective cases or Prader-Willi syndrome; children with an intracranial tumor; intracranial hypertension; untreated hypothyroidism, obesity.

CAUTIOUS USE Diabetes mellitus or family history of the disease, history of glucose intolerance; concurrent use of insulin or antihypertensive agents; Prader-Willi syndrome; skeletal abnormalities; history of upper airway obstruction, sleep apnea, or unidentified URI; concomitant or prior use of thyroid or androgens in prepubertal male; hypothyroidism; chronic renal failure; pregnancy (category B or category C depending on the brand), lactation.

ROUTE & DOSAGE

Note: Dosing will vary with specific products
Growth Hormone Deficiency
Adult: **Subcutaneous Humatrope** 0.006 mg/kg (0.018 international unit/kg) daily, may increase [max: 0.0125 mg/kg/day (0.0375 international unit/kg/day)]; **Nutropin, Nutropin AQ** 0.006 mg/kg daily (max: *younger than 35 y,* 0.025 mg/kg/day; *older than 35 y,* 0.0125 mg/kg/day) **Genotropin, Omnitrope** 0.04 mg/kg qwk divided into daily doses (max: 0.08 mg/kg/wk) **Norditropin** Initially 0.004 mg/kg/day (max: 0.016 mg/kg) *Child:* **Subcutaneous Genotropin** 0.16–0.24 mg/kg/wk divided into 6–7 daily doses; **Humatrope** 0.18 mg/kg/wk (0.54 international unit/kg/wk) divided into equal doses given on either 3 alternate days or 6 times/wk; **Norditropin** 0.024–0.034 mg/kg/day 6–7 times/wk; **Nutropin, Nutropin AQ** 0.3 mg/kg/wk (0.9 international unit/kg/wk) divided

S

Common adverse effects in *italic*, life-threatening effects <u>underlined</u>; generic names in **bold**; classifications in SMALL CAPS; ♣ Canadian drug name; ✪ Prototype drug

1393

into 6–7 daily doses **Omnitrope**
0.16–0.24 mg/kg/week divided
into 6–7 injections

**Inadequate Growth Hormone
Secretion**

Child: **Subcutaneous Nutropin,
Nutropin AQ** 0.3 mg/kg every
week **Accretropin** 0.18–0.3 mg/
kg/wk divided into daily doses
Genotropin 0.16–0.24 mg/kg/
wk divided into daily doses
**Subcutanous or IM Humatrop,
Saizen** 0.18 mg/kg/wk in
divided doses

Small for Gestational Age

Child: **Subcutaneous Omnitrope**
0.48 mg/kg/week in divided
doses

AIDS Wasting or Cachexia

Adult: **SC Serostim** *Weight greater
than 55 kg,* 6 mg each night;
weight 45–55 kg, 5 mg each
night; *weight 35–45 kg,* 4 mg
each night; *weight less than 35
kg,* 0.1 mg/kg each night
Child (6–17 y): **SC** 0.04–0.07
mg/kg/day

Short Bowel Syndrome

Adult: **SC Zorbtive** 0.1 mg/kg once
daily for 4 wk (max: 8 mg/day)

ADMINISTRATION

Subcutaneous

- Reconstitute each brand following
  its manufacturer's instructions
  (vary from brand to brand).
- Read and carefully follow directions for use supplied with
  the Nutropin AQ Pen™ cartridge if this is the product being used.
- Rotate injection sites; abdomen
  and thighs are preferred sites. Do
  not use buttocks until the child has
  been walking for a year or more

and the muscle is adequately
developed.

- Store lyophilized powder at
  2°–8° C (36°–46° F). After reconstitution, most preparations
  are stable for at least 14 days
  under refrigeration. DO NOT
  FREEZE.

ADVERSE EFFECTS (≥1%) **Body as
a Whole:** Pain, swelling at injection
site; myalgia. Fatalities reported in
patients with Prader-Willi syndrome
and one or more of the following:
Severe obesity, history of respiratory
impairment or sleep apnea, or unidentified respiratory infection, especially male patients. **Metabolic:** *Hypercalciuria;* oversaturation of bile
with cholesterol, hyperglycemia,
ketosis. **Endocrine:** High circulating
GH antibodies with resulting treatment failure, accelerated growth of
intracranial tumor.

INTERACTIONS Drug: ANABOLIC STER-
OIDS, **thyroid hormone,** ANDRO-
GENS, ESTROGENS may accelerate epi-
physeal closure; **ACTH,** CORTICOSTER-
OIDS may inhibit growth response to
somatropin.

PHARMACOKINETICS Metabolism:
In liver. **Elimination:** In urine. **Half-
Life:** 15–50 min.

NURSING IMPLICATIONS

Assessment & Drug Effects

- Monitor growth at designated
  intervals.
- Lab test: Periodic serum and urine
  calcium and plasma glucose. Test
  for circulating GH antibodies (an-
  tisomatropin antibodies) in patients
  who respond initially but later fail
  to respond to therapy.
- Hypercalciuria, a frequent adverse
  effect in the first 2–3 mo of therapy,
  may be symptomless; however, it
  may be accompanied by renal
  calculi, with these reportable

symptoms: Flank pain and colic, GI symptoms, urinary frequency, chills, fever, hematuria.

- Observe diabetics or those with family history of diabetes closely. Obtain regular fasting blood glucose and HbA1C.

- Examine patients with GH deficiency secondary to intracranial lesion frequently for progression or recurrence of underlying disease.

Patient & Family Education
- Be aware that during first 6 mo of successful treatment, linear growth rates may be increased 8–16 cm or more per year (average about 7 cm/y or approximately 3 in.). Additionally, subcutaneous fat diminishes but returns to pretreatment value later.

- Record accurate height measurements at regular intervals and report to prescriber if rate is less than expected.

- In general, growth response to somatropin is inversely proportional to duration of treatment.

- Bone age is typically assessed annually in all patients and especially those also receiving concurrent thyroid or androgen treatment, since these drugs may precipitate early epiphyseal closure. Take child for bone age assessment on appointed annual dates.

SORAFENIB
(sor-a-fe′nib)
Nexavar
Classifications: ANTINEOPLASTIC; TYROSINE KINASE INHIBITOR (TKI); MULTI-KINASE INHIBITOR
Therapeutic: ANTINEOPLASTIC
Prototype: Gefitinib
Pregnancy Category: D

AVAILABILITY 200 mg tablets

ACTION & *THERAPEUTIC EFFECT*
Sorafenib is a multi-kinase inhibitor targeting enzyme systems in both tumor cells and tumor vasculature. It appears to be cytostatic, requiring continued drug exposure for tumor growth inhibition. *Sorafenib inhibits enzymes responsible for uncontrolled tumor cellular proliferation and angiogenesis.*

USES Treatment of advanced renal cell cancer.
UNLABELED USES Treatment of advanced malignant melanoma. Treatment of metastatic hepatocellular cancer.

CONTRAINDICATIONS Active infection; severe renal impairment (less than 30 mL/min), or hemodialysis; pregnancy (category D), lactation.
CAUTIOUS USE Previous myelosuppressive therapy, either radiation or chemotherapy; mild or moderate renal disease; hepatic disease; heart failure, ventricular dysfunction, cardiac disease, peripheral edema; females of childbearing age. Safe use in children younger than 18 y is not established.

ROUTE & DOSAGE

Renal Cell Cancer
Adult: **PO** 400 mg b.i.d.

Dosage Adjustments for Skin Toxicity
Grade 2 (1st episode): Continue therapy and treat symptoms. If no improvement in 7 days, discontinue until toxicity resolves to at least grade 1, then resume with 400 mg/day or 400 mg every other day
Grade 2 (2nd or 3rd episode): Discontinue until toxicity resolves

S

Common adverse effects in *italic*, life-threatening effects underlined; generic names in **bold**; classifications in SMALL CAPS; ✦ Canadian drug name; ⊘ Prototype drug

1395

to at least grade 1, then resume with 400 mg/day or 400 mg every other day
Grade 2 (4th episode): Discontinue therapy
Grade 3 (1st or 2nd episode): Discontinue until toxicity resolves to at least grade 1, then resume with 400 mg/day or 400 mg every other day
Grade 3 (3rd episode): Discontinue therapy

ADMINISTRATION

Oral
- Tablets **must be** swallowed whole. They should not be crushed, broken, or chewed.
- Give on an empty stomach 1 h before or 2 h after eating.
- Store at 15°–30° C (59°–86° F). Protect from moisture.

ADVERSE EFFECTS (≥1%) **Body as a Whole:** Asthenia, bone pain, decreased appetite, *fatigue,* influenza-like illness, *joint pain,* mouth pain, muscle pain, pyrexia. **CNS:** Depression, *headache, sensory neuropathy.* **CV:** *Hypertension.* **GI:** *Abdominal pain, anorexia, constipation, diarrhea,* dyspepsia, dysphagia, mucositis, *nausea,* stomatitis, *vomiting.* **Hematologic:** Anemia, <u>hemorrhage</u>, <u>leukopenia,</u> *lymphopenia,* <u>neutropenia,</u> <u>thrombocytopenia,</u> <u>thrombotic events.</u> **Metabolic:** *Amylase elevation, hypophosphatemia, lipase elevation, weight loss.* **Musculoskeletal:** Arthralgia, myalgia. **Respiratory:** *Cough, dyspnea,* hoarseness. **Skin:** Acne, *alopecia, desquamation, dry skin,* erythema, exfoliative dermatitis, flushing, *hand-foot skin reaction, rash.* **Urogenital:** Erectile dysfunction.

INTERACTIONS Drug: Sorafenib may increase levels of drugs requiring glucuronidation by the UGT1A1 and UGT1A9 pathways (e.g., **irinotecan**). Due to thrombocytopenic effects, sorafenib can contribute to increased bleeding from NON-STEROIDAL ANTI-INFLAMMATORY DRUGS, PLATELET INHIBITORS (e.g., **aspirin, clopidogrel**), THROMBOLYTIC AGENTS, and **warfarin.** INDUCERS OF CYP3A4 (e.g., **carbamazepine, phenobarbital, phenytoin, rifampin**) may decrease the levels of sorafenib. **Food:** Food decreases the absorption of sorafenib. **Herbal: St. John's wort** may decrease the levels of sorafenib.

PHARMACOKINETICS Absorption: 38–49% absorbed. **Peak:** 3 h. **Distribution:** 99.5% protein bound. **Metabolism:** In the liver. **Elimination:** Primarily fecal (77%) with minor elimination in the urine (19%). **Half-Life:** 25–48 h.

NURSING IMPLICATIONS

Assessment & Drug Effects
- Monitor for and report S&S of skin toxicity (e.g., rash, erythema, dermatitis, paresthesia, swelling, or pain in hands or feet). Severe reactions may require temporary suspension of therapy or dose reduction.
- Monitor for S&S of bleeding, especially in those on anticoagulation therapy.
- Monitor BP weekly for the first 6 wk of therapy and periodically thereafter. New-onset hypertension has been associated with sorafenib.
- Lab tests: Periodic CBC with differential and platelet count, serum electrolytes, LFTs, lipase, amylase, and alkaline phosphatase.
- Monitor blood levels of warfarin with concurrent therapy.

Patient & Family Education
- Report any of the following to a health care provider: Skin rash;

Common adverse effects in *italic*, life-threatening effects <u>underlined</u>; generic names in **bold**; classifications in SMALL CAPS; ✦ Canadian drug name; ⊙ Prototype drug

redness, blisters, pain or swelling of the palms or hands or soles of feet; signs of bleeding; unexplained chest, shoulder, neck and jaw, or back pain.

- Do not take any prescription or nonprescription drugs without consulting the prescriber.
- Male and female patients should use effective birth control during treatment and for at least 2 wk following completion of treatment.

SOTALOL
(so-ta′lol)

Betapace, Betapace AF
Classifications: BETA-ADRENERGIC ANTAGONIST; CLASS II AND III ANTIARRHYTHMIC
Therapeutic: CLASS II AND III ANTIARRHYTHMIC
Prototype: Amiodarone
Pregnancy Category: B

AVAILABILITY Betapace: 80 mg, 120 mg, 160 mg, 240 mg tablets; **Betapace AF:** 80 mg, 120 mg, 160 mg tablets

ACTION & *THERAPEUTIC EFFECT* Has both class II and class III antiarrhythmic properties. Slows heart rate, decreases AV nodal conduction, and increases AV nodal refractoriness. Produces significant reduction in both systolic and diastolic blood pressure. *Antiarrhythmic properties are effective in controlling ventricular arrhythmias as well as atrial fibrillation/flutter. Regulates blood pressure values.*

USES Treatment of life-threatening ventricular arrhythmias (sustained ventricular tachycardia) and maintenance of normal sinus rhythm in patients with atrial fibrillation/flutter.
UNLABELED USES Hypertension, angina.

CONTRAINDICATIONS Hypersensitivity to sotalol; bronchial asthma, acute bronchospasm; sinus bradycardia, sick sinus syndrome; second and third degree heart block, long QT syndrome, cardiogenic shock, uncontrolled CHF; chronic bronchitis, emphysema; hypokalemia less than 4 mEq/L; creatinine clearance of less than 40 mL/min.
CAUTIOUS USE CHF, electrolyte disturbances, recent MI, DM; sick sinus rhythm, renal impairment; excessive diarrhea, or profuse sweating; pregnancy (category B), lactation.

ROUTE & DOSAGE

Ventricular Arrhythmias (Betapace)

Adult: **PO** Initial dose of 80 mg b.i.d. or 160 mg daily taken prior to meals, may increase every 3–4 days in 40–160 mg increments (most patients respond to 240–320 mg/day in 2 or 3 divided doses, doses greater than 640 mg/day have not been studied)

Renal Impairment Dosage Adjustment

CrCl greater than 60 mL/min: q12h; *30–60 mL/min:* q24h; *10–30 mL/min:* q36–48h; *less than 10 mL/min:* Individualize carefully

Atrial Fibrillation/Flutter (Betapace AF)

Adult: **PO** Initial dose of 80 mg b.i.d., may increase every 3–4 days (max: 240 mg/day in 1–2 divided doses)

Renal Impairment Dosage Adjustment

CrCl greater than 60 mL/min: q12h; *40–60 mL/min:* q24h; *less than 40 mL/min* contraindicated

S

Common adverse effects in *italic*, life-threatening effects underlined; generic names in **bold**; classifications in SMALL CAPS; ♣ Canadian drug name; ⊙ Prototype drug

1397

ADMINISTRATION

Oral

- Give on an empty stomach 1 h before or 2 h after meals. Do not give with milk or milk products.
- Drug should be initiated and doses increased only under close supervision, preferably in a hospital with cardiac rhythm monitoring and frequent assessment.
- Use smallest effective dose for patients with nonallergic bronchospasms.
- Do not discontinue drug abruptly. Gradually reduce dose over 1–2 wk.
- Store at room temperature, 15°–30° C (59°–86° F).

ADVERSE EFFECTS (≥1%) **CV:** AV block, hypotension, aggravation of CHF, although the incidence of heart failure may be lower than for other beta-blockers, <u>life-threatening ventricular arrhythmias, including polymorphous ventricular tachycardia or torsades de pointes</u>, *bradycardia, dyspnea, chest pain, palpitation*, bleeding (less than 2%). **CNS:** Headache, *fatigue, dizziness,* weakness, lethargy, depression, lassitude. **GI:** Nausea, vomiting, diarrhea, dyspepsia, dry mouth. **Urogenital:** Impotence, decreased libido. **Metabolic:** Hyperglycemia. **Special Senses:** Visual disturbances. **Respiratory:** Respiratory complaints. **Skin:** Rash.

INTERACTIONS Drug: Antagonizes the effects of BETA AGONISTS. **Amiodarone** may lead to symptomatic bradycardia and sinus arrest. The hypoglycemic effects of ORAL HYPOGLYCEMIC AGENTS may be potentiated. May cause resistance to **epinephrine** in anaphylactic reactions. Should be used with caution with other ANTIARRHYTHMIC AGENTS. **Food:** Absorption may be reduced by food, especially **milk** and MILK PRODUCTS.

PHARMACOKINETICS Absorption: Slowly and completely from GI tract. Negligible first-pass metabolism. Reduced by food, especially milk and milk products. **Peak:** 2–3 h. **Duration:** 24 h. **Distribution:** Drug is hydrophilic and will enter the CSF slowly (about 10%). Crosses placental barrier. Distributed in breast milk. Not appreciably protein bound. **Metabolism:** Does not undergo significant hepatic enzyme metabolism and no active metabolites have been identified. **Elimination:** In urine with 75% of the drug excreted unchanged within 72 h. **Half-Life:** 7–18 h.

NURSING IMPLICATIONS

Assessment & Drug Effects

- Monitor ECG at baseline and periodically thereafter (especially when doses are increased) because proarrhythmic events most often occur within 7 days of initiating therapy or increasing dose.
- Lab test: Baseline serum electrolytes. Correct electrolyte imbalances of hypokalemia or hypomagnesemia prior to initiating therapy.
- Monitor cardiac status throughout therapy. Exercise special caution when sotalol is used concurrently with other antiarrhythmics, digoxin, or calcium channel blockers.
- Monitor patients with bronchospastic disease (e.g., bronchitis, emphysema) carefully for inhibition of bronchodilation.
- Monitor diabetics for loss of glycemic control. Beta blockage reduces the release of endogenous insulin in response to hyperglycemia and may blunt symptoms of acute hypoglycemia (e.g., tachycardia, BP changes).

Patient & Family Education

- Be aware of risk for hypotension and syncope, especially with concurrent treatment with catecholamine-depleting drugs (e.g., reserpine, guanethidine).

- Take radial pulse daily and report marked bradycardia (pulse below 60 or other established parameter) to prescriber.
- Type 2 diabetics are at increased risk for hyperglycemia. All diabetics are at risk of possible masking of symptoms of hypoglycemia.
- Do not abruptly discontinue drug because of the risk of exacerbation of angina, arrhythmias, and possible myocardial infarction.

SPINOSAD

(spin' oh sad)

Nartroba
Classifications: SCABICIDE; PEDICULICIDE
Therapeutic: SCABICIDE; PEDICULICIDE
Prototype: Permethrin
Pregnancy Category: B

AVAILABILITY 0.9% topical suspension

ACTION & *THERAPEUTIC EFFECT*
Spinosad kills lice directly without measurable absorption into the human body. *Eliminates lice infestation in humans.*

USES Topical treatment of head lice infestations caused by *Pediculus capitis* in patients 4 y and older.

CONTRAINDICATIONS Contains benzyl alcohol and is not recommended for use in neonates and infants below the age of 6 months.
CAUTIOUS USE Inflammatory skin conditions (e.g., psoriasis, eczema); pregnancy (category B); lactation; children 4 y or older.

ROUTE & DOSAGE

Head Lice Infestation

Adult/Child (4 y and older):
Topical Once daily for 10 min; may repeat in 7 d if necessary.

ADMINISTRATION

Topical
- Shake well immediately prior to use.
- Apply sufficient amount (up to one bottle) to cover dry scalp, then apply to dry hair; leave on 10 min, then rinse hair and scalp thoroughly with warm water.
- Store at 15°–30° C (59°–86° F).

ADVERSE EFFECTS (≥1%) Body as a Whole: Application-site erythema and irritation. **Special Senses:** Ocular erythema.

NURSING IMPLICATIONS

Assessment & Drug Effects
- Assess hair and scalp for presence of lice. If still present after 7 days of treatment, a second treatment should be applied.

Patient & Family Education
- Avoid contact with eyes. If spinosad gets in or near the eyes, rinse thoroughly with water.
- Wash hands after applying spinosad.
- Use on children only under direct supervision of an adult.

SPIRONOLACTONE 🔘

(speer-on-oh-lak'tone)

Aldactone, Novospiroton ♣
Classifications: ELECTROLYTIC AND WATER BALANCE; ALDOSTERONE ANTAGONIST; POTASSIUM-SPARING DIURETIC
Therapeutic: POTASSIUM-SPARING DIURETIC; ANTIHYPERTENSIVE
Pregnancy Category: C and D in gestational hypertension

S

Common adverse effects in *italic*, life-threatening effects underlined; generic names in **bold**; classifications in SMALL CAPS; ♣ Canadian drug name; 🔘 Prototype drug

1399

AVAILABILITY 25 mg, 50 mg, 100 mg tablets

ACTION & *THERAPEUTIC EFFECT*
Specific pharmacologic antagonist of aldosterone that competes with aldosterone for cellular receptor sites in distal renal tubules. Promotes sodium and chloride excretion without loss of potassium. *A diuretic agent that promotes sodium and chloride excretion without concomitant loss of potassium. Lowers systolic and diastolic pressures in hypertensive patients. Effective in treatment of primary aldosteronism.*

USES Essential hypertension, refractory edema due to CHF, hepatic cirrhosis, nephrotic syndrome, hypokalemia, and idiopathic edema. May be used to potentiate actions of other diuretics and antihypertensive agents or for its potassium-sparing effect. Also used for treatment of (and as presumptive test for) primary aldosteronism.

UNLABELED USES Hirsutism in women with polycystic ovary syndrome or idiopathic hirsutism; adjunct in treatment of myasthenia gravis and familial periodic paralysis.

CONTRAINDICATIONS Anuria, acute renal insufficiency; renal failure; diabetic nephropathy; progressing impairment of kidney function, or worsening of liver disease; hyperkalemia; lactic acidosis; pregnancy (category D, if used in gestational hypertension); lactation.

CAUTIOUS USE BUN of 40 mg/dL or greater, mild or moderate renal impairment; liver disease; older adults; pregnancy (category C, if not used in gestational hypertension).

ROUTE & DOSAGE

Edema

Adult: **PO** 25–200 mg/day in divided doses, continued for at least 5 days (dose adjusted to optimal response; if no response, a thiazide or loop diuretic may be added)
Child: **PO** 3.3 mg/kg/day in single or divided doses, continued for at least 5 days (dose adjusted to optimal response)
Neonate: **PO** 1–3 mg/kg/day divided q12–24h

Hypertension

Adult: **PO** 25–100 mg/day in single or divided doses, continued for at least 2 wk (dose adjusted to optimal response)

Primary Aldosteronism: Diagnosis

Adult: **PO** Short Test: 400 mg/day for 4 days; long test: 400 mg/day for 3–4 wk

Primary Aldosteronism

Adult: **PO** 100–400 mg/day in divided doses

Hypokalemia

Adult: **PO** 25–100 mg daily

ADMINISTRATION

Oral

- Give with food to enhance absorption.
- Crush tablets and give with fluid of patient's choice if unable to swallow whole.
- Store in tight, light-resistant containers. Suspension is stable for 1 mo under refrigeration.

ADVERSE EFFECTS (≥1%) **CNS:** Lethargy, mental confusion, fatigue (with rapid weight loss), headache, drowsiness, ataxia.

Common adverse effects in *italic*, life-threatening effects underlined; generic names in **bold**; classifications in SMALL CAPS; ♥ Canadian drug name; ⊕ Prototype drug

Endocrine: Gynecomastia (both sexes), inability to achieve or maintain erection, androgenic effects (hirsutism, irregular menses, deepening of voice); parathyroid changes, decreased glucose tolerance, SLE. **GI:** Abdominal cramps, nausea, vomiting, anorexia, diarrhea. **Skin:** Maculopapular or erythematous rash, urticaria. **Metabolic:** Fluid and electrolyte imbalance (particularly hyperkalemia and hyponatremia); elevated BUN, mild acidosis, hyperuricemia, gout. **Body as a Whole:** Drug fever. **Hematologic:** Agranulocytosis. **CV:** Hypertension (post-sympathectomy patient).

DIAGNOSTIC TEST INTERFERENCE May produce marked increases in *plasma cortisol* determinations by *Mattingly fluorometric* method; these may persist for several days after termination of drug (spironolactone metabolite produces fluorescence). There is the possibility of false elevations in measurements of *digoxin serum levels* by *RIA* procedures.

INTERACTIONS Drug: Combinations of spironolactone and acidifying doses of **ammonium chloride** may produce systemic acidosis; use these combinations with caution. Diuretic effect of spironolactone may be antagonized by **aspirin** and other SALICYLATES. **Digoxin** should be monitored for decreased effect of CARDIAC GLYCOSIDES. Hyperkalemia may result with POTASSIUM SUPPLEMENTS, ACE INHIBITORS, ARBS, **heparin** may decrease **lithium** clearance resulting in increased tenacity; may alter anticoagulant response in **warfarin. Food:** Salt substitutes may increase risk of hyperkalemia.

PHARMACOKINETICS Absorption: ~73% from GI tract. **Onset:** Gradual. **Peak:** 2–3 days; maximum effect may take up to 2 wk. **Duration:** 2–3 days or longer. **Distribution:** Crosses placenta, distributed into breast milk. **Metabolism:** In liver and kidneys to active metabolites. **Elimination:** 40–57% in urine, 35–40% in bile. **Half-Life:** 1.3–2.4 h parent compound, 18–23 h metabolites.

NURSING IMPLICATIONS

Assessment & Drug Effects

- Check blood pressure before initiation of therapy and at regular intervals throughout therapy.
- Lab tests: Monitor serum electrolytes (sodium and potassium) especially during early therapy; monitor digoxin level when used concurrently.
- Assess for signs of fluid and electrolyte imbalance, and signs of digoxin toxicity.
- Monitor daily I&O and check for edema. Report lack of diuretic response or development of edema; both may indicate tolerance to drug.
- Weigh patient under standard conditions before therapy begins and daily throughout therapy. Weight is a useful index of need for dosage adjustment. For patients with ascites, prescriber may want measurements of abdominal girth.
- Observe for and report immediately the onset of mental changes, lethargy, or stupor in patients with liver disease.
- Adverse reactions are generally reversible with discontinuation of drug. Gynecomastia appears to be related to dosage level and duration of therapy; it may persist in some after drug is stopped.

Patient & Family Education

- Be aware that the maximal diuretic effect may not occur until third day of therapy and that diuresis may continue for 2–3 days after drug is withdrawn.
- Report signs of hyponatremia or hyperkalemia (see Appendix F),

S

Common adverse effects in *italic*, life-threatening effects <u>underlined</u>; generic names in **bold**; classifications in SMALL CAPS; ♣ Canadian drug name; ⊙ Prototype drug

1401

most likely to occur in patients with severe cirrhosis.

- Avoid replacing fluid losses with large amounts of free water (can result in dilutional hyponatremia).
- Weigh 2–3 times each wk. Report gains/loss of 5 lb or more.
- Do not drive or engage in potentially hazardous activities until response to the drug is known.
- Avoid excessive intake of high-potassium foods and salt substitutes.

STAVUDINE (D4T)

(sta′vu-deen)

Zerit

Classifications: ANTIRETROVIRAL; NUCLEOSIDE REVERSE TRANSCRIPTASE INHIBITOR (NRTI)

Therapeutic: ANTIRETROVIRAL; NRTI

Prototype: Lamivudine

Pregnancy Category: C

AVAILABILITY 15 mg, 20 mg, 30 mg, 40 mg capsules; 1 mg/mL oral solution

ACTION & *THERAPEUTIC EFFECT*
Synthetic analog of thymidine (a nucleoside in DNA) with antiviral action against HIV. Appears to act by being incorporated into growing DNA chains by viral transcriptase, thus terminating viral replication. *Inhibits the replication of HIV in human cells and decreases viral load.*

USES HIV treatment.

UNLABELED USES HIV prophylaxis.

CONTRAINDICATIONS Hypersensitivity to stavudine; lactic acidosis; lactation.

CAUTIOUS USE Previous hypersensitivity to zidovudine, didanosine, or zalcitabine; folic acid or B$_{12}$ deficiency; liver and renal insufficiency;

alcoholism; peripheral neuropathy; history of pancreatitis; pregnancy (category C).

ROUTE & DOSAGE

HIV Infection

Adult/Adolescent/Child: **PO** *Weight less than 60 kg,* 30 mg q12h; *weight 60 kg or greater,* 40 mg q12h
Child/Infant/Neonate (older than 14 days): **PO** *Weight less than 30 kg,* 1 mg/kg q12h
Neonate (younger than 13 days): **PO** 0.5 mg/kg q12h

Renal Impairment Dosage Adjustment

CrCl 25–50 mL/min: Reduce dose by 50%; *less than 25 mL/min:* Reduce dose by 50% and extend interval to q24h

Toxicity Dosage Adjustment

Adult/Adolescent: Weight greater than 60 kg, 20 mg q12h; weight less than 60 kg, 15 mg q12h
Child: Reduce dose by 50%

ADMINISTRATION

Oral

- Adhere strictly to 12-h interval between doses.
- Reconstitute powder by adding 202 mL of water to the container. Shake vigorously. Yields 200 mL of 1 mg/mL solution.
- Store at room temperature, 15°–30° C (59°–86° F).

ADVERSE EFFECTS (≥1%) **CNS:** *Peripheral neuropathy,* paresthesias. **GI:** *Anorexia, nausea, vomiting, diarrhea,* cramping, pancreatitis, abdominal pain, elevated liver function tests, abdominal pain. **Body as a Whole:** *Headache,* chills/fever, *myalgia.* **Hematologic:** Anemia, neutropenia. **Skin:** *Rash.* **Metabolic:** Lactic acidosis in pregnant women; hyperglycemia, lipodystrophy.

INTERACTIONS Drug: Didanosine may increase risk of pancreatitis and hepatotoxicity; **probenecid** can decrease elimination; **zalcitabine** increases risk of neuropathy; **zidovudine** may impact metabolism, avoid concurrent use. Use INTERFERONS, **ribavirin** cautiously.

PHARMACOKINETICS Absorption: Readily absorbed from GI tract; 82% reaches systemic circulation. **Peak:** Effect 6 wk. **Distribution:** Distributes into CSF; excreted in breast milk of animals. **Metabolism:** Unknown. **Elimination:** In urine. **Half-Life:** 1–1.6 h.

NURSING IMPLICATIONS

Assessment & Drug Effects

- Monitor for peripheral neuropathy and report numbness, tingling, or pain, which may indicate a need to interrupt stavudine.
- Lab tests: Monitor LFTs, CBC with differential, PT and INR, and kidney function periodically.
- Monitor for development of opportunistic infection.

Patient & Family Education

- Take drug exactly as prescribed.
- Report to prescriber any adverse drug effects that are bothersome.
- Report symptoms of peripheral neuropathy to prescriber immediately.

STREPTOMYCIN SULFATE

(strep-toe-mye'sin)

Classifications: AMINOGLYCOSIDE ANTIBIOTIC; ANTITUBERCULOSIS
Therapeutic: ANTIBIOTIC; ANTITUBERCULOSIS
Prototype: Gentamicin
Pregnancy Category: D

AVAILABILITY 400 mg/mL, 1 g injection

ACTION & *THERAPEUTIC EFFECT* Aminoglycoside antibiotic that works by inhibiting bacterial protein synthesis through irreversible binding to the 30S ribosomal subunit of susceptible bacteria. *Active against a variety of gram-positive, gram-negative, and acid-fast organisms.*

USES Only in combination with other antitubercular drugs in treatment of all forms of active tuberculosis caused by susceptible organisms. Used alone or in conjunction with tetracycline for tularemia, plague, and brucellosis. Also used with other antibiotics in treatment of subacute bacterial endocarditis due to *Enterococci* and *Streptococci* (viridans group) and *Haemophilus influenzae* and in treatment of peritonitis, respiratory tract infections, granuloma inguinale, and chancroid when other drugs have failed.

CONTRAINDICATIONS History of toxic reaction or hypersensitivity to streptomycin or another aminoglycosides; labyrinthine disease; concurrent or sequential use of neurotoxic and/or nephrotoxic drugs; myasthenia gravis; pregnancy (category D).
CAUTIOUS USE Impaired kidney function (given in reduced dosages); use in older adults and in prematures, neonates, and children.

ROUTE & DOSAGE

Tuberculosis
Adult: **IM** 15 mg/kg up to 1 g/ day as single dose
Geriatric: **IM** 10 mg/kg (max: 750 mg/day)
Child: **IM** 20–40 mg/kg/day up to 1 g/day as single dose
Infant: **IM** 10–15 mg/kg q12h
Neonate: **IM** 10–20 mg/kg q24h

S

Common adverse effects in *italic*, life-threatening effects underlined; generic names in **bold**; classifications in SMALL CAPS; ♣ Canadian drug name; ❷ Prototype drug

1403

Tularemia

Adult: **IM** 1–2 g/day in 1–2 divided doses for 7–10 days
Child: **IM** 20–40 mg/kg/day divided q6–12h

Plague

Adult: **IM** 2 g/day in 2–4 divided doses
Child: **IM** 30 mg/kg/day divided q8–12h

ADMINISTRATION

Intramuscular

- Give IM deep into large muscle mass to minimize possibility of irritation. Injections are painful.
- Avoid direct contact with drug; sensitization can occur. Use gloves during preparation of drug.
- Use commercially prepared IM solution undiluted; intended only for IM injection (contains a preservative, and therefore is not suitable for other routes).
- Store ampules at room temperature. Protect from light; exposure to light may slightly darken solution, with no apparent loss of potency.

ADVERSE EFFECTS (≥1%) CNS:

Paresthesias (peripheral, facial). **Body as a Whole:** Hypersensitivity angio-edema, drug fever, enlarged lymph nodes, <u>anaphylactic shock</u>, headache, inability to concentrate, lassitude, muscular weakness, *pain and irritation at IM site,* superinfections, neuromuscular blockade, arachnoid-itis. **GI:** Stomatitis, hepatotoxicity. **Hematologic:** Blood dyscrasias (<u>leukopenia</u>, neutropenia, pancytopenia, hemolytic or aplastic anemia, eosinophilia). **Special Senses:** *Labyrinthine damage,* auditory damage, optic nerve toxicity (scotomas). **Urogenital:** Nephrotoxicity. **CNS:** Encephalopathy, <u>CNS depression syndrome in infants (stupor, flaccidity, coma, paralysis, cardiac arrest)</u>. **Respiratory:** <u>Respiratory depression</u>. **Skin:** Skin rashes, pruritus, <u>exfoliative dermatitis</u>.

DIAGNOSTIC TEST INTERFERENCE

Streptomycin reportedly produces false-positive ***urinary glucose*** tests using ***copper sulfate methods (Benedict's solution, Clinitest)*** but not with ***glucose oxidase methods*** (e.g., ***Clinistix, TesTape***). False increases in protein content in ***urine*** and ***CSF*** using ***Folin-Ciocalteau reaction*** and decreased ***BUN*** readings with ***Berthelot reaction*** may occur from test interferences. ***C&S*** tests may be affected if patient is taking salts such as sodium and potassium chloride, sodium sulfate and tartrate, ammonium acetate, calcium and magnesium ions.

INTERACTIONS Drug:

May potentiate anticoagulant effects of **warfarin;** additive nephrotoxicity with **acyclovir, amphotericin B,** AMINOGLYCOSIDES, **carboplatin, cidofovir, cisplatin, cyclosporine, foscarnet, ganciclovir,** SALICYLATES, **tacrolimus, vancomycin.**

PHARMACOKINETICS Peak:

1–2 h. **Distribution:** Diffuses into most body tissues and extracellular fluids; crosses placenta; distributed into breast milk. **Elimination:** In urine. **Half-Life:** 2–3 h adults, 4–10 h newborns.

NURSING IMPLICATIONS

Assessment & Drug Effects

- Lab tests: Obtain C&S prior to and periodically during course of therapy. In patients with impaired kidney function, frequent determinations of serum drug concentrations (serum concentrations should not exceed 25 mcg/mL in these patients); and periodic LFTs and kidney function tests are advised.

- Be alert for and report immediately symptoms of ototoxicity (see Appendix F). Symptoms are most likely to occur in patients with impaired kidney function, patients receiving high doses (1.8–2 g/day) or other ototoxic or neurotoxic drugs, and older adults. Irreversible damage may occur if drug is not discontinued promptly.

- Early damage to vestibular portion of eighth cranial nerve (higher incidence than auditory toxicity) is initially manifested by moderately severe headache, nausea, vomiting, vertigo in upright position, difficulty in reading, unsteadiness, and positive Romberg sign.

- Be aware that auditory nerve damage is usually preceded by vestibular symptoms and high-pitched tinnitus, roaring noises, impaired hearing (especially to high-pitched sounds), sense of fullness in ears. Audiometric test should be done if these symptoms appear, and drug should be discontinued. Hearing loss can be permanent if damage is extensive. Tinnitus may persist several days to weeks after drug is stopped.

- Monitor I&O. Report oliguria or changes in I&O ratio (possible signs of diminishing kidney function). Sufficient fluids to maintain urinary output of 1500 mL/24 h are generally advised. Consult prescriber.

Patient & Family Education

- Report any unusual symptoms. Review adverse reactions with prescriber periodically, especially with prolonged therapy.

- Be aware of possibility of ototoxicity and its symptoms (see Appendix F).

- Report to prescriber immediately any of the following: Nausea, vomiting, vertigo, incoordination, tinnitus, fullness in ears, impaired hearing.

STREPTOZOCIN
(strep-toe-zoe′sin)
Zanosar
Classification: ANTINEOPLASTIC, ALKYLATING; NITROSOUREA
Therapeutic: ANTINEOPLASTIC
Prototype: Cyclophosphamide
Pregnancy Category: D

AVAILABILITY 1 g injection

ACTION & *THERAPEUTIC EFFECT*
Streptozocin inhibits DNA synthesis in cells and prevents progression of cells into mitosis, affecting all phases of the cell cycle (cell-cycle nonspecific). *Successful therapy with streptozocin (alone or in combination) produces a biochemical response evidenced by decreased secretion of hormones as well as measurable tumor regression. Thus, serial fasting insulin levels during treatment indicate response to this drug.*

USES Pancreatic cancer.
UNLABELED USES Carinoid.

CONTRAINDICATIONS Concurrent use with nephrotoxic drugs; pregnancy (category D), lactation.
CAUTIOUS USE Renal impairment; hepatic disease, hepatic impairment; patients with history of hypoglycemia; DM. Safe use in children is not established.

ROUTE & DOSAGE

Islet Cell Carcinoma of Pancreas
Adult/Adolescent/Child: **IV** 500 mg/m^2/day for 5 consecutive days q4– 6wk or 1 g/m^2/wk for 2 wk, then increase to 1.5 g/m^2/ wk x 4–6 wk

---

Common adverse effects in *italic*, life-threatening effects underlined; generic names in **bold**; classifications in SMALL CAPS; ♣ Canadian drug name; ⊙ Prototype drug

STREPTOZOCIN

Renal Impairment Dosage Adjustment
Dose reduction or discontinuation recommended

ADMINISTRATION

Intravenous
Use only under constant supervision by prescriber experienced in therapy with cytotoxic agents and only when the benefit to risk ratio is fully and thoroughly understood by patient and family.

▪ Wear gloves to protect against topical exposure, which may pose a carcinogen hazard, when handling streptozocin. ▪ If solution or powder comes in contact with skin or mucosa, promptly flush the area thoroughly with soap and water.

PREPARE: **IV Infusion:** Reconstitute with 9.5 mL D5W or NS, to yield 100 mg/mL. Solution will be pale gold. ▪ May be further diluted with up to 250 mL of the original diluent. ▪ Protect reconstituted solution and vials of drug from light.
ADMINISTER: **IV Infusion:** Give over 15–60 min. ▪ Inspect injection site frequently for signs of extravasation (patient complaints of stinging or burning at site, swelling around site, no blood return or questionable blood return). ▪ If extravasation occurs, area requires immediate attention to prevent necrosis. Remove needle, apply ice, and contact prescriber regarding further treatment to infiltrated tissue.
INCOMPATIBILITIES **Y-site:** Acyclovir, allopurinol, amphotericin, aztreonam, cefepime, daptomycin, pantoprazole, piperacillin/tazobactam, trastuzumab.

▪ Note: An antiemetic given routinely every 4 or 6 h and prophylactically 30 min before a treatment may provide sufficient control to maintain the treatment regimen (even if it reduces but does not completely eliminate nausea and vomiting).

▪ Discard reconstituted solutions after 12 h (contains no preservative and not intended for multidose use).

ADVERSE EFFECTS (≥1%) **CNS:** Confusion, lethargy, depression. **GI:** *Nausea, vomiting,* diarrhea, transient increase in AST, ALT, or alkaline phosphatase; hypoalbuminemia. **Hematologic:** *Mild* to moderate myelosuppression (leukopenia, thrombocytopenia, *anemia*). **Metabolic:** Glucose tolerance abnormalities (moderate and reversible); glycosuria without hyperglycemia, insulin shock (rare). **Urogenital:** Nephrotoxicity: Azotemia, anuria, proteinuria, hypophosphatemia, hyperchloremia; *Fanconi-like syndrome* (proximal renal tubular reabsorption defects, alkaline pH of urine, glucosuria, acetonuria, aminoaciduria); hypokalemia, hypocalcemia. **Other:** Local necrosis following extravasation.

INTERACTIONS Drug: MYELOSUPPRESSIVE AGENTS add to hematologic toxicity; nephrotoxic agents (e.g., AMINOGLYCOSIDES, **vancomycin, amphotericin B, cisplatin**) increase risk of nephrotoxicity; **phenytoin** may reduce cytotoxic effect on pancreatic beta cells.

PHARMACOKINETICS Absorption: Undetectable in plasma within 3 h. **Distribution:** Metabolite enters CSF. **Metabolism:** In liver and kidneys. **Elimination:** 70–80% of dose in urine, 1% in feces, and 5% in expired air. **Half-Life:** 35–40 min.

S

Common adverse effects in *italic*, life-threatening effects underlined; generic names in **bold**; classifications in SMALL CAPS; ♣ Canadian drug name; ⊘ Prototype drug

NURSING IMPLICATIONS

Assessment & Drug Effects

- Lab tests: Weekly CBC; LFTs prior to each course of therapy; serial urinalyses and kidney function tests; baseline and weekly serum electrolytes, then for 4 wk after termination of therapy.
- Ensure that repeat courses of streptozocin treatment are not given until patient's liver, kidney, and hematologic functions are within acceptable limits.
- Be alert to and report promptly early laboratory evidence of kidney dysfunction: Hypophosphatemia, mild proteinuria, and changes in I&O ratio and pattern.
- Mild adverse renal effects may be reversible following discontinuation of streptozocin, but nephrotoxicity may be irreversible, severe, or fatal.
- Be alert to symptoms of sepsis and superinfections or increased tendency to bleed (thrombocytopenia). Myelosuppression is severe in 10–20% of patients and may be cumulative and more severe if patient has had prior exposure to radiation or to other antineoplastics.
- Monitor and record temperature pattern to promptly recognize impending sepsis.

Patient & Family Education

- Inspect site at weekly intervals and report changes in tissue appearance if extravasation occurred during IV infusion.
- Report symptoms of hypoglycemia (see Appendix F) even though this drug has minimal, if any, diabetogenic action.
- Drink fluids liberally (2000–3000 mL/day). Hydration may protect against drug toxicity effects.
- Report S&S of nephrotoxicity (see Appendix F).
- Do not take aspirin or NSAIDs without consulting prescriber.

- Report to prescriber promptly any signs of bleeding: Hematuria, epistaxis, ecchymoses, petechial.
- Report symptoms that suggest anemia: Shortness of breath, pale mucous membranes and nail beds, exhaustion, rapid pulse.

SUCCINYLCHOLINE CHLORIDE ⊕

(suk-sin-ill-koe'leen)

Anectine, Quelicin
Classification: DEPOLARIZING SKELETAL MUSCLE RELAXANT
Therapeutic: SKELETAL MUSCLE RELAXANT
Pregnancy Category: C

AVAILABILITY 20 mg/mL, 50 mg/mL, 100 mg/mL injection

ACTION & *THERAPEUTIC EFFECT*
Synthetic, ultrashort-acting depolarizing neuromuscular blocking agent with high affinity for acetylcholine (ACh) receptor sites. *Initial transient contractions and fasciculations are followed by sustained flaccid skeletal muscle paralysis produced by state of accommodation that develops in adjacent excitable muscle membranes.*

USES To produce skeletal muscle relaxation as adjunct to anesthesia; to facilitate intubation and endoscopy, to increase pulmonary compliance in assisted or controlled respiration, and to reduce intensity of muscle contractions in pharmacologically induced or electroshock convulsions.

CONTRAINDICATIONS Hypersensitivity to succinylcholine; family history of malignant hyperthermia; burns; trauma.

CAUTIOUS USE Kidney, liver, pulmonary, metabolic, or cardiovascular disorders; myasthenia gravis; dehydration, electrolyte imbalance, patients taking digitalis, severe

S

Common adverse effects in *italic*, life-threatening effects underlined; generic names in **bold**; classifications in SMALL CAPS; ✦ Canadian drug name; ⊕ Prototype drug

1407

burns or trauma, fractures, spinal cord injuries, degenerative or dystrophic neuromuscular diseases, low plasma pseudocholinesterase levels (recessive genetic trait, but often associated with severe liver disease, severe anemia, dehydration, marked changes in body temperature, exposure to neurotoxic insecticides, certain drugs); collagen diseases, porphyria, intraocular surgery, glaucoma; during delivery with cesarean section; pregnancy (category C), lactation.

ROUTE & DOSAGE

Surgical and Anesthetic Procedures

Adult: **IV** 0.3–1.1 mg/kg administered over 10–30 sec, may give additional doses prn **IM** 2.5–4 mg/kg up to 150 mg
Child: **IV** 1–2 mg/kg administered over 10–30 sec, may give additional doses prn **IM** 2.5–4 mg/kg up to 150 mg

Prolonged Muscle Relaxation

Adult: **IV** 0.5–10 mg/min by continuous infusion

Obesity Dosage Adjustment

Dose based on IBW

ADMINISTRATION

Intramuscular

- Give IM injections deeply, preferably high into deltoid muscle.

Intravenous

Use only freshly prepared solutions; succinylcholine hydrolyzes rapidly with consequent loss of potency.

- Give initial small test dose (0.1 mg/kg) to determine individual drug sensitivity and recovery time.

PREPARE: **Direct:** Give undiluted. **Intermittent/Continuous:** Dilute 1 g in 500–1000 mL of D5W or NS.
ADMINISTER: **Direct:** Give a bolus dose over 10–30 sec. **Intermittent/Continuous (Preferred):** Give at a rate of 0.5–10 mg/min. Do not exceed 10 mg/min.
INCOMPATIBILITIES **Solution/additive: Aminophylline, ampicillin, cephalothin, diazepam, epinephrine, hydrocortisone, methicillin, methohexital, nitrofurantoin, oxacillin, oxytetracycline, sodium bicarbonate, thiopental, warfarin. Y-site: Thiopental.**

- Note: Expiration date and storage before and after reconstitution; varies with the manufacturer.

ADVERSE EFFECTS (≥1%) **CNS:** *Muscle fasciculations,* profound and prolonged muscle relaxation, muscle pain. **CV:** *Bradycardia,* tachycardia, hypotension, hypertension, arrhythmias, sinus arrest. **Respiratory:** <u>Respiratory depression</u>, bronchospasm, hypoxia, <u>apnea</u>. **Body as a Whole:** <u>Malignant hyperthermia</u>, increased IOP, excessive salivation, enlarged salivary glands. **Metabolic:** Myoglobinemia, hyperkalemia. **GI:** Decreased tone and motility of GI tract (large doses).

INTERACTIONS Drug: Aminoglycosides, colistin, cyclophosphamide, cyclopropane, echothiophate iodide, halothane, lidocaine, MAGNESIUM SALTS, **methotrimeprazine,** NARCOTIC ANALGESICS, ORGANOPHOSPHAMIDE INSECTICIDES, MAO INHIBITORS, PHENOTHIAZINES, **procaine, procainamide, quinidine, quinine, propranolol** may prolong neuromuscular blockade; DIGITALIS GLYCOSIDES may increase risk of cardiac arrhythmias.

PHARMACOKINETICS Onset: 0.5–1 min IV; 2–3 min IM. **Duration:** 2–3 min IV; 10–30 min IM. **Distribution:** Crosses placenta in small amounts. **Metabolism:** In plasma by pseudo-cholinesterases. **Elimination:** In urine.

NURSING IMPLICATIONS

Assessment & Drug Effects

- Lab tests: Obtain baseline serum electrolytes. Electrolyte imbalance (particularly potassium, calcium, magnesium) can potentiate effects of neuromuscular blocking agents.
- Be aware that transient apnea usually occurs at time of maximal drug effect (1–2 min); spontaneous respiration should return in a few seconds or, at most, 3 or 4 min.
- Have immediately available: Facilities for emergency endotracheal intubation, artificial respiration, and assisted or controlled respiration with oxygen.
- Monitor vital signs and keep airway clear of secretions.

Patient & Family Education

- Patient may experience postprocedural muscle stiffness and pain (caused by initial fasciculations following injection) for as long as 24–30 h.
- Be aware that hoarseness and sore throat are common even when pharyngeal airway has not been used.
- Report residual muscle weakness to prescriber.

SUCRALFATE
(soo-kral'fate)

Carafate, Sulcrate ✦
Classifications: ANTIULCER; GASTROADHESIVE
Therapeutic: ANTIULCER; GASTROPROTECTANT
Pregnancy Category: B

AVAILABILITY 1 g tablets; 1 g/10 mL suspension

ACTION & *THERAPEUTIC EFFECT*
Sucralfate and gastric acid react to form a viscous, adhesive, paste-like substance that resists further reaction with gastric acid. This "paste" adheres to the GI mucosa with a major portion binding electrostatically to the positively charged protein molecules in the damaged mucosa of an ulcer crater or an acute gastric erosion. *Absorbs bile, inhibits the enzyme pepsin, and blocks back diffusion of H^+ ions. These actions plus adherence of the paste-like complex protect damaged mucosa against further destruction from ulcerogenic secretions and drugs.*

USES Duodenal ulcer.
UNLABELED USES Short-term treatment of gastric ulcer, aspirin-induced erosions, suspension for chemotherapy-induced mucositis.
CAUTIOUS USE Chronic kidney failure or dialysis due to aluminum accumulation; chronic renal impairment; pregnancy (category B). Safe use in children not established.

ROUTE & DOSAGE

Duodenal Ulcer (Active disease)
Adult: **PO** 1 g q.i.d. 1 h a.c. and at bedtime
Duodenal Ulcer (Maintenance)
Adult: **PO** 1 g b.i.d.

ADMINISTRATION

Oral

- Use drug solubilized in an appropriate diluent by a pharmacist when given through nasogastric tube.

Common adverse effects in *italic*, life-threatening effects underlined; generic names in **bold**; classifications in SMALL CAPS; ✦ Canadian drug name; ❷ Prototype drug

1409

- Administer antacids prescribed for pain relief 30 min before or after sucralfate.
- Separate administration of quinolones, digoxin, phenytoin, tetracycline from that of sucralfate by 2 h to prevent sucralfate from binding to these compounds in the intestinal tract and reducing their bioavailability.
- Store in tight container at room temperature, 15°–30° C (59°–86° F). Stable for 2 y after manufacture.

ADVERSE EFFECTS (≥1%) **GI:** Nausea, gastric discomfort, *constipation,* diarrhea.

INTERACTIONS Drug: May decrease absorption of QUINOLONES (e.g., **ciprofloxacin, norfloxacin**), **digoxin, phenytoin, tetracycline.**

PHARMACOKINETICS Absorption: Minimally absorbed from GI tract (less than 5%). **Duration:** Up to 6 h (depends on contact time with ulcer crater). **Elimination:** 90% in feces.

NURSING IMPLICATIONS

Assessment & Drug Effects

- Be aware of drug interactions and schedule other medications accordingly.

Patient & Family Education

- Although healing has occurred within the first 2 wk of therapy, treatment is usually continued 4–8 wk.
- Be aware that constipation is a drug-related problem. Follow these measures unless contraindicated: Increase water intake to 8–10 glasses per day; increase physical exercise, increase dietary bulk. Consult prescriber: A suppository or bulk laxative (e.g., Metamucil) may be prescribed.

SULFACETAMIDE SODIUM
(sul-fa-see′ta-mide)

AK-Sulf, Bleph 10, Cetamide, Isopto Cetamide, Ophthacet, Sebizon, Sodium Sulamyd, Sulf-10

SULFACETAMIDE SODIUM/ SULFUR

Sulfacet, Rosula
Classification: SULFONAMIDE ANTIBIOTIC
Therapeutic: ANTIBIOTIC
Prototype: Silver sulfadiazide
Pregnancy Category: C; D near term

AVAILABILITY Sulfacetamide: 10% lotion; 1%, 10%, 15%, 30% solution; 10% ointment; **Sulfacetamide/Sulfur:** 10%/5% gel, lotion

ACTION & *THERAPEUTIC EFFECT* Highly soluble sulfonamide that exerts bacteriostatic effect by interfering with bacterial utilization of PABA, thereby inhibiting folic acid biosynthesis required for bacterial growth. *Effective against a wide range of gram-positive and gram-negative microorganisms.*

USES Ophthalmic preparations are used for conjunctivitis, corneal ulcers, and other superficial ocular infections and as adjunct to systemic sulfonamide therapy for trachoma. The topical lotion is used for scaly dermatoses, seborrheic dermatitis, seborrhea sicca, and other bacterial skin infections.

CONTRAINDICATIONS Hypersensitivity to sulfonamides; neonates, pregnancy (category D near term).
CAUTIOUS USE Application of lotion to denuded or debrided skin; pregnancy (category C except near term), lactation. Safe use in children is not established.

ROUTE & DOSAGE

Conjunctivitis

Adult: **Ophthalmic** 1–3 drops of 10%, 15%, or 30% solution into lower conjunctival sac q2–3h, may increase interval as patient responds or use 1.5–2.5 cm (½–1 in.) of 10% ointment q6h and at bedtime

Seborrhea, Rosacea

Adult: **Topical** Apply thin film to affected area 1–3 times/day

ADMINISTRATION

Instillation

- Be aware that ophthalmic preparations and skin lotion are not interchangeable.
- Check strength of medication prescribed.
- See patient instructions for instilling eye drops.
- Discard darkened solutions; results when left standing for a long time.
- Store at 8°–15° C (46°–59° F) in tightly closed containers unless otherwise directed.

ADVERSE EFFECTS (≥1%) **Special Senses:** *Temporary stinging or burning sensation,* retardation of corneal healing associated with long-term use of ophthalmic ointment. **Body as a Whole:** Hypersensitivity reactions (<u>Stevens-Johnson syndrome</u>, lupus-like syndrome), superinfections with nonsusceptible organisms.

INTERACTIONS Drug: Tetracaine and other LOCAL ANESTHETICS DERIVED FROM PABA may antagonize the antibacterial effects of SULFONAMIDES; SILVER PREPARATIONS may precipitate sulfacetamide from solution.

PHARMACOKINETICS Absorption: Minimal systemic absorption, but may be enough to cause

sensitization. **Metabolism:** In liver to inactive metabolites. **Elimination:** In urine.

NURSING IMPLICATIONS

Assessment & Drug Effects

- Discontinue if symptoms of hypersensitivity appear (erythema, skin rash, pruritus, urticaria).

Patient & Family Education

- Wash hands thoroughly with soap and running water (before and after instillation).
- Examine eye medication; discard if cloudy or dark in color.
- Avoid contaminating any part of eye dropper that is inserted in bottle.
- Tilt head back, pull down lower lid. At the same time, look up while drop is being instilled into conjunctival sac. Immediately apply gentle pressure just below the eyelid and next to nose for 1 min. Close eyes gently, so as not to squeeze out medication.
- Report purulent eye discharge to prescriber. Sulfacetamide sodium is inactivated by purulent exudates.

SULFADIAZINE

(sul-fa-dye'a-zeen)

Classification: SULFONAMIDE ANTIBIOTIC
Therapeutic: ANTIBIOTIC
Prototype: Silver sulfadiazide
Pregnancy Category: C; D near term

AVAILABILITY 500 mg tablets

ACTION & *THERAPEUTIC EFFECT*

Short-acting sulfonamide that exerts bacteriostatic effect by interfering with bacterial utilization of

Common adverse effects in *italic*, life-threatening effects <u>underlined</u>; generic names in **bold**; classifications in SMALL CAPS; ♣ Canadian drug name; ⦿ Prototype drug

1411

PABA, thereby inhibiting folic acid biosynthesis required for bacterial growth. *Effective against a wide range of gram-positive and gram-negative microorganisms.*

USES Used in combination with pyrimethamine for treatment of cerebral toxoplasmosis and chloroquineresistant malaria.

CONTRAINDICATIONS Hypersensitivity to sulfonamides or to any ingredients in the formulation; porphyria; pregnancy (category D near term), lactation.
CAUTIOUS USE Application of lotion to denuded or debrided skin; dehydration; hepatic disease; impaired renal function; pregnancy (category C except near term).

ROUTE & DOSAGE

Mild to Moderate Infections

Adult: **PO Loading Dose** 2–4 g loading dose **PO Maintenance Dose** 2–4 g/day in 4–6 divided doses
Child (older than 2 mo): **PO Loading Dose** 75 mg/kg **PO Maintenance Dose** 150 mg/kg/day in 4–6 divided doses (max: 6 g/day)

Rheumatic Fever Prophylaxis

Adult: **PO** *Weight less than 30 kg,* 500 mg/day; *weight greater than 30 kg,* 1 g/day

Toxoplasmosis

Adult: **PO** 2–8 g/day divided q6h
Child (older than 2 mo): **PO** 100–200 mg/kg/day divided q6h
Neonate: **PO** 50 mg/kg q12h × 12 mo

ADMINISTRATION

Oral

- Maintain sufficient fluid intake to produce urinary output of at least 1500 mL/24 h for children between 3000 and 4000 mL/24 h for adults. Concomitant administration of urinary alkalinizer may be prescribed to reduce possibility of crystalluria and stone formation.
- Store in tight, light-resistant containers.

ADVERSE EFFECTS (≥1%) **CNS:** Headache, peripheral neuritis, peripheral neuropathy, tinnitus, hearing loss, vertigo, insomnia, drowsiness, mental depression, acute psychosis, ataxia, convulsions, kernicterus (newborns). **GI:** *Nausea, vomiting, diarrhea,* abdominal pains, hepatitis, jaundice, pancreatitis, stomatitis. **Hematologic:** Acute hemolytic anemia (especially in patients with G6PD deficiency), aplastic anemia, methemoglobinemia, agranulocytosis, thrombocytopenia, leukopenia, eosinophilia, hypoprothrombinemia. **Body as a Whole:** Headache, *fever,* chills, arthralgia, malaise, allergic myocarditis, serum sickness, anaphylactoid reactions, lymphadenopathy, local reaction following IM injection, fixed drug eruptions, diuresis, overgrowth of nonsusceptible organisms, LE phenomenon. **Skin:** Pruritus, urticaria, rash, erythema multiforme including Stevens-Johnson syndrome, exfoliative dermatitis, alopecia, photosensitivity, vascular lesions. **Urogenital:** *Crystalluria,* hematuria, proteinuria, anuria, toxic nephrosis, reduction in sperm count. **Metabolic:** Goiter, hypoglycemia. **Special Senses:** Conjunctivitis, conjunctival or scleral infection, retardation of corneal healing (ophthalmic ointment).

INTERACTIONS Drug: PABA-CONTAINING LOCAL ANESTHETICS may antagonize sulfa's effects; ORAL ANTICOAGULANTS potentiate hypoprothrombinemia; may potentiate SULFONYLUREA-induced hypoglycemia. May decrease concentrations of **cyclosporine;** may increase levels of **phenytoin.**

PHARMACOKINETICS Absorption: Readily absorbed from GI tract. **Peak:** 3–6 h. **Distribution:** Distributed to most tissues, including CSF; crosses placenta. **Metabolism:** In liver. **Elimination:** In urine.

NURSING IMPLICATIONS

Assessment & Drug Effects

- Lab tests: Baseline and periodic urine C&S to determine drug effectiveness; with long-term therapy, CBC, Hct, and Hgb.
- Monitor hydration status.

Patient & Family Education

- Take drug exactly as prescribed. Do not alter schedule or dose; take total amount prescribed unless prescriber changes the regimen.
- Drink fluids liberally unless otherwise directed.
- Report early signs of blood dyscrasias (sore throat, pallor, fever) promptly to the prescriber.

SULFAMETHOXAZOLE-TRIMETHOPRIM (SMZ-TMP)

(sul-fa-meth′ox-a-zole-tri-meth′o-prim)

Bactrim, Bactrim DS, Co-Trim, Septra, Septra DS, Sulfatrom
Classifications: URINARY TRACT ANTI-INFECTIVE; SULFONAMIDE
Therapeutic: URINARY TRACT ANTI-INFECTIVE
Prototype: Trimethoprim
Pregnancy Category: C; D near term

AVAILABILITY 400 mg SMZ/80 mg TMP, 800 mg SMZ/160 mg TMP tablets; 200 mg SMZ/40 mg TMP suspension; 80 mg SMZ/16 mg TMP injection

ACTION & *THERAPEUTIC EFFECT*

Fixed combination of sulfamethoxazole (SMZ), an intermediate-acting anti-infective sulfonamide, and trimethoprim (TMP), a synthetic anti-infective. Both components of the combination are synthetic folate antagonist anti-infectives. Principal action is by enzyme inhibition that prevents bacterial synthesis of essential nucleic acids and proteins. *Effective against* Pneumocystis jiroveci *pneumonitis (formerly PCP), Shigellosis enteritis, and severely complicated UTIs due to most strains of the Enterobacteriaceae.*

USES Pneumocystis jiroveci pneumonitis (formerly PCP), shigellosis enteritis, and severe complicated UTIs. Also children with acute otitis media due to susceptible strains of *Haemophilus influenzae,* and acute episodes of chronic bronchitis or traveler's diarrhea in adults.

UNLABELED USES Isosporiasis; cholera; genital ulcers caused by *Haemophilus ducreyi;* prophylaxis for *P. jiroveci* pneumonia (formerly PCP) in neutropenic patients.

CONTRAINDICATIONS Hypersensitivity to SMZ, TMP, sulfonamides, or bisulfites, carbonic anhydrase inhibitors; group A beta-hemolytic streptococcal pharyngitis; megaloblastic anemia due to folate deficiency; G6PD deficiency; hyperkalemia; porphyria; pregnancy (category D near term), lactation.

CAUTIOUS USE Impaired kidney or liver function; bone marrow depression; possible folate

S

Common adverse effects in *italic,* life-threatening effects <u>underlined</u>; generic names in **bold**; classifications in SMALL CAPS; ♣ Canadian drug name; ⊘ Prototype drug

1413

deficiency; severe allergy or bronchial asthma; hypersensitivity to sulfonamide derivative drugs (e.g., acetazolamide, thiazides, tolbutamide); pregnancy (category C except near term); infants younger than 2 mo.

ROUTE & DOSAGE

(Weight-based doses are calculated on TMP component) Systemic Infections

Adult: **PO** 160 mg TMP/800 mg SMZ q12h **IV** 8–10 mg/kg/day TMP divided q6–12h
Child: **PO** *Older than 2 mo, weight less than 40 kg,* 4 mg/kg/day TMP q12h; *weight greater than 40 kg,* 160 mg TMP/800 mg SMZ q12h
Child/Infant (older than 2 mo): **IV** 6–10 mg/kg/day TMP divided q6–12h

Pneumocystis jiroveci Pneumonia (formerly PCP)

Adult: **PO** 15–20 mg/kg/day TMP divided q6h
Adult/Adolescent/Child: **IV** 15–20 mg/kg/day TMP divided q6h

Prophylaxis for Pneumocystis jiroveci Pneumonia (formerly PCP)

Adult: **PO** 160 mg TMP/800 mg SMZ q24h
Child: **PO** 150 mg/m² TMP/750 mg/m² SMZ b.i.d. 3 consecutive days/wk (max: 320 mg TMP/day)

Renal Impairment Dosage Adjustment

CrCl 10–30 mL/min: Reduce dose by 50%; *less than 10 mL/min:* Reduce dose by 50–75%

ADMINISTRATION

Oral

- Give with a full glass of desired fluid.
- Maintain adequate fluid intake (at least 1500 mL/day) during therapy.
- Store at 15°–30° C (59°–86° F) in dry place protected from light. Avoid freezing.

Intravenous

PREPARE: **Intermittent:** Add contents of the 5 mL ampule to 125 mL D5W. ▪ Use within 6 h. ▪ If less fluid is desired, dilute in 75 of 100 mL and use within 2 or 4 h, respectively.
ADMINISTER: **Intermittent:** Infuse over 60–90 min. ▪ Avoid rapid infusion.
INCOMPATIBILITIES **Solution/additive:** Stability in **dextrose** and **normal saline** is concentration dependent, **fluconazole, linezolid, verapamil. Y-site:** Alfentanil, amikacin, aminophylline, amphotericin B, ampicillin, ampicillin/sulbactam, atropine, azathioprine, benztropine, bumetanide, buprenorphine, butaorphanol, calcium, capsofungin, cefamandol, cefazolin, cefmetazole, cefonicid, cefotaxime, cefotetan, cefoxitin, ceftazidime, ceftizoxime, ceftriaxone, cefuroxime, cephalothin, chloramphenicol, chlorpromazine, cimetidine, cisatracurium, clindamycin, codeiene, cyanocobalamine, cyclosporine, dantrolene, dexamethasone, dexmedetomidine, diazepam, diazoxide, digoxin, diphenhydramine, dobutamine, dopamine, doxorubicin, doxycycline, ephedrine, ephienphrine, epirubicin, epoetin alfa, erythromycin, famotidine, fentanyl, fluconazole,

foscarnet, furosemide, ganciclovir, gentamicin, haloperidol, heparin, hydralazine, hydrocortisone, hydroxyzine, idarubicin, imipenem-cilastatin, indomethacin, isoproterenol, ketorolac, lidocaine, mannitol, metaraminol, methicillin, methyldopate, methylprednisolone, metoclopramide, mezlocillin, miconazole, milrinone acetate, minocycline, midazolam, nafcillin, nalbuphine, naloxone, netilmicin, nitroglycerin, nitroprusside, norepinephrine, ondansetron, oxacillin, oxytocin, papaverine, penicillin, pentamidine, pentazocine, pentobarbital, phenobarbital, phentolamine, phenylephrine, phenytoin, phytonadione, piperacillin, potassium, prochlorperazine, promethazine, propranolol, protamine, pyridoxine, quinidine, quinupristin/dalfopristin, ranitidine, ritodrine, succinylcholine, sufentanil, theophylline, ticarcillin, tobramycin, tolazoline, vancomycin, verapamil, vinorelbine.

- Store unopened ampule at 15°–30° C (50°–86° F)

ADVERSE EFFECTS (≥1%) Skin:
Mild to moderate rashes (including fixed drug eruptions), <u>toxic epidermal necrolysis</u>. **GI:** *Nausea, vomiting,* diarrhea, *anorexia,* hepatitis, <u>pseudomembranous enterocolitis</u>, stomatitis, glossitis, abdominal pain. **Urogenital:** Kidney failure, oliguria, anuria, crystalluria. **Hematologic:** <u>Agranulocytosis</u> (rare), <u>aplastic anemia</u> (rare), megaloblastic anemia, hypoprothrombinemia, <u>thrombocytopenia</u> (rare). **Body as a Whole:** Weakness, arthralgia, myalgia, photosensitivity, <u>allergic myocarditis</u>.

DIAGNOSTIC TEST INTERFERENCE
May elevate levels of ***serum creatinine, transaminase, bilirubin, alkaline phosphatase.***

INTERACTIONS Drug:
May enhance hypoprothrombinemic effects of ORAL ANTICOAGULANTS; may increase **methotrexate** toxicity. **Alcohol** may cause disulfiram reaction.

PHARMACOKINETICS Absorption:
Readily from GI tract. **Peak:** 1–4 h (oral). **Distribution:** Widely distributed, including CNS; crosses placenta; distributed into breast milk. **Metabolism:** In liver. **Elimination:** In urine. **Half-Life:** 8–10 h TMP, 10–13 h SMZ.

NURSING IMPLICATIONS
Assessment & Drug Effects
- Be aware that IV forms of the drug may contain sodium metabisulfite, which produces allergic-type reactions in susceptible patients: Hives, itching, wheezing, anaphylaxis. Susceptibility (low in general population) is seen most frequently in asthmatics or atopic nonasthmatic persons.
- Lab tests: Baseline and follow-up urinalysis; CBC with differential, platelet count, BUN and creatinine clearance with prolonged therapy.
- Monitor coagulation tests and prothrombin times in patient also receiving warfarin. Change in warfarin dosage may be indicated.
- Monitor I&O volume and pattern. Report significant changes to forestall renal calculi formation. Also report failure of treatment (i.e., continued UTI symptoms).
- Older adult patients are at risk for severe adverse reactions, especially if liver or kidney function is compromised or if certain other drugs

Common adverse effects in *italic*, life-threatening effects <u>underlined</u>; generic names in **bold**; classifications in SMALL CAPS; ♣ Canadian drug name; ⊘ Prototype drug

1415

are given. Most frequently observed: Thrombocytopenia (with concurrent thiazide diuretics); severe decrease in platelets (with or without purpura); bone marrow suppression; severe skin reactions.

- Be alert for overdose symptoms (no extensive experience has been reported): Nausea, vomiting, anorexia, headache, dizziness, mental depression, confusion, and bone marrow depression.

Patient & Family Education

- Report immediately to prescriber if rash appears. Other reportable symptoms are sore throat, fever, purpura, jaundice; all are early signs of serious reactions.
- Monitor for and report fixed eruptions to prescriber. This drug can cause fixed eruptions at the same sites each time the drug is administered. Every contact with drug may not result in eruptions; therefore, patient may overlook the relationship.
- Drink 2.5–3 L (1 L is approximately equal to 1 qt) daily, unless otherwise directed.

SULFASALAZINE
(sul-fa-sal'a-zeen)

Azulfidine, PMS Sulfasalazine ✦, PMS Sulfasalazine E.C. ✦, Salazopyrin ✦, SAS Enteric-500 ✦, S.A.S.-500 ✦

Classifications: ANTI-INFLAMMATORY; SULFONAMIDE
Therapeutic: GI ANTI-INFLAMMATORY; IMMUNOMODULATOR; DISEASE-MODIFYING ANTIRHEUMATIC DRUG (DMARD)
Prototype: Mesalamine
Pregnancy Category: C; D near term

AVAILABILITY 500 mg tablets; 500 mg sustained release tablets

ACTION & *THERAPEUTIC EFFECT*

Locally acting sulfonamide, believed to be converted by intestinal microflora to sulfapyridine (provides antibacterial action) and 5-aminosalicylic acid (5-ASA) or mesalamine, which may exert an anti-inflammatory effect. Inhibits prostaglandins known to cause diarrhea and affect mucosal transport as well as interference with absorption of fluids and electrolytes from colon. *Reduces* Clostridium *and* Escherichia coli *in the stools. Anti-inflammatory and immunomodulatory properties are effective in controlling the S&S of ulcerative colitis and rheumatoid arthritis.*

USES Ulcerative colitis and relatively mild regional enteritis; rheumatoid arthritis.
UNLABELED USES Granulomatous colitis, Crohn's disease, scleroderma.

CONTRAINDICATIONS Sensitivity to sulfasalazine, other sulfonamides and salicylates, trimethoprim; folate deficiency; megaloblastic anemia; renal failure, renal impairment; agranulocytosis; intestinal and urinary tract obstruction; porphyria; pregnancy (category D near term), lactation.
CAUTIOUS USE Severe allergy, or bronchial asthma; blood dyscrasias; hepatic or renal impairment; G6PD deficiency; older adults; pregnancy (category C except near term); children younger than 6 y.

ROUTE & DOSAGE

Ulcerative Colitis, Rheumatoid Arthritis
Adult: **PO** 1–2 g/day in 4 divided doses, may increase up to 8 g/day if needed
Child: **PO** 40–50 mg/kg/day in 4 divided doses (max: 75 mg/kg/day)

Juvenile Rheumatoid Arthritis

Child: **PO** 10 mg/kg/day, increase weekly by 10 mg/kg/day [usual dose: 15–25 mg/kg q12h (max: 2 g/day)]

ADMINISTRATION

Oral

- Give after eating to provide longer intestine transit time.
- Do not crush or chew sustained release tablets; **must be** swallowed whole.
- Use evenly divided doses over each 24-h period; do not exceed 8-h intervals between doses.
- Consult prescriber if GI intolerance occurs after first few doses. Symptoms are probably due to irritation of stomach mucosa and may be relieved by spacing total daily dose more evenly over 24 h or by administration of enteric-coated tablets.
- Store at 15°–30° C (59°–86° F) in tight, light-resistant containers.

ADVERSE EFFECTS (≥1%) Body as a Whole: *Nausea, vomiting, bloody diarrhea; anorexia,* arthralgia, rash, anemia, oligospermia (reversible), blood dyscrasias, liver injury, infectious mononucleosis–like reaction, *allergic reactions.*

INTERACTIONS Drug: Iron, ANTIBIOTICS may alter absorption of sulfasalazine.

PHARMACOKINETICS Absorption: 10–15% from GI tract unchanged; remaining drug is hydrolyzed in colon to sulfapyridine (most of which is absorbed) and 5-aminosalicylic acid (30% of which is absorbed). **Peak:** 1.5–6 h sulfasalazine; 6–24 h sulfapyridine. **Distribution:** Crosses placenta; distributed into breast milk. **Metabolism:** In intestines and liver. **Elimination:** All metabolites are excreted in urine. **Half-Life:** 5–10 h.

NURSING IMPLICATIONS

Assessment & Drug Effects

- Monitor for GI distress. GI symptoms that develop after a few days of therapy may indicate need for dosage adjustment. If symptoms persist, prescriber may withhold drug for 5–7 days and restart it at a lower dosage level.
- Be aware that adverse reactions generally occur within a few days to 12 wk after start of therapy; most likely to occur in patients receiving high doses (4 g or more).
- Lab tests: Measure RBC folate in patients on high doses (more than 2 g/day); a daily supplement may be prescribed.

Patient & Family Education

- Examine stools and report to prescriber if enteric-coated tablets have passed intact in feces. Some patients lack enzymes capable of dissolving coating; conventional tablet will be ordered.
- Be aware that drug may color alkaline urine and skin orange-yellow.
- Remain under close medical supervision. Relapses occur in about 40% of patients after initial satisfactory response. Response to therapy and duration of treatment are governed by endoscopic examinations.

SULINDAC

(sul-in′dak)

Clinoril
Classifications: ANALGESIC, NONSTEROIDAL ANTI-INFLAMMATORY DRUG (NSAID); ANTIPYRETIC
Therapeutic: NONNARCOTIC ANALGESIC, NSAID
Prototype: Ibuprofen
Pregnancy Category: C; D third trimester

S

Common adverse effects in *italic*, life-threatening effects underlined; generic names in **bold**; classifications in SMALL CAPS; ♣ Canadian drug name; ⓟ Prototype drug

AVAILABILITY 150 mg, 200 mg tablets

ACTION & *THERAPEUTIC EFFECT*
Anti-inflammatory action thought to result from inhibition of prostaglandin synthesis. *Exhibits anti-inflammatory, analgesic, and antipyretic properties.*

USES Acute and long-term symptomatic treatment of osteoarthritis, rheumatoid arthritis, ankylosing spondylitis; acute painful shoulder (acute subacromial bursitis or supraspinatus tendinitis); acute gouty arthritis.

CONTRAINDICATIONS Hypersensitivity to sulindac; hypersensitivity to aspirin (patients with "aspirin triad": Acute asthma, rhinitis, nasal polyps), other NSAIDs, or salicylates; significant kidney or liver dysfunction; CABG perioperative pain; pregnancy (category D third trimester); lactation.

CAUTIOUS USE History of upper GI tract disorders; anticoagulant therapy; CHF; moderate or mild renal impairment; compromised cardiac function, hypertension, hemophilia or other bleeding tendencies; pregnancy (category C first and second trimesters). Safety in children not established.

ROUTE & DOSAGE

Arthritis, Ankylosing Spondylitis, Acute Gouty Arthritis
Adult: **PO** 150–200 mg b.i.d. (max: 400 mg/day)

ADMINISTRATION
Oral
- Crush and give mixed with liquid or food if patient cannot swallow tablet.

- Administer with food, milk, or antacid (if prescribed) to reduce possibility of GI upset. Note: Food retards absorption and delays and lowers peak concentrations.

ADVERSE EFFECTS (≥1%) **CNS:** Drowsiness, *dizziness, headache,* anxiety, nervousness. **CV:** Palpitation, peripheral edema, CHF, (patients with marginal cardiac function). **Special Senses:** Blurred vision, amblyopia, vertigo, tinnitus, decreased hearing. **GI:** *Abdominal pain, dyspepsia, nausea, vomiting, constipation,* diarrhea, ulceration, flatulence, anorexia; stomatitis, sore or dry mucous membranes, dry mouth; GI bleeding, gastritis. **Hematologic:** Prolonged bleeding time, <u>aplastic anemia</u>, <u>thrombocytopenia</u>, <u>leukopenia</u>, eosinophilia. **Body as a Whole:** Angioneurotic edema, fever, chills, <u>anaphylaxis</u>. **Skin:** Stevens-Johnson syndrome, <u>toxic epidermal necrolysis syndrome</u>, rash, pruritus. **Urogenital:** Renal impairment.

DIAGNOSTIC TEST INTERFERENCE Abnormalities in *liver function tests* may occur.

INTERACTIONS Drug: Do not use with **ketorolac** or **cidofovir. Heparin,** ORAL ANTICOAGULANTS may prolong bleeding time; may increase **lithium** or **methotrexate** toxicity; **aspirin,** other NSAIDs add to ulcerogenic effects; **dimethylsulfoxide (DMSO)** may decrease effects of sulindac. **Herbal: Feverfew, garlic, ginger, ginkgo** may increase bleeding potential.

PHARMACOKINETICS Absorption: 90% from GI tract. **Peak:** 2 h without food, 3–4 h with food. **Duration:** 10–12 h. **Distribution:** Minimal passage across placenta; distributed into breast milk. **Metabolism:** In

Common adverse effects in *italic*, life-threatening effects <u>underlined</u>; generic names in **bold**; classifications in SMALL CAPS; ✦ Canadian drug name; ⊙ Prototype drug

liver to active metabolite. **Elimination:** 75% in urine, 25% in feces. **Half-Life:** 7.8 h sulindac, 16.4 h sulfide metabolite.

NURSING IMPLICATIONS

Assessment & Drug Effects

- Assess for and report promptly unexplained GI distress.
- Lab tests: Obtain baseline and periodic evaluations of Hgb, kidney and liver function.
- Schedule auditory and ophthalmic examinations in patients receiving prolonged or high-dose therapy.

Patient & Family Education

- Do not drive or engage in potentially hazardous activities until response to drug is known.
- Report any incidence of unexplained bleeding or bruising immediately to prescriber (e.g., bleeding gums, black and tarry stools, coffee-colored emesis).
- Report onset of skin rash, itching, hives, jaundice, swelling of feet or hands, sore throat or mouth, shortness of breath, or night cough to prescriber.
- Be aware that adverse GI effects are relatively common. Report abdominal pain, nausea, dyspepsia, diarrhea, or constipation.
- Avoid alcohol and aspirin as they may increase risk of GI ulceration and bleeding tendencies.
- Inform dentist or surgeon of drug regimen because bleeding time may be prolonged.

SUMATRIPTAN ⊕

(sum-a-trip′tan)
Imitrex, ALSUMA auto-injector, Sumavel DosePro
Classification: SEROTONIN 5-HT$_1$ RECEPTOR AGONIST

Therapeutic: ANTIMIGRAINE
Pregnancy Category: C

AVAILABILITY 25 mg, 50 mg tablets; 12 mg/mL injection; 5 mg, 20 mg nasal spray

ACTION & *THERAPEUTIC EFFECT*

Selective agonist for a serotonin receptor (probably 5-HT$_{1D}$) that causes vasoconstriction of cranial carotid arteries. *Relieves migraine headache. Also relieves photophobia, phonophobia, nausea and vomiting associated with migraine attacks.*

USES Treatment of acute migraine attacks with or without aura, cluster headache.

CONTRAINDICATIONS Hypersensitivity to sumatriptan; CAD; acute MI, angina, arteriosclerosis; cerebrovascular disease; colitis; concurrent use with MAO inhibitors; uncontrolled hypertension; intracranial bleeding; PVD; Raynaud's disease; stroke; severe hepatic disease; Wolff-Parkinson-White syndrome; basilar or hemiplegic migraine; older adults.

CAUTIOUS USE Impaired liver or kidney function; MAO inhibitors; pregnancy (category C). Safety and effectiveness in children are not established.

ROUTE & DOSAGE

Migraine or Cluster Headache

Adult: **Subcutaneous** 6 mg any time after onset of migraine. If headache returns, may repeat with 6 mg subcutaneously at least 1 h after first injection (max: 12 mg/24 h). **PO** 25 mg × 1 dose, if headache returns may repeat once after 2 h (max: 100 mg).

Common adverse effects in *italic*, life-threatening effects underlined; generic names in **bold**; classifications in SMALL CAPS; ♣ Canadian drug name; ⊕ Prototype drug

Intranasal 5, 10, or 20 mg in one nostril. If headache returns, may repeat once after 2 h (max: 40 mg/24 h).

ADMINISTRATION

Note: Do not give within 24 h of an ergot-containing drug.

Oral

- Give any time after symptoms of migraine appear.
- A second tablet may be given if symptoms return but no sooner than 2 h after the first tablet.
- Do not exceed 100 mg in a single oral dose or 300 mg/day.

Intranasal

- Note: A single dose is one spray into ONE nostril.

Subcutaneous

- A second injection may be given 1 h or longer following first injection if initial relief is not obtained or if migraine returns.
- Be aware that if adverse effects are dose limiting, a lower dose may be effective.
- Store all forms at room temperature, 15°–30° C (59°–86° F). Protect from light.

ADVERSE EFFECTS (≥1%) **CV:** Chest pressure and tightness, hypotension or hypertension, hypertensive crisis, syncope, peripheral cyanosis, thromboembolism, heart block, sinus bradycardia, atrial fibrillation, ventricular fibrillation, ventricular tachycardia, <u>coronary artery vasospasm</u>, angina, transient myocardial ischemia, <u>MI, cardiac arrest</u>. **CNS:** *Tingling, warming sensation, pressure, numbness,* headache, *dizziness, vertigo,* drowsiness, sedation, seizure, CNS hemorrhage, subarachnoid hemorrhage, stroke. **Body as a Whole:** Dizziness,

lightheadedness, myalgia, or muscle cramps, *pain on injection,* weakness, flushing and a sensation of warmth or burning after injection. **GI:** Abdominal pain, cramping, diarrhea, nausea, vomiting.

INTERACTIONS Drug: Dihydroergotamine, ERGOT ALKALOIDS may cause vasospasm and a slight elevation in blood pressure. MAO INHIBITORS increase sumatriptan levels and toxicity (especially the oral form); do not use concurrently or within 2 wk of stopping MAO INHIBITORS; use with other serotonin altering drugs increases risk of serotonin syndrome (see Appendix F). **Herbal: St. John's wort** may increase triptan toxicity.

PHARMACOKINETICS Onset: 10–30 min after subcutaneous administration. **Duration:** 1–2 h. **Distribution:** Widely distributed, 10–20% protein bound. May be excreted in breast milk. **Metabolism:** Hepatically to inactive metabolite. **Elimination:** 57% in urine, 38% in feces. **Half-Life:** 2 h.

NURSING IMPLICATIONS

Assessment & Drug Effects

- Monitor cardiovascular status carefully following first dose in patients at relatively high risk for coronary artery disease (e.g., postmenopausal women, men over 40 years old, persons with known CAD risk factors) or who have coronary artery vasospasms.
- Report to prescriber immediately chest pain or tightness in chest or throat that is severe or does not quickly resolve following a dose of sumatriptan.
- Monitor therapeutic effectiveness. Pain relief usually begins within 10 min of injection, with complete

relief in approximately 65% of all patients within 2 h.

Patient & Family Education

- Review patient information leaflet provided by the manufacturer carefully.
- Learn correct use of autoinjector for self-administration of subcutaneous dose.
- Pain or redness at injection site is common but usually disappears in less than 1 h.
- Notify prescriber immediately if symptoms of severe angina (e.g., severe or persistent pain or tightness in chest, back, neck, or throat) or hypersensitivity (e.g., wheezing, facial swelling, skin rash, or hives) occur.
- Do not take any other serotonin receptor agonist (Axert, Maxalt, Zomig, Amerge) within 24 h of taking sumatriptan.
- Check with prescriber before taking any new OTC or prescription drugs.
- Report any other adverse effects (e.g., tingling, flushing, dizziness) at next prescriber visit.

SUNITINIB

(sun-i-ti′nib)
Sutent
Classifications: ANTINEOPLASTIC; PROTEIN-TYROSINE KINASE INHIBITOR
Therapeutic: ANTINEOPLASTIC
Prototype: Gefitinib
Pregnancy Category: D

AVAILABILITY 12.5 mg, 25 mg, 50 mg capsules

ACTION & *THERAPEUTIC EFFECT*
An antineoplastic agent that is a selective inhibitor of receptor tyrosine kinases (RTKs) in solid tumors. Carcinogenic activity within these tumors is a result of tumor angiogenesis and proliferation. *Sunitinib*

causes tumor regression and decreased tumor growth.

USES Treatment of advanced renal cell cancer; treatment of gastrointestinal stromal tumors (GIST), pancreatic neuroendocrine tumor.

CONTRAINDICATIONS Hypersensitivity to sunitinib; uncontrolled hypertension, acute MI; fever, abnormal bleeding, sore throat; severe hepatic adverse reaction to sunitinib; pregnancy (category D), lactation.

CAUTIOUS USE Cardiac disease, CHF, history of hypertension, history of MI; CVA; hepatic impairment; females of childbearing age; children.

ROUTE & DOSAGE

Advanced Renal Cell Cancer
Adult: **PO** 50 mg/day for 4 wk, followed by 2 wk off treatment; repeat 6-wk cycle as needed.

GIST Tumor
Adult: **PO** 50 mg daily x 4 wk then 2 wk off

Dosage Adjustments with Concurrent Hepatic CYP3A4 Modifiers
CYP3A4 Inducers: Increase to maximum of 87.5 mg/day
CYP3A4 Inhibitors: Decrease to minimum of 37.5 mg/day

Pancreatic Tumor
Adult: **PO** 37.5 mg daily

ADMINISTRATION

Oral
- Incremental dosage changes of 12.5 mg are recommended.
- Store at 15°–30° C (59°–86° F).

ADVERSE EFFECTS (≥1%) **Body as a Whole:** *Alopecia, asthenia, dizziness, fatigue, fever,* hair color change, *headache, peripheral edema.* **CV:**

Common adverse effects in *italic,* life-threatening effects underlined; generic names in **bold**; classifications in SMALL CAPS; ♣ Canadian drug name; ☯ Prototype drug

1421

Hypertension, myocardial ischemia. **GI:** *Abdominal pain, altered taste, anorexia, constipation, diarrhea, dyspepsia, flatulence, glossodynia, nausea, mucositis, stomatitis, vomiting.* **Hematologic:** *Anemia,* bleeding, neutropenia, lymphopenia, thrombocytopenia. **Metabolic:** *Dehydration, elevated hepatic enzymes,* hepatotoxicity, hypothyroidism, *hyperbilirubinemia.* **Musculoskeletal:** *Arthralgia, back pain, myalgia/limb pain.* **Respiratory:** *Cough, dyspnea.* **Skin:** *Dry skin, hand-foot syndrome, rash, skin discoloration.*

INTERACTIONS Drug: Coadministration of CYP3A4 INDUCERS (e.g., **carbamazepine, dexamethasone, phenobarbital, phenytoin, rifabutin, rifampin, rifapentine**) may decrease plasma levels of sunitinib. Coadministration of CYP3A4 INHIBITORS (e.g., **atazanavir, clarithromycin, erythromycin, indinavir, itraconazole, ketoconazole, nefazodone, nelfinavir, ritonavir, saquinavir, telithromycin, voriconazole**) may increase plasma levels of sunitinib. **Food: Grapefruit** and **grapefruit juice** may increase the plasma levels of sunitinib. **Herbal: St. John's wort** may decrease the plasma levels of sunitinib.

PHARMACOKINETICS Peak: 6–12 h. **Distribution:** 95–98% protein bound. **Metabolism:** Extensive hepatic metabolism. **Elimination:** Primarily fecal (61%). **Half-Life:** 40–60 h.

NURSING IMPLICATIONS

Assessment & Drug Effects

- Monitor for and report S&S of bleeding (e.g., GI, GU, gingival, etc.).
- Monitor BP regularly and assess regularly for S&S of congestive heart failure. Withhold drug and notify prescriber if severe hypertension or signs of heart failure develop.

- Lab tests: At the beginning of each treatment cycle, CBC with differential and platelet count; periodic serum electrolytes; thyroid function tests with symptoms suggestive of hypothyroidism.

Patient & Family Education

- Do not use any prescription or nonprescription drugs without consulting a prescriber.
- Skin discoloration (yellow color) and/or loss of skin and hair pigmentation may occur with this drug.
- Report any of the following to a health care provider: Painful redness of palms and soles of feet; severe abdominal pain, vomiting, and diarrhea; signs of bleeding; chest pain or discomfort; shortness of breath; swelling of feet, legs, or hands; rapid weight gain.
- Women of childbearing age are advised not to become pregnant while taking sunitinib.

TACRINE
(tac'rine)

Cognex
Classifications: CHOLINESTERASE INHIBITOR; ANTIDEMENTIA
Therapeutic: ALZHEIMER'S
Pregnancy Category: C

AVAILABILITY 10 mg, 20 mg, 30 mg, 40 mg capsules

ACTION & *THERAPEUTIC EFFECT*
Cholinesterase inhibitor that elevates acetylcholine in the cerebral cortex by slowing degradation of acetylcholine release by the remaining intact neurons. Balances pathologic changes in neurons that result in deficiency of acetylcholine in early stages of Alzheimer's disease. *Slows manifestations of Alzheimer's disease.*

USES Improvement of memory in mild to moderate Alzheimer's dementia.

UNLABELED USES HIV infection (severe dementia), tardive dyskinesia, acute anticholinergic syndrome with possible advantage over physostigmine.

CONTRAINDICATIONS Hypersensitivity to tacrine; patients who develop jaundice while taking tacrine.

CAUTIOUS USE Anesthesia, sick sinus rhythm, bradycardia; history of ulcers, GI bleeding, abnormal liver function; liver disease; patients with asthma, hypotension, COPD, hyperthyroidism, seizure disorders; urinary tract obstruction, intestinal obstruction; pregnancy (category C), lactation. Safety and efficacy in children are not established.

ROUTE & DOSAGE

Alzheimer's Disease

Adult: **PO** 10 mg q.i.d. (taken between meals if tolerated), increase in 40 mg/day increments not sooner than q6wk (max: 160 mg/day)

Hepatic Impairment Dosage Adjustment

Dose-related hepatotoxic effects have been observed; use with caution or not at all in patients with history of past or current liver disease

ADMINISTRATION

Oral

- Give at least 1 h before meals; bioavailability reduced 30–40% when taken with food. Effectiveness depends on administration at regular intervals.
- Withhold drug and notify prescriber if ALT exceeds 3 × ULN. Note: Drug usually discontinued if ALT exceeds 5 × ULN.

- Store at room temperature, 15°–30° C (59°–86° F), away from moisture.

ADVERSE EFFECTS (≥1%) CNS: Agitation, dizziness and confusion, ataxia, insomnia, somnolence, hallucinations. **GI:** Nausea, *vomiting,* belching, *diarrhea,* abdominal discomfort, anorexia, *hepatotoxicity.* **Skin:** Purpura. **Urogenital:** Excessive micturition and incontinence with UTI infections. **Body as a Whole:** Diaphoresis.

INTERACTIONS Drug: Prolongs action of **succinylcholine** and possibly other NEUROMUSCULAR BLOCKING AGENTS due to inhibition of plasma pseudocholinesterase. Increases **theophylline** concentrations two-fold. **Cimetidine** increases concentration of tacrine by 64%. **Herbal: Echinacea** may increase risk of hepatotoxicity.

PHARMACOKINETICS Absorption: Approximately 17% absorbed from GI tract. Food decreases rate and extent of absorption by 30–40%. **Onset:** 30–90 min. **Peak:** 2 h. Steady state in 24–36 h. **Distribution:** Penetrates blood–brain barrier. Protein binding is 55%. **Metabolism:** Metabolized in the liver by cytochrome P450 system. At least three hydroxylated metabolites have been identified that may be biologically active. Females have lower activity in cytochrome P450 isoenzymes so plasma levels are approximately 50% higher than men with same dose. **Elimination:** Less than 3% of dose recovered in urine in 24 h. **Half-Life:** 3.5 h.

NURSING IMPLICATIONS

Assessment & Drug Effects

- Monitor for clinical improvement (defined as a 4-point improvement in Alzheimer's Disease Assessment Scale/Cognitive Subscale). Improvement has been

T

Common adverse effects in *italic*, life-threatening effects underlined; generic names in **bold**; classifications in SMALL CAPS; ♣ Canadian drug name; ⊘ Prototype drug

1423

observed after 1–4 wk; may take 6 mo for maximum benefit.

- Lab tests: Monitor serum transaminase (ALT) levels according to following schedule: Every 2 wk for first 16 wk, then monthly for 2 mo, then every 3 mo thereafter; resume weekly monitoring for 6 wk with each dose increase; continue weekly monitoring if ALT remains more than 2 times normal; if therapy is interrupted more than 4 wk then restarted, resume full ALT monitoring schedule.
- Monitor I&O because tacrine may cause bladder outflow obstruction.
- Monitor for seizure activity and take appropriate precautions.
- Monitor patients with history of angle-closure glaucoma for a worsening of this condition.
- Monitor for GI distress and bleeding, especially in patients with a history of peptic ulcer disease or on concurrent NSAID therapy.
- Supervise ambulation because dizziness occurs in more than 10% of patients.
- Monitor cardiovascular status including periodic ECG monitoring. Assess for fluid retention and worsening of CHF.
- Monitor periodically for development of drug-induced diabetes.

Patient & Family Education
- Be aware of adverse effects related to initiation of therapy or dosage increases (e.g., nausea, vomiting, diarrhea) as well as delayed effects (e.g., rash, GI bleeding, jaundice). Report adverse effects to the prescriber.
- Do not discontinue or reduce dosage of 80 mg/day or more abruptly because it may precipitate acute deterioration of cognitive function.
- Make sure to have regular follow-up and liver function tests.

- Tacrine may induce seizures, vertigo, and syncope. Use appropriate precautions.
- Understand that tacrine therapy is not a cure and will become ineffective at some point as the disease progresses.

TACROLIMUS
(tac-rol'i-mus)
Prograf, Protopic
Classifications: BIOLOGIC RESPONSE MODIFIER; IMMUNOSUPPRESSANT
Therapeutic: IMMUNOSUPPRESSANT
Prototype: Cyclosporine
Pregnancy Category: C

AVAILABILITY 0.5 mg, 1 mg, 5 mg capsules; 5 mg/mL injection; 0.1%, 0.03% ointment

ACTION & *THERAPEUTIC EFFECT*
Inhibits helper T-lymphocytes by selectively inhibiting secretion of interleukin-2, interleukin-3, and interleukin-gamma, thus reducing transplant rejection. *Inhibits antibody production (thus subduing immune response) by creating an imbalance in favor of suppressor T-lymphocytes.*

USES Prophylaxis for organ transplant rejection, moderate to severe atopic dermatitis (e.g., eczema).
UNLABELED USES Acute organ transplant rejection, severe plaque-type psoriasis, ulcerative colitis, nephrotic syndrome.

CONTRAINDICATIONS Hypersensitivity to tacrolimus or castor oil; postoperative oliguria or renal failure with CrCl greater than or equal to 4 mg/dL; potassium sparing diretics; lactation.
CAUTIOUS USE Renal or hepatic insufficiency, hyperkalemia, QT prolongation; CHF; diabetes

mellitus, gout, history of seizures, hypertension; cardiomyopathy, left ventricular dysfunction (e.g., heart failure); neoplastic disease, especially lymphoproliferative disorders; pregnancy (category C); **tacrolimus ointment:** children younger than 2 y.

ROUTE & DOSAGE

Rejection Prophylaxis (Dose Varies Based on Organ)

Adult: **PO** 0.075–0.2 mg/kg/day in divided doses q12h, start no sooner than 6 h after transplant; give first oral dose 8–12 h after discontinuing IV therapy **IV** 0.01–0.05 mg/kg/day as continuous IV infusion, start no sooner than 6 h after transplant, continue until patient can take oral therapy
Child: **PO** 0.15–0.2 mg/kg/day **IV** 0.03–0.05 mg/kg/day

Atopic Dermatitis

Adult: **Topical** Apply thin layer to affected area b.i.d., continue until clearing of symptoms
Child (2–15 y): **Topical** Apply thin layer of 0.03% ointment to affected area b.i.d., continue until clearing of symptoms

Renal Impairment Dosage Adjustment

Start with lower dose

Hemodialysis Dosage Adjustment

Supplementation not necessary

ADMINISTRATION

Oral

- Patient should be converted from IV to oral therapy as soon as possible.
- Give first oral dose 8–12 h after discontinuing IV infusion.

Topical

- Ensure that skin is clean and completely dry before application.
- Apply a thin layer to the affected area and rub in gently and completely.
- Do not apply occlusive dressing over the site.

Intravenous

PREPARE: **IV Infusion:** Dilute 5 mg/mL ampules with NS or D5W to a concentration of 0.004–0.02 mg/mL (4–20 mcg/mL). ▪ Lower concentrations are preferred for children.
ADMINISTER: **IV Infusion:** Give as continuous IV. ▪ PVC-free tubing is recommended, especially at lower concentrations.
INCOMPATIBILITIES **Y-site: Acyclovir, allopurinol, azathioprine, dantrolene, diazepam, diazoxide, esomeprazole, ganciclovir, lansoprazole, levothyroxine, omeprazole, phenytoin, thiopental.**

- Store ampules between 5° and 25° C (41° and 77° F); store capsules at room temperature, 15°–30° C (59°–86° F). ▪ Store the diluted infusion in glass or polyethylene containers and discard after 24 h.

ADVERSE EFFECTS (≥1%) **CNS:** *Headache, tremors, insomnia, paresthesia, hyperesthesia and/or sensations of warmth, circumoral numbness.* **CV:** *Mild to moderate hypertension.* **Endocrine:** Hirsutism, *hyperglycemia, hyperkalemia, hypokalemia, hypomagnesemia,* hyperuricemia, decreased serum cholesterol. **GI:** *Nausea, abdominal pain, gas,* appetite changes, *vomiting, anorexia, constipation,* diarrhea, ascites. **Hematologic:** Anemia, leukocytosis, thrombocytopenia purpura. **Urogenital:** UTI, oliguria, nephrotoxicity, nephropathy. **Respiratory:**

Common adverse effects in *italic*, life-threatening effects underlined; generic names in **bold**; classifications in SMALL CAPS; ♣ Canadian drug name; ⊕ Prototype drug

1425

Pleural effusion, atelectasis, dysp-nea. **Special Senses:** Blurred vision, photophobia. **Skin:** *Flushing, rash, pruritus, skin irritation,* alopecia, erythema, folliculitis, hyperesthesia, <u>exfoliative dermatitis</u>, hirsutism, photosensitivity, skin discoloration, skin ulcer, sweating. **Body as a Whole:** *Pain, fever, peripheral edema, increased risk of cancer.*

INTERACTIONS Drug: Use with **cyclosporine** increases risk of nephrotoxicity. **Metoclopramide, lansoprazole,** CALCIUM CHANNEL BLOCKER, ANTIFUNGAL AGENTS, MACROLIDE ANTIBIOTICS, **bromocriptine, cimetidine, cyclosporine, methylprednisolone, omeprazole** may increase levels; **caspofungin, rifampin** may decrease levels. NSAIDS may lead to oliguria or anuria. **Herbal: St. John's wort** decreases efficacy. **Food: Grapefruit juice** (greater than 1 qt/day) may increase plasma concentrations and adverse effects.

PHARMACOKINETICS Absorption: Erratic and incompletely from GI tract; absolute bioavailability approximately 14–25%; absorption reduced by food. **Peak:** PO 1–4 h. **Duration:** IV 12 h. **Distribution:** Within plasma, tacrolimus is found primarily in lipoprotein-deficient fraction; 75–97% protein bound; distributed into red blood cells; blood:plasma ratio reported greater than 4; animal studies have demonstrated high concentrations of tacrolimus in lung, kidney, heart, and spleen; distributed into breast milk. **Metabolism:** Extensively in liver (CYP3A4). **Elimination:** Metabolites primarily in bile. **Half-Life:** 8.7–11.3 h.

NURSING IMPLICATIONS

Assessment & Drug Effects
- Lab tests: Monitor serum tacrolimus, serum electrolytes, blood glucose, LFTs, uric acid, BUN, and creatinine clearance periodically.
- Monitor kidney function closely; report elevated serum creatinine or decreased urinary output.
- Monitor for neurotoxicity, and report tremors, changes in mental status, or other signs of toxicity.
- Monitor cardiovascular status and report hypertension.

Patient & Family Education
- Report promptly unexplained hunger, thirst, and frequent urination.
- Be aware of potential adverse effects.
- Minimize exposure to natural or artificial sunlight while using the ointment.
- Notify prescriber of S&S of neurotoxicity.

TADALAFIL
(ta-dal'a-fil)

Adcirca, Cialis

Classifications: IMPOTENCE; PHOSPHODIESTERASE (PDE) INHIBITOR; VASODILATOR

Therapeutic: IMPOTENCE; PULMONARY ANTIHYPERTENSIVE

Prototype: Sildenafil

Pregnancy Category: B

AVAILABILITY 2.5 mg, 5 mg, 10 mg, 20 mg tablets

ACTION & *THERAPEUTIC EFFECT*
Tadalafil is a selective phosphodiesterase (PDE) inhibitor. PDE is responsible for degradation of cyclic GMP in the corpus cavernosum of the penis. Cyclic GMP causes smooth muscle relaxation in lung tissue and the corpus cavernosum, thereby allowing inflow of blood into the penis. PDE-5 inhibitors reduce pulmonary vasodilation by sustaining levels of cyclic guanosine monophosphate (cGMP). Additionally, tadalafil

produces a reduction in the pulmonary to systemic vascular resistance ratio. *Tadalafil promotes sustained erection only in the presence of sexual stimulation. It produces a significant improvement in arterial oxygenation in pulmonary hypertension.*

USES Treatment of erectile dysfunction, pulmonary hypertension, BPH.

CONTRAINDICATIONS Hypersensitivity to tadalafil, vardenafil, or sildenafil; MI within last 90 days; Class 2 or greater heart failure within last 6 mo; unstable angina or angina during intercourse; uncontrolled cardiac arrhythmias; nitrate/nitrite therapy; hypotension, uncontrolled hypertension; retinitis pigmentosa; CVA within last 6 mo; left ventricular outflow obstruction, aortic stenosis; severe (Child-Pugh class C) hepatic cirrhosis; women, lactation.

CAUTIOUS USE CAD, risk factors for CVA; renal insufficiency; mild to moderate (Child-Pugh class A or B) hepatic disease; anatomic deformity of the penis; sickle cell anemia; multiple myeloma; leukemia; active bleeding or a peptic ulcer; hiatal hernia, GERD; sickle cell disease; retinitis pigmentosa; severe renal impairment; concurrent use with other medicines for penile dysfunction; older adults; pregnancy (category B). Safety and efficacy for PAH in children younger than 18 y have not been established.

ROUTE & DOSAGE

Erectile Dysfunction
Adult: **PO** 2.5 mg daily *OR* 10 mg prior to anticipated sexual activity. May increase to max dose 20 mg/day or reduce to 5 mg/day if needed. If taking ritonavir,

itraconazole, ketoconazole, or voriconazole, max dose 10 mg q72h.

Pulmonary Hypertension
Adult: **PO** 40 mg

BPH
Adult: **PO** 5 mg

Hepatic Impairment Dosage Adjustment
Mild to moderate impairment (Child-Pugh class A and B): Max 10 mg/day; not recommended with severe hepatic impairment

Renal Impairment Dosage Adjustment
CrCl 30–50 mL/min: Start at half normal dose; *less than 30 mL/min:* Max dose 5 mg and once daily dosing is not recommended

Hemodialysis Dosage Adjustment
Once daily dosing is not recommended

ADMINISTRATION

Oral
- If not on the daily dose regimen, tadalafil is taken approximately 1 h before expected intercourse, but preferably not after a heavy or high-fat meal.
- Store at 15°–30° C (59°–86° F).

ADVERSE EFFECTS (≥1%) **Body as a Whole:** Flushing, back pain, asthenia, facial edema, fatigue, pain, transient global amnesia. **CNS:** *Headache,* dizziness, insomnia, somnolence, vertigo, hypesthesia, paresthesia. **CV:** Angina, chest pain, hypertension, hypotension, MI, orthostatic hypotension, palpitations, syncope, sinus tachycardia. **GI:** Dyspepsia, nausea, vomiting, abdominal pain, abnormal liver function tests, diarrhea, loose stools, dysphagia, esophagitis, gastritis, GERD, xerostomia. **Metabolic:** Increased GGTP. **Musculoskeletal:** Arthralgia, myalgia, neck pain. **Respiratory:** Nasal congestion, dyspnea, epistaxis, pharyngitis. **Skin:** Rash, pruritus, sweating. **Special Senses:** Blurred vision, changes in color vision, conjunctivitis, eye pain, lacrimation, swelling of eyelids, sudden vision loss. **Urogenital:** Spontaneous penile erection.

INTERACTIONS Drug: May potentiate hypotensive effects of ETHANOL, NITRATES, **alfuzosin, doxazosin, prazosin, tamsulosin** (doses greater than 0.4 mg/day), **terazosin; erythromycin** (and other MACROLIDES), **indinavir, itraconazole, ketoconazole,** PROTEASE INHIBITORS, **ritonavir, saquinavir, voriconazole** may increase levels and toxicity of tadalafil; **barbiturates, bosentan, carbamazepine, dexamethasone, fosphenytoin, nevirapine, rifampin phenytoin, rifabutin, troglitazone** may reduce level and effectiveness of tadalafil. **Food: Grapefruit juice** may increase levels and toxicity of tadalafil.

PHARMACOKINETICS Absorption: Rapidly absorbed, 15% reaches systemic circulation. **Onset:** 30–45 min. **Peak:** 2 h. **Duration:** Up to 36 h. **Metabolism:** In liver by CYP3A4. **Elimination:** In feces (61%) and urine (39%). **Half-Life:** 17.5 h.

NURSING IMPLICATIONS

Assessment & Drug Effects
- Monitor CV status and report angina or other S&S of cardiac dysfunction.
- Lab tests: Baseline and periodic LFTs.

Patient & Family Education
- Do not take more than once per day.

- Note: With moderate renal insufficiency, the maximum recommended dose is 10 mg not more than once in every 48 h.
- Moderate use of alcohol when taking this drug.
- Do not take this drug without consulting prescriber if you are taking drugs called "alpha-blockers" or "nitrates" or any other drugs for high blood pressure, chest pain, or enlarged prostate.
- Report promptly any of the following: Palpitations, chest pain, back pain, difficulty breathing, or shortness of breath; dizziness or fainting; changes in vision; swollen eyelids; muscle aches; painful or prolonged erection (lasting longer than 4 h); skin rash, or itching.

TAMOXIFEN CITRATE Ⓟ
(ta-mox'i-fen)

Nolvadex, Nolvadex-D ♦, Tamofen ♦

Classifications: ANTINEOPLASTIC; SELECTIVE ESTROGEN RECEPTOR MODIFIER (SERM)
Therapeutic: ANTINEOPLASTIC; ANTIESTROGEN
Pregnancy Category: D

AVAILABILITY 10 mg, 20 mg tablets

ACTION & *THERAPEUTIC EFFECT*
Nonsteroidal gonad-stimulating drug with potent antiestrogenic as well as estrogenic activity on various tissues. Competes with estradiol at estrogen receptor (ER-positive) sites in target tissues such as breast, uterus, vagina, anterior pituitary. Estrogen is thought to increase breast cancer in ER-positive tumors. Tamoxifen has no effect on the development of ER-negative breast cancer disease. *Has effects on*

tumor with high concentration of estrogen receptors. Tamoxifen-receptor complexes move into the cell nucleus, decreasing DNA synthesis and estrogen responses.

USES Palliative treatment of advanced with metastatic estrogen receptors (ER)-positive breast cancer in postmenopausal women, adjunctively with surgery in the treatment of breast carcinoma with positive lymph nodes.
UNLABELED USES Investigationally to stimulate ovulation in selected anovulatory women desiring pregnancy.

CONTRAINDICATIONS Anticoagulant therapy including coumadin; preexisting endometrial hyperplasia; intramuscular injections if platelets less than 50,000/mm³; history of thromboembolic disease; pregnancy (category D), especially during first trimester; lactation; children.
CAUTIOUS USE Vision disturbances; cataracts; visual disturbance; leukopenia, bone marrow suppression; thrombocytopenia; hypercalcemia; hypercholesterolemia, lipid protein abnormalities; women with ductal cardinoma in situ (DCIS).

ROUTE & DOSAGE

Breast Carcinoma
Adult: **PO** 10–20 mg 1–2 times/day (morning and evening)
Stimulation of Ovulation
Adult: **PO** 5–40 mg b.i.d. for 4 days

ADMINISTRATION
Oral
▪ Doses greater than 20 mg/day should be given in divided doses a.m. and p.m.

▪ Store at 20°–25° C (68°–77° F); protect from light. Oral solution should be used within 3 mo of opening.

ADVERSE EFFECTS (≥1%) **Body as a Whole:** Increased bone pain, and transient local disease flair; loss of hair, weight gain, shortness of breath, photosensitivity, *hot flashes.* **CNS:** Depression, lightheadedness, dizziness, headache, mental confusion, sleepiness. **CV:** Thrombosis, pulmonary embolism, increased risk of stroke. **GI:** *Nausea and vomiting (about 25% of patients),* distaste for food, anorexia. **Hematologic:** Leukopenia, thrombocytopenia. **Metabolic:** Hypercalcemia. **Skin:** Skin rash or dryness. **Special Senses:** Retinopathy, decreased visual acuity, blurred vision. **Urogenital:** Changes in menstrual period, milk production and leaking from breasts, vaginal discharge and bleeding, pruritus vulvae, risk of uterine malignancies.

DIAGNOSTIC TEST INTERFERENCE Tamoxifen may produce transient increase in *serum calcium.*

INTERACTIONS Drug: May enhance hypoprothrombinemic effects of **warfarin;** may increase risk of thromboembolic events with CYTOTOXIC AGENTS; **bromocriptine** may elevate tamoxifen levels, SSRI ANTIDEPRESSANTS may decrease effectiveness of tamoxifen.

PHARMACOKINETICS Absorption: Slowly from GI tract. **Peak:** 3–6 h. **Metabolism:** In liver (CYP2D6), enterohepatically cycled. **Elimination:** Primarily in feces. **Half-Life:** 7 days.

NURSING IMPLICATIONS
Assessment & Drug Effects
▪ Administer analgesics for pain relief as necessitated by bone and tumor pain or local disease flair.

Common adverse effects in *italic*, life-threatening effects underlined; generic names in **bold**; classifications in SMALL CAPS; ✚ Canadian drug name; ⊘ Prototype drug

1429

- Be aware that local swelling and marked erythema over preexisting lesions or the development of new lesions may signal soft-tissue disease response to tamoxifen. These symptoms rapidly subside.
- Lab tests: Monitor CBC, including platelet counts, periodically. Transient leukopenia and thrombocytopenia (50,000–100,000/mm³) without hemorrhagic tendency have been reported. Monitor serum calcium periodically.

Patient & Family Education

- Do not change established dose schedule.
- Report to prescriber any of the following: Marked weakness or numbness in face or leg, especially on one side of the body; difficulty walking or loss of balance; unexplained sleepiness or mental confusion; edema; shortness of breath; or blurred vision.
- Understand the possibility of drug-induced menstrual irregularities before starting treatment.
- Report promptly any unexpected vaginal discharge or pain or pressure in your pelvis.
- Avoid prolonged sun exposure, especially if skin is unprotected. Apply sunscreen lotions (SPF 12 or greater) to all exposed skin surfaces.
- Avoid OTC drugs unless specifically prescribed by the prescriber; particularly OTC pain medicines.
- Report onset of tenderness or redness in an extremity.

TAMSULOSIN HYDROCHLORIDE ℗

(tam'su-lo-sin)

Flomax
Classification: ALPHA-ADRENERGIC RECEPTOR ANTAGONIST

Therapeutic: SMOOTH MUSCLE RELAXANT OF BLADDER OUTLET & PROSTATE GLAND
Pregnancy Category: B

AVAILABILITY 0.4 mg capsules

ACTION & *THERAPEUTIC EFFECT*
Antagonist of the $alpha_{1A}$-adrenergic receptors located in the prostate. This blockage can cause smooth muscles in the bladder outlet and the prostate gland to relax, resulting in improvement in urinary blood flow and a reduction in symptoms of BPH. *Effectiveness is indicated by improved voiding. Improves symptoms related to benign prostatic hypertrophy (BPH) related to bladder outlet obstruction.*

USES Benign prostatic hypertrophy.

CONTRAINDICATIONS Hypersensitivity to tamsulosin; women; lactation, children.
CAUTIOUS USE History of syncope, hypersensitivity to sulfonamides; hypotension; renal impairment, renal failure, renal disease; older adults; pregnancy (category B).

ROUTE & DOSAGE

Benign Prostatic Hypertrophy
Adult: **PO** 0.4 mg daily 30 min after a meal, may increase up to 0.8 mg daily

ADMINISTRATION

Oral
- Give 30 min after the same meal each day.
- Instruct to swallow capsules whole; not to crush, chew, or open.
- If dose is interrupted for several days, reinitiate at the lowest dose, 0.4 mg.
- Store at 20°–25° C (68°–77° F).

ADVERSE EFFECTS (≥1%) **Body as a Whole:** Asthenia, back or chest pain. **CNS:** *Headache, dizziness,* insomnia. **CV:** *Orthostatic hypotension (especially with first dose).* **GI:** Diarrhea, nausea. **Respiratory:** *Rhinitis,* pharyngitis, increased cough, sinusitis. **Urogenital:** Decreased libido, *abnormal ejaculation.* **Special Senses:** Amblyopia.

INTERACTIONS Drug: Cimetidine may decrease clearance of tamsulosin. **Sildenafil, vardenafil,** and **tadalafil,** and alcohol may enhance hypotensive effects.

PHARMACOKINETICS Absorption: Rapidly from GI tract. Greater than 90% bioavailability. **Peak:** 4–5 h fasting, 6–7 h fed. **Distribution:** Widely distributed in body tissues, including kidney and prostate. **Metabolism:** In the liver. **Elimination:** 76% in urine. **Half-Life:** 14–15 h.

NURSING IMPLICATIONS

Assessment & Drug Effects

▪ Monitor for signs of orthostatic hypotension; take BP lying down, then upon standing. Report a systolic pressure drop of 15 mm Hg or more or a HR 15 beats or more upon standing.

▪ Monitor patients on warfarin therapy closely.

Patient & Family Education

▪ Make position changes slowly to minimize orthostatic hypotension.

▪ Report dizziness, vertigo, or fainting to prescriber. Exercise caution with hazardous activities until response to drug is known.

▪ Be aware that concurrent use of cimetidine may increase the orthostatic hypotension adverse effect.

TAPENTADOL

(ta-pent'a-dol)

Nucynta

Classifications: CENTRALLY ACTING NARCOTIC ANALGESIC; MU-OPIOID RECEPTOR AGONIST; INHIBITOR OF NOREPINEPHRINE REUPTAKE
Therapeutic: CENTRALLY ACTING NARCOTIC ANALGESIC
Pregnancy Category: C
Controlled Substance: Schedule II

AVAILABILITY 50 mg, 75 mg, 100 mg tablets

ACTION & *THERAPEUTIC EFFECT* Tapentadol is a centrally acting synthetic analgesic that is a mu-opioid agonist and also thought to inhibit norepinephrine reuptake. *Effective in treatment of moderate to severe acute pain.*

USES Relief of acute moderate to severe pain.

CONTRAINDICATIONS Impaired pulmonary function (e.g., significant respiratory depression, acute or severe bronchial asthma, hypercapnia without monitoring or the absence of resuscitative equipment); paralytic ileus; concomitant use of MAOI or use within 14 days; head injury, intracranial pressure (ICP); severe hepatic or renal impairment; labor and delivery, lactation.
CAUTIOUS USE Debilitated patients; upper airway obstruction; cranial lesions without increased ICP; history of drug or alcohol abuse; history of seizures; mild or moderate renal or hepatic impairment; older adults; pregnancy (category C). Safe use in children younger than 18 y has not been established.

T

Common adverse effects in *italic,* life-threatening effects <u>underlined</u>; generic names in **bold**; classifications in SMALL CAPS; ♣ Canadian drug name; ⊘ Prototype drug

1431

ROUTE & DOSAGE

Acute Moderate to Severe Pain

Adult: **PO** 50–100 mg q 4–6 h. On first day of dosing only may give a second dose 1 h later if initial dose ineffective (may titrate to max total daily dose: 700 mg day 1 and 600 mg thereafter).

Hepatic Impairment Dosage Adjustment

Moderate impairment: Reduce initial dose to 50 mg q 8 h. *Severe impairment (Child-Pugh class C):* Not recommended for use.

ADMINISTRATION

Oral

- On day 1 of therapy, a second dose may be given as soon as 1 h after the initial dose if pain relief is inadequate. All subsequent doses should be at 4–6 h intervals.
- Do not exceed a total daily dose of 700 mg on day 1 or 600 mg on subsequent days.
- Do not administer if a paralytic ileus is suspected.
- Do not give within 14 days of an MAOI.
- Store at 15°–30° C (59°–86° F).

ADVERSE EFFECTS (≥1%) **Body as a Whole:** Fatigue. **CNS:** Abnormal dreams, anxiety, confusional state, *dizziness,* insomnia, lethargy, *somnolence,* tremor. **CV:** Hot flush. **GI:** Constipation, dry mouth, dyspepsia, *nausea, vomiting.* **Metabolic:** Decrease appetite. **Musculoskeletal:** Arthralgia. **Respiratory:** Nasopharyngitis, <u>respiratory depression</u>, upper respiratory tract infection. **Skin:** Hyperhidrosis, pruritus, rash. **Urogenital:** Urinary tract infection.

INTERACTIONS Drug: Other OPIOID AGONISTS, GENERAL ANESTHETICS, PHENOTHIAZINES, ANTIEMETICS, other TRANQUILIZERS, SEDATIVES, HYPNOTICS, or other CNS DEPRESSANTS may cause additive CNS depression. MAO INHIBITORS can raise **norepinephrine** levels resulting in adverse cardiovascular events.

PHARMACOKINETICS Absorption: 32% bioavailability. **Peak:** 1.25 h. **Metabolism:** In liver. **Elimination:** Primarily renal. **Half-Life:** 4 h.

NURSING IMPLICATIONS

Assessment & Drug Effects

- Monitor degree of pain relief, mental status, and level of alertness.
- Monitor vital signs. Withhold drug and notify prescriber for a respiratory rate of 12/min or less.
- If an opioid antagonist is required to reverse the action of tapentadol, continue to monitor respiratory status since respiratory depression may outlast duration of action of the opioid antagonists.
- Withhold drug and report promptly S&S of serotonin syndrome (see Appendix F).
- Monitor ambulation. Fall precautions may be warranted.

Patient & Family Education

- Avoid engaging in hazardous activities until reaction to drug is known.
- Avoid alcohol while taking this drug.
- Consult prescriber before taking OTC drugs.

TAZAROTENE

(ta-zar′o-teen)

Avage, Tazorac
Classifications: RETINOID; ANTIACNE

Therapeutic: ANTIACNE
Prototype: Isotretinoin
Pregnancy Category: X

AVAILABILITY 0.05%, 0.1% gel, cream

ACTION & *THERAPEUTIC EFFECT*
Retinoid prodrug that blocks epidermal cell proliferation and hyperplasia. Suppresses inflammation present in the epidermis of psoriasis. *Effectiveness indicated by improvement in acne or psoriasis.*

USES Topical treatment of plaque psoriasis on up to 20% of the body, mild to moderate acne, facial fine wrinkling, mottled hypo- and hyperpigmentation (blotchy skin discoloration), and benign facial lentigines.

CONTRAINDICATIONS Hypersensitivity to tazarotene; pregnancy (category X), women who are or may become pregnant; lactation.
CAUTIOUS USE Concurrent administration with drugs that are photosensitizers (e.g., thiazide diuretics, tetracyclines); retinoid hypersensitivity. Safety and efficacy in children younger than 12 y are not established.

ROUTE & DOSAGE

Plaque Psoriasis
Adult: **Topical** Apply thin film to affected area once daily in evening

Acne
Adult: **Topical** After cleansing and drying face, apply thin film to acne lesions once daily in evening

Fine Wrinkles
Adult: **Topical** Apply thin film of cream to affected area once daily

ADMINISTRATION
Topical
- Dry skin completely before application of a thin film of medication.
- Apply medication to no more than 20% of body surface in those with psoriasis.
- Apply only to affected areas; avoid contact with eyes and mucous membranes.

ADVERSE EFFECTS (≥1%) **Skin:** *Pruritus, burning/stinging, erythema, worsening of psoriasis, irritation, skin pain,* rash, desquamation of skin, irritant contact dermatitis, inflammation, fissuring, bleeding, dry skin, sunburn.

INTERACTIONS Drug: Increased risk of photosensitivity reactions with QUINOLONES (especially **sparfloxacin**), PHENOTHIAZINES, SULFONAMIDES, SULFONYLUREAS, TETRACYCLINES, THIAZIDE DIURETICS.

PHARMACOKINETICS Absorption: Rapidly absorbed through skin. **Distribution:** Active metabolite greater than 99% protein bound; crosses placenta, distributed into breast milk. **Metabolism:** Undergoes esterase hydrolysis to active metabolite AGN 190299. **Elimination:** In both urine and feces. **Half-Life:** 18 h.

NURSING IMPLICATIONS
Assessment & Drug Effects
- Monitor for photosensitivity in those concurrently using any of the following: Thiazides, tetracyclines, fluoroquinolones, phenothiazines, sulfonamides.

Patient & Family Education
- Understand fully the risk of serious fetal harm. Use reliable forms of effective contraception. Discontinue treatment and notify prescriber if pregnancy occurs.

T

Common adverse effects in *italic*, life-threatening effects <u>underlined</u>; generic names in **bold**; classifications in SMALL CAPS; ✚ Canadian drug name; ⊘ Prototype drug

1433

- Alert: Immediately rinse thoroughly with water if contact with eyes occurs.
- Avoid all unnecessary exposure to sunlight or artificial UV light. If brief exposure is necessary, cover as much skin surface as possible and use sunscreens (minimum SPF 15).
- Do not apply to sunburned skin.
- Discontinue medication and notify prescriber if any of the following occur: Pruritus, burning, skin redness, excessive peeling, worsening of psoriasis.
- Limit application of topicals with strong skin-drying effects to skin areas being treated with tazarotene.

TELAPREVIR

(tel-a′-pre-vir)

Incivek

Classifications: ANTIVIRAL; ANTIHEPATITIS AGENT; NS3/4A PROTEASE INHIBITOR

Therapeutic: ANTIVIRAL; PROTEASE INHIBITOR

Pregnancy Category: X (in combination with ribavirin and peginterferon alfa)

AVAILABILITY 375 mg tablets

ACTION & *THERAPEUTIC EFFECT*
Telaprevir is an inhibitor of a protease enzyme necessary for viral replication. *Prevents replication of the hepatitis C virus.*

USES Treatment of chronic hepatitis C virus (HCV) genotype 1 infection in adult patients with compensated liver disease. Must be used in combination with ribavirin and peginterferon alfa, and can be used in patients with or without cirrhosis, and in patients who are previously untreated or who have previously been treated with interferon

and ribavirin therapy, including prior null responders, partial responders, and relapsers.

CONTRAINDICATIONS Moderate to severe hepatic impairment (Child-Pugh class B or C); serious skin reactions (e.g., drug rash with eosinophilia and systemic symptoms, Stevens-Johnson syndrome); coadministration with alfuzosin, rifampin, ergot derivatives, cisapride, St. John's wort, HMG-CoA reductase inhibitors, pimozide, sildenafil or tadalafil when used for the treatment of pulmonary hypertension, midazolam oral, and triazolam; women who are or may become pregnant (category X in combination therapy); men whose female partners are pregnant; lactation.

CAUTIOUS USE Mild hepatic impairment; severe skin reactions (e.g., generalized rash, rash with vesicles or bullae or ulcerations), anemia; coinfection with HCV/HIV or HCV/HBV. Safety and efficacy in children have not been established.

ROUTE & DOSAGE

Chronic Hepatitis C Infection

Adult: **PO** 750 mg t.i.d. (every 7–9 h) in combination with ribavirin and peginterferon alfa for 12 wk; then D/C telaprevir and continue ribavirin and peginterferon alfa for 12–36 wk

Hepatic Impairment Dosage Adjustment

Moderate to severe impairment (Child-Pugh class B or C) patients with decompensated liver disease: Not recommended

ADMINISTRATION

Oral

- Ensure that patient has consumed food containing approximately

20 g of fat (not low fat) within 30 min of each dose.

- The dose of telaprevir must not be reduced or interrupted because it may increase the possibility of treatment failure. However, if a dose is missed by more than 4 h, skip the dose and resume the usual schedule.
- Note: Telaprevir **must** be discontinued if peginterferon alfa or ribavirin is discontinued.
- Store at 15°–30° C (59°–86° F).

ADVERSE EFFECTS (≥5%) **CNS:** Dysgeusia, *fatigue*. **GI:** Anal pruritus, anorectal discomfort, *diarrhea*, hemorrhoids, *nausea*, vomiting. **Hematological:** *Anemia.* **Metabolic:** Decreased lymphocyte count, decreased platelet count, elevated bilirubin, hyperuricemia. **Skin:** *Pruritus, rash.*

INTERACTIONS Drug: ANTICON-VULSANTS (e.g., **carbamazepine, phenobarbital, phenytoin**), HIV PROTEASE INHIBITORS (e.g., **atazanavir, darunavir, fosamprenavir, lopinavir, ritonavir**), CORTICOS-TEROIDS (e.g., **dexamethasone**), RIFAMYCINS (e.g., **rifampin**), and **efavirenz** may decrease the levels of telaprevir. AZOLE ANTIFUNGALS (e.g., **itraconazole, ketoconazole, posaconazole, voriconazole**), and MACROLIDES (e.g., **clarithromycin, erythromycin, telithromycin**) may increase the levels of telaprevir. Telaprevir may increase the levels of **alfuzosin**, ANTIARRHYTHIC AGENTS (e.g., **amiodarone, bepridil, flecainide, lidocaine, propafenone, quinidine**), ANTIDEPRESSANTS (e.g., **desipramine, trazodone**), **atazanavir**, AZOLE ANTIFUNGALS (e.g., **itraconazole, ketoconazole, posaconazole**), BENZODIAZEPINES (e.g., **alprazolam, midazolam, triazolam**), **bosentan**, CALCIUM CHANNEL BLOCKERS (e.g., **amlodipine, diltiazem, felodipine, nicardipine, nifedipine, nisoldipine, verapamil**), **carbamazepine, colchicine**, CORTICOSTEROIDS (e.g., **budesonide, fluticasone, methylprednisolone, prednisone**), **digoxin**, IMMUNOSUPPRESSIVE AGENTS (e.g., **cyclosporine, sirolimus, tacrolimus**), MACROLIDES (e.g., **clarithromycin, erythromycin, telithromycin**), PHOSPHODIESTERASE 5 INHIBITORS (e.g., **sildenafil, tadalafil, vardenafil**), **pimozide**, RIFAMYCINS (e.g., **rifampin**), **salmeterol**, and **tenofovir.** Telaprevir may decrease the levels of **darunavir, efavirenz, fosamprenavir, methadone**, ORAL CONTRACEPTIVES (e.g., **ethinyl estradiol, norethindrone**), **voriconazole**, and **zolpidem.** Telapreviar may either increase or decrease the levels of **escitalopram, phenobarbital, phenytoin**, and **warfarin.** Coadministration with ERGOT DERIVATIVES (e.g., **dihydroergotamine, ergonovine**) increases the risk of ergot toxicity. Coadministration with HMG-COA REDUCTASE INHIBITORS (e.g., **lovastatin, simvastatin**) may increase the risk of myopathy and rhabdomyolysis. **Food:** Co-administration with food enhances absorption. **Herbal:** St. John's wort may decrease the levels of telaprevir.

PHARMACOKINETICS Peak: 4–5 h. **Distribution:** 59–76% plasma protein bound. **Metabolism:** Extensive hepatic metabolism. **Elimination:** Primarily fecal (82%). **Half-Life:** 9–11 h.

NURSING IMPLICATIONS

Assessment & Drug Effects

- Monitor for and report promptly development of skin rash or other

Common adverse effects in *italic*, life-threatening effects underlined; generic names in **bold**; classifications in SMALL CAPS; ✦ Canadian drug name; ⊙ Prototype drug

1435

serious skin reactions (including symptoms of Stevens-Johnson syndrome), mucosal erosions or ulcerations, fever, or facial edema. Immediate discontinuation of drug may be warranted.

▪ Lab tests: HCV-RNA levels at 4 and 12 wk; CBC with differential at 2, 4, 8 and 12 wk; baseline and periodic LFTs and bilirubin; baseline Hgb, then q4wk; frequent serum electrolytes, serum creatinine, uric acid, and TSH.

Patient & Family Education

▪ Telaprevir must be used with both peginterferon alfa and ribavirin. If peginterferon alfa or ribavirin is discontinued for any reason, discontinue telaprevir.

▪ Use extreme care to avoid pregnancy during treatment and for 6 mo after discontinuation of treatment. Two reliable, nonhormonal methods of contraception should be initiated prior to starting treatment. Hormonal contraception may not be reliable during treatment and for up to 2 wk following discontinuation of telaprevir.

▪ Report promptly to prescriber any skin changes, itching, or fever.

TELAVANCIN HYDROCHLORIDE

(tel-a-van'sin)

Vibativ

Classifications: ANTIBIOTIC; LIPOGLYCOPROTEIN

Therapeutic: ANTIBIOTIC

Pregnancy Category: C

AVAILABILITY 250 mg, 750 mg lyophilized powder for reconstitution and injection

ACTION & *THERAPEUTIC EFFECT*
Telavancin is a lipoglycoprotein antibiotic derived from vancomycin with a mechanism of action related to inhibiting cell wall synthesis of bacteria. It binds to the bacterial membrane and disrupts the barrier function of the cell membrane of gram-positive bacteria. *Effective against a broad range of gram-positive bacteria.*

USES Treatment of complicated skin and skin structure infections caused by susceptible gram-positive bacteria.

UNLABELED USES Treatment of hospital-acquired nosocomial pneumonia caused by susceptible gram-positive bacteria.

CONTRAINDICATIONS End-stage renal disease or hemodialysis; uncompensated heart failure; QT prolongation.

CAUTIOUS USE Moderate or severe renal impairment; renal disease; elderly; ulcerative colitis; severe hepatic impairment; women of childbearing potential; pregnancy (category C), lactation. Safe use in children younger than 18 y has not been established.

ROUTE & DOSAGE

> **Complicated Skin and Skin Structure Infections**
>
> *Adult:* **IV** 10 mg/kg over 60 min q24h for 7–14 days
>
> **Renal Impairment Dosage Adjustment**
>
> *CrCl 30–50 mL/min:* Decrease to 7.5 mg/kg q 24 h; *10–29 mL/min:* Decrease to 10 mg/kg q 48 h

ADMINISTRATION

Intravenous

***PREPARE:* Infusion:** Reconstitute each 250 mg with 15 mL of D5W or NS to yield 15 mg/mL. Mix thoroughly to dissolve. For doses of 150–800 mg, must further dilute in 100–250 mL of IV solution. For doses less than 100 mg or greater than 800 mg, should be further diluted in D5W, NS, or LR to a final concentration in the range of 0.6–8 mg/mL.

ADMINISTER: Infusion: Infuse over 60 min or longer to minimize infusion-related reactions.

▪ Do not infuse through the same line with any other drugs or additives.

▪ Storage: Reconstituted vials or infusion solution should be used within 4 h if at room temperature or 72 h if refrigerated. **Important note:** The total time holding time for reconstituted vials plus infusion solution cannot exceed 4 h at room temperature or 72 h refrigerated.

ADVERSE EFFECTS (≥1%) **Body as a Whole:** Generalized pruritus, rigors, infusion site erythema and pain. **CNS:** Dizziness, *taste disturbance*. **GI:** Abdominal pain, *Clostridium difficile*-associated diarrhea, diarrhea, *nausea, vomiting*. **Metabolic:** Decreased appetite, *increased serum creatinine*. **Skin:** Pruritus, rash. **Urogenital:** *Foamy urine*.

DIAGNOSTIC TEST INTERFERENCE Telavancin may cause increases in *PT, INR, aPTT*, and *ACT*. Telavancin interferes with *urine qualitative dipstick protein assays*, as well as quantitative *dye methods* (e.g., *pyrogallol red-molybdate*).

PHARMACOKINETICS Distribution: 93% bound to albumin. **Metabolism:** Metabolized to 3-hydroxylated metabolites. **Elimination:** Primarily via the urine. **Half-Life:** 8–9 h.

NURSING IMPLICATIONS
Assessment & Drug Effects
▪ Lab tests: Baseline culture and sensitivity test before initiation of

therapy. Baseline and frequent (q48–72h) kidney function tests throughout therapy.

▪ Monitor for red man syndrome (i.e., flushing of upper body, urticaria, pruritus, or rash) during infusion. If syndrome develops, slow infusion immediately. If reaction does not cease, stop infusion and notify prescriber.

▪ Withhold drug and notify prescriber for CrCl of 50 mL/min or less.

▪ Monitor for and report promptly the onset of watery diarrhea with or without fever, passage of tarry or bloody stools, pus, or mucus.

▪ Monitor ECG with concurrent use of drugs known to prolong the QT interval.

Patient & Family Education
▪ Report promptly appearance of rash or itching during drug infusion.

▪ Report promptly loose stools or diarrhea even after completion of drug.

▪ Women should use effective means of contraception while on this drug. Notify prescriber if pregnancy occurs during treatment.

TELBIVUDINE
(tel-bi'vu-deen)
Tyzeka
Classifications: ANTIRETROVIRAL; NUCLEOSIDE REVERSE TRANSCRIPTASE INHIBITOR (NRTI)
Therapeutic: ANTIRETROVIRAL; NRTI
Prototype: Lamivudine
Pregnancy Category: B

AVAILABILITY 600 mg tablets

ACTION & THERAPEUTIC EFFECT
Telbivudine is a nucleoside analog

Common adverse effects in *italic*, life-threatening effects underlined; generic names in **bold**; classifications in SMALL CAPS; ♣ Canadian drug name; ⊕ Prototype drug

1437

with activity against hepatitis B virus (HBV) DNA polymerase. Its metabolite inhibits HBV DNA polymerase (reverse transcriptase) by competing with the natural nucleoside substrate. Incorporation into HBV viral DNA causes DNA chain termination, resulting in inhibition of HBV replication. *Effectiveness is measured by reducing the viral load and preventing infection of new hepatocytes.*

USES Hepatitis B infection.

CONTRAINDICATIONS Hypersensitivity to telbivudine; lactic acidosis; severe hepatomegaly with steatosis; lactation.

CAUTIOUS USE Moderate to severe renal impairment, hemodialysis; alcoholism; obesity in females; risk of hepatic disease; individuals with organ transplants; older adults; pregnancy (category B). Safe use in children younger than 16 y has not been established.

ROUTE & DOSAGE

Chronic Hepatitis B

Adults/Adolescents (16 y or older): **PO** 600 mg/day

Renal Impairment Dosage Adjustment

CrCl 50 mL/min or greater: No dosage adjustment
CrCl 30–49 mL/min: 600 mg q48h
CrCl less than 30 mL/min (not requiring dialysis): 600 mg q72h
CrCl less than 5–10 mL/min (ESRD): 600 mg q96h

ADMINISTRATION

Oral
- May be given without regard to food.
- Store at 15°–30° C (59°–86° F).

ADVERSE EFFECTS (≥1%) **Body as a Whole:** *Fatigue and malaise, headache, influenza-like syndrome, post-procedural pain,* pyrexia. **CNS:** Dizziness, insomnia. **GI:** *Abdominal pain, diarrhea and loose stools,* dyspepsia, *nausea, vomiting.* **Metabolic:** *Increased CPK levels,* <u>lactic acidosis and severe hepatomegaly with steatosis.</u> **Musculoskeletal:** Arthralgia, back pain, myalgia. **Respiratory:** *Cough, nasopharyngitis,* pharyngolaryngeal pain, *upper respiratory tract infection.* **Skin:** Rash.

INTERACTIONS Drug: Coadministration with drugs that alter renal function may alter plasma concentrations of telbivudine. Anti-HBV activity of telbivudine is additive with **adefovir** and is not antagonized by the HIV NRTIs **didanosine** and **stavudine.** Telbivudine is not antagonistic to anti-HIV activity of **abacavir, didanosine, emtricitabine, lamivudine, stavudine, tenofovir,** or **zidovudine.**

PHARMACOKINETICS Peak: 1–4 h. **Distribution:** Minimal protein binding; widely distributed in tissues. **Elimination:** Primarily unchanged in urine. **Half-Life:** 40–49 h.

NURSING IMPLICATIONS

Assessment & Drug Effects
- Monitor for and report S&S of lactic acidosis (e.g., anorexia, nausea, vomiting, bloating, abdominal pain, malaise, tachycardia or other arrhythmia, and difficulty in breathing).
- Withhold drug and notify prescriber of any of the following: Suspected lactic acidosis, steatosis, or markedly elevated liver enzymes.
- Lab tests: Monitor LFTs during and for several months after

discontinuation of telbivudine; monitor renal functions in the elderly and those taking drugs that may impair renal function; periodic serum bicarbonate.

Patient & Family Education
- Avoid all alcohol consumption while taking this drug.
- Report any of the following to a health care provider: Loss of appetite, nausea and vomiting, abdominal pain, palpitations, or difficulty breathing.

TELITHROMYCIN
(tel-i-thro-my'sin)
Ketek
Classifications: ANTIBIOTIC; KETOLIDE
Therapeutic: ANTIBIOTIC
Prototype: Erythromycin
Pregnancy Category: C

AVAILABILITY 300 mg, 400 mg tablets

ACTION & *THERAPEUTIC EFFECT*
Telithromycin binds to bacterial ribosomal RNA site of the 50S subunit; this action results in inhibition of RNA-dependent protein synthesis of bacteria, thus resulting in cell death. Telithromycin concentrates in phagocytes where it works against intracellular respiratory pathogens. *Its broad spectrum of activity is effective against respiratory pathogens, including erythromycin- and penicillin-resistant pneumococci.*

USES Treatment of community-acquired pneumonia
UNLABELED USES Sinusitis.

CONTRAINDICATIONS Macrolide antibiotic hypersensitivity; QT prolongation; ongoing proarrhythmic conditions such as hypokalemia, hypomagnesemia, significant bradycardia, myasthenia gravis unless no other therapeutic option is available; severe renal impairment or renal failure; viral infections.
CAUTIOUS USE History of GI disease; hepatic disease; history of hepatitis or jaundice; pregnancy (category C), lactation.

ROUTE & DOSAGE

Community-Acquired Pneumonia
Adult: **PO** 800 mg daily for 7–10 days

ADMINISTRATION

Oral
- Do not administer concurrently with simvastatin, lovastatin, atorvastatin, Class 1A (e.g., quinidine, procainamide) or Class III (e.g., dofetilide) antiarrhythmic agents.
- Store at 15°–30° C (59°–86° F). Keep container tightly closed. Protect from light.

ADVERSE EFFECTS (≥1%) **CNS:** Headache, dizziness. **CV:** Potential to cause QT_c prolongation. **GI:** *Diarrhea,* nausea, vomiting, loose stools, dysgeusia. **Metabolic:** Elevated LFTs, liver failure. **Musculoskeletal:** May exacerbate myasthenia gravis. **Special Senses:** Blurred vision, diplopia, difficulty focusing.

INTERACTIONS Drug: Pimozide or CLASS IA or CLASS III ANTIARRHYTHMICS may cause life-threatening arrhythmias; may increase concentrations of **atorvastatin, lovastatin, simvastatin,** BENZODIAZEPINES; **rifampin** decreases telithromycin levels; ERGOT DERIVATIVES (**ergotamine,**

Common adverse effects in *italic,* life-threatening effects underlined; generic names in **bold**; classifications in SMALL CAPS; ♣ Canadian drug name; ⊘ Prototype drug

1439

dihydroergotamine) may cause severe peripheral vasospasm; **theophylline** may exacerbate adverse GI effects. **Food: Grapefruit juice** (greater than 1 qt/day) may increase plasma concentrations and adverse effects.

PHARMACOKINETICS Absorption: 57% bioavailable. **Peak:** 1 h. **Metabolism:** 50% in liver (CYP3A4), 50% by CYP-independent mechanisms. **Elimination:** In urine and feces. **Half-Life:** 10 h.

NURSING IMPLICATIONS

Assessment & Drug Effects

- Monitor ECG in patients at risk for QT_c interval prolongation (i.e., bradycardia).
- Withhold drug and notify prescriber for S&S of QT_c interval prolongation such as dizziness or fainting.
- Monitor for and report promptly S&S of liver dysfunction: Fatigue, anorexia, nausea, clay-colored stools, etc.
- Lab tests: Baseline LFTs, BUN and creatinine, serum potassium.

Patient & Family Education

- Stop taking drug and notify prescriber for episodes of dizziness or fainting; report signs of jaundice (yellow color of the skin and/or eyes), unexplained fatigue, loss of appetite, nausea, dark urine, or clay-colored stool.
- Exercise caution when engaging in potentially hazardous activities; visual disturbances (e.g., blurred vision, difficulty focusing, double vision) are potential side effects of this drug. If visual problems occur, avoid quick changes in viewing between close and distant objects.

TELMISARTAN
(tel-mi-sar′tan)

Micardis
Classifications: ANGIOTENSIN II RECEPTOR ANTAGONIST; ANTIHYPERTENSIVE
Therapeutic: ANTIHYPERTENSIVE
Prototype: Losartan potassium
Pregnancy Category: C first trimester; D second and third trimester

AVAILABILITY 20 mg, 40 mg, 80 mg tablets

ACTION & *THERAPEUTIC EFFECT*
Angiotensin II receptor (type AT_1) antagonist. Selectively blocks the binding of angiotensin II to the AT_1 receptors in many tissues (e.g., vascular smooth muscles, adrenal glands). Blocks the vasoconstricting and aldosterone-secreting effects of angiotensin II, thus resulting in an antihypertensive effect. *Effectiveness is indicated by a reduction in BP.*

USES Treatment of hypertension, cardiovascular risk reduction.

CONTRAINDICATIONS Hypersensitivity to telmisartan or other angiotensin receptor antagonists (e.g., losartan, eprosartan, etc.); pregnancy (category D second and third trimester), lactation.

CAUTIOUS USE CAD hypertropic cardiomyopathy; CHF; oliguria; hypotension; renal artery stenosis; biliary obstruction; liver dysfunction; renal impairment; older adults; pregnancy (category C first trimester). Safety and efficacy in children younger than 18 y are not established.

ROUTE & DOSAGE

Hypertension

Adult: **PO** 40 mg daily, may increase to 80 mg/day

Common adverse effects in *italic*; life-threatening effects underlined; generic names in **bold**; classifications in SMALL CAPS; ♣ Canadian drug name; ☯ Prototype drug

CV Risk Reduction
Adult: **PO** 80 mg/day

ADMINISTRATION

Oral

- Do not remove tablets from blister pack until immediately before administration.
- Correct volume depletion prior to initial dose.
- Store at 15°–30° C (59°–86° F).

ADVERSE EFFECTS (≥ 1%) **Body as a Whole:** Back pain, flu-like syndrome, myalgia, fatigue. **CNS:** Dizziness, headache. **CV:** Hypotension, hypertension, chest pain, peripheral edema. **GI:** Diarrhea, dyspepsia, abdominal pain, nausea. **Respiratory:** Sinusitis, pharyngitis.

INTERACTIONS Drug: Telmisartan may increase **digoxin** or **lithium** levels.

PHARMACOKINETICS Absorption: Dose dependent, 42% of 40 mg dose is absorbed. **Peak:** 0.5–1 h. **Distribution:** Greater than 99% protein bound. **Metabolism:** Minimally metabolized. **Elimination:** Primarily in feces. **Half-Life:** 24 h.

NURSING IMPLICATIONS

Assessment & Drug Effects

- Monitor BP carefully after initial dose; and periodically thereafter. Monitor more frequently with preexisting biliary obstructive disorders or hepatic insufficiency.
- Monitor dialysis patients closely for orthostatic hypotension.
- Lab tests: Periodic Hgb, creatinine clearance, LFTs.
- Monitor concomitant digoxin levels throughout therapy.

Patient & Family Education

- Report pregnancy to prescriber immediately.

- Allow between 2–4 wk for maximum therapeutic response.

TEMAZEPAM
(te-maz'e-pam)

Restoril
Classifications: BENZODIAZEPINE; ANXIOLYTIC; SEDATIVE-HYPNOTIC
Therapeutic: ANTIANXIETY; SEDATIVE-HYPNOTIC
Prototype: Lorazepam
Pregnancy Category: X
Controlled Substance: Schedule IV

AVAILABILITY 7.5 mg, 15 mg, 30 mg capsules

ACTION & *THERAPEUTIC EFFECT*
Benzodiazepine derivative with hypnotic, anxiolytic, sedative effects. Principal effect is significant improvement in sleep parameters. Minimal change in REM sleep. *Reduces night awakenings and early morning awakenings; increases total sleep times, absence of rebound effects.*

USES To relieve insomnia.

CONTRAINDICATIONS Benzodiazepine hypersensitivity; ethanol intoxication; narrow-angle glaucoma; psychoses; pregnancy (category X), lactation.
CAUTIOUS USE Severely depressed patient or one with suicidal ideation; history of drug abuse or dependence, acute intoxication; alcoholism; COPD; liver or kidney dysfunction; sleep apnea; older adults. Safe use in children younger than 18 y not established.

ROUTE & DOSAGE

Insomnia
Adult: **PO** 7.5–30 mg at bedtime
Geriatric: **PO** 7.5 mg at bedtime

T

Common adverse effects in *italic*, life-threatening effects underlined; generic names in **bold**; classifications in SMALL CAPS; ♣ Canadian drug name; ◐ Prototype drug

1441

ADMINISTRATION

Oral

- Give 20–30 min before patient retires.
- Store at 15°–30° C (59°–86° F) in tight container unless otherwise specified by manufacturer.

ADVERSE EFFECTS (≥1%) **CNS:** *Drowsiness,* dizziness, lethargy, confusion, headache, euphoria, relaxed feeling, weakness. **GI:** Anorexia, diarrhea. **CV:** Palpitations.

INTERACTIONS Drug: Alcohol, CNS DEPRESSANTS, ANTICONVULSANTS potentiate CNS depression; **cimetidine** increases temazepam plasma levels, thus increasing its toxicity; may decrease antiparkinsonism effects of **levodopa;** may increase **phenytoin** levels; smoking decreases sedative effects. **Herbal: Kava, valerian** may potentiate sedation.

PHARMACOKINETICS Absorption: Readily from GI tract. **Onset:** 30–50 min. **Peak:** 2–3 h. **Duration:** 10–12 h. **Distribution:** Crosses placenta; distributed into breast milk. **Metabolism:** In liver to oxazepam. **Elimination:** In urine. **Half-Life:** 8–24 h.

NURSING IMPLICATIONS

Assessment & Drug Effects

- Be alert to signs of paradoxical reaction (excitement, hyperactivity, and disorientation) in older adults. Psychoactive drugs are the most frequent cause of acute confusion in this age group.
- CNS adverse effects are more apt to occur in the patient with hypoalbuminemia, liver disease, and in older adults. Report promptly incidence of bradycardia, drowsiness, dizziness, clumsiness, lack of coordination. Supervise ambulation, especially at night.

- Lab tests: Obtain LFTs and kidney function tests during long-term use.
- Be alert to S&S of overdose: Weakness, bradycardia, somnolence, confusion, slurred speech, ataxia, coma with reduced or absent reflexes, hypertension, and respiratory depression.

Patient & Family Education

- Be aware that improvement in sleep will not occur until after 2–3 doses of drug.
- Notify prescriber if dreams or nightmares interfere with rest. An alternate drug or reduced dose may be prescribed.
- Be aware that difficulty getting to sleep may continue. Drug effect is evidenced by the increased amount of rest once asleep.
- Consult prescriber if insomnia continues in spite of medication.
- Do not smoke after medication is taken.
- Do not use OTC drugs (especially for insomnia) without advice of prescriber.
- Consult prescriber before discontinuing drug especially after long-term use. Gradual reduction of dose may be necessary to avoid withdrawal symptoms.
- Avoid use of alcohol and other CNS depressants.
- Do not drive or engage in other potentially hazardous activities until response to drug is known. This drug may depress psychomotor skills and cause sedation.

TEMOZOLOMIDE

(tem-o-zol'o-mide)

Temodar
Classifications: ANTINEOPLASTIC; IMIDAZOTETRAZINE DERIVATIVE
Therapeutic: ANTINEOPLASTIC
Pregnancy Category: D

AVAILABILITY 5 mg, 20 mg, 100 mg, 140 mg, 180 mg, 250 mg capsules; 100 mg solution for injection

ACTION & *THERAPEUTIC EFFECT*
Cytotoxic agent with alkylating properties that are cell cycle nonspecific. Interferes with purine (e.g., guanine) metabolism and thus protein synthesis in rapidly proliferating cells. *Effectiveness is indicated by objective evidence of tumor regression.*

USES Adult patients with refractory anaplastic astrocytoma, glioblastoma multiforme with radiotherapy. **UNLABELED USES** Malignant melanoma.

CONTRAINDICATIONS Hypersensitivity to temozolomide, or dacarbazine; severe bone marrow suppression; pregnancy (category D), lactation. **CAUTIOUS USE** Bacterial or viral infection; severe hepatic or renal impairment; myelosuppression; prior radiotherapy or chemotherapy; older adults; children younger than 3 y.

ROUTE & DOSAGE

Astrocytoma

Adult: **PO/IV** 150 mg/m^2 daily days 1–5 per 28-day treatment cycle (may increase to 200 mg/m^2/day); subsequent doses are based on absolute neutrophil count on day 21 or at least 48 h before next scheduled cycle (see prescribing information for dosage adjustments based on neutrophil count)

Glioblastoma Multiforme

Adult: **PO** 75 mg/m^2 daily for 42 days with focal radiotherapy; after 4 wk, maintenance phase of 150–200 mg/m^2 on days 1–5 of 28-day cycle

ADMINISTRATION

Oral

- Give consistently with regard to food.
- Do not administer unless absolute neutrophil count greater than 1500 per microliter and platelet count greater than 100,000 per microliter.
- Do not open capsules. Avoid inhalation or contact with skin or mucous membranes, if accidentally opened/damaged.
- Store at room temperature, 15°–30° C (59°–86° F).

Intravenous

***PREPARE:* Infusion:** Bring the 100 mg vial to room temperature, then reconstitute with 41 mL sterile water to yield 2.5 mg/mL. Gently swirl to dissolve but do not shake, and do not further dilute. Must use within 14 h of reconstitution, including infusion time.
***ADMINISTER:* Infusion:** Infuse over 90 min. Flush line before and after with NS. Note: May be administered in the same line with NS but not with any other IV solution.

ADVERSE EFFECTS (≥1%) **Body as a Whole:** *Headache, fatigue, asthenia, fever,* back pain, myalgia, weight gain; viral infection. **CNS:** *Convulsions, hemiparesis, dizziness, abnormal coordination, amnesia, insomnia,* paresthesia, somnolence, paresis, ataxia, dysphasia, abnormal gait, confusion, anxiety, depression. **CV:** *Peripheral edema.* **GI:** *Nausea, vomiting, constipation, diarrhea,* abdominal pain, anorexia. **Hematologic:** Anemia, <u>neutropenia, thrombocytopenia,</u> leukopenia, lymphopenia. **Respiratory:** Upper respiratory tract infection, pharyngitis, sinusitis, cough. **Skin:** Rash, pruritus. **Metabolic:** Adrenal

T

Common adverse effects in *italic*, life-threatening effects <u>underlined</u>; generic names in **bold**; classifications in SMALL CAPS; ✦ Canadian drug name; ◉ Prototype drug

1443

hypercorticism. **Urogenital:** Urinary incontinence. **Special Senses:** Diplopia, abnormal vision.

INTERACTIONS Drug: Valproic acid may decrease **temozolomide** levels.

PHARMACOKINETICS Absorption: Rapidly. **Peak:** 1 h. **Metabolism:** Spontaneously metabolized to active metabolite MTIC. **Elimination:** Primarily in urine. **Half-Life:** 1.8 h.

NURSING IMPLICATIONS

Assessment & Drug Effects

- Monitor for S&S of toxicity: Infection, bleeding episodes, jaundice, rash, CNS disturbances.
- Lab tests: CBC with differential on day 22 and weekly until absolute neutrophil count (ANC) greater than 1500 per microliter and platelet count greater than 100,000 per microliter; periodic LFT and routine serum chemistry, including serum calcium.

Patient & Family Education

- Take consistently with respect to meals.
- Report to prescriber signs of infection, bleeding, discoloration of skin or skin rash, dizziness, lack of balance, or other bothersome side effects promptly.
- Exercise caution with hazardous activities until response to drug is known.
- Use effective methods of contraception; avoid pregnancy.

TEMSIROLIMUS

(tem-si-ro-li'mus)

Torisel

Classifications: BIOLOGIC RESPONSE MODIFIER; ANTINEOPLASTIC; PROTEIN-TYROSINE KINASE INHIBITOR (TKI)
Therapeutic: ANTINEOPLASTIC
Prototype: Gefitinib
Pregnancy Category: D

AVAILABILITY 25 mg/mL concentrated solution

ACTION & *THERAPEUTIC EFFECT* Inhibits an intracellular protein that controls cell division in renal carcinoma and other tumor cells. *Results in arrest of growth in tumor cells.*

USES Treatment of advanced renal cell carcinoma.
UNLABELED USES Astrocytoma, mantle cell lymphoma (MCL).

CONTRAINDICATIONS Live vaccines; intracranial bleeding; pregnancy (category D); lactation.
CAUTIOUS USE Hypersensitivity to temsirolimus, sirolimus, polysorbate 80, or antihistamines; renal impairment; diabetes mellitus; history of hyperlipemia; respiratory disorders including interstitial lung disease; perioperative period due to potential for abnormal wound healing; CNS tumors (primary or by metastasis); hepatic impairment. Safe use in children not established.

ROUTE & DOSAGE

Renal Cell Carcinoma

Adult: **IV** 25 mg qwk

Dosage Adjustment

Regimen with a strong CYP3A4 inhibitor: 12.5 mg/wk
Regimen with a strong CYP3A4 inducer: 50 mg based on tolerability

Hepatic Impairment Adjustment

Dose reduction required

ADMINISTRATION

Intravenous

Patients should receive prophylactic IV diphenhydramine 25 to

50 mg (or similar antihistamine) 30 min before each dose.

PREPARE: **IV Infusion:** Inject 1.8 mL of supplied diluent into the 25 mg/mL vial to yield 10 mg/mL.▪ Withdraw the required dose and inject rapidly into a 250 mL DEHP-free container of NS. Invert to mix.

ADMINISTER: **IV Infusion:** Use DEHP-free infusion line with a 5 micron or less in-line filter. ▪ Infuse over 30–60 min. ▪ Complete infusion within 6 h of preparation.

INCOMPATIBILITIES **Solution/additive:** Do not add other drugs or agents to temsirolimus IV solutions.

▪ Store at 2°–8° C (36°–46° F). Protect from light. The 10 mg/mL drug solution is stable for up to 24 h at 15°–30° C (59°–86° F).

ADVERSE EFFECTS (≥1%) **Body as a Whole:** Allergic/hypersensitivity reactions, *asthenia,* chest pain, chills, *edema,* impaired wound healing, infections, pain, pyrexia. **CNS:** Depression, dysgeusia, headache, insomnia. **CV:** Hypertension, thrombophlebitis, venous thromboembolism. **GI:** Abdominal pain, *anorexia,* constipation, diarrhea, fatal bowel perforation, *mucositis, nausea,* vomiting. **Hematologic:** Decrease hemoglobin, *leukocytopenia, lymphopenia,* neutropenia, <u>thrombocytopenia</u>. **Metabolic:** *Elevated alkaline phosphatase, elevated AST, elevated serum creatinine,* hypokalemia, *hypophosphatemia, hyperbilirubinemia, hypercholesterolemia, hyperglycemia, hypertriglyceridemia,* weight loss. **Musculoskeletal:** Arthralgia, back pain, myalgia. **Respiratory:** Cough, dyspnea, epistaxis, interstitial lung disease, pharyngitis, pneumonia, rhinitis, upper respiratory tract infection. **Skin:** Acne, dry skin, nail disorder, pruritus, *rash.* **Special Senses:** Conjunctivitis. **Urogenital:** Urinary tract infection.

INTERACTIONS Drug: AZOLE ANTIFUNGAL AGENTS **(fluconazole, itraconazole, ketoconazole, posaconazole, voriconazole), cyclosporine,** INHIBITORS OF CYP3A4 (HIV PROTEASE INHIBITORS, **clarithromycin, diltiazem**), **mycophenolate mofetil,** and **sunitinib** increase the plasma levels of **temsirolimus.** INDUCERS OF CYP3A4 (**dexamethasone, rifampin, rifabutin, phenytoin**) decrease the plasma level of temsirolimus. **Food: Grapefruit** and **grapefruit juice** increase the plasma level of temsirolimus. **Herbal: St. John's wort** decreases the plasma level of temsirolimus.

PHARMACOKINETICS Peak: 0.5–2 h. **Metabolism:** In liver. **Elimination:** Primarily in stool. **Half-Life:** 17.3 h.

NURSING IMPLICATIONS

Assessment & Drug Effects

▪ Withhold drug and notify prescriber for absolute neutrophil count less than 1000/mm³ or platelet count less than 75,000/mm³.

▪ Monitor for infusion-related reactions during and for at least 1 h after completion of infusion.

▪ Slow or stop infusion for infusion-related reactions. If infusion is restarted after 30–60 min of observation, slow rate to up to 60 min and continue observation.

▪ Monitor respiratory status and report promptly dyspnea, cough, S&S of hypoxia, fever.

▪ Lab tests: Baseline and periodic CBC with differential and platelet count, lipid profile, LFTs, alkaline phosphatase, kidney function tests, serum electrolytes, plasma glucose, ABGs.

Common adverse effects in *italic*, life-threatening effects <u>underlined</u>; generic names in **bold**; classifications in SMALL CAPS; ✚ Canadian drug name; ⊘ Prototype drug

1445

- Monitor diabetics for loss of glycemic control.

Patient & Family Education
- Avoid live vaccines and close contact with those who have received live vaccines.
- Use effective contraceptive measures to prevent pregnancy.
- Men with partners of childbearing age should use reliable contraception throughout treatment and for 3 mo after the last dose of temsirolimus.
- Report promptly any of the following: S&S of infection, difficulty breathing, abdominal pain, blood in stools, abnormal wound healing, S&S of hypersensitivity (see Appendix F).

TENECTEPLASE RECOMBINANT
(ten-ect′e-plase)

TNKase
Classification: THROMBOLYTIC ENZYME, TISSUE PLASMINOGEN ACTIVATOR (t-PA)
Therapeutic: THROMBOLYTIC ENZYME
Prototype: Alteplase
Pregnancy Category: C

AVAILABILITY 50 mg vial

ACTION & *THERAPEUTIC EFFECT*
Tenecteplase is a thrombolytic agent that activates plasminogen, a substance created by endothelial cells in response to arterial wall injury that contributes to clot formation. Plasminogen is converted to plasmin which breaks down the fibrin mesh that binds the clot together, thus dissolving the clot. *Effective in producing thrombolysis of a clot involved in a myocardial infarction.*

USES Reduction of mortality associated with acute myocardial infarction (AMI).

CONTRAINDICATIONS Active internal bleeding; history of CVA; intracranial or intraspinal surgery with 2 mo; intracranial neoplasm; arteriovenous malformation, or aneurysm; known bleeding diathesis; brain tumor; increased intracranial pressure; coagulopathy; head trauma; stroke; surgery; severe uncontrolled hypertension; lactation.

CAUTIOUS USE Recent major surgery, previous puncture of compressible vessels, CVA, recent GI or GU bleeding, recent trauma; hypertension, mitral valve stenosis, acute pericarditis, bacterial endocarditis; severe liver or kidney disease; hemorrhagic ophthalmic conditions; septic thrombophlebitis or occluded, infected AV cannula; advanced age; pregnancy (category C). Safety and efficacy in children are not established.

ROUTE & DOSAGE

Acute Myocardial Infarction
Adult: **IV** Infuse dose over 5 sec, *weight less than 60 kg, 30 mg; weight 60–70 kg, 35 mg; weight 70–80 kg, 40 mg; weight 80–90 kg, 45 mg; weight greater than 90 kg, 50 mg*

ADMINISTRATION

Intravenous

***PREPARE:* Direct:** Read and follow instructions supplied with TwinPak™ Dual Cannula Device.
- Withdraw 10 mL of sterile water for injection from the supplied vial; inject entire contents into the TNKase vial directing the diluent stream into the powder.
- Gently swirl until dissolved but do not shake. The resulting solution contains 5 mg/mL.
- Withdraw the appropriate dose

Common adverse effects in *italic*, life-threatening effects underlined; generic names in **bold**; classifications in SMALL CAPS; ♣ Canadian drug name; ⊘ Prototype drug

and discard any unused solution. ▪ Follow directions supplied with TwinPak™ for proper handling of syringe.

ADMINISTER: Direct: Dextrose-containing IV line **must be** flushed before and after bolus with NS. ▪ Give as a single bolus dose over 5 sec. ▪ The total dose given should not exceed 50 mg.

INCOMPATIBILITIES Solution/additive: Dextrose solutions.

▪ Store unopened TwinPak™ at or below 30° C (86° F) or under refrigeration at 2°–8° C (36°–46° F).

ADVERSE EFFECTS (≥1%) **Hematologic:** <u>Major bleeding</u>, *hematoma,* GI bleed, bleeding at puncture site, hematuria, pharyngeal, epistaxis.

DIAGNOSTIC TEST INTERFERENCE Unreliable results for ***coagulation test I*** and measures of ***fibrinolytic activity.***

PHARMACOKINETICS Metabolism: In liver. **Half-Life:** 90–130 min.

NURSING IMPLICATIONS

Assessment & Drug Effects

▪ Avoid IM injections and unnecessary handling or invasive procedures for the first few hours after treatment.
▪ Monitor for S&S of bleeding. Should bleeding occur, withhold concomitant heparin and antiplatelet therapy; notify prescriber.
▪ Monitor cardiovascular and neurologic status closely. Persons at increased risk for life-threatening cardiac events include those with: A high potential for bleeding, recent surgery, severe hypertension, mitral stenosis and atrial fibrillation, anticoagulant therapy, and advanced age.
▪ Lab tests: Baseline and 1 h after administration of drug determine cardiac enzymes, circulating myoglobin, cardiac troponin-1, creatine kinase-MB; Hgb and Hct post-infusion; frequent aPTT, PT, and thrombin time.
▪ Coagulation parameters may not predict bleeding episodes.

Patient & Family Education

▪ Notify prescriber of the following immediately: A sudden, severe headache; any sign of bleeding; signs or symptoms of hypersensitivity (see Appendix F).
▪ Stay as still as possible and do not attempt to get out of bed until directed to do so.

TENOFOVIR DISOPROXIL FUMARATE

(ten-o-fo'vir di-so-prox'il fum'a-rate)

Viread

Classifications: ANTIRETROVIRAL; NUCLEOSIDE REVERSE TRANSCRIPTASE INHIBITOR (NRTI)

Therapeutic: ANTIRETROVIRAL; NRTI

Prototype: Zidovudine

Pregnancy Category: B

AVAILABILITY 300 mg tablets

ACTION & *THERAPEUTIC EFFECT* Tenofovir is a potent inhibitor of retroviruses, including HIV-1. The active form of tenofovir persists in HIV-infected cells for prolonged periods, thus, it results in sustained inhibition of HIV replication. *It reduces the viral load (plasma HIV-RNA), and CD4 counts.*

USES In combination with other antiretrovirals for the treatment of HIV; chronic hepatitis B infection.

CONTRAINDICATIONS Hypersensitivity to tenofovir; lactic acidosis, severe hepatomegaly, concurrent

Common adverse effects in *italic,* life-threatening effects <u>underlined;</u> generic names in **bold;** classifications in SMALL CAPS; ♣ Canadian drug name; ☻ Prototype drug

1447

administration of nephrotoxic agents, acute renal failure; lactation. **CAUTIOUS USE** Hepatic dysfunction, alcoholism, know risks for hepatic dysfunction; renal impairment; obesity; pathologic bone fractures; older adults; pregnancy (category B); children.

ROUTE & DOSAGE

HIV Infection
Adult: **PO** 300 mg once daily with meal

Chronic Hepatitis B
Adult: **PO** 300 mg daily

Renal Impairment Dosage Adjustment
CrCl 30–49 mL/min: Dose q48h; *10–29 mL/min:* Dose twice weekly

Hemodialysis Dosage Adjustment
Dose weekly or after 12 h of dialysis

ADMINISTRATION

Oral
- Give at the same time each day with a meal.
- Give 2 h before or 1 h after didanosine (if ordered concurrently).
- Store at room temperature; excursions to 15°–30° C (59°–80° F) are permitted.

ADVERSE EFFECTS (≥1%) **Body as a Whole:** Asthenia. **CNS:** Headache. **GI:** *Nausea,* vomiting, diarrhea, flatulence, abdominal pain, anorexia. **Hematologic:** Neutropenia. **Metabolic:** Increased *creatine kinase,* AST, ALT, serum amylase, triglycerides, serum glucose.

INTERACTIONS Drug: May increase **didanosine** toxicity; **acyclovir, amphotericin B, cidofovir, foscarnet, ganciclovir, probenecid, valacyclovir, valganciclovir** may increase tenofovir toxicity by decreasing its renal elimination. **Food:** Food increases absorption.

PHARMACOKINETICS Absorption: Bioavailability 25% fasting, 40% with high fat meal. **Peak:** 1 h. **Distribution:** Less than 7% protein bound. **Metabolism:** Not metabolized by CYP450 enzyme system. **Elimination:** Renally eliminated. **Half-Life:** 11–14 h.

NURSING IMPLICATIONS

Assessment & Drug Effects
- Lab tests: Monitor baseline and periodic renal function and LFTs; monitor periodically serum electrolytes, and ABGs if lactic acidosis is suspected.
- Monitor for S&S of bone abnormalities (e.g., bone pain, stress fractures).
- Monitor closely patients receiving other nephrotoxic agents for changes in serum creatinine and phosphorus. Withhold drug and notify prescriber for creatinine clearance less than 60 mL/min.
- Withhold drug and notify prescriber if patient develops clinical or lab findings suggestive of lactic acidosis or pronounced hepatotoxicity (e.g., hepatomegaly and steatosis even in the absence of marked transaminase elevations).

Patient & Family Education
- Take this drug exactly as prescribed. Do not miss any doses. If you miss a dose, take it as soon as possible and then take your next dose at its regular time. If it is almost time for your next dose, do not take the missed dose. Wait and take the next dose at the regular time. Do not double the next dose.

- Report any of the following to prescriber: Unexplained anorexia, nausea, vomiting, abdominal pain, fatigue, dark urine.

TERAZOSIN
(ter-ay′zoe-sin)

Classifications: ALPHA-ADRENERGIC RECEPTOR ANTAGONIST; ANTIHYPERTENSIVE
Therapeutic: ANTIHYPERTENSIVE; BPH AGENT
Prototype: Prazosin
Pregnancy Category: C

AVAILABILITY 1 mg, 2 mg, 5 mg, 10 mg capsules

ACTION & *THERAPEUTIC EFFECT*
Selectively blocks alpha₁-adrenergic receptors in vascular smooth muscle in many tissues, including vascular smooth muscle, the bladder neck, and the prostate. Promotes vasodilation, thus producing relaxation that leads to reduction of peripheral vascular resistance and lowered BP as well as increased urine flow. *Effectiveness is measured in lowering of blood pressure values and controlling the symptoms of benign prostate hypertrophy.*

USES To treat hypertension alone or in combination with other antihypertensive agents (beta-adrenergic blocking agents, diuretics). To treat benign prostatic hypertrophy (BPH) and urinary flow obstruction.

CONTRAINDICATIONS Hypersensitivity to terazosin.
CAUTIOUS USE Patients with BPH; prostate cancer; history of hypotensive episodes; angina; renal impairment, renal disease, renal failure; elderly; pregnancy (category C), lactation. Safe use in children is not established.

ROUTE & DOSAGE

Hypertension, Benign Prostatic Hypertrophy
Adult: **PO** Start with 1 mg at bedtime, then 1–5 mg/day (max: 20 mg/day)

ADMINISTRATION
Oral
- Give initial dose at bedtime to reduce the potential for severe hypotensive effect. After the initial dose, give any time of day.
- Store at 15°–30° C (59°–86° F) in tightly closed container away from heat and strong light. Do not freeze.

ADVERSE EFFECTS (≥1%) CNS: *Asthenia (weakness), dizziness, headache,* drowsiness, weakness. **CV:** Postural hypotension, palpitation, *first-dose phenomenon (syncope).* **Special Senses:** Blurred vision. **GI:** Nausea. **Body as a Whole:** Weight gain, pain in extremities, peripheral edema. **Respiratory:** Nasal congestion, sinusitis, dyspnea. **Urogenital:** Impotence.

INTERACTIONS Drug: Antihypertensive effects may be attenuated by NSAIDS. **Sildenafil, vardenafil,** and **tadalafil** may enhance hypotensive effects.

PHARMACOKINETICS Absorption: Readily from GI tract. **Peak:** 1–2 h. **Metabolism:** In liver. **Elimination:** 60% in feces, 40% in urine. **Half-Life:** 9–12 h.

NURSING IMPLICATIONS
Assessment & Drug Effects
- Be alert for possible first-dose phenomenon (precipitous decline in BP with consciousness disturbance). This is rare; occurs within 90–120 min of initial dose.

Common adverse effects in *italic*, life-threatening effects underlined; generic names in **bold**; classifications in SMALL CAPS; ✤ Canadian drug name; ⊘ Prototype drug

1449

- Monitor BP at end of dosing interval (just before next dose) to determine level of antihypertensive control.
- Be aware that drug-induced decrease in BP appears to be more position dependent (i.e., greater in the erect position) during the first few hours after dosing than at end of 24 h.
- A greatly diminished hypotensive response at end of 24 h indicates need for change in dosage (increased dose or twice daily regimen). Report to prescriber.

Patient & Family Education
- Avoid situations that would result in injury should syncope (loss of consciousness) occur after first dose. If faintness develops, lie down promptly.
- Make position changes slowly (i.e., change in direction or from recumbent to upright posture). Dangle legs and move ankles a minute or so before standing when arising. Orthostatic hypotension (greatest shortly after dosing) can pose a problem with ambulation.
- Do not drive or engage in potentially hazardous activities for at least 12 h after first dose, after dosage increase, or when treatment is resumed after interruption of therapy.
- Monitor weight: Report sudden gain of more than 0.5–1 kg (1–2 lb) accompanied by edema in extremities to prescriber. Dose adjustment may be indicated.
- Do not alter established drug regimen. Consult prescriber if drug is omitted for several days. Drug will be started with the initial dosing regimen.
- Keep scheduled appointments for assessment of BP control and other clinically significant tests.

- Do not take OTC medications, particularly those that may contain an adrenergic agent (e.g., remedies for coughs, colds, allergy) without first consulting prescriber.

TERBINAFINE HYDROCHLORIDE ℞

(ter-bin'a-feen)

Lamisil, Lamisil DermaGel
Classifications: ANTIBIOTIC; ANTIFUNGAL
Therapeutic: ANTIFUNGAL
Pregnancy Category: B

AVAILABILITY 250 mg tablets; 1% cream; 1.12% gel

ACTION & THERAPEUTIC EFFECT Synthetic antifungal agent that inhibits sterol biosynthesis in fungi and ultimately causes fungal cell death. *Effective as an antifungal.*

USES Topical treatment of superficial mycoses such as interdigital tinea pedis, tinea cruris, and tinea corporis oral treatment of onychomycosis due to tinea unguium.

CONTRAINDICATIONS Hypersensitivity to terbinafine; alcoholism; hepatic disease; hepatitis; jaundice; renal impairment; renal failure; lactation.
CAUTIOUS USE History of depression; pregnancy (category B). Safety and efficacy in children younger than 12 y are not established.

ROUTE & DOSAGE

Tinea Pedis, Tinea Cruris, or Tinea Corporis

Adult: **Topical** Apply daily or b.i.d. to affected and immediately surrounding areas until clinical signs and symptoms are significantly improved (1–7 wk)

T

Onychomycosis
Adult: **PO** 250 mg daily × 6 wk for fingernails or × 12 wk for toenails

ADMINISTRATION

Topical
- Apply externally. Avoid application to mucous membranes and avoid contact with eyes.
- Do not use occlusive dressings unless specifically directed to do so by prescriber.
- Store at 15°–30° C (59°–86° F).

ADVERSE EFFECTS (≥1%) **Skin:** Pruritus, local burning, dryness, rash, vesiculation, redness, contact dermatitis at application site. **CNS:** *Headache.* **GI:** Diarrhea, dyspepsia, abdominal pain, liver test abnormalities, liver failure (rare). **Hematologic:** Neutropenia (rare). **Special Senses:** Taste disturbances, vision impairment.

INTERACTIONS Drug: May increase **theophylline** levels; may decrease **cyclosporine** levels; **rifampin** may decrease **terbinafine** levels.

PHARMACOKINETICS Absorption: 70% PO; approximately 3.5% of topical dose is absorbed systemically. **Elimination:** In urine. **Half-Life:** 36 h.

NURSING IMPLICATIONS

Assessment & Drug Effects
- Monitor for and report increased skin irritation.

Patient & Family Education
- Learn correct technique for application of cream.
- Notify prescriber if drug causes increased skin irritation or sensitivity.

- Be aware that medication **must be** used for full treatment time to be effective.

TERBUTALINE SULFATE
(ter-byoo′te-leen)

Classifications: BETA-ADRENERGIC RECEPTOR AGONIST; BRONCHODILATOR
Therapeutic: RESPIRATORY SMOOTH MUSCLE RELAXANT; BRONCHODILATOR
Prototype: Albuterol
Pregnancy Category: C

AVAILABILITY 2.5 mg, 5 mg tablets; 1 mg/mL injection

ACTION & *THERAPEUTIC EFFECT* Synthetic adrenergic stimulant with selective $beta_2$-receptor activity in bronchial smooth muscles, inhibits histamine release from mast cells, and increases ciliary motility. *Relieves bronchospasm in chronic obstructive pulmonary disease (COPD) and significantly increases vital capacity. Increases uterine relaxation (thereby preventing or abolishing high intrauterine pressure).*

USES Orally or subcutaneously as a bronchodilator in bronchial asthma and for reversible airway obstruction associated with bronchitis and emphysema.

UNLABELED USES To delay delivery in preterm labor (FDA warns avoid use for this purpose).

CONTRAINDICATIONS Known hypersensitivity to sympathomimetic amines; severe hypertension and coronary artery disease; tachycardia with digitalis intoxication; within 14 days of MAO inhibitor therapy; angle-closure glaucoma; acute or maintenance tocolysis; prolonged use during preterm labor.

CAUTIOUS USE Angina, stroke, hypertension; DM; thyrotoxicosis;

Common adverse effects in *italic*, life-threatening effects underlined; generic names in **bold**; classifications in SMALL CAPS; ✦ Canadian drug name; ⊘ Prototype drug

1451

history of seizure disorders; MAOI therapy; cardiac arrhythmias; QT prolongation; thyroid disease; older adults; kidney and liver dysfunction; pregnancy (category C). Use caution in second and third trimester (may inhibit uterine contractions and labor).

ROUTE & DOSAGE

Bronchodilator

Adult: **PO** 5 mg t.i.d. at 6 h intervals (max: 15 mg/day) **Subcutaneous** 0.25 mg q15–30min up to 0.5 mg in 4 h *Adolescent (12–15 y):* **PO** 2.5 mg t.i.d. at 6 h intervals (max: 7.5 mg/day) **Subcutaneous** 0.25 mg q15–30min up to 0.5 mg in 4 h

ADMINISTRATION

Oral

- Give with fluid of patient's choice; tablets may be crushed.
- Be certain about recommended doses: PO preparation, 2.5 mg; subcutaneous, 0.25 mg. A decimal point error can be fatal.
- Give with food if GI symptoms occur.

Subcutaneous

- Give subcutaneous injection into lateral deltoid area.
- Store all forms at 15°–30° C (59°–86° F); protect from light. Do not freeze.

ADVERSE EFFECTS (≥1%) **CNS:** *Nervousness, tremor,* headache, *light-headedness,* drowsiness, fatigue, seizures. **CV:** *Tachycardia,* hypotension or hypertension, *palpitation,* maternal and fetal tachycardia. **GI:** Nausea, vomiting. **Body as a Whole:** Sweating, muscle cramps.

DIAGNOSTIC TEST INTERFERENCE

Terbutaline may increase **blood glucose** and free **fatty acids.**

INTERACTIONS Drug: Epinephrine, other SYMPATHOMIMETIC BRONCHODILATORS may add to effects; MAO INHIBITORS, TRICYCLIC ANTIDEPRESSANTS potentiate action on vascular system; effects of both BETA-ADRENERGIC BLOCKERS and terbutaline antagonized.

PHARMACOKINETICS Absorption: 33–50% from GI tract. **Onset:** 30 min PO; less than 15 min Subcutaneous; 5–30 min inhaled. **Peak:** 2–3 h PO; 30–60 min Subcutaneous; 1–2 h inhaled. **Duration:** 4–8 h PO; 1.5–4 h Subcutaneous; 3–4 h inhaled. **Distribution:** Into breast milk. **Metabolism:** In liver. **Elimination:** Primarily in urine, 3% in feces. **Half-Life:** 3–4 h.

NURSING IMPLICATIONS

Assessment & Drug Effects

- Assess vital signs: Baseline pulse and BP and before each dose. If significantly altered from baseline level, consult prescriber. Cardiovascular adverse effects are more apt to occur when drug is given by subcutaneous route or it is used by a patient with cardiac arrhythmia.
- Most adverse effects are transient, however, rapid heart rate may persist for a relatively long time.
- Aerosolized drug produces minimal cardiac stimulation or tremors.
- Be aware that muscle tremor is a fairly common adverse effect that appears to subside with continued use.
- Monitor patient being treated for premature labor for CV S&S for 12 h after drug is discontinued. Report tachycardia promptly.
- Monitor I&O ratio. Fluid restriction may be necessary. Consult prescriber.

Patient & Family Education

- Inhalator therapy: Review instructions for use of inhalator (included in the package).
- Learn how to take your own pulse and the limits of change that indicate need to notify the prescriber.
- Consult prescriber if breathing difficulty is not relieved or if it becomes worse within 30 min after an oral dose.
- Consult prescriber if symptomatic relief wanes; tolerance can develop with chronic use.
- Do not self-dose this drug, particularly during long-term therapy. In the face of waning response, increasing the dose will not improve the clinical condition and may cause overdosage. Understand that decreasing relief with continued treatment indicates need for another bronchodilator, not an increase in dose.
- Do not puncture container, use or store it near heat or open flame, or expose to temperatures above 49° C (120° F), which may cause bursting. Contents of the aerosol (inhalator) are under pressure.
- Do not use any other aerosol bronchodilator while being treated with aerosol terbutaline. Do not self-medicate with an OTC aerosol.
- Do not use OTC drugs without prescriber approval. Many cold and allergy remedies, for example, contain a sympathomimetic agent that when combined with terbutaline may cause harmful adverse effects.

TERCONAZOLE

(ter-con′a-zole)
Terazol 7, Terazol 3
Classification: AZOLE ANTIFUNGAL
Therapeutic: VAGINAL ANTIFUNGAL
Prototype: Fluconazole
Pregnancy Category: C

AVAILABILITY 0.4%, 0.8% vaginal cream; 80 mg vaginal suppositories

ACTION & THERAPEUTIC EFFECT
Terconazole is thought to exert antifungal activity by disruption of normal fungal cell membrane permeability. *Exhibits fungicidal activity against Candida albicans.*

USES Local treatment of vulvovaginal candidiasis.

CONTRAINDICATIONS Hypersensitivity to terconazole or azole antifungals; use of tampons; lactation.
CAUTIOUS USE Pregnancy (category C). Safety and efficacy in children younger than 18 y are not established.

ROUTE & DOSAGE

Candidiasis
Adult: **Intravaginal** One suppository (2.5 g) each night × 3 days; one applicator full of 0.4% cream each night × 7 days; one applicator full of 0.8% cream each night × 3 days

ADMINISTRATION

Intravaginal
- Insert applicator high into the vagina (except during pregnancy).
- Wash applicator before and after each use.
- Store away from direct heat and light.

ADVERSE EFFECTS (≥1%) **CNS:** *Headache.* **Urogenital:** Vaginal itching, burning, irritation. **Body as a Whole:** Rash, flu-like syndrome (fever, chills, headache, hypotension).

INTERACTIONS Drug: May inactivate **nonoxynol-9** spermicides.

PHARMACOKINETICS Absorption: Slow minimal absorption from

Common adverse effects in *italic*, life-threatening effects <u>underlined</u>; generic names in **bold**; classifications in SMALL CAPS; ♣ Canadian drug name; ⊕ Prototype drug

1453

vagina. **Onset:** Within 3 days. **Metabolism:** In liver. **Elimination:** Half in urine, half in feces. **Half-Life:** 4–11 h.

NURSING IMPLICATIONS

Assessment & Drug Effects

- Do not use if patient has a history of allergic reaction to other antifungal agents, such as miconazole.
- Monitor for sensitization and irritation; these may indicate need to discontinue drug.

Patient & Family Education

- Use correct application technique.
- Do not use tampons concurrently with terconazole.
- Learn potential adverse reactions, including sensitization and allergic response.
- Be aware that terconazole may interact with diaphragms and latex condoms; avoid concurrent use within 72 h.
- Refrain from sexual intercourse while using terconazole.
- Wear only cotton underwear; change daily.

TERIPARATIDE

(ter-i-par'a-tide)

Forteo
Classification: PARATHYROID HORMONE AGONIST
Therapeutic: PARATHYROID HORMONE AGONIST
Pregnancy Category: C

AVAILABILITY 750 mcg/3 mL injection

ACTION & *THERAPEUTIC EFFECT*
Parathyroid hormone (PTH) is the primary regulator of calcium and phosphate metabolism in bone and kidney. Biological actions of PTH and teriparatide are similar in bone

and kidneys. *Stimulates new bone formation by preferential stimulation of osteoblastic activity over osteoclastic activity; improves bone microarchitecture, and increases bone mass and strength by stimulating new bone formation.*

USES Treatment of osteoporosis in postmenopausal women at high risk for fracture; increase bone mass in men with primary or hypogonadal osteoporosis who are at high risk for fracture.

CONTRAINDICATIONS Hypersensitivity to teriparatide; osteosarcoma; Paget's disease; unexplained elevations of alkaline phosphatase; bone metastases or a history of skeletal malignancies; metabolic bone diseases other than osteoporosis; preexisting hypercalcemia; prior history of radiation therapy involving the skeleton; pediatric patients or young adults with open epiphyses; lactation.

CAUTIOUS USE Active or recent urolithiasis, hypercalciuria; hypotension; concurrent use of digitalis; hepatic, renal, and cardiac disease; pregnancy (category C); children younger than 18 y.

ROUTE & DOSAGE

Osteoporosis
Adult: **Subcutaneous** 20 mcg daily

ADMINISTRATION

Subcutaneous

- Do not administer to anyone with hypercalcemia. Consult prescriber.
- Rotate subcutaneous injection sites.

ADVERSE EFFECTS (≥1%) **Body as a Whole:** *Pain,* asthenia, neck pain. **CNS:** Headache, dizziness, depression, insomnia, vertigo. **CV:** Hypertension, angina, syncope.

Common adverse effects in *italic,* life-threatening effects underlined; generic names in **bold**; classifications in SMALL CAPS; ♣ Canadian drug name; ⊕ Prototype drug

GI: Nausea, constipation, dyspepsia, vomiting. **Metabolic:** *Transient increase in calcium levels,* increase in serum uric acid, antibodies to teriparatide after 12 mo therapy. **Musculoskeletal:** *Arthralgia,* leg cramps. **Respiratory:** Rhinitis, cough, pharyngitis, dyspnea, pneumonia. **Skin:** Rash, sweating.

INTERACTIONS Drug: May increase risk of **digoxin** toxicity.

PHARMACOKINETICS Absorption: Extensively absorbed from subcutaneous site. **Onset:** 2 h for calcium concentration increase. **Peak:** Max calcium concentrations 4–6 h. **Duration:** 16–24 h. **Metabolism:** Parathyroid hormone is metabolized by nonspecific enzymes. **Elimination:** Primarily in urine. **Half-Life:** 1 h Subcutaneous.

NURSING IMPLICATIONS

Assessment & Drug Effects
- Monitor cardiovascular status including BP and subjective reports of angina.
- Lab tests: Monitor periodically serum calcium, alkaline phosphatase, uric acid and bone density levels.
- Concurrent drugs: Monitor closely for digoxin toxicity with concurrent use.

Patient & Family Education
- Report unexplained leg cramps and bone pain.
- Learn correct technique for subcutaneous injection.

TESAMORELIN ACETATE
(tes′ a moe rel′ in as′ e tate)

Egrifta
Classifications: GROWTH HORMONE RELEASING FACTOR ANALOG; GROWTH HORMONE MODIFIER

Therapeutic: GROWTH HORMONE MODIFIER
Pregnancy Category: X

AVAILABILITY 1 mg powder for reconstitution and injection

ACTION & *THERAPEUTIC EFFECT* An analog of human growth hormone-releasing factor (GHRF) which acts on the pituitary. *It stimulates the synthesis and pulsatile release of endogenous growth hormone (GH) thereby increasing IGF-1 and insulin-like growth factor-binding protein 3 (IGFBP-3) levels.*

USES Reduction of excess abdominal fat in HIV-infected patients with lipodystrophy.

CONTRAINDICATIONS Active malignancy; hypersensitivity to tesamorelin and/or mannitol; suppression of the hypothalamic-pituitary from hypophysectomy, hypopituitarism, pituitary tumor/surgery, head irradiation or trauma; pregnancy (category X); lactation.

CAUTIOUS USE Acute illness or trauma; injection site reactions; nonmalignant neoplasms; renal or hepatic impairment; conditions resulting in or associated with fluid retention (e.g., edema, carpal tunnel syndrome); glucose intolerance or diabetes; retinopathy; age 65 y and older. Safety and effectiveness in children have not been established.

ROUTE & DOSAGE

Lipodystrophy
Adult: **SC** 2 mg once daily

ADMINISTRATION

Subcutaneous
- *Preparation:* Add 2.2 mL of sterile water into the tesamorelin vial;

T

Common adverse effects in *italic,* life-threatening effects <u>underlined</u>; generic names in **bold**; classifications in SMALL CAPS; ♦ Canadian drug name; ⊙ Prototype drug

1455

inject slowly against side of vial to avoid foaming. Keep vial upright with syringe in place and roll vial gently (do not shake) for 30 sec, until mixed.

- Withdraw entire contents of vial (2.2 mL) immediately after reconstitution and inject into abdomen. Rotate injection sites.
- Store dry vials at 2°–8° C (36°–46° F). Protect from light. Reconstituted solution should be discarded if not used immediately.

ADVERSE EFFECTS (≥1%) **Body as a Whole:** Chest pain, pain, peripheral edema. **CNS:** Depression, hypesthesia, insomnia, pareshesia, peripheral neuropathy. **CV:** Hypertension, palpitations. **GI:** Dyspepsia, nausea, upper abdominal pain, vomiting. **Injection site:** *Erythema*, hemorrhage, irritation, pain, pruritis, rash, reaction, swelling, urticaria. **Metabolic:** Increased blood creatine phosphate. **Musculoskeletal:** *Arthralgia*, carpal tunnel syndrome, joint stiffness and swelling, muscle spasm, muscle pain, musculoskeletal pain, musculoskeletal stiffness, *myalgia, pain in extremity*. **Skin:** Hot flush, night sweats, pruritis, rash, urticarial.

INTERACTIONS Drug: Tesamorelin may decrease the enzymatic activation of **cortisone** and **prednisone**. Tesamorelin may alter the metabolism of compounds requiring CYP-450 enzymes; careful monitoring is suggested.

PHARMACOKINETICS Peak: 15 min. **Half-Life:** 26–38 min.

NURSING IMPLICATIONS

Assessment & Drug Effects

- Evaluate abdominal girth with periodic measurements at the level of the umbilicus.

- Assess for injections site reactions (i.e., erythema, pruritus, pain, irritation, and bruising).
- Monitor prediabetics and diabetics for loss of glycemic control.
- Lab tests: Baseline and frequent IGF-1 levels; baseline and periodic FBG and HgA1$_C$.

Patient & Family Education

- Seek immediate medical attention for any of the following: Rash or hives; swelling of face or throat; shortness of breath or trouble breathing; rapid heartbeat; feeling of faintness or fainting.
- Notify your prescriber if you experience swelling, an increase in joint pain, or pain or numbness in your hands or wrist (carpal tunnel syndrome).
- Women should discontinue tesamorelin and notify prescriber if pregnancy occurs.
- If diabetic, monitor fasting and postprandial blood glucose as directed.

TESTOSTERONE ⓟ
(tess-toss′ter-one)

Androderm, AndroGel, Axiron, Foresta, Testim, Testopel

TESTOSTERONE CYPIONATE
Andro-Cyp, Depo-Testosterone, Depotest

TESTOSTERONE ENANTHATE
Delatestryl, Malogex ♦

Classifications: ANDROGEN/ANABOLIC STEROID; ANTINEOPLASTIC
Therapeutic: ANTINEOPLASTIC; ANABOLIC STEROID
Pregnancy Category: X
Controlled Substance: Schedule III

AVAILABILITY Testosterone: 75 mg implantable pellets; 2.5 mg/24 h, 4 mg/24 h, 5 mg/24 h, 6 mg/24 h,

T

transdermal patch; 1% gel; 2.5 g, 5 g gel packets; 30 mg buccal patch; **Testosterone Cypionate:** 100 mg/mL, 200 mg/mL injection; **Testosterone Enanthate:** 100 mg/mL, 200 mg/mL injection

ACTION & *THERAPEUTIC EFFECT*

Synthetic steroid compound with both androgenic and anabolic activity. Controls development and maintenance of secondary sexual characteristics. **Androgenic activity:** Responsible for the growth spurt of the adolescent, onset of puberty, and for growth termination by epiphyseal closure. **Anabolic activity:** Increases protein metabolism and decreases its catabolism. Large doses suppress spermatogenesis, thereby causing testicular atrophy. *Antagonizes effects of estrogen excess on female breast and endometrium. Responsible for the growth spurt of the adolescent male and onset of puberty.*

USES Androgen replacement therapy, delayed puberty (male), palliation of female mammary cancer (1–5 y postmenopausal), and to treat postpartum breast engorgement. Available in fixed combination with estrogens in many preparations.

CONTRAINDICATIONS Hypersensitivity or toxic reactions to androgens; serious cardiac, liver, or kidney disease; hypercalcemia; known or suspected prostatic or breast cancer in male; benign prostatic hypertrophy with obstruction; patients easily stimulated sexually; asthenic males who may react adversely to androgenic overstimulation; conditions aggravated by fluid retention; hypertension; older adults; pregnancy (category X), possibility of virilization of external genitalia of female fetus, lactation.

CAUTIOUS USE Cardiac, liver, and kidney disease; prepubertal males; DM; history of MI; CAD; BPH; geriatric patients, acute intermittent porphyria.

ROUTE & DOSAGE

Male Hypogonadism

Adult: **IM Cypionate, Enanthate** 50–400 mg q2–4wk **Topical** Start with 6 mg/day system applied daily, if scrotal area inadequate, use 4 mg/day system; **Androderm** Apply to torso; **AndroGel** Apply one packet to upper arms, shoulders, or abdomen once daily; **Striant** Apply one patch to the gum region just above the incisor tooth q12h

Delayed Puberty

Adult: **IM Cypionate, Enanthate** 50–200 mg q2–4wk **SC** 2–6 pellets inserted q3–6mo

Metastatic Breast Cancer

Adult: **IM Cypionate, Enanthate** 200–400 mg q2–4wk **IM** (propionate) 50–100 mg 3 × per wk

ADMINISTRATION

Buccal

- Apply buccal patch to gum just above the incisor tooth.

Transdermal

- Apply transdermal system on clean, dry scrotal skin. Dry shave scrotal hair for optimal skin contact. Do not use chemical depilatories. Wear patch for 22–24 h.
- Topical gel preparations may be applied to shoulders and upper arms.
- Store at 15°–30° C (59°–86° F).

Intramuscular

- Give IM injections deep into gluteal musculature.

Common adverse effects in *italic*, life-threatening effects <u>underlined</u>; generic names in **bold**; classifications in SMALL CAPS; ✦ Canadian drug name; ❂ Prototype drug

1457

- Store IM formulations prepared in oil at room temperature. Warming and shaking vial will redisperse precipitated crystals.

ADVERSE EFFECTS (≥1%) CNS: Excitation, insomnia. **CV:** Skin flushing and vascularization. **GI:** Nausea, vomiting, anorexia, diarrhea, gastric pain, jaundice. **Hematologic:** Leukopenia. **Metabolic:** Hypercalcemia, hypercholesterolemia, *sodium and water retention (especially in older adults) with edema.* **Renal:** Renal calculi (especially in the immobilized patient), bladder irritability. **Urogenital:** *Increased libido.* **Skin:** *Acne,* injection site irritation and sloughing. **Body as a Whole:** Hypersensitivity to testosterone, anaphylactoid reactions (rare). **Hematologic:** Precipitation of acute intermittent porphyria. **Endocrine:** Female—suppression of ovulation, lactation, or menstruation; hoarseness or deepening of voice (often irreversible); hirsutism; oily skin; clitoral enlargement; regression of breasts; male-pattern baldness (in disseminated breast cancer); flushing, sweating; vaginitis with pruritus, drying, bleeding; menstrual irregularities. Male—prepubertal-premature epiphyseal closure, phallic enlargement, priapism. Postpubertal—testicular atrophy, decreased ejaculatory volume, azoospermia, oligospermia (after prolonged administration or excessive dosage), impotence, epididymitis, priapism, *gynecomastia.*

DIAGNOSTIC TEST INTERFERENCE
Testosterone alters *glucose tolerance* tests; decreases *thyroxine-binding globulin concentration* (resulting in decreased *total T_4* serum levels and increased *resin of T_3 and T_4*). Increases *creatinine* and *creatinine* excretion (lasting up to 2 wk after therapy is discontinued) and alters response to *metyrapone test.* It suppresses *clotting factors II, V, VII, X* and decreases excretion of *17-ketosteroids.* May increase or decrease *serum cholesterol.*

INTERACTIONS Drug: ORAL ANTICOAGULANTS may potentiate hypoprothrombinemia. May decrease **insulin** requirements.

PHARMACOKINETICS Absorption: Cypionate and **enanthate** are slowly absorbed from lipid tissue. **Duration:** 2–4 wk **cypionate** and **enanthate. Distribution:** 98% bound to sex hormone-binding globulin. **Metabolism:** Primarily in liver. **Elimination:** 90% in urine, 6% in feces. **Half-Life:** 10–100 min.

NURSING IMPLICATIONS

Assessment & Drug Effects
- Check I&O and weigh patient daily during dose adjustment period. Weight gain (due to sodium and water retention) suggests need for decreased dosage. When dosage is stabilized, urge patient to check weight at least twice weekly and to report increases, particularly if accompanied by edema in dependent areas. Dose adjustment and diuretic therapy may be started.
- Lab tests: Periodic serum cholesterol, serum electrolytes as well as LFTs throughout therapy.
- Monitor serum calcium closely. Androgenic therapy is usually terminated if serum calcium rises above 14 mg/dL.
- Report S&S of hypercalcemia (see Appendix F) promptly. The immobilized patient is particularly prone to develop hypercalcemia, which indicates progression of bone metastasis in patients with metastatic breast cancer. Treatment includes withdrawing testosterone and checking calcium, phosphate, and BUN levels daily.

- Instruct diabetic to report sweating, tremor, anxiety, vertigo. Testosterone-induced anabolic action enhances hypoglycemia (hyperinsulinism). Dosage adjustment of antidiabetic agent may be required.
- Observe patients on concomitant anticoagulant treatment for signs of overdosage (e.g., ecchymoses, petechiae). Report promptly to prescriber; anticoagulant dose may need to be reduced.
- Monitor prepubertal or adolescent males throughout therapy to avoid precocious sexual development and premature epiphyseal closure. Skeletal stimulation may continue 6 mo beyond termination of therapy.

Patient & Family Education
- Review directions for application of transdermal patches.
- Report soreness at injection site, because a postinjection site boil may be an associated adverse reaction.
- Report priapism (sustained and often painful erections occurring especially in early replacement therapy), reduced ejaculatory volume, and gynecomastia to prescriber. Symptoms indicate necessity for temporary withdrawal or discontinuation of testosterone therapy.
- Notify prescriber promptly if pregnancy is suspected or planned. Masculinization of the fetus is most likely to occur if testosterone (androgen) therapy is provided during first trimester of pregnancy.
- Androgens may cause virilism in women at dosage required to treat carcinoma. Report increase in libido (early sign of toxicity), growth of facial hair, deepening of voice, male-pattern baldness. The onset of hoarseness can easily be overlooked unless its significance as an early and possibly irreversible sign of virilism is appreciated. Reevaluation of treatment plan is indicated.

TETRACAINE HYDROCHLORIDE
(tet′ra-kane)
Pontocaine
Classification: LOCAL ANESTHETIC (ESTER TYPE)
Therapeutic: LOCAL ANESTHETIC
Prototype: Procaine HCl
Pregnancy Category: C

AVAILABILITY 0.2%, 0.3%, 1% injection; 20 mg powder; 2% solution; 1%, 2% cream; 2% gel; 1% ointment; 0.5% ophth solution

ACTION & *THERAPEUTIC EFFECT*
A potent and toxic local anesthetic that depresses initial depolarization phase of the action potential, thus preventing propagation and conduction of the nerve impulse. *Effectiveness indicated by loss of sensation and motor activity in circumscribed body areas close to injection or application site.*

USES Spinal anesthesia (high, low, saddle block) and topically to produce surface anesthesia. **Eye:** To anesthetize conjunctiva and cornea prior to superficial procedures (including tonometry, gonioscopy, removal of foreign bodies or sutures, corneal scraping). **Nose and Throat:** To abolish laryngeal and esophageal reflexes prior to bronchoscopy, esophagoscopy. **Skin:** To relieve pruritus, pain, burning.

CONTRAINDICATIONS Debilitated patients; prolonged use of ophthalmic preparations; known hypersensitivity to tetracaine or other local anesthetics of ester type (e.g., procaine, chloroprocaine, cocaine), sulfite,

T

or to PABA or its derivatives; co-agulopathy; anticoagulant therapy; thrombocytopenia; increased bleeding time; infection at application or injection site.

CAUTIOUS USE Shock; cachexia, cardiac decompensation; QT prolongation; older adults; pregnancy (category C), lactation; children younger than 16 y.

ROUTE & DOSAGE

> **Local Anesthesia**
>
> *Adult:* **Topical** Before procedure, 1–2 drops of 0.5% solution or 1.25–2.5 cm of ointment in lower conjunctival fornix or 0.5% solution or ointment to nose or throat **Spinal** 1% solution diluted with equal volume of 10% dextrose injected in subarachnoid space

ADMINISTRATION

Topical

- Avoid use of solutions that are cloudy, discolored, or crystallized.
- When tetracaine is used on mucosa of larynx, trachea, or esophagus, the manufacturer recommends adding 0.06 mL of a 0.1% epinephrine solution to each mL tetracaine solution to slow absorption of the anesthetic.
- Store ophthalmic solution and ointment at 15°–30° C (59°–86° F); refrigerate topical. Avoid freezing. Use tight, light-resistant containers.

ADVERSE EFFECTS (≥1%) Body as a Whole: <u>Anaphylactic reactions</u>, convulsions, faintness, syncope. **CNS:** Postspinal headache, headache, spinal nerve paralysis, anxiety, nervousness, seizures. **CV:** Bradycardia, arrhythmias, hypotension. **Special Senses:** Stinging; corneal erosion, retardation or prevention of healing of corneal abrasion, transient pitting and sloughing of corneal surface, dry corneal epithelium; dry mucous membranes, prolonged depression of cough reflex.

INTERACTIONS Drug: May antagonize effects of SULFONAMIDES.

PHARMACOKINETICS Onset: 1 min eye; 3 min mucosal surface; 3 min spinal. **Duration:** Up to 15 min eye; 30–60 min mucosal surface; 1.5–3 h spinal. **Metabolism:** In liver and plasma. **Elimination:** In urine.

NURSING IMPLICATIONS

Assessment & Drug Effects

- Recovery from anesthesia to the pharyngeal area is complete when patient has feeling in the hard and soft palates and when muscles in the faucial (tonsillar) pillars contract with stimulation.
- Do not give food or liquids until these normal pharyngeal responses are present (usually about 1 h after anesthetic administration). The first small amount of liquid (water) should be given under supervision of care provider.
- Be aware that increased blood concentration of the drug may result from excess application of tetracaine to the skin (to relieve pruritus or burning), application to debrided or infected skin surfaces, or too rapid injection rate.
- High blood concentrations of tetracaine can lead to adverse systemic effects involving CNS and CV systems: Convulsions, respiratory arrest, dysrhythmias, cardiac arrest.

Patient & Family Education

- Do not use ophthalmic drug longer than prescribed period. Prolonged use to eye surface may cause corneal epithelial erosions and retard healing of corneal surface.

- Natural barriers to eye infection and injury are removed by the anesthesia. Do not rub eye after drug instillation until anesthetic effect has dissipated (evidenced by return of blink reflex). Patching for temporary protection of the corneal epithelium may be ordered.
- Wash or disinfect hands before and after self-administration of solutions or ointment.

TETRACYCLINE HYDROCHLORIDE ℗ᵣ

(tet-ra-sye′kleen)

Novotetra ♣, Sumycin

Classifications: ANTIBIOTIC; TETRA-CYCLINE

Therapeutic: ANTIBIOTIC

Pregnancy Category: D

AVAILABILITY 100 mg, 125 mg, 250 mg, 500 mg capsules; 125 mg/5 mL suspension

ACTION & *THERAPEUTIC EFFECT*
Tetracyclines exert antiacne action by suppressing growth of *Propionibacterium acnes* within sebaceous follicles. *Effective against a variety of gram-positive and gram-negative bacteria and against most chlamydiae, mycoplasmas, rickettsiae, and certain protozoa (e.g., amebae).* Exerts antiacne action against Propionibacterium acnes.

USES Chlamydial infections (e.g., lymphogranuloma venereum, psittacosis, trachoma, inclusion conjunctivitis, nongonococcal urethritis); mycoplasmal infections (e.g., *Mycoplasma pneumoniae*); rickettsial infections (e.g., Q fever, Rocky Mountain spotted fever, typhus); spirochetal infections: Relapsing fever *(Borrelia),* leptospirosis, syphilis (penicillin-hypersensitive patients); amebiases; uncommon gram-negative bacterial

infections [e.g., brucellosis, shigellosis, cholera, gonorrhea (penicillin-hypersensitive patients), granuloma inguinale, tularemia]; gram-positive infections (e.g., tetanus). Also used orally (solution) for inflammatory acne vulgaris.

UNLABELED USES Actinomycosis, acute exacerbations of chronic bronchitis; Lyme disease; pericardial effusion (metastatic); acute PID; sexually transmitted epididymoorchitis; with quinine for multi-drug-resistant strains of *Plasmodium falciparum* malaria; anti-infective prophylaxis for rape victims; recurrent cystic thyroid nodules; melioidosis; and as fluorescence test for malignancy.

CONTRAINDICATIONS Hypersensitivity to tetracyclines; severe renal or hepatic impairment, common bile duct obstruction; UV exposure; pregnancy (category D), lactation.

CAUTIOUS USE History of kidney or liver dysfunction; myasthenia gravis; history of allergy, asthma, hay fever, urticaria; undernourished patients; infants, children younger than 8 y.

ROUTE & DOSAGE

Systemic Infection
Adult: **PO** 250–500 mg b.i.d.–q.i.d. (1–2 g/day)
Child (older than 8 y): **PO** 25–50 mg/kg/day in 2–4 divided doses
Acne
Adult/Child (older than 8 y): **PO** 500–1000 mg/day in 4 divided doses

ADMINISTRATION

Oral

- Give with a full glass of water on an empty stomach at least 1 h

Common adverse effects in *italic*, life-threatening effects underlined; generic names in **bold**; classifications in SMALL CAPS; ♣ Canadian drug name; ℗ Prototype drug

1461

before or 2 h after meals (food, milk, and milk products can reduce absorption by 50% or more).

- Do not give immediately before bed.
- Give with food if patient is having GI symptoms (e.g., nausea, vomiting, anorexia); do not give with foods high in calcium such as milk or milk products.
- Shake suspension well before pouring to ensure uniform distribution of drug. Use calibrated liquid measure to dispense.
- Check expiration date for all tetracyclines. Fanconi-like syndrome (renal tubular dysfunction) and also an LE-like syndrome have been attributed to outdated tetracycline preparations.
- Tetracycline decomposes with age, exposure to light, and when improperly stored under conditions of extreme humidity, heat, or cold. The resultant product may be toxic.
- Store at 15°–30° C (59°–86° F) in tightly covered container in dry place. Protect from light.

ADVERSE EFFECTS (≥1%) CNS: Headache, intracranial hypertension (rare). **Special Senses:** Pigmentation of conjunctiva due to drug deposit. **GI:** Reported mostly for oral administration, but also may occur with parenteral tetracycline (*nausea, vomiting,* epigastric distress, heartburn, *diarrhea,* bulky loose stools, steatorrhea, *abdominal discomfort, flatulence,* dry mouth); dysphagia, retrosternal pain, esophagitis, esophageal ulceration with oral administration, abnormally high liver function test values, decrease in serum cholesterol, fatty degeneration of liver [jaundice, increasing nitrogen retention (azotemia), hyperphosphatemia, acidosis, irreversible shock]; foul-smelling stools

or vaginal discharge, stomatitis, glossitis; black hairy tongue (lingua nigra), diarrhea: Staphylococcal enterocolitis. **Body as a Whole:** Drug fever, angioedema, serum sickness, anaphylaxis. **Urogenital:** Particularly in patients with kidney disease; increase in BUN/serum creatinine, renal impairment even with therapeutic doses; Fanconi-like syndrome (outdated tetracycline) (characterized by polyuria, polydipsia, nausea, vomiting, glycosuria, proteinuria acidosis, aminoaciduria); vulvovaginitis, pruritus vulvae or ani (possibly hypersensitivity). **Skin:** Dermatitis, *phototoxicity:* Discoloration of nails, onycholysis (loosening of nails); cheilosis; fixed drug eruptions particularly on genitalia; thrombocytopenic purpura; urticaria, rash, exfoliative dermatitis; with topical applications: Skin irritation, dry scaly skin, transient stinging or burning sensation, slight yellowing of skin at application site, acute contact dermatitis. **Other:** Pancreatitis, local reactions: Pain and irritation (IM site), Jarisch-Herxheimer reaction.

DIAGNOSTIC TEST INTERFERENCE TETRACYCLINES may cause false increases in **urinary catecholamines** (by *fluorometric methods*), and false decreases in **urinary urobilinogen.** Parenteral TETRACYCLINES containing **ascorbic acid** reportedly may produce false-positive **urinary glucose** determinations by **copper reduction methods** (e.g., **Benedict's reagent, Clinitest**); TETRACYCLINES may cause false-negative results with **glucose oxidase methods** (e.g., **Clinistix, TesTape**).

INTERACTIONS Drug: ANTACIDS, **calcium**, and **magnesium** bind tetracycline in gut and decrease absorption. ORAL ANTICOAGULANTS potentiate hypoprothrombinemia. ANTIDIARRHEAL AGENTS with **kaolin** and pectin may decrease absorption. Effectiveness of ORAL CONTRACEPTIVES decreased. **Methoxyflurane** may produce fatal nephrotoxicity. **Food:** Dairy products and **iron, zinc** supplements decrease tetracycline absorption.

PHARMACOKINETICS Absorption: 75–80% of dose absorbed. **Peak:** 2–4 h. **Distribution:** Widely distributed, preferentially binds to rapid growing tissues; crosses placenta; enters breast milk. **Metabolism:** Not metabolized; enterohepatic cycling. **Elimination:** 50–60% in urine within 72 h. **Half-Life:** 6–12 h.

NURSING IMPLICATIONS

Assessment & Drug Effects

- Lab tests: Baseline and periodic C&S, kidney function tests, LFTs, and hematopoietic function tests, particularly during high-dose, long-term therapy. Determine serum tetracycline levels in patients at-risk for hepatotoxicity (occurs most frequently in patients receiving other hepatotoxic drugs or with history of renal or hepatic impairment).
- Report GI symptoms (e.g., nausea, vomiting, diarrhea) to prescriber. These are generally dose-dependent, occurring mostly in patients receiving 2 g/day or more and during prolonged therapy. Frequently, symptoms are controlled by reducing dosage or administering with compatible foods.
- Be alert to evidence of superinfections (see Appendix F). Regularly inspect tongue and mucous membrane of mouth for candidiasis (thrush). Suspect superinfection if patient complains of irritation or soreness of mouth, tongue, throat, vagina, or anus, or persistent itching of any area, diarrhea, or foul-smelling excreta or discharge.

- Withhold drug and notify prescriber if superinfection develops. Superinfections occur most frequently in patients receiving prolonged therapy, the debilitated, or those who have diabetes.
- Monitor I&O in patients receiving parenteral tetracycline. Report oliguria or any changes in appearance of urine or in I&O.

Patient & Family Education

- Report onset of diarrhea to prescriber. It is important to determine whether diarrhea is due to irritating drug effect or superinfections or pseudomembranous colitis (caused by overgrowth of toxin-producing bacteria: *Clostridium difficile*) (see Appendix F). The latter two conditions can be life threatening and require immediate withdrawal of tetracycline and prompt initiation of symptomatic and supportive therapy.
- Reduce incidence of superinfection (see Appendix F) by meticulous care of mouth, skin, and perineal area. Rinse mouth of food debris after eating; floss daily and use a soft-bristled toothbrush.
- Avoid direct exposure to sunlight during and for several days after therapy is terminated to reduce possibility of photosensitivity reaction (appearing like an exaggerated sunburn).
- Exercise caution with potentially hazardous activities until reaction to drug is known.
- Report immediately sudden onset of painful or difficult swallowing (dysphagia) to prescriber. Esophagitis and esophageal ulceration have been associated with bedtime administration of tetracycline capsules or tablets with insufficient fluid, particularly to patients with hiatal hernia or esophageal problems.

T

Common adverse effects in *italic*, life-threatening effects underlined; generic names in **bold**; classifications in SMALL CAPS; ✦ Canadian drug name; ⊘ Prototype drug

- Response to acne therapy usually requires 2–8 wk, maximal results may not be apparent for up to 12 wk.

TETRAHYDROZOLINE HYDROCHLORIDE

(tet-ra-hye-drozz′a-leen)

Mallazine, Murine Plus, Optigene, Soothe, Tyzine, Visine
Classifications: EYE AND NOSE PREPARATION; VASOCONSTRICTOR; DECONGESTANT
Therapeutic: NASAL DECONGESTANT; OCULAR VASOCONSTRICTOR
Prototype: Naphazoline
Pregnancy Category: C

AVAILABILITY 0.05% ophth solution; 0.05%, 0.1% nasal solution

ACTION & *THERAPEUTIC EFFECT*
Alpha-adrenergic agonist that causes intense vasoconstriction when applied topically to mucous membranes, and when applied as eyedrops. *Ophthalmic solution is effective for allergic reactions of the eye; nasal solution is antiinflammatory and also decreases allergic congestion.*

USES Symptomatic relief of minor eye irritation and allergies and for nasopharyngeal congestion of allergic or inflammatory origin.

CONTRAINDICATIONS Hypersensitivity to tetrahydrozoline; use of ophthalmic preparation in glaucoma or other serious eye diseases; use within 14 days of MAO inhibitor therapy.
CAUTIOUS USE Hypertension; cardiovascular disease; hyperthyroidism; DM; pregnancy (category C), lactation. Use in children younger than 2 y; use of 0.1% or higher strengths in children younger than 6 y.

ROUTE & DOSAGE

Decongestant

Adult: **Ophthalmic** See Appendix A-1 **Nasal** 2–4 drops of 0.1% solution or spray in each nostril q3h prn
Child: **Nasal** 2–6 y, 2–4 drops of 0.05% solution or spray in each nostril q3h prn; *6 y or older,* same as adult

ADMINISTRATION

Instillation

- Make sure interval between doses is at least 4–6 h since drug action lasts 4–8 h.
- Place patient in upright position when using nasal spray. (If patient is reclining, a stream rather than a spray may be ejected, with consequent overdosage.)
- Use lateral, head-low position to administer nasal drops.

ADVERSE EFFECTS (≥1%) **Special Senses:** *Transient stinging,* irritation, *sneezing,* dryness, headache, tremors, drowsiness, light-headedness, insomnia, palpitation. **Body as a Whole:** With overdose: Marked drowsiness, sweating, <u>coma</u>, hypotension, <u>shock</u>, bradycardia.

PHARMACOKINETICS Absorption: May be absorbed from nasal mucosa. **Duration:** 4–8 h.

NURSING IMPLICATIONS

Patient & Family Education

- Discontinue medication and consult prescriber if relief is not obtained within 48 h or if symptoms persist or increase.
- Do not exceed recommended dosage. Rebound congestion and rhinitis may occur with frequent

or prolonged use of nasal preparation.

THALIDOMIDE

(tha-lid′o-mide)

Thalomid

Classifications: IMMUNOMODULA-TOR; TUMOR NECROSIS FACTOR (TNF) MODIFIER

Therapeutic: IMMUNOSUPPRESSIVE; TNF MODIFIER

Pregnancy Category: X

AVAILABILITY 50 mg capsules

ACTION & *THERAPEUTIC EFFECT*
Anti-inflammatory effects may be due to its inhibition of neutrophil chemotaxis and decrease of monocyte phagocytosis. Immunosuppressive effect may result from suppression of excessive tumor necrosis factor-alpha (TNF-alpha) production. Also, reduces helper T cells and increases suppressor T cells. *Effectiveness indicated by control of cutaneous manifestations of erythema nodosum leprosum and/or improvement in Crohn's disease.*

USES Acute and maintenance treatment of cutaneous manifestations of moderate to severe erythema nodosum leprosum. Refractory Crohn's disease.

UNLABELED USES Stimulate appetite in patients with HIV-associated cachexia, lupus, multiple myeloma.

CONTRAINDICATIONS Hypersensitivity to thalidomide; peripheral neuropathy; pregnancy (category X), lactation.

CAUTIOUS USE Liver and kidney disease; CHF or hypertension; constipation or other GI disorders; neurologic disorders or history of neuritis; children younger than 11 y.

ROUTE & DOSAGE

Erythema Nodosum Leprosum

Adult: **PO** 100–300 mg daily (max: 400 mg/day) × at least 2 wk

Child (11–17 y): **PO** 100 mg daily

Refractory Crohn's Disease

Adult: **PO** 50–100 mg daily (doses up to 300 mg studied)

ADMINISTRATION

Oral

- Give at bedtime and at least 1 h after the evening meal.
- Give this drug only to persons who understand and have signed the required consent form.
- Verify, prior to administration, that this drug was prescribed and dispensed only by persons registered by the STEPS (System for Thalidomide Education and Prescribing Safety) program.
- Store at 15°–30° C (59°–86° F); protect from light.

ADVERSE EFFECTS (≥1%) **Body as a Whole:** Asthenia, back pain, chills, facial edema, *fever,* malaise, pain. **CNS:** Drowsiness, *somnolence,* peripheral neuropathy (possibly irreversible), *dizziness,* orthostatic hypotension, headache, agitation, insomnia, nervousness, paresthesia, tremor, vertigo, seizures. **CV:** Bradycardia, peripheral edema, hyperlipidemia. **GI:** Abdominal pain, anorexia, constipation, *diarrhea,* dry mouth, flatulence, abnormal liver function tests, nausea, oral moniliasis. **Hematologic:** Neutropenia, anemia, <u>leukopenia</u>, lymphadenopathy. **Respiratory:** Pharyngitis, rhinitis, sinusitis. **Skin:** *Rash,* acne, nail disorder, fungal dermatitis,

Common adverse effects in *italic*; life-threatening effects <u>underlined</u>; generic names in **bold**; classifications in SMALL CAPS; ♣ Canadian drug name; ⊕ Prototype drug

1465

pruritus, sweating, <u>toxic epidermal necrolysis</u>. **Body as a Whole:** Hypersensitivity reaction (rash, fever, tachycardia, hypotension), HIV viral load increase, infection. **Urogenital:** Teratogenicity, albuminuria, hematuria, impotence.

INTERACTIONS Drug: Enhances sedation associated with BARBITURATES, **alcohol, chlorpromazine, reserpine.**

PHARMACOKINETICS Absorption: Slowly absorbed from GI tract. **Peak:** 2.9–5.7 h. **Distribution:** Crosses placenta; present in ejaculate in males. **Metabolism:** Does not appear to be hepatically metabolized. **Half-Life:** 6–7.5 h.

NURSING IMPLICATIONS

Assessment & Drug Effects

- Lab tests: Monitor WBC with differential prior to therapy and periodically thereafter.
- Monitor carefully for and immediately report S&S of peripheral neuropathy. Discontinue drug and notify prescriber if peripheral neuropathy is suspected.

Patient & Family Education

- Do not share this medication with anyone else under any circumstances.
- Use effective methods of birth control (both women and men); starting 1 mo before, during, and 1 mo following discontinuation of thalidomide therapy. Men **MUST** use condoms when engaging in sexual activity to prevent birth defects.
- Exercise caution while driving or engaging in potentially hazardous activities because drug may cause dizziness.
- Report pain, numbness, or tingling in the hands or feet to prescriber immediately.

THEOPHYLLINE 🅿ⓡ

(thee-off′i-lin)

Elixophyllin, Lanophyllin, PMS Theophylline ♣, Pulmophylline ♣, Theo-24, Theolair, Theophylline Ethylenediamine, Theospan-SR, Uniphyl

Classifications: BRONCHODILATOR (RESPIRATORY SMOOTH MUSCLE RELAXANT); XANTHINE
Therapeutic: BRONCHODILATOR
Pregnancy Category: C

AVAILABILITY 100 mg, 125 mg, 200 mg, 250 mg, 300 mg tablets; 100 mg, 200 mg capsules; 80 mg/15 mL, 150 mg/15 mL liquid; 100 mg, 200 mg, 250 mg, 300 mg, 450 mg, 500 mg, 600 mg sustained release tablets; 50 mg, 75 mg, 100 mg, 125 mg, 200 mg, 250 mg, 260 mg, 300 mg sustained release capsules; 200 mg, 400 mg, 800 mg injection

ACTION & *THERAPEUTIC EFFECT* Xanthine derivative that relaxes smooth muscle by direct action, particularly of bronchi and pulmonary vessels, and stimulates medullary respiratory center with resulting increase in vital capacity. *Effective for relief of bronchospasm in asthmatics, chronic bronchitis, and emphysema.*

USES Prophylaxis and symptomatic relief of bronchial asthma, as well as bronchospasm associated with chronic bronchitis and emphysema. Also used for emergency treatment of paroxysmal cardiac dyspnea and edema of CHF.

UNLABELED USES Treatment of apnea and bradycardia of premature infants and to reduce severe bronchospasm associated with cystic fibrosis and acute descending respiratory infection.

CONTRAINDICATIONS Hypersensitivity to xanthines; CAD or angina pectoris when myocardial stimulation might be harmful; severe renal or liver impairment.

CAUTIOUS USE Compromised cardiac or circulatory function, hypertension; acute pulmonary edema; multiple organ failure; CHF; hyperthyroidism; peptic ulcer; prostatic hypertrophy; glaucoma; DM; older adults; pregnancy (category C), lactation; children, and neonates.

ROUTE & DOSAGE

Bronchospasm

Adult/Child: **PO/IV Loading Dose** 5 mg/kg
Adult: **PO/IV Maintenance Dose*** Nonsmoker, 0.4 mg/kg/h; Smoker, 0.6 mg/kg/h; with CHF or cirrhosis, 0.2 mg/kg/h
Child: **PO/IV Maintenance Dose*** 1–9 y, 0.8 mg/kg/h; 10–12 y, 0.6 mg/kg/h
Infant: **PO/IV Maintenance Dose*** 0.5–0.7 mg/kg/h
Neonate: **PO/IV Maintenance Dose*** 0.13 mg/kg/h
*IV by continuous infusion, PO divided q6h (immediate release) or q8–12h (sustained release)

Obesity Dosage Adjustment

Dose based on IBW.

ADMINISTRATION

Note: All doses based on ideal body weight.

Oral

- Wait 4–6 h after the last IV dose, when switching from IV to oral dosing.
- Give with a full glass of water and after meals to minimize gastric irritation.
- Give sustained release forms and enteric-coated tablets whole. Chewable tablets **must be** chewed

thoroughly before swallowing. Sustained release granules from capsules can be taken on an empty stomach or mixed with applesauce or water.

- Note: Timing of dose is critical. Be certain patient understands necessity to adhere to the correct intervals between doses.

Intravenous

PREPARE: **Direct/Intermittent:** Dilute, as needed, to a maximum concentration of 20 mg/mL. **IV Infusion:** Typically diluted to 0.8 mg/mL with D5W.

ADMINISTER: **Direct/Intermittent:** Typically infused over 20–30 min. DO NOT EXCEED 20 mg/min. **IV Infusion:** Infuse at a rate based on patient's weight.

INCOMPATIBILITIES **Solution/additive: Ascorbic acid, ceftriaxone, cimetidine, hetastarch. Y-site: Hetastarch, phenytoin.**

ADVERSE EFFECTS (≥1%) **CNS:** Stimulation (irritability, restlessness, insomnia, dizziness, headache, tremor, hyperexcitability, muscle twitching, <u>drug-induced seizures</u>). **CV:** Palpitation, *tachycardia,* extrasystoles, flushing, marked hypotension, <u>circulatory failure</u>. **GI:** *Nausea,* vomiting, anorexia, epigastric or abdominal pain, diarrhea, activation of peptic ulcer. **Urogenital:** Transient urinary frequency, albuminuria, kidney irritation. **Respiratory:** Tachypnea, <u>respiratory arrest</u>. **Body as a Whole:** Fever, dehydration.

DIAGNOSTIC TEST INTERFERENCE False-positive elevations of *serum uric acid* (*Bittner* or **colorimetric** methods). *Probenecid* may cause false high *serum theophylline* readings, and spectrophotometric methods of determining *serum theophylline* are affected by a furosemide, sulfathiazole, phenylbutazone, probenecid, theobromine.

Common adverse effects in *italic*, life-threatening effects <u>underlined</u>; generic names in **bold**; classifications in SMALL CAPS; ✤ Canadian drug name; ⊙ Prototype drug

1467

INTERACTIONS Drug: Increases **lithium** excretion, lowering lithium levels; **cimetidine,** high-dose **allopurinol** (600 mg/day), **tacrine,** QUINOLONES, MACROLIDE ANTIBIOTICS, and **zileuton** can significantly increase theophylline levels; **tobacco** use significantly decreases levels. **Herbal: St. John's wort** may decrease theophylline efficacy. **Daidzein** (in soy), **black pepper** increase serum concentrations and adverse effects.

PHARMACOKINETICS Absorption: Most products are 100% absorbed from GI tract. **Peak:** IV 30 min; uncoated tablet 1 h; sustained release 4–6 h. **Duration:** 4–8 h; varies with age, smoking, and liver function. **Distribution:** Crosses placenta. **Metabolism:** Extensively in liver. **Elimination:** Parent drug and metabolites excreted by kidneys; excreted in breast milk.

NURSING IMPLICATIONS

Assessment & Drug Effects

- Monitor vital signs. Improvement in respiratory status is the expected outcome.
- Observe and report early signs of possible toxicity: Anorexia, nausea, vomiting, dizziness, shakiness, restlessness, abdominal discomfort, irritability, palpitation, tachycardia, marked hypotension, cardiac arrhythmias, seizures.
- Monitor for tachycardia, which may be worse in patients with severe cardiac disease. Conversely, theophylline toxicity may be masked in patients with tachycardia.
- Lab tests: Monitor plasma level of theophylline. Be aware that therapeutic plasma level ranges from 10–20 mcg/mL (a narrow therapeutic range). Levels exceeding 20 mcg/mL are associated with toxicity.

- Monitor drug levels closely in heavy smokers. Cigarette smoking induces hepatic microsomal enzyme activity, decreasing serum half-life and increasing body clearance of theophylline. An increase of dosage from 50–100% is usual in heavy smokers.
- Monitor plasma drug level closely in patients with heart failure, kidney or liver dysfunction, alcoholism, high fever. Plasma clearance of xanthines may be reduced.
- Take necessary safety precautions and forewarn older adult patients of possible dizziness during early therapy.
- Monitor patients on sustained release preparations for S&S of overdosage. Continued slow absorption leads to high plasma concentrations for a prolonged period.
- Note: Neonates of mothers using this drug have exhibited slight tachycardia, jitteriness, and apnea.
- Monitor **CLOSELY** for adverse effects in infants younger than 6 mo and prematures; theophylline metabolism is prolonged as is the half-life in this age group.

Patient & Family Education

- Take medication at the same time every day.
- Avoid charcoal-broiled foods (high in polycyclic carbon content); may increase theophylline elimination and reduce the half-life as much as 50%.
- Limit caffeine intake because it may increase incidence of adverse effects.
- Cigarette smoking may significantly lower theophylline plasma concentration.
- Be aware that a low-carbohydrate, high-protein diet increases theophylline elimination, and a high-carbohydrate, low-protein diet decreases it.

- Drink fluids liberally (2000–3000 mL/day) if not contraindicated to decrease viscosity of airway secretions.
- Avoid self-dosing with OTC medications, especially cough suppressants, which may cause retention of secretions and CNS depression.

THIAMINE HYDROCHLORIDE (VITAMIN B₁)

(thye′a-min)

Classification: VITAMIN B₁
Therapeutic: VITAMIN B₁ REPLACEMENT THERAPY
Pregnancy Category: A; C if dose is above RDA

AVAILABILITY 50 mg, 100 mg, 250 mg tablets; 20 mg enteric-coated tablet; 100 mg/mL injection

ACTION & *THERAPEUTIC EFFECT*
Water-soluble B₁ vitamin and member of B-complex group used for thiamine replacement therapy. Functions as an essential coenzyme in carbohydrate metabolism and has a role in conversion of tryptophan to nicotinamide. *Effectiveness is evidenced by improvement of clinical manifestations of thiamine deficiency (e.g., anorexia, depression, loss of memory, etc.).*

USES Treatment and prophylaxis of beriberi, to correct anorexia due to thiamine deficiency states, and in treatment of neuritis associated with pregnancy, pellagra, and alcoholism, including Wernicke-Korsakoff syndrome. Therapy generally includes other members of vitamin B complex, since thiamine deficiency rarely occurs alone. Severe deficiency is characterized by ophthalmoplegia, polyneuropathy, muscle wasting ("dry" beriberi), edema, serous effusions, and CHF ("wet" beriberi).

CAUTIOUS USE Pregnancy (category A; category C if above RDA).

ROUTE & DOSAGE

Thiamine Deficiency
Adult: **IV/IM** 50–100 mg t.i.d., then 5–10 mg PO for 1 mo
Child: **IV/IM** 10–25 mg t.i.d. then, 5–10 mg PO for 1 mo

Alcohol Withdrawal
Adult: **IV/IM** 100 mg/day until PO 50–100 mg/day as tolerated

Wernicke's Encephalopathy
Adult: **IV/IM** 100 mg/day then 50–100 mg/day until on normal diet

Dietary Supplement
Adult: **PO** 15–30 mg/day
Child: **PO** 10–50 mg/day

ADMINISTRATION

Oral
- Do not crush or chew enteric-coated tablets. These **must be** swallowed whole.

Intramuscular
- Give deep IM into a large muscle; may be painful. Rotate sites and apply cold compresses to area if necessary for relief of discomfort.

Intravenous
Note: Intradermal test dose is recommended prior to administration in suspected thiamine sensitivity. Deaths have occurred following IV use.

PREPARE: **Direct:** Give undiluted. **IV Infusion:** Diluted in 1000 mL of most IV solutions.
ADMINISTER: **Direct:** Give at a rate of 100 mg over 5 min. **IV Infusion:** Give at the ordered rate.

T

INCOMPATIBILITIES **Solution/additive: Amobarbital, diazepam, furosemide, phenobarbital.**

▪ Preserve in tight, light-resistant, nonmetallic containers. Thiamine is unstable in alkaline solutions (e.g., solutions of acetates, barbiturates, bicarbonates, carbonates, citrates) and neutral solutions.

ADVERSE EFFECTS (≥1%) **Body as a Whole:** Feeling of warmth, weakness, sweating, restlessness, tightness of throat, angioneurotic edema, anaphylaxis. **Respiratory:** Cyanosis, pulmonary edema. **CV:** Cardiovascular collapse, slight fall in BP following rapid IV administration. **GI:** GI hemorrhage, nausea. **Skin:** Urticaria, pruritus.

INTERACTIONS Drug: No clinically significant interactions established.

PHARMACOKINETICS Absorption: Limited from GI tract. **Distribution:** Widely distributed, including into breast milk. **Elimination:** In urine.

NURSING IMPLICATIONS

Assessment & Drug Effects

▪ Record patient's dietary history carefully as an essential part of vitamin replacement therapy. Collaborate with prescriber, dietitian, patient, and responsible family member in developing a diet teaching plan that can be sustained by patient.

Patient & Family Education

▪ Food–drug relationships: Learn about rich dietary sources of thiamine (e.g., yeast, pork, beef, liver, wheat and other whole grains, nutrient-added breakfast cereals, fresh vegetables, especially peas and dried beans).

▪ Body requirement of thiamine is directly proportional to carbohydrate intake and metabolic rate; requirement increases when diet consists predominantly of carbohydrates.

THIOGUANINE (TG, 6-THIOGUANINE)
(thye-oh-gwah'neen)
Lanvis ♣, Tabloid
Classifications: ANTINEOPLASTIC; ANTIMETABOLITE; PURINE ANTAGONIST
Therapeutic: ANTINEOPLASTIC
Prototype: Mercaptopurine
Pregnancy Category: D

AVAILABILITY 40 mg tablets

ACTION & *THERAPEUTIC EFFECT*
Antimetabolite and purine antagonist with immunosuppressive activity. Drug is incorporated into the DNA and RNA of human bone marrow cells. *Delays myelosuppression; has potential mutagenic and carcinogenic properties.*

USES In combination with other antineoplastics for remission induction in acute myelogenous leukemia and as treatment of chronic myelogenous leukemia. Has little advantage over mercaptopurine.

CONTRAINDICATIONS Patients with prior resistance to this drug; severe bone marrow depression; pregnancy (category D), lactation.
CAUTIOUS USE Hepatic disease.

ROUTE & DOSAGE

Leukemia
Adult: **PO** 2 mg/kg/day, may increase to 3 mg/kg/day if no response after 4 wk

ADMINISTRATION

Oral
▪ Withhold drug and notify prescriber if toxicity develops. There is no

known antagonist; prompt discontinuation of the drug is essential to avoid irreversible myelosuppression from toxicity.
- Store at 15°–30° C (59°–86° F) in airtight container.

ADVERSE EFFECTS (≥1%) Hematologic: <u>Leukopenia, thrombocytopenia,</u> anemia. **GI:** Jaundice, nausea, vomiting, anorexia, stomatitis, diarrhea. **Urogenital:** *Hyperuricemia.* **Other:** Hepatotoxicity (risk increased with long-term use).

INTERACTIONS Drug: Severe hepatotoxicity with **busulfan;** may decrease immune response to VACCINES; increase risk of bleeding with ANTICOAGULANTS; NSAIDS, SALICYLATES; PLATELET INHIBITORS, THROMBOLYTIC AGENTS; effects may be reversed by **filgrastim, sargramostim.**

PHARMACOKINETICS Absorption: Variable and incomplete absorption from GI tract. **Peak:** 8 h. **Distribution:** Crosses placenta. **Metabolism:** In liver. **Elimination:** In urine. **Half-Life:** 11 h.

NURSING IMPLICATIONS

Assessment & Drug Effects
- Lab tests: Monitor blood counts weekly (CBC with differential and platelet count); periodic LFTs with long-term use.
- Determine hematologic parameters for withholding drug.
- Monitor I&O ratio and report oliguria.
- Observe patient's skin and sclera for jaundice. It should be reported promptly as a symptom of toxicity; drug will be discontinued promptly.
- Expect that the drop in leukocyte count may be slow over a period of 2–4 wk. Treatment is interrupted if there is a rapid fall within a few days.

Patient & Family Education
- Maintenance doses are continued throughout remissions.

THIORIDAZINE HYDROCHLORIDE
(thye-o-rid'a-zeen)

Novoridazine ◆
Classification: ANTIPSYCHOTIC; PHENOTHIAZINE
Therapeutic: ANTIPSYCHOTIC
Prototype: Chlorpromazine
Pregnancy Category: C

AVAILABILITY 10 mg, 15 mg, 25 mg, 50 mg, 100 mg, 150 mg, 200 mg tablets; 30 mg/mL, 100 mg/mL solution; 25 mg/5 mL suspension

ACTION & *THERAPEUTIC EFFECT*
Thioridazine blocks postsynaptic dopamine receptors in the mesolimbic system of the brain. The decrease in dopamine neurotransmission has been found to correlate to antipsychotic effects. *Effective in reducing excitement, hypermotility, abnormal initiative, affective tension, and agitation by inhibiting psychomotor functions. Also effective as an antipsychotic agent, and for behavioral disorders in children.*

USES Management of nonpsychotic behavioral disturbances of senility, manifestations of psychotic disorders, alcohol withdrawal; symptomatic treatment of organic brain disease. Short-term treatment of moderate to marked depression and for management of hyperkinetic behavior syndrome (attention deficit disorder).

CONTRAINDICATIONS Hypersensitivity to phenothiazines. Severe CNS depression; CV disease; family history of QT prolongation; suicidal ideation; lactation.

Common adverse effects in *italic*, life-threatening effects <u>underlined</u>; generic names in **bold**; classifications in SMALL CAPS; ◆ Canadian drug name; 🔵 Prototype drug

1471

CAUTIOUS USE Premature ventricular contractions; previously diagnosed breast cancer; patients exposed to extremes in heat or to organophosphorus insecticides; history of suicidal ideation; Parkinson's disease; seizure disorders; closed-angle glaucoma; respiratory disorders; pregnancy (category C); children younger than 2 y.

ROUTE & DOSAGE

Psychotic Disorders

Adult: **PO** 50–100 mg t.i.d., may increase up to 800 mg/day as needed or tolerated
Geriatric: **PO** 10 mg t.i.d., may increase up to 200 mg/day
Child (older than 2 y): **PO** 0.5–3 mg/kg/day in divided doses; if hospitalized, may start at 25 mg t.i.d.

Moderate to Marked Depression

Adult: **PO** 25 mg t.i.d., may increase up to 200 mg/day in divided doses

Dementia Behavior

Geriatric: **PO** 10–25 mg 1–2 times/day, may increase q4–7days (max: 400 mg/day in divided doses)

Pharmacogenetic Dosage Adjustment

Poor CYP2D6 metabolizers: Start with 40% of dose

ADMINISTRATION

Oral

- Give with fluid of patient's choice; tablet may be crushed.
- Schedule phenothiazine at least 1 h before or 1 h after an antacid or antidiarrheal medication.
- Dilute liquid concentrate just prior to administration with ½ glass of fruit juice, milk, water, carbonated beverage, or soup.
- Add increases in dose to the first dose of the day to prevent sleep disturbance.
- Store at 15°–30° C (59°–86° F) in tightly covered, light-resistant containers unless otherwise indicated.

ADVERSE EFFECTS (≥1%) **CNS:** *Sedation,* dizziness, drowsiness, lethargy, extrapyramidal syndrome, nocturnal confusion, hyperactivity. **Special Senses:** Nasal congestion, blurred vision, pigmentary retinopathy. **GI:** Xerostomia, *constipation,* <u>paralytic ileus</u>. **Urogenital:** Amenorrhea, breast engorgement, gynecomastia, galactorrhea, *urinary retention*. **CV:** Ventricular dysrhythmias, hypotension, prolonged QT_c interval.

INTERACTIONS Drug: Alcohol, ANXIOLYTICS, SEDATIVE-HYPNOTICS, other CNS DEPRESSANTS add to CNS depression; additive adverse effects with other PHENOTHIAZINES; **amiodarone, amoxapine, arsenic trioxide, bepridil, clarithromycin, daunorubicin, diltiazem, disopyramide, dofetilide, dolasetron, doxorubicin, encainide, erythromycin, flecainide, fluoxetine, fluvoxamine gatifloxacin, grepafloxacin, haloperidol, ibutilide, indapamide, local anesthetics, maprotiline, moxifloxacin, octreotide, paroxetine, pentamidine, pimozide, procainamide, probucol, quinidine, risperidone, sotalol, sertraline, sparfloxacin, terodiline, tocainide, tricyclic antidepressants, venlafaxine, verapamil, ziprasidone** can prolong QT_c interval resulting in arrhythmias. **Herbal: Kava** may increase risk and severity of dystonic reactions.

PHARMACOKINETICS Absorption: Well absorbed from GI tract. **Onset:** Days to weeks. **Distribution:** Crosses placenta; distributed into breast milk. **Metabolism:** In liver (CYP-2D6). **Elimination:** In urine. **Half-Life:** 26–36 h.

NURSING IMPLICATIONS

Assessment & Drug Effects

- Monitor for changes in behavior that may indicate increased possibility of suicide ideation.
- Orthostatic hypotension may occur in early therapy. Female patients appear to be more susceptible than males.
- Be aware that patients may be unable to adjust to extremes of temperature because drug effects heat regulatory center in the hypothalamus. Patient may complain of being cold even at average room temperature; older adults are particularly susceptible.
- Monitor I&O ratio and bowel elimination pattern. Check for abdominal distention and pain. Encourage adequate fluid intake as prophylaxis for constipation and xerostomia. The depressed patient may not seek help for either symptom or for urinary retention.
- Lab tests: Obtain periodic CBC and LFTs during therapy.

Patient & Family Education

- Exercise care not to spill drug on skin because of danger of contact dermatitis. Wash skin well in soap and water if liquid drug is spilled.
- Take drug as prescribed and do not alter dosing regimen or stop medication without consulting prescriber.
- Avoid alcohol during phenothiazine therapy. Concomitant use enhances CNS depression effects.
- Be aware that marked drowsiness generally subsides with continued therapy or reduction in dosage.
- Do not drive or engage in potentially hazardous activities until response to drug is known.
- Make position changes slowly, particularly from lying down to upright posture; dangle legs a few minutes before standing.
- Vasodilation produced by hot showers or baths or by long exposure to environmental heat may accentuate hypotensive effect.
- Report the onset of any change in visual acuity, brownish coloring of vision, or impairment of night vision to prescriber. An ophthalmic consultation may be indicated.
- Note: Thioridazine may color urine pink-red to reddish brown.
- Do not use any OTC drugs unless approved by the prescriber.

THIOTEPA

(thye-oh-tep'a)
Classifications: ANTINEOPLASTIC; ALKYLATING AGENT
Therapeutic: ANTINEOPLASTIC
Prototype: Cyclophosphamide
Pregnancy Category: D

AVAILABILITY 15 mg, 30 mg injection

ACTION & THERAPEUTIC EFFECT Cell-cycle nonspecific alkylating agent that selectively reacts with DNA phosphate groups to produce chromosome cross-linkage and consequent blocking of nucleoprotein synthesis. Highly toxic hematopoietic agent. *Myelosuppression is cumulative and unpredictable and may be delayed.*

USES To produce remissions in malignant lymphomas, including Hodgkin's disease, and adenocarcinoma of breast and ovary. Also in chronic granulocytic and lymphocytic leukemia, superficial papillary carcinoma of urinary bladder, bronchogenic carcinoma, and in malignant effusions secondary to neoplastic disease of serosal cavities.

UNLABELED USES Prevention of pterygium recurrences following postoperative beta-irradiation; leukemia, malignant meningeal neoplasms.

CONTRAINDICATIONS Hypersensitivity to thiotepa; acute leukemia; acute infection; pregnancy (category D), lactation.

CAUTIOUS USE Chronic lymphocytic leukemia; myelosuppression produced by radiation; bone marrow invasion by tumor cells; impaired kidney or liver function.

ROUTE & DOSAGE

Malignant Lymphomas

Adult: **IV** 0.3–0.4 mg/kg q1–4wk **Intracavitary** 0.6–0.8 mg/kg instilled through same tubing used for paracentesis at intervals of at least 1 wk **Intravesicular** 60 mg in 30–60 mL of distilled water instilled into bladder to be retained for 2 h once/wk for 4 wk **Intrathecal** 1–11.5 mg/m^2 1–2 times/wk
Child: **IV** 25–65 mg/m^2/dose every 21 days

ADMINISTRATION

Intravenous

Use only under constant supervision by prescribers experienced in therapy with cytotoxic agents.
▪ Avoid exposure of skin and respiratory tract to particles of thiotepa during solution preparation.

PREPARE: **Direct:** Reconstitute each 15 mg vial with 1.5 mL sterile water for injection (supplied) to yield 10 mg/mL. ▪ Filter solution through a 0.22 micron filter to eliminate haze. Use immediately.

ADMINISTER: **Direct:** Give 60 mg or fraction thereof over 1 min.

INCOMPATIBILITIES Solution/additive: Cisplatin. Y-site: Cisplatin, filgrastim, minocycline, vinorelbine.

▪ Store powder for injection and reconstituted solutions at 2°–8° C (35°–46° F); protect from light. ▪ Solutions reconstituted with sterile water only are stable for 8 h under refrigeration.

ADVERSE EFFECTS (≥1%) **GI:** Anorexia, nausea, vomiting, stomatitis, ulceration of intestinal mucosa. **Hematologic:** Leukopenia, thrombocytopenia, anemia, pancytopenia. **Skin:** Hives, rash, pruritus. **Urogenital:** Amenorrhea, interference with spermatogenesis. **Body as a Whole:** Headache, febrile reactions, pain and weeping of injection site, hyperuricemia, slowed or lessened response in heavily irradiated area, sensation of throat tightness. **Other:** Reported with intravesical administration (lower abdominal pain, hematuria, hemorrhagic chemical cystitis, vesical irritability).

INTERACTIONS Drug: May prolong muscle paralysis with **mivacurium;** ANTICOAGULANTS, NSAIDS, SALICYLATES, ANTIPLATELET AGENTS may increase risk of bleeding.

PHARMACOKINETICS Absorption: Rapidly cleared from plasma. **Onset:** Gradual response over several wk. **Metabolism:** In liver. **Elimination:** 60% in urine within 24–72 h.

NURSING IMPLICATIONS

Assessment & Drug Effects
▪ Monitor closely because most patients will manifest some evidence of toxicity.

- Be aware that because of cumulative effects, maximum myelosuppression may be delayed 3–4 wk after termination of therapy.
- Withhold drug and notify prescriber if leukocyte count falls to 3000/mm³ or below or if platelet count falls below 150,000/mm³.
- Lab tests: Determine Hgb level, WBC with differential, and thrombocyte (i.e., platelet) counts at least weekly during therapy and for at least 3 wk after therapy is discontinued; periodic LFTs, kidney function tests, and serum uric acid.

Patient & Family Education
- Be aware of possibility of amenorrhea (usually reversible in 6–8 mo).
- Report onset of fever, bleeding, a cold or illness, no matter how mild to prescriber; medical supervision may be necessary.

THIOTHIXENE HYDROCHLORIDE

(thye-oh-thix'een)

Navane

Classification: ANTIPSYCHOTIC; PHENOTHIAZINE
Therapeutic: ANTIPSYCHOTIC
Prototype: Chlorpromazine
Pregnancy Category: C

AVAILABILITY 1 mg, 2 mg, 5 mg, 10 mg, 20 mg capsules; 5 mg/mL solution

ACTION & THERAPEUTIC EFFECT
Mechanism of action is related to blockade of postsynaptic dopamine receptors in the mesolimbic region of the brain. Additionally, blockade of alpha$_1$-adrenergic receptors in the CNS produces sedation and muscle relaxation. *Possesses antipsychotic, sedative, adrenolytic, and antiemetic activity.*

USES Manifestations of psychotic disorders.
UNLABELED USES Antidepressant.

CONTRAINDICATIONS Hypersensitivity to thioxanthenes and phenothiazines; comatose states; CNS depression; circulatory collapse; blood dyscrasias; lactation.
CAUTIOUS USE History of convulsive disorders; alcohol withdrawal; glaucoma; prostatic hypertrophy; cardiovascular disease; patients who might be exposed to organophosphorus insecticides or to extreme heat; previously diagnosed breast cancer; pregnancy (category C); children younger than 12 y.

ROUTE & DOSAGE

Psychotic Disorders

Adult: **PO** 2 mg t.i.d., may increase up to 15 mg/day as needed or tolerated (max: 60 mg/day) **IM** 4 mg b.i.d. to q.i.d. (max: 30 mg/day)

Dementia Behavior

Geriatric: **PO** 1–2 mg 1–2 times/day, may increase q4–7days (max: 30 mg/day in divided doses)

ADMINISTRATION

Oral
- Avoid contact between oral concentrate and skin or clothing to prevent contact dermatitis. If concentrate spills, wash skin promptly with water.
- Give oral concentrate (contains 7% alcohol) diluted in a cupful of water, fruit juice, carbonated beverage, milk, or soup.
- Capsules may be opened and mixed with food or water for ease of administration.

Intramuscular
- Give IM injection deep into upper outer quadrant of buttock. Aspirate

Common adverse effects in *italic*, life-threatening effects underlined; generic names in **bold**; classifications in SMALL CAPS; ♣ Canadian drug name; ⓟ Prototype drug

1475

carefully before injection. Rotate injection sites.
- Store at 15°–30° C (59°–86° F) in light-resistant containers unless otherwise indicated.

ADVERSE EFFECTS (≥1%) **CNS:** *Drowsiness,* insomnia, dizziness, cerebral edema, convulsions, *extrapyramidal symptoms (dose related),* paradoxical exaggeration of psychotic symptoms; <u>sudden death, neuroleptic malignant syndrome</u>, tardive dyskinesia, depressed cough reflex. **GI:** Xerostomia, constipation. **CV:** Tachycardia, *orthostatic hypotension* (especially with IM). **Urogenital:** Impotence, gynecomastia, galactorrhea, amenorrhea. **Skin:** Rash, contact dermatitis, photosensitivity. **Special Senses:** Blurred vision, pigmentary retinopathy. **Metabolic:** Decreased serum uric acid levels.

INTERACTIONS Drug: **Alcohol,** ANXIOLYTICS, SEDATIVE-HYPNOTICS, other CNS DEPRESSANTS add to CNS depression; additive adverse effects with other PHENOTHIAZINES; **Herbal: Kava** may increase risk and severity of dystonic reactions.

PHARMACOKINETICS Absorption: Slowly absorbed from GI tract. **Onset:** Days to weeks PO; 1–6 h IM. **Duration:** Up to 12 h. **Distribution:** May remain in body for several weeks; crosses placenta. **Metabolism:** In liver. **Elimination:** In bile and feces. **Half-Life:** 34 h.

NURSING IMPLICATIONS

Assessment & Drug Effects
- Monitor for therapeutic response. Although therapeutic response can be observed 1–6 h following IM injection, it may be days or several weeks before there is a response with oral drug.

- Keep patient recumbent for at least 1 h following IM because of possibility of orthostatic hypotension. Check BP periodically.
- Monitor BP for excessive hypotensive response when thiothixene is added to drug regimen of patient on hypertensive treatment until therapy is stabilized.
- Monitor response when patient is changed from IM to PO forms (capsules, concentrate). Dosage adjustment may be necessary.
- Lab tests: Periodic blood chemistry and LFTs with prolonged therapy.
- Report extrapyramidal effects (pseudoparkinsonism, akathisia, dystonia) to prescriber; dose adjustment or short-term therapy with an antiparkinsonism agent may provide relief.
- Be alert to first symptoms of tardive dyskinesia (see Appendix F). Withhold drug and notify prescriber.

Patient & Family Education
- Make position changes slowly, particularly from lying down to upright because of danger of lightheadedness; sit a few minutes before walking.
- Do not drive or engage in potentially hazardous activities until response to drug is known.
- Avoid alcohol and other depressants during therapy.
- Take drug as prescribed; do not alter dosing regimen or stop medication without consulting prescriber. Abrupt discontinuation can cause delirium.
- Do not use any OTC drugs without approval of prescriber.
- Avoid excessive exposure to sunlight to prevent a photosensitivity reaction. If sun exposure is expected, protect skin with sunscreen lotion (SPF 12 or above).
- Schedule periodic eye exams and report blurred vision to prescriber.

THROMBIN
(throm'bin)

Recothrom, Thrombinar, Thrombostat

Classification: COAGULATOR; TOPICAL HEMOSTATIC
Therapeutic: COAGULATOR; HEMOSTATIC
Pregnancy Category: C

AVAILABILITY 1000, 5000, 10,000, 20,000, 50,000 unit vials

ACTION & *THERAPEUTIC EFFECT*
Plasma protein prepared from prothrombin of bovine origin. Induces clotting of whole blood or fibrinogen solution without addition of other substances. *Facilitates conversion of fibrinogen to fibrin resulting in clotting of whole blood.*

USES When oozing of blood from capillaries and small venules is accessible, as in dental extraction, plastic surgery, grafting procedures, and epistaxis; also to shorten bleeding time at puncture sites in heparinized patient (i.e., following hemodialysis).

CONTRAINDICATIONS Known hypersensitivity to any of thrombin components or to material of bovine origin; parenteral use; entry or infiltration into large blood vessels.
CAUTIOUS USE Pregnancy (category C). Safety and efficacy in children and infants are not established.

ROUTE & DOSAGE

Oozing Blood
Adult: **Topical** 100–2000 NIH units/mL, depending on extent of bleeding, may be used as solution, in dry form, by mixing thrombin with blood plasma to form a fibrin "glue," or in conjunction with absorbable gelatin sponge

ADMINISTRATION
Topical
- Ensure that sponge recipient area is free of blood before applying thrombin.
- Prepare solutions in sterile distilled water or isotonic saline.
- Use solutions within a few hours of preparation. If several hours are to elapse between time of preparation and use, solution should be refrigerated, or preferably frozen, and used within 48 h.
- Store lyophilized preparation at 2°–8° C (36°–46° F).

ADVERSE EFFECTS (≥1%) **Body as a Whole:** Sensitivity, allergic and febrile reactions, <u>intravascular clotting and death when thrombin is allowed to enter large blood vessels.</u>

THYROID
(thye'roid)

Armour Thyroid, Thyrar

Classification: THYROID HORMONE
Therapeutic: THYROID HORMONE REPLACEMENT
Prototype: Levothyroxine sodium
Pregnancy Category: A

AVAILABILITY 15 mg (¼ grain), 30 mg (½ grain), 60 mg (1 grain), 90 mg (1½ grain), 120 mg (2 grain), 180 mg (3 grain), 240 mg (4 grain), 300 mg (5 grain) tablets

ACTION & *THERAPEUTIC EFFECT*
Animal thyroid gland containing active thyroid hormones, l-thyroxine (T_4) and l-triiodothyronine (T_3). T_4 is largely converted to T_3, which exerts the principal effects. Influences growth and maturation of various tissues (including skeletal and CNS) at critical periods. Promotes a generalized increase in metabolic rate of body

T

tissues. *Effectiveness indicated by diuresis, followed by sense of well-being, increased pulse rate, increased pulse pressure, increased appetite, increased psychomotor activity, loss of constipation, normalization of skin texture and hair, and increased T_3 and T_4 serum levels.*

USES Replacement or substitution therapy in primary hypothyroidism (cretinism, myxedema, simple goiter, deficiency states in pregnancy and older adults) and secondary hypothyroidism caused by surgery, excess radiation, or antithyroid drug therapy. May be given as adjunct to antithyroid agents when it is desirable to limit release of thyrotropic hormones and to prevent goitrogenesis and hypothyroidism.

CONTRAINDICATIONS Thyrotoxicosis; acute MI not associated with hypothyroidism, cardiovascular disease; morphologic hypogonadism; nephrosis; uncorrected hypoadrenalism.

CAUTIOUS USE Angina pectoris, hypertension, older adults who may have occult cardiac disease; renal insufficiency; DM; history of hyperthyroidism; malabsorption states; pregnancy (category A).

ROUTE & DOSAGE

Mild to Moderate Hypothyroidism

Adult: **PO** 60 mg/day, may increase q30days to 60–180 mg/day

Severe Hypothyroidism

Adult: **PO** 15 mg/day, increased q2wk to 60 mg/day, then may increase q30days if needed

Child: **PO** 15 mg/day, may increase by 15 mg q2wk if needed

ADMINISTRATION

Oral

- Give preferably on an empty stomach.
- Initiate dosage generally at low level and systematically increase in small increments to desired maintenance dose.
- Store in dark bottle to minimize spontaneous deiodination. Keep desiccated thyroid dry.

ADVERSE EFFECTS (≥1%) **Endocrine:** Hyperthyroidism, thyroid storm [high temperature (as high as 41° C [106° F])], tachycardia, vomiting, <u>shock, coma</u>. **Special Senses:** Staring expression in eyes. **CV:** CHF, angina, cardiac arrhythmias, palpitation, tachycardia. **Body as a Whole:** Weight loss, tremors, headache, nervousness, fever, insomnia, warm and moist skin, heat intolerance, leg cramps, menstrual irregularities, <u>shock</u>, changes in appetite. **GI:** Diarrhea or abdominal cramps. **Metabolic:** Hyperglycemia (usually offset by increased tissue oxidation of sugar).

DIAGNOSTIC TEST INTERFERENCE Thyroid increases *basal metabolic rate;* may increase *blood glucose levels, creatine phosphokinase, AST, LDH, PBI.* It may decrease *serum uric acid, cholesterol, thyroid-stimulating hormone (TSH), iodine 131* uptake. Many medications may produce false results in *thyroid function tests.*

INTERACTIONS Drug: ORAL ANTICOAGULANTS potentiate hypoprothrombinemia; may increase requirements for **insulin,** SULFONYLUREAS; **epinephrine** may precipitate coronary insufficiency; **cholestyramine** may decrease thyroid absorption.

PHARMACOKINETICS Absorption: Variably absorbed from GI tract.

T

Peak: 1–3 wk. **Distribution:** Does not readily cross placenta; minimal amounts in breast milk. **Metabolism:** Deiodination in thyroid gland. **Elimination:** In urine and feces. **Half-Life:** T_3, 1–2 days; T_4, 6–7 days.

NURSING IMPLICATIONS

Assessment & Drug Effects

- Observe patient carefully during initial treatment for untoward reactions such as angina, palpitations, cardiac pain.
- Be alert for symptoms of overdosage (see ADVERSE EFFECTS) that may occur 1–3 wk after therapy is started.
- Monitor response until regimen is stabilized to prevent iatrogenic hyperthyroidism. In drug-induced hyperthyroidism, there may also be increased bone loss. Such a patient is vulnerable to pathologic fractures.
- Monitor vital signs: Assess pulse before each dose during period of dosage adjustment. Consult prescriber if rate is 100 or more or if there has been a marked change in rate or rhythm.
- Lab tests: Monitor thyroid function q3mo during dose adjustment period. Monitor prothrombin time if patient is receiving concurrent anticoagulant therapy. A decrease in requirement usually develops within 1–4 wk after starting treatment with thyroid.

Patient & Family Education

- Be aware that replacement therapy for hypothyroidism is life-long; continued follow-up care is important.
- Monitor pulse rate and report increases greater than parameter set by prescriber.
- Report to prescriber onset of chest pain or other signs of aggravated CV disease (dyspnea, tachycardia).
- Report evidence of any unexplained bleeding to prescriber when taking concomitant anticoagulant.
- Use monthly height and weight measurement to monitor growth in juvenile undergoing treatment.

TIAGABINE HYDROCHLORIDE

(ti-a′ga-been)

Gabitril Filmtabs
Classifications: ANTICONVULSANT; GABA INHIBITOR
Therapeutic: ANTICONVULSANT
Prototype: Valproic acid sodium (sodium valproate)
Pregnancy Category: C

AVAILABILITY 2 mg, 4 mg, 12 mg, 16 mg, 20 mg tablets

ACTION & *THERAPEUTIC EFFECT*
Potent and selective inhibitor of GABA uptake into presynaptic neurons; allows more GABA to bind to the surfaces of postsynaptic neurons in the CNS. *Effectiveness indicated by reduction in seizure activity.*

USES Adjunctive therapy for partial seizures.

CONTRAINDICATIONS Hypersensitivity to tiagabine; lactation.
CAUTIOUS USE Liver function impairment; history of spike and wave discharge on EEG; status epilepticus; pregnancy (category C); children younger than 12 y.

ROUTE & DOSAGE

Seizures
Adult: **PO** Start with 4 mg daily, may increase dose by 4–8 mg/day qwk (max: 56 mg/day in 2–4 divided doses)
Adolescent (12–18 y): **PO** Start with 4 mg daily, after 2 wk may

T

Common adverse effects in *italic*, life-threatening effects <u>underlined</u>; generic names in **bold**; classifications in SMALL CAPS; ♣ Canadian drug name; ⊚ Prototype drug

1479

increase dose by 4–8 mg/day qwk (max: 32 mg/day in 2–4 divided doses)

ADMINISTRATION

Oral
- Give with food.
- Make dosage increases, when needed, at weekly intervals.
- Store at 15°–30° C (59°–86° F) in a tightly closed container and protect from light.

ADVERSE EFFECTS (≥1%) Body as a Whole: Infection, flu-like syndrome, pain, myasthenia, allergic reactions, chills, malaise, arthralgia. **CNS:** *Dizziness, asthenia, tremor, somnolence, nervousness,* difficulty concentrating, ataxia, depression, insomnia, abnormal gait, hostility, confusion, speech disorder, difficulty with memory, paresthesias, emotional lability, agitation, dysarthria, euphoria, hallucinations, hyperkinesia, hypertonia, hypotonia, myoclonus, twitching, vertigo. Risk of new-onset seizures. **CV:** Vasodilation, hypertension, palpitations, tachycardia, syncope, edema, peripheral edema. **GI:** Abdominal pain, diarrhea, nausea, vomiting, increased appetite, mouth ulcers. **Respiratory:** Pharyngitis, cough, bronchitis, dyspnea, epistaxis, pneumonia. **Skin:** Rash, pruritus, alopecia, dry skin, sweating, ecchymoses. **Special Senses:** Amblyopia, nystagmus, tinnitus. **Urogenital:** Dysmenorrhea, dysuria, metrorrhagia, incontinence, vaginitis, UTI.

INTERACTIONS Drug: Carbamazepine, phenytoin, phenobarbital decrease levels of tiagabine. Use with ANTIDEPRESSANTS, ANTIPSYCHOTICS, STIMULANTS, and NARCOTICS may increase seizure risk. **Herbal: Ginkgo** may decrease anticonvulsant effectiveness. **Evening primrose oil** may affect seizure threshold.

PHARMACOKINETICS Absorption: Rapidly absorbed; 90% bioavailability. **Peak:** 45 min. **Distribution:** 96% protein bound. **Metabolism:** In liver, probably by cytochrome P450 3A isoform. **Elimination:** 25% in urine, 63% in feces. **Half-Life:** 7–9 h (4–7 h with other enzyme-inducing drugs).

NURSING IMPLICATIONS

Assessment & Drug Effects
- Lab tests: Measure plasma levels of tiagabine before and after changes are made in the drug regimen.
- Be aware that concurrent use of other anticonvulsants may decrease effectiveness of tiagabine or increase the potential for adverse effects.
- Monitor carefully for S&S of CNS depression.

Patient & Family Education
- Do not stop taking drug abruptly; may cause sudden onset of seizures.
- Exercise caution while engaging in potentially hazardous activities because drug may cause dizziness.
- Use caution when taking other prescription or OTC drugs that can cause drowsiness.
- Report any of the following to the prescriber: Rash or hives; red, peeling skin; dizziness; drowsiness; depression; GI distress; nervousness or tremors; difficulty concentrating or talking.

TICAGRELOR
(tye′ ka grel′)

Brilinta
Classifications: ANTIPLATELET; PLATELET P2Y12 ADP RECEPTOR ANTAGONIST; PLATELET AGGREGATION INHIBITOR; ANTITHROMBOTIC

Common adverse effects in *italic*, life-threatening effects <u>underlined</u>; generic names in **bold**; classifications in SMALL CAPS; ✤ Canadian drug name; 🅟 Prototype drug

Therapeutic: PLATELET AGGREGA-TION INHIBITOR; ANTITHROMBOTIC
Prototype: Clopidogrel
Pregnancy Category: C

AVAILABILITY 90 mg tablets

ACTION & *THERAPEUTIC EFFECT*
Ticagrelor and its major metabolite reversibly interact with the platelet ADP receptor to prevent signal transduction. *Prevents platelet activation and clot formation.*

USES Prophylaxis to prevent arterial thromboembolism in patients with acute coronary syndrome (ACS) (unstable angina, acute myocardial infarction) and patients undergoing percutaneous coronary intervention (PCI).

CONTRAINDICATIONS Severe hepatic impairment; history of intracranial hemorrhage; active bleeding (e.g., peptic ulcer).
CAUTIOUS USE Mild-to-moderate hepatic impairment; older adult; history of bleeding disorders, percutaneous invasive procedures, and use of drugs that increase risk of bleeding (e.g., anticoagulant and fibrinolytic therapy, high dose aspirin, long-term NSAIDs); compromised pulmonary function; pregnancy (category C); lactation. Safety and effectiveness in children have not been established.

ROUTE & DOSAGE

Arterial Thromboembolism Prophylaxis

Adult: **PO** 180 mg loading dose (with 325 mg aspirin) followed by 90 mg b.i.d. (with 75–100 mg aspirin once daily)

ADMINISTRATION
Oral
- May be given without regard to food.
- Store at 15°–30° C (59°–86° F).

ADVERSE EFFECTS (≥3%) **Body as a Whole:** Fatigue, noncardiac chest pain. **CNS:** Dizziness, headache. **CV:** Atrial fibrillation, chest pain, hypertension, hypotension, *increased bleeding*. **GI:** Diarrhea, nausea. **Metabolic:** Increased serum creatinine, increased uric acid levels. **Musculoskeletal:** Back pain. **Respiratory:** Cough, *dyspnea.*

INTERACTIONS Drug: Strong CYP3A4 inhibitors (e.g., **ketoconazole, itraconazole, voriconazole, clarithromycin, nefazodone, ritonavir, saquinavir, nelfinavir, indinavir, atazanavir,** and **telithromycin**) can increase ticagrelor levels, while potent CYP3A4 inducers (e.g. **rifampin, dexamethasone, phenytoin, carbamazepine** and **phenobarbital**) can decrease ticagrelor levels. Ticagrelor can increase the levels of **simvastatin** and **lovastatin.**

PHARMACOKINETICS Absorption: 30–42% Bioavailable. **Onset:** 1.5–5 h. **Distribution:** Greater than 99% plasma protein bound. **Metabolism:** CYP3A4 metabolism to active compound. **Elimination:** Fecal (58%) and renal (26%). **Half-Life:** 7 h (ticagrelor); 9 h (active metabolite).

NURSING IMPLICATIONS
Assessment & Drug Effects
- Monitor cardiac status with ECG and frequent BP measurements. Report immediately development of hypotension in any one who has recently undergone coronary angiography, PCI, CABG, or other surgical procedures.

Common adverse effects in *italic*, life-threatening effects underlined; generic names in **bold**; classifications in SMALL CAPS; ♣ Canadian drug name; ⊘ Prototype drug

1481

- Monitor for and report promptly S&S of bleeding or dyspnea.

Patient & Family Education
- Report promptly to prescriber any of the following: Unexplained, prolonged, or excessive bleeding; blood in stool or urine; shortness of breath.
- Daily doses of aspirin **should not** exceed 100 mg. Do not take any other drugs that contain aspirin. Consult prescriber if taking a non-steroidal anti-inflammatory drug (NSAID).
- Inform all healthcare providers that you are taking ticagrelor.

TICARCILLIN DISODIUM/ CLAVULANATE POTASSIUM
(tye-kar-sill'in/clav-yoo'la-nate)
Timentin
Classifications: ANTIBIOTIC; EXTENDED-SPECTRUM PENICILLIN
Therapeutic: ANTIBIOTIC
Prototype: Piperacillin/tazobactum
Pregnancy Category: B

AVAILABILITY 3.1 g injection

ACTION & *THERAPEUTIC EFFECT*
Extended-spectrum penicillin and fixed combination of ticarcillin disodium with the potassium salt of clavulanic acid, a beta-lactamase inhibitor. Used alone, clavulanic acid antibacterial activity is weak, but in combination with ticarcillin prevents degradation by beta-lactamase and extends ticarcillin spectrum of activity. *Combination drug extends ticarcillin spectrum of activity against many strains of beta-lactamase–producing bacteria (synergistic effect).*

USES Infections of lower respiratory tract and urinary tract and skin and skin structures, infections of bone and joint, and septicemia caused by susceptible organisms. Also mixed infections and as presumptive therapy before identification of causative organism.

CONTRAINDICATIONS Hypersensitivity to penicillins or to cephalosporins, coagulopathy.
CAUTIOUS USE DM; GI disease; asthma; history of allergies; renal impairment; pregnancy (category B), lactation.

ROUTE & DOSAGE

Moderate to Severe Infections
Adult (weight greater than 60 kg): **IV** 3.1 g q4–6h
Child (older than 3 mo): **IV** 200–300 mg/kg/day divided q4–6h (based on ticarcillin)
Infant (younger than 3 mo): **IV** 200–300 mg/kg/day divided q6–8h (based on ticarcillin)

Renal Impairment Dosage Adjustment
CrCl 30–60 mL/min: Give 2 g q4h or 3.1 g q8h; *10–30 mL/min:* Give 2 g q8h or 3.1 g q12h; *less than 10 mL/min:* Give 2 g q12h

Hemodialysis Dosage Adjustment
2 g q12h, supplement with 3.1 g after dialysis

ADMINISTRATION

Intravenous
Note: Verify correct IV concentration and rate of infusion for administration to infants and children with prescriber.

PREPARE: Intermittent: Reconstitute by adding 13 mL sterile water for injection or NS injection to the 3.1 g vial to yield 200 mg/mL

Common adverse effects in *italic*, life-threatening effects underlined; generic names in **bold**; classifications in SMALL CAPS; ♣ Canadian drug name; ⊕ Prototype drug

ticarcillin with 6.7 mg/mL clavulanic acid. Shake until dissolved.
▪ Further dilute the required does in NS, D5W, or LR to concentrations between 10–100 mg/mL.
▪ DO NOT use if discoloration or particulate matter is present.
ADMINISTER: **Intermittent:** Give over 30 min.
INCOMPATIBILITIES **Solution/additive:** AMINOGLYCOSIDES, **sodium bicarbonate. Y-site:** AMINOGLYCOSIDES, **amphotericin B cholesteryl complex, azithromycin, vancomycin.**

▪ Store vial with sterile powder at 21°–24° C (69°–75° F) or colder.
▪ If exposed to higher temperature, powder will darken, indicating degradation of clavulanate potassium and loss of potency. Discard vial. ▪ See package insert for information about storage and stability of reconstituted and diluted IV solutions of drug.

ADVERSE EFFECTS (≥1%) **Body as a Whole:** Hypersensitivity reactions, pain, burning, swelling at injection site; phlebitis, thrombophlebitis; superinfections. **CNS:** Headache, blurred vision, mental deterioration, convulsions, hallucinations, seizures, giddiness, neuromuscular hyperirritability. **GI:** *Diarrhea, nausea,* vomiting, disturbances of taste or smell, stomatitis, flatulence. **Hematologic:** Eosinophilia, thrombocytopenia, leukopenia, neutropenia, hemolytic anemia. **Metabolic:** Hypernatremia, transient increases in serum AST, ALT, BUN, and alkaline phosphatase; increases in serum LDH, bilirubin, and creatinine and decreased serum uric acid.

DIAGNOSTIC TEST INTERFERENCE May interfere with test methods used to determine *urinary* *proteins* except for tests for urinary protein that use *bromphenol blue. Positive direct antiglobulin (Coombs') test* results, apparently caused by clavulanic acid, have been reported. This test may interfere with *transfusion cross-matching procedures.*

INTERACTIONS Drugs: May increase risk of bleeding with ANTICOAGULANTS; **probenecid** decreases elimination of ticarcillin.

PHARMACOKINETICS Distribution: Widely distributed with highest concentrations in urine and bile; crosses placenta; distributed into breast milk. **Metabolism:** In liver. **Elimination:** In urine. **Half-Life:** 1.1–1.2 h ticarcillin, 1.1–1.5 h clavulanate.

NURSING IMPLICATIONS

Assessment & Drug Effects
▪ Lab tests: Obtain baseline C&S tests before initiating therapy; drug may be started pending results. Monitor LFTs and kidney functions, CBC, platelet count, and serum electrolytes during prolonged treatment.
▪ Be aware that serious and sometimes fatal anaphylactoid reactions have been reported in patients with penicillin hypersensitivity or history of sensitivity to multiple allergens. Reported incidence is low with this combination drug.
▪ Monitor cardiac status because of high sodium content of drug.
▪ Overdose symptoms: This drug may cause neuromuscular hyperirritability or seizures.

Patient & Family Education
▪ Report urticaria, rashes, or pruritus to prescriber immediately.
▪ Report frequent loose stools, diarrhea, or other possible signs of pseudomembranous colitis (see Appendix F) to prescriber.

T

Common adverse effects in *italic*, life-threatening effects underlined; generic names in **bold;** classifications in SMALL CAPS; ♣ Canadian drug name; ⊘ Prototype drug

1483

TICLOPIDINE HYDROCHLORIDE
(ti-clo'pi-deen)
Ticlid
Classifications: ANTICOAGULANT; PLATELET AGGREGATION INHIBITOR; ADP RECEPTOR ANTAGONIST
Therapeutic: PLATELET AGGREGATION INHIBITOR
Prototype: Clopidogrel
Pregnancy Category: B

AVAILABILITY 250 mg tablets

ACTION & *THERAPEUTIC EFFECT*
Platelet aggregation inhibitor that interferes with platelet membrane functioning and therefore platelet interactions. *Ticlopidine interferes with ADP-induced binding of fibrinogen to the platelet membrane at specific receptor sites. Platelet adhesion and platelet aggregation are therefore inhibited. Prevents release of platelet constituents and prolongs bleeding time.*

USES Reduction of the risk of thrombotic stroke in patients intolerant to aspirin.
UNLABELED USES Prevention of venous thromboembolic disorders; maintenance of bypass graft patency and of vascular access sites in hemodialysis patients; improvement of exercise performance in patients with ischemic heart disease and intermittent claudication; prevention of postoperative deep venous thrombosis (DVT).

CONTRAINDICATIONS Hypersensitivity to ticlopidine; hematopoietic disease, coagulopathy; neutropenia thrombotic thrombocytopenic purpura (TTP) leukemia; pathologic bleeding; severe liver impairment; lactation.
CAUTIOUS USE Hepatic function impairment, renal impairment; patients at risk for bleeding from trauma, surgery, or a bleeding disorder; GI bleeding; pregnancy (category B). Safe use in children younger than 18 y not established.

ROUTE & DOSAGE

Stroke Prevention
Adult: **PO** 250 mg b.i.d. with food

ADMINISTRATION
Oral
- Give with food or just after eating to minimize GI irritation.
- Do not give within 2 h of an antacid.
- Store at 15°–30° C (59°–86° F).

ADVERSE EFFECTS (≥1%) **CNS:** Dizziness. **GI:** Nausea, vomiting, abdominal cramps; dyspepsia, flatulence, anorexia; abnormal liver function tests (few cases of hepatotoxicity reported). **Hematologic:** Neutropenia (resolves in 1–3 wk), thrombocytopenia, leukopenia, agranulocytosis (usually within first 3 mo), and pancytopenia; hemorrhage (ecchymosis, epistaxis, menorrhagia, GI bleeding), thrombotic thrombocytopenia purpura (usually within first month). **Skin:** Urticaria, maculopapular rash, erythema nodosum (generally occur within the first 3 mo of therapy, with most occurring within the first 3–6 wk).

DIAGNOSTIC TEST INTERFERENCE Increases *total serum cholesterol* by 8–10% within 4 wk of beginning therapy. *Lipoprotein ratios* remain unchanged. Elevates *alkaline phosphatase* and *serum transaminases.*

INTERACTIONS Drug: ANTACIDS decrease bioavailability of ticlopidine. ANTICOAGULANTS increase risk of bleeding. **Cimetidine** decreases clearance of ticlopidine. CORTICOSTEROIDS counteract increased bleeding time associated with ticlopidine. May decrease **cyclosporine** levels (one case report). Increases **theophylline** half-life by 42%, possibly increasing **theophylline** serum levels. May increase **phenytoin** levels. **Food:** Food may increase bioavailability of ticlopidine. **Herbal:** Evening primrose oil increases bleeding risk.

PHARMACOKINETICS Absorption: 90% absorbed from GI tract; increased absorption when taken with food. **Onset:** Antiplatelet activity, 24–48 h; maximal effect at 3–5 days. **Peak:** Peak serum levels at 2h. **Duration:** Bleeding times return to baseline within 4–10 days. **Distribution:** 90% bound to plasma proteins. **Metabolism:** Rapidly and extensively metabolized in liver. **Elimination:** Only 1% excreted unchanged; 60% of metabolites excreted in urine, 23% in feces. **Half-Life:** 12.6 h; terminal half-life is 4–5 days with repeated dosing.

NURSING IMPLICATIONS

Assessment & Drug Effects
- Lab tests: Periodic platelet count and bleeding time. Monitor CBC with differentials q2wk from second week to end of third month of therapy and thereafter if S&S of infection develop.
- Report promptly laboratory values indicative of neutropenia, thrombocytopenia, or agranulocytosis.
- Monitor for signs of bleeding (e.g., ecchymosis, epistaxis, hematuria, GI bleeding).

Patient & Family Education
- Report promptly to prescriber any of the following: Nausea, diarrhea, rash, sore throat, or other signs of infection, signs of bleeding, or signs of cholestasis (e.g.,

yellow skin or sclera, dark urine or clay-colored stools).
- Understand risk of GI bleeding; do not take aspirin along with ticlopidine.
- Do not take antacids within 2 h of ticlopidine.
- Keep appointments for regularly scheduled blood tests.

TIGECYCLINE
(ti-ge-cy′cline)
Tygacil
Classifications: ANTIBIOTIC; GLYCYLCYCLINE
Therapeutic: ANTIBIOTIC
Prototype: Tetracycline
Pregnancy Category: D

AVAILABILITY 50 mg injection

ACTION & *THERAPEUTIC EFFECT* Tigecycline inhibits protein production in bacteria by binding to the 30S ribosomal subunit and blocking entry of transfer RNA molecules into ribosome of bacteria. This prevents formation of peptide chains within bacteria, thus interfering with their growth. *Tigecycline is active against a broad spectrum of bacterial pathogens and is bacteriostatic.*

USES Treatment of complicated skin and skin structure infections, community acquired pneumonia, and complicated intra-abdominal infections.

CONTRAINDICATIONS Hypersensitivity to tigecycline; viral infections; pregnancy (category D) and during tooth development of the fetus.
CAUTIOUS USE Severe hepatic impairment (Child-Pugh class C); hypersensitivity to tetracycline; intestinal perforations, intra-abdominal infections; GI disorders; lactation; children younger than 18 y.

Common adverse effects in *italic*, life-threatening effects underlined; generic names in **bold;** classifications in SMALL CAPS; ◆ Canadian drug name; ⊘ Prototype drug

1485

ROUTE & DOSAGE

Community Acquired Pneumonia, Complicated Skin/ Intra-abdominal Infections

Adult: **IV** 100 mg initially, followed by 50 mg q12h over 30–60 min × 5–14 days

Hepatic Impairment Dosage Adjustment

Child-Pugh class C: Initial dose 100 mg, followed by 25 mg q12h

ADMINISTRATION

- Note that dosage adjustment is required with severe hepatic impairment.

Intravenous

PREPARE: **Intermittent:** Reconstitute each 50 mg with 5.3 mL of NS or D5W to yield 10 mg/mL. • Swirl gently to dissolve; reconstituted solution should be yellow to orange in color. • After reconstitution, immediately withdraw exactly 5 mL from each vial and add to 100 mL of NS or D5W for infusion. • The maximum concentration in the IV bag should be 1 mg/mL (two 50 mg doses).

ADMINISTER: **Intermittent:** Give over 30–60 min; when using Y-site, flush IV line with NS or D5W before/after infusion.

INCOMPATIBILITIES **Y-site: Amphotericin B, chlorpromazine, methylprednisolone, voriconazole.**

- Store in the IV bag at room temperature for up to 6 h, or refrigerated at 2°–8° C (36°–46° F) for up to 24 h.

ADVERSE EFFECTS (≥1%) CNS: As-thenia, dizziness, headache, insomnia. **CV:** Hypertension, hypotension, peripheral edema, phlebitis. **GI:** Abdominal pain, constipation, diarrhea, dyspepsia, *nausea, vomiting.* **Hematologic/Lymphatic:** Abnormal healing, anemia, infection, leukocytosis, thrombocythemia. **Metabolic/ Nutritional:** Alkaline phosphatase increased, ALT increased, amylase increased, AST increased, bilirubinemia, BUN increased, hyperglycemia, hypokalemia, hypoproteinemia, lactic dehydrogenase increased. **Musculoskeletal:** Back pain. **Respiratory:** Dyspnea, increased cough, pulmonary physical findings. **Skin:** Pruritus, rash, sweating. **Body as a Whole:** Abscess, fever, local reaction to injection, pain.

INTERACTIONS Drug: Increased concentrations of **warfarin** required close monitoring of INR. Efficacy of ORAL CONTRACEPTIVES may be decreased when used in combination with tigecycline.

PHARMACOKINETICS Distribution: 71–89% protein bound. **Metabolism:** Negligible. **Elimination:** Fecal (major) and renal. **Half-Life:** 27 h (single dose); 42 h (multiple doses).

NURSING IMPLICATIONS

Assessment & Drug Effects

- Monitor for hypersensitivity reaction in those with reported tetracycline allergy.
- Monitor for and report S&S of superinfection (see Appendix F) or pseudomembranous enterocolitis (see Appendix F).
- Lab tests: C&S prior to initiation of therapy; periodic serum electrolytes, LFTs and kidney function tests; PT and INR with concurrent anticoagulant therapy.
- Monitor diabetics for loss of glycemic control.

Common adverse effects in *italic*, life-threatening effects underlined; generic names in **bold**; classifications in SMALL CAPS; ♣ Canadian drug name; ⊘ Prototype drug

Patient & Family Education

- Avoid direct exposure to sunlight during and for several days after therapy is terminated to reduce risk of photosensitivity reaction.
- Report to prescriber loose stools or diarrhea either during or shortly after termination of therapy.
- Use a barrier contraceptive in addition to oral contraceptives if trying to avoid pregnancy.

TILUDRONATE DISODIUM

(til-u'dro-nate)

Skelid

Classification: BONE METABOLISM REGULATOR (BISPHOSPHONATE)
Therapeutic: BONE METABOLISM REGULATOR
Prototype: Etidronate disodium
Pregnancy Category: C

AVAILABILITY 240 mg tablets

ACTION & *THERAPEUTIC EFFECT* This diphosphate inhibits osteoclastic activity that leads to resorption of the bone matrix. Acts primarily by inhibiting normal or abnormal bone resorption, thus reducing bone formation. *Effectiveness indicated by decreasing levels of alkaline phosphatase.*

USES Treatment of Paget's disease.

CONTRAINDICATIONS Hypersensitivity to diphosphonates (e.g., alendronate, etidronate, pamidronate, tiludronate); severe kidney failure (CrCl less than 30 mL/min); lactation.
CAUTIOUS USE Hypocalcemia, renal impairment; active UGI problems (e.g., gastritis, dysphagia, ulcer, esophageal disease); vitamin D deficiency; CHF; pregnancy (category C). Safety and efficacy in children are not established.

ROUTE & DOSAGE

Paget's Disease
Adult: **PO** 400 mg/day with 6–8 oz of water × 3 mo

ADMINISTRATION

Oral

- Give with 6–8 oz of plain water 2 h before or after food.
- Do not give within 2 h of drugs containing calcium, aspirin, or indomethacin. Give aluminum- or magnesium-containing antacids no sooner than 2 h after tiludronate.
- Store in manufacturer's packaging at 15°–30° C (59°–86° F).

ADVERSE EFFECTS (≥1%) **Body as a Whole:** *Pain,* flu-like syndrome, edema. **CNS:** Headache, dizziness, paresthesias. **CV:** Chest pain. **GI:** *Nausea, diarrhea,* dyspepsia, vomiting, flatulence. **Special Senses:** Cataract, conjunctivitis, glaucoma. **Respiratory:** Rhinitis, sinusitis, coughing, pharyngitis. **Skin:** Rash. **Metabolic:** Hyperparathyroidism, vitamin D deficiency, **Musculoskeletal:** Arthralgia, arthrosis.

INTERACTIONS Drug: Absorption decreased by CALCIUM-, ALUMINUM- or MAGNESIUM-CONTAINING ANTACIDS, **aspirin.** Absorption increased by **indomethacin.**

PHARMACOKINETICS Absorption: Poorly absorbed from GI tract. **Steady-State:** 30 days. **Metabolism:** Not metabolized. **Elimination:** Primarily in urine. **Half-Life:** 150 h.

NURSING IMPLICATIONS

Assessment & Drug Effects

- Monitor for S&S of upper GI dysfunction or ulceration.
- Lab tests: Periodic serum calcium and serum phosphate.

T

Common adverse effects in *italic*, life-threatening effects underlined; generic names in **bold;** classifications in SMALL CAPS; ◆ Canadian drug name; ⊘ Prototype drug

1487

Patient & Family Education

- Do not remove tablets from foil strips until time to be taken.
- Wait at least 2 h after taking tiludronate to take aluminum- and magnesium-containing antacids.
- Consult prescriber to determine appropriate daily intake of vitamin D and calcium.

TIMOLOL MALEATE

(tye′moe-lole)

Betimol, Blocadren, Istalol, Timoptic, Timoptic XE

Classifications: BETA-ADRENERGIC ANTAGONIST; EYE PREPARATION; MIOTIC; ANTIHYPERTENSIVE; ANTIANGINAL
Therapeutic: ANTIGLAUCOMA; ANTIHYPERTENSIVE; ANTIANGINAL
Prototype: Propranolol
Pregnancy Category: C

AVAILABILITY 5 mg, 10 mg, 20 mg tablets; 0.25%, 0.5% ophth solution or gel

ACTION & *THERAPEUTIC EFFECT* Nonselective beta-adrenergic antagonist that demonstrates antihypertensive, antiarrhythmic, and antianginal properties, and suppresses plasma renin activity. When applied topically, lowers elevated and normal intraocular pressure (IOP) by reducing formation of aqueous humor and possibly by increasing outflow. *Topically, lowers elevated and normal intraocular pressure (IOP). Orally, therapeutically useful for mild hypertension, angina, and migraine headaches.*

USES Topically (ophthalmic solution) to reduce elevated IOP in chronic, open-angle glaucoma, aphakic glaucoma, secondary glaucoma, and ocular hypertension. May be used alone or in conjunction with epinephrine, pilocarpine, or a carbonic anhydrase inhibitor such as acetazolamide. Oral preparation is used as monotherapy or in combination with a thiazide diuretic to prevent reinfarction after MI and to treat mild hypertension.
UNLABELED USES Prophylactic management of stable, uncomplicated angina pectoris and migraine headaches.

CONTRAINDICATIONS Bronchospasm; severe COPD; bronchial asthma; heart failure; abrupt discontinuation, acute bronchospasm, AV block, bradycardia, cardiogenic shock, acute pulmonary edema, compromised left ventricular dysfunction, Raynaud's disease.
CAUTIOUS USE Bronchitis, patients subject to bronchospasm, asthma; sinus bradycardia, greater than first-degree heart block, heart failure; renal impairment; hepatic disease; vasospastic angina, peripheral vascular disease; pheochromocytoma; thyrotoxicosis, hyperthyroidism; right ventricular failure secondary to pulmonary hypertension, COPD; stroke, cerebrovascular disease; depression; psoriasis; myasthenia gravis; older adults; pregnancy (category C). Safe use in children not established.

ROUTE & DOSAGE

Glaucoma
See Appendix A-1.
Hypertension, Post-MI Reinfarction
Adult: **PO** 10 mg b.i.d., may increase to 60 mg/day in 2 divided doses
Angina
Adult: **PO** 15–45 mg in 3 divided doses

ADMINISTRATION

Oral

- Give with fluid of patient's choice; tablet may be crushed.
- Dosage increases for hypertension should be made at weekly intervals.

ADVERSE EFFECTS (≥1%) **CNS:** Fatigue, lethargy, weakness, somnolence, anxiety, headache, dizziness, confusion, psychic dissociation, depression. **CV:** Palpitation, bradycardia, hypotension, syncope, AV conduction disturbances, CHF, aggravation of peripheral vascular insufficiency. **Special Senses:** *Eye irritation* including conjunctivitis, blepharitis, keratitis, superficial punctate keratopathy. **GI:** Anorexia, dyspepsia, nausea. **Skin:** Rash, urticaria. **Respiratory:** Difficulty in breathing, bronchospasm. **Body as a Whole:** Fever. **Metabolic:** Hypoglycemia, hypokalemia.

INTERACTIONS Drug: ANTIHYPERTENSIVE AGENTS, DIURETICS, SELECTIVE SEROTONIN REUPTAKE INHIBITORS potentiate hypotensive effects; NSAIDS may antagonize hypotensive effects.

PHARMACOKINETICS Absorption: 90% absorbed from GI tract; 50% reaches systemic circulation; some systemic absorption from topical application. **Peak:** 1–2 h PO; 1–5 h topical. **Distribution:** Distributed into breast milk. **Metabolism:** 80% metabolized in liver to inactive metabolites. **Elimination:** In urine.

NURSING IMPLICATIONS

Assessment & Drug Effects

- Check pulse before administering timolol, topical or oral. If there are extremes (rate or rhythm), withhold medication and notify prescriber.

- Assess pulse rate and BP at regular intervals and more aften in patients with severe heart disease.
- Note: Some patients develop tolerance during long-term therapy.

Patient & Family Education

- Be aware that drug may cause slight reduction in resting heart rate. Learn how to assess pulse rate and report significant changes. Consult prescriber for parameters.
- Do not stop drug abruptly; angina may be exacerbated. Dosage is reduced over a period of 1–2 wk.
- Report promptly to prescriber difficulty breathing. Drug withdrawal may be indicated.

TINIDAZOLE

(tin′i-da-zole)

Tindamax

Classification: AZOLE ANTIBIOTIC

Therapeutic: ANTIBIOTIC; ANTI-PROTOZOAL; AMEBICIDE

Prototype: Metronidazole

Pregnancy Category: X first trimester and C second and third trimester

AVAILABILITY 250 mg, 500 mg tablets

ACTION & *THERAPEUTIC EFFECT* Tinidazole is effective against dividing and nondividing cells of targeted bacteria and protozoa. It inhibits formation of their DNA helix and thus inhibits DNA synthesis of these organisms. This leads to bacterial and protozoal cell death. *Demonstrates activity against infections caused by protozoa and anaerobic bacteria.*

USES Treatment of trichomoniasis, giardiasis, amebiasis, bacterial vaginosis, and amebic liver abscess.

CONTRAINDICATIONS Hypersensitivity to tinidazole or other azole

Common adverse effects in *italic*, life-threatening effects underlined; generic names in **bold;** classifications in SMALL CAPS; ♣ Canadian drug name; ⊙ Prototype drug

1489

antibiotics; pregnancy (category X first trimester), lactation within 72 h of tinidazole use.

CAUTIOUS USE CNS diseases, liver dysfunction, alcoholism, ethanol intoxication; hematologic disease; neurologic disease; bone marrow depression; dialysis; candidiasis; pregnancy (category C second and third trimester); children younger than 3 y.

ROUTE & DOSAGE

Giardiasis

Adult: **PO** 2 g as single dose
Child (3 y or older): **PO** 50 mg/kg (up to 2 g) as single dose

Intestinal Amebiasis

Adult: **PO** 2 g once daily for 3 days
Child (3 y or older): **PO** 50 mg/kg/day (up to 2 g/day) once daily for 3 days

Amebic Liver Abscess

Adult: **PO** 2 g once daily for 3– 5 days
Child (3 y or older): **PO** 50 mg/kg/day (up to 2 g/day) once daily for 3–5 days

Trichomoniasis

Adult: **PO** 2 g as single dose

Bacterial Vaginosis

Adult: **PO** 2 g daily × 2 days OR 1 g daily × 5 days

Hemodialysis Dosage Adjustment

If dose given on dialysis day, give supplemental dose (½ regular dose) post-dialysis

ADMINISTRATION

Oral

- Give with food to minimize GI distress; may be crushed in artificial cherry syrup if tablets cannot be swallowed whole.
- If given on a dialysis day, add a 50% dose of tinidazole at the end of hemodialysis.
- Separate the dosing of cholestyramine and tinidazole by 2–4 h when used concurrently.
- Do not give within 2 wk of the last dose of disulfiram.
- Store at 15°–30° C (59°–86° F). Protect from light.

ADVERSE EFFECTS (≥1%) **Body as a Whole:** Weakness, fatigue, malaise. **CNS:** Dizziness, headache. **GI:** Metallic/bitter taste, nausea, anorexia, dyspepsia, cramps, epigastric discomfort, vomiting, constipation.

INTERACTIONS Drug: May increase INR with **warfarin; alcohol** may cause **disulfiram**-like reaction; may increase the half-life of **fosphenytoin, phenytoin;** may increase levels and toxicity of **lithium, fluorouracil, cyclosporine, tacrolimus; cholestyramine** may decrease absorption of tinidazole; **cimetidine** or **ketoconazole** may increase levels.

PHARMACOKINETICS Peak: 2 h. **Distribution:** Crosses blood–brain barrier and placenta and is excreted in breast milk. **Metabolism:** In the liver by CYP3A4. **Elimination:** Primarily in urine. **Half-Life:** 12–14 h.

NURSING IMPLICATIONS

Assessment & Drug Effects

- Withhold drug and notify prescriber for S&S of CNS dysfunction (e.g., seizures, numbness or paresthesia of extremities). Drug should be discontinued if abnormal neurologic signs appear.
- Lab tests: Baseline LTFs; CBC with differential, if retreatment is required.

- Monitor INR/PT frequently with concomitant oral anticoagulants. Continue monitoring for at least 8 days after discontinuation of tinidazole.
- Monitor serum lithium levels with concurrent use.
- Monitor for phenytoin toxicity with concurrent IV phenytoin.

Patient & Family Education
- Stop taking the drug and report promptly: Convulsions, numbness, tingling, pain, or weakness in the hands or feet; dizziness or unsteadiness; fever.
- Harmless urine discoloration may occur while taking this drug.

TIOCONAZOLE
(ti-o-con′a-zole)

Vagistat-1
Classifications: AZOLE ANTIFUNGAL
Therapeutic: ANTIFUNGAL
Prototype: Fluconazole
Pregnancy Category: C

AVAILABILITY 6.5% vaginal ointment

ACTION & *THERAPEUTIC EFFECT*
Broad-spectrum antifungal agent that inhibits growth of human pathogenic yeasts by disrupting normal fungal cell membrane permeability. *Effective against* Candida albicans, *other species of* Candida, *and* Torulopsis glabrata.

USES Local treatment of vulvovaginal candidiasis.

CONTRAINDICATIONS Hypersensitivity to tioconazole or other imidazole antifungal agents; lactation.
CAUTIOUS USE Diabetes mellitus; HIV infections; immunosuppression; pregnancy (category C); children younger than 12 y.

ROUTE & DOSAGE

Candidiasis
Adult: **Intravaginal** One applicator full at bedtime × 1 day

ADMINISTRATION

Instillation
- Insert applicator high into the vagina (except during pregnancy).
- Wash applicator before and after each use.
- Store away from direct heat and light.

ADVERSE EFFECTS (≥1%) Urogenital: Mild erythema, burning, discomfort, rash, itching.

INTERACTIONS Drug: May inactivate spermicidal effects of **nonoxynol-9.**

PHARMACOKINETICS Absorption: Minimal absorption from vagina.

NURSING IMPLICATIONS

Assessment & Drug Effects
- Do not use for patient with a history of allergic reaction to other antifungal agents, such as miconazole.
- Monitor for sensitization and irritation; these may be an indication to discontinue drug.

Patient & Family Education
- Learn correct application technique.
- Understand potential adverse reactions, including sensitization and allergic response.
- Tioconazole may interact with diaphragms and latex condoms; avoid concurrent use within 72 h.
- Refrain from sexual intercourse while using tioconazole.
- Wear only cotton underwear; change daily.

T

Common adverse effects in *italic*, life-threatening effects underlined; generic names in **bold**; classifications in SMALL CAPS; ♣ Canadian drug name; ⊘ Prototype drug

TIOTROPIUM BROMIDE

(ti-o-tro′pi-um)

Spiriva

Classifications: ANTICHOLINER-GIC; ANTIMUSCARINIC; ANTISPAS-MODIC; BRONCHODILATOR

Therapeutic: BRONCHODILATOR; ANTISPASMODIC

Prototype: Atropine

Pregnancy Category: C

AVAILABILITY 18 mcg capsules with powder for inhalation

ACTION & *THERAPEUTIC EFFECT* A long-acting, antispasmodic agent. In the bronchial airways, it exhibits inhibition of muscarinic receptors of the smooth muscle resulting in bronchodilation. *Bronchodilation after inhalation of tiotropium is predominantly a site-specific effect.*

USES Maintenance treatment of bronchospasm associated with chronic obstructive pulmonary disease (COPD); reducing COPD exacerbations.

CONTRAINDICATIONS Hypersensitivity to tiotropium, atropine, or ipratropium; acute bronchospasm.

CAUTIOUS USE Decreased renal function; BPH, urinary bladder neck obstruction; narrow-angle glaucoma; older adults; pregnancy (category C), lactation; children younger than 18 y.

ROUTE & DOSAGE

COPD

Adult: **Inhaled** Inhale 18 mcg once daily using inhaler device provided.

ADMINISTRATION

Inhalation

- Place capsule in HandiHaler® and press button to puncture.

Instruct patient to exhale deeply then put the mouthpiece to the lips and breathe in the dose deeply and slowly; remove HandiHaler® and hold breath for at least 10 sec, and then exhale slowly; rinse mouth with water to minimize dry mouth.

- Ensure that drug does not contact the eyes.
- Store at 15°–30° C (59°–86° F).

ADVERSE EFFECTS (≥1%) Body as a Whole: Nonspecific chest pain, dependent edema, infection, moniliasis, flu-like syndrome, cough, allergic reactions. **CNS:** Dysphonia, paresthesia, depression. **GI:** Abdominal pain, constipation, *dry mouth, dyspepsia,* vomiting, reflux, stomatitis. **Metabolic:** Hypercholesterolemia, hyperglycemia. **Musculoskeletal:** Myalgia, skeletal pain. **Respiratory:** Epistaxis, *pharyngitis, rhinitis,* laryngitis, *sinusitis, upper respiratory tract infection.* **Skin:** Rash. **Special Senses:** Cataract. **Urogenital:** Urinary tract infection.

INTERACTIONS Drug: May cause additive anticholinergic effects with other ANTICHOLINERGIC AGENTS.

PHARMACOKINETICS Absorption: 19.5% absorbed from the lungs. **Peak:** 5 min. **Metabolism:** Less than 25% of dose is metabolized in liver by CYP2D6 and 3A4. **Elimination:** 14% of dose excreted in urine; remaining is excreted in feces as nonabsorbed drug. **Half-Life:** 5–6 days.

NURSING IMPLICATIONS

Assessment & Drug Effects

- Withhold drug and notify prescriber if S&S of angioedema occurs.
- Monitor for anticholinergic effects (e.g., tachycardia, urinary retention).

Patient & Family Education

- Do not allow powdered medication to contact the eyes, as this may cause blurring of vision and pupil dilation.
- Tiotropium bromide is intended as a once-daily maintenance treatment. It is not useful for treatment of acute episodes of bronchospasm (i.e., rescue therapy).
- Withhold drug and notify prescriber if swelling around the face, mouth, or neck occurs.
- Report any of the following: Constipation, increased heart rate, blurred vision, urinary difficulty.

TIPRANAVIR

(ti-pra′na̞-vir)

Aptivus
Classifications: ANTIRETROVIRAL; PROTEASE INHIBITOR
Therapeutic: ANTIRETROVIRAL
Prototype: Saquinavir
Pregnancy Category: C

AVAILABILITY 250 mg capsules; 100 mg/mL oral solution

ACTION & *THERAPEUTIC EFFECT* A non-peptide protease inhibitor that inhibits virus-specific processing of the viral polyproteins in HIV-1 infected cells, thus preventing the formation of mature viral particles. *Helps decrease viral load of HIV-1 strains resistant to other protease inhibitors.*

USES Treatment of HIV-1 infection in combination with other agents.

CONTRAINDICATIONS Known hypersensitivity to tipranavir; moderate to severe (Child-Pugh class B and C, respectively) hepatic impairment; pancreatitis; lactation.

CAUTIOUS USE Hypersensitivity to sulfonamides; patients with chronic hepatitis B or hepatitis C coinfection; hemophilia; coagulopathy; elevated liver enzymes; diabetes mellitus or hyperglycemia; hyperlipidemia; pregnancy (category C); children.

ROUTE & DOSAGE

HIV-1 Infection
Adult: **PO** 500 mg (with 200 mg ritonavir) b.i.d.
Adolescent/Child (over 2): **PO** 14 mg/kg OR 375 mg/m² OR 500 mg b.i.d. (depending on ritonavir dose)

ADMINISTRATION

Oral
- Coadminister with ritonavir. Give with food.
- Store at 15°–30° C (59°–86° F). Once opened, use contents of bottle within 60 days.

ADVERSE EFFECTS (≥1%) **Body as a Whole:** Fatigue, pyrexia. **CNS:** Asthenia, depression, headache, insomnia. **GI:** Abdominal pain, *diarrhea*, nausea, vomiting. **Hematologic:** Decreased white blood cell levels, risk of hemorrhage. **Hepatic:** *Elevated liver enzymes (amylase, ALT, AST).* **Metabolic:** *Increased cholesterol, increased triglycerides.* **Respiratory:** Bronchitis, cough. **Skin:** Rash.

INTERACTIONS Drug: Do not use with **sildenafil, alfuzosin, salmeterol.** Adjust **tipranavir** dosing when used with **bosentan, tadalafil, colchicine.** **Aluminum-** and **magnesium-**based ANTACIDS may decrease tipranavir absorption. AZOLE ANTIFUNGAL AGENTS, **clarithromycin,**

Common adverse effects in *italic*, life-threatening effects underlined; generic names in **bold**; classifications in SMALL CAPS; ♣ Canadian drug name; ⊙ Prototype drug

1493

erythromycin, and other inhibitors of CYP3A4 may increase tipranavir levels. **Efavirenz, loperamide,** NRTIS, and RIFAMYCINS (e.g., **rifampin**) may decrease tipranavir levels. Tipranavir increases **rifabutin** levels. Coadministration of tipranavir and **tenofovir** decreases the levels of both compounds. Tipranavir increases the concentration of BENZODIAZEPINES, **desipramine,** ERGOT ALKALOIDS, and numerous ANTIARRHYTHMIC AGENTS **(amiodarone, flecainide, propafenone, quinidine).** Tipranavir may decrease **ethinyl estradiol** levels by 50%. Combination use of tipranavir and HMG COA REDUCTASE INHIBITORS increases the risk of myopathy. Tipranavir capsules contain **alcohol** that can produce disulfiram-like reactions with **metronidazole** and **disulfiram.** Adjust **midazolam** dose. **Food:** Food enhances the bioavailability of tipranavir, avoid large doses of vitamin E. **Herbal: St. John's wort** decreases the levels of tipranavir.

PHARMACOKINETICS Peak: 3 h. **Distribution:** Greater than 99.9% protein bound. **Metabolism:** Extensive hepatic oxidation to inactive metabolites (when given alone); minimal metabolism (when given with ritonavir). **Elimination:** Fecal (primary) and renal (minimal). **Half-Life:** 6 h.

NURSING IMPLICATIONS

Assessment & Drug Effects
- Monitor for and report immediately S&S of liver toxicity (see Appendix F).
- Monitor for S&S of adverse drug reactions and toxicity from concurrently administered drugs. Many drugs interact with tipranavir.
- Monitor diabetics for loss of glycemic control.

- Use barrier contraceptive if using hormonal contraceptive.
- Lab tests: Baseline and frequent LFTs, especially in those with hepatitis B or C; periodic lipid profile and fasting plasma glucose.
- Monitor blood levels of anticoagulants with concurrent therapy.

Patient & Family Education
- If a dose is missed, take it as soon as possible and then return to the normal schedule. Never double a dose.
- Inform prescriber of all medications and herbal products you are taking.
- Protect against sunlight exposure to minimize risk of photosensitivity.
- Report any of the following to prescriber: Fatigue, weakness, loss of appetite, nausea, jaundice, dark urine, or clay colored stools.

TIROFIBAN HYDROCHLORIDE
(tir-o-fi′ban)

Aggrastat
Classifications: ANTICOAGULANT; ANTIPLATELET; GLYCOPROTEIN (GP) IIB/IIIA RECEPTOR INHIBITOR
Therapeutic: ANTIPLATELET
Prototype: Abciximab
Pregnancy Category: B

AVAILABILITY 250 mcg/mL, 50 mcg/mL injection

ACTION & *THERAPEUTIC EFFECT*
Antiplatelet agent that binds to the glycoprotein IIb/IIIa receptor of platelets inhibiting platelet aggregation. *Effectiveness indicated by minimizing thrombotic events during treatment of acute coronary syndrome.*

USES Acute coronary syndromes (unstable angina, MI).

Common adverse effects in *italic*, life-threatening effects <u>underlined</u>; generic names in **bold**; classifications in SMALL CAPS; ♣ Canadian drug name; ⊙ Prototype drug

CONTRAINDICATIONS Hypersensitivity to tirofiban; active internal bleeding within 30 days; acute pericarditis; aortic dissection; intracranial aneurysm, intracranial mass, coagulopathy; history of aneurysm or AV malformation; history of intracranial hemorrhage or neoplasm; active abnormal bleeding; retinal bleeding; hemorrhagic retinopathy; major surgery or trauma within 3 days; stroke within 30 days; history of hemorrhagic stroke; thrombocytopenia following administration of tirofiban; history of thrombopenia due to prior exposure to tirofiban; severe rash during use of tirofiban; within 4 h of PCI; lactation.

CAUTIOUS USE Platelet count less than 150,000 mm³; severe renal insufficiency; pregnancy (category B). Safety and efficacy in children younger than 18 y are unknown.

ROUTE & DOSAGE

Acute Coronary Syndromes

Adult: **IV** 0.4 mcg/kg/min for 30 min, then 0.1 mcg/kg/min for 12–24 h after angioplasty or arteriectomy

Renal Impairment Dosage Adjustment

CrCl less than 30 mL/min: Reduce rate of infusion 50%

ADMINISTRATION

Intravenous

PREPARE: **IV Infusion** *Dilution of 250 mcg/mL concentrate to 50 mcg/mL:* Withdraw 100 mL from a 500-mL bag of NS or D5W and replace with 100 mL of tirofiban HCl injection. ▪ If a 250-mL IV bag is used, withdraw 50 mL of IV solution and replace with 50 mL of tirofiban injection. ▪ Either preparation yields 50 mcg/mL. ▪ Mix well before infusing. ▪ Note: Commercially premixed IV tirofiban solutions are available.

ADMINISTER: **IV Infusion:** An initial loading dose of 0.4 mcg/kg/min for 30 min is usually followed by a maintenance infusion of 0.1 mcg/kg/min.

INCOMPATIBILITIES **Y-site: Diazepam, tenecteplase.**

▪ Discard unused IV solution 24 h following start of infusion. ▪ Store unopened containers at 15°–30° C (59°–86° F). ▪ Do not freeze and protect from light.

ADVERSE EFFECTS (≥1%) Body as a Whole: Edema, swelling, pelvic pain, vasovagal reaction, leg pain. **CNS:** Dizziness. **CV:** Bradycardia, <u>coronary artery dissection</u>. **GI:** GI bleeding. **Hematologic:** *Bleeding* (<u>major bleeding</u>), anemia, <u>thrombocytopenia</u>. **Skin:** Sweating.

INTERACTIONS Drug: Increased risk of bleeding with ANTICOAGULANTS, NSAIDS, SALICYLATES, ANTIPLATELET AGENTS. **Herbal: Feverfew, garlic, ginger, ginkgo, horse chestnut** may increase risk of bleeding.

PHARMACOKINETICS Duration: 4–8 h after stopping infusion. **Distribution:** 65% protein bound. **Metabolism:** Minimally metabolized. **Elimination:** 65% in urine, 25% in feces. **Half-Life:** 2 h.

T

NURSING IMPLICATIONS

Assessment & Drug Effects

▪ Lab tests: Monitor platelet count, Hgb and Hct before treatment, (within 6 h of infusing loading dose), and frequently throughout treatment; monitor aPTT and ACT.
▪ Withhold drug and notify prescriber if thrombocytopenia (platelets less than 100,000) is confirmed.

Common adverse effects in *italic*, life-threatening effects <u>underlined</u>; generic names in **bold**; classifications in SMALL CAPS; ♣ Canadian drug name; ❷ Prototype drug

1495

- Monitor carefully for and immediately report S&S of internal or external bleeding.
- Minimize unnecessary invasive procedures and devices to reduce the risk of bleeding.

Patient & Family Education
- Report unexplained pelvic or abdominal pain.

TIZANIDINE HYDROCHLORIDE
(ti-zan'i-deen)
Zanaflex
Classification: CENTRAL ACTING SKELETAL MUSCLE RELAXANT
Therapeutic: SKELETAL MUSCLE RELAXANT; ANTISPASMODIC
Prototype: Cyclobenzaprine
Pregnancy Category: C

AVAILABILITY 4 mg tablets; 2 mg, 4 mg, 6 mg capsules

ACTION & *THERAPEUTIC EFFECT*
Centrally acting alpha-adrenergic agonist that reduces spasticity by increasing presynaptic inhibition of motor neurons. Greatest effect on polysynaptic afferent reflex activity at the spinal cord level. *Site of action is the spinal cord; reduces skeletal muscle spasms. Effectiveness indicated by decreased muscle tone.*

USES Acute and intermittent management of increased muscle tone associated with spasticity.

CONTRAINDICATIONS Hypersensitivity to tizanidine. Safety in labor and delivery is unknown.

CAUTIOUS USE Patients with hepatic impairment, hepatic disease; renal insufficiency (CrCl less than 25 mL/min), or renal failure; psychosis; women taking oral contraceptives; older adults because of renal impairment; pregnancy (category C),

lactation. Safety and efficacy in children are not established.

ROUTE & DOSAGE

Spasticity
Adult: **PO** Start with 4 mg and gradually increase to 8 mg q6–8h prn (max: 3 doses or 36 mg/ 24 h)

Renal Impairment Dosage Adjustment
CrCl less than 25 mL/min: Use lower dose

ADMINISTRATION

Oral
- Dose increases are usually made gradually in 2- to 4-mg increments.
- Store at 15°–30° C (59°–86° F).

ADVERSE EFFECTS (≥1%) Body as a Whole: *Asthenia (tiredness),* flu-like syndrome, fever, myasthenia, back pain, infection. **CNS:** *Somnolence, dizziness,* dyskinesia, nervousness, depression, anxiety, paresthesia. **CV:** *Hypotension, bradycardia.* **GI:** *Dry mouth,* constipation, abnormal liver function tests, vomiting, abdominal pain, diarrhea, dyspepsia. **Respiratory:** Pharyngitis, rhinitis. **Skin:** Rash, sweating, skin ulcer. **Urogenital:** *UTI,* urinary frequency. **Special Senses:** Speech disorder, blurred vision.

INTERACTIONS Drug: ORAL CONTRACEPTIVES decrease clearance of **tizanidine. Alcohol** and other CNS DEPRESSANTS increase CNS depression. **Fluvoxamine, ciprofloxacin** increase tizanidine levels and toxicity. **Herbal: Kava, valerian** may potentiate sedation.

PHARMACOKINETICS Absorption: Rapidly absorbed from GI

tract; 40% bioavailability. **Peak:** 1–2 h. **Duration:** 3–6 h. **Distribution:** Crosses placenta, distributed into breast milk. **Metabolism:** In the liver. **Elimination:** 60% in urine, 20% in feces. **Half-Life:** 2.5 h.

NURSING IMPLICATIONS

Assessment & Drug Effects

- Lab tests: Monitor LFTs during the first 6 mo of treatment (baseline, 1, 3, and 6 mo) and periodically thereafter.
- Monitor cardiovascular status and report orthostatic hypotension or bradycardia.
- Monitor closely older adults, those with renal impairment, and women taking oral contraceptives for adverse effects because drug clearance is reduced.

Patient & Family Education

- Exercise caution with potentially hazardous activities requiring alertness since sedation is a common adverse effect. Effects are additive with alcohol or other CNS depressants.
- Make position changes slowly because of the risk of orthostatic hypotension.
- Report unusual sensory experiences; hallucinations and delusions have occurred with tizanidine use.

TOBRAMYCIN SULFATE

(toe-bra-mye'sin)

AKTob, TobraDex, Tobrex, TOBI

Classification: AMINOGLYCOSIDE ANTIBIOTIC
Therapeutic: ANTIBIOTIC
Prototype: Gentamicin sulfate
Pregnancy Category: D

AVAILABILITY 10 mg/mL, 40 mg/mL injection; 300 mg/5 mL inhalation solution; 0.1%, 0.3% ophth solution; 0.1%, 0.3% ophth ointment

ACTION & *THERAPEUTIC EFFECT*

Broad-spectrum, aminoglycoside antibiotic that binds irreversibly to one of two aminoglycoside binding sites on the 30S ribosomal subunit of the bacteria, thus inhibiting protein synthesis, resulting in bacterial cell death. *Effective in treatment of gram-negative bacteria. Exhibits greater antibiotic activity against Pseudomonas aeruginosa than other aminoglycosides.*

USES Treatment of severe infections caused by susceptible organisms.

CONTRAINDICATIONS History of hypersensitivity to tobramycin and other aminoglycoside antibiotics; pregnancy (category D).
CAUTIOUS USE Impaired kidney function; renal disease; dehydration; hearing impairment; myasthenia gravis; parkinsonism; older adults; lactation; premature, neonatal infants.

ROUTE & DOSAGE

Moderate to Severe Infections

Adult: **IV/IM** 3 mg/kg/day divided q8h up to 5 mg/kg/day OR 4–7 mg/kg/day single dose
Topical 1–2 drops in affected eye q1–4h
Child: **IM/IV** *Younger than 5 y,* 2.5 mg/kg q8h **IV/IM** *5 y or older,* 2–2.5 mg/kg/day divided q8h
Neonate: **IM/IV** 2.5 mg/kg q12–24h

Cystic Fibrosis

Adult: Child: **IM/IV** 2.5–3.5 mg/kg q6–8h **Nebulized** 300 mg

T

---

inhaled b.i.d. × 28 days, may repeat after 28-day drug-free period

Renal Impairment Dosage Adjustment

Increase interval.

Hemodialysis Dosage Adjustment

Administer dose after dialysis and monitor levels

Obesity Dosage Adjustment

Dose based on IBW; in morbid obesity, use dosing weight of IBW + 0.4 (Weight − IBW)

ADMINISTRATION

Note: All doses should be based on ideal body weight.

Inhalation

- Administer over 10–15 min using a handheld reusable nebulizer with a compressor (use only those supplied). Patient should sit upright and breathe normally through the nebulizer mouthpiece.

Intramuscular

- Give deep IM into a large muscle. Rotate injection sites.

Intravenous ─────────

Note: Verify correct IV concentration and rate of infusion to neonates, infants, or children with prescriber.

PREPARE: **Intermittent** ▪ Dilute each dose in 50–100 mL or more of D5W, NS or D5/NS. ▪ Final concentration should not exceed 1 mg/mL.

ADMINISTER: **Intermittent:** Infuse diluted solution over 20–60 min. ▪ Avoid rapid infusion.

INCOMPATIBILITIES **Solution/ additive: Alcohol 5% in dextrose, cefamandole, cefepime, cefoxitin, clindamycin, heparin. Y-site: Allopurinol,** **amphotericin B cholesteryl complex, azithromycin, heparin, hetastarch, indomethacin, propofol, sargramostim.**

- Store at 15°–30° C (59°–86° F) prior to reconstitution. After reconstitution, solution may be refrigerated and used within 96 h. ▪ If kept at room temperature, use within 24 h.

ADVERSE EFFECTS (≥1%) **CNS:** Neurotoxicity (including ototoxicity), *nephrotoxicity,* increased AST, ALT, LDH, serum bilirubin; anemia, fever, rash, pruritus, urticaria, nausea, vomiting, headache, lethargy, superinfections; hypersensitivity. **Special Senses:** *Burning, stinging of eye after drug instillation;* lid itching and edema.

INTERACTIONS Drug: ANESTHETICS, SKELETAL MUSCLE RELAXANTS add to neuromuscular blocking effects; **acyclovir, amphotericin B, bacitracin, capreomycin,** CEPHALOSPORINS, **colistin, cisplatin, carboplatin, methoxyflurane, polymyxin B, vancomycin, furosemide, ethacrynic acid** increased risk of ototoxicity, nephrotoxicity.

PHARMACOKINETICS Peak: 30–90 min IM. **Duration:** Up to 8 h. **Distribution:** Crosses placenta; accumulates in renal cortex. **Elimination:** In urine. **Half-Life:** 2–3 h in adults.

NURSING IMPLICATIONS

Assessment & Drug Effects

- Weigh patient before treatment for calculation of dosage.
- Obtain bacterial C&S tests prior to and during therapy.
- Observe patient receiving tobramycin closely because of the high potential for toxicity, even in conventional doses.
- Lab tests: Baseline and periodic kidney function; monitor serum drug concentrations to minimize

Common adverse effects in *italic*, life-threatening effects underlined; generic names in **bold**; classifications in SMALL CAPS; ♣ Canadian drug name; ☯ Prototype drug

rise of toxicity. Prolonged peak serum concentrations greater than 10 mcg/mL or trough concentrations greater than 2 mcg/mL are not recommended.

- Monitor auditory, and vestibular functions closely, particularly in patients with known or suspected renal impairment and patients receiving high doses.
- Be aware that drug-induced auditory changes are irreversible (partial or total); usually bilateral. Partial or bilateral deafness may continue to develop even after therapy discontinued.
- Monitor I&O. Report oliguria, changes in I&O ratio, and cloudy or frothy urine (may indicate proteinuria). Keep patient well hydrated to prevent chemical irritation in renal tubules; older adults are especially susceptible to renal toxicity.
- Be aware that prolonged use of ophthalmic solution may encourage superinfection with nonsusceptible organisms including fungi.
- Report overdose symptoms for eye medication: Increased lacrimation, keratitis, edema and itching of eyelids.

Patient & Family Education
- Report symptoms of superinfections (see Appendix F) to prescriber. Prompt treatment with an antibiotic or antifungal medication may be necessary.
- Report S&S of hearing loss, tinnitus, or vertigo to prescriber.

TOCILIZUMAB
(to-si-ly'zu-mab)

Actemra
Classification: BIOLOGIC RESPONSE MODIFIER; INTERLEUKIN-6 INHIBITOR; DISEASE-MODIFYING ANTIRHEUMATIC DRUG (DMARD)

Therapeutic: DISEASE-MODIFYING ANTIRHEUMATIC DRUG (DMARD)
Pregnancy Category: C

AVAILABILITY 80 mg/4 mL, 200 mg/10 mL, 400 mg/20 mL solution for injection

ACTION & THERAPEUTIC EFFECT
Binds to interleukin 6 (IL-6) receptors inhibiting IL-6 mediated signaling and suppressing systemic inflammatory responses; also reduces IL-6 production in joints affected by inflammatory processes such as rheumatoid arthritis. *Reduction of joint inflammation reduces joint pain and improves joint function.*

USES Treatment of moderate to severe rheumatoid arthritis or juvenile rheumatoid arthritis with previous TNF antagonist therapy.

CONTRAINDICATIONS Active or local infection; history of anaphylaxis with tocilizumab; severe neutropenia, severe thrombocytopenia, or severe hepatic impairment.
CAUTIOUS USE History of chronic or recurring infections; history of hypersensitivity; mild hepatic impairment; thrombocytopenia; neutropenia; risk for GI perforation; demyelinating disorders; concurrent biologic DMARDs (e.g., TNF antagonists, IL-1R antagonists, anti-CD20 monoclonal antibodies, selective co-stimulation modulators); preexisting or recent-onset demyelinating disorders; older adults; pregnancy (category C); lactation. Safety and efficacy in children less than 2 y not established.

ROUTE & DOSAGE

Rheumatoid Arthritis
Adult: **IV** 4 mg/kg given over 1 h q4wk. May increase up to 8 mg/kg q4wk

Juvenile Rheumatoid Arthritis

Child/Adolescent (over 2 years and over 30 kg) **IV** 8 mg/kg q2wks *Over 2 years and under 30 kg:* **IV** 12 mg/kg

Absolute Neutrophil Count (ANC) Adjustment

ANC = 500–1000/mm³: Stop tocilizumab and restart at 4 mg/kg with an increase to 8 mg/kg as appropriate when ANC is greater than 1000 cells/mm³. *ANC less than 500/mm³:* Discontinue tocilizumab.

Platelet Count Dosage Adjustment

Platelet count 50,000–100,000/mm³: Stop tocilizumab and restart at 4 mg/kg with an increase to 8 mg/kg as appropriate when platelet count is greater than 100,000/mm³. *Platelet count less than 50,000/mm³:* Discontinue tocilizumab.

Hepatic Impairment Dosage Adjustment

Do not initiate tocilizumab if AST/ALT is greater than 1.5 × ULN. Mild impairment due to tocilizumab (AST and/or ALT 1–3 × ULN): Modify dose of concomitant DMARDs, if appropriate. For persistent increases in this range, reduce tocilizumab dose to 4 mg/kg or interrupt tocilizumab until AST/ALT have normalized. *Moderate impairment due to tocilizumab (AST and/or ALT greater than 3–5 × ULN confirmed by repeat testing):* Interrupt tocilizumab until AST/ALT is less than 3 × ULN, then follow recommendations for 1–3 × ULN. For persistent increases greater than 3 × ULN, discontinue tocilizumab. *Severe impairment due to tocilizumab (AST and/or ALT greater than 5 × ULN):* Discontinue tocilizumab.

ADMINISTRATION

Intravenous

***PREPARE:* IV Infusion:** From a 100 mL bag of NS, remove a volume of solution equal to the volume of tocilizumab necessary to deliver the required dose. Slowly add the required dose of tocilizumab to the IV bag then gently invert to mix. Solution should reach room temperature prior to infusion. Discard unused drug remaining in vial.

***ADMINISTER:* IV Infusion:** Give over 60 min. Do NOT give IV push or a bolus dose.

INCOMPATIBILITIES **Solution/additive:** Do not mix with another drug. **Y-site:** Do not infuse tocilizumab concurrently with any drug in the same IV line.

▪ Store IV solution refrigerated or at room temperature for up to 24 h. Protect from light. Store unopened vials refrigerated in original packaging to protect from light.

ADVERSE EFFECTS (≥ 1%) Body as a Whole: Infusion reactions, <u>serious infections (includes invasive fungal infections, tuberculosis, and bacterial, viral, and other infections caused by opportunistic pathogens)</u>, upper abdominal pain. **CV:** Hypertension. **GI:** Headache, dizziness, gastritis, mouth ulceration. **Hematological:** Neutropenia, <u>thrombocytopenia</u>. **Metabolic:** *Elevations of ALT and AST levels*, increased LDL levels. **Respiratory:** Bronchitis, nasopharyngitis, upper respiratory tract infection. **Skin:** Rash.

INTERACTIONS Drug: Tocilizumab has the potential to affect expression of multiple CYP enzymes. These effects may persist for several weeks after stopping tocilizumab. Caution should be used when tocilizumab is combined with CYP3A4 substrate drugs that have a narrow therapeutic index.

PHARMACOKINETICS Half-Life: 11–13 days.

NURSING IMPLICATIONS

Assessment & Drug Effects
- Monitor for signs of serious allergic reactions during and shortly following each infusion.
- Monitor for and promptly report S&S of TB or hepatitis B infection in carriers of HBV.
- Lab tests: Prior to initiation of therapy, screen for latent TB; baseline and frequent (q4–8wk) WBC with differential, and platelet count; baseline and periodic LFTs; periodic lipid profile.
- Monitor for S&S of polyneuropathy.

Patient & Family Education
- Do not accept vaccinations with live vaccines while taking this drug.
- You should be tested for TB prior to taking this drug.
- Promptly report S&S of infection including: Chills, fever, cough, shortness of breath, diarrhea, stomach or abdominal pain, burning on urination, sores anywhere on your body, unexplained or excessive fatigue.
- Carriers of hepatitis B should report promptly signs of activation of the virus (e.g., clay-colored stools, dark urine, jaundice, unexplained fatigue).
- Report promptly if you become pregnant.

TOLAZAMIDE
(tole-az′a-mide)
Tolinase
Classification: ANTIDIABETIC, SULFONYLUREA
Therapeutic: ANTIDIABETIC
Prototype: Glyburide
Pregnancy Category: C

AVAILABILITY 250 mg, 500 mg tablets

ACTION & THERAPEUTIC EFFECT
Sulfonylurea hypoglycemic that lowers blood glucose primarily by stimulating pancreatic beta cells to secrete insulin. Antidiabetic action is a result of stimulation of the pancreas to secrete more insulin in the presence of blood sugar; it requires functioning beta cells. *Effectiveness is measured in decreasing the serum blood level to within normal limits and decreasing HbA1C value to 6.5 or lower.*

USES Type 2 diabetes mellitus

CONTRAINDICATIONS Known sensitivity to sulfonylureas and to sulfonamides; type 1 diabetes complicated by ketoacidosis; infection; trauma; lactation.

CAUTIOUS USE Renal disease; renal failure, renal impairment; older adults; pregnancy (category C). Safety in children not established.

ROUTE & DOSAGE

Type 2 Diabetes Mellitus
Adult: **PO** 100 mg–250 mg daily or b.i.d. a.c., may adjust dose by 100–250 mg/day at weekly intervals (max: 1 g/day)

ADMINISTRATION

Oral
- Give in the morning with or before meals.

T

Common adverse effects in *italic*, life-threatening effects underlined; generic names in **bold**; classifications in SMALL CAPS; ♣ Canadian drug name; ☻ Prototype drug

1501

- Divide dose of more than 500 mg and give b.i.d.
- Store at 15°–30° C (59°–86° F) in a tightly closed container unless otherwise directed. Keep drug out of the reach of children.

ADVERSE EFFECTS (≥1%) **GI:** Nausea, vomiting, cholestatic jaundice. **Metabolic:** Hypoglycemia. **CNS:** Vertigo. **Skin:** Photosensitivity. **Hematologic:** Agranulocytosis.

INTERACTIONS Drug: Alcohol elicits disulfiram-type reaction in some patients; ORAL ANTICOAGULANTS, **chloramphenicol, clofibrate, phenylbutazone,** MAO INHIBITORS, SALICYLATES, **probenecid,** SULFONAMIDES may potentiate hypoglycemic actions; THIAZIDES may antagonize hypoglycemic effects; **cimetidine** may increase tolazamide levels, causing hypoglycemia. **Herbal: Ginseng, karela** may potentiate hypoglycemic effects.

PHARMACOKINETICS Absorption: Slowly from GI tract. **Onset:** 60 min. **Peak:** 4–6 h. **Duration:** 10–15 h (up to 20 h in some patients). **Distribution:** Distributed in highest concentrations in liver, kidneys, and intestines; crosses placenta; distributed into breast milk. **Metabolism:** Extensively in liver. **Elimination:** 85% in urine, 15% in feces. **Half-Life:** 7 h.

NURSING IMPLICATIONS

Assessment & Drug Effects
- Monitor closely for daytime and nighttime hypoglycemia since tolazamide is long acting.
- Lab tests: Periodic HbA1C and frequent FBS; periodic CBC.
- Monitor for and report signs of an allergic reaction (e.g., rash, urticaria, pruritus).

Patient & Family Education
- Check blood glucose daily or as ordered by prescriber. Important

to continue close medical supervision for first 6 wk of treatment.
- Learn the S&S of hypoglycemia and hyperglycemia and check blood glucose level when either is suspected.
- Report to prescriber if you experience frequent and/or severe episodes of hypoglycemia.
- Do not take OTC preparations unless approved or prescribed by prescriber.
- Understand that alcohol can precipitate a disulfiram-type reaction.
- Many drugs interact with tolazamide. Monitor blood glucose more frequently whenever a new drug is added to your regimen.

TOLBUTAMIDE
(tole-byoo′ta-mide)

Mobenol ♦, Novobutamide ♦, Orinase

TOLBUTAMIDE SODIUM

Orinase Diagnostic
Classification: ANTIDIABETIC, SULFONYLUREA
Therapeutic: ANTIDIABETIC
Prototype: Glyburide
Pregnancy Category: C

AVAILABILITY 500 mg tablets; 1 g vial

ACTION & *THERAPEUTIC EFFECT*
Short-acting sulfonylurea that lowers blood glucose concentration by stimulating pancreatic beta cells to synthesize and release insulin. No action demonstrated if functional beta cells are absent. *Lowers blood glucose concentration by stimulating pancreatic beta cells to synthesize and release insulin.*

USES Management of mild to moderately severe, stable type 2 diabetes that is not controlled by diet and

Common adverse effects in *italic*, life-threatening effects underlined; generic names in **bold**; classifications in SMALL CAPS; ♦ Canadian drug name; ⓟ Prototype drug

weight reduction alone. Also used in treatment of patients who are unresponsive to other sulfonylureas and adjunctively with insulin to stabilize certain cases of labile diabetes. Used as diagnostic agent to rule out pancreatic islet cell adenoma or diabetes.

CONTRAINDICATIONS Hypersensitivity to sulfonylureas or to sulfonamides; history of repeated episodes of diabetic ketoacidosis; type 1 diabetes as sole therapy; diabetic coma; severe stress, infection, trauma, or major surgery; severe renal insufficiency, liver or endocrine disease; lactation.

CAUTIOUS USE Cardiac, thyroid, pituitary, or adrenal dysfunction; severe hepatic disease, renal disease, renal impairment, renal failure; history of peptic ulcer; alcoholism; infection; debilitated, malnourished, or uncooperative patient, older adults; pregnancy (category C). Safe use in children not established.

ROUTE & DOSAGE

Type 2 Diabetes
Adult: **PO** 250 mg to 3 g/day in 1–2 divided doses

Diagnosis of Functioning Insulinoma
Adult: **IV** 1 g over 2–3 min

ADMINISTRATION

Oral
- Tablets may be crushed and given with full glass of water if patient desires.
- Do not give at bedtime because of danger of nocturnal hypoglycemia, unless specifically prescribed.
- Store below 40° C (104° F), preferably between 15°–30° C (59°–86° F)

in well-closed container. Avoid freezing.

ADVERSE EFFECTS (≥1%) **GI:** Nausea, epigastric fullness, heartburn, anorexia, constipation, diarrhea, cholestatic jaundice (rare). **Hematologic:** <u>Agranulocytosis</u>, <u>thrombocytopenia, leukopenia</u>, hemolytic anemia, <u>aplastic anemia</u>, pancytopenia. **Metabolic:** Hepatic porphyria, disulfiram-like reactions, SIADH, hypoglycemia without loss of consciousness or neurologic symptoms (unusual fatigue, tremulousness, hunger, drowsiness, GI distress, sweating, anxiety, headache) severe hypoglycemia (visual disturbances, ataxia, paresthesias, confusion, tachycardia, seizures, <u>coma</u>). **Skin:** Allergic skin reactions: Pruritus, erythema, urticaria, morbilliform or maculopapular eruptions; porphyria cutanea tarda, photosensitivity. **Special Senses:** Taste alterations. **CNS:** Headache.

DIAGNOSTIC TEST INTERFERENCE The SULFONYLUREAS may produce abnormal *thyroid function test* results and reduced *RAI uptake* (after long-term administration). A tolbutamide metabolite may cause false-positive *urinary protein* values when turbidity procedures are used (such as heat and *acetic acid* or *sulfosalicylic acid*); *Ames reagent* strips reportedly not affected.

INTERACTIONS Drug: Phenylbutazone increases hypoglycemic effects; THIAZIDE DIURETICS may attenuate hypoglycemic effects; **alcohol** may produce disulfiram reaction; BETA-BLOCKERS may mask symptoms of a hypoglycemic reaction. **Herbal: Ginseng, karela** may potentiate hypoglycemic effects.

PHARMACOKINETICS Absorption: Readily from GI tract. **Peak:** 3–5 h.

T

Common adverse effects in *italic*, life-threatening effects <u>underlined</u>; generic names in **bold**; classifications in SMALL CAPS; ♣ Canadian drug name; ☻ Prototype drug

1503

Distribution: Into extracellular fluids. **Metabolism:** Principally in liver. **Elimination:** 75–85% in urine; some in feces. **Half-Life:** 7 h.

NURSING IMPLICATIONS

Assessment & Drug Effects

- Supervise closely during initial period of therapy until dosage is established. One or 2 wk of therapy may be required before full therapeutic effect is achieved.
- Monitor closely during adjustment period, watching for S&S of impending hypoglycemia (see Appendix F). Detection of a hypoglycemic reaction in a diabetic patient also receiving a beta-blocker, especially older adults, is difficult.
- Evaluate nondefinitive vague complaints; hypoglycemic symptoms may be especially vague in older adults. Observe patient carefully, especially 2–3 h after eating, check urine for sugar and ketone bodies and capillary blood glucose.
- Lab tests: Baseline LFTs and kidney function tests; periodic HbA1C, serum electrolytes, CBC with differential, and platelet counts.
- Report repetitive complaints of headache and weakness a few hours after eating; may signal incipient hypoglycemia.
- Be aware that pruritus and rash, frequently reported adverse effects, may clear spontaneously; if these persist, drug will be discontinued.

Patient & Family Education

- Learn the S&S of hyperglycemia (see Appendix A) and check blood glucose level when hyperglycemia is suspected.
- Hypoglycemia is frequently caused by overdosage of hypoglycemic drug, inadequate or irregular food intake, nausea, vomiting, diarrhea, and added exercise without caloric supplement or dose adjustment.

Learn the S&S of hypoglycemia (see Appendix F) and check blood glucose level whenever hypoglycemia is suspected.

- Report any illness promptly. Prescriber may want to evaluate need for insulin.
- Do not self-medicate with OTC drugs unless approved or prescribed by prescriber.
- Be aware that alcohol, even in moderate amounts, can precipitate a disulfiram-type reaction (see Appendix F). A hypoglycemic response after ingesting alcohol requires emergency treatment.
- Protect exposed skin areas from the sun with a sunscreen lotion (SPF 12–15) because of potential photosensitivity (especially in the alcoholic).
- Monitor blood glucose daily or as directed by prescriber.
- Be alert to added danger of loss of control (hyperglycemia) when a drug that affects the hypoglycemic action of sulfonylureas (see DRUG INTERACTIONS) is withdrawn or added to the tolbutamide regimen. Monitor blood glucose carefully.
- Use or add barrier contraceptive if using hormonal contraceptives.

TOLCAPONE ⊘

(tol′ca-pone)

Tasmar

Classification: ANTICHOLINERGIC; CATECHOLAMINE-O-METHYLTRANSFERASE (COMT) INHIBITOR
Therapeutic: ANTIPARKINSON
Pregnancy Category: C

AVAILABILITY 100 mg, 200 mg tablets

ACTION & THERAPEUTIC EFFECT
Selective and reversible inhibitor of catecholamine-O-methyltransferase

Common adverse effects in *italic*, life-threatening effects underlined; generic names in **bold**; classifications in SMALL CAPS; ♥ Canadian drug name; ⊘ Prototype drug

(COMT). COMT is the enzyme responsible for metabolizing catecholamines and, therefore, levodopa. *Concurrent administration of tolcapone and levodopa increases the amount of levodopa available to control Parkinson's disease by increasing dopaminergic brain stimulation.*

USES Idiopathic Parkinson's disease as adjunct to levodopa/carbidopa.

CONTRAINDICATIONS Hypersensitivity to tolcapone; liver disease; MAOI therapy.
CAUTIOUS USE History of hypersensitivity to other COMT inhibitors (e.g., entacapone, nitecapone); anorexia nervosa; hematuria; hypo-tension; syncopy; renal disease; renal impairment; pregnancy (category C), lactation.

ROUTE & DOSAGE

Parkinson's Disease
Adult: **PO** 100 mg t.i.d. (max: 200 mg t.i.d.)

ADMINISTRATION

Oral
- Give with food if GI upset occurs.
- Give only in conjunction with levodopa/carbidopa therapy.
- Therapy with tolcapone should NOT be initiated if patient has ALT/AST greater than 2 × ULN or has known liver disease.
- Store at 20°–25° C (68°–77° F) in a tightly closed container.

ADVERSE EFFECTS (≥1%) **Body as a Whole:** Muscle cramps, orthostatic complaints, fatigue, falling, balance difficulties, hyperkinesia, stiffness, arthritis, hypokinesia. **CNS:** *Dyskinesia, sleep disorder, dystonia, excessive dreaming,* somnolence, confusion, dizziness, headache, hallucination, syncope, paresthesias. **CV:** Chest pain, hypotension. **GI:** *Nausea,* anorexia, diarrhea, vomiting, constipation, <u>fulminant liver failure, severe hepatocellular injury,</u> dry mouth, abdominal pain, dyspepsia, flatulence. **Respiratory:** URI, dyspnea, sinus congestion. **Skin:** Sweating. **Urogenital:** UTI, urine discoloration, micturition disorder.

INTERACTIONS Drug: Will increase **levodopa** levels when taken simultaneously; CNS DEPRESSANTS may cause additive sedation; do not give with non-selective MAOIS **(isocarboxazid, phenelzine, or tranylcypromine furazolidone, linezolid, procarbazine).** **Food:** Decreases levels.

PHARMACOKINETICS Absorption: Rapidly from GI tract, bioavailability 65%. **Peak:** 2 h. **Distribution:** Over 99% protein bound. **Metabolism:** Extensively metabolized by COMT and glucuronidation. **Elimination:** 60% in urine, 40% in feces; clearance reduced by 50% in patients with liver disease. **Half-Life:** 2–3 h.

NURSING IMPLICATIONS

Assessment & Drug Effects
- Lab tests: Monitor LFTs monthly for first 3 mo, every 6 wk for the next 3 mo, and periodically thereafter.
- Withhold drug and notify prescriber if liver dysfunction is suspected.
- Monitor PT and INR carefully when given concurrently with warfarin.
- Monitor carefully for and immediately report S&S of hepatic impairment (e.g., jaundice, dark urine).

Patient & Family Education
- Do not engage in hazardous activities until response to drug is known. Avoid use of alcohol or sedative drugs while on tolcapone.

T

Common adverse effects in *italic*, life-threatening effects <u>underlined</u>; generic names in **bold**; classifications in SMALL CAPS; ✚ Canadian drug name; ⊘ Prototype drug

1505

- Rise slowly from a sitting or lying position to avoid a rapid drop in BP with possible weakness or fainting.
- Nausea is a common possible adverse effect especially at the beginning of therapy.
- Do not suddenly stop taking this drug. Doses **must be** gradually reduced over time.
- Notify prescriber promptly about any of following: Increased loss of muscle control, fainting, yellowing of skin or eyes, darkening of urine, severe diarrhea, hallucinations.

TOLMETIN SODIUM

(tole′met-in)

Tolectin, Tolectin DS

Classification: ANALGESIC, NONSTEROIDAL ANTI-INFLAMMATORY (NSAID); ANTIPYRETIC

Therapeutic: NONNARCOTIC ANALGESIC; NSAID; DISEASE-MODIFYING ANTIRHEUMATIC DRUG (DMARD)

Prototype: Ibuprofen

Pregnancy Category: C first and second trimester; D third trimester

AVAILABILITY 200 mg, 600 mg tablets; 400 mg capsules

ACTION & *THERAPEUTIC EFFECT*
Nonsteroidal anti-inflammatory analgesic that competitively inhibits both cyclooxygenase (COX) isoenzymes, COX-1 and COX-2, by blocking arachidonate binding to prostaglandin sites, and thus inhibits prostaglandin synthesis. *Possesses analgesic, antiinflammatory, antipyretic and antirheumatic activity.*

USES In acute flares and management of chronic rheumatoid arthritis. May be used alone or in combination with gold or corticosteroids.

CONTRAINDICATIONS History of intolerance or hypersensitivity to tolmetin, aspirin, and other NSAIDs; active peptic ulcer, CABG perioperative pain; in patients with functional class IV rheumatoid arthritis (severely incapacitated, bedridden, or confined to a wheelchair). Safety during pregnancy (category D third trimester), lactation.

CAUTIOUS USE History of upper GI tract disease; impaired kidney function; SLE; compromised cardiac function; pregnancy (category C first and second trimester); children younger than 2 y.

ROUTE & DOSAGE

Arthritis
Adult: **PO** 400 mg t.i.d. (max: 2 g/day)
Child (2 y or older): **PO** 20 mg/kg/day in 3–4 divided doses (max: 30 mg/kg/day)

ADMINISTRATION

Oral

- Schedule to include a morning dose (on arising) and a bedtime dose.
- Give with fluid of patient's choice; crush tablet or empty capsule to mix with water or food if patient cannot swallow tablet/capsule.
- Store at 15°–30° C (59°–86° F) in tightly capped and light-resistant container unless otherwise instructed.

ADVERSE EFFECTS (≥1%) **CNS:** *Headache, dizziness, vertigo, lightheadedness,* mood elevation or depression, tension, nervousness, weakness, drowsiness, insomnia, tinnitus. **CV:** Mild edema (about 7% patients), sodium and water retention, mild to moderate hypertension. **GI:** Epigastric or abdominal pain, dyspepsia, *nausea,* vomiting,

---

Common adverse effects in *italic*, life-threatening effects underlined; generic names in **bold**; classifications in SMALL CAPS; ♣ Canadian drug name; ◉ Prototype drug

heartburn, constipation, peptic ulcer, GI bleeding. **Hematologic:** Transient and small decreases in hemoglobin and hematocrit, purpura, petechiae, granulocytopenia, leukopenia. **Urogenital:** Hematuria, proteinuria, increased BUN. **Skin:** Toxic epidermal necrolysis, morbilliform eruptions, urticaria, pruritus. **Body as a Whole:** Anaphylaxis (especially after drug is discontinued and then reinstituted).

DIAGNOSTIC TEST INTERFERENCE Tolmetin prolongs *bleeding time,* inhibits *platelet aggregation,* elevates *BUN, alkaline phosphatase,* and *AST* levels; may decrease *hemoglobin* and *hematocrit* values. Metabolites may produce false-positive results for *proteinuria* [with tests that rely on acid precipitation (e.g., *sulfosalicylic acid*)].

INTERACTIONS Drug: ORAL ANTICOAGULANTS, **heparin** may prolong bleeding time; may increase **lithium** toxicity; **aspirin,** other NSAIDs add to ulcerogenic effects; may increase **methotrexate** toxicity. **Herbal: Feverfew, garlic, ginger, ginkgo** may increase bleeding potential.

PHARMACOKINETICS Absorption: Rapidly from GI tract. **Peak:** 30–60 min. **Distribution:** Crosses blood–brain barrier and placenta; distributed into breast milk. **Metabolism:** In liver. **Elimination:** In urine. **Half-Life:** 60–90 min.

NURSING IMPLICATIONS

Assessment & Drug Effects

- Monitor patients with kidney damage closely. Evaluate I&O ratio and encourage patient to increase fluid intake to at least 8 full glasses per day.
- Lab tests: Obtain periodic kidney function tests (routine urinalysis, creatinine clearance, and serum creatinine) for patient on long-term therapy.
- Check self-medicating habits of the patient. Sodium bicarbonate alkalinizes the urine, which increases urinary excretion of tolmetin and may reduce degree and duration of effectiveness.

Patient & Family Education

- Take drug with meals or milk if GI disturbances occur. Notify prescriber if symptoms persist; dosage reduction may be necessary, or antacid added.
- Monitor weight and report an increase greater than 2 kg (4 lb)/wk with impaired kidney or cardiac function; check for swelling in ankles, tibiae, hands, and feet.
- Inform surgeon or dentist before treatment if you are taking tolmetin because of possible enhanced bleeding.
- Report promptly signs of abnormal bleeding (ecchymosis, epistaxis, melena, petechiae), itching, skin rash, persistent headache, edema.
- Avoid potentially hazardous activities until response to drug is known because dizziness and drowsiness are common adverse effects.

TOLNAFTATE
(tole-naf′tate)

Aftate, Pitrex ✦, Tinactin
Classification: ANTIFUNGAL ANTIBIOTIC
Therapeutic: ANTIFUNGAL
Pregnancy Category: C

AVAILABILITY 1% cream, solution, gel, powder, spray

ACTION & *THERAPEUTIC EFFECT*
Synthetic topical antifungal agent. Tolnaftate distorts hyphae and

Common adverse effects in *italic*, life-threatening effects underlined; generic names in **bold**; classifications in SMALL CAPS; ✦ Canadian drug name; ❂ Prototype drug

1507

stunts mycelial growth on susceptible fungi. *Fungistatic or fungicidal as well as anti-infective against bacteria, protozoa, and viruses.*

USES Tinea pedis (athlete's foot), tinea cruris (jock itch), tinea corporis (body ringworm); also tinea capitis and tinea unguium if infection is superficial, plantar or palmar lesions adjunctively with keratolytic agents, and tinea versicolor (caused by *Malassezia furfur*).

CONTRAINDICATIONS Skin irritations prior to therapy, nail and scalp infections; immunosuppressed patients, diabetes mellitus, peripheral vascular disease.

CAUTIOUS USE Excoriated skin; pregnancy (category C), lactation. Safe use in children younger than 2 y is not established.

ROUTE & DOSAGE

Tinea Infestations

Adult: Child: **Topical** Apply 0.5–1 cm (¼–½ in.) of cream or 3 drops of solution b.i.d. in morning and evening; powder may be used prophylactically in normally moist areas

ADMINISTRATION

Topical

- Cleanse site thoroughly with water and dry completely before applying. Massage thin layer gently into skin. Make sure area is not wet from excess drug after application.
- Shake aerosol powder container well before use.
- Note: Cream and powder are not recommended for nail or scalp infection.
- Use liquids (solutions) for scalp infection or to treat hairy areas.
- Store cream, gel, powder, and topical solution in light-resistant containers at 15°–30° C (59°–86° F);

store aerosol container at 2°–30° C (38°–86° F). Avoid freezing and exposure to light.

ADVERSE EFFECTS (≥1%) **Skin:** Local irritation, stinging of skin from aerosol formulation.

NURSING IMPLICATIONS

Patient & Family Education

- Expect relief from pruritus, soreness, and burning within 24–72 h after start of treatment.
- Continue treatment for 2–3 wk after disappearance of all symptoms to prevent recurrence.
- Return to prescriber for reevaluation in absence of improvement within 4 wk.
- Note: If skin has thickened as a result of the infection, desired clinical response may be delayed for 4–6 wk.
- Avoid contact with eyes of all drug forms.
- Place container in warm water to liquify contents if solution solidifies. Potency is unaffected.

TOLTERODINE TARTRATE

(tol-ter′o-deen tar′trate)

Detrol, Detrol LA
Classifications: ANTICHOLINERGIC; MUSCARINIC RECEPTOR ANTAGONIST
Therapeutic: ANTIMUSCARINIC; BLADDER ANTISPASMODIC
Prototype: Oxybutynin
Pregnancy Category: C

AVAILABILITY 1 mg, 2 mg tablets; 2 mg, 4 mg sustained release

ACTION & THERAPEUTIC EFFECT
Selective muscarinic urinary bladder receptor antagonist. Reduces urinary incontinence, urgency, and frequency. *Controls urinary*

bladder incontinence by controlling contractions.

USES Overactive bladder, urinary incontinence, urinary urgency.

CONTRAINDICATIONS Hypersensitivity to tolterodine; uncontrolled narrow-angle glaucoma; gastric retention; urinary retention; lactation. **CAUTIOUS USE** Cardiovascular disease; liver disease; controlled narrow-angle glaucoma; urinary retention; severe hepatic impairment; obstructive GI disease; obstructive uropathy; paralytic ileus or intestinal atony; renal impairment; ulcerative colitis; pregnancy (category C).

ROUTE & DOSAGE

Overactive Bladder

Adult: **PO** 2 mg b.i.d. or 4 mg (sustained release) daily

Hepatic Impairment Dosage Adjustment

Reduce dose by 50%

Renal Impairment Dosage Adjustment

CrCl less than 30 mL/min: Reduce dose 50%

ADMINISTRATION

Oral

- Do not crush or chew sustained release tablets. These **must be** swallowed whole.
- Doses greater than 1 mg b.i.d. are not recommended for those with significantly reduced liver function or kidney function or concurrently receiving macrolide antibiotics, azole antifungal agents, or other cytochrome P450 3A4 inhibitors.
- Store at 20°–25° C (68°–77° F) in a tightly closed container.

ADVERSE EFFECTS (≥1%) **Body as a Whole:** Back pain, fatigue, flu-like syndrome, falls, arthralgia, weight gain. **CNS:** Headache, paresthesias, vertigo, dizziness, nervousness, somnolence. **CV:** Chest pain, hypertension. **GI:** *Dry mouth,* dyspepsia, constipation, abdominal pain, diarrhea, flatulence, nausea, vomiting. **Urogenital:** Dysuria, micturition frequency, urinary retention, UTI. **Respiratory:** Bronchitis, cough, pharyngitis, rhinitis, sinusitis, URI. **Skin:** Pruritus, rash, erythema, dry skin. **Special Senses:** Dry eyes, vision abnormalities.

INTERACTIONS Drug: Additive anticholinergic effects with **amantadine, amoxapine, bupropion, clozapine, cyclobenzaprine, disopyramide, maprotiline, olanzapine, orphenadrine,** SEDATING H$_1$-BLOCKERS, PHENOTHIAZINES, TRICYCLIC ANTIDEPRESSANTS, Increased effects with **clarithromycin, cyclosporine, erythromycin, itraconazole,** or **ketoconazole. Food: Grapefruit juice** may increase **tolterodine** levels in some patients.

PHARMACOKINETICS Absorption: 77% absorbed, decreased with food. **Peak:** 1–2 h. **Distribution:** 96% protein bound. **Metabolism:** In liver by CYP2D6 active metabolite. **Elimination:** 77% in urine, 17% in feces. **Half-Life:** 1.9–3.7 h.

NURSING IMPLICATIONS

Assessment & Drug Effects

- Monitor voiding pattern and report promptly urinary retention.
- Monitor vital signs carefully (HR and BP), especially in those with cardiovascular disease.

Patient & Family Education

- Notify prescriber promptly if you experience eye pain, rapid heartbeat, difficulty breathing, skin rash or hives, confusion, or incoordination.

T

Common adverse effects in *italic*, life-threatening effects <u>underlined</u>; generic names in **bold**; classifications in SMALL CAPS; ✚ Canadian drug name; ◉ Prototype drug

1509

- Report blurred vision, sensitivity to light, and dry mouth (all common adverse effects) to prescriber if bothersome.
- Avoid the use of alcohol or OTC antihistamines.

TOLVAPTAN
(tol-vap'tan)
Samsca
Classification: ELECTROLYTE & WATER BALANCE AGENT; DIURETIC; VASOPRESSIN ANTAGONIST
Therapeutic: VASOPRESSIN ANTAGONIST; DIURETIC
Prototype: Conivaptan
Pregnancy Category: C

AVAILABILITY 15 mg, 30 mg tablets

ACTION & *THERAPEUTIC EFFECT*
Tolvaptan is a selective vasopressin V_2-receptor antagonist, thus antagonizing the effect of vasopressin and causing an increase in urine water excretion, decrease in urine osmolality, and increase in serum sodium concentrations. *Effectiveness is measured by increase in serum sodium level toward lower limit of normal, and/or decrease in sign and symptoms of hyponatremia.*

USES Treatment of hypervolemic and euvolemic hyponatremia including patients with heart failure, cirrhosis, and syndrome of inappropriate antidiuretic hormone (SIADH).

CONTRAINDICATIONS Rapid correction of serum sodium (e.g. greater than 12 mEq/L/24 h); cognitively impaired; serious neurologic symptoms; hypovolemic hyponatremia; hypertonic saline; concurrent administration with strong CYP3A inhibitors; lactation.
CAUTIOUS USE Must be initiated or reinitiated in a hospital setting; cirrhosis; pregnancy (category C).

Safe use in children younger than 18 y not established.

ROUTE & DOSAGE

> **Hyponatremia**
> *Adult:* **PO** Initially 15 mg once daily; may be adjusted up to 60 mg once daily

ADMINISTRATION
Oral
- Doses may be increased at 24 h intervals or greater.
- Do not administer if patient is unable to sense or respond to thirst.
- Store at 15°–30°C (59°–86°F).

ADVERSE EFFECTS (≥1%) **Body as a Whole:** *Asthenia,* pyrexia, *thirst.* **GI:** *Constipation, dry mouth, nausea.* **Metabolic:** Anorexia, *hyperglycemia.* **Urogenital:** *Pollakiuria, polyuria.*

INTERACTIONS Drug: Strong CYP3A INHIBITORS (e.g., **clarithromycin, ketoconazole, itraconazole, telithromycin, saquinavir, nelfinavir, ritonavir, nefazodone**) may increase tolvaptan levels. Moderate CYP3A INHIBITORS (e.g., **erythromycin, fluconazole, aprepitant, diltiazem, verapamil**) may increase tolvaptan levels. P-GLYCOPROTEIN INHIBITORS (e.g., **cyclosporine**) may require tolvaptan dosage reduction. CYP3A INDUCERS (e.g., **rifabutin, rifapentine, rifampin, barbiturates, phenytoin, carbamazepine**) decrease the levels of tolvaptan. Tolvaptan increases the levels of **digoxin**. BETA-BLOCKERS, ANGIOTENSIN RECEPTOR BLOCKERS, ANGIOTENSIN-CONVERTING ENZYME INHIBITORS and POTASSIUM-SPARING DIURETICS may cause hyperkalemia. **Food:** Administration with

grapefruit juice may increase tolvaptan levels. **Herbal: St. John's Wort** may decrease the levels of tolvaptan.

PHARMACOKINETICS Absorption: Approximately 40% absorbed. **Peak:** 2–4 h. **Distribution:** 99% plasma protein bound. **Metabolism:** In liver through CYP3A4. **Elimination:** Eliminated entirely by non-renal routes. **Half-Life:** 12 h.

NURSING IMPLICATIONS

Assessment & Drug Effects

- Monitor vital signs frequently throughout therapy. Monitor ECG as warranted.
- Monitor weight and I&O closely as copious diuresis is expected.
- Fluid restriction should be avoided during the first 24 h of therapy.
- Lab tests: Baseline and frequent serum electrolytes especially serum sodium level, and volume status; periodic blood glucose.
- Monitor mental status throughout treatment. Report promptly changes in mental status (e.g., lethargy, confusion, disorientation, hallucinations, seizures).
- Report promptly symptoms of osmotic demyelination syndrome (ODS): Dysarthria, mutism, dysphagia, lethargy, affective changes, spasticity, seizures, or coma.
- Monitor digoxin levels closely with concurrent administration.

Patient & Family Education

- Continue to drink fluid in response to thirst until otherwise directed. Fluid restrictions are usually required after the first 24 h of therapy.
- Report promptly any of the following: Trouble speaking or swallowing, drowsiness, mood changes, confusion, involuntary body movements, or muscle weakness in arms or legs.

TOPIRAMATE
(to-pir'a-mate)
Topamax
Classifications: ANTICONVULSANT; GAMMA-AMINOBUTYRATE (GABA) ENHANCER
Therapeutic: ANTICONVULSANT; ANTIEPILEPTIC
Pregnancy Category: D

AVAILABILITY 25 mg, 100 mg, 200 mg tablets; 15 mg, 25 mg, 50 mg capsules

ACTION & *THERAPEUTIC EFFECT*
Sulfamate-substituted monosaccharide with a broad spectrum of anticonvulsant activity. Exhibits sodium channel-blocking action, as well as enhancing the ability of GABA to induce a flux of chloride ions into the neurons, thus potentiating the activity of this inhibitory neurotransmitter (GABA). *Effectiveness indicated by a decrease in seizure activity.*

USES Adjunctive therapy for partial-onset seizures in adults and children age 2–16 y; generalized tonic-clonic seizures; migraine prophylaxis.
UNLABELED USES Cluster headache, bulimia nervosa, neuropathic pain, infantile spasms, weight loss.

CONTRAINDICATIONS Hypersensitivity to topiramate; pregnancy (category D). Effect on labor and delivery is unknown.
CAUTIOUS USE Moderate and severe renal impairment, hepatic impairment; COPD; severe pulmonary disease; lactation; children younger than 2 y.

ROUTE & DOSAGE

Partial-Onset Seizures

Adult: **PO** Initiate with 25 mg b.i.d., increase by 50 mg/wk to

T

Common adverse effects in *italic*; life-threatening effects underlined; generic names in **bold**; classifications in SMALL CAPS; ♣ Canadian drug name; ◐ Prototype drug

1511

efficacy **PO Maintenance Dose**
200–400 mg/day divided b.i.d.
(max: 1600 mg/day)

Child (2–16 y): **PO** Initiate with
1–3 mg/kg at bedtime × 1 wk,
then increase by 1–3 mg/kg/day
in 2 divided doses q1–2wk to a
target range of 5–9 mg/kg/day

Generalized Tonic-Clonic

Child: **PO** Initiate with 1–3 mg/kg
at bedtime; titrate to 6 mg/kg/
day by the end of 8 wk

Migraine Prophylaxis

Adult: **PO** Initiate with 25 mg
b.i.d., increase by 25 mg/wk to
200 mg/day or max tolerated
dose

Renal Impairment Dosage Adjustment

CrCl less than 70 mL/min:
Decrease dose by 50%

ADMINISTRATION

Oral

- Make dosage increments of 50 mg
  at weekly intervals to the recommended dose, usually 400 mg/day.
- Do not break tablets unless absolutely necessary because of bitter
  taste.
- Store at 15°–30° C (59°–86° F) in a
  tightly closed container. Protect
  from light and moisture.

ADVERSE EFFECTS (≥1%) **Body as a Whole:** *Fatigue, speech problems,* weight loss; decreased sweating and hyperthermia in children; metabolic acidosis. **CNS:** *Somnolence, dizzi-ness, ataxia, psychomotor slowing, confusion, nystagmus, paresthesia, memory difficulty, difficulty con-centrating, nervousness,* depression, anxiety, tremor. **GI:** Anorexia. **Special Senses:** Angle closure glaucoma (rare).

INTERACTIONS **Drug:** Increased CNS depression with **alcohol** and other CNS DEPRESSANTS; may increase **phenytoin** concentrations; may decrease ORAL CONTRACEPTIVE, **val-proate** concentrations; may in-crease risk of kidney stone forma-tion with other CARBONIC ANHYDRASE INHIBITORS. **Carbamazepine, phe-nytoin, valproate** may decrease topiramate concentrations. **Herbal: Ginkgo** may decrease anticonvul-sant effectiveness.

PHARMACOKINETICS **Absorption:** Rapidly absorbed from GI tract; 80% bioavailability. **Peak:** 2 h. **Dis-tribution:** 13–17% protein bound. **Metabolism:** Minimally metabo-lized in the liver. **Elimination:** Pri-marily in urine. **Half-Life:** 21 h.

NURSING IMPLICATIONS

Assessment & Drug Effects

- Monitor mental status and report
  significant cognitive impairment.
- Lab tests: Periodically monitor
  CBC with Hgb and Hct.

Patient & Family Education

- Do not stop drug abruptly; discontinue gradually to minimize
  seizures.
- To minimize risk of kidney
  stones, drink at least 6–8 full
  glasses of water each day.
- Exercise caution with potentially
  hazardous activities. Sedation is
  common, especially with concurrent use of alcohol or other CNS
  depressants.
- Use or add barrier contraceptive if
  using hormonal contraceptives.
- Be aware that psychomotor slowing and speech/language problems may develop while on
  topiramate therapy.
- Report adverse effects that interfere with activities of daily
  living.

TOPOTECAN HYDROCHLORIDE 🅿️

(toe-po-tee′can)

Hycamtin
Classifications: ANTINEOPLAS-
TIC; CAMPTOTHECIN; DNA TOPO-
ISOMERASE I INHIBITOR
Therapeutic: ANTINEOPLASTIC
Pregnancy Category: D

AVAILABILITY 4 mg injection

ACTION & *THERAPEUTIC EFFECT*
Antitumor mechanism is related to
inhibition of the activity of topoisomerase I, an enzyme required
for DNA replication. Topoisomerase I is essential for the relaxation
of supercoiled double-stranded
DNA that enables replication and
transcription to proceed. *Topote-*
can permits uncoiling of DNA
stands but prevents recoiling of
the two strands of DNA, resulting
in a permanent break in the DNA
strands.

USES Metastatic ovarian cancer, cervical cancer, small cell lung cancer.

CONTRAINDICATIONS Previous hypersensitivity to topotecan, irinotecan, or other camptothecin analogs;
acute infection; severe bone marrow depression; severe thrombocytopenia; pregnancy (category D),
lactation.
CAUTIOUS USE Myelosuppression;
severe renal impairment or renal
failure; history of bleeding disorders; previous cytotoxic or radiation therapy.

ROUTE & DOSAGE

**Metastatic Ovarian Cancer and
Small Cell Lung Cancer**

Adult: **IV** 1.5 mg/m² daily for
5 days starting on day 1 of a

21-day course. Four courses
of therapy recommended.
Subsequent doses can be adjusted
by 0.25 mg/m² depending on
toxicity.

Cervial Cancer

Adult: 0.75 mg/m² days 1–3
every 3 weeks

**Renal Impairment Dosage
Adjustment**

CrCl 20–39 mL/min: Use
0.75 mg/m²

Hemodialysis Dosage Adjustment
Supplementation not needed

ADMINISTRATION

Intravenous
▪ Initiate therapy only if baseline
neutrophil count 1500/mm³ or
higher and platelet count
100,000/mm³ or higher. ▪ Do not
give subsequent doses until neutrophils 1000/mm³ or higher,
platelets 100,000/mm³ or higher,
and Hgb greater than 9.0 mg/dL.
▪ Note: Dosage adjustments to
0.75 mg/m² are recommended
with moderate renal impairment.

***PREPARE:* IV Infusion:** Reconstitute
each 4-mg vial with 4 mL sterile
water for injection to yield 1 mg/
mL. ▪ Withdraw the required dose
and inject into 50–100 mL of NS or
D5W. ▪ If skin contacts drug during preparation, wash immediately with soap and water.
***ADMINISTER:* IV Infusion:** Give
over 30 min immediately after
preparation.
***INCOMPATIBILITIES* Y-site: Dexa-**
methasone, fluorouracil,
mitomycin.

▪ Store vials at 20°–25° C (68°–77° F);
protect from light. Reconstituted
vials are stable for 24 h.

Common adverse effects in *italic*, life-threatening effects <u>underlined</u>; generic names
in **bold**; classifications in SMALL CAPS; ✤ Canadian drug name; 🅿️ Prototype drug

1513

ADVERSE EFFECTS (≥1%) **Body as a Whole:** *Asthenia, fever, fatigue.* **GI:** *Nausea, vomiting, diarrhea, constipation, abdominal pain, stomatitis, anorexia,* transient elevations in liver function tests. **Hematologic:** <u>Leukopenia, *neutropenia*</u>, anemia, <u>thrombocytopenia</u>. **Respiratory:** *Dyspnea.* **Skin:** *Alopecia.*

INTERACTIONS Drug: Increased risk of bleeding with ANTICOAGULANTS, NSAIDS, SALICYLATES, ANTIPLATELET AGENTS.

PHARMACOKINETICS Distribution: 35% bound to plasma proteins. **Metabolism:** Undergoes pH-dependent hydrolysis. **Elimination:** ~30% in urine. **Half-Life:** 2–3 h.

NURSING IMPLICATIONS

Assessment & Drug Effects

- Lab tests: Obtain CBC counts with differential frequently; periodically monitor ALT.
- Assess for GI distress, respiratory distress, neurosensory symptoms, and S&S of infection throughout therapy.

Patient & Family Education

- Learn common adverse effects and measures to control or minimize when possible. Immediately report any distressing adverse effects to prescriber.
- Avoid pregnancy during therapy.

TOREMIFENE CITRATE

(tor-em'i-feen ci'trate)

Fareston
Classification: ANTINEOPLASTIC; SELECTIVE ESTROGEN RECEPTOR MODIFIER (SERM)
Therapeutic: ANTINEOPLASTIC; ANTI-ESTROGEN
Prototype: Tamoxifen
Pregnancy Category: D

AVAILABILITY 60 mg tablets

ACTION & *THERAPEUTIC EFFECT* Nonsteroidal antiestrogen chemical derivative of tamoxifen. Antitumor activity thought to be due to ability to compete with estrogen for binding sites in cancer cells. *Depresses growth in estrogen receptor–positive tumors in postmenopausal women.*

USES Metastatic breast cancer in postmenopausal women who are estrogen receptor positive.

CONTRAINDICATIONS Hypersensitivity to toremifene; history of thromboembolic disease; congenital/acquired QT prolongations; pregnancy (category D); lactation.
CAUTIOUS USE Preexisting endometrial hyperplasia; bone metastases (may result in hypercalcemia); older adults; geriatric patients; leukopenia and thrombocytopenia; liver disease; history of thrombolytic disease.

ROUTE & DOSAGE

Breast Cancer
Adult: **PO** 60 mg daily

ADMINISTRATION

Oral

- Withhold drug and notify prescriber if severe hypercalcemia develops.
- Store at 15°–30° C (59°–86° F) in a tightly closed container and protect from light.

ADVERSE EFFECTS (≥1%) **Body as a Whole:** *Hot flashes, sweating,* edema. **CNS:** Dizziness. **GI:** *Nausea,* vomiting, abnormal liver function tests. **Respiratory:** <u>Pulmonary embolism</u>. **Urogenital:** *Vaginal discharge,* vaginal bleeding. **Special Senses:** Cataracts, dry eyes, corneal keratopathy **CV:** Arrythmia.

Common adverse effects in *italic*, life-threatening effects <u>underlined</u>; generic names in **bold**; classifications in SMALL CAPS; ♣ Canadian drug name; ⊙ Prototype drug

INTERACTIONS Drug: THIAZIDE DIURETICS increase risk of hypercalcemia; increased PT on **warfarin; carbamazepine, phenobarbital, phenytoin** may increase toremifene metabolism.

PHARMACOKINETICS Absorption: Rapidly absorbed from GI tract. **Peak:** 3 h. **Distribution:** Greater than 99% protein bound; crosses placenta. **Metabolism:** In liver by cytochrome P450 3A4. **Elimination:** Primarily in feces. **Half-Life:** 5 days.

NURSING IMPLICATIONS

Assessment & Drug Effects

- Lab tests: Periodically monitor CBC with differential, serum calcium, LFTs and kidney functions.
- Monitor patients carefully with bone metastases or those on drugs that decrease calcium excretion (e.g., thiazide diuretics) for S&S of hypercalcemia (see Appendix F).
- Monitor PT and INR carefully when given concurrently with warfarin.

Patient & Family Education

- Report to prescriber promptly any of the following: Unexplained weakness or fatigue, musculoskeletal pain or calf pain and tenderness, sudden chest pain, vaginal bleeding.
- Schedule periodic eye exams with long-term therapy.

TORSEMIDE
(tor′se-mide)

Demadex
Classification: ELECTROLYTE AND WATER BALANCE; LOOP DIURETIC
Therapeutic: DIURETIC; ANTI-HYPERTENSIVE
Prototype: Furosemide
Pregnancy Category: B

AVAILABILITY 5 mg, 10 mg, 20 mg, 100 mg tablets; 10 mg/mL injection

ACTION & *THERAPEUTIC EFFECT*
Long-acting potent sulfonamide "loop" diuretic that inhibits reabsorption of sodium and chloride primarily in the loop of Henle and renal tubules. Binds to the sodium/potassium/chloride carrier in the loop of Henle and in the renal tubules. *Long-acting potent "loop" diuretic and antihypertensive agent.*

USES Management of edema associated with CHF, chronic kidney failure, hepatic cirrhosis; hypertension.

CONTRAINDICATIONS Hypersensitivity to torsemide or sulfonamides; anuria, fluid and electrolyte depletion states; acute MI; hepatic coma.
CAUTIOUS USE Renal impairment; ventricular arrhythmias; gout or hyperuricemia; diabetes mellitus or history of pancreatitis; liver disease; hearing impairment; pregnancy (category B); lactation.

ROUTE & DOSAGE

Edema of CHF, Chronic Kidney Failure

Adult: **PO/IV** 10–20 mg once daily, may increase up to 200 mg/day as needed

Hepatic Cirrhosis

Adult: **PO/IV** 5–10 mg once daily administered with an aldosterone antagonist or potassium-sparing diuretic, may increase up to 40 mg/day as needed

Hypertension

Adult: **PO** 2.5–5 mg once daily, may increase to 10 mg/day if no response after 4–6 wk

T

Common adverse effects in *italic*, life-threatening effects underlined; generic names in **bold**; classifications in SMALL CAPS; ♣ Canadian drug name; ⦿ Prototype drug

ADMINISTRATION

■ Note: With hepatic cirrhosis, use an aldosterone antagonist concomitantly to prevent hypokalemia and metabolic alkalosis.

Oral

■ Be aware that oral and IV doses are therapeutically equivalent; patients may be switched between the two forms with no change in dosage.

Intravenous

PREPARE: **Direct:** Given undiluted.
ADMINISTER: **Direct:** Give slowly over 2 min.
INCOMPATIBILITIES **Solution/additive: Dobutamine**.

■ Store at 15°–30° C (59°–86° F).

ADVERSE EFFECTS (≥1%) **CNS:** Headache, dizziness, fatigue, insomnia. **CV:** Orthostatic hypotension. **Endocrine:** *Hypokalemia,* hyponatremia, hyperuricemia. **GI:** Nausea, diarrhea. **Skin:** Rash, pruritus. **Body as a Whole:** Muscle cramps, rhinitis.

INTERACTIONS Drug: NSAIDS may reduce diuretic effects. Also see furosemide for potential drug interactions such as increased risk of **digoxin** toxicity due to hypokalemia, prolonged neuromuscular blockade with NEUROMUSCULAR BLOCKING AGENTS, and decreased **lithium** elimination with increased toxicity. **Herbal: Ginseng** may decrease efficacy.

PHARMACOKINETICS Absorption: Readily from GI tract. **Onset:** IV 10 min; PO 60 min. **Peak:** IV within 60 min; PO 60–120 min. **Duration:** 6–8 h. **Metabolism:** In liver (CYP system). **Elimination:** 80% in bile; 20% in urine. **Half-Life:** 210 min.

NURSING IMPLICATIONS

Assessment & Drug Effects

■ Monitor BP often and assess for orthostatic hypotension; assess respiratory status for S&S of pulmonary edema.

■ Monitor ECG, as electrolyte imbalances predispose to cardiac arrhythmias.

■ Lab tests: Monitor serum electrolytes, uric acid, blood glucose, BUN, and creatinine periodically throughout the course of therapy.

■ Monitor I&O with daily weights. Assess for improvement in edema.

■ Monitor diabetics for loss of glycemic control.

■ Monitor coagulation parameters and lithium levels in patients on concurrent anticoagulant and/or lithium therapy.

Patient & Family Education

■ Check weight at least weekly and report abrupt gains or losses to prescriber.

■ Understand the risk of orthostatic hypotension.

■ Report symptoms of hypokalemia (see Appendix F) or hearing loss immediately to prescriber.

■ Monitor blood glucose for loss of glycemic control if diabetic.

TRAMADOL HYDROCHLORIDE

(tra′ma-dol)

Rybix, Ryzolt, Ultram, Ultram ER, Zydol ◆
Classifications: ANALGESIC; NARCOTIC (OPIATE AGONIST)
Therapeutic: NARCOTIC ANALGESIC
Prototype: Morphine sulfate
Pregnancy Category: C
Controlled Substance: Schedule IV (in some states)

AVAILABILITY 50 mg tablets; 50 mg orally disintegrating tablets; 100 mg, 200 mg, 300 mg extended-release tablet

ACTION & *THERAPEUTIC EFFECT*
Centrally acting opiate receptor

agonist that inhibits the uptake of norepinephrine and serotonin, suggesting both opioid and nonopioid mechanisms of pain relief. May produce opioid-like effects, but causes less respiratory depression than morphine. *Effective agent for control of moderate to moderately severe pain.*

USES Management of moderate or moderately severe pain.

CONTRAINDICATIONS Hypersensitivity to tramadol or other opioid analgesics; severe respiratory depression; severe or acute asthmas; patients on MAO Inhibitors; substance abuse; alcohol intoxication; lactation.

CAUTIOUS USE Debilitated patients; chronic respiratory disorders; respiratory depression; older adults; liver disease; renal impairment; myxedema, hypothyroidism, or hypoadrenalism; GI disease; acute abdominal conditions; increased ICP or head injury, increased intracranial pressure; history of seizures; patients older than 75 y; pregnancy (category C); children younger than 16 y.

ROUTE & DOSAGE

Pain

Adult: **PO** immediate release: 25 mg daily, titrated up to dose of 100 mg/day (max: 400 mg/day)
Geriatric (over 75 years): **PO** 50–100 mg q4–6h prn (max: 300 mg/day), may start with 25 mg/day if not well tolerated, and increase by 25 mg q3days up to 200 mg/day

Renal Impairment Dosage Adjustment

CrCl less than 30 mL/min: Decrease to 50–100 mg q12h

Hepatic Impairment Dosage Adjustment

Cirrhosis: Decrease to 50–100 mg q12h

ADMINISTRATION

Oral

- Extended release tablets should be swallowed whole. They should not be crushed or chewed.
- Store at 15°–30° C (59°–86° F).

ADVERSE EFFECTS (≥1%) **CNS:** Drowsiness, *dizziness, vertigo, fatigue, headache, somnolence,* restlessness, euphoria, confusion, anxiety, coordination disturbance, sleep disturbances, seizures. **CV:** Palpitations, vasodilation. **GI:** *Nausea, constipation,* vomiting, xerostomia, dyspepsia, diarrhea, abdominal pain, anorexia, flatulence. **Body as a Whole:** Sweating, <u>anaphylactic reaction</u> (even with first dose), withdrawal syndrome (anxiety, sweating, nausea, tremors, diarrhea, piloerection, panic attacks, paresthesia, hallucinations) with abrupt discontinuation. **Skin:** Rash. **Special Senses:** Visual disturbances. **Urogenital:** Urinary retention/frequency, menopausal symptoms.

DIAGNOSTIC TEST INTERFERENCE Increased ***creatinine, liver enzymes;*** decreased ***hemoglobin; proteinuria.***

INTERACTIONS Drug: Carbamazepine significantly decreases tramadol levels (may need up to twice usual dose). Tramadol may increase adverse effects of MAO INHIBITORS. TRICYCLIC ANTIDEPRESSANTS, **cyclobenzaprine,** PHENOTHIAZINES, SELECTIVE SEROTONIN-REUPTAKE INHIBITORS (SSRIS), MAO INHIBITORS may enhance seizure risk with tramadol. May increase CNS adverse effects when used with other CNS

T

Common adverse effects in *italic*, life-threatening effects <u>underlined</u>; generic names in **bold**; classifications in SMALL CAPS; ✚ Canadian drug name; ⊙ Prototype drug

1517

DEPRESSANTS. **Herbal: St. John's wort** may increase sedation.

PHARMACOKINETICS Absorption: Rapidly absorbed from GI tract; 75% reaches systemic circulation. **Onset:** 30–60 min. **Peak:** 2 h. **Duration:** 3–7 h. **Distribution:** Approximately 20% bound to plasma proteins; probably crosses blood–brain barrier; crosses placenta; 0.1% excreted into breast milk. **Metabolism:** Extensively in liver by cytochrome P450 system. **Elimination:** Primarily in urine. **Half-Life:** 6–7 h.

NURSING IMPLICATIONS

Assessment & Drug Effects

- Assess for level of pain relief and administer prn dose as needed but not to exceed the recommended total daily dose.
- Monitor vital signs and assess for orthostatic hypotension or signs of CNS depression.
- Withhold drug and notify prescriber if S&S of hypersensitivity occur.
- Assess bowel and bladder function; report urinary frequency or retention.
- Use seizure precautions for patients who have a history of seizures or who are concurrently using drugs that lower the seizure threshold.
- Monitor ambulation and take appropriate safety precautions.

Patient & Family Education

- Exercise caution with potentially hazardous activities until response to drug is known.
- Do not exceed the total number of mg prescribed for a 24 h period.
- Understand potential adverse effects and report problems with bowel and bladder function, CNS impairment, and any other bothersome adverse effects to prescriber.

TRANDOLAPRIL
(tran-do'la-pril)
Mavik
Classifications: ANGIOTENSIN-CONVERTING ENZYME (ACE) INHIBITOR; ANTIHYPERTENSIVE
Therapeutic: ANTIHYPERTENSIVE; ACE INHIBITOR
Prototype: Enalapril
Pregnancy Category: D

AVAILABILITY 1 mg, 2 mg, 4 mg tablets

ACTION & *THERAPEUTIC EFFECT*
Inhibits ACE and interrupts conversion by renin which leads to the formation of angiotensin II from angiotensin I. Inhibition of ACE leads to vasodilation as well as to decreased aldosterone. Decreased aldosterone leads to diuresis and a slight increase in serum potassium. *Lowers blood pressure by specific inhibition of ACE. Unlike other ACE inhibitors, all racial groups respond to trandolapril, including low-renin hypertensives.*

USES Treatment of hypertension, reduction of CV morbidity/mortality post MI.
UNLABELED USES CHF.

CONTRAINDICATIONS Hypersensitivity to trandolapril or ACE inhibitors; history of angioedema related to previous treatment with an ACE inhibitor; pregnancy (category D), lactation.
CAUTIOUS USE Renal impairment, hepatic insufficiency; patients prone to hypotension (e.g., CHF, ischemic heart disease, aortic stenosis, CVA, dehydration); SLE, scleroderma. Safety and effectiveness in children younger than 18 y are not established.

ROUTE & DOSAGE

Hypertension

Adult: **PO** 1 mg in nonblack patients, 2 mg in black patients once daily, may increase weekly to 2–4 mg once daily (max: 8 mg/day). Lower dose used in patients also taking a diuretic.

Post MI

Adult: **PO** 1 mg daily for 2 days then increase to 4 mg daily for 2–4 years

Renal Impairment Dosage Adjustment

CrCl less than 30 mL/min: Start with 0.5 mg once daily

Hepatic Impairment Dosage Adjustment

Hepatic cirrhosis: Start with 0.5 mg once daily

ADMINISTRATION

Oral

- Note: If concurrently ordered diuretic cannot be discontinued 2–3 days before beginning trandolapril therapy, initial dose is usually reduced to 0.5 mg.
- Dosage adjustments are typically made at intervals of at least 1 wk.
- Store at 15°–30° C (59°–86° F).

ADVERSE EFFECTS (≥1%) **Body as a Whole:** Fatigue, angioedema, fever, malaise. **CNS:** Dizziness, drowsiness. **CV:** Hypotension, bradycardia, edema, palpitations. **GI:** Diarrhea. **Respiratory:** Cough. **Skin:** Rash, pruritus. **Metabolic:** Hyperkalemia, increased liver enzymes.

INTERACTIONS Drug: DIURETICS may enhance hypotensive effects. POTASSIUM-SPARING DIURETICS (**amiloride, spironolactone, triamterene**),

POTASSIUM SUPPLEMENTS, POTASSIUM-CONTAINING SALT SUBSTITUTES may increase risk of hyperkalemia. May increase serum levels and toxicity of **lithium.** NSAIDs may reduce the therapeutic response.

PHARMACOKINETICS Absorption: Rapidly absorbed from GI tract and converted to active form, trandolaprilat, in liver; 70% of dose reaches systemic circulation as trandolaprilat. **Peak:** 4–10 h. **Distribution:** 80% protein bound; crosses placenta, secreted into breast milk of animals (human secretion unknown). **Metabolism:** In liver to trandolaprilat. **Elimination:** 33% in urine, 66% in feces. **Half-Life:** 6 h trandolapril, 22.5 h trandolaprilat.

NURSING IMPLICATIONS

Assessment & Drug Effects

- Monitor BP carefully for 1–3 h following initial dose, especially in patients using concurrent diuretics, on salt restriction, or volume depleted.
- Lab tests: Baseline LFTs and kidney function tests; periodic serum potassium and sodium.
- Monitor serum lithium levels frequently with concurrent lithium therapy and assess for S&S of lithium toxicity; increase caution when diuretic therapy is also used.

Patient & Family Education

- Discontinue drug and immediately report S&S of angioedema of face or extremities to prescriber. Seek emergency help for swelling of the tongue or any other sign of potential airway obstruction.
- Be aware that lightheadedness can occur, especially during early therapy. Excess fluid loss of any

T

Common adverse effects in *italic*, life-threatening effects underlined; generic names in **bold**; classifications in SMALL CAPS; ♣ Canadian drug name; ⊙ Prototype drug

1519

kind will increase risk of hypotension and syncope.

TRANYLCYPROMINE SULFATE

(tran-ill-sip'roe-meen)

Parnate

Classification: ANTIDEPRESSANT; MONOAMINE OXIDASE INHIBITOR (MAOI)

Therapeutic: ANTIDEPRESSANT; MAOI

Prototype: Phenelzine

Pregnancy Category: C

AVAILABILITY 10 mg tablets

ACTION & *THERAPEUTIC EFFECT*

Potent MAO with a antidepressant activity that arises from the increased availability of monoamines resulting from the inhibition of the enzyme MAO. This leads to increased concentration of neurotransmitters, such as epinephrine, norepinephrine, and dopamine in the CNS. *Drug of last choice for severe depression unresponsive to other MAO inhibitors.*

USES Severe depression.

UNLABELED USES Orthostatic hypotension, panic disorder, social anxiety disorder.

CONTRAINDICATIONS Confirmed or suspected cerebrovascular defect, cardiovascular disease; hepatic disease or abnormal LFT; hypertension, pheochromocytoma, history of severe or recurrent headaches; recent acute MI; angina; renal failure; suicidal ideation; anuria; lactation.

CAUTIOUS USE Bipolar disorder; Parkinson's disease; psychosis; schizophrenia, anxiety/agitation; CHF; DM; seizure disorders; hyperthyroidism; history of suicidal attempts; renal impairment; older adults; pregnancy

(category C); children and adolescents with major depressive disorder or other psychiatric disorders.

ROUTE & DOSAGE

Severe Depression

Adult: **PO** 30 mg/day in 2 divided doses, may increase by 10 mg/day at 2–3 wk intervals (max: 60 mg/day)

ADMINISTRATION

Oral

- Contraindicated drugs should be discontinued 1–2 weeks before starting therapy.
- Crush tablet and give with fluid or mix with food if patient cannot swallow pill.
- Note: Doses given in the late evening may cause insomnia.

ADVERSE EFFECTS (≥1%) **CNS:** Vertigo, dizziness, tremors, muscle twitching, headache, blurred vision. **CV:** *Orthostatic hypotension,* arrhythmias, hypertensive crisis. **GI:** Dry mouth, anorexia, constipation, diarrhea, abdominal discomfort. **Skin:** Rash. **Urogenital:** Impotence. **Body as a Whole:** Peripheral edema, sweating.

INTERACTIONS Drug: TRICYCLIC ANTIDEPRESSANTS, SSRIS, AMPHETAMINES, **ephedrine, reserpine, guanethidine, buspirone, methyldopa, dopamine, levodopa, tryptophan** may precipitate hypertensive crisis, and must be discontinued before **tranylcypromine** treatment. **Alcohol** and other CNS DEPRESSANTS add to CNS depressant effects; **meperidine** can cause fatal cardiovascular collapse; ANESTHETICS exaggerate hypotensive and CNS depressant effects; **metrizamide** increases risk of seizures; DIURETICS and other ANTIHYPERTENSIVE AGENTS add to hypotensive effects. **Food: Tyramine-**containing foods may precipitate hypertensive crisis (e.g., aged cheeses, processed cheeses, sour cream, wine, champagne, beer, pickled herring, anchovies, caviar, shrimp, liver, dry sausage, figs, raisins, overripe bananas or avocados, chocolate, soy sauce, bean curd, yeast extracts, yogurt, papaya products, meat tenderizers, broad beans). **Herbal: Ginseng, ephedra, ma huang, St. John's wort** may lead to hypertensive crisis; **ginseng** may lead to manic episodes.

PHARMACOKINETICS Absorption: Completely absorbed from GI tract. **Onset:** 10 days. **Metabolism:** Rapidly metabolized in liver to active metabolite. **Elimination:** Primarily in urine. **Half-Life:** 2.5 h (but may take 120 h for urinary tryptamine levels to return to normal).

NURSING IMPLICATIONS

Assessment & Drug Effects

- Monitor BP closely. Severe hypertensive reactions are known to occur with MAO inhibitors.
- Monitor for changes in behavior that could indicate increased suicidality.
- Expect therapeutic response within 3 days, but full antidepressant effects may not be obtained until 2–3 wk of drug therapy.

Patient & Family Education

- Do not eat tyramine-containing foods (see FOOD–DRUG INTERACTIONS).
- Be aware that excessive use of caffeine-containing beverages (chocolate, coffee, tea, cola) can contribute to development of rapid heartbeat, arrhythmias, and hypertension.
- Make position changes slowly, particularly from recumbent to upright posture.

- Avoid potentially hazardous activities until response to drug is known.
- Avoid alcohol or other CNS depressants because of their possible additive effects.

TRASTUZUMAB

(tra-stu′zu-mab)

Herceptin
Classifications: IMMUNOMODULATOR; MONOCLONAL ANTIBODY; ANTINEOPLASTIC; ANTI-HUMAN EPIDERMAL GROWTH FACTOR (ANTI-HER)
Therapeutic: ANTINEOPLASTIC; IMMUNOMODULATOR; ANTI-HER
Pregnancy Category: D

AVAILABILITY 21 mg/mL injection

ACTION & *THERAPEUTIC EFFECT* Recombinant DNA monoclonal antibody (I_gG_1 kappa) that selectively binds to the human epidermal growth factor receptor-2 protein (HER_2). *Inhibits growth of human tumor cells that overexpress HER_2 proteins.*

USES Metastatic breast cancer in those whose tumors overexpress the HER_2 protein. HER_2-positive breast cancer after surgery.

CONTRAINDICATIONS Concurrent administration of anthracycline or radiation; pregnancy (category D); lactation during and for 6 mo following administration of trastuzumab.
CAUTIOUS USE Preexisting cardiac dysfunction; pulmonary disease; previous administration of cardiotoxic therapy (e.g., anthracycline or radiation); hypersensitivity to benzyl alcohol; older adults.

ROUTE & DOSAGE

Metastatic Breast Cancer
Adult: **IV** 4 mg/kg, then 2 mg/kg qwk

Common adverse effects in *italic*, life-threatening effects underlined; generic names in **bold**; classifications in SMALL CAPS; ✿ Canadian drug name; ◉ Prototype drug

1521

ADMINISTRATION

Intravenous

PREPARE: **IV Infusion:** Reconstitute each vial with 20 mL of supplied diluent (bacteriostatic water) to produce a multidose vial containing 21 mg/mL. ▪ Note: For patients with a hypersensitivity to benzyl alcohol, reconstitute with sterile water for injection; this solution **must be** used immediately with any unused portion discarded. ▪ Withdraw the ordered dose and add to a 250 mL of NS and invert bag to mix. ▪ Do not give or mix with dextrose solutions.

ADMINISTER: **IV Infusion:** Infuse loading dose (4 mg/kg) over 90 min; infuse subsequent doses (2 mg/kg) over 30 min. ▪ Do not give IV push or as a bolus dose.

INCOMPATIBILITIES **Solution/additive: Dextrose** solution; do not mix or coadminister with other drugs.

▪ Store unopened vials and reconstituted vials at 2°–8° C (36°–46° F).
▪ Discard reconstituted vials 28 days after reconstitution.

ADVERSE EFFECTS (≥1%) **Body as a Whole:** *Pain, asthenia, fever, chills,* flu syndrome, allergic reaction, bone pain, arthralgia, <u>hypersensitivity (anaphylaxis, urticaria, bronchospasm, angioedema, or hypotension)</u>, increased incidence of infections, infusion reaction (*chills, fever,* nausea, vomiting, pain, rigors, headache, dizziness, dyspnea, hypotension, rash). **CNS:** *Headache, insomnia, dizziness, paresthesias,* depression, peripheral neuritis, neuropathy. **CV:** <u>CHF</u>, cardiac dysfunction (dyspnea, cough, paroxysmal nocturnal dyspnea, peripheral edema, S3 gallop, reduced ejection fraction), tachycardia, edema, cardiotoxicity. **GI:** *Diarrhea, abdominal pain,* *nausea, vomiting,* anorexia. **Hematologic:** *Anemia,* <u>leukopenia</u>. **Respiratory:** *Cough, dyspnea,* rhinitis, pharyngitis, sinusitis. **Skin:** *Rash,* herpes simplex, acne.

INTERACTIONS Drug: **Paclitaxel** may increase trastuzumab levels and toxicity.

PHARMACOKINETICS Half-Life: 5.8 days.

NURSING IMPLICATIONS

Assessment & Drug Effects

▪ Lab tests: Periodically monitor CBC with differential, platelet count, and Hgb and Hct.
▪ Monitor for chills and fever during the first IV infusion; these adverse events usually respond to prompt treatment without the need to discontinue the infusion. Notify prescriber immediately.
▪ Monitor carefully cardiovascular status at baseline and throughout course of therapy, assessing for S&S of heart failure (e.g., dyspnea, increased cough, PND, edema, S3 gallop). Those with preexisting cardiac dysfunction are at high risk for cardiotoxicity.

Patient & Family Education

▪ Report promptly any unusual symptoms (e.g., chills, nausea, fever) during infusion.
▪ Report promptly any of the following: Shortness of breath, swelling of feet or legs, persistent cough, difficulty sleeping, loss of appetite, abdominal bloating.

TRAVOPROST

(tra'-vo-prost)

Travatan
Pregnancy Category: C
See Appendix A-1.

TRAZODONE HYDROCHLORIDE

(tray'zoe-done)

Oleptro

Classifications: ANTIDEPRESSANT
Therapeutic: ANTIDEPRESSANT
Prototype: Imipramine
Pregnancy Category: C

AVAILABILITY 50 mg, 100 mg, 150 mg, 300 mg tablets; 150 mg, 300 mg extended release tablet

ACTION & *THERAPEUTIC EFFECT*
Centrally acting antidepressant that potentiates serotonin effects by selectively blocking its reuptake at presynaptic membranes in CNS. Produces varying degrees of sedation in normal and mentally depressed patients. *Increases total sleep time, decreases number and duration of awakenings in depressed patient, and decreases REM sleep. Has antianxiety effect in severely depressed patient.*

USES Major depression.
UNLABELED USES Adjunctive treatment of alcohol dependence, anxiety, panic disorder, insomnia.

CONTRAINDICATIONS Initial recovery phase of MI; ventricular ectopy; electroshock therapy; suicidal ideation.
CAUTIOUS USE Bipolar disorder, older adults; history of suicidal tendencies; cardiac arrhythmias or disease; hepatic disease, renal impairment; pregnancy (category C), lactation. Safe use in children not established.

ROUTE & DOSAGE

Depression

Adult: **PO Immediate release:** 150 mg/day in divided doses, may increase by 50 mg/day q3–4days (max: 400–600 mg/day)
PO Extended release: 150 mg daily may increase by 75 mg/day at 3 day intervals (max: 375 mg/day)

Pharmacogenetic Dosage Adjustment

Poor CYP2D6 metabolizer: Start with 80% of normal dose

ADMINISTRATION

Oral

- Ensure that extended-release tablets are swallowed whole. They should not be crushed or chewed.
- Give drug with food; increases amount of absorption by 20% and appears to decrease incidence of dizziness or light-headedness. Maintain the same schedule for food-drug intake throughout treatment period to prevent variations in serum concentration.
- Store in tightly closed, light-resistant container at 15°–30° C (59°–86° F).

ADVERSE EFFECTS (≥1%) **CNS:** *Drowsiness,* light-headedness, tiredness, dizziness, insomnia, headache, agitation, impaired memory and speech, disorientation. **CV:** *Hypotension (including orthostatic hypotension),* hypertension, syncope, shortness of breath, chest pain, tachycardia, palpitations, bradycardia, PVCs, ventricular tachycardia (short episodes of 3–4 beats). **Special Senses:** Nasal and sinus congestion, blurred vision, eye irritation, sweating or clamminess, tinnitus. **GI:** *Dry mouth,* anorexia, constipation, abdominal distress, nausea, vomiting, dysgeusia, flatulence, diarrhea. **Urogenital:** Hematuria, increased frequency, delayed urine flow, early or absent

T

Common adverse effects in *italic*, life-threatening effects underlined; generic names in **bold**; classifications in SMALL CAPS; ♦ Canadian drug name; ☻ Prototype drug

1523

menses, male priapism, ejaculation inhibition. **Hematologic:** Anemia. **Musculoskeletal:** Skeletal aches and pains, muscle twitches. **Skin:** Skin eruptions, rash, pruritus, acne, photosensitivity. **Body as a Whole:** Weight gain or loss.

INTERACTIONS Drug: ANTIHYPERTENSIVE AGENTS may potentiate hypotensive effects; **alcohol** and other CNS DEPRESSANTS add to depressant effects; may increase **digoxin** or **phenytoin** levels; MAO INHIBITORS may precipitate hypertensive crisis; **ketoconazole, indinavir, ritonavir, saquinavir** may increase levels and toxicity. Use of other serotonergic agents may increase risk of serotonin syndrome. **Herbal: Ginkgo** may increase sedation.

PHARMACOKINETICS Absorption: Readily from GI tract. **Onset:** 1–2 wk. **Peak:** 1–2 h. **Distribution:** Distributed into breast milk. **Metabolism:** In liver (CYP2D6). **Elimination:** 75% in urine, 25% in feces. **Half-Life:** 5–9 h.

NURSING IMPLICATIONS

Assessment & Drug Effects

- Monitor BP and heart rate and rhythm. Report to prescriber development of tachycardia, bradycardia, or palpitations.
- Monitor for orthostatic hypotension, especially in the elderly or those taking concurrent antihypertensive drugs.
- Monitor children and adolescents for changes in behavior that indicate increased suicidality.
- Observe patient's level of activity. If it appears to be increasing toward sleeplessness and agitation with changes in reality orientation, report to prescriber. Manic episodes have been reported.
- Be aware that overdose is characterized by an extension

of common adverse effects: Vomiting, lethargy, drowsiness, and exaggerated anticholinergic effects.

Patient & Family Education

- Expect therapeutic response to begin in 1 wk; may require 2–4 wk to reach maximum levels. Adhere to regimen.
- Do not alter dose or intervals between doses.
- Consult prescriber if drowsiness becomes a distressing adverse effect. Dose regimen may be changed so that largest dose is at bedtime.
- Limit or abstain from alcohol use. The depressant effects of CNS depressants and alcohol may be potentiated by this drug.
- Do not self-medicate with OTC drugs for colds, allergy, or insomnia treatment without advice of prescriber. Many of these drugs contain CNS depressants.
- Male patient should report inappropriate or prolonged penile erections. The drug may be discontinued.

TREPROSTINIL SODIUM

(tre-pros′tin-il)

Remodulin, Tyvaso
Classification: PROSTAGLANDIN; PULMONARY ANTIHYPERTENSIVE; VASODILATOR
Therapeutic: PULMONARY ANTI-HYPERTENSIVE
Prototype: Epoprostenol
Pregnancy Category: B

AVAILABILITY 1 mg/mL, 2.5 mg/mL, 5 mg/mL, 10 mg/mL injection; 1.74 mg/2.9 mL solution for inhalation

ACTION & *THERAPEUTIC EFFECT*
Causes direct vasodilation of the

pulmonary and systemic arterial vascular beds, and inhibition of platelet aggregation. The vasodilatory effects reduce right and left ventricular afterload, and increase cardiac output and stroke volume. Also improves dyspnea, fatigue, and signs and symptoms of pulmonary arterial hypertension (PAH). *Vasodilation of the arteries in the pulmonary system results in lowering of pulmonary arterial hypertension (PAH).*

USES Treatment of pulmonary arterial hypertension (PAH).
UNLABELED USES Severe intermittent claudication.

CONTRAINDICATIONS Severe hepatic insufficiency; hypersensitivity to treprostinil.
CAUTIOUS USE Mild or moderate hepatic insufficiency; bleeding disorders; elderly; renal disease, renal impairment, renal failure; pregnancy (category B), lactation. Safety and efficacy in children younger than 16 y are not established.

ROUTE & DOSAGE

Pulmonary Arterial Hypertension

Adult/Adolescent: **Subcutaneous/IV** 1.25 ng/kg/min. If dose is not tolerated, reduce to 0.625 ng/kg/min. Then increase rate by no more than 1.25 ng/kg/min/wk for first 4 wk, then by 2.5 ng/kg/min/wk until achieve desired response. There is little experience with doses greater than 40 ng/kg/min. See package insert for cross taper schedule switching from epoprostenol to treprostinil.
Adult: **Inhalation** 3 breaths (18 mcg) via inhalation system 4 times/day, may increase by 3 breaths at 1–2 wk intervals.

ADMINISTRATION

Inhalation

- Give treprostinil inhalation solution undiluted using the Tyvaso Inhalation System. A single breath delivers approximately 6 mcg of treprostinil.

Subcutaneous

- Administer treprostinil undiluted by continuous infusion through a subcutaneous catheter. Use an infusion pump designed for subcutaneous delivery.
- Avoid abrupt withdrawal or sudden large reductions in dosage as these may lead to worsening of PAH symptom.

Intravenous

May be given as IV infusion via a central line when the subcutaneous route is not feasible.

PREPARE: Infusion: Dilute with sterile water for injection or NS.
- Follow specific directions for preparation provided by the manufacturer.

ADMINISTER: Infusion: Given by continuous infusion via a central venous catheter. Interruptions in drug delivery MUST be avoided.
- Have backup delivery system readily available.

- Store at 15°–25° C (59°–77° F).

ADVERSE EFFECTS (≥1%) **Body as a Whole:** *Jaw pain,* flushing, syncope. **CNS:** *Headache,* dizziness. **CV:** *Vasodilation,* edema, hypotension. **GI:** *Diarrhea, nausea, vomiting.* **Skin:** *Rash,* pruritus. **Other:** *Infusion site reactions (erythema, hematoma, induration, pruritus, rash, injection site pain).*

INTERACTIONS Drug: NSAIDS, ANTICOAGULANTS may increase risk of bleeding; ANTIHYPERTENSIVE AGENTS, DIURETICS, VASODILATORS may exacerbate hypotension; **ephedrine, pseudoephedrine** may antagonize

Common adverse effects in *italic*, life-threatening effects underlined; generic names in **bold**; classifications in SMALL CAPS; ♣ Canadian drug name; ⊘ Prototype drug

1525

antihypertensive effects. **Herbal: Ephedra, ma huang** may antagonize antihypertensive effects.

PHARMACOKINETICS Absorption: Completely absorbed from subcutaneous site. **Onset:** Steady state reached in 10 h. **Metabolism:** Extensively in liver by unknown enzyme system. **Elimination:** 79% in urine, 13% in feces. **Half-Life:** 2–4 h.

NURSING IMPLICATIONS

Assessment & Drug Effects

▪ Monitor for therapeutic effectiveness indicated by less dyspnea and fatigue, increased activity tolerance, and improved hemodynamic parameters.

▪ Monitor for and report symptoms of excessive response to the drug including: Headache, nausea, emesis, restlessness, anxiety and infusion site pain or reaction (e.g., erythema, induration or rash). If these occur, the rate of subcutaneous infusion should be slowed.

▪ Monitor BP closely, especially if taking concurrent antihypertensive drugs (e.g., diuretics, vasodilators).

▪ Lab tests: Baseline and periodic LFTs and renal function tests. Monitor periodically coagulation parameters (more often if on concurrent anticoagulation therapy).

Patient & Family Education

▪ Note: Therapy with this drug may be needed for prolonged periods, possibly years.

▪ Report any of the following: Headache, nausea, vomiting, restlessness, anxiety, and infusion site pain.

TRETINOIN
(tret′i-noyn)

Atralin, Avita, Refissa, Renova, Retin-A, Retin-A Micro, Retinoic Acid, Vesanoid

Classifications: ANTINEOPLASTIC; ANTIACNE (RETINOID); ANTIPSORIATIC **Therapeutic:** ANTINEOPLASTIC; ANTIACNE; ANTIPSORIATIC **Prototype:** Isotretinoin **Pregnancy Category:** D (oral form); C (topical form)

AVAILABILITY 0.025%, 0.05%, 0.1% cream; 0.025%, 0.01% gel; 0.05% liquid; 10 mg capsules

ACTION & *THERAPEUTIC EFFECT*
Antiacne activity: Contact irritant containing retinoic acid and vitamin A acid. Reverses retention hyperkeratosis and comedo formation, primary events in acne pathology. Suggests that keratinocytes in the sebaceous follicle become less adherent and turnover of follicular epithelial cells is increased; two processes that promote easy extrusion of the comedo and prevent it from reformation. **Antineoplastic activity:** Tretinoin induces cellular differentiation in malignant cells. As vitamin A (retinol) derivatives, retinoids are important regulators of cell reproduction, and cell proliferation and differentiation. Tretinoin represents a new class of anticancer drugs, differentiating agents. It is used in the treatment of acute promyelotic leukemia (APL). Tretinoin offers a less toxic means to induce complete remission than conventional chemotherapy. APL results from changes in the alpha-retinoic acid receptor (RAR) found on the long arm of chromosome 17. *Effective in early treatment and control of acne vulgaris grades I–III. Effective in treatment of APL.*

USES Topical treatment of acne vulgaris grades I–III, especially during early stages when number of comedones is greatest; adjunctively in management of associated comedones and in treatment of flat warts; oral for remission induction treatment of acute promyelocytic leukemia; cream as adjunctive therapy for mitigation of fine wrinkles. **UNLABELED USES** Psoriasis, senile keratosis, ichthyosis vulgaris, keratosis palmaris and plantaris, basal cell carcinoma, photodamaged skin (photoaging), and other skin conditions. **Orphan drug:** For squamous metaplasia of conjunctiva or cornea with mucous deficiency and keratinization.

CONTRAINDICATIONS Hypersensitivity to retinoid; eczema; exposure to sunlight or ultraviolet rays, sunburn; pregnancy (**oral:** category D); lactation.
CAUTIOUS USE Patient in an occupation necessitating considerable sun exposure or weather extremes; hepatic disease; pregnancy (**topical:** category C); children younger than 12 y.

ROUTE & DOSAGE

Acne
Adult: **Topical** Apply once/day at bedtime

Acute Promyelocytic Leukemia
Adult: **PO** 45 mg/m²/day

Antiwrinkle Cream
Adult: **Topical** (0.05% cream) Apply to face once daily at bedtime

ADMINISTRATION

Topical
- Cleanse using a mild bland soap, and thoroughly dry areas being treated before applying drug. Avoid use of medicated, drying, or abrasive soaps and cleansers.
- Wash hands before and after treatment. Apply lightly over affected areas. Do not apply to non-affected skin area.

- Avoid contact of drug with eyes, mouth, angles of nose, open wounds, mucous membranes.
- Store gel and liquid formulations below 30° C (86° F) and solution below 27° C (80° F).

ADVERSE EFFECTS (≥1%) **Body as a Whole: Note:** Listed adverse effects occur primarily with oral administration; only skin effects with topical administration. *Bone pain, malaise, shivering, hemorrhage, peripheral edema, pain, chest discomfort, weight gain or loss,* <u>DIC</u>. **CNS:** *Dizziness, paresthesias, anxiety, insomnia, depression, headache, fever, weakness, fatigue,* cerebral hemorrhage, intracranial hypertension, hallucinations. **CV:** *Arrhythmias, flushing, hypotension, hypertension,* CHF. **Special Senses:** Visual disturbances, ocular disturbances, change in visual acuity, earache. **GI:** *Nausea, vomiting, abdominal pain, diarrhea, constipation, dyspepsia,* <u>GI hemorrhage</u>. **Respiratory:** *Dyspnea, respiratory insufficiency, pneumonia, rales, pleural effusion, wheezing.* **Skin:** Local inflammatory reactions, transient stinging or warmth on site, *redness, scaling, severe erythema,* blistering, crusting and peeling, temporary hypopigmentation or hyperpigmentation, *increased sweating.* **Urogenital:** Renal insufficiency, dysuria, acute kidney failure.

INTERACTIONS Drug: TOPICAL ACNE MEDICATIONS (including **sulfur, resorcinol, benzoyl peroxide,** and **salicylic acid**) may increase inflammation and peeling; topical products containing **alcohol** or **menthol** may cause stinging.

PHARMACOKINETICS Absorption: Minimally absorbed from intact skin, Topical; 60% absorbed, PO. **Elimination:** About 0.1% of topical dose is excreted in urine within

Common adverse effects in *italic*, life-threatening effects <u>underlined</u>; generic names in **bold**; classifications in SMALL CAPS; ♣ Canadian drug name; ☻ Prototype drug

1527

24 h; 63% excreted in urine and 31% in feces, PO. **Half-Life:** 45 min, Topical; 2–2.5 h, PO.

NURSING IMPLICATIONS

Assessment & Drug Effects

- Be aware that treatment to dark-skinned individuals may cause unsightly postinflammatory hyperpigmentation; that is reversible with termination of drug treatment.
- Clinical response should be evident in 2–3 wk; complete and satisfactory response (in 75% of the patients) may require 3–4 mo.
- Be aware that erythema and desquamation during the first 1–3 wk of treatment do not represent exacerbation of the skin problem but a probable response to the drug from deep previously unseen lesions.

Patient & Family Education

- As treatment is continued, lesions gradually disappear, leaving an inflammatory background; scaling and redness decrease after 8–10 wk of therapy.
- Wash face no more often than 2–3 times daily.
- Be aware that drug is not curative; relapses commonly occur within 3–6 wk after treatment has been discontinued.
- Avoid exposure to sun; when cannot be avoided, use a SPF 15 or higher sunscreen.
- Do not self-medicate with additional acne treatment because of danger of drug interactions.

TRIAMCINOLONE

(trye-am-sin′oh-lone)
Atolone, Kenacort, Kenalog-E

TRIAMCINOLONE ACETONIDE

Azmacort, Cenocort A₂, Kenalog, Nasacort HFA, Triam-A, Triamonide, Tri-kort, Trilog, Tri-Nasal

TRIAMCINOLONE DIACETATE

Kenacort

TRIAMCINOLONE HEXACETONIDE

Aristospan
Classification: ADRENAL CORTICOSTEROID; GLUCOCORTICOID
Therapeutic: ANTI-INFLAMMATORY; IMMUNOSUPPRESSANT
Prototype: Prednisone
Pregnancy Category: C

AVAILABILITY **Triamcinolone:** 4 mg, 8 mg tablets; 4 mg/5 mL syrup; **Triamcinolone acetonide:** 3 mg/mL, 10 mg/mL, 40 mg/mL injection; 100 mcg aerosol; 55 mcg inhaler; 55 mcg spray; 0.5 mg/mL nasal spray; 0.025%, 0.1%, 0.5% cream, ointment, lotion; 10.3% topical spray; **Triamcinolone diacetate:** 4 mg tablet; **Triamcinolone hexacetonide:** 5 mg/mL, 20 mg/mL injection

ACTION & *THERAPEUTIC EFFECT* Immediate-acting synthetic fluorinated adrenal corticosteroid with glucocorticoid properties. Possesses minimal sodium and water retention properties in therapeutic doses. *Anti-inflammatory and immunosuppressant drug that is effective in the treatment of bronchial asthma.*

USES An anti-inflammatory or immunosuppressant agent. Orally inhaled: Bronchial asthma in patient who has not responded to conventional inhalation treatment. Therapeutic doses do not appear to suppress HPA (hypothalamic-pituitary-adrenal) axis.

CONTRAINDICATIONS Hypersensitivity to corticosteroids or benzyl alcohol; kidney dysfunction; glaucoma; acute bronchospasm; fungal infection.

CAUTIOUS USE Coagulopathy, hemophilia, diabetes mellitus, GI disease; CHF; herpes infection; infection; IBD; myasthenia gravis; MI; ocular exposure, ocular infection; osteoporosis; peptic ulcer disease; PVD; skin abrasion; pregnancy (category C), lactation; children younger than 6 y.

ROUTE & DOSAGE

Inflammation, Immunosuppression

Adult: IM/Subcutaneous 4–48 mg/day in divided doses Intra-articular/Intradermal 4–48 mg/day Inhaled 2–4 inhalations q.i.d. Topical See Appendix A
Child: IM/Subcutaneous 3.3–50 mg/m²/day in divided doses Intra-articular/Intradermal 3.3–50 mg/m²/day

Acetonide

Adult: IM 60 mg, may repeat with 20–100 mg q6wk Intradermal 1 mg per injection site (max: 30 mg total) Intra-articular 2.5–4.0 mg Inhalation See Appendix A
Child: IM 6–12 y, 0.03–0.2 mg q1–7days Inhalation See Appendix A

Hexacetonide

Adult: Intralesional Up to 0.5 mg/in² of skin Intra-articular 2–20 mg q3–4wk

ADMINISTRATION

Inhalation
▪ Follow manufacturer's directions for specific oral or nasal inhaler and instruct patient on proper administration technique.

Subcutaneous/Intramuscular
▪ Do not give triamcinolone injection IV.

▪ IM injections should only be given into a large, well developed muscle such as the gluteal muscle.
▪ Store at 15°–30° C (59°–86° F). Protect from light.

ADVERSE EFFECTS (≥1%) CNS: Euphoria, headache, insomnia, confusion, psychosis. CV: CHF, edema. GI: Nausea, vomiting, peptic ulcer. Musculoskeletal: Muscle weakness, delayed wound healing, muscle wasting, osteoporosis, aseptic necrosis of bone, spontaneous fractures. Endocrine: Cushingoid features, growth suppression in children, carbohydrate intolerance, hyperglycemia. Special Senses: Cataracts. Hematologic: Leukocytosis. Metabolic: Hypokalemia. Skin: Burning, itching, folliculitis, hypertrichosis, hypopigmentation.

INTERACTIONS Drug: BARBITURATES, phenytoin, rifampin increase steroid metabolism—may need increased doses of triamcinolone; amphotericin B, DIURETICS add to potassium loss; ambenonium, neostigmine, pyridostigmine may cause severe muscle weakness in patients with myasthenia gravis; may inhibit antibody response to VACCINES, TOXOIDS.

PHARMACOKINETICS Absorption: Readily absorbed from all routes. Onset: 24–48 h PO, IM. Peak: 1–2 h PO; 8–10 h IM. Duration: 2.25 days PO; 1–6 wk IM. Metabolism: In liver. Elimination: In urine. Half-Life: 2–5 h; HPA suppression, 18–36 h.

NURSING IMPLICATIONS

Assessment & Drug Effects
▪ Notify prescriber if wheezing occurs immediately following a dose of inhaled triamcinolone.
▪ Do not use occlusive dressing over topical application unless specifically ordered to do so.

Common adverse effects in *italic*, life-threatening effects underlined; generic names in **bold**; classifications in SMALL CAPS; ♣ Canadian drug name; ✪ Prototype drug

1529

- Monitor growth in children receiving prolonged, systemic triamcinolone therapy.
- Monitor for signs of negative nitrogen balance (e.g., muscle atrophy), especially in older or debilitated patients receiving prolonged therapy.
- Lab tests: Periodic serum electrolytes and blood glucose.
- Report to prescriber immediately if a local infection develops at site of topical application.
- Report symptoms of hypercortisolism or Cushing's syndrome (see Appendix F), hyperglycemia (see Appendix F), and glucosuria (e.g., polyuria). These may arise from systemic absorption after topical application, especially in children and if used over extensive areas for prolonged periods or if occlusive dressings are used.

Patient & Family Education
- Report promptly any of the following: Sore throat, fever, swelling of feet or ankles, or muscle weakness.
- Adhere to drug regimen; do not increase or decrease established regimen and do not discontinue abruptly.
- Asthmatics should report promptly worsening of asthma symptoms following oral inhalation.

TRIAMTERENE
(trye-am'ter-een)
Dyrenium
Classification: ELECTROLYTE AND WATER BALANCE AGENT; POTASSIUM-SPARING DIURETIC
Therapeutic: POTASSIUM-SPARING DIURETIC
Prototype: Spironolactone
Pregnancy Category: D

AVAILABILITY 50 mg, 100 mg capsules

ACTION & *THERAPEUTIC EFFECT*
Has weak diuretic action with a potassium-sparing effect. Promotes excretion of sodium, chloride, and carbonate. Blocks potassium excretion by direct action on distal renal tubule rather than by inhibiting aldosterone. *Has a diuretic action and a potassium-sparing effect.*

USES Treatment of edema, ascites, hypokalemia, hyperaldosteronism.

CONTRAINDICATIONS Hypersensitivity to triamterene; anuria, severe or progressive kidney disease or dysfunction; severe liver disease; diabetic neuropathy; elevated serum potassium; severe electrolyte or acid-base imbalance; pregnancy (category D).
CAUTIOUS USE Impaired kidney or liver function; gout; history of gouty arthritis; DM; history of kidney stones; older adults; lactation.

ROUTE & DOSAGE

Edema, Ascites, Hypokalemia
Adult: **PO** 50–100 mg b.i.d. (max: 300 mg/day)
Renal Impairment Dosage Adjustment
CrCl less than 10 mL/min: Avoid use

ADMINISTRATION

Oral
- Empty capsule and give with fluid or mix with food, if patient cannot swallow capsule.
- Give drug with or after meals to prevent or minimize nausea.
- Schedule doses to prevent interruption of sleep from diuresis

(e.g., with or after breakfast if a single dose is taken, or no later than 6 p.m. if more than one dose is prescribed). Consult prescriber.

- Withdraw drug gradually in patients on prolonged or high-dose therapy in order to prevent rebound increased urinary excretion of potassium.
- Store in tight, light-resistant containers at 15°–30° C (59°–86° F) unless otherwise directed.

ADVERSE EFFECTS (≥1%) **GI:** Diarrhea, nausea, vomiting, and other GI disturbances. **CNS:** Dizziness, headache, dry mouth, <u>anaphylaxis</u>, weakness, muscle cramps. **Skin:** Pruritus, rash, photosensitivity. **CV:** Hypotension (large doses). **Metabolic:** *Hyperkalemia* and other electrolyte imbalances, elevated BUN, elevated uric acid (patients predisposed to gouty arthritis), hyperchloremic acidosis. **Hematologic:** Blood dyscrasias: Granulocytopenia, eosinophilia, megaloblastic anemia in patients with reduced folic acid stores (e.g., hepatic cirrhosis).

DIAGNOSTIC TEST INTERFERENCE Pale blue fluorescence in urine interferes with *fluorometric assay* of *quinidine* and *lactic dehydrogenase activity.* Treatment of edema, ascites, hypokalemia, hyperaldosteronism.

INTERACTIONS Drug: May increase **lithium** levels, thus increasing its toxicity; **indomethacin** may decrease renal elimination of triamterene; ANGIOTENSIN-CONVERTING ENZYME (ACE) INHIBITORS, other POTASSIUM-SPARING DIURETICS may cause hyperkalemia. **Food:** High potassium foods may increase risk of hyperkalemia.

PHARMACOKINETICS Absorption: Rapidly but variably from GI tract.

Onset: 2–4 h. **Duration:** 7–9 h. **Metabolism:** In liver to active and inactive metabolites. **Elimination:** In urine. **Half-Life:** 100–150 min.

NURSING IMPLICATIONS

Assessment & Drug Effects

- Monitor BP during periods of dosage adjustment. Hypotensive reactions, although rare, have been reported. Take care with ambulation, particularly for older adults.
- Weigh patient under standard conditions, prior to drug initiation and daily during therapy.
- Diuretic response usually occurs on first day of therapy; maximum effect may not occur for several days.
- Monitor and report oliguria and unusual changes in I&O ratio. Consult prescriber regarding allowable fluid intake.
- Be alert for S&S of kidney stone formation; reported in patients taking high doses or who have low urine volume and increased urine acidity.
- Lab tests: Baseline and periodic serum potassium and other electrolytes; periodic kidney function tests with known or suspected renal insufficiency; periodic blood studies with prolonged therapy or with cirrhosis to monitor for megaloblastic anemia.
- Observe for S&S of hyperkalemia (see Appendix F), particularly in patients with renal insufficiency, on high-dose or prolonged therapy, older adults, and those with diabetes.
- Monitor diabetics closely for loss of glycemic control.

Patient & Family Education

- Do not use salt substitutes; unlike most diuretics, triamterene promotes potassium retention.

Common adverse effects in *italic*, life-threatening effects <u>underlined</u>; generic names in **bold**; classifications in SMALL CAPS; ♣ Canadian drug name; ☺ Prototype drug

1531

- Do not restrict salt; there is a possibility of low-salt syndrome (hyponatremia). Consult prescriber.
- Report significant fatigue or weakness, malaise, fever, sore throat, or mouth and unusual bleeding or bruising to prescriber.
- Be aware that drug may cause photosensitivity; avoid exposure to sun and sunlamps.
- Drug may impart a harmless pale blue fluorescence to urine.

TRIAZOLAM
(trye-ay′zoe-lam)

Halcion
Classifications: SEDATIVE-HYPNOTIC; ANXIOLYTIC; BENZODIAZEPINE
Therapeutic: SEDATIVE-HYPNOTIC; ANTIANXIETY
Prototype: Lorazepam
Pregnancy Category: X
Controlled Substance: Schedule IV

AVAILABILITY 0.125 mg, 0.25 mg tablets

ACTION & THERAPEUTIC EFFECT
Blockade of cortical and limbic arousal results in hypnotic activity. *Decreases sleep latency and number of nocturnal awakenings, decreases total nocturnal wake time, and increases duration of sleep.*

USES Short-term management of insomnia characterized by difficulty in falling asleep, frequent wakeful periods. Following long-term use, tolerance or adaptation may develop.

CONTRAINDICATIONS Hypersensitivity to triazolam and benzodiazepines; ethanol intoxication; suicidal ideations; pregnancy (category X), lactation.
CAUTIOUS USE Depression; bipolar disorder; dementia; psychosis; myasthenia gravis; Parkinson's

disease; debilitated patients; patients with suicidal tendency; impaired kidney or liver function; chronic pulmonary insufficiency; sleep apnea; older adults.

ROUTE & DOSAGE

Insomnia
Adult: **PO** 0.125–0.25 mg at bedtime (max: 0.5 mg/day)
Geriatric: **PO** 0.0625–0.125 mg at bedtime

ADMINISTRATION

Oral
- Give immediately before bed; onset of drug action is rapid.
- Do not exceed recommended doses.
- Store at 15°–30° C (59°–86° F).

ADVERSE EFFECTS (≥1%) **CNS:** *Drowsiness,* lightheadedness, headache, dizziness, ataxia, visual disturbances, confusional states, *memory impairment, "rebound insomnia," anterograde amnesia,* paradoxical reactions, minor changes in EEG patterns. **GI:** Nausea, vomiting, constipation.

INTERACTIONS Drug: Alcohol, CNS DEPRESSANTS, ANTICONVULSANTS, **nefazodone,** BENZODIAZEPINES potentiate CNS depression; **cimetidine** increases triazolam plasma levels, thus increasing its toxicity; may decrease antiparkinsonism effects of **levodopa. Herbal: Kava, valerian** may potentiate sedation. **St. John's wort** may decrease efficacy. **Food: Grapefruit juice** (greater than 1 qt/day) may increase plasma concentrations and adverse effects.

PHARMACOKINETICS Absorption: Readily from GI tract. **Onset:** 15–30 min. **Peak:** 1–2 h. **Duration:** 6–8 h. **Distribution:** Crosses placenta;

Common adverse effects in *italic*, life-threatening effects underlined; generic names in **bold**; classifications in SMALL CAPS; ♣ Canadian drug name; ◎ Prototype drug

distributed into breast milk. **Elimination:** In urine. **Half-Life:** 2–3 h.

NURSING IMPLICATIONS

Assessment & Drug Effects

- Be aware that signs of developing tolerance or adaptation (with long-term use) include increased daytime anxiety, increased wakefulness during last one third of the night.
- Lab tests: Obtain periodic blood counts, urinalysis, and blood chemistries during long-term use.
- Evaluate smoking habit. As with other benzodiazepines, smoking may decrease hypnotic effects.
- Monitor for symptoms of overdosage: Slurred speech, somnolence, confusion, impaired coordination, and coma.

Patient & Family Education

- Do not drive or engage in potentially hazardous activities until response to drug is known.
- Avoid use of alcohol or other CNS depressants while on this drug; they may increase sedative effects.
- Do not stop taking drug suddenly, especially if you are subject to seizures. Withdrawal symptoms may occur and range from mild dysphoria to more serious symptoms (e.g., tremors, abdominal and muscle cramps, convulsions). Consult prescriber for schedule to discontinue therapy.
- Do not increase dose without prescriber's advice because of toxic potential of drug.

TRIFLUOPERAZINE HYDROCHLORIDE

(trye-floo-oh-per'a-zeen)

Novoflurazine ♦, **Solazine ♦**, **Terfluzine ♦**

Classification: ANTIPSYCHOTIC; PHENOTHIAZIDE

Therapeutic: ANTIPSYCHOTIC
Prototype: Chlorpromazine
Pregnancy Category: C

AVAILABILITY 1 mg, 2 mg, 5 mg, 10 mg tablets

ACTION & *THERAPEUTIC EFFECT*
Phenothiazine with antipsychotic effects thought to be related to blockade of postsynaptic dopamine receptors in the brain. *Effectiveness indicated by increase in mental and physical activity.*

USES Management of schizophrenia, short term for anxiety.

CONTRAINDICATIONS Hypersensitivity to phenothiazines or sulfites; comatose states; CNS depression; ethanol intoxication; blood dyscrasias; hematologic disease, bone marrow depression; dementia in elderly; preexisting liver disease; lactation.

CAUTIOUS USE Previously detected breast cancer; history of QT prolongation; significant cardiac disease or pulmonary disease; compromised respiratory function; seizure disorders; impaired liver function; pregnancy (category C); children younger than 6 y.

ROUTE & DOSAGE

Schizophrenia

Adult: **PO** 1–2 mg b.i.d., may increase up to 20 mg/day
Child (6–12 y): **PO** 1 mg 1–2 times/day, may increase up to 15 mg/day

Anxiety

Adult: **PO** 1–2 mg 1–2 times per day (max: 6 mg/day)

ADMINISTRATION

Oral

- Dilute oral concentrate just before administration with about 60–120

Common adverse effects in *italic*, life-threatening effects underlined; generic names in **bold**; classifications in SMALL CAPS; ♦ Canadian drug name; ❷ Prototype drug

1533

mL suitable diluent (e.g., water, fruit juices, carbonated beverage, milk, soups, puddings).

- Crush tablet and give with fluid or mix with food if patient will not or cannot swallow pill.
- Store in light-resistant container at 15°–30° C (59°–86° F) unless otherwise directed.

ADVERSE EFFECTS (≥1%) **CNS:** *Drowsiness,* insomnia, dizziness, agitation, *extrapyramidal effects,* neuroleptic malignant syndrome. **Special Senses:** Nasal congestion, *dry mouth,* blurred vision, pigmentary retinopathy. **Hematologic:** Agranulocytosis. **Skin:** Photosensitivity, skin rash, sweating. **GI:** Constipation. **CV:** Tachycardia, *hypotension.* **Respiratory:** Depressed cough reflex. **Endocrine:** Gynecomastia, galactorrhea.

INTERACTIONS Drug: Alcohol and other CNS DEPRESSANTS add to CNS depression. **Herbal: Kava** may increase risk and severity of dystonic reactions.

PHARMACOKINETICS Absorption: Well absorbed from GI tract. **Onset:** Rapid onset. **Peak:** 2–3 h. **Duration:** Up to 12 h. **Metabolism:** In liver. **Elimination:** In bile and feces.

NURSING IMPLICATIONS

Assessment & Drug Effects

- Monitor HR and BP. Hypotension is a common adverse effect.
- Hypotension and extrapyramidal effects (especially akathisia and dystonia) are most likely to occur in patients receiving high doses or in older adults. Withhold drug and notify prescriber if patient has dysphagia, neck muscle spasm, or if tongue protrusion occurs.
- Monitor I&O ratio and bowel elimination pattern. Check for

abdominal distention and pain. Encourage adequate fluid intake as prophylaxis for constipation and xerostomia.

- Lab tests: Periodic CBC and serum prolactin, especially with prolonged therapy.
- Agitation, jitteriness, and sometimes insomnia may simulate original psychotic symptoms. These adverse effects may disappear spontaneously.
- Expect maximum therapeutic response within 2–3 wk after initiation of therapy.

Patient & Family Education

- Take drug as prescribed; do not alter dosing regimen or stop medication without consulting prescriber.
- Consult prescriber about use of any OTC drugs during therapy.
- Do not take alcohol and other depressants during therapy.
- Avoid potentially hazardous activities such as driving or operating machinery, until response to drug is known.
- Cover as much skin surface as possible with clothing when you **must be** in direct sunlight. Use an SPF higher than 12 sunscreen on exposed skin.
- Urine may be discolored or reddish brown and this is harmless.

TRIFLURIDINE
(trye-flure′i-deen)

Viroptic
Classification: ANTIVIRAL
Therapeutic: ANTIVIRAL
Pregnancy Category: C

AVAILABILITY 1% ophth solution

ACTION & *THERAPEUTIC EFFECT*
Pyrimidine nucleoside that appears to inhibit viral DNA synthesis and

viral replication. *Active against herpes simplex virus (HSV) types 1 and 2, vaccinia virus, and certain strains of adenovirus.*

USES Topically to eyes for treatment of primary keratoconjunctivitis and recurring epithelial keratitis caused by herpes simplex virus types 1 and 2. Also for other herpetic ophthalmic infections including stromal keratitis, uveitis, and for infections caused by vaccinia and *Adenovirus,* but clinical effectiveness has not been established.

CONTRAINDICATIONS Lactation.
CAUTIOUS USE Dry eye syndrome; pregnancy (category C); children younger than 6 y.

ROUTE & DOSAGE

Viral Infections of Eye

Adult: **Ophthalmic** 1 drop 1% ophthalmic solution into affected eye q2h during waking hours until healing (reepithelialization) has occurred (max: 9 drops/day); when healing appears to be complete, dosage reduced to 1 drop q4h during waking hours for an additional 7 days (max: 5 drops/day); continuous administration beyond 21 days not recommended

ADMINISTRATION

Instillation
▪ Wait several minutes between applications when used concurrently with other eye drops.
▪ Store refrigerated at 2°–8° C (36°–46° F) unless otherwise directed.

ADVERSE EFFECTS (≥1%) **Special Senses:** Mild transient burning or stinging, mild irritation of conjunctiva or cornea, photophobia, edema of eyelids and cornea, punctal occlusion, superficial punctate keratopathy, epithelial keratopathy, stromal edema, keratitis sicca, hyperemia, increased intraocular pressure.

PHARMACOKINETICS Absorption: Following topical application to eye, trifluridine penetrates cornea and aqueous humor (inflammation enhances penetration). Systemic absorption does not appear to be significant.

NURSING IMPLICATIONS

Assessment & Drug Effects
▪ Expect epithelial eye infections to respond to therapy within 2–7 days, with complete healing occurring in 1–2 wk.

Patient & Family Education
▪ Inform prescriber of progress and keep follow-up appointments. Herpetic eye infections have a tendency to recur and can lead to corneal damage if not adequately treated.

TRIHEXYPHENIDYL HYDROCHLORIDE
(trye-hex-ee-fen′i-dill)

Aparkane ✦, **Apo-Trihex** ✦, **Novohexidyl** ✦
Classifications: CENTRALLY ACTING CHOLINERGIC RECEPTOR ANTAGONIST; ANTIPARKINSON AGENT; ANTISPASMODIC
Therapeutic: ANTIPARKINSON; ANTISPASMODIC
Prototype: Benztropine
Pregnancy Category: C

AVAILABILITY 2 mg, 5 mg tablets; 2 mg/5 mL elixir

ACTION & *THERAPEUTIC EFFECT*
Thought to act by blocking excess of acetylcholine at certain cerebral

Common adverse effects in *italic*, life-threatening effects underlined; generic names in **bold**; classifications in SMALL CAPS; ✦ Canadian drug name; ❂ Prototype drug

1535

synaptic sites. Relaxes smooth muscle by direct effect and by atropinelike blocking action on the parasympathetic nervous system. *Diminishes the characteristic tremor of Parkinson's disease.*

USES Symptomatic treatment of all forms of parkinsonism (arteriosclerotic, idiopathic, postencephalitic). Also to prevent or control drug-induced extrapyramidal disorders.
UNLABELED USES Huntington's chorea, spasmodic torticollis.

CONTRAINDICATIONS Narrow-angle glaucoma; tardive dyskinesia; lactation.
CAUTIOUS USE History of drug hypersensitivities; arteriosclerosis; hypertension; cardiac disease, kidney or liver disorders; myasthenia gravis; alcoholism; obstructive diseases of GI or genitourinary tracts; older adults with prostatic hypertrophy; pregnancy (category C). Safe use in children is not established.

ROUTE & DOSAGE

Parkinsonism
Adult: **PO** 1 mg day 1, 2 mg day 2, then increase by 2 mg q3–5days up to 6–10 mg/day in 3 or more divided doses (max: 15 mg/day)
Extrapyramidal Effects
Adult: **PO** 5–15 mg/day in divided doses

ADMINISTRATION

Oral
- Give before or after meals, depending on how patient reacts. Older adults and patients prone to excessive salivation (e.g., postencephalitic parkinsonism) may prefer to take drug after meals. If drug causes

excessive mouth dryness, it may be better given before meals, unless it causes nausea.
- Do not crush or chew sustained release capsules. These **must be** swallowed whole.
- Store at 15°–30° C (59°–86° F) in tight container unless otherwise directed.

ADVERSE EFFECTS (≥1%) **GI:** *Dry mouth, nausea,* constipation. **Special Senses:** *Blurred vision,* mydriasis, photophobia, angle-closure glaucoma. **Urogenital:** Urinary hesitancy or retention. **CNS:** *Dizziness, nervousness,* insomnia, drowsiness, confusion, agitation, delirium, psychotic manifestations, euphoria. **CV:** Tachycardia, palpitations, hypotension, orthostatic hypotension. **Body as a Whole:** Hypersensitivity reactions.

INTERACTIONS Drug: Reduces therapeutic effects of **chlorpromazine, haloperidol,** PHENOTHIAZINES; increases bioavailability of **digoxin;** MAO INHIBITORS potentiate actions of trihexyphenidyl. **Herbal: Betel nut** may increase risk of extrapyramidal symptoms.

PHARMACOKINETICS Absorption: Readily from GI tract. **Onset:** Within 1 h. **Peak:** 2–3 h. **Duration:** 6–12 h. **Elimination:** In urine.

NURSING IMPLICATIONS

Assessment & Drug Effects
- Be aware that incidence and severity of adverse effects are usually dose related and may be minimized by dosage reduction. Older adults appear more sensitive to usual adult doses.
- Monitor vital signs. Pulse is a particularly sensitive indicator of response to drug. Report tachycardia, palpitations, paradoxical bradycardia, or fall in BP.

- Assess for and report severe CNS stimulation (see ADVERSE EFFECTS) that occurs with high doses, and in patients with arteriosclerosis, or those with history of hypersensitivity to other drugs.
- In patients with severe rigidity, tremors may appear to be accentuated during therapy as rigidity diminishes.
- Monitor daily I&O if patient develops urinary hesitancy or retention. Voiding before taking drug may relieve problem.
- Check for abdominal distention and bowel sounds if constipation is a problem.

Patient & Family Education
- Learn measures to relieve drug-induced dry mouth; rinse mouth frequently with water and suck ice chips, sugarless gum, or hard candy. Maintain adequate total daily fluid intake.
- Avoid excessive heat because drug suppresses perspiration and, therefore, heat loss.
- Do not to engage in potentially hazardous activities requiring alertness and skill. Drug causes dizziness, drowsiness, and blurred vision. Help walking may be indicated.

TRIMETHOBENZAMIDE HYDROCHLORIDE

(trye-meth-oh-ben′za-mide)

Tigan
Classification: ANTIEMETIC
Therapeutic: ANTIEMETIC
Prototype: Prochlorperazine
Pregnancy Category: C

AVAILABILITY 300 mg capsules; 100 mg/mL injection

ACTION & *THERAPEUTIC EFFECT*
Primary locus of action is thought to be the chemoreceptor trigger zone (CTZ) in medulla. *Less effective than phenothiazine antiemetics but produces fewer adverse effects.*

USES Control of nausea and vomiting.

CONTRAINDICATIONS Uncomplicated vomiting in viral illness; parenteral use in children or infants; rectal administration in prematures and newborns; known sensitivity to benzocaine (in suppository) or to similar local anesthetics.

CAUTIOUS USE In presence of high fever, dehydration, electrolyte imbalance; pregnancy (category C), lactation; children.

ROUTE & DOSAGE

Nausea and Vomiting
Adult: **PO** 300 mg t.i.d. or q.i.d. **IM** 200 mg t.i.d. or q.i.d. prn
Geriatric: May require dose reduction or changes in interval.

ADMINISTRATION

Oral
- Empty capsule and give with water or mix with food if patient cannot swallow capsule.

Intramuscular
- Give IM deep into upper outer quadrant of buttock.
- Minimize possibility of irritation and pain by avoiding escape of solution along needle track. Use Z-track technique. Rotate injection sites.

ADVERSE EFFECTS (≥1%) **Body as a Whole:** Hypersensitivity reactions (including allergic skin eruptions), muscle cramps, pain, stinging, burning, redness, irritation

T

Common adverse effects in *italic*, life-threatening effects underlined; generic names in **bold**; classifications in SMALL CAPS; ♣ Canadian drug name; ⊙ Prototype drug

1537

at IM site; local irritation following rectal administration. **CNS:** Pseudoparkinsonism. **CV:** Hypotension. **GI:** Diarrhea, exaggeration of nausea, acute hepatitis, jaundice.

INTERACTIONS Drug: Alcohol and other CNS DEPRESSANTS add to depressant activity; PHENOTHIAZINES may precipitate extrapyramidal syndrome.

PHARMACOKINETICS Onset: 10–40 min PO; 15 min IM. **Duration:** 3–4 h PO; 2–3 h IM. **Elimination:** 30–50% of dose excreted unchanged in urine within 48–72 h.

NURSING IMPLICATIONS

Assessment & Drug Effects

- Monitor BP. Hypotension may occur particularly in surgical patients receiving drug parenterally.
- Report promptly and stop drug therapy if an acute febrile illness accompanies or begins during therapy.
- Antiemetic effect of drug may obscure diagnoses of GI or other pathologic conditions or signs of toxicity from other drugs.

Patient & Family Education

- Report promptly to prescriber onset of rash or other signs of hypersensitivity (see Appendix F). Discontinue drug immediately.
- Do not drive or engage in potentially hazardous activities until response to drug is known.
- Do not drink alcohol or alcoholic beverages during therapy with this drug.

TRIMETHOPRIM ⓟ

(trye-meth′oh-prim)

Primsol
Classification: URINARY TRACT ANTI-INFECTIVE

Therapeutic: URINARY TRACT ANTI-INFECTIVE
Pregnancy Category: C

AVAILABILITY 100 mg, 200 mg tablets; 50 mg/5 mL liquid

ACTION & THERAPEUTIC EFFECT
Anti-infective and folic acid antagonist with slow bactericidal action. Binds and interferes with bacterial cell growth. *Effective against most common UTI pathogens. Most pathogens causing UTI are in normal vaginal and fecal flora. Effective in treatment of acute otitis media.*

USES Initial episodes of acute uncomplicated UTIs, acute otitis media in children.

UNLABELED USES Treatment and prophylaxis of chronic and recurrent UTI in both men and women; treatment in conjunction with dapsone of initial episodes of *Pneumocystis carinii* pneumonia; treatment of travelers' diarrhea.

CONTRAINDICATIONS Megaloblastic anemia secondary to folate deficiency; creatinine clearance less than 15 mL/min, impaired kidney or liver function; possible folate deficiency; children with fragile X chromosome associated with mental retardation.

CAUTIOUS USE Renal disease; mild or moderate renal impairment; pregnancy (category C), lactation; children younger than 6 mo.

ROUTE & DOSAGE

Urinary Tract Infection
Adult: **PO** 100 mg b.i.d. or 200 mg once/day
Child: **PO** 2–3 mg/kg q12h × 10 days
Acute Otitis Media
Child (older than 6 mo): **PO** 10 mg/kg divided q12h × 10 days

Common adverse effects in *italic*, life-threatening effects underlined; generic names in **bold**; classifications in SMALL CAPS; ♣ Canadian drug name; ⓟ Prototype drug

Travelers' Diarrhea
Adult: **PO** 200 mg b.i.d.

ADMINISTRATION

Oral
- Give with 240 mL (8 oz) of fluid if not contraindicated.
- Store at 15°–30° C (59°–86° F) in dry, light-protected place.

ADVERSE EFFECTS (≥1%) **GI:** Epigastric discomfort, nausea, vomiting, glossitis, abnormal taste sensation. **Hematologic:** Neutropenia, *megaloblastic anemia,* methemoglobinemia, <u>leukopenia, thrombocytopenia</u> (rare). **Skin:** *Rash, pruritus,* <u>exfoliative dermatitis</u>, photosensitivity. **Body as a Whole:** Fever. **Metabolic:** Increased serum transaminases (ALT, AST), bilirubin, creatinine, BUN.

DIAGNOSTIC TEST INTERFERENCE
Interferes with serum ***methotrexate assays*** that use a competitive binding protein technique with a bacterial dihydrofolate reductase as the binding protein. May cause falsely elevated ***creatinine*** values when ***Jaffe reaction*** is used.

INTERACTIONS Drug:
May inhibit **phenytoin** metabolism causing increased levels.

PHARMACOKINETICS Absorption:
Almost completely absorbed from GI tract. **Peak:** 1–4 h. **Distribution:** Widely distributed, including lung, saliva, middle ear fluid, bile, bone, CSF; crosses placenta; appears in breast milk. **Metabolism:** In liver. **Elimination:** 80% in urine unchanged. **Half-Life:** 8–11 h.

NURSING IMPLICATIONS

Assessment & Drug Effects
- Lab tests: Obtain C&S tests before trimethoprim therapy is initiated.

Obtain periodic urine cultures, BUN, creatinine clearance, CBC, Hgb, and Hct.
- Reinforce necessity to adhere to established drug regimen. Recurrent infection after terminating prophylactic treatment of UTI may occur even after 6 mo of therapy.
- Assess urinary pattern during treatment. Altered pattern (frequency, urgency, nocturia, retention, polyuria) may reflect emerging drug resistance, necessitating change of drug regimen. Periodically check for bladder distention.
- Be alert for toxic effects on bone marrow, particularly in older adults, malnourished, alcoholic, pregnant, or debilitated patients. Recognize and report signs of infection or anemia.
- Drug-induced rash, a common adverse effect, is usually maculopapular, pruritic, or morbilliform and appears 7–14 days after start of therapy with daily doses of 200 mg or less.

Patient & Family Education
- Drink fluids liberally (2000–3000 mL/day, if not contraindicated) to help flush out urinary bacteria.
- Report pain and hematuria to prescriber immediately.
- Do not postpone voiding even though increases in fluid intake may cause more frequent urination.
- Do not use douches or sprays during treatment periods; practice careful perineal hygiene to prevent reinfection.
- Report to prescriber promptly any symptoms of a blood disorder (fever, sore throat, pallor, purpura, ecchymosis).
- Consult prescriber if severe traveler's diarrhea does not respond to 3–5 days therapy (i.e.,

Common adverse effects in *italic*, life-threatening effects <u>underlined</u>; generic names in **bold**; classifications in SMALL CAPS; ♣ Canadian drug name; ◒ Prototype drug

1539

persistence of symptoms of severe nausea, abdominal pain, diarrhea with mucus or blood, and dehydration).

- Drug-induced rash, a common adverse effect, may appear 7–14 days after start of therapy. Report rash to prescriber for evaluation.

TRIMIPRAMINE MALEATE
(tri-mip′ra-meen)

Surmontil
Classification: TRICYCLIC ANTIDEPRESSANT
Therapeutic: TRICYCLIC ANTIDEPRESSANT (TCA)
Prototype: Imipramine
Pregnancy Category: C

AVAILABILITY 25 mg, 50 mg, 100 mg capsules

ACTION & *THERAPEUTIC EFFECT*
Tricyclic antidepressant (TCA) useful in depression associated with anxiety and sleep disturbances. Recent studies suggest strong, active H_2-receptor antagonism is a characteristic of TCAs. *More effective in alleviation of endogenous depression than other depressive states.*

USES Treatment of major depression.

CONTRAINDICATIONS Hypersensitivity to tricyclic antidepressants; prostatic hypertrophy; during recovery period after MI; AV block; QT prolongation; bundle-branch block; ileus; MAOI therapy; suicide ideation.

CAUTIOUS USE Schizophrenia, electroshock therapy, psychosis, bipolar disease; Parkinson's disease; seizure disorders; increased intraocular pressure; history of urinary retention; history of narrow-angle glaucoma; hyperthyroidism; suicidal tendency; cardiovascular, liver, thyroid, kidney

disease; pregnancy (category C), lactation. Safe use in children not established.

ROUTE & DOSAGE

Depression
Adult: **PO** 75–100 mg/day in divided doses, may increase gradually up to 300 mg/day if needed **PO Maintenance Dose** Usually 50–150 mg/day
Geriatric: **PO** 25 mg at bedtime, may increase q3days (max: 100 mg/day)

Pharmacogenetic Dosage Adjustment
Poor CYP2D6 metabolizer: Start with 30% of dose

ADMINISTRATION

Oral
- Give with food to decrease gastric distress.
- Store in tightly closed container at 15°–30° C (59°–86° F) unless otherwise specified.

ADVERSE EFFECTS (≥1%) **CNS:** Seizures, tremor, confusion, *sedation.* **Special Senses:** Blurred vision. **CV:** Tachycardia, *orthostatic hypotension,* hypertension. **GI:** *Xerostomia, constipation,* paralytic ileus. **Urogenital:** *Urinary retention.* **Skin:** Photosensitivity, sweating.

INTERACTIONS Drug: May decrease some antihypertensive response to ANTIHYPERTENSIVES; CNS DEPRESSANTS, **alcohol,** HYPNOTICS, BARBITURATES, SEDATIVES potentiate CNS depression; may increase hypoprothrombinemic effect of ORAL ANTICOAGULANTS; **levodopa,** SYMPATHOMIMETICS (e.g., **epinephrine, norepinephrine**) increase possibility of sympathetic hyperactivity with hypertension and hyperpyrexia; with MAO INHIBITORS, possibility of severe reactions, toxic psychosis, cardiovascular instability; **methylphenidate** increases plasma TCA levels; THYROID AGENTS may increase possibility of arrhythmias; **cimetidine** may increase plasma TCA levels. **Herbal: Ginkgo** may decrease seizure threshold; **St. John's wort** may cause **serotonin** syndrome.

PHARMACOKINETICS Absorption: Rapidly absorbed from GI tract. **Peak:** 2 h. **Metabolism:** In liver (CYP2D6). **Elimination:** In urine and feces. **Half-Life:** 9.1 h.

NURSING IMPLICATIONS

Assessment & Drug Effects

- Assess vital signs (BP and pulse rate) during adjustment period of tricyclic antidepressant (TCA) therapy. If BP falls more than 20 mm Hg or if there is a sudden increase in pulse rate, withhold medication and notify prescriber.
- Orthostatic hypotension may be sufficiently severe to require protective assistance when patient is ambulating. Instruct patient to change position from recumbency to standing slowly and in stages.
- Monitor for changes in behavior that may indicate increased incidence of suicidality.
- Report fine tremors, a distressing extrapyramidal adverse effect, to prescriber.
- Monitor bowel elimination pattern and I&O ratio. Severe constipation and urinary retention are potential problems, especially in older adults. Advise increased fluid intake to at least 1500 mL/day (if allowed).
- Inspect oral membranes daily with high-dose therapy. Urge

outpatient to report symptoms of stomatitis or xerostomia.
- Regulate environmental temperature and patient's clothing carefully; drug may cause intolerance to heat or cold.

Patient & Family Education

- Be aware that your ability to perform tasks requiring alertness and skill may be impaired.
- Do not use OTC drugs unless approved by prescriber.
- Understand that the actions of both alcohol and trimipramine are increased when used together during therapy and for up to 2 wk after the TCA is discontinued. Consult prescriber about safe amounts of alcohol, if any, that can be taken.
- Be aware that the effects of barbiturates and other CNS depressants may also be enhanced by trimipramine.
- Expect that therapeutic response will be delayed because TCAs have a "lag period" of 2–4 wk. Increased dosage does not shorten period but rather increases incidence of adverse reactions. Keep prescriber advised and do not interrupt therapy.

TRIPTORELIN PAMOATE

(trip-tor'e-lyn)

Trelstar Depot, Trelstar LA
Classification: GONADOTROPIN-RELEASING HORMONE (GNRH) AGONIST ANALOG
Therapeutic: GNRH AGONIST ANALOG
Prototype: Leuprolide acetate
Pregnancy Category: X

AVAILABILITY 3.75 mg, 11.25 mg, 22.5 mg injection

Common adverse effects in *italic*, life-threatening effects underlined; generic names in **bold**; classifications in SMALL CAPS; ♣ Canadian drug name; ⊕ Prototype drug

1541

ACTION & *THERAPEUTIC EFFECT*
Synthetic luteinizing releasing hormone agonist (LHRH or GnRH) with greater potency than naturally occurring luteinizing hormone. Potent inhibitor of gonadotropin secretion that causes decreased formation of testosterone. *In men, the level of serum testosterone is equivalent to a surgically castrated man.*

USES Palliative treatment of advanced prostate cancer.
UNLABELED USES Breast cancer, endometriosis, infertility, hirsutism.

CONTRAINDICATIONS Hypersensitivity to triptorelin, other LHRH agonists, or LHRH; dysfunctional uterine bleeding; pregnancy (category X), lactation.
CAUTIOUS USE Prostatic carcinoma; hepatic or renal dysfunction; patients with impending spinal cord compression or severe urogenital disorder; premenstrual syndrome; renal insufficiency. Safety and effectiveness in children have not been established.

ROUTE & DOSAGE

Prostate Cancer
Adult: **IM** 3.75 mg qmo OR 11.25 mg q12w or 22.5 mg q24w

ADMINISTRATION

Intramuscular
▪ Give deep into a large muscle.

ADVERSE EFFECTS (≥1%) **Body as a Whole:** *Hot flushes,* pain, leg pain, fatigue. **CV:** Hypertension. **GI:** Diarrhea, vomiting. **Hematologic:** Anemia. **Musculoskeletal:** Skeletal pain. **CNS:** Headache, dizziness, insomnia, impotence, emotional lability. **Skin:** Pruritus. **Urogenital:** Urinary retention, UTI. **Other:** Pain at injection site.

DIAGNOSTIC TEST INTERFERENCE May interfere with tests for ***pituitary-gonadal function.***

PHARMACOKINETICS Peak: 1–3 h. **Duration:** 1 mo. **Metabolism:** Unknown. **Elimination:** Eliminated by liver and kidneys. **Half-Life:** 3 h.

NURSING IMPLICATIONS

Assessment & Drug Effects
▪ Monitor for S&S of disease flare, especially during the first 1–2 wk of therapy: Increased bone pain, blood in urine, urinary obstruction, or symptoms of spinal compression.
▪ Lab tests: Periodic serum testosterone, PSA, acid phosphatase levels; urinary and serum calcium; urinary calcium/creatinine ratio; lipid profile in those at risk for atherosclerosis.

Patient & Family Education
▪ Disease flare (see ASSESSMENT & DRUG EFFECTS) is a common, temporary adverse effect of therapy; however, symptoms may become serious enough to report to the prescriber.
▪ Notify prescriber promptly of the following: S&S of an allergic reaction (itching, hives, swelling of face, arms, or legs; tingling in mouth or throat, tightness in chest or trouble breathing); weakness or loss of muscle control; rapid weight gain.

TROPICAMIDE
(troe-pik′a-mide)
Pregnancy Category: C
See Appendix A-1.

TROSPIUM CHLORIDE
(tro-spi′um)
Sanctura, Sanctura XR
Classifications: ANTICHOLINERGIC; ANTIMUSCARINIC; ANTISPASMODIC

Common adverse effects in *italic*, life-threatening effects underlined; generic names in **bold**; classifications in SMALL CAPS; ♣ Canadian drug name; ● Prototype drug

Therapeutic: URINARY SMOOTH MUS-
CLE RELAXANT
Prototype: Oxybutynin
Pregnancy Category: C

AVAILABILITY 20 mg tablets; 60 mg
extended release capsule

ACTION & *THERAPEUTIC EFFECT*
Antagonizes the effect of acetylcholine on muscarinic receptors
in smooth muscle. Its parasympa-
tholytic action reduces tonus of
the smooth muscle of the bladder.
*Decreases urinary frequency, ur-
gency, and urge incontinence in
patients with overactive bladders.*

USES Treatment of overactive (neurogenic) bladder, urinary incontinence.

CONTRAINDICATIONS Hypersensitivity to trospium; patients with
or at risk for urinary retention; uncontrolled narrow-angle glaucoma;
gastric tension, GI obstruction, ileus, pyloric stenosis, toxic megacolon, severe ulcerative colitis.

CAUTIOUS USE Significant bladder
obstruction, closed-angle glaucoma; BPH; ulcerative colitis, GERD,
intestinal atony; myasthenia gravis,
autonomic neuropathy; moderate or severe hepatic dysfunction;
severe renal insufficiency, renal
failure; older adults; pregnancy
(category C), lactation. Safety in
children has not been established.

ROUTE & DOSAGE

Overactive Bladder

Adult: **PO** 20 mg twice daily OR
60 mg (extended release) daily
Geriatric (75 y or older): **PO**
20 mg once daily at bedtime if
anti-cholinergic adverse effects
are intolerable

**Renal Impairment Dosage
Adjustment**

CrCl less than 30 mL/min: 20 mg
once daily at bedtime

ADMINISTRATION

Oral
- Give at least 1 h before meals or
on an empty stomach.
- Store at 20°–25° C (66°–77° F).

ADVERSE EFFECTS (≥1%) **Body
as a Whole:** Fatigue. **CNS:** Headache. **GI:** *Dry mouth, constipation,*
abdominal pain, dyspepsia, flatulence. **Special Senses:** Dry eyes,
blurred vision. **Urogenital:** Urinary
retention, urinary tract infection.

INTERACTIONS Drug: Increased
anticholinergic adverse effects with
ANTICHOLINERGIC AGENTS.

PHARMACOKINETICS Absorption:
Less than 10% absorbed orally.
Peak: 5–6 h. **Elimination:** Primarily
in feces. **Half-Life:** 20 h.

NURSING IMPLICATIONS

Assessment & Drug Effects
- Monitor bowel and bladder function. Report urinary hesitancy or
significant constipation.
- Withhold drug and notify
prescriber if urinary retention
develops.
- Monitor for and report worsening
of GI symptoms in those with
GERD.

Patient & Family Education
- Report promptly any of the following: Signs of an allergic reaction, (e.g., itching or hives),
blurred vision or difficulty focusing, confusion, dizziness, difficulty passing urine.
- Moderate intake of tea, coffee,
caffeinated sodas, and alcohol to
minimize side effects of this drug.

T

Common adverse effects in *italic*, life-threatening effects underlined; generic names
in **bold**; classifications in SMALL CAPS; ✦ Canadian drug name; ❂ Prototype drug

1543

- Avoid situations in which overheating is likely, as drug may impair sweating, which is a normal cooling mechanism.
- Do not engage in hazardous activities until response to the drug is known.

ULIPRISTAL
(u-li-pris'tal)
Ella
Classification: PROGESTERONE AGONIST/ANTAGONIST; POSTCOITAL CONTRACEPTIVE
Therapeutic: POSTCOITAL CONTRACEPTIVE
Pregnancy Category: X

AVAILABILITY 30 mg tablet

ACTION & *THERAPEUTIC EFFECT*
A selective progesterone receptor modulator with antagonistic and partial agonistic effects. It binds to the progesterone receptor preventing the binding of progesterone. *If taken immediately before ovulation, it delays follicular rupture and inhibits ovulation. It may also alter the endometrium and interfere with implantation of a fertilized ovum.*

USES Postcoital contraception after unprotected intercourse or a known or suspected contraceptive failure.

CONTRAINDICATIONS Known or suspected pregnancy (category X); lactation.

ROUTE & DOSAGE

Postcoital Contraception
Adult: **PO** 30 mg within 120 h of unprotected intercourse or known or suspected contraceptive failure; may repeat within 3 h if patient vomits initial dose

ADMINISTRATION
Oral
- Give within 120 h of unprotected intercourse.
- Store at controlled room temperature 20°–25° C (68°–77° F).

ADVERSE EFFECTS (≥5%) Body as a Whole: *Abdominal and upper abdominal pain.* **CNS:** Fatigue, dizziness, *headache.* **GI:** *Nausea.* **Urogenital:** Dysmenorrhea.

INTERACTIONS Drug: Co-administration of CYP3A4 inducers (e.g., **carbamazepine, phenobarbital, phenytoin, rifampin, topiramate**) can decrease the levels of ulipristal. Co-administration of strong CYP3A INHIBITORS (e.g., **itraconazole, ketoconazole**) can increase the levels of ulipristal. **Food: Grapefruit juice** can increase the levels of ulipristal. **Herbal: St. John's wort** can decrease the levels of ulipristal.

PHARMACOKINETICS Peak: 1 h. **Distribution:** 94% plasma protein bound. **Metabolism:** Hepatic oxidation by CYP3A4 to active metabolite. **Half-Life:** 32.4 h.

NURSING IMPLICATIONS
Assessment & Drug Effects
- Monitor for vomiting. Drug may be re-administered if patient vomits within 3 h of initial dose.

Patient & Family Education
- Ulipristal is not intended for repeated use. It should not replace conventional means of contraception.
- Certain concurrently taken drugs may interfere with the contraceptive effects of ulipristal. Consult prescriber.
- Ulipristal may interfere with the length of the next menstrual cycle

(i.e., cycle may occur sooner or later than expected).

- Ulipristal does not protect against sexually transmitted diseases.

USTEKINUMAB
(us-te-kin'u-mab)

Stelara

Classification: BIOLOGIC RESPONSE MODIFIER; MONOCLONAL ANTIBODY; INTERLEUKIN-12 AND INTERLEUKIN-23 RECEPTOR ANTAGONIST
Therapeutic: IMMUNOSUPPRESSANT
Prototype: Basiliximab
Pregnancy Category: B

AVAILABILITY 45 mg/0.5 mL, 90 mg/mL solution for injection

ACTION & *THERAPEUTIC EFFECT*
A human monoclonal antibody that disrupts IL-12 and IL-23 mediated signaling and cytokine cascades. *Inhibits inflammatory and immune responses associated with plaque psoriasis thereby improving psoriasis.*

USES Treatment of moderate to severe plaque psoriasis.

CONTRAINDICATIONS Clinically significant, active infection (e.g., active TB); reversible posterior leukoencephalopathy syndrome (RPLS).
CAUTIOUS USE Chronic infection or history of recurrent infection (e.g., latent TB); known malignancy; phototherapy; older adults; pregnancy (category B); lactation. Safety and efficacy under 18 y not established.

ROUTE & DOSAGE

Psoriasis

Adults 100 kg or less: **Subcutaneous** 45 mg initially and 4wk later, followed by 45 mg q12wk

Patients over 100 kg: **Subcutaneous** 90 mg initially and 4 wk later, followed by 90 mg q12wk

ADMINISTRATION

Subcutaneous

- Solution may contain a few small translucent or white particles. Do NOT shake.
- Note: Needle cover on prefilled syringe is natural rubber and should not be handled by latex-sensitive persons.
- When using a single-use vial, withdraw dose using a 27 gauge syringe with a one-half inch needle.
- Rotate injection sites and do not inject into an area that is tender, bruised, red, or irritated.
- Store unopened vials upright in refrigerator. Discard unused portions in single-use vials.

ADVERSE EFFECTS (≥1%) CNS: Depression, dizziness, fatigue, headache. **GI:** Diarrhea, pharyngolaryngeal pain. **Musculoskeletal:** Back pain, myalgia. **Respiratory:** Nasopharyngitis, upper respiratory tract infection. **Skin:** Injection-site erythema, pruritis.

INTERACTIONS Drug: LIVE VACCINES should not be administered concurrently with ustekinumab therapy. Closely monitor use with narrow therapeutic index drugs (e.g., **warfarin, cyclosporine**).

PHARMACOKINETICS Peak: 7–13.5 days. **Metabolism:** Degraded to smaller proteins. **Half-Life:** 14.9–45.6 days.

NURSING IMPLICATIONS

Assessment & Drug Effects
- Monitor for and promptly report S&S of TB or other infection.
- Lab tests: Prior to initiation of therapy, screen for latent TB.

Common adverse effects in *italic*, life-threatening effects underlined; generic names in **bold**; classifications in SMALL CAPS; ◆ Canadian drug name; 🅿 Prototype drug

1545

- Monitor neurologic status and report promptly: Seizures, problems with vision, headaches, or confusion.
- Note: BCG vaccines should not be given during treatment, for one year prior to initiating treatment, or one year following discontinuation of treatment.

Patient & Family Education
- Do not accept vaccinations with live vaccines while taking this drug. Note that non-live vaccines may not be effective if given during a course of ustekinumab.
- You should be tested for TB prior to taking this drug.
- Report promptly S&S of infection including: Chills, fever, cough, shortness of breath, diarrhea, stomach or abdominal pain, burning on urination, sores anywhere on your body, unexplained or excessive fatigue.
- Report immediately seizures, problems with vision, headaches, or confusion.

VACCINIA IMMUNE GLOBULIN (VIG-IV)
(vac-cin'i-a)
Classifications: BIOLOGIC RESPONSE MODIFIER; IMMUNOGLOBULIN
Therapeutic: IMMUNOGLOBULIN
Prototype: Immune globulin
Pregnancy Category: C

AVAILABILITY 50 mg/mL injection

ACTION & THERAPEUTIC EFFECT
Vaccinia immune globulin, VIG (VIG-IV) is a purified human immunoglobulin G (IgG). It is derived from adult human plasma collected from donors who received booster immunizations with the smallpox vaccine. VIG (VIG-IV) contains high titers of antivaccinia antibodies. *VIG is effective in the treatment of smallpox vaccine adverse reactions secondary to continued vaccinia virus replication after vaccination.*

USES Prevention of serious complications of smallpox vaccine; treatment of progressive vaccinia; severe generalized vaccinia; eczema vaccinatum; vaccinia infection in patients with skin conditions (e.g., burns, impetigo, varicella-zoster, poison ivy, or eczematous skin lesions); treatment or modification of aberrant infections induced by vaccinia virus.

CONTRAINDICATIONS Predisposition to acute renal failure (i.e., preexisting renal insufficiency, DM, volume depletion, sepsis proteinemia, patients older than 65 y); AIDS; chronic skin conditions; bone marrow suppression; chemotherapy; radiation therapy; corticosteroid therapy, eczema; hematologic disease, thrombosis; hypotension; herpes infection; postvaccinal encephalitis; aseptic meningitis syndrome (AMS); pulmonary edema; lactation.

CAUTIOUS USE Renal impairment; autoimmune disease; cardiomyopathy, impaired cardiac output, cardiac disease, history of hypercoagulation; pregnancy (category C). Safe use in children not established.

ROUTE & DOSAGE

Vaccinia
Adults: **IV** 100–500 mg/kg
Renal Impairment Dosage Adjustment
Maximum dose: 400 mg/kg

ADMINISTRATION

Intravenous
PREPARE: IV Infusion: No dilution required. Use solution as supplied.

ADMINISTER: **IV Infusion:** Begin infusion within 6 h of entering the vial. Complete infusion within 12 h of entering the vial. ▪ Use inline filter (0.22 microns), infusion pump, and dedicated IV line [may infuse into a preexisting catheter if it contains NS, D2.5W, D5W, D10W, or D20W (or any combination of these)]. ▪ Infuse at 1 mL/kg/h the first 30 min, increase to 2 mL/kg/h the next 30 min, and then increase to 3 mL/kg/h until infused.

ADVERSE EFFECTS (≥1%) **Body as a Whole:** Injection site reaction. **CNS:** Dizziness, *headache.* **GI:** Abdominal pain, nausea, vomiting. **Musculoskeletal:** Arthralgia, back pain. **Respiratory:** Upper respiratory infection. **Skin:** Erythema, flushing.

INTERACTIONS Drug: May interfere with the immune response to LIVE VIRUS VACCINES. Vaccination with LIVE VIRUS VACCINES should be deferred until approximately 6 mo after administration of VIG-IV.

PHARMACOKINETICS Half-Life: 22 days.

NURSING IMPLICATIONS

Assessment & Drug Effects
▪ Monitor vital signs continuously during infusion, especially after infusion rate changes.
▪ Slow infusion rate for any of the following: Flushing, chills, muscle cramps, back pain, fever, nausea, vomiting, arthralgia, and wheezing.
▪ Discontinue infusion, institute supportive measures, and notify prescriber for any of the following: Increase in heart rate, increase in respiratory rate, shortness of breath, rales or other signs of anaphylaxis.
▪ Have loop diuretic available for management of fluid overload.

Patient & Family Education
▪ Promptly report any discomfort that develops while drug is being infused.

VALACYCLOVIR HYDROCHLORIDE
(val-a-cy'clo-vir)
Valtrex
Classification: ANTIVIRAL
Therapeutic: ANTIVIRAL
Prototype: Acyclovir
Pregnancy Category: B

AVAILABILITY 500 mg, 1 g tablets

ACTION & *THERAPEUTIC EFFECT*
An antiviral agent hydrolyzed in the intestinal wall or liver to acyclovir, which interferes with viral DNA synthesis. *Active against herpes simplex virus types 1 (HSV-1) and 2 (HSV-2), varicella zoster virus, and cytomegalovirus. Inhibits viral replication.*

USES Herpes zoster (shingles) in immunocompetent adults. Treatment and suppression of recurrent genital herpes; suppression of recurrent herpes in HIV-positive patients; treatment of cold sores.

CONTRAINDICATIONS Hypersensitivity to, or intolerance of valacyclovir or acyclovir.
CAUTIOUS USE Renal impairment, patients receiving nephrotoxic drugs, advanced HIV disease, allogeneic bone marrow transplant and renal transplant recipients, treatment of disseminated herpes zoster, immunocompromised patients, pregnancy (category B); lactation; children younger than 2 y.

V

ROUTE & DOSAGE

Herpes Zoster

Adult: **PO** 1 g (2 × 500 mg) t.i.d. for 7 days, start within 48 h of onset of zoster rash

Renal Impairment Dosage Adjustment

CrCl 30–49 mL/min: 1 g q12h; *10–29 mL/min:* 1 g q24h; *less than 10 mL/min:* 500 mg q24h

Treatment of Recurrent Genital Herpes

Adult: **PO** 500 mg b.i.d. × 3 days

Renal Impairment Dosage Adjustment

CrCl 29 mL/min or less: 500 mg daily

Suppression of Recurrent Genital Herpes

Adult: **PO** 1 g daily; concurrent HIV infection 500 mg b.i.d.

Treatment of Cold Sores

Adult/Child: **PO** 2 g 12 h × 2 doses

Chickenpox

Adolescent/Child (older than 2 y): **PO** 20 mg/kg t.i.d. × 5 days (max: 1 g t.i.d.)

ADMINISTRATION

Oral

- Start drug as soon as possible after diagnosis of herpes zoster, preferably within 48 h of onset of rash.
- Note: Dosage reduction is recommended for patients with renal impairment.
- Give valacyclovir after hemodialysis.
- Store at 15°–30° C (59°–86° F).

ADVERSE EFFECTS (≥1%) **CNS:** *Headache,* weakness, somnolence, dizziness, fatigue, lethargy, confusion. **GI:** *Nausea, vomiting, diarrhea,* abdominal pain, dyspepsia, flatulence. **Urogenital:** Glomerulonephritis, renal tubular damage, acute renal failure. **Skin:** Rash, urticaria, pruritus.

INTERACTIONS Drug: Probenecid, cimetidine decrease valacyclovir elimination. **Zidovudine** may cause increased drowsiness and lethargy.

PHARMACOKINETICS Absorption: Rapidly absorbed from GI tract; 54% reaches systemic circulation as acyclovir. **Peak:** 1.5 h. **Distribution:** 13.5–17.9% bound to plasma proteins; distributes into plasma, cerebrospinal fluid, saliva, and major body organs; crosses placenta; excreted in breast milk. **Metabolism:** Rapidly converted to acyclovir during first pass through intestine and liver. **Elimination:** 40–50% in urine. **Half-Life:** 2.5–3.3 h.

NURSING IMPLICATIONS

Assessment & Drug Effects

- Monitor kidney function in patients with kidney impairment or those receiving potentially nephrotoxic drugs.
- Monitor for S&S of hypersensitivity; if present, withhold drug and notify prescriber.

Patient & Family Education

- Be aware of potential adverse effects and do not discontinue drug until full course is completed.
- Note: Post-herpes pain is likely to be present for several months after completion of therapy.

VALGANCICLOVIR HYDROCHLORIDE

(val-gan-ci′clo-vir)

Valcyte
Classification: ANTIVIRAL

V

Common adverse effects in *italic*, life-threatening effects underlined; generic names in **bold**; classifications in SMALL CAPS; ✚ Canadian drug name; ⊙ Prototype drug

Therapeutic: ANTIVIRAL
Prototype: Acyclovir
Pregnancy Category: C

AVAILABILITY 450 mg tablets, powder for oral solution

ACTION & *THERAPEUTIC EFFECT* Rapidly converted to ganciclovir by intestinal and hepatic enzymes. In cells infected with cytomegalovirus (CMV), ganciclovir is converted to ganciclovir triphosphate that inhibits viral DNA synthesis. *Effective antiviral that prevents replication of viral CMV DNA.*

USES Treatment of CMV retinitis; prevention of CMV disease in high-risk kidney, kidney-pancreas, and heart transplant patients (not effective in liver transplants).
UNLABELED USES Esophagitis, herpesvirus infection.

CONTRAINDICATIONS Hypersensitivity to valganciclovir, ganciclovir, or acyclovir; dental work; antimicrobial resistance; neutropenia, thrombocytopenia; females of childbearing age; lactation.
CAUTIOUS USE Impaired kidney function; dental disease; anemia; leukopenia; bone marrow depression; irradiation; older adults; pregnancy (category C). Safety and efficacy in children are not established.

ROUTE & DOSAGE

Cytomegalovirus Prophylaxis
Adult/Adolescent: **PO** 900 mg once daily, starting within 10 days of transplantation until 100–200 days post-transplantation
Child/Infant (over 4 mo): See package insert for dose calculation formula (max: 900 mg)

Cytomegalovirus Retinitis Treatment
Adult: **PO** 900 mg b.i.d. × 21 days then 900 mg daily

Renal Impairment Dosage Adjustment
CrCl 40–59 mL/min: 450 mg b.i.d. (induction) or daily (maintenance); 25–39 mL/min: 450 mg daily (induction) or q2days (maintenance); 10–24 mL/min: 450 mg q2days (induction) or twice weekly (maintenance)

ADMINISTRATION
Oral
- Exercise caution in handling tablets, powder for oral solution, and the oral solution. Do not crush or break tablets. Avoid direct contact of any form of the drug with skin or mucous membranes. Note that adults should receive tablets, not the oral solution.
- Give with food.
- Do not give to patients on hemodialysis.
- Store tablets at 25°–30° C (77°–86° F). Store oral solution at 2°–8° C (36°–46° F) for no longer than 49 days.

ADVERSE EFFECTS (≥1%) **Body as a Whole:** *Fever,* local and systemic infections, hypersensitivity reactions. **CNS:** *Headache, insomnia,* peripheral neuropathy, paresthesia, convulsions, psychosis, confusion, hallucinations, agitation. **GI:** *Diarrhea, nausea, vomiting, abdominal pain.* **Hematologic:** *Neutropenia, anemia,* thrombocytopenia, pancytopenia, bone marrow suppression, aplastic anemia. **Special Senses:** *Retinal detachment.*

INTERACTIONS Drug: ANTINEOPLASTIC AGENTS, **amphotericin B,**

V

Common adverse effects in *italic*, life-threatening effects underlined; generic names in **bold**; classifications in SMALL CAPS; ♣ Canadian drug name; ⊙ Prototype drug

1549

didanosine, trimethoprim-sulfamethoxazole (TMP-SMZ), dapsone, pentamidine, probenecid, zidovudine may increase bone marrow suppression and other toxic effects of valganciclovir; may increase risk of nephrotoxicity from cyclosporine; ANTIRETROVIRAL AGENTS may decrease valganciclovir levels; valganciclovir may increase levels and toxicity of ANTIRETROVIRAL AGENTS; may increase risk of seizures due to imipenem-cilastatin.

PHARMACOKINETICS Absorption: Well absorbed from GI tract, 60% reaches systemic circulation as ganciclovir. **Onset:** 3–8 days. **Peak:** 1–3h. **Duration:** Clinical relapse can occur 14 days to 3.5 mo after stopping therapy; positive blood and urine cultures recur 12–60 days after therapy. **Distribution:** Distributes throughout body including CSF, eye, lungs, liver, and kidneys; crosses placenta in animals; not known if distributed into breast milk. **Metabolism:** Metabolized in intestinal wall to ganciclovir, ganciclovir is not metabolized. **Elimination:** 94–99% of dose is excreted unchanged in urine. **Half-Life:** 4 h.

NURSING IMPLICATIONS

Assessment & Drug Effects

- Withhold drug and notify prescriber for any of the following: Absolute neutrophil count less than 500 cells/mm³, platelet count less than 25,000/mm³, hemoglobin less than 8 g/dL, declining creatinine clearance.
- Monitor for S&S of bronchospasm in asthma patients; notify prescriber immediately.
- Lab tests: Baseline and frequent serum creatinine or creatinine clearance, CBC with differential, platelet count, Hct and Hgb.

Patient & Family Education

- Schedule ophthalmologic follow-up examinations at least every 4–6 wk while being treated with valganciclovir.
- Keep all scheduled appointments for laboratory tests.
- Do not drive or engage in potentially hazardous activities until response to drug is known.
- Report any of the following immediately: Unexpected bleeding, infection.
- Use effective methods of contraception (barrier and other types) during and for at least 90 days following treatment.
- Discontinue drug and notify prescriber immediately in the event of pregnancy.

VALPROIC ACID (DIVALPROEX SODIUM, SODIUM VALPROATE) ⊙

(val-proe'ic)

Depacon, Depakene, Depakote, Depakote ER, Depakote Sprinkle, Epival ✦, Stavzor
Classifications: ANTICONVULSANT; GAMMA-AMINOBUTYRIC ACID (GABA) INHIBITOR
Therapeutic: ANTICONVULSANT
Pregnancy Category: D

AVAILABILITY 250 mg capsules; 125 mg sprinkle capsules; 125 mg, 250 mg, 500 mg delayed release tablets; 500 mg sustained release tablets; 250 mg/5 mL syrup; 100 mg/mL injection

ACTION & THERAPEUTIC EFFECT
Anticonvulsant with increased bioavailability of the inhibitory neurotransmitter gamma-aminobutyric acid (GABA) to brain neurons. It may also suppress repetitive neuronal firing through inhibition of voltage-sensitive

sodium channels. *Depresses abnormal neuron discharges in the CNS, thus decreasing seizure activity.*

USES Alone or with other anticonvulsants in management of absence (petit mal) and mixed seizures; mania; migraine headache prophylaxis. **UNLABELED USES** Status epilepticus refractory to IV diazepam, petit mal variant seizures, febrile seizures in children, other types of seizures including psychomotor (temporal lobe), myoclonic, akinetic and tonic-clonic seizures, photosensitivity seizures, and those refractory to other anticonvulsants.

CONTRAINDICATIONS Hypersensitivity to valproate sodium; hepatic disease or significant hepatic function impairment; urea cycle disorders. Hyperammonemia, encephalopathy; suicidal ideations; thrombocytopenia, patient with bleeding disorders or liver dysfunction or disease, cirrhosis, hepatitis; pancreatitis; congenital metabolic disorders, those with severe seizures, or on multiple anticonvulsant drugs; AIDS; fluid restriction or decreased food intake; pregnancy (category D); lactation. **CAUTIOUS USE** History of suicidal tendencies; history of kidney disease, renal impairment or failure; history of mild or moderate liver impairment; congenital metabolic disorders, those with severe epilepsy, use as sole anticonvulsant drug; HIV; hypoalbuminemia; organic brain syndrome or mental retardation; use as sore agent; older adults; children younger than 10 y.

ROUTE & DOSAGE

Note: May need to increase dose when converting from immediate release to extended release products

Management of Seizures

Adult/Child (10 y or older): **PO/IV** 10–15 mg/kg/day in divided doses when total daily dose greater than 250 mg, increase at 1 wk intervals by 5–10 mg/kg/ day until seizures are controlled or adverse effects develop (max: 60 mg/kg/day) *Conversion of PO to IV:* Give normal dose in divided doses q6h

Migraine Headache Prophylaxis

Adult: **PO** 250 mg b.i.d. (max: 1000 mg/day) or **Depakote ER** 500 mg daily x 1 wk, may increase to 1000 mg daily

Mania

Adult: **PO** 750 mg/day administered in divided doses OR 25 mg/kg/day using extended release tabs

Hepatic Impairment Dosage Adjustment

Dose reduction recommended

Renal Impairment Dosage Adjustment

Severe impairment may require close monitoring

ADMINISTRATION

Oral

- Give tablets and capsules whole; instruct patient to swallow whole and not to chew. Instruct to swallow sprinkle capsules whole or sprinkle entire contents on teaspoonful of soft food, and instruct to not chew food.
- Avoid using a carbonated drink as diluent for the syrup because it will release drug from delivery vehicle; free drug painfully irritates oral and pharyngeal membranes.
- Reduce gastric irritation by administering drug with food.

V

Common adverse effects in *italic*, life-threatening effects underlined; generic names in **bold;** classifications in SMALL CAPS; ♣ Canadian drug name; ⊙ Prototype drug

1551

Intravenous

PREPARE: **IV Infusion:** Dilute each dose in 50 mL or more of D5W, NS, or LR.
ADMINISTER: **IV Infusion:** Give a single dose over at least 60 min (20 mg/min or less). ▪ Avoid rapid infusion.
INCOMPATIBILITIES **Solution/additive:** Should avoid mixing with other drugs.

ADVERSE EFFECTS (≥1%) **CNS:** Breakthrough seizures, *sedation, drowsiness,* dizziness, increased alertness, hallucinations, emotional upset, aggression; <u>deep coma, death (with overdose)</u>. **GI:** *Nausea, vomiting, indigestion (transient),* hypersalivation, anorexia with weight loss, increased appetite with weight gain, abdominal cramps, diarrhea, constipation, <u>liver failure, pancreatitis</u>. **Hematologic:** *Prolonged bleeding time,* leukopenia, lymphocytosis, thrombocytopenia, hypofibrinogenemia, <u>bone marrow depression</u>, anemia. **Skin:** Skin rash, photosensitivity, transient hair loss, curliness or waviness of hair. **Endocrine:** Irregular menses, secondary amenorrhea. **Metabolic:** Hyperammonemia (usually asymptomatic) hyperammonemic encephalopathy in patients with urea cycle disorders. **Respiratory:** Pulmonary edema (with overdose).

DIAGNOSTIC TEST INTERFERENCE Valproic acid produces false-positive results for ***urine ketones,*** elevated ***AST, ALT, LDH,*** and ***serum alkaline phosphatase,*** prolonged ***bleeding time,*** altered ***thyroid function tests.***

INTERACTIONS Drug: Alcohol and other CNS DEPRESSANTS potentiate depressant effects; other ANTICONVULSANTS, BARBITURATES increase or decrease anticonvulsant and BARBITURATE levels; **haloperidol, loxapine, maprotiline,** MAOIS, PHENOTHIAZINES, THIOXANTHENES, TRICYCLIC ANTIDEPRESSANTS can increase CNS depression or lower seizure threshold; **aspirin, dipyridamole, warfarin** increase risk of spontaneous bleeding; **clonazepam** may precipitate absence seizures; SALICYLATES, **cimetidine, isoniazid** may increase valproic acid levels and toxicity. **Mefloquine** can decrease valproic acid levels; **meropenem** may decrease valproic acid levels; **cholestyramine** may decrease absorption. **Herbal: Ginkgo** may decrease anticonvulsant effectiveness.

PHARMACOKINETICS Absorption: Readily from GI tract. **Peak:** 1–4 h valproic acid; 3–5 h divalproex. **Therapeutic Range:** 50–100 mcg/mL. **Distribution:** Crosses placenta; distributed into breast milk. **Metabolism:** In liver. **Elimination:** Primarily in urine; small amount in feces and expired air. **Half-Life:** 5–20 h.

NURSING IMPLICATIONS

Assessment & Drug Effects

▪ Monitor for therapeutic effectiveness achieved with serum levels of valproic acid at 50–100 mcg/mL.
▪ Monitor patient alertness especially with multiple drug therapy for seizure control.
▪ Monitor patient carefully during dose adjustments and promptly report presence of adverse effects. Increased dosage is associated with frequency of adverse effects.
▪ Lab tests: Baseline platelet count, bleeding time, coagulation parameters, and serum ammonia, LFT, then repeat at least q2mo, especially during the first 6 mo of therapy.
▪ Multiple drugs for seizure control increase the risk of hyperammonemia, marked by lethargy, anorexia, asterixis, increased seizure frequency, and vomiting. Report such

Common adverse effects in *italic*, life-threatening effects <u>underlined</u>; generic names in **bold**; classifications in SMALL CAPS; ✦ Canadian drug name; ⊙ Prototype drug

symptoms promptly to prescriber. If they persist with decreased dosage, the drug will be discontinued.

Patient & Family Education

- Do not discontinue therapy abruptly; such action could result in loss of seizure control. Consult prescriber before you stop or alter dosage regimen.
- Note to diabetic patients: Drug may cause a false-positive test for urine ketones. Notify prescriber if this occurs.
- Notify prescriber promptly if spontaneous bleeding or bruising occurs (e.g., petechiae, ecchymotic areas, otorrhagia, epistaxis, melena).
- Withhold dose and notify prescriber for following symptoms: Visual disturbances, rash, jaundice, light-colored stools, protracted vomiting, diarrhea. Fatal liver failure has occurred in patients receiving this drug.
- Avoid alcohol and self-medication with other depressants during therapy.
- Consult prescriber before using any OTC drugs during anticonvulsant therapy, especially drugs containing aspirin or sedatives and medications for hay fever or other allergies.
- Do not drive or engage in potentially hazardous activities until response to drug is known.

VALRUBICIN
(val-roo′bi-sin)

Valstar
Classification: ANTINEOPLASTIC; (ANTIBIOTIC) ANTHRACYCLINE
Therapeutic: ANTINEOPLASTIC
Prototype: Doxorubicin hydrochloride
Pregnancy Category: C

AVAILABILITY 40 mg/mL solution

ACTION & THERAPEUTIC EFFECT A cytotoxic agent that inhibits the incorporation of nucleosides in DNA and RNA, resulting in extensive chromosomal damage. It arrests the cell cycle of the HIV virus in the G2 phase by interfering with normal DNA action of topoisomerase II, which is responsible for separating DNA strands and resealing them. *Valrubicin has higher antitumor efficacy and lower toxicity than doxorubicin.*

USES Intravesical therapy of BCG-refractory carcinoma *in situ* of the urinary bladder.

CONTRAINDICATIONS Hypersensitivity to valrubicin, doxorubicin, anthracyclines, or castor oil; patients with a perforated bladder, concurrent UTI, active infection; severe irritable bladder symptoms; severe myelosuppression; lactation.
CAUTIOUS USE Within 2 wk of a transureteral resection; compromised bladder mucosa; mild to moderate myelosuppression; history of bleeding disorders; GI disorders, renal impairment; pregnancy (category C).

ROUTE & DOSAGE

BCG-Refractory Bladder Carcinoma *in situ*
Adult: **Intravesically** 800 mg once/wk × 6 wk

ADMINISTRATION

Instillation
Avoid skin reactions by using gloves during preparation/administration.

- Use only glass, polypropylene, or polyolefin containers and tubing.

PREPARE: Slowly warm 4 vials (5 mL each) to room temperature. ▪ When a precipitate is initially present, warm vials in hands until solution clears. ▪ Add

V

Common adverse effects in *italic*, life-threatening effects underlined; generic names in **bold**; classifications in SMALL CAPS; ◆ Canadian drug name; ⊙ Prototype drug

1553

contents of 4 vials to 55 mL of 0.9% NaCl injection to yield 75 mL of diluted solution.

ADMINISTER: Aseptically insert a urethral catheter and drain the bladder. ▪ Use gravity drainage to instill valrubicin slowly over several min. ▪ Withdraw catheter; instruct patient not to void for 2 h. ▪ Note: Do not leave a clamped catheter in place.

▪ Refrigerate. Do not freeze.

ADVERSE EFFECTS (≥1%) **Body as a Whole:** Abdominal pain, asthenia, back pain, fever, headache, malaise, myalgia. **CNS:** Dizziness. **CV:** Vasodilation. **GI:** Diarrhea, flatulence, nausea, vomiting. **Urogenital:** *Urinary frequency, urgency, dysuria, bladder spasm, hematuria, bladder pain, incontinence, cystitis, UTI,* nocturia, local burning, urethral pain, pelvic pain, gross hematuria, urinary retention. **Respiratory:** Pneumonia. **Skin:** Rash. **Other:** Anemia, hyperglycemia, peripheral edema.

PHARMACOKINETICS Absorption: Not absorbed. **Distribution:** Penetrates bladder wall. **Metabolism:** Not metabolized. **Elimination:** Almost completely excreted by voiding the instillate.

NURSING IMPLICATIONS

Assessment & Drug Effects

▪ Therapeutic effectiveness: Indicated by regression of the bladder tumor.
▪ Notify prescriber if bladder spasms with spontaneous discharge of valrubicin occur during/shortly after instillation.

Patient & Family Education

▪ Expect red-tinged urine during the first 24 h after administration.
▪ Report prolonged passage of red-colored urine or prolonged bladder irritation.

▪ Drink plenty of fluids during 48 h period following administration.
▪ Use reliable contraception during therapy period (approximately 6 wk).

VALSARTAN
(val-sar'tan)

Diovan

Classifications: RENIN ANGIOTENSIN SYSTEM ANTAGONIST; ANTIHYPERTENSIVE

Therapeutic: ANTIHYPERTENSIVE; ANGIOTENSIN II RECEPTOR (AT_1) ANTAGONIST

Prototype: Losartan
Pregnancy Category: D

AVAILABILITY 40 mg, 80 mg, 160 mg capsules

ACTION & *THERAPEUTIC EFFECT*
An angiotensin II receptor (type AT_1 receptor subtype) antagonist that blocks the angiotensin converting enzyme (ACE). It inhibits the binding of angiotensin II to the AT_1 subtype receptors found in many tissues (e.g., vascular smooth muscle, adrenal glands). Angiotensin II is a potent vasoconstrictor and primary vasoactive hormoné of the renin–angiotensin–aldosterone system (RAAS). *Blocking angiotensin II receptors results in vasodilation as well as decreasing the aldosterone-secreting effects of angiotensin II. These actions result in the antihypertensive effect of valsartan.*

USES Treatment of hypertension, heart failure.

CONTRAINDICATIONS Hypersensitivity to valsartan or losartan; severe heart failure with compromised renal function; volume depletion; history of ACE inhibitor–induced

Common adverse effects in *italic*, life-threatening effects <u>underlined</u>; generic names in **bold**; classifications in SMALL CAPS; ♣ Canadian drug name; ⊙ Prototype drug

angioedema; children with CrCl of 30 mL/min per 1.73 mm³; pregnancy (category D), lactation.

CAUTIOUS USE Severe renal impairment; mild to moderate hepatic impairment; biliary stenosis; renal artery stenosis; hypovolemia; hyperkalemia; transient hypotension; use prior to surgery; history of angioedema; congestive heart failure; children younger than 6 y.

ROUTE & DOSAGE

Hypertension

Adult: **PO** 80 mg daily (max: 320 mg daily)
Adolescent/Child (6–16 y): **PO** 1.3 mg/kg (up to 40 mg/day)

Heart Failure

Adult: **PO** Start with 40 mg b.i.d. and titrate up to 160 mg b.i.d.

Hemodialysis Dosage Adjustment

Adjustment not needed

ADMINISTRATION

Oral

- Give on an empty stomach.
- Correct volume depletion prior to initiation of therapy to prevent hypotension.
- Reduce dosage with severe hepatic or renal impairment.
- Note: Daily dose may be titrated up to 320 mg.
- Store at 15°–30° C (59°–86° F).

ADVERSE EFFECTS (≥1%) **Body as a Whole:** Arthralgia. **CNS:** Headache, dizziness. **GI:** Diarrhea, nausea. **Respiratory:** Cough, sinusitis. **Metabolic:** Hyperkalemia.

PHARMACOKINETICS Absorption: Rapidly from GI tract, 25% bioavailability. **Onset:** Blood pressure decreased in 2 wk. **Peak:** Plasma levels, 2–4 h; blood pressure effect 4 wk. **Distribution:** 99% protein bound. **Metabolism:** In the liver. **Elimination:** Primarily in feces. **Half-Life:** 6 h.

NURSING IMPLICATIONS

Assessment & Drug Effects

- Monitor BP periodically; take trough readings, just prior to the next scheduled dose, when possible.
- Lab tests: Periodic LFTs, BUN and creatinine, serum potassium, and CBC with differential.

Patient & Family Education

- Inform prescriber immediately if you become pregnant.
- Note: Maximum pressure lowering effect is usually evident between 2 and 4 wk after initiation of therapy.
- Notify prescriber of episodes of dizziness, especially those that occur when making position changes.
- Do not use potassium supplements or salt substitutes containing potassium without consulting the prescribing prescriber.

VANCOMYCIN HYDROCHLORIDE
(van-koe-mye'sin)

Vancocin
Classifications: ANTIBIOTIC; GLYCOPEPTIDE
Therapeutic: ANTIBIOTIC
Pregnancy Category: B

AVAILABILITY 125 mg, 250 mg capsules; 500 mg, 1 g, 5 g, 10 g injection

ACTION & THERAPEUTIC EFFECT Bactericidal action is due to inhibition of cell-wall biosynthesis and alteration of bacterial cell-membrane permeability and ribonucleic acid (RNA) synthesis. *Active against many gram-positive organisms.*

V

Common adverse effects in *italic*, life-threatening effects underlined; generic names in **bold**; classifications in SMALL CAPS; ◆ Canadian drug name; ☻ Prototype drug

1555

USES Parenterally for potentially life-threatening infections in patients allergic, nonsensitive, or resistant to other less toxic antimicrobial drugs. Used orally only in *Clostridium difficile* colitis (not effective by oral route for treatment of systemic infections).

CONTRAINDICATIONS Hypersensitivity to vancomycin, allergy to corn or corn products, previous hearing loss. **CAUTIOUS USE** Impaired kidney function, renal failure, renal impairment, hearing impairment; colitis, inflammatory disorders of the intestine; older adults; pregnancy (category B), lactation; children, neonates.

ROUTE & DOSAGE

Systemic Infections

Adult/Adolescent: **IV** 500 mg q6h or 1 g q12h or 15 mg/kg q12h
Child/Infant (older than 1 mo): **IV** 30–40 mg/kg/day divided q6–8h
Neonate: **IV** 10–15 mg/kg q8–24h

Clostridium difficile Colitis

Adult: **PO** 125–500 mg q6h
Child: **PO** 40 mg/kg/day divided q6h (max: 2 g/day)
Neonate: **PO** 10 mg/kg/day in divided doses

Surgical Prophylaxis (in Patients Allergic to Beta-Lactams)

Adult/Adolescent/Child (weight at least 27 kg): **IV** 10–15 mg/kg starting 1 h before surgery
Child (weight less than 27 kg): **IV** 20 mg/kg starting 1 h before surgery

Renal Impairment Dosage Adjustment

CrCl 40–60 mL/min: Dose q24h; *less than 40 mL/min:* Extend interval based on monitoring levels

Hemodialysis Dosage Adjustment

Not dialyzed

ADMINISTRATION

Oral

- May be given with or without food.
- Note: Some parenteral products may be administered orally; check manufacturer's package insert.

Intravenous

PREPARE: Intermittent: Reconstitute 500 mg vial or 1 g vial with 10 mL or 20 mL, respectively, of sterile water for injection to yield 50 mg/mL. ▪ Further dilute each 500 mg with at least 100 mL and each 1 g with at least 200 mL of D5W, NS, or LR.

ADMINISTER: Intermittent: Give a single dose at a rate of 10 mg/min or over NOT LESS than 60 min (whichever is longer). ▪ Avoid rapid infusion, which may cause sudden hypotension. ▪ Monitor IV site closely; necrosis and tissue sloughing will result from extravasation.

INCOMPATIBILITIES Solution/additive: Aminophylline, BARBITURATES, **aztreonam** (high concentration), **calcium chloride, chloramphenicol, chlorothiazide, dexamethasone, erythromycin, heparin, methicillin, sodium bicarbonate, warfarin. Y-site: Albumin, amphotericin B cholesteryl, aztreonam, bivalirudin, cefazolin, cefepime, cefotaxime, cefotetan, cefoxitin, ceftazidime, ceftriaxone, cefuroxime, drotrecogin, foscarnet, heparin, idarubicin, lansoprazole, nafcillin, omeprazole, piperacillin/tazobactam, sargramostim,**

V

ticarcillin, ticarcillin/clavu-
lanate, warfarin.

▪ Store oral and parenteral solutions in refrigerator for up to 14 days; after further dilution, parenteral solution is stable 24 h at room temperature.

ADVERSE EFFECTS (≥1%) **Special Senses:** Ototoxicity (auditory portion of eighth cranial nerve). **Urogenital:** Nephrotoxicity leading to uremia. **Body as a Whole:** Hypersensitivity reactions (chills, fever, skin rash, urticaria, shock-like state), anaphylactoid reaction with vascular collapse, superinfections, severe pain, thrombophlebitis at injection site, generalized tingling following rapid IV infusion. **Hematologic:** Transient leukopenia, eosinophilia. **GI:** Nausea, warmth. **Other:** Injection reaction that includes *hypotension accompanied by flushing and erythematous rash on face and upper body* ("red-neck syndrome") following rapid IV infusion.

INTERACTIONS Drug: Adds to toxicity of ototoxic and nephrotoxic drugs (AMINOGLYCOSIDES), **amphotericin B, colistin, capreomycin; cidofovir; cisplatin; cyclosporine; foscarnet; ganciclovir; IV pentamidine; polymyxin B; streptozocin; tacrolimus**). **Cholestyramine, colestipol** can decrease absorption of oral vancomycin; may increase risk of lactic acidosis with **metformin.**

PHARMACOKINETICS Absorption: Not absorbed. **Peak:** 30 min after end of infusion. **Distribution:** Diffuses into pleural, ascitic, pericardial, and synovial fluids; small amount penetrates CSF if meninges are inflamed; crosses placenta. **Elimination:** 80–90% of IV dose in urine within 24 h; PO dose excreted in feces. **Half-Life:** 4–8 h.

NURSING IMPLICATIONS
Assessment & Drug Effects

▪ Monitor BP and heart rate continuously through period of drug administration.

▪ Lab tests: Periodic urinalysis, LFTs, kidney functions, and hematologic studies. Monitor serial tests of vancomycin blood levels (peak and trough) in patients with borderline kidney function, in infants and neonates, and in patients older than 60 y.

▪ Assess hearing. Drug may cause damage to auditory branch (not vestibular branch) of eighth cranial nerve, with consequent deafness, which may be permanent.

▪ Be aware that serum levels of 60–80 mcg/mL are associated with ototoxicity. Tinnitus and high-tone hearing loss may precede deafness, which may progress even after drug is withdrawn. Older adults and those on high doses are especially susceptible.

▪ Monitor I&O: Report changes in I&O ratio and pattern. Oliguria or cloudy or pink urine may be a sign of nephrotoxicity (also manifested by transient elevations in BUN, albumin, and hyaline and granular casts in urine).

Patient & Family Education

▪ Notify prescriber promptly of ringing in ears.

▪ Adhere to drug regimen (i.e., do not increase, decrease, or interrupt dosage. The full course of prescribed drug therapy **must be** completed).

V

VARDENAFIL HYDROCHLORIDE
(var-den′a-fil hy-dro-chlo′ride)
Levitra, Staxyn
Classifications: IMPOTENCE AGENT; PHOSPHODIESTERASE (PDE) INHIBITOR; VASODILATOR

Common adverse effects in *italic*, life-threatening effects underlined; generic names in **bold**; classifications in SMALL CAPS; ♣ Canadian drug name; ⊚ Prototype drug

1557

Therapeutic: IMPOTENCE; PDE TYPE 5 INHIBITOR
Prototype: Sildenafil
Pregnancy Category: B

AVAILABILITY 2.5 mg, 5 mg, 10 mg, 20 mg tablets; 10 mg orally disintegrating tablet

ACTION & *THERAPEUTIC EFFECT*
Phosphodiesterases-5 (PDE5) is an enzyme that speeds up the degradation of cyclic guanosine monophosphate (cGMP), an enzyme needed to cause and maintain increased blood flow into the penis necessary for an erection. Vardenafil is a PDE5 inhibitor. *It enhances erectile function by increasing the amount of cGMP in the penis.*

USES Treatment of erectile dysfunction.

CONTRAINDICATIONS Hypersensitivity to vardenafil or sildenafil; QT prolongation, renal failure, severe renal impairment; retinitis pigmentosa; lactation.

CAUTIOUS USE CAD, MI, or stroke within 6 mo; hypotension, or hypertension; risk factors for CVA; anatomic deformity of the penis; subaortic stenosis; sickle cell anemia; leukemia; multiple myeloma; leukemia; coagulopathy; active bleeding or a peptic ulcer; coagulopathy; GERD; hepatitis, cirrhosis; older adults; pregnancy (category B).

ROUTE & DOSAGE

Erectile Dysfunction

Adult: **PO** 10 mg approximately 60 min before sexual activity. May increase to max 20 mg/day if needed. If taking ritonavir, max dose is 2.5 mg/72 h. If taking erythromycin, indinavir, itraconazole, ketoconazole, max dose is 2.5–5 mg/24 h.
Geriatric: **PO** Start with 5 mg 60 min before sexual activity (max: 20 mg/day)

Hepatic Impairment Dosage Adjustment

Moderate impairment: Reduce dose to 5 mg (max: 10 mg/day)

ADMINISTRATION

Oral
- Take approximately 1 h before expected intercourse, but preferably not after a heavy or high-fat meal.
- Store at 15°–30° C (59°–86° F).

ADVERSE EFFECTS (≥1%) **Body as a Whole:** *Flushing,* flu-like syndrome, back pain, anaphylactoid reactions, asthenia, facial edema, pain, paresthesias. **CNS:** *Headache,* dizziness, insomnia, somnolence, vertigo. **CV:** Angina, hypertension, hypotension, MI, orthostatic hypotension, palpitations, syncope, sinus tachycardia. **GI:** Dyspepsia, nausea, vomiting, abdominal pain, abnormal liver function tests, diarrhea, dysphagia, esophagitis, gastritis, GERD, xerostomia. **Metabolic:** Increased creatine kinase. **Musculoskeletal:** Arthralgia, myalgia, hypertonia, hyperesthesia. **Respiratory:** Rhinitis, sinusitis, dyspnea, epistaxis, pharyngitis. **Skin:** Photosensitivity, rash, pruritus, sweating. **Special Senses:** Tinnitus, sudden vision loss, blurred vision, changes in color vision. **Urogenital:** Ejaculation dysfunction.

INTERACTIONS Drug: May potentiate hypotensive effects of NITRATES, **alfuzosin, doxazosin, prazosin, tamsulosin, terazosin; amiodarone, dofetilide, procainamide, quinidine, sotalol** may increase QT_c interval leading to arrhythmias;

V

erythromycin (and other MAC-ROLIDES), **indinavir, itraconazole, ketoconazole,** PROTEASE INHIBITORS, **ritonavir, voriconazole** may increase level and toxicity of vardenafil.

PHARMACOKINETICS Absorption: Rapidly absorbed, 15% reaches systemic circulation. **Onset:** Within 1 h. **Peak:** 0.5–2 h. **Metabolism:** In liver by CYP3A4. **Elimination:** Primarily in feces (90–95%). **Half-Life:** 4–5 h.

NURSING IMPLICATIONS

Assessment & Drug Effects

- Monitor CV status and report angina or other S&S of cardiac dysfunction.
- Lab tests: Baseline and periodic LFTs.

Patient & Family Education

- Do not take more than once a day and never take more than the prescribed dose.
- Do not take this drug without consulting prescriber if you are taking drugs called "alpha-blockers" or "nitrates" or any other drugs for high blood pressure, chest pain, or enlarged prostate.
- Report promptly any of the following: Palpitations, chest pain, back pain, difficulty breathing, or shortness of breath; dizziness or fainting; changes in vision; dizziness; swollen eyelids; muscle aches; painful or prolonged erection (lasting longer than 4 h); skin rash, or itching.

VARENICLINE
(var-en′i-cline)

Chantix
Classifications: SMOKING DETERRENT; NICOTINIC RECEPTOR AGONIST
Therapeutic: SMOKING CESSATION
Prototype: Nicotine
Pregnancy Category: C

AVAILABILITY 0.5 mg, 1 mg tablet

ACTION & *THERAPEUTIC EFFECT*
Nicotine increases dopamine release in the brain and cravings for nicotine are stimulated by low levels of dopamine during periods of abstinence. Varenicline is a partial agonist at nicotinic acetylcholine receptors (nAChRs), the sites responsible for the dopamine effects of nicotine. It partially stimulates these receptors to produce a modest level of dopamine but blocks nicotine from binding to many of the nicotinic receptor sites. *By blocking nicotinic receptors, it reduces effects of nicotine in cases where patient relapses and uses tobacco.*

USES Adjunct for smoking cessation in patients experiencing nicotine withdrawal.

CONTRAINDICATIONS Suicidal ideation; chronic depression; serious psychiatric disease; lactation.
CAUTIOUS USE History of suicidal tendencies, depression; renal disease, older adults; pregnancy (category C). Safe use in children younger than 18 y not known.

ROUTE & DOSAGE

Smoking Cessation
Adult: **PO** Begin with 0.5 mg/day for 3 days, increase to 0.5 mg b.i.d. for 4 days, then increase to 1 mg b.i.d. on day 8. Treat for 12 wk and may repeat an additional 12 wk.
Renal Impairment Dosage Adjustment
CrCl 50 mL/min or less: Titrate to 0.5 mg b.i.d. (max)

ADMINISTRATION
Oral
- Give after a meal with a full glass of water.

V

Common adverse effects in *italic*, life-threatening effects underlined; generic names in **bold**; classifications in SMALL CAPS; ♦ Canadian drug name; ❷ Prototype drug

1559

- Dose titration over 8 days (from 0.5 mg to 2 mg daily) is recommended to minimize adverse effects.
- Store at 15°–30° C (59°–86° F).

ADVERSE EFFECTS (≥1%) **Body as a Whole:** Fatigue, flushing, gingivitis, headache, influenza-like symptoms, lethargy, malaise, thirst. **CNS:** *Abnormal dreams,* anorexia, anxiety, asthenia, disturbance in attention, depression, dizziness, drowsiness, emotional lability, *insomnia,* irritability, *nightmares,* restlessness, sensory disturbance, *sleep disorder, suicidality.* **CV:** Chest pain, hypertension. **GI:** Abdominal pain, *constipation,* diarrhea, dyspepsia, *flatulence,* gastroesophageal reflux, *nausea, vomiting.* **Metabolic:** Abnormal liver function test, appetite stimulation, weight gain. **Musculoskeletal:** Arthralgia, back pain, muscle cramps, musculoskeletal pain, myalgia. **Respiratory:** Dyspnea, epistaxis, respiratory disorder, rhinorrhea. **Skin:** Hyperhidrosis, pruritus, rash. **Special Senses:** Dysgeusia, xerostomia. **Urogenital:** Menstrual irregularity, polyuria.

INTERACTIONS Drug: Cimetidine increases systemic exposure to varenicline by 29%.

PHARMACOKINETICS Absorption: Complete absorption from GI tract. **Peak:** 3–4 h. **Distribution:** Less than 20% protein bound. **Metabolism:** Minimal. **Elimination:** Primarily eliminated unchanged in the urine. **Half-Life:** 24 h.

NURSING IMPLICATIONS

Assessment & Drug Effects
- Monitor smoking cessation behavior and adverse effects.
- Monitor BP for new-onset hypertension.
- Monitor diabetics for loss of glycemic control.

- Monitor for increased suicidality, or increase in agitation or aggression.

Patient & Family Education
- Report persistent nausea, vomiting, or insomnia to a health care provider.
- Report new-onset of depressed mood, suicidal ideation, or changes in emotion and behavior resulting from the use of varenicline.

VARICELLA VACCINE
(var-i-cel′la)

Varivax
Pregnancy Category: C
See Appendix J.

VASOPRESSIN INJECTION ℗
(vay-soe-press′in)

Classifications: PITUITARY HORMONE; ANTIDIURETIC HORMONE (ADH)
Therapeutic: ADH REPLACEMENT
Pregnancy Category: C

AVAILABILITY 20 pressor units/mL injection

ACTION & *THERAPEUTIC EFFECT* Polypeptide hormone extracted from animal posterior pituitaries, and possesses pressor and antidiuretic (ADH) properties. Produces concentrated urine by increasing tubular reabsorption of water (ADH activity), thus reabsorbing up to 90% of water in renal tubules. Causes contraction of smooth muscles of the GI tract as well as the vascular system, especially capillaries, arterioles and venules. *Effective in reversing diuresis caused by diabetes insipidus. When given intravenously, it is effective as an adjunct in treating massive GI bleeding.*

USES Antidiuretic to treat diabetes insipidus, to dispel gas shadows in

Common adverse effects in *italic*, life-threatening effects underlined; generic names in **bold**; classifications in SMALL CAPS; ♣ Canadian drug name; ℗ Prototype drug

abdominal roentgenography, and as prevention and treatment of postoperative abdominal distention.

UNLABELED USES Test for differential diagnosis of nephrogenic, psychogenic, and neurohypophyseal diabetes insipidus; test to elevate ability of kidney to concentrate urine, and provocative test for pituitary release of corticotropin and growth hormone; emergency and adjunct pressor agent in the control of massive GI hemorrhage (e.g., esophageal varices).

CONTRAINDICATIONS Chronic nephritis accompanied by nitrogen retention; ischemic heart disease, PVCs, advanced arteriosclerosis; lactation.

CAUTIOUS USE Epilepsy; migraine; asthma; heart failure, angina pectoris; any state in which rapid addition to extracellular fluid may be hazardous; vascular disease; preoperative and postoperative polyuric patients, kidney disease; goiter with cardiac complications; older adult patients, labor and delivery, pregnancy (category C), children.

ROUTE & DOSAGE

Diabetes Insipidus

Adult: **IM/Subcutaneous** 5–10 units aqueous solution 2–4 times/day (5–60 units/day) or 1.25–2.5 units in oil q2–3days **Intranasal** Apply to cotton pledget or intranasal spray
Child: **IM/Subcutaneous** 2.5–10 units aqueous solution 2–4 times/day

Abdominal Distention, Abdominal Radiographic Procedures

Adult: **IM/Subcutaneous** 5 units with 5–10 units q3–4h prn or 5–15 units 2 h and 30 min prior to procedure

GI Hemorrhage

Adult: **IV** 20 units bolus then 0.2–0.4 units/min up to 0.9 units/min

ADMINISTRATION

Intramuscular/Subcutaneous

- Give 1–2 glasses of water with vasopressin to reduce adverse effects such as skin blanching, abdominal cramps and nausea.
- Give IM injection deeply into a large muscle.
- With subcutaneous injection, exercise caution not to inject intradermally.

Intravenous

PREPARE: **Direct/IV Infusion:** Dilute with NS or D5W to a concentration of 0.1–1 units/mL.
ADMINISTER: **Direct:** Give rapid bolus dose. **IV Infusion:** Titrate dose and rate to patient's response.
- Ensure patency prior to injection or infusion as extravasation may cause severe vasoconstriction with tissue necrosis and gangrene.

ADVERSE EFFECTS (≥1%) **Skin:** Rash, urticaria. **Body as a Whole:** Anaphylaxis; *tremor;* sweating, bronchoconstriction, *circumoral and facial pallor,* angioneurotic edema, *pounding in head, water intoxication* (especially with tannate), gangrene at injection site with intraarterial infusion. **GI:** *Eructations, passage of gas, nausea, vomiting,* heartburn, abdominal cramps, increased bowel movements secondary to excessive use. **CV:** Angina (in patient with coronary vascular disease); cardiac arrest, hypertension, bradycardia, minor arrhythmias, premature atrial contraction, heart block, peripheral vascular collapse, coronary insufficiency, MI; cardiac arrhythmia, pulmonary edema, bradycardia (with intraarterial

Common adverse effects in *italic*, life-threatening effects underlined; generic names in **bold**; classifications in SMALL CAPS; ✦ Canadian drug name; ⊙ Prototype drug

1561

infusion). **Urogenital:** Uterine cramps. **Respiratory:** Congestion, rhinorrhea, irritation, mucosal ulceration and pruritus, postnasal drip. **Special Senses:** Conjunctivitis.

DIAGNOSTIC TEST INTERFERENCE
Vasopressin increases *plasma cortisol* levels.

INTERACTIONS Drug: **Alcohol, demeclocycline, epinephrine, heparin, lithium, phenytoin** may decrease antidiuretic effects of vasopressin; **guanethidine, neostigmine** increase vasopressor actions; **chlorpropamide, clofibrate, carbamazepine,** THIAZIDE DIURETICS may increase antidiuretic activity.

PHARMACOKINETICS Duration:
2–8 h in aqueous solution, 48–72 h in oil, 30–60 min IV infusion. **Distribution:** Extracellular fluid. **Metabolism:** In liver and kidneys. **Elimination:** In urine. **Half-Life:** 10–20 min.

NURSING IMPLICATIONS

Assessment & Drug Effects
- Monitor infants and children closely. They are more susceptible to volume disturbances (such as sudden reversal of polyuria) than adults.
- Establish baseline data of BP, weight, I&O pattern and ratio. Monitor BP and weight throughout therapy. (Dose used to stimulate diuresis has little effect on BP.) Report sudden changes in pattern to prescriber.
- Be alert to the fact that even small doses of vasopressin may precipitate MI or coronary insufficiency, especially in older adult patients. Keep emergency equipment and drugs (antiarrhythmics) readily available.
- Check patient's alertness and orientation frequently during therapy. Lethargy and confusion associated with headache may

signal onset of water intoxication, which, although insidious in rate of development, can lead to convulsions and terminal coma.
- Monitor urine output, specific gravity, and serum osmolality while patient is hospitalized.
- Withhold vasopressin, restrict fluid intake, and notify prescriber if urine-specific gravity is less than 1.015.

Patient & Family Education
- Be prepared for possibility of anginal attack and have coronary vasodilator available (e.g., nitroglycerin) if there is a history of coronary artery disease. Report to prescriber.
- With diabetes insipidus, measure and record data related to polydipsia and polyuria. Keep an accurate record of output. Understand that treatment should diminish intense thirst and restore undisturbed normal sleep.
- Avoid concentrated fluids (e.g., undiluted syrups), since these increase urine volume.

VECURONIUM BROMIDE
(vek-yoo-roe'nee-um)

Classifications: NONDEPOLARIZING SKELETAL MUSCLE RELAXANT; ACETYLCHOLINE RECEPTOR ANTAGONIST
Therapeutic: SKELETAL MUSCLE RELAXANT
Prototype: Atracurium
Pregnancy Category: C

AVAILABILITY 4 mg, 10 mg, 20 mg vials

ACTION & *THERAPEUTIC EFFECT*
Intermediate-acting nondepolarizing skeletal muscle relaxant that inhibits neuromuscular transmission by competitively binding with acetylcholine at receptors located on motor endplate receptors. *Effective as a skeletal muscle relaxant.*

V

USES Adjunct for general anesthesia to produce skeletal muscle relaxation during surgery. Especially useful for patients with severe kidney disease, limited cardiac reserve, and history of asthma or allergy. Also to facilitate endotracheal intubation.

UNLABELED USES Continuous infusion for facilitation of mechanical ventilation.

CONTRAINDICATIONS Hypersensitivity to bromide; malignant hyperthermia.

CAUTIOUS USE Severe liver disease; impaired acid–base, fluid and electrolyte balance; severe obesity; adrenal or neuromuscular disease (myasthenia gravis, Eaton-Lambert syndrome); patients with slow circulation time (cardiovascular disease, old age, edematous states); obesity, pregnancy (category C), lactation; children younger than 1 y and older than 7 wk.

ROUTE & DOSAGE

Skeletal Muscle Relaxation

Adult/Child (10 y or older): **IV** 0.04–0.1 mg/kg initially, then after 25–40 min, 0.01–0.15 mg/kg q12–15min or 0.001 mg/kg/min by continuous infusion for prolonged procedures
Child (1–10 y)/Infant: **IV** Varies greatly; may require higher initial dose

Obesity Dosage Adjustment

Dose based on IBW

ADMINISTRATION

Note: Vecuronium is administered only by qualified clinicians.

Intravenous

PREPARE: Direct: Reconstitute the 10 or 20 mg vial with 10 or 20 mL,

respectively, of sterile water for injection (supplied). **Continuous:** Further dilute the reconsituted solution in up to 100 mL D5W, NS, or LR to yield 0.1–0.2 mg/mL. **ADMINISTER: Direct:** Give a bolus dose over 30 sec. **Continuous:** Give at the required rate.

INCOMPATIBILITIES Y-site: Amphotericin B cholesteryl complex, diazepam, etomidate, furosemide, thiopental.

■ Refrigerate after reconstitution below 30° C (86° F), unless otherwise directed. Discard solution after 24 h.

ADVERSE EFFECTS (≥1%) **Body as a Whole:** Skeletal muscle weakness, malignant hyperthermia. **Respiratory:** Respiratory depression.

INTERACTIONS Drug: GENERAL ANESTHETICS increase neuromuscular blockade and duration of action; AMINOGLYCOSIDES, **bacitracin, polymyxin B, clindamycin, lidocaine, parenteral magnesium, quinidine, quinine, trimethaphan, verapamil** increase neuromuscular blockade; DIURETICS may increase or decrease neuromuscular blockade; **lithium** prolongs duration of neuromuscular blockade; NARCOTIC ANALGESICS increase possibility of additive respiratory depression; **succinylcholine** increases onset and depth of neuromuscular blockade; **phenytoin** may cause resistance to or reversal of neuromuscular blockade.

PHARMACOKINETICS Onset: Less than 1 min. **Peak:** 3–5 min. **Duration:** 25–40 min. **Distribution:** Well distributed to tissues and extracellular fluids; crosses placenta; distribution into breast milk unknown. **Metabolism:** Rapid nonenzymatic degradation in bloodstream. **Elimination:** 30–35% in urine, 30–35% in bile. **Half-Life:** 30–80 min.

V

Common adverse effects in *italic*, life-threatening effects underlined; generic names in **bold**; classifications in SMALL CAPS; ♣ Canadian drug name; ⊘ Prototype drug

1563

NURSING IMPLICATIONS

Assessment & Drug Effects

- Lab tests: Baseline serum electrolytes, acid–base balance, LFTs, and kidney functions.
- Use peripheral nerve stimulator during and following drug administration to avoid risk of overdosage and to identify residual paralysis during recovery period. This is especially indicated when cautious use of drug is specified.
- Monitor vital signs at least q15min until stable, then q30min for the next 2 h. Also monitor airway patency until assured that patient has fully recovered from drug effects. Note rate, depth, and pattern of respirations. Obese patients and patients with myasthenia gravis or other neuromuscular disease may have ventilation problems.
- Evaluate patients for recovery from neuromuscular blocking (curare-like) effects as evidenced by ability to breathe naturally or take deep breaths and cough, to keep eyes open, and to lift head keeping mouth closed and by adequacy of hand grip strength. Notify prescriber if recovery is delayed.
- Note: Recovery time may be delayed in patients with cardiovascular disease, edematous states, and in older adults.

VENLAFAXINE ⊕

(ven-la-fax'een)

Effexor, Effexor XR

DESVENLAFAXINE

Pristiq

Classifications: ANTIDEPRESSANT; SEROTONIN NOREPINEPHRINE REUPTAKE INHIBITOR (SNRI)

Therapeutic: ANTIDEPRESSANT; SNRI

Pregnancy Category: C

AVAILABILITY 25 mg, 37.5 mg, 50 mg, 75 mg, 100 mg tablets; 37.5 mg, 75 mg, 150 mg sustained release capsules; **Desvenlafaxine:** 50 mg, 100 mg extended release tablet

ACTION & THERAPEUTIC EFFECT Potent inhibitor of neuronal serotonin and norepinephrine reuptake. *Antidepressant effect presumed to be due to potentiation of neurotransmitter activity in the CNS.*

USES Depression, generalized anxiety disorder; social anxiety disorder. **UNLABELED USES** Obsessive-compulsive disorder.

CONTRAINDICATIONS Hypersensitivity to venlafaxine, desvenlafaxine, or other SNRI drugs; concurrent administration with MAO inhibitors; abrupt discontinuation; suicide ideation; lactation.

CAUTIOUS USE Renal and hepatic impairment, renal failure; anorexia nervosa, history of mania, history of suicidal tendencies especially in individuals younger than 24 y; elevated intraocular pressure, acute closed-angle glaucoma; cardiac disorders, recent MI, heart failure; hypertension; hyperthyroidism; CNS depression; history of seizures or seizure disorders; older adults; pregnancy (category C). Safety in children younger than 18 y is not established.

ROUTE & DOSAGE

Depression

Adult: **PO** 25–125 mg t.i.d.
Desvenlafaxine: 50 mg daily, may increase dose (max: 400 mg/day)
Geriatric: **PO (venlafaxine and desvenlafaxine)** Start with lower doses in older adults

Anxiety

Adult: **PO** Start with 37.5 mg sustained release daily and increase to 75–225 mg sustained release per day

Renal Impairment Dosage Adjustment

Venlafaxine: CrCl 10–70 mL/min: Reduce total daily dose by 25–50%; less than 10 mL/min: Reduce total daily dose by 50% **Desvenlafaxine:** CrCl 3–50 mL/min: Max dose is 50 mg/day; less than 30 mL/min: 50 mg every other day

Pharmacogenetic Dosage Adjustment

Poor CYP2D6 metabolizer: Start with 70% of dose

ADMINISTRATION

Oral

- Give with food. Extended-release *capsules* **must be** swallowed whole or carefully opened and sprinkled on a spoonful of applesauce. The applesauce mixture should be swallowed without chewing and followed with a glass of water. Extended-release *tablets* should be swallowed whole and not divided, crushed, or chewed.
- Dosage increments of up to 75 mg/day are usually made at 4 days or longer intervals.
- Allow 14 days interval after discontinuing an MAO inhibitor before starting venlafaxine or desvenlafaxine.
- Do not abruptly withdraw drug after 1 wk or more of therapy.
- Store at room temperature, 15°–30° C (59°–86° F).

ADVERSE EFFECTS (≥1%) **CV:** *Increased blood pressure and heart rate*, palpitations. **CNS:** *Dizziness*, fatigue, headache, anxiety, insomnia, *somnolence*, <u>suicidality</u>. **Endocrine:** Small but statistically significant increase in serum cholesterol, weight loss (approximately 3 lb). **GI:** *Nausea, vomiting, dry mouth*, constipation. **Urogenital:** Sexual dysfunction, erectile failure, delayed orgasm, anorgasmia, impotence, abnormal ejaculation. **Special Senses:** Blurred vision. **Body as a Whole:** *Sweating*, asthenia.

INTERACTIONS Drug: Cimetidine, MAO INHIBITORS, **desipramine, haloperidol** may increase levels and toxicity. Should not use in combination with MAO INHIBITORS: Do not start until greater than 14 days after stopping MAO INHIBITOR; do not start MAO INHIBITOR until 7 days after stopping venlafaxine/desvenlafaxine. **Trazodone** may lead to **serotonin** syndrome. **Herbal: St. John's wort, sour date nut** may cause **serotonin** syndrome.

PHARMACOKINETICS Absorption: Well absorbed from GI tract. **Onset:** 2 wk. **Peak:** Venlafaxine 1–2 h; metabolite 3–4 h. **Duration:** Extensively tissue bound. **Metabolism:** Undergoes substantial first-pass metabolism to its major active metabolite. **Elimination:** ~60% in urine as parent compound and metabolites. **Half-Life:** Venlafaxine 3–4 h, desvenlafaxine ~11 h.

NURSING IMPLICATIONS

Assessment & Drug Effects

- Monitor for worsening of depression or emergence of suicidal ideation.
- Monitor cardiovascular status periodically with measurements of HR and BP.
- Lab tests: Periodic lipid profile.
- Monitor neurologic status and report excessive anxiety, nervousness, and insomnia.
- Monitor weight periodically and report excess weight loss.

V

Common adverse effects in *italic*, life-threatening effects <u>underlined</u>; generic names in **bold;** classifications in SMALL CAPS; ♣ Canadian drug name; ⊘ Prototype drug

1565

- Assess safety, as dizziness and sedation are common.

Patient & Family Education
- Be aware of potential adverse effects and notify prescriber of those that are bothersome.
- Report promptly worsening mental status, especially thoughts of suicide.
- Do not drive or engage in potentially hazardous activities until response to drug is known.
- Avoid using alcohol while on venlafaxine.
- Do not use herbal medications without consulting prescriber.

VERAPAMIL HYDROCHLORIDE ℗

(ver-ap′a-mill)

Calan, Calan SR, Covera-HS, Isoptin, Isoptin SR, Verelan, Verelan PM

Classifications: CALCIUM CHANNEL BLOCKER; ANTIHYPERTENSIVE; CLASS IV ANTIARRHYTHMIC

Therapeutic: CLASS IV ANTIARRHYTHMIC; ANTIHYPERTENSIVE; ANTIANGINAL

Pregnancy Category: C

AVAILABILITY 40 mg, 80 mg, 120 mg tablets; 120 mg, 180 mg, 240 mg sustained release tablets; 100 mg, 120 mg, 180 mg, 200 mg, 240 mg, 300 mg sustained release capsules; 5 mg/2 mL injection

ACTION & *THERAPEUTIC EFFECT* Inhibits calcium ion influx through slow channels into cells of myocardial and arterial smooth muscle. Dilates coronary arteries and arterioles and inhibits coronary artery spasm. Decreases and slows SA and AV node conduction without affecting normal arterial action potential or intraventricular conduction. Dilates peripheral arterioles, causing decreased total peripheral resistance, and this results in lowering the BP. *Decreases angina attacks by dilating coronary arteries and inhibiting coronary vasospasms. Decreases nodal conduction, resulting in an antiarrhythmic effect. Decreased total peripheral vascular resistance; and therefore, reduction in BP.*

USES Supraventricular tachyarrhythmias; Prinzmetal's (variant) angina, chronic stable angina; unstable, crescendo or preinfarctive angina and essential hypertension.

UNLABELED USES Paroxysmal supraventricular tachycardia, atrial fibrillation; prophylaxis of migraine headache; and as alternate therapy in manic depression.

CONTRAINDICATIONS Severe hypotension (systolic less than 90 mm Hg), cardiogenic shock, cardiomegaly, digitalis toxicity, second- or third-degree AV block; Wolff-Parkinson-White syndrome including atrial flutter and fibrillation; accessory AV pathway, left ventricular dysfunction, severe CHF, sinus node disease, sick sinus syndrome (except in patients with functioning ventricular pacemaker); lactation.

CAUTIOUS USE Duchenne's muscular dystrophy; hepatic and renal impairment; MI followed by coronary occlusion, aortic stenosis; GI obstruction, GERD, hiatal hernia, ileus; pregnancy (category C); **extended release tablets:** children younger than 18 y.

ROUTE & DOSAGE

Angina

Adult: **PO** 80 mg q6–8h, may increase up to 320–480 mg/day in divided doses (Note: **Covera-HS must be** given once daily at bedtime)

Hypertension

Adult: **PO** 80 mg t.i.d. or 90–240 mg sustained release 1–2 times/day up to 480 mg/day

---

Common adverse effects in *italic*, life-threatening effects underlined; generic names in **bold**; classifications in SMALL CAPS; ♣ Canadian drug name; ℗ Prototype drug

(Note: Covera-HS **must be** given once daily at bedtime)

Supraventricular Tachycardia, Atrial Fibrillation

Adult/Adolescent (older than 15 y): **IV** 5–10 mg after 30 min may give 10 mg (max total dose: 20 mg)
Child/Adolescent (up to 15 y): **IV** 0.1–0.3 mg/kg (do not exceed 5 mg)
Infant: **IV** 0.1–0.2 mg/kg may repeat after 30 min

Renal Impairment Dosage Adjustment

CrCl less than 10 mL/min: Give 50–75% of dose

Hemodialysis Dosage Adjustment

Supplemental dose not necessary

Hepatic Impairment Dosage Adjustment

Cirrhosis: Use 20–50% of normal dose

ADMINISTRATION

Oral

- Give with food to reduce gastric irritation.
- Capsules can be opened and contents sprinkled on food. Do NOT dissolve or chew capsule contents.
- Do not withdraw abruptly; may increase and extend duration of pain in the angina patient.

Intravenous

PREPARE: **IV Direct:** Given undiluted or diluted in 5 mL of sterile water for injection. ▪ Inspect parenteral drug preparation before administration. Make sure solution is clear and colorless.
ADMINISTER: **Direct:** Give a single dose over 2–3 min.

INCOMPATIBILITIES **Solution/additive: Albumin, aminophylline, amphotericin B, hydralazine, trimethoprim/sulfamethoxazole. Y-site: Albumin, amphotericin B cholesteryl complex, ampicillin, lansoprazole, mezlocillin, nafcillin, oxacillin, propofol, sodium bicarbonate.**

▪ Store at 15°–30° C (59°–86° F) and protect from light.

ADVERSE EFFECTS (≥1%) **CNS:** Dizziness, vertigo, *headache,* fatigue, sleep disturbances, depression, syncope. **CV:** *Hypotension,* congestive heart failure, bradycardia, severe tachycardia, peripheral edema, <u>AV block</u>. **GI:** Nausea, abdominal discomfort, *constipation,* elevated liver enzymes. **Body as a Whole:** Flushing, pulmonary edema, muscle fatigue, diaphoresis. **Skin:** Pruritus.

DIAGNOSTIC TEST INTERFERENCE May cause false positive for urine detection of **methadone.**

INTERACTIONS Drug: BETA-BLOCKERS increase risk of CHF, bradycardia, or heart block; significantly increased levels of **digoxin** and **carbamazepine** and toxicity; potentiates hypotensive effects of HYPOTENSIVE AGENTS; levels of **lithium** and **cyclosporine** may be increased, increasing their toxicity; **calcium salts** (IV) may antagonize verapamil effects. **Food: Grapefruit juice** may increase verapamil levels. **Herbal: Hawthorne** may have additive hypotensive effects. **St. John's wort** may decrease efficacy.

PHARMACOKINETICS Absorption: 90% absorbed, but only 25–30% reaches systemic circulation (first pass metabolism). **Peak:** 1–2 h PO; 4–8 h sustained release; 5 min IV.

V

Common adverse effects in *italic*, life-threatening effects <u>underlined</u>; generic names in **bold**; classifications in SMALL CAPS; ♣ Canadian drug name; ☯ Prototype drug

1567

Distribution: Widely distributed, including CNS; crosses placenta; present in breast milk. **Metabolism:** In liver (CYP3A4). **Elimination:** 70% in urine; 16% in feces. **Half-Life:** 2–8 h.

NURSING IMPLICATIONS

Assessment & Drug Effects

- Establish baseline data and periodically monitor BP and pulse with oral administration.
- Lab tests: Baseline and periodic LFTs and kidney functions.
- Following IV infusion, instruct patient to remain in recumbent position for at least 1 h after dose is given to diminish subjective effects of transient asymptomatic hypotension that may accompany infusion.
- Monitor for AV block or excessive bradycardia when IV infusion is given concurrently with digitalis.
- Monitor I&O ratio during IV and early oral maintenance therapy. Renal impairment prolongs duration of action, increasing potential for toxicity and incidence of adverse effects. Advise patient to report gradual weight gain and evidence of edema.
- Monitor ECG continuously during IV administration. Essential because drug action may be prolonged and incidence of adverse reactions is highest during IV administration in older adults, patients with impaired kidney function, and patients of small stature.
- Check BP shortly before administration of next dose to evaluate degree of control during early treatment for hypertension.

Patient & Family Education

- Monitor radial pulse before each dose, notify prescriber of an irregular pulse or one slower than established guideline.
- Do not drive or engage in potentially hazardous activities until response to drug is known.
- Decrease intake of caffeine-containing beverage (i.e., coffee, tea, chocolate).
- Change positions slowly from lying down to standing to prevent falls because of drug-related vertigo until tolerance to reduced BP is established.
- Notify prescriber of easy bruising, petechiae, unexplained bleeding.
- Do not use OTC drugs, especially aspirin, unless they are specifically prescribed by prescriber.

VILAZODONE
(vil az′ oh done)
Vibryd
Classifications: SELECTIVE SEROTONIN REUPTAKE INHIBITOR (SSRI); PSYCHOTHERAPEUTIC AGENT; ANTIDEPRESSANT
Therapeutic: ANTIDEPRESSANT; SSRI
Prototype: Fluoxetine
Pregnancy Category: C

AVAILABILITY 10 mg, 20 mg, and 40 mg tablets

ACTION & *THERAPEUTIC EFFECT* A selective serotonin reuptake inhibitor (SSRI). Antidepressant effect is presumed to be linked to inhibition of CNS neuronal uptake of the neurotransmitter, serotonin. *Improves mood in those with major depressive disorder.*

USES Treatment of major depressive disorder.

CONTRAINDICATIONS Concomitant use of MAOIs or MAOIs within the preceding 14 days.
CAUTIOUS USE Suicidal ideation; potential precursors to suicidal impulses (e.g., anxiety, agitation, panic attacks, insomnia, irritability,

hostility, aggressiveness, impulsivity, psychomotor restlessness, hypomania, mania); history of seizure disorder; concurrent use of NSAIDs, aspirin, or other drugs that affect coagulation or bleeding; pregnancy (category C); lactation. The safety and efficacy in children have not been established.

ROUTE & DOSAGE

Major Depressive Disorder

Adult: **PO** Initial dose of 10 mg once daily for 7 days, increase to 20 mg once daily for 7 days, then to 40 mg once daily

ADMINISTRATION

Oral

▪ Give with food to enhance absorption.

▪ Store at 15°–30° C (59°–86° F).

ADVERSE EFFECTS (≥1%) **Body as a Whole:** Arthralgia, decreased appetite, feeling jittery, increased appetite. **CNS:** Abnormal dreams, akathisia, *dizziness*, fatigue, insomnia, libido decreased, migraine, abnormal orgasm, paresthesia, restless leg syndrome, restlessness, sedation, somnolence, tremor. **CV:** Palpitations. **GI:** *Diarrhea*, dry mouth, dyspepsia, flatulence, gastroenteritis, *nausea*, vomiting. **Skin:** Hyperhidrosis, night sweats. **Special Senses:** Blurred vision, dry eye. **Urogenital:** Delayed ejaculation, erectile dysfunction.

INTERACTIONS Drug: Inducers of CYP3A4 (e.g., **phenytoin**) may reduce the levels of vilazodone. Coadministration or moderate (e.g., **erythromycin**) and strong (e.g., **ketoconazole**) increase the levels of vilazodone and require a dosage reduction to 20 mg once daily. MONOAMINE OXIDASE INHIBITORS increase the levels of vilazodone and should not be used in combination with vilazodone. Coadministration of serotonergic agents (e.g, **busipirone, tramadol, thyptophan**, SELECTIVE SEROTONIN REUPTAKE INHIBITORS, SEROTONIN-NOR-EPINEPHRINE REUPTAKE INHIBITORS, TRIPTANS) increase the risk of serotonin syndrome. Vilazodone can increase the risk of bleeding if used in combination with **aspirin** or NSAIDs. **Food:** Ingestion of food enhances the absorption of vilazodone.

PHARMACOKINETICS Absorption: Bioavailability 72%. **Peak:** 4–5 h. **Distribution:** 96–99% plasma protein bound. **Metabolism:** Hepatic oxidative metabolism. **Elimination:** Renal and fecal. **Half-Life:** 25 h.

NURSING IMPLICATIONS

Assessment & Drug Effects

▪ Monitor closely and report promptly any of the following: Worsening of clinical symptoms; emergence of suicidal ideation; agitation, irritability, or unusual changes in behavior; signs of serotonin syndrome or neuroleptic malignant syndrome (see Appendix F).

▪ Supervise patients closely who are high suicide risks; be especially vigilant for changes in behavior and suicidal ideation in children and adolescents.

▪ Monitor those at risk for volume depletion (e.g., diuretics use) for S&S of hyponatremia.

▪ Lab tests: Periodic serum sodium.

Patient & Family Education

▪ Report promptly worsening of condition, suicidal thoughts or thoughts of self-harm.

▪ Do not abruptly stop taking this drug without consulting prescriber.

Common adverse effects in *italic*, life-threatening effects underlined; generic names in **bold**; classifications in SMALL CAPS; ♣ Canadian drug name; ◐ Prototype drug

1569

- Notify prescriber if you become pregnant or intend to breast feed.
- Do not drive or engage in other hazardous activities until response to drug is known.
- Consult prescriber prior to using nonsteroidal anti-inflammatory drugs (NSAIDs), aspirin, or other drugs that affect coagulation.

VINBLASTINE SULFATE
(vin-blast′een)

Classifications: ANTINEOPLASTIC; MITOTIC INHIBITOR; VINCA ALKALOID
Therapeutic: ANTINEOPLASTIC
Prototype: Vincristine
Pregnancy Category: D

AVAILABILITY 10 mg powder for injection; 1 mg/mL vial

ACTION & *THERAPEUTIC EFFECT* Cell cycle–specific drug that interferes with microtubules that form mitotic spindle fibers required to complete the process of mitosis. Has an effect on cell energy production needed for mitosis and interferes with nucleic acid synthesis. *Interrupts the cell cycle in metaphase, thus preventing cell replication.*

USES Palliative treatment of Hodgkin's disease and non-Hodgkin's lymphomas, choriocarcinoma, lymphosarcoma, neuroblastoma, mycosis fungoides, advanced testicular germinal cell cancer, histiocytosis, and other malignancies resistant to other chemotherapy. Used singly or in combination with other chemotherapeutic drugs.

CONTRAINDICATIONS Severe bone marrow suppression, leukopenia, bacterial infection, adynamic ileus; older adult patients with cachexia or skin ulcers; men and women of childbearing potential; pregnancy (category D), lactation.

CAUTIOUS USE Malignant cell infiltration of bone marrow; obstructive jaundice, hepatic impairment; history of gout; use of small amount of drug for long periods; use in eyes.

ROUTE & DOSAGE

Antineoplastic

Adult: **IV** 7–10 mg/m² once weekly, dose varies based on protocol and may increase incrementally up to 18.5 mg/m² if tolerated
Child: **IV** 2.5 mg/m² may increase up to 12.5 mg/m² if tolerated

Hepatic Impairment Dosage Adjustment

Bilirubin 1.5–3 mg/dL: Reduce dose 50%; *bilirubin over 3 mg/dL:* Reduce dose 75%

ADMINISTRATION

Intravenous

PREPARE: Direct: Add 10 mL NS to 10 mg of drug to yield 1 mg/mL. Do not use other diluents. ■ Avoid contact with eyes. Severe irritation and persisting corneal changes may occur. Flush immediately and thoroughly with copious amounts of water. Wash both eyes; do not assume one eye escaped contamination.

ADMINISTER: Direct: Drug is usually injected into tubing of running IV infusion of NS or D5W over period of 1 min. ■ Stop injection promptly if extravasation occurs. Use applications of moderate heat and local injection of hyaluronidase to help disperse extravasated drug. ■ Observe injection site for sloughing. ■ Restart infusion in another vein.

INCOMPATIBILITIES **Solution/additive: Furosemide, heparin. Y-site: Cefepime, furosemide, lansoprazole.**

V

- Refrigerate reconstituted solution in tight, light-resistant containers up to 30 days without loss of potency.

ADVERSE EFFECTS (≥1%) **Body as a Whole:** Fever, weight loss, muscular pains, weakness, parotid gland pain and tenderness, tumor site pain, Raynaud's phenomenon. **CNS:** Mental depression, peripheral neuritis, numbness and paresthesias of tongue and extremities, loss of deep tendon reflexes, headache, convulsions. **GI:** Vesiculation of mouth, stomatitis, pharyngitis, anorexia, *nausea, vomiting,* diarrhea, ileus, abdominal pain, constipation, rectal bleeding, <u>hemorrhagic enterocolitis</u>, bleeding of old peptic ulcer. **Hematologic:** <u>Leukopenia</u>, thrombocytopenia, and anemia. **Skin:** *Alopecia (reversible),* vesiculation, photosensitivity, phlebitis, cellulitis, and sloughing following extravasation (at injection site). **Urogenital:** Urinary retention, *hyperuricemia,* aspermia. **Respiratory:** <u>Bronchospasm</u>.

INTERACTIONS Drug: Mitomycin may cause acute shortness of breath and severe bronchospasm; may decrease **phenytoin** levels; ALFA INTERFERONS, **erythromycin, itraconazole** may increase vinblastine toxicity.

PHARMACOKINETICS Distribution: Concentrates in liver, platelets, and leukocytes; poor penetration of blood–brain barrier. **Metabolism:** Partially in liver. **Elimination:** In feces and urine. **Half-Life:** 24 h.

NURSING IMPLICATIONS

Assessment & Drug Effects

- Lab tests: Monitor WBC count. Recovery from leukopenic nadir occurs usually within 7–14 days. With high doses, total leukocyte count may not return to normal for 3 wk.

- Do not administer drug unless WBC count has returned to at least 4000/mm³, even if 7 days have passed.
- Monitor for unexplained bruising or bleeding, which should be promptly reported, even though thrombocyte reduction seldom occurs unless patient has had prior treatment with other anti-neoplastics.
- Adverse reactions seldom persist beyond 24 h with exception of epilation, leukopenia, and neurological adverse effects.
- Monitor bowel elimination pattern and bowel sounds to recognize severe constipation or paralytic ileus. A stool softener may be necessary.
- Inspect skin surfaces over pressure areas daily if patient is not ambulating. Note condition of skin of older adults especially.
- Report promptly if oral mucosa tissue breakdown is noted.

Patient & Family Education

- Be aware that temporary mental depression sometimes occurs on second or third day after treatment begins.
- Avoid exposure to infection, injury to skin or mucous membranes, and excessive physical stress, especially during leukocyte nadir period.
- Notify prescriber promptly about onset of symptoms of agranulocytosis (see Appendix F). Do not delay seeking appropriate treatment.
- Avoid exposure to sunlight unless protected with sunscreen lotion (SPF greater than 12) and clothing.

VINCRISTINE SULFATE ⊙
(vin-kris′teen)

Vincasar PFS
Classifications: ANTINEOPLASTIC; MITOTIC INHIBITOR; VINCA ALKALOID
Therapeutic: ANTINEOPLASTIC
Pregnancy Category: D

Common adverse effects in *italic*, life-threatening effects <u>underlined</u>; generic names in **bold**; classifications in SMALL CAPS; ♣ Canadian drug name; ⊙ Prototype drug

1571

AVAILABILITY 1 mg/mL injection

ACTION & *THERAPEUTIC EFFECT*
Cell cycle–specific vinca alkaloid
arrests mitosis at metaphase by inhibition of mitotic spindle function,
thereby inhibiting cell division. *Induction of metaphase arrest in 50%
of cells results in inhibition of cancer cell proliferation.*

USES Acute lymphoblastic and other leukemias, Hodgkin's disease,
lymphosarcoma, neuroblastoma,
Wilms' tumor, lung and breast cancer, reticular cell carcinoma, and osteogenic and other sarcomas.

UNLABELED USES Idiopathic thrombocytopenic purpura, alone or
adjunctively with other antineoplastics, breast cancer, colorectal
cancer, thymoma.

CONTRAINDICATIONS Obstructive jaundice; active infection; adynamic ileus; radiation of the liver;
patient with demyelinating form
of Charcot-Marie-Tooth syndrome;
men and women of childbearing
potential; pregnancy (category D),
lactation.

CAUTIOUS USE Leukopenia; preexisting neuromuscular or neurologic
disease; hypertension; hepatic or
biliary tract disease; older adults.

ROUTE & DOSAGE

ALL
Adult: **IV** 1.4 mg/m² (max: 2 mg/ m²) at weekly intervals
Child: **IV** *Weight greater than 10 kg,* 1.5–2 mg/m² at weekly intervals; *weight less than 10 kg,* 0.05 mg/kg initial weekly dose, then titrate

**Malignant Glioma/Hodgkin's
Disease/non-Hodgkin's
Lymphoma**
Adult/Adolescent/Child: **IV**
1.4 mg/m² on days 1 and 8 every
28 days

Neuroblastoma
Child: **IV** 1 mg/m²/day for
72 hours

Multiple Myeloma
Adult/Adolescent/Child: **IV**
0.4 mg/day × 4 days

**Hepatic Impairment Dosage
Adjustment**
Bilirubin 1.5–3 mg/dL: Use 50% of
dose; *3–5 mg/dL:* Use 25% of dose;
greater than 5 mg/dL: Skip dose

ADMINISTRATION

Intravenous

PREPARE: **Direct:** No dilution is
required. Administer as supplied.
▪ Avoid contact with eyes. Severe
irritation and persisting corneal
changes may occur. Flush immediately and thoroughly with copious amounts of water. Wash both
eyes; do not assume one eye escaped contamination.
ADMINISTER: **Direct:** Drug is usually injected into tubing of running infusion over a 1 min period.
▪ Stop injection promptly if extravasation occurs. Use applications of moderate heat and local
injection of hyaluronidase to help
disperse extravasated drug. ▪ Restart infusion in another vein. Observe injection site for sloughing.
INCOMPATIBILITIES **Solution/ad-
ditive: Furosemide. Y-site: Amphotericin B colloidal, cefepime, diazepam, furosemide,
gemtuzumab, idarubicin, lansoprazole, pantoprazole, phenytoin, sodium bicarbonate.**

▪ Store solution in the refrigerator.

ADVERSE EFFECTS (≥1%) **CNS:** *Peripheral neuropathy,* neuritic pain,
paresthesias, especially of hands and

V

feet; foot and hand drop, sensory loss, athetosis, ataxia, loss of deep tendon reflexes, muscle atrophy, dysphagia, weakness in larynx and extrinsic eye muscles, ptosis, diplopia, mental depression. **Special Senses:** Optic atrophy with blindness; transient cortical blindness, ptosis, diplopia, photophobia. **GI:** Stomatitis, pharyngitis, anorexia, nausea, vomiting, diarrhea, abdominal cramps, *severe constipation (upper-colon impaction), paralytic ileus (especially in children),* rectal bleeding; hepatotoxicity. **Urogenital:** Urinary retention, polyuria, dysuria, SIADH (high urinary sodium excretion, hyponatremia, dehydration, hypotension); uric acid nephropathy. **Skin:** Urticaria, rash, *alopecia,* cellulitis and phlebitis following extravasation (at injection site). **Body as a Whole:** Convulsions with hypertension, malaise, fever, headache, pain in parotid gland area, weight loss. **Metabolic:** Hyperuricemia, hyperkalemia. **CV:** Hypertension, hypotension. **Respiratory:** Bronchospasm.

INTERACTIONS Drug: Mitomycin may cause acute shortness of breath and severe bronchospasm; may decrease **digoxin, phenytoin** levels.

PHARMACOKINETICS Distribution: Concentrates in liver, platelets, and leukocytes; poor penetration of blood–brain barrier. **Metabolism:** Partially in liver (CYP3A4). **Elimination:** Primarily in feces. **Half-Life:** 10–155 h.

NURSING IMPLICATIONS

Assessment & Drug Effects
- Monitor I&O ratio and pattern, BP, and temperature daily.
- Monitor for and report steady weight gain.
- Lab tests: Monitor serum electrolytes and CBC with differential.
- Be aware that neuromuscular adverse effects, most apt to appear in the patient with preexisting neuromuscular disease, usually disappear after 6 wk of treatment. Children are especially susceptible to neuromuscular adverse effects.
- Assess for hand muscular weakness, and check deep tendon reflexes (depression of Achilles reflex is the earliest sign of neuropathy). Also observe for and report promptly: Mental depression, ptosis, double vision, hoarseness, paresthesias, neuritic pain, and motor difficulties.
- Provide special protection against infection or injury during leukopenic days. Leukopenia occurs in a significant number of patients; leukocyte count in children usually reaches nadir on fourth day and begins to rise on fifth day after drug administration.
- Avoid use of rectal thermometer or intrusive tubing to prevent injury to rectal mucosa.
- Monitor ability to ambulate and supply support as needed.
- Start a prophylactic regimen against constipation and paralytic ileus at beginning of treatment (paralytic ileus is most likely to occur in young children).

Patient & Family Education
- Notify prescriber promptly of stomach, bone, or joint pain, and swelling of lower legs and ankles.
- Report changes in bowel habit as soon as manifested.
- Report a steady gain or sudden weight change to prescriber.

V

VINORELBINE TARTRATE
(vin-o-rel′been)

Navelbine
Classifications: ANTINEOPLASTIC; MITOTIC INHIBITOR; VINCA ALKALOID
Therapeutic: ANTINEOPLASTIC
Prototype: Vincristine
Pregnancy Category: D

Common adverse effects in *italic,* life-threatening effects underlined; generic names in **bold;** classifications in SMALL CAPS; ♣ Canadian drug name; ⊘ Prototype drug

1573

AVAILABILITY 10 mg/mL injection

ACTION & *THERAPEUTIC EFFECT*
A semisynthetic vinca alkaloid with antineoplastic activity. Inhibits polymerization of tubules into microtubules, which disrupts mitotic spindle formation. *Arrests mitosis at metaphase, thereby inhibiting cell division in cancer cells.*

USES Non–small-cell lung cancer.
UNLABELED USES Breast cancer, ovarian cancer, Hodgkin's disease.

CONTRAINDICATIONS Hypersensitivity to vinorelbine, infection; severe bone marrow suppression; granulocyte counts greater than or equal to 1000 cells/mm^3; pulmonary toxicity to drug; constipation, ileus; pregnancy (category D), lactation.
CAUTIOUS USE Hypersensitivity to vincristine or vinblastine; leukopenia or other indicator(s) of bone marrow suppression; chickenpox or herpes zoster infection; hepatic insufficiency, severe liver disease; pulmonary disease; preexisting neurologic or neuromuscular disorders; older adults. Safety and efficacy in children are not established.

ROUTE & DOSAGE

Non–Small-Cell Lung Cancer
Adult: **IV** 25–30 mg/m^2 weekly; may require toxicity adjustment
Hepatic Impairment Dosage Adjustment
Bilirubin 2.1–3 mg/dL: Use 50% of dose; *greater than 3 mg/dL:* Use 25% of dose

ADMINISTRATION

Intravenous
Use caution to prevent contact with skin, mucous membranes, or eyes during preparation.

PREPARE: **Direct:** Dilute each 10 mg in a syringe with either 2 or 5 mL of D5W or NS to yield 3 mg/mL or 1.5 mg/L, respectively. **IV Infusion:** Dilute the required dose in an IV bag with D5W, NS, or LR to a final concentration of 0.5–2 mg/mL (example: 10 mg diluted in 19 mL yields 0.5 mg/mL).
ADMINISTER: **IV Infusion:** Give diluted solution over 6–10 min into the side port closest to an IV bag with free-flowing IV solution; follow by flushing with at least 75–125 mL of IV solution over 10 min. ▪Take every precaution to avoid extravasation. If suspected, discontinue IV immediately and begin in a different site.
INCOMPATIBILITIES **Solution/additive:** Acyclovir, aminophylline, amphotericin B, ampicillin, cefazolin, ceforanide, cefotaxime, cefotetan, ceftazidime, ceftriaxone, cefuroxime, fluorouracil, furosemide, ganciclovir, methylprednisolone, mitomycin, piperacillin, sodium bicarbonate, thiotepa, trimethoprim–sulfamethoxazole. **Y-site:** Acyclovir, allopurinol, aminophylline, amphotericin B, amphotericin B cholesteryl complex, ampicillin, cefazolin, cefotetan, ceftriaxone, cefuroxime, fluorouracil, furosemide, ganciclovir, heparin, lansoprazole, methylprednisolone, mitomycin, piperacillin, sodium bicarbonate, thiotepa, trimethoprim-sulfamethoxazole.

▪ Store at 2°–8° C (36°–46° F).

ADVERSE EFFECTS (≥1%) **CNS:** *Decreased deep tendon reflexes, paresthesia, fatigue, asthenia, peripheral neuropathy,* myalgia, jaw pain. **Hematologic:** *Anemia,* neutropenia, granulocytopenia, thrombocytopenia. **GI:** Paralytic ileus, *constipation, nausea, vomiting, diarrhea,* stomatitis, mucositis, hepatotoxicity *(elevated LFT).* **Body as a Whole:** *Pain on injection,* venous pain, thrombophlebitis, *alopecia,* myalgia, muscle weakness.

INTERACTIONS Drug: Increased severity of granulocytopenia in combination with **cisplatin;** increased risk of acute pulmonary reactions in combination with **mitomycin; paclitaxel** may increase neuropathy.

PHARMACOKINETICS Distribution: 60–80% bound to plasma proteins (including platelets and lymphocytes); sequestered in tissues, especially lung, spleen, liver, and kidney, and released slowly. **Metabolism:** In liver (CYP3A4). **Elimination:** Primarily in bile and feces (50%), 10% in urine. **Half-Life:** 42–45 h.

NURSING IMPLICATIONS

Assessment & Drug Effects

- Withhold drug and notify prescriber if the granulocyte count is less than 1000 cells/mm³.
- Monitor for neurologic dysfunction including paresthesia, decreased deep tendon reflexes, weakness, constipation, and paralytic ileus.
- Lab tests: Monitor CBC with differential throughout therapy and on the day of treatment prior to each infusion. Monitor LFTs and kidney functions, and serum electrolytes periodically.
- Monitor for S&S of infection, especially during period of granulocyte nadir 7–10 days after dosing.

Patient & Family Education

- Be aware of potential and inevitable adverse effects.
- Women should use reliable forms of contraception to prevent pregnancy.
- Notify prescriber of distressing adverse effects, especially symptoms of leukopenia (e.g., chills,

fever, cough) and peripheral neuropathy (e.g., pain, numbness, tingling in extremities).
- Report changes in bowel habits as soon as manifested.

VITAMIN A

(vye′ta-min A)

Aquasol A, Del-Vi-A
Classification: VITAMIN SUPPLEMENT
Therapeutic: VITAMIN A REPLACEMENT
Pregnancy Category: A (X if greater than RDA)

AVAILABILITY 5000 international unit tablets; 10,000 international unit, 15,000 international unit, 25,000 international unit capsules; 50,000 international units/mL injection

ACTION & *THERAPEUTIC EFFECT* Vitamin A, a fat-soluble vitamin, acts as a cofactor in mucopolysaccharide synthesis, cholesterol synthesis, and the metabolism of hydroxysteroids. *Essential for normal growth and development of bones and teeth, for integrity of epithelial and mucosal surfaces, and for synthesis of visual purple necessary for visual adaptation to the dark. Has antioxidant properties.*

USES Vitamin A deficiency and as dietary supplement during periods of increased requirements, such as pregnancy, lactation, infancy, and infections. Used as replacement therapy in conditions that affect absorption, mobilization, or storage of vitamin A (e.g., steatorrhea, severe biliary obstruction, liver cirrhosis, total gastrectomy). Used in skin disorders [e.g., folliculosis keratosis (Darier's disease), psoriasis]; however, other retinoids are being preferentially selected. Also used as a screening test for fat malabsorption.

V

Common adverse effects in *italic*, life-threatening effects <u>underlined</u>; generic names in **bold**; classifications in SMALL CAPS; ♣ Canadian drug name; ⊘ Prototype drug

1575

CONTRAINDICATIONS History of sensitivity to vitamin A, hypervitaminosis A, oral administration to patients with malabsorption syndrome. Safe use in amounts exceeding 6000 international units during pregnancy (category X if greater than RDA) is not established.

CAUTIOUS USE Women on oral contraceptives, hepatic disease, hepatic dysfunction, hepatitis; renal disease; pregnancy (category A within RDA limit), lactation; low-birth weight infants, children.

ROUTE & DOSAGE

Severe Deficiency

Adult/Child (older than 8 y): **PO** 500,000 international units/day for 3 days followed by 50,000 international units/day for 2 wk, then 10,000–20,000 international units/day for 2 mo **IM** 100,000 international units/day for 3 days followed by 50,000 international units/day for 2 wk
Child: **PO/IM** *Younger than 1 y,* 10,000 international units/kg/day for 3 days followed by 7500–15,000 international units/day for 10 days; *1–8 y,* 10,000 international units/kg/day for 3 days followed by 17,000–35,000 international units/day for 2 wk

Dietary Supplement

Child: **PO** *Younger than 4 y,* 10,000 international units/day; *4–8 y,* 15,000 international units/day

ADMINISTRATION

Oral

- Give on an empty stomach or following food or milk if GI upset occurs.
- Store in tight, light-resistant containers.

Intramuscular

- Use IM route only if oral route not feasible.
- Inject deeply into a large muscle.

ADVERSE EFFECTS (≥1%) **CNS:** Irritability, headache, intracranial hypertension (pseudotumor cerebri), increased intracranial pressure, bulging fontanelles, papilledema, exophthalmos, miosis, nystagmus. **Metabolic:** Hypervitaminosis A syndrome (malaise, lethargy, abdominal discomfort, anorexia, vomiting), hypercalcemia. **Musculoskeletal:** Slow growth; deep, tender, hard lumps (subperiosteal thickening) over radius, tibia, occiput; migratory arthralgia; retarded growth; premature closure of epiphyses. **Skin:** Gingivitis, lip fissures, excessive sweating, drying or cracking of skin, pruritus, increase in skin pigmentation, massive desquamation, brittle nails, alopecia. **Urogenital:** Hypomenorrhea. **GI:** Hepatosplenomegaly, jaundice. **Endocrine:** Polydipsia, polyurea. **Hematologic:** <u>Leukopenia</u>, hypoplastic anemias, vitamin A plasma levels greater than 1200 international units/dL, elevations of sedimentation rate and prothrombin time. **Body as a Whole:** <u>Anaphylaxis, death</u> (after IV use).

DIAGNOSTIC TEST INTERFERENCE Vitamin A may falsely increase *serum cholesterol* determinations *(Zlatkis-Zak reaction);* may falsely elevate *bilirubin* determination (with *Ehrlich's reagent*).

INTERACTIONS Drug: Mineral oil, cholestyramine may decrease absorption of vitamin A.

PHARMACOKINETICS Absorption: Readily absorbed from GI tract in presence of bile salts, pancreatic lipase, and dietary fat. **Distribution:** Stored mainly in liver; small amounts also found in kidney and body fat; distributed into breast milk. **Metabolism:** In liver. **Elimination:** In feces and urine.

V

NURSING IMPLICATIONS

Assessment & Drug Effects

- Take dietary and drug history (e.g., intake of fortified foods, dietary supplements, self-administration or prescription drug sources). Women taking oral contraceptives tend to have significantly higher plasma vitamin A levels.
- Monitor therapeutic effectiveness. Vitamin A deficiency is often associated with protein malnutrition as well as other vitamin deficiencies. It may manifest as night blindness, restriction of growth and development, epithelial alterations, susceptibility to infection, abnormal dryness of skin, mouth, and eyes (xerophthalmia) progressing to keratomalacia (ulceration and necrosis of cornea and conjunctiva), and urinary tract calculi.

Patient & Family Education

- Avoid use of mineral oil while on vitamin A therapy.
- Notify prescriber of symptoms of overdosage (e.g., nausea, vomiting, anorexia, drying and cracking of skin or lips, headache, loss of hair).

VITAMIN B₁
See Thiamine HCl.

VITAMIN B₂
See Riboflavin.

VITAMIN B₃
See Niacin.

VITAMIN B₆
See Pyridoxine.

VITAMIN B₉
See Folic acid.

VITAMIN B₁₂
See Cyanocobalamin.

VITAMIN B₁₂ₐ
See Hydroxocobalamin.

VITAMIN C
See Ascorbic acid.

VITAMIN D
See Calcitriol, Ergocalciferol.

VITAMIN E (TOCOPHEROL)
(vit′a-min E)

Aquasol E, Vita-Plus E, Vitec
Classification: VITAMIN SUPPLEMENT
Therapeutic: VITAMIN E SUPPLEMENT
Pregnancy Category: A within RDA

AVAILABILITY 100 international unit, 200 international unit, 400 international unit, 500 international unit, 800 international unit tablets; 100 international unit, 200 international unit, 400 international unit, 1000 international unit capsules; 15 international units/0.3 mL, 15 international units/30 mL liquid

ACTION & *THERAPEUTIC EFFECT* A group of naturally occurring fat-soluble substances known as tocopherols. Alpha tocopherol, comprising 90% of the tocopherols, is the most biologically potent. An antioxidant, it prevents peroxidation, a process that gives rise to free radicals (highly reactive chemical structures that damage cell membranes and alter nuclear proteins). *Prevents cell membrane and protein damage, protects against blood clot formation by decreasing platelet aggregation, enhances vitamin A utilization, and promotes normal growth, development, and tone of muscles.*

USES To treat and prevent hemolytic anemia due to vitamin E deficiency in premature neonates; to prevent retrolental fibroplasia secondary to oxygen treatment in neonates, and

V

in treatment of diseases with secondary erythrocyte membrane abnormalities (e.g., sickle cell anemia, and G6PD deficiency and as supplement in malabsorption syndromes). Used in patients on diets containing large amounts of polyunsaturated fats for long periods and in the patient who abruptly discontinues such a diet. Also used topically for dry or chapped skin and minor skin disorders.

UNLABELED USES Muscular dystrophy and a number of other conditions with no conclusive evidence of value. A component of many multivitamin formulations and of topical deodorant preparations as an antioxidant.

CONTRAINDICATIONS Bleeding disorders; thrombocytopenia.

CAUTIOUS USE Large doses may exacerbate iron deficiency anemia; pregnancy (category A within RDA).

ROUTE & DOSAGE

Vitamin E Deficiency

Adult: **PO** 60–75 international units/day
Child: **PO** 1 international unit/ kg/day

Prophylaxis for Vitamin E Deficiency

Adult: **PO** 12–15 international units/day
Child: **PO** 7–10 international units/day
Neonate: **PO** 5 international units/day

ADMINISTRATION

Oral

- Give on an empty stomach or following food or milk if GI upset occurs.
- Ensure that capsules are swallowed whole. They should not be crushed or chewed.

- Store in tight containers protected from light.

ADVERSE EFFECTS (≥1%) **Body as a Whole:** Skeletal muscle weakness, headache, fatigue (with excessive doses). **GI:** Nausea, diarrhea, intestinal cramps. **Urogenital:** Gonadal dysfunction. **Metabolic:** Increased serum creatine kinase, cholesterol, triglycerides; decreased serum thyroxine and triiodothyronine; increased urinary estrogens, androgens; creatinuria. **Skin:** Sterile abscess, thrombophlebitis, contact dermatitis. **Special Senses:** Blurred vision.

INTERACTIONS Herbal: Mineral oil, cholestyramine may decrease absorption of vitamin E; may enhance anticoagulant activity of **warfarin.**

PHARMACOKINETICS Absorption: 20–60% absorbed from GI tract if fat absorption is normal; enters blood via lymph. **Distribution:** Stored mainly in adipose tissue; crosses placenta. **Metabolism:** In liver. **Elimination:** Primarily in bile.

NURSING IMPLICATIONS

Patient & Family Education

- Natural sources of vitamin E are found in wheat germ (the richest source) as well as in vegetable oils (sunflower, corn, soybean, cottonseed), green leafy vegetables, nuts, dairy products, eggs, cereals, meat, and liver.

VORICONAZOLE

(vor-i-con'a-zole)

Vfend
Classifications: ANTIBIOTIC; AZOLE ANTIFUNGAL
Therapeutic: ANTIFUNGAL
Prototype: Fluconazole
Pregnancy Category: D

Common adverse effects in *italic*, life-threatening effects underlined; generic names in **bold;** classifications in SMALL CAPS; ♣ Canadian drug name; ● Prototype drug

AVAILABILITY 50 mg, 200 mg tablets; 200 mg injection, 45 mg/mL oral suspension

ACTION & *THERAPEUTIC EFFECT*
Inhibits fungal cytochrome P450 enzymes used for an essential step in fungal ergosterol biosynthesis. The subsequent loss of ergosterol in the fungal cell wall is thought to be responsible for the antifungal activity. *Voriconazole is active against Aspergillus and* Candida.

USES Treatment of invasive aspergillosis, esophageal candidiasis, candidemia in nonneutropenic patients and disseminated skin infections, and abdomen, kidney, bladder wall, and wound infections due to *Candida.*

CONTRAINDICATIONS Known hypersensitivity to voriconazole; **IV form:** Should be avoided in moderate or severe renal impairment (CrCl less than 50 mL/min) and severe Child-Pugh class C hepatic impairment. History of galactose intolerance; Lapp lactase deficiency or glucose-galactose malabsorption; sunlight (UV) exposure; pregnancy (category D); lactation.
CAUTIOUS USE Mild to moderate hepatic cirrhosis, hepatitis, Child-Pugh class A and B hepatic disease; renal disease. **PO & IV form:** Mild or moderate renal impairment; ocular disease; hypersensitivity to other azole antifungal agents such as fluconazole. Safety and efficacy have not been established in children younger than 12 y.

ROUTE & DOSAGE

Aspergillosis

Adult: **IV** 6 mg/kg q12h day 1, then 3–4 mg/kg q12h. Treatment continues until 7–14 days after symptom resolution. **PO** *Weight greater than 40 kg,* 400 mg q12h day 1, then 200 mg q12h. May increase to 300 mg q12h if

inadequate response. *Weight less than 40 kg,* 400 mg q12h day 1, then 100 mg q12h. May increase to 150 mg q12h if inadequate response.

Esophageal Candidiasis

Adult: **PO** *Weight greater than 40 kg,* 200 mg q12h for a minimum of 14 days and for at least 7 days after resolution of symptoms; *weight less than 40 kg,* 100 mg q12h for a minimum of 14 days and for at least 7 days after resolution of symptoms

Dose Adjustment for Concomitant Fosphenytoin or Phenytoin

Adult: **IV** 6 mg/kg q12h day 1, then 5 mg/kg q12h. **PO** *Weight greater than 40 kg,* 400 mg q12h day 1, then 400 mg q12h; *weight less than 40 kg,* 400 mg q12h day 1, then 200 mg q12h

Renal Impairment Dosage Adjustment

CrCl less than 50 mL/min: Switch to PO therapy after loading dose; hemodialysis does not require supplemental dose

Hepatic Impairment Dosage Adjustment

Child-Pugh class A or B: Reduce maintenance dose by 50%; *Child-Pugh class C:* Avoid drug use

ADMINISTRATION

Oral
- Give at least 1 h before or 1 h following a meal.
- Store tablets at 15°–30° C (59°–86° F).

Intravenous

PREPARE: **Intermittent:** Use a 20 mL syringe to reconstitute each

V

Common adverse effects in *italic,* life-threatening effects underlined; generic names in **bold;** classifications in SMALL CAPS; ✤ Canadian drug name; ◎ Prototype drug

1579

200 mg powder vial with exactly 19 mL of sterile water for injection to yield 10 mg/mL. Discard vial if a vacuum does not pull the diluent into vial. Shake until completely dissolved. ▪ Calculate the required dose of voriconazole based on patient's weight. ▪ From an IV infusion bag of NS, D5W, D5/NS, D5/.45NS, LR or other suitable solution, withdraw and discard a volume of IV solution equal to the required dose. ▪ Inject the required dose of voriconazole into the IV bag. The IV solution should have a final voriconazole concentration of 0.5–5 mg/mL. ▪ Infuse immediately.

ADMINISTER: Intermittent: Infuse over 1–2 h at a maximum rate of 3 mg/kg/h. ▪ DO NOT give a bolus dose.

***INCOMPATIBILITIES* Solution/additive:** Do not dilute with **sodium bicarbonate;** do not mix with any other drugs. **Y-site:** Do not infuse with other drugs.

▪ Store unreconstituted vials at 15°–30° C (59°–86° F).

ADVERSE EFFECTS (≥1%) Body as a Whole: Peripheral edema, fever, chills. **CNS:** Headache, hallucinations, dizziness. **CV:** Tachycardia, hypotension, hypertension, vasodilation. **GI:** Nausea, vomiting, abdominal pain, abnormal LFTs, diarrhea, cholestatic jaundice, dry mouth. **Metabolic:** Increased alkaline phosphatase, AST, ALT, hypokalemia, hypomagnesemia. **Skin:** Rash, pruritus. **Special Senses:** *Abnormal vision (enhanced brightness, blurred vision, or color vision changes),* photophobia.

INTERACTIONS Drug: Due to significant increased toxicity or decreased activity, the following drugs are contraindicated with voriconazole: BARBITURATES, **carbamazepine, efavirenz,** ERGOT ALKALOIDS, **pimozide,**

quinidine, rifabutin, sirolimus; fosphenytoin, phenytoin, rifampin, ritonavir may significantly decrease voriconazole levels. PROTEASE INHIBITORS (except **indinavir**) may increase voriconazole toxicity; voriconazole may increase the toxicity of BENZODIAZEPINES, **cyclosporine,** PROTEASE INHIBITORS (except **indinavir**), NONNUCLEOSIDE REVERSE TRANSCRIPTASE INHIBITORS, **omeprazole, tacrolimus, vinblastine, vincristine, warfarin, fentanyl, oxycodone,** NSAIDS; NONNUCLEOSIDE REVERSE TRANSCRIPTASE INHIBITORS may increase or decrease voriconazole levels. **Food:** Absorption reduced with high-fat meals. **Herbal: St. John's wort** may decrease efficacy.

PHARMACOKINETICS Absorption: 96% absorbed. Has a nonlinear pharmacokinetic profile, a small change in dose may cause a large change in serum levels. Steady state not achieved until day 5–6 if no loading dose is given. **Peak:** 1–2 h. **Metabolism:** In liver by (and inhibits) CYP3A4, 2C9 and 2C19. **Elimination:** Primarily in urine. **Half-Life:** 6 h–6 days depending on dose.

NURSING IMPLICATIONS

Assessment & Drug Effects

▪ Visual acuity, visual field, and color perception should be monitored if treatment continues beyond 28 days.
▪ Withhold drug and notify prescriber if skin rash develops.
▪ Monitor cardiovascular status especially with preexisting CV disease.
▪ Lab tests: Monitor baseline and periodic LFTs including bilirubin; patients who develop abnormal LFTs during therapy should be monitored for the development of more severe hepatic injury. Monitor frequently renal function tests, especially serum creatinine. Monitor periodic CBC with platelet count,

Common adverse effects in *italic*, life-threatening effects underlined; generic names in **bold;** classifications in SMALL CAPS; ♣ Canadian drug name; ⊘ Prototype drug

Hct and Hgb, serum electrolytes, alkaline phosphatase, blood glucose, and lipid profile.

- Concurrent drugs: Monitor PT/INR closely with warfarin as dose adjustments of warfarin may be needed. Monitor frequently blood glucose levels with sulfonylurea drugs as reduction in the sulfonylurea dosage may be needed. Monitor for and report any of the following: S&S of rhabdomyolysis in patient receiving a statin drug; prolonged sedation in patient receiving a benzodiazepine; S&S of heart block, bradycardia, or CHF in patient receiving a calcium channel blocker.

Patient & Family Education

- Use reliable means of birth control to prevent pregnancy. If you suspect you are pregnant, contact prescriber immediately.
- Do not drive at night while taking voriconazole as the drug may cause blurred vision and photophobia.
- Do not drive or engage in other potentially hazardous activities until reaction to drug is known.
- Avoid strong, direct sunlight while taking voriconazole.

WARFARIN SODIUM
(war'far-in)

Coumadin Sodium, Warfilone ♣
Classification: ANTICOAGULANT
Therapeutic: ANTICOAGULANT
Pregnancy Category: X

AVAILABILITY 1 mg, 2 mg, 2.5 mg, 3 mg, 4 mg, 5 mg, 6 mg, 7.5 mg, 10 mg tablets; 2.5 mg/mL injection

ACTION & *THERAPEUTIC EFFECT*
Indirectly interferes with blood clotting by depressing hepatic synthesis of vitamin K-dependent coagulation factors: II, VII, IX, and X. *Deters further extension of existing thrombi and prevents new clots from forming.*

USES Prophylaxis and treatment of deep vein thrombosis and its extension, pulmonary embolism; treatment of atrial fibrillation with embolization. Also used as adjunct in treatment of coronary occlusion, cerebral transient ischemic attacks (TIAs), and as a prophylactic in patients with prosthetic cardiac valves. Used extensively as rodenticide.

CONTRAINDICATIONS Hemorrhagic tendencies, vitamin C or K deficiency, hemophilia, coagulation factor deficiencies, dyscrasias; active bleeding; open wounds, active peptic ulcer, visceral carcinoma, esophageal varices, malabsorption syndrome; uncontrolled hypertension, cerebral vascular disease; heparin-induced thrombocytopenia (HIT); pericarditis with acute MI; severe hepatic or renal disease; continuous tube drainage of any orifice; subacute bacterial endocarditis; recent surgery of brain, spinal cord, or eye; regional or lumbar block anesthesia; threatened abortion; unreliable patients; pregnancy (category X).

CAUTIOUS USE Alcoholism, allergic disorders, during menstruation, older adults, senility, psychosis; debilitated patients. Endogenous factors that may increase prothrombin time response (enhance anticoagulant effect): Carcinoma, CHF, collagen diseases, hepatic and renal insufficiency, diarrhea, fever, pancreatic disorders, malnutrition, vitamin K deficiency. Endogenous factors that may decrease prothrombin time response (decrease anticoagulant response): Edema, hypothyroidism, hyperlipidemia, hypercholesterolemia, chronic alcoholism, hereditary resistance to coumarin therapy.

W

ROUTE & DOSAGE

Anticoagulant

Adult: **PO/IV** Usual dose 2–10 mg daily with dose adjusted to maintain a PT 1.2–2 × control or INR of 2–3
Child: **PO** 0.1–0.3 mg/kg/day, adjust to maintain INR of 2–3

Pharmacogenetic Dosage Adjustment

Variations in CYP2C9 or VKORC1 may require dose adjustments

ADMINISTRATION

Note: Antidote for bleeding—anticoagulant effect usually is reversed by omitting 1 or more doses of warfarin and by administration of specific antidote phytonadione (vitamin K_1) 2.5–10 mg orally. Prescriber may advise patient to carry vitamin K_1 at all times, but not to take it until after consultation. If bleeding persists or progresses to a severe level, vitamin K 15–25 mg IV is given, or a fresh whole blood transfusion may be necessary.

Oral

- Give tablet whole or crushed with fluid of patient's choice.

Intravenous

PREPARE: **Direct:** Add 2.7 mL of sterile water for injection to the 5 mg vial.
ADMINISTER: **Direct:** Give required dose over 1–2 min.
INCOMPATIBILITIES **Solution/additive: Ammonium chloride, 5% dextrose, lactated Ringer's, atropine, calcium chloride, calcium gluconate, chloramphenicol, chlorothiazide, chlortetracycline, erythromycin, methicillin, nitrofurantoin, oxacillin, oxytetracycline, penicillin, pentobarbital, phenobarbital, promethazine, sodium bicarbonate, succinyl chloride, vitamin B with C. Y-site: Aminophylline, ammonium chloride, ceftazidime, cephalothin, cimetidine, ciprofloxacin, dobutamine, esmolol, gentamicin, labetalol, metronidazole, promazine, lactated Ringer's, vancomycin.**

- Store at 15°–30° C (59°–86° F). Discard discolored or precipitated solutions. Protect all preparations from light and moisture.

ADVERSE EFFECTS (≥1%) **Body as a Whole:** Major or minor hemorrhage from any tissue or organ; hypersensitivity (dermatitis, urticaria, pruritus, fever). **GI:** Anorexia, nausea, vomiting, abdominal cramps, diarrhea, steatorrhea, stomatitis. **Other:** Increased serum transaminase levels, hepatitis, jaundice, burning sensation of feet, transient hair loss. **Overdosage:** Internal or external bleeding, paralytic ileus; skin necrosis of toes (purple toes syndrome), tip of nose, buttocks, thighs, calves, female breast, abdomen, and other fat-rich areas.

DIAGNOSTIC TEST INTERFERENCE Warfarin (coumarins) may cause alkaline urine to be red-orange; may enhance *uric acid* excretion, cause elevation of *serum transaminases,* and may increase *lactic dehydrogenase* activity.

INTERACTIONS Drug: In addition to the drugs listed below, many other drugs have been reported to alter the expected response to warfarin; however, clinical importance of these reports has not been substantiated. The addition or withdrawal of any drug to an established drug regimen should be made cautiously, with more

Common adverse effects in *italic,* life-threatening effects <u>underlined</u>; generic names in **bold;** classifications in SMALL CAPS; ✚ Canadian drug name; ⊘ Prototype drug

W

frequent INR determinations than usual and with careful observation of the patient and dose adjustment as indicated. The following may enhance the anticoagulant effects of warfarin: **Acetohexamide, acetaminophen,** ALKYLATING AGENTS, **allopurinol,** AMINOGLYCOSIDES, **aminosalicylic acid, amiodarone,** ANABOLIC STEROIDS, ANTIBIOTICS (ORAL), ANTIMETABOLITES, ANTIPLATELET DRUGS, **aspirin, asparaginase, capecitabine, celecoxib, chloramphenicol, chlorpropamide, chymotrypsin, cimetidine, clofibrate, co-trimoxazole, danazol, dextran, dextrothyroxine, diazoxide, disulfiram, erythromycin, ethacrynic acid, fluconazole, glucagons, guanethidine,** HEPATOTOXIC DRUGS, **influenza vaccine, isoniazid, itraconazole, ketoconazole,** MAO INHIBITORS, **meclofenamate, mefenamic acid, methyldopa, methylphenidate, metronidazole, miconazole, mineral oil, nalidixic acid, neomycin (oral),** NONSTEROIDAL ANTI-INFLAMMATORY DRUGS, **oxandrolone, plicamycin,** POTASSIUM PRODUCTS, **propoxyphene, propylthiouracil, quinidine, quinine, rofecoxib, salicylates, streptokinase, sulindac,** SULFONAMIDES, SULFONYLUREAS, TETRACYCLINES, THIAZIDES, THYROID DRUGS, **tolbutamide,** TRICYCLIC ANTIDEPRESSANTS, **urokinase, vitamin E, zileuton.** The following may increase or decrease the anticoagulant effects of warfarin: **Alcohol** (acute intoxication may increase, chronic alcoholism may decrease effects), **chloral hydrate,** DIURETICS. The following may decrease the anticoagulant effects of warfarin: BARBITURATES, **carbamazepine, cholestyramine,** CORTICOSTEROIDS, **corticotropin, glutethimide, griseofulvin,** LAXATIVES, **mercaptopurine,** ORAL CONTRACEPTIVES, **rifampin, spironolactone, vitamin C, vitamin K. Herbal: Boldo, capsicum, celery, chamomile, chondroitin, clove, coenzyme Q10, danshen, devil's claw, dong quai, echinacea, evening primrose oil, fenugreek, feverfew, fish oil, garlic, ginger, ginkgo, glucosamine, horse chestnut, licorice root, passionflower herb, turmeric, willow bark** may increase risk of bleeding; **ginseng, green tea, seaweed, soy, St. John's wort** may decrease effectiveness of warfarin. **Food: Cranberry juice** may increase INR. Green leafy vegetables may affect efficacy. **Avocado** may decrease effectiveness of warfarin.

PHARMACOKINETICS Absorption: Well absorbed from GI tract. **Onset:** 2–7 days. **Peak:** 0.5–3 days. **Distribution:** 97% protein bound; crosses placenta. **Metabolism:** In liver (CY-P2C9). **Elimination:** In urine and bile. **Half-Life:** 0.5–3 days.

NURSING IMPLICATIONS

Assessment & Drug Effects

- Determine PT/INP prior to initiation of therapy and then daily until maintenance dosage is established.
- Obtain a COMPLETE medication history prior to start of therapy and whenever altered responses to therapy require interpretation; extremely IMPORTANT since many drugs interfere with the activity of anticoagulant drugs (see INTERACTIONS).
- Dose is typically adjusted to maintain PT at 1½–2½ times the control (12–15 sec), or 15–35% of normal prothrombin activity, or an INR of 2–4 depending on diagnosis.
- Lab tests: For maintenance dosage, PT/INR determinations at

W

Common adverse effects in *italic*; life-threatening effects <u>underlined</u>; generic names in **bold**; classifications in SMALL CAPS; ✚ Canadian drug name; ⊙ Prototype drug

1583

1–4-wk intervals depending on patient's response; periodic urinalyses, stool guaiac, and LFTs. Blood samples for PT/INR should be drawn at 12–18 h after last dose (optimum).

- Note: Patients at greatest risk of hemorrhage include those whose PT/INR are difficult to regulate, who have an aortic valve prosthesis, who are receiving long-term anticoagulant therapy, and older adult and debilitated patients.

Patient & Family Education

- Understand that bleeding can occur even though PT/INR are within therapeutic range. Stop drug and notify prescriber immediately if bleeding or signs of bleeding appear: Blood in urine, bright red or black tarry stools, vomiting of blood, bleeding with tooth brushing, blue or purple spots on skin or mucous membrane, round pinpoint purplish red spots (often occur in ankle areas), nosebleed, bloody sputum; chest pain; abdominal or lumbar pain or swelling, profuse menstrual bleeding, pelvic pain; severe or continuous headache, faintness or dizziness; prolonged oozing from any minor injury (e.g., nicks from shaving).
- Stop drug and report immediately any symptoms of hepatitis (dark urine, itchy skin, jaundice, abdominal pain, light stools) or hypersensitivity reaction (see Appendix F).
- Take drug at same time each day, and do NOT alter dose.
- Risk of bleeding is increased for up to 1 mo after receiving the influenza vaccine.
- Fever, prolonged hot weather, malnutrition, and diarrhea lengthen PT/INR (enhanced anticoagulant effect).
- A high-fat diet, sudden increase in vitamin K–rich foods (cabbage, cauliflower, broccoli, asparagus, lettuce, turnip greens, onions, spinach, kale, fish, liver), coffee or green tea (caffeine), or by tube feedings with high vitamin K content shorten PT/INR.
- Avoid excess intake of alcohol.
- Use a soft toothbrush and floss teeth gently with waxed floss.
- Use barrier contraceptive measures; if you become pregnant while on anticoagulant therapy the fetus is at great potential risk of congenital malformations.
- Do not take any other prescription or OTC drug unless specifically approved by prescriber or pharmacist.

XYLOMETAZOLINE HYDROCHLORIDE
(zye-loe-met-az'oh-leen)

Otrivin
Classifications: NASAL DECONGESTANT; VASOCONSTRICTOR
Therapeutic: NASAL DECONGESTANT
Prototype: Naphazoline
Pregnancy Category: C

AVAILABILITY 0.05%, 0.1% nasal solution

ACTION & *THERAPEUTIC EFFECT*
Markedly constricts dilated arterioles of nasal membrane. *Decreases fluid exudate and mucosal engorgement associated with rhinitis and may open up obstructed eustachian tubes.*

USES Temporary relief of nasal congestion associated with common cold, sinusitis, acute and chronic rhinitis, and hay fever and other allergies.

CONTRAINDICATIONS Sensitivity to adrenergic substances; angle-closure glaucoma; concurrent therapy with MAO inhibitors or tricyclic antidepressants; lactation.

Common adverse effects in *italic*, life-threatening effects underlined; generic names in **bold**; classifications in SMALL CAPS; ♣ Canadian drug name; ⊕ Prototype drug

CAUTIOUS USE Hypertension; hyperthyroidism; heart disease, including angina; advanced arteriosclerosis, older adults, pregnancy (category C). **PO:** children younger than 12 y, **Nasal:** children younger than 2 y and infants.

ROUTE & DOSAGE

Nasal Congestion

Adult/Child (12 y or older): **Nasal** 1–2 sprays or 1–2 drops of 0.1% solution in each nostril q8–10h (max: 3 doses/day)
Child (2–12 y): **Nasal** 1 spray or 2–3 drops of 0.05% solution in each nostril q8–10h (max: 3 doses/day)

ADMINISTRATION

Instillation

- Have patient clear each nostril gently before administering spray or drops.
- Store at 15°–30° C (59°–86° F) in a tight, light-resistant container.

ADVERSE EFFECTS (≥1%) **All:** Usually mild and infrequent; local stinging, burning, dryness and ulceration, sneezing, headache, insomnia, drowsiness. **With Excessive Use:** *Rebound nasal congestion* and vasodilation, tremulousness, hypertension, palpitations, tachycardia, arrhythmia, somnolence, sedation, coma.

INTERACTIONS Drug: May cause increase BP with **guanethidine, methyldopa,** MAO INHIBITORS; PHENOTHIAZINES may decrease effectiveness of nasal decongestant.

PHARMACOKINETICS Onset: 5–10 min. **Duration:** 5–6 h.

NURSING IMPLICATIONS

Assessment & Drug Effects

- Evaluate for development of rebound congestion (see ADVERSE EFFECTS).

Patient & Family Education

- Prevent contamination of nasal solution and spread of infection by rinsing dropper and tip of nasal spray in hot water after each use; restrict use to the individual patient.
- Note: Prolonged use can cause rebound congestion and chemical rhinitis. Do NOT exceed prescribed dosage and report to prescriber if drug fails to provide relief within 3–4 days.
- Do NOT self-medicate with OTC drugs, sprays, or drops without prescriber's approval.
- Note: Excessive use by a child may lead to CNS depression.

ZAFIRLUKAST 🅟

(za-fir-lu'kast)

Accolate
Classifications: RESPIRATORY SMOOTH MUSCLE RELAXANT; LEUKOTRIENE RECEPTOR ANTAGONIST (LTRA); BRONCHODILATOR
Therapeutic: BRONCHODILATOR; LTRA
Pregnancy Category: B

AVAILABILITY 10 mg, 20 mg tablets

ACTION & *THERAPEUTIC EFFECT*
Selective leukotriene receptor antagonist (LTRA) that inhibits binding of leukotriene D_4 and E_4, thus inhibiting inflammation and bronchoconstriction. Leukotriene production and receptor affinity have been correlated with the pathogenesis of asthma. *Zafirlukast helps to prevent the signs and symptoms of asthma, including airway edema, smooth muscle constriction, and altered cellular activity due to inflammation.*

USES Prophylaxis and chronic treatment of asthma in adults and children older than 5 y (not for acute bronchospasm).

Common adverse effects in *italic*, life-threatening effects <u>underlined</u>; generic names in **bold**; classifications in SMALL CAPS; ♦ Canadian drug name; 🅟 Prototype drug

1585

CONTRAINDICATIONS Hypersensitivity to zafirlukast; acute asthma attacks, including status asthmaticus, acute bronchospasm; lactation.

CAUTIOUS USE Hepatic impairment, hepatic disease; corticosteroid withdrawal or reduction in dose; patients 65 y or older, pregnancy (category B); children younger than 5 y.

ROUTE & DOSAGE

Asthma

Adult: **PO** 20 mg b.i.d. 1 h before or 2 h after meals
Child (older than 5 y): **PO** 10 mg b.i.d.

ADMINISTRATION

Oral

- Give 1 h before or 2 h after meals.
- Store at 20°–25° C (68°–77° F); protect from light and moisture.

ADVERSE EFFECTS (≥1%) **Body as a Whole:** Generalized pain, asthenia, myalgia, fever, back pain. **CNS:** *Headache,* dizziness. **GI:** Nausea, diarrhea, abdominal pain, vomiting, dyspepsia; liver dysfunction, increased liver function tests, <u>hepatic failure</u>. **Other:** <u>Churg-Strauss syndrome</u> (fever, muscle aches and pains, weight loss).

INTERACTIONS Drug: May increase prothrombin time (PT) in patients on **warfarin. Erythromycin** decreases bioavailability of zafirlukast.

PHARMACOKINETICS Absorption: Rapidly from GI tract, bioavailability significantly reduced by food. **Onset:** 1 wk. **Peak:** 3 h. **Distribution:** Greater than 99% protein bound; secreted into breast milk. **Metabolism:** In liver (CYP2C9). **Elimination:** 90% in feces, 10% in urine. **Half-Life:** 10 h.

NURSING IMPLICATIONS

Assessment & Drug Effects

- Assess respiratory status and airway function regularly.
- Lab tests: Periodic LFTs.
- Monitor closely PT and INR with concurrent warfarin therapy.
- Monitor closely phenytoin level with concurrent phenytoin therapy.

Patient & Family Education

- Taking medication regularly, even during symptom-free periods.
- Note: Drug is not intended to treat acute episodes of asthma.
- Report S&S of hepatic toxicity (see Appendix F) or flu-like symptoms to prescriber. Follow-up lab work is very important.
- Notify prescriber immediately if condition worsens while using prescribed doses of all antiasthmatic medications.

ZALEPLON

(zal'ep-lon)

Sonata

Classifications: ANXIOLYTIC; SEDATIVE-HYPNOTIC; NONBENZODIAZEPINE
Therapeutic: SEDATIVE-HYPNOTIC; ANTIANXIETY
Prototype: Zolpidem
Pregnancy Category: C
Controlled Substance: Schedule IV

AVAILABILITY 5 mg, 10 mg capsules

ACTION & *THERAPEUTIC EFFECT*
Short-acting nonbenzodiazepine with sedative-hypnotic, muscle relaxant, and anticonvulsant activity. *Reduces difficulty in initially falling asleep. Preserves deep sleep (stage 3 through stage 4) at hypnotic dose with minimal-to-absent rebound insomnia when discontinued.*

USES Short-term treatment of insomnia.

Common adverse effects in *italic*, life-threatening effects <u>underlined</u>; generic names in **bold**; classifications in SMALL CAPS; ♣ Canadian drug name; ♦ Prototype drug

CONTRAINDICATIONS Hypersensitivity to zaleplon, or tartrazine dye (Yellow 5); suicidal ideation; lactation.
CAUTIOUS USE Hypersensitivity to salicylates; concurrent use of other CNS depressants (e.g., benzodiazepines, alcohol); chronic depression; history of drug abuse; COPD; respiratory insufficiency; hepatic or renal impairment; pulmonary disease; pregnancy (category C). Safe use in children not established.

ROUTE & DOSAGE

Insomnia

Adult: **PO** 10 mg at bedtime (max: 20 mg at bedtime)
Geriatric: **PO** 5 mg at bedtime (max: 10 mg at bedtime)

ADMINISTRATION

Oral

- Give immediately before bedtime; not while patient is still ambulating.
- Ensure that extended release tablets are swallowed whole and are not crushed or chewed.
- Sublingual tablets should not be given with water and should not be swallowed.
- Oral spray container must be primed before first use. Ensure that a full dose (5 mg) is sprayed directly into the mouth over the tongue. For a 10 mg dose, administer a second spray.
- Store at 20°–25° C (68°–77° F).

ADVERSE EFFECTS (≥1%) **Body as a Whole:** Asthenia, fever, *headache,* migraine, myalgia, back pain. **CNS:** Amnesia, dizziness, paresthesia, somnolence, tremor, vertigo, depression, hypertonia, nervousness, difficulty concentrating. **GI:** Abdominal pain, dyspepsia, nausea, constipation, dry mouth. **Respiratory:** Bronchitis. **Skin:** Pruritus, rash. **Special Senses:** Eye pain, hyperacusis, conjunctivitis. **Urogenital:** Dysmenorrhea.

INTERACTIONS Drug: Alcohol, imipramine, thioridazine may cause additive CNS impairment; **rifampin** increases metabolism of **zaleplon; cimetidine** increases serum levels of **zaleplon. Herbal: Valerian, melatonin** may produce additive sedative effects. **Food:** High-fat meals may delay absorption.

PHARMACOKINETICS Absorption: Rapidly and completely absorbed, 30% reaches systemic circulation. **Onset:** 15–20 min. **Peak:** 1 h. **Duration:** 3–4 h. **Distribution:** 60% protein bound. **Metabolism:** Extensively in liver (CYP3A4) to inactive metabolites. **Elimination:** 70% in urine, 17% in feces. **Half-Life:** 1 h.

NURSING IMPLICATIONS

Assessment & Drug Effects

- Monitor behavior and notify prescriber for significant changes. Use extra caution with preexisting clinical depression.
- Provide safe environment and monitor ambulation after drug is ingested.
- Monitor respiratory status with preexisting compromised pulmonary function.

Patient & Family Education

- Exercise caution when walking; avoid all hazardous activities after taking zaleplon.
- Do not take in combination with alcohol or any other sleep medication.
- Note: Exhibits altered effectiveness if taken with/immediately after high-fat meal.
- Do not use longer than 2–3 wk.
- Expect possible mild/brief rebound insomnia after discontinuing regimen.
- Report use of OTC medications to prescriber (e.g., cimetidine).
- Report pregnancy to prescriber immediately.

Common adverse effects in *italic*, life-threatening effects underlined; generic names in **bold**; classifications in SMALL CAPS; ♣ Canadian drug name; ⊘ Prototype drug

Z

1587

ZANAMIVIR

(zan'a-mi-vir)

Relenza
Classification: ANTIVIRAL
Therapeutic: ANTIINFLUENZA
Pregnancy Category: C

AVAILABILITY 5 mg blister for inhalation

ACTION & *THERAPEUTIC EFFECT*
Inhibitor of influenza A and B viral enzyme; does not permit the release of newly formed viruses from the surface of the infected cells. *Prevents viral spread across the mucus lining of the respiratory tract, and inhibits the replication of influenza A and B virus. Relieves flu-like symptoms.*

USES Uncomplicated acute influenza in patients symptomatic less than 2 days; prophylaxis for influenza.

CONTRAINDICATIONS Hypersensitivity to zanamivir or milk protein; severe renal impairment, renal failure; COPD; severe asthma.
CAUTIOUS USE Renal impairment; cardiac disease; severe metabolic disease; older adults; pregnancy (category C), lactation. **Acute influenza:** Safety and efficacy in children younger than 7 y are unknown. **Influenza prophylaxis:** Safe use in children younger than 5 y is unknown.

ROUTE & DOSAGE

Influenza Treatment
Adult/Child (older than 7 y):
Inhaled 2 inhalations (one 5 mg blister/inhalation) b.i.d. × 5 days
Influenza Prophylaxis
Adult/Child (older than 5 y):
Inhaled 2 inhalations daily for 10 days (household prophylaxis) or for 28 days (community outbreak)

ADMINISTRATION

Inhalation
- Most effective if initiated within 48 h of onset of flu-like symptoms.
- Give any scheduled inhaled bronchodilator before zanamivir.
- Store at 25° C (77° F).

ADVERSE EFFECTS (≥1%) **Body as a Whole:** Headache, abnormal behavior. **CNS:** Dizziness. **GI:** Nausea, diarrhea, vomiting. **Respiratory:** Nasal symptoms, bronchitis, cough, sinusitis; ear, nose, throat infection.

INTERACTIONS Drug: Do not use with LIVE VACCINES.

PHARMACOKINETICS Absorption: 4–17% of inhaled dose is systemically absorbed. **Peak:** 1–2 h. **Distribution:** Less than 10% protein bound. **Metabolism:** Not metabolized. **Elimination:** In urine. **Half-Life:** 2.5–5.1 h.

NURSING IMPLICATIONS

Patient & Family Education
- Start within 48 h of onset of flu-like symptoms for most effective response.
- Use any scheduled inhaled bronchodilator first; then use zanamivir.

ZICONOTIDE

(zi-con'o-tide)

Prialt
Classifications: MISCELLANEOUS ANALGESIC; N-TYPE CALCIUM CHANNEL ANTAGONIST
Therapeutic: MISCELLANEOUS ANALGESIC
Pregnancy Category: C

AVAILABILITY 100 mcg/mL, 500 mcg/5mL, 500 mcg/20mL injection

ACTION & *THERAPEUTIC EFFECT*
Ziconotide binds to N-type calcium channels located on the afferent

Common adverse effects in *italic*, life-threatening effects underlined; generic names in **bold**; classifications in SMALL CAPS; ♣ Canadian drug name; ● Prototype drug

nerves in the dorsal horn in the spinal cord. It is thought that these binding blocks of N-type calcium channels lead to a blockade of excitatory neurotransmitter release in the afferent nerve endings. *Ziconotide is effective in controlling severe chronic pain that is intractable to other analgesics.*

USES Management of severe chronic pain in patients for whom intrathecal (IT) therapy is warranted.
UNLABELED USES Spasticity associated with spinal cord trauma.

CONTRAINDICATIONS Hypersensitivity to ziconotide; preexisting history of psychosis; epidural or intravenous administration; sepsis; depression with suicidal ideation; cognitive impairment; bipolar disorder; schizophrenia; dementia; presence of infection at the injection site, uncontrolled bleeding, or spinal canal obstruction that impairs circulation of CSF; coagulopathy; seizures; lactation.
CAUTIOUS USE Renal, hepatic, and cardiac impairment; older adults; pregnancy (category C). Safe use in children and infants not established.

ROUTE & DOSAGE

Severe Chronic Pain
Adult: **Intrathecal** Initial 0.1 mcg/h; may titrate up 0.1 mcg/h q2–3days to 0.8 mcg/h (19.2 mcg/day)

ADMINISTRATION

Intrathecal
- May be administered undiluted (25 mcg/mL in 20 mL vial) or diluted using the 100 mcg/mL vials. Diluted ziconotide is prepared with NS without preservatives.
- Administer using an implanted variable-rate microinfusion device or an external microinfusion device and catheter.

- Note: Due to serious adverse events, 19.2 mcg/day (0.8 mcg/h) is the maximum recommended dose.
- Doses should normally be titrated upward by no more than 2.4 mcg/day (0.1 mcg/h) at intervals of 2–3 times/wk.
- Refrigerate all ziconotide solutions after preparation and begin infusion within 24 h. Discard any unused portion left in a vial.

ADVERSE EFFECTS (≥1%) **Body as a Whole:** Accidental injury, back pain, catheter complication, catheter-site pain, cellulitis, chest pain, chills, *fever*, flu syndrome, infection, malaise, neck pain, neck rigidity, *pain*, pump-site complication, pump-site mass, pump-site pain, viral infection. **CNS:** Abnormal dreams, *abnormal gait*, agitation, *anxiety, aphasia, asthenia, ataxia,* CSF abnormal, *confusion,* depression, difficulty concentrating, *dizziness,* dry mouth, *dysesthesia,* emotional lability, *headache,* hostility, hyperesthesia, *hypertonia,* incoordination, insomnia, *memory impairment,* mental slowing, meningitis, *nervousness,* neuralgia, paranoid reaction, *paresthesia,* reflexes decreased, *somnolence, speech disorder,* stupor, abnormal thinking, tremor, twitching, *vertigo.* **CV:** Hypertension, hypotension, postural hypotension, syncope, tachycardia, vasodilation. **GI:** Abdominal pain, *anorexia,* constipation, *diarrhea,* dyspepsia, gastrointestinal disorder, *nausea, vomiting.* **GU:** Dysuria, urinary incontinence, *urinary retention,* urinary tract infection, impaired urination. **Hematologic:** Anemia, ecchymosis. **Metabolic/Nutritional:** Creatinine phosphokinase increased, dehydration, edema, hypokalemia, peripheral edema, weight loss. **Musculoskeletal:** Arthralgia, arthritis, leg cramps, myalgia, myasthenia.

Common adverse effects in *italic*, life-threatening effects underlined; generic names in **bold**; classifications in SMALL CAPS; ✦ Canadian drug name; ☉ Prototype drug

1589

Respiratory: Bronchitis, cough increased, dyspnea, lung disorder, pharyngitis, pneumonia, rhinitis, sinusitis. **Skin:** Cutaneous surgical complication, dry skin, pruritus, rash, skin disorder, sweating. **Special Senses:** *Abnormal vision,* diplopia, *nystagmus,* photophobia, taste perversion, tinnitus.

INTERACTIONS Drug: Ethanol and other CNS DEPRESSANTS may increase drowsiness, dizziness, and confusion.

PHARMACOKINETICS Distribution: 50% protein bound. **Metabolism:** Hydrolyzed by peptidases. **Half-Life:** 4.6 h.

NURSING IMPLICATIONS

Assessment & Drug Effects

- Monitor for and report S&S of meningitis, cognitive impairment, hallucinations, changes in mood or consciousness, or other psychiatric symptoms.
- Lab tests: Serum creatine kinase every other week for first month and monthly thereafter.

Patient & Family Education

- Report any of the following to prescriber: Muscle pain, soreness, or weakness, confusion, unusual behavior, symptoms of depression or suicidal thoughts, fever, headache, stiff neck, nausea or vomiting, seizures.
- Note: Taking this drug with other depressants (e.g., alcohol, sedatives, tranquilizers) will increase the risk of side effects.

ZIDOVUDINE (FORMERLY AZIDOTHYMIDINE, AZT)

(zye-doe'vyoo-deen)

Retrovir

Classifications: ANTIVIRAL; NUCLEOSIDE REVERSE TRANSCRIPTASE INHIBITOR

Therapeutic: ANTIVIRAL; NRTI
Prototype: Lamivudine
Pregnancy Category: C

AVAILABILITY 300 mg tablets; 100 mg capsules; 50 mg/5 mL syrup; 10 mg/mL injection

ACTION & *THERAPEUTIC EFFECT*
Appears to act by being incorporated into growing DNA chains by viral reverse transcriptase, thereby terminating viral replication. *Zidovudine has antiviral action against HIV, LAV (lymphadenopathy-associated virus), and ARV (AIDS-associated retrovirus).*

USES Treatment of HIV (along with other antiretroviral agents), prevention of perinatal transfer of HIV during pregnancy.
UNLABELED USES Postexposure chemoprophylaxis, thrombocytopenia.

CONTRAINDICATIONS Life-threatening allergic reactions to any of the components of the drug; lactic acidosis; lactation.
CAUTIOUS USE Impaired renal or hepatic function, alcoholism; anemia; chemotherapy; radiation therapy; bone marrow depression; pregnancy (category C).

ROUTE & DOSAGE

HIV Infection

Adult/Adolescent/Child (at least 30 kg): **PO** 300 mg b.i.d. OR 200 mg t.i.d.
IV 1 mg/kg given 5–6 times daily
Infant (over 4 weeks)/Child/Adolescent (25–30 kg): **PO** 250 mg b.i.d./day in divided doses
Child (19–25 kg): **PO** 200 mg b.i.d.
Infant (at least 4 weeks of age)/Child (13–19 kg): **PO** 150 mg b.i.d.

Common adverse effects in *italic,* life-threatening effects underlined; generic names in **bold;** classifications in SMALL CAPS; ♥ Canadian drug name; ☺ Prototype drug

Infant (at least 4 weeks of age)/ Child (7–13 kg): 200 mg/day

Prevention of Maternal-Fetal Transmission

Neonate: **PO** *Greater than 34 wk:* 2 mg/kg q6h for 6 wk beginning within 12 h after birth; **IV** *Full term:* 1.5 mg/kg q6h × 6 wk *Maternal:* **PO** 100 mg 5 times daily OR 300 mg b.i.d. from 14–34 wk gestation until delivery **IV** During labor, 2 mg/kg loading dose, then 1 mg/kg/h until clamping umbilical cord

Toxicity Dosage Adjustment
Hemoglobin falls below 7.5 g/dL or falls 25% from baseline: Interrupt therapy. ANC falls below 750 cells/mm3 or decreases 50% from baseline: Interrupt therapy.

ADMINISTRATION

Oral

▪ May be given with or without food.

Intravenous

PREPARE: **Intermittent:** Withdraw required dose from vial and dilute with D5W to a concentration not to exceed 4 mg/mL.

ADMINISTER: **Intermittent for HIV infection:** Give calculated dose at a constant rate over 60 min; avoid rapid infusion. **IV Infusions for Prevention of Maternal-Fetal Transmission:** Give maternal loading dose over 1 h, then continuous infusion at 1 mg/kg/h. **Intermittent Infusion for Prevention of Maternal Transmission of HIV to Neonate:** Give calculated dose at a constant rate over 30 min.

INCOMPATIBILITIES **Solution/additive:** Meropenem. **Y-site:** Gemtuzumab, lansoprazole, meropenem.

▪ Store at 15°–25° C (59°–77° F) and protect from light. Store diluted IV solutions refrigerated for 24 h.

ADVERSE EFFECTS (≥1%) **Body as a Whole:** *Fever,* dyspnea, *malaise,* weakness, *myalgia,* myopathy. **CNS:** *Headache,* insomnia, dizziness, paresthesias, mild confusion, anxiety, restlessness, agitation. **GI:** *Nausea,* diarrhea, *vomiting, anorexia,* GI pain. **Hematologic:** <u>*Bone marrow depression, granulocytopenia, anemia.*</u> **Respiratory:** *Cough, wheezing.* **Skin:** *Rash,* itching, diaphoresis.

INTERACTIONS Drug: **Acetaminophen ganciclovir, interferon-alfa** may enhance bone marrow suppression; **atovaquone, amphotericin B, aspirin, dapsone, doxorubicin, fluconazole, flucytosine, indomethacin, interferon alfa, methadone, pentamidine, vincristine, valproic acid** may increase risk of AZT toxicity; **probenecid** will decrease AZT elimination, resulting in increased serum levels and thus toxicity. **Nelfinavir, rifampin, ritonavir** may decrease zidovudine (AZT) concentrations; other ANTIRETROVIRAL AGENTS may cause lactic acidosis and severe hepatomegaly with steatosis; **stavudine, doxorubicin** may antagonize AZT effects.

PHARMACOKINETICS Absorption: Readily from GI tract; 60–70% reaches systemic circulation (first-pass metabolism). **Peak:** 0.5–1.5 h. **Distribution:** Crosses blood–brain barrier and placenta. **Metabolism:** In liver. **Elimination:** 63–95% in urine. **Half-Life:** 1 h.

NURSING IMPLICATIONS

Assessment & Drug Effects

▪ Evaluate patient at least weekly during the first month of therapy.
▪ Lab tests: Baseline and frequent (at least q2wk) blood counts, CD4

Common adverse effects in *italic*, life-threatening effects <u>underlined</u>; generic names in **bold**; classifications in SMALL CAPS; ✦ Canadian drug name; ❂ Prototype drug

1591

(T_4) lymphocyte count, Hgb, and granulocyte count to detect hematologic toxicity.

- Myelosuppression results in anemia, which commonly occurs after 4–6 wk of therapy, and granulocytopenia in 6–8 wk. Frequently, both respond to dosage adjustment. Significant anemia (Hgb less than 7.5 g/dL or reduction greater than 25% of baseline value), or granulocyte count less than 750/mm^3 (or reduction greater than 50% of baseline) may require temporary interruption of therapy and transfusions.
- Monitor for common adverse effects, especially severe headache, nausea, insomnia, and myalgia.

Patient & Family Education
- Contact prescriber promptly if health status worsens or any unusual symptoms develop.
- Understand that this drug is not a cure for HIV infection; you will continue to be at risk for opportunistic infections.
- Do not share drug with others; take drug exactly as prescribed.
- Drug does NOT reduce the risk of transmission of HIV infection through body fluids.

ZILEUTON
(zi-leu′ton)

Zyflo, Zyflo CR
Classifications: RESPIRATORY SMOOTH MUSCLE RELAXANT; BRONCHODILATOR; LEUKOTRIENE RECEPTOR ANTAGONIST (LTRA)
Therapeutic: BRONCHODILATOR; LTRA
Prototype: Zafirlukast
Pregnancy Category: C

AVAILABILITY 600 mg immediate release and controlled release tablets

ACTION & *THERAPEUTIC EFFECT*
Inhibits 5-lipoxygenase, the enzyme needed to start the conversion of arachidonic acid to leukotrienes, which are important inflammatory agents that induce bronchoconstriction and mucus production. *Zileuton helps to prevent the signs and symptoms of asthma including airway edema, smooth muscle constriction, and altered cellular activity due to inflammation.*

USES Prophylaxis and chronic treatment of asthma in adults and children younger than 12 y.

CONTRAINDICATIONS Hypersensitivity to zileuton or zafirlukast, active liver disease, status asthmaticus; QT prolongation; lactation.
CAUTIOUS USE Hepatic insufficiency; alcoholism; fever; infection; history of QT prolongation; older females; older adults; pregnancy (category C). Safety and effectiveness in children younger than 12 y are not established.

ROUTE & DOSAGE

Asthma

Adult/Child (older than 12 y): **PO** 1200 mg controlled release tablets b.i.d. or 600 mg immediate release tablets q.i.d.

ADMINISTRATION

Oral
- Ensure that controlled release tablets are swallowed whole. They should not be crushed or chewed.
- Store at room temperature, 15°–30° C (59°–86° F); protect from light.

ADVERSE EFFECTS (≥1%) **Body as a Whole:** Pain, asthenia, myalgia, arthralgia, fever, malaise, neck pain/

Z

rigidity. **CNS:** *Headache,* dizziness, insomnia, nervousness, somnolence. **CV:** Chest pain. **GI:** Abdominal pain, *dyspepsia,* nausea, constipation, flatulence, vomiting, elevated liver function tests, asymptomatic hepatitis. **Skin:** Pruritus. **Other:** Conjunctivitis, hypertonia, lymphadenopathy, vaginitis, UTI, leukopenia.

INTERACTIONS Drug: May double **theophylline** levels and increase toxicity. Increases hypoprothrombinemic effects of **warfarin.** May increase levels of BETA-BLOCKERS (especially **propranolol**), leading to hypotension and bradycardia.

PHARMACOKINETICS Absorption: Rapidly from GI tract. **Peak:** 1.7 h. **Duration:** 5–8 h. **Distribution:** 93% protein bound; secreted in the breast milk of rats. **Metabolism:** In liver primarily via glucuronide conjugation. **Elimination:** Primarily in urine (94%). **Half-Life:** 2.5 h.

NURSING IMPLICATIONS

Assessment & Drug Effects

- Assess respiratory status and airway function regularly.
- Lab tests: Periodic CBC and routine blood chemistry; monthly LFTs for 3 mo, then every 2–3 mo for rest of first year, then periodically.
- Monitor closely each of the following with concurrent drug therapy: With theophylline, theophylline levels; with warfarin, PT and INR; with phenytoin, phenytoin level; with propranolol, HR and BP for excessive beta blockade.

Patient & Family Education

- Take medication regularly even during symptom-free periods.
- Drug is not intended to treat acute episodes of asthma.

- Report to prescriber promptly S&S of hepatic toxicity (see Appendix F) or flu-like symptoms. Follow-up lab work is very important.
- Notify prescriber if condition worsens while using prescribed doses of all antiasthmatic medications.

ZIPRASIDONE HYDROCHLORIDE
(zip-ra-si′done)
Geodon
Classification: ATYPICAL ANTIPSYCHOTIC
Therapeutic: ANTIPSYCHOTIC
Prototype: Clozapine
Pregnancy Category: C

AVAILABILITY 20 mg, 40 mg, 60 mg, 80 mg capsules; 20 mg/mL injection

ACTION & THERAPEUTIC EFFECT
Exerts antischizophrenic effects through dopamine (D_2) and serotonin (5-HT_{2A}) receptor antagonism. Exerts antidepressant effects through 5-HT_{1A} agonism, 5-HT_{1D} antagonism, and serotonin/norepinephrine reuptake inhibition. *Improves signs and symptoms of schizophrenia, schizoaffective disorder, and psychotic depression.*

USES Treatment of schizophrenia, acute bipolar mania, acute psychosis, agitation.
UNLABELED USES Tourette's syndrome.

CONTRAINDICATIONS Hypersensitivity to ziprasidone; history of QT prolongation including congenital long QT syndrome or with other drugs known to prolong the QT interval; AV block, bundle branch block, cardiac arrhythmias, congenital heart disease, recent MI or uncompensated heart failure; bradycardia, hypokalemia or hypomagnesemia; neuroleptic malignant syndrome and tardive dyskinesia; dehydration or hypovolemia; UV exposure and tanning beds; lactation.

CAUTIOUS USE History of seizures, CVA, dementia, Parkinson's disease, or Alzheimer disease; known cardiovascular disease, conduction abnormalities, cerebrovascular disease; hepatic impairment; seizure disorder, seizures; breast cancer; risk factors for elevated core body temperature; esophageal motility disorders and risk of aspiration pneumonia; schizophrenia; suicide potential; pregnancy (category C); children older than 7 y for use in Tourette's syndrome only. Safety and efficacy in children or adolescents (except for treatment of Tourette's syndrome) are not established.

ROUTE & DOSAGE

Schizophrenia

Adult: **PO** Start with 20 mg b.i.d. with food, may increase q2days up to 80 mg b.i.d. if needed

Acute Episodes of Agitation/ Acute Psychosis

Adult: **IM** 10 mg q2h or 20 mg q4h up to max of 40 mg/day

Acute Mania/Bipolar Disorder

Adult: **PO** Start with 40 mg b.i.d. with food; may increase q2days up to 80 mg b.i.d. if needed

ADMINISTRATION

Note: CONTRAINDICATIONS for this drug. Do NOT administer to anyone with a history of cardiac arrhythmias or other cardiac disease, hypokalemia, hypomagnesemia, prolonged QT/QT$_c$ interval, or to anyone on other drugs known to prolong the QT$_c$ interval.

- Withhold drug and consult prescriber if any of the foregoing conditions are present.

Oral
- Give with food.
- Make dosage adjustments at intervals of 2 days or more.

Intramuscular
- Give deep IM into a large muscle.
- Store at 15°–30° C (59°–86° F).

ADVERSE EFFECTS (≥1%) **Body as a Whole:** Asthenia, myalgia, weight gain, flu-like syndrome, face edema, chills, hypothermia. **CNS:** *Somnolence,* akathisia, dizziness, extrapyramidal effects, dystonia, hypertonia, agitation, tremor, dyskinesias, hostility, paresthesia, confusion, vertigo, hypokinesia, hyperkinesias, abnormal gait, oculogyric crisis, hypesthesia, ataxia, amnesia, cogwheel rigidity, delirium, hypotonia, akinesia, dysarthria, withdrawal syndrome, buccoglossal syndrome, choreoathetosis, diplopia, incoordination, neuropathy. **CV:** Tachycardia, postural hypotension, prolonged QT$_c$ interval, hypertension. **GI:** *Nausea,* constipation, dyspepsia, diarrhea, dry mouth, anorexia, abdominal pain, vomiting. **Metabolic:** Hyperglycemia, diabetes mellitus. **Respiratory:** Rhinitis, increased cough, dyspnea. **Skin:** Rash, fungal dermatitis, photosensitivity. **Special Senses:** Abnormal vision.

INTERACTIONS Drug: Carbamazepine may decrease **ziprasidone** levels; **ketoconazole** may increase **ziprasidone** levels; may enhance hypotensive effects of ANTIHYPERTENSIVE AGENTS; may antagonize effects of **levodopa;** increased risk of arrhythmias and heart block due to prolonged QT$_c$ interval with ANTIARRHYTHMIC AGENTS, **amoxapine, arsenic trioxide, chlorpromazine, clarithromycin, daunorubicin, diltiazem, dolasetron, doxorubicin, droperidol, erythromycin, halofantrine,**

indapamide, levomethadyl, LOCAL ANESTHETICS, **maprotiline, mefloquine, mesoridazine, octreotide, pentamidine, pimozide, probucol, gatifloxacin, grepafloxacin, levofloxacin, moxifloxacin, sparfloxacin,** TRICYCLIC ANTIDEPRESSANTS, **tacrolimus, thioridazine, troleandomycin;** additive CNS depression with SEDATIVE-HYPNOTICS, ANXIOLYTICS, **ethanol,** OPIATE AGONISTS.

PHARMACOKINETICS Absorption: Well absorbed with 60% reaching systemic circulation. **Peak:** 6–8 h. **Metabolism:** In liver (CYP3A4). **Elimination:** Feces and urine. **Half-Life:** 7 h.

NURSING IMPLICATIONS

Assessment & Drug Effects

- Lab tests: Baseline and periodic ECG, serum potassium and serum magnesium, especially with concomitant diuretic therapy. Periodically monitor blood glucose.
- Monitor diabetics for loss of glycemic control.
- Monitor for S&S of torsade de pointes (e.g., dizziness, palpitations, syncope), tardive dyskinesia (see Appendix F) especially in older adult women and with prolonged therapy, and the appearance of an unexplained rash. Withhold drug and report to prescriber immediately if any of these develop.
- Monitor for signs and symptoms of suicidality.
- Monitor I&O ratio and pattern: Notify prescriber if diarrhea, vomiting or any other conditions develops which may cause electrolyte imbalance.
- Monitor BP lying, sitting, and standing. Report orthostatic hypotension to prescriber.
- Monitor cognitive status and take appropriate precautions.

- Monitor for loss of seizure control, especially with a history of seizures or dementia.

Patient & Family Education

- Carefully monitor blood glucose levels if diabetic.
- Be aware that therapeutic effect may not be evident for several weeks.
- Report any of the following to a health care provider immediately: Palpitations, faintness or loss of consciousness, rash, abnormal muscle movements, vomiting or diarrhea.
- Do not drive or engage in potentially hazardous activities until response to drug is known.
- Make position changes slowly and in stages to prevent dizziness upon arising.
- Avoid strenuous exercise, exposure to extreme heat, or other activities that may cause dehydration.

ZOLEDRONIC ACID
(zo-le-dron'ic)
Aclasta ◆, Reclast, Zometa
Classifications: BISPHOSPHONATE; BONE METABOLISM REGULATOR
Therapeutic: BONE METABOLISM REGULATOR
Prototype: Etidronate disodium
Pregnancy Category: D

AVAILABILITY 4 mg/5 mL, 5 mg/100 mL injection

ACTION & *THERAPEUTIC EFFECT*
Zoledronic acid inhibits various stimulatory factors of osteoclastic activity produced by bone tumors. It also induces osteoclast apoptosis. *Zoledronic acid blocks osteoclastic resorption of bone, thus reducing the amount of calcium released from bone.*

Z

Common adverse effects in *italic*, life-threatening effects <u>underlined</u>; generic names in **bold**; classifications in SMALL CAPS; ◆ Canadian drug name; ⊘ Prototype drug

1595

USES Treatment of hypercalcemia of malignancy, multiple myeloma, and bony metastases from solid tumors, Paget's disease (Reclast), postmenopausal or glucocorticoid-induced osteoporosis (Reclast).

CONTRAINDICATIONS Hypersensitivity to zoledronic acid or other bisphosphonates; preexisting hypocalcemia; serum creatinine of 0.5 mg/dL; pregnancy (category D); lactation. **CAUTIOUS USE** Aspirin-sensitive asthma; cancer chemotherapy; renal and/or hepatic impairment; dental work; multiple myeloma; older adults. Safety and effectiveness of zoledronic acid in children have not been established.

ROUTE & DOSAGE

Hypercalcemia of Malignancy (Zometa)

Adult: **IV** 4 mg over a minimum of 15 min. May consider retreatment if serum calcium has not returned to normal, may repeat after 7 days

Multiple Myeloma and Bony Metastases from Solid Tumors (Zometa)

Adult: **IV** 4 mg over a minimum of 15 min q3–4wk

Osteoporosis (Reclast)

Adult: **IV** 5 mg infusion once per year

Osteoporosis Prophylaxis in Postmenopausal Women

Adult: **IV** 5 mg every other year

Paget's Disease (Reclast)

Adult: **IV** 5 mg dose, retreatment may be necessary

Renal Impairment Dosage Adjustment (Zometa)

CrCl 50–60 mL/min: 3.5 mg; 40–49 mL/min: 3.3 mg; 30–39 mL/min: 3 mg; *less than 30 mL/min:* Do not use

Renal Impairment Dosage Adjustment (Reclast)

Do not use if CrCl less than 35 mL/min

ADMINISTRATION

Intravenous

Do not administer to anyone who is dehydrated or suspected of being dehydrated. Consult prescriber.

- Do not administer zoledronic acid unless patient is adequately rehydrated.
- Do not administer until serum creatinine values have been evaluated by the prescriber.

PREPARE: **IV Infusion:** *Injection Concentrate:* Withdraw required dose from the 4 mg/5 mL vial and dilute in 100 mL of D5W or NS. DO NOT use lactated Ringer's solution. ■ If not used immediately, refrigerate. The total time between reconstitution and end of infusion must not exceed 24 h.

ADMINISTER: **IV Infusion:** Infuse a single dose over NO LESS than 15 min.

INCOMPATIBILITIES **Solution/additive and Y-site:** Do not mix or infuse with **calcium**-containing solutions (e.g., **lactated Ringer's**).

- Store at 2°–8° C (36°–46° F) following dilution. ■ **Must be** completely infused within 24 h of reconstitution.

ADVERSE EFFECTS (≥1%) **Body as a Whole:** *Fever,* flu-like syndrome, redness and swelling at injection site, asthenia, chest pain, leg edema, mucositis, rigors. **CNS:** *Insomnia, anxiety, confusion, agitation,* headache, somnolence. **CV:**

Z

Hypotension. **GI:** *Nausea, vomiting, constipation, abdominal pain, anorexia,* dysphagia. **Hematologic:** *Anemia,* granulocytopenia, thrombocytopenia, pancytopenia. **Metabolic:** *Hypophosphatemia, hypokalemia, hypomagnesemia,* hypocalcemia, dehydration. **Musculoskeletal:** Skeletal pain, arthralgias, osteonecrosis of the jaw in cancer patients. **Respiratory:** *Dyspnea, cough,* pleural effusion. **Skin:** Alopecia, dermatitis. **Urogenital:** Renal deterioration (increase in S_{cr}).

INTERACTIONS Drug: LOOP DIURETICS may increase risk of hypocalcemia; **thalidomide** and other NEPHROTOXIC DRUGS may increase risk of renal toxicity.

PHARMACOKINETICS Onset: 4–10 days. **Duration:** 3–4 wk. **Metabolism:** Not metabolized. **Elimination:** In urine. **Half-Life:** 146 h.

NURSING IMPLICATIONS

Assessment & Drug Effects

- Lab tests: Baseline kidney function tests prior to each dose and periodically thereafter; periodic ionized calcium or corrected serum calcium levels, serum phosphate and magnesium, electrolytes, CBC with differential, Hct and Hgb.
- Notify prescriber immediately of deteriorating renal function as indicated by rising serum creatinine levels over baseline value.
- Withhold zoledronic acid and notify prescriber if serum creatinine is not within 10% of the baseline value.
- Monitor closely patient's hydration status. Note that loop diuretics should be used with caution due to the risk of hypocalcemia.
- Monitor for S&S of bronchospasm in aspirin-sensitive asthma patients; notify prescriber immediately.

Patient & Family Education

- Maintain adequate daily fluid intake. Consult with prescriber for guidelines.
- Report unexplained weakness, tiredness, irritation, muscle pain, insomnia, or flu-like symptoms.
- Use reliable means of birth control to prevent pregnancy. If you suspect you are pregnant, contact prescriber immediately.

ZOLMITRIPTAN
(zol-mi-trip′tan)
Zomig, Zomig ZMT, Zomig Nasal Spray
Classifications: SEROTONIN 5-HT$_1$ RECEPTOR AGONIST; ERGOT ALKALOID
Therapeutic: ANTIMIGRAINE
Prototype: Sumatriptan
Pregnancy Category: C

AVAILABILITY 2.5 mg, 5 mg tablets orally disintegrating tablets; 5 mg nasal spray

ACTION & *THERAPEUTIC EFFECT*
Selective serotonin (5-HT$_{1B/1D}$) receptor agonist. The agonist effects at 5-HT$_{1B/1D}$ reverse the vasodilation of cranial blood vessels and inhibit release of pro-inflammatory neuropeptides. *Vasoconstricts dilated cranial blood vessels and decreased neuropeptide release relieve the pain of a migraine headache.*

USES Acute migraine headaches with or without aura.

CONTRAINDICATIONS Hypersensitivity to zolmitriptan; ischemic heart disease (angina pectoris, arteriosclerosis, ECG changes, history of MI or Prinzmetal's angina); cardiac arrhythmias, symptomatic Wolff-Parkinson-White syndrome, uncontrolled hypertension; hemiplegia or basilar migraine; concurrent administration of ergotamine or sumatriptan; PKU.

Z

Common adverse effects in *italic*, life-threatening effects underlined; generic names in **bold**; classifications in SMALL CAPS; ♦ Canadian drug name; ❷ Prototype drug

1597

CAUTIOUS USE Men older than 40 y; postmenopausal women; patients with other cardiac risk factors, such as diabetes, obesity, cigarette smoking, high cholesterol levels, strong family history of CAD; concurrent administration of MAOIs; GI disease, PVD, ischemic colitis, Raynaud's disease, cerebrovascular disease, stroke, intracranial bleeding; renal failure or renal disease; adults older than 65 y; pregnancy (category C), lactation; children younger than 18 y.

ROUTE & DOSAGE

Acute Migraine
Adult: **PO** 2.5–5 mg, may repeat in 2 h if necessary (max: 10 mg/24 h) **Nasal Spray** One spray into one nostril

ADMINISTRATION

Oral
- Give any time after symptoms of migraine appear. Give 2.5 mg or less by breaking a 5 mg tablet in half. If headache returns, may repeat q2h up to 10 mg in 24 h.
- Do NOT give zolmitriptan within 24 h of an ergot-containing drug or other 5-HT$_1$ agonist.
- Discard unused tablets that have been removed from the packaging.

Intranasal
- Unit-dose spray device delivers a 5 mg dose. Do not exceed the maximum dose of 10 mg in 24 h.
- Store at 2°–25° C (36°–77° F) and protect from light.

ADVERSE EFFECTS (≥1%) **Body as a Whole:** Asthenia, fatigue, malaise, pain, pressure sensation, paresthesia, throat pressure, warm/cold sensations, hypesthesia. **CNS:** Somnolence, dizziness, drowsiness, headache, hypesthesia, decreased mental acuity, euphoria, tremor. **CV:** Coronary artery vasospasm, transient myocardial ischemia, MI, ventricular tachycardia, ventricular fibrillation, chest pain/tightness/heaviness, palpitations. **GI:** Dry mouth, nausea, vomiting. **Respiratory:** Dyspnea. **Skin:** Flushing. **Other:** Hot flushes.

INTERACTIONS Drug: Dihydroergotamine, methysergide, other 5-HT$_1$ AGONISTS may cause prolonged vasospastic reactions; SSRIS have rarely caused weakness, hyperreflexia, and incoordination; MAOIS should not be used with 5-HT$_1$ AGONISTS; **cimetidine** increases half-life of zolmitriptan. **Herbal: St. John's wort** may increase triptan toxicity.

PHARMACOKINETICS Absorption: Rapidly absorbed, 40% bioavailability. **Peak:** 2–3 h. **Distribution:** 25% protein bound. **Metabolism:** In liver to active metabolite. **Elimination:** Primarily in urine (65%), 30% in feces. **Half-Life:** 3 h.

NURSING IMPLICATIONS

Assessment & Drug Effects
- Monitor for therapeutic effectiveness: Relief or reduction of migraine pain within 1–4 h.
- Monitor cardiovascular status carefully following first dose in patients at risk for CAD (e.g., postmenopausal women, men older than 40 y, persons with known CAD risk factors) or coronary artery vasospasms.
- Periodic cardiovascular evaluation is recommended with long-term use.
- Report to prescriber immediately chest pain, nausea, or tightness in chest or throat that is severe or does not quickly resolve.

Z

Patient & Family Education

- Carefully review patient information insert and guidelines for taking drug.
- Do NOT take zolmitriptan during the aura phase, but as early as possible after onset of migraine.
- Do not remove orally disintegrating tablet from blister until just prior to dosing.
- Concurrent oral contraceptive use may increase incidence of adverse effects.
- Contact prescriber immediately if any of the following occur after zolmitriptan use: Symptoms of angina (e.g., severe or persistent pain or tightness in chest or throat, sudden nausea), hypersensitivity (e.g., wheezing, facial swelling, skin rash, hives), fainting, or abdominal pain.
- Report any other adverse effects (e.g., tingling, flushing, dizziness) at next prescriber visit.

✓ZOLPIDEM ☻

(zol′-pi-dem)

Ambien, Ambien CR, Edluar, Intermezzo Tovalt ODT, Zolpimist
Classifications: ANXIOLYTIC; SEDATIVE-HYPNOTIC, NON-BENZODIAZEPINE
Therapeutic: SEDATIVE-HYPNOTIC; ANTIANXIETY
Pregnancy Category: C
Controlled Substance: Schedule IV

AVAILABILITY 5 mg, 10 mg tablets; 6.25 mg, 12.5 mg extended release tablets; 5 mg, 10 mg sublingual tablets

ACTION & *THERAPEUTIC EFFECT*

An agonist that binds the gamma-aminobutyric acid (GABA)-A receptor chloride channel, thus inhibiting its action potential in the cortical region of the brain. *Effective as a sedative.*

USES Short-term treatment of insomnia.

CONTRAINDICATIONS Suicidal ideation; labor or obstetric delivery.
CAUTIOUS USE Depressed patients, hepatic/renal impairment, older adults, alcohol or drug abuse; patients with compromised respiratory status, COPD, sleep apnea; chronic depression; pregnancy (category C); children younger than 18 y.

ROUTE & DOSAGE

Short-Term Treatment of Insomnia

Adult: **PO** 10 mg (immediate release/sublingual) OR 12.5 mg (extended release) at bedtime
Spray 1–2 sprays before bedtime
Geriatric: **PO** 5 mg (immediate release/sublingual) or 6.25 mg (extended release) at bedtime
Spray 1 spray before bedtime (max: 2 sprays)

Hepatic Impairment Dosage Adjustment

5 mg (immediate release) or 6.25 mg (extended release) at bedtime

ADMINISTRATION

Oral

- Give immediately before bedtime; for more rapid sleep onset, do NOT give with or immediately after a meal.
- Ensure that sublingual tablets are not swallowed.
- Extended release tablets should be swallowed whole. Ensure that they are not crushed or chewed.
- Store at room temperature, 15°–30° C (59°–86° F).

Z

Common adverse effects in *italic*, life-threatening effects <u>underlined</u>; generic names in **bold**; classifications in SMALL CAPS; ♣ Canadian drug name; ☻ Prototype drug

1599

ADVERSE EFFECTS (≥1%) **CNS:** Headache on awakening, drowsiness or fatigue, lethargy, drugged feeling, depression, anxiety, irritability, dizziness, double vision. Confusion and falls reported in elderly. Doses greater than 10 mg may be associated with anterograde amnesia or memory impairment. **GI:** Dyspepsia, nausea, vomiting. **Other:** Myalgia.

INTERACTIONS Drug: CNS DEPRESSANTS, **alcohol,** PHENOTHIAZINES by augmenting CNS depression. **Food:** Extent and rate of absorption of zolpidem are significantly decreased.

PHARMACOKINETICS Absorption: Readily from GI tract. 70% reaches systemic circulation. **Onset:** 7–27 min. **Peak:** 0.5–2.3 h. **Duration:** 6–8 h. **Distribution:** Highly protein bound. Lowest concentrations in CNS, highest concentrations in glandular tissue and fat. Crosses placenta. **Metabolism:** In the liver to 3 inactive metabolites. **Elimination:** 79–96% in the bile, urine, and feces. **Half-Life:** 1.7–2.5 h.

NURSING IMPLICATIONS

Assessment & Drug Effects

- Assess respiratory function in patients with compromised respiratory status. Report immediately to prescriber significantly depressed respiratory rate (less than 12/min).
- Monitor patients for S&S of depression (see Appendix F); zolpidem may increase level of depression.
- Monitor closely older adult or debilitated patients for impaired cognitive or motor function and unusual sensitivity to the drug's effects.

Patient & Family Education

- Avoid taking alcohol or other CNS depressants while on zolpidem.
- Do not drive or engage in other potentially hazardous activities until response to drug is known.
- Report vision changes to prescriber.
- Note: Onset of drug is more rapid when taken on an empty stomach.

ZONISAMIDE ⊕

(zon-i'sa-mide)
Zonegran
Classifications: ANTICONVULSANT; SULFONAMIDE
Therapeutic: ANTICONVULSANT
Pregnancy Category: C

AVAILABILITY 25 mg, 50 mg, 100 mg capsules

ACTION & *THERAPEUTIC EFFECT* A broad-spectrum anticonvulsant that facilitates dopaminergic and serotonergic neurotransmission but does not potentiate the activity of gamma-aminobutyric acid (GABA) in the synapses of the CNS neurons. *Suppresses focal spike discharges and electroshock seizures. Effective against a variety of seizure types.*

USES Adjunctive therapy for partial seizures in adults.
UNLABELED USES Bipolar disorder.

CONTRAINDICATIONS Hypersensitivity to sulfonamides or zonisamide; lactation.
CAUTIOUS USE Renal or hepatic insufficiency, dehydration, hypovolemia; renal impairment; older adults; pregnancy (category C); children younger than 16 y.

Z

ROUTE & DOSAGE

Partial Seizures
Adult: **PO** Start at 100 mg daily, may increase after 2 wk to 200 mg/day, may then increase q2wk, if necessary (max: 400 mg/day in 1–2 divided doses)

ADMINISTRATION

Oral
- Do not crush or break capsules; ensure capsules are swallowed whole with adequate fluid.
- Withdraw drug gradually when discontinued to minimize seizure potential.
- Store at 25° C (77° F); room temperature permitted. Protect from light and moisture.

ADVERSE EFFECTS (≥1%) **Body as a Whole:** Flu-like syndrome, weight loss. **CNS:** Agitation, irritability, anxiety, ataxia, confusion, depression, difficulty concentrating, difficulty with memory, *dizziness,* fatigue, *headache,* insomnia, mental slowing, nervousness, nystagmus, paresthesia, schizophrenic behavior, *somnolence,* tiredness, tremor, convulsion, abnormal gait, hyperesthesia, incoordination. **GI:** Abdominal pain, *anorexia,* constipation, diarrhea, dyspepsia, nausea, dry mouth, flatulence, gingivitis, gum hyperplasia, gastritis, stomatitis, cholelithiasis, glossitis, melena, rectal hemorrhage, ulcerative stomatitis, ulcer, dysphagia. **Metabolic:** Oligohidrosis, sometimes resulting in heat stroke and hyperthermia in children. **Respiratory:** Rhinitis, pharyngitis, cough. **Skin:** Ecchymosis, rash, pruritus. **Special Senses:** Difficulties in verbal expression, diplopia, speech abnormalities, taste perversion, amblyopia, tinnitus. **Urogenital:** Kidney stones.

INTERACTIONS Drug: **Phenytoin, carbamazepine, phenobarbital, valproic acid** may decrease half-life of zonisamide.

PHARMACOKINETICS Peak: 2–6 h. **Distribution:** 40% protein bound, extensively binds to erythrocytes. **Metabolism:** Acetylated in liver by CYP3A4. **Elimination:** Primarily in urine. **Half-Life:** 63–105 h.

NURSING IMPLICATIONS

Assessment & Drug Effects
- Withhold drug and notify prescriber if an unexplained rash or S&S of hypersensitivity appear (see Appendix F).
- Monitor for and report S&S of CNS impairment (somnolence, excessive fatigue, cognitive deficits, speech or language problems, incoordination, gait disturbances); oligohidrosis (lack of sweating) and hyperthermia in pediatric patients.
- Lab tests: Periodic BUN and serum creatinine, and CBC with differential.

Patient & Family Education
- Do not abruptly stop taking this medication.
- Increase daily fluid intake to minimize risk of renal stones. Notify prescriber immediately of S&S of renal stones: Sudden back or abdominal pain, and blood in urine.
- Report any of the following: Dizziness, excess drowsiness, frequent headaches, malaise, double vision, lack of coordination, persistent nausea, sore throat, fever, mouth ulcers, or easy bruising.
- Exercise special caution with concurrent use of alcohol or CNS depressants.
- Do not drive or engage in other potentially hazardous activities until response to drug is known.

Z

Common adverse effects in *italic*, life-threatening effects <u>underlined</u>; generic names in **bold;** classifications in SMALL CAPS; ✦ Canadian drug name; ⊙ Prototype drug

1601

APPENDIXES

◆ APPENDIX A

(Generic names are in **bold**)

APPENDIX A-1
OCULAR MEDICATIONS:

BETA-ADRENERGIC BLOCKERS Prototype for classification: Propranolol HCl Use: Intraocular hypertension and chronic open-angle glaucoma.

Betaxolol HCl Betoptic, Betoptic S, 0.25%, 0.5% solution	*Adult:* **Topical** 1–2 drops in affected eye twice daily.
Carteolol HCl Ocupress, 1% solution	*Adult:* **Topical** 1 drop b.i.d.
Levobunolol Betagan, 0.25%, 0.5% solution	*Adult:* **Topical** 1–2 drops 1–2 times/day.
Metipranolol HCl OptiPranolol, 0.3% solution	*Adult:* **Topical** 1 drop b.i.d.
Timolol maleate Betimol, Istalol, Timoptic, Timoptic XE, 0.25%, 0.5% solution	*Adult:* **Topical** 1 drop of 0.25–0.5% solution b.i.d.; may decrease to daily. Apply gel daily. Apply Istalol solution once daily.

Adverse Effects/Clinical Implications: May cause *mild ocular stinging* and discomfort; tearing; may also have the adverse effects of systemic beta-blockers. May precipitate thyroid storm in patients with hyperthyroidism. Patients with impaired cardiac function and the elderly should report to prescriber signs and symptoms of CHF (see Appendix G). Monitor BP for hypotension and heart rate for bradycardia.

MIOTICS Prototype for classification: Pilocarpine HCl Use: Open-angle and angle-closure glaucomas; to reduce IOP and to protect the lens during surgery and laser iridotomy; to counteract effects of mydriatics and cycloplegics following surgery or ophthalmoscopic examination.

Apraclonidine HCl Iopidine, 0.5%, 1% solution	**Intraoperative and Postsurgical Increase in IOP:** *Adult:* **Topical** 1 drop of 1% solution in affected eye 1 h before surgery and 1 drop in same eye immediately after surgery. **Open-Angle Glaucoma:** *Adult:* **Topical** 1 drop of 0.5% solution in affected eye q12h.
Brimonidine tartrate Alphagan P, 0.1%, 0.15% solution	**Glaucoma:** *Adult:* **Topical** 1 drop in affected eye(s) t.i.d. approximately 8 h apart.

Brinzolamide Azopt, 1% suspension	**Ocular Hypertension or Open-Angle Glaucoma:** *Adult:* **Topical** 1 drop in affected eye(s) t.i.d.
Carbachol Carbastat, Miostat, 0.01% solution	*Adult:* **Intraocular** 0.5 ml of 0.01% solution injected into anterior chamber of eye.
Dorzolamide Trusopt, 2% solution	**Ocular Hypertension or Open-Angle Glaucoma:** *Adult, Child:* **Topical** 1 drop in affected eye(s) t.i.d.
Pilocarpine HCl Isopto Carpine, Pilopine HS, 1%, 2%, 4% solution	**Acute Glaucoma:** *Adult:* **Topical** 1 drop of 1–2% solution in affected eye q5–10min for 3–6 doses, then 1 drop q1–3h until IOP is reduced. **Chronic Glaucoma:** *Adult:* **Topical** 1 drop of 0.5–4% solution in affected eye q4–12h or 1 ocular system (Ocusert) q7days **Miotic:** *Adult:* **Topical** 1 drop of 1% solution in affected eye.

Adverse Effects/Clinical Implications: Ocular: Ciliary spasm with brow ache, twitching of eyelids, eye pain with change in eye focus, miosis, *diminished vision in poorly illuminated areas,* blurred vision, reduced visual acuity, sensitivity, contact allergy, lacrimation, follicular conjunctivitis, conjunctival irritation, cataract, retinal detachment. **CNS:** *Headache, drowsiness,* depression, syncope. **GI:** Abnormal taste, dry mouth. **Clinical Implications:** Wait 15 min after instillation before inserting soft contact lenses to avoid staining lenses. Use with MAO inhibitors may increase risk of hypertensive emergency. May increase the effects of beta-blockers and other antihypertensives on blood pressure and heart rate. TCAs may reduce the effects of **brimonidine. Brinzolamide** is a carbonic anhydrase inhibitor (prototype: Acetazolamide) and is a sulfonamide. It should not be used by patients with sulfa allergies. Reconstituted solutions of **echothiophate** remain stable for 1 mo at room temperature. Expiration date should appear on label. The length of time solutions remain stable under refrigeration varies with manufacturer. **Echothiophate** therapy is generally discontinued 2–6 wk before surgery. If necessary, alternate therapy is substituted. Medication should be given in the evening. Give at least 5 min apart from other topical ophthalmic drugs. The patient with brown or hazel eyes may require a stronger ophthalmic solution or more frequent instillation of **physostigmine** for desired effects than the patient with blue eyes.

PROSTAGLANDINS Prototype for classification: Latanoprost
Use: Open-angle glaucoma and intraocular hypertension.

Bimatoprost Lumigan, `0.03% solution	*Adult:* **Topical** 1 drop in affected eye(s) once daily in the evening.

| **Latanoprost** Xalatan, 0.005% solution | *Adult:* **Topical** 1 drop (1.5 mcg) in affected eye(s) once daily in the evening. |
| **Travoprost** Travatan, 0.004% solution | *Adult:* **Topical** 1 drop in affected eye(s) once daily in the evening. |

Adverse Effects: Ocular: *Conjunctival hyperemia, growth of eyelashes, ocular pruritus,* ocular dryness, visual disturbance, ocular burning, foreign body sensation, eye pain, pigmentation of the periocular skin, blepharitis, cataract, superficial punctate keratitis, eyelid erythema, ocular irritation, eyelash darkening, eye discharge, tearing, photophobia, allergic conjunctivitis, increases in iris pigmentation (brown pigment), conjunctival edema. **Body as a Whole:** Headaches, abnormal liver function tests, asthenia, and hirsutism. **Clinical Implications:** Should instill in the evening. Wait 15 min after instillation before inserting soft contact lenses to avoid staining the lenses. Give at least 5 min apart from other topical ophthalmic drugs.

MYDRIATIC Prototype for classification: Homatropine HBr
Use: Mydriatic for ocular examination and as cycloplegic to measure errors of refraction. Also inflammatory conditions of uveal tract, ciliary spasm, as a cycloplegic and mydriatic in preoperative and postoperative conditions, and as an optical aid in select patients with axial lens opacities.

Atropine HCl AK-Pentolate, Cyclogyl, 0.5%, 1%, 2% solution	**Cycloplegic Refraction:** *Adult:* **Topical** 1 drop of 1% solution in eye 40–50 min before procedure, followed by 1 drop in 5 min; may need 2% solution in patients with darkly pigmented eyes. *Child:* **Topical** 1 drop of 0.5–1% solution in eye 40–50 min before procedure, followed by 1 drop in 5 min; may need 2% solution in patients with darkly pigmented eyes.
Homatropine HBr AK-Homatropine, Homatrine, Isopto Homatropine, 2%, 5% solution	**Cycloplegic Refraction:** *Adult:* **Topical** 1–2 drops of 2% or 5% solution in eye repeated in 5–10 min if necessary. **Ocular Inflammation:** *Adult:* **Topical** 1–2 drops of 2% or 5% solution in eye up to q3–4h.
Phenylephrine HCl AK-Dilate Ophthalmic, Mydfrin, 0.12%, 2.5%, 10% solution	**Ophthalmoscopy:** *Adult:* **Topical** 1 drop of 2.5% or 10% solution before examination. *Child:* **Topical** 1 drop of 2.5% solution before examination. **Vasoconstrictor:** *Adult:* **Topical** 2 drops of 0.12% solution q3–4h as necessary.

Tropicamide
Mydral, Mydriacyl,
Tropicacyl, 0.5%,
1% solution

Refraction: *Adult:* **Topical** 1-2 drops of 1% solution in each eye, repeat in 5 min; if patient is not seen within 20-30 min, an additional drop may be instilled. **Examination of Fundus:** *Adult:* **Topical** 1-2 drops of 0.5% solution in each eye 15-20 min prior to examination; may repeat q30min if necessary.

Contraindicated in: Primary (narrow-angle) glaucoma or predisposition to glaucoma; children younger than 6 y. **Cautious Use in:** Increased IOP, infants, children, pregnancy (category C), the elderly or debilitated; hypertension; hyperthyroidism; diabetes; cardiac disease. **Adverse Effects:** Increased IOP, *blurred vision, photophobia.* **Prolonged Use:** Local irritation, congestion, edema, eczema, follicular conjunctivitis. **Excessive Dosage/Systemic Absorption:** Symptoms of atropine poisoning (flushing, dry skin, mouth, nose; decreased sweating; fever, rash, rapid/irregular pulse; abdominal and bladder distention; hallucinations, confusion). **CNS:** Psychotic reaction, behavior disturbances, ataxia, incoherent speech, restlessness, hallucinations, somnolence, disorientation, failure to recognize people, grand mal seizures. **Clinical Implications:** Carefully monitor **cyclopentolate** patients with seizure disorders, since systemic absorption may precipitate a seizure. Photophobia associated with mydriasis may require patient to wear dark glasses. Since drug causes blurred vision, supervision of activity may be indicated.

VASOCONSTRICTOR; DECONGESTANT Prototype for classification:
Naphazoline HCl Use: Ocular vasoconstrictor.

Naphazoline HCl
AK-Con, Allerest,
Clear Eyes, Comfort,
Degest-2, Naphcon,
0.012%, 0.03%, 0.1%
solution

Adult: **Topical** 1-3 drops of 0.1% solution q3-4h prn or 1-2 drops of a 0.012-0.03% solution q4h prn.

**Tetrahydrozoline
HCl**
Opti-Clear,
Visine Original,
0.05% solution

Adult: **Topical** 1-2 drops of 0.05% solution in eye b.i.d. or t.i.d.

Contraindicated in: Narrow-angle glaucoma; concomitant use with MAO INHIBITORS or TRICYCLIC ANTIDEPRESSANTS **Cautious Use in:** Hypertension, cardiac irregularities, advanced arteriosclerosis; diabetes; hyperthyroidism; elderly patients. **Adverse Effects:** Pupillary dilation, increased intraocular pressure, rebound redness of the eye, headache, hypertension, nausea, weakness, sweating. **Overdosage:** Drowsiness, hypothermia, bradycardia, shocklike hypotension, coma.

CORTICOSTEROID, ANTI-INFLAMMATORY Prototype for classification: Hydrocortisone Use: Inflammation. **Unlabeled Use:** Anterior uveitis.

Dexamethasone sodium phosphate
Maxidex
Ophthalmic, 0.1% suspension

Adult: **Topical** 1–2 drops in conjunctival sac up to 4–6 times/day; may instill hourly for severe disease.

Difluprednate
Durezol, 0.05% suspension

Adult: **Topical** 1 drop in conjunctival sac q.i.d. for the first 24 h; q.i.d. for 2 wk then b.i.d. for 2 wk.

Fluorometholone
Flarex, Fluor-Op,
FML Forte,
FML Liquifilm,
0.1%, 0.25% suspension; 0.1% ointment

Adult/Child (older than 2 y): **Topical** 1–2 drops of suspension in conjunctival sac q.h. for the first 24–48 h; then b.i.d. to q.i.d.; or a thin strip of ointment q4h for the first 24–48 h; then 1–3 times/day.

Loteprednol etabonate
Alrex, Lotemax, 0.2%, 0.5% suspension

Adult: **Topical** 1–2 drops in conjunctival sac q.i.d. during initial treatment, may increase to q1h if necessary.

Prednisolone sodium phosphate
Inflamase Mild,
Pred Mild, Prednisol,
Inflamase Forte,
0.11%, 0.12%, 0.9% suspension

Adult: **Topical** 1–2 drops in conjunctival sac q.h. during the day; then q2h at night; may decrease to 1 drop t.i.d. or q.i.d.

Rimexolone
Vexol, 1% solution

Postoperative Ocular Inflammation: *Adult:* **Topical** 1–2 drops q.i.d. beginning 24 h after surgery, continue through first 2 wk postoperatively. **Anterior Uveitis:** *Adult:* **Topical** 1–2 drops in affected eye every hour while awake for first week, then q2h for second week, then taper frequency until uveitis resolves.

Contraindicated in: Ocular fungal diseases, *herpes simplex* keratitis, ocular infections, ocular mycobacterial infections, viral disease of cornea or conjunctiva such as vaccinia, varicella. **Adverse Effects: Ocular:** Blurred vision, photophobia, conjunctival edema, corneal edema, erosion, eye discharge, dryness, irritation, pain; prolonged use: Glaucoma, ocular hypertension, damage to optic nerve, defects in visual acuity and visual fields, posterior subcapsular cataract formation, secondary ocular infections. **Other:** Headache, taste perversion. **Clinical Implications:** Shake all products well before use.

OCULAR ANTIHISTAMINES Prototype for classification: Emedastine Use: Relief of signs and symptoms of allergic conjunctivitis.

Azelastine HCl
OPTIVAR, 0.05%
solution

Adult/Child (older than 3 y): **Topical**
1 drop in affected eye(s) b.i.d.

Bepotastine
Bepreve, 1.5% solution

Adult: **Topical** 1 drop in affected eye(s) b.i.d.

Cromolyn sodium
Crolom, Opticrom, 4%
solution

Adult: **Topical** 1-2 drops in each eye
4-6 times/day.

**Emedastine
difumarate**
Emadine, 0.05% solution

Adult/Child (older than 3 y): **Topical**
1 drop in affected eye(s) up to q.i.d.

**Epinastine
hydrochloride**
Elestat, 0.05% solution

Adult/Child (older than 3 y): **Topical**
1 drop in affected eye(s) up to b.i.d.

Ketotifen fumarate
Zaditor, 0.025% solution

Adult: **Topical** 1 drop in affected eye(s)
q8-12h.

Lodoxamide
Alomide, 0.1% solution

Adult/Child (older than 2 mo): **Topical**
1-2 drop in affected eye(s) q.i.d. for up to
3 mo.

Nedocromil sodium
Alocril, 2% solution

Adult/Child (older than 3 y): **Topical**
1-2 drops in affected eye(s) b.i.d.

Olopatadine HCl
Patanol, 0.1% solutionPata-
day, 0.2% solution Patan-
ase, 665 mcg nasal spray

Adult/Child (older than 3 y): **Topical**
1-2 drops in affected eye(s) b.i.d. at least
6-8 h apart. *Adult/Adolescent:* **Intranasal**
1 spray in each nostril b.i.d.

Pemirolast potassium
Alamast, 0.1% solution

Adult: **Topical** 1-2 drops in affected eye(s)
q.i.d.

Adverse Effects: Ocular: Allergic reactions, *burning, stinging,* discharge,
dry eyes, eye pain, eyelid disorder, itching, keratitis, lacrimation disorder,
mydriasis, photophobia, rash. **CNS:** Drowsiness, fatigue, headache. **Other:**
Dry mouth, cold syndrome, pharyngitis, rhinitis, sinusitis, taste perversion.
Clinical Implications: Wait 10 min after instilling **emedastine** before inserting soft contact lenses; do not use **olopatadine** with soft contact lenses.

OCULAR NONSTEROIDAL ANTI-INFLAMMATORY DRUGS:
Prototype for classification: Ibuprofen Use: Treatment of ocular pain
and inflammation associated with cataract surgery.

Bromfenac
Xibrom, 0.09%
solution

Adult: **Topical** 1 drop into affected eye(s) b.i.d.
beginning 24 h after cataract surgery and continuing for 14 days.

Ketorolac
Acular, Acuvail,
0.45%, 0.5% solution

Adult: **Topical** 1 drop to affected eye(s) 4 times/
day beginning 24 h after surgery and continuing
for 14 days.

Nepafenac
Nevanac, 0.1%
suspension

Adult/Child (older than 10 y): **Topical** 1 drop
into affected eye(s) t.i.d. beginning 24 h after
cataract surgery and continuing for 14 days.

Adverse Effects: Ocular: Conjunctival hyperemia, ocular hypertension, foreign body sensation, decreased visual acuity, headache, iritis, ocular inflammation (e.g., edema, erythema), ocular irritation (burning/stinging), ocular pruritus, ocular pain, photophobia, lacrimation, abnormal sensation in the eye, delayed wound healing, keratitis, lid margin crusting, corneal erosion, corneal perforation, corneal thinning, and epithelial breakdown. Continued use can lead to ulceration or perforation. **Clinical Implications: Nepafenac** suspension **must be** shaken well prior to use.

OCULAR ANTIBIOTIC, QUINOLONE
Prototype for classification: Ciprofloxacin Use: Treatment of ocular infection.

Besifloxacin Besivance, 0.6% suspension	*Adult/Adolescent/Child (over 1 year):* **Topical** 1 drop in affected eye(s) t.i.d. × 7 days
Ciprofloxacin Ciloxin, 0.3% solution	*Adult/Adolescent/Child (older than 1 y):* **Topical** 1–2 drops in affected eye(s) q2h × 2 days then q4h × 5 days.
Gatifloxacin Zymar, 0.3% solution	*Adult/Adolescent/Child (older than 1 y):* **Topical** 1 drop in affected eye(s) q2h × 2 days then 1 drop in affected eye(s) up to 4 times/day.
Moxifloxacin Vigamox, 0.5% solution	*Adult/Child (older than 1 y):* **Topical** 1 drop in affected eye(s) t.i.d. × 7 days.

Adverse Effects: Ocular: Conjunctival redness, blurred vision, irritation, pain, pruritus. **CNS:** Headache.

APPENDIX A-2
LOW MOLECULAR WEIGHT HEPARINS:

ANTICOAGULANT, LOW MOLECULAR WEIGHT HEPARIN Prototype for classification: Enoxaparin Use: Prevention and treatment of DVT following hip or knee replacement or abdominal surgery, unstable angina, acute coronary syndromes.

Dalteparin sodium Fragmin, 10,000 international units/mL, 25,000 international units/mL	**DVT Prophylaxis, Abdominal Surgery:** *Adult:* **Subcutaneous** 2500 international units daily starting 1–2 h prior to surgery and continuing for 5–10 days postoperatively. **DVT Prophylaxis, Total Hip Arthroplasty:** *Adult:* **Subcutaneous** 2500–5000 international units daily starting 1–2 h prior to surgery and continuing for 5–14 days postoperatively. **Acute Thromboembolism:** *Adult:* **Subcutaneous** 120 international units/kg b.i.d. for at least 5 days. **Recurrent Thromboembolism:** *Adult:* **Subcutaneous** 5000 international units b.i.d. for 3–6 mo. **Unstable Angina/Non–Q-Wave MI:** *Adult:* **Subcutaneous** 120 international units/kg (max: 10,000 international units) q12h.

DVT Prophylaxis with risk of PE: *Adult:* **Subcutaneous** 5000 international units once daily for 12-14 days. **Extended Treatment of VTE or Proximal DVT:** *Adult:* **Subcutaneous** 200 international units/kg daily for 1 mo then 150 international units/kg daily for 2-6 mo.

Enoxaparin
Lovenox,
100 mg/mL

Prevention of DVT after Hip or Knee Surgery: *Adult:* **Subcutaneous** 30 mg subcutaneously b.i.d. for 10-14 days starting 12-24 h post-surgery. **Prevention of DVT after Abdominal Surgery:** *Adult:* **Subcutaneous** 40 mg daily starting 2 h before surgery and continuing for 7-10 days (max: 12 days). **Treatment of DVT and Pulmonary Embolus:** *Adult:* **Subcutaneous** 1 mg/kg b.i.d.; monitor anti-Xa activity to determine appropriate dose. **Acute Coronary Syndrome:** *Adult:* **Subcutaneous** 1 mg/kg q12h × 2-8 days. Give concurrently with aspirin 100-325 mg/day.

Contraindicated in: Hypersensitivity to ardeparin, other low molecular weight heparins, pork products, or parabens; active major bleeding, thrombocytopenia that is positive for antiplatelet antibodies with ardeparin; uncontrolled hypertension; nursing mothers. **Tizaparin** contraindicated in patients over 90 years old with CrCl less than 60 mL/min. **Cautious Use in:** Hypersensitivity to heparin; history of heparin-induced thrombocytopenia; bacterial endocarditis; severe and uncontrolled hypertension, cerebral aneurysm or hemorrhagic stroke, bleeding disorders, recent GI bleeding or associated GI disorders (e.g., ulcerative colitis), thrombocytopenia, or platelet disorders; severe liver or renal disease, diabetic retinopathy, hypertensive retinopathy, invasive procedures; pregnancy (category C). **Adverse Effects: Body as a Whole:** Allergic reactions (rash, urticaria), arthralgia, pain and inflammation at injection site, peripheral edema, fever. **CNS:** *CVA*, dizziness, headache, insomnia. **CV:** Chest pain. **GI:** Nausea, vomiting. **Hematologic:** *Hemorrhage*, thrombocytopenia, ecchymoses, anemia. **Respiratory:** Dyspnea. **Skin:** Rash, pruritus. **Drug Interactions: Aspirin,** NSAIDs, **warfarin** can increase risk of hemorrhage **Clinical Implications:** Alternate injection sites using the abdomen, anterior thigh, or outer aspect of upper arms. **Lab Tests:** CBC with platelet count, urinalysis, and stool for occult blood should be tested throughout therapy. Routine coagulation tests are not required. Carefully monitor for and immediately report S&S of excessive anticoagulation (e.g., bleeding at venipuncture sites or surgical site) or hemorrhage (e.g., drop in BP or Hct). Patients on oral anticoagulants, platelet inhibitors, or with impaired renal function **must be** very carefully monitored for hemorrhage. Patient should be sitting or lying supine for injection. Inject deep subcutaneously with entire length of needle inserted into skin fold. Hold skin fold gently throughout injection and do not rub site after injection.

APPENDIX A-3
INHALED CORTICOSTEROIDS (ORAL AND NASAL INHALATIONS):

CORTICOSTEROID, ANTIINFLAMMATORY Prototype for classification: Hydrocortisone Use: Oral inhalation to treat steroid-dependent asthma, nasal inhalation for the management of the symptoms of seasonal or perennial rhinitis.

Beclomethasone dipropionate
Beconase AQ,
QVAR,
Vancenase AQ

Asthma: *Adult:* **Oral Inhaler** 2 inhalations t.i.d. or q.i.d. up to 20 inhalations/day; may try to reduce systemic steroids after 1 wk of concomitant therapy; QVAR 40–80 mcg b.i.d. (max: 320 mcg/day). *Child (6–12 y):* **Oral Inhaler** 1–2 inhalations t.i.d. or q.i.d. up to 10 inhalations/day; QVAR 5–11 y, 40–80 mcg b.i.d. (max: 160 mcg/day). **Allergic Rhinitis:** *Adult:* **Nasal Inhaler** 1 spray per nostril b.i.d. to q.i.d. Child (older than 6 y): 1–2 sprays daily.

Budesonide
Pulmicort,
Rhinocort Aqua

Asthma, Maintenance Therapy: *Adult:* **Oral Inhalation** 1 or 2 inhalations (200 mcg/inhalation) daily–b.i.d. (max: 800 mcg b.i.d.). *Child (6 y or older):* **Oral Inhalation** 1 inhalation (200 mcg/inhalation) daily–b.i.d. (max: 400 mcg b.i.d.). *Child (12 mo–8 y):* **Nebulization** 0.5 mg/day in 1–2 divided doses. **Rhinitis:** *Adult/Child (6 y or older):* **Intranasal** 2 sprays per nostril in the morning and evening or 4 sprays per nostril in the morning. Each actuation releases 32 mcg from the nasal adapter.

Ciclesonide
Omnaris, 50 mcg/spray 80 mcg,
160 mcg inhaled solution

Rhinitis: *Adult/Child (older than 6 y):* **Intranasal** 1–2 sprays per nostril once daily (200 mcg/day). **Asthma:** *Adult/Child (older than 12 y):* **Inhaled** 80 mcg b.i.d.; may increase to 160 mcg b.i.d.

Dexamethasone
Decadron,
Decaspray, 0.04% solution

Adult: **Oral Inhalation** Up to 3 inhalations t.i.d. or q.i.d. (max: 12 inhalations/day). **Intranasal** 2 sprays per nostril b.i.d. or t.i.d. (max: 12 sprays/day). *Child:* **Oral Inhalation** Up to 2 inhalations q.i.d. (max: 8 inhalations/day). **Intranasal** 1 or 2 sprays per nostril b.i.d. (max: 8 sprays/day).

Flunisolide
0.025% solution

Allergic Rhinitis: *Adult:* **Intranasal** 2 sprays orally, or intranasally per nostril, b.i.d.; may increase to t.i.d., if needed. *Child:* **Intranasal** 6–14 y, 1 spray orally, or intranasally per nostril t.i.d. or 2 sprays b.i.d.

Fluticasone
Flonase,
Flovent
Flovent HFA, 44 mcg,
110 mcg, 220 mcg
aerosol,
Veramyst, 27.5 mcg/
actuation

Seasonal Allergic Rhinitis: *Adult:* **Intranasal** 100 mcg (1 inhalation) per nostril 1-2 times daily (max: 4 times daily). **Inhalation** 1-2 inhalations b.i.d. *Child (4 y or older):* **Intranasal** 1 spray per nostril once daily. May increase to 2 sprays per nostril once daily if inadequate response, then decrease to 1 spray per nostril once daily when control is achieved. *Adult/Adolescent/Child (12 y or older):* **Intranasal (Veramyst)** 2 sprays per nostril daily then reduce to 1 spray daily. *Child (2-11 y):* **Intranasal (Veramyst)** 1 spray per nostril daily, may increase to 2 sprays per nostril daily if necessary. *Adult/Adolescent:* **Inhaled (Advair)** 1-2 inhalations q12h.

Mometasone furoate
Asmanex, Nasonex,
Twisthaler, 220 mcg/
inhalation, 50 mcg/
inhalation

Adult: **Intranasal** 2 sprays (50 mcg each) in each nostril once daily. *Child (2 y or older):* **Intranasal** 1 spray in each nostril once daily. *Adult/Child (older than 12 y):* **Powder for Inhalation** 1 inhalation (220 mcg) once daily (max: 1 inhalation b.i.d.). *Child (4-11 y):* **Powder for Inhalation** 100 mcg daily.

Triamcinolone acetonide
Azmacort, 100 mcg/
inhalation

Adult: **Inhalation** 2 puffs 3-4 times/day (max: 16 puffs/day) or 4 puffs b.i.d. **Nasal Spray** 2 sprays/nostril once daily (max: 8 sprays/day). *Child (6-12 y):* **Inhalation** 1-2 sprays t.i.d. or q.i.d. (max: 12 sprays/day) or 2-4 sprays b.i.d.

Contraindicated in: Nonasthmatic bronchitis, primary treatment of status asthmaticus, acute attack of asthma. **Cautious Use in:** Patients receiving systemic corticosteroids; use with extreme caution if at all in respiratory tuberculosis, untreated fungal, bacterial, or viral infections, and ocular herpes simplex; nasal inhalation therapy for nasal septal ulcers, nasal trauma, or surgery. **Adverse Effects: Oral Inhalation:** *Candidal infection of oropharynx* and occasionally larynx, hoarseness, dry mouth, sore throat, sore mouth. **Nasal (Inhaler):** *Transient nasal irritation, burning, sneezing,* epistaxis, bloody mucous, nasopharyngeal itching, dryness, crusting, and ulceration; headache, nausea, vomiting. **Other:** With excessive doses, symptoms of hypercorticism. Increase risk of adverse effects if Advair is used with other long-acting beta-agonists. **Clinical Implications:** Note that oral inhalation and nasal inhalation products are not to be used interchangeably. **Oral Inhaler:** Emphasize the following: (1) Shake inhaler well before using. (2) After exhaling fully, place mouthpiece well into mouth with lips closed firmly around it. (3) Inhale slowly through mouth while activating the inhaler. (4) Hold breath 5–10 sec, if possible, then exhale slowly. (5) Wait 1 min between puffs. Clean inhaler daily. Separate parts as directed in package insert, rinse them with warm water, and dry them thoroughly. Rinsing mouth and gargling with warm water after each oral inhalation removes residual medication from oropharyngeal area. Mouth care may also delay or prevent onset of oral dryness, hoarseness, and candidiasis. **Nasal Inhaler:** Directions for use of nasal inhaler provided by manufacturer should be carefully reviewed with patient. Emphasize the following points: (1) Gently blow nose to clear nostrils. (2) Shake inhaler well before using. (3) If 2 sprays in each nostril are prescribed, direct one spray toward upper, and the other toward lower part of nostril. (4) Wash cap and plastic nosepiece daily with warm water; dry thoroughly. Inhaled steroids do not provide immediate symptomatic relief and are not prescribed for this purpose.

APPENDIX A-4
TOPICAL CORTICOSTEROIDS:

CORTICOSTEROID, ANTI-INFLAMMATORY Prototype for classification: Hydrocortisone Use: As a topical corticosteroid, the drug is used for the relief of the inflammatory and pruritic manifestations of corticosteroid-responsive dermatoses.

Hydrocortisone
Aeroseb-HC, Alphaderm, Cetacort, Cortaid, Cortenema, Dermolate, Hytone, Rectacort, Synacort, Caldecort, 0.5%, 1%, 2.5% cream, lotion, ointment, spray

Adult: **Topical** Apply a small amount to the affected area 1-4 times/day. **PR** Insert 1% cream, 10% foam, 10-25 mg suppository, or 100 mg enema nightly.

Hydrocortisone acetate
Anusol HC, Carmol HC, Coli foam, Cortaid, Cort-Dome, Corticaine, Cortifoam, Epifoam, 0.5%, 1% ointment, cream

Alclometasone dipropionate
Aclovate, 0.05% cream, ointment

Adult: **Topical** 0.05% cream or ointment applied sparingly b.i.d. or t.i.d.; may use occlusive dressing for resistant dermatoses.

Amcinonide
Cyclocort, 0.1% cream, lotion, ointment

Adult: **Topical** Apply thin film b.i.d. or t.i.d.

Betamethasone dipropionate
Diprolene, Diprolene AF, Diprosone, Maxivate, 0.05% cream, gel, lotion, ointment

Adult: **Topical** Apply thin film b.i.d.

Betamethasone valerate
Luxiq, Valisone, Psorion, Beta-Val, 0.1% cream, ointment, lotion; 0.12% aerosol foam

Adult: **Topical** Apply sparingly b.i.d.

Clobetasol propionate
Clobex, Cormax, Embeline, Olux, Temovate, 0.05% cream, gel, ointment, lotion, aerosol foam

Adult: **Topical** Apply sparingly b.i.d. (max: 50 g/wk), or b.i.d. 3 day/wk or 1–2 times/wk for up to 6 mo.

Clocortolone pivalate
Cloderm, 0.1% cream

Adult: **Topical** Apply thin layer 1–4 times/day.

Desonide
DesOwen, Tridesilon, 0.05% cream, ointment, lotion

Adult: **Topical** Apply thin layer b.i.d. to q.i.d.

Desoximetasone
Topicort, Topicort-LP, 0.05% cream, ointment

Adult: **Topical** Apply thin layer b.i.d.

Diclofenac
Flector, Pennsaid, 1.3% topical patch; 1% topical gel

Adult: **Topical** Up to 4 g 4 times/day on affected joint (max: 16 g).

Diflorasone diacetate
Florone, Florone E, Maxiflor, Psorcon E, Psorcon, 0.05% cream, ointment

Adult: **Topical** Apply thin layer of ointment 1–3 times/day or cream 2–4 times/day.

Fluocinolone acetonide
Fluoderm, Synalar, 0.025% ointment, cream; 0.2% cream; 0.01% cream, solution, shampoo, oil; 0.59 mg ophthalmic insert

Adult: **Topical** Apply thin layer b.i.d. to q.i.d.

Fluocinonide
Vanos, 0.05% cream, ointment, solution, gel; 0.1% cream

Adult: **Topical** Apply thin layer b.i.d. to q.i.d.

Flurandrenolide
Cordran, Cordran SP, 0.05% cream, lotion; 4 mcg/sq cm tape

Adult: **Topical** Apply thin layer b.i.d. or t.i.d.; apply tape 1–2 times/day at 12 h intervals. *Child:* **Topical** Apply thin layer 1–2 times/day; apply tape once/day.

Fluticasone
Cutivate, 0.005%, 0.05% cream; 0.005% ointment

Adult/Child (older than 3 mo): **Topical** Apply a thin film of cream or ointment to affected area once or twice daily.

Halcinonide
Halog, 0.1% cream, ointment, solution

Adult: **Topical** Apply thin layer b.i.d. or t.i.d. *Child:* **Topical** Apply thin layer once/day.

Halobetasol
Ultravate, 0.05% cream, ointment

Adult: **Topical** Apply sparingly b.i.d.

Mometasone furoate
Elocon, 0.1% cream, lotion, ointment

Adult: **Topical** Apply a thin film of cream or ointment or a few drops of lotion to affected area once/day.

Triamcinolone
Kenalog,Triderm,
0.025%, 0.5%, 0.1%
cream, ointment; 0.025%,
0.1% lotion

Adult: **Topical** Apply sparingly b.i.d. or t.i.d.

Contraindicated in: Topical steroids contraindicated in presence of varicella, vaccinia, on surfaces with compromised circulation, and in children younger than 2 y. **Cautious Use in:** Children; diabetes mellitus; stromal *herpes simplex;* glaucoma, tuberculosis of eye; osteoporosis; untreated fungal, bacterial, or viral infections **Adverse Effects: Skin:** Skin thinning and atrophy, *acne,impaired wound healing;* petechiae, ecchymosis, easy bruising; suppression of skin test reaction; hypopigmentation or hyperpigmentation, hirsutism, acneiform eruptions, subcutaneous fat atrophy; allergic dermatitis, urticaria, angioneurotic edema, increased sweating. **Clinical Implications:** Administer retention enema preferably after a bowel movement. The enema should be retained at least w1 h or all night if possible. If an occlusive dressing is to be used, apply medication sparingly, rub until it disappears, and then reapply, leaving a thin coat over lesion. Completely cover area with transparent plastic or other occlusive device or vehicle. Avoid covering a weeping or exudative lesion. Usually, occlusive dressings are not applied to face, scalp, scrotum, axilla, and groin. Inspect skin carefully between applications for ecchymotic, petechial, and purpuric signs, maceration, secondary infection, skin atrophy, striae or miliaria; if present, stop medication and notify prescriber. Warn patient not to self-dose with OTC topical preparations of a corticosteroid more than 7 days. They should not be used for children younger than 2 y. If symptoms do not abate, consult prescriber. Usually, topical preparations are applied after a shower or bath when skin is damp or wet. Cleansing and application of prescribed preparation should be done with extreme gentleness because of fragility, easy bruisability, and poor-healing skin. Hazard of systemic toxicity is higher in small children because of the greater ratio of skin surface area to body weight. Apply sparingly. Urge patient on long-term therapy with topical corticosterone to check expiration date.

Schedule I

High potential for abuse and of no currently accepted medical use. Examples: heroin, LSD, marijuana, mescaline, peyote. Not obtainable by prescription but may be legally procured for research, study, or instructional use.

Schedule II

High abuse potential and high liability for severe psychological or physical dependence. Prescription required and cannot be renewed.[a] Includes opium derivatives, other opioids, and short-acting barbiturates. Examples: amphetamine, cocaine, meperidine, morphine, secobarbital.

Schedule III

Potential for abuse is less than that for drugs in Schedules I and II. Moderate to low physical dependence and high psychological dependence. Includes certain stimulants and depressants not included in the above schedules and preparations containing limited quantities of certain opioids. Examples: chlorphentermine, glutethimide, mazindol, paregoric, phendimetrazine. Prescription required.[a,b]

Schedule IV

Lower potential for abuse than Schedule III drugs. Examples: certain psychotropics (tranquilizers), chloral hydrate, chlordiazepoxide, diazepam, meprobamate, phenobarbital. Prescription required.[a,b]

Schedule V

Abuse potential less than that for Schedule IV drugs. Preparations contain limited quantities of certain narcotic drugs; generally intended for antitussive and antidiarrheal purposes and may be distributed without a prescription provided that:

1. Such distribution is made only by a pharmacist.
2. Not more than 240 mL or not more than 48 solid dosage units of any substance containing opium, nor more than 120 mL or not more than 24 solid dosage units of any other controlled substance may be distributed at retail to the same purchaser in any given 48-hour period without a valid prescription order.
3. The purchaser is at least 18 years old.
4. The pharmacist knows the purchaser or requests suitable identification.

5. The pharmacist keeps an official written record of: name and address of purchaser, name and quantity of controlled substance purchased, date of sale, initials of dispensing pharmacist. This record is to be made available for inspection and copying by U.S. officers authorized by the Attorney General.

6. Other federal, state, or local law does not require a prescription order.

Under jurisdiction of the Federal Controlled Substances Act:
[a]Except when dispensed directly by a practitioner, other than a pharmacist, to an ultimate user, no controlled substance in Schedule II may be dispensed without a *written* prescription, except that in emergency situations such drug may be dispensed upon oral prescription and a written prescription must be obtained within the time frame prescribed by law. No prescription for a controlled substance in Schedule II may be refilled.
[b]Refillable up to 5 times within 6 mo, but only if so indicated by prescriber.

The FDA requires that all prescription drugs absorbed systemically or known to be potentially harmful to the fetus be classified according to one of five pregnancy categories (A, B, C, D, X). The identifying letter signifies the level of risk to the fetus and is to appear in the precautions section of the package insert. The categories described by the FDA are as follows:

Category A

Controlled studies in women fail to demonstrate a risk to the fetus in the first trimester (and there is no evidence of risk in later trimesters), and the possibility of fetal harm appears remote.

Category B

Either animal-reproduction studies have not demonstrated a fetal risk but there are no controlled studies in pregnant women, or animal-reproduction studies have shown an adverse effect (other than a decrease in fertility) that was not confirmed in controlled studies in women in the first trimester (and there is no evidence of a risk in later trimesters).

Category C

Either studies in animals have revealed adverse effects on the fetus (teratogenic or embryocidal effects or other) and there are no controlled studies in women, or studies in women and animals are not available. Drugs should be given only if the potential benefit justifies the potential risk to the fetus.

Category D

There is positive evidence of human fetal risk, but the benefits from use in pregnant women may be acceptable despite the risk (e.g., if the drug is needed in a life-threatening situation or for a serious disease for which safer drugs cannot be used or are ineffective). There will be an appropriate statement in the "warnings" section of the labeling.

Category X

Studies in animals or human beings have demonstrated fetal abnormalities or there is evidence of fetal risk based on human experience, or both, and the risk of the use of the drug in pregnant women clearly outweighs any possible benefit. The drug is contraindicated in women who are or may become pregnant. There will be an appropriate statement in the "contraindications" section of the labeling.

Some oral dosage forms should not be crushed or chewed. These dosage forms have been specially designed to release the drug slowly over several hours, to protect the drug from the low pH of the stomach, and/or to protect the stomach from the irritating effects of the drug.

Drugs may have an **enteric coating** which is designed to allow the drug to pass through the stomach intact with the drug being released in the intestines. This protects the stomach from the irritating effects of the drug, protects the drug from being destroyed by the acid pH of the stomach, and can delay the onset of action.

Modified release formulations are designed to release the drug over an extended period of time. These formulations can include multiple-layer compressed tablets where drug is released as each layer dissolves, mixed-release pellets that dissolve at different time intervals, and special tablets that are themselves inert but are designed to release drug slowly from the formulation. Some modified release dosage forms are scored and may be broken in half without affecting the release mechanism but still should not be crushed or chewed. Some mixed-release capsule formulations can be opened and the contents sprinkled on food. However, the pellets should not be crushed or chewed. Some modified release formulations can be identified by common abbreviations used in their brand names. These abbreviations include: CR (controlled release), CRT (controlled release tablet), LA (long acting), SR (sustained release), TR (time release), SA (sustained action), and ER or XL or XR (extended release).

Occasionally, drugs should not be crushed because they are oral mucosa irritants, are extremely bitter, or contain dyes that may stain teeth or mucosal tissue. Many medications that are combinations (containing multiple ingredients) should not be split.

The table contains a partial listing of drugs found in the Guide that should not be crushed or chewed. A liquid dosage form may be available for many of these drugs. However, the dose or frequency of administration may be different from the slow-release product. Check with your pharmacist for liquid availability and dosing conversions.

	Generic Name	**Comments**
Accutane	isotretinoin	mucous membrane irritant
AcipHex	rabeprazole	slow release
Actiq Oralet	fentanyl citrate	lozenge product
Actonel	risendronate	irritant
Adalat CC	nifedipine	slow release
Adderall XR	amphetamine	slow release
Advicor	niacin/lovastatin	slow release

	Generic Name	Comments
Afinitor	everolimus	mucous membrane irritant
Aggrenox	aspirin/dipyridamole	slow release; may be opened and contents taken without crushing
Allegra D	fexofenadine/ pseudoephedrine	slow release
Alophen	bisacodyl	enteric coated
Altoprev	lovastatin	slow release
Amrix	cyclobenzaprine	slow release
Aptivus	tipranavir	taste
Arthrotec	diclofenac/misoprostol	enteric coated
Asacol	mesalamine	slow release
Augmentin XR	amoxicillin/clavulanic acid	slow release; may be scored and broken
Avinza	morphine	slow release; capsule may be opened
Avodart	dutasteride	skin contact may cause tumor production
Azor	amlodipine/olmesartan	combination product
Azulfidine En-tabs	sulfasalazine	enteric coated
Bayer Extra Strength Enteric 500	aspirin, enteric coated	enteric coated; slow release
Bayer Low Adult 81 mg	aspirin, enteric coated	enteric coated
Bayer Caplet	aspirin, enteric coated	enteric coated
Biaxin XL	clarithromycin	slow release
Bisacodyl	bisacodyl	enteric coated
Bisco-Lax	bisacodyl	enteric coated
Boniva	ibandronate	irritant
Calan SR	verapamil	slow release; may break tablet
Cama Arthritis Strength	aspirin, magnesium oxide, aluminum hydroxide	special tablet formulation
Cardizem, Cardizem CD, Cardizem SR, Cardizem LA	diltiazem	slow release; capsules may be opened and contents taken without chewing or crushing
Ceftin	cefuroxime	taste; use liquid formulation

	Generic Name	Comments
Cellcept	mycophenolate	teratogenic potential
Chloral Hydrate	chloral hydrate	liquid-filled capsule
Chlor-Trimeton Repetab	chlorpheniramine	slow release
Choledyl SA	oxtriphylline	slow release
Cipro	ciprofloxacin	taste
Compazine Spansule	prochlorperazine	slow release; capsules may be opened and contents taken without chewing or crushing
Concerta	methylphenidate	slow release
Constant T	theophylline	slow release; capsules may be opened and contents taken without chewing or crushing
Cotazym S	pancrelipase	enteric coated; capsules may be opened and contents taken without chewing or crushing
Covera-HS	verapamil	slow release
Crixivan	indinavir	taste
Cymbalta	duloxetine	enteric coated
Deconamine SR	chlorpheniramine, pseudoephedrine	slow release
Depakene	valproic acid	slow release; mucous membrane irritant
Depakote	valproate disodium	slow release
Desoxyn Gradumets	methamphetamine	slow release
Detrol LA	tolterodine	slow release
Dexedrine Spansule	dextroamphetamine	slow release
Diamox Sequels	acetazolamide	slow release
Dilacor XR	diltiazem	slow release
Dilatrate-SR	isosorbide dinitrate	slow release
Disophrol Chronotab	dexbrompheniramine, pseudoephedrine	slow release
Donnatal Extentab	atropine, scopolamine, hyoscyamine, phenobarbital	slow release

	Generic Name	Comments
Donnazyme	pancreatin, pepsin, bile salts, atropine, scopolamine, hyoscyamine, phenobarbital	slow release
Drixoral	dexbrompheniramine, pseudoephedrine	slow release
Droxia	hydroxyurea	skin toxicity when exposed to powder
Dulcolax	bisacodyl	enteric coated
Duratuss	phenylephrine, guaifenesin	slow release
Easprin	aspirin	enteric coated
Ecotrin	aspirin	enteric coated
E.E.S. 400	erythromycin ethylsuccinate	enteric coated
Effexor XR	venlafaxine	slow release
Elixophyllin SR	theophylline	slow release; capsules may be opened and contents taken without chewing or crushing
Embeda	morphine	slow release; do not give via NG tube
E-Mycin	erythromycin	enteric coated
Ergostat	ergotamine	sublingual tablet
Eryc	erythromycin	enteric coated; capsules may be opened and contents taken without chewing or crushing
Ery-Tab	erythromycin	enteric coated
Erythrocin Stearate	erythromycin	enteric coated
Erythromycin Base	erythromycin	enteric coated
Eskalith CR	lithium	slow release
Evista	raloxifene	taste
Fedahist Timecaps	chlorpheniramine, pseudoephedrine	slow release
Feldene	piroxicam	mucous membrane irritant
Feosol	ferrous sulfate	enteric coated

	Generic Name	Comments
Feosol Spansule	ferrous sulfate	slow release; capsules may be opened and contents taken without chewing or crushing
Fergon	ferrous gluconate	slow release; capsules may be opened and contents taken without chewing or crushing
Ferro-Sequels	ferrous fumarate, docusate	slow release
Flomax	tamsulosin	slow release
Fosamax	alendronate	mucous membrane irritant
Gleevec	imatinib	taste
Glucophage XR	metformin	slow release
Glucotrol XL	glipizide	slow release
Gris-Peg	griseofulvin ultramicrosize	crushing may result in precipitation of drug as larger particles
Imdur	isosorbide mononitrate	slow release
Inderal LA	propranolol	slow release
Indocin SR	indomethacin	slow release; capsules may be opened and contents taken without chewing or crushing
Intelence	etravirine	may be dissolved in water
Isoptin SR	verapamil	slow release
Iso-Bid	isosorbide dinitrate	slow release
Isosorbide Dinitrate SR	isosorbide dinitrate	slow release
Janumet	sitagliptin/metformin	combination product
Kadian	morphine	slow release
Kaon CL 10	potassium chloride	slow release
Klor-Con	potassium chloride	slow release
Klotrix	potassium chloride	slow release
K-Tab	potassium chloride	slow release
Levsinex Timecaps	hyoscyamine	slow release
Letairis	ambrisentan	slow release
Lithobid	lithium	slow release
Lovaza	fish oils	irritant

	Generic Name	Comments
Meprospan	meprobamate	slow release; capsules may be opened and contents taken without chewing or crushing
Mestinon Timespan	pyridostigmine	slow release
Micro K	potassium chloride	slow release
Moxatag	amoxicillin	slow release
MS Contin	morphine	slow release
Mucinex	guaifenesin	slow release
Nexium	esomeprazole	slow release
Nico-400	niacin	slow release
Nicobid	niacin	slow release
Nitro Bid	nitroglycerin	slow release; capsules may be opened and contents taken without chewing or crushing
Nitroglyn	nitroglycerin	slow release; capsules may be opened and contents taken without chewing or crushing
Norflex	orphenadrine	slow release
Norpace CR	disopyramide	slow release
Norvir	ritonavir	decreases bioavailability
Novafed A	pseudoephedrine, chlorpheniramine	slow release
Oracea	doxycycline	slow release
Oramorph SR	morphine	slow release
Pancrease	pancrelipase	enteric coated
Papaverine Sustained Action	papaverine	slow release
Perdiem	psyllium hydrophilic mucilloid	wax coated
Phazyme, Phazyme 95	simethicone	slow release
Plendil	felodipine	slow release
Prevacid	lansoprazole	slow release
Prilosec	omeprazole	slow release
Procainamide HCl SR	procainamide	slow release
Procan SR	procainamide	slow release

	Generic Name	Comments
Protonix	pantoprazole	slow release
Procardia XL	nifedipine	slow release
Pronestyl SR	procainamide	slow release
Proventil Repetabs	albuterol	slow release
Ranexa	ranolazine	slow release
Ritalin SR	methylphenidate	slow release
Robimycin Robitab	erythromycin	enteric coated
Rondec TR	pseudoephedrine, carbinoxamine	slow release
Roxanol SR	morphine	slow release
Sinemet CR	levodopa, carbidopa	slow release; tablet is scored and may be broken in half
Slo-Bid Gyrocaps	theophylline	slow release; capsules may be opened and contents taken without chewing or crushing
Slo-Phyllin Gyrocaps	theophylline	slow release; capsules may be opened and contents taken without chewing or crushing
Slow-Fe	ferrous sulfate	slow release
Slow-K	potassium chloride	slow release
Sorbitrate SA	isosorbide dinitrate	slow release
Strattera	atomoxetine	slow release
Sudafed 12 hour	pseudoephedrine	slow release
Tarka	trandolapril, verapamil	slow release
Tasigna	nilotinib	may increase toxicity risk
Teldrin	chlorpheniramine	slow release; capsules may be opened and contents taken without chewing or crushing
Tepanil Ten-Tab	diethylpropion	slow release
Tessalon Perles	benzonatate	slow release
Theo-24	theophylline	slow release
Thorazine Spansule	chlorpromazine	slow release
Toprol XL	metoprolol	slow release
Trental	pentoxifylline	slow release
Treximet	naproxen/sumatriptan	decreases efficacy

	Generic Name	Comments
Trilafon Repetabs	perphenazine	slow release
Triptone Caplets	scopolamine	slow release
Uniphyl	theophylline	slow release
Valcyte	valganciclovir	irritant; teratogen
Verelan	verapamil	slow release; capsules may be opened and contents taken without chewing or crushing
Wellbutrin SR, Wellbutrin XL	bupropion	slow release; mucous membrane irritant
Zolinza	vorinostat	irritant
ZORprin	aspirin	slow release
Zyban	bupropion	slow release

Note: This listing is not comprehensive. Please check with your pharmacist for additional questions.

Information from The Institute for Safe Medication Practices available at: http://www.ismp.org/tools/DoNotCrush.pdf

Acanya (ANTIACNE) *gel:* benzoyl peroxide 2.5%/clindamycin (see p. 339) 1.2%.

Accuretic (ANTIHYPERTENSIVE) *tablet:* quinapril (see p. 1293) 10 mg/hydrochlorothiazide (see p. 728) 12.5 mg; 20 mg quinapril (see p. 1293)/12.5 mg hydrochlorothiazide; 20 mg quinapril/25 mg hydrochlorothiazide.

Activella (HORMONE REPLACEMENT THERAPY) *tablet:* estradiol (see p. 571) 1 mg/norethindrone acetate (see p. 1079) 0.5 mg.

Actonel with Calcium (BISPHOSPHONATE) *tablet:* risedronate (see p. 1334) 35 mg/1250 calcium carbonate (see p. 221).

ACTOplus Met (ANTIDIABETIC) *tablet:* pioglitazone (see p. 1208) 15 mg/metformin (see p. 947) 500 mg; pioglitazone 15 mg/metformin 850 mg.

Advair Diskus (BRONCHODILATOR) *Inhalation powder:* fluticasone propionate (see p. 658) 100 mcg/salmeterol (see p. 1355) 50 mcg; fluticasone propionate 250 mcg/salmeterol 50 mcg; fluticasone propionate 115 mcg/salmeterol 21 mcg; fluticasone propionate 230 mcg/salmeterol 21 mcg; fluticasone propionate 45 mcg/salmeterol 21 mcg.

Advicor (ANTILIPEMIC) *tablets, sustained release:* niacin (see p. 1054) 500 mg/lovastatin (see p. 898) 20 mg; niacin 1000 mg/lovastatin 20 mg.

Aggrenox (ANTIPLATELET) *extended release capsule:* dipyridamole (see p. 472) 200 mg/aspirin (see p. 119) 25 mg.

Aldactazide 25/25 (DIURETIC) *tablet:* spironolactone (see p. 1399) 25 mg/hydrochlorothiazide (see p. 728) 25 mg.

Aldactazide 50/50 (DIURETIC) *tablet:* spironolactone (see p. 1399) 50 mg/hydrochlorothiazide (see p. 728) 50 mg.

Allegra D 12 hour (ANTIHISTAMINE, DECONGESTANT) *tablet, extended release:* fexofenadine (see p. 627) 60 mg/pseudoephedrine (see p. 1281) 120 mg.

Allegra D 24 hour (ANTIHISTAMINE, DECONGESTANT) *tablet, extended release:* fexofenadine (see p. 627) 180 mg/pseudoephedrine (see p. 1281) 240 mg.

Anexsia (NARCOTIC ANALGESIC [schedule III]) *tablet:* hydrocodone (see p. 730) 5 mg/acetaminophen (see p. 11) 500 mg.

Anexsia 5/325 (NARCOTIC ANALGESIC [schedule III]) *tablet:* hydrocodone (see p. 730) 5 mg/acetaminophen (see p. 11) 325 mg.

Anexsia 7.5/325 (NARCOTIC ANALGESIC [schedule III]) *tablet:* hydrocodone (see p. 730) 7.5 mg/acetaminophen (see p. 11) 325 mg.

Anexsia 7.5/650 (NARCOTIC ANALGESIC [schedule III]) *tablet:* hydrocodone (see p. 730) 7.5 mg/acetaminophen (see p. 11) 650 mg.

Angeliq (HORMONE) *tablet:* drospirenone 0.5 mg/estradiol (see p. 571) 1 mg.

Apresazide 25/25 (ANTIHYPERTENSIVE) *capsule:* hydralazine hydrochloride (see p. 727) 25 mg/hydrochlorothiazide (see p. 728) 25 mg.

Apresazide 50/50 (ANTIHYPERTENSIVE) *capsule:* hydralazine hydrochloride (see p. 727) 50 mg/hydrochlorothiazide (see p. 728) 50 mg.

Apresodex (ANTIHYPERTENSIVE) *tablet:* hydralazine hydrochloride (see p. 727) 25 mg/hydrochlorothiazide (see p. 728) 15 mg.

Aralen Phosphate with Primaquine Phosphate (ANTIMALARIAL) *tablet:* chloroquine phosphate (see p. 301) 500 mg (300 mg base)/primaquine phosphate (see p. 1248) 79 mg (45 mg base).

Arthrotec 50 (NSAID) *tablet:* diclofenac sodium (see p. 446) 50 mg/misoprostol (see p. 1001) 200 mcg.

Arthrotec 75 (NSAID) *tablet:* diclofenac sodium (see p. 446) 75 mg/misoprostol (see p. 1001) 200 mcg.

Atacand HCT (ANTIHYPERTENSIVE) *tablet:* candesartan (see p. 228) 32 mg/hydrochlorothiazide (see p. 728) 12.5 mg; candesartan 16 mg/hydrochlorothiazide 12.5 mg.

Atripla (ANTIRETROVIRAL) *tablet:* 600 mg efavirenz (see p. 519)/200 mg emtricitabine (see p. 525)/300 mg tenofovir (see p. 1447).

Augmentin (ANTIBIOTIC) *tablet:* amoxicillin (see p. 83) 250 mg/clavulanic acid 125 mg; amoxicillin 500 mg/clavulanic acid 125 mg; amoxicillin 875 mg/clavulanic acid 125 mg; amoxicillin 1000 mg/clavulanic acid 125 mg; *chewable tablet:* amoxicillin 125 mg/clavulanic acid 31.25 mg; amoxicillin 200 mg/clavulanic acid 28.5 mg; amoxicillin 250 mg/clavulanic acid 62.5 mg; amoxicillin 400 mg/clavulanic acid 57 mg; *suspension (per 5 mL):* amoxicillin 125 mg/clavulanic acid 31.25 mg; amoxicillin 200 mg/clavulanic acid 28.5 mg; amoxicillin 250 mg/clavulanic acid 62.5 mg; amoxicillin 400 mg/clavulanic acid 57 mg; amoxicillin 600 mg/clavulanic acid 42.9 mg.

Auralgan Otic (OTIC PREPARATION: DECONGESTANT, ANALGESIC) *solution:* acetic acid 0.1%, antipyrine 5.4%, benzocaine (see p. 164) 1.4%, u-polycosanol 410 0.01%.

Avalide (ANTIHYPERTENSIVE) *tablet:* irbesartan (see p. 801) 150 mg/hydrochlorothiazide (see p. 728) 12.5 mg; irbesartan 300 mg/hydrochlorothiazide 12.5 mg; irbesartan 300 mg/hydrochlorothiazide 25 mg.

Avandamet (HYPOGLYCEMIC AGENT) *tablet:* 1 mg rosiglitazone maleate (see p. 1352)/500 mg metformin HCl (see p. 947); 2 mg rosiglitazone/500 mg metformin; 4 mg rosiglitazone/500 mg metformin; 2 mg rosiglitazone/1000 mg metformin; 4 mg rosiglitazone/1000 mg metformin.

Avandaryl (HYPOGLYCEMIC AGENT) *tablet:* rosiglitazone (see p. 1352) 4 mg/glimepiride (see p. 697) 1 mg; rosiglitazone 4 mg/glimepiride 2 mg; rosiglitazone 4 mg/glimepiride 4 mg.

Azo Gantanol (URINARY ANTI-INFECTIVE, ANALGESIC) *tablet:* sulfamethoxazole (see p. 1413) 500 mg, phenazopyridine hydrochloride (see p. 1183) 100 mg.

Azo Gantrisin (URINARY ANTI-INFECTIVE, ANALGESIC) *tablet:* 500 mg/phenazopyridine hydrochloride (see p. 1183) 50 mg.

Azor (ANTIHYPERTENSIVE) *tablet:* amlodipine (see p. 78) 10 mg/olmesartan (see p. 1094) 20 mg; amlodipine 10 mg/olmesartan 40 mg; amlodipine 5 mg/olmesartan 20 mg; amlodipine 5 mg/olmesartan 40 mg.

B-A-C (ANALGESIC) acetaminophen (see p. 11) 650 mg/caffeine (see p. 214) 40 mg/butalbital 50 mg.

Bacticort Ophthalmic (ANTI-INFLAMMATORY) *suspension:* hydrocortisone (see p. 731) 1%/neomycin sulfate (see p. 1047) 0.35%/polymyxin B (see p. 1220) 10,000 units.

Bactrim (URINARY TRACT AGENT) *tablet:* sulfamethoxazole (see p. 1413) 400 mg/trimethoprim (see p. 1538) 80 mg.

Bactrim DS (URINARY TRACT AGENT) *tablet:* sulfamethoxazole (see p. 1413) 800 mg/trimethoprim (see p. 1538) 160 mg.

Benicar HCT (ANTIHYPERTENSIVE) *tablet:* 20 mg olmesartan medoxomil (see p. 1094)/12.5 mg hydrochlorothiazide (see p. 728); 40 mg olmesartan medoxomil/12.5 mg hydrochlorothiazide; 40 olmesartan medoxomil/25 mg hydrochlorothiazide.

Betoptic Pilo Suspension (ANTIGLAUCOMA) *suspension:* betaxolol (see p. 171) 0.25%/pilocarpine (see p. 1202) 1.75%.

Beyaz (ORAL CONTRACEPTIVE) *tablet:* drosperinone 3 mg/ethinyl estradiol 0.02 mg/levomefolate calcium 0.451 mg.

BiDil (ANTIHYPERTENSIVE) *tablet:* isosorbide dinitrate (see p. 813) 20 mg/hydralazine (see p. 727) 37.5 mg.

Blephamide (OPHTHALMIC STEROID, SULFONAMIDE) *suspension:* prednisolone acetate (see p. 1242) 0.2%/sulfacetamide sodium (see p. 1410) 10%.

Blephamide S.O.P. (OPHTHALMIC STEROID, SULFONAMIDE) *ointment:* prednisolone acetate (see p. 1242) 0.2%/sulfacetamide sodium (see p. 1410) 10%.

Brevicon (MONOPHASIC ORAL CONTRACEPTIVE [ESTROGEN, PROGESTIN]) *tablet:* ethinyl estradiol (see p. 591) 35 mcg/norethindrone (see p. 1079) 0.5 mg.

Bromfed (DECONGESTANT, ANTIHISTAMINE) *sustained release capsule:* pseudoephedrine hydrochloride (see p. 1281) 120 mg/brompheniramine maleate (see p. 195) 12 mg.

Bromfed-PD (DECONGESTANT, ANTIHISTAMINE) *sustained release capsule:* pseudoephedrine hydrochloride (see p. 1281) 60 mg/brompheniramine maleate (see p. 195) 6 mg.

Bronchial Capsules (ANTIASTHMATIC) *capsule:* theophylline (see p. 1466) 150 mg/guaifenesin (see p. 713) 90 mg.

Caduet (ANTIHYPERTENSIVE/ANTILIPEMIC) *tablet:* 2.5 mg amlodipine (see p. 78)/10 mg atorvastatin (see p. 128); 2.5 mg amlodipine/20 mg atorvastatin; 2.5 mg amlodipine/40 mg atorvastatin; 5 mg amlodipine/10 mg atorvastatin; 10 mg amlodipine/10 mg atorvastatin; 5 mg amlodipine/20 mg atorvastatin; 10 mg amlodipine/20 mg atorvastatin; 5 mg amlodipine/40 mg atorvastatin; 10 mg amlodipine/40 mg atorvastatin; 5 mg amlodipine/80 mg atorvastatin; 10 mg amlodipine/80 mg atorvastatin.

Cafergot Suppositories (ANTIMIGRAINE) *suppository:* ergotamine tartrate (see p. 552) 2 mg/caffeine (see p. 214) 100 mg.

Cam-ap-es (ANTIHYPERTENSIVE) *suspension, tablet:* hydrochlorothiazide (see p. 728) 15 mg/reserpine (see p. 1314) 0.1 mg/hydralazine hydrochloride (see p. 727) 25 mg.

Capital with Codeine (NARCOTIC ANALGESIC [schedule V]) *suspension (per 5 mL):* codeine phosphate (see p. 359) 12 mg/acetaminophen (see p. 11) 120 mg.

Capozide 25/15 (ANTIHYPERTENSIVE) *tablet:* captopril (see p. 233) 25 mg, hydrochlorothiazide (see p. 728) 15 mg.

Capozide 25/25 (ANTIHYPERTENSIVE) *tablet:* captopril (see p. 233) 25 mg/hydrochlorothiazide (see p. 728) 25 mg.

Capozide 50/15 (ANTIHYPERTENSIVE) *tablet:* captopril (see p. 233) 50 mg/hydrochlorothiazide (see p. 728) 15 mg.

Capozide 50/25 (ANTIHYPERTENSIVE) *tablet:* captopril (see p. 233) 50 mg/hydrochlorothiazide (see p. 728) 25 mg.

Carisoprodol Compound (SKELETAL MUSCLE RELAXANT, ANALGESIC) *tablet:* carisoprodol (see p. 245) 200 mg/aspirin (see p. 119) 325 mg.

Carmol HC (ANTI-INFLAMMATORY) *cream:* hydrocortisone acetate (see p. 731) 1%/urea 10%.

Celestone-Soluspan (GLUCOCORTICOID) *injection (suspension) (per mL):* betamethasone acetate (see p. 169) 3 mg/betamethasone sodium phosphate (see p. 169) 3 mg.

Cetacaine (TOPICAL ANESTHETIC) *gel, liquid, ointment, aerosol:* benzocaine (see p. 164) 14%/tetracaine hydrochloride (see p. 1459) 2%/butamben 2%/benzalkonium chloride (see p. 163) 0.5%.

Cheracol Syrup (NARCOTIC ANTITUSSIVE, EXPECTORANT [schedule V]) *syrup (per 5 mL):* codeine phosphate (see p. 359) 10 mg/guaifenesin (see p. 713) 100 mg/alcohol 4.75%.

Cipro HC Otic (ANTI-INFECTIVE/ANTI-INFLAMMATORY) *topical:* ciprofloxacin (see p. 326) 2 mg/dexamethasone (see p. 428) 10 mg otic suspension.

Ciprodex Otic (ANTI-INFECTIVE/ANTI- INFLAMMATORY) *topical:* ciprofloxacin (see p. 326) 0.3%/dexamethasone (see p. 428) 0.1% otic suspension.

Claritin D (ANTIHISTAMINE, DECONGESTANT) loratadine (see p. 894), 5 mg/pseudoephedrine (see p. 1281) 120 mg; loratadine 10 mg/pseudoephedrine 240 mg.

Clarinex D 24 hr (ANTIHISTAMINE, DECONGESTANT) *tablet:* desloratadine (see p. 425) 5 mg/pseudoephedrine (see p. 1281) 240 mg.

Climara Pro (HORMONE REPLACEMENT THERAPY) *transdermal patch:* estradiol (see p. 571) 0.045 mg/levonorgestrel acetate 0.015 mg.

Codiclear DH Syrup (ANTITUSSIVE [schedule III]) *syrup (per 5 mL):* hydrocodone (see p. 730) 5 mg/guaifenesin (see p. 713) 100 mg/alcohol 10%.

Codimal DH (ANTITUSSIVE [schedule III]) *syrup (per 5 mL):* phenylephrine hydrochloride (see p. 1193) 5 mg/pyrilamine maleate 8.33 mg/hydrocodone bitartrate (see p. 730) 1.66 mg.

Codimal PH (ANTITUSSIVE [schedule III]) *syrup (per 5 mL):* codeine (see p. 359) 10 mg/pyrilamine maleate 8.33 mg/phenylephrine (see p. 1193) 5 mg.

Co-Gesic (NARCOTIC ANALGESIC [schedule V]) *tablet:* hydrocodone (see p. 730) 5 mg/acetaminophen (see p. 11) 500 mg.

Coly-Mycin S Otic (OTIC: STEROID, ANTIBIOTIC) *suspension (per mL):* hydrocortisone acetate (see p. 731) 1%/neomycin sulfate (see p. 1047) 3.3 mg/colistin sulfate 3 mg/thonzonium bromide 0.05%.

Combigan (GLAUCOMA) *ophthalmic solution:* brimonidine (see p. 193) 0.2%/timolol (see p. 1488) 0.5%.

CombiPatch (HORMONE REPLACEMENT THERAPY) *transdermal patch:* estradiol (see p. 571) 0.05 mg/norethindrone acetate (see p. 1079) 0.14 mg; estradiol 0.05 mg/norethindrone acetate 0.25 mg.

Combivir (ANTIVIRAL) *tablet:* zidovudine (see p. 1590) 300 mg/lamivudine (see p. 841) 150 mg.

Complera (ANTIRETROVIRAL) *tablet:* rilpivirine 25 mg/emtricitabine 200 mg/tenofovir 300 mg.

Cortisporin (OPHTHALMIC STEROID, ANTIBIOTIC) *suspension (per mL):* hydrocortisone (see p. 731) 1%/neomycin sulfate (see p. 1047) (equivalent to 0.35% neomycin base)/polymyxin B sulfate (see p. 1220) 10,000 units.

Cortisporin Ointment (OPHTHALMIC STEROID, ANTIBIOTIC) *ointment:* hydrocortisone (see p. 731) 1%/neomycin sulfate (see p. 1047) (equivalent to 0.35% neomycin base)/bacitracin zinc (see p. 150) 400 units, polymyxin B sulfate (see p. 1220) 10,000 units/g.

Corzide 40/5 (ANTIHYPERTENSIVE) *tablet:* nadolol (see p. 1023) 40 mg/bendroflumethiazide 5 mg.

Corzide 80/5 (ANTIHYPERTENSIVE) *tablet:* nadolol (see p. 1023) 80 mg/bendroflumethiazide 5 mg.

Cosopt (OPHTHALMIC, GLAUCOMA) *ophthalmic solution:* dorzolamide (see p. 492) 2%/timolol (see p. 1488) 0.5%.

Cotrim (ANTI-INFECTIVE) *tablet:* trimethoprim (see p. 1538) 80 mg/sulfamethoxazole (see p. 1413) 400 mg.

Cotrim DS (Double Strength) (ANTI-INFECTIVE) *tablet:* trimethoprim (see p. 1538) 160 mg/sulfamethoxazole (see p. 1413) 800 mg.

Cotrim Pediatric (ANTI-INFECTIVE) *suspension:* trimethoprim (see p. 1538) 40 mg/sulfamethoxazole (see p. 1413) 200 mg/5 mL.

Cyclomydril (OPHTHALMIC DECONGESTANT) *ophthalmic solution:* atropine hydrochloride (see p. 381) 0.2%/phenylephrine hydrochloride (see p. 1193) 1%.

Decadron with Xylocaine (GLUCOCORTICOID) *injection (per mL):* dexamethasone sodium phosphate (see p. 428) 4 mg/lidocaine hydrochloride (see p. 871) 10 mg.

Deconamine (DECONGESTANT, ANTIHISTAMINE) *syrup (per 5 mL):* pseudoephedrine hydrochloride (see p. 1281) 30 mg/chlorpheniramine maleate (see p. 305) 2 mg; *tablet:* pseudoephedrine hydrochloride 60 mg/chlorpheniramine maleate 4 mg.

Deconamine SR (DECONGESTANT, ANTIHISTAMINE) *sustained release capsule:* pseudoephedrine hydrochloride (see p. 1281) 120 mg/chlorpheniramine maleate (see p. 305) 8 mg.

Demulen 1/50 (ORAL CONTRACEPTIVE) *tablet:* ethinyl estradiol (see p. 591) 50 mcg/norethindrone (see p. 1079) 1 mg.

Depo-Testadiol (ESTROGEN, ANDROGEN) *injection (per mL):* estradiol cypionate (see p. 571) 2 mg/testosterone cypionate (see p. 1456) 50 mg.

Dilaudid Cough Syrup (NARCOTIC ANTITUSSIVE [schedule II]) *syrup:* hydromorphone (see p. 735) 1 mg/guaifenesin (see p. 713) 100 mg/alcohol 5%.

Dilor G (ANTIASTHMATIC) *liquid (per 5 mL):* dyphylline (see p. 512) 100 mg/guaifenesin (see p. 713) 100 mg.

Diovan HCT (ANTIHYPERTENSIVE) *tablet:* hydrochlorothiazide (see p. 728) 12.5 mg/valsartan (see p. 1554) 80 mg; hydrochlorothiazide 12.5 mg/valsartan 160 mg; hydrochlorothiazide 25 mg/valsartan 160 mg; hydrochlorothiazide 12.5 mg/valsartan 320 mg; hydrochlorothiazide 25 mg/valsartan 320 mg.

Donnatal (GASTROINTESTINAL ANTICHOLINERGIC, SEDATIVE) *capsule, tablet, elixir:* atropine sulfate (see p. 134) 0.0194 mg/scopolamine hydrobromide (see p. 1362) 0.0065 mg/hyoscyamine hydrobromide or sulfate (see p. 746) 0.1037 mg/phenobarbital (see p. 1186) 16.2 mg. The elixir contains alcohol 23%/5 mL.

Donnatal Extentab (GASTROINTESTINAL ANTICHOLINERGIC, SEDATIVE) *tablet:* atropine sulfate (see p. 134) 0.0582 mg/scopolamine hydrobromide (see p. 1362) 0.0195 mg/hyoscyamine sulfate (see p. 746) 0.3111 mg/phenobarbital (see p. 1186) 48.6 mg.

Duac (ANTIACNE) *gel:* clindamycin (see p. 339) 1%/benzoyl peroxide 5%.

Duetact (ANTIDIABETIC) *tablet:* pioglitazone (see p. 1208) 30 mg/glimepiride (see p. 697) 2 mg; pioglitazone 30 mg/glimepiride 4 mg.

Dulera (CORTICOSTEROID/BETA-AGONIST) *inhaler:* mometasone 100 mcg/fomoterol 5 mcg; mometasone 200 mcg/fomoterol 5 mcg.

DuoNeb (BETA-AGONIST/ANTICHOLINERGIC BRONCHODILATOR) *inhalation solution:* 3 mg albuterol sulfate (see p. 31)/0.5 mg ipratropium bromide (see p. 799) per 3 mL.

Dyazide (DIURETIC) *capsule:* triamterene (see p. 1530) 37.5 mg/hydrochlorothiazide (see p. 728) 25 mg.

Embeda (ANALGESIC [schedule II]) *tablet:* morphine (see p. 1011) 20 mg/naltrexone (see p. 1031) 0.8 mg; morphine 30 mg/naltrexone 1.2 mg; morphine 50 mg/naltrexone 2 mg; morphine 60 mg/naltrexone 2.4 mg; morphine 80 mg/naltrexone 3.2 mg; morphine 100 mg/naltrexone 4 mg.

Endocet (NARCOTIC ANALGESIC [schedule II]) *tablet:* oxycodone (see p. 1120) 7.5 mg/acetaminophen (see p. 11) 325 mg; oxycodone 7.5 mg/acetaminophen 500 mg; oxycodone 10 mg/acetaminophen 325 mg; oxycodone 10 mg/acetaminophen 650 mg.

Epiduo (ANTIACNE) *gel:* adapalene (see p. 24) 0.1%/benzoyl peroxide 2.5%.

Epzicom (ANTIRETROVIRAL AGENT) *tablet:* abacavir (see p. 1) 600 mg/lamivudine (see p. 841) 300 mg.

Estratest (ESTROGEN, ANDROGEN) *tablet:* esterified estrogens (see p. 581) 1.25 mg/methyltestosterone (see p. 973) 2.5 mg.

Estratest H.S. (ESTROGEN, ANDROGEN) *tablet:* esterified estrogens (see p. 581) 0.625 mg/methyltestosterone (see p. 973) 1.25 mg.

Exforge (ANTIHYPERTENSIVE) *tablet:* amlodipine (see p. 78) 5 mg/valsartan (see p. 1554) 160 mg; amlodipine 10 mg/valsartan 160 mg; amlodipine 5 mg/valsartan 320 mg; amlodipine 10 mg/valsartan 320 mg.

Exforge HCT (ANTIHYPERTENSIVE) *tablet:* amlodipine (see p. 78) 5 mg/valsartan (see p. 1554) 160 mg/hydrochlorothiazide (see p. 728) 12.5 mg; amlodipine 5 mg/valsartan 160 mg/hydrochlorothiazide 25 mg; amlodipine 10 mg/valsartan 160 mg/hydrochlorothiazide 12.5 mg; amlodipine 10 mg/valsartan 160 mg/hydrochlorothiazide 25 mg.

Femhrt (HORMONES) *tablet:* ethinyl estradiol (see p. 591) 5 mcg/norethindrone acetate (see p. 1079) 1 mg; ethinyl estradiol 2.5 mcg/norethindrone acetate 0.5 mg.

Fioricet (NONNARCOTIC AGONIST ANALGESIC) *tablet:* acetaminophen (see p. 11) 325 mg/butalbital 50 mg/caffeine (see p. 214) 40 mg.

Fiorinal (NONNARCOTIC AGONIST ANALGESIC [schedule III]) *capsule, tablet:* aspirin (see p. 119) 325 mg/butalbital 50 mg/caffeine (see p. 214) 40 mg.

Fiorinal with Codeine (NARCOTIC AGONIST ANALGESIC [schedule III]) *capsule:* codeine phosphate (see p. 359) 30 mg/aspirin (see p. 119) 325 mg/caffeine (see p. 214) 40 mg/butalbital 50 mg.

Fluress (OPHTHALMIC ANESTHETIC) *ophthalmic solution:* benoxinate hydrochloride 0.4%/fluorescein sodium (see p. 645) 0.25%.

Fosamax Plus D (BISPHOSPHONATE) *tablet:* alendronate (see p. 36) 70 mg/vitamin D (see p. 1577) 2800 international units.

Glucovance (ANTIDIABETIC) *tablet:* glyburide (see p. 701) 1.25 mg/metformin (see p. 947) 250 mg; glyburide 2.5 mg/metformin 500 mg; glyburide 5 mg/metformin 500 mg.

Helidac (ANTIULCER, ANTIBIOTIC) *tablet:* bismuth subsalicylate (see p. 179) 262.4 mg/metronidazole (see p. 981) 250 mg/tetracycline (see p. 1461) 500 mg.

Hycodan (ANTITUSSIVE [schedule III]) *tablet, syrup:* hydrocodone bitartrate (see p. 730) 5 mg/homatropine methylbromide 1.5 mg.

Hycotuss Expectorant (ANTITUSSIVE [schedule III]) guaifenesin (see p. 713) 100 mg/hydrocodone (see p. 730) 5 mg.

Hydrocet (NARCOTIC ANALGESIC [schedule III]) *capsule:* hydrocodone (see p. 730) 5 mg/acetaminophen (see p. 11) 500 mg.

Hyzaar (ANTIHYPERTENSIVE) *tablet:* losartan (see p. 897) 50 mg/hydrochlorothiazide (see p. 728) 12.5 mg/losartan 100 mg/hydrochlorothiazide 12.5 mg/losartan 100 mg/hydrochlorothiazide 25 mg.

Inderide 40/25 (ANTIHYPERTENSIVE) *tablet:* propranolol hydrochloride (see p. 1271) 40 mg/hydrochlorothiazide (see p. 728) 25 mg.

Jalyn (BPH AGENT) *capsule:* dutasteride 0.5 mg/tamsulosin 0.4 mg.

Janumet (ANTIDIABETIC) *tablet:* sitagliptin (see p. 1382) 50 mg/metformin (see p. 947) 500 mg; sitagliptin 50 mg/metformin 1000 mg.

Jusvisync (ANTIDIABETIC) *tablet:* sitagliptin 100 mg/simvastatin 10mg; sitagliptin 100 mg/simvastatin 20 mg; sitagliptin 100 mg/simvastatin 40 mg.

Kombiglyze XR (ANTIDIABETIC) *extended release tablet:* saxagliptin 5 mg/metformin 500 mg; saxagliptin 2.5 mg/metformin 1000 mg; saxagliptin 5 mg/metformin 1000 mg.

Levlen (ORAL CONTRACEPTIVE) *tablet:* ethinyl estradiol (see p. 591) 30 mcg/levonorgestrel 0.15 mg.

LidoSite (LOCAL ANESTHETIC) *transdermal patch:* lidocaine (see p. 871) 100 mg/epinephrine (see p. 537) 1.05 mg.

Limbitrol (PSYCHOTHERAPEUTIC [schedule IV]) *tablet:* chlordiazepoxide (see p. 299) 5 mg/amitriptyline (see p. 76) 12.5 mg.

Limbitrol DS (PSYCHOTHERAPEUTIC [schedule IV]) *tablet:* chlordiazepoxide (see p. 299) 10 mg/amitriptyline (see p. 76) 25 mg.

Loestrin 1/20 (ORAL CONTRACEPTIVE) *tablet:* ethinyl estradiol (see p. 591) 20 mcg/norethindrone acetate (see p. 1079) 1 mg.

Loestrin 1/20 Fe (ORAL CONTRACEPTIVE) *tablet:* ethinyl estradiol (see p. 591) 20 mcg/norethindrone acetate (see p. 1079) 1 mg/ferrous fumarate 75 mg in last 7 tablets.

Loestrin 1.5/30 (ORAL CONTRACEPTIVE) *tablet:* ethinyl estradiol (see p. 591) 30 mcg/norethindrone acetate (see p. 1079) 1.5 mg.

Loestrin 1.5/30 Fe (ORAL CONTRACEPTIVE) *tablet:* ethinyl estradiol (see p. 591) 30 mcg/norethindrone acetate (see p. 1079) 1.5 mg/ferrous fumarate 75 mg in last 7 tablets.

Lomotil (ANTIDIARRHEAL) *tablet:* diphenoxylate (see p. 471) 2.5 mg/atropine (see p. 134) 0.025 mg.

Lo/Ovral (ORAL CONTRACEPTIVE) *tablet:* ethinyl estradiol (see p. 591) 30 mcg/norgestrel 0.3 mg.

Lopressor HCT 50/25 (ANTIHYPERTENSIVE) *tablet:* metoprolol tartrate (see p. 978) 50 mg/hydrochlorothiazide (see p. 728) 25 mg.

Lopressor HCT 100/25 (ANTIHYPERTENSIVE) *tablet:* metoprolol tartrate (see p. 978) 100 mg/hydrochlorothiazide (see p. 728) 25 mg.

Lopressor HCT 100/50 (ANTIHYPERTENSIVE) *tablet:* metoprolol tartrate (see p. 978) 100 mg/hydrochlorothiazide (see p. 728) 50 mg.

Lorcet (NARCOTIC ANALGESIC [schedule III]) *tablet:* hydrocodone (see p. 730) 5 mg/acetaminophen (see p. 11) 500 mg.

Lorcet 10/650 (NARCOTIC ANALGESIC [schedule III]) *tablet:* hydrocodone (see p. 730) 10 mg/acetaminophen (see p. 11) 650 mg.

Lorcet-HD (NARCOTIC ANALGESIC [schedule III]) *tablet:* hydrocodone (see p. 730) 5 mg/acetaminophen (see p. 11) 500 mg.

Lortab 5 (NARCOTIC ANALGESIC [schedule III]) *tablet:* hydrocodone (see p. 730) 5 mg/acetaminophen (see p. 11) 500 mg.

Lortab 7.5/500 (NARCOTIC ANALGESIC [schedule III]) *tablet:* hydrocodone (see p. 730) 7.5 mg/acetaminophen (see p. 11) 500 mg.

LoSeasonique (ORAL CONTRACEPTIVE) *tablet:* ethinyl estradiol (see p. 591) 0.01 mg/ethinyl estradiol 0.02 mg/levonorgestrel (see p. 866) 0.1 mg.

Lotensin HCT 20/25 (ANTIHYPERTENSIVE) *tablet:* hydrochlorothiazide (see p. 728) 25 mg/benazepril (see p. 160) 20 mg.

Lotensin HCT 20/12.5 (ANTIHYPERTENSIVE) *tablet:* hydrochlorothiazide (see p. 728) 12.5/benazepril (see p. 160) 20 mg.

Lotensin HCT 10/12.5 (ANTIHYPERTENSIVE) *tablet:* hydrochlorothiazide (see p. 728) 12.5 mg/benazepril (see p. 160) 10 mg.

Lotrel (ANTIHYPERTENSIVE) *tablet:* amlodipine (see p. 78) 2.5 mg/benazepril (see p. 160) 10 mg; amlodipine 5 mg/benazepril 10 mg; amlodipine 5 mg/benazepril 20 mg; amlodipine 10 mg/benazepril 20 mg.

Lotrisone (CORTICOSTEROID, ANTIFUNGAL) *cream:* betamethasone (see p. 169) (as dipropionate) 0.05%/clotrimazole (see p. 355) 1%.

Malarone (ANTIMALARIAL) *tablet:* atovaquone (see p. 131) 250 mg/proguanil HCl (see p. 131) 100 mg; atovaquone 62.5 mg/proguanil HCl 25 mg.

Maxitrol (OPHTHALMIC STEROID, ANTIBIOTIC) *ophthalmic ointment, ophthalmic suspension:* dexamethasone (see p. 428) 0.1%/neomycin sulfate (see p. 1047) (equivalent to 0.35% neomycin base)/polymyxin B sulfate (see p. 1220) 10,000 units.

Maxzide (DIURETIC) *tablet:* triamterene (see p. 1530) 75 mg/hydrochlorothiazide (see p. 728) 50 mg.

Maxzide 25 (DIURETIC) *tablet:* triamterene (see p. 1530) 37.5 mg/hydrochlorothiazide (see p. 728) 25 mg.

Metaglip (HYPOGLYCEMIC AGENT) *tablet:* glipizide (see p. 698) 2.5 mg/metformin HCl (see p. 947) 250 mg; glipizide 2.5 mg/metformin 500 mg; glipizide 5 mg/metformin 500 mg.

Micardis HCT (ANTIHYPERTENSIVE) *tablet:* telmisartan (see p. 1440) 40 mg/hydrochlorothiazide (see p. 728) 12.5 mg; telmisartan 80 mg/hydrochlorothiazide 12.5 mg; telmisartan 80 mg/hydrochlorothiazide 25 mg.

Minizide (ANTIHYPERTENSIVE) *capsule:* polythiazide 0.5 mg/prazosin hydrochloride (see p. 1241) 1 mg; polythiazide 0.5 mg/

prazosin hydrochloride 2 mg; polythiazide 0.5 mg/prazosin hydrochloride 5 mg.

Modicon (ORAL CONTRACEPTIVE) *tablet:* ethinyl estradiol (see p. 591) 35 mcg/norethindrone (see p. 1079) 0.5 mg.

Moduretic (DIURETIC) *tablet:* amiloride hydrochloride (see p. 66) 5 mg/hydrochlorothiazide (see p. 728) 50 mg.

Mycitracin (OPHTHALMIC ANTIBIOTIC) *ophthalmic ointment:* polymyxin B sulfate (see p. 1220) 10,000 units/neomycin sulfate (see p. 1047) 3.5 mg/bacitracin (see p. 150) 500 units/g.

Mycodone (NARCOTIC ANALGESIC [schedule III]) *syrup (per 5 mL):* homatropine methylbromide 1.5 mg/hydrocodone (see p. 730) 5 mg.

Mycolog II (CORTICOSTEROID, ANTIFUNGAL) *cream, ointment:* triamcinolone acetonide (see p. 1528) 0.1%/nystatin (see p. 1085) 100,000 units/g.

Neo-Cortef (CORTICOSTEROID ANTIBIOTIC) *water-soluble cream, topical ointment:* hydrocortisone acetate (see p. 731) 1%/neomycin sulfate (see p. 1047) 0.5%.

Neosporin (OPHTHALMIC ANTIBIOTIC) *ophthalmic drops:* polymyxin B sulfate (see p. 1220) 10,000 units/neomycin sulfate (see p. 1047) 1.75 mg/gramicidin 0.025 mg/mL; *ophthalmic ointment:* polymyxin B sulfate 10,000 units/neomycin sulfate 3.5 mg/bacitracin zinc (see p. 150) 400 units/g.

Neosporin G.U. Irrigant (ANTIBIOTIC) *solution:* neomycin sulfate (see p. 1047) 40 mg/polymyxin B sulfate (see p. 1220) 200,000 units/mL.

Neutra-Phos (PHOSPHORUS REPLACEMENT) *capsule, powder:* phosphorus 250 mg/potassium (see p. 1225) 278 mg/sodium (see p. 1383) 164 mg/combination of monobasic, dibasic, sodium, and potassium phosphate.

Norco (NARCOTIC AGONIST ANALGESIC [schedule III]) *tablet:* hydrocodone bitartrate (see p. 730) 10 mg/acetaminophen (see p. 11) 325 mg; hydrocodone bitartrate 7.5 mg/acetaminophen 325 mg.

Nordette (ORAL CONTRACEPTIVE) *tablet:* ethinyl estradiol (see p. 591) 30 mcg/levonorgestrel (see p. 866) 0.15 mg.

Norethin 1/35 E (ORAL CONTRACEPTIVE) *tablet:* ethinyl estradiol (see p. 591) 35 mcg/norethindrone (see p. 1079) 1 mg.

Norethin 1/50 M (ORAL CONTRACEPTIVE) *tablet:* mestranol 50 mcg/norethindrone (see p. 1079) 1 mg.

Norgesic (SKELETAL MUSCLE RELAXANT) *tablet:* orphenadrine citrate (see p. 1105) 25 mg/aspirin (see p. 119) 385 mg/caffeine (see p. 214) 30 mg.

Norgesic Forte (SKELETAL MUSCLE RELAXANT, ANALGESIC) *tablet:* orphenadrine citrate (see p. 1105) 50 mg/aspirin (see p. 119) 770 mg/caffeine (see p. 214) 60 mg.

Norinyl 1+35 (ORAL CONTRACEPTIVE) *tablet:* ethinyl estradiol (see p. 591) 35 mcg/norethindrone (see p. 1079) 1 mg.

Norinyl 1+50 (ORAL CONTRACEPTIVE) *tablet:* mestranol 50 mcg/norethindrone (see p. 1079) 1 mg.

Ortho Evra (CONTRACEPTIVE) *transdermal patch:* norelgestromin 0.15 mg/ethinyl estradiol (see p. 591) 0.02 mg.

Paremyd (MYDRIATIC) *ophthalmic solution:* 1% hydroxyamphetamine hydrobromide, 0.25% tropicamide (see p. 1542).

Percocet (NARCOTIC ANALGESIC [schedule II]) *tablet:* oxycodone (see p. 1120) 2.5 mg/acetaminophen (see p. 11) 325 mg;

oxycodone 7.5 mg/acetaminophen 325 mg; oxycodone 10 mg/acetaminophen 325 mg; oxycodone 10 mg/acetaminophen 650 mg.

Percodan (NARCOTIC ANALGESIC [schedule II]) *tablet:* oxycodone hydrochloride (see p. 1120) 4.5 mg/oxycodone terephthalate 0.38 mg/aspirin (see p. 119) 325 mg.

Phrenilin (NONNARCOTIC AGONIST ANALGESIC) *tablet:* acetaminophen (see p. 11) 325 mg/butalbital 50 mg.

Phrenilin Forte (NONNARCOTIC AGONIST ANALGESIC) *capsule:* acetaminophen (see p. 11) 650 mg/butalbital 50 mg.

Polysporin Ointment (ANTI-INFECTIVE [OPHTHALMIC]) *ophthalmic ointment:* polymyxin B sulfate (see p. 1220) 10,000 units/bacitracin zinc (see p. 150) 500 units/g.

Prandimet (HYPOGLYCEMIC AGENT) *tablet:* repaglinide (see p. 1313) 1 mg/metformin (see p. 947) 500 mg; repaglinide 2 mg/metformin 500 mg.

Pravigard Pac (LIPID-LOWERING AGENT) *tablet:* pravastatin (see p. 1238) 20 mg/buffered aspirin (see p. 119) 81 mg; pravastatin 20 mg/buffered aspirin 325 mg; pravastatin 40 mg/aspirin 81 mg; pravastatin 40 mg/buffered aspirin 325 mg; pravastatin 80 mg/buffered aspirin 81 mg; pravastatin 80 mg/buffered aspirin 325 mg.

Premarin with Methyltestosterone (ESTROGEN, ANDROGEN) *tablet:* conjugated estrogens (see p. 579) 0.625 mg/methyltestosterone (see p. 973) 5 mg.

Premphase (ESTROGEN, PROGESTERONE) *tablet:* conjugated estrogens (see p. 579) 0.625 mg/medroxyprogesterone acetate (see p. 924) 5 mg.

Prempro (ESTROGEN, PROGESTIN) *tablet:* conjugated estrogens (see p. 579) 0.3 mg/medroxyprogesterone (see p. 924) 1.5 mg; conjugated estrogen 0.45 mg/medroxyprogesterone 1.5 mg; conjugated estrogen 0.625 mg/medroxyprogesterone 2.5 mg; conjugated estrogen 0.625 mg/medroxyprogesterone 5 mg.

Prevacid NapraPAC (PROTON PUMP INHIBITOR/ANTI-INFLAMMATORY) *capsules and tablets:* lansoprazole (see p. 845) 15 mg capsule/naproxen sodium (see p. 1035) 375 mg table; lansoprazole 15 mg capsule/naproxen sodium 500 mg tablet.

Prevpac (ANTIBIOTIC/ANTISECRETORY) *capsules and tablets:* amoxicillin (see p. 83) 500 mg capsules, clarithromycin (see p. 335) 500 mg tablets, lansoprazole (see p. 845) 30 mg capsules.

Prinzide (ANTIHYPERTENSIVE) *tablet:* hydrochlorothiazide (see p. 728) 12.5 mg/lisinopril (see p. 885) 10 mg; hydrochlorothiazide 12.5 mg/lisinopril 20 mg; hydrochlorothiazide 25 mg/lisinopril 20 mg.

Probenecid and Colchicine (ANTIGOUT) *tablet:* probenecid (see p. 1251) 500 mg/colchicine (see p. 361) 0.5 mg.

Pyridium Plus (ANALGESIC) *tablet:* phenazopyridine hydrochloride (see p. 1183) 150 mg/hyoscyamine hydrobromide (see p. 746) 0.3 mg/butalbital 15 mg.

Rebetron (INTERFERON, ANTIVIRAL) ribavirin (see p. 1322) tablet polythiazide 2 mg/reserpine (see p. 1314) 0.25 mg.

Rifamate (ANTITUBERCULOSIS) *capsule:* isoniazid (see p. 809) 150 mg/rifampin (see p. 1326) 300 mg.

Rifater (ANTITUBERCULOSIS) *tablet:* rifampin (see p. 1326) 120 mg/isoniazid (see p. 809) 50 mg/

pyrazinamide (see p. 1284) 300 mg.

Rimactane/INH Dual Pack (ANTITUBERCULOSIS) *pack:* thirty isoniazid (see p. 809) 300 mg tablets, sixty rifampin (see p. 1326) 300 mg capsules.

Robitussin A-C (ANTITUSSIVE, EXPECTORANT [schedule V]) *syrup (per 5 mL):* codeine phosphate (see p. 359) 10 mg/guaifenesin (see p. 713) 100 mg/alcohol 3.5%.

Rondec (DECONGESTANT, ANTIHISTAMINE) *tablet:* pseudoephedrine hydrochloride (see p. 1281) 60 mg/carbinoxamine maleate 4 mg; *drops (per mL):* pseudoephedrine hydrochloride 25 mg/carbinoxamine maleate 2 mg; *syrup (per mL):* pseudoephedrine hydrochloride 60 mg/carbinoxamine maleate 4 mg.

Roxicet (NARCOTIC ANALGESIC [schedule II]) *tablet:* oxycodone (see p. 1120) 5 mg/acetaminophen (see p. 11) 325 mg; *syrup (per 5 mL):* oxycodone 5 mg/acetaminophen 325 mg.

Simcor (ANTILIPIDEMIC) *extended release tablet:* niacin (see p. 1054) 1000 mg/simvastatin (see p. 1377) 20 mg; niacin 1000 mg/simvastatin 40 mg, niacin 500 mg/simvastatin 20 mg, niacin 500 mg/simvastatin 40 mg, niacin 750 mg/simvastatin 20 mg.

Soma Compound (SKELETAL MUSCLE RELAXANT) *tablet:* carisoprodol (see p. 245) 200 mg/aspirin (see p. 119) 325 mg.

Soma Compound with Codeine (SKELETAL MUSCLE RELAXANT [schedule III]) *tablet:* carisoprodol (see p. 245) 200 mg, aspirin (see p. 119) 325 mg/codeine phosphate (see p. 359) 16 mg.

Stalevo (ANTIPARKINSON AGENT) *tablet:* carbidopa (see p. 238) 12.5 mg/levodopa (see p. 861) 50 mg/entacapone (see p. 532) 200 mg;

carbidopa 18.75 mg/levodopa 75 mg/entacapone 200 mg; carbidopa 25 mg/levodopa 100 mg/entacapone 200 mg; carbidopa 31.25 mg/levodopa 125 mg/entacapone 200 mg; carbidopa 37.5 mg/levodopa 150 mg/entacapone 200 mg.

Suboxone (ANALGESIC) *sublingual tablet:* buprenorphine (see p. 200) 2 mg/naloxone (see p. 1029) 0.5 mg; buprenorphine 8 mg/naloxone 2 mg.

Symbicort (MINERALOCORTICOID/BRONCHODILATOR) *inhaler:* budesonide (see p. 196) 0.08 mg/formoterol (see p. 664) 0.045 mg; budesonide 0.16 mg/formoterol 0.045 mg.

Symbyax (ATYPICAL ANTIPSYCHOTIC/SSRI) *capsule:* olanzapine (see p. 1092) 6 mg/fluoxetine (see p. 649) 25 mg; olanzapine 6 mg/fluoxetine 50 mg; olanzapine 12 mg/fluoxetine 25 mg; olanzapine 12 mg/fluoxetine 50 mg.

Synalgos-DC (NARCOTIC AGONIST ANALGESIC [schedule III]) *capsule:* dihydrocodeine bitartrate 16 mg/aspirin (see p. 119) 356.4 mg/caffeine (see p. 214) 30 mg.

Synera (LOCAL ANESTHETIC) *transdermal patch:* lidocaine (see p. 871) 70 mg/tetracycline (see p. 1461) 70 mg.

Syntest D.S. (ESTROGEN, ANDROGEN) *tablet:* esterified estrogens (see p. 581) 1.25 mg/methyltestosterone (see p. 973) 2.5 mg.

Syntest H.S. (ESTROGEN, ANDROGEN) *tablet:* esterified estrogens (see p. 581) 0.625 mg/methyltestosterone (see p. 973) 1.25 mg.

Taclonex Scalp (NSAID) *topical suspension:* betamethasone (see p. 169) 0.064%/calcipotriene (see p. 216) 0.005%.

Talacen (NARCOTIC AGONIST-ANTAGONIST ANALGESIC [schedule IV]) *tablet:* pentazocine hydrochloride (see p. 1173) 25 mg/acetaminophen (see p. 11) 625 mg.

Talwin NX (NARCOTIC ANALGESIC [schedule IV]) *tablet:* pentazocine (see p. 1173) 50 mg/naloxone (see p. 1029) 0.5 mg.

Tanafed DP (DECONGESTANT, ANTIHISTAMINE) *suspension:* pseudoephedrine (see p. 1281) 75 mg/chlorpheniramine tannate 4.5 mg.

Tarka (ANTIHYPERTENSIVE) *tablet:* trandolapril (see p. 1518) 2 mg/verapamil HCl (see p. 1566) 180 mg; trandolapril 4 mg/verapamil HCl 240 mg; trandolapril 1 mg/verapamil HCl 240 mg, trandolapril 2 mg/verapamil HCl 240 mg.

Tekamlo (ANTIHYPERTENSIVE) *tablet:* aliskiren 150 mg/amlodipine 10 mg; aliskiren 150 mg/amlodipine 5mg; aliskiren 300 mg/amlodipine 10 mg; aliskiren 300 mg/amlodipine 5 mg.

Tekurna HCT (ANTIHYPERTENSIVE) *tablet:* aliskiren (see p. 40) 150 mg/hydrochlorothiazide (see p. 728) 12.5 mg.

Tenoretic 50 (ANTIHYPERTENSIVE) *tablet:* chlorthalidone (see p. 311) 25 mg/atenolol (see p. 125) 50 mg.

Tenoretic 100 (ANTIHYPERTENSIVE) *tablet:* chlorthalidone (see p. 311) 25 mg/atenolol (see p. 125) 100 mg.

Terra-Cortril Suspension (OCULAR STEROID AND ANTIBIOTIC) *suspension:* hydrocortisone acetate (see p. 731) 1.5%/oxytetracycline 0.5%.

Teveten HCT (ANTIHYPERTENSIVE) *tablet:* eprosartan mesylate (see p. 547) 600 mg/hydrochlorothiazide (see p. 728) 12.5 mg; eprosartan 600 mg/hydrochlorothiazide 25 mg.

Timolide (ANTIHYPERTENSIVE) *tablet:* hydrochlorothiazide (see p. 728) 25 mg/timolol maleate (see p. 1488) 10 mg.

Tobradex (ANTI-INFECTIVE) *ophthalmic suspension:* dexamethasone (see p. 428) 0.1%/tobramycin (see p. 1497) 0.3%; *ophthalmic ointment:* dexamethasone 0.1%/tobramycin 0.3%.

Treximet (ANALGESIC) *tablet:* naproxen (see p. 1035) 500 mg/sumatriptan (see p. 1419) 85 mg.

Triacin-C Cough Syrup (ANTITUSSIVE [schedule V]) *syrup:* pseudoephedrine (see p. 1281) 30 mg/triprolidine 1.25 mg/codeine (see p. 359) 10 mg.

Tribenzor (ANTIHYPERTENSIVE) *tablet:* olmesartan 20 mg/amlodipine 5 mg/hydrochlorothiazide 12.5 mg; olmesartan 40 mg/amlodipine 10 mg/hydrochlorothiazide 12.5 mg; olmesartan 40 mg/amlodipine 10 mg/hydrochlorothiazide 25 mg; olmesartan 40 mg/amlodipine 5 mg/hydrochlorothiazide 12.5 mg; olmesartan 40 mg/amlodipine 5 mg/hydrochlorothiazide 25 mg.

Tri-Hydroserpine (ANTIHYPERTENSIVE) *tablet:* hydrochlorothiazide (see p. 728) 15 mg/reserpine (see p. 1314) 0.1 mg/hydralazine hydrochloride (see p. 727) 25 mg.

Tri-Levlen (ORAL CONTRACEPTIVE) *tablet:* ethinyl estradiol (see p. 591) 30 mcg/levonorgestrel (see p. 866) 0.05 mg × 6 days, ethinyl estradiol 40 mcg/levonorgestrel 0.075 mg × 5 days, ethinyl estradiol 30 mcg/levonorgestrel 0.125 mg × 10 days.

Tri-Luma Cream (STEROID) *cream:* 4% hydroquinone (see p. 737)/0.05% tretinoin (see p. 1526)/0.01% fluocinolone acetonide (see p. 645).

Tri-Norinyl (ORAL CONTRACEPTIVE) *tablet:* ethinyl estradiol (see p. 591) 35 mcg/norethindrone (see p. 1079) 0.5 mg × 7 d, ethinyl estradiol 35 mcg/norethindrone 1 mg × 9 d, ethinyl estradiol 35 mcg/norethindrone 0.5 mg × 5 d.

Triphasil (ORAL CONTRACEPTIVE) *tablet:* ethinyl estradiol (see p. 591) 30 mcg/levonorgestrel (see p. 866) 0.05 mg × 6 days, ethinyl estradiol 40 mcg/levonorgestrel 0.075 mg × 5 days, ethinyl estradiol 30 mcg/levonorgestrel 0.125 mg × 10 days.

Triple Antibiotic (OPHTHALMIC ANTIBIOTIC) *ophthalmic ointment:* hydrocortisone (see p. 731) 1%/neomycin sulfate (see p. 1047) 0.5%/bacitracin zinc (see p. 150) 400 units/polymyxin B sulfate (see p. 1220) 10,000 units/g.

Trizivir (REVERSE TRANSCRIPTASE INHIBITOR) *tablet:* abacavir (see p. 1) 300 mg/lamivudine (see p. 841) 150 mg/zidovudine (see p. 1590) 300 mg.

Truvada (NUCLEOSIDE REVERSE TRANSCRIPTASE INHIBITOR) *tablet:* emtricitabine (see p. 525) 200 mg/tenofovir disoproxil fumarate (see p. 1447) 300 mg.

Tussicap (ANTITUSSIVE [schedule III]) *extended release capsule:* chlorpheniramine (see p. 305) 8 mg/hydrocodone (see p. 730) 10 mg.

Tussigon (ANTITUSSIVE [schedule III]) *tablet:* homatropine methylbromide 1.5 mg/hydrocodone (see p. 730) 5 mg.

Tussionex (ANTITUSSIVE [schedule III]) *tablet:* chlorpheniramine (see p. 305) 8 mg/hydrocodone (see p. 730) 10 mg.

Twinrix (VACCINE) *injection:* hepatitis A vaccine (see p. 723) 720 ELU/hepatitis B recombinant vaccine (see p. 724) 20 mcg per single dose vial.

Twynsta (ANTIHYPERTENSIVE) *tablet:* telmisartan 40 mg/amlodipine 5 mg; telmisartan 40 mg/amlodipine 10 mg; telmisartan 80 mg/amlodipine 5 mg; telmisartan 80 mg/amlodipine 10 mg.

Tylenol with Codeine Elixir (NARCOTIC AGONIST ANALGESIC [schedule V]) *elixir (per 5 mL):* acetaminophen (see p. 11) 120 mg/codeine phosphate (see p. 359) 12 mg/alcohol 7%.

Tylenol with Codeine No. 1 (NARCOTIC AGONIST ANALGESIC [schedule III]) *tablet:* acetaminophen (see p. 11) 300 mg/codeine phosphate (see p. 359) 7.5 mg.

Tylenol with Codeine No. 2 (NARCOTIC AGONIST ANALGESIC [schedule III]) *tablet:* acetaminophen (see p. 11) 300 mg/codeine phosphate (see p. 359) 15 mg.

Tylenol with Codeine No. 3 (NARCOTIC AGONIST ANALGESIC [schedule III]) *tablet:* acetaminophen (see p. 11) 300 mg/codeine phosphate (see p. 359) 30 mg.

Tylenol with Codeine No. 4 (NARCOTIC AGONIST ANALGESIC [schedule III]) *tablet:* acetaminophen (see p. 11) 300 mg/codeine phosphate (see p. 359) 60 mg.

Tylox (NARCOTIC AGONIST ANALGESIC [schedule II]) *capsule:* oxycodone hydrochloride (see p. 1120) 5 mg/acetaminophen (see p. 11) 500 mg.

Ultracet (ANALGESIC/ANTIPYRETIC) *tablet:* tramadol (see p. 1516) 37.5 mg/acetaminophen (see p. 11) 325 mg.

Uniretic (ANTIHYPERTENSIVE) *tablet:* moexipril (see p. 1008) 7.5 mg/hydrochlorothiazide (see p. 728) 12.5 mg; moexipril 15 mg/hydrochlorothiazide 12.5 mg, moexipril 15 mg/hydrochlorothiazide 25 mg.

Urised (URINARY ANTI-INFECTIVE) *tablet:* methenamine (see p. 953) 40.8 mg/phenyl salicylate 18.1 mg/atropine sulfate (see p. 134) 0.03 mg/hyoscyamine (see p. 746) 0.03 mg/benzoic acid 4.5 mg/methylene blue 5.4 mg.

Valturna (ANTIHYPERTENSIVE) *tablet:* aliskiren (see p. 40) 150 mg/valsartan (see p. 1554)

160 mg; aliskiren 300 mg/valsartan 320 mg.

Vaseretic (ANTIHYPERTENSIVE) *tablet:* enalapril maleate (see p. 526) 10 mg/hydrochlorothiazide (see p. 728) 25 mg.

Vasocidin (OPHTHALMIC CORTICOSTEROID, ANTI-INFECTIVE) *ophthalmic solution:* prednisolone sodium phosphate (see p. 1243) 0.23%/sulfacetamide sodium (see p. 1410) 10%.

Vasocon-A (OPHTHALMIC DECONGESTANT) *ophthalmic solution:* naphazoline hydrochloride (see p. 1034) 0.05%/antazoline phosphate 0.5%.

Vicodin (NARCOTIC AGONIST ANALGESIC [schedule III]) *tablet:* hydrocodone bitartrate (see p. 730) 5 mg/acetaminophen (see p. 11) 500 mg.

Vicodin ES (NARCOTIC AGONIST ANALGESIC [schedule III]) *tablet:* hydrocodone (see p. 730) 7.5 mg/acetaminophen (see p. 11) 750 mg.

Vicodin HP (NARCOTIC AGONIST ANALGESIC [schedule III]) *tablet:* hydrocodone (see p. 730) 10 mg/acetaminophen (see p. 11) 660 mg.

Vicoprofen (NARCOTIC AGONIST ANALGESIC [schedule III]) *tablet:* hydrocodone bitartrate (see p. 730) 7.5 mg/ibuprofen (see p. 749) 200 mg.

Vimovo (ANALGESIC) *tablet:* naproxen 375 mg/esomeprazole 20 mg; naproxen 500 mg/esomperazole 20 mg.

Vytorin (ANTILIPEMIC AGENT) *tablet:* ezetimibe (see p. 606) 10 mg/simvastatin (see p. 1377) 10 mg; ezetimibe 10 mg/simvastatin 20 mg; ezetimibe 10 mg/simvastatin 40 mg; ezetimibe 10 mg/simvastatin 80 mg.

Xerese (TOPICAL ANTIINFECTIVE) *cream:* acyclovir 5%/hydrocortisone 1%.

Wygesic (NARCOTIC AGONIST ANALGESIC [schedule IV]) *tablet:* 65 mg/acetaminophen (see p. 11) 650 mg.

Yasmin (ORAL CONTRACEPTIVE) *tablet:* ethinyl estradiol (see p. 591) 30 mcg/drospirenone 3 mg.

Zestoretic (ANTIHYPERTENSIVE) *tablet:* hydrochlorothiazide (see p. 728) 12.5 mg/lisinopril (see p. 885) 10 mg; hydrochlorothiazide 12.5 mg/lisinopril 20 mg; hydrochlorothiazide 25 mg/lisinopril 20 mg.

Ziac (ANTIHYPERTENSIVE) *tablet:* bisoprolol (see p. 181) 2.5 mg/hydrochlorothiazide (see p. 728) 6.25 mg; bisoprolol 5 mg/hydrochlorothiazide 6.25 mg; bisoprolol 10 mg/hydrochlorothiazide 6.25 mg.

Zylet (OPHTHALMIC ANTIBIOTIC) *solution:* loteprednol etabonate (see p. 898) 0.5%/tobramycin (see p. 1497) 0.3%.

Zyrtec-D (ANTIHISTAMINE/DECONGESTANT) *tablet, sustained release:* cetirizine (see p. 287) 5 mg/pseudoephedrine (see p. 1281) 120 mg.

acute coronary syndrome an acute ischemic event with or without marked ST segment elevation.

acute dystonia extrapyramidal symptom manifested by abnormal posturing, grimacing, spastic torticollis (neck torsion), and oculogyric (eyeball movement) crisis.

adverse effect unintended, unpredictable, and nontherapeutic response to drug action. Adverse effects occur at doses used therapeutically or for prophylaxis or diagnosis. They generally result from drug toxicity, idiosyncrasies, or hypersensitivity reactions caused by the drug itself or by ingredients added during manufacture (e.g., preservatives, dyes, or vehicles).

afterload resistance that ventricles must work against to eject blood into the aorta during systole.

agranulocytosis sudden drop in leukocyte count; often followed by a severe infection manifested by high fever, chills, prostration, and ulcerations of mucous membrane such as in the mouth, rectum, or vagina.

akathisia extrapyramidal symptom manifested by a compelling need to move or pace, without specific pattern, and an inability to be still.

analeptic restorative medication that enhances excitation of the CNS without affecting inhibitory impulses.

anaphylactoid reaction excessive allergic response manifested by wheezing, chills, generalized pruritic urticaria, diaphoresis, sense of uneasiness, agitation, flushing, palpitations, coughing, difficulty breathing, and cardiovascular collapse.

anticholinergic actions inhibition of parasympathetic response manifested by dry mouth, decreased peristalsis, constipation, blurred vision, and urinary retention.

bioavailability fraction of active drug that reaches its action sites after administration by any route. Following an IV dose, bioavailability is 100%; however, such factors as first-pass effect, enterohepatic cycling, and biotransformation reduce bioavailability of an orally administered drug.

blood dyscrasia pathological condition manifested by fever, sore mouth or throat, unexplained fatigue, easy bruising or bleeding.

cardiotoxicity impairment of cardiac function manifested by one or more of the following: hypotension, arrhythmias, precordial pain, dyspnea, electrocardiogram (ECG) abnormalities, cardiac dilation, congestive failure.

cholinergic response stimulation of the parasympathetic response manifested by lacrimation, diaphoresis, salivation, abdominal cramps, diarrhea, nausea, and vomiting.

circulatory overload excessive vascular volume manifested by increased central venous pressure (CVP), elevated blood pressure, tachycardia, distended neck veins, peripheral edema, dyspnea, cough, and pulmonary rales.

CNS stimulation excitement of the CNS manifested by hyperactivity, excitement, nervousness, insomnia, and tachycardia.

CNS toxicity impairment of CNS function manifested by ataxia, tremor, incoordination, paresthesias, numbness, impairment of pain or touch sensation, drowsiness, confusion, headache, anxiety, tremors, and behavior changes.

congestive heart failure (CHF) impaired pumping ability of the heart manifested by paroxysmal nocturnal dyspnea, cough, fatigue or dyspnea on exertion, tachycardia, peripheral or pulmonary edema, and weight gain.

Cushing's syndrome fatty swellings in the interscapular area (buffalo hump) and in the facial area (moon face), distention of the abdomen, ecchymoses following even minor trauma, impotence, amenorrhea, high blood pressure, general weakness, loss of muscle mass, osteoporosis, and psychosis.

dehydration decreased intracellular or extracellular fluid manifested by elevated temperature, dry skin and mucous membranes, decrease tissue turgor, sunken eyes, furrowed tongue, low blood pressure, diminished or irregular pulse, muscle or abdominal cramps, thick secretions, hard feces and impaction, scant urinary output, urine specific gravity above 1.030, an elevated hemoglobin.

disulfiram-type   reaction Antabuse-type reaction manifested by facial flushing, pounding headache, sweating, slurred speech, abdominal cramps, nausea, vomiting, tachycardia, fever, palpitations, drop in blood pressure, dyspnea, and sense of chest constriction. Symptoms may last up to 24 hours.

enzyme induction stimulation of microsomal enzymes by a drug resulting in its accelerated metabolism and decreased activity. If reactive intermediates are formed, drug-mediated toxicity may be exacerbated.

first-pass effect reduced bioavailability of an orally administered drug due to metabolism in GI epithelial cells and liver or to biliary excretion. Effect may be avoided by use of sublingual tablets or rectal suppositories.

fixed drug eruption drug-induced circumscribed skin lesion that persists or recurs in the same site. Residual pigmentation may remain following drug withdrawal.

gamma-glutamyl   transpeptidase screening test for possible liver damage and/or suspected alcohol abuse.

half-life (t$_{1/2}$) time required for concentration of a drug in the body to decrease by 50%. Half-life also represents the time necessary to reach steady state or to decline from steady state after a change (i.e., starting or stopping) in the dosing regimen. Half-life may be affected by a disease state and age of the drug user.

heart failure left- and/or right-sided failure associated with systolic and/or diastolic dysfunction.

heat stroke a life-threatening condition manifested by absence of sweating; red, dry, hot skin; dilated pupils; dyspnea; full bounding pulse; temperature above 40° C (105° F); and mental confusion.

hepatotoxicity impairment of liver function manifested by jaundice, dark urine, pruritus, light-colored stools, eosinophilia, itchy skin or rash, and persistently high elevations of alanine amino-transferase (ALT) and aspartate aminotransferase (AST).

hyperammonemia elevated level of ammonia or ammonium in the blood manifested by lethargy, decreased appetite, vomiting, asterixis (flapping tremor), weak pulse, irritability, decreased responsiveness, and seizures.

hypercalcemia elevated serum calcium manifested by deep bone and flank pain, renal calculi, anorexia, nausea, vomiting, thirst, constipation, muscle hypotonicity, pathologic fracture, bradycardia, lethargy, and psychosis.

hyperglycemia elevated blood glucose manifested by flushed, dry skin, low blood pressure and elevated pulse, tachypnea, Kussmaul's respirations, polyuria, polydipsia; polyphagia, lethargy, and drowsiness.

hyperkalemia excessive potassium in blood, which may produce life-threatening cardiac arrhythmias, including bradycardia and heart block, unusual fatigue, weakness or heaviness of limbs, general muscle weakness, muscle cramps, paresthesias, flaccid paralysis of extremities, shortness of breath, nervousness, confusion, diarrhea, and GI distress.

hypermagnesemia excessive magnesium in blood, which may produce cathartic effect, profound thirst, flushing, sedation, confusion, depressed deep tendon reflexes (DTRs), muscle weakness, hypotension, and depressed respirations.

hypernatremia excessive sodium in blood, which may produce confusion, neuromuscular excitability, muscle weakness, seizures, thirst, dry and flushed skin, dry mucous membranes, pyrexia, agitation, and oliguria or anuria.

hypersensitivity reactions excessive and abnormal sensitivity to given agent manifested by urticaria, pruritus, wheezing, edema, redness, and anaphylaxis.

hyperthyroidism excessive secretion by the thyroid glands, which increases basal metabolic rate, resulting in warm, flushed, moist skin; tachycardia, exophthalmos; infrequent lid blinking; lid edema; weight loss despite increased appetite; frequent urination; menstrual irregularity; breathlessness; hypoventilation; congestive heart failure; excessive sweating.

hyperuricemia excessive uric acid in blood, resulting in pain in flank; stomach; or joints, and changes in intake and output ratio and pattern.

hypocalcemia abnormally low calcium level in blood, which may result in depression; psychosis; hyperreflexia; diarrhea; cardiac arrhythmias; hypotension; muscle spasms; paresthesias of feet, fingers, tongue; positive Chvostek's sign. Severe deficiency (tetany) may result in carpopedal spasms, spasms of face muscle, laryngospasm, and generalized convulsions.

hypoglycemia abnormally low glucose level in the blood, which may result in acute fatigue, restlessness, malaise, marked irritability and weakness, cold sweats, excessive hunger, headache, dizziness, confusion, slurred speech, loss of consciousness, and death.

hypokalemia abnormally low level of potassium in blood, which may result in malaise, fatigue, paresthesias, depressed reflexes, muscle weakness and cramps, rapid, irregular pulse, arrhythmias, hypotension, vomiting, paralytic ileus, mental confusion, depression, delayed thought process, abdominal distention, polyuria, shallow breathing, and shortness of breath.

hypomagnesemia abnormally low level of magnesium in blood, resulting in nausea, vomiting, cardiac arrhythmias, and neuromuscular symptoms (tetany, positive Chvostek's and Trousseau's signs, seizures, tremors, ataxia, vertigo, nystagmus, muscular fasciculations).

hyponatremia a decreased serum concentration (less than 125 mEq) of sodium that results in intracellular swelling. The resulting signs and symptoms include nausea, malaise, headache, lethargy, obtundation, seizures, and coma. There is significant variability in the symptomatology of hyponatremia manifested in patients.

hypophosphatemia abnormally low level of phosphates in blood, resulting in muscle weakness, anorexia, malaise, absent deep tendon reflexes, bone pain, paresthesias, tremors, negative calcium balance, osteomalacia, osteoporosis.

hypothyroidism condition caused by thyroid hormone deficiency that lowers basal metabolic rate and may result in periorbital edema, lethargy, puffy hands and feet, cool, pale skin, vertigo, nocturnal cramps, decreased GI motility, constipation, hypotension, slow pulse, depressed muscular activity, and enlarged thyroid gland.

hypoxia insufficient oxygenation in the blood manifested by dyspnea, tachypnea, headache, restlessness, cyanosis, tachycardia, dysrhythmias, confusion, decreased level of consciousness, and euphoria or delirium.

international normalizing ratio measurement that normalizes for the differences obtained from various laboratory readings in the value for thromboplastin blood level.

leukopenia abnormal decrease in number of white blood cells, usually below 5000 per cubic millimeter, resulting in fever, chills, sore mouth or throat, and unexplained fatigue.

liver toxicity manifested by anorexia, nausea, fatigue, lethargy, itching, jaundice, abdominal pain, dark-colored urine, and flu-like symptoms.

metabolic acidosis decrease in pH value of the extracellular fluid caused by either an increase in hydrogen ions or a decrease in bicarbonate ions. It may result in one or more of the following: lethargy, headache, weakness, abdominal pain, nausea, vomiting, dyspnea, hyperpnea progressing to Kussmaul breathing, dehydration, thirst, weakness, flushed face, full bounding pulse, progressive drowsiness, mental confusion, combativeness.

metabolic alkalosis increase in pH value of the extracellular fluid caused by either a loss of acid from the body (e.g., through vomiting) or an increased level of bicarbonate ions (e.g., through ingestion of sodium bicarbonate). It may result in muscle weakness, irritability, confusion, muscle twitching, slow and shallow respirations, and convulsive seizures.

microsomal enzymes drug-metabolizing enzymes located in the endoplasmic reticulum of the liver and other tissues chiefly responsible for oxidative drug metabolism (e.g., cytochrome P450).

myopathy any disease or abnormal condition of striated muscles manifested by muscle weakness, myalgia, diaphoresis, fever, and reddish-brown urine (myoglobinuria) or oliguria.

nephrotoxicity impairment of the nephrons of the kidney manifested by one or more of the following: oliguria, urinary frequency, hematuria, cloudy urine, rising BUN and serum creatinine, fever, graft tenderness or enlargement.

neuroleptic malignant syndrome (NMS) potentially fatal complication associated with antipsychotic drugs manifested by hyperpyrexia, altered mental status, muscle rigidity, irregular pulse, fluctuating BP, diaphoresis, and tachycardia.

orphan drug (as defined by the Orphan Drug Act, an amendment of the Federal Food, Drug, and Cosmetic Act which took effect in January 1983): drug or biological product used in the treatment, diagnosis, or prevention of a rare disease. A rare disease or condition is one that affects fewer than 200,000 persons in the United States, or affects more than 200,000 persons but for which there is no reasonable expectation that drug research and development costs can be recovered from sales within the United States.

ototoxicity impairment of the ear manifested by one or more of the following: headache, dizziness or vertigo, nausea and vomiting with motion, ataxia, nystagmus.

pharmacogenetic genetic variation affecting response to different drugs.

prodrug inactive drug form that becomes pharmacologically active through biotransformation.

protein binding reversible interaction between protein and drug resulting in a drug-protein complex (bound drug) which is in equilibrium with free (active) drug in plasma and tissues. Since only free drug can diffuse to action sites, factors that influence drug-binding (e.g., displacement of bound drug by another drug, or decreased albumin concentration) may potentiate pharmacologic effect.

pseudomembranous enterocolitis life-threatening superinfection characterized by severe diarrhea and fever.

pseudoparkinsonism extrapyramidal symptom manifested by slowing of volitional movement (akinesia), mask facies, rigidity and tremor at rest (especially of upper extremities); and pill rolling motion.

pulmonary edema excessive fluid in the lung tissue manifested by one or more of the following: shortness of breath, cyanosis, persistent productive cough (frothy sputum may be blood tinged), expiratory rales, restlessness, anxiety, increased heart rate, sense of chest pressure.

renal insufficiency reduced capacity of the kidney to perform its functions as manifested by one or more of the following: dysuria, oliguria, hematuria, swelling of lower legs and feet.

serotonin syndrome manifested by restlessness, myoclonus, mental status changes, hyperreflexia, diaphoresis, shivering, and tremor.

Somogyi effect rebound phenomenon clinically manifested by fasting hyperglycemia and worsening of diabetic control due to unnecessarily large p.m. insulin doses. Hormonal response to unrecognized hypoglycemia (i.e., release of epinephrine, glucagon, growth hormone, cortisol) causes insensitivity to insulin. Increasing the amount of insulin required to treat the hyperglycemia intensifies the hypoglycemia.

superinfection new infection by an organism different from the initial infection being treated by antimicrobial therapy manifested by one or more of the following: black, hairy tongue; glossitis, stomatitis; anal itching; loose, foul-smelling stools; vaginal itching or discharge; sudden fever; cough.

tachyphylaxis rapid decrease in response to a drug after administration of a few doses. Initial drug response cannot be restored by an increase in dose.

tardive dyskinesia extrapyramidal symptom manifested by involuntary rhythmic, bizarre movements of face, jaw, mouth, tongue, and sometimes extremities.

vasovagal symptoms transient vascular and neurogenic reaction marked by pallor, nausea, vomiting, bradycardia, and rapid fall in arterial blood pressure.

water intoxication (dilutional hyponatremia) less than normal concentration of sodium in the blood resulting from excess extracellular and intracellular fluid and producing one or more of the following: lethargy, confusion, headache, decreased skin turgor, tremors, convulsions, coma, anorexia, nausea, vomiting, diarrhea, sternal fingerprinting, weight gain, edema, full bounding pulse, jugular vein distention, rales, signs and symptoms of pulmonary edema.

ABGs	arterial blood gases
ABW	adjusted body weight
a.c.	before meals (*ante cibum*)
ACD	acid–citrate–dextrose
ACE	angiotensin-converting enzyme
ACh	acetylcholine
ACIP	Advisory Committee on Immunization Practices
ACLS	advanced cardiac life support
ACS	acute coronary syndrome
ACT	activated clotting time
ACTH	adrenocorticotropic hormone
AD	Alzheimer's disease
ADD	attention deficit disorder
ADH	antidiuretic hormone
ADLs	activities of daily living
ad lib	as desired (*ad libitum*)
ADP	adenosine diphosphate
ADT	alternate-day drug (administration)
AED	antiepilectic drug
AF	atrial fibrillation
AFL	atrial flutter
AIDS	acquired immunodeficiency syndrome
AIP	acute intermittent porphoria
alpha1-PI	alpha1-proteinase inhibitor
ALS	amyotrophic lateral sclerosis
ALT	alanine aminotransferase (formerly SGPT)
AMI	acute myocardial infarction
AML	acute myelogenous leukemia
AMP	adenosine monophosphate
ANA	antinuclear antibody(ies)
ANC	absolute neutrophil count
ANH	atrial natriuretic hormone
ANLL	acute nonlymphocytic leukemia
aPTT	activated partial thromboplastin time
ARC	AIDS-related complex
ARDS	adult respiratory distress syndrome
ASHD	arteriosclerotic heart disease
AST	aspartate aminotransferase (formerly SGOT)
AT$_1$	angiotensin II receptor subtype I
AT$_2$	angiotensin II receptor subtype II
ATP	adenosine triphosphate
AV	atrioventricular
b.i.d.	two times a day
BMD	bone mineral density
BMI	body mass index
BMR	basal metabolic rate
BP	blood pressure

BPH	benign prostatic hypertrophy
bpm	beats per minute
BSA	body surface area
BSE	breast self-exam
BSP	bromsulphalein
BT	bleeding time
BUN	blood urea nitrogen
C	centigrade, Celsius
CAD	coronary artery disease
cAMP	cyclic adenosine monophosphate
CBC	complete blood count
CCR5	cellular chemokine coreceptor-5
CDC	Centers for Disease Control and Prevention
CF	cystic fibrosis
cGMP	cyclic guanosine monophosphate
CHF	congestive heart failure
CKD	chronic kidney disease
CLL	chronic lymphocytic leukemia
cm	centimeter
CML	chronic myeloid leukemia
CMV	cytomegalovirus-I
CNS	central nervous system
Coll	collyrium (eye wash)
COMT	catecholamine-O-methyl transferase
COPD	chronic obstructive pulmonary disease
COX-1	cyclooxygenase-1
COX-2	cyclooxygenase-2
CPK	creatinine phosphokinase
CPR	cardiopulmonary resuscitation
CrCl	creatinine clearance
CRF	chronic renal failure
CRFD	chronic renal failure disease
C&S	culture and sensitivity
CSF	cerebrospinal fluid
CSP	cellulose sodium phosphate
CSSSI	complicated skin and skin structure infections
CT	clotting time
CTZ	chemoreceptor trigger zone
CV	cardiovascular
CVA	cerebrovascular accident
CVP	central venous pressure
CYP	cytochrome P450 system of enzymes
CYP3A4	cytochrome 3A4
D5W	5% dextrose in water
D&C	dilation and curettage
DIC	disseminated intravascular coagulation
DKA	diabetic ketoacidosis
dL	deciliter (100 mL or 0.1 liter)
DM	diabetes mellitus
DMARD	disease-modifying antirheumatic drug

DNA	deoxyribonucleic acid
DPD	dihydropyrimidine dehydrogenase
DTRs	deep tendon reflexes
DVT	deep venous thrombosis
ECG, EKG	electrocardiogram
ECT	electroconvulsive therapy
EEG	electroencephalogram
EENT	eye, ear, nose, throat
e.g.	for example (*exempli gratia*)
EGFR	epidermal growth factor receptor
EIB	exercise-induced bronchoconstriction
ENT	ear, nose, throat
EPS	extrapyramidal symptoms (or syndrome)
ER	estrogen receptor
ESRF	end-stage renal failure
F	Fahrenheit
FBS	fasting blood sugar
FDA	Food and Drug Administration
FSH	follicle-stimulating hormone
FTI	free thyroxine index
5-FU	5-fluorouracil
FUO	fever of unknown origin
g	gram
G6PD	glucose-6-phosphate dehydrogenase
GABA	gamma-aminobutyric acid
GERD	gastroesophageal reflux disease
GFR	glomerular filtration rate
GGT	gamma-glutamyl transferase
GGTP	gamma-glutamyl transpeptidase
GH	growth hormone
GI	gastrointestinal
GIST	gastrointestinal stomal tumor
GM-CSF	granulocyte-macrophage colony-stimulating factor
GnRH	gonadotropic releasing hormone
GPIIb/IIIa	glycoprotein IIb/IIIa
GTT	glucose tolerance test
GU	genitourinary
h	hour
HACA	human antichimeric antibody
HbA1C	glycosylated hemoglobin
hBNP	human B-type natriuretic peptide
HBV	viral hepatitis B
HCG	human chorionic gonadotropin
Hct	hematocrit
HDD-CKD	hemodialysis-dependant chronic kidney disease
HDL-C	high-density-lipoprotein cholesterol
HER	human epidermal growth factor
HF	heart failure
Hgb	hemoglobin
5-HIAA	5-hydroxyindoleacetic acid

HIT	heparin-induced thrombocytopenia
HIV	human immunodeficiency virus
HMG-CoA	3-hydroxy-3-methyl-glutaryl coenzyme A
HPA	hypothalamic–pituitary–adrenocortical (axis)
HPV	human papillomavirus
HR	heart rate
HSV-1	herpes simplex virus type 1
HSV-2	herpes simplex virus type 2
5-HT	5-hydroxytryptamine (serotonin receptor)
IBD	inflammatory bowel disease
IBW	ideal body weight
IC	intracoronary
ICP	intracranial pressure
ICU	intensive care unit
ID	intradermal
IFN	interferon
Ig	immunoglobulin
IGF-1	insulin-like growth factor 1
IL	interleukin
IM	intramuscular
INR	international normalized ratio
IOP	intraocular pressure
IPPB	intermittent positive pressure breathing
iPTH	idiopathic parathyroid hormone
IT	intrathecal
ITP	idiopathic thrombocytopenic purpura
IV	intravenous
JRA	juvenile rheumatoid arthritis
kg	kilogram
KGF	keratinocyte growth factor
17-KGS	17-ketogenic steroids
17-KS	17-ketosteroids
KVO	keep vein open
L	liter
LDH	lactic dehydrogenase
LDL	low density lipoprotein
LDL-C	low-density-lipoprotein cholesterol
LE	lupus erythematosus
LFT	liver function test
LH	luteinizing hormone
LR	lactated Ringer's
LSD	lysergic acid diethylamide
LTRA	leukotriene receptor antagonist
LVEDP	left ventricular end diastolic pressure
LVEF	left ventricular ejection fraction
M	molar (strength of a solution)
m^2	square meter (of body surface area)
MAO	monoamine oxidase
MAOI	monoamine oxidase inhibitor
MBD	minimal brain dysfunction

MCH	mean corpuscular hemoglobin
MCHC	mean corpuscular hemoglobin concentration
mCi	millicurie
MCL	mantle cell lymphoma
mcg	microgram (1/1000 of a milligram)
MDI	metered dose inhaler
MDR	minimum daily requirements
MDS	muscular dystrophy syndrome
mEq	milliequivalent
mg	milligram
MI	myocardial infarction
MIC	minimum inhibitory concentration
min	minute
mL	milliliter (0.001 liter)
mm	millimeter
mo	month
MPS I	mucopolysaccharidosis I
MRSA	methicillin-resistant *Staphylococcus aureus*
MS	multiple sclerosis
mu-m	micrometer
N	normal (strength of a solution)
NADH	reduced form of nicotine adenine dinucleotide
NAPA	*N*-acetyl procainamide
nb	note well (*nota bene*)
NDD	non-hemodialysis dependent
NDD-CKD	non–hemodialysis-dependent chronic kidney disease
ng	nanogram (1/1000 of a microgram)
NMS	neuroleptic malignant syndrome
NNRTI	nonnucleoside reverse transcriptase inhibitor
NON-PVC	nonpolyvinyl chloride IV bag or tubing
NPN	nonprotein nitrogen
NPO	nothing by mouth
NRTI	nucleoside reverse transcriptase inhibitor
NS	normal saline
NSAID	nonsteroidal anti-inflammatory drug
NSCLC	non–small-cell lung cancer
NSR	normal sinus rhythm
NYHA Class I, II, III, IV	New York Heart Association classes of heart failure
OAB	overactive bladder
OC	oral contraceptive
ODT	oral disintegrating tablet
17-OHCS	17-hydroxycorticosteroids
OTC	over the counter (nonprescription)
P450	cytochrome P450 system of enzymes
PABA	*para*-aminobenzoic acid
PAS	*para*-aminosalicylic acid
PAWP	pulmonary artery wedge pressure
PBI	protein-bound iodine
PBP	penicillin-binding protein

p.c.	after meals (*post cibum*)
PCI	percutaneous coronary intervention
PCP	*Pneumocystis carinii* pneumonia
PCWP	pulmonary capillary wedge pressure
PDD	peritoneal dialysis dependent
PDD-CKD	peritoneal dialysis–dependent chronic kidney disease
PDE	phosphodiesterase
PDE5	phosphodiesterase type-5
PE	pulmonary embolism
PERLA	pupils equal, react to light and accommodation
PG	prostaglandin
PGE$_2$	prostaglandin E$_2$
pH	hydrogen ion concentration
Ph	Philadelphia (chromosome)
PI	protease inhibitor
PID	pelvic inflammatory disease
PJP	*Pneumocystis jirovecii* pneumonia
PKU	phenylketonuria
PMDD	premenstrual dysphoric disorder
PML	progressive multifocal leukoencephalopathy
PND	paroxysmal nocturnal dyspnea
PNE	primary nocturnal enuresis
PO	by mouth or orally (*per os*)
PPI	proton pump inhibitor
PPM	parts per million
PR	rectally (*per rectum*)
prn	when required (*pro re nata*)
PSA	prostate-specific antigen
PSP	phenolsulfonphthalein
PSVT	paroxysmal supraventricular tachycardia
PT	prothrombin time
PTCL	peripheral T-cell lymphocyte
PTH	parathyroid hormone
PTT	partial thromboplastin time
PUD	peptic ulcer disease
PVC	polyvinyl chloride IV bag or tubing
PVC	premature ventricular contraction
PVD	peripheral vascular disease
PZI	protamine zinc insulin
q.i.d.	four times daily
RA	rheumatoid arthritis
RAAS	renin angiotension aldosterone system
RAI	radioactive iodine
RAR	retinoic acid receptor
RAST	radioallergosorbent test
RBC	red blood (cell) count
RDA	recommended (daily) dietary allowance
RDS	respiratory distress syndrome
REM	rapid eye movement

rem	radiation equivalent man
RES	reticuloendothelial system
RIA	radioimmunoassay
RNA	ribonucleic acid
ROM	range of motion
RSV	respiratory syncytial virus
RT	reverse transcriptase
RT₃U	total serum thyroxine concentration
S&S	signs and symptoms
SA	sinoatrial
SARA	selective aldosterone receptor antagonist
SBE	subacute bacterial endocarditis
SC	subcutaneous
S$_{cr}$	serum creatinine
sec	second
SERMs	selective estrogen receptor modulators
SGGT	serum gamma-glutamyl transferase
SGOT	serum glutamic–oxaloacetic transaminase (*see* AST)
SGPT	serum glutamic–pyruvic transaminase (*see* ALT)
SIADH	syndrome of inappropriate antidiuretic hormone
SI Units	International System of Units
SK	streptokinase
SL	sublingual
SLE	systemic lupus erythematosus
SMA	sequential multiple analysis
SNRI	serotonin norepinephrine reuptake inhibitor
SOS	if necessary (*si opus cit*)
sp	species
SPF	sun protection factor
sq	square
SR	sedimentation rate
SRS-A	slow-reactive substance of anaphylaxis
SSRI	selective serotonin reuptake inhibitor
stat	immediately
STD	sexually transmitted disease
STEMI	ST elevated MI
SVT	supraventricular tachyarrhythmias
SW	sterile water
t$_{1/2}$	half-life
T$_3$	triiodothyronine
T$_4$	thyroxine
TCA	tricyclic antidepressant
TG	total triglycerides
TIA	transient ischemic attack
t.i.d.	three times a day (*ter in die*)
TKI	tyrosine kinase inhibitor
TNF	tumor necrosis factor
tPA	tissue plasminogen activator
TPN	total parenteral nutrition
TPR	temperature, pulse, respirations

TSH	thyroid-stimulating hormone
TSS	toxic shock syndrome
TT	thrombin time
TTP	thrombotic cyclopenic purpura
UA	urinary analysis
ULN	upper limit of normal
URI	upper respiratory infection
USP	United States Pharmacopeia
USPHS	United States Public Health Service
UTI	urinary tract infection
UV-A, UVA	ultraviolet A wave
VDRL	Venereal Disease Research Laboratory
VEGF	vascular endothelial growth factor
VLDL	very low density lipoprotein
VMA	vanillylmandelic acid
VREF	vancomycin-resistant *Enterococcus faecium*
VRSA	vancomycin-resistant *Staphylococcus aureus*
VS	vital signs
wk	week
WBC	white blood (cell) count
WBCT	whole blood clotting time
y	year

As patient interest in dietary supplements and other natural products increases, there is an increased need for information on this topic. These products are not standardized or stringently regulated by FDA guidelines; therefore, caution should be used when discussing these products. Consumers should note that since rigid quality control standards are not required for these products, substantial variability can occur in both potency and purity of a given product, especially between different commercial companies.

Many of these products have limited research on safety; thus, side effects and potential drug interactions are not well understood. Dietary supplements may either increase or decrease the level of a drug in the patient's body.

This table provides basic information on some of the most commonly sold dietary supplements. For additional information, a specialty resource on herbal and/or dietary supplements should be consulted.

Name	Common Use	Significant Safety Concerns
Bilberry	Eye health	Long-term, high-dose use can cause liver problems
Black cohosh	Menopausal symptoms	Should be avoided in pregnant patients
Cranberry	Urinary tract infections	Considered safe at usual doses; high dose may increase bleeding risk
Echinacea	Infections	May cause allergic reactions; should be used only short-term
Eleuthera (Siberian ginseng)	Energy	Avoid use with digoxin
Evening primrose	Menopausal symptoms	May affect seizure threshold
Garlic	Cholesterol	Significant drug interactions with drugs metabolized by CYP system
Ginger	Nausea	Overdoses may cause cardiac arrhythmias
Ginkgo	Memory enhancement	Potential increased bleeding risk
Ginseng (American ginseng)	Energy	Should not be used with MAO inhibitors; may affect anticoagulants

Name	Common Use	Significant Safety Concerns
Glucosamine	Osteoarthritis	Considered safe at usual doses; at higher doses, possible interaction with warfarin and other coumarin anticoagulants
Green tea	Energy, weight loss	High doses may cause cardiovascular side effects
Horny goat weed	Sexual function	Should be avoided in pregnant/lactating women
Horse chestnut	Congestive heart failure	Potential hepatotoxicity
Milk thistle	Liver function	May affect CYP metabolism and interact with drugs metabolized by this system
Saw palmetto	Benign prostatic hyperplasia	Adverse effects appear mild
Soy	Menopausal symptoms	GI-related side effects may be significant for some patients
St. John's wort	Depression	Significant drug interactions with several drugs metabolized by CYP system
Valerian	Sleep disorder	Potentially hepatotoxic
Yohimbe	Sexual function	Do not use with drugs affecting serotonin system

There are a number of medications whose names look or sound similar when pronounced. The Institute of Safe Medication Practices, the Food and Drug Administration, the Joint Commission, and other patient safety groups have reported multiple medication errors related to these medications. The list below comes from the FDA approved list of established drug names that require use of Tall Man Letters (capitalized letters to more easily distinguish between similar medication names). More extensive lists of medications are available at http://www.ismp.org/tools/confuseddrugnames.pdf.

acetoHEXAMIDE	acetaZOLAMIDE	
buPROPion	busPIRone	
chlorproMAZINE	chlorproPAMIDE	
clomiPHENE	clomiPRAMINE	
cycloSPORINE	cycloSERINE	
DAUNOrubicin	DOXOrubicin	
dimenhyDRINATE	diphenhydrAMINE	
DOBUTamine	DOPamine	
glipiZIDE	glyBURIDE	
hydrALAZINE	hydrOXYzine	
medoxyPROGESTERone	methylPREDNISolone	methylTESTOSTERone
niCARdipine	NIFEdipine	
predniSONE	prednisoLONE	
sulfADIAZINE	sulfiSOXAZOLE	
TOLAZamide	TOLBUTamide	
vinBLAStine	vinCRIStine	

The Center for Disease Control and Prevention (CDC) contains detailed listings of what vaccinations are recommended for various age groups as well as patients with multiple concurrent disease states. The full recommendations are available at: http://www.cdc.gov/vaccines/recs/schedules/default.htm.

Dosing information is provided for some common vaccines. There are many other vaccines and vaccine formulations. Please consult the package insert for specific details regarding each of the following:

BCG (BACILLUS CALMETTE-GUERIN) VACCINE
Tice, TheraCys

USES To protect groups with excessive rates of TB infection.

Adult/Adolescent/Child (over 1 mo): **Percutaneous** 0.2–0.3 mL of vaccine **Intradermal** 0.1 mL
Child (younger than 1 mo): **Percutaneous** Reduce adult dose by half, may need to reactivate at 1 y of age

CONTRAINDICATIONS Hypersensitivity to any vaccine component; active or past mycobacterial infection; immunosuppressed patients.

HAEMOPHILUS B CONJUGATE VACCINE (HIB)
ActHIB, Hiberix, Liquid Pedvax-HIB

USES To provide active immunity to *H. influenzae* type b (Hib) infection.

Infant: **IM** 0.5 mL at 2, 4, and 6 mo (ActHIB)

Infant/Child: **IM** 0.5 mL at 2 and 4 mo and again at 12 to 15 mo (Liquid PedvaxHIB)
Child (15–59 m): **IM** 0.5 mL as booster (Hiberix)

CONTRAINDICATIONS Hypersensitivity to any vaccine components; fever of unknown etiology.

HEPATITIS A VACCINE
Havrix, Vaqta

USES To provide active immunity to hepatitis A.

Adult: **IM** 1 mL dose then 1 mL 6–12 mo later
Child (1–18 y): **IM** 0.5 mL then 0.5 mL 1 mo later, then 0.5 mL booster dose 6–12 mo later

CONTRAINDICATIONS Hypersensitivity to any vaccine component; neomycin hypersensitivity.

HEPATITIS B IMMUNE GLOBULIN
HepaGam B, HyperHep, Nabi-HB

USES To provide passive immunity to hepatitis B infections in individuals exposed to HBV or Hbs Ag-positive materials; to prevent hepatitis B recurrence in post-liver transplant (HepaGam B).

Adult/Child: **IM** 0.06 mL/kg as soon as possible post-exposure then repeat 28–30 day post-exposure
Neonate: **IM** 0.5 mL within 24 h of birth then repeat at 3 and 6 mo
Adult: **HepaGam B IV** 20,000 U/dose daily every 2 wk (2–12 wk) then monthly

CONTRAINDICATIONS Hypersensitivity to any vaccine component.

HEPATITIS B VACCINE (RECOMBINANT)
Engerix-B, Recombivax HB

USES To provide active immunity to individuals at high risk for exposure.

Adult (at least 20 y old): **IM/SC** 10 mcg then repeat at 1 and 6 mo (Recombivax); 20 mcg then repeat at 1 and 6 mo (Engerix-B)
Adult (less than 20 y)/Adolescent/ Child/Infant: **IM/SC** 5 mcg then repeat at 1 and 6 mo (Recombivax)
Adult/Adolescent/Child (over 11 y): **IM** 10–20 mcg then repeat at 1 and 6 mo (Engerix-B) *Neonate* **IM/SC** 5 mcg within 12 h of birth then repeat at 1 and 6 mo (Recombivax); 10 mcg within 12 h of birth then repeat at 1 and 6 mo (Engerix-B)

CONTRAINDICATIONS Hypersensitivity to any vaccine component; yeast hypersensitivity.

HUMAN PAPILLOMAVIRUS BIVALENT (RECOMBINANT)
Cervarix

USES To prevent disease associated with HPV exposure.

Adult/Adolescent/Child (over 10 y): **IM** 0.5 mL dose at 0, 1, and 6 mo

CONTRAINDICATIONS Hypersensitivity to any vaccine component; yeast hypersensitivity.

HUMAN PAPILLOMAVIRUS QUADRIVALENT VACCINE (RECOMBINANT)
Gardasil

USES To prevent HPV related disease; to prevent genital warts, prevent anal cancer.

Adult/Child (over 9 y): **IM** 0.5 mL followed by doses at 2 and 6 mo

CONTRAINDICATIONS Hypersensitivity to any vaccine component; yeast hypersensitivity.

INFLUENZA VACCINE
Afluria, Fluarix, FluLaval, Fluvirin, Fluzone, Flumist

USES To prevent influenza.

Adult (49 y and younger): **Intranasal** 0.2 mL dose **IM** 0.5 mL
Child (over 9 y old): **Intranasal** 00.1 mL per nostril
Child (2–8 y): **IM** 0.2 mL dose 0.5 mL

CONTRAINDICATIONS Hypersensitivity to egg, kanamycin, neomycin or polymyxin.

MENINGOCOCCAL DIPHTHERIA TOXOID CONJUGATE
Menveo, Menactra

USES To provide prophylaxis against meningococcal infection.

Adult (less than 56 y)/Adolescent/ Child (over 2 y): **IM** 0.5 mL injection
Infant/Child (over 9 mo): Menactra only: 0.5 mL injection (2 dose series)

CONTRAINDICATIONS Hypersensitivity to any vaccine component; history of Guillain-Barré syndrome, latex hypersensitivity.

PNEUMOCOCCAL 13-VALENT VACCINE, DIPHTHERIA
Prevnar 13

USES To provide routine prophylaxis for infants and children as part of the primary childhood immunization schedule.

Infant (6 wk or less): **IM** 0.5 mL dose repeated at 4–8 wk, then 4–8 wk after the second dose and again at 12–15 mo of age (total of 4 doses)

Infant (7–11 mo): **IM** If no previous doses given, 0.5 mL followed by a second dose in 4 wk, then a third dose after the 1 y birthday (total of 3 doses)

Child (12–23 mo): **IM** If no previous doses given, 0.5 mL followed by a second dose after 8 wk (total of 2 doses)

Child (2–6 y): **IM** If no previous doses given, 0.5 mL single dose

CONTRAINDICATIONS Hypersensitivity to any vaccine component.

ROTAVIRUS VACCINE
Rotarix, RotaTeq

USES To prevent gastroenteritis caused by rotavirus infection.

Infant (6-32 wk) (RotaTeq only): **PO** 2 mL at 2, 4 and 6 mo

Rotarix only: **PO** 1 ml at 2 and 4 mo

CONTRAINDICATIONS Hypersensitivity to any vaccine component, history of intussisception, or severe combined immunodeficiency.

VARICELLA VACCINE
Varivax

USES To provide vaccination against varicella.

Adult/Adolescent: **SC** 0.5 mL followed by a second dose 4–8 wk later

Child (1–12 y): **SC** 0.5 mL single dose

CONTRAINDICATIONS Hypersensitivity to any vaccine component; AIDS; bone marrow suppression; concurrent chemotherapy or corticosteroid therapy; gelatin hypersensitivity; hypogammaglobulinemia; current leukemia or lymphoma; neomycin hypersensitivity; pregnancy; tuberculosis; concurrent radiation therapy.

ZOSTER VACCINE LIVE
Zostavax

USES To prevent herpes zoster in people over 50 y old.

Adult (at least 50 y): **SC** Administer single-dose vial.

CONTRAINDICATIONS Hypersensitivity to any vaccine component; AIDS; gelatin hypersensitivity; immunosuppression; leukemia; lymphoma; neomycin hypersensitivity; pregnancy.

BIBLIOGRAPHY

American Academy of Pediatrics Committee on Drugs. The transfer of drugs and other chemicals into human milk. *Pediatrics.* 2001;108: 776–89.

American Hospital Formulary Service (AHFS) Drug Information. 2011. Bethesda, MD: American Society of Health-System Pharmacists. 2010.

Bindler R, Howry L. *Prentice Hall Pediatric Drug Guide with Nursing Implications.* Upper Saddle River, NJ: Prentice Hall Health. 2005.

Clinical Pharmacology. http://www.gsm.com. Gold Standard Media. 2011.

Drug Facts and Comparisons. http://factsandcomparisons.com. Version 4.0 online. St. Louis: Wolters Kluwer Health, Inc. 2011.

Food and Drug Administration. http://www.fda.gov. 2011.

King Guide to Parenteral Admixtures. 35th ed. Napa, CA: King Guide Publications, Inc. 2006.

Kirschheiner J, Nickchen K, Bauer M, et al. Pharmacogenetics of antidepressants and antipsychotics: the contribution of allelic variations to the phenotype of drug response. Preview. *Mol Psychiatry.* 2004;9:442–73.

Lacy CF, Armstrong LL, Goldman MP, Lance LL. *Drug Information Handbook.* 20th ed. Hudson, OH: Lexi-Comp. 2011.

Phelps SJ, Hak EB, Crill CM. *Pediatric Injectable Drugs (Teddy Bear Book).* 8th ed. American Society of Health-System Pharmacists (ASHP). 2007.

Tatro DS. *Drug Interaction Facts: Herbal Supplements and Food.* 1st ed (Paperback). Facts and Comparisons. 2004.

Trissel LA. *Handbook of Injectable Drugs.* 16th ed. Bethesda, MD: American Society of Health-System Pharmacists. 2010.

INDEX

Drug categories are in SMALL CAPS. Prototypes in **bold.** ·
Generic drug names are given in parentheses.

Drug categories are in SMALL CAPS. Prototypes in **bold.**
Generic drug names are given in parentheses.

1663

Drug categories are in SMALL CAPS. Prototypes in **bold.**
Generic drug names are given in parentheses.

1665

Drug categories are in SMALL CAPS. Prototypes in **bold.**
Generic drug names are given in parentheses.

Drug categories are in SMALL CAPS. Prototypes in **bold.**
Generic drug names are given in parentheses.

Drug categories are in SMALL CAPS. Prototypes in **bold.**
Generic drug names are given in parentheses.

Drug categories are in SMALL CAPS. Prototypes in **bold.**
Generic drug names are given in parentheses.

Drug categories are in SMALL CAPS. Prototypes in **bold.**
Generic drug names are given in parentheses.

Drug categories are in SMALL CAPS. Prototypes in **bold.**
Generic drug names are given in parentheses.

INDEX

Drug categories are in SMALL CAPS. Prototypes in **bold.**
Generic drug names are given in parentheses.

Drug categories are in SMALL CAPS. Prototypes in **bold.**
Generic drug names are given in parentheses.

1673

Drug categories are in SMALL CAPS. Prototypes in **bold.**
Generic drug names are given in parentheses.

Drug categories are in SMALL CAPS. Prototypes in **bold.**
Generic drug names are given in parentheses.

1675

Drug categories are in SMALL CAPS. Prototypes in **bold**.
Generic drug names are given in parentheses.

Drug categories are in SMALL CAPS. Prototypes in **bold.**
Generic drug names are given in parentheses.

1677

Drug categories are in SMALL CAPS. Prototypes in **bold.**
Generic drug names are given in parentheses.

1679

Drug categories are in SMALL CAPS. Prototypes in **bold.**
Generic drug names are given in parentheses.

Drug categories are in SMALL CAPS. Prototypes in **bold.**
Generic drug names are given in parentheses.

Drug categories are in SMALL CAPS. Prototypes in **bold.**
Generic drug names are given in parentheses.

Drug categories are in SMALL CAPS. Prototypes in **bold.**
Generic drug names are given in parentheses.

1683

Drug categories are in SMALL CAPS. Prototypes in **bold.**
Generic drug names are given in parentheses.

Drug categories are in SMALL CAPS. Prototypes in **bold.**
Generic drug names are given in parentheses.

1685

Drug categories are in SMALL CAPS. Prototypes in **bold.**
Generic drug names are given in parentheses.

Drug categories are in SMALL CAPS. Prototypes in **bold.**
Generic drug names are given in parentheses.

1687

1688

Drug categories are in SMALL CAPS. Prototypes in **bold.**
Generic drug names are given in parentheses.

Drug categories are in SMALL CAPS. Prototypes in **bold.**
Generic drug names are given in parentheses.

1689

Drug categories are in SMALL CAPS. Prototypes in **bold.**
Generic drug names are given in parentheses.

Drug categories are in SMALL CAPS. Prototypes in **bold.**
Generic drug names are given in parentheses.

1691

Drug categories are in SMALL CAPS. Prototypes in **bold.**
Generic drug names are given in parentheses.

Drug categories are in SMALL CAPS. Prototypes in **bold.**
Generic drug names are given in parentheses.

Drug categories are in SMALL CAPS. Prototypes in **bold.**
Generic drug names are given in parentheses.

Drug categories are in SMALL CAPS. Prototypes in **bold.**
Generic drug names are given in parentheses.

1695

Drug categories are in SMALL CAPS. Prototypes in **bold.**
Generic drug names are given in parentheses.

Drug categories are in SMALL CAPS. Prototypes in **bold.**
Generic drug names are given in parentheses.

1697

1698

Drug categories are in SMALL CAPS. Prototypes in **bold.**
Generic drug names are given in parentheses.

Drug categories are in SMALL CAPS. Prototypes in **bold.**
Generic drug names are given in parentheses.

1699

Drug categories are in SMALL CAPS. Prototypes in **bold.**
Generic drug names are given in parentheses.

Drug categories are in SMALL CAPS. Prototypes in **bold.**
Generic drug names are given in parentheses.

1701

Drug categories are in SMALL CAPS. Prototypes in **bold.**
Generic drug names are given in parentheses.

Drug categories are in SMALL CAPS. Prototypes in **bold.**
Generic drug names are given in parentheses.

1703

Drug categories are in SMALL CAPS. Prototypes in **bold.**
Generic drug names are given in parentheses.

Drug categories are in SMALL CAPS. Prototypes in **bold.**
Generic drug names are given in parentheses.

1705

Drug categories are in SMALL CAPS. Prototypes in **bold.**
Generic drug names are given in parentheses.

Drug categories are in SMALL CAPS. Prototypes in **bold.**
Generic drug names are given in parentheses.

1707

Drug categories are in SMALL CAPS. Prototypes in **bold.**
Generic drug names are given in parentheses.

Drug categories are in SMALL CAPS. Prototypes in **bold**.
Generic drug names are given in parentheses.

1709

Drug categories are in SMALL CAPS. Prototypes in **bold.**
Generic drug names are given in parentheses.

Drug categories are in SMALL CAPS. Prototypes in **bold.**
Generic drug names are given in parentheses.

1711

Drug categories are in SMALL CAPS. Prototypes in **bold.**
Generic drug names are given in parentheses.

Drug categories are in SMALL CAPS. Prototypes in **bold.**
Generic drug names are given in parentheses.

1713

Drug categories are in SMALL CAPS. Prototypes in **bold.**
Generic drug names are given in parentheses.

Drug categories are in SMALL CAPS. Prototypes in **bold.**
Generic drug names are given in parentheses.

1715

Infergen (interferon alfacon-1), 790–791

Inflamase Forte (prednisolone sodium phosphate), 1243–1244, 1607

Inflamase Mild (prednisolone sodium phosphate), 1243–1244, 1607

infliximab, 775–777

influenza vaccine, 1659

Infumorph (morphine sulfate), 1011–1014

INH (isoniazid), 809–811

INHALED CORTICOSTEROIDS
beclomethasone dipropionate, 1611
budesonide, 1611
ciclesonide, 1611
dexamethasone, 1611
flunisolide, 1611
fluticasone, 1612
mometasone furoate, 1612
triamcinolone acetonide, 1612

InnoPran XL (propranolol hydrochloride), 1271–1275

INOTROPICS
digoxin, 456–458
dobutamine hydrochloride, 477–479
dopamine hydrochloride, 487–489
milrinone lactate, 995–996
norepinephrine bitartrate, 1077–1079

Inspra (eplerenone), 542–544

Insta-Char (activated charcoal), 293–294

INSULINS, INTERMEDIATE-ACTING
insulin, isophane, 785–787

INSULINS, LONG-ACTING, BASAL
insulin detemir, 778–780
insulin glargine, 780–781

INSULINS, RAPID-ACTING
insulin aspart, 777–778
insulin glulisine, 781–782
insulin lispro, 787

INSULIN, SHORT-ACTING
insulin, 783–785

Intal (cromolyn sodium), 369–371

INTEGRASE INHIBITOR
raltegravir, 1302–1303

Integrilin (eptifibatide), 548–549

INTEGRIN INHIBITOR
natalizumab, 1038–1039

Intelence (etravirine), 601–602, 1623

INTERFERONS. *See* IMMUNOMODULATORS, INTERFERON

INTERLEUKIN INHIBITORS
denileukin diftitox, 420–421
tocilizumab, 1499–1501

INTERLEUKIN RECEPTOR ANTAGONISTS
anakinra, 101–103
basiliximab, 154–156
ustekinumab, 1545–1546

Intermezzo (zolpidem), 1599–1600

Intralipid (fat emulsion, intravenous), 611–613

Intron A (interferon alfa-2b), 788–790

Intuniv (guanfacine hydrochloride), 714–716

Invanz (ertapenem sodium), 558–559

Invega (paliperidone), 1131–1133

Invega Sustenna (paliperidone palmitate), 1131–1133

Invirase (saquinavir mesylate), 1357–1358

iodoquinol, 796–797

Ionsys (fentanyl citrate), 621–624

Iopidine (apraclonidine hydrochloride), 1603

ipecac syrup, 797–798

Ipecac Syrup (ipecac syrup), 797–798

ipilimumab, 798–799

ipratropium bromide, 799–801

Iquix (levofloxacin), 864–866

irbesartan, 801–802

Ircon-FA (ferrous fumarate), 624–626

Iressa (gefitinib), 688–689

irinotecan hydrochloride, 802–803

IRON PREPARATIONS
ferrous fumarate, 624–626
ferrous gluconate, 624–626
ferrous sulfate, 624–626
iron dextran, 803–805

Drug categories are in SMALL CAPS. Prototypes in **bold.**
Generic drug names are given in parentheses.

Drug categories are in SMALL CAPS. Prototypes in **bold.**
Generic drug names are given in parentheses.

1719

Drug categories are in SMALL CAPS. Prototypes in **bold.**
Generic drug names are given in parentheses.

Drug categories are in SMALL CAPS. Prototypes in **bold.**
Generic drug names are given in parentheses.

1721

Drug categories are in SMALL CAPS. Prototypes in **bold.**
Generic drug names are given in parentheses.

Drug categories are in SMALL CAPS. Prototypes in **bold.**
Generic drug names are given in parentheses.

1723

Drug categories are in SMALL CAPS. Prototypes in **bold.**
Generic drug names are given in parentheses.
1725

Drug categories are in SMALL CAPS. Prototypes in **bold.**
Generic drug names are given in parentheses.

Drug categories are in SMALL CAPS. Prototypes in **bold.**
Generic drug names are given in parentheses.
1727

Drug categories are in SMALL CAPS. Prototypes in **bold.**
Generic drug names are given in parentheses.

Drug categories are in SMALL CAPS. Prototypes in **bold.**
Generic drug names are given in parentheses.

1729

Drug categories are in SMALL CAPS. Prototypes in **bold.**
Generic drug names are given in parentheses.

Drug categories are in SMALL CAPS. Prototypes in **bold.**
Generic drug names are given in parentheses.

1731

Drug categories are in SMALL CAPS. Prototypes in **bold.**
Generic drug names are given in parentheses.

Drug categories are in SMALL CAPS. Prototypes in **bold.**
Generic drug names are given in parentheses.

1733

Drug categories are in SMALL CAPS. Prototypes in **bold.**
Generic drug names are given in parentheses.

Drug categories are in SMALL CAPS. Prototypes in **bold.**
Generic drug names are given in parentheses.
1735

Drug categories are in SMALL CAPS. Prototypes in **bold.**
Generic drug names are given in parentheses.

Drug categories are in SMALL CAPS. Prototypes in **bold.**
Generic drug names are given in parentheses.
1737

Drug categories are in SMALL CAPS. Prototypes in **bold.**
Generic drug names are given in parentheses.

Drug categories are in SMALL CAPS. Prototypes in **bold.**
Generic drug names are given in parentheses.

1739

Drug categories are in SMALL CAPS. Prototypes in **bold.**
Generic drug names are given in parentheses.

Drug categories are in SMALL CAPS. Prototypes in **bold.**
Generic drug names are given in parentheses.

1741

Drug categories are in SMALL CAPS. Prototypes in **bold.**
Generic drug names are given in parentheses.

Drug categories are in SMALL CAPS. Prototypes in **bold.**
Generic drug names are given in parentheses.

1743

Drug categories are in SMALL CAPS. Prototypes in **bold.**
Generic drug names are given in parentheses.

1745

Drug categories are in SMALL CAPS. Prototypes in **bold.**
Generic drug names are given in parentheses.

Drug categories are in SMALL CAPS. Prototypes in **bold.**
Generic drug names are given in parentheses.

Drug categories are in SMALL CAPS. Prototypes in **bold.**
Generic drug names are given in parentheses.

Drug categories are in SMALL CAPS. Prototypes in **bold.**
Generic drug names are given in parentheses.

1749

Drug categories are in SMALL CAPS. Prototypes in **bold.**
Generic drug names are given in parentheses.

Drug categories are in SMALL CAPS. Prototypes in **bold.**
Generic drug names are given in parentheses.

1751

Drug categories are in SMALL CAPS. Prototypes in **bold.**
Generic drug names are given in parentheses.

Drug categories are in SMALL CAPS. Prototypes in **bold.**
Generic drug names are given in parentheses.

1753

Drug categories are in SMALL CAPS. Prototypes in **bold.**
Generic drug names are given in parentheses.

1755

Drug categories are in SMALL CAPS. Prototypes in **bold.**
Generic drug names are given in parentheses.
1757

Drug categories are in SMALL CAPS. Prototypes in **bold.**
Generic drug names are given in parentheses.

Drug categories are in SMALL CAPS. Prototypes in **bold.**
Generic drug names are given in parentheses.

Drug categories are in SMALL CAPS. Prototypes in **bold.**
Generic drug names are given in parentheses.

Drug categories are in SMALL CAPS. Prototypes in **bold.**
Generic drug names are given in parentheses.

1761

COMMON DRUG IV-SITE COMPATIBILITY CHART

	AMINOPHYLLINE	DOBUTAMINE	DOPAMINE	HEPARIN	MEPERIDINE	MORPHINE	NITROGLYCERIN	ONDANSETRON	POTASSIUM CL
acyclovir	C	—	—	C	I/C	I/C	C	I	C
alteplase		—	C	—			—	C	C
amikacin	C	C	C	—	C	C	C		C
amino acids (TPN)	I/C	C	I/C	C	C	C	C	—	C
aminophylline	C	—	C	C	C	C	C		C
amiodarone	—	C	C	—	C	C	C	—	C
ampicillin	—	—	—	I/C	I/C	I/C	I/C	—	I/C
ampicillin/sulbactam	I/C	—	I/C	I/C	I/C	I/C	I/C	—	I/C
aztreonam	C	C	C	C	C	C	C	C	C
bretylium	C	C	C	C	C	C	C	C	C
bumetanide	C	C	C	C	C	C	C	C	C
calcium chloride	C	C	C	C	C	C	C	C	C
cefazolin	C	—	—	C	C	C	C	C	C
cefotaxime	C	I/C	C	C	C	C	C	I/C	C
cefotetan	C	—	C	C	I/C	C	C	C	C
cefoxitin	C	—	—	C	C	C	C	C	C
ceftazidime	C	—	C	C	C	C	C	I/C	C
ceftizoxime	C	I/C	C	C	I/C	C	C	C	C
ceftriaxone	C	—	C	C	C	C	C	I/C	C
cefuroxime	C	—	—	C	C	C	C	C	C
chloramphenicol	C	—	C	C	I/C	C	C	—	C
cimetidine	C	C	C	C	C	C	C	C	C
ciprofloxacin	—	C	C	—	C	C		C	C
clindamycin	C	I/C	—	C	C	C	C	C	C
dexamethasone	C	—	C	C	I/C	C	C	C	C
diazepam	—	I/C	—	—	—	I/C	—	I/C	—
digoxin	C	C	C	C	C	C	C	C	C

COMMON DRUG IV-SITE COMPATIBILITY CHART

	AMINOPHYLLINE	DOBUTAMINE	DOPAMINE	HEPARIN	MEPERIDINE	MORPHINE	NITROGLYCERIN	ONDANSETRON	POTASSIUM CL
diltiazem	I/C	C	C	I/C	C	C	C		C
diphenhydramine	—	C	C	I/C	C	C	C	C	C
dobutamine	—	C	C	I/C	C	C	C	C	C
dopamine	C	C	C	C	C	C	C	C	C
doxycycline	C	C	C	—	C	C	C	C	C
enalaprilat	C	C	C	C	C	C	C	C	C
epinephrine	—	C	C	C	C	C	C	C	C
eptifibatide	C	C		C	C	C	C	C	C
erythromycin	C	C	C	I/C	C	C	C	C	C
esmolol	C	C	C	I/C	C	C	C	C	C
famotidine	C	C	C	C	C	C	C	C	C
filgrastim	C	C	C	—	C	C		C	C
fluconazole	C	C	C	C	C	C	C	C	C
foscarnet	C	—	C	C		C			C
furosemide	C	I/C	I/C	I/C	I/C	I/C	I/C	I	C
ganciclovir	—	—	—	C	—	—		—	C
gentamicin	C	C	C	—	C	C	C	C	C
heparin	C	I/C	C	C	I/C	C	C	C	C
hydrocortisone	C	—	C	C	I/C	C	C	C	C
hydromorphone	C	C	C	C	C	C	C	C	C
imipenem/cilastatin	I/C	I/C	C	C	I/C	C	C		C
inamrinone	I/C	I/C	I/C	—	I/C	—	I/C	—	C
insulin	C	I/C	I/C	I/C	I/C	I/C	C	I/C	C
isoproterenol	—	C	C	C	C	C	C	C	C
labetolol	C	C	C	I/C	C	C	C	C	C
lidocaine	C	C	C	C	C	C	C	C	C
lorazepam	C	C	C	C	—	C	C	—	C